HANDBOOK of MATHEMATICAL, SCIENTIFIC, and ENGINEERING

FORMULAS, TABLES, FUNCTIONS, GRAPHS, TRANSFORMS

Staff of Research and Education Association,
Dr. M. Fogiel, Director

Research and Education Association
505 Eighth Avenue
New York, N. Y. 10018

HANDBOOK of MATHEMATICAL, SCIENTIFIC, and ENGINEERING FORMULAS, TABLES, FUNCTIONS, GRAPHS, TRANSFORMS

Printed in the United States of America

Library of Congress Catalog Card Number 80-52490

International Standard Book Number 0-87891-521-4

Revised Printing, 1986

PREFACE

This handbook is designed to be a particularly useful reference for persons working in mathematics, science, engineering and other technical fields.

The handbook includes the most-often used formulas, tables, transforms, functions and graphs which are needed as tools in solving problems. The entire field of special functions is also covered. The information has been arranged in complete but condensed form to enable the user of the handbook to easily find what he or she may need. For this purpose, a very extensive table of contents is presented at the beginning of the book. The user can simply refer to the chapter or topic that interests him/her and see the numerous items of reference information available in the handbook under that topic. Each topic includes parts from elementary to advanced. The very detailed index at the end of the book also allows the user to retrieve information in a minimum of time.

Unusually comprehensive compilations of integrals and Laplace transforms make the book useful for solving problems in mathematics and the various sciences. Scientific data which is often of interest to scientists and engineers have also been included.

Special thanks are due to staff members of R E A who patiently selected the material to be included and arranged it to be easily accessible and readable.

Max Fogiel, Ph. D.
Program Director

LIST OF CHAPTERS

(DETAILED TABLE OF CONTENTS FOLLOWS THIS LIST)

SCIENCE AND ENGINEERING DATA

CONTENTS

GEOMETRY 136

5. LOGS & EXPONENTIALS 190

6. TRIGONOMETRY 207

10. ORTHOGONAL COORDINATE SYSTEMS 297

11. SERIES 304

12. DERIVATIVES 312

13. INTEGRALS 316

16. SPECIAL FUNCTIONS 593

SCIENCE AND ENGINEERING DATA

MATHEMATICAL SIGNS AND SYMBOLS

$\pm (\mp)$	plus or minus (minus or plus)
$:$	divided by, ratio sign
$::$	proportional sign
$<$	less than
$\nless$	not less than
$>$	greater than
$\ngtr$	not greater than
$\cong$	approximately equals, congruent
$\sim$	similar to
$\rightleftharpoons$	equivalent to
$\neq$	not equal to
$\doteq$	approaches, is approximately equal to
$\propto$	varies as
∞	infinity
$\therefore$	therefore
$\sqrt{\ }$	square root
$\sqrt[3]{\ }$	cube root
$\sqrt[n]{\ }$	*n*th root
$\angle$	angle
$\perp$	perpendicular to
$\parallel$	parallel to
$\lvert x \rvert$	numerical value of x
log or $\log_{10}$	common logarithm or Briggsian logarithm
$\log_e$ or ln	natural logarithm or hyperbolic logarithm or Napierian logarithm
e	base (2.718) of natural system of logarithms
a°	an angle a degrees
a'	a prime, an angle a minutes
a''	a double prime, an angle a seconds, a second
sin	sine
cos	cosine
tan	tangent
ctn or cot	cotangent
sec	secant
csc	cosecant
vers	versed sine
covers	coversed sine
exsec	exsecant
$\sin^{-1}$	anti sine or angle whose sine is
sinh	hyperbolic sine
cosh	hyperbolic cosine
tanh	hyperbolic tangent
$\sinh^{-1}$	anti hyperbolic sine or angle whose hyperbolic sine is
$f(x)$ or $\phi(x)$	function of x
Δx	increment of x
Σ	summation of
dx	differential of x
dy/dx or y'	derivative of y with respect to x
d^2y/dx^2 or y''	second derivative of y with respect to x
d^ny/dx^n	*n*th derivative of y with respect to x
$\partial y/\partial x$	partial derivative of y with respect to x
$\partial^n y/\partial x^n$	*n*th partial derivative of y with respect to x
$\dfrac{\partial^n y}{\partial x\, \partial y}$	*n*th partial derivative with respect to x and y
$\int$	integral of
$\int_a^b$	integral between the limits a and b
$\dot{y}$	first derivative of y with respect to time
$\ddot{y}$	second derivative of y with respect to time
Δ or ∇^2	the "Laplacian" $\left(\dfrac{\partial^2}{\partial x^2} + \dfrac{\partial^2}{\partial y^2} + \dfrac{\partial^2}{\partial z^2}\right)$
δ	sign of a variation
$\oint$	sign of integration around a closed path

GREEK ALPHABET

Alpha	= Α, α	= A, a
Beta	= Β, β	= B, b
Gamma	= Γ, γ	= G, g
Delta	= Δ, δ	= D, d
Epsilon	= Ε, ϵ	= E, e
Zeta	= Ζ, ζ	= Z, z
Eta	= Η, η	= E, e
Theta	= Θ, θ	= Th, th
Iota	= Ι, ι	= I, i
Kappa	= Κ, κ	= K, k
Lambda	= Λ, λ	= L, l
Mu	= Μ, μ	= M, m
Nu	= Ν, ν	= N, n
Xi	= Ξ, ξ	= X, x
Omicron	= Ο, ο	= O, o
Pi	= Π, π	= P, p
Rho	= Ρ, ρ	= R, r
Sigma	= Σ, σ	= S, s
Tau	= Τ, τ	= T, t
Upsilon	= Υ, υ	= U, u
Phi	= Φ, φ	= Ph, ph
Chi	= Χ, χ	= Ch, ch
Psi	= Ψ, ψ	= Ps, ps
Omega	= Ω, ω	= O, o

CHAPTER 1

CONSTANTS

π CONSTANTS

π = 3.14159 26535 89793 23846 26433 83279 50288 41971 69399 37511
$1/\pi$ = 0.31830 98861 83790 67153 77675 26745 02872 40689 19291 48091
π^2 = 9.86960 44010 89358 61883 44909 99876 15113 53136 99407 24079
$\log_e \pi$ = 1.14472 98858 49400 17414 34273 51353 05871 16472 94812 91531
$\log_{10} \pi$ = 0.49714 98726 94133 85435 12682 88290 89887 36516 78324 38044
$\log_{10} \sqrt{2\pi}$ = 0.39908 99341 79057 52478 25035 91507 69595 02099 34102 92128

NUMBERS CONTAINING π

	Number	Logarithm		Number	Logarithm
π	3.1415 927	0.4971 499	$2\pi^2$	19.7392 088	1.2953 297
2π	6.2831 853	0.7981 799	$\pi/180$	0.0174 533	8.2418 774 − 10
3π	9.4247 780	0.9742 711	$180/\pi$	57.2957 795	1.7581 226
4π	12.5663 706	1.0992 099	$4\pi^2$	39.4784 176	1.5963 597
8π	25.1327 412	1.4002 399	$1/\pi^2$	0.1013 212	9.0057 003 − 10
$\pi/2$	1.5707 963	0.1961 199	$1/(2\pi^2)$	0.0506 606	8.7046 703 − 10
$\pi/3$	1.0471 976	0.0200 286	$1/(4\pi^2)$	0.0253 303	8.4036 403 − 10
$\pi/4$	0.7853 982	9.8950 899 − 10	$\sqrt{\pi}$	1.7724 539	0.2485 749
$\pi/6$	0.5235 988	9.7189 986 − 10	$\frac{\sqrt{\pi}}{2}$	0.8862 269	9.9475 449 − 10
$\pi/8$	0.3926 991	9.5940 599 − 10	$\frac{\sqrt{\pi}}{4}$	0.4431 135	9.6465 149 − 10
$2\pi/3$	2.0943 951	0.3210 586	$\sqrt{\frac{\pi}{2}}$	1.2533 141	0.0980 599
$4\pi/3$	4.1887 902	0.6220 886	$\sqrt{\frac{2}{\pi}}$	0.7978 846	9.9019 401 − 10
$1/\pi$	0.3183 099	9.5028 501 − 10	π^3	31.0062 767	1.4914 496
$2/\pi$	0.6366 198	9.8038 801 − 10	$\sqrt[3]{\pi}$	1.4645 919	0.1657 166
$4/\pi$	1.2732 395	0.1049 101	$1/\sqrt[3]{\pi}$	0.6827 841	9.8342 834 − 10
$1/(2\pi)$	0.1591 549	9.2018 201 − 10	$\sqrt[3]{\pi^2}$	2.1450 294	0.3314 332
$1/(4\pi)$	0.0795 775	8.9007 901 − 10	$1/\sqrt{\pi}$	0.5641 896	9.7514 251 − 10
$1/(6\pi)$	0.0530 516	8.7246 989 − 10	$1/\sqrt{2\pi}$	0.3989 423	9.6009 101 − 10
$1/(8\pi)$	0.0397 887	8.5997 601 − 10	$2/\sqrt{\pi}$	1.1283 792	0.0524 551
π^2	9.8696 044	0.9942 997			

MULTIPLES OF $\frac{\pi}{2}$

n	$n\frac{\pi}{2}$	n	$n\frac{\pi}{2}$	n	$n\frac{\pi}{2}$	n	$n\frac{\pi}{2}$
1	1.57079 63268	26	40.84070 44967	51	80.11061 26665	76	119.38502 08364
2	3.14159 26536	27	42.41150 08235	52	81.68140 89933	77	120.95131 71632
3	4.71238 89804	28	43.98229 71503	53	83.25220 53201	78	122.52211 34900
4	6.28318 53072	29	45.55309 34771	54	84.82300 16469	79	124.09290 98168
5	7.85398 16340	30	47.12388 98038	55	86.39379 79737	80	125.66370 61436
6	9.42477 79608	31	48.69468 61306	56	87.96459 43005	81	127.23450 24704
7	10.99557 42876	32	50.26548 24574	57	89.53539 06273	82	128.80529 87972
8	12.56637 06144	33	51.83627 87842	58	91.10618 69541	83	130.37609 51240
9	14.13716 69412	34	53.40707 51110	59	92.67698 32809	84	131.94689 14508
10	15.70796 32679	35	54.97787 14378	60	94.24777 96077	85	133.51768 77776
11	17.27875 95947	36	56.54866 77646	61	95.81857 59345	86	135.08848 41044
12	18.84955 59215	37	58.11946 40914	62	97.38937 22613	87	136.65928 04312
13	20.42035 22483	38	59.69026 04182	63	98.96016 85881	88	138.23007 67580
14	21.99114 85751	39	61.26105 67450	64	100.53096 49149	89	139.80087 30847
15	23.56194 49019	40	62.83185 30718	65	102.10176 12417	90	141.37166 94115
16	25.13274 12287	41	64.40264 93986	66	103.67255 75685	91	142.94246 57383
17	26.70533 75555	42	65.97344 57254	67	105.24335 38953	92	144.51326 20651
18	28.27433 38823	43	67.54424 20522	68	106.81415 02221	93	146.08405 83919
19	29.84513 02091	44	69.11503 83790	69	108.38494 65488	94	147.65485 47187
20	31.41592 65359	45	70.68583 47058	70	109.95574 28765	95	149.22565 10455
21	32.98672 28627	46	72.25663 10326	71	111.52653 92024	96	150.79644 73723
22	34.55751 91895	47	73.82742 73594	72	113.09733 55292	97	152.36724 36991
23	36.12831 55163	48	75.39822 36862	73	114.66813 18560	98	153.93804 00259
24	37.69911 18431	49	76.96902 00129	74	116.23892 81828	99	155.50883 63527
25	39.26990 81699	50	78.53981 63397	75	117.80972 45096	100	157.07963 26795

CONSTANTS INVOLVING e

e = 2.71828 18284 59045 23536 02874 71352 66249 77572 47093 69996

$1/e$ = 0.36787 94411 71442 32159 55237 70161 46086 74458 11131 03177

e^2 = 7.38905 60989 30650 22723 04274 60575 00781 31803 15570 55185

$M = \log_{10} e$ = 0.43429 44819 03251 82765 11289 18916 60508 22943 97005 80367

$1/M = \log_e 10$ = 2.30258 50929 94045 68401 79914 54684 36420 76011 01488 62877

$\log_{10} M$ = 9.63778 43113 00536 78912 29674 98645 − 10

π^e AND e^π CONSTANTS

π^e = 22.45915 77183 61045 47342 71522

e^π = 23.14069 26327 79269 00572 90864

$e^{-\pi}$ = 0.04321 39182 63772 24977 44177

$e^{\frac{1}{2}\pi}$ = 4.81047 73809 65351 65547 30357

$i^i = e^{-\frac{1}{2}\pi}$ = 0.20787 95763 50761 90854 69556

NUMERICAL CONSTANTS

$\sqrt{2}$ = 1.41421 35623 73095 04880 16887 24209 69807 85696 71875 37695

$\sqrt[3]{2}$ = 1.25992 10498 94873 16476 72106 07278 22835 05702 51464 70151

$\log_e 2$ = 0.69314 71805 59945 30941 72321 21458 17656 80755 00134 36026

$\log_{10} 2$ = 0.30102 99956 63981 19521 37388 94724 49302 67681 89881 46211

$\sqrt{3}$ = 1.73205 08075 68877 29352 74463 41505 87236 69428 05253 81039

$\sqrt[3]{3}$ = 1.44224 95703 07408 38232 16383 10780 10958 83918 69253 49935

$\log_e 3$ = 1.09861 22886 68109 69139 52452 36922 52570 46474 90557 82275

$\log_{10} 3$ = 0.47712 12547 19662 43729 50279 03255 11530 92001 28864 19070

OTHER CONSTANTS

Euler's Constant γ = 0.57721 56649 01532 86061

$\log_e \gamma$ = −0.54953 93129 81644 82234

Golden Ratio ϕ = 1.61803 39887 49894 84820 45868 34365 63811 77203 09180

TABLE OF MATHEMATICAL CONSTANTS[1]

n(prime)	$\sqrt{n}$
2	1. 4142
3	1. 7320
5	2. 2360
7	2. 6457
11	3. 3166
13	3. 6055
17	4. 1231
19	4. 3588
23	4. 7958
29	5. 3851
31	5. 5677
37	6. 0827
41	6. 4031
43	6. 5574
47	6. 8556
53	7. 2801
59	7. 6811
61	7. 8102
67	8. 1853
71	8. 4261
73	8. 5440
79	8. 8881
83	9. 1104
89	9. 4339
97	9. 8488

$10^{1/2}$		3. 1622
$10^{1/3}$		2. 1544
$10^{1/4}$		1. 7782
$10^{1/5}$		1. 5848
$100^{1/3}$		4. 6415
$100^{1/5}$		2. 5118
$1000^{1/4}$		5. 6234
$1000^{1/5}$		3. 9810
$2^{1/3}$		1. 2599
$3^{1/3}$		1. 4422
$2^{1/4}$		1. 1892
$3^{1/4}$		1. 3160
$2^{-1/2}$	(− 1)	7. 0710
$3^{-1/2}$	(− 1)	5. 7735
$5^{-1/2}$	(− 1)	4. 4721

$e^{\pi/2}$		4. 8104
$e^{\pi/4}$		2. 1932
$e^{-\pi/2}$	(− 1)	2. 0787
$e^{-\pi/4}$	(− 1)	4. 5593
$e^{1/2}$		1. 6487
$e^{-1/2}$	(− 1)	6. 0653
$e^{1/3}$		1. 3956
$e^{-1/3}$	(− 1)	7. 1653

n		e^n
1		2. 7182
2		7. 3890
3	(1)	2. 0085
4	(1)	5. 4598
5	(2)	1. 4841
6	(2)	4. 0342
7	(3)	1. 0966
8	(3)	2. 9809
9	(3)	8. 1030
10	(4)	2. 2026

n		e^{-n}
1	(− 1)	3. 6787
2	(− 1)	1. 3533
3	(− 2)	4. 9787
4	(− 2)	1. 8315
5	(− 3)	6. 7379
6	(− 3)	2. 4787
7	(− 4)	9. 1188
8	(− 4)	3. 3546
9	(− 4)	1. 2340
10	(− 5)	4. 5399

n		$e^{n\pi}$
1	(1)	2. 3140
2	(2)	5. 3549
3	(4)	1. 2391
4	(5)	2. 8675
5	(6)	6. 6356
6	(8)	1. 5355
7	(9)	3. 5533
8	(10)	8. 2226
9	(12)	1. 9027
10	(13)	4. 4031
e^e	(1)	1. 5154
e^γ		1. 7810

n		$e^{-n\pi}$
1	(− 2)	4. 3213
2	(− 3)	1. 8674
3	(− 5)	8. 0699
4	(− 6)	3. 4873
5	(− 7)	1. 5070
6	(− 9)	6. 5124
7	(−10)	2. 8142
8	(−11)	1. 2161
9	(−13)	5. 2554
10	(−14)	2. 2711
e^{-e}	(− 2)	6. 5988
$e^{-\gamma}$	(− 1)	5. 6145

n	$\ln n$
2	0. 6931
3	1. 0986
4	1. 3862
5	1. 6094
6	1. 7917
7	1. 9459
8	2. 0794
9	2. 1972
10	2. 3025
11	2. 3978
13	2. 5649
17	2. 8332
19	2. 9444
23	3. 1354
29	3. 3672
31	3. 4339
37	3. 6109
41	3. 7135
43	3. 7612

n		$\log_{10} n$
2	(−1)	3. 0102
3	(−1)	4. 7712
4	(−1)	6. 0205
5	(−1)	6. 9897
6	(−1)	7. 7815
7	(−1)	8. 4509
8	(−1)	9. 0308
9	(−1)	9. 5424
10		1. 0000
11		1. 0413
13		1. 1139
17		1. 2304
19		1. 2787
23		1. 3617
29		1. 4623
31		1. 4913
37		1. 5682
41		1. 6127
43		1. 6334

[1] From "Handbook of Mathematical Functions", National Bureau of Standards

TABLE OF MATHEMATICAL CONSTANTS

n	$\ln n$	n	$\log_{10} n$
47	3. 8501	47	1. 6720
53	3. 9702	53	1. 7242
59	4. 0775	59	1. 7708
61	4. 1108	61	1. 7853
67	4. 2046	67	1. 8260
71	4. 2626	71	1. 8512
73	4. 2904	73	1. 8633
79	4. 3694	79	1. 8976
83	4. 4188	83	1. 9190
89	4. 4886	89	1. 9493
97	4. 5747	97	1. 9867
$\ln \pi$	1. 1447	$\log_{10}\pi$	(−1) 4. 9714
$\ln\sqrt{2\pi}$	(−1) 9. 1893	$\log_{10}e$	(−1) 4. 3429

n	$n \ln 10$	n	$n\pi$
1	2. 3025	1	3. 1415
2	4. 6051	2	6. 2831
3	6. 9077	3	9. 4247
4	9. 2103	4	(1) 1. 2566
5	(1) 1. 1512	5	(1) 1. 5707
6	(1) 1. 3815	6	(1) 1. 8849
7	(1) 1. 6118	7	(1) 2. 1991
8	(1) 1. 8420	8	(1) 2. 5132
9	(1) 2. 0723	9	(1) 2. 8274

n	π^n	n	π^{-n}
1	3. 1415	1	(−1) 3. 1830
2	9. 8696	2	(−1) 1. 0132
3	(1) 3. 1006	3	(−2) 3. 2251
4	(1) 9. 7409	4	(−2) 1. 0265
5	(2) 3. 0601	5	(−3) 3. 2677
6	(2) 9. 6138	6	(−3) 1. 0401
7	(3) 3. 0202	7	(−4) 3. 3109
8	(3) 9. 4885	8	(−4) 1. 0539
9	(4) 2. 9809	9	(−5) 3. 3546
10	(4) 9. 3648	10	(−5) 1. 0678

$\pi/2$	1. 5707	$3\pi/2$	4. 7123
$\pi/3$	1. 0471	$4\pi/3$	4. 1887
$\pi/4$	(−1) 7. 8539	$\pi(2)^{1/2}$	4. 4428
$\pi^{1/2}$	1. 7724	$\pi^{-1/2}$	(−1) 5. 6418
$\pi^{1/3}$	1. 4645	$\pi^{-1/3}$	(−1) 6. 8278
$\pi^{1/4}$	1. 3313	$\pi^{-1/4}$	(−1) 7. 5112
$\pi^{2/3}$	2. 1450	$\pi^{-2/3}$	(−1) 4. 6619
$\pi^{3/4}$	2. 3597	$\pi^{-3/4}$	(−1) 4. 2377
$\pi^{3/2}$	5. 5683	$\pi^{-3/2}$	(−1) 1. 7958
π^{e}	(1) 2. 2459	π^{-e}	(−2) 4. 4525
$(2\pi)^{1/2}$	2. 5066	$(2\pi)^{-1/2}$	(−1) 3. 9894
$(\pi/2)^{1/2}$	1. 2533	$(2/\pi)^{1/2}$	(−1) 7. 9788
$\pi(2)^{-1/2}$	2. 2214	$2^{1/2}/\pi$	(−1) 4. 5015
$1r$	57. 2957	$1'$	0. 0002
$1°$	0. 0174	$1''$	0. 0000

$\Gamma(1/2)$	1. 7724	$1/\Gamma(1/2)$	0. 5641
$\Gamma(1/3)$	2. 6789	$1/\Gamma(1/3)$	0. 3732
$\Gamma(2/3)$	1. 3541	$1/\Gamma(2/3)$	0. 7384
$\Gamma(1/4)$	3. 6256	$1/\Gamma(1/4)$	0. 2758
$\Gamma(3/4)$	1. 2254	$1/\Gamma(3/4)$	0. 8160
$\Gamma(4/3)$	0. 8929	$1/\Gamma(4/3)$	1. 1198
$\Gamma(5/3)$	0. 9027	$1/\Gamma(5/3)$	1. 1077
$\Gamma(5/4)$	0. 9064	$1/\Gamma(5/4)$	1. 1032
$\Gamma(7/4)$	0. 9190	$1/\Gamma(7/4)$	1. 0880
$\ln \Gamma(1/3)$	0. 9854	$\ln \Gamma(4/3)$	−0. 1131
$\ln \Gamma(2/3)$	0. 3031	$\ln \Gamma(5/3)$	−0. 1023
$\ln \Gamma(1/4)$	1. 2880	$\ln \Gamma(5/4)$	−0. 0982
$\ln \Gamma(3/4)$	0. 2032	$\ln \Gamma(7/4)$	−0. 0844

CHAPTER 2

ALGEBRA

LAWS OF ALGEBRAIC OPERATIONS

(a) **Commutative law:** $a + b = b + a, \quad ab = ba;$
(b) **Associative law:** $a + (b + c) = (a + b) + c, \quad a(bc) = (ab)c;$
(c) **Distributive law:** $c(a + b) = ca + cb.$

SPECIAL PRODUCTS AND FACTORS

$$(x+y)^2 = x^2 + 2xy + y^2$$
$$(x-y)^2 = x^2 - 2xy + y^2$$
$$(x+y)^3 = x^3 + 3x^2y + 3xy^2 + y^3$$
$$(x-y)^3 = x^3 - 3x^2y + 3xy^2 - y^3$$
$$(x+y)^4 = x^4 + 4x^3y + 6x^2y^2 + 4xy^3 + y^4$$
$$(x-y)^4 = x^4 - 4x^3y + 6x^2y^2 - 4xy^3 + y^4$$
$$(x+y)^5 = x^5 + 5x^4y + 10x^3y^2 + 10x^2y^3 + 5xy^4 + y^5$$
$$(x-y)^5 = x^5 - 5x^4y + 10x^3y^2 - 10x^2y^3 + 5xy^4 - y^5$$
$$(x+y)^6 = x^6 + 6x^5y + 15x^4y^2 + 20x^3y^3 + 15x^2y^4 + 6xy^5 + y^6$$
$$(x-y)^6 = x^6 - 6x^5y + 15x^4y^2 - 20x^3y^3 + 15x^2y^4 - 6xy^5 + y^6$$

The results above are special cases of the *binomial formula.*

$$x^2 - y^2 = (x-y)(x+y)$$
$$x^3 - y^3 = (x-y)(x^2 + xy + y^2)$$
$$x^3 + y^3 = (x+y)(x^2 - xy + y^2)$$
$$x^4 - y^4 = (x-y)(x+y)(x^2 + y^2)$$
$$x^5 - y^5 = (x-y)(x^4 + x^3y + x^2y^2 + xy^3 + y^4)$$
$$x^5 + y^5 = (x+y)(x^4 - x^3y + x^2y^2 - xy^3 + y^4)$$

$$x^6 - y^6 = (x-y)(x+y)(x^2+xy+y^2)(x^2-xy+y^2)$$
$$x^4 + x^2y^2 + y^4 = (x^2+xy+y^2)(x^2-xy+y^2)$$
$$x^4 + 4y^4 = (x^2+2xy+2y^2)(x^2-2xy+2y^2)$$

Some generalizations of the above are given by the following results where n is a positive integer

$$\begin{aligned} x^{2n+1} - y^{2n+1} &= (x-y)(x^{2n} + x^{2n-1}y + x^{2n-2}y^2 + \cdots + y^{2n}) \\ &= (x-y)\left(x^2 - 2xy\cos\frac{2\pi}{2n+1} + y^2\right)\left(x^2 - 2xy\cos\frac{4\pi}{2n+1} + y^2\right) \\ &\qquad \cdots\left(x^2 - 2xy\cos\frac{2n\pi}{2n+1} + y^2\right) \end{aligned}$$

$$\begin{aligned} x^{2n+1} + y^{2n+1} &= (x+y)(x^{2n} - x^{2n-1}y + x^{2n-2}y^2 - \cdots + y^{2n}) \\ &= (x+y)\left(x^2 + 2xy\cos\frac{2\pi}{2n+1} + y^2\right)\left(x^2 + 2xy\cos\frac{4\pi}{2n+1} + y^2\right) \\ &\qquad \cdots\left(x^2 + 2xy\cos\frac{2n\pi}{2n+1} + y^2\right) \end{aligned}$$

$$\begin{aligned} x^{2n} - y^{2n} &= (x-y)(x+y)(x^{n-1} + x^{n-2}y + x^{n-3}y^2 + \cdots)(x^{n-1} - x^{n-2}y + x^{n-3}y^2 - \cdots) \\ &= (x-y)(x+y)\left(x^2 - 2xy\cos\frac{\pi}{n} + y^2\right)\left(x^2 - 2xy\cos\frac{2\pi}{n} + y^2\right) \\ &\qquad \cdots\left(x^2 - 2xy\cos\frac{(n-1)\pi}{n} + y^2\right) \end{aligned}$$

$$\begin{aligned} x^{2n} + y^{2n} &= \left(x^2 + 2xy\cos\frac{\pi}{2n} + y^2\right)\left(x^2 + 2xy\cos\frac{3\pi}{2n} + y^2\right) \\ &\qquad \cdots\left(x^2 + 2xy\cos\frac{(2n-1)\pi}{2n} + y^2\right) \end{aligned}$$

POWERS AND ROOTS

$a^x \times a^y = a^{(x+y)}.$ $\qquad a^0 = 1$ [if $a \neq 0$] $\qquad (ab)^x = a^x b^x.$

$\dfrac{a^x}{a^y} = a^{(x-y)}.$ $\qquad a^{-x} = \dfrac{1}{a^x}.$ $\qquad \left(\dfrac{a}{b}\right)^x = \dfrac{a^x}{b^x}.$

$(a^x)^y = a^{xy}.$ $\qquad a^{\frac{1}{x}} = \sqrt[x]{a}.$ $\qquad \sqrt[x]{ab} = \sqrt[x]{a}\sqrt[x]{b}.$

$\sqrt[x]{\sqrt[y]{a}} = \sqrt[xy]{a}.$ $\qquad a^{\frac{x}{y}} = \sqrt[y]{a^x}.$ $\qquad \sqrt[x]{\dfrac{a}{b}} = \dfrac{\sqrt[x]{a}}{\sqrt[x]{b}}$

PROPORTION

If $\dfrac{a}{b} = \dfrac{c}{d}$, then $\dfrac{a+b}{b} = \dfrac{c+d}{d}$,

$\dfrac{a-b}{b} = \dfrac{c-d}{d}$, $\qquad \dfrac{a-b}{a+b} = \dfrac{c-d}{c+d}$.

SUM OF ARITHMETIC PROGRESSION TO n TERMS[1]

$$a+(a+d)+(a+2d)+\ldots+(a+(n-1)d)$$
$$=na+\frac{1}{2}n(n-1)d=\frac{n}{2}(a+l),$$

last term in series $=l=a+(n-1)d$

SUM OF GEOMETRIC PROGRESSION TO n TERMS

$$s_n=a+ar+ar^2+\ldots+ar^{n-1}=\frac{a(1-r^n)}{1-r}$$

$$\lim_{n\to\infty} s_n=a/(1-r) \qquad (-1<r<1)$$

ARITHMETIC MEAN OF n QUANTITIES A

$$A=\frac{a_1+a_2+\ldots+a_n}{n}$$

GEOMETRIC MEAN OF n QUANTITIES G

$$G=(a_1a_2\ldots a_n)^{1/n} \qquad (a_k>0, k=1,2,\ldots,n)$$

HARMONIC MEAN OF n QUANTITIES H

$$\frac{1}{H}=\frac{1}{n}\left(\frac{1}{a_1}+\frac{1}{a_2}+\cdots+\frac{1}{a_n}\right) \qquad (a_k>0, k=1,2,\ldots,n)$$

GENERALIZED MEAN

$$M(t)=\left(\frac{1}{n}\sum_{k=1}^{n} a_k^t\right)^{1/t}$$

$M(t)=0(t<0$, some a_k zero)

$$\lim_{t\to\infty} M(t)=\max. \qquad (a_1, a_2, \ldots, a_n)=\max. a$$

$$\lim_{t\to-\infty} M(t)=\min. \qquad (a_1, a_2, \ldots, a_n)=\min. a$$

$$\lim_{t\to 0} M(t)=G$$

$$M(1)=A$$

$$M(-1)=H$$

---

[1]From "Handbook of Mathematical Functions", National Bureau of Standards

SOLUTION OF QUADRATIC EQUATIONS

Given $az^2+bz+c=0$,

$$z_{1,2}=-\left(\frac{b}{2a}\right)\pm\frac{1}{2a}\,q^{\frac{1}{2}},\ q=b^2-4ac,$$

$$z_1+z_2=-b/a,\ z_1z_2=c/a$$

If $q>0$, two real roots,
$q=0$, two equal roots,
$q<0$, pair of complex conjugate roots.

SOLUTION OF CUBIC EQUATIONS

Given $z^3+a_2z^2+a_1z+a_0=0$, let

$$q=\frac{1}{3}\,a_1-\frac{1}{9}\,a_2^2;\ r=\frac{1}{6}\,(a_1a_2-3a_0)-\frac{1}{27}\,a_2^3.$$

If $q^3+r^2>0$, one real root and a pair of complex conjugate roots,

$q^3+r^2=0$, all roots real and at least two are equal,

$q^3+r^2<0$, all roots real (irreducible case).

Let

$$s_1=[r+(q^3+r^2)^{\frac{1}{2}}]^{\frac{1}{3}},\ s_2=[r-(q^3+r^2)^{\frac{1}{2}}]^{\frac{1}{3}}$$

then

$$z_1=(s_1+s_2)-\frac{a_2}{3}$$

$$z_2=-\frac{1}{2}\,(s_1+s_2)-\frac{a_2}{3}+\frac{i\sqrt{3}}{2}\,(s_1-s_2)$$

$$z_3=-\frac{1}{2}\,(s_1+s_2)-\frac{a_2}{3}-\frac{i\sqrt{3}}{2}\,(s_1-s_2).$$

If z_1, z_2, z_3 are the roots of the cubic equation

$$z_1+z_2+z_3=-a_2$$

$$z_1z_2+z_1z_3+z_2z_3=a_1$$

$$z_1z_2z_3=-a_0$$

TRIGONOMETRIC SOLUTION OF THE CUBIC EQUATION

The form $x^3 + ax + b = 0$ with $ab \neq 0$ can always be solved by transforming it to the trignometric identity

$$4\cos^3\theta - 3\cos\theta - \cos(3\theta) \equiv 0.$$

Let $x = m\cos\theta$, then

$$x^3 + ax + b \equiv m^3\cos^3\theta + am\cos\theta + b \equiv 4\cos^3\theta - 3\cos\theta - \cos(3\theta) \equiv 0.$$

Hence

$$\frac{4}{m^3} = -\frac{3}{am} = \frac{-\cos(3\theta)}{b},$$

from which follows that

$$m = 2\sqrt{-\frac{a}{3}}, \quad \cos(3\theta) = \frac{3b}{am}.$$

Any solution θ_1 which satisfies $\cos(3\theta) = \dfrac{3b}{am}$, will also have the solutions

$$\theta_1 + \frac{2\pi}{3} \quad \text{and} \quad \theta_1 + \frac{4\pi}{3}.$$

The roots of the cubic $x^3 + ax + b = 0$ are

$$2\sqrt{-\frac{a}{3}}\cos\theta_1, \quad 2\sqrt{-\frac{a}{3}}\cos\left(\theta_1 + \frac{2\pi}{3}\right), \quad 2\sqrt{-\frac{a}{3}}\cos\left(\theta_1 + \frac{4\pi}{3}\right).$$

SOLUTION OF QUARTIC EQUATIONS

Given $z^4+a_3z^3+a_2z^2+a_1z+a_0=0$, find the real root u_1 of the cubic equation

$$u^3-a_2u^2+(a_1a_3-4a_0)u-(a_1^2+a_0a_3^2-4a_0a_2)=0$$

and determine the four roots of the quartic as solutions of the two quadratic equations

$$v^2+\left[\frac{a_3}{2}\mp\left(\frac{a_3^2}{4}+u_1-a_2\right)^{\frac{1}{2}}\right]v+\frac{u_1}{2}\mp\left[\left(\frac{u_1}{2}\right)^2-a_0\right]^{\frac{1}{2}}=0$$

If all roots of the cubic equation are real, use the value of u_1 which gives real coefficients in the quadratic equation and select signs so that if

$$z^4+a_3z^3+a_2z^2+a_1z+a_0=(z^2+p_1z+q_1)(z^2+p_2z+q_2),$$

then

$$p_1+p_2=a_3, p_1p_2+q_1+q_2=a_2, p_1q_2+p_2q_1=a_1, q_1q_2=a_0.$$

If z_1, z_2, z_3, z_4 are the roots,

$$\Sigma z_i=-a_3, \Sigma z_iz_jz_k=-a_1,$$

$$\Sigma z_iz_j=a_2, z_1z_2z_3z_4=a_0.$$

PARTIAL FRACTIONS

This section applies only to rational algebraic fractions with numerator of lower degree than the denominator. Improper fractions can be reduced to proper fractions by long division.

Every fraction may be expressed as the sum of component fractions whose denominators are factors of the denominator of the original fraction.

Let $N(x)$ = numerator, a polynomial of the form

$$N(x) = n_0 + n_1x + n_2x^2 + \cdots + n_ix^i$$

NON-REPEATED LINEAR FACTORS

$$\frac{N(x)}{(x-a)G(x)} = \frac{A}{x-a} + \frac{F(x)}{G(x)}$$

$$A = \left[\frac{N(x)}{G(x)}\right]_{x=a}$$

$F(x)$ determined by methods discussed in the following sections.

REPEATED LINEAR FACTORS

$$\frac{N(x)}{x^mG(x)} = \frac{A_0}{x^m} + \frac{A_1}{x^{m-1}} + \cdots + \frac{A_{m-1}}{x} + \frac{F(x)}{G(x)}$$

$$N(x) = n_0 + n_1x + n_2x^2 + n_3x^3 + \ldots$$

$$F(x) = f_0 + f_1x + f_2x^2 + \cdots,\ G(x) = g_0 + g_1x + g_2x^2 + \cdots$$

$$A_0 = \frac{n_0}{g_0},\quad A_1 = \frac{n_1 - A_0g_1}{g_0},\quad A_2 = \frac{n_2 - A_0g_2 - A_1g_1}{g_0}$$

General term:

$$A_0 = \frac{n_0}{g_0},\ A_k = \frac{1}{g_0}\left[n_k - \sum_{i=0}^{k-1} A_ig_{k-i}\right] k \geq 1$$

$$m^* = 1\begin{cases} f_0 = n_1 - A_0g_1 \\ f_1 = n_2 - A_0g_2 \\ f_j = n_{j+1} - A_0g_{j+1} \end{cases}$$

$$m = 2 \begin{cases} f_0 = n_2 - A_0 g_2 - A_1 g_1 \\ f_1 = n_3 - A_0 g_3 - A_1 g_2 \\ f_j = n_{j+2} - [A_0 g_{j+2} + A_1 g_{j+1}] \end{cases}$$

$$m = 3 \begin{cases} f_0 = n_3 - A_0 g_3 - A_1 g_2 - A_2 g_1 \\ f_1 = n_3 - A_0 g_4 - A_1 g_3 - A_2 g_2 \\ f_j = n_{j+3} - [A_0 g_{j+3} + A_1 g_{j+2} + A_2 g_{j+1}] \end{cases}$$

$$\text{any } m\text{: } f_j = n_{m+j} - \sum_{i=0}^{m-1} A_i g_{m+j-i}$$

$$\frac{N(x)}{(x-a)^m G(x)} = \frac{A_0}{(x-a)^m} + \frac{A_1}{(x-a)^{m-1}} + \cdots + \frac{A_{m-1}}{(x-a)} + \frac{F(x)}{G(x)}$$

Change to form $\dfrac{N'(y)}{y^m G'(y)}$ by substitution of $x = y + a$. Resolve into partial fractions in terms of y as described above. Then express in terms of x by substitution $y = x - a$.

REPEATED LINEAR FACTORS

Alternative method of determining coefficients:

$$\frac{N(x)}{(x-a)^m G(x)} = \frac{A_0}{(x-a)^m} + \cdots + \frac{A_k}{(x-a)^{m-k}} + \cdots + \frac{A_{m-1}}{x-a} + \frac{F(x)}{G(x)}$$

$$A_k = \frac{1}{k!}\left\{D_x^k\left[\frac{N(x)}{G(x)}\right]\right\}_{x=a}$$

where D_x^k is the differentiating operator, and the derivative of zero order is defined as:

$$D_x^0 u = u.$$

FACTORS OF HIGHER DEGREE

Factors of higher degree have the corresponding numerators indicated.

$$\frac{N(x)}{(x^2 + h_1 x + h_0)G(x)} = \frac{a_1 x + a_0}{x^2 + h_1 x + h_0} + \frac{F(x)}{G(x)}$$

$$\frac{N(x)}{(x^2 + h_1 x + h_0)^2 G(x)} = \frac{a_1 x + a_0}{(x^2 + h_1 x + h_0)^2} + \frac{b_1 x + b_0}{(x^2 + h_1 x + h_0)} + \frac{F(x)}{G(x)}$$

$$\frac{N(x)}{(x^3 + h_2 x^2 + h_1 x + h_0)G(x)} = \frac{a_2 x^2 + a_1 x + a_0}{x^3 + h_2 x^2 + h_1 x + h_0} + \frac{F(x)}{G(x)}$$

etc.

Problems of this type are determined first by solving for the coefficients due to linear factors as shown above, and then determining the remaining coefficients by the general methods given below.

GENERAL METHODS FOR EVALUATING COEFFICIENTS

1. $$\frac{N(x)}{D(x)} = \frac{N(x)}{G(x)H(x)L(x)} = \frac{A(x)}{G(x)} + \frac{B(x)}{H(x)} + \frac{C(x)}{L(x)} + \cdots$$

Multiply both sides of equation by $D(x)$ to clear fractions. Then collect terms, equate like powers of x, and solve the resulting simultaneous equations for the unknown coefficients.

2. Clear fractions as above. Then let x assume certain convenient values ($x = 1, 0, -1, \ldots$). Solve the resulting equations for the unknown coefficients.

3. $$\frac{N(x)}{G(x)H(x)} = \frac{A(x)}{G(x)} + \frac{B(x)}{H(x)}$$

Then

$$\frac{N(x)}{G(x)H(x)} - \frac{A(x)}{G(x)} = \frac{B(x)}{H(x)}$$

If $A(x)$ can be determined, such as by Method 1, then $B(x)$ can be found as above.

MATRICES

DEFINITION 1. Matrix of order m×n. A *matrix*, $A_{m\times n}$, *of order* $m \times n$ (read "m by n") is a rectangular array of $m \times n$ numbers (or functions) called *elements* arranged in m rows and n columns subject to a set of rules of operation (to be defined) and is denoted by the symbol

$$A_{m\times n} = \begin{bmatrix} a_{11} & a_{12} & \cdots & a_{1n} \\ a_{21} & a_{22} & \cdots & a_{2n} \\ \cdot & & & \\ \cdot & & & \\ \cdot & & & \\ a_{m1} & a_{m2} & \cdots & a_{mn} \end{bmatrix} \qquad [2.1]$$

where a_{ij} is the element of $A_{m\times n}$ located in the ith row and the jth column.

We shall use upper-case letters to represent matrices. When the order is immaterial in the discussion, a single letter A, without subscripts, will be used to denote a matrix. We should point out also that when the order is the only information required, expression [2.1] may then be abbreviated as

$$A = [a_{ij}]_{m\times n}$$

which gives essentially the same information as that in Equation [2.1]. The element a_{ij} is called the typical element or the (i,j) element. As in the case of a determinant, the first subscript i of a_{ij} indicates its row position, and the second subscript, j, shows its column position.

The reader should note that a determinant is always a square array; whereas, a matrix, say $A_{m\times n}$, is, in general, rectangular in shape unlesss $m = n$; that is, unless the number of rows equal the number of columns. In this case, $A_{m \times n}$ becomes a square matrix and a single-subscripted letter A_n may be used to indicate such information.

DEFINITION 2. Real matrix. A matrix $R = [r_{ij}]_{m \times n}$ is called a *real matrix* if and only if r_{ij} is real for every pair of i and j; that is, R is a real matrix if and only if *all* its elements are real.

In general, the elements of a matrix may be real or complex (or even functions). If one of its elements is complex, the matrix is a "complex" matrix.

DEFINITION 3. Conjugate of a matrix. Let $A = [a_{ij}]_{m \times n}$ be a matrix of complex elements of order mn. If every element $a_{ij} = \alpha_{ij} + \beta_{ij}\sqrt{-1}$ (α, β real) of A is replaced by its complex conjugate $\bar{a}_{ij} = \alpha_{ij} - \beta_{ij}\sqrt{-1}$, the resulting matrix, denoted by $\bar{A}$, is defined as the *conjugate of A*.

DEFINITION 4. Row and column matrices. A *row matrix* of m elements is a matrix of order $1 \times m$; and a *column matrix* of n elements is a

DEFINITION 5. Scalar matrix. A *scalar matrix* is defined as a diagonal matrix such that all the elements in the major diagonal are equal to a scalar. In other words, a square matrix $K = [k_{ij}]_n$ of order n is called a diagonal matrix if and only if $k_{ii} = \alpha$, a scalar for all i; and $k_{ij} = 0$ for all $i \neq j$.

DEFINITION 6. Submatrix of a matrix. A *submatrix* of a given matrix is the resulting matrix that is obtained by deleting certain rows and/or columns of the given matrix.

DEFINITION 7. Rank of a matrix. The *rank* of a matrix A of order $m \times n$ is defined to be the order of the determinant of a highest-order nonsingular submatrix A_r of A. We denote the rank of A by $R(A) = r$ and say that matrix A is of rank r.

DEFINITION 8. Elementary transformations of a matrix. The following three operations are called the *elementary transformations* of a matrix:

(1) Interchange of any two rows (or columns).

(2) Multiplication of all the elements of any row (or column) by a nonzero constant.

(3) Addition of the elements of a row (or column) multiplied by a nonzero constant to the corresponding elements of another row (or column).

THEOREM 1. The rank of a matrix will remain unaltered under (repeated applications of) any or all of the three operations defined in Definition 8

DEFINITION 9. Equality of matrices. Two matrices, $A = [a_{ij}]_{m \times n}$ and $B = [b_{ij}]_{p \times q}$, are said to be *equal* if and only if

(a) A and B are of the same order: $m = p$, $n = q$; and

(b) $a_{ij} = b_{ij}$ for every pair of i and j.
In short, A and B are equal if and only if they are of the same order and their corresponding elements are equal one by one.

DEFINITION 10. Conformability for addition. Two matrices A and B are said to be *conformable* for addition if they both have the same order.

DEFINITION 11. Addition of matrices. If two matrices A and B are conformable for addition and if $A = [a_{ij}]_{m \times n}$ and $B = [b_{ij}]_{m \times n}$, then the *sum*, $A + B$, is defined as $A + B = [(a_{ij} + b_{ij})]_{m \times n}$.

DEFINITION 12. Negative of a matrix. If $A = [a_{ij}]_{m \times n}$, then the *negative* of A, denoted by $-A$, is defined as $-A = [-a_{ij}]_{m \times n}$; that is, the negative of a matrix is obtained by taking the negative of every element of the matrix.

DEFINITION 13. Zero (null) matrix. A matrix of order $m \times n$ is called a *zero matrix* (or *null matrix*) if all its elements are equal to zero.

DEFINITION 14. Subtraction of matrices. If $A = [a_{ij}]_{m \times n}$ and $B = [b_{ij}]_{m \times n}$, then the *difference*, $A - B$, is defined as $A = B = A + (-B) = [(a_{ij} - b_{ij})]_{m \times n}$.

THEOREM 2. If A, B, and C are three matrices of the same order, then both the commutative and associative laws for addition hold for the three matrices; that is,

$$A + B = B + A \quad \text{(commutative law)}$$

$$A + (B + C) = (A + B) + C \quad \text{(associative law)}$$

DEFINITION 15. Scalar multiple of a matrix. *The product of a matrix* $A = [a_{ij}]_{m \times n}$ *and a scalar* k is the matrix $kA = [ka_{ij}]_{m \times n}$; that is, when a matrix A is multiplied by a scalar k, every element of A is multiplied by k.

DEFINITION 16. Conformability for multiplication. Two matrices, $A = [a_{ij}]_{m \times n}$ and $B = [b_{ij}]_{p \times q}$, are said to be *conformable for multiplication* in the order AB if and only if $n = p$. In other words, A and B are conformable for multiplication in the order AB if and only if the number of columns in A is equal to the number of rows in B.

DEFINITION 17. Product of two matrices. *The product* C *of two matrices*, $A = [a_{ij}]_{m \times n}$ and $B = [b_{ij}]_{n \times q}$, which are conformable for multiplication in the order AB is defined by

$$A \times B = C = [c_{ij}]_{m \times q}$$

where

$$c_{ij} = a_{i1}b_{1j} + a_{i2}b_{2j} + \cdots + a_{in}b_{nj} = \sum_{k=1}^{n} a_{ik}b_{kj}$$

In other words, we say that the (i,j) element c_{ij} of the product matrix $C = A \times B$ is the sum of the products of the elements in the ith row of A and the corresponding elements in the jth column of B.

In general, matrix multiplication is not commutative

$$\mathbf{AB} \neq \mathbf{BA}$$

Matrix multiplication is associative

$$\mathbf{A(BC)} = \mathbf{(AB)C}$$

The distributive law for multiplication and addition holds as in the case of scalars,

$$\mathbf{(A + B)C} = \mathbf{AC} + \mathbf{BC}$$
$$\mathbf{C(A + B)} = \mathbf{CA} + \mathbf{CB}$$

In some applications, the term-by-term product of two matrices **A** and **B** of identical order is defined as

$$\mathbf{C} = \mathbf{A} * \mathbf{B}$$

where

$$c_{ij} = a_{ij}b_{ij}$$

$(\mathbf{ABC})' = \mathbf{C}'\mathbf{B}'\mathbf{A}'$
$(\mathbf{ABC})^H = \mathbf{C}^H\mathbf{B}^H\mathbf{A}^H$

If both **A** and **B** are symmetric, then $(\mathbf{AB})' = \mathbf{BA}$. Note that the product of two symmetric matrices is generally not symmetric.

DEFINITION 18. Transpose of a matrix. The matrix A^T of order $n \times m$, obtained by interchanging the rows and columns of an $m \times n$ matrix A, is defined as the *transpose* of A. In symbolic terms, if $A = [a_{ij}]_{m \times n}$, then $A^T = [a_{ji}]_{n \times m}$.

Property 1: The transpose of a matrix always exists.

Property 2: The orders of a matrix and its transpose are the same if and only if the matrix is a square matrix.

Property 3: The transpose of the transpose of a matrix is the matrix itself; that is, $(A^T)^T = A$.

Property 4: The transpose of the scalar multiple of a matrix is equal to the scalar multiple of its transpose; that is, $(kA)^T = kA^T$.

Property 5: The transpose of the sum of two matrices is equal to the sum of their transposes. Thus, if A and B are two matrices of the same order, we have $(A + B)^T = A^T + B^T$.

Property 6: The transpose of the product of two matrices is equal to the product of the two transposes *in the reverse order*. That is, if A and B are conformable for multiplication, we write $(AB)^T = B^T A^T$.

DETERMINANTS

1. A determinant $|\mathbf{A}|$ or $\det(\mathbf{A})$ is a scalar function of a square matrix defined in such a way that

$$|\mathbf{A}|\,|\mathbf{B}| = |\mathbf{AB}|$$

and

$$\begin{vmatrix} a_{11} & a_{12} \\ a_{21} & a_{22} \end{vmatrix} = a_{11}a_{22} - a_{12}a_{21}$$

$|\mathbf{A}| = |\mathbf{A}'|$

$$\begin{vmatrix} a_{11} & a_{12} & a_{13} \\ a_{21} & a_{22} & a_{23} \\ a_{31} & a_{32} & a_{33} \end{vmatrix} = a_{11}a_{22}a_{33} + a_{12}a_{23}a_{31} + a_{13}a_{21}a_{32} - a_{13}a_{22}a_{31} - a_{11}a_{23}a_{32} - a_{12}a_{21}a_{33}$$

$$\begin{vmatrix} a_{11} & a_{12} & \cdots & a_{1n} \\ a_{21} & a_{22} & \cdots & a_{2n} \\ & & \vdots & \\ a_{n1} & a_{n2} & \cdots & a_{nm} \end{vmatrix} = \sum(-1)^{\delta} a_{1i_1} a_{2i_2} \cdots a_{ni_n}$$

where the sum is over all permutations

$$i_1 \neq i_2 \neq \cdots i_n$$

and δ denotes the number of exchanges necessary to bring the sequence $(i_1, i_2, \ldots i_n)$ back into the natural order $(1, 2, \ldots n)$.

2. If two rows (columns) in a matrix are exchanged, the determinant will change its sign.

3. A determinant does not change its value if a linear combination of other rows (columns) is added to any given row (column).

$\gamma_1, \gamma_3, \gamma_4$ arbitrary.

4. If the i'th row (column) equals (a constant times) the j'th row (column) of a matrix, its determinant is equal to zero, $(i \neq j)$.

5. If, in a matrix $\mathbf{A}$, each element of a row (column) is multiplied by a constant γ, the determinant is multiplied by γ.

6. $|\gamma\mathbf{A}| = \gamma^n |\mathbf{A}|$ assuming that $\mathbf{A}$ is of order $(n \times n)$.

7. The cofactor of a square matrix $\mathbf{A}$, $\mathrm{cof}_{ij}(\mathbf{A})$ is the determinant of a matrix obtained by striking the i'th row and j'th column of $\mathbf{A}$ and choosing positive (negative) sign if $i + j$ is even (odd).

8. **(Laplace Development)**

$$|\mathbf{A}| = a_{i1}\mathrm{cof}_{i1}(\mathbf{A}) + a_{i2}\mathrm{cof}_{i2}(\mathbf{A}) + \cdots + a_{in}\mathrm{cof}_{in}(\mathbf{A})$$
$$= a_{1j}\mathrm{cof}_{1j}(\mathbf{A}) + a_{2j}\mathrm{cof}_{2j}(\mathbf{A}) + \cdots + a_{nj}\mathrm{cof}_{nj}(\mathbf{A})$$

for any row i or any column j.

DEFINITION 19. Singular and nonsingular matrices. A *singular matrix* is a square matrix whose determinant is equal to zero, whereas a *nonsingular matrix* is a square matrix whose determinant is different from zero.

DEFINITION 20. Symmetric matrix. A *square* matrix $S_n = [s_{ij}]_n$ of order n is called a *symmetric* matrix if and only if $s_{ij} = s_{ji}$ for every pair of i and j.

From this definition it is seen that a matrix A is symmetric if and only if $A = A^T$.

DEFINITION 21. Diagonal matrix. A *square* matrix $D_n = [d_{ij}]_n$ of order n is called a *diagonal* matrix if and only if $d_{ij} = 0$ for $i \neq j$.

DEFINITION 22. Unit (or identity) matrix. A *square* matrix $U_n = [u_{ij}]_n$ of order n is defined as a *unit matrix* if and only if $u_{ii} = 1$ for all i, and $u_{ij} = 0$ for all i and j where $i \neq j$. (Note that I_n is used to denote an identity matrix. Thus $I_n \equiv U_n$, and the two symbols represent the same matrix.)

DEFINITION 23. Inverse. The *inverse* of a given matrix A is defined as a matrix A^{-1} such that the products of the two matrices in both orders yield the identity matrix I; that is,

$$A \cdot A^{-1} = A^{-1} \cdot A = I$$

Property 1: The inverse of a matrix, if it exists, is unique.

Property 2: The inverse of a matrix exists if and only if (a) the matrix is square and (b) the determinant of the matrix is different from zero; that is, the matrix is nonsingular.

Of course, the second condition implies the first since, by definition, both singular and nonsingular matrices must be square matrices.[1]

Property 3: If A is a nonsingular matrix (so that A^{-1} exists), then

$$(A^T)^{-1} = (A^{-1})^T$$

Property 4: If A and B are two nonsingular matrices of the same order, then

$$(A \cdot B)^{-1} = B^{-1} \cdot A^{-1}$$

CHAPTER 3

ELEMENTARY ANALYSIS

SQUARES, SQUARE ROOT, CUBES AND CUBE ROOT

n	n^2	$\sqrt{n}$	$\sqrt{10n}$	n^3	$\sqrt[3]{n}$	$\sqrt[3]{10n}$	$\sqrt[3]{100n}$
1	1	1.000 000	3.162 278	1	1.000 000	2.154 435	4.641 589
2	4	1.414 214	4.472 136	8	1.259 921	2.714 418	5.848 035
3	9	1.732 051	5.477 226	27	1.442 250	3.107 233	6.694 330
4	16	2.000 000	6.324 555	64	1.587 401	3.419 952	7.368 063
5	25	2.236 068	7.071 068	125	1.709 976	3.684 031	7.937 005
6	36	2.449 490	7.745 967	216	1.817 121	3.914 868	8.434 327
7	49	2.645 751	8.366 600	343	1.912 931	4.121 285	8.879 040
8	64	2.828 427	8.944 272	512	2.000 000	4.308 869	9.283 178
9	81	3.000 000	9.486 833	729	2.080 084	4.481 405	9.654 894
10	100	3.162 278	10.00000	1 000	2.154 435	4.641 589	10.00000
11	121	3.316 625	10.48809	1 331	2.223 980	4.791 420	10.32280
12	144	3.464 102	10.95445	1 728	2.289 428	4.932 424	10.62659
13	169	3.605 551	11.40175	2 197	2.351 335	5.065 797	10.91393
14	196	3.741 657	11.83216	2 744	2.410 142	5.192 494	11.18689
15	225	3.872 983	12.24745	3 375	2.466 212	5.313 293	11.44714
16	256	4.000 000	12.64911	4 096	2.519 842	5.428 835	11.69607
17	289	4.123 106	13.03840	4 913	2.571 282	5.539 658	11.93483
18	324	4.242 641	13.41641	5 832	2.620 741	5.646 216	12.16440
19	361	4.358 899	13.78405	6 859	2.668 402	5.748 897	12.38562
20	400	4.472 136	14.14214	8 000	2.714 418	5.848 035	12.59921
21	441	4.582 576	14.49138	9 261	2.758 924	5.943 922	12.80579
22	484	4.690 416	14.83240	10 648	2.802 039	6.036 811	13.00591
23	529	4.795 832	15.16575	12 167	2.843 867	6.126 926	13.20006
24	576	4.898 979	15.49193	13 824	2.884 499	6.214 465	13.38866
25	625	5.000 000	15.81139	15 625	2.924 018	6.299 605	13.57209
26	676	5.099 020	16.12452	17 576	2.962 496	6.382 504	13.75069
27	729	5.196 152	16.43168	19 683	3.000 000	6.463 304	13.92477
28	784	5.291 503	16.73320	21 952	3.036 589	6.542 133	14.09460
29	841	5.385 165	17.02939	24 389	3.072 317	6.619 106	14.26043
30	900	5.477 226	17.32051	27 000	3.107 233	6.694 330	14.42250
31	961	5.567 764	17.60682	29 791	3.141 381	6.767 899	14.58100
32	1 024	5.656 854	17.88854	32 768	3.174 802	6.839 904	14.73613
33	1 089	5.744 563	18.16590	35 937	3.207 534	6.910 423	14.88806
34	1 156	5.830 952	18.43909	39 304	3.239 612	6.979 532	15.03695
35	1 225	5.916 080	18.70829	42 875	3.271 066	7.047 299	15.18294
36	1 296	6.000 000	18.97367	46 656	3.301 927	7.113 787	15.32619
37	1 369	6.082 763	19.23538	50 653	3.332 222	7.179 054	15.46680
38	1 444	6.164 414	19.49359	54 872	3.361 975	7.243 156	15.60491
39	1 521	6.244 998	19.74842	59 319	3.391 211	7.306 144	15.74061
40	1 600	6.324 555	20.00000	64 000	3.419 952	7.368 063	15.87401
41	1 681	6.403 124	20.24846	68 921	3.448 217	7.428 959	16.00521
42	1 764	6.480 741	20.49390	74 088	3.476 027	7.488 872	16.13429
43	1 849	6.557 439	20.73644	79 507	3.503 398	7.547 842	16.26133
44	1 936	6.633 250	20.97618	85 184	3.530 348	7.605 905	16.38643
45	2 025	6.708 204	21.21320	91 125	3.556 893	7.663 094	16.50964
46	2 116	6.782 330	21.44761	97 336	3.583 048	7.719 443	16.63103
47	2 209	6.855 655	21.67948	103 823	3.608 826	7.774 980	16.75069
48	2 304	6.928 203	21.90890	110 592	3.634 241	7.829 735	16.86865
49	2 401	7.000 000	22.13594	117 649	3.659 306	7.883 735	16.98499
50	2 500	7.071 068	22.36068	125 000	3.684 031	7.937 005	17.09976

SQUARES, SQUARE ROOT, CUBES AND CUBE ROOT

n	n^2	$\sqrt{n}$	$\sqrt{10n}$	n^3	$\sqrt[3]{n}$	$\sqrt[3]{10n}$	$\sqrt[3]{100n}$
50	2 500	7.071 068	22.36068	125 000	3.684 031	7.937 005	17.09976
51	2 601	7.141 428	22.58318	132 651	3.708 430	7.989 570	17.21301
52	2 704	7.211 103	22.80351	140 608	3.732 511	8.041 452	17.32478
53	2 809	7.280 110	23.02173	148 877	3.756 286	8.092 672	17.43513
54	2 916	7.348 469	23.23790	157 464	3.779 763	8.143 253	17.54411
55	3 025	7.416 198	23.45208	166 375	3.802 952	8.193 213	17.65174
56	3 136	7.483 315	23.66432	175 616	3.825 862	8.242 571	17.75808
57	3 249	7.549 834	23.87467	185 193	3.848 501	8.291 344	17.86316
58	3 364	7.615 773	24.08319	195 112	3.870 877	8.339 551	17.96702
59	3 481	7.681 146	24.28992	205 379	3.892 996	8.387 207	18.06969
60	3 600	7.745 967	24.49490	216 000	3.914 868	8.434 327	18.17121
61	3 721	7.810 250	24.69818	226 981	3.936 497	8.480 926	18.27160
62	3 844	7.874 008	24.89980	238 328	3.957 892	8.527 019	18.37091
63	3 969	7.937 254	25.09980	250 047	3.979 057	8.572 619	18.46915
64	4 096	8.000 000	25.29822	262 144	4.000 000	8.617 739	18.56636
65	4 225	8.062 258	25.49510	274 625	4.020 726	8.662 391	18.66256
66	4 356	8.124 038	25.69047	287 496	4.041 240	8.706 588	18.75777
67	4 489	8.185 353	25.88436	300 763	4.061 548	8.750 340	18.85204
68	4 624	8.246 211	26.07681	314 432	4.081 655	8.793 659	18.94536
69	4 761	8.306 624	26.26785	328 509	4.101 566	8.836 556	19.03778
70	4 900	8.366 600	26.45751	343 000	4.121 285	8.879 040	19.12931
71	5 041	8.426 150	26.64583	357 911	4.140 818	8.921 121	19.21997
72	5 184	8.485 281	26.83282	373 248	4.160 168	8.962 809	19.30979
73	5 329	8.544 004	27.01851	389 017	4.179 339	9.004 113	19.39877
74	5 476	8.602 325	27.20294	405 224	4.198 336	9.045 042	19.48695
75	5 625	8.660 254	27.38613	421 875	4.217 163	9.085 603	19.57434
76	5 776	8.717 798	27.56810	438 976	4.235 824	9.125 805	19.66095
77	5 929	8.774 964	27.74887	456 533	4.254 321	9.165 656	19.74681
78	6 084	8.831 761	27.92848	474 552	4.272 659	9.205 164	19.83192
79	6 241	8.888 194	28.10694	493 039	4.290 840	9.244 335	19.91632
80	6 400	8.944 272	28.28427	512 000	4.308 869	9.283 178	20.00000
81	6 561	9.000 000	28.46050	531 441	4.326 749	9.321 698	20.08299
82	6 724	9.055 385	28.63564	551 368	4.344 481	9.359 902	20.16530
83	6 889	9.110 434	28.80972	571 787	4.362 071	9.397 796	20.24694
84	7 056	9.165 151	28.98275	592 704	4.379 519	9.435 388	20.32793
85	7 225	9.219 544	29.15476	614 125	4.396 830	9.472 682	20.40828
86	7 396	9.273 618	29.32576	636 056	4.414 005	9.509 685	20.48800
87	7 569	9.327 379	29.49576	658 503	4.431 048	9.546 403	20.56710
88	7 744	9.380 832	29.66479	681 472	4.447 960	9.582 840	20.64560
89	7 921	9.433 981	29.83287	704 969	4.464 745	9.619 002	20.72351
90	8 100	9.486 833	30.00000	729 000	4.481 405	9.654 894	20.80084
91	8 281	9.539 392	30.16621	753 571	4.497 941	9.690 521	20.87759
92	8 464	9.591 663	30.33150	778 688	4.514 357	9.725 888	20.95379
93	8 649	9.643 651	30.49590	804 357	4.530 655	9.761 000	21.02944
94	8 836	9.695 360	30.65942	830 584	4.546 836	9.795 861	21.10454
95	9 025	9.746 794	30.82207	857 375	4.562 903	9.830 476	21.17912
96	9 216	9.797 959	30.98387	884 736	4.578 857	9.864 848	21.25317
97	9 409	9.848 858	31.14482	912 673	4.594 701	9.898 983	21.32671
98	9 604	9.899 495	31.30495	941 192	4.610 436	9.932 884	21.39975
99	9 801	9.949 874	31.46427	970 299	4.626 065	9.966 555	21.47229
100	10 000	10.00000	31.62278	1 000 000	4.641 589	10.00000	21.54435

SQUARES, SQUARE ROOT, CUBES AND CUBE ROOT

n	n^2	$\sqrt{n}$	$\sqrt{10n}$	n^3	$\sqrt[3]{n}$	$\sqrt[3]{10n}$	$\sqrt[3]{100n}$
100	10 000	10.00000	31.62278	1 000 000	4.641 589	10.00000	21.54435
101	10 201	10.04988	31.78050	1 030 301	4.657 010	10.03322	21.61592
102	10 404	10.09950	31.93744	1 061 208	4.672 329	10.06623	21.68703
103	10 609	10.14889	32.09361	1 092 727	4.687 548	10.09902	21.75767
104	10 816	10.19804	32.24903	1 124 864	4.702 669	10.13159	21.82786
105	11 025	10.24695	32.40370	1 157 625	4.717 694	10.16396	21.89760
106	11 236	10.29563	32.55764	1 191 016	4.732 623	10.19613	21.96689
107	11 449	10.34408	32.71085	1 225 043	4.747 459	10.22809	22.03575
108	11 664	10.39230	32.86335	1 259 712	4.762 203	10.25986	22.10419
109	11 881	10.44031	33.01515	1 295 029	4.776 856	10.29142	22.17220
110	12 100	10.48809	33.16625	1 331 000	4.791 420	10.32280	22.23980
111	12 321	10.53565	33.31666	1 367 631	4.805 896	10.35399	22.30699
112	12 544	10.58301	33.46640	1 404 928	4.820 285	10.38499	22.37378
113	12 769	10.63015	33.61547	1 442 897	4.834 588	10.41580	22.44017
114	12 996	10.67708	33.76389	1 481 544	4.848 808	10.44644	22.50617
115	13 225	10.72381	33.91165	1 520 875	4.862 944	10.47690	22.57179
116	13 456	10.77033	34.05877	1 560 896	4.876 999	10.50718	22.63702
117	13 689	10.81665	34.20526	1 601 613	4.890 973	10.53728	22.70189
118	13 924	10.86278	34.35113	1 643 032	4.904 868	10.56722	22.76638
119	14 161	10.90871	34.49638	1 685 159	4.918 685	10.59699	22.83051
120	14 400	10.95445	34.64102	1 728 000	4.932 424	10.62659	22.89428
121	14 641	11.00000	34.78505	1 771 561	4.946 087	10.65602	22.95770
122	14 884	11.04536	34.92850	1 815 848	4.959 676	10.68530	23.02078
123	15 129	11.09054	35.07136	1 860 867	4.973 190	10.71441	23.08350
124	15 376	11.13553	35.21363	1 906 624	4.986 631	10.74337	23.14589
125	15 625	11.18034	35.35534	1 953 125	5.000 000	10.77217	23.20794
126	15 876	11.22497	35.49648	2 000 376	5.013 298	10.80082	23.26967
127	16 129	11.26943	35.63706	2 048 383	5.026 526	10.82932	23.33107
128	16 384	11.31371	35.77709	2 097 152	5.039 684	10.85767	23.39214
129	16 641	11.35782	35.91657	2 146 689	5.052 774	10.88587	23.45290
130	16 900	11.40175	36.05551	2 197 000	5.065 797	10.91393	23.51335
131	17 161	11.44552	36.19392	2 248 091	5.078 753	10.94184	23.57348
132	17 424	11.48913	36.33180	2 299 968	5.091 643	10.96961	23.63332
133	17 689	11.53256	36.46917	2 352 637	5.104 469	10.99724	23.69285
134	17 956	11.57584	36.60601	2 406 104	5.117 230	11.02474	23.75208
135	18 225	11.61895	36.74235	2 460 375	5.129 928	11.05209	23.81102
136	18 496	11.66190	36.87818	2 515 456	5.142 563	11.07932	23.86966
137	18 769	11.70470	37.01351	2 571 353	5.155 137	11.10641	23.92803
138	19 044	11.74734	37.14835	2 628 072	5.167 649	11.13336	23.98610
139	19 321	11.78983	37.28270	2 685 619	5.180 101	11.16019	24.04390
140	19 600	11.83216	37.41657	2 744 000	5.192 494	11.18689	24.10142
141	19 881	11.87434	37.54997	2 803 221	5.204 828	11.21346	24.15867
142	20 164	11.91638	37.68289	2 863 288	5.217 103	11.23991	24.21565
143	20 449	11.95826	37.81534	2 924 207	5.229 322	11.26623	24.27236
144	20 736	12.00000	37.94733	2 985 984	5.241 483	11.29243	24.32881
145	21 025	12.04159	38.07887	3 048 625	5.253 588	11.31851	24.38499
146	21 316	12.08305	38.20995	3 112 136	5.265 637	11.34447	24.44092
147	21 609	12.12436	38.34058	3 176 523	5.277 632	11.37031	24.49660
148	21 904	12.16553	38.47077	3 241 792	5.289 572	11.39604	24.55202
149	22 201	12.20656	38.60052	3 307 949	5.301 459	11.42165	24.60719
150	22.500	12.24745	38.72983	3 375 000	5.313 293	11.44714	24.66212

SQUARES, SQUARE ROOT, CUBES AND CUBE ROOT

n	n^2	$\sqrt{n}$	$\sqrt{10n}$	n^3	$\sqrt[3]{n}$	$\sqrt[3]{10n}$	$\sqrt[3]{100n}$
150	22 500	12.24745	38.72983	3 375 000	5.313 293	11.44714	24.66212
151	22 801	12.28821	38.85872	3 442 951	5.325 074	11.47252	24.71680
152	23 104	12.32883	38.98718	3 511 808	5.336 803	11.49779	24.77125
153	23 409	12.36932	39.11521	3 581 577	5.348 481	11.52295	24.82545
154	23 716	12.40967	39.24283	3 652 264	5.360 108	11.54800	24.87942
155	24 025	12.44990	39.37004	3 723 875	5.371 685	11.57295	24.93315
156	24 336	12.49000	39.49684	3 796 416	5.383 213	11.59778	24.98666
157	24 649	12.52996	39.62323	3 869 893	5.394 691	11.62251	25.03994
158	24 964	12.56981	39.74921	3 944 312	5.406 120	11.64713	25.09299
159	25 281	12.60952	39.87480	4 019 679	5.417 502	11.67165	25.14581
160	25 600	12.64911	40.00000	4 096 000	5.428 835	11.69607	25.19842
161	25 921	12.68858	40.12481	4 173 281	5.440 122	11.72039	25.25081
162	26 244	12.72792	40.24922	4 251 528	5.451 362	11.74460	25.30298
163	26 569	12.76715	40.37326	4 330 747	5.462 556	11.76872	25.35494
164	26 896	12.80625	40.49691	4 410 944	5.473 704	11.79274	25.40668
165	27 225	12.84523	40.62019	4 492 125	5.484 807	11.81666	25.45822
166	27 556	12.88410	40.74310	4 574 296	5.495 865	11.84048	25.50954
167	27 889	12.92285	40.86563	4 657 463	5.506 878	11.86421	25.56067
168	28 224	12.96148	40.98780	4 741 632	5.517 848	11.88784	25.61158
169	28 561	13.00000	41.10961	4 826 809	5.528 775	11.91138	25.66230
170	28 900	13.03840	41.23106	4 913 000	5.539 658	11.93483	25.71282
171	29 241	13.07670	41.35215	5 000 211	5.550 499	11.95819	25.76313
172	29 584	13.11488	41.47288	5 088 448	5.561 298	11.98145	25.81326
173	29 929	13.15295	41.59327	5 177 717	5.572 055	12.00463	25.86319
174	30 276	13.19091	41.71331	5 268 024	5.582 770	12.02771	25.91292
175	30 625	13.22876	41.83300	5 359 375	5.593 445	12.05071	25.96247
176	30 976	13.26650	41.95235	5 451 776	5.604 079	12.07362	26.01183
177	31 329	13.30413	42.07137	5 545 233	5.614 672	12.09645	26.06100
178	31 684	13.34166	42.19005	5 639 752	5.625 226	12.11918	26.10999
179	32 041	13.37909	42.30839	5 735 339	5.635 741	12.14184	26.15879
180	32 400	13.41641	42.42641	5 832 000	5.646 216	12.16440	26.20741
181	32 761	13.45362	42.54409	5 929 741	5.656 653	12.18689	26.25586
182	33 124	13.49074	42.66146	6 028 568	5.667 051	12.20929	26.30412
183	33 489	13.52775	42.77850	6 128 487	5.677 411	12.23161	26.35221
184	33 856	13.56466	42.89522	6 229 504	5.687 734	12.25385	26.40012
185	34 225	13.60147	43.01163	6 331 625	5.698 019	12.27601	26.44786
186	34 596	13.63818	43.12772	6 434 856	5.708 267	12.29809	26.49543
187	34 969	13.67479	43.24350	6 539 203	5.718 479	12.32009	26.54283
188	35 344	13.71131	43.35897	6 644 672	5.728 654	12.34201	26.59006
189	35 721	13.74773	43.47413	6 751 269	5.738 794	12.36386	26.63712
190	36 100	13.78405	43.58899	6 859 000	5.748 897	12.38562	26.68402
191	36 481	13.82027	43.70355	6 967 871	5.758 965	12.40731	26.73075
192	36 864	13.85641	43.81780	7 077 888	5.768 998	12.42893	26.77732
193	37 249	13.89244	43.93177	7 189 057	5.778 997	12.45047	26.82373
194	37 636	13.92839	44.04543	7 301 384	5.788 960	12.47194	26.86997
195	38 025	13.96424	44.15880	7 414 875	5.798 890	12.49333	26.91606
196	38 416	14.00000	44.27189	7 529 536	5.808 786	12.51465	26.96199
197	38 809	14.03567	44.38468	7 645 373	5.818 648	12.53590	27.00777
198	39 204	14.07125	44.49719	7 762 392	5.828 477	12.55707	27.05339
199	39 601	14.10674	44.60942	7 880 599	5.838 272	12.57818	27.09886
200	40 000	14.14214	44.72136	8 000 000	5.848 035	12.59921	27.14418

SQUARES, SQUARE ROOT, CUBES AND CUBE ROOT

n	n^2	$\sqrt{n}$	$\sqrt{10n}$	n^3	$\sqrt[3]{n}$	$\sqrt[3]{10n}$	$\sqrt[3]{100n}$
200	40 000	14.14214	44.72136	8 000 000	5.848 035	12.59921	27.14418
201	40 401	14.17745	44.83302	8 120 601	5.857 766	12.62017	27.18934
202	40 804	14.21267	44.94441	8 242 408	5.867 464	12.64107	27.23436
203	41 209	14.24781	45.05552	8 365 427	5.877 131	12.66189	27.27922
204	41 616	14.28286	45.16636	8 489 664	5.886 765	12.68265	27.32394
205	42 025	14.31782	45.27693	8 615 125	5.896 369	12.70334	27.36852
206	42 436	14.35270	45.38722	8 741 816	5.905 941	12.72396	27.41295
207	42 849	14.38749	45.49725	8 869 743	5.915 482	12.74452	27.45723
208	43 264	14.42221	45.60702	8 998 912	5.924 992	12.76501	27.50138
209	43 681	14.45683	45.71652	9 129 329	5.934 472	12.78543	27.54538
210	44 100	14.49138	45.82576	9 261 000	5.943 922	12.80579	27.58924
211	44 521	14.52584	45.93474	9 393 931	5.953 342	12.82609	27.63296
212	44 944	14.56022	46.04346	9 528 128	5.962 732	12.84632	27.67655
213	45 369	14.59452	46.15192	9 663 597	5.972 093	12.86648	27.72000
214	45 796	14.62874	46.26013	9 800 344	5.981 424	12.88659	27.76331
215	46 225	14.66288	46.36809	9 938 375	5.990 726	12.90663	27.80649
216	46 656	14.69694	46.47580	10 077 696	6.000 000	12.92661	27.84953
217	47 089	14.73092	46.58326	10 218 313	6.009 245	12.94653	27.89244
218	47 524	14.76482	46.69047	10 360 232	6.018 462	12.96638	27.93522
219	47 961	14.79865	46.79744	10 503 459	6.027 650	12.98618	27.97787
220	48 400	14.83240	46.90416	10 648 000	6.036 811	13.00591	28.02039
221	48 841	14.86607	47.01064	10 793 861	6.045 944	13.02559	28.06278
222	49 284	14.89966	47.11688	10 941 048	6.055 049	13.04521	28.10505
223	49 729	14.93318	47.22288	11 089 567	6.064 127	13.06477	28.14718
224	50 176	14.96663	47.32864	11 239 424	6.073 178	13.08427	28.18919
225	50 625	15.00000	47.43416	11 390 625	6.082 202	13.10371	28.23108
226	51 076	15.03330	47.53946	11 543 176	6.091 199	13.12309	28.27284
227	51 529	15.06652	47.64452	11 697 083	6.100 170	13.14242	28.31448
228	51 984	15.09967	47.74935	11 852 352	6.109 115	13.16169	28.35600
229	52 441	15.13275	47.85394	12 008 989	6.118 033	13.18090	28.39739
230	52 900	15.16575	47.95832	12 167 000	6.126 926	13.20006	28.43867
231	53 361	15.19868	48.06246	12 326 391	6.135 792	13.21916	28.47983
232	53 824	15.23155	48.16638	12 487 168	6.144 634	13.23821	28.52086
233	54 289	15.26434	48.27007	12 649 337	6.153 449	13.25721	28.56178
234	54 756	15.29706	48.37355	12 812 904	6.162 240	13.27614	28.60259
235	55 225	15.32971	48.47680	12 977 875	6.171 006	13.29503	28.64327
236	55 696	15.36229	48.57983	13 144 256	6.179 747	13.31386	28.68384
237	56 169	15.39480	48.68265	13 312 053	6.188 463	13.33264	28.72430
238	56 644	15.42725	48.78524	13 481 272	6.197 154	13.35136	28.76464
239	57 121	15.45962	48.88763	13 651 919	6.205 822	13.37004	28.80487
240	57 600	15.49193	48.98979	13 824 000	6.214 465	13.38866	28.84499
241	58 081	15.52417	49.09175	13 997 521	6.223 084	13.40723	28.88500
242	58 564	15.55635	49.19350	14 172 488	6.231 680	13.42575	28.92489
243	59 049	15.58846	49.29503	14 348 907	6.240 251	13.44421	28.96468
244	59 536	15.62050	49.39636	14 526 784	6.248 800	13.46263	29.00436
245	60 025	15.65248	49.49747	14 706 125	6.257 325	13.48100	29.04393
246	60 516	15.68439	49.59839	14 886 936	6.265 827	13.49931	29.08339
247	61 009	15.71623	49.69909	15 069 223	6.274 305	13.51758	29.12275
248	61 504	15.74802	49.79960	15 252 992	6.282 761	13.53580	29.16199
249	62 001	15.77973	49.89990	15 438 249	6.291 195	13.55397	29.20114
250	62 500	15.81139	50.00000	15 625 000	6.299 605	13.57209	29.24018

SQUARES, SQUARE ROOT, CUBES AND CUBE ROOT

n	n^2	$\sqrt{n}$	$\sqrt{10n}$	n^3	$\sqrt[3]{n}$	$\sqrt[3]{10n}$	$\sqrt[3]{100n}$
250	62 500	15.81139	50.00000	15 625 000	6.299 605	13.57209	29.24018
251	63 001	15.84298	50.09990	15 813 251	6.307 994	13.59016	29.27911
252	63 504	15.87451	50.19960	16 003 008	6.316 360	13.60818	29.31794
253	64 009	15.90597	50.29911	16 194 277	6.324 704	13.62616	29.35667
254	64 516	15.93738	50.39841	16 387 064	6.333 026	13.64409	29.39530
255	65 025	15.96872	50.49752	16 581 375	6.341 326	13.66197	29.43383
256	65 536	16.00000	50.59644	16 777 216	6.349 604	13.67981	29.47225
257	66 049	16.03122	50.69517	16 974 593	6.357 861	13.69760	29.51058
258	66 564	16.06238	50.79370	17 173 512	6.366 097	13.71534	29.54880
259	67 081	16.09348	50.89204	17 373 979	6.374 311	13.73304	29.58693
260	67 600	16.12452	50.99020	17 576 000	6.382 504	13.75069	29.62496
261	68 121	16.15549	51.08816	17 779 581	6.390 677	13.76830	29.66289
262	68 644	16.18641	51.18594	17 984 728	6.398 828	13.78586	29.70073
263	69 169	16.21727	51.28353	18 191 447	6.406 959	13.80337	29.73847
264	69 696	16.24808	51.38093	18 399 744	6.415 069	13.82085	29.77611
265	70 225	16.27882	51.47815	18 609 625	6.423 158	13.83828	29.81366
266	70 756	16.30951	51.57519	18 821 096	6.431 228	13.85566	29.85111
267	71 289	16.34013	51.67204	19 034 163	6.439 277	13.87300	29.88847
268	71 824	16.37071	51.76872	19 248 832	6.447 306	13.89030	29.92574
269	72 361	16.40122	51.86521	19 465 109	6.455 315	13.90755	29.96292
270	72 900	16.43168	51.96152	19 683 000	6.463 304	13.92477	30.00000
271	73 441	16.46208	52.05766	19 902 511	6.471 274	13.94194	30.03699
272	73 984	16.49242	52.15362	20 123 648	6.479 224	13.95906	30.07389
273	74 529	16.52271	52.24940	20 346 417	6.487 154	13.97615	30.11070
274	75 076	16.55295	52.34501	20 570 824	6.495 065	13.99319	30.14742
275	75 625	16.58312	52.44044	20 796 875	6.502 957	14.01020	30.18405
276	76 176	16.61325	52.53570	21 024 576	6.510 830	14.02716	30.22060
277	76 729	16.64332	52.63079	21 253 933	6.518 684	14.04408	30.25705
278	77 284	16.67333	52.72571	21 484 952	6.526 519	14.06096	30.29342
279	77 841	16.70329	52.82045	21 717 639	6.534 335	14.07780	30.32970
280	78 400	16.73320	52.91503	21 952 000	6.542 133	14.09460	30.36589
281	78 961	16.76305	53.00943	22 188 041	6.549 912	14.11136	30.40200
282	79 524	16.79286	53.10367	22 425 768	6.557 672	14.12808	30.43802
283	80 089	16.82260	53.19774	22 665 187	6.565 414	14.14476	30.47395
284	80 656	16.85230	53.29165	22 906 304	6.573 138	14.16140	30.50981
285	81 225	16.88194	53.38539	23 149 125	6.580 844	14.17800	30.54557
286	81 796	16.91153	53.47897	23 393 656	6.588 532	14.19456	30.58126
287	82 369	16.94107	53.57238	23 639 903	6.596 202	14.21109	30.61686
288	82 944	16.97056	53.66563	23 887 872	6.603 854	14.22757	30.65238
289	83 521	17.00000	53.75872	24 137 569	6.611 489	14.24402	30.68781
290	84 100	17.02939	53.85165	24 389 000	6.619 106	14.26043	30.72317
291	84 681	17.05872	53.94442	24 642 171	6.626 705	14.27680	30.75844
292	85 264	17.08801	54.03702	24 897 088	6.634 287	14.29314	30.79363
293	85 849	17.11724	54.12947	25 153 757	6.641 852	14.30944	30.82875
294	86 436	17.14643	54.22177	25 412 184	6.649 400	14.32570	30.86378
295	87 025	17.17556	54.31390	25 672 375	6.656 930	14.34192	30.89873
296	87 616	17.20465	54.40588	25 934 336	6.664 444	14.35811	30.93361
297	88 209	17.23369	54.49771	26 198 073	6.671 940	14.37426	30.96840
298	88 804	17.26268	54.58938	26 463 592	6.679 420	14.39037	31.00312
299	89 401	17.29162	54.68089	26 730 899	6.686 883	14.40645	31.03776
300	90 000	17.32051	54.77226	27 000 000	6.694 330	14.42250	31.07233

SQUARES, SQUARE ROOT, CUBES AND CUBE ROOT

n	n^2	$\sqrt{n}$	$\sqrt{10n}$	n^3	$\sqrt[3]{n}$	$\sqrt[3]{10n}$	$\sqrt[3]{100n}$
300	90 000	17 32051	54.77226	27 000 000	6.694 330	14.42250	31.07233
301	90 601	17.34935	54.86347	27 270 901	6.701 759	14.43850	31.10681
302	91 204	17.37815	54.95453	27 543 608	6.709 173	14.45447	31.14122
303	91 809	17.40690	55.04544	27 818 127	6.716 570	14.47041	31.17556
304	92 416	17.43560	55.13620	28 094 464	6.723 951	14.48631	31.20982
305	93 025	17.46425	55.22681	28 372 625	6.731 315	14.50218	31.24400
306	93 636	17.49286	55.31727	28 652 616	6.738 664	14.51801	31.27811
307	94 249	17.52142	55.40758	28 934 443	6.745 997	14.53381	31.31214
308	94 864	17.54993	55.49775	29 218 112	6.753 313	14.54957	31.34610
309	95 481	17.57840	55.58777	29 503 629	6.760 614	14.56530	31.37999
310	96 100	17.60682	55.67764	29 791 000	6.767 899	14.58100	31.41381
311	96 721	17.63519	55.76737	30 080 231	6.775 169	14.59666	31.44755
312	97 344	17.66352	55.85696	30 371 328	6.782 423	14.61229	31.48122
313	97 969	17.69181	55.94640	30 664 297	6.789 661	14.62788	31.51482
314	98 596	17.72005	56.03570	30 959 144	6.796 884	14.64344	31.54834
315	99 225	17.74824	56.12486	31 255 875	6.804 092	14.65897	31.58180
316	99 856	17.77639	56.21388	31 554 496	6.811 285	14.67447	31.61518
317	100 489	17.80449	56.30275	31 855 013	6.818 462	14.68993	31.64850
318	101 124	17.83255	56.39149	32 157 432	6.825 624	14.70536	31.68174
319	101 761	17.86057	56.48008	32 461 759	6.832 771	14.72076	31.71492
320	102 400	17.88854	56.56854	32 768 000	6.839 904	14.73613	31.74802
321	103 041	17.91647	56.65686	33 076 161	6.847 021	14.75146	31.78106
322	103 684	17.94436	56.74504	33 386 248	6.854 124	14.76676	31.81403
323	104 329	17.97220	56.83309	33 698 267	6.861 212	14.78203	31.84693
324	104 976	18.00000	56.92100	34 012 224	6.868 285	14.79727	31.87976
325	105 625	18.02776	57.00877	34 328 125	6.875 344	14.81248	31.91252
326	106 276	18.05547	57.09641	34 645 976	6.882 389	14.82766	31.94522
327	106 929	18.08314	57.18391	34 965 783	6.889 419	14.84280	31.97785
328	107 584	18.11077	57.27128	35 287 552	6.896 434	14.85792	32.01041
329	108 241	18.13836	57.35852	35 611 289	6.903 436	14.87300	32.04291
330	108 900	18.16590	57.44563	35 937 000	6.910 423	14.88806	32.07534
331	109 561	18.19341	57.53260	36 264 691	6.917 396	14.90308	32.10771
332	110 224	18.22087	57.61944	36 594 368	6.924 356	14.91807	32.14001
333	110 889	18.24829	57.70615	36 926 037	6.931 301	14.93303	32.17225
334	111 556	18.27567	57.79273	37 259 704	6.938 232	14.94797	32.20442
335	112 225	18.30301	57.87918	37 595 375	6.945 150	14.96287	32.23653
336	112 896	18.33030	57.96551	37 933 056	6.952 053	14.97774	32.26857
337	113 569	18.35756	58.05170	38 272 753	6.958 943	14.99259	32.30055
338	114 244	18.38478	58.13777	38 614 472	6.965 820	15.00740	32.33247
339	114 921	18.41195	58.22371	38 958 219	6.972 683	15.02219	32.36433
340	115 600	18.43909	58.30952	39 304 000	6.979 532	15.03695	32.39612
341	116 281	18.46619	58.39521	39 651 821	6.986 368	15.05167	32.42785
342	116 964	18.49324	58.48077	40 001 688	6.993 191	15.06637	32.45952
343	117 649	18.52026	58.56620	40 353 607	7.000 000	15.08104	32.49112
344	118 336	18.54724	58.65151	40 707 584	7.006 796	15.09568	32.52267
345	119 025	18.57418	58.73670	41 063 625	7.013 579	15.11030	32.55415
346	119 716	18.60108	58.82176	41 421 736	7.020 349	15.12488	32.58557
347	120 409	18.62794	58.90671	41 781 923	7.027 106	15.13944	32.61694
348	121 104	18.65476	58.99152	42 144 192	7.033 850	15.15397	32.64824
349	121 801	18.68154	59.07622	42 508 549	7.040 581	15.16847	32.67948
350	122 500	18.70829	59.16080	42 875 000	7.047 299	15.18294	32.71066

SQUARES, SQUARE ROOT, CUBES AND CUBE ROOT

n	n^2	$\sqrt{n}$	$\sqrt{10n}$	n^3	$\sqrt[3]{n}$	$\sqrt[3]{10n}$	$\sqrt[3]{100n}$
350	122 500	18.70829	59.16080	42 875 000	7.047 299	15.18294	32.71066
351	123 201	18.73499	59.24525	43 243 551	7.054 004	15.19739	32.74179
352	123 904	18.76166	59.32959	43 614 208	7.060 697	15.21181	32.77285
353	124 609	18.78829	59.41380	43 986 977	7.067 377	15.22620	32.80386
354	125 316	18.81489	59.49790	44 361 864	7.074 044	15.24057	32.83480
355	126 025	18.84144	59.58188	44 738 875	7.080 699	15.25490	32.86569
356	126 736	18.86796	59.66574	45 118 016	7.087 341	15.26921	32.89652
357	127 449	18.89444	59.74948	45 499 293	7.093 971	15.28350	32.92730
358	128 164	18.92089	59.83310	45 882 712	7.100 588	15.29775	32.95801
359	128 881	18.94730	59.91661	46 268 279	7.107 194	15.31198	32.98867
360	129 600	18.97367	60.00000	46 656 000	7.113 787	15.32619	33.01927
361	130 321	19.00000	60.08328	47 045 881	7.120 367	15.34037	33.04982
362	131 044	19.02630	60.16644	47 437 928	7.126 936	15.35452	33.08031
363	131 769	19.05256	60.24948	47 832 147	7.133 492	15.36864	33.11074
364	132 496	19.07878	60.33241	48 228 544	7.140 037	15.38274	33.14112
365	133 225	19.10497	60.41523	48 627 125	7.146 569	15.39682	33.17144
366	133 956	19.13113	60.49793	49 027 896	7.153 090	15.41087	33.20170
367	134 689	19.15724	60.58052	49 430 863	7.159 599	15.42489	33.23191
368	135 424	19.18333	60.66300	49 836 032	7.166 096	15.43889	33.26207
369	136 161	19.20937	60.74537	50 243 409	7.172 581	15.45286	33.29217
370	136 900	19.23538	60.82763	50 653 000	7.179 054	15.46680	33.32222
371	137 641	19.26136	60.90977	51 064 811	7.185 516	15.48073	33.35221
372	138 384	19.28730	60.99180	51 478 848	7.191 966	15.49462	33.38215
373	139 129	19.31321	61.07373	51 895 117	7.198 405	15.50849	33.41204
374	139 876	19.33908	61.15554	52 313 624	7.204 832	15.52234	33.44187
375	140 625	19.36492	61.23724	52 734 375	7.211 248	15.53616	33.47165
376	141 376	19.39072	61.31884	53 157 376	7.217 652	15.54996	33.50137
377	142 129	19.41649	61.40033	53 582 633	7.224 045	15.56373	33.53105
378	142 884	19.44222	61.48170	54 010 152	7.230 427	15.57748	33.56067
379	143 641	19.46792	61.56298	54 439 939	7.236 797	15.59121	33.59024
380	144 400	19.49359	61.64414	54 872 000	7.243 156	15.60491	33.61975
381	145 161	19.51922	61.72520	55 306 341	7.249 505	15.61858	33.64922
382	145 924	19.54482	61.80615	55 742 968	7.255 842	15.63224	33.67863
383	146 689	19.57039	61.88699	56 181 887	7.262 167	15.64587	33.70800
384	147 456	19.59592	61.96773	56 623 104	7.268 482	15.65947	33.73731
385	148 225	19.62142	62.04837	57 066 625	7.274 786	15.67305	33.76657
386	148 996	19.64688	62.12890	57 512 456	7.281 079	15.68661	33.79578
387	149 769	19.67232	62.20932	57 960 603	7.287 362	15.70014	33.82494
388	150 544	19.69772	62.28965	58 411 072	7.293 633	15.71366	33.85405
389	151 321	19.72308	62.36986	58 863 869	7.299 894	15.72714	33.88310
390	152 100	19.74842	62.44998	59 319 000	7.306 144	15.74061	33.91211
391	152 881	19.77372	62.52999	59 776 471	7.312 383	15.75405	33.94107
392	153 664	19.79899	62.60990	60 236 288	7.318 611	15.76747	33.96999
393	154 449	19.82423	62.68971	60 698 457	7.324 829	15.78087	33.99885
394	155 236	19.84943	62.76942	61 162 984	7.331 037	15.79424	34.02766
395	156 025	19.87461	62.84903	61 629 875	7.337 234	15.80759	34.05642
396	156 816	19.89975	62.92853	62 099 136	7.343 420	15.82092	34.08514
397	157 609	19.92486	63.00794	62 570 773	7.349 597	15.83423	34.11381
398	158 404	19.94994	63.08724	63 044 792	7.355 762	15.84751	34.14242
399	159 201	19.97498	63.16645	63 521 199	7.361 918	15.86077	34.17100
400	160 000	20.00000	63.24555	64 000 000	7.368 063	15.87401	34.19952

SQUARES, SQUARE ROOT, CUBES AND CUBE ROOT

n	n^2	$\sqrt{n}$	$\sqrt{10n}$	n^3	$\sqrt[3]{n}$	$\sqrt[3]{10n}$	$\sqrt[3]{100n}$
400	160 000	20.00000	63.24555	64 000 000	7.368 063	15.87401	34.19952
401	160 801	20.02498	63.32456	64 481 201	7.374 198	15.88723	34.22799
402	161 604	20.04994	63.40347	64 964 808	7.380 323	15.90042	34.25642
403	162 409	20.07486	63.48228	65 450 827	7.386 437	15.91360	34.28480
404	163 216	20.09975	63.56099	65 939 264	7.392 542	15.92675	34.31314
405	164 025	20.12461	63.63961	66 430 125	7.398 636	15.93988	34.34143
406	164 836	20.14944	63.71813	66 923 416	7.404 721	15.95299	34.36967
407	165 649	20.17424	63.79655	67 419 143	7.410 795	15.96607	34.39786
408	166 464	20.19901	63.87488	67 917 312	7.416 860	15.97914	34.42601
409	167 281	20.22375	63.95311	68 417 929	7.422 914	15.99218	34.45412
410	168 100	20.24846	64.03124	68 921 000	7.428 959	16.00521	34.48217
411	168 921	20.27313	64.10928	69 426 531	7.434 994	16.01821	34.51018
412	169 744	20.29778	64.18723	69 934 528	7.441 019	16.03119	34.53815
413	170 569	20.32240	64.26508	70 444 997	7.447 034	16.04415	34.56607
414	171 396	20.34699	64.34283	70 957 944	7.453 040	16.05709	34.59395
415	172 225	20.37155	64.42049	71 473 375	7.459 036	16.07001	34.62178
416	173 056	20.39608	64.49806	71 991 296	7.465 022	16.08290	34.64956
417	173 889	20.42058	64.57554	72 511 713	7.470 999	16.09578	34.67731
418	174 724	20.44505	64.65292	73 034 632	7.476 966	16.10864	34.70500
419	175 561	20.46949	64.73021	73 560 059	7.482 924	16.12147	34.73266
420	176 400	20.49390	64.80741	74 088 000	7.488 872	16.13429	34.76027
421	177 241	20.51828	64.88451	74 618 461	7.494 811	16.14708	34.78783
422	178 084	20.54264	64.96153	75 151 448	7.500 741	16.15986	34.81535
423	178 929	20.56696	65.03845	75 686 967	7.506 661	16.17261	34.84283
424	179 776	20.59126	65.11528	76 225 024	7.512 572	16.18534	34.87027
425	180 625	20.61553	65.19202	76 765 625	7.518 473	16.19806	34.89766
426	181 476	20.63977	65.26868	77 308 776	7.524 365	16.21075	34.92501
427	182 329	20.66398	65.34524	77 854 483	7.530 248	16.22343	34.95232
428	183 184	20.68816	65.42171	78 402 752	7.536 122	16.23608	34.97958
429	184 041	20.71232	65.49809	78 953 589	7.541 987	16.24872	35.00680
430	184 900	20.73644	65.57439	79 507 000	7.547 842	16.26133	35.03398
431	185 761	20.76054	65.65059	80 062 991	7.553 689	16.27393	35.06112
432	186 624	20.78461	65.72671	80 621 568	7.559 526	16.28651	35.08821
433	187 489	20.80865	65.80274	81 182 737	7.565 355	16.29906	35.11527
434	188 356	20.83267	65.87868	81 746 504	7.571 174	16.31160	35.14228
435	189 225	20.85665	65.95453	82 312 875	7.576 985	16.32412	35.16925
436	190 096	20.88061	66.03030	82 881 856	7.582 787	16.33662	35.19618
437	190 969	20.90454	66.10598	83 453 453	7.588 579	16.34910	35.22307
438	191 844	20.92845	66.18157	84 027 672	7.594 363	16.36156	35.24991
439	192 721	20.95233	66.25708	84 604 519	7.600 139	16.37400	35.27672
440	193 600	20.97618	66.33250	85 184 000	7.605 905	16.38643	35.30348
441	194 481	21.00000	66.40783	85 766 121	7.611 663	16.39883	35.33021
442	195 364	21.02380	66.48308	86 350 888	7.617 412	16.41122	35.35689
443	196 249	21.04757	66.55825	86 938 307	7.623 152	16.42358	35.38354
444	197 136	21.07131	66.63332	87 528 384	7.628 884	16.43593	35.41014
445	198 025	21.09502	66.70832	88 121 125	7.634 607	16.44826	35.43671
446	198 916	21.11871	66.78323	88 716 536	7.640 321	16.46057	35.46323
447	199 809	21.14237	66.85806	89 314 623	7.646 027	16.47287	35.48971
448	200 704	21.16601	66.93280	89 915 392	7.651 725	16.48514	35.51616
449	201 601	21.18962	67.00746	90 518 849	7.657 414	16.49740	35.54257
450	202 500	21.21320	67.08204	91 125 000	7.663 094	16.50964	35.56893

SQUARES, SQUARE ROOT, CUBES AND CUBE ROOT

n	n^2	$\sqrt{n}$	$\sqrt{10n}$	n^3	$\sqrt[3]{n}$	$\sqrt[3]{10n}$	$\sqrt[3]{100n}$
450	202 500	21.21320	67.08204	91 125 000	7.663 094	16.50964	35.56893
451	203 401	21.23676	67.15653	91 733 851	7.668 766	16.52186	35.59526
452	204 304	21.26029	67.23095	92 345 408	7.674 430	16.53406	35.62155
453	205 209	21.28380	67.30527	92 959 677	7.680 086	16.54624	35.64780
454	206 116	21.30728	67.37952	93 576 664	7.685 733	16.55841	35.67401
455	207 025	21.33073	67.45369	94 196 375	7.691 372	16.57056	35.70018
456	207 936	21.35416	67.52777	94 818 816	7.697 002	16.58269	35.72632
457	208 849	21.37756	67.60178	95 443 993	7.702 625	16.59480	35.75242
458	209 764	21.40093	67.67570	96 071 912	7.708 239	16.60690	35.77848
459	210 681	21.42429	67.74954	96 702 579	7.713 845	16.61897	35.80450
460	211 600	21.44761	67.82330	97 336 000	7.719 443	16.63103	35.83048
461	212 521	21.47091	67.89698	97 972 181	7.725 032	16.64308	35.85642
462	213 444	21.49419	67.97058	98 611 128	7.730 614	16.65510	35.88233
463	214 369	21.51743	68.04410	99 252 847	7.736 188	16.66711	35.90820
464	215 296	21.54066	68.11755	99 897 344	7.741 753	16.67910	35.93404
465	216 225	21.56386	68.19091	100 544 625	7.747 311	16.69108	35.95983
466	217 156	21.58703	68.26419	101 194 696	7.752 861	16.70303	35.98559
467	218 089	21.61018	68.33740	101 847 563	7.758 402	16.71497	36.01131
468	219 024	21.63331	68.41053	102 503 232	7.763 936	16.72689	36.03700
469	219 961	21.65641	68.48357	103 161 709	7.769 462	16.73880	36.06265
470	220 900	21.67948	68.55655	103 823 000	7.774 980	16.75069	36.08826
471	221 841	21.70253	68.62944	104 487 111	7.780 490	16.76256	36.11384
472	222 784	21.72556	68.70226	105 154 048	7.785 993	16.77441	36.13938
473	223 729	21.74856	68.77500	105 823 817	7.791 488	16.78625	36.16488
474	224 676	21.77154	68.84766	106 496 424	7.796 975	16.79807	36.19035
475	225 625	21.79449	68.92024	107 171 875	7.802 454	16.80988	36.21578
476	226 576	21.81742	68.99275	107 850 176	7.807 925	16.82167	36.24118
477	227 529	21.84033	69.06519	108 531 333	7.813 389	16.83344	36.26654
478	228 484	21.86321	69 13754	109 215 352	7.818 846	16.84519	36.29187
479	229 441	21.88607	69.20983	109 902 239	7.824 294	16.85693	36.31716
480	230 400	21.90890	69 28203	110 592 000	7.829 735	16.86865	36.34241
481	231 361	21.93171	69.35416	111 284 641	7.835 169	16.88036	36.36763
482	232 324	21.95450	69.42622	111 980 168	7.840 595	16.89205	36.39282
483	233 289	21.97726	69 49820	112 678 587	7.846 013	16.90372	36.41797
484	234 256	22.00000	69.57011	113 379 904	7.851 424	16.91538	36.44308
485	235 225	22.02272	69.64194	114 084 125	7.856 828	16.92702	36.46817
486	236 196	22.04541	69.71370	114 791 256	7.862 224	16.93865	36.49321
487	237 169	22.06808	69.78539	115 501 303	7.867 613	16.95026	36.51822
488	238 144	22.09072	69.85700	116 214 272	7.872 994	16.96185	36.54320
489	239 121	22.11334	69.92853	116 930 169	7.878 368	16.97343	36.56815
490	240 100	22.13594	70.00000	117 649 000	7.883 735	16.98499	36.59306
491	241 081	22.15852	70.07139	118 370 771	7.889 095	16.99654	36.61793
492	242 064	22.18107	70.14271	119 095 488	7.894 447	17.00807	36.64278
493	243 049	22.20360	70.21396	119 823 157	7.899 792	17.01959	36.66758
494	244 036	22.22611	70.28513	120 553 784	7.905 129	17.03108	36.69236
495	245 025	22.24860	70.35624	121 287 375	7.910 460	17.04257	36.71710
496	246 016	22.27106	70.42727	122 023 936	7.915 783	17.05404	36.74181
497	247 009	22.29350	70.49823	122 763 473	7.921 099	17.06549	36.76649
498	248 004	22.31591	70.56912	123 505 992	7.926 408	17.07693	36.79113
499	249 001	22.33831	70.63993	124 251 499	7.931 710	17.08835	36.81574
500	250 000	22.36068	70.71068	125 000 000	7.937 005	17.09976	36.84031

SQUARES, SQUARE ROOT, CUBES AND CUBE ROOT

n	n^2	$\sqrt{n}$	$\sqrt{10n}$	n^3	$\sqrt[3]{n}$	$\sqrt[3]{10n}$	$\sqrt[3]{100n}$
500	250 000	22.36068	70.71068	125 000 000	7.937 005	17.09976	36.84031
501	251 001	22.38303	70.78135	125 751 501	7.942 293	17.11115	36.86486
502	252 004	22.40536	70.85196	126 506 008	7.947 574	17.12253	36.88937
503	253 009	22.42766	70.92249	127 263 527	7.952 848	17.13389	36.91385
504	254 016	22.44994	70.99296	128 024 064	7.958 114	17.14524	36.93830
505	255 025	22.47221	71.06335	128 787 625	7.963 374	17.15657	36.96271
506	256 036	22.49444	71.13368	129 554 216	7.968 627	17.16789	36.98709
507	257 049	22.51666	71.20393	130 323 843	7.973 873	17.17919	37.01144
508	258 064	22.53886	71.27412	131 096 512	7.979 112	17.19048	37.03576
509	259 081	22.56103	71.34424	131 872 229	7.984 344	17.20175	37.06004
510	260 100	22.58318	71.41428	132 651 000	7.989 570	17.21301	37.08430
511	261 121	22.60531	71.48426	133 432 831	7.994 788	17.22425	37.10852
512	262 144	22.62742	71.55418	134 217 728	8.000 000	17.23548	37.13271
513	263 169	22.64950	71.62402	135 005 697	8.005 205	17.24669	37.15687
514	264 196	22.67157	71.69379	135 796 744	8.010 403	17.25789	37.18100
515	265 225	22.69361	71.76350	136 590 875	8.015 595	17.26908	37.20509
516	266 256	22.71563	71.83314	137 388 096	8.020 779	17.28025	37.22916
517	267 289	22.73763	71.90271	138 188 413	8.025 957	17.29140	37.25319
518	268 324	22.75961	71.97222	138 991 832	8.031 129	17.30254	37.27720
519	269 361	22.78157	72.04165	139 798 359	8.036 293	17.31367	37.30117
520	270 400	22.80351	72.11103	140 608 000	8.041 452	17.32478	37.32511
521	271 441	22.82542	72.18033	141 420 761	8.046 603	17.33588	37.34902
522	272 484	22.84732	72.24957	142 236 648	8.051 748	17.34696	37.37290
523	273 529	22.86919	72.31874	143 055 667	8.056 886	17.35804	37.39675
524	274 576	22.89105	72.38784	143 877 824	8.062 018	17.36909	37.42057
525	275 625	22.91288	72.45688	144 703 125	8.067 143	17.38013	37.44436
526	276 676	22.93469	72.52586	145 531 576	8.072 262	17.39116	37.46812
527	277 729	22.95648	72.59477	146 363 183	8.077 374	17.40218	37.49185
528	278 784	22.97825	72.66361	147 197 952	8.082 480	17.41318	37.51555
529	279 841	23.00000	72.73239	148 035 889	8.087 579	17.42416	37.53922
530	280 900	23.02173	72.80110	148 877 000	8.092 672	17.43513	37.56286
531	281 961	23.04344	72.86975	149 721 291	8.097 759	17.44609	37.58647
532	283 024	23.06513	72.93833	150 568 768	8.102 839	17.45704	37.61005
533	284 089	23.08679	73.00685	151 419 437	8.107 913	17.46797	37.63360
534	285 156	23.10844	73.07530	152 273 304	8.112 980	17.47889	37.65712
535	286 225	23.13007	73.14369	153 130 375	8.118 041	17.48979	37.68061
536	287 296	23.15167	73.21202	153 990 656	8.123 096	17.50068	37.70407
537	288 369	23.17326	73.28028	154 854 153	8.128 145	17.51156	37.72751
538	289 444	23.19483	73.34848	155 720 872	8.133 187	17.52242	37.75091
539	290 521	23.21637	73.41662	156 590 819	8.138 223	17.53327	37.77429
540	291 600	23.23790	73.48469	157 464 000	8.143 253	17.54411	37.79763
541	292 681	23.25941	73.55270	158 340 421	8.148 276	17.55493	37.82095
542	293 764	23.28089	73.62065	159 220 088	8.153 294	17.56574	37.84424
543	294 849	23.30236	73.68853	160 103 007	8.158 305	17.57654	37.86750
544	295 936	23.32381	73.75636	160 989 184	8.163 310	17.58732	37.89073
545	297 025	23.34524	73.82412	161 878 625	8.168 309	17.59809	37.91393
546	298 116	23.36664	73.89181	162 771 336	8.173 302	17.60885	37.93711
547	299 209	23.38803	73.95945	163 667 323	8.178 289	17.61959	37.96025
548	300 304	23.40940	74.02702	164 566 592	8.183 269	17.63032	37.98337
549	301 401	23.43075	74.09453	165 469 149	8.188 244	17.64104	38.00646
550	302 500	23.45208	74.16198	166 375 000	8.193 213	17.65174	38.02952

SQUARES, SQUARE ROOT, CUBES AND CUBE ROOT

n	n^2	$\sqrt{n}$	$\sqrt{10n}$	n^3	$\sqrt[3]{n}$	$\sqrt[3]{10n}$	$\sqrt[3]{100n}$
550	302 500	23.45208	74.16198	166 375 000	8.193 213	17.65174	38.02952
551	303 601	23.47339	74.22937	167 284 151	8.198 175	17.66243	38.05256
552	304 704	23.49468	74.29670	168 196 608	8.203 132	17.67311	38.07557
553	305 809	23.51595	74.36397	169 112 377	8.208 082	17.68378	38.09854
554	306 916	23.53720	74.43118	170 031 464	8.213 027	17.69443	38.12149
555	308 025	23.55844	74.49832	170 953 875	8.217 966	17.70507	38.14442
556	309 136	23.57965	74.56541	171 879 616	8.222 899	17.71570	38.16731
557	310 249	23.60085	74.63243	172 808 693	8.227 825	17.72631	38.19018
558	311 364	23.62202	74.69940	173 741 112	8.232 746	17.73691	38.21302
559	312 481	23.64318	74.76630	174 676 879	8.237 661	17.74750	38.23584
560	313 600	23.66432	74.83315	175 616 000	8.242 571	17.75808	38.25862
561	314 721	23.68544	74.89993	176 558 481	8.247 474	17.76864	38.28138
562	315 844	23.70654	74.96666	177 504 328	8.252 372	17.77920	38.30412
563	316 969	23.72762	75.03333	178 453 547	8.257 263	17.78973	38.32682
564	318 096	23.74868	75.09993	179 406 144	8.262 149	17.80026	38.34950
565	319 225	23.76973	75.16648	180 362 125	8.267 029	17.81077	38.37215
566	320 356	23.79075	75.23297	181 321 496	8.271 904	17.82128	38.39478
567	321 489	23.81176	75.29940	182 284 263	8.276 773	17.83177	38.41737
568	322 624	23.83275	75.36577	183 250 432	8.281 635	17.84224	38.43995
569	323 761	23.85372	75.43209	184 220 009	8.286 493	17.85271	38.46249
570	324 900	23.87467	75.49834	185 193 000	8.291 344	17.86316	38.48501
571	326 041	23.89561	75.56454	186 169 411	8.296 190	17.87360	38.50750
572	327 184	23.91652	75.63068	187 149 248	8.301 031	17.88403	38.52997
573	328 329	23.93742	75.69676	188 132 517	8.305 865	17.89444	38.55241
574	329 476	23.95830	75.76279	189 119 224	8.310 694	17.90485	38.57482
575	330 625	23.97916	75.82875	190 109 375	8.315 517	17.91524	38.59721
576	331 776	24.00000	75.89466	191 102 976	8.320 335	17.92562	38.61958
577	332 929	24.02082	75.96052	192 100 033	8.325 148	17.93599	38.64191
578	334 084	24.04163	76.02631	193 100 552	8.329 954	17.94634	38.66422
579	335 241	24.06242	76.09205	194 104 539	8.334 755	17.95669	38.68651
580	336 400	24.08319	76.15773	195 112 000	8.339 551	17.96702	38.70877
581	337 561	24.10394	76.22336	196 122 941	8.344 341	17.97734	38.73100
582	338 724	24.12468	76.28892	197 137 368	8.349 126	17.98765	38.75321
583	339 889	24.14539	76.35444	198 155 287	8.353 905	17.99794	38.77539
584	341 056	24.16609	76.41989	199 176 704	8.358 678	18.00823	38.79755
585	342 225	24.18677	76.48529	200 201 625	8.363 447	18.01850	38.81968
586	343 396	24.20744	76.55064	201 230 056	8.368 209	18.02876	38.84179
587	344 569	24.22808	76.61593	202 262 003	8.372 967	18.03901	38.86387
588	345 744	24.24871	76.68116	203 297 472	8.377 719	18.04925	38.88593
589	346 921	24.26932	76.74634	204 336 469	8.382 465	18.05947	38.90796
590	348 100	24.28992	76.81146	205 379 000	8.387 207	18.06969	38.92996
591	349 281	24.31049	76.87652	206 425 071	8.391 942	18.07989	38.95195
592	350 464	24.33105	76.94154	207 474 688	8.396 673	18.09008	38.97390
593	351 649	24.35159	77.00649	208 527 857	8.401 398	18.10026	38.99584
594	352 836	24.37212	77.07140	209 584 584	8.406 118	18.11043	39.01774
595	354 025	24.39262	77.13624	210 644 875	8.410 833	18.12059	39.03963
596	355 216	24.41311	77.20104	211 708 736	8.415 542	18.13074	39.06149
597	356 409	24.43358	77.26578	212 776 173	8.420 246	18.14087	39.08332
598	357 604	24.45404	77.33046	213 847 192	8.424 945	18.15099	39.10513
599	358 801	24.47448	77.39509	214 921 799	8.429 638	18.16111	39.12692
600	360 000	24.49490	77.45967	216 000 000	8.434 327	18.17121	39.14868

n	n^2	$\sqrt{n}$	$\sqrt{10n}$	n^3	$\sqrt[3]{n}$	$\sqrt[3]{10n}$ *	$\sqrt[3]{100n}$
600	360 000	24.49490	77.45967	216 000 000	8.434 327	18.17121	39.14868
601	361 201	24.51530	77.52419	217 081 801	8.439 010	18.18130	39.17041
602	362 404	24.53569	77.58866	218 167 208	8.443 688	18.19137	39.19213
603	363 609	24.55606	77.65307	219 256 227	8.448 361	18.20144	39.21382
604	364 816	24.57641	77.71744	220 348 864	8.453 028	18.21150	39.23548
605	366 025	24.59675	77.78175	221 445 125	8.457 691	18.22154	39.25712
606	367 236	24.61707	77.84600	222 545 016	8.462 348	18.23158	39.27874
607	368 449	24.63737	77.91020	223 648 543	8.467 000	18.24160	39.30033
608	369 664	24.65766	77.97435	224 755 712	8.471 647	18.25161	39.32190
609	370 881	24.67793	78.03845	225 866 529	8.476 289	18.26161	39.34345
610	372 100	24.69818	78.10250	226 981 000	8.480 926	18.27160	39.36497
611	373 321	24.71841	78.16649	228 099 131	8.485 558	18.28158	39.38647
612	374 544	24.73863	78.23043	229 220 928	8.490 185	18.29155	39.40795
613	375 769	24.75884	78.29432	230 346 397	8.494 807	18.30151	39.42940
614	376 996	24.77902	78.35815	231 475 544	8.499 423	18.31145	39.45083
615	378 225	24.79919	78.42194	232 608 375	8.504 035	18.32139	39.47223
616	379 456	24.81935	78.48567	233 744 896	8.508 642	18.33131	39.49362
617	380 689	24.83948	78.54935	234 885 113	8.513 243	18.34123	39.51498
618	381 924	24.85961	78.61298	236 029 032	8.517 840	18.35113	39.53631
619	383 161	24.87971	78.67655	237 176 659	8.522 432	18.36102	39.55763
620	384 400	24.89980	78.74008	238 328 000	8.527 019	18.37091	39.57892
621	385 641	24.91987	78.80355	239 483 061	8.531 601	18.38078	39.60018
622	386 884	24.93993	78.86698	240 641 848	8.536 178	18.39064	39.62143
623	388 129	24.95997	78.93035	241 804 367	8.540 750	18.40049	39.64265
624	389 376	24.97999	78.99367	242 970 624	8.545 317	18.41033	39.66385
625	390 625	25.00000	79.05694	244 140 625	8.549 880	18.42016	39.68503
626	391 876	25.01999	79.12016	245 314 376	8.554 437	18.42998	39.70618
627	393 129	25.03997	79.18333	246 491 883	8.558 990	18.43978	39.72731
628	394 384	25.05993	79.24645	247 673 152	8.563 538	18.44958	39.74842
629	395 641	25.07987	79.30952	248 858 189	8.568 081	18.45937	39.76951
630	396 900	25.09980	79.37254	250 047 000	8.572 619	18.46915	39.79057
631	398 161	25.11971	79.43551	251 239 591	8.577 152	18.47891	39.81161
632	399 424	25.13961	79.49843	252 435 968	8.581 681	18.48867	39.83263
633	400 689	25.15949	79.56130	253 636 137	8.586 205	18.49842	39.85363
634	401 956	25.17936	79.62412	254 840 104	8.590 724	18.50815	39.87461
635	403 225	25.19921	79.68689	256 047 875	8.595 238	18.51788	39.89556
636	404 496	25.21904	79.74961	257 259 456	8.599 748	18.52759	39.91649
637	405 769	25.23886	79.81228	258 474 853	8.604 252	18.53730	39.93740
638	407 044	25.25866	79.87490	259 694 072	8.608 753	18.54700	39.95829
639	408 321	25.27845	79.93748	260 917 119	8.613 248	18.55668	39.97916
640	409 600	25.29822	80.00000	262 144 000	8.617 739	18.56636	40.00000
641	410 881	25.31798	80.06248	263 374 721	8.622 225	18.57602	40.02082
642	412 164	25.33772	80.12490	264 609 288	8.626 706	18.58568	40.04162
643	413 449	25.35744	80.18728	265 847 707	8.631 183	18.59532	40.06240
644	414 736	25 37716	80.24961	267 089 984	8.635 655	18.60495	40.08316
645	416 025	25.39685	80.31189	268 336 125	8.640 123	18.61458	40.10390
646	417 316	25.41653	80.37413	269 586 136	8.644 585	18.62419	40.12461
647	418 609	25.43619	80.43631	270 840 023	8.649 044	18.63380	40.14530
648	419 904	25.45584	80.49845	272 097 792	8.653 497	18.64340	40.16598
649	421 201	25.47548	80.56054	273 359 449	8.657 947	18.65298	40.18663
650	422 500	25.49510	80.62258	274 625 000	8.662 391	18.66256	40.20726

SQUARES, SQUARE ROOT, CUBES AND CUBE ROOT

n	n^2	$\sqrt{n}$	$\sqrt{10n}$	n^3	$\sqrt[3]{n}$	$\sqrt[3]{10n}$	$\sqrt[3]{100n}$
650	422 500	25.49510	80.62258	274 625 000	8.662 391	18.66256	40.20726
651	423 801	25.51470	80.68457	275 894 451	8.666 831	18.67212	40.22787
652	425 104	25.53429	80.74652	277 167 808	8.671 266	18.68168	40.24845
653	426 409	25.55386	80.80842	278 445 077	8.675 697	18.69122	40.26902
654	427 716	25.57342	80.87027	279 726 264	8.680 124	18.70076	40.28957
655	429 025	25.59297	80.93207	281 011 375	8.684 546	18.71029	40.31009
656	430 336	25.61250	80.99383	282 300 416	8.688 963	18.71980	40.33059
657	431 649	25.63201	81.05554	283 593 393	8.693 376	18.72931	40.35108
658	432 964	25.65151	81.11720	284 890 312	8.697 784	18.73881	40.37154
659	434 281	25.67100	81.17881	286 191 179	8.702 188	18.74830	40.39198
660	435 600	25.69047	81.24038	287 496 000	8.706 588	18.75777	40.41240
661	436 921	25.70992	81.30191	288 804 781	8.710 983	18.76724	40.43280
662	438 244	25.72936	81.36338	290 117 528	8.715 373	18.77670	40.45318
663	439 569	25.74879	81.42481	291 434 247	8.719 760	18.78615	40.47354
664	440 896	25.76820	81.48620	292 754 944	8.724 141	18.79559	40.49388
665	442 225	25.78759	81.54753	294 079 625	8.728 519	18.80502	40.51420
666	443 556	25.80698	81.60882	295 408 296	8.732 892	18.81444	40.53449
667	444 889	25.82634	81.67007	296 740 963	8.737 260	18.82386	40.55477
668	446 224	25.84570	81.73127	298 077 632	8.741 625	18.83326	40.57503
669	447 561	25.86503	81.79242	299 418 309	8.745 985	18.84265	40.59526
670	448 900	25.88436	81.85353	300 763 000	8.750 340	18.85204	40.61548
671	450 241	25.90367	81.91459	302 111 711	8.754 691	18.86141	40.63568
672	451 584	25.92296	81.97561	303 464 448	8.759 038	18.87078	40.65585
673	452 929	25.94224	82.03658	304 821 217	8.763 381	18.88013	40.67601
674	454 276	25.96151	82.09750	306 182 024	8.767 719	18.88948	40.69615
675	455 625	25.98076	82.15838	307 546 875	8.772 053	18.89882	40.71626
676	456 976	26.00000	82.21922	308 915 776	8.776 383	18.90814	40.73636
677	458 329	26.01922	82.28001	310 288 733	8.780 708	18.91746	40.75644
678	459 684	26.03843	82.34076	311 665 752	8.785 030	18.92677	40.77650
679	461 041	26.05763	82.40146	313 046 839	8.789 347	18.93607	40.79653
680	462 400	26.07681	82.46211	314 432 000	8.793 659	18.94536	40.81655
681	463 761	26.09598	82.52272	315 821 241	8.797 968	18.95465	40.83655
682	465 124	26.11513	82.58329	317 214 568	8.802 272	18.96392	40.85653
683	466 489	26.13427	82.64381	318 611 987	8.806 572	18.97318	40.87649
684	467 856	26.15339	82.70429	320 013 504	8.810 868	18.98244	40.89643
685	469 225	26.17250	82.76473	321 419 125	8.815 160	18.99169	40.91635
686	470 596	26.19160	82.82512	322 828 856	8.819 447	19.00092	40.93625
687	471 969	26.21068	82.88546	324 242 703	8.823 731	19.01015	40.95613
688	473 344	26.22975	82.94577	325 660 672	8.828 010	19.01937	40.97599
689	474 721	26.24881	83.00602	327 082 769	8.832 285	19.02858	40.99584
690	476 100	26.26785	83.06624	328 509 000	8.836 556	19.03778	41.01566
691	477 481	26.28688	83.12641	329 939 371	8.840 823	19.04698	41.03546
692	478 864	26.30589	83.18654	331 373 888	8.845 085	19.05616	41.05525
693	480 249	26.32489	83.24662	332 812 557	8.849 344	19.06533	41.07502
694	481 636	26.34388	83.30666	334 255 384	8.853 599	19.07450	41.09476
695	483 025	26.36285	83.36666	335 702 375	8.857 849	19.08366	41.11449
696	484 416	26.38181	83.42661	337 153 536	8.862 095	19.09281	41.13420
697	485 809	26.40076	83.48653	338 608 873	8.866 338	19.10195	41.15389
698	487 204	26.41969	83.54639	340 068 392	8.870 576	19.11108	41.17357
699	488 601	26.43861	83.60622	341 532 099	8.874 810	19.12020	41.19322
700	490 000	26.45751	83.66600	343 000 000	8.879 040	19.12931	41.21285

SQUARES, SQUARE ROOT, CUBES AND CUBE ROOT

n	n^2	$\sqrt{n}$	$\sqrt{10n}$	n^3	$\sqrt[3]{n}$	$\sqrt[3]{10n}$	$\sqrt[3]{100n}$
700	490 000	26.45751	83.66600	343 000 000	8.879 040	19.12931	41.21285
701	491 401	26.47640	83.72574	344 472 101	8.883 266	19.13842	41.23247
702	492 804	26.49528	83.78544	345 948 408	8.887 488	19.14751	41.25207
703	494 209	26.51415	83.84510	347 428 927	8.891 706	19.15660	41.27164
704	495 616	26.53300	83.90471	348 913 664	8.895 920	19.16568	41.29120
705	497 025	26.55184	83.96428	350 402 625	8.900 130	19.17475	41.31075
706	498 436	26.57066	84.02381	351 895 816	8.904 337	19.18381	41.33027
707	499 849	26.58947	84.08329	353 393 243	8.908 539	19.19286	41.34977
708	501 264	26.60827	84.14274	354 894 912	8.912 737	19.20191	41.36926
709	502 681	26.62705	84.20214	356 400 829	8.916 931	19.21095	41.38873
710	504 100	26.64583	84.26150	357 911 000	8.921 121	19.21997	41.40818
711	505 521	26.66458	84.32082	359 425 431	8.925 308	19.22899	41.42761
712	506 944	26.68333	84.38009	360 944 128	8.929 490	19.23800	41.44702
713	508 369	26.70206	84.43933	362 467 097	8.933 669	19.24701	41.46642
714	509 796	26.72078	84.49852	363 994 344	8.937 843	19.25600	41.48579
715	511 225	26.73948	84.55767	365 525 875	8.942 014	19.26499	41.50515
716	512 656	26.75818	84.61678	367 061 696	8.946 181	19.27396	41.52449
717	514 089	26.77686	84.67585	368 601 813	8.950 344	19.28293	41.54382
718	515 524	26.79552	84.73488	370 146 232	8.954 503	19.29189	41.56312
719	516 961	26.81418	84.79387	371 694 959	8.958 658	19.30084	41.58241
720	518 400	26.83282	84.85281	373 248 000	8.962 809	19.30979	41.60168
721	519 841	26.85144	84.91172	374 805 361	8.966 957	19.31872	41.62093
722	521 284	26.87006	84.97058	376 367 048	8.971 101	19.32765	41.64016
723	522 729	26.88866	85.02941	377 933 067	8.975 241	19.33657	41.65938
724	524 176	26.90725	85.08819	379 503 424	8.979 377	19.34548	41.67857
725	525 625	26.92582	85.14693	381 078 125	8.983 509	19.35438	41.69775
726	527 076	26.94439	85.20563	382 657 176	8.987 637	19.36328	41.71692
727	528 529	26.96294	85.26429	384 240 583	8.991 762	19.37216	41.73606
728	529 984	26.98148	85.32292	385 828 352	8.995 883	19.38104	41.75519
729	531 441	27.00000	85.38150	387 420 489	9.000 000	19.38991	41.77430
730	532 900	27.01851	85.44004	389 017 000	9.004 113	19.39877	41.79339
731	534 361	27.03701	85.49854	390 617 891	9.008 223	19.40763	41.81247
732	535 824	27.05550	85.55700	392 223 168	9.012 329	19.41647	41.83152
733	537 289	27.07397	85.61542	393 832 837	9.016 431	19.42531	41.85056
734	538 756	27.09243	85.67380	395 446 904	9.020 529	19.43414	41.86959
735	540 225	27.11088	85.73214	397 065 375	9.024 624	19.44296	41.88859
736	541 696	27.12932	85.79044	398 688 256	9.028 715	19.45178	41.90758
737	543 169	27.14774	85.84870	400 315 553	9.032 802	19.46058	41.92655
738	544 644	27.16616	85.90693	401 947 272	9.036 886	19.46938	41.94551
739	546 121	27.18455	85.96511	403 583 419	9.040 966	19.47817	41.96444
740	547 600	27.20294	86.02325	405 224 000	9.045 042	19.48695	41.98336
741	549 081	27.22132	86.08136	406 869 021	9.049 114	19.49573	42.00227
742	550 564	27.23968	86.13942	408 518 488	9.053 183	19.50449	42.02115
743	552 049	27.25803	86.19745	410 172 407	9.057 248	19.51325	42.04002
744	553 536	27.27636	86.25543	411 830 784	9.061 310	19.52200	42.05887
745	555 025	27.29469	86.31338	413 493 625	9.065 368	19.53074	42.07771
746	556 516	27.31300	86.37129	415 160 936	9.069 422	19.53948	42.09653
747	558 009	27.33130	86.42916	416 832 723	9.073 473	19.54820	42.11533
748	559 504	27.34959	86.48699	418 508 992	9.077 520	19.55692	42.13411
749	561 001	27.36786	86.54479	420 189 749	9.081 563	19.56563	42.15288
750	562 500	27.38613	86.60254	421 875 000	9.085 603	19.57434	42.17163

SQUARES, SQUARE ROOT, CUBES AND CUBE ROOT

n	n^2	$\sqrt{n}$	$\sqrt{10n}$	n^3	$\sqrt[3]{n}$	$\sqrt[3]{10n}$	$\sqrt[3]{100n}$
750	562 500	27.38613	86.60254	421 875 000	9.085 603	19.57434	42.17163
751	564 001	27.40438	86.66026	423 564 751	9.089 639	19.58303	42.19037
752	565 504	27.42262	86.71793	425 259 008	9.093 672	19.59172	42.20909
753	567 009	27.44085	86.77557	426 957 777	9.097 701	19.60040	42.22779
754	568 516	27.45906	86.83317	428 661 064	9.101 727	19.60908	42.24647
755	570 025	27.47726	86.89074	430 368 875	9.105 748	19.61774	42.26514
756	571 536	27.49545	86.94826	432 081 216	9.109 767	19.62640	42.28379
757	573 049	27.51363	87.00575	433 798 093	9.113 782	19.63505	42.30243
758	574 564	27.53180	87.06320	435 519 512	9.117 793	19.64369	42.32105
759	576 081	27.54995	87.12061	437 245 479	9.121 801	19.65232	42.33965
760	577 600	27.56810	87.17798	438 976 000	9.125 805	19.66095	42.35824
761	579 121	27.58623	87.23531	440 711 081	9.129 806	19.66957	42.37681
762	580 644	27.60435	87.29261	442 450 728	9.133 803	19.67818	42.39536
763	582 169	27.62245	87.34987	444 194 947	9.137 797	19.68679	42.41390
764	583 696	27.64055	87.40709	445 943 744	9.141 787	19.69538	42.43242
765	585 225	27.65863	87.46428	447 697 125	9.145 774	19.70397	42.45092
766	586 756	27.67671	87.52143	449 455 096	9.149 758	19.71256	42.46941
767	588 289	27.69476	87.57854	451 217 663	9.153 738	19.72113	42.48789
768	589 824	27.71281	87.63561	452 984 832	9.157 714	19.72970	42.50634
769	591 361	27.73085	87.69265	454 756 609	9.161 687	19.73826	42.52478
770	592 900	27.74887	87.74964	456 533 000	9.165 656	19.74681	42.54321
771	594 441	27.76689	87.80661	458 314 011	9.169 623	19.75535	42.56162
772	595 984	27.78489	87.86353	460 099 648	9.173 585	19.76389	42.58001
773	597 529	27.80288	87.92042	461 889 917	9.177 544	19.77242	42.59839
774	599 076	27.82086	87.97727	463 684 824	9.181 500	19.78094	42.61675
775	600 625	27.83882	88.03408	465 484 375	9.185 453	19.78946	42.63509
776	602 176	27.85678	88.09086	467 288 576	9.189 402	19.79797	42.65342
777	603 729	27.87472	88.14760	469 097 433	9.193 347	19.80647	42.67174
778	605 284	27.89265	88.20431	470 910 952	9.197 290	19.81496	42.69004
779	606 841	27.91057	88.26098	472 729 139	9.201 229	19.82345	42.70832
780	608 400	27.92848	88.31761	474 552 000	9.205 164	19.83192	42.72659
781	609 961	27.94638	88.37420	476 379 541	9.209 096	19.84040	42.74484
782	611 524	27.96426	88.43076	478 211 768	9.213 025	19.84886	42.76307
783	613 089	27.98214	88.48729	480 048 687	9.216 950	19.85732	42.78129
784	614 656	28.00000	88.54377	481 890 304	9.220 873	19.86577	42.79950
785	616 225	28.01785	88.60023	483 736 625	9.224 791	19.87421	42.81769
786	617 796	28.03569	88.65664	485 587 656	9.228 707	19.88265	42.83586
787	619 369	28.05352	88.71302	487 443 403	9.232 619	19.89107	42.85402
788	620 944	28.07134	88.76936	489 303 872	9.236 528	19.89950	42.87216
789	622 521	28.08914	88.82567	491 169 069	9.240 433	19.90791	42.89029
790	624 100	28.10694	88.88194	493 039 000	9.244 335	19.91632	42.90840
791	625 681	28.12472	88.93818	494 913 671	9.248 234	19.92472	42.92650
792	627 264	28.14249	88.99438	496 793 088	9.252 130	19.93311	42.94458
793	628 849	28.16026	89.05055	498 677 257	9.256 022	19.94150	42.96265
794	630 436	28.17801	89.10668	500 566 184	9.259 911	19.94987	42.98070
795	632 025	28.19574	89.16277	502 459 875	9.263 797	19.95825	42.99874
796	633 616	28.21347	89.21883	504 358 336	9.267 680	19.96661	43.01676
797	635 209	28.23119	89.27486	506 261 573	9.271 559	19.97497	43.03477
798	636 804	28.24889	89.33085	508 169 592	9.275 435	19.98332	43.05276
799	638 401	28.26659	89.38680	510 082 399	9.279 308	19.99166	43.07073
800	640 000	28.28427	89.44272	512 000 000	9.283 178	20.00000	43.08869

SQUARES, SQUARE ROOT, CUBES AND CUBE ROOT

n	n^2	$\sqrt{n}$	$\sqrt{10n}$	n^3	$\sqrt[3]{n}$	$\sqrt[3]{10n}$	$\sqrt[3]{100n}$
800	640 000	28.28427	89.44272	512 000 000	9.283 178	20.00000	43.08869
801	641 601	28.30194	89.49860	513 922 401	9.287 044	20.00833	43.10664
802	643 204	28.31960	89.55445	515 849 608	9.290 907	20.01665	43.12457
803	644 809	28.33725	89.61027	517 781 627	9.294 767	20.02497	43.14249
804	646 416	28.35489	89.66605	519 718 464	9.298 624	20.03328	43.16039
805	648 025	28.37252	89.72179	521 660 125	9.302 477	20.04158	43.17828
806	649 636	28.39014	89.77750	523 606 616	9.306 328	20.04988	43.19615
807	651 249	28.40775	89.83318	525 557 943	9.310 175	20.05816	43.21400
808	652 864	28.42534	89.88882	527 514 112	9.314 019	20.06645	43.23185
809	654 481	28.44293	89.94443	529 475 129	9.317 860	20.07472	43.24967
810	656 100	28.46050	90.00000	531 441 000	9.321 698	20.08299	43.26749
811	657 721	28.47806	90.05554	533 411 731	9.325 532	20.09125	43.28529
812	659 344	28.49561	90.11104	535 387 328	9.329 363	20.09950	43.30307
813	660 969	28.51315	90.16651	537 367 797	9.333 192	20.10775	43.32084
814	662 596	28.53069	90.22195	539 353 144	9.337 017	20.11599	43.33859
815	664 225	28.54820	90.27735	541 343 375	9.340 839	20.12423	43.35633
816	665 856	28.56571	90.33272	543 338 496	9.344 657	20.13245	43.37406
817	667 489	28.58321	90.38805	545 338 513	9.348 473	20.14067	43.39177
818	669 124	28.60070	90.44335	547 343 432	9.352 286	20.14889	43.40947
819	670 761	28.61818	90.49862	549 353 259	9.356 095	20.15710	43.42715
820	672 400	28.63564	90.55385	551 368 000	9.359 902	20.16530	43.44481
821	674 041	28.65310	90.60905	553 387 661	9.363 705	20.17349	43.46247
822	675 684	28.67054	90.66422	555 412 248	9.367 505	20.18168	43.48011
823	677 329	28.68798	90.71935	557 441 767	9.371 302	20.18986	43.49773
824	678 976	28.70540	90.77445	559 476 224	9.375 096	20.19803	43.51534
825	680 625	28.72281	90.82951	561 515 625	9.378 887	20.20620	43.53294
826	682 276	28.74022	90.88454	563 559 976	9.382 675	20.21436	43.55052
827	683 929	28.75761	90.93954	565 609 283	9.386 460	20.22252	43.56809
828	685 584	28.77499	90.99451	567 663 552	9.390 242	20.23066	43.58564
829	687 241	28.79236	91.04944	569 722 789	9.394 021	20.23880	43.60318
830	688 900	28.80972	91.10434	571 787 000	9.397 796	20.24694	43.62071
831	690 561	28.82707	91.15920	573 856 191	9.401 569	20.25507	43.63822
832	692 224	28.84441	91.21403	575 930 368	9.405 339	20.26319	43.65572
833	693 889	28.86174	91.26883	578 009 537	9.409 105	20.27130	43.67320
834	695 556	28.87906	91.32360	580 093 704	9.412 869	20.27941	43.69067
835	697 225	28.89637	91.37833	582 182 875	9.416 630	20.28751	43.70812
836	698 896	28.91366	91.43304	584 277 056	9.420 387	20.29561	43.72556
837	700 569	28.93095	91.48770	586 376 253	9.424 142	20.30370	43.74299
838	702 244	28.94823	91.54234	588 480 472	9.427 894	20.31178	43.76041
839	703 921	28 96550	91.59694	590 589 719	9.431 642	20.31986	43.77781
840	705 600	28.98275	91.65151	592 704 000	9.435 388	20.32793	43.79519
841	707 281	29.00000	91.70605	594 823 321	9.439 131	20.33599	43.81256
842	708 964	29.01724	91.76056	596 947 688	9.442 870	20.34405	43.82992
843	710 649	29.03446	91.81503	599 077 107	9.446 607	20.35210	43.84727
844	712 336	29.05168	91.86947	601 211 584	9.450 341	20.36014	43.86460
845	714 025	29.06888	91.92388	603 351 125	9.454 072	20.36818	43.88191
846	715 716	29.08608	91.97826	605 495 736	9.457 800	20.37621	43.89922
847	717 409	29.10326	92.03260	607 645 423	9.461 525	20.38424	43.91651
848	719 104	29.12044	92.08692	609 800 192	9.465 247	20.39226	43.93378
849	720 801	29.13760	92.14120	611 960 049	9.468 966	20.40027	43.95105
850	722 500	29.15476	92.19544	614 125 000	9.472 682	20.40828	43.96830

SQUARES, SQUARE ROOT, CUBES AND CUBE ROOT

n	n^2	$\sqrt{n}$	$\sqrt{10n}$	n^3	$\sqrt[3]{n}$	$\sqrt[3]{10n}$	$\sqrt[3]{100n}$
850	722 500	29.15476	92.19544	614 125 000	9.472 682	20.40828	43.96830
851	724 201	29.17190	92.24966	616 295 051	9.476 396	20.41628	43.98553
852	725 904	29.18904	92.30385	618 470 208	9.480 106	20.42427	44.00275
853	727 609	29.20616	92.35800	620 650 477	9.483 814	20.43226	44.01996
854	729 316	29.22328	92.41212	622 835 864	9.487 518	20.44024	44.03716
855	731 025	29.24038	92.46621	625 026 375	9.491 220	20.44821	44.05434
856	732 736	29.25748	92.52027	627 222 016	9.494 919	20.45618	44.07151
857	734 449	29.27456	92.57429	629 422 793	9.498 615	20.46415	44.08866
858	736 164	29.29164	92.62829	631 628 712	9.502 308	20.47210	44.10581
859	737 881	29.30870	92.68225	633 839 779	9.505 998	20.48005	44.12293
860	739 600	29.32576	92.73618	636 056 000	9.509 685	20.48800	44.14005
861	741 321	29.34280	92.79009	638 277 381	9.513 370	20.49593	44.15715
862	743 044	29.35984	92.84396	640 503 928	9.517 052	20.50387	44.17424
863	744 769	29.37686	92.89779	642 735 647	9.520 730	20.51179	44.19132
864	746 496	29.39388	92.95160	644 972 544	9.524 406	20.51971	44.20838
865	748 225	29.41088	93.00538	647 214 625	9.528 079	20.52762	44.22543
866	749 956	29.42788	93.05912	649 461 896	9.531 750	20.53553	44.24246
867	751 689	29.44486	93.11283	651 714 363	9.535 417	20.54343	44.25949
868	753 424	29.46184	93.16652	653 972 032	9.539 082	20.55133	44.27650
869	755 161	29.47881	93.22017	656 234 909	9.542 744	20.55922	44.29349
870	756 900	29.49576	93.27379	658 503 000	9.546 403	20.56710	44.31048
871	758 641	29.51271	93.32738	660 776 311	9.550 059	20.57498	44.32745
872	760 384	29.52965	93.38094	663 054 848	9.553 712	20.58285	44.34440
873	762 129	29.54657	93.43447	665 338 617	9.557 363	20.59071	44.36135
874	763 876	29.56349	93.48797	667 627 624	9.561 011	20.59857	44.37828
875	765 625	29.58040	93.54143	669 921 875	9.564 656	20.60643	44.39520
876	767 376	29.59730	93.59487	672 221 376	9.568 298	20.61427	44.41211
877	769 129	29.61419	93.64828	674 526 133	9.571 938	20.62211	44.42900
878	770 884	29.63106	93.70165	676 836 152	9.575 574	20.62995	44.44588
879	772 641	29.64793	93.75500	679 151 439	9.579 208	20.63778	44.46275
880	774 400	29.66479	93.80832	681 472 000	9.582 840	20.64560	44.47960
881	776 161	29.68164	93.86160	683 797 841	9.586 468	20.65342	44.49644
882	777 924	29.69848	93.91486	686 128 968	9.590 094	20.66123	44.51327
883	779 689	29.71532	93.96808	688 465 387	9.593 717	20.66904	44.53009
884	781 456	29.73214	94.02127	690 807 104	9.597 337	20.67684	44.54689
885	783 225	29.74895	94.07444	693 154 125	9.600 955	20.68463	44.56368
886	784 996	29.76575	94.12757	695 506 456	9.604 570	20.69242	44.58046
887	786 769	29.78255	94.18068	697 864 103	9.608 182	20.70020	44.59723
888	788 544	29.79933	94.23375	700 227 072	9.611 791	20.70798	44.61398
889	790 321	29.81610	94.28680	702 595 369	9.615 398	20.71575	44.63072
890	792 100	29.83287	94.33981	704 969 000	9.619 002	20.72351	44.64745
891	793 881	29.84962	94.39280	707 347 971	9.622 603	20.73127	44.66417
892	795 664	29.86637	94.44575	709 732 288	9.626 202	20.73902	44.68087
893	797 449	29.88311	94.49868	712 121 957	9.629 797	20.74677	44.69756
894	799 236	29.89983	94.55157	714 516 984	9.633 391	20.75451	44.71424
895	801 025	29.91655	94.60444	716 917 375	9.636 981	20.76225	44.73090
896	802 816	29.93326	94.65728	719 323 136	9.640 569	20.76998	44.74756
897	804 609	29.94996	94.71008	721 734 273	9.644 154	20.77770	44.76420
898	806 404	29.96665	94.76286	724 150 792	9.647 737	20.78542	44.78083
899	808 201	29.98333	94.81561	726 572 699	9.651 317	20.79313	44.79744
900	810 000	30.00000	94.86833	729 000 000	9.654 894	20.80084	44.81405

SQUARES, SQUARE ROOT, CUBES AND CUBE ROOT

n	n^2	$\sqrt{n}$	$\sqrt{10n}$	n^3	$\sqrt[3]{n}$	$\sqrt[3]{10n}$	$\sqrt[3]{100n}$
900	810 000	30.00000	94.86833	729 000 000	9.654 894	20.80084	44.81405
901	811 801	30.01666	94.92102	731 432 701	9.658 468	20.80854	44.83064
902	813 604	30.03331	94.97368	733 870 808	9.662 040	20.81623	44.84722
903	815 409	30.04996	95.02631	736 314 327	9.665 610	20.82392	44.86379
904	817 216	30.06659	95.07891	738 763 264	9.669 176	20.83161	44.88034
905	819 025	30.08322	95.13149	741 217 625	9.672 740	20.83929	44.89688
906	820 836	30.09983	95.18403	743 677 416	9.676 302	20.84696	44.91341
907	822 649	30.11644	95.23655	746 142 643	9.679 860	20.85463	44.92993
908	824 464	30.13304	95.28903	748 613 312	9.683 417	20.86229	44.94644
909	826 281	30.14963	95.34149	751 089 429	9.686 970	20.86994	44.96293
910	828 100	30.16621	95.39392	753 571 000	9.690 521	20.87759	44.97941
911	829 921	30.18278	95.44632	756 058 031	9.694 069	20.88524	44.99588
912	831 744	30.19934	95.49869	758 550 528	9.697 615	20.89288	45.01234
913	833 569	30.21589	95.55103	761 048 497	9.701 158	20.90051	45.02879
914	835 396	30.23243	95.60335	763 551 944	9.704 699	20.90814	45.04522
915	837 225	30.24897	95.65563	766 060 875	9.708 237	20.91576	45.06164
916	839 056	30.26549	95.70789	768 575 296	9.711 772	20.92338	45.07805
917	840 889	30.28201	95.76012	771 095 213	9.715 305	20.93099	45.09445
918	842 724	30.29851	95.81232	773 620 632	9.718 835	20.93860	45.11084
919	844 561	30.31501	95.86449	776 151 559	9.722 363	20.94620	45.12721
920	846 400	30.33150	95.91663	778 688 000	9.725 888	20.95379	45.14357
921	848 241	30.34798	95.96874	781 229 961	9.729 411	20.96138	45.15992
922	850 084	30.36445	96.02083	783 777 448	9.732 931	20.96896	45.17626
923	851 929	30.38092	96.07289	786 330 467	9.736 448	20.97654	45.19259
924	853 776	30.39737	96.12492	788 889 024	9.739 963	20.98411	45.20891
925	855 625	30.41381	96.17692	791 453 125	9.743 476	20.99168	45.22521
926	857 476	30.43025	96.22889	794 022 776	9.746 986	20.99924	45.24150
927	859 329	30.44667	96.28084	796 597 983	9.750 493	21.00680	45.25778
928	861 184	30.46309	96.33276	799 178 752	9.753 998	21.01435	45.27405
929	863 041	30.47950	96.38465	801 765 089	9.757 500	21.02190	45.29030
930	864 900	30.49590	96.43651	804 357 000	9.761 000	21.02944	45.30655
931	866 761	30.51229	96.48834	806 954 491	9.764 497	21.03697	45.32278
932	868 624	30.52868	96.54015	809 557 568	9.767 992	21.04450	45.33900
933	870 489	30.54505	96.59193	812 166 237	9.771 485	21.05203	45.35521
934	872 356	30.56141	96.64368	814 780 504	9.774 974	21.05954	45.37141
935	874 225	30.57777	96.69540	817 400 375	9.778 462	21.06706	45.38760
936	876 096	30.59412	96.74709	820 025 856	9.781 946	21.07456	45.40377
937	877 969	30.61046	96.79876	822 656 953	9.785 429	21.08207	45.41994
938	879 844	30.62679	96.85040	825 293 672	9.788 909	21.08956	45.43609
939	881 721	30.64311	96.90201	827 936 019	9.792 386	21.09706	45.45223
940	883 600	30.65942	96.95360	830 584 000	9.795 861	21.10454	45.46836
941	885 481	30.67572	97.00515	833 237 621	9.799 334	21.11202	45.48448
942	887 364	30.69202	97.05668	835 896 888	9.802 804	21.11950	45.50058
943	889 249	30.70831	97.10819	838 561 807	9.806 271	21.12697	45.51668
944	891 136	30.72458	97.15966	841 232 384	9.809 736	21.13444	45.53276
945	893 025	30.74085	97.21111	843 908 625	9.813 199	21.14190	45.54883
946	894 916	30.75711	97.26253	846 590 536	9.816 659	21.14935	45.56490
947	896 809	30.77337	97.31393	849 278 123	9.820 117	21.15680	45.58095
948	898 704	30.78961	97.36529	851 971 392	9.823 572	21.16424	45.59698
949	900 601	30.80584	97.41663	854 670 349	9.827 025	21.17168	45.61301
950	902 500	30.82207	97.46794	857 375 000	9.830 476	21.17912	45.62903

SQUARES, SQUARE ROOT, CUBES AND CUBE ROOT

n	n^2	$\sqrt{n}$	$\sqrt{10n}$	n^3	$\sqrt[3]{n}$	$\sqrt[3]{10n}$	$\sqrt[3]{100n}$
950	902 500	30.82207	97.46794	857 375 000	9.830 476	21.17912	45.62903
951	904 401	30.83829	97.51923	860 085 351	9.833 924	21.18655	45.64503
952	906 304	30.85450	97.57049	862 801 408	9.837 369	21.19397	45.66102
953	908 209	30.87070	97.62172	865 523 177	9.840 813	21.20139	45.67701
954	910 116	30.88689	97.67292	868 250 664	9.844 254	21.20880	45.69298
955	912 025	30.90307	97.72410	870 983 875	9.847 692	21.21621	45.70894
956	913 936	30.91925	97.77525	873 722 816	9.851 128	21.22361	45.72489
957	915 849	30.93542	97.82638	876 467 493	9.854 562	21.23101	45.74082
958	917 764	30.95158	97.87747	879 217 912	9.857 993	21.23840	45.75675
959	919 681	30.96773	97.92855	881 974 079	9.861 422	21.24579	45.77267
960	921 600	30.98387	97.97959	884 736 000	9.864 848	21.25317	45.78857
961	923 521	31.00000	98.03061	887 503 681	9.868 272	21.26055	45.80446
962	925 444	31.01612	98.08160	890 277 128	9.871 694	21.26792	45.82035
963	927 369	31.03224	98.13256	893 056 347	9.875 113	21.27529	45.83622
964	929 296	31.04835	98.18350	895 841 344	9.878 530	21.28265	45.85208
965	931 225	31.06445	98.23441	898 632 125	9.881 945	21.29001	45.86793
966	933 156	31.08054	98.28530	901 428 696	9.885 357	21.29736	45.88376
967	935 089	31.09662	98.33616	904 231 063	9.888 767	21.30470	45.89959
968	937 024	31.11270	98.38699	907 039 232	9.892 175	21.31204	45.91541
969	938 961	31.12876	98.43780	909 853 209	9.895 580	21.31938	45.93121
970	940 900	31.14482	98.48858	912 673 000	9.898 983	21.32671	45.94701
971	942 841	31.16087	98.53933	915 498 611	9.902 384	21.33404	45.96279
972	944 784	31.17691	98.59006	918 330 048	9.905 782	21.34136	45.97857
973	946 729	31.19295	98.64076	921 167 317	9.909 178	21.34868	45.99433
974	948 676	31.20897	98.69144	924 010 424	9.912 571	21.35599	46.01008
975	950 625	31.22499	98.74209	926 859 375	9.915 962	21.36329	46.02582
976	952 576	31.24100	98.79271	929 714 176	9.919 351	21.37059	46.04155
977	954 529	31.25700	98.84331	932 574 833	9.922 738	21.37789	46.05727
978	956 484	31.27299	98.89388	935 441 352	9.926 122	21.38518	46.07298
979	958 441	31.28898	98.94443	938 313 739	9.929 504	21.39247	46.08868
980	960 400	31.30495	98.99495	941 192 000	9.932 884	21.39975	46.10436
981	962 361	31.32092	99.04544	944 076 141	9.936 261	21.40703	46.12004
982	964 324	31.33688	99.09591	946 966 168	9.939 636	21.41430	46.13571
983	966 289	31.35283	99.14636	949 862 087	9.943 009	21.42156	46.15136
984	968 256	31.36877	99.19677	952 763 904	9.946 380	21.42883	46.16700
985	970 225	31.38471	99.24717	955 671 625	9.949 748	21.43608	46.18264
986	972 196	31.40064	99.29753	958 585 256	9.953 114	21.44333	46.19826
987	974 169	31.41656	99.34787	961 504 803	9.956 478	21.45058	46.21387
988	976 144	31.43247	99.39819	964 430 272	9.959 839	21.45782	46.22948
989	978 121	31.44837	99.44848	967 361 669	9.963 198	21.46506	46.24507
990	980 100	31.46427	99.49874	970 299 000	9.966 555	21.47229	46.26065
991	982 081	31.48015	99.54898	973 242 271	9.969 910	21.47952	46.27622
992	984 064	31.49603	99.59920	976 191 488	9.973 262	21.48674	46.29178
993	986 049	31.51190	99.64939	979 146 657	9.976 612	21.49396	46.30733
994	988 036	31.52777	99.69955	982 107 784	9.979 960	21.50117	46.32287
995	990 025	31.54362	99.74969	985 074 875	9.983 305	21.50838	46.33840
996	992 016	31.55947	99.79980	988 047 936	9.986 649	21.51558	46.35392
997	994 009	31.57531	99.84989	991 026 973	9.989 990	21.52278	46.36943
998	996 004	31.59114	99.89995	994 011 992	9.993 329	21.52997	46.38492
999	998 001	31.60696	99.94999	997 002 999	9.996 666	21.53716	46.40041
1000	1 000 000	31.62278	100.00000	1 000 000.000	10.000 000	21.54435	46.41589

[1]POWERS AND ROOTS n^k

Floating decimal notation:

9^{10} = 34867 84401

= (9)3.4867 84401

k				
1	$n^1=$	2	3	4
2	$n^2=$	4	9	16
3	$n^3=$	8	27	64
4	$n^4=$	16	81	256
5	$n^5=$	32	243	1024
6	$n^6=$	64	729	4096
7	$n^7=$	128	2187	16384
8	$n^8=$	256	6561	65536
9	$n^9=$	512	19683	2 62144
10	$n^{10}=$	1024	59049	10 48576
24	$n^{24}=$	167 77216	(11)2.8242 95365	(14)2.8147 49767
1/2	$n^{1/2}=$	1.4142 13562	1.7320 50808	2.0000 00000
1/3	$n^{1/3}=$	1.2599 21050	1.4422 49570	1.5874 01052
1/4	$n^{1/4}=$	1.1892 07115	1.3160 74013	1.4142 13562
1/5	$n^{1/5}=$	1.1486 98355	1.2457 30940	1.3195 07911

k					
1	5	6	7	8	9
2	25	36	49	64	81
3	125	216	343	512	729
4	625	1296	2401	4096	6561
5	3125	7776	16807	32768	59049
6	15625	46656	1 17649	2 62144	5 31441
7	78125	2 79936	8 23543	20 97152	47 82969
8	3 90625	16 79616	57 64801	167 77216	430 46721
9	19 53125	100 77696	403 53607	1342 17728	3874 20489
10	97 65625	604 66176	2824 75249	(9)1.0737 41824	(9)3.4867 84401
24	(16)5.9604 64478	(18)4.7383 81338	(20)1.9158 12314	(21)4.7223 66483	(22)7.9766 44308
1/2	2.2360 67977	2.4494 89743	2.6457 51311	2.8284 27125	3.0000 00000
1/3	1.7099 75947	1.8171 20593	1.9129 31183	2.0000 00000	2.0800 83823
1/4	1.4953 48781	1.5650 84580	1.6265 76562	1.6817 92831	1.7320 50808
1/5	1.3797 29662	1.4309 69081	1.4757 73162	1.5157 16567	1.5518 45574

k					
1	10	11	12	13	14
2	100	121	144	169	196
3	1000	1331	1728	2197	2744
4	10000	14641	20736	28561	38416
5	1 00000	1 61051	2 48832	3 71293	5 37824
6	10 00000	17 71561	29 85984	48 26809	75 29536
7	100 00000	194 87171	358 31808	627 48517	1054 13504
8	1000 00000	2143 58881	4299 81696	8157 30721	(9)1.4757 89056
9	(9)1.0000 00000	(9)2.3579 47691	(9)5.1597 80352	(10)1.0604 49937	(10)2.0661 04678
10	(10)1.0000 00000	(10)2.5937 42460	(10)6.1917 36422	(11)1.3785 84918	(11)2.8925 46550
24	(24)1.0000 00000	(24)9.8497 32676	(25)7.9496 84720	(26)5.4280 07704	(27)3.2141 99700
1/2	3.1622 77660	3.3166 24790	3.4641 01615	3.6055 51275	3.7416 57387
1/3	2.1544 34690	2.2239 80091	2.2894 28485	2.3513 34688	2.4101 42264
1/4	1.7782 79410	1.8211 60287	1.8612 09718	1.8988 28922	1.9343 36420
1/5	1.5848 93192	1.6153 94266	1.6437 51830	1.6702 77652	1.6952 18203

k					
1	15	16	17	18	19
2	225	256	289	324	361
3	3375	4096	4913	5832	6859
4	50625	65536	83521	1 04976	1 30321
5	7 59375	10 48576	14 19857	18 89568	24 76099
6	113 90625	167 77216	241 37569	340 12224	470 45881
7	1708 59375	2684 35456	4103 38673	6122 20032	8938 71739
8	(9)2.5628 90625	(9)4.2949 67296	(9)6.9757 57441	(10)1.1019 96058	(10)1.6983 56304
9	(10)3.8443 35938	(10)6.8719 47674	(11)1.1858 78765	(11)1.9835 92904	(11)3.2268 76978
10	(11)5.7665 03906	(12)1.0995 11628	(12)2.0159 93900	(12)3.5704 67227	(12)6.1310 66258
24	(28)1.6834 11220	(28)7.9228 16251	(29)3.3944 86713	(30)1.3382 58845	(30)4.8987 62931
1/2	3.8729 83346	4.0000 00000	4.1231 05626	4.2426 40687	4.3588 98944
1/3	2.4662 12074	2.5198 42100	2.5712 81591	2.6207 41394	2.6684 01649
1/4	1.9679 89671	2.0000 00000	2.0305 43185	2.0597 67144	2.0877 97630
1/5	1.7187 71928	1.7411 01127	1.7623 40348	1.7826 02458	1.8019 83127

k					
1	20	21	22	23	24
2	400	441	484	529	576
3	8000	9261	10648	12167	13824
4	1 60000	1 94481	2 34256	2 79841	3 31776
5	32 00000	40 84101	51 53632	64 36343	79 62624
6	640 00000	857 66121	1133 79904	1480 35889	1911 02976
7	(9)1.2800 00000	(9)1.8010 88541	(9)2.4943 57888	(9)3.4048 25447	(9)4.5864 71424
8	(10)2.5600 00000	(10)3.7822 85936	(10)5.4875 87354	(10)7.8310 98528	(11)1.1007 53142
9	(11)5.1200 00000	(11)7.9428 00466	(12)1.2072 69218	(12)1.8011 52661	(12)2.6418 07540
10	(13)1.0240 00000	(13)1.6679 88098	(13)2.6559 92279	(13)4.1426 51121	(13)6.3403 38097
24	(31)1.6777 21600	(31)5.4108 19838	(32)1.6525 10926	(32)4.8025 07640	(33)1.3337 35777
1/2	4.4721 35955	4.5825 75695	4.6904 15760	4.7958 31523	4.8989 79486
1/3	2.7144 17617	2.7589 24176	2.8020 39331	2.8438 66980	2.8844 99141
1/4	2.1147 42527	2.1406 95143	2.1657 36771	2.1899 38703	2.2133 63839
1/5	1.8205 64203	1.8384 16287	1.8556 00736	1.8721 71231	1.8881 75023

[1] **From "Handbook of Mathematical Functions", National Bureau of Standards**

k	25	26	27	28	29
1	25	26	27	28	29
2	625	676	729	784	841
3	15625	17576	19683	21952	24389
4	3 90625	4 56976	5 31441	6 14656	7 07281
5	97 65625	118 81376	143 48907	172 10368	205 11149
6	2441 40625	3089 15776	3874 20489	4818 90304	5948 23321
7	(9) 6.1035 15625	(9) 8.0318 10176	(10) 1.0460 35320	(10) 1.3492 92851	(10) 1.7249 87631
8	(11) 1.5258 78906	(11) 2.0882 70646	(11) 2.8242 95365	(11) 3.7780 19983	(11) 5.0024 64130
9	(12) 3.8146 97266	(12) 5.4295 03679	(12) 7.6255 97485	(13) 1.0578 45595	(13) 1.4507 14598
10	(13) 9.5367 43164	(14) 1.4116 70957	(14) 2.0589 11321	(14) 2.9619 67667	(14) 4.2070 72333
24	(33) 3.5527 13679	(33) 9.1066 85770	(34) 2.2528 39954	(34) 5.3925 32264	(35) 1.2518 49008
1/2	5.0000 00000	5.0990 19514	5.1961 52423	5.2915 02622	5.3851 64807
1/3	2.9240 17738	2.9624 96068	3.0000 00000	3.0365 88972	3.0723 16826
1/4	2.2360 67977	2.2581 00864	2.2795 07057	2.3003 26634	2.3205 95787
1/5	1.9036 53939	1.9186 45192	1.9331 82045	1.9472 94361	1.9610 09057

k	30	31	32	33	34
1	30	31	32	33	34
2	900	961	1024	1089	1156
3	27000	29791	32768	35937	39304
4	8 10000	9 23521	10 48576	11 85921	13 36336
5	243 00000	286 29151	335 54432	391 35393	454 35424
6	7290 00000	8875 03681	(9) 1.0737 41824	(9) 1.2914 67969	(9) 1.5448 04416
7	(10) 2.1870 00000	(10) 2.7512 61411	(10) 3.4359 73837	(10) 4.2618 44298	(10) 5.2523 35014
8	(11) 6.5610 00000	(11) 8.5289 10374	(12) 1.0995 11628	(12) 1.4064 08618	(12) 1.7857 93905
9	(13) 1.9683 00000	(13) 2.6439 62216	(13) 3.5184 37209	(13) 4.6411 48440	(13) 6.0716 99277
10	(14) 5.9049 00000	(14) 8.1962 82870	(15) 1.1258 99907	(15) 1.5315 78985	(15) 2.0643 77754
24	(35) 2.8242 95365	(35) 6.2041 26610	(36) 1.3292 27996	(36) 2.7818 55434	(36) 5.6950 03680
1/2	5.4772 25575	5.5677 64363	5.6568 54249	5.7445 62647	5.8309 51895
1/3	3.1072 32506	3.1413 80652	3.1748 02104	3.2075 34330	3.2396 11801
1/4	2.3403 47319	2.3596 11062	2.3784 14230	2.3967 81727	2.4147 36403
1/5	1.9743 50486	1.9873 40755	2.0000 00000	2.0123 46617	2.0243 97459

k	35	36	37	38	39
1	35	36	37	38	39
2	1225	1296	1369	1444	1521
3	42875	46656	50653	54872	59319
4	15 00625	16 79616	18 74161	20 85136	23 13441
5	525 21875	604 66176	693 43957	792 35168	902 24199
6	(9) 1.8382 65625	(9) 2.1767 82336	(9) 2.5657 26409	(9) 3.0109 36384	(9) 3.5187 43761
7	(10) 6.4339 29688	(10) 7.8364 16410	(10) 9.4931 87713	(11) 1.1441 55826	(11) 1.3723 10067
8	(12) 2.2518 75391	(12) 2.8211 09907	(12) 3.5124 79454	(12) 4.3477 92138	(12) 5.3520 09260
9	(13) 7.8815 63867	(14) 1.0155 99567	(14) 1.2996 17398	(14) 1.6521 61013	(14) 2.0872 83612
10	(15) 2.7585 47354	(15) 3.6561 58440	(15) 4.8085 84372	(15) 6.2782 11848	(15) 8.1404 06085
24	(37) 1.1419 13124	(37) 2.2452 25771	(37) 4.3335 25711	(37) 8.2187 60383	(38) 1.5330 29700
1/2	5.9160 79783	6.0000 00000	6.0827 62530	6.1644 14003	6.2449 97998
1/3	3.2710 66310	3.3019 27249	3.3322 21852	3.3619 75407	3.3912 11443
1/4	2.4322 99279	2.4494 89743	2.4663 25715	2.4828 23796	2.4989 99399
1/5	2.0361 68005	2.0476 72511	2.0589 24137	2.0699 35054	2.0807 16549

k	40	41	42	43	44
1	40	41	42	43	44
2	1600	1681	1764	1849	1936
3	64000	68921	74088	79507	85184
4	25 60000	28 25761	31 11696	34 18801	37 48096
5	1024 00000	1158 56201	1306 91232	1470 08443	1649 16224
6	(9) 4.0960 00000	(9) 4.7501 04241	(9) 5.4890 31744	(9) 6.3213 63049	(9) 7.2563 13856
7	(11) 1.6384 00000	(11) 1.9475 42739	(11) 2.3053 93332	(11) 2.7181 86111	(11) 3.1927 78097
8	(12) 6.5536 00000	(12) 7.9849 25229	(12) 9.6826 51996	(13) 1.1688 20028	(13) 1.4048 22363
9	(14) 2.6214 40000	(14) 3.2738 19344	(14) 4.0667 13838	(14) 5.0259 26119	(14) 6.1812 18395
10	(16) 1.0485 76000	(16) 1.3422 65931	(16) 1.7080 19812	(16) 2.1611 48231	(16) 2.7197 36094
24	(38) 2.8147 49767	(38) 5.0911 10945	(38) 9.0778 49315	(39) 1.5967 72093	(39) 2.7724 53276
1/2	6.3245 55320	6.4031 24237	6.4807 40698	6.5574 38524	6.6332 49581
1/3	3.4199 51893	3.4482 17240	3.4760 26645	3.5033 98060	3.5303 48335
1/4	2.5148 66859	2.5304 39534	2.5457 29895	2.5607 49602	2.5755 09577
1/5	2.0912 79105	2.1016 32478	2.1117 85765	2.1217 47461	2.1315 25513

k	45	46	47	48	49
1	45	46	47	48	49
2	2025	2116	2209	2304	2401
3	91125	97336	1 03823	1 10592	1 17649
4	41 00625	44 77456	48 79681	53 08416	57 64801
5	1845 28125	2059 62976	2293 45007	2548 03968	2824 75249
6	(9) 8.3037 65625	(9) 9.4742 96896	(10) 1.0779 21533	(10) 1.2230 59046	(10) 1.3841 28720
7	(11) 3.7366 94531	(11) 4.3581 76572	(11) 5.0662 31205	(11) 5.8706 83423	(11) 6.7822 30728
8	(13) 1.6815 12539	(13) 2.0047 61223	(13) 2.3811 28666	(13) 2.8179 28043	(13) 3.3232 93057
9	(14) 7.5668 06426	(14) 9.2219 01627	(15) 1.1191 30473	(15) 1.3526 05461	(15) 1.6284 13598
10	(16) 3.4050 62892	(16) 4.2420 74748	(16) 5.2599 13224	(16) 6.4925 06211	(16) 7.9792 26630
24	(39) 4.7544 50505	(39) 8.0572 70802	(40) 1.3500 46075	(40) 2.2376 37322	(40) 3.6703 36822
1/2	6.7082 03932	6.7823 29983	6.8556 54600	6.9282 03230	7.0000 00000
1/3	3.5568 93304	3.5830 47871	3.6088 26080	3.6342 41186	3.6593 05710
1/4	2.5900 20064	2.6042 90687	2.6183 30499	2.6321 48026	2.6457 51311
1/5	2.1411 27368	2.1505 60013	2.1598 30012	2.1689 43542	2.1779 06425

$n^{\frac{1}{2}}\begin{bmatrix}(-4)3\\8\end{bmatrix}$ $n^{\frac{1}{3}}\begin{bmatrix}(-4)1\\7\end{bmatrix}$ $n^{\frac{1}{4}}\begin{bmatrix}(-5)9\\8\end{bmatrix}$ $n^{\frac{1}{5}}\begin{bmatrix}(-5)7\\6\end{bmatrix}$

The numbers in square brackets at the bottom, of the page mean that the maximum error in a linear interpolate is $a \times 10^{-p}$ (p in parentheses), and that to interpolate to the full tabular accuracy m points must be used in Lagrange's and Aitkens methods for the respective functions $n^{1/r}$.

POWERS AND ROOTS n^k

k					
1	50	51	52	53	54
2	2500	2601	2704	2809	2916
3	1 25000	1 32651	1 40608	1 48877	1 57464
4	62 50000	67 65201	73 11616	78 90481	85 03056
5	3125 00000	3450 25251	3802 04032	4181 95493	4591 65024
6	(10) 1.5625 00000	(10) 1.7596 28780	(10) 1.9770 60966	(10) 2.2164 36113	(10) 2.4794 91130
7	(11) 7.8125 00000	(11) 8.9741 06779	(12) 1.0280 71703	(12) 1.1747 11140	(12) 1.3389 25210
8	(13) 3.9062 50000	(13) 4.5767 94457	(13) 5.3459 72853	(13) 6.2259 69041	(13) 7.2301 96134
9	(15) 1.9531 25000	(15) 2.3341 65173	(15) 2.7799 05884	(15) 3.2997 63592	(15) 3.9043 05912
10	(16) 9.7656 25000	(17) 1.1904 24238	(17) 1.4455 51059	(17) 1.7488 74704	(17) 2.1083 25193
24	(40) 5.9604 64478	(40) 9.5870 33090	(41) 1.5278 48342	(41) 2.4133 53110	(41) 3.7796 38253
1/2	7.0710 67812	7.1414 28429	7.2111 02551	7.2801 09889	7.3484 69228
1/3	3.6840 31499	3.7084 29769	3.7325 11157	3.7562 85754	3.7797 63150
1/4	2.6591 47948	2.6723 45118	2.6853 49614	2.6981 67876	2.7108 06011
1/5	2.1867 24148	2.1954 01897	2.2039 44575	2.2123 56822	2.2206 43035

k					
1	55	56	57	58	59
2	3025	3136	3249	3364	3481
3	1 66375	1 75616	1 85193	1 95112	2 05379
4	91 50625	98 34496	105 56001	113 16496	121 17361
5	5032 84375	5507 31776	6016 92057	6563 56768	7149 24299
6	(10) 2.7680 64063	(10) 3.0840 97946	(10) 3.4296 44725	(10) 3.8068 69254	(10) 4.2180 53364
7	(12) 1.5224 35234	(12) 1.7270 94850	(12) 1.9548 97493	(12) 2.2079 84168	(12) 2.4886 51485
8	(13) 8.3733 93789	(13) 9.6717 31157	(14) 1.1142 91571	(14) 1.2806 30817	(14) 1.4683 04376
9	(15) 4.6053 66584	(15) 5.4161 69448	(15) 6.3514 61955	(15) 7.4276 58740	(15) 8.6629 95819
10	(17) 2.5329 51621	(17) 3.0330 54891	(17) 3.6203 33315	(17) 4.3080 42069	(17) 5.1111 67533
24	(41) 5.8708 98173	(41) 9.0471 67858	(42) 1.3835 55344	(42) 2.1002 54121	(42) 3.1655 43453
1/2	7.4161 98487	7.4833 14774	7.5498 34435	7.6157 73106	7.6811 45748
1/3	3.8029 52461	3.8258 62366	3.8485 01131	3.8708 76641	3.8929 96416
1/4	2.7232 69815	2.7355 64800	2.7476 96205	2.7596 69021	2.7714 88002
1/5	2.2288 07384	2.2368 53829	2.2447 86134	2.2526 07878	2.2603 22470

k					
1	60	61	62	63	64
2	3600	3721	3844	3969	4096
3	2 16000	2 26981	2 38328	2 50047	2 62144
4	129 60000	138 45841	147 76336	157 52961	167 77216
5	7776 00000	8445 96301	9161 32832	9924 36543	(9) 1.0737 41824
6	(10) 4.6656 00000	(10) 5.1520 37436	(10) 5.6800 23558	(10) 6.2523 50221	(10) 6.8719 47674
7	(12) 2.7993 60000	(12) 3.1427 42836	(12) 3.5216 14606	(12) 3.9389 80639	(12) 4.3980 46511
8	(14) 1.6796 16000	(14) 1.9170 73130	(14) 2.1834 01056	(14) 2.4815 57803	(14) 2.8147 49767
9	(16) 1.0077 69600	(16) 1.1694 14609	(16) 1.3537 08655	(16) 1.5633 81416	(16) 1.8014 39851
10	(17) 6.0466 17600	(17) 7.1334 29117	(17) 8.3929 93659	(17) 9.8493 02919	(18) 1.1529 21505
24	(42) 4.7383 81338	(42) 7.0455 68477	(43) 1.0408 79722	(43) 1.5281 75339	(43) 2.2300 74520
1/2	7.7459 66692	7.8102 49676	7.8740 07874	7.9372 53933	8.0000 00000
1/3	3.9148 67641	3.9364 97183	3.9578 91610	3.9790 57208	4.0000 00000
1/4	2.7831 57684	2.7946 82393	2.8060 66263	2.8173 13247	2.8284 27125
1/5	2.2679 33155	2.2754 43032	2.2828 55056	2.2901 72049	2.2973 96710

k					
1	65	66	67	68	69
2	4225	4356	4489	4624	4761
3	2 74625	2 87496	3 00763	3 14432	3 28509
4	178 50625	189 74736	201 51121	213 81376	226 67121
5	(9) 1.1602 90625	(9) 1.2523 32576	(9) 1.3501 25107	(9) 1.4539 33568	(9) 1.5640 31349
6	(10) 7.5418 89063	(10) 8.2653 95002	(10) 9.0458 38217	(10) 9.8867 48262	(11) 1.0791 81631
7	(12) 4.9022 27891	(12) 5.4551 60701	(12) 6.0607 11605	(12) 6.7229 88818	(12) 7.4463 53253
8	(14) 3.1864 48129	(14) 3.6004 06063	(14) 4.0606 76776	(14) 4.5716 32397	(14) 5.1379 83744
9	(16) 2.0711 91284	(16) 2.3762 68001	(16) 2.7206 53440	(16) 3.1087 10030	(16) 3.5452 08784
10	(18) 1.3462 74334	(18) 1.5683 36881	(18) 1.8228 37805	(18) 2.1139 22820	(18) 2.4461 94061
24	(43) 3.2353 44710	(43) 4.6671 78950	(43) 6.6956 88867	(43) 9.5546 30685	(44) 1.3563 70007
1/2	8.0622 57748	8.1240 38405	8.1853 52772	8.2462 11251	8.3066 23863
1/3	4.0207 25759	4.0412 40021	4.0615 48100	4.0816 55102	4.1015 65930
1/4	2.8394 11514	2.8502 69883	2.8610 05553	2.8716 21711	2.8821 21417
1/5	2.3045 31620	2.3115 79249	2.3185 41963	2.3254 22030	2.3322 21626

k					
1	70	71	72	73	74
2	4900	5041	5184	5329	5476
3	3 43000	3 57911	3 73248	3 89017	4 05224
4	240 10000	254 11681	268 73856	283 98241	299 86576
5	(9) 1.6807 00000	(9) 1.8042 29351	(9) 1.9349 17632	(9) 2.0730 71593	(9) 2.2190 06624
6	(11) 1.1764 90000	(11) 1.2810 02839	(11) 1.3931 40695	(11) 1.5133 42263	(11) 1.6420 64902
7	(12) 8.2354 30000	(12) 9.0951 20158	(13) 1.0030 61300	(13) 1.1047 39852	(13) 1.2151 28027
8	(14) 5.7648 01000	(14) 6.4575 35312	(14) 7.2220 41363	(14) 8.0646 00919	(14) 8.9919 47402
9	(16) 4.0353 60700	(16) 4.5848 50072	(16) 5.1998 69781	(16) 5.8871 58671	(16) 6.6540 41078
10	(18) 2.8247 52490	(18) 3.2552 43551	(18) 3.7439 06243	(18) 4.2976 25830	(18) 4.9239 90397
24	(44) 1.9158 12314	(44) 2.6927 76876	(44) 3.7668 63772	(44) 5.2450 38047	(44) 7.2704 49690
1/2	8.3666 00265	8.4261 49773	8.4852 81374	8.5440 03745	8.6023 25267
1/3	4.1212 85300	4.1408 17749	4.1601 67646	4.1793 39196	4.1983 36454
1/4	2.8925 07608	2.9027 83108	2.9129 50630	2.9230 12786	2.9329 72088
1/5	2.3389 42837	2.3455 87669	2.3521 58045	2.3586 55818	2.3650 82769

$n^{\frac{1}{2}}\left[\begin{matrix}(-5)9\\6\end{matrix}\right]$ $n^{\frac{1}{3}}\left[\begin{matrix}(-5)4\\6\end{matrix}\right]$ $n^{\frac{1}{4}}\left[\begin{matrix}(-5)3\\5\end{matrix}\right]$ $n^{\frac{1}{5}}\left[\begin{matrix}(-5)2\\5\end{matrix}\right]$

POWERS AND ROOTS n^k

k					
1	75	76	77	78	79
2	5625	5776	5929	6084	6241
3	4 21875	4 38976	4 56533	4 74552	4 93039
4	316 40625	333 62176	351 53041	370 15056	389 50081
5	(9) 2.3730 46875	(9) 2.5355 25376	(9) 2.7067 84157	(9) 2.8871 74368	(9) 3.0770 56399
6	(11) 1.7797 85156	(11) 1.9269 99286	(11) 2.0842 23801	(11) 2.2519 96007	(11) 2.4308 74555
7	(13) 1.3348 38867	(13) 1.4645 19457	(13) 1.6048 52327	(13) 1.7565 56885	(13) 1.9203 90899
8	(15) 1.0011 29150	(15) 1.1130 34787	(15) 1.2357 36292	(15) 1.3701 14371	(15) 1.5171 08810
9	(16) 7.5084 68628	(16) 8.4590 64385	(16) 9.5151 69445	(17) 1.0686 89209	(17) 1.1985 15960
10	(18) 5.6313 51471	(18) 6.4288 88932	(18) 7.3266 80473	(18) 8.3357 75831	(18) 9.4682 76083
24	(45) 1.0033 91278	(45) 1.3788 79182	(45) 1.8870 23915	(45) 2.5719 97041	(45) 3.4918 06676
1/2	8.6602 54038	8.7177 97887	8.7749 64387	8.8317 60866	8.8881 94417
1/3	4.2171 63326	4.2358 23584	4.2543 20865	4.2726 58682	4.2908 40427
1/4	2.9428 30956	2.9525 91724	2.9622 56638	2.9718 27866	2.9813 07501
1/5	2.3714 40610	2.3777 30992	2.3839 55503	2.3901 15677	2.3962 12991

k					
1	80	81	82	83	84
2	6400	6561	6724	6889	7056
3	5 12000	5 31441	5 51368	5 71787	5 92704
4	409 60000	430 46721	452 12176	474 58321	497 87136
5	(9) 3.2768 00000	(9) 3.4867 84401	(9) 3.7073 98432	(9) 3.9390 40643	(9) 4.1821 19424
6	(11) 2.6214 40000	(11) 2.8242 95365	(11) 3.0400 66714	(11) 3.2694 03734	(11) 3.5129 80316
7	(13) 2.0971 52000	(13) 2.2876 79245	(13) 2.4928 54706	(13) 2.7136 05099	(13) 2.9509 03466
8	(15) 1.6777 21600	(15) 1.8530 20189	(15) 2.0441 40859	(15) 2.2522 92232	(15) 2.4787 58911
9	(17) 1.3421 77280	(17) 1.5009 46353	(17) 1.6761 95504	(17) 1.8694 02553	(17) 2.0821 57485
10	(19) 1.0737 41824	(19) 1.2157 66546	(19) 1.3744 80313	(19) 1.5516 04119	(19) 1.7490 12288
24	(45) 4.7223 66483	(45) 6.3626 85441	(45) 8.5414 66801	(46) 1.1425 47375	(46) 1.5230 10388
1/2	8.9442 71910	9.0000 00000	9.0553 85138	9.1104 33579	9.1651 51390
1/3	4.3088 69380	4.3267 48711	4.3444 81486	4.3620 70671	4.3795 19140
1/4	2.9906 97562	3.0000 00000	3.0092 16698	3.0183 49479	3.0274 00104
1/5	2.4022 48868	2.4082 24685	2.4141 41771	2.4200 01407	2.4258 04834

k					
1	85	86	87	88	89
2	7225	7396	7569	7744	7921
3	6 14125	6 36056	6 58503	6 81472	7 04969
4	522 00625	547 00816	572 89761	599 69536	627 42241
5	(9) 4.4370 53125	(9) 4.7042 70176	(9) 4.9842 09207	(9) 5.2773 19168	(9) 5.5840 59449
6	(11) 3.7714 95156	(11) 4.0456 72351	(11) 4.3362 62010	(11) 4.6440 40868	(11) 4.9698 12910
7	(13) 3.2057 70883	(13) 3.4792 78222	(13) 3.7725 47949	(13) 4.0867 55964	(13) 4.4231 33490
8	(15) 2.7249 05250	(15) 2.9921 79271	(15) 3.2821 16715	(15) 3.5963 45248	(15) 3.9365 88806
9	(17) 2.3161 69463	(17) 2.5732 74173	(17) 2.8554 41542	(17) 3.1647 83818	(17) 3.5035 64037
10	(19) 1.9687 44043	(19) 2.2130 15789	(19) 2.4842 34142	(19) 2.7850 09760	(19) 3.1181 71993
24	(46) 2.0232 71747	(46) 2.6789 39031	(46) 3.5355 91351	(46) 4.6514 04745	(46) 6.1004 25945
1/2	9.2195 44457	9.2736 18495	9.3273 79053	9.3808 31520	9.4339 81132
1/3	4.3968 29672	4.4140 04962	4.4310 47622	4.4479 60181	4.4647 45096
1/4	3.0363 70277	3.0452 61646	3.0540 75810	3.0628 14314	3.0714 78656
1/5	2.4315 53252	2.4372 47818	2.4428 89656	2.4484 79851	2.4540 19455

k					
1	90	91	92	93	94
2	8100	8281	8464	8649	8836
3	7 29000	7 53571	7 78688	8 04357	8 30584
4	656 10000	685 74961	716 39296	748 05201	780 74896
5	(9) 5.9049 00000	(9) 6.2403 21451	(9) 6.5908 15232	(9) 6.9568 83693	(9) 7.3390 40224
6	(11) 5.3144 10000	(11) 5.6786 92520	(11) 6.0635 50013	(11) 6.4699 01834	(11) 6.8986 97811
7	(13) 4.7829 69000	(13) 5.1676 10194	(13) 5.5784 66012	(13) 6.0170 08706	(13) 6.4847 75942
8	(15) 4.3046 72100	(15) 4.7025 25276	(15) 5.1321 88731	(15) 5.5958 18097	(15) 6.0956 89385
9	(17) 3.8742 04890	(17) 4.2792 98001	(17) 4.7216 13633	(17) 5.2041 10830	(17) 5.7299 48022
10	(19) 3.4867 84401	(19) 3.8941 61181	(19) 4.3438 84542	(19) 4.8398 23072	(19) 5.3861 51141
24	(46) 7.9766 44308	(47) 1.0399 04400	(47) 1.3517 85726	(47) 1.7522 28603	(47) 2.2650 01461
1/2	9.4868 32981	9.5393 92014	9.5916 63047	9.6436 50761	9.6953 59715
1/3	4.4814 04747	4.4979 41445	4.5143 57435	4.5306 54896	4.5468 35944
1/4	3.0800 70288	3.0885 90619	3.0970 41015	3.1054 22799	3.1137 37258
1/5	2.4595 09486	2.4649 50932	2.4703 44749	2.4756 91866	2.4809 93182

k					
1	95	96	97	98	99
2	9025	9216	9409	9604	9801
3	8 57375	8 84736	9 12673	9 41192	9 70299
4	814 50625	849 34656	885 29281	922 36816	960 59601
5	(9) 7.7378 09375	(9) 8.1537 26976	(9) 8.5873 40257	(9) 9.0392 07968	(9) 9.5099 00499
6	(11) 7.3509 18906	(11) 7.8275 77897	(11) 8.3297 20049	(11) 8.8584 23809	(11) 9.4148 01494
7	(13) 6.9833 72961	(13) 7.5144 74781	(13) 8.0798 28448	(13) 8.6812 55332	(13) 9.3206 53479
8	(15) 6.6342 04313	(15) 7.2138 95790	(15) 7.8374 33594	(15) 8.5076 30226	(15) 9.2274 46944
9	(17) 6.3024 94097	(17) 6.9253 39958	(17) 7.6023 10587	(17) 8.3374 77621	(17) 9.1351 72475
10	(19) 5.9873 69392	(19) 6.6483 26360	(19) 7.3742 41269	(19) 8.1707 28069	(19) 9.0438 20750
24	(47) 2.9198 90243	(47) 3.7541 32467	(47) 4.8141 72219	(47) 6.1578 03365	(47) 7.8567 81408
1/2	9.7467 94345	9.7979 58971	9.8488 57802	9.8994 94937	9.9498 74371
1/3	4.5629 02635	4.5788 56970	4.5947 00892	4.6104 36292	4.6260 65009
1/4	3.1219 85641	3.1301 69160	3.1382 88993	3.1463 46284	3.1543 42146
1/5	2.4862 49570	2.4914 61879	2.4966 30932	2.5017 57527	2.5068 42442

$n^{\frac{1}{2}}\begin{bmatrix}(-5)5\\5\end{bmatrix}$ $n^{\frac{1}{3}}\begin{bmatrix}(-5)2\\5\end{bmatrix}$ $n^{\frac{1}{4}}\begin{bmatrix}(-5)1\\5\end{bmatrix}$ $n^{\frac{1}{5}}\begin{bmatrix}(-6)9\\5\end{bmatrix}$

k					
1	100	101	102	103	104
2	10000	10201	10404	10609	10816
3	10 00000	10 30301	10 61208	10 92727	11 24864
4	1000 00000	1040 60401	1082 43216	1125 50881	1169 85856
5	(10) 1.0000 00000	(10) 1.0510 10050	(10) 1.1040 80803	(10) 1.1592 74074	(10) 1.2166 52902
6	(12) 1.0000 00000	(12) 1.0615 20151	(12) 1.1261 62419	(12) 1.1940 52297	(12) 1.2653 19018
7	(14) 1.0000 00000	(14) 1.0721 35352	(14) 1.1486 85668	(14) 1.2298 73865	(14) 1.3159 31779
8	(16) 1.0000 00000	(16) 1.0828 56706	(16) 1.1716 59381	(16) 1.2667 70081	(16) 1.3685 69050
9	(18) 1.0000 00000	(18) 1.0936 85273	(18) 1.1950 92569	(18) 1.3047 73184	(18) 1.4233 11812
10	(20) 1.0000 00000	(20) 1.1046 22125	(20) 1.2189 94420	(20) 1.3439 16379	(20) 1.4802 44285
24	(48) 1.0000 00000	(48) 1.2697 34649	(48) 1.6084 37249	(48) 2.0327 94106	(48) 2.5633 04165
1/2	(1) 1.0000 00000	(1) 1.0049 87562	(1) 1.0099 50494	(1) 1.0148 89157	(1) 1.0198 03903
1/3	4.6415 88834	4.6570 09508	4.6723 28728	4.6875 48148	4.7026 69375
1/4	3.1622 77660	3.1701 53880	3.1779 71828	3.1857 32501	3.1934 36868
1/5	2.5118 86432	2.5168 90229	2.5218 54548	2.5267 80083	2.5316 67508

k					
1	105	106	107	108	109
2	11025	11236	11449	11664	11881
3	11 57625	11 91016	12 25043	12 59712	12 95029
4	1215 50625	1262 47696	1310 79601	1360 48896	1411 58161
5	(10) 1.2762 81563	(10) 1.3382 25578	(10) 1.4025 51731	(10) 1.4693 28077	(10) 1.5386 23955
6	(12) 1.3400 95641	(12) 1.4185 19112	(12) 1.5007 30352	(12) 1.5868 74323	(12) 1.6771 00111
7	(14) 1.4071 00423	(14) 1.5036 30259	(14) 1.6057 81476	(14) 1.7138 24269	(14) 1.8280 39121
8	(16) 1.4774 55444	(16) 1.5938 48075	(16) 1.7181 86180	(16) 1.8509 30210	(16) 1.9925 62642
9	(18) 1.5513 28216	(18) 1.6894 78959	(18) 1.8384 59212	(18) 1.9990 04627	(18) 2.1718 93279
10	(20) 1.6288 94627	(20) 1.7908 47697	(20) 1.9671 51357	(20) 2.1589 24997	(20) 2.3673 63675
24	(48) 3.2250 99944	(48) 4.0489 34641	(48) 5.0723 66953	(48) 6.3411 80737	(48) 7.9110 83175
1/2	(1) 1.0246 95077	(1) 1.0295 63014	(1) 1.0344 08043	(1) 1.0392 30485	(1) 1.0440 30651
1/3	4.7176 93980	4.7326 23491	4.7474 59398	4.7622 03156	4.7768 56181
1/4	3.2010 85873	3.2086 80436	3.2162 21453	3.2237 09795	3.2311 46315
1/5	2.5365 17482	2.5413 30642	2.5461 07613	2.5508 49001	2.5555 55397

k					
1	110	111	112	113	114
2	12100	12321	12544	12769	12996
3	13 31000	13 67631	14 04928	14 42897	14 81544
4	1464 10000	1518 07041	1573 51936	1630 47361	1688 96016
5	(10) 1.6105 10000	(10) 1.6850 58155	(10) 1.7623 41683	(10) 1.8424 35179	(10) 1.9254 14582
6	(12) 1.7715 61000	(12) 1.8704 14552	(12) 1.9738 22685	(12) 2.0819 51753	(12) 2.1949 72624
7	(14) 1.9487 17100	(14) 2.0761 60153	(14) 2.2106 81407	(14) 2.3526 05480	(14) 2.5022 68791
8	(16) 2.1435 88810	(16) 2.3045 37770	(16) 2.4759 63176	(16) 2.6584 44193	(16) 2.8525 86422
9	(18) 2.3579 47691	(18) 2.5580 36924	(18) 2.7730 78757	(18) 3.0040 41938	(18) 3.2519 48521
10	(20) 2.5937 42460	(20) 2.8394 20986	(20) 3.1058 48208	(20) 3.3945 67390	(20) 3.7072 21314
24	(48) 9.8497 32676	(49) 1.2239 15658	(49) 1.5178 62893	(49) 1.8788 09051	(49) 2.3212 20685
1/2	(1) 1.0488 08848	(1) 1.0535 65375	(1) 1.0583 00524	(1) 1.0630 14581	(1) 1.0677 07825
1/3	4.7914 19857	4.8058 95534	4.8202 84528	4.8345 88127	4.8488 07586
1/4	3.2385 31840	3.2458 67180	3.2531 53123	3.2603 90439	3.2675 79877
1/5	2.5602 27376	2.5648 65499	2.5694 70314	2.5740 42354	2.5785 82140

k					
1	115	116	117	118	119
2	13225	13456	13689	13924	14161
3	15 20875	15 60896	16 01613	16 43032	16 85159
4	1749 00625	1810 63936	1873 88721	1938 77776	2005 33921
5	(10) 2.0113 57188	(10) 2.1003 41658	(10) 2.1924 48036	(10) 2.2877 57757	(10) 2.3863 53660
6	(12) 2.3130 60766	(12) 2.4363 96323	(12) 2.5651 64202	(12) 2.6995 54153	(12) 2.8397 60855
7	(14) 2.6600 19880	(14) 2.8262 19734	(14) 3.0012 42116	(14) 3.1854 73901	(14) 3.3793 15418
8	(16) 3.0590 22863	(16) 3.2784 14892	(16) 3.5114 53276	(16) 3.7588 59203	(16) 4.0213 85347
9	(18) 3.5178 76292	(18) 3.8029 61275	(18) 4.1084 00333	(18) 4.4354 53859	(18) 4.7854 48563
10	(20) 4.0455 57736	(20) 4.4114 35079	(20) 4.8068 28389	(20) 5.2338 35554	(20) 5.6946 83790
24	(49) 2.8625 17619	(49) 3.5236 41704	(49) 4.3297 28675	(49) 5.3109 00627	(49) 6.5031 99444
1/2	(1) 1.0723 80529	(1) 1.0770 32961	(1) 1.0816 65383	(1) 1.0862 78049	(1) 1.0908 71211
1/3	4.8629 44131	4.8769 98961	4.8909 73246	4.9048 68131	4.9186 84734
1/4	3.2747 22171	3.2818 18035	3.2888 68168	3.2958 73252	3.3028 33952
1/5	2.5830 90178	2.5875 66964	2.5920 12982	2.5964 28703	2.6008 14587

k					
1	120	121	122	123	124
2	14400	14641	14884	15129	15376
3	17 28000	17 71561	18 15848	18 60867	19 06624
4	2073 60000	2143 58881	2215 33456	2288 86641	2364 21376
5	(10) 2.4883 20000	(10) 2.5937 42460	(10) 2.7027 08163	(10) 2.8153 05684	(10) 2.9316 25062
6	(12) 2.9859 84000	(12) 3.1384 28377	(12) 3.2973 03959	(12) 3.4628 25992	(12) 3.6352 15077
7	(14) 3.5831 80800	(14) 3.7974 98336	(14) 4.0227 10830	(14) 4.2592 75970	(14) 4.5076 66696
8	(16) 4.2998 16960	(16) 4.5949 72986	(16) 4.9077 07213	(16) 5.2389 09443	(16) 5.5895 06703
9	(18) 5.1597 80352	(18) 5.5599 17313	(18) 5.9874 02800	(18) 6.4438 58615	(18) 6.9309 88312
10	(20) 6.1917 36422	(20) 6.7274 99949	(20) 7.3046 31415	(20) 7.9259 46096	(20) 8.5944 25506
24	(49) 7.9496 84720	(49) 9.7017 23378	(50) 1.1820 50242	(50) 1.4378 80104	(50) 1.7463 06393
1/2	(1) 1.0954 45115	(1) 1.1000 00000	(1) 1.1045 36102	(1) 1.1090 53651	(1) 1.1135 52873
1/3	4.9324 24149	4.9460 87443	4.9596 75664	4.9731 89833	4.9866 30952
1/4	3.3097 50920	3.3166 24790	3.3234 56186	3.3302 45713	3.3369 93965
1/5	2.6051 71085	2.6094 98635	2.6137 97668	2.6180 68602	2.6223 11847

$n^{\frac{1}{2}}\begin{bmatrix}(-5)3\\4\end{bmatrix}$ $n^{\frac{1}{3}}\begin{bmatrix}(-5)1\\5\end{bmatrix}$ $n^{\frac{1}{4}}\begin{bmatrix}(-6)8\\5\end{bmatrix}$ $n^{\frac{1}{5}}\begin{bmatrix}(-6)5\\4\end{bmatrix}$

k	125	126	127	128	129
1	125	126	127	128	129
2	15625	15876	16129	16384	16641
3	19 53125	20 00376	20 48383	20 97152	21 46689
4	2441 40625	2520 47376	2601 44641	2684 35456	2769 22881
5	(10) 3.0517 57813	(10) 3.1757 96938	(10) 3.3038 36941	(10) 3.4359 73837	(10) 3.5723 05165
6	(12) 3.8146 97266	(12) 4.0015 04141	(12) 4.1958 72915	(12) 4.3980 46511	(12) 4.6082 73663
7	(14) 4.7683 71582	(14) 5.0418 95218	(14) 5.3287 58602	(14) 5.6294 99534	(14) 5.9446 73025
8	(16) 5.9604 64478	(16) 6.3527 87975	(16) 6.7675 23424	(16) 7.2057 59404	(16) 7.6686 28202
9	(18) 7.4505 80597	(18) 8.0045 12848	(18) 8.5947 54749	(18) 9.2233 72037	(18) 9.8925 30381
10	(20) 9.3132 25746	(21) 1.0085 68619	(21) 1.0915 33853	(21) 1.1805 91621	(21) 1.2761 36419
24	(50) 2.1175 82368	(50) 2.5638 52774	(50) 3.0994 83316	(50) 3.7414 44192	(50) 4.5097 56022
1/2	(1) 1.1180 33989	(1) 1.1224 97216	(1) 1.1269 42767	(1) 1.1313 70850	(1) 1.1357 81669
1/3	5.0000 00000	5.0132 97935	5.0265 25695	5.0396 84200	5.0527 74347
1/4	3.3437 01525	3.3503 68959	3.3569 96823	3.3635 85661	3.3701 36005
1/5	2.6265 27804	2.6307 16865	2.6348 79413	2.6390 15822	2.6431 26458

k	130	131	132	133	134
1	130	131	132	133	134
2	16900	17161	17424	17689	17956
3	21 97000	22 48091	22 99968	23 52637	24 06104
4	2856 10000	2944 99921	3035 95776	3129 00721	3224 17936
5	(10) 3.7129 30000	(10) 3.8579 48965	(10) 4.0074 64243	(10) 4.1615 79589	(10) 4.3204 00342
6	(12) 4.8268 09000	(12) 5.0539 13144	(12) 5.2898 52801	(12) 5.5349 00854	(12) 5.7893 36459
7	(14) 6.2748 51700	(14) 6.6206 26219	(14) 6.9826 05697	(14) 7.3614 18136	(14) 7.7577 10855
8	(16) 8.1573 07210	(16) 8.6730 20347	(16) 9.2170 39521	(16) 9.7906 86120	(17) 1.0395 33255
9	(19) 1.0604 49937	(19) 1.1361 65665	(19) 1.2166 49217	(19) 1.3021 61254	(19) 1.3929 74561
10	(21) 1.3785 84918	(21) 1.4883 77022	(21) 1.6059 76966	(21) 1.7318 74468	(21) 1.8665 85912
24	(50) 5.4280 07704	(50) 6.5239 57088	(50) 7.8302 26935	(50) 9.3851 10346	(51) 1.1233 50184
1/2	(1) 1.1401 75425	(1) 1.1445 52314	(1) 1.1489 12529	(1) 1.1532 56259	(1) 1.1575 83690
1/3	5.0657 97019	5.0787 53078	5.0916 43370	5.1044 68722	5.1172 29947
1/4	3.3766 48375	3.3831 23282	3.3895 61224	3.3959 62690	3.4023 28159
1/5	2.6472 11681	2.6512 71840	2.6553 07280	2.6593 18337	2.6633 05339

k	135	136	137	138	139
1	135	136	137	138	139
2	18225	18496	18769	19044	19321
3	24 60375	25 15456	25 71353	26 28072	26 85619
4	3321 50625	3421 02016	3522 75361	3626 73936	3733 01041
5	(10) 4.4840 33438	(10) 4.6525 87418	(10) 4.8261 72446	(10) 5.0049 00317	(10) 5.1888 84470
6	(12) 6.0534 45141	(12) 6.3275 18888	(12) 6.6118 56251	(12) 6.9067 62437	(12) 7.2125 49413
7	(14) 8.1721 50940	(14) 8.6054 25688	(14) 9.0582 43063	(14) 9.5313 32163	(15) 1.0025 44368
8	(17) 1.1032 40377	(17) 1.1703 37894	(17) 1.2409 79300	(17) 1.3153 23839	(17) 1.3935 36672
9	(19) 1.4893 74509	(19) 1.5916 59535	(19) 1.7001 41641	(19) 1.8151 46897	(19) 1.9370 15974
10	(21) 2.0106 55587	(21) 2.1646 56968	(21) 2.3291 94048	(21) 2.5049 02718	(21) 2.6924 52204
24	(51) 1.3427 97252	(51) 1.6030 01028	(51) 1.9111 44882	(51) 2.2756 11258	(51) 2.7061 70815
1/2	(1) 1.1618 95004	(1) 1.1661 90379	(1) 1.1704 69991	(1) 1.1747 34012	(1) 1.1789 82612
1/3	5.1299 27840	5.1425 63181	5.1551 36735	5.1676 49252	5.1801 01467
1/4	3.4086 58099	3.4149 52970	3.4212 13222	3.4274 39296	3.4336 31623
1/5	2.6672 68608	2.6712 08461	2.6751 25206	2.6790 19145	2.6828 90577

k	140	141	142	143	144
1	140	141	142	143	144
2	19600	19881	20164	20449	20736
3	27 44000	28 03221	28 63288	29 24207	29 85984
4	3841 60000	3952 54161	4065 86896	4181 61601	4299 81696
5	(10) 5.3782 40000	(10) 5.5730 83670	(10) 5.7735 33923	(10) 5.9797 10894	(10) 6.1917 36422
6	(12) 7.5295 36000	(12) 7.8580 47975	(12) 8.1984 18171	(12) 8.5509 86579	(12) 8.9161 00448
7	(15) 1.0541 35040	(15) 1.1079 84764	(15) 1.1641 75380	(15) 1.2227 91081	(15) 1.2839 18465
8	(17) 1.4757 89056	(17) 1.5622 58518	(17) 1.6531 29040	(17) 1.7485 91246	(17) 1.8488 42589
9	(19) 2.0661 04678	(19) 2.2027 84510	(19) 2.3474 43237	(19) 2.5004 85481	(19) 2.6623 33328
10	(21) 2.8925 46550	(21) 3.1059 26159	(21) 3.3333 69396	(21) 3.5756 94238	(21) 3.8337 59992
24	(51) 3.2141 99700	(51) 3.8129 28871	(51) 4.5177 29930	(51) 5.3464 42484	(51) 6.3197 48715
1/2	(1) 1.1832 15957	(1) 1.1874 34209	(1) 1.1916 37529	(1) 1.1958 26074	(1) 1.2000 00000
1/3	5.1924 94102	5.2048 27863	5.2171 03446	5.2293 21532	5.2414 82788
1/4	3.4397 90628	3.4459 16727	3.4520 10326	3.4580 71824	3.4641 01615
1/5	2.6867 39790	2.6905 67070	2.6943 72696	2.6981 56943	2.7019 20077

k	145	146	147	148	149
1	145	146	147	148	149
2	21025	21316	21609	21904	22201
3	30 48625	31 12136	31 76523	32 41792	33 07949
4	4420 50625	4543 71856	4669 48881	4797 85216	4928 84401
5	(10) 6.4097 34063	(10) 6.6338 29098	(10) 6.8641 48551	(10) 7.1008 21197	(10) 7.3439 77575
6	(12) 9.2941 14391	(12) 9.6853 90482	(13) 1.0090 29837	(13) 1.0509 21537	(13) 1.0942 52659
7	(15) 1.3476 46587	(15) 1.4140 67010	(15) 1.4832 73860	(15) 1.5553 63875	(15) 1.6304 36461
8	(17) 1.9540 87551	(17) 2.0645 37835	(17) 2.1804 12575	(17) 2.3019 38535	(17) 2.4293 50327
9	(19) 2.8334 26948	(19) 3.0142 25239	(19) 3.2052 06485	(19) 3.4068 69032	(19) 3.6197 31988
10	(21) 4.1084 69075	(21) 4.4007 68850	(21) 4.7116 53533	(21) 5.0421 66167	(21) 5.3934 00662
24	(51) 7.4616 01544	(51) 8.7997 13625	(52) 1.0366 11527	(52) 1.2197 79049	(52) 1.4337 40132
1/2	(1) 1.2041 59458	(1) 1.2083 04597	(1) 1.2124 35565	(1) 1.2165 52506	(1) 1.2206 55562
1/3	5.2535 87872	5.2656 37428	5.2776 32088	5.2895 72473	5.3014 59192
1/4	3.4701 00082	3.4760 67602	3.4820 04545	3.4879 11275	3.4937 88147
1/5	2.7056 62363	2.7093 84058	2.7130 85417	2.7167 66686	2.7204 28110

$n^{\frac{1}{2}}\begin{bmatrix}(-5)2\\4\end{bmatrix}$ $n^{\frac{1}{3}}\begin{bmatrix}(-6)9\\5\end{bmatrix}$ $n^{\frac{1}{4}}\begin{bmatrix}(-6)5\\4\end{bmatrix}$ $n^{\frac{1}{5}}\begin{bmatrix}(-6)3\\4\end{bmatrix}$

k					
1	150	151	152	153	154
2	22500	22801	23104	23409	23716
3	33 75000	34 42951	35 11808	35 81577	36 52264
4	5062 50000	5198 85601	5337 94816	5479 81281	5624 48656
5	(10) 7.5937 50000	(10) 7.8502 72575	(10) 8.1136 81203	(10) 8.3841 13599	(10) 8.6617 09302
6	(13) 1.1390 62500	(13) 1.1853 91159	(13) 1.2332 79543	(13) 1.2827 69381	(13) 1.3339 03233
7	(15) 1.7085 93750	(15) 1.7899 40650	(15) 1.8745 84905	(15) 1.9626 37152	(15) 2.0542 10978
8	(17) 2.5628 90625	(17) 2.7028 10381	(17) 2.8493 69056	(17) 3.0028 34843	(17) 3.1634 84906
9	(19) 3.8443 35938	(19) 4.0812 43676	(19) 4.3310 40965	(19) 4.5943 37310	(19) 4.8717 66756
10	(21) 5.7665 03906	(21) 6.1626 77950	(21) 6.5831 82267	(21) 7.0293 36085	(21) 7.5025 20804
24	(52) 1.6834 11220	(52) 1.9744 52704	(52) 2.3133 75387	(52) 2.7076 61312	(52) 3.1659 00782
1/2	(1) 1.2247 44871	(1) 1.2288 20573	(1) 1.2328 82801	(1) 1.2369 31688	(1) 1.2409 67365
1/3	5.3132 92846	5.3250 74022	5.3368 03297	5.3484 81241	5.3601 08411
1/4	3.4996 35512	3.5054 53712	3.5112 43086	3.5170 03963	3.5227 36670
1/5	2.7240 69927	2.7276 92374	2.7312 95679	2.7348 80069	2.7384 45765

k					
1	155	156	157	158	159
2	24025	24336	24649	24964	25281
3	37 23875	37 96416	38 69893	39 44312	40 19679
4	5772 00625	5922 40896	6075 73201	6232 01296	6391 28961
5	(10) 8.9466 09688	(10) 9.2389 57978	(10) 9.5388 99256	(10) 9.8465 80477	(11) 1.0162 15048
6	(13) 1.3867 24502	(13) 1.4412 77445	(13) 1.4976 07183	(13) 1.5557 59715	(13) 1.6157 81926
7	(15) 2.1494 22977	(15) 2.2483 92813	(15) 2.3512 43278	(15) 2.4581 00350	(15) 2.5690 93263
8	(17) 3.3316 05615	(17) 3.5074 92789	(17) 3.6914 51946	(17) 3.8837 98553	(17) 4.0848 58288
9	(19) 5.1639 88703	(19) 5.4716 88751	(19) 5.7955 79555	(19) 6.1364 01714	(19) 6.4949 24678
10	(21) 8.0041 82490	(21) 8.5358 34451	(21) 9.0990 59901	(21) 9.6955 14709	(22) 1.0326 93024
24	(52) 3.6979 47627	(52) 4.3150 94990	(52) 5.0302 74186	(52) 5.8582 79483	(52) 6.8160 22003
1/2	(1) 1.2449 89960	(1) 1.2489 99600	(1) 1.2529 96409	(1) 1.2569 80509	(1) 1.2609 52021
1/3	5.3716 85355	5.3832 12612	5.3946 90712	5.4061 20176	5.4175 01515
1/4	3.5284 41525	3.5341 18843	3.5397 68931	3.5453 92093	3.5509 88625
1/5	2.7419 92987	2.7455 21947	2.7490 32856	2.7525 25920	2.7560 01343

k					
1	160	161	162	163	164
2	25600	25921	26244	26569	26896
3	40 96000	41 73281	42 51528	43 30747	44 10944
4	6553 60000	6718 98241	6887 47536	7059 11761	7233 94816
5	(11) 1.0485 76000	(11) 1.0817 56168	(11) 1.1157 71008	(11) 1.1506 36170	(11) 1.1863 67498
6	(13) 1.6777 21600	(13) 1.7416 27430	(13) 1.8075 49033	(13) 1.8755 36958	(13) 1.9456 42697
7	(15) 2.6843 54560	(15) 2.8040 20163	(15) 2.9282 29434	(15) 3.0571 25241	(15) 3.1908 54023
8	(17) 4.2949 67296	(17) 4.5144 72463	(17) 4.7437 31683	(17) 4.9831 14143	(17) 5.2330 00598
9	(19) 6.8719 47674	(19) 7.2683 00665	(19) 7.6848 45327	(19) 8.1224 76053	(19) 8.5821 20981
10	(22) 1.0995 11628	(22) 1.1701 96407	(22) 1.2449 44943	(22) 1.3239 63597	(22) 1.4074 67841
24	(52) 7.9228 16251	(52) 9.2007 03274	(53) 1.0674 81480	(53) 1.2373 78329	(53) 1.4330 20335
1/2	(1) 1.2649 11064	(1) 1.2688 57754	(1) 1.2727 92206	(1) 1.2767 14533	(1) 1.2806 24847
1/3	5.4288 35233	5.4401 21825	5.4513 61778	5.4625 55571	5.4737 03675
1/4	3.5565 58820	3.5621 02966	3.5676 21345	3.5731 14235	3.5785 81908
1/5	2.7594 59323	2.7629 00056	2.7663 23734	2.7697 30547	2.7731 20681

k					
1	165	166	167	168	169
2	27225	27556	27889	28224	28561
3	44 92125	45 74296	46 57463	47 41632	48 26809
4	7412 00625	7593 33136	7777 96321	7965 94176	8157 30721
5	(11) 1.2229 81031	(11) 1.2604 93006	(11) 1.2989 19856	(11) 1.3382 78216	(11) 1.3785 84918
6	(13) 2.0179 18702	(13) 2.0924 18390	(13) 2.1691 96160	(13) 2.2483 07402	(13) 2.3298 08512
7	(15) 3.3295 65858	(15) 3.4734 14527	(15) 3.6225 57587	(15) 3.7771 56436	(15) 3.9373 76386
8	(17) 5.4937 83665	(17) 5.7658 68114	(17) 6.0496 71170	(17) 6.3456 22812	(17) 6.6541 66092
9	(19) 9.0647 43047	(19) 9.5713 41070	(20) 1.0102 95085	(20) 1.0660 64632	(20) 1.1245 54070
10	(22) 1.4956 82603	(22) 1.5888 42618	(22) 1.6871 92792	(22) 1.7909 88583	(22) 1.9004 96377
24	(53) 1.6581 15050	(53) 1.9168 76411	(53) 2.2140 90189	(53) 2.5551 87425	(53) 2.9463 26763
1/2	(1) 1.2845 23258	(1) 1.2884 09873	(1) 1.2922 84798	(1) 1.2961 48140	(1) 1.3000 00000
1/3	5.4848 06552	5.4958 64660	5.5068 78446	5.5178 48353	5.5287 74814
1/4	3.5840 24634	3.5894 42676	3.5948 36294	3.6002 05744	3.6055 51275
1/5	2.7764 94317	2.7798 51635	2.7831 92813	2.7865 18023	2.7898 27436

k					
1	170	171	172	173	174
2	28900	29241	29584	29929	30276
3	49 13000	50 00211	50 88448	51 77717	52 68024
4	8352 10000	8550 36081	8752 13056	8957 45041	9166 36176
5	(11) 1.4198 57000	(11) 1.4621 11699	(11) 1.5053 66456	(11) 1.5496 38921	(11) 1.5949 46946
6	(13) 2.4137 56900	(13) 2.5002 11004	(13) 2.5892 30305	(13) 2.6808 75333	(13) 2.7752 07686
7	(15) 4.1033 86730	(15) 4.2753 60818	(15) 4.4534 76124	(15) 4.6379 14326	(15) 4.8288 61374
8	(17) 6.9757 57441	(17) 7.3108 66998	(17) 7.6599 78934	(17) 8.0235 91785	(17) 8.4022 18792
9	(20) 1.1858 78765	(20) 1.2501 58257	(20) 1.3175 16377	(20) 1.3880 81379	(20) 1.4619 86070
10	(22) 2.0159 93900	(22) 2.1377 70619	(22) 2.2661 28168	(22) 2.4013 80785	(22) 2.5438 55761
24	(53) 3.3944 86713	(53) 3.9075 68945	(53) 4.4945 13878	(53) 5.1654 29935	(53) 5.9317 37979
1/2	(1) 1.3038 40481	(1) 1.3076 69683	(1) 1.3114 87705	(1) 1.3152 94644	(1) 1.3190 90596
1/3	5.5396 58257	5.5504 99103	5.5612 97766	5.5720 54656	5.5827 70172
1/4	3.6108 73137	3.6161 71571	3.6214 46817	3.6266 99110	3.6319 28683
1/5	2.7931 21220	2.7963 99540	2.7996 62559	2.8029 10436	2.8061 43329

$n^{\frac{1}{2}}\left[{(-5)2 \atop 4}\right]$ $n^{\frac{1}{3}}\left[{(-6)7 \atop 4}\right]$ $n^{\frac{1}{4}}\left[{(-6)4 \atop 4}\right]$ $n^{\frac{1}{5}}\left[{(-6)3 \atop 4}\right]$

POWERS AND ROOTS n^k

k					
1	175	176	177	178	179
2	30625	30976	31329	31684	32041
3	53 59375	54 51776	55 45233	56 39752	57 35339
4	9378 90625	9595 12576	9815 06241	(9) 1.0038 75856	(9) 1.0266 25681
5	(11) 1.6413 08594	(11) 1.6887 42134	(11) 1.7372 66047	(11) 1.7868 99024	(11) 1.8376 59969
6	(13) 2.8722 90039	(13) 2.9721 86155	(13) 3.0749 60902	(13) 3.1806 80262	(13) 3.2894 11344
7	(15) 5.0265 07568	(15) 5.2310 47634	(15) 5.4426 80797	(15) 5.6616 10867	(15) 5.8880 46307
8	(17) 8.7963 88245	(17) 9.2066 43835	(17) 9.6335 45011	(18) 1.0077 66734	(18) 1.0539 60289
9	(20) 1.5393 67943	(20) 1.6203 69315	(20) 1.7051 37467	(20) 1.7938 24787	(20) 1.8865 88917
10	(22) 2.6938 93900	(22) 2.8518 49994	(22) 3.0180 93317	(22) 3.1930 08121	(22) 3.3769 94162
24	(53) 6.8063 32613	(53) 7.8037 62212	(53) 8.9404 29702	(54) 1.0234 81638	(54) 1.1707 73122
1/2	(1) 1.3228 75656	(1) 1.3266 49916	(1) 1.3304 13470	(1) 1.3341 66406	(1) 1.3379 08816
1/3	5.5934 44710	5.6040 78661	5.6146 72408	5.6252 26328	5.6357 40794
1/4	3.6371 35763	3.6423 20574	3.6474 83337	3.6526 24271	3.6577 43589
1/5	2.8093 61392	2.8125 64777	2.8157 53634	2.8189 28111	2.8220 88352

k					
1	180	181	182	183	184
2	32400	32761	33124	33489	33856
3	58 32000	59 29741	60 28568	61 28487	62 29504
4	(9) 1.0497 60000	(9) 1.0732 83121	(9) 1.0971 99376	(9) 1.1215 13121	(9) 1.1462 28736
5	(11) 1.8895 68000	(11) 1.9426 42449	(11) 1.9969 02864	(11) 2.0523 69011	(11) 2.1090 60874
6	(13) 3.4012 22400	(13) 3.5161 82833	(13) 3.6343 63213	(13) 3.7558 35291	(13) 3.8806 72009
7	(15) 6.1222 00320	(15) 6.3642 90927	(15) 6.6145 41048	(15) 6.8731 78582	(15) 7.1404 36496
8	(18) 1.1019 96058	(18) 1.1519 36658	(18) 1.2038 46471	(18) 1.2577 91681	(18) 1.3138 40315
9	(20) 1.9835 92904	(20) 2.0850 05351	(20) 2.1910 00577	(20) 2.3017 58775	(20) 2.4174 66180
10	(22) 3.5704 67227	(22) 3.7738 59685	(22) 3.9876 21050	(22) 4.2122 18559	(22) 4.4481 37771
24	(54) 1.3382 58845	(54) 1.5285 71637	(54) 1.7446 70074	(54) 1.9898 76639	(54) 2.2679 20111
1/2	(1) 1.3416 40786	(1) 1.3453 62405	(1) 1.3490 73756	(1) 1.3527 74926	(1) 1.3564 65997
1/3	5.6462 16173	5.6566 52826	5.6670 51108	5.6774 11371	5.6877 33960
1/4	3.6628 41501	3.6679 18217	3.6729 73940	3.6780 08871	3.6830 23210
1/5	2.8252 34501	2.8283 66697	2.8314 85080	2.8345 89786	2.8376 80950

k					
1	185	186	187	188	189
2	34225	34596	34969	35344	35721
3	63 31625	64 34856	65 39203	66 44672	67 51269
4	(9) 1.1713 50625	(9) 1.1968 83216	(9) 1.2228 30961	(9) 1.2491 98336	(9) 1.2759 89841
5	(11) 2.1669 98656	(11) 2.2262 02782	(11) 2.2866 93897	(11) 2.3484 92872	(11) 2.4116 20799
6	(13) 4.0089 47514	(13) 4.1407 37174	(13) 4.2761 17588	(13) 4.4151 66599	(13) 4.5579 63311
7	(15) 7.4165 52901	(15) 7.7017 71144	(15) 7.9963 39889	(15) 8.3005 13206	(15) 8.6145 50658
8	(18) 1.3720 62287	(18) 1.4325 29433	(18) 1.4953 15559	(18) 1.5604 96483	(18) 1.6281 50074
9	(20) 2.5383 15230	(20) 2.6645 04745	(20) 2.7962 40096	(20) 2.9337 33387	(20) 3.0772 03640
10	(22) 4.6958 83176	(22) 4.9559 78826	(22) 5.2289 68979	(22) 5.5154 18768	(22) 5.8159 14881
24	(54) 2.5829 82606	(54) 2.9397 51775	(54) 3.3434 78670	(54) 3.8000 41874	(54) 4.3160 18526
1/2	(1) 1.3601 47051	(1) 1.3638 18170	(1) 1.3674 79433	(1) 1.3711 30920	(1) 1.3747 72708
1/3	5.6980 19215	5.7082 67473	5.7184 79065	5.7286 54316	5.7387 93548
1/4	3.6880 17151	3.6929 90888	3.6979 44609	3.7028 78502	3.7077 92751
1/5	2.8407 58702	2.8438 23174	2.8468 74493	2.8499 12786	2.8529 38178

k					
1	190	191	192	193	194
2	36100	36481	36864	37249	37636
3	68 59000	69 67871	70 77888	71 89057	73 01384
4	(9) 1.3032 10000	(9) 1.3308 63361	(9) 1.3589 54496	(9) 1.3874 88001	(9) 1.4164 68496
5	(11) 2.4760 99000	(11) 2.5419 49020	(11) 2.6091 92632	(11) 2.6778 51842	(11) 2.7479 48882
6	(13) 4.7045 88100	(13) 4.8551 22627	(13) 5.0096 49854	(13) 5.1682 54055	(13) 5.3310 20832
7	(15) 8.9387 17390	(15) 9.2732 84218	(15) 9.6185 27720	(15) 9.9747 30326	(16) 1.0342 18041
8	(18) 1.6983 56304	(18) 1.7711 97286	(18) 1.8467 57322	(18) 1.9251 22953	(18) 2.0063 83000
9	(20) 3.2268 76978	(20) 3.3829 86816	(20) 3.5457 74059	(20) 3.7154 87299	(20) 3.8923 83020
10	(22) 6.1310 66258	(22) 6.4615 04818	(22) 6.8078 86193	(22) 7.1708 90487	(22) 7.5512 23059
24	(54) 4.8987 62931	(54) 5.5564 93542	(54) 6.2983 89130	(54) 7.1346 95065	(54) 8.0768 40718
1/2	(1) 1.3784 04875	(1) 1.3820 27496	(1) 1.3856 40646	(1) 1.3892 44399	(1) 1.3928 38828
1/3	5.7488 97079	5.7589 65220	5.7689 98281	5.7789 96565	5.7889 60372
1/4	3.7126 87538	3.7175 63041	3.7224 19436	3.7272 56899	3.7320 75599
1/5	2.8559 50791	2.8589 50746	2.8619 38162	2.8649 13156	2.8678 75844

k					
1	195	196	197	198	199
2	38025	38416	38809	39204	39601
3	74 14875	75 29536	76 45373	77 62392	78 80599
4	(9) 1.4459 00625	(9) 1.4757 89056	(9) 1.5061 38481	(9) 1.5369 53616	(9) 1.5682 39201
5	(11) 2.8195 06219	(11) 2.8925 46550	(11) 2.9670 92808	(11) 3.0431 68160	(11) 3.1207 96010
6	(13) 5.4980 37127	(13) 5.6693 91238	(13) 5.8451 72831	(13) 6.0254 72956	(13) 6.2103 84060
7	(16) 1.0721 17240	(16) 1.1112 00683	(16) 1.1514 99048	(16) 1.1930 43645	(16) 1.2358 66428
8	(18) 2.0906 28617	(18) 2.1779 53338	(18) 2.2684 53124	(18) 2.3622 26418	(18) 2.4593 74192
9	(20) 4.0767 25804	(20) 4.2687 88542	(20) 4.4688 52654	(20) 4.6772 08307	(20) 4.8941 54641
10	(22) 7.9496 15318	(22) 8.3668 25543	(22) 8.8036 39729	(22) 9.2608 72448	(22) 9.7393 67736
24	(54) 9.1375 69069	(55) 1.0331 07971	(55) 1.1673 18660	(55) 1.3181 49187	(55) 1.4875 57746
1/2	(1) 1.3964 24004	(1) 1.4000 00000	(1) 1.4035 66885	(1) 1.4071 24728	(1) 1.4106 73598
1/3	5.7988 89998	5.8087 85734	5.8186 47867	5.8284 76683	5.8382 72461
1/4	3.7368 75706	3.7416 57387	3.7464 20805	3.7511 66123	3.7558 93499
1/5	2.8708 26340	2.8737 64756	2.8766 91203	2.8796 05790	2.8825 08624

$n^{\frac{1}{2}}\begin{bmatrix}(-5)1\\4\end{bmatrix}$ $n^{\frac{1}{3}}\begin{bmatrix}(-6)5\\4\end{bmatrix}$ $\iota^{\frac{1}{4}}\begin{bmatrix}(-6)3\\4\end{bmatrix}$ $n^{\frac{1}{5}}\begin{bmatrix}(-6)2\\4\end{bmatrix}$

k					
1	200	201	202	203	204
2	40000	40401	40804	41209	41616
3	80 00000	81 20601	82 42408	83 65427	84 89664
4	(9)1.6000 00000	(9)1.6322 40801	(9)1.6649 66416	(9)1.6981 81681	(9)1.7318 91456
5	(11)3.2000 00000	(11)3.2808 04010	(11)3.3632 32160	(11)3.4473 08812	(11)3.5330 58570
6	(13)6.4000 00000	(13)6.5944 16060	(13)6.7937 28964	(13)6.9980 36889	(13)7.2074 39483
7	(16)1.2800 00000	(16)1.3254 77628	(16)1.3723 33251	(16)1.4206 01489	(16)1.4703 17655
8	(18)2.5600 00000	(18)2.6642 10032	(18)2.7721 13166	(18)2.8838 21022	(18)2.9994 48015
9	(20)5.1200 00000	(20)5.3550 62165	(20)5.5996 68596	(20)5.8541 56674	(20)6.1188 73951
10	(23)1.0240 00000	(23)1.0763 67495	(23)1.1311 33056	(23)1.1883 93805	(23)1.2482 50286
24	(55)1.6777 21600	(55)1.8910 60303	(55)2.1302 61246	(55)2.3983 07745	(55)2.6985 09916
1/2	(1)1.4142 13562	(1)1.4177 44688	(1)1.4212 67040	(1)1.4247 80685	(1)1.4282 85686
1/3	5.8480 35476	5.8577 66003	5.8674 64308	5.8771 30659	5.8867 65317
1/4	3.7606 03093	3.7652 95059	3.7699 69549	3.7746 26716	3.7792 66709
1/5	2.8853 99812	2.8882 79458	2.8911 47666	2.8940 04537	2.8968 50171

k					
1	205	206	207	208	209
2	42025	42436	42849	43264	43681
3	86 15125	87 41816	88 69743	89 98912	91 29329
4	(9)1.7661 00625	(9)1.8008 14096	(9)1.8360 36801	(9)1.8717 73696	(9)1.9080 29761
5	(11)3.6205 06281	(11)3.7096 77038	(11)3.8005 96178	(11)3.8932 89288	(11)3.9877 82200
6	(13)7.4220 37877	(13)7.6419 34698	(13)7.8672 34089	(13)8.0980 41718	(13)8.3344 64799
7	(16)1.5215 17765	(16)1.5742 38548	(16)1.6285 17456	(16)1.6843 92677	(16)1.7419 03143
8	(18)3.1191 11418	(18)3.2429 31408	(18)3.3710 31135	(18)3.5035 36769	(18)3.6405 77569
9	(20)6.3941 78406	(20)6.6804 38701	(20)6.9780 34449	(20)7.2873 56480	(20)7.6088 07119
10	(23)1.3108 06573	(23)1.3761 70372	(23)1.4444 53131	(23)1.5157 70148	(23)1.5902 40688
24	(55)3.0345 38594	(55)3.4104 62581	(55)3.8307 89523	(55)4.3005 10765	(55)4.8251 50531
1/2	(1)1.4317 82106	(1)1.4352 70009	(1)1.4387 49457	(1)1.4422 20510	(1)1.4456 83229
1/3	5.8963 68540	5.9059 40584	5.9154 81700	5.9249 92137	5.9344 72140
1/4	3.7838 89674	3.7884 95756	3.7930 85099	3.7976 57844	3.8022 14131
1/5	2.8996 84668	2.9025 08125	2.9053 20638	2.9081 22302	2.9109 13212

k					
1	210	211	212	213	214
2	44100	44521	44944	45369	45796
3	92 61000	93 93931	95 28128	96 63597	98 00344
4	(9)1.9448 10000	(9)1.9821 19441	(9)2.0199 63136	(9)2.0583 46161	(9)2.0972 73616
5	(11)4.0841 01000	(11)4.1822 72021	(11)4.2823 21848	(11)4.3842 77323	(11)4.4881 65538
6	(13)8.5766 12100	(13)8.8245 93963	(13)9.0785 22318	(13)9.3385 10698	(13)9.6046 74252
7	(16)1.8010 88541	(16)1.8619 89326	(16)1.9246 46732	(16)1.9891 02779	(16)2.0554 00290
8	(18)3.7822 85936	(18)3.9287 97478	(18)4.0802 51071	(18)4.2367 88919	(18)4.3985 56620
9	(20)7.9428 00466	(20)8.2897 62679	(20)8.6501 32270	(20)9.0243 60396	(20)9.4129 11168
10	(23)1.6679 88098	(23)1.7491 39925	(23)1.8338 28041	(23)1.9221 88764	(23)2.0143 62990
24	(55)5.4108 19838	(55)6.0642 75557	(55)6.7929 85105	(55)7.6051 97251	(55)8.5100 19601
1/2	(1)1.4491 37675	(1)1.4525 83905	(1)1.4560 21978	(1)1.4594 51952	(1)1.4628 73884
1/3	5.9439 21953	5.9533 41813	5.9627 31958	5.9720 92620	5.9814 24030
1/4	3.8067 54096	3.8112 77876	3.8157 85604	3.8202 77414	3.8247 53435
1/5	2.9136 93459	2.9164 63134	2.9192 22328	2.9219 71130	2.9247 09627

k					
1	215	216	217	218	219
2	46225	46656	47089	47524	47961
3	99 38375	100 77696	102 18313	103 60232	105 03459
4	(9)2.1367 50625	(9)2.1767 82336	(9)2.2173 73921	(9)2.2585 30576	(9)2.3002 57521
5	(11)4.5940 13844	(11)4.7018 49846	(11)4.8117 01409	(11)4.9235 96656	(11)5.0375 63971
6	(13)9.8771 29764	(14)1.0155 99567	(14)1.0441 39206	(14)1.0733 44071	(14)1.1032 26510
7	(16)2.1235 82899	(16)2.1936 95064	(16)2.2657 82076	(16)2.3398 90075	(16)2.4160 66056
8	(18)4.5657 03233	(18)4.7383 81338	(18)4.9167 47106	(18)5.1009 60363	(18)5.2911 84663
9	(20)9.8162 61952	(21)1.0234 90369	(21)1.0669 34122	(21)1.1120 09359	(21)1.1587 69441
10	(23)2.1104 96320	(23)2.2107 39197	(23)2.3152 47045	(23)2.4241 80403	(23)2.5377 05076
24	(55)9.5175 03342	(56)1.0638 73589	(56)1.1885 94216	(56)1.3272 59512	(56)1.4813 53665
1/2	(1)1.4662 87830	(1)1.4696 93846	(1)1.4730 91986	(1)1.4764 82306	(1)1.4798 64859
1/3	5.9907 26415	6.0000 00000	6.0092 45007	6.0184 61655	6.0276 50160
1/4	3.8292 13796	3.8336 58625	3.8380 88048	3.8425 02187	3.8469 01167
1/5	2.9274 37906	2.9301 56052	2.9328 64149	2.9355 62280	2.9382 50529

k					
1	220	221	222	223	224
2	48400	48841	49284	49729	50176
3	106 48000	107 93861	109 41048	110 89567	112 39424
4	(9)2.3425 60000	(9)2.3854 43281	(9)2.4289 12656	(9)2.4729 73441	(9)2.5176 30976
5	(11)5.1536 32000	(11)5.2718 29651	(11)5.3921 86096	(11)5.5147 30773	(11)5.6394 93386
6	(14)1.1337 99040	(14)1.1650 74353	(14)1.1970 65313	(14)1.2297 84962	(14)1.2632 46519
7	(16)2.4943 57888	(16)2.5748 14320	(16)2.6574 84996	(16)2.7424 20466	(16)2.8296 72201
8	(18)5.4875 87354	(18)5.6903 39647	(18)5.8996 16690	(18)6.1155 97640	(18)6.3384 65731
9	(21)1.2072 69218	(21)1.2575 65062	(21)1.3097 14905	(21)1.3637 78274	(21)1.4198 16324
10	(23)2.6559 92279	(23)2.7792 18787	(23)2.9075 67090	(23)3.0412 25550	(23)3.1803 88565
24	(56)1.6525 10926	(56)1.8425 30003	(56)2.0533 89736	(56)2.2872 66205	(56)2.5465 51362
1/2	(1)1.4832 39697	(1)1.4866 06875	(1)1.4899 66443	(1)1.4933 18452	(1)1.4966 62955
1/3	6.0368 10737	6.0459 43596	6.0550 48947	6.0641 26994	6.0731 77944
1/4	3.8512 85107	3.8556 54127	3.8600 08345	3.8643 47878	3.8686 72841
1/5	2.9409 28975	2.9435 97699	2.9462 56780	2.9489 06295	2.9515 46323

$n^{\frac{1}{2}}\begin{bmatrix}(-5)1\\4\end{bmatrix}$ $n^{\frac{1}{3}}\begin{bmatrix}(-6)5\\4\end{bmatrix}$ $n^{\frac{1}{4}}\begin{bmatrix}(-6)2\\4\end{bmatrix}$ $n^{\frac{1}{5}}\begin{bmatrix}(-6)2\\4\end{bmatrix}$

k					
1	225	226	227	228	229
2	50625	51076	51529	51984	52441
3	113 90625	115 43176	116 97083	118 52352	120 08989
4	(9)2.5628 90625	(9)2.6087 57776	(9)2.6552 37841	(9)2.7023 36256	(9)2.7500 58481
5	(11)5.7665 03906	(11)5.8957 92574	(11)6.0273 89899	(11)6.1613 26664	(11)6.2976 33921
6	(14)1.2974 63379	(14)1.3324 49122	(14)1.3682 17507	(14)1.4047 82479	(14)1.4421 58168
7	(16)2.9192 92603	(16)3.0113 35015	(16)3.1058 53741	(16)3.2029 04053	(16)3.3025 42205
8	(18)6.5684 08356	(18)6.8056 17134	(18)7.0502 87992	(18)7.3026 21240	(18)7.5628 21649
9	(21)1.4778 91880	(21)1.5380 69472	(21)1.6004 15374	(21)1.6649 97643	(21)1.7318 86158
10	(23)3.3252 56730	(23)3.4760 37007	(23)3.6329 42900	(23)3.7961 94626	(23)3.9660 19301
24	(56)2.8338 73334	(56)3.1521 18526	(56)3.5044 55686	(56)3.8943 62082	(56)4.3256 51988
1/2	(1)1.5000 00000	(1)1.5033 29638	(1)1.5066 51917	(1)1.5099 66887	(1)1.5132 74595
1/3	6.0822 01996	6.0911 99349	6.1001 70200	6.1091 14744	6.1180 33173
1/4	3.8729 83346	3.8772 79507	3.8815 61435	3.8858 29238	3.8900 83026
1/5	2.9541 76939	2.9567 98218	2.9594 10235	2.9620 13062	2.9646 06773

k					
1	230	231	232	233	234
2	52900	53361	53824	54289	54756
3	121 67000	123 26391	124 87168	126 49337	128 12904
4	(9)2.7984 10000	(9)2.8473 96321	(9)2.8970 22976	(9)2.9472 95521	(9)2.9982 19536
5	(11)6.4363 43000	(11)6.5774 85502	(11)6.7210 93304	(11)6.8671 98564	(11)7.0158 33714
6	(14)1.4803 58890	(14)1.5193 99151	(14)1.5592 93647	(14)1.6000 57265	(14)1.6417 05089
7	(16)3.4048 25447	(16)3.5098 12038	(16)3.6175 61260	(16)3.7281 33428	(16)3.8415 89909
8	(18)7.8310 98528	(18)8.1076 65809	(18)8.3927 42123	(18)8.6865 50888	(18)8.9893 20386
9	(21)1.8011 52661	(21)1.8728 70802	(21)1.9471 16173	(21)2.0239 66357	(21)2.1035 00970
10	(23)4.1426 51121	(23)4.3263 31552	(23)4.5173 09521	(23)4.7158 41612	(23)4.9221 92271
24	(56)4.8025 07640	(56)5.3295 12896	(56)5.9116 89798	(56)6.5545 38287	(56)7.2640 79321
1/2	(1)1.5165 75089	(1)1.5198 68415	(1)1.5231 54621	(1)1.5264 33752	(1)1.5297 05854
1/3	6.1269 25675	6.1357 92440	6.1446 33651	6.1534 49494	6.1622 40148
1/4	3.8943 22905	3.8985 48980	3.9027 61357	3.9069 60138	3.9111 45426
1/5	2.9671 91438	2.9697 67129	2.9723 33915	2.9748 91866	2.9774 41049

k					
1	235	236	237	238	239
2	55225	55696	56169	56644	57121
3	129 77875	131 44256	133 12053	134 81272	136 51919
4	(9)3.0498 00625	(9)3.1020 44416	(9)3.1549 56561	(9)3.2085 42736	(9)3.2628 08641
5	(11)7.1670 31469	(11)7.3208 24822	(11)7.4772 47050	(11)7.6363 31712	(11)7.7981 12652
6	(14)1.6842 52395	(14)1.7277 14658	(14)1.7721 07551	(14)1.8174 46947	(14)1.8637 48924
7	(16)3.9579 93129	(16)4.0774 06593	(16)4.1998 94895	(16)4.3255 23735	(16)4.4543 59928
8	(18)9.3012 83852	(18)9.6226 79559	(18)9.9537 50902	(19)1.0294 74649	(19)1.0645 92023
9	(21)2.1858 01705	(21)2.2709 52376	(21)2.3590 38964	(21)2.4501 49664	(21)2.5443 74934
10	(23)5.1366 34007	(23)5.3594 47607	(23)5.5909 22344	(23)5.8313 56201	(23)6.0810 56093
24	(56)8.0469 01671	(56)8.9102 12697	(56)9.8618 93410	(57)1.0910 55818	(57)1.2065 61943
1/2	(1)1.5329 70972	(1)1.5362 29150	(1)1.5394 80432	(1)1.5427 24862	(1)1.5459 62483
1/3	6.1710 05793	6.1797 46606	6.1884 62762	6.1971 54435	6.2058 21795
1/4	3.9153 17320	3.9194 75921	3.9236 21327	3.9277 53635	3.9318 72942
1/5	2.9799 81531	2.9825 13380	2.9850 36660	2.9875 51438	2.9900 57776

k					
1	240	241	242	243	244
2	57600	58081	58564	59049	59536
3	138 24000	139 97521	141 72488	143 48907	145 26784
4	(9)3.3177 60000	(9)3.3734 02561	(9)3.4297 42096	(9)3.4867 84401	(9)3.5445 35296
5	(11)7.9626 24000	(11)8.1299 00172	(11)8.2999 75872	(11)8.4728 86094	(11)8.6486 66122
6	(14)1.9110 29760	(14)1.9593 05941	(14)2.0085 94161	(14)2.0589 11321	(14)2.1102 74534
7	(16)4.5864 71424	(16)4.7219 27319	(16)4.8607 97870	(16)5.0031 54510	(16)5.1490 69863
8	(19)1.1007 53142	(19)1.1379 84484	(19)1.1763 13085	(19)1.2157 66546	(19)1.2563 73046
9	(21)2.6418 07540	(21)2.7425 42606	(21)2.8466 77665	(21)2.9543 12707	(21)3.0655 50233
10	(23)6.3403 38097	(23)6.6095 27681	(23)6.8889 59948	(23)7.1789 79877	(23)7.4799 42569
24	(57)1.3337 35777	(57)1.4736 99791	(57)1.6276 79087	(57)1.7970 10300	(57)1.9831 51223
1/2	(1)1.5491 93338	(1)1.5524 17470	(1)1.5556 34919	(1)1.5588 45727	(1)1.5620 49935
1/3	6.2144 65012	6.2230 84253	6.2316 79684	6.2402 51469	6.2487 99770
1/4	3.9359 79343	3.9400 72930	3.9441 53798	3.9482 22039	3.9522 77742
1/5	2.9925 55740	2.9950 45390	2.9975 26790	3.0000 00000	3.0024 65081

k					
1	245	246	247	248	249
2	60025	60516	61009	61504	62001
3	147 06125	148 86936	150 69223	152 52992	154 38249
4	(9)3.6030 00625	(9)3.6621 86256	(9)3.7220 98081	(9)3.7827 42016	(9)3.8441 24001
5	(11)8.8273 51531	(11)9.0089 78190	(11)9.1935 82260	(11)9.3812 00200	(11)9.5718 68762
6	(14)2.1627 01125	(14)2.2162 08635	(14)2.2708 14818	(14)2.3265 37650	(14)2.3833 95322
7	(16)5.2986 17757	(16)5.4518 73241	(16)5.6089 12601	(16)5.7698 13371	(16)5.9346 54351
8	(19)1.2981 61350	(19)1.3411 60817	(19)1.3854 01412	(19)1.4309 13716	(19)1.4777 28934
9	(21)3.1804 95308	(21)3.2992 55611	(21)3.4219 41489	(21)3.5486 66016	(21)3.6795 45044
10	(23)7.7922 13506	(23)8.1161 68802	(23)8.4521 95477	(23)8.8006 91719	(23)9.1620 67161
24	(57)2.1876 91225	(57)2.4123 62509	(57)2.6590 52293	(57)2.9298 15956	(57)3.2268 91257
1/2	(1)1.5652 47584	(1)1.5684 38714	(1)1.5716 23365	(1)1.5748 01575	(1)1.5779 73384
1/3	6.2573 24746	6.2658 26556	6.2743 05357	6.2827 61305	6.2911 94552
1/4	3.9563 20998	3.9603 51896	3.9643 70523	3.9683 76966	3.9723 71312
1/5	3.0049 22094	3.0073 71096	3.0098 12147	3.0122 45305	3.0146 70627

$n^{\frac{1}{2}}\left[\begin{matrix}(-6)9\\4\end{matrix}\right]$ $n^{\frac{1}{3}}\left[\begin{matrix}(-6)3\\4\end{matrix}\right]$ $n^{\frac{1}{4}}\left[\begin{matrix}(-6)2\\4\end{matrix}\right]$ $n^{\frac{1}{5}}\left[\begin{matrix}(-6)1\\4\end{matrix}\right]$

k					
1	250	251	252	253	254
2	62500	63001	63504	64009	64516
3	156 25000	158 13251	160 03008	161 94277	163 87064
4	(9) 3.9062 50000	(9) 3.9691 26001	(9) 4.0327 58016	(9) 4.0971 52081	(9) 4.1623 14256
5	(11) 9.7656 25000	(11) 9.9625 06263	(12) 1.0162 55020	(12) 1.0365 79476	(12) 1.0572 27821
6	(14) 2.4414 06250	(14) 2.5005 89072	(14) 2.5609 62650	(14) 2.6225 46076	(14) 2.6853 58665
7	(16) 6.1035 15625	(16) 6.2764 78570	(16) 6.4536 25879	(16) 6.6350 41571	(16) 6.8208 11010
8	(19) 1.5258 78906	(19) 1.5753 96121	(19) 1.6263 13722	(19) 1.6786 65517	(19) 1.7324 85997
9	(21) 3.8146 97266	(21) 3.9542 44264	(21) 4.0983 10578	(21) 4.2470 23759	(21) 4.4005 14431
10	(23) 9.5367 43164	(23) 9.9251 53103	(24) 1.0327 74266	(24) 1.0744 97011	(24) 1.1177 30666
24	(57) 3.5527 13679	(57) 3.9099 33001	(57) 4.3014 31179	(57) 4.7303 41643	(57) 5.2000 70108
1/2	(1) 1.5811 38830	(1) 1.5842 97952	(1) 1.5874 50787	(1) 1.5905 97372	(1) 1.5937 37745
1/3	6.2996 05249	6.3079 93549	6.3163 59598	6.3247 03543	6.3330 25531
1/4	3.9763 53644	3.9803 24047	3.9842 82604	3.9882 29397	3.9921 64507
1/5	3.0170 88168	3.0194 97986	3.0219 00136	3.0242 94671	3.0266 81647

k					
1	255	256	257	258	259
2	65025	65536	66049	66564	67081
3	165 81375	167 77216	169 74593	171 73512	173 73979
4	(9) 4.2282 50625	(9) 4.2949 67296	(9) 4.3624 70401	(9) 4.4307 66096	(9) 4.4998 60561
5	(12) 1.0782 03909	(12) 1.0995 11628	(12) 1.1211 54893	(12) 1.1431 37653	(12) 1.1654 63886
6	(14) 2.7494 19969	(14) 2.8147 49767	(14) 2.8813 68075	(14) 2.9492 95144	(14) 3.0185 51463
7	(16) 7.0110 20921	(16) 7.2057 59404	(16) 7.4051 15953	(16) 7.6091 81472	(16) 7.8180 48289
8	(19) 1.7878 10335	(19) 1.8446 74407	(19) 1.9031 14800	(19) 1.9631 68820	(19) 2.0248 74507
9	(21) 4.5589 16354	(21) 4.7223 66483	(21) 4.8910 05036	(21) 5.0649 75555	(21) 5.2444 24973
10	(24) 1.1625 23670	(24) 1.2089 25820	(24) 1.2569 88294	(24) 1.3067 63693	(24) 1.3583 06068
24	(57) 5.7143 17018	(57) 6.2771 01735	(57) 6.8927 88615	(57) 7.5661 15089	(57) 8.3022 21920
1/2	(1) 1.5968 71942	(1) 1.6000 00000	(1) 1.6031 21954	(1) 1.6062 37840	(1) 1.6093 47694
1/3	6.3413 25705	6.3496 04208	6.3578 61180	6.3660 96760	6.3743 11088
1/4	3.9960 88015	4.0000 00000	4.0039 00541	4.0077 89716	4.0116 67601
1/5	3.0290 61117	3.0314 33133	3.0337 97748	3.0361 55014	3.0385 04982

k					
1	260	261	262	263	264
2	67600	68121	68644	69169	69696
3	175 76000	177 79581	179 84728	181 91447	183 99744
4	(9) 4.5697 60000	(9) 4.6404 70641	(9) 4.7119 98736	(9) 4.7843 50561	(9) 4.8575 32416
5	(12) 1.1881 37600	(12) 1.2111 62837	(12) 1.2345 43669	(12) 1.2582 84198	(12) 1.2823 88558
6	(14) 3.0891 57760	(14) 3.1611 35005	(14) 3.2345 04412	(14) 3.3092 87440	(14) 3.3855 05793
7	(16) 8.0318 10176	(16) 8.2505 62364	(16) 8.4744 01560	(16) 8.7034 25966	(16) 8.9377 35293
8	(19) 2.0882 70646	(19) 2.1533 96777	(19) 2.2202 93209	(19) 2.2890 01029	(19) 2.3595 62117
9	(21) 5.4295 03679	(21) 5.6203 65588	(21) 5.8171 68207	(21) 6.0200 72706	(21) 6.2292 43990
10	(24) 1.4116 70957	(24) 1.4669 15418	(24) 1.5240 98070	(24) 1.5832 79122	(24) 1.6445 20413
24	(57) 9.1066 85770	(57) 9.9855 54265	(58) 1.0945 38372	(58) 1.1993 27974	(58) 1.3136 94086
1/2	(1) 1.6124 51550	(1) 1.6155 49442	(1) 1.6186 41406	(1) 1.6217 27474	(1) 1.6248 07681
1/3	6.3825 04299	6.3906 76528	6.3988 27910	6.4069 58577	6.4150 68660
1/4	4.0155 34273	4.0193 89807	4.0232 34278	4.0270 67760	4.0308 90325
1/5	3.0408 47703	3.0431 83226	3.0455 11602	3.0478 32879	3.0501 47105

k					
1	265	266	267	268	269
2	70225	70756	71289	71824	72361
3	186 09625	188 21096	190 34163	192 48832	194 65109
4	(9) 4.9315 50625	(9) 5.0064 11536	(9) 5.0821 21521	(9) 5.1586 86976	(9) 5.2361 14321
5	(12) 1.3068 60916	(12) 1.3317 05469	(12) 1.3569 26446	(12) 1.3825 28110	(12) 1.4085 14752
6	(14) 3.4631 81426	(14) 3.5423 36546	(14) 3.6229 93611	(14) 3.7051 75334	(14) 3.7889 04684
7	(16) 9.1774 30780	(16) 9.4226 15213	(16) 9.6733 92942	(16) 9.9298 69894	(17) 1.0192 15360
8	(19) 2.4320 19157	(19) 2.5064 15647	(19) 2.5827 95915	(19) 2.6612 05132	(19) 2.7416 89318
9	(21) 6.4448 50765	(21) 6.6670 65620	(21) 6.8960 65094	(21) 7.1320 29753	(21) 7.3751 44266
10	(24) 1.7078 85453	(24) 1.7734 39455	(24) 1.8412 49380	(24) 1.9113 83974	(24) 1.9839 13808
24	(58) 1.4384 70548	(58) 1.5745 60235	(58) 1.7229 40472	(58) 1.8846 68868	(58) 2.0608 89564
1/2	(1) 1.6278 82060	(1) 1.6309 50643	(1) 1.6340 13464	(1) 1.6370 70554	(1) 1.6401 21947
1/3	6.4231 58289	6.4312 27591	6.4392 76696	6.4473 05727	6.4553 14811
1/4	4.0347 02045	4.0385 02994	4.0422 93240	4.0460 72854	4.0498 41906
1/5	3.0524 54329	3.0547 54599	3.0570 47961	3.0593 34462	3.0616 14147

k					
1	270	271	272	273	274
2	72900	73441	73984	74529	75076
3	196 83000	199 02511	201 23648	203 46417	205 70824
4	(9) 5.3144 10000	(9) 5.3935 80481	(9) 5.4736 32256	(9) 5.5545 71841	(9) 5.6364 05776
5	(12) 1.4348 90700	(12) 1.4616 60310	(12) 1.4888 27974	(12) 1.5163 98113	(12) 1.5443 75183
6	(14) 3.8742 04890	(14) 3.9610 99441	(14) 4.0496 12088	(14) 4.1397 66847	(14) 4.2315 88000
7	(17) 1.0460 35320	(17) 1.0734 57949	(17) 1.1014 94488	(17) 1.1301 56349	(17) 1.1594 55112
8	(19) 2.8242 95365	(19) 2.9090 71041	(19) 2.9960 65007	(19) 3.0853 26834	(19) 3.1769 07007
9	(21) 7.6255 97485	(21) 7.8835 82520	(21) 8.1492 96820	(21) 8.4229 42256	(21) 8.7047 25200
10	(24) 2.0589 11321	(24) 2.1364 50863	(24) 2.2166 08735	(24) 2.2994 63236	(24) 2.3850 94705
24	(58) 2.2528 39954	(58) 2.4618 57897	(58) 2.6893 89450	(58) 2.9369 97176	(58) 3.2063 69049
1/2	(1) 1.6431 67673	(1) 1.6462 07763	(1) 1.6492 42250	(1) 1.6522 71164	(1) 1.6552 94536
1/3	6.4633 04070	6.4712 73627	6.4792 23603	6.4871 54117	6.4950 65288
1/4	4.0536 00464	4.0573 48596	4.0610 86370	4.0648 13851	4.0685 31106
1/5	3.0638 87063	3.0661 53254	3.0684 12765	3.0706 65640	3.0729 11923

$n^{\frac{1}{2}}\left[\begin{matrix}(-6)8\\4\end{matrix}\right]$ $n^{\frac{1}{3}}\left[\begin{matrix}(-6)3\\4\end{matrix}\right]$ $n^{\frac{1}{4}}\left[\begin{matrix}(-6)2\\4\end{matrix}\right]$ $n^{\frac{1}{5}}\left[\begin{matrix}(-6)1\\4\end{matrix}\right]$

k					
1	275	276	277	278	279
2	75625	76176	76729	77284	77841
3	207 96875	210 24576	212 53933	214 84952	217 17639
4	(9)5.7191 40625	(9)5.8027 82976	(9)5.8873 39441	(9)5.9728 16656	(9)6.0592 21281
5	(12)1.5727 63672	(12)1.6015 68101	(12)1.6307 93025	(12)1.6604 43030	(12)1.6905 22737
6	(14)4.3251 00098	(14)4.4203 27960	(14)4.5172 96630	(14)4.6160 31624	(14)4.7165 58437
7	(17)1.1894 02527	(17)1.2200 10517	(17)1.2512 91180	(17)1.2832 56792	(17)1.3159 19804
8	(19)3.2708 56949	(19)3.3672 29027	(19)3.4660 76569	(19)3.5674 53881	(19)3.6714 16253
9	(21)8.9948 56609	(21)9.2935 52114	(21)9.6010 32097	(21)9.9175 21788	(22)1.0243 25135
10	(24)2.4735 85568	(24)2.5650 20383	(24)2.6594 85891	(24)2.7570 71057	(24)2.8578 67126
24	(58)3.4993 28001	(58)3.8178 42160	(58)4.1640 35828	(58)4.5402 01230	(58)4.9488 11121
1/2	(1)1.6583 12395	(1)1.6613 24773	(1)1.6643 31698	(1)1.6673 33200	(1)1.6703 29309
1/3	6.5029 57234	6.5108 30071	6.5186 83915	6.5265 18879	6.5343 35077
1/4	4.0722 38199	4.0759 35196	4.0796 22161	4.0832 99156	4.0869 66245
1/5	3.0751 51657	3.0773 84885	3.0796 11650	3.0818 31992	3.0840 45954

k					
1	280	281	282	283	284
2	78400	78961	79524	80089	80656
3	219 52000	221 88041	224 25768	226 65187	229 06304
4	(9)6.1465 60000	(9)6.2348 39521	(9)6.3240 66576	(9)6.4142 47921	(9)6.5053 90336
5	(12)1.7210 36800	(12)1.7519 89905	(12)1.7833 86774	(12)1.8152 32162	(12)1.8475 30855
6	(14)4.8189 03040	(14)4.9230 91634	(14)5.0291 50704	(14)5.1371 07017	(14)5.2469 87629
7	(17)1.3492 92851	(17)1.3833 88749	(17)1.4182 20498	(17)1.4538 01286	(17)1.4901 44487
8	(19)3.7780 19983	(19)3.8873 22385	(19)3.9993 81806	(19)4.1142 57639	(19)4.2320 10342
9	(22)1.0578 45595	(22)1.0923 37590	(22)1.1278 25669	(22)1.1643 34912	(22)1.2018 90937
10	(24)2.9619 67667	(24)3.0694 68629	(24)3.1804 68387	(24)3.2950 67801	(24)3.4133 70262
24	(58)5.3925 32264	(58)5.8742 39885	(58)6.3970 33126	(58)6.9642 51599	(58)7.5794 93086
1/2	(1)1.6733 20053	(1)1.6763 05461	(1)1.6792 85562	(1)1.6822 60384	(1)1.6852 29955
1/3	6.5421 32620	6.5499 11620	6.5576 72186	6.5654 14427	6.5731 38451
1/4	4.0906 23489	4.0942 70950	4.0979 08689	4.1015 36766	4.1051 55240
1/5	3.0862 53577	3.0884 54901	3.0906 49967	3.0928 38815	3.0950 21484

k					
1	285	286	287	288	289
2	81225	81796	82369	82944	83521
3	231 49125	233 93656	236 39903	238 87872	241 37569
4	(9)6.5975 00625	(9)6.6905 85616	(9)6.7846 52161	(9)6.8797 07136	(9)6.9757 57441
5	(12)1.8802 87678	(12)1.9135 07486	(12)1.9471 95170	(12)1.9813 55655	(12)2.0159 93900
6	(14)5.3588 19883	(14)5.4726 31410	(14)5.5884 50138	(14)5.7063 04287	(14)5.8262 22372
7	(17)1.5272 63667	(17)1.5651 72583	(17)1.6038 85190	(17)1.6434 15635	(17)1.6837 78266
8	(19)4.3527 01450	(19)4.4763 93589	(19)4.6031 50495	(19)4.7330 37028	(19)4.8661 19188
9	(22)1.2405 19913	(22)1.2802 48566	(22)1.3211 04192	(22)1.3631 14664	(22)1.4063 08445
10	(24)3.5354 81753	(24)3.6615 10900	(24)3.7915 69031	(24)3.9257 70232	(24)4.0642 31407
24	(58)8.2466 32480	(58)8.9698 42039	(58)9.7536 13040	(59)1.0602 77893	(59)1.1522 54005
1/2	(1)1.6881 94302	(1)1.6911 53453	(1)1.6941 07435	(1)1.6970 56275	(1)1.7000 00000
1/3	6.5808 44365	6.5885 32275	6.5962 02284	6.6038 54498	6.6114 89018
1/4	4.1087 64171	4.1123 63618	4.1159 53637	4.1195 34288	4.1231 05626
1/5	3.0971 98013	3.0993 68441	3.1015 32807	3.1036 91148	3.1058 43502

k					
1	290	291	292	293	294
2	84100	84681	85264	85849	86436
3	243 89000	246 42171	248 97088	251 53757	254 12184
4	(9)7.0728 10000	(9)7.1708 71761	(9)7.2699 49696	(9)7.3700 50801	(9)7.4711 82096
5	(12)2.0511 14900	(12)2.0867 23682	(12)2.1228 25311	(12)2.1594 24885	(12)2.1965 27536
6	(14)5.9482 33210	(14)6.0723 65916	(14)6.1986 49909	(14)6.3271 14912	(14)6.4577 90956
7	(17)1.7249 87631	(17)1.7670 58482	(17)1.8100 05773	(17)1.8538 44669	(17)1.8985 90541
8	(19)5.0024 64130	(19)5.1421 40181	(19)5.2852 16858	(19)5.4317 64881	(19)5.5818 56191
9	(22)1.4507 14598	(22)1.4963 62793	(22)1.5432 83323	(22)1.5915 07110	(22)1.6410 65720
10	(24)4.2070 72333	(24)4.3544 15727	(24)4.5063 87302	(24)4.6631 15833	(24)4.8247 33217
24	(59)1.2518 49008	(59)1.3596 64428	(59)1.4763 46962	(59)1.6025 91698	(59)1.7391 45550
1/2	(1)1.7029 38637	(1)1.7058 72211	(1)1.7088 00749	(1)1.7117 24277	(1)1.7146 42820
1/3	6.6191 05948	6.6267 05387	6.6342 87437	6.6418 52195	6.6493 99761
1/4	4.1266 67707	4.1302 20588	4.1337 64325	4.1372 98970	4.1408 24580
1/5	3.1079 89906	3.1101 30396	3.1122 65011	3.1143 93785	3.1165 16755

k					
1	295	296	297	298	299
2	87025	87616	88209	88804	89401
3	256 72375	259 34336	261 98073	264 63592	267 30899
4	(9)7.5733 50625	(9)7.6765 63456	(9)7.7808 27681	(9)7.8861 50416	(9)7.9925 38801
5	(12)2.2341 38434	(12)2.2722 62783	(12)2.3109 05821	(12)2.3500 72824	(12)2.3897 69101
6	(14)6.5907 08381	(14)6.7258 97838	(14)6.8633 90289	(14)7.0032 17015	(14)7.1454 09613
7	(17)1.9442 58973	(17)1.9908 65760	(17)2.0384 26916	(17)2.0869 58671	(17)2.1364 77474
8	(19)5.7355 63969	(19)5.8929 62649	(19)6.0541 27940	(19)6.2191 36838	(19)6.3880 67649
9	(22)1.6919 91371	(22)1.7443 16944	(22)1.7980 75998	(22)1.8533 02778	(22)1.9100 32227
10	(24)4.9913 74544	(24)5.1631 78155	(24)5.3402 85715	(24)5.5228 42278	(24)5.7109 96358
24	(59)1.8868 10930	(59)2.0464 49657	(59)2.2189 87131	(59)2.4054 16789	(59)2.6068 04847
1/2	(1)1.7175 56404	(1)1.7204 65053	(1)1.7233 68794	(1)1.7262 67650	(1)1.7291 61647
1/3	6.6569 30232	6.6644 43703	6.6719 40272	6.6794 20032	6.6868 83077
1/4	4.1443 41207	4.1478 48904	4.1513 47726	4.1548 37723	4.1583 18947
1/5	3.1186 33956	3.1207 45423	3.1228 51191	3.1249 51295	3.1270 45768

$n^{\frac{1}{2}}\left[{(-6)7 \atop 4}\right]$ $n^{\frac{1}{3}}\left[{(-6)2 \atop 4}\right]$ $n^{\frac{1}{4}}\left[{(-6)1 \atop 4}\right]$ $n^{\frac{1}{5}}\left[{(-7)8 \atop 4}\right]$

k	300	301	302	303	304
1	300	301	302	303	304
2	90000	90601	91204	91809	92416
3	270 00000	272 70901	275 43608	278 18127	280 94464
4	(9)8.1000 00000	(9)8.2085 41201	(9)8.3181 69616	(9)8.4288 92481	(9)8.5407 17056
5	(12)2.4300 00000	(12)2.4707 70902	(12)2.5120 87224	(12)2.5539 54422	(12)2.5963 77985
6	(14)7.2900 00000	(14)7.4370 20414	(14)7.5865 03417	(14)7.7384 81898	(14)7.8929 89074
7	(17)2.1870 00000	(17)2.2385 43144	(17)2.2911 24032	(17)2.3447 60015	(17)2.3994 68679
8	(19)6.5610 00000	(19)6.7380 14865	(19)6.9191 94576	(19)7.1046 22846	(19)7.2943 84783
9	(22)1.9683 00000	(22)2.0281 42474	(22)2.0895 96762	(22)2.1527 00722	(22)2.2174 92974
10	(24)5.9049 00000	(24)6.1047 08848	(24)6.3105 82221	(24)6.5226 83188	(24)6.7411 78641
24	(59)2.8242 95365	(59)3.0591 15639	(59)3.3125 81949	(59)3.5861 05682	(59)3.8811 99856
1/2	(1)1.7320 50808	(1)1.7349 35157	(1)1.7378 14720	(1)1.7406 89519	(1)1.7435 59577
1/3	6.6943 29501	6.7017 59395	6.7091 72852	6.7165 69962	6.7239 50814
1/4	4.1617 91450	4.1652 55283	4.1687 10496	4.1721 57138	4.1755 95260
1/5	3.1291 34645	3.1312 17958	3.1332 95743	3.1353 68030	3.1374 34853

k	305	306	307	308	309
1	305	306	307	308	309
2	93025	93636	94249	94864	95481
3	283 72625	286 52616	289 34443	292 18112	295 03629
4	(9)8.6536 50625	(9)8.7677 00496	(9)8.8828 74001	(9)8.9991 78496	(9)9.1166 21361
5	(12)2.6393 63441	(12)2.6829 16352	(12)2.7270 42318	(12)2.7717 46977	(12)2.8170 36001
6	(14)8.0500 58494	(14)8.2097 24036	(14)8.3720 19917	(14)8.5369 80688	(14)8.7046 41242
7	(17)2.4552 67841	(17)2.5121 75555	(17)2.5702 10115	(17)2.6293 90052	(17)2.6897 34144
8	(19)7.4885 66914	(19)7.6872 57199	(19)7.8905 45052	(19)8.0985 21360	(19)8.3112 78504
9	(22)2.2840 12909	(22)2.3523 00703	(22)2.4223 97331	(22)2.4943 44579	(22)2.5681 85058
10	(24)6.9662 39372	(24)7.1980 40151	(24)7.4367 59806	(24)7.6825 81303	(24)7.9356 91828
24	(59)4.1994 86063	(59)4.5427 01868	(59)4.9127 08679	(59)5.3115 00125	(59)5.7412 10972
1/2	(1)1.7464 24920	(1)1.7492 85568	(1)1.7521 41547	(1)1.7549 92877	(1)1.7578 39583
1/3	6.7313 15497	6.7386 64101	6.7459 96712	6.7533 13417	6.7606 14302
1/4	4.1790 24910	4.1824 46136	4.1858 58988	4.1892 63512	4.1926 59756
1/5	3.1394 96244	3.1415 52236	3.1436 02859	3.1456 48146	3.1476 88127

k	310	311	312	313	314
1	310	311	312	313	314
2	96100	96721	97344	97969	98596
3	297 91000	300 80231	303 71328	306 64297	309 59144
4	(9)9.2352 10000	(9)9.3549 51841	(9)9.4758 54336	(9)9.5979 24961	(9)9.7211 71216
5	(12)2.8629 15100	(12)2.9093 90023	(12)2.9564 66553	(12)3.0041 50513	(12)3.0524 47762
6	(14)8.8750 36810	(14)9.0482 02970	(14)9.2241 75645	(14)9.4029 91105	(14)9.5846 85972
7	(17)2.7512 61411	(17)2.8139 91124	(17)2.8779 42801	(17)2.9431 36216	(17)3.0095 91395
8	(19)8.5289 10374	(19)8.7515 12395	(19)8.9791 81540	(19)9.2120 16356	(19)9.4501 16981
9	(22)2.6439 62216	(22)2.7217 20355	(22)2.8015 04640	(22)2.8833 61119	(22)2.9673 36732
10	(24)8.1962 82870	(24)8.4645 50303	(24)8.7406 94478	(24)9.0249 20304	(24)9.3174 37339
24	(59)6.2041 26610	(59)6.7026 93132	(59)7.2395 28072	(59)7.8174 31800	(59)8.4393 99655
1/2	(1)1.7606 81686	(1)1.7635 19209	(1)1.7663 52173	(1)1.7691 80601	(1)1.7720 04515
1/3	6.7678 99452	6.7751 68952	6.7824 22886	6.7896 61336	6.7968 84386
1/4	4.1960 47767	4.1994 27591	4.2027 99273	4.2061 62861	4.2095 18398
1/5	3.1497 22833	3.1517 52295	3.1537 76544	3.1557 95609	3.1578 09519

k	315	316	317	318	319
1	315	316	317	318	319
2	99225	99856	1 00489	1 01124	1 01761
3	312 55875	315 54496	318 55013	321 57432	324 61759
4	(9)9.8456 00625	(9)9.9712 20736	(10)1.0098 03912	(10)1.0226 06338	(10)1.0355 30112
5	(12)3.1013 64197	(12)3.1509 05753	(12)3.2010 78401	(12)3.2518 88154	(12)3.3033 41058
6	(14)9.7692 97220	(14)9.9568 62178	(15)1.0147 41853	(15)1.0341 00433	(15)1.0537 65797
7	(17)3.0773 28624	(17)3.1463 68448	(17)3.2167 31675	(17)3.2884 39376	(17)3.3615 12894
8	(19)9.6935 85167	(19)9.9425 24297	(20)1.0197 03941	(20)1.0457 23722	(20)1.0723 22613
9	(22)3.0534 79328	(22)3.1418 37678	(22)3.2324 61493	(22)3.3254 01435	(22)3.4207 09136
10	(24)9.6184 59882	(24)9.9282 07062	(25)1.0246 90293	(25)1.0574 77656	(25)1.0912 06214
24	(59)9.1086 34822	(59)9.8285 62028	(60)1.0602 84208	(60)1.1435 38734	(60)1.2330 37808
1/2	(1)1.7748 23935	(1)1.7776 38883	(1)1.7804 49381	(1)1.7832 55450	(1)1.7860 57110
1/3	6.8040 92116	6.8112 84608	6.8184 61941	6.8256 24197	6.8327 71452
1/4	4.2128 65931	4.2162 05502	4.2195 37156	4.2228 60938	4.2261 76889
1/5	3.1598 18306	3.1618 21997	3.1638 20622	3.1658 14209	3.1678 02787

k	320	321	322	323	324
1	320	321	322	323	324
2	1 02400	1 03041	1 03684	1 04329	1 04976
3	327 68000	330 76161	333 86248	336 98267	340 12224
4	(10)1.0485 76000	(10)1.0617 44768	(10)1.0750 37186	(10)1.0884 54024	(10)1.1019 96058
5	(12)3.3554 43200	(12)3.4082 00706	(12)3.4616 19738	(12)3.5157 06498	(12)3.5704 67227
6	(15)1.0737 41824	(15)1.0940 32426	(15)1.1146 41556	(15)1.1355 73199	(15)1.1568 31381
7	(17)3.4359 73837	(17)3.5118 44089	(17)3.5891 45809	(17)3.6679 01432	(17)3.7481 33676
8	(20)1.0995 11628	(20)1.1273 01953	(20)1.1557 04950	(20)1.1847 32163	(20)1.2143 95311
9	(22)3.5184 37209	(22)3.6186 39268	(22)3.7213 69940	(22)3.8266 84885	(22)3.9346 40808
10	(25)1.1258 99907	(25)1.1615 83205	(25)1.1982 81121	(25)1.2360 19218	(25)1.2748 23622
24	(60)1.3292 27996	(60)1.4325 86248	(60)1.5436 21862	(60)1.6628 78568	(60)1.7909 36736
1/2	(1)1.7888 54382	(1)1.7916 47287	(1)1.7944 35844	(1)1.7972 20076	(1)1.8000 00000
1/3	6.8399 03787	6.8470 21278	6.8541 24002	6.8612 12036	6.8682 85455
1/4	4.2294 85054	4.2327 85474	4.2360 78192	4.2393 63249	4.2426 40687
1/5	3.1697 86385	3.1717 65030	3.1737 38749	3.1757 07571	3.1776 71523

$n^{\frac{1}{2}}\left[\begin{matrix}(-6)6\\4\end{matrix}\right]$ $n^{\frac{1}{3}}\left[\begin{matrix}(-6)2\\4\end{matrix}\right]$ $n^{\frac{1}{4}}\left[\begin{matrix}(-6)1\\4\end{matrix}\right]$ $n^{\frac{1}{5}}\left[\begin{matrix}(-7)7\\4\end{matrix}\right]$

k					
1	325	326	327	328	329
2	1 05625	1 06276	1 06929	1 07584	1 08241
3	343 28125	346 45976	349 65783	352 87552	356 11289
4	(10) 1.1156 64063	(10) 1.1294 58818	(10) 1.1433 81104	(10) 1.1574 31706	(10) 1.1716 11408
5	(12) 3.6259 08203	(12) 3.6820 35745	(12) 3.7388 56210	(12) 3.7963 75994	(12) 3.8546 01533
6	(15) 1.1784 20166	(15) 1.2003 43653	(15) 1.2226 05981	(15) 1.2452 11326	(15) 1.2681 63904
7	(17) 3.8298 65540	(17) 3.9131 20309	(17) 3.9979 21557	(17) 4.0842 93150	(17) 4.1722 59245
8	(20) 1.2447 06300	(20) 1.2756 77221	(20) 1.3073 20349	(20) 1.3396 48153	(20) 1.3726 73292
9	(22) 4.0452 95476	(22) 4.1587 07739	(22) 4.2749 37542	(22) 4.3940 45942	(22) 4.5160 95129
10	(25) 1.3147 21030	(25) 1.3557 38723	(25) 1.3979 04576	(25) 1.4412 47069	(25) 1.4857 95298
24	(60) 1.9284 15722	(60) 2.0759 76350	(60) 2.2343 23554	(60) 2.4042 09169	(60) 2.5864 34894
1/2	(1) 1.8027 75638	(1) 1.8055 47009	(1) 1.8083 14132	(1) 1.8110 77028	(1) 1.8138 35715
1/3	6.8753 44335	6.8823 88750	6.8894 18774	6.8964 34481	6.9034 35942
1/4	4.2459 10547	4.2491 72871	4.2524 27697	4.2556 75067	4.2589 15020
1/5	3.1796 30632	3.1815 84924	3.1835 34426	3.1854 79164	3.1874 19165

k					
1	330	331	332	333	334
2	1 08900	1 09561	1 10224	1 10889	1 11556
3	359 37000	362 64691	365 94368	369 26037	372 59704
4	(10) 1.1859 21000	(10) 1.2003 61272	(10) 1.2149 33018	(10) 1.2296 37032	(10) 1.2444 74114
5	(12) 3.9135 39300	(12) 3.9731 95811	(12) 4.0335 77618	(12) 4.0946 91317	(12) 4.1565 43539
6	(15) 1.2914 67969	(15) 1.3151 27813	(15) 1.3391 47769	(15) 1.3635 32209	(15) 1.3882 85542
7	(17) 4.2618 44298	(17) 4.3530 73062	(17) 4.4459 70594	(17) 4.5405 62254	(17) 4.6368 73711
8	(20) 1.4064 08618	(20) 1.4408 67184	(20) 1.4760 62237	(20) 1.5120 07231	(20) 1.5487 15819
9	(22) 4.6411 48440	(22) 4.7692 70378	(22) 4.9005 26628	(22) 5.0349 84078	(22) 5.1727 10837
10	(25) 1.5315 78985	(25) 1.5786 28495	(25) 1.6269 74840	(25) 1.6766 49698	(25) 1.7276 85420
24	(60) 2.7818 55434	(60) 2.9913 81825	(60) 3.2159 84959	(60) 3.4566 99320	(60) 3.7146 26935
1/2	(1) 1.8165 90212	(1) 1.8193 40540	(1) 1.8220 86716	(1) 1.8248 28759	(1) 1.8275 66688
1/3	6.9104 23230	6.9173 96417	6.9243 55573	6.9313 00768	6.9382 32074
1/4	4.2621 47595	4.2653 72832	4.2685 90770	4.2718 01446	4.2750 04899
1/5	3.1893 54454	3.1912 85058	3.1932 11001	3.1951 32308	3.1970 49006

k					
1	335	336	337	338	339
2	1 12225	1 12896	1 13569	1 14244	1 14921
3	375 95375	379 33056	382 72753	386 14472	389 58219
4	(10) 1.2594 45063	(10) 1.2745 50682	(10) 1.2897 91776	(10) 1.3051 69154	(10) 1.3206 83624
5	(12) 4.2191 40959	(12) 4.2824 90290	(12) 4.3465 98285	(12) 4.4114 71739	(12) 4.4771 17486
6	(15) 1.4134 12221	(15) 1.4389 16737	(15) 1.4648 03622	(15) 1.4910 77448	(15) 1.5177 42828
7	(17) 4.7349 30942	(17) 4.8347 60238	(17) 4.9363 88207	(17) 5.0398 41774	(17) 5.1451 48186
8	(20) 1.5862 01865	(20) 1.6244 79440	(20) 1.6635 62826	(20) 1.7034 66520	(20) 1.7442 05235
9	(22) 5.3137 76249	(22) 5.4582 50918	(22) 5.6062 06723	(22) 5.7577 16836	(22) 5.9128 55747
10	(25) 1.7801 15044	(25) 1.8339 72309	(25) 1.8892 91666	(25) 1.9461 08291	(25) 2.0044 58098
24	(60) 3.9909 41565	(60) 4.2868 93134	(60) 4.6038 12427	(60) 4.9431 16051	(60) 5.3063 11693
1/2	(1) 1.8303 00522	(1) 1.8330 30278	(1) 1.8357 55975	(1) 1.8384 77631	(1) 1.8411 95264
1/3	6.9451 49558	6.9520 53290	6.9589 43337	6.9658 19768	6.9726 82649
1/4	4.2782 01166	4.2813 90286	4.2845 72295	4.2877 47230	4.2909 15128
1/5	3.1989 61118	3.2008 68669	3.2027 71684	3.2046 70186	3.2065 64201

k					
1	340	341	342	343	344
2	1 15600	1 16281	1 16964	1 17649	1 18336
3	393 04000	396 51821	400 01688	403 53607	407 07584
4	(10) 1.3363 36000	(10) 1.3521 27096	(10) 1.3680 57730	(10) 1.3841 28720	(10) 1.4003 40890
5	(12) 4.5435 42400	(12) 4.6107 53398	(12) 4.6787 57435	(12) 4.7475 61510	(12) 4.8171 72660
6	(15) 1.5448 04416	(15) 1.5722 66909	(15) 1.6001 35043	(15) 1.6284 13598	(15) 1.6571 07395
7	(17) 5.2523 35014	(17) 5.3614 30158	(17) 5.4724 61847	(17) 5.5854 58641	(17) 5.7004 49439
8	(20) 1.7857 93905	(20) 1.8282 47684	(20) 1.8715 81952	(20) 1.9158 12314	(20) 1.9609 54607
9	(22) 6.0716 99277	(22) 6.2343 24602	(22) 6.4008 10274	(22) 6.5712 36236	(22) 6.7456 83848
10	(25) 2.0643 77754	(25) 2.1259 04689	(25) 2.1890 77114	(25) 2.2539 34029	(25) 2.3205 15244
24	(60) 5.6950 03680	(60) 6.1108 98859	(60) 6.5558 12822	(60) 7.0316 76479	(60) 7.5405 43015
1/2	(1) 1.8439 08891	(1) 1.8466 18531	(1) 1.8493 24201	(1) 1.8520 25918	(1) 1.8547 23699
1/3	6.9795 32047	6.9863 68028	6.9931 90657	7.0000 00000	7.0067 96121
1/4	4.2940 76026	4.2972 29958	4.3003 76961	4.3035 17071	4.3066 50321
1/5	3.2084 53751	3.2103 38860	3.2122 19552	3.2140 95850	3.2159 67776

k					
1	345	346	347	348	349
2	1 19025	1 19716	1 20409	1 21104	1 21801
3	410 63625	414 21736	417 81923	421 44192	425 08549
4	(10) 1.4166 95063	(10) 1.4331 92066	(10) 1.4498 32728	(10) 1.4666 17882	(10) 1.4835 48360
5	(12) 4.8875 97966	(12) 4.9588 44547	(12) 5.0309 19567	(12) 5.1038 30228	(12) 5.1775 83777
6	(15) 1.6862 21298	(15) 1.7157 60213	(15) 1.7457 29090	(15) 1.7761 32919	(15) 1.8069 76738
7	(17) 5.8174 63479	(17) 5.9365 30338	(17) 6.0576 79941	(17) 6.1809 42559	(17) 6.3063 48816
8	(20) 2.0070 24900	(20) 2.0540 39497	(20) 2.1020 14939	(20) 2.1509 68011	(20) 2.2009 15737
9	(22) 6.9242 35905	(22) 7.1069 76659	(22) 7.2939 91840	(22) 7.4853 68677	(22) 7.6811 95921
10	(25) 2.3888 61387	(25) 2.4590 13924	(25) 2.5310 15168	(25) 2.6049 08300	(25) 2.6807 37377
24	(60) 8.0845 95243	(60) 8.6661 53376	(60) 9.2876 83235	(60) 9.9518 04932	(61) 1.0661 30203
1/2	(1) 1.8574 17562	(1) 1.8601 07524	(1) 1.8627 93601	(1) 1.8654 75811	(1) 1.8681 54169
1/3	7.0135 79083	7.0203 48952	7.0271 05788	7.0338 49656	7.0405 80617
1/4	4.3097 76748	4.3128 96386	4.3160 09269	4.3191 15431	4.3222 14906
1/5	3.2178 35355	3.2196 98608	3.2215 57557	3.2234 12226	3.2252 62636

$n^{\frac{1}{2}}\begin{bmatrix}(-6)5\\4\end{bmatrix}$ $n^{\frac{1}{3}}\begin{bmatrix}(-6)2\\4\end{bmatrix}$ $n^{\frac{1}{4}}\begin{bmatrix}(-6)1\\4\end{bmatrix}$ $n^{\frac{1}{5}}\begin{bmatrix}(-7)6\\4\end{bmatrix}$

k					
1	350	351	352	353	354
2	1 22500	1 23201	1 23904	1 24609	1 25316
3	428 75000	432 43551	436 14208	439 86977	443 61864
4	(10)1.5006 25000	(10)1.5178 48640	(10)1.5352 20122	(10)1.5527 40288	(10)1.5704 09986
5	(12)5.2521 87500	(12)5.3276 48727	(12)5.4039 74828	(12)5.4811 73217	(12)5.5592 51349
6	(15)1.8382 65625	(15)1.8700 04703	(15)1.9021 99139	(15)1.9348 54146	(15)1.9679 74978
7	(17)6.4339 29688	(17)6.5637 16508	(17)6.6957 40971	(17)6.8300 35134	(17)6.9666 31421
8	(20)2.2518 75391	(20)2.3038 64494	(20)2.3569 00822	(20)2.4110 02402	(20)2.4661 87523
9	(22)7.8815 63867	(22)8.0865 64375	(22)8.2962 90893	(22)8.5108 38480	(22)8.7303 03831
10	(25)2.7585 47354	(25)2.8383 84096	(25)2.9202 94394	(25)3.0043 25983	(25)3.0905 27556
24	(61)1.1419 13124	(61)1.2228 43263	(61)1.3092 54042	(61)1.4014 99442	(61)1.4999 55202
1/2	(1)1.8708 28693	(1)1.8734 99400	(1)1.8761 66304	(1)1.8788 29423	(1)1.8814 88772
1/3	7.0472 98732	7.0540 04063	7.0606 96671	7.0673 76615	7.0740 43955
1/4	4.3253 07727	4.3283 93928	4.3314 73541	4.3345 46600	4.3376 13137
1/5	3.2271 08809	3.2289 50768	3.2307 88532	3.2326 22125	3.2344 51567

k					
1	355	356	357	358	359
2	1 26025	1 26736	1 27449	1 28164	1 28881
3	447 38875	451 18016	454 99293	458 82712	462 68279
4	(10)1.5882 30063	(10)1.6062 01370	(10)1.6243 24760	(10)1.6426 01090	(10)1.6610 31216
5	(12)5.6382 16722	(12)5.7180 76876	(12)5.7988 39394	(12)5.8805 11901	(12)5.9631 02066
6	(15)2.0015 66936	(15)2.0356 35368	(15)2.0701 85663	(15)2.1052 23260	(15)2.1407 53642
7	(17)7.1055 62624	(17)7.2468 61909	(17)7.3905 62819	(17)7.5366 99273	(17)7.6853 05573
8	(20)2.5224 74731	(20)2.5798 82840	(20)2.6384 30926	(20)2.6981 38340	(20)2.7590 24701
9	(22)8.9547 85297	(22)9.1843 82909	(22)9.4191 98407	(22)9.6593 35256	(22)9.9048 98676
10	(25)3.1789 48780	(25)3.2696 40316	(25)3.3626 53831	(25)3.4580 42022	(25)3.5558 58625
24	(61)1.6050 20092	(61)1.7171 17251	(61)1.8366 95605	(61)1.9642 31355	**(61)2.1002 29556**
1/2	(1)1.8841 44368	(1)1.8867 96226	(1)1.8894 44363	(1)1.8920 88793	(1)1.8947 29532
1/3	7.0806 98751	7.0873 41061	7.0939 70945	7.1005 88459	7.1071 93661
1/4	4.3406 73183	4.3437 26771	4.3467 73933	4.3498 14700	4.3528 49104
1/5	3.2362 76880	3.2380 98084	3.2399 15199	3.2417 28247	3.2435 37249

k					
1	360	361	362	363	364
2	1 29600	1 30321	1 31044	1 31769	1 32496
3	466 56000	470 45881	474 37928	478 32147	482 28544
4	(10)1.6796 16000	(10)1.6983 56304	(10)1.7172 52994	(10)1.7363 06936	(10)1.7555 19002
5	(12)6.0466 17600	(12)6.1310 66258	(12)6.2164 55837	(12)6.3027 94178	(12)6.3900 89166
6	(15)2.1767 82336	(15)2.2133 14919	(15)2.2503 57013	(15)2.2879 14287	(15)2.3259 92456
7	(17)7.8364 16410	(17)7.9900 66858	(17)8.1462 92387	(17)8.3051 28860	(17)8.4666 12541
8	(20)2.8211 09907	(20)2.8844 14136	(20)2.9489 57844	(20)3.0147 61776	(20)3.0818 46965
9	(23)1.0155 99567	(23)1.0412 73503	(23)1.0675 22740	(23)1.0943 58525	(23)1.1217 92295
10	(25)3.6561 58440	(25)3.7589 97346	(25)3.8644 32317	(25)3.9725 21445	(25)4.0833 23955
24	(61)2.2452 25771	(61)2.3997 87825	(61)2.5645 17652	(61)2.7400 53237	(61)2.9270 70667
1/2	(1)1.8973 66596	(1)1.9000 00000	(1)1.9026 29759	(1)1.9052 55888	(1)1.9078 78403
1/3	7.1137 86609	7.1203 67359	7.1269 35967	7.1334 92490	7.1400 36982
1/4	4.3558 77175	4.3588 98944	4.3619 14441	4.3649 23697	4.3679 26743
1/5	3.2453 42223	3.2471 43191	3.2489 40172	3.2507 33187	3.2525 22254

k					
1	365	366	367	368	369
2	1 33225	1 33956	1 34689	1 35424	1 36161
3	486 27125	490 27896	494 30863	498 36032	502 43409
4	(10)1.7748 90063	(10)1.7944 20994	(10)1.8141 12672	(10)1.8339 65978	(10)1.8539 81792
5	(12)6.4783 48728	(12)6.5675 80837	(12)6.6577 93507	(12)6.7489 94798	(12)6.8411 92813
6	(15)2.3645 97286	(15)2.4037 34586	(15)2.4434 10217	(15)2.4836 30086	(15)2.5244 00148
7	(17)8.6307 80093	(17)8.7976 68585	(17)8.9673 15496	(17)9.1397 58715	(17)9.3150 36546
8	(20)3.1502 34734	(20)3.2199 46702	(20)3.2910 04787	(20)3.3634 31207	(20)3.4372 48485
9	(23)1.1498 35678	(23)1.1785 00493	(23)1.2077 98757	(23)1.2377 42684	(23)1.2683 44691
10	(25)4.1969 00224	(25)4.3133 11804	(25)4.4326 21438	(25)4.5548 93078	(25)4.6801 91910
24	(61)3.1262 86296	(61)3.3384 59019	(61)3.5643 92671	(61)3.8049 38558	(61)4.0609 98114
1/2	(1)1.9104 97317	(1)1.9131 12647	(1)1.9157 24406	(1)1.9183 32609	(1)1.9209 37271
1/3	7.1465 69499	7.1530 90095	7.1595 98825	7.1660 95742	7.1725 80900
1/4	4.3709 23607	4.3739 14319	4.3768 98909	4.3798 77406	4.3828 49839
1/5	3.2543 07394	3.2560 88625	3.2578 65967	3.2596 39439	3.2614 09059

k					
1	370	371	372	373	374
2	1 36900	1 37641	1 38384	1 39129	1 39876
3	506 53000	510 64811	514 78848	518 95117	523 13624
4	(10)1.8741 61000	(10)1.8945 04488	(10)1.9150 13146	(10)1.9356 87864	(10)1.9565 29538
5	(12)6.9343 95700	(12)7.0286 11651	(12)7.1238 48902	(12)7.2201 15733	(12)7.3174 20471
6	(15)2.5657 26409	(15)2.6076 14922	(15)2.6500 71791	(15)2.6931 03168	(15)2.7367 15256
7	(17)9.4931 87713	(17)9.6742 51362	(17)9.8582 67064	(18)1.0045 27482	(18)1.0235 31506
8	(20)3.5124 79454	(20)3.5891 47255	(20)3.6672 75348	(20)3.7468 87507	(20)3.8280 07832
9	(23)1.2996 17398	(23)1.3315 73632	(23)1.3642 26429	(23)1.3975 89040	(23)1.4316 74929
10	(25)4.8085 84372	(25)4.9401 38174	(25)5.0749 22317	(25)5.2130 07120	(25)5.3544 64234
24	(61)4.3335 25711	(61)4.6235 31606	(61)4.9320 85051	(61)5.2603 17567	(61)5.6094 26383
1/2	(1)1.9235 38406	(1)1.9261 36028	(1)1.9287 30152	(1)1.9313 20792	(1)1.9339 07961
1/3	7.1790 54352	7.1855 16151	7.1919 66348	7.1984 04996	7.2048 32147
1/4	4.3858 16237	4.3887 76627	4.3917 31039	4.3946 79501	4.3976 22040
1/5	3.2631 74848	3.2649 36822	3.2666 95001	3.2684 49404	3.2702 00047

$n^{\frac{1}{2}}\left[{(-6)5 \atop 4}\right]$ $n^{\frac{1}{3}}\left[{(-6)2 \atop 4}\right]$ $n^{\frac{1}{4}}\left[{(-7)8 \atop 4}\right]$ $n^{\frac{1}{5}}\left[{(-7)5 \atop 4}\right]$

POWERS AND ROOTS n^k

k	375	376	377	378	379
1	375	376	377	378	379
2	1 40625	1 41376	1 42129	1 42884	1 43641
3	527 34375	531 57376	535 82633	540 10152	544 39939
4	(10) 1.9775 39063	(10) 1.9987 17338	(10) 2.0200 65264	(10) 2.0415 83746	(10) 2.0632 73688
5	(12) 7.4157 71484	(12) 7.5151 77189	(12) 7.6156 46046	(12) 7.7171 86558	(12) 7.8198 07278
6	(15) 2.7809 14307	(15) 2.8257 06623	(15) 2.8710 98559	(15) 2.9170 96519	(15) 2.9637 06958
7	(18) 1.0428 42865	(18) 1.0624 65690	(18) 1.0824 04157	(18) 1.1026 62484	(18) 1.1232 44937
8	(20) 3.9106 60744	(20) 3.9948 70996	(20) 4.0806 63671	(20) 4.1680 64190	(20) 4.2570 98312
9	(23) 1.4664 97779	(23) 1.5020 71494	(23) 1.5384 10204	(23) 1.5755 28264	(23) 1.6134 40260
10	(25) 5.4993 66671	(25) 5.6477 88819	(25) 5.7998 06469	(25) 5.9554 96838	(25) 6.1149 38586
24	(61) 5.9806 78067	(61) 6.3754 12334	(61) 6.7950 46060	(61) 7.2410 77507	(61) 7.7150 90756
1/2	(1) 1.9364 91673	(1) 1.9390 71943	(1) 1.9416 48784	(1) 1.9442 22210	(1) 1.9467 92233
1/3	7.2112 47852	7.2176 52160	7.2240 45124	7.2304 26792	7.2367 97216
1/4	4.4005 58684	4.4034 89461	4.4064 14397	4.4093 33520	4.4122 46858
1/5	3.2719 46950	3.2736 90130	3.2754 29605	3.2771 65392	3.2788 97510

k	380	381	382	383	384
1	380	381	382	383	384
2	1 44400	1 45161	1 45924	1 46689	1 47456
3	548 72000	553 06341	557 42968	561 81887	566 23104
4	(10) 2.0851 36000	(10) 2.1071 71592	(10) 2.1293 81378	(10) 2.1517 66272	(10) 2.1743 27194
5	(12) 7.9235 16800	(12) 8.0283 23766	(12) 8.1342 36862	(12) 8.2412 64822	(12) 8.3494 16423
6	(15) 3.0109 36384	(15) 3.0587 91355	(15) 3.1072 78481	(15) 3.1564 04427	(15) 3.2061 75907
7	(18) 1.1441 55826	(18) 1.1653 99506	(18) 1.1869 80380	(18) 1.2089 02895	(18) 1.2311 71548
8	(20) 4.3477 92138	(20) 4.4401 72119	(20) 4.5342 65051	(20) 4.6300 98090	(20) 4.7276 98745
9	(23) 1.6521 61013	(23) 1.6917 05577	(23) 1.7320 89250	(23) 1.7733 27568	(23) 1.8154 36318
10	(25) 6.2782 11848	(25) 6.4453 98249	(25) 6.6165 80933	(25) 6.7918 44587	(25) 6.9712 75461
24	(61) 8.2187 60383	(61) 8.7538 56362	(61) 9.3222 49236	(61) 9.9259 15535	(62) 1.0566 94349
1/2	(1) 1.9493 58869	(1) 1.9519 22130	(1) 1.9544 82029	(1) 1.9570 38579	(1) 1.9595 91794
1/3	7.2431 56443	7.2495 04524	7.2558 41507	7.2621 67440	7.2684 82371
1/4	4.4151 54436	4.4180 56280	4.4209 52418	4.4238 42876	4.4267 27679
1/5	3.2806 25976	3.2823 50807	3.2840 72019	3.2857 89631	3.2875 03659

k	385	386	387	388	389
1	385	386	387	388	389
2	1 48225	1 48996	1 49769	1 50544	1 51321
3	570 66625	575 12456	579 60603	584 11072	588 63869
4	(10) 2.1970 65063	(10) 2.2199 80802	(10) 2.2430 75336	(10) 2.2663 49594	(10) 2.2898 04504
5	(12) 8.4587 00491	(12) 8.5691 25894	(12) 8.6807 01551	(12) 8.7934 36423	(12) 8.9073 39521
6	(15) 3.2565 99689	(15) 3.3076 82595	(15) 3.3594 31500	(15) 3.4118 53332	(15) 3.4649 55074
7	(18) 1.2537 90880	(18) 1.2767 65482	(18) 1.3000 99991	(18) 1.3237 99093	(18) 1.3478 67524
8	(20) 4.8270 94889	(20) 4.9283 14759	(20) 5.0313 86963	(20) 5.1363 40480	(20) 5.2432 04667
9	(23) 1.8584 31532	(23) 1.9023 29497	(23) 1.9471 46755	(23) 1.9929 00106	(23) 2.0396 06615
10	(25) 7.1549 61399	(25) 7.3429 91859	(25) 7.5354 57941	(25) 7.7324 52413	(25) 7.9340 69734
24	(62) 1.1247 53901	(62) 1.1970 03202	(62) 1.2736 88303	(62) 1.3550 69013	(62) 1.4414 19629
1/2	(1) 1.9621 41687	(1) 1.9646 88270	(1) 1.9672 31557	(1) 1.9697 71560	(1) 1.9723 08292
1/3	7.2747 86349	7.2810 79420	7.2873 61631	7.2936 33030	7.2998 93662
1/4	4.4296 06853	4.4324 80423	4.4353 48416	4.4382 10856	4.4410 67768
1/5	3.2892 14120	3.2909 21030	3.2926 24406	3.2943 24265	3.2960 20622

k	390	391	392	393	394
1	390	391	392	393	394
2	1 52100	1 52881	1 53664	1 54449	1 55236
3	593 19000	597 76471	602 36288	606 98457	611 62984
4	(10) 2.3134 41000	(10) 2.3372 60016	(10) 2.3612 62490	(10) 2.3854 49360	(10) 2.4098 21570
5	(12) 9.0224 19900	(12) 9.1386 86663	(12) 9.2561 48959	(12) 9.3748 15985	(12) 9.4946 96984
6	(15) 3.5187 43761	(15) 3.5732 26485	(15) 3.6284 10392	(15) 3.6843 02682	(15) 3.7409 10612
7	(18) 1.3723 10067	(18) 1.3971 31556	(18) 1.4223 36874	(18) 1.4479 30954	(18) 1.4739 18781
8	(20) 5.3520 09260	(20) 5.4627 84383	(20) 5.5755 60545	(20) 5.6903 68650	(20) 5.8072 39997
9	(23) 2.0872 83612	(23) 2.1359 48694	(23) 2.1856 19734	(23) 2.2363 14879	(23) 2.2880 52559
10	(25) 8.1404 06085	(25) 8.3515 59392	(25) 8.5676 29356	(25) 8.7887 17476	(25) 9.0149 27082
24	(62) 1.5330 29700	(62) 1.6302 04837	(62) 1.7332 67559	(62) 1.8425 58176	(62) 1.9584 35730
1/2	(1) 1.9748 41766	(1) 1.9773 71993	(1) 1.9798 98987	(1) 1.9824 22760	(1) 1.9849 43324
1/3	7.3061 43574	7.3123 82812	7.3186 11420	7.3248 29445	7.3310 36930
1/4	4.4439 19178	4.4467 65109	4.4496 05586	4.4524 40634	4.4552 70277
1/5	3.2977 13494	3.2994 02898	3.3010 88848	3.3027 71361	3.3044 50453

k	395	396	397	398	399
1	395	396	397	398	399
2	1 56025	1 56816	1 57609	1 58404	1 59201
3	616 29875	620 99136	625 70773	630 44792	635 21199
4	(10) 2.4343 80063	(10) 2.4591 25786	(10) 2.4840 59688	(10) 2.5091 82722	(10) 2.5344 95840
5	(12) 9.6158 01247	(12) 9.7381 38111	(12) 9.8617 16962	(12) 9.9865 47232	(13) 1.0112 63840
6	(15) 3.7982 41493	(15) 3.8563 02692	(15) 3.9151 01634	(15) 3.9746 45798	(15) 4.0349 42722
7	(18) 1.5003 05390	(18) 1.5270 95866	(18) 1.5542 95349	(18) 1.5819 09028	(18) 1.6099 42146
8	(20) 5.9262 06289	(20) 6.0472 99629	(20) 6.1705 52534	(20) 6.2959 97930	(20) 6.4236 69163
9	(23) 2.3408 51484	(23) 2.3947 30653	(23) 2.4497 09356	(23) 2.5058 07176	(23) 2.5630 43996
10	(25) 9.2463 63362	(25) 9.4831 33387	(25) 9.7253 46143	(25) 9.9731 12562	(26) 1.0226 54554
24	(62) 2.0812 78965	(62) 2.2114 87364	(62) 2.3494 82217	(62) 2.4957 07762	(62) 2.6506 32365
1/2	(1) 1.9874 60691	(1) 1.9899 74874	(1) 1.9924 85885	(1) 1.9949 93734	(1) 1.9974 98436
1/3	7.3372 33921	7.3434 20462	7.3495 96597	7.3557 62368	7.3619 17821
1/4	4.4580 94538	4.4609 13443	4.4637 27013	4.4665 35273	4.4693 38246
1/5	3.3061 26138	3.3077 98433	3.3094 67354	3.3111 32914	3.3127 95131

$n^{\frac{1}{2}}\begin{bmatrix}(-6)4\\4\end{bmatrix}$ $n^{\frac{1}{3}}\begin{bmatrix}(-6)2\\4\end{bmatrix}$ $n^{\frac{1}{4}}\begin{bmatrix}(-7)8\\4\end{bmatrix}$ $n^{\frac{1}{5}}\begin{bmatrix}(-7)5\\4\end{bmatrix}$

k	400	401	402	403	404
1	400	401	402	403	404
2	1 60000	1 60801	1 61604	1 62409	1 63216
3	640 00000	644 81201	649 64808	654 50827	659 39264
4	(10) 2.5600 00000	(10) 2.5856 96160	(10) 2.6115 85282	(10) 2.6376 68328	(10) 2.6639 46266
5	(13) 1.0240 00000	(13) 1.0368 64160	(13) 1.0498 57283	(13) 1.0629 80336	(13) 1.0762 34291
6	(15) 4.0960 00000	(15) 4.1578 25282	(15) 4.2204 26278	(15) 4.2838 10755	(15) 4.3479 86537
7	(18) 1.6384 00000	(18) 1.6672 87938	(18) 1.6966 11364	(18) 1.7263 75734	(18) 1.7565 86561
8	(20) 6.5536 00000	(20) 6.6858 24632	(20) 6.8203 77683	(20) 6.9572 94209	(20) 7.0966 09706
9	(23) 2.6214 40000	(23) 2.6810 15678	(23) 2.7417 91829	(23) 2.8037 89566	(23) 2.8670 30321
10	(26) 1.0485 76000	(26) 1.0750 87287	(26) 1.1022 00315	(26) 1.1299 27195	(26) 1.1582 80250
24	(62) 2.8147 49767	(62) 2.9885 80393	(62) 3.1726 72718	(62) 3.3676 04703	(62) 3.5739 85306
1/2	(1) 2.0000 00000	(1) 2.0024 98439	(1) 2.0049 93766	(1) 2.0074 85990	(1) 2.0099 75124
1/3	7.3680 62997	7.3741 97940	7.3803 22692	7.3864 37295	7.3925 41792
1/4	4.4721 35955	4.4749 28423	4.4777 15674	4.4804 97729	4.4832 74611
1/5	3.3144 54017	3.3161 09590	3.3177 61862	3.3194 10850	3.3210 56568

k	405	406	407	408	409
1	405	406	407	408	409
2	1 64025	1 64836	1 65649	1 66464	1 67281
3	664 30125	669 23416	674 19143	679 17312	684 17929
4	(10) 2.6904 20063	(10) 2.7170 90690	(10) 2.7439 59120	(10) 2.7710 26330	(10) 2.7982 93296
5	(13) 1.0896 20125	(13) 1.1031 38820	(13) 1.1167 91362	(13) 1.1305 78742	(13) 1.1445 01958
6	(15) 4.4129 61508	(15) 4.4787 43609	(15) 4.5453 40843	(15) 4.6127 61269	(15) 4.6810 13009
7	(18) 1.7872 49411	(18) 1.8183 69905	(18) 1.8499 53723	(18) 1.8820 06598	(18) 1.9145 34321
8	(20) 7.2383 60113	(20) 7.3825 81816	(20) 7.5293 11653	(20) 7.6785 86919	(20) 7.8304 45371
9	(23) 2.9315 35846	(23) 2.9973 28217	(23) 3.0644 29843	(23) 3.1328 63463	(23) 3.2026 52157
10	(26) 1.1872 72017	(26) 1.2169 15256	(26) 1.2472 22946	(26) 1.2782 08293	(26) 1.3098 84732
24	(62) 3.7924 56055	(62) 4.0236 92707	(62) 4.2684 06980	(62) 4.5273 48373	(62) 4.8013 06073
1/2	(1) 2.0124 61180	(1) 2.0149 44168	(1) 2.0174 24100	(1) 2.0199 00988	(1) 2.0223 74842
1/3	7.3986 36223	7.4047 20630	7.4107 95055	7.4168 59539	7.4229 14120
1/4	4.4860 46344	4.4888 12948	4.4915 74446	4.4943 30860	4.4970 82211
1/5	3.3226 99030	3.3243 38251	3.3259 74245	3.3276 07026	3.3292 36609

k	410	411	412	413	414
1	410	411	412	413	414
2	1 68100	1 68921	1 69744	1 70569	1 71396
3	689 21000	694 26531	699 34528	704 44997	709 57944
4	(10) 2.8257 61000	(10) 2.8534 30424	(10) 2.8813 02554	(10) 2.9093 78376	(10) 2.9376 58882
5	(13) 1.1585 62010	(13) 1.1727 59904	(13) 1.1870 96652	(13) 1.2015 73269	(13) 1.2161 90777
6	(15) 4.7501 04241	(15) 4.8200 43207	(15) 4.8908 38207	(15) 4.9624 97602	(15) 5.0350 29817
7	(18) 1.9475 42739	(18) 1.9810 37758	(18) 2.0150 25341	(18) 2.0495 11510	(18) 2.0845 02344
8	(20) 7.9849 25229	(20) 8.1420 65185	(20) 8.3019 04405	(20) 8.4644 82535	(20) 8.6298 39705
9	(23) 3.2738 19344	(23) 3.3463 88791	(23) 3.4203 84615	(23) 3.4958 31287	(23) 3.5727 53638
10	(26) 1.3422 65931	(26) 1.3753 65793	(26) 1.4091 98461	(26) 1.4437 78322	(26) 1.4791 20006
24	(62) 5.0911 10945	(62) 5.3976 37632	(62) 5.7218 06738	(62) 6.0645 87127	(62) 6.4269 98328
1/2	(1) 2.0248 45673	(1) 2.0273 13493	(1) 2.0297 78313	(1) 2.0322 40143	(1) 2.0346 98995
1/3	7.4289 58841	7.4349 93742	7.4410 18861	7.4470 34238	7.4530 39914
1/4	4.4998 28522	4.5025 69814	4.5053 06108	4.5080 37426	4.5107 63788
1/5	3.3308 63008	3.3324 86236	3.3341 06308	3.3357 23237	3.3373 37037

k	415	416	417	418	419
1	415	416	417	418	419
2	1 72225	1 73056	1 73889	1 74724	1 75561
3	714 73375	719 91296	725 11713	730 34632	735 60059
4	(10) 2.9661 45063	(10) 2.9948 37914	(10) 3.0237 38432	(10) 3.0528 47618	(10) 3.0821 66472
5	(13) 1.2309 50201	(13) 1.2458 52572	(13) 1.2608 98926	(13) 1.2760 90304	(13) 1.2914 27752
6	(15) 5.1084 43334	(15) 5.1827 46700	(15) 5.2579 48522	(15) 5.3340 57471	(15) 5.4110 82280
7	(18) 2.1200 03984	(18) 2.1560 22627	(18) 2.1925 64534	(18) 2.2296 36023	(18) 2.2672 43475
8	(20) 8.7980 16532	(20) 8.9690 54129	(20) 9.1429 94106	(20) 9.3198 78576	(20) 9.4997 50162
9	(23) 3.6511 76861	(23) 3.7311 26518	(23) 3.8126 28542	(23) 3.8957 09245	(23) 3.9803 95318
10	(26) 1.5152 38397	(26) 1.5521 48631	(26) 1.5898 66102	(26) 1.6284 06464	(26) 1.6677 85638
24	(62) 6.8101 13045	(62) 7.2150 59801	(62) 7.6430 25690	(62) 8.0952 59269	(62) 8.5730 73581
1/2	(1) 2.0371 54879	(1) 2.0396 07805	(1) 2.0420 57786	(1) 2.0445 04830	(1) 2.0469 48949
1/3	7.4590 35926	7.4650 22314	7.4709 99115	7.4769 66370	7.4829 24114
1/4	4.5134 85215	4.5162 01729	4.5189 13349	4.5216 20097	4.5243 21992
1/5	3.3389 47722	3.3405 55305	3.3421 59799	3.3437 61218	3.3453 59575

k	420	421	422	423	424
1	420	421	422	423	424
2	1 76400	1 77241	1 78084	1 78929	1 79776
3	740 88000	746 18461	751 51448	756 86967	762 25024
4	(10) 3.1116 96000	(10) 3.1414 37208	(10) 3.1713 91106	(10) 3.2015 58704	(10) 3.2319 41018
5	(13) 1.3069 12320	(13) 1.3225 45065	(13) 1.3383 27047	(13) 1.3542 59332	(13) 1.3703 42991
6	(15) 5.4890 31744	(15) 5.5679 14722	(15) 5.6477 40136	(15) 5.7285 16974	(15) 5.8102 54284
7	(18) 2.3053 93332	(18) 2.3440 92098	(18) 2.3833 46338	(18) 2.4231 62680	(18) 2.4635 47816
8	(20) 9.6826 51996	(20) 9.8686 27732	(21) 1.0057 72154	(21) 1.0249 97814	(21) 1.0445 44274
9	(23) 4.0667 13838	(23) 4.1546 92275	(23) 4.2443 58492	(23) 4.3357 40751	(23) 4.4288 67722
10	(26) 1.7080 19812	(26) 1.7491 25448	(26) 1.7911 19284	(26) 1.8340 18338	(26) 1.8778 39914
24	(62) 9.0778 49315	(62) 9.6110 38126	(63) 1.0174 16609	(63) 1.0768 83734	(63) 1.1396 73784
1/2	(1) 2.0493 90153	(1) 2.0518 28453	(1) 2.0542 63858	(1) 2.0566 96380	(1) 2.0591 26028
1/3	7.4888 72387	7.4948 11226	7.5007 40668	7.5066 60749	7.5125 71508
1/4	4.5270 19056	4.5297 11307	4.5323 98767	4.5350 81455	4.5377 59390
1/5	3.3469 54883	3.3485 47155	3.3501 36405	3.3517 22644	3.3533 05887

$n^{\frac{1}{2}}\begin{bmatrix}(-6)4\\4\end{bmatrix}$ $n^{\frac{1}{3}}\begin{bmatrix}(-6)1\\4\end{bmatrix}$ $n^{\frac{1}{4}}\begin{bmatrix}(-7)7\\4\end{bmatrix}$ $n^{\frac{1}{5}}\begin{bmatrix}(-7)4\\4\end{bmatrix}$

k					
1	425	426	427	428	429
2	1 80625	1 81476	1 82329	1 83184	1 84041
3	767 65625	773 08776	778 54483	784 02752	789 53589
4	(10) 3.2625 39063	(10) 3.2933 53858	(10) 3.3243 86424	(10) 3.3556 37786	(10) 3.3871 08968
5	(13) 1.3865 79102	(13) 1.4029 68743	(13) 1.4195 13003	(13) 1.4362 12972	(13) 1.4530 69747
6	(15) 5.8929 61182	(15) 5.9766 46847	(15) 6.0613 20523	(15) 6.1469 91521	(15) 6.2336 69216
7	(18) 2.5045 08502	(18) 2.5460 51557	(18) 2.5881 83863	(18) 2.6309 12371	(18) 2.6742 44094
8	(21) 1.0644 16113	(21) 1.0846 17963	(21) 1.1051 54510	(21) 1.1260 30495	(21) 1.1472 50716
9	(23) 4.5237 68482	(23) 4.6204 72523	(23) 4.7190 09756	(23) 4.8194 10518	(23) 4.9217 05572
10	(26) 1.9226 01605	(26) 1.9683 21295	(26) 2.0150 17166	(26) 2.0627 07702	(26) 2.1114 11691
24	(63) 1.2059 63938	(63) 1.2759 40370	(63) 1.3497 98685	(63) 1.4277 44370	(63) 1.5099 93273
1/2	(1) 2.0615 52813	(1) 2.0639 76744	(1) 2.0663 97832	(1) 2.0688 16087	(1) 2.0712 31518
1/3	7.5184 72981	7.5243 65204	7.5302 48212	7.5361 22043	7.5419 86732
1/4	4.5404 32593	4.5431 01082	4.5457 64877	4.5484 23998	4.5510 78463
1/5	3.3548 86145	3.3564 63431	3.3580 37758	3.3596 09138	3.3611 77583

k					
1	430	431	432	433	434
2	1 84900	1 85761	1 86624	1 87489	1 88356
3	795 07000	800 62991	806 21568	811 82737	817 46504
4	(10) 3.4188 01000	(10) 3.4507 14912	(10) 3.4828 51738	(10) 3.5152 12512	(10) 3.5477 98274
5	(13) 1.4700 84430	(13) 1.4872 58127	(13) 1.5045 91951	(13) 1.5220 87018	(13) 1.5397 44451
6	(15) 6.3213 63049	(15) 6.4100 82528	(15) 6.4998 37227	(15) 6.5906 36787	(15) 6.6824 90916
7	(18) 2.7181 86111	(18) 2.7627 45570	(18) 2.8079 29682	(18) 2.8537 45729	(18) 2.9002 01058
8	(21) 1.1688 20028	(21) 1.1907 43340	(21) 1.2130 25623	(21) 1.2356 71901	(21) 1.2586 87259
9	(23) 5.0259 26119	(23) 5.1321 03797	(23) 5.2402 70690	(23) 5.3504 59329	(23) 5.4627 02704
10	(26) 2.1611 48231	(26) 2.2119 36737	(26) 2.2637 96938	(26) 2.3167 48890	(26) 2.3708 12974
24	(63) 1.5967 72093	(63) 1.6883 18906	(63) 1.7848 83700	(63) 1.8867 28946	(63) 1.9941 30189
1/2	(1) 2.0736 44135	(1) 2.0760 53949	(1) 2.0784 60969	(1) 2.0808 65205	(1) 2.0832 66666
1/3	7.5478 42314	7.5536 88825	7.5595 26299	7.5653 54772	7.5711 74278
1/4	4.5537 28292	4.5563 73502	4.5590 14114	4.5616 50145	4.5642 81614
1/5	3.3627 43107	3.3643 05720	3.3658 65436	3.3674 22267	3.3689 76223

k					
1	435	436	437	438	439
2	1 89225	1 90096	1 90969	1 91844	1 92721
3	823 12875	828 81856	834 53453	840 27672	846 04519
4	(10) 3.5806 10063	(10) 3.6136 48922	(10) 3.6469 15896	(10) 3.6804 12034	(10) 3.7141 38384
5	(13) 1.5575 65377	(13) 1.5755 50930	(13) 1.5937 02247	(13) 1.6120 20471	(13) 1.6305 06751
6	(15) 6.7754 09391	(15) 6.8694 02054	(15) 6.9644 78818	(15) 7.0606 49662	(15) 7.1579 24635
7	(18) 2.9473 03085	(18) 2.9950 59296	(18) 3.0434 77243	(18) 3.0925 64552	(18) 3.1423 28915
8	(21) 1.2820 76842	(21) 1.3058 45853	(21) 1.3299 99555	(21) 1.3545 43274	(21) 1.3794 82394
9	(23) 5.5770 34263	(23) 5.6934 87918	(23) 5.8120 98057	(23) 5.9328 99539	(23) 6.0559 27708
10	(26) 2.4260 09904	(26) 2.4823 60732	(26) 2.5398 86851	(26) 2.5986 09998	(26) 2.6585 52264
24	(63) 2.1073 76666	(63) 2.2267 71952	(63) 2.3526 34640	(63) 2.4852 99040	(63) 2.6251 15920
1/2	(1) 2.0856 65361	(1) 2.0880 61302	(1) 2.0904 54496	(1) 2.0928 44954	(1) 2.0952 32684
1/3	7.5769 84852	7.5827 86527	7.5885 79338	7.5943 63318	7.6001 38502
1/4	4.5669 08540	4.5695 30941	4.5721 48834	4.5747 62238	4.5773 71171
1/5	3.3705 27318	3.3720 75562	3.3736 20969	3.3751 63549	3.3767 03314

k					
1	440	441	442	443	444
2	1 93600	1 94481	1 95364	1 96249	1 97136
3	851 84000	857 66121	863 50888	869 38307	875 28384
4	(10) 3.7480 96000	(10) 3.7822 85936	(10) 3.8167 09250	(10) 3.8513 67000	(10) 3.8862 60250
5	(13) 1.6491 62240	(13) 1.6679 88098	(13) 1.6869 85488	(13) 1.7061 55581	(13) 1.7254 99551
6	(15) 7.2563 13856	(15) 7.3558 27511	(15) 7.4564 75858	(15) 7.5582 69224	(15) 7.6612 18006
7	(18) 3.1927 78097	(18) 3.2439 19933	(18) 3.2957 62329	(18) 3.3483 13266	(18) 3.4015 80795
8	(21) 1.4048 22363	(21) 1.4305 68690	(21) 1.4567 26950	(21) 1.4833 02777	(21) 1.5103 01873
9	(23) 6.1812 18395	(23) 6.3088 07924	(23) 6.4387 33117	(23) 6.5710 31302	(23) 6.7057 40315
10	(26) 2.7197 36094	(26) 2.7821 84294	(26) 2.8459 20038	(26) 2.9109 66867	(26) 2.9773 48700
24	(63) 2.7724 53276	(63) 2.9276 97132	(63) 3.0912 52385	(63) 3.2635 43677	(63) 3.4450 16313
1/2	(1) 2.0976 17696	(1) 2.1000 00000	(1) 2.1023 79604	(1) 2.1047 56518	(1) 2.1071 30751
1/3	7.6059 04922	7.6116 62611	7.6174 11603	7.6231 51930	7.6288 83626
1/4	4.5799 75651	4.5825 75695	4.5851 71321	4.5877 62546	4.5903 49388
1/5	3.3782 40276	3.3797 74445	3.3813 05834	3.3828 34454	3.3843 60316

k					
1	445	446	447	448	449
2	1 98025	1 98916	1 99809	2 00704	2 01601
3	881 21125	887 16536	893 14623	899 15392	905 18849
4	(10) 3.9213 90063	(10) 3.9567 57506	(10) 3.9923 63648	(10) 4.0282 09562	(10) 4.0642 96320
5	(13) 1.7450 18578	(13) 1.7647 13847	(13) 1.7845 86551	(13) 1.8046 37884	(13) 1.8248 69048
6	(15) 7.7653 32671	(15) 7.8706 23760	(15) 7.9771 01882	(15) 8.0847 77719	(15) 8.1936 62024
7	(18) 3.4555 73039	(18) 3.5102 98197	(18) 3.5657 64541	(18) 3.6219 80418	(18) 3.6789 54249
8	(21) 1.5377 30002	(21) 1.5655 92996	(21) 1.5938 96750	(21) 1.6226 47227	(21) 1.6518 50458
9	(23) 6.8428 98510	(23) 6.9825 44761	(23) 7.1247 18472	(23) 7.2694 59578	(23) 7.4168 08555
10	(26) 3.0450 89837	(26) 3.1142 14964	(26) 3.1847 49157	(26) 3.2567 17891	(26) 3.3301 47041
24	(63) 3.6361 37215	(63) 3.8373 95917	(63) 4.0493 05610	(63) 4.2724 04226	(63) 4.5072 55570
1/2	(1) 2.1095 02311	(1) 2.1118 71208	(1) 2.1142 37451	(1) 2.1166 01049	(1) 2.1189 62010
1/3	7.6346 06721	7.6403 21250	7.6460 27242	7.6517 24731	7.6574 13748
1/4	4.5929 31864	4.5955 09991	4.5980 83787	4.6006 53268	4.6032 18450
1/5	3.3858 83431	3.3874 03811	3.3889 21465	3.3904 36406	3.3919 48644

$n^{\frac{1}{2}}\begin{bmatrix}(-6)4\\4\end{bmatrix}$ $n^{\frac{1}{3}}\begin{bmatrix}(-6)1\\4\end{bmatrix}$ $n^{\frac{1}{4}}\begin{bmatrix}(-7)6\\4\end{bmatrix}$ $n^{\frac{1}{5}}\begin{bmatrix}(-7)4\\4\end{bmatrix}$

POWERS AND ROOTS n^k

k					
1	450	451	452	453	454
2	2 02500	2 03401	2 04304	2 05209	2 06116
3	911 25000	917 33851	923 45408	929 59677	935 76664
4	(10) 4.1006 25000	(10) 4.1371 96680	(10) 4.1740 12442	(10) 4.2110 73368	(10) 4.2483 80546
5	(13) 1.8452 81250	(13) 1.8658 75703	(13) 1.8866 53624	(13) 1.9076 16236	(13) 1.9287 64768
6	(15) 8.3037 65625	(15) 8.4150 99419	(15) 8.5276 74379	(15) 8.6415 01548	(15) 8.7565 92045
7	(18) 3.7366 94531	(18) 3.7952 09838	(18) 3.8545 08819	(18) 3.9146 00201	(18) 3.9754 92789
8	(21) 1.6815 12539	(21) 1.7116 39637	(21) 1.7422 37986	(21) 1.7733 13891	(21) 1.8048 73726
9	(23) 7.5668 06426	(23) 7.7194 94763	(23) 7.8749 15698	(23) 8.0331 11927	(23) 8.1941 26716
10	(26) 3.4050 62892	(26) 3.4814 92138	(26) 3.5594 61895	(26) 3.6389 99703	(26) 3.7201 33529
24	(63) 4.7544 50505	(63) 5.0146 08183	(63) 5.2883 77338	(63) 5.5764 37619	(63) 5.8795 01000
1/2	(1) 2.1213 20344	(1) 2.1236 76058	(1) 2.1260 29163	(1) 2.1283 79665	(1) 2.1307 27575
1/3	7.6630 94324	7.6687 66491	7.6744 30279	7.6800 85719	7.6857 32843
1/4	4.6057 79352	4.6083 35988	4.6108 88377	4.6134 36534	4.6159 80476
1/5	3.3934 58190	3.3949 65055	3.3964 69249	3.3979 70784	3.3994 69669

k					
1	455	456	457	458	459
2	2 07025	2 07936	2 08849	2 09764	2 10681
3	941 96375	948 18816	954 43993	960 71912	967 02579
4	(10) 4.2859 35063	(10) 4.3237 38010	(10) 4.3617 90480	(10) 4.4000 93570	(10) 4.4386 48376
5	(13) 1.9501 00453	(13) 1.9716 24532	(13) 1.9933 38249	(13) 2.0152 42855	(13) 2.0373 39605
6	(15) 8.8729 57063	(15) 8.9906 07868	(15) 9.1095 55800	(15) 9.2298 12275	(15) 9.3513 88785
7	(18) 4.0371 95464	(18) 4.0997 17188	(18) 4.1630 67001	(18) 4.2272 54022	(18) 4.2922 87452
8	(21) 1.8369 23936	(21) 1.8694 71038	(21) 1.9025 21619	(21) 1.9360 82342	(21) 1.9701 59941
9	(23) 8.3580 03909	(23) 8.5247 87931	(23) 8.6945 23800	(23) 8.8672 57127	(23) 9.0430 34128
10	(26) 3.8028 91778	(26) 3.8873 03297	(26) 3.9733 97377	(26) 4.0612 03764	(26) 4.1507 52665
24	(63) 6.1983 13235	(63) 6.5336 55383	(63) 6.8863 45396	(63) 7.2572 39774	(63) 7.6472 35292
1/2	(1) 2.1330 72901	(1) 2.1354 15650	(1) 2.1377 55833	(1) 2.1400 93456	(1) 2.1424 28529
1/3	7.6913 71681	7.6970 02263	7.7026 24618	7.7082 38778	7.7138 44772
1/4	4.6185 20218	4.6210 55778	4.6235 87171	4.6261 14413	4.6286 37519
1/5	3.4009 65915	3.4024 59532	3.4039 50532	3.4054 38923	3.4069 24718

k					
1	460	461	462	463	464
2	2 11600	2 12521	2 13444	2 14369	2 15296
3	973 36000	979 72181	986 11128	992 52847	998 97344
4	(10) 4.4774 56000	(10) 4.5165 17544	(10) 4.5558 34114	(10) 4.5954 06816	(10) 4.6352 36762
5	(13) 2.0596 29760	(13) 2.0821 14588	(13) 2.1047 95360	(13) 2.1276 73356	(13) 2.1507 49857
6	(15) 9.4742 96896	(15) 9.5985 48250	(15) 9.7241 54565	(15) 9.8511 27638	(15) 9.9794 79338
7	(18) 4.3581 76572	(18) 4.4249 30743	(18) 4.4925 59409	(18) 4.5610 72096	(18) 4.6304 78413
8	(21) 2.0047 61223	(21) 2.0398 93073	(21) 2.0755 62447	(21) 2.1117 76381	(21) 2.1485 41984
9	(23) 9.2219 01627	(23) 9.4039 07065	(23) 9.5890 98505	(23) 9.7775 24642	(23) 9.9692 34804
10	(26) 4.2420 74748	(26) 4.3352 01157	(26) 4.4301 63510	(26) 4.5269 93909	(26) 4.6257 24949
24	(63) 8.0572 70802	(63) 8.4883 29103	(63) 8.9414 38903	(63) 9.4176 76852	(63) 9.9181 69666
1/2	(1) 2.1447 61059	(1) 2.1470 91055	(1) 2.1494 18526	(1) 2.1517 43479	(1) 2.1540 65923
1/3	7.7194 42629	7.7250 32380	7.7306 14052	7.7361 87677	7.7417 53281
1/4	4.6311 56507	4.6336 71390	4.6361 82186	4.6386 88909	4.6411 91574
1/5	3.4084 07924	3.4098 88554	3.4113 66616	3.4128 42121	3.4143 15079

k					
1	465	466	467	468	469
2	2 16225	2 17156	2 18089	2 19024	2 19961
3	1005 44625	1011 94696	1018 47563	1025 03232	1031 61709
4	(10) 4.6753 25063	(10) 4.7156 72834	(10) 4.7562 81192	(10) 4.7971 51258	(10) 4.8382 84152
5	(13) 2.1740 26154	(13) 2.1975 03540	(13) 2.2211 83317	(13) 2.2450 66789	(13) 2.2691 55267
6	(16) 1.0109 22162	(16) 1.0240 36650	(16) 1.0372 92609	(16) 1.0506 91257	(16) 1.0642 33820
7	(18) 4.7007 88052	(18) 4.7720 10788	(18) 4.8441 56484	(18) 4.9172 35083	(18) 4.9912 56618
8	(21) 2.1858 66444	(21) 2.2237 57027	(21) 2.2622 21078	(21) 2.3012 66019	(21) 2.3408 99354
9	(24) 1.0164 27896	(24) 1.0362 70775	(24) 1.0564 57243	(24) 1.0769 92497	(24) 1.0978 81797
10	(26) 4.7263 89719	(26) 4.8290 21810	(26) 4.9336 55326	(26) 5.0403 24885	(26) 5.1490 65627
24	(64) 1.0444 09634	(64) 1.0996 69046	(64) 1.1577 24259	(64) 1.2187 10278	(64) 1.2827 68318
1/2	(1) 2.1563 85865	(1) 2.1587 03314	(1) 2.1610 18278	(1) 2.1633 30765	(1) 2.1656 40783
1/3	7.7473 10895	7.7528 60547	7.7584 02264	7.7639 36077	7.7694 62012
1/4	4.6436 90198	4.6461 84795	4.6486 75380	4.6511 61968	4.6536 44575
1/5	3.4157 85500	3.4172 53393	3.4187 18768	3.4201 81635	3.4216 42003

k					
1	470	471	472	473	474
2	2 20900	2 21841	2 22784	2 23729	2 24676
3	1038 23000	1044 87111	1051 54048	1058 23817	1064 96424
4	(10) 4.8796 81000	(10) 4.9213 42928	(10) 4.9632 71066	(10) 5.0054 66544	(10) 5.0479 30498
5	(13) 2.2934 50070	(13) 2.3179 52519	(13) 2.3426 63943	(13) 2.3675 85675	(13) 2.3927 19056
6	(16) 1.0779 21533	(16) 1.0917 55637	(16) 1.1057 37381	(16) 1.1198 68024	(16) 1.1341 48832
7	(18) 5.0662 31205	(18) 5.1421 69048	(18) 5.2190 80439	(18) 5.2969 75756	(18) 5.3758 65466
8	(21) 2.3811 28666	(21) 2.4219 61622	(21) 2.4634 05967	(21) 2.5054 69532	(21) 2.5481 60231
9	(24) 1.1191 30473	(24) 1.1407 43924	(24) 1.1627 27616	(24) 1.1850 87089	(24) 1.2078 27949
10	(26) 5.2599 13224	(26) 5.3729 03881	(26) 5.4880 74350	(26) 5.6054 61930	(26) 5.7251 04480
24	(64) 1.3500 46075	(64) 1.4206 98007	(64) 1.4948 85630	(64) 1.5727 77826	(64) 1.6545 51159
1/2	(1) 2.1679 48339	(1) 2.1702 53441	(1) 2.1725 56098	(1) 2.1748 56317	(1) 2.1771 54106
1/3	7.7749 80097	7.7804 90361	7.7859 92832	7.7914 87536	7.7969 74500
1/4	4.6561 23215	4.6585 97902	4.6610 68652	4.6635 35480	4.6659 98399
1/5	3.4230 99883	3.4245 55283	3.4260 08213	3.4274 58683	3.4289 06701

$n^{\frac{1}{2}}\begin{bmatrix}(-6)3\\4\end{bmatrix}$ $n^{\frac{1}{3}}\begin{bmatrix}(-6)1\\4\end{bmatrix}$ $n^{\frac{1}{4}}\begin{bmatrix}(-7)5\\4\end{bmatrix}$ $n^{\frac{1}{5}}\begin{bmatrix}(-7)3\\3\end{bmatrix}$

k					
1	475	476	477	478	479
2	2 25625	2 26576	2 27529	2 28484	2 29441
3	1071 71875	1078 50176	1085 31333	1092 15352	1099 02239
4	(10) 5.0906 64063	(10) 5.1336 68378	(10) 5.1769 44584	(10) 5.2204 93826	(10) 5.2643 17248
5	(13) 2.4180 65430	(13) 2.4436 26148	(13) 2.4694 02567	(13) 2.4953 96049	(13) 2.5216 07962
6	(16) 1.1485 81079	(16) 1.1631 66046	(16) 1.1779 05024	(16) 1.1927 99311	(16) 1.2078 50214
7	(18) 5.4557 60126	(18) 5.5366 70380	(18) 5.6186 06966	(18) 5.7015 80708	(18) 5.7856 02524
8	(21) 2.5914 86060	(21) 2.6354 55101	(21) 2.6800 75523	(21) 2.7253 55578	(21) 2.7713 03609
9	(24) 1.2309 55878	(24) 1.2544 76628	(24) 1.2783 96024	(24) 1.3027 19966	(24) 1.3274 54429
10	(26) 5.8470 40422	(26) 5.9713 08750	(26) 6.0979 49036	(26) 6.2270 01440	(26) 6.3585 06713
24	(64) 1.7403 90207	(64) 1.8304 87912	(64) 1.9250 45935	(64) 2.0242 75033	(64) 2.1283 95451
1/2	(1) 2.1794 49472	(1) 2.1817 42423	(1) 2.1840 32967	(1) 2.1863 21111	(1) 2.1886 06863
1/3	7.8024 53753	7.8079 25322	7.8133 89232	7.8188 45511	7.8242 94186
1/4	4.6684 57424	4.6709 12569	4.6733 63849	4.6758 11278	4.6782 54870
1/5	3.4303 52278	3.4317 95422	3.4332 36143	3.4346 74449	3.4361 10350

k					
1	480	481	482	483	484
2	2 30400	2 31361	2 32324	2 33289	2 34256
3	1105 92000	1112 84641	1119 80168	1126 78587	1133 79904
4	(10) 5.3084 16000	(10) 5.3527 91232	(10) 5.3974 44098	(10) 5.4423 75752	(10) 5.4875 87354
5	(13) 2.5480 39680	(13) 2.5746 92583	(13) 2.6015 68055	(13) 2.6286 67488	(13) 2.6559 92279
6	(16) 1.2230 59046	(16) 1.2384 27132	(16) 1.2539 55803	(16) 1.2696 46397	(16) 1.2855 00263
7	(18) 5.8706 83423	(18) 5.9568 34506	(18) 6.0440 66968	(18) 6.1323 92097	(18) 6.2218 21273
8	(21) 2.8179 28043	(21) 2.8652 37397	(21) 2.9132 40279	(21) 2.9619 45383	(21) 3.0113 61496
9	(24) 1.3526 05461	(24) 1.3781 79188	(24) 1.4041 81814	(24) 1.4306 19620	(24) 1.4574 98964
10	(26) 6.4925 06211	(26) 6.6290 41895	(26) 6.7681 56345	(26) 6.9098 92764	(26) 7.0542 94987
24	(64) 2.2376 37322	(64) 2.3522 41094	(64) 2.4724 57971	(64) 2.5985 50361	(64) 2.7307 92362
1/2	(1) 2.1908 90230	(1) 2.1931 71220	(1) 2.1954 49840	(1) 2.1977 26098	(1) 2.2000 00000
1/3	7.8297 35282	7.8351 68827	7.8405 94846	7.8460 13365	7.8514 24411
1/4	4.6806 94639	4.6831 30598	4.6855 62762	4.6879 91145	4.6904 15760
1/5	3.4375 43855	3.4389 74973	3.4404 03713	3.4418 30083	3.4432 54092

k					
1	485	486	487	488	489
2	2 35225	2 36196	2 37169	2 38144	2 39121
3	1140 84125	1147 91256	1155 01303	1162 14272	1169 30169
4	(10) 5.5330 80063	(10) 5.5788 55042	(10) 5.6249 13456	(10) 5.6712 56474	(10) 5.7178 85264
5	(13) 2.6835 43830	(13) 2.7113 23550	(13) 2.7393 32853	(13) 2.7675 73159	(13) 2.7960 45894
6	(16) 1.3015 18758	(16) 1.3177 03245	(16) 1.3340 55099	(16) 1.3505 75702	(16) 1.3672 66442
7	(18) 6.3123 65975	(18) 6.4040 37773	(18) 6.4968 48334	(18) 6.5908 09424	(18) 6.6859 32903
8	(21) 3.0614 97498	(21) 3.1123 62358	(21) 3.1639 65139	(21) 3.2163 14999	(21) 3.2694 21189
9	(24) 1.4848 26286	(24) 1.5126 08106	(24) 1.5408 51023	(24) 1.5695 61719	(24) 1.5987 46962
10	(26) 7.2014 07489	(26) 7.3512 75394	(26) 7.5039 44480	(26) 7.6594 61191	(26) 7.8178 72642
24	(64) 2.8694 70250	(64) 3.0148 82996	(64) 3.1673 42798	(64) 3.3271 75643	(64) 3.4947 21879
1/2	(1) 2.2022 71555	(1) 2.2045 40769	(1) 2.2068 07649	(1) 2.2090 72203	(1) 2.2113 34439
1/3	7.8568 28008	7.8622 24183	7.8676 12960	7.8729 94366	7.8783 68425
1/4	4.6928 36620	4.6952 53740	4.6976 67133	4.7000 76812	4.7024 82790
1/5	3.4446 75750	3.4460 95065	3.4475 12045	3.4489 26700	3.4503 39037

k					
1	490	491	492	493	494
2	2 40100	2 41081	2 42064	2 43049	2 44036
3	1176 49000	1183 70771	1190 95488	1198 23157	1205 53784
4	(10) 5.7648 01000	(10) 5.8120 04856	(10) 5.8594 98010	(10) 5.9072 81640	(10) 5.9553 56930
5	(13) 2.8247 52490	(13) 2.8536 94384	(13) 2.8828 73021	(13) 2.9122 89849	(13) 2.9419 46323
6	(16) 1.3841 28720	(16) 1.4011 63943	(16) 1.4183 73526	(16) 1.4357 58895	(16) 1.4533 21484
7	(18) 6.7822 30728	(18) 6.8797 14959	(18) 6.9783 97749	(18) 7.0782 91354	(18) 7.1794 08129
8	(21) 3.3232 93057	(21) 3.3779 40045	(21) 3.4333 71692	(21) 3.4895 97638	(21) 3.5466 27616
9	(24) 1.6284 13598	(24) 1.6585 68562	(24) 1.6892 18873	(24) 1.7203 71635	(24) 1.7520 34042
10	(26) 7.9792 26630	(26) 8.1435 71639	(26) 8.3109 56854	(26) 8.4814 32162	(26) 8.6550 48169
24	(64) 3.6703 36822	(64) 3.8543 91376	(64) 4.0472 72689	(64) 4.2493 84825	(64) 4.4611 49467
1/2	(1) 2.2135 94362	(1) 2.2158 51981	(1) 2.2181 07301	(1) 2.2203 60331	(1) 2.2226 11077
1/3	7.8837 35163	7.8890 94604	7.8944 46773	7.8997 91695	7.9051 29393
1/4	4.7048 85081	4.7072 83697	4.7096 78653	4.7120 69960	4.7144 57633
1/5	3.4517 49066	3.4531 56794	3.4545 62231	3.4559 65384	3.4573 66263

k					
1	495	496	497	498	499
2	2 45025	2 46016	2 47009	2 48004	2 49001
3	1212 87375	1220 23936	1227 63473	1235 05992	1242 51499
4	(10) 6.0037 25063	(10) 6.0523 87226	(10) 6.1013 44608	(10) 6.1505 98402	(10) 6.2001 49800
5	(13) 2.9718 43906	(13) 3.0019 84064	(13) 3.0323 68270	(13) 3.0629 98004	(13) 3.0938 74750
6	(16) 1.4710 62733	(16) 1.4889 84096	(16) 1.5070 87030	(16) 1.5253 73006	(16) 1.5438 43500
7	(18) 7.2817 60531	(18) 7.3853 61115	(18) 7.4902 22541	(18) 7.5963 57570	(18) 7.7037 79067
8	(21) 3.6044 71463	(21) 3.6631 39113	(21) 3.7226 40603	(21) 3.7829 86070	(21) 3.8441 85754
9	(24) 1.7842 13374	(24) 1.8169 17000	(24) 1.8501 52380	(24) 1.8839 27063	(24) 1.9182 48691
10	(26) 8.8318 56201	(26) 9.0119 08320	(26) 9.1952 57326	(26) 9.3819 56772	(26) 9.5720 60970
24	(64) 4.6830 06649	(64) 4.9154 15513	(64) 5.1588 55098	(64) 5.4138 25162	(64) 5.6808 47029
1/2	(1) 2.2248 59546	(1) 2.2271 05745	(1) 2.2293 49681	(1) 2.2315 91360	(1) 2.2338 30790
1/3	7.9104 59893	7.9157 83219	7.9210 99395	7.9264 08444	7.9317 10391
1/4	4.7168 41683	4.7192 22124	4.7215 98967	4.7239 72227	4.7263 41916
1/5	3.4587 64874	3.4601 61227	3.4615 55329	3.4629 47190	3.4643 36816

$n^{\frac{1}{2}}\begin{bmatrix}(-6)3\\3\end{bmatrix}$ $n^{\frac{1}{3}}\begin{bmatrix}(-6)1\\4\end{bmatrix}$ $n^{\frac{1}{4}}\begin{bmatrix}(-7)5\\4\end{bmatrix}$ $n^{\frac{1}{5}}\begin{bmatrix}(-7)3\\3\end{bmatrix}$

k					
1	500	501	502	503	504
2	2 50000	2 51001	2 52004	2 53009	2 54016
3	1250 00000	1257 51501	1265 06008	1272 63527	1280 24064
4	(10) 6.2500 00000	(10) 6.3001 50200	(10) 6.3506 01602	(10) 6.4013 55408	(10) 6.4524 12826
5	(13) 3.1250 00000	(13) 3.1563 75250	(13) 3.1880 02004	(13) 3.2198 81770	(13) 3.2520 16064
6	(16) 1.5625 00000	(16) 1.5813 44000	(16) 1.6003 77006	(16) 1.6196 00530	(16) 1.6390 16096
7	(18) 7.8125 00000	(18) 7.9225 33442	(18) 8.0338 92570	(18) 8.1465 90668	(18) 8.2606 41125
8	(21) 3.9062 50000	(21) 3.9691 89254	(21) 4.0330 14070	(21) 4.0977 35106	(21) 4.1633 63127
9	(24) 1.9531 25000	(24) 1.9885 63816	(24) 2.0245 73063	(24) 2.0611 60758	(24) 2.0983 35016
10	(26) 9.7656 25000	(26) 9.9627 04720	(27) 1.0163 35678	(27) 1.0367 63861	(27) 1.0575 60848
24	(64) 5.9604 64478	(64) 6.2532 44659	(64) 6.5597 79050	(64) 6.8806 84448	(64) 7.2166 04000
1/2	(1) 2.2360 67977	(1) 2.2383 02929	(1) 2.2405 35650	(1) 2.2427 66149	(1) 2.2449 94432
1/3	7.9370 05260	7.9422 93073	7.9475 73855	7.9528 47628	7.9581 14416
1/4	4.7287 08045	4.7310 70628	4.7334 29676	4.7357 85203	4.7381 37221
1/5	3.4657 24216	3.4671 09398	3.4684 92370	3.4698 73139	3.4712 51715

k					
1	505	506	507	508	509
2	2 55025	2 56036	2 57049	2 58064	2 59081
3	1287 87625	1295 54216	1303 23843	1310 96512	1318 72229
4	(10) 6.5037 75063	(10) 6.5554 43330	(10) 6.6074 18840	(10) 6.6597 02810	(10) 6.7122 96456
5	(13) 3.2844 06407	(13) 3.3170 54325	(13) 3.3499 61352	(13) 3.3831 29027	(13) 3.4165 58896
6	(16) 1.6586 25235	(16) 1.6784 29488	(16) 1.6984 30405	(16) 1.7186 29546	(16) 1.7390 28478
7	(18) 8.3760 57438	(18) 8.4928 53211	(18) 8.6110 42156	(18) 8.7306 38093	(18) 8.8516 54954
8	(21) 4.2299 09006	(21) 4.2973 83725	(21) 4.3657 98373	(21) 4.4351 64151	(21) 4.5054 92371
9	(24) 2.1361 04048	(24) 2.1744 76165	(24) 2.2134 59775	(24) 2.2530 63389	(24) 2.2932 95617
10	(27) 1.0787 32544	(27) 1.1002 84939	(27) 1.1222 24106	(27) 1.1445 56202	(27) 1.1672 87469
24	(64) 7.5682 08268	(64) 7.9361 96349	(64) 8.3212 97020	(64) 8.7242 69942	(64) 9.1459 06897
1/2	(1) 2.2472 20505	(1) 2.2494 44376	(1) 2.2516 66050	(1) 2.2538 85534	(1) 2.2561 02835
1/3	7.9633 74242	7.9686 27129	7.9738 73099	7.9791 12176	7.9843 44383
1/4	4.7404 85740	4.7428 30775	4.7451 72336	4.7475 10436	4.7498 45086
1/5	3.4726 28104	3.4740 02314	3.4753 74353	3.4767 44229	3.4781 11950

k					
1	510	511	512	513	514
2	2 60100	2 61121	2 62144	2 63169	2 64196
3	1326 51000	1334 32831	1342 17728	1350 05697	1357 96744
4	(10) 6.7652 01000	(10) 6.8184 17664	(10) 6.8719 47674	(10) 6.9257 92256	(10) 6.9799 52642
5	(13) 3.4502 52510	(13) 3.4842 11426	(13) 3.5184 37209	(13) 3.5529 31427	(13) 3.5876 95658
6	(16) 1.7596 28780	(16) 1.7804 32039	(16) 1.8014 39851	(16) 1.8226 53822	(16) 1.8440 75568
7	(18) 8.9741 06779	(18) 9.0980 07719	(18) 9.2233 72037	(18) 9.3502 14108	(18) 9.4785 48420
8	(21) 4.5767 94457	(21) 4.6490 81944	(21) 4.7223 66483	(21) 4.7966 59837	(21) 4.8719 73888
9	(24) 2.3341 65173	(24) 2.3756 80873	(24) 2.4178 51639	(24) 2.4606 86497	(24) 2.5041 94578
10	(27) 1.1904 24238	(27) 1.2139 72926	(27) 1.2379 40039	(27) 1.2623 32173	(27) 1.2871 56013
24	(64) 9.5870 33090	(65) 1.0048 50848	(65) 1.0531 22917	(65) 1.1036 12886	(65) 1.1564 18034
1/2	(1) 2.2583 17958	(1) 2.2605 30911	(1) 2.2627 41700	(1) 2.2649 50331	(1) 2.2671 56810
1/3	7.9895 69740	7.9947 88272	8.0000 00000	8.0052 04946	8.0104 03133
1/4	4.7521 76299	4.7545 04087	4.7568 28460	4.7591 49431	4.7614 67011
1/5	3.4794 77522	3.4808 40954	3.4822 02253	3.4835 61427	3.4849 18483

k					
1	515	516	517	518	519
2	2 65225	2 66256	2 67289	2 68324	2 69361
3	1365 90875	1373 88096	1381 88413	1389 91832	1397 98359
4	(10) 7.0344 30063	(10) 7.0892 25754	(10) 7.1443 40952	(10) 7.1997 76898	(10) 7.2555 34832
5	(13) 3.6227 31482	(13) 3.6580 40489	(13) 3.6936 24272	(13) 3.7294 84433	(13) 3.7656 22578
6	(16) 1.8657 06713	(16) 1.8875 48892	(16) 1.9096 03749	(16) 1.9318 72936	(16) 1.9543 58118
7	(18) 9.6083 89574	(18) 9.7397 52284	(18) 9.8726 51381	(19) 1.0007 10181	(19) 1.0143 11863
8	(21) 4.9483 20630	(21) 5.0257 12179	(21) 5.1041 60764	(21) 5.1836 78738	(21) 5.2642 78570
9	(24) 2.5483 85125	(24) 2.5932 67484	(24) 2.6388 51115	(24) 2.6851 45586	(24) 2.7321 60578
10	(27) 1.3124 18339	(27) 1.3381 26022	(27) 1.3642 86026	(27) 1.3909 05414	(27) 1.4179 91340
24	(65) 1.2116 39706	(65) 1.2693 83471	(65) 1.3297 59294	(65) 1.3928 81704	(65) 1.4588 69982
1/2	(1) 2.2693 61144	(1) 2.2715 63338	(1) 2.2737 63400	(1) 2.2759 61335	(1) 2.2781 57150
1/3	8.0155 94581	8.0207 79314	8.0259 57353	8.0311 28718	8.0362 93433
1/4	4.7637 81212	4.7660 92045	4.7683 99522	4.7707 03654	4.7730 04452
1/5	3.4862 73428	3.4876 26271	3.4889 77017	3.4903 25675	3.4916 72252

k					
1	520	521	522	523	524
2	2 70400	2 71441	2 72484	2 73529	2 74576
3	1406 08000	1414 20761	1422 36648	1430 55667	1438 77824
4	(10) 7.3116 16000	(10) 7.3680 21648	(10) 7.4247 53026	(10) 7.4818 11384	(10) 7.5391 97978
5	(13) 3.8020 40320	(13) 3.8387 39279	(13) 3.8757 21079	(13) 3.9129 87354	(13) 3.9505 39740
6	(16) 1.9770 60966	(16) 1.9999 83164	(16) 2.0231 26403	(16) 2.0464 92386	(16) 2.0700 82824
7	(19) 1.0280 71703	(19) 1.0419 91229	(19) 1.0560 71983	(19) 1.0703 15518	(19) 1.0847 23400
8	(21) 5.3459 72853	(21) 5.4287 74301	(21) 5.5126 95749	(21) 5.5977 50159	(21) 5.6839 50615
9	(24) 2.7799 05884	(24) 2.8283 91411	(24) 2.8776 27181	(24) 2.9276 23333	(24) 2.9783 90122
10	(27) 1.4455 51059	(27) 1.4735 91925	(27) 1.5021 21389	(27) 1.5311 47003	(27) 1.5606 76424
24	(65) 1.5278 48342	(65) 1.5999 46126	(65) 1.6752 98008	(65) 1.7540 44200	(65) 1.8363 30669
1/2	(1) 2.2803 50850	(1) 2.2825 42442	(1) 2.2847 31932	(1) 2.2869 19325	(1) 2.2891 04628
1/3	8.0414 51517	8.0466 02993	8.0517 47881	8.0568 86203	8.0620 17979
1/4	4.7753 01928	4.7775 96092	4.7798 86957	4.7821 74532	4.7844 58829
1/5	3.4930 16754	3.4943 59190	3.4956 99566	3.4970 37889	3.4983 74167

$n^{\frac{1}{2}}\left[{(-6)3 \atop 3}\right]$ $n^{\frac{1}{3}}\left[{(-7)9 \atop 4}\right]$ $n^{\frac{1}{4}}\left[{(-7)5 \atop 3}\right]$ $n^{\frac{1}{5}}\left[{(-7)3 \atop 3}\right]$

POWERS AND ROOTS n^k

k					
1	525	526	527	528	529
2	2 75625	2 76676	2 77729	2 78784	2 79841
3	1447 03125	1455 31576	1463 63183	1471 97952	1480 35889
4	(10) 7.5969 14063	(10) 7.6549 60898	(10) 7.7133 39744	(10) 7.7720 51866	(10) 7.8310 98528
5	(13) 3.9883 79883	(13) 4.0265 09432	(13) 4.0649 30045	(13) 4.1036 43385	(13) 4.1426 51121
6	(16) 2.0938 99438	(16) 2.1179 43961	(16) 2.1422 18134	(16) 2.1667 23707	(16) 2.1914 62443
7	(19) 1.0992 97205	(19) 1.1140 38524	(19) 1.1289 48957	(19) 1.1440 30117	(19) 1.1592 83632
8	(21) 5.7713 10327	(21) 5.8598 42634	(21) 5.9495 61001	(21) 6.0404 79020	(21) 6.1326 10416
9	(24) 3.0299 37922	(24) 3.0822 77226	(24) 3.1354 18647	(24) 3.1893 72923	(24) 3.2441 50910
10	(27) 1.5907 17409	(27) 1.6212 77821	(27) 1.6523 65627	(27) 1.6839 88903	(27) 1.7161 55831
24	(65) 1.9223 09365	(65) 2.0121 38448	(65) 2.1059 82534	(65) 2.2040 12944	(65) 2.3064 07963
1/2	(1) 2.2912 87847	(1) 2.2934 68988	(1) 2.2956 48057	(1) 2.2978 25059	(1) 2.3000 00000
1/3	8.0671 43230	8.0722 61977	8.0773 74241	8.0824 80041	8.0875 79399
1/4	4.7867 39859	4.7890 17632	4.7912 92160	4.7935 63454	4.7958 31523
1/5	3.4997 08406	3.5010 40614	3.5023 70797	3.5036 98962	3.5050 25117

k					
1	530	531	532	533	534
2	2 80900	2 81961	2 83024	2 84089	2 85156
3	1488 77000	1497 21291	1505 68768	1514 19437	1522 73304
4	(10) 7.8904 81000	(10) 7.9502 00552	(10) 8.0102 58458	(10) 8.0706 55992	(10) 8.1313 94434
5	(13) 4.1819 54930	(13) 4.2215 56493	(13) 4.2614 57499	(13) 4.3016 59644	(13) 4.3421 64628
6	(16) 2.2164 36113	(16) 2.2416 46498	(16) 2.2670 95390	(16) 2.2927 84590	(16) 2.3187 15911
7	(19) 1.1747 11140	(19) 1.1903 14290	(19) 1.2060 94747	(19) 1.2220 54187	(19) 1.2381 94297
8	(21) 6.2259 69041	(21) 6.3205 68882	(21) 6.4164 24056	(21) 6.5135 48814	(21) 6.6119 57543
9	(24) 3.2997 63592	(24) 3.3562 22076	(24) 3.4135 37598	(24) 3.4717 21518	(24) 3.5307 85328
10	(27) 1.7488 74704	(27) 1.7821 53922	(27) 1.8160 02002	(27) 1.8504 27569	(27) 1.8854 39365
24	(65) 2.4133 53110	(65) 2.5250 41417	(65) 2.6416 73716	(65) 2.7634 58943	(65) 2.8906 14446
1/2	(1) 2.3021 72887	(1) 2.3043 43724	(1) 2.3065 12519	(1) 2.3086 79276	(1) 2.3108 44002
1/3	8.0926 72335	8.0977 58868	8.1028 39019	8.1079 12808	8.1129 80255
1/4	4.7980 96379	4.8003 58033	4.8026 16494	4.8048 71774	4.8071 23882
1/5	3.5063 49267	3.5076 71420	3.5089 91583	3.5103 09762	3.5116 25964

k					
1	535	536	537	538	539
2	2 86225	2 87296	2 88369	2 89444	2 90521
3	1531 30375	1539 90656	1548 54153	1557 20872	1565 90819
4	(10) 8.1924 75063	(10) 8.2538 99162	(10) 8.3156 68016	(10) 8.3777 82914	(10) 8.4402 45144
5	(13) 4.3829 74158	(13) 4.4240 89951	(13) 4.4655 13725	(13) 4.5072 47208	(13) 4.5492 92133
6	(16) 2.3448 91175	(16) 2.3713 12214	(16) 2.3979 80870	(16) 2.4248 98998	(16) 2.4520 68460
7	(19) 1.2545 16778	(19) 1.2710 23346	(19) 1.2877 15727	(19) 1.3045 95661	(19) 1.3216 64900
8	(21) 6.7116 64765	(21) 6.8126 85137	(21) 6.9150 33455	(21) 7.0187 24655	(21) 7.1237 73809
9	(24) 3.5907 40649	(24) 3.6515 99233	(24) 3.7133 72966	(24) 3.7760 73864	(24) 3.8397 14083
10	(27) 1.9210 46247	(27) 1.9572 57189	(27) 1.9940 81282	(27) 2.0315 27739	(27) 2.0696 05891
24	(65) 3.0233 66304	(65) 3.1619 49669	(65) 3.3066 09101	(65) 3.4575 98937	(65) 3.6151 83652
1/2	(1) 2.3130 06701	(1) 2.3151 67381	(1) 2.3173 26045	(1) 2.3194 82701	(1) 2.3216 37353
1/3	8.1180 41379	8.1230 96201	8.1281 44739	8.1331 87014	8.1382 23044
1/4	4.8093 72829	4.8116 18626	4.8138 61283	4.8161 00810	4.8183 37217
1/5	3.5129 40196	3.5142 52463	3.5155 62774	3.5168 71134	3.5181 77550

k					
1	540	541	542	543	544
2	2 91600	2 92681	2 93764	2 94849	2 95936
3	1574 64000	1583 40421	1592 20088	1601 03007	1609 89184
4	(10) 8.5030 56000	(10) 8.5662 16776	(10) 8.6297 28770	(10) 8.6935 93280	(10) 8.7578 11610
5	(13) 4.5916 50240	(13) 4.6343 23276	(13) 4.6773 12993	(13) 4.7206 21151	(13) 4.7642 49516
6	(16) 2.4794 91130	(16) 2.5071 68892	(16) 2.5351 03642	(16) 2.5632 97285	(16) 2.5917 51736
7	(19) 1.3389 25210	(19) 1.3563 78371	(19) 1.3740 26174	(19) 1.3918 70426	(19) 1.4099 12945
8	(21) 7.2301 96134	(21) 7.3380 06986	(21) 7.4472 21864	(21) 7.5578 56412	(21) 7.6699 26419
9	(24) 3.9043 05912	(24) 3.9698 61779	(24) 4.0363 94250	(24) 4.1039 16032	(24) 4.1724 39972
10	(27) 2.1083 25193	(27) 2.1476 95223	(27) 2.1877 25684	(27) 2.2284 26405	(27) 2.2698 07345
24	(65) 3.7796 38253	(65) 3.9512 48669	(65) 4.1303 12169	(65) 4.3171 37789	(65) 4.5120 46770
1/2	(1) 2.3237 90008	(1) 2.3259 40670	(1) 2.3280 89345	(1) 2.3302 36040	(1) 2.3323 80758
1/3	8.1432 52850	8.1482 76449	8.1532 93862	8.1583 05107	8.1633 10204
1/4	4.8205 70514	4.8228 00711	4.8250 27819	4.8272 51847	4.8294 72806
1/5	3.5194 82029	3.5207 84576	3.5220 85199	3.5233 83903	3.5246 80696

k					
1	545	546	547	548	549
2	2 97025	2 98116	2 99209	3 00304	3 01401
3	1618 78625	1627 71336	1636 67323	1645 66592	1654 69149
4	(10) 8.8223 85063	(10) 8.8873 14946	(10) 8.9526 02568	(10) 9.0182 49242	(10) 9.0842 56280
5	(13) 4.8081 99859	(13) 4.8524 73960	(13) 4.8970 73605	(13) 4.9420 00584	(13) 4.9872 56698
6	(16) 2.6204 68923	(16) 2.6494 50782	(16) 2.6786 99262	(16) 2.7082 16320	(16) 2.7380 03927
7	(19) 1.4281 55563	(19) 1.4466 00127	(19) 1.4652 48496	(19) 1.4841 02543	(19) 1.5031 64156
8	(21) 7.7834 47819	(21) 7.8984 36694	(21) 8.0149 09274	(21) 8.1328 81938	(21) 8.2523 71216
9	(24) 4.2419 79061	(24) 4.3125 46435	(24) 4.3841 55373	(24) 4.4568 19302	(24) 4.5305 51798
10	(27) 2.3118 78588	(27) 2.3546 50354	(27) 2.3981 32989	(27) 2.4423 36978	(27) 2.4872 72937
24	(65) 4.7153 73024	(65) 4.9274 63602	(65) 5.1486 79188	(65) 5.3793 94612	(65) 5.6199 99369
1/2	(1) 2.3345 23506	(1) 2.3366 64289	(1) 2.3388 03113	(1) 2.3409 39982	(1) 2.3430 74903
1/3	8.1683 09170	8.1733 02026	8.1782 88788	8.1832 69477	8.1882 44110
1/4	4.8316 90704	4.8339 05553	4.8361 17361	4.8383 26138	4.8405 31895
1/5	3.5259 75582	3.5272 68570	3.5285 59664	3.5298 48871	3.5311 36198

$n^{\frac{1}{2}}\begin{bmatrix}(-6)3\\3\end{bmatrix}$ $n^{\frac{1}{3}}\begin{bmatrix}(-7)8\\4\end{bmatrix}$ $n^{\frac{1}{4}}\begin{bmatrix}(-7)4\\3\end{bmatrix}$ $n^{\frac{1}{5}}\begin{bmatrix}(-7)3\\3\end{bmatrix}$

k	550	551	552	553	554
1	550	551	552	553	554
2	3 02500	3 03601	3 04704	3 05809	3 06916
3	1663 75000	1672 84151	1681 96608	1691 12377	1700 31464
4	(10) 9.1506 25000	(10) 9.2173 56720	(10) 9.2844 52762	(10) 9.3519 14448	(10) 9.4197 43106
5	(13) 5.0328 43750	(13) 5.0787 63553	(13) 5.1250 17924	(13) 5.1716 08690	(13) 5.2185 37681
6	(16) 2.7680 64063	(16) 2.7983 98718	(16) 2.8290 09894	(16) 2.8598 99605	(16) 2.8910 69875
7	(19) 1.5224 35234	(19) 1.5419 17693	(19) 1.5616 13462	(19) 1.5815 24482	(19) 1.6016 52711
8	(21) 8.3733 93789	(21) 8.4959 66491	(21) 8.6201 06308	(21) 8.7458 30384	(21) 8.8731 56018
9	(24) 4.6053 66584	(24) 4.6812 77536	(24) 4.7582 98682	(24) 4.8364 44203	(24) 4.9157 28434
10	(27) 2.5329 51621	(27) 2.5793 83922	(27) 2.6265 80873	(27) 2.6745 53644	(27) 2.7233 13552
24	(65) 5.8708 98173	(65) 6.1325 11516	(65) 6.4052 76258	(65) 6.6896 46227	(65) 6.9860 92851
1/2	(1) 2.3452 07880	(1) 2.3473 38919	(1) 2.3494 68025	(1) 2.3515 95203	(1) 2.3537 20459
1/3	8.1932 12706	8.1981 75283	8.2031 31859	8.2080 82453	8.2130 27082
1/4	4.8427 34641	4.8449 34384	4.8471 31136	4.8493 24905	4.8515 15700
1/5	3.5324 21650	3.5337 05234	3.5349 86956	3.5362 66821	3.5375 44836

k	555	556	557	558	559
1	555	556	557	558	559
2	3 08025	3 09136	3 10249	3 11364	3 12481
3	1709 53875	1718 79616	1728 08693	1737 41112	1746 76879
4	(10) 9.4879 40063	(10) 9.5565 06650	(10) 9.6254 44200	(10) 9.6947 54050	(10) 9.7644 37536
5	(13) 5.2658 06735	(13) 5.3134 17697	(13) 5.3613 72419	(13) 5.4096 72760	(13) 5.4583 20583
6	(16) 2.9225 22738	(16) 2.9542 60240	(16) 2.9862 84438	(16) 3.0185 97400	(16) 3.0512 01206
7	(19) 1.6220 00119	(19) 1.6425 68693	(19) 1.6633 60432	(19) 1.6843 77349	(19) 1.7056 21474
8	(21) 9.0021 00663	(21) 9.1326 81934	(21) 9.2649 17605	(21) 9.3988 25608	(21) 9.5344 24040
9	(24) 4.9961 65868	(24) 5.0777 71156	(24) 5.1605 59106	(24) 5.2445 44689	(24) 5.3297 43038
10	(27) 2.7728 72057	(27) 2.8232 40762	(27) 2.8744 31422	(27) 2.9264 55937	(27) 2.9793 26358
24	(65) 7.2951 05803	(65) 7.6171 93672	(65) 7.9528 84664	(65) 8.3027 27311	(65) 8.6672 91224
1/2	(1) 2.3558 43798	(1) 2.3579 65225	(1) 2.3600 84744	(1) 2.3622 02362	(1) 2.3643 18084
1/3	8.2179 65765	8.2228 98519	8.2278 25361	8.2327 46311	8.2376 61384
1/4	4.8537 03532	4.8558 88409	4.8580 70341	4.8602 49337	4.8624 25407
1/5	3.5388 21007	3.5400 95340	3.5413 67840	3.5426 38514	3.5439 07368

k	560	561	562	563	564
1	560	561	562	563	564
2	3 13600	3 14721	3 15844	3 16969	3 18096
3	1756 16000	1765 58481	1775 04328	1784 53547	1794 06144
4	(10) 9.8344 96000	(10) 9.9049 30784	(10) 9.9757 43234	(11) 1.0046 93470	(11) 1.0118 50652
5	(13) 5.5073 17760	(13) 5.5566 66170	(13) 5.6063 67697	(13) 5.6564 24234	(13) 5.7068 37678
6	(16) 3.0840 97946	(16) 3.1172 89721	(16) 3.1507 78646	(16) 3.1845 66844	(16) 3.2186 56450
7	(19) 1.7270 94850	(19) 1.7487 99534	(19) 1.7707 37599	(19) 1.7929 11133	(19) 1.8153 22238
8	(21) 9.6717 31157	(21) 9.8107 65384	(21) 9.9515 45306	(22) 1.0094 08968	(22) 1.0238 41742
9	(24) 5.4161 69448	(24) 5.5038 39380	(24) 5.5927 68462	(24) 5.6829 72489	(24) 5.7744 67426
10	(27) 3.0330 54891	(27) 3.0876 53892	(27) 3.1431 35876	(27) 3.1995 13511	(27) 3.2567 99629
24	(65) 9.0471 67858	(65) 9.4429 71309	(65) 9.8553 39138	(66) 1.0284 93323	(66) 1.0732 44065
1/2	(1) 2.3664 31913	(1) 2.3685 43856	(1) 2.3706 53918	(1) 2.3727 62104	(1) 2.3748 68417
1/3	8.2425 70600	8.2474 73974	8.2523 71525	8.2572 63270	8.2621 49226
1/4	4.8645 98558	4.8667 68801	4.8689 36145	4.8711 00598	4.8732 62170
1/5	3.5451 74407	3.5464 39637	3.5477 03064	3.5489 64695	3.5502 24533

k	565	566	567	568	569
1	565	566	567	568	569
2	3 19225	3 20356	3 21489	3 22624	3 23761
3	1803 62125	1813 21496	1822 84263	1832 50432	1842 20009
4	(11) 1.0190 46006	(11) 1.0262 79667	(11) 1.0335 51771	(11) 1.0408 62454	(11) 1.0482 11851
5	(13) 5.7576 09935	(13) 5.8087 42917	(13) 5.8602 38543	(13) 5.9120 98737	(13) 5.9643 25433
6	(16) 3.2530 49613	(16) 3.2877 48491	(16) 3.3227 55254	(16) 3.3580 72083	(16) 3.3937 01172
7	(19) 1.8379 73032	(19) 1.8608 65646	(19) 1.8840 02229	(19) 1.9073 84943	(19) 1.9310 15967
8	(22) 1.0384 54763	(22) 1.0532 49956	(22) 1.0682 29264	(22) 1.0833 94648	(22) 1.0987 48085
9	(24) 5.8672 69410	(24) 5.9613 94749	(24) 6.0568 59926	(24) 6.1536 81599	(24) 6.2518 76604
10	(27) 3.3150 07217	(27) 3.3741 49428	(27) 3.4342 39578	(27) 3.4952 91148	(27) 3.5573 17788
24	(66) 1.1198 57461	(66) 1.1684 07534	(66) 1.2189 71112	(66) 1.2716 27927	(66) 1.3264 60719
1/2	(1) 2.3769 72865	(1) 2.3790 75451	(1) 2.3811 76180	(1) 2.3832 75058	(1) 2.3853 72088
1/3	8.2670 29409	8.2719 03838	8.2767 72529	8.2816 35499	8.2864 92764
1/4	4.8754 20869	4.8775 76704	4.8797 29685	4.8818 79820	4.8840 27117
1/5	3.5514 82586	3.5527 38859	3.5539 93358	3.5552 46087	3.5564 97054

k	570	571	572	573	574
1	570	571	572	573	574
2	3 24900	3 26041	3 27184	3 28329	3 29476
3	1851 93000	1861 69411	1871 49248	1881 32517	1891 19224
4	(11) 1.0556 00100	(11) 1.0630 27337	(11) 1.0704 93699	(11) 1.0779 99322	(11) 1.0855 44346
5	(13) 6.0169 20570	(13) 6.0698 86093	(13) 6.1232 23956	(13) 6.1769 36117	(13) 6.2310 24545
6	(16) 3.4296 44725	(16) 3.4659 04959	(16) 3.5024 84103	(16) 3.5393 84395	(16) 3.5766 08089
7	(19) 1.9548 97493	(19) 1.9790 31732	(19) 2.0034 20907	(19) 2.0280 67258	(19) 2.0529 73043
8	(22) 1.1142 91571	(22) 1.1300 27119	(22) 1.1459 56759	(22) 1.1620 82539	(22) 1.1784 06527
9	(24) 6.3514 61955	(24) 6.4524 54848	(24) 6.5548 72660	(24) 6.6587 32949	(24) 6.7640 53463
10	(27) 3.6203 33315	(27) 3.6843 51718	(27) 3.7493 87161	(27) 3.8154 53980	(27) 3.8825 66688
24	(66) 1.3835 55344	(66) 1.4430 00887	(66) 1.5048 89774	(66) 1.5693 17896	(66) 1.6363 84728
1/2	(1) 2.3874 67277	(1) 2.3895 60629	(1) 2.3916 52149	(1) 2.3937 41841	(1) 2.3958 29710
1/3	8.2913 44342	8.2961 90248	8.3010 30501	8.3058 65115	8.3106 94107
1/4	4.8861 71586	4.8883 13236	4.8904 52074	4.8925 88109	4.8947 21351
1/5	3.5577 46263	3.5589 93720	3.5602 39430	3.5614 83400	3.5627 25633

$n^{\frac{1}{2}}\begin{bmatrix}(-6)2\\3\end{bmatrix}$ $n^{\frac{1}{3}}\begin{bmatrix}(-7)8\\4\end{bmatrix}$ $n^{\frac{1}{4}}\begin{bmatrix}(-7)4\\3\end{bmatrix}$ $n^{\frac{1}{5}}\begin{bmatrix}(-7)2\\3\end{bmatrix}$

k	575	576	577	578	579
1	575	576	577	578	579
2	3 30625	3 31776	3 32929	3 34084	3 35241
3	1901 09375	1911 02976	1921 00033	1931 00552	1941 04539
4	(11) 1.0931 28906	(11) 1.1007 53142	(11) 1.1084 17190	(11) 1.1161 21191	(11) 1.1238 65281
5	(13) 6.2854 91211	(13) 6.3403 38097	(13) 6.3955 67189	(13) 6.4511 80481	(13) 6.5071 79976
6	(16) 3.6141 57446	(16) 3.6520 34744	(16) 3.6902 42268	(16) 3.7287 82318	(16) 3.7676 57206
7	(19) 2.0781 40532	(19) 2.1035 72012	(19) 2.1292 69789	(19) 2.1552 36180	(19) 2.1814 73522
8	(22) 1.1949 30806	(22) 1.2116 57479	(22) 1.2285 88668	(22) 1.2457 26512	(22) 1.2630 73169
9	(24) 6.8708 52133	(24) 6.9791 47080	(24) 7.0889 56614	(24) 7.2002 99239	(24) 7.3131 93651
10	(27) 3.9507 39976	(27) 4.0199 88718	(27) 4.0903 27966	(27) 4.1617 72960	(27) 4.2343 39124
24	(66) 1.7061 93459	(66) 1.7788 51122	(66) 1.8544 68735	(66) 1.9331 61432	(66) 2.0150 48620
1/2	(1) 2.3979 15762	(1) 2.4000 00000	(1) 2.4020 82430	(1) 2.4041 63056	(1) 2.4062 41883
1/3	8.3155 17494	8.3203 35292	8.3251 47517	8.3299 54185	8.3347 55313
1/4	4.8968 51807	4.8989 79486	4.9011 04396	4.9032 26546	4.9053 45944
1/5	3.5639 66137	3.5652 04916	3.5664 41976	3.5676 77321	3.5689 10958

k	580	581	582	583	584
1	580	581	582	583	584
2	3 36400	3 37561	3 38724	3 39889	3 41056
3	1951 12000	1961 22941	1971 37368	1981 55287	1991 76704
4	(11) 1.1316 49600	(11) 1.1394 74287	(11) 1.1473 39482	(11) 1.1552 45323	(11) 1.1631 91951
5	(13) 6.5635 67680	(13) 6.6203 45609	(13) 6.6775 15784	(13) 6.7350 80234	(13) 6.7930 40996
6	(16) 3.8068 69254	(16) 3.8464 20799	(16) 3.8863 14186	(16) 3.9265 51777	(16) 3.9671 35942
7	(19) 2.2079 84168	(19) 2.2347 70484	(19) 2.2618 34856	(19) 2.2891 79686	(19) 2.3168 07390
8	(22) 1.2806 30817	(22) 1.2984 01651	(22) 1.3163 87886	(22) 1.3345 91757	(22) 1.3530 15516
9	(24) 7.4276 58740	(24) 7.5437 13594	(24) 7.6613 77499	(24) 7.7806 69942	(24) 7.9016 10612
10	(27) 4.3080 42069	(27) 4.3828 97598	(27) 4.4589 21704	(27) 4.5361 30576	(27) 4.6145 40597
24	(66) 2.1002 54121	(66) 2.1889 06331	(66) 2.2811 38380	(66) 2.3770 88299	(66) 2.4768 99188
1/2	(1) 2.4083 18916	(1) 2.4103 94159	(1) 2.4124 67616	(1) 2.4145 39294	(1) 2.4166 09195
1/3	8.3395 50915	8.3443 41009	8.3491 25609	8.3539 04732	8.3586 78393
1/4	4.9074 62599	4.9095 76518	4.9116 87710	4.9137 96184	4.9159 01946
1/5	3.5701 42892	3.5713 73127	3.5726 01670	3.5738 28526	3.5750 53698

k	585	586	587	588	589
1	585	586	587	588	589
2	3 42225	3 43396	3 44569	3 45744	3 46921
3	2002 01625	2012 30056	2022 62003	2032 97472	2043 36469
4	(11) 1.1711 79506	(11) 1.1792 08128	(11) 1.1872 77958	(11) 1.1953 89135	(11) 1.2035 41802
5	(13) 6.8514 00112	(13) 6.9101 59631	(13) 6.9693 21611	(13) 7.0288 88116	(13) 7.0888 61216
6	(16) 4.0080 69065	(16) 4.0493 53544	(16) 4.0909 91786	(16) 4.1329 86212	(16) 4.1753 39256
7	(19) 2.3447 20403	(19) 2.3729 21177	(19) 2.4014 12178	(19) 2.4301 95893	(19) 2.4592 74822
8	(22) 1.3716 61436	(22) 1.3905 31810	(22) 1.4096 28949	(22) 1.4289 55185	(22) 1.4485 12870
9	(24) 8.0242 19400	(24) 8.1485 16404	(24) 8.2745 21928	(24) 8.4022 56487	(24) 8.5317 40805
10	(27) 4.6941 68349	(27) 4.7750 30613	(27) 4.8571 44372	(27) 4.9405 26815	(27) 5.0251 95334
24	(66) 2.5807 19397	(66) 2.6887 02707	(66) 2.8010 08521	(66) 2.9178 02055	(66) 3.0392 54545
1/2	(1) 2.4186 77324	(1) 2.4207 43687	(1) 2.4228 08288	(1) 2.4248 71131	(1) 2.4269 32220
1/3	8.3634 46607	8.3682 09391	8.3729 66760	8.3777 18728	8.3824 65312
1/4	4.9180 05007	4.9201 05372	4.9222 03051	4.9242 98052	4.9263 90382
1/5	3.5762 77194	3.5774 99018	3.5787 19175	3.5799 37670	3.5811 54508

k	590	591	592	593	594
1	590	591	592	593	594
2	3 48100	3 49281	3 50464	3 51649	3 52836
3	2053 79000	2064 25071	2074 74688	2085 27857	2095 84584
4	(11) 1.2117 36100	(11) 1.2199 72170	(11) 1.2282 50153	(11) 1.2365 70192	(11) 1.2449 32429
5	(13) 7.1492 42990	(13) 7.2100 35522	(13) 7.2712 40906	(13) 7.3328 61239	(13) 7.3948 98628
6	(16) 4.2180 53364	(16) 4.2611 30994	(16) 4.3045 74616	(16) 4.3483 86715	(16) 4.3925 69785
7	(19) 2.4886 51485	(19) 2.5183 28417	(19) 2.5483 08173	(19) 2.5785 93322	(19) 2.6091 86452
8	(22) 1.4683 04376	(22) 1.4883 32095	(22) 1.5085 98438	(22) 1.5291 05840	(22) 1.5498 56753
9	(24) 8.6629 95819	(24) 8.7960 42679	(24) 8.9309 02754	(24) 9.0675 97630	(24) 9.2061 49111
10	(27) 5.1111 67533	(27) 5.1984 61223	(27) 5.2870 94431	(27) 5.3770 85394	(27) 5.4684 52572
24	(66) 3.1655 43453	(66) 3.2968 52680	(66) 3.4333 72793	(66) 3.5753 01250	(66) 3.7228 42640
1/2	(1) 2.4289 91560	(1) 2.4310 49156	(1) 2.4331 05012	(1) 2.4351 59132	(1) 2.4372 11521
1/3	8.3872 06527	8.3919 42387	8.3966 72908	8.4013 98104	8.4061 17992
1/4	4.9284 80050	4.9305 67063	4.9326 51429	4.9347 33156	4.9368 12252
1/5	3.5823 69695	3.5835 83235	3.5847 95134	3.5860 05396	3.5872 14026

k	595	596	597	598	599
1	595	596	597	598	599
2	3 54025	3 55216	3 56409	3 57604	3 58801
3	2106 44875	2117 08736	2127 76173	2138 47192	2149 21799
4	(11) 1.2533 37006	(11) 1.2617 84067	(11) 1.2702 73753	(11) 1.2788 06208	(11) 1.2873 81576
5	(13) 7.4573 55187	(13) 7.5202 33037	(13) 7.5835 34304	(13) 7.6472 61125	(13) 7.7114 15640
6	(16) 4.4371 26336	(16) 4.4820 58890	(16) 4.5273 69980	(16) 4.5730 62153	(16) 4.6191 37969
7	(19) 2.6400 90170	(19) 2.6713 07098	(19) 2.7028 39878	(19) 2.7346 91167	(19) 2.7668 63643
8	(22) 1.5708 53651	(22) 1.5920 99031	(22) 1.6135 95407	(22) 1.6353 45318	(22) 1.6573 51322
9	(24) 9.3465 79225	(24) 9.4889 10223	(24) 9.6331 64580	(24) 9.7793 65002	(24) 9.9275 34420
10	(27) 5.5612 14639	(27) 5.6553 90493	(27) 5.7509 99254	(27) 5.8480 60271	(27) 5.9465 93118
24	(66) 3.8762 08928	(66) 4.0356 19703	(66) 4.2013 02448	(66) 4.3734 92798	(66) 4.5524 34829
1/2	(1) 2.4392 62184	(1) 2.4413 11123	(1) 2.4433 58345	(1) 2.4454 03852	(1) 2.4474 47650
1/3	8.4108 32585	8.4155 41899	8.4202 45948	8.4249 44747	8.4296 38310
1/4	4.9388 88725	4.9409 62581	4.9430 33830	4.9451 02478	4.9471 68534
1/5	3.5884 21030	3.5896 26411	3.5908 30176	3.5920 32329	3.5932 32875

$n^{\frac{1}{2}}\left[{(-6)2 \atop 3}\right]$ $n^{\frac{1}{3}}\left[{(-7)7 \atop 4}\right]$ $n^{\frac{1}{4}}\left[{(-7)4 \atop 3}\right]$ $n^{\frac{1}{5}}\left[{(-7)2 \atop 3}\right]$

k					
1	600	601	602	603	604
2	3 60000	3 61201	3 62404	3 63609	3 64816
3	2160 00000	2170 81801	2181 67208	2192 56227	2203 48864
4	(11) 1.2960 00000	(11) 1.3046 61624	(11) 1.3133 66592	(11) 1.3221 15049	(11) 1.3309 07139
5	(13) 7.7760 00000	(13) 7.8410 16360	(13) 7.9064 66885	(13) 7.9723 53744	(13) 8.0386 79117
6	(16) 4.6656 00000	(16) 4.7124 50833	(16) 4.7596 93065	(16) 4.8073 29308	(16) 4.8553 62187
7	(19) 2.7993 60000	(19) 2.8321 82950	(19) 2.8653 35225	(19) 2.8988 19573	(19) 2.9326 38761
8	(22) 1.6796 16000	(22) 1.7021 41953	(22) 1.7249 31805	(22) 1.7479 88202	(22) 1.7713 13811
9	(25) 1.0077 69600	(25) 1.0229 87314	(25) 1.0384 08947	(25) 1.0540 36886	(25) 1.0698 73542
10	(27) 6.0466 17600	(27) 6.1481 53756	(27) 6.2512 21860	(27) 6.3558 42422	(27) 6.4620 36194
24	(66) 4.7383 81338	(66) 4.9315 94142	(66) 5.1323 44384	(66) 5.3409 12849	(66) 5.5575 90288
1/2	(1) 2.4494 89743	(1) 2.4515 30134	(1) 2.4535 68829	(1) 2.4556 05832	(1) 2.4576 41145
1/3	8.4343 26653	8.4390 09789	8.4436 87734	8.4483 60500	8.4530 28104
1/4	4.9492 32004	4.9512 92896	4.9533 51218	4.9554 06978	4.9574 60182
1/5	3.5944 31819	3.5956 29165	3.5968 24918	3.5980 19083	3.5992 11665
1	605	606	607	608	609
2	3 66025	3 67236	3 68449	3 69664	3 70881
3	2214 45125	2225 45016	2236 48543	2247 55712	2258 66529
4	(11) 1.3397 43006	(11) 1.3486 22797	(11) 1.3575 46656	(11) 1.3665 14729	(11) 1.3755 27162
5	(13) 8.1054 45188	(13) 8.1726 54150	(13) 8.2403 08202	(13) 8.3084 09552	(13) 8.3769 60414
6	(16) 4.9037 94339	(16) 4.9526 28415	(16) 5.0018 67079	(16) 5.0515 13008	(16) 5.1015 68892
7	(19) 2.9667 95575	(19) 3.0012 92819	(19) 3.0361 33317	(19) 3.0713 19909	(19) 3.1068 55455
8	(22) 1.7949 11323	(22) 1.8187 83448	(22) 1.8429 32923	(22) 1.8673 62504	(22) 1.8920 74972
9	(25) 1.0859 21350	(25) 1.1021 82770	(25) 1.1186 60284	(25) 1.1353 56403	(25) 1.1522 73658
10	(27) 6.5698 24169	(27) 6.6792 27585	(27) 6.7902 67926	(27) 6.9029 66929	(27) 7.0173 46578
24	(66) 5.7826 77757	(66) 6.0164 86963	(66) 6.2593 40623	(66) 6.5115 72833	(66) 6.7735 29447
1/2	(1) 2.4596 74775	(1) 2.4617 06725	(1) 2.4637 36999	(1) 2.4657 65601	(1) 2.4677 92536
1/3	8.4576 90558	8.4623 47878	8.4670 00076	8.4716 47168	8.4762 89168
1/4	4.9595 10838	4.9615 58954	4.9636 04536	4.9656 47592	4.9676 88130
1/5	3.6004 02669	3.6015 92098	3.6027 79959	3.6039 66255	3.6051 50991
1	610	611	612	613	614
2	3 72100	3 73321	3 74544	3 75769	3 76996
3	2269 81000	2280 99131	2292 20928	2303 46397	2314 75544
4	(11) 1.3845 84100	(11) 1.3936 85690	(11) 1.4028 32079	(11) 1.4120 23414	(11) 1.4212 59840
5	(13) 8.4459 63010	(13) 8.5154 19568	(13) 8.5853 32326	(13) 8.6557 03525	(13) 8.7265 35419
6	(16) 5.1520 37436	(16) 5.2029 21356	(16) 5.2542 23383	(16) 5.3059 46261	(16) 5.3580 92747
7	(19) 3.1427 42836	(19) 3.1789 84949	(19) 3.2155 84711	(19) 3.2525 45058	(19) 3.2898 68947
8	(22) 1.9170 73130	(22) 1.9423 59804	(22) 1.9679 37843	(22) 1.9938 10121	(22) 2.0199 79533
9	(25) 1.1694 14609	(25) 1.1867 81840	(25) 1.2043 77960	(25) 1.2222 05604	(25) 1.2402 67433
10	(27) 7.1334 29117	(27) 7.2512 37043	(27) 7.3707 93114	(27) 7.4921 20352	(27) 7.6152 42041
24	(66) 7.0455 68477	(66) 7.3280 60494	(66) 7.6213 89047	(66) 7.9259 51097	(66) 8.2421 57465
1/2	(1) 2.4698 17807	(1) 2.4718 41419	(1) 2.4738 63375	(1) 2.4758 83681	(1) 2.4779 02339
1/3	8.4809 26088	8.4855 57944	8.4901 84749	8.4948 06516	8.4994 23260
1/4	4.9697 26156	4.9717 61679	4.9737 94704	4.9758 25239	4.9778 53291
1/5	3.6063 34171	3.6075 15802	3.6086 95885	3.6098 74428	3.6110 51433
1	615	616	617	618	619
2	3 78225	3 79456	3 80689	3 81924	3 83161
3	2326 08375	2337 44896	2348 85113	2360 29032	2371 76659
4	(11) 1.4305 41506	(11) 1.4398 68559	(11) 1.4492 41147	(11) 1.4586 59418	(11) 1.4681 23519
5	(13) 8.7978 30263	(13) 8.8695 90326	(13) 8.9418 17878	(13) 9.0145 15202	(13) 9.0876 84584
6	(16) 5.4106 65612	(16) 5.4636 67641	(16) 5.5171 01631	(16) 5.5709 70395	(16) 5.6252 76757
7	(19) 3.3275 59351	(19) 3.3656 19267	(19) 3.4040 51706	(19) 3.4428 59704	(19) 3.4820 46313
8	(22) 2.0464 49001	(22) 2.0732 21468	(22) 2.1002 99903	(22) 2.1276 87297	(22) 2.1553 86668
9	(25) 1.2585 66136	(25) 1.2771 04424	(25) 1.2958 85040	(25) 1.3149 10750	(25) 1.3341 84347
10	(27) 7.7401 81734	(27) 7.8669 63254	(27) 7.9956 10697	(27) 8.1261 48432	(27) 8.2586 01110
24	(66) 8.5704 33286	(66) 8.9112 18488	(66) 9.2649 68280	(66) 9.6321 53659	(67) 1.0013 26192
1/2	(1) 2.4799 19354	(1) 2.4819 34729	(1) 2.4839 48470	(1) 2.4859 60579	(1) 2.4879 71061
1/3	8.5040 34993	8.5086 41730	8.5132 43484	8.5178 40269	8.5224 32097
1/4	4.9798 78868	4.9819 01975	4.9839 22621	4.9859 40813	4.9879 56556
1/5	3.6122 26906	3.6134 00850	3.6145 73271	3.6157 44173	3.6169 13560
1	620	621	622	623	624
2	3 84400	3 85641	3 86884	3 88129	3 89376
3	2383 28000	2394 83061	2406 41848	2418 04367	2429 70624
4	(11) 1.4776 33600	(11) 1.4871 89809	(11) 1.4967 92295	(11) 1.5064 41206	(11) 1.5161 36694
5	(13) 9.1613 28320	(13) 9.2354 48713	(13) 9.3100 48072	(13) 9.3851 28716	(13) 9.4606 92969
6	(16) 5.6800 23558	(16) 5.7352 13651	(16) 5.7908 49901	(16) 5.8469 35190	(16) 5.9034 72413
7	(19) 3.5216 14606	(19) 3.5615 67677	(19) 3.6019 08638	(19) 3.6426 40623	(19) 3.6837 66786
8	(22) 2.1834 01056	(22) 2.2117 33527	(22) 2.2403 87173	(22) 2.2693 65108	(22) 2.2986 70474
9	(25) 1.3537 08655	(25) 1.3734 86521	(25) 1.3935 20822	(25) 1.4138 14463	(25) 1.4343 70376
10	(27) 8.3929 93659	(27) 8.5293 51293	(27) 8.6676 99511	(27) 8.8080 64101	(27) 8.9504 71145
24	(67) 1.0408 79722	(67) 1.0819 28109	(67) 1.1245 25305	(67) 1.1687 27115	(67) 1.2145 91262
1/2	(1) 2.4899 79920	(1) 2.4919 87159	(1) 2.4939 92783	(1) 2.4959 96795	(1) 2.4979 99199
1/3	8.5270 18983	8.5316 00940	8.5361 77980	8.5407 50116	8.5453 17363
1/4	4.9899 69859	4.9919 80728	4.9939 89170	4.9959 95191	4.9979 98799
1/5	3.6180 81437	3.6192 47808	3.6204 12677	3.6215 76049	3.6227 37928

$n^{\frac{1}{2}}\left[{(-6)2 \atop 3}\right]$ $n^{\frac{1}{3}}\left[{(-7)7 \atop 4}\right]$ $n^{\frac{1}{4}}\left[{(-7)3 \atop 3}\right]$ $n^{\frac{1}{5}}\left[{(-7)2 \atop 3}\right]$

k	625	626	627	628	629
2	3 90625	3 91876	3 93129	3 94384	3 95641
3	2441 40625	2453 14376	2464 91883	2476 73152	2488 58189
4	(11)1.5258 78906	(11)1.5356 67994	(11)1.5455 04106	(11)1.5553 87395	(11)1.5653 18009
5	(13)9.5367 43164	(13)9.6132 81641	(13)9.6903 10747	(13)9.7678 32838	(13)9.8458 50275
6	(16)5.9604 64478	(16)6.0179 14307	(16)6.0758 24838	(16)6.1341 99022	(16)6.1930 39823
7	(19)3.7252 90298	(19)3.7672 14356	(19)3.8095 42174	(19)3.8522 76986	(19)3.8954 22049
8	(22)2.3283 06437	(22)2.3582 76187	(22)2.3885 82943	(22)2.4192 29947	(22)2.4502 20469
9	(25)1.4551 91523	(25)1.4762 80893	(25)1.4976 41505	(25)1.5192 76407	(25)1.5411 88675
10	(27)9.0949 47018	(27)9.2415 18391	(27)9.3902 12238	(27)9.5410 55835	(27)9.6940 76765
24	(67)1.2621 77448	(67)1.3115 47419	(67)1.3627 65028	(67)1.4158 96309	(67)1.4710 09545
1/2	(1)2.5000 00000	(1)2.5019 99201	(1)2.5039 96805	(1)2.5059 92817	(1)2.5079 87241
1/3	8.5498 79733	8.5544 37239	8.5589 89894	8.5635 37711	8.5680 80703
1/4	5.0000 00000	5.0019 98801	5.0039 95209	5.0059 89230	5.0079 80871
1/5	3.6238 98318	3.6250 57224	3.6262 14650	3.6273 70600	3.6285 25079

k	630	631	632	633	634
2	3 96900	3 98161	3 99424	4 00689	4 01956
3	2500 47000	2512 39591	2524 35968	2536 36137	2548 40104
4	(11)1.5752 96100	(11)1.5853 21819	(11)1.5953 95318	(11)1.6055 16747	(11)1.6156 86259
5	(13)9.9243 65430	(14)1.0003 38068	(14)1.0082 89841	(14)1.0162 92101	(14)1.0243 45088
6	(16)6.2523 50221	(16)6.3121 33209	(16)6.3723 91794	(16)6.4331 28999	(16)6.4943 47861
7	(19)3.9389 80639	(19)3.9829 56055	(19)4.0273 51614	(19)4.0721 70657	(19)4.1174 16544
8	(22)2.4815 57803	(22)2.5132 45270	(22)2.5452 86220	(22)2.5776 84026	(22)2.6104 42089
9	(25)1.5633 81416	(25)1.5858 57766	(25)1.6086 20891	(25)1.6316 73988	(25)1.6550 20284
10	(27)9.8493 02919	(28)1.0006 76250	(28)1.0166 48403	(28)1.0328 49635	(28)1.0492 82860
24	(67)1.5281 75339	(67)1.5874 66692	(67)1.6489 59081	(67)1.7127 30535	(67)1.7788 61719
1/2	(1)2.5099 80080	(1)2.5119 71337	(1)2.5139 61018	(1)2.5159 49125	(1)2.5179 35662
1/3	8.5726 18882	8.5771 52262	8.5816 80854	8.5862 04672	8.5907 23728
1/4	5.0099 70139	5.0119 57040	5.0139 41581	5.0159 23768	5.0179 03608
1/5	3.6296 78090	3.6308 29638	3.6319 79727	3.6331 28361	3.6342 75544

k	635	636	637	638	639
2	4 03225	4 04496	4 05769	4 07044	4 08321
3	2560 47875	2572 59456	2584 74853	2596 94072	2609 17119
4	(11)1.6259 04006	(11)1.6361 70140	(11)1.6464 84814	(11)1.6568 48179	(11)1.6672 60390
5	(14)1.0324 49044	(14)1.0406 04209	(14)1.0488 10826	(14)1.0570 69138	(14)1.0653 79389
6	(16)6.5560 51429	(16)6.6182 42770	(16)6.6809 24963	(16)6.7441 01103	(16)6.8077 74299
7	(19)4.1630 92658	(19)4.2092 02402	(19)4.2557 49202	(19)4.3027 36504	(19)4.3501 67777
8	(22)2.6435 63838	(22)2.6770 52728	(22)2.7109 12241	(22)2.7451 45889	(22)2.7797 57209
9	(25)1.6786 63037	(25)1.7026 05535	(25)1.7268 51098	(25)1.7514 03077	(25)1.7762 64857
10	(28)1.0659 51028	(28)1.0828 57120	(28)1.1000 04149	(28)1.1173 95163	(28)1.1350 33244
24	(67)1.8474 36020	(67)1.9185 39634	(67)1.9922 61654	(67)2.0686 94164	(67)2.1479 32334
1/2	(1)2.5199 20634	(1)2.5219 04043	(1)2.5238 85893	(1)2.5258 66188	(1)2.5278 44932
1/3	8.5952 38034	8.5997 47604	8.6042 52449	8.6087 52582	8.6132 48015
1/4	5.0198 81108	5.0218 56273	5.0238 29110	5.0257 99626	5.0277 67827
1/5	3.6354 21280	3.6365 65574	3.6377 08430	3.6388 49851	3.6399 89842

k	640	641	642	643	644
2	4 09600	4 10881	4 12164	4 13449	4 14736
3	2621 44000	2633 74721	2646 09288	2658 47707	2670 89984
4	(11)1.6777 21600	(11)1.6882 31962	(11)1.6987 91629	(11)1.7094 00756	(11)1.7200 59497
5	(14)1.0737 41824	(14)1.0821 56687	(14)1.0906 24226	(14)1.0991 44686	(14)1.1077 18316
6	(16)6.8719 47674	(16)6.9366 24366	(16)7.0018 07530	(16)7.0675 00332	(16)7.1337 05955
7	(19)4.3980 46511	(19)4.4463 76219	(19)4.4951 60434	(19)4.5444 02713	(19)4.5941 06635
8	(22)2.8147 49767	(22)2.8501 27156	(22)2.8858 92999	(22)2.9220 50945	(22)2.9586 04673
9	(25)1.8014 39851	(25)1.8269 31507	(25)1.8527 43305	(25)1.8788 78757	(25)1.9053 41409
10	(28)1.1529 21505	(28)1.1710 63096	(28)1.1894 61202	(28)1.2081 19041	(28)1.2270 39868
24	(67)2.2300 74520	(67)2.3152 22362	(67)2.4034 80891	(67)2.4949 58638	(67)2.5897 67740
1/2	(1)2.5298 22128	(1)2.5317 97780	(1)2.5337 71892	(1)2.5357 44467	(1)2.5377 15508
1/3	8.6177 38760	8.6222 24830	8.6267 06237	8.6311 82992	8.6356 55108
1/4	5.0297 33719	5.0316 97308	5.0336 58602	5.0356 17605	5.0375 74325
1/5	3.6411 28406	3.6422 65548	3.6434 01272	3.6445 35581	3.6456 68481

k	645	646	647	648	649
2	4 16025	4 17316	4 18609	4 19904	4 21201
3	2683 36125	2695 86136	2708 40023	2720 97792	2733 59449
4	(11)1.7307 68006	(11)1.7415 26439	(11)1.7523 34949	(11)1.7631 93692	(11)1.7741 02824
5	(14)1.1163 45364	(14)1.1250 26079	(14)1.1337 60712	(14)1.1425 49513	(14)1.1513 92733
6	(16)7.2004 27598	(16)7.2676 68472	(16)7.3354 31806	(16)7.4037 20841	(16)7.4725 38836
7	(19)4.6442 75801	(19)4.6949 13833	(19)4.7460 24378	(19)4.7976 11105	(19)4.8496 77704
8	(22)2.9955 57891	(22)3.0329 14336	(22)3.0706 77773	(22)3.1088 51996	(22)3.1474 40830
9	(25)1.9321 34840	(25)1.9592 62661	(25)1.9867 28519	(25)2.0145 36093	(25)2.0426 89099
10	(28)1.2462 26972	(28)1.2656 83679	(28)1.2854 13352	(28)1.3054 19389	(28)1.3257 05225
24	(67)2.6880 24057	(67)2.7898 47292	(67)2.8953 61105	(67)3.0046 93247	(67)3.1179 75679
1/2	(1)2.5396 85020	(1)2.5416 53005	(1)2.5436 19468	(1)2.5455 84412	(1)2.5475 47841
1/3	8.6401 22598	8.6445 85472	8.6490 43742	8.6534 97422	8.6579 46522
1/4	5.0395 28767	5.0414 80939	5.0434 30845	5.0453 78492	5.0473 23886
1/5	3.6467 99973	3.6479 30063	3.6490 58755	3.6501 86051	3.6513 11957

$n^{\frac{1}{2}}\left[{(-6)2 \atop 3}\right]$ $n^{\frac{1}{3}}\left[{(-7)6 \atop 4}\right]$ $n^{\frac{1}{4}}\left[{(-7)3 \atop 3}\right]$ $n^{\frac{1}{5}}\left[{(-7)2 \atop 3}\right]$

k					
1	650	651	652	653	654
2	4 22500	4 23801	4 25104	4 26409	4 27716
3	2746 25000	2758 94451	2771 67808	2784 45077	2797 26264
4	(11) 1.7850 62500	(11) 1.7960 72876	(11) 1.8071 34108	(11) 1.8182 46353	(11) 1.8294 09767
5	(14) 1.1602 90625	(14) 1.1692 43442	(14) 1.1782 51439	(14) 1.1873 14868	(14) 1.1964 33987
6	(16) 7.5418 89063	(16) 7.6117 74809	(16) 7.6821 99379	(16) 7.7531 66091	(16) 7.8246 78277
7	(19) 4.9022 27891	(19) 4.9552 65401	(19) 5.0087 93995	(19) 5.0628 17457	(19) 5.1173 39593
8	(22) 3.1864 48129	(22) 3.2258 77776	(22) 3.2657 33685	(22) 3.3060 19800	(22) 3.3467 40094
9	(25) 2.0711 91284	(25) 2.1000 46432	(25) 2.1292 58363	(25) 2.1588 30929	(25) 2.1887 68021
10	(28) 1.3462 74334	(28) 1.3671 30227	(28) 1.3882 76452	(28) 1.4097 16597	(28) 1.4314 54286
24	(67) 3.2353 44710	(67) 3.3569 41134	(67) 3.4829 10364	(67) 3.6134 02582	(67) 3.7485 72888
1/2	(1) 2.5495 09757	(1) 2.5514 70164	(1) 2.5534 29067	(1) 2.5553 86468	(1) 2.5573 42371
1/3	8.6623 91053	8.6668 31029	8.6712 66460	8.6756 97359	8.6801 23736
1/4	5.0492 67033	5.0512 07939	5.0531 46611	5.0550 83054	5.0570 17274
1/5	3.6524 36476	3.6535 59612	3.6546 81368	3.6558 01749	3.6569 20758

k					
1	655	656	657	658	659
2	4 29025	4 30336	4 31649	4 32964	4 34281
3	2810 11375	2823 00416	2835 93393	2848 90312	2861 91179
4	(11) 1.8406 24506	(11) 1.8518 90729	(11) 1.8632 08592	(11) 1.8745 78253	(11) 1.8859 99870
5	(14) 1.2056 09052	(14) 1.2148 40318	(14) 1.2241 28045	(14) 1.2334 72490	(14) 1.2428 73914
6	(16) 7.8967 39288	(16) 7.9693 52487	(16) 8.0425 21255	(16) 8.1162 48987	(16) 8.1905 39094
7	(19) 5.1723 64234	(19) 5.2278 95232	(19) 5.2839 36465	(19) 5.3404 91834	(19) 5.3975 65263
8	(22) 3.3878 98573	(22) 3.4294 99272	(22) 3.4715 46257	(22) 3.5140 43626	(22) 3.5569 95508
9	(25) 2.2190 73565	(25) 2.2497 51522	(25) 2.2808 05891	(25) 2.3122 40706	(25) 2.3440 60040
10	(28) 1.4534 93185	(28) 1.4758 36999	(28) 1.4984 89470	(28) 1.5214 54385	(28) 1.5447 35566
24	(67) 3.8885 81447	(67) 4.0335 93654	(67) 4.1837 80288	(67) 4.3393 17689	(67) 4.5003 87920
1/2	(1) 2.5592 96778	(1) 2.5612 49695	(1) 2.5632 01124	(1) 2.5651 51068	(1) 2.5670 99531
1/3	8.6845 45603	8.6889 62971	8.6933 75853	8.6977 84260	8.7021 88202
1/4	5.0589 49277	5.0608 79069	5.0628 06656	5.0647 32044	5.0666 55239
1/5	3.6580 38399	3.6591 54676	3.6602 69592	3.6613 83152	3.6624 95358

k					
1	660	661	662	663	664
2	4 35600	4 36921	4 38244	4 39569	4 40896
3	2874 96000	2888 04781	2901 17528	2914 34247	2927 54944
4	(11) 1.8974 73600	(11) 1.9089 99602	(11) 1.9205 78035	(11) 1.9322 09058	(11) 1.9438 92828
5	(14) 1.2523 32576	(14) 1.2618 48737	(14) 1.2714 22659	(14) 1.2810 54605	(14) 1.2907 44838
6	(16) 8.2653 95002	(16) 8.3408 20153	(16) 8.4168 18005	(16) 8.4933 92032	(16) 8.5705 45724
7	(19) 5.4551 60701	(19) 5.5132 82121	(19) 5.5719 33519	(19) 5.6311 18918	(19) 5.6908 42360
8	(22) 3.6004 06063	(22) 3.6442 79482	(22) 3.6886 19990	(22) 3.7334 31842	(22) 3.7787 19327
9	(25) 2.3762 68001	(25) 2.4088 68738	(25) 2.4418 66433	(25) 2.4752 65311	(25) 2.5090 69633
10	(28) 1.5683 36881	(28) 1.5922 62236	(28) 1.6165 15579	(28) 1.6411 00901	(28) 1.6660 22237
24	(67) 4.6671 78950	(67) 4.8398 84834	(67) 5.0187 05901	(67) 5.2038 48947	(67) 5.3955 27431
1/2	(1) 2.5690 46516	(1) 2.5709 92026	(1) 2.5729 36066	(1) 2.5748 78638	(1) 2.5768 19745
1/3	8.7065 87691	8.7109 82739	8.7153 73356	8.7197 59553	8.7241 41343
1/4	5.0685 76246	5.0704 95071	5.0724 11720	5.0743 26200	5.0762 38514
1/5	3.6636 06215	3.6647 15727	3.6658 23896	3.6669 30727	3.6680 36224

k					
1	665	666	667	668	669
2	4 42225	4 43556	4 44889	4 46224	4 47561
3	2940 79625	2954 08296	2967 40963	2980 77632	2994 18309
4	(11) 1.9556 29506	(11) 1.9674 19251	(11) 1.9792 62223	(11) 1.9911 58582	(11) 2.0031 08487
5	(14) 1.3004 93622	(14) 1.3103 01221	(14) 1.3201 67903	(14) 1.3300 93933	(14) 1.3400 79578
6	(16) 8.6482 82584	(16) 8.7266 06135	(16) 8.8055 19912	(16) 8.8850 27470	(16) 8.9651 32376
7	(19) 5.7511 07918	(19) 5.8119 19686	(19) 5.8732 81781	(19) 5.9351 98350	(19) 5.9976 73560
8	(22) 3.8244 86766	(22) 3.8707 38511	(22) 3.9174 78948	(22) 3.9647 12498	(22) 4.0124 43612
9	(25) 2.5432 83699	(25) 2.5779 11848	(25) 2.6129 58458	(25) 2.6484 27948	(25) 2.6843 24776
10	(28) 1.6912 83660	(28) 1.7168 89291	(28) 1.7428 43292	(28) 1.7691 49870	(28) 1.7958 13275
24	(67) 5.5939 61683	(67) 5.7993 79113	(67) 6.0120 14426	(67) 6.2321 09844	(67) 6.4599 15340
1/2	(1) 2.5787 59392	(1) 2.5806 97580	(1) 2.5826 34314	(1) 2.5845 69597	(1) 2.5865 03431
1/3	8.7285 18735	8.7328 91741	8.7372 60372	8.7416 24639	8.7459 84552
1/4	5.0781 48670	5.0800 56673	5.0819 62528	5.0838 66242	5.0857 67819
1/5	3.6691 40389	3.6702 43226	3.6713 44740	3.6724 44934	3.6735 43810

k					
1	670	671	672	673	674
2	4 48900	4 50241	4 51584	4 52929	4 54276
3	3007 63000	3021 11711	3034 64448	3048 21217	3061 82024
4	(11) 2.0151 12100	(11) 2.0271 69581	(11) 2.0392 81091	(11) 2.0514 46790	(11) 2.0636 66842
5	(14) 1.3501 25107	(14) 1.3602 30789	(14) 1.3703 96893	(14) 1.3806 23690	(14) 1.3909 11451
6	(16) 9.0458 38217	(16) 9.1271 48592	(16) 9.2090 67120	(16) 9.2915 97433	(16) 9.3747 43182
7	(19) 6.0607 11605	(19) 6.1243 16705	(19) 6.1884 93105	(19) 6.2532 45073	(19) 6.3185 76905
8	(22) 4.0606 76776	(22) 4.1094 16509	(22) 4.1586 67366	(22) 4.2084 33934	(22) 4.2587 20834
9	(25) 2.7206 53440	(25) 2.7574 18478	(25) 2.7946 24470	(25) 2.8322 76038	(25) 2.8703 77842
10	(28) 1.8228 37805	(28) 1.8502 27799	(28) 1.8779 87644	(28) 1.9061 21773	(28) 1.9346 34665
24	(67) 6.6956 88867	(67) 6.9396 96605	(67) 7.1922 13208	(67) 7.4535 22063	(67) 7.7239 15552
1/2	(1) 2.5884 35821	(1) 2.5903 66769	(1) 2.5922 96279	(1) 2.5942 24354	(1) 2.5961 50997
1/3	8.7503 40123	8.7546 91362	8.7590 38280	8.7633 80887	8.7677 19196
1/4	5.0876 67266	5.0895 64588	5.0914 59790	5.0933 52878	5.0952 43858
1/5	3.6746 41374	3.6757 37627	3.6768 32575	3.6779 26219	3.6790 18565

$n^{\frac{1}{2}}\begin{bmatrix}(-6)2\\3\end{bmatrix}$ $n^{\frac{1}{3}}\begin{bmatrix}(-7)6\\4\end{bmatrix}$ $n^{\frac{1}{4}}\begin{bmatrix}(-7)3\\3\end{bmatrix}$ $n^{\frac{1}{5}}\begin{bmatrix}(-7)2\\3\end{bmatrix}$

POWERS AND ROOTS n^k

k	675	676	677	678	679
2	4 55625	4 56976	4 58329	4 59684	4 61041
3	3075 46875	3089 15776	3102 88733	3116 65752	3130 46839
4	(11) 2.0759 41406	(11) 2.0882 70646	(11) 2.1006 54722	(11) 2.1130 93799	(11) 2.1255 88037
5	(14) 1.4012 60449	(14) 1.4116 70957	(14) 1.4221 43247	(14) 1.4326 77595	(14) 1.4432 74277
6	(16) 9.4585 08032	(16) 9.5428 95666	(16) 9.6279 09783	(16) 9.7135 54097	(16) 9.7998 32341
7	(19) 6.3844 92922	(19) 6.4509 97470	(19) 6.5180 94923	(19) 6.5857 89678	(19) 6.6540 86159
8	(22) 4.3095 32722	(22) 4.3608 74290	(22) 4.4127 50263	(22) 4.4651 65402	(22) 4.5181 24502
9	(25) 2.9089 34587	(25) 2.9479 51020	(25) 2.9874 31928	(25) 3.0273 82142	(25) 3.0678 06537
10	(28) 1.9635 30847	(28) 1.9928 14890	(28) 2.0224 91415	(28) 2.0525 65092	(28) 2.0830 40639
24	(67) 8.0036 95322	(67) 8.2931 72571	(67) 8.5926 68325	(67) 8.9025 13744	(67) 9.2230 50418
1/2	(1) 2.5980 76211	(1) 2.6000 00000	(1) 2.6019 22366	(1) 2.6038 43313	(1) 2.6057 62844
1/3	8.7720 53215	8.7763 82955	8.7807 08428	8.7850 29644	8.7893 46612
1/4	5.0971 32735	5.0990 19514	5.1009 04200	5.1027 86801	5.1046 67319
1/5	3.6801 09614	3.6811 99371	3.6822 87840	3.6833 75023	3.6844 60923

k	680	681	682	683	684
2	4 62400	4 63761	4 65124	4 66489	4 67856
3	3144 32000	3158 21241	3172 14568	3186 11987	3200 13504
4	(11) 2.1381 37600	(11) 2.1507 42651	(11) 2.1634 03354	(11) 2.1761 19871	(11) 2.1888 92367
5	(14) 1.4539 33568	(14) 1.4646 55745	(14) 1.4754 41087	(14) 1.4862 89872	(14) 1.4972 02379
6	(16) 9.8867 48262	(16) 9.9743 05627	(17) 1.0062 50822	(17) 1.0151 35983	(17) 1.0240 86427
7	(19) 6.7229 88818	(19) 6.7925 02132	(19) 6.8626 30603	(19) 6.9333 78761	(19) 7.0047 51164
8	(22) 4.5716 32397	(22) 4.6256 93952	(22) 4.6803 14071	(22) 4.7354 97694	(22) 4.7912 49796
9	(25) 3.1087 10030	(25) 3.1500 97581	(25) 3.1919 74196	(25) 3.2343 44925	(25) 3.2772 14860
10	(28) 2.1139 22820	(28) 2.1452 16453	(28) 2.1769 26402	(28) 2.2090 57584	(28) 2.2416 14965
24	(67) 9.5546 30685	(67) 9.8976 17949	(68) 1.0252 38701	(68) 1.0619 32441	(68) 1.0998 82878
1/2	(1) 2.6076 80962	(1) 2.6095 97670	(1) 2.6115 12971	(1) 2.6134 26869	(1) 2.6153 39366
1/3	8.7936 59344	8.7979 67850	8.8022 72141	8.8065 72225	8.8108 68115
1/4	5.1065 45762	5.1084 22134	5.1102 96441	5.1121 68688	5.1140 38880
1/5	3.6855 45546	3.6866 28893	3.6877 10968	3.6887 91774	3.6898 71315

k	685	686	687	688	689
2	4 69225	4 70596	4 71969	4 73344	4 74721
3	3214 19125	3228 28856	3242 42703	3256 60672	3270 82769
4	(11) 2.2017 21006	(11) 2.2146 05952	(11) 2.2275 47370	(11) 2.2405 45423	(11) 2.2536 00278
5	(14) 1.5081 78889	(14) 1.5192 19683	(14) 1.5303 25043	(14) 1.5414 95251	(14) 1.5527 30592
6	(17) 1.0331 02539	(17) 1.0421 84703	(17) 1.0513 33304	(17) 1.0605 48733	(17) 1.0698 31378
7	(19) 7.0767 52393	(19) 7.1493 87060	(19) 7.2226 59802	(19) 7.2965 75282	(19) 7.3711 38193
8	(22) 4.8475 75389	(22) 4.9044 79523	(22) 4.9619 67284	(22) 5.0200 43794	(22) 5.0787 14215
9	(25) 3.3205 89142	(25) 3.3644 72953	(25) 3.4088 71524	(25) 3.4537 90130	(25) 3.4992 34094
10	(28) 2.2746 03562	(28) 2.3080 28446	(28) 2.3418 94737	(28) 2.3762 07610	(28) 2.4109 72291
24	(68) 1.1391 31118	(68) 1.1797 19551	(68) 1.2216 91886	(68) 1.2650 93189	(68) 1.3099 69927
1/2	(1) 2.6172 50466	(1) 2.6191 60171	(1) 2.6210 68484	(1) 2.6229 75410	(1) 2.6248 80950
1/3	8.8151 59819	8.8194 47349	8.8237 30714	8.8280 09925	8.8322 84991
1/4	5.1159 07022	5.1177 73120	5.1196 37179	5.1214 99204	5.1233 59200
1/5	3.6909 49595	3.6920 26615	3.6931 02381	3.6941 76894	3.6952 50159

k	690	691	692	693	694
2	4 76100	4 77481	4 78864	4 80249	4 81636
3	3285 09000	3299 39371	3313 73888	3328 12557	3342 55384
4	(11) 2.2667 12100	(11) 2.2798 81054	(11) 2.2931 07305	(11) 2.3063 91020	(11) 2.3197 32365
5	(14) 1.5640 31349	(14) 1.5753 97808	(14) 1.5868 30255	(14) 1.5983 28977	(14) 1.6098 94261
6	(17) 1.0791 81631	(17) 1.0885 99885	(17) 1.0980 86536	(17) 1.1076 41981	(17) 1.1172 66617
7	(19) 7.4463 53253	(19) 7.5222 25208	(19) 7.5987 58832	(19) 7.6759 58928	(19) 7.7538 30324
8	(22) 5.1379 83744	(22) 5.1978 57619	(22) 5.2583 41112	(22) 5.3194 39537	(22) 5.3811 58245
9	(25) 3.5452 08784	(25) 3.5917 19614	(25) 3.6387 72050	(25) 3.6863 71599	(25) 3.7345 23822
10	(28) 2.4461 94061	(28) 2.4818 78254	(28) 2.5180 30258	(28) 2.5546 55518	(28) 2.5917 59533
24	(68) 1.3563 70007	(68) 1.4043 42816	(68) 1.4539 39271	(68) 1.5052 11857	(68) 1.5582 14678
1/2	(1) 2.6267 85107	(1) 2.6286 87886	(1) 2.6305 89288	(1) 2.6324 89316	(1) 2.6343 87974
1/3	8.8365 55922	8.8408 22729	8.8450 85422	8.8493 44010	8.8535 98503
1/4	5.1252 17173	5.1270 73128	5.1289 27069	5.1307 79001	5.1326 28931
1/5	3.6963 22179	3.6973 92956	3.6984 62494	3.6995 30796	3.7005 97866

k	695	696	697	698	699
2	4 83025	4 84416	4 85809	4 87204	4 88601
3	3357 02375	3371 53536	3386 08873	3400 68392	3415 32099
4	(11) 2.3331 31506	(11) 2.3465 88611	(11) 2.3601 03845	(11) 2.3736 77376	(11) 2.3873 09372
5	(14) 1.6215 26397	(14) 1.6332 25673	(14) 1.6449 92380	(14) 1.6568 26809	(14) 1.6687 29251
6	(17) 1.1269 60846	(17) 1.1367 25068	(17) 1.1465 59689	(17) 1.1564 65112	(17) 1.1664 41746
7	(19) 7.8323 77878	(19) 7.9116 06476	(19) 7.9915 21031	(19) 8.0721 26484	(19) 8.1534 27808
8	(22) 5.4435 02625	(22) 5.5064 78107	(22) 5.5700 90158	(22) 5.6343 44286	(22) 5.6992 46038
9	(25) 3.7832 34325	(25) 3.8325 08763	(25) 3.8823 52840	(25) 3.9327 72312	(25) 3.9837 72980
10	(28) 2.6293 47856	(28) 2.6674 26099	(28) 2.7059 99930	(28) 2.7450 75074	(28) 2.7846 57313
24	(68) 1.6130 03502	(68) 1.6696 35809	(68) 1.7281 70846	(68) 1.7886 69670	(68) 1.8511 95210
1/2	(1) 2.6362 85265	(1) 2.6381 81192	(1) 2.6400 75756	(1) 2.6419 68963	(1) 2.6438 60813
1/3	8.8578 48911	8.8620 95243	8.8663 37511	8.8705 75722	8.8748 09888
1/4	5.1344 76863	5.1363 22801	5.1381 66751	5.1400 08719	5.1418 48708
1/5	3.7016 63707	3.7027 28321	3.7037 91713	3.7048 53884	3.7059 14839

$n^{\frac{1}{2}}\left[{(-6)2 \atop 3}\right]$ $n^{\frac{1}{3}}\left[{(-7)5 \atop 4}\right]$ $n^{\frac{1}{4}}\left[{(-7)3 \atop 3}\right]$ $n^{\frac{1}{5}}\left[{(-7)2 \atop 3}\right]$

k					
1	700	701	702	703	704
2	4 90000	4 91401	4 92804	4 94209	4 95616
3	3430 00000	3444 72101	3459 48408	3474 28927	3489 13664
4	(11) 2.4010 00000	(11) 2.4147 49428	(11) 2.4285 57824	(11) 2.4424 25357	(11) 2.4563 52195
5	(14) 1.6807 00000	(14) 1.6927 39349	(14) 1.7048 47593	(14) 1.7170 25026	(14) 1.7292 71945
6	(17) 1.1764 90000	(17) 1.1866 10284	(17) 1.1968 03010	(17) 1.2070 68593	(17) 1.2174 07449
7	(19) 8.2354 30000	(19) 8.3181 38089	(19) 8.4015 57130	(19) 8.4856 92210	(19) 8.5705 48443
8	(22) 5.7648 01000	(22) 5.8310 14800	(22) 5.8978 93105	(22) 5.9654 41624	(22) 6.0336 66104
9	(25) 4.0353 60700	(25) 4.0875 41375	(25) 4.1403 20960	(25) 4.1937 05461	(25) 4.2477 00937
10	(28) 2.8247 52490	(28) 2.8653 66504	(28) 2.9065 05314	(28) 2.9481 74939	(28) 2.9903 81460
24	(68) 1.9158 12314	(68) 1.9825 87808	(68) 2.0515 90555	(68) 2.1228 91511	(68) 2.1965 63787
1/2	(1) 2.6457 51311	(1) 2.6476 40459	(1) 2.6495 28260	(1) 2.6514 14717	(1) 2.6532 99832
1/3	8.8790 40017	8.8832 66120	8.8874 88205	8.8917 06283	8.8959 20362
1/4	5.1436 86724	5.1455 22771	5.1473 56856	5.1491 88981	5.1510 19154
1/5	3.7069 74581	3.7080 33112	3.7090 90435	3.7101 46554	3.7112 01473

k					
1	705	706	707	708	709
2	4 97025	4 98436	4 99849	5 01264	5 02681
3	3504 02625	3518 95816	3533 93243	3548 94912	3564 00829
4	(11) 2.4703 38506	(11) 2.4843 84461	(11) 2.4984 90228	(11) 2.5126 55977	(11) 2.5268 81878
5	(14) 1.7415 88647	(14) 1.7539 75429	(14) 1.7664 32591	(14) 1.7789 60432	(14) 1.7915 59251
6	(17) 1.2278 19996	(17) 1.2383 06653	(17) 1.2488 67842	(17) 1.2595 03986	(17) 1.2702 15509
7	(19) 8.6561 30972	(19) 8.7424 44971	(19) 8.8294 95643	(19) 8.9172 88218	(19) 9.0058 27960
8	(22) 6.1025 72335	(22) 6.1721 66150	(22) 6.2424 53419	(22) 6.3134 40059	(22) 6.3851 32023
9	(25) 4.3023 13497	(25) 4.3575 49302	(25) 4.4134 14568	(25) 4.4699 15561	(25) 4.5270 58605
10	(28) 3.0331 31015	(28) 3.0764 29807	(28) 3.1202 84099	(28) 3.1647 00218	(28) 3.2096 84551
24	(68) 2.2726 82709	(68) 2.3513 25887	(68) 2.4325 73275	(68) 2.5165 07242	(68) 2.6032 12640
1/2	(1) 2.6551 83609	(1) 2.6570 66051	(1) 2.6589 47160	(1) 2.6608 26939	(1) 2.6627 05391
1/3	8.9001 30453	8.9043 36564	8.9085 38706	8.9127 36887	8.9169 31117
1/4	5.1528 47377	5.1546 73657	5.1564 97998	5.1583 20404	5.1601 40881
1/5	3.7122 55193	3.7133 07718	3.7143 59051	3.7154 09195	3.7164 58153

k					
1	710	711	712	713	714
2	5 04100	5 05521	5 06944	5 08369	5 09796
3	3579 11000	3594 25431	3609 44128	3624 67097	3639 94344
4	(11) 2.5411 68100	(11) 2.5555 14814	(11) 2.5699 22191	(11) 2.5843 90402	(11) 2.5989 19616
5	(14) 1.8042 29351	(14) 1.8169 71033	(14) 1.8297 84600	(14) 1.8426 70356	(14) 1.8556 28606
6	(17) 1.2810 02839	(17) 1.2918 66404	(17) 1.3028 06635	(17) 1.3138 23964	(17) 1.3249 18825
7	(19) 9.0951 20158	(19) 9.1851 70136	(19) 9.2759 83244	(19) 9.3675 64864	(19) 9.4599 20408
8	(22) 6.4575 35312	(22) 6.5306 55967	(22) 6.6045 00070	(22) 6.6790 73748	(22) 6.7543 83171
9	(25) 4.5848 50072	(25) 4.6432 96392	(25) 4.7024 04050	(25) 4.7621 79582	(25) 4.8226 29584
10	(28) 3.2552 43551	(28) 3.3013 83735	(28) 3.3481 11683	(28) 3.3954 34042	(28) 3.4433 57523
24	(68) 2.6927 76876	(68) 2.7852 89985	(68) 2.8808 44702	(68) 2.9795 36544	(68) 3.0814 63889
1/2	(1) 2.6645 82519	(1) 2.6664 58325	(1) 2.6683 32813	(1) 2.6702 05985	(1) 2.6720 77843
1/3	8.9211 21404	8.9253 07760	8.9294 90191	8.9336 68708	8.9378 43321
1/4	5.1619 59433	5.1637 76065	5.1655 90782	5.1674 03588	5.1692 14489
1/5	3.7175 05928	3.7185 52523	3.7195 97942	3.7206 42186	3.7216 85260

k					
1	715	716	717	718	719
2	5 11225	5 12656	5 14089	5 15524	5 16961
3	3655 25875	3670 61696	3686 01813	3701 46232	3716 94959
4	(11) 2.6135 10006	(11) 2.6281 61743	(11) 2.6428 74999	(11) 2.6576 49946	(11) 2.6724 86755
5	(14) 1.8686 59654	(14) 1.8817 63808	(14) 1.8949 41374	(14) 1.9081 92661	(14) 1.9215 17977
6	(17) 1.3360 91653	(17) 1.3473 42887	(17) 1.3586 72965	(17) 1.3700 82331	(17) 1.3815 71425
7	(19) 9.5530 55319	(19) 9.6469 75069	(19) 9.7416 85162	(19) 9.8371 91134	(19) 9.9334 98549
8	(22) 6.8304 34553	(22) 6.9072 34149	(22) 6.9847 88261	(22) 7.0631 03234	(22) 7.1421 85457
9	(25) 4.8837 60705	(25) 4.9455 79651	(25) 5.0080 93183	(25) 5.0713 08122	(25) 5.1352 31343
10	(28) 3.4918 88904	(28) 3.5410 35030	(28) 3.5908 02813	(28) 3.6411 99232	(28) 3.6922 31336
24	(68) 3.1867 28051	(68) 3.2954 33372	(68) 3.4076 87302	(68) 3.5236 00491	(68) 3.6432 86875
1/2	(1) 2.6739 48391	(1) 2.6758 17632	(1) 2.6776 85568	(1) 2.6795 52201	(1) 2.6814 17536
1/3	8.9420 14037	8.9461 80866	8.9503 43817	8.9545 02899	8.9586 58122
1/4	5.1710 23488	5.1728 30591	5.1746 35801	5.1764 39125	5.1782 40566
1/5	3.7227 27165	3.7237 67905	3.7248 07483	3.7258 45902	3.7268 83164

k					
1	720	721	722	723	724
2	5 18400	5 19841	5 21284	5 22729	5 24176
3	3732 48000	3748 05361	3763 67048	3779 33067	3795 03424
4	(11) 2.6873 85600	(11) 2.7023 46653	(11) 2.7173 70087	(11) 2.7324 56074	(11) 2.7476 04790
5	(14) 1.9349 17632	(14) 1.9483 91937	(14) 1.9619 41202	(14) 1.9755 65742	(14) 1.9892 65868
6	(17) 1.3931 40695	(17) 1.4047 90586	(17) 1.4165 21548	(17) 1.4283 34031	(17) 1.4402 28488
7	(20) 1.0030 61300	(20) 1.0128 54013	(20) 1.0227 28558	(20) 1.0326 85505	(20) 1.0427 25426
8	(22) 7.2220 41363	(22) 7.3026 77432	(22) 7.3841 00187	(22) 7.4663 16199	(22) 7.5493 32081
9	(25) 5.1998 69781	(25) 5.2652 30428	(25) 5.3313 20335	(25) 5.3981 46612	(25) 5.4657 16426
10	(28) 3.7439 06243	(28) 3.7962 31139	(28) 3.8492 13282	(28) 3.9028 60000	(28) 3.9571 78693
24	(68) 3.7668 63772	(68) 3.8944 51981	(68) 4.0261 75870	(68) 4.1621 63488	(68) 4.3025 46659
1/2	(1) 2.6832 81573	(1) 2.6851 44316	(1) 2.6870 05769	(1) 2.6888 65932	(1) 2.6907 24809
1/3	8.9628 09493	8.9669 57022	8.9711 00718	8.9752 40590	8.9793 76646
1/4	5.1800 40128	5.1818 37817	5.1836 33637	5.1854 27593	5.1872 19688
1/5	3.7279 19273	3.7289 54232	3.7299 88042	3.7310 20708	3.7320 52232

$n^{\frac{1}{2}}\left[{(-6)2 \atop 3}\right]$ $n^{\frac{1}{3}}\left[{(-7)5 \atop 4}\right]$ $n^{\frac{1}{4}}\left[{(-7)2 \atop 3}\right]$ $n^{\frac{1}{5}}\left[{(-7)2 \atop 3}\right]$

k					
1	725	726	727	728	729
2	5 25625	5 27076	5 28529	5 29984	5 31441
3	3810 78125	3826 57176	3842 40583	3858 28352	3874 20489
4	(11) 2.7628 16406	(11) 2.7780 91098	(11) 2.7934 29038	(11) 2.8088 30403	(11) 2.8242 95365
5	(14) 2.0030 41895	(14) 2.0168 94137	(14) 2.0308 22911	(14) 2.0448 28533	(14) 2.0589 11321
6	(17) 1.4522 05374	(17) 1.4642 65143	(17) 1.4764 08256	(17) 1.4886 35172	(17) 1.5009 46353
7	(20) 1.0528 48896	(20) 1.0630 56494	(20) 1.0733 48802	(20) 1.0837 26405	(20) 1.0941 89891
8	(22) 7.6331 54495	(22) 7.7177 90147	(22) 7.8032 45793	(22) 7.8895 28230	(22) 7.9766 44308
9	(25) 5.5340 37009	(25) 5.6031 15647	(25) 5.6729 59691	(25) 5.7435 76552	(25) 5.8149 73700
10	(28) 4.0121 76831	(28) 4.0678 61960	(28) 4.1242 41696	(28) 4.1813 23730	(28) 4.2391 15828
24	(68) 4.4474 61095	(68) 4.5970 46501	(68) 4.7514 46686	(68) 4.9108 09683	(68) 5.0752 87861
1/2	(1) 2.6925 82404	(1) 2.6944 38717	(1) 2.6962 93753	(1) 2.6981 47513	(1) 2.7000 00000
1/3	8.9835 08896	8.9876 37347	8.9917 62009	8.9958 82891	9.0000 00000
1/4	5.1890 09928	5.1907 98317	5.1925 84860	5.1943 69560	5.1961 52423
1/5	3.7330 82616	3.7341 11864	3.7351 39979	3.7361 66963	3.7371 92819

k					
1	730	731	732	733	734
2	5 32900	5 34361	5 35824	5 37289	5 38756
3	3890 17000	3906 17891	3922 23168	3938 32837	3954 46904
4	(11) 2.8398 24100	(11) 2.8554 16783	(11) 2.8710 73590	(11) 2.8867 94695	(11) 2.9025 80275
5	(14) 2.0730 71593	(14) 2.0873 09669	(14) 2.1016 25868	(14) 2.1160 20512	(14) 2.1304 93922
6	(17) 1.5133 42263	(17) 1.5258 23368	(17) 1.5383 90135	(17) 1.5510 43035	(17) 1.5637 82539
7	(20) 1.1047 39852	(20) 1.1153 76882	(20) 1.1261 01579	(20) 1.1369 14545	(20) 1.1478 16384
8	(22) 8.0646 00919	(22) 8.1534 05006	(22) 8.2430 63558	(22) 8.3335 83612	(22) 8.4249 72255
9	(25) 5.8871 58671	(25) 5.9601 39059	(25) 6.0339 22524	(25) 6.1085 16788	(25) 6.1839 29635
10	(28) 4.2976 25830	(28) 4.3568 61652	(28) 4.4168 31288	(28) 4.4775 42805	(28) 4.5390 04352
24	(68) 5.2450 38047	(68) 5.4202 21655	(68) 5.6010 04807	(68) 5.7875 58467	(68) 5.9800 58576
1/2	(1) 2.7018 51217	(1) 2.7037 01167	(1) 2.7055 49852	(1) 2.7073 97274	(1) 2.7092 43437
1/3	9.0041 13346	9.0082 22937	9.0123 28782	9.0164 30890	9.0205 29268
1/4	5.1979 33452	5.1997 12653	5.2014 90029	5.2032 65584	5.2050 39324
1/5	3.7382 17550	3.7392 41158	3.7402 63647	3.7412 85019	3.7423 05277

k					
1	735	736	737	738	739
2	5 40225	5 41696	5 43169	5 44644	5 46121
3	3970 65375	3986 88256	4003 15553	4019 47272	4035 83419
4	(11) 2.9184 30506	(11) 2.9343 45564	(11) 2.9503 25626	(11) 2.9663 70867	(11) 2.9824 81466
5	(14) 2.1450 46422	(14) 2.1596 78335	(14) 2.1743 89986	(14) 2.1891 81700	(14) 2.2040 53804
6	(17) 1.5766 09120	(17) 1.5895 23255	(17) 1.6025 25420	(17) 1.6156 16095	(17) 1.6287 95761
7	(20) 1.1588 07703	(20) 1.1698 89115	(20) 1.1810 61234	(20) 1.1923 24678	(20) 1.2036 80067
8	(22) 8.5172 36620	(22) 8.6103 83890	(22) 8.7044 21297	(22) 8.7993 56123	(22) 8.8951 95697
9	(25) 6.2601 68916	(25) 6.3372 42543	(25) 6.4151 58496	(25) 6.4939 24819	(25) 6.5735 49620
10	(28) 4.6012 24153	(28) 4.6642 10512	(28) 4.7279 71812	(28) 4.7925 16516	(28) 4.8578 53170
24	(68) 6.1786 86185	(68) 6.3836 27605	(68) 6.5950 74542	(68) 6.8132 24254	(68) 7.0382 79698
1/2	(1) 2.7110 88342	(1) 2.7129 31993	(1) 2.7147 74392	(1) 2.7166 15541	(1) 2.7184 55444
1/3	9.0246 23926	9.0287 14871	9.0328 02112	9.0368 85658	9.0409 65517
1/4	5.2068 11253	5.2085 81374	5.2103 49693	5.2121 16213	5.2138 80938
1/5	3.7433 24423	3.7443 42461	3.7453 59393	3.7463 75222	3.7473 89950

k					
1	740	741	742	743	744
2	5 47600	5 49081	5 50564	5 52049	5 53536
3	4052 24000	4068 69021	4085 18488	4101 72407	4118 30784
4	(11) 2.9986 57600	(11) 3.0148 99446	(11) 3.0312 07181	(11) 3.0475 80984	(11) 3.0640 21033
5	(14) 2.2190 06624	(14) 2.2340 40489	(14) 2.2491 55728	(14) 2.2643 52671	(14) 2.2796 31649
6	(17) 1.6420 64902	(17) 1.6554 24002	(17) 1.6688 73550	(17) 1.6824 14035	(17) 1.6960 45947
7	(20) 1.2151 28027	(20) 1.2266 69186	(20) 1.2383 04174	(20) 1.2500 33628	(20) 1.2618 58184
8	(22) 8.9919 47402	(22) 9.0896 18667	(22) 9.1882 16974	(22) 9.2877 49854	(22) 9.3882 24890
9	(25) 6.6540 41078	(25) 6.7354 07432	(25) 6.8176 56995	(25) 6.9007 98142	(25) 6.9848 39318
10	(28) 4.9239 90397	(28) 4.9909 36907	(28) 5.0587 01490	(28) 5.1272 93019	(28) 5.1967 20453
24	(68) 7.2704 49690	(68) 7.5099 49065	(68) 7.7569 98844	(68) 8.0118 26396	(68) 8.2746 65623
1/2	(1) 2.7202 94102	(1) 2.7221 31518	(1) 2.7239 67694	(1) 2.7258 02634	(1) 2.7276 36339
1/3	9.0450 41696	9.0491 14206	9.0531 83053	9.0572 48245	9.0613 09792
1/4	5.2156 43874	5.2174 05023	5.2191 64391	5.2209 21982	5.2226 77799
1/5	3.7484 03580	3.7494 16115	3.7504 27557	3.7514 37909	3.7524 47174

k					
1	745	746	747	748	749
2	5 55025	5 56516	5 58009	5 59504	5 61001
3	4134 93625	4151 60936	4168 32723	4185 08992	4201 89749
4	(11) 3.0805 27506	(11) 3.0971 00583	(11) 3.1137 40441	(11) 3.1304 47260	(11) 3.1472 21220
5	(14) 2.2949 92992	(14) 2.3104 37035	(14) 2.3259 64109	(14) 2.3415 74551	(14) 2.3572 68694
6	(17) 1.7097 69779	(17) 1.7235 86028	(17) 1.7374 95190	(17) 1.7514 97764	(17) 1.7655 94252
7	(20) 1.2737 78485	(20) 1.2857 95177	(20) 1.2979 08907	(20) 1.3101 20327	(20) 1.3224 30094
8	(22) 9.4896 49717	(22) 9.5920 32018	(22) 9.6953 79533	(22) 9.7997 00049	(22) 9.9050 01408
9	(25) 7.0697 89039	(25) 7.1556 55886	(25) 7.2424 48511	(25) 7.3301 75636	(25) 7.4188 46054
10	(28) 5.2669 92834	(28) 5.3381 19291	(28) 5.4101 09038	(28) 5.4829 71376	(28) 5.5567 15695
24	(68) 8.5457 57129	(68) 8.8253 48404	(68) 9.1136 94019	(68) 9.4110 55807	(68) 9.7177 03069
1/2	(1) 2.7294 68813	(1) 2.7313 00057	(1) 2.7331 30074	(1) 2.7349 58866	(1) 2.7367 86437
1/3	9.0653 67701	9.0694 21981	9.0734 72639	9.0775 19683	9.0815 63122
1/4	5.2244 31847	5.2261 84131	5.2279 34653	5.2296 83419	5.2314 30432
1/5	3.7534 55355	3.7544 62453	3.7554 68472	3.7564 73415	3.7574 77282

$n^{\frac{1}{2}}\left[{(-6)2 \atop 3}\right]$ $n^{\frac{1}{3}}\left[{(-7)5 \atop 4}\right]$ $n^{\frac{1}{4}}\left[{(-7)2 \atop 3}\right]$ $n^{\frac{1}{5}}\left[{(-7)1 \atop 3}\right]$

k	750	751	752	753	754
2	5 62500	5 64001	5 65504	5 67009	5 68516
3	4218 75000	4235 64751	4252 59008	4269 57777	4286 61064
4	(11) 3.1640 62500	(11) 3.1809 71280	(11) 3.1979 47740	(11) 3.2149 92061	(11) 3.2321 04423
5	(14) 2.3730 46875	(14) 2.3889 09431	(14) 2.4048 56701	(14) 2.4208 89022	(14) 2.4370 06735
6	(17) 1.7797 85156	(17) 1.7940 70983	(17) 1.8084 52239	(17) 1.8229 29433	(17) 1.8375 03078
7	(20) 1.3348 38867	(20) 1.3473 47308	(20) 1.3599 56084	(20) 1.3726 65863	(20) 1.3854 77321
8	(23) 1.0011 29150	(23) 1.0118 57828	(23) 1.0226 86975	(23) 1.0336 17395	(23) 1.0446 49900
9	(25) 7.5084 68628	(25) 7.5990 52291	(25) 7.6906 06051	(25) 7.7831 38985	(25) 7.8766 60245
10	(28) 5.6313 51471	(28) 5.7068 88271	(28) 5.7833 35750	(28) 5.8607 03656	(28) 5.9390 01825
24	(69) 1.0033 91278	(69) 1.0359 96977	(69) 1.0696 16698	(69) 1.1042 80565	(69) 1.1400 19555
1/2	(1) 2.7386 12788	(1) 2.7404 37921	(1) 2.7422 61840	(1) 2.7440 84547	(1) 2.7459 06044
1/3	9.0856 02964	9.0896 39217	9.0936 71888	9.0977 00985	9.1017 26517
1/4	5.2331 75697	5.2349 19217	5.2366 60997	5.2384 01041	5.2401 39353
1/5	3.7584 80079	3.7594 81806	3.7604 82467	3.7614 82064	3.7624 80599

k	755	756	757	758	759
2	5 70025	5 71536	5 73049	5 74564	5 76081
3	4303 68875	4320 81216	4337 98093	4355 19512	4372 45479
4	(11) 3.2492 85006	(11) 3.2665 33993	(11) 3.2838 51564	(11) 3.3012 37901	(11) 3.3186 93186
5	(14) 2.4532 10180	(14) 2.4694 99699	(14) 2.4858 75634	(14) 2.5023 38329	(14) 2.5188 88128
6	(17) 1.8521 73686	(17) 1.8669 41772	(17) 1.8818 07855	(17) 1.8967 72453	(17) 1.9118 36089
7	(20) 1.3983 91133	(20) 1.4114 07980	(20) 1.4245 28546	(20) 1.4377 53520	(20) 1.4510 83592
8	(23) 1.0557 85305	(23) 1.0670 24433	(23) 1.0783 68109	(23) 1.0898 17168	(23) 1.1013 72446
9	(25) 7.9711 79054	(25) 8.0667 04711	(25) 8.1632 46588	(25) 8.2608 14132	(25) 8.3594 16865
10	(28) 6.0182 40186	(28) 6.0984 28762	(28) 6.1795 77667	(28) 6.2616 97112	(28) 6.3447 97401
24	(69) 1.1768 65520	(69) 1.2148 51214	(69) 1.2540 10313	(69) 1.2943 77441	(69) 1.3359 88198
1/2	(1) 2.7477 26333	(1) 2.7495 45417	(1) 2.7513 63298	(1) 2.7531 79980	(1) 2.7549 95463
1/3	9.1057 48491	9.1097 66916	9.1137 81798	9.1177 93146	9.1218 00968
1/4	5.2418 75936	5.2436 10795	5.2453 43934	5.2470 75356	5.2488 05067
1/5	3.7634 78075	3.7644 74495	3.7654 69862	3.7664 64176	3.7674 57442

k	760	761	762	763	764
2	5 77600	5 79121	5 80644	5 82169	5 83696
3	4389 76000	4407 11081	4424 50728	4441 94947	4459 43744
4	(11) 3.3362 17600	(11) 3.3538 11326	(11) 3.3714 74547	(11) 3.3892 07446	(11) 3.4070 10204
5	(14) 2.5355 25376	(14) 2.5522 50419	(14) 2.5690 63605	(14) 2.5859 65281	(14) 2.6029 55796
6	(17) 1.9269 99286	(17) 1.9422 62569	(17) 1.9576 26467	(17) 1.9730 91509	(17) 1.9886 58228
7	(20) 1.4645 19457	(20) 1.4780 61815	(20) 1.4917 11368	(20) 1.5054 68822	(20) 1.5193 34886
8	(23) 1.1130 34787	(23) 1.1248 05041	(23) 1.1366 84062	(23) 1.1486 72711	(23) 1.1607 71853
9	(25) 8.4590 64385	(25) 8.5597 66364	(25) 8.6615 32555	(25) 8.7643 72784	(25) 8.8682 96958
10	(28) 6.4288 88932	(28) 6.5139 82203	(28) 6.6000 87807	(28) 6.6872 16435	(28) 6.7753 78876
24	(69) 1.3788 79182	(69) 1.4230 88020	(69) 1.4686 53390	(69) 1.5156 15056	(69) 1.5640 13890
1/2	(1) 2.7568 09750	(1) 2.7586 22845	(1) 2.7604 34748	(1) 2.7622 45463	(1) 2.7640 54992
1/3	9.1258 05271	9.1298 06063	9.1338 03351	9.1377 97144	9.1417 87449
1/4	5.2505 33069	5.2522 59366	5.2539 83963	5.2557 06863	5.2574 28071
1/5	3.7684 49662	3.7694 40838	3.7704 30972	3.7714 20068	3.7724 08126

k	765	766	767	768	769
2	5 85225	5 86756	5 88289	5 89824	5 91361
3	4476 97125	4494 55096	4512 17663	4529 84832	4547 56609
4	(11) 3.4248 83006	(11) 3.4428 26035	(11) 3.4608 39475	(11) 3.4789 23510	(11) 3.4970 78323
5	(14) 2.6200 35500	(14) 2.6372 04743	(14) 2.6544 63877	(14) 2.6718 13255	(14) 2.6892 53231
6	(17) 2.0043 27157	(17) 2.0200 98833	(17) 2.0359 73794	(17) 2.0519 52580	(17) 2.0680 35734
7	(20) 1.5333 10275	(20) 1.5473 95706	(20) 1.5615 91900	(20) 1.5758 99582	(20) 1.5903 19480
8	(23) 1.1729 82361	(23) 1.1853 05111	(23) 1.1977 40987	(23) 1.2102 90879	(23) 1.2229 55680
9	(25) 8.9733 15059	(25) 9.0794 37150	(25) 9.1866 73373	(25) 9.2950 33948	(25) 9.4045 29178
10	(28) 6.8645 86020	(28) 6.9548 48857	(28) 7.0461 78477	(28) 7.1385 86072	(28) 7.2320 82938
24	(69) 1.6138 91907	(69) 1.6652 92289	(69) 1.7182 59425	(69) 1.7728 38934	(69) 1.8290 77701
1/2	(1) 2.7658 63337	(1) 2.7676 70501	(1) 2.7694 76485	(1) 2.7712 81292	(1) 2.7730 84925
1/3	9.1457 74274	9.1497 57625	9.1537 37512	9.1577 13940	9.1616 86919
1/4	5.2591 47590	5.2608 65424	5.2625 81576	5.2642 96052	5.2660 08854
1/5	3.7733 95151	3.7743 81144	3.7753 66108	3.7763 50045	3.7773 32958

k	770	771	772	773	774
2	5 92900	5 94441	5 95984	5 97529	5 99076
3	4565 33000	4583 14011	4600 99648	4618 89917	4636 84824
4	(11) 3.5153 04100	(11) 3.5336 01025	(11) 3.5519 69283	(11) 3.5704 09058	(11) 3.5889 20538
5	(14) 2.7067 84157	(14) 2.7244 06390	(14) 2.7421 20286	(14) 2.7599 26202	(14) 2.7778 24496
6	(17) 2.0842 23801	(17) 2.1005 17327	(17) 2.1169 16861	(17) 2.1334 22954	(17) 2.1500 36160
7	(20) 1.6048 52327	(20) 1.6194 98859	(20) 1.6342 59817	(20) 1.6491 35944	(20) 1.6641 27988
8	(23) 1.2357 36292	(23) 1.2486 33620	(23) 1.2616 48578	(23) 1.2747 82084	(23) 1.2880 35063
9	(25) 9.5151 69445	(25) 9.6269 65212	(25) 9.7399 27025	(25) 9.8540 65513	(25) 9.9693 91385
10	(28) 7.3266 80473	(28) 7.4223 90179	(28) 7.5192 23664	(28) 7.6171 92641	(28) 7.7163 08932
24	(69) 1.8870 23915	(69) 1.9467 27094	(69) 2.0082 38127	(69) 2.0716 09310	(69) 2.1368 94378
1/2	(1) 2.7748 87385	(1) 2.7766 88675	(1) 2.7784 88798	(1) 2.7802 87755	(1) 2.7820 85549
1/3	9.1656 56454	9.1696 22555	9.1735 85227	9.1775 44479	9.1815 00317
1/4	5.2677 19986	5.2694 29452	5.2711 37257	5.2728 43403	5.2745 47894
1/5	3.7783 14849	3.7792 95720	3.7802 75573	3.7812 54412	3.7822 32239

$n^{\frac{1}{2}}\left[{(-6)2 \atop 3}\right]$ $n^{\frac{1}{3}}\left[{(-7)5 \atop 3}\right]$ $n^{\frac{1}{4}}\left[{(-7)2 \atop 3}\right]$ $n^{\frac{1}{5}}\left[{(-7)1 \atop 3}\right]$

k					
1	775	776	777	778	779
2	6 00625	6 02176	6 03729	6 05284	6 06841
3	4654 84375	4672 88576	4690 97433	4709 10952	4727 29139
4	(11) 3.6075 03906	(11) 3.6261 59350	(11) 3.6448 87054	(11) 3.6636 87207	(11) 3.6825 59993
5	(14) 2.7958 15527	(14) 2.8138 99655	(14) 2.8320 77241	(14) 2.8503 48647	(14) 2.8687 14234
6	(17) 2.1667 57034	(17) 2.1835 86133	(17) 2.2005 24016	(17) 2.2175 71247	(17) 2.2347 28389
7	(20) 1.6792 36701	(20) 1.6944 62839	(20) 1.7098 07161	(20) 1.7252 70430	(20) 1.7408 53415
8	(23) 1.3014 08443	(23) 1.3149 03163	(23) 1.3285 20164	(23) 1.3422 60395	(23) 1.3561 24810
9	(26) 1.0085 91544	(26) 1.0203 64854	(26) 1.0322 60167	(26) 1.0442 78587	(26) 1.0564 21227
10	(28) 7.8165 84463	(28) 7.9180 31271	(28) 8.0206 61501	(28) 8.1244 87408	(28) 8.2295 21359
24	(69) 2.2041 48547	(69) 2.2734 28553	(69) 2.3447 92689	(69) 2.4183 00846	(69) 2.4940 14558
1/2	(1) 2.7838 82181	(1) 2.7856 77655	(1) 2.7874 71973	(1) 2.7892 65136	(1) 2.7910 57147
1/3	9.1854 52750	9.1894 01784	9.1933 47428	9.1972 89687	9.2012 28569
1/4	5.2762 50735	5.2779 51928	5.2796 51478	5.2813 49388	5.2830 45663
1/5	3.7832 09055	3.7841 84864	3.7851 59667	3.7861 33467	3.7871 06266

k					
1	780	781	782	783	784
2	6 08400	6 09961	6 11524	6 13089	6 14656
3	4745 52000	4763 79541	4782 11768	4800 48687	4818 90304
4	(11) 3.7015 05600	(11) 3.7205 24215	(11) 3.7396 16026	(11) 3.7587 81219	(11) 3.7780 19983
5	(14) 2.8871 74368	(14) 2.9057 29412	(14) 2.9243 79732	(14) 2.9431 25695	(14) 2.9619 67667
6	(17) 2.2519 96007	(17) 2.2693 74671	(17) 2.2868 64951	(17) 2.3044 67419	(17) 2.3221 82651
7	(20) 1.7565 56885	(20) 1.7723 81618	(20) 1.7883 28391	(20) 1.8043 97989	(20) 1.8205 91198
8	(23) 1.3701 14371	(23) 1.3842 30044	(23) 1.3984 72802	(23) 1.4128 43625	(23) 1.4273 43499
9	(26) 1.0686 89209	(26) 1.0810 83664	(26) 1.0936 05731	(26) 1.1062 56559	(26) 1.1190 37304
10	(28) 8.3357 75831	(28) 8.4432 63416	(28) 8.5519 96818	(28) 8.6619 88854	(28) 8.7732 52460
24	(69) 2.5719 97041	(69) 2.6523 13239	(69) 2.7350 29868	(69) 2.8202 15463	(69) 2.9079 40422
1/2	(1) 2.7928 48009	(1) 2.7946 37722	(1) 2.7964 26291	(1) 2.7982 13716	(1) 2.8000 00000
1/3	9.2051 64083	9.2090 96233	9.2130 25029	9.2169 50477	9.2208 72584
1/4	5.2847 40305	5.2864 33318	5.2881 24706	5.2898 14473	5.2915 02622
1/5	3.7880 78066	3.7890 48871	3.7900 18681	3.7909 87500	3.7919 55329

k					
1	785	786	787	788	789
2	6 16225	6 17796	6 19369	6 20944	6 22521
3	4837 36625	4855 87656	4874 43403	4893 03872	4911 69069
4	(11) 3.7973 32506	(11) 3.8167 18976	(11) 3.8361 79582	(11) 3.8557 14511	(11) 3.8753 23954
5	(14) 2.9809 06017	(14) 2.9999 41115	(14) 3.0190 73331	(14) 3.0383 03035	(14) 3.0576 30600
6	(17) 2.3400 11224	(17) 2.3579 53717	(17) 2.3760 10711	(17) 2.3941 82792	(17) 2.4124 70543
7	(20) 1.8369 08811	(20) 1.8533 51621	(20) 1.8699 20430	(20) 1.8866 16040	(20) 1.9034 39259
8	(23) 1.4419 73416	(23) 1.4567 34374	(23) 1.4716 27378	(23) 1.4866 53439	(23) 1.5018 13575
9	(26) 1.1319 49132	(26) 1.1449 93218	(26) 1.1581 70747	(26) 1.1714 82910	(26) 1.1849 30911
10	(28) 8.8858 00685	(28) 8.9996 46695	(28) 9.1148 03776	(28) 9.2312 85332	(28) 9.3491 04886
24	(69) 2.9982 77060	(69) 3.0912 99652	(69) 3.1870 84488	(69) 3.2857 09926	(69) 3.3872 56439
1/2	(1) 2.8017 85145	(1) 2.8035 69154	(1) 2.8053 52028	(1) 2.8071 33770	(1) 2.8089 14381
1/3	9.2247 91357	9.2287 06804	9.2326 18931	9.2365 27746	9.2404 33255
1/4	5.2931 89157	5.2948 74081	5.2965 57399	5.2982 39113	5.2999 19227
1/5	3.7929 22172	3.7938 88029	3.7948 52904	3.7958 16799	3.7967 79716

k					
1	790	791	792	793	794
2	6 24100	6 25681	6 27264	6 28849	6 30436
3	4930 39000	4949 13671	4967 93088	4986 77257	5005 66184
4	(11) 3.8950 08100	(11) 3.9147 67138	(11) 3.9346 01257	(11) 3.9545 10648	(11) 3.9744 95501
5	(14) 3.0770 56399	(14) 3.0965 80806	(14) 3.1162 04196	(14) 3.1359 26944	(14) 3.1557 49428
6	(17) 2.4308 74555	(17) 2.4493 95417	(17) 2.4680 33723	(17) 2.4867 90066	(17) 2.5056 65046
7	(20) 1.9203 90899	(20) 1.9374 71775	(20) 1.9546 82708	(20) 1.9720 24523	(20) 1.9894 98046
8	(23) 1.5171 08810	(23) 1.5325 40174	(23) 1.5481 08705	(23) 1.5638 15447	(23) 1.5796 61449
9	(26) 1.1985 15960	(26) 1.2122 39278	(26) 1.2261 02094	(26) 1.2401 05649	(26) 1.2542 51190
10	(28) 9.4682 76083	(28) 9.5888 12687	(28) 9.7107 28588	(28) 9.8340 37797	(28) 9.9587 54451
24	(69) 3.4918 06676	(69) 3.5994 45514	(69) 3.7102 60118	(69) 3.8243 39997	(69) 3.9417 77065
1/2	(1) 2.8106 93865	(1) 2.8124 72222	(1) 2.8142 49456	(1) 2.8160 25568	(1) 2.8178 00561
1/3	9.2443 35465	9.2482 34384	9.2521 30018	9.2560 22375	9.2599 11460
1/4	5.3015 97745	5.3032 74670	5.3049 50005	5.3066 23755	5.3082 95923
1/5	3.7977 41656	3.7987 02623	3.7996 62619	3.8006 21646	3.8015 79705

k					
1	795	796	797	798	799
2	6 32025	6 33616	6 35209	6 36804	6 38401
3	5024 59875	5043 58336	5062 61573	5081 69592	5100 82399
4	(11) 3.9945 56006	(11) 4.0146 92355	(11) 4.0349 04737	(11) 4.0551 93344	(11) 4.0755 58368
5	(14) 3.1756 72025	(14) 3.1956 95114	(14) 3.2158 19075	(14) 3.2360 44289	(14) 3.2563 71136
6	(17) 2.5246 59260	(17) 2.5437 73311	(17) 2.5630 07803	(17) 2.5823 63342	(17) 2.6018 40538
7	(20) 2.0071 04112	(20) 2.0248 43555	(20) 2.0427 17219	(20) 2.0607 25947	(20) 2.0788 70590
8	(23) 1.5956 47769	(23) 1.6117 75470	(23) 1.6280 45624	(23) 1.6444 59306	(23) 1.6610 17601
9	(26) 1.2685 39976	(26) 1.2829 73274	(26) 1.2975 52362	(26) 1.3122 78526	(26) 1.3271 53063
10	(29) 1.0084 89281	(29) 1.0212 46726	(29) 1.0341 49232	(29) 1.0471 98264	(29) 1.0603 95298
24	(69) 4.0626 65702	(69) 4.1871 02820	(69) 4.3151 87922	(69) 4.4470 23172	(69) 4.5827 13463
1/2	(1) 2.8195 74436	(1) 2.8213 47196	(1) 2.8231 18843	(1) 2.8248 89378	(1) 2.8266 58805
1/3	9.2637 97282	9.2676 79846	9.2715 59160	9.2754 35230	9.2793 08064
1/4	5.3099 66512	5.3116 35526	5.3133 02968	5.3149 68841	5.3166 33150
1/5	3.8025 36800	3.8034 92932	3.8044 48104	3.8054 02317	3.8063 55574

$n^{\frac{1}{2}}\left[\begin{matrix}(-6)2\\3\end{matrix}\right]$ $n^{\frac{1}{3}}\left[\begin{matrix}(-7)4\\3\end{matrix}\right]$ $n^{\frac{1}{4}}\left[\begin{matrix}(-7)2\\3\end{matrix}\right]$ $n^{\frac{1}{5}}\left[\begin{matrix}(-7)1\\3\end{matrix}\right]$

k					
1	800	801	802	803	804
2	6 40000	6 41601	6 43204	6 44809	6 46416
3	5120 00000	5139 22401	5158 49608	5177 81627	5197 18464
4	(11) 4.0960 00000	(11) 4.1165 18432	(11) 4.1371 13856	(11) 4.1577 86465	(11) 4.1785 36451
5	(14) 3.2768 00000	(14) 3.2973 31264	(14) 3.3179 65313	(14) 3.3387 02531	(14) 3.3595 43306
6	(17) 2.6214 40000	(17) 2.6411 62342	(17) 2.6610 08181	(17) 2.6809 78133	(17) 2.7010 72818
7	(20) 2.0971 52000	(20) 2.1155 71036	(20) 2.1341 28561	(20) 2.1528 25440	(20) 2.1716 62546
8	(23) 1.6777 21600	(23) 1.6945 72400	(23) 1.7115 71106	(23) 1.7287 18829	(23) 1.7460 16687
9	(26) 1.3421 77280	(26) 1.3573 52492	(26) 1.3726 80027	(26) 1.3881 61219	(26) 1.4037 97416
10	(29) 1.0737 41824	(29) 1.0872 39346	(29) 1.1008 89382	(29) 1.1146 93459	(29) 1.1286 53123
24	(69) 4.7223 66483	(69) 4.8660 92789	(69) 5.0140 05879	(69) 5.1662 22264	(69) 5.3228 61548
1/2	(1) 2.8284 27125	(1) 2.8301 94340	(1) 2.8319 60452	(1) 2.8337 25463	(1) 2.8354 89376
1/3	9.2831 77667	9.2870 44047	9.2909 07211	9.2947 67164	9.2986 23915
1/4	5.3182 95897	5.3199 57086	5.3216 16720	5.3232 74803	5.3249 31338
1/5	3.8073 07877	3.8082 59229	3.8092 09631	3.8101 59085	3.8111 07593

k					
1	805	806	807	808	809
2	6 48025	6 49636	6 51249	6 52864	6 54481
3	5216 60125	5236 06616	5255 57943	5275 14112	5294 75129
4	(11) 4.1993 64006	(11) 4.2202 69325	(11) 4.2412 52600	(11) 4.2623 14025	(11) 4.2834 53794
5	(14) 3.3804 88025	(14) 3.4015 37076	(14) 3.4226 90848	(14) 3.4439 49732	(14) 3.4653 14119
6	(17) 2.7212 92860	(17) 2.7416 38883	(17) 2.7621 11515	(17) 2.7827 11384	(17) 2.8034 39122
7	(20) 2.1906 40752	(20) 2.2097 60940	(20) 2.2290 23992	(20) 2.2484 30798	(20) 2.2679 82250
8	(23) 1.7634 65806	(23) 1.7810 67318	(23) 1.7988 22362	(23) 1.8167 32085	(23) 1.8347 97640
9	(26) 1.4195 89974	(26) 1.4355 40258	(26) 1.4516 49646	(26) 1.4679 19524	(26) 1.4843 51291
10	(29) 1.1427 69929	(29) 1.1570 45448	(29) 1.1714 81264	(29) 1.1860 78976	(29) 1.2008 40194
24	(69) 5.4840 46503	(69) 5.6499 03151	(69) 5.8205 60843	(69) 5.9961 52346	(69) 6.1768 13927
1/2	(1) 2.8372 52192	(1) 2.8390 13913	(1) 2.8407 74542	(1) 2.8425 34081	(1) 2.8442 92531
1/3	9.3024 77468	9.3063 27832	9.3101 75012	9.3140 19016	9.3178 59849
1/4	5.3265 86329	5.3282 39778	5.3298 91690	5.3315 42067	5.3331 90912
1/5	3.8120 55159	3.8130 01783	3.8139 47468	3.8148 92216	3.8158 36029

k					
1	810	811	812	813	814
2	6 56100	6 57721	6 59344	6 60969	6 62596
3	5314 41000	5334 11731	5353 87328	5373 67797	5393 53144
4	(11) 4.3046 72100	(11) 4.3259 69138	(11) 4.3473 45103	(11) 4.3688 00190	(11) 4.3903 34592
5	(14) 3.4867 84401	(14) 3.5083 60971	(14) 3.5300 44224	(14) 3.5518 34554	(14) 3.5737 32358
6	(17) 2.8242 95365	(17) 2.8452 80748	(17) 2.8663 95910	(17) 2.8876 41493	(17) 2.9090 18139
7	(20) 2.2876 79245	(20) 2.3075 22686	(20) 2.3275 13479	(20) 2.3476 52533	(20) 2.3679 40765
8	(23) 1.8530 20189	(23) 1.8714 00899	(23) 1.8899 40945	(23) 1.9086 41510	(23) 1.9275 03783
9	(26) 1.5009 46353	(26) 1.5177 06129	(26) 1.5346 32047	(26) 1.5517 25547	(26) 1.5689 88079
10	(29) 1.2157 66546	(29) 1.2308 59670	(29) 1.2461 21222	(29) 1.2615 52870	(29) 1.2771 56297
24	(69) 6.3626 85441	(69) 6.5539 10420	(69) 6.7506 36166	(69) 6.9530 13847	(69) 7.1611 98588
1/2	(1) 2.8460 49894	(1) 2.8478 06173	(1) 2.8495 61370	(1) 2.8513 15486	(1) 2.8530 68524
1/3	9.3216 97518	9.3255 32030	9.3293 63391	9.3331 91608	9.3370 16687
1/4	5.3348 38230	5.3364 84023	5.3381 28295	5.3397 71049	5.3414 12288
1/5	3.8167 78910	3.8177 20859	3.8186 61880	3.8196 01974	3.8205 41144

k					
1	815	816	817	818	819
2	6 64225	6 65856	6 67489	6 69124	6 70761
3	5413 43375	5433 38496	5453 38513	5473 43432	5493 53259
4	(11) 4.4119 48506	(11) 4.4336 42127	(11) 4.4554 15651	(11) 4.4772 69274	(11) 4.4992 03191
5	(14) 3.5957 38033	(14) 3.6178 51976	(14) 3.6400 74587	(14) 3.6624 06266	(14) 3.6848 47414
6	(17) 2.9305 26497	(17) 2.9521 67212	(17) 2.9739 40938	(17) 2.9958 48326	(17) 3.0178 90032
7	(20) 2.3883 79095	(20) 2.4089 68445	(20) 2.4297 09746	(20) 2.4506 03930	(20) 2.4716 51936
8	(23) 1.9465 28962	(23) 1.9657 18251	(23) 1.9850 72863	(23) 2.0045 94015	(23) 2.0242 82936
9	(26) 1.5864 21104	(26) 1.6040 26093	(26) 1.6218 04529	(26) 1.6397 57904	(26) 1.6578 87724
10	(29) 1.2929 33200	(29) 1.3088 85292	(29) 1.3250 14300	(29) 1.3413 21966	(29) 1.3578 10046
24	(69) 7.3753 49576	(69) 7.5956 30157	(69) 7.8222 07941	(69) 8.0552 54907	(69) 8.2949 47511
1/2	(1) 2.8548 20485	(1) 2.8565 71371	(1) 2.8583 21186	(1) 2.8600 69929	(1) 2.8618 17604
1/3	9.3408 38634	9.3446 57457	9.3484 73160	9.3522 85752	9.3560 95237
1/4	5.3430 52016	5.3446 90236	5.3463 26950	5.3479 62163	5.3495 95877
1/5	3.8214 79391	3.8224 16717	3.8233 53125	3.8242 88616	3.8252 23193

k					
1	820	821	822	823	824
2	6 72400	6 74041	6 75684	6 77329	6 78976
3	5513 68000	5533 87661	5554 12248	5574 41767	5594 76224
4	(11) 4.5212 17600	(11) 4.5433 12697	(11) 4.5654 88679	(11) 4.5877 45742	(11) 4.6100 84086
5	(14) 3.7073 98432	(14) 3.7300 59724	(14) 3.7528 31694	(14) 3.7757 14746	(14) 3.7987 09287
6	(17) 3.0400 66714	(17) 3.0623 79033	(17) 3.0848 27652	(17) 3.1074 13236	(17) 3.1301 36452
7	(20) 2.4928 54706	(20) 2.5142 13186	(20) 2.5357 28330	(20) 2.5574 01093	(20) 2.5792 32437
8	(23) 2.0441 40859	(23) 2.0641 69026	(23) 2.0843 68687	(23) 2.1047 41100	(23) 2.1252 87528
9	(26) 1.6761 95504	(26) 1.6946 82770	(26) 1.7133 51061	(26) 1.7322 01925	(26) 1.7512 36923
10	(29) 1.3744 80313	(29) 1.3913 34555	(29) 1.4083 74572	(29) 1.4256 02184	(29) 1.4430 19224
24	(69) 8.5414 66801	(69) 8.7949 98523	(69) 9.0557 33244	(69) 9.3238 66467	(69) 9.5995 98755
1/2	(1) 2.8635 64213	(1) 2.8653 09756	(1) 2.8670 54237	(1) 2.8687 97658	(1) 2.8705 40019
1/3	9.3599 01623	9.3637 04916	9.3675 05121	9.3713 02245	9.3750 96295
1/4	5.3512 28095	5.3528 58822	5.3544 88059	5.3561 15810	5.3577 42079
1/5	3.8261 56858	3.8270 89612	3.8280 21458	3.8289 52397	3.8298 82432

$n^{\frac{1}{2}}\begin{bmatrix}(-6)1\\3\end{bmatrix}$ $n^{\frac{1}{3}}\begin{bmatrix}(-7)4\\3\end{bmatrix}$ $n^{\frac{1}{4}}\begin{bmatrix}(-7)2\\3\end{bmatrix}$ $n^{\frac{1}{5}}\begin{bmatrix}(-7)1\\3\end{bmatrix}$

k	825	826	827	828	829
1	825	826	827	828	829
2	6 80625	6 82276	6 83929	6 85584	6 87241
3	5615 15625	5635 59976	5656 09283	5676 63552	5697 22789
4	(11) 4.6325 03906	(11) 4.6550 05402	(11) 4.6775 88770	(11) 4.7002 54211	(11) 4.7230 01921
5	(14) 3.8218 15723	(14) 3.8450 34462	(14) 3.8683 65913	(14) 3.8918 10486	(14) 3.9153 68592
6	(17) 3.1529 97971	(17) 3.1759 98465	(17) 3.1991 38610	(17) 3.2224 19083	(17) 3.2458 40563
7	(20) 2.6012 23326	(20) 2.6233 74732	(20) 2.6456 87631	(20) 2.6681 63000	(20) 2.6908 01827
8	(23) 2.1460 09244	(23) 2.1669 07529	(23) 2.1879 83671	(23) 2.2092 38964	(23) 2.2306 74714
9	(26) 1.7704 57626	(26) 1.7898 65619	(26) 1.8094 62496	(26) 1.8292 49863	(26) 1.8492 29338
10	(29) 1.4606 27542	(29) 1.4784 29001	(29) 1.4964 25484	(29) 1.5146 18886	(29) 1.5330 11121
24	(69) 9.8831 35853	(70) 1.0174 68882	(70) 1.0474 47415	(70) 1.0782 71392	(70) 1.1099 63591
1/2	(1) 2.8722 81323	(1) 2.8740 21573	(1) 2.8757 60769	(1) 2.8774 98914	(1) 2.8792 36010
1/3	9.3788 87277	9.3826 75196	9.3864 60060	9.3902 41873	9.3940 20643
1/4	5.3593 66869	5.3609 90182	5.3626 12021	5.3642 32391	5.3658 51293
1/5	3.8308 11564	3.8317 39795	3.8326 67128	3.8335 93565	3.8345 19107

k	830	831	832	833	834
1	830	831	832	833	834
2	6 88900	6 90561	6 92224	6 93889	6 95556
3	5717 87000	5738 56191	5759 30368	5780 09537	5800 93704
4	(11) 4.7458 32100	(11) 4.7687 44947	(11) 4.7917 40662	(11) 4.8148 19443	(11) 4.8379 81491
5	(14) 3.9390 40643	(14) 3.9628 27051	(14) 3.9867 28231	(14) 4.0107 44596	(14) 4.0348 76564
6	(17) 3.2694 03734	(17) 3.2931 09279	(17) 3.3169 57888	(17) 3.3409 50249	(17) 3.3650 87054
7	(20) 2.7136 05099	(20) 2.7365 73811	(20) 2.7597 08963	(20) 2.7830 11557	(20) 2.8064 82603
8	(23) 2.2522 92232	(23) 2.2740 92837	(23) 2.2960 77857	(23) 2.3182 48627	(23) 2.3406 06491
9	(26) 1.8694 02553	(26) 1.8897 71148	(26) 1.9103 36777	(26) 1.9311 01106	(26) 1.9520 65814
10	(29) 1.5516 04119	(29) 1.5703 99824	(29) 1.5894 00198	(29) 1.6086 07222	(29) 1.6280 22889
24	(70) 1.1425 47375	(70) 1.1760 46709	(70) 1.2104 86167	(70) 1.2458 90957	(70) 1.2822 86929
1/2	(1) 2.8809 72058	(1) 2.8827 07061	(1) 2.8844 41020	(1) 2.8861 73938	(1) 2.8879 05816
1/3	9.3977 96375	9.4015 69076	9.4053 38751	9.4091 05407	9.4128 69049
1/4	5.3674 68731	5.3690 84709	5.3706 99229	5.3723 12294	5.3739 23907
1/5	3.8354 43756	3.8363 67514	3.8372 90383	3.8382 12366	3.8391 33463

k	835	836	837	838	839
1	835	836	837	838	839
2	6 97225	6 98896	7 00569	7 02244	7 03921
3	5821 82875	5842 77056	5863 76253	5884 80472	5905 89719
4	(11) 4.8612 27006	(11) 4.8845 56188	(11) 4.9079 69238	(11) 4.9314 66355	(11) 4.9550 47742
5	(14) 4.0591 24550	(14) 4.0834 88973	(14) 4.1079 70252	(14) 4.1325 68806	(14) 4.1572 85056
6	(17) 3.3893 68999	(17) 3.4137 96782	(17) 3.4383 71101	(17) 3.4630 92659	(17) 3.4879 62162
7	(20) 2.8301 23115	(20) 2.8539 34109	(20) 2.8779 16611	(20) 2.9020 71648	(20) 2.9264 00254
8	(23) 2.3631 52801	(23) 2.3858 88916	(23) 2.4088 16204	(23) 2.4319 36041	(23) 2.4552 49813
9	(26) 1.9732 32589	(26) 1.9946 03133	(26) 2.0161 79163	(26) 2.0379 62403	(26) 2.0599 54593
10	(29) 1.6476 49211	(29) 1.6674 88220	(29) 1.6875 41959	(29) 1.7078 12493	(29) 1.7283 01904
24	(70) 1.3197 00592	(70) 1.3581 59133	(70) 1.3976 90431	(70) 1.4383 23072	(70) 1.4800 86372
1/2	(1) 2.8896 36655	(1) 2.8913 66459	(1) 2.8930 95228	(1) 2.8948 22965	(1) 2.8965 49672
1/3	9.4166 29685	9.4203 87319	9.4241 41957	9.4278 93606	9.4316 42272
1/4	5.3755 34071	5.3771 42790	5.3787 50067	5.3803 55904	5.3819 60304
1/5	3.8400 53677	3.8409 73010	3.8418 91464	3.8428 09040	3.8437 25741

k	840	841	842	843	844
1	840	841	842	843	844
2	7 05600	7 07281	7 08964	7 10649	7 12336
3	5927 04000	5948 23321	5969 47688	5990 77107	6012 11584
4	(11) 4.9787 13600	(11) 5.0024 64130	(11) 5.0262 99533	(11) 5.0502 20012	(11) 5.0742 25769
5	(14) 4.1821 19424	(14) 4.2070 72333	(14) 4.2321 44207	(14) 4.2573 35470	(14) 4.2826 46549
6	(17) 3.5129 80316	(17) 3.5381 47832	(17) 3.5634 65422	(17) 3.5889 33801	(17) 3.6145 53687
7	(20) 2.9509 03466	(20) 2.9755 82327	(20) 3.0004 37885	(20) 3.0254 71195	(20) 3.0506 83312
8	(23) 2.4787 58911	(23) 2.5024 64737	(23) 2.5263 68700	(23) 2.5504 72217	(23) 2.5747 76715
9	(26) 2.0821 57485	(26) 2.1045 72844	(26) 2.1272 02445	(26) 2.1500 48079	(26) 2.1731 11548
10	(29) 1.7490 12288	(29) 1.7699 45762	(29) 1.7911 04459	(29) 1.8124 90531	(29) 1.8341 06146
24	(70) 1.5230 10388	(70) 1.5671 25939	(70) 1.6124 64626	(70) 1.6590 58848	(70) 1.7069 41821
1/2	(1) 2.8982 75349	(1) 2.9000 00000	(1) 2.9017 23626	(1) 2.9034 46228	(1) 2.9051 67809
1/3	9.4353 87961	9.4391 30677	9.4428 70428	9.4466 07220	9.4503 41057
1/4	5.3835 63271	5.3851 64807	5.3867 64916	5.3883 63600	5.3899 60862
1/5	3.8446 41568	3.8455 56523	3.8464 70609	3.8473 83826	3.8482 96177

k	845	846	847	848	849
1	845	846	847	848	849
2	7 14025	7 15716	7 17409	7 19104	7 20801
3	6033 51125	6054 95736	6076 45423	6098 00192	6119 60049
4	(11) 5.0983 17006	(11) 5.1224 93927	(11) 5.1467 56733	(11) 5.1711 05628	(11) 5.1955 40816
5	(14) 4.3080 77870	(14) 4.3336 29862	(14) 4.3593 02953	(14) 4.3850 97573	(14) 4.4110 14153
6	(17) 3.6403 25800	(17) 3.6662 50863	(17) 3.6923 29601	(17) 3.7185 62742	(17) 3.7449 51016
7	(20) 3.0760 75301	(20) 3.1016 48230	(20) 3.1274 03172	(20) 3.1533 41205	(20) 3.1794 63412
8	(23) 2.5992 83630	(23) 2.6239 94403	(23) 2.6489 10487	(23) 2.6740 33342	(23) 2.6993 64437
9	(26) 2.1963 94667	(26) 2.2198 99265	(26) 2.2436 27182	(26) 2.2675 80274	(26) 2.2917 60407
10	(29) 1.8559 53494	(29) 1.8780 34778	(29) 1.9003 52223	(29) 1.9229 08072	(29) 1.9457 04586
24	(70) 1.7561 47601	(70) 1.8067 11101	(70) 1.8586 68111	(70) 1.9120 55324	(70) 1.9669 10351
1/2	(1) 2.9068 88371	(1) 2.9086 07914	(1) 2.9103 26442	(1) 2.9120 43956	(1) 2.9137 60457
1/3	9.4540 71946	9.4577 99893	9.4615 24903	9.4652 46982	9.4689 66137
1/4	5.3915 56705	5.3931 51133	5.3947 44148	5.3963 35753	5.3979 25951
1/5	3.8492 07664	3.8501 18288	3.8510 28051	3.8519 36956	3.8528 45003

$n^{\frac{1}{2}}\left[\begin{matrix}(-6)1\\3\end{matrix}\right]$ $n^{\frac{1}{3}}\left[\begin{matrix}(-7)4\\3\end{matrix}\right]$ $n^{\frac{1}{4}}\left[\begin{matrix}(-7)2\\3\end{matrix}\right]$ $n^{\frac{1}{5}}\left[\begin{matrix}(-7)1\\3\end{matrix}\right]$

k					
1	850	851	852	853	854
2	7 22500	7 24201	7 25904	7 27609	7 29316
3	6141 25000	6162 95051	6184 70208	6206 50477	6228 35864
4	(11) 5.2200 62500	(11) 5.2446 70884	(11) 5.2693 66172	(11) 5.2941 48569	(11) 5.3190 18279
5	(14) 4.4370 53125	(14) 4.4632 14922	(14) 4.4894 99979	(14) 4.5159 08729	(14) 4.5424 41610
6	(17) 3.7714 95156	(17) 3.7981 95899	(17) 3.8250 53982	(17) 3.8520 70146	(17) 3.8792 45135
7	(20) 3.2057 70883	(20) 3.2322 64710	(20) 3.2589 45993	(20) 3.2858 15835	(20) 3.3128 75345
8	(23) 2.7249 05250	(23) 2.7506 57268	(23) 2.7766 21986	(23) 2.8028 00907	(23) 2.8291 95545
9	(26) 2.3161 69463	(26) 2.3408 09335	(26) 2.3656 81932	(26) 2.3907 89174	(26) 2.4161 32995
10	(29) 1.9687 44043	(29) 1.9920 28744	(29) 2.0155 61006	(29) 2.0393 43165	(29) 2.0633 77578
24	(70) 2.0232 71747	(70) 2.0811 79034	(70) 2.1406 72719	(70) 2.2017 94325	(70) 2.2645 86409
1/2	(1) 2.9154 75947	(1) 2.9171 90429	(1) 2.9189 03904	(1) 2.9206 16373	(1) 2.9223 27839
1/3	9.4726 82372	9.4763 95693	9.4801 06107	9.4838 13619	9.4875 18234
1/4	5.3995 14744	5.4011 02137	5.4026 88131	5.4042 72729	5.4058 55935
1/5	3.8537 52195	3.8546 58534	3.8555 64021	3.8564 68659	3.8573 72448

k					
1	855	856	857	858	859
2	7 31025	7 32736	7 34449	7 36164	7 37881
3	6250 26375	6272 22016	6294 22793	6316 28712	6338 39779
4	(11) 5.3439 75506	(11) 5.3690 20457	(11) 5.3941 53336	(11) 5.4193 74349	(11) 5.4446 83702
5	(14) 4.5690 99058	(14) 4.5958 81511	(14) 4.6227 89409	(14) 4.6498 23191	(14) 4.6769 83300
6	(17) 3.9065 79694	(17) 3.9340 74574	(17) 3.9617 30523	(17) 3.9895 48298	(17) 4.0175 28654
7	(20) 3.3401 25639	(20) 3.3675 67835	(20) 3.3952 03059	(20) 3.4230 32440	(20) 3.4510 57114
8	(23) 2.8558 07421	(23) 2.8826 38067	(23) 2.9096 89021	(23) 2.9369 61833	(23) 2.9644 58061
9	(26) 2.4417 15345	(26) 2.4675 38185	(26) 2.4936 03491	(26) 2.5199 13253	(26) 2.5464 69474
10	(29) 2.0876 66620	(29) 2.1122 12686	(29) 2.1370 18192	(29) 2.1620 85571	(29) 2.1874 17279
24	(70) 2.3290 92589	(70) 2.3953 57569	(70) 2.4634 27165	(70) 2.5333 48329	(70) 2.6051 69182
1/2	(1) 2.9240 38303	(1) 2.9257 47768	(1) 2.9274 56234	(1) 2.9291 63703	(1) 2.9308 70178
1/3	9.4912 19958	9.4949 18797	9.4986 14756	9.5023 07842	9.5059 98059
1/4	5.4074 37751	5.4090 18180	5.4105 97225	5.4121 74889	5.4137 51174
1/5	3.8582 75391	3.8591 77490	3.8600 78746	3.8609 79161	3.8618 78737

k					
1	860	861	862	863	864
2	7 39600	7 41321	7 43044	7 44769	7 46496
3	6360 56000	6382 77381	6405 03928	6427 35647	6449 72544
4	(11) 5.4700 81600	(11) 5.4955 68250	(11) 5.5211 43859	(11) 5.5468 08634	(11) 5.5725 62780
5	(14) 4.7042 70176	(14) 4.7316 84264	(14) 4.7592 26007	(14) 4.7868 95851	(14) 4.8146 94242
6	(17) 4.0456 72351	(17) 4.0739 80151	(17) 4.1024 52818	(17) 4.1310 91119	(17) 4.1598 95825
7	(20) 3.4792 78222	(20) 3.5076 96910	(20) 3.5363 14329	(20) 3.5651 31636	(20) 3.5941 49993
8	(23) 2.9921 79271	(23) 3.0201 27039	(23) 3.0483 02952	(23) 3.0767 08602	(23) 3.1053 45594
9	(26) 2.5732 74173	(26) 2.6003 29381	(26) 2.6276 37144	(26) 2.6551 99523	(26) 2.6830 18593
10	(29) 2.2130 15789	(29) 2.2388 83597	(29) 2.2650 23218	(29) 2.2914 37189	(29) 2.3181 28064
24	(70) 2.6789 39031	(70) 2.7547 08410	(70) 2.8325 29097	(70) 2.9124 54150	(70) 2.9945 37938
1/2	(1) 2.9325 75660	(1) 2.9342 80150	(1) 2.9359 83651	(1) 2.9376 86164	(1) 2.9393 87691
1/3	9.5096 85413	9.5133 69910	9.5170 51555	9.5207 30354	9.5244 06312
1/4	5.4153 26084	5.4168 99621	5.4184 71787	5.4200 42587	5.4216 12022
1/5	3.8627 77475	3.8636 75378	3.8645 72447	3.8654 68684	3.8663 64090

k					
1	865	866	867	868	869
2	7 48225	7 49956	7 51689	7 53424	7 55161
3	6472 14625	6494 61896	6517 14363	6539 72032	6562 34909
4	(11) 5.5984 06506	(11) 5.6243 40019	(11) 5.6503 63527	(11) 5.6764 77238	(11) 5.7026 81359
5	(14) 4.8426 21628	(14) 4.8706 78457	(14) 4.8988 65178	(14) 4.9271 82242	(14) 4.9556 30101
6	(17) 4.1888 67708	(17) 4.2180 07544	(17) 4.2473 16109	(17) 4.2767 94186	(17) 4.3064 42558
7	(20) 3.6233 70568	(20) 3.6527 94533	(20) 3.6824 23067	(20) 3.7122 57354	(20) 3.7422 98583
8	(23) 3.1342 15541	(23) 3.1633 20065	(23) 3.1926 60799	(23) 3.2222 39383	(23) 3.2520 57468
9	(26) 2.7110 96443	(26) 2.7394 35177	(26) 2.7680 36913	(26) 2.7969 03785	(26) 2.8260 37940
10	(29) 2.3450 98423	(29) 2.3723 50863	(29) 2.3998 88003	(29) 2.4277 12485	(29) 2.4558 26970
24	(70) 3.0788 36164	(70) 3.1654 05907	(70) 3.2543 05644	(70) 3.3455 95291	(70) 3.4393 36231
1/2	(1) 2.9410 88234	(1) 2.9427 87794	(1) 2.9444 86373	(1) 2.9461 83973	(1) 2.9478 80595
1/3	9.5280 79435	9.5317 49727	9.5354 17196	9.5390 81845	9.5427 43681
1/4	5.4231 80095	5.4247 46809	5.4263 12167	5.4278 76171	5.4294 38824
1/5	3.8672 58668	3.8681 52418	3.8690 45344	3.8699 37445	3.8708 28725

k					
1	870	871	872	873	874
2	7 56900	7 58641	7 60384	7 62129	7 63876
3	6585 03000	6607 76311	6630 54848	6653 38617	6676 27624
4	(11) 5.7289 76100	(11) 5.7553 61669	(11) 5.7818 38275	(11) 5.8084 06126	(11) 5.8350 65434
5	(14) 4.9842 09207	(14) 5.0129 20014	(14) 5.0417 62975	(14) 5.0707 38548	(14) 5.0998 47189
6	(17) 4.3362 62010	(17) 4.3662 53332	(17) 4.3964 17315	(17) 4.4267 54753	(17) 4.4572 66443
7	(20) 3.7725 47949	(20) 3.8030 06652	(20) 3.8336 75898	(20) 3.8645 56899	(20) 3.8956 50871
8	(23) 3.2821 16715	(23) 3.3124 18794	(23) 3.3429 65383	(23) 3.3737 58173	(23) 3.4047 98862
9	(26) 2.8554 41542	(26) 2.8851 16769	(26) 2.9150 65814	(26) 2.9452 90885	(26) 2.9757 94205
10	(29) 2.4842 34142	(29) 2.5129 36706	(29) 2.5419 37390	(29) 2.5712 38943	(29) 2.6008 44135
24	(70) 3.5355 91351	(70) 3.6344 25075	(70) 3.7359 03403	(70) 3.8400 93943	(70) 3.9470 65953
1/2	(1) 2.9495 76241	(1) 2.9512 70913	(1) 2.9529 64612	(1) 2.9546 57341	(1) 2.9563 49100
1/3	9.5464 02709	9.5500 58934	9.5537 12362	9.5573 62998	9.5610 10846
1/4	5.4310 00130	5.4325 60090	5.4341 18707	5.4356 75984	5.4372 31924
1/5	3.8717 19185	3.8726 08827	3.8734 97651	3.8743 85661	3.8752 72857

$n^{\frac{1}{2}}\left[\begin{matrix}(-6)1\\3\end{matrix}\right]$ $n^{\frac{1}{3}}\left[\begin{matrix}(-7)4\\3\end{matrix}\right]$ $n^{\frac{1}{4}}\left[\begin{matrix}(-7)2\\3\end{matrix}\right]$ $n^{\frac{1}{5}}\left[\begin{matrix}(-7)1\\3\end{matrix}\right]$

POWERS AND ROOTS n^k

k					
1	875	876	877	878	879
2	7 65625	7 67376	7 69129	7 70884	7 72641
3	6699 21875	6722 21376	6745 26133	6768 36152	6791 51439
4	(11) 5.8618 16406	(11) 5.8886 59254	(11) 5.9155 94186	(11) 5.9426 21415	(11) 5.9697 41149
5	(14) 5.1290 89355	(14) 5.1584 65506	(14) 5.1879 76101	(14) 5.2176 21602	(14) 5.2474 02470
6	(17) 4.4879 53186	(17) 4.5188 15784	(17) 4.5498 55041	(17) 4.5810 71767	(17) 4.6124 66771
7	(20) 3.9269 59038	(20) 3.9584 82626	(20) 3.9902 22871	(20) 4.0221 81011	(20) 4.0543 58292
8	(23) 3.4360 89158	(23) 3.4676 30781	(23) 3.4994 25458	(23) 3.5314 74928	(23) 3.5637 80938
9	(26) 3.0065 78013	(26) 3.0376 44564	(26) 3.0689 96127	(26) 3.1006 34987	(26) 3.1325 63445
10	(29) 2.6307 55762	(29) 2.6609 76638	(29) 2.6915 09603	(29) 2.7223 57518	(29) 2.7535 23268
24	(70) 4.0568 90376	(70) 4.1696 39882	(70) 4.2853 88904	(70) 4.4042 13682	(70) 4.5261 92303
1/2	(1) 2.9580 39892	(1) 2.9597 29717	(1) 2.9614 18579	(1) 2.9631 06478	(1) 2.9647 93416
1/3	9.5646 55914	9.5682 98205	9.5719 37725	9.5755 74480	9.5792 08475
1/4	5.4387 86530	5.4403 39803	5.4418 91747	5.4434 42365	5.4449 91658
1/5	3.8761 59242	3.8770 44816	3.8779 29583	3.8788 13542	3.8796 96696

k					
1	880	881	882	883	884
2	7 74400	7 76161	7 77924	7 79689	7 81456
3	6814 72000	6837 97841	6861 28968	6884 65387	6908 07104
4	(11) 5.9969 53600	(11) 6.0242 58979	(11) 6.0516 57498	(11) 6.0791 49367	(11) 6.1067 34799
5	(14) 5.2773 19168	(14) 5.3073 72161	(14) 5.3375 61913	(14) 5.3678 88891	(14) 5.3983 53563
6	(17) 4.6440 40868	(17) 4.6757 94874	(17) 4.7077 29607	(17) 4.7398 45891	(17) 4.7721 44549
7	(20) 4.0867 55964	(20) 4.1193 75284	(20) 4.1522 17514	(20) 4.1852 83922	(20) 4.2185 75782
8	(23) 3.5963 45248	(23) 3.6291 69625	(23) 3.6622 55847	(23) 3.6956 05703	(23) 3.7292 20991
9	(26) 3.1647 83818	(26) 3.1972 98440	(26) 3.2301 09657	(26) 3.2632 19836	(26) 3.2966 31356
10	(29) 2.7850 09760	(29) 2.8168 19925	(29) 2.8489 56718	(29) 2.8814 23115	(29) 2.9142 22119
24	(70) 4.6514 04745	(70) 4.7799 32920	(70) 4.9118 60716	(70) 5.0472 74047	(70) 5.1862 60897
1/2	(1) 2.9664 79395	(1) 2.9681 64416	(1) 2.9698 48481	(1) 2.9715 31592	(1) 2.9732 13749
1/3	9.5828 39714	9.5864 68204	9.5900 93948	9.5937 16954	9.5973 37224
1/4	5.4465 39631	5.4480 86284	5.4496 31621	5.4511 75645	5.4527 18358
1/5	3.8805 79047	3.8814 60596	3.8823 41346	3.8832 21296	3.8841 00450

k					
1	885	886	887	888	889
2	7 83225	7 84996	7 86769	7 88544	7 90321
3	6931 54125	6955 06456	6978 64103	7002 27072	7025 95369
4	(11) 6.1344 14006	(11) 6.1621 87200	(11) 6.1900 54594	(11) 6.2180 16399	(11) 6.2460 72830
5	(14) 5.4289 56396	(14) 5.4596 97859	(14) 5.4905 78425	(14) 5.5215 98563	(14) 5.5527 58746
6	(17) 4.8046 26410	(17) 4.8372 92303	(17) 4.8701 43063	(17) 4.9031 79524	(17) 4.9364 02525
7	(20) 4.2520 94373	(20) 4.2858 40981	(20) 4.3198 16896	(20) 4.3540 23417	(20) 4.3884 61845
8	(23) 3.7631 03520	(23) 3.7972 55109	(23) 3.8316 77587	(23) 3.8663 72794	(23) 3.9013 42580
9	(26) 3.3303 46615	(26) 3.3643 68027	(26) 3.3986 98020	(26) 3.4333 39041	(26) 3.4682 93554
10	(29) 2.9473 56754	(29) 2.9808 30072	(29) 3.0146 45144	(29) 3.0488 05069	(29) 3.0833 12969
24	(70) 5.3289 11365	(70) 5.4753 17719	(70) 5.6255 74442	(70) 5.7797 78281	(70) 5.9380 28303
1/2	(1) 2.9748 94956	(1) 2.9765 75213	(1) 2.9782 54522	(1) 2.9799 32885	(1) 2.9816 10303
1/3	9.6009 54766	9.6045 69584	9.6081 81682	9.6117 91067	9.6153 97744
1/4	5.4542 59763	5.4557 99862	5.4573 38658	5.4588 76153	5.4604 12350
1/5	3.8849 78808	3.8858 56373	3.8867 33146	3.8876 09128	3.8884 84321

k					
1	890	891	892	893	894
2	7 92100	7 93881	7 95664	7 97449	7 99236
3	7049 69000	7073 47971	7097 32288	7121 21957	7145 16984
4	(11) 6.2742 24100	(11) 6.3024 70422	(11) 6.3308 12009	(11) 6.3592 49076	(11) 6.3877 81837
5	(14) 5.5840 59449	(14) 5.6155 01146	(14) 5.6470 84312	(14) 5.6788 09425	(14) 5.7106 76962
6	(17) 4.9698 12910	(17) 5.0034 11521	(17) 5.0371 99206	(17) 5.0711 76816	(17) 5.1053 45204
7	(20) 4.4231 33490	(20) 4.4580 39665	(20) 4.4931 81692	(20) 4.5285 60897	(20) 4.5641 78613
8	(23) 3.9365 88806	(23) 3.9721 13342	(23) 4.0079 18069	(23) 4.0440 04881	(23) 4.0803 75680
9	(26) 3.5035 64037	(26) 3.5391 52987	(26) 3.5750 62918	(26) 3.6112 96359	(26) 3.6478 55858
10	(29) 3.1181 71993	(29) 3.1533 85312	(29) 3.1889 56123	(29) 3.2248 87648	(29) 3.2611 83137
24	(70) 6.1004 25945	(70) 6.2670 75070	(70) 6.4380 82017	(70) 6.6135 55666	(70) 6.7936 07487
1/2	(1) 2.9832 86778	(1) 2.9849 62311	(1) 2.9866 36905	(1) 2.9883 10559	(1) 2.9899 83278
1/3	9.6190 01716	9.6226 02990	9.6262 01570	9.6297 97462	9.6333 90671
1/4	5.4619 47252	5.4634 80860	5.4650 13179	5.4665 44210	5.4680 73955
1/5	3.8893 58728	3.8902 32348	3.8911 05185	3.8919 77239	3.8928 48512

k					
1	895	896	897	898	899
2	8 01025	8 02816	8 04609	8 06404	8 08201
3	7169 17375	7193 23136	7217 34273	7241 50792	7265 72699
4	(11) 6.4164 10506	(11) 6.4451 35299	(11) 6.4739 56429	(11) 6.5028 74112	(11) 6.5318 88564
5	(14) 5.7426 87403	(14) 5.7748 41228	(14) 5.8071 38917	(14) 5.8395 80953	(14) 5.8721 67819
6	(17) 5.1397 05226	(17) 5.1742 57740	(17) 5.2090 03608	(17) 5.2439 43696	(17) 5.2790 78869
7	(20) 4.6000 36177	(20) 4.6361 34935	(20) 4.6724 76237	(20) 4.7090 61439	(20) 4.7458 91904
8	(23) 4.1170 32378	(23) 4.1539 76902	(23) 4.1912 11184	(23) 4.2287 37172	(23) 4.2665 56821
9	(26) 3.6847 43979	(26) 3.7219 63304	(26) 3.7595 16432	(26) 3.7974 05980	(26) 3.8356 34582
10	(29) 3.2978 45861	(29) 3.3348 79120	(29) 3.3722 86240	(29) 3.4100 70570	(29) 3.4482 35490
24	(70) 6.9783 51604	(70) 7.1679 04854	(70) 7.3623 86846	(70) 7.5619 20026	(70) 7.7666 29743
1/2	(1) 2.9916 55060	(1) 2.9933 25909	(1) 2.9949 95826	(1) 2.9966 64813	(1) 2.9983 32870
1/3	9.6369 81200	9.6405 69057	9.6441 54244	9.6477 36769	9.6513 16634
1/4	5.4696 02417	5.4711 29599	5.4726 55504	5.4741 80133	5.4757 03489
1/5	3.8937 19006	3.8945 88722	3.8954 57662	3.8963 25828	3.8971 93220

$n^{\frac{1}{2}}\left[{(-6)1 \atop 3}\right]$ $n^{\frac{1}{3}}\left[{(-7)3 \atop 3}\right]$ $n^{\frac{1}{4}}\left[{(-7)2 \atop 3}\right]$ $n^{\frac{1}{5}}\left[{(-7)1 \atop 3}\right]$

k	900	901	902	903	904
1	900	901	902	903	904
2	8 10000	8 11801	8 13604	8 15409	8 17216
3	7290 00000	7314 32701	7338 70808	7363 14327	7387 63264
4	(11) 6.5610 00000	(11) 6.5902 08636	(11) 6.6195 14688	(11) 6.6489 18373	(11) 6.6784 19907
5	(14) 5.9049 00000	(14) 5.9377 77981	(14) 5.9708 02249	(14) 6.0039 73291	(14) 6.0372 91596
6	(17) 5.3144 10000	(17) 5.3499 37961	(17) 5.3856 63628	(17) 5.4215 87881	(17) 5.4577 11602
7	(20) 4.7829 69000	(20) 4.8202 94103	(20) 4.8578 68593	(20) 4.8956 93857	(20) 4.9337 71289
8	(23) 4.3046 72100	(23) 4.3430 84987	(23) 4.3817 97471	(23) 4.4208 11553	(23) 4.4601 29245
9	(26) 3.8742 04890	(26) 3.9131 19573	(26) 3.9523 81319	(26) 3.9919 92832	(26) 4.0319 56837
10	(29) 3.4867 84401	(29) 3.5257 20735	(29) 3.5650 47949	(29) 3.6047 69527	(29) 3.6448 88981
24	(70) 7.9766 44308	(70) 8.1920 95066	(70) 8.4131 16465	(70) 8.6398 46120	(70) 8.8724 24888
1/2	(1) 3.0000 00000	(1) 3.0016 66204	(1) 3.0033 31484	(1) 3.0049 95840	(1) 3.0066 59276
1/3	9.6548 93846	9.6584 68409	9.6620 40328	9.6656 09608	9.6691 76254
1/4	5.4772 25575	5.4787 46393	5.4802 65946	5.4817 84235	5.4833 01264
1/5	3.8980 59841	3.8989 25692	3.8997 90774	3.9006 55089	3.9015 18640

k	905	906	907	908	909
1	905	906	907	908	909
2	8 19025	8 20836	8 22649	8 24464	8 26281
3	7412 17625	7436 77416	7461 42643	7486 13312	7510 89429
4	(11) 6.7080 19506	(11) 6.7377 17389	(11) 6.7675 13772	(11) 6.7974 08873	(11) 6.8274 02910
5	(14) 6.0707 57653	(14) 6.1043 71954	(14) 6.1381 34991	(14) 6.1720 47257	(14) 6.2061 09245
6	(17) 5.4940 35676	(17) 5.5305 60991	(17) 5.5672 88437	(17) 5.6042 18909	(17) 5.6413 53304
7	(20) 4.9721 02287	(20) 5.0106 88258	(20) 5.0495 30612	(20) 5.0886 30769	(20) 5.1279 90153
8	(23) 4.4997 52570	(23) 4.5396 83561	(23) 4.5799 24265	(23) 4.6204 76739	(23) 4.6613 43049
9	(26) 4.0722 76076	(26) 4.1129 53307	(26) 4.1539 91309	(26) 4.1953 92879	(26) 4.2371 60832
10	(29) 3.6854 09848	(29) 3.7263 35696	(29) 3.7676 70117	(29) 3.8094 16734	(29) 3.8515 79196
24	(70) 9.1109 96943	(70) 9.3557 09844	(70) 9.6067 14616	(70) 9.8641 65825	(71) 1.0128 22166
1/2	(1) 3.0083 21791	(1) 3.0099 83389	(1) 3.0116 44069	(1) 3.0133 03835	(1) 3.0149 62686
1/3	9.6727 40271	9.6763 01663	9.6798 60436	9.6834 16593	9.6869 70141
1/4	5.4848 17035	5.4863 31551	5.4878 44813	5.4893 56824	5.4908 67587
1/5	3.9023 81426	3.9032 43449	3.9041 04712	3.9049 65216	3.9058 24962

k	910	911	912	913	914
1	910	911	912	913	914
2	8 28100	8 29921	8 31744	8 33569	8 35396
3	7535 71000	7560 58031	7585 50528	7610 48497	7635 51944
4	(11) 6.8574 96100	(11) 6.8876 88662	(11) 6.9179 80815	(11) 6.9483 72778	(11) 6.9788 64768
5	(14) 6.2403 21451	(14) 6.2746 84371	(14) 6.3091 98504	(14) 6.3438 64346	(14) 6.3786 82398
6	(17) 5.6786 92520	(17) 5.7162 37462	(17) 5.7539 89035	(17) 5.7919 48148	(17) 5.8301 15712
7	(20) 5.1676 10194	(20) 5.2074 92328	(20) 5.2476 38000	(20) 5.2880 48659	(20) 5.3287 25761
8	(23) 4.7025 25276	(23) 4.7440 25511	(23) 4.7858 45856	(23) 4.8279 88426	(23) 4.8704 55345
9	(26) 4.2792 98001	(26) 4.3218 07241	(26) 4.3646 91421	(26) 4.4079 53433	(26) 4.4515 96186
10	(29) 3.8941 61181	(29) 3.9371 66396	(29) 3.9805 98576	(29) 4.0244 61484	(29) 4.0687 58914
24	(71) 1.0399 04400	(71) 1.0676 79852	(71) 1.0961 65476	(71) 1.1253 78622	(71) 1.1553 37042
1/2	(1) 3.0166 20626	(1) 3.0182 77655	(1) 3.0199 33774	(1) 3.0215 88986	(1) 3.0232 43292
1/3	9.6905 21083	9.6940 69425	9.6976 15172	9.7011 58327	9.7046 98896
1/4	5.4923 77104	5.4938 85378	5.4953 92410	5.4968 98203	5.4984 02760
1/5	3.9066 83951	3.9075 42186	3.9083 99668	3.9092 56397	3.9101 12376

k	915	916	917	918	919
1	915	916	917	918	919
2	8 37225	8 39056	8 40889	8 42724	8 44561
3	7660 60875	7685 75296	7710 95213	7736 20632	7761 51559
4	(11) 7.0094 57006	(11) 7.0401 49711	(11) 7.0709 43103	(11) 7.1018 37402	(11) 7.1328 32827
5	(14) 6.4136 53161	(14) 6.4487 77136	(14) 6.4840 54826	(14) 6.5194 86735	(14) 6.5550 73368
6	(17) 5.8684 92642	(17) 5.9070 79856	(17) 5.9458 78275	(17) 5.9848 88823	(17) 6.0241 12425
7	(20) 5.3696 70767	(20) 5.4108 85148	(20) 5.4523 70378	(20) 5.4941 27939	(20) 5.5361 59319
8	(23) 4.9132 48752	(23) 4.9563 70796	(23) 4.9998 23637	(23) 5.0436 09448	(23) 5.0877 30414
9	(26) 4.4956 22608	(26) 4.5400 35649	(26) 4.5848 38275	(26) 4.6300 33473	(26) 4.6756 24251
10	(29) 4.1134 94687	(29) 4.1586 72654	(29) 4.2042 96698	(29) 4.2503 70729	(29) 4.2968 98686
24	(71) 1.1860 58902	(71) 1.2175 62793	(71) 1.2498 67732	(71) 1.2829 93183	(71) 1.3169 59057
1/2	(1) 3.0248 96692	(1) 3.0265 49190	(1) 3.0282 00786	(1) 3.0298 51482	(1) 3.0315 01278
1/3	9.7082 36884	9.7117 72294	9.7153 05133	9.7188 35404	9.7223 63112
1/4	5.4999 06083	5.5014 08174	5.5029 09036	5.5044 08671	5.5059 07081
1/5	3.9109 67606	3.9118 22089	3.9126 75826	3.9135 28819	3.9143 81068

k	920	921	922	923	924
1	920	921	922	923	924
2	8 46400	8 48241	8 50084	8 51929	8 53776
3	7786 88000	7812 29961	7837 77448	7863 30467	7888 89024
4	(11) 7.1639 29600	(11) 7.1951 27941	(11) 7.2264 28071	(11) 7.2578 30210	(11) 7.2893 34582
5	(14) 6.5908 15232	(14) 6.6267 12833	(14) 6.6627 66681	(14) 6.6989 77284	(14) 6.7353 45154
6	(17) 6.0635 50013	(17) 6.1032 02520	(17) 6.1430 70880	(17) 6.1831 56033	(17) 6.2234 58922
7	(20) 5.5784 66012	(20) 5.6210 49521	(20) 5.6639 11351	(20) 5.7070 53019	(20) 5.7504 76044
8	(23) 5.1321 88731	(23) 5.1769 86608	(23) 5.2221 26266	(23) 5.2676 09936	(23) 5.3134 39864
9	(26) 4.7216 13633	(26) 4.7680 04666	(26) 4.8148 00417	(26) 4.8620 03971	(26) 4.9096 18435
10	(29) 4.3438 84542	(29) 4.3913 32298	(29) 4.4392 45985	(29) 4.4876 29665	(29) 4.5364 87434
24	(71) 1.3517 85726	(71) 1.3874 94035	(71) 1.4241 05308	(71) 1.4616 41363	(71) 1.5001 24518
1/2	(1) 3.0331 50178	(1) 3.0347 98181	(1) 3.0364 45290	(1) 3.0380 91506	(1) 3.0397 36831
1/3	9.7258 88262	9.7294 10859	9.7329 30906	9.7364 48410	9.7399 63373
1/4	5.5074 04268	5.5089 00236	5.5103 94986	5.5118 88520	5.5133 80842
1/5	3.9152 32576	3.9160 83344	3.9169 33373	3.9177 82664	3.9186 31220

$n^{\frac{1}{2}}\left[\begin{matrix}(-6)1\\3\end{matrix}\right]$ $n^{\frac{1}{3}}\left[\begin{matrix}(-7)3\\3\end{matrix}\right]$ $n^{\frac{1}{4}}\left[\begin{matrix}(-7)2\\3\end{matrix}\right]$ $n^{\frac{1}{5}}\left[\begin{matrix}(-7)1\\3\end{matrix}\right]$

k					
1	925	926	927	928	929
2	8 55625	8 57476	8 59329	8 61184	8 63041
3	7914 53125	7940 22776	7965 97983	7991 78752	8017 65089
4	(11) 7.3209 41406	(11) 7.3526 50906	(11) 7.3844 63302	(11) 7.4163 78819	(11) 7.4483 97677
5	(14) 6.7718 70801	(14) 6.8085 54739	(14) 6.8453 97481	(14) 6.8823 99544	(14) 6.9195 61442
6	(17) 6.2639 80491	(17) 6.3047 21688	(17) 6.3456 83465	(17) 6.3868 66776	(17) 6.4282 72579
7	(20) 5.7941 81954	(20) 5.8381 72283	(20) 5.8824 48572	(20) 5.9270 12369	(20) 5.9718 65226
8	(23) 5.3596 18307	(23) 5.4061 47534	(23) 5.4530 29826	(23) 5.5002 67478	(23) 5.5478 62795
9	(26) 4.9576 46934	(26) 5.0060 92617	(26) 5.0549 58649	(26) 5.1042 48220	(26) 5.1539 64537
10	(29) 4.5858 23414	(29) 4.6356 41763	(29) 4.6859 46668	(29) 4.7367 42348	(29) 4.7880 33055
24	(71) 1.5395 77607	(71) 1.5800 23988	(71) 1.6214 87554	(71) 1.6639 92748	(71) 1.7075 64573
1/2	(1) 3.0413 81265	(1) 3.0430 24811	(1) 3.0446 67470	(1) 3.0463 09242	(1) 3.0479 50131
1/3	9.7434 75802	9.7469 85700	9.7504 93072	9.7539 97922	9.7575 00256
1/4	5.5148 71952	5.5163 61854	5.5178 50550	5.5193 38042	5.5208 24332
1/5	3.9194 79042	3.9203 26131	3.9211 72488	3.9220 18115	3.9228 63013

k					
1	930	931	932	933	934
2	8 64900	8 66761	8 68624	8 70489	8 72356
3	8043 57000	8069 54491	8095 57568	8121 66237	8147 80504
4	(11) 7.4805 20100	(11) 7.5127 46311	(11) 7.5450 76534	(11) 7.5775 10991	(11) 7.6100 49907
5	(14) 6.9568 83693	(14) 6.9943 66816	(14) 7.0320 11329	(14) 7.0698 17755	(14) 7.1077 86613
6	(17) 6.4699 01834	(17) 6.5117 55505	(17) 6.5538 34559	(17) 6.5961 39965	(17) 6.6386 72697
7	(20) 6.0170 08706	(20) 6.0624 44376	(20) 6.1081 73809	(20) 6.1541 98588	(20) 6.2005 20299
8	(23) 5.5958 18097	(23) 5.6441 35714	(23) 5.6928 17990	(23) 5.7418 67282	(23) 5.7912 85959
9	(26) 5.2041 10830	(26) 5.2546 90349	(26) 5.3057 06367	(26) 5.3571 62174	(26) 5.4090 61086
10	(29) 4.8398 23072	(29) 4.8921 16715	(29) 4.9449 18334	(29) 4.9982 32309	(29) 5.0520 63054
24	(71) 1.7522 28603	(71) 1.7980 10997	(71) 1.8449 38512	(71) 1.8930 38514	(71) 1.9423 38996
1/2	(1) 3.0495 90136	(1) 3.0512 29260	(1) 3.0528 67504	(1) 3.0545 04870	(1) 3.0561 41358
1/3	9.7610 00077	9.7644 97390	9.7679 92199	9.7714 84510	9.7749 74326
1/4	5.5223 09423	5.5237 93317	5.5252 76015	5.5267 57521	5.5282 37837
1/5	3.9237 07185	3.9245 50630	3.9253 93351	3.9262 35348	3.9270 76625

k					
1	935	936	937	938	939
2	8 74225	8 76096	8 77969	8 79844	8 81721
3	8174 00375	8200 25856	8226 56953	8252 93672	8279 36019
4	(11) 7.6426 93506	(11) 7.6754 42012	(11) 7.7082 95650	(11) 7.7412 54643	(11) 7.7743 19218
5	(14) 7.1459 18428	(14) 7.1842 13723	(14) 7.2226 73024	(14) 7.2612 96855	(14) 7.3000 85746
6	(17) 6.6814 33731	(17) 6.7244 24045	(17) 6.7676 44623	(17) 6.8110 96450	(17) 6.8547 80516
7	(20) 6.2471 40538	(20) 6.2940 60906	(20) 6.3412 83012	(20) 6.3888 08471	(20) 6.4366 38904
8	(23) 5.8410 76403	(23) 5.8912 41008	(23) 5.9417 82182	(23) 5.9927 02345	(23) 6.0440 03931
9	(26) 5.4614 06437	(26) 5.5142 01584	(26) 5.5674 49905	(26) 5.6211 54800	(26) 5.6753 19691
10	(29) 5.1064 15018	(29) 5.1612 92682	(29) 5.2167 00561	(29) 5.2726 43202	(29) 5.3291 25190
24	(71) 1.9928 68584	(71) 2.0446 56558	(71) 2.0977 32860	(71) 2.1521 28115	(71) 2.2078 73640
1/2	(1) 3.0577 76970	(1) 3.0594 11708	(1) 3.0610 45573	(1) 3.0626 78566	(1) 3.0643 10689
1/3	9.7784 61652	9.7819 46493	9.7854 28852	9.7889 08735	9.7923 86145
1/4	5.5297 16964	5.5311 94905	5.5326 71663	5.5341 47239	5.5356 21636
1/5	3.9279 17180	3.9287 57017	3.9295 96137	3.9304 34540	3.9312 72229

k					
1	940	941	942	943	944
2	8 83600	8 85481	8 87364	8 89249	8 91136
3	8305 84000	8332 37621	8358 96888	8385 61807	8412 32384
4	(11) 7.8074 89600	(11) 7.8407 66014	(11) 7.8741 48685	(11) 7.9076 37840	(11) 7.9412 33705
5	(14) 7.3390 40224	(14) 7.3781 60819	(14) 7.4174 48061	(14) 7.4569 02483	(14) 7.4965 24617
6	(17) 6.8986 97811	(17) 6.9428 49330	(17) 6.9872 36074	(17) 7.0318 59042	(17) 7.0767 19239
7	(20) 6.4847 75942	(20) 6.5332 21220	(20) 6.5819 76381	(20) 6.6310 43076	(20) 6.6804 22962
8	(23) 6.0956 89385	(23) 6.1477 61168	(23) 6.2002 21751	(23) 6.2530 73621	(23) 6.3063 19276
9	(26) 5.7299 48022	(26) 5.7850 43259	(26) 5.8406 08890	(26) 5.8966 48424	(26) 5.9531 65396
10	(29) 5.3861 51141	(29) 5.4437 25707	(29) 5.5018 53574	(29) 5.5605 39464	(29) 5.6197 88134
24	(71) 2.2650 01461	(71) 2.3235 44328	(71) 2.3835 35733	(71) 2.4450 09921	(71) 2.5080 01911
1/2	(1) 3.0659 41943	(1) 3.0675 72330	(1) 3.0692 01851	(1) 3.0708 30507	(1) 3.0724 58299
1/3	9.7958 61087	9.7993 33566	9.8028 03585	9.8062 71149	9.8097 36263
1/4	5.5370 94855	5.5385 66899	5.5400 37771	5.5415 07472	5.5429 76005
1/5	3.9321 09204	3.9329 45467	3.9337 81020	3.9346 15863	3.9354 49998

k					
1	945	946	947	948	949
2	8 93025	8 94916	8 96809	8 98704	9 00601
3	8439 08625	8465 90536	8492 78123	8519 71392	8546 70349
4	(11) 7.9749 36506	(11) 8.0087 46471	(11) 8.0426 63825	(11) 8.0766 88796	(11) 8.1108 21612
5	(14) 7.5363 14998	(14) 7.5762 74161	(14) 7.6164 02642	(14) 7.6567 00979	(14) 7.6971 69710
6	(17) 7.1218 17673	(17) 7.1671 55356	(17) 7.2127 33302	(17) 7.2585 52528	(17) 7.3046 14055
7	(20) 6.7301 17701	(20) 6.7801 28967	(20) 6.8304 58437	(20) 6.8811 07796	(20) 6.9320 78738
8	(23) 6.3599 61228	(23) 6.4140 02003	(23) 6.4684 44140	(23) 6.5232 90191	(23) 6.5785 42722
9	(26) 6.0101 63360	(26) 6.0676 45895	(26) 6.1256 16600	(26) 6.1840 79101	(26) 6.2430 37043
10	(29) 5.6796 04376	(29) 5.7399 93016	(29) 5.8009 58921	(29) 5.8625 06988	(29) 5.9246 42154
24	(71) 2.5725 47511	(71) 2.6386 83331	(71) 2.7064 46809	(71) 2.7758 76218	(71) 2.8470 10693
1/2	(1) 3.0740 85230	(1) 3.0757 11300	(1) 3.0773 36511	(1) 3.0789 60864	(1) 3.0805 84360
1/3	9.8131 98931	9.8166 59156	9.8201 16944	9.8235 72299	9.8270 25224
1/4	5.5444 43371	5.5459 09574	5.5473 74614	5.5488 38494	5.5503 01217
1/5	3.9362 83427	3.9371 16151	3.9379 48170	3.9387 79487	3.9396 10103

$n^{\frac{1}{2}}\left[\begin{matrix}(-6)1\\3\end{matrix}\right]$ $n^{\frac{1}{3}}\left[\begin{matrix}(-7)3\\3\end{matrix}\right]$ $n^{\frac{1}{4}}\left[\begin{matrix}(-7)2\\3\end{matrix}\right]$ $n^{\frac{1}{5}}\left[\begin{matrix}(-8)9\\3\end{matrix}\right]$

k	950	951	952	953	954
1	950	951	952	953	954
2	9 02500	9 04401	9 06304	9 08209	9 10116
3	8573 75000	8600 85351	8628 01408	8655 23177	8682 50664
4	(11) 8.1450 62500	(11) 8.1794 11688	(11) 8.2138 69404	(11) 8.2484 35877	(11) 8.2831 11335
5	(14) 7.7378 09375	(14) 7.7786 20515	(14) 7.8196 03673	(14) 7.8607 59391	(14) 7.9020 88213
6	(17) 7.3509 18906	(17) 7.3974 68110	(17) 7.4442 62696	(17) 7.4913 03699	(17) 7.5385 92155
7	(20) 6.9833 72961	(20) 7.0349 92173	(20) 7.0869 38087	(20) 7.1392 12425	(20) 7.1918 16916
8	(23) 6.6342 04313	(23) 6.6902 77556	(23) 6.7467 65059	(23) 6.8036 69441	(23) 6.8609 93338
9	(26) 6.3024 94097	(26) 6.3624 53956	(26) 6.4229 20336	(26) 6.4838 96978	(26) 6.5453 87645
10	(29) 5.9873 69392	(29) 6.0506 93712	(29) 6.1146 20160	(29) 6.1791 53820	(29) 6.2442 99813
24	(71) 2.9198 90243	(71) 2.9945 55775	(71) 3.0710 49109	(71) 3.1494 12996	(71) 3.2296 91146
1/2	(1) 3.0822 07001	(1) 3.0838 28789	(1) 3.0854 49724	(1) 3.0870 69808	(1) 3.0886 89042
1/3	9.8304 75725	9.8339 23805	9.8373 69469	9.8408 12721	9.8442 53565
1/4	5.5517 62784	5.5532 23198	5.5546 82461	5.5561 40574	5.5575 97541
1/5	3.9404 40019	3.9412 69236	3.9420 97756	3.9429 25580	3.9437 52709

k	955	956	957	958	959
1	955	956	957	958	959
2	9 12025	9 13936	9 15849	9 17764	9 19681
3	8709 83875	8737 22816	8764 67493	8792 17912	8819 74079
4	(11) 8.3178 96006	(11) 8.3527 90121	(11) 8.3877 93908	(11) 8.4229 07597	(11) 8.4581 31418
5	(14) 7.9435 90686	(14) 7.9852 67356	(14) 8.0271 18770	(14) 8.0691 45478	(14) 8.1113 48029
6	(17) 7.5861 29105	(17) 7.6339 15592	(17) 7.6819 52663	(17) 7.7302 41368	(17) 7.7787 82760
7	(20) 7.2447 53295	(20) 7.2980 23306	(20) 7.3516 28698	(20) 7.4055 71230	(20) 7.4598 52667
8	(23) 6.9187 39397	(23) 6.9769 10280	(23) 7.0355 08664	(23) 7.0945 37239	(23) 7.1539 98708
9	(26) 6.6073 96124	(26) 6.6699 26228	(26) 6.7329 81792	(26) 6.7965 66675	(26) 6.8606 84761
10	(29) 6.3100 63299	(29) 6.3764 49474	(29) 6.4434 63575	(29) 6.5111 10874	(29) 6.5793 96686
24	(71) 3.3119 28238	(71) 3.3961 69948	(71) 3.4824 62966	(71) 3.5708 55021	(71) 3.6613 94899
1/2	(1) 3.0903 07428	(1) 3.0919 24967	(1) 3.0935 41660	(1) 3.0951 57508	(1) 3.0967 72513
1/3	9.8476 92005	9.8511 28046	9.8545 61691	9.8579 92945	9.8614 21813
1/4	5.5590 53362	5.5605 08040	5.5619 61578	5.5634 13977	5.5648 65240
1/5	3.9445 79145	3.9454 04889	3.9462 29943	3.9470 54307	3.9478 77983

k	960	961	962	963	964
1	960	961	962	963	964
2	9 21600	9 23521	9 25444	9 27369	9 29296
3	8847 36000	8875 03681	8902 77128	8930 56347	8958 41344
4	(11) 8.4934 65600	(11) 8.5289 10374	(11) 8.5644 65971	(11) 8.6001 32622	(11) 8.6359 10556
5	(14) 8.1537 26976	(14) 8.1962 82870	(14) 8.2390 16264	(14) 8.2819 27715	(14) 8.3250 17776
6	(17) 7.8275 77897	(17) 7.8766 27838	(17) 7.9259 33646	(17) 7.9754 96389	(17) 8.0253 17136
7	(20) 7.5144 74781	(20) 7.5694 39352	(20) 7.6247 48168	(20) 7.6804 03023	(20) 7.7364 05719
8	(23) 7.2138 95790	(23) 7.2742 31217	(23) 7.3350 07737	(23) 7.3962 28111	(23) 7.4578 95113
9	(26) 6.9253 39958	(26) 6.9905 36200	(26) 7.0562 77443	(26) 7.1225 67671	(26) 7.1894 10889
10	(29) 6.6483 26360	(29) 6.7179 05288	(29) 6.7881 38901	(29) 6.8590 32667	(29) 6.9305 92097
24	(71) 3.7541 32467	(71) 3.8491 18699	(71) 3.9464 05693	(71) 4.0460 46699	(71) 4.1480 96142
1/2	(1) 3.0983 86677	(1) 3.1000 00000	(1) 3.1016 12484	(1) 3.1032 24130	(1) 3.1048 34939
1/3	9.8648 48297	9.8682 72403	9.8716 94135	9.8751 13495	9.8785 30490
1/4	5.5663 15367	5.5677 64363	5.5692 12228	5.5706 58964	5.5721 04575
1/5	3.9487 00972	3.9495 23275	3.9503 44894	3.9511 65831	3.9519 86085

k	965	966	967	968	969
1	965	966	967	968	969
2	9 31225	9 33156	9 35089	9 37024	9 38961
3	8986 32125	9014 28696	9042 31063	9070 39232	9098 53209
4	(11) 8.6718 00006	(11) 8.7078 01203	(11) 8.7439 14379	(11) 8.7801 39766	(11) 8.8164 77595
5	(14) 8.3682 87006	(14) 8.4117 35962	(14) 8.4553 65205	(14) 8.4991 75293	(14) 8.5431 66790
6	(17) 8.0753 96961	(17) 8.1257 36940	(17) 8.1763 38153	(17) 8.2272 01684	(17) 8.2783 28619
7	(20) 7.7927 58067	(20) 7.8494 61884	(20) 7.9065 18994	(20) 7.9639 31230	(20) 8.0217 00432
8	(23) 7.5200 11535	(23) 7.5825 80180	(23) 7.6456 03867	(23) 7.7090 85431	(23) 7.7730 27719
9	(26) 7.2568 11131	(26) 7.3247 72454	(26) 7.3932 98939	(26) 7.4623 94697	(26) 7.5320 63859
10	(29) 7.0028 22742	(29) 7.0757 30190	(29) 7.1493 20074	(29) 7.2235 98067	(29) 7.2985 69880
24	(71) 4.2526 09649	(71) 4.3596 44069	(71) 4.4692 57504	(71) 4.5815 09331	(71) 4.6964 60232
1/2	(1) 3.1064 44913	(1) 3.1080 54054	(1) 3.1096 62361	(1) 3.1112 69837	(1) 3.1128 76483
1/3	9.8819 45122	9.8853 57396	9.8887 67316	9.8921 74886	9.8955 80110
1/4	5.5735 49061	5.5749 92425	5.5764 34668	5.5778 75794	5.5793 15803
1/5	3.9528 05659	3.9536 24554	3.9544 42771	3.9552 60312	3.9560 77177

k	970	971	972	973	974
1	970	971	972	973	974
2	9 40900	9 42841	9 44784	9 46729	9 48676
3	9126 73000	9154 98611	9183 30048	9211 67317	9240 10424
4	(11) 8.8529 28100	(11) 8.8894 91513	(11) 8.9261 68067	(11) 8.9629 57994	(11) 8.9998 61530
5	(14) 8.5873 40257	(14) 8.6316 96259	(14) 8.6762 35361	(14) 8.7209 58129	(14) 8.7658 65130
6	(17) 8.3297 20049	(17) 8.3813 77067	(17) 8.4333 00771	(17) 8.4854 92259	(17) 8.5379 52637
7	(20) 8.0798 28448	(20) 8.1383 17132	(20) 8.1971 68349	(20) 8.2563 83968	(20) 8.3159 65868
8	(23) 7.8374 33594	(23) 7.9023 05936	(23) 7.9676 47635	(23) 8.0334 61601	(23) 8.0997 50755
9	(26) 7.6023 10587	(26) 7.6731 39063	(26) 7.7445 53501	(26) 7.8165 58138	(26) 7.8891 57236
10	(29) 7.3742 41269	(29) 7.4506 18031	(29) 7.5277 06003	(29) 7.6055 11068	(29) 7.6840 39148
24	(71) 4.8141 72219	(71) 4.9347 08664	(71) 5.0581 34323	(71) 5.1845 15371	(71) 5.3139 19427
1/2	(1) 3.1144 82300	(1) 3.1160 87290	(1) 3.1176 91454	(1) 3.1192 94792	(1) 3.1208 97307
1/3	9.8989 82992	9.9023 83537	9.9057 81747	9.9091 77627	9.9125 71181
1/4	5.5807 54698	5.5821 92482	5.5836 29155	5.5850 64719	5.5864 99178
1/5	3.9568 93368	3.9577 08886	3.9585 23732	3.9593 37908	3.9601 51415

$n^{\frac{1}{2}}\left[{(-6)1 \atop 3}\right]$ $n^{\frac{1}{3}}\left[{(-7)3 \atop 3}\right]$ $n^{\frac{1}{4}}\left[{(-7)2 \atop 3}\right]$ $n^{\frac{1}{5}}\left[{(-8)9 \atop 3}\right]$

k					
1	975	976	977	978	979
2	9 50625	9 52576	9 54529	9 56484	9 58441
3	9268 59375	9297 14176	9325 74833	9354 41352	9383 13739
4	(11)9.0368 78906	(11)9.0740 10358	(11)9.1112 56118	(11)9.1486 16423	(11)9.1860 91505
5	(14)8.8109 56934	(14)8.8562 34109	(14)8.9016 97228	(14)8.9473 46861	(14)8.9931 83583
6	(17)8.5906 83010	(17)8.6436 84491	(17)8.6969 58191	(17)8.7505 05230	(17)8.8043 26728
7	(20)8.3759 15935	(20)8.4362 36063	(20)8.4969 28153	(20)8.5579 94115	(20)8.6194 35867
8	(23)8.1665 18037	(23)8.2337 66397	(23)8.3014 98806	(23)8.3697 18245	(23)8.4384 27713
9	(26)7.9623 55086	(26)8.0361 56004	(26)8.1105 64333	(26)8.1855 84443	(26)8.2612 20731
10	(29)7.7632 96209	(29)7.8432 88260	(29)7.9240 21353	(29)8.0055 01586	(29)8.0877 35096
24	(71)5.4464 15584	(71)5.5820 74443	(71)5.7209 68141	(71)5.8631 70383	(71)6.0087 56477
1/2	(1)3.1224 98999	(1)3.1240 99870	(1)3.1256 99922	(1)3.1272 99154	(1)3.1288 97569
1/3	9.9159 62413	9.9193 51328	9.9227 37928	9.9261 22218	9.9295 04202
1/4	5.5879 32533	5.5893 64785	5.5907 95938	5.5922 25992	5.5936 54950
1/5	3.9609 64254	3.9617 76427	3.9625 87934	3.9633 98776	3.9642 08956

k					
1	980	981	982	983	984
2	9 60400	9 62361	9 64324	9 66289	9 68256
3	9411 92000	9440 76141	9469 66168	9498 62087	9527 63904
4	(11)9.2236 81600	(11)9.2613 86943	(11)9.2992 07770	(11)9.3371 44315	(11)9.3751 96815
5	(14)9.0392 07968	(14)9.0854 20591	(14)9.1318 22030	(14)9.1784 12862	(14)9.2251 93666
6	(17)8.8584 23809	(17)8.9127 97600	(17)8.9674 49233	(17)9.0223 79843	(17)9.0775 90568
7	(20)8.6812 55332	(20)8.7434 54446	(20)8.8060 35147	(20)8.8689 99386	(20)8.9323 49119
8	(23)8.5076 30226	(23)8.5773 28811	(23)8.6475 26515	(23)8.7182 26396	(23)8.7894 31533
9	(26)8.3374 77621	(26)8.4143 59564	(26)8.4918 71037	(26)8.5700 16548	(26)8.6488 00628
10	(29)8.1707 28069	(29)8.2544 86732	(29)8.3390 17359	(29)8.4243 26266	(29)8.5104 19818
24	(71)6.1578 03365	(71)6.3103 89657	(71)6.4665 95666	(71)6.6265 03443	(71)6.7901 96812
1/2	(1)3.1304 95168	(1)3.1320 91953	(1)3.1336 87923	(1)3.1352 83081	(1)3.1368 77428
1/3	9.9328 83884	9.9362 61267	9.9396 36356	9.9430 09155	9.9463 79667
1/4	5.5950 82813	5.5965 09584	5.5979 35265	5.5993 59857	5.6007 83363
1/5	3.9650 18474	3.9658 27331	3.9666 35529	3.9674 43069	3.9682 49952

k					
1	985	986	987	988	989
2	9 70225	9 72196	9 74169	9 76144	9 78121
3	9556 71625	9585 85256	9615 04803	9644 30272	9673 61669
4	(11)9.4133 65506	(11)9.4516 50624	(11)9.4900 52406	(11)9.5285 71087	(11)9.5672 06906
5	(14)9.2721 65024	(14)9.3193 27515	(14)9.3666 81724	(14)9.4142 28234	(14)9.4619 67630
6	(17)9.1330 82548	(17)9.1888 56930	(17)9.2449 14862	(17)9.3012 57495	(17)9.3578 85987
7	(20)8.9960 86310	(20)9.0602 12933	(20)9.1247 30969	(20)9.1896 42406	(20)9.2549 49241
8	(23)8.8611 45015	(23)8.9333 69952	(23)9.0061 09466	(23)9.0793 66697	(23)9.1531 44799
9	(26)8.7282 27840	(26)8.8083 02773	(26)8.8890 30043	(26)8.9704 14296	(26)9.0524 60206
10	(29)8.5973 04423	(29)8.6849 86534	(29)8.7734 72653	(29)8.8627 69325	(29)8.9528 83144
24	(71)6.9577 61406	(71)7.1292 84708	(71)7.3048 56083	(71)7.4845 66822	(71)7.6685 10178
1/2	(1)3.1384 70965	(1)3.1400 63694	(1)3.1416 55614	(1)3.1432 46729	(1)3.1448 37039
1/3	9.9497 47896	9.9531 13846	9.9564 77521	9.9598 38925	9.9631 98061
1/4	5.6022 05785	5.6036 27123	5.6050 47381	5.6064 66560	5.6078 84662
1/5	3.9690 56179	3.9698 61752	3.9706 66671	3.9714 70939	3.9722 74555

k					
1	990	991	992	993	994
2	9 80100	9 82081	9 84064	9 86049	9 88036
3	9702 99000	9732 42271	9761 91488	9791 46657	9821 07784
4	(11)9.6059 60100	(11)9.6448 30906	(11)9.6838 19561	(11)9.7229 26304	(11)9.7621 51373
5	(14)9.5099 00499	(14)9.5580 27427	(14)9.6063 49004	(14)9.6548 65820	(14)9.7035 78465
6	(17)9.4148 01494	(17)9.4720 05181	(17)9.5294 98212	(17)9.5872 81759	(17)9.6453 56994
7	(20)9.3206 53479	(20)9.3867 57134	(20)9.4532 62227	(20)9.5201 70787	(20)9.5874 84852
8	(23)9.2274 46944	(23)9.3022 76320	(23)9.3776 36129	(23)9.4535 29591	(23)9.5299 59943
9	(26)9.1351 72475	(26)9.2185 55833	(26)9.3026 15040	(26)9.3873 54884	(26)9.4727 80183
10	(29)9.0438 20750	(29)9.1355 88830	(29)9.2281 94120	(29)9.3216 43400	(29)9.4159 43502
24	(71)7.8567 81408	(71)8.0494 77813	(71)8.2466 98779	(71)8.4485 45822	(71)8.6551 22630
1/2	(1)3.1464 26545	(1)3.1480 15248	(1)3.1496 03150	(1)3.1511 90251	(1)3.1527 76554
1/3	9.9665 54934	9.9699 09547	9.9732 61904	9.9766 12009	9.9799 59866
1/4	5.6093 01690	5.6107 17644	5.6121 32527	5.6135 46340	5.6149 59086
1/5	3.9730 77521	3.9738 79839	3.9746 81509	3.9754 82534	3.9762 82913

k					
1	995	996	997	998	999
2	9 90025	9 92016	9 94009	9 96004	9 98001
3	9850 74875	9880 47936	9910 26973	9940 11992	9970 02999
4	(11)9.8014 95006	(11)9.8409 57443	(11)9.8805 38921	(11)9.9202 39680	(11)9.9600 59960
5	(14)9.7524 87531	(14)9.8015 93613	(14)9.8508 97304	(14)9.9003 99201	(14)9.9500 99900
6	(17)9.7037 25094	(17)9.7623 87238	(17)9.8213 44612	(17)9.8805 98402	(17)9.9401 49800
7	(20)9.6552 06468	(20)9.7233 37689	(20)9.7918 80578	(20)9.8608 37206	(20)9.9302 09650
8	(23)9.6069 30436	(23)9.6844 44339	(23)9.7625 04937	(23)9.8411 15531	(23)9.9202 79441
9	(26)9.5588 95784	(26)9.6457 06561	(26)9.7332 17422	(26)9.8214 33300	(26)9.9103 59161
10	(29)9.5111 01305	(29)9.6071 23735	(29)9.7040 17769	(29)9.8017 90434	(29)9.9004 48802
24	(71)8.8665 35105	(71)9.0828 91413	(71)9.3043 02025	(71)9.5308 79767	(71)9.7627 39866
1/2	(1)3.1543 62059	(1)3.1559 46768	(1)3.1575 30681	(1)3.1591 13800	(1)3.1606 96126
1/3	9.9833 05478	9.9866 48849	9.9899 89983	9.9933 28884	9.9966 65555
1/4	5.6163 70767	5.6177 81384	5.6191 90939	5.6205 99434	5.6220 06871
1/5	3.9770 82648	3.9778 81740	3.9786 80191	3.9794 78001	3.9802 75173

$n^{\frac{1}{2}}\begin{bmatrix}(-6)1\\3\end{bmatrix}$ $n^{\frac{1}{3}}\begin{bmatrix}(-7)3\\3\end{bmatrix}$ $n^{\frac{1}{4}}\begin{bmatrix}(-7)1\\3\end{bmatrix}$ $n^{\frac{1}{5}}\begin{bmatrix}(-8)8\\3\end{bmatrix}$

SUMS OF POWERS OF INTEGERS, $\sum_{k=1}^{n} k^m$

$(m = 1, 2, 3, 4)$; $1 \leq n \leq 40$

n	Σk	Σk^2	Σk^3	Σk^4
1	1	1	1	1
2	3	5	9	17
3	6	14	36	98
4	10	30	100	354
5	15	55	225	979
6	21	91	441	2275
7	28	140	784	4676
8	36	204	1296	8772
9	45	285	2025	15333
10	55	385	3025	25333
11	66	506	4356	39974
12	78	650	6084	60710
13	91	819	8281	89271
14	105	1015	11025	127687
15	120	1240	14400	178312
16	136	1496	18496	243848
17	153	1785	23409	327369
18	171	2109	29241	432345
19	190	2470	36100	562666
20	210	2870	44100	722666
21	231	3311	53361	917147
22	253	3795	64009	1151403
23	276	4324	76176	1431244
24	300	4900	90000	1763020
25	325	5525	105625	2153645
26	351	6201	123201	2610621
27	378	6930	142884	3142062
28	406	7714	164836	3756718
29	435	8555	189225	4463999
30	465	9455	216225	5273999
31	496	10416	246016	6197520
32	528	11440	278784	7246096
33	561	12529	314721	8432017
34	595	13685	354025	9768353
35	630	14910	396900	11268978
36	666	16206	443556	12948594
37	703	17575	494209	14822755
38	741	19019	549081	16907891
39	780	20540	608400	19221332
40	820	22140	672400	21781332

SUMS OF POWERS OF THE FIRST n INTEGERS

$$\sum_{k=1}^{n} k = 1 + 2 + 3 + \cdots + n = \frac{n(n+1)}{2}$$

$$\sum_{k=1}^{n} k^2 = 1^2 + 2^2 + 3^2 + \cdots + n^2 = \frac{n(n+1)(2n+1)}{6}$$

$$\sum_{k=1}^{n} k^3 = \frac{n^2(n+1)^2}{4}$$

$$\sum_{k=1}^{n} k^4 = \frac{n}{30}(n+1)(2n+1)(3n^2+3n-1).$$

$$\sum_{k=1}^{n} k^5 = \frac{n^2}{12}(n+1)^2(2n^2+2n-1).$$

$$\sum_{k=1}^{n} k^6 = \frac{n}{42}(n+1)(2n+1)(3n^4+6n^3-3n+1).$$

$$\sum_{k=1}^{n} k^7 = \frac{n^2}{24}(n+1)^2(3n^4+6n^3-n^2-4n+2).$$

$$\sum_{k=1}^{n} k^8 = \frac{n}{90}(n+1)(2n+1)(5n^6+15n^5+5n^4-15n^3-n^2+9n-3).$$

$$\sum_{k=1}^{n} k^9 = \frac{n^2}{20}(n+1)^2(2n^6+6n^5+n^4-8n^3+n^2+6n-3).$$

$$\sum_{k=1}^{n} k^{10} = \frac{n}{66}(n+1)(2n+1)(3n^8+12n^7+8n^6-18n^5 - 10n^4+24n^3+2n^2-15n+5).$$

Note that

$\sum_{k=1}^{n} k^p = 1^p + 2^p + 3^p + \cdots + n^p$ is a function of n which can be conveniently generated by use of the *following proposition*

If $$\sum_{k=1}^{n} k^p = a_1 n^{p+1} + a_2 n^p + a_3 n^{p-1} + \cdots + a_{p+1} n$$

then

$$\sum_{k=1}^{n} k^{p+1} = \frac{p+1}{p+2}a_1 n^{p+2} + \frac{p+1}{p+1}a_2 n^{p+1} + \frac{p+1}{p}a_3 n^p$$

$$+ \cdots + \frac{p+1}{2}a_{p+1}n^2 + \left[1 - (p+1)\sum_{k=1}^{p+1}\frac{a_k}{(p+3-k)}\right]n$$

Example Since $\sum_{k=1}^{n} k = \frac{1}{2}n^2 + \frac{1}{2}n$, then

$\sum_{k=1}^{n} k^2 = \frac{1}{3}n^3 + \frac{1}{2}n^2 + \frac{1}{6}n$ and from this result

$\sum_{k=1}^{n} k^3 = \frac{1}{4}n^4 + \frac{1}{2}n^3 + \frac{1}{4}n^2$ etc.

SUMS OF RECIPROCAL POWERS*

n	$\zeta(n) = \sum_{k=1}^{\infty} k^{-n}$				$\sum_{k=1}^{\infty} (-1)^{k-1} k^{-n}$			
1	∞				0.69314	71805	59945	30942
2	1.64493	40668	48226	43637	0.82246	70334	24113	21824
3	1.20205	69031	59594	28540	0.90154	26773	69695	71405
4	1.08232	32337	11138	19152	0.94703	28294	97245	91758
5	1.03692	77551	43369	92633	0.97211	97704	46909	30594
6	1,01734	30619	84449	13971	0.98555	10912	97435	10410
7	1.00834	92773	81922	82684	0.99259	38199	22830	28267
8	1.00407	73561	97944	33938	0.99623	30018	52647	89923
9	1.00200	83928	26082	21442	0.99809	42975	41605	33077
10	1.00099	45751	27818	08534	0.99903	95075	98271	56564
11	1.00049	41886	04119	46456	0.99951	71434	98060	75414
12	1.00024	60865	53308	04830	0.99975	76851	43858	19085
13	1.00012	27133	47578	48915	0.99987	85427	63265	11549
14	1.00006	12481	35058	70483	0.99993	91703	45979	71817
15	1.00003	05882	36307	02049	0.99996	95512	13099	23808
16	1.00001	52822	59408	65187	0.99998	47642	14906	10644
17	1.00000	76371	97637	89976	0.99999	23782	92041	01198
18	1.00000	38172	93264	99984	0.99999	61878	69610	11348
19	1.00000	19082	12716	55394	0.99999	80935	08171	67511
20	1.00000	09539	62033	87280	0.99999	90466	11581	52212
21	1.00000	04769	32986	78781	0.99999	95232	58215	54282
22	1.00000	02384	50502	72773	0.99999	97616	13230	82255
23	1.00000	01192	19925	96531	0.99999	98808	01318	43950
24	1.00000	00596	08189	05126	0.99999	99403	98892	39463
25	1.00000	00298	03503	51465	0.99999	99701	98856	96283
26	1.00000	00149	01554	82837	0.99999	99850	99231	99657
27	1.00000	00074	50711	78984	0.99999	99925	49550	48496
28	1.00000	00037	25334	02479	0.99999	99962	74753	40011
29	1.00000	00018	62659	72351	0.99999	99981	37369	41811
30	1.00000	00009	31327	43242	0.99999	99990	68682	28145
31	1.00000	00004	65662	90650	0.99999	99995	34340	33145
32	1.00000	00002	32831	18337	0.99999	99997	67169	89595
33	1.00000	00001	16415	50173	0.99999	99998	83584	85805
34	1.00000	00000	58207	72088	0.99999	99999	41792	39905
35	1.00000	00000	29103	85044	0.99999	99999	70896	18953
36	1.00000	00000	14551	92189	0.99999	99999	85448	09143
37	1.00000	00000	07275	95984	0.99999	99999	92724	04461
38	1.00000	00000	03637	97955	0.99999	99999	96362	02193
39	1.00000	00000	01818	98965	0.99999	99999	98181	01084
40	1.00000	00000	00909	49478	0.99999	99999	99090	50538
41	1.00000	00000	00454	74738	0.99999	99999	99545	25268
42	1.00000	00000	00227	37368	0.99999	99999	99772	62633

For $n > 42$, $\sum_{k=1}^{\infty} k^{-(n+1)} = \frac{1}{2}\left[1 + \sum_{k=1}^{\infty} k^{-n}\right]$, $\sum_{k=1}^{\infty} (-1)^{k-1} k^{-(n+1)} = \frac{1}{2}\left[1 + \sum_{k=1}^{\infty} (-1)^{k-1} k^{-n}\right]$

*Note: By definition Riemann's Zeta Function is $\zeta(p)$ = Zeta $(p) = 1 + \frac{1}{2^p} + \frac{1}{3^p} + \frac{1}{4^p} + \ldots$

SUMS OF RECIPROCAL POWERS

n	$\sum_{k=0}^{\infty} (2k+1)^{-n}$				$\sum_{k=0}^{\infty} (-1)^k (2k+1)^{-n}$			
1	∞				0.78539	81633	97448	310
2	1.23370	05501	36169	82735	0.91596	55941	77219	015
3	1.05179	97902	64644	99972	0.96894	61462	59369	380
4	1.01467	80316	04192	05455	0.98894	45517	41105	336
5	1.00452	37627	95139	61613	0.99615	78280	77088	064
6	1.00144	70766	40942	12191	0.99868	52222	18438	135
7	1.00047	15486	52376	55476	0.99955	45078	90539	909
8	1.00015	51790	25296	11930	0.99984	99902	46829	656
9	1.00005	13451	83843	77259	0.99994	96841	87220	090
10	1.00001	70413	63044	82549	0.99998	31640	26196	877
11	1.00000	56660	51090	10935	0.99999	43749	73823	699
12	1.00000	18858	48583	11958	0.99999	81223	50587	882
13	1.00000	06280	55421	80232	0.99999	93735	83771	841
14	1.00000	02092	40519	21150	0.99999	97910	87248	734
15	1.00000	00697	24703	12929	0.99999	99303	40842	624
16	1.00000	00232	37157	37916	0.99999	99767	75950	903
17	1.00000	00077	44839	45587	0.99999	99922	57782	104
18	1.00000	00025	81437	55666	0.99999	99974	19086	745
19	1.00000	00008	60444	11452	0.99999	99991	39660	745
20	1.00000	00002	86807	69746	0.99999	99997	13213	274
21	1.00000	00000	95601	16531	0.99999	99999	04403	029
22	1.00000	00000	31866	77514	0.99999	99999	68134	064
23	1.00000	00000	10622	20241	0.99999	99999	89377	965
24	1.00000	00000	03540	72294	0.99999	99999	96459	311
25	1.00000	00000	01180	23874	0.99999	99999	98819	768
26	1.00000	00000	00393	41247	0.99999	99999	99606	589
27	1.00000	00000	00131	13740	0.99999	99999	99868	863
28	1.00000	00000	00043	71245	0.99999	99999	99956	288
29	1.00000	00000	00014	57081	0.99999	99999	99985	429
30	1.00000	00000	00004	85694	0.99999	99999	99995	143
31	1.00000	00000	00001	61898	0.99999	99999	99998	381
32	1.00000	00000	00000	53966	0.99999	99999	99999	460
33	1.00000	00000	00000	17989	0.99999	99999	99999	820
34	1.00000	00000	00000	05996	0.99999	99999	99999	940
35	1.00000	00000	00000	01999	0.99999	99999	99999	980
36	1.00000	00000	00000	00666	0.99999	99999	99999	993
37	1.00000	00000	00000	00222	0.99999	99999	99999	998
38	1.00000	00000	00000	00074	0.99999	99999	99999	999
39	1.00000	00000	00000	00025				
40	1.00000	00000	00000	00008				
41	1.00000	00000	00000	00003				
42	1.00000	00000	00000	00001				

FACTORIAL n, $n! = 1 \cdot 2 \cdot 3 \cdot \ldots \cdot n$

n	$n!$	n	$n!$	n	$n!$
0	1 (by definition)	40	8.15915×10^{47}	80	7.15695×10^{118}
1	1	41	3.34525×10^{49}	81	5.79713×10^{120}
2	2	42	1.40501×10^{51}	82	4.75364×10^{122}
3	6	43	6.04153×10^{52}	83	3.94552×10^{124}
4	24	44	2.65827×10^{54}	84	3.31424×10^{126}
5	120	45	1.19622×10^{56}	85	2.81710×10^{128}
6	720	46	5.50262×10^{57}	86	2.42271×10^{130}
7	5040	47	2.58623×10^{59}	87	2.10776×10^{132}
8	40,320	48	1.24139×10^{61}	88	1.85483×10^{134}
9	362,880	49	6.08282×10^{62}	89	1.65080×10^{136}
10	3,628,800	50	3.04141×10^{64}	90	1.48572×10^{138}
11	39,916,800	51	1.55112×10^{66}	91	1.35200×10^{140}
12	479,001,600	52	8.06582×10^{67}	92	1.24384×10^{142}
13	6,227,020,800	53	4.27488×10^{69}	93	1.15677×10^{144}
14	87,178,291,200	54	2.30844×10^{71}	94	1.08737×10^{146}
15	1,307,674,368,000	55	1.26964×10^{73}	95	1.03300×10^{148}
16	20,922,789,888,000	56	7.10999×10^{74}	96	9.91678×10^{149}
17	355,687,428,096,000	57	4.05269×10^{76}	97	9.61928×10^{151}
18	6,402,373,705,728,000	58	2.35056×10^{78}	98	9.42689×10^{153}
19	121,645,100,408,832.000	59	1.38683×10^{80}	99	9.33262×10^{155}
20	2,432,902,008,176,640,000	60	8.32099×10^{81}	100	9.33262×10^{157}
21	51,090,942,171,709,440,000	61	5.07580×10^{83}		
22	1,124,000,727,777,607,680,000	62	3.14700×10^{85}		
23	25,852,016,738,884,976,640,000	63	1.98261×10^{87}		
24	620,448,401,733,239,439,360,000	64	1.26887×10^{89}		
25	15,511,210,043,330,985,984,000,000	65	8.24765×10^{90}		
26	403,291,461,126,605,635,584,000,000	66	5.44345×10^{92}		
27	10,888,869,450,418,352,160,768,000,000	67	3.64711×10^{94}		
28	304,888,344,611,713,860,501,504,000,000	68	2.48004×10^{96}		
29	8,841,761,993,739,701,954,543,616,000,000	69	1.71122×10^{98}		
30	265,252,859,812,191,058,636,308,480,000,000	70	1.19786×10^{100}		
31	8.22284×10^{33}	71	8.50479×10^{101}		
32	2.63131×10^{35}	72	6.12345×10^{103}		
33	8.68332×10^{36}	73	4.47012×10^{105}		
34	2.95233×10^{38}	74	3.30789×10^{107}		
35	1.03331×10^{40}	75	2.48091×10^{109}		
36	3.71993×10^{41}	76	1.88549×10^{111}		
37	1.37638×10^{43}	77	1.45183×10^{113}		
38	5.23023×10^{44}	78	1.13243×10^{115}		
39	2.03979×10^{46}	79	8.94618×10^{116}		

RECIPROCALS OF FACTORIALS AND THEIR COMMON LOGARITHMS

This table presents the reciprocals of the factorials and their logarithms for numbers from 1 to 100.

n	$1/n!$	$\log(1/n!)$	n	$1/n!$	$\log(1/n!)$
1	1.	.00000	51	$.64470 \times 10^{-66}$	$\overline{67}.80935$
2	0.5	$\overline{1}.69897$	52	$.12398 \times 10^{-67}$	$\overline{68}.09335$
3	.16667	$\overline{1}.22185$	53	$.23392 \times 10^{-69}$	$\overline{70}.36908$
*4	$.41667 \times 10^{-1}$	$\overline{2}.61979$	54	$.43319 \times 10^{-71}$	$\overline{72}.63668$
5	$.83333 \times 10^{-2}$	$\overline{3}.92082$	55	$.78762 \times 10^{-73}$	$\overline{74}.89632$
6	$.13889 \times 10^{-2}$	$\overline{3}.14267$	56	$.14065 \times 10^{-74}$	$\overline{75}.14813$
7	$.19841 \times 10^{-3}$	$\overline{4}.29757$	57	$.24675 \times 10^{-76}$	$\overline{77}.39226$
8	$.24802 \times 10^{-4}$	$\overline{5}.39448$	58	$.42543 \times 10^{-78}$	$\overline{79}.62883$
9	$.27557 \times 10^{-5}$	$\overline{6}.44024$	59	$.72107 \times 10^{-80}$	$\overline{81}.85798$
10	$.27557 \times 10^{-6}$	$\overline{7}.44024$	60	$.12018 \times 10^{-81}$	$\overline{82}.07983$
11	$.25052 \times 10^{-7}$	$\overline{8}.39884$	61	$.19701 \times 10^{-83}$	$\overline{84}.29450$
12	$.20877 \times 10^{-8}$	$\overline{9}.31966$	62	$.31776 \times 10^{-85}$	$\overline{86}.50210$
13	$.16059 \times 10^{-9}$	$\overline{10}.20572$	63	$.50439 \times 10^{-87}$	$\overline{88}.70276$
14	$.11471 \times 10^{-10}$	$\overline{11}.05959$	64	$.78810 \times 10^{-89}$	$\overline{90}.89658$
15	$.76472 \times 10^{-12}$	$\overline{13}.88350$	65	$.12125 \times 10^{-90}$	$\overline{91}.08367$
16	$.47795 \times 10^{-13}$	$\overline{14}.67938$	66	$.18371 \times 10^{-92}$	$\overline{93}.26413$
17	$.28115 \times 10^{-14}$	$\overline{15}.44893$	67	$.27419 \times 10^{-94}$	$\overline{95}.43805$
18	$.15619 \times 10^{-15}$	$\overline{16}.19366$	68	$.40322 \times 10^{-96}$	$\overline{97}.60554$
19	$.82206 \times 10^{-17}$	$\overline{18}.91491$	69	$.58438 \times 10^{-98}$	$\overline{99}.76669$
20	$.41103 \times 10^{-18}$	$\overline{19}.61388$	70	$.83482 \times 10^{-100}$	$\overline{101}.92159$
21	$.19573 \times 10^{-19}$	$\overline{20}.29166$	71	$.11758 \times 10^{-101}$	$\overline{102}.07034$
22	$.88968 \times 10^{-21}$	$\overline{22}.94923$	72	$.16331 \times 10^{-103}$	$\overline{104}.21300$
23	$.38682 \times 10^{-22}$	$\overline{23}.58751$	73	$.22371 \times 10^{-105}$	$\overline{106}.34968$
24	$.16117 \times 10^{-23}$	$\overline{24}.20729$	74	$.30231 \times 10^{-107}$	$\overline{108}.48045$
25	$.64470 \times 10^{-25}$	$\overline{26}.80935$	75	$.40308 \times 10^{-109}$	$\overline{110}.60539$
26	$.24796 \times 10^{-26}$	$\overline{27}.39438$	76	$.53036 \times 10^{-111}$	$\overline{112}.72457$
27	$.91837 \times 10^{-28}$	$\overline{29}.96302$	77	$.68879 \times 10^{-113}$	$\overline{114}.83808$
28	$.32799 \times 10^{-29}$	$\overline{30}.51586$	78	$.88306 \times 10^{-115}$	$\overline{116}.94599$
29	$.11310 \times 10^{-30}$	$\overline{31}.05346$	79	$.11178 \times 10^{-116}$	$\overline{117}.04836$
30	$.37700 \times 10^{-32}$	$\overline{33}.57634$	80	$.13972 \times 10^{-118}$	$\overline{119}.14527$
31	$.12161 \times 10^{-33}$	$\overline{34}.08498$	81	$.17250 \times 10^{-120}$	$\overline{121}.23679$
32	$.38004 \times 10^{-35}$	$\overline{36}.57983$	82	$.21036 \times 10^{-122}$	$\overline{123}.32297$
33	$.11516 \times 10^{-36}$	$\overline{37}.06131$	83	$.25345 \times 10^{-124}$	$\overline{125}.40390$
34	$.33872 \times 10^{-38}$	$\overline{39}.52984$	84	$.30173 \times 10^{-126}$	$\overline{127}.47962$
35	$.96776 \times 10^{-40}$	$\overline{41}.98577$	85	$.35497 \times 10^{-128}$	$\overline{129}.55020$
36	$.26882 \times 10^{-41}$	$\overline{42}.42946$	86	$.41276 \times 10^{-130}$	$\overline{131}.61570$
37	$.72655 \times 10^{-43}$	$\overline{44}.86126$	87	$.47444 \times 10^{-132}$	$\overline{133}.67618$
38	$.19120 \times 10^{-44}$	$\overline{45}.28148$	88	$.53913 \times 10^{-134}$	$\overline{135}.73170$
39	$.49025 \times 10^{-46}$	$\overline{47}.69041$	89	$.60577 \times 10^{-136}$	$\overline{137}.78231$
40	$.12256 \times 10^{-47}$	$\overline{48}.08835$	90	$.67308 \times 10^{-138}$	$\overline{139}.82806$
41	$.29893 \times 10^{-49}$	$\overline{50}.47557$	91	$.73964 \times 10^{-140}$	$\overline{141}.86902$
42	$.71174 \times 10^{-51}$	$\overline{52}.85232$	92	$.80396 \times 10^{-142}$	$\overline{143}.90524$
43	$.16552 \times 10^{-52}$	$\overline{53}.21885$	93	$.86447 \times 10^{-144}$	$\overline{145}.93675$
44	$.37618 \times 10^{-54}$	$\overline{55}.57540$	94	$.91965 \times 10^{-146}$	$\overline{147}.96362$
45	$.83597 \times 10^{-56}$	$\overline{57}.92219$	95	$.96806 \times 10^{-148}$	$\overline{149}.98590$
46	$.18173 \times 10^{-57}$	$\overline{58}.25943$	96	$.10084 \times 10^{-149}$	$\overline{150}.00363$
47	$.38666 \times 10^{-59}$	$\overline{60}.58733$	97	$.10396 \times 10^{-151}$	$\overline{152}.01686$
48	$.80555 \times 10^{-61}$	$\overline{62}.90609$	98	$.10608 \times 10^{-153}$	$\overline{154}.02563$
49	$.16440 \times 10^{-62}$	$\overline{63}.21590$	99	$.10715 \times 10^{-155}$	$\overline{156}.03000$
50	$.32879 \times 10^{-64}$	$\overline{65}.51693$	100	$.10715 \times 10^{-157}$	$\overline{158}.03000$

* For example $\log \frac{1}{4!} = \overline{2}.61979 = .61979 - 2 = 8.61979 - 10$.

NUMBER OF PERMUTATIONS $P(n,m)$

This table contains the number of permutations of n distinct things taken m at a time, given by

$$P(n,m) = \frac{n!}{(n-m)!} = n(n-1)\cdots(n-m+1)$$

n \ m	0	1	2	3	4	5	6	7	8	9	10
0	1										
1	1	1									
2	1	2	2								
3	1	3	6	6							
4	1	4	12	24	24						
5	1	5	20	60	120	120					
6	1	6	30	120	360	720	720				
7	1	7	42	210	840	2520	5040	5040			
8	1	8	56	336	1680	6720	20160	40320	40320		
9	1	9	72	504	3024	15120	60480	1 81440	3 62880	3 62880	
10	1	10	90	720	5040	30240	1 51200	6 04800	18 14400	36 28800	36 28800
11	1	11	110	990	7920	55440	3 32640	16 63200	66 52800	199 58400	399 16800
12	1	12	132	1320	11880	95040	6 65280	39 91680	199 58400	798 33600	2395 00800
13	1	13	156	1716	17160	1 54440	12 35520	86 48640	518 91840	2594 59200	10378 36800
14	1	14	182	2184	24024	2 40240	21 62160	172 97280	1210 80960	7264 85760	36324 28800
15	1	15	210	2730	32760	3 60360	36 03600	324 32400	2594 59200	18162 14400	1 08972 86400

n \ m	11	12	13	14	15
8					
9					
10					
11	399 16800				
12	4790 01600	4790 01600			
13	31135 10400	62270 20800	62270 20800		
14	1 45297 15200	4 35891 45600	8 71782 91200	8 71782 91200	
15	5 44864 32000	21 79457 28000	65 38371 84000	130 76743 68000	130 76743 68000

THE BINOMINAL FORMULA AND BINOMIAL COEFFICIENTS

FACTORIAL n

If $n = 1, 2, 3, \ldots$ *factorial n* or *n factorial* is defined as

$$n! = 1 \cdot 2 \cdot 3 \cdot \cdots \cdot n$$

We also define *zero factorial* as

$$0! = 1$$

BINOMIAL FORMULA FOR POSITIVE INTEGRAL n

If $n = 1, 2, 3, \ldots$ then

3.1 $$(x+y)^n = x^n + nx^{n-1}y + \frac{n(n-1)}{2!}x^{n-2}y^2 + \frac{n(n-1)(n-2)}{3!}x^{n-3}y^3 + \cdots + y^n$$

This is called the *binomial formula.* It can be extended to other values of n and then is an infinite series.

BINOMIAL COEFFICIENTS

The result 3.1 can also be written

$$(x+y)^n = x^n + \binom{n}{1}x^{n-1}y + \binom{n}{2}x^{n-2}y^2 + \binom{n}{3}x^{n-3}y^3 + \cdots + \binom{n}{n}y^n$$

where the coefficients, called *binomial coefficients*, are given by

$$\binom{n}{k} = \frac{n(n-1)(n-2)\cdots(n-k+1)}{k!} = \frac{n!}{k!\,(n-k)!} = \binom{n}{n-k}$$

PROPERTIES OF BINOMIAL COEFFICIENTS

$$\binom{n}{k} + \binom{n}{k+1} = \binom{n+1}{k+1}$$

This leads to *Pascal's triangle*.

$$\binom{n}{0} + \binom{n}{1} + \binom{n}{2} + \cdots + \binom{n}{n} = 2^n$$

$$\binom{n}{0} - \binom{n}{1} + \binom{n}{2} - \cdots (-1)^n \binom{n}{n} = 0$$

$$\binom{n}{n} + \binom{n+1}{n} + \binom{n+2}{n} + \cdots + \binom{n+m}{n} = \binom{n+m+1}{n+1}$$

$$\binom{n}{0} + \binom{n}{2} + \binom{n}{4} + \cdots = 2^{n-1}$$

$$\binom{n}{1} + \binom{n}{3} + \binom{n}{5} + \cdots = 2^{n-1}$$

$$\binom{n}{0}^2 + \binom{n}{1}^2 + \binom{n}{2}^2 + \cdots + \binom{n}{n}^2 = \binom{2n}{n}$$

$$\binom{m}{0}\binom{n}{p} + \binom{m}{1}\binom{n}{p-1} + \cdots + \binom{m}{p}\binom{n}{0} = \binom{m+n}{p}$$

$$(1)\binom{n}{1} + (2)\binom{n}{2} + (3)\binom{n}{3} + \cdots + (n)\binom{n}{n} = n2^{n-1}$$

$$(1)\binom{n}{1} - (2)\binom{n}{2} + (3)\binom{n}{3} - \cdots (-1)^{n+1}(n)\binom{n}{n} = 0$$

MULTINOMIAL FORMULA

$$(x_1 + x_2 + \cdots + x_p)^n = \sum \frac{n!}{n_1!\, n_2! \cdots n_p!} x_1^{n_1} x_2^{n_2} \cdots x_p^{n_p}$$

where the sum, denoted by Σ, is taken over all nonnegative integers $n_1, n_2, \ldots, n_p$ for which $n_1 + n_2 + \cdots + n_p = n$.

BINOMIAL COEFFICIENTS[1] $\binom{n}{m}$

n \ m	0	1	2	3	4	5	6	7	8
1	1	1							
2	1	2	1						
3	1	3	3	1					
4	1	4	6	4	1				
5	1	5	10	10	5	1			
6	1	6	15	20	15	6	1		
7	1	7	21	35	35	21	7	1	
8	1	8	28	56	70	56	28	8	1
9	1	9	36	84	126	126	84	36	9
10	1	10	45	120	210	252	210	120	45
11	1	11	55	165	330	462	462	330	165
12	1	12	66	220	495	792	924	792	495
13	1	13	78	286	715	1287	1716	1716	1287
14	1	14	91	364	1001	2002	3003	3432	3003
15	1	15	105	455	1365	3003	5005	6435	6435
16	1	16	120	560	1820	4368	8008	11440	12870
17	1	17	136	680	2380	6188	12376	19448	24310
18	1	18	153	816	3060	8568	18564	31824	43758
19	1	19	171	969	3876	11628	27132	50388	75582
20	1	20	190	1140	4845	15504	38760	77520	1 25970
21	1	21	210	1330	5985	20349	54264	1 16280	2 03490
22	1	22	231	1540	7315	26334	74613	1 70544	3 19770
23	1	23	253	1771	8855	33649	1 00947	2 45157	4 90314
24	1	24	276	2024	10626	42504	1 34596	3 46104	7 35471
25	1	25	300	2300	12650	53130	1 77100	4 80700	10 81575
26	1	26	325	2600	14950	65780	2 30230	6 57800	15 62275
27	1	27	351	2925	17550	80730	2 96010	8 88030	22 20075
28	1	28	378	3276	20475	98280	3 76740	11 84040	31 08105
29	1	29	406	3654	23751	1 18755	4 75020	15 60780	42 92145
30	1	30	435	4060	27405	1 42506	5 93775	20 35800	58 52925
31	1	31	465	4495	31465	1 69911	7 36281	26 29575	78 88725
32	1	32	496	4960	35960	2 01376	9 06192	33 65856	105 18300
33	1	33	528	5456	40920	2 37336	11 07568	42 72048	138 84156
34	1	34	561	5984	46376	2 78256	13 44904	53 79616	181 56204
35	1	35	595	6545	52360	3 24632	16 23160	67 24520	235 35820
36	1	36	630	7140	58905	3 76992	19 47792	83 47680	302 60340
37	1	37	666	7770	66045	4 35897	23 24784	102 95472	386 08020
38	1	38	703	8436	73815	5 01942	27 60681	126 20256	489 03492
39	1	39	741	9139	82251	5 75757	32 62623	153 80937	615 23748
40	1	40	780	9880	91390	6 58008	38 38380	186 43560	769 04685
41	1	41	820	10660	101270	7 49398	44 96388	224 81940	955 48245
42	1	42	861	11480	111930	8 50668	52 45786	269 78328	1180 30185
43	1	43	903	12341	123410	9 62598	60 96454	322 24114	1450 08513
44	1	44	946	13244	135751	10 86008	70 59052	383 20568	1772 32627
45	1	45	990	14190	148995	12 21759	81 45060	453 79620	2155 53195
46	1	46	1035	15180	163185	13 70754	93 66819	535 24680	2609 32815
47	1	47	1081	16215	178365	15 33939	107 37573	628 91499	3144 57495
48	1	48	1128	17296	194580	17 12304	122 71512	736 29072	3773 48994
49	1	49	1176	18424	211876	19 06884	139 83816	859 00584	4509 78066
50	1	50	1225	19600	230300	21 18760	158 90700	998 84400	5368 78650

[1] From "Handbook of Mathematical Functions", National Bureau of Standards

BINOMIAL COEFFICIENTS $\binom{n}{m}$

$n \backslash m$	9	10	11	12	13
9	1				
10	10	1			
11	55	11	1		
12	220	66	12	1	
13	715	286	78	13	1
14	2002	1001	364	91	14
15	5005	3003	1365	455	105
16	11440	8008	4368	1820	560
17	24310	19448	12376	6188	2380
18	48620	43758	31824	18564	8568
19	92378	92378	75582	50388	27132
20	1 67960	1 84756	1 67960	1 25970	77520
21	2 93930	3 52716	3 52716	2 93930	2 03490
22	4 97420	6 46646	7 05432	6 46646	4 97420
23	8 17190	11 44066	13 52078	13 52078	11 44066
24	13 07504	19 61256	24 96144	27 04156	24 96144
25	20 42975	32 68760	44 57400	52 00300	52 00300
26	31 24550	53 11735	77 26160	96 57700	104 00600
27	46 86825	84 36285	130 37895	173 83860	200 58300
28	69 06900	131 23110	214 74180	304 21755	374 42160
29	100 15005	200 30010	345 97290	518 95935	678 63915
30	143 07150	300 45015	546 27300	864 93225	1197 59850
31	201 60075	443 52165	846 72315	1411 20525	2062 53075
32	280 48800	645 12240	1290 24480	2257 92840	3473 73600
33	385 67100	925 61040	1935 36720	3548 17320	5731 66440
34	524 51256	1311 28140	2860 97760	5483 54040	9279 83760
35	706 07460	1835 79396	4172 25900	8344 51800	14763 37800
36	941 43280	2541 86856	6008 05296	12516 77700	23107 89600
37	1244 03620	3483 30136	8549 92152	18524 82996	35624 67300
38	1630 11640	4727 33756	12033 22288	27074 75148	54149 50296
39	2119 15132	6357 45396	16760 56044	39107 97436	81224 25444
40	2734 38880	8476 60528	23118 01440	55868 53480	1 20332 22880
41	3503 43565	11210 99408	31594 61968	78986 54920	1 76200 76360
42	4458 91810	14714 42973	42805 61376	1 10581 16888	2 55187 31280
43	5639 21995	19173 34783	57520 04349	1 53386 78264	3 65768 48168
44	7089 30508	24812 56778	76693 39132	2 10906 82613	5 19155 26432
45	8861 63135	31901 87286	1 01505 95910	2 87600 21745	7 30062 09045
46	11017 16330	40763 50421	1 33407 83196	3 89106 17655	10 17662 30790
47	13626 49145	51780 66751	1 74171 33617	5 22514 00851	14 06768 48445
48	16771 06640	65407 15896	2 25952 00368	6 96685 34468	19 29282 49296
49	20544 55634	82178 22536	2 91359 16264	9 22637 34836	26 25967 83764
50	25054 33700	1 02722 78170	3 73537 38800	12 13996 51100	35 48605 18600

BINOMIAL COEFFICIENTS $\binom{n}{m}$

n\m	14	15	16	17	18	19
14	1					
15	15	1				
16	120	16	1			
17	680	136	17	1		
18	3060	816	153	18	1	
19	11628	3876	969	171	19	1
20	38760	15504	4845	1140	190	20
21	1 16280	54264	20349	5985	1330	210
22	3 19770	1 70544	74613	26334	7315	1540
23	8 17190	4 90314	2 45157	1 00947	33649	8855
24	19 61256	13 07504	7 35471	3 46104	1 34596	42504
25	44 57400	32 68760	20 42975	10 81575	4 80700	1 77100
26	96 57700	77 26160	53 11735	31 24550	15 62275	6 57800
27	200 58300	173 83860	130 37895	84 36285	46 86825	22 20075
28	401 16600	374 42160	304 21755	214 74180	131 23110	69 06900
29	775 58760	775 58760	678 63915	518 95935	345 97290	200 30010
30	1454 22675	1551 17520	1454 22675	1197 59850	864 93225	546 27300
31	2651 82525	3005 40195	3005 40195	2651 82525	2062 53075	1411 20525
32	4714 35600	5657 22720	6010 80390	5657 22720	4714 35600	3473 73600
33	8188 09200	10371 58320	11668 03110	11668 03110	10371 58320	8188 09200
34	13919 75640	18559 67520	22039 61430	23336 06220	22039 61430	18559 67520
35	23199 59400	32479 43160	40599 28950	45375 67650	45375 67650	40599 28950
36	37962 97200	55679 02560	73078 72110	85974 96600	90751 35300	85974 96600
37	61070 86800	93641 99760	1 28757 74670	1 59053 68710	1 76726 31900	1 76726 31900
38	96695 54100	1 54712 86560	2 22399 74430	2 87811 43380	3 35780 00610	3 53452 63800
39	1 50845 04396	2 51408 40660	3 77112 60990	5 10211 17810	6 23591 43990	6 89232 64410
40	2 32069 29840	4 02253 45056	6 28521 01650	8 87323 78800	11 33802 61800	13 12824 08400
41	3 52401 52720	6 34322 74896	10 30774 46706	15 15844 80450	20 21126 40600	24 46626 70200
42	5 28602 29080	9 86724 27616	16 65097 21602	25 46619 27156	35 36971 21050	44 67753 10800
43	7 83789 60360	15 15326 56696	26 51821 49218	42 11716 48758	60 83590 48206	80 04724 31850
44	11 49558 08528	22 99116 17056	41 67148 05914	68 63537 97976	102 95306 96964	140 88314 80056
45	16 68713 34960	34 48674 25584	64 66264 22970	110 30686 03890	171 58844 94940	243 83621 77020
46	23 98775 44005	51 17387 60544	99 14938 48554	174 96950 26860	281 89530 98830	415 42466 71960
47	34 16437 74795	75 16163 04549	150 32326 09098	274 11888 75414	456 86481 25690	697 31997 70790
48	48 23206 23240	109 32600 79344	225 48489 13647	424 44214 84512	730 98370 01104	1154 18478 96480
49	67 52488 72536	157 55807 02584	334 81089 92991	649 92703 98159	1155 42584 85616	1885 16848 97584
50	93 78456 56300	225 08295 75120	492 36896 95575	984 73793 91150	1805 35288 83775	3040 59433 83200

BINOMIAL COEFFICIENTS $\binom{n}{m}$

$n\backslash m$	20	21	22	23	24	25
20	1					
21	21	1				
22	231	22	1			
23	1771	253	23	1		
24	10626	2024	276	24	1	
25	53130	12650	2300	300	25	1
26	2 30230	65780	14950	2600	325	26
27	8 88030	2 96010	80730	17550	2925	351
28	31 08105	11 84040	3 76740	98280	20475	3276
29	100 15005	42 92145	15 60780	4 75020	1 18755	23751
30	300 45015	143 07150	58 52925	20 35800	5 93775	1 42506
31	846 72315	443 52165	201 60075	78 88725	26 29575	7 36281
32	2257 92840	1290 24480	645 12240	280 48800	105 18300	33 65856
33	5731 66440	3548 17320	1935 36720	925 61040	385 67100	138 84156
34	13919 75640	9279 83760	5483 54040	2860 97760	1311 28140	524 51256
35	32479 43160	23199 59400	14763 37800	8344 51800	4172 25900	1835 79396
36	73078 72110	55679 02560	37962 97200	23107 89600	12516 77700	6008 05296
37	1 59053 68710	1 28757 74670	93641 99760	61070 86800	35624 67300	18524 82996
38	3 35780 00610	2 87811 43380	2 22399 74430	1 54712 86560	96695 54100	54149 50296
39	6 89232 64410	6 23591 43990	5 10211 17810	3 77112 60990	2 51408 40660	1 50845 04396
40	13 78465 28820	13 12824 08400	11 33802 61800	8 87323 78800	6 28521 01650	4 02253 45056
41	26 91289 37220	26 91289 37220	24 46626 70200	20 21126 40600	15 15844 80450	10 30774 46706
42	51 37916 07420	53 82578 74440	51 37916 07420	44 67753 10800	35 36971 21050	25 46619 27156
43	96 05669 18220	105 20494 81860	105 20494 81860	96 05669 18220	80 04724 31850	60 83590 48206
44	176 10393 50070	201 26164 00080	210 40989 63720	201 26164 00080	176 10393 50070	140 88314 80056
45	316 98708 30126	377 36557 50150	411 67153 63800	411 67153 63800	377 36557 50150	316 98708 30126
46	560 82330 07146	694 35265 80276	789 03711 13950	823 34307 27600	789 03711 13950	694 35265 80276
47	976 24796 79106	1255 17595 87422	1483 38976 94226	1612 38018 41550	1612 38018 41550	1483 38976 94226
48	1673 56794 49896	2231 42392 66528	2738 56572 81648	3095 76995 35776	3224 76036 83100	3095 76995 35776
49	2827 75273 46376	3904 99187 16424	4969 98965 48176	5834 33568 17424	6320 53032 18876	6320 53032 18876
50	4712 92122 43960	6732 74460 62800	8874 98152 64600	10804 32533 66600	12154 86600 36300	12641 06064 37752

PRIMES

The following table contains all primes from 1 to 100,000.

	0	1	2	3	4	5	6	7	8	9	10	11	12	13	14	15	16	17	18	19	20	21	22	23	24
1	2	547	1229	1993	2749	3581	4421	5281	6143	7001	7927	8837	9739	10663	11677	12569	13513	14533	15413	16411	17393	18329	19427	20359	21391
2	3	557	1231	1997	2753	3583	4423	5297	6151	7013	7933	8839	9743	10667	11681	12577	13523	14537	15427	16417	17401	18341	19429	20369	21397
3	5	563	1237	1999	2767	3593	4441	5303	6163	7019	7937	8849	9749	10687	11689	12583	13537	14543	15439	16421	17417	18353	19433	20389	21401
4	7	569	1249	2003	2777	3607	4447	5309	6173	7027	7949	8861	9767	10691	11699	12589	13553	14549	15443	16427	17419	18367	19441	20393	21407
5	11	571	1259	2011	2789	3613	4451	5323	6197	7039	7951	8863	9769	10709	11701	12601	13567	14551	15451	16433	17431	18371	19447	20399	21419
6	13	577	1277	2017	2791	3617	4457	5333	6199	7043	7963	8867	9781	10711	11717	12611	13577	14557	15461	16447	17443	18379	19457	20407	21433
7	17	587	1279	2027	2797	3623	4463	5347	6203	7057	7993	8887	9787	10723	11719	12613	13591	14561	15467	16451	17449	18397	19463	20411	21467
8	19	593	1283	2029	2801	3631	4481	5351	6211	7069	8009	8893	9791	10729	11731	12619	13597	14563	15473	16453	17467	18401	19469	20431	21481
9	23	599	1289	2039	2803	3637	4483	5381	6217	7079	8011	8923	9803	10733	11743	12637	13613	14591	15493	16477	17471	18413	19471	20441	21487
10	29	601	1291	2053	2819	3643	4493	5387	6221	7103	8017	8929	9811	10739	11777	12641	13619	14593	15497	16481	17477	18427	19477	20443	21491
11	31	607	1297	2063	2833	3659	4507	5393	6229	7109	8039	8933	9817	10753	11779	12647	13627	14621	15511	16487	17483	18433	19483	20477	21493
12	37	613	1301	2069	2837	3671	4513	5399	6247	7121	8053	8941	9829	10771	11783	12653	13633	14627	15527	16493	17489	18439	19489	20479	21499
13	41	617	1303	2081	2843	3673	4517	5407	6257	7127	8059	8951	9833	10781	11789	12659	13649	14629	15541	16519	17491	18443	19501	20483	21503
14	43	619	1307	2083	2851	3677	4519	5413	6263	7129	8069	8963	9839	10789	11801	12671	13669	14633	15551	16529	17497	18451	19507	20507	21517
15	47	631	1319	2087	2857	3691	4523	5417	6269	7151	8081	8969	9851	10799	11807	12689	13679	14639	15559	16547	17509	18457	19531	20509	21521
16	53	641	1321	2089	2861	3697	4547	5419	6271	7159	8087	8971	9857	10831	11813	12697	13681	14653	15569	16553	17519	18461	19541	20521	21523
17	59	643	1327	2099	2879	3701	4549	5431	6277	7177	8089	8999	9859	10837	11821	12703	13687	14657	15581	16561	17539	18481	19543	20533	21529
18	61	647	1361	2111	2887	3709	4561	5437	6287	7187	8093	9001	9871	10847	11827	12713	13691	14669	15583	16567	17551	18493	19553	20543	21557
19	67	653	1367	2113	2897	3719	4567	5441	6299	7193	8101	9007	9883	10853	11831	12721	13693	14683	15601	16573	17569	18503	19559	20549	21559
20	71	659	1373	2129	2903	3727	4583	5443	6301	7207	8111	9011	9887	10859	11833	12739	13697	14699	15607	16603	17573	18517	19571	20551	21563
21	73	661	1381	2131	2909	3733	4591	5449	6311	7211	8117	9013	9901	10861	11839	12743	13709	14713	15619	16607	17579	18521	19577	20563	21569
22	79	673	1399	2137	2917	3739	4597	5471	6317	7213	8123	9029	9907	10867	11863	12757	13711	14717	15629	16619	17581	18523	19583	20593	21577
23	83	677	1409	2141	2927	3761	4603	5477	6323	7219	8147	9041	9923	10883	11867	12763	13721	14723	15641	16631	17597	18539	19597	20599	21587
24	89	683	1423	2143	2939	3767	4621	5479	6329	7229	8161	9043	9929	10889	11887	12781	13723	14731	15643	16633	17599	18541	19603	20611	21589
25	97	691	1427	2153	2953	3769	4637	5483	6337	7237	8167	9049	9931	10891	11897	12791	13729	14737	15647	16649	17609	18553	19609	20627	21599
26	101	701	1429	2161	2957	3779	4639	5501	6343	7243	8171	9059	9941	10903	11903	12799	13751	14741	15649	16651	17623	18583	19661	20639	21601
27	103	709	1433	2179	2963	3793	4643	5503	6353	7247	8179	9067	9949	10909	11909	12809	13757	14747	15661	16657	17627	18587	19681	20641	21611
28	107	719	1439	2203	2969	3797	4649	5507	6359	7253	8191	9091	9967	10937	11923	12821	13759	14753	15667	16661	17657	18593	19687	20663	21613
29	109	727	1447	2207	2971	3803	4651	5519	6361	7283	8209	9103	9973	10939	11927	12823	13763	14759	15671	16673	17659	18617	19697	20681	21617
30	113	733	1451	2213	2999	3821	4657	5521	6367	7297	8219	9109	10007	10949	11933	12829	13781	14767	15679	16691	17669	18637	19699	20693	21647
31	127	739	1453	2221	3001	3823	4663	5527	6373	7307	8221	9127	10009	10957	11939	12841	13789	14771	15683	16693	17681	18661	19709	20707	21649
32	131	743	1459	2237	3011	3833	4673	5531	6379	7309	8231	9133	10037	10973	11941	12853	13799	14779	15727	16699	17683	18671	19717	20717	21661
33	137	751	1471	2239	3019	3847	4679	5557	6389	7321	8233	9137	10039	10979	11953	12889	13807	14783	15731	16703	17707	18679	19727	20719	21673
34	139	757	1481	2243	3023	3851	4691	5563	6397	7331	8237	9151	10061	10987	11959	12893	13829	14797	15733	16729	17713	18691	19739	20731	21683
35	149	761	1483	2251	3037	3853	4703	5569	6421	7333	8243	9157	10067	10993	11969	12899	13831	14813	15737	16741	17729	18701	19751	20743	21701
36	151	769	1487	2267	3041	3863	4721	5573	6427	7349	8263	9161	10069	11003	11971	12907	13841	14821	15739	16747	17737	18713	19753	20747	21713
37	157	773	1489	2269	3049	3877	4723	5581	6449	7351	8269	9173	10079	11027	11981	12911	13859	14827	15749	16759	17747	18719	19759	20749	21727
38	163	787	1493	2273	3061	3881	4729	5591	6451	7369	8273	9181	10091	11047	11987	12917	13873	14831	15761	16763	17749	18731	19763	20753	21737
39	167	797	1499	2281	3067	3889	4733	5623	6469	7393	8287	9187	10093	11057	12007	12919	13877	14843	15767	16787	17761	18743	19777	20759	21739
40	173	809	1511	2287	3079	3907	4751	5639	6473	7411	8291	9199	10099	11059	12011	12923	13879	14851	15773	16811	17783	18749	19793	20771	21751
41	179	811	1523	2293	3083	3911	4759	5641	6481	7417	8293	9203	10103	11069	12037	12941	13883	14867	15787	16823	17789	18757	19801	20773	21757
42	181	821	1531	2297	3089	3917	4783	5647	6491	7433	8297	9209	10111	11071	12041	12953	13901	14869	15791	16829	17791	18773	19813	20789	21767
43	191	823	1543	2309	3109	3919	4787	5651	6521	7451	8311	9221	10133	11083	12043	12959	13903	14879	15797	16831	17807	18787	19819	20807	21773
44	193	827	1549	2311	3119	3923	4789	5653	6529	7457	8317	9227	10139	11087	12049	12967	13907	14887	15803	16843	17827	18793	19841	20809	21787
45	197	829	1553	2333	3121	3929	4793	5657	6547	7459	8329	9239	10141	11093	12071	12973	13913	14891	15809	16871	17837	18797	19843	20849	21799
46	199	839	1559	2339	3137	3931	4799	5659	6551	7477	8353	9241	10151	11113	12073	12979	13921	14897	15817	16879	17839	18803	19853	20857	21803
47	211	853	1567	2341	3163	3943	4801	5669	6553	7481	8363	9257	10159	11117	12097	12983	13931	14923	15823	16883	17851	18839	19861	20873	21817
48	223	857	1571	2347	3167	3947	4813	5683	6563	7487	8369	9277	10163	11119	12101	13001	13933	14929	15859	16889	17863	18859	19867	20879	21821
49	227	859	1579	2351	3169	3967	4817	5689	6569	7489	8377	9281	10169	11131	12107	13003	13963	14939	15877	16901	17881	18869	19889	20887	21839
50	229	863	1583	2357	3181	3989	4831	5693	6571	7499	8387	9283	10177	11149	12109	13007	13967	14947	15881	16903	17891	18899	19891	20897	21841

PRIMES

	0	1	2	3	4	5	6	7	8	9	10	11	12	13	14	15	16	17	18	19	20	21	22	23	24
51	233	877	1597	2371	3187	4001	4861	5701	6577	7507	8389	9293	10181	11159	12113	13009	13997	14951	15887	16921	17903	18911	19913	20899	21851
52	239	881	1601	2377	3191	4003	4871	5711	6581	7517	8419	9311	10193	11161	12119	13033	13999	14957	15889	16927	17909	18913	19919	20903	21859
53	241	883	1607	2381	3203	4007	4877	5717	6599	7523	8423	9319	10211	11171	12143	13037	14009	14969	15901	16931	17911	18917	19927	20921	21863
54	251	887	1609	2383	3209	4013	4889	5737	6607	7529	8429	9323	10223	11173	12149	13043	14011	14983	15907	16937	17921	18919	19937	20929	21871
55	257	907	1613	2389	3217	4019	4903	5741	6619	7537	8431	9337	10243	11177	12157	13049	14029	15013	15913	16943	17923	18947	19949	20939	21881
56	263	911	1619	2393	3221	4021	4909	5743	6637	7541	8443	9341	10247	11197	12161	13063	14033	15017	15919	16963	17929	18959	19961	20947	21893
57	269	919	1621	2399	3229	4027	4919	5749	6653	7547	8447	9343	10253	11213	12163	13093	14051	15031	15923	16979	17939	18973	19963	20959	21911
58	271	929	1627	2411	3251	4049	4931	5779	6659	7549	8461	9349	10259	11239	12197	13099	14057	15053	15937	16981	17957	18979	19973	20963	21929
59	277	937	1637	2417	3253	4051	4933	5783	6661	7559	8467	9371	10267	11243	12203	13103	14071	15061	15959	16987	17959	19001	19979	20981	21937
60	281	941	1657	2423	3257	4057	4937	5791	6673	7561	8501	9377	10271	11251	12211	13109	14081	15073	15971	16993	17971	19009	19991	20983	21943
61	283	947	1663	2437	3259	4073	4943	5801	6679	7573	8513	9391	10273	11257	12227	13121	14083	15077	15973	17011	17977	19013	19993	21001	21961
62	293	953	1667	2441	3271	4079	4951	5807	6689	7577	8521	9397	10289	11261	12239	13127	14087	15083	15991	17021	17981	19031	19997	21011	21977
63	307	967	1669	2447	3299	4091	4957	5813	6691	7583	8527	9403	10301	11273	12241	13147	14107	15091	16001	17027	17987	19037	20011	21013	21991
64	311	971	1693	2459	3301	4093	4967	5821	6701	7589	8537	9413	10303	11279	12251	13151	14143	15101	16007	17029	17989	19051	20021	21017	21997
65	313	977	1697	2467	3307	4099	4969	5827	6703	7591	8539	9419	10313	11287	12253	13159	14149	15107	16033	17033	18013	19069	20023	21019	22003
66	317	983	1699	2473	3313	4111	4973	5839	6709	7603	8543	9421	10321	11299	12263	13163	14153	15121	16057	17041	18041	19073	20029	21023	22013
67	331	991	1709	2477	3319	4127	4987	5843	6719	7607	8563	9431	10331	11311	12269	13171	14159	15131	16061	17047	18043	19079	20047	21031	22027
68	337	997	1721	2503	3323	4129	4993	5849	6733	7621	8573	9433	10333	11317	12277	13177	14173	15137	16063	17053	18047	19081	20051	21059	22031
69	347	1009	1723	2521	3329	4133	4999	5851	6737	7639	8581	9437	10337	11321	12281	13183	14177	15139	16067	17077	18049	19087	20063	21061	22037
70	349	1013	1733	2531	3331	4139	5003	5857	6761	7643	8597	9439	10343	11329	12289	13187	14197	15149	16069	17093	18059	19121	20071	21067	22039
71	353	1019	1741	2539	3343	4153	5009	5861	6763	7649	8599	9461	10357	11351	12301	13217	14207	15161	16073	17099	18061	19139	20089	21089	22051
72	359	1021	1747	2543	3347	4157	5011	5867	6779	7669	8609	9463	10369	11353	12323	13219	14221	15173	16087	17107	18077	19141	20101	21101	22063
73	367	1031	1753	2549	3359	4159	5021	5869	6781	7673	8623	9467	10391	11369	12329	13229	14243	15187	16091	17117	18089	19157	20107	21107	22067
74	373	1033	1759	2551	3361	4177	5023	5879	6791	7681	8627	9473	10399	11383	12343	13241	14249	15193	16097	17123	18097	19163	20113	21121	22073
75	379	1039	1777	2557	3371	4201	5039	5881	6793	7687	8629	9479	10427	11393	12347	13249	14251	15199	16103	17137	18119	19181	20117	21139	22079
76	383	1049	1783	2579	3373	4211	5051	5897	6803	7691	8641	9491	10429	11399	12373	13259	14281	15217	16111	17159	18121	19183	20123	21143	22091
77	389	1051	1787	2591	3389	4217	5059	5903	6823	7699	8647	9497	10433	11411	12377	13267	14293	15227	16127	17167	18127	19207	20129	21149	22093
78	397	1061	1789	2593	3391	4219	5077	5923	6827	7703	8663	9511	10453	11423	12379	13291	14303	15233	16139	17183	18131	19211	20143	21157	22109
79	401	1063	1801	2609	3407	4229	5081	5927	6829	7717	8669	9521	10457	11437	12391	13297	14321	15241	16141	17189	18133	19213	20147	21163	22111
80	409	1069	1811	2617	3413	4231	5087	5939	6833	7723	8677	9533	10459	11443	12401	13309	14323	15259	16183	17191	18143	19219	20149	21169	22123
81	419	1087	1823	2621	3433	4241	5099	5953	6841	7727	8681	9539	10463	11447	12409	13313	14327	15263	16187	17203	18149	19231	20161	21179	22129
82	421	1091	1831	2633	3449	4243	5101	5981	6857	7741	8689	9547	10477	11467	12413	13327	14341	15269	16189	17207	18169	19237	20173	21187	22133
83	431	1093	1847	2647	3457	4253	5107	5987	6863	7753	8693	9551	10487	11471	12421	13331	14347	15271	16193	17209	18181	19249	20177	21191	22147
84	433	1097	1861	2657	3461	4259	5113	6007	6869	7757	8699	9587	10499	11483	12433	13337	14369	15277	16217	17231	18191	19259	20183	21193	22153
85	439	1103	1867	2659	3463	4261	5119	6011	6871	7759	8707	9601	10501	11489	12437	13339	14387	15287	16223	17239	18199	19267	20201	21211	22157
86	443	1109	1871	2663	3467	4271	5147	6029	6883	7789	8713	9613	10513	11491	12451	13367	14389	15289	16229	17257	18211	19273	20219	21221	22159
87	449	1117	1873	2671	3469	4273	5153	6037	6899	7793	8719	9619	10529	11497	12457	13381	14401	15299	16231	17291	18217	19289	20231	21227	22171
88	457	1123	1877	2677	3491	4283	5167	6043	6907	7817	8731	9623	10531	11503	12473	13397	14407	15307	16249	17293	18223	19301	20233	21247	22189
89	461	1129	1879	2683	3499	4289	5171	6047	6911	7823	8737	9629	10559	11519	12479	13399	14411	15313	16253	17299	18229	19309	20249	21269	22193
90	463	1151	1889	2687	3511	4297	5179	6053	6917	7829	8741	9631	10567	11527	12487	13411	14419	15319	16267	17317	18233	19319	20261	21277	22229
91	467	1153	1901	2689	3517	4327	5189	6067	6947	7841	8747	9643	10589	11549	12491	13417	14423	15329	16273	17321	18251	19333	20269	21283	22247
92	479	1163	1907	2693	3527	4337	5197	6073	6949	7853	8753	9649	10597	11551	12497	13421	14431	15331	16301	17327	18253	19373	20287	21313	22259
93	487	1171	1913	2699	3529	4339	5209	6079	6959	7867	8761	9661	10601	11579	12503	13441	14437	15349	16319	17333	18257	19379	20297	21317	22271
94	491	1181	1931	2707	3533	4349	5227	6089	6961	7873	8779	9677	10607	11587	12511	13451	14447	15359	16333	17341	18269	19381	20323	21319	22273
95	499	1187	1933	2711	3539	4357	5231	6091	6967	7877	8783	9679	10613	11593	12517	13457	14449	15361	16339	17351	18287	19387	20327	21323	22277
96	503	1193	1949	2713	3541	4363	5233	6101	6971	7879	8803	9689	10627	11597	12527	13463	14461	15373	16349	17359	18289	19391	20333	21341	22279
97	509	1201	1951	2719	3547	4373	5237	6113	6977	7883	8807	9697	10631	11617	12539	13469	14479	15377	16361	17377	18301	19403	20341	21347	22283
98	521	1213	1973	2729	3557	4391	5261	6121	6983	7901	8819	9719	10639	11621	12541	13477	14489	15383	16363	17383	18307	19417	20347	21377	22291
99	523	1217	1979	2731	3559	4397	5273	6131	6991	7907	8821	9721	10651	11633	12547	13487	14503	15391	16369	17387	18311	19421	20353	21379	22303
100	541	1223	1987	2741	3571	4409	5279	6133	6997	7919	8831	9733	10657	11657	12553	13499	14519	15401	16381	17389	18313	19423	20357	21383	22307

PRIMES

	25	26	27	28	29	30	31	32	33	34	35	36	37	38	39	40	41	42	43	44	45	46	47	48	49
1	22343	23327	24317	25409	26407	27457	28513	29453	30577	31607	32611	33617	34651	35771	36787	37831	38923	39979	41113	42083	43063	44203	45317	46451	47533
2	22349	23333	24329	25411	26417	27479	28517	29473	30593	31627	32621	33619	34667	35797	36791	37847	38933	39983	41117	42089	43067	44207	45319	46457	47543
3	22367	23339	24337	25423	26423	27481	28537	29483	30631	31643	32633	33623	34673	35801	36793	37853	38953	39989	41131	42101	43093	44221	45329	46471	47563
4	22369	23357	24359	25439	26431	27487	28541	29501	30637	31649	32647	33629	34679	35803	36809	37861	38959	40009	41141	42131	43103	44249	45337	46477	47569
5	22381	23369	24371	25447	26437	27509	28547	29527	30643	31657	32653	33637	34687	35809	36821	37871	38971	40013	41143	42139	43117	44257	45341	46489	47581
6	22391	23371	24373	25453	26449	27527	28549	29531	30649	31663	32687	33641	34693	35831	36833	37879	38977	40031	41149	42157	43133	44263	45343	46499	47591
7	22397	23399	24379	25457	26459	27529	28559	29537	30661	31667	32693	33647	34703	35837	36847	37889	38993	40037	41161	42169	43151	44267	45361	46507	47599
8	22409	23417	24391	25463	26479	27539	28571	29567	30671	31687	32707	33679	34721	35839	36857	37897	39019	40039	41177	42179	43159	44269	45377	46511	47609
9	22433	23431	24407	25469	26489	27541	28573	29569	30677	31699	32713	33703	34729	35851	36871	37907	39023	40063	41179	42181	43177	44273	45389	46523	47623
10	22441	23447	24413	25471	26497	27551	28579	29573	30689	31721	32717	33713	34739	35863	36877	37951	39041	40087	41183	42187	43189	44279	45403	46549	47629
11	22447	23459	24419	25523	26501	27581	28591	29581	30697	31723	32719	33721	34747	35869	36887	37957	39043	40093	41189	42193	43201	44281	45413	46559	47639
12	22453	23473	24421	25537	26513	27583	28597	29587	30703	31727	32749	33739	34757	35879	36899	37963	39047	40099	41201	42197	43207	44293	45427	46567	47653
13	22469	23497	24439	25541	26539	27611	28603	29599	30707	31729	32771	33749	34759	35897	36901	37967	39079	40111	41203	42209	43223	44351	45433	46573	47657
14	22481	23509	24443	25561	26557	27617	28607	29611	30713	31741	32779	33751	34763	35899	36913	37987	39089	40123	41213	42221	43237	44357	45439	46589	47659
15	22483	23531	24469	25577	26561	27631	28619	29629	30727	31751	32783	33757	34781	35911	36919	37991	39097	40127	41221	42223	43261	44371	45481	46591	47681
16	22501	23537	24473	25579	26573	27647	28621	29633	30757	31769	32789	33767	34807	35923	36923	37993	39103	40129	41227	42227	43271	44381	45491	46601	47699
17	22511	23539	24481	25583	26591	27653	28627	29641	30763	31771	32797	33769	34819	35933	36929	37997	39107	40151	41231	42239	43283	44383	45497	46619	47701
18	22531	23549	24499	25589	26597	27673	28631	29663	30773	31793	32801	33773	34841	35951	36931	38011	39113	40153	41233	42257	43291	44389	45503	46633	47711
19	22541	23557	24509	25601	26627	27689	28643	29669	30781	31799	32803	33791	34843	35963	36943	38039	39119	40163	41243	42281	43313	44417	45523	46639	47713
20	22543	23561	24517	25603	26633	27691	28649	29671	30803	31817	32831	33797	34847	35969	36947	38047	39133	40169	41257	42283	43319	44449	45533	46643	47717
21	22549	23563	24527	25609	26641	27697	28657	29683	30809	31847	32833	33809	34849	35977	36973	38053	39139	40177	41263	42293	43321	44453	45541	46649	47737
22	22567	23567	24533	25621	26647	27701	28661	29717	30817	31849	32839	33811	34871	35983	36979	38069	39157	40189	41269	42299	43331	44483	45553	46663	47741
23	22571	23581	24547	25633	26669	27733	28663	29723	30829	31859	32843	33827	34877	35993	36997	38083	39161	40193	41281	42307	43391	44491	45557	46679	47743
24	22573	23593	24551	25639	26681	27737	28669	29741	30839	31873	32869	33829	34883	35999	37003	38113	39163	40213	41299	42323	43397	44497	45569	46681	47777
25	22613	23599	24571	25643	26683	27739	28687	29753	30841	31883	32887	33851	34897	36007	37013	38119	39181	40231	41333	42331	43399	44501	45587	46687	47779
26	22619	23603	24593	25657	26687	27743	28697	29759	30851	31891	32909	33857	34913	36011	37019	38149	39191	40237	41341	42337	43403	44507	45589	46691	47791
27	22621	23609	24611	25667	26693	27749	28703	29761	30853	31907	32911	33863	34919	36013	37021	38153	39199	40241	41351	42349	43411	44519	45599	46703	47797
28	22637	23623	24623	25673	26699	27751	28711	29789	30859	31957	32917	33871	34939	36017	37039	38167	39209	40253	41357	42359	43427	44531	45613	46723	47807
29	22639	23627	24631	25679	26701	27763	28723	29803	30869	31963	32933	33889	34949	36037	37049	38177	39217	40277	41381	42373	43441	44533	45631	46727	47809
30	22643	23629	24659	25693	26711	27767	28729	29819	30871	31973	32939	33893	34961	36061	37057	38183	39227	40283	41387	42379	43451	44537	45641	46747	47819
31	22651	23633	24671	25703	26713	27773	28751	29833	30881	31981	32941	33911	34963	36067	37061	38189	39229	40289	41389	42391	43457	44543	45659	46751	47837
32	22669	23663	24677	25717	26717	27779	28753	29837	30893	31991	32957	33923	34981	36073	37087	38197	39233	40343	41399	42397	43481	44549	45667	46757	47843
33	22679	23669	24683	25733	26723	27791	28759	29851	30911	32003	32969	33931	35023	36083	37097	38201	39239	40351	41411	42403	43487	44563	45673	46769	47857
34	22691	23671	24691	25741	26729	27793	28771	29863	30931	32009	32971	33937	35027	36097	37117	38219	39241	40357	41413	42407	43499	44579	45677	46771	47869
35	22697	23677	24697	25747	26731	27799	28789	29867	30937	32027	32983	33941	35051	36107	37123	38231	39251	40361	41443	42409	43517	44587	45691	46807	47881
36	22699	23687	24709	25759	26737	27803	28793	29873	30941	32029	32987	33961	35053	36109	37139	38237	39293	40387	41453	42433	43541	44617	45697	46811	47903
37	22709	23689	24733	25763	26759	27809	28807	29879	30949	32051	32993	33967	35059	36131	37159	38239	39301	40423	41467	42437	43543	44621	45707	46817	47911
38	22717	23719	24749	25771	26777	27817	28813	29881	30971	32057	32999	33997	35069	36137	37171	38261	39313	40427	41479	42443	43573	44623	45737	46819	47917
39	22721	23741	24763	25793	26783	27823	28817	29917	30977	32059	33013	34019	35081	36151	37181	38273	39317	40429	41491	42451	43577	44633	45751	46829	47933
40	22727	23743	24767	25799	26801	27827	28837	29921	30983	32063	33023	34031	35083	36161	37189	38281	39323	40433	41507	42457	43579	44641	45757	46831	47939
41	22739	23747	24781	25801	26813	27847	28843	29927	31013	32069	33029	34033	35089	36187	37199	38287	39341	40459	41513	42461	43591	44647	45763	46853	47947
42	22741	23753	24793	25819	26821	27851	28859	29947	31019	32077	33037	34039	35099	36191	37201	38299	39343	40471	41519	42463	43597	44651	45767	46861	47951
43	22751	23761	24799	25841	26833	27883	28867	29959	31033	32083	33049	34057	35107	36209	37217	38303	39359	40483	41521	42467	43607	44657	45779	46867	47963
44	22769	23767	24809	25847	26839	27893	28871	29983	31039	32089	33053	34061	35111	36217	37223	38317	39367	40487	41539	42473	43609	44683	45817	46877	47969
45	22777	23773	24821	25849	26849	27901	28879	29989	31051	32099	33071	34123	35117	36229	37243	38321	39371	40493	41543	42487	43613	44687	45821	46889	47977
46	22783	23789	24841	25867	26861	27917	28901	30011	31063	32117	33073	34127	35129	36241	37253	38327	39373	40499	41549	42491	43627	44699	45823	46901	47981
47	22787	23801	24847	25873	26863	27919	28909	30013	31069	32119	33083	34129	35141	36251	37273	38329	39383	40507	41579	42499	43633	44701	45827	46919	48017
48	22807	23813	24851	25889	26879	27941	28921	30029	31079	32141	33091	34141	35149	36263	37277	38333	39397	40519	41593	42509	43649	44711	45833	46933	48023
49	22811	23819	24859	25903	26881	27943	28927	30047	31081	32143	33107	34147	35153	36269	37307	38351	39409	40529	41597	42533	43651	44729	45841	46957	48029
50	22817	23827	24877	25913	26891	27947	28933	30059	31091	32159	33113	34157	35159	36277	37309	38371	39419	40531	41603	42557	43661	44741	45853	46993	48049

PRIMES

PRIMES

	25	26	27	28	29	30	31	32	33	34	35	36	37	38	39	40	41	42	43	44	45	46	47	48	49
51	22853	23831	24889	25919	26893	27953	28949	30071	31121	32173	33119	34159	35171	36293	37313	38377	39439	40543	41609	42569	43669	44753	45863	46997	48073
52	22859	23833	24907	25931	26903	27961	28961	30089	31123	32183	33149	34171	35201	36299	37321	38393	39443	40559	41611	42571	43691	44771	45869	47017	48079
53	22861	23857	24917	25933	26921	27967	28979	30091	31139	32189	33151	34183	35221	36307	37337	38431	39451	40577	41617	42577	43711	44773	45887	47041	48091
54	22871	23869	24919	25939	26927	27983	29009	30097	31147	32191	33161	34211	35227	36313	37339	38447	39461	40583	41621	42589	43717	44777	45893	47051	48109
55	22877	23873	24923	25943	26947	27997	29017	30103	31151	32203	33179	34213	35251	36319	37357	38449	39499	40591	41627	42611	43721	44789	45943	47057	48119
56	22901	23879	24943	25951	26951	28001	29021	30109	31153	32213	33181	34217	35257	36341	37361	38453	39503	40597	41641	42641	43753	44797	45949	47059	48121
57	22907	23887	24953	25969	26953	28019	29023	30113	31159	32233	33191	34231	35267	36343	37363	38459	39509	40609	41647	42643	43759	44809	45953	47087	48131
58	22921	23893	24967	25981	26959	28027	29027	30119	31177	32237	33199	34253	35279	36353	37369	38461	39511	40627	41651	42649	43777	44819	45959	47093	48157
59	22937	23899	24971	25997	26981	28031	29033	30133	31181	32251	33203	34259	35281	36373	37379	38501	39521	40637	41659	42667	43781	44839	45971	47111	48163
60	22943	23909	24977	25999	26987	28051	29059	30137	31183	32257	33211	34261	35291	36383	37397	38543	39541	40639	41669	42677	43783	44843	45979	47119	48179
61	22961	23911	24979	26003	26993	28057	29063	30139	31189	32261	33223	34267	35311	36389	37409	38557	39551	40693	41681	42683	43787	44851	45989	47123	48187
62	22963	23917	24989	26017	27011	28069	29077	30161	31193	32297	33247	34273	35317	36433	37423	38561	39563	40697	41687	42689	43789	44867	46021	47129	48193
63	22973	23929	25013	26021	27017	28081	29101	30169	31219	32299	33287	34283	35323	36451	37441	38567	39569	40699	41719	42697	43793	44879	46027	47137	48197
64	22993	23957	25031	26029	27031	28087	29123	30181	31223	32303	33289	34297	35327	36457	37447	38569	39581	40709	41729	42701	43801	44887	46049	47143	48221
65	23003	23971	25033	26041	27043	28097	29129	30187	31231	32309	33301	34301	35339	36467	37463	38593	39607	40739	41737	42703	43853	44893	46051	47147	48239
66	23011	23977	25037	26053	27059	28099	29131	30197	31237	32321	33311	34303	35353	36469	37483	38603	39619	40751	41759	42709	43867	44909	46061	47149	48247
67	23017	23981	25057	26083	27061	28109	29137	30203	31247	32323	33317	34313	35363	36473	37489	38609	39623	40759	41761	42719	43889	44917	46073	47161	48259
68	23021	23993	25073	26099	27067	28111	29147	30211	31249	32327	33329	34319	35381	36479	37493	38611	39631	40763	41771	42727	43891	44927	46091	47189	48271
69	23027	24001	25087	26107	27073	28123	29153	30223	31253	32341	33331	34327	35393	36493	37501	38629	39659	40771	41777	42737	43913	44939	46093	47207	48281
70	23029	24007	25097	26111	27077	28151	29167	30241	31259	32353	33343	34337	35401	36497	37507	38639	39667	40787	41801	42743	43933	44953	46099	47221	48299
71	23039	24019	25111	26113	27091	28163	29173	30253	31267	32359	33347	34351	35407	36523	37511	38651	39671	40801	41809	42751	43943	44959	46103	47237	48311
72	23041	24023	25117	26119	27103	28181	29179	30259	31271	32363	33349	34361	35419	36527	37517	38653	39679	40813	41813	42767	43951	44963	46133	47251	48313
73	23053	24029	25121	26141	27107	28183	29191	30269	31277	32369	33353	34367	35423	36529	37529	38669	39703	40819	41843	42773	43961	44971	46141	47269	48337
74	23057	24043	25127	26153	27109	28201	29201	30271	31307	32371	33359	34369	35437	36541	37537	38671	39709	40823	41849	42787	43963	44983	46147	47279	48341
75	23059	24049	25147	26161	27127	28211	29207	30293	31319	32377	33377	34381	35447	36551	37547	38677	39719	40829	41851	42793	43969	44987	46153	47287	48353
76	23063	24061	25153	26171	27143	28219	29209	30307	31321	32381	33391	34403	35449	36559	37549	38693	39727	40841	41863	42797	43973	45007	46171	47293	48371
77	23071	24071	25163	26177	27179	28229	29221	30313	31327	32401	33403	34421	35461	36563	37561	38699	39733	40847	41879	42821	43987	45013	46181	47297	48383
78	23081	24077	25169	26183	27191	28277	29231	30319	31333	32411	33409	34429	35491	36571	37567	38707	39749	40849	41887	42829	43991	45053	46183	47303	48397
79	23087	24083	25171	26189	27197	28279	29243	30323	31337	32413	33413	34439	35507	36583	37571	38711	39761	40853	41893	42839	43997	45061	46187	47309	48407
80	23099	24091	25183	26203	27211	28283	29251	30341	31357	32423	33427	34457	35509	36587	37573	38713	39769	40867	41897	42841	44017	45077	46199	47317	48409
81	23117	24097	25189	26209	27239	28289	29269	30347	31379	32429	33457	34469	35521	36599	37579	38723	39779	40879	41903	42853	44021	45083	46219	47339	48413
82	23131	24103	25219	26227	27241	28297	29287	30367	31387	32441	33461	34471	35527	36607	37589	38729	39791	40883	41911	42859	44027	45119	46229	47351	48437
83	23143	24107	25229	26237	27253	28307	29297	30389	31391	32443	33469	34483	35531	36629	37591	38737	39799	40897	41927	42863	44029	45121	46237	47353	48449
84	23159	24109	25237	26249	27259	28309	29303	30391	31393	32467	33479	34487	35533	36637	37607	38747	39821	40903	41941	42899	44041	45127	46261	47363	48463
85	23167	24113	25243	26251	27271	28319	29311	30403	31397	32479	33487	34499	35537	36643	37619	38749	39827	40927	41947	42901	44053	45131	46271	47381	48473
86	23173	24121	25247	26261	27277	28349	29327	30427	31469	32491	33493	34501	35543	36653	37633	38767	39829	40933	41953	42923	44059	45137	46273	47387	48479
87	23189	24133	25253	26263	27281	28351	29333	30431	31477	32497	33503	34511	35569	36671	37643	38783	39839	40939	41957	42929	44071	45139	46279	47389	48481
88	23197	24137	25261	26267	27283	28387	29339	30449	31481	32503	33521	34513	35573	36677	37649	38791	39841	40949	41959	42937	44087	45161	46301	47407	48487
89	23201	24151	25301	26293	27299	28393	29347	30467	31489	32507	33529	34519	35591	36683	37657	38803	39847	40961	41969	42943	44089	45179	46307	47417	48491
90	23203	24169	25303	26297	27329	28403	29363	30469	31511	32531	33533	34537	35593	36691	37663	38821	39857	40973	41981	42953	44101	45181	46309	47419	48497
91	23209	24179	25307	26309	27337	28409	29383	30491	31513	32533	33547	34543	35597	36697	37691	38833	39863	40993	41983	42961	44111	45191	46327	47431	48523
92	23227	24181	25309	26317	27361	28411	29387	30493	31517	32537	33563	34549	35603	36709	37693	38839	39869	41011	41999	42967	44119	45197	46337	47441	48527
93	23251	24197	25321	26321	27367	28429	29389	30497	31531	32561	33569	34583	35617	36713	37699	38851	39877	41017	42013	42979	44123	45233	46349	47459	48533
94	23269	24203	25339	26339	27397	28433	29399	30509	31541	32563	33577	34589	35671	36721	37717	38861	39883	41023	42017	42989	44129	45247	46351	47491	48539
95	23279	24223	25343	26347	27407	28439	29401	30517	31543	32569	33581	34591	35677	36739	37747	38867	39887	41039	42019	43003	44131	45259	46381	47497	48541
96	23291	24229	25349	26357	27409	28447	29411	30529	31547	32573	33587	34603	35729	36749	37781	38873	39901	41047	42023	43013	44159	45263	46399	47501	48563
97	23293	24239	25357	26371	27427	28463	29423	30539	31567	32579	33589	34607	35731	36761	37783	38891	39929	41051	42043	43019	44171	45281	46411	47507	48571
98	23297	24247	25367	26387	27431	28477	29429	30553	31573	32587	33599	34613	35747	36767	37799	38903	39937	41057	42061	43037	44179	45289	46439	47513	48589
99	23311	24251	25373	26393	27437	28493	29437	30557	31583	32603	33601	34631	35753	36779	37811	38917	39953	41077	42071	43049	44189	45293	46441	47521	48593
100	23321	24281	25391	26399	27449	28499	29443	30559	31601	32609	33613	34649	35759	36781	37813	38921	39971	41081	42073	43051	44201	45307	46447	47527	48611

PRIMES

PRIMES

	50	51	52	53	54	55	56	57	58	59	60	61	62	63	64	65	66	67	68	69	70	71	72	73	74
1	48619	49667	50767	51817	52937	54001	55109	56197	57193	58243	59369	60509	61637	62791	63823	65071	66107	67247	68389	69497	70663	71719	72859	73999	75083
2	48623	49669	50773	51827	52951	54011	55117	56207	57203	58271	59377	60521	61643	62801	63839	65089	66109	67261	68399	69499	70667	71741	72869	74017	75109
3	48647	49681	50777	51829	52957	54013	55127	56209	57221	58309	59387	60527	61651	62819	63841	65099	66137	67271	68437	69539	70687	71761	72871	74021	75133
4	48649	49697	50789	51839	52963	54037	55147	56237	57223	58313	59393	60539	61657	62827	63853	65101	66161	67273	68443	69557	70709	71777	72883	74027	75149
5	48661	49711	50821	51853	52967	54049	55163	56239	57241	58321	59399	60589	61667	62851	63857	65111	66169	67289	68447	69593	70717	71789	72889	74047	75161
6	48673	49727	50833	51859	52973	54059	55171	56249	57251	58337	59407	60601	61673	62861	63863	65119	66173	67307	68449	69623	70729	71807	72893	74051	75167
7	48677	49739	50839	51869	52981	54083	55201	56263	57259	58363	59417	60607	61681	62869	63901	65123	66179	67339	68473	69653	70753	71809	72901	74071	75169
8	48679	49741	50849	51871	52999	54091	55207	56267	57269	58367	59419	60611	61687	62873	63907	65129	66191	67343	68477	69661	70769	71821	72907	74077	75181
9	48731	49747	50857	51893	53003	54101	55213	56269	57271	58369	59441	60617	61703	62897	63913	65141	66221	67349	68483	69677	70783	71837	72911	74093	75193
10	48733	49757	50867	51899	53017	54121	55217	56299	57283	58379	59443	60623	61717	62903	63929	65147	66239	67369	68489	69691	70793	71843	72923	74099	75209
11	48751	49783	50873	51907	53047	54133	55219	56311	57287	58391	59447	60631	61723	62921	63949	65167	66271	67391	68491	69697	70823	71849	72931	74101	75211
12	48757	49787	50891	51913	53051	54139	55229	56333	57301	58393	59453	60637	61729	62927	63977	65171	66293	67399	68501	69709	70841	71861	72937	74131	75217
13	48761	49789	50893	51929	53069	54151	55243	56359	57329	58403	59467	60647	61751	62929	63997	65173	66301	67409	68507	69737	70843	71867	72949	74143	75223
14	48767	49801	50909	51941	53077	54163	55249	56369	57331	58411	59471	60649	61757	62939	64007	65179	66337	67411	68521	69739	70849	71879	72953	74149	75227
15	48779	49807	50923	51949	53087	54167	55259	56377	57347	58417	59473	60659	61781	62969	64013	65183	66343	67421	68531	69761	70853	71881	72959	74159	75239
16	48781	49811	50929	51971	53089	54181	55291	56383	57349	58427	59497	60661	61813	62971	64019	65203	66347	67427	68539	69763	70867	71887	72973	74161	75253
17	48787	49823	50951	51973	53093	54193	55313	56393	57367	58439	59509	60679	61819	62981	64033	65213	66359	67429	68543	69767	70877	71899	72977	74167	75269
18	48799	49831	50957	51977	53101	54217	55331	56401	57373	58441	59513	60689	61837	62983	64037	65239	66361	67433	68567	69779	70879	71909	72997	74177	75277
19	48809	49843	50969	51991	53113	54251	55333	56417	57383	58451	59539	60703	61843	62987	64063	65257	66373	67447	68581	69809	70891	71917	73009	74189	75289
20	48817	49853	50971	52009	53117	54269	55337	56431	57389	58453	59557	60719	61861	62989	64067	65267	66377	67453	68597	69821	70901	71933	73013	74197	75307
21	48821	49871	50989	52021	53129	54277	55339	56437	57397	58477	59561	60727	61871	63029	64081	65269	66383	67477	68611	69827	70913	71941	73019	74201	75323
22	48823	49877	50993	52027	53147	54287	55343	56443	57413	58481	59567	60733	61879	63031	64091	65287	66403	67481	68633	69829	70919	71947	73037	74203	75329
23	48847	49891	51001	52051	53149	54293	55351	56453	57427	58511	59581	60737	61909	63059	64109	65293	66413	67489	68639	69833	70921	71963	73039	74209	75337
24	48857	49919	51031	52057	53161	54311	55373	56467	57457	58537	59611	60757	61927	63067	64123	65309	66431	67493	68659	69847	70937	71971	73043	74219	75347
25	48859	49921	51043	52067	53171	54319	55381	56473	57467	58543	59617	60761	61933	63073	64151	65323	66449	67499	68669	69857	70949	71983	73061	74231	75353
26	48869	49927	51047	52069	53173	54323	55399	56477	57487	58549	59621	60763	61949	63079	64153	65327	66457	67511	68683	69859	70951	71987	73063	74257	75367
27	48871	49937	51059	52081	53189	54331	55411	56479	57493	58567	59627	60773	61961	63097	64157	65353	66463	67523	68687	69877	70957	71993	73079	74279	75377
28	48883	49939	51061	52103	53197	54347	55439	56489	57503	58573	59629	60779	61967	63103	64171	65357	66467	67531	68699	69899	70969	71999	73091	74287	75389
29	48889	49943	51071	52121	53201	54361	55441	56501	57527	58579	59651	60793	61979	63113	64187	65371	66491	67537	68711	69911	70979	72019	73121	74293	75391
30	48907	49957	51109	52127	53231	54367	55457	56503	57529	58601	59659	60811	61981	63127	64189	65381	66499	67547	68713	69929	70981	72031	73127	74297	75401
31	48947	49991	51131	52147	53233	54371	55469	56509	57557	58603	59663	60821	61987	63131	64217	65393	66509	67559	68729	69931	70991	72043	73133	74311	75403
32	48953	49993	51133	52153	53239	54377	55487	56519	57559	58613	59669	60859	61991	63149	64223	65407	66523	67567	68737	69941	70997	72047	73141	74317	75407
33	48973	49999	51137	52163	53267	54401	55501	56527	57571	58631	59671	60869	62003	63179	64231	65413	66529	67577	68743	69959	70999	72053	73181	74323	75431
34	48989	50021	51151	52177	53269	54403	55511	56531	57587	58657	59693	60887	62011	63197	64237	65419	66533	67579	68749	69991	71011	72073	73189	74353	75437
35	48991	50023	51157	52181	53279	54409	55529	56533	57593	58661	59699	60889	62017	63199	64271	65423	66541	67589	68767	69997	71023	72077	73237	74357	75479
36	49003	50033	51169	52183	53281	54413	55541	56543	57601	58679	59707	60899	62039	63211	64279	65437	66553	67601	68771	70001	71039	72089	73243	74363	75503
37	49009	50047	51193	52189	53299	54419	55547	56569	57637	58687	59723	60901	62047	63241	64283	65447	66569	67607	68777	70003	71059	72091	73259	74377	75511
38	49019	50051	51197	52201	53309	54421	55579	56591	57641	58693	59729	60913	62053	63247	64301	65449	66571	67619	68791	70009	71069	72101	73277	74381	75521
39	49031	50053	51199	52223	53323	54437	55589	56597	57649	58699	59743	60917	62057	63277	64303	65479	66587	67631	68813	70019	71081	72103	73291	74383	75527
40	49033	50069	51203	52237	53327	54443	55603	56599	57653	58711	59747	60919	62071	63281	64319	65497	66593	67651	68819	70039	71089	72109	73303	74411	75533
41	49037	50077	51217	52249	53353	54449	55609	56611	57667	58727	59753	60923	62081	63299	64327	65519	66601	67679	68821	70051	71119	72139	73309	74413	75539
42	49043	50087	51229	52253	53359	54469	55619	56629	57679	58733	59771	60937	62099	63311	64333	65521	66617	67699	68863	70061	71129	72161	73327	74419	75541
43	49057	50093	51239	52259	53377	54493	55621	56633	57689	58741	59779	60943	62119	63313	64373	65537	66629	67709	68879	70067	71143	72167	73331	74441	75553
44	49069	50101	51241	52267	53381	54497	55631	56659	57697	58757	59791	60953	62129	63317	64381	65539	66643	67723	68881	70079	71147	72169	73351	74449	75557
45	49081	50111	51257	52289	53401	54499	55633	56663	57709	58763	59797	60961	62131	63331	64399	65543	66653	67733	68891	70099	71153	72173	73361	74453	75571
46	49103	50119	51263	52291	53407	54503	55639	56671	57713	58771	59809	61001	62137	63337	64403	65551	66683	67741	68897	70111	71161	72211	73363	74471	75577
47	49109	50123	51283	52301	53411	54517	55661	56681	57719	58787	59833	61007	62141	63347	64433	65557	66697	67751	68899	70117	71167	72221	73369	74489	75583
48	49117	50129	51287	52313	53419	54521	55663	56687	57727	58789	59863	61027	62143	63353	64439	65563	66701	67757	68903	70121	71171	72223	73379	74507	75611
49	49121	50131	51307	52321	53437	54539	55667	56701	57731	58831	59879	61031	62171	63361	64451	65579	66713	67759	68909	70123	71191	72227	73387	74509	75617
50	49123	50147	51329	52361	53441	54541	55673	56711	57737	58889	59887	61043	62189	63367	64453	65581	66721	67763	68917	70139	71209	72229	73417	74521	75619

PRIMES

	50	51	52	53	54	55	56	57	58	59	60	61	62	63	64	65	66	67	68	69	70	71	72	73	74
51	49139	50153	51341	52363	53453	54547	55681	56713	57751	58897	59921	61051	62191	63377	64483	65587	66733	67777	68927	70141	71233	72251	73421	74527	75629
52	49157	50159	51343	52369	53479	54559	55691	56731	57773	58901	59929	61057	62201	63389	64489	65599	66739	67783	68947	70157	71237	72253	73433	74531	75641
53	49169	50177	51347	52379	53503	54563	55697	56737	57781	58907	59951	61091	62207	63391	64499	65609	66749	67789	68963	70163	71249	72269	73453	74551	75653
54	49171	50207	51349	52387	53507	54577	55711	56747	57787	58909	59957	61099	62213	63397	64513	65617	66751	67801	68993	70177	71257	72271	73459	74561	75659
55	49177	50221	51361	52391	53527	54581	55717	56767	57791	58913	59971	61121	62219	63409	64553	65629	66763	67807	69001	70181	71261	72277	73471	74567	75679
56	49193	50227	51383	52433	53549	54583	55721	56773	57793	58921	59981	61129	62233	63419	64567	65633	66791	67819	69011	70183	71263	72287	73477	74573	75683
57	49199	50231	51407	52453	53551	54601	55733	56779	57803	58937	59999	61141	62273	63421	64577	65647	66797	67829	69019	70199	71287	72307	73483	74587	75689
58	49201	50261	51413	52457	53569	54617	55763	56783	57809	58943	60013	61151	62297	63439	64579	65651	66809	67843	69029	70201	71293	72313	73517	74597	75703
59	49207	50263	51419	52489	53591	54623	55787	56807	57829	58963	60017	61153	62299	63443	64591	65657	66821	67853	69031	70207	71317	72337	73523	74609	75707
60	49211	50273	51421	52501	53593	54629	55793	56809	57839	58967	60029	61169	62303	63463	64601	65677	66841	67867	69061	70223	71327	72341	73529	74611	75709
61	49223	50287	51427	52511	53597	54631	55799	56813	57847	58979	60037	61211	62311	63467	64609	65687	66851	67883	69067	70229	71329	72353	73547	74623	75721
62	49253	50291	51431	52517	53609	54647	55807	56821	57853	58991	60041	61223	62323	63473	64613	65699	66853	67891	69073	70237	71333	72367	73553	74653	75731
63	49261	50311	51437	52529	53611	54667	55813	56827	57859	58997	60077	61231	62327	63487	64621	65701	66863	67901	69109	70241	71339	72379	73561	74687	75743
64	49277	50321	51439	52541	53617	54673	55817	56843	57881	59009	60083	61253	62347	63493	64627	65707	66877	67927	69119	70249	71341	72383	73571	74699	75767
65	49279	50329	51449	52543	53623	54679	55819	56857	57899	59011	60089	61261	62351	63499	64633	65713	66883	67931	69127	70271	71347	72421	73583	74707	75773
66	49297	50333	51461	52553	53629	54709	55823	56873	57901	59021	60091	61283	62383	63521	64661	65717	66889	67933	69143	70289	71353	72431	73589	74713	75781
67	49307	50341	51473	52561	53633	54713	55829	56891	57917	59023	60101	61291	62401	63527	64663	65719	66919	67939	69149	70297	71359	72461	73597	74717	75787
68	49331	50359	51479	52567	53639	54721	55837	56893	57923	59029	60103	61297	62417	63533	64667	65729	66923	67943	69151	70309	71363	72467	73607	74719	75793
69	49333	50363	51481	52571	53653	54727	55843	56897	57943	59051	60107	61331	62423	63541	64679	65731	66931	67957	69163	70313	71387	72469	73609	74729	75797
70	49339	50377	51487	52579	53657	54751	55849	56909	57947	59053	60127	61333	62459	63559	64693	65761	66943	67961	69191	70321	71389	72481	73613	74731	75821
71	49363	50383	51503	52583	53681	54767	55871	56911	57973	59063	60133	61339	62467	63577	64709	65777	66947	67967	69193	70327	71399	72493	73637	74747	75833
72	49367	50387	51511	52609	53693	54773	55889	56921	57977	59069	60139	61343	62473	63587	64717	65789	66949	67979	69197	70351	71411	72497	73643	74759	75853
73	49369	50411	51517	52627	53699	54779	55897	56923	57991	59077	60149	61357	62477	63589	64747	65809	66959	67987	69203	70373	71413	72503	73651	74761	75869
74	49391	50417	51521	52631	53717	54787	55901	56929	58013	59083	60161	61363	62483	63599	64763	65827	66973	67993	69221	70379	71419	72533	73673	74771	75883
75	49393	50423	51539	52639	53719	54799	55903	56941	58027	59093	60167	61379	62497	63601	64781	65831	66977	68023	69233	70381	71429	72547	73679	74779	75913
76	49409	50441	51551	52667	53731	54829	55921	56951	58031	59107	60169	61381	62501	63607	64783	65837	67003	68041	69239	70393	71437	72551	73681	74797	75931
77	49411	50459	51563	52673	53759	54833	55927	56957	58043	59113	60209	61403	62507	63611	64793	65839	67021	68053	69247	70423	71443	72559	73693	74821	75937
78	49417	50461	51577	52691	53773	54851	55931	56963	58049	59119	60217	61409	62533	63617	64811	65843	67033	68059	69257	70429	71453	72577	73699	74827	75941
79	49429	50497	51581	52697	53777	54869	55933	56983	58057	59123	60223	61417	62539	63629	64817	65851	67043	68071	69259	70439	71471	72613	73709	74831	75967
80	49433	50503	51593	52709	53783	54877	55949	56989	58061	59141	60251	61441	62549	63647	64849	65867	67049	68087	69263	70451	71473	72617	73721	74843	75979
81	49451	50513	51599	52711	53791	54881	55967	56993	58067	59149	60257	61463	62563	63649	64853	65881	67057	68099	69313	70457	71479	72623	73727	74857	75983
82	49459	50527	51607	52721	53813	54907	55987	56999	58073	59159	60259	61469	62581	63659	64871	65899	67061	68111	69317	70459	71483	72643	73751	74861	75989
83	49463	50539	51613	52727	53819	54917	55997	57037	58099	59167	60271	61471	62591	63667	64877	65921	67073	68113	69337	70481	71503	72647	73757	74869	75991
84	49477	50543	51631	52733	53831	54919	56003	57041	58109	59183	60289	61483	62597	63671	64879	65927	67079	68141	69341	70487	71527	72649	73771	74873	75997
85	49481	50549	51637	52747	53849	54941	56009	57047	58111	59197	60293	61487	62603	63689	64891	65929	67103	68147	69371	70489	71537	72661	73783	74887	76001
86	49499	50551	51647	52757	53857	54949	56039	57059	58129	59207	60317	61493	62617	63691	64901	65951	67121	68161	69379	70501	71549	72671	73819	74891	76003
87	49523	50581	51659	52769	53861	54959	56041	57073	58147	59209	60331	61507	62627	63697	64919	65957	67129	68171	69383	70507	71551	72673	73823	74897	76031
88	49529	50587	51673	52783	53881	54973	56053	57077	58151	59219	60337	61511	62633	63703	64921	65963	67139	68207	69389	70529	71563	72679	73847	74903	76039
89	49531	50591	51679	52807	53887	54979	56081	57089	58153	59221	60343	61519	62639	63709	64927	65981	67141	68209	69401	70537	71569	72689	73849	74923	76079
90	49537	50593	51683	52813	53891	54983	56087	57097	58169	59233	60353	61543	62653	63719	64937	65983	67153	68213	69403	70549	71593	72701	73859	74929	76081
91	49547	50599	51691	52817	53897	55001	56093	57107	58171	59239	60373	61547	62659	63727	64951	65993	67157	68219	69427	70571	71597	72707	73867	74933	76091
92	49549	50627	51713	52837	53899	55009	56099	57119	58189	59243	60383	61553	62683	63737	64969	66029	67169	68227	69431	70573	71633	72719	73877	74941	76099
93	49559	50647	51719	52859	53917	55021	56101	57131	58193	59263	60397	61559	62687	63743	64997	66037	67181	68239	69439	70583	71647	72727	73883	74959	76103
94	49597	50651	51721	52861	53923	55049	56113	57139	58199	59273	60413	61561	62701	63761	65003	66041	67187	68261	69457	70589	71663	72733	73897	75011	76123
95	49603	50671	51749	52879	53927	55051	56123	57143	58207	59281	60427	61583	62723	63773	65011	66047	67189	68279	69463	70607	71671	72739	73907	75013	76129
96	49613	50683	51767	52883	53939	55057	56131	57149	58211	59333	60443	61603	62731	63781	65027	66067	67211	68281	69467	70619	71693	72763	73939	75017	76147
97	49627	50707	51769	52889	53951	55061	56149	57163	58217	59341	60449	61609	62743	63793	65029	66071	67213	68311	69473	70621	71699	72767	73943	75029	76157
98	49633	50723	51787	52901	53959	55073	56167	57173	58229	59351	60457	61613	62753	63799	65033	66083	67217	68329	69481	70627	71707	72797	73951	75037	76159
99	49639	50741	51797	52903	53987	55079	56171	57179	58231	59357	60493	61627	62761	63803	65053	66089	67219	68351	69491	70639	71711	72817	73961	75041	76163
100	49663	50753	51803	52919	53993	55103	56179	57191	58237	59359	60497	61631	62773	63809	65063	66103	67231	68371	69493	70657	71713	72823	73973	75079	76207

PRIMES

	75	76	77	78	79	80	81	82	83	84	85	86	87	88	89	90	91	92	93	94	95
1	76213	77359	78487	79627	80737	81817	82903	84131	85243	86381	87557	88807	89867	90989	92177	93187	94351	95443	96587	97829	98953
2	76231	77369	78497	79631	80747	81839	82913	84137	85247	86389	87559	88811	89891	90997	92179	93199	94379	95461	96589	97841	98963
3	76243	77377	78509	79633	80749	81847	82939	84143	85259	86399	87583	88813	89897	91009	92189	93229	94397	95467	96601	97843	98981
4	76249	77383	78511	79657	80761	81853	82963	84163	85297	86413	87587	88817	89899	91019	92203	93239	94399	95471	96643	97847	98993
5	76253	77417	78517	79669	80777	81869	82981	84179	85303	86423	87589	88819	89909	91033	92219	93241	94421	95479	96661	97849	98999
6	76259	77419	78539	79687	80779	81883	82997	84181	85313	86441	87613	88843	89917	91079	92221	93251	94427	95483	96667	97859	99013
7	76261	77431	78541	79691	80783	81899	83003	84191	85331	86453	87623	88853	89923	91081	92227	93253	94433	95507	96671	97861	99017
8	76283	77447	78553	79693	80789	81901	83009	84199	85333	86461	87629	88861	89939	91097	92233	93257	94439	95527	96697	97871	99023
9	76289	77471	78569	79697	80803	81919	83023	84211	85361	86467	87631	88867	89959	91099	92237	93263	94441	95531	96703	97879	99041
10	76303	77477	78571	79699	80809	81929	83047	84221	85363	86477	87641	88873	89963	91121	92243	93281	94447	95539	96731	97883	99053
11	76333	77479	78577	79757	80819	81931	83059	84223	85369	86491	87643	88883	89977	91127	92251	93283	94463	95549	96737	97919	99079
12	76343	77489	78583	79769	80831	81937	83063	84229	85381	86501	87649	88897	89983	91129	92269	93287	94477	95561	96739	97927	99083
13	76367	77491	78593	79777	80833	81943	83071	84239	85411	86509	87671	88903	89989	91139	92297	93307	94483	95569	96749	97931	99089
14	76369	77509	78607	79801	80849	81953	83077	84247	85427	86531	87679	88919	90001	91141	92311	93319	94513	95581	96757	97943	99103
15	76379	77513	78623	79811	80863	81967	83089	84263	85429	86533	87683	88937	90007	91151	92317	93323	94529	95597	96763	97961	99109
16	76387	77521	78643	79813	80897	81971	83093	84299	85439	86539	87691	88951	90011	91153	92333	93329	94531	95603	96769	97967	99119
17	76403	77527	78649	79817	80909	81973	83101	84307	85447	86561	87697	88969	90017	91159	92347	93337	94541	95617	96779	97973	99131
18	76421	77543	78653	79823	80911	82003	83117	84313	85451	86573	87701	88993	90019	91163	92353	93371	94543	95621	96787	97987	99133
19	76423	77549	78691	79829	80917	82007	83137	84317	85453	86579	87719	88997	90023	91183	92357	93377	94547	95629	96797	98009	99137
20	76441	77551	78697	79841	80923	82009	83177	84319	85469	86587	87721	89003	90031	91193	92363	93383	94559	95633	96799	98011	99139
21	76463	77557	78707	79843	80929	82013	83203	84347	85487	86599	87739	89009	90053	91199	92369	93407	94561	95651	96821	98017	99149
22	76471	77563	78713	79847	80933	82021	83207	84349	85513	86627	87743	89017	90059	91229	92377	93419	94573	95701	96823	98041	99173
23	76481	77569	78721	79861	80953	82031	83219	84377	85517	86629	87751	89021	90067	91237	92381	93427	94583	95707	96827	98047	99181
24	76487	77573	78737	79867	80963	82037	83221	84389	85523	86677	87767	89041	90071	91243	92383	93463	94597	95713	96847	98057	99191
25	76493	77587	78779	79873	80989	82039	83227	84391	85531	86689	87793	89051	90073	91249	92387	93479	94603	95717	96851	98081	99223
26	76507	77591	78781	79889	81001	82051	83231	84401	85549	86693	87797	89057	90089	91253	92399	93481	94613	95723	96857	98101	99233
27	76511	77611	78787	79901	81013	82067	83233	84407	85571	86711	87803	89069	90107	91283	92401	93487	94621	95731	96893	98123	99241
28	76519	77617	78791	79903	81017	82073	83243	84421	85577	86719	87811	89071	90121	91291	92413	93491	94649	95737	96907	98129	99251
29	76537	77621	78797	79907	81019	82129	83257	84431	85597	86729	87833	89083	90127	91297	92419	93493	94651	95747	96911	98143	99257
30	76541	77641	78803	79939	81023	82139	83267	84437	85601	86743	87853	89087	90149	91303	92431	93497	94687	95773	96931	98179	99259
31	76543	77647	78809	79943	81031	82141	83269	84443	85607	86753	87869	89101	90163	91309	92459	93503	94693	95783	96953	98207	99277
32	76561	77659	78823	79967	81041	82153	83273	84449	85619	86767	87877	89107	90173	91331	92461	93523	94709	95789	96959	98213	99289
33	76579	77681	78839	79973	81043	82163	83299	84457	85621	86771	87881	89113	90187	91367	92467	93529	94723	95791	96973	98221	99317
34	76597	77687	78853	79979	81047	82171	83311	84463	85627	86783	87887	89119	90191	91369	92479	93553	94727	95801	96979	98227	99347
35	76603	77689	78857	79987	81049	82183	83339	84467	85639	86813	87911	89123	90197	91373	92489	93557	94747	95803	96989	98251	99349
36	76607	77699	78877	79997	81071	82189	83341	84481	85643	86837	87917	89137	90199	91381	92503	93559	94771	95813	96997	98257	99367
37	76631	77711	78887	79999	81077	82193	83357	84499	85661	86843	87931	89153	90203	91387	92507	93563	94777	95819	97001	98269	99371
38	76649	77713	78889	80021	81083	82207	83383	84503	85667	86851	87943	89189	90217	91393	92551	93581	94781	95857	97003	98297	99377
39	76651	77719	78893	80039	81097	82217	83389	84509	85669	86857	87959	89203	90227	91397	92557	93601	94789	95869	97007	98299	99391
40	76667	77723	78901	80051	81101	82219	83399	84521	85691	86861	87961	89209	90239	91411	92567	93607	94793	95873	97021	98317	99397
41	76673	77731	78919	80071	81119	82223	83401	84523	85703	86869	87973	89213	90247	91423	92569	93629	94811	95881	97039	98321	99401
42	76679	77743	78929	80077	81131	82231	83407	84533	85711	86923	87977	89227	90263	91433	92581	93637	94819	95891	97073	98323	99409
43	76697	77747	78941	80107	81157	82237	83417	84551	85717	86927	87991	89231	90271	91453	92593	93683	94823	95911	97081	98327	99431
44	76717	77761	78977	80111	81163	82241	83423	84559	85733	86929	88001	89237	90281	91457	92623	93701	94837	95917	97103	98347	99439
45	76733	77773	78979	80141	81173	82261	83431	84589	85751	86939	88003	89261	90289	91459	92627	93703	94841	95923	97117	98369	99469
46	76753	77783	78989	80147	81181	82267	83437	84629	85781	86951	88007	89269	90313	91463	92639	93719	94847	95929	97127	98377	99487
47	76757	77797	79031	80149	81197	82279	83443	84631	85793	86959	88019	89273	90353	91493	92641	93739	94849	95947	97151	98387	99497
48	76771	77801	79039	80153	81199	82301	83449	84649	85817	86969	88037	89293	90359	91499	92647	93761	94873	95957	97157	98389	99523
49	76777	77813	79043	80167	81203	82307	83459	84653	85819	86981	88069	89303	90371	91513	92657	93763	94889	95959	97159	98407	99527
50	76781	77839	79063	80173	81223	82339	83471	84659	85829	86993	88079	89317	90373	91529	92669	93787	94903	95971	97169	98411	99529

	75	76	77	78	79	80	81	82	83	84	85	86	87	88	89	90	91	92	93	94	95
51	76801	77849	79087	80177	81233	82349	83477	84673	85831	87011	88093	89329	90379	91541	92671	93809	94907	95987	97171	98419	99551
52	76819	77863	79103	80191	81239	82351	83497	84691	85837	87013	88117	89363	90397	91571	92681	93811	94933	95989	97177	98429	99559
53	76829	77867	79111	80207	81281	82361	83537	84697	85843	87037	88129	89371	90401	91573	92683	93827	94949	96001	97187	98443	99563
54	76831	77893	79133	80209	81283	82373	83557	84701	85847	87041	88169	89381	90403	91577	92693	93851	94951	96013	97213	98453	99571
55	76837	77899	79139	80221	81293	82387	83561	84713	85853	87049	88177	89387	90407	91583	92699	93871	94961	96017	97231	98459	99577
56	76847	77929	79147	80231	81299	82393	83563	84719	85889	87071	88211	89393	90437	91591	92707	93887	94993	96043	97241	98467	99581
57	76871	77933	79151	80233	81307	82421	83579	84731	85903	87083	88223	89399	90439	91621	92717	93889	94999	96053	97259	98473	99607
58	76873	77951	79153	80239	81331	82457	83591	84737	85909	87103	88237	89413	90469	91631	92723	93893	95003	96059	97283	98479	99611
59	76883	77969	79159	80251	81343	82463	83597	84751	85931	87107	88241	89417	90473	91639	92737	93901	95009	96079	97301	98491	99623
60	76907	77977	79181	80263	81349	82469	83609	84761	85933	87119	88259	89431	90481	91673	92753	93911	95021	96097	97303	98507	99643
61	76913	77983	79187	80273	81353	82471	83617	84787	85991	87121	88261	89443	90499	91691	92761	93913	95027	96137	97327	98519	99661
62	76919	77999	79193	80279	81359	82483	83621	84793	85999	87133	88289	89449	90511	91703	92767	93923	95063	96149	97367	98533	99667
63	76943	78007	79201	80287	81371	82487	83639	84809	86011	87149	88301	89459	90523	91711	92779	93937	95071	96157	97369	98543	99679
64	76949	78017	79229	80309	81373	82493	83641	84811	86017	87151	88321	89477	90527	91733	92789	93941	95083	96167	97373	98561	99689
65	76961	78031	79231	80317	81401	82499	83653	84827	86027	87179	88327	89491	90529	91753	92791	93949	95087	96179	97379	98563	99707
66	76963	78041	79241	80329	81409	82507	83663	84857	86029	87181	88337	89501	90533	91757	92801	93967	95089	96181	97381	98573	99709
67	76991	78049	79259	80341	81421	82529	83689	84859	86069	87187	88339	89513	90547	91771	92809	93971	95093	96199	97387	98597	99713
68	77003	78059	79273	80347	81439	82531	83701	84869	86077	87211	88379	89519	90583	91781	92821	93979	95101	96211	97397	98621	99719
69	77017	78079	79279	80363	81457	82549	83717	84871	86083	87221	88397	89521	90599	91801	92831	93983	95107	96221	97423	98627	99721
70	77023	78101	79283	80369	81463	82559	83719	84913	86111	87223	88411	89527	90617	91807	92849	93997	95111	96223	97429	98639	99733
71	77029	78121	79301	80387	81509	82561	83737	84919	86113	87251	88423	89533	90619	91811	92857	94007	95131	96233	97441	98641	99761
72	77041	78137	79309	80407	81517	82567	83761	84947	86117	87253	88427	89561	90631	91813	92861	94009	95143	96259	97453	98663	99767
73	77047	78139	79319	80429	81527	82571	83773	84961	86131	87257	88463	89563	90641	91823	92863	94033	95153	96263	97459	98669	99787
74	77069	78157	79333	80447	81533	62591	83777	84967	86137	87277	88469	89567	90647	91837	92867	94049	95177	96269	97463	98689	99793
75	77081	78163	79337	80449	81547	82601	83791	84977	86143	87281	88471	89591	90659	91841	92893	94057	95189	96281	97499	98711	99809
76	77093	78167	79349	80471	81551	82609	83813	84979	86161	87293	88493	89597	90677	91867	92899	94063	95191	96289	97501	98713	99817
77	77101	78173	79357	80473	81553	82613	83833	84991	86171	87299	88499	89599	90679	91873	92921	94079	95203	96293	97511	98717	99823
78	77137	78179	79367	80489	81559	82619	83843	85009	86179	87313	88513	89603	90697	91909	92927	94099	95213	96323	97523	98729	99829
79	77141	78191	79379	80491	81563	82633	83857	85021	86183	87317	88523	89611	90703	91921	92941	94109	95219	96329	97547	98731	99833
80	77153	78193	79393	80513	81569	82651	83869	85027	86197	87323	88547	89627	90709	91939	92951	94111	95231	96331	97549	98737	99839
81	77167	78203	79397	80527	81611	82657	83873	85037	86201	87337	88589	89633	90731	91943	92957	94117	95233	96337	97553	98773	99859
82	77171	78229	79399	80537	81619	82699	83891	85049	86209	87359	88591	89653	90749	91951	92959	94121	95239	96353	97561	98779	99871
83	77191	78233	79411	80557	81629	82721	83903	85061	86239	87383	88607	89657	90787	91957	92987	94151	95257	96377	97571	98801	99877
84	77201	78241	79423	80567	81637	82723	83911	85081	86243	87403	88609	89659	90793	91961	92993	94153	95261	96401	97577	98807	99881
85	77213	78259	79427	80599	81647	82727	83921	85087	86249	87407	88643	89669	90803	91967	93001	94169	95267	96419	97579	98809	99901
86	77237	78277	79433	80603	81649	82729	83933	85091	86257	87421	88651	89671	90821	91969	93047	94201	95273	96431	97583	98837	99907
87	77239	78283	79451	80611	81667	82757	83939	85093	86263	87427	88657	89681	90823	91997	93053	94207	95279	96443	97607	98849	99923
88	77243	78301	79481	80621	81671	82759	83969	85103	86269	87433	88661	89689	90833	92003	93059	94219	95287	96451	97609	98867	99929
89	77249	78307	79493	80627	81677	82763	83983	85109	86287	87443	88663	89753	90841	92009	93077	94229	95311	96457	97613	98869	99961
90	77261	78311	79531	80629	81689	82781	83987	85121	86291	87473	88667	89759	90847	92033	93083	94253	95317	96461	97649	98873	99971
91	77263	78317	79537	80651	81701	82787	84011	85133	86293	87481	88681	89767	90863	92041	93089	94261	95327	96469	97651	98887	99989
92	77267	78341	79549	80657	81703	82793	84017	85147	86297	87491	88721	89779	90887	92051	93097	94273	95339	96479	97673	98893	99991
93	77269	78347	79559	80669	81707	82799	84047	85159	86311	87509	88729	89783	90901	92077	93103	94291	95369	96487	97687	98897	
94	77279	78367	79561	80671	81727	82811	84053	85193	86323	87511	88741	89797	90907	92083	93113	94307	95383	96493	97711	98899	
95	77291	78401	79579	80677	81737	82813	84059	85199	86341	87517	88747	89809	90911	92107	93131	94309	95393	96497	97729	98909	
96	77317	78427	79589	80681	81749	82837	84061	85201	86351	87523	88771	89819	90917	92111	93133	94321	95401	96517	97771	98911	
97	77323	78437	79601	80683	81761	82847	84067	85213	86353	87539	88789	89821	90931	92119	93139	94327	95413	96527	97777	98927	
98	77339	78439	79609	80687	81769	82883	84089	85223	86357	87541	88793	89833	90947	92143	93151	94331	95419	96553	97787	98929	
99	77347	78467	79613	80701	81773	82889	84121	85229	86369	87547	88799	89839	90971	92153	93169	94343	95429	96557	97789	98939	
100	77351	78479	79621	80713	81799	82891	84127	85237	86371	87553	88801	89849	90977	92173	93179	94349	95441	96581	97813	98947	

FACTORS AND PRIMES

This table presents the prime factors of all factorable numbers and the mantissas of the common logarithms of all prime numbers from 1 to 2,000. The table runs across two facing pages. Thus, the factors of 258 are found on a line with 25 and under vertical column 8 to be $2 \cdot 3 \cdot 43$. If n is prime, the mantissa of its common logarithm is given. If n is not prime its prime factors are given.

n	0	1	2	3	4
0	....	0000000	**3010300**	**4771213**	2^2
1	$2 \cdot 5$	**0413927**	$2^2 \cdot 3$	**1139434**	$2 \cdot 7$
2	$2^2 \cdot 5$	$3 \cdot 7$	$2 \cdot 11$	**3617278**	$2^3 \cdot 3$
3	$2 \cdot 3 \cdot 5$	**4913617**	2^5	$3 \cdot 11$	$2 \cdot 17$
4	$2^3 \cdot 5$	**6127839**	$2 \cdot 3 \cdot 7$	**6334685**	$2^2 \cdot 11$
5	$2 \cdot 5^2$	$3 \cdot 17$	$2^2 \cdot 13$	**7242759**	$2 \cdot 3^3$
6	$2^2 \cdot 3 \cdot 5$	**7853298**	$2 \cdot 31$	$3^2 \cdot 7$	2^6
7	$2 \cdot 5 \cdot 7$	**8512583**	$2^3 \cdot 3^2$	**8633229**	$2 \cdot 37$
8	$2^4 \cdot 5$	3^4	$2 \cdot 41$	**9190781**	$2^2 \cdot 3 \cdot 7$
9	$2 \cdot 3^2 \cdot 5$	$7 \cdot 13$	$2^2 \cdot 23$	$3 \cdot 31$	$2 \cdot 47$
10	$2^2 \cdot 5^2$	**0043214**	$2 \cdot 3 \cdot 17$	**0128372**	$2^3 \cdot 13$
11	$2 \cdot 5 \cdot 11$	$3 \cdot 37$	$2^4 \cdot 7$	**0530784**	$2 \cdot 3 \cdot 19$
12	$2^3 \cdot 3 \cdot 5$	11^2	$2 \cdot 61$	$3 \cdot 41$	$2^2 \cdot 31$
13	$2 \cdot 5 \cdot 13$	**1172713**	$2^2 \cdot 3 \cdot 11$	$7 \cdot 19$	$2 \cdot 67$
14	$2^2 \cdot 5 \cdot 7$	$3 \cdot 47$	$2 \cdot 71$	$11 \cdot 13$	$2^4 \cdot 3^2$
15	$2 \cdot 3 \cdot 5^2$	**1789769**	$2^3 \cdot 19$	$3^2 \cdot 17$	$2 \cdot 7 \cdot 11$
16	$2^5 \cdot 5$	$7 \cdot 23$	$2 \cdot 3^4$	**2121876**	$2^2 \cdot 41$
17	$2 \cdot 5 \cdot 17$	$3^2 \cdot 19$	$2^2 \cdot 43$	**2380461**	$2 \cdot 3 \cdot 29$
18	$2^2 \cdot 3^2 \cdot 5$	**2576786**	$2 \cdot 7 \cdot 13$	$3 \cdot 61$	$2^3 \cdot 23$
19	$2 \cdot 5 \cdot 19$	**2810334**	$2^6 \cdot 3$	**2855573**	$2 \cdot 97$
20	$2^3 \cdot 5^2$	$3 \cdot 67$	$2 \cdot 101$	$7 \cdot 29$	$2^2 \cdot 3 \cdot 17$
21	$2 \cdot 3 \cdot 5 \cdot 7$	**3242825**	$2^2 \cdot 53$	$3 \cdot 71$	$2 \cdot 107$
22	$2^2 \cdot 5 \cdot 11$	$13 \cdot 17$	$2 \cdot 3 \cdot 37$	**3483049**	$2^5 \cdot 7$
23	$2 \cdot 5 \cdot 23$	$3 \cdot 7 \cdot 11$	$2^3 \cdot 29$	**3673559**	$2 \cdot 3^2 \cdot 13$
24	$2^4 \cdot 3 \cdot 5$	**3820170**	$2 \cdot 11^2$	3^5	$2^2 \cdot 61$
25	$2 \cdot 5^3$	**3996737**	$2^2 \cdot 3^2 \cdot 7$	$11 \cdot 23$	$2 \cdot 127$
26	$2^2 \cdot 5 \cdot 13$	$3^2 \cdot 29$	$2 \cdot 131$	**4199557**	$2^3 \cdot 3 \cdot 11$
27	$2 \cdot 3^3 \cdot 5$	**4329693**	$2^4 \cdot 17$	$3 \cdot 7 \cdot 13$	$2 \cdot 137$
28	$2^3 \cdot 5 \cdot 7$	**4487063**	$2 \cdot 3 \cdot 47$	**4517864**	$2^2 \cdot 71$
29	$2 \cdot 5 \cdot 29$	$3 \cdot 97$	$2^2 \cdot 73$	**4668676**	$2 \cdot 3 \cdot 7^2$
30	$2^2 \cdot 3 \cdot 5^2$	$7 \cdot 43$	$2 \cdot 151$	$3 \cdot 101$	$2^4 \cdot 19$
31	$2 \cdot 5 \cdot 31$	**4927604**	$2^3 \cdot 3 \cdot 13$	**4955443**	$2 \cdot 157$
32	$2^6 \cdot 5$	$3 \cdot 107$	$2 \cdot 7 \cdot 23$	$17 \cdot 19$	$2^2 \cdot 3^4$
33	$2 \cdot 3 \cdot 5 \cdot 11$	**5198280**	$2^2 \cdot 83$	$3^2 \cdot 37$	$2 \cdot 167$
34	$2^2 \cdot 5 \cdot 17$	$11 \cdot 31$	$2 \cdot 3^2 \cdot 19$	7^3	$2^3 \cdot 43$
35	$2 \cdot 5^2 \cdot 7$	$3^3 \cdot 13$	$2^5 \cdot 11$	**5477747**	$2 \cdot 3 \cdot 59$
36	$2^3 \cdot 3^2 \cdot 5$	19^2	$2 \cdot 181$	$3 \cdot 11^2$	$2^2 \cdot 7 \cdot 13$
37	$2 \cdot 5 \cdot 37$	$7 \cdot 53$	$2^2 \cdot 3 \cdot 31$	**5717088**	$2 \cdot 11 \cdot 17$
38	$2^2 \cdot 5 \cdot 19$	$3 \cdot 127$	$2 \cdot 191$	**5831988**	$2^7 \cdot 3$
39	$2 \cdot 3 \cdot 5 \cdot 13$	$17 \cdot 23$	$2^3 \cdot 7^2$	$3 \cdot 131$	$2 \cdot 197$
40	$2^4 \cdot 5^2$	**6031444**	$2 \cdot 3 \cdot 67$	$13 \cdot 31$	$2^2 \cdot 101$
41	$2 \cdot 5 \cdot 41$	$3 \cdot 137$	$2^2 \cdot 103$	$7 \cdot 59$	$2 \cdot 3^2 \cdot 23$
42	$2^2 \cdot 3 \cdot 5 \cdot 7$	**6242821**	$2 \cdot 211$	$3^2 \cdot 47$	$2^3 \cdot 53$
43	$2 \cdot 5 \cdot 43$	**6344773**	$2^4 \cdot 3^3$	**6364879**	$2 \cdot 7 \cdot 31$
44	$2^3 \cdot 5 \cdot 11$	$3^2 \cdot 7^2$	$2 \cdot 13 \cdot 17$	**6464037**	$2^2 \cdot 3 \cdot 37$
45	$2 \cdot 3^2 \cdot 5^2$	$11 \cdot 41$	$2^2 \cdot 113$	$3 \cdot 151$	$2 \cdot 227$
46	$2^2 \cdot 5 \cdot 23$	**6637009**	$2 \cdot 3 \cdot 7 \cdot 11$	**6655810**	$2^4 \cdot 29$
47	$2 \cdot 5 \cdot 47$	$3 \cdot 157$	$2^3 \cdot 59$	$11 \cdot 43$	$2 \cdot 3 \cdot 79$
48	$2^5 \cdot 3 \cdot 5$	$13 \cdot 37$	$2 \cdot 241$	$3 \cdot 7 \cdot 23$	$2^2 \cdot 11^2$
49	$2 \cdot 5 \cdot 7^2$	**6910815**	$2^2 \cdot 3 \cdot 41$	$17 \cdot 29$	$2 \cdot 13 \cdot 19$

FACTORS AND PRIMES

n	5	6	7	8	9
0	**6989700**	$2 \cdot 3$	**8450980**	2^3	3^2
1	$3 \cdot 5$	2^4	**2304489**	$2 \cdot 3^2$	**2787536**
2	5^2	$2 \cdot 13$	3^3	$2^2 \cdot 7$	**4623980**
3	$5 \cdot 7$	$2^2 \cdot 3^2$	**5682017**	$2 \cdot 19$	$3 \cdot 13$
4	$3^2 \cdot 5$	$2 \cdot 23$	**6720979**	$2^4 \cdot 3$	7^2
5	$5 \cdot 11$	$2^3 \cdot 7$	$3 \cdot 19$	$2 \cdot 29$	**7708520**
6	$5 \cdot 13$	$2 \cdot 3 \cdot 11$	**8260748**	$2^2 \cdot 17$	$3 \cdot 23$
7	$3 \cdot 5^2$	$2^2 \cdot 19$	$7 \cdot 11$	$2 \cdot 3 \cdot 13$	**8976271**
8	$5 \cdot 17$	$2 \cdot 43$	$3 \cdot 29$	$2^3 \cdot 11$	**9493900**
9	$5 \cdot 19$	$2^5 \cdot 3$	**9867717**	$2 \cdot 7^2$	$3^2 \cdot 11$
10	$3 \cdot 5 \cdot 7$	$2 \cdot 53$	**0293838**	$2^2 \cdot 3^3$	**0374265**
11	$5 \cdot 23$	$2^2 \cdot 29$	$3^2 \cdot 13$	$2 \cdot 59$	$7 \cdot 17$
12	5^3	$2 \cdot 3^2 \cdot 7$	**1038037**	2^7	$3 \cdot 43$
13	$3^3 \cdot 5$	$2^3 \cdot 17$	**1367206**	$2 \cdot 3 \cdot 23$	**1430148**
14	$5 \cdot 29$	$2 \cdot 73$	$3 \cdot 7^2$	$2^2 \cdot 37$	**1731863**
15	$5 \cdot 31$	$2^2 \cdot 3 \cdot 13$	**1958997**	$2 \cdot 79$	$3 \cdot 53$
16	$3 \cdot 5 \cdot 11$	$2 \cdot 83$	**2227165**	$2^3 \cdot 3 \cdot 7$	13^2
17	$5^2 \cdot 7$	$2^4 \cdot 11$	$3 \cdot 59$	$2 \cdot 89$	**2528530**
18	$5 \cdot 37$	$2 \cdot 3 \cdot 31$	$11 \cdot 17$	$2^2 \cdot 47$	$3^3 \cdot 7$
19	$3 \cdot 5 \cdot 13$	$2^2 \cdot 7^2$	**2944662**	$2 \cdot 3^2 \cdot 11$	**2988531**
20	$5 \cdot 41$	$2 \cdot 103$	$3^2 \cdot 23$	$2^4 \cdot 13$	$11 \cdot 19$
21	$5 \cdot 43$	$2^3 \cdot 3^3$	$7 \cdot 31$	$2 \cdot 109$	$3 \cdot 73$
22	$3^2 \cdot 5^2$	$2 \cdot 113$	**3560259**	$2^2 \cdot 3 \cdot 19$	**3598355**
23	$5 \cdot 47$	$2^2 \cdot 59$	$3 \cdot 79$	$2 \cdot 7 \cdot 17$	**3783979**
24	$5 \cdot 7^2$	$2 \cdot 3 \cdot 41$	$13 \cdot 19$	$2^3 \cdot 31$	$3 \cdot 83$
25	$3 \cdot 5 \cdot 17$	2^8	**4099331**	$2 \cdot 3 \cdot 43$	$7 \cdot 37$
26	$5 \cdot 53$	$2 \cdot 7 \cdot 19$	$3 \cdot 89$	$2^2 \cdot 67$	**4297523**
27	$5^2 \cdot 11$	$2^2 \cdot 3 \cdot 23$	**4424798**	$2 \cdot 139$	$3^2 \cdot 31$
28	$3 \cdot 5 \cdot 19$	$2 \cdot 11 \cdot 13$	$7 \cdot 41$	$2^5 \cdot 3^2$	17^2
29	$5 \cdot 59$	$2^3 \cdot 37$	$3^3 \cdot 11$	$2 \cdot 149$	$13 \cdot 23$
30	$5 \cdot 61$	$2 \cdot 3^2 \cdot 17$	**4871384**	$2^2 \cdot 7 \cdot 11$	$3 \cdot 103$
31	$3^2 \cdot 5 \cdot 7$	$2^2 \cdot 79$	**5010593**	$2 \cdot 3 \cdot 53$	$11 \cdot 29$
32	$5^2 \cdot 13$	$2 \cdot 163$	$3 \cdot 109$	$2^3 \cdot 41$	$7 \cdot 47$
33	$5 \cdot 67$	$2^4 \cdot 3 \cdot 7$	**5276299**	$2 \cdot 13^2$	$3 \cdot 113$
34	$3 \cdot 5 \cdot 23$	$2 \cdot 173$	**5403295**	$2^2 \cdot 3 \cdot 29$	**5428254**
35	$5 \cdot 71$	$2^2 \cdot 89$	$3 \cdot 7 \cdot 17$	$2 \cdot 179$	**5550944**
36	$5 \cdot 73$	$2 \cdot 3 \cdot 61$	**5646661**	$2^4 \cdot 23$	$3^2 \cdot 41$
37	$3 \cdot 5^3$	$2^3 \cdot 47$	$13 \cdot 29$	$2 \cdot 3^3 \cdot 7$	**5786392**
38	$5 \cdot 7 \cdot 11$	$2 \cdot 193$	$3^2 \cdot 43$	$2^2 \cdot 97$	**5899496**
39	$5 \cdot 79$	$2^2 \cdot 3^2 \cdot 11$	**5987905**	$2 \cdot 199$	$3 \cdot 7 \cdot 19$
40	$3^4 \cdot 5$	$2 \cdot 7 \cdot 29$	$11 \cdot 37$	$2^3 \cdot 3 \cdot 17$	**6117233**
41	$5 \cdot 83$	$2^5 \cdot 13$	$3 \cdot 139$	$2 \cdot 11 \cdot 19$	**6222140**
42	$5^2 \cdot 17$	$2 \cdot 3 \cdot 71$	$7 \cdot 61$	$2^2 \cdot 107$	$3 \cdot 11 \cdot 13$
43	$3 \cdot 5 \cdot 29$	$2^2 \cdot 109$	$19 \cdot 23$	$2 \cdot 3 \cdot 73$	**6424645**
44	$5 \cdot 89$	$2 \cdot 223$	$3 \cdot 149$	$2^6 \cdot 7$	**6522463**
45	$5 \cdot 7 \cdot 13$	$2^3 \cdot 3 \cdot 19$	**6599162**	$2 \cdot 229$	$3^3 \cdot 17$
46	$3 \cdot 5 \cdot 31$	$2 \cdot 233$	**6693169**	$2^2 \cdot 3^2 \cdot 13$	$7 \cdot 67$
47	$5^2 \cdot 19$	$2^2 \cdot 7 \cdot 17$	$3^2 \cdot 53$	$2 \cdot 239$	**6803355**
48	$5 \cdot 97$	$2 \cdot 3^5$	**6875290**	$2^3 \cdot 61$	$3 \cdot 163$
49	$3^2 \cdot 5 \cdot 11$	$2^4 \cdot 31$	$7 \cdot 71$	$2 \cdot 3 \cdot 83$	**6981005**

FACTORS AND PRIMES

n	0	1	2	3	4
50	$2^2 \cdot 5^3$	$3 \cdot 167$	$2 \cdot 251$	**7015680**	$2^3 \cdot 3^2 \cdot 7$
51	$2 \cdot 3 \cdot 5 \cdot 17$	$7 \cdot 73$	2^9	$3^3 \cdot 19$	$2 \cdot 257$
52	$2^3 \cdot 5 \cdot 13$	**7168377**	$2 \cdot 3^2 \cdot 29$	**7185017**	$2^2 \cdot 131$
53	$2 \cdot 5 \cdot 53$	$3^2 \cdot 59$	$2^2 \cdot 7 \cdot 19$	$13 \cdot 41$	$2 \cdot 3 \cdot 89$
54	$2^2 \cdot 3^3 \cdot 5$	**7331973**	$2 \cdot 271$	$3 \cdot 181$	$2^5 \cdot 17$
55	$2 \cdot 5^2 \cdot 11$	$19 \cdot 29$	$2^3 \cdot 3 \cdot 23$	$7 \cdot 79$	$2 \cdot 277$
56	$2^4 \cdot 5 \cdot 7$	$3 \cdot 11 \cdot 17$	$2 \cdot 281$	**7505084**	$2^2 \cdot 3 \cdot 47$
57	$2 \cdot 3 \cdot 5 \cdot 19$	**7566361**	$2^2 \cdot 11 \cdot 13$	$3 \cdot 191$	$2 \cdot 7 \cdot 41$
58	$2^2 \cdot 5 \cdot 29$	$7 \cdot 83$	$2 \cdot 3 \cdot 97$	$11 \cdot 53$	$2^3 \cdot 73$
59	$2 \cdot 5 \cdot 59$	$3 \cdot 197$	$2^4 \cdot 37$	**7730547**	$2 \cdot 3^3 \cdot 11$
60	$2^3 \cdot 3 \cdot 5^2$	**7788745**	$2 \cdot 7 \cdot 43$	$3^2 \cdot 67$	$2^2 \cdot 151$
61	$2 \cdot 5 \cdot 61$	$13 \cdot 47$	$2^2 \cdot 3^2 \cdot 17$	**7874605**	$2 \cdot 307$
62	$2^2 \cdot 5 \cdot 31$	$3^3 \cdot 23$	$2 \cdot 311$	$7 \cdot 89$	$2^4 \cdot 3 \cdot 13$
63	$2 \cdot 3^2 \cdot 5 \cdot 7$	**8000294**	$2^3 \cdot 79$	$3 \cdot 211$	$2 \cdot 317$
64	$2^7 \cdot 5$	**8068580**	$2 \cdot 3 \cdot 107$	**8082110**	$2^2 \cdot 7 \cdot 23$
65	$2 \cdot 5^2 \cdot 13$	$3 \cdot 7 \cdot 31$	$2^2 \cdot 163$	**8149132**	$2 \cdot 3 \cdot 109$
66	$2^2 \cdot 3 \cdot 5 \cdot 11$	**8202015**	$2 \cdot 331$	$3 \cdot 13 \cdot 17$	$2^3 \cdot 83$
67	$2 \cdot 5 \cdot 67$	$11 \cdot 61$	$2^5 \cdot 3 \cdot 7$	**8280151**	$2 \cdot 337$
68	$2^3 \cdot 5 \cdot 17$	$3 \cdot 227$	$2 \cdot 11 \cdot 31$	**8344207**	$2^2 \cdot 3^2 \cdot 19$
69	$2 \cdot 3 \cdot 5 \cdot 23$	**8394780**	$2^2 \cdot 173$	$3^2 \cdot 7 \cdot 11$	$2 \cdot 347$
70	$2^2 \cdot 5^2 \cdot 7$	**8457180**	$2 \cdot 3^3 \cdot 13$	$19 \cdot 37$	$2^6 \cdot 11$
71	$2 \cdot 5 \cdot 71$	$3^2 \cdot 79$	$2^3 \cdot 89$	$23 \cdot 31$	$2 \cdot 3 \cdot 7 \cdot 17$
72	$2^4 \cdot 3^2 \cdot 5$	$7 \cdot 103$	$2 \cdot 19^2$	$3 \cdot 241$	$2^2 \cdot 181$
73	$2 \cdot 5 \cdot 73$	$17 \cdot 43$	$2^2 \cdot 3 \cdot 61$	**8651040**	$2 \cdot 367$
74	$2^2 \cdot 5 \cdot 37$	$3 \cdot 13 \cdot 19$	$2 \cdot 7 \cdot 53$	**8709888**	$2^3 \cdot 3 \cdot 31$
75	$2 \cdot 3 \cdot 5^3$	**8756399**	$2^4 \cdot 47$	$3 \cdot 251$	$2 \cdot 13 \cdot 29$
76	$2^3 \cdot 5 \cdot 19$	**8813847**	$2 \cdot 3 \cdot 127$	$7 \cdot 109$	$2^2 \cdot 191$
77	$2 \cdot 5 \cdot 7 \cdot 11$	$3 \cdot 257$	$2^2 \cdot 193$	**8881795**	$2 \cdot 3^2 \cdot 43$
78	$2^2 \cdot 3 \cdot 5 \cdot 13$	$11 \cdot 71$	$2 \cdot 17 \cdot 23$	$3^3 \cdot 29$	$2^4 \cdot 7^2$
79	$2 \cdot 5 \cdot 79$	$7 \cdot 113$	$2^3 \cdot 3^2 \cdot 11$	$13 \cdot 61$	$2 \cdot 397$
80	$2^5 \cdot 5^2$	$3^2 \cdot 89$	$2 \cdot 401$	$11 \cdot 73$	$2^2 \cdot 3 \cdot 67$
81	$2 \cdot 3^4 \cdot 5$	**9090209**	$2^2 \cdot 7 \cdot 29$	$3 \cdot 271$	$2 \cdot 11 \cdot 37$
82	$2^2 \cdot 5 \cdot 41$	**9143432**	$2 \cdot 3 \cdot 137$	**9153998**	$2^3 \cdot 103$
83	$2 \cdot 5 \cdot 83$	$3 \cdot 277$	$2^6 \cdot 13$	$7^2 \cdot 17$	$2 \cdot 3 \cdot 139$
84	$2^3 \cdot 3 \cdot 5 \cdot 7$	29^2	$2 \cdot 421$	$3 \cdot 281$	$2^2 \cdot 211$
85	$2 \cdot 5^2 \cdot 17$	$23 \cdot 37$	$2^2 \cdot 3 \cdot 71$	**9309490**	$2 \cdot 7 \cdot 61$
86	$2^2 \cdot 5 \cdot 43$	$3 \cdot 7 \cdot 41$	$2 \cdot 431$	**9360108**	$2^5 \cdot 3^3$
87	$2 \cdot 3 \cdot 5 \cdot 29$	$13 \cdot 67$	$2^3 \cdot 109$	$3^2 \cdot 97$	$2 \cdot 19 \cdot 23$
88	$2^4 \cdot 5 \cdot 11$	**9449759**	$2 \cdot 3^2 \cdot 7^2$	**9459607**	$2^2 \cdot 13 \cdot 17$
89	$2 \cdot 5 \cdot 89$	$3^4 \cdot 11$	$2^2 \cdot 223$	$19 \cdot 47$	$2 \cdot 3 \cdot 149$
90	$2^2 \cdot 3^2 \cdot 5^2$	$17 \cdot 53$	$2 \cdot 11 \cdot 41$	$3 \cdot 7 \cdot 43$	$2^3 \cdot 113$
91	$2 \cdot 5 \cdot 7 \cdot 13$	**9595184**	$2^4 \cdot 3 \cdot 19$	$11 \cdot 83$	$2 \cdot 457$
92	$2^3 \cdot 5 \cdot 23$	$3 \cdot 307$	$2 \cdot 461$	$13 \cdot 71$	$2^2 \cdot 3 \cdot 7 \cdot 11$
93	$2 \cdot 3 \cdot 5 \cdot 31$	$7^2 \cdot 19$	$2^2 \cdot 233$	$3 \cdot 311$	$2 \cdot 467$
94	$2^2 \cdot 5 \cdot 47$	**9735896**	$2 \cdot 3 \cdot 157$	$23 \cdot 41$	$2^4 \cdot 59$
95	$2 \cdot 5^2 \cdot 19$	$3 \cdot 317$	$2^3 \cdot 7 \cdot 17$	**9790929**	$2 \cdot 3^2 \cdot 53$
96	$2^6 \cdot 3 \cdot 5$	31^2	$2 \cdot 13 \cdot 37$	$3^2 \cdot 107$	$2^2 \cdot 241$
97	$2 \cdot 5 \cdot 97$	**9872192**	$2^2 \cdot 3^5$	$7 \cdot 139$	$2 \cdot 487$
98	$2^2 \cdot 5 \cdot 7^2$	$3^2 \cdot 109$	$2 \cdot 491$	**9925535**	$2^3 \cdot 3 \cdot 41$
99	$2 \cdot 3^2 \cdot 5 \cdot 11$	**9960737**	$2^5 \cdot 31$	$3 \cdot 331$	$2 \cdot 7 \cdot 71$

FACTORS AND PRIMES

n	5	6	7	8	9
50	$5 \cdot 101$	$2 \cdot 11 \cdot 23$	$3 \cdot 13^2$	$2^2 \cdot 127$	**7067178**
51	$5 \cdot 103$	$2^2 \cdot 3 \cdot 43$	$11 \cdot 47$	$2 \cdot 7 \cdot 37$	$3 \cdot 173$
52	$3 \cdot 5^2 \cdot 7$	$2 \cdot 263$	$17 \cdot 31$	$2^4 \cdot 3 \cdot 11$	23^2
53	$5 \cdot 107$	$2^3 \cdot 67$	$3 \cdot 179$	$2 \cdot 269$	$7^2 \cdot 11$
54	$5 \cdot 109$	$2 \cdot 3 \cdot 7 \cdot 13$	**7379873**	$2^2 \cdot 137$	$3^2 \cdot 61$
55	$3 \cdot 5 \cdot 37$	$2^2 \cdot 139$	**7458552**	$2 \cdot 3^2 \cdot 31$	$13 \cdot 43$
56	$5 \cdot 113$	$2 \cdot 283$	$3^4 \cdot 7$	$2^3 \cdot 71$	**7551123**
57	$5^2 \cdot 23$	$2^6 \cdot 3^2$	**7611758**	$2 \cdot 17^2$	$3 \cdot 193$
58	$3^2 \cdot 5 \cdot 13$	$2 \cdot 293$	**7686381**	$2^2 \cdot 3 \cdot 7^2$	$19 \cdot 31$
59	$5 \cdot 7 \cdot 17$	$2^2 \cdot 149$	$3 \cdot 199$	$2 \cdot 13 \cdot 23$	**7774268**
60	$5 \cdot 11^2$	$2 \cdot 3 \cdot 101$	**7831887**	$2^5 \cdot 19$	$3 \cdot 7 \cdot 29$
61	$3 \cdot 5 \cdot 41$	$2^3 \cdot 7 \cdot 11$	**7902852**	$2 \cdot 3 \cdot 103$	**7916906**
62	5^4	$2 \cdot 313$	$3 \cdot 11 \cdot 19$	$2^2 \cdot 157$	$17 \cdot 37$
63	$5 \cdot 127$	$2^2 \cdot 3 \cdot 53$	$7^2 \cdot 13$	$2 \cdot 11 \cdot 29$	$3^2 \cdot 71$
64	$3 \cdot 5 \cdot 43$	$2 \cdot 17 \cdot 19$	**8109043**	$2^3 \cdot 3^4$	$11 \cdot 59$
65	$5 \cdot 131$	$2^4 \cdot 41$	$3^2 \cdot 73$	$2 \cdot 7 \cdot 47$	**8188854**
66	$5 \cdot 7 \cdot 19$	$2 \cdot 3^2 \cdot 37$	$23 \cdot 29$	$2^2 \cdot 167$	$3 \cdot 223$
67	$3^3 \cdot 5^2$	$2^2 \cdot 13^2$	**8305887**	$2 \cdot 3 \cdot 113$	$7 \cdot 97$
68	$5 \cdot 137$	$2 \cdot 7^3$	$3 \cdot 229$	$2^4 \cdot 43$	$13 \cdot 53$
69	$5 \cdot 139$	$2^3 \cdot 3 \cdot 29$	$17 \cdot 41$	$2 \cdot 349$	$3 \cdot 233$
70	$3 \cdot 5 \cdot 47$	$2 \cdot 353$	$7 \cdot 101$	$2^2 \cdot 3 \cdot 59$	**8506462**
71	$5 \cdot 11 \cdot 13$	$2^2 \cdot 179$	$3 \cdot 239$	$2 \cdot 359$	**8567289**
72	$5^2 \cdot 29$	$2 \cdot 3 \cdot 11^2$	**8615344**	$2^3 \cdot 7 \cdot 13$	3^6
73	$3 \cdot 5 \cdot 7^2$	$2^5 \cdot 23$	$11 \cdot 67$	$2 \cdot 3^2 \cdot 41$	**8686444**
74	$5 \cdot 149$	$2 \cdot 373$	$3^2 \cdot 83$	$2^2 \cdot 11 \cdot 17$	$7 \cdot 107$
75	$5 \cdot 151$	$2^2 \cdot 3^3 \cdot 7$	**8790959**	$2 \cdot 379$	$3 \cdot 11 \cdot 23$
76	$3^2 \cdot 5 \cdot 17$	$2 \cdot 383$	$13 \cdot 59$	$2^8 \cdot 3$	**8859263**
77	$5^2 \cdot 31$	$2^3 \cdot 97$	$3 \cdot 7 \cdot 37$	$2 \cdot 389$	$19 \cdot 41$
78	$5 \cdot 157$	$2 \cdot 3 \cdot 131$	**8959747**	$2^2 \cdot 197$	$3 \cdot 263$
79	$3 \cdot 5 \cdot 53$	$2^2 \cdot 199$	**9014583**	$2 \cdot 3 \cdot 7 \cdot 19$	$17 \cdot 47$
80	$5 \cdot 7 \cdot 23$	$2 \cdot 13 \cdot 31$	$3 \cdot 269$	$2^3 \cdot 101$	**9079485**
81	$5 \cdot 163$	$2^4 \cdot 3 \cdot 17$	$19 \cdot 43$	$2 \cdot 409$	$3^2 \cdot 7 \cdot 13$
82	$3 \cdot 5^2 \cdot 11$	$2 \cdot 7 \cdot 59$	**9175055**	$2^2 \cdot 3^2 \cdot 23$	**9185545**
83	$5 \cdot 167$	$2^2 \cdot 11 \cdot 19$	$3^3 \cdot 31$	$2 \cdot 419$	**9237620**
84	$5 \cdot 13^2$	$2 \cdot 3^2 \cdot 47$	$7 \cdot 11^2$	$2^4 \cdot 53$	$3 \cdot 283$
85	$3^2 \cdot 5 \cdot 19$	$2^3 \cdot 107$	**9329808**	$2 \cdot 3 \cdot 11 \cdot 13$	**9339932**
86	$5 \cdot 173$	$2 \cdot 433$	$3 \cdot 17^2$	$2^2 \cdot 7 \cdot 31$	$11 \cdot 79$
87	$5^3 \cdot 7$	$2^2 \cdot 3 \cdot 73$	**9429996**	$2 \cdot 439$	$3 \cdot 293$
88	$3 \cdot 5 \cdot 59$	$2 \cdot 443$	**9479236**	$2^3 \cdot 3 \cdot 37$	$7 \cdot 127$
89	$5 \cdot 179$	$2^7 \cdot 7$	$3 \cdot 13 \cdot 23$	$2 \cdot 449$	$29 \cdot 31$
90	$5 \cdot 181$	$2 \cdot 3 \cdot 151$	**9576073**	$2^2 \cdot 227$	$3^2 \cdot 101$
91	$3 \cdot 5 \cdot 61$	$2^2 \cdot 229$	$7 \cdot 131$	$2 \cdot 3^3 \cdot 17$	**9633155**
92	$5^2 \cdot 37$	$2 \cdot 463$	$3^2 \cdot 103$	$2^5 \cdot 29$	**9680157**
93	$5 \cdot 11 \cdot 17$	$2^3 \cdot 3^2 \cdot 13$	**9717396**	$2 \cdot 7 \cdot 67$	$3 \cdot 313$
94	$3^3 \cdot 5 \cdot 7$	$2 \cdot 11 \cdot 43$	**9763500**	$2^2 \cdot 3 \cdot 79$	$13 \cdot 73$
95	$5 \cdot 191$	$2^2 \cdot 239$	$3 \cdot 11 \cdot 29$	$2 \cdot 479$	$7 \cdot 137$
96	$5 \cdot 193$	$2 \cdot 3 \cdot 7 \cdot 23$	**9854265**	$2^3 \cdot 11^2$	$3 \cdot 17 \cdot 19$
97	$3 \cdot 5^2 \cdot 13$	$2^4 \cdot 61$	**9898946**	$2 \cdot 3 \cdot 163$	$11 \cdot 89$
98	$5 \cdot 197$	$2 \cdot 17 \cdot 29$	$3 \cdot 7 \cdot 47$	$2^2 \cdot 13 \cdot 19$	$23 \cdot 43$
99	$5 \cdot 199$	$2^2 \cdot 3 \cdot 83$	**9986952**	$2 \cdot 499$	$3^3 \cdot 37$

FACTORS AND PRIMES

n	0	1	2	3	4
100	$2^3 \cdot 5^3$	$7 \cdot 11 \cdot 13$	$2 \cdot 3 \cdot 167$	$17 \cdot 59$	$2^2 \cdot 251$
101	$2 \cdot 5 \cdot 101$	$3 \cdot 337$	$2^2 \cdot 11 \cdot 23$	**0056094**	$2 \cdot 3 \cdot 13^2$
102	$2^2 \cdot 3 \cdot 5 \cdot 17$	**0090257**	$2 \cdot 7 \cdot 73$	$3 \cdot 11 \cdot 31$	2^{10}
103	$2 \cdot 5 \cdot 103$	**0132587**	$2^3 \cdot 3 \cdot 43$	**0141003**	$2 \cdot 11 \cdot 47$
104	$2^4 \cdot 5 \cdot 13$	$3 \cdot 347$	$2 \cdot 521$	$7 \cdot 149$	$2^2 \cdot 3^2 \cdot 29$
105	$2 \cdot 3 \cdot 5^2 \cdot 7$	**0216027**	$2^2 \cdot 263$	$3^4 \cdot 13$	$2 \cdot 17 \cdot 31$
106	$2^2 \cdot 5 \cdot 53$	**0257154**	$2 \cdot 3^2 \cdot 59$	**0265333**	$2^3 \cdot 7 \cdot 19$
107	$2 \cdot 5 \cdot 107$	$3^2 \cdot 7 \cdot 17$	$2^4 \cdot 67$	$29 \cdot 37$	$2 \cdot 3 \cdot 179$
108	$2^3 \cdot 3^3 \cdot 5$	$23 \cdot 47$	$2 \cdot 541$	$3 \cdot 19^2$	$2^2 \cdot 271$
109	$2 \cdot 5 \cdot 109$	**0378248**	$2^2 \cdot 3 \cdot 7 \cdot 13$	**0386202**	$2 \cdot 547$
110	$2^2 \cdot 5^2 \cdot 11$	$3 \cdot 367$	$2 \cdot 19 \cdot 29$	**0425755**	$2^4 \cdot 3 \cdot 23$
111	$2 \cdot 3 \cdot 5 \cdot 37$	$11 \cdot 101$	$2^3 \cdot 139$	$3 \cdot 7 \cdot 53$	$2 \cdot 557$
112	$2^5 \cdot 5 \cdot 7$	$19 \cdot 59$	$2 \cdot 3 \cdot 11 \cdot 17$	**0503798**	$2^2 \cdot 281$
113	$2 \cdot 5 \cdot 113$	$3 \cdot 13 \cdot 29$	$2^2 \cdot 283$	$11 \cdot 103$	$2 \cdot 3^4 \cdot 7$
114	$2^2 \cdot 3 \cdot 5 \cdot 19$	$7 \cdot 163$	$2 \cdot 571$	$3^2 \cdot 127$	$2^3 \cdot 11 \cdot 13$
115	$2 \cdot 5^2 \cdot 23$	**0610753**	$2^7 \cdot 3^2$	**0618293**	$2 \cdot 577$
116	$2^3 \cdot 5 \cdot 29$	$3^3 \cdot 43$	$2 \cdot 7 \cdot 83$	**0655797**	$2^2 \cdot 3 \cdot 97$
117	$2 \cdot 3^2 \cdot 5 \cdot 13$	**0685569**	$2^2 \cdot 293$	$3 \cdot 17 \cdot 23$	$2 \cdot 587$
118	$2^2 \cdot 5 \cdot 59$	**0722499**	$2 \cdot 3 \cdot 197$	$7 \cdot 13^2$	$2^5 \cdot 37$
119	$2 \cdot 5 \cdot 7 \cdot 17$	$3 \cdot 397$	$2^3 \cdot 149$	**0766404**	$2 \cdot 3 \cdot 199$
120	$2^4 \cdot 3 \cdot 5^2$	**0795430**	$2 \cdot 601$	$3 \cdot 401$	$2^2 \cdot 7 \cdot 43$
121	$2 \cdot 5 \cdot 11^2$	$7 \cdot 173$	$2^2 \cdot 3 \cdot 101$	**0838608**	$2 \cdot 607$
122	$2^2 \cdot 5 \cdot 61$	$3 \cdot 11 \cdot 37$	$2 \cdot 13 \cdot 47$	**0874265**	$2^3 \cdot 3^2 \cdot 17$
123	$2 \cdot 3 \cdot 5 \cdot 41$	**0902581**	$2^4 \cdot 7 \cdot 11$	$3^2 \cdot 137$	$2 \cdot 617$
124	$2^3 \cdot 5 \cdot 31$	$17 \cdot 73$	$2 \cdot 3^3 \cdot 23$	$11 \cdot 113$	$2^2 \cdot 311$
125	$2 \cdot 5^4$	$3^2 \cdot 139$	$2^2 \cdot 313$	$7 \cdot 179$	$2 \cdot 3 \cdot 11 \cdot 19$
126	$2^2 \cdot 3^2 \cdot 5 \cdot 7$	$13 \cdot 97$	$2 \cdot 631$	$3 \cdot 421$	$2^4 \cdot 79$
127	$2 \cdot 5 \cdot 127$	$31 \cdot 41$	$2^3 \cdot 3 \cdot 53$	$19 \cdot 67$	$2 \cdot 7^2 \cdot 13$
128	$2^8 \cdot 5$	$3 \cdot 7 \cdot 61$	$2 \cdot 641$	**1082267**	$2^2 \cdot 3 \cdot 107$
129	$2 \cdot 3 \cdot 5 \cdot 43$	**1109262**	$2^2 \cdot 17 \cdot 19$	$3 \cdot 431$	$2 \cdot 647$
130	$2^2 \cdot 5^2 \cdot 13$	**1142773**	$2 \cdot 3 \cdot 7 \cdot 31$	**1149444**	$2^3 \cdot 163$
131	$2 \cdot 5 \cdot 131$	$3 \cdot 19 \cdot 23$	$2^5 \cdot 41$	$13 \cdot 101$	$2 \cdot 3^2 \cdot 73$
132	$2^3 \cdot 3 \cdot 5 \cdot 11$	**1209028**	$2 \cdot 661$	$3^3 \cdot 7^2$	$2^2 \cdot 331$
133	$2 \cdot 5 \cdot 7 \cdot 19$	11^3	$2^2 \cdot 3^2 \cdot 37$	$31 \cdot 43$	$2 \cdot 23 \cdot 29$
134	$2^2 \cdot 5 \cdot 67$	$3^2 \cdot 149$	$2 \cdot 11 \cdot 61$	$17 \cdot 79$	$2^6 \cdot 3 \cdot 7$
135	$2 \cdot 3^3 \cdot 5^2$	$7 \cdot 193$	$2^3 \cdot 13^2$	$3 \cdot 11 \cdot 41$	$2 \cdot 677$
136	$2^4 \cdot 5 \cdot 17$	**1338581**	$2 \cdot 3 \cdot 227$	$29 \cdot 47$	$2^2 \cdot 11 \cdot 31$
137	$2 \cdot 5 \cdot 137$	$3 \cdot 457$	$2^2 \cdot 7^3$	**1376705**	$2 \cdot 3 \cdot 229$
138	$2^2 \cdot 3 \cdot 5 \cdot 23$	**1401937**	$2 \cdot 691$	$3 \cdot 461$	$2^3 \cdot 173$
139	$2 \cdot 5 \cdot 139$	$13 \cdot 107$	$2^4 \cdot 3 \cdot 29$	$7 \cdot 199$	$2 \cdot 17 \cdot 41$
140	$2^3 \cdot 5^2 \cdot 7$	$3 \cdot 467$	$2 \cdot 701$	$23 \cdot 61$	$2^2 \cdot 3^3 \cdot 13$
141	$2 \cdot 3 \cdot 5 \cdot 47$	$17 \cdot 83$	$2^2 \cdot 353$	$3^2 \cdot 157$	$2 \cdot 7 \cdot 101$
142	$2^2 \cdot 5 \cdot 71$	$7^2 \cdot 29$	$2 \cdot 3^2 \cdot 79$	**1532049**	$2^4 \cdot 89$
143	$2 \cdot 5 \cdot 11 \cdot 13$	$3^3 \cdot 53$	$2^3 \cdot 179$	**1562462**	$2 \cdot 3 \cdot 239$
144	$2^5 \cdot 3^2 \cdot 5$	$11 \cdot 131$	$2 \cdot 7 \cdot 103$	$3 \cdot 13 \cdot 37$	$2^2 \cdot 19^2$
145	$2 \cdot 5^2 \cdot 29$	**1616674**	$2^2 \cdot 3 \cdot 11^2$	**1622656**	$2 \cdot 727$
146	$2^2 \cdot 5 \cdot 73$	$3 \cdot 487$	$2 \cdot 17 \cdot 43$	$7 \cdot 11 \cdot 19$	$2^3 \cdot 3 \cdot 61$
147	$2 \cdot 3 \cdot 5 \cdot 7^2$	**1676127**	$2^6 \cdot 23$	$3 \cdot 491$	$2 \cdot 11 \cdot 67$
148	$2^3 \cdot 5 \cdot 37$	**1705551**	$2 \cdot 3 \cdot 13 \cdot 19$	**1711412**	$2^2 \cdot 7 \cdot 53$
149	$2 \cdot 5 \cdot 149$	$3 \cdot 7 \cdot 71$	$2^2 \cdot 373$	**1740598**	$2 \cdot 3^2 \cdot 83$

n	5	6	7	8	9
100	$3 \cdot 5 \cdot 67$	$2 \cdot 503$	$19 \cdot 53$	$2^4 \cdot 3^2 \cdot 7$	**0038912**
101	$5 \cdot 7 \cdot 29$	$2^3 \cdot 127$	$3^2 \cdot 113$	$2 \cdot 509$	**0081742**
102	$5^2 \cdot 41$	$2 \cdot 3^3 \cdot 19$	$13 \cdot 79$	$2^2 \cdot 257$	$3 \cdot 7^3$
103	$3^2 \cdot 5 \cdot 23$	$2^2 \cdot 7 \cdot 37$	$17 \cdot 61$	$2 \cdot 3 \cdot 173$	**0166155**
104	$5 \cdot 11 \cdot 19$	$2 \cdot 523$	$3 \cdot 349$	$2^3 \cdot 131$	**0207755**
105	$5 \cdot 211$	$2^5 \cdot 3 \cdot 11$	$7 \cdot 151$	$2 \cdot 23^2$	$3 \cdot 353$
106	$3 \cdot 5 \cdot 71$	$2 \cdot 13 \cdot 41$	$11 \cdot 97$	$2^2 \cdot 3 \cdot 89$	**0289777**
107	$5^2 \cdot 43$	$2^2 \cdot 269$	$3 \cdot 359$	$2 \cdot 7^2 \cdot 11$	$13 \cdot 83$
108	$5 \cdot 7 \cdot 31$	$2 \cdot 3 \cdot 181$	**0362295**	$2^6 \cdot 17$	$3^2 \cdot 11^2$
109	$3 \cdot 5 \cdot 73$	$2^3 \cdot 137$	**0402066**	$2 \cdot 3^2 \cdot 61$	$7 \cdot 157$
110	$5 \cdot 13 \cdot 17$	$2 \cdot 7 \cdot 79$	$3^3 \cdot 41$	$2^2 \cdot 277$	**0449315**
111	$5 \cdot 223$	$2^2 \cdot 3^2 \cdot 31$	**0480532**	$2 \cdot 13 \cdot 43$	$3 \cdot 373$
112	$3^2 \cdot 5^3$	$2 \cdot 563$	$7^2 \cdot 23$	$2^3 \cdot 3 \cdot 47$	**0526939**
113	$5 \cdot 227$	$2^4 \cdot 71$	$3 \cdot 379$	$2 \cdot 569$	$17 \cdot 67$
114	$5 \cdot 229$	$2 \cdot 3 \cdot 191$	$31 \cdot 37$	$2^2 \cdot 7 \cdot 41$	$3 \cdot 383$
115	$3 \cdot 5 \cdot 7 \cdot 11$	$2^2 \cdot 17^2$	$13 \cdot 89$	$2 \cdot 3 \cdot 193$	$19 \cdot 61$
116	$5 \cdot 233$	$2 \cdot 11 \cdot 53$	$3 \cdot 389$	$2^4 \cdot 73$	$7 \cdot 167$
117	$5^2 \cdot 47$	$2^3 \cdot 3 \cdot 7^2$	$11 \cdot 107$	$2 \cdot 19 \cdot 31$	$3^2 \cdot 131$
118	$3 \cdot 5 \cdot 79$	$2 \cdot 593$	**0744507**	$2^2 \cdot 3^3 \cdot 11$	$29 \cdot 41$
119	$5 \cdot 239$	$2^2 \cdot 13 \cdot 23$	$3^2 \cdot 7 \cdot 19$	$2 \cdot 599$	$11 \cdot 109$
120	$5 \cdot 241$	$2 \cdot 3^2 \cdot 67$	$17 \cdot 71$	$2^3 \cdot 151$	$3 \cdot 13 \cdot 31$
121	$3^5 \cdot 5$	$2^6 \cdot 19$	**0852906**	$2 \cdot 3 \cdot 7 \cdot 29$	$23 \cdot 53$
122	$5^2 \cdot 7^2$	$2 \cdot 613$	$3 \cdot 409$	$2^2 \cdot 307$	**0895519**
123	$5 \cdot 13 \cdot 19$	$2^2 \cdot 3 \cdot 103$	**0923697**	$2 \cdot 619$	$3 \cdot 7 \cdot 59$
124	$3 \cdot 5 \cdot 83$	$2 \cdot 7 \cdot 89$	$29 \cdot 43$	$2^5 \cdot 3 \cdot 13$	**0965624**
125	$5 \cdot 251$	$2^3 \cdot 157$	$3 \cdot 419$	$2 \cdot 17 \cdot 37$	**1000257**
126	$5 \cdot 11 \cdot 23$	$2 \cdot 3 \cdot 211$	$7 \cdot 181$	$2^2 \cdot 317$	$3^3 \cdot 47$
127	$3 \cdot 5^2 \cdot 17$	$2^2 \cdot 11 \cdot 29$	**1061909**	$2 \cdot 3^2 \cdot 71$	**1068705**
128	$5 \cdot 257$	$2 \cdot 643$	$3^2 \cdot 11 \cdot 13$	$2^3 \cdot 7 \cdot 23$	**1102529**
129	$5 \cdot 7 \cdot 37$	$2^4 \cdot 3^4$	**1129400**	$2 \cdot 11 \cdot 59$	$3 \cdot 433$
130	$3^2 \cdot 5 \cdot 29$	$2 \cdot 653$	**1162756**	$2^2 \cdot 3 \cdot 109$	$7 \cdot 11 \cdot 17$
131	$5 \cdot 263$	$2^2 \cdot 7 \cdot 47$	$3 \cdot 439$	$2 \cdot 659$	**1202448**
132	$5^2 \cdot 53$	$2 \cdot 3 \cdot 13 \cdot 17$	**1228709**	$2^4 \cdot 83$	$3 \cdot 443$
133	$3 \cdot 5 \cdot 89$	$2^3 \cdot 167$	$7 \cdot 191$	$2 \cdot 3 \cdot 223$	$13 \cdot 103$
134	$5 \cdot 269$	$2 \cdot 673$	$3 \cdot 449$	$2^2 \cdot 337$	$19 \cdot 71$
135	$5 \cdot 271$	$2^2 \cdot 3 \cdot 113$	$23 \cdot 59$	$2 \cdot 7 \cdot 97$	$3^2 \cdot 151$
136	$3 \cdot 5 \cdot 7 \cdot 13$	$2 \cdot 683$	**1357685**	$2^3 \cdot 3^2 \cdot 19$	37^2
137	$5^3 \cdot 11$	$2^5 \cdot 43$	$3^4 \cdot 17$	$2 \cdot 13 \cdot 53$	$7 \cdot 197$
138	$5 \cdot 277$	$2 \cdot 3^2 \cdot 7 \cdot 11$	$19 \cdot 73$	$2^2 \cdot 347$	$3 \cdot 463$
139	$3^2 \cdot 5 \cdot 31$	$2^2 \cdot 349$	$11 \cdot 127$	$2 \cdot 3 \cdot 233$	**1458177**
140	$5 \cdot 281$	$2 \cdot 19 \cdot 37$	$3 \cdot 7 \cdot 67$	$2^7 \cdot 11$	**1489110**
141	$5 \cdot 283$	$2^3 \cdot 3 \cdot 59$	$13 \cdot 109$	$2 \cdot 709$	$3 \cdot 11 \cdot 43$
142	$3 \cdot 5^2 \cdot 19$	$2 \cdot 23 \cdot 31$	**1544240**	$2^2 \cdot 3 \cdot 7 \cdot 17$	**1550322**
143	$5 \cdot 7 \cdot 41$	$2^2 \cdot 359$	$3 \cdot 479$	$2 \cdot 719$	**1580608**
144	$5 \cdot 17^2$	$2 \cdot 3 \cdot 241$	**1604685**	$2^3 \cdot 181$	$3^2 \cdot 7 \cdot 23$
145	$3 \cdot 5 \cdot 97$	$2^4 \cdot 7 \cdot 13$	$31 \cdot 47$	$2 \cdot 3^6$	**1640553**
146	$5 \cdot 293$	$2 \cdot 733$	$3^2 \cdot 163$	$2^2 \cdot 367$	$13 \cdot 113$
147	$5^2 \cdot 59$	$2^2 \cdot 3^2 \cdot 41$	$7 \cdot 211$	$2 \cdot 739$	$3 \cdot 17 \cdot 29$
148	$3^3 \cdot 5 \cdot 11$	$2 \cdot 743$	**1723110**	$2^4 \cdot 3 \cdot 31$	**1728947**
149	$5 \cdot 13 \cdot 23$	$2^3 \cdot 11 \cdot 17$	$3 \cdot 499$	$2 \cdot 7 \cdot 107$	**1758016**

FACTORS AND PRIMES

n	0	1	2	3	4
150	$2^2 \cdot 3 \cdot 5^3$	$19 \cdot 79$	$2 \cdot 751$	$3^2 \cdot 167$	$2^5 \cdot 47$
151	$2 \cdot 5 \cdot 151$	**1792645**	$2^3 \cdot 3^3 \cdot 7$	$17 \cdot 89$	$2 \cdot 757$
152	$2^4 \cdot 5 \cdot 19$	$3^2 \cdot 13^2$	$2 \cdot 761$	**1826999**	$2^2 \cdot 3 \cdot 127$
153	$2 \cdot 3^2 \cdot 5 \cdot 17$	**1849752**	$2^2 \cdot 383$	$3 \cdot 7 \cdot 73$	$2 \cdot 13 \cdot 59$
154	$2^2 \cdot 5 \cdot 7 \cdot 11$	$23 \cdot 67$	$2 \cdot 3 \cdot 257$	**1883659**	$2^3 \cdot 193$
155	$2 \cdot 5^2 \cdot 31$	$3 \cdot 11 \cdot 47$	$2^4 \cdot 97$	**1911715**	$2 \cdot 3 \cdot 7 \cdot 37$
156	$2^3 \cdot 3 \cdot 5 \cdot 13$	$7 \cdot 223$	$2 \cdot 11 \cdot 71$	$3 \cdot 521$	$2^2 \cdot 17 \cdot 23$
157	$2 \cdot 5 \cdot 157$	**1961762**	$2^2 \cdot 3 \cdot 131$	$11^2 \cdot 13$	$2 \cdot 787$
158	$2^2 \cdot 5 \cdot 79$	$3 \cdot 17 \cdot 31$	$2 \cdot 7 \cdot 113$	**1994809**	$2^4 \cdot 3^2 \cdot 11$
159	$2 \cdot 3 \cdot 5 \cdot 53$	$37 \cdot 43$	$2^3 \cdot 199$	$3^3 \cdot 59$	$2 \cdot 797$
160	$2^6 \cdot 5^2$	**2043913**	$2 \cdot 3^2 \cdot 89$	$7 \cdot 229$	$2^2 \cdot 401$
161	$2 \cdot 5 \cdot 7 \cdot 23$	$3^2 \cdot 179$	$2^2 \cdot 13 \cdot 31$	**2076344**	$2 \cdot 3 \cdot 269$
162	$2^2 \cdot 3^4 \cdot 5$	**2097830**	$2 \cdot 811$	$3 \cdot 541$	$2^3 \cdot 7 \cdot 29$
163	$2 \cdot 5 \cdot 163$	$7 \cdot 233$	$2^5 \cdot 3 \cdot 17$	$23 \cdot 71$	$2 \cdot 19 \cdot 43$
164	$2^3 \cdot 5 \cdot 41$	$3 \cdot 547$	$2 \cdot 821$	$31 \cdot 53$	$2^2 \cdot 3 \cdot 137$
165	$2 \cdot 3 \cdot 5^2 \cdot 11$	$13 \cdot 127$	$2^2 \cdot 7 \cdot 59$	$3 \cdot 19 \cdot 29$	$2 \cdot 827$
166	$2^2 \cdot 5 \cdot 83$	$11 \cdot 151$	$2 \cdot 3 \cdot 277$	**2208922**	$2^7 \cdot 13$
167	$2 \cdot 5 \cdot 167$	$3 \cdot 557$	$2^3 \cdot 11 \cdot 19$	$7 \cdot 239$	$2 \cdot 3^3 \cdot 31$
168	$2^4 \cdot 3 \cdot 5 \cdot 7$	41^2	$2 \cdot 29^2$	$3^2 \cdot 11 \cdot 17$	$2^2 \cdot 421$
169	$2 \cdot 5 \cdot 13^2$	$19 \cdot 89$	$2^2 \cdot 3^2 \cdot 47$	**2286570**	$2 \cdot 7 \cdot 11^2$
170	$2^2 \cdot 5^2 \cdot 17$	$3^5 \cdot 7$	$2 \cdot 23 \cdot 37$	$13 \cdot 131$	$2^3 \cdot 3 \cdot 71$
171	$2 \cdot 3^2 \cdot 5 \cdot 19$	$29 \cdot 59$	$2^4 \cdot 107$	$3 \cdot 571$	$2 \cdot 857$
172	$2^3 \cdot 5 \cdot 43$	**2357809**	$2 \cdot 3 \cdot 7 \cdot 41$	**2362853**	$2^2 \cdot 431$
173	$2 \cdot 5 \cdot 173$	$3 \cdot 577$	$2^2 \cdot 433$	**2387986**	$2 \cdot 3 \cdot 17^2$
174	$2^2 \cdot 3 \cdot 5 \cdot 29$	**2407988**	$2 \cdot 13 \cdot 67$	$3 \cdot 7 \cdot 83$	$2^4 \cdot 109$
175	$2 \cdot 5^3 \cdot 7$	$17 \cdot 103$	$2^3 \cdot 3 \cdot 73$	**2437819**	$2 \cdot 877$
176	$2^5 \cdot 5 \cdot 11$	$3 \cdot 587$	$2 \cdot 881$	$41 \cdot 43$	$2^2 \cdot 3^2 \cdot 7^2$
177	$2 \cdot 3 \cdot 5 \cdot 59$	$7 \cdot 11 \cdot 23$	$2^2 \cdot 443$	$3^2 \cdot 197$	$2 \cdot 887$
178	$2^2 \cdot 5 \cdot 89$	$13 \cdot 137$	$2 \cdot 3^4 \cdot 11$	**2511513**	$2^3 \cdot 223$
179	$2 \cdot 5 \cdot 179$	$3^2 \cdot 199$	$2^8 \cdot 7$	$11 \cdot 163$	$2 \cdot 3 \cdot 13 \cdot 23$
180	$2^3 \cdot 3^2 \cdot 5^2$	**2555137**	$2 \cdot 17 \cdot 53$	$3 \cdot 601$	$2^2 \cdot 11 \cdot 41$
181	$2 \cdot 5 \cdot 181$	**2579185**	$2^2 \cdot 3 \cdot 151$	$7^2 \cdot 37$	$2 \cdot 907$
182	$2^2 \cdot 5 \cdot 7 \cdot 13$	$3 \cdot 607$	$2 \cdot 911$	**2607867**	$2^5 \cdot 3 \cdot 19$
183	$2 \cdot 3 \cdot 5 \cdot 61$	**2626883**	$2^3 \cdot 229$	$3 \cdot 13 \cdot 47$	$2 \cdot 7 \cdot 131$
184	$2^4 \cdot 5 \cdot 23$	$7 \cdot 263$	$2 \cdot 3 \cdot 307$	$19 \cdot 97$	$2^2 \cdot 461$
185	$2 \cdot 5^2 \cdot 37$	$3 \cdot 617$	$2^2 \cdot 463$	$17 \cdot 109$	$2 \cdot 3^2 \cdot 103$
186	$2^2 \cdot 3 \cdot 5 \cdot 31$	**2697464**	$2 \cdot 7^2 \cdot 19$	$3^4 \cdot 23$	$2^3 \cdot 233$
187	$2 \cdot 5 \cdot 11 \cdot 17$	**2720738**	$2^4 \cdot 3^2 \cdot 13$	**2725378**	$2 \cdot 937$
188	$2^3 \cdot 5 \cdot 47$	$3^2 \cdot 11 \cdot 19$	$2 \cdot 941$	$7 \cdot 269$	$2^2 \cdot 3 \cdot 157$
189	$2 \cdot 3^3 \cdot 5 \cdot 7$	$31 \cdot 61$	$2^2 \cdot 11 \cdot 43$	$3 \cdot 631$	$2 \cdot 947$
190	$2^2 \cdot 5^2 \cdot 19$	**2789821**	$2 \cdot 3 \cdot 317$	$11 \cdot 173$	$2^4 \cdot 7 \cdot 17$
191	$2 \cdot 5 \cdot 191$	$3 \cdot 7^2 \cdot 13$	$2^3 \cdot 239$	**2817150**	$2 \cdot 3 \cdot 11 \cdot 29$
192	$2^7 \cdot 3 \cdot 5$	$17 \cdot 113$	$2 \cdot 31^2$	$3 \cdot 641$	$2^2 \cdot 13 \cdot 37$
193	$2 \cdot 5 \cdot 193$	**2857823**	$2^2 \cdot 3 \cdot 7 \cdot 23$	**2862319**	$2 \cdot 967$
194	$2^2 \cdot 5 \cdot 97$	$3 \cdot 647$	$2 \cdot 971$	$29 \cdot 67$	$2^3 \cdot 3^5$
195	$2 \cdot 3 \cdot 5^2 \cdot 13$	**2902573**	$2^5 \cdot 61$	$3^2 \cdot 7 \cdot 31$	$2 \cdot 977$
196	$2^3 \cdot 5 \cdot 7^2$	$37 \cdot 53$	$2 \cdot 3^2 \cdot 109$	$13 \cdot 151$	$2^2 \cdot 491$
197	$2 \cdot 5 \cdot 197$	$3^3 \cdot 73$	$2^2 \cdot 17 \cdot 29$	**2951271**	$2 \cdot 3 \cdot 7 \cdot 47$
198	$2^2 \cdot 3^2 \cdot 5 \cdot 11$	$7 \cdot 283$	$2 \cdot 991$	$3 \cdot 661$	$2^6 \cdot 31$
199	$2 \cdot 5 \cdot 199$	$11 \cdot 181$	$2^3 \cdot 3 \cdot 83$	**2995073**	$2 \cdot 997$

FACTORS AND PRIMES

n	5	6	7	8	9
150	$5 \cdot 7 \cdot 43$	$2 \cdot 3 \cdot 251$	$11 \cdot 137$	$2^2 \cdot 13 \cdot 29$	$3 \cdot 503$
151	$3 \cdot 5 \cdot 101$	$2^2 \cdot 379$	$37 \cdot 41$	$2 \cdot 3 \cdot 11 \cdot 23$	$7^2 \cdot 31$
152	$5^2 \cdot 61$	$2 \cdot 7 \cdot 109$	$3 \cdot 509$	$2^3 \cdot 191$	$11 \cdot 139$
153	$5 \cdot 307$	$2^9 \cdot 3$	$29 \cdot 53$	$2 \cdot 769$	$3^4 \cdot 19$
154	$3 \cdot 5 \cdot 103$	$2 \cdot 773$	$7 \cdot 13 \cdot 17$	$2^2 \cdot 3^2 \cdot 43$	**1900514**
155	$5 \cdot 311$	$2^2 \cdot 389$	$3^2 \cdot 173$	$2 \cdot 19 \cdot 41$	**1928461**
156	$5 \cdot 313$	$2 \cdot 3^3 \cdot 29$	**1950690**	$2^5 \cdot 7^2$	$3 \cdot 523$
157	$3^2 \cdot 5^2 \cdot 7$	$2^3 \cdot 197$	$19 \cdot 83$	$2 \cdot 3 \cdot 263$	**1983821**
158	$5 \cdot 317$	$2 \cdot 13 \cdot 61$	$3 \cdot 23^2$	$2^2 \cdot 397$	$7 \cdot 227$
159	$5 \cdot 11 \cdot 29$	$2^2 \cdot 3 \cdot 7 \cdot 19$	**2033049**	$2 \cdot 17 \cdot 47$	$3 \cdot 13 \cdot 41$
160	$3 \cdot 5 \cdot 107$	$2 \cdot 11 \cdot 73$	**2060159**	$2^3 \cdot 3 \cdot 67$	**2065560**
161	$5 \cdot 17 \cdot 19$	$2^4 \cdot 101$	$3 \cdot 7^2 \cdot 11$	$2 \cdot 809$	**2092468**
162	$5^3 \cdot 13$	$2 \cdot 3 \cdot 271$	**2113876**	$2^2 \cdot 11 \cdot 37$	$3^2 \cdot 181$
163	$3 \cdot 5 \cdot 109$	$2^2 \cdot 409$	**2140487**	$2 \cdot 3^2 \cdot 7 \cdot 13$	$11 \cdot 149$
164	$5 \cdot 7 \cdot 47$	$2 \cdot 823$	$3^3 \cdot 61$	$2^4 \cdot 103$	$17 \cdot 97$
165	$5 \cdot 331$	$2^3 \cdot 3^2 \cdot 23$	**2193225**	$2 \cdot 829$	$3 \cdot 7 \cdot 79$
166	$3^2 \cdot 5 \cdot 37$	$2 \cdot 7^2 \cdot 17$	**2219356**	$2^2 \cdot 3 \cdot 139$	**2224563**
167	$5^2 \cdot 67$	$2^2 \cdot 419$	$3 \cdot 13 \cdot 43$	$2 \cdot 839$	$23 \cdot 73$
168	$5 \cdot 337$	$2 \cdot 3 \cdot 281$	$7 \cdot 241$	$2^3 \cdot 211$	$3 \cdot 563$
169	$3 \cdot 5 \cdot 113$	$2^5 \cdot 53$	**2296818**	$2 \cdot 3 \cdot 283$	**2301934**
170	$5 \cdot 11 \cdot 31$	$2 \cdot 853$	$3 \cdot 569$	$2^2 \cdot 7 \cdot 61$	**2327421**
171	$5 \cdot 7^3$	$2^2 \cdot 3 \cdot 11 \cdot 13$	$17 \cdot 101$	$2 \cdot 859$	$3^2 \cdot 191$
172	$3 \cdot 5^2 \cdot 23$	$2 \cdot 863$	$11 \cdot 157$	$2^6 \cdot 3^3$	$7 \cdot 13 \cdot 19$
173	$5 \cdot 347$	$2^3 \cdot 7 \cdot 31$	$3^2 \cdot 193$	$2 \cdot 11 \cdot 79$	$37 \cdot 47$
174	$5 \cdot 349$	$2 \cdot 3^2 \cdot 97$	**2422929**	$2^2 \cdot 19 \cdot 23$	$3 \cdot 11 \cdot 53$
175	$3^3 \cdot 5 \cdot 13$	$2^2 \cdot 439$	$7 \cdot 251$	$2 \cdot 3 \cdot 293$	**2452658**
176	$5 \cdot 353$	$2 \cdot 883$	$3 \cdot 19 \cdot 31$	$2^3 \cdot 13 \cdot 17$	$29 \cdot 61$
177	$5^2 \cdot 71$	$2^4 \cdot 3 \cdot 37$	**2496874**	$2 \cdot 7 \cdot 127$	$3 \cdot 593$
178	$3 \cdot 5 \cdot 7 \cdot 17$	$2 \cdot 19 \cdot 47$	**2521246**	$2^2 \cdot 3 \cdot 149$	**2526103**
179	$5 \cdot 359$	$2^2 \cdot 449$	$3 \cdot 599$	$2 \cdot 29 \cdot 31$	$7 \cdot 257$
180	$5 \cdot 19^2$	$2 \cdot 3 \cdot 7 \cdot 43$	$13 \cdot 139$	$2^4 \cdot 113$	$3^3 \cdot 67$
181	$3 \cdot 5 \cdot 11^2$	$2^3 \cdot 227$	$23 \cdot 79$	$2 \cdot 3^2 \cdot 101$	$17 \cdot 107$
182	$5^2 \cdot 73$	$2 \cdot 11 \cdot 83$	$3^2 \cdot 7 \cdot 29$	$2^2 \cdot 457$	$31 \cdot 59$
183	$5 \cdot 367$	$2^2 \cdot 3^3 \cdot 17$	$11 \cdot 167$	$2 \cdot 919$	$3 \cdot 613$
184	$3^2 \cdot 5 \cdot 41$	$2 \cdot 13 \cdot 71$	**2664669**	$2^3 \cdot 3 \cdot 7 \cdot 11$	43^2
185	$5 \cdot 7 \cdot 53$	$2^6 \cdot 29$	$3 \cdot 619$	$2 \cdot 929$	$11 \cdot 13^2$
186	$5 \cdot 373$	$2 \cdot 3 \cdot 311$	**2711443**	$2^2 \cdot 467$	$3 \cdot 7 \cdot 89$
187	$3 \cdot 5^4$	$2^2 \cdot 7 \cdot 67$	**2734643**	$2 \cdot 3 \cdot 313$	**2739268**
188	$5 \cdot 13 \cdot 29$	$2 \cdot 23 \cdot 41$	$3 \cdot 17 \cdot 37$	$2^5 \cdot 59$	**2762320**
189	$5 \cdot 379$	$2^3 \cdot 3 \cdot 79$	$7 \cdot 271$	$2 \cdot 13 \cdot 73$	$3^2 \cdot 211$
190	$3 \cdot 5 \cdot 127$	$2 \cdot 953$	**2803507**	$2^2 \cdot 3^2 \cdot 53$	$23 \cdot 83$
191	$5 \cdot 383$	$2^2 \cdot 479$	$3^3 \cdot 71$	$2 \cdot 7 \cdot 137$	$19 \cdot 101$
192	$5^2 \cdot 7 \cdot 11$	$2 \cdot 3^2 \cdot 107$	$41 \cdot 47$	$2^3 \cdot 241$	$3 \cdot 643$
193	$3^2 \cdot 5 \cdot 43$	$2^4 \cdot 11^2$	$13 \cdot 149$	$2 \cdot 3 \cdot 17 \cdot 19$	$7 \cdot 277$
194	$5 \cdot 389$	$2 \cdot 7 \cdot 139$	$3 \cdot 11 \cdot 59$	$2^2 \cdot 487$	**2898118**
195	$5 \cdot 17 \cdot 23$	$2^2 \cdot 3 \cdot 163$	$19 \cdot 103$	$2 \cdot 11 \cdot 89$	$3 \cdot 653$
196	$3 \cdot 5 \cdot 131$	$2 \cdot 983$	$7 \cdot 281$	$2^4 \cdot 3 \cdot 41$	$11 \cdot 179$
197	$5^2 \cdot 79$	$2^3 \cdot 13 \cdot 19$	$3 \cdot 659$	$2 \cdot 23 \cdot 43$	**2964458**
198	$5 \cdot 397$	$2 \cdot 3 \cdot 331$	**2981979**	$2^2 \cdot 7 \cdot 71$	$3^2 \cdot 13 \cdot 17$
199	$3 \cdot 5 \cdot 7 \cdot 19$	$2^2 \cdot 499$	**3003781**	$2 \cdot 3^3 \cdot 37$	**3008128**

TOTIENT FUNCTION $\phi(n)$

Introductory facts

$\phi(n)$ is the number of integers not exceeding and relatively prime to n.

σ_k is the sum of the k'th powers of divisors of n.

σ_0 is usually denoted by $d(n)$ and specifies the number of divisors of n.

For example if $n = 10$, then the divisors are 1, 2, 5, 10 and the sum of the first powers is 18, i.e. $\sigma_0 = d(10) = 4$; $\sigma_1 = 18$; $\phi(10) = 4$.

n	$\phi(n)$	σ_0	σ_1	n	$\phi(n)$	σ_0	σ_1	n	$\phi(n)$	σ_0	σ_1	n	$\phi(n)$	σ_0	σ_1
1	1	1	1	41	40	2	42	81	54	5	121	121	110	3	133
2	1	2	3	42	12	8	96	82	40	4	126	122	60	4	186
3	2	2	4	43	42	2	44	83	82	2	84	123	80	4	168
4	2	3	7	44	20	6	84	84	24	12	224	124	60	6	224
5	4	2	6	45	24	6	78	85	64	4	108	125	100	4	156
6	2	4	12	46	22	4	72	86	42	4	132	126	36	12	312
7	6	2	8	47	46	2	48	87	56	4	120	127	126	2	128
8	4	4	15	48	16	10	124	88	40	8	180	128	64	8	255
9	6	3	13	49	42	3	57	89	88	2	90	129	84	4	176
10	4	4	18	50	20	6	93	90	24	12	234	130	48	8	252
11	10	2	12	51	32	4	72	91	72	4	112	131	130	2	132
12	4	6	28	52	24	6	98	92	44	6	168	132	40	12	336
13	12	2	14	53	52	2	54	93	60	4	128	133	108	4	160
14	6	4	24	54	18	8	120	94	46	4	144	134	66	4	204
15	8	4	24	55	40	4	72	95	72	4	120	135	72	8	240
16	8	5	31	56	24	8	120	96	32	12	252	136	64	8	270
17	16	2	18	57	36	4	80	97	96	2	98	137	136	2	138
18	6	6	39	58	28	4	90	98	42	6	171	138	44	8	288
19	18	2	20	59	58	2	60	99	60	6	156	139	138	2	140
20	8	6	42	60	16	12	168	100	40	9	217	140	48	12	336
21	12	4	32	61	60	2	62	101	100	2	102	141	92	4	192
22	10	4	36	62	30	4	96	102	32	8	216	142	70	4	216
23	22	2	24	63	36	6	104	103	102	2	104	143	120	4	168
24	8	8	60	64	32	7	127	104	48	8	210	144	48	15	403
25	20	3	31	65	48	4	84	105	48	8	192	145	112	4	180
26	12	4	42	66	20	8	144	106	52	4	162	146	72	4	222
27	18	4	40	67	66	2	68	107	106	2	108	147	84	6	228
28	12	6	56	68	32	6	126	108	36	12	280	148	72	6	266
29	28	2	30	69	44	4	96	109	108	2	110	149	148	2	150
30	8	8	72	70	24	8	144	110	40	8	216	150	40	12	372
31	30	2	32	71	70	2	72	111	72	4	152	151	150	2	152
32	16	6	63	72	24	12	195	112	48	10	248	152	72	8	300
33	20	4	48	73	72	2	74	113	112	2	114	153	96	6	234
34	16	4	54	74	36	4	114	114	36	8	240	154	60	8	288
35	24	4	48	75	40	6	124	115	88	4	144	155	120	4	192
36	12	9	91	76	36	6	140	116	56	6	210	156	48	12	392
37	36	2	38	77	60	4	96	117	72	6	182	157	156	2	158
38	18	4	60	78	24	8	168	118	58	4	180	158	78	4	240
39	24	4	56	79	78	2	80	119	96	4	144	159	104	4	216
40	16	8	90	80	32	10	186	120	32	16	360	160	64	12	378

TOTIENT FUNCTION $\phi(n)$

n	$\phi(n)$	σ_0	σ_1	n	$\phi(n)$	σ_0	σ_1	n	$\phi(n)$	σ_0	σ_1	n	$\phi(n)$	σ_0	σ_1
161	132	4	192	211	210	2	212	261	168	6	390	311	310	2	312
162	54	10	363	212	104	6	378	262	130	4	396	312	96	16	840
163	162	2	164	213	140	4	288	263	262	2	264	313	312	2	314
164	80	6	294	214	106	4	324	264	80	16	720	314	156	4	474
165	80	8	288	215	168	4	264	265	208	4	324	315	144	12	624
166	82	4	252	216	72	16	600	266	108	8	480	316	156	6	560
167	166	2	168	217	180	4	256	267	176	4	360	317	316	2	318
168	48	16	480	218	108	4	330	268	132	6	476	318	104	8	648
169	156	3	183	219	144	4	296	269	268	2	270	319	280	4	360
170	64	8	324	220	80	12	504	270	72	16	720	320	128	14	762
171	108	6	260	221	192	4	252	271	270	2	272	321	212	4	432
172	84	6	308	222	72	8	456	272	128	10	558	322	132	8	576
173	172	2	174	223	222	2	224	273	144	8	448	323	288	4	360
174	56	8	360	224	96	12	504	274	136	4	414	324	108	15	847
175	120	6	248	225	120	9	403	275	200	6	372	325	240	6	434
176	80	10	372	226	112	4	342	276	88	12	672	326	162	4	492
177	116	4	240	227	226	2	228	277	276	2	278	327	216	4	440
178	88	4	270	228	72	12	560	278	138	4	420	328	160	8	630
179	178	2	180	229	228	2	230	279	180	6	416	329	276	4	384
180	48	18	546	230	88	8	432	280	96	16	720	330	80	16	864
181	180	2	182	231	120	8	384	281	280	2	282	331	330	2	332
182	72	8	336	232	112	8	450	282	92	8	576	332	164	6	588
183	120	4	248	233	232	2	234	283	282	2	284	333	216	6	494
184	88	8	360	234	72	12	546	284	140	6	504	334	166	4	504
185	144	4	228	235	184	4	288	285	144	8	480	335	264	4	408
186	60	8	384	236	116	6	420	286	120	8	504	336	96	20	992
187	160	4	216	237	156	4	320	287	240	4	336	337	336	2	338
188	92	6	336	238	96	8	432	288	96	18	819	338	156	6	549
189	108	8	320	239	238	2	240	289	272	3	307	339	224	4	456
190	72	8	360	240	64	20	744	290	112	8	540	340	128	12	756
191	190	2	192	241	240	2	242	291	192	4	392	341	300	4	384
192	64	14	508	242	110	6	399	292	144	6	518	342	108	12	780
193	192	2	194	243	162	6	364	293	292	2	294	343	294	4	400
194	96	4	294	244	120	6	434	294	84	12	684	344	168	8	660
195	96	8	336	245	168	6	342	295	232	4	360	345	176	8	576
196	84	9	399	246	80	8	504	296	144	8	570	346	172	4	522
197	196	2	198	247	216	4	280	297	180	8	480	347	346	2	348
198	60	12	468	248	120	8	480	298	148	4	450	348	112	12	840
199	198	2	200	249	164	4	336	299	264	4	336	349	348	2	350
200	80	12	465	250	100	8	468	300	80	18	868	350	120	12	744
201	132	4	272	251	250	2	252	301	252	4	352	351	216	8	560
202	100	4	306	252	72	18	728	302	150	4	456	352	160	12	756
203	168	4	240	253	220	4	288	303	200	4	408	353	352	2	354
204	64	12	504	254	126	4	384	304	144	10	620	354	116	8	720
205	160	4	252	255	128	8	432	305	240	4	372	355	280	4	432
206	102	4	312	256	128	9	511	306	96	12	702	356	176	6	630
207	132	6	312	257	256	2	258	307	306	2	308	357	192	8	576
208	96	10	434	258	84	8	528	308	120	12	672	358	178	4	540
209	180	4	240	259	216	4	304	309	204	4	416	359	358	2	360
210	48	16	576	260	96	12	588	310	120	8	576	360	96	24	1170

TOTIENT FUNCTION $\phi(n)$

n	$\phi(n)$	σ_0	σ_1	n	$\phi(n)$	σ_0	σ_1	n	$\phi(n)$	σ_0	σ_1	n	$\phi(n)$	σ_0	σ_1
361	342	3	381	411	272	4	552	461	460	2	462	511	432	4	592
362	180	4	546	412	204	6	728	462	120	16	1152	512	256	10	1023
363	220	6	532	413	348	4	480	463	462	2	464	513	324	8	800
364	144	12	784	414	132	12	936	464	224	10	930	514	256	4	774
365	288	4	444	415	328	4	504	465	240	8	768	515	408	4	624
366	120	8	744	416	192	12	882	466	232	4	702	516	168	12	1232
367	366	2	368	417	276	4	560	467	466	2	468	517	460	4	576
368	176	10	744	418	180	8	720	468	144	18	1274	518	216	8	912
369	240	6	546	419	418	2	420	469	396	4	544	519	344	4	696
370	144	8	684	420	96	24	1344	470	184	8	864	520	192	16	1260
371	312	4	432	421	420	2	422	471	312	4	632	521	520	2	522
372	120	12	896	422	210	4	636	472	232	8	900	522	168	12	1170
373	372	2	374	423	276	6	624	473	420	4	528	523	522	2	524
374	160	8	648	424	208	8	810	474	156	8	960	524	260	6	924
375	200	8	624	425	320	6	558	475	360	6	620	525	240	12	992
376	184	8	720	426	140	8	864	476	192	12	1008	526	262	4	792
377	336	4	420	427	360	4	496	477	312	6	702	527	480	4	576
378	108	16	960	428	212	6	756	478	238	4	720	528	160	20	1488
379	378	2	380	429	240	8	672	479	478	2	480	529	506	3	553
380	144	12	840	430	168	8	792	480	128	24	1512	530	208	8	972
381	252	4	512	431	430	2	432	481	432	4	532	531	348	6	780
382	190	4	576	432	144	20	1240	482	240	4	726	532	216	12	1120
383	382	2	384	433	432	2	434	483	264	8	768	533	480	4	588
384	128	16	1020	434	180	8	768	484	220	9	931	534	176	8	1080
385	240	8	576	435	224	8	720	485	384	4	588	535	424	4	648
386	192	4	582	436	216	6	770	486	162	12	1092	536	264	8	1020
387	252	6	572	437	396	4	480	487	486	2	488	537	356	4	720
388	192	6	686	438	144	8	888	488	240	8	930	538	268	4	810
389	388	2	390	439	438	2	440	489	324	4	656	539	420	6	684
390	96	16	1008	440	160	16	1080	490	168	12	1026	540	144	24	1680
391	352	4	432	441	252	9	741	491	490	2	492	541	540	2	542
392	168	12	855	442	192	8	756	492	160	12	1176	542	270	4	816
393	260	4	528	443	442	2	444	493	448	4	540	543	360	4	728
394	196	4	594	444	144	12	1064	494	216	8	840	544	256	12	1134
395	312	4	480	445	352	4	540	495	240	12	936	545	432	4	660
396	120	18	1092	446	222	4	672	496	240	10	992	546	144	16	1344
397	396	2	398	447	296	4	600	497	420	4	576	547	546	2	548
398	198	4	600	448	192	14	1016	498	164	8	1008	548	272	6	966
399	216	8	640	449	448	2	450	499	498	2	500	549	360	6	806
400	160	15	961	450	120	18	1209	500	200	12	1092	550	200	12	1116
401	400	2	402	451	400	4	504	501	332	4	672	551	504	4	600
402	132	8	816	452	224	6	798	502	250	4	756	552	176	16	1440
403	360	4	448	453	300	4	608	503	502	2	504	553	468	4	640
404	200	6	714	454	226	4	684	504	144	24	1560	554	276	4	834
405	216	10	726	455	288	8	672	505	400	4	612	555	288	8	912
406	168	8	720	456	144	16	1200	506	220	8	864	556	276	6	980
407	360	4	456	457	456	2	458	507	312	6	732	557	556	2	558
408	128	16	1080	458	228	4	690	508	252	6	896	558	180	12	1248
409	408	2	410	459	288	8	720	509	508	2	510	559	504	4	616
410	160	8	756	460	176	12	1008	510	128	16	1296	560	192	20	1488

TOTIENT FUNCTION $\phi(n)$

n	$\phi(n)$	σ_0	σ_1	n	$\phi(n)$	σ_0	σ_1	n	$\phi(n)$	σ_0	σ_1	n	$\phi(n)$	σ_0	σ_1
561	320	8	864	611	552	4	672	661	660	2	662	711	468	6	1040
562	280	4	846	612	192	18	1638	662	330	4	996	712	352	8	1350
563	562	2	564	613	612	2	614	663	384	8	1008	713	660	4	768
564	184	12	1344	614	306	4	924	664	328	8	1260	714	192	16	1728
565	448	4	684	615	320	8	1008	665	432	8	960	715	480	8	1008
566	282	4	852	616	240	16	1440	666	216	12	1482	716	356	6	1260
567	324	10	968	617	616	2	618	667	616	4	720	717	476	4	960
568	280	8	1080	618	204	8	1248	668	332	6	1176	718	358	4	1080
569	568	2	570	619	618	2	620	669	444	4	896	719	718	2	720
570	144	16	1440	620	240	12	1344	670	264	8	1224	720	192	30	2418
571	570	2	572	621	396	8	960	671	600	4	744	721	612	4	832
572	240	12	1176	622	310	4	936	672	192	24	2016	722	342	6	1143
573	380	4	768	623	528	4	720	673	672	2	674	723	480	4	968
574	240	8	1008	624	192	20	1736	674	336	4	1014	724	360	6	1274
575	440	6	744	625	500	5	781	675	360	12	1240	725	560	6	930
576	192	21	1651	626	312	4	942	676	312	9	1281	726	220	12	1596
577	576	2	578	627	360	8	960	677	676	2	678	727	726	2	728
578	272	6	921	628	312	6	1106	678	224	8	1368	728	288	16	1680
579	384	4	776	629	576	4	684	679	576	4	784	729	486	7	1093
580	224	12	1260	630	144	24	1872	680	256	16	1620	730	288	8	1332
581	492	4	672	631	630	2	632	681	452	4	912	731	672	4	792
582	192	8	1176	632	312	8	1200	682	300	8	1152	732	240	12	1736
583	520	4	648	633	420	4	848	683	682	2	684	733	732	2	734
584	288	8	1110	634	316	4	954	684	216	18	1820	734	366	4	1104
585	288	12	1092	635	504	4	768	685	544	4	828	735	336	12	1368
586	292	4	882	636	208	12	1512	686	294	8	1200	736	352	12	1512
587	586	2	588	637	504	6	798	687	456	4	920	737	660	4	816
588	168	18	1596	638	280	8	1080	688	336	10	1364	738	240	12	1638
589	540	4	640	639	420	6	936	689	624	4	756	739	738	2	740
590	232	8	1080	640	256	16	1530	690	176	16	1728	740	288	12	1596
591	392	4	792	641	640	2	642	691	690	2	692	741	432	8	1120
592	288	10	1178	642	212	8	1296	692	344	6	1218	742	312	8	1296
593	592	2	594	643	642	2	644	693	360	12	1248	743	742	2	744
594	180	16	1440	644	264	12	1344	694	346	4	1044	744	240	16	1920
595	384	8	864	645	336	8	1056	695	552	4	840	745	592	4	900
596	296	6	1050	646	288	8	1080	696	224	16	1800	746	372	4	1122
597	396	4	800	647	646	2	648	697	640	4	756	747	492	6	1092
598	264	8	1008	648	216	20	1815	698	348	4	1050	748	320	12	1512
599	598	2	600	649	580	4	720	699	464	4	936	749	636	4	864
600	160	24	1860	650	240	12	1302	700	240	18	1736	750	200	16	1872
601	600	2	602	651	360	8	1024	701	700	2	702	751	750	2	752
602	252	8	1056	652	324	6	1148	702	216	16	1680	752	368	10	1488
603	396	6	884	653	652	2	654	703	648	4	760	753	500	4	1008
604	300	6	1064	654	216	8	1320	704	320	14	1524	754	336	8	1260
605	440	6	798	655	520	4	792	705	368	8	1152	755	600	4	912
606	200	8	1224	656	320	10	1302	706	352	4	1062	756	216	24	2240
607	606	2	608	657	432	6	962	707	600	4	816	757	756	2	758
608	288	12	1260	658	276	8	1152	708	232	12	1680	758	378	4	1140
609	336	8	960	659	658	2	660	709	708	2	710	759	440	8	1152
610	240	8	1116	660	160	24	2016	710	280	8	1296	760	288	16	1800

TOTIENT FUNCTION $\phi(n)$

n	$\phi(n)$	σ_0	σ_1	n	$\phi(n)$	σ_0	σ_1	n	$\phi(n)$	σ_0	σ_1	n	$\phi(n)$	σ_0	σ_1
761	760	2	762	811	810	2	812	861	480	8	1344	911	910	2	912
762	252	8	1536	812	336	12	1680	862	430	4	1296	912	288	20	2480
763	648	4	880	813	540	4	1088	863	862	2	864	913	820	4	1008
764	380	6	1344	814	360	8	1368	864	288	24	2520	914	456	4	1374
765	384	12	1404	815	648	4	984	865	688	4	1044	915	480	8	1488
766	382	4	1152	816	256	20	2232	866	432	4	1302	916	456	6	1610
767	696	4	840	817	756	4	880	867	544	6	1228	917	780	4	1056
768	256	18	2044	818	408	4	1230	868	360	12	1792	918	288	16	2160
769	768	2	770	819	432	12	1456	869	780	4	960	919	918	2	920
770	240	16	1728	820	320	12	1764	870	224	16	2160	920	352	16	2160
771	512	4	1032	821	820	2	822	871	792	4	952	921	612	4	1232
772	384	6	1358	822	272	8	1656	872	432	8	1650	922	460	4	1386
773	772	2	774	823	822	2	824	873	576	6	1274	923	840	4	1008
774	252	12	1716	824	408	8	1560	874	396	8	1440	924	240	24	2688
775	600	6	992	825	400	12	1488	875	600	8	1248	925	720	6	1178
776	384	8	1470	826	348	8	1440	876	288	12	2072	926	462	4	1392
777	432	8	1216	827	826	2	828	877	876	2	878	927	612	6	1352
778	388	4	1170	828	264	18	2184	878	438	4	1320	928	448	12	1890
779	720	4	840	829	828	2	830	879	584	4	1176	929	928	2	930
780	192	24	2352	830	328	8	1512	880	320	20	2232	930	240	16	2304
781	700	4	864	831	552	4	1112	881	880	2	882	931	756	6	1140
782	352	8	1296	832	384	14	1778	882	252	18	2223	932	464	6	1638
783	504	8	1200	833	672	6	1026	883	882	2	884	933	620	4	1248
784	336	15	1767	834	276	8	1680	884	384	12	1764	924	466	4	1404
785	624	4	948	835	664	4	1008	885	464	8	1440	935	640	8	1296
786	260	8	1584	836	360	12	1680	886	442	4	1332	936	288	24	2730
787	786	2	788	837	540	8	1280	887	886	2	888	937	936	2	938
788	392	6	1386	838	418	4	1260	888	288	16	2280	938	396	8	1632
789	524	4	1056	839	838	2	840	889	756	4	1024	939	624	4	1256
790	312	8	1440	840	192	32	2880	890	352	8	1620	940	368	12	2016
791	672	4	912	841	812	3	871	891	540	10	1452	941	940	2	942
792	240	24	2340	842	420	4	1266	892	444	6	1568	942	312	8	1896
793	720	4	868	843	560	4	1128	893	828	4	960	943	880	4	1008
794	396	4	1194	844	420	6	1484	894	296	8	1800	944	464	10	1860
795	416	8	1296	845	624	6	1098	895	712	4	1080	945	432	16	1920
796	396	6	1400	846	276	12	1872	896	384	16	2040	946	420	8	1584
797	796	2	798	847	660	6	1064	897	528	8	1344	947	946	2	948
798	216	16	1920	848	416	10	1674	898	448	4	1350	948	312	12	2240
799	736	4	864	849	564	4	1136	899	840	4	960	949	864	4	1036
800	320	18	1953	850	320	12	1674	900	240	27	2821	950	360	12	1860
801	528	6	1170	851	792	4	912	901	832	4	972	951	632	4	1272
802	400	4	1206	852	280	12	2016	902	400	8	1512	952	384	16	2160
803	720	4	888	853	852	2	854	903	504	8	1408	953	952	2	954
804	264	12	1904	854	360	8	1488	904	448	8	1710	954	312	12	2106
805	528	8	1152	855	432	12	1560	905	720	4	1092	955	760	4	1152
806	360	8	1344	856	424	8	1620	906	300	8	1824	956	476	6	1680
807	536	4	1080	857	856	2	858	907	906	2	908	957	560	8	1440
808	400	8	1530	858	240	16	2016	908	452	6	1596	958	478	4	1440
809	808	2	810	859	858	2	860	909	600	6	1326	959	816	4	1104
810	216	20	2178	860	336	12	1848	910	288	16	2016	960	256	28	3048

TOTIENT FUNCTION $\phi(n)$

n	$\phi(n)$	σ_0	σ_1	n	$\phi(n)$	σ_0	σ_1	n	$\phi(n)$	σ_0	σ_1	n	$\phi(n)$	σ_0	σ_1
961	930	3	993	971	970	2	972	981	648	6	1430	991	990	2	992
962	432	8	1596	972	324	18	2548	982	490	4	1476	992	480	12	2016
963	636	6	1404	973	828	4	1120	983	982	2	984	993	660	4	1328
964	480	6	1694	974	486	4	1464	984	320	16	2520	994	420	8	1728
965	768	4	1164	975	480	12	1736	985	784	4	1188	995	792	4	1200
966	264	16	2304	976	480	10	1922	986	448	8	1620	996	328	12	2352
967	966	2	968	977	976	2	978	987	552	8	1536	997	996	2	998
968	440	12	1995	978	324	8	1968	988	432	12	1960	998	498	4	1500
969	576	8	1440	979	880	4	1080	989	924	4	1056	999	648	8	1520
970	384	8	1764	980	336	18	2394	990	240	24	2808	1000	400	16	2340

REPRESENTATION OF NUMBERS[1]

Any positive real number x can be uniquely represented in the scale of some integer $b>1$ as

$$x=(A_m \ldots A_1A_0 \cdot a_{-1}a_{-2} \ldots)_{(b)},$$

where every A_i and a_{-j} is one of the integers 0, 1, . . ., $b-1$, not all A_i, a_{-j} are zero, and $A_m>0$ if $x\geq 1$. There is a one-to-one correspondence between the number and the sequence

$$x=A_mb^m+ \ldots +A_1b+A_0+\sum_1^{\infty} a_{-j}b^{-j}$$

where the infinite series converges. The integer b is called the base or radix of the scale.

The sequence for x in the scale of b may terminate, i.e., $a_{-n-1}=a_{-n-2}= \ldots =0$ for some $n\geq 1$ so that

$$x=(A_m \ldots A_1A_0 \cdot a_{-1}a_{-2} \ldots a_{-n})_{(b)};$$

then x is said to be a finite b-adic number.

A sequence which does not terminate may have the property that the infinite sequence a_{-1}, a_{-2}, . . . becomes periodic from a certain digit $a_{-n}(n\geq 1)$ on; according as $n=1$ or $n>1$ the sequence is then said to be pure or mixed recurring.

A sequence which neither terminates nor recurs represents an irrational number.

NAMES OF SCALES

Base	Scale	Base	Scale
2	Binary	8	Octal
3	Ternary	9	Nonary
4	Quaternary	10	Decimal
5	Quinary	11	Undenary
6	Senary	12	Duodenary
7	Septenary	16	Hexadecimal

[1] From "Handbook of Mathematical Functions", National Bureau of Standards

GENERAL CONVERSION METHODS

Any number can be converted from the scale of b to the scale of some integer $\bar{b} \neq b$, $\bar{b} > 1$, by using arithmetic operations in either the b-scale or the $\bar{b}$-scale. Accordingly, there are four methods of conversion, depending on whether the number to be converted is an integer or a proper fraction.

INTEGERS $X = (A_m \ldots A_1 A_0)_{(b)}$

(I) b-scale arithmetic. Convert $\bar{b}$ to the b-scale and define

$$X/\bar{b} = X_1 + \bar{A}'_0/\bar{b},$$

$$X_1/\bar{b} = X_2 + \bar{A}'_1/\bar{b},$$

$$\vdots$$

$$X_{\bar{m}}/\bar{b} = 0 + \bar{A}'_{\bar{m}}/\bar{b},$$

where $\bar{A}'_0, \bar{A}'_1, \ldots, \bar{A}'_{\bar{m}}$ are the remainders and $X_1, X_2, \ldots, X_{\bar{m}}$ the quotients (in the b-scale) where $X, X_1, \ldots, X_{\bar{m}-1}$, respectively are divided by $\bar{b}$ in the b-scale. Then convert the remainders to the $\bar{b}$-scale,

$$(\bar{A}'_0)_{(\bar{b})} = \bar{A}_0,\ (\bar{A}'_1)_{(\bar{b})} = \bar{A}_1,\ \ldots,\ (\bar{A}'_{\bar{m}})_{(\bar{b})} = \bar{A}_{\bar{m}}$$

and obtain

$$X = (\bar{A}_{\bar{m}} \ldots \bar{A}_1 \bar{A}_0)_{(\bar{b})}.$$

(II) $\bar{b}$-scale arithmetic. Convert b and A_0, $A_1, \ldots, A_m$ to the $\bar{b}$-scale and define, using arithmetic operations in the $\bar{b}$-scale,

$$X_{m-1} = A_m b + A_{m-1},$$

$$X_{m-2} = X_{m-1} b + A_{m-2},$$

$$X_1 = X_2 b + A_1,$$

then

$$X = X_1 b + A_0.$$

PROPER FRACTIONS $x = (0.a_{-1}a_{-2}\cdots)_{(b)}$

To convert a proper fraction x, given to n digits in the b-scale, to the scale of $\bar{b} \neq b$ such that inverse conversion from the $\bar{b}$-scale may yield the same n rounded digits in the b-scale, the representation of x in the $\bar{b}$-scale must be obtained to $\bar{n}$ rounded digits where n satisfies $\bar{b}^{\bar{n}} > b^n$.

(III) b-scale arithmetic. Convert $\bar{b}$ to the b-scale and define

$$x\bar{b} = x_1 + \bar{a}'_{-1}$$
$$x_1\bar{b} = x_2 + \bar{a}'_{-2}$$
$$x_{\bar{n}-1}\bar{b} = x_{\bar{n}} + \bar{a}'_{-n}$$

where $\bar{a}_{-1}, \bar{a}'_{-2}, \ldots, \bar{a}_{-\bar{n}}$ are the integral parts and $x_1, x_2, \ldots, x_{\bar{n}}$ the fractional parts (in the b-scale) of the products $x\bar{b}, x_1\bar{b}, \ldots, x_{\bar{n}-1}\bar{b}$, respectively. Then convert the integral parts to the $\bar{b}$-scale,

$$(\bar{a}'_{-1})_{(\bar{b})} = \bar{a}_{-1},\ (\bar{a}'_{-2})_{(\bar{b})} = \bar{a}_{-2}, \ldots, (\bar{a}'_{-n})_{(\bar{b})} = \bar{a}_{-n},$$

and obtain

$$x = (0.\bar{a}_{-1}\bar{a}_{-2}\ldots\bar{a}_{-n})_{(\bar{b})}.$$

(IV) $\bar{b}$-scale arithmetic. Convert b and a_{-1}, $a_{-2}, \ldots, a_{-n}$ to the $\bar{b}$-scale and define, using arithmetic operations in the $\bar{b}$-scale,

$$x_{-n+1} = a_{-n}/b + a_{-n+1},$$
$$x_{-n+2} = x_{-n+1}/b + a_{-n+2},$$
$$x_{-1} = x_2/b + a_{-1};$$

then

$$x = x_{-1}/b.$$

$2^{\pm n}$ IN DECIMAL

2^n	n	2^{-n}
1	0	1.0
2	1	0.5
4	2	0.25
8	3	0.125
16	4	0.0625
32	5	0.03125
64	6	0.01562 5
128	7	0.00781 25
256	8	0.00390 625
512	9	0.00195 3125
1024	10	0.00097 65625
2048	11	0.00048 82812 5
4096	12	0.00024 41406 25
8192	13	0.00012 20703 125
16384	14	0.00006 10351 5625
32768	15	0.00003 05175 78125
65536	16	0.00001 52587 89062 5
1 31072	17	0.00000 76293 94531 25
2 62144	18	0.00000 38146 97265 625
5 24288	19	0.00000 19073 48632 8125
10 48576	20	0.00000 09536 74316 40625
20 97152	21	0.00000 04768 37158 20312 5
41 94304	22	0.00000 02384 18579 10156 25
83 88608	23	0.00000 01192 09289 55078 125
167 77216	24	0.00000 00596 04644 77539 0625
335 54432	25	0.00000 00298 02322 38769 53125
671 08864	26	0.00000 00149 01161 19384 76562 5
1342 17728	27	0.00000 00074 50580 59692 38281 25
2684 35456	28	0.00000 00037 25290 29846 19140 625
5368 70912	29	0.00000 00018 62645 14923 09570 3125
10737 41824	30	0.00000 00009 31322 57461 54785 15625
21474 83648	31	0.00000 00004 65661 28730 77392 57812 5
42949 67296	32	0.00000 00002 32830 64365 38696 28906 25
85899 34592	33	0.00000 00001 16415 32182 69348 14453 125
1 71798 69184	34	0.00000 00000 58207 66091 34674 07226 5625
3 43597 38368	35	0.00000 00000 29103 83045 67337 03613 28125
6 87194 76736	36	0.00000 00000 14551 91522 83668 51806 64062 5
13 74389 53472	37	0.00000 00000 07275 95761 41834 25903 32031 25
27 48779 06944	38	0.00000 00000 03637 97880 70917 12951 66015 625
54 97558 13888	39	0.00000 00000 01818 98940 35458 56475 83007 8125
109 95116 27776	40	0.00000 00000 00909 49470 17729 28237 91503 90625
219 90232 55552	41	0.00000 00000 00454 74735 08864 64118 95751 95312 5
439 80465 11104	42	0.00000 00000 00227 37367 54432 32059 47875 97656 25
879 60930 22208	43	0.00000 00000 00113 68683 77216 16029 73937 98828 125
1759 21860 44416	44	0.00000 00000 00056 84341 88608 08014 86968 99414 0625
3518 43720 88832	45	0.00000 00000 00028 42170 94304 04007 43484 49707 03125
7036 87441 77664	46	0.00000 00000 00014 21085 47152 02003 71742 24853 51562 5
14073 74883 55328	47	0.00000 00000 00007 10542 73576 01001 85871 12426 75781 25
28147 49767 10656	48	0.00000 00000 00003 55271 36788 00500 92935 56213 37890 625
56294 99534 21312	49	0.00000 00000 00001 77635 68394 00250 46467 78106 68945 3125
112589 99068 42624	50	0.00000 00000 00000 88817 84197 00125 23233 89053 34472 65625

2^x IN DECIMAL

x	2^x	x	2^x	x	2^x
0.001	1.00069 33874 62581	0.01	1.00695 55500 56719	0.1	1.07177 34625 36293
0.002	1.00138 72557 11335	0.02	1.01395 94797 90029	0.2	1.14869 83549 97035
0.003	1.00208 16050 79633	0.03	1.02101 21257 07193	0.3	1.23114 44133 44916
0.004	1.00277 64359 01078	0.04	1.02811 38266 56067	0.4	1.31950 79107 72894
0.005	1.00347 17485 09503	0.05	1.03526 49238 41377	0.5	1.41421 35623 73095
0.006	1.00416 75432 38973	0.06	1.04246 57608 41121	0.6	1.51571 65665 10398
0.007	1.00486 38204 23785	0.07	1.04971 66836 23067	0.7	1.62450 47927 12471
0.008	1.00556 05803 98468	0.08	1.05701 80405 61380	0.8	1.74110 11265 92248
0.009	1.00625 78234 97782	0.09	1.06437 01824 53360	0.9	1.86606 59830 73615

$10^{\pm n}$ IN OCTAL

10^n	n	10^{-n}	10^n	n	10^{-n}
1	0	1.000 000 000 000 000 000 00	112 402 762 000	10	0.000 000 000 006 676 337 66
12	1	0.063 146 314 631 463 146 31	1 351 035 564 000	11	0.000 000 000 000 537 657 77
144	2	0.005 075 341 217 270 243 66	16 432 451 210 000	12	0.000 000 000 000 043 136 32
1 750	3	0.000 406 111 564 570 651 77	221 411 634 520 000	13	0.000 000 000 000 003 411 35
23 420	4	0.000 032 155 613 530 704 15	2 657 142 036 440 000	14	0.000 000 000 000 000 264 11
303 240	5	0.000 002 476 132 610 706 64	34 327 724 461 500 000	15	0.000 000 000 000 000 022 01
3 641 100	6	0.000 000 206 157 364 055 37	434 157 115 760 200 000	16	0.000 000 000 000 000 001 63
46 113 200	7	0.000 000 015 327 745 152 75	5 432 127 413 542 400 000	17	0.000 000 000 000 000 000 14
575 360 400	8	0.000 000 001 257 143 561 06	67 405 553 164 731 000 000	18	0.000 000 000 000 000 000 01
7 346 545 000	9	0.000 000 000 104 560 276 41			

$n \log_{10} 2$, $n \log_2 10$ IN DECIMAL

n	$n \log_{10} 2$	$n \log_2 10$	n	$n \log_{10} 2$	$n \log_2 10$
1	0.30102 99957	3.32192 80949	6	1.80617 99740	19.93156 85693
2	0.60205 99913	6.64385 61898	7	2.10720 99696	23.25349 66642
3	0.90308 99870	9.96578 42847	8	2.40823 99653	26.57542 47591
4	1.20411 99827	13.28771 23795	9	2.70926 99610	29.89735 28540
5	1.50514 99783	16.60964 04744	10	3.01029 99566	33.21928 09489

ADDITION AND MULTIPLICATION TABLES

Binary Scale

Addition

$$0 + 0 = 0$$
$$0 + 1 = 1 + 0 = 1$$
$$1 + 1 = 10$$

Multiplication

$$0 \times 0 = 0$$
$$0 \times 1 = 1 \times 0 = 0$$
$$1 \times 1 = 1$$

Octal Scale

Addition

0	01	02	03	04	05	06	07
1	02	03	04	05	06	07	10
2	03	04	05	06	07	10	11
3	04	05	06	07	10	11	12
4	05	06	07	10	11	12	13
5	06	07	10	11	12	13	14
6	07	10	11	12	13	14	15
7	10	11	12	13	14	15	16

Multiplication

1	02	03	04	05	06	07
2	04	06	10	12	14	16
3	06	11	14	17	22	25
4	10	14	20	24	30	34
5	12	17	24	31	36	43
6	14	22	30	36	44	52
7	16	25	34	43	52	61

MATHEMATICAL CONSTANTS IN OCTAL SCALE

$$\pi = (3.11037\ 552421)_{(8)}$$

$$\pi^{-1} = (0.24276\ 301556)_{(8)}$$

$$\sqrt{\pi} = (1.61337\ 611067)_{(8)}$$

$$\ln \pi = (1.11206\ 404435)_{(8)}$$

$$\log_2 \pi = (1.51544\ 163223)_{(8)}$$

$$\sqrt{10} = (3.12305\ 407267)_{(8)}$$

$$\gamma = (0.44742\ 147707)_{(8)}$$

$$\ln \gamma = -(0.43127\ 233602)_{(8)}$$

$$\log_2 \gamma = -(0.62573\ 030645)_{(8)}$$

$$e = (2.55760\ 521305)_{(8)}$$

$$e^{-1} = (0.27426\ 530661)_{(8)}$$

$$\sqrt{e} = (1.51411\ 230704)_{(8)}$$

$$\log_{10} e = (0.33626\ 754251)_{(8)}$$

$$\log_2 e = (1.34252\ 166245)_{(8)}$$

$$\log_2 10 = (3.24464\ 741136)_{(8)}$$

$$\sqrt{2} = (1.32404\ 746320)_{(8)}$$

$$\ln 2 = (0.54271\ 027760)_{(8)}$$

$$\ln 10 = (2.23273\ 067355)_{(8)}$$

OCTAL-DECIMAL INTEGER CONVERSION TABLE

	0	1	2	3	4	5	6	7
0000	0000	0001	0002	0003	0004	0005	0006	0007
0010	0008	0009	0010	0011	0012	0013	0014	0015
0020	0016	0017	0018	0019	0020	0021	0022	0023
0030	0024	0025	0026	0027	0028	0029	0030	0031
0040	0032	0033	0034	0035	0036	0037	0038	0039
0050	0040	0041	0042	0043	0044	0045	0046	0047
0060	0048	0049	0050	0051	0052	0053	0054	0055
0070	0056	0057	0058	0059	0060	0061	0062	0063
0100	0064	0065	0066	0067	0068	0069	0070	0071
0110	0072	0073	0074	0075	0076	0077	0078	0079
0120	0080	0081	0082	0083	0084	0085	0086	0087
0130	0088	0089	0090	0091	0092	0093	0094	0095
0140	0096	0097	0098	0099	0100	0101	0102	0103
0150	0104	0105	0106	0107	0108	0109	0110	0111
0160	0112	0113	0114	0115	0116	0117	0118	0119
0170	0120	0121	0122	0123	0124	0125	0126	0127
0200	0128	0129	0130	0131	0132	0133	0134	0135
0210	0136	0137	0138	0139	0140	0141	0142	0143
0220	0144	0145	0146	0147	0148	0149	0150	0151
0230	0152	0153	0154	0155	0156	0157	0158	0159
0240	0160	0161	0162	0163	0164	0165	0166	0167
0250	0168	0169	0170	0171	0172	0173	0174	0175
0260	0176	0177	0178	0179	0180	0181	0182	0183
0270	0184	0185	0186	0187	0188	0189	0190	0191
0300	0192	0193	0194	0195	0196	0197	0198	0199
0310	0200	0201	0202	0203	0204	0205	0206	0207
0320	0208	0209	0210	0211	0212	0213	0214	0215
0330	0216	0217	0218	0219	0220	0221	0222	0223
0340	0224	0225	0226	0227	0228	0229	0230	0231
0350	0232	0233	0234	0235	0236	0237	0238	0239
0360	0240	0241	0242	0243	0244	0245	0246	0247
0370	0248	0249	0250	0251	0252	0253	0254	0255

	0	1	2	3	4	5	6	7
0400	0256	0257	0258	0259	0260	0261	0262	0263
0410	0264	0265	0266	0267	0268	0269	0270	0271
0420	0272	0273	0274	0275	0276	0277	0278	0279
0430	0280	0281	0282	0283	0284	0285	0286	0287
0440	0288	0289	0290	0291	0292	0293	0294	0295
0450	0296	0297	0298	0299	0300	0301	0302	0303
0460	0304	0305	0306	0307	0308	0309	0310	0311
0470	0312	0313	0314	0315	0316	0317	0318	0319
0500	0320	0321	0322	0323	0324	0325	0326	0327
0510	0328	0329	0330	0331	0332	0333	0334	0335
0520	0336	0337	0338	0339	0340	0341	0342	0343
0530	0344	0345	0346	0347	0348	0349	0350	0351
0540	0352	0353	0354	0355	0356	0357	0358	0359
0550	0360	0361	0362	0363	0364	0365	0366	0367
0560	0368	0369	0370	0371	0372	0373	0374	0375
0570	0376	0377	0378	0379	0380	0381	0382	0383
0600	0384	0385	0386	0387	0388	0389	0390	0391
0610	0392	0393	0394	0395	0396	0397	0398	0399
0620	0400	0401	0402	0403	0404	0405	0406	0407
0630	0408	0409	0410	0411	0412	0413	0414	0415
0640	0416	0417	0418	0419	0420	0421	0422	0423
0650	0424	0425	0426	0427	0428	0429	0430	0431
0660	0432	0433	0434	0435	0436	0437	0438	0439
0670	0440	0441	0442	0443	0444	0445	0446	0447
0700	0448	0449	0450	0451	0452	0453	0454	0455
0710	0456	0457	0458	0459	0460	0461	0462	0463
0720	0464	0465	0466	0467	0468	0469	0470	0471
0730	0472	0473	0474	0475	0476	0477	0478	0479
0740	0480	0481	0482	0483	0484	0485	0486	0487
0750	0488	0489	0490	0491	0492	0493	0494	0495
0760	0496	0497	0498	0499	0500	0501	0502	0503
0770	0504	0505	0506	0507	0508	0509	0510	0511

0000 to 0777 (Octal)	0000 to 0511 (Decimal)

Octal	Decimal
10000	**4096**
20000	**8192**
30000	**12288**
40000	**16384**
50000	**20480**
60000	**24576**
70000	**28672**

	0	1	2	3	4	5	6	7
1000	0512	0513	0514	0515	0516	0517	0518	0519
1010	0520	0521	0522	0523	0524	0525	0526	0527
1020	0528	0529	0530	0531	0532	0533	0534	0535
1030	0536	0537	0538	0539	0540	0541	0542	0543
1040	0544	0545	0546	0547	0548	0549	0550	0551
1050	0552	0553	0554	0555	0556	0557	0558	0559
1060	0560	0561	0562	0563	0564	0565	0566	0567
1070	0568	0569	0570	0571	0572	0573	0574	0575
1100	0576	0577	0578	0579	0580	0581	0582	0583
1110	0584	0585	0586	0587	0588	0589	0590	0591
1120	0592	0593	0594	0595	0596	0597	0598	0599
1130	0600	0601	0602	0603	0604	0605	0606	0607
1140	0608	0609	0610	0611	0612	0613	0614	0615
1150	0616	0617	0618	0619	0620	0621	0622	0623
1160	0624	0625	0626	0627	0628	0629	0630	0631
1170	0632	0633	0634	0635	0636	0637	0638	0639
1200	0640	0641	0642	0643	0644	0645	0646	0647
1210	0648	0649	0650	0651	0652	0653	0654	0655
1220	0656	0657	0658	0659	0660	0661	0662	0663
1230	0664	0665	0666	0667	0668	0669	0670	0671
1240	0672	0673	0674	0675	0676	0677	0678	0679
1250	0680	0681	0682	0683	0684	0685	0686	0687
1260	0688	0689	0690	0691	0692	0693	0694	0695
1270	0696	0697	0698	0699	0700	0701	0702	0703
1300	0704	0705	0706	0707	0708	0709	0710	0711
1310	0712	0713	0714	0715	0716	0717	0718	0719
1320	0720	0721	0722	0723	0724	0725	0726	0727
1330	0728	0729	0730	0731	0732	0733	0734	0735
1340	0736	0737	0738	0739	0740	0741	0742	0743
1350	0744	0745	0746	0747	0748	0749	0750	0751
1360	0752	0753	0754	0755	0756	0757	0758	0759
1370	0760	0761	0762	0763	0764	0765	0766	0767

	0	1	2	3	4	5	6	7
1400	0768	0769	0770	0771	0772	0773	0774	0775
1410	0776	0777	0778	0779	0780	0781	0782	0783
1420	0784	0785	0786	0787	0788	0789	0790	0791
1430	0792	0793	0794	0795	0796	0797	0798	0799
1440	0800	0801	0802	0803	0804	0805	0806	0807
1450	0808	0809	0810	0811	0812	0813	0814	0815
1460	0816	0817	0818	0819	0820	0821	0822	0823
1470	0824	0825	0826	0827	0828	0829	0830	0831
1500	0832	0833	0834	0835	0836	0837	0838	0839
1510	0840	0841	0842	0843	0844	0845	0846	0847
1520	0848	0849	0850	0851	0852	0853	0854	0855
1530	0856	0857	0858	0859	0860	0861	0862	0863
1540	0864	0865	0866	0867	0868	0869	0870	0871
1550	0872	0873	0874	0875	0876	0877	0878	0879
1560	0880	0881	0882	0883	0884	0885	0886	0887
1570	0888	0889	0890	0891	0892	0893	0894	0895
1600	0896	0897	0898	0899	0900	0901	0902	0903
1610	0904	0905	0906	0907	0908	0909	0910	0911
1620	0912	0913	0914	0915	0916	0917	0918	0919
1630	0920	0921	0922	0923	0924	0925	0926	0927
1640	0928	0929	0930	0931	0932	0933	0934	0935
1650	0936	0937	0938	0939	0940	0941	0942	0943
1660	0944	0945	0946	0947	0948	0949	0950	0951
1670	0952	0953	0954	0955	0956	0957	0958	0959
1700	0960	0961	0962	0963	0964	0965	0966	0967
1710	0968	0969	0970	0971	0972	0973	0974	0975
1720	0976	0977	0978	0979	0980	0981	0982	0983
1730	0984	0985	0986	0987	0988	0989	0990	0991
1740	0992	0993	0994	0995	0996	0997	0998	0999
1750	1000	1001	1002	1003	1004	1005	1006	1007
1760	1008	1009	1010	1011	1012	1013	1014	1015
1770	1016	1017	1018	1019	1020	1021	1022	1023

1000 to 1777 (Octal)	0512 to 1023 (Decimal)

OCTAL-DECIMAL INTEGER CONVERSION TABLE

2000 to 2777 (Octal) | 1024 to 1535 (Decimal)

Octal	Decimal
10000-	**4096**
20000-	**8192**
30000-	**12288**
40000-	**16384**
50000-	**20480**
60000-	**24576**
70000-	**28672**

	0	1	2	3	4	5	6	7
2000	1024	1025	1026	1027	1028	1029	1030	1031
2010	1032	1033	1034	1035	1036	1037	1038	1039
2020	1040	1041	1042	1043	1044	1045	1046	1047
2030	1048	1049	1050	1051	1052	1053	1054	1055
2040	1056	1057	1058	1059	1060	1061	1062	1063
2050	1064	1065	1066	1067	1068	1069	1070	1071
2060	1072	1073	1074	1075	1076	1077	1078	1079
2070	1080	1081	1082	1083	1084	1085	1086	1087
2100	1088	1089	1090	1091	1092	1093	1094	1095
2110	1096	1097	1098	1099	1100	1101	1102	1103
2120	1104	1105	1106	1107	1108	1109	1110	1111
2130	1112	1113	1114	1115	1116	1117	1118	1119
2140	1120	1121	1122	1123	1124	1125	1126	1127
2150	1128	1129	1130	1131	1132	1133	1134	1135
2160	1136	1137	1138	1139	1140	1141	1142	1143
2170	1144	1145	1146	1147	1148	1149	1150	1151
2200	1152	1153	1154	1155	1156	1157	1158	1159
2210	1160	1161	1162	1163	1164	1165	1166	1167
2220	1168	1169	1170	1171	1172	1173	1174	1175
2230	1176	1177	1178	1179	1180	1181	1182	1183
2240	1184	1185	1186	1187	1188	1189	1190	1191
2250	1192	1193	1194	1195	1196	1197	1198	1199
2260	1200	1201	1202	1203	1204	1205	1206	1207
2270	1208	1209	1210	1211	1212	1213	1214	1215
2300	1216	1217	1218	1219	1220	1221	1222	1223
2310	1224	1225	1226	1227	1228	1229	1230	1231
2320	1232	1233	1234	1235	1236	1237	1238	1239
2330	1240	1241	1242	1243	1244	1245	1246	1247
2340	1248	1249	1250	1251	1252	1253	1254	1255
2350	1256	1257	1258	1259	1260	1261	1262	1263
2360	1264	1265	1266	1267	1268	1269	1270	1271
2370	1272	1273	1274	1275	1276	1277	1278	1279

	0	1	2	3	4	5	6	7
2400	1280	1281	1282	1283	1284	1285	1286	1287
2410	1288	1289	1290	1291	1292	1293	1294	1295
2420	1296	1297	1298	1299	1300	1301	1302	1303
2430	1304	1305	1306	1307	1308	1309	1310	1311
2440	1312	1313	1314	1315	1316	1317	1318	1319
2450	1320	1321	1322	1323	1324	1325	1326	1327
2460	1328	1329	1330	1331	1332	1333	1334	1335
2470	1336	1337	1338	1339	1340	1341	1342	1343
2500	1344	1345	1346	1347	1348	1349	1350	1351
2510	1352	1353	1354	1355	1356	1357	1358	1359
2520	1360	1361	1362	1363	1364	1365	1366	1367
2530	1368	1369	1370	1371	1372	1373	1374	1375
2540	1376	1377	1378	1379	1380	1381	1382	1383
2550	1384	1385	1386	1387	1388	1389	1390	1391
2560	1392	1393	1394	1395	1396	1397	1398	1399
2570	1400	1401	1402	1403	1404	1405	1406	1407
2600	1408	1409	1410	1411	1412	1413	1414	1415
2610	1416	1417	1418	1419	1420	1421	1422	1423
2620	1424	1425	1426	1427	1428	1429	1430	1431
2630	1432	1433	1434	1435	1436	1437	1438	1439
2640	1440	1441	1442	1443	1444	1445	1446	1447
2650	1448	1449	1450	1451	1452	1453	1454	1455
2660	1456	1457	1458	1459	1460	1461	1462	1463
2670	1464	1465	1466	1467	1468	1469	1470	1471
2700	1472	1473	1474	1475	1476	1477	1478	1479
2710	1480	1481	1482	1483	1484	1485	1486	1487
2720	1488	1489	1490	1491	1492	1493	1494	1495
2730	1496	1497	1498	1499	1500	1501	1502	1503
2740	1504	1505	1506	1507	1508	1509	1510	1511
2750	1512	1513	1514	1515	1516	1517	1518	1519
2760	1520	1521	1522	1523	1524	1525	1526	1527
2770	1528	1529	1530	1531	1532	1533	1534	1535

3000 to 3777 (Octal) | 1536 to 2047 (Decimal)

	0	1	2	3	4	5	6	7
3000	1536	1537	1538	1539	1540	1541	1542	1543
3010	1544	1545	1546	1547	1548	1549	1550	1551
3020	1552	1553	1554	1555	1556	1557	1558	1559
3030	1560	1561	1562	1563	1564	1565	1566	1567
3040	1568	1569	1570	1571	1572	1573	1574	1575
3050	1576	1577	1578	1579	1580	1581	1582	1583
3060	1584	1585	1586	1587	1588	1589	1590	1591
3070	1592	1593	1594	1595	1596	1597	1598	1599
3100	1600	1601	1602	1603	1604	1605	1606	1607
3110	1608	1609	1610	1611	1612	1613	1614	1615
3120	1616	1617	1618	1619	1620	1621	1622	1623
3130	1624	1625	1626	1627	1628	1629	1630	1631
3140	1632	1633	1634	1635	1636	1637	1638	1639
3150	1640	1641	1642	1643	1644	1645	1646	1647
3160	1648	1649	1650	1651	1652	1653	1654	1655
3170	1656	1657	1658	1659	1660	1661	1662	1663
3200	1664	1665	1666	1667	1668	1669	1670	1671
3210	1672	1673	1674	1675	1676	1677	1678	1679
3220	1680	1681	1682	1683	1684	1685	1686	1687
3230	1688	1689	1690	1691	1692	1693	1694	1695
3240	1696	1697	1698	1699	1700	1701	1702	1703
3250	1704	1705	1706	1707	1708	1709	1710	1711
3260	1712	1713	1714	1715	1716	1717	1718	1719
3270	1720	1721	1722	1723	1724	1725	1726	1727
3300	1728	1729	1730	1731	1732	1733	1734	1735
3310	1736	1737	1738	1739	1740	1741	1742	1743
3320	1744	1745	1746	1747	1748	1749	1750	1751
3330	1752	1753	1754	1755	1756	1757	1758	1759
3340	1760	1761	1762	1763	1764	1765	1766	1767
3350	1768	1769	1770	1771	1772	1773	1774	1775
3360	1776	1777	1778	1779	1780	1781	1782	1783
3370	1784	1785	1786	1787	1788	1789	1790	1791

	0	1	2	3	4	5	6	7
3400	1792	1793	1794	1795	1796	1797	1798	1799
3410	1800	1801	1802	1803	1804	1805	1806	1807
3420	1808	1809	1810	1811	1812	1813	1814	1815
3430	1816	1817	1818	1819	1820	1821	1822	1823
3440	1824	1825	1826	1827	1828	1829	1830	1831
3450	1832	1833	1834	1835	1836	1837	1838	1839
3460	1840	1841	1842	1843	1844	1845	1846	1847
3470	1848	1849	1850	1851	1852	1853	1854	1855
3500	1856	1857	1858	1859	1860	1861	1862	1863
3510	1864	1865	1866	1867	1868	1869	1870	1871
3520	1872	1873	1874	1875	1876	1877	1878	1879
3530	1880	1881	1882	1883	1884	1885	1886	1887
3540	1888	1889	1890	1891	1892	1893	1894	1895
3550	1896	1897	1898	1899	1900	1901	1902	1903
3560	1904	1905	1906	1907	1908	1909	1910	1911
3570	1912	1913	1914	1915	1916	1917	1918	1919
3600	1920	1921	1922	1923	1924	1925	1926	1927
3610	1928	1929	1930	1931	1932	1933	1934	1935
3620	1936	1937	1938	1939	1940	1941	1942	1943
3630	1944	1945	1946	1947	1948	1949	1950	1951
3640	1952	1953	1954	1955	1956	1957	1958	1959
3650	1960	1961	1962	1963	1964	1965	1966	1967
3660	1968	1969	1970	1971	1972	1973	1974	1975
3670	1976	1977	1978	1979	1980	1981	1982	1983
3700	1984	1985	1986	1987	1988	1989	1990	1991
3710	1992	1993	1994	1995	1996	1997	1998	1999
3720	2000	2001	2002	2003	2004	2005	2006	2007
3730	2008	2009	2010	2011	2012	2013	2014	2015
3740	2016	2017	2018	2019	2020	2021	2022	2023
3750	2024	2025	2026	2027	2028	2029	2030	2031
3760	2032	2033	2034	2035	2036	2037	2038	2039
3770	2040	2041	2042	2043	2044	2045	2046	2047

OCTAL-DECIMAL INTEGER CONVERSION TABLE

	0	1	2	3	4	5	6	7
4000	2048	2049	2050	2051	2052	2053	2054	2055
4010	2056	2057	2058	2059	2060	2061	2062	2063
4020	2064	2065	2066	2067	2068	2069	2070	2071
4030	2072	2073	2074	2075	2076	2077	2078	2079
4040	2080	2081	2082	2083	2084	2085	2086	2087
4050	2088	2089	2090	2091	2092	2093	2094	2095
4060	2096	2097	2098	2099	2100	2101	2102	2103
4070	2104	2105	2106	2107	2108	2109	2110	2111
4100	2112	2113	2114	2115	2116	2117	2118	2119
4110	2120	2121	2122	2123	2124	2125	2126	2127
4120	2128	2129	2130	2131	2132	2133	2134	2135
4130	2136	2137	2138	2139	2140	2141	2142	2143
4140	2144	2145	2146	2147	2148	2149	2150	2151
4150	2152	2153	2154	2155	2156	2157	2158	2159
4160	2160	2161	2162	2163	2164	2165	2166	2167
4170	2168	2169	2170	2171	2172	2173	2174	2175
4200	2176	2177	2178	2179	2180	2181	2182	2183
4210	2184	2185	2186	2187	2188	2189	2190	2191
4220	2192	2193	2194	2195	2196	2197	2198	2199
4230	2200	2201	2202	2203	2204	2205	2206	2207
4240	2208	2209	2210	2211	2212	2213	2214	2215
4250	2216	2217	2218	2219	2220	2221	2222	2223
4260	2224	2225	2226	2227	2228	2229	2230	2231
4270	2232	2233	2234	2235	2236	2237	2238	2239
4300	2240	2241	2242	2243	2244	2245	2246	2247
4310	2248	2249	2250	2251	2252	2253	2254	2255
4320	2256	2257	2258	2259	2260	2261	2262	2263
4330	2264	2265	2266	2267	2268	2269	2270	2271
4340	2272	2273	2274	2275	2276	2277	2278	2279
4350	2280	2281	2282	2283	2284	2285	2286	2287
4360	2288	2289	2290	2291	2292	2293	2294	2295
4370	2296	2297	2298	2299	2300	2301	2302	2303

	0	1	2	3	4	5	6	7
4400	2304	2305	2306	2307	2308	2309	2310	2311
4410	2312	2313	2314	2315	2316	2317	2318	2319
4420	2320	2321	2322	2323	2324	2325	2326	2327
4430	2328	2329	2330	2331	2332	2333	2334	2335
4440	2336	2337	2338	2339	2340	2341	2342	2343
4450	2344	2345	2346	2347	2348	2349	2350	2351
4460	2352	2353	2354	2355	2356	2357	2358	2359
4470	2360	2361	2362	2363	2364	2365	2366	2367
4500	2368	2369	2370	2371	2372	2373	2374	2375
4510	2376	2377	2378	2379	2380	2381	2382	2383
4520	2384	2385	2386	2387	2388	2389	2390	2391
4530	2392	2393	2394	2395	2396	2397	2398	2399
4540	2400	2401	2402	2403	2404	2405	2406	2407
4550	2408	2409	2410	2411	2412	2413	2114	2415
4560	2416	2417	2418	2419	2420	2421	2422	2423
4570	2424	2425	2426	2427	2428	2429	2430	2431
4600	2432	2433	2434	2435	2436	2437	2438	2439
4610	2440	2441	2442	2443	2444	2445	2446	2447
4620	2448	2449	2450	2451	2452	2453	2454	2455
4630	2456	2457	2458	2459	2460	2461	2462	2463
4640	2464	2465	2466	2467	2468	2469	2470	2471
4650	2472	2473	2474	2475	2476	2477	2478	2479
4660	2480	2481	2482	2483	2484	2485	2486	2487
4670	2488	2489	2490	2491	2492	2493	2494	2495
4700	2496	2497	2498	2499	2500	2501	2502	2503
4710	2504	2505	2506	2507	2508	2509	2510	2511
4720	2512	2513	2514	2515	2516	2517	2518	2519
4730	2520	2521	2522	2523	2524	2525	2526	2527
4740	2528	2529	2530	2531	2532	2533	2534	2535
4750	2536	2537	2538	2539	2540	2541	2542	2543
4760	2544	2545	2546	2547	2548	2549	2550	2551
4770	2552	2553	2554	2555	2556	2557	2558	**2559**

4000 to 4777 (Octal)	2048 to 2559 (Decimal)

Octal	Decimal
10000-	**4096**
20000-	**8192**
30000-	**12288**
40000-	**16384**
50000-	**20480**
60000-	**24576**
70000-	**28672**

	0	1	2	3	4	5	6	7
5000	2560	2561	2562	2563	2564	2565	2566	2567
5010	2568	2569	2570	2571	2572	2573	2574	2575
5020	2576	2577	2578	2579	2580	2581	2582	2583
5030	2584	2585	2586	2587	2588	2589	2590	2591
5040	2592	2593	2594	2595	2596	2597	2598	2599
5050	2600	2601	2602	2603	2604	2605	2606	2607
5060	2608	2609	2610	2611	2612	2613	2614	2615
5070	2616	2617	2618	2619	2620	2621	2622	2623
5100	2624	2625	2626	2627	2628	2629	2630	2631
5110	2632	2633	2634	2635	2636	2637	2638	2639
5120	2640	2641	2642	2643	2644	2645	2646	2647
5130	2648	2649	2650	2651	2652	2653	2654	2655
5140	2656	2657	2658	2659	2660	2661	2662	2663
5150	2664	2665	2666	2667	2668	2669	2670	2671
5160	2672	2673	2674	2675	2676	2677	2678	2679
5170	2680	2681	2682	2683	2684	2685	2686	2687
5200	2688	2689	2690	2691	2692	2693	2694	2695
5210	2696	2697	2698	2699	2700	2701	2702	2703
5220	2704	2705	2706	2707	2708	2709	2710	2711
5230	2712	2713	2714	2715	2716	2717	2718	2719
5240	2720	2721	2722	2723	2724	2725	2726	2727
5250	2728	2729	2730	2731	2732	2733	2734	2735
5260	2736	2737	2738	2739	2740	2741	2742	2743
5270	2744	2745	2746	2747	2748	2749	2750	2751
5300	2752	2753	2754	2755	2756	2757	2758	2759
5310	2760	2761	2762	2763	2764	2765	2766	2767
5320	2768	2769	2770	2771	2772	2773	2774	2775
5330	2776	2777	2778	2779	2780	2781	2782	2783
5340	2784	2785	2786	2787	2788	2789	2790	2791
5350	2792	2793	2794	2795	2796	2797	2798	2799
5360	2800	2801	2802	2803	2804	2805	2806	2807
5370	2808	2809	2810	2811	2812	2813	2814	2815

	0	1	2	3	4	5	6	7
5400	2816	2817	2818	2819	2820	2821	2822	2823
5410	2824	2825	2826	2827	2828	2829	2830	2831
5420	2832	2833	2834	2835	2836	2837	2838	2839
5430	2840	2841	2842	2843	2844	2845	2846	2847
5440	2848	2849	2850	2851	2852	2853	2854	2855
5450	2856	2857	2858	2859	2860	2861	2862	2863
5460	2864	2865	2866	2867	2868	2869	2870	2871
5470	2872	2873	2874	2875	2876	2877	2878	2879
5500	2880	2881	2882	2883	2884	2885	2886	2887
5510	2888	2889	2890	2891	2892	2893	2894	2895
5520	2896	2897	2898	2899	2900	2901	2902	2903
5530	2904	2905	2906	2907	2908	2909	2910	2911
5540	2912	2913	2914	2915	2916	2917	2918	2919
5550	2920	2921	2922	2923	2924	2925	2926	2927
5560	2928	2929	2930	2931	2932	2933	2934	2935
5570	2936	2937	2938	2939	2940	2941	2942	2943
5600	2944	2945	2946	2947	2948	2949	2950	2951
5610	2952	2953	2954	2955	2956	2957	2958	2959
5620	2960	2961	2962	2963	2964	2965	2966	2967
5630	2968	2969	2970	2971	2972	2973	2974	2975
5640	2976	2977	2978	2979	2980	2981	2982	2983
5650	2984	2985	2986	2987	2988	2989	2990	2991
5660	2992	2993	2994	2995	2996	2997	2998	2999
5670	3000	3001	3002	3003	3004	3005	3006	3007
5700	3008	3009	3010	3011	3012	3013	3014	3015
5710	3016	3017	3018	3019	3020	3021	3022	3023
5720	3024	3025	3026	3027	3028	3029	3030	3031
5730	3032	3033	3034	3035	3036	3037	3038	3039
5740	3040	3041	3042	3043	3044	3045	3046	3047
5750	3048	3049	3050	3051	3052	3053	3054	3055
5760	3056	3057	**3058**	3059	3060	3061	3062	3063
5770	3064	3065	3066	3067	3068	3069	3070	3071

5000 to 5777 (Octal)	2560 to 3071 (Decimal)

OCTAL-DECIMAL INTEGER CONVERSION TABLE

6000 to 6777 (Octal) | 3072 to 3583 (Decimal)

Octal	Decimal
10000-	4096
20000-	8192
30000-	12288
40000-	16384
50000-	20480
60000-	24576
70000-	28672

	0	1	2	3	4	5	6	7
6000	3072	3073	3074	3075	3076	3077	3078	3079
6010	3080	3081	3082	3083	3084	3085	3086	3087
6020	3088	3089	3090	3091	3092	3093	3094	3095
6030	3096	3097	3098	3099	3100	3101	3102	3103
6040	3104	3105	3106	3107	3108	3109	3110	3111
6050	3112	3113	3114	3115	3116	3117	3118	3119
6060	3120	3121	3122	3123	3124	3125	3126	3127
6070	3128	3129	3130	3131	3132	3133	3134	3135
6100	3136	3137	3138	3139	3140	3141	3142	3143
6110	3144	3145	3146	3147	3148	3149	3150	3151
6120	3152	3153	3154	3155	3156	3157	3158	3159
6130	3160	3161	3162	3163	3164	3165	3166	3167
6140	3168	3169	3170	3171	3172	3173	3174	3175
6150	3176	3177	3178	3179	3180	3181	3182	3183
6160	3184	3185	3186	3187	3188	3189	3190	3191
6170	3192	3193	3194	3195	3196	3197	3198	3199
6200	3200	3201	3202	3203	3204	3205	3206	3207
6210	3208	3209	3210	3211	3212	3213	3214	3215
6220	3216	3217	3218	3219	3220	3221	3222	3223
6230	3224	3225	3226	3227	3228	3229	3230	3231
6240	3232	3233	3234	3235	3236	3237	3238	3239
6250	3240	3241	3442	3243	3244	3245	3246	3247
6260	3248	3249	3250	3251	3252	3253	3254	3255
6270	3256	3257	3258	3259	3260	3261	3262	3263
6300	3264	3265	3266	3267	3268	3269	3270	3871
6310	3272	3273	3274	3275	3276	3277	3278	3279
6320	3280	3281	3282	3283	3284	3285	3286	3287
6330	3288	3289	3290	3291	3292	3293	3294	3295
6340	3296	3297	3298	3299	3300	3301	3302	3003
6350	3304	3305	3306	3307	3308	3309	3310	3311
6360	3312	3313	3314	3315	3316	3317	3318	3319
6370	3320	3321	3322	3323	3324	3325	3326	3327

	0	1	2	3	4	5	6	7
6400	3328	3329	3330	3331	3332	3333	3334	3335
6410	3336	3337	3338	3339	3340	3341	3342	3343
6420	3344	3345	3346	3347	3348	3349	3350	3351
6430	3352	3353	3354	3355	3356	3357	3358	3359
6440	3360	3361	3362	3363	3364	3365	3366	3367
6450	3368	3369	3370	3371	3372	3373	3374	3375
6460	3376	3377	3378	3379	3380	3381	3382	3383
6470	3384	3385	3386	3387	3388	3389	3390	3391
6500	3392	3393	3394	3395	3396	3397	3398	3399
6510	3400	3401	3402	3403	3404	3405	3406	3407
6520	3408	3409	3410	3411	3412	3413	3414	3415
6530	3416	3417	3418	3419	3420	3421	3422	3423
6540	3424	3425	3426	3427	3428	3429	3430	3431
6550	3432	3433	3434	3435	3436	3437	3438	3439
6560	3440	3441	3442	3443	3444	3445	3446	3447
6570	3448	3449	3450	3451	3452	3453	3454	3455
6600	3456	3457	3458	3459	3460	3461	3462	3463
6610	3464	3465	3466	3467	3468	3469	3470	3471
6620	3472	3473	3474	3475	3476	3477	3478	3479
6630	3480	3481	3482	3483	3484	3485	3486	3487
6640	3488	3489	3490	3491	3492	3493	3494	3495
6650	3496	3497	3498	3499	3500	3501	3502	3503
6660	3504	3505	3506	3507	3508	3509	3510	3511
6670	3512	3513	3514	3515	3516	3517	3518	3519
6700	3520	3521	3522	3523	3524	3525	3526	3527
6710	3528	3529	3530	3531	3532	3533	3534	3535
6720	3536	3537	3538	3539	3540	3541	3542	3543
6730	3544	3545	3546	3547	3548	3549	3550	3551
6740	3552	3553	3554	3555	3556	3557	3558	3559
6750	3560	3561	3562	3563	3564	3655	3566	3567
6760	3568	3569	3570	3571	3572	3573	3574	3575
6770	3576	3577	3578	3579	3580	3581	3582	3583

7000 to 7777 (Octal) | 3584 to 4095 (Decimal)

	0	1	2	3	4	5	6	7
7000	3584	3585	3586	3587	3588	3589	3590	3591
7010	3592	3593	3594	3595	3596	3597	3598	3599
7020	3600	3601	3602	3603	3604	3605	3606	3607
7030	3608	3609	3610	3611	3612	3613	3614	3615
7040	3616	3617	3618	3619	3620	3621	3622	3623
7050	3624	3625	3626	3627	3628	3629	3630	3631
7060	3632	3633	3634	3635	3636	3637	3638	3639
7070	3640	3641	3642	3643	3644	3645	3646	3647
7100	3648	3649	3650	3651	3652	3653	3654	3655
7110	3656	3657	3658	3659	3660	3661	3662	3663
7120	3664	3665	3666	3667	3668	3669	3670	3671
7130	3672	3673	3674	3675	3676	3677	3678	3679
7140	3680	3681	3682	3683	3684	3685	3686	3687
7150	3688	3689	3690	3691	3692	3693	3694	3695
7160	3696	3697	3698	3699	3700	3701	3702	3703
7170	3704	3705	3706	3707	3708	3709	3710	3711
7200	3712	3713	3714	3715	3716	3717	3718	3719
7210	3720	3721	3722	3723	3724	3725	3726	3727
7220	3728	3729	3730	3731	3732	3733	3734	3735
7230	3736	3737	3738	3739	3740	3741	3742	3743
7240	3744	3745	3746	3747	3748	3749	3750	3751
7250	3752	3753	3754	3755	3756	3757	3758	3759
7260	3760	3761	3762	3763	3764	3765	3766	3767
7270	3768	3769	3770	3771	3772	3773	3774	3775
7300	3776	3777	3778	3779	3780	3781	3782	3783
7310	3784	3785	3786	3787	3788	3789	3790	3791
7320	3792	3893	3794	3795	3796	3797	3798	3799
7330	3800	3801	3802	3803	3804	3805	3806	3807
7340	3808	3809	3810	3811	3812	3813	3814	3815
7350	3816	3817	3818	3819	3820	3821	3822	3823
7360	3824	3825	3826	3827	3828	3829	3830	3831
7370	3832	3833	3834	3835	3836	3837	3838	3839

	0	1	2	3	4	5	6	7
7400	3840	3841	3482	3843	3844	3845	3846	3847
7410	3848	3849	3850	3851	3852	3853	3854	3855
7420	3856	3857	3858	3859	3860	3861	3862	3863
7430	3864	3865	3866	3867	3868	3869	3870	3871
7440	3872	3873	3874	3875	3876	3877	3878	3879
7450	3880	3881	3882	3883	3884	3885	3886	3887
7460	3888	3889	3890	3891	3892	3893	3894	3895
7470	3896	3897	3898	3899	3900	3901	3902	3903
7500	3904	3905	3906	3907	3908	3909	3910	3911
7510	3912	3913	3914	3915	3916	3917	3918	3919
7520	3920	3921	3922	3923	3924	3925	3926	3927
7530	3928	3929	3930	3931	3932	3933	3934	3935
7540	3936	3937	3938	3939	3940	3941	3942	3943
7550	3944	3945	3946	3947	3948	3949	3950	3951
7560	3952	3953	3954	3955	3956	3957	3958	3959
7570	3960	3961	3962	3963	3964	3965	3966	3967
7600	3968	3969	4970	3971	3972	3973	3974	3975
7610	3976	3977	3978	3979	3980	3981	3982	3983
7620	3984	3985	3986	3987	3988	3989	3990	3991
7630	3992	3993	3994	3995	3996	3997	3998	3999
7640	4000	4001	4002	4003	4004	4005	4006	4007
7650	4008	4009	4010	4011	4012	4013	4014	4015
7660	4016	4017	4018	4019	4020	4021	4022	4023
7670	4024	4025	4026	4027	4028	4029	4030	4031
7700	4032	4033	4034	4035	4036	4037	4038	4039
7710	4040	4041	4042	4043	4044	4045	4046	4047
7720	4048	4049	4050	4051	4052	4053	4054	4055
7730	4056	4057	4058	4059	4060	4061	4062	4063
7740	4064	4065	4066	4067	4068	4069	4070	4071
7750	4072	4073	4074	4075	4076	4077	4078	4079
7760	4080	4081	4082	4083	4084	4085	4086	4087
7770	4088	4089	4090	4091	4092	4093	4094	4095

OCTAL-DECIMAL FRACTION CONVERSION TABLE

This table covers the entries from $(.000)_8$ to $(.377)_8$. For entries from $(.400)_8$ to $(.777)_8$, cognizance should be made of the fact that $(.400)_8$ is $(.500)_{10}$. Hence if $(.637)_8$ is desired, find $(.237)_8$ in table, namely, $(.310456)_{10}$ and add $(.50000)_{10}$ for $(.400)_8$. Thus

$$
\begin{aligned}
(.637)_8 &= (.237)_8 + (.400)_8 \\
&= (.310456)_{10} + (.50000)_{10} \\
&= (.810456)_{10}.
\end{aligned}
$$

OCTAL	DEC.	OCTAL	DEC.	OCTAL	DEC.	OCTAL	DEC.
.000	.000000	.100	.125000	.200	.250000	.300	.375000
.001	.001953	.101	.126953	.201	.251953	.301	.376953
.002	.003906	.102	.128906	.202	.253906	.302	.378906
.003	.005859	.103	.130859	.203	.255859	.303	.380859
.004	.007812	.104	.132812	.204	.257812	.304	.382812
.005	.009765	.105	.134765	.205	.259765	.305	.384765
.006	.011718	.106	.136718	.206	.261718	.306	.386718
.007	.013671	.107	.138671	.207	.263671	.307	.388671
.010	.015625	.110	.140625	.210	.265625	.310	.390625
.011	.017578	.111	.142578	.211	.267578	.311	.392578
.012	.019531	.112	.144531	.212	.269531	.312	.394531
.013	.021484	.113	.146484	.213	.271484	.313	.396484
.014	.023437	.114	.148437	.214	.273437	.314	.398437
.015	.025390	.115	.150390	.215	.275390	.315	.400490
.016	.027343	.116	.152343	.216	.277343	.316	.402343
.017	.029296	.117	.154296	.217	.279296	.317	.404296
.020	.031250	.120	.156250	.220	.281250	.320	.406250
.021	.033203	.121	.158203	.221	.283203	.321	.408203
.022	.035156	.122	.160156	.222	.285156	.322	.410156
.023	.037109	.123	.162109	.223	.287109	.323	.412109
.024	.039062	.124	.164062	.224	.289062	.324	.414062
.025	.041015	.125	.166015	.225	.291015	.325	.416015
.026	.042968	.126	.167968	.226	.292968	.326	.417968
.027	.044921	.127	.169921	.227	.294921	.327	.419921
.030	.046875	.130	.171875	.230	.294875	.330	.421875
.031	.048828	.131	.173828	.231	.298828	.331	.423828
.032	.050781	.132	.175781	.232	.300781	.332	.425781
.033	.052734	.133	.177734	.233	.302734	.333	.427734
.034	.054687	.134	.179687	.234	.304687	.334	.429687
.035	.056640	.135	.181640	.235	.306640	.335	.431640
.036	.058593	.136	.183593	.236	.308593	.336	.433593
.037	.060546	.137	.185546	.237	.310546	.337	.435546
.040	.062500	.140	.187500	.240	.312500	.340	.437500
.041	.064453	.141	.189453	.241	.314453	.341	.439453
.042	.066406	.142	.191406	.242	.316406	.342	.441406
.043	.068359	.143	.193359	.243	.318359	.343	.443359
.044	.070312	.144	.195312	.244	.320312	.344	.445312
.045	.072265	.145	.197265	.245	.322265	.345	.447265
.046	.074218	.146	.199218	.246	.324218	.346	.449218
.047	.076171	.147	.201171	.247	.326171	.347	.451171
.050	.078125	.150	.203125	.250	.328125	.350	.453125
.051	.080078	.151	.205078	.251	.330078	.351	.455078
.052	.082031	.152	.207031	.252	.332031	.352	.457031
.053	.083984	.153	.208984	.253	.333984	.353	.458984
.054	.085937	.154	.210937	.254	.335937	.354	.460937
.055	.087890	.155	.212890	.255	.337890	.355	.462890
.056	.089843	.156	.214843	.256	.339843	.356	.464843
.057	.091796	.157	.216796	.257	.341796	.357	.466796
.060	.093750	.160	.218750	.260	.343750	.360	.468750
.061	.095703	.161	.220703	.261	.345703	.361	.470703
.062	.097656	.162	.222656	.262	.347656	.362	.472656
.063	.099609	.163	.224609	.263	.349609	.363	.474609
.064	.101562	.164	.226562	.264	.351562	.364	.476562
.065	.103515	.165	.228515	.265	.353515	.365	.478515
.066	.105468	.166	.230468	.266	.355468	.366	.480468
.067	.107421	.167	.232421	.267	.357421	.367	.482421
.070	.109375	.170	.234375	.270	.359375	.370	.484375
.071	.111328	.171	.236328	.271	.361328	.371	.486328
.072	.113281	.172	.238281	.272	.363281	.372	.488281
.073	.115234	.173	.240234	.273	.365234	.373	.490234
.074	.117187	.174	.242187	.274	.367187	.374	.492187
.075	.119140	.175	.244140	.275	.369140	.375	.494140
.076	.121093	.176	.246093	.276	.371093	.376	.496093
.077	.123046	.177	.248046	.277	.373046	.377	.498046

OCTAL-DECIMAL FRACTION CONVERSION TABLE

OCTAL	DEC.	OCTAL	DEC.	OCTAL	DEC.	OCTAL	DEC.
.000000	.000000	.000100	.000244	.000200	.000488	.000300	.000732
.000001	.000004	.000101	.000247	.000201	.000492	.000301	.000736
.000002	.000007	.000102	.000251	.000202	.000495	.000302	.000740
.000003	.000011	.000103	.000255	.000203	.000499	.000303	.000743
.000004	.000015	.000104	.000259	.000204	.000503	.000304	.000747
.000005	.000019	.000105	.000263	.000205	.000507	.000305	.000751
.000006	.000022	.000106	.000267	.000206	.000511	.000306	.000755
.000007	.000026	.000107	.000270	.000207	.000514	.000307	.000759
.000010	.000030	.000110	.000274	.000210	.000518	.000310	.000762
.000011	.000034	.000111	.000278	.000211	.000522	.000311	.000766
.000012	.000038	.000112	.000282	.000212	.000526	.000312	.000770
.000013	.000041	.000113	.000286	.000213	.000530	.000313	.000774
.000014	.000045	.000114	.000289	.000214	.000534	.000314	.000778
.000015	.000049	.000115	.000293	.000215	.000537	.000315	.000782
.000016	.000053	.000116	.000297	.000216	.000541	.000316	.000785
.000017	.000057	.000117	.000301	.000217	.000545	.000317	.000789
.000020	.000061	.000120	.000305	.000220	.000549	.000320	.000793
.000021	.000064	.000121	.000308	.000221	.000553	.000321	.000797
.000022	.000068	.000122	.000312	.000222	.000556	.000322	.000801
.000023	.000072	.000123	.000316	.000223	.000560	.000323	.000805
.000024	.000076	.000124	.000320	.000224	.000564	.000324	.000808
.000025	.000080	.000125	.000324	.000225	.000568	.000325	.000812
.000026	.000083	.000126	.000328	.000226	.000572	.000326	.000816
.000027	.000087	.000127	.000331	.000227	.000576	.000327	.000820
.000030	.000091	.000130	.000335	.000230	.000579	.000330	.000823
.000031	.000095	.000131	.000339	.000231	.000583	.000331	.000827
.000032	.000099	.000132	.000343	.000232	.000587	.000332	.000831
.000033	.000102	.000133	.000347	.000233	.000591	.000333	.000835
.000034	.000106	.000134	.000350	.000234	.000595	.000334	.000839
.000035	.000110	.000135	.000354	.000235	.000598	.000335	.000843
.000036	.000114	.000136	.000358	.000236	.000602	.000336	.000846
.000037	.000118	.000137	.000362	.000237	.000606	.000337	.000850
.000040	.000122	.000140	.000366	.000240	.000610	.000340	.000854
.000041	.000125	.000141	.000370	.000241	.000614	.000341	.000858
.000042	.000129	.000142	.000373	.000242	.000617	.000342	.000862
.000043	.000133	.000143	.000377	.000243	.000621	.000343	.000865
.000044	.000137	.000144	.000381	.000244	.000625	.000344	.000869
.000045	.000141	.000145	.000385	.000245	.000629	.000345	.000873
.000046	.000144	.000146	.000389	.000246	.000633	.000346	.000877
.000047	.000148	.000147	.000392	.000247	.000637	.000347	.000881
.000050	.000152	.000150	.000396	.000250	.000640	.000350	.000885
.000051	.000156	.000151	.000400	.000251	.000644	.000351	.000888
.000052	.000160	.000152	.000404	.000252	.000648	.000352	.000892
.000053	.000164	.000153	.000408	.000253	.000652	.000353	.000896
.000054	.000167	.000154	.000411	.000254	.000656	.000354	.000900
.000055	.000171	.000155	.000415	.000255	.000659	.000355	.000904
.000056	.000175	.000156	.000419	.000256	.000663	.000356	.000907
.000057	.000179	.000157	.000423	.000257	.000667	.000357	.000911
.000060	.000183	.000160	.000427	.000260	.000671	.000360	.000915
.000061	.000186	.000161	.000431	.000261	.000675	.000361	.000919
.000062	.000190	.000162	.000434	.000262	.000679	.000362	.000923
.000063	.000194	.000163	.000438	.000263	.000682	.000363	.000926
.000064	.000198	.000164	.000442	.000264	.000686	.000364	.000930
.000065	.000202	.000165	.000446	.000265	.000690	.000365	.000934
.000066	.000205	.000166	.000450	.000266	.000694	.000366	.000938
.000067	.000209	.000167	.000453	.000267	.000698	.000367	.000942
.000070	.000213	.000170	.000457	.000270	.000701	.000370	.000946
.000071	.000217	.000171	.000461	.000271	.000705	.000371	.000949
.000072	.000221	.000172	.000465	.000272	.000709	.000372	.000953
.000073	.000225	.000173	.000469	.000273	.000713	.000373	.000957
.000074	.000228	.000174	.000473	.000274	.000717	.000374	.000961
.000075	.000232	.000175	.000476	.000275	.000720	.000375	.000965
.000076	.000326	.000176	.000480	.000276	.000724	.000376	.000968
.000077	.000240	.000177	.000484	.000277	.000728	.000377	.000972

OCTAL-DECIMAL FRACTION CONVERSION TABLE

OCTAL	DEC.	OCTAL	DEC.	OCTAL	DEC.	OCTAL	DEC.
.000400	.000976	.000500	.001220	.000600	.001464	.000700	.001708
.000401	.000980	.000501	.001224	.000601	.001468	.000701	.001712
.000402	.000984	.000502	.001228	.000602	.001472	.000702	.001716
.000403	.000988	.000503	.001232	.000603	.001476	.000703	.001720
.000404	.000991	.000504	.001235	.000604	.001480	.000704	.001724
.000405	.000995	.000505	.001239	.000605	.001483	.000705	.001728
.000406	.000999	.000506	.001243	.000606	.001487	.000706	.001731
.000407	.001003	.000507	.001247	.000607	.001491	.000707	.001735
.000410	.001007	.000510	.001251	.000610	.001495	.000710	.001739
.000411	.001010	.000511	.001255	.000611	.001499	.000711	.001743
.000412	.001014	.000512	.001258	.000612	.001502	.000712	.001747
.000413	.001018	.000513	.001262	.000613	.001506	.000713	.001750
.000414	.001022	.000514	.001266	.000614	.001510	.000714	.001754
.000415	.001026	.000515	.001270	.000615	.001514	.000715	.001758
.000416	.001029	.000516	.001274	.000616	.001518	.000716	.001762
.000417	.001033	.000517	.001277	.000617	.001522	.000717	.001766
.000420	.001037	.000520	.001281	.000620	.001525	.000720	.001770
.000421	.001041	.000521	.001285	.000621	.001529	.000721	.001733
.000422	.001045	.000522	.001289	.000622	.001533	.000722	.001777
.000423	.001049	.000523	.001293	.000623	.001537	.000723	.001781
.000424	.001052	.000524	.001296	.000624	.001541	.000724	.001785
.000425	.001056	.000525	.001300	.000625	.001544	.000725	.001789
.000426	.001060	.000526	.001304	.000626	.001548	.000726	.001792
.000427	.001064	.000527	.001308	.000627	.001552	.000727	.001796
.000430	.001068	.000530	.001312	.000630	.001556	.000730	.001800
.000431	.001071	.000531	.001316	.000631	.001560	.000731	.001804
.000432	.001075	.000532	.001319	.000632	.001564	.000732	.001808
.000433	.001079	.000533	.001323	.000633	.001567	.000733	.001811
.000434	.001083	.000534	.001327	.000634	.001571	.000734	.001815
.000435	.001087	.000535	.001331	.000635	.001575	.000735	.001819
.000436	.001091	.000536	.001335	.000636	.001579	.000736	.001823
.000437	.001094	.000537	.001338	.000637	.001583	.000737	.001827
.000440	.001098	.000540	.001342	.000640	.001586	.000740	.001831
.000441	.001102	.000541	.001346	.000641	.001590	.000741	.001834
.000442	.001106	.000542	.001350	.000642	.001594	.000742	.001838
.000443	.001110	.000543	.001354	.000643	.001598	.000743	.001842
.000444	.001113	.000544	.001358	.000644	.001602	.000744	.001846
.000445	.001117	.000545	.001361	.000645	.001605	.000745	.001850
.000446	.001121	.000546	.001365	.000646	.001609	.000746	.001853
.000447	.001125	.000547	.001369	.000647	.001613	.000747	.001857
.000450	.001129	.000550	.001373	.000650	.001617	.000750	.001861
.000451	.001132	.000551	.001377	.000651	.001621	.000751	.001865
.000452	.001136	.000552	.001380	.000652	.001625	.000752	.001869
.000453	.001140	.000553	.001384	.000653	.001628	.000753	.001873
.000454	.001144	.000554	.001388	.000654	.001632	.000754	.001876
.000455	.001148	.000555	.001392	.000655	.001636	.000755	.001880
.000456	.001152	.000556	.001396	.000656	.001640	.000756	.001884
.000457	.001155	.000557	.001399	.000657	.001644	.000757	.001888
.000460	.001159	.000560	.001403	.000660	.001647	.000760	.001892
.000461	.001163	.000561	.001407	.000661	.001651	.000761	.001895
.000462	.001167	.000562	.001411	.000662	.001655	.000762	.001899
.000463	.001171	.000563	.001415	.000663	.001659	.000763	.001903
.000464	.001174	.000564	.001419	.000664	.001663	.000764	.001907
.000465	.001178	.000565	.001422	.000665	.001667	.000765	.001911
.000466	.001182	.000566	.001426	.000666	.001670	.000766	.001914
.000467	.001186	.000567	.001430	.000667	.001674	000767	.001918
.000470	.001190	.000570	.001434	.000670	.001678	.000770	.001922
.000471	.001194	.000571	.001438	.000671	.001682	.000771	.001926
.000472	.001197	.000572	.001441	.000672	.001686	.000772	.001930
.000473	.001201	.000573	.001445	.000673	.001689	.000773	.001934
.000474	.001205	.000574	.001449	.000674	.001693	.000774	.001937
.000475	.001209	.000575	.001453	.000675	.001697	.000775	.001941
.000476	.001213	.000576	.001457	.000676	.001701	.000776	.001945
.000477	.001216	.000577	.001461	.000677	.001705	.000777	.001949

I. HEXADECIMAL AND DECIMAL DIRECT CONVERSION TABLE

The following tables aid in converting hexadecimal (base 16) numbers to decimal, and the reverse. Note that the base 16 digits for the decimal values 10–15 are represented by the letters A–F, respectively.

This table provides direct conversion of decimal and hexadecimal numbers in these ranges:

HEXADECIMAL	DECIMAL
000 to FFF	0000 to 4095

For numbers outside the range of the table, add the following values to the table figures:

HEXADECIMAL	DECIMAL
1000	4096
2000	8192
3000	12288
4000	16384
5000	20480
6000	24576
7000	28672
8000	32768
9000	36864
A000	40960
B000	45056
C000	49152
D000	53248
E000	57344
F000	61440

	0	1	2	3	4	5	6	7	8	9	A	B	C	D	E	F
00_	0000	0001	0002	0003	0004	0005	0006	0007	0008	0009	0010	0011	0012	0013	0014	0015
01_	0016	0017	0018	0019	0020	0021	0022	0023	0024	0025	0026	0027	0028	0029	0030	0031
02_	0032	0033	0034	0035	0036	0037	0038	0039	0040	0041	0042	0043	0044	0045	0046	0047
03_	0048	0049	0050	0051	0052	0053	0054	0055	0056	0057	0058	0059	0060	0061	0062	0063
04_	0064	0065	0066	0067	0068	0069	0070	0071	0072	0073	0074	0075	0076	0077	0078	0079
05_	0080	0081	0082	0083	0084	0085	0086	0087	0088	0089	0090	0091	0092	0093	0094	0095
06_	0096	0097	0098	0099	0100	0101	0102	0103	0104	0105	0106	0107	0108	0109	0110	0111
07_	0112	0113	0114	0115	0116	0117	0118	0119	0120	0121	0122	0123	0124	0125	0126	0127
08_	0128	0129	0130	0131	0132	0133	0134	0135	0136	0137	0138	0139	0140	0141	0142	0143
09_	0144	0145	0146	0147	0148	0149	0150	0151	0152	0153	0154	0155	0156	0157	0158	0159
0A_	0160	0161	0162	0163	0164	0165	0166	0167	0168	0169	0170	0171	0172	0173	0174	0175
0B_	0176	0177	0178	0179	0180	0181	0182	0183	0184	0185	0186	0187	0188	0189	0190	0191
0C_	0192	0193	0194	0195	0196	0197	0198	0199	0200	0201	0202	0203	0204	0205	0206	0207
0D_	0208	0209	0210	0211	0212	0213	0214	0215	0216	0217	0218	0219	0220	0221	0222	0223
0E_	0224	0225	0226	0227	0228	0229	0230	0231	0232	0233	0234	0235	0236	0237	0238	0239
0F_	0240	0241	0242	0243	0244	0245	0246	0247	0248	0249	0250	0251	0252	0253	0254	0255
10_	0256	0257	0258	0259	0260	0261	0262	0263	0264	0265	0266	0267	0268	0269	0270	0271
11_	0272	0273	0274	0275	0276	0277	0278	0279	0280	0281	0282	0283	0284	0285	0286	0287
12_	0288	0289	0290	0291	0292	0293	0294	0295	0296	0297	0298	0299	0300	0301	0302	0303
13_	0304	0305	0306	0307	0308	0309	0310	0311	0312	0313	0314	0315	0316	0317	0318	0319
14_	0320	0321	0322	0323	0324	0325	0326	0327	0328	0329	0330	0331	0332	0333	0334	0335
15_	0336	0337	0338	0339	0340	0341	0342	0343	0344	0345	0346	0347	0348	0349	0350	0351
16_	0352	0353	0354	0355	0356	0357	0358	0359	0360	0361	0362	0363	0364	0365	0366	0367
17_	0368	0369	0370	0371	0372	0373	0374	0375	0376	0377	0378	0379	0380	0381	0382	0383
18_	0384	0385	0386	0387	0388	0389	0390	0391	0392	0393	0394	0395	0396	0397	0398	0399
19_	0400	0401	0402	0403	0404	0405	0406	0407	0408	0409	0410	0411	0412	0413	0414	0415
1A_	0416	0417	0418	0419	0420	0421	0422	0423	0424	0425	0426	0427	0428	0429	0430	0431
1B_	0432	0433	0434	0435	0436	0437	0438	0439	0440	0441	0442	0443	0444	0445	0446	0447
1C_	0448	0449	0450	0451	0452	0453	0454	0455	0456	0457	0458	0459	0460	0461	0462	0463
1D_	0464	0465	0466	0467	0468	0469	0470	0471	0472	0473	0474	0475	0476	0477	0478	0479
1E_	0480	0481	0482	0483	0484	0485	0486	0487	0488	0489	0490	0491	0492	0493	0494	0495
1F_	0496	0497	0498	0499	0500	0501	0502	0503	0504	0505	0506	0507	0508	0509	0510	0511

DIRECT CONVERSION TABLE

	0	1	2	3	4	5	6	7	8	9	A	B	C	D	E	F
20_	0512	0513	0514	0515	0516	0517	0518	0519	0520	0521	0522	0523	0524	0525	0526	0527
21_	0528	0529	0530	0531	0532	0533	0534	0535	0536	0537	0538	0539	0540	0541	0542	0543
22_	0544	0545	0546	0547	0548	0549	0550	0551	0552	0553	0554	0555	0556	0557	0558	0559
23_	0560	0561	0562	0563	0564	0565	0566	0567	0568	0569	0570	0571	0572	0573	0574	0575
24_	0576	0577	0578	0579	0580	0581	0582	0583	0584	0585	0586	0587	0588	0589	0590	0591
25_	0592	0593	0594	0595	0596	0597	0598	0599	0600	0601	0602	0603	0604	0605	0606	0607
26_	0608	0609	0610	0611	0612	0613	0614	0615	0616	0617	0618	0619	0620	0621	0622	0623
27_	0624	0625	0626	0627	0628	0629	0630	0631	0632	0633	0634	0635	0636	0637	0638	0639
28_	0640	0641	0642	0643	0644	0645	0646	0647	0648	0649	0650	0651	0652	0653	0654	0655
29_	0656	0657	0658	0659	0660	0661	0662	0663	0664	0665	0666	0667	0668	0669	0670	0671
2A_	0672	0673	0674	0675	0676	0677	0678	0679	0680	0681	0682	0683	0684	0685	0686	0687
2B_	0688	0689	0690	0691	0692	0693	0694	0695	0696	0697	0698	0699	0700	0701	0702	0703
2C_	0704	0705	0706	0707	0708	0709	0710	0711	0712	0713	0714	0715	0716	0717	0718	0719
2D_	0720	0721	0722	0723	0724	0725	0726	0727	0728	0729	0730	0731	0732	0733	0734	0735
2E_	0736	0737	0738	0739	0740	0741	0742	0743	0744	0745	0746	0747	0748	0749	0750	0751
2F_	0752	0753	0754	0755	0756	0757	0758	0759	0760	0761	0762	0763	0764	0765	0766	0767
30_	0768	0769	0770	0771	0772	0773	0774	0775	0776	0777	0778	0779	0780	0781	0782	0783
31_	0784	0785	0786	0787	0788	0789	0790	0791	0792	0793	0794	0795	0796	0797	0798	0799
32_	0800	0801	0802	0803	0804	0805	0806	0807	0808	0809	0810	0811	0812	0813	0814	0815
33_	0816	0817	0818	0819	0820	0821	0822	0823	0824	0825	0826	0827	0828	0829	0830	0831
34_	0832	0833	0834	0835	0836	0837	0838	0839	0840	0841	0842	0843	0844	0845	0846	0847
35_	0848	0849	0850	0851	0852	0853	0854	0855	0856	0857	0858	0859	0860	0861	0862	0863
36_	0864	0865	0866	0867	0868	0869	0870	0871	0872	0873	0874	0875	0876	0877	0878	0879
37_	0880	0881	0882	0883	0884	0885	0886	0887	0888	0889	0890	0891	0892	0893	0894	0895
38_	0896	0897	0898	0899	0900	0901	0902	0903	0904	0905	0906	0907	0908	0909	0910	0911
39_	0912	0913	0914	0915	0916	0917	0918	0919	0920	0921	0922	0923	0924	0925	0926	0927
3A_	0928	0929	0930	0931	0932	0933	0934	0935	0936	0937	0938	0939	0940	0941	0942	0943
3B_	0944	0945	0946	0947	0948	0949	0950	0951	0952	0953	0954	0955	0956	0957	0958	0959
3C_	0960	0961	0962	0963	0964	0965	0966	0967	0968	0969	0970	0971	0972	0973	0974	0975
3D_	0976	0977	0978	0979	0980	0981	0982	0983	0984	0985	0986	0987	0988	0989	0990	0991
3E_	0992	0993	0994	0995	0996	0997	0998	0999	1000	1001	1002	1003	1004	1005	1006	1007
3F_	1008	1009	1010	1011	1012	1013	1014	1015	1016	1017	1018	1019	1020	1021	1022	1023
40_	1024	1025	1026	1027	1028	1029	1030	1031	1032	1033	1034	1035	1036	1037	1038	1039
41_	1040	1041	1042	1043	1044	1045	1046	1047	1048	1049	1050	1051	1052	1053	1054	1055
42_	1056	1057	1058	1059	1060	1061	1062	1063	1064	1065	1066	1067	1068	1069	1070	1071
43_	1072	1073	1074	1075	1076	1077	1078	1079	1080	1081	1082	1083	1084	1085	1086	1087
44_	1088	1089	1090	1091	1092	1093	1094	1095	1096	1097	1098	1099	1100	1101	1102	1103
45_	1104	1105	1106	1107	1108	1109	1110	1111	1112	1113	1114	1115	1116	1117	1118	1119
46_	1120	1121	1122	1123	1124	1125	1126	1127	1128	1129	1130	1131	1132	1133	1134	1135
47_	1136	1137	1138	1139	1140	1141	1142	1143	1144	1145	1146	1147	1148	1149	1150	1151
48_	1152	1153	1154	1155	1156	1157	1158	1159	1160	1161	1162	1163	1164	1165	1166	1167
49_	1168	1169	1170	1171	1172	1173	1174	1175	1176	1177	1178	1179	1180	1181	1182	1183
4A_	1184	1185	1186	1187	1188	1189	1190	1191	1192	1193	1194	1195	1196	1197	1198	1199
4B_	1200	1201	1202	1203	1204	1205	1206	1207	1208	1209	1210	1211	1212	1213	1214	1215
4C_	1216	1217	1218	1219	1220	1221	1222	1223	1224	1225	1226	1227	1228	1229	1230	1231
4D_	1232	1233	1234	1235	1236	1237	1238	1239	1240	1241	1242	1243	1244	1245	1246	1247
4E_	1248	1249	1250	1251	1252	1253	1254	1255	1256	1257	1258	1259	1260	1261	1262	1263
4F_	1264	1265	1266	1267	1268	1269	1270	1271	1272	1273	1274	1275	1276	1277	1278	1279
50_	1280	1281	1282	1283	1284	1285	1286	1287	1288	1289	1290	1291	1292	1293	1294	1295
51_	1296	1297	1298	1299	1300	1301	1302	1303	1304	1305	1306	1307	1308	1309	1310	1311
52_	1312	1313	1314	1315	1316	1317	1318	1319	1320	1321	1322	1323	1324	1325	1326	1327
53_	1328	1329	1330	1331	1332	1333	1334	1335	1336	1337	1338	1339	1340	1341	1342	1343
54_	1344	1345	1346	1347	1348	1349	1350	1351	1352	1353	1354	1355	1356	1357	1358	1359
55_	1360	1361	1362	1363	1364	1365	1366	1367	1368	1369	1370	1371	1372	1373	1374	1375
56_	1376	1377	1378	1379	1380	1381	1382	1383	1384	1385	1386	1387	1388	1389	1390	1391
57_	1392	1393	1394	1395	1396	1397	1398	1399	1400	1401	1402	1403	1404	1405	1406	1407
58_	1408	1409	1410	1411	1412	1413	1414	1415	1416	1417	1418	1419	1420	1421	1422	1423
59_	1424	1425	1426	1427	1428	1429	1430	1431	1432	1433	1434	1435	1436	1437	1438	1439
5A_	1440	1441	1442	1443	1444	1445	1446	1447	1448	1449	1450	1451	1452	1453	1454	1455
5B_	1456	1457	1458	1459	1460	1461	1462	1463	1464	1465	1466	1467	1468	1469	1470	1471
5C_	1472	1473	1474	1475	1476	1477	1478	1479	1480	1481	1482	1483	1484	1485	1486	1487
5D_	1488	1489	1490	1491	1492	1493	1494	1495	1496	1497	1498	1499	1500	1501	1502	1503
5E_	1504	1505	1506	1507	1508	1509	1510	1511	1512	1513	1514	1515	1516	1517	1518	1519
5F_	1520	1521	1522	1523	1524	1525	1526	1527	1528	1529	1530	1531	1532	1533	1534	1535

DIRECT CONVERSION TABLE

	0	1	2	3	4	5	6	7	8	9	A	B	C	D	E	F
60_	1536	1537	1538	1539	1540	1541	1542	1543	1544	1545	1546	1547	1548	1549	1550	1551
61_	1552	1553	1554	1555	1556	1557	1558	1559	1560	1561	1562	1563	1564	1565	1566	1567
62_	1568	1569	1570	1571	1572	1573	1574	1575	1576	1577	1578	1579	1580	1581	1582	1583
63_	1584	1585	1586	1587	1588	1589	1590	1591	1592	1593	1594	1595	1596	1597	1598	1599
64_	1600	1601	1602	1603	1604	1605	1606	1607	1608	1609	1610	1611	1612	1613	1614	1615
65_	1616	1617	1618	1619	1620	1621	1622	1623	1624	1625	1626	1627	1628	1629	1630	1631
66_	1632	1633	1634	1635	1636	1637	1638	1639	1640	1641	1642	1643	1644	1645	1646	1647
67_	1648	1649	1650	1651	1652	1653	1654	1655	1656	1657	1658	1659	1660	1661	1662	1663
68_	1664	1665	1666	1667	1668	1669	1670	1671	1672	1673	1674	1675	1676	1677	1678	1679
69_	1680	1681	1682	1683	1684	1685	1686	1687	1688	1689	1690	1691	1692	1693	1694	1695
6A_	1696	1697	1698	1699	1700	1701	1702	1703	1704	1705	1706	1707	1708	1709	1710	1711
6B_	1712	1713	1714	1715	1716	1717	1718	1719	1720	1721	1722	1723	1724	1725	1726	1727
6C_	1728	1729	1730	1731	1732	1733	1734	1735	1736	1737	1738	1739	1740	1741	1742	1743
6D_	1744	1745	1746	1747	1748	1749	1750	1751	1752	1753	1754	1755	1756	1757	1758	1759
6E_	1760	1761	1762	1763	1764	1765	1766	1767	1768	1769	1770	1771	1772	1773	1774	1775
6F_	1776	1777	1778	1779	1780	1781	1782	1783	1784	1785	1786	1787	1788	1789	1790	1791
70_	1792	1793	1794	1795	1796	1797	1798	1799	1800	1801	1802	1803	1804	1805	1806	1807
71_	1808	1809	1810	1811	1812	1813	1814	1815	1816	1817	1818	1819	1820	1821	1822	1823
72_	1824	1825	1826	1827	1828	1829	1830	1831	1832	1833	1834	1835	1836	1837	1838	1839
73_	1840	1841	1842	1843	1844	1845	1846	1847	1848	1849	1850	1851	1852	1853	1854	1855
74_	1856	1857	1858	1859	1860	1861	1862	1863	1864	1865	1866	1867	1868	1869	1870	1871
75_	1872	1873	1874	1875	1876	1877	1878	1879	1880	1881	1882	1883	1884	1885	1886	1887
76_	1888	1889	1890	1891	1892	1893	1894	1895	1896	1897	1898	1899	1900	1901	1902	1903
77_	1904	1905	1906	1907	1908	1909	1910	1911	1912	1913	1914	1915	1916	1917	1918	1919
78_	1920	1921	1922	1923	1924	1925	1926	1927	1928	1929	1930	1931	1932	1933	1934	1935
79_	1936	1937	1938	1939	1940	1941	1942	1943	1944	1945	1946	1947	1948	1949	1950	1951
7A_	1952	1953	1954	1955	1956	1957	1958	1959	1960	1961	1962	1963	1964	1965	1966	1967
7B_	1968	1969	1970	1971	1972	1973	1974	1975	1976	1977	1978	1979	1980	1981	1982	1983
7C_	1984	1985	1986	1987	1988	1989	1990	1991	1992	1993	1994	1995	1996	1997	1998	1999
7D_	2000	2001	2002	2003	2004	2005	2006	2007	2008	2009	2010	2011	2012	2013	2014	2015
7E_	2016	2017	2018	2019	2020	2021	2022	2023	2024	2025	2026	2027	2028	2029	2030	2031
7F_	2032	2033	2034	2035	2036	2037	2038	2039	2040	2041	2042	2043	2044	2045	2046	2047
80_	2048	2049	2050	2051	2052	2053	2054	2055	2056	2057	2058	2059	2060	2061	2062	2063
81_	2064	2065	2066	2067	2068	2069	2070	2071	2072	2073	2074	2075	2076	2077	2078	2079
82_	2080	2081	2082	2083	2084	2085	2086	2087	2088	2089	2090	2091	2092	2093	2094	2095
83_	2096	2097	2098	2099	2100	2101	2102	2103	2104	2105	2106	2107	2108	2109	2110	2111
84_	2112	2113	2114	2115	2116	2117	2118	2119	2120	2121	2122	2123	2124	2125	2126	2127
85_	2128	2129	2130	2131	2132	2133	2134	2135	2136	2137	2138	2139	2140	2141	2142	2143
86_	2144	2145	2146	2147	2148	2149	2150	2151	2152	2153	2154	2155	2156	2157	2158	2159
87_	2160	2161	2162	2163	2164	2165	2166	2167	2168	2169	2170	2171	2172	2173	2174	2175
88_	2176	2177	2178	2179	2180	2181	2182	2183	2184	2185	2186	2187	2188	2189	2190	2191
89_	2192	2193	2194	2195	2196	2197	2198	2199	2200	2201	2202	2203	2204	2205	2206	2207
8A_	2208	2209	2210	2211	2212	2213	2214	2215	2216	2217	2218	2219	2220	2221	2222	2223
8B_	2224	2225	2226	2227	2228	2229	2230	2231	2232	2233	2234	2235	2236	2237	2238	2239
8C_	2240	2241	2242	2243	2244	2245	2246	2247	2248	2249	2250	2251	2252	2253	2254	2255
8D_	2256	2257	2258	2259	2260	2261	2262	2263	2264	2265	2266	2267	2268	2269	2270	2271
8E_	2272	2273	2274	2275	2276	2277	2278	2279	2280	2281	2282	2283	2284	2285	2286	2287
8F_	2288	2289	2290	2291	2292	2293	2294	2295	2296	2297	2298	2299	2300	2301	2302	2303
90_	2304	2305	2306	2307	2308	2309	2310	2311	2312	2313	2314	2315	2316	2317	2318	2319
91_	2320	2321	2322	2323	2324	2325	2326	2327	2328	2329	2330	2331	2332	2333	2334	2335
92_	2336	2337	2338	2339	2340	2341	2342	2343	2344	2345	2346	2347	2348	2349	2350	2351
93_	2352	2353	2354	2355	2356	2357	2358	2359	2360	2361	2362	2363	2364	2365	2366	2367
94_	2368	2369	2370	2371	2372	2373	2374	2375	2376	2377	2378	2379	2380	2381	2382	2383
95_	2384	2385	2386	2387	2388	2389	2390	2391	2392	2393	2394	2395	2396	2397	2398	2399
96_	2400	2401	2402	2403	2404	2405	2406	2407	2408	2409	2410	2411	2412	2413	2414	2415
97_	2416	2417	2418	2419	2420	2421	2422	2423	2424	2425	2426	2427	2428	2429	2430	2431
98_	2432	2433	2434	2435	2436	2437	2438	2439	2440	2441	2442	2443	2444	2445	2446	2447
99_	2448	2449	2450	2451	2452	2453	2454	2455	2456	2457	2458	2459	2460	2461	2462	2463
9A_	2464	2465	2466	2467	2468	2469	2470	2471	2472	2473	2474	2475	2476	2477	2478	2479
9B_	2480	2481	2482	2483	2484	2485	2486	2487	2488	2489	2490	2491	2492	2493	2494	2495
9C_	2496	2497	2498	2499	2500	2501	2502	2503	2504	2505	2506	2507	2508	2509	2510	2511
9D_	2512	2513	2514	2515	2516	2517	2518	2519	2520	2521	2522	2523	2524	2525	2526	2527
9E_	2528	2529	2530	2531	2532	2533	2534	2535	2536	2537	2538	2539	2540	2541	2542	2543
9F_	2544	2545	2546	2547	2548	2549	2550	2551	2552	2553	2554	2555	2556	2557	2558	2559

DIRECT CONVERSION TABLE

	0	1	2	3	4	5	6	7	8	9	A	B	C	D	E	F
A0__	2560	2561	2562	2563	2564	2565	2566	2567	2568	2569	2570	2571	2572	2573	2574	2575
A1__	2576	2577	2578	2579	2580	2581	2582	2583	2584	2585	2586	2587	2588	2589	2590	2591
A2__	2592	2593	2594	2595	2596	2597	2598	2599	2600	2601	2602	2603	2604	2605	2606	2607
A3__	2608	2609	2610	2611	2612	2613	2614	2615	2616	2617	2618	2619	2620	2621	2622	2623
A4__	2624	2625	2626	2627	2628	2629	2630	2631	2632	2633	2634	2635	2636	2637	2638	2639
A5__	2640	2641	2642	2643	2644	2645	2646	2647	2648	2649	2650	2651	2652	2653	2654	2655
A6__	2656	2657	2658	2659	2660	2661	2662	2663	2664	2665	2666	2667	2668	2669	2670	2671
A7__	2672	2673	2674	2675	2676	2677	2678	2679	2680	2681	2682	2683	2684	2685	2686	2687
A8__	2688	2689	2690	2691	2692	2693	2694	2695	2696	2697	2698	2699	2700	2701	2702	2703
A9__	2704	2705	2706	2707	2708	2709	2710	2711	2712	2713	2714	2715	2716	2717	2718	2719
AA__	2720	2721	2722	2723	2724	2725	2726	2727	2728	2729	2730	2731	2732	3733	2734	2735
AB__	2736	2737	2738	2739	2740	2741	2742	2743	2744	2745	2746	2747	2748	2749	2750	2751
AC__	2752	2753	2754	2755	2756	2757	2758	2759	2760	2761	2762	2763	2764	2765	2766	2767
AD__	2768	2769	2770	2771	2772	2773	2774	2775	2776	2777	2778	2779	2780	2781	2782	2783
AE__	2784	2785	2786	2787	2788	2789	2790	2791	2792	2793	2794	2795	2796	2797	2798	2799
AF__	2800	2801	2802	2803	2804	2805	2806	2807	2808	2809	2810	2811	2812	2813	2814	2815
B0__	2816	2817	2818	2819	2820	2821	2822	2823	2824	2825	2826	2827	2828	2829	2830	2831
B1__	2832	2833	2834	2835	2836	2837	2838	2839	2840	2841	2842	2843	2844	2845	2846	2847
B2__	2848	2849	2850	2851	2852	2853	2854	2855	2856	2857	2858	2859	2860	2861	2862	2863
B3__	2864	2865	2866	2867	2868	2869	2870	2871	2872	2873	2874	2875	2876	2877	2878	2879
B4__	2880	2881	2882	2883	2884	2885	2886	2887	2888	2889	2890	2891	2892	2893	2894	2895
B5__	2896	2897	2898	2899	2900	2901	2902	2903	2904	2905	2906	2907	2908	2909	2910	2911
B6__	2912	2913	2914	2915	2916	2917	2918	2919	2920	2921	2922	2923	2924	2925	2926	2927
B7__	2928	2929	2930	2931	2932	2933	2934	2935	2936	2937	2938	2939	2940	2941	2942	2943
B8__	2944	2945	2946	2947	2948	2949	2950	2951	2952	2953	2954	2955	2956	2957	2958	2959
B9__	2960	2961	2962	2963	2964	2965	2966	2967	2968	2969	2970	2971	2972	2973	2974	2975
BA__	2976	2977	2978	2979	2980	2981	2982	2983	2984	2985	2986	2987	2988	2989	2990	2991
BB__	2992	2993	2994	2995	2996	2997	2998	2999	3000	3001	3002	3003	3004	3005	3006	3007
BC__	3008	3009	3010	3011	3012	3013	3014	3015	3016	3017	3018	3019	3020	3021	3022	3023
BD__	3024	3025	3026	3027	3028	3029	3030	3031	3032	3033	3034	3035	3036	3037	3038	3039
BE__	3040	3041	3042	3043	3044	3045	3046	3047	3048	3049	3050	3051	3052	3053	3054	3055
BF__	3056	3057	3058	3059	3060	3061	3062	3063	3064	3065	3066	3067	3068	3069	3070	3071
C0__	3072	3073	3074	3075	3076	3077	3078	3079	3080	3081	3082	3083	3084	3085	3086	3087
C1__	3088	3089	3090	3091	3092	3093	3094	3095	3096	3097	3098	3099	3100	3101	3102	3103
C2__	3104	3105	3106	3107	3108	3109	3110	3111	3112	3113	3114	3115	3116	3117	3118	3119
C3__	3120	3121	3122	3123	3124	3125	3126	3127	3128	3129	3130	3131	3132	3133	3134	3135
C4__	3136	3137	3138	3139	3140	3141	3142	3143	3144	3145	3146	3147	3148	3149	3150	3151
C5__	3152	3153	3154	3155	3156	3157	3158	3159	3160	3161	3162	3613	3164	3165	3166	3167
C6__	3168	3169	3170	3171	3172	3173	3174	3175	3176	3177	3178	3179	3180	3181	3182	3183
C7__	3184	3185	3186	3187	3188	3189	3190	3191	3192	3193	3194	3195	3196	3197	3198	3199
C8__	3200	3201	3202	3203	3204	3205	3206	3207	3208	3209	3210	3211	3212	3213	3214	3215
C9__	3216	3217	3218	3219	3220	3221	3222	3223	3224	3225	3226	3227	3228	3229	3230	3231
CA__	3232	3233	3234	3235	3236	3237	3238	3239	3240	3241	3242	3243	3244	3245	3246	3247
CB__	3248	3249	3250	3251	3252	3253	3254	3255	3256	3257	3258	3259	3260	3261	3262	3263
CC__	3264	3265	3266	3267	3268	3269	3270	3271	3272	3273	3274	3275	3276	3277	3278	3279
CD__	3280	3281	3282	3283	3284	3285	3286	3287	3288	3289	3290	3291	3292	3293	3294	3295
CE__	3296	3297	3298	3299	3300	3301	3302	3303	3304	3305	3306	3307	3308	3309	3310	3311
CF__	3312	3313	3314	3315	3316	3317	3318	3319	3320	3321	3322	3323	3324	3325	3326	3327
D0__	3328	3329	3330	3331	3332	3333	3334	3335	3336	3337	3338	3339	3340	3341	3342	3343
D1__	3344	3345	3346	3347	3348	3349	3350	3351	3352	3353	3354	3355	3356	3357	3358	3359
D2__	3360	3361	3362	3363	3364	3365	3366	3367	3368	3369	3370	3371	3372	3373	3374	3375
D3__	3376	3377	3378	3379	3380	3381	3382	3383	3384	3385	3386	3387	3388	3389	3390	3391
D4__	3392	3393	3394	3395	3396	3397	3398	3399	3400	3401	3402	3403	3404	3405	3406	3407
D5__	3408	3409	3410	3411	3412	3413	3414	3415	3416	3417	3418	3419	3420	3421	3422	3423
D6__	3424	3425	3426	3427	3428	3429	3430	3431	3432	3433	3434	3435	3436	3437	3438	3439
D7__	3440	3441	3442	3443	3444	3445	3446	3447	3448	3449	3450	3451	3452	3453	3454	3455
D8__	3456	3457	3458	3459	3460	3461	3462	3463	3464	3465	3466	3467	3468	3469	3470	3471
D9__	3472	3473	3474	3475	3476	3477	3478	3479	3480	3481	3482	3483	3484	3485	3486	3487
DA__	3488	3489	3490	3491	3492	3493	3494	3495	3496	3497	3498	3499	3500	3501	3502	3503
DB__	3504	3505	3506	3507	3508	3509	3510	3511	3512	3513	3514	3515	3516	3517	3518	3519
DC__	3520	3521	3522	3523	3524	3525	3526	3527	3528	3529	3530	3531	3532	3533	3534	3535
DD__	3536	3537	3538	3539	3540	3541	3542	3543	3544	3545	3546	3547	3548	3549	3550	3551
DE__	3552	3553	3554	3555	3556	3557	3558	3559	3560	3561	3562	3563	3564	3565	3566	3567
DF__	3568	3569	3570	3571	3572	3573	3574	3575	3576	3577	3578	3579	3580	3581	3582	3583

DIRECT CONVERSION TABLE

	0	1	2	3	4	5	6	7	8	9	A	B	C	D	E	F
E0__	3584	3585	3586	3587	3588	3589	3590	3591	3592	3593	3594	3595	3596	3597	3598	3599
E1__	3600	3601	3602	3603	3604	3605	3606	3607	3608	3609	3610	3611	3612	3613	3614	3615
E2__	3616	3617	3618	3619	3620	3621	3622	3623	3624	3625	3626	3627	3628	3629	3630	3631
E3__	3632	3633	3634	3635	3636	3637	3638	3639	3640	3641	3642	3643	3644	3645	3646	3647
E4__	3648	3649	3650	3651	3652	3653	3654	3655	3656	3657	3658	3659	3660	3661	3662	3663
E5__	3664	3665	3666	3667	3668	3669	3670	3671	3672	3673	3674	3675	3676	3677	3678	3679
E6__	3680	3681	3682	3683	3684	3685	3686	3687	3688	3689	3690	3691	3692	3693	3694	3695
E7__	3696	3697	3698	3699	3700	3701	3702	3703	3704	3705	3706	3707	3708	3709	3710	3711
E8__	3712	3713	3714	3715	3716	3717	3718	3719	3720	3721	3722	3723	3724	3725	3726	3727
E9__	3728	3729	3730	3731	3732	3733	3734	3735	3736	3737	3738	3739	3740	3741	3742	3743
EA__	3744	3745	3746	3747	3748	3749	3750	3751	3752	3753	3754	3755	3756	3757	3758	3759
EB__	3760	3761	3762	3763	3764	3765	3766	3767	3768	3769	3770	3771	3772	3773	3774	3775
EC__	3776	3777	3778	3779	3780	3781	3782	3783	3784	3785	3786	3787	3788	3789	3790	3791
ED__	3792	3793	3794	3795	3796	3797	3798	3799	3800	3801	3802	3803	3804	3805	3806	3807
EE__	3808	3809	3810	3811	3812	3813	3814	3815	3816	3817	3818	3819	3820	3821	3822	3823
EF__	3824	3825	3826	3827	3828	3829	3830	3831	3832	3833	3834	3835	3836	3837	3838	3839
F0__	3840	3841	3842	3843	3844	3845	3846	3847	3848	3849	3850	3851	3852	3853	3854	3855
F1__	3856	3857	3858	3859	3860	3861	3862	3863	3864	3865	3866	3867	3868	3869	3870	3871
F2__	3872	3873	3874	3875	3876	3877	3878	3879	3880	3881	3882	3883	3884	3885	3886	3887
F3__	3888	3889	3890	3891	3892	3893	3894	3895	3896	3897	3898	3899	3900	3901	3902	3903
F4__	3904	3905	3906	3907	3908	3909	3910	3911	3912	3913	3914	3915	3916	3917	3918	3919
F5__	3920	3921	3922	3923	3924	3925	3926	3927	3928	3929	3930	3931	3932	3933	3934	3935
F6__	3936	3937	3938	3939	3940	3941	3942	3943	3944	3945	3946	3947	3948	3949	3950	3951
F7__	3952	3953	3954	3955	3956	3957	3958	3959	3960	3961	3962	3963	3964	3965	3966	3967
F8__	3968	3969	3970	3971	3972	3973	3974	3975	3976	3977	3978	3979	3980	3981	3982	3983
F9__	3984	3985	3986	3987	3988	3989	3990	3991	3992	3993	3994	3995	3996	3997	3998	3999
FA__	4000	4001	4002	4003	4004	4005	4006	4007	4008	4009	4010	4011	4012	4013	4014	4015
FB__	4016	4017	4018	4019	4020	4021	4022	4023	4024	4025	4026	4027	4028	4029	4030	4031
FC__	4032	4033	4034	4035	4036	4037	4038	4039	4040	4041	4042	4043	4044	4045	4046	4047
FD__	4048	4049	4050	4051	4052	4053	4054	4055	4056	4057	4058	4059	4060	4061	4062	4063
FE__	4064	4065	4066	4067	4068	4069	4070	4071	4072	4073	4074	4075	4076	4077	4078	4079
FF__	4080	4081	4082	4083	4084	4085	4086	4087	4088	4089	4090	4091	4092	4093	4094	4095

II. HEXADECIMAL AND DECIMAL INTEGER CONVERSION TABLE

8		7		6		5		4		3		2		1	
Hex	Decimal	Hex	Decimal	Hex	Decimal	Hex	Decimal	Hex	Decimal	Hex	Decimal	Hex	Decimal	Hex	Decimal
0	0	0	0	0	0	0	0	0	0	0	0	0	0	0	0
1	268,435,456	1	16,777,216	1	1,048,576	1	65,536	1	4,096	1	256	1	16	1	1
2	536,870,912	2	33,554,432	2	2,097,152	2	131,072	2	8,192	2	512	2	32	2	2
3	805,306,368	3	50,331,648	3	3,145,728	3	196,608	3	12,288	3	768	3	48	3	3
4	1,073,741,824	4	67,108,864	4	4,194,304	4	262,144	4	16,384	4	1,024	4	64	4	4
5	1,342,177,280	5	83,886,080	5	5,242,880	5	327,680	5	20,480	5	1,280	5	80	5	5
6	1,610,612,736	6	100,663,296	6	6,291,456	6	393,216	6	24,576	6	1,536	6	96	6	6
7	1,879,048,192	7	117,440,512	7	7,340,032	7	458,752	7	28,672	7	1,792	7	112	7	7
8	2,147,483,648	8	134,217,728	8	8,388,608	8	524,288	8	32,768	8	2,048	8	128	8	8
9	2,415,919,104	9	150,994,944	9	9,437,184	9	589,824	9	36,864	9	2,304	9	144	9	9
A	2,684,354,560	A	167,772,160	A	10,485,760	A	655,360	A	40,960	A	2,560	A	160	A	10
B	2,952,790,016	B	184,549,376	B	11,534,336	B	720,896	B	45,056	B	2,816	B	176	B	11
C	3,221,225,472	C	201,326,592	C	12,582,912	C	786,432	C	49,152	C	3,072	C	192	C	12
D	3,489,660,928	D	218,103,808	D	13,631,488	D	851,968	D	53,248	D	3,328	D	208	D	13
E	3,758,096,384	E	234,881,024	E	14,680,064	E	917,504	E	57,344	E	3,584	E	224	E	14
F	4,026,531,840	F	251,658,240	F	15,728,640	F	983,040	F	61,440	F	3,840	F	240	F	15
8		7		6		5		4		3		2		1	

INTEGER CONVERSION TABLE

TO CONVERT HEXADECIMAL TO DECIMAL

1. Locate the column of decimal numbers corresponding to the left-most digit or letter of the hexadecimal; select from this column and record the number that corresponds to the position of the hexadecimal digit or letter.
2. Repeat step 1 for the next (second from the left) position.
3. Repeat step 1 for the units (third from the left) position.
4. Add the numbers selected from the table to form the decimal number.

EXAMPLE

Conversion of Hexadecimal Value	D34
1. D	3328
2. 3	48
3. 4	4
4. Decimal	3380

To convert integer numbers greater than the capacity of table, use the techniques below:

HEXADECIMAL TO DECIMAL

Successive cumulative multiplication from left to right, adding units position.

Example: $D34_{16} = 3380_{10}$

D =	13
	×16
	208
3 =	+3
	211
	×16
	3376
4 =	+4
	3380

TO CONVERT DECIMAL TO HEXADECIMAL

1. (a) Select from the table the highest decimal number that is equal to or less than the number to be converted.
   (b) Record the hexadecimal of the column containing the selected number.
   (c) Subtract the selected decimal from the number to be converted.
2. Using the remainder from step 1(c) repeat all of step 1 to develop the second position of the hexadecimal (and a remainder).
3. Using the remainder from step 2 repeat all of step 1 to develop the units position of the hexadecimal.
4. Combine terms to form the hexadecimal number.

EXAMPLE

Conversion of Decimal Value	3380
1. D	−3328
	52
2. 3	−48
	4
3. 4	−4
4. Hexadecimal	D34

DECIMAL TO HEXADECIMAL

Divide and collect the remainder in reverse order.

Example: $3380_{10} = X_{16}$

	remainder
16 | 3380	
16 | 211	4
16 | 13	3
	D

$3380_{10} = D34_{16}$

INTEGER CONVERSION TABLE

POWERS OF 16 TABLE

Example: $268{,}435{,}456_{10} = (2.68435456 \times 10^8)_{10} = 1000\ 0000_{16} = (10^7)_{16}$

16^n	n
1	0
16	1
256	2
4 096	3
65 536	4
1 048 576	5
16 777 216	6
268 435 456	7
4 294 967 296	8
68 719 476 736	9
1 099 511 627 776	10 = A
17 592 186 044 416	11 = B
281 474 976 710 656	12 = C
4 503 599 627 370 496	13 = D
72 057 594 037 927 936	14 = E
1 152 921 504 606 846 976	15 = F
Decimal Values	

III. HEXADECIMAL AND DECIMAL FRACTION CONVERSION TABLE

1		2			3				4				
Hex	Decimal	Hex	Decimal		Hex	Decimal			Hex	Decimal Equivalent			
.0	.0000	.00	.0000	0000	.000	.0000	0000	0000	.0000	.0000	0000	0000	0000
.1	.0625	.01	.0039	0625	.001	.0002	4414	0625	.0001	.0000	1525	8789	0625
.2	.1250	.02	.0078	1250	.002	.0004	8828	1250	.0002	.0000	3051	7578	1250
.3	.1875	.03	.0117	1875	.003	.0007	3242	1875	.0003	.0000	4577	6367	1875
.4	.2500	.04	.0156	2500	.004	.0009	7656	2500	.0004	.0000	6103	5156	2500
.5	.3125	.05	.0195	3125	.005	.0012	2070	3125	.0005	.0000	7629	3945	3125
.6	.3750	.06	.0234	3750	.006	.0014	6484	3750	.0006	.0000	9155	2734	3750
.7	.4375	.07	.0273	4375	.007	.0017	0898	4375	.0007	.0001	0681	1523	4375
.8	.5000	.08	.0312	5000	.008	.0019	5312	5000	.0008	.0001	2207	0312	5000
.9	.5625	.09	.0351	5625	.009	.0021	9726	5625	.0009	.0001	3732	9101	5625
.A	.6250	.0A	.0390	6250	.00A	.0024	4140	6250	.000A	.0001	5258	7890	6250
.B	.6875	.0B	.0429	6875	.00B	.0026	8554	6875	.000B	.0001	6784	6679	6875
.C	.7500	.0C	.0468	7500	.00C	.0029	2968	7500	.000C	.0001	8310	5468	7500
.D	.8125	.0D	.0507	8125	.00D	.0031	7382	8125	.000D	.0001	9836	4257	8125
.E	.8750	.0E	.0546	8750	.00E	.0034	1796	8750	.000E	.0002	1362	3046	8750
.F	.9375	.0F	.0585	9375	.00F	.0036	6210	9375	.000F	.0002	2888	1835	9375
1		2			3				4				

TO CONVERT .ABC HEXADECIMAL TO DECIMAL

Find .A in position 1	.6250
Find .0B in position 2	.0429 6875
Find .00C in position 3	.0029 2968 7500
.ABC Hex is equal to	.6708 9843 7500

FRACTION CONVERSION TABLE

TO CONVERT .13 DECIMAL TO HEXADECIMAL

1. Find .1250 next lowest to	.1300	
subtract	– .1250	= .2 Hex
2. Find .0039 0625 next lowest to	.0050 0000	
	– .0039 0625	= .01
3. Find .0009 7656 2500	.0010 9375 0000	
	– .0009 7656 2500	= .004
4. Find .0001 0681 1523 4375	.0001 1718 7500 0000	
	– .0001 0681 1523 4375	= .0007
	.0000 1037 5976 5625	= .2147 Hex

5. 13 Decimal is approximately equal to ——— .2147 Hex

To convert fractions beyond the capacity of table, use techniques below:

HEXADECIMAL FRACTION TO DECIMAL

Convert the hexadecimal fraction to its decimal equivalent using the same technique as for integer numbers. Divide the results by 16^n (n is the number of fraction positions).

Example: $.8A7_{16} = .540771_{10}$

$8A7_{16} = 2215_{10}$

$16^3 = 4096$

$2215.000000 \div 4096 = .540771$

DECIMAL FRACTION TO HEXADECIMAL

Collect integer parts of product in the order of calculation.

Example: $.5408_{10} = .8A7_{16}$

```
          .5408
          ×  16
8 ← [ 8]  .6528
          ×  16
A ← [10]  .4448
          ×  16
7 ← [ 7]  .1168
```

HEXADECIMAL ADDITION AND SUBTRACTION TABLE

Example: 6 + 2 = 8, 8 − 2 = 6, and 8 − 6 = 2

	1	2	3	4	5	6	7	8	9	A	B	C	D	E	F
1	02	03	04	05	06	07	08	09	0A	0B	0C	0D	0E	0F	10
2	03	04	05	06	07	08	09	0A	0B	0C	0D	0E	0F	10	11
3	04	05	06	07	08	09	0A	0B	0C	0D	0E	0F	10	11	12
4	05	06	07	08	09	0A	0B	0C	0D	0E	0F	10	11	12	13
5	06	07	08	09	0A	0B	0C	0D	0E	0F	10	11	12	13	14
6	07	08	09	0A	0B	0C	0D	0E	0F	10	11	12	13	14	15
7	08	09	0A	0B	0C	0D	0E	0F	10	11	12	13	14	15	16
8	09	0A	0B	0C	0D	0E	0F	10	11	12	13	14	15	16	17
9	0A	0B	0C	0D	0E	0F	10	11	12	13	14	15	16	17	18
A	0B	0C	0D	0E	0F	10	11	12	13	14	15	16	17	18	19
B	0C	0D	0E	0F	10	11	12	13	14	15	16	17	18	19	1A
C	0D	0E	0F	10	11	12	13	14	15	16	17	18	19	1A	1B
D	0E	0F	10	11	12	13	14	15	16	17	18	19	1A	1B	1C
E	0F	10	11	12	13	14	15	16	17	18	19	1A	1B	1C	1D
F	10	11	12	13	14	15	16	17	18	19	1A	1B	1C	1D	1E

HEXADECIMAL MULTIPLICATION TABLE

Example: 2 × 4 = 08, F × 2 = 1E

	1	2	3	4	5	6	7	8	9	A	B	C	D	E	F
1	01	02	03	04	05	06	07	08	09	0A	0B	0C	0D	0E	0F
2	02	04	06	08	0A	0C	0E	10	12	14	16	18	1A	1C	1E
3	03	06	09	0C	0F	12	15	18	1B	1E	21	24	27	2A	2D
4	04	08	0C	10	14	18	1C	20	24	28	2C	30	34	38	3C
5	05	0A	0F	14	19	1E	23	28	2D	32	37	3C	41	46	4B
6	06	0C	12	18	1E	24	2A	30	36	3C	42	48	4E	54	5A
7	07	0E	15	1C	23	2A	31	38	3F	46	4D	54	5B	62	69
08	08	10	18	20	28	30	38	40	48	50	58	60	68	70	78
9	09	12	1B	24	2D	36	3F	48	51	5A	63	6C	75	7E	87
A	0A	14	1E	28	32	3C	46	50	5A	64	6E	78	82	8C	96
B	0B	16	21	2C	37	42	4D	58	63	6E	79	84	8F	9A	A5
C	0C	18	24	30	3C	48	54	60	6C	78	84	90	9C	A8	B4
D	0D	1A	27	34	41	4E	5B	68	75	82	8F	9C	A9	B6	C3
E	0E	1C	2A	38	46	54	62	70	7E	8C	9A	A8	B6	C4	D2
F	0F	1E	2D	3C	4B	5A	69	78	87	96	A5	B4	C3	D2	E1

CHAPTER 4

GEOMETRY

MENSURATION FORMULAS

TRIANGLES

In the following: K = area, r = radius of the inscribed circle, R = radius of the circumscribed circle.

RIGHT TRIANGLE

$A + B = C = 90°$

$c^2 = a^2 + b^2$ (*Pythagorean relation*)

$a = \sqrt{(c + b)(c - b)}$

$K = \frac{1}{2}ab$

$r = \dfrac{ab}{a + b + c}, \quad R = \frac{1}{2}c$

$h = \dfrac{ab}{c}, \quad m = \dfrac{b^2}{c}, \quad n = \dfrac{a^2}{c}$

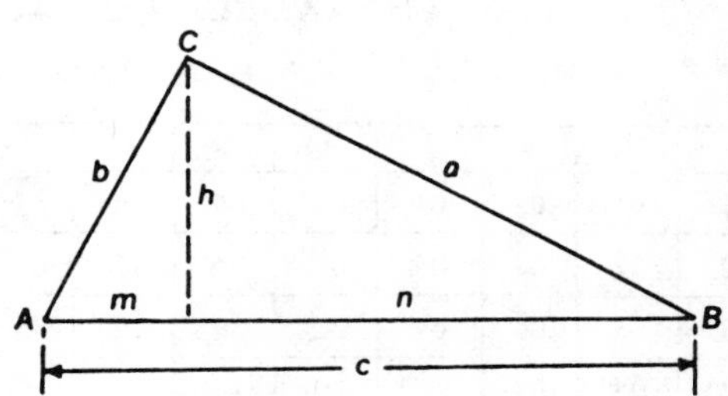

EQUILATERAL TRIANGLE

$A = B = C = 60°$

$K = \frac{1}{4}a^2\sqrt{3}$

$r = \frac{1}{6}a\sqrt{3}, \quad R = \frac{1}{3}a\sqrt{3}$

$h = \frac{1}{2}a\sqrt{3}$

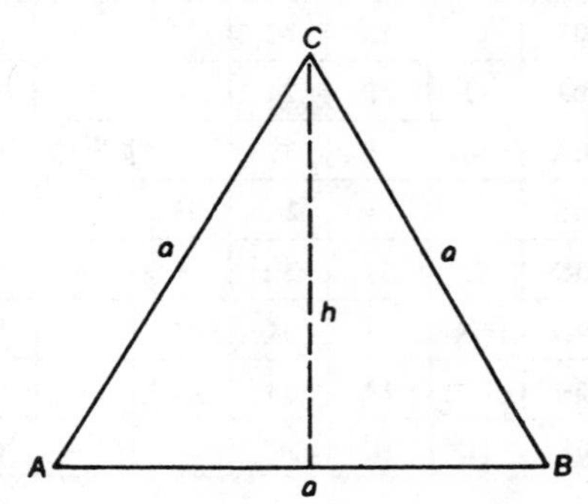

GENERAL TRIANGLE

Let $s = \frac{1}{2}(a + b + c)$, h_c = length of altitude on side c, t_c = length of bisector of angle C, m_c = length of median to side c.

$$A + B + C = 180°$$

$$c^2 = a^2 + b^2 - 2ab \cos C \quad \text{(law of cosines)}$$

$$K = \tfrac{1}{2}h_c c = \tfrac{1}{2}ab \sin C = \frac{c^2 \sin A \sin B}{2 \sin C} = rs = \frac{abc}{4R} = \sqrt{s(s-a)(s-b)(s-c)} \quad \text{(Heron's formula)}$$

$$r = c \sin \frac{A}{2} \sin \frac{B}{2} \sec \frac{C}{2} = \frac{ab \sin C}{2s} = (s - c) \tan \frac{C}{2} = \sqrt{\frac{(s-a)(s-b)(s-c)}{s}} = \frac{K}{s} = 4R \sin \frac{A}{2} \sin \frac{B}{2} \sin \frac{C}{2}$$

$$R = \frac{c}{2 \sin C} = \frac{abc}{4\sqrt{s(s-a)(s-b)(s-c)}} = \frac{abc}{4K}$$

$$h_c = a \sin B = b \sin A = \frac{2K}{c}$$

$$t_c = \frac{2ab}{a+b} \cos \frac{C}{2} = \sqrt{ab\left\{1 - \frac{c^2}{(a+b)^2}\right\}}$$

$$m_c = \sqrt{\frac{a^2}{2} + \frac{b^2}{2} - \frac{c^2}{4}}$$

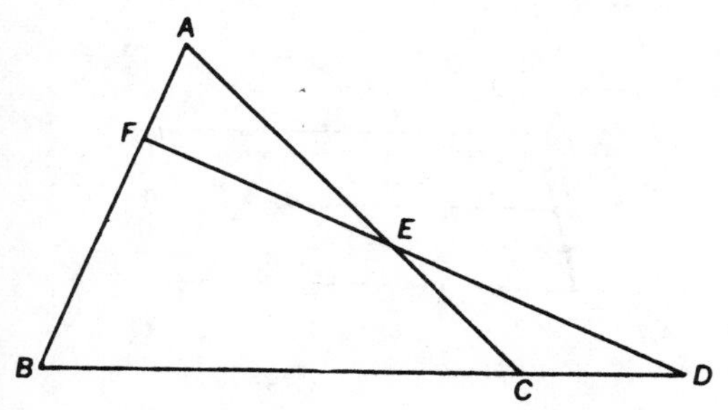

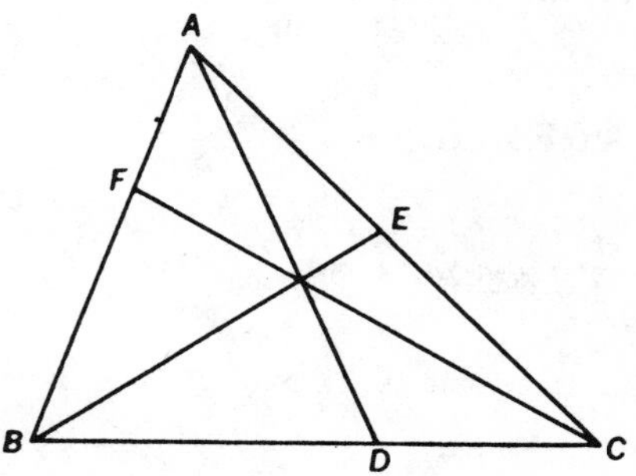

Menelaus' Theorem. A necessary and sufficient condition for points D, E, F on the respective side lines BC, CA, AB of a triangle ABC to be collinear is that

$$BD \cdot CE \cdot AF = -\, DC \cdot EA \cdot FB,$$

where all segments in the formula are directed segments.

Ceva's Theorem. A necessary and sufficient condition for AD, BE, CF, where D, E, F are points on the respective side lines BC, CA, AB of a triangle ABC, to be concurrent is that

$$BD \cdot CE \cdot AF = +\, DC \cdot EA \cdot FB,$$

where all segments in the formula are directed segments.

QUADRILATERALS

In the following: K = area, p and q are diagonals.

RECTANGLE

$A = B = C = D = 90°$
$K = ab, \quad p = \sqrt{a^2 + b^2}$

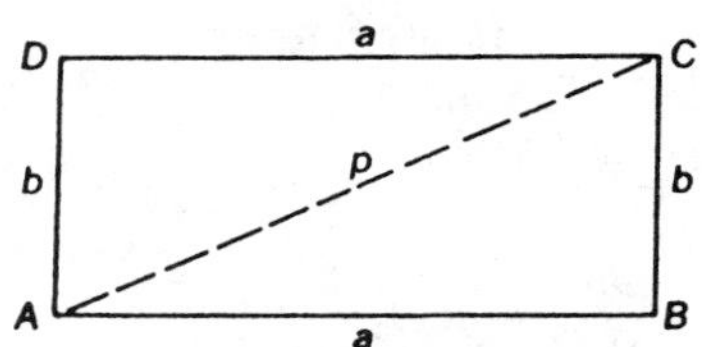

PARALLELOGRAM

$A = C, \quad B = D, \quad A + B = 180°$
$K = bh = ab \sin A = ab \sin B$
$h = a \sin A = a \sin B$
$p = \sqrt{a^2 + b^2 - 2ab \cos A}$
$q = \sqrt{a^2 + b^2 - 2ab \cos B} = \sqrt{a^2 + b^2 + 2ab \cos A}$

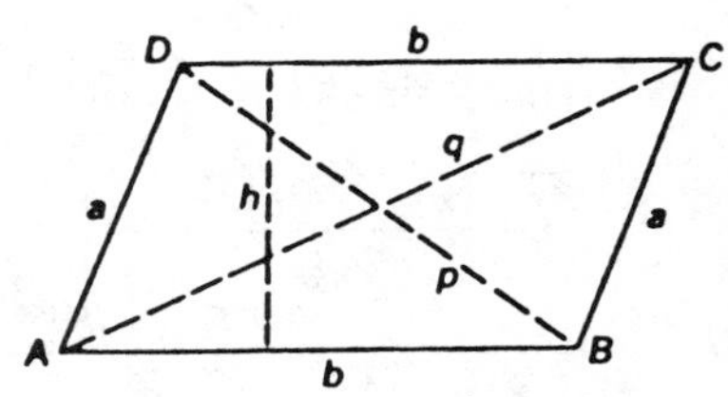

RHOMBUS

$p^2 + q^2 = 4a^2$
$K = \frac{1}{2}pq$

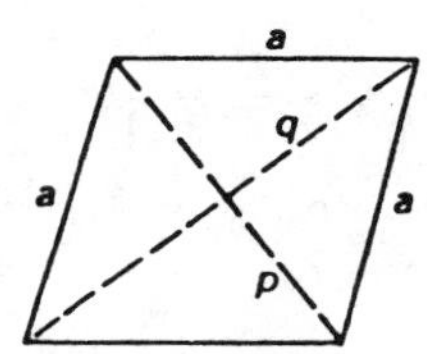

TRAPEZOID

$m = \frac{1}{2}(a + b)$
$K = \frac{1}{2}(a + b)h = mh$

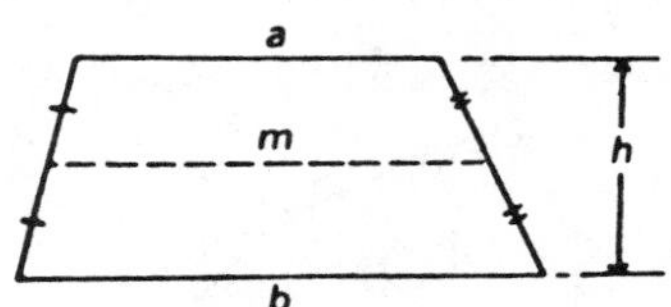

GENERAL QUADRILATERAL

Let $s = \frac{1}{2}(a + b + c + d)$.

$$K = \frac{1}{2}pq \sin \theta$$
$$= \frac{1}{4}(b^2 + d^2 - a^2 - c^2) \tan \theta$$
$$= \frac{1}{4}\sqrt{4p^2q^2 - (b^2 + d^2 - a^2 - c^2)^2}$$

(*Bretschneider's formula*)

$$= \sqrt{(s - a)(s - b)(s - c)(s - d) - abcd \cos^2\left(\frac{A + B}{2}\right)}$$

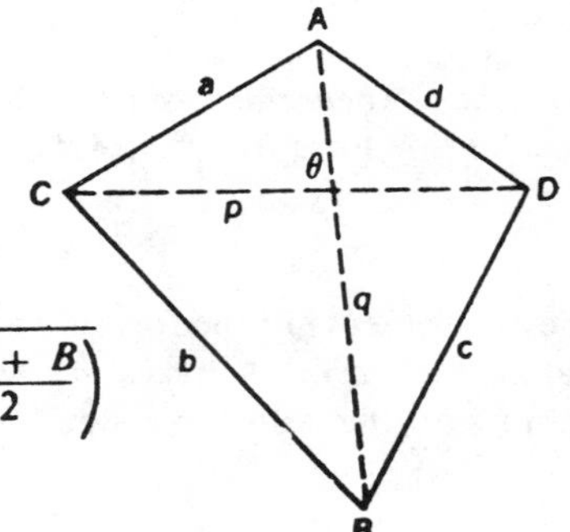

THEOREM. The diagonals of a quadrilateral with consecutive sides a, b, c, d are perpendicular if and only if $a^2 + c^2 = b^2 + d^2$.

REGULAR POLYGON OF n SIDES EACH OF LENGTH b

Area $= \frac{1}{4}nb^2 \cot \frac{\pi}{n} = \frac{1}{4}nb^2 \frac{\cos(\pi/n)}{\sin(\pi/n)}$

Perimeter $= nb$

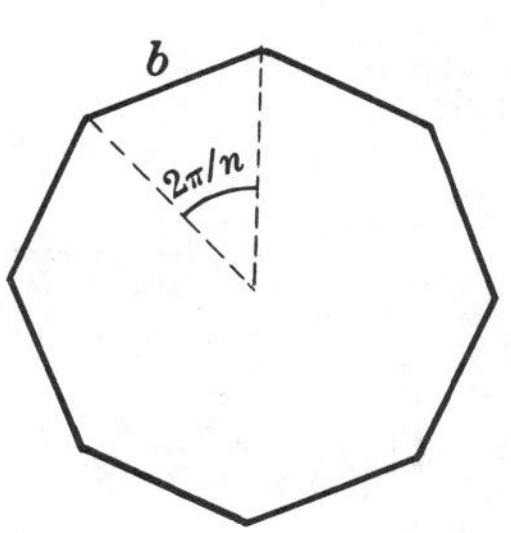

CIRCLE OF RADIUS r

Area $= \pi r^2$

Perimeter $= 2\pi r$

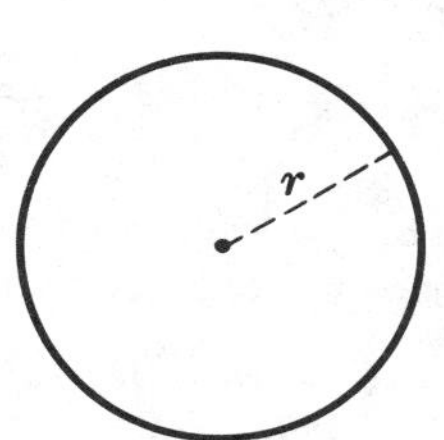

REGULAR POLYGON OF n SIDES INSCRIBED IN A CIRCLE OF RADIUS r

Area $= \frac{1}{2}nr^2 \sin \frac{2\pi}{n} = \frac{1}{2}nr^2 \sin \frac{360°}{n}$

Perimeter $= 2nr \sin \frac{\pi}{n} = 2nr \sin \frac{180°}{n}$

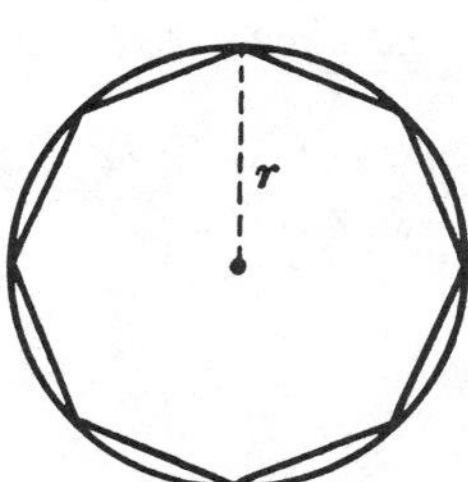

REGULAR POLYGON OF n SIDES CIRCUMSCRIBING A CIRCLE OF RADIUS r

Area $= nr^2 \tan \frac{\pi}{n} = nr^2 \tan \frac{180°}{n}$

Perimeter $= 2nr \tan \frac{\pi}{n} = 2nr \tan \frac{180°}{n}$

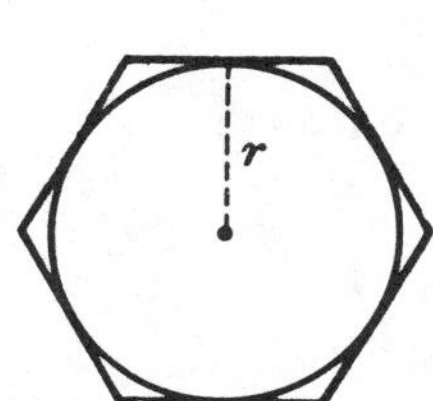

CYCLIC QUADRILATERAL

Let R = radius of the circumscribed circle.

$A + C = B + D = 180°$

(*Brahmagupta's formula*)

$$K = \sqrt{(s-a)(s-b)(s-c)(s-d)} = \frac{\sqrt{(ac+bd)(ad+bc)(ab+cd)}}{4R}$$

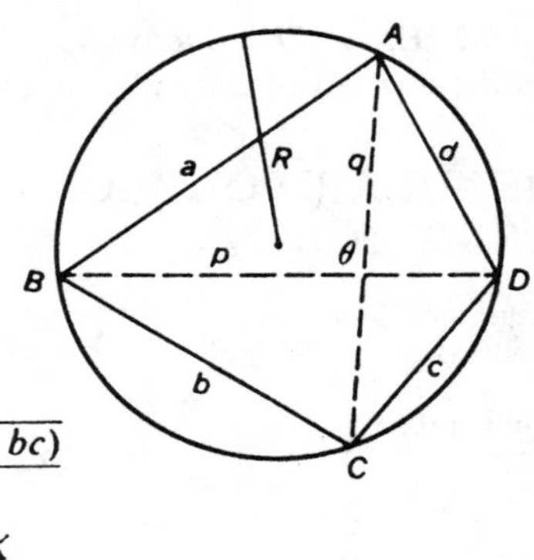

$$p = \sqrt{\frac{(ac+bd)(ab+cd)}{ad+bc}}, \quad q = \sqrt{\frac{(ac+bd)(ad+bc)}{ab+cd}}$$

$$R = \tfrac{1}{2}\sqrt{\frac{(ac+bd)(ad+bc)(ab+cd)}{(s-a)(s-b)(s-c)(s-d)}}, \quad \sin\theta = \frac{2K}{ac+bd}$$

Ptolemy's Theorem. **A convex quadrilateral with consecutive sides a, b, c, d and diagonals p and q is cyclic if and only if $ac + bd = pq$.**

CYCLIC-INSCRIPTABLE QUADRILATERAL

Let r = radius of the inscribed circle, R = radius of the circumscribed circle, m = distance between the centers of the inscribed and the circumscribed circles.

$$A + C = B + D = 180^\circ$$

$$a + c = b + d$$

$$K = \sqrt{abcd}$$

$$\frac{1}{(R-m)^2} + \frac{1}{(R+m)^2} = \frac{1}{r^2}$$

$$r = \frac{\sqrt{abcd}}{s}$$

$$R = \tfrac{1}{2}\sqrt{\frac{(ac+bd)(ad+bc)(ab+cd)}{abcd}}$$

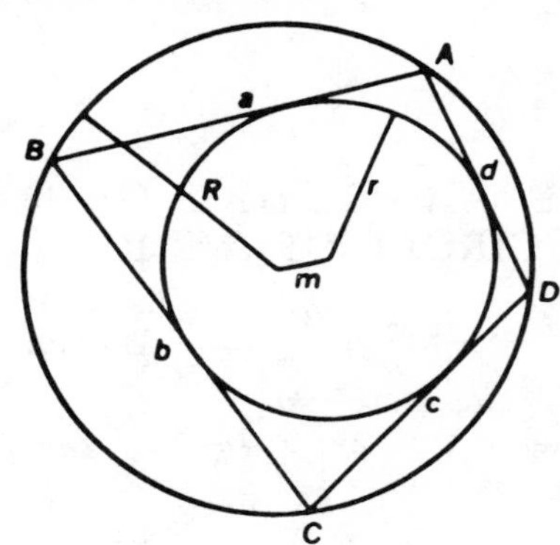

SECTOR OF CIRCLE OF RADIUS r

Area $= \frac{1}{2}r^2\theta$ $\quad$ [θ in radians]

Arc length $s = r\theta$

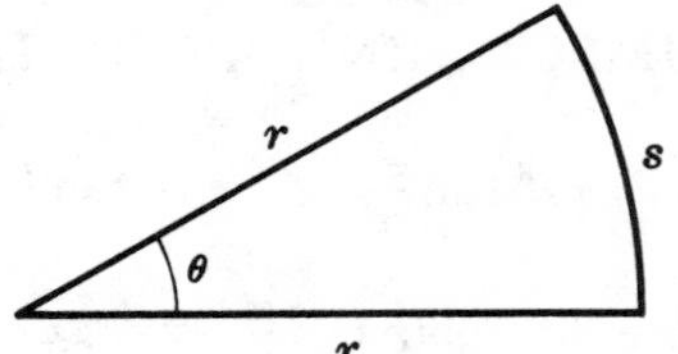

RADIUS OF CIRCLE INSCRIBED IN A TRIANGLE OF SIDES a, b, c

$$r = \frac{\sqrt{s(s-a)(s-b)(s-c)}}{s}$$

where $s = \frac{1}{2}(a+b+c)$ = semiperimeter

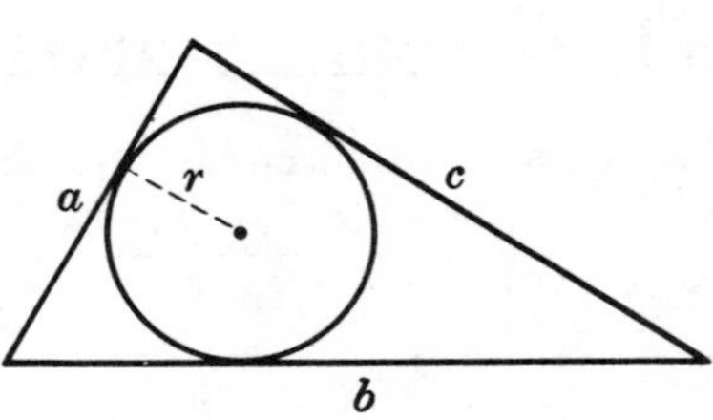

RADIUS OF CIRCLE CIRCUMSCRIBING A TRIANGLE OF SIDES a, b, c

$$R = \frac{abc}{4\sqrt{s(s-a)(s-b)(s-c)}}$$

where $s = \frac{1}{2}(a+b+c)$ = semiperimeter

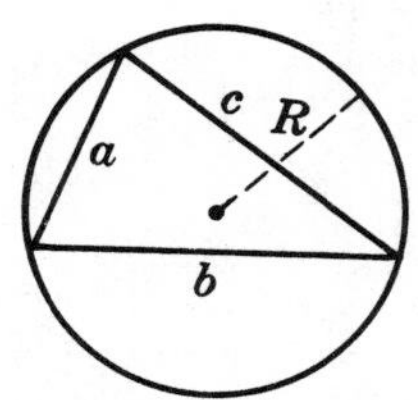

SEGMENT OF CIRCLE OF RADIUS r

Area of shaded part $= \frac{1}{2}r^2(\theta - \sin\theta)$

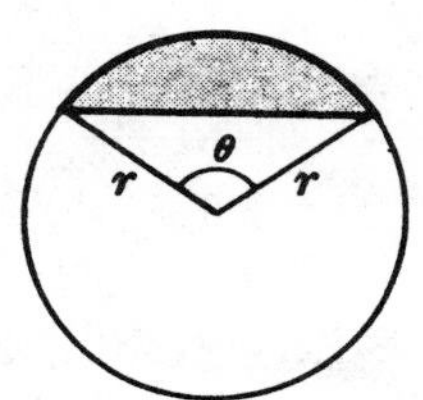

ELLIPSE OF SEMI-MAJOR AXIS a AND SEMI-MINOR AXIS b

Area $= \pi ab$

$$\text{Perimeter} = 4a\int_0^{\pi/2} \sqrt{1 - k^2 \sin^2\theta}\, d\theta$$

$$= 2\pi\sqrt{\tfrac{1}{2}(a^2+b^2)} \qquad \text{[approximately]}$$

where $k = \sqrt{a^2 - b^2}/a$.

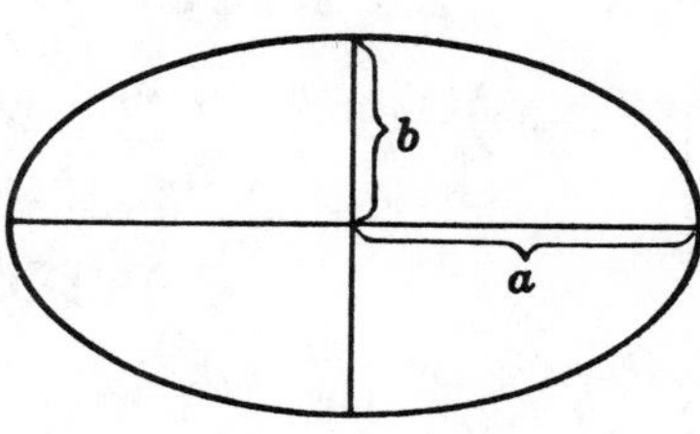

SEGMENT OF A PARABOLA

Area $= \frac{2}{3}ab$

$$\text{Arc length } ABC = \tfrac{1}{2}\sqrt{b^2+16a^2} + \frac{b^2}{8a}\ln\left(\frac{4a+\sqrt{b^2+16a^2}}{b}\right)$$

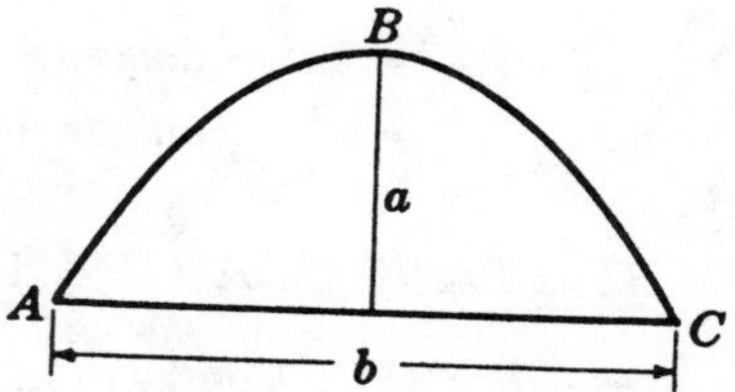

PLANAR AREAS BY APPROXIMATION

Divide the planar area K into n strips by equidistant parallel chords of lengths $y_0, y_1, y_2, \ldots, y_n$ (where y_0 and/or y_n may be zero), and let h denote the common distance between the chords.

Then, approximately:

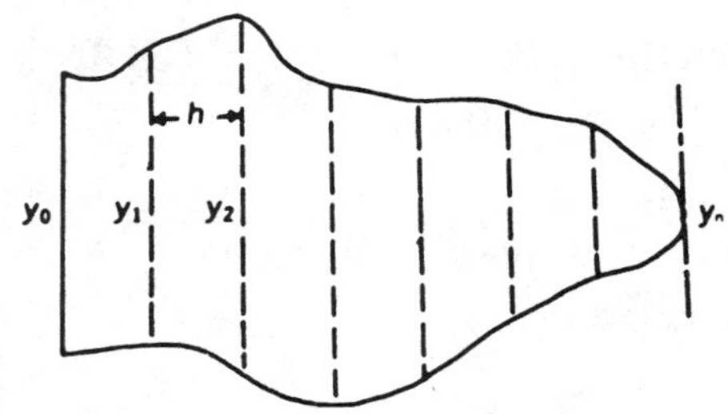

TRAPEZOIDAL RULE

$$K = h(\tfrac{1}{2}y_0 + y_1 + y_2 + \cdots + y_{n-1} + \tfrac{1}{2}y_n)$$

DURAND'S RULE

$$K = h(\tfrac{4}{10}y_0 + \tfrac{11}{10}y_1 + y_2 + y_3 + \cdots + y_{n-2} + \tfrac{11}{10}y_{n-1} + \tfrac{4}{10}y_n)$$

SIMPSON'S RULE (n even)

$$K = \tfrac{1}{3}h(y_0 + 4y_1 + 2y_2 + 4y_3 + 2y_4 + \cdots + 2y_{n-2} + 4y_{n-1} + y_n)$$

WEDDLE'S RULE (n = 6)

$$K = \tfrac{3}{10}h(y_0 + 5y_1 + y_2 + 6y_3 + y_4 + 5y_5 + y_6)$$

SOLIDS BOUNDED BY PLANES

In the following: S = lateral surface, T = total surface, V = volume.

CUBE

Let a = length of each edge.

$T = 6a^2$, diagonal of face $= a\sqrt{2}$

$V = a^3$, diagonal of cube $= a\sqrt{3}$

RECTANGULAR PARALLELEPIPED (OR BOX)

Let a, b, c be the lengths of its edges.

$T = 2(ab + bc + ca)$, $V = abc$

diagonal $= \sqrt{a^2 + b^2 + c^2}$

PRISM

S = (perimeter of right section) × (lateral edge)

V = (area of right section) × (lateral edge)

= (area of base) × (altitude)

TRUNCATED TRIANGULAR PRISM

V = (area of right section) × $\frac{1}{3}$(sum of the three lateral edges)

PYRAMID

S of regular pyramid = $\frac{1}{2}$(perimeter of base) × (slant height)
$V = \frac{1}{3}$(area of base) × (altitude)

FRUSTUM OF PYRAMID

Let B_1 = area of lower base, B_2 = area of upper base, h = altitude.

S of regular figure = $\frac{1}{2}$(sum of perimeters of bases) × (slant height)

$$V = \frac{1}{3}h(B_1 + B_2 + \sqrt{B_1 B_2})$$

PRISMATOID

A *prismatoid* is a polyhedron having for bases two polygons in parallel planes, and for lateral faces triangles or trapezoids with one side lying in one base, and the opposite vertex or side lying in the other base, of the polyhedron. Let B_1 = area of lower base, M = area of midsection, B_2 = area of upper base, h = altitude.

$$V = \frac{1}{6}h(B_1 + 4M + B_2) \quad \text{(the } \textit{prismoidal formula}\text{)}$$

Note: Since cubes, rectangular parallelepipeds, prisms, pyramids, and frustums of pyramids are all examples of prismatoids, the formula for the volume of a prismatoid subsumes most of the above volume formulae.

REGULAR POLYHEDRA

Let v = number of vertices, e = number of edges, f = number of faces, α = each dihedral angle, a = length of each edge, r = radius of the inscribed sphere, R = radius of the circumscribed sphere, A = area of each face, T = total area, V = volume.

$v - e + f = 2$ (the *Euler-Descartes formula*—actually holds for *any* convex polyhedron)

$$T = fA$$
$$V = \frac{1}{3}rfA = \frac{1}{3}rT$$

Name	Nature of Surface	T	V
Tetrahedron	4 equilateral triangles	$1.73205a^2$	$0.11785a^3$
Hexahedron (cube)	6 squares	$6.00000a^2$	$1.00000a^3$
Octahedron	8 equilateral triangles	$3.46410a^2$	$0.47140a^3$
Dodecahedron	12 regular pentagons	$20.64573a^2$	$7.66312a^3$
Icosahedron	20 equilateral triangles	$8.66025a^2$	$2.18169a^3$

Name	v	e	f	α	a	r
Tetrahedron	4	6	4	70° 32′	$1.633R$	$0.333R$
Hexahedron	8	12	6	90°	$1.155R$	$0.577R$
Octahedron	6	12	8	109° 28′	$1.414R$	$0.577R$
Dodecahedron	20	30	12	116° 34′	$0.714R$	$0.795R$
Icosahedron	12	30	20	138° 11′	$1.051R$	$0.795R$

Name	A	r	R	V
Tetrahedron	$\frac{1}{4}a^2\sqrt{3}$	$\frac{1}{12}a\sqrt{6}$	$\frac{1}{4}a\sqrt{6}$	$\frac{1}{12}a^3\sqrt{2}$
Hexahedron	a^2	$\frac{1}{2}a$	$\frac{1}{2}a\sqrt{3}$	a^3
Octahedron	$\frac{1}{4}a^2\sqrt{3}$	$\frac{1}{6}a\sqrt{6}$	$\frac{1}{2}a\sqrt{2}$	$\frac{1}{3}a^3\sqrt{2}$
Dodecahedron	$\frac{1}{4}a^2\sqrt{25+10\sqrt{5}}$	$\frac{1}{20}a\sqrt{250+110\sqrt{5}}$	$\frac{1}{4}a(\sqrt{15}+\sqrt{3})$	$\frac{1}{4}a^3(15+7\sqrt{5})$
Icosahedron	$\frac{1}{4}a^2\sqrt{3}$	$\frac{1}{12}a\sqrt{42+18\sqrt{5}}$	$\frac{1}{4}a\sqrt{10+2\sqrt{5}}$	$\frac{5}{12}a^3(3+\sqrt{5})$

SPHERE OF RADIUS r

Volume $= \frac{4}{3}\pi r^3$

Surface area $= 4\pi r^2$

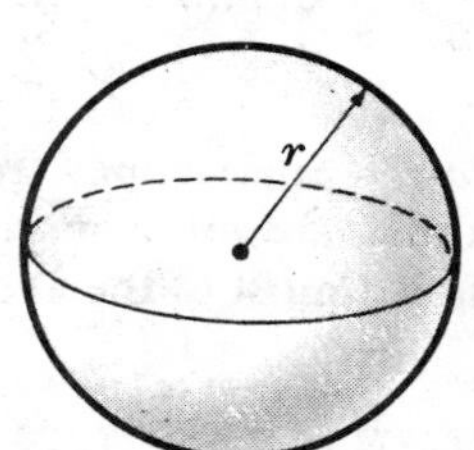

RIGHT CIRCULAR CYLINDER OF RADIUS r AND HEIGHT h

4.1 Volume $= \pi r^2 h$

4.2 Lateral surface area $= 2\pi rh$

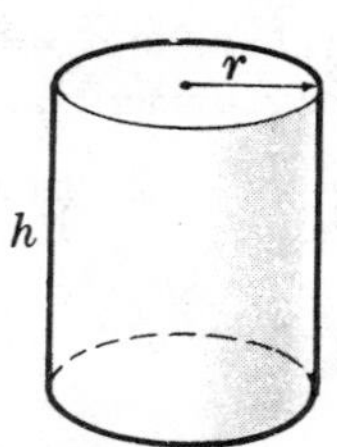

CIRCULAR CYLINDER OF RADIUS r AND SLANT HEIGHT ℓ

4.3 Volume $= \pi r^2 h = \pi r^2 \ell \sin\theta$

4.4 Lateral surface area $= 2\pi r\ell = \dfrac{2\pi rh}{\sin\theta} = 2\pi rh \csc\theta$

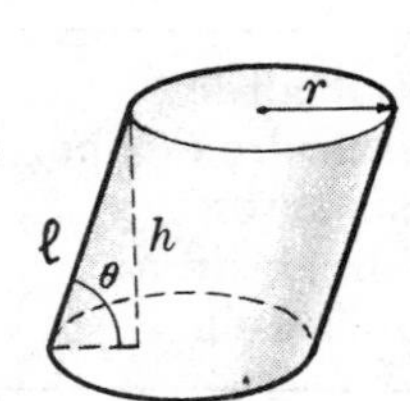

CYLINDER OF CROSS-SECTIONAL AREA A AND SLANT HEIGHT ℓ

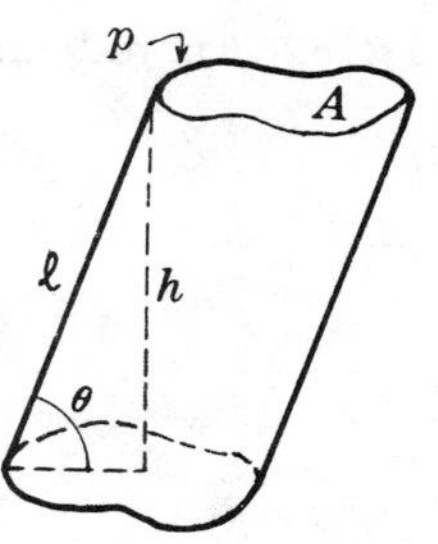

Volume $= Ah = A\ell \sin\theta$

Lateral surface area $= p\ell = \dfrac{ph}{\sin\theta} = ph \csc\theta$

Note that formulas 4.1 to 4.4 are special cases.

RIGHT CIRCULAR CONE OF RADIUS r AND HEIGHT h

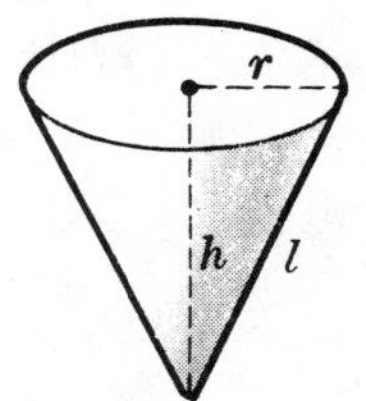

Volume $= \frac{1}{3}\pi r^2 h$

Lateral surface area $= \pi r\sqrt{r^2 + h^2} = \pi r l$

SPHERICAL CAP OF RADIUS r AND HEIGHT h

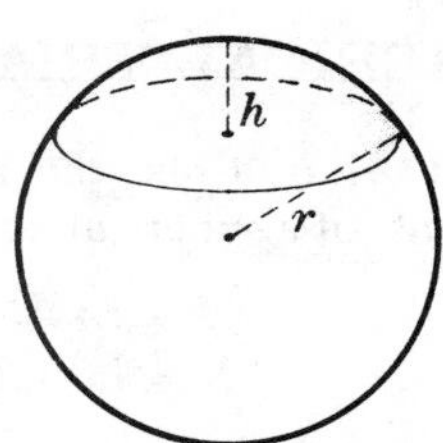

Volume (shaded in figure) $= \frac{1}{3}\pi h^2(3r - h)$

Surface area $= 2\pi r h$

FRUSTUM OF RIGHT CIRCULAR CONE OF RADII a, b AND HEIGHT h

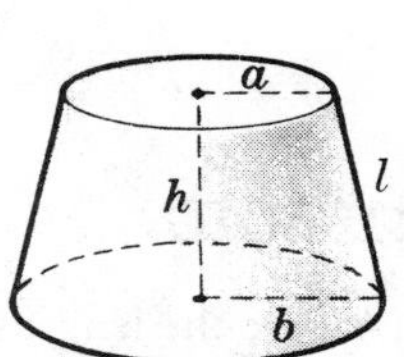

Volume $= \frac{1}{3}\pi h(a^2 + ab + b^2)$

Lateral surface area $= \pi(a + b)\sqrt{h^2 + (b - a)^2}$
$= \pi(a + b)l$

ZONE AND SEGMENT OF TWO BASES

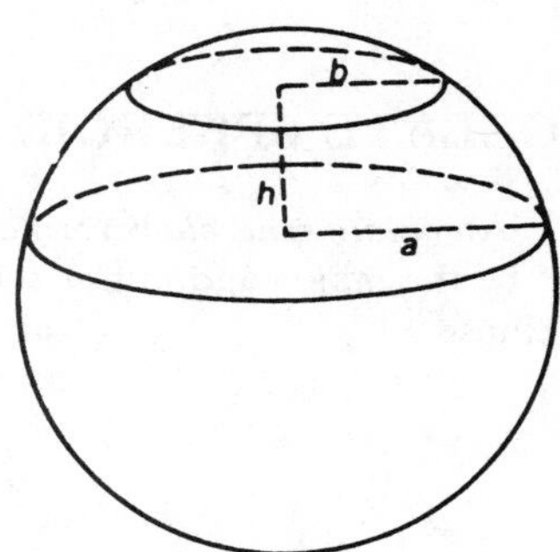

$S = 2\pi Rh = \pi Dh$

$V = \frac{1}{6}\pi h(3a^2 + 3b^2 + h^2)$

LUNE

$S = 2R^2\theta$, θ in radians

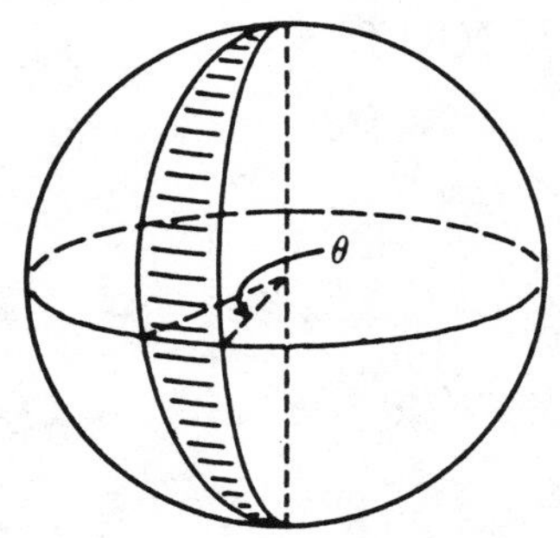

SPHERICAL SECTOR

$V = \frac{2}{3}\pi R^2 h = \frac{1}{6}\pi D^2 h$

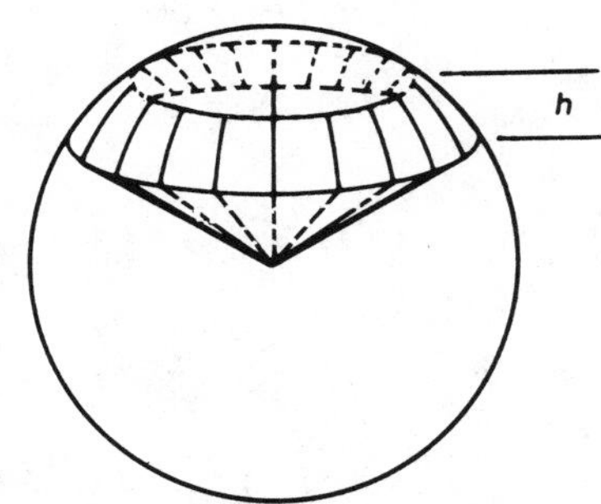

SPHERICAL TRIANGLE AND POLYGON

Let A, B, C be the angles, in radians, of the triangle; let θ = sum of angles, in radians, of a spherical polygon of n sides.

$$S = (A + B + C - \pi)R^2$$
$$S = [\theta - (n - 2)\pi]R^2$$

SPHEROIDS

ELLIPSOID

Let a, b, c be the lengths of the semiaxes.

$$V = \frac{4}{3}\pi abc$$

OBLATE SPHEROID

An *oblate spheroid* is formed by the rotation of an ellipse about its minor axis. Let a and b be the major and minor semiaxes, respectively, and ϵ the eccentricity, of the revolving ellipse.

$$S = 2\pi a^2 + \pi \frac{b^2}{\epsilon} \log_e \frac{1 + \epsilon}{1 - \epsilon}$$
$$V = \frac{4}{3}\pi a^2 b$$

PROLATE SPHEROID

A *prolate spheroid* is formed by the rotation of an ellipse about its major axis. Let a and b be the major and minor semiaxes, respectively, and ϵ the eccentricity, of the revolving ellipse.

$$S = 2\pi b^2 + 2\pi \frac{ab}{\epsilon} \sin^{-1} \epsilon$$

$$V = \tfrac{4}{3}\pi ab^2$$

CIRCULAR TORUS

A *circular torus* is formed by the rotation of a circle about an axis in the plane of the circle and not cutting the circle. Let r be the radius of the revolving circle and let R be the distance of its center from the axis of rotation.

$$S = 4\pi^2 Rr$$
$$V = 2\pi^2 Rr^2$$

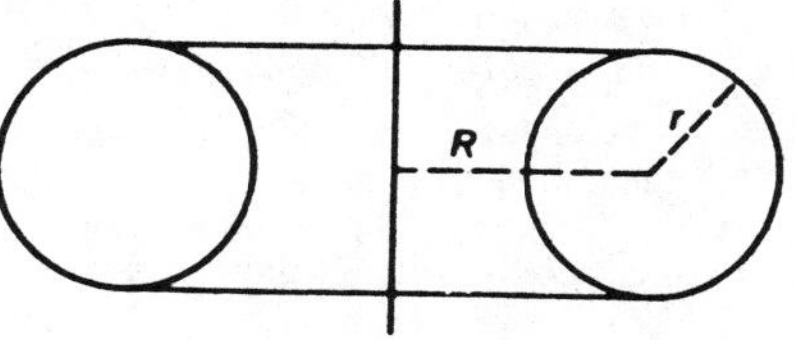

RECIPROCALS, CIRCUMFERENCE AND AREA OF CIRCLES

As a matter of convenience, the values of 1000 × (1/*n*) are given in the table. To obtain the actual value of the reciprocal, shift the decimal point three places to the left.

Circumferences and areas of circles are given for the values of *n* as the diameter.

n = dia	$1000\frac{1}{n}$	Circumference πn	Area $\frac{\pi n^2}{4}$	n = dia	$1000\frac{1}{n}$	Circumference πn	Area $\frac{\pi n^2}{4}$
0	∞	0.000000	.0000000	**50**	20.00000	157.0796	1963.495
1	1000.000	3.141593	.7853982	51	19.60784	160.2212	2042.821
2	500.0000	6.283185	3.141593	52	19.23077	163.3628	2123.717
3	333.3333	9.424778	7.068583	53	18.86792	166.5044	2206.183
4	250.0000	12.56637	12.56637	54	18.51852	169.6460	2290.221
5	200.0000	15.70796	19.63495	55	18.18182	172.7876	2375.829
6	166.6667	18.84956	28.27433	56	17.85714	175.9292	2463.009
7	142.8571	21.99115	38.48451	57	17.54386	179.0708	2551.759
8	125.0000	25.13274	50.26548	58	17.24138	182.2124	2642.079
9	111.1111	28.27433	63.61725	59	16.94915	185.3540	2733.971
10	100.0000	31.41593	78.53982	**60**	16.66667	188.4956	2827.433
11	90.90909	34.55752	95.03318	61	16.39344	191.6372	2922.467
12	83.33333	37.69911	113.0973	62	16.12903	194.7787	3019.071
13	76.92308	40.84070	132.7323	63	15.87302	197.9203	3117.245
14	71.42857	43.98230	153.9380	64	15.62500	201.0619	3216.991
15	66.66667	47.12389	176.7146	65	15.38462	204.2035	3318.307
16	62.50000	50.26548	201.0619	66	15.15152	207.3451	3421.194
17	58.82353	53.40708	226.9801	67	14.92537	210.4867	3525.652
18	55.55556	56.54867	254.4690	68	14.70588	213.6283	3631.681
19	52.63158	59.69026	283.5287	69	14.49275	216.7699	3739.281
20	50.00000	62.83185	314.1593	**70**	14.28571	219.9115	3848.451
21	47.61905	65.97345	346.3606	71	14.08451	223.0531	3959.192
22	45.45455	69.11504	380.1327	72	13.88889	226.1947	4071.504
23	43.47826	72.25663	415.4756	73	13.69863	229.3363	4185.387
24	41.66667	75.39822	452.3893	74	13.51351	232.4779	4300.840
25	40.00000	78.53982	490.8739	75	13.33333	235.6194	4417.865
26	38.46154	81.68141	530.9292	76	13.15789	238.7610	4536.460
27	37.03704	84.82300	572.5553	77	12.98701	241.9026	4656.626
28	35.71429	87.96459	615.7522	78	12.82051	245.0442	4778.362
29	34.48276	91.10619	660.5199	79	12.65823	248.1858	4901.670
30	33.33333	94.24778	706.8583	**80**	12.50000	251.3274	5026.548
31	32.25806	97.38937	754.7676	81	12.34568	254.4690	5152.997
32	31.25000	100.5310	804.2477	82	12.19512	257.6106	5281.017
33	30.30303	103.6726	855.2986	83	12.04819	260.7522	5410.608
34	29.41176	106.8142	907.9203	84	11.90476	263.8938	5541.769
35	28.57143	109.9557	962.1128	85	11.76471	267.0354	5674.502
36	27.77778	113.0973	1017.876	86	11.62791	270.1770	5808.805
37	27.02703	116.2389	1075.210	87	11.49425	273.3186	5944.679
38	26.31579	119.3805	1134.115	88	11.36364	276.4602	6082.123
39	25.64103	122.5221	1194.591	89	11.23596	279.6017	6221.139
40	25.00000	125.6637	1256.637	**90**	11.11111	282.7433	6361.725
41	24.39024	128.8053	1320.254	91	10.98901	285.8849	6503.882
42	23.80952	131.9469	1385.442	92	10.86957	289.0265	6647.610
43	23.25581	135.0885	1452.201	93	10.75269	292.1681	6792.909
44	22.72727	138.2301	1520.531	94	10.63830	295.3097	6939.778
45	22.22222	141.3717	1590.431	95	10.52632	298.4513	7088.218
46	21.73913	144.5133	1661.903	96	10.41667	301.5929	7238.229
47	21.27660	147.6549	1734.945	97	10.30928	304.7345	7389.811
48	20.83333	150.7964	1809.557	98	10.20408	307.8761	7542.964
49	20.40816	153.9380	1885.741	99	10.10101	311.0177	7697.687
50	20.00000	157.0796	1963.495	**100**	10.00000	314.1593	7853.982

RECIPROCALS, CIRCUMFERENCE AND AREA OF CIRCLES

n = dia	$1000\frac{1}{n}$	Circum-ference πn	Area $\frac{\pi n^2}{4}$	n = dia	$1000\frac{1}{n}$	Circum-ference πn	Area $\frac{\pi n^2}{4}$
100	10.00000	314.1593	7853.982	**150**	6.666 667	471.2389	17671.46
101	9.900 990	317.3009	8011.847	151	6.622 517	474.3805	17907.86
102	9.803 922	320.4425	8171.282	152	6.578 947	477.5221	18145.84
103	9.708 738	323.5840	8332.289	153	6.535 948	480.6637	18385.39
104	9.615 385	326.7256	8494.867	154	6.493 506	483.8053	18626.50
105	9.523 810	329.8672	8659.015	155	6.451 613	486.9469	18869.19
106	9.433 962	333.0088	8824.734	156	6.410 256	490.0885	19113.45
107	9.345 794	336.1504	8992.024	157	6.369 427	493.2300	19359.28
108	9.259 259	339.2920	9160.884	158	6.329 114	496.3716	19606.68
109	9.174 312	342.4336	9331.316	159	6.289 308	499.5132	19855.65
110	9.090 909	345.5752	9503.318	**160**	6.250 000	502.6548	20106.19
111	9.009 009	348.7168	9676.891	161	6.211 180	505.7964	20358.31
112	8.928 571	351.8584	9852.035	162	6.172 840	508.9380	20611.99
113	8.849 558	355.0000	10028.75	163	6.134 969	512.0796	20867.24
114	8.771 930	358.1416	10207.03	164	6.097 561	515.2212	21124.07
115	8.695 652	361.2832	10386.89	165	6.060 606	518.3628	21382.46
116	8.620 690	364.4247	10568.32	166	6.024 096	521.5044	21642.43
117	8.547 009	367.5663	10751.32	167	5.988 024	524.6460	21903.97
118	8.474 576	370.7079	10935.88	168	5.952 381	527.7876	22167.08
119	8.403 361	373.8495	11122.02	169	5.917 160	530.9292	22431.76
120	8.333 333	376.9911	11309.73	**170**	5.882 353	534.0708	22698.01
121	8.264 463	380.1327	11499.01	171	5.847 953	537.2123	22965.83
122	8.196 721	383.2743	11689.87	172	5.813 953	540.3539	23235.22
123	8.130 081	386.4159	11882.29	173	5.780 347	543.4955	23506.18
124	8.064 516	389.5575	12076.28	174	5.747 126	546.6371	23778.71
125	8.000 000	392.6991	12271.85	175	5.714 286	549.7787	24052.82
126	7.936 508	395.8407	12468.98	176	5.681 818	552.9203	24328.49
127	7.874 016	398.9823	12667.69	177	5.649 718	556.0619	24605.74
128	7.812 500	402.1239	12867.96	178	5.617 978	559.2035	24884.56
129	7.751 938	405.2655	13069.81	179	5.586 592	562.3451	25164.94
130	7.692 308	408.4070	13273.23	**180**	5.555 556	565.4867	25446.90
131	7.633 588	411.5486	13478.22	181	5.524 862	568.6283	25730.43
132	7.575 758	414.6902	13684.78	182	5.494 505	571.7699	26015.53
133	7.518 797	417.8318	13892.91	183	5.464 481	574.9115	26302.20
134	7.462 687	420.9734	14102.61	184	5.434 783	578.0530	26590.44
135	7.407 407	424.1150	14313.88	185	5.405 405	581.1946	26880.25
136	7.352 941	427.2566	14526.72	186	5.376 344	584.3362	27171.63
137	7.299 270	430.3982	14741.14	187	5.347 594	587.4778	27464.59
138	7.246 377	433.5398	14957.12	188	5.319 149	590.6194	27759.11
139	7.194 245	436.6814	15174.68	189	5.291 005	593.7610	28055.21
140	7.142 857	439.8230	15393.80	**190**	5.263 158	596.9026	28352.87
141	7.092 199	442.9646	15614.50	191	5.235 602	600.0442	28652.11
142	7.042 254	446.1062	15836.77	192	5.208 333	603.1858	28952.92
143	6.993 007	449.2477	16060.61	193	5.181 347	606.3274	29255.30
144	6.944 444	452.3893	16286.02	194	5.154 639	609.4690	29559.25
145	6.896 552	455.5309	16513.00	195	5.128 205	612.6106	29864.77
146	6.849 315	458.6725	16741.55	196	5.102 041	615.7522	30171.86
147	6.802 721	461.8141	16971.67	197	5.076 142	618.8938	30480.52
148	6.756 757	464.9557	17203.36	198	5.050 505	622.0353	30790.75
149	6.711 409	468.0973	17436.62	199	5.025 126	625.1769	31102.55
150	6.666 667	471.2389	17671.46	**200**	5.000 000	628.3185	31415.93

RECIPROCALS, CIRCUMFERENCE AND AREA OF CIRCLES

$n =$ dia	$1000\frac{1}{n}$	Circumference πn	Area $\frac{\pi n^2}{4}$	$n =$ dia	$1000\frac{1}{n}$	Circumference πn	Area $\frac{\pi n^2}{4}$
200	5.000 000	628.3185	31415.93	**250**	4.000 000	785.3982	49087.39
201	4.975 124	631.4601	31730.87	251	3.984 064	788.5398	49480.87
202	4.950 495	634.6017	32047.39	252	3.968 254	791.6813	49875.92
203	4.926 108	637.7433	32365.47	253	3.952 569	794.8229	50272.55
204	4.901 961	640.8849	32685.13	254	3.937 008	797.9645	50670.75
205	4.878 049	644.0265	33006.36	255	3.921 569	801.1061	51070.52
206	4.854 369	647.1681	33329.16	256	3.906 250	804.2477	51471.85
207	4.830 918	650.3097	33653.53	257	3.891 051	807.3893	51874.76
208	4.807 692	653.4513	33979.47	258	3.875 969	810.5309	52279.24
209	4.784 689	656.5929	34306.98	259	3.861 004	813.6725	52685.29
210	4.761 905	659.7345	34636.06	**260**	3.846 154	816.8141	53092.92
211	4.739 336	662.8760	34966.71	261	3.831 418	819.9557	53502.11
212	4.716 981	666.0176	35298.94	262	3.816 794	823.0973	53912.87
213	4.694 836	669.1592	35632.73	263	3.802 281	826.2389	54325.21
214	4.672 897	672.3008	35968.09	264	3.787 879	829.3805	54739.11
215	4.651 163	675.4424	36305.03	265	3.773 585	832.5221	55154.59
216	4.629 630	678.5840	36643.54	266	3.759 398	835.6636	55571.63
217	4.608 295	681.7256	36983.61	267	3.745 318	838.8052	55990.25
218	4.587 156	684.8672	37325.26	268	3.731 343	841.9468	56410.44
219	4.566 210	688.0088	37668.48	269	3.717 472	845.0884	56832.20
220	4.545 455	691.1504	38013.27	**270**	3.703 704	848.2300	57255.53
221	4.524 887	694.2920	38359.63	271	3.690 037	851.3716	57680.43
222	4.504 505	697.4336	38707.56	272	3.676 471	854.5132	58106.90
223	4.484 305	700.5752	39057.07	273	3.663 004	857.6548	58534.94
224	4.464 286	703.7168	39408.14	274	3.649 635	860.7964	58964.55
225	4.444 444	706.8583	39760.78	275	3.636 364	863.9380	59395.74
226	4.424 779	709.9999	40115.00	276	3.623 188	867.0796	59828.49
227	4.405 286	713.1415	40470.78	277	3.610 108	870.2212	60262.82
228	4.385 965	716.2831	40828.14	278	3.597 122	873.3628	60698.71
229	4.366 812	719.4247	41187.07	279	3.584 229	876.5044	61136.18
230	4.347 826	722.5663	41547.56	**280**	3.571 429	879.6459	61575.22
231	4.329 004	725.7079	41909.63	281	3.558 719	882.7875	62015.82
232	4.310 345	728.8495	42273.27	282	3.546 099	885.9291	62458.00
233	4.291 845	731.9911	42638.48	283	3.533 569	889.0707	62901.75
234	4.273 504	735.1327	43005.26	284	3.521 127	892.2123	63347.07
235	4.255 319	738.2743	43373.61	285	3.508 772	895.3539	63793.97
236	4.237 288	741.4159	43743.54	286	3.496 503	898.4955	64242.43
237	4.219 409	744.5575	44115.03	287	3.484 321	901.6371	64692.46
238	4.201 681	747.6991	44488.09	288	3.472 222	904.7787	65144.07
239	4.184 100	750.8406	44862.73	289	3.460 208	907.9203	65597.24
240	4.166 667	753.9822	45238.93	**290**	3.448 276	911.0619	66051.99
241	4.149 378	757.1238	45616.71	291	3.436 426	914.2035	66508.30
242	4.132 231	760.2654	45996.06	292	3.424 658	917.3451	66966.19
243	4.115 226	763.4070	46376.98	293	3.412 969	920.4866	67425.65
244	4.098 361	766.5486	46759.47	294	3.401 361	923.6282	67886.68
245	4.081 633	769.6902	47143.52	295	3.389 831	926.7698	68349.28
246	4.065 041	772.8318	47529.16	296	3.378 378	929.9114	68813.45
247	4.048 583	775.9734	47916.36	297	3.367 003	933.0530	69279.19
248	4.032 258	779.1150	48305.13	298	3.355 705	936.1946	69746.50
249	4.016 064	782.2566	48695.47	299	3.344 482	939.3362	70215.38
250	4.000 000	785.3982	49087.39	**300**	3.333 333	942.4778	70685.83

RECIPROCALS, CIRCUMFERENCE AND AREA OF CIRCLES

n = dia	$1000\frac{1}{n}$	Circum-ference πn	Area $\frac{\pi n^2}{4}$	n = dia	$1000\frac{1}{n}$	Circum-ference πn	Area $\frac{\pi n^2}{4}$
300	3.333 333	942.4778	70685.83	**350**	2.857 143	1099.557	96211.28
301	3.322 259	945.6194	71157.86	351	2.849 003	1102.699	96761.84
302	3.311 258	948.7610	71631.45	352	2.840 909	1105.841	97313.97
303	3.300 330	951.9026	72106.62	353	2.832 861	1108.982	97867.68
304	3.289 474	955.0442	72583.36	354	2.824 859	1112.124	98422.96
305	3.278 689	958.1858	73061.66	355	2.816 901	1115.265	98979.80
306	3.267 974	961.3274	73541.54	356	2.808 989	1118.407	99538.22
307	3.257 329	964.4689	74022.99	357	2.801 120	1121.549	100 098.2
308	3.246 753	967.6105	74506.01	358	2.793 296	1124.690	100 659.8
309	3.236 246	970.7521	74990.60	359	2.785 515	1127.832	101 222.9
310	3.225 806	973.8937	75476.76	**360**	2.777 778	1130.973	101 787.6
311	3.215 434	977.0353	75964.50	361	2.770 083	1134.115	102 353.9
312	3.205 128	980.1769	76453.80	362	2.762 431	1137.257	102 921.7
313	3.194 888	983.3185	76944.67	363	2.754 821	1140.398	103 491.1
314	3.184 713	986.4601	77437.12	364	2.747 253	1143.540	104 062.1
315	3.174 603	989.6017	77931.13	365	2.739 726	1146.681	104 634.7
316	3.164 557	992.7433	78426.72	366	2.732 240	1149.823	105 208.8
317	3.154 574	995.8849	78923.88	367	2.724 796	1152.965	105 784.5
318	3.144 654	999.0265	79422.60	368	2.717 391	1156.106	106 361.8
319	3.134 796	1002.168	79922.90	369	2.710 027	1159.248	106 940.6
320	3.125 000	1005.310	80424.77	**370**	2.702 703	1162.389	107 521.0
321	3.115 265	1008.451	80928.21	371	2.695 418	1165.531	108 103.0
322	3.105 590	1011.593	81433.22	372	2.688 172	1168.672	108 686.5
323	3.095 975	1014.734	81939.80	373	2.680 965	1171.814	109 271.7
324	3.086 420	1017.876	82447.96	374	2.673 797	1174.956	109 858.4
325	3.076 923	1021.018	82957.68	375	2.666 667	1178.097	110 446.6
326	3.067 485	1024.159	83468.98	376	2.659 574	1181.239	111 036.5
327	3.058 104	1027.301	83981.84	377	2.652 520	1184.380	111 627.9
328	3.048 780	1030.442	84496.28	378	2.645 503	1187.522	112 220.8
329	3.039 514	1033.584	85012.28	379	2.638 522	1190.664	112 815.4
330	3.030 303	1036.726	85529.86	**380**	2.631 579	1193.805	113 411.5
331	3.021 148	1039.867	86049.01	381	2.624 672	1196.947	114 009.2
332	3.012 048	1043.009	86569.73	382	2.617 801	1200.088	114 608.4
333	3.003 003	1046.150	87092.02	383	2.610 966	1203.230	115 209.3
334	2.994 012	1049.292	87615.88	384	2.604 167	1206.372	115 811.7
335	2.985 075	1052.434	88141.31	385	2.597 403	1209.513	116 415.6
336	2.976 190	1055.575	88668.31	386	2.590 674	1212.655	117 021.2
337	2.967 359	1058.717	89196.88	387	2.583 979	1215.796	117 628.3
338	2.958 580	1061.858	89727.03	388	2.577 320	1218.938	118 237.0
339	2.949 853	1065.000	90258.74	389	2.570 694	1222.080	118 847.2
340	2.941 176	1068.142	90792.03	**390**	2.564 103	1225.221	119 459.1
341	2.932 551	1071.283	91326.88	391	2.557 545	1228.363	120 072.5
342	2.923 977	1074.425	91863.31	392	2.551 020	1231.504	120 687.4
343	2.915 452	1077.566	92401.31	393	2.544 529	1234.646	121 304.0
344	2.906 977	1080.708	92940.88	394	2.538 071	1237.788	121 922.1
345	2.898 551	1083.849	93482.02	395	2.531 646	1240.929	122 541.7
346	2.890 173	1086.991	94024.73	396	2.525 253	1244.071	123 163.0
347	2.881 844	1090.133	94569.01	397	2.518 892	1247.212	123 785.8
348	2.873 563	1093.274	95114.86	398	2.512 563	1250.354	124 410.2
349	2.865 330	1096.416	95662.28	399	2.506 266	1253.495	125 036.2
350	2.857 143	1099.557	96211.28	**400**	2.500 000	1256.637	125 663.7

RECIPROCALS, CIRCUMFERENCE AND AREA OF CIRCLES

n = dia	$1000\frac{1}{n}$	Circumference πn	Area $\frac{\pi n^2}{4}$	n = dia	$1000\frac{1}{n}$	Circumference πn	Area $\frac{\pi n^2}{4}$
400	2.500 000	1256.637	125 663.7	**450**	2.222 222	1413.717	159 043.1
401	2.493 766	1259.779	126 292.8	451	2.217 295	1416.858	159 750.8
402	2.487 562	1262.920	126 923.5	452	2.212 389	1420.000	160 460.0
403	2.481 390	1266.062	127 555.7	453	2.207 506	1423.141	161 170.8
404	2.475 248	1269.203	128 189.5	454	2.202 643	1426.283	161 883.1
405	2.469 136	1272.345	128 824.9	455	2.197 802	1429.425	162 597.1
406	2.463 054	1275.487	129 461.9	456	2.192 982	1432.566	163 312.6
407	2.457 002	1278.628	130 100.4	457	2.188 184	1435.708	164 029.6
408	2.450 980	1281.770	130 740.5	458	2.183 406	1438.849	164 748.3
409	2.444 988	1284.911	131 382.2	459	2.178 649	1441.991	165 468.5
410	2.439 024	1288.053	132 025.4	**460**	2.173 913	1445.133	166 190.3
411	2.433 090	1291.195	132 670.2	461	2.169 197	1448.274	166 913.6
412	2.427 184	1294.336	133 316.6	462	2.164 502	1451.416	167 638.5
413	2.421 308	1297.478	133 964.6	463	2.159 827	1454.557	168 365.0
414	2.415 459	1300.619	134 614.1	464	2.155 172	1457.699	169 093.1
415	2.409 639	1303.761	135 265.2	465	2.150 538	1460.841	169 822.7
416	2.403 846	1306.903	135 917.9	466	2.145 923	1463.982	170 553.9
417	2.398 082	1310.044	136 572.1	467	2.141 328	1467.124	171 286.7
418	2 392 344	1313.186	137 227.9	468	2.136 752	1470.265	172 021.0
419	2.386 635	1316.327	137 885.3	469	2.132 196	1473.407	172 757.0
420	2.380 952	1319.469	138 544.2	**470**	2.127 660	1476.549	173 494.5
421	2.375 297	1322.611	139 204.8	471	2.123 142	1479.690	174 233.5
422	2.369 668	1325.752	139 866.8	472	2.118 644	1482.832	174 974.1
423	2.364 066	1328.894	140 530.5	473	2.114 165	1485.973	175 716.3
424	2.358 491	1332.035	141 195.7	474	2.109 705	1489.115	176 460.1
425	2.352 941	1335.177	141 862.5	475	2.105 263	1492.257	177 205.5
426	2.347 418	1338.318	142 530.9	476	2.100 840	1495.398	177 952.4
427	2.341 920	1341.460	143 200.9	477	2.096 436	1498.540	178 700.9
428	2.336 449	1344.602	143 872.4	478	2.092 050	1501.681	179 450.9
429	2.331 002	1347.743	144 545.5	479	2.087 683	1504.823	180 202.5
430	2.325 581	1350.885	145 220.1	**480**	2.083 333	1507.964	180 955.7
431	2.320 186	1354.026	145 896.3	481	2.079 002	1511.106	181 710.5
432	2.314 815	1357.168	146 574.1	482	2.074 689	1514.248	182 466.8
433	2.309 469	1360.310	147 253.5	483	2.070 393	1517.389	183 224.8
434	2.304 147	1363.451	147 934.5	484	2.066 116	1520.531	183 984.2
435	2.298 851	1366.593	148 617.0	485	2.061 856	1523.672	184 745.3
436	2.293 578	1369.734	149 301.0	486	2.057 613	1526.814	185 507.9
437	2.288 330	1372.876	149 986.7	487	2.053 388	1529.956	186 272.1
438	2.283 105	1376.018	150 673.9	488	2.049 180	1533.097	187 037.9
439	2.277 904	1379.159	151 362.7	489	2.044 990	1536.239	187 805.2
440	2.272 727	1382.301	152 053.1	**490**	2.040 816	1539.380	188 574.1
441	2.267 574	1385.442	152 745.0	491	2.036 660	1542.522	189 344.6
442	2.262 443	1388.584	153 438.5	492	2.032 520	1545.664	190 116.6
443	2.257 336	1391.726	154 133.6	493	2.028 398	1548.805	190 890.2
444	2.252 252	1394.867	154 830.3	494	2.024 291	1551.947	191 665.4
445	2.247 191	1398.009	155 528.5	495	2.020 202	1555.088	192 442.2
446	2.242 152	1401.150	156 228.3	496	2.016 129	1558.230	193 220.5
447	2.237 136	1404.292	156 929.6	497	2.012 072	1561.372	194 000.4
448	2.232 143	1407.434	157 632.6	498	2.008 032	1564.513	194 781.9
449	2.227 171	1410.575	158 337.1	499	2.004 008	1567.655	195 564.9
450	2.222 222	1413.717	159 043.1	**500**	2.000 000	1570.796	196 349.5

RECIPROCALS, CIRCUMFERENCE AND AREA OF CIRCLES

n = dia	$1000\frac{1}{n}$	Circumference πn	Area $\frac{\pi n^2}{4}$	n = dia	$1000\frac{1}{n}$	Circumference πn	Area $\frac{\pi n^2}{4}$
500	2.000 000	1570.796	196 349.5	**550**	1.818 182	1727.876	237 582.9
501	1.996 008	1573.938	197 135.7	551	1.814 882	1731.018	238 447.7
502	1.992 032	1577.080	197 923.5	552	1.811 594	1734.159	239 314.0
503	1.988 072	1580.221	198 712.8	553	1.808 318	1737.301	240 181.8
504	1.984 127	1583.363	199 503.7	554	1.805 054	1740.442	241 051.3
505	1.980 198	1586.504	200 296.2	555	1.801 802	1743.584	241 922.3
506	1.976 285	1589.646	201 090.2	556	1.798 561	1746.726	242 794.8
507	1.972 387	1592.787	201 885.8	557	1.795 332	1749.867	243 669.0
508	1.968 504	1595.929	202 683.0	558	1.792 115	1753.009	244 544.7
509	1.964 637	1599.071	203 481.7	559	1.788 909	1756.150	245 422.0
510	1.960 784	1602.212	204 282.1	**560**	1.785 714	1759.292	246 300.9
511	1.956 947	1605.354	205 084.0	561	1.782 531	1762.433	247 181.3
512	1.953 125	1608.495	205 887.4	562	1.779 359	1765.575	248 063.3
513	1.949 318	1611.637	206 692.4	563	1.776 199	1768.717	248 946.9
514	1.945 525	1614.779	207 499.1	564	1.773 050	1771.858	249 832.0
515	1.941 748	1617.920	208 307.2	565	1.769 912	1775.000	250 718.7
516	1.937 984	1621.062	209 117.0	566	1.766 784	1778.141	251 607.0
517	1.934 236	1624.203	209 928.3	567	1.763 668	1781.283	252 496.9
518	1.930 502	1627.345	210 741.2	568	1.760 563	1784.425	253 388.3
519	1.926 782	1630.487	211 555.6	569	1.757 469	1787.566	254 281.3
520	1.923 077	1633.628	212 371.7	**570**	1.754 386	1790.708	255 175.9
521	1.919 386	1636.770	213 189.3	571	1.751 313	1793.849	256 072.0
522	1.915 709	1639.911	214 008.4	572	1.748 252	1796.991	256 969.7
523	1.912 046	1643.053	214 829.2	573	1.745 201	1800.133	257 869.0
524	1.908 397	1646.195	215 651.5	574	1.742 160	1803.274	258 769.8
525	1.904 762	1649.336	216 475.4	575	1.739 130	1806.416	259 672.3
526	1.901 141	1652.478	217 300.8	576	1.736 111	1809.557	260 576.3
527	1.897 533	1655.619	218 127.8	577	1.733 102	1812.699	261 481.8
528	1.893 939	1658.761	218 956.4	578	1.730 104	1815.841	262 389.0
529	1.890 359	1661.903	219 786.6	579	1.727 116	1818.982	263 297.7
530	1.886 792	1665.044	220 618.3	**580**	1.724 138	1822.124	264 207.9
531	1.883 239	1668.186	221 451.7	581	1.721 170	1825.265	265 119.8
532	1.879 699	1671.327	222 286.5	582	1.718 213	1828.407	266 033.2
533	1.876 173	1674.469	223 123.0	583	1.715 266	1831.549	266 948.2
534	1.872 659	1677.610	223 961.0	584	1.712 329	1834.690	267 864.8
535	1.869 159	1680.752	224 800.6	585	1.709 402	1837.832	268 782.9
536	1.865 672	1683.894	225 641.8	586	1.706 485	1840.973	269 702.6
537	1.862 197	1687.035	226 484.5	587	1.703 578	1844.115	270 623.9
538	1.858 736	1690.177	227 328.8	588	1.700 680	1847.256	271 546.7
539	1.855 288	1693.318	228 174.7	589	1.697 793	1850.398	272 471.1
540	1.851 852	1696.460	229 022.1	**590**	1.694 915	1853.540	273 397.1
541	1.848 429	1699.602	229 871.1	591	1.692 047	1856.681	274 324.7
542	1.845 018	1702.743	230 721.7	592	1.689 189	1859.823	275 253.8
543	1.841 621	1705.885	231 573.9	593	1.686 341	1862.964	276 184.5
544	1.838 235	1709.026	232 427.6	594	1.683 502	1866.106	277 116.7
545	1.834 862	1712.168	233 282.9	595	1.680 672	1869.248	278 050.6
546	1.831 502	1715.310	234 139.8	596	1.677 852	1872.389	278 986.0
547	1.828 154	1718.451	234 998.2	597	1.675 042	1875.531	279 923.0
548	1.824 818	1721.593	235 858.2	598	1.672 241	1878.672	280 861.5
549	1.821 494	1724.734	236 719.8	599	1.669 449	1881.814	281 801.6
550	1.818 182	1727.876	237.582.9	**600**	1.666 667	1884.956	282 743.3

n = dia	$1000\frac{1}{n}$	Circumference πn	Area $\frac{\pi n^2}{4}$	n = dia	$1000\frac{1}{n}$	Circumference πn	Area $\frac{\pi n^2}{4}$
600	1.666 667	1884.956	282 743.3	**650**	1.538 462	2042.035	331 830.7
601	1.663 894	1888.097	283 686.6	651	1.536 098	2045.177	332 852.5
602	1.661 130	1891.239	284 631.4	652	1.533 742	2048.318	333 875.9
603	1.658 375	1894.380	285 577.8	653	1.531 394	2051.460	334 900.8
604	1.655 629	1897.522	286 525.8	654	1.529 052	2054.602	335 927.4
605	1.652 893	1900.664	287 475.4	655	1.526 718	2057.743	336 955.4
606	1.650 165	1903.805	288 426.5	656	1.524 390	2060.885	337 985.1
607	1.647 446	1906.947	289 379.2	657	1.522 070	2064.026	339 016.3
608	1.644 737	1910.088	290 333.4	658	1.519 757	2067.168	340 049.1
609	1.642 036	1913.230	291 289.3	659	1.517 451	2070.310	341 083.5
610	1.639 344	1916.372	292 246.7	**660**	1.515 152	2073.451	342 119.4
611	1.636 661	1919.513	293 205.6	661	1.512 859	2076.593	343 157.0
612	1.633 987	1922.655	294 166.2	662	1.510 574	2079.734	344 196.0
613	1.631 321	1925.796	295 128.3	663	1.508 296	2082.876	345 236.7
614	1.628 664	1928.938	296 092.0	664	1.506 024	2086.018	346 278.9
615	1.626 016	1932.079	297 057.2	665	1.503 759	2089.159	347 322.7
616	1.623 377	1935.221	298 024.0	666	1.501 502	2092.301	348 368.1
617	1.620 746	1938.363	298 992.4	667	1.499 250	2095.442	349 415.0
618	1.618 123	1941.504	299 962.4	668	1.497 006	2098.584	350 463.5
619	1.615 509	1944.646	300 933.9	669	1.494 768	2101.725	351 513.6
620	1.612 903	1947.787	301 907.1	**670**	1.492 537	2104.867	352 565.2
621	1.610 306	1950.929	302 881.7	671	1.490 313	2108.009	353 618.5
622	1.607 717	1954.071	303 858.0	672	1.488 095	2111.150	354 673.2
623	1.605 136	1957.212	304 835.8	673	1.485 884	2114.292	355 729.6
624	1.602 564	1960.354	305 815.2	674	1.483 680	2117.433	356 787.5
625	1.600 000	1963.495	306 796.2	675	1.481 481	2120.575	357 847.0
626	1.597 444	1966.637	307 778.7	676	1.479 290	2123.717	358 908.1
627	1.594 896	1969.779	308 762.8	677	1.477 105	2126.858	359 970.8
628	1.592 357	1972.920	309 748.5	678	1.474 926	2130.000	361 035.0
629	1.589 825	1976.062	310 735.7	679	1.472 754	2133.141	362 100.8
630	1.587 302	1979.203	311 724.5	**680**	1.470 588	2136.283	363 168.1
631	1.584 786	1982.345	312 714.9	681	1.468 429	2139.425	364 237.0
632	1.582 278	1985.487	313 706.9	682	1.466 276	2142.566	365 307.5
633	1.579 779	1988.628	314 700.4	683	1.464 129	2145.708	366 379.6
634	1.577 287	1991.770	315 695.5	684	1.461 988	2148.849	367 453.2
635	1.574 803	1994.911	316 692.2	685	1.459 854	2151.991	368 528.5
636	1.572 327	1998.053	317 690.4	686	1.457 726	2155.133	369 605.2
637	1.569 859	2001.195	318 690.2	687	1.455 604	2158.274	370 683.6
638	1.567 398	2004.336	319 691.6	688	1.453 488	2161.416	371 763.5
639	1.564 945	2007.478	320 694.6	689	1.451 379	2164.557	372 845.0
640	1.562 500	2010.619	321 699.1	**690**	1.449 275	2167.699	373 928.1
641	1.560 062	2013.761	322 705.2	691	1.447 178	2170.841	375 012.7
642	1.557 632	2016.902	323 712.8	692	1.445 087	2173.982	376 098.9
643	1.555 210	2020.044	324 722.1	693	1.443 001	2177.124	377 186.7
644	1.552 795	2023.186	325 732.9	694	1.440 922	2180.265	378 276.0
645	1.550 388	2026.327	326 745.3	695	1.438 849	2183.407	379 366.9
646	1.547 988	2029.469	327 759.2	696	1.436 782	2186.548	380 459.4
647	1.545 595	2032.610	328 774.7	697	1.434 720	2189.690	381 553.5
648	1.543 210	2035.752	329 791.8	698	1.432 665	2192.832	382 649.1
649	1.540 832	2038.894	330 810.5	699	1.430 615	2195.973	383 746.3
650	1.538 462	2042.035	331 830.7	**700**	1.428 571	2199.115	384 845.1

RECIPROCALS, CIRCUMFERENCE AND AREA OF CIRCLES

n = dia	$1000\frac{1}{n}$	Circum-ference πn	Area $\frac{\pi n^2}{4}$	n = dia	$1000\frac{1}{n}$	Circum-ference πn	Area $\frac{\pi n^2}{4}$
700	1.428 571	2199.115	384 845.1	**750**	1.333 333	2356.194	441 786.5
701	1.426 534	2202.256	385 945.4	751	1.331 558	2359.336	442 965.3
702	1.424 501	2205.398	387 047.4	752	1.329 787	2362.478	444 145.8
703	1.422 475	2208 540	388 150.8	753	1.328 021	2365.619	445 327.8
704	1.420 455	2211.681	389 255.9	754	1.326 260	2368.761	446 511.4
705	1.418 440	2214.823	390 362.5	755	1.324 503	2371.902	447 696.6
706	1.416 431	2217.964	391 470.7	756	1.322 751	2375.044	448 883.3
707	1.414 427	2221.106	392 580.5	757	1.321 004	2378.186	450 071.6
708	1.412 429	2224.248	393 691.8	758	1.319 261	2381.327	451 261.5
709	1.410 437	2227.389	394 804.7	759	1.317 523	2384.469	452 453.0
710	1.408 451	2230.531	395 919.2	**760**	1.315 789	2387.610	453 646.0
711	1.406 470	2233.672	397 035.3	761	1.314 060	2390.752	454 840.6
712	1.404 494	2236.814	398 152.9	762	1.312 336	2393.894	456 036.7
713	1.402 525	2239.956	399 272.1	763	1.310 616	2397.035	457 234.5
714	1.400 560	2243.097	400 392.8	764	1.308 901	2400.177	458 433.8
715	1.398 601	2246.239	401 515.2	765	1.307 190	2403.318	459 634.6
716	1.396 648	2249.380	402 639.1	766	1.305 483	2406.460	460 837.1
717	1.394 700	2252.522	403 764.6	767	1.303 781	2409.602	462 041.1
718	1.392 758	2255.664	404 891.6	768	1.302 083	2412.743	463 246.7
719	1.390 821	2258.805	406 020.2	769	1.300 390	2415.885	464 453.8
720	1.388 889	2261.947	407 150.4	**770**	1.298 701	2419.026	465 662.6
721	1.386 963	2265.088	408 282.2	771	1.297 017	2422.168	466 872.9
722	1.385 042	2268.230	409 415.5	772	1.295 337	2425.310	468 084.7
723	1.383 126	2271.371	410 550.4	773	1.293 661	2428.451	469 298.2
724	1.381 215	2274.513	411 686.9	774	1.291 990	2431.593	470 513.2
725	1.379 310	2277.655	412 824.9	775	1.290 323	2434.734	471 729.8
726	1.377 410	2280.796	413 964.5	776	1.288 660	2437.876	472 947.9
727	1.375 516	2283.938	415 105.7	777	1.287 001	2441.017	474 167.6
728	1.373 626	2287.079	416 248.5	778	1.285 347	2444.159	475 388.9
729	1.371 742	2290.221	417 392.8	779	1.283 697	2447.301	476 611.8
730	1.369 863	2293.363	418 538.7	**780**	1.282 051	2450.442	477 836.2
731	1.367 989	2296.504	419 686.1	781	1.280 410	2453.584	479 062.2
732	1.366 120	2299.646	420 835.2	782	1.278 772	2456.725	480 289.8
733	1.364 256	2302.787	421 985.8	783	1.277 139	2459.867	481 519.0
734	1.362 398	2305.929	423 138.0	784	1.275 510	2463.009	482 749.7
735	1.360 544	2309.071	424 291.7	785	1.273 885	2466.150	483 982.0
736	1.358 696	2312.212	425 447.0	786	1.272 265	2469.292	485 215.8
737	1.356 852	2315.354	426 603.9	787	1.270 648	2472.433	486 451.3
738	1.355 014	2318.495	427 762.4	788	1.269 036	2475.575	487 688.3
739	1.353 180	2321.637	428 922.4	789	1.267 427	2478.717	488 926.9
740	1.351 351	2324.779	430 084.0	**790**	1.265 823	2481.858	490 167.0
741	1.349 528	2327.920	431 247.2	791	1.264 223	2485.000	491 408.7
742	1.347 709	2331.062	432 412.0	792	1.262 626	2488.141	492 652.0
743	1.345 895	2334.203	433 578.3	793	1.261 034	2491.283	493 896.8
744	1.344 086	2337.345	434 746.2	794	1.259 446	2494.425	495 143.3
745	1.342 282	2340.487	435 915.6	795	1.257 862	2497.566	496 391.3
746	1.340 483	2343.628	437 086.6	796	1.256 281	2500.708	497 640.8
747	1.338 688	2346.770	438 259.2	797	1.254 705	2503.849	498 892.0
748	1.336 898	2349.911	439 433.4	798	1.253 133	2506.991	500 144.7
749	1.335 113	2353.053	440 609.2	799	1.251 564	2510.133	501 399.0
750	1.333 333	2356.194	441 786.5	**800**	1.250 000	2513.274	502 654.8

RECIPROCALS, CIRCUMFERENCE AND AREA OF CIRCLES

n = dia	$1000\frac{1}{n}$	Circum-ference πn	Area $\frac{\pi n^2}{4}$	n = dia	$1000\frac{1}{n}$	Circum-ference πn	Area $\frac{\pi n^2}{4}$
800	1.250 000	2513.274	502 654.8	**850**	1.176 471	2670.354	567 450.2
801	1.248 439	2516.416	503 912.2	851	1.175 088	2673.495	568 786.1
802	1.246 883	2519.557	505 171.2	852	1.173 709	2676.637	570 123.7
803	1.245 330	2522.699	506 431.8	853	1.172 333	2679.779	571 462.8
804	1.243 781	2525.840	507 693.9	854	1.170 960	2682.920	572 803.4
805	1.242 236	2528.982	508 957.6	855	1.169 591	2686.062	574 145.7
806	1.240 695	2532.124	510 222.9	856	1.168 224	2689.203	575 489.5
807	1.239 157	2535.265	511 489.8	857	1.166 861	2692.345	576 834.9
808	1.237 624	2538.407	512 758.2	858	1.165 501	2695.486	578 181.9
809	1.236 094	2541.548	514 028.2	859	1.164 144	2698.628	579 530.4
810	1.234 568	2544.690	515 299.7	**860**	1.162 791	2701.770	580 880.5
811	1.233 046	2547.832	516 572.9	861	1.161 440	2704.911	582 232.2
812	1.231 527	2550.973	517 847.6	862	1.160 093	2708.053	583 585.4
813	1.230 012	2554.115	519 123.8	863	1.158 749	2711.194	584 940.2
814	1.228 501	2557.256	520 401.7	864	1.157 407	2714.336	586 296.6
815	1.226 994	2560.398	521 681.1	865	1.156 069	2717.478	587 654.5
816	1.225 490	2563.540	522 962.1	866	1.154 734	2720.619	589 014.1
817	1.223 990	2566.681	524 244.6	867	1.153 403	2723.761	590 375.2
818	1.222 494	2569.823	525 528.8	868	1.152 074	2726.902	591 737.8
819	1.221 001	2572.964	526 814.5	869	1.150 748	2730.044	593 102.1
820	1.219 512	2576.106	528 101.7	**870**	1.149 425	2733.186	594 467.9
821	1.218 027	2579.248	529 390.6	871	1.148 106	2736.327	595 835.2
822	1.216 545	2582.389	530 681.0	872	1.146 789	2739.469	597 204.2
823	1.215 067	2585.531	531 973.0	873	1.145 475	2742.610	598 574.7
824	1.213 592	2588.672	533 266.5	874	1.144 165	2745.752	599 946.8
825	1.212 121	2591.814	534 561.6	875	1.142 857	2748.894	601 320.5
826	1.210 654	2594.956	535 858.3	876	1.141 553	2752.035	602 695.7
827	1.209 190	2598.097	537 156.6	877	1.140 251	2755.177	604 072.5
828	1.207 729	2601.239	538 456.4	878	1.138 952	2758.318	605 450.9
829	1.206 273	2604.380	539 757.8	879	1.137 656	2761.460	606 830.8
830	1.204 819	2607.522	541 060.8	**880**	1.136 364	2764.602	608 212.3
831	1.203 369	2610.663	542 365.3	881	1.135 074	2767.743	609 595.4
832	1.201 923	2613.805	543 671.5	882	1.133 787	2770.885	610 980.1
833	1.200 480	2616.947	544 979.1	883	1.132 503	2774.026	612 366.3
834	1.199 041	2620.088	546 288.4	884	1.131 222	2777.168	613 754.1
835	1.197 605	2623.230	547 599.2	885	1.129 944	2780.309	615 143.5
836	1.196 172	2626.371	548 911.6	886	1.128 668	2783.451	616 534.4
837	1.194 743	2629.513	550 225.6	887	1.127 396	2786.593	617 926.9
838	1.193 317	2632.655	551 541.1	888	1.126 126	2789.734	619 321.0
839	1.191 895	2635.796	552 858.3	889	1.124 859	2792.876	620 716.7
840	1.190 476	2638.938	554 176.9	**890**	1.123 596	2796.017	622 113.9
841	1.189 061	2642.079	555 497.2	891	1.122 334	2799.159	623 512.7
842	1.187 648	2645.221	556 819.0	892	1.121 076	2802.301	624 913.0
843	1.186 240	2648.363	558 142.4	893	1.119 821	2805.442	626 315.0
844	1.184 834	2651.504	559 467.4	894	1.118 568	2808.584	627 718.5
845	1.183 432	2654.646	560 793.9	895	1.117 318	2811.725	629 123.6
846	1.182 033	2657.787	562 122.0	896	1.116 071	2814.867	630 530.2
847	1.180 638	2660.929	563 451.7	897	1.114 827	2818.009	631 938.4
848	1.179 245	2664.071	564 783.0	898	1.113 586	2821.150	633 348.2
849	1.177 856	2667.212	566 115.8	899	1.112 347	2824.292	634 759.6
850	1.176 471	2670.354	567 450.2	**900**	1.111 111	2827.433	636 172.5

n = dia	$1000\frac{1}{n}$	Circumference πn	Area $\frac{\pi n^2}{4}$	n = dia	$1000\frac{1}{n}$	Circumference πn	Area $\frac{\pi n^2}{4}$
900	1.111 111	2827.433	636 172.5	**950**	1.052 632	2984.513	708 821.8
901	1.109 878	2830.575	637 587.0	951	1.051 525	2987.655	710 314.9
902	1.108 647	2833.717	639 003.1	952	1.050 420	2990.796	711 809.5
903	1.107 420	2836.858	640 420.7	953	1.049 318	2993.938	713 305.7
904	1.106 195	2840.000	641 839.9	954	1.048 218	2997.079	714 803.4
905	1.104 972	2843.141	643 260.7	955	1.047 120	3000.221	716 302.8
906	1.103 753	2846.283	644 683.1	956	1.046 025	3003.363	717 803.7
907	1.102 536	2849.425	646 107.0	957	1.044 932	3006.504	719 306.1
908	1.101 322	2852.566	647 532.5	958	1.043 841	3009.646	720 810.2
909	1.100 110	2855.708	648 959.6	959	1.042 753	3012.787	722 315.8
910	1.098 901	2858.849	650 388.2	**960**	1.041 667	3015.929	723 822.9
911	1.097 695	2861.991	651 818.4	961	1.040 583	3019.071	725 331.7
912	1.096 491	2865.133	653 250.2	962	1.039 501	3022.212	726 842.0
913	1.095 290	2868.274	654 683.6	963	1.038 422	3025.354	728 353.9
914	1.094 092	2871.416	656 118.5	964	1.037 344	3028.495	729 867.4
915	1.092 896	2874.557	657 555.0	965	1.036 269	3031.637	731 382.4
916	1.091 703	2877.699	658 993.0	966	1.035 197	3034.779	732 899.0
917	1.090 513	2880.840	660 432.7	967	1.034 126	3037.920	734 417.2
918	1.089 325	2883.982	661 873.9	968	1.033 058	3041.062	735 936.9
919	1.088 139	2887.124	663 316.7	969	1.031 992	3044.203	737 458.2
920	1.086 957	2890.265	664 761.0	**970**	1.030 928	3047.345	738 981.1
921	1.085 776	2893.407	666 206.9	971	1.029 866	3050.486	740 505.6
922	1.084 599	2896.548	667 654.4	972	1.028 807	3053.628	742 031.6
923	1.083 424	2899.690	669 103.5	973	1.027 749	3056.770	743 559.2
924	1.082 251	2902.832	670 554.1	974	1.026 694	3059.911	745 088.4
925	1.081 081	2905.973	672 006.3	975	1.025 641	3063.053	746 619.1
926	1.079 914	2909.115	673 460.1	976	1.024 590	3066.194	748 151.4
927	1.078 749	2912.256	674 915.4	977	1.023 541	3069.336	749 685.3
928	1.077 586	2915.398	676 372.3	978	1.022 495	3072.478	751 220.8
929	1.076 426	2918.540	677 830.8	979	1.021 450	3075.619	752 757.8
930	1.075 269	2921.681	679 290.9	**980**	1.020 408	3078.761	754 296.4
931	1.074 114	2924.823	680 752.5	981	1.019 368	3081.902	755 836.6
932	1.072 961	2927.964	682 215.7	982	1.018 330	3085.044	757 378.3
933	1.071 811	2931.106	683 680.5	983	1.017 294	3088.186	758 921.6
934	1.070 664	2934.248	685 146.8	984	1.016 260	3091.327	760 466.5
935	1.069 519	2937.389	686 614.7	985	1.015 228	3094.469	762 012.9
936	1.068 376	2940.531	688 084.2	986	1.014 199	3097.610	763 561.0
937	1.067 236	2943.672	689 555.2	987	1.013 171	3100.752	765 110.5
938	1.066 098	2946.814	691 027.9	988	1.012 146	3103.894	766 661.7
939	1.064 963	2949.956	692 502.1	989	1.011 122	3107.035	768 214.4
940	1.063 830	2953.097	693 977.8	**990**	1.010 101	3110.177	769 768.7
941	1.062 699	2956.239	695 455.2	991	1.009 082	3113.318	771 324.6
942	1.061 571	2959.380	696 934.1	992	1.008 065	3116.460	772 882.1
943	1.060 445	2962.522	698 414.5	993	1.007 049	3119.602	774 441.1
944	1.059 322	2965.663	699 896.6	994	1.006 036	3122.743	776 001.7
945	1.058 201	2968.805	701 380.2	995	1.005 025	3125.885	777 563.8
946	1.057 082	2971.947	702 865.4	996	1.004 016	3129.026	779 127.5
947	1.055 966	2975.088	704 352.1	997	1.003 009	3132.168	780 692.8
948	1.054 852	2978.230	705 840.5	998	1.002 004	3135.309	782 259.7
949	1.053 741	2981.371	707 330.4	999	1.001 001	3138.451	783 828.2
950	1.052 632	2984.513	708 821.8	**1000**	1.000 000	3141.593	785 398.2

FORMULAS FROM PLANE ANALYTIC GEOMETRY

DISTANCE d BETWEEN TWO POINTS $P_1(x_1, y_1)$ AND $P_2(x_2, y_2)$

$$d = \sqrt{(x_2 - x_1)^2 + (y_2 - y_1)^2}$$

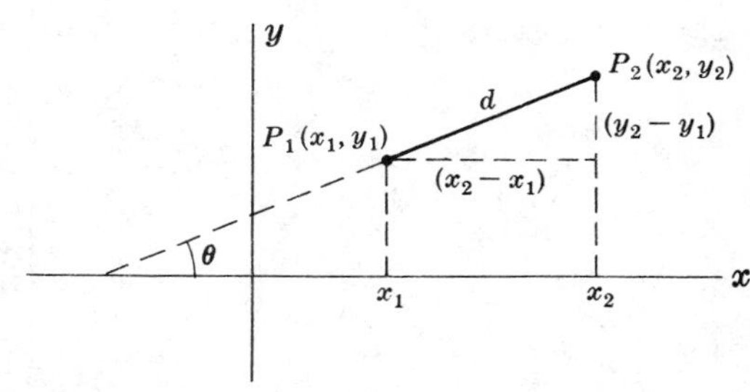

SLOPE m OF LINE JOINING TWO POINTS $P_1(x_1, y_1)$ AND $P_2(x_2, y_2)$

$$m = \frac{y_2 - y_1}{x_2 - x_1} = \tan\theta$$

EQUATION OF LINE JOINING TWO POINTS $P_1(x_1, y_1)$ AND $P_2(x_2, y_2)$

$$\frac{y - y_1}{x - x_1} = \frac{y_2 - y_1}{x_2 - x_1} = m \quad \text{or} \quad y - y_1 = m(x - x_1)$$

$$y = mx + b$$

where $b = y_1 - mx_1 = \dfrac{x_2y_1 - x_1y_2}{x_2 - x_1}$ is the *intercept* on the y axis, i.e. the y *intercept*

EQUATION OF LINE IN TERMS OF x INTERCEPT $a \neq 0$ AND y INTERCEPT $b \neq 0$

$$\frac{x}{a} + \frac{y}{b} = 1$$

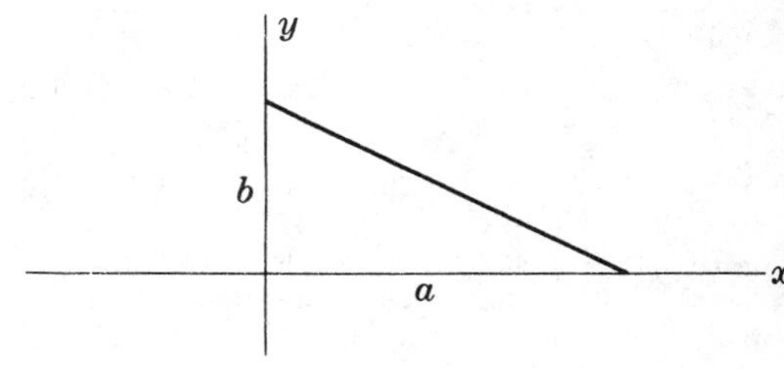

NORMAL FORM FOR EQUATION OF LINE

$$x \cos \alpha + y \sin \alpha = p$$

where p = perpendicular distance from origin O to line

and α = angle of inclination of perpendicular with positive x axis.

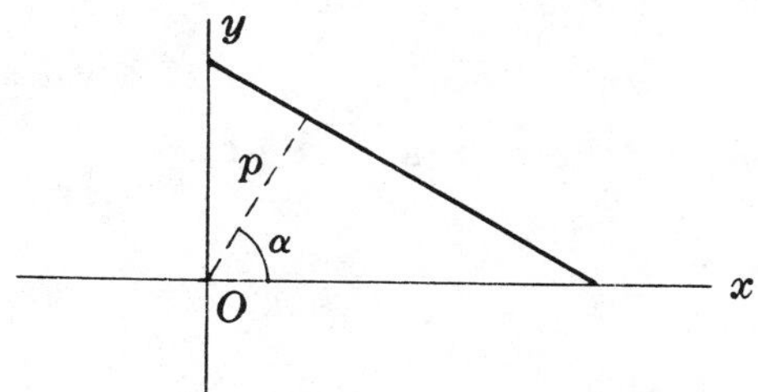

GENERAL EQUATION OF LINE

$$Ax + By + C = 0$$

DISTANCE FROM POINT (x_1, y_1) TO LINE $Ax + By + C = 0$

$$\frac{Ax_1 + By_1 + C}{\pm\sqrt{A^2 + B^2}}$$

where the sign is chosen so that the distance is nonnegative.

ANGLE ψ BETWEEN TWO LINES HAVING SLOPES m_1 AND m_2

$$\tan \psi = \frac{m_2 - m_1}{1 + m_1 m_2}$$

Lines are parallel or coincident if and only if $m_1 = m_2$.

Lines are perpendicular if and only if $m_2 = -1/m_1$.

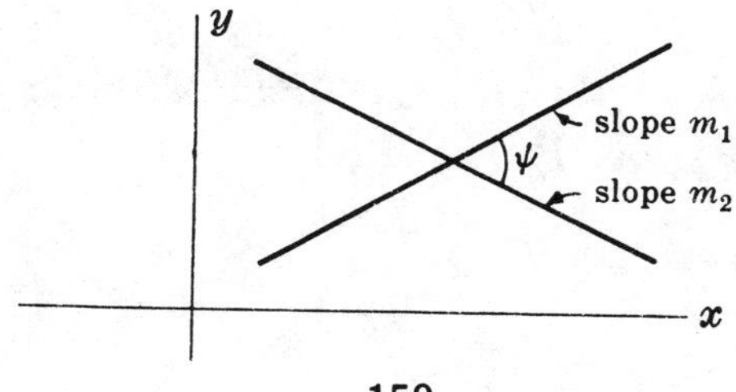

AREA OF TRIANGLE WITH VERTICLES AT (x_1, y_1), (x_2, y_2), (x_3, y_3)

$$\text{Area} = \pm\frac{1}{2}\begin{vmatrix} x_1 & y_1 & 1 \\ x_2 & y_2 & 1 \\ x_3 & y_3 & 1 \end{vmatrix}$$

$$= \pm\frac{1}{2}(x_1y_2 + y_1x_3 + y_3x_2 - y_2x_3 - y_1x_2 - x_1y_3)$$

where the sign is chosen so that the area is nonnegative.

If the area is zero the points all lie on a line.

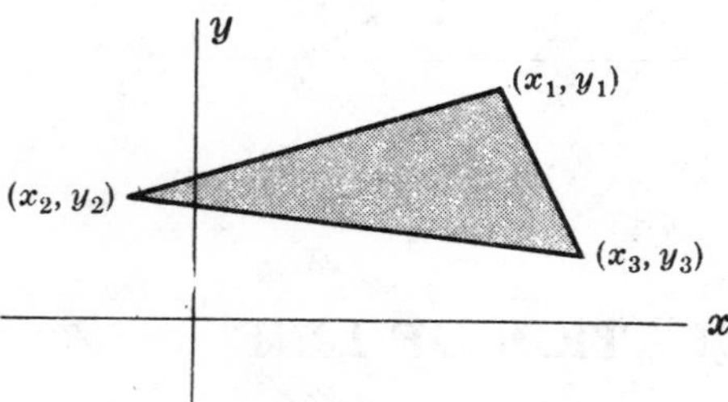

TRANSFORMATION OF COORDINATES INVOLVING PURE TRANSLATION

$$\begin{cases} x = x' + x_0 \\ y = y' + y_0 \end{cases} \quad \text{or} \quad \begin{cases} x' = x - x_0 \\ y' = y - y_0 \end{cases}$$

where (x, y) are old coordinates [i.e. coordinates relative to xy system], (x', y') are new coordinates [relative to $x'y'$ system] and (x_0, y_0) are the coordinates of the new origin O' relative to the old xy coordinate system.

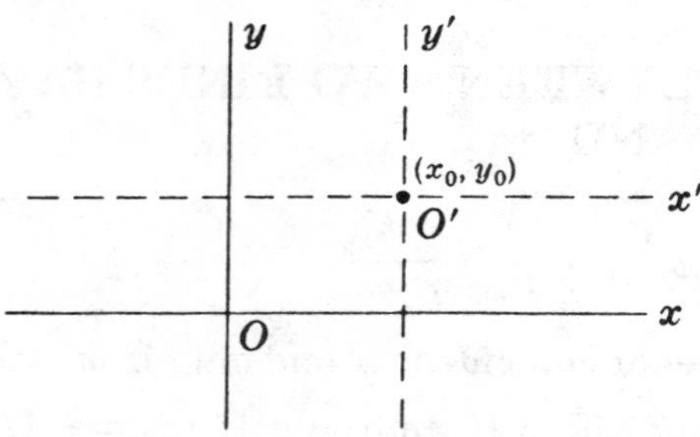

TRANSFORMATION OF COORDINATES INVOLVING PURE ROTATION

$$\begin{cases} x = x' \cos\alpha - y' \sin\alpha \\ y = x' \sin\alpha + y' \cos\alpha \end{cases} \quad \text{or} \quad \begin{cases} x' = x \cos\alpha + y \sin\alpha \\ y' = y \cos\alpha - x \sin\alpha \end{cases}$$

where the origins of the old $[xy]$ and new $[x'y']$ coordinate systems are the same but the x' axis makes an angle α with the positive x axis.

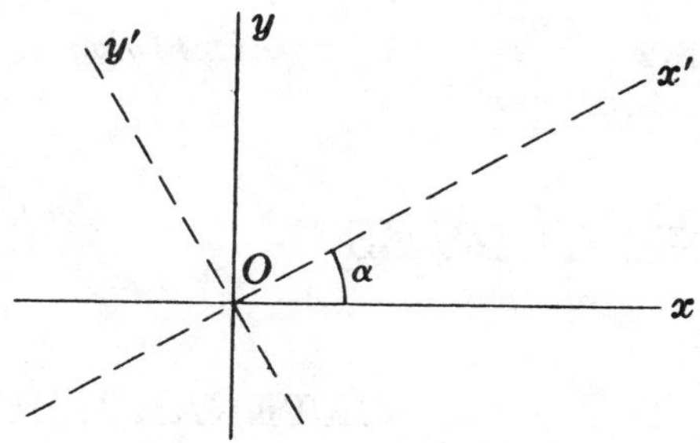

TRANSFORMATION OF COORDINATES INVOLVING TRANSLATION AND ROTATION

$$\begin{cases} x = x' \cos\alpha - y' \sin\alpha + x_0 \\ y = x' \sin\alpha + y' \cos\alpha + y_0 \end{cases}$$

or

$$\begin{cases} x' = (x - x_0)\cos\alpha + (y - y_0)\sin\alpha \\ y' = (y - y_0)\cos\alpha - (x - x_0)\sin\alpha \end{cases}$$

where the new origin O' of $x'y'$ coordinate system has coordinates (x_0, y_0) relative to the old xy coordinate system and the x' axis makes an angle α with the positive x axis.

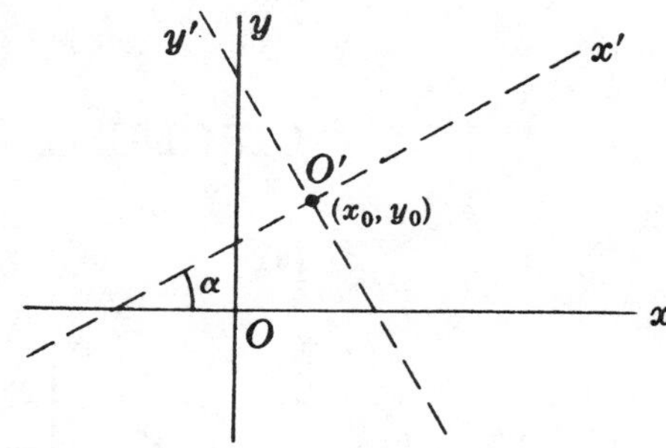

POLAR COORDINATES (r, θ)

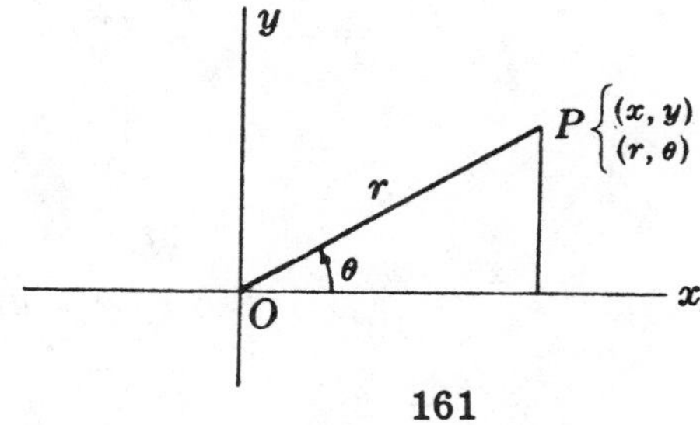

A point P can be located by rectangular coordinates (x, y) or polar coordinates (r, θ). The transformation between these coordinates is

$$\begin{cases} x = r\cos\theta \\ y = r\sin\theta \end{cases} \quad \text{or} \quad \begin{cases} r = \sqrt{x^2 + y^2} \\ \theta = \tan^{-1}(y/x) \end{cases}$$

PLANE CURVES

BIFOLIUM

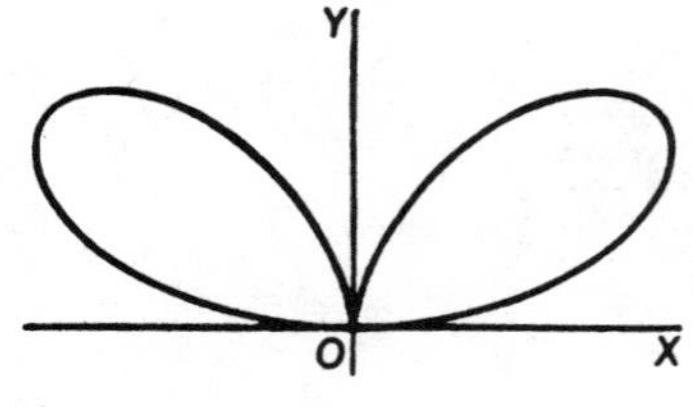

$$(x^2 + y^2)^2 = ax^2y$$
$$r = a\sin\theta\cos^2\theta$$

CARDIOID

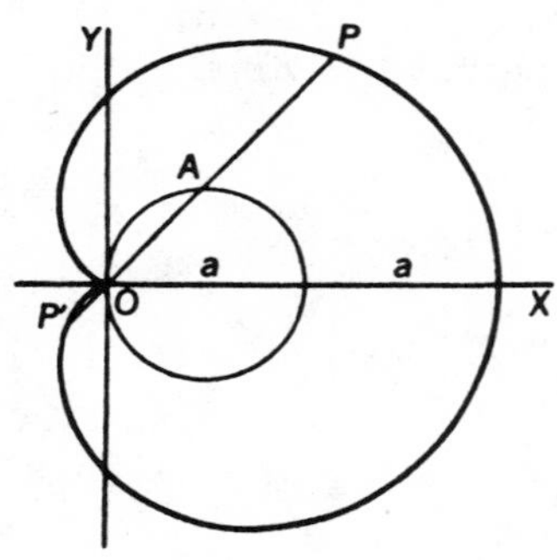

$$(x^2 + y^2 - ax)^2 = a^2(x^2 + y^2)$$
$$r = a(\cos\theta + 1)$$
or
$$r = a(\cos\theta - 1)$$
$$[P'A = AP = a]$$

CASSINIAN CURVES

See: Ovals of Cassini

CATENARY, HYPERBOLIC COSINE

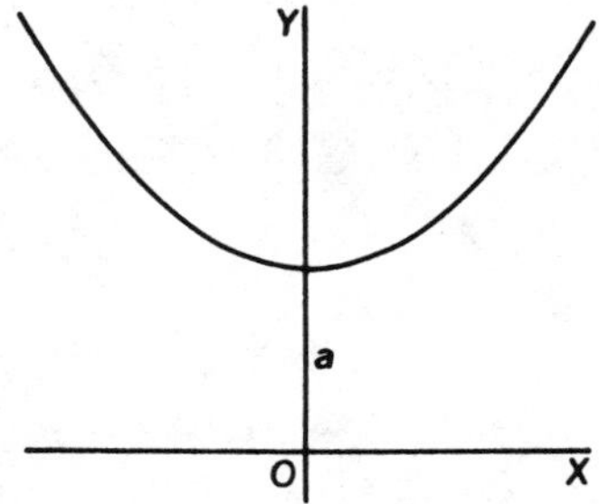

$$y = \frac{a}{2}\left(e^{x/a} + e^{-x/a}\right) = a\cosh\frac{x}{a}$$

CIRCLE

(a)

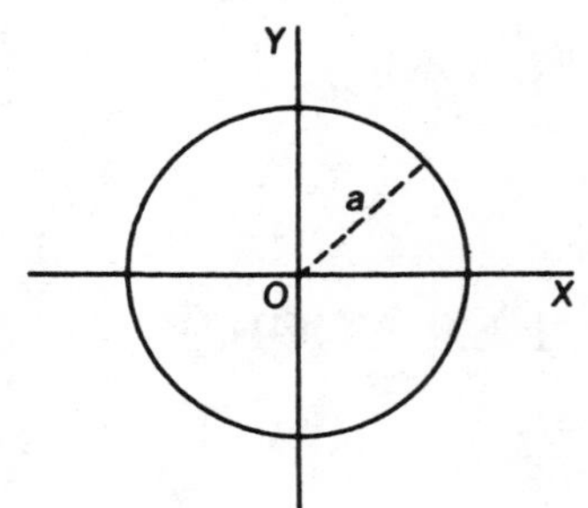

$$x^2 + y^2 = a^2$$
$$r = a$$

b)

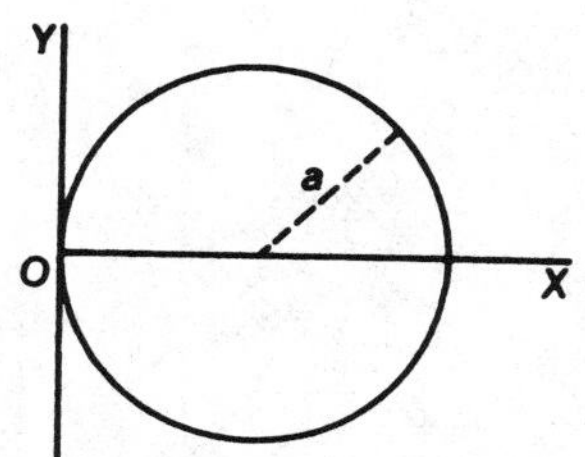

$$x^2 + y^2 = 2ax$$
$$r = 2a\cos\theta$$

c)

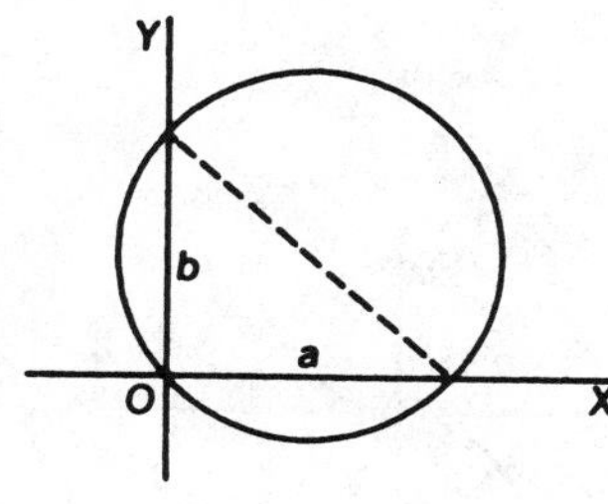

$$x^2 + y^2 = ax + by$$
$$r = a\cos\theta + b\sin\theta$$

CISSOID OF DIOCLES

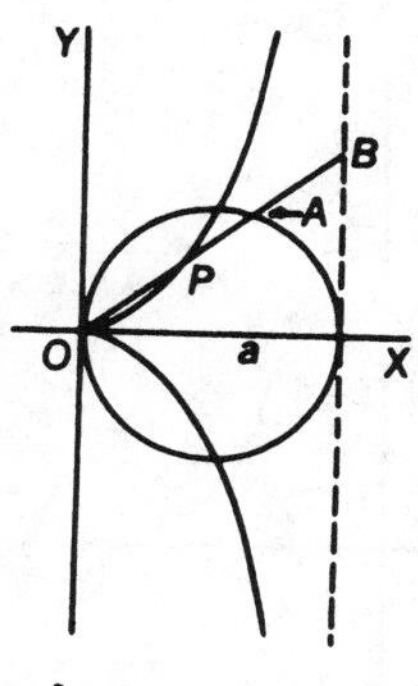

$$y^2(a - x) = x^3$$
$$r = a\sin\theta\tan\theta$$
$$[OP = AB]$$

COCHLEOID, OUI-JA BOARD CURVE

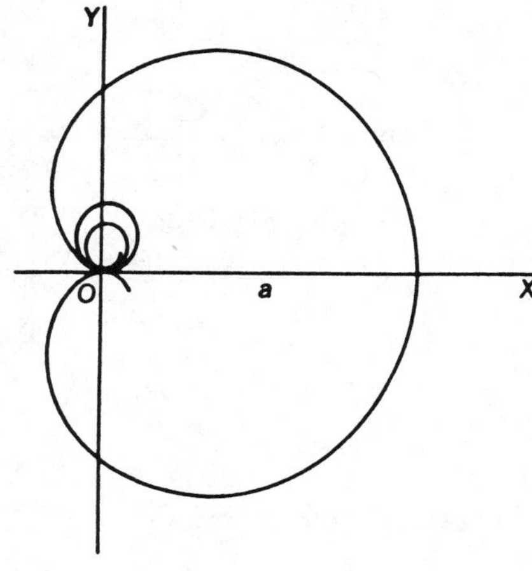

$$(x^2 + y^2)\tan^{-1}(y/x) = ay$$
$$r\theta = a\sin\theta$$

COMPANION TO THE CYCLOID

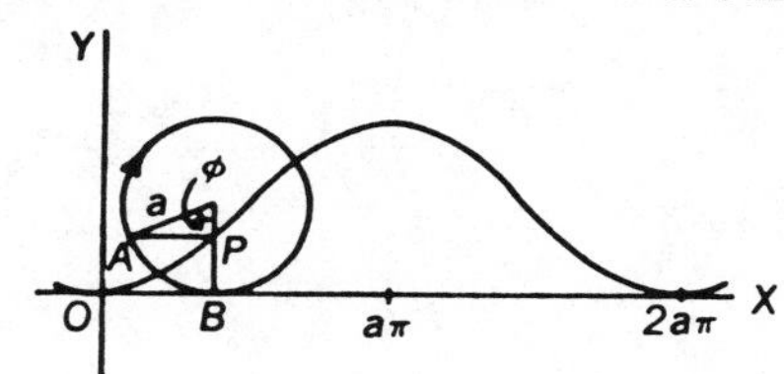

$$\begin{cases} x = a\phi \\ y = a(1 - \cos\phi) \end{cases}$$
$$[OB = \widehat{AB}]$$

(This is a sinusoid)

CONCHOID OF NICOMEDES

(a) $a < b$

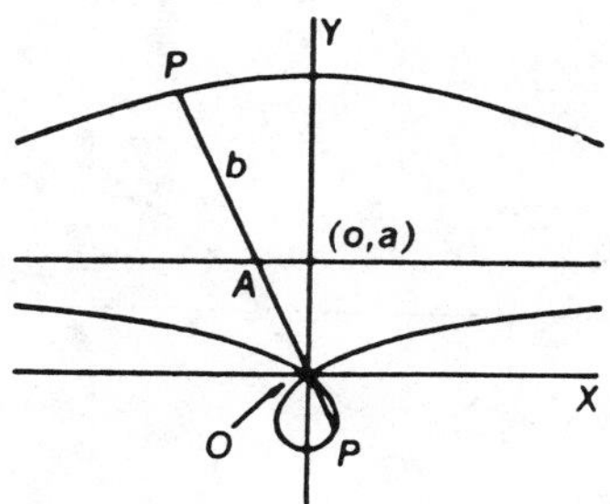

(b) $a > b$

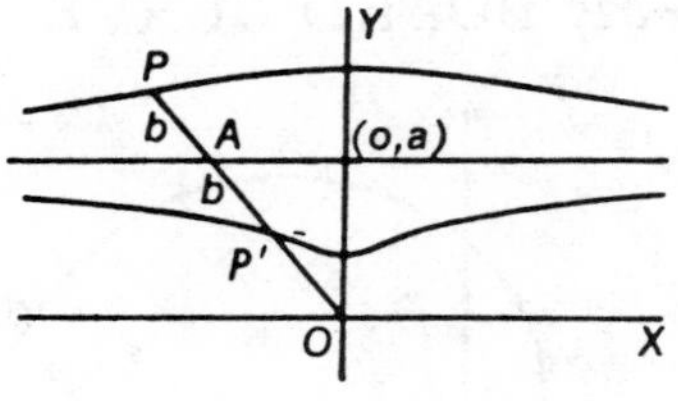

$$(y - a)^2(x^2 + y^2) = b^2y^2$$
$$r = a \csc\theta \pm b$$
$$[P'A = AP = b]$$

CONIC SECTIONS

See: Circle; Ellipse; Hyperbola; Parabola

COSECANT CURVE

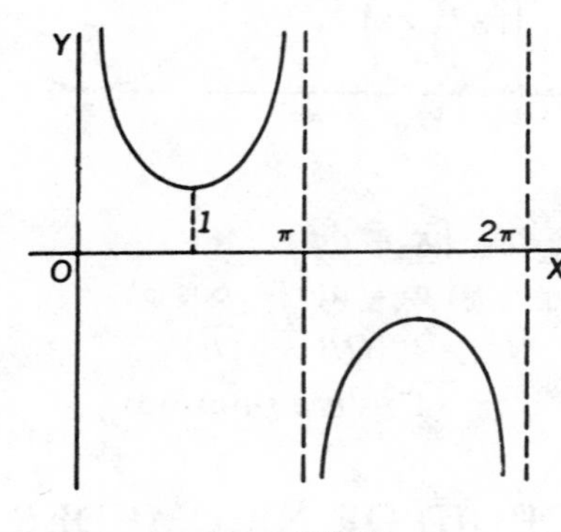

$y = \csc x$

COSINE CURVE

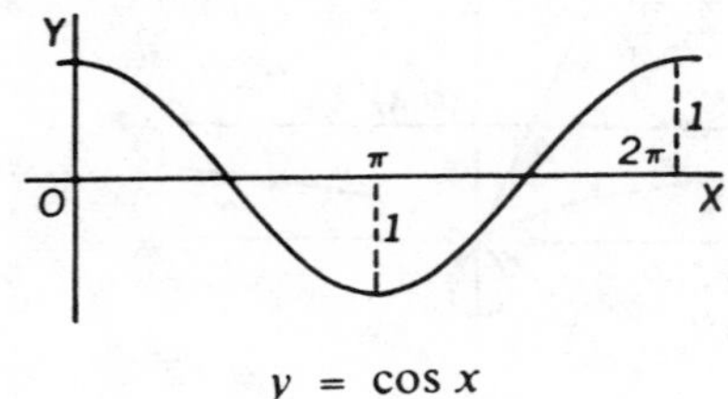

$y = \cos x$

COTANGENT CURVE

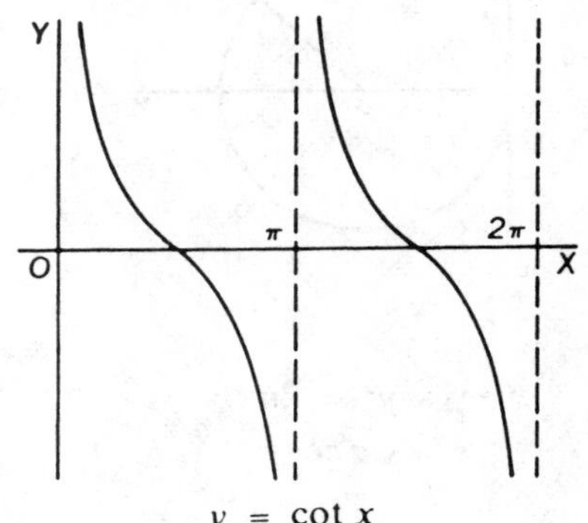

$y = \cot x$

CUBICAL PARABOLA (SPECIAL)

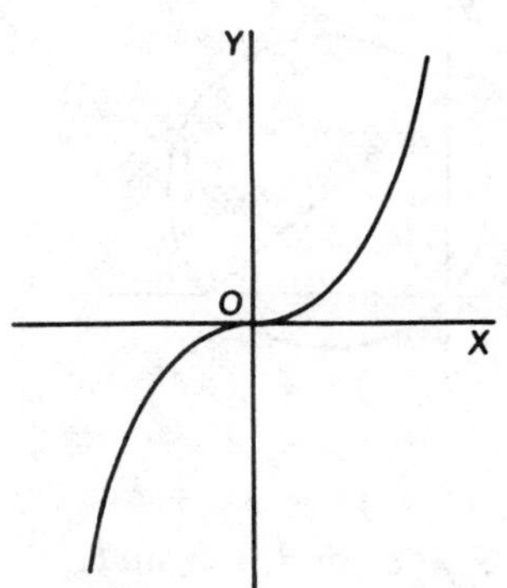

$$y = ax^3, \quad a > 0$$
$$r^2 = \frac{1}{a}\sec^2\theta\tan\theta, \quad a > 0$$

CUBICAL PARABOLA (GENERAL)

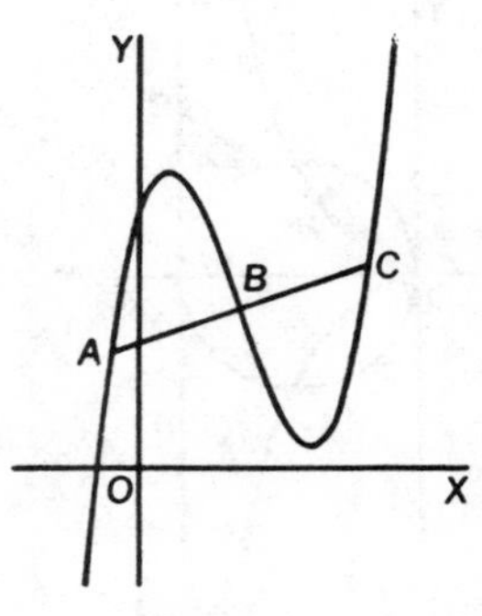

$$y = ax^3 + bx^2 + cx + d, \quad a >$$
$$[AB = BC]$$
(abscissa of $B = -b/3a$)

CURTATE CYCLOID, TROCHOIDS

See: Cycloid, curtate

CYCLOID (CUSP AT ORIGIN)

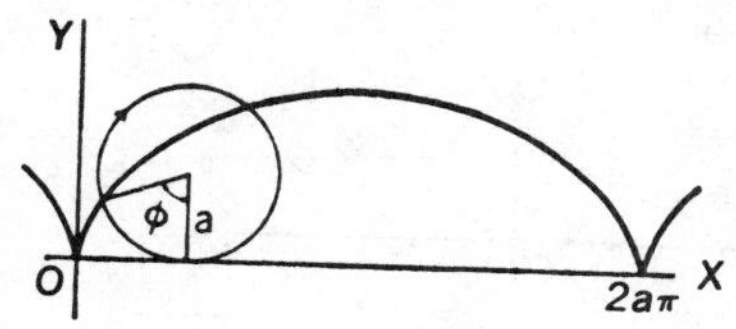

$$x = a \arccos \frac{a - y}{a} \mp \sqrt{2ay - y^2}$$

$$\begin{cases} x = a(\phi - \sin \phi) \\ y = a(1 - \cos \phi) \end{cases}$$

(For one arch: arc length $= 8a$; area $= 3\pi a^2$)

CYCLOID (VERTEX AT ORIGIN)

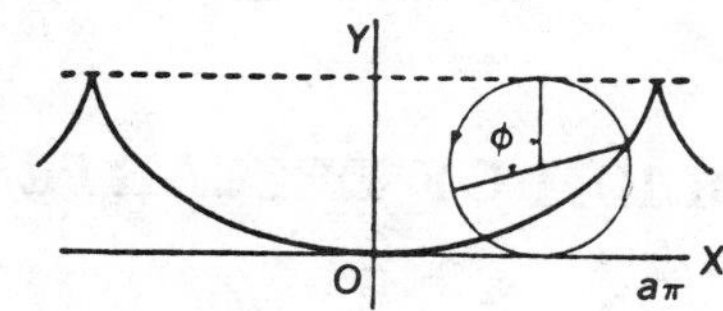

$$x = 2a \arcsin \sqrt{y/2a} + \sqrt{2ay - y^2}$$

$$\begin{cases} x = a(\phi + \sin \phi) \\ y = a(1 - \cos \phi) \end{cases}$$

CYCLOID, CURTATE

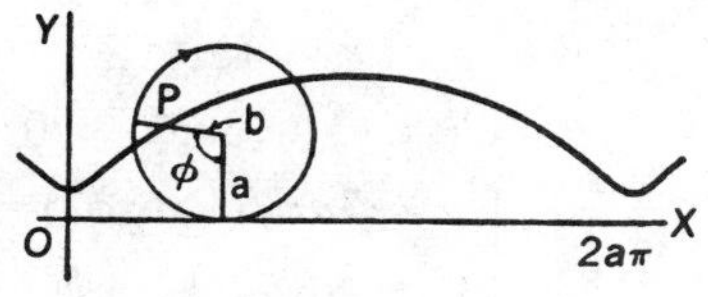

$$\begin{cases} x = a\phi - b \sin \phi \\ y = a - b \cos \phi \end{cases}$$

$$a > b$$

CYCLOID, PROLATE

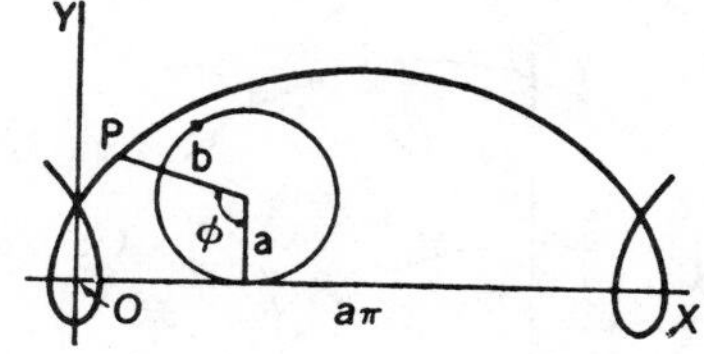

$$\begin{cases} x = a\phi - b \sin \phi \\ y = a - b \cos \phi \end{cases}$$

$$a < b$$

DELTOID

See: Hypocycloid of three cusps

ELLIPSE

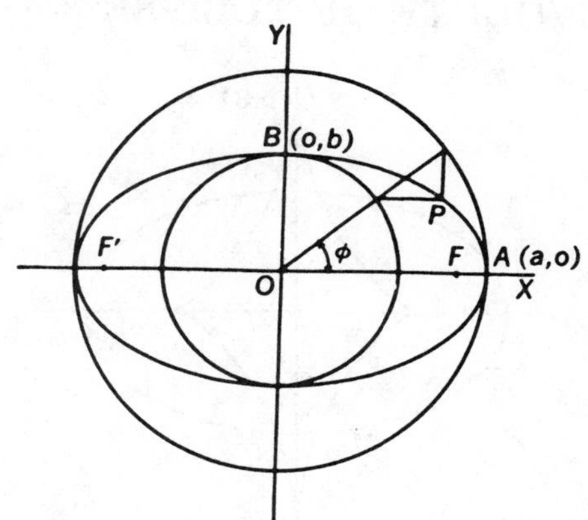

$$x^2/a^2 + y^2/b^2 = 1$$

$$\begin{cases} x = a \cos \phi \\ y = b \sin \phi \end{cases}$$

$$[BF' = BF = a, \quad PF' + PF = 2a]$$

EPICYCLOID

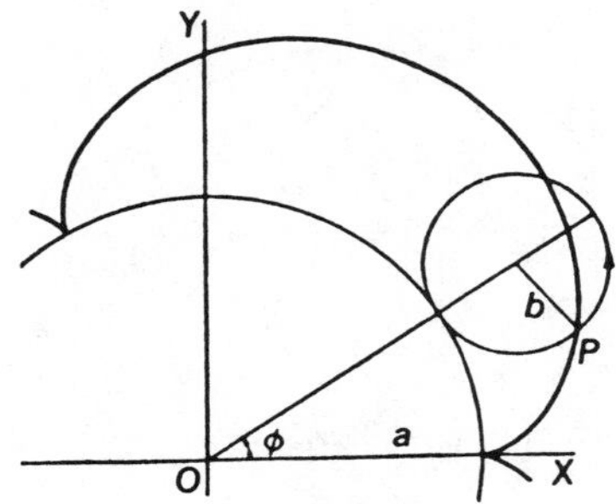

$$\begin{cases} x = (a + b)\cos\phi - b\cos\left(\dfrac{a+b}{b}\phi\right) \\ y = (a + b)\sin\phi - b\sin\left(\dfrac{a+b}{b}\phi\right) \end{cases}$$

EQUIANGULAR SPIRAL

See: Spiral, logarithmic or equiangular

EQUILATERAL HYPERBOLA

See: Hyperbola, equilateral or rectangular

EVOLUTE OF ELLIPSE

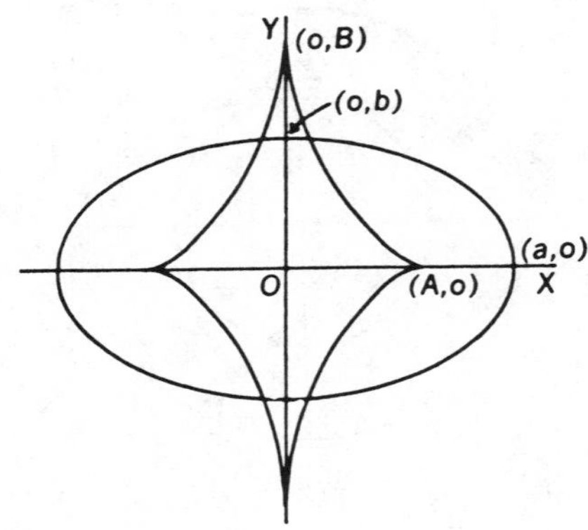

$$(ax)^{2/3} + (by)^{2/3} = (a^2 - b^2)^{2/3}$$

$$\begin{cases} x = A\cos^3\phi \\ y = B\sin^3\phi \end{cases}$$

$$[A = (a^2 - b^2)/a, \quad B = (a^2 - b^2)/b]$$

EXPONENTIAL CURVE

(1) $a > 0$

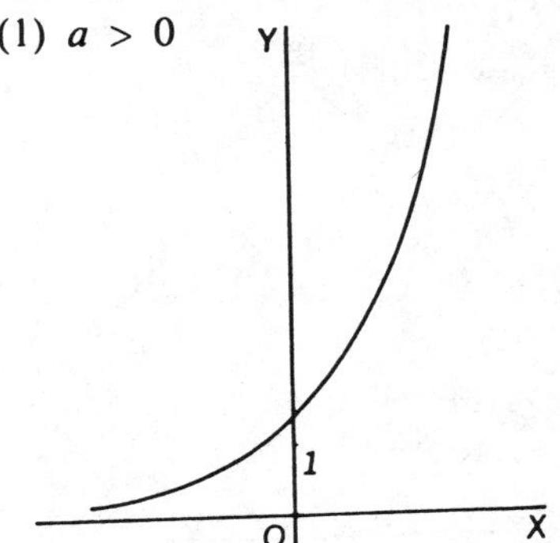

(2) $a < 0$

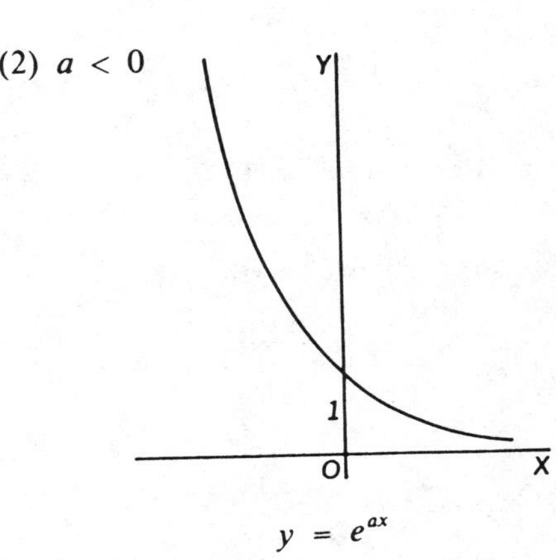

$$y = e^{ax}$$

FOLIUM OF DESCARTES

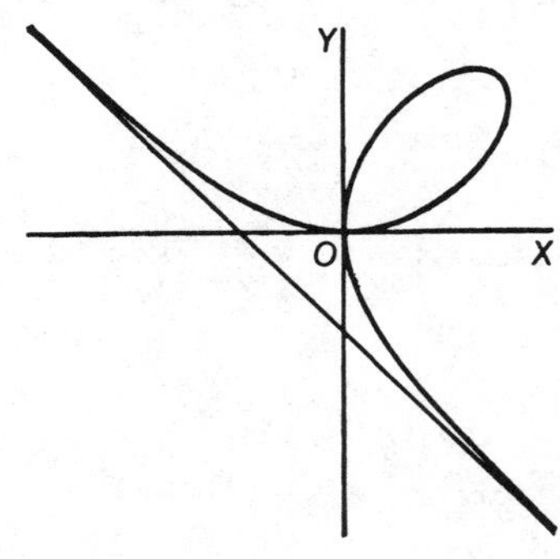

$$x^3 + y^3 - 3axy = 0$$

$$\begin{cases} x = 3a\phi/(1 + \phi^3) \\ y = 3a\phi^2/(1 + \phi^3) \end{cases}$$

$$r = \frac{3a\sin\theta\cos\theta}{\sin^3\theta + \cos^3\theta}$$

[asymptote: $x + y + a = 0$]

GAMMA FUNCTION

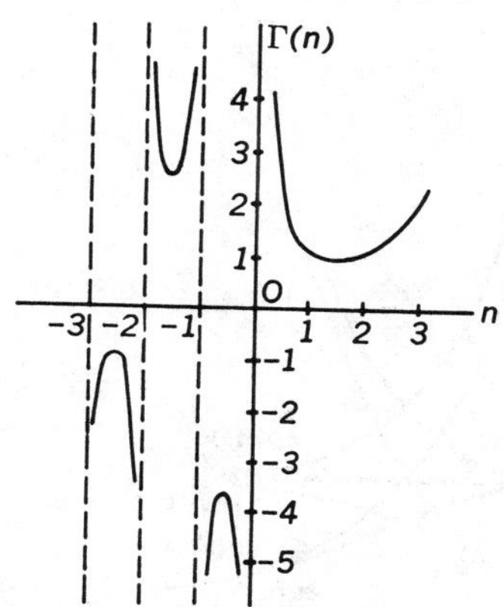

$$\Gamma(n) = \int_0^\infty x^{n-1} e^{-x} dx \quad (n > 0)$$

$$\Gamma(n) = \frac{\Gamma(n+1)}{n} \quad (0 > n \neq -1, -2, -3, \ldots)$$

HYPERBOLA

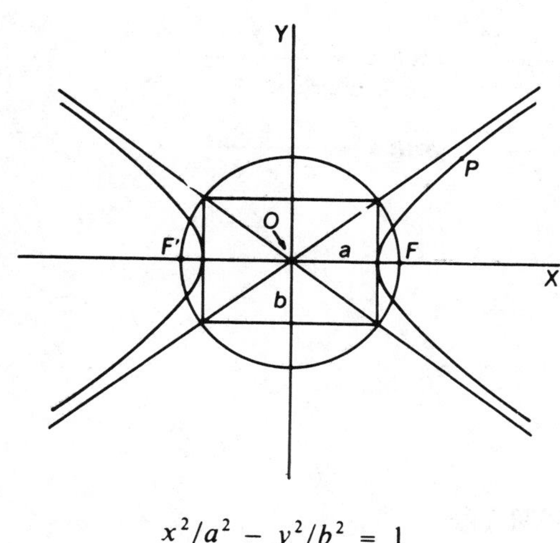

$x^2/a^2 - y^2/b^2 = 1$
$[F'P - FP = 2a]$

HYPERBOLA, EQUILATERAL OR RECTANGULAR

(1)

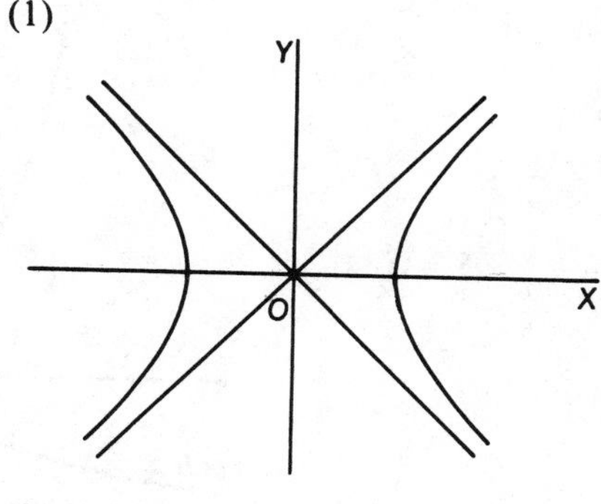

$x^2 - y^2 = a^2$

(2)

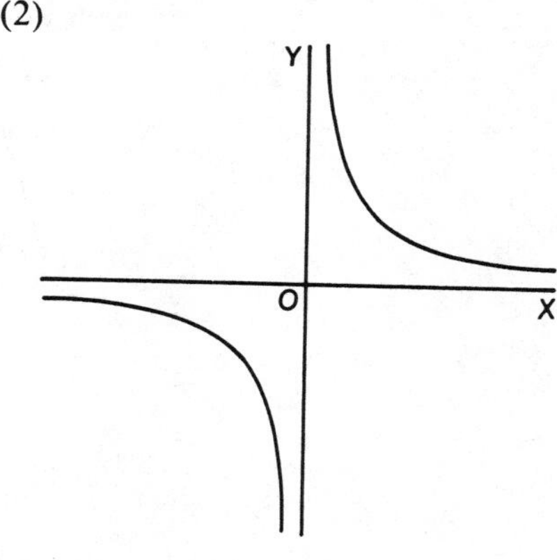

$xy = k, \quad k > 0$

(3)

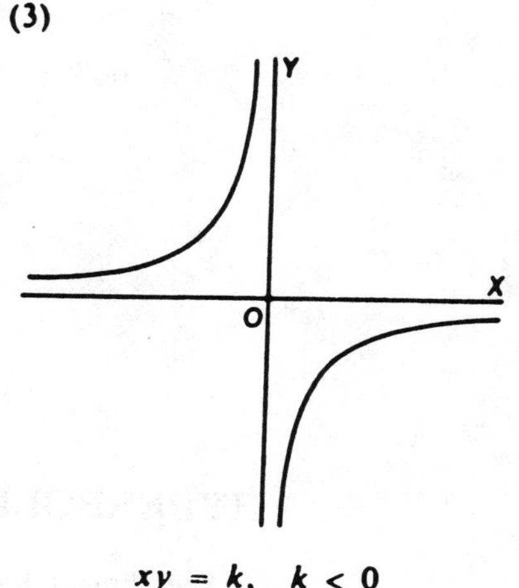

$\boldsymbol{xy = k, \quad k < 0}$

Hyperbolic functions

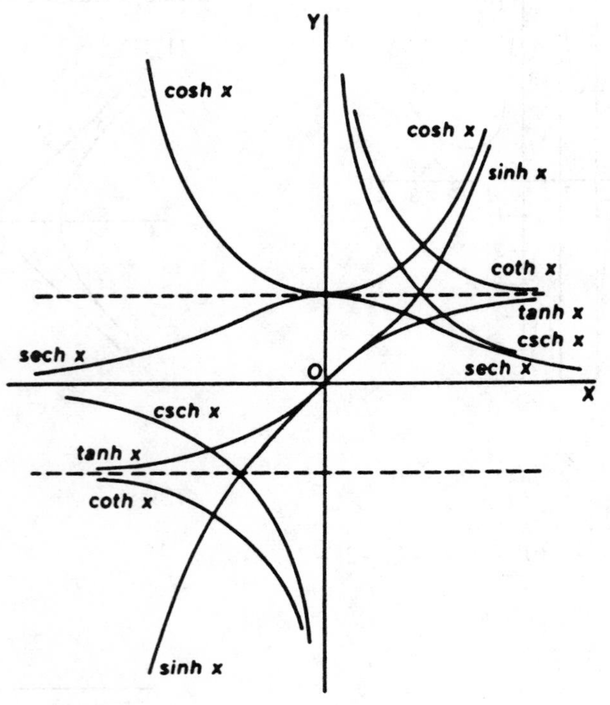

$$\sinh x = \frac{e^x - e^{-x}}{2} \qquad \operatorname{csch} x = \frac{2}{e^x - e^{-x}}$$

$$\cosh x = \frac{e^x + e^{-x}}{2} \qquad \operatorname{sech} x = \frac{2}{e^x + e^{-x}}$$

$$\tanh x = \frac{e^x - e^{-x}}{e^x + e^{-x}} \qquad \coth x = \frac{e^x + e^{-x}}{e^x - e^{-x}}$$

HYPERBOLIC SPIRAL

See: Spiral, hyperbolic or reciprocal

HYPOCYCLOID OF THREE CUSPS, DELTOID

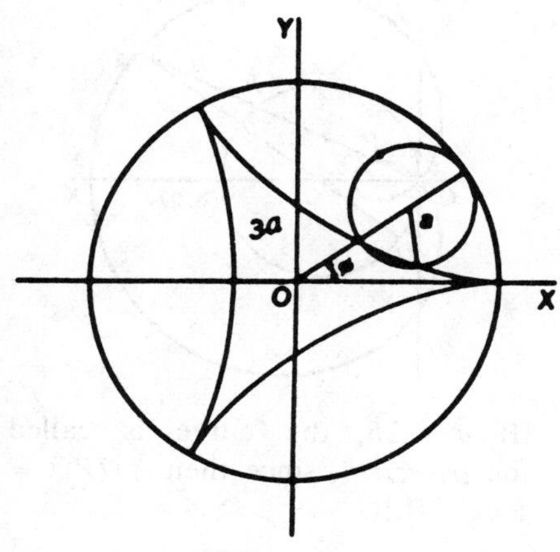

$$\begin{cases} x = 2a\cos\phi + a\cos 2\phi \\ y = 2a\sin\phi - a\sin 2\phi \end{cases}$$

HYPOCYCLOID OF FOUR CUSPS, ASTEROID

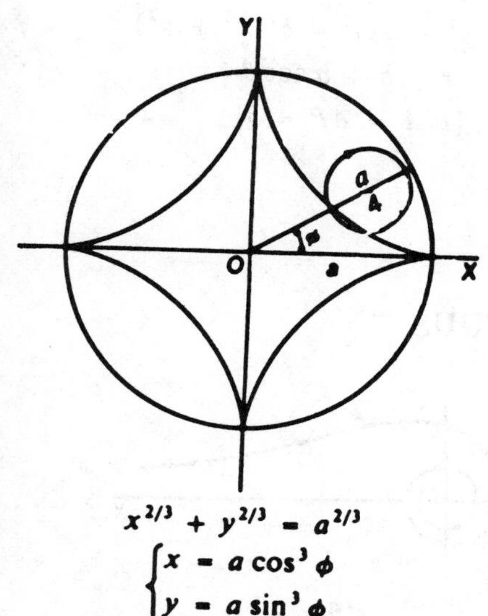

$$x^{2/3} + y^{2/3} = a^{2/3}$$

$$\begin{cases} x = a\cos^3\phi \\ y = a\sin^3\phi \end{cases}$$

INVERSE COSINE CURVE

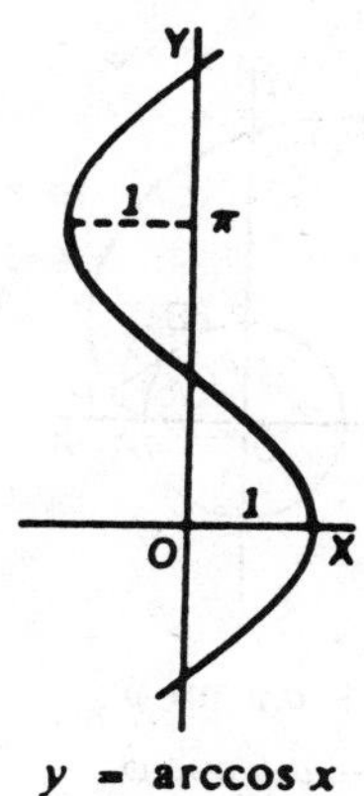

$y = \arccos x$

INVERSE SINE CURVE

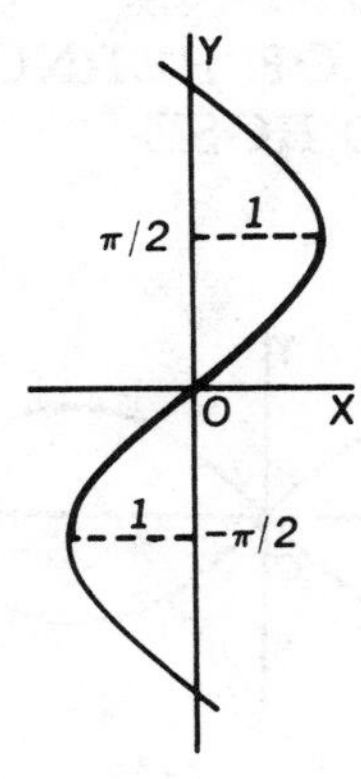

$y = \arcsin x$

INVERSE TANGENT CURVE

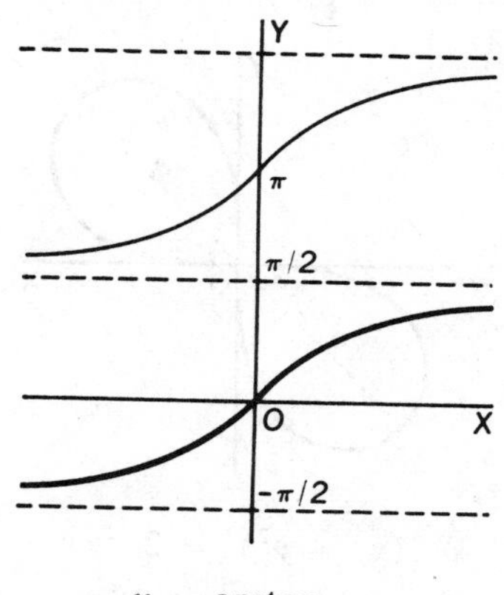

$y = \arctan x$

INVOLUTE OF CIRCLE

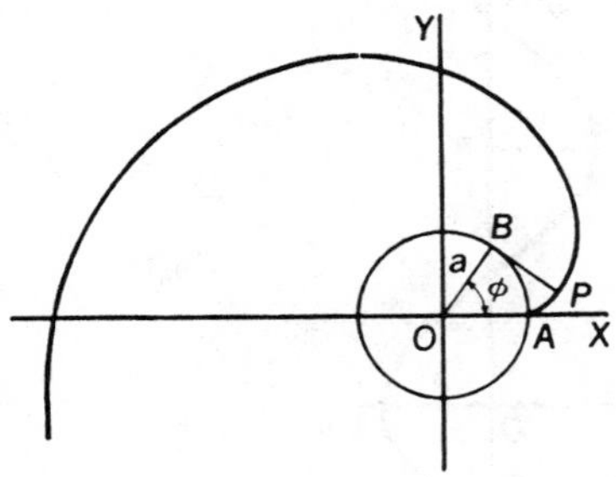

$$\begin{cases} x = a\cos\phi + a\phi\sin\phi \\ y = a\sin\phi - a\phi\cos\phi \end{cases}$$

$[BP = \widehat{BA}]$

LEMNISCATE OF BERNOULLI, TWO-LEAVED ROSE

(a)

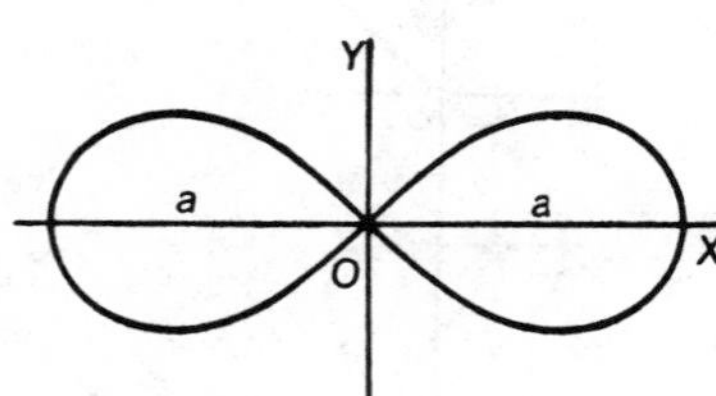

$$(x^2 + y^2)^2 = a^2(x^2 - y^2)$$
$$r^2 = a^2 \cos 2\theta$$

(b)

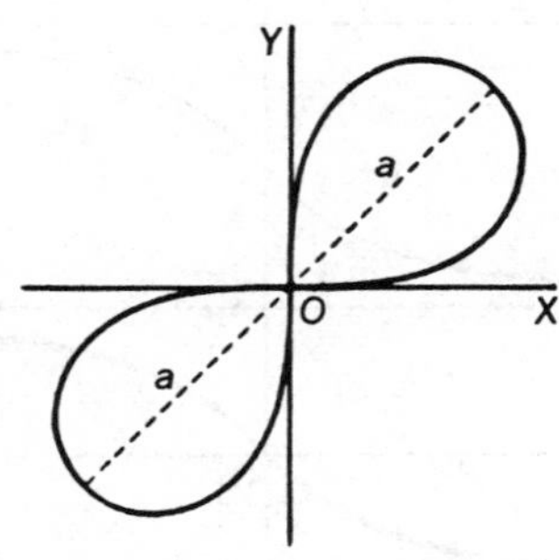

$$(x^2 + y^2)^2 = 2a^2 xy$$
$$r^2 = a^2 \sin 2\theta$$

LIMACON OF PASCAL

(1) $a > b$

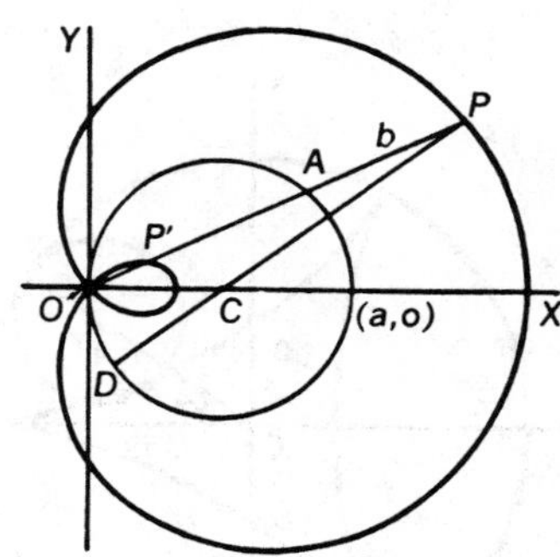

[If $a = 2b$, the curve is called the *trisectrix*, since then $\sphericalangle OPD = \frac{1}{3} \sphericalangle OCD$.]

(2) $a = b$
See: Cardioid

(3) $a < b$

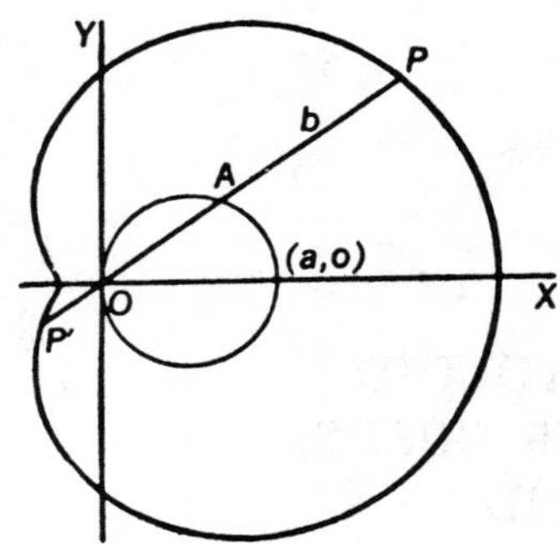

$$(x^2 + y^2 - ax)^2 = b^2(x^2 + y^2)$$
$$r = b + a\cos\theta$$
$$[P'A = AP = b]$$

LITUUS

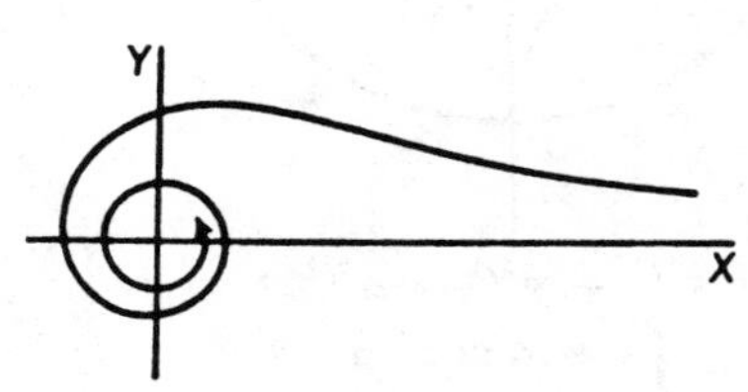

$$r^2\theta = a^2$$

LOGARITHMIC CURVE

(1) $a > 1$

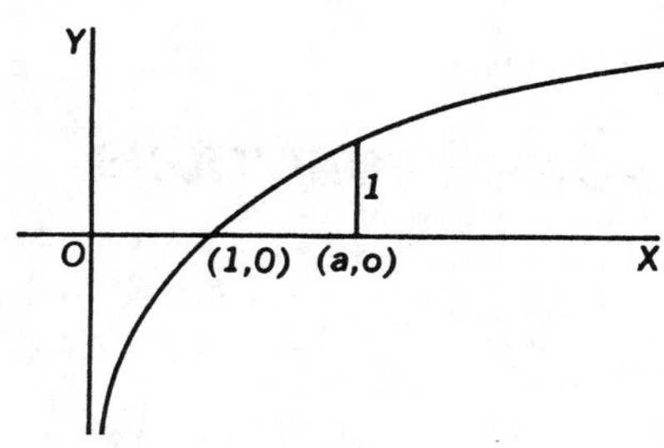

(2) $0 < a < 1$

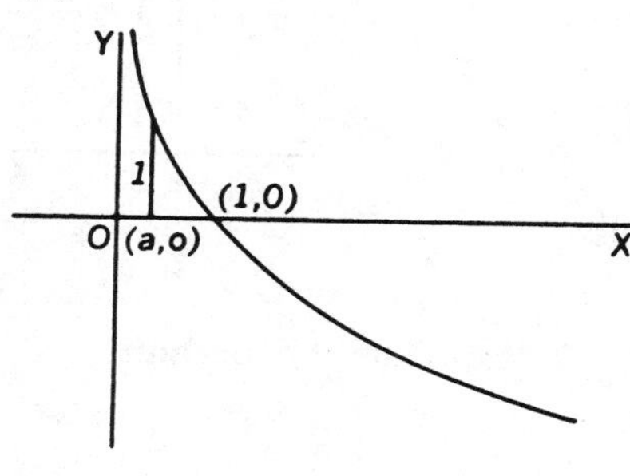

$$y = \log_a x$$

LOGARITHMIC SPIRAL

See: Spiral, logarithmic or equiangular

NEPHROID

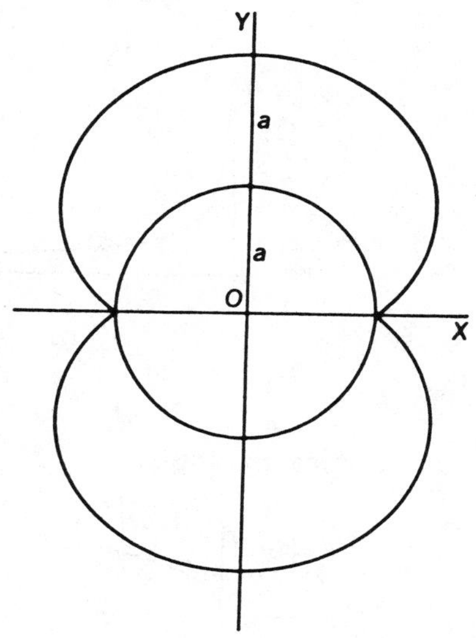

$$\begin{cases} x = \frac{1}{2}a(3\cos\phi - \cos 3\phi) \\ y = \frac{1}{2}a(3\sin\phi - \sin 3\phi) \end{cases}$$

[The nephroid is a 2-cusped epicycloid.]

OUI-JA BOARD CURVE

See: Cochleoid

OVALS OF CASSINI

(1) $b > k$

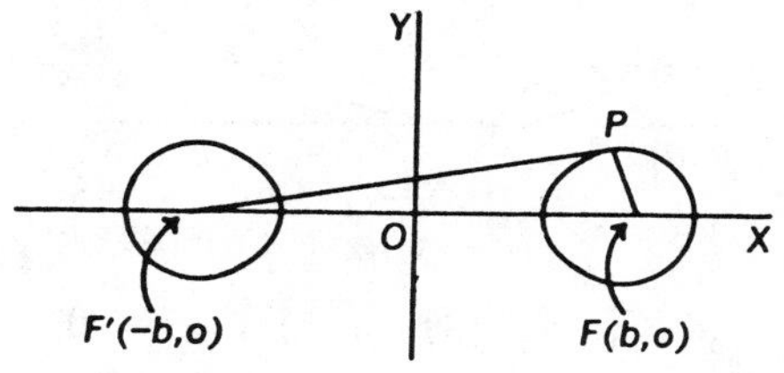

(2) $b = k$

See: Lemniscate of Bernoulli

(3) $b < k$

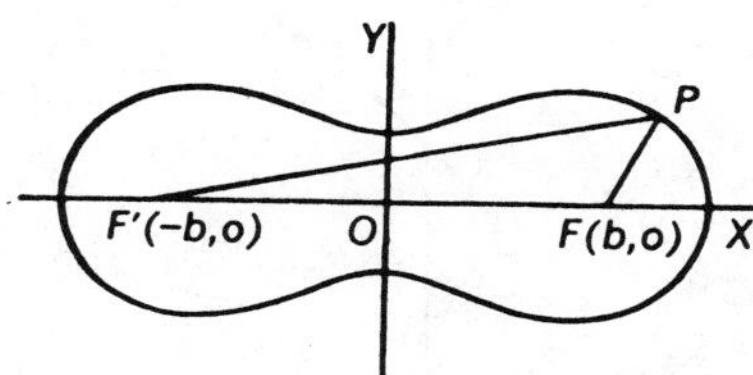

$$(x^2 + y^2 + b^2)^2 - 4b^2x^2 = k^4$$

$$r^4 + b^4 - 2r^2b^2\cos 2\theta = k^4$$

$$[F'P \cdot FP = k^2]$$

[These curves are sections of a torus on planes parallel to the axis of the torus.]

PARABOLA

(1)

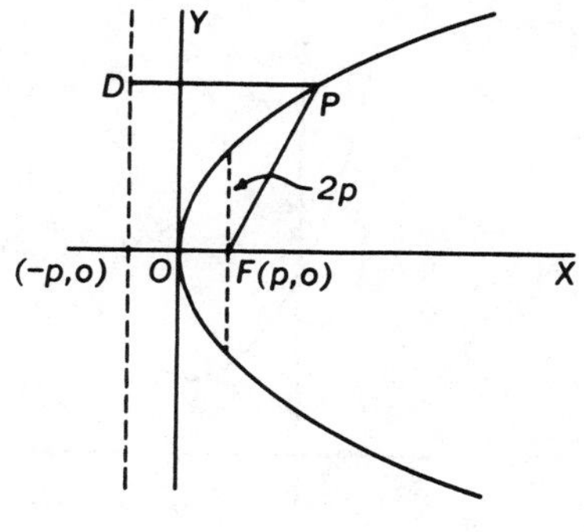

$$y^2 = 4px$$

$$[DP = FP]$$

(2)

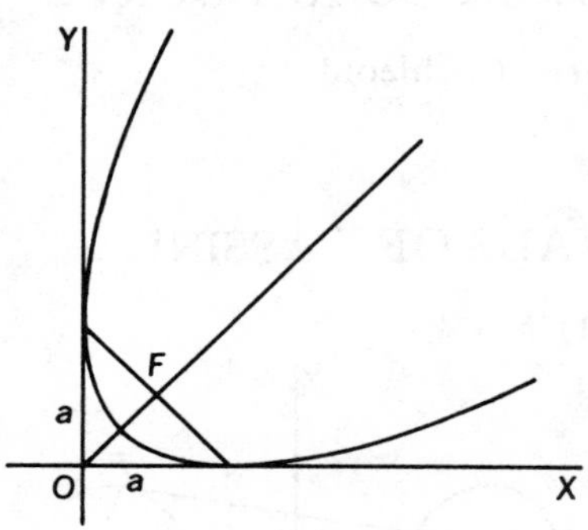

$$\pm x^{1/2} \pm y^{1/2} = a^{1/2}$$
$$(x - y)^2 - 2a(x + y) + a^2 = 0$$

(3)

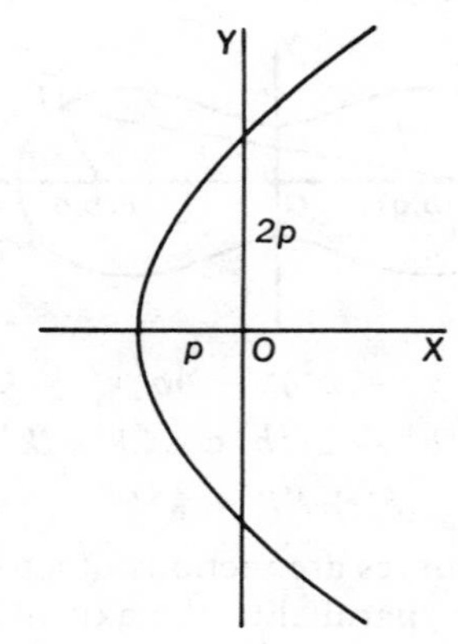

$$r = 2p/(1 - \cos\theta)$$

(4)

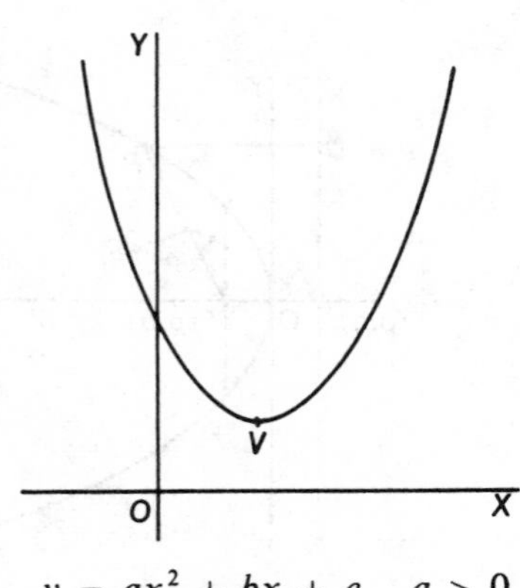

$$y = ax^2 + bx + c, \quad a > 0$$
[abscissa of vertex $= -b/2a$]

PARABOLIC SPIRAL

See: Spiral, parabolic

POWER FUNCTIONS

(1)

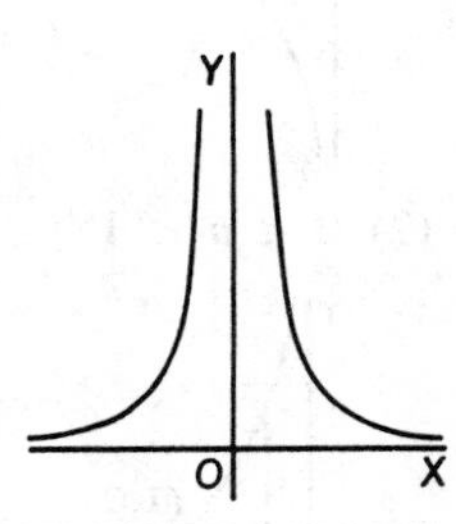

$$y = x^{-2}$$

(2) Equilateral hyperbola

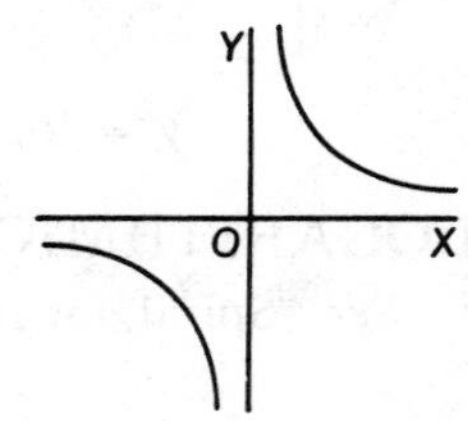

$$y = x^{-1}$$

(3)

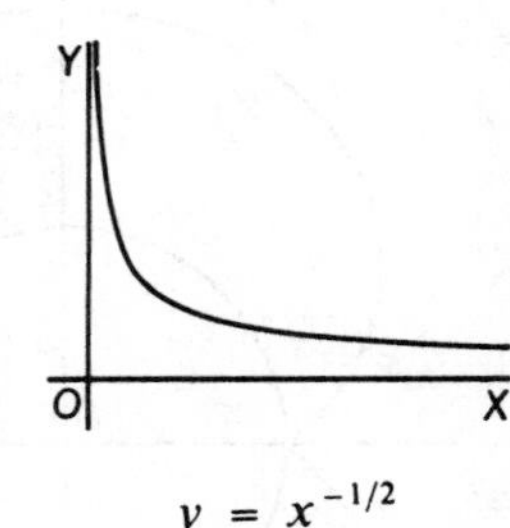

$$y = x^{-1/2}$$

(4) Cubical parabola

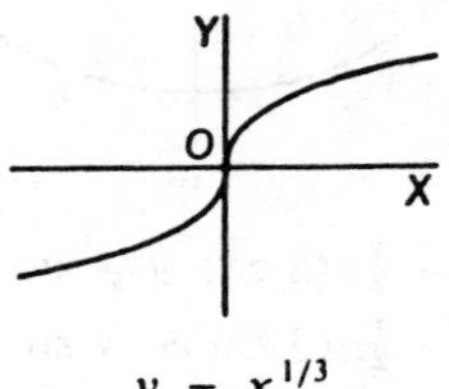

$$y = x^{1/3}$$

(5) Half of a parabola

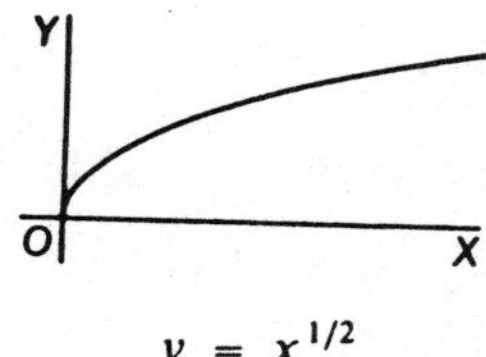

$y = x^{1/2}$

(6) Semicubical parabola

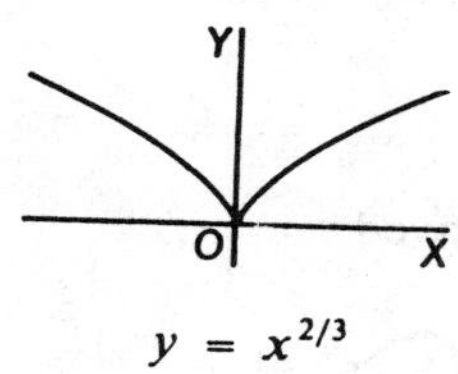

$y = x^{2/3}$

(7) Half of semicubical parabola

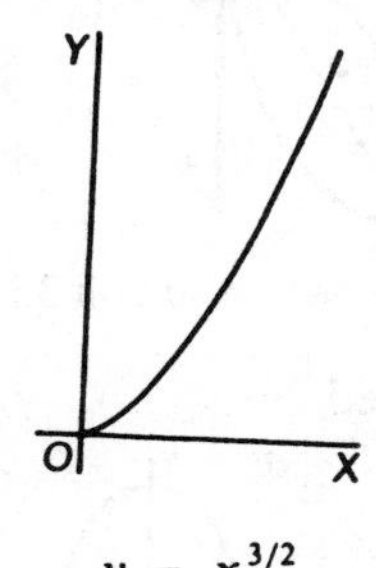

$y = x^{3/2}$

(8) Parabola

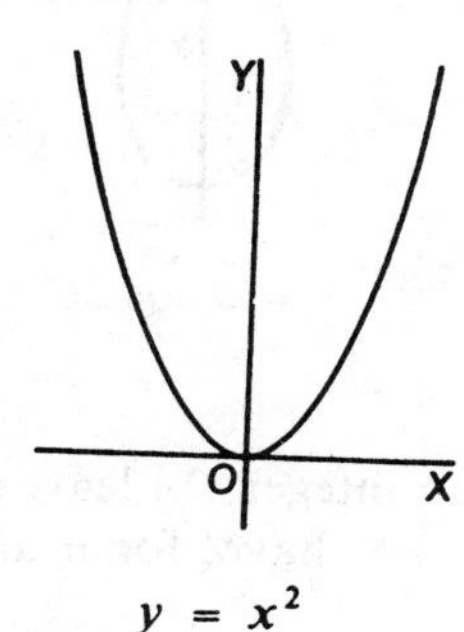

$y = x^2$

(9) Cubical parabola

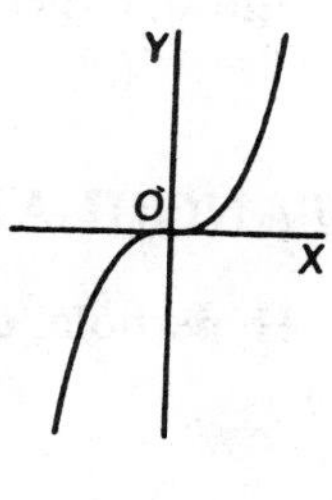

$y = x^3$

PROBABILITY CURVE

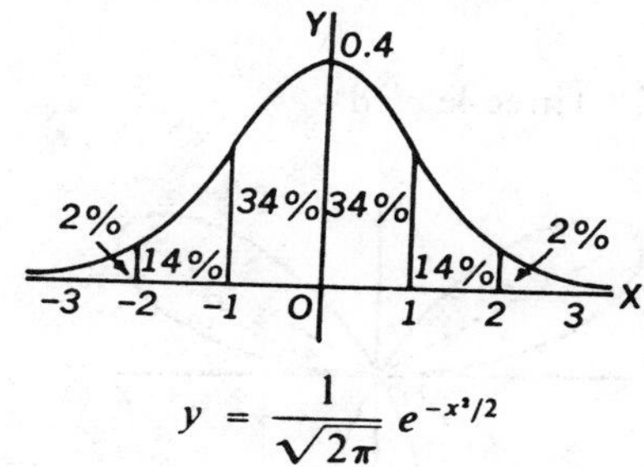

$$y = \frac{1}{\sqrt{2\pi}} e^{-x^2/2}$$

PROLATE CYCLOID

See: Cycloid, prolate

PURSUIT CURVE

See: Tractrix

QUADRATRIX OF HIPPIAS

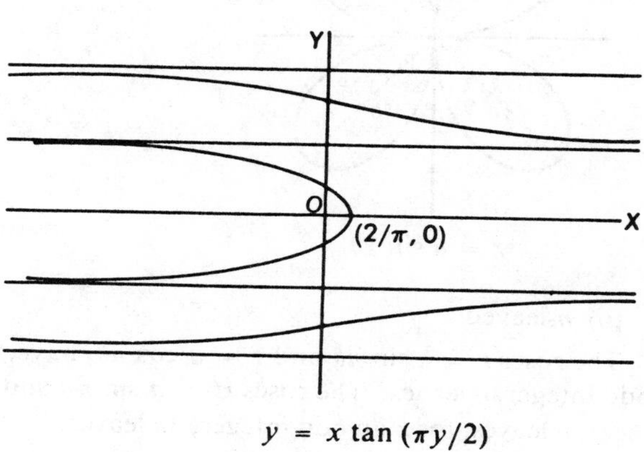

$y = x \tan(\pi y/2)$

RECIPROCAL SPIRAL

See: Spiral, hyperbolic or reciprocal

RECTANGULAR HYPERBOLA

See: Hyperbola, equilateral or rectangular

ROSE CURVES

(1) Two-leaved
See: Lemniscate of Bernoulli, Two-leaved rose

(2) Three-leaved

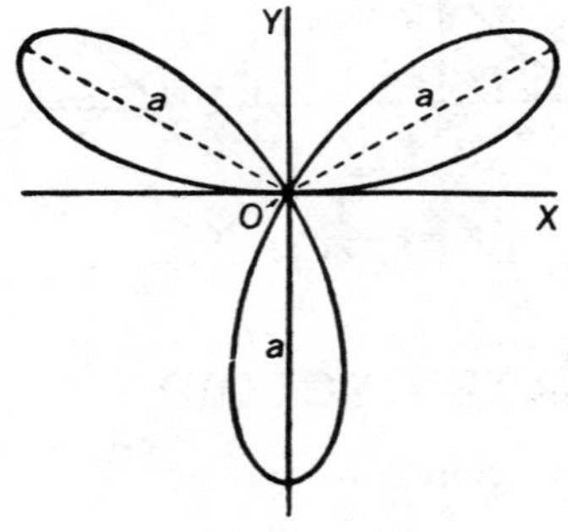

$r = a \sin 3\theta$

(3) Three-leaved

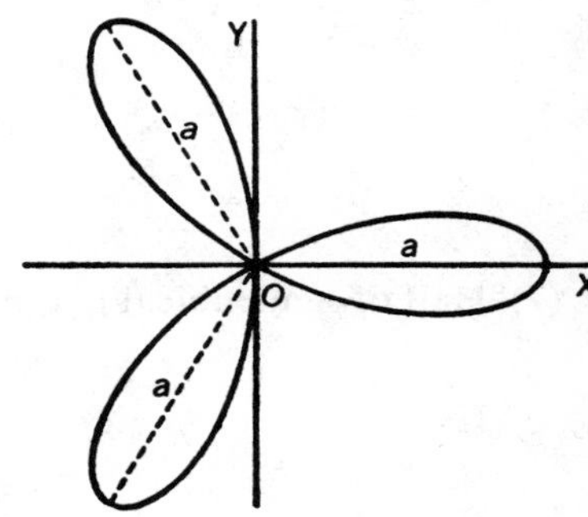

$r = a \cos 3\theta$

(4) Four-leaved

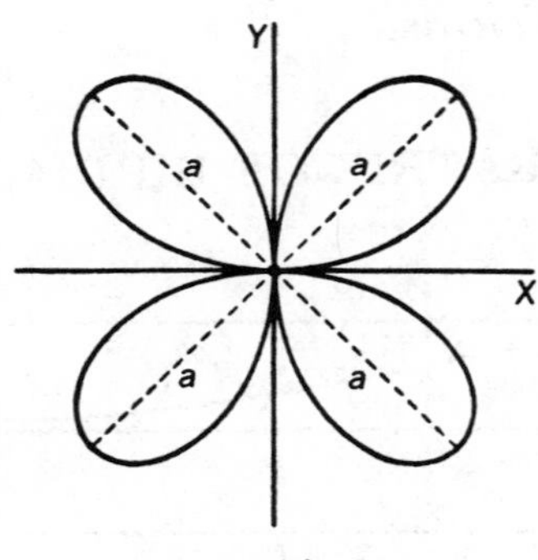

$r = a \sin 2\theta$

(5) Four-leaved

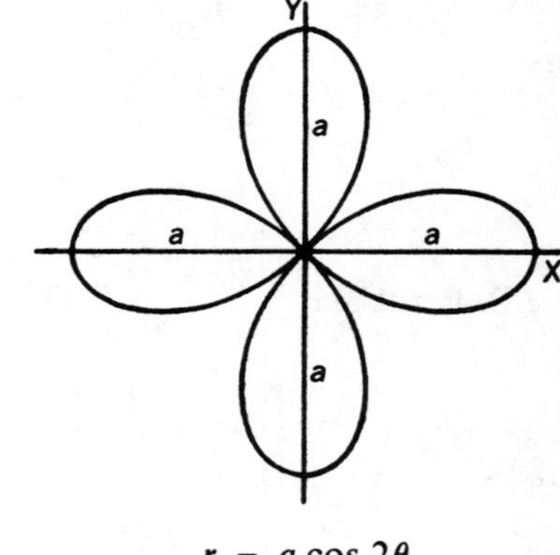

$r = a \cos 2\theta$

(6) n-leaved

The roses $r = a \sin n\theta$ and $r = a \cos n\theta$, have, for n an even integer, $2n$ leaves; for n a odd integer, n leaves. The roses $r^2 = a \sin n\theta$ and $r^2 = a \cos n\theta$, have, for n an even i teger, n leaves; for n an odd integer, $2n$ leaves.

SECANT CURVE

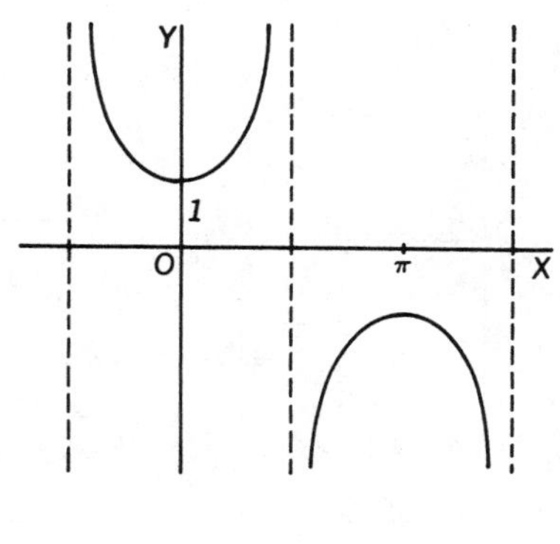

$$y = \sec x$$

SINE CURVE

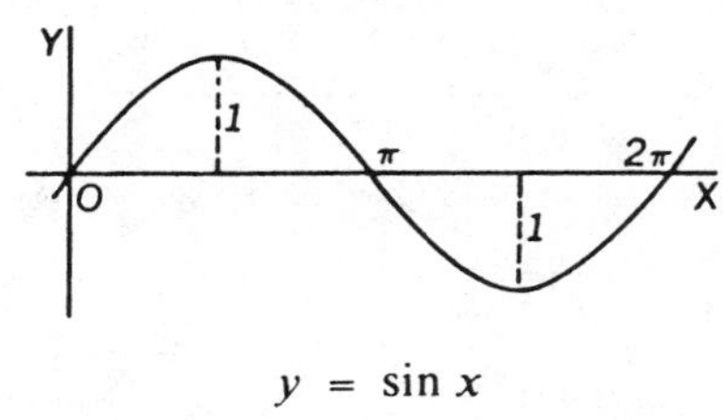

$$y = \sin x$$

SEMICUBICAL PARABOLA

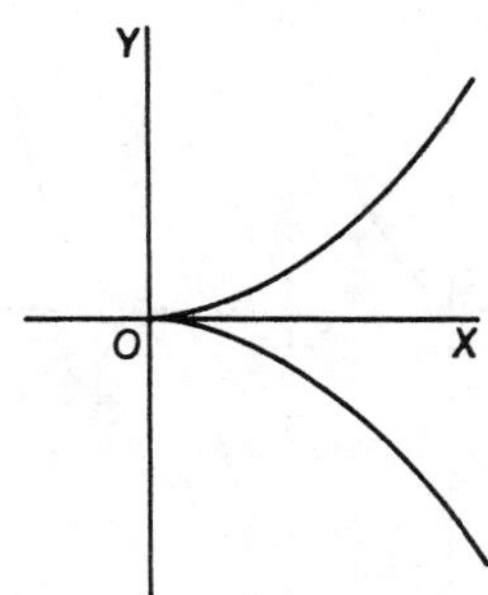

$$y^2 = ax^3$$

$$r = \frac{1}{a}\tan^2\theta\sec\theta$$

SINUSOID

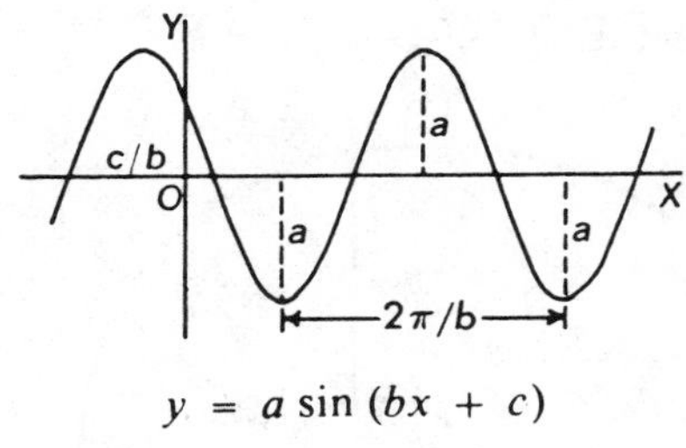

$$y = a\sin(bx + c)$$

SERPENTINE CURVE

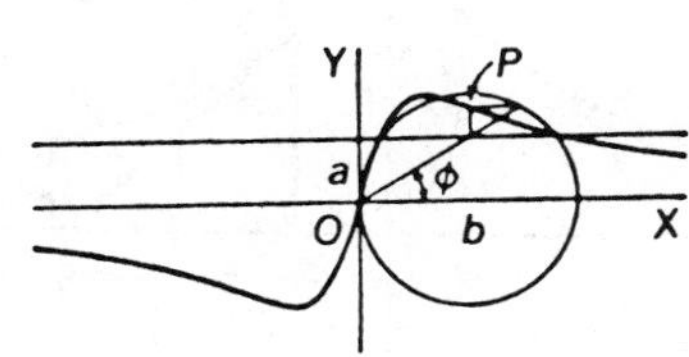

$$(a^2 + x^2)y = abx$$

$$\begin{cases} x = a\cot\phi \\ y = b\sin\phi\cos\phi \end{cases}$$

SPIRAL OF ARCHIMEDES

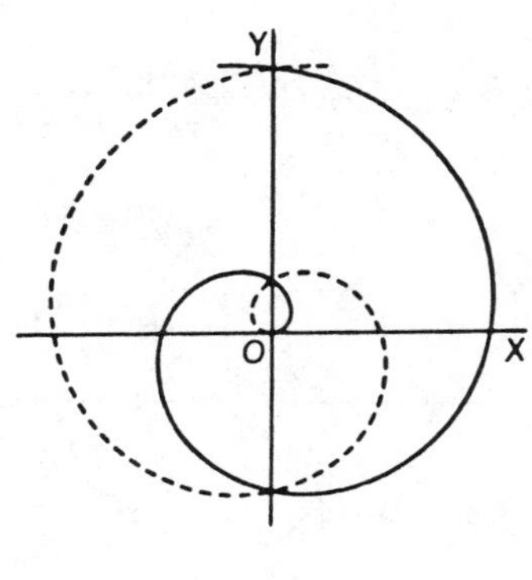

$$r = a\theta$$

SPIRAL, HYPERBOLIC OR RECIPROCAL

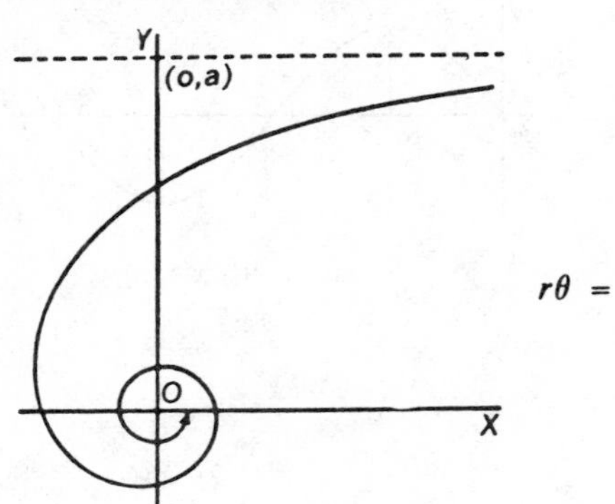

$$r\theta = a$$

STROPHOID

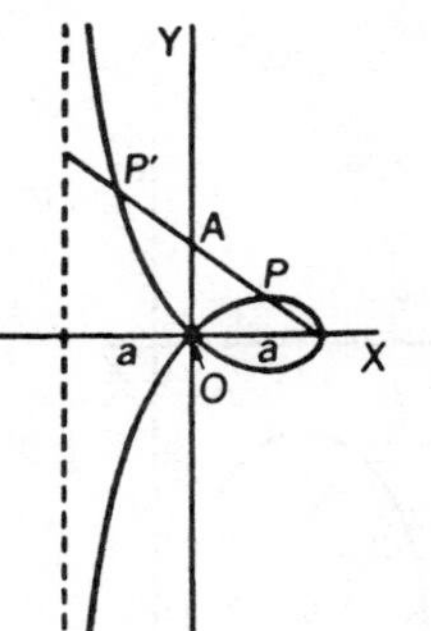

$$y^2 = x^2 \frac{a - x}{a + x}$$

$$r = a \cos 2\theta \sec \theta$$

$$[P'A = AP = OA]$$

SPIRAL, LOGARITHMIC OR EQUIANGULAR

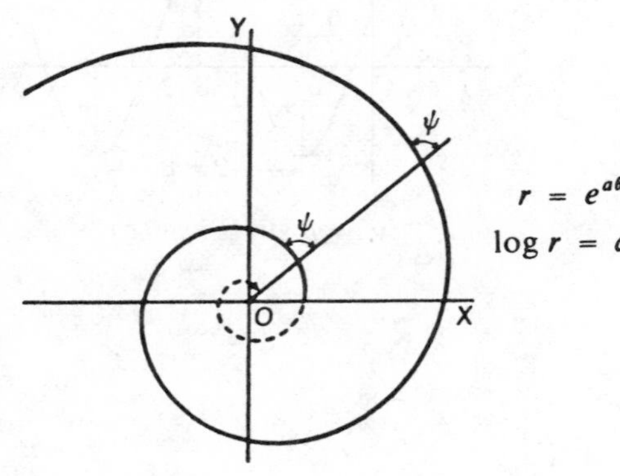

$$r = e^{a\theta}$$

$$\log r = a\theta$$

TANGENT CURVE

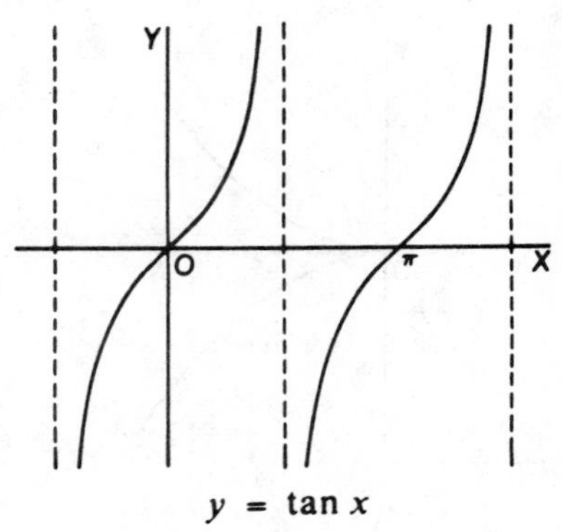

$$y = \tan x$$

SPIRAL, PARABOLIC

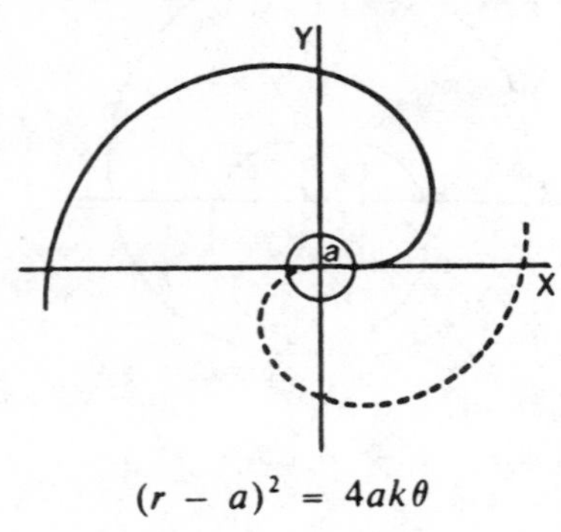

$$(r - a)^2 = 4ak\theta$$

TRACTRIX, PURSUIT CURVE

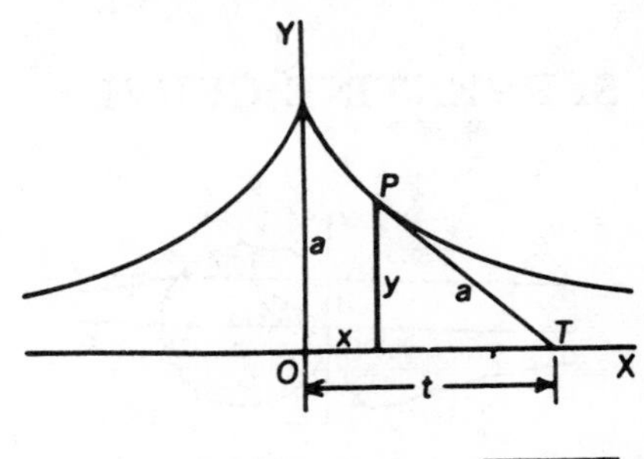

$$x = a \operatorname{sech}^{-1}(y/a) - \sqrt{a^2 - y^2}$$

$$\begin{cases} x = t - a \tanh (t/a) \\ y = a \operatorname{sech} (t/a) \end{cases}$$

$$[PT = a]$$

TRAJECTORY (A PARABOLA)

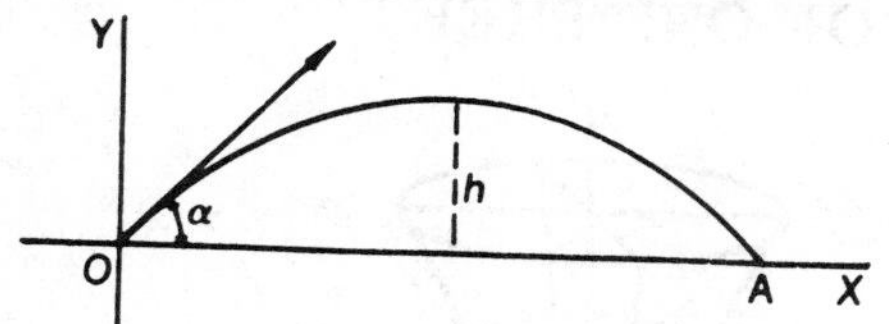

$$y = x \tan \alpha - gx^2/(2v_0^2 \cos^2 \alpha)$$

$$x = (v_0 \cos \alpha) t$$

$$y = (v_0 \sin \alpha) t - gt^2/2$$

TRIGONOMETRIC FUNCTIONS

See: Cosecant curve; Cosine curve; Cotangent curve; Secant curve Sine curve; Tangent curve

TRISECTRIX

See: Limaçon of Pascal (1)

WITCH OF AGNESI

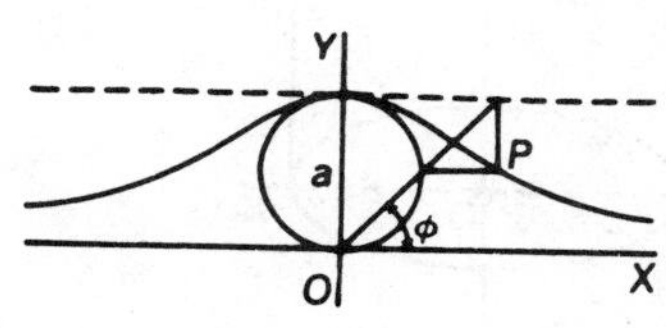

$$y = a^3/(x^2 + a^2)$$

$$\begin{cases} x = a \cot \phi \\ y = a \sin^2 \phi \end{cases}$$

QUADRIC SURFACES*

ELLIPSOID

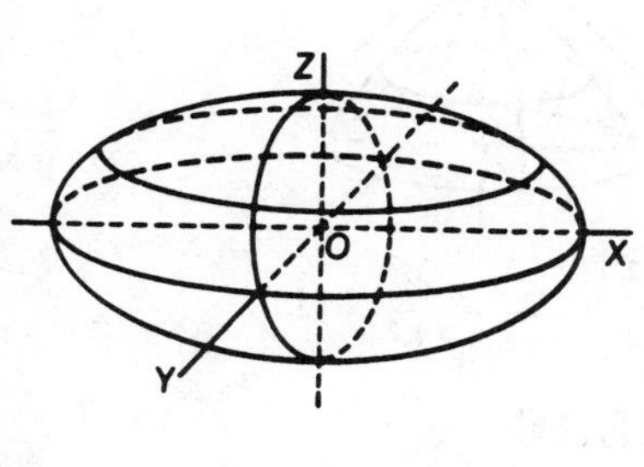

$$\frac{x^2}{a^2} + \frac{y^2}{b^2} + \frac{z^2}{c^2} = 1$$

ELLIPTIC CONE

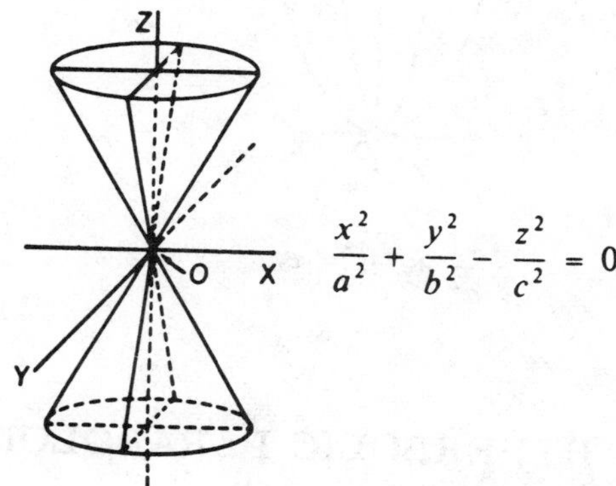

$$\frac{x^2}{a^2} + \frac{y^2}{b^2} - \frac{z^2}{c^2} = 0$$

*Each of the equations is given for the case where the origin is located at (0, 0, 0), the center of the quadric surface. If, however, the center of the surface is at (h, k, l), replace x by $x - h$, y by $y - k$, and z by $z - l$, and the particular standardized form will be that of the surface with center at (h, k, l). For example, the elliptic paraboloid would be

$$\frac{(x - h)^2}{a^2} + \frac{(y - k)^2}{b^2} = c(z - l).$$

ELLIPTIC CYLINDER

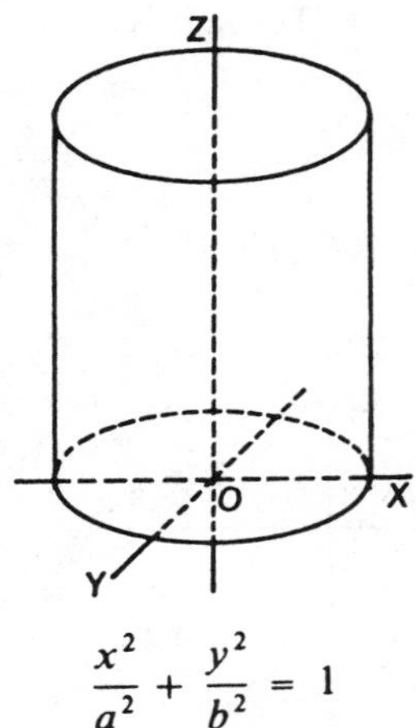

$$\frac{x^2}{a^2} + \frac{y^2}{b^2} = 1$$

HYPERBOLOID OF ONE SHEET

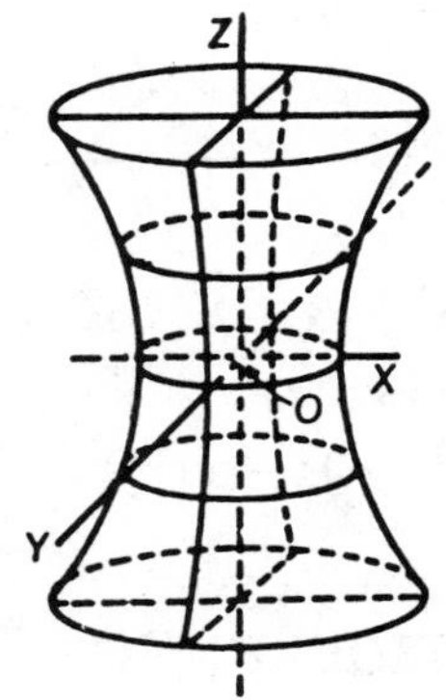

$$\frac{x^2}{a^2} + \frac{y^2}{b^2} - \frac{z^2}{c^2} = 1$$

ELLIPTIC PARABOLOID

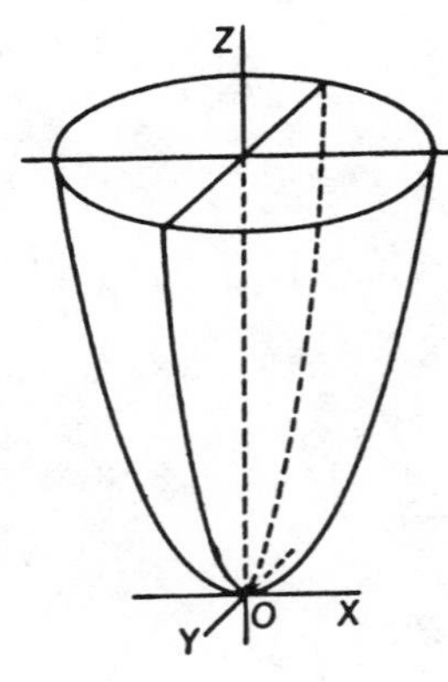

$$\frac{x^2}{a^2} + \frac{y^2}{b^2} = cz$$

HYPERBOLOID OF TWO SHEETS

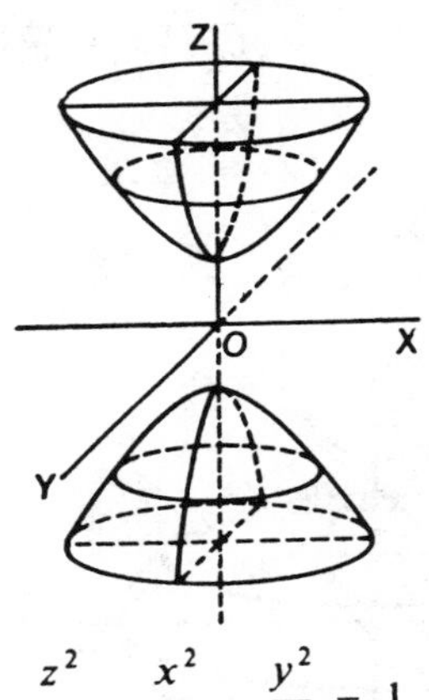

$$\frac{z^2}{c^2} - \frac{x^2}{a^2} - \frac{y^2}{b^2} = 1$$

HYPERBOLIC PARABOLOID

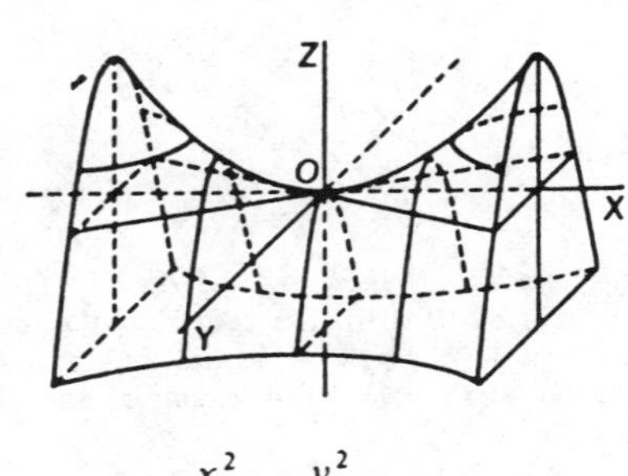

$$\frac{x^2}{a^2} - \frac{y^2}{b^2} = cz$$

SPHERE

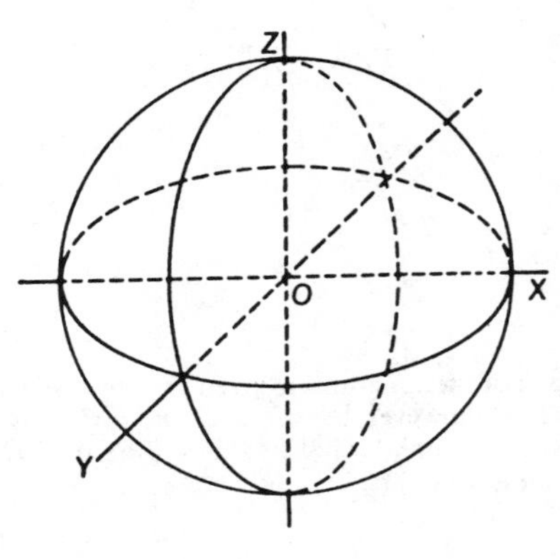

$$x^2 + y^2 + z^2 = a^2$$

PATTERNS OF REGULAR POLYHEDRA

TETRAHEDRON

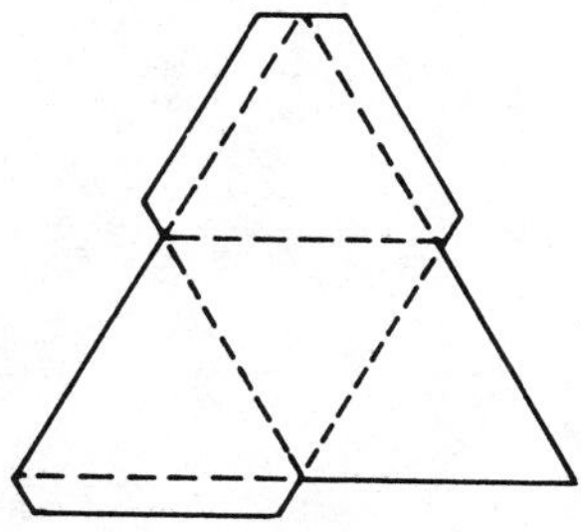

OCTAHEDRON

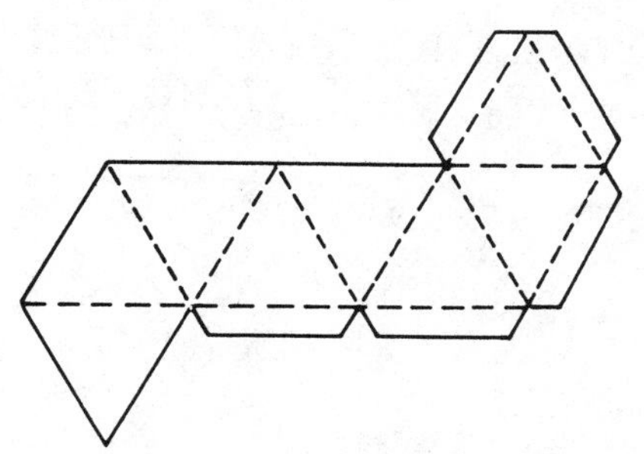

HEXAHEDRON, OR CUBE

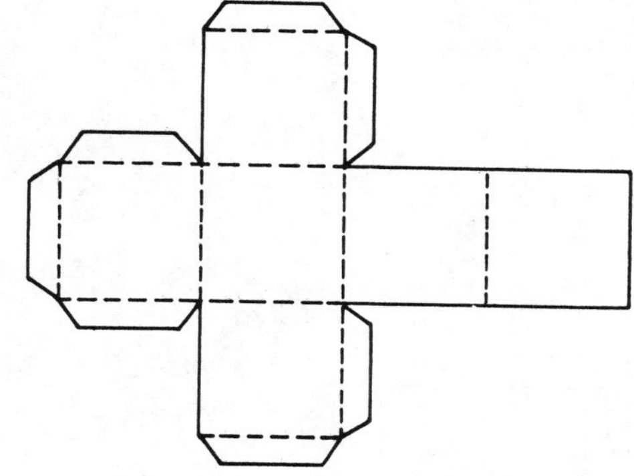

ICOSAHEDRON

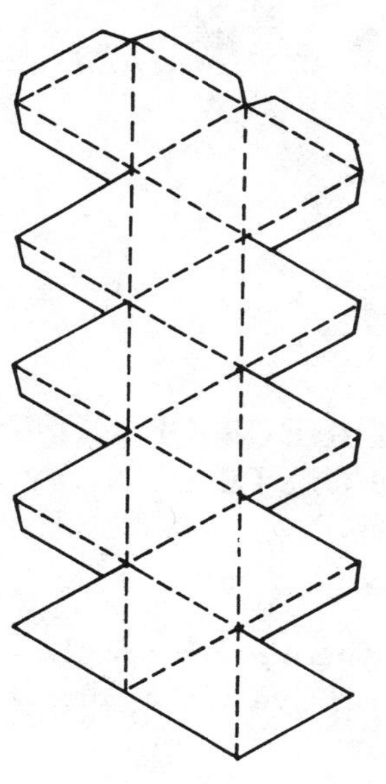

DODECAHEDRON

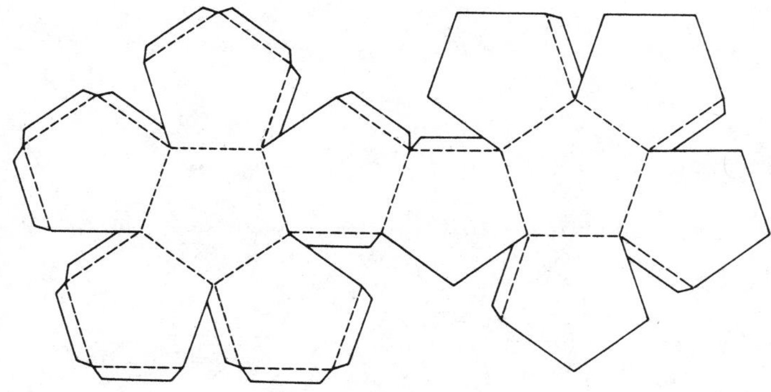

FORMULAS FROM SOLID ANALYTIC GEOMETRY

DISTANCE d BETWEEN TWO POINTS $P_1(x_1, y_1, z_1)$ AND $P_2(x_2, y_2, z_2)$

4.5 $$d = \sqrt{(x_2 - x_1)^2 + (y_2 - y_1)^2 + (z_2 - z_1)^2}$$

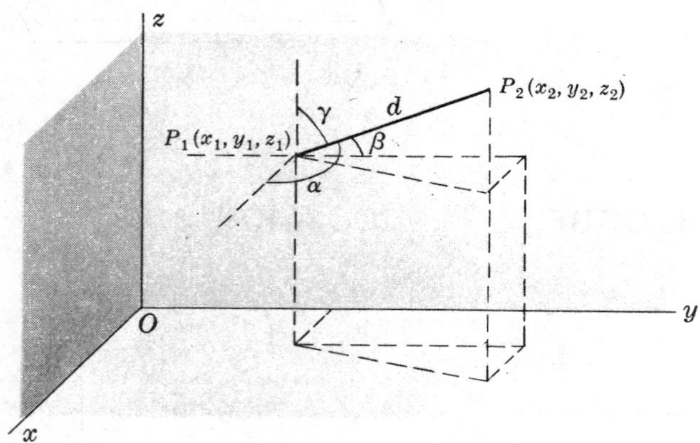

Fig. 1

DIRECTION COSINES OF LINE JOINING POINTS $P_1(x_1, y_1, z_1)$ AND $P_2(x_2, y_2, z_2)$

$$l = \cos\alpha = \frac{x_2 - x_1}{d}, \qquad m = \cos\beta = \frac{y_2 - y_1}{d}, \qquad n = \cos\gamma = \frac{z_2 - z_1}{d}$$

where α, β, γ are the angles which line P_1P_2 makes with the positive x, y, z axes respectively and d is given by 4.5 [See Fig. 1].

RELATIONSHIP BETWEEN DIRECTION COSINES

$$\cos^2\alpha + \cos^2\beta + \cos^2\gamma = 1 \qquad \text{or} \qquad l^2 + m^2 + n^2 = 1$$

DIRECTION NUMBERS

Numbers L, M, N which are proportional to the direction cosines l, m, n are called *direction numbers*. The relationship between them is given by

$$l = \frac{L}{\sqrt{L^2 + M^2 + N^2}}, \qquad m = \frac{M}{\sqrt{L^2 + M^2 + N^2}}, \qquad n = \frac{N}{\sqrt{L^2 + M^2 + N^2}}$$

EQUATIONS OF LINE JOINING $P_1(x_1, y_1, z_1)$ AND $P_2(x_2, y_2, z_2)$ IN STANDARD FORM

$$\frac{x - x_1}{x_2 - x_1} = \frac{y - y_1}{y_2 - y_1} = \frac{z - z_1}{z_2 - z_1} \quad \text{or} \quad \frac{x - x_1}{l} = \frac{y - y_1}{m} = \frac{z - z_1}{n}$$

These are also valid if l, m, n are replaced by L, M, N respectively.

EQUATIONS OF LINE JOINING $P_1(x_1, y_1, z_1)$ AND $P_2(x_2, y_2, z_2)$ IN PARAMETRIC FORM

$$x = x_1 + lt, \quad y = y_1 + mt, \quad z = z_1 + nt$$

These are also valid if l, m, n are replaced by L, M, N respectively.

ANGLE ϕ BETWEEN TWO LINES WITH DIRECTION COSINES ℓ_1, m_1, n_1 AND ℓ_2, m_2, n_2

$$\cos\phi = l_1 l_2 + m_1 m_2 + n_1 n_2$$

GENERAL EQUATION OF A PLANE

$$Ax + By + Cz + D = 0 \qquad [A, B, C, D \text{ are constants}]$$

EQUATION OF PLANE PASSING THROUGH POINTS (x_1, y_1, z_1), (x_2, y_2, z_2), (x_3, y_3, z_3)

$$\begin{vmatrix} x - x_1 & y - y_1 & z - z_1 \\ x_2 - x_1 & y_2 - y_1 & z_2 - z_1 \\ x_3 - x_1 & y_3 - y_1 & z_3 - z_1 \end{vmatrix} = 0$$

or

$$\begin{vmatrix} y_2 - y_1 & z_2 - z_1 \\ y_3 - y_1 & z_3 - z_1 \end{vmatrix}(x - x_1) + \begin{vmatrix} z_2 - z_1 & x_2 - x_1 \\ z_3 - z_1 & x_3 - x_1 \end{vmatrix}(y - y_1) +$$

$$\begin{vmatrix} x_2 - x_1 & y_2 - y_1 \\ x_3 - x_1 & y_3 - y_1 \end{vmatrix}(z - z_1) = 0$$

EQUATION OF PLANE IN INTERCEPT FORM

$$\frac{x}{a} + \frac{y}{b} + \frac{z}{c} = 1$$

where a, b, c are the intercepts on the x, y, z axes respectively.

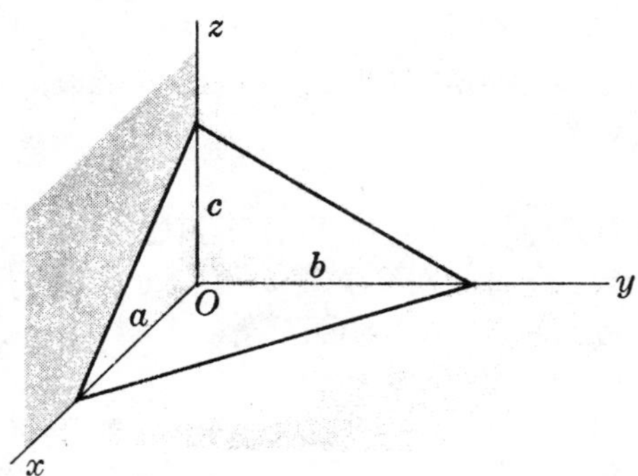

EQUATIONS OF LINE THROUGH (x_0, y_0, z_0) AND PERPENDICULAR TO PLANE $Ax + By + Cz + D = 0$

$$\frac{x - x_0}{A} = \frac{y - y_0}{B} = \frac{z - z_0}{C}$$

or $\quad x = x_0 + At, \quad y = y_0 + Bt, \quad z = z_0 + Ct$

Note that the direction numbers for a line perpendicular to the plane $Ax + By + Cz + D = 0$ are A, B, C.

DISTANCE FROM POINT (x , y , z) TO PLANE Ax + By + D = 0

$$\frac{Ax_0 + By_0 + Cz_0 + D}{\pm\sqrt{A^2 + B^2 + C^2}}$$

where the sign is chosen so that the distance is nonnegative.

NORMAL FORM FOR EQUATION OF PLANE

$$x \cos\alpha + y \cos\beta + z \cos\gamma = p$$

where p = perpendicular distance from O to plane at P and α, β, γ are angles between OP and positive x, y, z axes.

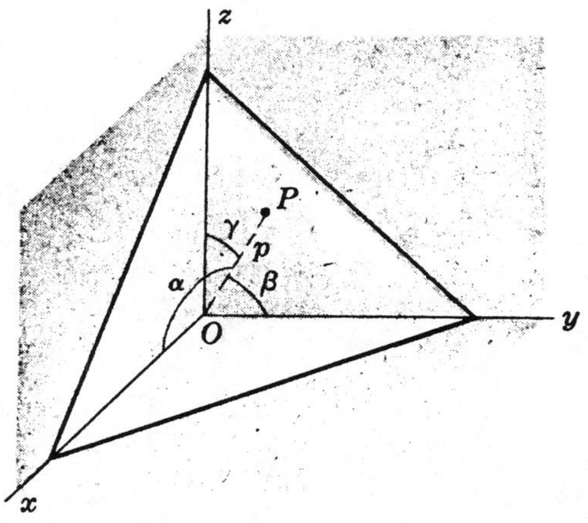

TRANSFORMATION OF COORDINATES INVOLVING PURE TRANSLATION

$$\begin{cases} x = x' + x_0 \\ y = y' + y_0 \\ z = z' + z_0 \end{cases} \quad \text{or} \quad \begin{cases} x' = x - x_0 \\ y' = y - y_0 \\ z' = z - z_0 \end{cases}$$

where (x, y, z) are old coordinates [i.e. coordinates relative to xyz system], (x', y', z') are new coordinates [relative to $x'y'z'$ system] and (x_0, y_0, z_0) are the coordinates of the new origin O' relative to the old xyz coordinate system.

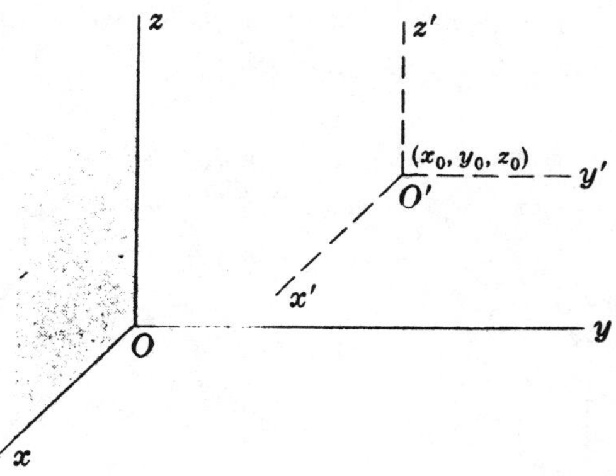

TRANSFORMATION OF COORDINATES INVOLVING PURE ROTATION

$$\begin{cases} x = l_1x' + l_2y' + l_3z' \\ y = m_1x' + m_2y' + m_3z' \\ z = n_1x' + n_2y' + n_3z' \end{cases}$$

$$\text{or}\quad \begin{cases} x' = l_1x + m_1y + n_1z \\ y' = l_2x + m_2y + n_2z \\ z' = l_3x + m_3y + n_3z \end{cases}$$

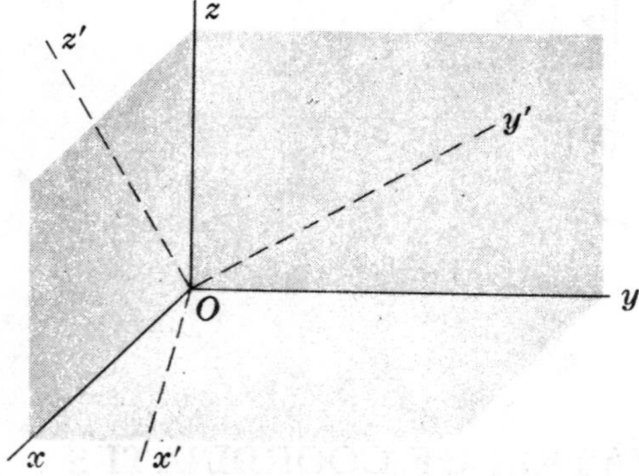

where the origins of the xyz and $x'y'z'$ systems are the same and l_1, m_1, n_1; l_2, m_2, n_2; l_3, m_3, n_3 are the direction cosines of the x', y', z' axes relative to the x, y, z axes respectively.

TRANSFORMATION OF COORDINATES INVOLVING TRANSLATION AND ROTATION

$$\begin{cases} x = l_1x' + l_2y' + l_3z' + x_0 \\ y = m_1x' + m_2y' + m_3z' + y_0 \\ z = n_1x' + n_2y' + n_3z' + z_0 \end{cases}$$

$$\text{or}\quad \begin{cases} x' = l_1(x - x_0) + m_1(y - y_0) + n_1(z - z_0) \\ y' = l_2(x - x_0) + m_2(y - y_0) + n_2(z - z_0) \\ z' = l_3(x - x_0) + m_3(y - y_0) + n_3(z - z_0) \end{cases}$$

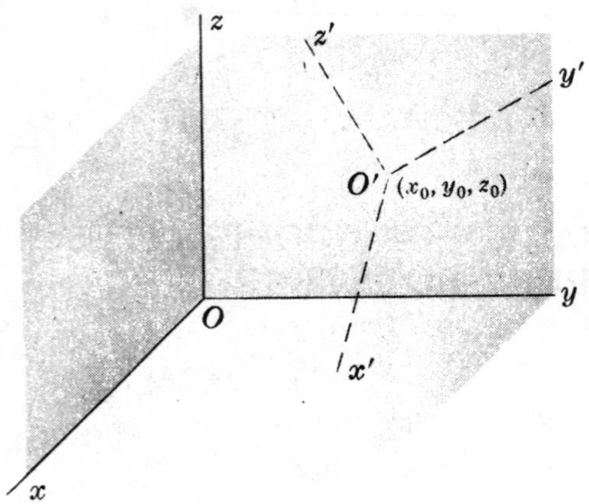

where the origin O' of the $x'y'z'$ system has coordinates (x_0, y_0, z_0) relative to the xyz system and l_1, m_1, n_1; l_2, m_2, n_2; l_3, m_3, n_3 are the direction cosines of the x', y', z' axes relative to the x, y, z axes respectively.

CYCLINDRICAL COORDINATES (r, θ, z)

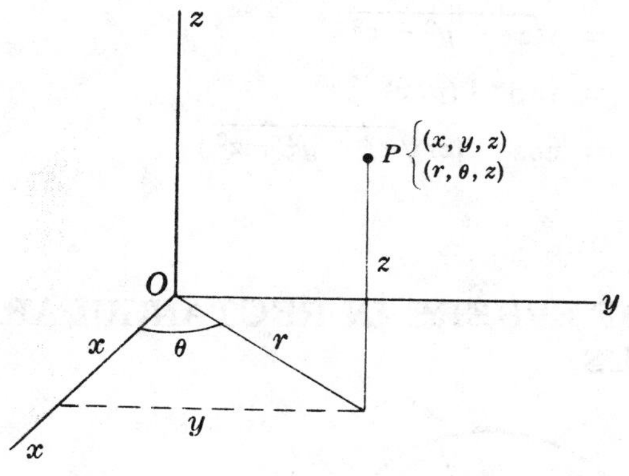

Fig. 2

A point P can be located by cylindrical coordinates (r, θ, z) (See Fig. 2) as well as rectangular coordinates (x, y, z). The transformation between these coordinates is

$$\begin{cases} x = r\cos\theta \\ y = r\sin\theta \\ z = z \end{cases} \quad \text{or} \quad \begin{cases} r = \sqrt{x^2 + y^2} \\ \theta = \tan^{-1}(y/x) \\ z = z \end{cases}$$

SPHERICAL COORDINATES (r, θ, ϕ)

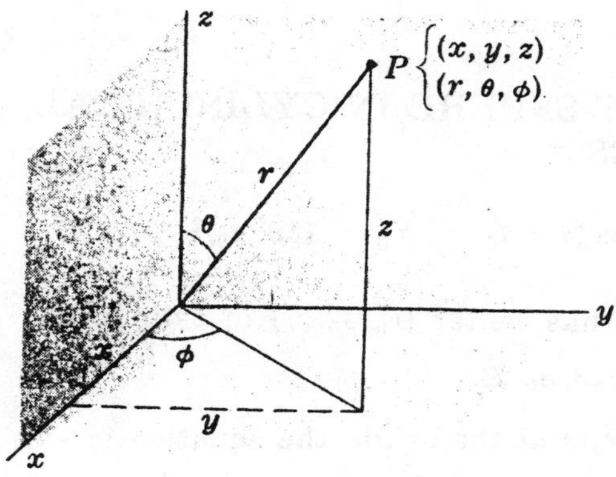

Fig. 3

A point P can be located by spherical coordinates (r, θ, ϕ) (See Fig. 3) as well as rectangular coordinates (x, y, z). The transformation between those coordinates is

$$\begin{cases} x = r \sin \theta \cos \phi \\ y = r \sin \theta \sin \phi \\ z = r \cos \theta \end{cases}$$

or

$$\begin{cases} r = \sqrt{x^2 + y^2 + z^2} \\ \phi = \tan^{-1} (y/x) \\ \theta = \cos^{-1} (z/\sqrt{x^2 + y^2 + z^2}) \end{cases}$$

EQUATION OF SPHERE IN RECTANGULAR COORDINATES

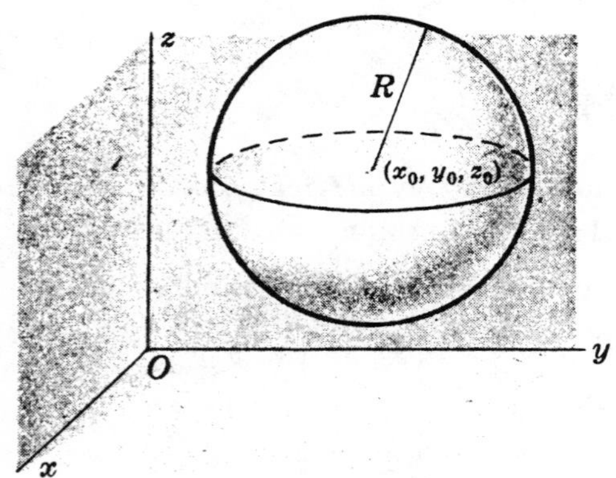

$$(x - x_0)^2 + (y - y_0)^2 + (z - z_0)^2 = R^2$$

where the sphere has center (x_0, y_0, z_0) and radius R.

EQUATION OF SPHERE IN CYLINDRICAL COORDINATES

$$r^2 - 2r_0 r \cos (\theta - \theta_0) + r_0^2 + (z - z_0)^2 = R^2$$

where the sphere has center (r_0, θ_0, z_0) in cylindrical coordinates and radius R.

If the center is at the origin the equation is

$$r^2 + z^2 = R^2$$

EQUATION OF SPHERE IN SPHERICAL COORDINATES

$$r^2 + r_0^2 - 2r_0 r \sin\theta \sin\theta_0 \cos(\phi - \phi_0) = R^2$$

where the sphere has center (r_0, θ_0, ϕ_0) in spherical coordinates and radius R.

If the center is at the origin the equation is

$$r = R$$

EQUATION OF ELLIPSOID WITH CENTER (x_0, y_0, z_0) AND SEMI-AXES a, b, c

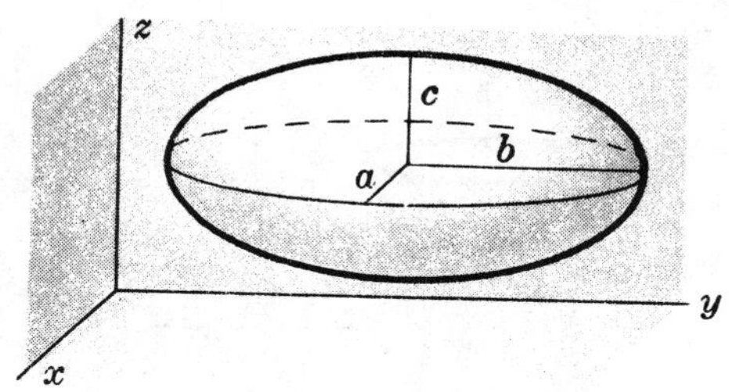

$$\frac{(x-x_0)^2}{a^2} + \frac{(y-y_0)^2}{b^2} + \frac{(z-z_0)^2}{c^2} = 1$$

ELLIPTIC CYLINDER WITH AXIS AS z AXIS

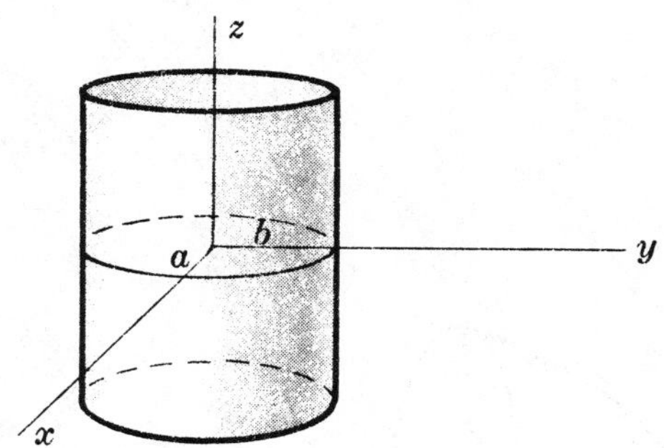

$$\frac{x^2}{a^2} + \frac{y^2}{b^2} = 1$$

where a, b are semi-axes of elliptic cross section.

If $b = a$ it becomes a circular cylinder of radius a.

ELLIPTIC CONE WITH AXIS AS z AXIS

$$\frac{x^2}{a^2} + \frac{y^2}{b^2} = \frac{z^2}{c^2}$$

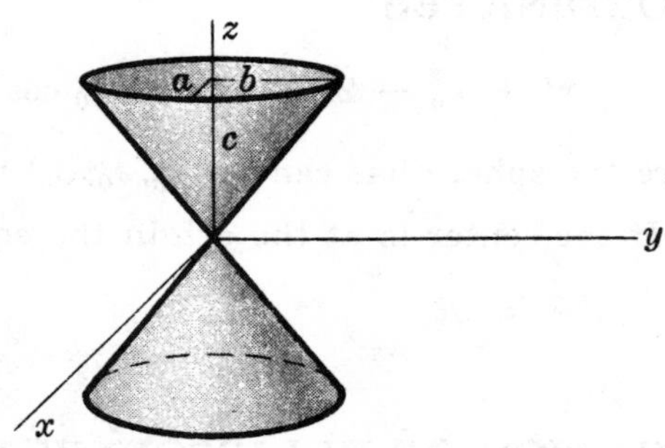

HYPERBOLOID OF ONE SHEET

$$\frac{x^2}{a^2} + \frac{y^2}{b^2} - \frac{z^2}{c^2} = 1$$

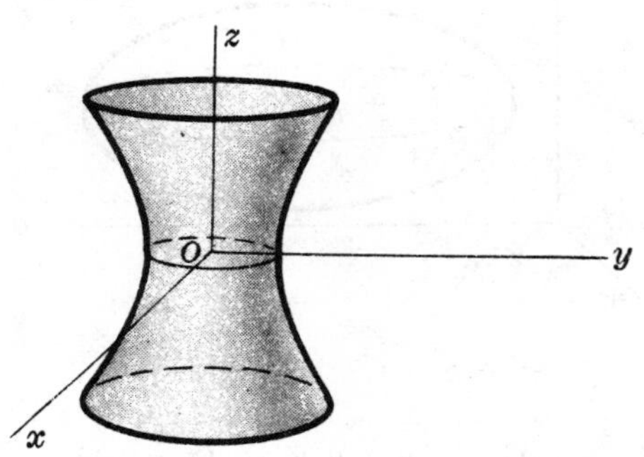

HYPERBOLOID OF TWO SHEETS

$$\frac{x^2}{a^2} - \frac{y^2}{b^2} - \frac{z^2}{c^2} = 1$$

Note orientation of axes

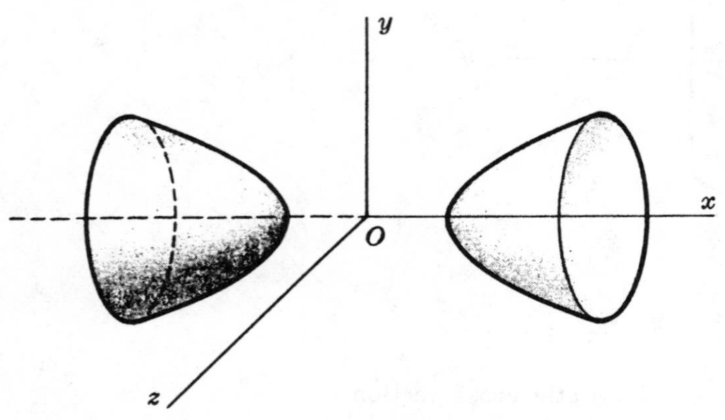

ELLIPTIC PARABOLOID

$$\frac{x^2}{a^2} + \frac{y^2}{b^2} = \frac{z}{c}$$

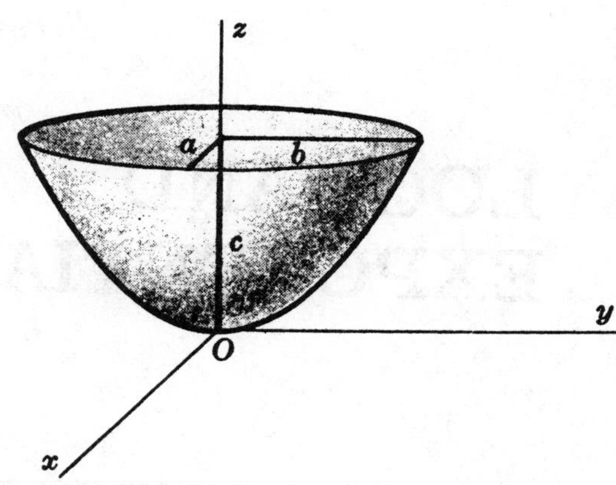

HYPERBOLIC PARABOLOID

$$\frac{x^2}{a^2} - \frac{y^2}{b^2} = \frac{z}{c}$$

Note orientation of axes

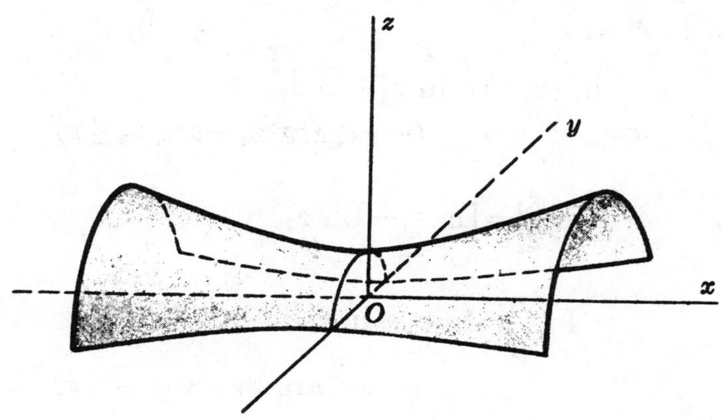

CHAPTER 5

LOGS AND EXPONENTIALS

LOGARITHMIC FUNCTION[1]

LOGARITHMIC IDENTITIES

$$\text{Ln}\,(z_1 z_2) = \text{Ln}\,z_1 + \text{Ln}\,z_2.$$

(i.e., every value of Ln $(z_1 z_2)$ is one of the values of Ln z_1+Ln z_2.)

$$\ln\,(z_1 z_2) = \ln z_1 + \ln z_2 \qquad (-\pi < \arg z_1 + \arg z_2 \leq \pi)$$

$$\text{Ln}\,\frac{z_1}{z_2} = \text{Ln}\,z_1 - \text{Ln}\,z_2$$

$$\ln \frac{z_1}{z_2} = \ln z_1 - \ln z_2 \qquad (-\pi < \arg z_1 - \arg z_2 \leq \pi)$$

$$\text{Ln}\,z^n = n\,\text{Ln}\,z \qquad (n \text{ integer})$$

$$\ln z^n = n \ln z \qquad (n \text{ integer}, \quad -\pi < n \arg z \leq \pi)$$

SPECIAL VALUES

$$\ln 1 = 0$$

[1] From "Handbook of Mathematical Functions", National Bureau of Standards

$$\ln 0 = -\infty$$

$$\ln(-1) = \pi i$$

$$\ln(\pm i) = \pm \tfrac{1}{2}\pi i$$

$\ln e = 1$, e is the real number such that

$$\int_1^e \frac{dt}{t} = 1$$

5.1 $$e = \lim_{n\to\infty}\left(1+\frac{1}{n}\right)^n = 2.71828\ 18284\ldots$$

(see 5.2)

LOGARITHMS TO GENERAL BASE

$$\log_a z = \ln z / \ln a$$

$$\log_a z = \frac{\log_b z}{\log_b a}$$

$$\log_a b = \frac{1}{\log_b a}$$

$$\log_e z = \ln z$$

$$\log_{10} z = \ln z/\ln 10 = \log_{10} e \ln z$$
$$= (.43429\quad 44819\ldots)\ln z$$

$$\ln z = \ln 10 \log_{10} z = (2.30258\ 50929\ldots)\log_{10} z$$

($\log_e x = \ln x$, called natural, Napierian, or hyperbolic logarithms; $\log_{10} x$, called common or Briggs logarithms.)

SERIES EXPANSIONS

$$\ln(1+z) = z - \tfrac{1}{2}z^2 + \tfrac{1}{3}z^3 - \ldots$$
$$(|z| \leq 1 \text{ and } z \neq -1)$$

$$\ln z = \left(\frac{z-1}{z}\right) + \frac{1}{2}\left(\frac{z-1}{z}\right)^2 + \frac{1}{3}\left(\frac{z-1}{z}\right)^3 + \ldots$$
$$(\mathscr{R}z \geq \tfrac{1}{2})$$

$$\ln z=(z-1)-\tfrac{1}{2}(z-1)^2+\tfrac{1}{3}(z-1)^3-\ldots$$
$$(|z-1|\leq 1, \quad z\neq 0)$$

$$\ln z=2\left[\left(\frac{z-1}{z+1}\right)+\frac{1}{3}\left(\frac{z-1}{z+1}\right)^3+\frac{1}{5}\left(\frac{z-1}{z+1}\right)^5+\ldots\right]$$
$$(\mathscr{R}z\geq 0, \quad z\neq 0)$$

$$\ln\left(\frac{z+1}{z-1}\right)=2\left(\frac{1}{z}+\frac{1}{3z^3}+\frac{1}{5z^5}+\ldots\right)$$
$$(|z|\geq 1, \quad z\neq\pm 1)$$

$$\ln(z+a)=\ln a+2\left[\left(\frac{z}{2a+z}\right)+\frac{1}{3}\left(\frac{z}{2a+z}\right)^3+\frac{1}{5}\left(\frac{z}{2a+z}\right)^5+\ldots\right]$$
$$(a>0, \quad \mathscr{R}z\geq -a\neq z)$$

LIMITING VALUES

$$\lim_{x\to\infty} x^{-\alpha}\ln x=0$$
$$(\alpha \text{ constant}, \quad \mathscr{R}\alpha>0)$$

$$\lim_{x\to 0} x^{\alpha}\ln x=0$$
$$(\alpha \text{ constant}, \quad \mathscr{R}\alpha>0)$$

$$\lim_{m\to\infty}\left(\sum_{k=1}^{m}\frac{1}{k}-\ln m\right)=\gamma \text{ (Euler's constant)}$$
$$=.57721\ 56649\ldots$$

INEQUALITIES

$$\frac{x}{1+x}<\ln(1+x)<x$$
$$(x>-1, \quad x\neq 0)$$

$$x<-\ln(1-x)<\frac{x}{1-x}$$
$$(x<1, \quad x\neq 0)$$

$$|\ln(1-x)| < \frac{3x}{2} \qquad (0 < x \leq .5828)$$

$$\ln x \leq x-1 \qquad (x>0)$$

$$\ln x \leq n(x^{1/n}-1) \text{ for any positive } n \qquad (x>0)$$

$$|\ln(1+z)| \leq -\ln(1-|z|) \qquad (|z|<1)$$

CONTINUED FRACTIONS

$$\ln(1+z) = \frac{z}{1+}\,\frac{z}{2+}\,\frac{z}{3+}\,\frac{4z}{4+}\,\frac{4z}{5+}\,\frac{9z}{6+}\cdots$$

(z in the plane cut from -1 to $-\infty$)

$$\ln\left(\frac{1+z}{1-z}\right) = \frac{2z}{1-}\,\frac{z^2}{3-}\,\frac{4z^2}{5-}\,\frac{9z^2}{7-}\cdots$$

POLYNOMIAL APPROXIMATIONS

$$\frac{1}{\sqrt{10}} \leq x \leq \sqrt{10}$$

$$\log_{10} x = a_1 t + a_3 t^3 + \epsilon(x), \quad t=(x-1)/(x+1)$$

$$|\epsilon(x)| \leq 6\times 10^{-4}$$

$$a_1 = .86304 \qquad a_3 = .36415$$

$$\frac{1}{\sqrt{10}} \leq x \leq \sqrt{10}$$

$$\log_{10} x = a_1 t + a_3 t^3 + a_5 t^5 + a_7 t^7 + a_9 t^9 + \epsilon(x)$$

$$t=(x-1)/(x+1)$$

$$|\epsilon(x)| \leq 10^{-7}$$

$a_1 = .86859\,1718$ $\qquad a_7 = .09437\,6476$

$a_3 = .28933\,5524$ $\qquad a_9 = .19133\,7714$

$a_5 = .17752\,2071$

$$0 \leq x \leq 1$$

$$\ln (1+x) = a_1 x + a_2 x^2 + a_3 x^3 + a_4 x^4 + a_5 x^5 + \epsilon(x)$$

$$|\epsilon(x)| \leq 1 \times 10^{-5}$$

$a_1 = .99949\,556$ $\qquad a_4 = -.13606\,275$

$a_2 = -.49190\,896$ $\qquad a_5 = .03215\,845$

$a_3 = .28947\,478$

$$0 \leq x \leq 1$$

$$\ln (1+x) = a_1 x + a_2 x^2 + a_3 x^3 + a_4 x^4 + a_5 x^5 + a_6 x^6 + a_7 x^7 + a_8 x^8 + \epsilon(x)$$

$$|\epsilon(x)| \leq 3 \times 10^{-8}$$

$a_1 =$	$.99999\,64239$	$a_5 =$	$.16765\,40711$
$a_2 =$	$-.49987\,41238$	$a_6 =$	$-.09532\,93897$
$a_3 =$	$.33179\,90258$	$a_7 =$	$.03608\,84937$
$a_4 =$	$-.24073\,38084$	$a_8 =$	$-.00645\,35442$

EXPONENTIAL FUNCTION

SERIES EXPANSION

$$e^z = \exp z = 1 + \frac{z}{1!} + \frac{z^2}{2!} + \frac{z^3}{3!} + \cdots \quad (z = x + iy)$$

where e is the real number defined in 5.1

FUNDAMENTAL PROPERTIES

$$\text{Ln} (\exp z) = z + 2k\pi i \quad (k \text{ any integer})$$

$$\ln (\exp z) = z \quad (-\pi < \mathscr{I} z \leq \pi)$$

$$\exp (\ln z) = \exp (\text{Ln } z) = z$$

$$\frac{d}{dz} \exp z = \exp z$$

DEFINITION OF GENERAL POWERS

$$\text{If } N=a^z, \text{ then } z=\text{Log}_a N$$

$$a^z=\exp(z \ln a)$$

$$\text{If } a=|a| \exp(i \arg a) \quad (-\pi<\arg a\leq\pi)$$

$$|a^z|=|a|^x e^{-y \arg a}$$

$$\arg(a^z)=y \ln |a|+x \arg a$$

$$\text{Ln } a^z=z \ln a \quad \text{for one of the values of Ln } a^z$$

$$\ln a^x=x \ln a \quad (a \text{ real and positive})$$

$$|e^z|=e^x$$

$$\arg(e^z)=y$$

$$a^{z_1}a^{z_2}=a^{z_1+z_2}$$

$$a^z b^z=(ab)^z \qquad (-\pi<\arg a+\arg b\leq\pi)$$

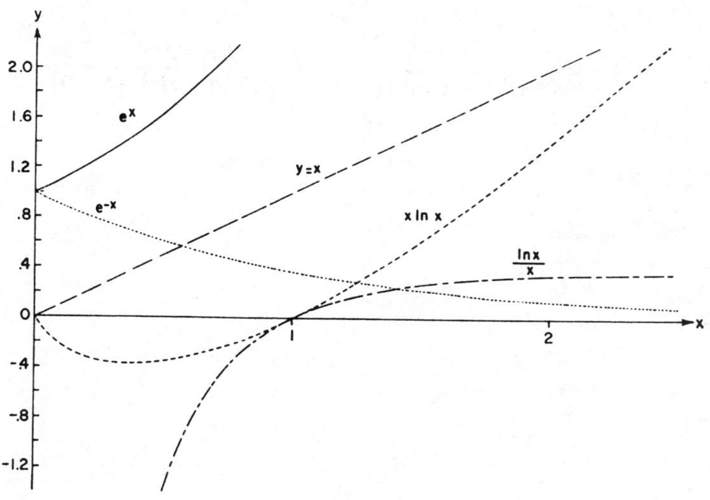

Logarithmic and exponential functions.

PERIODIC PROPERTY

$$e^{z+2\pi ki}=e^z \qquad (k \text{ any integer})$$

EXPONENTIAL IDENTITIES

$$e^{z_1}e^{z_2}=e^{z_1+z_2}$$

$$(e^{z_1})^{z_2}=e^{z_1 z_2} \qquad (-\pi<\mathscr{I}z_1\leq\pi)$$

The restriction $(-\pi<\mathscr{I}z_1\leq\pi)$ can be removed if z_2 is an integer.

LIMITING VALUES

$$\lim_{|z|\to\infty} z^\alpha e^{-z}=0 \quad (|\arg z|\leq\tfrac{1}{2}\pi-\epsilon<\tfrac{1}{2}\pi, \quad \alpha \text{ constant})$$

5.2 $$\lim_{m\to\infty}\left(1+\frac{z}{m}\right)^m=e^z$$

CONTINUED FRACTIONS

$$e^z=\frac{1}{1-}\,\frac{z}{1+}\,\frac{z}{2-}\,\frac{z}{3+}\,\frac{z}{2-}\,\frac{z}{5+}\,\frac{z}{2-}\cdots \qquad (|z|<\infty)$$

$$=1+\frac{z}{1-}\,\frac{z}{2+}\,\frac{z}{3-}\,\frac{z}{2+}\,\frac{z}{5-}\,\frac{z}{2+}\,\frac{z}{7-}\cdots \qquad (|z|<\infty)$$

$$=1+\frac{z}{(1-z/2)+}\,\frac{z^2/4\cdot3}{1+}\,\frac{z^2/4\cdot15}{1+}\,\frac{z^2/4\cdot35}{1+}\cdots\frac{z^2/4(4n^2-1)}{1+}\cdots \quad (|z|<\infty)$$

$$e^z-e_{n-1}(z)=\frac{z^n}{n!-}\,\frac{n!z}{(n+1)+}\,\frac{z}{(n+2)-}\,\frac{(n+1)z}{(n+3)+}\,\frac{2z}{(n+4)-}\,\frac{(n+2)z}{(n+5)+}\,\frac{3z}{(n+6)-}\cdots \quad (|z|<\infty)$$

SPECIAL VALUES

$e=2.71828\ 18284\ldots$

$e^0=1$

$e^\infty=\infty$

$e^{-\infty}=0$

$e^{\pm\pi i}=-1$

$e^{\pm\frac{\pi i}{2}}=\pm i$

$e^{2\pi k i}=1 \quad$ (k any integer)

EXPONENTIAL INEQUALITIES

If x is real and different from zero

$$e^{-\frac{x}{1-x}}<1-x<e^{-x} \quad (x<1)$$

$$e^x>1+x$$

$$e^x < \frac{1}{1-x} \quad (x<1)$$

$$\frac{x}{1+x} < (1-e^{-x}) < x \quad (x>-1)$$

$$x < (e^x - 1) < \frac{x}{1-x} \quad (x<1)$$

$$1+x > e^{\frac{x}{1+x}} \quad (x>-1)$$

$$e^x > 1 + \frac{x^n}{n!} \quad (n>0, \quad x>0)$$

$$e^x > \left(1+\frac{x}{y}\right)^y > e^{\frac{xy}{x+y}} \quad (x>0, \quad y>0)$$

$$e^{-x} < 1 - \frac{x}{2} \quad (0 < x \leq 1.5936)$$

$$\frac{1}{4}|z| < |e^z - 1| < \frac{7}{4}|z| \quad (0<|z|<1)$$

$$|e^z - 1| \leq e^{|z|} - 1 \leq |z| e^{|z|} \quad (\text{all } z)$$

$$e^{2a \arctan \frac{1}{z}} = 1 + \frac{2a}{z-a+} \, \frac{a^2+1}{3z+} \, \frac{a^2+4}{5z+} \, \frac{a^2+9}{7z+} \cdots$$

POLYNOMIAL APPROXIMATIONS

$$0 \leq x \leq \ln 2 = .693 \ldots$$

$$e^{-x} = 1 + a_1 x + a_2 x^2 + \epsilon(x)$$

$$|\epsilon(x)| \leq 3 \times 10^{-3}$$

$$a_1 = -.9664 \qquad a_2 = .3536$$

$$0 \leq x \leq \ln 2$$

$$e^{-x} = 1 + a_1 x + a_2 x^2 + a_3 x^3 + a_4 x^4 + \epsilon(x)$$

$$|\epsilon(x)| \leq 3 \times 10^{-5}$$

$a_1 = -.99986\,84$ $\qquad$ $a_3 = -.15953\,32$
$a_2 = .49829\,26$ $\qquad$ $a_4 = .02936\,41$

$$0 \leq x \leq \ln 2$$

$$e^{-x}=1+a_1x+a_2x^2+a_3x^3+a_4x^4+a_5x^5$$
$$+a_6x^6+a_7x^7+\epsilon(x)$$

$$|\epsilon(x)|\leq 2\times 10^{-10}$$

$$\begin{aligned} a_1 &= -.99999\,99995 & a_5 &= -.00830\,13598 \\ a_2 &= .49999\,99206 & a_6 &= .00132\,98820 \\ a_3 &= -.16666\,53019 & a_7 &= -.00014\,13161 \\ a_4 &= .04165\,73475 & & \end{aligned}$$

$$0\leq x\leq 1$$

$$10^x=(1+a_1x+a_2x^2+a_3x^3+a_4x^4)^2+\epsilon(x)$$

$$|\epsilon(x)|\leq 7\times 10^{-4}$$

$$\begin{aligned} a_1 &= 1.14991\,96 & a_3 &= .20800\,30 \\ a_2 &= .67743\,23 & a_4 &= .12680\,89 \end{aligned}$$

$$0\leq x\leq 1$$

$$10^x=(1+a_1x+a_2x^2+a_3x^3+a_4x^4+a_5x^5$$
$$+a_6x^6+a_7x^7)^2+\epsilon(x)$$

$$|\epsilon(x)|<5\times 10^{-8}$$

$$\begin{aligned} a_1 &= 1.\,15129\ 277603 & a_5 &= .\,01742\ 111988 \\ a_2 &= .\,66273\ 088429 & a_6 &= .\,00255\ 491796 \\ a_3 &= .\,25439\ 357484 & a_7 &= .\,00093\ 264267 \\ a_4 &= .\,07295\ 173666 & & \end{aligned}$$

FOUR PLACE COMMON LOGARITHMS

Log_{10} N or Log N

N	0	1	2	3	4	5	6	7	8	9	Proportional Parts 1	2	3	4	5	6	7	8	9
10	0000	0043	0086	0128	0170	0212	0253	0294	0334	0374	4	8	12	17	21	25	29	33	37
11	0414	0453	0492	0531	0569	0607	0645	0682	0719	0755	4	8	11	15	19	23	26	30	34
12	0792	0828	0864	0899	0934	0969	1004	1038	1072	1106	3	7	10	14	17	21	24	28	31
13	1139	1173	1206	1239	1271	1303	1335	1367	1399	1430	3	6	10	13	16	19	23	26	29
14	1461	1492	1523	1553	1584	1614	1644	1673	1703	1732	3	6	9	12	15	18	21	24	27
15	1761	1790	1818	1847	1875	1903	1931	1959	1987	2014	3	6	8	11	14	17	20	22	25
16	2041	2068	2095	2122	2148	2175	2201	2227	2253	2279	3	5	8	11	13	16	18	21	24
17	2304	2330	2355	2380	2405	2430	2455	2480	2504	2529	2	5	7	10	12	15	17	20	22
18	2553	2577	2601	2625	2648	2672	2695	2718	2742	2765	2	5	7	9	12	14	16	19	21
19	2788	2810	2833	2856	2878	2900	2923	2945	2967	2989	2	4	7	9	11	13	16	18	20
20	3010	3032	3054	3075	3096	3118	3139	3160	3181	3201	2	4	6	8	11	13	15	17	19
21	3222	3243	3263	3284	3304	3324	3345	3365	3385	3404	2	4	6	8	10	12	14	16	18
22	3424	3444	3464	3483	3502	3522	3541	3560	3579	3598	2	4	6	8	10	12	14	15	17
23	3617	3636	3655	3674	3692	3711	3729	3747	3766	3784	2	4	6	7	9	11	13	15	17
24	3802	3820	3838	3856	3874	3892	3909	3927	3945	3962	2	4	5	7	9	11	12	14	16
25	3979	3997	4014	4031	4048	4065	4082	4099	4116	4133	2	3	5	7	9	10	12	14	15
26	4150	4166	4183	4200	4216	4232	4249	4265	4281	4298	2	3	5	7	8	10	11	13	15
27	4314	4330	4346	4362	4378	4393	4409	4425	4440	4456	2	3	5	6	8	9	11	13	14
28	4472	4487	4502	4518	4533	4548	4564	4579	4594	4609	2	3	5	6	8	9	11	12	14
29	4624	4639	4654	4669	4683	4698	4713	4728	4742	4757	1	3	4	6	7	9	10	12	13
30	4771	4786	4800	4814	4829	4843	4857	4871	4886	4900	1	3	4	6	7	9	10	11	13
31	4914	4928	4942	4955	4969	4983	4997	5011	5024	5038	1	3	4	6	7	8	10	11	12
32	5051	5065	5079	5092	5105	5119	5132	5145	5159	5172	1	3	4	5	7	8	9	11	12
33	5185	5198	5211	5224	5237	5250	5263	5276	5289	5302	1	3	4	5	6	8	9	10	12
34	5315	5328	5340	5353	5366	5378	5391	5403	5416	5428	1	3	4	5	6	8	9	10	11
35	5441	5453	5465	5478	5490	5502	5514	5527	5539	5551	1	2	4	5	6	7	9	10	11
36	5563	5575	5587	5599	5611	5623	5635	5647	5658	5670	1	2	4	5	6	7	8	10	11
37	5682	5694	5705	5717	5729	5740	5752	5763	5775	5786	1	2	3	5	6	7	8	9	10
38	5798	5809	5821	5832	5843	5855	5866	5877	5888	5899	1	2	3	5	6	7	8	9	10
39	5911	5922	5933	5944	5955	5966	5977	5988	5999	6010	1	2	3	4	5	7	8	9	10
40	6021	6031	6042	6053	6064	6075	6085	6096	6107	6117	1	2	3	4	5	6	8	9	10
41	6128	6138	6149	6160	6170	6180	6191	6201	6212	6222	1	2	3	4	5	6	7	8	9
42	6232	6243	6253	6263	6274	6284	6294	6304	6314	6325	1	2	3	4	5	6	7	8	9
43	6335	6345	6355	6365	6375	6385	6395	6405	6415	6425	1	2	3	4	5	6	7	8	9
44	6435	6444	6454	6464	6474	6484	6493	6503	6513	6522	1	2	3	4	5	6	7	8	9
45	6532	6542	6551	6561	6571	6580	6590	6599	6609	6618	1	2	3	4	5	6	7	8	9
46	6628	6637	6646	6656	6665	6675	6684	6693	6702	6712	1	2	3	4	5	6	7	7	8
47	6721	6730	6739	6749	6758	6767	6776	6785	6794	6803	1	2	3	4	5	5	6	7	8
48	6812	6821	6830	6839	6848	6857	6866	6875	6884	6893	1	2	3	4	4	5	6	7	8
49	6902	6911	6920	6928	6937	6946	6955	6964	6972	6981	1	2	3	4	4	5	6	7	8
50	6990	6998	7007	7016	7024	7033	7042	7050	7059	7067	1	2	3	3	4	5	6	7	8
51	7076	7084	7093	7101	7110	7118	7126	7135	7143	7152	1	2	3	3	4	5	6	7	8
52	7160	7168	7177	7185	7193	7202	7210	7218	7226	7235	1	2	2	3	4	5	6	7	7
53	7243	7251	7259	7267	7275	7284	7292	7300	7308	7316	1	2	2	3	4	5	6	6	7
54	7324	7332	7340	7348	7356	7364	7372	7380	7388	7396	1	2	2	3	4	5	6	6	7
N	0	1	2	3	4	5	6	7	8	9	1	2	3	4	5	6	7	8	9

FOUR PLACE COMMON LOGARITHMS
$\log_{10} N$ or $\log N$

N	0	1	2	3	4	5	6	7	8	9	Proportional Parts 1	2	3	4	5	6	7	8	9
55	7404	7412	7419	7427	7435	7443	7451	7459	7466	7474	1	2	2	3	4	5	5	6	[illegible]
56	7482	7490	7497	7505	7513	7520	7528	7536	7543	7551	1	2	2	3	4	5	5	6	[illegible]
57	7559	7566	7574	7582	7589	7597	7604	7612	7619	7627	1	2	2	3	4	5	5	6	[illegible]
58	7634	7642	7649	7657	7664	7672	7679	7686	7694	7701	1	1	2	3	4	4	5	6	[illegible]
59	7709	7716	7723	7731	7738	7745	7752	7760	7767	7774	1	1	2	3	4	4	5	6	[illegible]
60	7782	7789	7796	7803	7810	7818	7825	7832	7839	7846	1	1	2	3	4	4	5	6	6
61	7853	7860	7868	7875	7882	7889	7896	7903	7910	7917	1	1	2	3	4	4	5	6	6
62	7924	7931	7938	7945	7952	7959	7966	7973	7980	7987	1	1	2	3	3	4	5	6	6
63	7993	8000	8007	8014	8021	8028	8035	8041	8048	8055	1	1	2	3	3	4	5	5	6
64	8062	8069	8075	8082	8089	8096	8102	8109	8116	8122	1	1	2	3	3	4	5	5	6
65	8129	8136	8142	8149	8156	8162	8169	8176	8182	8189	1	1	2	3	3	4	5	5	6
66	8195	8202	8209	8215	8222	8228	8235	8241	8248	8254	1	1	2	3	3	4	5	5	6
67	8261	8267	8274	8280	8287	8293	8299	8306	8312	8319	1	1	2	3	3	4	5	5	6
68	8325	8331	8338	8344	8351	8357	8363	8370	8376	8382	1	1	2	3	3	4	4	5	6
69	8388	8395	8401	8407	8414	8420	8426	8432	8439	8445	1	1	2	2	3	4	4	5	6
70	8451	8457	8463	8470	8476	8482	8488	8494	8500	8506	1	1	2	2	3	4	4	5	6
71	8513	8519	8525	8531	8537	8543	8549	8555	8561	8567	1	1	2	2	3	4	4	5	5
72	8573	8579	8585	8591	8597	8603	8609	8615	8621	8627	1	1	2	2	3	4	4	5	5
73	8633	8639	8645	8651	8657	8663	8669	8675	8681	8686	1	1	2	2	3	4	4	5	5
74	8692	8698	8704	8710	8716	8722	8727	8733	8739	8745	1	1	2	2	3	4	4	5	5
75	8751	8756	8762	8768	8774	8779	8785	8791	8797	8802	1	1	2	2	3	3	4	5	5
76	8808	8814	8820	8825	8831	8837	8842	8848	8854	8859	1	1	2	2	3	3	4	5	5
77	8865	8871	8876	8882	8887	8893	8899	8904	8910	8915	1	1	2	2	3	3	4	4	5
78	8921	8927	8932	8938	8943	8949	8954	8960	8965	8971	1	1	2	2	3	3	4	4	5
79	8976	8982	8987	8993	8998	9004	9009	9015	9020	9025	1	1	2	2	3	3	4	4	5
80	9031	9036	9042	9047	9053	9058	9063	9069	9074	9079	1	1	2	2	3	3	4	4	5
81	9085	9090	9096	9101	9106	9112	9117	9122	9128	9133	1	1	2	2	3	3	4	4	5
82	9138	9143	9149	9154	9159	9165	9170	9175	9180	9186	1	1	2	2	3	3	4	4	5
83	9191	9196	9201	9206	9212	9217	9222	9227	9232	9238	1	1	2	2	3	3	4	4	5
84	9243	9248	9253	9258	9263	9269	9274	9279	9284	9289	1	1	2	2	3	3	4	4	5
85	9294	9299	9304	9309	9315	9320	9325	9330	9335	9340	1	1	2	2	3	3	4	4	5
86	9345	9350	9355	9360	9365	9370	9375	9380	9385	9390	1	1	2	2	3	3	4	4	5
87	9395	9400	9405	9410	9415	9420	9425	9430	9435	9440	0	1	1	2	2	3	3	4	4
88	9445	9450	9455	9460	9465	9469	9474	9479	9484	9489	0	1	1	2	2	3	3	4	4
89	9494	9499	9504	9509	9513	9518	9523	9528	9533	9538	0	1	1	2	2	3	3	4	4
90	9542	9547	9552	9557	9562	9566	9571	9576	9581	9586	0	1	1	2	2	3	3	4	4
91	9590	9595	9600	9605	9609	9614	9619	9624	9628	9633	0	1	1	2	2	3	3	4	4
92	9638	9643	9647	9652	9657	9661	9666	9671	9675	9680	0	1	1	2	2	3	3	4	4
93	9685	9689	9694	9699	9703	9708	9713	9717	9722	9727	0	1	1	2	2	3	3	4	4
94	9731	9736	9741	9745	9750	9754	9759	9763	9768	9773	0	1	1	2	2	3	3	4	4
95	9777	9782	9786	9791	9795	9800	9805	9809	9814	9818	0	1	1	2	2	3	3	4	4
96	9823	9827	9832	9836	9841	9845	9850	9854	9859	9863	0	1	1	2	2	3	3	4	4
97	9868	9872	9877	9881	9886	9890	9894	9899	9903	9908	0	1	1	2	2	3	3	4	4
98	9912	9917	9921	9926	9930	9934	9939	9943	9948	9952	0	1	1	2	2	3	3	4	4
99	9956	9961	9965	9969	9974	9978	9983	9987	9991	9996	0	1	1	2	2	3	3	3	4
N	0	1	2	3	4	5	6	7	8	9	1	2	3	4	5	6	7	8	9

FOUR PLACE COMMON ANTILOGARITHMS
10^p or antilog p

p	0	1	2	3	4	5	6	7	8	9	PP 1	PP 2	PP 3	PP 4	PP 5	PP 6	PP 7	PP 8	PP 9
00	1000	1002	1005	1007	1009	1012	1014	1016	1019	1021	0	0	1	1	1	1	2	2	2
01	1023	1026	1028	1030	1033	1035	1038	1040	1042	1045	0	0	1	1	1	1	2	2	2
02	1047	1050	1052	1054	1057	1059	1062	1064	1067	1069	0	0	1	1	1	1	2	2	2
03	1072	1074	1076	1079	1081	1084	1086	1089	1091	1094	0	0	1	1	1	1	2	2	2
04	1096	1099	1102	1104	1107	1109	1112	1114	1117	1119	0	1	1	1	1	2	2	2	2
05	1122	1125	1127	1130	1132	1135	1138	1140	1143	1146	0	1	1	1	1	2	2	2	2
06	1148	1151	1153	1156	1159	1161	1164	1167	1169	1172	0	1	1	1	1	2	2	2	2
07	1175	1178	1180	1183	1186	1189	1191	1194	1197	1199	0	1	1	1	1	2	2	2	2
08	1202	1205	1208	1211	1213	1216	1219	1222	1225	1227	0	1	1	1	1	2	2	2	3
09	1230	1233	1236	1239	1242	1245	1247	1250	1253	1256	0	1	1	1	1	2	2	2	3
10	1259	1262	1265	1268	1271	1274	1276	1279	1282	1285	0	1	1	1	1	2	2	2	3
11	1288	1291	1294	1297	1300	1303	1306	1309	1312	1315	0	1	1	1	2	2	2	2	3
12	1318	1321	1324	1327	1330	1334	1337	1340	1343	1346	0	1	1	1	2	2	2	2	3
13	1349	1352	1355	1358	1361	1365	1368	1371	1374	1377	0	1	1	1	2	2	2	3	3
14	1380	1384	1387	1390	1393	1396	1400	1403	1406	1409	0	1	1	1	2	2	2	3	3
15	1413	1416	1419	1422	1426	1429	1432	1435	1439	1442	0	1	1	1	2	2	2	3	3
16	1445	1449	1452	1455	1459	1462	1466	1469	1472	1476	0	1	1	1	2	2	2	3	3
17	1479	1483	1486	1489	1493	1496	1500	1503	1507	1510	0	1	1	1	2	2	2	3	3
18	1514	1517	1521	1524	1528	1531	1535	1538	1542	1545	0	1	1	1	2	2	2	3	3
19	1549	1552	1556	1560	1563	1567	1570	1574	1578	1581	0	1	1	1	2	2	3	3	3
20	1585	1589	1592	1596	1600	1603	1607	1611	1614	1618	0	1	1	1	2	2	3	3	3
21	1622	1626	1629	1633	1637	1641	1644	1648	1652	1656	0	1	1	2	2	2	3	3	3
22	1660	1663	1667	1671	1675	1679	1683	1687	1690	1694	0	1	1	2	2	2	3	3	3
23	1698	1702	1706	1710	1714	1718	1722	1726	1730	1734	0	1	1	2	2	2	3	3	4
24	1738	1742	1746	1750	1754	1758	1762	1766	1770	1774	0	1	1	2	2	2	3	3	4
25	1778	1782	1786	1791	1795	1799	1803	1807	1811	1816	0	1	1	2	2	2	3	3	4
26	1820	1824	1828	1832	1837	1841	1845	1849	1854	1858	0	1	1	2	2	3	3	3	4
27	1862	1866	1871	1875	1879	1884	1888	1892	1897	1901	0	1	1	2	2	3	3	3	4
28	1905	1910	1914	1919	1923	1928	1932	1936	1941	1945	0	1	1	2	2	3	3	4	4
29	1950	1954	1959	1963	1968	1972	1977	1982	1986	1991	0	1	1	2	2	3	3	4	4
30	1995	2000	2004	2009	2014	2018	2023	2028	2032	2037	0	1	1	2	2	3	3	4	4
31	2042	2046	2051	2056	2061	2065	2070	2075	2080	2084	0	1	1	2	2	3	3	4	4
32	2089	2094	2099	2104	2109	2113	2118	2123	2128	2133	0	1	1	2	2	3	3	4	4
33	2138	2143	2148	2153	2158	2163	2168	2173	2178	2183	0	1	1	2	2	3	3	4	4
34	2188	2193	2198	2203	2208	2213	2218	2223	2228	2234	1	1	2	2	3	3	4	4	5
35	2239	2244	2249	2254	2259	2265	2270	2275	2280	2286	1	1	2	2	3	3	4	4	5
36	2291	2296	2301	2307	2312	2317	2323	2328	2333	2339	1	1	2	2	3	3	4	4	5
37	2344	2350	2355	2360	2366	2371	2377	2382	2388	2393	1	1	2	2	3	3	4	4	5
38	2399	2404	2410	2415	2421	2427	2432	2438	2443	2449	1	1	2	2	3	3	4	4	5
39	2455	2460	2466	2472	2477	2483	2489	2495	2500	2506	1	1	2	2	3	3	4	5	5
40	2512	2518	2523	2529	2535	2541	2547	2553	2559	2564	1	1	2	2	3	4	4	5	5
41	2570	2576	2582	2588	2594	2600	2606	2612	2618	2624	1	1	2	2	3	4	4	5	5
42	2630	2636	2642	2649	2655	2661	2667	2673	2679	2685	1	1	2	2	3	4	4	5	6
43	2692	2698	2704	2710	2716	2723	2729	2735	2742	2748	1	1	2	3	3	4	4	5	6
44	2754	2761	2767	2773	2780	2786	2793	2799	2805	2812	1	1	2	3	3	4	4	5	6
45	2818	2825	2831	2838	2844	2851	2858	2864	2871	2877	1	1	2	3	3	4	5	5	6
46	2884	2891	2897	2904	2911	2917	2924	2931	2938	2944	1	1	2	3	3	4	5	5	6
47	2951	2958	2965	2972	2979	2985	2992	2999	3006	3013	1	1	2	3	3	4	5	5	6
48	3020	3027	3034	3041	3048	3055	3062	3069	3076	3083	1	1	2	3	4	4	5	6	6
49	3090	3097	3105	3112	3119	3126	3133	3141	3148	3155	1	1	2	3	4	4	5	6	6
	0	1	2	3	4	5	6	7	8	9	1	2	3	4	5	6	7	8	9

FOUR PLACE COMMON ANTILOGARITHMS
10^p or antilog p

p	0	1	2	3	4	5	6	7	8	9	Proportional Parts 1	2	3	4	5	6	7	8	9
.50	3162	3170	3177	3184	3192	3199	3206	3214	3221	3228	1	1	2	3	4	4	5	6	[illegible]
.51	3236	3243	3251	3258	3266	3273	3281	3289	3296	3304	1	2	2	3	4	5	5	6	[illegible]
.52	3311	3319	3327	3334	3342	3350	3357	3365	3373	3381	1	2	2	3	4	5	5	6	[illegible]
.53	3388	3396	3404	3412	3420	3428	3436	3443	3451	3459	1	2	2	3	4	5	6	6	[illegible]
.54	3467	3475	3483	3491	3499	3508	3516	3524	3532	3540	1	2	2	3	4	5	6	6	[illegible]
.55	3548	3556	3565	3573	3581	3589	3597	3606	3614	3622	1	2	2	3	4	5	6	7	[illegible]
.56	3631	3639	3648	3656	3664	3673	3681	3690	3698	3707	1	2	3	3	4	5	6	7	[illegible]
.57	3715	3724	3733	3741	3750	3758	3767	3776	3784	3793	1	2	3	3	4	5	6	7	[illegible]
.58	3802	3811	3819	3828	3837	3846	3855	3864	3873	3882	1	2	3	4	4	5	6	7	[illegible]
.59	3890	3899	3908	3917	3926	3936	3945	3954	3963	3972	1	2	3	4	5	5	6	7	[illegible]
.60	3981	3990	3999	4009	4018	4027	4036	4046	4055	4064	1	2	3	4	5	6	6	7	[illegible]
.61	4074	4083	4093	4102	4111	4121	4130	4140	4150	4159	1	2	3	4	5	6	7	8	[illegible]
.62	4169	4178	4188	4198	4207	4217	4227	4236	4246	4256	1	2	3	4	5	6	7	8	[illegible]
.63	4266	4276	4285	4295	4305	4315	4325	4335	4345	4355	1	2	3	4	5	6	7	8	[illegible]
.64	4365	4375	4385	4395	4406	4416	4426	4436	4446	4457	1	2	3	4	5	6	7	8	[illegible]
.65	4467	4477	4487	4498	4508	4519	4529	4539	4550	4560	1	2	3	4	5	6	7	8	[illegible]
.66	4571	4581	4592	4603	4613	4624	4634	4645	4656	4667	1	2	3	4	5	6	7	9	[illegible]
.67	4677	4688	4699	4710	4721	4732	4742	4753	4764	4775	1	2	3	4	5	7	8	9	[illegible]
.68	4786	4797	4808	4819	4831	4842	4853	4864	4875	4887	1	2	3	4	6	7	8	9	[illegible]
.69	4898	4909	4920	4932	4943	4955	4966	4977	4989	5000	1	2	3	5	6	7	8	9	[illegible]
.70	5012	5023	5035	5047	5058	5070	5082	5093	5105	5117	1	2	4	5	6	7	8	9	[illegible]
.71	5129	5140	5152	5164	5176	5188	5200	5212	5224	5236	1	2	4	5	6	7	8	10	[illegible]
.72	5248	5260	5272	5284	5297	5309	5321	5333	5346	5358	1	2	4	5	6	7	9	10	[illegible]
.73	5370	5383	5395	5408	5420	5433	5445	5458	5470	5483	1	3	4	5	6	8	9	10	[illegible]
.74	5495	5508	5521	5534	5546	5559	5572	5585	5598	5610	1	3	4	5	6	8	9	10	[illegible]
.75	5623	5636	5649	5662	5675	5689	5702	5715	5728	5741	1	3	4	5	7	8	9	10	[illegible]
.76	5754	5768	5781	5794	5808	5821	5834	5848	5861	5875	1	3	4	5	7	8	9	11	[illegible]
.77	5888	5902	5916	5929	5943	5957	5970	5984	5998	6012	1	3	4	5	7	8	10	11	[illegible]
.78	6026	6039	6053	6067	6081	6095	6109	6124	6138	6152	1	3	4	6	7	8	10	11	[illegible]
.79	6166	6180	6194	6209	6223	6237	6252	6266	6281	6295	1	3	4	6	7	9	10	11	[illegible]
.80	6310	6324	6339	6353	6368	6383	6397	6412	6427	6442	1	3	4	6	7	9	10	12	[illegible]
.81	6457	6471	6486	6501	6516	6531	6546	6561	6577	6592	2	3	5	6	8	9	11	12	[illegible]
.82	6607	6622	6637	6653	6668	6683	6699	6714	6730	6745	2	3	5	6	8	9	11	12	[illegible]
.83	6761	6776	6792	6808	6823	6839	6855	6871	6887	6902	2	3	5	6	8	9	11	13	[illegible]
.84	6918	6934	6950	6966	6982	6998	7015	7031	7047	7063	2	3	5	6	8	10	11	13	[illegible]
.85	7079	7096	7112	7129	7145	7161	7178	7194	7211	7228	2	3	5	7	8	10	12	13	[illegible]
.86	7244	7261	7278	7295	7311	7328	7345	7362	7379	7396	2	3	5	7	8	10	12	13	[illegible]
.87	7413	7430	7447	7464	7482	7499	7516	7534	7551	7568	2	3	5	7	9	10	12	14	[illegible]
.88	7586	7603	7621	7638	7656	7674	7691	7709	7727	7745	2	4	5	7	9	11	12	14	[illegible]
.89	7762	7780	7798	7816	7834	7852	7870	7889	7907	7925	2	4	5	7	9	11	13	14	[illegible]
.90	7943	7962	7980	7998	8017	8035	8054	8072	8091	8110	2	4	6	7	9	11	13	15	[illegible]
.91	8128	8147	8166	8185	8204	8222	8241	8260	8279	8299	2	4	6	8	9	11	13	15	[illegible]
.92	8318	8337	8356	8375	8395	8414	8433	8453	8472	8492	2	4	6	8	10	12	14	15	[illegible]
.93	8511	8531	8551	8570	8590	8610	8630	8650	8670	8690	2	4	6	8	10	12	14	16	[illegible]
.94	8710	8730	8750	8770	8790	8810	8831	8851	8872	8892	2	4	6	8	10	12	14	16	[illegible]
.95	8913	8933	8954	8974	8995	9016	9036	9057	9078	9099	2	4	6	8	10	12	15	17	[illegible]
.96	9120	9141	9162	9183	9204	9226	9247	9268	9290	9311	2	4	6	8	11	13	15	17	[illegible]
.97	9333	9354	9376	9397	9419	9441	9462	9484	9506	9528	2	4	7	9	11	13	15	17	[illegible]
.98	9550	9572	9594	9616	9638	9661	9683	9705	9727	9750	2	4	7	9	11	13	16	18	[illegible]
.99	9772	9795	9817	9840	9863	9886	9908	9931	9954	9977	2	5	7	9	11	14	16	18	[illegible]
p	0	1	2	3	4	5	6	7	8	9	1	2	3	4	5	6	7	8	9

NATURAL OR NAPIERIAN LOGARITHMS
$\log_e x$ or $\ln x$

x	0	1	2	3	4	5	6	7	8	9
1.0	.00000	.00995	.01980	.02956	.03922	.04879	.05827	.06766	.07696	.08618
1.1	.09531	.10436	.11333	.12222	.13103	.13976	.14842	.15700	.16551	.17395
1.2	.18232	.19062	.19885	.20701	.21511	.22314	.23111	.23902	.24686	.25464
1.3	.26236	.27003	.27763	.28518	.29267	.30010	.30748	.31481	.32208	.32930
1.4	.33647	.34359	.35066	.35767	.36464	.37156	.37844	.38526	.39204	.39878
1.5	.40547	.41211	.41871	.42527	.43178	.43825	.44469	.45108	.45742	.46373
1.6	.47000	.47623	.48243	.48858	.49470	.50078	.50682	.51282	.51879	.52473
1.7	.53063	.53649	.54232	.54812	.55389	.55962	.56531	.57098	.57661	.58222
1.8	.58779	.59333	.59884	.60432	.60977	.61519	.62058	.62594	.63127	.63658
1.9	.64185	.64710	.65233	.65752	.66269	.66783	.67294	.67803	.68310	.68813
2.0	.69315	.69813	.70310	.70804	.71295	.71784	.72271	.72755	.73237	.73716
2.1	.74194	.74669	.75142	.75612	.76081	.76547	.77011	.77473	.77932	.78390
2.2	.78846	.79299	.79751	.80200	.80648	.81093	.81536	.81978	.82418	.82855
2.3	.83291	.83725	.84157	.84587	.85015	.85442	.85866	.86289	.86710	.87129
2.4	.87547	.87963	.88377	.88789	.89200	.89609	.90016	.90422	.90826	.91228
2.5	.91629	.92028	.92426	.92822	.93216	.93609	.94001	.94391	.94779	.95166
2.6	.95551	.95935	.96317	.96698	.97078	.97456	.97833	.98208	.98582	.98954
2.7	.99325	.99695	1.00063	1.00430	1.00796	1.01160	1.01523	1.01885	1.02245	1.02604
2.8	1.02962	1.03318	1.03674	1.04028	1.04380	1.04732	1.05082	1.05431	1.05779	1.06126
2.9	1.06471	1.06815	1.07158	1.07500	1.07841	1.08181	1.08519	1.08856	1.09192	1.09527
3.0	1.09861	1.10194	1.10526	1.10856	1.11186	1.11514	1.11841	1.12168	1.12493	1.12817
3.1	1.13140	1.13462	1.13783	1.14103	1.14422	1.14740	1.15057	1.15373	1.15688	1.16002
3.2	1.16315	1.16627	1.16938	1.17248	1.17557	1.17865	1.18173	1.18479	1.18784	1.19089
3.3	1.19392	1.19695	1.19996	1.20297	1.20597	1.20896	1.21194	1.21491	1.21788	1.22083
3.4	1.22378	1.22671	1.22964	1.23256	1.23547	1.23837	1.24127	1.24415	1.24703	1.24990
3.5	1.25276	1.25562	1.25846	1.26130	1.26413	1.26695	1.26976	1.27257	1.27536	1.27815
3.6	1.28093	1.28371	1.28647	1.28923	1.29198	1.29473	1.29746	1.30019	1.30291	1.30563
3.7	1.30833	1.31103	1.31372	1.31641	1.31909	1.32176	1.32442	1.32708	1.32972	1.33237
3.8	1.33500	1.33763	1.34025	1.34286	1.34547	1.34807	1.35067	1.35325	1.35584	1.35841
3.9	1.36098	1.36354	1.36609	1.36864	1.37118	1.37372	1.37624	1.37877	1.38128	1.38379
4.0	1.38629	1.38879	1.39128	1.39377	1.39624	1.39872	1.40118	1.40364	1.40610	1.40854
4.1	1.41099	1.41342	1.41585	1.41828	1.42070	1.42311	1.42552	1.42792	1.43031	1.43270
4.2	1.43508	1.43746	1.43984	1.44220	1.44456	1.44692	1.44927	1.45161	1.45395	1.45629
4.3	1.45862	1.46094	1.46326	1.46557	1.46787	1.47018	1.47247	1.47476	1.47705	1.47933
4.4	1.48160	1.48387	1.48614	1.48840	1.49065	1.49290	1.49515	1.49739	1.49962	1.50185
4.5	1.50408	1.50630	1.50851	1.51072	1.51293	1.51513	1.51732	1.51951	1.52170	1.52388
4.6	1.52606	1.52823	1.53039	1.53256	1.53471	1.53687	1.53902	1.54116	1.54330	1.54543
4.7	1.54756	1.54969	1.55181	1.55393	1.55604	1.55814	1.56025	1.56235	1.56444	1.56653
4.8	1.56862	1.57070	1.57277	1.57485	1.57691	1.57898	1.58104	1.58309	1.58515	1.58719
4.9	1.58924	1.59127	1.59331	1.59534	1.59737	1.59939	1.60141	1.60342	1.60543	1.60744

$\ln 10 = 2.30259$ $\quad 4 \ln 10 = 9.21034$ $\quad 7 \ln 10 = 16.11810$

$2 \ln 10 = 4.60517$ $\quad 5 \ln 10 = 11.51293$ $\quad 8 \ln 10 = 18.42068$

$3 \ln 10 = 6.90776$ $\quad 6 \ln 10 = 13.81551$ $\quad 9 \ln 10 = 20.72327$

NATURAL OR NAPIERIAN LOGARITHMS
$\log_e x$ or $\ln x$

x	0	1	2	3	4	5	6	7	8	9
5.0	1.60944	1.61144	1.61343	1.61542	1.61741	1.61939	1.62137	1.62334	1.62531	1.62728
5.1	1.62924	1.63120	1.63315	1.63511	1.63705	1.63900	1.64094	1.64287	1.64481	1.64673
5.2	1.64866	1.65058	1.65250	1.65441	1.65632	1.65823	1.66013	1.66203	1.66393	1.66582
5.3	1.66771	1.66959	1.67147	1.67335	1.67523	1.67710	1.67896	1.68083	1.68269	1.68455
5.4	1.68640	1.68825	1.69010	1.69194	1.69378	1.69562	1.69745	1.69928	1.70111	1.70293
5.5	1.70475	1.70656	1.70838	1.71019	1.71199	1.71380	1.71560	1.71740	1.71919	1.72098
5.6	1.72277	1.72455	1.72633	1.72811	1.72988	1.73166	1.73342	1.73519	1.73695	1.73871
5.7	1.74047	1.74222	1.74397	1.74572	1.74746	1.74920	1.75094	1.75267	1.75440	1.75613
5.8	1.75786	1.75958	1.76130	1.76302	1.76473	1.76644	1.76815	1.76985	1.77156	1.77326
5.9	1.77495	1.77665	1.77834	1.78002	1.78171	1.78339	1.78507	1.78675	1.78842	1.79009
6.0	1.79176	1.79342	1.79509	1.79675	1.79840	1.80006	1.80171	1.80336	1.80500	1.80665
6.1	1.80829	1.80993	1.81156	1.81319	1.81482	1.81645	1.81808	1.81970	1.82132	1.82294
6.2	1.82455	1.82616	1.82777	1.82938	1.83098	1.83258	1.83418	1.83578	1.83737	1.83896
6.3	1.84055	1.84214	1.84372	1.84530	1.84688	1.84845	1.85003	1.85160	1.85317	1.85473
6.4	1.85630	1.85786	1.85942	1.86097	1.86253	1.86408	1.86563	1.86718	1.86872	1.87026
6.5	1.87180	1.87334	1.87487	1.87641	1.87794	1.87947	1.88099	1.88251	1.88403	1.88555
6.6	1.88707	1.88858	1.89010	1.89160	1.89311	1.89462	1.89612	1.89762	1.89912	1.90061
6.7	1.90211	1.90360	1.90509	1.90658	1.90806	1.90954	1.91102	1.91250	1.91398	1.91545
6.8	1.91692	1.91839	1.91986	1.92132	1.92279	1.92425	1.92571	1.92716	1.92862	1.93007
6.9	1.93152	1.93297	1.93442	1.93586	1.93730	1.93874	1.94018	1.94162	1.94305	1.94448
7.0	1.94591	1.94734	1.94876	1.95019	1.95161	1.95303	1.95445	1.95586	1.95727	1.95869
7.1	1.96009	1.96150	1.96291	1.96431	1.96571	1.96711	1.96851	1.96991	1.97130	1.97269
7.2	1.97408	1.97547	1.97685	1.97824	1.97962	1.98100	1.98238	1.98376	1.98513	1.98650
7.3	1.98787	1.98924	1.99061	1.99198	1.99334	1.99470	1.99606	1.99742	1.99877	2.00013
7.4	2.00148	2.00283	2.00418	2.00553	2.00687	2.00821	2.00956	2.01089	2.01223	2.01357
7.5	2.01490	2.01624	2.01757	2.01890	2.02022	2.02155	2.02287	2.02419	2.02551	2.02683
7.6	2.02815	2.02946	2.03078	2.03209	2.03340	2.03471	2.03601	2.03732	2.03862	2.03992
7.7	2.04122	2.04252	2.04381	2.04511	2.04640	2.04769	2.04898	2.05027	2.05156	2.05284
7.8	2.05412	2.05540	2.05668	2.05796	2.05924	2.06051	2.06179	2.06306	2.06433	2.06560
7.9	2.06686	2.06813	2.06939	2.07065	2.07191	2.07317	2.07443	2.07568	2.07694	2.07819
8.0	2.07944	2.08069	2.08194	2.08318	2.08443	2.08567	2.08691	2.08815	2.08939	2.09063
8.1	2.09186	2.09310	2.09433	2.09556	2.09679	2.09802	2.09924	2.10047	2.10169	2.10291
8.2	2.10413	2.10535	2.10657	2.10779	2.10900	2.11021	2.11142	2.11263	2.11384	2.11505
8.3	2.11626	2.11746	2.11866	2.11986	2.12106	2.12226	2.12346	2.12465	2.12585	2.12704
8.4	2.12823	2.12942	2.13061	2.13180	2.13298	2.13417	2.13535	2.13653	2.13771	2.13889
8.5	2.14007	2.14124	2.14242	2.14359	2.14476	2.14593	2.14710	2.14827	2.14943	2.15060
8.6	2.15176	2.15292	2.15409	2.15524	2.15640	2.15756	2.15871	2.15987	2.16102	2.16217
8.7	2.16332	2.16447	2.16562	2.16677	2.16791	2.16905	2.17020	2.17134	2.17248	2.17361
8.8	2.17475	2.17589	2.17702	2.17816	2.17929	2.18042	2.18155	2.18267	2.18380	2.18493
8.9	2.18605	2.18717	2.18830	2.18942	2.19054	2.19165	2.19277	2.19389	2.19500	2.19611
9.0	2.19722	2.19834	2.19944	2.20055	2.20166	2.20276	2.20387	2.20497	2.20607	2.20717
9.1	2.20827	2.20937	2.21047	2.21157	2.21266	2.21375	2.21485	2.21594	2.21703	2.21812
9.2	2.21920	2.22029	2.22138	2.22246	2.22354	2.22462	2.22570	2.22678	2.22786	2.22894
9.3	2.23001	2.23109	2.23216	2.23324	2.23431	2.23538	2.23645	2.23751	2.23858	2.23965
9.4	2.24071	2.24177	2.24284	2.24390	2.24496	2.24601	2.24707	2.24813	2.24918	2.25024
9.5	2.25129	2.25234	2.25339	2.25444	2.25549	2.25654	2.25759	2.25863	2.25968	2.26072
9.6	2.26176	2.26280	2.26384	2.26488	2.26592	2.26696	2.26799	2.26903	2.27006	2.27109
9.7	2.27213	2.27316	2.27419	2.27521	2.27624	2.27727	2.27829	2.27932	2.28034	2.28136
9.8	2.28238	2.28340	2.28442	2.28544	2.28646	2.28747	2.28849	2.28950	2.29051	2.29152
9.9	2.29253	2.29354	2.29455	2.29556	2.29657	2.29757	2.29858	2.29958	2.30058	2.30158

EXPONENTIAL FUNCTIONS e^x

x	0	1	2	3	4	5	6	7	8	9
.0	1.0000	1.0101	1.0202	1.0305	1.0408	1.0513	1.0618	1.0725	1.0833	1.0942
.1	1.1052	1.1163	1.1275	1.1388	1.1503	1.1618	1.1735	1.1853	1.1972	1.2092
.2	1.2214	1.2337	1.2461	1.2586	1.2712	1.2840	1.2969	1.3100	1.3231	1.3364
.3	1.3499	1.3634	1.3771	1.3910	1.4049	1.4191	1.4333	1.4477	1.4623	1.4770
.4	1.4918	1.5068	1.5220	1.5373	1.5527	1.5683	1.5841	1.6000	1.6161	1.6323
.5	1.6487	1.6653	1.6820	1.6989	1.7160	1.7333	1.7507	1.7683	1.7860	1.8040
.6	1.8221	1.8404	1.8589	1.8776	1.8965	1.9155	1.9348	1.9542	1.9739	1.9937
.7	2.0138	2.0340	2.0544	2.0751	2.0959	2.1170	2.1383	2.1598	2.1815	2.2034
.8	2.2255	2.2479	2.2705	2.2933	2.3164	2.3396	2.3632	2.3869	2.4109	2.4351
.9	2.4596	2.4843	2.5093	2.5345	2.5600	2.5857	2.6117	2.6379	2.6645	2.6912
1.0	2.7183	2.7456	2.7732	2.8011	2.8292	2.8577	2.8864	2.9154	2.9447	2.9743
1.1	3.0042	3.0344	3.0649	3.0957	3.1268	3.1582	3.1899	3.2220	3.2544	3.2871
1.2	3.3201	3.3535	3.3872	3.4212	3.4556	3.4903	3.5254	3.5609	3.5966	3.6328
1.3	3.6693	3.7062	3.7434	3.7810	3.8190	3.8574	3.8962	3.9354	3.9749	4.0149
1.4	4.0552	4.0960	4.1371	4.1787	4.2207	4.2631	4.3060	4.3492	4.3929	4.4371
1.5	4.4817	4.5267	4.5722	4.6182	4.6646	4.7115	4.7588	4.8066	4.8550	4.9037
1.6	4.9530	5.0028	5.0531	5.1039	5.1552	5.2070	5.2593	5.3122	5.3656	5.4195
1.7	5.4739	5.5290	5.5845	5.6407	5.6973	5.7546	5.8124	5.8709	5.9299	5.9895
1.8	6.0496	6.1104	6.1719	6.2339	6.2965	6.3598	6.4237	6.4883	6.5535	6.6194
1.9	6.6859	6.7531	6.8210	6.8895	6.9588	7.0287	7.0993	7.1707	7.2427	7.3155
2.0	7.3891	7.4633	7.5383	7.6141	7.6906	7.7679	7.8460	7.9248	8.0045	8.0849
2.1	8.1662	8.2482	8.3311	8.4149	8.4994	8.5849	8.6711	8.7583	8.8463	8.9352
2.2	9.0250	9.1157	9.2073	9.2999	9.3933	9.4877	9.5831	9.6794	9.7767	9.8749
2.3	9.9742	10.074	10.176	10.278	10.381	10.486	10.591	10.697	10.805	10.913
2.4	11.023	11.134	11.246	11.359	11.473	11.588	11.705	11.822	11.941	12.061
2.5	12.182	12.305	12.429	12.554	12.680	12.807	12.936	13.066	13.197	13.330
2.6	13.464	13.599	13.736	13.874	14.013	14.154	14.296	14.440	14.585	14.732
2.7	14.880	15.029	15.180	15.333	15.487	15.643	15.800	15.959	16.119	16.281
2.8	16.445	16.610	16.777	16.945	17.116	17.288	17.462	17.637	17.814	17.993
2.9	18.174	18.357	18.541	18.728	18.916	19.106	19.298	19.492	19.688	19.886
3.0	20.086	20.287	20.491	20.697	20.905	21.115	21.328	21.542	21.758	21.977
3.1	22.198	22.421	22.646	22.874	23.104	23.336	23.571	23.807	24.047	24.288
3.2	24.533	24.779	25.028	25.280	25.534	25.790	26.050	26.311	26.576	26.843
3.3	27.113	27.385	27.660	27.938	28.219	28.503	28.789	29.079	29.371	29.666
3.4	29.964	30.265	30.569	30.877	31.187	31.500	31.817	32.137	32.460	32.786
3.5	33.115	33.448	33.784	34.124	34.467	34.813	35.163	35.517	35.874	36.234
3.6	36.598	36.966	37.338	37.713	38.092	38.475	38.861	39.252	39.646	40.045
3.7	40.447	40.854	41.264	41.679	42.098	42.521	42.948	43.380	43.816	44.256
3.8	44.701	45.150	45.604	46.063	46.525	46.993	47.465	47.942	48.424	48.911
3.9	49.402	49.899	50.400	50.907	51.419	51.935	52.457	52.985	53.517	54.055
4.	54.598	60.340	66.686	73.700	81.451	90.017	99.484	109.95	121.51	134.29
5.	148.41	164.02	181.27	200.34	221.41	244.69	270.43	298.87	330.30	365.04
6.	403.43	445.86	492.75	544.57	601.85	665.14	735.10	812.41	897.85	992.27
7.	1096.6	1212.0	1339.4	1480.3	1636.0	1808.0	1998.2	2208.3	2440.6	2697.3
8.	2981.0	3294.5	3641.0	4023.9	4447.1	4914.8	5431.7	6002.9	6634.2	7332.0
9.	8103.1	8955.3	9897.1	10938	12088	13360	14765	16318	18034	19930
10.	22026									

EXPONENTIAL FUNCTIONS e^{-x}

x	0	1	2	3	4	5	6	7	8	9
.0	1.00000	.99005	.98020	.97045	.96079	.95123	.94176	.93239	.92312	.91393
.1	.90484	.89583	.88692	.87810	.86936	.86071	.85214	.84366	.83527	.82696
.2	.81873	.81058	.80252	.79453	.78663	.77880	.77105	.76338	.75578	.74826
.3	.74082	.73345	.72615	.71892	.71177	.70469	.69768	.69073	.68386	.67706
.4	.67032	.66365	.65705	.65051	.64404	.63763	.63128	.62500	.61878	.61263
.5	.60653	.60050	.59452	.58860	.58275	.57695	.57121	.56553	.55990	.55433
.6	.54881	.54335	.53794	.53259	.52729	.52205	.51685	.51171	.50662	.50158
.7	.49659	.49164	.48675	.48191	.47711	.47237	.46767	.46301	.45841	.45384
.8	.44933	.44486	.44043	.43605	.43171	.42741	.42316	.41895	.41478	.41066
.9	.40657	.40252	.39852	.39455	.39063	.38674	.38289	.37908	.37531	.37158
1.0	.36788	.36422	.36060	.35701	.35345	.34994	.34646	.34301	.33960	.33622
1.1	.33287	.32956	.32628	.32303	.31982	.31664	.31349	.31037	.30728	.30422
1.2	.30119	.29820	.29523	.29229	.28938	.28650	.28365	.28083	.27804	.27527
1.3	.27253	.26982	.26714	.26448	.26185	.25924	.25666	.25411	.25158	.24908
1.4	.24660	.24414	.24171	.23931	.23693	.23457	.23224	.22993	.22764	.22537
1.5	.22313	.22091	.21871	.21654	.21438	.21225	.21014	.20805	.20598	.20393
1.6	.20190	.19989	.19790	.19593	.19398	.19205	.19014	.18825	.18637	.18452
1.7	.18268	.18087	.17907	.17728	.17552	.17377	.17204	.17033	.16864	.16696
1.8	.16530	.16365	.16203	.16041	.15882	.15724	.15567	.15412	.15259	.15107
1.9	.14957	.14808	.14661	.14515	.14370	.14227	.14086	.13946	.13807	.13670
2.0	.13534	.13399	.13266	.13134	.13003	.12873	.12745	.12619	.12493	.12369
2.1	.12246	.12124	.12003	.11884	.11765	.11648	.11533	.11418	.11304	.11192
2.2	.11030	.10970	.10861	.10753	.10646	.10540	.10435	.10331	.10228	.10127
2.3	.10026	.09926	.09827	.09730	.09633	.09537	.09442	.09348	.09255	.09163
2.4	.09072	.08982	.08892	.08804	.08716	.08629	.08543	.08458	.08374	.08291
2.5	.08208	.08127	.08046	.07966	.07887	.07808	.07730	.07654	.07577	.07502
2.6	.07427	.07353	.07280	.07208	.07136	.07065	.06995	.06925	.06856	.06788
2.7	.06721	.06654	.06587	.06522	.06457	.06393	.06329	.06266	.06204	.06142
2.8	.06081	.06020	.05961	.05901	.05843	.05784	.05727	.05670	.05613	.05558
2.9	.05502	.05448	.05393	.05340	.05287	.05234	.05182	.05130	.05079	.05029
3.0	.04979	.04929	.04880	.04832	.04783	.04736	.04689	.04642	.04596	.04550
3.1	.04505	.04460	.04416	.04372	.04328	.04285	.04243	.04200	.04159	.04117
3.2	.04076	.04036	.03996	.03956	.03916	.03877	.03839	.03801	.03763	.03725
3.3	.03688	.03652	.03615	.03579	.03544	.03508	.03474	.03439	.03405	.03371
3.4	.03337	.03304	.03271	.03239	.03206	.03175	.03143	.03112	.03081	.03050
3.5	.03020	.02990	.02960	.02930	.02901	.02872	.02844	.02816	.02788	.02760
3.6	.02732	.02705	.02678	.02652	.02625	.02599	.02573	.02548	.02522	.02497
3.7	.02472	.02448	.02423	.02399	.02375	.02352	.02328	.02305	.02282	.02260
3.8	.02237	.02215	.02193	.02171	.02149	.02128	.02107	.02086	.02065	.02045
3.9	.02024	.02004	.01984	.01964	.01945	.01925	.01906	.01887	.01869	.01850
4.	.018316	.016573	.014996	.013569	.012277	.011109	.010052	$.0^{2}90953$	$.0^{2}82297$	$.0^{2}74466$
5.	$.0^{2}67379$	$.0^{2}60967$	$.0^{2}55166$	$.0^{2}49916$	$.0^{2}45166$	$.0^{2}40868$	$.0^{2}36979$	$.0^{2}33460$	$.0^{2}30276$	$.0^{2}27394$
6.	$.0^{2}24788$	$.0^{2}22429$	$.0^{2}20294$	$.0^{2}18363$	$.0^{2}16616$	$.0^{2}15034$	$.0^{2}13604$	$.0^{2}12309$	$.0^{2}11138$	$.0^{2}10078$
7.	$.0^{3}91188$	$.0^{3}82510$	$.0^{3}74659$	$.0^{3}67554$	$.0^{3}61125$	$.0^{3}55308$	$.0^{3}50045$	$.0^{3}45283$	$.0^{3}40973$	$.0^{3}37074$
8.	$.0^{3}33546$	$.0^{3}30354$	$.0^{3}27465$	$.0^{3}24852$	$.0^{3}22487$	$.0^{3}20347$	$.0^{3}18411$	$.0^{3}16659$	$.0^{3}15073$	$.0^{3}13639$
9.	$.0^{3}12341$	$.0^{3}11167$	$.0^{3}10104$	$.0^{4}91424$	$.0^{4}82724$	$.0^{4}74852$	$.0^{4}67729$	$.0^{4}61283$	$.0^{4}55452$	$.0^{4}50175$
10.	$.0^{4}45400$									

CHAPTER 6

TRIGONOMETRY

DEFINITION OF TRIGONOMETRIC FUNCTIONS FOR A RIGHT TRIANGLE

Triangle ABC has a right angle (90°) at C and sides of length a, b, c. The trigonometric functions of angle A are defined as follows.

$$\textit{sine} \text{ of } A = \sin A = \frac{a}{c} = \frac{\text{opposite}}{\text{hypotenuse}}$$

$$\textit{cosine} \text{ of } A = \cos A = \frac{b}{c} = \frac{\text{adjacent}}{\text{hypotenuse}}$$

$$\textit{tangent} \text{ of } A = \tan A = \frac{a}{b} = \frac{\text{opposite}}{\text{adjacent}}$$

$$\textit{cotangent} \text{ of } A = \cot A = \frac{b}{a} = \frac{\text{adjacent}}{\text{opposite}}$$

$$\textit{secant} \text{ of } A = \sec A = \frac{c}{b} = \frac{\text{hypotenuse}}{\text{adjacent}}$$

$$\textit{cosecant} \text{ of } A = \csc A = \frac{c}{a} = \frac{\text{hypotenuse}}{\text{opposite}}$$

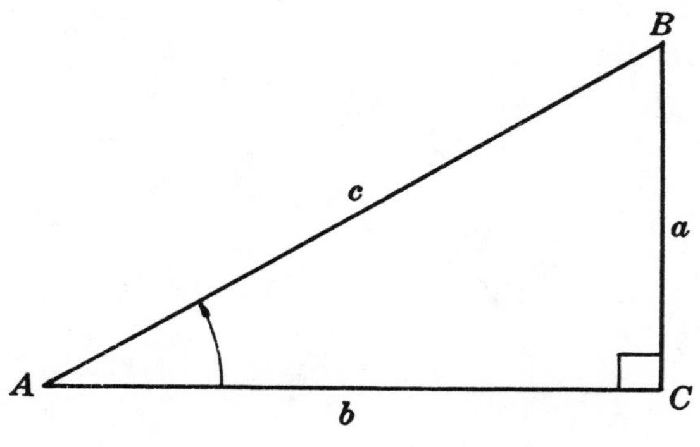

THE TRIGONOMETRIC FUNCTIONS OF AN ARBITRARY ANGLE

Let α be any angle in standard position and let $P(x,y)$ be any point on the terminal side of the angle. Denote the positive distance OP by r. Then

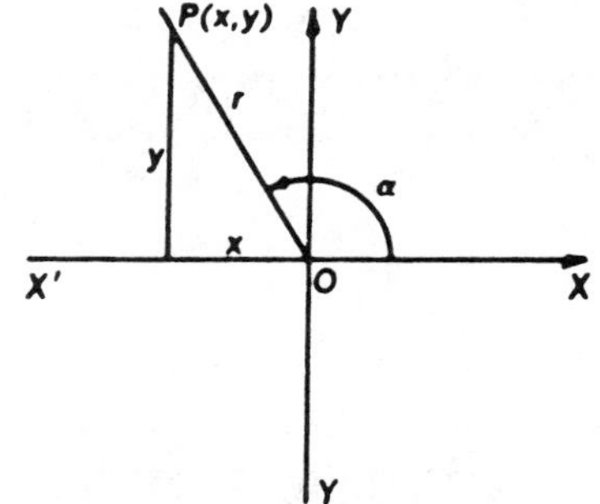

$\sin\alpha = y/r$	$\csc\alpha = r/y$
$\cos\alpha = x/r$	$\sec\alpha = r/x$
$\tan\alpha = y/x$	$\cot\alpha = \text{ctn}\,\alpha = x/y$
$\text{exsec}\,\alpha = \sec\alpha - 1$	$\text{covers}\,\alpha = 1 - \sin\alpha$
$\text{vers}\,\alpha = 1 - \cos\alpha$	$\text{hav}\,\alpha = \frac{1}{2}\,\text{vers}\,\alpha$

$\text{cis}\,\alpha = \cos\alpha + i\sin\alpha = e^{i\alpha}$, α in radians, $i = \sqrt{-1}$

GRAPHS OF TRIGONOMETRIC FUNCTIONS

In each graph x is in radians.

$y = \sin x$

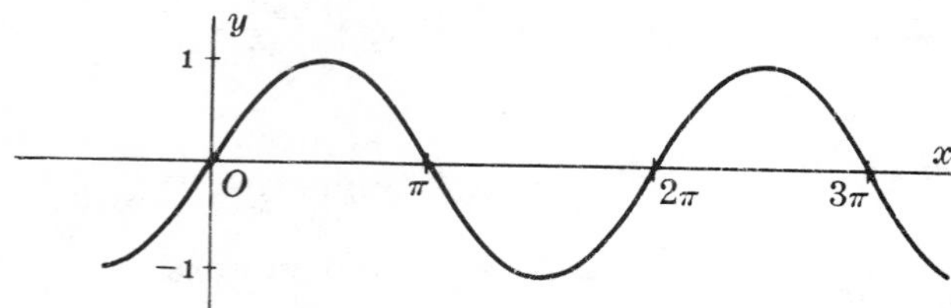

$y = \cos x$

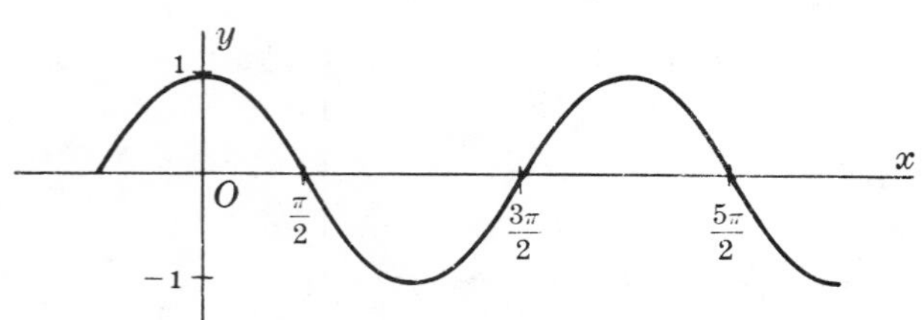

$y = \tan x$

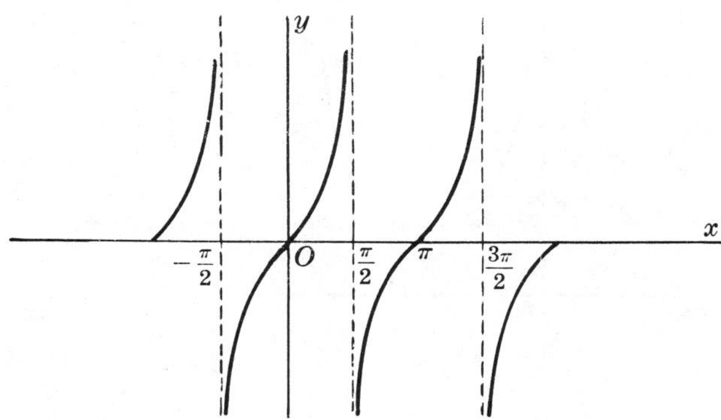
y
x
$-\frac{\pi}{2}$
O
$\frac{\pi}{2}$
π
$\frac{3\pi}{2}$

$y = \cot x$

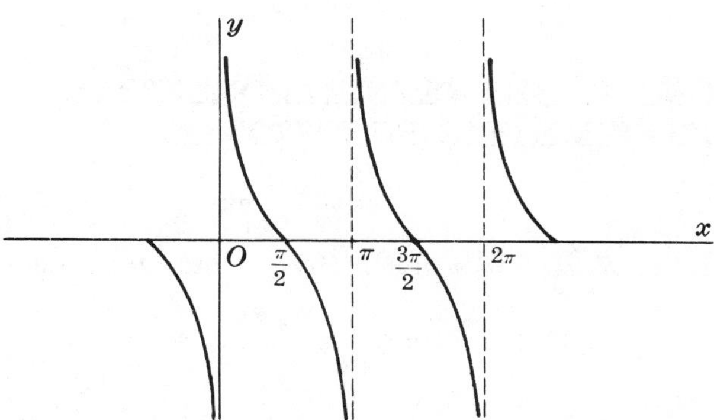
y
x
O
$\frac{\pi}{2}$
π
$\frac{3\pi}{2}$
2π

$y = \sec x$

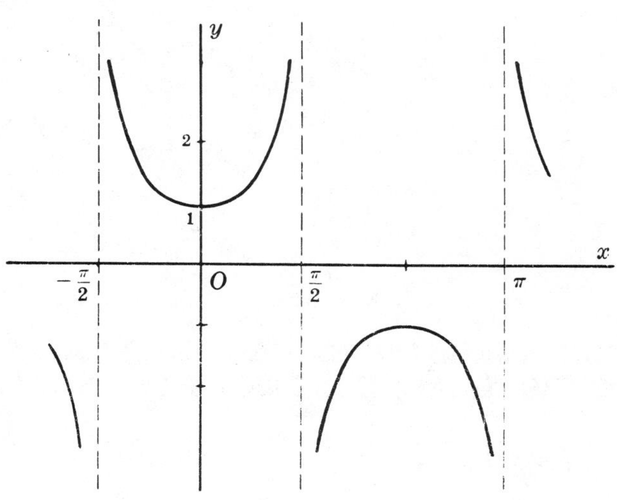
y
2
1
x
$-\frac{\pi}{2}$
O
$\frac{\pi}{2}$
π

$y = \csc x$

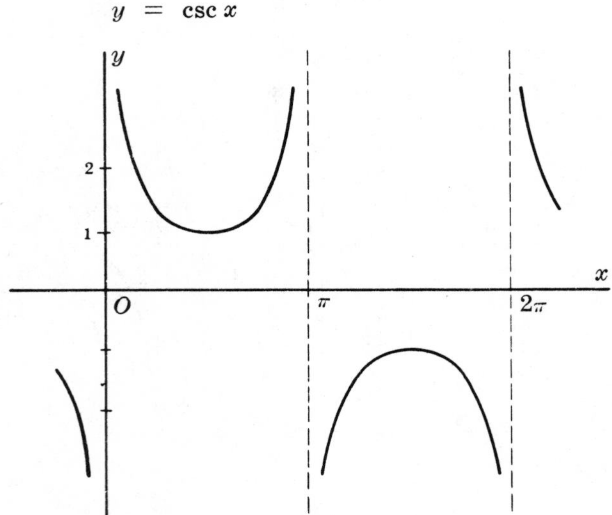

RELATIONSHIP BETWEEN EXPONENTIAL AND TRIGONOMETRIC FUNCTIONS

$$e^{i\theta} = \cos\theta + i\sin\theta, \qquad e^{-i\theta} = \cos\theta - i\sin\theta$$

These are called *Euler's identities*. Here i is the imaginary unit.

$$\sin\theta = \frac{e^{i\theta} - e^{-i\theta}}{2i}$$

$$\cos\theta = \frac{e^{i\theta} + e^{-i\theta}}{2}$$

$$\tan\theta = \frac{e^{i\theta} - e^{-i\theta}}{i(e^{i\theta} + e^{-i\theta})} = -i\left(\frac{e^{i\theta} - e^{-i\theta}}{e^{i\theta} + e^{-i\theta}}\right)$$

$$\cot\theta = i\left(\frac{e^{i\theta} + e^{-i\theta}}{e^{i\theta} - e^{-i\theta}}\right)$$

$$\sec\theta = \frac{2}{e^{i\theta} + e^{-i\theta}}$$

$$\csc\theta = \frac{2i}{e^{i\theta} - e^{-i\theta}}$$

RELATIONSHIP BETWEEN CIRCULAR (OR INVERSE CIRCULAR) FUNCTIONS

$$\left(0 \le x \le \frac{\pi}{2}\right)$$

	$\sin x = a$	$\cos x = a$	$\tan x = a$
$\sin x$	a	$(1 - a^2)^{1/2}$	$a(1 + a^2)^{-1/2}$
$\cos x$	$(1 - a^2)^{1/2}$	a	$(1 + a^2)^{-1/2}$
$\tan x$	$a(1 - a^2)^{-1/2}$	$a^{-1}(1 - a^2)^{1/2}$	a
$\csc x$	a^{-1}	$(1 - a^2)^{-1/2}$	$a^{-1}(1 + a^2)^{1/2}$
$\sec x$	$(1 - a^2)^{-1/2}$	a^{-1}	$(1 + a^2)^{1/2}$
$\cot x$	$a^{-1}(1 - a^2)^{1/2}$	$a(1 - a^2)^{-1/2}$	a^{-1}

	$\csc x = a$	$\sec x = a$	$\cot x = a$
$\sin x$	a^{-1}	$a^{-1}(a^2 - 1)^{1/2}$	$(1 + a^2)^{-1/2}$
$\cos x$	$a^{-1}(a^2 - 1)^{1/2}$	a^{-1}	$a(1 + a^2)^{-1/2}$
$\tan x$	$(a^2 - 1)^{-1/2}$	$(a^2 - 1)^{1/2}$	a^{-1}
$\csc x$	a	$a(a^2 - 1)^{-1/2}$	$(1 + a^2)^{1/2}$
$\sec x$	$a(a^2 - 1)^{-1/2}$	a	$a^{-1}(1 + a^2)^{1/2}$
$\cot x$	$(a^2 - 1)^{1/2}$	$(a^2 - 1)^{-1/2}$	a

Examples:

If $\sec x = a$, then $\tan x = (a^2 - 1)^{1/2}$

$\arctan a = \arccos (1 + a^2)^{-1/2}$

SIGNS OF THE TRIGONOMETRIC FUNCTIONS

Quadrant	sin	cos	tan	cot	sec	csc
I	+	+	+	+	+	+
II	+	−	−	−	−	+
III	−	−	+	+	−	−
IV	−	+	−	−	+	−

VARIATIONS OF THE TRIGONOMETRIC FUNCTIONS

Quadrant	sin	cos	tan	cot	sec	csc
I	$0 \to +1$	$+1 \to 0$	$0 \to +\infty$	$+\infty \to 0$	$+1 \to +\infty$	$+\infty \to +1$
II	$+1 \to 0$	$0 \to -1$	$-\infty \to 0$	$0 \to -\infty$	$-\infty \to -1$	$+1 \to +\infty$
III	$0 \to -1$	$-1 \to 0$	$0 \to +\infty$	$+\infty \to 0$	$-1 \to -\infty$	$-\infty \to -1$
IV	$-1 \to 0$	$0 \to +1$	$-\infty \to 0$	$0 \to -\infty$	$+\infty \to +1$	$-1 \to -\infty$

TRIGONOMETRIC FUNCTIONS OF SOME SPECIAL ANGLES

Angle	sin	cos	tan	cot	sec	csc
$0° = 0$	0	1	0	$\cdots$	1	$\cdots$
$15° = \frac{\pi}{12}$	$\frac{\sqrt{2}}{4}(\sqrt{3}-1)$	$\frac{\sqrt{2}}{4}(\sqrt{3}+1)$	$2-\sqrt{3}$	$2+\sqrt{3}$	$\sqrt{2}(\sqrt{3}-1)$	$\sqrt{2}(\sqrt{3}+1)$
$30° = \frac{\pi}{6}$	$1/2$	$\sqrt{3}/2$	$\sqrt{3}/3$	$\sqrt{3}$	$2\sqrt{3}/3$	2
$45° = \frac{\pi}{4}$	$\sqrt{2}/2$	$\sqrt{2}/2$	1	1	$\sqrt{2}$	$\sqrt{2}$
$60° = \frac{\pi}{3}$	$\sqrt{3}/2$	$1/2$	$\sqrt{3}$	$\sqrt{3}/3$	2	$2\sqrt{3}/3$
$75° = \frac{5\pi}{12}$	$\frac{\sqrt{2}}{4}(\sqrt{3}+1)$	$\frac{\sqrt{2}}{4}(\sqrt{3}-1)$	$2+\sqrt{3}$	$2-\sqrt{3}$	$\sqrt{2}(\sqrt{3}+1)$	$\sqrt{2}(\sqrt{3}-1)$
$90° = \frac{\pi}{2}$	1	0	$\cdots$	0	$\cdots$	1
$105° = \frac{7\pi}{12}$	$\frac{\sqrt{2}}{4}(\sqrt{3}+1)$	$-\frac{\sqrt{2}}{4}(\sqrt{3}-1)$	$-(2+\sqrt{3})$	$-(2-\sqrt{3})$	$-2(\sqrt{3}+1)$	$\sqrt{2}(\sqrt{3}-1)$
$120° = \frac{2\pi}{3}$	$\sqrt{3}/2$	$-1/2$	$-\sqrt{3}$	$-\sqrt{3}/3$	-2	$2\sqrt{3}/3$
$135° = \frac{3\pi}{4}$	$\sqrt{2}/2$	$-\sqrt{2}/2$	-1	-1	$-\sqrt{2}$	$\sqrt{2}$
$150° = \frac{5\pi}{6}$	$1/2$	$-\sqrt{3}/2$	$-\sqrt{3}/3$	$-\sqrt{3}$	$-2\sqrt{3}/3$	2
$165° = \frac{11\pi}{12}$	$\frac{\sqrt{2}}{4}(\sqrt{3}-1)$	$-\frac{\sqrt{2}}{4}(\sqrt{3}+1)$	$-(2-\sqrt{3})$	$-(2+\sqrt{3})$	$-\sqrt{2}(\sqrt{3}-1)$	$\sqrt{2}(\sqrt{3}+1)$
$180° = \pi$	0	-1	0	$\cdots$	-1	$\cdots$
$270° = \frac{3\pi}{2}$	-1	0	$\cdots$	0	$\cdots$	-1

RELATIONS AMONG THE FUNCTIONS

$$\sin x = \frac{1}{\csc x} \qquad \csc x = \frac{1}{\sin x}$$

$$\cos x = \frac{1}{\sec x} \qquad \sec x = \frac{1}{\cos x}$$

$$\tan x = \frac{1}{\cot x} = \frac{\sin x}{\cos x} \qquad \sin^2 x + \cos^2 x = 1$$

$$\cot x = \frac{1}{\tan x} = \frac{\cos x}{\sin x} \qquad 1 + \tan^2 x = \sec^2 x \qquad 1 + \cot^2 x = \csc^2 x$$

$$* \sin x = \pm\sqrt{1 - \cos^2 x} \qquad * \cos x = \pm\sqrt{1 - \sin^2 x}$$

$$* \tan x = \pm\sqrt{\sec^2 x - 1} \qquad * \sec x = \pm\sqrt{\tan^2 x + 1}$$

$$* \cot x = \pm\sqrt{\csc^2 x - 1} \qquad * \csc x = \pm\sqrt{\cot^2 x + 1}$$

$$\sin x = \cos(90° - x) = \sin(180° - x)$$

$$\cos x = \sin(90° - x) = -\cos(180° - x)$$

$$\tan x = \cot(90° - x) = -\tan(180° - x)$$

$$\cot x = \tan(90° - x) = -\cot(180° - x)$$

$$\csc x = \cot\frac{x}{2} - \cot x$$

*The sign in front of radical depends on quadrant in which x falls.

REDUCTION FORMULAS

$$\begin{aligned}
\sin\alpha &= +\cos(\alpha - 90^\circ) = -\sin(\alpha - 180^\circ) = -\cos(\alpha - 270^\circ)\\
\cos\alpha &= -\sin(\alpha - 90^\circ) = -\cos(\alpha - 180^\circ) = +\sin(\alpha - 270^\circ)\\
\tan\alpha &= -\cot(\alpha - 90^\circ) = +\tan(\alpha - 180^\circ) = -\cot(\alpha - 270^\circ)\\
\cot\alpha &= -\tan(\alpha - 90^\circ) = +\cot(\alpha - 180^\circ) = -\tan(\alpha - 270^\circ)\\
\sec\alpha &= -\csc(\alpha - 90^\circ) = -\sec(\alpha - 180^\circ) = +\csc(\alpha - 270^\circ)\\
\csc\alpha &= +\sec(\alpha - 90^\circ) = -\csc(\alpha - 180^\circ) = -\sec(\alpha - 270^\circ)
\end{aligned}$$

FURTHER REDUCTION FORMULAS

	sin	cos	tan	cot	sec	csc
$-\alpha$	$-\sin\alpha$	$+\cos\alpha$	$-\tan\alpha$	$-\cot\alpha$	$+\sec\alpha$	$-\csc\alpha$
$90^\circ + \alpha$	$+\cos\alpha$	$-\sin\alpha$	$-\cot\alpha$	$-\tan\alpha$	$-\csc\alpha$	$+\sec\alpha$
$90^\circ - \alpha$	$+\cos\alpha$	$+\sin\alpha$	$+\cot\alpha$	$+\tan\alpha$	$+\csc\alpha$	$+\sec\alpha$
$180^\circ + \alpha$	$-\sin\alpha$	$-\cos\alpha$	$+\tan\alpha$	$+\cot\alpha$	$-\sec\alpha$	$-\csc\alpha$
$180^\circ - \alpha$	$+\sin\alpha$	$-\cos\alpha$	$-\tan\alpha$	$-\cot\alpha$	$-\sec\alpha$	$+\csc\alpha$
$270^\circ + \alpha$	$-\cos\alpha$	$+\sin\alpha$	$-\cot\alpha$	$-\tan\alpha$	$+\csc\alpha$	$-\sec\alpha$
$270^\circ - \alpha$	$-\cos\alpha$	$-\sin\alpha$	$+\cot\alpha$	$+\tan\alpha$	$-\csc\alpha$	$-\sec\alpha$
$360^\circ + \alpha$	$+\sin\alpha$	$+\cos\alpha$	$+\tan\alpha$	$+\cot\alpha$	$+\sec\alpha$	$+\csc\alpha$
$360^\circ - \alpha$	$-\sin\alpha$	$+\cos\alpha$	$-\tan\alpha$	$-\cot\alpha$	$+\sec\alpha$	$-\csc\alpha$

FUNDAMENTAL IDENTITIES

Where a double sign appears in the following, the choice of sign depends upon the quadrant in which the angle terminates.

RECIPROCAL RELATIONS

$$\sin\alpha = \frac{1}{\csc\alpha}, \quad \cos\alpha = \frac{1}{\sec\alpha}, \quad \tan\alpha = \frac{1}{\cot\alpha}$$

$$\csc\alpha = \frac{1}{\sin\alpha}, \quad \sec\alpha = \frac{1}{\cos\alpha}, \quad \cot\alpha = \frac{1}{\tan\alpha}$$

PRODUCT RELATIONS

$$\sin\alpha = \tan\alpha\cos\alpha, \quad \cos\alpha = \cot\alpha\sin\alpha$$

$$\tan\alpha = \sin\alpha\sec\alpha, \quad \cot\alpha = \cos\alpha\csc\alpha$$

$$\sec\alpha = \csc\alpha\tan\alpha, \quad \csc\alpha = \sec\alpha\cot\alpha$$

QUOTIENT RELATIONS

$$\sin\alpha = \frac{\tan\alpha}{\sec\alpha}, \qquad \cos\alpha = \frac{\cot\alpha}{\csc\alpha}, \qquad \tan\alpha = \frac{\sin\alpha}{\cos\alpha}$$

$$\csc\alpha = \frac{\sec\alpha}{\tan\alpha}, \qquad \sec\alpha = \frac{\csc\alpha}{\cot\alpha}, \qquad \cot\alpha = \frac{\cos\alpha}{\sin\alpha}$$

PYTHAGOREAN RELATIONS

$$\sin^2\alpha + \cos^2\alpha = 1, \qquad 1 + \tan^2\alpha = \sec^2\alpha, \qquad 1 + \cot^2\alpha = \csc^2\alpha$$

ANGLE-SUM AND ANGLE-DIFFERENCE RELATIONS

$$\sin(\alpha + \beta) = \sin\alpha\cos\beta + \cos\alpha\sin\beta$$
$$\sin(\alpha - \beta) = \sin\alpha\cos\beta - \cos\alpha\sin\beta$$
$$\cos(\alpha + \beta) = \cos\alpha\cos\beta - \sin\alpha\sin\beta$$
$$\cos(\alpha - \beta) = \cos\alpha\cos\beta + \sin\alpha\sin\beta$$
$$\tan(\alpha + \beta) = \frac{\tan\alpha + \tan\beta}{1 - \tan\alpha\tan\beta}$$
$$\tan(\alpha - \beta) = \frac{\tan\alpha - \tan\beta}{1 + \tan\alpha\tan\beta}$$
$$\cot(\alpha + \beta) = \frac{\cot\beta\cot\alpha - 1}{\cot\beta + \cot\alpha}$$
$$\cot(\alpha - \beta) = \frac{\cot\beta\cot\alpha + 1}{\cot\beta - \cot\alpha}$$
$$\sin(\alpha + \beta)\sin(\alpha - \beta) = \sin^2\alpha - \sin^2\beta = \cos^2\beta - \cos^2\alpha$$
$$\cos(\alpha + \beta)\cos(\alpha - \beta) = \cos^2\alpha - \sin^2\beta = \cos^2\beta - \sin^2\alpha$$

DOUBLE-ANGLE RELATIONS

$$\sin 2\alpha = 2\sin\alpha\cos\alpha = \frac{2\tan\alpha}{1 + \tan^2\alpha}$$

$$\cos 2\alpha = \cos^2\alpha - \sin^2\alpha = 2\cos^2\alpha - 1 = 1 - 2\sin^2\alpha = \frac{1 - \tan^2\alpha}{1 + \tan^2\alpha}$$

$$\tan 2\alpha = \frac{2\tan\alpha}{1 - \tan^2\alpha}, \qquad \cot 2\alpha = \frac{\cot^2\alpha - 1}{2\cot\alpha}$$

MULTIPLE-ANGLE RELATIONS

$$\sin 3\alpha = 3\sin\alpha - 4\sin^3\alpha$$
$$\cos 3\alpha = 4\cos^3\alpha - 3\cos\alpha$$
$$\sin 4\alpha = 4\sin\alpha\cos\alpha - 8\sin^3\alpha\cos\alpha$$
$$\cos 4\alpha = 8\cos^4\alpha - 8\cos^2\alpha + 1$$
$$\sin 5\alpha = 5\sin\alpha - 20\sin^3\alpha + 16\sin^5\alpha$$
$$\cos 5\alpha = 16\cos^5\alpha - 20\cos^3\alpha + 5\cos\alpha$$

$$\sin 6\alpha = 32\cos^5\alpha \sin\alpha - 32\cos^3\alpha\sin\alpha + 6\cos\alpha\sin\alpha$$
$$\cos 6\alpha = 32\cos^6\alpha - 48\cos^4\alpha + 18\cos^2\alpha - 1$$
$$\sin n\alpha = 2\sin(n-1)\alpha\cos\alpha - \sin(n-2)\alpha$$
$$\cos n\alpha = 2\cos(n-1)\alpha\cos\alpha - \cos(n-2)\alpha$$
$$\tan 3\alpha = \frac{3\tan\alpha - \tan^3\alpha}{1 - 3\tan^2\alpha}$$
$$\tan 4\alpha = \frac{4\tan\alpha - 4\tan^3\alpha}{1 - 6\tan^2\alpha + \tan^4\alpha}$$
$$\tan n\alpha = \frac{\tan(n-1)\alpha + \tan\alpha}{1 - \tan(n-1)\alpha\tan\alpha}$$

FUNCTION-PRODUCT RELATIONS

$$\sin\alpha\sin\beta = \tfrac{1}{2}\cos(\alpha-\beta) - \tfrac{1}{2}\cos(\alpha+\beta)$$
$$\cos\alpha\cos\beta = \tfrac{1}{2}\cos(\alpha-\beta) + \tfrac{1}{2}\cos(\alpha+\beta)$$
$$\sin\alpha\cos\beta = \tfrac{1}{2}\sin(\alpha+\beta) + \tfrac{1}{2}\sin(\alpha-\beta)$$
$$\cos\alpha\sin\beta = \tfrac{1}{2}\sin(\alpha+\beta) - \tfrac{1}{2}\sin(\alpha-\beta)$$

FUNCTION-SUM AND FUNCTION-DIFFERENCE RELATIONS

$$\sin\alpha + \sin\beta = 2\sin\tfrac{1}{2}(\alpha+\beta)\cos\tfrac{1}{2}(\alpha-\beta)$$
$$\sin\alpha - \sin\beta = 2\cos\tfrac{1}{2}(\alpha+\beta)\sin\tfrac{1}{2}(\alpha-\beta)$$
$$\cos\alpha + \cos\beta = 2\cos\tfrac{1}{2}(\alpha+\beta)\cos\tfrac{1}{2}(\alpha-\beta)$$
$$\cos\alpha - \cos\beta = -2\sin\tfrac{1}{2}(\alpha+\beta)\sin\tfrac{1}{2}(\alpha-\beta)$$
$$\tan\alpha + \tan\beta = \frac{\sin(\alpha+\beta)}{\cos\alpha\cos\beta}, \qquad \tan\alpha - \tan\beta = \frac{\sin(\alpha-\beta)}{\cos\alpha\cos\beta}$$
$$\cot\alpha + \cot\beta = \frac{\sin(\alpha+\beta)}{\sin\alpha\sin\beta}, \qquad \cot\alpha - \cot\beta = \frac{\sin(\beta-\alpha)}{\sin\alpha\sin\beta}$$
$$\frac{\sin\alpha + \sin\beta}{\sin\alpha - \sin\beta} = \frac{\tan\frac{1}{2}(\alpha+\beta)}{\tan\frac{1}{2}(\alpha-\beta)}, \qquad \frac{\sin\alpha + \sin\beta}{\cos\alpha - \cos\beta} = \cot\tfrac{1}{2}(\beta-\alpha)$$
$$\frac{\sin\alpha + \sin\beta}{\cos\alpha + \cos\beta} = \tan\tfrac{1}{2}(\alpha+\beta), \qquad \frac{\sin\alpha - \sin\beta}{\cos\alpha + \cos\beta} = \tan\tfrac{1}{2}(\alpha-\beta)$$

HALF-ANGLE RELATIONS

$$\sin\frac{\alpha}{2} = \pm\sqrt{\frac{1-\cos\alpha}{2}}, \qquad \cos\frac{\alpha}{2} = \pm\sqrt{\frac{1+\cos\alpha}{2}}$$
$$\tan\frac{\alpha}{2} = \pm\sqrt{\frac{1-\cos\alpha}{1+\cos\alpha}} = \frac{1-\cos\alpha}{\sin\alpha} = \frac{\sin\alpha}{1+\cos\alpha}$$
$$\cot\frac{\alpha}{2} = \pm\sqrt{\frac{1+\cos\alpha}{1-\cos\alpha}} = \frac{1+\cos\alpha}{\sin\alpha} = \frac{\sin\alpha}{1-\cos\alpha}$$

POWER RELATIONS

$$\sin^2\alpha = \tfrac{1}{2}(1 - \cos 2\alpha), \qquad \sin^3\alpha = \tfrac{1}{4}(3\sin\alpha - \sin 3\alpha)$$
$$\sin^4\alpha = \tfrac{1}{8}(3 - 4\cos 2\alpha + \cos 4\alpha)$$
$$\cos^2\alpha = \tfrac{1}{2}(1 + \cos 2\alpha), \qquad \cos^3\alpha = \tfrac{1}{4}(3\cos\alpha + \cos 3\alpha)$$
$$\cos^4\alpha = \tfrac{1}{8}(3 + 4\cos 2\alpha + \cos 4\alpha)$$
$$\tan^2\alpha = \frac{1 - \cos 2\alpha}{1 + \cos 2\alpha}, \qquad \cot^2\alpha = \frac{1 + \cos 2\alpha}{1 - \cos 2\alpha}$$

THE TRIGONOMETRIC FUNCTIONS IN TERMS OF ONE ANOTHER

Function	$\sin\alpha$	$\cos\alpha$	$\tan\alpha$	$\cot\alpha$	$\sec\alpha$	$\csc\alpha$
$\sin\alpha$	$\sin\alpha$	$\pm\sqrt{1-\cos^2\alpha}$	$\frac{\tan\alpha}{\pm\sqrt{1+\tan^2\alpha}}$	$\frac{1}{\pm\sqrt{1+\cot^2\alpha}}$	$\frac{\pm\sqrt{\sec^2\alpha-1}}{\sec\alpha}$	$\frac{1}{\csc\alpha}$
$\cos\alpha$	$\pm\sqrt{1-\sin^2\alpha}$	$\cos\alpha$	$\frac{1}{\pm\sqrt{1+\tan^2\alpha}}$	$\frac{\cot\alpha}{\pm\sqrt{1+\cot^2\alpha}}$	$\frac{1}{\sec\alpha}$	$\frac{\pm\sqrt{\csc^2\alpha-1}}{\csc\alpha}$
$\tan\alpha$	$\frac{\sin\alpha}{\pm\sqrt{1-\sin^2\alpha}}$	$\frac{\pm\sqrt{1-\cos^2\alpha}}{\cos\alpha}$	$\tan\alpha$	$\frac{1}{\cot\alpha}$	$\pm\sqrt{\sec^2\alpha-1}$	$\frac{1}{\pm\sqrt{\csc^2\alpha-1}}$
$\cot\alpha$	$\frac{\pm\sqrt{1-\sin^2\alpha}}{\sin\alpha}$	$\frac{\cos\alpha}{\pm\sqrt{1-\cos^2\alpha}}$	$\frac{1}{\tan\alpha}$	$\cot\alpha$	$\frac{1}{\pm\sqrt{\sec^2\alpha-1}}$	$\pm\sqrt{\csc^2\alpha-1}$
$\sec\alpha$	$\frac{1}{\pm\sqrt{1-\sin^2\alpha}}$	$\frac{1}{\cos\alpha}$	$\pm\sqrt{1+\tan^2\alpha}$	$\frac{\pm\sqrt{1+\cot^2\alpha}}{\cot\alpha}$	$\sec\alpha$	$\frac{\csc\alpha}{\pm\sqrt{\csc^2\alpha-1}}$
$\csc\alpha$	$\frac{1}{\sin\alpha}$	$\frac{1}{\pm\sqrt{1-\cos^2\alpha}}$	$\frac{\pm\sqrt{1+\tan^2\alpha}}{\tan\alpha}$	$\pm\sqrt{1+\cot^2\alpha}$	$\frac{\sec\alpha}{\pm\sqrt{\sec^2\alpha-1}}$	$\csc\alpha$

Note. The choice of sign depends upon the quadrant in which the angle terminates.

PRINCIPAL VALUES FOR INVERSE TRIGONOMETRIC FUNCTIONS

Principal values for $x \geqq 0$	Principal values for $x < 0$
$0 \leqq \sin^{-1}x \leqq \pi/2$	$-\pi/2 \leqq \sin^{-1}x < 0$
$0 \leqq \cos^{-1}x \leqq \pi/2$	$\pi/2 < \cos^{-1}x \leqq \pi$
$0 \leqq \tan^{-1}x < \pi/2$	$-\pi/2 < \tan^{-1}x < 0$
$0 < \cot^{-1}x \leqq \pi/2$	$\pi/2 < \cot^{-1}x < \pi$
$0 \leqq \sec^{-1}x < \pi/2$	$\pi/2 < \sec^{-1}x \leqq \pi$
$0 < \csc^{-1}x \leqq \pi/2$	$-\pi/2 \leqq \csc^{-1}x < 0$

RELATIONS BETWEEN INVERSE TRIGONOMETRIC FUNCTIONS

In all cases it is assumed that principal values are used.

$$\sin^{-1} x + \cos^{-1} x = \pi/2$$

$$\tan^{-1} x + \cot^{-1} x = \pi/2$$

$$\sec^{-1} x + \csc^{-1} x = \pi/2$$

$$\csc^{-1} x = \sin^{-1}(1/x)$$

$$\sec^{-1} x = \cos^{-1}(1/x)$$

$$\cot^{-1} x = \tan^{-1}(1/x)$$

$$\sin^{-1}(-x) = -\sin^{-1} x$$

$$\cos^{-1}(-x) = \pi - \cos^{-1} x$$

$$\tan^{-1}(-x) = -\tan^{-1} x$$

$$\cot^{-1}(-x) = \pi - \cot^{-1} x$$

$$\sec^{-1}(-x) = \pi - \sec^{-1} x$$

$$\csc^{-1}(-x) = -\csc^{-1} x$$

GRAPHS OF INVERSE TRIGONOMETRIC FUNCTIONS

In each graph y is in radians.

Solid portions of curves correspond to principal values.

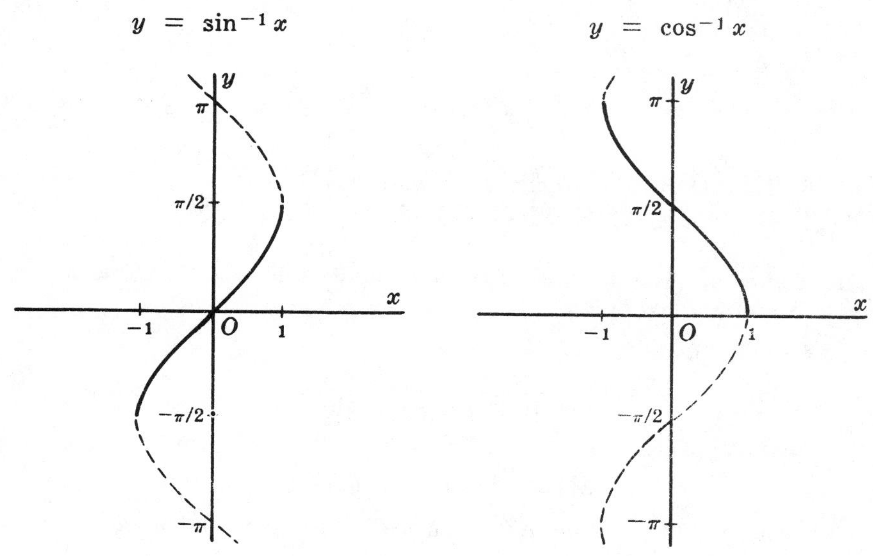

$y = \cot^{-1} x$

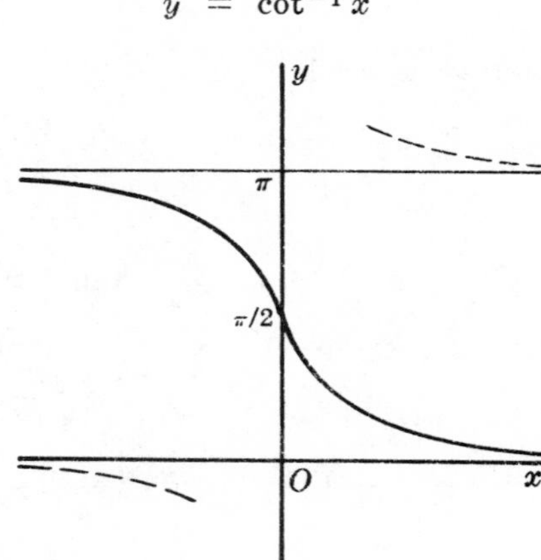

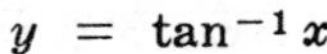

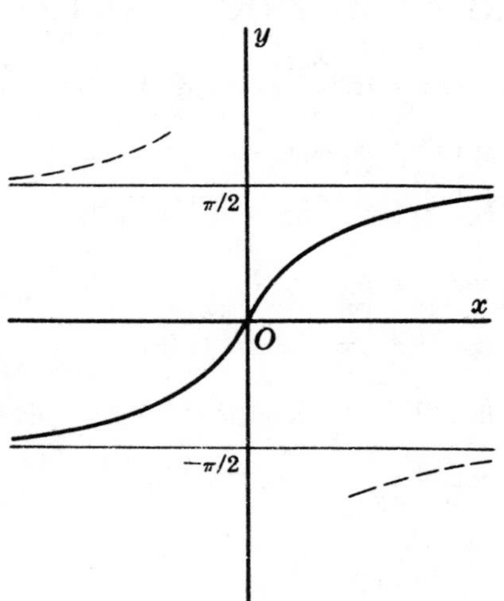

$y = \sec^{-1} x$

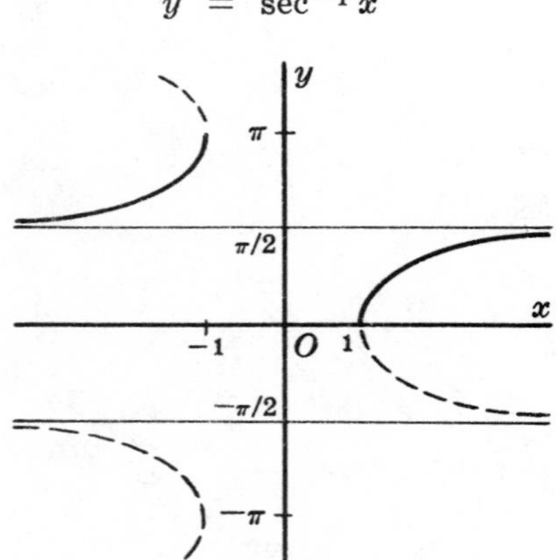

$y = \csc^{-1} x$

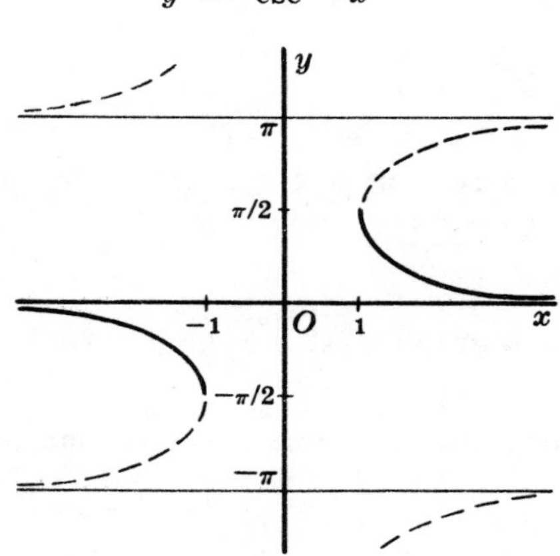

RELATIONSHIPS BETWEEN SIDES AND ANGLES OF A PLANE TRIANGLE

The following results hold for any plane triangle ABC with sides a, b, c and angles A, B, C.

Law of Sines

$$\frac{a}{\sin A} = \frac{b}{\sin B} = \frac{c}{\sin C}$$

Law of Cosines

$$c^2 = a^2 + b^2 - 2ab \cos C$$

with similar relations involving the other sides and angles.

Law of Tangents

$$\frac{a+b}{a-b} = \frac{\tan\frac{1}{2}(A+B)}{\tan\frac{1}{2}(A-B)}$$

with similar relations involving the other sides and angles.

$$\sin A = \frac{2}{bc}\sqrt{s(s-a)(s-b)(s-c)}$$

where $s = \frac{1}{2}(a+b+c)$ is the semiperimeter of the triangle. Similar relations involving angles B and C can be obtained.

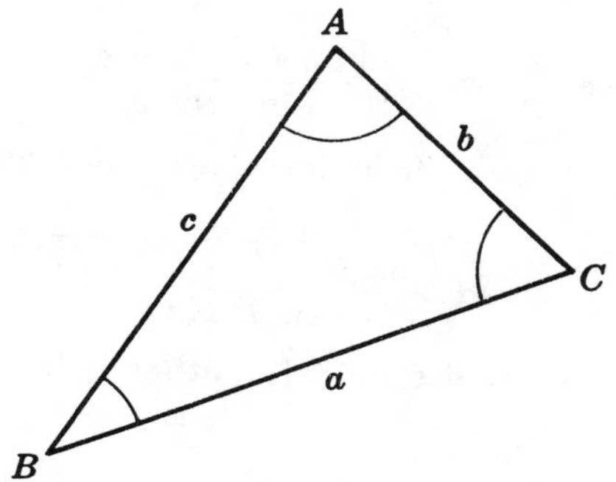

RELATIONSHIPS BETWEEN SIDES AND ANGLES OF A SPHERICAL TRIANGLE

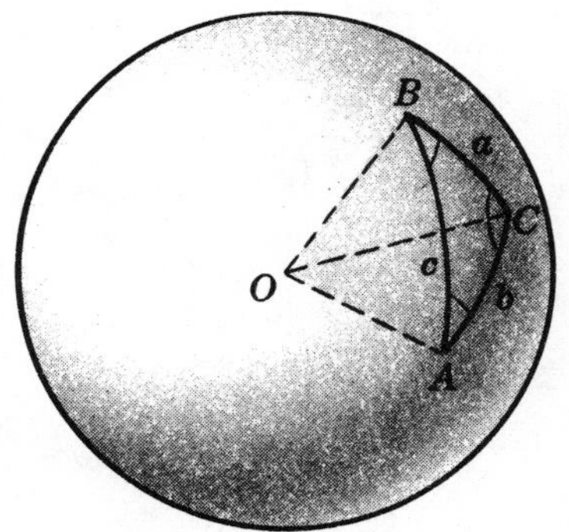

Fig. 1

Spherical triangle ABC is on the surface of a sphere as shown in Fig. 1. Sides a, b, c [which are arcs of great circles] are measured by their angles subtended at center O of the sphere. A, B, C are the angles opposite sides a, b, c respectively. Then the following results hold.

Law of Sines

$$\frac{\sin a}{\sin A} = \frac{\sin b}{\sin B} = \frac{\sin c}{\sin C}$$

6.1 **Law of Cosines**

$$\cos a = \cos b \cos c + \sin b \sin c \cos A$$

$$\cos A = -\cos B \cos C + \sin B \sin C \cos a$$

with similar results involving other sides and angles.

Law of Tangents

$$\frac{\tan \frac{1}{2}(A+B)}{\tan \frac{1}{2}(A-B)} = \frac{\tan \frac{1}{2}(a+b)}{\tan \frac{1}{2}(a-b)}$$

with similar results involving other sides and angles.

$$\cos \frac{A}{2} = \sqrt{\frac{\sin s \sin (s-c)}{\sin b \sin c}}$$

where $s = \frac{1}{2}(a+b+c)$. Similar results hold for other sides and angles.

$$\cos \frac{a}{2} = \sqrt{\frac{\cos (S-B) \cos (S-C)}{\sin B \sin C}}$$

where $S = \frac{1}{2}(A+B+C)$. Similar results hold for other sides and angles.

NAPIER'S RULES FOR RIGHT ANGLED SPHERICAL TRIANGLES

Except for right angle C, there are five parts of spherical triangle ABC which if arranged in the order as given in Fig. 2 would be a, b, A, c, B.

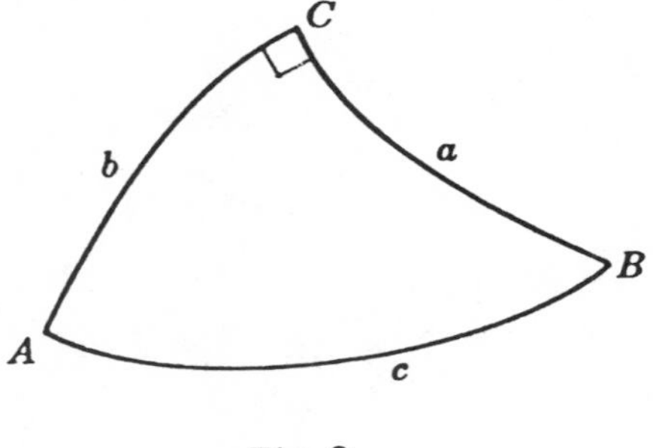

Fig. 2

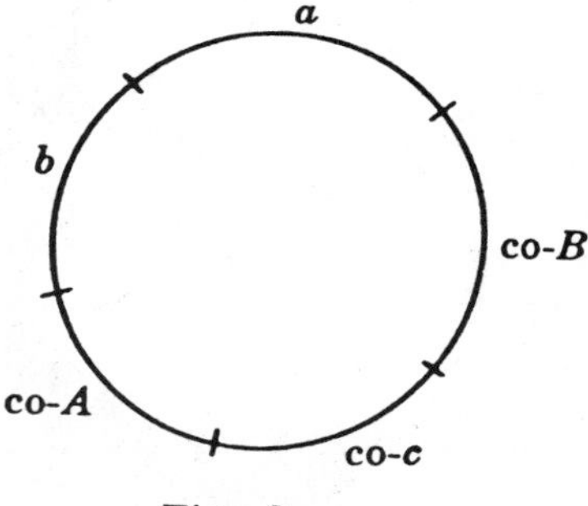

Fig. 3

Suppose these quantities are arranged in a circle as in Fig. 3 where we attach the prefix co [indicating *complement*] to hypotenuse c and angles A and B

Any one of the parts of this circle is called a *middle part*, the two neighboring parts are called *adjacent parts* and the two remaining parts are called *opposite parts*.

Then Napier's rules are:

i. The sine of any middle part equals the product of the tangents of the adjacent p

ii. The sine of any middle part equals the product of the cosines of the opposite pa

Example: Since co-$A = 90° - A$, co-$B = 90° - B$, we have

$$\sin a = \tan b \tan(\text{co-}B) \qquad \text{or} \qquad \sin a = \tan b \cot B$$

$$\sin(\text{co-}A) = \cos a \cos(\text{co-}B) \qquad \text{or} \qquad \cos A = \cos a \sin B$$

These can of course be obtained also from the results of 6.1.

DEGREES, MINUTES, AND SECONDS TO RADIANS

Units in degrees, minutes *or* seconds	Degrees to Radians	Minutes to Radians	Seconds to Radians
10	0.174 5329	0.002 9089	0.000 0485
20	0.349 0659	0.005 8178	0.000 0970
30	0.523 5988	0.008 7266	0.000 1454
40	0.698 1317	0.011 6355	0.000 1939
50	0.872 6646	0.014 5444	0.000 2424
60	1.047 1976	0.017 4533	0.000 2909
70	1.221 7305	(0.020 3622)	(0.000 3394)
80	1.396 2634	(0.023 2711)	(0.000 3879)
90	1.570 7963	(0.026 1799)	(0.000 4363)
100	1.745 3293	....	....
200	3.490 6585	....	....
300	5.235 9878	....	....

where n = 1, 2, 3, 4, etc. n (100°) = n (1.745 3293)

RADIANS TO DEGREES, MINUTES, AND SECONDS

Radians	1.0	0.1	0.01	0.001	0.0001
1	57° 17′ 44.8″	5° 43′ 46.5″	0° 34′ 22.6″	0° 03′ 26.3″	0° 00′ 20.6″
2	114° 35′ 29.6″	11° 27′ 33.0″	1° 08′ 45.3″	0° 06′ 52.5″	0° 00′ 41.3″
3	171° 53′ 14.4″	17° 11′ 19.4″	1° 43′ 07.9′	0° 10′ 18.8″	0° 01′ 01.9″
4	229° 10′ 59.2″	22° 55′ 05.9″	2° 17′ 30.6″	0° 13′ 45.1′	0° 01′ 22.5″
5	286° 28′ 44.0″	28° 38′ 52.4″	2° 51′ 53.2″	0° 17′ 11.3″	0° 01′ 43.1″
6	343° 46′ 28.8″	34° 22′ 38.9″	3° 26′ 15.9″	0° 20′ 37.6″	0° 02′ 03.8″
7	401° 04′ 13.6″	40° 06′ 25.4″	4° 00′ 38.5″	0° 24′ 03.9″	0° 02′ 24.4″
8	458° 21′ 58.4″	45° 50′ 11.8″	4° 35′ 01.2″	0° 27′ 30.1″	0° 02′ 45.0″
9	515° 39′ 43.3″	51° 33′ 58.3″	5° 09′ 23.8″	0° 30′ 56.4″	0° 03′ 05.6″

Example: If 3.214 is desired in degrees, minutes and seconds it is obtained as follows:

3 =	171° 53′ 14.4″
.2 =	11° 27′ 33.0″
.01 =	0° 34′ 22.6″
.004 =	0° 13′ 45.1″
3.214 =	184° 8′ 55.1″

MILS—RADIANS—DEGREES

1 mil = 0.00098175 radians = 0.05625° = 3.375′ = 202.5″
1000 mils = 0.98175 radians = 56.25°
6400 mils = 360° = 2π radians
1 radian = 1018.6 mils
1° = 17.777778 mils
1′ = 0.296296 mils
1″ = 0.0049383 mils

DEGREES—RADIANS

1 radian = 57° 17′ 44″ .80625

		log
1 radian	= 57.29577 95131 degrees	1.75812 26324
1 radian	= 3437.74677 07849 minutes	3.53627 38828
1 radian	= 206264.80625 seconds	5.31442 51332
1 degree	= 0.01745 32925 19943 radians	8.24187 73676–10
1 minute	= 0.00029 08882 08666 radians	6.46372 61172–10
1 second	= 0.00000 48481 36811 radians	4.68557 48668–10

DEGREES—RADIANS

The table gives in radians the angle which is expressed in degrees and minutes at the side and top. Angles expressed to the nearest minute and second can readily be converted to radians by adding to the equivalent of the whole number of degrees the equivalents of the minutes and seconds found on the third page of this table.

°	00′	10	20	30	40	50
0	0.00000	0.00291	0.00582	0.00873	0.01164	0.01454
1	0.01745	0.02036	0.02327	0.02618	0.02909	0.03200
2	0.03491	0.03782	0.04072	0.04363	0.04654	0.04945
3	0.05236	0.05527	0.05818	0.06109	0.06400	0.06690
4	0.06981	0.07272	0.07563	0.07854	0.08145	0.08436
5	0.08727	0.09018	0.09308	0.09599	0.09890	0.10181
6	0.10472	0.10763	0.11054	0.11345	0.11636	0.11926
7	0.12217	0.12508	0.12799	0.13090	0.13381	0.13672
8	0.13963	0.14254	0.14544	0.14835	0.15126	0.15417
9	0.15708	0.15999	0.16290	0.16581	0.16872	0.17162
10	0.17453	0.17744	0.18035	0.18326	0.18617	0.18908
11	0.19199	0.19490	0.19780	0.20071	0.20362	0.20653
12	0.20944	0.21235	0.21526	0.21817	0.22108	0.22398
13	0.22689	0.22980	0.23271	0.23562	0.23853	0.24144
14	0.24435	0.24725	0.25016	0.25307	0.25598	0.25889
15	0.26180	0.26471	0.26762	0.27053	0.27343	0.27634
16	0.27925	0.28216	0.28507	0.28798	0.29089	0.29380
17	0.29671	0.29961	0.30252	0.30543	0.30834	0.31125
18	0.31416	0.31707	0.31998	0.32289	0.32579	0.32870
19	0.33161	0.33452	0.33743	0.34034	0.34325	0.34616
20	0.34907	0.35197	0.35488	0.35779	0.36070	0.36361
21	0.36652	0.36943	0.37234	0.37525	0.37815	0.38106
22	0.38397	0.38688	0.38979	0.39270	0.39561	0.39852
23	0.40143	0.40433	0.40724	0.41015	0.41306	0.41597
24	0.41888	0.42179	0.42470	0.42761	0.43051	0.43342
25	0.43633	0.43924	0.44215	0.44506	0.44797	0.45088
26	0.45379	0.45669	0.45960	0.46251	0.46542	0.46833
27	0.47124	0.47415	0.47706	0.47997	0.48287	0.48578
28	0.48869	0.49160	0.49451	0.49742	0.50033	0.50324
29	0.50615	0.50905	0.51196	0.51487	0.51778	0.52069
30	0.52360	0.52651	0.52942	0.53233	0.53523	0.53814
31	0.54105	0.54396	0.54687	0.54978	0.55269	0.55560
32	0.55851	0.56141	0.56432	0.56723	0.57014	0.57305
33	0.57596	0.57887	0.58178	0.58469	0.58759	0.59050
34	0.59341	0.59632	0.59923	0.60214	0.60505	0.60796
35	0.61087	0.61377	0.61668	0.61959	0.62250	0.62541
36	0.62832	0.63123	0.63414	0.63705	0.63995	0.64286
37	0.64577	0.64868	0.65159	0.65450	0.65741	0.66032
38	0.66323	0.66613	0.66904	0.67195	0.67486	0.67777
39	0.68068	0.68359	0.68650	0.68941	0.69231	0.69522
40	0.69813	0.70104	0.70395	0.70686	0.70977	0.71268
41	0.71558	0.71849	0.72140	0.72431	0.72722	0.73013
42	0.73304	0.73595	0.73886	0.74176	0.74467	0.74758
43	0.75049	0.75340	0.75631	0.75922	0.76213	0.76504
44	0.76794	0.77085	0.77376	0.77667	0.77958	0.78249
45	0.78540	0.78831	0.79122	0.79412	0.79703	0.79994
46	0.80285	0.80576	0.80867	0.81158	0.81449	0.81740
47	0.82030	0.82321	0.82612	0.82903	0.83194	0.83485
48	0.83776	0.84067	0.84358	0.84648	0.84939	0.85230
49	0.85521	0.85812	0.86103	0.86394	0.86685	0.86976

DEGREES—RADIANS

°	00′	10	20	30	40	50
50	0.87266	0.87557	0.87848	0.88139	0.88430	0.88721
51	0.89012	0.89303	0.89594	0.89884	0.90175	0.90466
52	0.90757	0.91048	0.91339	0.91630	0.91921	0.92212
53	0.92502	0.92793	0.93084	0.93375	0.93666	0.93957
54	0.94248	0.94539	0.94830	0.95120	0.95411	0.95702
55	0.95993	0.96284	0.96575	0.96866	0.97157	0.97448
56	0.97738	0.98029	0.98320	0.98611	0.98902	0.99193
57	0.99484	0.99775	1.00066	1.00356	1.00647	1.00938
58	1.01229	1.01520	1.01811	1.02102	1.02393	1.02684
59	1.02974	1.03265	1.03556	1.03847	1.04138	1.04429
60	1.04720	1.05011	1.05302	1.05592	1.05883	1.06174
61	1.06465	1.06756	1.07047	1.07338	1.07629	1.07920
62	1.08210	1.08501	1.08792	1.09083	1.09374	1.09665
63	1.09956	1.10247	1.10538	1.10828	1.11119	1.11410
64	1.11701	1.11992	1.12283	1.12574	1.12865	1.13156
65	1.13446	1.13737	1.14028	1.14319	1.14610	1.14901
66	1.15192	1.15483	1.15774	1.16064	1.16355	1.16646
67	1.16937	1.17228	1.17519	1.17810	1.18101	1.18392
68	1.18682	1.18973	1.19264	1.19555	1.19846	1.20137
69	1.20428	1.20719	1.21009	1.21300	1.21591	1.21882
70	1.22173	1.22464	1.22755	1.23046	1.23337	1.23627
71	1.23918	1.24209	1.24500	1.24791	1.25082	1.25373
72	1.25664	1.25955	1.26245	1.26536	1.26827	1.27118
73	1.27409	1.27700	1.27991	1.28282	1.28573	1.28863
74	1.29154	1.29445	1.29736	1.30027	1.30318	1.30609
75	1.30900	1.31191	1.31481	1.31772	1.32063	1.32354
76	1.32645	1.32936	1.33227	1.33518	1.33809	1.34099
77	1.34390	1.34681	1.34972	1.35263	1.35554	1.35845
78	1.36136	1.36427	1.36717	1.37008	1.37299	1.37590
79	1.37881	1.38172	1.38463	1.38754	1.39045	1.39335
80	1.39626	1.39917	1.40208	1.40499	1.40790	1.41081
81	1.41372	1.41663	1.41953	1.42244	1.42535	1.42826
82	1.43117	1.43408	1.43699	1.43990	1.44281	1.44571
83	1.44862	1.45153	1.45444	1.45735	1.46026	1.46317
84	1.46608	1.46899	1.47189	1.47480	1.47771	1.48062
85	1.48353	1.48644	1.48935	1.49226	1.49517	1.49807
86	1.50098	1.50389	1.50680	1.50971	1.51262	1.51553
87	1.51844	1.52135	1.52425	1.52716	1.53007	1.53298
88	1.53589	1.53880	1.54171	1.54462	1.54753	1.55043
89	1.55334	1.55625	1.55916	1.56207	1.56498	1.56789
90	1.57080	1.57371	1.57661	1.57952	1.58243	1.58534
91	1.58825	1.59116	1.59407	1.59698	1.59989	1.60279
92	1.60570	1.60861	1.61152	1.61443	1.61734	1.62025
93	1.62316	1.62607	1.62897	1.63188	1.63479	1.63770
94	1.64061	1.64352	1.64643	1.64934	1.65225	1.65515
95	1.65806	1.66097	1.66388	1.66679	1.66970	1.67261
96	1.67552	1.67842	1.68133	1.68424	1.68715	1.69006
97	1.69297	1.69588	1.69879	1.70170	1.70460	1.70751
98	1.71042	1.71333	1.71624	1.71915	1.72206	1.72497
99	1.72788	1.73078	1.73369	1.73660	1.73951	1.74242
100	1.74533	1.74824	1.75115	1.75406	1.75696	1.75987
101	1.76278	1.76569	1.76860	1.77151	1.77442	1.77733
102	1.78024	1.78314	1.78605	1.78896	1.79187	1.79478
103	1.79769	1.80060	1.80351	1.80642	1.80932	1.81223
104	1.81514	1.81805	1.82096	1.82387	1.82678	1.82969
105	1.83260	1.83550	1.83841	1.84132	1.84423	1.84714
106	1.85005	1.85296	1.85587	1.85878	1.86168	1.86459
107	1.86750	1.87041	1.87332	1.87623	1.87914	1.88205
108	1.88496	1.88786	1.89077	1.89368	1.89659	1.89950
109	1.90241	1.90532	1.90823	1.91114	1.91404	1.91695
110	1.91986	1.92277	1.92568	1.92859	1.93150	1.93441

DEGREES—RADIANS

Deg.	Radians	Deg.	Radians	Min.	Radians	Sec.	Radians
90	1.57080	**150**	2.61799	**0**	0.00000	**0**	0.00000
91	1.58825	151	2.63545	1	0.00029	1	0.00000
92	1.60570	152	2.65290	2	0.00058	2	0.00001
93	1.62316	153	2.67035	3	0.00087	3	0.00001
94	1.64061	154	2.68781	4	0.00116	4	0.00002
95	1.65806	**155**	2.70526	**5**	0.00145	**5**	0.00002
96	1.67552	156	2.72271	6	0.00175	6	0.00003
97	1.69297	157	2.74017	7	0.00204	7	0.00003
98	1.71042	158	2.75762	8	0.00233	8	0.00004
99	1.72788	159	2.77507	9	0.00262	9	0.00004
100	1.74533	**160**	2.79253	**10**	0.00291	**10**	0.00005
101	1.76278	161	2.80998	11	0.00320	11	0.00005
102	1.78024	162	2.82743	12	0.00349	12	0.00006
103	1.79769	163	2.84489	13	0.00378	13	0.00006
104	1.81514	164	2.86234	14	0.00407	14	0.00007
105	1.83260	**165**	2.87979	**15**	0.00436	**15**	0.00007
106	1.85005	166	2.89725	16	0.00465	16	0.00008
107	1.86750	167	2.91470	17	0.00495	17	0.00008
108	1.88496	168	2.93215	18	0.00524	18	0.00009
109	1.90241	169	2.94961	19	0.00553	19	0.00009
110	1.91986	**170**	2.96706	**20**	0.00582	**20**	0.00010
111	1.93732	171	2.98451	21	0.00611	21	0.00010
112	1.95477	172	3.00197	22	0.00640	22	0.00011
113	1.97222	173	3.01942	23	0.00669	23	0.00011
114	1.98968	174	3.03687	24	0.00698	24	0.00012
115	2.00713	**175**	3.05433	**25**	0.00727	**25**	0.00012
116	2.02458	176	3.07178	26	0.00756	26	0.00013
117	2.04204	177	3.08923	27	0.00785	27	0.00013
118	2.05949	178	3.10669	28	0.00814	28	0.00014
119	2.07694	179	3.12414	29	0.00844	29	0.00014
120	2.09440	**180**	3.14159	**30**	0.00873	**30**	0.00015
121	2.11185	190	3.31613	31	0.00902	31	0.00015
122	2.12930	200	3.49066	32	0.00931	32	0.00016
123	2.14675	210	3.66519	33	0.00960	33	0.00016
124	2.16421	220	3.83972	34	0.00989	34	0.00016
125	2.18166	**230**	4.01426	**35**	0.01018	**35**	0.00017
126	2.19911	240	4.18879	36	0.01047	36	0.00017
127	2.21657	250	4.36332	37	0.01076	37	0.00018
128	2.23402	260	4.53786	38	0.01105	38	0.00018
129	2.25147	270	4.71239	39	0.01134	39	0.00019
130	2.26893	**280**	4.88692	**40**	0.01164	**40**	0.00019
131	2.28638	290	5.06145	41	0.01193	41	0.00020
132	2.30383	300	5.23599	42	0.01222	42	0.00020
133	2.32129	310	5.41052	43	0.01251	43	0.00021
134	2.33874	320	5.58505	44	0.01280	44	0.00021
135	2.35619	**330**	5.75959	**45**	0.01309	**45**	0.00022
136	2.37365	340	5.93412	46	0.01338	46	0.00022
137	2.39110	350	6.10865	47	0.01367	47	0.00023
138	2.40855	360	6.28319	48	0.01396	48	0.00023
139	2.42601	370	6.45772	49	0.01425	49	0.00024
140	2.44346	**380**	6.63225	**50**	0.01454	**50**	0.00024
141	2.46091	390	6.80678	51	0.01484	51	0.00025
142	2.47837	400	6.98132	52	0.01513	52	0.00025
143	2.49582	410	7.15585	53	0.01542	53	0.00026
144	2.51327	420	7.33038	54	0.01571	54	0.00026
145	2.53073	**430**	7.50492	**55**	0.01600	**55**	0.00027
146	2.54818	440	7.67945	56	0.01629	56	0.00027
147	2.56563	450	7.85398	57	0.01658	57	0.00028
148	2.58309	460	8.02851	58	0.01687	58	0.00028
149	2.60054	470	8.20305	59	0.01716	59	0.00029
150	2.61799	**480**	8.37758	**60**	0.01745	**60**	0.00029

DEGREES AND DECIMAL FRACTIONS TO RADIANS

The table below facilitates conversion of an angle expressed in degrees and decimal fractions into radians. To convert 25.78 into radians, find the equivalents, successively, of 20°, 5°, 0°.7, 0°.08 and add.

Deg.	Radians	Deg.	Radians	Deg.	Radians	Deg.	Radians	Deg.	Radians
10	0.174533	1	0.017453	0.1	0.001745	0.01	0.000175	0.001	0.000017
20	0.349066	2	.034907	.2	.003491	.02	.000349	.002	.000035
30	0.523599	3	.052360	.3	.005236	.03	.000524	.003	.000052
40	0.698132	4	.069813	.4	.006981	.04	.000698	.004	.000070
50	0.872665	5	.087266	.5	.008727	.05	.000873	.005	.000087
60	1.047198	6	.104720	.6	.010472	.06	.001047	.006	.000105
70	1.221730	7	.122173	.7	.012217	.07	.001222	.007	.000122
80	1.396263	8	.139626	.8	.013963	.08	.001396	.008	.000140
90	1.570796	9	.157080	.9	.015708	.09	.001571	.009	.000157

RADIANS TO DEGREES AND DECIMALS

Radians	Degrees	Radians	Degrees	Radians	Degrees	Radians	Degrees
1	57.2958	0.1	5.7296	0.01	0.5730	0.001	0.0573
2	114.5916	.2	11.4592	.02	1.1459	.002	.1146
3	171.8873	.3	17.1887	.03	1.7189	.003	.1719
4	229.1831	.4	22.9183	.04	2.2918	.004	.2292
5	286.4789	.5	28.6479	.05	2.8648	.005	.2865
6	343.7747	.6	34.3775	.06	3.4377	.006	.3438
7	401.0705	.7	40.1070	.07	4.0107	.007	.4011
8	458.3662	.8	45.8366	.08	4.5837	.008	.4584
9	515.6620	.9	51.5662	.09	5.1566	.009	.5157
10	572.9578	1.0	57.2958	.10	5.7296	.010	.5730

RADIANS—DEGREES

Multiples and Fractions of π Radians in Degrees

Radians	Radians	Deg.	Radians	Radians	Deg.	Radians	Radians	Deg.
π	3.1416	180	$\pi/2$	1.5708	90	$2\pi/3$	2.0944	120
2π	6.2832	360	$\pi/3$	1.0472	60	$3\pi/4$	2.3562	135
3π	9.4248	540	$\pi/4$	0.7854	45	$5\pi/6$	2.6180	150
4π	12.5664	720	$\pi/5$	0.6283	36	$7\pi/6$	3.6652	210
5π	15.7080	900	$\pi/6$	0.5236	30	$5\pi/4$	3.9270	225
6π	18.8496	1080	$\pi/7$	0.4488	25.714	$4\pi/3$	4.1888	240
7π	21.9911	1260	$\pi/8$	0.3927	22.5	$3\pi/2$	4.7124	270
8π	25.1327	1440	$\pi/9$	0.3491	20	$5\pi/3$	5.2360	300
9π	28.2743	1620	$\pi/10$	0.3142	18	$7\pi/4$	5.4978	315
10π	31.4159	1800	$\pi/12$	0.2618	15	$11\pi/6$	5.7596	330

CONVERSION OF ANGLES FROM ARC TO TIME

Arc	Time	Arc	Time	Arc	Time	Arc	Time
°	h m	°	h m	″	s	″	s
′	m s	′	m s				
0	0 00	20	1 20	0	0.00	8	0.53
1	0 04	30	2 00	1	0.07	9	0.60
2	0 08	40	2 40	2	0.13	10	0.67
3	0 12	50	3 20	3	0.20	20	1.33
4	0 16	60	4 00	4	0.27	30	2.00
5	0 20	70	4 40	5	0.33	40	2.67
6	0 24	80	5 20	6	0.40	50	3.33
7	0 28	90	6 00	7	0.47	60	4.00
8	0 32	100	6 40				
9	0 36	200	13 20				
10	0 40	300	20 00				

MINUTES AND SECONDS TO DECIMAL PARTS OF A DEGREE

MINUTES AND SECONDS TO DECIMAL PARTS OF A DEG.

Min.	Degrees	Sec.	Degrees
0	0.00000	**0**	0.00000
1	.01667	1	.00028
2	.03333	2	.00056
3	.05	3	.00083
4	.06667	4	.00111
5	.08333	5	.00139
6	.10	6	.00167
7	.11667	7	.00194
8	.13333	8	.00222
9	.15	9	.0025
10	0.16667	**10**	0.00278
11	.18333	11	.00306
12	.20	12	.00333
13	.21667	13	.00361
14	.23333	14	.00389
15	.25	15	.00417
16	.26667	16	.00444
17	.28333	17	.00472
18	.30	18	.005
19	.31667	19	.00528
20	0.33333	**20**	0.00556
21	.35	21	.00583
22	.36667	22	.00611
23	.38333	23	.00639
24	.40	24	.00667
25	.41667	25	.00694
26	.43333	26	.00722
27	.45	27	.0075
28	.46667	28	.00778
29	.48333	29	.00806
30	0.50	**30**	0.00833
31	.51667	31	.00861
32	.53333	32	.00889
33	.55	33	.00917
34	.56667	34	.00944
35	.58333	35	.00972
36	.60	36	.01
37	.61667	37	.01028
38	.63333	38	.01056
39	.65	39	.01083
40	0.66667	**40**	0.01111
41	.68333	41	.01139
42	.70	42	.01167
43	.71667	43	.01194
44	.73333	44	.01222
45	.75	45	.0125
46	.76667	46	.01278
47	.78333	47	.01306
48	.80	48	.01333
49	.81667	49	.01361
50	0.83333	**50**	0.01389
51	.85	51	.01417
52	.86667	52	.01444
53	.88333	53	.01472
54	.90	54	.015
55	.91667	55	.01528
56	.93333	56	.01556
57	.95	57	.01583
58	.96667	58	.01611
59	.98333	59	.01639
60	1.00	**60**	0.01667

DECIMAL PARTS OF A DEGREE TO MINUTES AND SECONDS

Deg.	′	″
0.00	0	00
.01	0	36
.02	1	12
.03	1	48
.04	2	24
.05	3	
.06	3	36
.07	4	12
.08	4	48
.09	5	24
0.10	6	
.11	6	36
.12	7	12
.13	7	48
.14	8	24
.15	9	
.16	9	36
.17	10	12
.18	10	48
.19	11	24
0.20	12	
.21	12	36
.22	13	12
.23	13	48
.24	14	24
.25	15	
.26	15	36
.27	16	12
.28	16	48
.29	17	24
0.30	18	
.31	18	36
.32	19	12
.33	19	48
.34	20	24
.35	21	
.36	21	36
.37	22	12
.38	22	48
.39	23	24
0.40	24	
.41	24	36
.42	25	12
.43	25	48
.44	26	24
.45	27	
.46	27	36
.47	28	12
.48	28	48
.49	29	24
0.50	30	
.51	30	36
.52	31	12
.53	31	48
.54	32	24
.55	83	
.56	33	36
.57	34	12
.58	34	48
.59	35	24
0.60	36	

Deg.	′	″
0.60	36	
.61	36	36
.62	37	12
.63	37	48
.64	38	24
.65	39	
.66	39	36
.67	40	12
.68	40	48
.69	41	24
0.70	42	
.71	42	36
.72	43	12
.73	43	48
.74	44	24
.75	45	
.76	45	36
.77	46	12
.78	46	48
.79	47	24
0.80	48	
.81	48	36
.82	49	12
.83	49	48
.84	50	24
.85	51	
.86	51	36
.87	52	12
.88	52	48
.89	53	24
0.90	54	
.91	54	36
.92	55	12
.93	55	48
94	56	24
.95	57	
.96	57	36
.97	58	12
.98	58	48
.99	59	24
1.00	60	

Deg.	Sec.
0.000	0.0
.001	3.6
.002	7.2
.003	10.8
.004	14.4
.005	18.
.006	21.6
.007	25.2
.008	28.8
.009	32.4
0.010	36.

NATURAL TRIGONOMETRIC FUNCTIONS TO FIVE PLACES

For degrees indicated at the top of the page use the top column headings. For degrees indicated at the bottom use the column indications at the bottom. With degrees at the left of each block (top or bottom), use the minute column at the left;with degrees at the right of each block, use the minute column at the right. Appropriate signs for the functions must be supplied in accordance with the quadrant in which the angle measure belongs. Linear interpolation may be used except in regions where the functions are rapidly changing. See "Use of Logarithm Tables" for further discussion in interpolation practice.

0° (180°) — (359)° **179°**

′	Sin	Tan	Cot	Cos	Sec	Csc	′
0	.00000	.00000	——	1.0000	1.0000	——	**60**
1	.00029	.00029	3437.7	1.0000	1.0000	3437.7	59
2	.00058	.00058	1718.9	1.0000	1.0000	1718.9	58
3	.00087	.00087	1145.9	1.0000	1.0000	1145.9	57
4	.00116	.00116	859.44	1.0000	1.0000	859.44	56
5	.00145	.00145	687.55	1.0000	1.0000	687.55	**55**
6	.00175	.00175	572.96	1.0000	1.0000	572.96	54
7	.00204	.00204	491.11	1.0000	1.0000	491.11	53
8	.00233	.00233	429.72	1.0000	1.0000	429.72	52
9	.00262	.00262	381.97	1.0000	1.0000	381.97	51
10	.00291	.00291	343.77	1.0000	1.0000	343.78	**50**
11	.00320	.00320	312.52	.99999	1.0000	312.52	49
12	.00349	.00349	286.48	.99999	1.0000	286.48	48
13	.00378	.00378	264.44	.99999	1.0000	264.44	47
14	.00407	.00407	245.55	.99999	1.0000	245.55	46
15	.00436	.00436	229.18	.99999	1.0000	229.18	**45**
16	.00465	.00465	214.86	.99999	1.0000	214.86	44
17	.00495	.00495	202.22	.99999	1.0000	202.22	43
18	.00524	.00524	190.98	.99999	1.0000	190.99	42
19	.00553	.00553	180.93	.99998	1.0000	180.93	41
20	.00582	.00582	171.89	.99998	1.0000	171.89	**40**
21	.00611	.00611	163.70	.99998	1.0000	163.70	39
22	.00640	.00640	156.26	.99998	1.0000	156.26	38
23	.00669	.00669	149.47	.99998	1.0000	149.47	37
24	.00698	.00698	143.24	.99998	1.0000	143.24	36
25	.00727	.00727	137.51	.99997	1.0000	137.51	**35**
26	.00756	.00756	132.22	.99997	1.0000	132.22	34
27	.00785	.00785	127.32	.99997	1.0000	127.33	33
28	.00814	.00815	122.77	.99997	1.0000	122.78	32
29	.00844	.00844	118.54	.99996	1.0000	118.54	31
30	.00873	.00873	114.59	.99996	1.0000	114.59	**30**
31	.00902	.00902	110.89	.99996	1.0000	110.90	29
32	.00931	.00931	107.43	.99996	1.0000	107.43	28
33	.00960	.00960	104.17	.99995	1.0000	104.18	27
34	.00989	.00989	101.11	.99995	1.0000	101.11	26
35	.01018	.01018	98.218	.99995	1.0001	98.223	**25**
36	.01047	.01047	95.489	.99995	1.0001	95.495	24
37	.01076	.01076	92.908	.99994	1.0001	92.914	23
38	.01105	.01105	90.463	.99994	1.0001	90.469	22
39	.01134	.01135	88.144	.99994	1.0001	88.149	21
40	.01164	.01164	85.940	.99993	1.0001	85.946	**20**
41	.01193	.01193	83.844	.99993	1.0001	83.849	19
42	.01222	.01222	81.847	.99993	1.0001	81.853	18
43	.01251	.01251	79.943	.99992	1.0001	79.950	17
44	.01280	.01280	78.126	.99992	1.0001	78.133	16
45	.01309	.01309	76.390	.99991	1.0001	76.397	**15**
46	.01338	.01338	74.729	.99991	1.0001	74.736	14
47	.01367	.01367	73.139	.99991	1.0001	73.146	13
48	.01396	.01396	71.615	.99990	1.0001	71.622	12
49	.01425	.01425	70.153	.99990	1.0001	70.160	11
50	.01454	.01455	68.750	.99989	1.0001	68.757	**10**
51	.01483	.01484	67.402	.99989	1.0001	67.409	9
52	.01513	.01513	66.105	.99989	1.0001	66.113	8
53	.01542	.01542	64.858	.99988	1.0001	64.866	7
54	.01571	.01571	63.657	.99988	1.0001	63.665	6
55	.01600	.01600	62.499	.99987	1.0001	62.507	**5**
56	.01629	.01629	61.383	.99987	1.0001	61.391	4
57	.01658	.01658	60.306	.99986	1.0001	60.314	3
58	.01687	.01687	59.266	.99986	1.0001	59.274	2
59	.01716	.01716	58.261	.99985	1.0001	58.270	1
60	.01745	.01746	57.290	.99985	1.0002	57.299	**0**
′	Cos	Cot	Tan	Sin	Csc	Sec	′

90° (270°) — (269°) **89°**

1° (181°) — (358°) **178°**

′	Sin	Tan	Cot	Cos	Sec	Csc	′
0	.01745	.01746	57.290	.99985	1.0002	57.299	**60**
1	.01774	.01775	56.351	.99984	1.0002	56.359	59
2	.01803	.01804	55.442	.99984	1.0002	55.451	58
3	.01832	.01833	54.561	.99983	1.0002	54.570	57
4	.01862	.01862	53.709	.99983	1.0002	53.718	56
5	.01891	.01891	52.882	.99982	1.0002	52.892	**55**
6	.01920	.01920	52.081	.99982	1.0002	52.090	54
7	.01949	.01949	51.303	.99981	1.0002	51.313	53
8	.01978	.01978	50.549	.99980	1.0002	50.558	52
9	.02007	.02007	49.816	.99980	1.0002	49.826	51
10	.02036	.02036	49.104	.99979	1.0002	49.114	**50**
11	.02065	.02066	48.412	.99979	1.0002	48.422	49
12	.02094	.02095	47.740	.99978	1.0002	47.750	48
13	.02123	.02124	47.085	.99977	1.0002	47.096	47
14	.02152	.02153	46.449	.99977	1.0002	46.460	46
15	.02181	.02182	45.829	.99976	1.0002	45.840	**45**
16	.02211	.02211	45.226	.99976	1.0002	45.237	44
17	.02240	.02240	44.639	.99975	1.0003	44.650	43
18	.02269	.02269	44.066	.99974	1.0003	44.077	42
19	.02298	.02298	43.508	.99974	1.0003	43.520	41
20	.02327	.02328	42.964	.99973	1.0003	42.976	**40**
21	.02356	.02357	42.433	.99972	1.0003	42.445	39
22	.02385	.02386	41.916	.99972	1.0003	41.928	38
23	.02414	.02415	41.411	.99971	1.0003	41.423	37
24	.02443	.02444	40.917	.99970	1.0003	40.930	36
25	.02472	.02473	40.436	.99969	1.0003	40.448	**35**
26	.02501	.02502	39.965	.99969	1.0003	39.978	34
27	.02530	.02531	39.506	.99968	1.0003	39.519	33
28	.02560	.02560	39.057	.99967	1.0003	39.070	32
29	.02589	.02589	38.618	.99966	1.0003	38.631	31
30	.02618	.02619	38.188	.99966	1.0003	38.202	**30**
31	.02647	.02648	37.769	.99965	1.0004	37.782	29
32	.02676	.02677	37.358	.99964	1.0004	37.371	28
33	.02705	.02706	36.956	.99963	1.0004	36.970	27
34	.02734	.02735	36.563	.99963	1.0004	36.576	26
35	.02763	.02764	36.178	.99962	1.0004	36.191	**25**
36	.02792	.02793	35.801	.99961	1.0004	35.815	24
37	.02821	.02822	35.431	.99960	1.0004	35.445	23
38	.02850	.02851	35.070	.99959	1.0004	35.084	22
39	.02879	.02881	34.715	.99959	1.0004	34.730	21
40	.02908	.02910	34.368	.99958	1.0004	34.382	**20**
41	.02938	.02939	34.027	.99957	1.0004	34.042	19
42	.02967	.02968	33.694	.99956	1.0004	33.708	18
43	.02996	.02997	33.366	.99955	1.0004	33.381	17
44	.03025	.03026	33.045	.99954	1.0005	33.060	16
45	.03054	.03055	32.730	.99953	1.0005	32.746	**15**
46	.03083	.03084	32.421	.99952	1.0005	32.437	14
47	.03112	.03114	32.118	.99952	1.0005	32.134	13
48	.03141	.03143	31.821	.99951	1.0005	31.836	12
49	.03170	.03172	31.528	.99950	1.0005	31.544	11
50	.03199	.03201	31.242	.99949	1.0005	31.258	**10**
51	.03228	.03230	30.960	.99948	1.0005	30.976	9
52	.03257	.03259	30.683	.99947	1.0005	30.700	8
53	.03286	.03288	30.412	.99946	1.0005	30.428	7
54	.03316	.03317	30.145	.99945	1.0006	30.161	6
55	.03345	.03346	29.882	.99944	1.0006	29.899	**5**
56	.03374	.03376	29.624	.99943	1.0006	29.641	4
57	.03403	.03405	29.371	.99942	1.0006	29.388	3
58	.03432	.03434	29.122	.99941	1.0006	29.139	2
59	.03461	.03463	28.877	.99940	1.0006	28.894	1
60	.03490	.03492	28.636	.99939	1.0006	28.654	**0**
′	Cos	Cot	Tan	Sin	Csc	Sec	′

91° (271°) — (268°) **88°**

NATURAL TRIGONOMETRIC FUNCTIONS TO FIVE PLACES

2° (182°) (357°) 177°

′	Sin	Tan	Cot	Cos	Sec	Csc	′
0	.03490	.03492	28.636	.99939	1.0006	28.654	60
1	.03519	.03521	28.399	.99938	1.0006	28.417	59
2	.03548	.03550	28.166	.99937	1.0006	28.184	58
3	.03577	.03579	27.937	.99936	1.0006	27.955	57
4	.03606	.03609	27.712	.99935	1.0007	27.730	56
5	.03635	.03638	27.490	.99934	1.0007	27.508	55
6	.03664	.03667	27.271	.99933	1.0007	27.290	54
7	.03693	.03696	27.057	.99932	1.0007	27.075	53
8	.03723	.03725	26.845	.99931	1.0007	26.864	52
9	.03752	.03754	26.637	.99930	1.0007	26.655	51
10	.03781	.03783	26.432	.99929	1.0007	26.451	50
11	.03810	.03812	26.230	.99927	1.0007	26.249	49
12	.03839	.03842	26.031	.99926	1.0007	26.050	48
13	.03868	.03871	25.835	.99925	1.0007	25.854	47
14	.03897	.03900	25.642	.99924	1.0008	25.661	46
15	.03926	.03929	25.452	.99923	1.0008	25.471	45
16	.03955	.03958	25.264	.99922	1.0008	25.284	44
17	.03984	.03987	25.080	.99921	1.0008	25.100	43
18	.04013	.04016	24.898	.99919	1.0008	24.918	42
19	.04042	.04046	24.719	.99918	1.0008	24.739	41
20	.04071	.04075	24.542	.99917	1.0008	24.562	40
21	.04100	.04104	24.368	.99916	1.0008	24.388	39
22	.04129	.04133	24.196	.99915	1.0009	24.216	38
23	.04159	.04162	24.026	.99913	1.0009	24.047	37
24	.04188	.04191	23.859	.99912	1.0009	23.880	36
25	.04217	.04220	23.695	.99911	1.0009	23.716	35
26	.04246	.04250	23.532	.99910	1.0009	23.553	34
27	.04275	.04279	23.372	.99909	1.0009	23.393	33
28	.04304	.04308	23.214	.99907	1.0009	23.235	32
29	.04333	.04337	23.058	.99906	1.0009	23.079	31
30	.04362	.04366	22.904	.99905	1.0010	22.926	30
31	.04391	.04395	22.752	.99904	1.0010	22.774	29
32	.04420	.04424	22.602	.99902	1.0010	22.624	28
33	.04449	.04454	22.454	.99901	1.0010	22.476	27
34	.04478	.04483	22.308	.99900	1.0010	22.330	26
35	.04507	.04512	22.164	.99898	1.0010	22.187	25
36	.04536	.04541	22.022	.99897	1.0010	22.044	24
37	.04565	.04570	21.881	.99896	1.0010	21.904	23
38	.04594	.04599	21.743	.99894	1.0011	21.766	22
39	.04623	.04628	21.606	.99893	1.0011	21.629	21
40	.04653	.04658	21.470	.99892	1.0011	21.494	20
41	.04682	.04687	21.337	.99890	1.0011	21.360	19
42	.04711	.04716	21.205	.99889	1.0011	21.229	18
43	.04740	.04745	21.075	.99888	1.0011	21.098	17
44	.04769	.04774	20.946	.99886	1.0011	20.970	16
45	.04798	.04803	20.819	.99885	1.0012	20.843	15
46	.04827	.04833	20.693	.99883	1.0012	20.717	14
47	.04856	.04862	20.569	.99882	1.0012	20.593	13
48	.04885	.04891	20.446	.99881	1.0012	20.471	12
49	.04914	.04920	20.325	.99879	1.0012	20.350	11
50	.04943	.04949	20.206	.99878	1.0012	20.230	10
51	.04972	.04978	20.087	.99876	1.0012	20.112	9
52	.05001	.05007	19.970	.99875	1.0013	19.995	8
53	.05030	.05037	19.855	.99873	1.0013	19.880	7
54	.05059	.05066	19.740	.99872	1.0013	19.766	6
55	.05088	.05095	19.627	.99870	1.0013	19.653	5
56	.05117	.05124	19.516	.99869	1.0013	19.541	4
57	.05146	.05153	19.405	.99867	1.0013	19.431	3
58	.05175	.05182	19.296	.99866	1.0013	19.322	2
59	.05205	.05212	19.188	.99864	1.0014	19.214	1
60	.05234	.05241	19.081	.99863	1.0014	19.107	0
′	Cos	Cot	Tan	Sin	Csc	Sec	′

° (272°) (267°) 87°

3° (183°) (356°) 176°

′	Sin	Tan	Cot	Cos	Sec	Csc	′
0	.05234	.05241	19.081	.99863	1.0014	19.107	60
1	.05263	.05270	18.976	.99861	1.0014	19.002	59
2	.05292	.05299	18.871	.99860	1.0014	18.898	58
3	.05321	.05328	18.768	.99858	1.0014	18.794	57
4	.05350	.05357	18.666	.99857	1.0014	18.692	56
5	.05379	.05387	18.564	.99855	1.0014	18.591	55
6	.05408	.05416	18.464	.99854	1.0015	18.492	54
7	.05437	.05445	18.366	.99852	1.0015	18.393	53
8	.05466	.05474	18.268	.99851	1.0015	18.295	52
9	.05495	.05503	18.171	.99849	1.0015	18.198	51
10	.05524	.05533	18.075	.99847	1.0015	18.103	50
11	.05553	.05562	17.980	.99846	1.0015	18.008	49
12	.05582	.05591	17.886	.99844	1.0016	17.914	48
13	.05611	.05620	17.793	.99842	1.0016	17.822	47
14	.05640	.05649	17.702	.99841	1.0016	17.730	46
15	.05669	.05678	17.611	.99839	1.0016	17.639	45
16	.05698	.05708	17.521	.99838	1.0016	17.549	44
17	.05727	.05737	17.431	.99836	1.0016	17.460	43
18	.05756	.05766	17.343	.99834	1.0017	17.372	42
19	.05785	.05795	17.256	.99833	1.0017	17.285	41
20	.05814	.05824	17.169	.99831	1.0017	17.198	40
21	.05844	.05854	17.084	.99829	1.0017	17.113	39
22	.05873	.05883	16.999	.99827	1.0017	17.028	38
23	.05902	.05912	16.915	.99826	1.0017	16.945	37
24	.05931	.05941	16.832	.99824	1.0018	16.862	36
25	.05960	.05970	16.750	.99822	1.0018	16.779	35
26	.05989	.05999	16.668	.99821	1.0018	16.698	34
27	.06018	.06029	16.587	.99819	1.0018	16.618	33
28	.06047	.06058	16.507	.99817	1.0018	16.538	32
29	.06076	.06087	16.428	.99815	1.0019	16.459	31
30	.06105	.06116	16.350	.99813	1.0019	16.380	30
31	.06134	.06145	16.272	.99812	1.0019	16.303	29
32	.06163	.06175	16.195	.99810	1.0019	16.226	28
33	.06192	.06204	16.119	.99808	1.0019	16.150	27
34	.06221	.06233	16.043	.99806	1.0019	16.075	26
35	.06250	.06262	15.969	.99804	1.0020	16.000	25
36	.06279	.06291	15.895	.99803	1.0020	15.926	24
37	.06308	.06321	15.821	.99801	1.0020	15.853	23
38	.06337	.06350	15.748	.99799	1.0020	15.780	22
39	.06366	.06379	15.676	.99797	1.0020	15.708	21
40	.06395	.06408	15.605	.99795	1.0021	15.637	20
41	.06424	.06438	15.534	.99793	1.0021	15.566	19
42	.06453	.06467	15.464	.99792	1.0021	15.496	18
43	.06482	.06496	15.394	.99790	1.0021	15.427	17
44	.06511	.06525	15.325	.99788	1.0021	15.358	16
45	.06540	.06554	15.257	.99786	1.0021	15.290	15
46	.06569	.06584	15.189	.99784	1.0022	15.222	14
47	.06598	.06613	15.122	.99782	1.0022	15.155	13
48	.06627	.06642	15.056	.99780	1.0022	15.089	12
49	.06656	.06671	14.990	.99778	1.0022	15.023	11
50	.06685	.06700	14.924	.99776	1.0022	14.958	10
51	.06714	.06730	14.860	.99774	1.0023	14.893	9
52	.06743	.06759	14.795	.99772	1.0023	14.829	8
53	.06773	.06788	14.732	.99770	1.0023	14.766	7
54	.06802	.06817	14.669	.99768	1.0023	14.703	6
55	.06831	.06847	14.606	.99766	1.0023	14.640	5
56	.06860	.06876	14.544	.99764	1.0024	14.578	4
57	.06889	.06905	14.482	.99762	1.0024	14.517	3
58	.06918	.06934	14.421	.99760	1.0024	14.456	2
59	.06947	.06963	14.361	.99758	1.0024	14.395	1
60	.06976	.06993	14.301	.99756	1.0024	14.336	0
′	Cos	Cot	Tan	Sin	Csc	Sec	′

93° (273°) (266°) 86°

NATURAL TRIGONOMETRIC FUNCTIONS TO FIVE PLACES

4° (184°) (355°) **175°**

′	Sin	Tan	Cot	Cos	Sec	Csc	′
0	.06976	.06993	14.301	.99756	1.0024	14.336	**60**
1	.07005	.07022	14.241	.99754	1.0025	14.276	59
2	.07034	.07051	14.182	.99752	1.0025	14.217	58
3	.07063	.07080	14.124	.99750	1.0025	14.159	57
4	.07092	.07110	14.065	.99748	1.0025	14.101	56
5	.07121	.07139	14.008	.99746	1.0025	14.044	**55**
6	.07150	.07168	13.951	.99744	1.0026	13.987	54
7	.07179	.07197	13.894	.99742	1.0026	13.930	53
8	.07208	.07227	13.838	.99740	1.0026	13.874	52
9	.07237	.07256	13.782	.99738	1.0026	13.818	51
10	.07266	.07285	13.727	.99736	1.0027	13.763	**50**
11	.07295	.07314	13.672	.99734	1.0027	13.708	49
12	.07324	.07344	13.617	.99731	1.0027	13.654	48
13	.07353	.07373	13.563	.99729	1.0027	13.600	47
14	.07382	.07402	13.510	.99727	1.0027	13.547	46
15	.07411	.07431	13.457	.99725	1.0028	13.494	**45**
16	.07440	.07461	13.404	.99723	1.0028	13.441	44
17	.07469	.07490	13.352	.99721	1.0028	13.389	43
18	.07498	.07519	13.300	.99719	1.0028	13.337	42
19	.07527	.07548	13.248	.99716	1.0028	13.286	41
20	.07556	.07578	13.197	.99714	1.0029	13.235	**40**
21	.07585	.07607	13.146	.99712	1.0029	13.184	39
22	.07614	.07636	13.096	.99710	1.0029	13.134	38
23	.07643	.07665	13.046	.99708	1.0029	13.084	37
24	.07672	.07695	12.996	.99705	1.0030	13.035	36
25	.07701	.07724	12.947	.99703	1.0030	12.985	**35**
26	.07730	.07753	12.898	.99701	1.0030	12.937	34
27	.07759	.07782	12.850	.99699	1.0030	12.888	33
28	.07788	.07812	12.801	.99696	1.0030	12.840	32
29	.07817	.07841	12.754	.99694	1.0031	12.793	31
30	.07846	.07870	12.706	.99692	1.0031	12.745	**30**
31	.07875	.07899	12.659	.99689	1.0031	12.699	29
32	.07904	.07929	12.612	.99687	1.0031	12.652	28
33	.07933	.07958	12.566	.99685	1.0032	12.606	27
34	.07962	.07987	12.520	.99683	1.0032	12.560	26
35	.07991	.08017	12.474	.99680	1.0032	12.514	**25**
36	.08020	.08046	12.429	.99678	1.0032	12.469	24
37	.08049	.08075	12.384	.99676	1.0033	12.424	23
38	.08078	.08104	12.339	.99673	1.0033	12.379	22
39	.08107	.08134	12.295	.99671	1.0033	12.335	21
40	.08136	.08163	12.251	.99668	1.0033	12.291	**20**
41	.08165	.08192	12.207	.99666	1.0034	12.248	19
42	.08194	.08221	12.163	.99664	1.0034	12.204	18
43	.08223	.08251	12.120	.99661	1.0034	12.161	17
44	.08252	.08280	12.077	.99659	1.0034	12.119	16
45	.08281	.08309	12.035	.99657	1.0034	12.076	**15**
46	.08310	.08339	11.992	.99654	1.0035	12.034	14
47	.08339	.08368	11.950	.99652	1.0035	11.992	13
48	.08368	.08397	11.909	.99649	1.0035	11.951	12
49	.08397	.08427	11.867	.99647	1.0035	11.909	11
50	.08426	.08456	11.826	.99644	1.0036	11.868	**10**
51	.08455	.08485	11.785	.99642	1.0036	11.828	9
52	.08484	.08514	11.745	.99639	1.0036	11.787	8
53	.08513	.08544	11.705	.99637	1.0036	11.747	7
54	.08542	.08573	11.664	.99635	1.0037	11.707	6
55	.08571	.08602	11.625	.99632	1.0037	11.668	**5**
56	.08600	.08632	11.585	.99630	1.0037	11.628	4
57	.08629	.08661	11.546	.99627	1.0037	11.589	3
58	.08658	.08690	11.507	.99625	1.0038	11.551	2
59	.08687	.08720	11.468	.99622	1.0038	11.512	1
60	.08716	.08749	11.430	.99619	1.0038	11.474	**0**
′	Cos	Cot	Tan	Sin	Csc	Sec	′

94° (274°) (265°) **85°**

5° (185°) (354°) **17**

′	Sin	Tan	Cot	Cos	Sec	Csc	′
0	.08716	.08749	11.430	.99619	1.0038	11.474	**60**
1	.08745	.08778	11.392	.99617	1.0038	11.436	59
2	.08774	.08807	11.354	.99614	1.0039	11.398	58
3	.08803	.08837	11.316	.99612	1.0039	11.360	57
4	.08831	.08866	11.279	.99609	1.0039	11.323	56
5	.08860	.08895	11.242	.99607	1.0039	11.286	**55**
6	.08889	.08925	11.205	.99604	1.0040	11.249	54
7	.08918	.08954	11.168	.99602	1.0040	11.213	53
8	.08947	.08983	11.132	.99599	1.0040	11.176	52
9	.08976	.09013	11.095	.99596	1.0041	11.140	51
10	.09005	.09042	11.059	.99594	1.0041	11.105	**50**
11	.09034	.09071	11.024	.99591	1.0041	11.069	49
12	.09063	.09101	10.988	.99588	1.0041	11.034	48
13	.09092	.09130	10.953	.99586	1.0042	10.998	47
14	.09121	.09159	10.918	.99583	1.0042	10.963	46
15	.09150	.09189	10.883	.99580	1.0042	10.929	**45**
16	.09179	.09218	10.848	.99578	1.0042	10.894	44
17	.09208	.09247	10.814	.99575	1.0043	10.860	43
18	.09237	.09277	10.780	.99572	1.0043	10.826	42
19	.09266	.09306	10.746	.99570	1.0043	10.792	41
20	.09295	.09335	10.712	.99567	1.0043	10.758	**4**
21	.09324	.09365	10.678	.99564	1.0044	10.725	3
22	.09353	.09394	10.645	.99562	1.0044	10.692	3
23	.09382	.09423	10.612	.99559	1.0044	10.659	3
24	.09411	.09453	10.579	.99556	1 0045	10.626	3
25	.09440	.09482	10.546	.99553	1.0045	10.593	**3**
26	.09469	.09511	10.514	.99551	1.0045	10.561	3
27	.09498	.09541	10.481	.99548	1.0045	10.529	3
28	.09527	.09570	10.449	.99545	1.0046	10.497	3
29	.09556	.09600	10.417	.99542	1.0046	10.465	3
30	.09585	.09629	10.385	.99540	1.0046	10.433	**3**
31	.09614	.09658	10.354	.99537	1.0047	10.402	2
32	.09642	.09688	10.322	.99534	1.0047	10.371	2
33	.09671	.09717	10.291	.99531	1.0047	10.340	2
34	.09700	.09746	10.260	.99528	1.0047	10.309	2
35	.09729	.09776	10.229	.99526	1.0048	10.278	**2**
36	.09758	.09805	10.199	.99523	1.0048	10.248	2
37	.09787	.09834	10.168	.99520	1.0048	10.217	2
38	.09816	.09864	10.138	.99517	1.0049	10.187	2
39	.09845	.09893	10.108	.99514	1.0049	10.157	2
40	.09874	.09923	10.078	.99511	1.0049	10.128	**2**
41	.09903	.09952	10.048	.99508	1.0049	10.098	1
42	.09932	.09981	10.019	.99506	1.0050	10.068	1
43	.09961	.10011	9.9893	.99503	1.0050	10.039	1
44	.09990	.10040	9.9601	.99500	1.0050	10.010	1
45	.10019	.10069	9.9310	.99497	1.0051	9.9812	**1**
46	.10048	.10099	9.9021	.99494	1.0051	9.9525	1
47	.10077	.10128	9.8734	.99491	1.0051	9.9239	1
48	.10106	.10158	9.8448	.99488	1.0051	9.8955	1
49	.10135	.10187	9.8164	.99485	1.0052	9.8672	1
50	.10164	.10216	9.7882	.99482	1.0052	9.8391	**1**
51	.10192	.10246	9.7601	.99479	1.0052	9.8112	
52	.10221	.10275	9.7322	.99476	1.0053	9.7834	
53	.10250	.10305	9.7044	.99473	1.0053	9.7558	
54	.10279	.10334	9.6768	.99470	1.0053	9.7283	
55	.10308	.10363	9.6493	.99467	1.0054	9.7010	
56	.10337	.10393	9.6220	.99464	1.0054	9.6739	
57	.10366	.10422	9.5949	.99461	1.0054	9.6469	
58	.10395	.10452	9.5679	.99458	1.0054	9.6200	
59	.10424	.10481	9.5411	.99455	1.0055	9.5933	
60	.10453	.10510	9.5144	.99452	1.0055	9.5668	
′	Cos	Cot	Tan	Sin	Csc	Sec	

95° (275°) (264°)

NATURAL TRIGONOMETRIC FUNCTIONS TO FIVE PLACES

6° (186°) (353°) **173°**

′	Sin	Tan	Cot	Cos	Sec	Csc	′
0	.10453	.10510	9.5144	.99452	1.0055	9.5668	**60**
1	.10482	.10540	9.4878	.99449	1.0055	9.5404	59
2	.10511	.10569	9.4614	.99446	1.0056	9.5141	58
3	.10540	.10599	9.4352	.99443	1.0056	9.4880	57
4	.10569	.10628	9.4090	.99440	1.0056	9.4620	56
5	.10597	.10657	9.3831	.99437	1.0057	9.4362	**55**
6	.10626	.10687	9.3572	.99434	1.0057	9.4105	54
7	.10655	.10716	9.3315	.99431	1.0057	9.3850	53
8	.10684	.10746	9.3060	.99428	1.0058	9.3596	52
9	.10713	.10775	9.2806	.99424	1.0058	9.3343	51
10	.10742	.10805	9.2553	.99421	1.0058	9.3092	**50**
11	.10771	.10834	9.2302	.99418	1.0059	9.2842	49
12	.10800	.10863	9.2052	.99415	1.0059	9.2593	48
13	.10829	.10893	9.1803	.99412	1.0059	9.2346	47
14	.10858	.10922	9.1555	.99409	1.0059	9.2100	46
15	.10887	.10952	9.1309	.99406	1.0060	9.1855	**45**
16	.10916	.10981	9.1065	.99402	1.0060	9.1612	44
17	.10945	.11011	9.0821	.99399	1.0060	9.1370	43
18	.10973	.11040	9.0579	.99396	1.0061	9.1129	42
19	.11002	.11070	9.0338	.99393	1.0061	9.0890	41
20	.11031	.11099	9.0098	.99390	1.0061	9.0652	**40**
21	.11060	.11128	8.9860	.99386	1.0062	9.0415	39
22	.11089	.11158	8.9623	.99383	1.0062	9.0179	38
23	.11118	.11187	8.9387	.99380	1.0062	8.9944	37
24	.11147	.11217	8.9152	.99377	1.0063	8.9711	36
25	.11176	.11246	8.8919	.99374	1.0063	8.9479	**35**
26	.11205	.11276	8.8686	.99370	1.0063	8.9248	34
27	.11234	.11305	8.8455	.99367	1.0064	8.9019	33
28	.11263	.11335	8.8225	.99364	1.0064	8.8790	32
29	.11291	.11364	8.7996	.99360	1.0064	8.8563	31
30	.11320	.11394	8.7769	.99357	1.0065	8.8337	**30**
31	.11349	.11423	8.7542	.99354	1.0065	8.8112	29
32	.11378	.11452	8.7317	.99351	1.0065	8.7888	28
33	.11407	.11482	8.7093	.99347	1.0066	8.7665	27
34	.11436	.11511	8.6870	.99344	1.0066	8.7444	26
35	.11465	.11541	8.6648	.99341	1.0066	8.7223	**25**
36	.11494	.11570	8.6427	.99337	1.0067	8.7004	24
37	.11523	.11600	8.6208	.99334	1.0067	8.6786	23
38	.11552	.11629	8.5989	.99331	1.0067	8.6569	22
39	.11580	.11659	8.5772	.99327	1.0068	8.6353	21
40	.11609	.11688	8.5555	.99324	1.0068	8.6138	**20**
41	.11638	.11718	8.5340	.99320	1.0068	8.5924	19
42	.11667	.11747	8.5126	.99317	1.0069	8.5711	18
43	.11696	.11777	8.4913	.99314	1.0069	8.5500	17
44	.11725	.11806	8.4701	.99310	1.0069	8.5289	16
45	.11754	.11836	8.4490	.99307	1.0070	8.5079	**15**
46	.11783	.11865	8.4280	.99303	1.0070	8.4871	14
47	.11812	.11895	8.4071	.99300	1.0070	8.4663	13
48	.11840	.11924	8.3863	.99297	1.0071	8.4457	12
49	.11869	.11954	8.3656	.99293	1.0071	8.4251	11
50	.11898	.11983	8.3450	.99290	1.0072	8.4047	**10**
51	.11927	.12013	8.3245	.99286	1.0072	8.3843	9
52	.11956	.12042	8.3041	.99283	1.0072	8.3641	8
53	.11985	.12072	8.2838	.99279	1.0073	8.3439	7
54	.12014	.12101	8.2636	.99276	1.0073	8.3238	6
55	.12043	.12131	8.2434	.99272	1.0073	8.3039	**5**
56	.12071	.12160	8.2234	.99269	1.0074	8.2840	4
57	.12100	.12190	8.2035	.99265	1.0074	8.2642	3
58	.12129	.12219	8.1837	.99262	1.0074	8.2446	2
59	.12158	.12249	8.1640	.99258	1.0075	8.2250	1
60	.12187	.12278	8.1443	.99255	1.0075	8.2055	**0**
′	Cos	Cot	Tan	Sin	Csc	Sec	′

96° (276°) (263°) **83°**

7° (187°) (352°) **172°**

′	Sin	Tan	Cot	Cos	Sec	Csc	′
0	.12187	.12278	8.1443	.99255	1.0075	8.2055	**60**
1	.12216	.12308	8.1248	.99251	1.0075	8.1861	59
2	.12245	.12338	8.1054	.99248	1.0076	8.1668	58
3	.12274	.12367	8.0860	.99244	1.0076	8.1476	57
4	.12302	.12397	8.0667	.99240	1.0077	8.1285	56
5	.12331	.12426	8.0476	.99237	1.0077	8.1095	**55**
6	.12360	.12456	8.0285	.99233	1.0077	8.0905	54
7	.12389	.12485	8.0095	.99230	1.0078	8.0717	53
8	.12418	.12515	7.9906	.99226	1.0078	8.0529	52
9	.12447	.12544	7.9718	.99222	1.0078	8.0342	51
10	.12476	.12574	7.9530	.99219	1.0079	8.0156	**50**
11	.12504	.12603	7.9344	.99215	1.0079	7.9971	49
12	.12533	.12633	7.9158	.99211	1.0079	7.9787	48
13	.12562	.12662	7.8973	.99208	1.0080	7.9604	47
14	.12591	.12692	7.8789	.99204	1.0080	7.9422	46
15	.12620	.12722	7.8606	.99200	1.0081	7.9240	**45**
16	.12649	.12751	7.8424	.99197	1.0081	7.9059	44
17	.12678	.12781	7.8243	.99193	1.0081	7.8879	43
18	.12706	.12810	7.8062	.99189	1.0082	7.8700	42
19	.12735	.12840	7.7882	.99186	1.0082	7.8522	41
20	.12764	.12869	7.7704	.99182	1.0082	7.8344	**40**
21	.12793	.12899	7.7525	.99178	1.0083	7.8168	39
22	.12822	.12929	7.7348	.99175	1.0083	7.7992	38
23	.12851	.12958	7.7171	.99171	1.0084	7.7817	37
24	.12880	.12988	7.6996	.99167	1.0084	7.7642	36
25	.12908	.13017	7.6821	.99163	1.0084	7.7469	**35**
26	.12937	.13047	7.6647	.99160	1.0085	7.7296	34
27	.12966	.13076	7.6473	.99156	1.0085	7.7124	33
28	.12995	.13106	7.6301	.99152	1.0086	7.6953	32
29	.13024	.13136	7.6129	.99148	1.0086	7.6783	31
30	.13053	.13165	7.5958	.99144	1.0086	7.6613	**30**
31	.13081	.13195	7.5787	.99141	1.0087	7.6444	29
32	.13110	.13224	7.5618	.99137	1.0087	7.6276	28
33	.13139	.13254	7.5449	.99133	1.0087	7.6109	27
34	.13168	.13284	7.5281	.99129	1.0088	7.5942	26
35	.13197	.13313	7.5113	.99125	1.0088	7.5776	**25**
36	.13226	.13343	7.4947	.99122	1.0089	7.5611	24
37	.13254	.13372	7.4781	.99118	1.0089	7.5446	23
38	.13283	.13402	7.4615	.99114	1.0089	7.5282	22
39	.13312	.13432	7.4451	.99110	1.0090	7.5119	21
40	.13341	.13461	7.4287	.99106	1.0090	7.4957	**20**
41	.13370	.13491	7.4124	.99102	1.0091	7.4795	19
42	.13399	.13521	7.3962	.99098	1.0091	7.4635	18
43	.13427	.13550	7.3800	.99094	1.0091	7.4474	17
44	.13456	.13580	7.3639	.99091	1.0092	7.4315	16
45	.13485	.13609	7.3479	.99087	1.0092	7.4156	**15**
46	.13514	.13639	7.3319	.99083	1.0093	7.3998	14
47	.13543	.13669	7.3160	.99079	1.0093	7.3840	13
48	.13572	.13698	7.3002	.99075	1.0093	7.3684	12
49	.13600	.13728	7.2844	.99071	1.0094	7.3527	11
50	.13629	.13758	7.2687	.99067	1.0094	7.3372	**10**
51	.13658	.13787	7.2531	.99063	1.0095	7.3217	9
52	.13687	.13817	7.2375	.99059	1.0095	7.3063	8
53	.13716	.13846	7.2220	.99055	1.0095	7.2909	7
54	.13744	.13876	7.2066	.99051	1.0096	7.2757	6
55	.13773	.13906	7.1912	.99047	1.0096	7.2604	**5**
56	.13802	.13935	7.1759	.99043	1.0097	7.2453	4
57	.13831	.13965	7.1607	.99039	1.0097	7.2302	3
58	.13860	.13995	7.1455	.99035	1.0097	7.2152	2
59	.13889	.14024	7.1304	.99031	1.0098	7.2002	1
60	.13917	.14054	7.1154	.99027	1.0098	7.1853	**0**
′	Cos	Cot	Tan	Sin	Csc	Sec	′

97° (277°) (262°) **82°**

NATURAL TRIGONOMETRIC FUNCTIONS TO FIVE PLACES

8° (188°) (351°) **171°**

′	Sin	Tan	Cot	Cos	Sec	Csc	′
0	.13917	.14054	7.1154	.99027	1.0098	7.1853	**60**
1	.13946	.14084	7.1004	.99023	1.0099	7.1705	59
2	.13975	.14113	7.0855	.99019	1.0099	7.1557	58
3	.14004	.14143	7.0706	.99015	1.0100	7.1410	57
4	.14033	.14173	7.0558	.99011	1.0100	7.1263	56
5	.14061	.14202	7.0410	.99006	1.0100	7.1117	**55**
6	.14090	.14232	7.0264	.99002	1.0101	7.0972	54
7	.14119	.14262	7.0117	.98998	1.0101	7.0827	53
8	.14148	.14291	6.9972	.98994	1.0102	7.0683	52
9	.14177	.14321	6.9827	.98990	1.0102	7.0539	51
10	.14205	.14351	6.9682	.98986	1.0102	7.0396	**50**
11	.14234	.14381	6.9538	.98982	1.0103	7.0254	49
12	.14263	.14410	6.9395	.98978	1.0103	7.0112	48
13	.14292	.14440	6.9252	.98973	1.0104	6.9971	47
14	.14320	.14470	6.9110	.98969	1.0104	6.9830	46
15	.14349	.14499	6.8969	.98965	1.0105	6.9690	**45**
16	.14378	.14529	6.8828	.98961	1.0105	6.9550	44
17	.14407	.14559	6.8687	.98957	1.0105	6.9411	43
18	.14436	.14588	6.8548	.98953	1.0106	6.9273	42
19	.14464	.14618	6.8408	.98948	1.0106	6.9135	41
20	.14493	.14648	6.8269	.98944	1.0107	6.8998	**40**
21	.14522	.14678	6.8131	.98940	1.0107	6.8861	39
22	.14551	.14707	6.7994	.98936	1.0108	6.8725	38
23	.14580	.14737	6.7856	.98931	1.0108	6.8589	37
24	.14608	.14767	6.7720	.98927	1.0108	6.8454	36
25	.14637	.14796	6.7584	.98923	1.0109	6.8320	**35**
26	.14666	.14826	6.7448	.98919	1.0109	6.8186	34
27	.14695	.14856	6.7313	.98914	1.0110	6.8052	33
28	.14723	.14886	6.7179	.98910	1.0110	6.7919	32
29	.14752	.14915	6.7045	.98906	1.0111	6.7787	31
30	.14781	.14945	6.6912	.98902	1.0111	6.7655	**30**
31	.14810	.14975	6.6779	.98897	1.0112	6.7523	29
32	.14838	.15005	6.6646	.98893	1.0112	6.7392	28
33	.14867	.15034	6.6514	.98889	1.0112	6.7262	27
34	.14896	.15064	6.6383	.98884	1.0113	6.7132	26
35	.14925	.15094	6.6252	.98880	1.0113	6.7003	**25**
36	.14954	.15124	6.6122	.98876	1.0114	6.6874	24
37	.14982	.15153	6.5992	.98871	1.0114	6.6745	23
38	.15011	.15183	6.5863	.98867	1.0115	6.6618	22
39	.15040	.15213	6.5734	.98863	1.0115	6.6490	21
40	.15069	.15243	6.5606	.98858	1.0116	6.6363	**20**
41	.15097	.15272	6.5478	.98854	1.0116	6.6237	19
42	.15126	.15302	6.5350	.98849	1.0116	6.6111	18
43	.15155	.15332	6.5223	.98845	1.0117	6.5986	17
44	.15184	.15362	6.5097	.98841	1.0117	6.5861	16
45	.15212	.15391	6.4971	.98836	1.0118	6.5736	**15**
46	.15241	.15421	6.4846	.98832	1.0118	6.5612	14
47	.15270	.15451	6.4721	.98827	1.0119	6.5489	13
48	.15299	.15481	6.4596	.98823	1.0119	6.5366	12
49	.15327	.15511	6.4472	.98818	1.0120	6.5243	11
50	.15356	.15540	6.4348	.98814	1.0120	6.5121	**10**
51	.15385	.15570	6.4225	.98809	1.0120	6.4999	9
52	.15414	.15600	6.4103	.98805	1.0121	6.4878	8
53	.15442	.15630	6.3980	.98800	1.0121	6.4757	7
54	.15471	.15660	6.3859	.98796	1.0122	6.4637	6
55	.15500	.15689	6.3737	.98791	1.0122	6.4517	**5**
56	.15529	.15719	6.3617	.98787	1.0123	6.4398	4
57	.15557	.15749	6.3496	.98782	1.0123	6.4279	3
58	.15586	.15779	6.3376	.98778	1.0124	6.4160	2
59	.15615	.15809	6.3257	.98773	1.0124	6.4042	1
60	.15643	.15838	6.3138	.98769	1.0125	6.3925	**0**
′	Cos	Cot	Tan	Sin	Csc	Sec	′

98° (278°) (261°) **81°**

9° (189°) (350°) **17**

′	Sin	Tan	Cot	Cos	Sec	Csc
0	.15643	.15838	6.3138	.98769	1.0125	6.3925
1	.15672	.15868	6.3019	.98764	1.0125	6.3807
2	.15701	.15898	6.2901	.98760	1.0126	6.3691
3	.15730	.15928	6.2783	.98755	1.0126	6.3574
4	.15758	.15958	6.2666	.98751	1.0127	6.3458
5	.15787	.15988	6.2549	.98746	1.0127	6.3343
6	.15816	.16017	6.2432	.98741	1.0127	6.3228
7	.15845	.16047	6.2316	.98737	1.0128	6.3113
8	.15873	.16077	6.2200	.98732	1.0128	6.2999
9	.15902	.16107	6.2085	.98728	1.0129	6.2885
10	.15931	.16137	6.1970	.98723	1.0129	6.2772
11	.15959	.16167	6.1856	.98718	1.0130	6.2659
12	.15988	.16196	6.1742	.98714	1.0130	6.2546
13	.16017	.16226	6.1628	.98709	1.0131	6.2434
14	.16046	.16256	6.1515	.98704	1.0131	6.2323
15	.16074	.16286	6.1402	.98700	1.0132	6.2211
16	.16103	.16316	6.1290	.98695	1.0132	6.2100
17	.16132	.16346	6.1178	.98690	1.0133	6.1990
18	.16160	.16376	6.1066	.98686	1.0133	6.1880
19	.16189	.16405	6.0955	.98681	1.0134	6.1770
20	.16218	.16435	6.0844	.98676	1.0134	6.1661
21	.16246	.16465	6.0734	.98671	1.0135	6.1552
22	.16275	.16495	6.0624	.98667	1.0135	6.1443
23	.16304	.16525	6.0514	.98662	1.0136	6.1335
24	.16333	.16555	6.0405	.98657	1.0136	6.1227
25	.16361	.16585	6.0296	.98652	1.0137	6.1120
26	.16390	.16615	6.0188	.98648	1.0137	6.1013
27	.16419	.16645	6.0080	.98643	1.0138	6.0906
28	.16447	.16674	5.9972	.98638	1.0138	6.0800
29	.16476	.16704	5.9865	.98633	1.0139	6.0694
30	.16505	.16734	5.9758	.98629	1.0139	6.0589
31	.16533	.16764	5.9651	.98624	1.0140	6.0483
32	.16562	.16794	5.9545	.98619	1.0140	6.0379
33	.16591	.16824	5.9439	.98614	1.0141	6.0274
34	.16620	.16854	5.9333	.98609	1.0141	6.0170
35	.16648	.16884	5.9228	.98604	1.0142	6.0067
36	.16677	.16914	5.9124	.98600	1.0142	5.9963
37	.16706	.16944	5.9019	.98595	1.0143	5.9860
38	.16734	.16974	5.8915	.98590	1.0143	5.9758
39	.16763	.17004	5.8811	.98585	1.0144	5.9656
40	.16792	.17033	5.8708	.98580	1.0144	5.9554
41	.16820	.17063	5.8605	.98575	1.0145	5.9452
42	.16849	.17093	5.8502	.98570	1.0145	5.9351
43	.16878	.17123	5.8400	.98565	1.0146	5.9250
44	.16906	.17153	5.8298	.98561	1.0146	5.9150
45	.16935	.17183	5.8197	.98556	1.0147	5.9049
46	.16964	.17213	5.8095	.98551	1.0147	5.8950
47	.16992	.17243	5.7994	.98546	1.0148	5.8850
48	.17021	.17273	5.7894	.98541	1.0148	5.8751
49	.17050	.17303	5.7794	.98536	1.0149	5.8652
50	.17078	.17333	5.7694	.98531	1.0149	5.8554
51	.17107	.17363	5.7594	.98526	1.0150	5.8456
52	.17136	.17393	5.7495	.98521	1.0150	5.8358
53	.17164	.17423	5.7396	.98516	1.0151	5.8261
54	.17193	.17453	5.7297	.98511	1.0151	5.8164
55	.17222	.17483	5.7199	.98506	1.0152	5.8067
56	.17250	.17513	5.7101	.98501	1.0152	5.7970
57	.17279	.17543	5.7004	.98496	1.0153	5.7874
58	.17308	.17573	5.6906	.98491	1.0153	5.7778
59	.17336	.17603	5.6809	.98486	1.0154	5.7683
60	.17365	.17633	5.6713	.98481	1.0154	5.7588
′	Cos	Cot	Tan	Sin	Csc	Sec

99° (279°) (260°)

NATURAL TRIGONOMETRIC FUNCTIONS TO FIVE PLACES

(190°) (349°) **169°**

Sin	Tan	Cot	Cos	Sec	Csc	′
.17365	.17633	5.6713	.98481	1.0154	5.7588	**60**
.17393	.17663	5.6617	.98476	1.0155	5.7493	59
.17422	.17693	5.6521	.98471	1.0155	5.7398	58
.17451	.17723	5.6425	.98466	1.0156	5.7304	57
.17479	.17753	5.6329	.98461	1.0156	5.7210	56
.17508	.17783	5.6234	.98455	1.0157	5.7117	**55**
.17537	.17813	5.6140	.98450	1.0157	5.7023	54
.17565	.17843	5.6045	.98445	1.0158	5.6930	53
.17594	.17873	5.5951	.98440	1.0158	5.6838	52
.17623	.17903	5.5857	.98435	1.0159	5.6745	51
.17651	.17933	5.5764	.98430	1.0160	5.6653	**50**
.17680	.17963	5.5671	.98425	1.0160	5.6562	49
.17708	.17993	5.5578	.98420	1.0161	5.6470	48
.17737	.18023	5.5485	.98414	1.0161	5.6379	47
.17766	.18053	5.5393	.98409	1.0162	5.6288	46
.17794	.18083	5.5301	.98404	1.0162	5.6198	**45**
.17823	.18113	5.5209	.98399	1.0163	5.6107	44
.17852	.18143	5.5118	.98394	1.0163	5.6017	43
.17880	.18173	5.5026	.98389	1.0164	5.5928	42
.17909	.18203	5.4936	.98383	1.0164	5.5838	41
.17937	.18233	5.4845	.98378	1.0165	5.5749	**40**
.17966	.18263	5.4755	.98373	1.0165	5.5660	39
.17995	.18293	5.4665	.98368	1.0166	5.5572	38
.18023	.18323	5.4575	.98362	1.0166	5.5484	37
.18052	.18353	5.4486	.98357	1.0167	5.5396	36
.18081	.18384	5.4397	.98352	1.0168	5.5308	**35**
.18109	.18414	5.4308	.98347	1.0168	5.5221	34
.18138	.18444	5.4219	.98341	1.0169	5.5134	33
.18166	.18474	5.4131	.98336	1.0169	5.5047	32
.18195	.18504	5.4043	.98331	1.0170	5.4960	31
.18224	.18534	5.3955	.98325	1.0170	5.4874	**30**
.18252	.18564	5.3868	.98320	1.0171	5.4788	29
.18281	.18594	5.3781	.98315	1.0171	5.4702	28
.18309	.18624	5.3694	.98310	1.0172	5.4617	27
.18338	.18654	5.3607	.98304	1.0173	5.4532	26
.18367	.18684	5.3521	.98299	1.0173	5.4447	**25**
.18395	.18714	5.3435	.98294	1.0174	5.4362	24
.18424	.18745	5.3349	.98288	1.0174	5.4278	23
.18452	.18775	5.3263	.98283	1.0175	5.4194	22
.18481	.18805	5.3178	.98277	1.0175	5.4110	21
.18509	.18835	5.3093	.98272	1.0176	5.4026	**20**
.18538	.18865	5.3008	.98267	1.0176	5.3943	19
.18567	.18895	5.2924	.98261	1.0177	5.3860	18
.18595	.18925	5.2839	.98256	1.0178	5.3777	17
.18624	.18955	5.2755	.98250	1.0178	5.3695	16
.18652	.18986	5.2672	.98245	1.0179	5.3612	**15**
.18681	.19016	5.2588	.98240	1.0179	5.3530	14
.18710	.19046	5.2505	.98234	1.0180	5.3449	13
.18738	.19076	5.2422	.98229	1.0180	5.3367	12
.18767	.19106	5.2339	.98223	1.0181	5.3286	11
.18795	.19136	5.2257	.98218	1.0181	5.3205	**10**
.18824	.19166	5.2174	.98212	1.0182	5.3124	9
.18852	.19197	5.2092	.98207	1.0183	5.3044	8
.18881	.19227	5.2011	.98201	1.0183	5.2963	7
.18910	.19257	5.1929	.98196	1.0184	5.2883	6
.18938	.19287	5.1848	.98190	1.0184	5.2804	**5**
.18967	.19317	5.1767	.98185	1.0185	5.2724	4
.18995	.19347	5.1686	.98179	1.0185	5.2645	3
.19024	.19378	5.1606	.98174	1.0186	5.2566	2
.19052	.19408	5.1526	.98168	1.0187	5.2487	1
.19081	.19438	5.1446	.98163	1.0187	5.2408	**0**
Cos	**Cot**	**Tan**	**Sin**	**Csc**	**Sec**	**′**

(280°) (259°) **79°**

11° (191°) (348°) **168°**

′	Sin	Tan	Cot	Cos	Sec	Csc	′
0	.19081	.19438	5.1446	.98163	1.0187	5.2408	**60**
1	.19109	.19468	5.1366	.98157	1.0188	5.2330	59
2	.19138	.19498	5.1286	.98152	1.0188	5.2252	58
3	.19167	.19529	5.1207	.98146	1.0189	5.2174	57
4	.19195	.19559	5.1128	.98140	1.0189	5.2097	56
5	.19224	.19589	5.1049	.98135	1.0190	5.2019	**55**
6	.19252	.19619	5.0970	.98129	1.0191	5.1942	54
7	.19281	.19649	5.0892	.98124	1.0191	5.1865	53
8	.19309	.19680	5.0814	.98118	1.0192	5.1789	52
9	.19338	.19710	5.0736	.98112	1.0192	5.1712	51
10	.19366	.19740	5.0658	.98107	1.0193	5.1636	**50**
11	.19395	.19770	5.0581	.98101	1.0194	5.1560	49
12	.19423	.19801	5.0504	.98096	1.0194	5.1484	48
13	.19452	.19831	5.0427	.98090	1.0195	5.1409	47
14	.19481	.19861	5.0350	.98084	1.0195	5.1333	46
15	.19509	.19891	5.0273	.98079	1.0196	5.1258	**45**
16	.19538	.19921	5.0197	.98073	1.0197	5.1183	44
17	.19566	.19952	5.0121	.98067	1.0197	5.1109	43
18	.19595	.19982	5.0045	.98061	1.0198	5.1034	42
19	.19623	.20012	4.9969	.98056	1.0198	5.0960	41
20	.19652	.20042	4.9894	.98050	1.0199	5.0886	**40**
21	.19680	.20073	4.9819	.98044	1.0199	5.0813	39
22	.19709	.20103	4.9744	.98039	1.0200	5.0739	38
23	.19737	.20133	4.9669	.98033	1.0201	5.0666	37
24	.19766	.20164	4.9594	.98027	1.0201	5.0593	36
25	.19794	.20194	4.9520	.98021	1.0202	5.0520	**35**
26	.19823	.20224	4.9446	.98016	1.0202	5.0447	34
27	.19851	.20254	4.9372	.98010	1.0203	5.0375	33
28	.19880	.20285	4.9298	.98004	1.0204	5.0302	32
29	.19908	.20315	4.9225	.97998	1.0204	5.0230	31
30	.19937	.20345	4.9152	.97992	1.0205	5.0159	**30**
31	.19965	.20376	4.9078	.97987	1.0205	5.0087	29
32	.19994	.20406	4.9006	.97981	1.0206	5.0016	28
33	.20022	.20436	4.8933	.97975	1.0207	4.9944	27
34	.20051	.20466	4.8860	.97969	1.0207	4.9873	26
35	.20079	.20497	4.8788	.97963	1.0208	4.9803	**25**
36	.20108	.20527	4.8716	.97958	1.0209	4.9732	24
37	.20136	.20557	4.8644	.97952	1.0209	4.9662	23
38	.20165	.20588	4.8573	.97946	1.0210	4.9591	22
39	.20193	.20618	4.8501	.97940	1.0210	4.9521	21
40	.20222	.20648	4.8430	.97934	1.0211	4.9452	**20**
41	.20250	.20679	4.8359	.97928	1.0212	4.9382	19
42	.20279	.20709	4.8288	.97922	1.0212	4.9313	18
43	.20307	.20739	4.8218	.97916	1.0213	4.9244	17
44	.20336	.20770	4.8147	.97910	1.0213	4.9175	16
45	.20364	.20800	4.8077	.97905	1.0214	4.9106	**15**
46	.20393	.20830	4.8007	.97899	1.0215	4.9037	14
47	.20421	.20861	4.7937	.97893	1.0215	4.8969	13
48	.20450	.20891	4.7867	.97887	1.0216	4.8901	12
49	.20478	.20921	4.7798	.97881	1.0217	4.8833	11
50	.20507	.20952	4.7729	.97875	1.0217	4.8765	**10**
51	.20535	.20982	4.7659	.97869	1.0218	4.8697	9
52	.20563	.21013	4.7591	.97863	1.0218	4.8630	8
53	.20592	.21043	4.7522	.97857	1.0219	4.8563	7
54	.20620	.21073	4.7453	.97851	1.0220	4.8496	6
55	.20649	.21104	4.7385	.97845	1.0220	4.8429	**5**
56	.20677	.21134	4.7317	.97839	1.0221	4.8362	4
57	.20706	.21164	4.7249	.97833	1.0222	4.8296	3
58	.20734	.21195	4.7181	.97827	1.0222	4.8229	2
59	.20763	.21225	4.7114	.97821	1.0223	4.8163	1
60	.20791	.21256	4.7046	.97815	1.0223	4.8097	**0**
′	**Cos**	**Cot**	**Tan**	**Sin**	**Csc**	**Sec**	**′**

101° (281°) (258°) **78°**

NATURAL TRIGONOMETRIC FUNCTIONS TO FIVE PLACES

12° (192°) (347°) 167°

′	Sin	Tan	Cot	Cos	Sec	Csc	′
0	.20791	.21256	4.7046	.97815	1.0223	4.8097	60
1	.20820	.21286	4.6979	.97809	1.0224	4.8032	59
2	.20848	.21316	4.6912	.97803	1.0225	4.7966	58
3	.20877	.21347	4.6845	.97797	1.0225	4.7901	57
4	.20905	.21377	4.6779	.97791	1.0226	4.7836	56
5	.20933	.21408	4.6712	.97784	1.0227	4.7771	55
6	.20962	.21438	4.6646	.97778	1.0227	4.7706	54
7	.20990	.21469	4.6580	.97772	1.0228	4.7641	53
8	.21019	.21499	4.6514	.97766	1.0228	4.7577	52
9	.21047	.21529	4.6448	.97760	1.0229	4.7512	51
10	.21076	.21560	4.6382	.97754	1.0230	4.7448	50
11	.21104	.21590	4.6317	.97748	1.0230	4.7384	49
12	.21132	.21621	4.6252	.97742	1.0231	4.7321	48
13	.21161	.21651	4.6187	.97735	1.0232	4.7257	47
14	.21189	.21682	4.6122	.97729	1.0232	4.7194	46
15	.21218	.21712	4.6057	.97723	1.0233	4.7130	45
16	.21246	.21743	4.5993	.97717	1.0234	4.7067	44
17	.21275	.21773	4.5928	.97711	1.0234	4.7004	43
18	.21303	.21804	4.5864	.97705	1.0235	4.6942	42
19	.21331	.21834	4.5800	.97698	1.0236	4.6879	41
20	.21360	.21864	4.5736	.97692	1.0236	4.6817	40
21	.21388	.21895	4.5673	.97686	1.0237	4.6755	39
22	.21417	.21925	4.5609	.97680	1.0238	4.6693	38
23	.21445	.21956	4.5546	.97673	1.0238	4.6631	37
24	.21474	.21986	4.5483	.97667	1.0239	4.6569	36
25	.21502	.22017	4.5420	.97661	1.0240	4.6507	35
26	.21530	.22047	4.5357	.97655	1.0240	4.6446	34
27	.21559	.22078	4.5294	.97648	1.0241	4.6385	33
28	.21587	.22108	4.5232	.97642	1.0241	4.6324	32
29	.21616	.22139	4.5169	.97636	1.0242	4.6263	31
30	.21644	.22169	4.5107	.97630	1.0243	4.6202	30
31	.21672	.22200	4.5045	.97623	1.0243	4.6142	29
32	.21701	.22231	4.4983	.97617	1.0244	4.6081	28
33	.21729	.22261	4.4922	.97611	1.0245	4.6021	27
34	.21758	.22292	4.4860	.97604	1.0245	4.5961	26
35	.21786	.22322	4.4799	.97598	1.0246	4.5901	25
36	.21814	.22353	4.4737	.97592	1.0247	4.5841	24
37	.21843	.22383	4.4676	.97585	1.0247	4.5782	23
38	.21871	.22414	4.4615	.97579	1.0248	4.5722	22
39	.21899	.22444	4.4555	.97573	1.0249	4.5663	21
40	.21928	.22475	4.4494	.97566	1.0249	4.5604	20
41	.21956	.22505	4.4434	.97560	1.0250	4.5545	19
42	.21985	.22536	4.4373	.97553	1.0251	4.5486	18
43	.22013	.22567	4.4313	.97547	1.0251	4.5428	17
44	.22041	.22597	4.4253	.97541	1.0252	4.5369	16
45	.22070	.22628	4.4194	.97534	1.0253	4.5311	15
46	.22098	.22658	4.4134	.97528	1.0253	4.5253	14
47	.22126	.22689	4.4075	.97521	1.0254	4.5195	13
48	.22155	.22719	4.4015	.97515	1.0255	4.5137	12
49	.22183	.22750	4.3956	.97508	1.0256	4.5079	11
50	.22212	.22781	4.3897	.97502	1.0256	4.5022	10
51	.22240	.22811	4.3838	.97496	1.0257	4.4964	9
52	.22268	.22842	4.3779	.97489	1.0258	4.4907	8
53	.22297	.22872	4.3721	.97483	1.0258	4.4850	7
54	.22325	.22903	4.3662	.97476	1.0259	4.4793	6
55	.22353	.22934	4.3604	.97470	1.0260	4.4736	5
56	.22382	.22964	4.3546	.97463	1.0260	4.4679	4
57	.22410	.22995	4.3488	.97457	1.0261	4.4623	3
58	.22438	.23026	4.3430	.97450	1.0262	4.4566	2
59	.22467	.23056	4.3372	.97444	1.0262	4.4510	1
60	.22495	.23087	4.3315	.97437	1.0263	4.4454	0
′	Cos	Cot	Tan	Sin	Csc	Sec	′

102° (282°) (257°) 77°

13° (193°) (346°) 166

′	Sin	Tan	Cot	Cos	Sec	Csc	′
0	.22495	.23087	4.3315	.97437	1.0263	4.4454	60
1	.22523	.23117	4.3257	.97430	1.0264	4.4398	59
2	.22552	.23148	4.3200	.97424	1.0264	4.4342	58
3	.22580	.23179	4.3143	.97417	1.0265	4.4287	57
4	.22608	.23209	4.3086	.97411	1.0266	4.4231	56
5	.22637	.23240	4.3029	.97404	1.0266	4.4176	55
6	.22665	.23271	4.2972	.97398	1.0267	4.4121	54
7	.22693	.23301	4.2916	.97391	1.0268	4.4066	53
8	.22722	.23332	4.2859	.97384	1.0269	4.4011	52
9	.22750	.23363	4.2803	.97378	1.0269	4.3956	51
10	.22778	.23393	4.2747	.97371	1.0270	4.3901	50
11	.22807	.23424	4.2691	.97365	1.0271	4.3847	49
12	.22835	.23455	4 2635	.97358	1.0271	4.3792	48
13	.22863	.23485	4.2580	.97351	1.0272	4.3738	47
14	.22892	.23516	4.2524	.97345	1.0273	4.3684	46
15	.22920	.23547	4.2468	.97338	1.0273	4.3630	45
16	.22948	.23578	4.2413	.97331	1.0274	4.3576	44
17	.22977	.23608	4.2358	.97325	1.0275	4.3522	43
18	.23005	.23639	4.2303	.97318	1.0276	4.3469	42
19	.23033	.23670	4.2248	.97311	1.0276	4.3415	41
20	.23062	.23700	4.2193	.97304	1.0277	4.3362	40
21	.23090	.23731	4.2139	.97298	1.0278	4.3309	39
22	.23118	.23762	4.2084	.97291	1.0278	4.3256	38
23	.23146	.23793	4.2030	.97284	1.0279	4.3203	37
24	.23175	.23823	4.1976	.97278	1.0280	4.3150	36
25	.23203	.23854	4.1922	.97271	1.0281	4.3098	35
26	.23231	.23885	4.1868	.97264	1.0281	4.3045	34
27	.23260	.23916	4.1814	.97257	1.0282	4.2993	33
28	.23288	.23946	4.1760	.97251	1.0283	4.2941	32
29	.23316	.23977	4.1706	.97244	1.0283	4.2889	31
30	.23345	.24008	4.1653	.97237	1.0284	4.2837	30
31	.23373	.24039	4.1600	.97230	1.0285	4.2785	29
32	.23401	.24069	4.1547	.97223	1.0286	4.2733	28
33	.23429	.24100	4.1493	.97217	1.0286	4.2681	27
34	.23458	.24131	4.1441	.97210	1.0287	4.2630	26
35	.23486	.24162	4.1388	.97203	1.0288	4.2579	25
36	.23514	.24193	4.1335	.97196	1.0288	4.2527	24
37	.23542	.24223	4.1282	.97189	1.0289	4.2476	23
38	.23571	.24254	4.1230	.97182	1.0290	4.2425	22
39	.23599	.24285	4.1178	.97176	1.0291	4.2375	21
40	.23627	.24316	4.1126	.97169	1.0291	4.2324	20
41	.23656	.24347	4.1074	.97162	1.0292	4.2273	19
42	.23684	.24377	4.1022	.97155	1.0293	4.2223	18
43	.23712	.24408	4.0970	.97148	1.0294	4.2173	17
44	.23740	.24439	4.0918	.97141	1.0294	4.2122	16
45	.23769	.24470	4.0867	.97134	1.0295	4.2072	15
46	.23797	.24501	4.0815	.97127	1.0296	4.2022	14
47	.23825	.24532	4.0764	.97120	1.0297	4.1973	13
48	.23853	.24562	4.0713	.97113	1.0297	4.1923	12
49	.23882	.24593	4.0662	.97106	1.0298	4.1873	11
50	.23910	.24624	4.0611	.97100	1.0299	4.1824	10
51	.23938	.24655	4.0560	.97093	1.0299	4.1774	9
52	.23966	.24686	4.0509	.97086	1.0300	4.1725	8
53	.23995	.24717	4.0459	.97079	1.0301	4.1676	7
54	.24023	.24747	4.0408	.97072	1.0302	4.1627	6
55	.24051	.24778	4.0358	.97065	1.0302	4.1578	5
56	.24079	.24809	4.0308	.97058	1.0303	4.1529	4
57	.24108	.24840	4.0257	.97051	1.0304	4.1481	3
58	.24136	.24871	4.0207	.97044	1.0305	4.1432	2
59	.24164	.24902	4.0158	.97037	1.0305	4.1384	1
60	.24192	.24933	4.0108	.97030	1.0306	4.1336	0
′	Cos	Cot	Tan	Sin	Csc	Sec	′

103° (283°) (256°) 7

NATURAL TRIGONOMETRIC FUNCTIONS TO FIVE PLACES

14° (194°) (345°) **165°**

′	Sin	Tan	Cot	Cos	Sec	Csc	′
0	.24192	.24933	4.0108	.97030	1.0306	4.1336	**60**
1	.24220	.24964	4.0058	.97023	1.0307	4.1287	59
2	.24249	.24995	4.0009	.97015	1.0308	4.1239	58
3	.24277	.25026	3.9959	.97008	1.0308	4.1191	57
4	.24305	.25056	3.9910	.97001	1.0309	4.1144	56
5	.24333	.25087	3.9861	.96994	1.0310	4.1096	**55**
6	.24362	.25118	3.9812	.96987	1.0311	4.1048	54
7	.24390	.25149	3.9763	.96980	1.0311	4.1001	53
8	.24418	.25180	3.9714	.96973	1.0312	4.0954	52
9	.24446	.25211	3.9665	.96966	1.0313	4.0906	51
10	.24474	.25242	3.9617	.96959	1.0314	4.0859	**50**
11	.24503	.25273	3.9568	.96952	1.0314	4.0812	49
12	.24531	.25304	3.9520	.96945	1.0315	4.0765	48
13	.24559	.25335	3.9471	.96937	1.0316	4.0718	47
14	.24587	.25366	3.9423	.96930	1.0317	4.0672	46
15	.24615	.25397	3.9375	.96923	1.0317	4.0625	**45**
16	.24644	.25428	3.9327	.96916	1.0318	4.0579	44
17	.24672	.25459	3.9279	.96909	1.0319	4.0532	43
18	.24700	.25490	3.9232	.96902	1.0320	4.0486	42
19	.24728	.25521	3.9184	.96894	1.0321	4.0440	41
20	.24756	.25552	3.9136	.96887	1.0321	4.0394	**40**
21	.24784	.25583	3.9089	.96880	1.0322	4.0348	39
22	.24813	.25614	3.9042	.96873	1.0323	4.0302	38
23	.24841	.25645	3.8995	.96866	1.0324	4.0256	37
24	.24869	.25676	3.8947	.96858	1.0324	4.0211	36
25	.24897	.25707	3.8900	.96851	1.0325	4.0165	**35**
26	.24925	.25738	3.8854	.96844	1.0326	4.0120	34
27	.24954	.25769	3.8807	.96837	1.0327	4.0075	33
28	.24982	.25800	3.8760	.96829	1.0327	4.0029	32
29	.25010	.25831	3.8714	.96822	1.0328	3.9984	31
30	.25038	.25862	3.8667	.96815	1.0329	3.9939	**30**
31	.25066	.25893	3.8621	.96807	1.0330	3.9894	29
32	.25094	.25924	3.8575	.96800	1.0331	3.9850	28
33	.25122	.25955	3.8528	.96793	1.0331	3.9805	27
34	.25151	.25986	3.8482	.96786	1.0332	3.9760	26
35	.25179	.26017	3.8436	.96778	1.0333	3.9716	**25**
36	.25207	.26048	3.8391	.96771	1.0334	3.9672	24
37	.25235	.26079	3.8345	.96764	1.0334	3.9627	23
38	.25263	.26110	3.8299	.96756	1.0335	3.9583	22
39	.25291	.26141	3.8254	.96749	1.0336	3.9539	21
40	.25320	.26172	3.8208	.96742	1.0337	3.9495	**20**
41	.25348	.26203	3.8163	.96734	1.0338	3.9451	19
42	.25376	.26235	3.8118	.96727	1.0338	3.9408	18
43	.25404	.26266	3.8073	.96719	1.0339	3.9364	17
44	.25432	.26297	3.8028	.96712	1.0340	3.9320	16
45	.25460	.26328	3.7983	.96705	1.0341	3.9277	**15**
46	.25488	.26359	3.7938	.96697	1.0342	3.9234	14
47	.25516	.26390	3.7893	.96690	1.0342	3.9190	13
48	.25545	.26421	3.7848	.96682	1.0343	3.9147	12
49	.25573	.26452	3.7804	.96675	1.0344	3.9104	11
50	.25601	.26483	3.7760	.96667	1.0345	3.9061	**10**
51	.25629	.26515	3.7715	.96660	1.0346	3.9018	9
52	.25657	.26546	3.7671	.96653	1.0346	3.8976	8
53	.25685	.26577	3.7627	.96645	1.0347	3.8933	7
54	.25713	.26608	3.7583	.96638	1.0348	3.8890	6
55	.25741	.26639	3.7539	.96630	1.0349	3.8848	**5**
56	.25769	.26670	3.7495	.96623	1.0350	3.8806	4
57	.25798	.26701	3.7451	.96615	1.0350	3.8763	3
58	.25826	.26733	3.7408	.96608	1.0351	3.8721	2
59	.25854	.26764	3.7364	.96600	1.0352	3.8679	1
60	.25882	.26795	3.7321	.96593	1.0353	3.8637	**0**
′	Cos	Cot	Tan	Sin	Csc	Sec	′

104° (284°) (255°) **75°**

15° (195°) (344°) **164°**

′	Sin	Tan	Cot	Cos	Sec	Csc	′
0	.25882	.26795	3.7321	.96593	1.0353	3.8637	**60**
1	.25910	.26826	3.7277	.96585	1.0354	3.8595	59
2	.25938	.26857	3.7234	.96578	1.0354	3.8553	58
3	.25966	.26888	3.7191	.96570	1.0355	3.8512	57
4	.25994	.26920	3.7148	.96562	1.0356	3.8470	56
5	.26022	.26951	3.7105	.96555	1.0357	3.8428	**55**
6	.26050	.26982	3.7062	.96547	1.0358	3.8387	54
7	.26079	.27013	3.7019	.96540	1.0358	3.8346	53
8	.26107	.27044	3.6976	.96532	1.0359	3.8304	52
9	.26135	.27076	3.6933	.96524	1.0360	3.8263	51
10	.26163	.27107	3.6891	.96517	1.0361	3.8222	**50**
11	.26191	.27138	3.6848	.96509	1.0362	3.8181	49
12	.26219	.27169	3.6806	.96502	1.0363	3.8140	48
13	.26247	.27201	3.6764	.96494	1.0363	3.8100	47
14	.26275	.27232	3.6722	.96486	1.0364	3.8059	46
15	.26303	.27263	3.6680	.96479	1.0365	3.8018	**45**
16	.26331	.27294	3.6638	.96471	1.0366	3.7978	44
17	.26359	.27326	3.6596	.96463	1.0367	3.7937	43
18	.26387	.27357	3.6554	.96456	1.0367	3.7897	42
19	.26415	.27388	3.6512	.96448	1.0368	3.7857	41
20	.26443	.27419	3.6470	.96440	1.0369	3.7817	**40**
21	.26471	.27451	3.6429	.96433	1.0370	3.7777	39
22	.26500	.27482	3.6387	.96425	1.0371	3.7737	38
23	.26528	.27513	3.6346	.96417	1.0372	3.7697	37
24	.26556	.27545	3.6305	.96410	1.0372	3.7657	36
25	.26584	.27576	3.6264	.96402	1.0373	3.7617	**35**
26	.26612	.27607	3.6222	.96394	1.0374	3.7577	34
27	.26640	.27638	3.6181	.96386	1.0375	3.7538	33
28	.26668	.27670	3.6140	.96379	1.0376	3.7498	32
29	.26696	.27701	3.6100	.96371	1.0377	3.7459	31
30	.26724	.27732	3.6059	.96363	1.0377	3.7420	**30**
31	.26752	.27764	3.6018	.96355	1.0378	3.7381	29
32	.26780	.27795	3.5978	.96347	1.0379	3.7341	28
33	.26808	.27826	3.5937	.96340	1.0380	3.7302	27
34	.26836	.27858	3.5897	.96332	1.0381	3.7263	26
35	.26864	.27889	3.5856	.96324	1.0382	3.7225	**25**
36	.26892	.27921	3.5816	.96316	1.0382	3.7186	24
37	.26920	.27952	3.5776	.96308	1.0383	3.7147	23
38	.26948	.27983	3.5736	.96301	1.0384	3.7108	22
39	.26976	.28015	3.5696	.96293	1.0385	3.7070	21
40	.27004	.28046	3.5656	.96285	1.0386	3.7032	**20**
41	.27032	.28077	3.5616	.96277	1.0387	3.6993	19
42	.27060	.28109	3.5576	.96269	1.0388	3.6955	18
43	.27088	.28140	3.5536	.96261	1.0388	3.6917	17
44	.27116	.28172	3.5497	.96253	1.0389	3.6879	16
45	.27144	.28203	3.5457	.96246	1.0390	3.6840	**15**
46	.27172	.28234	3.5418	.96238	1.0391	3.6803	14
47	.27200	.28266	3.5379	.96230	1.0392	3.6765	13
48	.27228	.28297	3.5339	.96222	1.0393	3.6727	12
49	.27256	.28329	3.5300	.96214	1.0394	3.6689	11
50	.27284	.28360	3.5261	.96206	1.0394	3.6652	**10**
51	.27312	.28391	3.5222	.96198	1.0395	3.6614	9
52	.27340	.28423	3.5183	.96190	1.0396	3.6575	8
53	.27368	.28454	3.5144	.96182	1.0397	3.6539	7
54	.27396	.28486	3.5105	.96174	1.0398	3.6502	6
55	.27424	.28517	3.5067	.96166	1.0399	3.6465	**5**
56	.27452	.28549	3.5028	.96158	1.0400	3.6427	4
57	.27480	.28580	3.4989	.96150	1.0400	3.6390	3
58	.27508	.28612	3.4951	.96142	1.0401	3.6353	2
59	.27536	.28643	3.4912	.96134	1.0402	3.6316	1
60	.27564	.28675	3.4874	.96126	1.0403	3.6280	**0**
′	Cos	Cot	Tan	Sin	Csc	Sec	′

105° (285°) (254°) **74°**

16° (196°) (343°) **163°**

′	Sin	Tan	Cot	Cos	Sec	Csc	′
0	.27564	.28675	3.4874	.96126	1.0403	3.6280	**60**
1	.27592	.28706	3.4836	.96118	1.0404	3.6243	59
2	.27620	.28738	3.4798	.96110	1.0405	3.6206	58
3	.27648	.28769	3.4760	.96102	1.0406	3.6169	57
4	.27676	.28801	3.4722	.96094	1.0406	3.6133	56
5	.27704	.28832	3.4684	.96086	1.0407	3.6097	**55**
6	.27731	.28864	3.4646	.96078	1.0408	3.6060	54
7	.27759	.28895	3.4608	.96070	1.0409	3.6024	53
8	.27787	.28927	3.4570	.96062	1.0410	3.5988	52
9	.27815	.28958	3.4533	.96054	1.0411	3.5951	51
10	.27843	.28990	3.4495	.96046	1.0412	3.5915	**50**
11	.27871	.29021	3.4458	.96037	1.0413	3.5879	49
12	.27899	.29053	3.4420	.96029	1.0413	3.5843	48
13	.27927	.29084	3.4383	.96021	1.0414	3.5808	47
14	.27955	.29116	3.4346	.96013	1.0415	3.5772	46
15	.27983	.29147	3.4308	.96005	1.0416	3.5736	**45**
16	.28011	.29179	3.4271	.95997	1.0417	3.5700	44
17	.28039	.29210	3.4234	.95989	1.0418	3.5665	43
18	.28067	.29242	3.4197	.95981	1.0419	3.5629	42
19	.28095	.29274	3.4160	.95972	1.0420	3.5594	41
20	.28123	.29305	3.4124	.95964	1.0421	3.5559	**40**
21	.28150	.29337	3.4087	.95956	1.0421	3.5523	39
22	.28178	.29368	3.4050	.95948	1.0422	3.5488	38
23	.28206	.29400	3.4014	.95940	1.0423	3.5453	37
24	.28234	.29432	3.3977	.95931	1.0424	3.5418	36
25	.28262	.29463	3.3941	.95923	1.0425	3.5383	**35**
26	.28290	.29495	3.3904	.95915	1.0426	3.5348	34
27	.28318	.29526	3.3868	.95907	1.0427	3.5313	33
28	.28346	.29558	3.3832	.95898	1.0428	3.5279	32
29	.28374	.29590	3.3796	.95890	1.0429	3.5244	31
30	.28402	.29621	3.3759	.95882	1.0429	3.5209	**30**
31	.28429	.29653	3.3723	.95874	1.0430	3.5175	29
32	.28457	.29685	3.3687	.95865	1.0431	3.5140	28
33	.28485	.29716	3.3652	.95857	1.0432	3.5106	27
34	.28513	.29748	3.3616	.95849	1.0433	3.5072	26
35	.28541	.29780	3.3580	.95841	1.0434	3.5037	**25**
36	.28569	.29811	3.3544	.95832	1.0435	3.5003	24
37	.28597	.29843	3.3509	.95824	1.0436	3.4969	23
38	.28625	.29875	3.3473	.95816	1.0437	3.4935	22
39	.28652	.29906	3.3438	.95807	1.0438	3.4901	21
40	.28680	.29938	3.3402	.95799	1.0439	3.4867	**20**
41	.28708	.29970	3.3367	.95791	1.0439	3.4833	19
42	.28736	.30001	3.3332	.95782	1.0440	3.4799	18
43	.28764	.30033	3.3297	.95774	1.0441	3.4766	17
44	.28792	.30065	3.3261	.95766	1.0442	3.4732	16
45	.28820	.30097	3.3226	.95757	1.0443	3.4699	**15**
46	.28847	.30128	3.3191	.95749	1.0444	3.4665	14
47	.28875	.30160	3.3156	.95740	1.0445	3.4632	13
48	.28903	.30192	3.3122	.95732	1.0446	3.4598	12
49	.28931	.30224	3.3087	.95724	1.0447	3.4565	11
50	.28959	.30255	3.3052	.95715	1.0448	3.4532	**10**
51	.28987	.30287	3.3017	.95707	1.0449	3.4499	9
52	.29015	.30319	3.2983	.95698	1.0450	3.4465	8
53	.29042	.30351	3.2948	.95690	1.0450	3.4432	7
54	.29070	.30382	3.2914	.95681	1.0451	3.4399	6
55	.29098	.30414	3.2879	.95673	1.0452	3.4367	**5**
56	.29126	.30446	3.2845	.95664	1.0453	3.4334	4
57	.29154	.30478	3.2811	.95656	1.0454	3.4301	3
58	.29182	.30509	3.2777	.95647	1.0455	3.4268	2
59	.29209	.30541	3.2743	.95639	1.0456	3.4236	1
60	.29237	.30573	3.2709	.95630	1.0457	3.4203	**0**
′	Cos	Cot	Tan	Sin	Csc	Sec	′

106° (286°) (253°) **73°**

17° (197°) (342°) **162°**

′	Sin	Tan	Cot	Cos	Sec	Csc	′
0	.29237	.30573	3.2709	.95630	1.0457	3.4203	**60**
1	.29265	.30605	3.2675	.95622	1.0458	3.4171	59
2	.29293	.30637	3.2641	.95613	1.0459	3.4138	58
3	.29321	.30669	3.2607	.95605	1.0460	3.4106	57
4	.29348	.30700	3.2573	.95596	1.0461	3.4073	56
5	.29376	.30732	3.2539	.95588	1.0462	3.4041	**55**
6	.29404	.30764	3.2506	.95579	1.0463	3.4009	54
7	.29432	.30796	3.2472	.95571	1.0463	3.3977	53
8	.29460	.30828	3.2438	.95562	1.0464	3.3945	52
9	.29487	.30860	3.2405	.95554	1.0465	3.3913	51
10	.29515	.30891	3.2371	.95545	1.0466	3.3881	**50**
11	.29543	.30923	3.2338	.95536	1.0467	3.3849	49
12	.29571	.30955	3.2305	.95528	1.0468	3.3817	48
13	.29599	.30987	3.2272	.95519	1.0469	3.3785	47
14	.29626	.31019	3.2238	.95511	1.0470	3.3754	46
15	.29654	.31051	3.2205	.95502	1.0471	3.3722	**45**
16	.29682	.31083	3.2172	.95493	1.0472	3.3691	44
17	.29710	.31115	3.2139	.95485	1.0473	3.3659	43
18	.29737	.31147	3.2106	.95476	1.0474	3.3628	42
19	.29765	.31178	3.2073	.95467	1.0475	3.3596	41
20	.29793	.31210	3.2041	.95459	1.0476	3.3565	**40**
21	.29821	.31242	3.2008	.95450	1.0477	3.3534	39
22	.29849	.31274	3.1975	.95441	1.0478	3.3502	38
23	.29876	.31306	3.1943	.95433	1.0479	3.3471	37
24	.29904	.31338	3.1910	.95424	1.0480	3.3440	36
25	.29932	.31370	3.1878	.95415	1.0480	3.3409	**35**
26	.29960	.31402	3.1845	.95407	1.0481	3.3378	34
27	.29987	.31434	3.1813	.95398	1.0482	3.3347	33
28	.30015	.31466	3.1780	.95389	1.0483	3.3317	32
29	.30043	.31498	3.1748	.95380	1.0484	3.3286	31
30	.30071	.31530	3.1716	.95372	1.0485	3.3255	**30**
31	.30098	.31562	3.1684	.95363	1.0486	3.3224	29
32	.30126	.31594	3.1652	.95354	1.0487	3.3194	28
33	.30154	.31626	3.1620	.95345	1.0488	3.3163	27
34	.30182	.31658	3.1588	.95337	1.0489	3.3133	26
35	.30209	.31690	3.1556	.95328	1.0490	3.3102	**25**
36	.30237	.31722	3.1524	.95319	1.0491	3.3072	24
37	.30265	.31754	3.1492	.95310	1.0492	3.3042	23
38	.30292	.31786	3.1460	.95301	1.0493	3.3012	22
39	.30320	.31818	3.1429	.95293	1.0494	3.2981	21
40	.30348	.31850	3.1397	.95284	1.0495	3.2951	**20**
41	.30376	.31882	3.1366	.95275	1.0496	3.2921	19
42	.30403	.31914	3.1334	.95266	1.0497	3.2891	18
43	.30431	.31946	3.1303	.95257	1.0498	3.2861	17
44	.30459	.31978	3.1271	.95248	1.0499	3.2831	16
45	.30486	.32010	3.1240	.95240	1.0500	3.2801	**15**
46	.30514	.32042	3.1209	.95231	1.0501	3.2772	14
47	.30542	.32074	3.1178	.95222	1.0502	3.2742	13
48	.30570	.32106	3.1146	.95213	1.0503	3.2712	12
49	.30597	.32139	3.1115	.95204	1.0504	3.2683	11
50	.30625	.32171	3.1084	.95195	1.0505	3.2653	**10**
51	.30653	.32203	3.1053	.95186	1.0506	3.2624	9
52	.30680	.32235	3.1022	.95177	1.0507	3.2594	8
53	.30708	.32267	3.0991	.95168	1.0508	3.2565	7
54	.30736	.32299	3.0961	.95159	1.0509	3.2535	6
55	.30763	.32331	3.0930	.95150	1.0510	3.2506	**5**
56	.30791	.32363	3.0899	.95142	1.0511	3.2477	4
57	.30819	.32396	3.0868	.95133	1.0512	3.2448	3
58	.30846	.32428	3.0838	.95124	1.0513	3.2419	2
59	.30874	.32460	3.0807	.95115	1.0514	3.2390	1
60	.30902	.32492	3.0777	.95106	1.0515	3.2361	**0**
′	Cos	Cot	Tan	Sin	Csc	Sec	′

107° (287°) (252°) **7**

(198°) (341°) **161°**

′	Sin	Tan	Cot	Cos	Sec	Csc	′
0	.30902	.32492	3.0777	.95106	1.0515	3.2361	**60**
1	.30929	.32524	3.0746	.95097	1.0516	3.2332	59
2	.30957	.32556	3.0716	.95088	1.0517	3.2303	58
3	.30985	.32588	3.0686	.95079	1.0518	3.2274	57
4	.31012	.32621	3.0655	.95070	1.0519	3.2245	56
5	.31040	.32653	3.0625	.95061	1.0520	3.2217	**55**
6	.31068	.32685	3.0595	.95052	1.0521	3.2188	54
7	.31095	.32717	3.0565	.95043	1.0522	3.2159	53
8	.31123	.32749	3.0535	.95033	1.0523	3.2131	52
9	.31151	.32782	3.0505	.95024	1.0524	3.2102	51
10	.31178	.32814	3.0475	.95015	1.0525	3.2074	**50**
11	.31206	.32846	3.0445	.95006	1.0526	3.2045	49
12	.31233	.32878	3.0415	.94997	1.0527	3.2017	48
13	.31261	.32911	3.0385	.94988	1.0528	3.1989	47
14	.31289	.32943	3.0356	.94979	1.0529	3.1960	46
15	.31316	.32975	3.0326	.94970	1.0530	3.1932	**45**
16	.31344	.33007	3.0296	.94961	1.0531	3.1904	44
17	.31372	.33040	3.0267	.94952	1.0532	3.1876	43
18	.31399	.33072	3.0237	.94943	1.0533	3.1848	42
19	.31427	.33104	3.0208	.94933	1.0534	3.1820	41
20	.31454	.33136	3.0178	.94924	1.0535	3.1792	**40**
21	.31482	.33169	3.0149	.94915	1.0536	3.1764	39
22	.31510	.33201	3.0120	.94906	1.0537	3.1736	38
23	.31537	.33233	3.0090	.94897	1.0538	3.1708	37
24	.31565	.33266	3.0061	.94888	1.0539	3.1681	36
25	.31593	.33298	3.0032	.94878	1.0540	3.1653	**35**
26	.31620	.33330	3.0003	.94869	1.0541	3.1625	34
27	.31648	.33363	2.9974	.94860	1.0542	3.1598	33
28	.31675	.33395	2.9945	.94851	1.0543	3.1570	32
29	.31703	.33427	2.9916	.94842	1.0544	3.1543	31
30	.31730	.33460	2.9887	.94832	1.0545	3.1515	**30**
31	.31758	.33492	2.9858	.94823	1.0546	3.1488	29
32	.31786	.33524	2.9829	.94814	1.0547	3.1461	28
33	.31813	.33557	2.9800	.94805	1.0548	3.1433	27
34	.31841	.33589	2.9772	.94795	1.0549	3.1406	26
35	.31868	.33621	2.9743	.94786	1.0550	3.1379	**25**
36	.31896	.33654	2.9714	.94777	1.0551	3.1352	24
37	.31923	.33686	2.9686	.94768	1.0552	3.1325	23
38	.31951	.33718	2.9657	.94758	1.0553	3.1298	22
39	.31979	.33751	2.9629	.94749	1.0554	3.1271	21
40	.32006	.33783	2.9600	.94740	1.0555	3.1244	**20**
41	.32034	.33816	2.9572	.94730	1.0556	3.1217	19
42	.32061	.33848	2.9544	.94721	1.0557	3.1190	18
43	.32089	.33881	2.9515	.94712	1.0558	3.1163	17
44	.32116	.33913	2.9487	.94702	1.0559	3.1137	16
45	.32144	.33945	2.9459	.94693	1.0560	3.1110	**15**
46	.32171	.33978	2.9431	.94684	1.0561	3.1083	14
47	.32199	.34010	2.9403	.94674	1.0563	3.1057	13
48	.32227	.34043	2.9375	.94665	1.0564	3.1030	12
49	.32254	.34075	2.9347	.94656	1.0565	3.1004	11
50	.32282	.34108	2.9319	.94646	1.0566	3.0977	**10**
51	.32309	.34140	2.9291	.94637	1.0567	3.0951	9
52	.32337	.34173	2.9263	.94627	1.0568	3.0925	8
53	.32364	.34205	2.9235	.94618	1.0569	3.0898	7
54	.32392	.34238	2.9208	.94609	1.0570	3.0872	6
55	.32419	.34270	2.9180	.94599	1.0571	3.0846	**5**
56	.32447	.34303	2.9152	.94590	1.0572	3.0820	4
57	.32474	.34335	2.9125	.94580	1.0573	3.0794	3
58	.32502	.34368	2.9097	.94571	1.0574	3.0768	2
59	.32529	.34400	2.9070	.94561	1.0575	3.0742	1
60	.32557	.34433	2.9042	.94552	1.0576	3.0716	**0**
′	Cos	Cot	Tan	Sin	Csc	Sec	′

(288°) (251°) **71°**

19° (199°) (340°) **160°**

′	Sin	Tan	Cot	Cos	Sec	Csc	′
0	.32557	.34433	2.9042	.94552	1.0576	3.0716	**60**
1	.32584	.34465	2.9015	.94542	1.0577	3.0690	59
2	.32612	.34498	2.8987	.94533	1.0578	3.0664	58
3	.32639	.34530	2.8960	.94523	1.0579	3.0638	57
4	.32667	.34563	2.8933	.94514	1.0580	3.0612	56
5	.32694	.34596	2.8905	.94504	1.0582	3.0586	**55**
6	.32722	.34628	2.8878	.94495	1.0583	3.0561	54
7	.32749	.34661	2.8851	.94485	1.0584	3.0535	53
8	.32777	.34693	2.8824	.94476	1.0585	3.0509	52
9	.32804	.34726	2.8797	.94466	1.0586	3.0484	51
10	.32832	.34758	2.8770	.94457	1.0587	3.0458	**50**
11	.32859	.34791	2.8743	.94447	1.0588	3.0433	49
12	.32887	.34824	2.8716	.94438	1.0589	3.0407	48
13	.32914	.34856	2.8689	.94428	1.0590	3.0382	47
14	.32942	.34889	2.8662	.94418	1.0591	3.0357	46
15	.32969	.34922	2.8636	.94409	1.0592	3.0331	**45**
16	.32997	.34954	2.8609	.94399	1.0593	3.0306	44
17	.33024	.34987	2.8582	.94390	1.0594	3.0281	43
18	.33051	.35020	2.8556	.94380	1.0595	3.0256	42
19	.33079	.35052	2.8529	.94370	1.0597	3.0231	41
20	.33106	.35085	2.8502	.94361	1.0598	3.0206	**40**
21	.33134	.35118	2.8476	.94351	1.0599	3.0181	39
22	.33161	.35150	2.8449	.94342	1.0600	3.0156	38
23	.33189	.35183	2.8423	.94332	1.0601	3.0131	37
24	.33216	.35216	2.8397	.94322	1.0602	3.0106	36
25	.33244	.35248	2.8370	.94313	1.0603	3.0081	**35**
26	.33271	.35281	2.8344	.94303	1.0604	3.0056	34
27	.33298	.35314	2.8318	.94293	1.0605	3.0031	33
28	.33326	.35346	2.8291	.94284	1.0606	3.0007	32
29	.33353	.35379	2.8265	.94274	1.0607	2.9982	31
30	.33381	.35412	2.8239	.94264	1.0608	2.9957	**30**
31	.33408	.35445	2.8213	.94254	1.0610	2.9933	29
32	.33436	.35477	2.8187	.94245	1.0611	2.9908	28
33	.33463	.35510	2.8161	.94235	1.0612	2.9884	27
34	.33490	.35543	2.8135	.94225	1.0613	2.9859	26
35	.33518	.35576	2.8109	.94215	1.0614	2.9835	**25**
36	.33545	.35608	2.8083	.94206	1.0615	2.9811	24
37	.33573	.35641	2.8057	.94196	1.0616	2.9786	23
38	.33600	.35674	2.8032	.94186	1.0617	2.9762	22
39	.33627	.35707	2.8006	.94176	1.0618	2.9738	21
40	.33655	.35740	2.7980	.94167	1.0619	2.9713	**20**
41	.33682	.35772	2.7955	.94157	1.0621	2.9689	19
42	.33710	.35805	2.7929	.94147	1.0622	2.9665	18
43	.33737	.35838	2.7903	.94137	1.0623	2.9641	17
44	.33764	.35871	2.7878	.94127	1.0624	2.9617	16
45	.33792	.35904	2.7852	.94118	1.0625	2.9593	**15**
46	.33819	.35937	2.7827	.94108	1.0626	2.9569	14
47	.33846	.35969	2.7801	.94098	1.0627	2.9545	13
48	.33874	.36002	2.7776	.94088	1.0628	2.9521	12
49	.33901	.36035	2.7751	.94078	1.0629	2.9498	11
50	.33929	.36068	2.7725	.94068	1.0631	2.9474	**10**
51	.33956	.36101	2.7700	.94058	1.0632	2.9450	9
52	.33983	.36134	2.7675	.94049	1.0633	2.9426	8
53	.34011	.36167	2.7650	.94039	1.0634	2.9403	7
54	.34038	.36199	2.7625	.94029	1.0635	2.9379	6
55	.34065	.36232	2.7600	.94019	1.0636	2.9355	**5**
56	.34093	.36265	2.7575	.94009	1.0637	2.9332	4
57	.34120	.36298	2.7550	.93999	1.0638	2.9308	3
58	.34147	.36331	2.7525	.93989	1.0640	2.9285	2
59	.34175	.36364	2.7500	.93979	1.0641	2.9261	1
60	.34202	.36397	2.7475	.93969	1.0642	2.9238	**0**
′	Cos	Cot	Tan	Sin	Csc	Sec	′

109° (289°) (250°) **70°**

NATURAL TRIGONOMETRIC FUNCTIONS TO FIVE PLACES

20° (200°) (339°) **159°**

′	Sin	Tan	Cot	Cos	Sec	Csc	′
0	.34202	.36397	2.7475	.93969	1.0642	2.9238	**60**
1	.34229	.36430	2.7450	.93959	1.0643	2.9215	59
2	.34257	.36463	2.7425	.93949	1.0644	2.9191	58
3	.34284	.36496	2.7400	.93939	1.0645	2.9168	57
4	.34311	.36529	2.7376	.93929	1.0646	2.9145	56
5	.34339	.36562	2.7351	.93919	1.0647	2.9122	**55**
6	.34366	.36595	2.7326	.93909	1.0649	2.9099	54
7	.34393	.36628	2.7302	.93899	1.0650	2.9075	53
8	.34421	.36661	2.7277	.93889	1.0651	2.9052	52
9	.34448	.36694	2.7253	.93879	1.0652	2.9029	51
10	.34475	.36727	2.7228	.93869	1.0653	2.9006	**50**
11	.34503	.36760	2.7204	.93859	1.0654	2.8983	49
12	.34530	.36793	2.7179	.93849	1.0655	2.8960	48
13	.34557	.36826	2.7155	.93839	1.0657	2.8938	47
14	.34584	.36859	2.7130	.93829	1.0658	2.8915	46
15	.34612	.36892	2.7106	.93819	1.0659	2.8892	**45**
16	.34639	.36925	2.7082	.93809	1.0660	2.8869	44
17	.34666	.36958	2.7058	.93799	1.0661	2.8846	43
18	.34694	.36991	2.7034	.93789	1.0662	2.8824	42
19	.34721	.37024	2.7009	.93779	1.0663	2.8801	41
20	.34748	.37057	2.6985	.93769	1.0665	2.8779	**40**
21	.34775	.37090	2.6961	.93759	1.0666	2.8756	39
22	.34803	.37123	2.6937	.93748	1.0667	2.8733	38
23	.34830	.37157	2.6913	.93738	1.0668	2.8711	37
24	.34857	.37190	2.6889	.93728	1.0669	2.8688	36
25	.34884	.37223	2.6865	.93718	1.0670	2.8666	**35**
26	.34912	.37256	2.6841	.93708	1.0671	2.8644	34
27	.34939	.37289	2.6818	.93698	1.0673	2.8621	33
28	.34966	.37322	2.6794	.93688	1.0674	2.8599	32
29	.34993	.37355	2.6770	.93677	1.0675	2.8577	31
30	.35021	.37388	2.6746	.93667	1.0676	2.8555	**30**
31	.35048	.37422	2.6723	.93657	1.0677	2.8532	29
32	.35075	.37455	2.6699	.93647	1.0678	2.8510	28
33	.35102	.37488	2.6675	.93637	1.0680	2.8488	27
34	.35130	.37521	2.6652	.93626	1.0681	2.8466	26
35	.35157	.37554	2.6628	.93616	1.0682	2.8444	**25**
36	.35184	.37588	2.6605	.93606	1.0683	2.8422	24
37	.35211	.37621	2.6581	.93596	1.0684	2.8400	23
38	.35239	.37654	2.6558	.93585	1.0685	2.8378	22
39	.35266	.37687	2.6534	.93575	1.0687	2.8356	21
40	.35293	.37720	2.6511	.93565	1.0688	2.8334	**20**
41	.35320	.37754	2.6488	.93555	1.0689	2.8312	19
42	.35347	.37787	2.6464	.93544	1.0690	2.8291	18
43	.35375	.37820	2.6441	.93534	1.0691	2.8269	17
44	.35402	.37853	2.6418	.93524	1.0692	2.8247	16
45	.35429	.37887	2.6395	.93514	1.0694	2.8225	**15**
46	.35456	.37920	2.6371	.93503	1.0695	2.8204	14
47	.35484	.37953	2.6348	.93493	1.0696	2.8182	13
48	.35511	.37986	2.6325	.93483	1.0697	2.8161	12
49	.35538	.38020	2.6302	.93472	1.0698	2.8139	11
50	.35565	.38053	2.6279	.93462	1.0700	2.8117	**10**
51	.35592	.38086	2.6256	.93452	1.0701	2.8096	9
52	.35619	.38120	2.6233	.93441	1.0702	2.8075	8
53	.35647	.38153	2.6210	.93431	1.0703	2.8053	7
54	.35674	.38186	2.6187	.93420	1.0704	2.8032	6
55	.35701	.38220	2.6165	.93410	1.0705	2.8010	**5**
56	.35728	.38253	2.6142	.93400	1.0707	2.7989	4
57	.35755	.38286	2.6119	.93389	1.0708	2.7968	3
58	.35782	.38320	2.6096	.93379	1.0709	2.7947	2
59	.35810	.38353	2.6074	.93368	1.0710	2.7925	1
60	.35837	.38386	2.6051	.93358	1.0711	2.7904	**0**
′	Cos	Cot	Tan	Sin	Csc	Sec	′

110° (290°) (249°) **69°**

21° (201°) (338°) **158°**

′	Sin	Tan	Cot	Cos	Sec	Csc	′
0	.35837	.38386	2.6051	.93358	1.0711	2.7904	**60**
1	.35864	.38420	2.6028	.93348	1.0713	2.7883	59
2	.35891	.38453	2.6006	.93337	1.0714	2.7862	58
3	.35918	.38487	2.5983	.93327	1.0715	2.7841	57
4	.35945	.38520	2.5961	.93316	1.0716	2.7820	56
5	.35973	.38553	2.5938	.93306	1.0717	2.7799	**55**
6	.36000	.38587	2.5916	.93295	1.0719	2.7778	54
7	.36027	.38620	2.5893	.93285	1.0720	2.7757	53
8	.36054	.38654	2.5871	.93274	1.0721	2.7736	52
9	.36081	.38687	2.5848	.93264	1.0722	2.7715	51
10	.36108	.38721	2.5826	.93253	1.0723	2.7695	**50**
11	.36135	.38754	2.5804	.93243	1.0725	2.7674	49
12	.36162	.38787	2.5782	.93232	1.0726	2.7653	48
13	.36190	.38821	2.5759	.93222	1.0727	2.7632	47
14	.36217	.38854	2.5737	.93211	1.0728	2.7612	46
15	.36244	.38888	2.5715	.93201	1.0730	2.7591	**45**
16	.36271	.38921	2.5693	.93190	1.0731	2.7570	44
17	.36298	.38955	2.5671	.93180	1.0732	2.7550	43
18	.36325	.38988	2.5649	.93169	1.0733	2.7529	42
19	.36352	.39022	2.5627	.93159	1.0734	2.7509	41
20	.36379	.39055	2.5605	.93148	1.0736	2.7488	**40**
21	.36406	.39089	2.5583	.93137	1.0737	2.7468	39
22	.36434	.39122	2.5561	.93127	1.0738	2.7447	38
23	.36461	.39156	2.5539	.93116	1.0739	2.7427	37
24	.36488	.39190	2.5517	.93106	1.0740	2.7407	36
25	.36515	.39223	2.5495	.93095	1.0742	2.7386	**35**
26	.36542	.39257	2.5473	.93084	1.0743	2.7366	34
27	.36569	.39290	2.5452	.93074	1.0744	2.7346	33
28	.36596	.39324	2.5430	.93063	1.0745	2.7325	32
29	.36623	.39357	2.5408	.93052	1.0747	2.7305	31
30	.36650	.39391	2.5386	.93042	1.0748	2.7285	**30**
31	.36677	.39425	2.5365	.93031	1.0749	2.7265	29
32	.36704	.39458	2.5343	.93020	1.0750	2.7245	28
33	.36731	.39492	2.5322	.93010	1.0752	2.7225	27
34	.36758	.39526	2.5300	.92999	1.0753	2.7205	26
35	.36785	.39559	2.5279	.92988	1.0754	2.7185	**25**
36	.36812	.39593	2.5257	.92978	1.0755	2.7165	24
37	.36839	.39626	2.5236	.92967	1.0757	2.7145	23
38	.36867	.39660	2.5214	.92956	1.0758	2.7125	22
39	.36894	.39694	2.5193	.92945	1.0759	2.7105	21
40	.36921	.39727	2.5172	.92935	1.0760	2.7085	**20**
41	.36948	.39761	2.5150	.92924	1.0761	2.7065	19
42	.36975	.39795	2.5129	.92913	1.0763	2.7046	18
43	.37002	.39829	2.5108	.92902	1.0764	2.7026	17
44	.37029	.39862	2.5086	.92892	1.0765	2.7006	16
45	.37056	.39896	2.5065	.92881	1.0766	2.6986	**15**
46	.37083	.39930	2.5044	.92870	1.0768	2.6967	14
47	.37110	.39963	2.5023	.92859	1.0769	2.6947	13
48	.37137	.39997	2.5002	.92849	1.0770	2.6927	12
49	.37164	.40031	2.4981	.92838	1.0771	2.6908	11
50	.37191	.40065	2.4960	.92827	1.0773	2.6888	**10**
51	.37218	.40098	2.4939	.92816	1.0774	2.6869	9
52	.37245	.40132	2.4918	.92805	1.0775	2.6849	8
53	.37272	.40166	2.4897	.92794	1.0777	2.6830	7
54	.37299	.40200	2.4876	.92784	1.0778	2.6811	6
55	.37326	.40234	2.4855	.92773	1.0779	2.6791	**5**
56	.37353	.40267	2.4834	.92762	1.0780	2.6772	4
57	.37380	.40301	2.4813	.92751	1.0782	2.6752	3
58	.37407	.40335	2.4792	.92740	1.0783	2.6733	2
59	.37434	.40369	2.4772	.92729	1.0784	2.6714	1
60	.37461	.40403	2.4751	.92718	1.0785	2.6695	**0**
′	Cos	Cot	Tan	Sin	Csc	Sec	′

111° (291°) (248°) **68°**

NATURAL TRIGONOMETRIC FUNCTIONS TO FIVE PLACES

22° (202°) (337°) **157°**

′	Sin	Tan	Cot	Cos	Sec	Csc	′
0	.37461	.40403	2.4751	.92718	1.0785	2.6695	**60**
1	.37488	.40436	2.4730	.92707	1.0787	2.6675	59
2	.37515	.40470	2.4709	.92697	1.0788	2.6656	58
3	.37542	.40504	2.4689	.92686	1.0789	2.6637	57
4	.37569	.40538	2.4668	.92675	1.0790	2.6618	56
5	.37595	.40572	2.4648	.92664	1.0792	2.6599	**55**
6	.37622	.40606	2.4627	.92653	1.0793	2.6580	54
7	.37649	.40640	2.4606	.92642	1.0794	2.6561	53
8	.37676	.40674	2.4586	.92631	1.0796	2.6542	52
9	.37703	.40707	2.4566	.92620	1.0797	2.6523	51
10	.37730	.40741	2.4545	.92609	1.0798	2.6504	**50**
11	.37757	.40775	2.4525	.92598	1.0799	2.6485	49
12	.37784	.40809	2.4504	.92587	1.0801	2.6466	48
13	.37811	.40843	2.4484	.92576	1.0802	2.6447	47
14	.37838	.40877	2.4464	.92565	1.0803	2.6429	46
15	.37865	.40911	2.4443	.92554	1.0804	2.6410	**45**
16	.37892	.40945	2.4423	.92543	1.0806	2.6391	44
17	.37919	.40979	2.4403	.92532	1.0807	2.6372	43
18	.37946	.41013	2.4383	.92521	1.0808	2.6354	42
19	.37973	.41047	2.4362	.92510	1.0810	2.6335	41
20	.37999	.41081	2.4342	.92499	1.0811	2.6316	**40**
21	.38026	.41115	2.4322	.92488	1.0812	2.6298	39
22	.38053	.41149	2.4302	.92477	1.0814	2.6279	38
23	.38080	.41183	2.4282	.92466	1.0815	2.6260	37
24	.38107	.41217	2.4262	.92455	1.0816	2.6242	36
25	.38134	.41251	2.4242	.92444	1.0817	2.6223	**35**
26	.38161	.41285	2.4222	.92432	1.0819	2.6205	34
27	.38188	.41319	2.4202	.92421	1.0820	2.6186	33
28	.38215	.41353	2.4182	.92410	1.0821	2.6168	32
29	.38241	.41387	2.4162	.92399	1.0823	2.6150	31
30	.38268	.41421	2.4142	.92388	1.0824	2.6131	**30**
31	.38295	.41455	2.4122	.92377	1.0825	2.6113	29
32	.38322	.41490	2.4102	.92366	1.0827	2.6095	28
33	.38349	.41524	2.4083	.92355	1.0828	2.6076	27
34	.38376	.41558	2.4063	.92343	1.0829	2.6058	26
35	.38403	.41592	2.4043	.92332	1.0830	2.6040	**25**
36	.38430	.41626	2.4023	.92321	1.0832	2.6022	24
37	.38456	.41660	2.4004	.92310	1.0833	2.6003	23
38	.38483	.41694	2.3984	.92299	1.0834	2.5985	22
39	.38510	.41728	2.3964	.92287	1.0836	2.5967	21
40	.38537	.41763	2.3945	.92276	1.0837	2.5949	**20**
41	.38564	.41797	2.3925	.92265	1.0838	2.5931	19
42	.38591	.41831	2.3906	.92254	1.0840	2.5913	18
43	.38617	.41865	2.3886	.92243	1.0841	2.5895	17
44	.38644	.41899	2.3867	.92231	1.0842	2.5877	16
45	.38671	.41933	2.3847	.92220	1.0844	2.5859	**15**
46	.38698	.41968	2.3828	.92209	1.0845	2.5841	14
47	.38725	.42002	2.3808	.92198	1.0846	2.5823	13
48	.38752	.42036	2.3789	.92186	1.0848	2.5805	12
49	.38778	.42070	2.3770	.92175	1.0849	2.5788	11
50	.38805	.42105	2.3750	.92164	1.0850	2.5770	**10**
51	.38832	.42139	2.3731	.92152	1.0852	2.5752	9
52	.38859	.42173	2.3712	.92141	1.0853	2.5734	8
53	.38886	.42207	2.3693	.92130	1.0854	2.5716	7
54	.38912	.42242	2.3673	.92119	1.0856	2.5699	6
55	.38939	.42276	2.3654	92107	1.0857	2.5681	**5**
56	.38966	.42310	2.3635	92096	1.0858	2.5663	4
57	.38993	.42345	2.3616	.92085	1.0860	2.5646	3
58	.39020	.42379	2.3597	.92073	1.0861	2.5628	2
59	.39046	.42413	2.3578	.92062	1.0862	2.5611	1
60	.39073	.42447	2.3559	.92050	1.0864	2.5593	**0**
′	Cos	Cot	Tan	Sin	Csc	Sec	′

112° **(292°)** (247°) **67°**

23° (203°) (336°) **156°**

′	Sin	Tan	Cot	Cos	Sec	Csc	′
0	.39073	.42447	2.3559	.92050	1.0864	2.5593	**60**
1	.39100	.42482	2.3539	.92039	1.0865	2.5576	59
2	.39127	.42516	2.3520	.92028	1.0866	2.5558	58
3	.39153	.42551	2.3501	.92016	1.0868	2.5541	57
4	.39180	.42585	2 3483	92005	1.0869	2.5523	56
5	.39207	.42619	2.3464	.91994	1.0870	2.5506	**55**
6	.39234	.42654	2.3445	.91982	1.0872	2.5488	54
7	.39260	.42688	2.3426	.91971	1.0873	2.5471	53
8	.39287	.42722	2.3407	.91959	1.0874	2.5454	52
9	.39314	.42757	2.3388	.91948	1.0876	2.5436	51
10	.39341	.42791	2.3369	.91936	1.0877	2.5419	**50**
11	.39367	.42826	2.3351	.91925	1.0878	2.5402	49
12	.39394	.42860	2.3332	.91914	1.0880	2.5384	48
13	.39421	.42894	2.3313	.91902	1.0881	2.5367	47
14	.39448	.42929	2.3294	.91891	1.0883	2.5350	46
15	.39474	.42963	2.3276	.91879	1.0884	2.5333	**45**
16	.39501	.42998	2.3257	.91868	1.0885	2.5316	44
17	.39528	.43032	2.3238	.91856	1.0887	2.5299	43
18	.39555	.43067	2.3220	.91845	1.0888	2.5282	42
19	.39581	.43101	2.3201	.91833	1.0889	2.5264	41
20	.39608	.43136	2.3183	.91822	1.0891	2.5247	**40**
21	.39635	.43170	2.3164	.91810	1.0892	2.5230	39
22	.39661	.43205	2.3146	.91799	1.0893	2.5213	38
23	.39688	.43239	2.3127	.91787	1.0895	2.5196	37
24	.39715	.43274	2.3109	.91775	1.0896	2.5180	36
25	.39741	.43308	2.3090	.91764	1.0898	2.5163	**35**
26	.39768	.43343	2.3072	.91752	1.0899	2.5146	34
27	.39795	.43378	2.3053	.91741	1.0900	2.5129	33
28	.39822	.43412	2.3035	.91729	1.0902	2.5112	32
29	.39848	.43447	2.3017	.91718	1.0903	2.5095	31
30	.39875	.43481	2.2998	.91706	1.0904	2.5078	**30**
31	.39902	.43516	2.2980	.91694	1.0906	2.5062	29
32	.39928	.43550	2.2962	.91683	1.0907	2.5045	28
33	.39955	.43585	2.2944	.91671	1.0909	2.5028	27
34	.39982	.43620	2.2925	.91660	1.0910	2.5012	26
35	.40008	.43654	2.2907	.91648	1.0911	2.4995	**25**
36	.40035	.43689	2.2889	.91636	1.0913	2.4978	24
37	.40062	.43724	2.2871	.91625	1.0914	2.4962	23
38	.40088	.43758	2.2853	.91613	1.0915	2.4945	22
39	.40115	.43793	2.2835	.91601	1.0917	2.4928	21
40	.40141	.43828	2.2817	.91590	1.0918	2.4912	**20**
41	.40168	.43862	2.2799	.91578	1.0920	2.4895	19
42	.40195	.43897	2.2781	.91566	1.0921	2.4879	18
43	.40221	.43932	2.2763	.91555	1.0922	2.4862	17
44	.40248	.43966	2.2745	.91543	1.0924	2.4846	16
45	.40275	.44001	2.2727	.91531	1.0925	2.4830	**15**
46	.40301	.44036	2.2709	.91519	1.0927	2.4813	14
47	.40328	.44071	2.2691	.91508	1.0928	2.4797	13
48	.40355	.44105	2.2673	.91496	1.0929	2.4780	12
49	.40381	.44140	2.2655	.91484	1.0931	2.4764	11
50	.40408	.44175	2.2637	.91472	1.0932	2.4748	**10**
51	.40434	.44210	2.2620	.91461	1.0934	2.4731	9
52	.40461	.44244	2.2602	.91449	1.0935	2.4715	8
53	.40488	.44279	2.2584	.91437	1.0936	2.4699	7
54	.40514	.44314	2.2566	.91425	1.0938	2.4683	6
55	.40541	.44349	2.2549	.91414	1.0939	2.4667	**5**
56	.40567	.44384	2.2531	.91402	1.0941	2.4650	4
57	.40594	.44418	2.2513	.91390	1.0942	2.4634	3
58	.40621	.44453	2.2496	.91378	1.0944	2.4618	2
59	.40647	.44488	2.2478	.91366	1.0945	2.4602	1
60	.40674	.44523	2.2460	.91355	1.0946	2.4586	**0**
′	Cos	Cot	Tan	Sin	Csc	Sec	′

113° **(293°)** **(246°)** **66°**

NATURAL TRIGONOMETRIC FUNCTIONS TO FIVE PLACES

24° (204°) (335°) **155°**

′	Sin	Tan	Cot	Cos	Sec	Csc	′
0	.40674	.44523	2.2460	.91355	1.0946	2.4586	**60**
1	.40700	.44558	2.2443	.91343	1.0948	2.4570	59
2	.40727	.44593	2.2425	.91331	1.0949	2.4554	58
3	.40753	.44627	2.2408	.91319	1.0951	2.4538	57
4	.40780	.44662	2.2390	.91307	1.0952	2.4522	56
5	.40806	.44697	2.2373	.91295	1.0953	2.4506	**55**
6	.40833	.44732	2.2355	.91283	1.0955	2.4490	54
7	.40860	.44767	2.2338	.91272	1.0956	2.4474	53
8	.40886	.44802	2.2320	.91260	1.0958	2.4458	52
9	.40913	.44837	2.2303	.91248	1.0959	2.4442	51
10	.40939	.44872	2.2286	.91236	1.0961	2.4426	**50**
11	.40966	.44907	2.2268	.91224	1.0962	2.4411	49
12	.40992	.44942	2.2251	.91212	1.0963	2.4395	48
13	.41019	.44977	2.2234	.91200	1.0965	2.4379	47
14	.41045	.45012	2.2216	.91188	1.0966	2.4363	46
15	.41072	.45047	2.2199	.91176	1.0968	2.4348	**45**
16	.41098	.45082	2.2182	.91164	1.0969	2.4332	44
17	.41125	.45117	2.2165	.91152	1.0971	2.4316	43
18	.41151	.45152	2.2148	.91140	1.0972	2.4300	42
19	.41178	.45187	2.2130	.91128	1.0974	2.4285	41
20	.41204	.45222	2.2113	.91116	1.0975	2.4269	**40**
21	.41231	.45257	2.2096	.91104	1.0976	2.4254	39
22	.41257	.45292	2.2079	.91092	1.0978	2.4238	38
23	.41284	.45327	2.2062	.91080	1.0979	2.4222	37
24	.41310	.45362	2.2045	.91068	1.0981	2.4207	36
25	.41337	.45397	2.2028	.91056	1.0982	2.4191	**35**
26	.41363	.45432	2.2011	.91044	1.0984	2.4176	34
27	.41390	.45467	2.1994	.91032	1.0985	2.4160	33
28	.41416	.45502	2.1977	.91020	1.0987	2.4145	32
29	.41443	.45538	2.1960	.91008	1.0988	2.4130	31
30	.41469	.45573	2.1943	.90996	1.0989	2.4114	**30**
31	.41496	.45608	2.1926	.90984	1.0991	2.4099	29
32	.41522	.45643	2.1909	.90972	1.0992	2.4083	28
33	.41549	.45678	2.1892	.90960	1.0994	2.4068	27
34	.41575	.45713	2.1876	.90948	1.0995	2.4053	26
35	.41602	.45748	2.1859	.90936	1.0997	2.4038	**25**
36	.41628	.45784	2.1842	.90924	1.0998	2.4022	24
37	.41655	.45819	2.1825	.90911	1.1000	2.4007	23
38	.41681	.45854	2.1808	.90899	1.1001	2.3992	22
39	.41707	.45889	2.1792	.90887	1.1003	2.3977	21
40	.41734	.45924	2.1775	.90875	1.1004	2.3961	**20**
41	.41760	.45960	2.1758	.90863	1.1006	2.3946	19
42	.41787	.45995	2.1742	.90851	1.1007	2.3931	18
43	.41813	.46030	2.1725	.90839	1.1009	2.3916	17
44	.41840	.46065	2.1708	.90826	1.1010	2.3901	16
45	.41866	.46101	2.1692	.90814	1.1011	2.3886	**15**
46	.41892	.46136	2.1675	.90802	1.1013	2.3871	14
47	.41919	.46171	2.1659	.90790	1.1014	2.3856	13
48	.41945	.46206	2.1642	.90778	1.1016	2.3841	12
49	.41972	.46242	2.1625	.90766	1.1017	2.3826	11
50	.41998	.46277	2.1609	.90753	1.1019	2.3811	**10**
51	.42024	.46312	2.1592	.90741	1.1020	2.3796	9
52	.42051	.46348	2.1576	.90729	1.1022	2.3781	8
53	.42077	.46383	2.1560	.90717	1.1023	2.3766	7
54	.42104	.46418	2.1543	.90704	1.1025	2.3751	6
55	.42130	.46454	2.1527	.90692	1.1026	2.3736	**5**
56	.42156	.46489	2.1510	.90680	1.1028	2.3721	4
57	.42183	.46525	2.1494	.90668	1.1029	2.3706	3
58	.42209	.46560	2.1478	.90655	1.1031	2.3692	2
59	.42235	.46595	2.1461	.90643	1.1032	2.3677	1
60	.42262	.46631	2.1445	.90631	1.1034	2.3662	**0**
′	Cos	Cot	Tan	Sin	Csc	Sec	′

114° (294°) (245°) **65°**

25° (205°) (334°) **15**

′	Sin	Tan	Cot	Cos	Sec	Csc	′
0	.42262	.46631	2.1445	.90631	1.1034	2.3662	
1	.42288	.46666	2.1429	.90618	1.1035	2.3647	
2	.42315	.46702	2.1413	.90606	1.1037	2.3633	
3	.42341	.46737	2.1396	.90594	1.1038	2.3618	
4	.42367	.46772	2.1380	.90582	1.1040	2.3603	
5	.42394	.46808	2.1364	.90569	1.1041	2.3588	
6	.42420	.46843	2.1348	.90557	1.1043	2.3574	
7	.42446	.46879	2.1332	.90545	1.1044	2.3559	
8	.42473	.46914	2.1315	.90532	1.1046	2.3545	
9	.42499	.46950	2.1299	.90520	1.1047	2.3530	
10	.42525	.46985	2.1283	.90507	1.1049	2.3515	
11	.42552	.47021	2.1267	.90495	1.1050	2.3501	
12	.42578	.47056	2.1251	.90483	1.1052	2.3486	
13	.42604	.47092	2.1235	.90470	1.1053	2.3472	
14	.42631	.47128	2.1219	.90458	1.1055	2.3457	
15	.42657	.47163	2.1203	.90446	1.1056	2.3443	
16	.42683	.47199	2.1187	.90433	1.1058	2.3428	
17	.42709	.47234	2.1171	.90421	1.1059	2.3414	
18	.42736	.47270	2.1155	.90408	1.1061	2.3400	
19	.42762	.47305	2.1139	.90396	1.1062	2.3385	
20	.42788	.47341	2.1123	.90383	1.1064	2.3371	
21	.42815	.47377	2.1107	.90371	1.1066	2.3356	
22	.42841	.47412	2.1092	.90358	1.1067	2.3342	
23	.42867	.47448	2.1076	.90346	1.1069	2.3328	
24	.42894	.47483	2.1060	.90334	1.1070	2.3314	
25	.42920	.47519	2.1044	.90321	1.1072	2.3299	
26	.42946	.47555	2.1028	.90309	1.1073	2.3285	
27	.42972	.47590	2.1013	.90296	1.1075	2.3271	
28	.42999	.47626	2.0997	.90284	1.1076	2.3257	
29	.43025	.47662	2.0981	.90271	1.1078	2.3242	
30	.43051	.47698	2.0965	.90259	1.1079	2.3228	
31	.43077	.47733	2.0950	.90246	1.1081	2.3214	
32	.43104	.47769	2.0934	.90233	1.1082	2.3200	
33	.43130	.47805	2.0918	.90221	1.1084	2.3186	
34	.43156	.47840	2.0903	.90208	1.1085	2.3172	
35	.43182	.47876	2.0887	.90196	1.1087	2.3158	
36	.43209	.47912	2.0872	.90183	1.1089	2.3144	
37	.43235	.47948	2.0856	.90171	1.1090	2.3130	
38	.43261	.47984	2.0840	.90158	1.1092	2.3115	
39	.43287	.48019	2.0825	.90146	1.1093	2.3101	
40	.43313	.48055	2.0809	.90133	1.1095	2.3088	
41	.43340	.48091	2.0794	.90120	1.1096	2.3074	
42	.43366	.48127	2.0778	.90108	1.1098	2.3060	
43	.43392	.48163	2.0763	.90095	1.1099	2.3046	
44	.43418	.48198	2.0748	.90082	1.1101	2.3032	
45	.43445	.48234	2.0732	.90070	1.1102	2.3018	
46	.43471	.48270	2.0717	.90057	1.1104	2.3004	
47	.43497	.48306	2.0701	.90045	1.1106	2.2990	
48	.43523	.48342	2.0686	.90032	1.1107	2.2976	
49	.43549	.48378	2.0671	.90019	1.1109	2.2962	
50	.43575	.48414	2.0655	.90007	1.1110	2.2949	
51	.43602	.48450	2.0640	.89994	1.1112	2.2935	
52	.43628	.48486	2.0625	.89981	1.1113	2.2921	
53	.43654	.48521	2.0609	.89968	1.1115	2.2907	
54	.43680	.48557	2.0594	.89956	1.1117	2.2894	
55	.43706	.48593	2.0579	.89943	1.1118	2.2880	
56	.43733	.48629	2.0564	.89930	1.1120	2.2866	
57	.43759	.48665	2.0549	.89918	1.1121	2.2853	
58	.43785	.48701	2.0533	.89905	1.1123	2.2839	
59	.43811	.48737	2.0518	.89892	1.1124	2.2825	
60	.43837	.48773	2.0503	.89879	1.1126	2.2812	
′	Cos	Cot	Tan	Sin	Csc	Sce	′

115° (295°) (244°) **6**

NATURAL TRIGONOMETRIC FUNCTIONS TO FIVE PLACES

(206°) (333°) **153°**

′	Sin	Tan	Cot	Cos	Sec	Csc	′
0	.43837	.48773	2.0503	.89879	1.1126	2.2812	**60**
1	.43863	.48809	2.0488	.89867	1.1128	2.2798	59
2	.43889	.48845	2.0473	.89854	1.1129	2.2785	58
3	.43916	.48881	2.0458	.89841	1.1131	2.2771	57
4	.43942	.48917	2.0443	.89828	1.1132	2.2757	56
5	.43968	.48953	2.0428	.89816	1.1134	2.2744	**55**
6	.43994	.48989	2.0413	.89803	1.1136	2.2730	54
7	.44020	.49026	2.0398	.89790	1.1137	2.2717	53
8	.44046	.49062	2.0383	.89777	1.1139	2.2703	52
9	.44072	.49098	2.0368	.89764	1.1140	2.2690	51
10	.44098	.49134	2.0353	.89752	1.1142	2.2677	**50**
11	.44124	.49170	2.0338	.89739	1.1143	2.2663	49
12	.44151	.49206	2.0323	.89726	1.1145	2.2650	48
13	.44177	.49242	2.0308	.89713	1.1147	2.2636	47
14	.44203	.49278	2.0293	.89700	1.1148	2.2623	46
15	.44229	.49315	2.0278	.89687	1.1150	2.2610	**45**
16	.44255	.49351	2.0263	.89674	1.1151	2.2596	44
17	.44281	.49387	2.0248	.89662	1.1153	2.2583	43
18	.44307	.49423	2.0233	.89649	1.1155	2.2570	42
19	.44333	.49459	2.0219	.89636	1.1156	2.2556	41
20	.44359	.49495	2.0204	.89623	1.1158	2.2543	**40**
21	.44385	.49532	2.0189	.89610	1.1159	2.2530	39
22	.44411	.49568	2.0174	.89597	1.1161	2.2517	38
23	.44437	.49604	2.0160	.89584	1.1163	2.2504	37
24	.44464	.49640	2.0145	.89571	1.1164	2.2490	36
25	.44490	.49677	2.0130	.89558	1.1166	2.2477	**35**
26	.44516	.49713	2.0115	.89545	1.1168	2.2464	34
27	.44542	.49749	2.0101	.89532	1.1169	2.2451	33
28	.44568	.49786	2.0086	.89519	1.1171	2.2438	32
29	.44594	.49822	2.0072	.89506	1.1172	2.2425	31
30	.44620	.49858	2.0057	.89493	1.1174	2.2412	**30**
31	.44646	.49894	2.0042	.89480	1.1176	2.2399	29
32	.44672	.49931	2.0028	.89467	1.1177	2.2385	28
33	.44698	.49967	2.0013	.89454	1.1179	2.2372	27
34	.44724	.50004	1.9999	.89441	1.1180	2.2359	26
35	.44750	.50040	1.9984	.89428	1.1182	2.2346	**25**
36	.44776	.50076	1.9970	.89415	1.1184	2.2333	24
37	.44802	.50113	1.9955	.89402	1.1185	2.2320	23
38	.44828	.50149	1.9941	.89389	1.1187	2.2308	22
39	.44854	.50185	1.9926	.89376	1.1189	2.2295	21
40	.44880	.50222	1.9912	.89363	1.1190	2.2282	**20**
41	.44906	.50258	1.9897	.89350	1.1192	2.2269	19
42	.44932	.50295	1.9883	.89337	1.1194	2.2256	18
43	.44958	.50331	1.9868	.89324	1.1195	2.2243	17
44	.44984	.50368	1.9854	.89311	1.1197	2.2230	16
45	.45010	.50404	1.9840	.89298	1.1198	2.2217	**15**
46	.45036	.50441	1.9825	.89285	1.1200	2.2205	14
47	.45062	.50477	1.9811	.89272	1.1202	2.2192	13
48	.45088	.50514	1.9797	.89259	1.1203	2.2179	12
49	.45114	.50550	1.9782	.89245	1.1205	2.2166	11
50	.45140	.50587	1.9768	.89232	1.1207	2.2153	**10**
51	.45166	.50623	1.9754	.89219	1.1208	2.2141	9
52	.45192	.50660	1.9740	.89206	1.1210	2.2128	8
53	.45218	.50696	1.9725	.89193	1.1212	2.2115	7
54	.45243	.50733	1.9711	.89180	1.1213	2.2103	6
55	.45269	.50769	1.9697	.89167	1.1215	2.2090	**5**
56	.45295	.50806	1.9683	.89153	1.1217	2.2077	4
57	.45321	.50843	1.9669	.89140	1.1218	2.2065	3
58	.45347	.50879	1.9654	.89127	1.1220	2.2052	2
59	.45373	.50916	1.9640	.89114	1.1222	2.2039	1
60	.45399	.50953	1.9626	.89101	1.1223	2.2027	**0**
′	Cos	Cot	Tan	Sin	Csc	Sec	′

(296°) (243°) **63°**

27° (207°) (332°) **152°**

′	Sin	Tan	Cot	Cos	Sec	Csc	′
0	.45399	.50953	1.9626	.89101	1.1223	2.2027	**60**
1	.45425	.50989	1.9612	.89087	1.1225	2.2014	59
2	.45451	.51026	1.9598	.89074	1.1227	2.2002	58
3	.45477	.51063	1.9584	.89061	1.1228	2.1989	57
4	.45503	.51099	1.9570	.89048	1.1230	2.1977	56
5	.45529	.51136	1.9556	.89035	1.1232	2.1964	**55**
6	.45554	.51173	1.9542	.89021	1.1233	2.1952	54
7	.45580	.51209	1.9528	.89008	1.1235	2.1939	53
8	.45606	.51246	1.9514	.88995	1.1237	2.1927	52
9	.45632	.51283	1.9500	.88981	1.1238	2.1914	51
10	.45658	.51319	1.9486	.88968	1.1240	2.1902	**50**
11	.45684	.51356	1.9472	.88955	1.1242	2.1890	49
12	.45710	.51393	1.9458	.88942	1.1243	2.1877	48
13	.45736	.51430	1.9444	.88928	1.1245	2.1865	47
14	.45762	.51467	1.9430	.88915	1.1247	2.1852	46
15	.45787	.51503	1.9416	.88902	1.1248	2.1840	**45**
16	.45813	.51540	1.9402	.88888	1.1250	2.1828	44
17	.45839	.51577	1.9388	.88875	1.1252	2.1815	43
18	.45865	.51614	1.9375	.88862	1.1253	2.1803	42
19	.45891	.51651	1.9361	.88848	1.1255	2.1791	41
20	.45917	.51688	1.9347	.88835	1.1257	2.1779	**40**
21	.45942	.51724	1.9333	.88822	1.1259	2.1766	39
22	.45968	.51761	1.9319	.88808	1.1260	2.1754	38
23	.45994	.51798	1.9306	.88795	1.1262	2.1742	37
24	.46020	.51835	1.9292	.88782	1.1264	2.1730	36
25	.46046	.51872	1.9278	.88768	1.1265	2.1718	**35**
26	.46072	.51909	1.9265	.88755	1.1267	2.1705	34
27	.46097	.51946	1.9251	.88741	1.1269	2.1693	33
28	.46123	.51983	1.9237	.88728	1.1270	2.1681	32
29	.46149	.52020	1.9223	.88715	1.1272	2.1669	31
30	.46175	.52057	1.9210	.88701	1.1274	2.1657	**30**
31	.46201	.52094	1.9196	.88688	1.1276	2.1645	29
32	.46226	.52131	1.9183	.88674	1.1277	2.1633	28
33	.46252	.52168	1.9169	.88661	1.1279	2.1621	27
34	.46278	.52205	1.9155	.88647	1.1281	2.1609	26
35	.46304	.52242	1.9142	.88634	1.1282	2.1596	**25**
36	.46330	.52279	1.9128	.88620	1.1284	2.1584	24
37	.46355	.52316	1.9115	.88607	1.1286	2.1572	23
38	.46381	.52353	1.9101	.88593	1.1288	2.1560	22
39	.46407	.52390	1.9088	.88580	1.1289	2.1549	21
40	.46433	.52427	1.9074	.88566	1.1291	2.1537	**20**
41	.46458	.52464	1.9061	.88553	1.1293	2.1525	19
42	.46484	.52501	1.9047	.88539	1.1294	2.1513	18
43	.46510	.52538	1.9034	.88526	1.1296	2.1501	17
44	.46536	.52575	1.9020	.88512	1.1298	2.1489	16
45	.46561	.52613	1.9007	.88499	1.1300	2.1477	**15**
46	.46587	.52650	1.8993	.88485	1.1301	2.1465	14
47	.46613	.52687	1.8980	.88472	1.1303	2.1453	13
48	.46639	.52724	1.8967	.88458	1.1305	2.1441	12
49	.46664	.52761	1.8953	.88445	1.1307	2.1430	11
50	.46690	.52798	1.8940	.88431	1.1308	2.1418	**10**
51	.46716	.52836	1.8927	.88417	1.1310	2.1406	9
52	.46742	.52873	1.8913	.88404	1.1312	2.1394	8
53	.46767	.52910	1.8900	.88390	1.1313	2.1382	7
54	.46793	.52947	1.8887	.88377	1.1315	2.1371	6
55	.46819	.52985	1.8873	.88363	1.1317	2.1359	**5**
56	.46844	.53022	1.8860	.88349	1.1319	2.1347	4
57	.46870	.53059	1.8847	.88336	1.1320	2.1336	3
58	.46896	.53096	1.8834	.88322	1.1322	2.1324	2
59	.46921	.53134	1.8820	.88308	1.1324	2.1312	1
60	.46947	.53171	1.8807	.88295	1.1326	2.1301	**0**
′	Cos	Cot	Tan	Sin	Csc	Sec	′

117° (297°) (242°) **62°**

NATURAL TRIGONOMETRIC FUNCTIONS TO FIVE PLACES

28° (208°) (331°) **151°**

′	Sin	Tan	Cot	Cos	Sec	Csc	′
0	.46947	.53171	1.8807	.88295	1.1326	2.1301	**60**
1	.46973	.53208	1.8794	.88281	1.1327	2.1289	59
2	.46999	.53246	1.8781	.88267	1.1329	2.1277	58
3	.47024	.53283	1.8768	.88254	1.1331	2.1266	57
4	.47050	.53320	1.8755	.88240	1.1333	2.1254	56
5	.47076	.53358	1.8741	.88226	1.1334	2.1242	**55**
6	.47101	.53395	1.8728	.88213	1.1336	2.1231	54
7	.47127	.53432	1.8715	.88199	1.1338	2.1219	53
8	.47153	.53470	1.8702	.88185	1.1340	2.1208	52
9	.47178	.53507	1.8689	.88172	1.1342	2.1196	51
10	.47204	.53545	1.8676	.88158	1.1343	2.1185	**50**
11	.47229	.53582	1.8663	.88144	1.1345	2.1173	49
12	.47255	.53620	1.8650	.88130	1.1347	2.1162	48
13	.47281	.53657	1.8637	.88117	1.1349	2.1150	47
14	.47306	.53694	1.8624	.88103	1.1350	2.1139	46
15	.47332	.53732	1.8611	.88089	1.1352	2.1127	**45**
16	.47358	.53769	1.8598	.88075	1.1354	2.1116	44
17	.47383	.53807	1.8585	.88062	1.1356	2.1105	43
18	.47409	.53844	1.8572	.88048	1.1357	2.1093	42
19	.47434	.53882	1.8559	.88034	1.1359	2.1082	41
20	.47460	.53920	1.8546	.88020	1.1361	2.1070	**40**
21	.47486	.53957	1.8533	.88006	1.1363	2.1059	39
22	.47511	.53995	1.8520	.87993	1.1365	2.1048	38
23	.47537	.54032	1.8507	.87979	1.1366	2.1036	37
24	.47562	.54070	1.8495	.87965	1.1368	2.1025	36
25	.47588	.54107	1.8482	.87951	1.1370	2.1014	**35**
26	.47614	.54145	1.8469	.87937	1.1372	2.1002	34
27	.47639	.54183	1.8456	.87923	1.1374	2.0991	33
28	.47665	.54220	1.8443	.87909	1.1375	2.0980	32
29	.47690	.54258	1.8430	.87896	1.1377	2.0969	31
30	.47716	.54296	1.8418	.87882	1.1379	2.0957	**30**
31	.47741	.54333	1.8405	.87868	1.1381	2.0946	29
32	.47767	.54371	1.8392	.87854	1.1383	2.0935	28
33	.47793	.54409	1.8379	.87840	1.1384	2.0924	27
34	.47818	.54446	1.8367	.87826	1.1386	2.0913	26
35	.47844	.54484	1.8354	.87812	1.1388	2.0901	**25**
36	.47869	.54522	1.8341	.87798	1.1390	2.0890	24
37	.47895	.54560	1.8329	.87784	1.1392	2.0879	23
38	.47920	.54597	1.8316	.87770	1.1393	2.0868	22
39	.47946	.54635	1.8303	.87756	1.1395	2.0857	21
40	.47971	.54673	1.8291	.87743	1.1397	2.0846	**20**
41	.47997	.54711	1.8278	.87729	1.1399	2.0835	19
42	.48022	.54748	1.8265	.87715	1.1401	2.0824	18
43	.48048	.54786	1.8253	.87701	1.1402	2.0813	17
44	.48073	.54824	1.8240	.87687	1.1404	2.0802	16
45	.48099	.54862	1.8228	.87673	1.1406	2.0791	**15**
46	.48124	.54900	1.8215	.87659	1.1408	2.0779	14
47	.48150	.54938	1.8202	.87645	1.1410	2.0768	13
48	.48175	.54975	1.8190	.87631	1.1412	2.0757	12
49	.48201	.55013	1.8177	.87617	1.1413	2.0747	11
50	.48226	.55051	1.8165	.87603	1.1415	2.0736	**10**
51	.48252	.55089	1.8152	.87589	1.1417	2.0725	9
52	.48277	.55127	1.8140	.87575	1.1419	2.0714	8
53	.48303	.55165	1.8127	.87561	1.1421	2.0703	7
54	.48328	.55203	1.8115	.87546	1.1423	2.0692	6
55	.48354	.55241	1.8103	.87532	1.1424	2.0681	**5**
56	.48379	.55279	1.8090	.87518	1.1426	2.0670	4
57	.48405	.55317	1.8078	.87504	1.1428	2.0659	3
58	.48430	.55355	1.8065	.87490	1.1430	2.0648	2
59	.48456	.55393	1.8053	.87476	1.1432	2.0637	1
60	.48481	.55431	1.8040	.87462	1.1434	2.0627	**0**
′	Cos	Cot	Tan	Sin	Csc	Sec	′

118° (298°) (241°) **61°**

29° (209°) (330)° **150°**

′	Sin	Tan	Cot	Cos	Sec	Csc	′
0	.48481	.55431	1.8040	.87462	1.1434	2.0627	**60**
1	.48506	.55469	1.8028	.87448	1.1435	2.0616	59
2	.48532	.55507	1.8016	.87434	1.1437	2.0605	58
3	.48557	.55545	1.8003	.87420	1.1439	2.0594	57
4	.48583	.55583	1.7991	.87406	1.1441	2.0583	56
5	.48608	.55621	1.7979	.87391	1.1443	2.0573	**55**
6	.48634	.55659	1.7966	.87377	1.1445	2.0562	54
7	.48659	.55697	1.7954	.87363	1.1446	2.0551	53
8	.48684	.55736	1.7942	.87349	1.1448	2.0540	52
9	.48710	.55774	1.7930	.87335	1.1450	2.0530	51
10	.48735	.55812	1.7917	.87321	1.1452	2.0519	**50**
11	.48761	.55850	1.7905	.87306	1.1454	2.0508	49
12	.48786	.55888	1.7893	.87292	1.1456	2.0498	48
13	.48811	.55926	1.7881	.87278	1.1458	2.0487	47
14	.48837	.55964	1.7868	.87264	1.1460	2.0476	46
15	.48862	.56003	1.7856	.87250	1.1461	2.0466	**45**
16	.48888	.56041	1.7844	.87235	1.1463	2.0455	44
17	.48913	.56079	1.7832	.87221	1.1465	2.0445	43
18	.48938	.56117	1.7820	.87207	1.1467	2.0434	42
19	.48964	.56156	1.7808	.87193	1.1469	2.0423	41
20	.48989	.56194	1.7796	.87178	1.1471	2.0413	**40**
21	.49014	.56232	1.7783	.87164	1.1473	2.0402	39
22	.49040	.56270	1.7771	.87150	1.1474	2.0392	38
23	.49065	.56309	1.7759	.87136	1.1476	2.0381	37
24	.49090	.56347	1.7747	.87121	1.1478	2.0371	36
25	.49116	.56385	1.7735	.87107	1.1480	2.0360	**35**
26	.49141	.56424	1.7723	.87093	1.1482	2.0350	34
27	.49166	.56462	1.7711	.87079	1.1484	2.0339	33
28	.49192	.56501	1.7699	.87064	1.1486	2.0329	32
29	.49217	.56539	1.7687	.87050	1.1488	2.0318	31
30	.49242	.56577	1.7675	.87036	1.1490	2.0308	**30**
31	.49268	.56616	1.7663	.87021	1.1491	2.0297	29
32	.49293	.56654	1.7651	.87007	1.1493	2.0287	28
33	.49318	.56693	1.7639	.86993	1.1495	2.0276	27
34	.49344	.56731	1.7627	.86978	1.1497	2.0266	26
35	.49369	.56769	1.7615	.86964	1.1499	2.0256	**25**
36	.49394	.56808	1.7603	.86949	1.1501	2.0245	24
37	.49419	.56846	1.7591	.86935	1.1503	2.0235	23
38	.49445	.56885	1.7579	.86921	1.1505	2.0225	22
39	.49470	.56923	1.7567	.86906	1.1507	2.0214	21
40	.49495	.56962	1.7556	.86892	1.1509	2.0204	**20**
41	.49521	.57000	1.7544	.86878	1.1510	2.0194	19
42	.49546	.57039	1.7532	.86863	1.1512	2.0183	18
43	.49571	.57078	1.7520	.86849	1.1514	2.0173	17
44	.49596	.57116	1.7508	.86834	1.1516	2.0163	16
45	.49622	.57155	1.7496	.86820	1.1518	2.0152	**15**
46	.49647	.57193	1.7485	.86805	1.1520	2.0142	14
47	.49672	.57232	1.7473	.86791	1.1522	2.0132	13
48	.49697	.57271	1.7461	.86777	1.1524	2.0122	12
49	.49723	.57309	1.7449	.86762	1.1526	2.0112	11
50	.49748	.57348	1.7437	.86748	1.1528	2.0101	**10**
51	.49773	.57386	1.7426	.86733	1.1530	2.0091	9
52	.49798	.57425	1.7414	.86719	1.1532	2.0081	8
53	.49824	.57464	1.7402	.86704	1.1533	2.0071	7
54	.49849	.57503	1.7391	.86690	1.1535	2.0061	6
55	.49874	.57541	1.7379	.86675	1.1537	2.0051	**5**
56	.49899	.57580	1.7367	.86661	1.1539	2.0040	4
57	.49924	.57619	1.7355	.86646	1.1541	2.0030	3
58	.49950	.57657	1.7344	.86632	1.1543	2.0020	2
59	.49975	.57696	1.7332	.86617	1.1545	2.0010	1
60	.50000	.57735	1.7321	.86603	1.1547	2.0000	**0**
′	Cos	Cot	Tan	Sin	Csc	Sec	′

119° (299°) (240°) **60°**

NATURAL TRIGONOMETRIC FUNCTIONS TO FIVE PLACES

30° (210°) (329°) **149°**

′	Sin	Tan	Cot	Cos	Sec	Csc	′
0	.50000	.57735	1.7321	.86603	1.1547	2.0000	**60**
1	.50025	.57774	1.7309	.86588	1.1549	1.9990	59
2	.50050	.57813	1.7297	.86573	1.1551	1.9980	58
3	.50076	.57851	1.7286	.86559	1.1553	1.9970	57
4	.50101	.57890	1.7274	.86544	1.1555	1.9960	56
5	.50126	.57929	1.7262	.86530	1.1557	1.9950	**55**
6	.50151	.57968	1.7251	.86515	1.1559	1.9940	54
7	.50176	.58007	1.7239	.86501	1.1561	1.9930	53
8	.50201	.58046	1.7228	.86486	1.1563	1.9920	52
9	.50227	.58085	1.7216	.86471	1.1565	1.9910	51
10	.50252	.58124	1.7205	.86457	1.1566	1.9900	**50**
11	.50277	.58162	1.7193	.86442	1.1568	1.9890	49
12	.50302	.58201	1.7182	.86427	1.1570	1.9880	48
13	.50327	.58240	1.7170	.86413	1.1572	1.9870	47
14	.50352	.58279	1.7159	.86398	1.1574	1.9860	46
15	.50377	.58318	1.7147	.86384	1.1576	1.9850	**45**
16	.50403	.58357	1.7136	.86369	1.1578	1.9840	44
17	.50428	.58396	1.7124	.86354	1.1580	1.9830	43
18	.50453	.58435	1.7113	.86340	1.1582	1.9821	42
19	.50478	.58474	1.7102	.86325	1.1584	1.9811	41
20	.50503	.58513	1.7090	.86310	1.1586	1.9801	**40**
21	.50528	.58552	1.7079	.86295	1.1588	1.9791	39
22	.50553	.58591	1.7067	.86281	1.1590	1.9781	38
23	.50578	.58631	1.7056	.86266	1.1592	1.9771	37
24	.50603	.58670	1.7045	.86251	1.1594	1.9762	36
25	.50628	.58709	1.7033	.86237	1.1596	1.9752	**35**
26	.50654	.58748	1.7022	.86222	1.1598	1.9742	34
27	.50679	.58787	1.7011	.86207	1.1600	1.9732	33
28	.50704	.58826	1.6999	.86192	1.1602	1.9722	32
29	.50729	.58865	1.6988	.86178	1.1604	1.9713	31
30	.50754	.58905	1.6977	.86163	1.1606	1.9703	**30**
31	.50779	.58944	1.6965	.86148	1.1608	1.9693	29
32	.50804	.58983	1.6954	.86133	1.1610	1.9684	28
33	.50829	.59022	1.6943	.86119	1.1612	1.9674	27
34	.50854	.59061	1.6932	.86104	1.1614	1.9664	26
35	.50879	.59101	1.6920	.86089	1.1616	1.9654	**25**
36	.50904	.59140	1.6909	.86074	1.1618	1.9645	24
37	.50929	.59179	1.6898	.86059	1.1620	1.9635	23
38	.50954	.59218	1.6887	.86045	1.1622	1.9625	22
39	.50979	.59258	1.6875	.86030	1.1624	1.9616	21
40	.51004	.59297	1.6864	.86015	1.1626	1.9606	**20**
41	.51029	.59336	1.6853	.86000	1.1628	1.9597	19
42	.51054	.59376	1.6842	.85985	1.1630	1.9587	18
43	.51079	.59415	1.6831	.85970	1.1632	1.9577	17
44	.51104	.59454	1.6820	.85956	1.1634	1.9568	16
45	.51129	.59494	1.6808	.85941	1.1636	1.9558	**15**
46	.51154	.59533	1.6797	.85926	1.1638	1.9549	14
47	.51179	.59573	1.6786	.85911	1.1640	1.9539	13
48	.51204	.59612	1.6775	.85896	1.1642	1.9530	12
49	.51229	.59651	1.6764	.85881	1.1644	1.9520	11
50	.51254	.59691	1.6753	.85866	1.1646	1.9511	**10**
51	.51279	.59730	1.6742	.85851	1.1648	1.9501	9
52	.51304	.59770	1.6731	.85836	1.1650	1.9492	8
53	.51329	.59809	1.6720	.85821	1.1652	1.9482	7
54	.51354	.59849	1.6709	.85806	1.1654	1.9473	6
55	.51379	.59888	1.6698	.85792	1.1656	1.9463	**5**
56	.51404	.59928	1.6687	.85777	1.1658	1.9454	4
57	.51429	.59967	1.6676	.85762	1.1660	1.9444	3
58	.51454	.60007	1.6665	.85747	1.1662	1.9435	2
59	.51479	.60046	1.6654	.85732	1.1664	1.9425	1
60	.51504	.60086	1.6643	.85717	1.1666	1.9416	**0**
′	Cos	Cot	Tan	Sin	Csc	Sec	′

120° (300°) (239°) **59°**

31° (211°) (328°) **148°**

′	Sin	Tan	Cot	Cos	Sec	Csc	′
0	.51504	.60086	1.6643	.85717	1.1666	1.9416	**60**
1	.51529	.60126	1.6632	.85702	1.1668	1.9407	59
2	.51554	.60165	1.6621	.85687	1.1670	1.9397	58
3	.51579	.60205	1.6610	.85672	1.1672	1.9388	57
4	.51604	.60245	1.6599	.85657	1.1675	1.9379	56
5	.51628	.60284	1.6588	.85642	1.1677	1.9369	**55**
6	.51653	.60324	1.6577	.85627	1.1679	1.9360	54
7	.51678	.60364	1.6566	.85612	1.1681	1.9351	53
8	.51703	.60403	1.6555	.85597	1.1683	1.9341	52
9	.51728	.60443	1.6545	.85582	1.1685	1.9332	51
10	.51753	.60483	1.6534	.85567	1.1687	1.9323	**50**
11	.51778	.60522	1.6523	.85551	1.1689	1.9313	49
12	.51803	.60562	1.6512	.85536	1.1691	1.9304	48
13	.51828	.60602	1.6501	.85521	1.1693	1.9295	47
14	.51852	.60642	1.6490	.85506	1.1695	1.9285	46
15	.51877	.60681	1.6479	.85491	1.1697	1.9276	**45**
16	.51902	.60721	1.6469	.85476	1.1699	1.9267	44
17	.51927	.60761	1.6458	.85461	1.1701	1.9258	43
18	.51952	.60801	1.6447	.85446	1.1703	1.9249	42
19	.51977	.60841	1.6436	.85431	1.1705	1.9239	41
20	.52002	.60881	1.6426	.85416	1.1707	1.9230	**40**
21	.52026	.60921	1.6415	.85401	1.1710	1.9221	39
22	.52051	.60960	1.6404	.85385	1.1712	1.9212	38
23	.52076	.61000	1.6393	.85370	1.1714	1.9203	37
24	.52101	.61040	1.6383	.85355	1.1716	1.9194	36
25	.52126	.61080	1.6372	.85340	1.1718	1.9184	**35**
26	.52151	.61120	1.6361	.85325	1.1720	1.9175	34
27	.52175	.61160	1.6351	.85310	1.1722	1.9166	33
28	.52200	.61200	1.6340	.85294	1.1724	1.9157	32
29	.52225	.61240	1.6329	.85279	1.1726	1.9148	31
30	.52250	.61280	1.6319	.85264	1.1728	1.9139	**30**
31	.52275	.61320	1.6308	.85249	1.1730	1.9130	29
32	.52299	.61360	1.6297	.85234	1.1732	1.9121	28
33	.52324	.61400	1.6287	.85218	1.1735	1.9112	27
34	.52349	.61440	1.6276	.85203	1.1737	1.9103	26
35	.52374	.61480	1.6265	.85188	1.1739	1.9094	**25**
36	.52399	.61520	1.6255	.85173	1.1741	1.9084	24
37	.52423	.61561	1.6244	.85157	1.1743	1.9075	23
38	.52448	.61601	1.6234	.85142	1.1745	1.9066	22
39	.52473	.61641	1.6223	.85127	1.1747	1.9057	21
40	.52498	.61681	1.6212	.85112	1.1749	1.9048	**20**
41	.52522	.61721	1.6202	.85096	1.1751	1.9039	19
42	.52547	.61761	1.6191	.85081	1.1753	1.9031	18
43	.52572	.61801	1.6181	.85066	1.1756	1.9022	17
44	.52597	.61842	1.6170	.85051	1.1758	1.9013	16
45	.52621	.61882	1.6160	.85035	1.1760	1.9004	**15**
46	.52646	.61922	1.6149	.85020	1.1762	1.8995	14
47	.52671	.61962	1.6139	.85005	1.1764	1.8986	13
48	.52696	.62003	1.6128	.84989	1.1766	1.8977	12
49	.52720	.62043	1.6118	.84974	1.1768	1.8968	11
50	.52745	.62083	1.6107	.84959	1.1770	1.8959	**10**
51	.52770	.62124	1.6097	.84943	1.1773	1.8950	9
52	.52794	.62164	1.6087	.84928	1.1775	1.8941	8
53	.52819	.62204	1.6076	.84913	1.1777	1.8933	7
54	.52844	.62245	1.6066	.84897	1.1779	1.8924	6
55	.52869	.62285	1.6055	.84882	1.1781	1.8915	**5**
56	.52893	.62325	1.6045	.84866	1.1783	1.8906	4
57	.52918	.62366	1.6034	.84851	1.1785	1.8897	3
58	.52943	.62406	1.6024	.84836	1.1788	1.8888	2
59	.52967	.62446	1.6014	.84820	1.1790	1.8880	1
60	.52992	.62487	1.6003	.84805	1.1792	1.8871	**0**
′	Cos	Cot	Tan	Sin	Csc	Sec	′

121° (301°) (238°) **58°**

NATURAL TRIGONOMETRIC FUNCTIONS TO FIVE PLACES

32° (212°) (327°) **147°**

′	Sin	Tan	Cot	Cos	Sec	Csc	′
0	.52992	.62487	1.6003	.84805	1.1792	1.8871	**60**
1	.53017	.62527	1.5993	.84789	1.1794	1.8862	59
2	.53041	.62568	1.5983	.84774	1.1796	1.8853	58
3	.53066	.62608	1.5972	.84759	1.1798	1.8844	57
4	.53091	.62649	1.5962	.84743	1.1800	1.8836	56
5	.53115	.62689	1.5952	.84728	1.1803	1.8827	**55**
6	.53140	.62730	1.5941	.84712	1.1805	1.8818	54
7	.53164	.62770	1.5931	.84697	1.1807	1.8810	53
8	.53189	.62811	1.5921	.84681	1.1809	1.8801	52
9	.53214	.62852	1.5911	.84666	1.1811	1.8792	51
10	.53238	.62892	1.5900	.84650	1.1813	1.8783	**50**
11	.53263	.62933	1.5890	.84635	1.1815	1.8775	49
12	.53288	.62973	1.5880	.84619	1.1818	1.8766	48
13	.53312	.63014	1.5869	.84604	1.1820	1.8757	47
14	.53337	.63055	1.5859	.84588	1.1822	1.8749	46
15	.53361	.63095	1.5849	.84573	1.1824	1.8740	**45**
16	.53386	.63136	1.5839	.84557	1.1826	1.8731	44
17	.53411	.63177	1.5829	.84542	1.1828	1.8723	43
18	.53435	.63217	1.5818	.84526	1.1831	1.8714	42
19	.53460	.63258	1.5808	.84511	1.1833	1.8706	41
20	.53484	.63299	1.5798	.84495	1.1835	1.8697	**40**
21	.53509	.63340	1.5788	.84480	1.1837	1.8688	39
22	.53534	.63380	1.5778	.84464	1.1830	1.8680	38
23	.53558	.63421	1.5768	.84448	1.1842	1.8671	37
24	.53583	.63462	1.5757	.84433	1.1844	1.8663	36
25	.53607	.63503	1.5747	.84417	1.1846	1.8654	**35**
26	.53632	.63544	1.5737	.84402	1.1848	1.8646	34
27	.53656	.63584	1.5727	.84386	1.1850	1.8637	33
28	.53681	.63625	1.5717	.84370	1.1852	1.8629	32
29	.53705	.63666	1.5707	.84355	1.1855	1.8620	31
30	.53730	.63707	1.5697	.84339	1.1857	1.8612	**30**
31	.53754	.63748	1.5687	.84324	1.1859	1.8603	29
32	.53779	.63789	1.5677	.84308	1.1861	1.8595	28
33	.53804	.63830	1.5667	.84292	1.1863	1.8586	27
34	.53828	.63871	1.5657	.84277	1.1866	1.8578	26
35	.53853	.63912	1.5647	.84261	1.1868	1.8569	**25**
36	.53877	.63953	1.5637	.84245	1.1870	1.8561	24
37	.53902	.63994	1.5627	.84230	1.1872	1.8552	23
38	.53926	.64035	1.5617	.84214	1.1875	1.8544	22
39	.53951	.64076	1.5607	.84198	1.1877	1.8535	21
40	.53975	.64117	1.5597	.84182	1.1879	1.8527	**20**
41	.54000	.64158	1.5587	.84167	1.1881	1.8519	19
42	.54024	.64199	1.5577	.84151	1.1883	1.8510	18
43	.54049	.64240	1.5567	.84135	1.1886	1.8502	17
44	.54073	.64281	1.5557	.84120	1.1888	1.8494	16
45	.54097	.64322	1.5547	.84104	1.1890	1.8485	**15**
46	.54122	.64363	1.5537	.84088	1.1892	1.8477	14
47	.54146	.64404	1.5527	.84072	1.1895	1.8468	13
48	.54171	.64446	1.5517	.84057	1.1897	1.8460	12
49	.54195	.64487	1.5507	.84041	1.1899	1.8452	11
50	.54220	.64528	1.5497	.84025	1.1901	1.8443	**10**
51	.54244	.64569	1.5487	.84009	1.1903	1.8435	9
52	.54269	.64610	1.5477	.83994	1.1906	1.8427	8
53	.54293	.64652	1.5468	.83978	1.1908	1.8419	7
54	.54317	.64693	1.5458	.83962	1.1910	1.8410	6
55	.54342	.64734	1.5448	.83946	1.1912	1.8402	**5**
56	.54366	.64775	1.5438	.83930	1.1915	1.8394	4
57	.54391	.64817	1.5428	.83915	1.1917	1.8385	3
58	.54415	.64858	1.5418	.83899	1.1919	1.8377	2
59	.54440	.64899	1.5408	.83883	1.1921	1.8369	1
60	.54464	.64941	1.5399	.83867	1.1924	1.8361	**0**
′	Cos	Cot	Tan	Sin	Csc	Sec	′

122° (302°) (237°) **57°**

33° (213°) (326°) **1**

′	Sin	Tan	Cot	Cos	Sec	Csc	′
0	.54464	.64941	1.5399	.83867	1.1924	1.8361	**6**
1	.54488	.64982	1.5389	.83851	1.1926	1.8353	5
2	.54513	.65024	1.5379	.83835	1.1928	1.8344	5
3	.54537	.65065	1.5369	.83819	1.1930	1.8336	5
4	.54561	.65106	1.5359	.83804	1.1933	1.8328	5
5	.54586	.65148	1.5350	.83788	1.1935	1.8320	**5**
6	.54610	.65189	1.5340	.83772	1.9137	1.8312	5
7	.54635	.65231	1.5330	.83756	1.1939	1.8303	5
8	.54659	.65272	1.5320	.83740	1.1942	1.8295	5
9	.54683	.65314	1.5311	.83724	1.1944	1.8287	5
10	.54708	.65355	1.5301	.83708	1.1946	1.8279	**5**
11	.54732	.65397	1.5291	.83692	1.1949	1.8271	4
12	.54756	.65438	1.5282	.83676	1.1951	1.8263	4
13	.54781	.65480	1.5272	.83660	1.1953	1.8255	4
14	.54805	.65521	1.5262	.83645	1.1955	1.8247	4
15	.54829	.65563	1.5253	.83629	1.1958	1.8238	**4**
16	.54854	.65604	1.5243	.83613	1.1960	1.8230	4
17	.54878	.65646	1.5233	.83597	1.1962	1.8222	4
18	.54902	.65688	1.5224	.83581	1.1964	1.8214	4
19	.54927	.65729	1.5214	.83565	1.1967	1.8206	4
20	.54951	.65771	1.5204	.83549	1.1969	1.8198	**4**
21	.54975	.65813	1.5195	.83533	1.1971	1.8190	3
22	.54999	.65854	1.5185	.83517	1.1974	1.8182	3
23	.55024	.65896	1.5175	.83501	1.1976	1.8174	3
24	.55048	.65938	1.5166	.83485	1.1978	1.8166	3
25	.55072	.65980	1.5156	.83469	1.1981	1.8158	**3**
26	.55097	.66021	1.5147	.83453	1.1983	1.8150	3
27	.55121	.66063	1.5137	.83437	1.1985	1.8142	3
28	.55145	.66105	1.5127	.83421	1.1987	1.8134	3
29	.55169	.66147	1.5118	.83405	1.1990	1.8126	3
30	.55194	.66189	1.5108	.83389	1.1992	1.8118	**3**
31	.55218	.66230	1.5099	.83373	1.1994	1.8110	2
32	.55242	.66272	1.5089	.83356	1.1997	1.8102	2
33	.55266	.66314	1.5080	.83340	1.1999	1.8094	2
34	.55291	.66356	1.5070	.83324	1.2001	1.8086	2
35	.55315	.66398	1.5061	.83308	1.2004	1.8078	**2**
36	.55339	.66440	1.5051	.83292	1.2006	1.8070	2
37	.55363	.66482	1.5042	.83276	1.2008	1.8062	2
38	.55388	.66524	1.5032	.83260	1.2011	1.8055	2
39	.55412	.66566	1.5023	.83244	1.2013	1.8047	2
40	.55436	.66608	1.5013	.83228	1.2015	1.8039	**2**
41	.55460	.66650	1.5004	.83212	1.2018	1.8031	1
42	.55484	.66692	1.4994	.83195	1.2020	1.8023	1
43	.55509	.66734	1.4985	.83179	1.2022	1.8015	1
44	.55533	.66776	1.4975	.83163	1.2025	1.8007	1
45	.55557	.66818	1.4966	.83147	1.2027	1.8000	**1**
46	.55581	.66860	1.4957	.83131	1.2029	1.7992	1
47	.55605	.66902	1.4947	.83115	1.2032	1.7984	1
48	.55630	.66944	1.4938	.83098	1.2034	1.7976	1
49	.55654	.66986	1.4928	.83082	1.2036	1.7968	1
50	.55678	.67028	1.4919	.83066	1.2039	1.7960	**1**
51	.55702	.67071	1.4910	.83050	1.2041	1.7953	
52	.55726	.67113	1.4900	.83034	1.2043	1.7945	
53	.55750	.67155	1.4891	.83017	1.2046	1.7937	
54	.55775	.67197	1.4882	.83001	1.2048	1.7929	
55	.55799	.67239	1.4872	.82985	1.2050	1.7922	
56	.55823	.67282	1.4863	.82969	1.2053	1.7914	
57	.55847	.67324	1.4854	.82953	1.2055	1.7906	
58	.55871	.67366	1.4844	.82936	1.2057	1.7898	
59	.55895	.67409	1.4835	.82920	1.2060	1.7891	
60	.55919	.67451	1.4826	.82904	1.2062	1.7883	
′	Cos	Cot	Tan	Sin	Csc	Sec	′

123° (303°) (236°) **5**

NATURAL TRIGONOMETRIC FUNCTIONS TO FIVE PLACES

34° (214°) (325°) **145°**

′	Sin	Tan	Cot	Cos	Sec	Csc	′
0	.55919	.67451	1.4826	.82904	1.2062	1.7883	**60**
1	.55943	.67493	1.4816	.82887	1.2065	1.7875	59
2	.55968	.67536	1.4807	.82871	1.2067	1.7868	58
3	.55992	.67578	1.4798	.82855	1.2069	1.7860	57
4	.56016	.67620	1.4788	.82839	1.2072	1.7852	56
5	.56040	.67663	1.4779	.82822	1.2074	1.7844	**55**
6	.56064	.67705	1.4770	.82806	1.2076	1.7837	54
7	.56088	.67748	1.4761	.82790	1.2079	1.7829	53
8	.56112	.67790	1.4751	.82773	1.2081	1.7821	52
9	.56136	.67832	1.4742	.82757	1.2084	1.7814	51
10	.56160	.67875	1.4733	.82741	1.2086	1.7806	**50**
11	.56184	.67917	1.4724	.82724	1.2088	1.7799	49
12	.56208	.67960	1.4715	.82708	1.2091	1.7791	48
13	.56232	.68002	1.4705	.82692	1.2093	1.7783	47
14	.56256	.68045	1.4696	.82675	1.2096	1.7776	46
15	.56280	.68088	1.4687	.82659	1.2098	1.7768	**45**
16	.56305	.68130	1.4678	.82643	1.2100	1.7761	44
17	.56329	.68173	1.4669	.82626	1.2103	1.7753	43
18	.56353	.68215	1.4659	.82610	1.2105	1.7745	42
19	.56377	.68258	1.4650	.82593	1.2108	1.7738	41
20	.56401	.68301	1.4641	.82577	1.2110	1.7730	**40**
21	.56425	.68343	1.4632	.82561	1.2112	1.7723	39
22	.56449	.68386	1.4623	.82544	1.2115	1.7715	38
23	.56473	.68429	1.4614	.82528	1.2117	1.7708	37
24	.56497	.68471	1.4605	.82511	1.2120	1.7700	36
25	.56521	.68514	1.4596	.82495	1.2122	1.7693	**35**
26	.56545	.68557	1.4586	.82478	1.2124	1.7685	34
27	.56569	.68600	1.4577	.82462	1.2127	1.7678	33
28	.56593	.68642	1.4568	.82446	1.2129	1.7670	32
29	.56617	.68685	1.4559	.82429	1.2132	1.7663	31
30	.56641	.68728	1.4550	.82413	1.2134	1.7655	**30**
31	.56665	.68771	1.4541	.82396	1.2136	1.7648	29
32	.56689	.68814	1.4532	.82380	1.2139	1.7640	28
33	.56713	.68857	1.4523	.82363	1.2141	1.7633	27
34	.56736	.68900	1.4514	.82347	1.2144	1.7625	26
35	.56760	.68942	1.4505	.82330	1.2146	1.7618	**25**
36	.56784	.68985	1.4496	.82314	1.2149	1.7610	24
37	.56808	.69028	1.4487	.82297	1.2151	1.7603	23
38	.56832	.69071	1.4478	.82281	1.2154	1.7596	22
39	.56856	.69114	1.4469	.82264	1.2156	1.7588	21
40	.56880	.69157	1.4460	.82248	1.2158	1.7581	**20**
41	.56904	.69200	1.4451	.82231	1.2161	1.7573	19
42	.56928	.69243	1.4442	.82214	1.2163	1.7566	18
43	.56952	.69286	1.4433	.82198	1.2166	1.7559	17
44	.56976	.69329	1.4424	.82181	1.2168	1.7551	16
45	.57000	.69372	1.4415	.82165	1.2171	1.7544	**15**
46	.57024	.69416	1.4406	.82148	1.2173	1.7537	14
47	.57047	.69459	1.4397	.82132	1.2176	1.7529	13
48	.57071	.69502	1.4388	.82115	1.2178	1.7522	12
49	.57095	.69545	1.4379	.82098	1.2181	1.7515	11
50	.57119	.69588	1.4370	.82082	1.2183	1.7507	**10**
51	.57143	.69631	1.4361	.82065	1.2185	1.7500	9
52	.57167	.69675	1.4352	.82048	1.2188	1.7493	8
53	.57191	.69718	1.4344	.82032	1.2190	1.7485	7
54	.57215	.69761	1.4335	.82015	1.2193	1.7478	6
55	.57238	.69804	1.4326	.81999	1.2195	1.7471	**5**
56	.57262	.69847	1.4317	.81982	1.2198	1.7463	4
57	.57286	.69891	1.4308	.81965	1.2200	1.7456	3
58	.57310	.69934	1.4299	.81949	1.2203	1.7449	2
59	.57334	.69977	1.4290	.81932	1.2205	1.7442	1
60	.57358	.70021	1.4281	.81915	1.2208	1.7434	**0**
′	Cos	Cot	Tan	Sin	Csc	Sec	′

124° (304°) (235°) **55°**

35° (215°) (324°) **144°**

′	Sin	Tan	Cot	Cos	Sec	Csc	′
0	.57358	.70021	1.4281	.81915	1.2208	1.7434	**60**
1	.57381	.70064	1.4273	.81899	1.2210	1.7427	59
2	.57405	.70107	1.4264	.81882	1.2213	1.7420	58
3	.57429	.70151	1.4255	.81865	1.2215	1.7413	57
4	.57453	.70194	1.4246	.81848	1.2218	1.7406	56
5	.57477	.70238	1.4237	.81832	1.2220	1.7398	**55**
6	.57501	.70281	1.4229	.81815	1.2223	1.7391	54
7	.57524	.70325	1.4220	.81798	1.2225	1.7384	53
8	.57548	.70368	1.4211	.81782	1.2228	1.7377	52
9	.57572	.70412	1.4202	.81765	1.2230	1.7370	51
10	.57596	.70455	1.4193	.81748	1.2233	1.7362	**50**
11	.57619	.70499	1.4185	.81731	1.2235	1.7355	49
12	.57643	.70542	1.4176	.81714	1.2238	1.7348	48
13	.57667	.70586	1.4167	.81698	1.2240	1.7341	47
14	.57691	.70629	1.4158	.81681	1.2243	1.7334	46
15	.57715	.70673	1.4150	.81664	1.2245	1.7327	**45**
16	.57738	.70717	1.4141	.81647	1.2248	1.7320	44
17	.57762	.70760	1.4132	.81631	1.2250	1.7312	43
18	.57786	.70804	1.4124	.81614	1.2253	1.7305	42
19	.57810	.70848	1.4115	.81597	1.2255	1.7298	41
20	.57833	.70891	1.4106	.81580	1.2258	1.7291	**40**
21	.57857	.70935	1.4097	.81563	1.2260	1.7284	39
22	.57881	.70979	1.4089	.81546	1.2263	1.7277	38
23	.57904	.71023	1.4080	.81530	1.2265	1.7270	37
24	.57928	.71066	1.4071	.81513	1.2268	1.7263	36
25	.57952	.71110	1.4063	.81496	1.2271	1.7256	**35**
26	.57976	.71154	1.4054	.81479	1.2273	1.7249	34
27	.57999	.71198	1.4045	.81462	1.2276	1.7242	33
28	.58023	.71242	1.4037	.81445	1.2278	1.7235	32
29	.58047	.71285	1.4028	.81428	1.2281	1.7228	31
30	.58070	.71329	1.4019	.81412	1.2283	1.7221	**30**
31	.58094	.71373	1.4011	.81395	1.2286	1.7213	29
32	.58118	.71417	1.4002	.81378	1.2288	1.7206	28
33	.58141	.71461	1.3994	.81361	1.2291	1.7199	27
34	.58165	.71505	1.3985	.81344	1.2293	1.7192	26
35	.58189	.71549	1.3976	.81327	1.2296	1.7185	**25**
36	.58212	.71593	1.3968	.81310	1.2299	1.7179	24
37	.58236	.71637	1.3959	.81293	1.2301	1.7172	23
38	.58260	.71681	1.3951	.81276	1.2304	1.7165	22
39	.58283	.71725	1.3942	.81259	1.2306	1.7158	21
40	.58307	.71769	1.3934	.81242	1.2309	1.7151	**20**
41	.58330	.71813	1.3925	.81225	1.2311	1.7144	19
42	.58354	.71857	1.3916	.81208	1.2314	1.7137	18
43	.58378	.71901	1.3908	.81191	1.2317	1.7130	17
44	.58401	.71946	1.3899	.81174	1.2319	1.7123	16
45	.58425	.71990	1.3891	.81157	1.2322	1.7116	**15**
46	.58449	.72034	1.3882	.81140	1.2324	1.7109	14
47	.58472	.72078	1.3874	.81123	1.2327	1.7102	13
48	.58496	.72122	1.3865	.81106	1.2329	1.7095	12
49	.58519	.72167	1.3857	.81089	1.2332	1.7088	11
50	.58543	.72211	1.3848	.81072	1.2335	1.7081	**10**
51	.58567	.72255	1.3840	.81055	1.2337	1.7075	9
52	.58590	.72299	1.3831	.81038	1.2340	1.7068	8
53	.58614	.72344	1.3823	.81021	1.2342	1.7061	7
54	.58637	.72388	1.3814	.81004	1.2345	1.7054	6
55	.58661	.72432	1.3806	.80987	1.2348	1.7047	**5**
56	.58684	.72477	1.3798	.80970	1.2350	1.7040	4
57	.58708	.72521	1.3789	.80953	1.2353	1.7033	3
58	.58731	.72565	1.3781	.80936	1.2355	1.7027	2
59	.58755	.72610	1.3772	.80919	1.2358	1.7020	1
60	.58779	.72654	1.3764	.80902	1.2361	1.7013	**0**
′	Cos	Cot	Tan	Sin	Csc	Sec	′

125° (305°) (234°) **54°**

NATURAL TRIGONOMETRIC FUNCTIONS TO FIVE PLACES

36° (216°) (323°) **143°**

′	Sin	Tan	Cot	Cos	Sec	Csc	′
0	.58779	.72654	1.3764	.80902	1.2361	1.7013	**60**
1	.58802	.72699	1.3755	.80885	1.2363	1.7006	59
2	.58826	.72743	1.3747	.80867	1.2366	1.6999	58
3	.58849	.72788	1.3739	.80850	1.2369	1.6993	57
4	.58873	.72832	1.3730	.80833	1.2371	1.6986	56
5	.58896	.72877	1.3722	.80816	1.2374	1.6979	**55**
6	.58920	.72921	1.3713	.80799	1.2376	1.6972	54
7	.58943	.72966	1.3705	.80782	1.2379	1.6966	53
8	.58967	.73010	1.3697	.80765	1.2382	1.6959	52
9	.58990	.73055	1.3688	.80748	1.2384	1.6952	51
10	.59014	.73100	1.3680	.80730	1.2387	1.6945	**50**
11	.59037	.73144	1.3672	.80713	1.2390	1.6939	49
12	.59061	.73189	1.3663	.80696	1.2392	1.6932	48
13	.59084	.73234	1.3655	.80679	1.2395	1.6925	47
14	.59108	.73278	1.3647	.80662	1.2397	1.6918	46
15	.59131	.73323	1.3638	.80644	1.2400	1.6912	**45**
16	.59154	.73368	1.3630	.80627	1.2403	1.6905	44
17	.59178	.73413	1.3622	.80610	1.2405	1.6898	43
18	.59201	.73457	1.3613	.80593	1.2408	1.6892	42
19	.59225	.73502	1.3605	.80576	1.2411	1.6885	41
20	.59248	.73547	1.3597	.80558	1.2413	1.6878	**40**
21	.59272	.73592	1.3588	.80541	1.2416	1.6871	39
22	.59295	.73637	1.3580	.80524	1.2419	1.6865	38
23	.59318	.73681	1.3572	.80507	1.2421	1.6858	37
24	.59342	.73726	1.3564	.80489	1.2424	1.6852	36
25	.59365	.73771	1.3555	.80472	1.2427	1.6845	**35**
26	.59389	.73816	1.3547	.80455	1.2429	1.6838	34
27	.59412	.73861	1.3539	.80438	1.2432	1.6832	33
28	.59436	.73906	1.3531	.80420	1.2435	1.6825	32
29	.59459	.73951	1.3522	.80403	1.2437	1.6818	31
30	.59482	.73996	1.3514	.80386	1.2440	1.6812	**30**
31	.59506	.74041	1.3506	.80368	1.2443	1.6805	29
32	.59529	.74086	1.3498	.80351	1.2445	1.6799	28
33	.59552	.74131	1.3490	.80334	1.2448	1.6792	27
34	.59576	.74176	1.3481	.80316	1.2451	1.6785	26
35	.59599	.74221	1.3473	.80299	1.2453	1.6779	**25**
36	.59622	.74267	1.3465	.80282	1.2456	1.6772	24
37	.59646	.74312	1.3457	.80264	1.2459	1.6766	23
38	.59669	.74357	1.3449	.80247	1.2462	1.6759	22
39	.59693	.74402	1.3440	.80230	1.2464	1.6753	21
40	.59716	.74447	1.3432	.80212	1.2467	1.6746	**20**
41	.59739	.74492	1.3424	.80195	1.2470	1.6739	19
42	.59763	.74538	1.3416	.80178	1.2472	1.6733	18
43	.59786	.74583	1.3408	.80160	1.2475	1.6726	17
44	.59809	.74628	1.3400	.80143	1.2478	1.6720	16
45	.59832	.74674	1.3392	.80125	1.2480	1.6713	**15**
46	.59856	.74719	1.3384	.80108	1.2483	1.6707	14
47	.59879	.74764	1.3375	.80091	1.2486	1.6700	13
48	.59902	.74810	1.3367	.80073	1.2489	1.6694	12
49	.59926	.74855	1.3359	.80056	1.2491	1.6687	11
50	.59949	.74900	1.3351	.80038	1.2494	1.6681	**10**
51	.59972	.74946	1.3343	.80021	1.2497	1.6674	9
52	.59995	.74991	1.3335	.80003	1.2499	1.6668	8
53	.60019	.75037	1.3327	.79986	1.2502	1.6661	7
54	.60042	.75082	1.3319	.79968	1.2505	1.6655	6
55	.60065	.75128	1.3311	.79951	1.2508	1.6649	**5**
56	.60089	.75173	1.3303	.79934	1.2510	1.6642	4
57	.60112	.75219	1.3295	.79916	1.2513	1.6636	3
58	.60135	.75264	1.3287	.79899	1.2516	1.6629	2
59	.60158	.75310	1.3278	.79881	1.2519	1.6623	1
60	.60182	.75355	1.3270	.79864	1.2521	1.6616	**0**
′	Cos	Cot	Tan	Sin	Csc	Sec	′

126° (306°) (233°) **53°**

37° (217°) (322°) **142°**

′	Sin	Tan	Cot	Cos	Sec	Csc	′
0	.60182	.75355	1.3270	.79864	1.2521	1.6616	**60**
1	.60205	.75401	1.3262	.79846	1.2524	1.6610	59
2	.60228	.75447	1.3254	.79829	1.2527	1.6604	58
3	.60251	.75492	1.3246	.79811	1.2530	1.6597	57
4	.60274	.75538	1.3238	.79793	1.2532	1.6591	56
5	.60298	.75584	1.3230	.79776	1.2535	1.6584	**55**
6	.60321	.75629	1.3222	.79758	1.2538	1.6578	54
7	.60344	.75675	1.3214	.79741	1.2541	1.6572	53
8	.60367	.75721	1.3206	.79723	1.2543	1.6565	52
9	.60390	.75767	1.3198	.79706	1.2546	1.6559	51
10	.60414	.75812	1.3190	.79688	1.2549	1.6553	**50**
11	.60437	.75858	1.3182	.79671	1.2552	1.6546	49
12	.60460	.75904	1.3175	.79653	1.2554	1.6540	48
13	.60483	.75950	1.3167	.79635	1.2557	1.6534	47
14	.60506	.75996	1.3159	.79618	1.2560	1,6527	46
15	.60529	.76042	1.3151	.79600	1.2563	1.6521	**45**
16	.60553	.76088	1.3143	.79583	1.2566	1.6515	44
17	.60576	.76134	1.3135	.79565	1.2568	1.6508	43
18	.60599	.76180	1.3127	.79547	1.2571	1.6502	42
19	.60622	.76226	1.3119	.79530	1.2574	1.6496	41
20	.60645	.76272	1.3111	.79512	1.2577	1.6489	**40**
21	.60668	.76318	1.3103	.79494	1.2579	1.6483	39
22	.60691	.76364	1.3095	.79477	1.2582	1.6477	38
23	.60714	.76410	1.3087	.79459	1.2585	1.6471	37
24	.60738	.76456	1.3079	.79441	1.2588	1.6464	36
25	.60761	.76502	1.3072	.79424	1.2591	1.6458	**35**
26	.60784	.76548	1.3064	.79406	1.2593	1.6452	34
27	.60807	.76594	1.3056	.79388	1.2596	1.6446	33
28	.60830	.76640	1.3048	.79371	1.2599	1.6439	32
29	.60853	.76686	1.3040	.79353	1.2602	1.6433	31
30	.60876	.76733	1.3032	.79335	1.2605	1.6427	**30**
31	.60899	.76779	1.3024	.79318	1.2608	1.6421	29
32	.60922	.76825	1.3017	.79300	1.2610	1.6414	28
33	.60945	.76871	1.3009	.79282	1.2613	1.6408	27
34	.60968	.76918	1.3001	.79264	1.2616	1.6402	26
35	.60991	.76964	1.2993	.79247	1.2619	1.6396	**25**
36	.61015	.77010	1.2985	.79229	1.2622	1.6390	24
37	.61038	.77057	1.2977	.79211	1.2624	1.6383	23
38	.61061	.77103	1.2970	.79193	1.2627	1.6377	22
39	.61084	.77149	1.2962	.79176	1.2630	1.6371	21
40	.61107	.77196	1.2954	.79158	1.2633	1.6365	**20**
41	.61130	.77242	1.2946	.79140	1.2636	1.6359	19
42	.61153	.77289	1.2938	.79122	1.2639	1.6353	18
43	.61176	.77335	1.2931	.79105	1.2641	1.6346	17
44	.61199	.77382	1.2923	.79087	1.2644	1.6340	16
45	.61222	.77428	1.2915	.79069	1.2647	1.6334	**15**
46	.61245	.77475	1.2907	.79051	1.2650	1.6328	14
47	.61268	.77521	1.2900	.79033	1.2653	1.6322	13
48	.61291	.77568	1.2892	.79016	1.2656	1.6316	12
49	.61314	.77615	1.2884	.78998	1.2659	1.6310	11
50	.61337	.77661	1.2876	.78980	1.2661	1.6303	**10**
51	.61360	.77708	1.2869	.78962	1.2664	1.6297	9
52	.61383	.77754	1.2861	.78944	1.2667	1.6291	8
53	.61406	.77801	1.2853	.78926	1.2670	1.6285	7
54	.61429	.77848	1.2846	.78908	1.2673	1.6279	6
55	.61451	.77895	1.2838	.78891	1.2676	1.6273	**5**
56	.61474	.77941	1.2830	.78873	1.2679	1.6267	4
57	.61497	.77988	1.2822	.78855	1.2682	1.6261	3
58	.61520	.78035	1.2815	.78837	1.2684	1.6255	2
59	.61543	.78082	1.2807	.78819	1.2687	1.6249	1
60	.61566	.78129	1.2799	.78801	1.2690	1.6243	**0**
′	Cos	Cot	Tan	Sin	Csc	Sec	′

127° (307°) (232°) **52°**

NATURAL TRIGONOMETRIC FUNCTIONS TO FIVE PLACES

38° (218°) — (321°) **141°**

′	Sin	Tan	Cot	Cos	Sec	Csc	′
0	.61566	.78129	1.2799	.78801	1.2690	1.6243	**60**
1	.61589	.78175	1.2792	.78783	1.2693	1.6237	59
2	.61612	.78222	1.2784	.78765	1.2696	1.6231	58
3	.61635	.78269	1.2776	.78747	1.2699	1.6225	57
4	.61658	.78316	1.2769	.78729	1.2702	1.6219	56
5	.61681	.78363	1.2761	.78711	1.2705	1.6213	**55**
6	.61704	.78410	1.2753	.78694	1.2708	1.6207	54
7	.61726	.78457	1.2746	.78676	1.2710	1.6201	53
8	.61749	.78504	1.2738	.78658	1.2713	1.6195	52
9	.61772	.78551	1.2731	.78640	1.2716	1.6189	51
10	.61795	.78598	1.2723	.78622	1.1719	1.6183	**50**
11	.61818	.78645	1.2715	.78604	1.2722	1.6177	49
12	.61841	.78692	1.2708	.78586	1.2725	1.6171	48
13	.61864	.78739	1.2700	.78568	1.2728	1.6165	47
14	.61887	.78786	1.2693	.78550	1.2731	1.6159	46
15	.61909	.78834	1.2685	.78532	1.2734	1.6153	**45**
16	.61932	.78881	1.2677	.78514	1.2737	1.6147	44
17	.61955	.78928	1.2670	.78496	1.2740	1.6141	43
18	.61978	.78975	1.2662	.78478	1.2742	1.6135	42
19	.62001	.79022	1.2655	.78460	1.2745	1.6129	41
20	.62024	.79070	1.2647	.78442	1.2748	1.6123	**40**
21	.62046	.79117	1.2640	.78424	1.2751	1.6117	39
22	.62069	.79164	1.2632	.78405	1.2754	1.6111	38
23	.62092	.79212	1.2624	.78387	1.2757	1.6105	37
24	.62115	.79259	1.2617	.78369	1.2760	1.6099	36
25	.62138	.79306	1.2609	.78351	1.2763	1.6093	**35**
26	.62160	.79354	1.2602	.78333	1.2766	1.6087	34
27	.62183	.79401	1.2594	.78315	1.2769	1.6082	33
28	.62206	.79449	1.2587	.78297	1.2772	1.6076	32
29	.62229	.79496	1.2579	.78279	1.2775	1.6070	31
30	.62251	.79544	1.2572	.78261	1.2778	1.6064	**30**
31	.62274	.79591	1.2564	.78243	1.2781	1.6058	29
32	.62297	.79639	1.2557	.78225	1.2784	1.6052	28
33	.62320	.79686	1.2549	.78206	1.2787	1.6046	27
34	.62342	.79734	1.2542	.78188	1.2790	1.6040	26
35	.62365	.79781	1.2534	.78170	1.2793	1.6035	**25**
36	.62388	.79829	1.2527	.78152	1.2796	1.6029	24
37	.62411	.79877	1.2519	.78134	1.2799	1.6023	23
38	.62433	.79924	1.2512	.78116	1.2802	1.6017	22
39	.62456	.79972	1.2504	.78098	1.2804	1.6011	21
40	.62479	.80020	1.2497	.78079	1.2807	1.6005	**20**
41	.62502	.80067	1.2489	.78061	1.2810	1.6000	19
42	.62524	.80115	1.2482	.78043	1.2813	1.5994	18
43	.62547	.80163	1.2475	.78025	1.2816	1.5988	17
44	.62570	.80211	1.2467	.78007	1.2819	1.5982	16
45	.62592	.80258	1.2460	.77988	1.2822	1.5976	**15**
46	.62615	.80306	1.2452	.77970	1.2825	1.5971	14
47	.62638	.80354	1.2445	.77952	1.2828	1.5965	13
48	.62660	.80402	1.2437	.77934	1.2831	1.5959	12
49	.62683	.80450	1.2430	.77916	1.2834	1.5953	11
50	.62706	.80498	1.2423	.77897	1.2837	1.5948	**10**
51	.62728	.80546	1.2415	.77879	1.2840	1.5942	9
52	.62751	.80594	1.2408	.77861	1.2843	1.5936	8
53	.62774	.80642	1.2401	.77843	1.2846	1.5930	7
54	.62796	.80690	1.2393	.77824	1.2849	1.5925	6
55	.62819	.80738	1.2386	.77806	1.2852	1.5919	**5**
56	.62842	.80786	1.2378	.77788	1.2855	1.5913	4
57	.62864	.80834	1.2371	.77769	1.2859	1.5907	3
58	.62887	.80882	1.2364	.77751	1.2862	1.5902	2
59	.62909	.80930	1.2356	.77733	1.2865	1.5896	1
60	.62932	.80978	1.2349	.77715	1.2868	1.5890	**0**
′	Cos	Cot	Tan	Sin	Csc	Sec	′

128° (308°) — (231°) **51°**

39° (219°) — (320°) **140°**

′	Sin	Tan	Cot	Cos	Sec	Csc	′
0	.62932	.80978	1.2349	.77715	1.2868	1.5890	**60**
1	.62955	.81027	1.2342	.77696	1.2871	1.5884	59
2	.62977	.81075	1.2334	.77678	1.2874	1.5879	58
3	.63000	.81123	1.2327	.77660	1.2877	1.5873	57
4	.63022	.81171	1.2320	.77641	1.2880	1.5867	56
5	.63045	.81220	1.2312	.77623	1.2883	1.5862	**55**
6	.63068	.81268	1.2305	.77605	1.2886	1.5856	54
7	.63090	.81316	1.2298	.77586	1.2889	1.5850	53
8	.63113	.81364	1.2290	.77568	1.2892	1.5845	52
9	.63135	.81413	1.2283	.77550	1.2895	1.5839	51
10	.63158	.81461	1.2276	.77531	1.2898	1.5833	**50**
11	.63180	.81510	1.2268	.77513	1.2901	1.5828	49
12	.63203	.81558	1.2261	.77494	1.2904	1.5822	48
13	.63225	.81606	1.2254	.77476	1.2907	1.5816	47
14	.63248	.81655	1.2247	.77458	1.2910	1.5811	46
15	.63271	.81703	1.2239	.77439	1.2913	1.5805	**45**
16	.63293	.81752	1.2232	.77421	1.2916	1.5800	44
17	.63316	.81800	1.2225	.77402	1.2919	1.5794	43
18	.63338	.81849	1.2218	.77384	1.2923	1.5788	42
19	.63361	.81898	1.2210	.77366	1.2926	1.5783	41
20	.63383	.81946	1.2203	.77347	1.2929	1.5777	**40**
21	.63406	.81995	1.2196	.77329	1.2932	1.5771	39
22	.63428	.82044	1.2189	.77310	1.2935	1.5766	38
23	.63451	.82092	1.2181	.77292	1.2938	1.5760	37
24	.63473	.82141	1.2174	.77273	1.2941	1.5755	36
25	.63496	.82190	1.2167	.77255	1.2944	1.5749	**35**
26	.63518	.82238	1.2160	.77236	1.2947	1.5744	34
27	.63540	.82287	1.2153	.77218	1.2950	1.5738	33
28	.63563	.82336	1.2145	.77199	1.2953	1.5732	32
29	.63585	.82385	1.2138	.77181	1.2957	1.5727	31
30	.63608	.82434	1.2131	.77162	1.2960	1.5721	**30**
31	.63630	.82483	1.2124	.77144	1.2963	1.5716	29
32	.63653	.82531	1.2117	.77125	1.2966	1.5710	28
33	.63675	.82580	1.2109	.77107	1.2969	1.5705	27
34	.63698	.82629	1.2102	.77088	1.2972	1.5699	26
35	.63720	.82678	1.2095	.77070	1.2975	1.5694	**25**
36	.63742	.82727	1.2088	.77051	1.2978	1.5688	24
37	.63765	.82776	1.2081	.77033	1.2981	1.5683	23
38	.63787	.82825	1.2074	.77014	1.2985	1.5677	22
39	.63810	.82874	1.2066	.76996	1.2988	1.5672	21
40	.63832	.82923	1.2059	.76977	1.2991	1.5666	**20**
41	.63854	.82972	1.2052	.76959	1.2994	1.5661	19
42	.63877	.83022	1.2045	.76940	1.2997	1.5655	18
43	.63899	.83071	1.2038	.76921	1.3000	1.5650	17
44	.63922	.83120	1.2031	.76903	1.3003	1.5644	16
45	.63944	.83169	1.2024	.76884	1.3007	1.5639	**15**
46	.63966	.83218	1.2017	.76866	1.3010	1.5633	14
47	.63989	.83268	1.2009	.76847	1.3013	1.5628	13
48	.64011	.83317	1.2002	.76828	1.3016	1.5622	12
49	.64033	.83366	1.1995	.76810	1.3019	1.5617	11
50	.64056	.83415	1.1988	.76791	1.3022	1.5611	**10**
51	.64078	.83465	1.1981	.76772	1.3026	1.5606	9
52	.64100	.83514	1.1974	.76754	1.3029	1.5601	8
53	.64123	.83564	1.1967	.76735	1.3032	1.5595	7
54	.64145	.83613	1.1960	.76717	1.3035	1.5590	6
55	.64167	.83662	1.1953	.76698	1.3038	1.5584	**5**
56	.64190	.83712	1.1946	.76679	1.3041	1.5579	4
57	.64212	.83761	1.1939	.76661	1.3045	1.5573	3
58	.64234	.83811	1.1932	.76642	1.3048	1.5568	2
59	.64256	.83860	1.1925	.76623	1.3051	1.5563	1
60	.64279	.83910	1 1918	.76604	1.3054	1.5557	**0**
′	Cos	Cot	Tan	Sin	Csc	Sec	′

129° (309°) — (230°) **50°**

NATURAL TRIGONOMETRIC FUNCTIONS TO FIVE PLACES

40° (220°) — (319°) **139°**

′	Sin	Tan	Cot	Cos	Sec	Csc	′
0	.64279	.83910	1.1918	.76604	1.3054	1.5557	**60**
1	.64301	.83960	1.1910	.76586	1.3057	1.5552	59
2	.64323	.84009	1.1903	.76567	1.3060	1.5546	58
3	.64346	.84059	1.1896	.76548	1.3064	1.5541	57
4	.64368	.84108	1.1889	.76530	1.3067	1.5536	56
5	.64390	.84158	1.1882	.76511	1.3070	1.5530	**55**
6	.64412	.84208	1.1875	.76492	1.3073	1.5525	54
7	.64435	.84258	1.1868	.76473	1.3076	1.5520	53
8	.64457	.84307	1.1861	.76455	1.3080	1.5514	52
9	.64479	.84357	1.1854	.76436	1.3083	1.5509	51
10	.64501	.84407	1.1847	.76417	1.3086	1.5504	**50**
11	.64524	.84457	1.1840	.76398	1.3089	1.5498	49
12	.64546	.84507	1.1833	.76380	1.3093	1.5493	48
13	.64568	.84556	1.1826	.76361	1.3096	1.5488	47
14	.64590	.84606	1.1819	.76342	1.3099	1.5482	46
15	.64612	.84656	1.1812	.76323	1.3102	1.5477	**45**
16	.64635	.84706	1.1806	.76304	1.3105	1.5472	44
17	.64657	.84756	1.1799	.76286	1.3109	1.5466	43
18	.64679	.84806	1.1792	.76267	1.3112	1.5461	42
19	.64701	.84856	1.1785	.76248	1.3115	1.5456	41
20	.64723	.84906	1.1778	.76229	1.3118	1.5450	**40**
21	.64746	.84956	1.1771	.76210	1.3122	1.5445	39
22	.64768	.85006	1.1764	.76192	1.3125	1.5440	38
23	.64790	.85057	1.1757	.76173	1.3128	1.5435	37
24	.64812	.85107	1.1750	.76154	1.3131	1.5429	36
25	.64834	.85157	1.1743	.76135	1.3135	1.5424	**35**
26	.64856	.85207	1.1736	.76116	1.3138	1.5419	34
27	.64878	.85257	1.1729	.76097	1.3141	1.5413	33
28	.64901	.85308	1.1722	.76078	1.3144	1.5408	32
29	.64923	.85358	1.1715	.76059	1.3148	1.5403	31
30	.64945	.85408	1.1708	.76041	1.3151	1.5398	**30**
31	.64967	.85458	1.1702	.76022	1.3154	1.5392	29
32	.64989	.85509	1.1695	.76003	1.3157	1.5387	28
33	.65011	.85559	1.1688	.75984	1.3161	1.5382	27
34	.65033	.85609	1.1681	.75965	1.3164	1.5377	26
35	.65055	.85660	1.1674	.75946	1.3167	1.5372	**25**
36	.65077	.85710	1.1667	.75927	1.3171	1.5366	24
37	.65100	.85761	1.1660	.75908	1.3174	1.5361	23
38	.65122	.85811	1.1653	.75889	1.3177	1.5356	22
39	.65144	.85862	1.1647	.75870	1.3180	1.5351	21
40	.65166	.85912	1.1640	.75851	1.3184	1.5345	**20**
41	.65188	.85963	1.1633	.75832	1.3187	1.5340	19
42	.65210	.86014	1.1626	.75813	1.3190	1.5335	18
43	.65232	.86064	1.1619	.75794	1.3194	1.5330	17
44	.65254	.86115	1.1612	.75775	1.3197	1.5325	16
45	.65276	.86166	1.1606	.75756	1.3200	1.5320	**15**
46	.65298	.86216	1.1599	.75738	1.3203	1.5314	14
47	.65320	.86267	1.1592	.75719	1.3207	1.5309	13
48	.65342	.86318	1.1585	.75700	1.3210	1.5304	12
49	.65364	.86368	1.1578	.75680	1.3213	1.5299	11
50	.65386	.86419	1.1571	.75661	1.3217	1.5294	**10**
51	.65408	.86470	1.1565	.75642	1.3220	1.5289	9
52	.65430	.86521	1.1558	.75623	1.3223	1.5283	8
53	.65452	.86572	1.1551	.75604	1.3227	1.5278	7
54	.65474	.86623	1.1544	.75585	1.3230	1.5273	6
55	.65496	.86674	1.1538	.75566	1.3233	1.5268	**5**
56	.65518	.86725	1.1531	.75547	1.3237	1.5263	4
57	.65540	.86776	1.1524	.75528	1.3240	1.5258	3
58	.65562	.86827	1.1517	.75509	1.3243	1.5253	2
59	.65584	.86878	1.1510	.75490	1.3247	1.5248	1
60	.65606	.86929	1.1504	.75471	1.3250	1.5243	**0**
′	Cos	Cot	Tan	Sin	Csc	Sec	′

130° (310°) — (229°) **49°**

41° (221°) — (318°) **138°**

′	Sin	Tan	Cot	Cos	Sec	Csc	′
0	.65606	.86929	1.1504	.75471	1.3250	1.5243	**60**
1	.65628	.86980	1.1497	.75452	1.3253	1.5237	59
2	.65650	.87031	1.1490	.75433	1.3257	1.5232	58
3	.65672	.87082	1.1483	.75414	1.3260	1.5227	57
4	.65694	.87133	1.1477	.75395	1.3264	1.5222	56
5	.65716	.87184	1.1470	.75375	1.3267	1.5217	**55**
6	.65738	.87236	1.1463	.75356	1.3270	1.5212	54
7	.65759	.87287	1.1456	.75337	1.3274	1.5207	53
8	.65781	.87338	1.1450	.75318	1.3277	1.5202	52
9	.65803	.87389	1.1443	.75299	1.3280	1.5197	51
10	.65825	.87441	1.1436	.75280	1.3284	1.5192	**50**
11	.65847	.87492	1.1430	.75261	1.3287	1.5187	49
12	.65869	.87543	1.1423	.75241	1.3291	1.5182	48
13	.65891	.87595	1.1416	.75222	1.3294	1.5177	47
14	.65913	.87646	1.1410	.75203	1.3297	1.5172	46
15	.65935	.87698	1.1403	.75184	1.3301	1.5167	**45**
16	.65956	.87749	1.1396	.75165	1.3304	1.5162	44
17	.65978	.87801	1.1389	.75146	1.3307	1.5156	43
18	.66000	.87852	1.1383	.75126	1.3311	1.5151	42
19	.66022	.87904	1.1376	.75107	1.3314	1.5146	41
20	.66044	.87955	1.1369	.75088	1.3318	1.5141	**40**
21	.66066	.88007	1.1363	.75069	1.3321	1.5136	39
22	.66088	.88059	1.1356	.75050	1.3325	1.5131	38
23	.66109	.88110	1.1349	.75030	1.3328	1.5126	37
24	.66131	.88162	1.1343	.75011	1.3331	1.5121	36
25	.66153	.88214	1.1336	.74992	1.3335	1.5116	**35**
26	.66175	.88265	1.1329	.74973	1.3338	1.5111	34
27	.66197	.88317	1.1323	.74953	1.3342	1.5107	33
28	.66218	.88369	1.1316	.74934	1.3345	1.5102	32
29	.66240	.88421	1.1310	.74915	1.3348	1.5097	31
30	.66262	.88473	1.1303	.74896	1.3352	1.5092	**30**
31	.66284	.88524	1.1296	.74876	1.3355	1.5087	29
32	.66306	.88576	1.1290	.74857	1.3359	1.5082	28
33	.66327	.88628	1.1283	.74838	1.3362	1.5077	27
34	.66349	.88680	1.1276	.74818	1.3366	1.5072	26
35	.66371	.88732	1.1270	.74799	1.3369	1.5067	**25**
36	.66393	.88784	1.1263	.74780	1.3373	1.5062	24
37	.66414	.88836	1.1257	.74760	1.3376	1.5057	23
38	.66436	.88888	1.1250	.74741	1.3380	1.5052	22
39	.66458	.88940	1.1243	.74722	1.3383	1.5047	21
40	.66480	.88992	1.1237	.74703	1.3386	1.5042	**20**
41	.66501	.89045	1.1230	.74683	1.3390	1.5037	19
42	.66523	.89097	1.1224	.74664	1.3393	1.5032	18
43	.66545	.89149	1.1217	.74644	1.3397	1.5027	17
44	.66566	.89201	1.1211	.74625	1.3400	1.5023	16
45	.66588	.89253	1.1204	.74606	1.3404	1.5018	**15**
46	.66610	.89306	1.1197	.74586	1.3407	1.5013	14
47	.66632	.89358	1.1191	.74567	1.3411	1.5008	13
48	.66653	.89410	1.1184	.74548	1.3414	1.5003	12
49	.66675	.89463	1.1178	.74528	1.3418	1.4998	11
50	.66697	.89515	1.1171	.74509	1.3421	1.4993	**10**
51	.66718	.89567	1.1165	.74489	1.3425	1.4988	9
52	.66740	.89620	1.1158	.74470	1.3428	1.4984	8
53	.66762	.89672	1.1152	.74451	1.3432	1.4979	7
54	.66783	.89725	1.1145	.74431	1.3435	1.4974	6
55	.66805	.89777	1.1139	.74412	1.3439	1.4969	**5**
56	.66827	.89830	1.1132	.74392	1.3442	1.4964	4
57	.66848	.89883	1.1126	.74373	1.3446	1.4959	3
58	.66870	.89935	1.1119	.74353	1.3449	1.4954	2
59	.66891	.89988	1.1113	.74334	1.3453	1.4950	1
60	.66913	.90040	1.1106	.74314	1.3456	1.4945	**0**
′	Cos	Cot	Tan	Sin	Csc	Sec	′

131° (311°) — (228°) **48°**

NATURAL TRIGONOMETRIC FUNCTIONS TO FIVE PLACES

42° (222°) — (317°) **137°**

′	Sin	Tan	Cot	Cos	Sec	Csc	′
0	.66913	.90040	1.1106	.74314	1.3456	1.4945	**60**
1	.66935	.90093	1.1100	.74295	1.3460	1.4940	59
2	.66956	.90146	1.1093	.74276	1.3463	1.4935	58
3	.66978	.90199	1.1087	.74256	1.3467	1.4930	57
4	.66999	.90251	1.1080	.74237	1.3470	1.4925	56
5	.67021	.90304	1.1074	.74217	1.3474	1.4921	**55**
6	.67043	.90357	1.1067	.74198	1.3478	1.4916	54
7	.67064	.90410	1.1061	.74178	1.3481	1.4911	53
8	.67086	.90463	1.1054	.74159	1.3485	1.4906	52
9	.67107	.90516	1.1048	.74139	1.3488	1.4901	51
10	.67129	.90569	1.1041	.74120	1.3492	1.4897	**50**
11	.67151	.90621	1.1035	.74100	1.3495	1.4892	49
12	.67172	.90674	1.1028	.74080	1.3499	1.4887	48
13	.67194	.90727	1.1022	.74061	1.3502	1.4882	47
14	.67215	.90781	1.1016	.74041	1.3506	1.4878	46
15	.67237	.90834	1.1009	.74022	1.3510	1.4873	**45**
16	.67258	.90887	1.1003	.74002	1.3513	1.4868	44
17	.67280	.90940	1.0996	.73983	1.3517	1.4863	43
18	.67301	.90993	1.0990	.73963	1.3520	1.4859	42
19	.67323	.91046	1.0983	.73944	1.3524	1.4854	41
20	.67344	.91099	1.0977	.73924	1.3527	1.4849	**40**
21	.67366	.91153	1.0971	.73904	1.3531	1.4844	39
22	.67387	.91206	1.0964	.73885	1.3535	1.4840	38
23	.67409	.91259	1.0958	.73865	1.3538	1.4835	37
24	.67430	.91313	1.0951	.73846	1.3542	1.4830	36
25	.67452	.91366	1.0945	.73826	1.3545	1.4825	**35**
26	.67473	.91419	1.0939	.73806	1.3549	1.4821	34
27	.67495	.91473	1.0932	.73787	1.3553	1.4816	33
28	.67516	.91526	1.0926	.73767	1.3556	1.4811	32
29	.67538	.91580	1.0919	.73747	1.3560	1.4807	31
30	.67559	.91633	1.0913	.73728	1.3563	1.4802	**30**
31	.67580	.91687	1.0907	.73708	1.3567	1.4797	29
32	.67602	.91740	1.0900	.73688	1.3571	1.4792	28
33	.67623	.91794	1.0894	.73669	1.3574	1.4788	27
34	.67645	.91847	1.0888	.73649	1.3578	1.4783	26
35	.67666	.91901	1.0881	.73629	1.3582	1.4778	**25**
36	.67688	.91955	1.0875	.73610	1.3585	1.4774	24
37	.67709	.92008	1.0869	.73590	1.3589	1.4769	23
38	.67730	.92062	1.0862	.73570	1.3592	1.4764	22
39	.67752	.92116	1.0856	.73551	1.3596	1.4760	21
40	.67773	.92170	1.0850	.73531	1.3600	1.4755	**20**
41	.67795	.92224	1.0843	.73511	1.3603	1.4750	19
42	.67816	.92277	1.0837	.73491	1.3607	1.4746	18
43	.67837	.92331	1.0831	.73472	1.3611	1.4741	17
44	.67859	.92385	1.0824	.73452	1.3614	1.4737	16
45	.67880	.92439	1.0818	.73432	1.3618	1.4732	**15**
46	.67901	.92493	1.0812	.73413	1.3622	1.4727	14
47	.67923	.92547	1.0805	.73393	1.3625	1.4723	13
48	.67944	.92601	1.0799	.73373	1.3629	1.4718	12
49	.67965	.92655	1.0793	.73353	1.3633	1.4713	11
50	.67987	.92709	1.0786	.73333	1.3636	1.4709	**10**
51	.68008	.92763	1.0780	.73314	1.3640	1.4704	9
52	.68029	.92817	1.0774	.73294	1.3644	1.4700	8
53	.68051	.92872	1.0768	.73274	1.3647	1.4695	7
54	.68072	.92926	1.0761	.73254	1.3651	1.4690	6
55	.68093	.92980	1.0755	.73234	1.3655	1.4686	**5**
56	.68115	.93034	1.0749	.73215	1.3658	1.4681	4
57	.68136	.93088	1.0742	.73195	1.3662	1.4677	3
58	.68157	.93143	1.0736	.73175	1.3666	1.4672	2
59	.68179	.93197	1.0730	.73155	1.3670	1.4667	1
60	.68200	.93252	1.0724	.73135	1.3673	1.4663	**0**
′	Cos	Cot	Tan	Sin	Csc	Sec	′

132° (312°) — (227°) **47°**

43° (223°) — (316°) **136°**

′	Sin	Tan	Cot	Cos	Sec	Csc	′
0	.68200	.93252	1.0724	.73135	1.3673	1.4663	**60**
1	.68221	.93306	1.0717	.73116	1.3677	1.4658	59
2	.68242	.93360	1.0711	.73096	1.3681	1.4654	58
3	.68264	.93415	1.0705	.73076	1.3684	1.4649	57
4	.68285	.93469	1.0699	.73056	1.3688	1.4645	56
5	.68306	.93524	1.0692	.73036	1.3692	1.4640	**55**
6	.68327	.93578	1.0686	.73016	1.3696	1.4635	54
7	.68349	.93633	1.0680	.72996	1.3699	1.4631	53
8	.68370	.93688	1.0674	.72976	1.3703	1.4626	52
9	.68391	.93742	1.0668	.72957	1.3707	1.4622	51
10	.68412	.93797	1.0661	.72937	1.3711	1.4617	**50**
11	.68434	.93852	1.0655	.72917	1.3714	1.4613	49
12	.68455	.93906	1.0649	.72897	1.3718	1.4608	48
13	.68476	.93961	1.0643	.72877	1.3722	1.4604	47
14	.68497	.94016	1.0637	.72857	1.3726	1.4599	46
15	.68518	.94071	1.0630	.72837	1.3729	1.4595	**45**
16	.68539	.94125	1.0624	.72817	1.3733	1.4590	44
17	.68561	.94180	1.0618	.72797	1.3737	1.4586	43
18	.68582	.94235	1.0612	.72777	1.3741	1.4581	42
19	.68603	.94290	1.0606	.72757	1.3744	1.4577	41
20	.68624	.94345	1.0599	.72737	1.3748	1.4572	**40**
21	.68645	.94400	1.0593	.72717	1.3752	1.4568	39
22	.68666	.94455	1.0587	.72697	1.3756	1.4563	38
23	.68688	.94510	1.0581	.72677	1.3759	1.4559	37
24	.68709	.94565	1.0575	.72657	1.3763	1.4554	36
25	.68730	.94620	1.0569	.72637	1.3767	1.4550	**35**
26	.68751	.94676	1.0562	.72617	1.3771	1.4545	34
27	.68772	.94731	1.0556	.72597	1.3775	1.4541	33
28	.68793	.94786	1.0550	.72577	1.3778	1.4536	32
29	.68814	.94841	1.0544	.72557	1.3782	1.4532	31
30	.68835	.94896	1.0538	.72537	1.3786	1.4527	**30**
31	.68857	.94952	1.0532	.72517	1.3790	1.4523	29
32	.68878	.95007	1.0526	.72497	1.3794	1.4518	28
33	.68899	.95062	1.0519	.72477	1.3797	1.4514	27
34	.68920	.95118	1.0513	.72457	1.3801	1.4510	26
35	.68941	.95173	1.0507	.72437	1.3805	1.4505	**25**
36	.68962	.95229	1.0501	.72417	1.3809	1.4501	24
37	.68983	.95284	1.0495	.72397	1.3813	1.4496	23
38	.69004	.95340	1.0489	.72377	1.3817	1.4492	22
39	.69025	.95395	1.0483	.72357	1.3820	1.4487	21
40	.69046	.95451	1.0477	.72337	1.3824	1.4483	**20**
41	.69067	.95506	1.0470	.72317	1.3828	1.4479	19
42	.69088	.95562	1.0464	.72297	1.3832	1.4474	18
43	.69109	.95618	1.0458	.72277	1.3836	1.4470	17
44	.69130	.95673	1.0452	.72257	1.3840	1.4465	16
45	.69151	.95729	1.0446	.72236	1.3843	1.4461	**15**
46	.69172	.95785	1.0440	.72216	1.3847	1.4457	14
47	.69193	.95841	1.0434	.72196	1.3851	1.4452	13
48	.69214	.95897	1.0428	.72176	1.3855	1.4448	12
49	.69235	.95952	1.0422	.72156	1.3859	1.4443	11
50	.69256	.96008	1.0416	.72136	1.3863	1.4439	**10**
51	.69277	.96064	1.0410	.72116	1.3867	1.4435	9
52	.69298	.96120	1.0404	.72095	1.3871	1.4430	8
53	.69319	.96176	1.0398	.72075	1.3874	1.4426	7
54	.69340	.96232	1.0392	.72055	1.3878	1.4422	6
55	.69361	.96288	1.0385	.72035	1.3882	1.4417	**5**
56	.69382	.96344	1.0379	.72015	1.3886	1.4413	4
57	.69403	.96400	1.0373	.71995	1.3890	1.4409	3
58	.69424	.96457	1.0367	.71974	1.3894	1.4404	2
59	.69445	.96513	1.0361	.71954	1.3898	1.4400	1
60	.69466	.96569	1.0355	.71934	1.3902	1.4396	**0**
′	Cos	Cot	Tan	Sin	Csc	Sec	′

133° (313°) — (226°) **46°**

NATURAL TRIGONOMETRIC FUNCTIONS TO FIVE PLACES

44° (224°) (315°) **135°**

′	Sin	Tan	Cot	Cos	Sec	Csc	′
0	.69466	.96569	1.0355	.71934	1.3902	1.4396	**60**
1	.69487	.96625	1.0349	.71914	1.3906	1.4391	59
2	.69508	.96681	1.0343	.71894	1.3909	1.4387	58
3	.69529	.96738	1.0337	.71873	1.3913	1.4383	57
4	.69549	.96794	1.0331	.71853	1.3917	1.4378	56
5	.69570	.96850	1.0325	.71833	1.3921	1.4374	**55**
6	.69591	.96907	1.0319	.71813	1.3925	1.4370	54
7	.69612	.96963	1.0313	.71792	1.3929	1.4365	53
8	.69633	.97020	1.0307	.71772	1.3933	1.4361	52
9	.69654	.97076	1.0301	.71752	1.3937	1.4357	51
10	.69675	.97133	1.0295	.71732	1.3941	1.4352	**50**
11	.69696	.97189	1.0289	.71711	1.3945	1.4348	49
12	.69717	.97246	1.0283	.71691	1.3949	1.4344	48
13	.69737	.97302	1.0277	.71671	1.3953	1.4340	47
14	.69758	.97359	1.0271	.71650	1.3957	1.4335	46
15	.69779	.97416	1.0265	.71630	1.3961	1.4331	**45**
16	.69800	.97472	1.0259	.71610	1.3965	1.4327	44
17	.69821	.97529	1.0253	.71590	1.3969	1.4322	43
18	.69842	.97586	1.0247	.71569	1.3972	1.4318	42
19	.69862	.97643	1.0241	.71549	1.3976	1.4314	41
20	.69883	.97700	1.0235	.71529	1.3980	1.4310	**40**
21	.69904	.97756	1.0230	.71508	1.3984	1.4305	39
22	.69925	.97813	1.0224	.71488	1.3988	1.4301	38
23	.69946	.97870	1.0218	.71468	1.3992	1.4297	37
24	.69966	.97927	1.0212	.71447	1.3996	1.4293	36
25	.69987	.97984	1.0206	.71427	1.4000	1.4288	**35**
26	.70008	.98041	1.0200	.71407	1.4004	1.4284	34
27	.70029	.98098	1.0194	.71386	1.4008	1.4280	33
28	.70049	.98155	1.0188	.71366	1.4012	1.4276	32
29	.70070	.98213	1.0182	.71345	1.4016	1.4271	31
30	.70091	.98270	1.0176	.71325	1.4020	1.4267	**30**
31	.70112	.98327	1.0170	.71305	1.4024	1.4263	29
32	.70132	.98384	1.0164	.71284	1.4028	1.4259	28
33	.70153	.98441	1.0158	.71264	1.4032	1.4255	27
34	.70174	.98499	1.0152	.71243	1.4036	1.4250	26
35	.70195	.98556	1.0147	.71223	1.4040	1.4246	**25**
36	.70215	.98613	1.0141	.71203	1.4044	1.4242	24
37	.70236	.98671	1.0135	.71182	1.4048	1.4238	23
38	.70257	.98728	1.0129	.71162	1.4052	1.4234	22
39	.70277	.98786	1.0123	.71141	1.4057	1.4229	21
40	.70298	.98843	1.0117	.71121	1.4061	1.4225	**20**
41	.70319	.98901	1.0111	.71100	1.4065	1.4221	19
42	.70339	.98958	1.0105	.71080	1.4069	1.4217	18
43	.70360	.99016	1.0099	.71059	1.4073	1.4213	17
44	.70381	.99073	1.0094	.71039	1.4077	1.4208	16
45	.70401	.99131	1.0088	.71019	1.4081	1.4204	**15**
46	.70422	.99189	1.0082	.70998	1.4085	1.4200	14
47	.70443	.99247	1.0076	.70978	1.4089	1.4196	13
48	.70463	.99304	1.0070	.70957	1.4093	1.4192	12
49	.70484	.99362	1.0064	.70937	1.4097	1.4188	11
50	.70505	.99420	1.0058	.70916	1.4101	1.4183	**10**
51	.70525	.99478	1.0052	.70896	1.4105	1.4179	9
52	.70546	.99536	1.0047	.70875	1.4109	1.4175	8
53	.70567	.99594	1.0041	.70855	1.4113	1.4171	7
54	.70587	.99652	1.0035	.70834	1.4118	1.4167	6
55	.70608	.99710	1.0029	.70813	1.4122	1.4163	**5**
56	.70628	.99768	1.0023	.70793	1.4126	1.4159	4
57	.70649	.99826	1.0017	.70772	1.4130	1.4154	3
58	.70670	.99884	1.0012	.70752	1.4134	1.4150	2
59	.70690	.99942	1.0006	.70731	1.4138	1.4146	1
60	.70711	1.0000	1.0000	.70711	1.4142	1.4142	**0**
′	Cos	Cot	Tan	Sin	Csc	Sec	′

134° (314°) (225°) **45°**

NATURAL TRIGONOMETRIC FUNCTIONS
FOR ANGLES IN RADIANS

x	Sin	Tan	Cot	Cos	x	Sin	Tan	Cot	Cos
.00	.00000	.00000	∞	1.00000	**.50**	.47943	.54630	1.8305	.87758
01	.01000	.01000	99.997	0.99995	.51	.48818	.55936	1.7878	.87274
.02	.02000	.02000	49.993	.99980	.52	.49688	.57256	1.7465	.86782
.03	.03000	.03001	33.323	.99955	.53	.50553	.58592	1.7067	.86281
.04	.03999	.04002	24.987	.99920	.54	.51414	.59943	1.6683	.85771
.05	.04998	.05004	19.983	.99875	.55	.52269	.61311	1.6310	.85252
.06	.05996	.06007	16.647	.99820	.56	.53119	.62695	1.5950	.84726
.07	.06994	.07011	14.262	.99755	.57	.53963	.64097	1.5601	.84190
.08	.07991	.08017	12.473	.99680	.58	.54802	.65517	1.5263	.83646
.09	.08988	.09024	11.081	.99595	.59	.55636	.66956	1.4935	.83094
.10	.09983	.10033	9.9666	.99500	**.60**	.56464	.68414	1.4617	.82534
.11	.10978	.11045	9.0542	.99396	.61	.57287	.69892	1.4308	.81965
.12	.11971	.12058	8.2933	.99281	.62	.58104	.71391	1.4007	.81388
.13	.12963	.13074	7.6489	.99156	.63	.58914	.72911	1.3715	.80803
.14	.13954	.14092	7.0961	.99022	.64	.59720	.74454	1.3431	.80210
.15	.14944	.15114	6.6166	.98877	.65	.60519	.76020	1.3154	.79608
.16	.15932	.16138	6.1966	.98723	.66	.61312	.77610	1.2885	.78999
.17	.16918	.17166	5.8256	.98558	.67	.62099	.79225	1.2622	.78382
.18	.17903	.18197	5.4954	.98384	.68	.62879	.80866	1.2366	.77757
.19	.18886	.19232	5.1997	.98200	.69	.63654	.82534	1.2116	.77125
.20	.19867	.20271	4.9332	.98007	**.70**	.64422	.84229	1.1872	.76484
.21	.20846	.21314	4.6917	.97803	.71	.65183	.85953	1.1634	.75836
.22	.21823	.22362	4.4719	.97590	.72	.65938	.87707	1.1402	.75181
.23	.22798	.23414	4.2709	.97367	.73	.66687	.89492	1.1174	.74517
.24	.23770	.24472	4.0864	.97134	.74	.67429	.91309	1.0952	.73847
.25	.24740	.25534	3.9163	.96891	.75	.68164	.93160	1.0734	.73169
.26	.25708	.26602	3.7591	.96639	.76	.68892	.95045	1.0521	.72484
.27	.26673	.27676	3.6133	.96377	.77	.69614	.96967	1.0313	.71791
.28	.27636	.28755	3.4776	.96106	.78	.70328	.98926	1.0109	.71091
.29	.28595	.29841	3.3511	.95824	.79	.71035	1.0092	.99084	.70385
.30	.29552	.30934	3.2327	.95534	**.80**	.71736	1.0296	.97121	.69671
.31	.30506	.32033	3.1218	.95233	.81	.72429	1.0505	.95197	.68950
.32	.31457	.33139	3.0176	.94924	.82	.73115	1.0717	.93309	.68222
.33	.32404	.34252	2.9195	.94604	.83	.73793	1.0934	.91455	.67488
.34	.33349	.35374	2.8270	.94275	.84	.74464	1.1156	.89635	.66746
.35	.34290	.36503	2.7395	.93937	.85	.75128	1.1383	.87848	.65998
.36	.35227	.37640	2.6567	.93590	.86	.75784	1.1616	.86091	.65244
.37	.36162	.38786	2.5782	.93233	.87	.76433	1.1853	.84365	.64483
.38	.37092	.39941	2.5037	.92866	.88	.77074	1.2097	.82668	.63715
.39	.38019	.41105	2.4328	.92491	.89	.77707	1.2346	.80998	.62941
.40	.38942	.42279	2.3652	.92106	**.90**	.78333	1.2602	.79355	.62161
.41	.39861	.43463	2.3008	.91712	.91	.78950	1.2864	.77738	.61375
.42	.40776	.44657	2.2393	.91309	.92	.79560	1.3133	.76146	.60582
.43	.41687	.45862	2.1804	.90897	.93	.80162	1.3409	.74578	.59783
.44	.42594	.47078	2.1241	.90475	.94	.80756	1.3692	.73034	.58979
.45	.43497	.48306	2.0702	.90045	.95	.81342	1.3984	.71511	.58168
.46	.44395	.49545	2.0184	.89605	.96	.81919	1.4284	.70010	.57352
.47	.45289	.50797	1.9686	.89157	.97	.82489	1.4592	.68531	.56530
.48	.46178	.52061	1.9208	.88699	.98	.83050	1.4910	.67071	.55702
.49	.47063	.53339	1.8748	.88233	.99	.83603	1.5237	.65631	.54869
.50	.47943	.54630	1.8305	.87758	**1.00**	.84147	1.5574	.64209	.54030
x	Sin	Tan	Cot	Cos	x	Sin	Tan	Cot	Cos

NATURAL TRIGONOMETRIC FUNCTIONS FOR ANGLES IN RADIANS

x	Sin	Tan	Cot	Cos	x	Sin	Tan	Cot	Cos
1.00	.84147	1.5574	.64209	.54030	1.50	.99749	14.101	.07091	.07074
1.01	.84683	1.5922	.62806	.53186	1.51	.99815	16.428	.06087	.06076
1.02	.85211	1.6281	.61420	.52337	1.52	.99871	19.670	.05084	.05077
1.03	.85730	1.6652	.60051	.51482	1.53	.99917	24.498	.04082	.04079
1.04	.86240	1.7036	.58699	.50622	1.54	.99953	32.461	.03081	.03079
1.05	.86742	1.7433	.57362	.49757	1.55	.99978	48.078	.02080	.02079
1.06	.87236	1.7844	.56040	.48887	1.56	.99994	92.620	.01080	.01080
1.07	.87720	1.8270	.54734	.48012	1.57	1.00000	1255.8	.00080	.00080
1.08	.88196	1.8712	.53441	.47133	1.58	.99996	−108.65	−.00920	−.00920
1.09	.88663	1.9171	.52162	.46249	1.59	.99982	−52.067	−.01921	−.01920
1.10	.89121	1.9648	.50897	.45360	1.60	.99957	−34.233	−.02921	−.02920
1.11	.89570	2.0143	.49644	.44466	1.61	.99923	−25.495	−.03922	−.03919
1.12	.90010	2.0660	.48404	.43568	1.62	.99879	−20.307	−.04924	−.04918
1.13	.90441	2.1198	.47175	.42666	1.63	.99825	−16.871	−.05927	−.05917
1.14	.90863	2.1759	.45959	.41759	1.64	.99761	−14.427	−.06931	−.06915
1.15	.91276	2.2345	.44753	.40849	1.65	.99687	−12.599	−.07397	−.07912
1.16	.91680	2.2958	.43558	.39934	1.66	.99602	−11.181	−.08944	−.08909
1.17	.92075	2.3600	.42373	.39015	1.67	.99508	−10.047	−.09953	−.09904
1.18	.92461	2.4273	.41199	.38092	1.68	.99404	− 9.1208	−.10964	−.10899
1.19	.92837	2.4979	.40034	.37166	1.69	.99290	− 8.3492	−.11977	−.11892
1.20	.93204	2.5722	.38878	.36236	1.70	.99166	− 7.6966	−.12993	−.12884
1.21	.93562	2.6503	.37731	.35302	1.71	.99033	− 7.1373	−.14011	−.13875
1.22	.93910	2.7328	.36593	.34365	1.72	.98889	− 6.6524	−.15032	−.14865
1.23	.94249	2.8198	.35463	.33424	1.73	.98735	− 6.2281	−.16056	−.15853
1.24	.94578	2.9119	.34341	.32480	1.74	.98572	− 5.8535	−.17084	−.16840
1.25	.94898	3.0096	.33227	.31532	1.75	.98399	− 5.5204	−.18115	−.17825
1.26	.95209	3.1133	.32121	.30582	1.76	.98215	− 5.2221	−.19149	−.18808
1.27	.95510	3.2236	.31021	.29628	1.77	.98022	− 4.9534	−.20188	−.19789
1.28	.95802	3.3413	.29928	.28672	1.78	.97820	− 4.7101	−.21231	−.20768
1.29	.96084	3.4672	.28842	.27712	1.79	.97607	− 4.4887	−.22278	−.21745
1.30	.96356	3.6021	.27762	.26750	1.80	.97385	− 4.2863	−.23330	−.22720
1.31	.96618	3.7471	.26687	.25785	1.81	.97153	− 4.1005	−.24387	−.23693
1.32	.96872	3.9033	.25619	.24818	1.82	.96911	− 3.9294	−.25449	−.24663
1.33	.97115	4.0723	.24556	.23848	1.83	.96659	− 3.7712	−.26517	−.25631
1.34	.97348	4.2556	.23498	.22875	1.84	.96398	− 3.6245	−.27590	−.26596
1.35	.97572	4.4552	.22446	.21901	1.85	.96128	− 3.4881	−.28669	−.27559
1.36	.97786	4.6734	.21398	.20924	1.86	.95847	− 3.3608	−.29755	−.28519
1.37	.97991	4.9131	.20354	.19945	1.87	.95557	− 2.2419	−.30846	−.29476
1.38	.98185	5.1774	.19315	.18964	1.88	.95258	− 3.1304	−.31945	−.30430
1.39	.98370	5.4707	.18279	.17981	1.89	.94949	− 3.0257	−.33051	−.31381
1.40	.98545	5.7979	.17248	.16997	1.90	.94630	− 2.9271	−.34164	−.32329
1.41	.98710	6.1654	.16220	.16010	1.91	.94302	− 2.8341	−.35284	−.33274
1.42	.98865	6.5811	.15195	.15023	1.92	.93965	− 2.7463	−.36413	−.34215
1.43	.99010	7.0555	.14173	.14033	1.93	.93618	− 2.6632	−.37549	−.35153
1.44	.99146	7.6018	.13155	.13042	1.94	.93262	− 2.5843	−.38695	−.36087
1.45	.99271	8.2381	.12139	.12050	1.95	.92896	− 2.5095	−.39849	−.37018
1.46	.99387	8.9886	.11125	.11057	1.96	.92521	− 2.4383	−.41012	−.37945
1.47	.99492	9.8874	.10114	.10063	1.97	.92137	− 2.3705	−.42185	−.38868
1.48	.99588	10.983	.09105	.09067	1.98	.91744	− 2.3058	−.43368	−.39788
1.49	.99674	12.350	.08097	.08071	1.99	.91341	− 2.2441	−.44562	−.40703
1.50	.99749	14.101	.07091	.07074	2.00	.90930	− 2.1850	−.45766	−.41615
x	Sin	Tan	Cot	Cos	x	Sin	Tan	Cot	Cos

NATURAL TRIGONOMETRIC FUNCTIONS
SECANTS AND COSECANTS
FOR ANGLES IN RADIANS

x	sec x	csc x	x	sec x	csc x	x	sec x	csc x
0.00	1.00000	∞	0.55	1.17299	1.91319	1.10	2.20460	1.12207
0.01	1.00005	100.00167	0.56	1.18028	1.88258	1.11	2.24890	1.11645
0.02	1.00020	50.00333	0.57	1.18779	1.85311	1.12	2.29525	1.11099
0.03	1.00045	33.33833	0.58	1.19551	1.82474	1.13	2.34379	1.10569
0.04	1.00080	25.00667	0.59	1.20346	1.79739	1.14	2.39467	1.10055
0.05	1.00125	20.00834	0.60	1.21163	1.77103	1.15	2.44806	1.09557
0.06	1.00180	16.67667	0.61	1.22004	1.74560	1.16	2.50413	1.09075
0.07	1.00246	14.29739	0.62	1.22868	1.72107	1.17	2.56311	1.08607
0.08	1.00321	12.51334	0.63	1.23758	1.69738	1.18	2.62519	1.08154
0.09	1.00406	11.12613	0.64	1.24673	1.67449	1.19	2.69063	1.07716
0.10	1.00502	10.01669	0.65	1.25615	1.65238	1.20	2.75970	1.07292
0.11	1.00608	9.10927	0.66	1.26584	1.63101	1.21	2.83271	1.06881
0.12	1.00724	8.35337	0.67	1.27580	1.61034	1.22	2.90997	1.06485
0.13	1.00851	7.71402	0.68	1.28605	1.59035	1.23	2.99188	1.06102
0.14	1.00988	7.16624	0.69	1.29660	1.57100	1.24	3.07885	1.05732
0.15	1.01136	6.69173	0.70	1.30746	1.55227	1.24	3.17136	1.05376
0.16	1.01294	6.27675	0.71	1.31863	1.53413	1.26	3.26993	1.05032
0.17	1.01463	5.91078	0.72	1.33013	1.51657	1.27	3.37518	1.04701
0.18	1.01642	5.58567	0.73	1.34197	1.49954	1.28	3.48778	1.04382
0.19	1.01833	5.29496	0.74	1.35415	1.48305	1.29	3.60853	1.04076
0.20	1.02034	5.03349	0.75	1.36670	1.46705	1.30	3.73833	1.03782
0.21	1.02246	4.79709	0.76	1.37962	1.45154	1.31	3.87822	1.03500
0.22	1.02470	4.58233	0.77	1.39293	1.43650	1.32	4.02941	1.03230
0.23	1.02705	4.38640	0.78	1.40664	1.42191	1.33	4.19329	1.02971
0.24	1.02951	4. 20694	0.79	1.42077	1.40775	1.34	4.37153	1.02724
0.25	1.03209	4.04197	0.80	1.43532	1.39401	1.35	4.56607	1.02488
0.26	1.03478	3.88983	0.81	1.45033	1.38067	1.36	4.77923	1.02264
0.27	1.03759	3.74909	0.82	1.46580	1.36772	1.37	5.01379	1.02050
0.28	1.04052	3.61853	0.83	1.48175	1.35514	1.38	5.27313	1.01848
0.29	1.04358	3.49709	0.84	1.49821	1.34293	1.39	5.56133	1.01657
0.30	1.04675	3.38386	0.85	1.51519	1.33106	1.40	5.88349	1.01477
0.31	1.05005	3.27806	0.86	1.53271	1.31954	1.41	6.24593	1.01307
0.32	1.05348	3.17898	0.87	1.55080	1.30834	1.42	6.65666	1.01148
0.33	1.05704	3.08601	0.88	1.56949	1.29746	1.43	7.12598	1.00999
0.34	1.06072	2.99862	0.89	1.58878	1.28688	1.44	7.66732	1.00862
0.35	1.06454	2.91632	0.90	1.60873	1.27661	1.45	8.29856	1.00734
0.36	1.06849	2.83870	0.91	1.62934	1.26662	1.46	9.04406	1.00617
0.37	1.07258	2.76537	0.92	1.65065	1.25691	1.47	9.93782	1.00510
0.38	1.07682	2.69600	0.93	1.67271	1.24747	1.48	11.02881	1.00414
0.39	1.08119	2.63027	0.94	1.69552	1.23830	1.49	12.39028	1.00327
0.40	1.08570	2.56793	0.95	1.71915	1.22938	1.50	14.13683	1.00251
0.41	1.09037	2.50872	0.96	1.74362	1.22072	1.51	16.45850	1.00185
0.42	1.09518	2.45242	0.97	1.76897	1.21229	1.52	19.69493	1.00129
0.43	1.10015	2.39882	0.98	1.79526	1.20410	1.53	24.51881	1.00083
0.44	1.10528	2.34775	0.99	1.82252	1.19614	1.54	32.47654	1.00047
0.45	1.11056	2.29903	1.00	1.85082	1.18840	1.55	48.08888	1.00022
0.46	1.11601	2.25252	1.01	1.88019	1.18087	1.56	92.62589	1.00006
0.47	1.12162	2.20806	1.02	1.91071	1.17356	1.57	+1255.76599	1.00000
0.48	1.12740	2.16554	1.03	1.94243	1.16645	1.58	− 108.65381	1.00004
0.49	1.13336	2.12483	1.04	1.97542	1.15955	1.59	− 52.07657	1.00018
0.50	1.13949	2.08583	1.05	2.00976	1.15284	1.60	− 34.24714	1.00043
0.51	1.14581	2.04844	1.06	2.04552	1.14632			
0.52	1.15231	2.01256	1.07	2.08279	1.13999			
0.53	1.15901	1.97811	1.08	2.12166	1.13384			
0.54	1.16590	1.94501	1.09	2.16223	1.12787			
0.55	1.17299	1.91319	1.10	2.20460	1.12207			

NATURAL TRIGONOMETRIC FUNCTIONS
SINE, TANGENT, COTANGENT, COSINE
FOR ANGLES IN πx RADIANS

x		Sin (πx)	Tan (πx)	Cot (πx)	Cos (πx)
.00 or	1.00	.00000	.00000	inf	1.00000
.01	.99	.03141	.03143	31.821	.99951
.02	.98	.06279	.06291	15.895	.99803
.03	.97	.09411	.09453	10.579	.99556
.04	.96	.12533	.12633	7.9158	.99211
.05	.95	.15643	.15838	6.3138	.98769
.06	.94	.18738	.19076	5.2422	.98229
.07	.93	.21814	.22353	4.4737	.97592
.08	.92	.24869	.25676	3.8947	.96858
.09	.91	.27899	.29053	3.4420	.96029
.10	.90	.30902	.32492	3.0777	.95106
.11	.89	.33874	.36002	2.7776	.94088
.12	.88	.36812	.39593	2.5257	.92978
.13	.87	.39715	.43274	2.3109	.91775
.14	.86	.42578	.47056	2.1251	.90483
.15	.85	.45399	.50953	1.9626	.89101
.16	.84	.48175	.54975	1.8190	.87631
.17	.83	.50904	.59140	1.6909	.86074
.18	.82	.53583	.63462	1.5757	.84433
.19	.81	.56208	.67960	1.4715	.82708
.20	.80	.58779	.72654	1.3764	.80902
.21	.79	.61291	.77568	1.2892	.79016
.22	.78	.63742	.82727	1.2088	.77051
.23	.77	.66131	.88162	1.1343	.75011
.24	.76	.68455	.93906	1.0649	.72897
.25	.75	.70711	1.0000	1.0000	.70711
.26	.74	.72897	1.0649	.93906	.68455
.27	.73	.75011	1.1343	.88162	.66131
.28	.72	.77051	1.2088	.82727	.63742
.29	.71	.79016	1.2892	.77568	.61291
.30	.70	.80902	1.3764	.72654	.58779
.31	.69	.82708	1.4715	.67960	.56208
.32	.68	.84433	1.5757	.63462	.53583
.33	.67	.86074	1.6909	.59140	.50904
.34	.66	.87631	1.8190	.54975	.48175
.35	.65	.89101	1.9626	.50953	.45399
.36	.64	.90483	2.1251	.47056	.42578
.37	.63	.91775	2.3109	.43274	.39715
.38	.62	.92978	2.5257	.39593	.36812
.39	.61	.94088	2.7776	.36002	.33874
.40	.60	.95106	3.0777	.32492	.30902
.41	.59	.96029	3.4420	.29053	.27899
.42	.58	.96858	3.8947	.25676	.24869
.43	.57	.97592	4.4737	.22353	.21814
.44	.56	.98229	5.2422	.19076	.18738
.45	.55	.98769	6.3138	.15838	.15643
.46	.54	.99211	7.9158	.12633	.12533
.47	.53	.99556	10.579	.09453	.09411
.48	.52	.99803	15.895	.06291	.06279
.49	.51	.99951	31.821	.03143	.03141
.50	.50	1.0000	inf	.00000	.00000

These functions are useful in the solution of wave equations such as the displacement equation of a sound wave in the form:

$$y = A \sin 2\pi n x$$

without the necessity of reducing the angular rotation either to radians or to degrees in order to find the value of the function.

The algebraic sign of the function follows the familiar Quadrant Law for the particular function desired. Thus a numerical value of 9.13π radians becomes (by the subtraction of the greatest multiple of 2π radians) 1.13π radians. This is the same as $.13\pi$ radians in the 3rd Quadrant, which would give, from the tables above, a value of the sine function of $-.39715$.

Submitted by J. A. Blythe Jr.

SINE AND COSINE FUNCTIONS FOR SPECIAL MULTIPLES OF π RADIANS

k	$\sin \frac{2\pi k}{m}$	$\cos \frac{2\pi k}{m}$	$\sin \frac{2\pi k}{m}$	$\cos \frac{2\pi k}{m}$	$\sin \frac{2\pi k}{m}$	$\cos \frac{2\pi k}{m}$
	$m = 3$		$m = 4$		$m = 5$	
1	0.86603	−0.50000	1.00000	+0.00000	0.95106	+0.30902
2			0.00000	−1.00000	0.58779	−0.80902
	$m = 6$		$m = 7$		$m = 8$	
1	0.86603	+0.50000	0.78183	+0.62349	0.70711	0.70711
2	0.86603	−0.50000	0.97493	−0.22252	1.00000	+0.00000
3	0.00000	−1.00000	0.43388	−0.90097	0.70711	−0.70711
4					0.00000	−1.00000
	$m = 9$		$m = 10$		$m = 11$	
1	0.64279	0.76604	0.58779	0.80902	0.54064	0.84125
2	0.98481	+0.17365	0.95106	+0.30902	0.90963	+0.41542
3	0.86603	−0.50000	0.95106	−0.30902	0.98982	−0.14231
4	0.34202	−0.93969	0.58779	−0.80902	0.75575	−0.65486
5			0.00000	−1.00000	0.28173	−0.95949
	$m = 12$		$m = 13$		$m = 14$	
1	0.50000	0.86603	0.46472	0.88546	0.43388	0.90097
2	0.86603	0.50000	0.82298	0.56806	0.78183	0.62349
3	1.00000	+0.00000	0.99271	+0.12054	0.97493	+0.22252
4	0.86603	−0.50000	0.93502	−0.35460	0.97493	−0.22252
5	0.50000	−0.86603	0.66312	−0.74851	0.78183	−0.62349
6	0.00000	−1.00000	0.23932	−0.97094	0.43388	−0.90097
7					0.00000	−1.00000
	$m = 15$		$m = 16$		$m = 17$	
1	0.40674	0.91355	0.38268	0.92388	0.36124	0.93247
2	0.74314	0.66913	0.70711	0.70711	0.67370	0.73901
3	0.95106	+0.30902	0.92388	0.38268	0.89516	0.44574
4	0.99452	−0.10453	1.00000	+0.00000	0.99573	+0.09227
5	0.86603	−0.50000	0.92388	−0.38268	0.96183	−0.27366
6	0.58779	−0.80902	0.70711	−0.70711	0.79802	−0.60263
7	0.20791	−0.97815	0.38268	−0.92388	0.52643	−0.85022
8			0.00000	−1.00000	0.18375	−0.98297
	$m = 18$		$m = 19$		$m = 20$	
1	0.34202	0.93969	0.32470	0.94582	0.30902	0.95106
2	0.64279	0.76604	0.61421	0.78914	0.58779	0.80902
3	0.86603	0.50000	0.83717	0.54695	0.80902	0.58779
4	0.98481	+0.17365	0.96940	+0.24549	0.95106	0.30902
5	0.98481	−0.17365	0.99658	−0.08258	1.00000	+0.00000
6	0.86603	−0.50000	0.91577	−0.40170	0.95106	−0.30902
7	0.64279	−0.76604	0.73572	−0.67728	0.80902	−0.58779
8	0.34202	−0.93969	0.47595	−0.87947	0.58779	−0.80902
9	0.00000	−1.00000	0.16459	−0.98636	0.30902	−0.95106
10					0.00000	−1.00000
	$m = 21$		$m = 22$		$m = 23$	
1	0.29476	0.95557	0.28173	0.95949	0.26980	0.96292
2	0.56332	0.82624	0.54064	0.84125	0.51958	0.85442
3	0.78183	0.62349	0.75575	0.65486	0.73084	0.68255
4	0.93087	0.36534	0.90963	0.41542	0.88789	0.46007
5	0.99720	+0.07473	0.98982	+0.14231	0.97908	+0.20346
6	0.97493	−0.22252	0.98982	−0.14231	0.99767	−0.06824
7	0.86603	−0.50000	0.90963	−0.41542	0.94226	−0.33488
8	0.68017	−0.73305	0.75575	−0.65486	0.81697	−0.57668
9	0.43388	−0.90097	0.54064	−0.84125	0.63109	−0.77571
10	0.14904	−0.98883	0.28173	−0.95949	0.39840	−0.91721
11			0.00000	−1.00000	0.13617	−0.99069
	$m = 24$		$m = 25$			
1	0.25882	0.96593	0.24869	0.96858		
2	0.50000	0.86603	0.48175	0.87631		
3	0.70711	0.70711	0.68455	0.72897		
4	0.86603	0.50000	0.84433	0.53583		
5	0.96593	0.25882	0.95106	0.30902		
6	1.00000	+0.00000	0.99803	+0.06279		
7	0.96593	−0.25882	0.98229	−0.18738		
8	0.86603	−0.50000	0.90483	−0.42578		
9	0.70711	−0.70711	0.77051	−0.63742		
10	0.50000	−0.86603	0.58779	−0.80902		
11	0.25882	−0.96593	0.36812	−0.92978		
12	0.00000	−1.00000	0.12533	−0.99211		

CHAPTER 7

HYPERBOLICS

HYPERBOLIC FUNCTIONS[1]

DEFINITIONS

$$\sinh z = \frac{e^z - e^{-z}}{2} \qquad (z = x + iy)$$

$$\cosh z = \frac{e^z + e^{-z}}{2}$$

$$\tanh z = \sinh z / \cosh z$$

$$\operatorname{csch} z = 1/\sinh z$$

$$\operatorname{sech} z = 1/\cosh z$$

$$\coth z = 1/\tanh z$$

Hyperbolic formulas can be derived from trigonometric identities by replacing z by iz

$$\sinh z = -i \sin iz$$

$$\cosh z = \cos iz$$

$$\tanh z = -i \tan iz$$

$$\operatorname{csch} z = i \csc iz$$

$$\operatorname{sech} z = \sec iz$$

$$\coth z = i \cot iz$$

---

[1] From "Handbook of Mathematical Functions", National Bureau of Standards

GRAPHS OF HYPERBOLIC FUNCTIONS

$y = \sinh x$

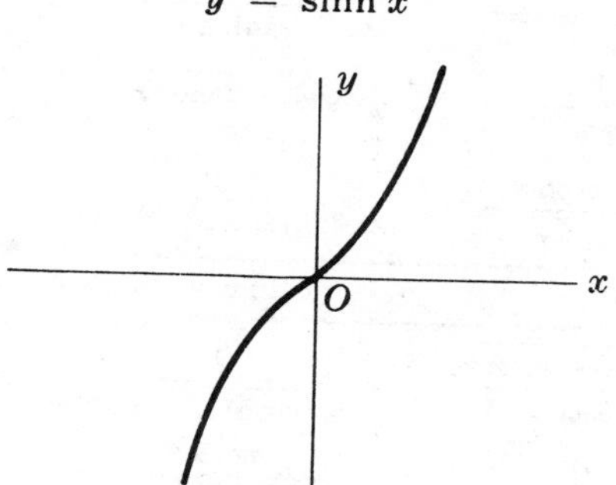

$y = \cosh x$

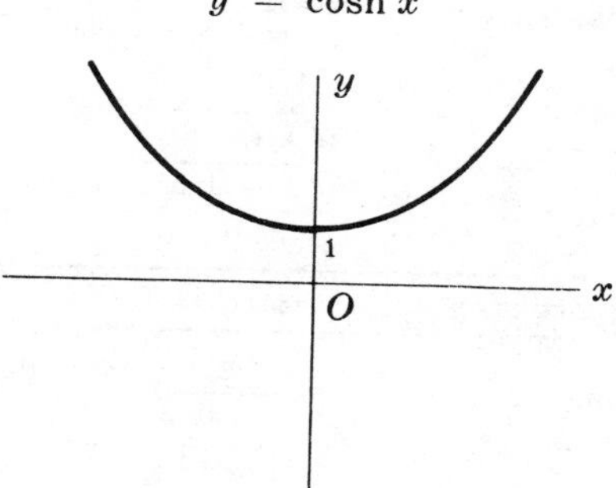

$y = \tanh x$

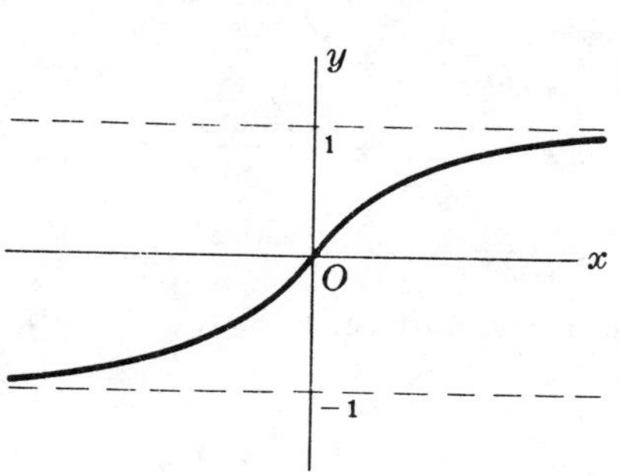

$y = \coth x$

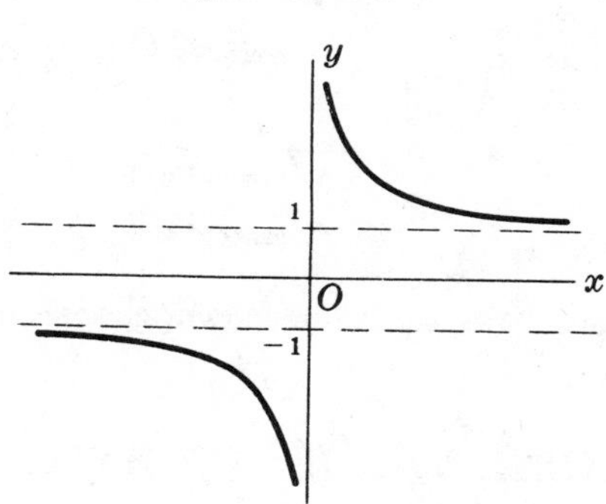

$y = \operatorname{sech} x$

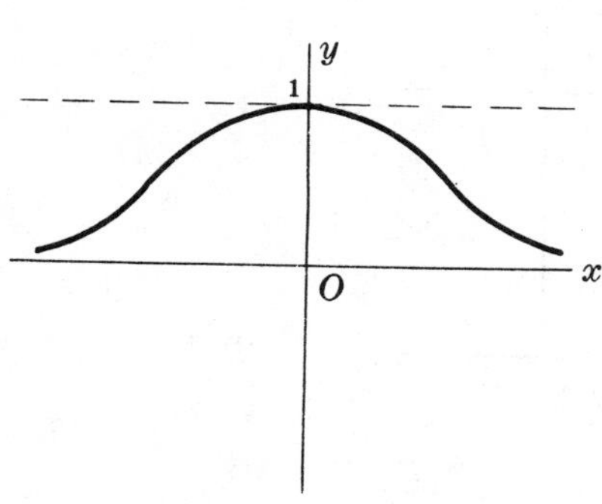

$y = \operatorname{csch} x$

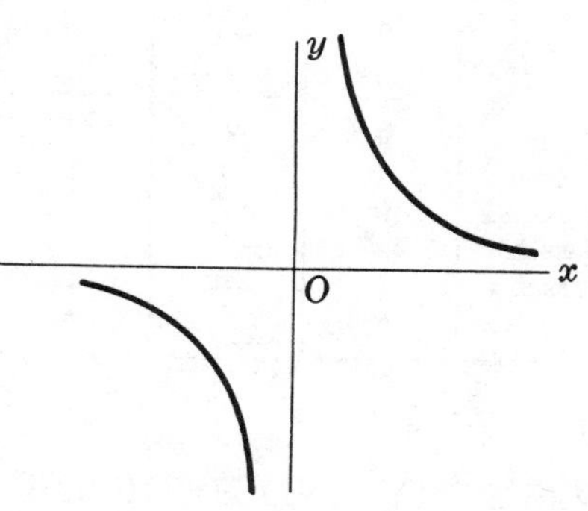

HYPERBOLIC FUNCTIONS IN TERMS OF ONE ANOTHER

Function	$\sinh x$	$\cosh x$	$\tanh x$
$\sinh x =$	$\sinh x$	$\pm\sqrt{\cosh^2 x - 1}$	$\dfrac{\tanh x}{\sqrt{1 - \tanh^2 x}}$
$\cosh x =$	$\sqrt{1 + \sinh^2 x}$	$\cosh x$	$\dfrac{1}{\sqrt{1 - \tanh^2 x}}$
$\tanh x =$	$\dfrac{\sinh x}{\sqrt{1 + \sinh^2 x}}$	$\pm\dfrac{\sqrt{\cosh^2 x - 1}}{\cosh x}$	$\tanh x$

Function	$\sinh x$	$\cosh x$	$\tanh x$
$\operatorname{cosech} x =$	$\dfrac{1}{\sinh x}$	$\pm\dfrac{1}{\sqrt{\cosh^2 x - 1}}$	$\dfrac{\sqrt{1-\tanh^2 x}}{\tanh x}$
$\operatorname{sech} x =$	$\dfrac{1}{\sqrt{1+\sinh^2 x}}$	$\dfrac{1}{\cosh x}$	$\sqrt{1-\tanh^2 x}$
$\coth x =$	$\dfrac{\sqrt{1+\sinh^2 x}}{\sinh x}$	$\dfrac{\pm \cosh x}{\sqrt{\cosh^2 x - 1}}$	$\dfrac{1}{\tanh x}$

Function	$\operatorname{cosech} x$	$\operatorname{sech} x$	$\coth x$
$\sinh x =$	$\dfrac{1}{\operatorname{cosech} x}$	$\pm\dfrac{\sqrt{1-\operatorname{sech}^2 x}}{\operatorname{sech} x}$	$\dfrac{\pm 1}{\sqrt{\coth^2 x - 1}}$
$\cosh x =$	$\pm\dfrac{\sqrt{\operatorname{cosech}^2 x + 1}}{\operatorname{cosech} x}$	$\dfrac{1}{\operatorname{sech} x}$	$\pm\dfrac{\coth x}{\sqrt{\coth^2 x - 1}}$
$\tanh x =$	$\dfrac{1}{\sqrt{\operatorname{cosech}^2 x + 1}}$	$\pm\sqrt{1-\operatorname{sech}^2 x}$	$\dfrac{1}{\coth x}$
$\operatorname{cosech} x =$	$\operatorname{cosech} x$	$\pm\dfrac{\operatorname{sech} x}{\sqrt{1-\operatorname{sech}^2 x}}$	$\pm\dfrac{\sqrt{\coth^2 x - 1}}{1}$
$\operatorname{sech} x =$	$\pm\dfrac{\operatorname{cosec} x}{\sqrt{\operatorname{cosech}^2 x + 1}}$	$\operatorname{sech} x$	$\pm\dfrac{\sqrt{\coth^2 x - 1}}{\coth x}$
$\coth x =$	$\sqrt{\operatorname{cosech}^2 x + 1}$	$\pm\dfrac{1}{\sqrt{1-\operatorname{sech}^2 x}}$	$\coth x$

Whenever two signs are shown, choose + sign if x is positive, − sign if x is negative.

SPECIAL VALUES OF HYPERBOLIC FUNCTIONS

x	0	$\frac{\pi}{2}i$	πi	$\frac{3\pi}{2}i$	∞
$\sinh x$	0	i	0	$-i$	∞
$\cosh x$	1	0	−1	0	∞
$\tanh x$	0	∞i	0	$-\infty i$	1
$\operatorname{csch} x$	∞	$-i$	∞	i	0
$\operatorname{sech} x$	1	∞	−1	∞	0
$\coth x$	∞	0	∞	0	1

SYMMETRY AND PERIODICITY

$$\sinh(-u) = -\sinh u, \qquad \operatorname{csch}(-u) = -\operatorname{csch} u$$
$$\cosh(-u) = \cosh u, \qquad \operatorname{sech}(-u) = \operatorname{sech} u$$
$$\tanh(-u) = -\tanh u, \qquad \coth(-u) = -\coth u$$

$\sinh u$, $\cosh u$, $\operatorname{cosech} u$ and $\operatorname{sech} u$ are periodic with a period $2\pi i$; $\tanh u$ and $\coth u$ are periodic with a period πi.

RECIPROCAL RELATIONS

$$\operatorname{csch} u = \frac{1}{\sinh u}, \qquad \operatorname{sech} u = \frac{1}{\cosh u}, \qquad \coth u = \frac{1}{\tanh u}$$

PRODUCT RELATIONS

$$\sinh u = \tanh u \cosh u \qquad \cosh u = \coth u \sinh u$$
$$\tanh u = \sinh u \operatorname{sech} u \qquad \coth u = \cosh u \operatorname{cosech} u$$
$$\operatorname{sech} u = \operatorname{cosech} u \tanh u \qquad \operatorname{cosech} u = \operatorname{sech} u \coth u$$

QUOTIENT RELATIONS

$$\sinh u = \frac{\tanh u}{\operatorname{sech} u} \qquad \cosh u = \frac{\coth u}{\operatorname{cosech} u} \qquad \tanh u = \frac{\sinh u}{\cosh u}$$
$$\operatorname{cosech} u = \frac{\operatorname{sech} u}{\tanh u} \qquad \operatorname{sech} u = \frac{\operatorname{cosech} u}{\coth u} \qquad \coth u = \frac{\cosh u}{\sinh u}$$

ANGLE-SUM AND ANGLE-DIFFERENCE RELATIONS

$$\sinh (u + v) = \sinh u \cosh v + \cosh u \sinh v$$
$$\sinh (u - v) = \sinh u \cosh v - \cosh u \sinh v$$
$$\cosh (u + v) = \cosh u \cosh v + \sinh u \sinh v$$
$$\cosh (u - v) = \cosh u \cosh v - \sinh u \sinh v$$
$$\tanh (u + v) = \frac{\tanh u + \tanh v}{1 + \tanh u \tanh v} = \frac{\sinh 2u + \sinh 2v}{\cosh 2u + \cosh 2v}$$
$$\tanh (u - v) = \frac{\tanh u - \tanh v}{1 - \tanh u \tanh v} = \frac{\sinh 2u - \sinh 2v}{\cosh 2u - \cosh 2v}$$
$$\coth (u + v) = \frac{1 + \coth u \coth v}{\coth u + \coth v} = \frac{\sinh 2u - \sinh 2v}{\cosh 2u - \cosh 2v}$$
$$\coth (u - v) = \frac{1 - \coth u \coth v}{\coth u - \coth v} = \frac{\sinh 2u + \sinh 2v}{\cosh 2u - \cosh 2v}$$

MULTIPLE ANGLE RELATIONS

$$\sinh 2u = 2 \sinh u \cosh u = \frac{2 \tanh u}{1 - \tanh^2 u}$$
$$\cosh 2u = \cosh^2 u + \sinh^2 u = 2 \cosh^2 u - 1 = 1 + 2 \sinh^2 u = \frac{1 + \tanh^2 u}{1 - \tanh^2 u}$$
$$\tanh 2u = \frac{2 \tanh u}{1 + \tanh^2 u}$$
$$\coth 2u = \frac{\coth^2 u + 1}{2 \coth u}$$

$$\sinh 3u = 3 \sinh u + 4 \sinh^3 u = \sinh u (4 \cosh^2 u - 1)$$
$$\cosh 3u = 4 \cosh^3 u - 3 \cosh u = \cosh u (1 + 4 \sinh^2 u)$$

$$\tanh 3u = \frac{3 \tanh u + \tanh^3 u}{1 + 3 \tanh^2 u}$$

$$\coth 3u = \frac{3 \coth u + \coth^3 u}{1 + 3 \coth^2 u}$$

$$
\begin{aligned}
\sinh 4u &= 4 \sinh u \cosh u(2 \cosh^2 u - 1) \\
&= 4 \sinh u \cosh u(1 + 2 \sinh^2 u) \\
&= 4 \sinh u \cosh u(\cosh^2 u + \sinh^2 u) \\
\cosh 4u &= 1 + 8 \cosh^2 u(\cosh^2 u - 1) \\
&= 1 + 8 \sinh^2 u(\sinh^2 u + 1) \\
&= \cosh^4 u + 6 \sinh^2 u \cosh^2 u + \sinh^4 u \\
\tanh 4u &= \frac{4 \tanh u(1 + \tanh^2 u)}{1 + 6 \tanh^2 u + \tanh^4 u} \\
\coth 4u &= \frac{\coth^4 u + 6 \coth^2 u + 1}{4 \coth u(\coth^2 u + 1)} \\
\sinh 5u &= \sinh u(16 \sinh^4 u + 20 \sinh^2 u + 5) \\
&= \sinh u(16 \cosh^4 u - 12 \cosh^2 u + 1) \\
\cosh 5u &= \cosh u(16 \cosh^4 u - 20 \cosh^2 u + 5) \\
&= \cosh u(16 \sinh^4 u + 12 \sinh^2 u + 1) \\
\sinh 6u &= 2 \sinh u \cosh u(16 \cosh^4 u - 16 \cosh^2 u + 3) \\
&= 2 \sinh u \cosh u(16 \sinh^4 u + 16 \sinh^2 u + 3) \\
\cosh 6u &= 32 \cosh^6 u - 48 \cosh^4 u + 18 \cosh^2 u - 1 \\
&= 32 \sinh^6 u + 48 \sinh^4 u + 18 \sinh^2 u + 1 \\
\sinh nu &= \sinh u \Bigg[(2 \cosh u)^{n-1} - \frac{(n-2)}{1!} \cdot (2 \cosh u)^{n-3} + \frac{(n-3)(n-4)}{2!} \\
&\qquad \cdot (2 \cosh u)^{n-5} - \frac{(n-4)(n-5)(n-6)}{3!} (2 \cosh u)^{n-7} + \cdots \Bigg]. \\
\cosh nu &= \frac{1}{2} \Bigg[(2 \cosh u)^n - \frac{n}{1!} (2 \cosh u)^{n-2} + \frac{n(n-3)}{2!} (2 \cosh u)^{n-4} \\
&\qquad - \frac{n(n-4)(n-5)}{3!} (2 \cosh u)^{n-6} + \cdots \Bigg].
\end{aligned}
$$

HALF ANGLE RELATIONS

$$
\begin{aligned}
\sinh \tfrac{1}{2}u &= \pm \sqrt{\tfrac{1}{2}(\cosh u - 1)} \\
\cosh \tfrac{1}{2}u &= \sqrt{\tfrac{1}{2}(\cosh u + 1)} \\
\tanh \frac{u}{2} &= \frac{\sinh u}{1 + \cosh u} = \frac{\cosh u - 1}{\sinh u} = \pm \sqrt{\frac{\cosh u - 1}{\cosh u + 1}} \\
\coth \frac{u}{2} &= \frac{1 + \cosh u}{\sinh u} = \frac{\sinh u}{\cosh u - 1} = \pm \sqrt{\frac{\cosh u + 1}{\cosh u - 1}}
\end{aligned}
$$

Choose + sign if u is positive, otherwise choose the − sign.

RELATIONS BETWEEN SQUARES OF FUNCTIONS

$$\cosh^2 u - \sinh^2 u = 1, \qquad \tanh^2 u + \operatorname{sech}^2 u = 1$$
$$\coth^2 u - \operatorname{csch}^2 u = 1, \qquad \operatorname{csch}^2 u - \operatorname{sech}^2 u = \operatorname{csch}^2 u \operatorname{sech}^2 u$$

FUNCTION SUM AND FUNCTION DIFFERENCE RELATIONS

$$\sinh u + \sinh v = 2 \sinh \tfrac{1}{2}(u+v) \cosh \tfrac{1}{2}(u-v)$$
$$\sinh u - \sinh v = 2 \cosh \tfrac{1}{2}(u+v) \sinh \tfrac{1}{2}(u-v)$$
$$\cosh u + \cosh v = 2 \cosh \tfrac{1}{2}(u+v) \cosh \tfrac{1}{2}(u-v)$$
$$\cosh u - \cosh v = 2 \sinh \tfrac{1}{2}(u+v) \sinh \tfrac{1}{2}(u-v)$$

$$\tanh u + \tanh v = (1 + \tanh u \tanh v) \tanh (u+v) = \frac{\sinh (u+v)}{\cosh u \cosh v}$$

$$\tanh u - \tanh v = (1 - \tanh u \tanh v) \tanh (u-v) = \frac{\sinh (u-v)}{\cosh u \cosh v}$$

$$\coth u + \coth v = \frac{1 + \coth u \coth v}{\coth (u+v)} = \frac{\sinh (u+v)}{\sinh u \sinh v}$$

$$\coth u - \coth v = \frac{1 - \coth u \coth v}{\coth (u-v)} = \frac{\sinh (u-v)}{\sinh u \sinh v}$$

$$\sinh u + \cosh u = \frac{1 + \tanh \frac{1}{2}u}{1 - \tanh \frac{1}{2}u} = e^{u} \qquad \cosh u - \sinh u = \frac{1 - \tanh \frac{1}{2}u}{1 + \tanh \frac{1}{2}u} = e^{-u}$$

FUNCTION PRODUCT RELATIONS

$$\sinh u \cosh v = \tfrac{1}{2} \sinh (u+v) + \tfrac{1}{2} \sinh (u-v)$$
$$\cosh u \sinh v = \tfrac{1}{2} \sinh (u+v) - \tfrac{1}{2} \sinh (u-v)$$
$$\cosh u \cosh v = \tfrac{1}{2} \cosh (u+v) + \tfrac{1}{2} \cosh (u-v)$$
$$\sinh u \sinh v = \tfrac{1}{2} \cosh (u+v) - \tfrac{1}{2} \cosh (u-v)$$
$$\sinh (u+v) \sinh (u-v) = \sinh^2 u - \sinh^2 v = \cosh^2 u - \cosh^2 v$$
$$\cosh (u+v) \cosh (u-v) = \sinh^2 u + \cosh^2 v = \cosh^2 u + \sinh^2 v$$

POWER RELATIONS

$$\sinh^2 u = \tfrac{1}{2}(\cosh 2u - 1)$$
$$\cosh^2 u = \tfrac{1}{2}(\cosh 2u + 1)$$
$$\sinh^3 u = \tfrac{1}{4}(-3 \sinh u + \sinh 3u)$$
$$\cosh^3 u = \tfrac{1}{4}(3 \cosh u + \cosh 3u)$$
$$\sinh^4 u = \tfrac{1}{8}(3 - 4 \cosh 2u + \cosh 4u)$$
$$\cosh^4 u = \tfrac{1}{8}(3 + 4 \cosh 2u + \cosh 4u)$$
$$\sinh^5 u = \tfrac{1}{16}(10 \sinh u - 5 \sinh 3u + \sinh 5u)$$
$$\cosh^5 u = \tfrac{1}{16}(10 \cosh u + 5 \cosh 3u + \cosh 5u)$$
$$\sinh^6 u = \tfrac{1}{32}(-10 + 15 \cosh 2u - 6 \cosh 4u + \cosh 6u)$$
$$\cosh^6 u = \tfrac{1}{32}(10 + 15 \cosh 2u + 6 \cosh 4u + \cosh 6u)$$
$$(\cosh u \pm \sinh u)^n = \cosh nu \pm \sinh nu$$

RELATIONS WITH CIRCULAR FUNCTIONS

$$\sinh iu = i \sin u, \qquad \sinh u = -i \sin iu$$
$$\cosh iu = \cos u, \qquad \cosh u = \cos iu$$
$$\tanh iu = i \tan u, \qquad \tanh u = -i \tan iu$$
$$\operatorname{cosech} iu = -i \operatorname{cosec} u \qquad \operatorname{cosech} u = i \operatorname{cosec} iu$$
$$\operatorname{sech} iu = \sec u \qquad \operatorname{sech} u = \sec iu$$
$$\coth iu = -i \cot u \qquad \coth u = i \coth iu$$

HYPERBOLIC FUNCTIONS OF COMPLEX ARGUMENT

$$\sinh(u + iv) = \sinh u \cos v + i \cosh u \sin v$$
$$\sinh(u - iv) = \sinh u \cos v - i \cosh u \sin v$$
$$\cosh(u + iv) = \cosh u \cos v + i \sinh u \sin v$$
$$\cosh(u - iv) = \cosh u \cos v - i \sinh u \sin v$$
$$\tanh(u + iv) = \frac{\sinh 2u + i \sin 2v}{\cosh 2u + \cos 2v}$$
$$\tanh(u - iv) = \frac{\sinh 2u - i \sin 2v}{\cosh 2u + \cos 2v}$$
$$\coth(u + iv) = \frac{\sinh 2u - i \sin 2v}{\cosh 2u - \cos 2v}$$
$$\coth(u - iv) = \frac{\sinh 2u + i \sin 2v}{\cosh 2u - \cos 2v}$$

$$\sinh(u + \tfrac{1}{2}\pi i) = i \cosh u, \qquad \cosh(u + \tfrac{1}{2}\pi i) = i \sinh u$$
$$\sinh(u + \pi i) = -\sinh u, \qquad \cosh(u + \pi i) = -\cosh u$$
$$\sinh(u + 2\pi i) = \sinh u, \qquad \cosh(u + 2\pi i) = \cosh u$$

SERIES FOR HYPERBOLIC FUNCTIONS

(see series expansions for sinh nu and cosh nu under multiple angle relations).

$$\sinh x = x + \frac{x^3}{3!} + \frac{x^5}{5!} + \frac{x^7}{7!} + \cdots + \frac{x^{2n+1}}{(2n+1)!} + \cdots \qquad |x| < \infty$$

$$\sinh ax = \frac{2}{\pi} \sinh \pi a \left[\frac{\sin x}{a^2 + 1^2} - \frac{2 \sin 2x}{a^2 + 2^2} + \frac{3 \sin 3x}{a^2 + 3^2} - + \cdots\right]$$
$$= \frac{2}{\pi} \sinh \pi a \sum_{n=1}^{\infty} (-1)^{n+1} \frac{n \sin nx}{n^2 + a^2}, \qquad |x| < \pi$$

$$\cosh x = 1 + \frac{x^2}{2!} + \frac{x^4}{4!} + \frac{x^6}{6!} + \cdots + \frac{x^{2n}}{(2n)!} + \cdots \qquad |x| < \infty$$

$$\cosh ax = \frac{2a}{\pi} \sinh \pi a \left[\frac{1}{2a^2} - \frac{\cos x}{a^2 + 1^2} + \frac{\cos 2x}{a^2 + 2^2} - \frac{\cos 3x}{a^2 + 3^2} + - \cdots\right]$$
$$= \frac{\sinh \pi a}{a\pi} + \frac{2a}{\pi} \sinh \pi a \sum_{n=1}^{\infty} (-1)^n \frac{\cos nx}{n^2 + a^2}, \qquad |x| < \pi$$

$$\tanh x = x - \frac{1}{3} x^3 + \frac{2}{15} x^5 - \frac{17}{315} x^7 + \frac{62}{2835} x^9 - \cdots$$
$$+ \frac{2^{2n}(2^{2n} - 1) B_{2n} x^{2n-1}}{(2n)!} \pm \cdots \text{ (1)} \qquad |x| < \frac{\pi}{2}$$

$$\tanh x = 1 - 2e^{-2x} + 2e^{-4x} - 2e^{-6x} + - \cdots$$
$$= 1 + 2 \sum_{n=1}^{\infty} (-1)^n e^{-2nx}, \qquad \mathrm{Re}\,(x) > 0$$

$$\tanh x = 2x \left[\frac{1}{\left(\frac{\pi}{2}\right)^2 + x^2} + \frac{1}{\left(\frac{3\pi}{2}\right)^2 + x^2} + \frac{1}{\left(\frac{5\pi}{2}\right)^2 + x^2} + \cdots\right]$$

$$= 2x \sum_{n=0}^{\infty} \frac{1}{\left(n + \frac{1}{2}\right)^2 \pi^2 + x^2}$$

$$\coth x = \frac{1}{x} + \frac{x}{3} - \frac{x^3}{45} + \frac{2x^5}{945} - \frac{x^7}{4725} + \cdots + \frac{2^{2n}}{(2n)!} B_{2n} x^{2n-1} \pm \cdots \text{(1)}$$

$$0 < |x| < \pi$$

$$\coth x = 1 + 2e^{-2x} + 2e^{-4x} + 2e^{-6x} + \cdots$$

$$= 1 + 2 \sum_{n=1}^{\infty} e^{-2nx} \qquad \mathrm{Re}\,(x) > 0$$

$$\coth x = \frac{1}{x} + 2x \left[\frac{1}{\pi^2 + x^2} + \frac{1}{(2\pi)^2 + x^2} + \frac{1}{(3\pi)^2 + x^2} + \cdots\right]$$

$$= \frac{1}{x} + 2x \sum_{n=1}^{\infty} \frac{1}{(n\pi)^2 + x^2}$$

$$\operatorname{sech} x = 1 - \frac{1}{2!} x^2 + \frac{5}{4!} x^4 - \frac{61}{6!} x^6 + \frac{1385}{8!} x^8 - \cdots + \frac{E_{2n} x^{2n}}{(2n)!} \pm \cdots \text{(2)}$$

$$|x| < \frac{\pi}{2}$$

$$\operatorname{sech} x = 2e^{-x} - 2e^{-3x} + 2e^{-5x} - 2e^{-7x} + - \cdots$$

$$= 2 \sum_{n=0}^{\infty} (-1)^n e^{-(2n+1)x}, \quad \mathrm{Re}\,(x) > 0$$

$$\operatorname{sech} x = 4\pi \left[\frac{1}{\pi^2 + 4x^2} - \frac{3}{(3\pi)^2 + 4x^2} + \frac{5}{(5\pi)^2 + 4x^2} - + \cdots\right]$$

$$= 4\pi \sum_{n=0}^{\infty} \frac{(-1)^n (2n+1)}{(2n+1)^2 \pi^2 + 4x^2}$$

$$\operatorname{cosech} x = \frac{1}{x} - \frac{x}{6} + \frac{7x^3}{360} - \frac{31x^5}{15{,}120} + \cdots - \frac{2(2^{2n-1} - 1)}{(2n)!} B_{2n} x^{2n-1} + \cdots \text{(1)}$$

$$0 < |x| < \pi$$

$$\operatorname{cosech} x = 2e^{-x} + 2e^{-3x} + 2e^{-5x} + \cdots$$

$$= 2 \sum_{n=0}^{\infty} e^{-(2n+1)x}, \quad \mathrm{Re}\,(x) > 0$$

$$\operatorname{cosech} x = \frac{1}{x} - \frac{2x}{\pi^2 + x^2} + \frac{2x}{(2\pi)^2 + x^2} - \frac{2x}{(3\pi)^2 + x^2} + - \cdots$$

$$= \frac{1}{x} + 2x \sum_{n=1}^{\infty} \frac{(-1)^n}{(n\pi)^2 + x^2}$$

(1) B_{2n} denotes Bernoulli numbers.

INVERSE HYPERBOLIC FUNCTIONS

DEFINITIONS

If $x = \sinh y$, then $y = \sinh^{-1} x$. Other inverse functions are denned similarly.

DOMAIN AND RANGE

Function	Domain	Range	Remarks
$\sinh^{-1} x$	$(-\infty, +\infty)$	$(-\infty, +\infty)$	odd function
$\cosh^{-1} x$	$[1, +\infty)$	$(-\infty, +\infty)$	even function double valued
$\tanh^{-1} x$	$(-1, 1)$	$(-\infty, +\infty)$	odd function
$\operatorname{cosech}^{-1} x$	$(-\infty, 0), (0, \infty)$	$(0, -\infty), (\infty, 0)$	odd function two branches, pole at $x = 0$
$\operatorname{sech}^{-1} x$	$(0, 1]$	$(-\infty, +\infty)$	double valued
$\coth^{-1} x$	$(-\infty, -1), (1, \infty)$	$(0, -\infty), (\infty, 0)$	odd function two branches

GRAPHS OF INVERSE FUNCTIONS

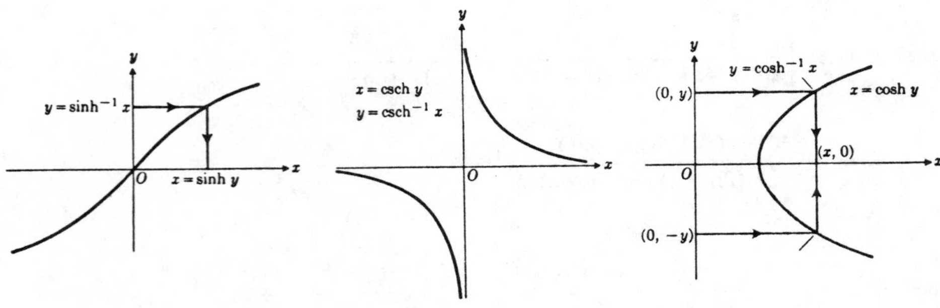

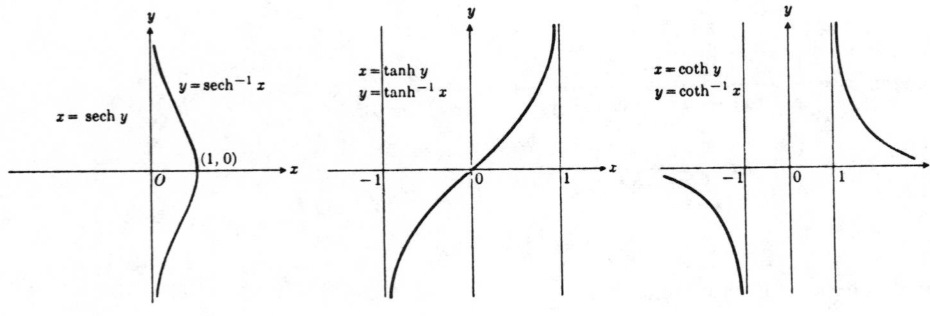

INVERSE HYPERBOLIC FUNCTIONS IN TERMS OF ONE ANOTHER

Function	$\sinh^{-1} x$	$\cosh^{-1} x$ + if $x > 0$ − if $x < 0$	$\tanh^{-1} x$
$\sinh^{-1} x =$	$\sinh^{-1} x$	$\pm \cosh^{-1} \sqrt{x^2 + 1}$	$\tanh^{-1} \frac{x}{\sqrt{1 + x^2}}$
$\cosh^{-1} x =$	$\pm \sinh^{-1} \sqrt{x^2 - 1}$	$\cosh^{-1} x$	$\pm \tanh^{-1} \frac{\sqrt{x^2 - 1}}{x}$
$\tanh^{-1} x =$	$\sinh^{-1} \frac{x}{\sqrt{1 - x^2}}$	$\pm \cosh^{-1} \frac{1}{\sqrt{1 - x^2}}$	$\tanh^{-1} x$
$\operatorname{cosech}^{-1} x =$	$\sinh^{-1} \frac{1}{x}$	$\pm \cosh^{-1} \frac{\sqrt{x^2 + 1}}{x}$	$\tanh^{-1} \frac{1}{\sqrt{1 + x^2}}$
$\operatorname{sech}^{-1} x =$	$\pm \sinh^{-1} \frac{\sqrt{1 - x^2}}{x}$	$\cosh^{-1} \frac{1}{x}$	$\pm \tanh^{-1} \sqrt{1 - x^2}$
$\coth^{-1} x =$	$\sinh^{-1} \frac{1}{\sqrt{x^2 - 1}}$	$\pm \cosh^{-1} \frac{x}{\sqrt{x^2 - 1}}$	$\tanh^{-1} \frac{1}{x}$

Function	$\operatorname{cosech}^{-1} x$	$\operatorname{sech}^{-1} x$ + if $x > 0$ − if $x < 0$	$\coth^{-1} x$
$\sinh^{-1} x =$	$\operatorname{cosech}^{-1} \frac{1}{x}$	$\pm \operatorname{sech}^{-1} \frac{1}{\sqrt{1 + x^2}}$	$\coth^{-1} \frac{\sqrt{1 + x^2}}{x}$
$\cosh^{-1} x =$	$\pm \operatorname{cosech}^{-1} \frac{1}{\sqrt{x^2 - 1}}$	$\operatorname{sech}^{-1} \frac{1}{x}$	$\pm \coth^{-1} \frac{x}{\sqrt{x^2 - 1}}$
$\tanh^{-1} x =$	$\operatorname{cosech}^{-1} \frac{\sqrt{1 - x^2}}{x}$	$\pm \operatorname{sech}^{-1} \sqrt{1 - x^2}$	$\coth^{-1} \frac{1}{x}$
$\operatorname{cosech}^{-1} x =$	$\operatorname{cosech}^{-1} x$	$\pm \operatorname{sech}^{-1} \frac{x}{\sqrt{x^2 + 1}}$	$\coth^{-1} \sqrt{1 + x^2}$
$\operatorname{sech}^{-1} x =$	$\pm \operatorname{cosech}^{-1} \frac{x}{\sqrt{1 - x^2}}$	$\operatorname{sech}^{-1} x$	$\pm \coth^{-1} \frac{1}{\sqrt{1 - x^2}}$
$\coth^{-1} x =$	$\operatorname{cosech}^{-1} \sqrt{x^2 - 1}$	$\operatorname{sech}^{-1} \frac{\sqrt{x^2 - 1}}{x}$	$\coth^{-1} x$

FUNDAMENTAL IDENTITIES

RELATIONS WITH LOGARITHMIC FUNCTIONS

$\sinh^{-1} x = \log_e (x + \sqrt{x^2 + 1})$

$\cosh^{-1} x = \log_e (x \pm \sqrt{x^2 - 1}), \quad x \geqq 1.$ The plus sign is used for the principal value.

$$\tanh^{-1} x = \tfrac{1}{2} \log_e \left(\frac{1+x}{1-x}\right), \quad x^2 < 1$$

$$\operatorname{csch}^{-1} x = \log_e \left(\frac{1 \pm \sqrt{1+x^2}}{x}\right).$$ The plus sign is used if $x > 0$, the minus sign if $x < 0$.

$$\operatorname{sech}^{-1} x = \log_e \left(\frac{1 \pm \sqrt{1-x^2}}{x}\right), \quad 0 < x \leq 1.$$ The plus sign is used for the principal values.

$$\coth^{-1} x = \tfrac{1}{2} \log_e \left(\frac{x+1}{x-1}\right), \quad x^2 > 1$$

RELATIONS WITH CIRCULAR FUNCTIONS

$$\sinh^{-1} x = -i \sin^{-1} (ix) \qquad \sinh^{-1} (ix) = i \sin^{-1} x$$
$$\cosh^{-1} x = \pm i \cos^{-1} x \qquad \cosh^{-1} (ix) = \pm i \cos^{-1} (ix)$$
$$\tanh^{-1} x = -i \tan^{-1} (ix) \qquad \tanh^{-1} (ix) = i \tan^{-1} x$$
$$\operatorname{cosech}^{-1} x = i \operatorname{cosec}^{-1} (ix) \qquad \operatorname{cosech}^{-1} (ix) = -i \operatorname{cosec}^{-1} x$$
$$\operatorname{sech}^{-1} x = \pm i \sec^{-1} x \qquad \operatorname{sech}^{-1} (ix) = \pm i \sec^{-1} (ix)$$
$$\coth^{-1} x = i \coth^{-1} (ix) \qquad \coth^{-1} (ix) = -i \cot^{-1} (x)$$

FUNCTION SUM AND FUNCTION DIFFERENCE RELATIONS

$$\sinh^{-1} x + \sinh^{-1} y = \sinh^{-1} (x\sqrt{1+y^2} + y\sqrt{1+x^2})$$
$$\sinh^{-1} x - \sinh^{-1} y = \sinh^{-1} (x\sqrt{1+y^2} - y\sqrt{1+x^2})$$
$$\cosh^{-1} x + \cosh^{-1} y = \cosh^{-1} (xy + \sqrt{(x^2-1)(y^2-1)})$$
$$\cosh^{-1} x - \cosh^{-1} y = \cosh^{-1} (xy - \sqrt{(x^2-1)(y^2-1)})$$
$$\tanh^{-1} x + \tanh^{-1} y = \tanh^{-1} \left(\frac{x+y}{1+xy}\right)$$
$$\tanh^{-1} x - \tanh^{-1} y = \tanh^{-1} \left(\frac{x-y}{1-xy}\right)$$
$$\sinh^{-1} x + \cosh^{-1} y = \sinh^{-1} (xy + \sqrt{(1+x^2)(y^2-1)})$$
$$= \cosh^{-1} (y\sqrt{1+x^2} + x\sqrt{y^2-1})$$
$$\sinh^{-1} x - \cosh^{-1} y = \sinh^{-1} (xy - \sqrt{(1+x^2)(y^2-1)})$$
$$= \cosh^{-1} (y\sqrt{1+x^2} - x\sqrt{y^2-1})$$
$$\tanh^{-1} x + \coth^{-1} y = \tanh^{-1} \left(\frac{xy+1}{y+x}\right)$$
$$= \coth^{-1} \left(\frac{y+x}{xy+1}\right)$$
$$\tanh^{-1} x - \coth^{-1} y = \tanh^{-1} \left(\frac{xy-1}{y-x}\right)$$
$$= \coth^{-1} \left(\frac{y-x}{xy-1}\right)$$

SERIES EXPANSIONS

$$\sinh^{-1} x = x - \frac{1}{2 \cdot 3} x^3 + \frac{1 \cdot 3}{2 \cdot 4 \cdot 5} x^5 - \frac{1 \cdot 3 \cdot 5}{2 \cdot 4 \cdot 6 \cdot 7} x^7 + \cdots + (-1)^n \cdot \frac{1 \cdot 3 \cdot 5 \ldots (2n-1)}{2 \cdot 4 \cdot 6 \ldots 2n(2n+1)} x^{2n+1} \pm \cdots \qquad |x| < 1$$

$$\sinh^{-1} x = \ln (2x) + \frac{1}{2} \cdot \frac{1}{2x^2} - \frac{1 \cdot 3}{2 \cdot 4} \cdot \frac{1}{4x^4} + \frac{1 \cdot 3 \cdot 5}{2 \cdot 4 \cdot 6} \cdot \frac{1}{6x^6} - \cdots$$

$$= \ln (2x) + \sum_{n=1}^{\infty} (-1)^{n+1} \frac{(2n)!x^{-2n}}{2^{2n}(n!)^2 2n}, \qquad |x| > 1$$

$$\cosh^{-1} x = \pm \left[\ln (2x) - \frac{1}{2 \cdot 2x^2} - \frac{1 \cdot 3}{2 \cdot 4 \cdot 4x^4} - \frac{1 \cdot 3 \cdot 5}{2 \cdot 4 \cdot 6 \cdot 6x^6} - \cdots \right] \qquad x > 1$$

$$\operatorname{cosech}^{-1} x = \frac{1}{x} - \frac{1}{2} \cdot \frac{1}{3x^3} + \frac{1 \cdot 3}{2 \cdot 4} \cdot \frac{1}{5x^5} - \frac{1 \cdot 3 \cdot 5}{2 \cdot 4 \cdot 6} \cdot \frac{1}{7x^7} + - \cdots$$

$$= \sum_{n=0}^{\infty} (-1)^n \frac{(2n)!x^{-2n-1}}{2^{2n}(n!)^2(2n+1)}, \qquad |x| > 1$$

$$\operatorname{cosech}^{-1} x = \ln \frac{2}{x} + \frac{1}{2} \cdot \frac{x^2}{2} - \frac{1 \cdot 3}{2 \cdot 4} \cdot \frac{x^4}{4} + \frac{1 \cdot 3 \cdot 5}{2 \cdot 4 \cdot 6} \cdot \frac{x^6}{6} - + \cdots$$

$$= \ln \frac{2}{x} + \sum_{n=1}^{\infty} \frac{(-1)^{n+1}(2n)!x^{2n}}{2^{2n}(n!)^2 2n}, \qquad 0 < x < 1$$

$$\operatorname{sech}^{-1} x = \ln \frac{2}{x} - \frac{1}{2}\frac{x^2}{2} - \frac{1 \cdot 3}{2 \cdot 4} \cdot \frac{x^4}{4} - \frac{1 \cdot 3 \cdot 5}{2 \cdot 4 \cdot 6} \cdot \frac{x^6}{6} - - \cdots$$

$$= \ln \frac{2}{x} - \sum_{n=1}^{\infty} \frac{(2n)!x^{2n}}{2^{2n}(n!)^2 2n}, \qquad 0 < x < 1$$

$${}^{*}\ \tanh^{-1} x = x + \frac{x^3}{3} + \frac{x^5}{5} + \frac{x^7}{7} + \cdots + \frac{x^{2n+1}}{2n+1} + \cdots \qquad |x| < 1$$

$${}^{*}\ \coth^{-1} x = \frac{1}{x} + \frac{1}{3x^3} + \frac{1}{5x^5} + \frac{1}{7x^7} + \cdots + \frac{1}{(2n+1)x^{2n+1}} + \cdots \qquad |x| > 1$$

GUDERMANNIAN FUNCTION

This function is useful because it relates circular and hyperbolic functions without the use of functions of imaginary argument.

DEFINITION

$$\gamma = \text{the gudermannian of } x, \text{ written as gd } x$$

$$= 2 \tan^{-1} e^x - \frac{\pi}{2}$$

* Can also be written as arg tanh x and arg coth x respectively.

$$x = \text{gd}^{-1}\,\gamma$$
$$= \ln \tan\left(\frac{\pi}{4} + \frac{\gamma}{2}\right)$$
$$= \ln(\sec\gamma + \tan\gamma)$$

SPECIAL VALUES OF GUDERMANNIAN FUNCTION

x	0	∞	$-\infty$
gd x	0	$\frac{1}{2}\pi$	$-\frac{1}{2}\pi$

RELATIONS WITH HYPERBOLIC AND CIRCULAR FUNCTIONS

$$\sinh x = \tan(\text{gd}\,x) \qquad \text{cosech}\,x = \cot(\text{gd}\,x)$$
$$\cosh x = \sec(\text{gd}\,x) \qquad \text{sech}\,(x) = \cos(\text{gd}\,x)$$
$$\tanh x = \sin(\text{gd}\,x) \qquad \coth(x) = \text{cosec}\,(\text{gd}\,x)$$

DERIVATIVES

$$\frac{d}{dx}(\text{gd}\,x) = \text{sech}\,x$$
$$\frac{d}{dx}(\text{gd}^{-1}\,\gamma) = \sec\gamma$$

FUNDAMENTAL IDENTITIES

$$\tanh(\tfrac{1}{2}x) = \tan(\tfrac{1}{2}\text{gd}\,x)$$
$$e^x = \cosh x + \sinh x = \sec(\text{gd}\,x) + \tan(\text{gd}\,x)$$
$$= \tan\left(\frac{\pi}{4} + \frac{1}{2}\,\text{gd}\,x\right) = \frac{1 + \sin(\text{gd}\,x)}{\cos(\text{gd}\,x)} = \frac{1 + \tan(\frac{1}{2}\text{gd}\,x)}{1 - \tan(\frac{1}{2}\text{gd}\,x)}$$
$$\text{gd}\,x = 2\tan^{-1}\left(\tanh\frac{1}{2}\,x\right) = \int_0^x \frac{dt}{\cosh t}$$
$$\text{gd}^{-1}\,\gamma = \int_0^\gamma \frac{dt}{\cos t}$$
$$i\,\text{gd}^{-1}\,\gamma = \text{gd}\,i\gamma, \text{ where } i = \sqrt{-1}.$$

If

$$\alpha + i\beta = \text{gd}(x + iy)$$

then

$$\tan\alpha = \frac{\sinh x}{\cos y} \qquad \tanh\beta = \frac{\sin y}{\cosh x}$$
$$\tanh x = \frac{\sin\alpha}{\cosh\beta} \qquad \tan y = \frac{\sinh\beta}{\cosh\alpha}$$

SERIES EXPANSIONS

$$\text{gd}\, x = x - \frac{1}{6}x^3 + \frac{1}{24}x^5 - \frac{61}{5040}x^7 + - \cdots$$

$$= \sum_{n=0}^{\infty} \frac{E_{2n}}{(2n+1)!} x^{2n+1}, \quad |x| < 1 \; *$$

$$\text{gd}\, x = \frac{\pi}{2} - \text{sech}\, x - \frac{1}{2}\frac{\text{sech}^3 x}{3} - \frac{1 \cdot 3}{2 \cdot 4}\frac{\text{sech}^5 x}{5} + - \cdots$$

$$= \frac{\pi}{2} - \sum_{n=0}^{\infty} \frac{(2n)!}{2^{2n}(n!)^2} \frac{\text{sech}^{2n+1} x}{(2n+1)} \qquad x \text{ large}$$

$$\text{gd}\, x = \frac{2}{1}\tanh\frac{x}{2} - \frac{2}{3}\tanh^3\frac{x}{2} + \frac{2}{5}\tanh^5\frac{x}{2} - \frac{2}{7}\tanh^7\frac{x}{2} + - \cdots$$

$$= 2\sum_{n=0}^{\infty} \frac{(-1)^n}{2n+1} \tanh^{2n+1}\frac{x}{2}$$

$$\text{gd}^{-1}\, \gamma = \gamma + \frac{1}{6}\gamma^3 + \frac{1}{24}\gamma^5 + \frac{61}{5040}\gamma^7 + \cdots$$

$$= \sum_{n=0}^{\infty} \frac{(-1)^n E_{2n} \gamma^{2n+1}}{(2n+1)!}, \qquad |\gamma| < \frac{\pi}{2} \; *$$

$$\text{gd}^{-1}\, \gamma = \frac{2}{1}\tan\frac{\gamma}{2} + \frac{2}{3}\tan^3\frac{\gamma}{2} + \frac{2}{5}\tan^5\frac{\gamma}{2} + \cdots$$

$$= 2\sum_{n=0}^{\infty} \frac{1}{2n+1} \tan^{2n+1}\frac{\gamma}{2}$$

*E_n are Euler's numbers.

HYPERBOLIC FUNCTIONS AND THEIR COMMON LOGARITHMS

The logarithms given below show the mantissa only. The proper characteristic must be added.

x	Sinh x		Cosh x		Tanh x		Coth x	
	Value	$\log_{10}$	Value	$\log_{10}$	Value	$\log_{10}$	Value	$\log_{10}$
0.00	0.00000	$-\infty$	1.00000	.00000	0.00000	$-\infty$	∞	∞
0.01	.01000	.00001	1.00005	.00002	.01000	.99999	100.003	.00001
0.02	.02000	.30106	1.00020	.00009	.02000	.30097	50.007	.69903
0.03	.03000	.47719	1.00045	.00020	.02999	.47699	33.343	.52301
0.04	.04001	.60218	1.00080	.00035	.03998	.60183	25.013	.39817
0.05	0.05002	.69915	1.00125	.00054	0.04996	.69861	20.017	.30139
0.06	.06004	.77841	1.00180	.00078	.05993	.77763	16.687	.22237
0.07	.07006	.84545	1.00245	.00106	.06989	.84439	14.309	.15561
0.08	.08009	.90355	1.00320	.00139	.07983	.90216	12.527	.09784
0.09	.09012	.95483	1.00405	.00176	.08976	.95307	11.141	.04693
0.10	0.10017	.00072	1.00500	.00217	0.09967	.99856	10.0333	.00144
0.11	.11022	.04227	1.00606	.00262	.10956	.03965	9.1275	.96035
0.12	.12029	.08022	1.00721	.00312	.11943	.07710	8.3733	.92290
0.13	.13037	.11517	1.00846	.00366	.12927	.11151	7.7356	.88849
0.14	.14046	.14755	1.00982	.00424	.13909	.14330	7.1895	.85670
0.15	0.15056	.17772	1.01127	.00487	0.14889	.17285	6.7166	.82715
0.16	.16068	.20597	1.01283	.00554	.15865	.20044	6.3032	.79956
0.17	.17082	.23254	1.01448	.00625	.16838	.22629	5.9389	.77371
0.18	.18097	.25762	1.01624	.00700	.17808	.25062	5.6154	.74938
0.19	.19115	.28136	1.01810	.00779	.18775	.27357	5.3263	.72643
0.20	0.20134	.30392	1.02007	.00863	0.19738	.29529	5.0665	.70471
0.21	.21155	.32541	1.02213	.00951	.20697	.31590	4.8317	.68410
0.22	.22178	.34592	1.02430	.01043	.21652	.33549	4.6186	.66451
0.23	.23203	.36555	1.02657	.01139	.22603	.35416	4.4242	.64584
0.24	.24231	.38437	1.02894	.01239	.23550	.37198	4.2464	.62802
0.25	0.25261	.40245	1.03141	.01343	0.24492	.38902	4.0830	.61098
0.26	.26294	.41986	1.03399	.01452	.25430	.40534	3.9324	.59466
0.27	.27329	.43663	1.03667	.01564	.26362	.42099	3.7933	.57901
0.28	.28367	.45282	1.03946	.01681	.27291	.43601	3.6643	.56399
0.29	.29408	.46847	1.04235	.01801	.28213	.45046	3.5444	.54954
0.30	0.30452	.48362	1.04534	.01926	0.29131	.46436	3.4327	.53564
0.31	.31499	.49830	1.04844	.02054	.30044	.47775	3.3285	.52225
0.32	.32549	.51254	1.05164	.02187	.30951	.49067	3.2309	.50933
0.33	.33602	.52637	1.05495	.02323	.31852	.50314	3.1395	.49686
0.34	.34659	.53981	1.05836	.02463	.32748	.51518	3.0536	.48482
0.35	0.35719	.55290	1.06188	.02607	0.33638	.52682	2.9729	.47318
0.36	.36783	.56564	1.06550	.02755	.34521	.53809	2.8968	.46191
0.37	.37850	.57807	1.06923	.02907	.35399	.54899	2.8249	.45101
0.38	.38921	.59019	1.07307	.03063	.36271	.55956	2.7570	.44044
0.39	.39996	.60202	1.07702	.03222	.37136	.56980	2.6928	.43020
0.40	0.41075	.61358	1.08107	.03385	0.37995	.57973	2.6319	.42027
0.41	.42158	.62488	1.08523	.03552	.38847	.58936	2.5742	.41064
0.42	.43246	.63594	1.08950	.03723	.39693	.59871	2.5193	.40129
0.43	.44337	.64677	1.09388	.03897	.40532	.60780	2.4672	.39220
0.44	.45434	.65738	1.09837	.04075	.41364	.61663	2.4175	.38337
0.45	0.46534	.66777	1.10297	.04256	0.42190	.62521	2.3702	.37479
0.46	.47640	.67797	1.10768	.04441	.43008	.63355	2.3251	.36645
0.47	.48750	.68797	1.11250	.04630	.43820	.64167	2.2821	.35833
0.48	.49865	.69779	1.11743	.04822	.44624	.64957	2.2409	.35043
0.49	.50984	.70744	1.12247	.05018	.45422	.65726	2.2016	.34274

HYPERBOLIC FUNCTIONS AND THEIR COMMON LOGARITHMS

x	Sinh x		Cosh x		Tanh x		Coth x	
	Value	$\log_{10}$	Value	$\log_{10}$	Value	$\log_{10}$	Value	$\log_{10}$
0.50	0.52110	.71692	1.12763	.05217	0.46212	.66475	2.1640	.33525
0.51	.53240	.72624	1.13289	.05419	.46995	.67205	2.1279	.32795
0.52	.54375	.73540	1.13827	.05625	.47770	.67916	2.0934	.32084
0.53	.55516	.74442	1.14377	.05834	.48538	.68608	2.0602	.31392
0.54	.56663	.75330	1.14938	.06046	.49299	.69284	2.0284	.30716
0.55	0.57815	.76204	1.15510	.06262	0.50052	.69942	1.9979	.30058
0.56	.58973	.77065	1.16094	.06481	.50798	.70584	1.9686	.29416
0.57	.60137	.77914	1.16690	.06703	.51536	.71211	1.9404	.28789
0.58	.61307	.78751	1.17297	.06929	.52267	.71822	1.9133	.28178
0.59	.62483	.79576	1.17916	.07157	.52990	.72419	1.8872	.27581
0.60	0.63665	.80390	1.18547	.07389	0.53705	.73001	1.8620	.26999
0.61	.64854	.81194	1.19189	.07624	.54413	.73570	1.8378	.26430
0.62	.66049	.81987	1.19844	.07861	.55113	.74125	1.8145	.25875
0.63	.67251	.82770	1.20510	.08102	.55805	.74667	1.7919	.25333
0.64	.68459	.83543	1.21189	.08346	.56490	.75197	1.7702	.24803
0.65	0.69675	.84308	1.21879	.08593	0.57167	.75715	1.7493	.24285
0.66	.70897	.85063	1.22582	.08843	.57836	.76220	1.7290	.23780
0.67	.72126	.85809	1.23297	.09095	.58498	.76714	1.7095	.23286
0.68	.73363	.86548	1.24025	.09351	.59152	.77197	1.6906	.22803
0.69	.74607	.87278	1.24765	.09609	.59798	.77669	1.6723	.22331
0.70	0.75858	.88000	1.25517	.09870	0.60437	.78130	1.6546	.21870
0.71	.77117	.88715	1.26282	.10134	.61068	.78581	1.6375	.21419
0.72	.78384	.89423	1.27059	.10401	.61691	.79022	1.6210	.20978
0.73	.79659	.90123	1.27849	.10670	.62307	.79453	1.6050	.20547
0.74	.80941	.90817	1.28652	.10942	.62915	.79875	1.5895	.20125
0.75	0.82232	.91504	1.29468	.11216	0.63515	.80288	1.5744	.19712
0.76	.83530	.92185	1.30297	.11493	.64108	.80691	1.5599	.19309
0.77	.84838	.92859	1.31139	.11773	.64693	.81086	1.5458	.18914
0.78	.86153	.93527	1.31994	.12055	.65271	.81472	1.5321	.18528
0.79	.87478	.94190	1.32862	.12340	.65841	.81850	1.5188	.18150
0.80	0.88811	.94846	1.33743	.12627	0.66404	.82219	1.5059	.17781
0.81	.90152	.95498	1.34638	.12917	.66959	.82581	1.4935	.17419
0.82	.91503	.96144	1.35547	.13209	.67507	.82935	1.4813	.17065
0.83	.92863	.96784	1.36468	.13503	.68048	.83281	1.4696	.16719
0.84	.94233	.97420	1.37404	.13800	.68581	.83620	1.4581	.16380
0.85	0.95612	.98051	1.38353	.14099	0.69107	.83952	1.4470	.16048
0.86	.97000	.98677	1.39316	.14400	.69626	.84277	1.4362	.15723
0.87	.98398	.99299	1.40293	.14704	.70137	.84595	1.4258	.15405
0.88	.99806	.99916	1.41284	.15009	.70642	.84906	1.4156	.15094
0.89	1.01224	.00528	1.42289	.15317	.71139	.85211	1.4057	.14789
0.90	1.02652	.01137	1.43309	.15627	0.71630	.85509	1.3961	.14491
0.91	1.04090	.01741	1.44342	.15939	.72113	.85801	1.3867	.14199
0.92	1.05539	.02341	1.45390	.16254	.72590	.86088	1.3776	.13912
0.93	1.06998	.02937	1.46453	.16570	.73059	.86368	1.3687	.13632
0.94	1.08468	.03530	1.47530	.16888	.73522	.86642	1.3601	.13358
0.95	1.09948	.04119	1.48623	.17208	0.73978	.86910	1.3517	.13090
0.96	1.11440	.04704	1.49729	.17531	.74428	.87173	1.3436	.12827
0.97	1.12943	.05286	1.50851	.17855	.74870	.87431	1.3356	.12569
0.98	1.14457	.05864	1.51988	.18181	.75307	.87683	1.3279	.12317
0.99	1.15983	.06439	1.53141	.18509	.75736	.87930	1.3204	.12070

HYPERBOLIC FUNCTIONS AND THEIR COMMON LOGARITHMS

x	Sinh x		Cosh x		Tanh x		Coth x	
	Value	$\log_{10}$	Value	$\log_{10}$	Value	$\log_{10}$	Value	$\log_{10}$
1.00	1.17520	.07011	1.54308	.18839	0.76159	.88172	1.3130	.11828
1.01	1.19069	.07580	1.55491	.19171	.76576	.88409	1.3059	.11591
1.02	1.20630	.08146	1.56689	.19504	.76987	.88642	1.2989	.11358
1.03	1.22203	.08708	1.57904	.19839	.77391	.88869	1.2921	.11131
1.04	1.23788	.09268	1.59134	.20176	.77789	.89092	1.2855	.10908
1.05	1.25386	.09825	1.60379	.20515	0.78181	.89310	1.2791	.10690
1.06	1.26996	.10379	1.61641	.20855	.78566	.89524	1.2728	.10476
1.07	1.28619	.10930	1.62919	.21197	.78946	.89733	1.2667	.10267
1.08	1.30254	.11479	1.64214	.21541	.79320	.89938	1.2607	.10062
1.09	1.31903	.12025	1.65525	.21886	.79688	.90139	1.2549	.09861
1.10	1.33565	.12569	1.66852	.22233	0.80050	.90336	1.2492	.09664
1.11	1.35240	.13111	1.68196	.22582	.80406	.90529	1.2437	.09471
1.12	1.36929	.13649	1.69557	.22931	.80757	.90718	1.2383	.09282
1.13	1.38631	.14186	1.70934	.23283	.81102	.90903	1.2330	.09097
1.14	1.40347	.14720	1.72329	.23636	.81441	.91085	1.2279	.08915
1.15	1.42078	.15253	1.73741	.23990	0.81775	.91262	1.2229	.08738
1.16	1.43822	.15783	1.75171	.24346	.82104	.91436	1.2180	.08564
1.17	1.45581	.16311	1.76618	.24703	.82427	.91607	1.2132	.08393
1.18	1.47355	.16836	1.78083	.25062	.82745	.91774	1.2085	.08226
1.19	1.49143	.17360	1.79565	.25422	.83058	.91938	1.2040	.08062
1.20	1.50946	.17882	1.81066	.25784	0.83365	.92099	1.1995	.07901
1.21	1.52764	.18402	1.82584	.26146	.83668	.92256	1.1952	.07744
1.22	1.54598	.18920	1.84121	.26510	.83965	.92410	1.1910	.07590
1.23	1.56447	.19437	1.85676	.26876	.84258	.92561	1.1868	.07439
1.24	1.58311	.19951	1.87250	.27242	.84546	.92709	1.1828	.07291
1.25	1.60192	.20464	1.88842	.27610	0.84828	.92854	1.1789	.07146
1.26	1.62088	.20975	1.90454	.27979	.85106	.92996	1.1750	.07004
1.27	1.64001	.21485	1.92084	.28349	.85380	.93135	1.1712	.06865
1.28	1.65930	.21993	1.93734	.28721	.85648	.93272	1.1676	.06728
1.29	1.67876	.22499	1.95403	.29093	.85913	.93406	1.1640	.06594
1.30	1.69838	.23004	1.97091	.29467	0.86172	.93537	1.1605	.06463
1.31	1.71818	.23507	1.98800	.29842	.86428	.93665	1.1570	.06335
1.32	1.73814	.24009	2.00528	.30217	.86678	.93791	1.1537	.06209
1.33	1.75828	.24509	2.02276	.30594	.86925	.93914	1.1504	.06086
1.34	1.77860	.25008	2.04044	.30972	.87167	.94035	1.1472	.05965
1.35	1.79909	.25505	2.05833	.31352	0.87405	.94154	1.1441	.05846
1.36	1.81977	.26002	2.07643	.31732	.87639	.94270	1.1410	.05730
1.37	1.84062	.26496	2.09473	.32113	.87869	.94384	1.1381	.05616
1.38	1.86166	.26990	2.11324	.32495	.88095	.94495	1.1351	.05505
1.39	1.88289	.27482	2.13196	.32878	.88317	.94604	1.1323	.05396
1.40	1.90430	.27974	2.15090	.33262	0.88535	.94712	1.1295	.05288
1.41	1.92591	.28464	2.17005	.33647	.88749	.94817	1.1268	.05183
1.42	1.94770	.28952	2.18942	.34033	.88960	.94919	1.1241	.05081
1.43	1.96970	.29440	2.20900	.34420	.89167	.95020	1.1215	.04980
1.44	1.99188	.29926	2.22881	.34807	.89370	.95119	1.1189	.04881
1.45	2.01427	.30412	2.24884	.35196	0.89569	.95216	1.1165	.04784
1.46	2.03686	.30896	2.26910	.35585	.89765	.95311	1.1140	.04689
1.47	2.05965	.31379	2.28958	.35976	.89958	.95404	1.1116	.04596
1.48	2.08265	.31862	2.31029	.36367	.90147	.95495	1.1093	.04505
1.49	2.10586	.32343	2.33123	.36759	.90332	.95584	1.1070	.04416

HYPERBOLIC FUNCTIONS AND THEIR COMMON LOGARITHMS

x	Sinh x		Cosh x		Tanh x		Coth x	
	Value	$\log_{10}$	Value	$\log_{10}$	Value	$\log_{10}$	Value	$\log_{10}$
1.50	2.12928	.32823	2.35241	.37151	0.90515	.95672	1.1048	.04328
1.51	2.15291	.33303	2.37382	.37545	.90694	.95758	1.1026	.04242
1.52	2.17676	.33781	2.39547	.37939	.90870	.95842	1.1005	.04158
1.53	2.20082	.34258	2.41736	.38334	.91042	.95924	1.0984	.04076
1.54	2.22510	.34735	2.43949	.38730	.91212	.96005	1.0963	.03995
1.55	2.24961	.35211	2.46186	.39126	0.91379	.96084	1.0943	.03916
1.56	2.27434	.35686	2.48448	.39524	.91542	.96162	1.0924	.03838
1.57	2.29930	.36160	2.50735	.39921	.91703	.96238	1.0905	.03762
1.58	2.32449	.36633	2.53047	.40320	.91860	.96313	1.0886	.03687
1.59	2.34991	.37105	2.55384	.40719	.92015	.96386	1.0868	.03614
1.60	2.37557	.37577	2.57746	.41119	0.92167	.96457	1.0850	.03543
1.61	2.40146	.38048	2.60135	.41520	.92316	.96528	1.0832	.03472
1.62	2.42760	.38518	2.62549	.41921	.92462	.96597	1.0815	.03403
1.63	2.45397	.38987	2.64990	.42323	.92606	.96664	1.0798	.03336
1.64	2.48059	.39456	2.67457	.42725	.92747	.96730	1.0782	.03270
1.65	2.50746	.39923	2.69951	.43129	0.92886	.96795	1.0766	.03205
1.66	2.53459	.40391	2.72472	.43532	.93022	.96858	1.0750	.03142
1.67	2.56196	.40857	2.75021	.43937	.93155	.96921	1.0735	.03079
1.68	2.58959	.41323	2.77596	.44341	.93286	.96982	1.0720	.03018
1.69	2.61748	.41788	2.80200	.44747	.93415	.97042	1.0705	.02958
1.70	2.64563	.42253	2.82832	.45153	.93541	.97100	1.0691	.02900
1.71	2.67405	.42717	2.85491	.45559	.93665	.97158	1.0676	.02842
1.72	2.70273	.43180	2.88180	.45966	.93786	.97214	1.0663	.02786
1.73	2.73168	.43643	2.90897	.46374	.93906	.97269	1.0649	.02731
1.74	2.76091	.44105	2.93643	.46782	.94023	.97323	1.0636	.02677
1.75	2.79041	.44567	2.96419	.47191	0.94138	.97376	1.0623	.02624
1.76	2.82020	.45028	2.99224	.47600	.94250	.97428	1.0610	.02572
1.77	2.85026	.45488	3.02059	.48009	.94361	.97479	1.0598	.02521
1.78	2.88061	.45948	3.04925	.48419	.94470	.97529	1.0585	.02471
1.79	2.91125	.46408	3.07821	.48830	.94576	.97578	1.0574	.02422
1.80	2.94217	.46867	3.10747	.49241	0.94681	.97626	1.0562	.02374
1.81	2.97340	.47325	3.13705	.49652	.94783	.97673	1.0550	.02327
1.82	3.00492	.47783	3.16694	.50064	.94884	.97719	1.0539	.02281
1.83	3.03674	.48241	3.19715	.50476	.94983	.97764	1.0528	.02236
1.84	3.06886	.48698	3.22768	.50889	.95080	.97809	1.0518	.02191
1.85	3.10129	.49154	3.25853	.51302	0.95175	.97852	1.0507	.02148
1.86	3.13403	.49610	3.28970	.51716	.95268	.97895	1.0497	.02105
1.87	3.16709	.50066	3.32121	.52130	.95359	.97936	1.0487	.02064
1.88	3.20046	.50521	3.35305	.52544	.95449	.97977	1.0477	.02023
1.89	3.23415	.50976	3.38522	.52959	.95537	.98017	1.0467	.01983
1.90	3.26816	.51430	3.41773	.53374	0.95624	.98057	1.0458	.01943
1.91	3.30250	.51884	3.45058	.53789	.95709	.98095	1.0448	.01905
1.92	3.33718	.52338	3.48378	.54205	.95792	.98133	1.0439	.01867
1.93	3.37218	.52791	3.51733	.54621	.95873	.98170	1.0430	.01830
1.94	3.40752	.53244	3.55123	.55038	.95953	.98206	1.0422	.01794
1.95	3.44321	.53696	3.58548	.55455	0.96032	.98242	1.0413	.01758
1.96	3.47923	.54148	3.62009	.55872	.96109	.98276	1.0405	.01724
1.97	3.51561	.54600	3.65507	.56290	.96185	.98311	1.0397	.01689
1.98	3.55234	.55051	3.69041	.56707	.96259	.98344	1.0389	.01656
1.99	3.58942	.55502	3.72611	.57126	.96331	.98377	1.0381	.01623

HYPERBOLIC FUNCTIONS AND THEIR COMMON LOGARITHMS

x	Sinh x		Cosh x		Tanh x		Coth x	
	Value	$\log_{10}$	Value	$\log_{10}$	Value	$\log_{10}$	Value	$\log_{10}$
2.00	3.62686	.55953	3.76220	.57544	0.96403	.98409	1.0373	.01591
2.01	3.66466	.56403	3.79865	.57963	.96473	.98440	1.0366	.01560
2.02	3.70283	.56853	3.83549	.58382	.96541	.98471	1.0358	.01529
2.03	3.74138	.57303	3.87271	.58802	.96609	.98502	1.0351	.01498
2.04	3.78029	.57753	3.91032	.59221	.96675	.98531	1.0344	.01469
2.05	3.81958	.58202	3.94832	.59641	0.96740	.98560	1.0337	.01440
2.06	3.85926	.58650	3.98671	.60061	.96803	.98589	1.0330	.01411
2.07	3.89932	.59099	4.02550	.60482	.96865	.98617	1.0324	.01383
2.08	3.93977	.59547	4.06470	.60903	.96926	.98644	1.0317	.01356
2.09	3.98061	.59995	4.10430	.61324	.96986	.98671	1.0311	.01329
2.10	4.02186	.60443	4.14431	.61745	0.97045	.98697	1.0304	.01303
2.11	4.06350	.60890	4.18474	.62167	.97103	.98723	1.0298	.01277
2.12	4.10555	.61337	4.22558	.62589	.97159	.98748	1.0292	.01252
2.13	4.14801	.61784	4.26685	.63011	.97215	.98773	1.0286	.01227
2.14	4.19089	.62231	4.30855	.63433	.97269	.98798	1.0281	.01202
2.15	4.23419	.62677	4.35067	.63856	0.97323	.98821	1.0275	.01179
2.16	4.27791	.63123	4.39323	.64278	.97375	.98845	1.0270	.01155
2.17	4.32205	.63569	4.43623	.64701	.97426	.98868	1.0264	.01132
2.18	4.36663	.64015	4.47967	.65125	.97477	.98890	1.0259	.01110
2.19	4.41165	.64460	4.52356	.65548	.97526	.98912	1.0254	.01088
2.20	4.45711	.64905	4.56791	.65972	0.97574	.98934	1.0249	.01066
2.21	4.50301	.65350	4.61271	.66396	.97622	.98955	1.0244	.01045
2.22	4.54936	.65795	4.65797	.66820	.97668	.98975	1.0239	.01025
2.23	4.59617	.66240	4.70370	.67244	.97714	.98996	1.0234	.01004
2.24	4.64344	.66684	4.74989	.67668	.97759	.99016	1.0229	.00984
2.25	4.69117	.67128	4.79657	.68093	0.97803	.99035	1.0225	.00965
2.26	4.73937	.67572	4.84372	.68518	.97846	.99054	1.0220	.00946
2.27	4.78804	.68016	4.89136	.68943	.97888	.99073	1.0216	.00927
2.28	4.83720	.68459	4.93948	.69368	.97929	.99091	1.0211	.00909
2.29	4.88684	.68903	4.98810	.69794	.97970	.99109	1.0207	.00891
2.30	4.93696	.69346	5.03722	.70219	0.98010	.99127	1.0203	.00873
2.31	4.98758	.69789	5.08684	.70645	.98049	.99144	1.0199	.00856
2.32	5.03870	.70232	5.13697	.71071	.98087	.99161	1.0195	.00839
2.33	5.09032	.70675	5.18762	.71497	.98124	.99178	1.0191	.00822
2.34	5.14245	.71117	5.23878	.71923	.98161	.99194	1.0187	.00806
2.35	5.19510	.71559	5.29047	.72349	0.98197	.99210	1.0184	.00790
2.36	5.24827	.72002	5.34269	.72776	.98233	.99226	1.0180	.00774
2.37	5.30196	.72444	5.39544	.73203	.98267	.99241	1.0176	.00759
2.38	5.35618	.72885	5.44873	.73630	.98301	.99256	1.0173	.00744
2.39	5.41093	.73327	5.50256	.74056	.98335	.99271	1.0169	.00729
2.40	5.46623	.73769	5.55695	.74484	0.98367	.99285	1.0166	.00715
2.41	5.52207	.74210	5.61189	.74911	.98400	.99299	1.0163	.00701
2.42	5.57847	.74652	5.66739	.75338	.98431	.99313	1.0159	.00687
2.43	5.63542	.75093	5.72346	.75766	.98462	.99327	1.0156	.00673
2.44	5.69294	.75534	5.78010	.76194	.98492	.99340	1.0153	.00660
2.45	5.75103	.75975	5.83732	.76621	0.98522	.99353	1.0150	.00647
2.46	5.80969	.75415	5.89512	.77049	.98551	.99366	1.0147	.00634
2.47	5.86893	.76856	5.95352	.77477	.98579	.99379	1.0144	.00621
2.48	5.92876	.77296	6.01250	.77906	.98607	.99391	1.0141	.00609
2.49	5.98918	.77737	6.07209	.78334	.98635	.99403	1.0138	.00597

HYPERBOLIC FUNCTIONS AND THEIR COMMON LOGARITHMS

x	Sinh x		Cosh x		Tanh x		Coth x	
	Value	$\log_{10}$	Value	$\log_{10}$	Value	$\log_{10}$	Value	$\log_{10}$
2.50	6.05020	.78177	6.13229	.78762	0.98661	.99415	1.0136	.00585
2.51	6.11183	.78617	6.19310	.79191	.98688	.99426	1.0133	.00574
2.52	6.17407	.79057	6.25453	.79619	.98714	.99438	1.0130	.00562
2.53	6.23692	.79497	6.31658	.80048	.98739	.99449	1.0128	.00551
2.54	6.30040	.79937	6.37927	.80477	.98764	.99460	1.0125	.00540
2.55	6.36451	.80377	6.44259	.80906	0.98788	.99470	1.0123	.00530
2.56	6.42926	.80816	6.50656	.81335	.98812	.99481	1.0120	.00519
2.57	6.49464	.81256	6.57118	.81764	.98835	.99491	1.0118	.00509
2.58	6.56068	.81695	6.63646	.82194	.98858	.99501	1.0115	.00499
2.59	6.62738	.82134	6.70240	.82623	.98881	.99511	1.0113	.00489
2.60	6.69473	.82573	6.76901	.83052	0.98903	.99521	1.0111	.00479
2.61	6.76276	.83012	6.83629	.83482	.98924	.99530	1.0109	.00470
2.62	6.83146	.83451	6.90426	.83912	.98946	.99540	1.0107	.00460
2.63	6.90085	.83890	6.97292	.84341	.98966	.99549	1.0104	.00451
2.64	6.97092	.84329	7.04228	.84771	.98987	.99558	1.0102	.00442
2.65	7.04169	.84768	7.11234	.85201	0.99007	.99566	1.0100	.00434
2.66	7.11317	.85206	7.18312	.85631	.99026	.99575	1.0098	.00425
2.67	7.18536	.85645	7.25461	.86061	.99045	.99583	1.0096	.00417
2.68	7.25827	.86083	7.32683	.86492	.99064	.99592	1.0094	.00408
2.69	7.33190	.86522	7.39978	.86922	.99083	.99600	1.0093	.00400
2.70	7.40626	.86960	7.47347	.87352	0.99101	.99608	1.0091	.00392
2.71	7.48137	.87398	7.54791	.87783	.99118	.99615	1.0089	.00385
2.72	7.55722	.87836	7.62310	.88213	.99136	.99623	1.0087	.00377
2.73	7.63383	.88274	7.69905	.88644	.99153	.99631	1.0085	.00369
2.74	7.71121	.88712	7.77578	.89074	.99170	.99638	1.0084	.00362
2.75	7.78935	.89150	7.85328	.89505	0.99186	.99645	1.0082	.00355
2.76	7.86828	.89588	7.93157	.89936	.99202	.99652	1.0080	.00348
2.77	7.94799	.90026	8.01065	.90367	.99218	.99659	1.0079	.00341
2.78	8.02849	.90463	8.09053	.90798	.99233	.99666	1.0077	.00334
2.79	8.10980	.90901	8.17122	.91229	.99248	.99672	1.0076	.00328
2.80	8.19192	.91339	8.25273	.91660	0.99263	.99679	1.0074	.00321
2.81	8.27486	.91776	8.33506	.92091	.99278	.99685	1.0073	.00315
2.82	8.35862	.92213	8.41823	.92522	.99292	.99691	1.0071	.00309
2.83	8.44322	.92651	8.50224	.92953	.99306	.99698	1.0070	.00302
2.84	8.52867	.93088	8.58710	.93385	.99320	.99704	1.0069	.00296
2.85	8.61497	.93525	8.67281	.93816	0.99333	.99709	1.0067	.00291
2.86	8.70213	.93963	8.75940	.94247	.99346	.99715	1.0066	.00285
2.87	8.79016	.94400	8.84686	.94679	.99359	.99721	1.0065	.00279
2.88	8.87907	.94837	8.93520	.95110	.99372	.99726	1.0063	.00274
2.89	8.96887	.95274	9.02444	.95542	.99384	.99732	1.0062	.00268
2.90	9.05956	.95711	9.11458	.95974	0.99396	.99737	1.0061	.00263
2.91	9.15116	.96148	9.20564	.96405	.99408	.99742	1.0060	.00258
2.92	9.24368	.96584	9.29761	.96837	.99420	.99747	1.0058	.00253
2.93	9.33712	.97021	9.39051	.97269	.99431	.99752	1.0057	.00248
2.94	9.43149	.97458	9.48436	.97701	.99443	.99757	1.0056	.00243
2.95	9.52681	.97895	9.57915	.98133	0.99454	.99762	1.0055	.00238
2.96	9.62308	.98331	9.67490	.98565	.99464	.99767	1.0054	.00233
2.97	9.72031	.98768	9.77161	.98997	.99475	.99771	1.0053	.00229
2.98	9.81851	.99205	9.86930	.99429	.99485	.99776	1.0052	.00224
2.99	9.91770	.99641	9.96798	.99861	.99496	.99780	1.0051	.00220

HYPERBOLIC FUNCTIONS AND THEIR COMMON LOGARITHMS

x	Sinh x		Cosh x		Tanh x		Coth x	
	Value	$\log_{10}$	Value	$\log_{10}$	Value	$\log_{10}$	Value	$\log_{10}$
3.0	10.0179	.00078	10.0677	.00293	0.99505	.99785	1.0050	.00215
3.1	11.0765	.04440	11.1215	.04616	.99595	.99824	1.0041	.00176
3.2	12.2459	.08799	12.2866	.08943	.99668	.99856	1.0033	.00144
3.3	13.5379	.13155	13.5748	.13273	.99728	.99882	1.0027	.00118
3.4	14.9654	.17509	14.9987	.17605	.99777	.99903	1.0022	.00097
3.5	16.5426	.21860	16.5728	.21940	0.99818	.99921	1.0018	.00079
3.6	18.2855	.26211	18.3128	.26275	.99851	.99935	1.0015	.00065
3.7	20.2113	.30559	20.2360	.30612	.99878	.99947	1.0012	.00053
3.8	22.3394	.34907	22.3618	.34951	.99900	.99957	1.0010	.00043
3.9	24.6911	.39254	24.7113	.39290	.99918	.99964	1.0008	.00036
4.0	27.2899	.43600	27.3082	.43629	0.99933	.99971	1.0007	.00029
4.1	30.1619	.47946	30.1784	.47970	.99945	.99976	1.0005	.00024
4.2	33.3357	.52291	33.3507	.52310	.99955	.99980	1.0004	.00020
4.3	36.8431	.56636	36.8567	.56652	.99963	.99984	1.0004	.00016
4.4	40.7193	.60980	40.7316	.60993	.99970	.99987	1.0003	.00013
4.5	45.0030	.65324	45.0141	.65335	0.99975	.99989	1.0002	.00011
4.6	49.7371	.69668	49.7472	.69677	.99980	.99991	1.0002	.00009
4.7	54.9690	.74012	54.9781	.74019	.99983	.99993	1.0002	.00007
4.8	60.7511	.78355	60.7593	.78361	.99986	.99994	1.0001	.00006
4.9	67.1412	.82699	67.1486	.82704	.99989	.99995	1.0001	.00005
5.0	74.2032	.87042	74.2099	.87046	0.99991	.99996	1.0001	.00004
5.1	82.008	.91386	82.014	.91389	.99993	.99997	1.0001	.00003
5.2	90.633	.95729	90.639	.95731	.99994	.99997	1.0001	.00003
5.3	100.17	.00072	100.17	.00074	.99995	.99998	1.0000	.00002
5.4	110.70	.04415	110.71	.04417	.99996	.99998	1.0000	.00002
5.5	122.34	.08758	122.35	.08760	0.99997	.99999	1.0000	.00001
5.6	135.21	.13101	135.22	.13103	.99997	.99999	1.0000	.00001
5.7	149.43	.17444	149.44	.17445	.99998	.99999	1.0000	.00001
5.8	165.15	.21787	165.15	.21788	.99998	.99999	1.0000	.00001
5.9	182.52	.26130	182.52	.26131	.99998	.99999	1.0000	.00001
6.0	201.71	.30473	201.72	.30474	0.99999	.00000	1.0000	.00000
6.1	222.93	.34817	222.93	.34817	.99999	.00000	1.0000	.00000
6.2	246.37	.39159	246.38	.39161	.99999	.00000	1.0000	.00000
6.3	272.29	.43503	272.29	.43503	.99999	.00000	1.0000	.00000
6.4	300.92	.47845	300.92	.47845	.99999	.00000	1.0000	.00000
6.5	332.57	.52188	332.57	.52188	1.0000	.00000	1.0000	.00000
6.6	367.55	.56532	367.55	.56532	1.0000	.00000	1.0000	.00000
6.7	406.20	.60874	406.20	.60874	1.0000	.00000	1.0000	.00000
6.8	448.92	.65217	448.92	.65217	1.0000	.00000	1.0000	.00000
6.9	496.14	.69560	496.14	.69560	1.0000	.00000	1.0000	.00000
7.0	548.32	.73903	548.32	.73903	1.0000	.00000	1.0000	.00000
7.1	605.98	.78246	605.98	.78246	1.0000	.00000	1.0000	.00000
7.2	669.72	.82589	669.72	.82589	1.0000	.00000	1.0000	.00000
7.3	740.15	.86932	740.15	.86932	1.0000	.00000	1.0000	.00000
7.4	817.99	.91275	817.99	.91275	1.0000	.00000	1.0000	.00000
7.5	904.02	.95618	904.02	.95618	1.0000	.00000	1.0000	.00000
7.6	999.10	.99961	999.10	.99961	1.0000	.00000	1.0000	.00000
7.7	1104.2	.04305	1104.2	.04305	1.0000	.00000	1.0000	.00000
7.8	1220.3	.08647	1220.3	.08647	1.0000	.00000	1.0000	.00000
7.9	1348.6	.12988	1348.6	.12988	1.0000	.00000	1.0000	.00000

HYPERBOLIC FUNCTIONS AND THEIR COMMON LOGARITHMS

x	Sinh x		Cosh x		Tanh x		Coth x	
	Value	$\log_{10}$	Value	$\log_{10}$	Value	$\log_{10}$	Value	$\log_{10}$
8.0	1490.5	.17333	1490.5	.17333	1.0000	.00000	1.0000	.00000
8.1	1647.2	.21675	1647.2	.21675	1.0000	.00000	1.0000	.00000
8.2	1820.5	.26019	1820.5	.26019	1.0000	.00000	1.0000	.00000
8.3	2011.9	.30360	2011.9	.30360	1.0000	.00000	1.0000	.00000
8.4	2223.5	.34704	2223.5	.34704	1.0000	.00000	1.0000	.00000
8.5	2457.4	.39048	2457.4	.39048	1.0000	.00000	1.0000	.00000
8.6	2715.8	.43390	2715.8	.43390	1.0000	.00000	1.0000	.00000
8.7	3001.5	.47734	3001.5	.47734	1.0000	.00000	1.0000	.00000
8.8	3317.1	.52076	3317.1	.52076	1.0000	.00000	1.0000	.00000
8.9	3666.0	.56419	3666.0	.56419	1.0000	.00000	1.0000	.00000
9.0	4051.5	.60762	4051.5	.60762	1.0000	.00000	1.0000	.00000
9.1	4477.6	.65105	4477.6	.65105	1.0000	.00000	1.0000	.00000
9.2	4948.6	.69448	4948.6	.69448	1.0000	.00000	1.0000	.00000
9.3	5469.0	.73791	5469.0	.73791	1.0000	.00000	1.0000	.00000
9.4	6044.2	.78134	6044.2	.78134	1.0000	.00000	1.0000	.00000
9.5	6679.9	.82477	6679.9	.82477	1.0000	.00000	1.0000	.00000
9.6	7382.4	.86820	7382.4	.86820	1.0000	.00000	1.0000	.00000
9.7	8158.8	.91163	8158.8	.91163	1.0000	.00000	1.0000	.00000
9.8	9016.9	.95506	9016.9	.95506	1.0000	.00000	1.0000	.00000
9.9	9965.2	.99849	9965.2	.99849	1.0000	.00000	1.0000	.00000
10.0	11013.2	.04191	11013.2	.04191	1.0000	.00000	1.0000	.00000

EXPONENTIAL AND HYPERBOLIC FUNCTIONS FOR πx

x	$e^{\pi x}$	$e^{-\pi x}$	$\sinh \pi x$	$\cosh \pi x$	$\tanh \pi x$
0.00	1.00000	1.00000	0.00000	1.00000	0.00000
0.01	1.03191	0.96907	0.03142	1.00049	0.03141
0.02	1.06485	0.93910	0.06287	1.00197	0.06275
0.03	1.09883	0.91006	0.09439	1.00444	0.09397
0.04	1.13390	0.88191	0.12599	1.00791	0.12501
0.05	1.17009	0.85464	0.15773	1.01236	0.15580
0.06	1.20743	0.82820	0.18961	1.01782	0.18629
0.07	1.24597	0.80259	0.22169	1.02428	0.21643
0.08	1.28573	0.77777	0.25398	1.03175	0.24617
0.09	1.32676	0.75371	0.28653	1.04024	0.27544
0.10	1.36911	0.73040	0.31935	1.04976	0.30422
0.11	1.41280	0.70781	0.35249	1.06031	0.33245
0.12	1.45789	0.68592	0.38598	1.07191	0.36009
0.13	1.50442	0.66471	0.41986	1.08456	0.38712
0.14	1.55243	0.64415	0.45414	1.09829	0.41350
0.15	1.60198	0.62423	0.48887	1.11310	0.43920
0.16	1.65310	0.60492	0.52409	1.12901	0.46420
0.17	1.70586	0.58621	0.55982	1.14604	0.48849
0.18	1.76030	0.56808	0.59611	1.16419	0.51204
0.19	1.81648	0.55051	0.63298	1.18350	0.53484
0.20	1.87446	0.53349	0.67048	1.20397	0.55689
0.21	1.93428	0.51699	0.70865	1.22563	0.57819
0.22	1.99601	0.50100	0.74751	1.24850	0.59872
0.23	2.05971	0.48550	0.78710	1.27261	0.61850
0.24	2.12545	0.47049	0.82748	1.29797	0.63752
0.25	2.19328	0.45594	0.86867	1.32461	0.65579
0.26	2.26328	0.44184	0.91072	1.35256	0.67333
0.27	2.33551	0.42817	0.95367	1.38184	0.69014
0.28	2.41005	0.41493	0.99756	1.41249	0.70624
0.29	2.48696	0.40210	1.04243	1.44453	0.72164
0.30	2.56633	0.38966	1.08834	1.47800	0.73636
0.31	2.64824	0.37761	1.13531	1.51292	0.75041
0.32	2.73275	0.36593	1.18341	1.54934	0.76382
0.33	2.81997	0.35461	1.23268	1.58729	0.77659
0.34	2.90997	0.34365	1.28316	1.62681	0.78876
0.35	3.00284	0.33302	1.33491	1.66793	0.80034
0.36	3.09867	0.32272	1.38798	1.71070	0.81135
0.37	3.19756	0.31274	1.44241	1.75515	0.82182
0.38	3.29961	0.30307	1.49827	1.80134	0.83176
0.39	3.40492	0.29369	1.55561	1.84931	0.84119
0.40	3.51359	0.28461	1.61449	1.89910	0.85013
0.41	3.62572	0.27581	1.67496	1.95076	0.85862
0.42	3.74143	0.26728	1.73708	2.00436	0.86665
0.43	3.86084	0.25901	1.80091	2.05993	0.87426
0.44	3.98406	0.25100	1.86653	2.11753	0.88147
0.45	4.11121	0.24324	1.93398	2.17722	0.88828
0.46	4.24241	0.23571	2.00335	2.23906	0.89473
0.47	4.37781	0.22842	2.07469	2.30312	0.90082
0.48	4.51753	0.22136	2.14808	2.36944	0.90658
0.49	4.66170	0.21451	2.22359	2.43811	0.91202
0.50	4.81048	0.20788	2.30130	2.50918	0.91715
0.51	4.96400	0.20145	2.38128	2.58273	0.92200
0.52	5.12243	0.19522	2.46360	2.65882	0.92658
0.53	5.28591	0.18918	2.54836	2.73754	0.93089
0.54	5.45460	0.18333	2.63564	2.81897	0.93497

EXPONENTIAL AND HYPERBOLIC FUNCTIONS FOR πx

x	$e^{\pi x}$	$e^{-\pi x}$	$\sinh \pi x$	$\cosh \pi x$	$\tanh \pi x$
0.55	5.62869	0.17766	2.72551	2.90317	0.93880
0.56	5.80832	0.17217	2.81808	2.99024	0.94242
0.57	5.99369	0.16684	2.91343	3.08027	0.94584
0.58	6.18498	0.16168	3.01165	3.17333	0.94905
0.59	6.38237	0.15668	3.11284	3.26953	0.95208
0.60	6.58606	0.15184	3.21711	3.36895	0.95493
0.61	6.79625	0.14714	3.32456	3.47170	0.95762
0.62	7.01315	0.14259	3.43528	3.57787	0.96015
0.63	7.23698	0.13818	3.54940	3.68758	0.96253
0.64	7.46794	0.13391	3.66702	3.80092	0.96477
0.65	7.70628	0.12976	3.78826	3.91802	0.96688
0.66	7.95222	0.12575	3.91323	4.03899	0.96887
0.67	8.20601	0.12186	4.04208	4.16394	0.97073
0.68	8.46790	0.11809	4.17491	4.29300	0.97249
0.69	8.73815	0.11444	4.31186	4.42630	0.97415
0.70	9.01703	0.11090	4.45306	4.56396	0.97570
0.71	9.30480	0.10747	4.59867	4.70614	0.97716
0.72	9.60176	0.10415	4.74881	4.85296	0.97854
0.73	9.90820	0.10093	4.90364	5.00456	0.97983
0.74	10.22442	0.09781	5.06331	5.16111	0.98105
0.75	10.55072	0.09478	5.22797	5.32275	0.98219
0.76	10.88745	0.09185	5.39780	5.48965	0.98327
0.77	11.23492	0.08901	5.57295	5.66196	0.98428
0.78	11.59347	0.08626	5.75361	5.83986	0.98523
0.79	11.96347	0.08359	5.93994	6.02353	0.98612
0.80	12.34528	0.08100	6.13214	6.21314	0.98696
0.81	12.73928	0.07850	6.33039	6.40889	0.98775
0.82	13.14585	0.07607	6.53489	6.61096	0.98849
0.83	13.56539	0.07372	6.74584	6.81955	0.98919
0.84	13.99833	0.07144	6.96344	7.03488	0.98985
0.85	14.44508	0.06923	7.18793	7.25715	0.99046
0.86	14.90609	0.06709	7.41950	7.48659	0.99104
0.87	15.38181	0.06501	7.65840	7.72341	0.99158
0.88	15.87271	0.06300	7.90486	7.96786	0.99209
0.89	16.37929	0.06105	8.15912	8.22017	0.99257
0.90	16.90202	0.05916	8.41243	8.48059	0.99302
0.91	17.44145	0.05733	8.69206	8.74939	0.99345
0.92	17.99808	0.05556	8.97126	9.02682	0.99384
0.93	18.57248	0.05384	9.25932	9.31316	0.99422
0.94	19.16522	0.05218	9.55652	9.60870	0.99457
0.95	19.77687	0.05056	9.86315	9.91372	0.99490
0.96	20.40804	0.04900	10.17952	10.22852	0.99521
0.97	21.05935	0.04748	10.50594	10.55342	0.99550
0.98	21.73146	0.04602	10.84272	10.88874	0.99577
0.99	22.42501	0.04459	11.19021	11.23480	0.99603
1.00	23.14069	0.04321	11.54874	11.59195	0.99627

INVERSE HYPERBOLIC FUNCTIONS

x	$\sinh^{-1} x$	$\tanh^{-1} x$	$\operatorname{cosech}^{-1} x$	$\operatorname{sech}^{-1} x$
0.0	0.00000	0.00000	∞	∞
0.01	0.01000	0.01000	5.29834	5.29829
0.02	0.02000	0.02000	4.60527	4.60507
0.03	0.03000	0.03001	4.19993	4.19948
0.04	0.03999	0.04002	3.91242	3.91162
0.05	0.04998	0.05004	3.68950	3.68825
0.06	0.05996	0.06007	3.50746	3.50566
0.07	0.06994	0.07011	3.35363	3.35118
0.08	0.07991	0.08017	3.22047	3.21727
0.09	0.08988	0.09024	3.10311	3.09906
0.10	0.09983	0.10034	2.99822	2.99322
0.11	0.10978	0.11045	2.90343	2.89738
0.12	0.11971	0.12058	2.81699	2.80979
0.13	0.12964	0.13074	2.73757	2.72912
0.14	0.13955	0.14093	2.66412	2.65432
0.15	0.14944	0.15114	2.59585	2.58459
0.16	0.15933	0.16139	2.53207	2.51927
0.17	0.16919	0.17167	2.47225	2.45780
0.18	0.17904	0.18198	2.41595	2.39975
0.19	0.18888	0.19234	2.36278	2.34473
0.20	0.19869	0.20273	2.31244	2.29243
0.21	0.20849	0.21317	2.26464	2.24258
0.22	0.21826	0.22366	2.21916	2.19495
0.23	0.22802	0.23419	2.17579	2.14933
0.24	0.23775	0.24477	2.13436	2.10554
0.25	0.24747	0.25541	2.09471	2.06344
0.26	0.25716	0.26611	2.05671	2.02288
0.27	0.26682	0.27686	2.02023	1.98374
0.28	0.27646	0.28768	1.98516	1.94591
0.29	0.28608	0.29857	1.95141	1.90930
0.30	0.29567	0.30952	1.91890	1.87382
0.31	0.30524	0.32055	1.88753	1.83939
0.32	0.31478	0.33165	1.85725	1.80594
0.33	0.32429	0.34283	1.82799	1.77340
0.34	0.33377	0.35409	1.79968	1.74172
0.35	0.34322	0.36544	1.77228	1.71083
0.36	0.35265	0.37689	1.74573	1.68070
0.37	0.36204	0.38842	1.71999	1.65127
0.38	0.37140	0.40006	1.69502	1.62250
0.39	0.38073	0.41180	1.67078	1.59436
0.40	0.39004	0.42365	1.64723	1.56680
0.41	0.39930	0.43561	1.62434	1.53979
0.42	0.40854	0.44769	1.60209	1.51331
0.43	0.41774	0.45990	1.58043	1.48731
0.44	0.42691	0.47223	1.55935	1.46178
0.45	0.43605	0.48470	1.53882	1.43669
0.46	0.44515	0.49731	1.51881	1.41200
0.47	0.45422	0.51007	1.49931	1.38771
0.48	0.46325	0.52298	1.48029	1.36379
0.49	0.47225	0.53606	1.46174	1.34021
0.50	0.48121	0.54931	1.44364	1.31696

INVERSE HYPERBOLIC FUNCTIONS

x	$\sinh^{-1} x$	$\tanh^{-1} x$	$\operatorname{cosech}^{-1} x$	$\operatorname{sech}^{-1} x$
0.50	0.48121	0.54931	1.44364	1.31696
0.51	0.49014	0.56273	1.42596	1.29401
0.52	0.49903	0.57634	1.40870	1.27136
0.53	0.50788	0.59015	1.39183	1.24898
0.54	0.51670	0.60416	1.37535	1.22686
0.55	0.52548	0.61838	1.35924	1.20497
0.56	0.53422	0.63283	1.34348	1.18331
0.57	0.54293	0.64752	1.32807	1.16186
0.58	0.55160	0.66246	1.31299	1.14060
0.59	0.56023	0.67767	1.29824	1.11952
0.60	0.56882	0.69315	1.28380	1.09861
0.61	0.57738	0.70892	1.26965	1.07785
0.62	0.58590	0.72501	1.25580	1.05723
0.63	0.59438	0.74142	1.24223	1.03673
0.64	0.60282	0.75817	1.22894	1.01635
0.65	0.61122	0.77530	1.21591	0.99606
0.66	0.61959	0.79281	1.20314	0.97585
0.67	0.62792	0.81074	1.19062	0.95572
0.68	0.63620	0.82911	1.17834	0.93564
0.69	0.64446	0.84796	1.16629	0.91560
0.70	0.65267	0.86730	1.15448	0.89559
0.71	0.66084	0.88718	1.14288	0.87559
0.72	0.66897	0.90764	1.13151	0.85558
0.73	0.67707	0.92873	1.12034	0.83555
0.74	0.68513	0.95048	1.10938	0.81549
0.75	0.69315	0.97296	1.09861	0.79537
0.76	0.70113	0.99622	1.08804	0.77517
0.77	0.70907	1.02033	1.07766	0.75487
0.78	0.71697	1.04537	1.06746	0.73445
0.79	0.72484	1.07143	1.05744	0.71388
0.80	0.73267	1.09861	1.04759	0.69315
0.81	0.74046	1.12703	1.03792	0.67221
0.82	0.74821	1.15682	1.02840	0.65103
0.83	0.75592	1.18814	1.01905	0.62958
0.84	0.76360	1.22117	1.00986	0.60781
0.85	0.77124	1.25615	1.00082	0.58568
0.86	0.77884	1.29334	0.99193	0.56313
0.87	0.78640	1.33308	0.98319	0.54008
0.88	0.79393	1.37577	0.97459	0.51647
0.89	0.80142	1.42193	0.96613	0.49220
0.90	0.80887	1.47222	0.95780	0.46715
0.91	0.81628	1.52752	0.94961	0.44116
0.92	0.82366	1.58903	0.94154	0.41406
0.93	0.83100	1.65839	0.93361	0.38560
0.94	0.83830	1.73805	0.92580	0.35542
0.95	0.84557	1.83178	0.91810	0.32304
0.96	0.85280	1.94591	0.91053	0.28768
0.97	0.86000	2.09230	0.90307	0.24807
0.98	0.86716	2.29756	0.89573	0.20169
0.99	0.87428	2.64665	0.88850	0.14201
1.00	0.88137	∞	0.88137	0.00000

INVERSE HYPERBOLIC FUNCTIONS

x	$\sinh^{-1} x$	$\cosh^{-1} x$	$\operatorname{cosech}^{-1} x$	$\coth^{-1} x$
1.00	0.88137	0.00000	0.88137	∞
1.01	0.88843	0.14130	0.87436	2.65165
1.02	0.89545	0.19967	0.86744	2.30756
1.03	0.90243	0.24434	0.86063	2.10730
1.04	0.90938	0.28191	0.85391	1.96591
1.05	0.91629	0.31492	0.84730	1.85679
1.06	0.92317	0.34470	0.84078	1.76806
1.07	0.93002	0.37202	0.83435	1.69340
1.08	0.93683	0.39738	0.82801	1.62905
1.09	0.94360	0.42114	0.82177	1.57255
1.10	0.95035	0.44357	0.81561	1.52226
1.11	0.95706	0.46485	0.80954	1.47698
1.12	0.96373	0.48513	0.80355	1.43584
1.13	0.97038	0.50453	0.79764	1.39817
1.14	0.97699	0.52316	0.79182	1.36346
1.15	0.98357	0.54110	0.78607	1.33129
1.16	0.99011	0.55840	0.78041	1.30134
1.17	0.99663	0.57514	0.77482	1.27334
1.18	1.00311	0.59135	0.76930	1.24706
1.19	1.00956	0.60708	0.76386	1.22232
1.20	1.01597	0.62236	0.75849	1.19895
1.21	1.02236	0.63724	0.75319	1.17682
1.22	1.02871	0.65173	0.74796	1.15582
1.23	1.03504	0.66586	0.74279	1.13584
1.24	1.04133	0.67966	0.73770	1.11680
1.25	1.04759	0.69315	0.73267	1.09861
1.26	1.05382	0.70634	0.72770	1.08122
1.27	1.06003	0.71924	0.72280	1.06456
1.28	1.06620	0.73189	0.71796	1.04857
1.29	1.07234	0.74428	0.71318	1.03321
1.30	1.07845	0.75643	0.70846	1.01844
1.31	1.08453	0.76836	0.70380	1.00422
1.32	1.09059	0.78007	0.69920	0.99050
1.33	1.09661	0.79157	0.69465	0.97727
1.34	1.10261	0.80288	0.69016	0.96448
1.35	1.10857	0.81400	0.68572	0.95212
1.36	1.11451	0.82494	0.68134	0.94016
1.37	1.12042	0.83570	0.67701	0.92857
1.38	1.12630	0.84630	0.67273	0.91734
1.39	1.13216	0.85673	0.66851	0.90645
1.40	1.13798	0.86701	0.66433	0.89588
1.41	1.14378	0.87715	0.66020	0.88561
1.42	1.14955	0.88714	0.65612	0.87563
1.43	1.15530	0.89699	0.65209	0.86593
1.44	1.16101	0.90670	0.64811	0.85649
1.45	1.16670	0.91629	0.64417	0.84730
1.46	1.17237	0.92575	0.64028	0.83835
1.47	1.17801	0.93509	0.63643	0.82962
1.48	1.18362	0.94432	0.63263	0.82111
1.49	1.18920	0.95343	0.62886	0.81282
1.50	1.19476	0.96242	0.62515	0.80472

INVERSE HYPERBOLIC FUNCTIONS

x	$\sinh^{-1} x$	$\cosh^{-1} x$	$\operatorname{cosech}^{-1} x$	$\coth^{-1} x$
1.50	1.19476	0.96242	0.62515	0.80472
1.51	1.20030	0.97131	0.62147	0.79681
1.52	1.20581	0.98010	0.61783	0.78909
1.53	1.21129	0.98879	0.61424	0.78155
1.54	1.21675	0.99737	0.61068	0.77418
1.55	1.22218	1.00587	0.60716	0.76697
1.56	1.22759	1.01426	0.60368	0.75991
1.57	1.23298	1.02257	0.60024	0.75301
1.58	1.23834	1.03079	0.59684	0.74626
1.59	1.24367	1.03892	0.59347	0.73965
1.60	1.24898	1.04697	0.59014	0.73317
1.61	1.25427	1.05493	0.58685	0.72682
1.62	1.25954	1.06282	0.58359	0.72061
1.63	1.26478	1.07063	0.58036	0.71451
1.64	1.26999	1.07836	0.57717	0.70853
1.65	1.27519	1.08601	0.57401	0.70267
1.66	1.28036	1.09360	0.57089	0.69692
1.67	1.28551	1.10111	0.56780	0.69128
1.68	1.29064	1.10855	0.56474	0.68574
1.69	1.29574	1.11592	0.56171	0.68030
1.70	1.30082	1.12323	0.55871	0.67496
1.71	1.30588	1.13047	0.55574	0.66972
1.72	1.31092	1.13765	0.55281	0.66457
1.73	1.31593	1.14476	0.54990	0.65951
1.74	1.32093	1.15182	0.54702	0.65453
1.75	1.32590	1.15881	0.54417	0.64964
1.76	1.33085	1.16574	0.54135	0.64483
1.77	1.33578	1.17262	0.53856	0.64011
1.78	1.34069	1.17944	0.53579	0.63546
1.79	1.34557	1.18620	0.53305	0.63088
1.80	1.35044	1.19291	0.53034	0.62638
1.81	1.35529	1.19957	0.52766	0.62195
1.82	1.36011	1.20617	0.52500	0.61759
1.83	1.36492	1.21272	0.52237	0.61330
1.84	1.36970	1.21922	0.51976	0.60908
1.85	1.37447	1.22567	0.51718	0.60492
1.86	1.37921	1.23207	0.51462	0.60082
1.87	1.38394	1.23842	0.51208	0.59679
1.88	1.38864	1.24473	0.50957	0.59281
1.89	1.39333	1.25098	0.50709	0.58890
1.90	1.39800	1.25720	0.50462	0.58504
1.91	1.40265	1.26336	0.50218	0.58123
1.92	1.40728	1.26949	0.49977	0.57748
1.93	1.41188	1.27557	0.49737	0.57379
1.94	1.41648	1.28160	0.49500	0.57014
1.95	1.42105	1.28760	0.49265	0.56655
1.96	1.42560	1.29355	0.49032	0.56301
1.97	1.43014	1.29946	0.48801	0.55951
1.98	1.43466	1.30533	0.48572	0.55606
1.99	1.43915	1.31117	0.48346	0.55266
2.00	1.44364	1.31696	0.48121	0.54931

INVERSE HYPERBOLIC FUNCTIONS

x	$\sinh^{-1} x$	$\cosh^{-1} x$	$\operatorname{cosech}^{-1} x$	$\coth^{-1} x$
2.00	1.44364	1.31696	0.48121	0.54931
2.10	1.48748	1.37286	0.45982	0.51805
2.20	1.52966	1.42542	0.44019	0.49041
2.30	1.57028	1.47504	0.42213	0.46578
2.40	1.60944	1.52208	0.40547	0.44365
2.50	1.64723	1.56680	0.39004	0.42365
2.60	1.68374	1.60944	0.37571	0.40547
2.70	1.71905	1.65019	0.36239	0.38885
2.80	1.75323	1.68924	0.34996	0.37361
2.90	1.78634	1.72671	0.33834	0.35956
3.00	1.81845	1.76275	0.32745	0.34657
3.10	1.84960	1.79746	0.31723	0.33452
3.20	1.87986	1.83094	0.30763	0.32331
3.30	1.90927	1.86328	0.29857	0.31285
3.40	1.93788	1.89456	0.29003	0.30307
3.50	1.96572	1.92485	0.28196	0.29389
3.60	1.99284	1.95421	0.27432	0.28527
3.70	2.01926	1.98270	0.26708	0.27716
3.80	2.04503	2.01037	0.26021	0.26950
3.90	2.07017	2.03727	0.25368	0.26226
4.00	2.09471	2.06344	0.24747	0.25541
4.10	2.11869	2.08892	0.24155	0.24892
4.20	2.14211	2.11375	0.23590	0.24275
4.30	2.16502	2.13796	0.23051	0.23689
4.40	2.18742	2.16158	0.22536	0.23131
4.50	2.20935	2.18464	0.22043	0.22599
4.60	2.23081	2.20717	0.21571	0.22092
4.70	2.25184	2.22920	0.21119	0.21607
4.80	2.27244	2.25073	0.20685	0.21143
4.90	2.29264	2.27180	0.20269	0.20699
5.00	2.31244	2.29243	0.19869	0.20273
5.10	2.33186	2.31263	0.19484	0.19865
5.20	2.35093	2.33243	0.19114	0.19473
5.30	2.36964	2.35183	0.18758	0.19097
5.40	2.38801	2.37086	0.18414	0.18735
5.50	2.40606	2.38953	0.18083	0.18386
5.60	2.42379	2.40784	0.17764	0.18051
5.70	2.44122	2.42583	0.17455	0.17727
5.80	2.45836	2.44349	0.17157	0.17415
5.90	2.47521	2.46084	0.16869	0.17114
6.00	2.49178	2.47789	0.16590	0.16824
6.10	2.50809	2.49465	0.16321	0.16543
6.20	2.52414	2.51113	0.16060	0.16271
6.30	2.53994	2.52734	0.15807	0.16008
6.40	2.55549	2.54329	0.15562	0.15754
6.50	2.57081	2.55898	0.15325	0.15508
6.60	2.58591	2.57443	0.15094	0.15269
6.70	2.60078	2.58964	0.14871	0.15038
6.80	2.61543	2.60462	0.14653	0.14813
6.90	2.62988	2.61938	0.14442	0.14596
7.00	2.64412	2.63392	0.14238	0.14384

GUDERMANNIAN FUNCTION

x	0	1	2	3	4	5	6	7	8	9
0.0	0.00000	0.01000	0.02000	0.03000	0.03999	0.04998	0.05996	0.06994	0.07991	0.08988
0.1	0.09983	0.10978	0.11971	0.12964	0.13954	0.14944	0.15932	0.16919	0.17904	0.18887
0.2	0.19868	0.20847	0.21825	0.22800	0.23773	0.24744	0.25712	0.26678	0.27641	0.28602
0.3	0.29560	0.30515	0.31467	0.32417	0.33363	0.34307	0.35247	0.36184	0.37117	0.38047
0.4	0.38974	0.39897	0.40817	0.41733	0.42645	0.43554	0.44459	0.45359	0.46256	0.47149
0.5	0.48038	0.48923	0.49803	0.50680	0.51552	0.52420	0.53284	0.54143	0.54997	0.55848
0.6	0.56694	0.57535	0.58372	0.59204	0.60031	0.60854	0.61672	0.62486	0.63294	0.64098
0.7	0.64897	0.65692	0.66481	0.67266	0.68045	0.68820	0.69590	0.70355	0.71115	0.71870
0.8	0.72620	0.73366	0.74106	0.74841	0.75571	0.76297	0.77017	0.77732	0.78443	0.79148
0.9	0.79848	0.80544	0.81234	0.81919	0.82599	0.83275	0.83945	0.84611	0.85271	0.85926
1.0	0.86577	0.87223	0.87863	0.88499	0.89130	0.89756	0.90377	0.90993	0.91604	0.92211
1.1	0.92813	0.93410	0.94002	0.94589	0.95172	0.95750	0.96323	0.96892	0.97455	0.98015
1.2	0.98569	0.99119	0.99665	1.00205	1.00742	1.01274	1.01801	1.02324	1.02842	1.03356
1.3	1.03866	1.04371	1.04872	1.05368	1.05860	1.06348	1.06832	1.07312	1.07787	1.08258
1.4	1.08725	1.09188	1.09647	1.10101	1.10552	1.10999	1.11441	1.11880	1.12315	1.12746
1.5	1.13173	1.13596	1.14015	1.14431	1.14843	1.15251	1.15655	1.16056	1.16453	1.16846
1.6	1.17236	1.17622	1.18005	1.18384	1.18760	1.19132	1.19500	1.19866	1.20228	1.20586
1.7	1.20941	1.21293	1.21642	1.21987	1.22330	1.22668	1.23004	1.23337	1.23666	1.23993
1.8	1.24316	1.24636	1.24954	1.25268	1.25579	1.25888	1.26193	1.26496	1.26795	1.27092
1.9	1.27386	1.27677	1.27966	1.28251	1.28534	1.28815	1.29092	1.29367	1.29639	1.29909
2.0	1.30176	1.30441	1.30703	1.30962	1.31219	1.31473	1.31726	1.31975	1.32222	1.32467
2.1	1.32710	1.32950	1.33188	1.33423	1.33656	1.33887	1.34116	1.34343	1.34567	1.34789
2.2	1.35009	1.35227	1.35443	1.35656	1.35868	1.36077	1.36285	1.36490	1.36694	1.36895
2.3	1.37095	1.37292	1.37488	1.37682	1.37873	1.38063	1.38251	1.38438	1.38622	1.38805
2.4	1.38986	1.39165	1.39342	1.39518	1.39691	1.39864	1.40034	1.40203	1.40370	1.40535
2.5	1.40699	1.40862	1.41022	1.41181	1.41339	1.41495	1.41649	1.41802	1.41954	1.42104
2.6	1.42252	1.42399	1.42545	1.42689	1.42832	1.42973	1.43113	1.43251	1.43388	1.43524
2.7	1.43659	1.43792	1.43924	1.44054	1.44183	1.44311	1.44438	1.44564	1.44688	1.44811
2.8	1.44933	1.45053	1.45173	1.45291	1.45408	1.45524	1.45638	1.45752	1.45864	1.45976
2.9	1.46086	1.46195	1.46303	1.46410	1.46516	1.46621	1.46725	1.46828	1.46930	1.47031
3.0	1.47130	1.47229	1.47327	1.47424	1.47520	1.47615	1.47709	1.47802	1.47894	1.47986
3.1	1.48076	1.48165	1.48254	1.48342	1.48428	1.48514	1.48600	1.48684	1.48767	1.48850
3.2	1.48932	1.49013	1.49093	1.49172	1.49251	1.49329	1.49406	1.49482	1.49558	1.49632
3.3	1.49706	1.49780	1.49852	1.49924	1.49995	1.50066	1.50135	1.50204	1.50273	1.50340
3.4	1.50407	1.50474	1.50539	1.50605	1.50669	1.50733	1.50796	1.50858	1.50920	1.50981
3.5	1.51042	1.51102	1.51161	1.51220	1.51279	1.51336	1.51393	1.51450	1.51506	1.51561
3.6	1.51616	1.51671	1.51724	1.51778	1.51830	1.51883	1.51934	1.51985	1.52036	1.52086
3.7	1.52136	1.52185	1.52234	1.52282	1.52330	1.52377	1.52424	1.52470	1.52516	1.52561
3.8	1.52606	1.52651	1.52695	1.52738	1.52782	1.52824	1.52867	1.52909	1.52950	1.52991
3.9	1.53032	1.53072	1.53112	1.53151	1.53190	1.53229	1.53267	1.53305	1.53343	1.53380
4.0	1.53417	1.53453	1.53489	1.53525	1.53561	1.53596	1.53630	1.53664	1.53698	1.53732
4.1	1.53765	1.53798	1.53831	1.53863	1.53895	1.53927	1.53958	1.53989	1.54020	1.54051
4.2	1.54081	1.54111	1.54140	1.54169	1.54198	1.54227	1.54255	1.54283	1.54311	1.54339
4.3	1.54366	1.54393	1.54420	1.54446	1.54472	1.54498	1.54524	1.54550	1.54575	1.54600
4.4	1.54624	1.54649	1.54673	1.54697	1.54721	1.54744	1.54767	1.54790	1.54813	1.54836
4.5	1.54858	1.54880	1.54902	1.54924	1.54945	1.54966	1.54987	1.55008	1.55029	1.55049
4.6	1.55069	1.55089	1.55109	1.55129	1.55148	1.55167	1.55186	1.55205	1.55224	1.55242
4.7	1.55261	1.55279	1.55297	1.55314	1.55332	1.55349	1.55367	1.55384	1.55400	1.55417
4.8	1.55434	1.55450	1.55466	1.55482	1.55498	1.55514	1.55530	1.55545	1.55560	1.55575
4.9	1.55590	1.55605	1.55620	1.55634	1.55649	1.55663	1.55677	1.55691	1.55705	1.55719

GUDERMANNIAN FUNCTION

x	0	1	2	3	4	5	6	7	8	9
5.0	1.55732	1.55745	1.55759	1.55772	1.55785	1.55798	1.55811	1.55823	1.55836	1.55848
5.1	1.55860	1.55872	1.55884	1.55896	1.55908	1.55920	1.55931	1.55943	1.55954	1.55965
5.2	1.55976	1.55987	1.55998	1.56009	1.56020	1.56030	1.56041	1.56051	1.56061	1.56071
5.3	1.56081	1.56091	1.56101	1.56111	1.56120	1.56130	1.56139	1.56149	1.56158	1.56167
5.4	1.56176	1.56185	1.56194	1.56203	1.56212	1.56220	1.56229	1.56237	1.56246	1.56254
5.5	1.56262	1.56270	1.56278	1.56286	1.56294	1.56302	1.56310	1.56318	1.56325	1.56333
5.6	1.56340	1.56347	1.56355	1.56362	1.56369	1.56376	1.56383	1.56390	1.56397	1.56404
5.7	1.56410	1.56417	1.56424	1.56430	1.56437	1.56443	1.56449	1.56456	1.56462	1.56468
5.8	1.56474	1.56480	1.56486	1.56492	1.56498	1.56504	1.56509	1.56515	1.56521	1.56526
5.9	1.56532	1.56537	1.56543	1.56548	1.56553	1.56558	1.56564	1.56569	1.56574	1.56579
6.0	1.56584	1.56589	1.56594	1.56599	1.56603	1.56608	1.56613	1.56617	1.56622	1.56627
6.1	1.56631	1.56636	1.56640	1.56644	1.56649	1.56653	1.56657	1.56661	1.56666	1.56670
6.2	1.56674	1.56678	1.56682	1.56686	1.56690	1.56694	1.56697	1.56701	1.56705	1.56709
6.3	1.56712	1.56716	1.56720	1.56723	1.56727	1.56730	1.56734	1.56737	1.56741	1.56744
6.4	1.56747	1.56751	1.56754	1.56757	1.56760	1.56764	1.56767	1.56770	1.56773	1.56776
6.5	1.56779	1.56782	1.56785	1.56788	1.56791	1.56794	1.56796	1.56799	1.56802	1.56805
6.6	1.56808	1.56810	1.56813	1.56816	1.56818	1.56821	1.56823	1.56826	1.56828	1.56831
6.7	1.56833	1.56836	1.56838	1.56841	1.56843	1.56845	1.56848	1.56850	1.56852	1.56855
6.8	1.56857	1.56859	1.56861	1.56863	1.56866	1.56868	1.56870	1.56872	1.56874	1.56876
6.9	1.56878	1.56880	1.56882	1.56884	1.56886	1.56888	1.56890	1.56892	1.56894	1.56895
7.0	1.56897	1.56899	1.56901	1.56903	1.56904	1.56906	1.56908	1.56910	1.56911	1.56913
7.1	1.56915	1.56916	1.56918	1.56919	1.56921	1.56923	1.56924	1.56926	1.56927	1.56929
7.2	1.56930	1.56932	1.56933	1.56935	1.56936	1.56938	1.56939	1.56940	1.56942	1.56943
7.3	1.56945	1.56946	1.56947	1.56949	1.56950	1.56951	1.56952	1.56954	1.56955	1.56956
7.4	1.56957	1.56959	1.56960	1.56961	1.56962	1.56963	1.56965	1.56966	1.56967	1.56968
7.5	1.56969	1.56970	1.56971	1.56972	1.56973	1.56974	1.56975	1.56976	1.56978	1.56979
7.6	1.56980	1.56981	1.56982	1.56983	1.56983	1.56984	1.56985	1.56986	1.56987	1.56988
7.7	1.56989	1.56990	1.56991	1.56992	1.56993	1.56993	1.56994	1.56995	1.56996	1.56997
7.8	1.56998	1.56999	1.56999	1.57000	1.57001	1.57002	1.57002	1.57003	1.57004	1.57005
7.9	1.57005	1.57006	1.57007	1.57008	1.57008	1.57009	1.57010	1.57010	1.57011	1.57012
8.0	1.57013	1.57013	1.57014	1.57015	1.57015	1.57016	1.57016	1.57017	1.57018	1.57018
8.1	1.57019	1.57020	1.57020	1.57021	1.57021	1.57022	1.57022	1.57023	1.57024	1.57024
8.2	1.57025	1.57025	1.57026	1.57026	1.57027	1.57027	1.57028	1.57028	1.57029	1.57029
8.3	1.57030	1.57030	1.57031	1.57031	1.57032	1.57032	1.57033	1.57033	1.57034	1.57034
8.4	1.57035	1.57035	1.57036	1.57036	1.57036	1.57037	1.57037	1.57038	1.57038	1.57039
8.5	1.57039	1.57039	1.57040	1.57040	1.57041	1.57041	1.57041	1.57042	1.57042	1.57042
8.6	1.57043	1.57043	1.57044	1.57044	1.57044	1.57045	1.57045	1.57045	1.57046	1.57046
8.7	1.57046	1.57047	1.57047	1.57047	1.57048	1.57048	1.57048	1.57049	1.57049	1.57049
8.8	1.57049	1.57050	1.57050	1.57050	1.57051	1.57051	1.57051	1.57052	1.57052	1.57052
8.9	1.57052	1.57053	1.57053	1.57053	1.57053	1.57054	1.57054	1.57054	1.57054	1.57055
9.0	1.57055	1.57055	1.57055	1.57056	1.57056	1.57056	1.57056	1.57057	1.57057	1.57057
9.1	1.57057	1.57058	1.57058	1.57058	1.57058	1.57058	1.57059	1.57059	1.57059	1.57059
9.2	1.57059	1.57060	1.57060	1.57060	1.57060	1.57060	1.57061	1.57061	1.57061	1.57061
9.3	1.57061	1.57062	1.57062	1.57062	1.57062	1.57062	1.57062	1.57063	1.57063	1.57063
9.4	1.57063	1.57063	1.57063	1.57064	1.57064	1.57064	1.57064	1.57064	1.57064	1.57065
9.5	1.57065	1.57065	1.57065	1.57065	1.57065	1.57065	1.57066	1.57066	1.57066	1.57066
9.6	1.57066	1.57066	1.57066	1.57066	1.57067	1.57067	1.57067	1.57067	1.57067	1.57067
9.7	1.57067	1.57067	1.57068	1.57068	1.57068	1.57068	1.57068	1.57068	1.57068	1.57068
9.8	1.57069	1.57069	1.57069	1.57069	1.57069	1.57069	1.57069	1.57069	1.57069	1.57069
9.9	1.57070	1.57070	1.57070	1.57070	1.57070	1.57070	1.57070	1.57070	1.57070	1.57070

INVERSE GUDERMANNIAN FUNCTION

x	0	1	2	3	4	5	6	7	8	9
0.0	0.00000	0.01000	0.02000	0.03000	0.04001	0.05002	0.06004	0.07006	0.08009	0.09012
0.1	0.10017	0.11022	0.12029	0.13037	0.14046	0.15057	0.16069	0.17082	0.18098	0.19115
0.2	0.20135	0.21156	0.22180	0.23206	0.24234	0.25265	0.26298	0.27334	0.28373	0.29415
0.3	0.30460	0.31509	0.32561	0.33616	0.34675	0.35737	0.36804	0.37874	0.38949	0.40028
0.4	0.41111	0.42199	0.43292	0.44390	0.45493	0.46600	0.47714	0.48833	0.49957	0.51087
0.5	0.52224	0.53366	0.54515	0.55671	0.56834	0.58003	0.59180	0.60364	0.61555	0.62755
0.6	0.63962	0.65178	0.66402	0.67636	0.68878	0.70129	0.71390	0.72661	0.73942	0.75233
0.7	0.76535	0.77848	0.79172	0.80508	0.81856	0.83217	0.84590	0.85976	0.87376	0.88790
0.8	0.90218	0.91660	0.93118	0.94592	0.96082	0.97589	0.99113	1.00654	1.02215	1.03794
0.9	1.05392	1.07011	1.08651	1.10313	1.11997	1.13704	1.15435	1.17192	1.18974	1.20783
1.0	1.22619	1.24485	1.26380	1.28306	1.30265	1.32258	1.34285	1.36349	1.38451	1.40593
1.1	1.42776	1.45003	1.47275	1.49594	1.51963	1.54384	1.56860	1.59394	1.61987	1.64645
1.2	1.67370	1.70166	1.73037	1.75987	1.79022	1.82147	1.85367	1.88689	1.92120	1.95667
1.3	1.99340	2.03147	2.07100	2.11210	2.15491	2.19959	2.24630	2.29524	2.34666	2.40080
1.4	2.45800	2.51861	2.58307	2.65193	2.72583	2.80558	2.89219	2.98695	3.09160	3.20843
1.5	3.34068	3.49307	3.67286	3.89217	4.17343	4.56609	5.22169	7.82865		

CHAPTER 8

COMPLEX NUMBERS

DEFINITIONS INVOLVING COMPLEX NUMBERS

A *complex number* is generally written as $a + bi$ where a and b are real numbers and i, called the *imaginary unit*, has the property that $i^2 = -1$. The real numbers a and b are called the *real* and *imaginary parts* of $a + bi$ respectively.

The complex numbers $a + bi$ and $a - bi$ are called *complex conjugates* of each other.

EQUALITY OF COMPLEX NUMBERS

$$a + bi = c + di \quad \text{if and only if} \quad a = c \text{ and } b = d$$

ADDITION OF COMPLEX NUMBERS

$$(a + bi) + (c + di) = (a + c) + (b + d)i$$

SUBTRACTION OF COMPLEX NUMBERS

$$(a + bi) - (c + di) = (a - c) + (b - d)i$$

MULTIPLICATION OF COMPLEX NUMBERS

$$(a + bi)(c + di) = (ac - bd) + (ad + bc)i$$

DIVISION OF COMPLEX NUMBERS

$$\frac{a + bi}{c + di} = \frac{a + bi}{c + di} \cdot \frac{c - di}{c - di} = \frac{ac + bd}{c^2 + d^2} + \left(\frac{bc - ad}{c^2 + d^2}\right)i$$

Note that the above operations are obtained by using the ordinary rules of algebra and replacing i^2 by -1 wherever it occurs.

GRAPH OF A COMPLEX NUMBER

A complex number $a + bi$ can be plotted as a point (a, b) on an xy plane called an *Argand diagram* or *Gaussian plane*. For example in Fig. 1 P represents the complex number $-3 + 4i$.

A complex number can also be interpreted as a *vector* from O to P.

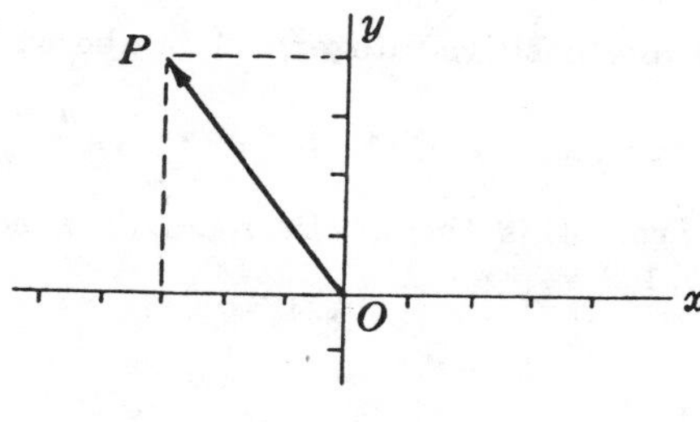

Fig. 1

POLAR FORM OF A COMPLEX NUMBER

In Fig. 2 point P with coordinates (x, y) represents the complex number $x + iy$. Point P can also be represented by *polar coordinates* (r, θ). Since $x = r\cos\theta$, $y = r\sin\theta$ we have

$$x + iy = r(\cos\theta + i\sin\theta)$$

called the *polar form* of the complex number. We often call $r = \sqrt{x^2 + y^2}$ the *modulus* and θ the *amplitude* of $x + iy$.

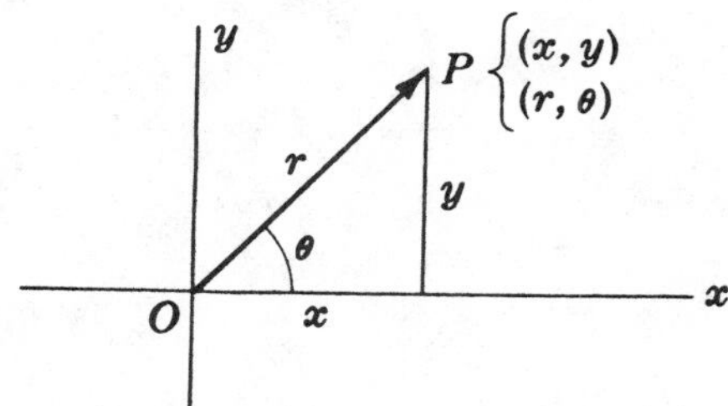

Fig. 2

MULTIPLICATION AND DIVISION OF COMPLEX NUMBERS IN POLAR FORM

$$[r_1(\cos\theta_1 + i\sin\theta_1)][r_2(\cos\theta_2 + i\sin\theta_2)] = r_1 r_2[\cos(\theta_1 + \theta_2) + i\sin(\theta_1 + \theta_2)]$$

$$\frac{r_1(\cos\theta_1 + i\sin\theta_1)}{r_2(\cos\theta_2 + i\sin\theta_2)} = \frac{r_1}{r_2}[\cos(\theta_1 - \theta_2) + i\sin(\theta_1 - \theta_2)]$$

DE MOIVRE'S THEOREM

If p is any real number, De Moivre's theorem states that

8.1 $$[r(\cos\theta + i\sin\theta)]^p = r^p(\cos p\theta + i\sin p\theta)$$

ROOTS OF COMPLEX NUMBERS

If $p = 1/n$ where n is any positive integer, 8.1 can be written

$$[r(\cos\theta + i\sin\theta)]^{1/n} = r^{1/n}\left[\cos\frac{\theta + 2k\pi}{n} + i\sin\frac{\theta + 2k\pi}{n}\right]$$

where k is any integer. From this the n nth roots of a complex number can be obtained by putting $k = 0, 1, 2, \ldots, n-1$.

CHAPTER 9

VECTOR ANALYSIS

VECTORS AND SCALARS

Various quantities in physics such as temperature, volume and speed can be specified by a real number. Such quantities are called *scalars.*

Other quantities such as force, velocity and momentum require for their specification a direction as well as magnitude. Such quantities are called *vectors.* A vector is represented by an arrow or directed line segment indicating direction. The magnitude of the vector is determined by the length of the arrow, using an appropriate unit.

NOTATION FOR VECTORS

A vector is denoted by a bold faced letter such as $\mathbf{A}$ [Fig. 1]. The magnitude is denoted by $|\mathbf{A}|$ or A. The tail end of the arrow is called the *initial point* while the head is called the *terminal point.*

COMPONENTS OF A VECTOR

A vector $\mathbf{A}$ can be represented with initial point at the origin of a rectangular coordinate system. If $\mathbf{i}, \mathbf{j}, \mathbf{k}$ are unit vectors in the directions of the positive x, y, z axes, then

$$\mathbf{A} = A_1\mathbf{i} + A_2\mathbf{j} + A_3\mathbf{k}$$

where $A_1\mathbf{i}, A_2\mathbf{j}, A_3\mathbf{k}$ are called *component vectors* of $\mathbf{A}$ in the $\mathbf{i}, \mathbf{j}, \mathbf{k}$ directions and A_1, A_2, A_3 are called the *components* of $\mathbf{A}$.

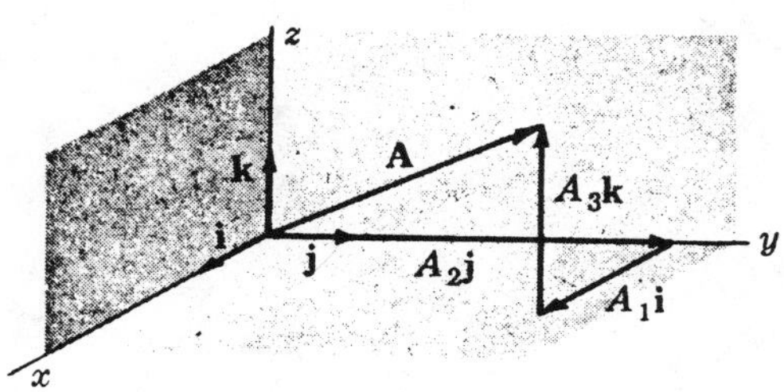

FUNDAMENTAL DEFINITIONS

1. **Equality of vectors.** Two vectors are equal if they have the same magnitude and direction. Thus $\mathbf{A} = \mathbf{B}$ in Fig. 1

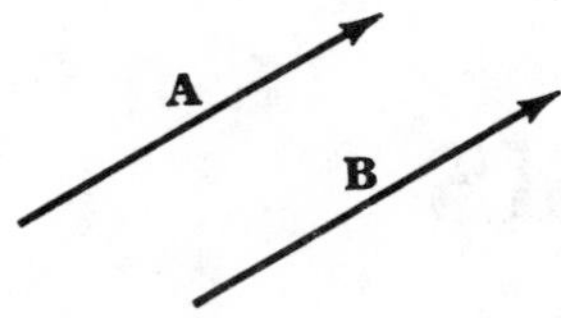

Fig. 1

2. **Multiplication of a vector by a scalar.** If m is any real number (scalar), then $m\mathbf{A}$ is a vector whose magnitude is $|m|$ times the magnitude of $\mathbf{A}$ and whose direction is the same as or opposite to $\mathbf{A}$ according as $m > 0$ or $m < 0$. If $m = 0$, then $m\mathbf{A} = \mathbf{0}$ is called the *zero* or *null vector*.

3. **Sums of vectors.** The sum or resultant of $\mathbf{A}$ and $\mathbf{B}$ is a vector $\mathbf{C} = \mathbf{A} + \mathbf{B}$ formed by placing the initial point of $\mathbf{B}$ on the terminal point of $\mathbf{A}$ and joining the initial point of $\mathbf{A}$ to the terminal point of $\mathbf{B}$ [Fig. 2(b)]. This definition is equivalent to the parallelogram law for vector addition as indicated in Fig. 2(c). The vector $\mathbf{A} - \mathbf{B}$ is defined as $\mathbf{A} + (-\mathbf{B})$.

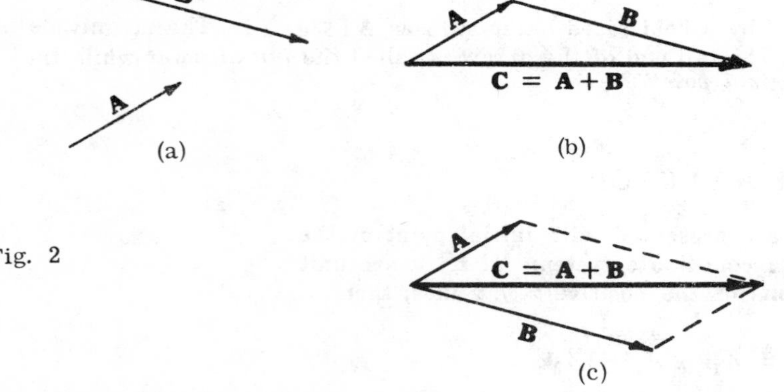

Fig. 2

Extensions to sums of more than two vectors are immediate. Thus Fig. 3 shows how to obtain the sum $\mathbf{E}$ of the vectors $\mathbf{A}$, $\mathbf{B}$, $\mathbf{C}$ and $\mathbf{D}$.

Fig. 3

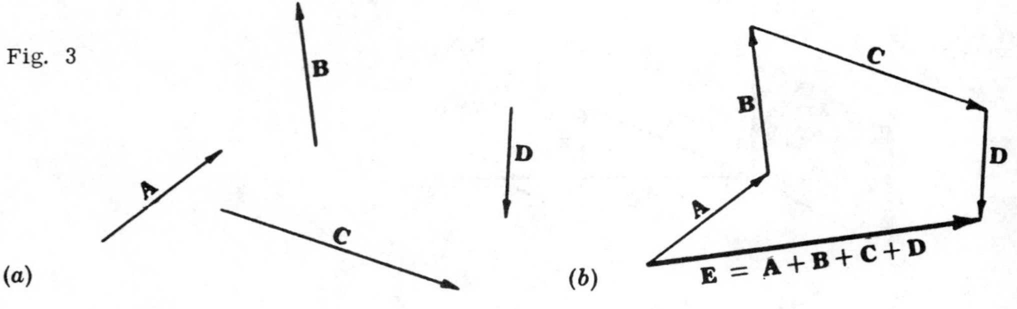

4. **Unit vectors.** A *unit vector* is a vector with unit magnitude. If $\mathbf{A}$ is a vector, then a unit vector in the direction of $\mathbf{A}$ is $\mathbf{a} = \mathbf{A}/A$ where $A > 0$.

LAWS OF VECTOR ALGEBRA

If $\mathbf{A}, \mathbf{B}, \mathbf{C}$ are vectors and m, n are scalars, then

$\mathbf{A} + \mathbf{B} = \mathbf{B} + \mathbf{A}$	Commutative law for addition
$\mathbf{A} + (\mathbf{B} + \mathbf{C}) = (\mathbf{A} + \mathbf{B}) + \mathbf{C}$	Associative law for addition
$m(n\mathbf{A}) = (mn)\mathbf{A} = n(m\mathbf{A})$	Associative law for scalar multiplication
$(m+n)\mathbf{A} = m\mathbf{A} + n\mathbf{A}$	Distributive law
$m(\mathbf{A} + \mathbf{B}) = m\mathbf{A} + m\mathbf{B}$	Distributive law

DOT OR SCALAR PRODUCT

$$\mathbf{A} \cdot \mathbf{B} = AB \cos\theta \qquad 0 \leqq \theta \leqq \pi$$

where θ is the angle between $\mathbf{A}$ and $\mathbf{B}$.

Fundamental results are

$$\mathbf{A} \cdot \mathbf{B} = \mathbf{B} \cdot \mathbf{A} \qquad \text{Commutative law}$$

$$\mathbf{A} \cdot (\mathbf{B} + \mathbf{C}) = \mathbf{A} \cdot \mathbf{B} + \mathbf{A} \cdot \mathbf{C} \qquad \text{Distributive law}$$

$$\mathbf{A} \cdot \mathbf{B} = A_1B_1 + A_2B_2 + A_3B_3$$

where $\mathbf{A} = A_1\mathbf{i} + A_2\mathbf{j} + A_3\mathbf{k}$, $\mathbf{B} = B_1\mathbf{i} + B_2\mathbf{j} + B_3\mathbf{k}$.

CROSS OR VECTOR PRODUCT

$$\mathbf{A} \times \mathbf{B} = AB \sin\theta\, \mathbf{u} \qquad 0 \leqq \theta \leqq \pi$$

where θ is the angle between $\mathbf{A}$ and $\mathbf{B}$ and $\mathbf{u}$ is a unit vector perpendicular to the plane of $\mathbf{A}$ and $\mathbf{B}$ such that $\mathbf{A}, \mathbf{B}, \mathbf{u}$ form a *right-handed system* [i.e. a right-threaded screw rotated through an angle less than 180° from $\mathbf{A}$ to $\mathbf{B}$ will advance in the direction of $\mathbf{u}$ as in Fig. 4].

Fundamental results are

$$\mathbf{A} \times \mathbf{B} = \begin{vmatrix} \mathbf{i} & \mathbf{j} & \mathbf{k} \\ A_1 & A_2 & A_3 \\ B_1 & B_2 & B_3 \end{vmatrix}$$

$$= (A_2B_3 - A_3B_2)\mathbf{i} + (A_3B_1 - A_1B_3)\mathbf{j} + (A_1B_2 - A_2B_1)\mathbf{k}$$

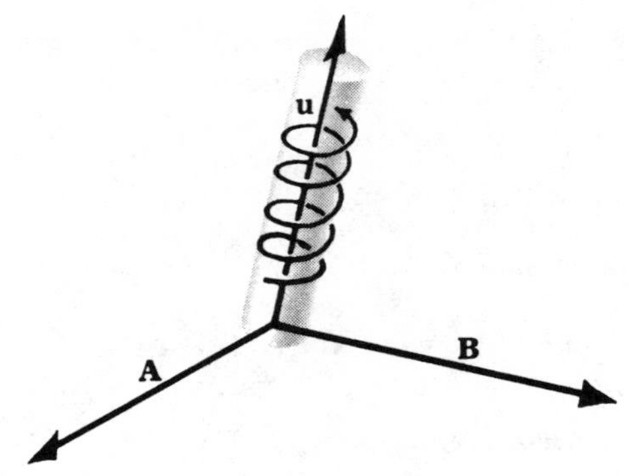

Fig. 4

$$\mathbf{A}\times\mathbf{B} \;=\; -\mathbf{B}\times\mathbf{A}$$

$$\mathbf{A}\times(\mathbf{B}+\mathbf{C}) \;=\; \mathbf{A}\times\mathbf{B} + \mathbf{A}\times\mathbf{C}$$

$|\mathbf{A}\times\mathbf{B}|$ = area of parallelogram having sides **A** and **B**

MISCELLANEOUS FORMULAS INVOLVING DOT AND CROSS PRODUCTS

$$\mathbf{A}\cdot(\mathbf{B}\times\mathbf{C}) \;=\; \begin{vmatrix} A_1 & A_2 & A_3 \\ B_1 & B_2 & B_3 \\ C_1 & C_2 & C_3 \end{vmatrix} \;=\; A_1B_2C_3 + A_2B_3C_1 + A_3B_1C_2 - A_3B_2C_1 - A_2B_1C_3 - A_1B_3C_2$$

$|\mathbf{A}\cdot(\mathbf{B}\times\mathbf{C})|$ = volume of parallelepiped with sides **A, B, C**

$$\mathbf{A}\times(\mathbf{B}\times\mathbf{C}) \;=\; \mathbf{B}(\mathbf{A}\cdot\mathbf{C}) - \mathbf{C}(\mathbf{A}\cdot\mathbf{B})$$

$$(\mathbf{A}\times\mathbf{B})\times\mathbf{C} \;=\; \mathbf{B}(\mathbf{A}\cdot\mathbf{C}) - \mathbf{A}(\mathbf{B}\cdot\mathbf{C})$$

$$(\mathbf{A}\times\mathbf{B})\cdot(\mathbf{C}\times\mathbf{D}) \;=\; (\mathbf{A}\cdot\mathbf{C})(\mathbf{B}\cdot\mathbf{D}) - (\mathbf{A}\cdot\mathbf{D})(\mathbf{B}\cdot\mathbf{C})$$

$$\begin{aligned}(\mathbf{A}\times\mathbf{B})\times(\mathbf{C}\times\mathbf{D}) &= \mathbf{C}\{\mathbf{A}\cdot(\mathbf{B}\times\mathbf{D})\} - \mathbf{D}\{\mathbf{A}\cdot(\mathbf{B}\times\mathbf{C})\} \\ &= \mathbf{B}\{\mathbf{A}\cdot(\mathbf{C}\times\mathbf{D})\} - \mathbf{A}\{\mathbf{B}\cdot(\mathbf{C}\times\mathbf{D})\}\end{aligned}$$

THE DEL OPERATOR

The operator *del* is defined by

$$\nabla \;=\; \mathbf{i}\frac{\partial}{\partial x} + \mathbf{j}\frac{\partial}{\partial y} + \mathbf{k}\frac{\partial}{\partial z}$$

In the results below we assume that $U = U(x, y, z)$, $V = V(x, y, z)$, $\mathbf{A} = \mathbf{A}(x, y, z)$ and $\mathbf{B} = \mathbf{B}(x, y, z)$ have partial derivatives.

DIFFERENTIATION OF VECTORS

If $\mathbf{V}_1 = a_1\mathbf{i} + b_1\mathbf{j} + c_1\mathbf{k}$, and $\mathbf{V}_2 = a_2\mathbf{i} + b_2\mathbf{j} + c_2\mathbf{k}$, and if $\mathbf{V}_1$ and $\mathbf{V}_2$ are functions of the scalar t, then

$$\frac{d}{dt}(\mathbf{V}_1 + \mathbf{V}_2 + \cdots) = \frac{d\mathbf{V}_1}{dt} + \frac{d\mathbf{V}_2}{dt} + \cdots,$$

$$\text{where } \frac{d\mathbf{V}_1}{dt} = \frac{da_1}{dt}\mathbf{i} + \frac{db_1}{dt}\mathbf{j} + \frac{dc_1}{dt}\mathbf{k}, \text{ etc.}$$

$$\frac{d}{dt}(\mathbf{V}_1 \cdot \mathbf{V}_2) = \frac{d\mathbf{V}_1}{dt} \cdot \mathbf{V}_2 + \mathbf{V}_1 \cdot \frac{d\mathbf{V}_2}{dt}$$

$$\frac{d}{dt}(\mathbf{V}_1 \times \mathbf{V}_2) = \frac{d\mathbf{V}_1}{dt} \times \mathbf{V}_2 + \mathbf{V}_1 \times \frac{d\mathbf{V}_2}{dt}$$

$$\mathbf{V} \cdot \frac{d\mathbf{V}}{dt} = v \cdot \frac{dv}{dt}$$

In particular, if $\mathbf{V}$ is a vector of constant length then the right hand side of the last equation is identically zero showing that $\mathbf{V}$ is perpendicular to its derivative.

The derivatives of the triple products are

$$\frac{d}{dt}[\mathbf{V}_1\mathbf{V}_2\mathbf{V}_3] = \left[\left(\frac{d\mathbf{V}_1}{dt}\right)\mathbf{V}_2\mathbf{V}_3\right] + \left[\mathbf{V}_1\left(\frac{d\mathbf{V}_2}{dt}\right)\mathbf{V}_3\right] + \left[\mathbf{V}_1\mathbf{V}_2\left(\frac{d\mathbf{V}_3}{dt}\right)\right]$$

and
$$\frac{d}{dt}\left\{\mathbf{V}_1 \times (\mathbf{V}_2 \times \mathbf{V}_3)\right\} = \left(\frac{d\mathbf{V}_1}{dt}\right) \times (\mathbf{V}_2 \times \mathbf{V}_3) + \mathbf{V}_1 \times \left(\left(\frac{d\mathbf{V}_2}{dt}\right) \times \mathbf{V}_3\right) + \mathbf{V}_1 \times \left(\mathbf{V}_2 \times \left(\frac{d\mathbf{V}_3}{dt}\right)\right).$$

FORMULAS OF VECTOR ANALYSIS

	Rectangular coordinates	Cylindrical coordinates	Spherical coordinates
Conversion to rectangular coordinates		$x = r\cos\varphi \quad y = r\sin\varphi \quad z = z$	$x = r\cos\varphi\sin\theta \quad y = r\sin\varphi\sin\theta \quad z = r\cos\theta$
Gradient	$\nabla\phi = \frac{\partial\phi}{\partial x}\mathbf{i} + \frac{\partial\phi}{\partial y}\mathbf{j} + \frac{\partial\phi}{\partial z}\mathbf{k}$	$\nabla\phi = \frac{\partial\phi}{\partial r}\mathbf{r} + \frac{1}{r}\frac{\partial\phi}{\partial\varphi}\boldsymbol{\phi} + \frac{\partial\phi}{\partial z}\mathbf{k}$	$\nabla\phi = \frac{\partial\phi}{\partial r}\mathbf{r} + \frac{1}{r}\frac{\partial\phi}{\partial\theta}\boldsymbol{\theta} + \frac{1}{r\sin\theta}\frac{\partial\phi}{\partial\varphi}\boldsymbol{\phi}$
Divergence	$\nabla\cdot\mathbf{A} = \frac{\partial A_x}{\partial x} + \frac{\partial A_y}{\partial y} + \frac{\partial A_z}{\partial z}$	$\nabla\cdot\mathbf{A} = \frac{1}{r}\frac{\partial(rA_r)}{\partial r} + \frac{1}{r}\frac{\partial A\varphi}{\partial\varphi} + \frac{\partial A_z}{\partial z}$	$\nabla\cdot\mathbf{A} = \frac{1}{r^2}\frac{\partial(r^2A_r)}{\partial r} + \frac{1}{r\sin\theta}\frac{\partial(A_\theta\sin\theta)}{\partial\theta} + \frac{1}{r\sin\theta}\frac{\partial A_\varphi}{\partial\varphi}$
Curl	$\nabla\times\mathbf{A} = \begin{vmatrix} \mathbf{i} & \mathbf{j} & \mathbf{k} \\ \frac{\partial}{\partial x} & \frac{\partial}{\partial y} & \frac{\partial}{\partial z} \\ A_x & A_y & A_z \end{vmatrix}$	$\nabla\times\mathbf{A} = \begin{vmatrix} \frac{1}{r}\mathbf{r} & \boldsymbol{\phi} & \frac{1}{r}\mathbf{k} \\ \frac{\partial}{\partial r} & \frac{\partial}{\partial\varphi} & \frac{\partial}{\partial z} \\ A_r & rA_\varphi & A_z \end{vmatrix}$	$\nabla\times\mathbf{A} = \begin{vmatrix} \frac{\mathbf{r}}{r^2\sin\theta} & \frac{\boldsymbol{\theta}}{r\sin\theta} & \frac{\boldsymbol{\phi}}{r} \\ \frac{\partial}{\partial r} & \frac{\partial}{\partial\theta} & \frac{\partial}{\partial\varphi} \\ A_r & rA_\theta & rA_\varphi\sin\theta \end{vmatrix}$
Laplacian	$\nabla^2\phi = \frac{\partial^2\phi}{\partial x^2} + \frac{\partial^2\phi}{\partial y^2} + \frac{\partial^2\phi}{\partial z^2}$	$\nabla^2\phi = \frac{1}{r}\frac{\partial}{\partial r}\left(r\frac{\partial\phi}{\partial r}\right) + \frac{1}{r^2}\frac{\partial^2\phi}{\partial\varphi^2} + \frac{\partial^2\phi}{\partial z^2}$	$\nabla^2\phi = \frac{1}{r^2}\frac{\partial}{\partial r}\left(r^2\frac{\partial\phi}{\partial r}\right) + \frac{1}{r^2\sin\theta}\frac{\partial}{\partial\theta}\left(\sin\theta\frac{\partial\phi}{\partial\theta}\right) + \frac{1}{r^2\sin^2\theta}\frac{\partial^2\phi}{\partial\varphi^2}$

CHAPTER 10

ORTHOGONAL COORDINATE SYSTEMS

CYLINDRICAL COORDINATES (r, θ, z)

$$x = r\cos\theta, \quad y = r\sin\theta, \quad z = z$$

$$h_1^2 = 1, \quad h_2^2 = r^2, \quad h_3^2 = 1$$

$$\nabla^2\Phi = \frac{\partial^2\Phi}{\partial r^2} + \frac{1}{r}\frac{\partial\Phi}{\partial r} + \frac{1}{r^2}\frac{\partial^2\Phi}{\partial\theta^2} + \frac{\partial^2\Phi}{\partial z^2}$$

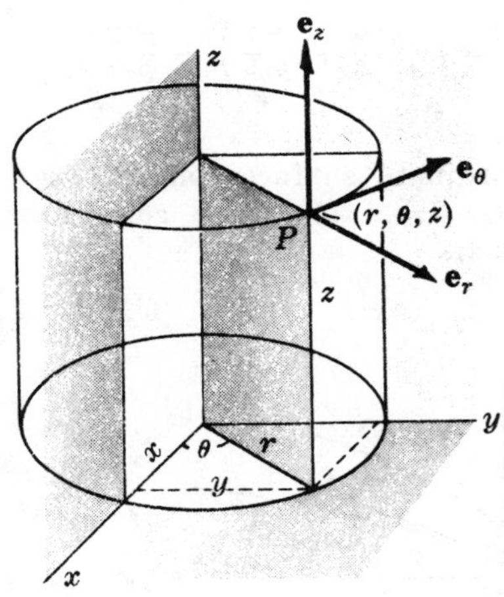

Fig. 1. Cylindrical coordinates.

SPHERICAL COORDINATES (r, θ, φ)

$$x = r\sin\theta\cos\phi, \quad y = r\sin\theta\sin\phi, \quad z = r\cos\theta$$

$$h_1^2 = 1, \quad h_2^2 = r^2, \quad h_3^2 = r^2\sin^2\theta$$

$$\nabla^2\Phi = \frac{1}{r^2}\frac{\partial}{\partial r}\left(r^2\frac{\partial\Phi}{\partial r}\right) + \frac{1}{r^2\sin\theta}\frac{\partial}{\partial\theta}\left(\sin\theta\frac{\partial\Phi}{\partial\theta}\right) + \frac{1}{r^2\sin^2\theta}\frac{\partial^2\Phi}{\partial\phi^2}$$

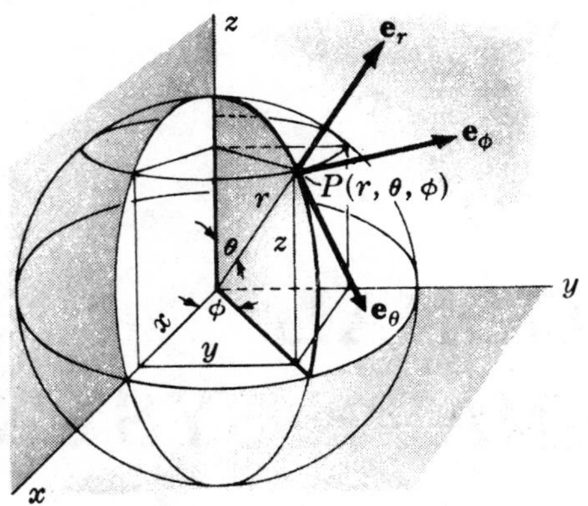

Fig. 2. Spherical coordinates.

PARABOLIC CYLINDRICAL COORDINATES (u, ν, z)

$$x = \tfrac{1}{2}(u^2 - v^2), \quad y = uv, \quad z = z$$

$$h_1^2 = h_2^2 = u^2 + v^2, \quad h_3^2 = 1$$

$$\nabla^2\Phi = \frac{1}{u^2 + v^2}\left(\frac{\partial^2\Phi}{\partial u^2} + \frac{\partial^2\Phi}{\partial v^2}\right) + \frac{\partial^2\Phi}{\partial z^2}$$

The traces of the coordinate surfaces on the xy plane are shown in Fig. 3. They are confocal parabolas with a common axis.

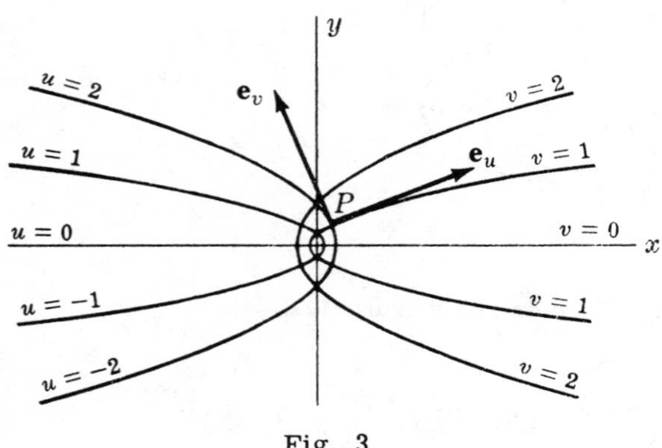

Fig. 3

PARABOLOIDAL COORDINATES (u, ν, φ)

$$x = uv \cos\phi, \quad y = uv \sin\phi, \quad z = \tfrac{1}{2}(u^2 - v^2)$$

where

$$u \geqq 0, \quad v \geqq 0, \quad 0 \leqq \phi < 2\pi$$

$$h_1^2 = h_2^2 = u^2 + v^2, \quad h_3^2 = u^2v^2$$

$$\nabla^2\Phi = \frac{1}{u(u^2+v^2)}\frac{\partial}{\partial u}\left(u\frac{\partial\Phi}{\partial u}\right) + \frac{1}{v(u^2+v^2)}\frac{\partial}{\partial v}\left(v\frac{\partial\Phi}{\partial v}\right) + \frac{1}{u^2v^2}\frac{\partial^2\Phi}{\partial\phi^2}$$

Two sets of coordinate surfaces are obtained by revolving the parabolas of Fig. 3 about the x axis which is then relabeled the z axis.

ELLIPTIC CYLINDRICAL COORDINATES (u, v, z)

$$x = a\cosh u\cos v, \quad y = a\sinh u\sin v, \quad z = z$$

where

$$u \geqq 0, \quad 0 \leqq v < 2\pi, \quad -\infty < z < \infty$$

$$h_1^2 = h_2^2 = a^2(\sinh^2 u + \sin^2 v), \quad h_3^2 = 1$$

$$\nabla^2\Phi = \frac{1}{a^2(\sinh^2 u + \sin^2 v)}\left(\frac{\partial^2\Phi}{\partial u^2} + \frac{\partial^2\Phi}{\partial v^2}\right) + \frac{\partial^2\Phi}{\partial z^2}$$

The traces of the coordinate surfaces on the xy plane are shown in Fig. 4. They are confocal ellipses and hyperbolas.

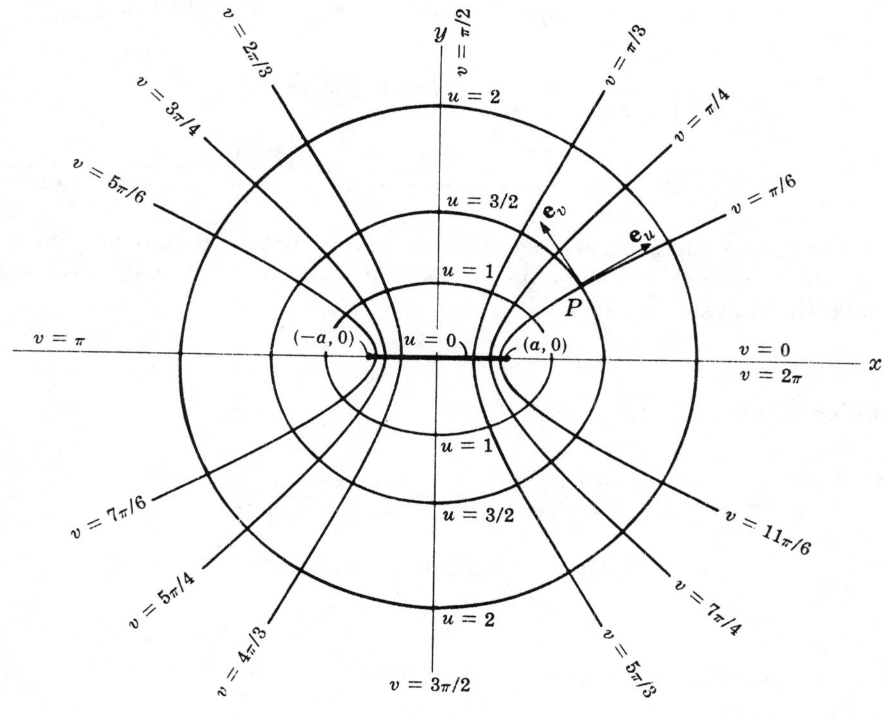

Fig. 4. Elliptic cylindrical coordinates.

PROLATE SPHEROIDAL COORDINATES $(\xi,\ \eta,\ \phi)$

$$x = a \sinh \xi \sin \eta \cos \phi, \quad y = a \sinh \xi \sin \eta \sin \phi, \quad z = a \cosh \xi \cos \eta$$

where
$$\xi \geqq 0, \quad 0 \leqq \eta \leqq \pi, \quad 0 \leqq \phi < 2\pi$$

$$h_1^2 = h_2^2 = a^2(\sinh^2 \xi + \sin^2 \eta), \quad h_3^2 = a^2 \sinh^2 \xi \sin^2 \eta$$

$$\nabla^2\Phi = \frac{1}{a^2(\sinh^2 \xi + \sin^2 \eta) \sinh \xi} \frac{\partial}{\partial \xi}\left(\sinh \xi \frac{\partial \Phi}{\partial \xi}\right) + \frac{1}{a^2(\sinh^2 \xi + \sin^2 \eta) \sin \eta} \frac{\partial}{\partial \eta}\left(\sin \eta \frac{\partial \Phi}{\partial \eta}\right) + \frac{1}{a^2 \sinh^2 \xi \sin^2 \eta} \frac{\partial^2 \Phi}{\partial \phi^2}$$

Two sets of coordinate surfaces are obtained by revolving the curves of Fig. 4 about the x axis which is relabeled the z axis. The third set of coordinate surfaces consists of planes passing through this axis.

OBLATE SPHEROIDAL COORDINATES $(\xi,\ \eta,\ \phi)$

$$x = a \cosh \xi \cos \eta \cos \phi, \quad y = a \cosh \xi \cos \eta \sin \phi, \quad z = a \sinh \xi \sin$$

where
$$\xi \geqq 0, \quad -\pi/2 \leqq \eta \leqq \pi/2, \quad 0 \leqq \phi < 2\pi$$

$$h_1^2 = h_2^2 = a^2(\sinh^2 \xi + \sin^2 \eta), \quad h_3^2 = a^2 \cosh^2 \xi \cos^2 \eta$$

$$\nabla^2\Phi = \frac{1}{a^2(\sinh^2 \xi + \sin^2 \eta) \cosh \xi} \frac{\partial}{\partial \xi}\left(\cosh \xi \frac{\partial \Phi}{\partial \xi}\right) + \frac{1}{a^2(\sinh^2 \xi + \sin^2 \eta) \cos \eta} \frac{\partial}{\partial \eta}\left(\cos \eta \frac{\partial \Phi}{\partial \eta}\right) + \frac{1}{a^2 \cosh^2 \xi \cos^2 \eta} \frac{\partial^2}{\partial \phi}$$

Two sets of coordinate surfaces are obtained by revolving the curves of Fig. 4 about the y axis which is relabeled the z axis. The third set of coordinate surfaces are planes passing through this axis.

BIPOLAR COORDINATES $(u,\ v,\ z)$

$$x = \frac{a \sinh v}{\cosh v - \cos u}, \quad y = \frac{a \sin u}{\cosh v - \cos u}, \quad z = z$$

where
$$0 \leqq u < 2\pi, \quad -\infty < v < \infty, \quad -\infty < z < \infty$$

or

$$x^2 + (y - a \cot u)^2 = a^2 \csc^2 u, \quad (x - a \coth v)^2 + y^2 = a^2 \operatorname{csch}^2 v, \quad z =$$

$$h_1^2 = h_2^2 = \frac{a^2}{(\cosh v - \cos u)^2}, \quad h_3^2 = 1$$

$$\nabla^2\Phi = \frac{(\cosh v - \cos u)^2}{a^2}\left(\frac{\partial^2\Phi}{\partial u^2} + \frac{\partial^2\Phi}{\partial v^2}\right) + \frac{\partial^2\Phi}{\partial z^2}$$

The traces of the coordinate surfaces on the xy plane are shown in Fig. 5.

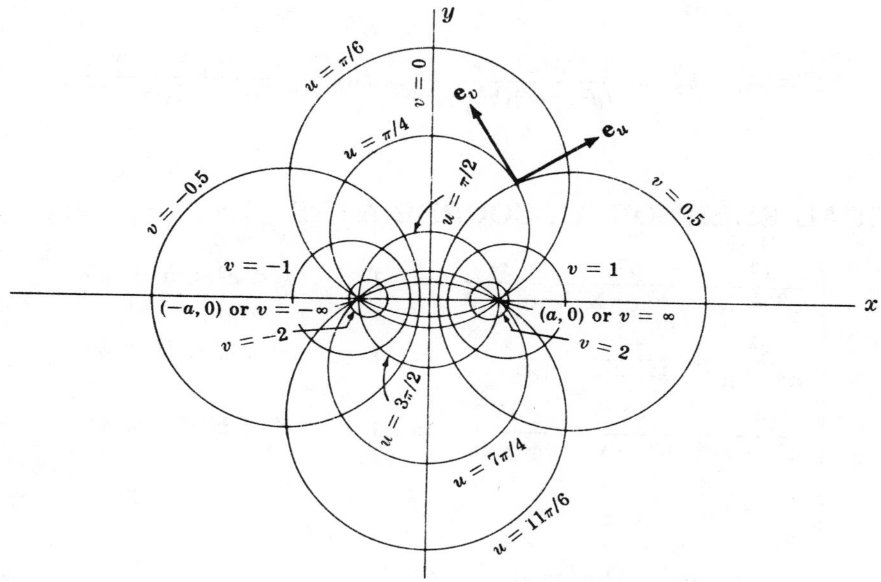

Fig. 5. Bipolar coordinates.

TOROIDAL COORDINATES (u, ν, φ)

$$x = \frac{a \sinh v \cos\phi}{\cosh v - \cos u}, \quad y = \frac{a \sinh v \sin\phi}{\cosh v - \cos u}, \quad z = \frac{a \sin u}{\cosh v - \cos u}$$

$$h_1^2 = h_2^2 = \frac{a^2}{(\cosh v - \cos u)^2}, \quad h_3^2 = \frac{a^2 \sinh^2 v}{(\cosh v - \cos u)^2}$$

$$\nabla^2\Phi = \frac{(\cosh v - \cos u)^3}{a^2}\frac{\partial}{\partial u}\left(\frac{1}{\cosh v - \cos u}\frac{\partial\Phi}{\partial u}\right)$$

$$+ \frac{(\cosh v - \cos u)^3}{a^2 \sinh v}\frac{\partial}{\partial v}\left(\frac{\sinh v}{\cosh v - \cos u}\frac{\partial\Phi}{\partial v}\right) + \frac{(\cosh v - \cos u)^2}{a^2 \sinh^2 v}\frac{\partial^2\Phi}{\partial\phi^2}$$

The coordinate surfaces are obtained by revolving the curves of Fig. 5 about the y axis which is relabeled the z axis.

CONICAL COORDINATES (λ, μ, ν)

$$x = \frac{\lambda\mu\nu}{ab}, \quad y = \frac{\lambda}{a}\sqrt{\frac{(\mu^2-a^2)(\nu^2-a^2)}{a^2-b^2}}, \quad z = \frac{\lambda}{b}\sqrt{\frac{(\mu^2-b^2)(\nu^2-b^2)}{b^2-a^2}}$$

$$h_1^2 = 1, \quad h_2^2 = \frac{\lambda^2(\mu^2-\nu^2)}{(\mu^2-a^2)(b^2-\mu^2)}, \quad h_3^2 = \frac{\lambda^2(\mu^2-\nu^2)}{(\nu^2-a^2)(\nu^2-b^2)}$$

CONFOCAL ELLIPSOIDAL COORDINATES (λ, μ, ν)

$$\begin{cases} \dfrac{x^2}{a^2-\lambda} + \dfrac{y^2}{b^2-\lambda} + \dfrac{z^2}{c^2-\lambda} = 1 & \lambda < c^2 < b^2 < a^2 \\ \dfrac{x^2}{a^2-\mu} + \dfrac{y^2}{b^2-\mu} + \dfrac{z^2}{c^2-\mu} = 1 & c^2 < \mu < b^2 < a^2 \\ \dfrac{x^2}{a^2-\nu} + \dfrac{y^2}{b^2-\nu} + \dfrac{z^2}{c^2-\nu} = 1 & c^2 < b^2 < \nu < a^2 \end{cases}$$

or

$$\begin{cases} x^2 = \dfrac{(a^2-\lambda)(a^2-\mu)(a^2-\nu)}{(a^2-b^2)(a^2-c^2)} \\ y^2 = \dfrac{(b^2-\lambda)(b^2-\mu)(b^2-\nu)}{(b^2-a^2)(b^2-c^2)} \\ z^2 = \dfrac{(c^2-\lambda)(c^2-\mu)(c^2-\nu)}{(c^2-a^2)(c^2-b^2)} \end{cases}$$

$$\begin{cases} h_1^2 = \dfrac{(\mu-\lambda)(\nu-\lambda)}{4(a^2-\lambda)(b^2-\lambda)(c^2-\lambda)} \\ h_2^2 = \dfrac{(\nu-\mu)(\lambda-\mu)}{4(a^2-\mu)(b^2-\mu)(c^2-\mu)} \\ h_3^2 = \dfrac{(\lambda-\nu)(\mu-\nu)}{4(a^2-\nu)(b^2-\nu)(c^2-\nu)} \end{cases}$$

CONFOCAL PARABOLOIDAL COORDINATES (λ, μ, ν)

$$\begin{cases} \dfrac{x^2}{a^2-\lambda} + \dfrac{y^2}{b^2-\lambda} = z - \lambda & -\infty < \lambda < b^2 \\ \dfrac{x^2}{a^2-\mu} + \dfrac{y^2}{b^2-\mu} = z - \mu & b^2 < \mu < a^2 \\ \dfrac{x^2}{a^2-\nu} + \dfrac{y^2}{b^2-\nu} = z - \nu & a^2 < \nu < \infty \end{cases}$$

or

$$\begin{cases} x^2 = \dfrac{(a^2-\lambda)(a^2-\mu)(a^2-\nu)}{b^2-a^2} \\[2ex] y^2 = \dfrac{(b^2-\lambda)(b^2-\mu)(b^2-\nu)}{a^2-b^2} \\[2ex] z = \lambda + \mu + \nu - a^2 - b^2 \end{cases}$$

$$\begin{cases} h_1^2 = \dfrac{(\mu-\lambda)(\nu-\lambda)}{4(a^2-\lambda)(b^2-\lambda)} \\[2ex] h_2^2 = \dfrac{(\nu-\mu)(\lambda-\mu)}{4(a^2-\mu)(b^2-\mu)} \\[2ex] h_3^2 = \dfrac{(\lambda-\nu)(\mu-\nu)}{16(a^2-\nu)(b^2-\nu)} \end{cases}$$

CHAPTER 11

SERIES

SERIES OF CONSTANTS

ARITHMETIC SERIES

$$a + (a+d) + (a+2d) + \cdots + \{a + (n-1)d\} = \tfrac{1}{2}n\{2a + (n-1)d\}$$
$$= \tfrac{1}{2}n(a+l)$$

where $l = a + (n-1)d$ is the last term.

Some special cases are

$$1 + 2 + 3 + \cdots + n = \tfrac{1}{2}n(n+1)$$

$$1 + 3 + 5 + \cdots + (2n-1) = n^2$$

GEOMETRIC SERIES

$$a + ar + ar^2 + ar^3 + \cdots + ar^{n-1} = \frac{a(1-r^n)}{1-r} = \frac{a - rl}{1-r}$$

where $l = ar^{n-1}$ is the last term and $r \neq 1$.

If $-1 < r < 1$, then

$$a + ar + ar^2 + ar^3 + \cdots = \frac{a}{1-r}$$

ARITHMETIC-GEOMETRIC SERIES

$$a + (a+d)r + (a+2d)r^2 + \cdots + \{a + (n-1)d\}r^{n-1} =$$

$$\frac{a(1-r^n)}{1-r} + \frac{rd\{1 - nr^{n-1} + (n-1)r^n}{(1-r)^2}$$

where $r \neq 1$.

If $-1 < r < 1$, then

$$a + (a+d)r + (a+2d)r^2 + \cdots = \frac{a}{1-r} + \frac{rd}{(1-r)^2}$$

SUMS OF POWERS OF POSITIVE INTEGERS

$$1^p + 2^p + 3^p + \cdots + n^p = \frac{n^{p+1}}{p+1} + \tfrac{1}{2}n^p + \frac{B_1 p n^{p-1}}{2!} - \frac{B_2 p(p-1)(p-2)n^{p-3}}{4!} + \cdots$$

where the series terminates at n^2 or n according as p is odd or even, and B_k are the *Bernoulli numbers*.

Some special cases are

$$1 + 2 + 3 + \cdots + n = \frac{n(n+1)}{2}$$

$$1^2 + 2^2 + 3^2 + \cdots + n^2 = \frac{n(n+1)(2n+1)}{6}$$

$$1^3 + 2^3 + 3^3 + \cdots + n^3 = \frac{n^2(n+1)^2}{4} = (1+2+3+\cdots+n)^2$$

$$1^4 + 2^4 + 3^4 + \cdots + n^4 = \frac{n(n+1)(2n+1)(3n^2+3n-1)}{30}$$

If $S_k = 1^k + 2^k + 3^k + \cdots + n^k$ where k and n are positive integers, then

$$\binom{k+1}{1}S_1 + \binom{k+1}{2}S_2 + \cdots + \binom{k+1}{k}S_k = (n+1)^{k+1} - (n+1)$$

SERIES INVOLVING RECIPROCALS OF POWERS OF POSITIVE INTEGERS

$$1 - \frac{1}{2} + \frac{1}{3} - \frac{1}{4} + \frac{1}{5} - \cdots = \ln 2$$

$$1 - \frac{1}{3} + \frac{1}{5} - \frac{1}{7} + \frac{1}{9} - \cdots = \frac{\pi}{4}$$

$$1 - \frac{1}{4} + \frac{1}{7} - \frac{1}{10} + \frac{1}{13} - \cdots = \frac{\pi\sqrt{3}}{9} + \frac{1}{3}\ln 2$$

$$1 - \frac{1}{5} + \frac{1}{9} - \frac{1}{13} + \frac{1}{17} - \cdots = \frac{\pi\sqrt{2}}{8} + \frac{\sqrt{2}\ln(1+\sqrt{2})}{4}$$

$$\frac{1}{2} - \frac{1}{5} + \frac{1}{8} - \frac{1}{11} + \frac{1}{14} - \cdots = \frac{\pi\sqrt{3}}{9} - \frac{1}{3}\ln 2$$

$$\frac{1}{1^2} + \frac{1}{2^2} + \frac{1}{3^2} + \frac{1}{4^2} + \cdots = \frac{\pi^2}{6}$$

$$\frac{1}{1^4} + \frac{1}{2^4} + \frac{1}{3^4} + \frac{1}{4^4} + \cdots = \frac{\pi^4}{90}$$

$$\frac{1}{1^6} + \frac{1}{2^6} + \frac{1}{3^6} + \frac{1}{4^6} + \cdots = \frac{\pi^6}{945}$$

$$\frac{1}{1^2} - \frac{1}{2^2} + \frac{1}{3^2} - \frac{1}{4^2} + \cdots = \frac{\pi^2}{12}$$

$$\frac{1}{1^4} - \frac{1}{2^4} + \frac{1}{3^4} - \frac{1}{4^4} + \cdots = \frac{7\pi^4}{720}$$

$$\frac{1}{1^6} - \frac{1}{2^6} + \frac{1}{3^6} - \frac{1}{4^6} + \cdots = \frac{31\pi^6}{30{,}240}$$

$$\frac{1}{1^2} + \frac{1}{3^2} + \frac{1}{5^2} + \frac{1}{7^2} + \cdots = \frac{\pi^2}{8}$$

$$\frac{1}{1^4} + \frac{1}{3^4} + \frac{1}{5^4} + \frac{1}{7^4} + \cdots = \frac{\pi^4}{96}$$

$$\frac{1}{1^6} + \frac{1}{3^6} + \frac{1}{5^6} + \frac{1}{7^6} + \cdots = \frac{\pi^6}{960}$$

$$\frac{1}{1^3} - \frac{1}{3^3} + \frac{1}{5^3} - \frac{1}{7^3} + \cdots = \frac{\pi^3}{32}$$

$$\frac{1}{1^3} + \frac{1}{3^3} - \frac{1}{5^3} - \frac{1}{7^3} + \cdots = \frac{3\pi^3\sqrt{2}}{128}$$

$$\frac{1}{1\cdot 3} + \frac{1}{3\cdot 5} + \frac{1}{5\cdot 7} + \frac{1}{7\cdot 9} + \cdots = \frac{1}{2}$$

$$\frac{1}{1\cdot 3} + \frac{1}{2\cdot 4} + \frac{1}{3\cdot 5} + \frac{1}{4\cdot 6} + \cdots = \frac{3}{4}$$

$$\frac{1}{1^2\cdot 3^2} + \frac{1}{3^2\cdot 5^2} + \frac{1}{5^2\cdot 7^2} + \frac{1}{7^2\cdot 9^2} + \cdots = \frac{\pi^2 - 8}{16}$$

$$\frac{1}{1^2\cdot 2^2\cdot 3^2} + \frac{1}{2^2\cdot 3^2\cdot 4^2} + \frac{1}{3^2\cdot 4^2\cdot 5^2} + \cdots = \frac{4\pi^2 - 39}{16}$$

$$\frac{1}{a} - \frac{1}{a+d} + \frac{1}{a+2d} - \frac{1}{a+3d} + \cdots = \int_0^1 \frac{u^{a-1}\,du}{1+u^d}$$

$$\frac{1}{1^{2p}} + \frac{1}{2^{2p}} + \frac{1}{3^{2p}} + \frac{1}{4^{2p}} + \cdots = \frac{2^{2p-1}\pi^{2p}B_p}{(2p)!}$$

$$\frac{1}{1^{2p}} + \frac{1}{3^{2p}} + \frac{1}{5^{2p}} + \frac{1}{7^{2p}} + \cdots = \frac{(2^{2p}-1)\pi^{2p}B_p}{2(2p)!}$$

$$\frac{1}{1^{2p}} - \frac{1}{2^{2p}} + \frac{1}{3^{2p}} - \frac{1}{4^{2p}} + \cdots = \frac{(2^{2p-1}-1)\pi^{2p}B_p}{(2p)!}$$

$$\frac{1}{1^{2p+1}} - \frac{1}{3^{2p+1}} + \frac{1}{5^{2p+1}} - \frac{1}{7^{2p+1}} + \cdots = \frac{\pi^{2p+1}E_p}{2^{2p+2}(2p)!}$$

MISCELLANEOUS SERIES

$$\frac{1}{2} + \cos\alpha + \cos 2\alpha + \cdots + \cos n\alpha = \frac{\sin\,(n+\frac{1}{2})\alpha}{2\sin\,(\alpha/2)}$$

$$\sin\alpha + \sin 2\alpha + \sin 3\alpha + \cdots + \sin n\alpha = \frac{\sin\,[\frac{1}{2}(n+1)]\alpha\,\sin\frac{1}{2}n\alpha}{\sin\,(\alpha/2)}$$

$$1 + r\cos\alpha + r^2\cos 2\alpha + r^3\cos 3\alpha + \cdots = \frac{1 - r\cos\alpha}{1 - 2r\cos\alpha + r^2}, \qquad |r| < 1$$

$$r\sin\alpha + r^2\sin 2\alpha + r^3\sin 3\alpha + \cdots = \frac{r\sin\alpha}{1 - 2r\cos\alpha + r^2}, \qquad |r| < 1$$

$$1 + r\cos\alpha + r^2\cos 2\alpha + \cdots + r^n \cos n\alpha$$

$$= \frac{r^{n+2}\cos n\alpha - r^{n+1}\cos(n+1)\alpha - r\cos\alpha + 1}{1 - 2r\cos\alpha + r^2}$$

$$r\sin\alpha + r^2\sin 2\alpha + \cdots + r^n\sin n\alpha = \frac{r\sin\alpha - r^{n+1}\sin(n+1)\alpha + r^{n+2}\sin n\alpha}{1 - 2r\cos\alpha + r^2}$$

THE EULER-MACLAURIN SUMMATION FORMULA

$$\sum_{k=1}^{n-1} F(k) = \int_0^n F(k)\,dk - \frac{1}{2}\{F(0) + F(n)\}$$

$$+ \frac{1}{12}\{F'(n) - F'(0)\} - \frac{1}{720}\{F'''(n) - F'''(0)\}$$

$$+ \frac{1}{30{,}240}\{F^{(v)}(n) - F^{(v)}(0)\} - \frac{1}{1{,}209{,}600}\{F^{(vii)}(n) - F^{(vii)}(0)\}$$

$$+ \cdots (-1)^{p-1}\frac{B_p}{(2p)!}\{F^{(2p-1)}(n) - F^{(2p-1)}(0)\} + \cdots$$

THE POISSON SUMMATION FORMULA

$$\sum_{k=-\infty}^{\infty} F(k) = \sum_{m=-\infty}^{\infty}\left\{\int_{-\infty}^{\infty} e^{2\pi imx}F(x)\,dx\right\}$$

TAYLOR SERIES

TAYLOR SERIES FOR FUNCTIONS OF ONE VARIABLE

$$f(x) = f(a) + f'(a)(x-a) + \frac{f''(a)(x-a)^2}{2!} +$$

$$\cdots + \frac{f^{(n-1)}(a)(x-a)^{n-1}}{(n-1)!} + R_n$$

where R_n, the remainder after n terms, is given by either of the following forms:

Lagrange's form $$R_n = \frac{f^{(n)}(\xi)(x-a)^n}{n!}$$

Cauchy's form $$R_n = \frac{f^{(n)}(\xi)(x-\xi)^{n-1}(x-a)}{(n-1)!}$$

The value ξ, which may be different in the two forms, lies between a and x. The result holds if $f(x)$ has continuous derivatives of order n at least.

If $\lim_{n\to\infty} R_n = 0$, the infinite series obtained is called the *Taylor series* for

$f(x)$ about $x = a$. If $a = 0$ the series is often called a *Maclaurin series*. These series, often called power series, generally converge for all values of x in some interval called the *interval of convergence* and diverge for all x outside this interval.

BINOMIAL SERIES

$$\begin{aligned}(a+x)^n &= a^n + na^{n-1}x + \frac{n(n-1)}{2!}a^{n-2}x^2 + \frac{n(n-1)(n-2)}{3!}a^{n-3}x^3 + \cdots \\ &= a^n + \binom{n}{1}a^{n-1}x + \binom{n}{2}a^{n-2}x^2 + \binom{n}{3}a^{n-3}x^3 + \cdots\end{aligned}$$

Special cases are

$$(a+x)^2 = a^2 + 2ax + x^2$$

$$(a+x)^3 = a^3 + 3a^2x + 3ax^2 + x^3$$

$$(a+x)^4 = a^4 + 4a^3x + 6a^2x^2 + 4ax^3 + x^4$$

$$(1+x)^{-1} = 1 - x + x^2 - x^3 + x^4 - \cdots \qquad -1 < x < 1$$

$$(1+x)^{-2} = 1 - 2x + 3x^2 - 4x^3 + 5x^4 - \cdots \qquad -1 < x < 1$$

$$(1+x)^{-3} = 1 - 3x + 6x^2 - 10x^3 + 15x^4 - \cdots \qquad -1 < x < 1$$

$$(1+x)^{-1/2} = 1 - \frac{1}{2}x + \frac{1\cdot 3}{2\cdot 4}x^2 - \frac{1\cdot 3\cdot 5}{2\cdot 4\cdot 6}x^3 + \cdots \qquad -1 < x \leqq 1$$

$$(1+x)^{1/2} = 1 + \frac{1}{2}x - \frac{1}{2\cdot 4}x^2 + \frac{1\cdot 3}{2\cdot 4\cdot 6}x^3 - \cdots \qquad -1 < x \leqq 1$$

$$(1+x)^{-1/3} = 1 - \frac{1}{3}x + \frac{1\cdot 4}{3\cdot 6}x^2 - \frac{1\cdot 4\cdot 7}{3\cdot 6\cdot 9}x^3 + \cdots \qquad -1 < x \leqq 1$$

$$(1+x)^{1/3} = 1 + \frac{1}{3}x - \frac{2}{3\cdot 6}x^2 + \frac{2\cdot 5}{3\cdot 6\cdot 9}x^3 - \cdots \qquad -1 < x \leqq 1$$

SERIES FOR EXPONENTIAL AND LOGARITHMIC FUNCTIONS

$$e^x = 1 + x + \frac{x^2}{2!} + \frac{x^3}{3!} + \cdots \qquad -\infty < x < \infty$$

$$a^x = e^{x\ln a} = 1 + x\ln a + \frac{(x\ln a)^2}{2!} + \frac{(x\ln a)^3}{3!} + \cdots \qquad -\infty < x < \infty$$

$$\ln(1+x) = x - \frac{x^2}{2} + \frac{x^3}{3} - \frac{x^4}{4} + \cdots \qquad -1 < x \leqq 1$$

$$\frac{1}{2}\ln\left(\frac{1+x}{1-x}\right) = x + \frac{x^3}{3} + \frac{x^5}{5} + \frac{x^7}{7} + \cdots \qquad -1 < x < 1$$

$$\ln x = 2\left\{\left(\frac{x-1}{x+1}\right) + \frac{1}{3}\left(\frac{x-1}{x+1}\right)^3 + \frac{1}{5}\left(\frac{x-1}{x+1}\right)^5 + \cdots\right\} \qquad x > 0$$

$$\ln x = \left(\frac{x-1}{x}\right) + \frac{1}{2}\left(\frac{x-1}{x}\right)^2 + \frac{1}{3}\left(\frac{x-1}{x}\right)^3 + \cdots \qquad x \geqq \tfrac{1}{2}$$

SERIES FOR TRIGONOMETRIC FUNCTIONS

$$\sin x = x - \frac{x^3}{3!} + \frac{x^5}{5!} - \frac{x^7}{7!} + \cdots \qquad -\infty < x < \infty$$

$$\cos x = 1 - \frac{x^2}{2!} + \frac{x^4}{4!} - \frac{x^6}{6!} + \cdots \qquad -\infty < x < \infty$$

$$\tan x = x + \frac{x^3}{3} + \frac{2x^5}{15} + \frac{17x^7}{315} + \cdots + \frac{2^{2n}(2^{2n}-1)B_n x^{2n-1}}{(2n)!} + \cdots \qquad |x| < \frac{\pi}{2}$$

$$\cot x = \frac{1}{x} - \frac{x}{3} - \frac{x^3}{45} - \frac{2x^5}{945} - \cdots - \frac{2^{2n}B_n x^{2n-1}}{(2n)!} - \cdots \qquad 0 < |x| < \pi$$

$$\sec x = 1 + \frac{x^2}{2} + \frac{5x^4}{24} + \frac{61x^6}{720} + \cdots + \frac{E_n x^{2n}}{(2n)!} + \cdots \qquad |x| < \frac{\pi}{2}$$

$$\csc x = \frac{1}{x} + \frac{x}{6} + \frac{7x^3}{360} + \frac{31x^5}{15{,}120} + \cdots + \frac{2(2^{2n-1}-1)B_n x^{2n-1}}{(2n)!} + \cdots \qquad 0 < |x| < \pi$$

$$\sin^{-1} x = x + \frac{1}{2}\frac{x^3}{3} + \frac{1 \cdot 3}{2 \cdot 4}\frac{x^5}{5} + \frac{1 \cdot 3 \cdot 5}{2 \cdot 4 \cdot 6}\frac{x^7}{7} + \cdots \qquad |x| < 1$$

$$\cos^{-1} x = \frac{\pi}{2} - \sin^{-1} x = \frac{\pi}{2} - \left(x + \frac{1}{2}\frac{x^3}{3} + \frac{1 \cdot 3}{2 \cdot 4}\frac{x^5}{5} + \cdots\right) \qquad |x| < 1$$

$$\tan^{-1} x = \begin{cases} x - \dfrac{x^3}{3} + \dfrac{x^5}{5} - \dfrac{x^7}{7} + \cdots & |x| < 1 \\ \pm\dfrac{\pi}{2} - \dfrac{1}{x} + \dfrac{1}{3x^3} - \dfrac{1}{5x^5} + \cdots & [+ \text{ if } x \geqq 1,\ - \text{ if } x \leqq -1] \end{cases}$$

$$\cot^{-1} x = \frac{\pi}{2} - \tan^{-1} x$$

$$= \begin{cases} \dfrac{\pi}{2} - \left(x - \dfrac{x^3}{3} + \dfrac{x^5}{5} - \cdots\right) & |x| < 1 \\ p\pi + \dfrac{1}{x} - \dfrac{1}{3x^3} + \dfrac{1}{5x^5} - \cdots & [p = 0 \text{ if } x > 1,\ p = 1 \text{ if } x < -1] \end{cases}$$

$$\sec^{-1} x = \cos^{-1}(1/x) = \frac{\pi}{2} - \left(\frac{1}{x} + \frac{1}{2 \cdot 3x^3} + \frac{1 \cdot 3}{2 \cdot 4 \cdot 5x^5} + \cdots\right) \qquad |x| > 1$$

$$\csc^{-1} x = \sin^{-1}(1/x) = \frac{1}{x} + \frac{1}{2 \cdot 3x^3} + \frac{1 \cdot 3}{2 \cdot 4 \cdot 5x^5} + \cdots \qquad |x| > 1$$

SERIES FOR HYPERBOLIC FUNCTIONS

$$\sinh x = x + \frac{x^3}{3!} + \frac{x^5}{5!} + \frac{x^7}{7!} + \cdots \qquad -\infty < x < \infty$$

$$\cosh x = 1 + \frac{x^2}{2!} + \frac{x^4}{4!} + \frac{x^6}{6!} + \cdots \qquad -\infty < x < \infty$$

$$\tanh x = x - \frac{x^3}{3} + \frac{2x^5}{15} - \frac{17x^7}{315} + \cdots \frac{(-1)^{n-1}2^{2n}(2^{2n}-1)B_n x^{2n-1}}{(2n)!} + \cdots \qquad |x| < \frac{\pi}{2}$$

$$\coth x = \frac{1}{x} + \frac{x}{3} - \frac{x^3}{45} + \frac{2x^5}{945} + \cdots \frac{(-1)^{n-1}2^{2n}B_n x^{2n-1}}{(2n)!} + \cdots \qquad 0 < |x| < \pi$$

$$\operatorname{sech} x = 1 - \frac{x^2}{2} + \frac{5x^4}{24} - \frac{61x^6}{720} + \cdots \frac{(-1)^n E_n x^{2n}}{(2n)!} + \cdots \qquad |x| < \frac{\pi}{2}$$

$$\operatorname{csch} x = \frac{1}{x} - \frac{x}{6} + \frac{7x^3}{360} - \frac{31x^5}{15{,}120} + \cdots \frac{(-1)^n 2(2^{2n-1}-1)B_n x^{2n-1}}{(2n)!} + \cdots \qquad 0 < |x| < \pi$$

$$\sinh^{-1} x = \begin{cases} x - \dfrac{x^3}{2 \cdot 3} + \dfrac{1 \cdot 3x^5}{2 \cdot 4 \cdot 5} - \dfrac{1 \cdot 3 \cdot 5x^7}{2 \cdot 4 \cdot 6 \cdot 7} + \cdots & |x| < 1 \\ \pm\left(\ln|2x| + \dfrac{1}{2 \cdot 2x^2} - \dfrac{1 \cdot 3}{2 \cdot 4 \cdot 4x^4} + \dfrac{1 \cdot 3 \cdot 5}{2 \cdot 4 \cdot 6 \cdot 6x^6} - \cdots\right) & \begin{bmatrix} + \text{ if } x \geqq 1 \\ - \text{ if } x \leqq -1 \end{bmatrix} \end{cases}$$

$$\cosh^{-1} x = \pm\left\{\ln(2x) - \left(\frac{1}{2 \cdot 2x^2} + \frac{1 \cdot 3}{2 \cdot 4 \cdot 4x^4} + \frac{1 \cdot 3 \cdot 5}{2 \cdot 4 \cdot 6 \cdot 6x^6} + \cdots\right)\right\}$$

$$\begin{bmatrix} + \text{ if } \cosh^{-1} x > 0,\ x \geqq 1 \\ - \text{ if } \cosh^{-1} x < 0,\ x \geqq 1 \end{bmatrix}$$

$$\tanh^{-1} x = x + \frac{x^3}{3} + \frac{x^5}{5} + \frac{x^7}{7} + \cdots \qquad |x| < 1$$

$$\coth^{-1} x = \frac{1}{x} + \frac{1}{3x^3} + \frac{1}{5x^5} + \frac{1}{7x^7} + \cdots \qquad |x| > 1$$

MISCELLANEOUS SERIES

$$e^{\sin x} = 1 + x + \frac{x^2}{2} - \frac{x^4}{8} - \frac{x^5}{15} + \cdots \qquad -\infty < x < \infty$$

$$e^{\cos x} = e\left(1 - \frac{x^2}{2} + \frac{x^4}{6} - \frac{31x^6}{720} + \cdots\right) \qquad -\infty < x < \infty$$

$$e^{\tan x} = 1 + x + \frac{x^2}{2} + \frac{x^3}{2} + \frac{3x^4}{8} + \cdots \qquad |x| < \frac{\pi}{2}$$

$$e^x \sin x = x + x^2 + \frac{2x^3}{3} - \frac{x^5}{30} - \frac{x^6}{90} + \cdots + \frac{2^{n/2}\sin(n\pi/4)\, x^n}{n!} + \cdots \qquad -\infty < x < \infty$$

$$e^x \cos x = 1 + x - \frac{x^3}{3} - \frac{x^4}{6} + \cdots + \frac{2^{n/2}\cos(n\pi/4)\, x^n}{n!} + \cdots \qquad -\infty < x < \infty$$

$$\ln|\sin x| = \ln|x| - \frac{x^2}{6} - \frac{x^4}{180} - \frac{x^6}{2835} - \cdots - \frac{2^{2n-1}B_n x^{2n}}{n(2n)!} + \cdots \qquad 0 < |x| < \pi$$

$$\ln|\cos x| = -\frac{x^2}{2} - \frac{x^4}{12} - \frac{x^6}{45} - \frac{17x^8}{2520} - \cdots - \frac{2^{2n-1}(2^{2n}-1)B_n x^{2n}}{n(2n)!} + \cdots \qquad |x| < \frac{\pi}{2}$$

$$\ln|\tan x| = \ln|x| + \frac{x^2}{3} + \frac{7x^4}{90} + \frac{62x^6}{2835} + \cdots + \frac{2^{2n}(2^{2n-1}-1)B_n x^{2n}}{n(2n)!} + \cdots \quad 0 < |x| < \frac{\pi}{2}$$

$$\frac{\ln(1+x)}{1+x} = x - (1+\tfrac{1}{2})x^2 + (1+\tfrac{1}{2}+\tfrac{1}{3})x^3 - \cdots \quad |x| < 1$$

REVERSION OF POWER SERIES

If

$$y = c_1x + c_2x^2 + c_3x^3 + c_4x^4 + c_5x^5 + c_6x^6 + \cdots$$

then

$$x = C_1y + C_2y^2 + C_3y^3 + C_4y^4 + C_5y^5 + C_6y^6 + \cdots$$

where

$$c_1C_1 = 1$$

$$c_1^3C_2 = -c_2$$

$$c_1^5C_3 = 2c_2^2 - c_1c_3$$

$$c_1^7C_4 = 5c_1c_2c_3 - 5c_2^3 - c_1^2c_4$$

$$c_1^9C_5 = 6c_1^2c_2c_4 + 3c_1^2c_3^2 - c_1^3c_5 + 14c_2^4 - 21c_1c_2^2c_3$$

$$c_1^{11}C_6 = 7c_1^3c_2c_5 + 84c_1c_2^3c_3 + 7c_1^3c_3c_4 - 28c_1^2c_2c_3^2 - c_1^4c_6 - 28c_1^2c_2^2c_4 - 42c_2^5$$

TAYLOR SERIES FOR FUNCTIONS OF TWO VARIABLES

$$f(x,y) = f(a,b) + (x-a)f_x(a,b) + (y-b)f_y(a,b) + \frac{1}{2!}\{(x-a)^2f_{xx}(a,b) + 2(x-a)(y-b)f_{xy}(a,b) + (y-b)^2f_{yy}(a,b)\} + \cdots$$

where $f_x(a,b)$, $f_y(a,b)$, ... denote partial derivatives with respect to x, y, ... evaluated at $x = a$, $y = b$.

CHAPTER 12

DERIVATIVES

DEFINITION OF A DERIVATIVE

If $y = f(x)$, the derivative of y or $f(x)$ with respect to x is defined as

$$\frac{dy}{dx} = \lim_{h \to 0} \frac{f(x+h) - f(x)}{h} = \lim_{\Delta x \to 0} \frac{f(x+\Delta x) - f(x)}{\Delta x}$$

where $h = \Delta x$. The derivative is also denoted by y', df/dx or $f'(x)$. The process of taking a derivative is called *differentiation.*

GENERAL RULES OF DIFFERENTIATION

In the following, u, v, w are functions of x; a, b, c, n are constants [restricted if indicated]; $e = 2.71828\ldots$ is the natural base of logarithms; $\ln u$ is the natural logarithm of u [i.e. the logarithm to the base e] where it is assumed that $u > 0$ and all angles are in radians.

$$\frac{d}{dx}(c) = 0$$

$$\frac{d}{dx}(cx) = c$$

$$\frac{d}{dx}(cx^n) = ncx^{n-1}$$

$$\frac{d}{dx}(u \pm v \pm w \pm \cdots) = \frac{du}{dx} \pm \frac{dv}{dx} \pm \frac{dw}{dx} \pm \cdots$$

$$\frac{d}{dx}(cu) = c\frac{du}{dx}$$

$$\frac{d}{dx}(uv) = u\frac{dv}{dx} + v\frac{du}{dx}$$

$$\frac{d}{dx}(uvw) = uv\frac{dw}{dx} + uw\frac{dv}{dx} + vw\frac{du}{dx}$$

$$\frac{d}{dx}\left(\frac{u}{v}\right) = \frac{v(du/dx) - u(dv/dx)}{v^2}$$

$$\frac{d}{dx}(u^n) = nu^{n-1}\frac{du}{dx}$$

$$\frac{dy}{dx} = \frac{dy}{du}\frac{du}{dx} \qquad \text{(Chain rule)}$$

$$\frac{du}{dx} = \frac{1}{dx/du}$$

$$\frac{dy}{dx} = \frac{dy/du}{dx/du}$$

HIGHER DERIVATIVES

The second, third and higher derivatives are defined as follows.

$$\text{Second derivative} = \frac{d}{dx}\left(\frac{dy}{dx}\right) = \frac{d^2y}{dx^2} = f''(x) = y''$$

$$\text{Third derivative} = \frac{d}{dx}\left(\frac{d^2y}{dx^2}\right) = \frac{d^3y}{dx^3} = f'''(x) = y'''$$

$$n\text{th derivative} = \frac{d}{dx}\left(\frac{d^{n-1}y}{dx^{n-1}}\right) = \frac{d^ny}{dx^n} = f^{(n)}(x) = y^{(n)}$$

LEIBNITZ'S RULE FOR HIGHER DERIVATIVES OF PRODUCTS

Let D^p stand for the operator $\frac{d^p}{dx^p}$ so that $D^pu = \frac{d^pu}{dx^p}$ = the pth derivative of u. Then

$$D^n(uv) = uD^nv + \binom{n}{1}(Du)(D^{n-1}v) + \binom{n}{2}(D^2u)(D^{n-2}v) + \cdots + vD^nu$$

where $\binom{n}{1}, \binom{n}{2}, \ldots$ are the binomial coefficients.

As special cases we have

$$\frac{d^2}{dx^2}(uv) = u\frac{d^2v}{dx^2} + 2\frac{du}{dx}\frac{dv}{dx} + v\frac{d^2u}{dx^2}$$

$$\frac{d^3}{dx^3}(uv) = u\frac{d^3v}{dx^3} + 3\frac{du}{dx}\frac{d^2v}{dx^2} + 3\frac{d^2u}{dx^2}\frac{dv}{dx} + v\frac{d^3u}{dx^3}$$

DERIVATIVES OF EXPONENTIAL AND LOGARITHMIC FUNCTIONS

$$\frac{d}{dx}\sin u = \cos u\frac{du}{dx}$$

$$\frac{d}{dx}\cos u = -\sin u\frac{du}{dx}$$

$$\frac{d}{dx}\tan u = \sec^2 u\frac{du}{dx}$$

$$\frac{d}{dx}\cot u = -\csc^2 u\frac{du}{dx}$$

$$\frac{d}{dx}\sec u = \sec u \tan u\frac{du}{dx}$$

$$\frac{d}{dx}\csc u = -\csc u \cot u\frac{du}{dx}$$

$$\frac{d}{dx}\sin^{-1}u = \frac{1}{\sqrt{1-u^2}}\frac{du}{dx} \qquad \left[-\frac{\pi}{2} < \sin^{-1}u < \frac{\pi}{2}\right]$$

$$\frac{d}{dx}\cos^{-1}u = \frac{-1}{\sqrt{1-u^2}}\frac{du}{dx} \qquad [0 < \cos^{-1}u < \pi]$$

$$\frac{d}{dx}\tan^{-1}u = \frac{1}{1+u^2}\frac{du}{dx} \qquad \left[-\frac{\pi}{2} < \tan^{-1}u < \frac{\pi}{2}\right]$$

$$\frac{d}{dx}\cot^{-1}u = \frac{-1}{1+u^2}\frac{du}{dx} \qquad [0 < \cot^{-1}u < \pi]$$

$$\frac{d}{dx}\sec^{-1}u = \frac{1}{|u|\sqrt{u^2-1}}\frac{du}{dx} = \frac{\pm 1}{u\sqrt{u^2-1}}\frac{du}{dx} \qquad \left[\begin{array}{l} + \text{ if } 0 < \sec^{-1}u < \pi/2 \\ - \text{ if } \pi/2 < \sec^{-1}u < \pi \end{array}\right]$$

$$\frac{d}{dx}\csc^{-1}u = \frac{-1}{|u|\sqrt{u^2-1}}\frac{du}{dx} = \frac{\mp 1}{u\sqrt{u^2-1}}\frac{du}{dx} \qquad \left[\begin{array}{l} - \text{ if } 0 < \csc^{-1}u < \pi/2 \\ + \text{ if } -\pi/2 < \csc^{-1}u < 0 \end{array}\right]$$

DERIVATIVES OF TRIGONOMETRIC AND INVERSE TRIGONOMETRIC FUNCTIONS

$$\frac{d}{dx}\log_a u = \frac{\log_a e}{u}\frac{du}{dx} \qquad a \neq 0, 1$$

$$\frac{d}{dx}\ln u = \frac{d}{dx}\log_e u = \frac{1}{u}\frac{du}{dx}$$

$$\frac{d}{dx}a^u = a^u \ln a \frac{du}{dx}$$

$$\frac{d}{dx}e^u = e^u \frac{du}{dx}$$

$$\frac{d}{dx}u^v = \frac{d}{dx}e^{v \ln u} = e^{v \ln u}\frac{d}{dx}[v \ln u] = vu^{v-1}\frac{du}{dx} + u^v \ln u \frac{dv}{dx}$$

DERIVATIVES OF HYPERBOLIC AND INVERSE HYPERBOLIC FUNCTIONS

$$\frac{d}{dx}\sinh u = \cosh u \frac{du}{dx}$$

$$\frac{d}{dx}\cosh u = \sinh u \frac{du}{dx}$$

$$\frac{d}{dx}\tanh u = \operatorname{sech}^2 u \frac{du}{dx}$$

$$\frac{d}{dx}\coth u = -\operatorname{csch}^2 u \frac{du}{dx}$$

$$\frac{d}{dx}\operatorname{sech} u = -\operatorname{sech} u \tanh u \frac{du}{dx}$$

$$\frac{d}{dx}\operatorname{csch} u = -\operatorname{csch} u \coth u \frac{du}{dx}$$

$$\frac{d}{dx}\sinh^{-1}u = \frac{1}{\sqrt{u^2+1}}\frac{du}{dx}$$

$$\frac{d}{dx}\cosh^{-1}u = \frac{\pm 1}{\sqrt{u^2-1}}\frac{du}{dx} \qquad \begin{bmatrix} + \text{ if } \cosh^{-1}u > 0,\ u > 1 \\ - \text{ if } \cosh^{-1}u < 0,\ u > 1 \end{bmatrix}$$

$$\frac{d}{dx}\tanh^{-1}u = \frac{1}{1-u^2}\frac{du}{dx} \qquad [-1 < u < 1]$$

$$\frac{d}{dx}\coth^{-1}u = \frac{1}{1-u^2}\frac{du}{dx} \qquad [u > 1 \text{ or } u < -1]$$

$$\frac{d}{dx}\operatorname{sech}^{-1}u = \frac{\mp 1}{u\sqrt{1-u^2}}\frac{du}{dx} \qquad \begin{bmatrix} - \text{ if } \operatorname{sech}^{-1}u > 0,\ 0 < u < 1 \\ + \text{ if } \operatorname{sech}^{-1}u < 0,\ 0 < u < 1 \end{bmatrix}$$

$$\frac{d}{dx}\operatorname{csch}^{-1}u = \frac{-1}{|u|\sqrt{1+u^2}}\frac{du}{dx} = \frac{\mp 1}{u\sqrt{1+u^2}}\frac{du}{dx} \qquad [- \text{ if } u > 0,\ + \text{ if } u < 0]$$

CHAPTER 13

INTEGRALS

GENERAL FORMULAS

$$\int f(x)\,dx = xf(o) + \frac{x^2}{1.2} f'(o) + \frac{x^3}{1.2.3} f''(o) + \cdots$$

$$\int U dx = xU - \frac{x^2}{1.2} U' + \frac{x^3}{1.2.3} U'' - \cdots$$

$$\int U dx = xU - \int xU'\,dx.$$

$$\int U'V\,dx = UV - \int UV'\,dx.$$

$$\int UV\,dx = U_1V - \int U_1V'\,dx.$$

$$\int UV\,dx = UV_1 - U'V_2 + U''V_3 - \cdots$$

$$\int x^m V\,dx = x^m V_1 - mx^{m-1} V_2 + m(m-1)x^{m-2} V_3 - \cdots$$

$$\int U^m U'\,dx = \frac{U^{m+1}}{m+1}.$$

$$\int (aU + b)^m U'\,dx = \frac{(aU+b)^{m+1}}{a(m+1)}.$$

$$\int (U+V)^m\,dx = \int U(U+V)^{m-1}\,dx + \int V(U+V)^{m-1}\,dx.$$

$$\int (U'V + UV')\,dx = UV.$$

$$\int \frac{dx}{U(x^2 - a^2)} = \frac{1}{2a} \int \frac{dx}{U(x-a)} - \frac{1}{2a} \int \frac{dx}{U(x+a)} .$$

$$\int \frac{dx}{(U - x)^2} = \frac{1}{U - x} + \int \frac{U'}{(U - x)^2} dx.$$

$$\int \frac{dx}{U(U \pm V)} = \mp \int \frac{dx}{V(U \pm V)} \pm \int \frac{dx}{UV} .$$

$$\int \frac{U}{(U+a)(U+b)} dx = \frac{a}{a-b} \int \frac{dx}{U+a} - \frac{b}{a-b} \int \frac{dx}{U+b} .$$

$$\int \frac{U}{(U + V)^m} dx = \int \frac{dx}{(U + V)^{m-1}} - \int \frac{V}{(U+V)^m} dx.$$

$$\int \frac{U^{2m}}{U^{2m} - a^2} dx = \frac{1}{2} \int \frac{U^m}{U^m - a} dx + \frac{1}{2} \int \frac{U^m}{U^m + a} dx.$$

$$\int \frac{U^{2m}}{1 - U^{2m}} dx = - x + \int \frac{dx}{1 - U^{2m}} .$$

$$\int \frac{U'}{U} dx = \log U = - \frac{1}{m} \log \frac{1}{U^m} .$$

$$\int \frac{U'}{U^m} dx = - \frac{1}{(m - 1)\, U^{m-1}} .$$

$$\int \frac{U'}{aU + b} dx = \frac{1}{a} \log (aU + b).$$

$$\int \frac{U'}{U^2 + 1} dx = \text{arc tg } U = - \text{arc tg } \frac{1}{U} .$$

$$\int \frac{U'}{a^2 + U^2} dx = \frac{1}{a} \text{ arc tg } \frac{U}{a} .$$

$$\int \frac{U'}{b^2 + a^2U^2} dx = \frac{1}{ab} \text{ arc tg } \frac{aU}{b} .$$

$$\int \frac{U'}{a^2U^2 - b^2} dx = \frac{1}{2ab} \log \frac{aU - b}{aU + b} .$$

$$\int \frac{U'}{aU^2 + bU} dx = \frac{1}{b} \log \frac{U}{aU + b} .$$

$$\int \frac{U'}{aU^2 - bU} dx = \frac{1}{b} \log \frac{aU - b}{U} .$$

$$\int \frac{U'}{(aU + b)^2} dx = \frac{U}{b(aU + b)} .$$

$$\int \frac{U'}{(a-U)^2}\,dx = \frac{1}{a-U} = \frac{U}{a\,(a-U)} = \frac{1}{a-b}\,\frac{U-b}{a-U}.$$

$$\int \frac{U'}{(aU+b)\,(mU+n)}\,dx = \frac{1}{an-bm}\log\frac{aU+b}{mU+n}.$$

$$\int \frac{U'}{U\,(1+U^2)}\,dx = \log\frac{U}{\sqrt{1+U^2}}$$

$$\int \frac{U'}{(a^2+U^2)(1+U^2)}\,dx = \frac{1}{a\,(1-a^2)}\left\{\operatorname{arc\,tg}\frac{U}{a} - a\operatorname{arc\,tg} U\right\}.$$

$$\int \frac{U'}{\sqrt{U}}\,dx = 2\sqrt{U}.$$

$$\int \frac{U'}{\sqrt{aU+b}}\,dx = \frac{2}{a}\sqrt{aU+b}.$$

$$\int \frac{U'}{\sqrt{U^2+a}}\,dx = \log\,(U+\sqrt{U^2+a}).$$

$$\int \frac{U'}{\sqrt{1-U^2}}\,dx = \operatorname{arc\,sin} U.$$

$$\int \frac{U'}{\sqrt{a^2-U^2}}\,dx = \operatorname{arc\,sin}\frac{U}{a}.$$

$$\int \frac{U'}{\sqrt{U^2+aU}}\,dx = 2\log\,(\sqrt{U}+\sqrt{U+a}).$$

$$\int \frac{U'}{\sqrt{U-aU^2}}\,dx = -\frac{1}{\sqrt{a}}\operatorname{arc\,sin}\,(1-2aU).$$

$$\int \frac{U'}{\sqrt{(U+a)(U+b)}}\,dx = 2\log\,(\sqrt{U+a}+\sqrt{U+b}).$$

$$\int \frac{U'}{U\sqrt{U^2+a^2}}\,dx = \frac{1}{a}\log\frac{U}{a+\sqrt{U^2+a^2}}.$$

$$\int \frac{U'}{U\sqrt{a^2-U^2}}\,dx = \frac{1}{a}\log\frac{U}{a+\sqrt{a^2-U^2}}.$$

$$\int \frac{U'}{U\sqrt{2aU-a^2}}\,dx = \frac{1}{a}\operatorname{arc\,sin}\frac{U-a}{U}.$$

$$\int \frac{U'}{U\sqrt{U^2-1}}\,dx = \operatorname{arc\,séc} U.$$

$$\int \frac{U'}{(U - a^2)\sqrt{U}}\,dx = \frac{1}{a}\log\frac{\sqrt{U} - a}{\sqrt{U} + a}.$$

$$\int \frac{U'}{U\sqrt{U^4 - 1}}\,dx = -\frac{1}{2}\text{arc sin}\,\frac{1}{U^2}.$$

$$\int \frac{U'}{(b - U)^{\frac{3}{2}}\sqrt{U - a}}\,dx = \frac{2}{b - a}\sqrt{\frac{U - a}{b - U}}.$$

$$\int \frac{U'}{aU^{\frac{p}{q}} + U}\,dx = \frac{q}{q - p}\log\left(a + U^{\frac{q-p}{q}}\right).$$

$$\int \frac{UU'}{U^4 + 1}\,dx = \frac{1}{2}\text{arc tg}\,U^2.$$

$$\int \frac{UU'}{U^4 - 1}\,dx = \frac{1}{4}\log\frac{U^2 - 1}{U^2 + 1}.$$

$$\int \frac{UU'}{(1 - U^2)^2}\,dx = \frac{U^2}{2(1 - U^2)} = \frac{1 + U^2}{4(1 - U^2)}.$$

$$\int \frac{UU'}{\sqrt{1 - U^4}}\,dx = \frac{1}{2}\text{arc sin}\,U^2.$$

$$\int \frac{UU''}{U'^2\sqrt{1 + U'^2}}\,dx = -\frac{U\sqrt{1 + U'^2}}{U'} + \int\sqrt{1 + U'^2}\,dx.$$

$$\int \frac{U^2U'}{1 + U^2}\,dx = U - \text{arc tg}\,U.$$

$$\int \frac{U^2U'}{(1 + U^2)(a^2 + b^2U^2)}\,dx = \frac{1}{b^2 - a^2}\left\{\text{arc tg}\,U - \frac{a}{b}\text{arc tg}\,\frac{b}{a}U\right\}.$$

$$\int \frac{U^{m-1}U'}{U^{2m} + 1}\,dx = \frac{1}{m}\text{arc tg}\,U^m.$$

$$\int \frac{U^{m-1}U'}{U^{2m} - a^2}\,dx = \frac{1}{2am}\log\frac{U^m - a}{U^m + a}.$$

$$\int \frac{U^{m-1}U'}{(a - U^m)^2}\,dx = \frac{1}{2am}\,\frac{a + U^m}{a - U^m}.$$

$$\int \frac{U'\sqrt{a^2 - U^2}}{U^2}\,dx = -\frac{\sqrt{a^2 - U^2}}{U} - \text{arc sin}\,\frac{U}{a}.$$

$$\int \frac{UU'U''}{\sqrt{1+U'^2}}\,dx = U\sqrt{1+U'^2} - \int U'\sqrt{1+U'^2}\,dx.$$

$$\int \frac{VU'}{U^2}\,dx = -\frac{V}{U} + \int \frac{V'}{U}\,dx.$$

$$\int \frac{xU' + U}{(xU+a)^2}\,dx = \frac{1}{a-b}\,\frac{Ux+b}{Ux+a}.$$

$$\int \frac{U - xU'}{(aU+bx)^2}\,dx = \frac{1}{an-bm}\,\frac{mU+nx}{aU+bx}.$$

$$\int \frac{U - xU'}{(aU+bx)(mU+nx)}\,dx = \frac{1}{an-bm}\log\frac{mU+nx}{aU+bx}.$$

$$\int \frac{U'V - UV'}{V^2}\,dx = \frac{U}{V}$$

$$\int \frac{U'V - UV'}{UV}\,dx = \log\frac{U}{V}.$$

$$\int \frac{U'V - UV'}{U^2+V^2}\,dx = \operatorname{arc\,tg}\frac{U}{V}.$$

$$\int \frac{U'V - UV'}{V^2-U^2}\,dx = \frac{1}{2}\log\frac{U+V}{U-V}.$$

$$\int \frac{U'V - UV'}{(U+V)^2}\,dx = \frac{U-V}{2(U+V)} = -\frac{V}{U+V}.$$

$$\int \frac{U'V - UV'}{(U-V)^2}\,dx = \frac{V+U}{2(V-U)} = \frac{V}{V-U}.$$

$$\int \frac{U'V - UV'}{V\sqrt{V^2-U^2}}\,dx = \operatorname{arc\,sin}\frac{U}{V}.$$

$$\int \frac{U'V + UV'}{\sqrt{U^2V^2+a}}\,dx = \log\left\{UV + \sqrt{U^2V^2+a}\right\}.$$

$$\int \frac{2U - xU'}{(U+a^2x^2)\sqrt{U}}\,dx = \frac{2}{a}\operatorname{arc\,tg}\frac{ax}{\sqrt{U}}.$$

$$\int \frac{2U - xU'}{(U-a^2x^2)\sqrt{U}}\,dx = \frac{1}{a}\log\frac{\sqrt{U}+ax}{\sqrt{U}-ax}.$$

$$\int \frac{U'V^2 + V'U^2}{UV(U+V)}\,dx = \log\frac{UV}{U+V}.$$

ELEMENTARY ALGEBRAIC FUNCTIONS

$$\frac{1}{ax + b}$$

$$\int \frac{dx}{x} = \log x = \log\left(\frac{x}{a} + \frac{x}{b}\right).$$

$$\int \frac{dx}{x + a} = \log (x + a).$$

$$\int \frac{dx}{ax + b} = \frac{1}{a} \log (ax + b).$$

$$\int \frac{x}{x + a} dx = x - a \log (x + a).$$

$$\int \frac{x}{ax + b} dx = \frac{x}{a} - \frac{b}{a^2} \log (ax + b).$$

$$\int \frac{x}{b - ax} dx = - \frac{1}{a} \log (b - ax) = - \frac{x}{a} - \frac{b}{a^2} \log (b - ax).$$

$$\int \frac{mx + n}{ax + b} dx = \frac{m}{a} x + \frac{an - bm}{a^2} \log (ax + b).$$

$$\int \frac{x^2}{ax + b} dx = \frac{x^2}{2a} - \frac{bx}{a^2} + \frac{b^2}{a^3} \log (ax + b).$$

$$\int \frac{mx^2 + nx}{ax + b} dx = \frac{mx^2}{2a} + \frac{an - bm}{a^2} x - \frac{b\,(an - bm)}{a^3} \log (ax + b).$$

$$\int \frac{(x - a)\,(x - b)}{x - c} dx = \frac{x^2}{2} + (c - a - b)\, x + (c - a)\,(c - b) \log (x - c).$$

$$\int \frac{mx^2 + nx + p}{ax + b} dx = \frac{mx^2}{2a} + \frac{an - bm}{a^2} x + \frac{a^2p - abn + b^2m}{a^3} \log (ax + b).$$

$$\int \frac{x^3}{ax + b} dx = \frac{x^3}{3a} - \frac{bx^2}{2a^2} + \frac{b^2}{a^3} x - \frac{b^3}{a^4} \log (ax + b).$$

$$\int \frac{mx^3 + n}{ax + b} dx = \frac{mx^3}{3a} - \frac{bm}{2a^2} x^2 + \frac{b^2m}{a^3} x + \frac{a^3n - b^3m}{a^4} \log (ax + b).$$

$$\int \frac{mx^3 + nx}{ax + b} dx = \frac{mx^3}{3a} - \frac{bm}{2a^2} x^2 + \frac{a^2n - b^2m}{a^3} - \frac{b\,(a^2n + b^2m)}{a^4} \log (ax + b).$$

$$\int \frac{mx^3 + nx + p}{ax + b}\, dx = \frac{mx^3}{3a} - \frac{bm}{2a^2} x^2 + \frac{a^2 n + b^2 m}{a^3} x + \frac{a^2(ap - bn) - b^3 m}{a^4} \log (ax + b).$$

$$\int \frac{mx^3 + nx^2}{ax + b}\, dx = \frac{mx^3}{3a} + \frac{an - bm}{a^2} \left\{ \frac{x^2}{2} - \frac{b}{a} x + \frac{b^2}{a^2} \log (ax + b) \right\}.$$

$$\int \frac{x(x - a)^2}{x + a}\, dx = \frac{x^3}{3} - \frac{3a}{2} x^2 + 4a^2 x - 4a^3 \log (x + a).$$

$$\int \frac{mx^3 + nx^2 + px + q}{ax + b}\, dx = \frac{mx^3}{3a} + \frac{an - bm}{2a^2} x^2 + \frac{a^2 p - b(an - bm)}{a^3} x + \frac{a^2(aq - bp) + b^2(an - bm)}{a^4} \log (ax + b).$$

$$\int \frac{x^4}{ax + b}\, dx = \frac{x^4}{4a} - \frac{bx^3}{3a^2} + \frac{b^2 x^2}{2a^3} - \frac{b^3 x}{a^4} + \frac{b^4}{a^5} \log (ax + b).$$

$$\int \frac{(ax^2 + b)^2}{x}\, dx = \frac{1}{4} (ax^2 + b)^2 + \frac{b}{2} (ax^2 + b) + \frac{b^2}{2} \log x^2.$$

$$\int \frac{x^5}{ax + b}\, dx = \frac{1}{a^6} \left\{ \frac{\Phi^5}{5} - \frac{5b\Phi^4}{4} + \frac{10b^2\Phi^3}{3} - 5b^3\Phi^2 + 5b^4\Phi - b^5 \log \Phi \right\}.$$

$$\int \frac{x^6}{a + bx}\, dx = \frac{1}{b^7} \left\{ \frac{\Phi^6}{6} - \frac{6a\Phi^5}{5} + \frac{15a^2\Phi^4}{4} - \frac{20a^3\Phi^3}{3} + \frac{15a^4\Phi^2}{2} - 6a^5\Phi + a^6 \log \Phi \right\}.$$

$$\int \frac{dx}{ax^4 + bx^3} = \frac{a}{b^2 x} - \frac{1}{2bx^2} - \frac{a^2}{b^3} \log \frac{ax + b}{x}.$$

$$\int \frac{dx}{x^4 (ax + b)} = -\frac{a^2}{b^3 x} + \frac{a}{2b^2 x^2} - \frac{1}{3bx^3} + \frac{a^3}{b^4} \log \frac{ax + b}{x}.$$

$$\int \frac{dx}{x^5 (a + bx)} = -\frac{1}{4ax^4} + \frac{b}{3a^2 x^3} - \frac{b^2}{2a^3 x^2} + \frac{b^3}{a^4 x} - \frac{b^4}{a^5} \log \frac{a + bx}{x}.$$

$$\int \frac{dx}{x^6 (a + bx)} = -\frac{1}{5ax^5} + \frac{b}{4a^2 x^4} - \frac{b^2}{3a^3 x^3} + \frac{b^3}{2a^4 x^2} - \frac{b^4}{a^5 x} + \frac{b^5}{a^6} \log \frac{\Phi}{x}.$$

$$\int \frac{dx}{x^7 (a + bx)} = -\frac{1}{6ax^6} + \frac{b}{5a^2 x^5} - \frac{b^2}{4a^3 x^4} + \frac{b^3}{3a^4 x^3} - \frac{b^4}{2a^5 x^2} + \frac{b^5}{a^6 x} - \frac{b^6}{a^7} \log \frac{a + bx}{x}.$$

$$\frac{1}{ax^2 + b}$$

$$\int \frac{dx}{x^2 + a^2} = \frac{1}{a} \text{arc tg} \frac{x}{a} = \frac{1}{a} \text{arc sin} \frac{x}{\sqrt{x^2 + a^2}} = -\frac{1}{a} \text{arc sin} \frac{a}{\sqrt{x^2 + a^2}}$$

$$\int \frac{dx}{x^2 - a^2} = \frac{1}{2a} \log \frac{x - a}{x + a} = \frac{1}{2a} \log \frac{a - x}{a + x}.$$

$$\int \frac{dx}{ax^2 - b} = \frac{1}{2\sqrt{ab}} \log \frac{x\sqrt{a} - \sqrt{b}}{x\sqrt{a} + \sqrt{b}} = \frac{1}{\sqrt{ab}} \log \frac{x\sqrt{a} - \sqrt{b}}{\sqrt{ax^2 - b}}$$

$$= -\frac{1}{\sqrt{ab}} \log \frac{x\sqrt{a} + \sqrt{b}}{\sqrt{ax^2 - b}}.$$

$$\int \frac{dx}{a^2 - x^2} = \frac{1}{2a} \log \frac{a + x}{a - x} = \frac{1}{2a} \log \frac{x + a}{x - a}.$$

$$\int \frac{dx}{b - ax^2} = \frac{1}{2\sqrt{ab}} \log \frac{\sqrt{b} + x\sqrt{a}}{\sqrt{b} - x\sqrt{a}} = \frac{1}{\sqrt{ab}} \log \frac{\sqrt{b} - x\sqrt{a}}{\sqrt{b - ax^2}}.$$

$$\int \frac{dx}{ax^2 + b} = \frac{1}{\sqrt{ab}} \text{arc tg}\, x \sqrt{\frac{a}{b}} = -\frac{1}{\sqrt{ab}} \text{art tg} \frac{\sqrt{b}}{x\sqrt{a}}$$

$$= \frac{1}{2\sqrt{ab}} \text{art tg} \frac{2x\sqrt{ab}}{b - ax^2}$$

$$= \frac{1}{\sqrt{ab}} \text{arc sin}\, x \sqrt{\frac{a}{ax^2 + b}} = \frac{1}{2\sqrt{ab}} \text{arc sin} \frac{2x\sqrt{ab}}{ax^2 + b}$$

$$= \frac{1}{2\sqrt{ab}} \text{arc cos} \frac{b - ax^2}{b + ax^2}$$

$$= \frac{1}{\sqrt{ab}} \text{arc cos} \sqrt{\frac{b}{ax^2 + b}} = \frac{1}{\sqrt{ab}} \text{arc cotg} \sqrt{\frac{b}{ax^2}}$$

$$= \frac{1}{\sqrt{ab}} \text{arc séc.} \sqrt{\frac{ax^2 + b}{b}}$$

$$= \frac{1}{2\sqrt{ab}} \text{arc séc.} \frac{b + ax^2}{b - ax^2} = \frac{1}{2\sqrt{-ab}} \log \frac{\sqrt{b} + x\sqrt{-a}}{\sqrt{b} - x\sqrt{-a}}$$

$$= \frac{1}{2\sqrt{-ab}} \log \frac{\sqrt{-b} - x\sqrt{a}}{\sqrt{-b} + x\sqrt{a}}.$$

$$\int \frac{x}{ax^2 + b}\, dx = \frac{1}{2a} \log (ax^2 + b).$$

$$\int \frac{x + 1}{x^2 + 1}\, dx = \frac{1}{2} \log (x^2 + 1) + \text{arc tg}\, x.$$

$$\int \frac{mx - n}{x^2 - c^2}\,dx = \frac{m}{2}\log(x^2 - c^2) - \frac{n}{2c}\log\frac{x - c}{x + c}.$$

$$\int \frac{mx + n}{ax^2 + b}\,dx = \frac{m}{2a}\log(ax^2 + b) + \frac{n}{\sqrt{ab}}\operatorname{arc\,tg} x\sqrt{\frac{a}{b}}.$$

$$\int \frac{mx + n}{ax^2 - b}\,dx = \frac{m}{2a}\log(ax^2 - b) + \frac{n}{2\sqrt{ab}}\log\frac{x\sqrt{a} - \sqrt{b}}{x\sqrt{a} + \sqrt{b}}.$$

$$\int \frac{x^2}{a^2 + x^2}\,dx = x - a\operatorname{arc\,tg}\frac{x}{a} = x + a\arcsin\frac{a}{\sqrt{a^2 + x^2}}.$$

$$\int \frac{x^2}{x^2 - a^2}\,dx = x + \frac{a}{2}\log\frac{x - a}{x + a}.$$

$$\int \frac{x^2}{ax^2 + b}\,dx = \frac{x}{a} - \frac{b}{a}\int\frac{dx}{ax^2 + b} = \frac{x}{a} - \frac{\sqrt{b}}{a\sqrt{a}}\operatorname{arc\,tg} x\sqrt{\frac{a}{b}}.$$

$$\int \frac{x^2}{ax^2 - b}\,dx = \frac{x}{a} + \frac{\sqrt{b}}{2a\sqrt{a}}\log\frac{x\sqrt{a} - \sqrt{b}}{x\sqrt{a} + \sqrt{b}}.$$

$$\int \frac{x^2}{b - ax^2}\,dx = -\frac{x}{a} + \frac{\sqrt{b}}{2a\sqrt{a}}\log\frac{x\sqrt{a} + \sqrt{b}}{x\sqrt{a} - \sqrt{b}}.$$

$$\int \frac{mx^2 + n}{ax^2 + b}\,dx = \frac{m}{a}x + \frac{an - bm}{a\sqrt{ab}}\operatorname{arc\,tg} x\sqrt{\frac{a}{b}}.$$

$$\int \frac{mx^2 + n}{b - ax^2}\,dx = -\frac{m}{a}x + \frac{an + bm}{2a\sqrt{ab}}\log\frac{\sqrt{b} + x\sqrt{a}}{\sqrt{b} - x\sqrt{a}}.$$

$$\int \frac{mx^2 + nx}{ax^2 + b}\,dx = \frac{m}{a}x + \frac{n}{2a}\log(ax^2 + b) - \frac{m}{a}\sqrt{\frac{b}{a}}\operatorname{arc\,tg} x\sqrt{\frac{a}{b}}.$$

$$\int \frac{mx^2 + nx}{b - ax^2}\,dx = -\frac{m}{a}x - \frac{n}{2a}\log(b - ax^2) + \frac{m}{2a}\sqrt{\frac{b}{a}}\log\frac{x\sqrt{a} + \sqrt{b}}{x\sqrt{a} - \sqrt{b}}.$$

$$\int \frac{x^3}{ax^2 + b}\,dx = \frac{x^2}{2a} - \frac{b}{2a^2}\log(ax^2 + b).$$

$$\int \frac{x^4}{ax^2 + b}\,dx = \frac{x^3}{3a} - \frac{bx}{a^2} + \frac{b^2}{a^2}\int\frac{dx}{ax^2 + b}.$$

$$\int \frac{x^4}{ax^2 - b}\,dx = \frac{x^3}{3a} + \frac{bx}{a^2} + \frac{b\sqrt{b}}{2a^2\sqrt{a}}\log\frac{x\sqrt{a} - \sqrt{b}}{x\sqrt{a} + \sqrt{b}}.$$

$$\int \frac{x^4}{b - ax^2}\,dx = -\frac{x^3}{3a} - \frac{bx}{a^2} + \frac{b\sqrt{b}}{2a^2\sqrt{a}}\log\frac{\sqrt{b} + x\sqrt{a}}{\sqrt{b} - x\sqrt{a}}.$$

$$\int \frac{x^5}{ax^2 + b}\,dx = \frac{x^4}{4a} - \frac{bx^2}{2a^2} + \frac{b^2}{2a^3}\log(ax^2 + b).$$

$$\int \frac{x^6}{a+bx^2}\,dx = \frac{x^5}{5b} - \frac{ax^3}{3b^2} + \frac{a^2x}{b^3} - \frac{a^3}{b^3}\int \frac{dx}{\Phi}.$$

$$\int \frac{x^8}{a+bx^2}\,dx = \frac{x^7}{7b} - \frac{ax^5}{5b^2} + \frac{a^2x^5}{3b^3} - \frac{a^3x}{b^4} + \frac{a^4}{b^4}\int \frac{dx}{\Phi}.$$

$$\int \frac{dx}{x^2(ax^2+b)} = -\frac{1}{bx} - \frac{a}{b}\int \frac{dx}{ax^2+b}.$$

$$\int \frac{dx}{x^3(ax^2+b)} = -\frac{1}{2bx^2} + \frac{a}{2b^2}\log\frac{ax^2+b}{x^2}.$$

$$\int \frac{dx}{x^4(a+bx^2)} = -\frac{1}{3ax^3} + \frac{b}{a^2x} + \frac{b^2}{a^2}\int \frac{dx}{a+bx^2}.$$

$$\int \frac{dx}{x^5(x^2-a)} = \frac{1}{2a^2x^2} + \frac{1}{4ax^4} + \frac{1}{2a^3}\log\frac{x^2-a}{x^2}.$$

$$\int \frac{dx}{x^6(a+bx^2)} = -\frac{1}{5ax^5} + \frac{b}{3a^2x^3} - \frac{b^2}{a^3x} - \frac{b^3}{a^3}\int \frac{dx}{a+bx^2}.$$

$$\boxed{\frac{1}{ax^2+bx}}$$

$$\int \frac{dx}{ax^2+bx} = \frac{1}{b}\log\frac{x}{ax+b}$$

$$\int \frac{dx}{bx-ax^2} = \frac{1}{b}\log\frac{x}{ax-b}$$

$$\int \frac{dx}{ax^2-bx} = \frac{1}{b}\log\frac{ax-b}{x}.$$

$$\int \frac{mx+n}{ax^2+bx}\,dx = \frac{m}{2a}\log(ax^2+bx) + \frac{2an-bm}{2ab}\log\frac{ax}{ax+b}.$$

$$\int \frac{ax+b}{x-x^2}\,dx = \log\frac{x^b}{(x-1)^{a+b}}.$$

$$\boxed{\frac{1}{(ax+b)^2}}$$

$$\int \frac{dx}{(x+a)^2} = -\frac{1}{x+a} = \frac{x}{a(x+a)} = -\frac{1}{2a}\frac{a-x}{a+x}.$$

$$\int \frac{dx}{(ax+b)^2} = -\frac{1}{a(ax+b)} = \frac{x}{b(ax+b)} = \frac{mx+n}{(bm-an)(ax+b)}.$$

$$\int \frac{x}{(ax+b)^2}\,dx = \frac{b}{a^2(ax+b)} + \frac{1}{a^2}\log(ax+b)$$

$$= -\frac{x}{a(ax+b)} - \frac{1}{a^2}\log\frac{b}{ax+b}.$$

$$\int \frac{mx+n}{(ax+b)^2}\,dx = \frac{bm-an}{a^2(ax+b)} + \frac{m}{a^2}\log(ax+b).$$

$$\int \frac{mx^2+n}{(ax+b)^2}\,dx = \frac{max^2+2bmx-an}{a^2(ax+b)} - \frac{2bm}{a^3}\log(ax+b).$$

$$\int \frac{mx^2+nx}{(ax+b)^2}\,dx = \frac{mx(ax+2b)+bn}{a^2(ax+b)} + \frac{an-2bm}{a^3}\log(ax+b).$$

$$\int \frac{mx^2+nx+p}{(ax+b)^2}\,dx = \frac{m}{a^2}x - \frac{b(bm-an)+a^2p}{a^3(ax+b)} + \frac{an-2bm}{a^3}\log(ax+b).$$

$$\int \frac{x^3}{(ax+b)^2}\,dx = \frac{\Phi^2}{2a^4} - \frac{3b\Phi}{a^4} + \frac{b^3}{a^4\Phi} + \frac{3b^2}{a^4}\log\Phi.$$

$$\int \frac{x^4}{(ax+b)^2}\,dx = \frac{1}{a^5}\left\{\frac{\Phi^3}{3} - 2b\,\Phi^2 + 6b^2\,\Phi - \frac{b^4}{\Phi} - 4b^3\log\Phi\right\}.$$

$$\int \frac{x^6}{(a+bx)^2}\,dx = \frac{1}{b^7}\left\{\frac{\Phi^5}{5} - \frac{3a\Phi^4}{2} + 5a^2\Phi^3 - 10a^3\Phi^2\right.$$

$$\left. + 15a^4\Phi - \frac{a^6}{\Phi} - 6a^5\log\Phi\right\}.$$

$$\int \frac{dx}{x(ax+b)^2} = \frac{1}{b(ax+b)} - \frac{1}{b^2}\log\frac{ax+b}{x}.$$

$$\int \frac{dx}{x^2(x-a)^2} = \frac{a-2x}{a^2x(x-a)} + \frac{2}{a^3}\log\frac{x}{x-a}.$$

$$\int \frac{dx}{x^2(ax+b)^2} = -\frac{b+2ax}{b^2x(ax+b)} + \frac{2a}{b^3}\log\frac{ax+b}{x}.$$

$$\int \frac{dx}{x^3(ax+b)^2} = \frac{a^2}{b^3(ax+b)} - \frac{1}{2b^2x^2} + \frac{2a}{b^3x} + \frac{3a^2}{b^4}\log\frac{x}{ax+b}.$$

$$\int \frac{dx}{x^4 (a + bx)^2} = -\frac{b^3}{a^4 \Phi} - \frac{1}{3a^2x^3} + \frac{b}{a^3x^2} - \frac{3b^2}{a^4x} + \frac{4b^3}{a^5} \log \frac{\Phi}{x}.$$

$$\int \frac{dx}{x^5 (a + bx)^2} = \frac{b^4}{a^5 (a + bx)} - \frac{1}{4a^2x^4} + \frac{2b}{3a^3x^3} - \frac{3b^2}{2a^4x^2} + \frac{4b^3}{a^5x} - \frac{5b^4}{a^6} \log \frac{a + bx}{x}.$$

$$\int \frac{dx}{x^6 (a + bx)^2} = -\frac{b^5}{a^6 \Phi} - \frac{1}{5a^2x^5} + \frac{b}{2a^3x^4} - \frac{b^2}{a^4x^3} + \frac{2b^3}{a^5x^2} - \frac{5b^4}{a^6x} + \frac{6b^5}{a^7} \log \frac{\Phi}{x}.$$

$$\boxed{\frac{1}{(ax + b)(cx + d)}}$$

$$\int \frac{dx}{(x + a)(x + b)} = \frac{1}{b - a} \log \frac{x + a}{x + b}.$$

$$\int \frac{dx}{ax + b)(mx + n)} = \frac{1}{an - bm} \log \frac{ax + b}{mx + n}.$$

$$\int \frac{x}{(x + a)(x + b)}\, dx = \frac{1}{a - b} \log \frac{(x + a)^a}{(x + b)^b}.$$

$$\int \frac{x}{(ax + b)(mx + n)}\, dx = -\frac{b}{an - bm} \int \frac{dx}{ax + b} + \frac{n}{an - bm} \int \frac{dx}{mx + n}.$$

$$\int \frac{mx + n}{(ax + b)(px + q)}\, dx = \frac{bm - an}{bp - aq} \int \frac{dx}{ax + b} + \frac{np - mq}{bp - aq} \int \frac{dx}{px + q}.$$

$$\int \frac{dx}{x (x + a)(x + b)} = \frac{1}{ab} \log x + \frac{1}{a (a - b)} \log (x + a) + \frac{1}{b (b - a)} \log (x + b).$$

$$\boxed{\frac{1}{(ax + b)^2 + (mx + n)^2}}$$

$$\int \frac{dx}{(ax + b)^2 + (mx + n)^2} = \frac{1}{an - bm} \operatorname{arc\,tg} \frac{ax + b}{mx + n}.$$

$$\int \frac{dx}{(x - a)^2 + (b - x)^2} = \frac{1}{b - a} \operatorname{arc\,tg} \frac{2x - a - b}{b - a} = \frac{1}{b - a} \operatorname{art\,tg} \frac{x - a}{b - x}.$$

$$\frac{1}{(ax + b)^2 + c^2}$$

$$\int \frac{dx}{(ax+b)^2 + c^2} = \frac{1}{ac} \operatorname{arc\,tg} \frac{ax+b}{c}.$$

$$\int \frac{dx}{(ax+b)^2 - c^2} = \frac{1}{2ac} \log \frac{ax+b-c}{ax+b+c}.$$

$$\int \frac{mx+n}{(x+a)^2 + b^2}\, dx = \frac{m}{2} \log \left\{ (x+a)^2 + b^2 \right\}$$

$$- \frac{n - am}{b} \arccos \frac{x+a}{\sqrt{(x+a)^2 + b^2}}.$$

$$\frac{1}{ax^2 + bx + c}$$

$$\int \frac{dx}{ax^2 + bx + c} = \frac{1}{\sqrt{b^2 - 4ac}} \log \frac{2ax + b - \sqrt{b^2 - 4ac}}{2ax + b + \sqrt{b^2 - 4ac}}. \qquad b^2 > 4ac$$

$$= \frac{2}{\sqrt{b^2 - 4ac}} \log \frac{2ax + b - \sqrt{b^2 - 4ac}}{\sqrt{ax^2 + bx + c}}. \qquad \text{»}$$

$$= \frac{2}{\sqrt{4ac - b^2}} \operatorname{arc\,tg} \frac{2ax + b}{\sqrt{4ac - b^2}}. \qquad b^2 < 4ac$$

$$= \frac{1}{\sqrt{4ac - b^2}} \arcsin \frac{(2ax + b)\sqrt{4ac - b^2}}{2a\,(ax^2 + bx + c)}. \qquad \text{»}$$

$$= \frac{2}{\sqrt{4ac - b^2}} \arcsin \frac{2ax + b}{2\sqrt{a\,(ax^2 + bx + c)}}. \qquad \text{»}$$

$$= -\frac{2}{2ax + b}. \qquad b^2 = 4ac.$$

$$\int \frac{dx}{c + bx - ax^2} = \frac{1}{\sqrt{4ac + b^2}} \log \frac{\sqrt{4ac + b^2} + 2ax - b}{\sqrt{4ac + b^2} - 2ax + b}$$

$$= \frac{1}{\sqrt{4ac + b^2}} \log \frac{2ax - b + \sqrt{4ac + b^2}}{2ax - b - \sqrt{4ac + b^2}}.$$

$$\int \frac{x}{x^2 + 2bx + a^2} dx = \frac{1}{2} \log (x^2 + 2bx + a^2) - \frac{b}{\sqrt{a^2 - b^2}} \text{arc tg} \frac{x + b}{\sqrt{a^2 - b^2}}.$$

$$\int \frac{x}{ax^2 + bx + c} dx = \frac{1}{2a} \log (ax^2 + bx + c) - \frac{b}{2a} \int \frac{dx}{ax^2 + bx + c}.$$

$$= \frac{1}{2a} \log (ax^2 + bx + c) - \frac{b}{2a\sqrt{b^2 - 4ac}} \log \frac{2ax + b - \sqrt{b^2 - 4ac}}{2ax + b + \sqrt{b^2 - 4ac}}. \qquad b^2 > 4ac$$

$$= \frac{1}{2a} \log (ax^2 + bx + c) - \frac{b}{a\sqrt{4ac - b^2}} \text{arc tg} \frac{2ax + b}{\sqrt{4ac - b^2}}. \qquad b^2 < 4ac$$

$$= \frac{b}{a(2ax + b)} + \frac{1}{a} \log (2ax + b). \qquad b^2 = 4ac.$$

$$\int \frac{x}{c + bx - ax^2} dx = \frac{1}{2a} \log (c + bx - ax^2) + \frac{b}{2a\sqrt{b^2 + 4ac}} \log \frac{\sqrt{b^2 + 4ac} + 2ax - b}{\sqrt{b^2 + 4ac} - 2ax + b}.$$

$$\int \frac{mx + n}{ax^2 + bx + c} dx = \frac{m}{2a} \log (ax^2 + bx + c) + \frac{2an - bm}{2a\sqrt{b^2 - 4ac}} \log \frac{2ax + b - \sqrt{b^2 - 4ac}}{2ax + b + \sqrt{b^2 - 4ac}}.$$

$$b^2 > 4ac.$$

$$= \frac{m}{2a} \log (ax^2 + bx + c) + \frac{2an - bm}{a\sqrt{4ac - b^2}} \text{arc tg} \frac{2ax + b}{\sqrt{4ac - b^2}}.$$

$$b^2 < 4ac.$$

$$= \frac{bm - 2an}{a(2ax + b)} + \frac{m}{a} \log (2ax + b). \qquad b^2 = 4ac.$$

$$\int \frac{mx + n}{c + bx - ax^2} dx = -\frac{m}{2a} \log (c + bx - ax^2) + \frac{2an + bm}{2a\sqrt{b^2 + 4ac}} \log \frac{b - 2ax - \sqrt{b^2 + 4ac}}{b - 2ax + \sqrt{b^2 + 4ac}}.$$

$$\int \frac{x^2}{ax^2+bx+c}\,dx = \frac{x}{a} - \frac{b}{2a^2}\log(ax^2+bx+c) + \frac{b^2-2ac}{2a^2}\int \frac{dx}{ax^2+bx+c}.$$

$$= \frac{x}{a} - \frac{b}{2a^2}\log(ax^2+bx+c) + \frac{b^2-2ac}{2a^2\sqrt{b^2-4ac}}\log\frac{2ax+b-\sqrt{b^2-4ac}}{2ax+b+\sqrt{b^2-4ac}}. \quad b^2 > 4ac.$$

$$= \frac{x}{a} - \frac{b}{2a^2}\log(ax^2+bx+c) + \frac{b^2-2ac}{a^2\sqrt{4ac-b^2}}\operatorname{arc\,tg}\frac{2ax+b}{\sqrt{4ac-b^2}}. \quad b^2 < 4ac.$$

$$= \frac{2x(ax+b)}{a(2ax+b)} - \frac{b}{a^2}\log(2ax+b). \quad b^2 = 4ac.$$

$$\int \frac{mx^2+n}{ax^2+bx+c}\,dx = \frac{m}{a}x - \frac{bm}{2a^2}\log(ax^2+bx+c) + \frac{2a(an-cm)+b^2m}{2a^2\sqrt{b^2-4ac}}\log\frac{2ax+b-\sqrt{b^2-4ac}}{2ax+b+\sqrt{b^2-4ac}}.$$

$$b^2 > 4ac$$

$$= \frac{m}{a}x - \frac{bm}{2a^2}\log(ax^2+bx+c) + \frac{2a(an-cm)+b^2m}{a^2\sqrt{4ac-b^2}}\operatorname{arc\,tg}\frac{2ax+b}{\sqrt{4ac-b^2}}. \quad b^2 < 4ac$$

$$= \frac{2(amx^2+bmx-an)}{a(2ax+b)} - \frac{bm}{a^2}\log(2ax+b). \quad b^2 = 4ac.$$

$$\int \frac{mx^2+n}{c+bx-ax^2}\,dx = -\frac{m}{a}x - \frac{bm}{2a^2}\log(c+bx-ax^2) + \frac{2a(an+cm)+b^2m}{2a^2\sqrt{4ac+b^2}}\log\frac{\sqrt{4ac+b^2}+2ax-b}{\sqrt{4ac+b^2}-2ax+b}.$$

$$\int \frac{mx^2+nx}{ax^2+bx+c}\,dx = \frac{m}{a}x + \frac{an-bm}{2a^2}\log(ax^2+bx+c)$$

$$+ \frac{b^2m-2acm-abn}{2a^2\sqrt{b^2-4ac}}\log\frac{2ax+b-\sqrt{b^2-4ac}}{2ax+b+\sqrt{b^2-4ac}}. \quad b^2 > 4ac.$$

$$= \frac{m}{a}x + \frac{an-bm}{2a^2}\log(ax^2+bx+c) + \frac{b^2m-2acm-abn}{a^2\sqrt{4ac-b^2}}\operatorname{arc\,tg}\frac{2ax+b}{\sqrt{4ac-b^2}}. \quad b^2 < 4ac.$$

$$= \frac{m}{a} x + \frac{b(an - bm) + 2acm}{a^2(2ax + b)} + \frac{an - bm}{a^2} \log(2ax + b).$$

$$b^2 = 4ac.$$

$$\int \frac{mx^2 + nx + p}{ax^2 + bx + c} dx = \frac{m}{a} x + \frac{an - bm}{2a^2} \log(ax^2 + bx + c)$$

$$+ \frac{2a(ap - cm) - b(an - bm)}{2a^2\sqrt{b^2 - 4ac}} \log \frac{2ax + b - \sqrt{b^2 - 4ac}}{2ax + b + \sqrt{b^2 - 4ac}}.$$

$$b^2 > 4ac.$$

$$= \frac{m}{a} x + \frac{an - bm}{2a^2} \log(ax^2 + bx + c)$$

$$+ \frac{2a(ap - cm) - b(an - bm)}{a^2\sqrt{4ac - b^2}} \operatorname{arc\,tg} \frac{2ax + b}{\sqrt{4ac - b^2}}.$$

$$b^2 < 4ac.$$

$$= \frac{m}{a} x + \frac{b(an - bm) - 2a(ap - cm)}{a^2(2ax + b)}$$

$$+ \frac{an - bm}{a^2} \log(2ax + b). \quad b^2 = 4ac.$$

$$\int \frac{mx^3 + nx^2 + px + q}{ax^2 + bx + c} dx = \frac{mx^2}{2a} + \frac{an - bm}{a^2} x$$

$$+ \frac{a(ap - cm) - b(an - bm)}{2a^3} \log(ax^2 + bx + c)$$

$$+ \frac{2a^3q - ab(ap - cm) + (an - bm)(b^2 - 2ac)}{2a^3\sqrt{b^2 - 4ac}} \log \frac{2ax + b - \sqrt{b^2 - 4ac}}{2ax + b + \sqrt{b^2 - 4ac}}.$$

$$b^2 > 4ac.$$

$$= \frac{mx^2}{2a} + \frac{an - bm}{a^2} x$$

$$+ \frac{a(ap - cm) - b(an - bm)}{2a^3} \log(ax^2 + bx + c)$$

$$+ \frac{2a^3q - ab(ap - cm) + (an - bm)(b^2 - 2ac)}{a^3\sqrt{4ac - b^2}} \operatorname{arc\,tg} \frac{2ax + b}{\sqrt{4ac - b^2}}.$$

$$b^2 < 4ac.$$

$$= (2amx - 5bm + 4an) \frac{2ax + b}{8a^3}$$

$$+ \frac{b^3m - 2ab^2n + 4a^2bp - 8a^3q}{4a^3(2ax + b)}$$

$$+\frac{3b^2m+4a^2p-4abn}{4a^3}\log(2ax+b). \qquad b^2=4ac.$$

$$\int\frac{x^4}{ax^2+bx+c}dx=\frac{x^3}{3a}-\frac{bx^2}{2a^2}+\frac{b^2-ac}{a^3}x+\frac{(2ac-b^2)b}{2a^4}\log\Phi$$

$$+\frac{2a^2c^2-4ab^2c+b^4}{2a^4}\int\frac{dx}{\Phi}.$$

$$\int\frac{x^5}{ax^2+bx+c}dx=\frac{x^4}{4a}-\frac{bx^3}{3a^2}+\left(\frac{b^2}{a^3}-\frac{c}{a^2}\right)\frac{x^2}{2}-\left(\frac{b^3}{a^4}-\frac{2bc}{a^3}\right)x$$

$$+\left(\frac{b^4}{2a^5}-\frac{3b^2c}{2a^4}+\frac{c^2}{2a^3}\right)\log\Phi\quad-\left(\frac{b^5}{2a^5}-\frac{5b^3c}{2a^4}+\frac{5bc^2}{2a^3}\right)\int\frac{dx}{\Phi}.$$

$$\int\frac{dx}{x(ax^2+bx+c)}=\frac{1}{2c}\log\frac{x^2}{ax^2+bx+c}-\frac{b}{2c}\int\frac{dx}{ax^2+bx+c}.$$

$$=\frac{1}{2c}\log\frac{x^2}{ax^2+bx+c}$$

$$-\frac{b}{2c\sqrt{b^2-4ac}}\log\frac{2ax+b-\sqrt{b^2-4ac}}{2ax+b+\sqrt{b^2-4ac}}. \qquad b^2>4ac.$$

$$=\frac{1}{2c}\log\frac{x^2}{ax^2+bx+c}$$

$$-\frac{b}{c\sqrt{4ac-b^2}}\operatorname{arc\,tg}\frac{2ax+b}{\sqrt{4ac-b^2}}. \qquad b^2<4ac.$$

$$\int\frac{dx}{x^2(ax^2+bx+c)}=-\frac{1}{cx}-\frac{b}{2c^2}\log\frac{x^2}{\Phi}+\frac{b^2-2ac}{2c^2}\int\frac{dx}{\Phi}.$$

$$\int\frac{dx}{x^3(ax^2+bx+c)}=-\frac{1}{2cx^2}+\frac{b}{c^2x}+\left(\frac{b^2}{2c^3}-\frac{a}{2c^2}\right)\log\frac{x^2}{\Phi}+\left(\frac{3ab}{2c^2}-\frac{b^3}{2c^3}\right)\int\frac{dx}{\Phi}.$$

$$\int\frac{dx}{x^4(a+bx+cx^2)}=\frac{3bx-2a}{6a^2x^3}+\frac{ac-b^2}{a^3x}+\frac{2abc-b^3}{2a^4}\log\frac{x^2}{\Phi}$$

$$+\frac{2a^2c^2-4ab^2c+b^4}{2a^4}\int\frac{dx}{\Phi}.$$

$$\boxed{\frac{1}{ax^3+b}}$$

$$\int\frac{dx}{ax^3+b}=\frac{k}{6b}\log\frac{(x+k)^2}{x^2-kx+k^2}+\frac{k}{b\sqrt{3}}\operatorname{arc\,tg}\frac{x\sqrt{3}}{2k-x}. \qquad k=\left(\frac{b}{a}\right)^{\frac{1}{3}}.$$

$$\int \frac{dx}{x^3 + a^3} = \frac{1}{6a^2} \log \frac{(x+a)^2}{x^2 - ax + a^2} + \frac{1}{a^2\sqrt{3}} \text{arc tg} \frac{2x - a}{a\sqrt{3}}.$$

$$= \frac{1}{6a^2} \log \frac{(x+a)^2}{x^2 - ax + a^2} + \frac{1}{a^2\sqrt{3}} \text{arc tg} \frac{x\sqrt{3}}{2a - x}.$$

$$\int \frac{dx}{ax^3 - b} = \frac{1}{6ak^2} \log \frac{(x-k)^2}{x^2 + kx + k^2}$$

$$+ \frac{1}{ak^2\sqrt{3}} \text{arc tg} \frac{k\sqrt{3}}{2x + k} \qquad k = \left(\frac{b}{a}\right)^{\frac{1}{3}}$$

$$\int \frac{dx}{b - ax^3} = -\frac{1}{3bk} \log \frac{1 - kx}{\sqrt{1 + kx + k^2x^2}}$$

$$+ \frac{1}{bk\sqrt{3}} \text{arc tg} \frac{kx\sqrt{3}}{2 + kx}. \qquad k = \left(\frac{b}{a}\right)^{\frac{1}{3}}.$$

$$\int \frac{x}{a^3 + x^3} dx = -\frac{1}{6a} \log \frac{(x+a)^2}{x^2 - ax + a^2} + \frac{1}{a\sqrt{3}} \text{arc tg} \frac{2x - a}{a\sqrt{3}}.$$

$$= \frac{1}{6a} \log \frac{x^2 - ax + a^2}{(x+a)^2} + \frac{1}{a\sqrt{3}} \text{arc tg} \frac{x\sqrt{3}}{2a - x}.$$

$$\int \frac{x}{ax^3 + b} dx = -\frac{1}{6ak} \log \frac{(x+k)^2}{x^2 - kx + k^2}$$

$$+ \frac{1}{ak\sqrt{3}} \text{arc tg} \frac{x\sqrt{3}}{2k - x}. \qquad k = \left(\frac{b}{a}\right)^{\frac{1}{3}}.$$

$$\int \frac{x}{ax^3 - b} dx = \frac{1}{3a\,k} \left\{ \log (x - k) - \frac{1}{2} \log (x^2 + kx + k^2) \right.$$

$$\left. + \sqrt{3}\, \text{arc tg} \frac{2x + k}{k\sqrt{3}} \right\}. \qquad k = \left(\frac{b}{a}\right)^{\frac{1}{3}}.$$

$$\int \frac{x + a}{x^3 + a^3} dx = \frac{2}{a\sqrt{3}} \text{arc tg} \frac{2x - a}{a\sqrt{3}}.$$

$$\int \frac{mx + n}{ax^3 + b} dx = m \int \frac{x}{ax^3 + b} dx + n \int \frac{dx}{ax^3 + b}.$$

$$\int \frac{x^2}{ax^3 + b} dx = \frac{1}{3a} \log (ax^3 + b).$$

$$\int \frac{x^3}{ax^3 + b} dx = \frac{x}{a} - \frac{b}{a} \int \frac{dx}{ax^3 + b}.$$

$$\int \frac{x^6}{a + bx^3} dx = \frac{x^4}{4b} - \frac{ax}{b^2} + \frac{a^2}{b^2} \int \frac{dx}{\Phi}.$$

$$\int \frac{x^7}{a+bx^3}\,dx = \frac{x^5}{5b} - \frac{ax^2}{2b^2} + \frac{a^2}{b^2}\int \frac{x}{a+bx^3}\,dx.$$

$$\int \frac{dx}{ax^4+bx} = \frac{1}{3b}\log\frac{x^3}{ax^3+b}.$$

$$\int \frac{dx}{x^3(a+bx^3)} = -\frac{1}{2ax^2} - \frac{b}{a}\int \frac{dx}{a+bx^3}.$$

$$\int \frac{dx}{x^4(a+bx^3)} = -\frac{1}{3ax^3} + \frac{b}{3a^2}\log\frac{\Phi}{x^3}.$$

$$\int \frac{dx}{x^5(a+bx^3)} = -\frac{1}{4ax^4} + \frac{b}{a^2x} + \frac{b^2}{a^2}\int \frac{x}{a+bx^3}\,dx.$$

$$\int \frac{dx}{x^6(a+bx^3)} = -\frac{1}{5ax^5} + \frac{b}{2a^2x^2} + \frac{b^2}{a^2}\int \frac{dx}{\Phi}.$$

•

$$\int \frac{dx}{ax^3+bx^2} = -\frac{1}{bx} + \frac{a}{b^2}\log\frac{ax+b}{x}.$$

$$\frac{1}{(a+bx)^2(cx+d)}$$

$$\int \frac{dx}{(a+x)^2(b+x)} = \frac{1}{(a-b)(x+a)} + \frac{1}{(a-b)^2}\log\frac{x+b}{x+a}.$$

$$\int \frac{x}{(x+a)^2(x+b)}\,dx = -\frac{a}{(a-b)(x+a)} - \frac{b}{(a-b)^2}\log\frac{x+b}{x+a}.$$

$$\int \frac{x^2}{(x-a)^2(x+a)}\,dx = -\frac{a}{2(x-a)} + \frac{3}{4}\log(x-a) + \frac{1}{4}\log(x+a).$$

$$\frac{1}{(ax + b)^3}$$

$$\int \frac{dx}{(ax+b)^3} = -\frac{1}{2a(ax+b)^2}.$$

$$\int \frac{x}{(ax+b)^3}\,dx = \frac{x^2}{2b(ax+b)^2} = -\frac{2ax+b}{2a^2(ax+b)^2}.$$

$$\int \frac{x^2}{(x-a)^3}\,dx = -\frac{a^2}{2(x-a)^2} - \frac{2a}{x-a} + \log(x-a).$$

$$\int \frac{x^2}{(ax+b)^3}\,dx = \frac{4abx+3b^2}{2a^3(ax+b)^2} + \frac{1}{a^3}\log(ax+b).$$

$$\int \frac{x^2+a^2}{(x-a)^3}\,dx = -\frac{a(2x-a)}{(x-a)^2} + \log(x-a).$$

$$\int \frac{mx^2+nx+p}{(x-a)^3}\,dx = \frac{2am+n}{a-x} - \frac{p+a^2m+an}{2(x-a)^2} + m\log(x-a).$$

$$\int \frac{x^3}{(x-a)^3}\,dx = \Phi - \frac{3a^2}{\Phi} - \frac{a^3}{2\Phi^2} + 3a\log\Phi.$$

$$\int \frac{x^3}{(ax+b)^3}\,dx = \frac{\Phi}{a^4} - \frac{3b^2}{a^4\Phi} + \frac{b^3}{2a^4\Phi^2} - \frac{3b}{a^4}\log\Phi.$$

$$\int \frac{x^5}{(ax+b)^3}\,dx = \frac{1}{a^6}\left\{\frac{\Phi^3}{3} - \frac{5b\Phi^2}{2} + 10\,b^2\Phi - \frac{5b^4}{\Phi} + \frac{b^5}{2\Phi^2} - 10b^3\log\Phi\right\}.$$

$$\int \frac{x^6}{(a+bx)^3}\,dx = \frac{1}{b^7}\left\{\frac{\Phi^4}{4} - 2a\Phi^3 + \frac{15a^2\Phi^2}{2} - 20a^3\Phi\right.$$

$$\left. + \frac{6a^5}{\Phi} - \frac{a^6}{2\Phi^2} + 15a^4\log\Phi\right\}.$$

$$\int \frac{dx}{x(x-a)^3} = \frac{2x-3a}{2a^2(x-d)^2} + \frac{1}{a^3}\log\frac{x-a}{x}.$$

$$\int \frac{dx}{x(ax+b)^3} = \frac{1}{b^2\Phi} + \frac{1}{2b\Phi^2} + \frac{1}{b^3}\log\frac{x}{\Phi}.$$

$$\int \frac{dx}{x^3(a+bx)^3} = \frac{b^2}{2a^3\Phi^2} + \frac{3b^2}{a^4\Phi} - \frac{1}{2a^3x^2} + \frac{3b}{a^4x} - \frac{6b^2}{a^5}\log\frac{\Phi}{x}.$$

$$\int \frac{dx}{x^4 (a + bx)^3} = - \frac{b^3}{2a^4\Phi^2} - \frac{4b^3}{a^5\Phi} - \frac{1}{3a^3x^3} + \frac{3b}{2a^4x^2} - \frac{6b^2}{a^5x} + \frac{10b^3}{a^6} \log \frac{\Phi}{x} .$$

$$\int \frac{dx}{x^5 (a + bx)^3} = \frac{b^4}{2a^5\Phi^2} + \frac{5b^4}{a^6\Phi} - \frac{1}{4a^3x^4} + \frac{b}{a^4x^3} - \frac{3b^2}{a^5x^2} + \frac{10b^3}{a^6x} - \frac{15b^4}{a^7} \log \frac{\Phi}{x} .$$

$$\boxed{\frac{1}{(ax^2 + b)(cx + d)}}$$

$$\int \frac{dx}{(x^2 + a)(x + b)} = \frac{1}{b^2 + a} \log \frac{x + b}{\sqrt{x^2 + a}} + \frac{b}{b^2 + a} \int \frac{dx}{x^2 + a} .$$

$$\int \frac{x}{(x^2 + a)(x + b)} \, dx = \frac{b}{a + b^2} \log \frac{\sqrt{x^2 + a}}{x + b} + \frac{\sqrt{a}}{a + b^2} \operatorname{arc\,tg} \frac{x}{\sqrt{a}} .$$

$$\int \frac{dx}{x (x^2 + a)(x + b)} = \frac{1}{ab} \int (x + b)(x^2 + a)\, dx - \frac{1}{b (b^2 + a)} \int x (x^2 + a)\, dx$$

$$- \frac{1}{a (b^2 + a)} \int x (x + b)(bx + a)\, dx.$$

$$\boxed{\frac{1}{(ax + b)(cx + d)(ex + f)}}$$

$$\int \frac{dx}{(x - a)(x - b)(2x - a - b)} = \frac{1}{(a - b)^2} \log \frac{(x - a)(x - b)}{(2x - a - b)^2} .$$

$$\int \frac{dx}{(x - a)(x - b)(3x - 2a - b)} = \frac{1}{2 (a - b)^2} \log \frac{(x - a)^2 (x - b)}{(3x - 2a - b)^3} .$$

$$\int \frac{dx}{(x + a)(x + b)(x + c)} = \frac{1}{(b - a)(c - a)} \log (x + a) + \frac{1}{(a - b)(c - b)} \log (x + b)$$

$$+ \frac{1}{(b - c)(a - c)} \log (x + c).$$

$$\int \frac{x}{(x + a)(x + b)(x + c)} \, dx = - \frac{a \log (x + a)}{(b - a)(c - a)} - \frac{b \log (x + b)}{(a - b)(c - b)}$$

$$- \frac{c \log (x + c)}{(a - c)(b - c)} .$$

$$\boxed{\frac{1}{ax^3 + bx^2 + cx + d}}$$

$$\int \frac{dx}{ax^3 + bx^2 + cx + d} = \frac{\log (x - x_1)}{3ax_1^2 + 2bx_1 + c} + \frac{\log (x - x_2)}{3ax_2^2 + 2bx_2 + c} + \frac{\log (x - x_3)}{3ax_3^2 + 2bx_3 + c}.$$

x_1, x_2, x_3 being the roots of $X^3 = 0$.

$$= \frac{1}{3ax_1^2 + 2bx_1 + c}\left[\log (x - x_1) - \frac{1}{2} \log\left\{(x - \alpha)^2 + \beta^2\right\} + \frac{\alpha - x_1}{\beta} \operatorname{arc\,tg} \frac{x - \alpha}{\beta}\right].$$

$x_1, \quad x_2 = \alpha + \beta\sqrt{-1}, \quad x_3$

$= \alpha - \beta\sqrt{-1}$ being the roots of $X^3 = 0$.

$$\int \frac{x}{ax^3 + bx^2 + cx + d}\, dx = \frac{x_1 \log (x - x_1)}{3ax_1^2 + 2bx_1 + c} + \frac{x_2 \log (x - x_2)}{3ax_2^2 + 2bx_2 + c} + \frac{x_3 \log (x - x_3)}{3ax_3^2 + 2bx_3 + c}.$$

x_1, x_2, x_3 being the roots of $X^3 = 0$.

$$= \frac{1}{3ax_1^2 + 2bx_1 + c}\left[x_1 \log (x - x_1) - \frac{1}{2} x_1 \log\left\{(x - \alpha)^2 + \beta^2\right\} + \frac{\alpha^2 - \alpha x_1 + \beta^2}{\beta} \operatorname{arc\,tg} \frac{x - \alpha}{\beta}\right].$$

$x_1, \quad x_2 = \alpha + \beta\sqrt{-1}, \quad x_3$

$= \alpha - \beta\sqrt{-1}$ being the roots of $X^3 = 0$.

$$\int \frac{x^2}{ax^3 + bx^2 + cx + d}\, dx = \frac{1}{3a} \log \Phi - \frac{2b}{3a}\int \frac{x}{\Phi}\, dx - \frac{c}{3a}\int \frac{dx}{\Phi}.$$

$$\int \frac{dx}{x\,(ax^3 + bx^2 + cx + d)} = \frac{1}{3d} \log \frac{x^3}{\Phi} - \frac{b}{3d}\int \frac{x}{\Phi}\, dx - \frac{2c}{3d}\int \frac{dx}{\Phi}.$$

$$\frac{1}{ax^4 + b}$$

$$\int \frac{dx}{x^4 + a^4} = \frac{1}{4a^3\sqrt{2}} \log \frac{x^2 + ax\sqrt{2} + a^2}{x^2 - ax\sqrt{2} + a^2} + \frac{1}{2a^3\sqrt{2}} \operatorname{arc\,tg} \frac{ax\sqrt{2}}{a^2 - x^2}.$$

$$\int \frac{dx}{x^4 - a^4} = \frac{1}{4a^3} \log \frac{x-a}{x+a} - \frac{1}{2a^3} \operatorname{arc\,tg} \frac{x}{a}.$$

$$\int \frac{dx}{ax^4 + b} = \frac{k}{4b\sqrt{2}} \log \frac{x^2 + kx\sqrt{2} + k^2}{x^2 - kx\sqrt{2} + k^2} + \frac{k}{2b\sqrt{2}} \operatorname{arc\,tg} \frac{kx\sqrt{2}}{k^2 - x^2}. \quad k = \left(\frac{b}{a}\right)^{\frac{1}{4}}.$$

$$\int \frac{dx}{ax^4 - b} = \frac{k}{4b} \log \frac{k-x}{k+x} - \frac{k}{2b} \operatorname{arc\,tg} \frac{x}{k}. \qquad k = \left(\frac{b}{a}\right)^{\frac{1}{4}}$$

$$\int \frac{dx}{a^4 - x^4} = \frac{1}{4a^3} \log \frac{x+a}{x-a} + \frac{1}{2a^3} \operatorname{arc\,tg} \frac{x}{a}.$$

$$\int \frac{dx}{b^4 - a^4x^4} = \frac{1}{4ab^3} \log \frac{b+ax}{b-ax} + \frac{1}{2ab^3} \operatorname{arc\,tg} \frac{ax}{b}.$$

$$\int \frac{x}{ax^4 + b}\, dx = \frac{1}{2\sqrt{ab}} \operatorname{arc\,tg} x^2 \sqrt{\frac{a}{b}}.$$

$$\int \frac{x}{ax^4 - b}\, dx = \frac{1}{4\sqrt{ab}} \log \frac{x^2\sqrt{a} - \sqrt{b}}{x^2\sqrt{a} + \sqrt{b}}.$$

$$\int \frac{x}{b - ax^4}\, dx = \frac{1}{4\sqrt{ab}} \log \frac{\sqrt{b} + x^2\sqrt{a}}{\sqrt{b} - x^2\sqrt{a}}.$$

$$\int \frac{mx + n}{x^4 - 1}\, dx = \frac{m}{4} \log \frac{x^2 - 1}{x^2 + 1} + \frac{n}{4} \log \frac{x-1}{x+1} + \frac{n}{2} \operatorname{arc\,tg} x.$$

$$\int \frac{x^2}{x^4 + a^4}\, dx = \frac{1}{4a\sqrt{2}} \log \frac{x^2 - ax\sqrt{2} + a^2}{x^2 + ax\sqrt{2} + a^2} + \frac{1}{2a\sqrt{2}} \operatorname{arc\,tg} \frac{ax\sqrt{2}}{a^2 - x^2}.$$

$$\int \frac{x^2}{x^4 - a^4}\, dx = \frac{1}{4a} \log \frac{x-a}{x+a} + \frac{1}{2a} \operatorname{arc\,tg} \frac{x}{a}.$$

$$\int \frac{x^2}{ax^4 + b}\, dx = \frac{1}{4ak\sqrt{2}} \log \frac{x^2 - kx\sqrt{2} + k^2}{x^2 + kx\sqrt{2} + k^2}$$

$$+ \frac{1}{2ak\sqrt{2}} \operatorname{arc\,tg} \frac{kx\sqrt{2}}{k^2 - x^2}. \quad k = \left(\frac{b}{a}\right)^{\frac{1}{4}}.$$

$$\int \frac{x^2}{ax^4 - b}\, dx = \frac{1}{4ak} \log \frac{x-k}{x+k} + \frac{1}{2ak} \operatorname{arc\,tg} \frac{x}{k}. \quad k = \left(\frac{b}{a}\right)^{\frac{1}{4}}.$$

$$\int \frac{x^3}{ax^4 + b}\,dx = \frac{1}{4a}\log(ax^4 + b).$$

$$\int \frac{x^4}{x^4 - a^2}\,dx = \frac{1}{2}\int \frac{x^2}{x^2 + a}\,dx + \frac{1}{2}\int \frac{x^2}{x^2 - a}\,dx$$

$$\int \frac{x^4}{ax^4 + b}\,dx = \frac{x}{a} - \frac{b}{a}\int \frac{dx}{ax^4 + b}.$$

$$\int \frac{x^6}{a + bx^4}\,dx = \frac{x^3}{3b} - \frac{a}{b}\int \frac{x^2}{\Phi}\,dx.$$

$$\int \frac{dx}{x(ax^4 + b)} = \frac{1}{2b}\log\frac{x^2}{\sqrt{ax^4 + b}}.$$

$$\int \frac{dx}{x^2(a + bx^4)} = -\frac{1}{ax} - \frac{b}{a}\int \frac{x^2}{a + bx^4}\,dx.$$

$$\int \frac{dx}{x^3(a + bx^4)} = -\frac{1}{2ax^2} - \frac{b}{a}\int \frac{x}{\Phi}\,dx.$$

$$\int \frac{dx}{x^4(a + bx^4)} = -\frac{1}{3ax^3} - \frac{b}{a}\int \frac{dx}{a + bx^4}.$$

$\dfrac{1}{(ax + b)^4}$

$$\int \frac{dx}{(ax + b)^4} = -\frac{1}{3a(ax + b)^3}.$$

$$\int \frac{x}{(ax + b)^4}\,dx = -\frac{1}{2a^2(ax + b)^2} + \frac{b}{3a^2(ax + b)^3} = \frac{1}{2b^2}\left(\frac{x}{\Phi}\right)^2 - \frac{a}{3b^2}\left(\frac{x}{\Phi}\right)^3$$

$$\int \frac{x^2}{(ax + b)^4}\,dx = -\frac{1}{a^3\Phi} + \frac{b}{a^3\Phi^2} - \frac{b^2}{3a^3\Phi^3}.$$

$$\int \frac{x^3}{(ax + b)^4}\,dx = \frac{1}{a^4}\left\{\frac{3b}{\Phi} - \frac{3b^2}{2\Phi^2} + \frac{b^3}{3\Phi^3} + \log\Phi\right\}.$$

$$\int \frac{x^4}{(ax + b)^4}\,dx = \frac{1}{a^5}\left\{\Phi - \frac{6b^2}{\Phi} + \frac{2b^3}{\Phi^2} - \frac{b^4}{3\Phi^3} - 4b\log\Phi\right\}$$

$$\int \frac{x^5}{(ax + b)^4}\,dx = \frac{1}{a^6}\left\{\frac{\Phi^2}{2} - 5b\Phi + \frac{10b^3}{\Phi} - \frac{5b^4}{2\Phi^2} + \frac{b^5}{3\Phi^3} + 10b^2\log\Phi\right\}$$

$$\int \frac{x^6}{(a + bx)^4}\,dx = \frac{1}{b^7}\left\{\frac{\Phi^3}{3} - 3a\Phi^2 + 15a^2\Phi - \frac{15a^4}{\Phi}\right.$$

$$+\frac{3a^5}{\Phi^2}-\frac{a^6}{3\Phi^3}-20a^3\log\Phi\Big\}.$$

$$\int\frac{dx}{x(ax+b)^4}=\frac{1}{b^3\Phi}+\frac{1}{2b^2\Phi^2}+\frac{1}{3b\Phi^3}+\frac{1}{b^4}\log\frac{x}{\Phi}.$$

$$\int\frac{dx}{x^2(a+bx)^4}=-\frac{b}{3a^2\Phi^3}-\frac{b}{a^3\Phi^2}-\frac{3b}{a^4\Phi}-\frac{1}{a^4x}+\frac{4b}{a^5}\log\frac{\Phi}{x}.$$

$$\int\frac{dx}{x^3(a+bx)^4}=\frac{b^2}{3a^3\Phi^3}+\frac{3b^2}{2a^4\Phi^2}+\frac{6b^2}{a^5\Phi}-\frac{1}{2a^4x^2}+\frac{4b}{a^5x}-\frac{10b^2}{a^6}\log\frac{\Phi}{x}.$$

$$\int\frac{dx}{x^4(a+bx)^4}=-\frac{b^3}{3a^4\Phi^3}-\frac{2b^3}{a^5\Phi^2}-\frac{10b^3}{a^6\Phi}-\frac{1}{3a^4x^3}+\frac{2b}{a^5x^2}-\frac{10b^2}{a^6x}+\frac{20b^3}{a^7}\log\frac{\Phi}{x}.$$

$$\frac{1}{(ax^2+b)^2}$$

$$\int\frac{dx}{(x^2+a^2)^2}=\frac{x}{2a^2(x^2+a^2)}+\frac{1}{2a^3}\operatorname{arc\,tg}\frac{x}{a}.$$

$$\int\frac{dx}{(ax^2+b)^2}=\frac{x}{2b(ax^2+b)}+\frac{1}{2b}\int\frac{dx}{ax^2+b}.$$

$$\int\frac{dx}{(x^2-a)^2}=-\frac{x}{2a(x^2-a)}+\frac{1}{2a\sqrt{a}}\log\frac{x+\sqrt{a}}{\sqrt{x^2-a}}.$$

$$\int\frac{dx}{(ax^2-b)^2}=-\frac{x}{2b(ax^2-b)}+\frac{1}{4b\sqrt{ab}}\log\frac{x\sqrt{a}+\sqrt{b}}{x\sqrt{a}-\sqrt{b}}.$$

$$\int\frac{dx}{(b-ax^2)^2}=\frac{x}{2b(b-ax^2)}+\frac{1}{4b\sqrt{ab}}\log\frac{\sqrt{b}+x\sqrt{a}}{\sqrt{b}-x\sqrt{a}}.$$

$$\int\frac{x}{(ax^2+b)^2}dx=-\frac{1}{2a(ax^2+b)}=\frac{x^2}{2b(ax^2+b)}=-\frac{1}{4ab}\cdot\frac{b-ax^2}{b+ax^2}$$

$$=\frac{1}{2(bn-am)}\cdot\frac{nx^2+m}{ax^2+b}.$$

$$\int\frac{x}{(b-ax^2)^2}dx=\frac{1}{2a(b-ax^2)}=\frac{1}{2(bn+am)}\cdot\frac{m+nx^2}{b-ax^2}.$$

$$\int\frac{mx+n}{(x^2+1)^2}dx=\frac{nx-m}{2(x^2+1)}+\frac{n}{2}\operatorname{arc\,tg}x.$$

$$\int \frac{x^2}{(ax^2+b)^2}\,dx = -\frac{x}{2a(ax^2+b)} + \frac{1}{2a}\int \frac{dx}{ax^2+b}.$$

$$= -\frac{x}{2a(ax^2+b)} + \frac{1}{2a\sqrt{ab}} \operatorname{arc\,tg} x\sqrt{\frac{a}{b}}.$$

$$\int \frac{x^2}{(ax^2-b)^2}\,dx = -\frac{x}{2a(ax^2-b)} + \frac{1}{4a\sqrt{ab}} \log \frac{x\sqrt{a}-\sqrt{b}}{x\sqrt{a}+\sqrt{b}}.$$

$$\int \frac{x^2}{(b-ax^2)^2}\,dx = \frac{x}{2a(b-ax^2)} - \frac{1}{4a\sqrt{ab}} \log \frac{\sqrt{b}+x\sqrt{a}}{\sqrt{b}-x\sqrt{a}}.$$

$$\int \frac{a^2-x^2}{(a^2+x^2)^2}\,dx = \frac{x}{a^2+x^2}.$$

$$\int \frac{mx^2+n}{(ax^2+b)^2}\,dx = \frac{(an-bm)x}{2ab(ax^2+b)} + \frac{an+bm}{2ab\sqrt{ab}} \operatorname{arc\,tg} x\sqrt{\frac{a}{b}}$$

$$\int \frac{mx^2+n}{(ax^2-b)^2}\,dx = -\frac{(an+bm)x}{2ab(ax^2-b)} + \frac{bm-an}{4ab\sqrt{ab}} \log \frac{x\sqrt{a}-\sqrt{b}}{x\sqrt{a}+\sqrt{b}}.$$

$$\int \frac{mx^2+n}{(b-ax^2)^2}\,dx = \frac{(an+bm)x}{2ab(b-ax^2)} + \frac{an-bm}{4ab\sqrt{ab}} \log \frac{\sqrt{b}+x\sqrt{a}}{\sqrt{b}-x\sqrt{a}}.$$

$$\int \frac{x^3}{(ax^2+b)^2}\,dx = \frac{b}{2a^2(ax^2+b)} + \frac{1}{2a^2}\log(ax^2+b).$$

$$\int \frac{x^4}{(1-ax^2)^2}\,dx = \frac{x(3-2ax^2)}{2a^2(1-ax^2)} - \frac{3}{4a^2\sqrt{a}} \log \frac{1+x\sqrt{a}}{1-x\sqrt{a}}$$

$$\int \frac{x^4}{(x^2+a^2)^2}\,dx = -\frac{x^3}{2(x^2+a^2)} + \frac{3}{2}x - \frac{3a}{2} \operatorname{arc\,tg} \frac{x}{a}.$$

$$\int \frac{x^4}{(ax^2+b)^2}\,dx = \frac{x}{a^2} + \frac{bx}{2a^2\Phi} - \frac{3b}{2a^2}\int \frac{dx}{\Phi}$$

$$\int \frac{x^6}{(a+bx^2)^2}\,dx = \frac{x}{b^3}\left\{\frac{bx^2}{3} - 2a - \frac{a^2}{2\Phi}\right\} + \frac{5a^2}{2b^3}\int \frac{dx}{\Phi}.$$

$$\int \frac{x^8}{(a+bx^2)^2}\,dx = \frac{x^5}{5b^2} - \frac{2ax^3}{3b^3} + \frac{3a^2x}{b^4} + \frac{a^3x}{2b^4\Phi} - \frac{7a^3}{2b^4}\int \frac{dx}{\Phi}.$$

$$\int \frac{dx}{x(x^2-a^2)^2} = \frac{1}{2a^2(x^2-a^2)} + \frac{1}{2a^4}\log\frac{x^2-a^2}{x^2}.$$

$$\int \frac{dx}{x(ax^2+b)^2} = \frac{1}{2b(ax^2+b)} + \frac{1}{2b^2}\log\frac{x^2}{ax^2+b} = -\frac{ax^2}{2b^2\Phi} + \frac{1}{b^2}\log\frac{x}{\sqrt{\Phi}}.$$

$$\int \frac{dx}{x^2(a^2+x^2)^2} = -\frac{1}{a^4x} - \frac{x}{2a^4(a^2+x^2)} - \frac{3}{2a^5}\operatorname{arc\,tg}\frac{x}{a}.$$

$$\int \frac{dx}{x^2(x^2-a^2)^2} = -\frac{1}{a^4x} + \frac{1}{2a^5}\log\frac{x+a}{x-a} + \frac{1}{a^2}\int\frac{dx}{(x^2-a^2)^2}.$$

$$\int \frac{dx}{x^2(x^2-b)^2} = \frac{3x^2-2b}{2b^2x(b-x^2)} + \frac{3}{4b^2\sqrt{b}}\log\frac{x+\sqrt{b}}{x-\sqrt{b}}.$$

$$\int \frac{dx}{x^2(a+bx^2)^2} = -\frac{bx}{2a^2(a+bx^2)} - \frac{1}{a^2x} - \frac{3b}{2a^2}\int\frac{dx}{a+bx^2}.$$

$$\int \frac{dx}{x^4(a+bx^2)^2} = -\frac{1}{3a^2x^3} + \frac{2b}{a^3x} + \frac{b^2x}{2a^3\Phi} + \frac{5b^2}{2a^3}\int\frac{dx}{\Phi}.$$

$$\frac{1}{(ax^2+bx+c)^2}$$

$$\int \frac{dx}{(ax^2+bx+c)^2} = \frac{2ax+b}{(4ac-b^2)\Phi} + \frac{2a}{4ac-b^2}\int\frac{dx}{\Phi}.$$

$$\int \frac{x}{(ax^2+bx+c)^2}\,dx = -\frac{bx+2c}{(4ac-b^2)(ax^2+bx+c)} - \frac{b}{4ac-b^2}\int\frac{dx}{ax^2+bx+c}.$$

$$= -\frac{1}{2a(ax^2+bx+c)} - \frac{b}{2a}\int\frac{dx}{(ax^2+bx+c)^2}.$$

$$\int \frac{mx+n}{(ax^2+bx+c)^2}\,dx = \frac{(2an-bm)x+bn-2cm}{(4ac-b^2)\Phi} + \frac{2an-bm}{4ac-b^2}\int\frac{dx}{\Phi}.$$

$$\int \frac{x^2}{(ax^2+bx+c)^2}\,dx = \frac{(b^2-2ac)x+bc}{a(4ac-b^2)\Phi} + \frac{2c}{4ac-b^2}\int\frac{dx}{\Phi} = -\frac{x}{a\Phi} + \frac{c}{a}\int\frac{dx}{\Phi^2}.$$

$$\int \frac{mx^2+n}{(ax^2+bx+c)^2}\,dx = \frac{b(an+cm)+(mb^2-2acm+2a^2n)x}{a(4ac-b^2)\Phi}$$

$$+\frac{2(an+cm)}{4ac-b^2}\int\frac{dx}{\Phi}.$$

$$\int \frac{x^3}{(ax^2+bx+c)^2}\,dx = \frac{(3ac-b^2)bx+(2ac-b^2)c}{a^2(4ac-b^2)\Phi}$$

$$+\frac{1}{2a^2}\log\Phi - \frac{b(6ac-b^2)}{2a^2(4ac-b^2)}\int\frac{dx}{\Phi}.$$

$$\int \frac{x^4}{(ax^2+bx+c)^2}\,dx = \frac{x}{a^2} + 2\left(\frac{c}{a^2} - \frac{b^2}{a^3}\right)\frac{x}{\Phi} + \frac{b^3}{2a^4\Phi} - \frac{b}{a^3}\log\Phi$$

$$- \left\{\frac{b^4}{2a^4} - \frac{3b^2c}{a^3} + \frac{3c^2}{a^2}\right\}\int\frac{dx}{\Phi^2}.$$

$$\int \frac{dx}{x\,(ax^2+bx+c)^2} = \frac{2ac - abx - b^2}{c\,(4ac-b^2)\,\Phi} + \frac{1}{2c^2}\log\frac{x^2}{\Phi} - \frac{b\,(6ac-b^2)}{2c^2(4ac-b^2)}\int\frac{dx}{\Phi}.$$

$$\int \frac{dx}{x^2\,(a+bx+cx^2)^2} = -\frac{1}{a^2x} + \left[\frac{b^3}{a^2} - \frac{3bc}{a} + \left(\frac{b^2}{a^2} - \frac{2c}{a}\right)cx\right]\frac{1}{(4ac-b^2)\Phi}$$

$$+ \frac{b}{a^2}\log\frac{\Phi}{x^2} - \left(\frac{b^4}{a^3} - \frac{6b^2c}{a^2} + \frac{6c^2}{a}\right)\frac{1}{4ac-b^2}\int\frac{dx}{\Phi}.$$

$$\boxed{\frac{1}{(ax^2+b)(cx^2+d)}}$$

$$\int \frac{dx}{(x^2+a)(x^2+b)} = \frac{1}{b-a}\int\frac{dx}{x^2+a} - \frac{1}{b-a}\int\frac{dx}{x^2+b}.$$

$$\int \frac{dx}{(1+a^2x^2)(1+b^2x^2)} = \frac{1}{a^2-b^2}\,(a \text{ arc tg } ax - b \text{ arc tg } bx).$$

$$\int \frac{x}{(x^2+a)(x^2+b)}\,dx = \frac{1}{2\,(b-a)}\log\frac{x^2+a}{x^2+b}.$$

$$\int \frac{x^2}{(1+a^2x^2)(1+b^2x^2)}\,dx = \frac{1}{a\,(b^2-a^2)}\left\{\text{arc tg } ax - \frac{a}{b}\text{ arc tg } bx\right\}.$$

$$\int \frac{x^2}{(x^2+a)(x^2+b)}\,dx = \frac{a}{a-b}\int\frac{dx}{x^2+a} - \frac{b}{a-b}\int\frac{dx}{x^2+b}.$$

$$\int \frac{x^3}{(x^2-a)(x^2-b)}\,dx = \frac{1}{2\,(a-b)}\log\frac{(x^2-a)^a}{(x^2-b)^b}.$$

$$\int \frac{dx}{x\,(x^2+a)(x^2+b)} = \frac{1}{2b\,(a-b)}\log\frac{x^2}{x^2+b} - \frac{1}{2a\,(a-b)}\log\frac{x^2}{x^2+a}.$$

$$\frac{1}{ax^4 + bx^2 + c}$$

$$\int \frac{dx}{ax^4 + bx^2 + c} = \frac{2a}{\sqrt{b^2 - 4ac}} \left\{ \int \frac{dx}{2ax^2 + A} - \int \frac{dx}{2ax^2 + B} \right\}.$$

in which $A = b - \sqrt{b^2 - 4ac}$,

$B = b + \sqrt{b^2 - 4ac}, \quad b^2 > 4ac.$

$$= \frac{1}{4ak^3} \left\{ \frac{1}{2 \cos \frac{\alpha}{2}} \log \frac{x^2 + 2kx \cos \frac{\alpha}{2} + k^2}{x^2 - 2kx \cos \frac{\alpha}{2} + k^2} + \frac{1}{\sin \frac{\alpha}{2}} \operatorname{arc\,tg} \frac{2kx \sin \frac{\alpha}{2}}{k^2 - x^2} \right\}.$$

in which $k = \left(\frac{c}{a}\right)^{\frac{1}{4}}, \quad \cos \alpha = -\frac{b}{2\sqrt{ac}}, \quad b^2 < 4ac.$

$$= \frac{x}{2c + bx^2} + \frac{1}{\sqrt{2bc}} \operatorname{arc\,tg} x \sqrt{\frac{b}{2c}}. \qquad b^2 = 4ac.$$

$$\int \frac{x}{ax^4 + bx^2 + c}\, dx = \frac{1}{2\sqrt{b^2 - 4ac}} \log \frac{2ax^2 + b - \sqrt{b^2 - 4ac}}{2ax^2 + b + \sqrt{b^2 - 4ac}}. \qquad b^2 > 4ac.$$

$$= \frac{1}{\sqrt{4ac - b^2}} \operatorname{arc\,tg} \frac{2ax^2 + b}{\sqrt{4ac - b^2}}. \qquad b^2 < 4ac.$$

$$= -\frac{1}{2ax^2 + b}. \qquad b^2 = 4ac.$$

$$\int \frac{x^2}{ax^4 + bx^2 + c}\, dx = \frac{A - b}{A} \int \frac{dx}{2ax^2 + b - A} + \frac{A + b}{A} \int \frac{dx}{2ax^2 + b + A}.$$

in which $A = \sqrt{b^2 - 4ac}, \quad b^2 > 4ac.$

$$= \frac{1}{8ak \cos \frac{\varepsilon}{2}} \log \frac{x^2 - 2kx \cos \frac{\varepsilon}{2} + k^2}{x^2 + 2kx \cos \frac{\varepsilon}{2} + k^2} + \frac{1}{4ak \sin \frac{\varepsilon}{2}} \operatorname{arc\,tg} \frac{2kx \sin \frac{\varepsilon}{2}}{k^2 - x^2}.$$

in which $k = \left(\frac{c}{a}\right)^{\frac{1}{4}}, \quad \cos \varepsilon = -\frac{b}{2\sqrt{ac}}, \quad b^2 < 4ac.$

$$= -\frac{2cx}{b(2c + bx^2)} + \frac{2c}{b\sqrt{2bc}} \operatorname{arc\,tg} x \sqrt{\frac{b}{2c}}. \qquad b^2 = 4ac.$$

$$\int \frac{bx - 2ax^3}{ax^4 - bx^2 + c}\, dx = \frac{1}{2} \log \frac{1}{ax^4 - bx^2 + c}.$$

$$\int \frac{x^4}{ax + bx^2 + c}\,dx = \frac{x}{a} - \frac{c}{a}\int \frac{dx}{ax^4 + bx^2 + c} - \frac{b}{a}\int \frac{x^2}{ax^4 + bx^2 + c}\,dx.$$

$$\int \frac{dx}{x^2(a + bx^2 + cx^4)} = -\frac{1}{ax} - \frac{b}{a}\int \frac{dx}{\Phi} - \frac{c}{a}\int \frac{x^2}{\Phi}\,dx.$$

$$\int \frac{dx}{x^4(a + bx^2 + cx^4)} = -\frac{1}{3ax^3} + \frac{b}{a^2x} + \frac{b^2 - ac}{a^2}\int \frac{dx}{\Phi} + \frac{bc}{a^2}\int \frac{x^2}{\Phi}\,dx.$$

$$\boxed{\frac{1}{(ax + b)^2(cx^2 + d)}}$$

$$\int \frac{dx}{(x + b)^2(x^2 + a)} = -\frac{1}{(b^2 + a)(x + b)}$$

$$+ \frac{1}{(b^2 + a)^2}\left\{ b \log \frac{(x + b)^2}{x^2 + a} + \frac{b^2 - a}{\sqrt{a}} \operatorname{arc\,tg} \frac{x}{\sqrt{a}} \right\}.$$

$$\int \frac{x}{(b + x)^2(a + x^2)}\,dx = \frac{b}{(a + b^2)(b + x)}$$

$$+ \frac{1}{(a + b^2)^2}\left\{ \frac{a - b^2}{2} \log \frac{(b + x)^2}{a + x} + 2ab\int \frac{dx}{a + x^2} \right\}.$$

$$\int \frac{x^2}{(x + a)^2(x^2 + a^2)}\,dx = -\frac{1}{2(x + a)} + \frac{1}{4a} \log \frac{x^2 + a^2}{(x + a)^2}.$$

$$\boxed{\frac{1}{(ax + b)^2(cx + d)^2}}$$

$$\int \frac{dx}{(x + a)^2(x + b)^2} = -\frac{1}{(a - b)^2}\left(\frac{1}{x + a} + \frac{1}{x + b}\right) - \frac{2}{(a - b)^3} \log \frac{x + b}{x + a}.$$

$$\int \frac{dx}{(x - a)^2(x - b)^2} = -\frac{2x - a - b}{(a - b)^2(x - a)(x - b)} + \frac{2}{(a - b)^3} \log \frac{x - b}{x - a}$$

$$\int \frac{x}{(x + a)^2(x + b)^2}\,dx = \frac{1}{(a - b)^2}\left\{ \frac{a}{x + a} + \frac{b}{x + b} \right\} + \frac{a + b}{(a - b)^3} \log \frac{x + b}{x + a}.$$

•

$$\int \frac{dx}{(x - a)^2(x - b)(x - c)} = \frac{1}{(a - b)(a - c)(x - a)} + \frac{2a - b - c}{(a - b)^2(a - c)^2} \log(x - a)$$

$$+\frac{\log(x-b)}{(a-b)^2(b-c)}-\frac{\log(x-c)}{(a-c)^2(b-c)}.$$

•

$$\int\frac{dx}{(x-a)^2(x-b)^3}=\frac{(x-a)^2-5(x-a)(x-b)-2(x-b)^2}{2(a-b)^3(x-a)(x-b)^2}$$
$$-\frac{3}{(a-b)^4}\log\frac{x-a}{x-b}.$$

$$\boxed{\frac{1}{(ax+b)^5}}$$

$$\int\frac{x^2}{(ax+b)^5}dx=\frac{1}{a^3}\left\{-\frac{1}{2\phi^2}+\frac{2b}{3\phi^3}-\frac{b^2}{4\phi^4}\right\}$$

$$\int\frac{x^3}{(ax+b)^5}dx=\frac{1}{a^4}\left\{-\frac{1}{\phi}+\frac{3b}{2\phi^2}-\frac{b^2}{\phi^3}+\frac{b^3}{4\phi^4}\right\}.$$

$$\int\frac{x^4}{(ax+b)^5}dx=\frac{1}{a^5}\left\{\frac{4b}{\phi}-\frac{3b^2}{\phi^2}+\frac{4b^3}{3\phi^3}-\frac{b^4}{4\phi^4}+\log\phi\right\}.$$

$$\int\frac{x^5}{(ax+b)^5}dx=\frac{1}{a^6}\left\{\phi-\frac{10b^2}{\phi}+\frac{5b^3}{\phi^2}-\frac{5b^4}{3\phi^3}+\frac{b^5}{4\phi^4}-5b\log\phi\right\}.$$

$$\int\frac{x^6}{(a+bx)^5}dx=\frac{1}{b^7}\left\{\frac{\Phi^2}{2}-6a\phi+\frac{20a^3}{\Phi}-\frac{15a^4}{2\Phi^2}+\frac{2a^5}{\phi^3}-\frac{a^6}{4\Phi^4}+15a^2\log\Phi\right\}.$$

$$\int\frac{dx}{x(a+bx)^5}=\frac{1}{4a\phi^4}+\frac{1}{3a^2\phi^3}+\frac{1}{2a^3\phi^2}+\frac{1}{a^4\phi}-\frac{1}{a^5}\log\frac{\Phi}{x}.$$

$$\int\frac{dx}{x^2(a+bx)^5}=-\frac{b}{4a^2\phi^4}-\frac{2b}{3a^3\phi^3}-\frac{3b}{2a^4\phi^2}-\frac{4b}{a^5\phi}-\frac{1}{a^5x}+\frac{5b}{a^6}\log\frac{\Phi}{x}.$$

$$\int\frac{dx}{x^3(a+bx)^5}=\frac{b^2}{4a^3\phi^4}+\frac{b^2}{a^4\phi^3}+\frac{3b^2}{a^5\phi^2}+\frac{10b^2}{a^6\phi}-\frac{1}{2a^5x^2}+\frac{5b}{a^6x}-\frac{15b^2}{a^7}\log\frac{\Phi}{x}.$$

$$\frac{1}{(ax^5 + b)}$$

$$\int \frac{dx}{ax^5+b} = \frac{2k}{5a}\left\{\frac{1}{2}\log(x+k) - P_0\cos\frac{\pi}{5} + P_1\cos\frac{2\pi}{5} + Q_0\sin\frac{\pi}{5} + Q_1\sin\frac{2\pi}{5}\right\},$$

$$\int \frac{x}{ax^5+b}\,dx = \frac{2k^2}{5a}\left\{-\frac{1}{2}\log(x+k) - P_0\cos\frac{2\pi}{5} + P_1\cos\frac{\pi}{5} + Q_0\sin\frac{2\pi}{5} - Q_1\sin\frac{\pi}{5}\right\},$$

$$\int \frac{x^2}{ax^5+b}\,dx = \frac{2k^3}{5a}\left\{\frac{1}{2}\log(x+k) + P_0\cos\frac{2\pi}{5} - P_1\cos\frac{\pi}{5} + Q_0\sin\frac{2\pi}{5} - Q_1\sin\frac{\pi}{5}\right\},$$

$$\int \frac{x^3}{ax^5+b}\,dx = \frac{2k^4}{5a}\left\{-\frac{1}{2}\log(x+k) + P_0\cos\frac{\pi}{5} - P_1\cos\frac{2\pi}{5} + Q_0\sin\frac{\pi}{5} + Q_1\sin\frac{2\pi}{5}\right\},$$

in which $k = \left(\frac{b}{a}\right)^{\frac{1}{5}}$,

$$P_0 = \frac{1}{2}\log\left(x^2 - 2kx\cos\frac{\pi}{5} + k^2\right), \quad Q_0 = \text{arc tg}\,\frac{x\sin\frac{\pi}{5}}{k - x\cos\frac{\pi}{5}},$$

$$P_1 = \frac{1}{2}\log\left(x^2 + 2kx\cos\frac{2\pi}{5} + k^2\right), \quad Q_1 = \text{arc tg}\,\frac{x\sin\frac{2\pi}{5}}{k + x\cos\frac{2\pi}{5}}.$$

$$\int \frac{x^4}{x^5+a^5}\,dx = \frac{1}{5}\log(x^5 + a^5).$$

$$\int \frac{x^5}{ax^5+b}\,dx = \frac{x}{a} - \frac{b}{a}\int\frac{dx}{ax^5+b}.$$

$$\int \frac{x^6}{a+bx^5}\,dx = \frac{x^2}{2b} - \frac{a}{b}\int\frac{x}{\Phi}\,dx.$$

$$\int \frac{x^7}{a+bx^5}\,dx = \frac{x^3}{3b} - \frac{a}{b}\int \frac{x^2}{a+bx^5}\,dx.$$

$$\int \frac{x^8}{a+bx^5}\,dx = \frac{x^4}{4b} - \frac{a}{b}\int \frac{x^3}{\Phi}\,dx.$$

$$\int \frac{dx}{x^2(a+bx^5)} = -\frac{1}{ax} - \frac{b}{a}\int \frac{x^3}{\Phi}\,dx.$$

$$\int \frac{dx}{x^3(a+bx^5)} = -\frac{1}{2ax^2} - \frac{b}{a}\int \frac{x^2}{a+bx^5}\,dx.$$

$$\boxed{\frac{1}{a + bx^6}}$$

$$\int \frac{dx}{a+bx^6} = \frac{k}{a}\left\{\frac{1}{4\sqrt{3}}\log\frac{x^2+kx\sqrt{3}+k^2}{x^2-kx\sqrt{3}+k^2} + \frac{1}{6}\,\text{arc tg}\,\frac{kx}{k^2-x^2} + \frac{1}{3}\,\text{arc tg}\,\frac{x}{k}\right\}.$$

$$k = \left(\frac{a}{b}\right)^{\frac{1}{6}}$$

$$\int \frac{dx}{a-bx^6} = \frac{k}{2a}\left\{\frac{1}{6}\log\frac{x^2+kx+k^2}{x^2-kx+k^2} + \frac{1}{3}\log\frac{x+k}{x-k} + \frac{1}{\sqrt{3}}\,\text{arc tg}\,\frac{kx\sqrt{3}}{k^2-x^2}\right\}.$$

$$k = \left(\frac{a}{b}\right)^{\frac{1}{6}}.$$

$$\int \frac{x}{a+bx^6}\,dx = \frac{1}{2}\int \frac{dz}{a+bz^3}. \qquad z = x^2.$$

$$\int \frac{x^2}{a^6+x^6}\,dx = \frac{1}{3a^3}\,\text{arc tg}\left(\frac{x}{a}\right)^3.$$

$$\int \frac{x^2}{a^2-x^6}\,dx = \frac{1}{6a}\log\frac{x^3+a}{x^3-a}$$

$$\int \frac{x^2}{a+bx^6}\,dx = \frac{1}{3}\int \frac{dz}{a+bz^2} \qquad z = x^3.$$

$$\int \frac{x^3}{a+bx^6}\,dx = \frac{1}{2}\int \frac{z}{a+bz^3}\,dz. \qquad z = x^2$$

$$\int \frac{x^4}{a+bx^6}\,dx = \frac{k^5}{a}\left\{-\frac{1}{4\sqrt{3}}\log\frac{x^2+kx\sqrt{3}+k^2}{x^2-kx\sqrt{3}+k^2}+\frac{1}{6}\operatorname{arc\,tg}\frac{kx}{k^2-x^2}\right.$$

$$\left.+\frac{1}{3}\operatorname{arc\,tg}\frac{x}{k}\right\},$$

in which $k=\left(\frac{a}{b}\right)^{\frac{1}{6}}$

$$\boxed{\frac{1}{(ax^3+b)^2}}$$

$$\int \frac{dx}{(a+bx^3)^2} = \frac{x}{3a(a+bx^3)} + \frac{2}{3a}\int \frac{dx}{a+bx^3}.$$

$$\int \frac{x}{(a+bx^3)^2}\,dx = \frac{x^2}{3a\Phi} + \frac{1}{3a}\int \frac{x}{\Phi}\,dx.$$

$$\int \frac{x^2}{(a+bx^3)^2}\,dx = -\frac{1}{3b(a+bx^3)} = \frac{x^3}{3a(a+bx^3)}$$

$$\int \frac{x^3}{(a+bx^3)^2}\,dx = -\frac{x}{3b\Phi} + \frac{1}{3b}\int \frac{dx}{\Phi}.$$

$$\int \frac{x^4}{(a+bx^3)^2}\,dx = -\frac{x^2}{3b\Phi} + \frac{2}{3b}\int \frac{x}{\Phi}\,dx.$$

$$\int \frac{x^5}{(x^3+a)^2}\,dx = \frac{a}{3\Phi} + \frac{1}{3}\log \Phi.$$

$$\int \frac{x^6}{(a+bx^3)^2}\,dx = \frac{x}{b^2} + \frac{ax}{3b^2\Phi} - \frac{4a}{3b^2}\int \frac{dx}{\Phi}.$$

$$\int \frac{x^7}{(a+bx^3)^2}\,dx = \frac{x^2}{2b^2} + \frac{ax^2}{3b^2\Phi} - \frac{5a}{3b^2}\int \frac{x}{\Phi}\,dx.$$

$$\int \frac{dx}{x(a+bx^3)^2} = \frac{1}{3a(a+bx^3)} - \frac{1}{3a^2}\log\frac{a+bx^3}{x^3}.$$

$$\int \frac{dx}{x^2(a+bx^3)^2} = -\frac{1}{a^2x} - \frac{bx^2}{3a^2\Phi} - \frac{4b}{3a^2}\int \frac{x}{\Phi}\,dx.$$

$$\int \frac{dx}{x^3(a+bx^3)^2} = -\frac{1}{2a^2x^2} - \frac{bx}{3a^2\Phi} - \frac{5b}{3a^2}\int\frac{dx}{\Phi}.$$

$\frac{1}{(ax^2 + b)^3}$

$$\int \frac{dx}{(a^2+x^2)^3} = \frac{x(5a^2+3x^2)}{8a^4\Phi^2} + \frac{3}{8a^5}\operatorname{arc\,tg}\frac{x}{a}.$$

$$\int \frac{dx}{(a^2-x^2)^3} = \frac{x(5a^2-3x^2)}{8a^4\Phi^2} + \frac{3}{16a^5}\log\frac{a+x}{a-x}.$$

$$\int \frac{dx}{(a+bx^2)^3} = \frac{x}{4a\Phi^2} + \frac{3x}{8a^2\Phi} + \frac{3}{8a^2}\int\frac{dx}{\Phi}.$$

$$\int \frac{x}{(a+bx^2)^3}\,dx = -\frac{1}{4b(a+bx^2)^2}.$$

$$\int \frac{x^2}{(a^2+x^2)^3}\,dx = \frac{x}{8a^2\Phi} - \frac{x}{4\Phi^2} + \frac{1}{8a^3}\operatorname{arc\,tg}\frac{x}{a}.$$

$$\int \frac{x^2}{(x^2-a^2)^3}\,dx = \frac{x}{8a^2\Phi} - \frac{x^3}{4a^2\Phi^2} - \frac{1}{16a^3}\log\frac{x-a}{x+a}.$$

$$\int \frac{x^2}{(a+bx^2)^3}\,dx = \frac{x}{8ab\Phi} - \frac{x}{4b\Phi^2} + \frac{1}{8ab}\int\frac{dx}{\Phi}.$$

$$\int \frac{x^3}{(a+bx^2)^3}\,dx = \frac{x^4}{4a\Phi^2} = -\frac{a+2bx^2}{4b^2\Phi^2}.$$

$$\int \frac{x^4}{(a+bx^2)^3}\,dx = \frac{ax}{4b^2\Phi^2} - \frac{5x}{8b^2\Phi} + \frac{3}{8b^2}\int\frac{dx}{\Phi}.$$

$$\int \frac{x^3}{(x^2-a)^3}\,dx = -\int\frac{(1+az^2)^2}{z}\,dz. \qquad x^2 - a = \frac{1}{z^2}.$$

$$\int \frac{x^6}{(a+bx^2)^3}\,dx = \frac{x}{b^3}\left\{1 + \frac{9a}{8\Phi} - \frac{a^2}{4\Phi^2}\right\} - \frac{15a}{8b^3}\int\frac{dx}{\Phi}.$$

$$\int \frac{dx}{x^2(a+bx^2)^3} = -\frac{1}{a^3x} - \frac{bx}{a^2}\left\{\frac{1}{4\Phi^2} + \frac{7}{8a\Phi}\right\} - \frac{15b}{8a^3}\int\frac{dx}{\Phi}.$$

$$\frac{1}{(a + bx + cx^2)^3}$$

$$\int \frac{dx}{(a + bx + cx^2)^3} = \frac{2cx + b}{4ac - b^2} \left\{ \frac{1}{2\Phi^2} + \frac{3c}{(4ac - b^2)\Phi} \right\} + \frac{6c^2}{(4ac - b^2)^2} \int \frac{dx}{\Phi}.$$

$$\int \frac{x}{(a + bx + cx^2)^3}\, dx = -\frac{1}{4c\Phi^2} - \frac{b}{2c} \int \frac{dx}{\Phi^3}.$$

$$= -\frac{bx + 2a}{2(4ac - b^2)\Phi^2} - \frac{3b(2cx + b)}{2(4ac - b^2)^2\Phi} - \frac{3bc}{(4ac - b^2)^2} \int \frac{dx}{\Phi}.$$

$$\int \frac{x^2}{(a + bx + cx^2)^3}\, dx = \frac{(b^2 - 2ac)x + ab}{2c(4ac - b^2)\Phi^2} + \frac{(2ac + b^2)(2cx + b)}{2c(4ac - b^2)^2\Phi}$$

$$+ \frac{2ac + b^2}{(4ac - b^2)^2} \int \frac{dx}{\Phi}.$$

$$\int \frac{x^3}{(a + bx + cx^2)^3}\, dx = \frac{(3ac - b^2)bx + (2ac - b^2)a}{c^2(4ac - b^2)\Phi} + \frac{1}{2c^3} \log \Phi$$

$$- \frac{b(6ac - b^2)}{2c^2(4ac - b^2)} \int \frac{dx}{\Phi}$$

$$\int \frac{x^4}{(a + bx + cx^2)^3}\, dx = -\left\{ \frac{ax}{c^2} + \frac{bx^2}{2c^2} + \frac{x^3}{c} \right\} \frac{1}{\Phi^2} + \frac{a^2}{c^2} \int \frac{dx}{\Phi^3}.$$

$$\int \frac{dx}{x(a + bx + cx^2)^3} = \frac{1}{2a^2\Phi}\left(1 + \frac{a}{2\Phi}\right) - \frac{b(2cx + b)}{2a^2(4ac - b^2)\Phi} \left\{ 1 + 3q + \frac{a}{2\Phi} \right\}$$

$$+ \frac{1}{2a^3} \log \frac{x^2}{\Phi} - \frac{b}{2a^3} \left\{ 1 + 2q + 6q^2 \right\} \int \frac{dx}{\Phi},$$

in which $q = \dfrac{ac}{4ac - b^2}$.

$$\frac{1}{a + bx^3 + cx^6}$$

$$\int \frac{dx}{a + bx^3 + cx^6} = \frac{2c}{k} \int \frac{dx}{2cx^3 + A} - \frac{2c}{k} \int \frac{dx}{2cx^3 + B},$$

$$\int \frac{x}{a + bx^3 + cx^6}\, dx = \frac{2c}{k} \int \frac{x}{2cx^3 + A}\, dx - \frac{2c}{k} \int \frac{x}{2cx^3 + B}\, dx,$$

$$\int \frac{x^2}{x^6 - 10x^3 + 9}\, dx = \frac{1}{24} \log \frac{x^3 - 9}{x^3 - 1}.$$

$$\int \frac{x^2}{ax^6 + bx^3 + c}\, dx = \frac{1}{3\sqrt{b^2 - 4ac}} \log \frac{2ax^3 + b - \sqrt{b^2 - 4ac}}{2ax^3 + b + \sqrt{b^2 - 4ac}}.$$

$$= \frac{2}{3\sqrt{4ac - b^2}} \operatorname{arc\,tg} \frac{2ax^3 + b}{\sqrt{4ac - b^2}}.$$

$$\int \frac{x^3}{a + bx^3 + cx^6}\, dx = \frac{k - b}{k} \int \frac{dx}{2cx^3 + A} + \frac{k + b}{k} \int \frac{dx}{2cx^3 + B},$$

$$\int \frac{x^4}{a + bx^3 + cx^6}\, dx = \frac{k - b}{k} \int \frac{x}{2cx^3 + A}\, dx + \frac{k + b}{k} \int \frac{x}{2cx^3 + B}\, dx,$$

in which $k = \sqrt{b^2 - 4ac}$, $A = b - k$, $B = b + k$, $b^2 > 4ac$.

$$\frac{1}{(a + bx)^6}$$

$$\int \frac{x^2}{(a + bx)^6}\, dx = \frac{1}{b^3} \left\{ -\frac{1}{3\Phi^3} + \frac{a}{2\Phi^4} - \frac{a^2}{5\Phi^5} \right\}.$$

$$\int \frac{x^3}{(a + bx)^6}\, dx = \frac{1}{b^4} \left\{ -\frac{1}{2\Phi^2} + \frac{a}{\Phi^3} - \frac{3a^2}{4\Phi^4} + \frac{a^3}{5\Phi^5} \right\}.$$

$$\int \frac{x^4}{(a + bx)^6}\, dx = \frac{1}{b^5} \left\{ -\frac{1}{\Phi} + \frac{2a}{\Phi^2} - \frac{2a^2}{\Phi^3} + \frac{a^3}{\Phi^4} - \frac{a^4}{5\Phi^5} \right\}.$$

$$\int \frac{x^5}{(a + bx)^6}\, dx = \frac{1}{b^6} \left\{ \frac{5a}{\Phi} - \frac{5a^2}{\Phi^2} + \frac{10a^3}{3\Phi^3} - \frac{5a^4}{4\Phi^4} + \frac{a^5}{5\Phi^5} + \log \Phi \right\}.$$

$$\int \frac{x^6}{(a + bx)^6}\, dx = \frac{1}{b^7} \left\{ \Phi - \frac{15a^2}{\Phi} + \frac{10a^3}{\Phi^2} - \frac{5a^4}{\Phi^3} + \frac{3a^5}{2\Phi^4} - \frac{a^6}{5\Phi^5} - 6a \log \Phi \right\}.$$

$$\int \frac{dx}{x^2 (a + bx)^6} = -\frac{b}{5a^2\Phi^5} - \frac{b}{2a^3\Phi^4} - \frac{b}{a^4\Phi^3} - \frac{2b}{a^5\Phi^2} - \frac{5b}{a^6\Phi} - \frac{1}{a^6 x} + \frac{6b}{a^7} \log$$

$$\frac{1}{(a + bx)^7}$$

$$\int \frac{x}{(a+bx)^7}\,dx = \frac{1}{b^2}\left\{-\frac{1}{5\Phi^5} + \frac{a}{6\Phi^6}\right\}.$$

$$\int \frac{x^2}{(a+bx)^7}\,dx = \frac{1}{b^3}\left\{-\frac{1}{4\Phi^4} + \frac{2a}{5\Phi^5} - \frac{a^2}{6\Phi^6}\right\}.$$

$$\int \frac{x^3}{(a+bx)^7}\,dx = \frac{1}{b^4}\left\{-\frac{1}{3\Phi^3} + \frac{3a}{4\Phi^4} - \frac{3a^2}{5\Phi^5} + \frac{a^3}{6\Phi^6}\right\}.$$

$$\int \frac{x^4}{(a+bx)^7}\,dx = \frac{1}{b^5}\left\{-\frac{1}{2\Phi^2} + \frac{4a}{3\Phi^3} - \frac{3a^2}{2\Phi^4} + \frac{4a^3}{5\Phi^5} - \frac{a^4}{6\Phi^6}\right\}.$$

$$\int \frac{x^5}{(a+bx)^7}\,dx = \frac{1}{b^6}\left\{-\frac{1}{\Phi} + \frac{5a}{2\Phi^2} - \frac{10a^2}{3\Phi^3} + \frac{5a^3}{2\Phi^4} - \frac{a^4}{\Phi^5} + \frac{a^5}{6\Phi^6}\right\}.$$

$$\int \frac{x^6}{(a+bx)^7}\,dx = \frac{1}{b^7}\left\{\frac{6a}{\Phi} - \frac{15a^2}{2\Phi^2} + \frac{20a^3}{3\Phi^3} - \frac{15a^4}{4\Phi^4} + \frac{6a^5}{5\Phi^5} - \frac{a^6}{6\Phi^6} + \log\Phi\right\}.$$

$$\int \frac{dx}{x\,(a+bx)^7} = \frac{1}{6a\Phi^6} + \frac{1}{5a^2\Phi^5} + \frac{1}{4a^3\Phi^4} + \frac{1}{3a^4\Phi^3} + \frac{1}{2a^5\Phi^2} + \frac{1}{a^6\Phi} - \frac{1}{a^7}\log\frac{\Phi}{x}.$$

$$\frac{1}{(a + bx^2)^4}$$

$$\int \frac{dx}{(a^2+x^2)^4} = \frac{x}{6a^2\Phi^3} + \frac{5x}{24a^2\Phi^2} + \frac{5x}{16a^6\Phi} + \frac{5}{16a^7}\,\text{arc tg}\,\frac{x}{a}.$$

$$\int \frac{dx}{(a+bx^2)^4} = \frac{x}{6a\Phi^3} + \frac{5x}{24a^2\Phi^2} + \frac{5x}{16a^3\Phi} + \frac{5}{16a^3}\int\frac{dx}{\Phi}.$$

$$\int \frac{x^2}{(1+x^2)^4}\,dx = \frac{x}{16\Phi} + \frac{x}{24\Phi^2} - \frac{x}{6\Phi^3} + \frac{1}{16}\,\text{arc tg}\,x$$

$$= \frac{3x^5 + 8x^3 - 3x}{48\Phi^3} + \frac{1}{16}\,\text{arc tg}\,x.$$

$$\int \frac{x^3}{(a+bx^2)^4}\,dx = -\frac{1}{a^2}\int \frac{z^2-b}{z^7}\,dz. \qquad a+bx^2 = z^2x^2.$$

$$\int \frac{x^4}{(a+bx^2)^4}\,dx = \frac{x}{b^2}\left\{\frac{a}{6\Phi^3} - \frac{7}{24\Phi^2} + \frac{1}{16a\Phi}\right\} + \frac{1}{16ab^2}\int \frac{dx}{\Phi}.$$

$$\int \frac{x^6}{(a+bx^2)^4}\,dx = \frac{x}{b^3}\left\{-\frac{a^2}{6\Phi^3} + \frac{13a}{24\Phi^2} - \frac{11}{16\Phi}\right\} + \frac{5}{16b^3}\int \frac{dx}{\Phi}.$$

$$\int \frac{x^8}{(a+bx^2)^4}\,dx = \frac{x}{b^4} + \frac{x}{b^4}\left\{\frac{a^3}{6\Phi^3} - \frac{19a^2}{24\Phi^2} + \frac{29a}{16\Phi}\right\} - \frac{35a}{16b^4}\int \frac{dx}{\Phi}.$$

$$\frac{1}{(a+bx^4)^2}$$

$$\int \frac{dx}{(a+bx^4)^2} = \frac{x}{4a\Phi} + \frac{3}{4a}\int \frac{dx}{\Phi}.$$

$$\int \frac{x}{(ax^4+b)^2}\,dx = \frac{x^2}{4b\Phi} + \frac{1}{2b}\int \frac{x}{\Phi}\,dx.$$

$$\int \frac{x^2}{(a+bx^4)^2}\,dx = \frac{x^3}{4a\Phi} + \frac{1}{4a}\int \frac{x^2}{\Phi}\,dx.$$

$$\int \frac{x^4}{(a+bx^4)^2}\,dx = -\frac{x}{4b\Phi} + \frac{1}{4b}\int \frac{dx}{\Phi}.$$

$$\int \frac{x^6}{(a+bx^4)^2}\,dx = -\frac{x^3}{4b\Phi} + \frac{3}{4b}\int \frac{x^2}{\Phi}\,dx.$$

$$\int \frac{x^8}{(a+bx^4)^2}\,dx = \frac{x}{b^2} + \frac{ax}{4b^2\Phi} - \frac{5a}{4b^2}\int \frac{dx}{\Phi}.$$

$$\frac{1}{(a+bx+cx^2)^4}$$

$$\int \frac{dx}{(a+bx+cx^2)^4} = \frac{2cx+b}{4ac-b^2}\left\{\frac{1}{3\Phi^3} + \frac{5c}{3(4ac-b^2)\Phi^2} + \frac{10c^2}{(4ac-b^2)^2\Phi}\right\} + \frac{20c^3}{(4ac-b^2)^3}\int \frac{dx}{\Phi}.$$

$$\int \frac{x}{(a+bx+cx^2)^4}\,dx = -\frac{bx+2a}{3(4ac-b^2)\Phi^3}$$

$$-\frac{5b(2cx+b)}{3(4ac-b^2)^2}\left\{\frac{1}{2\Phi^2}+\frac{3c}{(4ac-b^2)\Phi}\right\}-\frac{10bc^2}{(4ac-b^2)^3}\int\frac{dx}{\Phi}.$$

$$\int \frac{x^2}{(a+bx+cx^2)^4}\,dx = \frac{(b^2-2ac)x+ab}{3c(4ac-b^2)\Phi^3}$$

$$+\frac{2(ac+b^2)(2cx+b)}{3c(4ac-b^2)^2}\left\{\frac{1}{2\Phi^2}+\frac{3c}{(4ac-b^2)\Phi}\right\}$$

$$+\frac{4c(ac+b^2)}{(4ac-b^2)^3}\int\frac{dx}{\Phi}.$$

$$\int \frac{x^3}{(a+bx+cx^2)^4}\,dx = \left\{-\frac{x^2}{c}-\frac{b^3x}{3c^2(4ac-b^2)}\right.$$

$$\left.-\frac{a(4ac+b^2)}{3c^2(4ac-b^2)}\right\}\frac{1}{4\Phi^3}-\frac{b(6ac+b^2)}{6c^2(4ac-b^2)}\int\frac{dx}{\Phi^3}$$

$$\int \frac{x^4}{(a+bx+cx^2)^4}\,dx = -\left\{\frac{x^3}{3c}+\frac{ax}{5c^2}-\frac{ab}{15c^3}\right\}\frac{1}{\Phi^3}+\frac{a^2c+ab^2}{5c^3}\int\frac{dx}{\Phi^4}.$$

$$\int \frac{x^5}{(a+bx+cx^2)^4}\,dx = -\left\{\frac{x^4}{2c}+\frac{bx^3}{6c^2}+\frac{ax^2}{2c^2}+\frac{a^2}{6c^3}\right\}\frac{1}{\Phi^3}-\frac{a^2b}{2c^3}\int\frac{dx}{\Phi^4}.$$

•

$$\int \frac{dx}{(a+bx^2+cx^4)^2} = -\frac{bcx^3+(b^2-2ac)x}{2a(4ac-b^2)\Phi}-\frac{bc}{2a(4ac-b^2)}\int\frac{x^2}{\Phi}\,dx$$

$$+\frac{6ac-b^2}{2a(4ac-b^2)}\int\frac{dx}{\Phi}.$$

$$\int \frac{x^6}{(a+bx^2+cx^4)^2}\,dx = \frac{(b^2-2ac)x^3+abx}{2c(4ac-b^2)\Phi}-\frac{ab}{2c(4ac-b^2)}\int\frac{dx}{\Phi}$$

$$+\frac{6ac-b^2}{2c(4ac-b^2)}\int\frac{x^2}{\Phi}\,dx.$$

$$\frac{1}{(a + bx)^8}$$

$$\int \frac{x^2}{(a+bx)^8}\,dx = \frac{1}{b^3}\left\{-\frac{1}{5\Phi^5} + \frac{a}{3\Phi^6} - \frac{a^2}{7\Phi^7}\right\}.$$

$$\int \frac{x^3}{(a+bx)^8}\,dx = \frac{1}{b^4}\left\{-\frac{1}{4\Phi^4} + \frac{3a}{5\Phi^5} - \frac{a^2}{2\Phi^6} + \frac{c^3}{7\Phi^7}\right\}.$$

$$\int \frac{x^4}{(a+bx)^8}\,dx = \frac{1}{b^5}\left\{-\frac{1}{3\Phi^3} + \frac{a}{\Phi^4} - \frac{6a^2}{5\Phi^5} + \frac{2a^3}{3\Phi^6} - \frac{a^4}{7\Phi^7}\right\}.$$

$$\int \frac{x^5}{(a+bx)^8}\,dx = \frac{1}{b^6}\left\{-\frac{1}{2\Phi^2} + \frac{5a}{3\Phi^3} - \frac{5a^2}{2\Phi^4} + \frac{2a^3}{\Phi^5} - \frac{5a^4}{6\Phi^6} + \frac{a^5}{7\Phi^7}\right\}.$$

$$\int \frac{x^6}{(a+bx)^8}\,dx = \frac{1}{b^7}\left\{-\frac{1}{\Phi} + \frac{3a}{\Phi^2} - \frac{5a^2}{\Phi^3} + \frac{5a^3}{\Phi^4} - \frac{3a^4}{\Phi^5} + \frac{a^5}{\Phi^6} - \frac{a^6}{7\Phi^7}\right\}.$$

$$\frac{1}{(a + bx^3)^3}$$

$$\int \frac{dx}{(a+bx^3)^3} = \frac{x}{a}\left\{\frac{1}{6\Phi^2} + \frac{5}{18a\Phi}\right\} + \frac{5}{9a^2}\int \frac{dx}{\Phi}.$$

$$\int \frac{x}{(a+bx^3)^3}\,dx = \frac{x^2}{a}\left\{\frac{1}{6\Phi^2} + \frac{2}{9a\Phi}\right\} + \frac{2}{9a^2}\int \frac{x}{\Phi}\,dx.$$

$$\int \frac{x^3}{(a+bx^3)^3}\,dx = -\frac{x}{b}\left\{\frac{1}{6\Phi^2} - \frac{1}{18a\Phi}\right\} + \frac{1}{9ab}\int \frac{dx}{\Phi}.$$

$$\int \frac{x^4}{(a+bx^3)^3}\,dx = -\frac{x^2}{b}\left\{\frac{1}{6\Phi^2} - \frac{1}{9a\Phi}\right\} + \frac{1}{9ab}\int \frac{x}{\Phi}\,dx.$$

SQUARE ROOTS

$\sqrt{X}$

$$\int \sqrt{ax+b}\,dx = \frac{2}{3a}(ax+b)^{\frac{3}{2}}.$$

$$\int \frac{dx}{\sqrt{ax+b}} = \frac{2}{a}\sqrt{ax+b}.$$

$$\int \sqrt{x^2+a^2}\,dx = \frac{x}{2}\sqrt{x^2+a^2} + \frac{a^2}{2}\log\left(x+\sqrt{x^2+a^2}\right).$$

$$\int \sqrt{x^2-a^2}\,dx = \frac{x}{2}\sqrt{x^2-a^2} - \frac{a^2}{2}\log\left(x+\sqrt{x^2-a^2}\right)$$

$$= \frac{(x+\sqrt{x^2-a^2})^4 - a^4}{8(x+\sqrt{x^2-a^2})^2} - \frac{a^2}{2}\log\left(x+\sqrt{x^2-a^2}\right).$$

$$\int \sqrt{a^2-x^2}\,dx = \frac{x}{2}\sqrt{a^2-x^2} + \frac{a^2}{2}\text{arc sin}\,\frac{x}{a}.$$

$$\int \sqrt{ax^2+b}\,dx = \frac{x}{2}\sqrt{ax^2+b} + \frac{b}{2}\int\frac{dx}{\sqrt{ax^2+b}}$$

$$= \frac{x}{2}\sqrt{ax^2+b} + \frac{b}{2\sqrt{a}}\log\left(x\sqrt{a}+\sqrt{ax^2+b}\right).$$

$$\int \sqrt{b-ax^2}\,dx = \frac{x}{2}\sqrt{b-ax^2} + \frac{b}{2\sqrt{a}}\text{arc sin}\,x\sqrt{\frac{a}{b}}.$$

$$\int \frac{dx}{\sqrt{ax^2+b}} = \frac{1}{\sqrt{a}}\log\left(x\sqrt{a}+\sqrt{ax^2+b}\right) = \frac{1}{2\sqrt{a}}\log\frac{\sqrt{ax^2+b}+x\sqrt{a}}{\sqrt{ax^2+b}-x\sqrt{a}}$$

$$= -\frac{1}{\sqrt{a}}\log\left(\sqrt{ax^2+b}-x\sqrt{a}\right).$$

$$\int \frac{dx}{\sqrt{b-ax^2}} = \frac{1}{\sqrt{a}}\text{arc sin}\,x\sqrt{\frac{a}{b}} = \frac{1}{\sqrt{a}}\text{arc cos}\sqrt{\frac{b-ax^2}{b}}$$

$$= \frac{2}{a}\text{arc tg}\sqrt{\frac{b+ax}{b-ax}}.$$

$$\int \frac{dx}{\sqrt{x^2+a}} = \log\left(x+\sqrt{x^2+a}\right) = \frac{1}{2}\log\frac{x+\sqrt{x^2+a}}{x-\sqrt{x^2+a}}$$

$$= \log \frac{x - \sqrt{a} + \sqrt{x^2 + a}}{x + \sqrt{a} - \sqrt{x^2 + a}}.$$

$$\int \frac{dx}{\sqrt{x^2 - a^2}} = \log\left(x + \sqrt{x^2 - a^2}\right) = \log\left\{\frac{x}{a} + \sqrt{\frac{x^2}{a^2} - 1}\right\}$$

$$= -2 \log\left(\sqrt{x + a} - \sqrt{x - a}\right).$$

$$\int \sqrt{ax^2 + bx}\, dx = \frac{2ax + b}{4a}\sqrt{ax^2 + bx}$$

$$- \frac{b^2}{8a\sqrt{a}} \log\left(2ax + b + 2\sqrt{a}\ \sqrt{ax^2 + bx}\right).$$

$$\int \sqrt{bx - ax^2}\, dx = \frac{2ax - b}{4a}\sqrt{bx - ax^2} + \frac{b^2}{8a\sqrt{a}} \text{arc sin} \frac{2ax - b}{b}.$$

$$\int \frac{dx}{\sqrt{x^2 + bx}} = \log\left(x + \frac{b}{2} + \sqrt{x^2 + bx}\right).$$

$$\int \frac{dx}{\sqrt{ax - x^2}} = \text{arc sin} \frac{2x - a}{a} = 2 \text{ arc sin} \sqrt{\frac{x}{a}} = \text{arc cos} \frac{a - 2x}{a}.$$

$$\int \frac{dx}{\sqrt{x - ax^2}} = -\frac{2}{\sqrt{a}} \text{arc sin} \sqrt{1 - ax} = -\frac{1}{\sqrt{a}} \text{arc sin}\,(1 - 2ax).$$

$$\int \frac{dx}{\sqrt{ax^2 + bx}} = \frac{2}{\sqrt{a}} \log\left(\sqrt{ax} + \sqrt{ax + b}\right) = \frac{1}{\sqrt{a}} \log \frac{\sqrt{\Phi} + x\sqrt{a}}{\sqrt{\Phi} - x\sqrt{a}}$$

$$= \frac{1}{\sqrt{a}} \log\left\{2ax + b + 2\sqrt{a}\sqrt{\Phi}\right\}.$$

$$\int \frac{dx}{\sqrt{bx - ax^2}} = \frac{1}{\sqrt{a}} \text{arc sin} \frac{2ax - b}{b} = \frac{2}{\sqrt{a}} \text{arc sin} \sqrt{\frac{ax}{b}}$$

$$= \frac{2}{\sqrt{a}} \text{arc tg} \sqrt{\frac{ax}{b - ax}} = \frac{2}{\sqrt{a}} \text{arc cos} \sqrt{\frac{b - ax}{b}}.$$

$$\int \sqrt{ax^2 + bx + c}\, dx = \frac{2ax + b}{4a}\sqrt{\Phi} + \frac{4ac - b^2}{8a} \int \frac{dx}{\sqrt{\Phi}} = \frac{2ax + b}{4a}\sqrt{\Phi}$$

$$+ \frac{4ac - b^2}{8a\sqrt{a}} \log\left(2ax + b + 2\sqrt{a\Phi}\right).$$

$$\int \sqrt{c + bx - ax^2}\, dx = \frac{2ax - b}{4a}\sqrt{\Phi} + \frac{4ac + b^2}{8a\sqrt{a}} \text{arc sin} \frac{2ax - b}{\sqrt{4ac + b^2}}.$$

$$\int \sqrt{(x - a)(b - x)}\, dx = \frac{2x - a - b}{4}\sqrt{\Phi} + \frac{(b - a)^2}{8} \text{arc sin} \frac{2x - a - b}{b - a}.$$

$$\int \frac{dx}{\sqrt{(x-a)(x-b)}} = 2 \log \left\{ \sqrt{x-a} + \sqrt{x-b} \right\}$$

$$= -2 \log \left\{ \sqrt{a-x} + \sqrt{b-x} \right\} = \log \frac{\sqrt{\Phi} + x - a}{\sqrt{\Phi} - x + a} \,..$$

$$\int \frac{dx}{\sqrt{(x-a)(b-x)}} = 2 \operatorname{arc\,tg} \sqrt{\frac{x-a}{b-x}} = -2 \operatorname{arc\,tg} \sqrt{\frac{b-x}{x-a}}$$

$$= -2 \operatorname{arc\,sin} \sqrt{\frac{b-x}{b-a}} \,.$$

$$\int \frac{dx}{\sqrt{(ax+b)(n-ax)}} = \frac{n}{a} \operatorname{arc\,tg} \sqrt{\frac{ax+b}{n-ax}} = \frac{n}{a} \operatorname{arc\,sin} \sqrt{\frac{ax+b}{b+n}} \,.$$

$$\int \frac{dx}{\sqrt{(b-ax)(mx+n)}} = \frac{2}{\sqrt{am}} \operatorname{arc\,tg} \sqrt{\frac{a(mx+n)}{m(b-ax)}} \,.$$

$$\int \frac{dx}{\sqrt{ax^2+bx+c}} = \frac{1}{\sqrt{a}} \log \left(b + 2ax + 2\sqrt{a\Phi} \right)$$

$$= -\frac{1}{\sqrt{a}} \log \left(b + 2ax - 2\sqrt{a\Phi} \right).$$

$$\int \frac{dx}{\sqrt{c+bx-ax^2}} = \frac{1}{\sqrt{a}} \operatorname{arc\,sin} \frac{2ax-b}{\sqrt{b^2+4ac}} = \frac{1}{\sqrt{a}} \operatorname{arc\,tg} \frac{2ax-b}{2\sqrt{a\Phi}}$$

$$= -\frac{2}{\sqrt{a}} \operatorname{arc\,tg} \frac{\sqrt{c}+\sqrt{\Phi}}{x\sqrt{a}} = \frac{2}{\sqrt{a}} \operatorname{arc\,tg} \sqrt{\frac{\sqrt{b^2+4ac}+2ax-b}{\sqrt{b^2+4ac}-2ax+b}}$$

$$= -\frac{2}{\sqrt{a}} \operatorname{arc\,tg} \sqrt{\frac{\alpha-x}{x-\beta}} = -\frac{2}{\sqrt{a}} \operatorname{arc\,sin} \sqrt{\frac{\alpha-x}{\alpha-\beta}} \,.$$

α, β being the roots of $+\, bx - ax^2 = 0$.

$$\sqrt{\frac{X_1}{X_2}}$$

$$\int \sqrt{\frac{x}{ax+b}}\, dx = \frac{1}{a} \sqrt{ax^2+bx} - \frac{b}{a\sqrt{a}} \log \left(\sqrt{ax} + \sqrt{ax+b} \right).$$

$$\int \sqrt{\frac{x}{b-ax}}\, dx = -\frac{1}{a} \sqrt{bx-ax^2} - \frac{b}{a\sqrt{a}} \operatorname{arc\,sin} \sqrt{\frac{b-ax}{b}}$$

$$\int \sqrt{\frac{b-ax}{x}}\, dx = \sqrt{bx-ax^2} - \frac{b}{\sqrt{a}} \operatorname{arc\,sin} \sqrt{\frac{b-ax}{b}} \,.$$

$$\int\sqrt{\frac{ax+b}{x}}\,dx=\sqrt{ax^2+bx}+\frac{b}{\sqrt{a}}\log\left(\sqrt{ax}+\sqrt{ax+b}\right).$$

$$\int\sqrt{\frac{ax+b}{mx+n}}\,dx=\frac{1}{m}\sqrt{amx^2+(an+bm)x+bn}$$

$$+\frac{bm-an}{m\sqrt{am}}\log\left\{\sqrt{amx+an}+\sqrt{amx+bm}\right\}.$$

$$\int\sqrt{\frac{ax+b}{n-mx}}\,dx=-\frac{1}{m}\sqrt{bn+(an-bm)x-amx^2}$$

$$+\frac{an+bm}{m\sqrt{am}}\text{arc sin}\sqrt{\frac{bm+amx}{an+bm}}.$$

$$\int\sqrt{\frac{b-ax}{mx+n}}\,dx=\frac{1}{m}\sqrt{bn+(bm-an)x-amx^2}$$

$$-\frac{an+bm}{m\sqrt{am}}\text{arc sin}\sqrt{\frac{bm-amx}{an+bm}}.$$

$$\int\sqrt{\frac{ax^2+b}{mx^2+n}}\,dx=\int\frac{ax^2+b}{\sqrt{(ax^2+b)(mx^2+n)}}\,dx,\quad\text{elliptic integral.}$$

$x^m\sqrt{X}$

$$\int x^m\sqrt{ax^2+bx}\,dx=\frac{x^{m+1}}{m+1}\sqrt{\Phi}-\frac{b}{2(m+1)}\int\frac{x^{m+1}}{\sqrt{\Phi}}\,dx$$

$$-\frac{a}{m+1}\int\frac{x^{m+2}}{\sqrt{\Phi}}\,dx.$$

$$\int\frac{\sqrt{ax^2+bx}}{x^n}\,dx=-\frac{\sqrt{ax^2+bx}}{(n-1)x^{n-1}}+\frac{b}{2(n-1)}\int\frac{dx}{x^{n-1}\sqrt{ax^2+bx}}$$

$$+\frac{a}{n-1}\int\frac{dx}{x^{n-2}\sqrt{ax^2+bx}}$$

$$\int x^m\sqrt{ax^2+bx+c}\,dx=\frac{x^{m+1}}{m+1}\sqrt{\Phi}-\frac{b}{2(m+1)}\int\frac{x^{m+1}}{\sqrt{\Phi}}\,dx$$

$$-\frac{a}{m+1}\int\frac{x^{m+2}}{\sqrt{\Phi}}\,dx.$$

$$\int\frac{\sqrt{ax^2+bx+c}}{x^n}\,dx=-\frac{\sqrt{\Phi}}{(n-1)x^{n-1}}+\frac{b}{2(n-1)}\int\frac{dx}{x^{n-1}\sqrt{\Phi}}$$

$$+\frac{a}{n-1}\int\frac{dx}{x^{n-2}\sqrt{\Phi}}.$$

$$\int\frac{dx}{x^m\sqrt{ax^2+bx+c}}=-\int\frac{z^{m-1}}{\sqrt{cz^2-bz+a}}dz. \qquad xz=1.$$

$$=-\frac{\sqrt{\Phi}}{c(m-1)x^{m-1}}-\frac{b(2m-3)}{2c(m-1)}\int\frac{dx}{x^{m-1}\sqrt{\Phi}}$$

$$-\frac{a(m-2)}{c(m-1)}\int\frac{dx}{x^{m-2}\sqrt{\Phi}}.$$

$$\int\frac{x^p}{(mx+n)\sqrt{ax^2+b}}dx=\frac{1}{m}\int\frac{x^{p-1}}{\sqrt{\Phi}}dx-\frac{n}{m^2}\int\frac{x^{p-2}}{\sqrt{\Phi}}dx$$

$$+\frac{n^2}{m^3}\int\frac{x^{p-3}}{\sqrt{\Phi}}dx-\dots\pm\frac{n^{p-1}}{m^p}\int\frac{dx}{\sqrt{\Phi}}\mp\frac{n^p}{m^p}\int\frac{dx}{(mx+n)\sqrt{\Phi}}.$$

$$\int\frac{dx}{x^p(mx+n)\sqrt{ax^2+bx+c}}=\frac{1}{n}\int\frac{dx}{x^p\sqrt{\Phi}}-\frac{m}{n^2}\int\frac{dx}{x^{p-1}\sqrt{\Phi}}$$

$$+\frac{m^2}{n^3}\int\frac{dx}{x^{p-2}\sqrt{\Phi}}-\dots\pm\frac{m^{p-1}}{n^p}\int\frac{dx}{x\sqrt{\Phi}}$$

$$\mp\frac{m^p}{n^p}\int\frac{dx}{(mx+n)\sqrt{\Phi}}.$$

$$\int\frac{x^{m-1}}{\sqrt{a^{2m}-x^{2m}}}dx=\frac{1}{m}\arcsin\left(\frac{x}{a}\right)^m.$$

$$\int\frac{x^{m-1}}{\sqrt{1-(ax^m+b)^2}}dx=\frac{1}{ma}\arcsin(ax^m+b).$$

$$\int\frac{x^{m-1}}{\sqrt{(x^m+a)(x^m+b)}}dx=\frac{2}{m}\log\left\{\sqrt{x^m+a}+\sqrt{x^m+b}\right\}.$$

$x\sqrt{X}$

$$\int x\sqrt{\frac{ax+b}{mx+n}}dx=a\int\frac{x^2}{\sqrt{(ax+b)(mx+n)}}dx$$

$$+b\int\frac{x}{\sqrt{(ax+b)(mx+n)}}dx.$$

$$\int x\sqrt{ax+b}\,dx=\frac{2(3ax-2b)}{15a^2}(ax+b)^{\frac{3}{2}}.$$

$$\int x\sqrt{\frac{a-bx^2}{m-nx^2}}\,dx = -\frac{1}{n\sqrt{n}}\int\sqrt{an-bm+bz^2}\,dz. \qquad m-nx^2=z^2.$$

$$\int x\sqrt{ax^2+b}\,dx = \frac{1}{3a}\left(ax^2+b\right)^{\frac{3}{2}}.$$

$$\int x\sqrt{ax^2+bx}\,dx = \frac{1}{3a}\left(ax^2+bx\right)^{\frac{3}{2}} - \frac{b}{2a}\int\sqrt{ax^2+bx}\,dx.$$

$$\int x\sqrt{ax^2+bx+c}\,dx = \frac{1}{3a}\,\Phi^{\frac{3}{2}} - \frac{b}{2a}\int\sqrt{\Phi}\,dx.$$

$$\boxed{\frac{\sqrt{X}}{x}}$$

$$\int\frac{1}{x}\sqrt{\frac{ax+b}{mx+n}}\,dx = a\int\frac{dx}{\sqrt{(mx+n)(ax+b)}} + b\int\frac{dx}{x\sqrt{(mx+n)(ax+b)}}.$$

$$\int\frac{\sqrt{ax+b}}{x}\,dx = 2\sqrt{\Phi} + \sqrt{b}\log\frac{\sqrt{\Phi}-\sqrt{b}}{\sqrt{\Phi}+\sqrt{b}}.$$

$$\int\frac{\sqrt{ax-b}}{x}\,dx = 2\sqrt{\Phi} - 2\sqrt{b}\ \text{arc tg}\sqrt{\frac{ax-b}{b}}.$$

$$\int\frac{1}{x}\sqrt{\frac{a^2-x^2}{x^2-b^2}}\,dx = \text{arc tg}\sqrt{\frac{a^2-x^2}{x^2-b^2}} - \frac{a}{b}\,\text{arc tg}\,\frac{b}{a}\sqrt{\frac{a^2-x^2}{x^2-b^2}}.$$

$$\int\frac{\sqrt{ax^2+b}}{x}\,dx = \sqrt{ax^2+b} + b\int\frac{dx}{x\sqrt{ax^2+b}}.$$

$$\int\frac{\sqrt{ax^2+bx}}{x}\,dx = \sqrt{ax^2+bx} + \frac{b}{2}\int\frac{dx}{\sqrt{ax^2+bx}}.$$

$$\int\frac{\sqrt{ax^2+bx+c}}{x}\,dx = \sqrt{\Phi} + \frac{b}{2}\int\frac{dx}{\sqrt{\Phi}} + c\int\frac{dx}{x\sqrt{\Phi}}.$$

$$\boxed{\frac{x}{\sqrt{X}}}$$

$$\int\frac{x}{\sqrt{ax+b}}\,dx = \frac{2ax-4b}{3a^2}\sqrt{ax+b}.$$

$$\int \frac{x}{\sqrt{ax^2+b}}\,dx = \frac{1}{a}\sqrt{ax^2+b}\,.$$

$$\int \frac{x}{\sqrt{ax^2+bx}}\,dx = \frac{1}{a}\sqrt{ax^2+bx} - \frac{b}{2a}\int \frac{dx}{\sqrt{ax^2+bx}}\,.$$

$$\int \frac{x}{\sqrt{(ax+b)(mx+n)}}\,dx = \frac{\sqrt{\Phi}}{am} - \frac{an+bm}{am\sqrt{am}}\log\left\{\sqrt{m(ax+b)}+\sqrt{a(mx+n)}\right\}.$$

$$\int \frac{x}{\sqrt{(ax+b)(n-mx)}}\,dx = -\frac{\sqrt{\Phi}}{am} + \frac{an-bm}{2am\sqrt{am}}\arcsin\frac{2amx+bm-an}{an+bm}\,.$$

$$\int \frac{x}{\sqrt{ax^2+bx+c}}\,dx = \frac{\sqrt{\Phi}}{a} - \frac{b}{2a}\int\frac{dx}{\sqrt{\Phi}} = \frac{\sqrt{\Phi}}{a} - \frac{b}{2a\sqrt{a}}\log\left(2ax+b+2\sqrt{a\Phi}\right).$$

$$\int \frac{x}{\sqrt{c+bx-ax^2}}\,dx = -\frac{\sqrt{\Phi}}{a} + \frac{b}{2a\sqrt{a}}\arcsin\frac{2ax-b}{\sqrt{b^2+4ac}}\,.$$

$$\int \frac{x}{\sqrt{a^2x^4+b}}\,dx = \frac{1}{2a}\log\left(ax^2+\sqrt{a^2x^4+b}\right).$$

$$\int \frac{x}{\sqrt{(ax^2+b)(ax^2+c)}}\,dx = \frac{1}{a}\log\left\{\sqrt{ax^2+b}+\sqrt{ax^2+c}\right\}.$$

$$\int \frac{x}{\sqrt{ax^4+bx^2+c}}\,dx = \frac{1}{2}\int\frac{dz}{\sqrt{az^2+bz+c}}\,. \qquad x^2 = z.$$

$$\boxed{\frac{1}{x\sqrt{X}}}$$

$$\int \frac{dx}{x\sqrt{ax+b}} = \frac{1}{\sqrt{b}}\log\frac{\sqrt{\Phi}-\sqrt{b}}{\sqrt{\Phi}+\sqrt{b}} = \frac{2}{\sqrt{b}}\log\frac{\sqrt{\Phi}-\sqrt{b}}{\sqrt{x}}$$

$$= -\frac{2}{\sqrt{b}}\log\frac{\sqrt{\Phi}+\sqrt{b}}{\sqrt{x}}\,.$$

$$\int \frac{dx}{x\sqrt{ax-b}} = \frac{2}{\sqrt{b}}\operatorname{arc\,tg}\sqrt{\frac{ax-b}{b}} = \frac{2}{\sqrt{b}}\arcsin\sqrt{\frac{ax-b}{ax}}$$

$$= \frac{2}{\sqrt{b}}\arccos\sqrt{\frac{b}{ax}} = \frac{1}{\sqrt{b}}\arccos\frac{2b-ax}{ax}\,.$$

$$\int \frac{dx}{x\sqrt{ax^2+b}} = \frac{1}{\sqrt{b}}\log\frac{\sqrt{ax^2+b}-\sqrt{b}}{x} = \frac{1}{\sqrt{b}}\log\frac{x}{\sqrt{ax^2+b}+\sqrt{b}}$$

$$= \frac{1}{2\sqrt{b}} \log \frac{\sqrt{ax^2+b}-\sqrt{b}}{\sqrt{ax^2+b}+\sqrt{b}}.$$

$$\int \frac{dx}{x\sqrt{ax^2-b}} = \frac{1}{\sqrt{b}} \operatorname{arc\,tg} \frac{\sqrt{ax^2-b}}{\sqrt{b}} = \frac{1}{\sqrt{b}} \operatorname{arc\,sin} \frac{\sqrt{ax^2-b}}{x\sqrt{a}}.$$

$$\int \frac{dx}{x\sqrt{ax^2+bx}} = -\frac{2}{bx}\sqrt{ax^2+bx}.$$

$$\int \frac{dx}{x\sqrt{(mx+n)(ax+b)}} = -2\int \frac{dz}{n-bz^2}. \qquad mx+n = z^2(ax+b).$$

$$\int \frac{dx}{x\sqrt{ax^2+bx+c}} = -\frac{1}{\sqrt{c}} \log \frac{2c+bx+2\sqrt{c\Phi}}{x} = \frac{1}{\sqrt{c}} \log \frac{2c+bx-2\sqrt{c\Phi}}{x}$$

$$= \frac{1}{\sqrt{c}} \log \frac{x\sqrt{a}-\sqrt{\Phi}+\sqrt{c}}{x\sqrt{a}-\sqrt{\Phi}-\sqrt{c}}.$$

$$\int \frac{dx}{x\sqrt{ax^2+bx-c}} = \frac{1}{\sqrt{c}} \operatorname{arc\,sin} \frac{bx-2c}{x\sqrt{b^2+4ac}} = \frac{1}{\sqrt{c}} \operatorname{arc\,tg} \frac{bx-2c}{2\sqrt{c\Phi}}.$$

$$\int \frac{dx}{x\sqrt{a^2+x^3}} = \frac{1}{3a} \log \frac{\sqrt{a^2+x^3}-a}{\sqrt{a^2+x^3}+a}.$$

$$\int \frac{dx}{x\sqrt{a^2+x^4}} = \frac{1}{2a} \log \frac{x^2}{a+\sqrt{a^2+x^4}}$$

$$\int \frac{dx}{x\sqrt{a^2-x^4}} = \frac{1}{2a} \log \frac{x^2}{a+\sqrt{a^2-x^4}}.$$

$$\int \frac{dx}{x\sqrt{(a+bx^2)(a+cx^2)}} = -\frac{1}{a} \log \left\{ \sqrt{\frac{a}{x^2}+b} + \sqrt{\frac{a}{x^2}+c} \right\}.$$

$$\int \frac{dx}{x\sqrt{(x^2-a^2)(b^2-x^2)}} = \frac{1}{ab} \operatorname{arc\,sin} \frac{b}{x}\sqrt{\frac{x^2-a^2}{b^2-a^2}} = \frac{1}{ab} \operatorname{arc\,tg} \frac{b}{a}\sqrt{\frac{x^2-a^2}{b^2-x^2}}.$$

$$\frac{\sqrt{X}}{ax+b}$$

$$\int \frac{\sqrt{x}}{ax+b}\,dx = \frac{2}{a}\sqrt{x} - \frac{b}{a}\int \frac{dx}{(ax+b)\sqrt{x}}.$$

$$\int \frac{\sqrt{x}}{x-a}\,dx = 2\sqrt{x} - \sqrt{a} \log \frac{\sqrt{x}+\sqrt{a}}{\sqrt{x}-\sqrt{a}}.$$

$$\int \frac{\sqrt{ax+b}}{mx+n}\,dx = \frac{2}{m}\sqrt{\Phi} - \frac{2}{m}\sqrt{\frac{an-bm}{m}}\,\text{arc tg}\sqrt{\frac{m\Phi}{an-bm}}\,.$$

$$\frac{ax+b}{\sqrt{X}}$$

$$\int \frac{mx+n}{\sqrt{ax+b}}\,dx = \frac{2}{3a^2}(3an - 2bm + amx)\sqrt{ax+b}\,.$$

$$\int \frac{a-x}{\sqrt{2ax-x^2}}\,dx = \sqrt{2ax-x^2}\,.$$

$$\int \frac{mx+n}{\sqrt{ax^2+bx+c}}\,dx = \frac{m}{a}\sqrt{\Phi} + \frac{2an-bm}{2a}\int \frac{dx}{\sqrt{\Phi}}\,.$$

$$\frac{1}{(ax+b)\sqrt{X}}$$

$$\int \frac{dx}{(ax+b)\sqrt{x}} = \frac{2}{\sqrt{ab}}\,\text{arc tg}\sqrt{\frac{ax}{b}}\,.$$

$$\int \frac{dx}{(ax-b)\sqrt{x}} = \frac{1}{\sqrt{ab}}\log\frac{2\sqrt{abx}-b-ax}{ax-b}\,.$$

$$\int \frac{dx}{(mx+n)\sqrt{ax+b}} = \frac{1}{\sqrt{m(bm-an)}}\log\frac{\sqrt{m(ax+b)}-\sqrt{bm-an}}{\sqrt{m(ax+b)}+\sqrt{bm-an}}\,.$$

$$bm - an > 0.$$

$$\int \frac{dx}{(n-mx)\sqrt{b-ax}} = \frac{1}{\sqrt{m}\sqrt{bm-an}}\log\frac{\sqrt{m(b-ax)}+\sqrt{bm-an}}{\sqrt{m(b-ax)}-\sqrt{bm-an}}\,.$$

$$bm - an > 0.$$

$$= -\frac{2}{\sqrt{m}\sqrt{an-bm}}\,\text{arc tg}\sqrt{\frac{m(b-ax)}{an-bm}}\,.$$

$$an - bm > 0.$$

$$\int \frac{dx}{(ax+b)\sqrt{a^2x^2-b^2}} = \frac{1}{ab}\sqrt{\frac{ax-b}{ax+b}}\,.$$

$$\int\frac{dx}{(x+b)\sqrt{x^2-a^2}} = -2\int\frac{dz}{(a+b)z^2+a-b}. \qquad \sqrt{x^2-a^2}=(x-a)z.$$

$$\int\frac{dx}{(x+a)\sqrt{b^2-x^2}} = 2\int\frac{dz}{(a+b)z^2+a-b}. \qquad \sqrt{b^2-x^2}=(b-x)z.$$

$$=\frac{1}{\sqrt{b^2-a^2}}\log\frac{\sqrt{(a+b)(b+x)}-\sqrt{(b-a)(b-x)}}{\sqrt{(a+b)(b+x)}+\sqrt{(b-a)(b-x)}}.$$

$$a^2 < b^2.$$

$$=\frac{2}{\sqrt{a^2-b^2}}\operatorname{arc\,tg}\sqrt{\frac{a+b}{a-b}\,\frac{b+x}{b-x}}. \qquad a^2 > b^2.$$

$$\int\frac{dx}{(mx+n)\sqrt{ax^2+b}} = \frac{1}{\sqrt{bm^2+an^2}}\log\frac{bm-anx-\sqrt{bm^2+an^2}\sqrt{ax^2+b}}{mx+n}.$$

$$bm^2+an^2 > 0.$$

$$=-\frac{1}{\sqrt{bm^2+an^2}}\log\frac{bm-anx+\sqrt{bm^2+an^2}\sqrt{ax^2+b}}{mx+n}.$$

$$bm^2+an^2 > 0.$$

$$=-\frac{1}{\sqrt{-(bm^2+an^2)}}\arcsin\frac{bm-anx}{(mx+n)\sqrt{-ab}}.$$

$$bm^2+an^2 < 0.$$

$$\int\frac{dx}{(mx+n)\sqrt{ax^2+bx}} = \frac{1}{\sqrt{an^2-bmn}}\log\frac{(bm-2an)x-bn-2\sqrt{(an^2-bmn)\Phi}}{mx+n}.$$

$$an^2-bmn > 0.$$

$$=\frac{1}{\sqrt{bmn-an^2}}\operatorname{arc\,tg}\frac{(bm-2an)x-bn}{2\sqrt{(bmn-an^2)\Phi}}. \qquad bmn > an^2.$$

$$\int\frac{dx}{(x-a)\sqrt{(x-a)(x-b)}} = \frac{2}{b-a}\sqrt{\frac{x-b}{x-a}}.$$

$$\int\frac{dx}{(x-a)\sqrt{(x-b)(c-x)}} = \frac{2}{\sqrt{(b-a)(c-a)}}\operatorname{arc\,tg}\sqrt{\frac{(x-b)(c-a)}{(c-x)(b-a)}}.$$

$$\int \frac{dx}{(ax+b)\sqrt{ax^2+2bx+c}} = \frac{1}{ak} \log \frac{\sqrt{\Phi}-k}{ax+b}. \qquad k^2 = \frac{ac-b^2}{a}.$$

$$= \frac{1}{ak} \operatorname{arc\,tg} \frac{\sqrt{\Phi}}{k}. \qquad k^2 = \frac{b^2-ac}{a}.$$

$$\int \frac{dx}{(bx+c)\sqrt{ax^2+2bx+c}} = \frac{1}{ck} \log \frac{bx+c}{\sqrt{\Phi}-k}. \qquad k^2 = \frac{ac-b^2}{c}.$$

$$= \frac{1}{ck} \operatorname{arc\,tg} \frac{kx}{\sqrt{\Phi}}. \qquad k^2 = \frac{b^2-ac}{c}.$$

$$\int \frac{dx}{(x+n)\sqrt{ax^2+bx+c}} = -\frac{1}{\sqrt{an^2+c-bn}} \log \left\{ \frac{\sqrt{\Phi}+\sqrt{an^2+c-bn}}{x+n} + \frac{b-2an}{2\sqrt{an^2+c-bn}} \right\}. \quad an^2+c-bn>0.$$

$$= -\frac{1}{\sqrt{bn-an^2-c}} \arcsin \frac{(b-2an)x+2(c-bn+an^2)}{(x+n)\sqrt{b^2-4ac}}.$$

$$an^2+c-bn<0.$$

$$= -\frac{2\sqrt{\Phi}}{(b-2an)(x+n)}. \qquad an^2+c-bn=0.$$

$$\int \frac{dx}{(x-n)\sqrt{ax^2+bx+c}} = -\int \frac{dz}{\sqrt{(an^2+bn+c)z^2+(2an+b)z+a}}.$$

$$x-n=\frac{1}{z}.$$

$$= -\frac{1}{\sqrt{an^2+bn+c}} \log \left\{ \frac{\sqrt{\Phi}+\sqrt{an^2+bn+c}}{x-n} + \frac{b+2an}{2\sqrt{an^2+bn+c}} \right\}$$

$$\cdot \; an^2+bn+c>0.$$

$$= \frac{1}{\sqrt{-(an^2+bn+c)}} \arcsin \frac{2c+bn+(b+2an)x}{(x-n)\sqrt{b^2-4ac}}.$$

$$an^2+bn+c<0.$$

$$\int \frac{dx}{(mx+n)\sqrt{ax^2+bx+c}} = -\int \frac{dz}{\sqrt{a(1-nz)^2+bm(1-nz)z+cm^2z^2}}.$$

$$mx+n=\frac{1}{z}$$

$$= \frac{1}{\sqrt{an^2 + cm^2 - bmn}} \log \frac{2cm - bn + (bm - 2an)x - 2\sqrt{(an^2 + cm^2 - bmn)\Phi}}{mx + n}.$$

$$an^2 + cm^2 - bmn > 0$$

$$= \frac{1}{\sqrt{bmn - an^2 - cm^2}} \operatorname{arc\,tg} \frac{2cm - bn + (bm - 2an)x}{2\sqrt{(bmn - an^2 - cm^2)\Phi}}.$$

$$an^2 + cm^2 - bmn < 0$$

$$\boxed{\frac{\sqrt{X}}{ax^2 + b}}$$

$$\int \frac{\sqrt{x}}{ax^2 + b}\,dx = 2\int \frac{z^2}{az^4 + b}\,dz. \qquad x = z^2.$$

$$= \frac{1}{2ak\sqrt{2}} \log \frac{x - k\sqrt{2x} + k^2}{x + k\sqrt{2x} + k^2} + \frac{1}{ak\sqrt{2}} \operatorname{arc\,tg} \frac{k\sqrt{2x}}{k^2 - x}. \qquad k = \left(\frac{b}{a}\right)^{\frac{1}{4}}$$

$$\int \frac{\sqrt{x}}{ax^2 - b}\,dx = \frac{1}{2ak} \log \frac{k - \sqrt{x}}{k + \sqrt{x}} + \frac{1}{ak} \operatorname{arc\,tg} \frac{\sqrt{x}}{k}. \qquad k = \left(\frac{b}{a}\right)^{\frac{1}{4}}.$$

$$\int \frac{\sqrt{ax^2 + b}}{mx^2 + n}\,dx = \frac{a}{m}\int \frac{dx}{\sqrt{ax^2 + b}} + \frac{bm - an}{m}\int \frac{dx}{(mx^2 + n)\sqrt{ax^2 + b}}.$$

$$\boxed{\frac{1}{(ax^2 + b)\sqrt{X}}}$$

$$\int \frac{dx}{(ax^2 + b)\sqrt{x}} = \frac{k}{2b\sqrt{2}} \log \frac{x + k\sqrt{2x} + k^2}{x - k\sqrt{2x} + k^2} + \frac{k}{b\sqrt{2}} \operatorname{arc\,tg} \frac{k\sqrt{2x}}{k^2 - x}. \qquad k = \left(\frac{b}{a}\right)^{\frac{1}{4}}$$

$$\int \frac{dx}{(ax^2 - b)\sqrt{x}} = -\frac{k}{2b} \log \frac{k + \sqrt{x}}{k - \sqrt{x}} - \frac{k}{b} \operatorname{arc\,tg} \frac{\sqrt{x}}{k}. \qquad k = \left(\frac{b}{a}\right)^{\frac{1}{4}}.$$

$$\int \frac{dx}{(x^2 + b^2)\sqrt{x^2 + a^2}} = \frac{1}{2b\sqrt{b^2 - a^2}} \log \frac{x\sqrt{b^2 - a^2} + b\sqrt{x^2 + a^2}}{x\sqrt{b^2 - a^2} - b\sqrt{x^2 + a^2}}. \qquad b^2 > a^2.$$

$$= \frac{1}{b\sqrt{a^2 - b^2}} \operatorname{arc\,tg} \frac{x\sqrt{a^2 - b^2}}{b\sqrt{x^2 + a^2}}. \qquad b^2 < a^2.$$

$$\int \frac{dx}{(n - mx^2)\sqrt{ax^2 + b}} = \frac{1}{2\sqrt{n(bm + an)}} \log \frac{x\sqrt{bm + an} + \sqrt{n(ax^2 + b)}}{x\sqrt{bm + an} - \sqrt{n(ax^2 + b)}} .$$

$$bm + an > 0.$$

$$= \frac{1}{\sqrt{-n(bm + an)}} \text{arc tg} \frac{x\sqrt{-(bm + an)}}{\sqrt{n(ax^2 + b)}} .$$

$$bm + an < 0.$$

$$\int \frac{dx}{(mx^2 + n)\sqrt{ax^2 + b}} = \frac{1}{2\sqrt{n(an - bm)}} \log \frac{\sqrt{n(ax^2 + b)} + x\sqrt{an - bm}}{\sqrt{n(ax^2 + b)} - x\sqrt{an - bm}} .$$

$$an - bm > 0.$$

$$= \frac{1}{2\sqrt{n(an - bm)}} \log \frac{x\sqrt{an - bm} + \sqrt{n(ax^2 + b)}}{x\sqrt{an - bm} - \sqrt{n(ax^2 + b)}} .$$

$$an - bm > 0.$$

$$= \frac{1}{\sqrt{n(bm - an)}} \text{arc tg} \frac{x\sqrt{bm - an}}{\sqrt{n(ax^2 + b)}} . \qquad bm - an > 0.$$

$$= \frac{x}{n\sqrt{ax^2 + b}} . \qquad an = bm.$$

$$\int \frac{dx}{(mx^2 + n)\sqrt{ax^2 - b}} = \frac{1}{2\sqrt{n(bm + an)}} \log \frac{x\sqrt{bm + an} + \sqrt{n(ax^2 - b)}}{x\sqrt{bm + an} - \sqrt{n(ax^2 - b)}} .$$

$$bm + an > 0.$$

$$= \frac{1}{\sqrt{-n(bm + an)}} \text{arc tg} \frac{x\sqrt{-(bm + an)}}{\sqrt{n(ax^2 - b)}} .$$

$$bm + an < 0.$$

$$\int \frac{dx}{(n - mx^2)\sqrt{ax^2 - b}} = \frac{1}{2\sqrt{n(an - bm)}} \log \frac{x\sqrt{an - bm} + \sqrt{n(ax^2 - b)}}{x\sqrt{an - bm} - \sqrt{n(ax^2 - b)}} .$$

$$an - bm > 0.$$

$$= \frac{1}{\sqrt{n(bm - an)}} \text{arc tg} \frac{x\sqrt{bm - an}}{\sqrt{n(ax^2 - b)}} . \qquad an - bm < 0.$$

$$\int \frac{dx}{(mx^2+n)\sqrt{b-ax^2}} = \frac{1}{2\sqrt{-n(bm+an)}} \log \frac{x\sqrt{-(bm+an)}+\sqrt{n(b-ax^2)}}{x\sqrt{-(bm+an)}-\sqrt{n(b-ax^2)}}.$$

$$bm+an<0.$$

$$= \frac{1}{\sqrt{n(bm+an)}} \operatorname{arc\,tg} \frac{x\sqrt{bm+an}}{\sqrt{n(b-ax^2)}}. \qquad bm+an>0.$$

$$\int \frac{dx}{(n-mx^2)\sqrt{b-ax^2}} = \frac{1}{2\sqrt{n(bm-an)}} \log \frac{x\sqrt{bm-an}+\sqrt{n(b-ax^2)}}{x\sqrt{bm-an}-\sqrt{n(b-ax^2)}}.$$

$$bm-an>0.$$

$$= \frac{1}{\sqrt{n(an-bm)}} \operatorname{arc\,tg} \frac{x\sqrt{an-bm}}{\sqrt{n(b-ax^2)}}. \qquad an-bm>0.$$

$$\boxed{\frac{\sqrt{X}}{(ax+b)^2}}$$

$$\int \frac{\sqrt{x}}{(x-a)^2}\,dx = -\frac{\sqrt{x}}{x-a} + \frac{1}{2\sqrt{a}} \log \frac{\sqrt{x}-\sqrt{a}}{\sqrt{x}+\sqrt{a}}.$$

$$\int \frac{\sqrt{x}}{(ax+b)^2}\,dx = -\frac{\sqrt{x}}{a(ax+b)} + \frac{1}{a\sqrt{ab}} \operatorname{arc\,tg} \sqrt{\frac{ax}{b}}.$$

$$\int \frac{(x+a)^2}{\sqrt{a^2-x^2}}\,dx = -\left(\frac{x}{2}+2a\right)\sqrt{a^2-x^2} + \frac{3a^2}{2} \arcsin \frac{x}{a}.$$

$$\int \frac{dx}{(x-a)^2\sqrt{x}} = -\frac{\sqrt{x}}{a(x-a)} + \frac{1}{a\sqrt{a}} \log \frac{\sqrt{x}+\sqrt{a}}{\sqrt{x-a}}.$$

$$\int \frac{dx}{(ax+b)^2\sqrt{x}} = \frac{\sqrt{x}}{b(ax+b)} + \frac{1}{2b}\int \frac{dx}{(ax+b)\sqrt{x}}.$$

$$\int \frac{dx}{(mx+n)^2\sqrt{ax+b}} = \frac{1}{\sqrt{m}}\int \frac{dz}{z^2\sqrt{az+bm-an}}. \qquad mx+n=z.$$

●

$$\int \frac{\sqrt{x}}{(ax^2+b)^2}\,dx = \frac{x\sqrt{x}}{2b(ax^2+b)} + \frac{1}{4b}\int \frac{\sqrt{x}}{ax^2+b}\,dx.$$

$$\int \frac{dx}{(ax^2+b)^2\sqrt{x}} = \frac{\sqrt{x}}{2b(ax^2+b)} + \frac{3}{4b}\int \frac{dx}{(ax^2+b)\sqrt{x}}.$$

$x^2\sqrt{X}$

$$\int x^2 \sqrt{\frac{ax+b}{mx+n}}\,dx = \frac{1}{m^3\sqrt{m}}\left\{ n^2 \int \sqrt{\frac{az+k}{z}}\,dz - 2n\int \sqrt{az^2+kz}\,dz \right.$$

$$\left. + \int z\sqrt{az^2+kz}\,dz. \right\} \cdot \quad mx+n=z. \quad k=bm-an.$$

$$\int x^2\sqrt{ax+b}\,dx = \frac{2}{a^3}\left\{ \frac{\Phi^2}{7} - \frac{2b}{5}\Phi + \frac{b^2}{3} \right\}\Phi^{\frac{3}{2}}.$$

$$\int x^2\sqrt{ax^2+b}\,dx = \frac{x(2ax^2+b)}{8a}\sqrt{ax^2+b} - \frac{b^2}{8a\sqrt{a}}\log(x\sqrt{a}+\sqrt{ax^2+b})$$

$$= \frac{x(2ax^2+b)}{8a}\sqrt{ax^2+b} - \frac{b^2}{16a\sqrt{a}}\log\left(2ax^2+b+2x\sqrt{a}\sqrt{ax^2+b}\right).$$

$$\int x^2\sqrt{a^2-x^2}\,dx = \frac{x(2x^2-a^2)}{8}\sqrt{a^2-x^2} + \frac{a^4}{8}\arcsin\frac{x}{a}.$$

$$\int x^2\sqrt{b-ax^2}\,dx = \frac{x(2ax^2-b)}{8a}\sqrt{b-ax^2} + \frac{b^2}{8a\sqrt{a}}\arcsin x\sqrt{\frac{a}{b}}.$$

$$\int x^2\sqrt{ax^2+bx}\,dx = \left\{ \frac{x}{4a} - \frac{5b}{24a^2} \right\}\left(ax^2+bx\right)^{\frac{3}{2}} + \frac{5b^2}{16a^2}\int \sqrt{ax^2+bx}\,dx.$$

$$\int x^2\sqrt{ax^2+bx+c}\,dx = \frac{6ax-5b}{24a^2}\Phi^{\frac{3}{2}} - \frac{4ac-5b^2}{16a^2}\int \sqrt{\Phi}\,dx.$$

$$\int x^2\sqrt{a+bx^3}\,dx = \frac{2}{9b}(a+bx^3)^{\frac{3}{2}}.$$

$$\int \frac{\sqrt{ax+b}}{x^2}\,dx = -\frac{\sqrt{\Phi}}{x} + \frac{a}{2\sqrt{b}}\log\frac{\sqrt{\Phi}-\sqrt{b}}{\sqrt{\Phi}+\sqrt{b}}$$

$$\int \frac{\sqrt{ax-b}}{x^2}\,dx = -\frac{\sqrt{\Phi}}{x} + \frac{a}{\sqrt{b}}\,\text{arc tg}\sqrt{\frac{ax-b}{b}}\,.$$

$$\int \frac{\sqrt{a^2-x^2}}{x^2}\,dx = -\frac{\sqrt{a^2-x^2}}{x} - \text{arc sin}\,\frac{x}{a}\,.$$

$$\int \frac{\sqrt{ax^2+b}}{x^2}\,dx = -\frac{\sqrt{ax^2+b}}{x} + a\int\frac{dx}{\sqrt{ax^2+b}}\,.$$

$$\int \frac{\sqrt{ax^2+bx}}{x^2}\,dx = -\frac{\sqrt{ax^2+bx}}{x} + \frac{b}{2}\int\frac{dx}{x\sqrt{ax^2+bx}} + a\int\frac{dx}{\sqrt{ax^2+bx}}\,.$$

$$\int \frac{\sqrt{ax^2+bx+c}}{x^2}\,dx = -\frac{\sqrt{\Phi}}{x} + \frac{b}{2}\int\frac{dx}{x\sqrt{\Phi}} + a\int\frac{dx}{\sqrt{\Phi}}\,.$$

●

$$\int \frac{x^2}{\sqrt{ax+b}}\,dx = \left\{\frac{\Phi^2}{5} - \frac{2b\Phi}{3} + b^2\right\}\frac{2\sqrt{\Phi}}{a^3}$$

$$\int \frac{x^2}{\sqrt{x^2-a^2}}\,dx = \frac{x}{2}\sqrt{x^2-a^2} + \frac{a^2}{2}\log\left(x+\sqrt{x^2-a^2}\right).$$

$$\int \frac{x^2}{\sqrt{a^2-x^2}}\,dx = -\frac{x}{2}\sqrt{a^2-x^2} + \frac{a^2}{2}\,\text{arc sin}\,\frac{x}{a}\,.$$

$$\int \frac{x^2}{\sqrt{ax^2+b}}\,dx = \frac{x\sqrt{ax^2+b}}{2a} - \frac{b}{2a}\int\frac{dx}{\sqrt{ax^2+b}}$$

$$= \frac{x\sqrt{ax^2+b}}{2a} - \frac{b}{2a\sqrt{a}}\log\left(x\sqrt{a}+\sqrt{ax^2+b}\right).$$

$$\int \frac{x^2}{\sqrt{b-ax^2}}\,dx = -\frac{x\sqrt{b-ax^2}}{2a} - \frac{b}{2a\sqrt{a}}\,\text{arc sin}\sqrt{\frac{b-ax^2}{b}}$$

$$= -\frac{x\sqrt{b-ax^2}}{2a} + \frac{b}{2a\sqrt{a}}\,\text{arc sin}\,x\sqrt{\frac{a}{b}}\,.$$

$$\int \frac{x^2}{\sqrt{ax^2+bx}}\,dx = \left\{\frac{x}{2a} - \frac{3b}{4a^2}\right\}\sqrt{ax^2+bx} + \frac{3b^2}{8a^2}\int\frac{dx}{\sqrt{ax^2+bx}}\,.$$

$$\int \frac{x^2}{\sqrt{(ax+b)(mx+n)}}\,dx = \frac{2amx-3(an+bm)}{4a^2m^2}\sqrt{\Phi}$$

$$+\frac{3(an+bm)^2-4abmn}{8a^2m^2}\int\frac{dx}{\sqrt{\Phi}}\,.$$

$$\int \frac{x^2}{\sqrt{(ax+b)(n-mx)}}\,dx = \frac{3(bm-an)-2amx}{4a^2m^2}\sqrt{\Phi} + \frac{3(a^2n^2+b^2m^2)-2abmn}{8a^2m^2\sqrt{am}} \arcsin \frac{2amx+bm-an}{an+bm}.$$

$$\int \frac{x^2}{\sqrt{ax^2+bx+c}}\,dx = \frac{2ax-3b}{4a^2}\sqrt{\Phi} + \frac{3b^2-4ac}{8a^2}\int \frac{dx}{\sqrt{\Phi}}.$$

$$\int \frac{x^2}{\sqrt{a+bx^3}}\,dx = \frac{2}{3b}\sqrt{a+bx^3}.$$

$$\int \frac{x^2}{\sqrt{a^6+x^6}}\,dx = \frac{1}{3}\log\left\{ x^3+\sqrt{a^6+x^6} \right\}.$$

•

$$\int \frac{dx}{x^2\sqrt{ax+b}} = -\frac{\sqrt{ax+b}}{bx} - \frac{a}{2b\sqrt{b}}\log\frac{\sqrt{ax+b}-\sqrt{b}}{\sqrt{ax+b}+\sqrt{b}}.$$

$$\int \frac{dx}{x^2\sqrt{ax-b}} = \frac{\sqrt{ax-b}}{bx} + \frac{a}{b\sqrt{b}}\,\mathrm{arc\,tg}\sqrt{\frac{ax-b}{b}}.$$

$$\int \frac{dx}{x^2\sqrt{ax^2-b}} = \frac{\sqrt{ax^2-b}}{bx}.$$

$$\int \frac{dx}{x^2\sqrt{ax^2+bx}} = -\frac{2}{3bx}\left(\frac{1}{x}-\frac{2a}{b}\right)\sqrt{ax^2+bx}.$$

$$\int \frac{dx}{x^2\sqrt{ax^2+bx+c}} = -\frac{\sqrt{\Phi}}{cx} - \frac{b}{2c}\int \frac{dx}{x\sqrt{\Phi}}.$$

$x^3\sqrt{X}$

$$\int x^3\sqrt{ax+b}\,dx = \frac{2}{a^4}\left\{ \frac{\Phi^3}{9} - \frac{3b}{7}\Phi^2 + \frac{3b^2}{5}\Phi - \frac{b^3}{3} \right\}\Phi^{\frac{3}{2}}.$$

$$\int x^3\sqrt{ax^2+b}\,dx = \frac{1}{15a^2}(3a^2x^4+abx^2-2b^2)\sqrt{ax^2+b}.$$

$$\int x^3\sqrt{2ax-x^2}\,dx = -\left\{ \frac{x^2}{5} + \frac{7}{20}ax + \frac{7a^2}{12} \right\}\left(2ax-x^2\right)^{\frac{3}{2}} + \frac{7}{4}a^3\int \sqrt{2ax-x^2}\,dx.$$

$$\int \frac{\sqrt{ax+b}}{x^3}\,dx = -\frac{\Phi^{\frac{3}{2}}}{2bx^2} + \frac{a\sqrt{\Phi}}{4bx} - \frac{a^2}{8b}\int \frac{dx}{x\sqrt{\Phi}}$$

$$\int \frac{\sqrt{ax^2+b}}{x^3}\,dx = -\frac{\sqrt{ax^2+b}}{2x^2} + \frac{a}{2}\int \frac{dx}{x\sqrt{ax^2+b}}\cdot$$

$$\int \frac{\sqrt{ax^2+bx}}{x^3}\,dx = -\frac{1}{2x^2}\sqrt{ax^2+bx} + \frac{b}{4}\int \frac{dx}{x^2\sqrt{ax^2+bx}} + \frac{a}{2}\int \frac{dx}{x\sqrt{ax^2+bx}}\cdot$$

$$\int \frac{\sqrt{ax^2+bx+c}}{x^3}\,dx = -\frac{bx+2c}{4cx^2}\sqrt{\Phi} + \frac{4ac-b^2}{8c}\int \frac{dx}{x\sqrt{\Phi}}\cdot$$

●

$$\int \frac{x^3}{\sqrt{ax+b}}\,dx = \frac{2}{35a^4}(5a^3x^3 - 6a^2bx^2 + 8ab^2x - 16b^3)\sqrt{\Phi}$$

$$\int \frac{x^3}{\sqrt{ax^2+b}}\,dx = \frac{ax^2-2b}{3a^2}\sqrt{ax^2+b}\cdot$$

$$\int \frac{x^3}{\sqrt{2ax-x^2}}\,dx = -\frac{1}{6}(2x^2+5ax+15a^2)\sqrt{2ax-x^2} + \frac{5}{2}a^3 \operatorname{arc\,sin}\frac{x-a}{a}\cdot$$

$$\int \frac{x^3}{\sqrt{ax^2+bx+c}}\,dx = \left\{\frac{x^2}{3a} - \frac{5bx}{12a^2} + \frac{5b^2}{8a^3} - \frac{2c}{3a^2}\right\}\sqrt{\Phi} + \left\{\frac{3bc}{4a^2} - \frac{5b^3}{16a^3}\right\}\int \frac{dx}{\sqrt{\Phi}}\cdot$$

$$\int \frac{x^3}{\sqrt{ax^4+b}}\,dx = \frac{1}{2a}\sqrt{ax^4+b}\cdot$$

●

$$\int \frac{dx}{x^3\sqrt{ax+b}} = \frac{3ax-2b}{4b^2x^2}\sqrt{ax+b} + \frac{3a^2}{8b^2}\int \frac{dx}{x\sqrt{ax+b}}\cdot$$

$$\int \frac{dx}{x^3\sqrt{ax^2+b}} = -\frac{\sqrt{ax^2+b}}{2bx^2} - \frac{a}{2b}\int \frac{dx}{x\sqrt{ax^2+b}}\cdot$$

$$\int \frac{dx}{x^3\sqrt{a^2-x^2}} = -\frac{\sqrt{a^2-x^2}}{2a^2x^2} + \frac{1}{2a^3}\log\frac{a-\sqrt{a^2-x^2}}{x}\cdot$$

$$\int \frac{dx}{x^3\sqrt{ax^2+bx}} = -\frac{2}{5bx}\left\{\frac{1}{x^2} - \frac{4a}{3bx} + \frac{8a^2}{3b^2}\right\}\sqrt{ax^2+bx}.$$

$$\int \frac{dx}{x^3\sqrt{ax^2+bx+c}} = \frac{3bx-2c}{4c^2x^2}\sqrt{\Phi} + \frac{3b^2-4ac}{8c^2}\int \frac{dx}{x\sqrt{\Phi}}\cdot$$

$x^4 \sqrt{X}$

$$\int \frac{\sqrt{ax^2+b}}{x^4}\,dx = -\frac{(ax^2+b)^{\frac{3}{2}}}{3bx^3}.$$

$$\int \frac{x^4}{\sqrt{a^2-x^2}}\,dx = -\frac{3a^2+2x^2}{8}\,x\sqrt{a^2-x^2} + \frac{3a^4}{8}\,\text{arc sin}\,\frac{x}{a}.$$

$$\int \frac{x^4}{\sqrt{ax^4+bx^2+c}}\,dx = \frac{x\sqrt{\Phi}}{3a} - \frac{c}{3a}\int\frac{dx}{\sqrt{\Phi}} - \frac{2b}{3a}\int\frac{x^2}{\sqrt{\Phi}}\,dx.$$

•

$$\int \frac{dx}{x^4\sqrt{ax+b}} = \left\{-\frac{1}{3bx^3} + \frac{5a}{12b^2x^2} - \frac{5a^2}{8b^3x}\right\}\sqrt{\Phi} - \frac{5a^3}{16b^3}\int\frac{dx}{x\sqrt{\Phi}}.$$

$$\int \frac{dx}{x^4\sqrt{ax^2+b}} = \frac{2ax^2-b}{3b^2x^3}\sqrt{ax^2+b} = -\frac{1}{b^2}\int(z^2-a)\,dz. \qquad ax^2+b = x^2z^2.$$

$$\int \frac{dx}{x^4\sqrt{x^2-a^2}} = \frac{2x^2+a^2}{3a^4x^3}\sqrt{x^2-a^2}.$$

•

$$\int \frac{mx^2+nx+p}{\sqrt{ax^2+bx+c}}\,dx = \frac{2amx+4an-3bm}{4a^2}\sqrt{\Phi}$$

$$+\frac{4a(2ap-mc)-b(4an-3bm)}{8a^2}\int\frac{dx}{\sqrt{\Phi}}.$$

•

$$\int \frac{dx}{x(mx+n)\sqrt{ax+b}} = \frac{1}{n}\int\frac{dx}{x\sqrt{ax+b}} - \frac{m}{n}\int\frac{dx}{(mx+n)\sqrt{ax+b}}.$$

$$\int \frac{dx}{x(mx+n)\sqrt{ax^2+b}} = \frac{1}{n}\int\frac{dx}{x\sqrt{ax^2+b}} - \frac{m}{n}\int\frac{dx}{(mx+n)\sqrt{ax^2+b}}.$$

$$\int \frac{x}{(mx+n)\sqrt{ax+b}}\,dx = \frac{1}{m}\int\frac{dx}{\sqrt{\Phi}} - \frac{n}{m}\int\frac{dx}{(mx+n)\sqrt{\Phi}}.$$

$$x^n (mx^y + n)\sqrt{X}$$

$$\int \frac{x}{(mx+n)\sqrt{ax^2+b}}\,dx = \frac{1}{m}\int \frac{dx}{\sqrt{ax^2+b}} - \frac{n}{m}\int \frac{dx}{(mx+n)\sqrt{ax^2+b}}.$$

$$= \frac{1}{\sqrt{bm^2+an^2}} \log \frac{bm - anx - \sqrt{bm^2+an^2}\sqrt{ax^2+b}}{mx+n}.$$

$$bm^2 + an^2 > 0.$$

$$= \frac{1}{\sqrt{-(bm^2+an^2)}} \operatorname{arc\,tg} \frac{bm - anx}{\sqrt{-(bm^2+an^2)}\sqrt{ax^2+b}}$$

$$bm^2 + an^2 < 0.$$

$$\int \frac{x}{(mx+n)\sqrt{ax^2+bx+c}}\,dx = \frac{1}{m}\int \frac{dx}{\sqrt{\Phi}} - \frac{n}{m}\int \frac{dx}{(mx+n)\sqrt{\Phi}}.$$

$$\int \frac{x^2}{(mx+n)\sqrt{ax^2+bx+c}}\,dx = \frac{1}{m}\int \frac{x}{\sqrt{\Phi}}\,dx - \frac{n}{m^2}\int \frac{dx}{\sqrt{\Phi}} + \frac{n^2}{m^2}\int \frac{dx}{(mx+n)\sqrt{\Phi}}.$$

$$\int \frac{dx}{x^2(mx+n)\sqrt{ax^2+b}} = \frac{1}{n}\int \frac{dx}{x^2\sqrt{\Phi}} - \frac{m}{n^2}\int \frac{dx}{x\sqrt{\Phi}} + \frac{m^2}{n^2}\int \frac{dx}{(mx+n)\sqrt{\Phi}}.$$

$$\int \frac{x\sqrt{ax^2+b}}{mx^2+n}\,dx = \frac{a}{m}\int \frac{x}{\sqrt{\Phi}}\,dx + \frac{bm-an}{m}\int \frac{x}{(mx^2+n)\sqrt{\Phi}}\,dx.$$

$$\int \frac{x}{(mx^2+n)\sqrt{ax^2+b}}\,dx = \frac{1}{\sqrt{bm^2-amn}} \log \frac{m\sqrt{ax^2+b} - \sqrt{bm^2-amn}}{\sqrt{mx^2+n}}.$$

$$bm - an > 0.$$

$$= \frac{1}{\sqrt{amn-bm^2}} \operatorname{arc\,tg} \frac{m\sqrt{ax^2+b}}{\sqrt{amn-bm^2}}.$$

$$bm - an < 0.$$

$$\int \frac{x^2\sqrt{ax^2+b}}{mx^2+n}\,dx = \frac{a}{m}\int \frac{x^2}{\sqrt{\Phi}}\,dx + \frac{bm^2-an}{m^2}\int \frac{dx}{\sqrt{\Phi}} - \frac{bmn-an^2}{m^2}\int \frac{dx}{(mx^2+n)\sqrt{\Phi}}$$

$$\int \frac{x^2}{(mx^2+n)\sqrt{ax^2+b}}\,dx = \frac{1}{m}\int \frac{dx}{\sqrt{ax^2+b}} - \frac{n}{m}\int \frac{dx}{(mx+n)\sqrt{ax^2+b}}.$$

$$\int \frac{x^3}{(mx^2+n)\sqrt{ax^2+b}}\,dx = \frac{1}{m}\int \frac{x}{\sqrt{\Phi}}\,dx - \frac{n}{m}\int \frac{x}{(mx+n)\sqrt{\Phi}}\,dx.$$

FRACTIONAL ROOTS

$$x^n (X)^{p/q}$$

$$\int (a + bx)^{\frac{3}{2}}\, dx = \frac{2}{5b} (a + bx)^{\frac{5}{2}} \cdot$$

$$\int (a + bx)^{\frac{5}{2}}\, dx = \frac{2}{7b} (a + bx)^{\frac{7}{2}} \cdot$$

$$\int (a + bx)^{\frac{1}{3}}\, dx = \frac{3}{4b} (a + bx)^{\frac{4}{3}}.$$

$$\int (a + bx)^{\frac{2}{3}}\, dx = \frac{3}{5b} (a + bx)^{\frac{5}{3}} \cdot$$

$$\int (a + bx^2)^{\frac{3}{2}}\, dx = a \int \sqrt{\Phi}\, dx + b \int x^2 \sqrt{\Phi}\, dx.$$

$$\int (a^2 - x^2)^{\frac{3}{2}}\, dx = \frac{5a^2 - 2x^2}{8} x \sqrt{a^2 - x^2} + \frac{3a^4}{8} \text{arc sin} \frac{x}{a} \cdot$$

$$\int (a^2 - x^2)^{\frac{5}{2}}\, dx = \frac{x}{48} \left\{ 8x^4 - 26a^2x^2 + 33a^4 \right\} \sqrt{\Phi} + \frac{5a^6}{16} \text{arc sin} \frac{x}{a} \cdot$$

$$\int (a^2 - x^2)^{\frac{7}{2}}\, dx = \frac{x \sqrt{\Phi}}{8} \left\{ \Phi^3 + \frac{7}{6} a^2 \Phi^2 + \frac{35}{24} a^4 \Phi + \frac{105}{48} a^6 \right\} + \frac{105}{384} \text{arc sin} \frac{x}{a} \cdot$$

$$\int (ax^2 + bx)^{\frac{3}{2}}\, dx = \left\{ \frac{\Phi}{a} - \frac{3b^2}{8a^2} \right\} \frac{b + 2ax}{8} \sqrt{\Phi} + \frac{3b^4}{128a^2} \int \frac{dx}{\sqrt{\Phi}} \cdot$$

$$\int (ax^2 + bx + c)^{\frac{3}{2}}\, dx = \frac{2ax + b}{8a} \Phi^{\frac{3}{2}} + \frac{3\,(4ac - b^2)}{16a} \int \sqrt{\Phi}\, dx.$$

$$\int (a + bx + cx^2)^{\frac{5}{2}}\, dx = \frac{2cx + b}{12c} \Phi^{\frac{5}{2}} + \frac{5\,(4ac - b^2)}{24c} \int \Phi^{\frac{3}{2}}\, dx.$$

$$\int (a + bx^3)^{\frac{2}{3}}\, dx = \frac{az^2}{3\,(z^3 - b)} - \frac{2a}{3} \int \frac{z}{z^3 - b}\, dz. \qquad a + bx^3 = z^3x^3.$$

$$\int x\,(a + bx)^{\frac{3}{2}}\, dx = \frac{2}{b^2} \left\{ \frac{\Phi}{7} - \frac{a}{5} \right\} \Phi^{\frac{5}{2}} \cdot$$

$$\int x\,(a + x)^{\frac{1}{3}}\, dx = \frac{3}{28} (4x - 3a)(a + x)^{\frac{4}{3}} \cdot$$

$$\int x\,(a+bx^2)^{\frac{3}{2}}\,dx = \frac{1}{5b}\,\Phi^{\frac{5}{2}}.$$

$$\int x\,(a+bx^2)^{\frac{1}{3}}\,dx = \frac{3}{8b}\,(a+bx^2)^{\frac{4}{3}}.$$

$$\int x\,(ax^2+bx)^{\frac{3}{2}}\,dx = \frac{\Phi^{\frac{5}{2}}}{5a} - \frac{b}{2a}\int \Phi^{\frac{3}{2}}\,dx.$$

$$\int x\,(ax^2+bx+c)^{\frac{3}{2}}\,dx = \frac{\Phi^{\frac{5}{2}}}{5a} - \frac{b}{2a}\int \Phi^{\frac{3}{2}}\,dx.$$

$$\int x\,(a-bx^3)^{\frac{1}{3}}\,dx = -a\int \frac{z^3}{(z^3+b)^2}\,dz. \qquad a-bx^3 = z^3x^3.$$

$$\int x^2\,(a+bx)^{\frac{3}{2}}\,dx = \frac{2}{b^3}\left\{\frac{\Phi^2}{9} - \frac{2}{7}\,a\Phi + \frac{a^2}{5}\right\}\Phi^{\frac{5}{2}}.$$

$$\int x^2\,(a+bx^2)^{\frac{3}{2}}\,dx = \frac{1}{6b}\,x\,\Phi^{\frac{5}{2}} - \frac{a}{6b}\int \Phi^{\frac{3}{2}}\,dx$$

$$= \frac{x}{48b}\,(8b^2x^4+14abx^2+9a^2)\sqrt{\Phi} + \frac{a^3}{16b}\int\frac{dx}{\sqrt{\Phi}}.$$

$$\int x^2\,(ax^2+bx+c)^{\frac{3}{2}}\,dx = \left\{x - \frac{7b}{10a}\right\}\frac{\Phi^{\frac{5}{2}}}{6a} + \frac{7b^2-4ac}{24a^2}\int \Phi^{\frac{3}{2}}\,dx.$$

$$\int x^2\,(a+bx+cx^2)^{\frac{5}{2}}\,dx = \left(x - \frac{9b}{14c}\right)\frac{\Phi^{\frac{7}{2}}}{8c} + \frac{9b^2-4ac}{32c^2}\int \Phi^{\frac{5}{2}}\,dx.$$

$$\int x^3\,(a+x)^{\frac{2}{3}}\,dx = 3\left\{\frac{1}{14}\,\Phi^3 - \frac{3a}{11}\,\Phi^2 + \frac{3a^2}{8}\,\Phi - \frac{a^3}{5}\right\}\Phi^{\frac{5}{3}}.$$

$$\int x^3\,(a+x^2)^{\frac{1}{3}}\,dx = \frac{3}{56}\,(4x^2-3a)\,(a+x^2)^{\frac{4}{3}}.$$

$$\int x^3\,(ax^2+bx+c)^{\frac{3}{2}}\,dx = \left\{x^2 - \frac{3bx}{4a} + \frac{21b^2}{40a^2} - \frac{2c}{5a}\right\}\frac{\Phi^{\frac{5}{2}}}{7a} + \left\{\frac{bc}{4a^2} - \frac{3b^3}{16a^3}\right\}\int \Phi^{\frac{3}{2}}\,dx.$$

$$\int x^3\,(a+bx^4)^{\frac{1}{3}}\,dx = \frac{3}{16b}\,(a+bx^4)^{\frac{4}{3}}.$$

$$\int x^4\,(a+bx^3)^{\frac{1}{3}}\,dx = -a^2\int\frac{dz}{(z^3-b)^2} - a^2b\int\frac{dz}{(z^3-b)^3}. \qquad a+bx^3 = z^3x^3.$$

$$\int x^5\,(a+bx^2)^{\frac{2}{3}}\,dx = \frac{3}{2a^3}\left\{\frac{1}{11}\,\Phi^2 - \frac{a}{4}\,\Phi + \frac{b^2}{5}\right\}\Phi^{\frac{5}{3}}.$$

$$\frac{1}{x^n (X)^{p/q}}$$

$$\int \frac{dx}{(a+bx)^{\frac{3}{2}}} = -\frac{2}{b\sqrt{a+bx}}.$$

$$\int \frac{dx}{(a+bx)^{\frac{5}{2}}} = -\frac{2}{3b\,(a+bx)^{\frac{3}{2}}}.$$

$$\int \frac{dx}{(a+bx)^{\frac{1}{3}}} = \frac{3}{2b}(a+bx)^{\frac{2}{3}}.$$

$$\int \frac{dx}{(a+bx^2)^{\frac{3}{2}}} = \frac{x}{a\sqrt{\Phi}}.$$

$$\int \frac{dx}{(a+bx^2)^{\frac{5}{2}}} = \frac{x\,(3a+2bx^2)}{3a^2\,(a+bx^2)^{\frac{3}{2}}}.$$

$$\int \frac{dx}{(a^2+x^2)^{\frac{5}{2}}} = \frac{x\,(2x^2+3a^2)}{3a^4\,(a^2+x^2)^{\frac{3}{2}}}.$$

$$\int \frac{dx}{(ax^2+bx)^{\frac{3}{2}}} = -\frac{2\,(2ax+b)}{b^2\sqrt{\Phi}}.$$

$$\int \frac{dx}{(ax^2+bx)^{\frac{5}{2}}} = \frac{2}{3b^4}\,\frac{8a^2x^2+8abx-b^2}{(ax^2+bx)^{\frac{3}{2}}}\,(2ax+b).$$

$$\int \frac{dx}{(ax^2+bx+c)^{\frac{3}{2}}} = \frac{2\,(2ax+b)}{(4ac-b^2)\sqrt{\Phi}}.$$

$$\int \frac{dx}{(a+bx+cx^2)^{\frac{5}{2}}} = \frac{2\,(2cx+b)}{3\,(4ac-b^2)\sqrt{\Phi}}\left\{\frac{1}{\Phi}+\frac{8c}{4ac-b^2}\right\}.$$

$$\int \frac{dx}{(a+bx^3)^{\frac{1}{3}}} = -\int \frac{z}{z^3-b}\,dz. \qquad a+bx^3 = z^3x^3.$$

$$\int \frac{dx}{x\,(a+bx)^{\frac{3}{2}}} = \frac{2}{a\sqrt{a+bx}} + \frac{1}{a}\int \frac{dx}{x\sqrt{a+bx}}.$$

$$\int \frac{dx}{x\,(a+bx)^{\frac{5}{2}}} = \frac{2\,(4a+3bx)}{3a^2\Phi^{\frac{3}{2}}} + \frac{1}{a^2}\int \frac{dx}{x\sqrt{\Phi}}.$$

$$\int \frac{dx}{x(a+bx)^{\frac{1}{3}}} = 3\int \frac{z}{z^3 - a}\,dz. \qquad a + bx = z^3.$$

$$\int \frac{dx}{x(a+bx)^{\frac{2}{3}}} = 3\int \frac{dz}{z^3 - a}. \qquad a + bx = z^3.$$

$$\int \frac{dx}{x(a+bx^2)^{\frac{3}{2}}} = \frac{1}{a\sqrt{\Phi}} + \frac{1}{a}\int \frac{dx}{x\sqrt{\Phi}}.$$

$$\int \frac{dx}{x(a+bx^2)^{\frac{5}{2}}} = \frac{1}{a^2\sqrt{\Phi}} + \frac{1}{3a\Phi^{\frac{3}{2}}} + \frac{1}{2a^2\sqrt{a}}\log\frac{\sqrt{\Phi}-\sqrt{a}}{\sqrt{\Phi}+\sqrt{a}}.$$

$$\int \frac{dx}{x(ax^2+bx)^{\frac{3}{2}}} = -\frac{2}{3bx\sqrt{\Phi}} - \frac{4a}{3b}\int \frac{dx}{\Phi^{\frac{3}{2}}}.$$

$$\int \frac{dx}{x(ax^2+bx+c)^{\frac{3}{2}}} = \frac{4ac - 2b^2 - 2abx}{c(4ac-b^2)\sqrt{\Phi}} + \frac{1}{c}\int \frac{dx}{x\sqrt{\Phi}}.$$

$$\int \frac{dx}{x(a+bx^3)^{\frac{2}{3}}} = \int \frac{dz}{z^3 - a}. \qquad a + bx^3 = z^3.$$

$$\int \frac{dx}{x^2(a+bx)^{\frac{3}{2}}} = -\frac{a+3bx}{a^2x\sqrt{a+bx}} - \frac{3b}{2a^2}\int \frac{dx}{x\sqrt{a+bx}}.$$

$$\int \frac{dx}{x^2(a+bx^2)^{\frac{3}{2}}} = -\frac{a+2bx^2}{a^2x\sqrt{\Phi}}.$$

$$\int \frac{dx}{x^2(ax^2+bx)^{\frac{3}{2}}} = \frac{2(2ax-b)}{5b^2x^2\sqrt{\Phi}} + \frac{8a^2}{5b^2}\int \frac{dx}{\Phi^{\frac{3}{2}}}.$$

$$\int \frac{dx}{x^2(ax^2+bx+c)^{\frac{3}{2}}} = -\frac{4a(2ax+b)}{c(4ac-b^2)\sqrt{\Phi}} - \frac{1}{cx\sqrt{\Phi}} - \frac{3b}{2c}\int \frac{dx}{x\Phi^{\frac{3}{2}}}.$$

$$\int \frac{dx}{x^3(a+bx^3)^{\frac{1}{3}}} = -\frac{1}{2ax^2}(a+bx^3)^{\frac{2}{3}}.$$

$$\int \frac{dx}{x^3(a+bx^2)^{\frac{3}{2}}} = \frac{1}{ax^2\sqrt{\Phi}} - \frac{3\sqrt{\Phi}}{2a^2x^2} + \frac{3b}{2a^2\sqrt{a}}\log\frac{\sqrt{a}+\sqrt{\Phi}}{x}.$$

$$\int \frac{dx}{x^3(ax^2+bx)^{\frac{3}{2}}} = \frac{2}{7bx\sqrt{\Phi}}\left\{-\frac{1}{x^2} + \frac{8a}{5bx} - \frac{16a^2}{5b^2}\right\} - \frac{64a^3}{35b^3}\int \frac{dx}{\Phi^{\frac{3}{2}}}.$$

$$\int \frac{dx}{x^3(ax^2+bx+c)^{\frac{3}{2}}} = \left\{ \frac{5b}{2c} - \frac{1}{x} \right\} \frac{1}{2cx\sqrt{\Phi}} + \frac{5ab(2ax+b)}{c^2(4ac-b^2)\sqrt{\Phi}} + \frac{3(5b^2-4ac)}{8c^2} \int \frac{dx}{x\Phi^{\frac{3}{2}}}.$$

$$\int \frac{dx}{x^3(a+bx^3)^{\frac{4}{3}}} = -\frac{a+3bx^3}{2a^2x^2\Phi^{\frac{1}{3}}}.$$

$$\int \frac{dx}{x^4(a+bx^2)^{\frac{3}{2}}} = -\frac{\Phi^2 - 6bx^2\Phi - 3b^2x^4}{3a^3x^3\sqrt{\Phi}}.$$

$$\frac{(X)^{p/q}}{x^n}$$

$$\int \frac{(a+bx)^{\frac{3}{2}}}{x}\,dx = \left\{ \frac{2}{3}\Phi + 2a \right\} \sqrt{\Phi} + a^2 \int \frac{dx}{x\sqrt{\Phi}}.$$

$$\int \frac{(a+bx)^{\frac{5}{2}}}{x}\,dx = \left\{ \frac{\Phi^2}{5} + \frac{a\Phi}{3} + a^2 \right\} 2\sqrt{\Phi} + a^3 \int \frac{dx}{x\sqrt{\Phi}}.$$

$$\int \frac{(a+bx^2)^{\frac{3}{2}}}{x}\,dx = \frac{1}{3}(4a+bx^2)\sqrt{\Phi} + a^2 \int \frac{dx}{x\sqrt{\Phi}}.$$

$$\int \frac{(ax^2+bx+c)^{\frac{3}{2}}}{x}\,dx = \frac{3c+\Phi}{3}\sqrt{\Phi} + \frac{b}{2}\int \sqrt{\Phi}\,dx + \frac{bc}{2}\int \frac{dx}{\sqrt{\Phi}} + c^2 \int \frac{dx}{x\sqrt{\Phi}}.$$

$$\int \frac{(a+bx)^{\frac{3}{2}}}{x^2}\,dx = -\frac{\Phi^{\frac{5}{2}}}{ax} + \frac{3b}{2a}\int \frac{\Phi^{\frac{3}{2}}}{x}\,dx = 2b\int \frac{z^4}{(z^2-a)^2}\,dz. \qquad a+bx=z^2.$$

$$\int \frac{(a+bx^2)^{\frac{3}{2}}}{x^2}\,dx = -\frac{\Phi^{\frac{5}{2}}}{ax} + \frac{4b}{a}\int \Phi^{\frac{3}{2}}\,dx.$$

$$\int \frac{(ax^2+bx+c)^{\frac{3}{2}}}{x^2}\,dx = \frac{3bx-2\Phi}{2x}\sqrt{\Phi} + \frac{3b^2}{4}\int \frac{dx}{\sqrt{\Phi}} + 3a\int \sqrt{\Phi}\,dx + \frac{3bc}{2}\int \frac{dx}{x\sqrt{\Phi}}.$$

$$\int \frac{(a+bx)^{\frac{3}{2}}}{x^3}\,dx = 2b^2\int \frac{z^4}{(z^2-a)^3}\,dz. \qquad a+bx=z^2.$$

$$\int \frac{(a+bx^2)^{\frac{3}{2}}}{x^3}\,dx = \frac{2bx^2 - a}{2x^2}\sqrt{\Phi} + \frac{3ab}{2}\int \frac{dx}{x\sqrt{\Phi}}.$$

$$\int \frac{(ax^2+bx+c)^{\frac{3}{2}}}{x^3}\,dx = -\left\{\frac{\Phi}{2x^2} + \frac{3b}{4x} - \frac{3a}{2}\right\}\sqrt{\Phi}$$

$$+\left\{\frac{3b^2}{8} + \frac{3ac}{2}\right\}\int \frac{dx}{x\sqrt{\Phi}} + \frac{3ab}{2}\int \frac{dx}{\sqrt{\Phi}}.$$

$$\int \frac{(a+bx^2)^{\frac{3}{2}}}{x^5}\,dx = -\frac{\Phi^{\frac{3}{2}}}{4x^4} + \frac{3b}{4}\int \frac{\sqrt{\Phi}}{x^3}\,dx.$$

$$\boxed{\frac{x^n}{(X)^{p/q}}}$$

$$\int \frac{x}{(a+bx)^{\frac{3}{2}}}\,dx = \frac{2(2a+bx)}{b^2\sqrt{a+bx}}.$$

$$\int \frac{x}{(a+bx)^{\frac{5}{2}}}\,dx = -\frac{2(2a+3bx)}{3b^2(a+bx)^{\frac{3}{2}}}.$$

$$\int \frac{x}{(a+bx)^{\frac{1}{3}}}\,dx = \frac{3}{10b^2}(2bx-3a)(a+bx)^{\frac{2}{3}}.$$

$$\int \frac{x}{(a+bx)^{\frac{2}{3}}}\,dx = \frac{3}{4b^2}(bx-3a)(a+bx)^{\frac{1}{3}}.$$

$$\int \frac{x}{(a+bx)^{\frac{1}{5}}}\,dx = \frac{4bx-5a}{36b^2}\Phi^{\frac{4}{5}}.$$

$$\int \frac{x}{(a+bx^2)^{\frac{3}{2}}}\,dx = -\frac{1}{b\sqrt{a+bx^2}}.$$

$$\int \frac{x}{(ax^2+bx)^{\frac{3}{2}}}\,dx = \frac{2x}{b\sqrt{ax^2+bx}}.$$

$$\int \frac{x}{(ax^2+bx+c)^{\frac{3}{2}}}\,dx = -\frac{2(2c+bx)}{(4ac-b^2)\sqrt{\Phi}}.$$

$$\int \frac{x}{(a+bx+cx^2)^{\frac{5}{2}}}\,dx = -\frac{2(bx+2a)}{3(4ac-b^2)\Phi^{\frac{3}{2}}} - \frac{8b(2cx+b)}{3(4ac-b^2)^2\sqrt{\Phi}}.$$

$$\int \frac{x}{(a+bx^4)^{\frac{3}{2}}}\,dx = \frac{x^2}{2a\sqrt{a+bx^4}}.$$

$$\int \frac{x^2}{(a+bx)^{\frac{3}{2}}}\,dx = \left\{\frac{\Phi^2}{3} - 2a\Phi - a^2\right\}\frac{2}{b^3\sqrt{\Phi}}.$$

$$\int \frac{x^2}{(a+bx)^{\frac{5}{2}}}\,dx = \frac{2}{3b^3}\,\frac{3\Phi^2 + 6a\Phi - a^2}{\Phi^{\frac{3}{2}}}.$$

$$\int \frac{x^2}{(a+bx)^{\frac{1}{3}}}\,dx = 3\left\{\frac{x^2}{8b} - \frac{3ax}{20b^2} + \frac{9a^2}{40b^3}\right\}(a+bx)^{\frac{2}{3}}.$$

$$\int \frac{x^2}{(a+bx^2)^{\frac{3}{2}}}\,dx = -\frac{x}{b\sqrt{\Phi}} + \frac{1}{b}\int\frac{dx}{\sqrt{\Phi}}.$$

$$\int \frac{x^2}{(a^2-x^2)^{\frac{3}{2}}}\,dx = \frac{x}{\sqrt{\Phi}} + \text{arc sin}\,\frac{\sqrt{\Phi}}{a}.$$

$$\int \frac{x^2}{(a+bx^2)^{\frac{5}{2}}}\,dx = \frac{x^3}{3a\,(a+bx^2)^{\frac{3}{2}}}.$$

$$\int \frac{x^2}{(ax^2+bx)^{\frac{3}{2}}}\,dx = -\frac{2x}{a\sqrt{\Phi}} + \frac{1}{a}\int\frac{dx}{\sqrt{\Phi}}.$$

$$\int \frac{x^2}{(ax^2+bx+c)^{\frac{3}{2}}}\,dx = -\frac{(4ac-2b^2)\,x - 2bc}{a\,(4ac-b^2)\sqrt{\Phi}} + \frac{1}{a}\int\frac{dx}{\sqrt{\Phi}}.$$

$$\int \frac{x^2}{(a+bx+cx^2)^{\frac{5}{2}}}\,dx = \frac{(2b^2-4ac)\,x + 2ab}{3c\,(4ac-b^2)\,\Phi^{\frac{3}{2}}} + \frac{2\,(4ac+b^2)\,(2cx+b)}{3c\,(4ac-b^2)^2\sqrt{\Phi}}.$$

$$\int \frac{x^2}{(a+bx^3)^{\frac{3}{2}}}\,dx = -\frac{2}{3b\sqrt{a+bx^3}}.$$

$$\int \frac{x^2}{(a+bx^3)^{\frac{2}{3}}}\,dx = \frac{1}{b}(a+bx^3)^{\frac{1}{3}}.$$

$$\int \frac{x^2}{(a+bx^3)^{\frac{4}{3}}}\,dx = -\frac{1}{b\,\Phi^{\frac{1}{3}}}.$$

$$\int \frac{x^2}{(a+bx^4)^{\frac{3}{4}}}\,dx = -\frac{1}{4b^{\frac{3}{4}}}\log\frac{\Phi^{\frac{1}{4}} - b^{\frac{1}{4}}x}{\Phi^{\frac{1}{4}} + b^{\frac{1}{4}}x} + \frac{1}{2b^{\frac{3}{4}}}\,\text{arc tg}\,\frac{\Phi^{\frac{1}{4}}}{b^{\frac{1}{4}}\,x}.$$

$$\int \frac{x^3}{(a+bx)^{\frac{3}{2}}}\,dx = \left\{\frac{\Phi^3}{5} - a\Phi^2 + 3a^2\Phi - a^3\right\}\frac{2}{b^4\sqrt{\Phi}}\cdot$$

$$\int \frac{x^3}{(a+bx)^{\frac{5}{2}}}\,dx = \left\{\frac{\Phi^3}{3} - 3a\Phi^2 - 3a^2\Phi + \frac{a^3}{3}\right\}\frac{2}{b^4\Phi^{\frac{3}{2}}}\cdot$$

$$\int \frac{x^3}{(a+bx)^{\frac{1}{3}}}\,dx = 3\left\{\frac{x^3}{11b} - \frac{9}{11.8}\cdot\frac{ax^2}{b^2} + \frac{9.6}{11.8.5}\cdot\frac{a^2x}{b^3} - \frac{9.6.3}{11.8.5.2}\,\frac{a^3}{b^4}\right\}(a+bx)^{\frac{2}{3}}\cdot$$

$$\int \frac{x^3}{(a+bx)^{\frac{7}{2}}}\,dx = \frac{2\sqrt{\Phi}}{7b}\left\{x^3 - \frac{6}{5}\,\frac{ax^2}{b} + \frac{8}{5}\,\frac{a^2x}{b^2} - \frac{16}{5}\,\frac{a^3}{b^3}\right\}\cdot$$

$$\int \frac{x^3}{(a+bx^2)^{\frac{3}{2}}}\,dx = \frac{2a+bx^2}{b^2\sqrt{\Phi}}\,.$$

$$\int \frac{x^3}{(a^2-x^2)^{\frac{1}{3}}}\,dx = -\frac{3}{20}\,(2x^2+3a^2)\,(a^2-x^2)^{\frac{2}{3}}.$$

$$\int \frac{x^3}{(ax^2+bx+c)^{\frac{3}{2}}}\,dx = \left\{\frac{x^2}{a} + \frac{10ac-3b^2}{a^2(4ac-b^2)}\,bx + \frac{c\,(8ac-3b^2)}{a^2(4ac-b^2)}\right\}\frac{1}{\sqrt{\Phi}} - \frac{3b}{2a^2}\int\frac{dx}{\sqrt{\Phi}}$$

$$\int \frac{x^3}{(a+bx^3)^{\frac{4}{3}}}\,dx = -\frac{x}{b\,\Phi^{\frac{1}{3}}} - \frac{1}{2b\sqrt{b}}\log\frac{\Phi^{\frac{1}{3}} - x\sqrt{b}}{\Phi^{\frac{1}{3}} + x\sqrt{b}}\,.$$

$$\int \frac{x^4}{(a^2-x^2)^{\frac{3}{2}}}\,dx = \frac{x^3}{\sqrt{a^2-x^2}} - 3\int\frac{x^2}{\sqrt{a^2-x^2}}\,dx.$$

$$\frac{x^{\pm 3/2}}{X}$$

$$\int \frac{dx}{(a+bx)\,x^{\frac{3}{2}}} = -\frac{2}{a\sqrt{x}} - \frac{b}{a}\int\frac{dx}{(a+bx)\sqrt{x}}\cdot$$

$$\int \frac{x^{\frac{3}{2}}}{a+bx}\,dx = \frac{bx-3a}{3b^2}\,2\sqrt{x} + \frac{a^2}{b^2}\int\frac{dx}{(a+bx)\sqrt{x}}\cdot$$

$$\int \frac{dx}{(x-a)\,x^{\frac{3}{2}}} = \frac{2}{a\sqrt{x}} + \frac{1}{a\sqrt{a}}\log\frac{\sqrt{x}-\sqrt{a}}{\sqrt{x}+\sqrt{a}}\cdot$$

$$\int \frac{x^{\frac{3}{2}}}{x-a}\,dx = \frac{2}{3}\,x^{\frac{3}{2}} + 2a\sqrt{x} + a\sqrt{a}\log\frac{\sqrt{x}-\sqrt{a}}{\sqrt{x}+\sqrt{a}}\cdot$$

$$\int \frac{dx}{(a+bx)^2\,x^{\frac{3}{2}}} = -\frac{2}{a\,(a+bx)\sqrt{x}} - \frac{3b}{a}\int\frac{dx}{(a+bx)^2\sqrt{x}}\cdot$$

$$\int \frac{x^{\frac{3}{2}}}{(a+bx)^2}\,dx = \frac{2x^{\frac{3}{2}}}{b\,(a+bx)} - \frac{3a}{b}\int\frac{\sqrt{x}}{(a+bx)^2}\,dx.$$

$$\int \frac{dx}{(x-a)^2\,x^{\frac{3}{2}}} = \frac{3x-2a}{a^2\,(x-a)\sqrt{x}} + \frac{3}{2a^2\sqrt{a}}\log\frac{\sqrt{x}+\sqrt{a}}{\sqrt{x}-\sqrt{a}}$$

$$= 2\int\frac{dz}{z^2\,(z^2-a)^2}\cdot \qquad x = z^2.$$

$$\int \frac{x^{\frac{3}{2}}}{(x-a)^2}\,dx = \frac{2x-3a}{x-a}\sqrt{x} + \frac{3}{2}\sqrt{a}\log\frac{\sqrt{x}-\sqrt{a}}{\sqrt{x}+\sqrt{a}}\cdot$$

$$\int \frac{x^{\frac{3}{2}}}{a+bx^2}\,dx = \frac{2}{a}\sqrt{x} - \int\frac{dx}{(a+bx^2)\sqrt{x}}\cdot$$

$$\int \frac{x^{\frac{3}{2}}}{(a+bx^2)^2}\,dx = -\frac{\sqrt{x}}{2b\,(a+bx^2)} + \frac{1}{4b}\int\frac{dx}{(a+bx^2)\sqrt{x}}$$

$$\int \frac{dx}{(x-a)\,x^{\frac{2}{3}}} = \frac{k^2}{2}\log\frac{(x^{\frac{1}{3}}-k)^2}{x^{\frac{2}{3}}+kx^{\frac{1}{3}}+k^2} - \frac{3}{k^2\sqrt{3}}\,\text{arc tg}\,\frac{2x^{\frac{1}{3}}+k}{k\sqrt{3}}\cdot \qquad k = a^{\frac{1}{3}}\cdot$$

•

$$\int\left(\frac{x-a}{x-b}\right)^{\frac{1}{3}}dx = (x-a)^{\frac{1}{3}}\,(x-b)^{\frac{2}{3}} + \frac{1}{2}\,(a-b)\log\left\{(x-b)^{\frac{1}{3}}-(x-a)^{\frac{1}{3}}\right\}$$

$$-\frac{a-b}{\sqrt{3}}\,\text{arc tg}\,\frac{2(x-a)^{\frac{1}{3}}+(x-b)^{\frac{1}{3}}}{\sqrt{3}\,(x-b)^{\frac{1}{3}}}.$$

$$\int\left(\frac{x-a}{x-b}\right)^{\frac{2}{3}}dx = (x-a)^{\frac{2}{3}}(x-b)^{\frac{1}{3}} + (a-b)\left\{(x-b)^{\frac{1}{3}}-(x-a)^{\frac{1}{3}}\right\}$$

$$+\frac{2}{\sqrt{3}}(a-b)\,\text{arc tg}\,\frac{2(x-a)^{\frac{1}{3}}+(x-b)^{\frac{1}{3}}}{(x-b)^{\frac{1}{3}}\sqrt{3}}\cdot$$

$$\int\frac{1}{x}\left(\frac{x-a}{x-b}\right)^{\frac{1}{3}}dx = 3\int\frac{dz}{1-z^3} - 3a\int\frac{dz}{a-bz^3}\cdot \qquad x = \frac{a-bz^3}{1-z^3}\cdot$$

$$\int \frac{dx}{\left\{(x-a)(x-b)\right\}^{\frac{3}{2}}} = \frac{2(a+b-2x)}{(a-b)^2 \sqrt{\Phi}}.$$

$$\int \frac{mx+n}{\left\{(x-a)(x-b)\right\}^{\frac{3}{2}}} dx = \frac{2n(a+b-2x) + 2m\left\{2ab-(a+b)x\right\}}{(a-b)^2 \sqrt{\Phi}}.$$

MULTIPLE FUNCTIONS

$$(X)^{\pm 1}$$

$$\int \frac{a+bx}{x^m}\,dx = -\frac{\Phi^2}{a(m-1)x^{m-1}} + \frac{b(3-m)}{a(m-1)}\int \frac{\Phi}{x^{m-1}}\,dx.$$

$$\int \frac{x^m}{a+bx}\,dx = \frac{x^m}{bm} - \frac{a}{b}\int \frac{x^{m-1}}{a+bx}\,dx = \frac{x^m}{bm} - \frac{ax^{m-1}}{(m-1)b^2}$$

$$+ \frac{a^2x^{m-2}}{(m-2)b^3} - \cdots \pm \frac{a^m}{b^{m+1}} \log(a+bx).$$

$$\int \frac{x^m}{a+bx^2}\,dx = \frac{x^{m-1}}{(m-1)b} - \frac{a}{b}\int \frac{x^{m-2}}{a+bx^2}\,dx.$$

$$\int \frac{x^{2m}}{a+bx^2}\,dx = \frac{x^{2m-1}}{(2m-1)b} - \frac{ax^{2m-3}}{(2m-3)b^2} + \frac{a^2x^{2m-5}}{(2m-5)b^3} - \cdots \pm \frac{a^m}{b^m}\int \frac{dx}{a+bx^2}.$$

$$\int \frac{x^{2m+1}}{a+bx^2}\,dx = \frac{x^{2m}}{2mb} - \frac{ax^{2m-2}}{(2m-2)b^2} + \frac{a^2x^{2m-4}}{(2m-4)b^3} - \cdots - \frac{a^m}{2b^m}\log(a+bx^2).$$

$$\int \frac{x^m}{a+bx+cx^2}\,dx = \frac{x^{m-1}}{c(m-1)} - \frac{a}{c}\int \frac{x^{m-2}}{\Phi}\,dx - \frac{b}{c}\int \frac{x^{m-1}}{\Phi}\,dx.$$

$$\int \frac{dx}{x^m(a+bx)} = -\frac{1}{a(m-1)x^{m-1}} - \frac{b}{a}\int \frac{dx}{x^{m-1}(a+bx)} = -\int \frac{z^{m-1}}{b+az}\,dz.$$

$$xz = 1.$$

$$= -\frac{1}{(m-1)ax^{m-1}} + \frac{b}{(m-2)a^2x^{m-2}}$$

$$- \frac{b^2}{(m-3)a^3x^{m-3}} + \cdots \pm \frac{b^{m-1}}{a^m}\log\frac{a+bx}{x}.$$

$$\int \frac{dx}{x^m(a+bx^2)} = -\frac{1}{a(m-1)x^{m-1}} - \frac{b}{a}\int \frac{dx}{x^{m-2}(a+bx^2)}.$$

$$\int \frac{dx}{x^{2m}(a+bx^2)} = -\frac{1}{(2m-1)ax^{2m-1}} + \frac{b}{(2m-3)a^2x^{2m-3}}$$

$$- \frac{b^2}{(2m-5)a^3x^{2m-5}} + \cdots \pm \frac{b^m}{a^m}\int \frac{dx}{a+bx^2}.$$

$$\int \frac{dx}{x^{2m+1}(a+bx^2)} = -\frac{1}{2max^{2m}} + \frac{b}{(2m-2)\,a^2x^{2m-2}}$$

$$-\frac{b^2}{(2m-4)\,a^3x^{2m-4}} + \ldots \pm \frac{b^m}{a^m} \log \frac{\sqrt{a+bx^2}}{x}.$$

$$\int \frac{dx}{x^m(a+bx+cx^2)} = -\frac{1}{a\,(m-1)\,x^{m-1}} - \frac{b}{a}\int \frac{dx}{x^{m-1}\Phi} - \frac{c}{a}\int \frac{dx}{x^{m-2}\Phi}.$$

$$\int \frac{x^n}{a+bx^n}\,dx = \frac{x}{b} - \frac{a}{b}\int \frac{dx}{a+bx^n}.$$

$$\int \frac{x^{n-1}}{a+bx^n}\,dx = \frac{1}{bn}\log(a+bx^n).$$

$$\int \frac{x^{2n-1}}{a+bx^n}\,dx = \frac{x^n}{bn} - \frac{a}{b^2n}\log(a+bx^n).$$

$$\int \frac{x^{\frac{n}{2}}}{a^2-x^n}\,dx = \frac{1}{2}\int \frac{dx}{a-x^{\frac{n}{2}}} - \frac{1}{2}\int \frac{dx}{a+x^{\frac{n}{2}}}.$$

$$\int \frac{x^{\frac{n}{2}-1}}{a+bx^n}\,dx = \frac{2}{n\sqrt{ab}}\,\text{arc tg}\, x^{\frac{n}{2}}\sqrt{\frac{b}{a}}.$$

$$\int \frac{x^{\frac{n}{2}-1}}{a-bx^n}\,dx = \frac{2}{n\sqrt{ab}}\log\frac{\sqrt{a}+x^{\frac{n}{2}}\sqrt{b}}{\sqrt{a}-x^{\frac{n}{2}}\sqrt{b}}.$$

$$\int \frac{x^m}{a+bx^n}\,dx = \frac{x^{m-n+1}}{b\,(m-n+1)} - \frac{a}{b}\int \frac{x^{m-n}}{a+bx^n}\,dx$$

$$= \frac{1}{a}\left(\frac{a}{b}\right)^{\frac{m+1}{n}}\int \frac{z^m}{1+z^n}\,dz. \qquad x = \left(\frac{a}{b}\right)^{\frac{1}{n}} z.$$

$$\int \frac{x^m}{a-bx^n}\,dx = \frac{1}{a}\left(\frac{a}{b}\right)^{\frac{m+1}{n}}\int \frac{z^m}{1-z^n}\,dz. \qquad x = \left(\frac{a}{b}\right)^{\frac{1}{n}} z.$$

$$\int \frac{x^m}{a+bx^n+cx^{2n}}\,dx = \frac{1}{c\,(\alpha-\beta)}\left[\int \frac{x^m}{x^n-\alpha}\,dx - \int \frac{x^m}{x^n-\beta}\,dx\right].$$

α, β being the real roots of $a+bx^n+cx^{2n}=0$.

$$= \frac{1}{a}\left(\frac{a}{c}\right)^{\frac{m+1}{2n}}\int \frac{z^m}{1-2z^n\cos n\theta+z^{2n}}\,dz. \qquad z = \left(\frac{c}{a}\right)^{\frac{1}{2n}} x.$$

where $\cos n\theta = -\dfrac{b}{2\sqrt{ac}}$; the roots being imaginary.

$$= \frac{2c}{\sqrt{b^2 - 4ac}} \left[\int \frac{x^m}{2cx^n + b - \sqrt{b^2 - 4ac}} dx - \int \frac{x^m}{2cx^n + b + \sqrt{b^2 - 4ac}} dx \right] \quad b^2 - 4ac > 0,\ m < n$$

$$= \frac{1}{ncq^{2n-m-1} \sin \varepsilon} \sum_{k=0}^{k=n-1} \left[-\sin (n - m - 1)\, \varepsilon_k . \frac{1}{2} \log (x^2 - 2qx \cos \varepsilon_k + q^2) + \cos (n - m - 1)\, \varepsilon_k . \operatorname{arc\,tg} \frac{x \sin \varepsilon_k}{q - x \cos \varepsilon_k} \right]$$

where $q = \left(\frac{a}{c}\right)^{\frac{1}{2n}}, \quad \cos \varepsilon = -\frac{b}{2\sqrt{ac}},$

$$\varepsilon_k = \frac{2k\pi + \varepsilon}{n}, \qquad b^2 - 4ac < 0, \qquad m < 2n.$$

$$\int \frac{x^{m-1}}{ax^{2m} + bx^m + c} dx = \frac{1}{m\sqrt{b^2 - 4ac}} \log \frac{2ax^m + b - \sqrt{b^2 - 4ac}}{2ax^m + b + \sqrt{b^2 - 4ac}} . \quad b^2 - 4ac > 0.$$

$$= \frac{2}{m\sqrt{4ac - b^2}} \operatorname{arc\,tg} \frac{2ax^m + b}{\sqrt{4ac - b^2}} . \qquad b^2 - 4ac < 0.$$

$$\int \frac{dx}{a + bx^n} = \frac{1}{a} \left(\frac{a}{b}\right)^{\frac{1}{n}} \int \frac{dz}{1 + z^n} . \qquad x = \left(\frac{a}{b}\right)^{\frac{1}{n}} z.$$

$$\int \frac{dx}{x(a + bx^n)} = \frac{1}{an} \log \frac{x^n}{a + bx^n} .$$

$$\int \frac{dx}{x(x^n - a)} = \frac{1}{an} \log \frac{x^n - a}{x^n} .$$

$$\int \frac{dx}{x^2 (a + bx^n)} = -\frac{1}{ax} - \frac{b}{a} \int \frac{x^{n-2}}{a + bx^n} dx.$$

$$\int \frac{dx}{x^3 (a + bx^n)} = -\frac{1}{2ax^2} - \frac{b}{a} \int \frac{x^{n-3}}{a + bx^n} dx.$$

$$\int \frac{dx}{x^m (a + bx^n)} = -\frac{1}{a(m - 1)\, x^{m-1}} - \frac{b}{a} \int \frac{x^{n-m}}{\Phi} dx.$$

$$\int \frac{dx}{x^n (x^n - a)} = \frac{1}{a(n - 1)\, x^{n-1}} + \frac{1}{a} \int \frac{dx}{x^n - a}$$

$$\int \frac{dx}{x^{n+1}(a+bx^n)} = -\frac{1}{anx^n} + \frac{b}{a^2n}\log\frac{\Phi}{x^n}.$$

$$\int \frac{dx}{x^{2n+1}(x^n-a)} = \frac{1}{n}\int\frac{dz}{z^3(z-a)}. \qquad x^n = z.$$

$$\int \frac{dx}{x^{n-m}(a+bx^n)} = -\frac{1}{a(n-m-1)x^{n-m-1}} - \frac{b}{a}\int\frac{x^m}{\Phi}dx.$$

$$(X)^{\pm 2}$$

$$\int \frac{(a+bx)^2}{x^m}dx = -\frac{\Phi^3}{a(m-1)x^{m-1}} + \frac{b(4-m)}{a(m-1)}\int\frac{\Phi^2}{x^{m-1}}dx.$$

$$\int \frac{x^m}{(a+bx)^2}dx = \frac{x^m}{b(m-1)(a+bx)} - \frac{am}{b(m-1)}\int\frac{x^{m-1}}{(a+bx)^2}dx.$$

$$\int \frac{x^m}{(a+bx^2)^2}dx = \frac{x^{m-1}}{b(m-3)\Phi} - \frac{a(m-1)}{b(m-3)}\int\frac{x^{m-2}}{\Phi^2}dx$$

$$= -\frac{x^{m-1}}{2b\Phi} + \frac{m-1}{2b}\int\frac{x^{m-2}}{\Phi}dx.$$

$$\int \frac{x^m}{(a+bx+cx^2)^2}dx = \frac{x^{m-1}}{c(m-3)\Phi} - \frac{a(m-1)}{c(m-3)}\int\frac{x^{m-2}}{\Phi}dx - \frac{b(m-2)}{c(m-3)}\int\frac{x^{m-1}}{\Phi^2}dx.$$

$$\int \frac{dx}{x^m(a+bx)^2} = \frac{1}{ax^{m-1}\Phi} + \frac{m}{a}\int\frac{dx}{x^m\Phi}.$$

$$\int \frac{dx}{x^m(a+bx^2)^2} = -\frac{1}{a(m-1)x^{m-1}\Phi} - \frac{b(m+1)}{a(m-1)}\int\frac{dx}{x^{m-2}\Phi^2}.$$

$$\int \frac{dx}{x^m(a+bx+cx^2)^2} = -\frac{1}{a(m-1)x^{m-1}\Phi} - \frac{bm}{a(m-1)}\int\frac{dx}{x^{m-1}\Phi^2}$$

$$-\frac{c(m+1)}{a(m-1)}\int\frac{dx}{x^{m-2}\Phi^2}.$$

$$\int \frac{x}{(a+bx^n)^2}dx = \frac{x^2}{an\Phi} + \frac{n-2}{an}\int\frac{x}{\Phi}dx.$$

$$\int \frac{x^2}{(a+bx^n)^2}dx = \frac{x^3}{an\Phi} + \frac{n-3}{an}\int\frac{x^2}{\Phi}dx.$$

$$\int \frac{x^n}{(a+bx^n)^2}\,dx = -\frac{x}{bn\Phi} + \frac{1}{bn}\int\frac{dx}{\Phi}.$$

$$\int \frac{x^{n-1}}{(a+bx^n)^2}\,dx = -\frac{1}{bn\Phi} = \frac{x^n}{an\Phi} = \frac{1}{n(ap-bq)}\cdot\frac{q+px^n}{a+bx^n}.$$

$$\int \frac{x^{n-1}}{(x^n-a)^2}\,dx = -\frac{1}{2an}\cdot\frac{x^n+a}{x^n-a}.$$

$$\int \frac{x^{2n-1}}{(a+bx^n)^2}\,dx = \frac{a}{nb^2\Phi} + \frac{1}{nb^2}\log\Phi = -\int\frac{dz}{z^{n+1}(z^n-b)}. \qquad a+bx^n = z^n x^n.$$

$$\int \frac{dx}{(a+bx^n)^2} = \frac{x}{an\Phi} + \frac{n-1}{an}\int\frac{dx}{\Phi}.$$

$$\int \frac{dx}{x.(a+bx^n)^2} = -\frac{1}{a^2}\int\frac{z^n-b}{z^{n+1}}\,dz. \qquad a+bx^n = x^n z^n.$$

$$\int \frac{dx}{x^{n-1}(a+bx^n)^2} = -\frac{1}{a(n-2)\Phi x^{n-2}} - \frac{2b(n-1)}{a(n-2)}\int\frac{x}{\Phi^2}\,dx.$$

$$\int \frac{dx}{x^m(a+bx^n)^2} = -\frac{1}{a(m-1)\Phi x^{m-1}} - \frac{b(m+n-1)}{a(m-1)}\int\frac{dx}{x^{m-n}\Phi^2}.$$

$(X)^{-3}$

$$\int \frac{x^m}{(x-a)^3}\,dx = \int\frac{(z+a)^m}{z^3}\,dz. \qquad x-a=z.$$

$$\int \frac{x^m}{(a+bx^n)^3}\,dx = \frac{x^{m-n+1}}{b(m-3n+1)\Phi^2} - \frac{a(m-n+1)}{b(m-3n+1)}\int\frac{x^{m-n}}{\Phi^3}\,dx.$$

$$\int \frac{x^{n-1}}{(a+bx^n)^3}\,dx = -\frac{1}{a^2}\int\frac{z^n-b}{z^{2n+1}}\,dz. \qquad a+bx^n = z^n x^n.$$

$$\int \frac{x^{2n-1}}{(a+bx^n)^3}\,dx = \frac{x^{2n}}{2an\Phi^2}.$$

$$\int \frac{x^{3n-1}}{(x^n-b)^3}\,dx = -\int\frac{(1+bz^n)^2}{z}\,dz. \qquad x^n - b - \frac{1}{z^n}.$$

$$\int \frac{dx}{x^m(a+bx^n)^3} = -\frac{1}{a(m-1)\Phi^2 x^{m-1}} - \frac{b(m+2n-1)}{a(m-1)}\int\frac{dx}{x^{m-n}\Phi^3}.$$

$$(X)^{\pm 1/2}$$

$$\int x^m \sqrt{a+bx}\,dx = \left\{ \frac{\Phi^m}{2m+3} - \frac{m}{1}\frac{a\Phi^{m-1}}{2m+1} + \frac{m(m-1)}{1.2}\frac{a^2\Phi^{m-2}}{2m-1} - \frac{m(m-1)(m-2)}{1.2.3}\frac{a^3\Phi^{m-3}}{2m-3} + \cdots \pm Na^m \right\} \frac{2\Phi^{\frac{3}{2}}}{b^{m+1}}.$$

$$\int x^{m+\frac{1}{2}} \sqrt{b+cx}\,dx = \frac{x^{m+1}}{m+1}\sqrt{bx+cx^2} - \frac{b}{2(m+1)}\int \frac{x^{m+1}}{\sqrt{bx+cx^2}}\,dx - \frac{c}{m+1}\int \frac{x^{m+2}}{\sqrt{bx+cx^2}}\,dx.$$

$$= \left\{ \frac{x^{m+1}}{(m+2)c} - \frac{(2m+1)\,bx^{m-2}}{(m+2)(m+1)2c^2} + \frac{(2m+1)(2m-1)b^2x^{m-3}}{(m+2)(m+1)m4c^3} - \frac{(2m+1)(2m-1)(2m-3)b^3x^{m-4}}{(m+2)(m+1)m(m-1)8c^4} + \cdots \pm N \right\} (bx+cx^2)^{\frac{3}{2}} \mp \frac{3bN}{2}\int \sqrt{bx+cx^2}\,dx.$$

$$\int x^m \sqrt{a+bx^2}\,dx = \frac{x^{m+1}}{m+1}\sqrt{\Phi} - \frac{b}{m+1}\int \frac{x^{m+2}}{\sqrt{\Phi}}\,dx.$$

$$\int x^{2m} \sqrt{a+bx^2}\,dx = -a^{m+1}\int \frac{z^2}{(z^2-b)^{m+2}}\,dz. \qquad a+bx^2 = z^2x^2.$$

$$\int x^{2m+1} \sqrt{a+bx^2}\,dx = \frac{1}{b^{m+1}}\int z^2(z^2-a)^m\,dz. \qquad a+bx^2 = z^2.$$

$$\int x^m \sqrt{bx+cx^2}\,dx = \frac{x^{m+1}}{m+1}\sqrt{\Phi} - \frac{b}{2(m+1)}\int \frac{x^{m+1}}{\sqrt{\Phi}}\,dx - \frac{c}{m+1}\int \frac{x^{m+2}}{\sqrt{\Phi}}\,dx.$$

$$= \left\{ \frac{x^{m-1}}{(m+2)c} - \frac{(2m+1)\,bx^{m-2}}{(m+2)(m+1)2c^2} + \frac{(2m+1)(2m-1)b^2x^{m-3}}{(m+2)(m+1)m4c^3} - \frac{(2m+1)(2m-1)(2m-3)b^3x^{m-4}}{(m+2)(m+1)m(m-1)8c^4} + \cdots \pm N \right\} \Phi^{\frac{3}{2}} \mp \frac{3bN}{2}\int \sqrt{\Phi}\,dx.$$

$$\int x^m \sqrt{a + bx + cx^2}\, dx = \frac{x^{m+1}}{m+1}\sqrt{\Phi} - \frac{b}{2(m+1)}\int \frac{x^{m+1}}{\sqrt{\Phi}}\, dx - \frac{c}{m+1}\int \frac{x^{m+2}}{\sqrt{\Phi}}\, dx.$$

$$\int x\sqrt{a + bx^n}\, dx = \frac{2x^{2-n}\,\Phi^{\frac{3}{2}}}{b(n+4)} - \frac{2a(2-n)}{b(n+4)}\int \frac{\sqrt{\Phi}}{x^{n-1}}\, dx.$$

$$\int x^2\sqrt{a + bx^n}\, dx = \frac{2x^{3-n}\,\Phi^{\frac{3}{2}}}{b(n+6)} - \frac{2a(3-n)}{b(n+6)}\int \frac{\sqrt{\Phi}}{x^{n-2}}\, dx.$$

$$\int x^{n-1}\sqrt{a + bx^n}\, dx = \frac{2}{3bn}\Phi^{\frac{3}{2}}.$$

$$\int x^{2n-1}\sqrt{a + bx^n}\, dx = \frac{1}{n}\int z\sqrt{a + bz}\, dz. \qquad x^n = z.$$

$$\int \frac{\sqrt{a + bx}}{x^m}\, dx = -\frac{\Phi^{\frac{3}{2}}}{a(m-1)x^{m-1}} + \frac{b(5-2m)}{2a(m-1)}\int \frac{\sqrt{\Phi}}{x^{m-1}}\, dx.$$

$$= \left\{ -\frac{1}{(m-1)ax^{m-1}} + \frac{1}{m-1}\,\frac{2m-5}{2m-4}\,\frac{b}{a^2x^{m-2}} \right.$$

$$\left. - \frac{1}{m-1}\,\frac{2m-5}{2m-4}\,\frac{2m-7}{2m-6}\,\frac{b^2}{a^3x^{m-3}} + \ldots \pm \frac{N}{x} \right\} \Phi^{\frac{3}{2}} \mp \frac{bN}{2}\int \frac{\sqrt{\Phi}}{x}\, dx.$$

$$\int \frac{\sqrt{a + bx^2}}{x^m}\, dx = -\frac{\Phi^{\frac{3}{2}}}{(m-1)ax^{m-1}} - \frac{b(m-4)}{a(m-1)}\int \frac{\sqrt{\Phi}}{x^{m-2}}\, dx$$

$$= -\frac{\sqrt{\Phi}}{(n-1)x^{m-1}} + \frac{b}{m-1}\int \frac{dx}{x^{m-2}\sqrt{\Phi}}.$$

$$\int \frac{\sqrt{bx + cx^2}}{x^m}\, dx = -\frac{\sqrt{\Phi}}{(m-1)x^{m-1}} + \frac{b}{2(m-1)}\int \frac{dx}{x^{m-1}\sqrt{\Phi}} + \frac{c}{m-1}\int \frac{dx}{x^{m-2}\sqrt{\Phi}}.$$

$$= \left\{ -\frac{2}{(2m-3)bx^m} + \frac{(m-3)\,4c}{(2m-3)(2m-5)b^2x^{m-1}} \right.$$

$$- \frac{(m-3)(m-4)\,8c^2}{(2m-3)(2m-5)(2m-7)b^3x^{m-2}}$$

$$+ \frac{(m-3)(m-4)(m-5)\,16c^3}{(2m-3)(2m-5)(2m-7)(2m-9)b^4x^{m-3}}$$

$$\left. - \ldots \pm \frac{N}{x^4} \right\} \Phi^{\frac{3}{2}} \pm cN\int \frac{\sqrt{\Phi}}{x^3}\, dx.$$

$$\int \frac{\sqrt{a + bx + cx^2}}{x^m}\, dx = -\frac{\sqrt{\Phi}}{(m-1)x^{m-1}} + \frac{b}{2(m-1)}\int \frac{dx}{x^{m-1}\sqrt{\Phi}} =$$

$$+\frac{c}{m-1}\int\frac{dx}{x^{m-2}\sqrt{\Phi}}.$$

$$\int\frac{\sqrt{a+bx^n}}{x}\,dx=\frac{2}{n}\sqrt{\Phi}+\frac{\sqrt{a}}{n}\log\frac{\sqrt{\Phi}-\sqrt{a}}{\sqrt{\Phi}+\sqrt{a}}.$$

$$\int\frac{\sqrt{bx^n-a}}{x}\,dx=\frac{2}{n}\sqrt{\Phi}-\frac{2\sqrt{a}}{n}\operatorname{arc\,tg}\sqrt{\frac{\Phi}{a}}.$$

$$\int\frac{\sqrt{bx^n+a}}{x^2}\,dx=\frac{\Phi^{\frac{3}{2}}}{ax}-\frac{b\,(3n-2)}{2a}\int x^{n-2}\sqrt{\Phi}\,dx.$$

$$\int\frac{\sqrt{a+bx^n}}{x^n}\,dx=-\frac{\sqrt{\Phi}}{(n-1)\,x^{n-1}}+\frac{nb}{2\,(n-1)}\int\frac{dx}{\sqrt{\Phi}}.$$

$$\int\frac{\sqrt{a+bx^n}}{x^{n+1}}\,dx=-\frac{\sqrt{\Phi}}{nx^n}+\frac{b}{2}\int\frac{dx}{x\sqrt{\Phi}}=\frac{2b}{n}\int\frac{z^2}{(z^2-a)^2}\,dz. \qquad a+bx^n=z^2.$$

$$\int\frac{\sqrt{a+bx^{2n}}}{x^{3n+1}}\,dx=-\frac{\Phi^{\frac{3}{2}}}{3anx^{3n}}.$$

$$\int\frac{\sqrt{ax^n+b}}{x^m}\,dx=a\int\frac{x^{n-m}}{\sqrt{\Phi}}\,dx+b\int\frac{dx}{\sqrt{\Phi}}.$$

$$\int\frac{\sqrt{a+bx^n+cx^m}}{x^p}\,dx=c\int\frac{x^{m-p}}{\sqrt{\Phi}}\,dx+b\int\frac{x^{n-p}}{\sqrt{\Phi}}\,dx+a\int\frac{dx}{x^p\sqrt{\Phi}}.$$

$$\int\frac{x^m}{\sqrt{a+bx}}\,dx=\frac{2x^m\sqrt{\Phi}}{(2m+1)\,b}-\frac{2am}{(2m+1)\,b}\int\frac{x^{m-1}}{\sqrt{\Phi}}\,dx.$$

$$=\left\{\frac{\Phi^m}{2m+1}-\frac{m}{1}\,\frac{a\Phi^{m-1}}{2m-1}+\frac{m(m-1)}{1.2}\,\frac{a^2\Phi^{m-2}}{2m-3}\right.$$

$$\left.-\frac{m\,(m-1)\,(m-2)}{1.2.3}\,\frac{a^3\Phi^{m-3}}{2m-5}+\ldots\pm a^m\right\}\frac{2\sqrt{\Phi}}{b^{m+1}}$$

$$\int\frac{x^m}{\sqrt{a+bx^2}}\,dx=\frac{x^{m-1}\sqrt{\Phi}}{bm}-\frac{a\,(m-1)}{bm}\int\frac{x^{m-2}}{\sqrt{\Phi}}\,dx.$$

$$\int\frac{x^m}{\sqrt{a^2-x^2}}\,dx=-\frac{x^{m-1}\sqrt{\Phi}}{m}+\frac{(m-1)\,a^2}{m}\int\frac{x^{m-2}}{\sqrt{\Phi}}\,dx.$$

$$\int\frac{x^{2m}}{\sqrt{a^2-x^2}}\,dx=-\sqrt{a^2-x^2}\left\{\frac{x^{2m-1}}{2m}+\frac{2m-1}{2m\,(2m-2)}a^2x^{2m-3}\right.$$

$$+\frac{(2m-1)(2m-3)}{2m(2m-2)(2m-4)}a^4x^{2m-5}+\ldots\frac{(2m-1)(2m-3)\ldots3}{2m(2m-2)\ldots2}a^{2m-2}x\Big\}$$

$$+\frac{(2m-1)(2m-3)\ldots3}{2m(2m-2)\ldots2}a^{2m}\arc\sin\frac{x}{a}\cdot$$

$$\int\frac{x^{2m+1}}{\sqrt{a^2-x^2}}dx=-\sqrt{a^2-x^2}\Big\{\frac{x^{2m}}{2m+1}+\frac{2m}{(2m+1)(2m-1)}a^2x^{2m-2}$$

$$+\frac{2m(2m-2)}{(2m+1)(2m-1)(2m-3)}a^4x^{2m-4}+\ldots+\frac{2m(2m-2)\ldots2}{(2m+1)(2m-1)\ldots1}a^{2m}\Big\}.$$

$$\int\frac{x^m}{\sqrt{ax-x^2}}dx=2a^m\int\frac{z^{2m}}{\sqrt{1-z^2}}dz. \qquad x=az^2.$$

$$=-2a^m\int\frac{dz}{(1+z^2)^{m+1}}. \qquad \sqrt{ax-x^2}=zx.$$

$$=-\frac{x^{m-1}}{m}\sqrt{\Phi}+\frac{a(2m-1)}{2m}\int\frac{x^{m-1}}{\sqrt{\Phi}}dx.$$

$$\int\frac{x^m}{\sqrt{bx+cx^2}}dx=\frac{x^{m-1}\sqrt{\Phi}}{cm}-\frac{b(2m-1)}{2cm}\int\frac{x^{m-1}}{\sqrt{\Phi}}dx.$$

$$=\Big\{\frac{x^{m-1}}{mc}-\frac{(2m-1)bx^{m-2}}{m(m-1)2c^2}+\frac{(2m-1)(2m-3)b^2x^{m-3}}{m(m-1)(m-2)4c^3}-\ldots$$

$$\pm Nx^o\Big\}\sqrt{\Phi}\pm\frac{bN}{2}\int\frac{dx}{\sqrt{\Phi}}.$$

$$\int\frac{x^m}{\sqrt{a+bx+cx^2}}dx=\frac{x^{m-1}\sqrt{\Phi}}{cm}-\frac{b(2m-1)}{2cm}\int\frac{x^{m-1}}{\sqrt{\Phi}}dx-\frac{a(m-1)}{cm}\int\frac{x^{m-2}}{\sqrt{\Phi}}dx.$$

$$\int\frac{x}{\sqrt{a+bx^n}}dx=-\frac{2x^2\sqrt{\Phi}}{an}+\frac{n+4}{an}\int x\sqrt{\Phi}\,dx.$$

$$\int\frac{x^2}{\sqrt{a+bx^n}}dx=-\frac{2x^3\sqrt{\Phi}}{an}+\frac{n+6}{an}\int x^2\sqrt{\Phi}\,dx.$$

$$\int\frac{x^{n-1}}{\sqrt{a+bx^n}}dx=\frac{2}{bn}\sqrt{\Phi}.$$

$$\int\frac{x^{2n-1}}{\sqrt{a+bx^n}}dx=\frac{2x^n}{bn}\sqrt{\Phi}-\frac{4}{3b^2n}\Phi^{\frac{3}{2}}.$$

$$\int\frac{x^{n-1}}{\sqrt{a+x^{2n}}}dx=\frac{1}{n}\log\left(x^n+\sqrt{\Phi}\right).$$

$$\int \frac{x^{n-1}}{\sqrt{a^2 - b^2x^{2n}}}\, dx = \frac{1}{bn} \arcsin \frac{b}{a} x^n = \frac{1}{bn} \arccos \frac{\sqrt{a^2 - b^2x^{2n}}}{a}.$$

$$\int \frac{dx}{x^m \sqrt{a + bx}} = 2b^{m-1} \int \frac{dz}{(z^2 - a)^m}. \qquad a + bx = z^2.$$

$$= - \frac{\sqrt{\Phi}}{a\,(m-1)\,x^{m-1}} - \frac{b\,(2m-3)}{2a\,(m-1)} \int \frac{dx}{x^{m-1}\sqrt{\Phi}}.$$

$$= \Big\{ - \frac{1}{(m-1)\,ax^{m-1}} + \frac{1}{(m-1)\,a^2} \cdot \frac{2m-3}{2m-4}\,\frac{b}{x^{m-2}}$$

$$- \frac{1}{(m-1)\,a^3}\,\frac{2m-3}{2m-4}\,\frac{2m-5}{2m-6}\,\frac{b^2}{x^{m-3}}.$$

$$+ \frac{1}{(m-1)\,a^4}\,\frac{2m-3}{2m-4}\,\frac{2m-5}{2m-6}\,\frac{2m-7}{2m-8}\,\frac{b^3}{x^{m-4}}$$

$$- \cdots \pm \frac{N}{x} \Big\} \sqrt{\Phi} \pm \frac{Nb}{2} \int \frac{dx}{x\sqrt{\Phi}}.$$

$$\int \frac{dx}{x^m \sqrt{a + bx^2}} = - \frac{\sqrt{\Phi}}{(m-1)\,ax^{m-1}} - \frac{b\,(m-2)}{a\,(m-1)} \int \frac{dx}{x^{m-2}\sqrt{\Phi}}.$$

$$\int \frac{dx}{x^m \sqrt{bx + cx^2}} = - \frac{2\sqrt{\Phi}}{b\,(2m-1)\,x^m} - \frac{2c\,(m-1)}{b\,(2m-1)} \int \frac{dx}{x^{m-1}\sqrt{\Phi}}.$$

$$= \Big\{ - \frac{2}{(2m-1)\,bx^m} + \frac{(m-1)\,4c}{(2m-1)\,(2m-3)\,b^2x^{m-1}}$$

$$- \frac{(m-1)(m-2)\,8c^2}{(2m-1)\,(2m-3)\,(2m-5)\,b^3x^{m-2}}$$

$$+ \frac{(m-1)\,(m-2)\,(m-3)\,16c^3}{(2m-1)\,(2m-3)\,(2m-5)\,(2m-7)\,b^4x^{m-3}} - \cdots \pm \frac{N}{x} \Big\} \sqrt{\Phi}.$$

$$\int \frac{dx}{x^m \sqrt{a + bx + cx^2}} = 2\sqrt{a} \int \frac{(c - az^2)^{m-1}}{(2az - b)^m}\, dz. \qquad x = \frac{2az - b}{c - az^2}.$$

$$= - \frac{\sqrt{\Phi}}{a\,(m-1)\,x^{m-1}} - \frac{b\,(2m-3)}{2a\,(m-1)} \int \frac{dx}{x^{m-1}\sqrt{\Phi}} - \frac{c\,(m-2)}{a\,(m-1)} \int \frac{dx}{x^{m-2}\sqrt{\Phi}}.$$

$$\int \frac{dx}{\sqrt{a + bx^n}} = - \frac{2x\sqrt{\Phi}}{an} + \frac{n+2}{an} \int \sqrt{\Phi}\, dx.$$

$$\int \frac{dx}{x\sqrt{a + bx^n}} = \frac{1}{n\sqrt{a}} \log \frac{\sqrt{\Phi} - \sqrt{a}}{\sqrt{\Phi} + \sqrt{a}} = \frac{2}{n\sqrt{a}} \log \frac{\sqrt{\Phi} - \sqrt{a}}{x^{\frac{n}{2}}}$$

$$= -\frac{2}{n\sqrt{a}} \log \frac{\sqrt{\Phi} + \sqrt{a}}{x^{\frac{n}{2}}}.$$

$$\int \frac{dx}{x\sqrt{x^n - a}} = \frac{1}{n\sqrt{a}} \text{arc sin} \frac{x^n - 2a}{x^n} = \frac{2}{n} \text{arc séc} \sqrt{x^n}.$$

$$\int \frac{dx}{x\sqrt{bx^n - a}} = \frac{2}{n\sqrt{a}} \text{arc tg} \sqrt{\frac{bx^n - a}{a}} = \frac{1}{n\sqrt{a}} \text{arc sin} \frac{bx^n - 2a}{bx^n}.$$

$$\int \frac{dx}{x^m\sqrt{a + bx^n}} = -\frac{\sqrt{\Phi}}{a(m-1)x^{m-1}} - \frac{b(2m - n - 2)}{2a(m-1)} \int \frac{dx}{x^{m-n}\sqrt{\Phi}}.$$

$$\int \frac{dx}{x^{n+1}\sqrt{a + bx^n}} = -\frac{\sqrt{\Phi}}{nax^n} - \frac{b}{2a} \int \frac{dx}{x\sqrt{\Phi}} = -\frac{\sqrt{\Phi}}{nax^n} + \frac{b}{na\sqrt{a}} \log \frac{\sqrt{\Phi} + \sqrt{a}}{x^{\frac{n}{2}}}.$$

$$\int \frac{dx}{x^{\frac{n}{2}+1}\sqrt{a + bx^n}} = -\frac{2\sqrt{\Phi}}{nax^{\frac{n}{2}}}.$$

$(X)^{\pm 3/2}$

$$\int x^m (a + bx)^{\frac{3}{2}} dx = \frac{2x^m \Phi^{\frac{5}{2}}}{b(2m+5)} - \frac{2am}{b(2m+5)} \int x^{m-1} \Phi^{\frac{3}{2}} dx.$$

$$\int x^m (a + bx^2)^{\frac{3}{2}} dx = \frac{x^{m+1}}{m+1} \Phi^{\frac{3}{2}} - \frac{3b}{m+1} \int x^{m+2} \sqrt{\Phi}\, dx.$$

$$\int x^{n-1} (a + bx^n)^{\frac{3}{2}} dx = \frac{2}{5bn} \Phi^{\frac{5}{2}}.$$

$$\int x^{2n-1} (a + bx^n)^{\frac{3}{2}} dx = \frac{1}{n} \int z(a + bz)^{\frac{3}{2}} dz. \qquad x^n = z.$$

$$\int \frac{(a + bx)^{\frac{3}{2}}}{x^m} dx = -\frac{\Phi^{\frac{5}{2}}}{a(m-1)x^{m-1}} + \frac{b(7 - 2m)}{2a(m-1)} \int \frac{\Phi^{\frac{3}{2}}}{x^{m-1}} dx.$$

$$\int \frac{(a + bx^2)^{\frac{3}{2}}}{x^m} dx = -\frac{\Phi^{\frac{3}{2}}}{(m-1)x^{m-1}} + \frac{3b}{m-1} \int \frac{\sqrt{\Phi}}{x^{m-2}} dx.$$

$$\int \frac{(a + bx^n)^{\frac{3}{2}}}{x} dx = \frac{2}{3n} \Phi^{\frac{3}{2}} + \frac{2a}{n}\sqrt{\Phi} + \frac{a\sqrt{a}}{n} \log \frac{\sqrt{\Phi} - \sqrt{a}}{\sqrt{\Phi} + \sqrt{a}}.$$

$$\int \frac{(a + bx^n)^{\frac{3}{2}}}{x^2} dx = -\frac{\Phi^{\frac{5}{2}}}{ax} + \frac{b(5n-2)}{2a} \int x^{n-2} \Phi^{\frac{3}{2}} dx.$$

$$\int \frac{(a+bx^n)^{\frac{3}{2}}}{x^m} dx = -\frac{\Phi^{\frac{5}{2}}}{a(m-1)x^{m-1}} - \frac{b(2m-5n-2)}{2a(m-1)} \int \frac{\Phi^{\frac{3}{2}}}{x^{m-n}} dx$$

$$= b^3 \int \frac{x^{3n-m}}{\Phi^{\frac{3}{2}}} dx + 3ab^2 \int \frac{x^{2n-m}}{\Phi^{\frac{3}{2}}} dx + 3a^2b \int \frac{x^{n-m}}{\Phi^{\frac{3}{2}}} dx + a^3 \int \frac{dx}{x^m \Phi^{\frac{3}{2}}}.$$

$$\int \frac{x^m}{(a+bx)^{\frac{3}{2}}} dx = \frac{2x^{m+1}}{a\sqrt{\Phi}} - \frac{2m+1}{a} \int \frac{x^m}{\sqrt{\Phi}} dx = \left\{ \frac{\Phi^m}{2m-1} - \frac{m}{1} \frac{a\Phi^{m-1}}{2m-3} \right.$$

$$\left. + \frac{m(m-1)}{1.2} \frac{a^2\Phi^{m-2}}{2m-5} - \dots \pm a^m \right\} \frac{2}{b^{m+1}\sqrt{\Phi}}$$

$$\int \frac{x^m}{(a+bx^2)^{\frac{3}{2}}} dx = \frac{x^{m+1}}{a\sqrt{\Phi}} - \frac{m}{a} \int \frac{x^m}{\sqrt{\Phi}} dx.$$

$$\int \frac{x^{2m}}{(a+bx^2)^{\frac{3}{2}}} dx = -a^{m-1} \int \frac{dz}{z^2(z^2-b)^m}. \qquad a+bx^2 = z^2x^2.$$

$$\int \frac{x^{2m+1}}{(a+bx^2)^{\frac{3}{2}}} dx = \frac{1}{b^{m+1}} \int \frac{(z^2-a)^m}{z^2} dz. \qquad a+bx^2 = z^2.$$

$$\int \frac{x}{(a+bx^n)^{\frac{3}{2}}} dx = \frac{2x^2}{an\sqrt{\Phi}} + \frac{n-4}{an} \int \frac{x}{\sqrt{\Phi}} dx.$$

$$\int \frac{x^2}{(a+bx^n)^{\frac{3}{2}}} dx = \frac{2x^3}{an\sqrt{\Phi}} + \frac{n-6}{an} \int \frac{x^2}{\sqrt{\Phi}} dx.$$

$$\int \frac{x^m}{(a+bx^n)^{\frac{3}{2}}} dx = \frac{2x^{m+1}}{an\sqrt{\Phi}} - \frac{2m-n+2}{an} \int \frac{x^m}{\sqrt{\Phi}} dx.$$

$$\int \frac{x^{n-1}}{(a+bx^n)^{\frac{3}{2}}} dx = -\frac{2}{bn\sqrt{\Phi}} = \frac{x^n}{an\sqrt{\Phi}}.$$

$$\int \frac{dx}{x^m(a+bx)^{\frac{3}{2}}} = -\frac{1}{(m-1)ax^{m-1}\sqrt{\Phi}} - \frac{b(2m-1)}{2a(m-1)} \int \frac{dx}{x^{m-1}\Phi^{\frac{3}{2}}}.$$

$$= \left\{ -\frac{1}{(m-1)ax^{m-1}} + \frac{1}{(m-1)a^2} \frac{2m-1}{2m-4} \frac{b}{x^{m-2}} \right.$$

$$\left. - \frac{1}{(m-1)a^3} \frac{2m-1}{2m-4} \frac{2m-3}{2m-6} \frac{b^2}{x^{m-3}} + \dots \pm \frac{N}{x} \right\} \frac{1}{\sqrt{\Phi}} \pm \frac{3bN}{2} \int \frac{dx}{x\Phi^{\frac{3}{2}}}.$$

$$\int \frac{dx}{x^m(bx+cx^2)^{\frac{3}{2}}} = -\frac{2\sqrt{\Phi}}{b(2m+1)x^m\Phi} - \frac{2c(m+1)}{b(2m+1)} \int \frac{dx}{x^{m-1}\Phi^{\frac{3}{2}}}.$$

$$\int \frac{dx}{x(a+bx^n)^{\frac{3}{2}}} = \frac{2x}{an\sqrt{\Phi}} + \frac{n-2}{an}\int \frac{dx}{\sqrt{\Phi}}.$$

$$\int \frac{dx}{x^m(a+bx^n)^{\frac{3}{2}}} = \frac{2}{anx^{m-1}\sqrt{\Phi}} + \frac{2m+n-2}{an}\int \frac{dx}{x^m\sqrt{\Phi}}.$$

$(X)^{2/3,\ \pm 1/3,\ -5/2}$

$$\int \frac{(a+bx)^{\frac{2}{3}}}{x^m}\,dx = -\frac{\Phi^{\frac{5}{3}}}{a(m-1)\,x^{m-1}} + \frac{b(8-3m)}{3a(m-1)}\int \frac{\Phi^{\frac{2}{3}}}{x^{m-1}}\,dx.$$

$$\int \frac{(a+bx)^{\frac{1}{3}}}{x^m}\,dx = -\frac{\Phi^{\frac{4}{3}}}{a(m-1)\,x^{m-1}} + \frac{b(7-3m)}{3a(m-1)}\int \frac{\Phi^{\frac{1}{3}}}{x^{m-1}}\,dx.$$

$$\int \frac{x^m}{(a+bx)^{\frac{1}{3}}}\,dx = \frac{3x^m\Phi^{\frac{2}{3}}}{(3m+2)\,b} - \frac{3am}{(3m+2)\,b}\int \frac{x^{m-1}}{\Phi^{\frac{1}{3}}}\,dx.$$

$$\int \frac{x^{m+n}}{(a+bx^n)^{\frac{5}{2}}}\,dx = -\frac{2(m+1)}{3bn}\left\{\frac{x^{m+1}}{(m+1)\,\Phi^{\frac{3}{2}}} - \frac{2x^{m+1}}{an\Phi^{\frac{1}{2}}} + \frac{2m-n+2}{an}\int \frac{x^m}{\sqrt{\Phi}}\,dx\right\}$$

$(X)^{+p}$

$$\int x^2 (a+bx)^p\,dx = \frac{\Phi^{p+1}}{b(p+3)}\left\{x^2 - \frac{2ax}{b(p+2)} + \frac{2a^2}{b^2(p+1)(p+2)}\right\}.$$

$$\int x^3 (a+bx)^p\,dx = \frac{\Phi^{p+1}}{b(p+4)}\left\{x^3 - \frac{3ax^2}{b(p+3)} + \frac{6a^2x}{b^2(p+2)(p+3)} - \frac{6a^3}{b^3(p+1)(p+2)(p+3)}\right\}.$$

$$\int x^m (a+bx)^p\,dx = \frac{x^m\Phi^{p+1}}{b(m+p+1)} - \frac{am}{b(m+p+1)}\int x^{m-1}\Phi^p\,dx$$

$$= \frac{x^{m+1}\Phi^p}{m+p+1} + \frac{ap}{m+p+1}\int x^m\Phi^{p-1}\,dx.$$

$$\int x^m (x+a)^p\,dx = \frac{1}{p+1} x^m \Phi^{p+1} - \frac{m}{(p+1)(p+2)} x^{m-1}\Phi^{p+2}$$

$$+ \frac{m(m-1)}{(p+1)(p+2)(p+3)} x^{m-2}\Phi^{p+3}$$

$$\frac{m(m-1)(m-2)}{(p+1)(p+2)(p+3)(p+4)} x^{m-3}\Phi^{p+4} + \dots \pm N x^o \Phi^{p+m+1}.$$

$$\int (a+bx^2)^p\,dx = \frac{x\Phi^p}{2p+1} + \frac{2ap}{2p+1}\int \Phi^{p-1}\,dx.$$

$$\int x^2 (a+bx^2)^p\,dx = \frac{x\Phi^{p+1}}{b(3+2p)} - \frac{a}{b(3+2p)}\int \Phi^p\,dx.$$

$$\int (bx+cx^2)^p\,dx = \int \left(cz^2 - \frac{b^2}{4c}\right)^p dz. \qquad x = z - \frac{b}{2c}\cdot$$

$$\int x^m (a+bx+cx^2)^p\,dx = \frac{x^{m-1}\Phi^{p+1}}{c(m+2p+1)} - \frac{a(m-1)}{c(m+2p+1)}\int x^{m-2}\Phi^p\,dx$$

$$- \frac{b(m+p)}{c(m+2p+1)}\int x^{m-1}\Phi^p\,dx.$$

$$\int x^5 (a+bx^3)^p\,dx = \frac{1}{3b^2}\left\{\frac{\Phi^{p+2}}{p+2} - \frac{a\Phi^{p+1}}{p+1}\right\}.$$

$$\int x^m (a+bx^4)^p\,dx = \frac{x^{m+1}\Phi^p}{m+1} - \frac{4bp}{m+1}\int x^{m+4}\Phi^{p-1}\,dx.$$

$$\int (a+bx^n)^p\,dx = a\int \Phi^{p-1}\,dx + b\int x^n\Phi^{p-1}\,dx = x\,\Phi^p - bnp\int x^n\Phi^{p-1}\,dx.$$

$$\int x (a+bx^n)^p\,dx = \frac{x^2\Phi^p}{np+2} + \frac{anp}{np+2}\int x\Phi^{p-1}\,dx.$$

$$\int x^2 (a+bx^n)^p\,dx = \frac{x^3\Phi^p}{np+3} + \frac{anp}{np+3}\int x^2\Phi^{p-1}\,dx.$$

$$\int x^m (a+bx^n)^p\,dx = \frac{x^{m+1}\Phi^p}{m+np+1} + \frac{anp}{m+np+1}\int x^m\Phi^{p-1}\,dx.$$

$$= \frac{x^{m-n+1}\Phi^{p+1}}{b.(np+m+1)} - \frac{a(m-n+1)}{b(np+m+1)}\int x^{m-n}\Phi^p\,dx.$$

$$= \frac{x^{m-n+1}\Phi^{p+1}}{bn(p+1)} - \frac{m-n+1}{bn(p+1)}\int x^{m-n}\Phi^{p+1}\,dx.$$

$$= \frac{x^{m+1}\Phi^p}{m+1} - \frac{bnp}{m+1}\int x^{m+n}\Phi^{p-1}\,dx.$$

$$= \frac{x^{m+1}\Phi^{p+1}}{a(m+1)} - \frac{b(np+m+n+1)}{a(m+1)}\int x^{m+n}\Phi^p\,dx.$$

$$= -\frac{x^{m+1}\,\Phi^{p+1}}{an(p+1)} + \frac{np+m+n+1}{an(p+1)}\int x^m\Phi^{p+1}\,dx.$$

$$\int (ax^r + bx^{r+n})^p\,dx = \frac{x\Phi^p}{pr+np+1} + \frac{anp}{pr+np+1}\int x^r\,\Phi^{p-1}\,dx.$$

$$\int x^m\,(ax^r + bx^{r+n})^p\,dx = \frac{x^{m+1}\Phi}{m+pr+1} - \frac{pnb}{m+pr+1}\int x^{m+r+n}\,\Phi^{p-1}\,dx.$$

$$= \frac{x^{m-r-n+1}\,\Phi^{p+1}}{(m+pr+np+1)\,b} - \frac{(m+pr-n+1)\,a}{(m+pr+np+1)\,b}\int x^{m-n}\,\Phi^p\,dx.$$

$$= \frac{x^{m+1}\,\Phi^p}{m+pr+np+1} + \frac{anp}{m+pr+np+1}\int x^{m+r}\,\Phi^{p-1}\,dx.$$

$$\int (a + bx^n + cx^m)^p\,dx = a\int \Phi^{p-1}dx + b\int x^n\Phi^{p-1}\,dx + c\int x^m\Phi^{p-1}\,dx.$$

$$\int x^m\,(a + bx^n + cx^{2n})^p\,dx = \frac{1}{m+1}\Big\{ x^{m+1}\,\Phi^p - bnp\int x^{m+n}\,\Phi^{p-1}\,dx$$

$$- 2cnp\int x^{m+2n}\,\Phi^{p-1}\,dx \Big\}.$$

$$= \frac{1}{bn(p+1)}\Big\{ x^{m-n+1}\,\Phi^{p+1} - (m-n+1)\int x^{m-n}\,\Phi^{p+1}\,dx$$

$$- 2cn\,(p+1)\int x^{m+n}\,\Phi^p\,dx \Big\}.$$

$$= \frac{1}{2cn(p+1)}\Big\{ x^{m-2n+1}\,\Phi^{p+1}$$

$$- (m-2n+1)\int x^{m-2n}\,\Phi^{p+1}\,dx - bn\,(p+1)\int x^{m-n}\,\Phi^p\,dx \Big\}.$$

$$= \frac{1}{m+np+1}\Big\{ x^{m+1}\,\Phi^p + anp\int x^m\Phi^{p-1}dx$$

$$- cnp\int x^{m+2n}\Phi^{p-1}\,dx \Big\}.$$

$$= \frac{1}{m+2np+1}\Big\{ x^{m+1}\Phi^p + 2anp\int x^m\,\Phi^{p-1}dx$$

$$+ bnp \int x^{m+n}\Phi^{p-1}\,dx \Big\}.$$

$$= \frac{1}{b(m+np+1)}\Big\{ x^{m-n+1}\Phi^{p+1} - a(m-n+1)\int x^{m-n}\Phi^{p}\,dx$$

$$- c(m+2np+n+1)\int x^{m+n}\Phi^{p}dx\Big\}.$$

$$= \frac{1}{bn(p+1)}\Big\{ - x^{m-n+1}\Phi^{p+1} + (m+2np+n+1)\int x^{m-n}\Phi^{p+1}\,dx$$

$$- 2an(p+1)\int x^{m-n}\Phi^{p}dx\Big\}.$$

$$= \frac{1}{cn(p+1)}\Big\{ x^{m-2n+1}\Phi^{p+1} + an(p+1)\int x^{m-2n}\Phi^{p}\,dx$$

$$-(m+np-n+1)\int x^{m-2n}\Phi^{p+1}\,dx\Big\}.$$

$$= \frac{1}{an(p+1)}\Big\{ - x^{m+1}\Phi^{p+1} + (m+np+n+1)\int x^{m}\Phi^{p+1}\,dx$$

$$+ cn(p+1)\int x^{m+2n}\Phi^{p}\,dx\Big\}.$$

$$= \frac{1}{2an(p+1)}\Big\{ - x^{m+1}\Phi^{p+1} + (m+2np+2n+1)\int x^{m}\Phi^{p+1}dx$$

$$- bn(p+1)\int x^{m+n}\Phi^{p}\,dx\Big\}.$$

$$= \frac{1}{a(m+1)}\Big\{ x^{m+1}\Phi^{p+1} - b(m+np+n+1)\int x^{m+n}\Phi^{p}dx$$

$$- c(m+2np+2n+1)\int x^{m+2n}\Phi^{p}\,dx\Big\}.$$

$$= \frac{1}{c(m+2np+1)}\Big\{ x^{m-2n+1}\Phi^{p+1} - b(m+np-n+1)\int x^{m-n}\Phi^{p}dx$$

$$- a(m-2n+1)\int x^{m-2n}\Phi^{p}dx\Big\}.$$

$$\int \frac{(a+bx)^{p}}{x}\,dx = \frac{\Phi^{p}}{p} + a\int \frac{\Phi^{p-1}}{x}\,dx.$$

$$\int \frac{(a+bx)^{p}}{x^{2}}\,dx = -\frac{\Phi^{p+1}}{ax} + \frac{bp}{a}\int \frac{\Phi^{p}}{x}\,dx.$$

$$\int \frac{(a+bx)^p}{x^3}\,dx = -\frac{\Phi^{p+1}}{2a}\left\{\frac{1}{x^2}+\frac{b\,(p-1)}{ax}\right\}+\frac{b^2p\,(p-1)}{2a^2}\int\frac{\Phi^p}{x}\,dx.$$

$$\int \frac{(a+bx)^p}{x^m}\,dx = -\frac{\Phi^{p+1}}{a\,(m-1)\,x^{m-1}}+\frac{b\,(p-m+2)}{a\,(m-1)}\int\frac{\Phi^p}{x^{m-1}}\,dx.$$

$$\int \frac{(a+bx)^p}{x^{p+q}}\,dx = -\frac{b^p}{(q-1)\,x^{q-1}}-\frac{p}{1}\,\frac{ab^{p-1}}{qx^q}-\frac{p\,(p-1)}{1.2}\,\frac{a^2b^{p-2}}{(q+1)\,x^{q+1}}-\dots.$$

$$\int \frac{(a+bx)^p}{x^{p+1}}\,dx = b^p\log x-\frac{p}{1}\,a\,\frac{b^{p-1}}{x}-\frac{p\,(p-1)}{1.2}\,a^2\,\frac{b^{p-2}}{2x^2}$$

$$-\frac{p\,(p-1)\,(p-2)}{1.2.3}\,a^3\,\frac{b^{p-3}}{3x^3}-\dots$$

$$\int \frac{(a-x)^p}{x^{p+2}}\,dx = -\frac{\Phi^{p+1}}{a\,(p+1)\,x^{p+1}}.$$

$$\int \frac{(a+bx^2)^{p-1}}{x^p}\,dx = -\frac{\Phi^{p-1}}{(p-1)\,x^{p-1}}+2b\int\frac{\Phi^{p-2}}{x^{p-2}}\,dx.$$

$$\int \frac{(a+bx+cx^2)^p}{x}\,dx = \frac{\Phi^p}{2p}+\frac{b}{2}\int\Phi^{p-1}\,dx+a\int\frac{\Phi^{p-1}}{x}\,dx.$$

$$\int \frac{(a+bx+cx^2)^p}{x^m}\,dx = -\frac{\Phi^{p+1}}{a\,(m-1)\,x^{m-1}}+\frac{b\,(p-m+2)}{a\,(m-1)}\int\frac{\Phi^p}{x^{m-1}}\,dx$$

$$+\frac{c\,(2p-m+3)}{a\,(m-1)}\int\frac{\Phi^p}{x^{m-2}}\,dx.$$

$$\int \frac{(a+bx^n)^p}{x^{2n+1}}\,dx = \frac{b^2}{n}\int\frac{z^p}{(z-a)^3}\,dz. \qquad a+bx^n=z.$$

$$= -\frac{\Phi^p}{2nx^{2n}}+\frac{bp}{2}\int\frac{\Phi^{p-1}}{x^{n+1}}\,dx.$$

$$\int \frac{(a+bx^n)^p}{x^{pn+1}}\,dx = -\int\frac{z^{pn+n-1}}{z^n-b}\,dz. \qquad a+bx^n=z^nx^n.$$

$$\int \frac{(ax^r+bx^{r+n})^p}{x^m}\,dx = -\frac{\Phi^{p+1}}{(m-pr-1)\,ax^{m+r-1}}$$

$$-\frac{(m-n-pr-np-1)\,b}{(m-pr-1)\,a}\int\frac{\Phi^p}{x^{m-n}}\,dx.$$

$$= -\frac{\Phi^p}{(m-pr-np-1)\,x^{m-1}}-\frac{pna}{m-pr-np-1}\int\frac{\Phi^{p-1}}{x^{m-r}}\,dx.$$

$$\int \frac{(a + bx^n + cx^{2n})^p}{x^m}\,dx = -\frac{\Phi^{p+1}}{a(m-1)x^{m-1}} - \frac{b\,(m - pn - n - 1)}{a\,(m-1)} \int \frac{\Phi^p}{x^{m-n}}\,dx$$

$$-\frac{c\,(m - 2pn - 2n - 1)}{a\,(m-1)} \int \frac{\Phi^p}{x^{m-2n}}\,dx.$$

$(X)^{-p}$

$$\int \frac{x}{(a+bx)^p}\,dx = \frac{a}{b^2\,(p-1)\,\Phi^{p-1}} - \frac{1}{b^2\,(p-2)\,\Phi^{p-2}}.$$

$$\int \frac{x^2}{(a+bx)^p}\,dx = \frac{1}{b^3}\left\{-\frac{1}{(p-3)\,\Phi^{p-3}} + \frac{2a}{(p-2)\,\Phi^{p-2}} - \frac{a^2}{(p-1)\,\Phi^{p-1}}\right\}.$$

$$\int \frac{x^3}{(a+bx)^p}\,dx = \frac{1}{b^4}\left\{-\frac{1}{(p-4)\,\Phi^{p-4}} + \frac{3a}{(p-3)\,\Phi^{p-3}} - \frac{3a^2}{(p-2)\,\Phi^{p-2}} + \frac{a^3}{(p-1)\,\Phi^{p-1}}\right\}.$$

$$\int \frac{x^m}{(a+bx)^p}\,dx = \frac{1}{b^{m+1}} \int \frac{(z-a)^m}{z^p}\,dz. \qquad a + bx = z.$$

$$= \frac{x^{m+1}}{a\,(p-1)\,\Phi^{p-1}} + \frac{p-m-2}{a\,(p-1)} \int \frac{x^m}{\Phi^{p-1}}\,dx.$$

$$= -\frac{1}{b^{m+1}\,\Phi^{p-m-1}}\left\{\frac{1}{p-m-1} - \frac{m}{p-m}\cdot\frac{a}{\Phi} + \frac{m(m-1)}{1.2\,(p-m+1)}\left(\frac{a}{\Phi}\right)^2\right.$$

$$- \frac{m(m-1)(m-2)}{1.2.3\,(p-m+2)}\left(\frac{a}{\Phi}\right)^3 + \cdots$$

$$\left.\pm \frac{1}{p-1}\left(\frac{a}{\Phi}\right)^m\right\}. \qquad p > m + 1.$$

$$\int \frac{x^m}{(x-a)^p}\,dx = -\frac{x^m}{(p-1)\,\Phi^{p-1}} + \frac{m}{p-1} \int \frac{x^{m-1}}{\Phi^{p-1}}\,dx.$$

$$\int \frac{x^p}{(a+bx)^p}\,dx = \frac{\Phi}{b^{p+1}}\left\{1 - \frac{p}{1}\,\frac{a}{\Phi}\log\Phi - \frac{p\,(p-1)}{1.2}\left(\frac{a}{\Phi}\right)^2 + \frac{1}{2}\,\frac{p\,(p-1)\,(p-2)}{1.2.3}\left(\frac{a}{\Phi}\right)^3\right.$$

$$\left.-\frac{1}{3}\,\frac{p(p-1)(p-2)(p-3)}{1.2.3.4}\left(\frac{a}{\Phi}\right)^4 + \cdots \pm \frac{1}{1-p}\left(\frac{a}{\Phi}\right)^p\right\}.$$

$$\int \frac{dx}{(a+bx^2)^p} = \frac{x}{2a\,(p-1)\,\Phi^{p-1}} + \frac{2p-3}{2a\,(p-1)} \int \frac{dx}{\Phi^{p-1}}.$$

$$\int \frac{dx}{(a^2+x^2)^p} = \frac{x}{2a^2(p-1)\,\Phi^{p-1}} + \frac{2p-3}{2a^2(p-1)}\int \frac{dx}{\Phi^{p-1}}.$$

$$= \frac{1}{2p-1}\sum_{k=o}^{k=p-2} \frac{(2p-1)(2p-3)\cdots(2p-1-2k)}{(p-1)(p-2)\cdots(p-1-k)} \cdot \frac{x}{(2a^2)^{k+1}\Phi^{p-k-1}}$$

$$+ \frac{(2p-3)(2p-5)\cdots 5.3.1}{(p-1)(p-2)\cdots 2.1\,(2a^2)^{p-1}}\,\frac{1}{a}\,\text{arc tg}\,\frac{x}{a}$$

$$= \frac{1}{(p-1)\,2a^2}\cdot\frac{x}{\Phi^{p-1}} + \frac{2p-3}{(p-1)(p-2)(2a^2)^2}\cdot\frac{x}{\Phi^{p-2}}$$

$$+ \frac{(2p-3)(2p-5)}{(p-1)(p-2)(p-3)(2a^2)^3}\cdot\frac{x}{\Phi^{p-3}} + \cdots$$

$$+ \frac{(2p-3)(2p-5)\cdots 5.3}{(p-1)(p-2)\cdots 2.1\,(2a^2)^{p-1}}\cdot\frac{x}{\Phi}$$

$$+ \frac{(2p-3)(2p-5)\cdots 5.3.1}{(p-1)(p-2)\cdots 2.1\,(2a^2)^{p-1}}\cdot\frac{1}{a}\,\text{arc tg}\,\frac{x}{a}.$$

$$\int \frac{dx}{(a^2-x^2)^p} = \frac{x}{2a^2(p-1)\,\Phi^{p-1}} + \frac{2p-3}{2a^2(p-1)}\int \frac{dx}{\Phi^{p-1}}.$$

$$= \frac{1}{2p-1}\sum_{k=o}^{k=p-2} \frac{(2p-1)(2p-3)\cdots(2p-1-2k)}{(p-1)(p-2)\cdots(p-1-k)}\,\frac{x}{(2a^2)^{k+1}\,\Phi^{p-1-k}}$$

$$+ \frac{(2p-3)(2p-5)\cdots 3.1}{(p-1)(p-2)\cdots 2.1}\,\frac{a}{(2a^2)^p}\log\frac{a+x}{a-x}.$$

$$\int \frac{dx}{(x^2-a^2)^p} = \frac{1}{(2a)^{2p-1}}\int \frac{(z-1)^{2p-2}}{z^p}\,dz. \qquad x-a = z\,(x+a).$$

$$= -\frac{x}{2a^2(p-1)\,\Phi^{p-1}} - \frac{2p-3}{2a^2(p-1)}\int \frac{dx}{\Phi^{p-1}}.$$

$$\int \frac{x^2}{(a+bx^2)^p}\,dx = -\frac{x}{2b(p-1)\,\Phi^{p-1}} + \frac{1}{2b(p-1)}\int \frac{dx}{\Phi^{p-1}}.$$

$$\int \frac{x^m}{(a+bx^2)^p}\,dx = \frac{1}{b}\int \frac{x^{m-2}}{\Phi^{p-1}}\,dx - \frac{a}{b}\int \frac{x^{m-2}}{\Phi^p}\,dx$$

$$= -\frac{x^{m-1}}{2b(p-1)\,\Phi^{p-1}} + \frac{m-1}{2b(p-1)}\int \frac{x^{m-2}}{\Phi^{p-1}}\,dx.$$

$$= -\frac{x^{m-1}}{b(2p-m-1)\,\Phi^{p-1}} + \frac{a(m-1)}{b(2p-m-1)}\int \frac{x^{m-2}}{\Phi^p}\,dx.$$

$$\int \frac{x^m}{(a^2 + x^2)^p}\,dx = -\frac{x^{m-1}}{2(p-1)\Phi^{p-1}} + \frac{m-1}{2(p-1)}\int \frac{x^{m-2}}{\Phi^{p-1}}\,dx.$$

$$\int \frac{dx}{(bx + cx^2)^p} = \int \frac{dz}{\left(cz^2 - \frac{b^2}{4c}\right)^p}. \qquad x = z - \frac{b}{2c}.$$

$$\int \frac{dx}{(a + bx + cx^2)^p} = 2^{2p-1} c^{p-1} \int \frac{dz}{(z^2 + 4ac - b^2)^p}. \qquad 2cx + b = z.$$

$$= \frac{2^{2p-1} c^{p-1}}{(4ac - b^2)^{p-\frac{1}{2}}} \int \frac{dz}{(z^2 + 1)^p}. \qquad x + \frac{b}{2c} = \frac{z}{2c}\sqrt{4ac - b^2}.$$

$$= \frac{2cx + b}{(p-1)(4ac - b^2)\Phi^{p-1}} + \frac{2c(2p-3)}{(p-1)4ac - b^2)}\int \frac{dx}{\Phi^{p-1}}.$$

$$\int \frac{dx}{(a + bx - cx^2)^p} = 2^{2p-1}c^{p-1} \int \frac{dz}{(4ac + b^2 - z^2)^p}. \qquad 2cx - b = z.$$

$$\int \frac{x}{(a + bx + cx^2)^p}\,dx = -\frac{1}{2c(p-1)\Phi^{p-1}} - \frac{b}{2c}\int \frac{dx}{\Phi^{p-1}}$$

$$= -\frac{bx + 2a}{(p-1)(4ac - b^2)\Phi^{p-1}} - \frac{b(2p-3)}{(p-1)(4ac - b^2)}\int \frac{dx}{\Phi^{p-1}}.$$

$$\int \frac{x^2}{(a + bx + cx^2)^p}\,dx = \frac{(b^2 - 2ac)x + ab}{c(p-1)(4ac - b^2)\Phi^{p-1}} + \frac{2ac + b^2(p-2)}{c(p-1)(4ac - b^2)}\int \frac{dx}{\Phi^{p-1}}.$$

$$\int \frac{x^3}{(a + bx + cx^2)^p}\,dx = \frac{1}{c}\int \frac{x}{\Phi^{p-1}}\,dx - \frac{b}{c}\int \frac{x^2}{\Phi^p}\,dx - \frac{a}{c}\int \frac{x}{\Phi^p}\,dx.$$

$$\int \frac{x^m}{(a + bx + cx^2)^p}\,dx = -\frac{x^{m-1}}{c(2p-m-1)\Phi^{p-1}} - \frac{b(p-m)}{c(2p-m-1)}\int \frac{x^{m-1}}{\Phi^p}\,dx$$

$$+ \frac{a(m-1)}{c(2p-m-1)}\int \frac{x^{m-2}}{\Phi^p}\,dx.$$

$$\int \frac{dx}{(a + bx^3)^p} = \frac{x}{3a(p-1)\Phi^{p-1}} + \frac{3p-4}{3a(p-1)}\int \frac{dx}{\Phi^{p-1}}.$$

$$\int \frac{x}{(a + bx^3)^p}\,dx = \frac{x^2}{3a(p-1)\Phi^{p-1}} + \frac{3p-5}{3a(p-1)}\int \frac{x}{\Phi^{p-1}}\,dx.$$

$$\int \frac{x^3}{(a + bx^3)^p}dx = -\frac{x}{3b(p-1)\Phi^{p-1}} + \frac{1}{3b(p-1)}\int \frac{dx}{\Phi^{p-1}}.$$

$$\int \frac{x^m}{(a+bx^3)^p}\,dx = \frac{x^{m+1}}{3a\,(p-1)\,\Phi^{p-1}} - \frac{m-3p+4}{3a\,(p-1)} \int \frac{x^m}{\Phi^{p-1}}\,dx.$$

$$\int \frac{dx}{(a+bx^4)^p} = \frac{x}{4a\,(p-1)\,\Phi^{p-1}} + \frac{4p-5}{4a\,(p-1)} \int \frac{dx}{\Phi^{p-1}}.$$

$$\int \frac{x^2}{(a+bx^4)^p}\,dx = \frac{x^3}{4a\,(p-1)\,\Phi^{p-1}} + \frac{4p-7}{4a\,(p-1)} \int \frac{x^2}{\Phi^{p-1}}\,dx.$$

$$\int \frac{x^m}{(a+bx^4)^p}\,dx = \frac{x^{m+1}}{4a\,(p-1)\,\Phi^{p-1}} + \frac{4p-m-5}{4a\,(p-1)} \int \frac{x^m}{\Phi^{p-1}}\,dx$$

$$= -\frac{x^{m-3}}{4b\,(p-1)\,\Phi^{p-1}} + \frac{m-3}{4b\,(p-1)} \int \frac{x^{m-4}}{\Phi^{p-1}}\,dx$$

$$\int \frac{x^2}{(a+bx^2+cx^4)^p}\,dx = \frac{2cx^3+bx}{2\,(p-1)\,(4ac-b^2)\,\Phi^{p-1}} + \frac{c\,(4p-7)}{(p-1)\,(4ac-b^2)} \int \frac{x^2}{\Phi^{p-1}}\,dx$$

$$- \frac{b}{2\,(p-1)\,(4ac-b^2)} \int \frac{dx}{\Phi^{p-1}}.$$

$$\int \frac{x^m}{(a+bx^2+cx^4)^p}\,dx = \frac{1}{c} \int \frac{x^{m-4}}{\Phi^{p-1}}\,dx - \frac{b}{c} \int \frac{x^{m-2}}{\Phi^p}\,dx - \frac{a}{c} \int \frac{x^{m-4}}{\Phi^p}\,dx.$$

$$= \frac{(2cx^2+b)\,x^{m-1}}{2\,(p-1)\,(4ac-b^2)\,\Phi^{p-1}} + \frac{c\,(4p-m-5)}{(p-1)\,(4ac-b^2)} \int \frac{x^m}{\Phi^{p-1}}\,dx$$

$$- \frac{b\,(m-1)}{2\,(p-1)(4ac-b^2)} \int \frac{x^{m-2}}{\Phi^{p-1}}\,dx.$$

$$\int \frac{x^m}{(a+bx^5)^p}\,dx = -\frac{x^{m-4}}{5b\,(p-1)\,\Phi^{p-1}} + \frac{m-4}{5b\,(p-1)} \int \frac{x^{m-5}}{\Phi^{p-1}}\,dx.$$

$$\int \frac{x^m}{(a+bx^n)^p}\,dx = \frac{1}{b} \int \frac{x^{m-n}}{\Phi^{p-1}}\,dx - \frac{a}{b} \int \frac{x^{m-n}}{\Phi^p}\,dx.$$

$$\int \frac{x^{pn-1}}{(a+bx^n)^p}\,dx = -\int \frac{z^{n-pn-1}}{z^n-b}\,dz. \qquad a+bx^n = z^n x^n.$$

$$\int \frac{x^{pn-n-1}}{(a+bx^n)^p}\,dx = \frac{x^{n\,(p-1)}}{an\,(p-1)\,\Phi^{p-1}}.$$

$$\int \frac{dx}{(ax^r+bx^{r+n})^p} = \frac{1}{(p-1)\,nax^{r-1}\,\Phi^{p-1}} + \frac{pr+np-n-1}{(p-1)\,na} \int \frac{dx}{x^r\,\Phi^{p-1}}.$$

$$\int \frac{x^m}{(ax^r+bx^{r+n})^p}\,dx = -\frac{x^{m-r\cdot n+1}}{(p-1)\,nb\,\Phi^{p-1}} + \frac{m-pr-n+1}{(p-1)\,nb} \int \frac{x^{m-r-n}}{\Phi^{p-1}}\,dx.$$

$$= \frac{x^{m-r-n+1}}{(m - pr - np + 1)\, b\Phi^{p-1}} - \frac{(m - pr - n + 1)\, a}{(m - pr - np + 1)\, b} \int \frac{x^{m-n}}{\Phi^p}\, dx.$$

$$= \frac{x^{m-r+1}}{(p - 1)\, na\Phi^{p-1}} - \frac{m + n - pr - np + 1}{(p - 1)\, na} \int \frac{x^{m-r}}{\Phi^{p-1}}\, dx.$$

$$\int \frac{x^m}{(a + bx^n + cx^{2n})^p}\, dx = \frac{1}{c} \int \frac{x^{m-2n}}{\Phi^{p-1}}\, dx - \frac{b}{c} \int \frac{x^{m-n}}{\Phi^p}\, dx - \frac{a}{c} \int \frac{x^{m-2n}}{\Phi^p} dx.$$

$$= \frac{(2cx^n + b)\, x^{m-n+1}}{n\,(p - 1)\,(4ac - b^2)\,\Phi^{p-1}} + \frac{2c\,(2pn - 2n - m - 1)}{n\,(p - 1)\,(4ac - b^2)} \int \frac{x^m}{\Phi^{p-1}}\, dx$$

$$- \frac{b\,(m - n + 1)}{n\,(p - 1)\,(4ac - b^2)} \int \frac{x^{m-n}}{\Phi^{p-1}}\, dx.$$

$$\int \frac{dx}{x\,(a + bx)^p} = \frac{1}{a\,(p - 1)\,\Phi^{p-1}} + \frac{1}{a} \int \frac{dx}{x\Phi^{p-1}}.$$

$$= \sum_{k=o}^{k=p-2} \frac{1}{(p - k - 1)\, a^{k+1}\, \Phi^{p-k-1}} - \frac{1}{a^p} \log \frac{\Phi}{x}.$$

$$= \frac{1}{a^p} \log x - \frac{b}{a^{p+1}} \int \left\{ \left(\frac{a}{\Phi}\right)^p + \left(\frac{a}{\Phi}\right)^{p-1} + \cdots + \frac{a}{\Phi} \right\} dx.$$

$$\int \frac{dx}{x^2\,(a + bx)^p} = - b \sum_{k=o}^{k=p-2} \frac{k + 1}{(p - k - 1)\, a^{k+2}\, \Phi^{p-k-1}} - \frac{1}{a^p x} + \frac{bp}{a^{p+1}} \log \frac{\Phi}{x}.$$

$$\int \frac{dx}{x^m\,(a + bx)^p} = - \int \frac{z^{m+p-2}}{(az + b)^p}\, dz. \qquad xz = 1.$$

$$= \frac{(-b)^{m-1}}{a^m} \left\{ \frac{1}{(p-1)\,\Phi^{p-1}} + \frac{m}{(p-2)\,a\Phi^{p-2}} + \frac{m\,(m + 1)}{1.2(p-3)a^2\Phi^{p-3}} \right.$$

$$\left. + \frac{m(m+1)(m+2)}{1.2.3(p-4)\,a^3\Phi^{p-4}} + \cdots \frac{m(m+1)\cdots(m+p-3)}{1.2.3\cdots(p-2)\,a^{p-2}\Phi} \right\}$$

$$+ \frac{1}{a^p} \left\{ - \frac{1}{(m-1)\, x^{m-1}} + \frac{pb}{(m-2)\,ax^{m-2}} - \frac{p\,(p + 1)\, b^2}{1.2(m-3)\,a^2x^{m-3}} \right.$$

$$\left. + \frac{p(p+1)(p+2)b^3}{1.2.3(m-4)\, a^3x^{m-4}} \cdots \pm \frac{p(p+1)\cdots(p+m-3)b^{m-2}}{1.2.3\cdots(m-2)\, a^{m-2}x} \right\}$$

$$+ (-1)^m \frac{m\,(m + 1)\cdots(m + p - 2)}{1.2.3\cdots(p - 1)} \frac{b^{m-1}}{a^{m+p-1}} \log \frac{\Phi}{x}.$$

$$= \frac{1}{a\,(p - 1)\, x^{m-1}\Phi^{p-1}} + \frac{m + p - 2}{a\,(p - 1)} \int \frac{dx}{x^m\Phi^{p-1}}.$$

$$= -\frac{1}{(m-1)\,ax^{m-1}\Phi^{p-1}} - \frac{b\,(m+p-2)}{a\,(m-1)}\int\frac{dx}{x^{m-1}\Phi^{p}}.$$

$$\int\frac{dx}{x^2\,(a+bx^2)^p} = \frac{1}{a}\int\frac{dx}{x^2\,\Phi^{p-1}} - \frac{b}{a}\int\frac{dx}{\Phi^p}.$$

$$\int\frac{dx}{x^4\,(a+bx^2)^p} = \frac{1}{a}\int\frac{dx}{x^4\,\Phi^{p-1}} - \frac{b}{a}\int\frac{dx}{x^2\Phi^p}.$$

$$\int\frac{dx}{x^m\,(a+bx^2)^p} = \frac{1}{a}\int\frac{dx}{x^m\Phi^{p-1}} - \frac{b}{a}\int\frac{dx}{x^{m-2}\Phi^p} = -\frac{1}{a\,(m-1)\,x^{m-1}\Phi^{p-1}}$$

$$-\frac{b\,(m+2p-3)}{a\,(m-1)}\int\frac{dx}{x^{m-2}\Phi^p}.$$

$$\int\frac{dx}{x\,(a+bx+cx^2)^p} = \frac{1}{2a\,(p-1)\,\Phi^{p-1}} + \frac{1}{a}\int\frac{dx}{x\Phi^{p-1}} - \frac{b}{2a}\int\frac{dx}{\Phi^p}.$$

$$\int\frac{dx}{x^2\,(a+bx+cx^2)^p} = -\frac{1}{ax\Phi^{p-1}} - \frac{bp}{a}\int\frac{dx}{x\Phi^p} - \frac{c\,(2p-1)}{a}\int\frac{dx}{\Phi^p}.$$

$$\int\frac{dx}{x^3\,(a+bx+cx^2)^p} = -\frac{1}{2ax^2\Phi^{p-1}} - \frac{b\,(p+1)}{2a}\int\frac{dx}{x^2\Phi} - \frac{cp}{a}\int\frac{dx}{x\Phi^p}$$

$$\int\frac{dx}{x^m\,(a+bx+cx^2)^p} = \frac{1}{a}\int\frac{dx}{x^m\Phi^{p-1}} - \frac{b}{a}\int\frac{dx}{x^{m-1}\Phi^p} - \frac{c}{a}\int\frac{dx}{x^{m-2}\Phi^p}.$$

$$= -\frac{1}{a\,(m-1)\,x^{m-1}\Phi^{p-1}} - \frac{b\,(p+m-2)}{a\,(m-1)}\int\frac{dx}{x^{m-1}\Phi^p}$$

$$-\frac{c\,(2p+m-3)}{a\,(m-1)}\int\frac{dx}{x^{m-2}\Phi^p}$$

$$\int\frac{dx}{x^2\,(a+bx^3)^p} = \frac{1}{a}\int\frac{dx}{x^2\Phi^{p-1}} - \frac{b}{a}\int\frac{x}{\Phi^p}\,dx.$$

$$\int\frac{dx}{x^3\,(a+bx^3)^p} = \frac{1}{a}\int\frac{dx}{x^3\Phi^{p-1}} - \frac{b}{a}\int\frac{dx}{\Phi^p}.$$

$$\int\frac{dx}{x^m\,(a+bx^3)^p} = \frac{1}{3a\,(p-1)\,x^{m-1}\Phi^{p-1}} + \frac{m+3p-4}{3a\,(p-1)}\int\frac{dx}{x^m\Phi^{p-1}}$$

$$\int\frac{dx}{x^m\,(a+bx^4)^p} = \frac{1}{a}\int\frac{dx}{x^m\Phi^{p-1}} - \frac{b}{a}\int\frac{dx}{x^{m-4}\Phi^p}.$$

$$\int \frac{dx}{(a+bx^2+cx^4)^p} = -\frac{bcx^3+(b^2-2ac)\,x}{2a(p-1)(4ac-b^2)\,\Phi^{p-1}} - \frac{bc\,(4p-7)}{2a\,(p-1)\,(4ac-b^2)} \int \frac{x^2}{\Phi^{p-1}}\,dx.$$

$$+\frac{2ac\,(4p-5)-b^2\,(2p-3)}{2a\,(p-1)\,(4ac-b^2)} \int \frac{dx}{\Phi^{p-1}}.$$

$$\int \frac{dx}{x^2\,(a+bx^2+cx^4)^p} = \frac{2cx^2+b}{2(p-1)(4ac-b^2)x^3\Phi^{p-1}} + \frac{c\,(4p-3)}{(p-1)(4ac-b^2)} \int \frac{dx}{x^2\Phi^{p-1}}$$

$$+\frac{3b}{2\,(p-1)\,(4ac-b^2)} \int \frac{dx}{x^4\Phi^{p-1}}.$$

$$\int \frac{dx}{x^m\,(a+bx^2+cx^4)^p} = \frac{1}{a}\int \frac{dx}{x^m\Phi^{p-1}} - \frac{b}{a}\int \frac{dx}{x^{m-2}\Phi^p} - \frac{c}{a}\int \frac{dx}{x^{m-4}\Phi^p}$$

$$= \frac{bcx^2+b^2-2ac}{2a\,(p-1)\,(b^2-4ac)\,x^{m-1}\Phi^{p-1}} + \frac{bc\,(4p+m-7)}{2a\,(p-1)\,(b^2-4ac)} \int \frac{dx}{x^{m-2}\Phi^{p-1}}$$

$$+\frac{b^2\,(2p+m-3)-2ac\,(4p+m-5)}{2a\,(p-1)\,(b^2-4ac)} \int \frac{dx}{x^m\Phi^{p-1}}$$

$$\int \frac{dx}{x^m\,(a+bx^n)^p} = \frac{1}{a}\int \frac{dx}{x^m\Phi^{p-1}} - \frac{b}{a}\int \frac{dx}{x^{m-n}\Phi^p}.$$

$$\int \frac{dx}{x^m\,(ax^r+bx^{r+n})^p} = \frac{1}{(p-1)\,nax^{m+r-1}\Phi^{p-1}} + \frac{m-n+pr+np-1}{(p-1)\,na} \int \frac{dx}{x^{m+r}\Phi^{p-1}}.$$

$$= -\frac{1}{(m+pr-1)\,ax^{m+r-1}\Phi^{p-1}} - \frac{(m-n+np+pr-1)\,b}{(m+pr-1)\,a} \int \frac{dx}{x^{m-n}\Phi^p}$$

$$\int \frac{dx}{x^m\,(a+bx^n+cx^{2n})^p} = \frac{1}{a}\int \frac{dx}{x^m\Phi^{p-1}} - \frac{b}{a}\int \frac{dx}{x^{m-n}\Phi^p} - \frac{c}{a}\int \frac{dx}{x^{m-2n}\Phi^p}$$

$$= -\frac{bcx^n+b^2-2ac}{an\,(p-1)\,(4ac-b^2)\,x^{m-1}\Phi^{p-1}} - \frac{bc\,(2pn-3n+m-1)}{an\,(p-1)\,(4ac-b^2)} \int \frac{dx}{x^{m-n}\Phi^{p-1}}$$

$$+\frac{2\,(2pn-2n+m-1)\,ac-(pn-n+m-1)\,b^2}{an\,(p-1)\,(4ac-b^2)} \int \frac{dx}{x^m\Phi^{p-1}}.$$

$(X)^{+p/q}$

$$\int (a+bx)^{\frac{p}{q}}\,dx = \frac{q}{b\,(p+q)}\,\Phi^{\frac{p}{q}+1}$$

$$\int x\left(a+bx\right)^{\frac{p}{q}}dx=\frac{q\Phi^{\frac{p}{q}+1}}{b^2}\left\{\frac{\Phi}{p+2q}-\frac{a}{p+q}\right\}.$$

$$\int x^2\left(a+bx\right)^{\frac{p}{q}}dx=\frac{q}{b^3}\int z^{p+q-1}(z^q-a)^2\,dz.\qquad a+bx=z^q.$$

$$\int x^m(a+bx)^{\frac{p}{q}}dx=\frac{qx^m\,\Phi^{\frac{p}{q}+1}}{b\,(mq+p+q)}-\frac{amq}{b\,(mq+p+q)}\int x^{m-1}\,\Phi^{\frac{p}{q}}\,dx.$$

$$=\left\{\frac{\Phi^m}{qm+p+q}-\frac{m}{1}\frac{a\Phi^{m-1}}{qm+p}+\frac{m\,(m-1)}{1.2}\frac{a^2\Phi^{m-2}}{qm+p-q}\right.$$

$$\left.-\frac{m\,(m-1)\,(m-2)}{1.2.3.}\frac{a^3\Phi^{m-3}}{qm+p-2q}+\cdots\pm\mathrm{N}a^m\right\}\frac{q\Phi^{\frac{p+q}{q}}}{b^{m+1}}.$$

$$\int x\,(a+bx^2)^{\frac{p}{q}}dx=\frac{q}{2b\,(p+q)}\,\Phi^{\frac{p}{q}+1}.$$

$$\int x^3\,(a+bx^2)^{\frac{p}{q}}dx=\frac{q}{2b^2}\int z^{p+q-1}\,(z^q-a)\,dz.\qquad a+bx^2=z^q.$$

$$\int x^2\,(a+bx^3)^{\frac{p}{q}}dx=\frac{q}{3b\,(p+q)}\,\Phi^{\frac{p}{q}+1}.$$

$$\int x\,(a+bx^n)^{\frac{p}{q}}dx=\frac{qx^{2-n}\,\Phi^{\frac{p}{q}+1}}{b\,(np+2q)}+\frac{aq\,(n-2)}{b\,(np+2q)}\int\frac{\Phi^{\frac{p}{q}}}{x^{n-1}}\,dx$$

$$=\frac{qx^2\Phi^{\frac{p}{q}}}{np+2q}+\frac{anp}{np+2q}\int x\,\Phi^{\frac{p}{q}-1}\,dx.$$

$$\int x^2\,(a+bx^n)^{\frac{p}{q}}dx=\frac{q\Phi^{\frac{p}{q}+1}}{b\,(np+3q)\,x^{n-3}}+\frac{aq\,(n-3)}{b\,(np+3q)}\int\frac{\Phi^{\frac{p}{q}}}{x^{n-2}}\,dx$$

$$=\frac{qx^3\Phi^{\frac{p}{q}}}{np+3q}+\frac{anp}{np+3q}\int x^2\Phi^{\frac{p}{q}-1}\,dx.$$

$$\int x^n\,(a+bx^n)^{\frac{p}{q}}dx=\frac{qx\Phi^{\frac{p}{q}+1}}{bn\,(p+q)}-\frac{q}{bn\,(p+q)}\int\Phi^{\frac{p}{q}+1}\,dx$$

$$=\frac{qx\Phi^{\frac{p}{q}+1}}{b\,(np+nq+q)}-\frac{aq}{b\,(np+nq+q)}\int\Phi^{\frac{p}{q}}\,dx.$$

$$\int x^{n-1}\,(a+bx^n)^{\frac{p}{q}}dx=\frac{q\Phi^{\frac{p}{q}+1}}{bn\,(p+q)}.$$

$$\int x^{2n-1}(a+bx^n)^{\frac{p}{q}}dx = \frac{1}{n}\int z(a+bz)^{\frac{p}{q}}\,dz. \qquad z = x^n.$$

$$= \frac{q}{b^2 n}\int z^{p+q-1}(z^q - a)\,dz. \qquad a + bx^n = z^q.$$

$$= \frac{qx^n\Phi^{\frac{p}{q}+1}}{bn(p+q)} - \frac{q^2\Phi^{\frac{p}{q}+2}}{b^2 n(p+q)(p+2q)}.$$

$$\int x^{kn-1}(a+bx^n)^{\frac{p}{q}}\,dx = \frac{q}{nb^k}\int z^{p+q-1}(z^q-a)^{k-1}\,dz. \qquad a + bx^n = z^q.$$

$$\int x^m(a+bx^n)^{\frac{p}{q}}\,dx = \frac{q}{nb^{\frac{m+1}{n}}}\int z^{p+q-1}(z^q-a)^{\frac{m+1}{n}-1}\,dz. \qquad a + bx^n = z^q$$

$$= -\frac{q}{nb^{\frac{m+1}{n}}}\int \frac{(1-az^q)^{\frac{m+1}{n}-1}}{z^{\frac{q(m+1)}{n}+p+1}}\,dz. \qquad a + bx^n = \frac{1}{z^q}.$$

$$= -\frac{qa^{\frac{m+1}{n}+\frac{p}{q}}}{n}\int \frac{z^{p+q-1}}{(z^q-b)^{\frac{m+1}{n}+\frac{p}{q}+1}}\,dz. \qquad a + bx^n = z^q x^n.$$

$$= -a^{\frac{m+1}{n}+\frac{p}{q}}\int \frac{z^{\frac{np}{q}+n-1}}{(z^n-b)^{\frac{m+1}{n}+\frac{p}{q}+1}}\,dz. \qquad a + bx^n = z^n x^n.$$

$$= \frac{qx^{m+1}\Phi^{\frac{p}{q}}}{np+mq+q} + \frac{anp}{np+mq+q}\int x^m\Phi^{\frac{p}{q}-1}\,dx.$$

$$= \frac{qx^{m-n+1}\Phi^{\frac{p}{q}+1}}{b(np+mq+q)} - \frac{aq(m-n+1)}{b(np+mq+q)}\int x^{m-n}\Phi^{\frac{p}{q}}\,dx.$$

$$= \frac{qx^{m-n+1}\Phi^{\frac{p}{q}+1}}{bn(p+q)} - \frac{q(m-n+1)}{bn(p+q)}\int x^{m-n}\Phi^{\frac{p}{q}+1}\,dx.$$

$$= \frac{x^{m+1}\Phi^{\frac{p}{q}}}{m+1} - \frac{bnp}{q(m+1)}\int x^{m+n}\Phi^{\frac{p}{q}-1}\,dx.$$

$$= \frac{x^{m+1}\Phi^{\frac{p}{q}+1}}{a(m+1)} - \frac{b(np+mq+nq+q)}{aq(m+1)}\int x^{m+n}\Phi^{\frac{p}{q}}\,dx.$$

$$= -\frac{qx^{m+1}\Phi^{\frac{p}{q}+1}}{an(p+q)} + \frac{np+mq+nq+q}{an(p+q)}\int x^m\Phi^{\frac{p}{q}+1}\,dx.$$

$$\int \frac{(a+bx)^{\frac{p}{q}}}{x}\,dx = q\int \frac{z^{p+q-1}}{z^q-a}\,dz. \qquad a + bx = z^q.$$

$$\int \frac{(a+bx)^{\frac{p}{q}}}{x^2}\,dx = -\frac{\Phi^{\frac{p}{q}+1}}{ax} + \frac{bp}{aq}\int \frac{\Phi^{\frac{p}{q}}}{x}\,dx = bq\int \frac{z^{p+q-1}}{(z^q-a)^2}\,dz. \qquad a + bx = z^q.$$

$$\int \frac{(a+bx)^{\frac{p}{q}}}{x^m}\,dx = -\frac{\Phi^{\frac{p}{q}+1}}{a(m-1)x^{m-1}} + \frac{b(p-mq+2q)}{aq(m-1)}\int \frac{\Phi^{\frac{p}{q}}}{x^{m-1}}\,dx.$$

$$= \Bigg\{ -\frac{1}{(m-1)ax^{m-1}} + \frac{1}{(m-1)a}\,\frac{(qm-p-2q)b}{(m-2)qax^{m-2}}$$

$$-\frac{1}{(m-1)a}\,\frac{(qm-b-2q)b}{(m-2)qa}\cdot\frac{(qm-b-3q)b}{(m-3)qax^{m-3}}$$

$$+\frac{1}{(m-1)a}\cdot\frac{(qm-p-2q)b}{(m-2)qa}\cdot\frac{(qm-p-3q)b}{(m-3)qa}\cdot\frac{(qm-p-4q)b}{(m-4)qax^{m-4}}$$

$$-\cdots\pm\frac{\mathrm{N}}{x}\Bigg\}\,\Phi^{\frac{p}{q}+1} \mp \frac{pb\mathrm{N}}{q}\int\frac{\Phi^{\frac{p}{q}}}{x}\,dx.$$

$$\int \frac{(a+bx^2)^{\frac{p}{q}}}{x}\,dx = \frac{q}{2}\int\frac{z^{p+q-1}}{z^q-a}\,dz. \qquad a+bx^2=z^q.$$

$$\int \frac{(a+bx^2)^{\frac{p}{q}}}{x^2}\,dx = -\frac{\Phi^{\frac{p}{q}+1}}{ax} + \frac{b(2p+q)}{aq}\int \Phi^{\frac{p}{q}}\,dx.$$

$$\int \frac{(a+bx^n)^{\frac{p}{q}}}{x}\,dx = \frac{q}{n}\int\frac{z^{p+q-1}}{z^q-a}\,dz. \qquad a+bx^n=z^q.$$

$$\int \frac{(a+bx^n)^{\frac{p}{q}}}{x^2}\,dx = -\frac{\Phi^{\frac{p}{q}+1}}{ax} - \frac{b(q-np-nq)}{aq}\int x^{n-2}\,\Phi^{\frac{p}{q}}\,dx.$$

$$\int \frac{(a+bx^n)^{\frac{p}{q}}}{x^n}\,dx = -\frac{\Phi^{\frac{p}{q}}}{(n-1)x^{n-1}} + \frac{bnp}{q(n-1)}\int \Phi^{\frac{p}{q}-1}\,dx$$

$$= -\frac{\Phi^{\frac{p}{q}+1}}{a(n-1)x^{n-1}} + \frac{b(np+q)}{aq(n-1)}\int \Phi^{\frac{p}{q}}\,dx.$$

$$\int \frac{(a+bx^n)^{\frac{p}{q}}}{x^m}\,dx = \frac{9b^{\frac{m-1}{n}}}{n}\int\frac{z^{p+q-1}}{(z^q-a)^{\frac{m-1}{n}+1}}\,dz. \qquad a+bx^n=z^q.$$

$$= -\frac{q}{na^{\frac{m-1}{n}-\frac{p}{q}}}\int z^{p+q-1}\,(z^q-b)^{\frac{m-1}{n}-\frac{p}{q}-1}\,dz. \qquad a+bx^n=z^qx^n.$$

$$= \frac{q\Phi^{\frac{p}{q}}}{(np-mq+q)x^{m-1}} + \frac{anp}{np-mq+q}\int\frac{\Phi^{\frac{p}{q}-1}}{x^m}\,dx.$$

$$= \frac{q\Phi^{\frac{p}{q}+1}}{b(np-mq+q)x^{m+n-1}} + \frac{aq(m+n-1)}{b(np-mq+q)}\int\frac{\Phi^{\frac{p}{q}}}{x^{m+n}}\,dx.$$

$$= \frac{q\Phi^{\frac{p}{q}+1}}{bn(p+q)x^{m+n-1}} + \frac{q(m+n-1)}{bn(p+q)}\int\frac{\Phi^{\frac{p}{q}+1}}{x^{m+n}}\,dx.$$

$$= -\frac{\Phi^{\frac{p}{q}}}{(m-1)x^{m-1}} + \frac{bnp}{q(m-1)}\int\frac{\Phi^{\frac{p}{q}-1}}{x^{m-n}}\,dx$$

$$= -\frac{\Phi^{\frac{p}{q}+1}}{a(m-1)x^{m-1}} + \frac{b(np-mq+nq+q)}{aq(m-1)}\int\frac{\Phi^{\frac{p}{q}}}{x^{m-n}}\,dx.$$

$$= -\frac{q\Phi^{\frac{p}{q}+1}}{an(p+q)x^{m-1}} + \frac{np-mq+nq+q}{an(p+q)}\int\frac{\Phi^{\frac{p}{q}+1}}{x^{m}}\,dx.$$

$(X)^{-p/q}$

$$\int\frac{dx}{(a+bx)^{\frac{p}{q}}} = -\frac{q}{b(p-q)\Phi^{\frac{p}{q}-1}}.$$

$$\int\frac{x}{(a+bx)^{\frac{p}{q}}}\,dx = \left\{\frac{\Phi}{2q-p} - \frac{a}{q-p}\right\}\frac{q}{b^2\Phi^{\frac{p}{q}-1}}.$$

$$\int\frac{x^2}{(a+bx)^{\frac{p}{q}}}\,dx = \left\{\frac{\Phi^2}{3q-p} - \frac{2a\Phi}{2q-p} + \frac{a^2}{q-p}\right\}\frac{q}{b^3\Phi^{\frac{p}{q}-1}}.$$

$$\int\frac{x^m}{(a+bx)^{\frac{p}{q}}}\,dx = \frac{qx^{m+1}}{a(p-q)\Phi^{\frac{p}{q}-1}} + \frac{p-mq-2q}{a(p-q)}\int\frac{x^m}{\Phi^{\frac{p}{q}-1}}\,dx.$$

$$= \left\{\frac{\Phi^m}{qm-p+q} - \frac{m}{1}\,\frac{a\Phi^{m-1}}{qm-p} + \frac{m(m-1)}{1.2}\,\frac{a^2\Phi^{m-2}}{qm-p-q}\right.$$

$$\left. - \cdots \pm Na^m\right\}\frac{q}{b^{m+1}\Phi^{\frac{p}{q}-1}}.$$

$$\int\frac{x}{(a+bx^2)^{\frac{p}{q}}}\,dx = \frac{q}{2b(q-p)\Phi^{\frac{p}{q}-1}}.$$

$$\int\frac{x^3}{(a+bx^2)^{\frac{p}{q}}}\,dx = \frac{q}{2b^2}\int z^{q-p-1}(z^q-a)\,dz. \qquad a+bx^2=z^q.$$

$$\int\frac{x^2}{(a+bx^3)^{\frac{p}{q}}}\,dx = -\frac{q}{3b(p-q)\Phi^{\frac{p}{q}-1}}.$$

$$\int \frac{dx}{(a+bx^n)^{\frac{p}{q}}} = \frac{qx}{an(p-q)\,\Phi^{\frac{p}{q}-1}} - \frac{nq+q-np}{an(p-q)} \int \frac{dx}{\Phi^{\frac{p}{q}-1}}.$$

$$\int \frac{x}{(a+bx^n)^{\frac{p}{q}}}\,dx = \frac{qx^2}{an(p-q)\,\Phi^{\frac{p}{q}-1}} - \frac{2q+nq-np}{an(p-q)} \int \frac{x}{\Phi^{\frac{p}{q}-1}}\,dx.$$

$$\int \frac{x^2}{(a+bx^n)^{\frac{p}{q}}}\,dx = \frac{qx^3}{an(p-q)\,\Phi^{\frac{p}{q}-1}} - \frac{3q+nq-np}{an(p-q)} \int \frac{x^2}{\Phi^{\frac{p}{q}-1}}\,dx.$$

$$\int \frac{x^m}{(a+bx^n)^{\frac{p}{q}}}\,dx = \frac{q}{nb^{\frac{m+1}{n}}} \int \frac{(z^q-a)^{\frac{m+1}{n}-1}}{z^{p-q+1}}\,dz. \qquad a+bx^n = z^q.$$

$$= -\frac{qa^{\frac{m+1}{n}-\frac{p}{q}}}{n} \int \frac{dz}{z^{p-q+1}\,(z_q-b)^{\frac{m+1}{n}-\frac{p}{q}+1}} \qquad a+bx^n = z^q x^n.$$

$$\int \frac{x^{p-1}}{(a+bx^n)^{\frac{p}{n}}}\,dx = \int \frac{z^{n-p-1}}{b-z^n}\,dz. \qquad a+bx^n = z^n x^n.$$

$$\int \frac{x^{p-n-1}}{(a+bx^n)^{\frac{p}{n}}}\,dx = \frac{x^{p-n}}{a(p-n)\,\Phi^{\frac{p}{n}-1}}.$$

$$\int \frac{x^n}{(a+bx^n)^{\frac{p}{q}}}\,dx = -\frac{qx}{bn(p-q)\,\Phi^{\frac{p}{q}-1}} - \frac{q}{bn(p-q)} \int \frac{dx}{\Phi^{\frac{p}{q}-1}}$$

$$= \frac{qx}{b(qn-np+q)\,\Phi^{\frac{p}{q}-1}} - \frac{aq}{b(qn-np+q)} \int \frac{dx}{\Phi^{\frac{p}{q}}}.$$

$$\int \frac{x^{2n-1}}{(a+bx^n)^{\frac{p}{q}}}\,dx = \frac{1}{n} \int \frac{z}{(a+bz)^{\frac{p}{q}}}\,dz. \qquad x^n = z.$$

$$= -\frac{q}{b^2 n} \int z^{p-2q-1}(1-az^q)\,dz. \qquad a+bx^n = \frac{1}{z^q}.$$

$$= -\frac{qx^n}{bn(p-q)\,\Phi^{\frac{p}{q}-1}} - \frac{q^2}{b^2 n(p-q)(p-2q)\,\Phi^{\frac{p}{q}-2}}.$$

$$\int \frac{x^{3n-1}}{(a+bx^n)^{\frac{p}{q}}}\,dx = \frac{1}{n} \int \frac{z^2}{(a+bz)^{\frac{p}{q}}}\,dz. \qquad x^n = z.$$

$$\int \frac{dx}{x(a+bx)^{\frac{p}{q}}} = q \int \frac{z^{q-p-1}}{z^q-a}\,dz. \qquad a+bx = z^q.$$

$$= -q\int \frac{z^{p-1}}{1-az^q}\,dz. \qquad a+bx=\frac{1}{z^q}.$$

$$\int \frac{dx}{x^2(a+bx)^{\frac{p}{q}}} = -bq\int \frac{z^{p+q-1}}{(1-az^q)^2}\,dz. \qquad a+bx=\frac{1}{z^q}.$$

$$\int \frac{dx}{x^3(a+bx)^{\frac{p}{q}}} = -b^2q\int \frac{z^{p+2q-1}}{(1-az^q)^3}\,dz. \qquad a+bx=\frac{1}{z^q}.$$

$$\int \frac{dx}{x^m(a+bx)^{\frac{p}{q}}} = \Big\{-\frac{1}{(m-1)ax^{m-1}} + \frac{1}{(m-1)a}\,\frac{(qm+p-2q)b}{(m-2)qax^{m-2}}$$

$$-\frac{1}{(m-1)a}\cdot\frac{(qm+p-2q)b}{(m-2)qa}\cdot\frac{(qm+p-3q)b}{(m-3)qax^{m-3}}$$

$$+\frac{1}{(m-1)a}\,\frac{(qm+p-2q)b}{(m-2)qa}\,\frac{(qm+p-3q)b}{(m-3)qa}\,\frac{(qm+p-4q)b}{(m-4)qax^{m-4}}$$

$$-\dots\pm\frac{\mathrm{N}}{x}\Big\}\frac{1}{\Phi^{\frac{p-q}{q}}}\pm\frac{pb\mathrm{N}}{q}\int\frac{dx}{x\Phi^{\frac{p}{q}}}.$$

$$\int \frac{dx}{x(a+bx^2)^{\frac{p}{q}}} = -\frac{q}{2}\int \frac{z^{p-1}}{1-az^q}\,dz. \qquad a+bx^2=\frac{1}{z^q}.$$

$$\int \frac{dx}{x^3(a+bx^2)^{\frac{p}{q}}} = \frac{bq}{2}\int \frac{z^{q-p-1}}{(z^q-a)^2}\,dz. \qquad a+bx^2=z^q.$$

$$\int \frac{dx}{x^m(a+bx^n)^{\frac{p}{q}}} = \frac{qb^{\frac{m-1}{n}}}{n}\int \frac{dz}{z^{p-q+1}(z^q-a)^{\frac{m-1}{n}+1}} \qquad a+bx^n=z^q.$$

$$= -\frac{q}{na^{\frac{m+1}{n}+\frac{p}{q}}}\int \frac{(z^q-b)^{\frac{m-1}{n}+\frac{p}{q}-1}}{z^{p-q+1}}\,dz. \quad a+bx^n=z^qx^n.$$

$$= -\frac{q}{(np+mq-q)x^{m-1}\Phi^{\frac{p}{q}}} + \frac{anp}{np+mq-q}\int\frac{dx}{x^m\Phi^{\frac{p}{q}+1}}$$

$$= -\frac{q}{b(np+mq-q)x^{m+n-1}\Phi^{\frac{p}{q}-1}} - \frac{aq(m+n-1)}{b(np+mq-q)}\int\frac{dx}{x^{m+n}\Phi^{\frac{p}{q}}}$$

$$= -\frac{q}{bn(p-q)x^{m+n-1}\Phi^{\frac{p}{q}-1}} - \frac{q(m+n-1)}{bn(p-q)}\int\frac{dx}{x^{m+n}\Phi^{\frac{p}{q}-1}}.$$

$$= -\frac{1}{(m-1)\,x^{m-1}\Phi^{\frac{p}{q}}} - \frac{bnp}{q\,(m-1)}\int\frac{dx}{x^{m-n}\Phi^{\frac{p}{q}+1}}\,.$$

$$= -\frac{1}{a\,(m-1)\,x^{m-1}\Phi^{\frac{p}{q}-1}} - \frac{b\,(np+mq-nq-q)}{aq\,(m-1)}\int\frac{dx}{x^{m-n}\Phi^{\frac{p}{q}}}\,.$$

$$= \frac{q}{an\,(p-q)\,x^{m-1}\Phi^{\frac{p}{q}-1}} + \frac{np+mq-nq-q}{an\,(p-q)}\int\frac{dx}{x^{m}\Phi^{\frac{p}{q}-1}}\,.$$

$$\int\frac{dx}{x^{p}\,(a+bx^{n})^{\frac{p}{n}}} = \frac{1}{a}\int\frac{\Phi^{\frac{n-p}{p}}}{x^{p}}\,dx - \frac{b}{a}\int\frac{x^{n-p}}{\Phi^{\frac{p}{n}}}\,dx.$$

$$\int\frac{dx}{x^{n}\,(a+bx^{n})^{\frac{p}{q}}} = -\frac{1}{(n-1)\,x^{n-1}\,\Phi^{\frac{p}{q}}} - \frac{bnp}{q\,(n-1)}\int\frac{dx}{\Phi^{\frac{p}{q}+1}}$$

$$= -\frac{1}{a\,(n-1)\,x^{n-1}\,\Phi^{\frac{p}{q}-1}} - \frac{b\,(np-q)}{aq\,(n-1)}\int\frac{dx}{\Phi^{\frac{p}{q}}}$$

$(X)^{p+1/2}$

$$\int(a+bx^{2})^{p+\frac{1}{2}}\,dx = \left\{\frac{\Phi^{p}}{2p+2} + \frac{(2p+1)\,a\Phi^{p-1}}{(2p+2)\,2p} + \frac{(2p+1)\,(2p-1)\,a^{2}\Phi^{p-2}}{(2p+2)\,2p\,(2p-2)}\right.$$

$$\left. + \frac{(2p+1)\,(2p-1)\,(2p-3)\,a^{3}\Phi^{p-3}}{(2p+2)\,2p\,(2p-2)\,(2p-4)} + \cdots + \mathrm{N}\Phi^{\circ}\right\}x\sqrt{\Phi} + \mathrm{N}a\int\frac{dx}{\sqrt{\Phi}}$$

$$\int(a^{2}-x^{2})^{p+\frac{1}{2}}\,dx = \frac{x\Phi^{p+\frac{1}{2}}}{2p+2} + \frac{(2p+1)\,a^{2}}{2p+2}\int\Phi^{p-\frac{1}{2}}\,dx.$$

$$\int x^{m}\,(a+bx^{2})^{p+\frac{1}{2}}\,dx = \frac{x^{m-1}\Phi^{p+\frac{3}{2}}}{b\,(m+2p+2)} - \frac{a\,(m-1)}{b\,(m+2p+2)}\int x^{m-2}\Phi^{p+\frac{1}{2}}\,dx.$$

$$\int x^{2m}\,(a+bx^{2})^{p+\frac{1}{2}}\,dx = -\,a^{m+p+1}\int\frac{z^{2p+2}}{(z^{2}-b)^{m+p+2}}\,dz. \qquad a+bx^{2} = z^{2}x^{2}.$$

$$\int x^{2m+1}\,(a+bx^{2})^{p+\frac{1}{2}}\,dx = \frac{1}{b^{m+1}}\int z^{2p+2}\,(z^{2}-a)^{m}\,dz. \qquad a+bx^{2} = z^{2}.$$

$$\int(a+bx+cx^{2})^{p+\frac{1}{2}}\,dx = \frac{(2cx+b)\,\Phi^{p+\frac{1}{2}}}{4c\,(p+1)} + \frac{(2p+1)\,(4ac-b^{2})}{8c\,(p+1)}\int\Phi^{p-\frac{1}{2}}\,dx.$$

$$\int\frac{(a+bx^2)^{p+\frac{1}{2}}}{x}dx=\left\{\frac{\Phi^p}{2p+1}+\frac{a\Phi^{p-1}}{2p-1}+\frac{a^2\Phi^{p-2}}{2p-3}+\cdots\frac{a^p}{1}\right\}\sqrt{\Phi}+a^{p+1}\int\frac{dx}{x\sqrt{\Phi}}.$$

$$\int\frac{(a+bx^2)^{p+\frac{1}{2}}}{x^m}dx=-\frac{\Phi^{p+\frac{1}{2}}}{(m-1)x^{m-1}}+\frac{b(2p+1)}{m-1}\int\frac{\Phi^{p-\frac{1}{2}}}{x^{m-2}}dx.$$

$(X)^{-(p+1/2)}$

$$\int\frac{dx}{(a+bx)^{p+\frac{1}{2}}}=-\frac{2}{b(2p-1)\Phi^{p-\frac{1}{2}}}.$$

$$\int\frac{x}{(a+bx)^{p+\frac{1}{2}}}dx=-\left\{x+\frac{2a}{b(2p-1)}\right\}\frac{2}{b(2p-3)\Phi^{p-\frac{1}{2}}}.$$

$$\int\frac{x^2}{(a+bx)^{p+\frac{1}{2}}}dx=\left\{x^2+\frac{4ax}{(2p-3)b}+\frac{4a^2}{(2p-1)(2p-3)b^2}\right\}\frac{2}{b(2p-5)\Phi^{p-\frac{1}{2}}}.$$

$$\int\frac{x^m}{(a+bx)^{p+\frac{1}{2}}}dx=\frac{2x^m}{b(2m-2p+1)\Phi^{p-\frac{1}{2}}}-\frac{2am}{b(2m-2p+1)}\int\frac{x^{m-1}}{\Phi^{p+\frac{1}{2}}}dx.$$

$$\int\frac{dx}{(a+bx^2)^{p+\frac{1}{2}}}=\left\{\frac{1}{a(2p-1)\Phi^{p-1}}+\frac{2p-2}{(2p-1)(2p-3)a^2\Phi^{p-2}}\right.$$
$$\left.+\frac{(2p-2)(2p-4)}{(2p-1)(2p-3)(2p-5)a^3\Phi^{p-3}}+\cdots+\frac{N}{\Phi^0}\right\}\frac{x}{\sqrt{\Phi}}.$$

$$\int\frac{dx}{(a^2+x^2)^{p+\frac{1}{2}}}=\frac{x}{a^2(2p-1)\Phi^{p-\frac{1}{2}}}+\frac{2p-2}{a^2(2p-1)}\int\frac{dx}{\Phi^{p-\frac{1}{2}}}.$$

$$\int\frac{dx}{(a+bx+cx^2)^{p+\frac{1}{2}}}=\frac{2(2cx+b)}{(2p-1)(4ac-b^2)\Phi^{p-\frac{1}{2}}}+\frac{8c(p-1)}{(2p-1)(4ac-b^2)}\int\frac{dx}{\Phi^{p-\frac{1}{2}}}.$$

$$\int\frac{x}{(a+bx+cx^2)^{p+\frac{1}{2}}}dx=-\frac{1}{(2p-1)c\Phi^{p-\frac{1}{2}}}-\frac{b}{2c}\int\frac{dx}{\Phi^{p+\frac{1}{2}}}.$$

$$\int\frac{x^2}{(a+bx+cx^2)^{p+\frac{1}{2}}}dx=\frac{(2b^2-4ac)x+2ab}{c(2p-1)(4ac-b^2)\Phi^{p-\frac{1}{2}}}+\frac{4ac+b^2(2p-3)}{c(4ac-b^2)(2p-1)}\int\frac{dx}{\Phi^{p-}}$$

$$\int \frac{x^m}{(a+bx+cx^2)^{p+\frac{1}{2}}}dx = -\frac{x^{m-1}}{c(2p-m)\Phi^{p-\frac{1}{2}}} - \frac{b(2p-2m+1)}{2c(2p-m)}\int \frac{x^{m-1}}{\Phi^{p+\frac{1}{2}}}dx$$

$$+\frac{a(m-1)}{c(2p-m)}\int \frac{x^{m-2}}{\Phi^{p+\frac{1}{2}}}dx.$$

$$\int \frac{dx}{x(a+bx)^{p+\frac{1}{2}}} = \frac{2}{a(2p-1)\Phi^{p-\frac{1}{2}}} + \frac{1}{a}\int \frac{dx}{x\Phi^{p-\frac{1}{2}}}$$

$$\int \frac{dx}{x^2(a+bx)^{p+\frac{1}{2}}} = -\frac{1}{ax\Phi^{p-\frac{1}{2}}} - \frac{b(2p+1)}{2a}\int \frac{dx}{x\Phi^{p+\frac{1}{2}}}.$$

$$\int \frac{dx}{x^3(a+bx)^{p+\frac{1}{2}}} = -\frac{2a-b(2p+3)x}{4a^2x^2\Phi^{p-\frac{1}{2}}} + \frac{(2p+1)(2p+3)b^2}{8a^2}\int \frac{dx}{x\Phi^{p+\frac{1}{2}}}.$$

$$\int \frac{dx}{x^m(a+bx)^{p+\frac{1}{2}}} = -\frac{1}{a(m-1)x^{m-1}\Phi^{p-\frac{1}{2}}} - \frac{b(2m+2p-3)}{2a(m-1)}\int \frac{dx}{x^{m-1}\Phi^{p+\frac{1}{2}}}.$$

$$\int \frac{dx}{x(a+bx^2)^{p+\frac{1}{2}}} = \left\{\frac{1}{(2p-1)a\Phi^{p-1}} + \frac{1}{(2p-3)a^2\Phi^{p-2}} + \frac{1}{(2p-5)a^3\Phi^{p-3}}\right.$$

$$\left. + \cdots \frac{1}{a^p}\right\}\frac{1}{\sqrt{\Phi}} + \frac{1}{a^p}\int \frac{dx}{x\sqrt{\Phi}}.$$

$$\int \frac{dx}{x(bx+cx^2)^{p+\frac{1}{2}}} = -\frac{2}{(2p+1)bx\Phi^{p-\frac{1}{2}}} - \frac{4pc}{(2p+1)b}\int \frac{dx}{\Phi^{p+\frac{1}{2}}}.$$

$$\int \frac{dx}{x^2(bx+cx^2)^{p+\frac{1}{2}}} = \frac{2}{(2p+3)bx\Phi^{p-\frac{1}{2}}}\left\{-\frac{1}{x}+\frac{2c}{b}\right\} + \frac{8pc^2}{(2p+3)b^2}\int \frac{dx}{\Phi^{p+\frac{1}{2}}}.$$

$$\int \frac{dx}{x^3(bx+cx^2)^{p+\frac{1}{2}}} = \frac{2}{(2p+5)bx\Phi^{p-\frac{1}{2}}}\left\{-\frac{1}{x^2}+\frac{4(p+1)c}{(2p+3)bx} - \frac{8(p+1)c^2}{(2p+3)b^2}\right\}$$

$$-\frac{32p(p+1)c^3}{(2p+3)(2p+5)b^3}\int \frac{dx}{\Phi^{p+\frac{1}{2}}}.$$

$$\int \frac{dx}{x^m(bx+cx^2)^{p+\frac{1}{2}}} = -\frac{2}{(2p+2m-1)bx^m\Phi^{p-\frac{1}{2}}} - \frac{2(2p+m-1)c}{(2p+2m-1)b}\int \frac{dx}{x^{m-1}\Phi^{p+\frac{1}{2}}}.$$

$$\int \frac{dx}{x(a+bx+cx^2)^{p+\frac{1}{2}}} = \frac{1}{(2p-1)\,a\Phi^{p-\frac{1}{2}}} + \frac{1}{a}\int \frac{dx}{x\Phi^{p-\frac{1}{2}}} - \frac{b}{2a}\int \frac{dx}{\Phi^{p+\frac{1}{2}}}.$$

$$\int \frac{dx}{x^2(a+bx+cx^2)^{p+\frac{1}{2}}} = -\frac{1}{ax\Phi^{p-\frac{1}{2}}} - \frac{b(2p+1)}{2a}\int \frac{dx}{x\Phi^{p+\frac{1}{2}}} - \frac{2pc}{a}\int \frac{dx}{\Phi^{p+\frac{1}{2}}}.$$

$$\int \frac{dx}{x^m(a+bx+cx^2)^{p+\frac{1}{2}}} = \frac{1}{a}\int \frac{dx}{x^m\Phi^{p-\frac{1}{2}}} - \frac{b}{a}\int \frac{dx}{x^{m-1}\Phi^{p+\frac{1}{2}}} - \frac{c}{a}\int \frac{dx}{x^{m-2}\Phi^{p+\frac{1}{2}}}.$$

$$= -\frac{1}{(m-1)\,ax^{m-1}\Phi^{p-\frac{1}{2}}} - \frac{b(2p+2m-3)}{2a(m-1)}\int \frac{dx}{x^{m-1}\Phi^{p+\frac{1}{2}}}$$

$$-\frac{c(2p+m-2)}{a(m-1)}\int \frac{dx}{x^{m-2}\Phi^{p+\frac{1}{2}}}.$$

$$(X)^{\pm(p-1/2)}$$

$$\int (a+bx)^{p-\frac{1}{2}}\,dx = \frac{2}{b(2p+1)}\Phi^{p+\frac{1}{2}}.$$

$$\int x(a+bx)^{p-\frac{1}{2}}\,dx = \frac{2\Phi^{p+\frac{1}{2}}}{b(2p+3)}\left\{x - \frac{2a}{b(2p+1)}\right\}.$$

$$\int x^2(a+bx)^{p-\frac{1}{2}}\,dx = \frac{2\Phi^{p+\frac{1}{2}}}{b(2p+5)}\left\{x^2 - \frac{4ax}{b(2p+3)} + \frac{8a^2}{b^2(2p+1)(2p+3)}\right\}.$$

$$\int x^m(a+bx)^{p-\frac{1}{2}}\,dx = \frac{2x^m\Phi^{p+\frac{1}{2}}}{b(2m+2p+1)} - \frac{2am}{b(2m+2p+1)}\int x^{m-1}\,\Phi^{p-\frac{1}{2}}\,dx.$$

$$\int x(a+bx+cx^2)^{p-\frac{1}{2}}\,dx = \frac{\Phi^{p+\frac{1}{2}}}{c(2p+1)} - \frac{b}{2c}\int \Phi^{p-\frac{1}{2}}\,dx.$$

$$\int x^2(a+bx+cx^2)^{p-\frac{1}{2}}\,dx = \frac{x\Phi^{p+\frac{1}{2}}}{2c(p+1)} - \frac{b(2p+3)}{4c(p+1)}\int x\Phi^{p-\frac{1}{2}}\,dx$$

$$-\frac{a}{2c(p+1)}\int \Phi^{p-\frac{1}{2}}\,dx.$$

$$= \left\{ x - \frac{b(2p+3)}{2c(2p+1)} \right\} \frac{\Phi^{p+\frac{1}{2}}}{2c(p+1)} + \frac{(2p+3)\,b^2 - 4ac}{8c^2(p+1)} \int \Phi^{p-\frac{1}{2}}\,dx.$$

$$\int x^3 (a+bx+cx^2)^{p-\frac{1}{2}}\,dx = \frac{x^2\Phi^{p+\frac{1}{2}}}{c(2p+3)} - \frac{b(2p+5)}{2c(2p+3)} \int x^2 \Phi^{p-\frac{1}{2}}\,dx$$

$$- \frac{2a}{c(2p+3)} \int x\Phi^{p-\frac{1}{2}}\,dx.$$

$$\int x^m (a+bx+cx^2)^{p-\frac{1}{2}}\,dx = \frac{x^{m-1}\,\Phi^{p+\frac{1}{2}}}{c(2p+m)} - \frac{b(2p+2m-1)}{2c(2p+m)} \int x^{m-1}\Phi^{p-\frac{1}{2}}\,dx$$

$$- \frac{a(m-1)}{c(2p+m)} \int x^{m-2}\Phi^{p-\frac{1}{2}}\,dx.$$

$$\int \frac{(a+bx)^{p-\frac{1}{2}}}{x}\,dx = \frac{2\Phi^{p-\frac{1}{2}}}{2p-1} + a \int \frac{\Phi^{p-\frac{3}{2}}}{x}\,dx = \frac{2}{\sqrt{\Phi}} \sum_{k=o}^{k=p-1} \frac{a^k\,\Phi^{p-k}}{2p-2k-1} + a^p \int \frac{dx}{x\sqrt{\Phi}}.$$

$$\int \frac{(a+bx)^{p-\frac{1}{2}}}{x^2}\,dx = -\frac{\Phi^{p+\frac{1}{2}}}{ax} + \frac{b(2p-1)}{2a} \int \frac{\Phi^{p-\frac{1}{2}}}{x}\,dx.$$

$$\int \frac{(a+bx)^{p-\frac{1}{2}}}{x^3}\,dx = -\frac{\Phi^{p+\frac{1}{2}}}{2ax} \left\{ \frac{1}{x} + \frac{b(2p-3)}{2a} \right\} + \frac{b^2(2p-1)(2p-3)}{8a^2} \int \frac{\Phi^{p-\frac{1}{2}}}{x}\,dx.$$

$$\int \frac{(a+bx)^{p-\frac{1}{2}}}{x^m}\,dx = -\frac{\Phi^{p+\frac{1}{2}}}{a(m-1)\,x^{m-1}} + \frac{b(2p-2m+3)}{2a(m-1)} \int \frac{\Phi^{p-\frac{1}{2}}}{x^{m-1}}\,dx.$$

$$\int \frac{(a+bx+cx^2)^{p-\frac{1}{2}}}{x}\,dx = \frac{\Phi^{p-\frac{1}{2}}}{2p-1} + a \int \frac{\Phi^{p-\frac{3}{2}}}{x}\,dx + \frac{b}{2} \int \Phi^{p-\frac{3}{2}}\,dx.$$

$$\int \frac{(a+bx+cx^2)^{p-\frac{1}{2}}}{x^2}\,dx = -\frac{\Phi^{p-\frac{1}{2}}}{x} + \frac{b(2p-1)}{2} \int \frac{\Phi^{p-\frac{3}{2}}}{x}\,dx + c(2p-1) \int \Phi^{p-\frac{3}{2}}\,dx.$$

$$\int \frac{(a+bx+cx^2)^{p-\frac{1}{2}}}{x^m}\,dx = -\frac{\Phi^{p-\frac{1}{2}}}{(m-1)\,x^{m-1}} + \frac{b(2p-1)}{2(m-1)} \int \frac{\Phi^{p-\frac{3}{2}}}{x^{m-1}}\,dx$$

$$+ \frac{c(2p-1)}{m-1} \int \frac{x^{p-\frac{3}{2}}}{x^{m-2}}\,dx.$$

$$\int \frac{x^{2m+1}}{(a+bx^2)^{p-\frac{1}{2}}}\,dx = \frac{1}{b^{m+1}} \int \frac{(z^2-a)^m}{z^{2p-2}}\,dz. \qquad a+bx^2 = z^2.$$

$$\int \frac{x^{2p}}{(a+bx^2)^{p-\frac{1}{2}}}\,dx = -a \int \frac{dz}{z^{2p-2}(z^2-b)^2}. \qquad a+bx^2 = z^2x^2.$$

$$\int \frac{dx}{x^{2m}(a+bx^2)^{p-\frac{1}{2}}} = -\frac{1}{a^{m+p-1}} \int \frac{(z^2-b)^{m+p-2}}{z^{2p-2}}\,dz. \qquad a+bx^2 = z^2x^2.$$

$(X)^{p \pm 1/3}$

$$\int (a+bx)^{p+\frac{1}{3}}\,dx = \frac{3}{b(3p+4)}\,\Phi^{p+\frac{4}{3}}.$$

$$\int x^m (a+bx)^{p+\frac{1}{3}}\,dx = \frac{3x^m\,\Phi^{p+\frac{4}{3}}}{b(3m+3p+4)} - \frac{3am}{b(3m+3p+4)} \int x^{m-1}\,\Phi^{p+\frac{1}{3}}\,dx.$$

$$\int \frac{(a+bx)^{p+\frac{1}{3}}}{x}\,dx = \frac{3\Phi^{p+\frac{1}{3}}}{3p+1} + a \int \frac{\Phi^{p-\frac{2}{3}}}{x}\,dx.$$

$$\int \frac{(a+bx)^{p+\frac{1}{3}}}{x^m}\,dx = \frac{\Phi^{p+\frac{4}{3}}}{a(m-1)\,x^{m-1}} + \frac{b(3p-3m+7)}{3a(m-1)} \int \frac{\Phi^{p+\frac{1}{3}}}{x^{m-1}}\,dx.$$

$$\int (a+bx)^{p-\frac{1}{3}}\,dx = \frac{3}{b(3p+2)}\,\Phi^{p+\frac{2}{3}}.$$

$$\int x\,(a+bx)^{p-\frac{1}{3}}\,dx = \frac{3\Phi^{p+\frac{2}{3}}}{b(3p+5)} \left\{ x - \frac{3a}{b(3p+2)} \right\}.$$

$$\int x^2 (a+bx)^{p-\frac{1}{3}}\,dx = \frac{3\Phi^{p+\frac{2}{3}}}{b(3p+8)} \left\{ x^2 - \frac{6ax}{b(3p+5)} + \frac{18a^2}{b^2(3p+2)(3p+5)} \right\}.$$

$$\int x^m (a+bx)^{p-\frac{1}{3}}\,dx = \frac{3x^m\Phi^{p+\frac{2}{3}}}{b(3m+3p+2)} - \frac{3am}{b(3m+3p+2)} \int x^{m-1}\,\Phi^{p-\frac{1}{3}}\,dx.$$

$$\int \frac{(a+bx)^{p-\frac{1}{3}}}{x}\,dx = \frac{3\Phi^{p-\frac{1}{3}}}{3p-1} + a \int \frac{\Phi^{p-\frac{4}{3}}}{x}\,dx = \frac{3}{\Phi^{\frac{1}{3}}} \sum_{k=o}^{k=p-1} \frac{a^k\,\Phi^{p-k}}{3p-3k-1} + a^p \int \frac{dx}{x\Phi^{\frac{1}{3}}}.$$

$$\int \frac{(a+bx)^{p-\frac{1}{3}}}{x^m}\,dx = -\frac{\Phi^{p+\frac{2}{3}}}{a\,(m-1)\,x^{m-1}} + \frac{b\,(3p-3m+5)}{3a\,(m-1)} \int \frac{\Phi^{p-\frac{1}{3}}}{x^{m-1}}\,dx.$$

$$\int \frac{(a+bx)^{p-\frac{1}{3}}}{x^2}\,dx = -\frac{\Phi^{p+\frac{2}{3}}}{ax} + \frac{b\,(3p-1)}{3a} \int \frac{\Phi^{p-\frac{1}{3}}}{x}\,dx.$$

$$\int \frac{(a+bx)^{p-\frac{1}{3}}}{x^m}\,dx = -\frac{\Phi^{p+\frac{2}{3}}}{a\,(m-1)\,x^{m-1}} + \frac{b\,(3p-3m+5)}{3a\,(m-1)} \int \frac{\Phi^{p-\frac{1}{3}}}{x^{m-1}}\,dx.$$

$(X)^{p\,-\,2/3}$

$$\int (a+bx)^{p-\frac{2}{3}}\,dx = \frac{3}{b\,(3p+1)}\,\Phi^{p+\frac{1}{3}}.$$

$$\int x\,(a+bx)^{p-\frac{2}{3}}\,dx = \frac{3\Phi^{p+\frac{1}{3}}}{b\,(3p+4)} \left\{ x - \frac{3a}{b\,(3p+1)} \right\}.$$

$$\int x^2\,(a+bx)^{p-\frac{2}{3}}\,dx = \frac{3\Phi^{p+\frac{1}{3}}}{b\,(3p+7)} \left\{ x^2 - \frac{6ax}{b\,(3p+4)} + \frac{18a^2}{b^2\,(3p+1)\,(3p+4)} \right\}.$$

$$\int x^m\,(a+bx)^{p-\frac{2}{3}}\,dx = \frac{3x^m\Phi^{p+\frac{1}{3}}}{b\,(3m+3p+1)} - \frac{3am}{b\,(3m+3p+1)} \int x^{m-1}\,\Phi^{p-\frac{2}{3}}\,dx.$$

$$\int \frac{(a+bx)^{p-\frac{2}{3}}}{x}\,dx = \frac{3\Phi^{p-\frac{2}{3}}}{3p-2} + a \int \frac{\Phi^{p-\frac{5}{3}}}{x}\,dx = \frac{3}{\Phi^{\frac{2}{3}}} \sum_{k=0}^{k=p-1} \frac{a^k\,\Phi^{p-k}}{3p-3k-2} + a^p \int \frac{dx}{x\Phi^{\frac{3}{2}}}.$$

$$\int \frac{(a+bx)^{p-\frac{2}{3}}}{x^2}\,dx = -\frac{\Phi^{p+\frac{1}{3}}}{ax} + \frac{b\,(3p-2)}{3a} \int \frac{\Phi^{p-\frac{2}{3}}}{x}\,dx.$$

$$\int \frac{(a+bx)^{p-\frac{2}{3}}}{x^m}\,dx = -\frac{\Phi^{p+\frac{1}{3}}}{a\,(m-1)\,x^{m-1}} + \frac{b\,(3p-3m+4)}{3a\,(m-1)} \int \frac{\Phi^{p-\frac{2}{3}}}{x^{m-1}}\,dx.$$

In that which follows, $\Phi = c_0x^3 + c_1x^2 + c_2x + c_3$.

$$\left\{\begin{aligned}\int\frac{dx}{\Phi^{p+1}} &= \left\{\frac{1}{3}U_0x^2 + (N_0U_0 + \frac{1}{3}U_1)x + N_0U_1 + \frac{N_1U_0}{3c_0}\right\}\frac{1}{p\Phi^p} + \frac{3p-2}{3p}U_0\int\frac{x}{\Phi^p}dx \\ &\quad + \frac{(3p-1)U_1 - 3N_0U_0}{3p}\int\frac{dx}{\Phi^p}. \\ \int\frac{x}{\Phi^{p+1}}dx &= \left\{\frac{1}{3}V_0x^2 + (N_0V_0 + \frac{1}{3}V_1)x + N_0V_1 + \frac{N_1V_0}{3c_0}\right\}\frac{1}{p\Phi^p} + \frac{3p-2}{3p}V_0\int\frac{x}{\Phi^p}dx \\ &\quad + \frac{(3p-1)V_1 - 3N_0V_0}{3p}\int\frac{dx}{\Phi^p}.\end{aligned}\right.$$

where $N_0 = \dfrac{c_1}{9c_0}$, $\quad N_1 = \dfrac{2}{3}c_2 - \dfrac{2c_1^2}{9c_0}$, $\quad N_2 = 3c_0c_3 - \dfrac{5}{3}c_1c_2 + \dfrac{4c_1^3}{9c_0}$,

$$N_3 = -6c_0c_1c_3 - 2c_0c_2^2 + 4c_1^2c_2 - \frac{8c_1^4}{9c_0}, \qquad D = N_1N_3 - N_2^2,$$

$$u_0 = \frac{9c_0^2N_1}{D}, \quad u_1 = -\frac{3c_0N_2}{D}, \quad V_0 = -\frac{3c_0(N_2 + 2c_1N_1)}{D}, \quad V_1 = \frac{N_3 + 2c_1N}{D}$$

$$\int\frac{x^2}{\Phi^{p+1}}dx = -\frac{1}{3pc_0\Phi^p} - \frac{2c_1}{3c_0}\int\frac{x}{\Phi^{p+1}}dx - \frac{c_2}{3c_0}\int\frac{dx}{\Phi^{p+1}}.$$

$$\int\frac{x^m}{\Phi^{p+1}}dx = \frac{1}{c_0}\int\frac{x^{m-3}}{\Phi^p}dx - \frac{c_1}{c_0}\int\frac{x^{m-1}}{\Phi^{p+1}}dx - \frac{c_2}{c_0}\int\frac{x^{m-2}}{\Phi^{p+1}}dx - \frac{c_3}{c_0}\int\frac{x^{m-3}}{\Phi^{p+1}}dx.$$

$$= -\frac{x^{m-2}}{(3p-m+2)c_0\Phi^p} - \frac{2p-m+2}{3p-m+2}\cdot\frac{c_1}{c_0}\int\frac{x^{m-1}}{\Phi^{p+1}}dx$$

$$-\frac{p-m+2}{3p-m+2}\cdot\frac{c_2}{c_0}\int\frac{x^{m-2}}{\Phi^{p+1}}dx + \frac{m-2}{3p-m+2}\cdot\frac{c_3}{c_2}\int\frac{x^{m-3}}{\Phi^{p+1}}dx.$$

$$\int\frac{dx}{x\Phi^{p+1}} = \frac{1}{3pc_3\Phi^p} - \frac{c_1}{3c_3}\int\frac{x}{\Phi^{p+1}}dx - \frac{2c_2}{3c_3}\int\frac{dx}{\Phi^{p+1}} + \frac{1}{c_3}\int\frac{dx}{x\Phi^p}.$$

$$\int\frac{dx}{x^2\Phi^{p+1}} = -\frac{1}{c_3x\Phi^p} - \frac{(3p+1)c_0}{c_3}\int\frac{x}{\Phi^{p+1}}dx - \frac{(2p+1)c_1}{c_3}\int\frac{dx}{\Phi^{p+1}}$$

$$-\frac{(p+1)c_2}{c_3}\int\frac{dx}{x\Phi^{p+1}}.$$

$$\int\frac{dx}{x^m\Phi^{p+1}}=\frac{1}{c_3}\int\frac{dx}{x^m\Phi^p}-\frac{c_2}{c_3}\int\frac{dx}{x^{m-1}\Phi^{p+1}}-\frac{c_1}{c_3}\int\frac{dx}{x^{m-2}\Phi^{p+1}}-\frac{c_0}{c_3}\int\frac{dx}{x^{m-3}\Phi^{p+1}}.$$

$$=-\frac{1}{(m-1)c_3x^{m-1}\Phi^p}-\frac{(3p+m-1)c_0}{(m-1)c_3}\int\frac{dx}{x^{m-3}\Phi^{p+1}}-\frac{(2p+m-1)c_1}{(m-1)c_3}\int\frac{dx}{x^{m-2}\Phi^{p+1}}$$

$$-\frac{(p+m-1)c_2}{(m-1)c_3}\int\frac{dx}{x^{m-1}\Phi^{p+1}}$$

In that which follows, $\Phi=c_0x^n+c_1x^{n-1}+\cdots c_n=f(x)$, $\quad f^k(x)=\frac{d^k\Phi}{dx^k}$.

$$\int\frac{dx}{x^m\Phi^{p+1}}=\frac{1}{c_n}\int\frac{dx}{x^m\Phi^p}-\sum_{k=1}^{k=n}\frac{c_{n-k}}{c_n}\int\frac{dx}{x^{m-k}\Phi^{p+1}}.$$

$$=-\frac{1}{(m-1)\,c_nx^{m-1}\Phi^p}-\sum_{k=1}^{k=n}\frac{kp+m-1}{m-1}\cdot\frac{c_{n-k}}{c_n}\int\frac{dx}{x^{m-k}\Phi^{p+1}}$$

$$\int\frac{dx}{(x-a)^2\,\Phi^{p+1}}=-\frac{1}{f(a)\,(x-a)\,\Phi^p}-(p+1)\,\frac{f'(a)}{f(a)}\int\frac{dx}{(x-a)\,\Phi^{p+1}}$$

$$-\sum_{k=2}^{k=n}\frac{kp+1}{k!}\cdot\frac{f^k(a)}{f(a)}\int\frac{(x-a)^{k-2}}{\Phi^{p+1}}\,dx.$$

$$\int\frac{dx}{(x-a)^m\,\Phi^{p+1}}=-\frac{1}{(m-1)\,f(a)\,(x-a)^{m-1}\,\Phi^p}$$

$$-\sum_{k=1}^{k=n}\frac{kp+m-1}{k!\,(m-1)}\cdot\frac{f^k(a)}{f(a)}\int\frac{dx}{(x-a)^{m-k}\,\Phi^{p+1}}.$$

$$\int\frac{x^{n-1}}{\Phi}\,dx=\frac{1}{nc_0}\log\Phi-\sum_{k=n}^{k=n-1}\frac{(n-k)\,c_k}{nc_0}\int\frac{x^{n-k-1}}{\Phi}\,dx.$$

$$\int\frac{x^{n-1}}{\Phi^{p+1}}\,dx=\frac{1}{npc_0\,\Phi^p}-\sum_{k=1}^{k=n-1}\frac{(n-k)\,c_k}{nc_0}\int\frac{x^{n-k-1}}{\Phi^{p+1}}\,dx.$$

$$\int\frac{x^m}{\Phi^{p+1}}\,dx=\frac{1}{c_0}\int\frac{x^{m-n}}{\Phi^p}\,dx-\frac{1}{c_0}\sum_{k=1}^{k=n}c_k\int\frac{x^{m-k}}{\Phi^{p+1}}\,dx.$$

$$=-\frac{x^{m-n+1}}{(np-m+n-1)\,c_0\,\Phi^p}$$

$$-\sum_{k=1}^{k=n}\frac{(n-k)\,p-m+n-1}{np-m+n-1}\cdot\frac{c_k}{c_0}\int\frac{x^{m-k}}{\Phi^{p+1}}\,dx.$$

IRRATIONAL MONOMIALS

$$\int \frac{dx}{\sqrt{x} + x^{\frac{1}{3}}} = 6 \int \frac{z^3}{z+1}\, dz. \qquad x = z^6.$$

$$\int \frac{dx}{\sqrt{x} + x^{\frac{1}{6}}} = 2\sqrt{x} - 6x^{\frac{1}{6}} + 6 \operatorname{arc\,tg} x^{\frac{1}{6}} \cdot$$

$$\int \frac{1 + x^{\frac{1}{3}}}{1 + x^{\frac{1}{4}}}\, dx = 12 \int \frac{z^{15} + z^{11}}{z^3 + 1}\, dz. \qquad x = z^{12}.$$

$$\int \frac{1 + x^{\frac{1}{4}}}{1 + x^{\frac{1}{3}}}\, dx = \frac{1}{2^{\frac{5}{2}}} \log \frac{x^{\frac{1}{6}} - x^{\frac{1}{12}}\sqrt{2} + 1}{x^{\frac{1}{6}} + x^{\frac{1}{12}}\sqrt{2} + 1} + 2 \operatorname{arc\,tg} \frac{x^{\frac{1}{12}}\sqrt{2}}{1 - x^{\frac{1}{6}}}$$

$$\int \frac{1 + \sqrt{x}}{1 - \sqrt{x}}\, dx = -x - 4\sqrt{x} - 4 \log (\sqrt{x} - 1) \cdot$$

$$\int \frac{1 - \sqrt{x}}{1 - x^{\frac{1}{3}}}\, dx = 6 \int \frac{z^5 - z^8}{1 - z^2}\, dz$$

$$= 6 \int \left\{ z^6 + z^4 - z^3 + z^2 - z + 1 - \frac{1}{1+z} \right\} dz. \qquad x = z^6.$$

$$\int \frac{1 + \sqrt{x}}{1 + x^{\frac{1}{3}}}\, dx = 6 \int \frac{z^8 + z^5}{z^2 + 1}\, dz$$

$$= 6 \int \left\{ z^6 - z^4 + z^3 + z^2 - z - 1 + \frac{z+1}{z^2+1} \right\} dz. \qquad x = z^6$$

$$\int \frac{\sqrt{x} - 1}{1 + x^{\frac{1}{3}}}\, dx = 6 \int \frac{z^8 - z^5}{z^2 + 1}\, dz$$

$$= 6 \int \left\{ z^6 - z^4 - z^3 + z^2 + z - 1 + \frac{1 - z}{z^2 + 1} \right\} dz. \qquad x = z^6.$$

$$\int \frac{\sqrt{x} - a}{x^{\frac{1}{3}} - \sqrt{x}}\, dx = 6 \int \frac{z^6 - az^3}{1 - z}\, dz. \qquad x = z^6.$$

$$\int \frac{\sqrt{x}}{1 - \sqrt{x}}\, dx = -x - 2\sqrt{x} - 2 \log (1 - \sqrt{x}) \cdot$$

$$\int \frac{x^{\frac{1}{4}}}{1+\sqrt{x}}\,dx = \frac{4}{3}x^{\frac{3}{4}} - 4x^{\frac{1}{4}} + 4 \operatorname{arc\,tg} x^{\frac{1}{4}}.$$

$$\int \frac{dx}{x^{\frac{3}{2}} - x} = 2\int \frac{dz}{z^2 - z}. \qquad x = z^2.$$

$$\int \frac{x - \sqrt{x} + x^{\frac{2}{3}}}{1 - \sqrt{x}}\,dx = -6\int z^8\,dz - 6\int \frac{z^9 - 1}{z^3 - 1}\,dz - 6\int \frac{dz}{z^3 - 1}. \qquad x = z^6.$$

SUMS AND DIFFERENCES OF BINOMIALS

$$\int \frac{x^{m-1}}{(ax^m + b)^2 + (px^m + q)^2}\, dx = \frac{1}{m(aq - bp)} \operatorname{arc\,tg} \frac{ax^m + b}{px^m + q}.$$

$$\int \frac{x^{m-1}}{\sqrt{ax^m + b} + \sqrt{ax^m + c}}\, dx = \frac{2}{3am(b - c)} \left\{ (ax^m + b)^{\frac{3}{2}} \mp (ax^m + c)^{\frac{3}{2}} \right\}.$$

$$\int \frac{x}{(1 + x)^{\frac{1}{3}} - \sqrt{1 + x}}\, dx = -\left\{ \frac{2}{3}\Phi^{\frac{5}{6}} + \frac{3}{4}\Phi^{\frac{2}{3}} + \frac{6}{7}\Phi^{\frac{1}{2}} + \Phi^{\frac{1}{3}} + \frac{6}{5}\Phi^{\frac{1}{6}} + \frac{3}{2} \right\} \Phi^{\frac{2}{3}}.$$

$$\int \frac{x}{(1 - x^2)^{\frac{1}{3}} + \sqrt{1 - x^2}}\, dx = -3 \int \frac{z^3}{z + 1}\, dz. \qquad 1 - x^2 = z^6.$$

$$\int \frac{dx}{\sqrt{x + a} + \sqrt{x + b}} = \frac{2}{3(a - b)} \left\{ (x + a)^{\frac{3}{2}} - (x + b)^{\frac{3}{2}} \right\}.$$

$$\int \frac{dx}{\sqrt{x + a} - \sqrt{x - a}} = \frac{1}{3a} \left\{ (x + a)^{\frac{3}{2}} + (x - a)^{\frac{3}{2}} \right\}.$$

$$\int \frac{dx}{\sqrt{1 + x^2} - \sqrt{1 - x^2}} = -\frac{\sqrt{1 + x^2} + \sqrt{1 - x^2}}{2x}$$

$$+ \frac{1}{2} \log (x + \sqrt{1 + x^2}) - \frac{1}{2} \arcsin x.$$

$$\int \frac{x}{(a + x)^{\frac{1}{4}} + (a + x)^{\frac{1}{2}}}\, dx = 4 \int \frac{(z - 1)^6 - a^2 (z - 1)^2}{z}\, dz. \qquad (a + x)^{\frac{1}{4}} = z - 1.$$

PRODUCTS AND QUOTIENTS OF BINOMIALS

$$\int (ax^m + b)(a'x^n + b')\,dx = \frac{aa'}{m+n+1}x^{m+n+1} + \frac{ab'}{m+1}x^{m+1} + \frac{ba'}{n+1}x^{n+1} + bb'x.$$

$$\int (ax+b)^p (a'x+b')^q\,dx = \frac{(ax+b)^p(a'x+b')^{q+1}}{a'(p+q+1)}$$

$$- \frac{p(ab'-a'b)}{a'(p+q+1)}\int (ax+b)^{p-1}(a'x+b')^q\,dx.$$

$$\int \frac{dx}{(1-x^n)(2x^n-1)^{\frac{1}{2n}}} = \int \frac{z^{2n-2}}{1-z^{2n}}\,dz. \qquad (2x^n-1)^{\frac{1}{2n}} = zx.$$

$$\int \frac{dx}{(1+x^n)(1+2x^n)^{\frac{1}{2n}}} = \int \frac{dz}{1+z^{2n}}. \qquad (1+2x^n)^{\frac{1}{2n}} = \frac{x}{z}.$$

$$\int \frac{x^m}{(x^m+a)(x^m+b)}\,dx = \frac{a}{a-b}\int \frac{dx}{x^m+a} - \frac{b}{a-b}\int \frac{dx}{x^m+b}.$$

$$\int \frac{x^{2m-1}}{(x^m-a)(x^m-b)}\,dx = \frac{1}{m(a-b)}\log\frac{(x^m-a)^a}{(x^m-b)^b}.$$

$$\int \frac{dx}{x\sqrt{(a+bx^m)(a+cx^m)}} = -\frac{2}{am}\log\left\{\sqrt{\frac{a}{x^m}+b} + \sqrt{\frac{a}{x^m}+c}\right\}.$$

$$\int \frac{x^{m-1}}{\sqrt{(ax^m+b)(ax^m+c)}}\,dx = \frac{2}{am}\log\left\{\sqrt{ax^m+b} + \sqrt{ax^m+c}\right\}.$$

$$\int \frac{x^{m-1}}{(ax^m+b)(px^m+q)}\,dx = \frac{1}{m(aq-bp)}\log\frac{ax^m+b}{px^m+q}.$$

$$\int \frac{x-4}{(x-2)^{\frac{4}{3}}\sqrt{x-1}}\,dx = 6\frac{\sqrt{x-1}}{(x-2)^{\frac{1}{3}}}.$$

$$\int \frac{x(2-x^2)}{(1+x^2)^{\frac{5}{2}}\sqrt{1-x^2}}\,dx = -\frac{1}{2}\sqrt{\frac{1-x^2}{(1+x^2)^3}}.$$

$$\int \frac{dx}{(1+nx^2)^p\sqrt{a+bx^2}} = \frac{1}{(2p-2)(an-b)}\left\{\frac{nx\sqrt{\Phi}}{(1+nx^2)^{p-1}}\right.$$

$$+ (2p-3)(an-2b)\int\frac{dx}{(1+nx^2)^{p-1}\sqrt{\Phi}} + b(2p-4)\int\frac{dx}{(1+nx^2)^{p-2}\sqrt{\Phi}}\Bigg\}.$$

$$\int\frac{dx}{x^p(mx+n)\sqrt{ax+b}} = \frac{1}{n}\int\frac{dx}{x^p\sqrt{\Phi}} - \frac{m}{n^2}\int\frac{dx}{x^{p-1}\sqrt{\Phi}} + \frac{m^2}{n^3}\int\frac{dx}{x^{p-2}\sqrt{\Phi}}$$

$$-\cdots \pm \frac{m^{p-1}}{n^p}\int\frac{dx}{x\sqrt{\Phi}} \mp \frac{m^p}{n^p}\int\frac{dx}{(mx+n)\sqrt{\Phi}}.$$

$$\int\frac{dx}{(x^m+a)(x^m+b)} = \frac{1}{a-b}\int\frac{dx}{x^m+b} - \frac{1}{a-b}\int\frac{dx}{x^m+a}.$$

$$\int\frac{dx}{(x-a)^{\frac{3}{4}}(x-b)^{\frac{1}{4}}} = \log\frac{(x-b)^{\frac{1}{4}}+(x-a)^{\frac{1}{4}}}{(x-b)^{\frac{1}{4}}-(x-a)^{\frac{1}{4}}} + 2\operatorname{arc\,tg}\left(\frac{x-a}{x-b}\right)^{\frac{1}{4}}.$$

$$\int\frac{x^p}{(x-a_1)(x-a_2)\cdots(x-a_n)}\,dx = \frac{a_1^p\log(x-a_1)}{(a_1-a_2)(a_1-a_3)\cdots(a_1-a_n)}$$

$$+\frac{a_2^p\log(x-a_2)}{(a_2-a_1)(a_2-a_3)\cdots(a_2-a_n)}+\cdots$$

$$+\frac{a_n^p\log(x-a_n)}{(a_n-a_1)(a_n-a_2)\cdots(a_n-a_{n-1})}.$$

$$\int\frac{dx}{\sqrt{ax+b}\,(mx+n)^{\frac{3}{2}}} = \frac{2}{an-bm}\sqrt{\frac{ax+b}{mx+n}}.$$

$$\int\frac{x^p}{(mx+n)\sqrt{ax^2+b}}\,dx = \frac{1}{m}\int\frac{x^{p-1}}{\sqrt{\Phi}}\,dx - \frac{n}{m^2}\int\frac{x^{p-2}}{\sqrt{\Phi}}\,dx + \frac{n^2}{m^3}\int\frac{x^{p-3}}{\sqrt{\Phi}}\,dx - \cdots$$

$$\mp\frac{n^{p-1}}{m^p}\int\frac{dx}{\sqrt{\Phi}} \mp \frac{n^p}{m^p}\int\frac{dx}{(mx+n)\sqrt{\Phi}}.$$

$$\int\frac{x^{m-1}}{(px^m+q)^{\frac{3}{2}}\sqrt{ax^m+b}}\,dx = \frac{2}{m(aq-bp)}\sqrt{\frac{ax^m+b}{px^m+q}}.$$

$$\int\frac{x^{m-1}}{(ax^m-1)(ax^m+1)^{\frac{3}{2}}}\,dx = \frac{1}{am}\left\{\frac{1}{\sqrt{ax^m+1}} + \frac{1}{\sqrt{2}}\log\frac{\sqrt{ax^m+1}-\sqrt{2}}{\sqrt{ax^m-1}}\right\}.$$

$$\int\frac{x^{m-1}}{(1+x^m)(1-x^m)^{\frac{1}{m}}}\,dx = -\int\frac{z^{m-2}}{2-z^m}\,dz. \qquad 1-x^m = z^m.$$

$$\int \frac{x^{m-2}}{(1-x^m)(2x^m-1)^{\frac{1}{2m}}}\,dx = 2\int \frac{z^{2m-1}}{1-z^{2m}}\,dz. \qquad 2x^m - 1 = z^{2m}.$$

$$\int \frac{\sqrt{x-a}}{(x-b)^{\frac{3}{2}}}\,dx = -2\left(\frac{x-a}{x-b}\right)^{\frac{1}{2}} + \log \frac{\sqrt{x-b}+\sqrt{x-a}}{\sqrt{x-b}-\sqrt{x-a}}.$$

$$\int \frac{\sqrt{1-x}}{(1+x)^{\frac{3}{2}}}\,dx = -2\left(\frac{1-x}{1+x}\right)^{\frac{1}{2}} + \text{arc cos } x.$$

$$\int \frac{mx+n}{(ax+b)^p}\,dx = -\frac{m}{a^2(p-2)\,\Phi^{p-2}} - \frac{an-bm}{a^2(p-1)\,\Phi^{p-1}}.$$

$$\int \frac{(ax+b)^m}{px+q}\,dx = \frac{a^m}{p^{m+1}}\left\{\frac{1}{m}(px+q)^m - \frac{m}{m-1}\,\frac{aq-bp}{a}(px+q)^{m-1}\right.$$

$$\left. + \frac{m(m-1)}{1.2\,(m-2)}\,\frac{(aq-bp)^2}{a^2}(px+q)^{m-2} - \ldots \pm \frac{(aq-bp)^m}{a^m}\log(px+q)\right\}.$$

$$\int \left(\frac{ax+b}{mx+n}\right)^p dx = \frac{1}{m^{p+1}}\int \frac{(az+bm-an)^p}{z^p}\,dz. \qquad mx+n=z.$$

$$\int \frac{(ax+b)^m}{(a'x+b')^n}\,dx = -\frac{(ax+b)^{m+1}}{(n-1)(a'b-ab')(a'x+b')^{n-1}}$$

$$-\frac{a(n-m-2)}{(n-1)(a'b-ab')}\int \frac{(ax+b)^m}{(a'x+b')^{n-1}}\,dx.$$

$$\int \frac{(a+bx)^m}{(a-bx)^n}\,dx = \frac{(2a)^m}{b(n-1)(a-bx)^{n-1}} - \frac{m}{1}\,\frac{(2a)^{m-1}}{b(n-2)(a-bx)^{n-2}}$$

$$+\frac{m(m-1)}{1.2}\,\frac{(2a)^{m-2}}{b(n-3)(a-bx)^{n-3}} - \ldots$$

$$\int \frac{(ax^2-c)^n}{(b-ax)^{n+1}}\,dx = -\frac{1}{a^{n+1}}\int \frac{(z^2-2bz+b^2-ac)^n}{z^{n+1}}\,dz. \qquad b-ax=z.$$

TRIGONOMETRIC FUNCTIONS

x^a cir X

$$\int \sin (ax + b)\, dx = -\frac{1}{a} \cos (ax + b).$$

$$\int \cos (px + q)\, dx = \frac{1}{p} \sin (px + q).$$

$$\int \operatorname{tg} (ax + b)\, dx = -\frac{1}{a} \log \cos (ax + b).$$

$$\int \operatorname{cotg} (ax + b)\, dx = \frac{1}{a} \log \sin (ax + b).$$

$$\int \text{séc}\, x\, dx = \log \operatorname{tg} \left(\frac{\pi}{4} + \frac{x}{2}\right).$$

$$\int \text{cosééc}\, x\, dx = \log \operatorname{tg} \frac{x}{2}.$$

$$\int x \sin x\, dx = \sin x - x \cos x.$$

$$\int x \cos x\, dx = \cos x + x \sin x.$$

$$\int x \cos ax\, dx = \frac{1}{a} x \sin ax + \frac{1}{a^2} \cos ax.$$

$$\int x^2 \sin x\, dx = 2x \sin x - (x^2 - 2) \cos x.$$

$$\int x^2 \cos x\, dx = 2x \cos x + (x^2 - 2) \sin x.$$

$$\int x^3 \sin x\, dx = (3x^2 - 6) \sin x - (x^3 - 6x) \cos x.$$

$$\int x^3 \cos x\, dx = (3x^2 - 6) \cos x + (x^3 - 6x) \sin x.$$

$$\int x^4 \sin x \, dx = (4x^3 - 24x) \sin x - (x^4 - 12x^2 + 24) \cos x.$$

$$\int x^4 \cos x \, dx = (4x^3 - 24x) \cos x + (x^4 - 12x^2 + 24) \sin x.$$

$$\int x^n \sin x \, dx = - x^n \cos x + n \int x^{n-1} \cos x \, dx$$

$$= - x^n \cos x + n x^{n-1} \sin x - n(n-1) \int x^{n-2} \sin x \, dx.$$

$$\int x^n \sin ax \, dx = \frac{x^{n-1}}{a^2} (n \sin ax - ax \cos ax) - \frac{n(n-1)}{a^2} \int x^{n-2} \sin ax \, dx.$$

$$\int x^m \cos x \, dx = x^m \sin x - m \int x^{m-1} \sin x \, dx.$$

$$= x^m \sin x + m x^{m-1} \cos x - m(m-1) \int x^{m-2} \cos x \, dx.$$

$$= \sin x \left\{ x^m - m(m-1) x^{m-2} \right.$$

$$\left. + m(m-1)(m-2)(m-3) x^{m-4} - \cdots \right\}$$

$$+ \cos x \left\{ m x^{m-1} - m(m-1)(m-2) x^{m-2} + \cdots \right\}.$$

$$\int x^m \cos ax \, dx = \frac{x^m \sin ax}{a} - \frac{m}{a} \int x^{m-1} \sin ax \, dx.$$

$$= \frac{x^{m-1}}{a^2} (m \cos ax + ax \sin ax) - \frac{m(m-1)}{a^2} \int x^{m-2} \cos ax \, dx.$$

$$\frac{\text{cir X}}{x^9}$$

$$\int \frac{\sin x}{x^2} dx = - \frac{\sin x}{x} + \int \frac{\cos x}{x} dx.$$

$$\int \frac{\cos x}{x^2} dx = - \frac{\cos x}{x} - \int \frac{\sin x}{x} dx.$$

$$\int \frac{\sin x}{x^3} dx = - \frac{\sin x}{2x^2} - \frac{\cos x}{2x} - \frac{1}{2} \int \frac{\sin x}{x} dx.$$

$$\int \frac{\cos x}{x^3} dx = -\frac{\cos x}{2x^2} + \frac{\sin x}{2x} - \frac{1}{2}\int \frac{\cos x}{x} dx.$$

$$\int \frac{\sin x}{x^4} dx = -\frac{\sin x}{3x^3} - \frac{\cos x}{6x^2} + \frac{\sin x}{6x} - \frac{1}{6}\int \frac{\cos x}{x} dx.$$

$$\int \frac{\cos x}{x^4} dx = -\frac{\cos x}{3x^3} + \frac{\sin x}{6x^2} + \frac{\cos x}{6x} + \frac{1}{6}\int \frac{\sin x}{x} dx.$$

$$\int \frac{\sin x}{x^m} dx = -\frac{\sin x}{(m-1)\,x^{m-1}} + \frac{1}{m-1}\int \frac{\cos x}{x^{m-1}} dx.$$

$$= -\frac{\sin x}{(m-1)\,x^{m-1}} - \frac{\cos x}{(m-1)\,(m-2)\,x^{m-2}} - \frac{1}{(m-1)\,(m-2)}\int \frac{\sin x}{x^{m-2}} dx.$$

$$\int \frac{\sin ax}{x^m} dx = -\frac{ax \cos ax + (m-2) \sin ax}{(m-1)\,(m-2)\,x^{m-1}} - \frac{a^2}{(m-1)\,(m-2)}\int \frac{\sin ax}{x^{m-2}} dx.$$

$$\int \frac{\cos x}{x^m} dx = -\frac{\cos x}{(m-1)\,x^{m-1}} - \frac{1}{m-1}\int \frac{\sin x}{x^{m-1}} dx.$$

$$\int \frac{\cos ax}{x^m} dx = \frac{ax \sin ax - (m-2) \cos ax}{(m-1)\,(m-2)\,x^{m-1}} - \frac{a^2}{(m-1)\,(m-2)}\int \frac{\cos ax}{x^{m-2}} dx.$$

(cir X) ^{Q}

$$\int \text{Sin}^2 x\, dx = -\frac{1}{2}(\sin x \cos x - x) = \frac{x}{2} - \frac{1}{4}\sin 2x.$$

$$\int \text{Sin}^2 mx\, dx = \frac{x}{2} - \frac{\sin 2mx}{4m}.$$

$$\int \cos^2 x\, dx = \frac{1}{2}(\sin x \cos x + x) = \frac{1}{2}\left(x + \frac{\sin 2x}{2}\right).$$

$$\int \cos^2 mx\, dx = \frac{x}{2} + \frac{\sin 2m\,x}{4m}.$$

$$\int \text{tg}^2 x\, dx = \text{tg}\, x - x.$$

$$\int \text{cotg}^2 x\, dx = -\,\text{cotg}\, x - x.$$

$$\int \text{séc}^2 x\, dx = \text{tg}\, x.$$

$$\int \cos^3 x\,dx = \frac{1}{3}\sin x\cos^2 x + \frac{2}{3}\sin x = \frac{1}{12}\sin 3x + \frac{3}{4}\sin x.$$

$$\int \operatorname{tg}^3 x\,dx = \frac{1}{2}\operatorname{tg}^2 x + \log\cos x.$$

$$\int \operatorname{cotg}^3 x\,dx = -\frac{1}{2}\operatorname{cotg}^2 x - \log\sin x.$$

$$\int \sin^4 x\,dx = \frac{1}{32}\sin 4x - \frac{1}{4}\sin 2x + \frac{3}{8}x = -\frac{\cos x}{4}\left(\sin^3 x + \frac{3}{2}\sin x\right) + \frac{3}{8}x.$$

$$\int \cos^4 x\,dx = \frac{1}{32}\sin 4x + \frac{1}{4}\sin 2x + \frac{3}{8}x = \frac{1}{4}\cos^3 x\sin x + \frac{3}{8}(\cos x\sin x + x)$$

$$= \frac{\sin x\cos x}{6}\left(\cos^4 x + \frac{5}{4}\cos^2 x + \frac{15}{8}\right) + \frac{5}{16}x.$$

$$\int \operatorname{tg}^4 x\,dx = \frac{1}{3}\operatorname{tg}^3 x - \operatorname{tg} x + x.$$

$$\int \text{séc}^4\,dx = \operatorname{tg} x + \frac{1}{3}\operatorname{tg}^3 x.$$

$$\int \sin^n x\,dx = -\frac{\sin^{n-1} x\cos x}{n} + \frac{n-1}{n}\int \sin^{n-2} x\,dx.$$

$$\int \sin^{2n} x\,dx = -\frac{\cos x}{2n}\left\{\sin^{2n-1} x + \frac{2n-1}{2n-2}\sin^{2n-3} x + \frac{(2n-1)(2n-3)}{(2n-2)(2n-4)}\sin^{2n-5} x + \right.$$

$$\left.\cdots\frac{(2n-1)\cdots 3.1}{(2n-2)\cdots 4.2}\sin x\right\} + \frac{(2n-1)\cdots 3.1}{2n\cdots 4.2}x.$$

$$= \frac{(-1)^n}{2^{2n-1}}\left\{\frac{\sin 2n\,x}{2n} - 2n\frac{\sin(2n-2)\,x}{2n-2}\right.$$

$$+ \frac{2n\,(2n-1)}{1.2}\,\frac{\sin(2n-4)\,x}{2n-4} - \cdots$$

$$+ (-1)^{n-1}\frac{2n\,(2n-1)\cdots(n+2)}{(n-1)\cdots 2.1}\,\frac{\sin 2x}{2}$$

$$\left. + (-1)^n\frac{2n\,(2n-1)\cdots(n+1)}{n\cdots 2.1}\cdot\frac{x}{2}\right\}.$$

$$\int \sin^{2n+1} x\,dx = -\frac{\cos x}{2n+1}\left\{\sin^{2n} x + \frac{2n}{2n-1}\sin^{2n-2} x + \frac{2n\,(2n-2)}{(2n-1)(2n-3)}\sin^{2n-4} x\right.$$

$$+\cdots+\frac{2n\cdots 4.2}{(2n-1)\cdots 3.1}\Big\}.$$

$$=\frac{(-1)^{n+1}}{2^{2n}}\Big\{\frac{\cos(2n+1)x}{2n+1}-(2n+1)\frac{\cos(2n-1)x}{2n-1}+\cdots$$

$$+(-1)^n\frac{(2n+1)\cdots(n+2)}{n\cdots 2.1}\cos x\Big\}.$$

$$\int\cos^n x\,dx=\frac{1}{n}\cos^{n-1}x\sin x+\frac{n-1}{n}\int\cos^{n-2}x\,dx.$$

$$\int\cos^{2n}x\,dx=\frac{1}{2^{2n-1}}\Big\{\frac{\sin 2nx}{2n}+2n\frac{\sin(2n-2)x}{2n-2}+\frac{2n(2n-1)}{1.2}\frac{\sin(2n-4)x}{2n-4}+\cdots$$

$$+\frac{2n(2n-1)\cdots(n+1)}{1.2\cdots n}\frac{x}{2}\Big\}.$$

$$=\frac{\sin x}{2n}\Big\{\cos^{2n-1}x+\frac{2n-1}{2n-2}\cos^{2n-3}x+\frac{(2n-1)(2n-3)}{(2n-2)(2n-4)}\cos^{2n-5}x+\cdots$$

$$+\frac{1.3\cdots(2n-1)}{2.4\cdots(2n-2)}\cos x\Big\}+\frac{1.3\cdots(2n-1)}{2.4\cdots 2n}x.$$

$$\int\cos^{2n+1}x\,dx=\frac{1}{2^{2n}}\Big\{\frac{\sin(2n+1)x}{2n+1}+(2n+1)\frac{\sin(2n-1)x}{2n-1}$$

$$+\frac{(2n+1)(2n-1)}{1.2}\frac{\sin(2n-3)x}{2n-3}+\cdots+\frac{(2n+1)\cdots(n+2)}{1.2\cdots n}\sin x\Big\}.$$

$$=\frac{\sin x}{2n+1}\Big\{\cos^{2n}x+\frac{2n}{2n-1}\cos^{2n-2}x+\cdots+\frac{2.4\cdots 2n}{1.3\cdots(2n-1)}\Big\}.$$

$$\int\operatorname{tg}^n x\,dx=\frac{1}{n-1}\operatorname{tg}^{n-1}x-\int\operatorname{tg}^{n-2}x\,dx.$$

$$\int\operatorname{tg}^{2n}x\,dx=\frac{1}{2n-1}\operatorname{tg}^{2n-1}x-\frac{1}{2n-3}\operatorname{tg}^{2n-3}x+\frac{1}{2n-5}\operatorname{tg}^{2n-5}x-\cdots$$

$$+(-1)^{n-1}\operatorname{tg}x+(-1)^n x.$$

$$\int\operatorname{tg}^{2n+1}x\,dx=\frac{1}{2n}\operatorname{tg}^{2n}x-\frac{1}{2n-2}\operatorname{tg}^{2n-2}x+\frac{1}{2n-4}\operatorname{tg}^{2n-4}x-\cdots$$

$$+(-1)^{n-1}\frac{1}{2}\operatorname{tg}^2 x+(-1)^{n-1}\log\cos x.$$

$$\int \operatorname{cotg}^n x\,dx = -\frac{1}{n-1}\operatorname{cotg}^{n-1} x - \int \operatorname{cotg}^{n-2} x\,dx.$$

$$\int \operatorname{cotg}^{2n} x\,dx = -\frac{\operatorname{cotg}^{2n-1} x}{2n-1} + \frac{\operatorname{cotg}^{2n-3} x}{2n-3} - \frac{\operatorname{cotg}^{2n-5} x}{2n-5} + \cdots$$
$$- (-1)^{n-1}\operatorname{cotg} x + (-1)^{n-1} x.$$

$$\int \operatorname{cotg}^{2n+1} x\,dx = -\frac{\operatorname{cotg}^{2n} x}{2n} + \frac{\operatorname{cotg}^{2n-2} x}{2n-2} - \frac{\operatorname{cotg}^{2n-4} x}{2n-4} + \cdots$$
$$+ (-1)^n \frac{\operatorname{cotg}^2 x}{2} + (-1)^n \log \sin x.$$

$$\frac{A}{\text{cir } X}$$

$$\int \frac{dx}{\sin(ax+b)} = \frac{1}{a}\log \operatorname{tg}\frac{1}{2}(ax+b).$$

$$\int \frac{dx}{\cos x} = \log \operatorname{tg}\left(\frac{\pi}{4}+\frac{x}{2}\right) = \frac{1}{2}\log\frac{1+\sin x}{1-\sin x}.$$

$$\int \frac{dx}{\operatorname{tg} x} = \log \sin x.$$

$$\int \frac{dx}{\operatorname{cotg} x} = -\log \cos x.$$

$$\int \frac{\sin(x+a)}{\sin x}\,dx = x\cos a + \sin a \log \sin x.$$

$$\int \frac{\sin 2x}{\sin x}\,dx = 2\sin x.$$

$$\int \frac{\sin 2x}{\cos x}\,dx = -2\cos x.$$

$$\int \frac{\cos 2x}{\sin x}\,dx = 2\cos x + \log \operatorname{tg}\frac{x}{2}.$$

$$\int \frac{\cos 2x}{\cos x}\,dx = 2\sin x - \log \operatorname{tg}\left(\frac{\pi}{4}+\frac{x}{2}\right).$$

$$\int \frac{\sin 3x}{\sin x}\,dx = x + 2\sin x \cos x.$$

$$\int \frac{\sin 3x}{\cos x}\,dx = 2\sin^2 x + \log \cos x.$$

$$\int \frac{\cos 3x}{\sin x}\,dx = -2\sin^2 x + \log \sin x.$$

$$\int \frac{\cos 3x}{\cos x}\,dx = 2\sin x \cos x - x.$$

$$\int \frac{\sin x}{\sin 2x}\,dx = \frac{1}{2}\log \operatorname{tg}\left(\frac{\pi}{4} + \frac{x}{2}\right)$$

$$\int \frac{\cos x}{\sin 2x}\,dx = \frac{1}{2}\log\ \operatorname{tg}\frac{x}{2}\cdot$$

$$\int \frac{\cos x}{\cos 2x}\,dx = \frac{1}{2\sqrt{2}}\log\frac{1+\sqrt{2}\sin x}{1-\sqrt{2}\sin x}\cdot$$

$$\int \frac{\sin x}{\sin 3x}\,dx = \frac{1}{2\sqrt{3}}\log\frac{\sin\left(\frac{\pi}{3}+x\right)}{\sin\left(\frac{\pi}{3}-x\right)}\cdot$$

$$\int \frac{\sin x}{\cos 3x}\,dx = \frac{1}{3}\log\cos x - \frac{1}{6}\log\left(\cos^2 x - \frac{3}{4}\right)\cdot$$

$$\int \frac{\cos x}{\sin 3x}\,dx = \frac{1}{3}\log\sin x - \frac{1}{6}\log\left(\sin^2 x - \frac{3}{4}\right)\cdot$$

$$\int \frac{\cos x}{\cos 3x}\,dx = \frac{1}{2\sqrt{3}}\log\frac{\cos\left(\frac{\pi}{3}-x\right)}{\cos\left(\frac{\pi}{3}+x\right)}\cdot$$

$$\int \frac{\sin 2x}{\cos 3x}\,dx = \frac{1}{2\sqrt{3}}\log\left\{\operatorname{cotg}\left(15^0 - \frac{x}{2}\right)\operatorname{cotg}\left(15^0 + \frac{x}{2}\right)\right\}\cdot$$

$$\int \frac{\sin^2(x+a)}{\sin x}\,dx = -\cos^2 a\cos x + \sin 2a \sin x + \sin^2 a\left\{\log\operatorname{tg}\frac{x}{2} + \cos x\right\}\cdot$$

$$\int \frac{\cos^2 x}{\sin x}\,dx = \cos x + \log\operatorname{tg}\frac{x}{2}\cdot$$

$$\int \frac{\cos^3 x}{\sin x}\,dx = \frac{1}{2}\cos^2 x + \log \sin x.$$

$$\int \frac{\sin^3 x}{\cos x}\,dx = \frac{1}{2}\cos^2 x - \log \cos x = \frac{1}{\cos^3 x}\left\{ \sin^2 x - \frac{2}{3} \right\}.$$

$$\int \frac{\cos^4 x}{\sin x}\,dx = \frac{1}{3}\cos^3 x + \cos x + \log \operatorname{tg} \frac{x}{2}\cdot$$

$$\int \frac{\sin^4 x}{\cos x}\,dx = -\frac{\sin^2 x + 3}{3}\sin x + \log \operatorname{tg}\left(\frac{\pi}{4} + \frac{x}{2}\right)\cdot$$

$$\int \frac{\cos^n x}{\sin x}\,dx = \frac{1}{n-1}\cos^{n-1} x + \int \frac{\cos^{n-2} x}{\sin x}\,dx.$$

$$\int \frac{\sin^m x}{\cos x}\,dx = -\frac{\sin^{m-1} x}{m-1} + \int \frac{\sin^{m-2} x}{\cos x}\,dx.$$

$$\int \frac{\sin^{2n} x}{\cos x}\,dx = -\sin x - \frac{\sin^3 x}{3} - \frac{\sin^5 x}{5} - \cdots - \frac{\sin^{2n-1} x}{2n-1} + \int \frac{dx}{\cos x}\cdot$$

$$\int \frac{\sin^{2n+1} x}{\cos x}\,dx = -\frac{\sin^2 x}{2} - \frac{\sin^4 x}{4} - \frac{\sin^6 x}{6} - \cdots - \frac{\sin^{2n} x}{2n} + \int \operatorname{tg} x\,dx.$$

$$\int \frac{\cos^2 x}{\sin 3x}\,dx = \frac{1}{3}\log \operatorname{tg} \frac{3}{2}x.$$

$$\int \frac{\sin^2 x}{\cos 3x}\,dx = \frac{1}{3}\log \operatorname{tg}\left(\frac{\pi}{4} + \frac{3}{2}x\right)\cdot$$

$$\boxed{\frac{A}{\operatorname{cir}^2 X}}$$

$$\int \frac{dx}{\sin^2 (px+q)} = -\frac{1}{p}\operatorname{cotg}(px+q).$$

$$\int \frac{dx}{\cos^2 (px+q)} = \frac{1}{p}\operatorname{tg}(px+q).$$

$$\int \frac{x}{\sin^2 x}\,dx = -x \operatorname{cotg} x + \log \sin x.$$

$$\int \frac{x}{\cos^2 x}\,dx = x \operatorname{tg} x + \log \cos x.$$

$$\int \frac{\cos x}{\sin^2 x}\, dx = -\ \text{coséc}\, x.$$

$$\int \frac{\cos\ (px+q)}{\sin^2\ (px+q)}\, dx = -\frac{1}{p}\, \text{coséc}\, (px+q).$$

$$\int \frac{\sin x}{\cos^2 x}\, dx = \frac{1}{\cos x} = \text{séc}\, x.$$

$$\int \frac{\sin\ (px+q)}{\cos^2\ (px+q)}\, dx = \frac{1}{p}\, \text{séc}\, (px+q).$$

$$\int \frac{\sin 2x}{\sin^2 x}\, dx = 2 \log \sin x.$$

$$\int \frac{\cos 2x}{\sin^2 x}\, dx = -\ 2x - \text{cotg}\, x.$$

$$\int \frac{\sin 2x}{\cos^2 x}\, dx = -\ 2 \log \cos x.$$

$$\int \frac{\cos 2x}{\cos^2 x}\, dx = 2x - \text{tg} x.$$

$$\int \frac{\sin 3x}{\sin^2 x}\, dx = 3 \log \text{tg}\, \frac{x}{2} + 4 \cos x.$$

$$\int \frac{\cos 3x}{\sin^2 x}\, dx = -\ 4 \sin x - \frac{1}{\sin x}\, .$$

$$\int \frac{\sin 3x}{\cos^2 x}\, dx = -\ 4 \cos x - \frac{1}{\cos x}$$

$$\int \frac{\cos 3x}{\cos^2 x}\, dx = 4 \sin x - 3 \log \text{tg} \left(\frac{\pi}{4} + \frac{x}{2}\right) \cdot$$

$$\int \frac{\cos nx}{\sin^2 nx}\, dx = -\frac{1}{n}\, \text{coséc}\, nx = -\ \frac{1}{n \sin nx} \cdot$$

$$\int \frac{\sin nx}{\cos^2 nx}\, dx = \frac{1}{n}\, \text{séc}\, nx.$$

$$\int \frac{\cos^2 x}{\sin^2 x}\, dx = -\ \text{cotg}\, x - x.$$

$$\int \frac{\cos^3 x}{\sin^2 x}\,dx = -\frac{1+\sin^2 x}{\sin x}.$$

$$\int \frac{\sin^3 x}{\cos^2 x}\,dx = \cos x + \text{séc}\, x.$$

$$\int \frac{\cos^4 x}{\sin^2 x}\,dx = -\frac{\cos^3 x}{\sin x} - \frac{3}{2}\sin x \cos x - \frac{3}{2}x.$$

$$\int \frac{\cos^n x}{\sin^2 x}\,dx = \frac{1}{n-2}\frac{\cos^{n-1} x}{\sin x} + \frac{n-1}{n-2}\int \frac{\cos^{n-2} x}{\sin^2 x}\,dx = -\frac{\cos^{n-1} x}{\sin x} - n\int \cos^n x\,dx.$$

$$\int \frac{\sin^m x}{\cos^2 x}\,dx = -\frac{\sin^{m-1} x}{(m-2)\cos x} + \frac{m-1}{m-2}\int \frac{\sin^{m-2} x}{\cos^2 x}\,dx = \frac{\sin^{m+1} x}{\cos x} - m\int \sin^m x\,dx.$$

$$\frac{A}{\text{cir}^3 X}$$

$$\int \frac{dx}{\sin^3 x} = -\frac{\cos x}{2\sin^2 x} + \frac{1}{2}\log \text{tg}\,\frac{x}{2}.$$

$$\int \frac{dx}{\cos^3 x} = \frac{\sin x}{2\cos^2 x} + \frac{1}{2}\log \text{tg}\left(\frac{\pi}{4}+\frac{x}{2}\right).$$

$$\int \frac{x}{\sin^3 x}\,dx = -\frac{\sin x + x\cos x}{2\sin^2 x} + \frac{1}{2}\int \frac{x}{\sin x}\,dx.$$

$$\int \frac{x}{\cos^3 x}\,dx = \frac{x\sin x - \cos x}{2\cos^2 x} + \frac{1}{2}\int \frac{x}{\cos x}\,dx.$$

$$\int \frac{\sin x}{\cos^3 x}\,dx = \frac{1}{2}\text{tg}^2\, x.$$

$$\int \frac{\sin 2x}{\sin^3 x}\,dx = -\frac{2}{\sin x}.$$

$$\int \frac{\cos 2x}{\sin^3 x}\,dx = -\frac{\cos x}{2\sin^2 x} - \frac{3}{2}\log \text{tg}\,\frac{x}{2}.$$

$$\int \frac{\cos 2x}{\cos^3 x}\,dx = -\frac{\sin x}{2\cos^2 x} + \frac{3}{2}\log \text{tg}\left(\frac{\pi}{4}+\frac{x}{2}\right).$$

$$\int \frac{\sin 3x}{\sin^3 x}\,dx = -3\,\text{cotg}\, x - 4x.$$

$$\int \frac{\cos 3x}{\sin^3 x}\, dx = -\frac{1}{2 \sin^2 x} - 4 \log \sin x.$$

$$\int \frac{\sin 3x}{\cos^3 x}\, dx = -\frac{1}{2 \cos^2 x} - 4 \log \cos x.$$

$$\int \frac{\cos 3x}{\cos^3 x}\, dx = 4x - 3 \operatorname{tg} x.$$

$$\int \frac{\cos^2 x}{\sin^3 x}\, dx = -\frac{\cos x}{2 \sin^2 x} - \frac{1}{2} \log \operatorname{tg} \frac{x}{2}.$$

$$\int \frac{\cos^3 x}{\sin^3 x}\, dx = -\frac{1}{2} \operatorname{cotg}^2 x - \log \sin x.$$

$$\int \frac{\cos^4 x}{\sin^3 x}\, dx = \frac{1}{\sin^2 x}\left(\cos^5 x - \frac{3}{2} \cos x\right) - \frac{3}{2} \log \operatorname{tg} \frac{x}{2}.$$

$$\int \frac{\cos^n x}{\sin^3 x}\, dx = -\frac{\cos^{n+1} x}{2 \sin^2 x} - \frac{n-1}{2} \int \frac{\cos^n x}{\sin x}\, dx$$

$$= \frac{\cos^{n-1} x}{(n-3) \sin^2 x} + \frac{n-1}{n-3} \int \frac{\cos^{n-2} x}{\sin^3 x}\, dx.$$

$$\int \frac{\sin^m x}{\cos^3 x}\, dx = -\frac{\sin^{m-1} x}{(m-3) \cos^2 x} + \frac{m-1}{m-3} \int \frac{\sin^{m-2} x}{\cos^3 x}\, dx$$

$$= \frac{\sin^{m+1} x}{2 \cos^2 x} - \frac{m-1}{2} \int \frac{\sin^m x}{\cos x}\, dx.$$

$$\frac{A}{\operatorname{cir}^4 X}$$

$$\int \frac{dx}{\sin^4 x} = -\frac{\cos x}{3 \sin^3 x} - \frac{2}{3} \operatorname{cotg} x = \frac{\cos x}{3 \sin^3 x}(1 + 2 \sin^2 x).$$

$$\int \frac{dx}{\cos^4 x} = \frac{\sin x}{3 \cos^3 x} + \frac{2}{3} \operatorname{tg} x = \operatorname{tg} x + \frac{1}{3} \operatorname{tg}^3 x.$$

$$\int \frac{x}{\sin^4 x}\, dx = -\frac{\sin x + 2x \cos x}{6 \sin^3 x} - \frac{2}{3}(x \operatorname{cotg} x - \log \sin x).$$

$$\int \frac{x}{\cos^4 x}\, dx = \frac{2\, x \sin x - \cos x}{6 \cos^3 x} + \frac{2}{3}(x \operatorname{tg} x + \log \cos x).$$

$$\int \frac{\sin 2x}{\sin^4 x}\,dx = -\frac{1}{\sin^2 x}\cdot$$

$$\int \frac{\cos 2x}{\sin^4 x}\,dx = -\frac{\cos x}{3\sin^3 x} + \frac{4}{3}\operatorname{cotg} x.$$

$$\int \frac{\cos 2x}{\cos^4 x}\,dx = \frac{1}{2}\operatorname{tg} x\left(4 - \frac{1}{\cos^2 x}\right)\cdot$$

$$\int \frac{\sin 3x}{\sin^4 x}\,dx = -\frac{3\operatorname{cotg} x}{2\sin x} - \frac{5}{2}\log\operatorname{tg}\frac{x}{2}\cdot$$

$$\int \frac{\cos 3x}{\sin^4 x}\,dx = -\frac{1}{3\sin^3 x} + \frac{4}{\sin x}\cdot$$

$$\int \frac{\sin 3x}{\cos^4 x}\,dx = \frac{4}{\cos x} - \frac{1}{3\cos^3 x}\cdot$$

$$\int \frac{\cos 3x}{\cos^4 x}\,dx = -\frac{3\sin x}{2\cos^2 x} + \frac{5}{2}\log\operatorname{tg}\left(\frac{\pi}{4} + \frac{x}{2}\right)\cdot$$

$$\int \frac{\cos^2 x}{\sin^4 x}\,dx = -\frac{\cos x}{3\sin^3 x} + \frac{1}{3}\operatorname{cotg} x.$$

$$\int \frac{\cos^3 x}{\sin^4 x}\,dx = -\frac{\cos^2 x}{3\sin^3 x} + \frac{2}{3\sin x}\cdot$$

$$\int \frac{\cos^4 x}{\sin^4 x}\,dx = -\frac{1}{3}\operatorname{cotg}^3 x + \operatorname{cotg} x + x.$$

$$\int \frac{\sin^m x}{\cos^4 x}\,dx = -\frac{\sin^{m-1} x}{(m-4)\cos^3 x} + \frac{m-1}{m-4}\int \frac{\sin^{m-2} x}{\cos^4 x}\,dx$$

$$= \frac{\sin^{m+1} x}{3\cos^3 x} - \frac{m-2}{3}\int \frac{\sin^m x}{\cos^2 x}\,dx.$$

$$\boxed{\frac{A}{\operatorname{cir}^n X}}$$

$$\int \frac{dx}{\sin^n x} = -\frac{\cos x}{(n-1)\sin^{n-1} x} + \frac{n-2}{n-1}\int \frac{dx}{\sin^{n-2} x}\cdot$$

$$\int \frac{dx}{\sin^{2n+2} x} = - \operatorname{cotg} x - \frac{n_1}{3} \operatorname{cotg}^3 x - \frac{n_2}{5} \operatorname{cotg}^5 x - \dots - \frac{1}{2n+1} \operatorname{cotg}^{2n+1} x.$$

$$\int \frac{dx}{\cos^n x} = \frac{\sin x}{(n-1)\cos^{n-1} x} + \frac{n-2}{n-1} \int \frac{dx}{\cos^{n-2} x}$$

$$\int \frac{dx}{\cos^{2n+2} x} = \operatorname{tg} x + \frac{n_1}{3} \operatorname{tg}^3 x + \frac{n_2}{5} \operatorname{tg}^5 x + \dots + \frac{1}{2n+1} \operatorname{tg}^{2n+1} x.$$

$$\int \frac{dx}{\operatorname{tg}^n x} = - \frac{1}{(n-1)\operatorname{tg}^{n-1} x} - \int \frac{dx}{\operatorname{tg}^{n-2} x}.$$

$$\int \frac{x}{\sin^n x} dx = - \frac{\sin x + (n-2) x \cos x}{(n-2)(n-1)\sin^{n-1} x} + \frac{n-2}{n-1} \int \frac{x}{\sin^{n-2} x} dx.$$

$$\int \frac{x}{\cos^n x} dx = \frac{(n-2) x \sin x - \cos x}{(n-2)(n-1)\cos^{n-1} x} + \frac{n-2}{n-1} \int \frac{x}{\cos^{n-2} x} dx.$$

$$\int \frac{\cos x}{\sin^m x} dx = - \frac{1}{(m-1)\sin^{m-1} x}.$$

$$\int \frac{\sin x}{\cos^n x} dx = \frac{1}{(n-1)\cos^{n-1} x}.$$

$$\int \frac{\sin 2x}{\sin^n x} dx = - \frac{2}{(n-2)\sin^{n-2} x}$$

$$\int \frac{\sin 2x}{\cos^n x} dx = \frac{2}{(n-2)\cos^{n-2} x}.$$

$$\int \frac{\cos 3x}{\sin^m x} dx = - \frac{1}{(m-1)\sin^{m-1} x} + \frac{4}{(m-3)\sin^{m-3} x}.$$

$$\int \frac{\sin 3x}{\cos^n x} dx = \frac{4}{(n-3)\cos^{n-3} x} - \frac{1}{(n-1)\cos^{n-1} x}.$$

$$\int \frac{\sin mx}{\sin^n x} dx = 2 \int \frac{\cos (m-1) x}{\sin^{n-1} x} dx + \int \frac{\sin (m-2) x}{\sin^n x} dx.$$

$$\int \frac{\sin ax}{\cos^n x} dx = 2 \int \frac{\sin (a-1) x}{\cos^{n-1} x} dx - \int \frac{\sin (a-2) x}{\sin^n x} dx.$$

$$\int \frac{\cos mx}{\sin^n x} dx = - 2 \int \frac{\sin (m-1) x}{\sin^{n-1} x} dx + \int \frac{\cos (m-2) x}{\sin^n x} dx.$$

$$\int \frac{\cos ax}{\cos^n x}\,dx = 2\int \frac{\cos(a-1)\,x}{\cos^{n-1} x}\,dx - \int \frac{\cos(a-2)\,x}{\cos^n x}\,dx.$$

$$\int \frac{\cos^2 x}{\sin^m x}\,dx = -\frac{\cos x}{(m-2)\sin^{m-1} x} - \frac{1}{m-2}\int \frac{dx}{\sin^m x}$$

$$= -\frac{\cos x}{(m-1)\sin^{m-1} x} - \frac{1}{m-1}\int \frac{dx}{\sin^{m-2} x}$$

$$\int \frac{\cos^3 x}{\sin^m x}\,dx = -\frac{\cos^2 x}{(m-3)\sin^{m-1} x} + \frac{2}{(m-1)(m-3)\sin^{m-1} x}$$

$$= -\frac{\cos^2 x}{(m-1)\sin^{m-1} x} + \frac{2}{(m-1)(m-3)\sin^{m-3} x}.$$

$$\int \frac{\cos^4 x}{\sin^m x}\,dx = -\frac{1}{(m-4)\sin^{m-1} x}\left(\cos^3 x - \frac{3}{m-2}\cos x\right)$$

$$+ \frac{3}{(m-2)(m-4)}\int \frac{dx}{\sin^m x}.$$

$$= -\frac{\cos^3 x}{(m-1)\sin^{m-1} x} - \frac{3}{m-1}\int \frac{\cos^2 x}{\sin^{m-2} x}\,dx.$$

$$\int \frac{\sin^m x}{\cos^n x}\,dx = -\int \frac{\cos^m z}{\sin^n z}\,dz. \qquad z = \frac{\pi}{2} - x.$$

$$= \frac{\sin^{m+1} x}{(n-1)\cos^{n-1} x} + \frac{n-m-2}{n-1}\int \frac{\sin^m x}{\cos^{n-2} x}\,dx.$$

$$= -\frac{\sin^{m-1} x}{(m-n)\cos^{n-1} x} + \frac{m-1}{m-n}\int \frac{\sin^{m-2} x}{\cos^n x}\,dx.$$

$$= \frac{\sin^{m-1} x}{(n-1)\cos^{n-1} x} - \frac{m-1}{n-1}\int \frac{\sin^{m-2} x}{\cos^{n-2} x}\,dx.$$

$$\int \frac{\cos^m x}{\sin^{m+2} x}\,dx = -\frac{1}{m+1}\operatorname{cotg}^{m+1} x.$$

$$\int \frac{\cos^n x}{\sin^m x}\,dx = -\frac{\cos^{n+1} x}{(m-1)\sin^{m-1} x} + \frac{m-n-2}{m-1}\int \frac{\cos^n x}{\sin^{m-2} x}\,dx.$$

$$= \frac{\cos^{n-1} x}{(n-m)\sin^{m-1} x} + \frac{n-1}{n-m}\int \frac{\cos^{n-2} x}{\sin^m x}\,dx.$$

$$= -\frac{\cos^{n-1} x}{(m-1)\sin^{m-1} x} - \frac{n-1}{m-1}\int \frac{\cos^{n-2} x}{\sin^{m-2} x}\,dx.$$

$\mathrm{Sin}^m X \ \mathrm{Cos}^n X$

$$\int \sin x \cos x \, dx = \frac{1}{2} \sin^2 x = -\frac{1}{4} \cos 2x.$$

$$\int \sin ax \cos bx \, dx = -\frac{\cos (a+b) x}{2(a+b)} - \frac{\cos (a-b) x}{2(a-b)}.$$

$$\int \sin^2 x \cos x \, dx = \frac{1}{3} \sin^3 x = -\frac{1}{4} \left(\frac{1}{3} \sin 3x - \sin x\right) = -\cos (\sin x).$$

$$\int \cos 3x \sin^2 x \, dx = \frac{1}{5} \sin x \left(2 \cos x \cos 3x + 3 \sin x \sin 3x\right) - \frac{2}{15} \sin 3x.$$

$$\int \cos 4x \sin^2 x \, dx = \frac{1}{6} \sin 4x \sin^2 x - \frac{1}{24} (2 \sin 2x - \cos 4x).$$

$$\int \sin^3 x \cos x \, dx = \frac{1}{8} \left\{ \frac{1}{4} \cos 4x - \cos 2x \right\} = \frac{1}{4} \sin^4 x.$$

$$\int \mathrm{Sin}^4 x \cos x \, dx = \frac{1}{16} \left\{ \frac{1}{5} \sin 5x - \sin 3x + 2 \sin x \right\}.$$

$$\int \sin^m x \cos x \, dx = \frac{1}{m+1} \sin^{m+1} x.$$

$$\int \sin^m x \cos ax \, dx = \frac{\sin^m x \sin ax}{m+a} - \frac{m}{m+a} \int \sin^{m-1} x \sin (a-1) x \, dx.$$

$$= \frac{\sin^{m-1} x}{a^2 - m^2} \left(m \cos x \cos ax + a \sin x \sin ax\right)$$

$$- \frac{m(m-1)}{a^2 - m^2} \int \sin^{m-2} x \cos ax \, dx.$$

$$\int \sin^{m-1} x \cos (m+1) x \, dx = \frac{1}{m} \cos mx \sin^m x.$$

$$\int \sin x \cos^2 x \, dx = -\frac{1}{3} \cos^3 x = -\frac{1}{4} \left(\frac{1}{3} \cos 3x + \cos x\right).$$

$$\int \sin 3x \cos^2 x \, dx = -\frac{1}{5} \cos x \left(2 \sin x \sin 3x + 3 \cos x \cos 3x\right) + \frac{2}{15} \cos 3x.$$

$$\int \sin 4x \cos^2 x\, dx = -\frac{1}{6} \cos x \left(\sin x \sin 4x + 2 \cos x \cos 4x\right) + \frac{1}{24} \cos 4x.$$

$$\int \sin^2 x \cos^2 x\, dx = \frac{1}{4} \int \sin^2 2x\, dx = \frac{1}{8} x - \frac{1}{32} \sin 4x$$

$$= \frac{1}{4} \sin^3 x \cos x - \frac{1}{8} \sin x \cos x + \frac{1}{8} x.$$

$$\int \sin^3 x \cos^2 x\, dx = \frac{1}{16} \left\{ \frac{1}{5} \cos 5x - \frac{1}{3} \cos 3x - 2 \cos x \right\}$$

$$= \frac{1}{5} \sin^4 x \cos x - \frac{1}{15} \sin^2 x \cos x - \frac{2}{15} \cos x$$

$$= -\frac{\cos^3 x}{5} \left(\sin^2 x + \frac{2}{3}\right) = -\frac{1}{3} \cos^3 x + \frac{1}{5} \cos^5 x.$$

$$\int \sin^4 x \cos^2 x\, dx = \frac{1}{32} \left\{ \frac{1}{6} \sin 6x - \frac{1}{2} \sin 4x - \frac{1}{2} \sin 2x + 2x \right\}.$$

$$\int \sin^m x \cos^2 x\, dx = \frac{1}{m+2} \sin^{m+1} x \cos x + \frac{1}{m+2} \int \sin^m x\, dx.$$

$$\int \sin x \cos^3 x\, dx = -\frac{1}{4} \cos^4 x = -\frac{1}{8} \left(\frac{1}{4} \cos 4x + \cos 2x\right).$$

$$\int \sin^2 x \cos^3 x\, dx = -\frac{1}{16} \left\{ \frac{1}{5} \sin 5x + \frac{1}{3} \sin 3x - 2 \sin x \right\} = \frac{1}{3} \sin^3 x - \frac{1}{5} \sin^5 x.$$

$$\int \sin^3 x \cos^3 x\, dx = \frac{1}{32} \left\{ \frac{1}{6} \cos 6x - \frac{3}{2} \cos 2x \right\}$$

$$= \frac{1}{6} \sin^4 x \cos^2 x + \frac{1}{12} \sin^4 x = -\frac{1}{4} \cos^4 x + \frac{1}{6} \cos^6 x.$$

$$\int \sin^4 x \cos^3 x\, dx = \frac{1}{64} \left\{ \frac{1}{7} \sin 7x - \frac{1}{5} \sin 5x - \sin 3x + 3 \sin x \right\}$$

$$= \frac{1}{5} \sin^5 x - \frac{1}{7} \sin^7 x.$$

$$\int \sin^m x \cos^3 x\, dx = \frac{1}{m+3} \sin^{m+1} x \cos^2 x + \frac{2}{m+3} \int \sin^m x \cos x\, dx$$

$$= \frac{\sin^{m+1} x}{m+1} \left(\cos^2 x + \frac{2}{m+3}\right).$$

$$\int \sin x \cos^4 x\, dx = -\frac{1}{5}\cos^5 x = -\frac{1}{16}\left(\frac{1}{5}\cos 5x + \cos 3x + 2\cos x\right)$$

$$\int \sin^2 x \cos^4 x\, dx = -\frac{1}{32}\left\{\frac{1}{6}\sin 6x + \frac{1}{2}\sin 4x - \frac{1}{2}\sin 2x - 2x\right\}.$$

$$= \frac{\sin^3 x \cos^3 x}{6} + \frac{\sin^3 x \cos x}{8} - \frac{1}{16}(\sin x \cos x - x).$$

$$\int \sin^3 x \cos^4 x\, dx = \frac{1}{64}\left\{\frac{1}{7}\cos 7x + \frac{1}{5}\cos 5x - \cos 3x - 3\cos x\right\}$$

$$= -\frac{1}{5}\sin^5 x + \frac{1}{7}\sin^7 x.$$

$$= \frac{1}{7}\sin^4 x \cos^3 x - \frac{3}{35}\sin^4 x \cos x + \frac{1}{35}\sin^2 x \cos x + \frac{2}{35}\cos x.$$

$$\int \sin^4 x \cos^4 x\, dx = \frac{1}{128}\left\{\frac{1}{8}\sin 8x - \sin 4x + 3x\right\}.$$

$$\int \sin^m x \cos^4 x\, dx = \frac{1}{m+4}\sin^{m+1} x \cos^3 x + \frac{3}{m+4}\int \sin^m x \cos^2 x\, dx.$$

$$\int \sin^4 x \cos^5 x\, dx = \int (z^4 - 2z^6 + z^8)\, dz. \qquad \sin x = z.$$

$$\int \sin^m x \cos^5 x\, dx = \frac{1}{m+5}\sin^{m+1} x \cos^4 x + \frac{4}{m+5}\int \sin^m x \cos^3 x\, dx.$$

$$\int \sin^m x \cos^6 x\, dx = \frac{1}{m+6}\sin^{m+1} x \cos^5 x + \frac{5}{m+6}\int \sin^m x \cos^4 x\, dx.$$

$$\int \sin x \cos^n x\, dx = -\frac{1}{n+1}\cos^{n+1} x.$$

$$\int \sin ax \cos^n x\, dx = -\frac{\cos^n x \cos ax}{n+a} + \frac{n}{n+a}\int \cos^{n-1} x \sin (a-1)x\, dx.$$

$$= \frac{\cos^{n-1} x}{n^2 - a^2}\left(n \sin x \sin ax + a \cos x \cos ax\right)$$

$$+ \frac{n(n-1)}{n^2 - a^2}\int \cos^{n-2} x \sin ax\, dx.$$

$$\int \sin^2 x \cos^n x\, dx = -\frac{\sin x \cos^{n+1} x}{n+2} + \frac{1}{n+2}\int \cos^n x\, dx.$$

$$\int \sin^3 x \cos^n x \, dx = -\frac{\cos^{n+1} x}{n+3}\left(\sin^2 x + \frac{2}{n+1}\right).$$

$$\int \sin^4 x \cos^n x \, dx = -\frac{\cos^{n+1} x}{n+4}\left\{\sin^3 x + \frac{3}{n+2}\sin x\right\}$$

$$+ \frac{3}{(n+4)(n+2)}\int \cos^n x \, dx.$$

$$\int \sin^m x \cos^n x \, dx = \int z^m (1-z^2)^{\frac{n-1}{2}} dz. \qquad z = \sin x.$$

$$= \int z^m (1+z^2)^{k-1} dz. \qquad z = \operatorname{tg} x \qquad m+n = -2k.$$

$$= \frac{\sin^{m+1} x \cos^{n-1} x}{m+1} + \frac{n-1}{m+1}\int \sin^{m+2} x \cos^{n-2} x \, dx.$$

$$= -\frac{\sin^{m-1} x \cos^{n+1} x}{n+1} + \frac{m-1}{n+1}\int \sin^{m-2} x \cos^{n+2} x \, dx.$$

$$= \frac{\sin^{m+1} x \cos^{n-1} x}{m+n} + \frac{n-1}{m+n}\int \sin^m x \cos^{n-2} x \, dx.$$

$$= -\frac{\sin^{m-1} x \cos^{n+1} x}{m+n} + \frac{m-1}{m+n}\int \sin^{m-2} x \cos^n x \, dx.$$

$$= \frac{\sin^{m-1} x \cos^{n-1} x}{m+n}\left\{\sin^2 x - \frac{m-1}{m+n-2}\right\}$$

$$+ \frac{(m-1)(n-1)}{(m+n)(m+n-2)}\int \sin^{m-2} x \cos^{n-2} x \, dx.$$

$$= -\frac{\sin^{m+1} x \cos^{n+1} x}{n+1} + \frac{m+n+2}{n+1}\int \sin^m x \cos^{n+2} x \, dx.$$

$$= \frac{\sin^{m+1} x \cos^{n+1} x}{m+1} + \frac{m+n+2}{m+1}\int \sin^{m+2} x \cos^n x \, dx.$$

$$= \frac{\sin^{m+1} x}{m+n}\left\{\cos^{n-1} x + \frac{n-1}{n+m-2}\cos^{n-3} x\right.$$

$$\left. + \frac{(n-1)(n-3)}{(n+m-2)(n+m-4)}\cos^{n-5} x + \cdots\right\}$$

$$+ \frac{(n-1)(n-3)\cdots 3.1}{(n+m)(n+m-2)\cdots(n+2)}\int \sin^m x \, dx. \quad \text{For } n \text{ even.}$$

$$= -\frac{\cos^{n+1} x}{m+n}\left\{\sin^{m-1} x + \frac{m-1}{n+m-2}\sin^{m-3}\right.$$

$$+\frac{(m-1)(m-3)}{(n+m-2)(n+m-4)}\sin^{m-5}x+\cdots\Big\}$$

$$+\frac{(m-1)(m-3)\cdots 3.1}{(n+m)(n+m-2)\cdots(n+2)}\int\cos^n x\,dx. \qquad \text{For } m \text{ even.}$$

$$\boxed{\frac{1}{\sin^m X \cos^n X}}$$

$$\int\frac{dx}{\sin x\cos x}=\log\operatorname{tg}x.$$

$$\int\frac{dx}{\sin^2 x\cos x}=-\frac{1}{\sin x}+\log\operatorname{tg}\left(\frac{\pi}{4}+\frac{x}{2}\right).$$

$$\int\frac{dx}{\sin^3\cos x}=-\frac{1}{2\sin^2 x}+\log\operatorname{tg}x.$$

$$\int\frac{dx}{\sin^4 x\cos x}=-\frac{1}{3\sin^3 x}-\frac{1}{\sin x}+\log\operatorname{tg}\left(\frac{\pi}{4}+\frac{x}{2}\right).$$

$$\int\frac{dx}{\sin^m x\cos x}=-\frac{1}{(m-1)\sin^{m-1}x}+\int\frac{dx}{\sin^{m-2}x\cos x}$$

$$\int\frac{dx}{\sin x\cos^2 x}=\frac{1}{\cos x}+\log\operatorname{tg}\frac{x}{2}.$$

$$\int\frac{dx}{\sin^2 x\cos^2 x}=\int\frac{dx}{\cos^2 x}+\int\frac{dx}{\sin^2 x}=\operatorname{tg}x-\operatorname{cotg}x=-2\operatorname{cotg}2x.$$

$$\int\frac{dx}{\sin^3 x\cos^2 x}=\frac{1}{\sin^2 x\cos x}-\frac{3\cos x}{2\sin^2 x}+\frac{3}{2}\log\operatorname{tg}\frac{x}{2}$$

$$=-\frac{1}{2\cos x\sin^2 x}+\frac{3}{2\cos x}+\frac{3}{2}\log\operatorname{tg}\frac{x}{2}.$$

$$\int\frac{dx}{\sin^4 x\cos^2 x}=-\frac{1}{3\cos x\sin^3 x}-\frac{8}{3}\operatorname{cotg}2x.$$

$$\int\frac{dx}{\sin^m x\cos^2 x}=\frac{1}{\sin^{m-1}x\cos x}+m\int\frac{dx}{\sin^m x}$$

$$=-\frac{1}{(m-1)\sin^{m-1}x\cos x}+\frac{m}{m-1}\int\frac{dx}{\sin^{m-2}x\cos^2 x}.$$

$$\int \frac{dx}{\sin x \cos^3 x} = \frac{1}{2\cos^2 x} + \log \operatorname{tg} x.$$

$$\int \frac{dx}{\sin^2 x \cos^3 x} = \frac{1}{2\sin x \cos^2 x} - \frac{3}{2\sin x} + \frac{3}{2}\log \operatorname{tg}\left(\frac{\pi}{4} + \frac{x}{2}\right).$$

$$\int \frac{dx}{\sin^3 x \cos^3 x} = \frac{1}{2\sin^2 x \cos^2 x} - \frac{2}{\sin^2 x} + 2\log \operatorname{tg} x = -\frac{2\cos 2x}{\sin^2 2x} + 2\log \operatorname{tg} x.$$

$$\int \frac{dx}{\sin^4 x \cos^3 x} = \frac{1}{2\cos^2 x \sin^3 x} - \frac{5}{6\sin^3 x} - \frac{5}{2\sin x} + \frac{5}{2}\log \operatorname{tg}\left(\frac{\pi}{4} + \frac{x}{2}\right).$$

$$\int \frac{dx}{\sin^m x \cos^3 x} = \frac{1}{2\sin^{m-1} x \cos^2 x} + \frac{m+1}{2}\int \frac{dx}{\sin^m x \cos x}$$

$$= -\frac{1}{(m-1)\sin^{m-1} x \cos^2 x} + \frac{m+1}{m-1}\int \frac{dx}{\sin^{m-2} x \cos^3 x}.$$

$$\int \frac{dx}{\sin x \cos^4 x} = \frac{1}{3\cos^3 x} + \frac{1}{\cos x} + \log \operatorname{tg}\frac{x}{2}.$$

$$\int \frac{dx}{\sin^2 x \cos^4 x} = \frac{1}{3\sin x \cos^3 x} - \frac{8}{3}\operatorname{cotg} 2x$$

$$= \frac{1}{3\sin x \cos^3 x} + \frac{4}{3\sin x \cos x} - \frac{8\cos x}{3\sin x}.$$

$$\int \frac{dx}{\sin^3 x \cos^4 x} = -\frac{1}{2\cos^3 x \sin^2 x} + \frac{5}{6\cos^3 x} + \frac{5}{2\cos x} + \frac{5}{2}\log \operatorname{tg}\frac{x}{2}.$$

$$\int \frac{dx}{\sin^4 x \cos^4 x} = -\frac{8}{3}\left\{\frac{1}{\sin^2 2x} + 2\right\}\operatorname{cotg} 2x.$$

$$\int \frac{dx}{\sin^m x \cos^4 x} = \frac{1}{3\sin^{m-1} x \cos^3 x} + \frac{m+2}{3}\int \frac{dx}{\sin^m x \cos^2 x}$$

$$= -\frac{1}{(m-1)\sin^{m-1} x \cos^3 x} + \frac{m+2}{m-1}\int \frac{dx}{\sin^{m-2} x \cos^4 x}.$$

$$\int \frac{dx}{\sin^2 x \cos^n x} = \frac{1 - n\cos^2 x}{(n-1)\cos^{n-1} x \sin x} + \frac{n(n-2)}{n-1}\int \frac{dx}{\cos^{n-2} x}.$$

$$\int \frac{dx}{\sin^3 x \cos^n x} = \frac{2 - (n+1)\cos^2 x}{2(n-1)\cos^{n-1} x \sin^2 x} + \frac{n+1}{2}\int \frac{dx}{\cos^{n-2} x \sin x}.$$

$$\int \frac{dx}{\sin^4 x \cos^n x} = -\frac{1}{3\cos^{n-1} x \sin^3 x} + \frac{n+2}{3} \int \frac{dx}{\cos^n x \sin^2 x}.$$

$$\int \frac{dx}{\sin^m x \cos^n x} = \int \frac{dx}{\sin^{m-2} x \cos^n x} + \int \frac{dx}{\sin^m x \cos^{n-2} x}.$$

$$= \frac{1}{(n-1)\sin^{m-1} x \cos^{n-1} x} + \frac{m+n-2}{n-1} \int \frac{dx}{\sin^m x \cos^{n-2} x}.$$

$$= -\frac{1}{(m-1)\sin^{m-1} x \cos^{n-1} x} + \frac{m+n-2}{m-1} \int \frac{dx}{\sin^{m-2} x \cos^n x}.$$

$$= \frac{(m-1)\sin^2 x - (n-1)\cos^2 x}{(n-1)(m-1)\cos^{n-1} x \sin^{m-1} x}$$

$$+ \frac{(n+m-2)(n+m-4)}{(n-1)(m-1)} \int \frac{dx}{\sin^{m-2} x \cos^{n-2} x}.$$

$$\int \frac{dx}{\sin^m x \cos^m x} = 2^{m-1} \int \frac{dz}{\sin^m z}. \qquad z = 2x.$$

$$\frac{A}{(Cir_1 X + Cir_2 X)^n}$$

$$\int \frac{dx}{\sin \alpha + \sin x} = \frac{1}{\cos \alpha} \log \frac{\sin \frac{x+\alpha}{2}}{\cos \frac{x-\alpha}{2}}.$$

$$\int \frac{dx}{a \sin x + b \cos x} = \frac{1}{\sqrt{a^2+b^2}} \log \operatorname{tg} \frac{x+\alpha}{2}. \qquad \frac{b}{\sin \alpha} = \frac{a}{\cos \alpha} = \sqrt{a^2+b^2}.$$

$$= \frac{1}{\sqrt{a^2+b^2}} \log \frac{b \operatorname{tg} \frac{x}{2} - a + \sqrt{a^2+b^2}}{b \operatorname{tg} \frac{x}{2} - a - \sqrt{a^2+b^2}}$$

$$= \frac{1}{\sqrt{a^2+b^2}} \log \frac{b \sin x - a \cos x + \sqrt{a^2+b^2}}{b \cos x + a \sin x}.$$

$$\int \frac{m \sin x + n \cos x + k}{a \sin x + b \cos x}\, dx = \frac{nb+ma}{a^2+b^2} \left\{ x - \int \frac{k}{a \sin x + b \cos x}\, dx \right\}$$

$$+ \frac{an-bm}{a^2+b^2} \log (a \sin x + b \cos x).$$

$$\int \frac{dx}{1 + \cos \alpha \sin x} = \frac{2}{\sin \alpha} \operatorname{arc\,tg} \left\{ \frac{\cos \alpha + \operatorname{tg} \frac{x}{2}}{\sin \alpha} \right\}$$

$$\int \frac{dx}{(a + b \sin x)^2} = \frac{b \cos x}{(a^2 - b^2)(a + b \sin x)} + \frac{a}{a^2 - b^2} \int \frac{dx}{a + b \sin x}.$$

$$= \frac{1}{a^2 - b^2} \left\{ \frac{b \cos x}{a + b \sin x} + \frac{2a}{\sqrt{a^2 - b^2}} \operatorname{arc\,tg} \left(\frac{a \operatorname{tg} \frac{x}{2} + b}{\sqrt{a^2 - b^2}} \right) \right\}. \quad a > b.$$

$$= -\frac{1}{b^2 - a^2} \left\{ \frac{b \cos x}{a + b \sin x} + \frac{a}{\sqrt{b^2 - a^2}} \log \frac{a \operatorname{tg} \frac{x}{2} + b - \sqrt{b^2 - a^2}}{a \operatorname{tg} \frac{x}{2} + b + \sqrt{b^2 - a^2}} \right\}. \quad a < b.$$

$$\int \frac{dx}{(a + b \cos x)^2} = -\frac{b \sin x}{(a^2 - b^2)(a + b \cos x)} + \frac{a}{a^2 - b^2} \int \frac{dx}{a + b \cos x}.$$

$$= -\frac{1}{a^2 - b^2} \left\{ \frac{b \sin x}{a + b \cos x} - \frac{2a}{\sqrt{a^2 - b^2}} \operatorname{arc\,tg} \left(\sqrt{\frac{a - b}{a + b}} \operatorname{tg} \frac{x}{2} \right) \right\}. \quad a > b.$$

$$= \frac{1}{b^2 - a^2} \left\{ \frac{b \sin x}{a + b \cos x} - \frac{a}{\sqrt{b^2 - a^2}} \log \frac{\sqrt{b - a} \operatorname{tg} \frac{x}{2} + \sqrt{b + a}}{\sqrt{b - a} \operatorname{tg} \frac{x}{2} - \sqrt{b + a}} \right\}. \quad a < b.$$

$$\int \frac{\cos x}{(a + b \cos x)^2} dx = \frac{1}{a^2 - b^2} \left\{ \frac{a \sin x}{a + b \cos x} - b \int \frac{dx}{a + b \cos x} \right\}.$$

$$= -\frac{2b}{(a^2 - b^2)^{\frac{3}{2}}} \operatorname{arc\,tg} \left\{ \sqrt{\frac{a - b}{a + b}} \operatorname{tg} \frac{x}{2} \right\} + \frac{a}{a^2 - b^2} \frac{\sin x}{a + b \cos x}.$$

$$\int \frac{dx}{(1 + e \cos x)^2} = \frac{2}{(1 - e)^{\frac{3}{2}}} \operatorname{arc\,tg} \left\{ \sqrt{\frac{1 - e}{1 + e}} \operatorname{tg} \frac{x}{2} \right\} - \frac{e}{1 - e^2} \frac{\sin x}{1 + e \cos x}.$$

$$\int \frac{dx}{(1 + e \cos mx)^2} = \frac{1}{m(1 - e^2)^{\frac{3}{2}}} \operatorname{arc\,tg} \left\{ \sqrt{\frac{1 - e}{1 + e}} \operatorname{tg} \frac{mx}{2} \right\}$$

$$- \frac{e}{m(1 - e^2)} \frac{\operatorname{tg} \frac{mx}{2}}{1 + e + (1 - e) \operatorname{tg}^2 \frac{mx}{2}}.$$

$$\int \frac{\sin x}{(a - b \cos x)^2} dx = -\frac{1}{b(a - b \cos x)}.$$

$$\int \frac{dx}{(a + b \cos x)^3} = \frac{1}{(b^2 - a^2)^2} \left\{ \frac{(b^2 - a^2) b \sin x}{2\Phi^2} - \frac{3ab \sin x}{2\Phi} \right\} \frac{a^2 + \frac{1}{2} b^2}{(b^2 - a^2)^2} \int \frac{dx}{\Phi}.$$

$$= -b \sin x \frac{4a^2 - b^2 + 3ab \cos x}{2c^4 (a + b \cos x)}$$

$$+ \frac{2a^2 + b^2}{c^5} \operatorname{arc\,tg} \left(\frac{a - b}{c} \operatorname{tg} \frac{x}{2} \right). \qquad a^2 - b^2 = c^2$$

$$\int \frac{dx}{(a + b \sin x)^n} = \frac{b \cos x}{(n-1)(a^2 - b^2) \Phi^{n-1}} + \frac{(2n-3)a}{(n-1)(a^2 - b^2)} \int \frac{dx}{\Phi^{n-1}}$$

$$- \frac{n-2}{(n-1)(a^2 - b^2)} \int \frac{dx}{\Phi^{n-2}}.$$

$$\int \frac{dx}{(a + b \cos x)^n} = - \frac{b \sin x}{(n-1)(a^2 - b^2) \Phi^{n-1}} + \frac{(2n-3)a}{(n-1)(a^2 - b^2)} \int \frac{dx}{\Phi^{n-1}}$$

$$- \frac{n-2}{(n-1)(a^2 - b^2)} \int \frac{dx}{\Phi^{n-2}}$$

$$\int \frac{\cos x}{(a + b \cos x)^n} dx = - \frac{a}{b} \int \frac{dx}{\Phi^n} + \frac{1}{b} \int \frac{dx}{\Phi^{n-1}} = \frac{a \sin x}{(n-1)(a^2 - b^2) \Phi^{n-1}}$$

$$- \frac{1}{(n-1)(a^2 - b^2)} \int \frac{(n-1)b - (n-2) a \cos x}{\Phi^{n-1}} dx.$$

$$\int \frac{\cos^2 x}{(a + b \cos x)^n} dx = \frac{a^2}{b^2} \int \frac{dx}{\Phi^n} - \frac{a}{b^2} \int \frac{dx}{\Phi^{n-1}} + \frac{1}{b} \int \frac{\cos x}{\Phi^{n-1}} dx.$$

$$\int \frac{a' + b' \cos x}{(a + b \cos x)^n} dx = \frac{(ab' - ba) \sin x}{(n-1)(a^2 - b^2)(a + b \cos x)^{n-1}}$$

$$+ \frac{1}{(n-1)(a^2 - b^2)} \int \frac{(n-1)(aa' - bb') + (n-2)(ab' - ba') \cos x}{(a + b \cos x)^{n-1}} dx.$$

$$\int \frac{\sin^2 x}{(a + b \cos x)^n} dx = \frac{\sin x}{b(n-1)(a + b \cos x)^{n-1}} - \frac{1}{b(n-1)} \int \frac{\cos x}{(a + b \cos x)^{n-1}} dx.$$

$$\frac{A}{\operatorname{Cir} X (a + b \operatorname{cir}^m X)}$$

$$\int \frac{dx}{\sin x (a + b \cos x)} = \frac{b}{a^2 - b^2} \log (a + b \cos x) + \frac{1}{b + a} \log \sin \frac{x}{2} + \frac{1}{b - a} \log \cos \frac{x}{2}.$$

$$= \frac{b}{a^2 - b^2} \log (a + b \cos x) + \frac{\log (1 - \cos x)}{2(a + b)} - \frac{\log (1 + \cos x)}{2(a - b)}.$$

$$= \frac{b}{a^2 - b^2} \log (a + b \cos x) + \frac{a}{a^2 - b^2} \log \operatorname{tg} \frac{x}{2} + \frac{b}{a^2 - b^2} \log \sin x.$$

$$= \frac{1}{a^2 - b^2} \log \left\{ \left(a \operatorname{cos\acute{e}c} x + b \operatorname{cotg} x\right)^b \operatorname{tg}^a \frac{x}{2} \right\}.$$

$$\int \frac{dx}{(a + b \cos x) \cos x} = \frac{1}{a} \int \frac{dx}{\cos x} - \frac{b}{a} \int \frac{dx}{a + b \cos x}.$$

$$\int \frac{\alpha + \beta \cos x}{\sin x\,(a + b \cos x)}\, dx = \frac{b\alpha - a\beta}{a^2 - b^2} \log (a + b \cos x) - \frac{\alpha - \beta}{a - b} \log \cos \frac{x}{2}$$

$$+ \frac{\alpha + \beta}{a + b} \log \sin \frac{x}{2}$$

$$\int \frac{\alpha + \beta \cos x}{(a + b \cos x) \cos x}\, dx = \frac{\alpha}{a} \log \operatorname{tg} \left(\frac{\pi}{4} + \frac{x}{2}\right) + \frac{a\beta - b\alpha}{a} \int \frac{dx}{a + b \cos x}$$

SQUARE ROOTS OF TRIGONOMETRIC FUNCTIONS

$$\int\sqrt{1-a^2\sin^2 x}\,\sin x\,dx = -\frac{1}{2}\sqrt{\Phi}\cos x - \frac{1-a^2}{2a}\log\left\{a\cos x+\sqrt{\Phi}\right\}.$$

$$\int\sqrt{1-a^2\sin^2 x}\,\cos x\,dx = \frac{1}{2}\sqrt{1-a^2\sin^2 x}\,\sin x + \frac{1}{2a}\text{arc sin}\,(a\sin x).$$

$$\int(1-a^2\sin^2 x)^{\frac{3}{2}}\sin x\,dx = -\frac{1}{4}\Phi^{\frac{3}{2}}\cos x + \frac{3}{4}(1-a^2)\int\sqrt{\Phi}\,\sin x\,dx.$$

$$\int\sqrt{1+\cos x}\,dx = 2\sqrt{2}\sin\frac{x}{2} = 2\sqrt{1-\cos x}\,.$$

$$\int\sqrt{1-\cos x}\,dx = -2\sqrt{2}\cos\frac{x}{2} = -2\sqrt{1+\cos x}\,.$$

$$\int\frac{dx}{\sqrt{\sin x\cos^{\frac{3}{2}} x}} = 2\sqrt{\text{tg}\,x}\left(1+\frac{1}{5}\text{tg}^2 x\right)$$

$$\int\frac{dx}{\sqrt{\sin 2x\cos^3 x}} = \sqrt{2\text{tg}\,x}\left(1+\frac{1}{5}\text{tg}^2 x\right).$$

$$\int\frac{dx}{\cos^2 x\sqrt{a+b\,\text{tg}\,x}} = \frac{2}{b}\sqrt{a+b\,\text{tg}\,x}\,.$$

$$\int\frac{dx}{\sqrt{a^2+b^2\,\text{tg}^2 x}} = \frac{1}{\sqrt{a^2-b^2}}\text{arc sin}\left\{\sqrt{1-\frac{b^2}{a^2}}\sin x\right\}. \qquad a > b.$$

$$= \frac{1}{\sqrt{b^2-a^2}}\log\left\{\sqrt{b^2-a^2}\sin x+\sqrt{a^2\cos^2 x+b^2\sin^2 x}\right\}. \quad a < b.$$

$$\int\frac{\text{tg}\,x}{\sqrt{a^2+b^2\,\text{tg}^2 x}}\,dx = \frac{1}{\sqrt{b^2-a^2}}\text{arc cos}\left\{\sqrt{1-\frac{a^2}{b^2}}\cos x\right\}. \qquad a < b$$

$$= -\frac{1}{\sqrt{a^2-b^2}}\log\left\{\sqrt{a^2-b^2}\cos x+\sqrt{a^2\cos^2 x+b^2\sin^2 x}\right\}. \qquad a > b.$$

$$\int\frac{\cos^3 x}{\sqrt{\sin x}}\,dx = \frac{2}{5}\left(5-\sin^2 x\right)\sqrt{\sin x}.$$

$$\int\frac{\sin^3 x}{\sqrt{\cos x}}\,dx = \frac{2}{5}\left(\cos^2 x-5\right)\sqrt{\cos x}.$$

$$\int \frac{x \cos (a^2 - x^2)}{\sqrt{\sin (a^2 - x^2)}} dx = -\sqrt{\sin (a^2 - x^2)}.$$

$$\int \sqrt{1 - \frac{\sin^2 \alpha}{\cos^2 x}}\, dx = \text{arc sin} \left(\frac{\sin x}{\cos \alpha}\right) - \sin \alpha \, \text{arc sin} (\text{tg}\, \alpha \, \text{tg}\, x).$$

$$\int \frac{\sqrt{a^2 + b^2 \sin^2 x}}{\sin x} dx = b \, \text{arc cos} \left\{ \frac{b \cos x}{\sqrt{a^2 + b^2}} \right\}$$

$$- a \log \left(a \, \text{cotg}\, x + \sqrt{b^2 + a^2 \, \text{coséc}^2 x}\right).$$

INVERSE TRIGONOMETRIC FUNCTIONS

$$\int \text{arc sin } x\, dx = x \text{ arc sin } x + \sqrt{1 - x^2}.$$

$$\int \text{arc cos } x\, dx = x \text{ arc cos } x - \sqrt{1 - x^2}.$$

$$\int \text{arc tg } x\, dx = x \text{ arc tg } x - \log \sqrt{1 + x^2}.$$

$$\int \text{arc cotg } x\, dx = x \text{ arc cotg } x + \log \sqrt{1 + x^2}.$$

$$\int \text{arc séc } x\, dx = x \text{ arc séc } x - \log (x + \sqrt{x^2 - 1}).$$

$$\int \text{arc coséc } x\, dx = x \text{ arc coséc } x + \log (x + \sqrt{x^2 - 1}).$$

$$\int \text{U arc sin } x\, dx = \text{U}_1 \text{ arc sin } x - \int \frac{\text{U}_1}{\sqrt{1 - x^2}}\, dx.$$

$$\int \text{U arc cos } x\, dx = \text{U}_1 \text{ arc cos } x + \int \frac{\text{U}_1}{\sqrt{1 - x^2}}\, dx.$$

$$\int \text{U arc tg } x\, dx = \text{U}_1 \text{ arc tg } x - \int \frac{\text{U}_1}{1 + x^2}\, dx.$$

$$\int \text{U arc cotg } x\, dx = \text{U}_1 \text{ arc cotg } x + \int \frac{\text{U}_1}{1 + x^2}\, dx.$$

$$\int \text{U arc séc } x\, dx = \text{U}_1 \text{ arc séc } x - \int \frac{\text{U}_1}{x\sqrt{x^2 - 1}}\, dx.$$

$$\int \text{U arc coséc } x\, dx = \text{U}_1 \text{ arc coséc } x + \int \frac{\text{U}_1}{x\sqrt{x^2 - 1}}\, dx.$$

$$\int x \text{ arc sin } x\, dx = \frac{1}{4} (2x^2 - 1) \text{ arc sin } x + \frac{1}{4} x \sqrt{1 - x^2}.$$

$$\int x \text{ arc tg } x\, dx = -\frac{1}{2} x + \frac{1}{2} (x^2 + 1) \text{ arc tg } x.$$

$$\int x^2 \operatorname{arc\,tg} x \, dx = \frac{1}{3} x^3 \operatorname{arc\,tg} x - \frac{1}{6} x^2 + \frac{1}{3} \log \sqrt{1 + x^2}.$$

$$\int \frac{\arcsin x}{x^2} dx = -\frac{1}{x} \arcsin x + \log \frac{x}{1 + \sqrt{1 - x^2}}.$$

$$\int \frac{\operatorname{arc\,tg} x}{x^2} dx = -\frac{1}{x} \operatorname{arc\,tg} x + \log \frac{x}{\sqrt{1 + x^2}}.$$

$$\int x^m \arcsin x \, dx = \frac{x^{m+1}}{m+1} \arcsin x - \frac{1}{m+1} \int \frac{x^{m+1}}{\sqrt{1 - x^2}} dx.$$

$$\int x^m \arccos x \, dx = \frac{x^{m+1}}{m+1} \arccos x + \frac{1}{m+1} \int \frac{x^{m+1}}{\sqrt{1 - x^2}} dx.$$

$$\int x^m \operatorname{arc\,tg} x \, dx = \frac{x^{m+1}}{m+1} \operatorname{arc\,tg} x - \frac{1}{m+1} \int \frac{x^{m+1}}{1 + x^2} dx.$$

$$\int x^m \operatorname{arc\,cotg} x \, dx = \frac{x^{m+1}}{m+1} \operatorname{arc\,cotg} x + \frac{1}{m+1} \int \frac{x^{m+1}}{1 + x^2} dx.$$

$$\int x^m \operatorname{arc\,séc} x \, dx = \frac{x^{m+1}}{m+1} \operatorname{arc\,séc} x - \frac{1}{m+1} \int \frac{x^m}{\sqrt{x^2 - 1}} dx.$$

$$\int x^m \operatorname{arc\,coséc} x \, dx = \frac{x^{m+1}}{m+1} \operatorname{arc\,coséc} x + \frac{1}{m+1} \int \frac{x^m}{\sqrt{x^2 - 1}} dx.$$

$$\int \frac{\arcsin x}{\sqrt{1 - x^2}} dx = \frac{1}{2} \left(\arcsin x\right)^2 = x - \sqrt{1 - x^2} \arcsin x.$$

$$\int \frac{x \arcsin x}{\sqrt{1 - x^2}} dx = x - \sqrt{1 - x^2} \arcsin x.$$

$$\int \frac{x^2 \arcsin x}{\sqrt{1 - x^2}} dx = \frac{1}{4} x^2 - \frac{1}{2} x \sqrt{1 - x^2} \arcsin x + \frac{1}{4} \left(\arcsin x\right)^2.$$

$$\int \frac{x^3 \arcsin x}{\sqrt{1 - x^2}} dx = \frac{2}{3} x + \frac{1}{9} x^3 - \frac{1}{3} (x^2 + 2) \sqrt{1 - x^2} \arcsin x.$$

$$\int \frac{dx}{\sqrt{1 - x^2} \arcsin x} = \log \arcsin x.$$

$$\int \frac{\arccos x}{\sqrt{1 - x^2}} dx = -\frac{1}{2} \left(\arccos x\right)^2.$$

$$\int \frac{dx}{\sqrt{1-x^2}\operatorname{arc}\cos x} = -\log \operatorname{arc}\cos x.$$

$$\int \frac{\operatorname{arc}\operatorname{tg} x}{1+x^2}\,dx = \frac{1}{2}\left(\operatorname{arc}\operatorname{tg} x\right)^2.$$

$$\int \frac{x \operatorname{arc}\operatorname{tg} x}{1+x^2}\,dx = \frac{1}{2}\operatorname{arc}\operatorname{tg} x \log(1+x^2) - \frac{1}{2}\int \frac{\log(1+x^2)}{1+x^2}\,dx.$$

$$\int \frac{x^2 \operatorname{arc}\operatorname{tg} x}{1+x^2}\,dx = -\log\sqrt{1+x^2} + \left(x - \frac{1}{2}\operatorname{arc}\operatorname{tg} x\right)\operatorname{arc}\operatorname{tg} x.$$

$$\int \frac{x^3 \operatorname{arc}\operatorname{tg} x}{1+x^2}\,dx = -\frac{1}{2}x + \frac{1}{2}(1+x^2)\operatorname{arc}\operatorname{tg} x - \int \frac{x \operatorname{arc}\operatorname{tg} x}{1+x^2}\,dx.$$

$$\int \frac{x^4 \operatorname{arc}\operatorname{tg} x}{1+x^2}\,dx = -\frac{1}{6}x^2 + \frac{2}{3}\log(1+x^2) + \left(\frac{x^3}{3} - x\right)\operatorname{arc}\operatorname{tg} x + \frac{1}{2}\left(\operatorname{arc}\operatorname{tg} x\right)^2.$$

$$\int \frac{dx}{(1+x^2)\operatorname{arc}\operatorname{tg} x} = \log \operatorname{arc}\operatorname{tg} x.$$

$$\int \frac{\operatorname{arc}\operatorname{cotg} x}{1+x^2}\,dx = -\frac{1}{2}\left(\operatorname{arc}\operatorname{cotg} x\right)^2.$$

$$\int \frac{dx}{(1+x^2)\operatorname{arc}\operatorname{cotg} x} = -\log \operatorname{arc}\operatorname{cotg} x.$$

$$\int \frac{\operatorname{arc}\operatorname{séc} x}{x\sqrt{x^2-1}}\,dx = \frac{1}{2}\left(\operatorname{arc}\operatorname{séc} x\right)^2.$$

$$\int \frac{dx}{x\sqrt{x^2-1}\operatorname{arc}\operatorname{séc} x} = \log \operatorname{arc}\operatorname{séc} x.$$

$$\int \frac{\operatorname{arc}\operatorname{coséc} x}{x\sqrt{x^2-1}}\,dx = -\frac{1}{2}\left(\operatorname{arc}\operatorname{coséc} x\right)^2.$$

$$\int \frac{dx}{x\sqrt{x^2-1}\operatorname{arc}\operatorname{coséc} x} = -\log \operatorname{arc}\operatorname{coséc} x.$$

$$\int \left(\operatorname{arc}\sin x\right)^n dx = x\,\Phi^n + n\Phi^{n-1}\sqrt{1-x^2} - n(n-1)\int \Phi^{n-2}\,dx.$$

$$= x\left\{\Phi^n - n(n-1)\Phi^{n-2} + \cdots\right\} + \sqrt{1-x^2}\left\{n\,\Phi^{n-1}\right.$$

$$\left. - n(n-1)(n-2)\Phi^{n-3} + \cdots\right\}.$$

$$\int \text{arc sin} \sqrt{\frac{x}{a+x}}\, dx = (a+x)\, \text{arc sin} \sqrt{\frac{x}{a+x}} - \sqrt{ax}$$

$$= (a+x)\, \text{arc tg} \sqrt{\frac{x}{a}} - \sqrt{ax}\,.$$

$$\int x\, \text{arc sin} \frac{1}{2} \sqrt{\frac{2a-x}{a}}\, dx = \frac{x^2}{2}\, \text{arc sin} \frac{1}{2} \sqrt{\frac{2a-x}{a}} + \frac{a^2}{2}\, \text{arc sin} \frac{x}{2a}$$

$$- \frac{x}{8} \sqrt{4a^2 - x^2}\,.$$

$$\int \frac{dx}{1 + \left(\text{arc sin} \frac{x}{a}\right)^2} = \sqrt{a^2 - x^2}\, \text{arc tg} \left(\text{arc sin} \frac{x}{a}\right).$$

$$\int \frac{\text{arc sin } x}{(1-x^2)^{\frac{3}{2}}}\, dx = \frac{x\, \text{arc sin } x}{\sqrt{1-x^2}} + \frac{1}{2} \log (1 - x^2).$$

$$\int \frac{x\, \text{arc sin } x}{(1-x^2)^{\frac{3}{2}}}\, dx = \frac{1}{2} \log \frac{1-x}{1+x} + \frac{\text{arc sin } x}{\sqrt{1-x^2}}.$$

$$\int \frac{\text{arc sin } (1-2x)}{\sqrt{x-x^2}}\, dx = -\frac{1}{2} \left\{ \text{arc sin } (1-2x) \right\}^2.$$

$$\int \frac{a-2x}{\sqrt{ax-x^2}}\, \text{arc sin} \sqrt{\frac{a-x}{a+x}}\, dx = 2 \sqrt{ax - x^2}\, \text{arc sin} \sqrt{\frac{a-x}{a+x}}$$

$$+ a \sqrt{2} \log (a + x).$$

$$\int \frac{\text{arc tg } x}{(1+x^2)^{\frac{3}{2}}}\, dx = \frac{1 + x\, \text{arc tg } x}{\sqrt{1+x^2}}.$$

$$\int \frac{x(3-x^2)\, \text{arc tg } x}{(1-x^2)^{\frac{3}{2}}}\, dx = \frac{1+x^2}{\sqrt{1-x^2}}\, \text{arc tg } x - \text{arc sin } x.$$

$$\int \frac{x}{\sqrt{1-x^2}}\, \text{arc séc} \sqrt{\frac{1+x^2}{1-x^2}}\, dx = -\sqrt{1-x^2}\, \text{arc séc} \sqrt{\frac{1+x^2}{1-x^2}} + \sqrt{2}\, \text{arc tg } x.$$

LOGARITHMIC FUNCTIONS

$$\int \log x\,dx = x\,(\log x - 1).$$

$$\int (\log x)^2\,dx = x\left\{(\log x)^2 - 2\log x + 2\right\}.$$

$$\int (\log x)^n\,dx = \int z^n e^z\,dz. \qquad \log x = z.$$

$$= x\,(\log x)^n - n\int (\log x)^{n-1}\,dx.$$

$$\int \frac{dx}{\log x} = \log\log x + \frac{\log x}{1} + \frac{1}{2}\frac{(\log x)^2}{1.2} + \frac{1}{3}\frac{(\log x)^3}{1.2.3} + \frac{1}{4}\frac{(\log x)^4}{1.2.3.4} + \ldots.$$

$$\int \frac{dx}{\log \frac{1}{x}} = \log\log x - \frac{\log x}{1} + \frac{1}{2}\frac{(\log x)^2}{1.2} - \frac{1}{3}\frac{(\log x)^3}{1.2.3} + \frac{1}{4}\frac{(\log x)^4}{1.2.3.4} - \ldots.$$

$$\int \mathrm{U}\log x\,dx = \mathrm{U}_1\log x - \int \frac{\mathrm{U}_1}{x}\,dx.$$

$$\int \mathrm{U}\,(\log x)^n\,dx = \mathrm{U}_1\,(\log x)^n - n\int \frac{\mathrm{U}_1\,(\log x)^{n-1}}{x}\,dx.$$

$$= \mathrm{U}_1\,(\log x)^{n-} - n\,\mathrm{U}_2\,(\log x)^{n-1} + n\,(n-1)\,\mathrm{U}_3\,(\log x)^{n-2} - \ldots.$$

$$\int \frac{\mathrm{U}}{(\log x)^n}\,dx = -\frac{\mathrm{U}x}{(n-1)\,(\log x)^{n-1}} - \frac{\mathrm{U}'x}{(n-1)\,(n-2)\,(\log x)^{n-2}}$$

$$-\frac{\mathrm{U}''x}{(n-1)\,(n-2)\,(n-3)\,(\log x)^{n-3}} - \ldots.$$

$$\int x^m\log x\,dx = \frac{x^{m+1}}{m+1}\left(\log x - \frac{1}{m+1}\right).$$

$$\int x^m\,(\log x)^2\,dx = x^{m+1}\left\{\frac{(\log x)^2}{m+1} - \frac{2\log x}{(m+1)^2} + \frac{2}{(m+1)^3}\right\}.$$

$$\int x^m\,(\log x)^3\,dx = \frac{x^{m+1}}{m+1}\left\{(\log x)^3 - \frac{3}{m+1}\,(\log x)^2\right.$$

$$\left. + \frac{6}{(m+1)^2}\log x - \frac{6}{(m+1)^3}\right\}.$$

$$\int x^m \left(\log x\right)^n dx = \int z^n e^{(m+1)z} dz. \qquad \log x = z.$$

$$= \frac{1}{(m+1)^{n+1}} \int \left(\log z\right)^n dz. \qquad x^{m+1} = z.$$

$$= \frac{x^{m+1}(\log x)^n}{m+1} - \frac{n}{m+1} \int x^m \left(\log x\right)^{n-1} dx.$$

$$= \frac{x^{m+1}}{m+1} \left\{ \left(\log x\right)^n - \frac{n}{m+1} \left(\log x\right)^{n-1} + \frac{n(n-1)}{(m+1)^2} \left(\log x\right)^{n-2} \right.$$

$$\left. - \frac{n(n-1)(n-2)}{(m+1)^3} \left(\log x\right)^{n-3} + \cdots \pm \frac{n(n-1)\cdots 2.1}{(m+1)^n} \right\}.$$

$$\int \frac{dx}{x \log x} = \log \log x.$$

$$\int \frac{dx}{x (\log x)^n} = - \frac{1}{(n-1)(\log x)^{n-1}}.$$

$$\int \frac{\log x}{x} dx = \frac{1}{2} \left(\log x\right)^2.$$

$$\int \frac{(\log x)^n}{x} dx = \frac{1}{n+1} \left(\log x\right)^{n+1}.$$

$$\int \frac{x^m}{\log x} dx = \log \log x + (m+1) \log x + \frac{(m+1)^2 (\log x)^2}{1.2^2} + \frac{(m+1)^3 (\log x)^3}{1.2.3^2}$$

$$+ \frac{(m+1)^4 (\log x)^4}{1.2.3.4^2} + \cdots$$

$$\int \frac{x^m}{(\log x)^n} dx = - \frac{x^{m+1}}{(n-1)(\log x)^{n-1}} + \frac{m+1}{n-1} \int \frac{x^m}{(\log x)^{n-1}} dx.$$

$$= - x^{m+1} \left\{ \frac{1}{(n-1)(\log x)^{n-1}} + \frac{m+1}{(n-1)(n-2)(\log x)^{n-2}} \right.$$

$$\left. + \frac{(m+1)^2}{(n-1)(n-2)(n-3)(\log x)^{n-3}} + \cdots + \frac{(m+1)^{n-2}}{(n-1)\cdots 2.1 \log x} \right\}$$

$$+ \frac{(m+1)^{n-1}}{(n-1)\cdots 2.1} \int \frac{x^m}{\log x} dx.$$

$$\int \left(a + bx\right) \log x \, dx = \frac{(a+bx)^2}{2b} \log x - \frac{a^2}{2b} \log x - ax - \frac{1}{4} bx^2.$$

$$\int \left(a + bx\right)^m \log x \, dx = \frac{(a + bx)^{m+1}}{(m + 1)\, b} \log x - \frac{1}{(m + 1)\; b} \int \frac{(a + bx)^{m+1}}{x} \, dx.$$

$$\int \frac{\log x}{a + bx} \, dx = \frac{1}{b} \log x \log (a + bx) - \frac{1}{b} \int \frac{\log (a + bx)}{x} \, dx.$$

$$= \frac{1}{b} \log x \log \frac{a + bx}{a} - \frac{x}{a} + \frac{bx^2}{(2a)^2} - \frac{b^2 x^3}{(3a)^2\, a} + \frac{b^3 x^4}{(4a)^2\, a^2} - \cdots$$

$$= \frac{1}{b} \log x \log (a + bx) - \frac{1}{2b} \left(\log bx\right)^2 + \frac{a}{b^2 x} - \frac{a^2}{2^2 b^3 x^2} + \frac{a^3}{3^2 b^4 x^3} - \cdots$$

$$\int \frac{\log x}{(a + bx)^m} \, dx = - \frac{\log x}{(m - 1)\, b\, (a + bx)^{m-1}} + \frac{1}{(m - 1)\; b} \int \frac{dx}{x\; (a + bx)^{m-1}} \cdot$$

$$= - \frac{\log x}{(m - 1)\; b\; \Phi^{m-1}} + \frac{1}{(m - 1)\, ab} \left\{ \frac{1}{(m - 2)\, \Phi^{m-2}} + \frac{1}{(m - 3)\, a\, \Phi^{m-3}} \right.$$

$$\left. + \frac{1}{(m - 4)\; a^2 \Phi^{m-4}} + \cdots + \frac{1}{2a^{m-4}\, \Phi^2} + \frac{1}{a^{m-3}\, \Phi} \right\}$$

$$+ \frac{1}{(m - 1)\; a^{m-1} b} \log \frac{x}{\Phi} \cdot$$

$$\int x^m \log (a + bx) \, dx = \frac{x^{m+1}}{m + 1} \log (a + bx) - \frac{b}{m + 1} \int \frac{x^{m+1}}{a + bx} \, dx.$$

$$\int x \log (1 - x^4) \, dx = \frac{1}{2} (1 + x^2) \log (1 + x^2) - \frac{1}{2} (1 - x^2) \log (1 - x^2) - x^2.$$

$$\int x^2 \log (x^2 - 1) \, dx = \frac{1}{3} x^3 \log (x^2 - 1) - \frac{2}{3} x - \frac{2}{9} x^3 - \frac{1}{3} \log \frac{x - 1}{x + 1} \cdot$$

$$\int x^3 \log (1 + x^2) \, dx = \frac{x^4}{4} \log (1 + x^2) - \frac{1}{8} x^4 + \frac{1}{4} x^2 - \frac{1}{2} \log (1 + x^2).$$

$$\int \frac{\log (a + bx)}{x} \, dx = \log a \log x + \frac{bx}{a} - \frac{b^2 x^2}{2^2 a^2} + \frac{b^3 x^3}{3^2 a^3} - \cdots.$$

$$= \frac{1}{2} \left(\log bx\right)^2 - \frac{a}{bx} + \frac{a^2}{2^2 b^2 x^2} - \frac{a^3}{3^2 b^3 x^3} + \cdots.$$

$$\int \log \left\{ \sqrt{x - a} + \sqrt{x - b} \right\} dx = \frac{1}{2} \left(2x - a - b\right) \log \left\{ \sqrt{x - a} + \sqrt{x - b} \right\}$$

$$- \frac{1}{2} \sqrt{(x - a)\,(x - b)} \,.$$

$$\int \frac{\log \log x}{x}\, dx = \left\{ \log \log x - 1 \right\} \log x.$$

$$\int \frac{1}{x} \log \frac{1}{1-x}\, dx = x + \frac{1}{4} x^2 + \frac{1}{9} x^3 + \frac{1}{16} x^4 + \cdots$$

$$\int \frac{\log (1-x)}{x\sqrt{x}}\, dx = -\frac{2}{\sqrt{x}} \log (1-x) + 2 \log \frac{1-\sqrt{x}}{1+\sqrt{x}} \cdot$$

$$\int \frac{\log x}{(a^2 + b^2x^2)^{\frac{3}{2}}}\, dx = \frac{1}{a^2b} \left\{ \frac{bx \log x}{\sqrt{a^2 + b^2x^2}} - \log \frac{a+bx}{a-bx} \right\}.$$

$$\int \frac{x}{(1+x^2) \log (1+x^2)}\, dx = \frac{1}{2} \log \left\{ \log (1+x^2) \right\} \cdot$$

$$\int \mathrm{V} \log \mathrm{U}\, dx = \mathrm{V}_1 \log \mathrm{U} - \int \frac{\mathrm{U}' \mathrm{V}_1}{\mathrm{U}}\, dx.$$

EXPONENTIAL FUNCTIONS

$$\int e^x \, dx = e^x.$$

$$\int e^{mx+n} \, dx = \frac{1}{m} e^{mx+n}.$$

$$\int \frac{dx}{e^{ax}} = -\frac{1}{ae^{ax}}.$$

$$\int x e^{ax} \, dx = \frac{ax - 1}{a^2} e^{ax}.$$

$$\int x^m e^{ax} \, dx = \frac{x^m e^{ax}}{a} - \frac{m}{a} \int x^{m-1} e^{ax} \, dx$$

$$= \frac{e^{ax}}{a} \left\{ x^m - \frac{m}{a} x^{m-1} + \frac{m(m-1)}{a^2} x^{m-2} - \ldots \right\}.$$

$$\int \frac{dx}{x^m e^x} = -\frac{1}{(m-1) x^{m-1} e^x} - \frac{1}{m-1} \int \frac{dx}{x^{m-1} e^x}.$$

$$\int \frac{e^{ax}}{x} \, dx = \log x + \frac{ax}{1} + \frac{1}{2} \frac{(ax)^2}{1.2} + \frac{1}{3} \frac{(ax)^3}{1.2.3} + \ldots.$$

$$\int \frac{e^{ax}}{x^m} \, dx = -\frac{e^{ax}}{(m-1) x^{m-1}} + \frac{a}{m-1} \int \frac{e^{ax}}{x^{m-1}} \, dx$$

$$= \frac{e^{ax}}{a} \left\{ \frac{1}{x^m} + \frac{m}{a x^{m+1}} + \frac{m(m+1)}{a^2 x^{m+2}} + \ldots \right\}.$$

$$\int \frac{dx}{a + be^x} = \frac{x}{a} - \frac{1}{a} \log (a + be^x).$$

$$\int \frac{dx}{e^x + e^{-x}} = \text{arc tg } e^x.$$

$$\int \frac{dx}{e^{ax} + e^{-ax}} = \frac{1}{a} \text{arc tg } e^{ax}.$$

$$\int \frac{dx}{a^2 e^x + b^2 e^{-x}} = \frac{1}{ab} \text{arc tg} \left(\frac{a}{b} e^x \right).$$

$$\int \frac{xe^x}{(1+x)^2}\,dx = \frac{e^x}{1+x}.$$

$$\int \frac{(ax-1)(bx-1)\,e^{(a+b)x}}{x^2}\,dx = ab\,e^{(a+b)x}\left\{\frac{1}{a+b} - \frac{1}{abx}\right\}.$$

$$\int \frac{x}{(e^x-1)^2}\,dx = -\frac{1}{2}\,\frac{e^x+1}{e^x-1}.$$

$$\int \frac{x}{e^{2x}}\,dx = -\frac{x}{2\,e^{2x}} - \frac{1}{4\,e^{2x}}.$$

$$\int \frac{x^3}{e^x}\,dx = -\frac{1}{e^x}\,(x^3 + 3x^2 + 6x + 6).$$

$$\int \frac{e^x-1}{e^x+1}\,dx = 2\log\left(e^{\frac{x}{2}} + e^{-\frac{x}{2}}\right).$$

$$\int \frac{a + be^x + ce^{2x}}{m+e^x}\,dx = \frac{ax}{m} + ce^x - \frac{a - bm + cm^2}{m}\log(m + e^x).$$

$$\int e^{m\operatorname{arc\,sin} x}\,dx = \frac{x + m\sqrt{1-x^2}}{1+m^2}\,e^{m\operatorname{arc\,sin} x}.$$

$$\int \frac{e^{a\operatorname{arc\,tg} x}}{(1+x^2)^{\frac{3}{2}}}\,dx = \frac{(a+x)\,e^{a\operatorname{arc\,tg} x}}{(1+a^2)\sqrt{1+x^2}}.$$

$$\int \frac{x\,e^{\operatorname{arc\,sin} x}}{\sqrt{1-x^2}}\,dx = \frac{1}{2}\left(x + \sqrt{1-x^2}\right)e^{\operatorname{arc\,sin} x}.$$

$$\int \frac{x\,e^{a\operatorname{arc\,tg} x}}{(1+x^2)^{\frac{3}{2}}}\,dx = \frac{(ax-1)\,e^{a\operatorname{arc\,tg} x}}{(1+a^2)\sqrt{1+x^2}}.$$

EXPONENTIAL TRIGONOMETRIC FUNCTIONS

$$\int e^{ax} \sin x \, dx = \frac{e^{ax}(a \sin x - \cos x)}{1 + a^2}.$$

$$\int e^{ax} \sin bx \, dx = \frac{e^{ax}(a \sin bx - b \cos bx)}{a^2 + b^2}.$$

$$\int e^{ax} \cos x \, dx = \frac{e^{ax}(\sin x + a \cos x)}{1 + a^2}.$$

$$\int e^{ax} \cos bx \, dx = \frac{e^{ax}(a \cos bx + b \sin bx)}{a^2 + b^2}.$$

$$\int x e^{x} \sin x \, dx = \frac{1}{2} x e^{x} (\sin x - \cos x) + \frac{1}{2} e^{x} \cos x.$$

$$\int x e^{x} \cos x \, dx = \frac{1}{2} x e^{x} \left(\sin x + \cos x\right) - \frac{1}{2} e^{x} \sin x.$$

$$\int x^m e^{x} \sin x \, dx = \frac{1}{2} x^m e^{x} (\sin x - \cos x) - \frac{m}{2} \int x^{m-1} e^{x} \sin x \, dx$$
$$+ \frac{m}{2} \int x^{m-1} e^{x} \cos x \, dx.$$

$$\int x^m e^{x} \cos x \, dx = \frac{1}{2} x^m e^{x} \left(\sin x + \cos x\right) - \frac{m}{2} \int x^{m-1} e^{x} \sin x \, dx$$
$$- \frac{m}{2} \int x^{m-1} e^{x} \cos x \, dx.$$

$$\int x^m e^{ax} \sin bx \, dx = x^m e^{ax} \frac{a \sin bx - b \cos bx}{a^2 + b^2}$$
$$- \frac{m}{a^2 + b^2} \int x^{m-1} e^{ax} (a \sin bx - b \cos bx) \, dx.$$
$$= e^{ax} \left[\frac{1}{\rho} x^m \sin (bx - \alpha) - \frac{m}{\rho^2} x^{m-1} \sin (bx - 2\alpha) + \cdots \right.$$
$$\left. \pm \frac{m (m - 1) \cdots 1}{\rho^{m+1}} \sin \left\{ bx - (m + 1) \alpha \right\} \right].$$

where $\quad a + b \sqrt{-1} = \rho (\cos \alpha + \sqrt{-1} \sin \alpha).$

$$\int x^m e^{ax} \cos bx\, dx = x^m e^{ax} \frac{a \cos bx + b \sin bx}{a^2 + b^2}$$

$$- \frac{m}{a^2 + b^2} \int x^{m-1} e^{ax} (a \cos bx + b \sin bx)\, dx.$$

$$= e^{ax} \left[\frac{1}{\rho} x^m \cos (bx - \alpha) - \frac{m}{\rho^2} x^{m-1} \cos (bx - 2\alpha) + \cdots \right.$$

$$\left. \pm \frac{m(m-1)\cdots 1}{\rho^{m+1}} \cos \left\{ bx - (m+1)\alpha \right\} \right].$$

where $\qquad a + b\sqrt{-1} = \rho (\cos \alpha + \sqrt{-1} \sin \alpha).$

$$\int e^{ax} \sin^2 x\, dx = \frac{e^{ax} \sin x (a \sin x - 2 \cos x)}{4 + a^2} + \frac{2e^{ax}}{a(4 + a^2)}.$$

$$= \frac{e^{ax}}{a(4 + a^2)} \left(a^2 \sin^2 x - a \sin 2x + 2 \right).$$

$$\int e^{ax} \cos^2 x\, dx = \frac{e^{ax} \cos x (2 \sin x + a \cos x)}{4 + a^2} + \frac{2e^{ax}}{a(4 + a^2)}.$$

$$= \frac{e^{ax}}{a(a^2 + 4)} (a^2 \cos^2 x + a \sin 2x + 2).$$

$$\int e^{ax} \operatorname{tg}^2 x\, dx = \frac{e^{ax}}{a} \left(a \operatorname{tg} x - 1 \right) - a \int e^{ax} \operatorname{tg} x\, dx.$$

$$\int e^{ax} \sin^3 x\, dx = \frac{e^{ax} \sin^2 x (a \sin x - 3 \cos x)}{a^2 + 9} + \frac{6 e^{ax} (a \sin x - \cos x)}{(a^2 + 9)(a^2 + 1)}.$$

$$\int e^{ax} \cos^3 x\, dx = \frac{e^{ax} \cos^2 x (3 \sin x + a \cos x)}{9 + a^2} + \frac{6 e^{ax} (\sin x + a \cos x)}{(1 + a^2)(9 + a^2)}.$$

$$\int e^{ax} \operatorname{tg}^3 x\, dx = \frac{e^{ax}}{2} \left(\operatorname{tg}^2 x - a \operatorname{tg} x + 1 \right) + \frac{1}{2} \left(a^2 - 2 \right) \int e^{ax} \operatorname{tg} x\, dx.$$

$$\int e^{ax} \sin^n bx\, dx = \frac{e^{ax} \sin^{n-1} bx (a \sin bx - nb \cos bx)}{a^2 + n^2 b^2}$$

$$+ \frac{n(n-1) b^2}{a^2 + n^2 b^2} \int e^{ax} \sin^{n-2} bx\, dx.$$

$$\int e^{ax} \cos^n bx\, dx = \frac{e^{ax} \cos^{n-1} bx\,(a \cos bx + nb \sin bx)}{a^2 + n^2b^2}$$

$$+ \frac{n\,(n-1)\,b^2}{a^2 + n^2b^2} \int e^{ax} \cos^{n-2} bx\, dx.$$

$$\int e^{ax} \operatorname{tg}^n x\, dx = \frac{e^{ax} \operatorname{tg}^{n-1} x}{n-1} - \frac{a}{n-1} \int e^{ax} \operatorname{tg}^{n-1} x\, dx - \int e^{ax} \operatorname{tg}^{n-2} x\, dx.$$

$$\int \frac{\sin x}{e^x}\, dx = -\frac{\sin x + \cos x}{2e^x}.$$

$$\int \frac{\cos x}{e^x}\, dx = \frac{\sin x - \cos x}{2e^x}.$$

$$\int \frac{x \sin x}{e^x}\, dx = -\frac{x\,(\sin x + \cos x)}{2e^x} - \frac{\cos x}{2e^x}.$$

$$\int \frac{x \cos x}{e^x}\, dx = \frac{x\,(\sin x - \cos x)}{2e^x} + \frac{\sin x}{2e^x}.$$

$$\int \frac{e^{ax}}{\sin^n x}\, dx = -\frac{e^{ax} \{ a \sin x + (n-2) \cos x \}}{(n-1)\,(n-2) \sin^{n-1} x} + \frac{a^2 + (n-2)^2}{(n-1)\,(n-2)} \int \frac{e^{ax}}{\sin^{n-2} x}\, dx.$$

$$\int \frac{e^{ax}}{\cos^n x}\, dx = -\frac{e^{ax} \{ a \cos x - (n-2) \sin x \}}{(n-1)\,(n-2) \cos^{n-1} x} + \frac{a^2 + (n-2)^2}{(n-1)\,(n-2)} \int \frac{e^{ax}}{\cos^{n-2} x}\, dx.$$

$$\int e^{ax} \sin x \cos x\, dx = \frac{e^{ax}\,(a \sin 2x - 2 \cos 2x)}{2\,(a^2 + 4)}.$$

$$\int e^{ax} \sin x \cos bx\, dx = \frac{e^{ax}}{c} \{ (a \sin x - \cos x) \cos (bx - \beta) - b \sin x \sin (bx - \beta) \}.$$

where $1 + a^2 - b^2 = c \cos \beta, \qquad 2ab = c \sin \beta.$

$$\int e^{ax} \cos^m x \sin^n x\, dx = \frac{e^{ax} \cos^{m-1} x \sin^n x \{ a \cos x + (m+n) \sin x \}}{(m+n)^2 + a^2}.$$

$$- \frac{na}{(m+n)^2 + a^2} \int e^{ax} \cos^{m-1} x \sin^{n-1} x\, dx$$

$$+ \frac{(m-1)\,(m+n)}{(m+n)^2 + a^2} \int e^{ax} \cos^{m-2} x \sin^n x\, dx.$$

$$= \frac{e^{ax} \cos^m x \sin^{n-1} x \{ a \sin x - (m+n) \cos x \}}{(m+n)^2 + a^2}$$

$$+\frac{ma}{(m+n)^2+a^2}\int e^{ax}\cos^{m-1}x\sin^{n-1}x\,dx$$

$$+\frac{(n-1)(m+n)}{(m+n)^2+a^2}\int e^{ax}\cos^{m}x\sin^{n-2}x\,dx.$$

$$=\frac{e^{ax}\cos^{m-1}x\sin^{n-1}x\,(a\sin x\cos x+m\sin^2 x-n\cos^2 x)}{(m+n)^2+a^2}$$

$$+\frac{m(m-1)}{(m+n)^2+a^2}\int e^{ax}\cos^{m-2}x\sin^{n}x\,dx$$

$$+\frac{n(n-1)}{(m+n)^2+a^2}\int e^{ax}\cos^{m}x\sin^{n-2}x\,dx.$$

$$=\frac{e^{ax}\cos^{m-1}x\sin^{n-1}x\,(a\cos x\sin x+m\sin^2 x-\cos^2 x)}{(m+n)^2+a^2}$$

$$+\frac{m(m-1)}{(m+n)^2+a^2}\int e^{ax}\cos^{m-2}x\sin^{n-2}x\,dx$$

$$+\frac{(n-m)(n+m-1)}{(m+n)^2+a^2}\int e^{ax}\cos^{m}x\sin^{n-2}x\,dx.$$

DIVERSE TRIGONOMETRIC, LOGARITHMIC, AND EXPONENTIAL FUNCTIONS

$$\int \sin \log x \, dx = \frac{x}{2}\left(\sin \log x - \cos \log x\right).$$

$$\int \sin x \log \cos x \, dx = \cos x \, (1 - \log \cos x).$$

$$\int \frac{\cos (\log x)}{x} \, dx = \sin (\log x).$$

$$\int \cos x \log \sin x \, dx = \sin x \, (\log \sin x - 1).$$

$$\int \text{coséc}\, x \, \text{cotg}\, x \, dx = - \text{coséc}\, x.$$

$$\int \text{séc}\, x \, \text{tg}\, x \, dx = \text{séc}\, x.$$

$$\int \frac{e^{mx}}{\sqrt{a + be^{mx}}} \, dx = \frac{2}{bm} \sqrt{a + be^{mx}}.$$

$$\int \frac{dx}{\sqrt{e^x + a^2}} = \frac{2}{a} \log \left\{ \sqrt{e^x + a^2} - a \right\} - \frac{x}{a}.$$

$$\int \sqrt{1 + e^{ax}} \, dx = \frac{2}{a} \sqrt{1 + e^{ax}} + \frac{1}{a} \log \frac{\sqrt{1 + e^{ax}} + 1}{\sqrt{1 + e^{ax}} - 1}.$$

$$\int \frac{dx}{\sqrt{e^{2x} - 1}} = \text{arc cos}\, e^{-x}.$$

$$\int x^{nx} \, dx = x \left\{ 1 - \frac{nx}{2^2} + \frac{n^2x^2}{3^3} - \frac{n^3x^3}{4^4} + \cdots \right\} + \frac{nx^2 \log x}{1} \left\{ \frac{1}{2} - \frac{nx}{3^2} + \frac{n^2x^2}{4^3} - \frac{n^3x^3}{5^4} + \cdots \right\}$$

$$+ \frac{n^2x^3 (\log x)^2}{1.2} \left\{ \frac{1}{3} - \frac{nx}{4^2} + \frac{n^2x^2}{5^3} - \frac{n^3x^3}{6^4} + \cdots \right\}$$

$$+ \frac{n^3x^4 (\log x)^3}{1.2.3} \left\{ \frac{1}{4} - \frac{nx}{5^2} + \frac{n^2x^2}{6^3} - \frac{n^3x^3}{7^4} + \cdots \right\} + \cdots$$

$$\int x^{m+nx}\,dx = \int x^m \left\{ 1 + nx \log x + \frac{1}{2}\left(nx \log x\right)^2 + \cdots \right\} dx.$$

$$\int \frac{x^m}{\sqrt{\log \frac{1}{x}}}\,dx = \frac{x^{m+1}}{(m+1)\sqrt{\log \frac{1}{x}}} \left\{ 1 + \frac{1}{(2m+2)\log x} + \frac{1.3}{\{(2m+2)\log x\}^2} + \frac{1.3.5}{\{(2m+2)\log x\}^3} + \cdots \right\}$$

$$\int e^{\cos x} \sin x\,dx = -\,e^{\cos x}.$$

a^u

$$\int a^{mx}\,dx = \frac{a^{mx}}{m \log a}.$$

$$\int U a^x\,dx = \frac{Ua^x}{\log a} - \frac{1}{\log a}\int U' a^x\,dx = U_1 a^x - \log a \int U_1 a^x\,dx.$$

$$= \frac{Ua^x}{\log a} - \frac{U' a^x}{(\log a)^2} + \frac{U'' a^x}{(\log a)^3} - \cdots$$

$$= U_1 a^x - U_2 a^x \log a + U_3 a^x \left(\log a\right)^2 - \cdots$$

$$\int x a^x\,dx = \frac{xa^x}{\log a} - \frac{a^x}{(\log a)^2}.$$

$$\int x^2 a^x\,dx = \frac{a^x}{\log a}\left\{x^2 - \frac{2x}{\log a} + \frac{2}{(\log a)^2}\right\}.$$

$$\int x^m a^x\,dx = \frac{a^x}{\log a}\left\{x^m - \frac{mx^{m-1}}{\log a} + \frac{m(m-1)x^{m-2}}{(\log a)^2}\right.$$

$$\left. - \frac{m(m-1)(m-2)x^{m-3}}{(\log a)^3} + \cdots \pm \frac{m(m-1)\cdots 2.1}{(\log a)^m}\right\}.$$

$$\int x^m a^{nx}\,dx = \int\left\{1 + \frac{nx\log x}{1} + \frac{n^2x^2(\log x)^2}{1.2} + \frac{n^3x^3(\log x)^3}{1.2.3} + \cdots\right\} x^m\,dx.$$

$$\int \frac{a^x}{x}\,dx = \log x + \frac{x\log a}{1} + \frac{1}{2}\frac{x^2(\log a)^2}{1.2} + \frac{1}{3}\frac{x^3(\log a)^3}{1.2.3} + \frac{1}{4}\frac{x^4(\log a)^4}{1.2.3.4} + \cdots$$

$$\int \frac{a^x}{x^2}\,dx = -\frac{a^x}{x} + \log a\int\frac{a^x}{x}\,dx.$$

$$\int \frac{a^x}{x^m}\,dx = -\frac{a^x}{(m-1)x^{m-1}} - \frac{a^x\log a}{(m-1)(m-2)x^{m-2}} - \frac{a^x(\log a)^2}{(m-1)(m-2)(m-3)x^{m-3}} - \cdots$$

$$- \frac{a^x(\log a)^{m-2}}{(m-1)\cdots 2.1\,x} + \frac{(\log a)^{m-1}}{(m-1)\cdots 2.1}\int\frac{a^x}{x}\,dx.$$

$$\int \frac{a^x}{\sqrt{x}}\,dx = \frac{a^x}{\sqrt{x}}\left\{\frac{1}{\log a} + \frac{1}{2x(\log a)^2} + \frac{1.3}{2^2x^2(\log a)^3} + \frac{1.3\,5}{2^3x^3(\log a)^4} + \cdots\right\}.$$

$$= \frac{a^x}{\sqrt{x}} \left\{ \frac{2x}{1} - \frac{2^2x^2 \log a}{1.3} + \frac{2^3x^3 (\log a)^2}{1.3.5} - \frac{2^4x^4 (\log a)^3}{1.3.5.7} + \cdots \right\}.$$

$$\int \frac{a^x}{1-x}\,dx = a^x \left\{ \frac{1}{(1-x) \log a} - \frac{1}{(1-x)^2 (\log a)^2} + \frac{1.2}{(1-x)^3 (\log a)^3} \right.$$

$$\left. - \frac{1.2.3}{(1-x)^4 (\log a)^4} + \cdots \right\}.$$

CHAPTER 14

DIFFERENTIAL EQUATIONS

DIFFERENTIAL EQUATIONS	METHOD OF SOLUTION
Separation of variables $f_1(x)\,g_1(y)\,dx + f_2(x)\,g_2(y)\,dy = 0$	$\displaystyle\int \frac{f_1(x)}{f_2(x)}\,dx + \int \frac{g_2(y)}{g_1(y)}\,dy = c$
Exact equation $M(x,y)dx + N(x,y)dy = 0$ where $\partial M/\partial y = \partial N/\partial x$	$\displaystyle\int M\,\partial x + \int\left(N - \frac{\partial}{\partial y}\int M\,\partial x\right) dy = c$ where ∂x indicates that the integration is to be performed with respect to x keeping y constant.
Linear first order equation $\dfrac{dy}{dx} + P(x)y = Q(x)$	$ye^{\int P dx} = \int Qe^{\int P dx}\,dx + c$
Bernoulli's equation $\dfrac{dy}{dx} + P(x)y = Q(x)y^n$	$ve^{(1-n)\int P dx} = (1-n)\int Qe^{(1-n)\int P dx}\,dx + c$ where $v = y^{1-n}$. If $n = 1$, the solution is $\ln y = \int (Q - P)\,dx + c$
Homogeneous equation $\dfrac{dy}{dx} = F\left(\dfrac{y}{x}\right)$	$\ln x = \displaystyle\int \frac{dv}{F(v) - v} + c$ where $v = y/x$. If $F(v) = v$, the solution is $y = cx$

Reducible to homogeneous $(a_1x + b_1y + c_1)\,dx + (a_2x + b_2y + c_2)\,dy = 0$ $\frac{a_1}{a_2} \neq \frac{b_1}{b_2}$	Set $u = a_1x + b_1y + c_1$ $v = a_2x + b_2y + c_2$ Eliminate x and y and the equation becomes homogenous
Reducible to separable $(a_1x + b_1y + c_1)\,dx + (a_2x + b_2y + c_2)\,dy = 0$ $\frac{a_1}{a_2} = \frac{b_1}{b_2}$	Set $u = a_1x + b_1y$ Eliminate x or y and equation becomes separable
$y\,F(xy)\,dx + x\,G(xy)dy = 0$	$\ln x = \int \frac{G(v)\,dv}{v\{G(v) - F(v)\}} + c$ where $v = xy$. If $G(v) = F(v)$, the solution is $xy = c$.
Linear, homogeneous second order equation $\frac{d^2y}{dx^2} + b\frac{dy}{dx} + cy = 0$ a, b are real constants	Let m_1, m_2 be the roots of $m^2 + bm + c = 0$. Then there are 3 cases: Let m_1, m_2 be the roots of $m^2 + bm + c = 0$. Then there are 3 cases: Case 1. m_1, m_2 real and distinct: $y = c_1e^{m_1x} + c_2e^{m_2x}$ Case 2. m_1, m_2 real and equal: $y = c_1e^{m_1x} + c_2xe^{m_1x}$ Case 3. $m_1 = p + qi$, $m_2 = p - qi$: $y = e^{px}(c_1\cos qx + c_2\sin qx)$ where $p = -b/2$, $q = \sqrt{4c - b^2}/2$
Linear, nonhomogeneous second order equation $\frac{d^2y}{dx^2} + b\frac{dy}{dx} + cy = R(x)$ b, c are real constants	There are 3 cases corresponding to those immediately above: Case 1. $y = c_1e^{m_1x} + c_2e^{m_2x} + \frac{e^{m_1x}}{m_1 - m_2}\int e^{-m_1x}R(x)\,dx + \frac{e^{m_2x}}{m_2 - m_1}\int e^{-m_2x}R(x)\,dx$

	Case 2. $$y = c_1 e^{m_1 x} + c_2 x e^{m_1 x} + x e^{m_1 x} \int e^{-m_1 x} R(x)\,dx - e^{m_1 x} \int x e^{-m_1 x} R(x)\,dx$$ Case 3. $$y = e^{px}(c_1 \cos qx + c_2 \sin qx) + \frac{e^{px} \sin qx}{q} \int e^{-px} R(x) \cos qx\,dx - \frac{e^{px} \cos qx}{q} \int e^{-px} R(x) \sin qx\,dx$$
Euler or Cauchy equation $$x^2 \frac{d^2 y}{dx^2} + bx \frac{dy}{dx} + cy = S(x)$$	Putting $x = e^t$, the equation becomes $$\frac{d^2 y}{dt^2} + (b-1) \frac{dy}{dt} + cy = S(e^t)$$ and can then be solved as a linear second order equation.
Bessel's equation $$x^2 \frac{d^2 y}{dx^2} + x \frac{dy}{dx} + (\lambda^2 x^2 - n^2)y = 0$$	$$y = c_1 J_n(\lambda x) + c_2 Y_n(\lambda x)$$
Transformed Bessel's equation $$x^2 \frac{d^2 y}{dx^2} + (2p+1)x \frac{dy}{dx} + (\alpha^2 x^{2r} + \beta^2)y = 0$$	$$y = x^{-p} \left\{ c_1 J_{q/r}\left(\frac{\alpha}{r} x^r\right) + c_2 Y_{q/r}\left(\frac{\alpha}{r} x^r\right) \right\}$$ where $q = \sqrt{p^2 - \beta^2}$.
Legendre's equation $$(1-x^2)\frac{d^2 y}{dx^2} - 2x \frac{dy}{dx} + n(n+1)y = 0$$	$$y = c_1 P_n(x) + c_2 Q_n(x)$$

CHAPTER 15

PROBABILITY FUNCTIONS[1]

PROBABILITY FUNCTION: DEFINITIONS AND PROPERTIES

UNIVARIATE CUMULATIVE DISTRIBUTION FUNCTIONS

A real-valued function $F(x)$ is termed a (univariate) cumulative distribution function (c.d.f.) or simply distribution function if

i) $F(x)$ is non-decreasing, i.e., $F(x_1) \leq F(x_2)$ for $x_1 \leq x_2$

---

a. We follow the customary convention of denoting a random variable by a capital letter, i.e., X, and using the corresponding lower case letter, i.e., x, for a particular value that the random variable assumes.

b. For statistical applications it is often convenient to have tabulated the "upper tail area," $1-F(x)$, or the c.d.f. for $|X|$, $F(x)-F(-x)$, instead of simply the c.d.f. $F(x)$. We use the notation P to indicate the c.d.f. of X, $Q=1-P$ to indicate the "upper tail area" and $A=P-Q$ to denote the c.d.f. of $|X|$. In particular we use $P(x)$, $Q(x)$, and $A(x)$ to denote the corresponding functions for the normal or Gaussian probability function, see 15.20 to 15.22. When these distributions depend on other parameters, say θ_1 and θ_2, we indicate this by writing $P(x|\theta_1, \theta_2)$, $Q(x|\theta_1, \theta_2)$, or $A(x|\theta_1, \theta_2)$. For example the chi-square distribution depends on the parameter ν and the tabulated function is written $Q(\chi^2|\nu)$.

---

[1]From "Handbook of Mathematical Functions", National Bureau of Standards

ii) $F(x)$ is everywhere continuous from the right, i.e., $F(x)=\lim_{\epsilon\to 0+} F(x+\epsilon)$

iii) $F(-\infty)=0$, $F(\infty)=1$.

The function $F(x)$ signifies the probability of the event "$X\leq x$" where X is a random variable, i.e., $Pr\{X\leq x\}=F(x)$, and thus describes the c.d.f. of X. The two principal types of distribution functions are termed *discrete* and *continuous*.

Discrete Distributions: Discrete distributions are characterized by the random variable X taking on an enumerable number of values $\ldots, x_{-1}, x_0, x_1, \ldots$ with point probabilities

$$p_n=Pr\{X=x_n\}\geq 0$$

which need only be subject to the restriction

$$\sum_n p_n=1.$$

The corresponding distribution function can then be written

$$F(x)=Pr\{X\leq x\}=\sum_{x_n\leq x} p_n$$

where the summation is over all values of x for which $x_n\leq x$. The set $\{x_n\}$ of values for which $p_n>0$ is termed the domain of the random variable X. A discrete distribution of a random variable is called a *lattice distribution* if there exist numbers a and $b\neq 0$ such that every possible value of X can be represented in the form $a+bn$ where n takes on only integral values. A summary of some properties of certain discrete distributions is presented in 15.1 to 15.6.

Continuous Distributions. Continuous distributions are characterized by $F(x)$ being absolutely continuous. Hence $F(x)$ possesses a derivative $F'(x)=f(x)$ and the c.d.f. can be written

$$F(x)=Pr\{X\leq x\}=\int_{-\infty}^{x} f(t)dt.$$

The derivative $f(x)$ is termed the *probability density function* (p.d.f.) or *frequency function*, and the values of x for which $f(x)>0$ make up the domain of the random variable X. A summary of some properties of certain selected continuous distributions is presented in 15.1 to 15.6.

MULTIVARIATE PROBABILITY FUNCTIONS

The real-valued function $F(x_1, x_2, \ldots x_n)$ defines an n-variate cumulative distribution function if

i) $F(x_1, x_2, \ldots x_n)$ is a non-decreasing function for each x_i

ii) $F(x_1, x_2, \ldots x_n)$ is continuous from the right in each x_i; i.e., $F(x_1, x_2, \ldots x_n) = \lim_{\epsilon \to 0+} F(x_1, \ldots, x_i+\epsilon, \ldots, x_n)$

iii) $F(x_1, x_2, \ldots x_n)=0$ when any $x_i=-\infty$; $F(\infty, \infty, \ldots, \infty)=1$. The function $F(x_1, x_2, \ldots, x_n)$ signifies the probability of the event $X_1 \leq x_1, X_2 \leq x_2, \ldots, X_n \leq x_n$ where $X_1, X_2, \ldots X_n$ is a set of n random variables.

Thus $Pr\{X_1 \leq x_1, \quad X_2 \leq x_2, \quad \ldots, \quad X_n \leq x_n\} = F(x_1, x_2, \ldots x_n)$. The two principal types of n-variate distribution functions termed *discrete* and *continuous*, are defined in a manner similar to the corresponding cases for the univariate distribution function.

RELATION OF THE CHARACTERISTIC FUNCTION TO MOMENTS ABOUT THE ORIGIN

$$\phi^{(n)}(0) = \left[\frac{d^n}{dt^n}\phi(t)\right]_{t=0} = i^n \mu_n'$$

CHARACTERISTICS OF DISTRIBUTION FUNCTIONS: MOMENTS, CHARACTERISTIC FUNCTIONS, CUMULANTS

	Continuous distributions	Discrete distributions
n^{th} moment about origin	$\mu_n' = \int_{-\infty}^{\infty} x^n f(x)dx$	$\mu_n' = \sum_s x_s^n p_s$
mean	$m = \mu_1' = \int_{-\infty}^{\infty} x f(x)dx$	$m = \mu_1' = \sum_s x_s p_s$
variance	$\sigma^2 = \mu_2' - m^2 = \int_{-\infty}^{\infty} (x-m)^2 f(x)dx$	$\sigma^2 = \mu_2' - m^2 = \sum_s (x_s - m)^2 p_s$
n^{th} central moment	$\mu_n = \int_{-\infty}^{\infty} (x-m)^n f(x)dx$	$\mu_n = \sum_s (x_s - m)^n p_s$
expected value operator for the function $g(x)$	$E[g(X)] = \int_{-\infty}^{\infty} g(x) f(x)dx$	$E[g(X)] = \sum_s g(x_s) p_s$
characteristic function of X	$\phi(t) = E(e^{itX}) = \int_{-\infty}^{\infty} e^{itx} f(x)dx$	$\phi(t) = E(e^{itX}) = \sum_s e^{itx_s} p_s$
characteristic function of $g(X)$	$\phi_g(t) = E(e^{itg(X)}) = \int_{-\infty}^{\infty} e^{itg(x)} f(x)dx$	$\phi_g(t) = E(e^{itg(X)}) = \sum_s e^{itg(x_s)} p_s$
inversion formula	$f(x) = \frac{1}{2\pi}\int_{-\infty}^{\infty} e^{-itx}\phi(t)dt$	$p_n = \frac{b}{2\pi}\int_{-\pi/b}^{\pi/b} e^{-itx_n}\phi(t)dt$ (lattice distributions only)

CUMULANT FUNCTION

$$\ln \phi(t) = \sum_{n=0}^{\infty} \kappa_n \frac{(it)^n}{n!}$$

κ_n is called the nth cumulant.

$$\kappa_1 = m,\ \kappa_2 = \sigma^2,\ \kappa_3 = \mu_3,\ \kappa_4 = \mu_4 - 3\mu_2^2$$

RELATION OF CENTRAL MOMENTS TO MOMENTS ABOUT THE ORIGIN

$$\mu_n = \sum_{j=0}^{n} \binom{n}{j} (-1)^{n-j} \mu_j' m^{n-j}$$

COEFFICIENTS OF SKEWNESS AND EXCESS

$$\gamma_1=\frac{\kappa_3}{\kappa_2^{3/2}}=\frac{\mu_3}{\sigma^3} \qquad \text{(skewness)}$$

$$\gamma_2=\frac{\kappa_4}{\kappa_2^2}=\frac{\mu_4}{\sigma^4}-3 \qquad \text{(excess)}$$

Occasionally coefficients of skewness and excess (or kurtosis) are given by

$$\beta_1=\gamma_1^2=\left(\frac{\mu_3}{\sigma^3}\right)^2 \qquad \text{(skewness)}$$

$$\beta_2=\gamma_2+3=\frac{\mu_4}{\sigma^4}$$

(excess or kurtosis)

SOME ONE-DIMENSIONAL DISCRETE DISTRIBUTION FUNCTIONS

Name	Domain	Point Probabilities	Restrictions on Parameters	Mean	Variance	Skewness γ_1	Excess γ_2	Characteristic function	Cumulants
15.1 Single point or degenerate	$x=c$ (c a constant)	$p=1$	$-\infty<c<+\infty$	c	0	----------	----------	$e^{i\lambda t}$	$\kappa_1=\lambda$, $\kappa_r=0$ for $r>1$
15.2 Binomial	$x_s=s$, for $s=0, 1, 2, \ldots, n$	$\binom{n}{s} p^s(1-p)^{n-s}$	$0<p<1$ ($q=1-p$)	np	npq	$\frac{q-p}{\sqrt{npq}}$	$\frac{1-6pq}{npq}$	$(q+pe^{it})^n$	$\kappa_1=np$ $\kappa_{r+1}=pq\frac{d\kappa_r}{dp}$ for $r\geq1$
15.3 Hypergeometric	$x_s=s$, for $s=0, 1, \ldots \min(n, N_1)$	$\frac{\binom{N_1}{s}\binom{N_2}{n-s}}{\binom{N_1+N_2}{n}}$	N_1 and N_2 integers, and $n\leq N_1+N_2$ ($N=N_1+N_2$, $p=N_1/N$ and $q=1-p=N_2/N$)	np	$npq\left(\frac{N-n}{N-1}\right)$	$\frac{q-p}{\sqrt{npq}}\left(\frac{N-1}{N-n}\right)^{\frac{1}{2}}\left(\frac{N-2n}{N-2}\right)$	Complicated	$\frac{\binom{N_2}{n}}{\binom{N}{n}}F(-n, -N_1; N_2-n+1; e^{it})$	Complicated
15.4 Poisson	$x_s=s$, for $s=0, 1, 2, \ldots, \infty$	$\frac{e^{-m}m^s}{s!}$	$0<m<\infty$	m	m	$m^{-\frac{1}{2}}$	m^{-1}	$e^{m(e^{it}-1)}$	$\kappa_r=m$ for $r=1, 2, \ldots$
15.5 Negative binomial	$x_s=s$, for $s=0, 1, 2, \ldots, \infty$	$\binom{n+s-1}{s}p^n(1-p)^s$	$n\geq0$ and $0<p<1$ ($p=1/Q$, and $1-p=P/Q$)	nP	nPQ	$\frac{Q+P}{\sqrt{nPQ}}$	$\frac{1+6PQ}{nPQ}$	$(Q-Pe^{it})^{-n}$	$\kappa_1=nP$ $\kappa_{r+1}=PQ\frac{d\kappa_r}{dQ}$ for $r\geq1$
15.6 Geometric	$x_s=s$, for $s=0, 1, 2, \ldots, \infty$	$p(1-p)^s$	$0<p<1$	$\frac{1-p}{p}$	$\frac{1-p}{p^2}$	$\frac{2-p}{\sqrt{1-p}}$	$6+\frac{p^2}{1-p}$	$p[1-(1-p)e^{it}]^{-1}$	$\kappa_1=\frac{1-p}{p}$, $\kappa_{r+1}=-(1-p)\frac{d\kappa_r}{dp}$, $r\geq1$

	Name	Domain	Probability Density Function $f(x)$	Restrictions on Parameters	Mean	Variance	Skewness γ_1	Excess γ_2	Characteristic function	Cumulants
15.7	Error function	$-\infty<x<\infty$	$\frac{h}{\sqrt{\pi}} e^{-h^2x^2}$	$0<h<\infty$	0	$\frac{1}{2h^2}$	0	0	$e^{\frac{-t^2}{4h^2}}$	$\kappa_1=0,\ \kappa_2=\frac{1}{2h^2}$ $\kappa_n=0$ for $n>2$
15.8	Normal	$-\infty<x<\infty$	$\frac{1}{\sigma\sqrt{2\pi}} e^{-\frac{1}{2}\left(\frac{x-m}{\sigma}\right)^2}$	$-\infty<m<\infty$ $0<\sigma<\infty$	m	σ^2	0	0	$e^{imt-\frac{\sigma^2t^2}{2}}$	$\kappa_1=m,\ \kappa_2=\sigma^2,\ \kappa_n=0$ for $n>2$
15.9	Cauchy	$-\infty<x<\infty$	$\frac{1}{\pi\beta}\frac{1}{1+\left(\frac{x-\alpha}{\beta}\right)^2}$	$-\infty<\alpha<\infty$ $0<\beta<\infty$	not defined	not defined	not defined	not defined	$e^{i\alpha t-\beta\|t\|}$	not defined
15.10	Exponential	$\alpha\le x<\infty$	$\frac{1}{\beta} e^{-\left(\frac{x-\alpha}{\beta}\right)}$	$-\infty<\alpha<\infty$ $0<\beta<\infty$	$\alpha+\beta$	β^2	2	6	$e^{i\alpha t}(1-i\beta t)^{-1}$	$\kappa_1=\alpha+\beta,\ \kappa_n=\beta^n\Gamma(n)$ for $n>1$
15.11	Laplace, or double exponential	$-\infty<x<\infty$	$\frac{1}{2\beta} e^{-\left\|\frac{x-\alpha}{\beta}\right\|}$	$-\infty<\alpha<\infty$ $0<\beta<\infty$	α	$2\beta^2$	0	3	$e^{i\alpha t}(1+\beta^2t^2)^{-1}$	$\kappa_1=\alpha,\ \kappa_2=2\beta^2$ $\kappa_{2n+1}=0,\ \kappa_{2n}=\frac{(2n)!}{n}\beta^{2n}$ for $n=1, 2, \ldots$
15.12	Extreme-Value,[1] (Fisher-Tippett Type I or doubly exponential)	$-\infty<x<\infty$	$\frac{1}{\beta}\exp(-y-e^{-y})$ with $y=\frac{x-\alpha}{\beta}$	$-\infty<\alpha<\infty$ $0<\beta<\infty$	$\alpha+\gamma\beta$	$\frac{(\pi\beta)^2}{6}$	1.3	2.4	$\Gamma(1-i\beta t)e^{i\alpha t}$	$\kappa_1=\gamma,\ \kappa_2=\frac{(\pi\beta)^2}{6}$ $\kappa_n=\beta^n\Gamma(n)\sum_{r=1}^{\infty}\frac{1}{r^n}$ for $n>2$
15.13	Pearson Type III	$\alpha\le x<\infty$	$\frac{1}{\beta\Gamma(p)} y^{p-1}e^{-y}$ with $y=\frac{x-\alpha}{\beta}$	$-\infty<\alpha<\infty$ $0<\beta<\infty$ $0<p<\infty$	$\alpha+p\beta$	$p\beta^2$	$\frac{2}{\sqrt{p}}$	$6/p$	$e^{i\alpha t}(1-i\beta t)^{-p}$	$\kappa_1=\alpha+\beta p,\ \kappa_n=\beta^n p\Gamma(n)$ for $n>1$
15.14	Gamma distribution	$0\le x<\infty$	$\frac{1}{\Gamma(p)} x^{p-1}e^{-x}$	$0<p<\infty$	p	p	$\frac{2}{\sqrt{p}}$	$6/p$	$(1-it)^{-p}$	$\kappa_1=p,\ \kappa_n=p\Gamma(n)$ for $n>1$
15.15	Beta distribution	$0\le x\le 1$	$\frac{1}{B(a,b)} x^{a-1}(1-x)^{b-1}$	$1\le a<\infty$ $1\le b<\infty$	$\frac{a}{a+b}$	$\frac{ab}{(a+b)^2(a+b+1)}$	$\frac{2(a-b)}{(a+b+2)}$	See footnote 2	$M(a, a+b, it)$	
15.16	Rectangular, or uniform	$m-\frac{h}{2}\le x\le m+\frac{h}{2}$	$\frac{1}{h}$	$-\infty<m<\infty$ $0<h<\infty$	m	$\frac{h^2}{12}$	0	-1.2	$\frac{2}{ht}\sin\left(\frac{ht}{2}\right)e^{imt}$	$\kappa_1=m,\ \kappa_{2n+1}=0$ $\kappa_{2n}=\frac{h^{2n}B_{2n}}{2n}$ B_{2n} (Bernoulli numbers), $B_2=\frac{1}{6},\ B_4=-\frac{1}{30}, \ldots$

[1] γ (Euler's constant) $=.57721\ 56649\ldots$

[2] $\gamma_2=\sqrt{\frac{a+b+1}{ab}}\left\{\frac{3(a+b+1)[2(a+b)^2+ab(a+b-6)]}{ab(a+b+2)(a+b+3)}-3\right\}.$

INEQUALITIES FOR DISTRIBUTION FUNCTIONS

($F(x)$ denotes the c.d.f. of the random variable X and t denotes a positive constant; further m is always assumed to be finite and all expectations are assumed to exist.)

	Inequality	Conditions
	$Pr\{g(X)\geq t\}\leq E[g(X)]/t$	$g(X)\geq 0$
	$Pr\{X\geq t\}\leq m/t$ $F(t)\geq 1-\frac{m}{t}$	$Pr\{X<0\}=0$ $E(X)=m$
15.17	$Pr\{\lvert X-m\rvert\geq t\sigma\}\leq 1/t^2$ $F(m+t\sigma)-F(m-t\sigma)\geq 1-\frac{1}{t^2}$	$E(X)=m$ [3] $E(X-m)^2=\sigma^2$
	$Pr\{\lvert\overline{X}-\overline{m}\rvert\geq t\overline{\sigma}\}\leq\frac{1}{nt^2}$	$E(X_i)=m_i$ $E(X_i-m_i)^2=\sigma_i^2$ $E([X_i-m_i][X_j-m_j])=0\,(i\neq j)$ $\overline{X}=\sum_{i=1}^{n}\frac{X_i}{n}$, $\overline{m}=\sum_{i=1}^{n}\frac{m_i}{n}$, $\overline{\sigma}=\left[\sum_{i=1}^{n}\frac{\sigma_i^2}{n}\right]^{\frac{1}{2}}$
	$Pr\{\lvert X-m\rvert\geq t\sigma\}\leq\frac{4}{9}\left\{\frac{1+\left(\frac{m-x_0}{\sigma}\right)^2}{\left(t-\left\lvert\frac{m-x_0}{\sigma}\right\rvert\right)^2}\right\}$ $F(m+t\sigma)-F(m-t\sigma)\geq 1-\frac{4}{9}\left\{\frac{1+\left(\frac{m-x_0}{\sigma}\right)^2}{\left(t-\left\lvert\frac{m-x_0}{\sigma}\right\rvert\right)^2}\right\}$	$E(X-m)^2=\sigma^2$ $F(x)$ is a continuous c.d.f. $F(x)$ is unimodal at x_0
15.18	$Pr\{\lvert X-m\rvert\geq t\sigma\}\leq 4/9t^2$ $F(m+t\sigma)-F(m-t\sigma)\geq 1-\frac{4}{9t^2}$	$E(X-m)^2=\sigma^2$ $F(x)$ is a continuous c.d.f. $F(x)$ is unimodal at x_0 [6] $m=x_0$
15.19	$Pr\{\lvert X-m\rvert\geq t\sigma\}\leq\frac{\mu_4-\sigma^4}{\mu_4+t^4\sigma^4-2t^2\sigma^4}$ $F(m+t\sigma)-F(m-t\sigma)\geq 1-\frac{\mu_4-\sigma^4}{\mu_4+t^4\sigma^4-2t^2\sigma^4}$	$E(X-m)^2=\sigma^2$ $E(X-m)^4=\mu_4$

3 x_0 is such that $F'(x_0)>F'(x)$ for $x\neq x_0$.

NORMAL OR GAUSSIAN PROBABILITY FUNCTION

$$Z(x)=\frac{1}{\sqrt{2\pi}}\,e^{-x^2/2}$$

15.20 $$P(x)=\frac{1}{\sqrt{2\pi}}\int_{-\infty}^{x} e^{-t^2/2}dt=\int_{-\infty}^{x} Z(t)dt$$

15.21 $$Q(x)=\frac{1}{\sqrt{2\pi}}\int_{x}^{\infty} e^{-t^2/2}dt=\int_{x}^{\infty} Z(t)dt$$

15.22 $$A(x)=\frac{1}{\sqrt{2\pi}}\int_{-x}^{x} e^{-t^2/2}dt=\int_{-x}^{x} Z(t)dt$$

$$P(x)+Q(x)=1$$
$$P(-x)=Q(x)$$
$$A(x)=2P(x)-1$$

PROBABILITY INTEGRAL WITH MEANS m AND VARIANCE σ^2

A random variable X is said to be normally distributed with mean m and variance σ^2 if the probability that X is less than or equal to x is given by

15.23

$$Pr\{X\leq x\}=\frac{1}{\sigma\sqrt{2\pi}}\int_{-\infty}^{x} e^{-\frac{(t-m)^2}{2\sigma^2}}\,dt$$

$$=\frac{1}{\sqrt{2\pi}}\int_{-\infty}^{(x-m)/\sigma} e^{-t^2/2}\,dt=P\left(\frac{x-m}{\sigma}\right).$$

The corresponding probability density function is

$$\frac{\partial}{\partial x}P\left(\frac{x-m}{\sigma}\right)=\frac{1}{\sigma}\,Z\left(\frac{x-m}{\sigma}\right)=\frac{1}{\sigma\sqrt{2\pi}}\,e^{-\frac{(x-m)^2}{2\sigma^2}}$$

and is symmetric around m, i.e.

$$Z\left(\frac{m+x}{\sigma}\right)=Z\left(\frac{m-x}{\sigma}\right).$$

The inflexion points of the probability density function are at $m\pm\sigma$.

POWER SERIES

$$P(x)=\frac{1}{2}+\frac{1}{\sqrt{2\pi}}\sum_{n=0}^{\infty}\frac{(-1)^n x^{2n+1}}{n!2^n(2n+1)}$$

$$P(x)=\frac{1}{2}+Z(x)\sum_{n=0}^{\infty}\frac{x^{2n+1}}{1\cdot 3\cdot 5\ldots(2n+1)}$$

ASYMPTOTIC EXPANSIONS ($x>0$)

$$Q(x)=\frac{Z(x)}{x}\left\{1-\frac{1}{x^2}+\frac{1\cdot 3}{x^4}+\ldots\right.$$

$$\left.+\frac{(-1)^n 1\cdot 3\ldots(2n-1)}{x^{2n}}\right\}+R_n$$

where

$$R_n=(-1)^{n+1}1\cdot 3\ldots(2n+1)\int_x^{\infty}\frac{Z(t)}{t^{2n+2}}\,dt$$

which is less in absolute value than the first neglected term.

$$Q(x)\sim\frac{Z(x)}{x}\left\{1-\frac{a_1}{x^2+2}+\frac{a_2}{(x^2+2)(x^2+4)}\right.$$

$$\left.-\frac{a_3}{(x^2+2)(x^2+4)(x^2+6)}+\ldots\right\}$$

where $a_1=1$, $a_2=1$, $a_3=5$, $a_4=9$, $a_5=129$ and the general term is

$$a_n=c_0 1\cdot 3\ldots(2n-1)+2c_1 1\cdot 3\ldots(2n-3)$$
$$+2^2c_2 1\cdot 3\ldots(2n-5)+\ldots+2^{n-1}c_{n-1}$$

and c_s is the coefficient of t^{n-s} in the expansion of $t(t-1)\ .\ .\ .\ (t-n+1)$.

CONTINUED FRACTION EXPANSIONS

$$Q(x)=Z(x)\left\{\frac{1}{x+}\,\frac{1}{x+}\,\frac{2}{x+}\,\frac{3}{x+}\,\frac{4}{x+}\cdots\right\}\qquad (x>0)$$

$$Q(x)=\frac{1}{2}-Z(x)\left\{\frac{x}{1-}\,\frac{x^2}{3+}\,\frac{2x^2}{5-}\,\frac{3x^2}{7+}\,\frac{4x^2}{9-}\cdots\right\}\qquad (x\geq 0)$$

POLYNOMIAL AND RATIONAL APPROXIMATIONS FOR P(x) AND Z(x)

$$0\leq x<\infty$$

$$P(x)=1-Z(x)(a_1t+a_2t^2+a_3t^3)+\epsilon(x),\qquad t=\frac{1}{1+px}$$

$$|\epsilon(x)|<1\times 10^{-5}$$

$p=.33267$ $\quad a_1=.43618\ 36$

$a_2=-.12016\ 76$

$a_3=.93729\ 80$

$$P(x)=1-Z(x)(b_1t+b_2t^2+b_3t^3+b_4t^4+b_5t^5)+\epsilon(x),$$

$$t=\frac{1}{1+px}$$

$$|\epsilon(x)|<7.5\times 10^{-8}$$

$p=.23164\ 19$

$b_1=\ .31938\ 1530$ $\qquad b_4=-1.82125\ 5978$

$b_2=-.35656\ 3782$ $\qquad b_5=\ 1.33027\ 4429$

$b_3=\ 1.78147\ 7937$

$$P(x)=1-\frac{1}{2}(1+c_1x+c_2x^2+c_3x^3+c_4x^4)^{-4}+\epsilon(x)$$

$$|\epsilon(x)|<2.5\times10^{-4}$$

$$c_1=.196854 \qquad c_3=.000344$$

$$c_2=.115194 \qquad c_4=.019527$$

$$P(x)=1-\frac{1}{2}\,(1+d_1x+d_2x^2+d_3x^3$$

$$+d_4x^4+d_5x^5+d_6x^6)^{-16}+\epsilon(x)$$

$$|\epsilon(x)|<1.5\times10^{-7}$$

$$d_1=.04986\ 73470 \qquad d_4=.00003\ 80036$$

$$d_2=.02114\ 10061 \qquad d_5=.00004\ 88906$$

$$d_3=.00327\ 76263 \qquad d_6=.00000\ 53830$$

$$Z(x)=(a_0+a_2x^2+a_4x^4+a_6x^6)^{-1}+\epsilon(x)$$

$$|\epsilon(x)|<2.7\times10^{-3}$$

$$a_0=2.490895 \qquad a_4=-.024393$$

$$a_2=1.466003 \qquad a_6=\ \ .178257$$

$$Z(x)=(b_0+b_2x^2+b_4x^4+b_6x^6+b_8x^8+b_{10}x^{10})^{-1}+\epsilon(x)$$

$$|\epsilon(x)|<2.3\times10^{-4}$$

$$b_0=2.50523\ 67 \qquad b_6=\ \ .13064\ 69$$

$$b_2=1.28312\ 04 \qquad b_8=-.02024\ 90$$

$$b_4=\ \ .22647\ 18 \qquad b_{10}=\ \ .00391\ 32$$

RATIONAL APPROXIMATIONS FOR x_p WHERE $Q(x_p) = p$

$$0<p\leq.5$$

$$x_p=t-\frac{a_0+a_1t}{1+b_1t+b_2t^2}+\epsilon(p), \qquad t=\sqrt{\ln\frac{1}{p^2}}$$

$$|\epsilon(p)|<3\times10^{-3}$$

$$a_0=2.30753 \qquad b_1=.99229$$

$$a_1=\ \ .27061 \qquad b_2=.04481$$

15.24

$$x_p = t - \frac{c_0 + c_1 t + c_2 t^2}{1 + d_1 t + d_2 t^2 + d_3 t^3} + \epsilon(p), \qquad t = \sqrt{\ln \frac{1}{p^2}}$$

$$|\epsilon(p)| < 4.5 \times 10^{-4}$$

$$\begin{array}{ll} c_0 = 2.515517 & d_1 = 1.432788 \\ c_1 = .802853 & d_2 = .189269 \\ c_2 = .010328 & d_3 = .001308 \end{array}$$

BOUNDS USEFUL AS APPROXIMATIONS TO THE NORMAL DISTRIBUTION FUNCTION

$$P(x) \leq \begin{cases} P_1(x) = \frac{1}{2} + \frac{1}{2}\left(1 - e^{-2x^2/\pi}\right)^{\frac{1}{2}} & (x > 0) \\ P_2(x) = 1 - \frac{(4 + x^2)^{\frac{1}{2}} - x}{2}(2\pi)^{-\frac{1}{2}} e^{-x^2/2} & (x > 1.4) \end{cases}$$

$$P(x) \geq \begin{cases} P_3(x) = \frac{1}{2} + \frac{1}{2}\left(1 - e^{-2x^2/\pi} - \frac{2(\pi - 3)}{3\pi^2} x^4 e^{-x^2/2}\right)^{\frac{1}{2}} & (x > 0) \\ P_4(x) = 1 - \frac{1}{x}(2\pi)^{-\frac{1}{2}} e^{-x^2/2} & (x > 2.2) \end{cases}$$

See Fig. 1 for error curves.

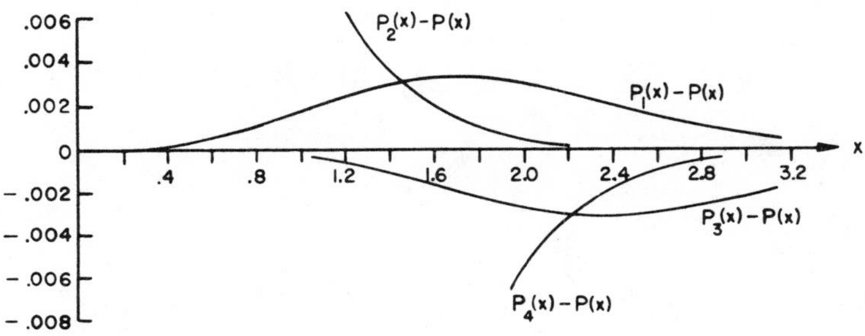

Fig. 1. *Error curves for bounds on normal distribution.*

DERIVATIVES OF THE NORMAL PROBABILITY DENSITY FUNCTION

$$Z^{(m)}(x)=\frac{d^m}{dx^m}Z(x)$$

DIFFERENTIAL EQUATION

$$Z^{(m+2)}(x)+xZ^{(m+1)}(x)+(m+1)Z^{(m)}(x)=0$$

VALUE AT x = 0

$$Z^{(m)}(0)=\begin{cases}\dfrac{(-1)^{m/2}m!}{\sqrt{2\pi}2^{m/2}\left(\dfrac{m}{2}\right)!} & \text{for } m=2r,\ r=0,1,\ldots\\ 0 & \text{for odd } m>0\end{cases}$$

RELATION OF P(x) AND Z(m)(x) TO OTHER FUNCTIONS

Function	*Relation*	
;rror function	$\operatorname{erf} x=2P(x\sqrt{2})-1$	$(x\geq 0)$
ncomplete gamma function (special case)	$\dfrac{\gamma\left(\frac{1}{2}, x\right)}{\Gamma\left(\frac{1}{2}\right)}=[2P(\sqrt{2x})-1]$	$(x\geq 0)$
Iermite polynomial	$He_n(x)=(-1)^n\dfrac{Z^{(n)}(x)}{Z(x)}$	
"	$H_n(x)=(-1)^n 2^{n/2}\dfrac{Z^{(n)}(x\sqrt{2})}{Z(x\sqrt{2})}$	
Ih function	$Hh_{-n}(x)=(-1)^{n-1}\sqrt{2\pi}Z^{(n-1)}(x)$	$(n>0)$
"	$Hh_n(x)=\dfrac{(-1)^n}{n!}Hh_{-1}(x)\dfrac{d^n}{dx^n}\left(\dfrac{Q(x)}{Z(x)}\right)$	$(n>0)$
etrachoric function	$\tau_n(x)=\dfrac{(-1)^{n-1}}{\sqrt{n!}}Z^{(n-1)}(x)$	
onfluent hypergeometric function (special case)	$M\left(\frac{1}{2}, \frac{3}{2}, -\frac{x^2}{2}\right)=\dfrac{\sqrt{2\pi}}{x}\left\{P(x)-\frac{1}{2}\right\}$	$(x>0)$
"	$M\left(1, \frac{3}{2}, \frac{x^2}{2}\right)=\dfrac{1}{xZ(x)}\left\{P(x)-\frac{1}{2}\right\}$	$(x>0)$
"	$M\left(\frac{2m+1}{2}, \frac{1}{2}, -\frac{x^2}{2}\right)=\dfrac{Z^{(2m)}(x)}{Z^{(2m)}(0)}$	$(x\geq 0)$
"	$M\left(\frac{2m+2}{2}, \frac{3}{2}, -\frac{x^2}{2}\right)=\dfrac{Z^{(2m-1)}(x)}{xZ^{(2m)}(0)}$	$(x\geq 0)$
arabolic cylinder function	$U\left(-n-\frac{1}{2}, x\right)=e^{-\frac{1}{4}x^2}(-1)^n\dfrac{Z^{(n)}(x)}{Z(x)}$	$(n>0)$

EPEATED INTEGRALS OF THE NORMAL ROBABILITY INTEGRAL

$$I_n(x)=\int_x^\infty I_{n-1}(t)dt \qquad (n\geq 0)$$

where $I_{-1}(x)=Z(x)$

$$I_{-n}(x)=\left(-\frac{d}{dx}\right)^{n-1}Z(x)=(-1)^{n-1}Z^{(n-1)}(x)$$

$$\left(\frac{d^2}{dx^2}+x\frac{dx}{dn}-n\right)I_n(x)=0 \qquad (n\geq -1)$$

$$(n+1)I_{n+1}(x)+xI_n(x)-I_{n-1}(x)=0 \qquad (n>-1)$$

$$I_n(x)=\int_x^\infty \frac{(t-x)^n}{n!} Z(t)dt=e^{-x^2/2}\int_0^\infty \frac{t^n}{n!} Z(t)dt \qquad (n>-1)$$

$$I_n(0)=I_{-n}(0)=\frac{1}{\left(\frac{n}{2}\right)!2^{\frac{n+2}{2}}} \qquad (n \text{ even})$$

ASYMPTOTIC EXPANSIONS OF AN ARBITRARY PROBABILITY DENSITY FUNCTION AND DISTRIBUTION FUNCTION

Let Y_i $(i=1, 2, \ldots, n)$ be n

independent random variables with mean m_i, variance σ_i^2, and higher cumulants $\kappa_{r,i}$. Then asymptotic expansions with respect to n for the probability density and cumulative distribution function of

$$X=\frac{\sum_{i=1}^{m}(Y_i-m_i)}{\left(\sum_{i=1}^{m}\sigma_i^2\right)^{\frac{1}{2}}} \text{ are}$$

$$f(x)\sim Z(x)-\left[\frac{\gamma_1}{6}Z^{(3)}(x)\right]+\left[\frac{\gamma_2}{24}Z^{(4)}(x)+\frac{\gamma_1^2}{72}Z^{(6)}(x)\right]$$
$$-\left[\frac{\gamma_3}{120}Z^{(5)}(x)+\frac{\gamma_1\gamma_2}{144}Z^{(7)}(x)+\frac{\gamma_1^3}{1296}Z^{(9)}(x)\right]$$
$$+\left[\frac{\gamma_4}{720}Z^{(6)}(x)+\frac{\gamma_2^2}{1152}Z^{(8)}(x)+\frac{\gamma_1\gamma_3}{720}Z^{(8)}(x)\right.$$
$$\left.+\frac{\gamma_1^2\gamma_2}{1728}Z^{(10)}(x)+\frac{\gamma_1^4}{31104}Z^{(12)}(x)\right]+\ldots$$

$$F(x)\sim P(x)-\left[\frac{\gamma_1}{6}Z^{(2)}(x)\right]+\left[\frac{\gamma_2}{24}Z^{(3)}(x)+\frac{\gamma_1^2}{72}Z^{(5)}(x)\right]$$
$$-\left[\frac{\gamma_3}{120}Z^{(4)}(x)+\frac{\gamma_1\gamma_2}{144}Z^{(6)}(x)+\frac{\gamma_1^3}{1296}Z^{(8)}(x)\right]$$
$$+\left[\frac{\gamma_4}{720}Z^{(5)}(x)+\frac{\gamma_2^2}{1152}Z^{(7)}(x)+\frac{\gamma_1\gamma_3}{720}Z^{(7)}(x)\right.$$

$$+\frac{\gamma_1^2\gamma_2}{1728} Z^{(9)}(x)+\frac{\gamma_1^4}{31104} Z^{(11)}(x)\Big]+\ldots$$

where

$$\gamma_{r-2}=\frac{1}{n^{\frac{r}{2}-1}}\frac{\left(\frac{1}{n}\sum_{i=1}^{n}\kappa_{r,i}\right)}{\left(\frac{1}{n}\sum_{i=1}^{n}\sigma_i^2\right)^{r/2}}$$

Terms in brackets are terms of the same order with respect to n. When the Y_i have the same distribution, then $m_i=m$, $\sigma_i^2=\sigma^2$, $\kappa_{r,i}=\kappa_r$ and

$$\gamma_{r-2}=\frac{1}{n^{\frac{1}{2}r-1}}\left(\frac{\kappa_r}{\sigma^r}\right)$$

ASYMPTOTIC EXPANSION FOR THE INVERSE FUNCTION OF AN ARBITRARY DISTRIBUTION FUNCTION

Let the cumulative distribution function of $Y=\sum_{i=1}^{n} Y_i$ be denoted by $F(y)$. Then the (Cornish-Fisher) asymptotic expansion with respect to n for the value of y_p such that $F(y_p)=1-p$ is

15.25 $$y_p\sim m+\sigma w$$

where

$$\begin{aligned}w=x&+[\gamma_1 h_1(x)]\\&+[\gamma_2 h_2(x)+\gamma_1^2 h_{11}(x)]\\&+[\gamma_3 h_3(x)+\gamma_1\gamma_2 h_{12}(x)+\gamma_1^3 h_{111}(x)]\\&+[\gamma_4 h_4(x)+\gamma_2^2 h_{22}(x)+\gamma_1\gamma_3 h_{13}(x)+\gamma_1^2\gamma_2 h_{112}(x)\\&\qquad+\gamma_1^4 h_{1111}(x)]+\ldots\end{aligned}$$

and

$$Q(x)=p,\qquad \gamma_{r-2}=\frac{\kappa_r}{\kappa_2^{r/2}},\qquad r=3,4,\ldots$$

15.26

$$h_1(x)=\frac{1}{6}He_2(x)$$

$$h_2(x)=\frac{1}{24}He_3(x)$$

$$h_{11}(x)=-\frac{1}{36}[2He_3(x)+He_1(x)]$$

$$h_3(x)=\frac{1}{120}[He_4(x)]$$

$$h_{12}(x)=-\frac{1}{24}[He_4(x)+He_2(x)]$$

$$h_{111}(x)=\frac{1}{324}[12He_4(x)+19He_2(x)]$$

$$h_4(x)=\frac{1}{720}He_5(x)$$

$$h_{22}(x)=-\frac{1}{384}[3He_5(x)+6He_3(x)+2He_1(x)]$$

$$h_{13}(x)=-\frac{1}{180}[2He_5(x)+3He_3(x)]$$

$$h_{112}(x)=\frac{1}{288}[14He_5(x)+37He_3(x)+8He_1(x)]$$

$$h_{1111}(x)=-\frac{1}{7776}[252He_5(x)+832He_3(x)+227He_1(x)]$$

Terms in brackets in 15.25 are terms of the same order with respect to n. The $He_n(x)$ are the Hermite polynomials.

15.27

$$He_n(x)=(-1)^n\frac{Z^{(n)}(x)}{Z(x)}=n!\sum_{m=0}^{\left[\frac{n}{2}\right]}\frac{(-1)^m}{2^m m!(n-2m)!}x^{n-2m}$$

In the following auxiliary table, the polynomial functions $h_1(x)$, $h_2(x)$ . . . $h_{1111}(x)$ are tabulated for

$p=.25, .1, .05, .025, .01, .005, .0025, .001, .0005.$

15.25

AUXILIARY COEFFICIENTS FOR USE WITH CORNISH-FISHER ASYMPTOTIC EXPANSION

	p								
	.25	.10	.05	.025	.01	.005	.0025	.001	.0005
x	.67449	1.28155	1.64485	1.95996	2.32635	2.57583	2.80703	3.09022	3.29053
$h_1(x)$	−.09084	.10706	.28426	.47358	.73532	.93915	1.14657	1.42491	1.63793
$h_2(x)$	−.07153	−.07249	−.02018	.06872	.23379	.39012	.57070	.84331	1.07320
$h_{11}(x)$	.07663	.06106	−.01878	−.14607	−.37634	−.59171	−.83890	−1.21025	−1.52234
$h_3(x)$	.00398	−.03464	−.04928	−.04410	−.00152	.06010	.14841	.30746	.46059
$h_{12}(x)$	.00282	.14644	.17532	.10210	−.17621	−.53531	−1.02868	−1.89355	−2.71243
$h_{111}(x)$	−.01428	−.11629	−.11900	−.02937	.25195	.59757	1.06301	1.86787	2.62337
$h_4(x)$	.00998	.00227	−.01082	−.02357	−.03176	−.02621	−.00666	.04591	.10950
$h_{22}(x)$	−.03285	.00776	.05985	.09659	.07888	−.01226	−.19116	−.59060	−1.03555
$h_{13}(x)$	−.05126	.01086	.09462	.16106	.16058	.05366	−.17498	−.70464	−1.30531
$h_{112}(x)$	.14764	−.10858	−.39517	−.55856	−.32621	.35696	1.60445	4.29304	7.23307
$h_{1111}(x)$	−.06898	.09585	.25623	.31624	.07286	−.46534	−1.39199	−3.32708	−5.40702

BIVARIATE NORMAL PROBABILITY FUNCTION

$$g(x, y, \rho) = [2\pi\sqrt{1-\rho^2}]^{-1} \exp -\frac{1}{2}\left(\frac{x^2-2\rho xy+y^2}{1-\rho^2}\right)$$

$$g(x, y, \rho) = (1-\rho^2)^{-\frac{1}{2}} Z(x) Z\left(\frac{y-\rho x}{\sqrt{1-\rho^2}}\right)$$

$$L(h, k, \rho) = \int_h^\infty dx \int_k^\infty g(x, y, \rho)dy$$

$$= \int_h^\infty Z(x)dx \int_w^\infty Z(w)\,dw, \qquad w = \left(\frac{k-\rho x}{\sqrt{1-\rho^2}}\right)$$

$$L(-h, -k, \rho) = \int_{-\infty}^h dx \int_{-\infty}^k g(x, y, \rho)dy$$

$$L(-h, k, -\rho) = \int_{-\infty}^h dx \int_k^\infty g(x, y, \rho)dy$$

$$L(h, -k, -\rho) = \int_h^\infty dx \int_{-\infty}^k g(x, y, \rho)dy$$

$$L(h, k, \rho) = L(k, h, \rho)$$

15.28 $$L(-h, k, \rho) + L(h, k, -\rho) = Q(k)$$

$$L(-h, -k, \rho) - L(h, k, \rho) = P(k) - Q(h)$$

$$2[L(h, k, \rho) + L(h, k, -\rho) + P(h) - Q(k)] - 1 = \int_{-h}^h dx \int_{-k}^k g(x, y, \rho)dy$$

PROBABILITY FUNCTION WITH MEANS m_x, m_y, VARIANCES σ^2_x, σ^2_y AND CORRELATION ρ

The random variables X, Y are said to be distributed as a bivariate Normal distribution with means and variances (m_x, m_y) and (σ_x^2, σ_y^2) and correlation ρ if the joint probability that X is less than or equal to h and Y less than or equal to k is given by

$$Pr\{X \leq h, Y \leq k\} = \frac{1}{\sigma_x \sigma_y} \int_{-\infty}^{\frac{h-m_x}{\sigma_x}} \int_{-\infty}^{\frac{k-m_y}{\sigma_y}} g(s, t, \rho) ds\, dt$$

$$= L\left(-\left(\frac{h-m_x}{\sigma_x}\right), -\left(\frac{k-m_y}{\sigma_y}\right), \rho\right)$$

The probability density function is

$$\frac{1}{2\pi\sigma_x\sigma_y\sqrt{1-\rho^2}} \exp \frac{-Q}{2(1-\rho^2)} = \frac{1}{\sigma_x\sigma_y} g\left(\frac{x-m_x}{\sigma_x}, \frac{y-m_y}{\sigma_y}, \rho\right)$$

where

$$Q = \frac{(x-m_x)^2}{\sigma_x^2} - \frac{2\rho(x-m_x)(y-m_y)}{\sigma_x\sigma_y} + \frac{(y-m_y)^2}{\sigma_y^2}$$

CIRCULAR NORMAL PROBABILITY DENSITY FUNCTION

$$\frac{1}{\sigma^2} g\left(\frac{x-m_x}{\sigma}, \frac{y-m_y}{\sigma}, 0\right) =$$

$$\frac{1}{2\pi\sigma^2} \exp - \frac{(x-m_x)^2 + (y-m_y)^2}{2\sigma^2}$$

SPECIAL VALUES OF L(h, k, ρ)

$$L(h, k, 0) = Q(h)Q(k)$$

$$L(h, k, -1) = 0 \qquad (h+k \geq 0)$$

$$L(h, k, -1) = P(h) - Q(k) \qquad (h+k \leq 0)$$

$$L(h, k, 1) = Q(h) \qquad (k \leq h)$$

$$L(h, k, 1) = Q(k) \qquad (k \geq h)$$

$$L(0, 0, \rho) = \frac{1}{4} + \frac{\arcsin \rho}{2\pi}$$

L(h, k, ρ) AS A FUNCTION OF L(h, O, ρ)

15.29

$$L(h,k,\rho)=L\left(h,0,\frac{(\rho h-k)(\operatorname{sgn} h)}{\sqrt{h^2-2\rho hk+k^2}}\right)$$
$$+L\left(k,0,\frac{(\rho k-h)(\operatorname{sgn} k)}{\sqrt{h^2-2\rho hk+k^2}}\right)$$
$$-\begin{cases}0 & \text{if } hk>0 \text{ or } hk=0 \\ & \text{and } h+k\geq 0 \\ \frac{1}{2} & \text{otherwise}\end{cases}$$

where sgn $h=1$ if $h\geq 0$ and sgn $h=-1$ if $h<0$.

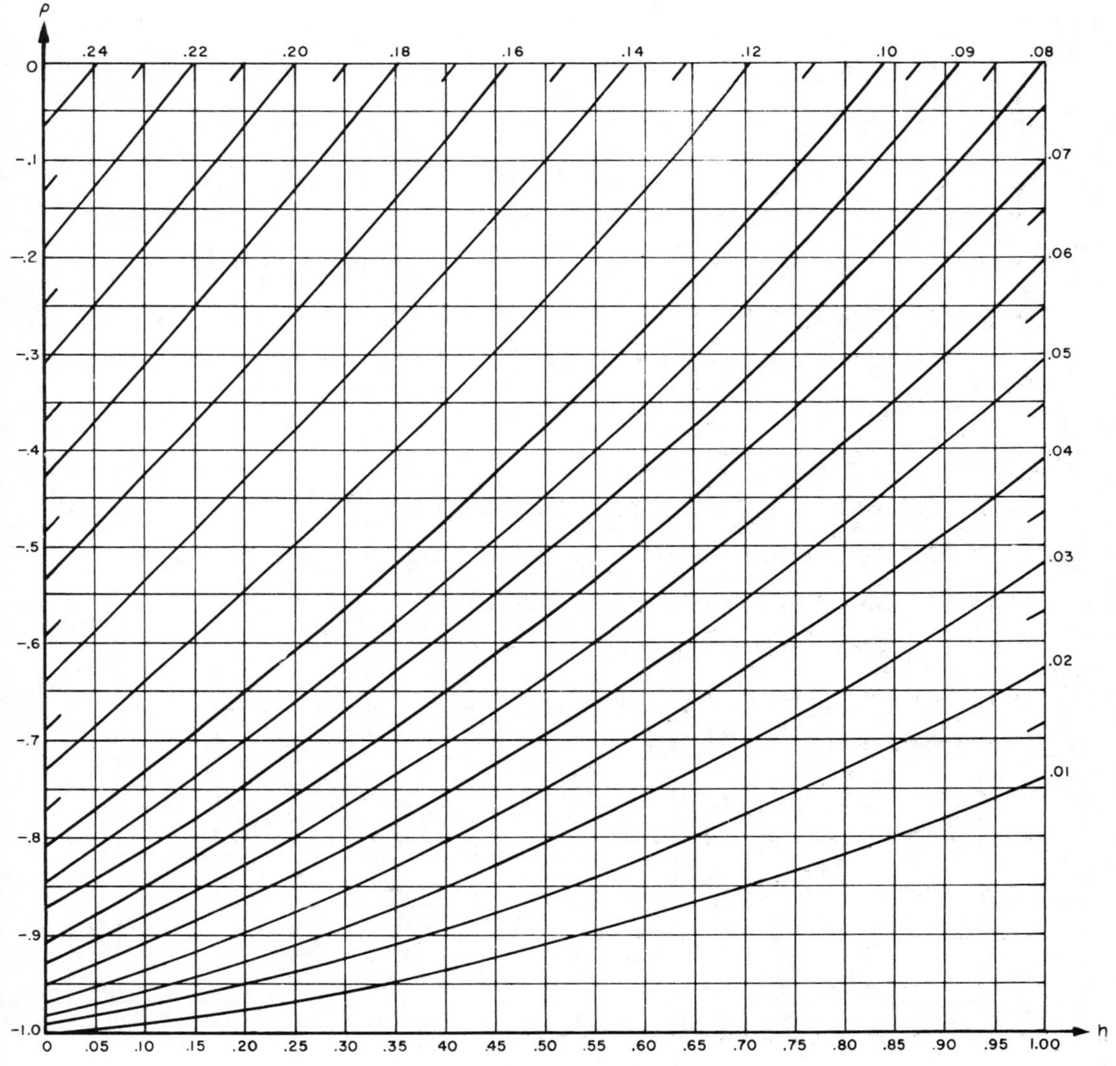

Fig. 2 $L(h, 0, \rho)$ *for* $0 \leq h \leq 1$ *and* $-1 \leq \rho \leq 0$.

Values for $h<0$ can be obtained using $L(h, 0, -\rho) = \frac{1}{2} - L(-h, 0, \rho)$.

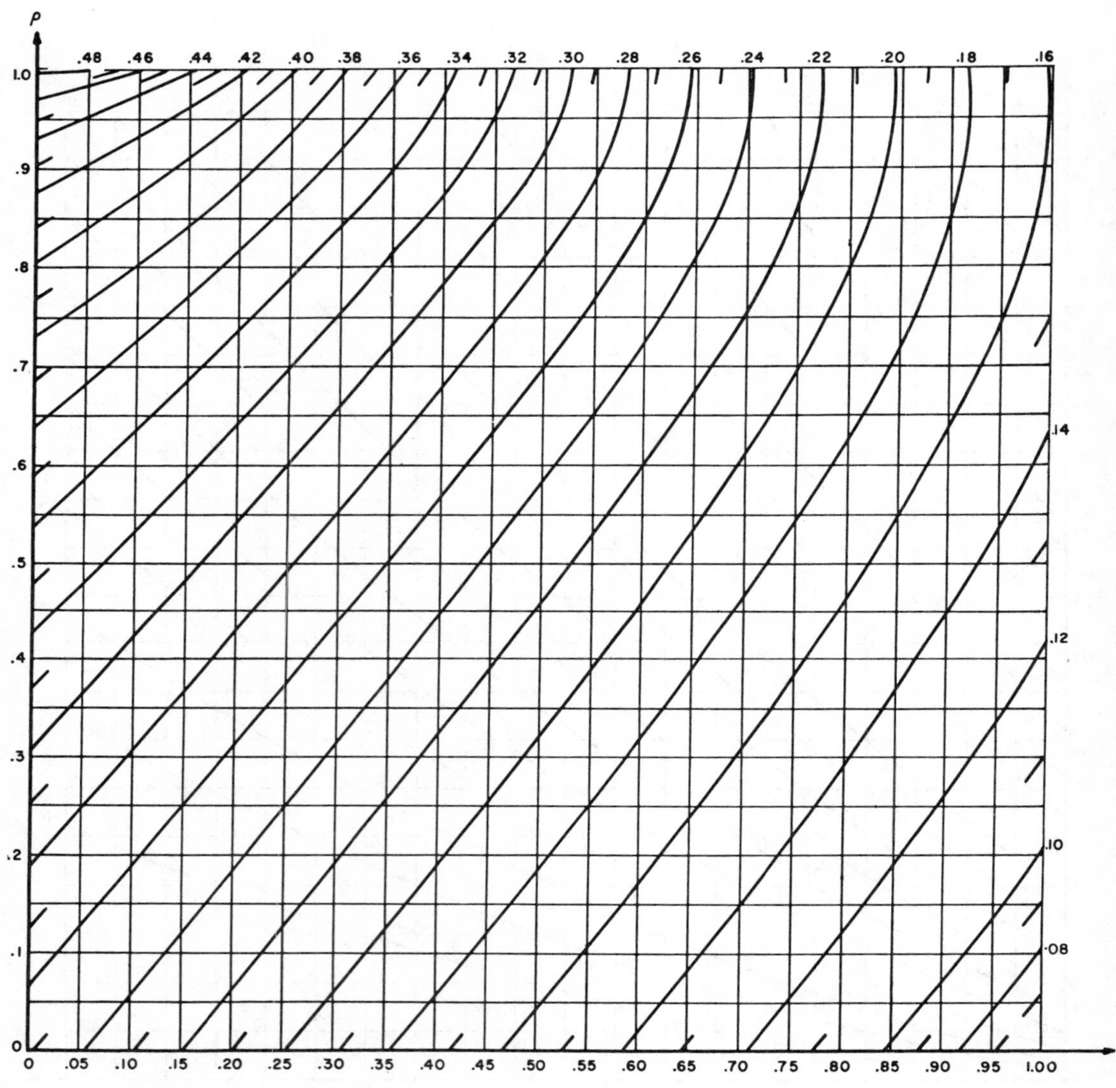

Fig. 3 $L(h,\ 0,\ \rho)$ *for* $0 \leq h \leq 1$ *and* $0 \leq \rho \leq 1$.

Values for $h<0$ can be obtained using $L(h,\ 0,\ -\rho) = \frac{1}{2} - L(-h,\ 0,\ \rho)$.

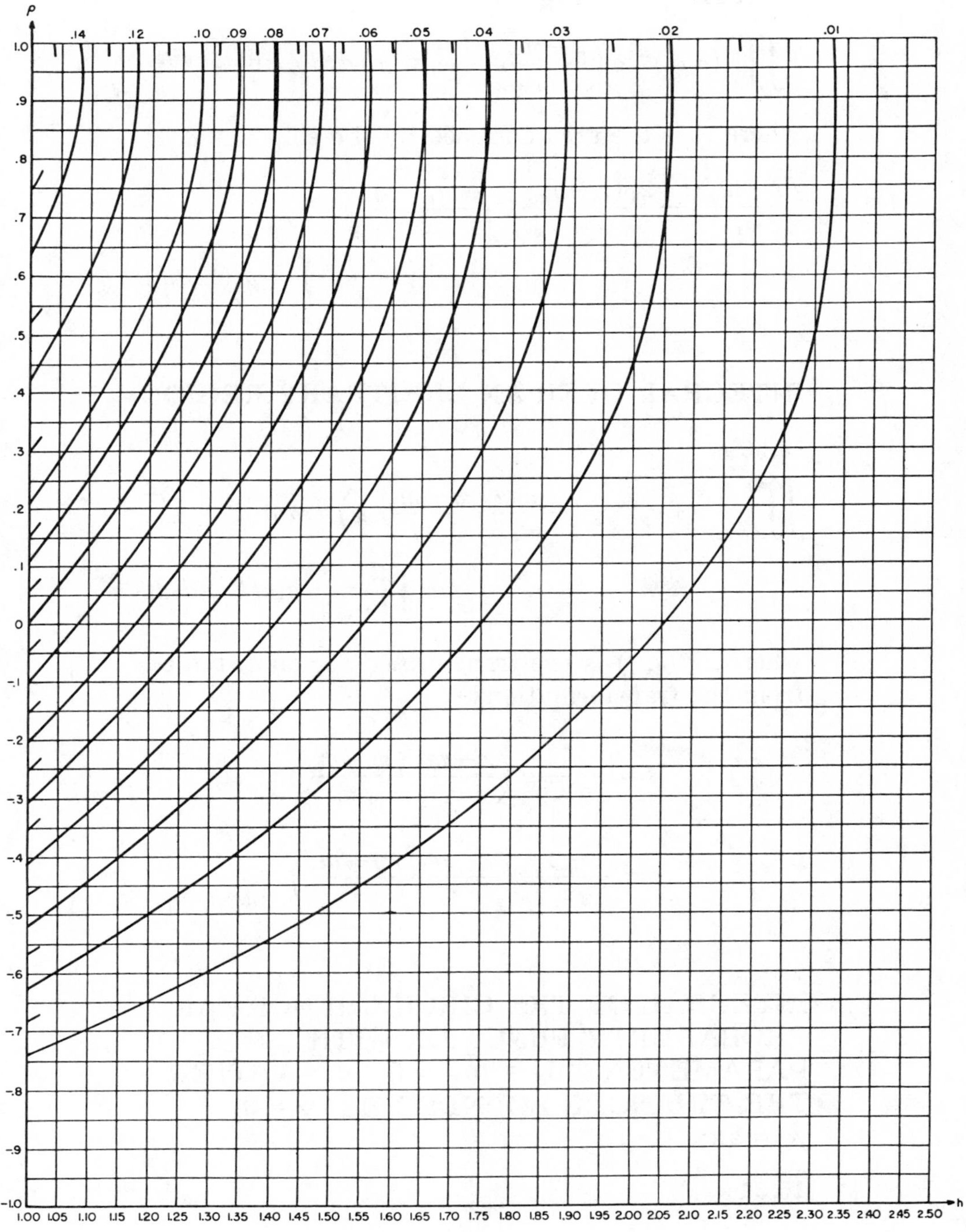

$L(h,\ 0,\ \rho)$ *for* $h \geq 1$ *and* $-1 \leq \rho \leq 1.$

Values for $h<0$ can be obtained using $L(h,\ 0,\ -\rho) = \frac{1}{2} - L(-h,\ 0,\ \rho.)$

INTEGRAL OVER AN ELLIPSE WITH CENTER AT (m_x, m_y)

$$\iint_A (\sigma_x\sigma_y)^{-1} g\left(\frac{x-m_x}{\sigma_x}, \frac{y-m_y}{\sigma_y}, \rho\right) dxdy = 1 - e^{-a^2/2}$$

where A is the area enclosed by the ellipse

$$\left(\frac{x-m_x}{\sigma_x}\right)^2 - \frac{2\rho(x-m_x)(y-m_y)}{\sigma_x\sigma_y} + \left(\frac{y-m_y}{\sigma_y}\right)^2 = a^2(1-\rho^2)$$

INTEGRAL OVER AN ARBITRARY REGION

15.30

$$\iint_{A(x,y)} (\sigma_x\sigma_y)^{-1} g\left(\frac{x-m_x}{\sigma_x}, \frac{y-m_y}{\sigma_y}, \rho\right) dxdy = \iint_{A^*(s,t)} g(s, t, o) dsdt$$

where $A^*(s, t)$ is the transformed region obtained from the transformation

$$s = \frac{1}{\sqrt{2+2\rho}}\left(\frac{x-m_x}{\sigma_x} + \frac{y-m_y}{\sigma_y}\right)$$

$$t = \frac{-1}{\sqrt{2-2\rho}}\left(\frac{x-m_x}{\sigma_x} - \frac{y-m_y}{\sigma_y}\right)$$

INTEGRAL OF THE CIRCULAR NORMAL PROBABILITY FUNCTION WITH PARAMETERS $m_x = m_y = 0$, $\sigma = 1$ OVER THE TRIANGLE BOUNDED BY $y = 0$, $y = ax$, $x = h$

15.31

$$V(h, ah) = \frac{1}{2\pi}\int_0^h \int_0^{ax} e^{-\frac{1}{2}(x^2+y^2)} dxdy$$

$$= \frac{1}{4} + L(h, 0, \rho) - L(0, 0, \rho) - \frac{1}{2} Q(h)$$

where

$$\rho = -\frac{a}{\sqrt{1+a^2}}$$

INTEGRAL OF CIRCULAR NORMAL DISTRIBUTION OVER AN OFFSET CIRCLE WITH RADIUS Rσ AND CENTER A DISTANCE rσ FROM (m_x, m_y)

15.32

$$\int_A \int \sigma^{-2} g\left(\frac{x-m_x}{\sigma}, \frac{y-m_y}{\sigma}, 0\right) dxdy = P(R^2|2, r^2)$$

where $P(R^2|2, r^2)$ is the c.d.f. of the non-central χ^2 distribution (see 15.46) with $\nu=2$ degrees of freedom and noncentrality parameter r^2.

APPROXIMATION TO $P(R^2|2, r^2)$

15.33

	Approximation	*Condition*
	$\frac{2R^2}{4+R^2} \exp - \frac{2r^2}{4+R^2}$	$R<1$
15.34	$P(x_1)$	$R>1$
15.35	$P(x_2)$	$R>5$

$$x_1 = \frac{[R^2/(2+r^2)]^{1/3} - \left[1 - \frac{2}{9}\frac{2+2r^2}{(2+r^2)^2}\right]}{\left[\frac{2}{9}\frac{2+2r^2}{(2+r^2)^2}\right]^{\frac{1}{2}}}$$

$x_2 = R - \sqrt{r^2-1}$ R, r both large

INEQUALITY

$$Q(h) - \frac{1-\rho^2}{\rho h - k} Z(k) \left[Q\left(\frac{h-\rho k}{\sqrt{1-\rho^2}}\right)\right] < L(h, k, \rho) < Q(h)$$

where

$$\rho h - k > 0, \qquad 0 < \rho < 1.$$

SERIES EXPANSION

$$L(h, k, \rho) = Q(h)Q(k) + \sum_{n=0}^{\infty} \frac{Z^{(n)}(h)Z^{(n)}(k)}{(n+1)!} \rho^{n+1}$$

CHI-SQUARE PROBABILITY FUNCTION

15.36

$$P(\chi^2|\nu) = \left[2^{\nu/2}\Gamma\left(\frac{\nu}{2}\right)\right]^{-1} \int_0^{\chi^2} (t)^{\frac{\nu}{2}-1} e^{-\frac{t}{2}} dt$$

$$(0 \leq \chi^2 < \infty)$$

$$Q(\chi^2|\nu) = 1 - P(\chi^2|\nu) \qquad (0 \leq \chi^2 < \infty)$$

$$= \left[2^{\nu/2}\,\Gamma\left(\frac{\nu}{2}\right)\right]^{-1} \int_{\chi^2}^{\infty} (t)^{\frac{\nu}{2}-1} e^{-\frac{t}{2}}\, dt$$

RELATION TO NORMAL DISTRIBUTION

Let $X_1, X_2, \ldots, X_\nu$ be independent and identically distributed random variables each following a normal distribution with mean zero and unit variance. Then $X^2 = \sum_{i=1}^{\nu} X_i^2$ is said to follow the chi-square distribution with ν degrees of freedom and the probability that $X^2 \leq \chi^2$ is given by $P(\chi^2|\nu)$.

CUMULANTS

$$\kappa_{n+1} = 2^n n!\nu \qquad (n = 0, 1, \ldots)$$

SERIES EXPANSIONS

$$Q(\chi^2|\nu) = 2Q(\chi) + 2Z(\chi) \sum_{r=1}^{\frac{\nu-1}{2}} \frac{\chi^{2r-1}}{1 \cdot 3 \cdot 5 \ldots (2r-1)}$$

(ν odd) and $\chi=\sqrt{\chi^2}$

$$Q(\chi^2|\nu)=\sqrt{2\pi}Z(\chi)\left\{1+\sum_{r=1}^{\frac{\nu-2}{2}}\frac{\chi^{2r}}{2\cdot 4\ldots(2r)}\right\}$$

(ν even)

$$P(\chi^2|\nu)=\left(\frac{1}{2}\chi^2\right)^{\nu/2}\frac{e^{-\chi^2/2}}{\Gamma\left(\frac{\nu+2}{2}\right)}$$

$$\left\{1+\sum_{r=1}^{\infty}\frac{\chi^{2r}}{(\nu+2)(\nu+4)\cdots(\nu+2r)}\right\}$$

$$P(\chi^2|\nu)=\frac{1}{\Gamma\left(\frac{\nu}{2}\right)}\sum_{n=0}^{\infty}\frac{(-1)^n(\chi^2/2)^{\frac{\nu}{2}+n}}{n!\left(\frac{\nu}{2}+n\right)}$$

RECURRENCE AND DIFFERENTIAL RELATIONS

15.37 $$Q(\chi^2|\nu+2)=Q(\chi^2|\nu)+\frac{(\chi^2/2)^{\nu/2}e^{-\chi^2/2}}{\Gamma\left(\frac{\nu}{2}+1\right)}$$

$$\frac{\partial^m Q(\chi^2|\nu)}{\partial(\chi^2)^m}=\frac{1}{2^m}\sum_{j=0}^{m}\binom{m}{j}(-1)^{m+j}Q(\chi^2|\nu-2j)$$

CONTINUED FRACTIONS

$$Q(\chi^2|\nu)=\frac{(\chi^2)^{\nu/2}e^{-\chi^2/2}}{2^{\nu/2}\Gamma(\nu/2)}$$

$$\left\{\frac{1}{\chi^2/2+}\,\frac{1-\nu/2}{1+}\,\frac{1}{\chi^2/2+}\,\frac{2-\nu/2}{1+}\,\frac{2}{\chi^2/2+}\cdots\right\}$$

ASYMPTOTIC DISTRIBUTION FOR LARGE ν

$$P(\chi^2|\nu)\sim P(x)\qquad\text{where } x=\frac{\chi^2-\nu}{\sqrt{2\nu}}$$

ASYMPTOTIC EXPANSIONS FOR LARGE χ^2

$$Q(\chi^2|\nu)\sim\frac{(\chi^2)^{\frac{\nu}{2}-1}e^{-\chi^2/2}}{2^{\nu/2}\Gamma(\nu/2)}\sum_{j=0}^{\infty}(-1)^j\frac{\Gamma\left(1-\frac{\nu}{2}+j\right)}{\Gamma\left(1-\frac{\nu}{2}\right)}\frac{2^{j+1}}{(\chi^2)^j}$$

APPROXIMATIONS TO THE CHI-SQUARE DISTRIBUTION FOR LARGE ν

15.38

	Approximation	*Condition*
$Q(\chi^2\|\nu)\approx Q(x_1)$,	$x_1=\sqrt{2\chi^2}-\sqrt{2\nu-1}$	$(\nu>100)$

15.39

$$Q(\chi^2|\nu)\approx Q(x_2),\qquad x_2=\frac{(\chi^2/\nu)^{1/3}-\left(1-\frac{2}{9\nu}\right)}{\sqrt{2/9\nu}}\qquad(\nu>30)$$

15.40

$$Q(\chi^2|\nu)\approx Q(x_2+h_\nu),\qquad h_\nu=\frac{60}{\nu}h_{60}\qquad(\nu>30)$$

Values of h_{60}

x	h_{60}	x	h_{60}	x	h_{60}
−3.5	−.0118	−1.0	+.0006	+1.5	−.0005
−3.0	−.0067	−.5	.0006	2.0	+.0002
−2.5	−.0033	.0	+.0002	2.5	.0017
−2.0	−.0010	+.5	−.0003	3.0	.0043
−1.5	+.0001	1.0	−.0006	3.5	.0082

APPROXIMATIONS FOR THE INVERSE FUNCTION FOR LARGE ν

If $Q(\chi_p^2|\nu)=p$ and $Q(x_p)=1-P(x_p)=p$, then

	Approximation	*Condition*
15.41	$\chi_p^2\approx\frac{1}{2}\left\{x_p+\sqrt{2\nu-1}\right\}^2$	$(\nu>100)$
15.42	$\chi_p^2\approx\nu\left\{1-\frac{2}{9\nu}+x_p\sqrt{\frac{2}{9\nu}}\right\}^3$	$(\nu>30)$
15.43	$\chi_p^2\approx\nu\cdot\left\{1-\frac{2}{9\nu}+(x_p-h_\nu)\sqrt{\frac{2}{9\nu}}\right\}^3$	$(\nu>30)$

where h_ν is given by 15.40

RELATION TO OTHER FUNCTIONS

15.44 Incomplete gamma function

$$\frac{\gamma(a,x)}{\Gamma(a)}=P(\chi^2|\nu), \qquad \nu=2a,\ \chi^2=2x$$

$$\frac{\Gamma(a,x)}{\Gamma(a)}=Q(\chi^2|\nu)$$

Pearson's incomplete gamma function

$$I(u,p)=\frac{1}{\Gamma(p+1)}\int_0^{u\sqrt{p+1}} t^p e^{-t}dt=P(\chi^2|\nu)$$

$$\nu=2(p+1),\ \chi^2=2u\sqrt{p+1}$$

15.45 Poisson distribution

$$Q(\chi^2|\nu)=\sum_{j=0}^{c-1} e^{-m}\frac{m^j}{j!}, \qquad c=\frac{\nu}{2},\ m=\frac{\chi^2}{2},\ (\nu \text{ even})$$

$$Q(\chi^2|\nu)-Q(\chi^2|\nu-2)=e^{-m}\frac{m^{c-1}}{(c-1)!}$$

Pearson Type III

$$\left[\frac{ab}{e}\right]^{ab}\int_{-a}^{x}\left(1+\frac{t}{a}\right)^{ab} e^{-bt}dt=P(\chi^2|\nu)$$

$$\nu=2ab+2,\ \chi^2=2b(x+a)$$

Incomplete moments of Normal distribution

$$\int_0^x t^n Z(t)dt=\begin{cases}(n-1)!!\,\dfrac{P(\chi^2|\nu)}{2} & (n \text{ even})\\[2ex] \dfrac{(n-1)!!}{\sqrt{2\pi}}P(\chi^2|\nu) & (n \text{ odd})\end{cases} \qquad \chi^2=x^2,\ \nu=n+1$$

Generalized Laguerre Polynomials

$$n!L_n^{(\alpha)}(x)=\frac{\sum_{j=0}^{n+1}(-1)^{n+j}\binom{n+1}{j}Q(\chi^2|\nu+2-2j)}{2^n[Q(\chi^2|\nu+2)-Q(\chi^2|\nu)]} \qquad x=\chi^2/2,\ \alpha=\nu/2$$

NON-CENTRAL χ^2 DISTRIBUTION FUNCTION

15.46

$$P(\chi'^2|\nu, \lambda) = \sum_{j=0}^{\infty} e^{-\lambda/2} \frac{(\lambda/2)^j}{j!} P(\chi'^2|\nu+2j)$$

where $\lambda \geq 0$ is termed the non-centrality parameter.

RELATION OF NON-CENTRAL χ^2 DISTRIBUTION WITH $\nu = 2$ TO THE INTEGRAL OF CIRCULAR NORMAL DISTRIBUTION ($\sigma^2 = 1$) OVER AN OFFSET CIRCLE HAVING RADIUS R AND CENTER A DISTANCE $r = \sqrt{\lambda}$ FROM THE ORIGIN [15.32 - 15.35]

$$\iint_A g(x, y, 0)\, dx dy = P(\chi^2 = R^2|\nu=2, \lambda)$$

$$= 1 - \sum_{j=0}^{\infty} \frac{e^{-\lambda/2}\lambda^j}{2^j j!} Q(R^2|2+2j)$$

APPROXIMATIONS TO THE NON-CENTRAL χ^2 DISTRIBUTION

$$a = \nu + \lambda \qquad b = \frac{\lambda}{\nu+\lambda}$$

χ^2 distribution

$$P(\chi'^2|\nu, \lambda) \approx P\left(\frac{\chi^2}{1+b}\middle|\nu^*\right), \qquad \nu^* = \frac{a}{1+b}$$

Normal distribution

$$P(\chi'^2|\nu, \lambda) \approx P(x), \qquad x = \frac{(\chi'^2/a)^{1/3} - \left[1 - \frac{2}{9}\left(\frac{1+b}{a}\right)\right]}{\sqrt{\frac{2}{9}\left(\frac{1+b}{a}\right)}}$$

Normal distribution

$$P(\chi'^2|\nu, \lambda) \approx P(x), \qquad x = \left[\frac{2\chi'^2}{1+b}\right]^{\frac{1}{2}} - \left[\frac{2a}{1+b} - 1\right]^{\frac{1}{2}}$$

APPROXIMATIONS TO THE INVERSE FUNCTION OF NON-CENTRAL χ^2 DISTRIBUTION

If $Q(\chi_p'^2|\nu, \lambda) = p$, $Q(\chi_p^2|\nu^*) = p$, and $Q(x_p) = p$ then

χ^2 distribution

$$\chi_p'^2 \approx (1+b)\chi_p^2$$

Normal distribution

$$\chi_p'^2 \approx \frac{1+b}{2}\left[x_p + \sqrt{\frac{2a}{1+b} - 1}\right]^2$$

Normal distribution

$$\chi_p'^2 \approx a\left[x_p\sqrt{\frac{2}{9}\left(\frac{1+b}{a}\right)} + 1 - \frac{2}{9}\left(\frac{1+b}{a}\right)\right]^3$$

PROPERTIES OF CHI-SQUARE, NON-CENTRAL CHI-SQUARE, AND RELATED QUANTITIES

$$a=\nu+\lambda \qquad b=\frac{\lambda}{\nu+\lambda}$$

$$\psi(z)=\frac{d}{dz}\ln\Gamma(z), \qquad \psi^{s}(z)=\frac{d^{s}}{dz^{s}}\psi(z)$$

	Variable	Mean	Variance	Coefficient of skewness (γ_1)	Coefficient of excess (γ_2)
15.47	χ^2	ν	2ν	$\frac{2^{3/2}}{\sqrt{\nu}}$	$12\nu^{-1}$
15.48	$\sqrt{2\chi^2}$	$(2\nu-1)^{\frac{1}{2}}\{1+[16\nu(\nu-1)]^{-1}\}+O(\nu^{-7/2})$	$1-\frac{1}{4\nu}-\frac{1}{8\nu^2}+\frac{5}{64\nu^3}-O(\nu^{-4})$	$\frac{1}{\sqrt{2\nu}}\left[1+\frac{5}{8\nu}-\frac{1}{128\nu^2}\right]+O(\nu^{-7/2})$	$\frac{3}{2^2}\frac{1}{\nu^2}\left[1+\frac{3}{2\nu}\right]+O(\nu^{-4})$
	$(\chi^2/\nu)^{1/3}$	$1-\frac{2}{3^2\nu}+\frac{80}{3^7\nu^3}+O(\nu^{-4})$	$\frac{2}{3^2\nu}-\frac{104}{3^7\nu^3}+O(\nu^{-4})$	$\frac{2^{7/2}}{3^3\nu^{3/2}}\left[1+\frac{8}{3^2\nu}\right]+O(\nu^{-7/2})$	$-\frac{4}{9\nu}\left[1+\frac{16}{9\nu}\right]+O(\nu^{-3})$
	$\ln(\chi^2/\nu)$	$\psi\left(\frac{\nu}{2}\right)-\ln\left(\frac{\nu}{2}\right)=-\frac{1}{\nu}-\frac{1}{3\nu^2}+O(\nu^{-4})$	$\psi'\left(\frac{\nu}{2}\right)=\frac{2}{\nu-1}\left[1-\frac{1}{3(\nu-1)^2}\right]+O((\nu-1)^{-5})$	$\frac{\psi''\left(\frac{\nu}{2}\right)}{\psi'\left(\frac{\nu}{2}\right)^{3/2}}=-\sqrt{\frac{2}{\nu-1}}\left[1-\frac{1}{2(\nu-1)^2}\right]+O((\nu-1)^{-9/2})$	$\frac{\psi^{(3)}\left(\frac{\nu}{2}\right)}{\psi'\left(\frac{\nu}{2}\right)^2}=\frac{4}{\nu-1}\left[1+\frac{4}{3(\nu-1)^2}\right]+O((\nu-1)^{-5})$
	χ'^2	a	$2a(1+b)$	$\left(\frac{2}{1+b}\right)^{3/2}(1+2b)a^{-\frac{1}{2}}$	$\frac{12}{a}\frac{(1+3b)}{(1+b)^2}$
	$\sqrt{2\chi'^2}$	$[2a-(1+b)]^{\frac{1}{2}}+O(a^{-3/2})$	$(1+b)-\frac{a^{-1}}{4}[8b+(1+b)(1-7b)]+O(a^{-2})$	$\frac{a^{-\frac{1}{2}}(1-b)(1+3b)}{2^{\frac{1}{2}}(1+b)^{3/2}}+O(a^{-1})$	$\frac{3b(b+2)}{(1+b)^2}+O(a^{-2})$
	$(\chi'^2/a)^{1/3}$	$1-\frac{2}{3^2}\frac{1+b}{a}-\frac{40}{3^4}\frac{b^2}{a^2}+O(a^{-3})$	$\frac{2}{9}a^{-1}(1+b)+\frac{16}{27}a^{-2}b^2+O(a^{-3})$	$\left(\frac{2}{1+b}\right)^{3/2}b^2a^{-\frac{1}{2}}+O(a^{-3/2})$	$-\frac{4}{3^2}\frac{(1+3b+12b^2-44b^3)}{a(1+b)^2}-O(a^{-2})$

INCOMPLETE BETA FUNCTION

$$I_x(a,b)=\frac{1}{B(a,b)}\int_0^x t^{a-1}(1-t)^{b-1}dt \qquad (0\leq x\leq 1)$$

$$I_x(a,b)=1-I_{1-x}(b,a)$$

RELATION TO THE CHI-SQUARE DISTRIBUTION

If X_1^2 and X_2^2 are independent random variables following chi-square distributions 15.36 with ν_1 and ν_2 degrees of freedom respectively, then $\frac{X_1^2}{X_1^2+X_2^2}$ is said to follow a beta distribution with ν_1 and ν_2 degrees of freedom and has the distribution function

$$P\left\{\frac{X_1^2}{X_1^2+X_2^2}\leq x\right\}=\frac{1}{B(a,b)}\int_0^x t^{a-1}(1-t)^{b-1}dt$$

$$=I_x(a,b) \qquad a=\frac{\nu_1}{2},\ b=\frac{\nu_2}{2}$$

SERIES EXPANSIONS $(0<x<1)$

$$I_x(a,b)=\frac{x^a(1-x)^b}{aB(a,b)}\left\{1+\sum_{n=0}^{\infty}\frac{B(a+1,n+1)}{B(a+b,n+1)}x^{n+1}\right\}$$

$$I_x(a,b)=\frac{x^a(1-x)^{b-1}}{aB(a,b)}\left\{1+\sum_{n=0}^{\infty}\frac{B(a+1,n+1)}{B(b-n-1,n+1)}\left(\frac{x}{1-x}\right)^{n+1}\right\}$$

$$=\frac{x^a(1-x)^{b-1}}{aB(a,b)}\left\{1+\sum_{n=0}^{s-2}\frac{B(a+1,n+1)}{B(b-n-1,n+1)}\left(\frac{x}{1-x}\right)^{n+1}\right\}+I_x(a+s,b-s)$$

15.49

$$1-I_x(a,b)=I_{1-x}(b,a)$$

$$=\frac{(1-x)^b}{B(a,b)}\sum_{i=0}^{a-1}(-1)^i\binom{a-1}{i}\frac{(1-x)^i}{b+i}\quad\text{(integer a)}$$

15.50

$$1-I_x(a,b)=I_{1-x}(b,a)$$

$$=(1-x)^{a+b-1}\sum_{i=0}^{a-1}\binom{a+b-1}{i}\left(\frac{x}{1-x}\right)^i\quad\text{(integer a)}$$

CONTINUED FRACTIONS

$$I_x(a,b)=\frac{x^a(1-x)^b}{aB(a,b)}\left\{\frac{1}{1+}\,\frac{d_1}{1+}\,\frac{d_2}{1+}\cdots\right\}$$

$$d_{2m+1}=-\frac{(a+m)(a+b+m)}{(a+2m)(a+2m+1)}\,x$$

$$d_{2m}=\frac{m(b-m)}{(a+2m-1)(a+2m)}\,x$$

Best results are obtained when $x<\frac{a-1}{a+b-2}$. Also the $4m$ and $4m+1$ convergents are less than $I_x(a,b)$ and the $4m+2$, $4m+3$ convergents are greater than $I_x(a,b)$.

$$I_x(a,b)=\frac{x^a(1-x)^{b-1}}{aB(a,b)}\left[\frac{e_1}{1+}\,\frac{e_2}{1+}\,\frac{e_3}{1+}\cdots\right]$$

$$x<1\qquad\qquad e_1=1$$

$$e_{2m}=-\frac{(a+m-1)(b-m)}{(a+2m-2)(a+2m-1)}\,\frac{x}{1-x}$$

$$e_{2m+1}=\frac{m(a+b-1+m)}{(a+2m-1)(a+2m)}\,\frac{x}{1-x}$$

RECURRENCE RELATIONS

$$I_x(a,b)=xI_x(a-1,b)+(1-x)I_x(a,b-1)$$

$$I_x(a,b)=\frac{1}{x}\{I_x(a+1,b)-(1-x)I_x(a+1,b-1)\}$$

$$I_x(a,b)=\frac{b}{a(1-x)+b}\{I_x(a,b+1)$$
$$+(1-x)I_x(a+1,b-1)\}$$

$$I_x(a,b)=\frac{1}{a+b}\,\{aI_x(a+1,b)+bI_x(a,b+1)\}$$

$$I_x(a,a)=\frac{1}{2}\,I_{1-x'}\left(a,\frac{1}{2}\right),\qquad x'=4\left(x-\frac{1}{2}\right)^2\ x\leqslant\frac{1}{2}$$

$$I_x(a,b)=\frac{\Gamma(a+b)}{\Gamma(a+1)\Gamma(b)}\,x^a(1-x)^{b-1}+I_x(a+1,b-1)$$

$$I_x(a,b)=\frac{\Gamma(a+b)}{\Gamma(a+1)\Gamma(b)}\,x^a(1-x)^b+I_x(a+1,b)$$

ASYMPTOTIC EXPANSIONS

$$1-I_x(a,b)=I_{1-x}(b,a)\sim\frac{\Gamma(b,y)}{\Gamma(b)}-\frac{1}{24N^2}\left\{\frac{y^be^{-y}}{(b-2)!}\,(b+1+y)\right\}$$
$$+\frac{1}{5760N^4}\left\{\frac{y^be^{-y}}{(b-2)!}\,[(b-3)(b-2)(5b+7)(b+1+y)\right.$$
$$\left.-(5b-7)\,(b+3+y)y^2]\right\}$$
$$y=-N\ln x,\qquad N=a+\frac{b}{2}-\frac{1}{2}$$

$$I_x(a,b)\sim\frac{\Gamma(a,w)}{\Gamma(a)}+\frac{e^{-w}w^a}{\Gamma(a)}\left\{\frac{(a-1-w)}{2b}+\frac{1}{(2b)^2}\left(\frac{a^3}{2}-\frac{5}{3}\,a^2+\frac{3}{2}\,a-\frac{1}{3}\right.\right.$$
$$\left.\left.-w\left[\frac{3}{2}\,a^2-\frac{11}{6}\,a+\frac{1}{3}\right]+w^2\left(\frac{3}{2}\,a-\frac{1}{6}\right)-\frac{1}{2}\,w^3\right)\right\}$$
$$w=b\left(\frac{x}{1-x}\right)$$

$$I_x(a,b) \sim P(y) - Z(y)\left[a_1 + \frac{a_2(y-a_1)}{1+a_2} + \frac{a_3(1+y^2/2)}{1+a_2} + \cdots\right]$$

$$a_1 = \frac{2}{3}(b-a)[(a+b-2)(a-1)(b-1)]^{-1/2}$$

$$a_2 = \frac{1}{12}\left[\frac{1}{a-1} + \frac{1}{b-1} - \frac{13}{a+b-1}\right]$$

$$a_3 = -\frac{8}{15}\left[a_1\left(a_2 + \frac{3}{a+b-2}\right)\right]$$

$$y^2 = 2\left[(a+b-1)\ln\frac{a+b-1}{a+b-2} + (a-1)\ln\frac{a-1}{(a+b-1)x} + (b-1)\ln\frac{b-1}{(a+b-1)(1-x)}\right]$$

and y is taken negative when $x < \dfrac{a-1}{a+b-2}$

APPROXIMATIONS

15.51 If $(a+b-1)(1-x) \leq .8$

$I_x(a, b) = Q(\chi^2|\nu) + \epsilon$,

$|\epsilon| < 5 \times 10^{-3}$ if $a+b > 6$

$\chi^2 = (a+b-1)(1-x)(3-x) - (1-x)(b-1)$,

$\nu = 2b$

15.52 If $(a+b-1)(1-x) \geq .8$

$I_x(a, b) = P(y) + \epsilon$,

$|\epsilon| < 5 \times 10^{-3}$ if $a+b > 6$

$$y = \frac{3\left[w_1\left(1-\frac{1}{9b}\right) - w_2\left(1-\frac{1}{9a}\right)\right]}{\left[\frac{w_1^2}{b} + \frac{w_2^2}{a}\right]^{\frac{1}{2}}},$$

$w_1 = (bx)^{1/3}$, $w_2 = [a(1-x)]^{1/3}$

APPROXIMATION TO THE INVERSE FUNCTION

15.53 If $I_{x_p}(a, b)=p$ and $Q(y_p)=p$ then

$$x_p \approx \frac{a}{a+be^{2w}}$$

$$w=\frac{y_p(h+\lambda)^{\frac{1}{2}}}{h}-\left(\frac{1}{2b-1}-\frac{1}{2a-1}\right)\left(\lambda+\frac{5}{6}-\frac{2}{3h}\right)$$

$$h=2\left(\frac{1}{2a-1}+\frac{1}{2b-1}\right)^{-1}, \qquad \lambda=\frac{y_p^2-3}{6}$$

RELATIONS TO OTHER FUNCTIONS AND DISTRIBUTIONS

Hypergeometric function

$$\frac{1}{B(a, b)}\,\frac{x^a}{a}\,F(a, 1-b; a+1; x)=I_x(a, b)$$

15.54 Binomial distribution

$$\sum_{s=a}^{n}\binom{n}{s}p^s(1-p)^{n-s}=I_p(a, n-a+1)$$

Binomial distribution

$$\binom{n}{a}p^a(1-p)^{n-a}=I_p(a, n-a+1)-I_p(a+1, n-a)$$

Negative binomial distribution

$$\sum_{s=a}^{n}\binom{n+s-1}{s}p^n q^s=I_q(a, n)$$

15.55 Student's distribution

$$\frac{1}{2}[1-A(t|\nu)]=\frac{1}{2}\,I_x\left(\frac{\nu}{2}, \frac{1}{2}\right), \qquad x=\frac{\nu}{\nu+t^2}$$

15.56 F-(variance-ratio) distribution

$$Q(F|\nu_1, \nu_2)=I_x\left(\frac{\nu_2}{2}, \frac{\nu_1}{2}\right), \qquad x=\frac{\nu_2}{\nu_2+\nu_1 F}$$

F-(VARIANCE-RATIO) DISTRIBUTION FUNCTION

$$P(F|\nu_1, \nu_2)=\frac{\nu_1^{\frac{1}{2}\nu_1}\nu_2^{\frac{1}{2}\nu_2}}{B\left(\frac{1}{2}\nu_1, \frac{1}{2}\nu_2\right)}\int_0^F t^{\frac{1}{2}(\nu_1-2)}(\nu_2+\nu_1 t)^{-\frac{1}{2}(\nu_1+\nu_2)}dt \qquad (F\geq 0)$$

$$Q(F|\nu_1, \nu_2)=1-P(F|\nu_1, \nu_2)=I_x\left(\frac{\nu_2}{2}, \frac{\nu_1}{2}\right)$$

where

$$x=\frac{\nu_2}{\nu_2+\nu_1 F}$$

RELATION TO THE CHI-SQUARE DISTRIBUTION

If X_1^2 and X_2^2 are independent random variables following chi-square distributions **15.36** with ν_1 and ν_2 degrees of freedom respectively, then the distribution of $F=\frac{X_1^2/\nu_1}{X_2^2/\nu_2}$ is said to follow the variance ratio or F-distribution with ν_1 and ν_2 degrees of freedom. The corresponding distribution function is $P(F|\nu_1, \nu_2)$.

STATISTICAL PROPERTIES

mean: $$m=\frac{\nu_2}{\nu_2-2} \qquad (\nu_2>2)$$

variance: $$\sigma^2=\frac{2\nu_2^2(\nu_1+\nu_2-2)}{\nu_1(\nu_2-2)^2(\nu_2-4)} \qquad (\nu_2>4)$$

third central moment:

$$\mu_3=\left(\frac{\nu_2}{\nu_1}\right)^3\frac{8\nu_1(\nu_1+\nu_2-2)(2\nu_1+\nu_2-2)}{(\nu_2-2)^3(\nu_2-4)(\nu_2-6)} \qquad (\nu_2>6)$$

moments about the origin:

$$\mu_n'=\left(\frac{\nu_2}{\nu_1}\right)^n\frac{\Gamma\left(\frac{\nu_1+2n}{2}\right)}{\Gamma\left(\frac{\nu_1}{2}\right)}\frac{\Gamma\left(\frac{\nu_1-2n}{2}\right)}{\Gamma\left(\frac{\nu_2}{2}\right)} \qquad (\nu_2>2n)$$

characteristic function:

$$\phi(t)=E(e^{iFt})=M\left(\frac{\nu_1}{2}, -\frac{\nu_2}{2}, -\frac{\nu_2}{\nu_1}\,it\right)$$

SERIES EXPANSIONS

$$x=\frac{\nu_2}{\nu_2+\nu_1 F}$$

15.57

$$Q(F|\nu_1, \nu_2)=x^{\nu_2/2}\left[1+\frac{\nu_2}{2}(1-x)+\frac{\nu_2(\nu_2+2)}{2\cdot 4}(1-x)^2+\ldots\right.$$

$$\left.+\frac{\nu_2(\nu_2+2)\ldots(\nu_2+\nu_1-4)}{2\cdot 4\ldots(\nu_1-2)}(1-x)^{\frac{\nu_1-2}{2}}\right] \qquad (\nu_1 \text{ even})$$

$$Q(F|\nu_1, \nu_2)=1-(1-x)^{\nu_1/2}\left[1+\frac{\nu_1}{2}x+\frac{\nu_1(\nu_1+2)}{2\cdot 4}x^2+\ldots\right.$$

$$\left.+\frac{\nu_1(\nu_1+2)\ldots(\nu_2+\nu_1-4)}{2\cdot 4\ldots(\nu_2-2)}x^{\frac{\nu_2-2}{2}}\right] \qquad (\nu_2 \text{ even})$$

$$Q(F|\nu_1, \nu_2)=x^{\frac{\nu_1+\nu_2-2}{2}}\left[1+\frac{\nu_1+\nu_2-2}{2}\left(\frac{1-x}{x}\right)\right.$$

$$+\frac{(\nu_1+\nu_2-2)(\nu_1+\nu_2-4)}{2\cdot 4}\left(\frac{1-x}{x}\right)^2+\ldots$$

$$\left.+\frac{(\nu_1+\nu_2-2)\ldots(\nu_2+2)}{2\cdot 4\ldots(\nu_1-2)}\left(\frac{1-x}{x}\right)^{\frac{\nu_1-2}{2}}\right] \qquad (\nu_1 \text{ even})$$

$$Q(F|\nu_1, \nu_2)=1-(1-x)^{\frac{\nu_1+\nu_2-2}{2}}\left[1+\frac{\nu_1+\nu_2-2}{2}\left(\frac{x}{1-x}\right)\right.$$

$$\left.+\ldots+\frac{(\nu_1+\nu_2-2)\ldots(\nu_1+2)}{2\cdot 4\ldots(\nu_2-2)}\left(\frac{x}{1-x}\right)^{\frac{\nu_2-2}{2}}\right]$$

$$(\nu_2 \text{ even})$$

15.58

$$Q(F|\nu_1, \nu_2)=1-A(t|\nu_2)+\beta(\nu_1, \nu_2) \qquad (\nu_1, \nu_2 \text{ odd})$$

$$A(t|\nu_2)=\begin{cases}\frac{2}{\pi}\left\{\theta+\sin\theta\left[\cos\theta+\frac{2}{3}\cos^3\theta+\ldots+\right.\right.\\ \qquad\left.\left.\frac{2\cdot 4\ldots(\nu_2-3)}{3.5\ldots(\nu_2-2)}\cos^{\nu_2-2}\theta\right]\right\}\text{for } \nu_2>1\\ \frac{2\theta}{\pi}\text{ for } \nu_2=1\end{cases}$$

$$\beta(\nu_1,\nu_2)=\begin{cases}\frac{2}{\sqrt{\pi}}\frac{\left(\frac{\nu_2-1}{2}\right)!}{\left(\frac{\nu_2-2}{2}\right)!}\sin\theta\cos^{\nu_2}\theta\left\{1+\right.\\ \qquad\frac{\nu_2+1}{3}\sin^2\theta+\cdots+\\ \left.\frac{(\nu_2+1)(\nu_2+3)\ldots(\nu_1+\nu_2-4)\sin^{\nu_1-3}\theta}{3\cdot 5\ldots(\nu_1-2)}\right\}\\ \qquad\text{for } \nu_2>1\\ 0\text{ for } \nu_1=1\end{cases}$$

where

$$\theta=\arctan\sqrt{\frac{\nu_1}{\nu_2}F}$$

REFLEXIVE RELATION

If $F_p(\nu_1, \nu_2)$ and $F_{1-p}(\nu_2, \nu_1)$ satisfy

$$Q(F_p(\nu_1, \nu_2)|\nu_1, \nu_2)=p$$

$$Q(F_{1-p}(\nu_2, \nu_1)|\nu_2, \nu_1)=1-p$$

15.59 then

$$F_p(\nu_1, \nu_2)=\frac{1}{F_{1-p}(\nu_2, \nu_1)}$$

RELATION TO STUDENT'S t-DISTRIBUTION (SEE STUDENT'S t-DISTRIBUTION)

$$Q(F|\nu_1=1, \nu_2)=1-A(t|\nu_2)\qquad t=\sqrt{F}$$

LIMITING FORMS

$$\lim_{\nu_2\to\infty} Q(F|\nu_1,\nu_2)=Q(\chi^2|\nu_1), \qquad \chi^2=\nu_1 F$$

$$\lim_{\nu_1\to\infty} Q(F|\nu_1,\nu_2)=P(\chi^2|\nu_2), \qquad \chi^2=\frac{\nu_2}{F}$$

APPROXIMATIONS

$$Q(F|\nu_1,\nu_2)\approx Q(x), \qquad x=\frac{F-\dfrac{\nu_2}{\nu_2-2}}{\dfrac{\nu_2}{\nu_2-2}\sqrt{\dfrac{2(\nu_1+\nu_2-2)}{\nu_1(\nu_2-4)}}}$$

(ν_1 and ν_2 large)

15.60

$$Q(F|\nu_1,\nu_2)\approx Q(x), \qquad x=\frac{\sqrt{(2\nu_2-1)\dfrac{\nu_1}{\nu_2}F}-\sqrt{2\nu_1-1}}{\sqrt{1+\dfrac{\nu_1}{\nu_2}F}}$$

15.61

$$Q(F|\nu_1,\nu_2)\approx Q(x), \qquad x=\frac{F^{1/3}\left(1-\dfrac{2}{9\nu_2}\right)-\left(1-\dfrac{2}{9\nu_1}\right)}{\sqrt{\dfrac{2}{9\nu_1}+F^{2/3}\dfrac{2}{9\nu_2}}}$$

APPROXIMATION TO THE INVERSE FUNCTION

15.62 If $Q(F_p|\nu_1,\nu_2)=p$, then

$F_p\approx e^{2w}$ where w is given by 15.53, with

$$\nu_1=2b,\ \nu_2=2a$$

NON-CENTRAL F-DISTRIBUTION FUNCTION

$$P(F'|\nu_1,\nu_2,\lambda)=\int_0^{F'} p(t|\nu_1,\nu_2,\lambda)dt=1-Q(F'|\nu_1,\nu_2,\lambda)$$

where

$$p(t|\nu_1,\nu_2,\lambda)=\sum_{j=0}^{\infty} e^{-\lambda/2}\frac{(\lambda/2)^j}{j!}\frac{(\nu_1+2j)^{\frac{\nu_1+2j}{2}}\nu_2^{\nu_2/2}}{B\left(\frac{\nu_1+2j}{2},\frac{\nu_2}{2}\right)}$$

$$\times t^{\frac{\nu_1+2j-2}{2}}[\nu_2+(\nu_1+2j)t]^{-(\nu_1+2j+\nu_2)/2}$$

and $\lambda\geq 0$ is termed the non-centrality parameter.

RELATION OF NON-CENTRAL F-DISTRIBUTION FUNCTION TO OTHER FUNCTIONS

F-distribution

$$P(F'|\nu_1,\nu_2,\lambda)=\sum_{j=0}^{\infty} e^{-\lambda/2}\frac{(\lambda/2)^j}{j!}P(F'|\nu_1+2j,\nu_2)$$

$$P(F'|\nu_1,\nu_2,\lambda=0)=P(F'|\nu_1,\nu_2)$$

Non-central t-distribution

$$P(F'|\nu_1=1,\nu_2,\lambda)=P(t'|\nu,\delta),\ t'=\sqrt{F'},\ \nu=\nu_2,\ \delta=\sqrt{\lambda}$$

Incomplete Beta function

$$P(F'|\nu_1,\nu_2)=\sum_{j=0}^{\infty} e^{-\lambda/2}\frac{(\lambda/2)^j}{j!}I_x\left(\frac{\nu_1}{2}+j,\frac{\nu_2}{2}\right),$$

$$x=\frac{\nu_1 F'}{\nu_1 F'+\nu_2}$$

Confluent hypergeometric function

$$P(F'|\nu_1,\nu_2,\lambda)=\sum_{i=0}^{\frac{\nu_2}{2}-1}\frac{2e^{-\lambda/2}}{(\nu_1+\nu_2)B\left(\frac{\nu_1}{2}+i+1,\frac{\nu_2}{2}-i\right)}\times$$

$$x^{\frac{\nu_1}{2}+1}(1-x)^{\frac{\nu_2}{2}-i-1}M\left(\frac{\nu_1+\nu_2}{2},\frac{\nu_1}{2}+i+1,\frac{\lambda x}{2}\right)$$

$$\left(\nu_2\text{ even and } x=\frac{\nu_2}{\nu_1 F'+\nu_2}\right)$$

SERIES EXPANSION

$$P(F'|\nu_1, \nu_2, \lambda) = e^{-\frac{\lambda}{2}(1-x)} x^{\frac{1}{2}(\nu_1+\nu_2-2)} \sum_{i=0}^{\frac{\nu_2}{2}-1} T_i \qquad (\nu_2 \text{ even})$$

where

$$T_0 = 1$$

$$T_1 = \frac{1}{2}(\nu_1 + \nu_2 - 2 + \lambda x)\frac{1-x}{x}$$

$$T_i = \frac{1-x}{2i}[(\nu_1 + \nu_2 - 2i + \lambda x)T_{i-1} + \lambda(1-x)T_{i-2}]$$

$$x = \frac{\nu_2}{\nu_1 F' + \nu_2}$$

LIMITING FORM

$$\lim_{\nu_2 \to \infty} P(F'|\nu_1, \nu_2, \lambda) = P(\chi'^2|\nu, \lambda), \qquad \chi'^2 = \nu_1 F', \ \nu = \nu_1$$

$$\lim_{\nu_1 \to \infty} P(F'|\nu_1, \nu_2, \lambda) = Q(\chi^2|\nu), \qquad \chi^2 = \frac{\nu_2(1+c^2)}{F'}$$

where $\lambda/\nu_1 \to c^2$ as $\nu_1 \to \infty$.

APPROXIMATIONS TO THE NON-CENTRAL F-DISTRIBUTION

$$P(F'|\nu_1, \nu_2, \lambda) \approx P(x_1), \qquad (\nu_1 \text{ and } \nu_2 \text{ large})$$

where

$$x_1 = \frac{F' - \frac{\nu_2(\nu_1+\lambda)}{\nu_1(\nu_2-2)}}{\frac{\nu_2}{\nu_1}\left[\frac{2}{(\nu_2-2)(\nu_2-4)}\left\{\frac{(\nu_1+\lambda)^2}{\nu_2-2} + \nu_1 + 2\lambda\right\}\right]^{\frac{1}{2}}}$$

$$P(F'|\nu_1, \nu_2, \lambda) \approx P(F|\nu_1, \nu_2),$$

$$F = \frac{\nu_1}{\nu_1+\lambda} F', \ \nu_1 = \frac{(\nu_1+\lambda)^2}{\nu_1+2\lambda}$$

15.63

$$P(F'|\nu_1,\nu_2,\lambda)\approx P(x_2),$$

$$x_2=\frac{\left[\frac{\nu_1F'}{(\nu_1+\lambda)}\right]^{1/3}\left[1-\frac{2}{9\nu_2}\right]-\left[1-\frac{2(\nu_1+2\lambda)}{9(\nu_1+\lambda)^2}\right]}{\left[\frac{2}{9}\frac{\nu_1+2\lambda}{(\nu_1+\lambda)^2}+\frac{2}{9\nu_2}\left(\frac{\nu_1}{\nu_1+\lambda}F'\right)^{2/3}\right]^{\frac{1}{2}}}$$

STUDENT'S t-DISTRIBUTION

If X is a random variable following a normal distribution with mean zero and variance unity, and χ^2 is a random variable following an independent chi-square distribution with ν degrees of freedom, then the distribution of the ratio $\frac{X}{\sqrt{\chi^2/\nu}}$ is called Student's t-distribution with ν degrees of freedom. The probability that $\frac{X}{\sqrt{\chi^2/\nu}}$ will be less in absolute value than a fixed constant t is

$$A(t|\nu)=P_r\left\{\left|\frac{X}{\sqrt{\chi^2/\nu}}\right|\leq t\right\}$$

$$=\left[\sqrt{\nu}B\left(\frac{1}{2},\frac{\nu}{2}\right)\right]^{-1}\int_{-t}^{t}\left(1+\frac{x^2}{\nu}\right)^{-\frac{\nu+1}{2}}dx$$

$$=1-I_x\left(\frac{\nu}{2},\frac{1}{2}\right),\qquad (0\leq t<\infty)$$

where

$$x=\frac{\nu}{\nu+t^2}$$

STATISTICAL PROPERTIES

mean: $m=0$

variance: $\sigma^2=\frac{\nu}{\nu-2}$ $\qquad (\nu>2)$

skewness: $\gamma_1=0$

excess: $\gamma_2=\dfrac{6}{\nu-4}$ $(\nu>4)$

moments:

$$\mu_{2n}=\frac{1\cdot 3\ .\ .\ .\ (2n-1)\nu^n}{(\nu-2)(\nu-4)\ .\ .\ .\ (\nu-2n)} \qquad (\nu>2n)$$

$$\mu_{2n+1}=0$$

characteristic function:

$$\phi(t)=E\left[\exp\left(it\ \frac{X}{\sqrt{\chi^2/\nu}}\right)\right]=\frac{\left(\frac{|t|}{2\sqrt{\nu}}\right)^{\nu/2}}{\pi\Gamma(\nu/2)}\ Y_{\frac{\nu}{2}}\left(\frac{|t|}{\sqrt{\nu}}\right)$$

SERIES EXPANSIONS

$$\left(\theta=\arctan\frac{t}{\sqrt{\nu}}\right)$$

$$A(t|\nu)=\begin{cases}\dfrac{2}{\pi}\left\{\theta+\sin\theta\left[\cos\theta+\dfrac{2}{3}\cos^3\theta+\ .\ .\ .\right.\right.\\ \qquad\left.\left.+\dfrac{2\cdot 4\ .\ .\ .\ (\nu-3)}{1\cdot 3\ .\ .\ .\ (\nu-2)}\cos^{\nu-2}\theta\right]\right\} \\ \qquad\qquad (\nu>1 \text{ and odd}) \\ \dfrac{2}{\pi}\theta \qquad (\nu=1)\end{cases}$$

$$A(t|\nu)=\sin\theta\left\{1+\frac{1}{2}\cos^2\theta+\frac{1\cdot 3}{2\cdot 4}\cos^4\theta+\ .\ .\ .\right.$$
$$\left.+\frac{1\cdot 3\cdot 5\ .\ .\ .\ (\nu-3)}{2\cdot 4\cdot 6\ .\ .\ .\ (\nu-2)}\cos^{\nu-2}\theta\right\} \qquad (\nu \text{ even})$$

ASYMPTOTIC EXPANSION FOR THE INVERSE FUNCTION

If $A(t_p|\nu)=1-2p$ and $Q(x_p)=p$, then

$$t_p\sim x_p+\frac{g_1(x_p)}{\nu}+\frac{g_2(x_p)}{\nu^2}+\frac{g_3(x_p)}{\nu^3}+\ .\ .\ .$$

$$g_1(x)=\frac{1}{4}(x^3+x)$$

$$g_2(x)=\frac{1}{96}\,(5x^5+16x^3+3x)$$

$$g_3(x)=\frac{1}{384}\,(3x^7+19x^5+17x^3-15x)$$

$$g_4(x)=\frac{1}{92160}\,(79x^9+776x^7+1482x^5-1920x^3-945x)$$

LIMITING DISTRIBUTION

$$\lim_{\nu\to\infty} A(t|\nu)=\frac{1}{\sqrt{2\pi}}\int_{-t}^{t} e^{-x^2/2}dx=A(t)$$

APPROXIMATION FOR LARGE VALUES OF t AND $\nu \leq 5$

$$A(t|\nu)\approx 1-2\left\{\frac{a_\nu}{t^\nu}+\frac{b_\nu}{t^{\nu+1}}\right\}$$

ν	1	2	3	4	5
a_ν	.3183	.4991	1.1094	3.0941	9.948
b_ν	.0000	.0518	−.0460	−2.756	−14.05

APPROXIMATION FOR LARGE ν

$$A(t|\nu)\approx 2P(x)-1, \qquad x=\frac{t\left(1-\frac{1}{4\nu}\right)}{\sqrt{1+\frac{t^2}{2\nu}}}$$

NON-CENTRAL t-DISTRIBUTION

$$P(t'|\nu,\delta)=$$

$$\frac{1}{\sqrt{\nu}B\left(\frac{1}{2},\frac{\nu}{2}\right)}\int_{-\infty}^{t'}\left(\frac{\nu}{\nu+x^2}\right)^{\frac{\nu+1}{2}} e^{-\frac{1}{2}\frac{\nu\delta^2}{\nu+x^2}}Hh_\nu\left(\frac{-\delta x}{\sqrt{\nu+x^2}}\right)dx$$

$$=1-\sum_{j=0}^{\infty} e^{-\delta^2/2}\,\frac{(\delta^2/2)^j}{2j!}\,I_x\left(\frac{\nu}{2},\frac{1}{2}+j\right), \qquad x=\frac{\nu}{\nu+t'^2}$$

where δ is termed the non-centrality parameter.

APPROXIMATION TO THE NON-CENTRAL t-DISTRIBUTION

$$P(t'|\nu,\delta)\approx P(x) \qquad \text{where } x=\frac{t'\left(1-\frac{1}{4\nu}\right)-\delta}{\left(1+\frac{t'^2}{2\nu}\right)^{\frac{1}{2}}}$$

METHODS OF GENERATING RANDOM NUMBERS AND THEIR APPLICATIONS

Random digits are digits generated by repeated independent drawings from the population 0, 1, 2, . . ., 9 where the probability of selecting any digit is one-tenth. This is equivalent to putting 10 balls, numbered from 0 to 9, into an urn and drawing one ball at a time, replacing the ball after each drawing. The recorded set of numbers forms a collection of random digits. Any group of n successive random digits is known as a *random number*.

Several lengthy tables of random digits are available. However, the use of random numbers in electronic computers has resulted in a need for random numbers to be generated in a completely deterministic way. The numbers so generated are termed pseudo-random numbers. The quality of pseudo-random numbers is determined by subjecting the numbers to several statistical tests The purpose of these statistical tests is to detect any properties of the pseudo-random numbers which are different from the (conceptual) properties of random numbers.

Experience has shown that the congruence method is the most preferable device for generating random numbers on a computer. Let the sequence of pseudo-random numbers be denoted by $\{X_n\}$, $n=0, 1, 2, \ldots$. Then the congruence method of generating pseudo-random numbers is

$$X_{n+1}=aX_n+b(\text{mod } T)$$

where b and T are relatively prime. The choice

of T is determined by the capacity and base of the computer; a and b are chosen so that: (1) the resulting sequence $\{X_n\}$ possesses the desired statistical properties of random numbers, (2) the period of the sequence is as long as possible, and (3) the speed of generation is fast. A guide for choosing a and b is to make the correlation between the numbers be near zero, e.g., the correlation between X_n and X_{n+s} is

$$\rho_s = \frac{1-6\frac{b_s}{T}\left(1-\frac{b_s}{T}\right)}{a_s} + e$$

where

$$a_s = a^s \pmod{T}$$
$$b_s = (1+a+a^2+ \ldots +a^{s-1})b \pmod{T}$$
$$|e| < a_s/T$$

which occur in

$$X_{n+s} = a_s X_n + b_s \pmod{T}$$

When a is chosen so that $a \approx T^{1/2}$, the correlation $\rho_1 \approx T^{-1/2}$.

The sequence defined by the multiplicative congruence method will have a full period of T numbers if

b is relatively prime to T
$a = 1 \pmod{p}$ if p is a prime factor of T
$a = 1 \pmod{4}$ if 4 is a factor of T.

Consequently if $T = 2^q$, b need only be odd, and $a = 1 \pmod{4}$. When $T = 10^q$, b need only be not divisible by 2 or 5, and $a = 1 \pmod{20}$. The most convenient choices for a are of the form $a = 2^s + 1$ (for binary computers) and $a = 10^s + 1$ (for decimal computers). This results in the fastest generation of random numbers as the operations only require a shift operation plus two additions. Also any number can serve as the starting point to generate a sequence of random digits.

Below are listed various congruence schemes and their properties.

CONGRUENCE METHODS FOR GENERATING RANDOM NUMBERS

$X_{n+1}=aX_n+b(\text{mod } T)$, T and b relatively prime

a	b	T	Period	X_0	Special cases for which random numbers have passed statistical tests for randomness*
$1+t^s$	odd	$T=t^q$	t^q	$0\leq X_0<T$	$T=2^{35}$, X_0 unknown; $a=2^7+1$, $b=1$; $T=2^{47}$, $a=2^9+1$, $b=29741\ 09625\ 8473$, $X_0=76293\ 94531\ 25$.
$r2^s\pm1$ (r odd, $s\geq2$)	0	$T=t^q$	t^{q-s}	relatively prime to T	$T=2^{40}, 2^{42}$, $X_0=1$; $a=5^{17}(s=2)$ $T=2^{36}$, $X_0=1$; $T=2^{39}$, $X_0=1-2^{-39}$, .5478126193; $a=5^{13}(s=2)$ $T=2^{35}$, $X_0=1$; $a=5^{15}(s=2)$
$r2^s\pm1$ (r odd, $s\geq2$)	0	$T=t^q\pm1$	(varies)	relatively prime to T	$T=2^{35}+1$, $X_0=10{,}987{,}654{,}321$; $a=23$; period $\approx10^6$ $T=10^8+1$, $X_0=47{,}594{,}118$; $a=23$; period $\approx5.8\times10^6$
7^{4s+1}	0	$T=10^q$	$5\cdot10^{q-3}$	relatively prime to T	$T=10^{10}$, $X_0=1$; $a=7$ $T=10^{11}$, $X_0=1$; $a=7^{13}$
3^{4s+1} ($s=0, 2, 3, 4$)	0	$T=10^q$	$5\cdot10^{q-2}$	relatively prime to T	

* X_0 given is the starting point for random numbers when statistical tests were made.

When the numbers are generated using a congruence scheme, the least significant digits have short periods. Hence the entire word length cannot be used. If one desired random numbers with as many digits as possible, one would have to modify the congruence schemes. One way is to generate the numbers mod $T \pm 1$. This unfortunately reduces the period.

GENERATION OF RANDOM DEVIATES

Let $\{X\}$ be a generated sequence of independent random numbers having the domain $(0, T)$. Then $\{U\}=\{T^{-1}X\}$ is a sequence of random deviates (numbers) from a uniform distribution on the interval (0, 1). This is usually a necessary preliminary step in the generation of random deviates having a given cumulative distribution function $F(y)$ or probability density function $f(y)$. Below are summarized some general techniques for producing arbitrary random deviates. (In what follows $\{U\}$ will always denote a sequence of random deviates from a uniform distribution on the interval (0, 1).)

1. Inverse Method

The solutions $\{y\}$ of the equations $\{u=F(y)\}$ form a sequence of independent random deviates with cumulative distribution function $F(y)$. (If $F(y)$ has a discontinuity at $y=y_0$, then whenever u is such that $F(y_0-0)<u<F(y_0)$, select y_0 as the corresponding deviate.) Generally the inverse method is not practical unless the inverse function $y=F^{-1}(u)$ can be obtained explicitly or can be conveniently approximated.

2. Generating a Discrete Random Variable

Let Y be a discrete random variable with point probabilities $p_i=Pr\{Y=y_i\}$ for $i=1, 2, \ldots$. The direct way to generate Y is to generate $\{U\}$ and put $Y=y_i$ if

$$p_1+p_2+ \ldots +p_{i-1}<U<p_1+p_2+ \ldots +p_i.$$

However, this method requires complicated machine programs that take too long.

An alternative way due to Marsaglia, is simple, fast, and seems to be well suited to high-speed computations. Let p_i for $i=1, 2, \ldots, n$ be expressed by k decimal digits as $p_i=.\delta_{1i}\delta_{2i} \ldots \delta_{ki}$ where the δ's are the decimal digits. (If the domain of the random variable is infinite, it is necessary to truncate the probability distribution at p_n.) Define

$$P_0=0,\ P_r=10^{-r}\sum_{i=1}^{n}\delta_{ri} \text{ for } r=1, 2, \ldots, k, \text{ and}$$

$$\Pi_s=\sum_{r=0}^{s}10^r P_r,\ s=1, 2, \ldots, k.$$

Number the computer memory locations by 0, 1, 2, . . . , Π_k-1. The memory locations are divided into k mutually exclusive sets such that the sth set consists of memory locations Π_{s-1}, $\Pi_{s-1}+1, \ldots, \Pi_s-1$. The information stored in the memory locations of the sth set consists of y_1 in δ_{s1} locations, y_2 in δ_{s2} locations, . . . , y_n in δ_{sn} locations.

Denote the decimal expansion of the uniform deviates generated by the computer by $u=\cdot d_1d_2d_3 \ldots$ and finally let $a\{m\}$ be the contents of memory location m. Then if

$$\sum_{i=0}^{s-1}P_i \leq U < \sum_{i=0}^{s}P_i$$

put

$$y=a\left\{d_1d_2 \ldots d_s+\Pi_{s-1}-10^s\sum_{i=1}^{s-1}P_i\right\}.$$

This method is perhaps the best all-around method for generating random deviates from a discrete distribution. In order to illustrate this method consider the problem of generating deviates from the binomial distribution with point probabilities

$$p_1=\binom{n}{i}p^i(1-p)^{n-i}$$

for $n=5$ and $p=.20$. The point probabilities to 4 D are

Value of Random Variable	*Point Probabilities*
0	$p_0=0.3277$
1	$p_1=\ .4096$
2	$p_2=\ .2048$
3	$p_3=\ .0512$
4	$p_4=\ .0064$
5	$p_5=\ .0003$

and thus $P_0=0$, $P_1=.9$, $P_2=.07$, $P_3=.027$, $P_4=.0030$ from which $\Pi_0=0$, $\Pi_1=9$, $\Pi_2=16$, $\Pi_3=43$, $\Pi_4=73$. The 73 memory locations are divided into 4 mutually exclusive sets such that

Set	*Memory Locations*
1	0, 1, . . . , 8
2	9, 10, . . . , 15
3	16, . . . , 42
4	43, . . . , 72

Among the nine memory locations of set 1, zero is stored $\delta_{10}=3$ times, 1 is stored $\delta_{11}=4$ times, 2 is stored $\delta_{12}=2$ times; the seven locations of set 2 store 0 $\delta_{20}=2$ times and 3 $\delta_{23}=5$ times; etc. A summary of the memory locations is set out below:

	Value of Random Variable					
	0	1	2	3	4	5
Frequency (set 1)	3	4	2	0	0	0
Frequency (set 2)	2	0	0	5	0	0
Frequency (set 3)	7	9	4	1	6	0
Frequency (set 4)	7	6	8	2	4	3

Then to generate the random variables if

$0\leq u<.9$	put	$y=a\{d_1\}$
$.9\leq u<.97$		$y=a\{d_1d_2-81\}$
$.97\leq u<.997$		$y=a\{d_1d_2d_3-954\}$
$.997\leq u<1.000$		$y=a\{d_1d_2d_3d_4-9927\}$

3. Generating a Continuous Random Variable

The method for generating deviates from a

discrete distribution can be adapted to random variables having a continuous distribution. Let $F(y)$ be the cumulative distribution function and assume that the domain of the random variable is (a,b) where the interval is finite. (If the domain is infinite, it must be truncated at (say) the points a and b.) Divide the interval $(b-a)$ into n sub-intervals of length Δ $(n\Delta=b-a)$ such that the boundary of the ith interval is (y_{i-1}, y_i) where $y_i=a+i\Delta$ for $i=0, 1, \ldots, n$. Now define a discrete distribution having domain

$$\left\{z_i=\frac{y_i+y_{i-1}}{2}\right\}$$

with point probabilities $p_i=F(y_i)-F(y_{i-1})$. Finally, let W be a random variable having a uniform distribution on $\left(-\frac{\Delta}{2}, \frac{\Delta}{2}\right)$. This can be done by setting $W=\Delta\left(U-\frac{1}{2}\right)$. Then random deviates from the distribution function $F(y)$, can be generated (approximately) by setting $y=z+w$ $=z+\Delta\left(u-\frac{1}{2}\right)$. This is simply an approximate decomposition of the continuous random variable into the sum of a discrete and continuous random variable. The discrete variable can be generated quickly by the method described previously. The smaller the value of Δ the better will be the approximation. Each number can be generated by using the leading digits of U to generate the discrete random variable Z and the remaining digits forming a uniformly distributed deviate having (0,1) domain.

4. Acceptance-Rejection Methods

In what follows the random variable Y will be assumed to have finite domain (a, b). If the domain is infinite, it must be truncated for computational purposes at (say) the points a and b. Then the resulting random deviates will only have this truncated domain.

a) Let f be the maximum of $f(y)$. Then the procedure for generating random deviates is: (1) generate a pair of uniform deviates U_1, U_2; (2) compute a point $y=a+(b-a)u_2$ in (a, b); (3) if $u_1<f(y)/f$ accept y as the random deviate, otherwise reject the pair (u_1, u_2) and start again. The acceptance ratio of deviates actually produced is $[(b-a)f]^{-1}$. Hence the acceptance ratio decreases as the domain increases. One way to increase the acceptance ratio is to divide the interval (a, b) into mutually exclusive sub-intervals and then carry out the acceptance-rejection process. For this purpose let the interval (a, b) be divided into k sub-intervals such that the end points of the jth interval are (ξ_{j-1}, ξ_j) with $\xi_0=a$, $\xi_k=b$ and $\int_{\xi_{j-1}}^{\xi_j} f(y)dy=p_j$; further let the maximum of $f(y)$ in the jth interval be f_j. Then to generate random deviates from $f(y)$, generate n pairs of deviates $(u_{1s}, u_{2s})s=1, 2, \ldots, n$. Assign $[np_j]$ such pairs to the jth interval and compute $y_j=\xi_{j-1}+(\xi_j-\xi_{j-1})u_{2s}$. If $u_{1s}<f(y_j)/f_j$ accept y_j as a deviate. The acceptance ratio of this method is

$$\sum_{j=1}^{k} p_j[(\xi_j-\xi_{j-1})f_j]^{-1}$$

b) Let $F(y)$ be such that $f(y)=f_1(y)f_2(y)$ where the domain of y is (a, b). Let f_1 and f_2 be the maximum of $f_1(y)$ and $f_2(y)$ respectively. Then the procedure for generating random deviates having the probability density function $f(y)$ is: (1) generate U_1, U_2, U_3; (2) define $z=a+(b-a)u_3$; (3) if both $u_1<\frac{f_1(z)}{f_1}$ and $u_2<\frac{f_2(z)}{f_2}$, take z as the random deviate; otherwise take another sample of three uniform deviates. The acceptance ratio of this method is $[(b-a)f_1f_2]^{-1}$ and can be increased by dividing (a, b) into sub-intervals as in the previous case.

c) Let the probability density function of Y be

$$f(y)=\int_{\alpha}^{\beta} g(y, t)dt, \ (\alpha \leq t \leq \beta), \ (a \leq y \leq b).$$

Let g be the maximum of $g(y, t)$. Then the procedure for generating random deviates having the probability density function $f(y)$ is: (1) generate U_1, U_2, U_3; (2) define $s=\alpha+(\beta-\alpha)u_2$; $z=a+(b-a)u_3$; (3) if $u_1 < \frac{g(z, s)}{g}$, take z as the random deviate; otherwise take another sample of three. The acceptance ratio for this method is $[(b-a)g]^{-1}$ and can be increased by dividing the domain of t and y into sub-domains.

5. Composition Method

Let $g_z(y)$ be a probability density function which depends on the parameter z; further let $H(z)$ be the cumulative distribution function for z. In order to generate random deviates Y having the frequency function

$$f(y)=\int_{-\infty}^{\infty} g_z(y)dH(z)$$

one draws a deviate having the cumulative distribution function $H(z)$; then draws a second sample having the probability density function $g_z(y)$.

6. Generation of Random Deviates From Well Known Distributions

a. Normal distribution

(1) *Inverse method:* The inverse method depends on having a convenient approximation to the inverse function $x=P^{-1}(u)$ where

$$u=(2\pi)^{-1/2} \int_{-\infty}^{x} e^{-t^2/2}dt.$$

Two ways of performing this operation are to i) use 15.24 with $t=\left(\ln \frac{1}{u^2}\right)^{1/2}$ or ii) approximate $x=P^{-1}(u)$ piecewise using Chebyshev polynomials.

(2) *Sum of uniform deviates:* Let U_1, U_2, . . ., U_n be a sequence of n uniform deviates. Then

$$X_n=\left(\sum_{i=1}^{n} U_i-\frac{n}{2}\right)\left(\frac{n}{12}\right)^{-1/2}$$

will be distributed asymptotically as a normal random deviate. When $n=12$, the maximum errors made in the normal deviate are 9×10^{-3} for $|X|<2$, 9×10^{-1} for $2<|X|<3$. An improvement can be made by taking a polynomial function of X_n (say)

$$X_n^*=X_n \sum_{s=0}^{k} a_{2s}X_n^{2s}$$

as the normal deviate where a_{2s} are suitable coefficients. These coefficients may be calculated using (say) Chebyshev polynomials or simply by making the asymptotic random deviate agree with the correc normal deviate at certain specified points. When $n=12$, the maximum error in the normal deviate is 8×10^{-4} using the coefficients

$a_0=9.8746$ $\quad a_6=(-7)-5.102$
$a_2=(-3)3.9439$ $\quad a_8=(-7)1.141$
$a_4=(-5)7.474$

(3) *Direct method:* Generate a pair of uniform deviates $(U_1,\ U_2)$. Then

$$X_1=(-2\ \ln U_1)^{1/2}\cos 2\pi U_2,$$

$X_2=(-2\ \ln U_1)^{1/2}\sin 2\pi U_2$ will be a pair of independent normal random deviates with mean zero and unit variance. This procedure can be modified by calculating $\cos 2\pi U$ and $\sin 2\pi U$ using an acceptance rejection method; e.g., (1) generate (U_1, U_2); (2) if $(2U_1-1)^2+(2U_2-1)^2\leq1$ generate a third uniform deviate U_3, otherwise reject the pair and start over; (3) calculate $y_1=(-\ln u_3)^{1/2}\frac{u_1^2-u_2^2}{u_1^2+u_2^2}$, $y_2=\pm2(-\ln u_3)^{1/2}\frac{u_1u_2}{u_1^2+u_2^2}$ ($\pm$ random). Both y_1 and y_2 are the desired random deviates.

(4) *Acceptance-rejection method:* 1) Generate a pair of uniform deviates $(U_1,\ U_2)$; 2) compute $x=-\ln u_1$; 3) if $e^{-½(x-1)^2}\geq u_2$ (or equivalently

$(x-1)^2 \leq -2(\ln u_2)$ accept x, otherwise reject the pair and start over. The quantity will be the required normal deviate with mean zero and unit variance.

b. Bivariate normal distribution

Let $\{X_1, X_2\}$ be a pair of independent normal deviates with mean zero and unit variance. Then $\{X_1, \rho X_1 + (1-\rho^2)^{1/2} X_2\}$ represent a pair of deviates from a bivariate normal distribution with zero means, unit variances, and correlation coefficient ρ.

c. Exponential distribution

(1) *Inverse method:* Since $F(x) = e^{-x/\theta}$, $X = -\theta \ln U$ will be a deviate from the exponential distribution with parameter θ.

(2) *Acceptance-rejection method:* 1) Generate a pair of independent uniform deviates (U_0, U_1); 2) if $U_1 < U_0$ generate a third value U_2; 3) if $U_1 + U_2 < U_0$ generate a fourth value U_3, etc.; 4) continue generating uniform deviates until an n is obtained such that $U_1 + U_2 + \ldots + U_{n-1} < U_0 < U_1 + \ldots + U_n$; 5) if n is even reject the procedure and start a fresh trial with a new value of U_0, otherwise if n is odd take $X = \theta U_0$ as the desired deviate; 6) in general if t is the number of trials until an acceptable sequence is obtained $X = \theta(t + U_0)$. The random deviates produced in this way follow an exponential distribution with parameter θ. One can expect to generate approximately six uniform deviates for every exponential deviate.

(3) *Discrete Distribution Method:* Let Y and n be discrete random variables with point probabilities

$$Pr\{Y=r\} = (e-1)e^{-(r+1)} \qquad r=0, 1, 2, \ldots$$

$$Pr\{n=s\} = [s!(e-1)]^{-1} \qquad s=1, 2, 3, \ldots .$$

Then $X = Y + \min(U_1, U_2, \ldots, U_n)$ will follow an exponential distribution. The average value of n is 1.58 so that one needs, on the average, only 1.58 u's from which the minimum is selected.

USE AND EXTENSION OF THE TABLES

USE OF PROBABILITY FUNCTION INEQUALITIES

Example 1. Let X be a random variable with finite mean and variance equal to m and σ^2, respectively. Use the inequalities for probability functions 15.17, 18, 19 to place lower bounds on

$$A(t)=F(t)-F(-t)=P\left\{\frac{|X-m|}{\sigma}\leq t\right\}$$

for $t=1(1)4$.

Lower bounds on $A(t)=F(t)-F(-t)$

$t=1$	2	3	4	Remarks
0	.7500	.8889	.9375	no knowledge of $F(t)$; **15.17**
.5556	.8889	.9506	.9722	$F(t)$ is unimodal and continuous; 15.18
0	.8182	.9697	.9912	$F(t)$ is such that $\mu_4=3$; 15.19

It is of interest to note that the standard normal distribution is unimodal, has mean zero, unit variance $\mu_4=3$, is continuous, and such that

$$A(t)=P(t)-P(-t)$$
$$=.6827,\ .9545,\ .9973,\text{ and }.9999$$

for $t=1$, 2, 3 and 4 respectively.

INTERPOLATION FOR P(x) IN TABLE 1

Example 2. Compute $P(x)$ for $x=2.576$ to fifteen decimal places using a Taylor expansion.

Writing $x=x_0+\theta$ we have

$$P(x)=P(x_0)+Z(x_0)\theta+Z^{(1)}(x_0)\frac{\theta^2}{2!}$$

$$+Z^{(2)}(x_0)\frac{\theta^3}{3!}+Z^{(3)}(x_0)\frac{\theta^4}{4!}+\ldots$$

Taking $x_0=2.58$ and $\theta=-4\times10^{-3}$ we calculate the successive terms to 16D

+.99505	99842	42230		
—	5	72204	35976	6
—		2952	57449	6
—		8	63097	8
—			1439	4
—				9
.99500	24676	84265	7	

The result correct to 17D is

$$P(2.576)=.99500\quad 24676\quad 84264\quad 98$$

CALCULATION FOR ARBITRARY MEAN AND VARIANCE

Example 3. Find the value to 5D of

$$P\{X\leq .50\}=\frac{1}{2\sqrt{2\pi}}\int_{-\infty}^{.5}e^{-1/2\left(\frac{t-1}{2}\right)^2}dt$$

using 15.23 and Table 1

This represents the probability of the random variable being less than or equal to .5 for a normal distribution with mean $m=1$ and variance $\sigma^2=4$. Using 15.23 we have

$$P\{X\leq .5\}=P\left(\frac{.5-1}{2}\right)=P(-.25)$$

Since $P(-x)=1-P(x)$, we have

$$P(-.25)=1-P(.25)=1-.59871=.40129$$

where a two-term Taylor series was used for interpolation. Note that when interpolating for $P(x)$ for a value of x midway between the tabulated values we can write $x=x_0+.01$ and a two-term Taylor series is $P(x)=P(x_0)+Z(x_0)10^{-2}$. Thus one need only multiply $Z(x_0)$ by 10^{-2} and add the result to $P(x_0)$.

CALCULATION OF P(x) FOR x APPROXIMATE

Example 4. Using **Table 1**, find $P(x)$ for $x=1.96$, when there is a possible error in x of $\pm 5\times 10^{-3}$.

This is an example where the argument is only known approximately. The question arises as to how many decimal places one should retain in $P(x)$. If Δx and $\Delta P(x)$ denote the error in x and the resulting error in $P(x)$, respectively, then

$$\Delta P(x) \approx Z(x)\Delta x$$

Hence $\Delta P(1.960)=3\times 10^{-4}$ which indicates that $P(1.960)$ need only be calculated to 4D. Therefore $P(1.960)=.9750$.

INVERSE INTERPOLATION FOR P(x)

Example 5. Find the value of x for which $P(x)=.97500\ 00000\ 00000$ using **Table 1** and determining as many decimal places as is consistent with the tabulated function.

For inverse interpolation the tabulated function $P(x)$ may be regarded as having a possible error of $.5\times 10^{-15}$. Hence

$$\Delta x \approx \frac{\Delta P(x)}{Z(x)} = \frac{.5\times 10^{-15}}{Z(x)}$$

Let $P(x_0)$ correspond to the closest tabulated value of $P(x)$. Then a convenient formula for inverse interpolation is

$$x = x_0 + t + \frac{x_0 t^2}{2} + \frac{2x_0^2+1}{6}\, t^3$$

where

$$t = \frac{P(x) - P(x_0)}{Z(x_0)}$$

If only the first two terms (i.e., $x = x_0 + t$) are used, the error in x will be bounded by $\frac{x}{8} \times 10^{-4}$ and the true value will always be greater than the value thus calculated.

With respect to this example, $\Delta x \approx 10^{-14}$ and thus the interpolated value of x may be in error by one unit in the fourteenth place. The closest value to $P(x) = .97500\ 00000\ 00000$ is $P(x_0) = .97500\ 21048\ 51780$ with $x_0 = 1.96$. Hence using the preceding inverse interpolation formulas with

$$t = -.00003\ 60167\ 31129$$

and carrying fifteen decimals we have the successive terms

+	1.96000	00000	00000
−	.00003	60167	31129
+		12	71261
−			68
			0
+	1.95996	39845	40064

EDGEWORTH ASYMPTOTIC EXPANSION

Example 6. Find the Edgeworth asymptotic expansion 15.25 for the c.d.f. of chi-square.

Method 1. Expansion for χ^2

Let

$$Q(\chi^2|\nu) = 1 - F(t)$$

where

$$t = \frac{\chi^2 - \nu}{(2\nu)^{\frac{1}{2}}}$$

Since the values of γ_1 and γ_2 15.47 are

$$\gamma_1 = 2\sqrt{2}/\nu^{\frac{1}{2}}$$

$$\gamma_2 = 12/\nu,$$

we obtain, by using the first two bracketed terms of 15.25

$$F(t) \sim P(t) - \frac{1}{\nu^{\frac{1}{2}}}\left[\frac{\sqrt{2}}{3} Z^{(2)}(t)\right] + \frac{1}{\nu}\left[\frac{1}{2} Z^{(3)}(t) + \frac{1}{9} Z^{(5)}(t)\right]$$

The Edgeworth expansion is an asymptotic expansion in terms of derivatives of the normal distribution function. It is often possible to transform a random variable so that the distribution of the transformed random variable more closely approximates the normal distribution function than does the distribution of the original random variable. Hence for the same number of terms, greater accuracy may be achieved by using the transformed variable in the expansion. Since the distribution of $\sqrt{2\chi^2}$ is more closely approximated by a normal distribution than χ^2 itself (as judged by a comparison of the values of γ_1 and γ_2), we would expect that the Edgeworth asymptotic expansion of $\sqrt{2\chi^2}$ would be superior to that of χ^2.

Method 2. Expansion for $\sqrt{2\chi^2}$. Let

$$Q(\chi^2|\nu) = 1 - F(t) = 1 - F\left(\frac{\sqrt{2\chi^2} - (2\nu-1)^{\frac{1}{2}}}{\left(1 - \frac{1}{4\nu}\right)^{\frac{1}{2}}}\right)$$

where $(2\nu-1)^{\frac{1}{2}}$ and $1 - \frac{1}{4\nu}$ are the mean and variance to terms of order ν^{-2} of $\sqrt{2\chi^2}$ (see 15.48) The values of γ_1 and γ_2 for $\sqrt{2\chi^2}$ are

$$\gamma_1 \approx \frac{1}{\sqrt{2\nu}}\left[1 + \frac{5}{8\nu}\right] \qquad \gamma_2 \approx \frac{3}{4\nu^2}$$

Thus we obtain

$$F(t) \sim P(t) - \frac{1}{\nu^{\frac{1}{2}}}\left[\frac{\sqrt{2}}{12}\left(1 + \frac{5}{8\nu}\right) Z^{(2)}(t)\right]$$

$$+\frac{1}{\nu}\left[\frac{1}{32\nu}Z^{(3)}(t)+\frac{1}{144}\left(1+\frac{5}{8\nu}\right)^2 Z^{(5)}(t)\right]$$

For numerical examples using these expansions see **Example 12.**

CALCULATION OF L(h , k, ρ)

Example 7. Find $L(.5, .4, .8)$. Using **15.29**

$$\sqrt{h^2-2\rho hk+k^2}=\sqrt{.09}=.3$$

$$L(.5, .4, .8)=L(.5, 0, 0)+L(.4, 0, -.6)$$

Reference to Fig. 2 yields

$$L(.5, 0, 0)+L(.4, 0, -.6)=.16+.08=.24$$

The answer to 3D is $L(.5, .4, .8)=.250$.

CALCULATION OF THE BIVARIATE NORMAL PROBABILITY FUNCTION

Example 8. Let X and Y follow a bivariate normal distribution with parameters $m_x=3$, $m_y=2$, $\sigma_x=4$, $\sigma_y=2$, and $\rho=-.125$. Find the value of $P_r\{X\geq 2, Y\geq 4\}$ using **15.29** and Figs. 2 and 3

Since $P_r\{X\geq h, Y\geq k\}=L\left(\frac{h-m_x}{\sigma_x}, \frac{k-m_y}{\sigma_y}, \rho\right)$ we have $P\{X\geq 2, Y\geq 4\}=L(-.25, 1, -.125)$. Using **15.29**

$$L(-.25, 1, -.125)=L(-.25, 0, .969)$$
$$+L(1, 0, .125)-\frac{1}{2}$$

Fig. 2 only gives values for $h>0$, however, using the relationship **15.28** with $k=0$, $L(-h, 0, \rho)=\frac{1}{2}-L(h, 0, -\rho)$ and thus $L(-.25, 0, .969)=\frac{1}{2}-L(.25, 0, -.969)$. Therefore $L(-.25, 1, -.125)=-L(.25, 0, -.969)+L(1, 0, .125)=-.01+.09=.08$.

The answer to 3D is $L(-.25, 1., -.125)=.080$.

INTEGRAL OF A BIVARIATE NORMAL DISTRIBUTION OVER A POLYGON

Example 9. Let the random variables X and Y have a bivariate normal distribution with parameters $m_x=5$, $\sigma_x=2$, $m_y=9$, $\sigma_y=4$, and $\rho=.5$. Find the probability that the point (X, Y) be inside the triangle whose vertices are $A=(7, 8)$, $B=(9, 13)$, and $C=(2, 9)$.

When obtaining the integral of a bivariate normal distribution over a polygon, it is first necessary to use 15.30 in order to transform the variates so that one deals with a circular normal distribution. The polygon in the region of the transformed variables is then divided into configurations such that the integral over any selected configuration can be easily obtained. Below are listed some of the most useful configurations.

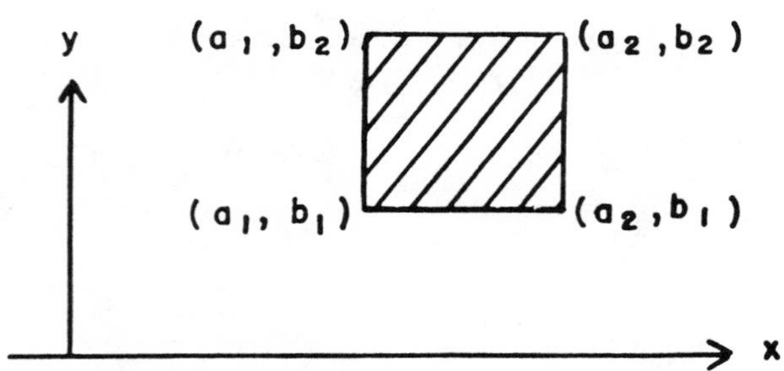

$$\int_{a_1}^{a_2}\int_{b_1}^{b_2} g(x, y, 0)dxdy=[P(a_2)-P(a_1)]\,[P(b_2)-P(b_1)]$$

$$\int_0^{\infty}\int_0^{ax} g(x, y, 0)dxdy=\frac{\arctan a}{2\pi}$$

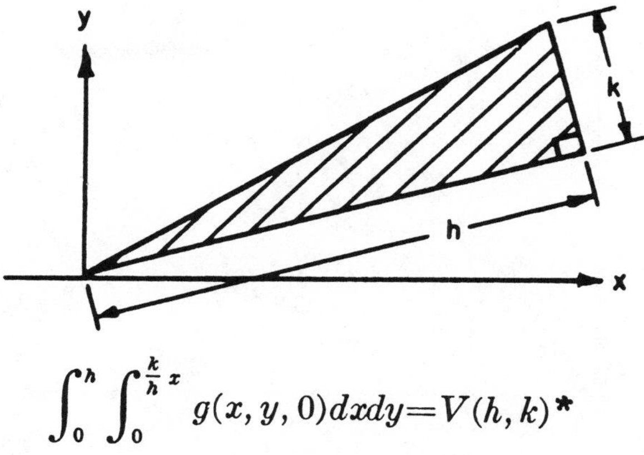

$$\int_0^h \int_0^{\frac{k}{h}x} g(x, y, 0)dxdy = V(h, k)^*$$

For the following two configurations we define

$$h = \frac{|t_2 s_1 - t_1 s_2|}{[(s_2 - s_1)^2 + (t_2 - t_1)^2]^{\frac{1}{2}}}$$

$$k_1 = \frac{|s_1(s_2 - s_1) + t_1(t_2 - t_1)|}{[(s_2 - s_1)^2 + (t_2 - t_1)^2]^{\frac{1}{2}}}$$

$$k_2 = \frac{|s_2(s_2 - s_1) + t_2(t_2 - t_1)|}{[(s_2 - s_1)^2 + (t_2 - t_1)^2]^{\frac{1}{2}}}$$

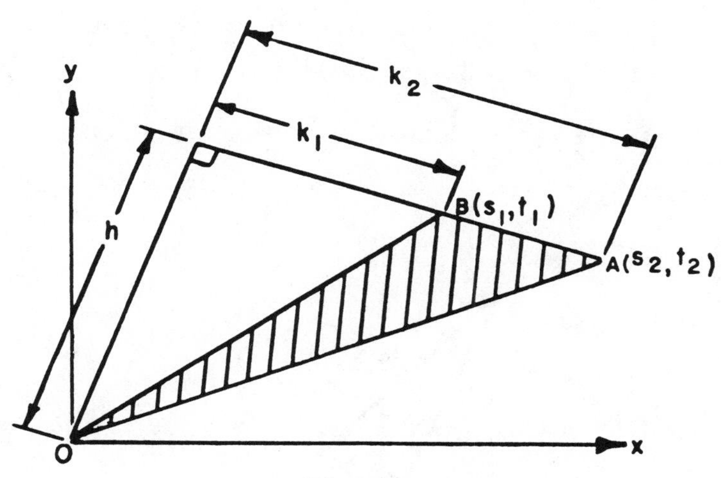

Fig. 4

$$\iint_{\Delta AOB} g(x, y, 0)dxdy = V(h, k_2) - V(h, k_1)$$

---

* See 15.31 for definition of $V(h, k)$.

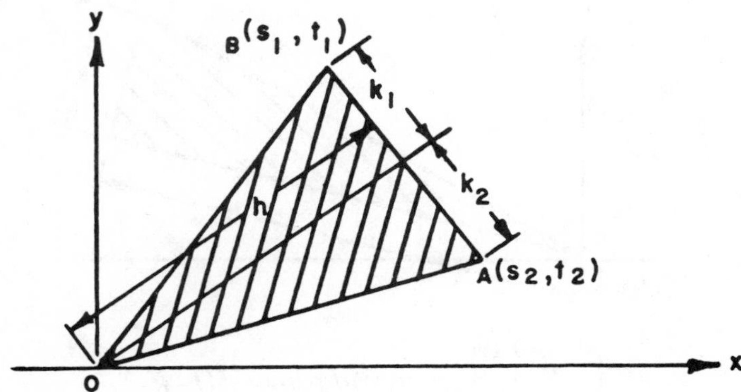

Fig. 5

$$\iint_{\Delta AOB} g(x, y, 0)\,dx dy = V(h, k_2) + V(h, k_1)$$

Using the circularizing transformation 15.30 for our example results in

$$s = \frac{1}{\sqrt{3}}\left(\frac{x-5}{2} + \frac{y-9}{4}\right)$$

$$t = -\frac{1}{1}\left(\frac{x-5}{2} - \frac{y-9}{4}\right)$$

The vertices of the triangle in the (s, t) coordinates become $A = (\sqrt{3}/4, \ -5/4)$, $B = (\sqrt{3}, \ -1)$ and $C = \left(-\frac{\sqrt{3}}{2}, \frac{3}{2}\right)$. These points are plotted below.

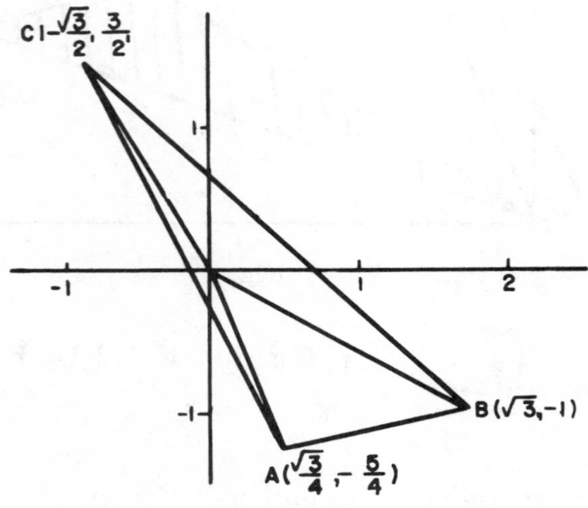

From the figure it is seen that the desired probability is the sum of the probabilities that the point having the transformed variables as coordinates is inside the triangles *AOB*, *AOC*, and *BOC*.

For these three triangles we have

	h	k_1	k_2
ΔAOB	$\frac{2}{7}\sqrt{21}$	$\sqrt{7}/14$	$\frac{4}{7}\sqrt{7}$
ΔAOC	$\frac{1}{74}\sqrt{111}$	$\frac{8}{37}\sqrt{37}$	$\frac{21}{74}\sqrt{37}$
ΔBOC	$\frac{1}{13}\sqrt{39}$	$\frac{7}{13}\sqrt{13}$	$\frac{6}{13}\sqrt{13}$

From the graph it is seen that the probability over *AOB* may be found in the same manner as that over Fig. 4, and over *AOC* and *BOC* the probabilities may be found as that over Fig. 5.

Hence

$$\iint_{\Delta} g(x, y, .5)dxdy = \iint_{\Delta ABC} g(s, t, 0)dsdt$$

$$= \iint_{\Delta AOB} g(s, t, 0)dsdt + \iint_{\Delta AOC} g(s, t, 0)dsdt$$

$$+ \iint_{\Delta BOC} g(s, t, 0)dsdt$$

and consequently using 15.31 and Fig. 2

$$\iint_{\Delta AOB} g(s, t, 0)dsdt = V\left(\frac{2}{7}\sqrt{21}, \frac{4\sqrt{7}}{7}\right) - V\left(\frac{2}{7}\sqrt{21}, \frac{\sqrt{7}}{14}\right)$$

$$= \left[\frac{1}{4} + L(1.31, 0, -.76) - L(0, 0, -.76) - \frac{1}{2}Q(1.31)\right]$$

$$- \left[\frac{1}{4} + L(1.31, 0, -.14) - L(0, 0, -.14) - \frac{1}{2}Q(1.31)\right]$$

$$=L(1.31,\ 0,-.76)-L(0,\ 0,-.76)$$

$$-L(1.31,\ 0,-.14)+L(0,\ 0,-.14)$$

$$=.00-.11-.04+.23=.08$$

$$\iint\limits_{\Delta AOC} g(s,t,0)dsdt=V\left(\frac{\sqrt{111}}{74},\frac{8\sqrt{37}}{37}\right)+V\left(\frac{\sqrt{111}}{74},\frac{21\sqrt{37}}{74}\right)$$

$$=\left[\frac{1}{4}+L(.14,\ 0,-.99)-L(0,\ 0,-.99)-\frac{1}{2}\,Q(.14)\right]$$

$$+\left[\frac{1}{4}+L(.14,\ 0,-1)-L(0,\ 0,-1)-\frac{1}{2}\,Q(.14)\right]$$

$$=.01+.02=.03$$

$$\iint\limits_{\Delta BOC} g(s,t,0)dsdt=V\left(\frac{\sqrt{39}}{13},\frac{7\sqrt{13}}{13}\right)+V\left(\frac{\sqrt{39}}{13},\frac{6\sqrt{13}}{13}\right)$$

$$=\left[\frac{1}{4}+L(.48,\ 0,-.97)-L(0,\ 0,-.97)-\frac{1}{2}\,Q(.48)\right]$$

$$+\left[\frac{1}{4}+L(.48,\ 0,-.96)-L(0,\ 0,-.96)-\frac{1}{2}\,Q(.48)\right]$$

$$=.05+.04=.09$$

Thus adding all parts, the probability that X and Y are in triangle ABC is $=.08+.03+.09=.20$. The answer to 3D is .211.

CALCULATION OF A CIRCULAR NORMAL DISTRIBUTION OVER AN OFFSET CIRCLE

Example 10. Let X and Y have a circular normal distribution with $\sigma=1000$. Find the probability that the point (X, Y) falls within a circle having a radius equal to 540 whose center is displaced 1210 from the mean of the circular normal distribution.

In units of σ, the radius and displacement from the center are, respectively, $R=\frac{540}{1000}=.54$ and $r=\frac{1210}{1000}=1.21$. The problem is thus reduced to finding the probability of X and Y falling in a circle of radius $R=.54$ displaced $r=1.21$ from the center of the distribution where $\sigma=1$.

Since $R<1$, the approximation 15.33 is used. This results in

$$P(R^2|2, r^2)=\frac{2(.54)^2}{4+(.54)^2}\exp\frac{-2(1.21)^2}{4+(.54)^2}$$

$$=(.1359)e^{-.6823}=.06869$$

The answer to 5D is .06870.

INTERPOLATION FOR $Q(\chi^2|\nu)$

Example 11. Find $Q(25.298|20)$ using the interpolation formula given with Table 2.

Taking $\chi^2=25$, $\theta=.298$ and applying the interpolation formula results in

$$Q(25.298|20)=\frac{1}{8}\{Q(25|16)\theta^2+Q(25|18)(4\theta-2\theta^2)$$
$$+Q(25|20)(8-4\theta+\theta^2)\}$$
$$=\frac{1}{8}\{(.06982)(.088804)$$
$$+(.12492)(1.014392)$$
$$+(.20143)(6.896804)\}$$
$$=.19027$$

A less accurate interpolate may be obtained by setting θ^2 equal to zero in the above formula. This results in the value .19003. The correct value to 6D is $Q(25.298|20)=.190259$.

On the other hand if $\chi^2=25.298$ is assumed to have an error of $\pm5\times10^{-4}$, then how large an error arises in $Q(\chi^2|\nu)$? Denoting the error in χ^2 by $\Delta\chi^2$ and the resulting error in $Q(\chi^2|\nu)$ by $\Delta Q(\chi^2|\nu)$, we then have the approximate relationship

$$\Delta Q(\chi^2|\nu)\approx\frac{\partial Q(\chi^2|\nu)}{\partial\chi^2}\Delta\chi^2$$

Using 15.37 we can write

$$\frac{\partial Q(\chi^2|\nu)}{\partial \chi^2}=\frac{1}{2}\left[Q(\chi^2|\nu-2)-Q(\chi^2|\nu)\right]$$

and

$$\Delta Q(\chi^2|\nu)\approx\frac{1}{2}\left[Q(\chi^2|\nu-2)-Q(\chi^2|\nu)\right]\Delta\chi^2$$

For practical purposes it is sufficient to evaluate the derivative to one or two significant figures. Consequently we can write

$$\frac{\partial Q(\chi^2|\nu)}{\partial \chi^2}\approx\frac{\partial Q(\chi_0^2|\nu)}{\partial \chi^2}$$

where χ_0^2 is the closest value to χ^2 for which Q is tabulated. Hence

$$\Delta Q(\chi^2|\nu)\approx\frac{1}{2}\left[Q(\chi_0^2|\nu-2)-Q(\chi_0^2|\nu)\right]\Delta\chi^2$$

For this example $\Delta\chi^2=\pm 5\times 10^{-4}$ and $\chi_0^2=25$. This results in

$$\Delta Q(\chi^2|\nu)=\frac{1}{2}(-.076)(\pm 5)10^{-4}=\pm 2\times 10^{-5}$$

as the possible error in $Q(\chi^2|\nu)$.

CALCULATION OF $Q(x^2|\nu)$ OUTSIDE THE RANGE OF TABLE 2

Example 12. Find the value of $Q(84|72)$.

Since this value is outside the range of Table 2 we can approximate $Q(84|72)$ by (1) using the Edgeworth expansion for $Q(\chi^2|\nu)$ given in **Example 6**, (2) the cube root approximation 15.39, (3) the improved cube root approximation 15.40 or (4) the square root approximation 15.38. The results of using all four methods are presented below:

1. Edgeworth expansion

The successive terms of the Edgeworth expansion for the distribution of chi-square result in

$$1-Q(84|72)=.841345$$
$$.000000$$
$$.001120$$
$$\overline{.842465}$$

Hence $Q(84|72)=.15754$.

The successive terms of the Edgeworth expansion for the distribution of $\sqrt{2\chi^2}$ result in

$$1-Q(84|72)=.842544$$
$$-.000034$$
$$-.000138$$
$$\overline{.842372}$$

Hence $Q(84|72)=.15764$.

2. Cube root approximation 15.39

Using the cube root approximation we have

$$Q(84|72)=Q(x)$$

where

$$x=\frac{\left(\frac{84}{72}\right)^{1/3}\left[1-\frac{2}{9(72)}\right]}{\left[\frac{2}{9(72)}\right]^{\frac{1}{2}}}=1.0046$$

This results in $Q(84|72)=Q(1.0046)=1-P(1.0046)=.15754$.

3. Improved cube root approximation 15.40

The improved cube root approximation involves calculating a correction factor h, to x. Linearly interpolating for h_{60} (which appears below 15.40) with $x=1.0046$ results in $h_{60}=-.0006$ and hence

$$h_{72}=\frac{60}{72}(-.0006)=-.00049$$

Thus

$$Q(84|72)=Q(1.0046-.0005)=Q(1.0041)$$
$$=1-P(1.0041)=.15766$$

4. Square root approximation 15.38

Using the square root approximation we have $Q(84|72)=Q(x)$ where

$$x=\sqrt{2(84)}-\sqrt{2(72)-1}=1.0032.$$

This results in

$$Q(84|72)=Q(1.0032)=1-P(1.0032)=.15788$$

The value correct to 6D is $Q(84|72)=.157653$. Generally the improved cube root approximation will be correct with a maximum error of a few units in the fifth decimal and is recommended for calculations which are outside the range of Table 2.

CALCULATION OF χ^2 FOR $Q(\chi^2|\nu)$ OUTSIDE THE RANGE OF TABLE 3

Example 13. Find the value of χ^2 for which $Q(\chi^2|144)=.01$.

Since $\nu=144$ is outside the range of Table 3, we can compute it by using (1) the Cornish-Fisher asymptotic expansion 15.26, for χ^2, (2) the cube approximation 15.42, (3) the improved cube approximation 15.43, or (4) the square approximation 15.41. We shall compute the value by all four methods.

1. Cornish-Fisher asymptotic expansion 15.26

The Cornish-Fisher asymptotic expansion for χ^2 with $\nu=144$ can be written as

$$\chi^2\sim 144+12\sqrt{2}x+4h_1(x)+\frac{4\sqrt{2}}{12}[3h_2(x)+2h_{11}(x)]$$

$$+\frac{8}{12^2}[6h_3(x)+3h_{12}(x)+2h_{111}(x)]+\frac{16\sqrt{2}}{12^3}[30h_4(x)$$

$$+9h_{22}(x)+12h_{13}(x)+6h_{112}(x)+4h_{1111}(x)]$$

Hence using the auxiliary table following 15.27

with $p=.01$ we have

$$\begin{array}{r} 144.0000 \\ 39.4794 \\ 2.9413 \\ -.0242 \\ -.0019 \\ +.0002 \\ \hline \chi^2=186.395 \end{array}$$

2. Cube approximation 15.42

Taking $\chi_{.01}=2.32635$ we have

$$\chi^2=144\left\{\left[1-\frac{2}{9(144)}\right]+(2.32635)\sqrt{\frac{2}{9(144)}}\right\}^3=186.405$$

3. Improved cube approximation 15.43

From the table for h_{60} we obtain using linear interpolation with $x=2.33$ (approximately)

$$h_{60}=.0012 \text{ and thus } h_{144}=\frac{60}{144}(.0012)=.00049$$

Hence

$$\chi^2=144\left[1-\frac{2}{9(144)}+(2.32635-.00049)\sqrt{\frac{2}{9(144)}}\right]^3=186.394$$

4. Square approximation 15.41

$$\chi^2=\frac{1}{2}\left[2.32635+\sqrt{2(144)-1}\right]^2=185.616$$

The correct answer to 3D is $\chi^2=186.394$. Generally the improved cube approximation will give results correct in the second or third decimal for $\nu>30$.

CALCULATION OF THE INCOMPLETE GAMMA FUNCTION

Example 14. Find the value of

$$\gamma(2.5,.9)=\int_0^{.9} t^{1.5}e^{-t}dt$$

making use of 15.44 and Table 2

Using 15.44 we have

$$\gamma(2.5, .9) = \Gamma(2.5) P(1.8|5) = \Gamma(2.5)[1 - Q(1.8|5)]$$

$$\gamma(2.5, .9) = \frac{3}{4}\sqrt{\pi}[1 - .87607] = .16475$$

POISSON DISTRIBUTION

Example 15. Find the value of m for which

$$\sum_{i=0}^{3} e^{-m} \frac{m^i}{i!} = .99$$

using 15.45 and Table 3.

From Table 3 with $\nu = 2c = 8$ and $Q = .99$ we have $\chi^2 = 1.646482$. Hence $m = \chi^2/2 = .823241$.

INVERSE OF THE INCOMPLETE BETA FUNCTION

Example 16. Find the value of x for which $I_x(10, 6) = .10$ using Table 4 and 15.52 . Using 15.52 we have

$$I_x(10, 6) = Q(F|12, 20) = .10 \text{ where } x = \frac{20}{20 + 12F}$$

From Table 4 the upper 10 percent point of F with 12 and 20 degrees of freedom is $F = 1.89$. Hence

$$x = \frac{20}{20 + 12(1.89)} = .469$$

The correct value to 4D is $x = .4683$.

CALCULATION OF $I_x(a, b)$ FOR a OR b SMALL INTEGERS

Example 17. Calculate $I_{.10}(3, 20)$.

Values of $I_x(a, b)$ for small integral a or b can conveniently be calculated using **15.49** or **15.50**.

Using **15.49** we have

$$1-I_{.90}(20,3)=\frac{(.9)^{20}}{B(3,20)}\left\{\sum_{i=0}^{2}(-1)^i\binom{2}{i}\frac{.9^i}{20+i}\right\}$$

$$=\frac{.121576}{.216450\times10^{-3}}(.110390\times10^{-2})=.620040$$

BINOMIAL DISTRIBUTION

Example 18. Find the value of p which satisfies

$$\sum_{s=0}^{20}\binom{50}{s}p^s q^{50-s}=.95, \qquad q=1-p$$

using **15.54** and **Table 4**.

Combining **15.54** and **15.56** we have

$$\sum_{s=a}^{n}\binom{n}{s}p^s q^{n-s}=Q(F|\nu_1,\nu_2)$$

where

$$\nu_1=2(n-a+1),\ \nu_2=2(a),\text{ and } p=\frac{a}{a+(n-a+1)F}$$

Hence

$$\sum_{s=0}^{20}\binom{50}{s}p^s q^{50-s}=1-\sum_{s=21}^{50}\binom{50}{s}p^s q^{50-s}$$

$$=1-Q(F|60,42)=.95$$

Harmonic interpolation on ν_2 in the table for which $Q(F|\nu_1,\nu_2)=.05$ results in $F=1.624$ for $\nu_1=60$, $\nu_2=42$, and thus $p=\frac{42}{42+60(1.624)}=.301$. The correct answer to 4D is $p=.3003$.

APPROXIMATING THE INCOMPLETE BETA FUNCTION

Example 19. Find $I_{.60}(16, 10.5)$ using **15.52**.

Values of $I_x(a, b)$ can conveniently be calculated with good accuracy using the approximation given by **15.51** or **15.52**. For this example

$(a+b-1)(1-x)=10.20$ which is greater than .8 and hence 15.52 will be used. Thus

$$w_1=[(10.5)(.60)]^{1/3}=1.8469, w_2=[16(.4)]^{1/3}=1.8566$$

$$y=\frac{3[(1.8469)(.98942)-(1.8566)(.99306)]}{\left[\frac{(1.8469)^2}{10.5}+\frac{(1.8566)^2}{16}\right]^{\frac{1}{2}}}=-.0668$$

and interpolating in Table 1 gives

$$P(-.0668)=1-P(.0668)=.47336$$

The answer correct to 5D is $I_{.60}(16, 10.5)=.47332$.

INTERPOLATION FOR F IN TABLE 4

Example 20. Find the value of F for which

$$Q(F|7, 20)=.05 \text{ using Table 4.}$$

Interpolation in Table 4 is approximately linear when the reciprocals of the degrees of freedom (ν_1, ν_2) are used as the interpolating variable. For this example it is only necessary to interpolate with respect to $1/\nu_1$. Thus linear interpolation on $1/\nu_1$ results in $F=2.51$ which is the correct interpolate.

CALCULATION OF F FOR $Q(F|\nu_1, \nu_2)>.50$

Example 21. Find the value of F for which $Q(F|4, 8)=.90$ using 15.59 and Table 4. Table 4 only tabulates values of F for which $Q(F|\nu_1, \nu_2)=p$ where $p=.500$, .250, .100, .050, .025, .010, .005, .001. However making, use of Table 4 we can find the values of F_p for which $p=.75$, .9, .95, .975, .99, .995, .999. For this example we have

$$F_{.90}(4,8)=\frac{1}{F_{.10}(8,4)}$$

and referring to the table for which $Q(F|\nu_1, \nu_2)=.10$ gives $F_{.10}(8,4)=3.95$ and thus $F_{.90}(4,8)=\frac{1}{3.95}$ $=.253$.

CALCULATION OF $Q(F|\nu_1,\nu_2)$ FOR SMALL INTEGRAL ν_1 OR ν_2

Example 22. Compute $Q(2.5|4, 15)$ using 15.57.

Values of $Q(F|\nu_1, \nu_2)$ can be readily computed for small ν_1 or ν_2 using the expansions 15.57 to 15.58 inclusive. We have using 15.57

$$x=\frac{15}{15+4(2.50)}=.60$$

and

$$Q(2.50|4,15)=(.6)^{7.5}\left[1+\frac{15}{2}(.4)\right]=.086\,735$$

APPROXIMATING $Q(F|\nu_1,\nu_2)$

Example 23. Calculate $Q(1.714|10,40)$ using 15.61.

The approximation given by 15.61 will result in a maximum error of .0005. For this example we have

$$x=\frac{(1.714)^{1/3}\left(1-\frac{2}{9(40)}\right)-\left(1-\frac{2}{9(10)}\right)}{\left[\frac{2}{9(10)}+(1.714)^{2/3}\,\frac{2}{9(40)}\right]^{\frac{1}{2}}}=1.2222$$

Interpolating in Table 1 results in

$$Q(1.714|10,40)\approx Q(1.2222)=1-P(1.2222)=.1108$$

The correct value to 5D is $Q(1.714|10, 40)=.11108$.

On the other hand the approximation given by 15.60 which is usually less accurate results in

$$x=\frac{\sqrt{[2(40)-1]\left(\frac{10}{40}\right)(1.714)}-\sqrt{2(10)-1}}{\sqrt{1+\frac{10}{40}(1.714)}}=1.2210$$

and interpolating in Table 1 gives

$$Q(1.714|10, 40)\approx Q(1.2210)=1-P(1.2210)=.1112$$

CALCULATION OF F OUTSIDE THE RANGE OF TABLE 4

Example 24. Find the value of F for which $Q(F|10, 20)\approx .0001$ using **15.62** and **15.53**.

For this problem we have $a=\frac{\nu_2}{2}=10$, $b=\frac{\nu_1}{2}=5$, $p=.0001$. The value of the normal deviate which cuts off .0001 in the tail of the distribution is $y=3.7190$ (i.e., $Q(3.7190)=.0001$). Hence substituting in **15.53** gives

$$h=2\left[\frac{1}{19}+\frac{1}{9}\right]^{-1}=12.2143$$

$$\lambda=\frac{3.7190^2-3}{6}=1.8052$$

$$w=3.7190\,\frac{(12.2143+1.8052)^{\frac{1}{2}}}{12.2143}-\left(\frac{1}{9}-\frac{1}{19}\right)\left[1.8052+.8333-\frac{2}{3(12.2143)}\right]$$

$$w=.9889$$

and thus $F\approx e^{2w}=7.23$. The correct answer is $F=7.180$.

APPROXIMATING THE NON-CENTRAL F-DISTRIBUTION

Example 25. Compute $P(3.71|3, 10, 4)$ using the approximation **15.63** to the non-central F-distribution.

Using **15.63** with $\nu_1=3$, $\nu_2=10$, $\lambda=4$, $F'=3.71$ we have

$$x=\frac{\left[\left(\frac{3}{3+4}\right)(3.71)\right]^{1/3}\left[1-\frac{2}{9(10)}\right]-\left[1-\frac{2}{9}\frac{(3+8)}{(3+4)^2}\right]}{\left[\frac{2}{9}\frac{3+8}{(3+4)^2}+\frac{2}{9(10)}\left[\left(\frac{3}{3+4}\right)(3.71)\right]^{2/3}\right]^{\frac{1}{2}}}$$

$$=.675$$

and interpolating in **Table 1** gives

$$P(3.71|3,10,4)\approx P(.675)=.750$$

The exact answer is $P(3.71|3,10,4)=.745$.

Table 1

NORMAL PROBABILITY FUNCTION AND DERIVATIVES

x	$P(x)$	$Z(x)$	$Z^{(1)}(x)$
0.00	0.50000 00000 00000	0.39894 22804 01433	0.00000 00000 00000
0.02	0.50797 83137 16902	0.39886 24999 23666	-0.00797 72499 98473
0.04	0.51595 34368 52831	0.39862 32542 04605	-0.01594 49301 68184
0.06	0.52392 21826 54107	0.39822 48301 95607	-0.02389 34898 11736
0.08	0.53188 13720 13988	0.39766 77055 11609	-0.03181 34164 40929
0.10	0.53982 78372 77029	0.39695 25474 77012	-0.03969 52547 47701
0.12	0.54775 84260 20584	0.39608 02117 93656	-0.04752 96254 15239
0.14	0.55567 00048 05907	0.39505 17408 34611	-0.05530 72437 16846
0.16	0.56355 94628 91433	0.39386 83615 68541	-0.06301 89378 50967
0.18	0.57142 37159 00901	0.39253 14831 20429	-0.07065 56669 61677
0.20	0.57925 97094 39103	0.39104 26939 75456	-0.07820 85387 95091
0.22	0.58706 44226 48215	0.38940 37588 33790	-0.08566 88269 43434
0.24	0.59483 48716 97796	0.38761 66151 25014	-0.09302 79876 30003
0.26	0.60256 81132 01761	0.38568 33691 91816	-0.10027 76759 89872
0.28	0.61026 12475 55797	0.38360 62921 53479	-0.10740 97618 02974
0.30	0.61791 14221 88953	0.38138 78154 60524	-0.11441 63446 38157
0.32	0.62551 58347 23320	0.37903 05261 52702	-0.12128 97683 68865
0.34	0.63307 17360 36028	0.37653 71618 33254	-0.12802 26350 23306
0.36	0.64057 64332 17991	0.37391 06053 73128	-0.13460 78179 34326
0.38	0.64802 72924 24163	0.37115 38793 59466	-0.14103 84741 56597
0.40	0.65542 17416 10324	0.36827 01403 03323	-0.14730 80561 21329
0.42	0.66275 72731 51751	0.36526 26726 22154	-0.15341 03225 01305
0.44	0.67003 14463 39407	0.36213 48824 13092	-0.15933 93482 61761
0.46	0.67724 18897 49653	0.35889 02910 33545	-0.16508 95338 75431
0.48	0.68438 63034 83778	0.35553 25285 05997	-0.17065 56136 82879
0.50	0.69146 24612 74013	0.35206 53267 64299	-0.17603 26633 82150
0.52	0.69846 82124 53034	0.34849 25127 58974	-0.18121 61066 34667
0.54	0.70540 14837 84302	0.34481 80014 39333	-0.18620 17207 77240
0.56	0.71226 02811 50973	0.34104 57886 30353	-0.19098 56416 32997
0.58	0.71904 26911 01436	0.33717 99438 22381	-0.19556 43674 16981
0.60	0.72574 68822 49927	0.33322 46028 91800	-0.19993 47617 35080
0.62	0.73237 11065 31017	0.32918 39607 70765	-0.20409 40556 77874
0.64	0.73891 37003 07139	0.32506 22640 84082	-0.20803 98490 13813
0.66	0.74537 30853 28664	0.32086 38037 71172	-0.21177 01104 88974
0.68	0.75174 77695 46430	0.31659 29077 10893	-0.21528 31772 43407
0.70	0.75803 63477 76927	0.31225 39333 66761	-0.21857 77533 56733
0.72	0.76423 75022 20749	0.30785 12604 69853	-0.22165 29075 38294
0.74	0.77035 00028 35210	0.30338 92837 56300	-0.22450 80699 79662
0.76	0.77637 27075 62401	0.29887 24057 75953	-0.22714 30283 89724
0.78	0.78230 45624 14267	0.29430 50297 88325	-0.22955 79232 34894
0.80	0.78814 46014 16604	0.28969 15527 61483	-0.23175 32422 09186
0.82	0.79389 19464 14187	0.28503 63584 89007	-0.23372 98139 60986
0.84	0.79954 58067 39551	0.28034 38108 39621	-0.23548 88011 05281
0.86	0.80510 54787 48192	0.27561 82471 53457	-0.23703 16925 51973
0.88	0.81057 03452 23288	0.27086 39717 98338	-0.23836 02951 82537
0.90	0.81593 98746 53241	0.26608 52498 98755	-0.23947 67249 08879
0.92	0.82121 36203 85629	0.26128 63012 49553	-0.24038 33971 49589
0.94	0.82639 12196 61376	0.25647 12944 25620	-0.24108 30167 60083
0.96	0.83147 23925 33162	0.25164 43410 98117	-0.24157 85674 54192
0.98	0.83645 69406 72308	0.24680 94905 67043	-0.24187 33007 55702
1.00	0.84134 47460 68543	0.24197 07245 19143	-0.24197 07245 19143
	$\left[\begin{matrix}(-5)1\\10\end{matrix}\right]$	$\left[\begin{matrix}(-5)2\\10\end{matrix}\right]$	$\left[\begin{matrix}(-5)3\\10\end{matrix}\right]$

$Z(x)=\frac{1}{\sqrt{2\pi}}e^{-\frac{1}{2}x^2}$ $P(x)=\int_{-\infty}^{x}Z(t)dt$ $Z^{(n)}(x)=\frac{d^n}{dx^n}Z(x)$ $He_n(x)=(-1)^nZ^{(n)}(x)/Z(x)$

Table 1

NORMAL PROBABILITY FUNCTION AND DERIVATIVES

x	$Z^{(2)}(x)$	$Z^{(3)}(x)$	$Z^{(4)}(x)$	$Z^{(5)}(x)$	$Z^{(6)}(x)$
0.00	-0.39894 22804	0.00000 000	1.19682 684	0.00000 000	-5.98413 421
0.02	-0.39870 29549	0.02392 856	1.19563 029	-0.11962 684	-5.97575 893
0.04	-0.39798 54570	0.04780 928	1.19204 400	-0.23891 887	-5.95066 325
0.06	-0.39679 12208	0.07159 445	1.18607 800	-0.35754 249	-5.90893 742
0.08	-0.39512 26322	0.09523 664	1.17774 897	-0.47516 649	-5.85073 151
0.10	-0.39298 30220	0.11868 881	1.16708 019	-0.59146 327	-5.77625 460
0.12	-0.39037 66567	0.14190 445	1.15410 144	-0.70610 997	-5.68577 399
0.14	-0.38730 87267	0.16483 771	1.13884 890	-0.81878 968	-5.57961 395
0.16	-0.38378 53315	0.18744 353	1.12136 503	-0.92919 252	-5.45815 435
0.18	-0.37981 34631	0.20967 776	1.10169 839	-1.03701 674	-5.32182 895
0.20	-0.37540 09862	0.23149 727	1.07990 350	-1.14196 980	-5.17112 356
0.22	-0.37055 66169	0.25286 011	1.05604 063	-1.24376 938	-5.00657 387
0.24	-0.36528 98981	0.27372 555	1.03017 556	-1.34214 434	-4.82876 317
0.26	-0.35961 11734	0.29405 426	1.00237 941	-1.43683 568	-4.63831 979
0.28	-0.35353 15588	0.31380 836	0.97272 834	-1.52759 737	-4.43591 441
0.30	-0.34706 29121	0.33295 156	0.94130 327	-1.61419 723	-4.22225 716
0.32	-0.34021 78003	0.35144 923	0.90818 965	-1.69641 762	-3.99809 459
0.34	-0.33300 94659	0.36926 849	0.87347 711	-1.77405 617	-3.76420 646
0.36	-0.32545 17909	0.38637 828	0.83725 919	-1.84692 643	-3.52140 244
0.38	-0.31755 92592	0.40274 947	0.79963 298	-1.91485 840	-3.27051 871
0.40	-0.30934 69179	0.41835 488	0.76069 880	-1.97769 904	-3.01241 439
0.42	-0.30083 03372	0.43316 939	0.72055 987	-2.03531 269	-2.74796 802
0.44	-0.29202 55692	0.44716 995	0.67932 193	-2.08758 144	-2.47807 382
0.46	-0.28294 91055	0.46033 566	0.63709 291	-2.13440 537	-2.20363 810
0.48	-0.27361 78339	0.47264 779	0.59398 256	-2.17570 278	-1.92557 548
0.50	-0.26404 89951	0.48408 982	0.55010 207	-2.21141 033	-1.64480 520
0.52	-0.25426 01373	0.49464 748	0.50556 372	-2.24148 307	-1.36224 740
0.54	-0.24426 90722	0.50430 874	0.46048 050	-2.26589 443	-1.07881 949
0.56	-0.23409 38293	0.51306 383	0.41496 574	-2.28463 613	-0.79543 249
0.58	-0.22375 26107	0.52090 525	0.36913 279	-2.29771 801	-0.51298 749
0.60	-0.21326 37459	0.52782 777	0.32309 457	-2.30516 783	-0.23237 218
0.62	-0.20264 56463	0.53382 841	0.27696 332	-2.30703 091	+0.04554 255
0.64	-0.19191 67607	0.53890 643	0.23085 017	-2.30336 981	0.31990 583
0.66	-0.18109 55308	0.54306 327	0.18486 483	-2.29426 388	0.58988 999
0.68	-0.17020 03472	0.54630 259	0.13911 528	-2.27980 875	0.85469 355
0.70	-0.15924 95060	0.54863 016	0.09370 741	-2.26011 583	1.11354 405
0.72	-0.14826 11670	0.55005 386	0.04874 473	-2.23531 162	1.36570 074
0.74	-0.13725 33120	0.55058 359	+0.00432 808	-2.20553 714	1.61045 709
0.76	-0.12624 37042	0.55023 127	-0.03944 465	-2.17094 715	1.84714 311
0.78	-0.11524 98497	0.54901 073	-0.08247 882	-2.13170 944	2.07512 746
0.80	-0.10428 89590	0.54693 765	-0.12468 324	-2.08800 401	2.29381 943
0.82	-0.09337 79110	0.54402 952	-0.16597 047	-2.04002 228	2.50267 061
0.84	-0.08253 32179	0.54030 551	-0.20625 697	-1.98796 617	2.70117 643
0.86	-0.07177 09916	0.53578 644	-0.24546 336	-1.93204 726	2.88887 745
0.88	-0.06110 69120	0.53049 467	-0.28351 458	-1.87248 587	3.06536 044
0.90	-0.05055 61975	0.52445 403	-0.32034 003	-1.80951 008	3.23025 923
0.92	-0.04013 35759	0.51768 968	-0.35587 378	-1.74335 486	3.38325 538
0.94	-0.02985 32587	0.51022 810	-0.39005 463	-1.67426 103	3.52407 854
0.96	-0.01972 89163	0.50209 689	-0.42282 627	-1.60247 436	3.65250 673
0.98	-0.00977 36558	0.49332 478	-0.45413 732	-1.52824 456	3.76836 628
1.00	0.00000 00000	0.48394 145	-0.48394 145	-1.45182 435	3.87153 159
	$\left[\begin{matrix}(-5)6\\6\end{matrix}\right]$	$\left[\begin{matrix}(-4)1\\6\end{matrix}\right]$	$\left[\begin{matrix}(-4)3\\6\end{matrix}\right]$	$\left[\begin{matrix}(-4)7\\6\end{matrix}\right]$	$\left[\begin{matrix}(-3)2\\7\end{matrix}\right]$

$P(-x)=1-P(x)$ $\quad$ $Z(-x)=Z(x)$ $\quad$ $Z^{(n)}(-x)=(-1)^n Z^{(n)}(x)$

Table 1

NORMAL PROBABILITY FUNCTION AND DERIVATIVES

x	$P(x)$	$Z(x)$	$Z^{(1)}(x)$
1.00	0.84134 47460 68543	0.24197 07245 19143	−0.24197 07245 19143
1.02	0.84613 57696 27265	0.23713 19520 19380	−0.24187 45910 59767
1.04	0.85083 00496 69019	0.23229 70047 43366	−0.24158 88849 33101
1.06	0.85542 77003 36091	0.22746 96324 57386	−0.24111 78104 04829
1.08	0.85992 89099 11231	0.22265 34987 51761	−0.24046 57786 51902
1.10	0.86433 39390 53618	0.21785 21770 32551	−0.23963 73947 35806
1.12	0.86864 31189 57270	0.21306 91467 75718	−0.23863 74443 88804
1.14	0.87285 68494 37202	0.20830 77900 47108	−0.23747 08806 53704
1.16	0.87697 55969 48657	0.20357 13882 90759	−0.23614 28104 17281
1.18	0.88099 98925 44800	0.19886 31193 87276	−0.23465 84808 76986
1.20	0.88493 03297 78292	0.19418 60549 83213	−0.23302 32659 79856
1.22	0.88876 75625 52166	0.18954 31580 91640	−0.23124 26528 71801
1.24	0.89251 23029 25413	0.18493 72809 63305	−0.22932 22283 94499
1.26	0.89616 53188 78700	0.18037 11632 27080	−0.22726 76656 66121
1.28	0.89972 74320 45558	0.17584 74302 97662	−0.22508 47107 81008
1.30	0.90319 95154 14390	0.17136 85920 47807	−0.22277 91696 62150
1.32	0.90658 24910 06528	0.16693 70417 41714	−0.22035 68950 99062
1.34	0.90987 73275 35548	0.16255 50552 25534	−0.21782 37740 02216
1.36	0.91308 50380 52915	0.15822 47903 70383	−0.21518 57149 03721
1.38	0.91620 66775 84986	0.15394 82867 62634	−0.21244 86357 32434
1.40	0.91924 33407 66229	0.14972 74656 35745	−0.20961 84518 90043
1.42	0.92219 61594 73454	0.14556 41300 37348	−0.20670 10646 53034
1.44	0.92506 63004 65673	0.14145 99652 24839	−0.20370 23499 23768
1.46	0.92785 49630 34106	0.13741 65392 82282	−0.20062 81473 52131
1.48	0.93056 33766 66669	0.13343 53039 51002	−0.19748 42498 47483
1.50	0.93319 27987 31142	0.12951 75956 65892	−0.19427 63934 98838
1.52	0.93574 45121 81064	0.12566 46367 89088	−0.19101 02479 19414
1.54	0.93821 98232 88188	0.12187 75370 32402	−0.18769 14070 29899
1.56	0.94062 00594 05207	0.11815 72950 59582	−0.18432 53802 92948
1.58	0.94294 65667 62246	0.11450 48002 59292	−0.18091 75844 09682
1.60	0.94520 07083 00442	0.11092 08346 79456	−0.17747 33354 87129
1.62	0.94738 38615 45748	0.10740 60751 13484	−0.17399 78416 83844
1.64	0.94949 74165 25897	0.10396 10953 28764	−0.17049 61963 39173
1.66	0.95154 27737 33277	0.10058 63684 27691	−0.16697 33715 89966
1.68	0.95352 13421 36280	0.09728 22693 31467	−0.16343 42124 76865
1.70	0.95543 45372 41457	0.09404 90773 76887	−0.15988 34315 40708
1.72	0.95728 37792 08671	0.09088 69790 16283	−0.15632 56039 08007
1.74	0.95907 04910 21193	0.08779 60706 10906	−0.15276 51628 62976
1.76	0.96079 60967 12518	0.08477 63613 08022	−0.14920 63959 02119
1.78	0.96246 20196 51483	0.08182 77759 92143	−0.14565 34412 66014
1.80	0.96406 96808 87074	0.07895 01583 00894	−0.14211 02849 41609
1.82	0.96562 04975 54110	0.07614 32736 96207	−0.13858 07581 27097
1.84	0.96711 58813 40836	0.07340 68125 81657	−0.13506 85351 50249
1.86	0.96855 72370 19248	0.07074 03934 56983	−0.13157 71318 29989
1.88	0.96994 59610 38800	0.06814 35661 01045	−0.12810 99042 69964
1.90	0.97128 34401 83998	0.06561 58147 74677	−0.12467 00480 71886
1.92	0.97257 10502 96163	0.06315 65614 35199	−0.12126 05979 55581
1.94	0.97381 01550 59548	0.06076 51689 54565	−0.11788 44277 71856
1.96	0.97500 21048 51780	0.05844 09443 33451	−0.11454 42508 93565
1.98	0.97614 82356 58492	0.05618 31419 03868	−0.11124 26209 69659
2.00	0.97724 98680 51821	0.05399 09665 13188	−0.10798 19330 26376
	$\begin{bmatrix}(-5)1\\10\end{bmatrix}$	$\begin{bmatrix}(-6)9\\10\end{bmatrix}$	$\begin{bmatrix}(-5)2\\10\end{bmatrix}$

$Z(x)=\frac{1}{\sqrt{2\pi}}e^{-\frac{1}{2}x^2}$ $\quad P(x)=\int_{-\infty}^{x} Z(t)dt$ $\quad Z^{(n)}(x)=\frac{d^n}{dx^n}Z(x)$ $\quad He_n(x)=(-1)^n Z^{(n)}(x)/Z(x)$

Table 1

NORMAL PROBABILITY FUNCTION AND DERIVATIVES

x	$Z^{(2)}(x)$	$Z^{(3)}(x)$	$Z^{(4)}(x)$	$Z^{(5)}(x)$	$Z^{(6)}(x)$
1.00	0.00000 00000	0.48394 145	−0.48394 145	−1.45182 435	3.87153 159
1.02	0.00958 01309	0.47397 745	−0.51219 739	−1.37346 846	3.96192 478
1.04	0.01895 54356	0.46346 412	−0.53886 899	−1.29343 272	4.03951 497
1.06	0.02811 52466	0.45243 346	−0.56392 521	−1.21197 312	4.10431 754
1.08	0.03704 95422	0.44091 805	−0.58734 012	−1.12934 487	4.15639 308
1.10	0.04574 89572	0.42895 094	−0.60909 290	−1.04580 155	4.19584 622
1.12	0.05420 47909	0.41656 552	−0.62916 776	−0.96159 420	4.22282 430
1.14	0.06240 90139	0.40379 549	−0.64755 390	−0.87697 050	4.23751 585
1.16	0.07035 42718	0.39067 467	−0.66424 543	−0.79217 397	4.24014 894
1.18	0.07803 38880	0.37723 697	−0.67924 129	−0.70744 317	4.23098 941
1.20	0.08544 18642	0.36351 629	−0.69254 515	−0.62301 100	4.21033 894
1.22	0.09257 28784	0.34954 639	−0.70416 524	−0.53910 399	4.17853 305
1.24	0.09942 22822	0.33536 083	−0.71411 427	−0.45594 161	4.13593 896
1.26	0.10598 60955	0.32099 285	−0.72240 928	−0.37373 571	4.08295 339
1.28	0.11226 09995	0.30647 534	−0.72907 143	−0.29268 993	4.02000 029
1.30	0.11824 43285	0.29184 071	−0.73412 591	−0.21299 916	3.94752 847
1.32	0.12393 40598	0.27712 083	−0.73760 168	−0.13484 911	3.86600 921
1.34	0.12932 88019	0.26234 695	−0.73953 132	−0.05841 584	3.77593 384
1.36	0.13442 77819	0.24754 965	−0.73995 087	+0.01613 459	3.67781 128
1.38	0.13923 08305	0.23275 873	−0.73889 953	0.08864 645	3.57216 556
1.40	0.14373 83670	0.21800 319	−0.73641 957	0.15897 463	3.45953 335
1.42	0.14795 13818	0.20331 117	−0.73255 600	0.22698 486	3.34046 152
1.44	0.15187 14187	0.18870 986	−0.72735 645	0.29255 386	3.21550 469
1.46	0.15550 05559	0.17422 548	−0.72087 087	0.35556 954	3.08522 283
1.48	0.15884 13858	0.15988 325	−0.71315 137	0.41593 103	2.95017 891
1.50	0.16189 69946	0.14570 730	−0.70425 193	0.47354 871	2.81093 657
1.52	0.16467 09400	0.13172 067	−0.69422 823	0.52834 425	2.66805 791
1.54	0.16716 72298	0.11794 528	−0.68313 742	0.58025 051	2.52210 132
1.56	0.16939 02982	0.10440 190	−0.67103 785	0.62921 147	2.37361 937
1.58	0.17134 49831	0.09111 010	−0.65798 890	0.67518 208	2.22315 681
1.60	0.17303 65021	0.07808 827	−0.64405 073	0.71812 810	2.07124 871
1.62	0.17447 04284	0.06535 359	−0.62928 410	0.75802 588	1.91841 857
1.64	0.17565 26667	0.05292 202	−0.61375 011	0.79486 211	1.76517 671
1.66	0.17658 94284	0.04080 829	−0.59751 005	0.82863 352	1.61201 862
1.68	0.17728 72076	0.02902 592	−0.58062 516	0.85934 661	1.45942 351
1.70	0.17775 27562	0.01758 718	−0.56315 647	0.88701 729	1.30785 296
1.72	0.17799 30597	+0.00650 315	−0.54516 459	0.91167 051	1.15774 966
1.74	0.17801 53128	−0.00421 632	−0.52670 954	0.93333 988	1.00953 633
1.76	0.17782 68955	−0.01456 254	−0.50785 061	0.95206 725	0.86361 469
1.78	0.17743 53495	−0.02452 804	−0.48864 614	0.96790 228	0.72036 463
1.80	0.17684 83546	−0.03410 647	−0.46915 342	0.98090 203	0.58014 345
1.82	0.17607 37061	−0.04329 263	−0.44942 853	0.99113 045	0.44328 526
1.84	0.17511 92921	−0.05208 243	−0.42952 621	0.99865 794	0.31010 045
1.86	0.17399 30717	−0.06047 285	−0.40949 971	1.00356 087	0.18087 536
1.88	0.17270 30539	−0.06846 193	−0.38940 073	1.00592 110	+0.05587 197
1.90	0.17125 72766	−0.07604 873	−0.36927 924	1.00582 548	−0.06467 219
1.92	0.16966 37866	−0.08323 327	−0.34918 347	1.00336 537	−0.18054 414
1.94	0.16793 06209	−0.09001 655	−0.32915 976	0.99863 613	−0.29155 530
1.96	0.16606 57874	−0.09640 044	−0.30925 250	0.99173 666	−0.39754 137
1.98	0.16407 72476	−0.10238 771	−0.28950 408	0.98276 891	−0.49836 204
2.00	0.16197 28995	−0.10798 193	−0.26995 483	0.97183 740	−0.59390 063
	$\left[\begin{matrix}(-5)4\\6\end{matrix}\right]$	$\left[\begin{matrix}(-5)7\\6\end{matrix}\right]$	$\left[\begin{matrix}(-4)2\\6\end{matrix}\right]$	$\left[\begin{matrix}(-4)4\\6\end{matrix}\right]$	$\left[\begin{matrix}(-3)1\\7\end{matrix}\right]$

$P(-x)=1-P(x) \qquad Z(-x)=Z(x) \qquad Z^{(n)}(-x)=(-1)^n Z^{(n)}(x)$

Table 1

NORMAL PROBABILITY FUNCTION AND DERIVATIVES

x	$P(x)$	$Z(x)$	$Z^{(1)}(x)$
2.00	0.97724 98680 51821	0.05399 09665 13188	-0.10798 19330 26376
2.02	0.97830 83062 32353	0.05186 35766 82821	-0.10476 44248 99298
2.04	0.97932 48371 33930	0.04980 00877 35071	-0.10159 21789 79544
2.06	0.98030 07295 90623	0.04779 95748 82077	-0.09846 71242 57079
2.08	0.98123 72335 65062	0.04586 10762 71055	-0.09539 10386 43794
2.10	0.98213 55794 37184	0.04398 35959 80427	-0.09236 55515 58897
2.12	0.98299 69773 52367	0.04216 61069 61770	-0.08939 21467 58953
2.14	0.98382 26166 27834	0.04040 75539 22860	-0.08647 21653 94921
2.16	0.98461 36652 16075	0.03870 68561 47456	-0.08360 68092 78504
2.18	0.98537 12692 24011	0.03706 29102 47806	-0.08079 71443 40218
2.20	0.98609 65524 86502	0.03547 45928 46231	-0.07804 41042 61709
2.22	0.98679 06161 92744	0.03394 07631 82449	-0.07534 84942 65037
2.24	0.98745 45385 64054	0.03246 02656 43697	-0.07271 09950 41882
2.26	0.98808 93745 81453	0.03103 19322 15008	-0.07013 21668 05919
2.28	0.98869 61557 61447	0.02965 45848 47341	-0.06761 24534 51938
2.30	0.98927 58899 78324	0.02832 70377 41601	-0.06515 21868 05683
2.32	0.98982 95613 31281	0.02704 80995 46882	-0.06275 15909 48766
2.34	0.99035 81300 54642	0.02581 65754 71588	-0.06041 07866 03515
2.36	0.99086 25324 69428	0.02463 12693 06382	-0.05812 97955 63063
2.38	0.99134 36809 74484	0.02349 09853 58201	-0.05590 85451 52519
2.40	0.99180 24640 75404	0.02239 45302 94843	-0.05374 68727 07623
2.42	0.99223 97464 49447	0.02134 07148 99923	-0.05164 45300 57813
2.44	0.99265 63690 44652	0.02032 83557 38226	-0.04960 11880 01271
2.46	0.99305 31492 11376	0.01935 62767 31737	-0.04761 64407 60073
2.48	0.99343 08808 64453	0.01842 33106 46862	-0.04568 98104 04218
2.50	0.99379 03346 74224	0.01752 83004 93569	-0.04382 07512 33921
2.52	0.99413 22582 84668	0.01667 01008 37381	-0.04200 86541 10200
2.54	0.99445 73765 56918	0.01584 75790 25361	-0.04025 28507 24416
2.56	0.99476 63918 36444	0.01505 96163 27377	-0.03855 26177 98086
2.58	0.99505 99842 42230	0.01430 51089 94150	-0.03690 71812 04906
2.60	0.99533 88119 76281	0.01358 29692 33686	-0.03531 57200 07583
2.62	0.99560 35116 51879	0.01289 21261 07895	-0.03377 73704 02686
2.64	0.99585 46986 38964	0.01223 15263 51278	-0.03229 12295 67374
2.66	0.99609 29674 25147	0.01160 01351 13703	-0.03085 63594 02449
2.68	0.99631 88919 90825	0.01099 69366 29406	-0.02947 17901 66807
2.70	0.99653 30261 96960	0.01042 09348 14423	-0.02813 65239 98941
2.72	0.99673 59041 84109	0.00987 11537 94751	-0.02684 95383 21723
2.74	0.99692 80407 81350	0.00934 66383 67612	-0.02560 97891 27258
2.76	0.99710 99319 23774	0.00884 64543 98237	-0.02441 62141 39135
2.78	0.99728 20550 77299	0.00836 96891 54653	-0.02326 77358 49935
2.80	0.99744 48696 69572	0.00791 54515 82980	-0.02216 32644 32344
2.82	0.99759 88175 25811	0.00748 28725 25781	-0.02110 17005 22701
2.84	0.99774 43233 08458	0.00707 11048 86019	-0.02008 19378 76295
2.86	0.99788 17949 59596	0.00667 93237 39203	-0.01910 28658 94119
2.88	0.99801 16241 45106	0.00630 67263 96266	-0.01816 33720 21246
2.90	0.99813 41866 99616	0.00595 25324 19776	-0.01726 23440 17350
2.92	0.99824 98430 71324	0.00561 59835 95991	-0.01639 86721 00294
2.94	0.99835 89387 65843	0.00529 63438 65311	-0.01557 12509 64014
2.96	0.99846 18047 88262	0.00499 28992 13612	-0.01477 89816 72293
2.98	0.99855 87580 82660	0.00470 49575 26934	-0.01402 07734 30263
3.00	0.99865 01019 68370	0.00443 18484 11938	-0.01329 55452 35814
	$\left[\begin{matrix}(-6)5\\10\end{matrix}\right]$	$\left[\begin{matrix}(-6)8\\10\end{matrix}\right]$	$\left[\begin{matrix}(-6)7\\10\end{matrix}\right]$

$Z(x)=\frac{1}{\sqrt{2\pi}}e^{-\frac{1}{2}x^2}$ $\quad P(x)=\int_{-\infty}^{x}Z(t)dt$ $\quad Z^{(n)}(x)=\frac{d^n}{dx^n}Z(x)$ $\quad He_n(x)=(-1)^nZ^{(n)}(x)/Z(x)$

Table 1

NORMAL PROBABILITY FUNCTION AND DERIVATIVES

x	$Z^{(2)}(x)$	$Z^{(3)}(x)$	$Z^{(4)}(x)$	$Z^{(5)}(x)$	$Z^{(6)}(x)$
.00	0.16197 28995	-0.10798 193	-0.26995 483	0.97183 740	-0.59390 063
.02	0.15976 05616	-0.11318 748	-0.25064 297	0.95904 873	-0.68406 360
.04	0.15744 79574	-0.11800 948	-0.23160 454	0.94451 117	-0.76878 007
.06	0.15504 27011	-0.12245 372	-0.21287 345	0.92833 417	-0.84800 114
.08	0.15255 22841	-0.12652 667	-0.19448 137	0.91062 795	-0.92169 927
.10	0.14998 40623	-0.13023 543	-0.17645 779	0.89150 307	-0.98986 750
.12	0.14734 52442	-0.13358 762	-0.15882 997	0.87107 003	-1.05251 862
.14	0.14464 28800	-0.13659 143	-0.14162 297	0.84943 890	-1.10968 436
.16	0.14188 38519	-0.13925 550	-0.12485 967	0.82671 890	-1.16141 446
.18	0.13907 48644	-0.14158 892	-0.10856 076	0.80301 811	-1.20777 570
.20	0.13622 24365	-0.14360 115	-0.09274 478	0.77844 311	-1.24885 097
.22	0.13333 28941	-0.14530 204	-0.07742 816	0.75309 866	-1.28473 823
.24	0.13041 23633	-0.14670 170	-0.06262 527	0.72708 743	-1.31554 947
.26	0.12746 67648	-0.14781 055	-0.04834 844	0.70050 969	-1.34140 971
.28	0.12450 18090	-0.14863 922	-0.03460 801	0.67346 314	-1.36245 589
.30	0.12152 29919	-0.14919 851	-0.02141 241	0.64604 257	-1.37883 587
.32	0.11853 55915	-0.14949 939	-0.00876 819	0.61833 976	-1.39070 730
.34	0.11554 46652	-0.14955 294	+0.00331 989	0.59044 323	-1.39823 661
.36	0.11255 50482	-0.14937 032	0.01484 882	0.56243 808	-1.40159 796
.38	0.10957 13521	-0.14896 273	0.02581 724	0.53440 589	-1.40097 220
.40	0.10659 79642	-0.14834 137	0.03622 539	0.50642 453	-1.39654 584
.42	0.10363 90478	-0.14751 744	0.04607 505	0.47856 812	-1.38851 010
.44	0.10069 85430	-0.14650 207	0.05536 942	0.45090 689	-1.37705 991
.46	0.09778 01675	-0.14530 633	0.06411 307	0.42350 717	-1.36239 299
.48	0.09488 74192	-0.14394 118	0.07231 187	0.39643 129	-1.34470 892
.50	0.09202 35776	-0.14241 744	0.07997 287	0.36973 759	-1.32420 833
.52	0.08919 17075	-0.14074 579	0.08710 428	0.34348 039	-1.30109 199
.54	0.08639 46618	-0.13893 674	0.09371 533	0.31771 001	-1.27556 010
.56	0.08363 50852	-0.13700 058	0.09981 624	0.29247 277	-1.24781 146
.58	0.08091 54185	-0.13494 742	0.10541 808	0.26781 102	-1.21804 284
.60	0.07823 79028	-0.13278 711	0.11053 277	0.24376 323	-1.18644 824
.62	0.07560 45843	-0.13052 927	0.11517 293	0.22036 399	-1.15321 833
.64	0.07301 73197	-0.12818 326	0.11935 186	0.19764 415	-1.11853 985
.66	0.07047 77809	-0.12575 818	0.12308 341	0.17563 084	-1.08259 509
.68	0.06798 74610	-0.12326 282	0.12638 196	0.15434 760	-1.04556 139
.70	0.06554 76800	-0.12070 569	0.12926 232	0.13381 449	-1.00761 072
.72	0.06315 95904	-0.11809 501	0.13173 965	0.11404 817	-0.96890 932
.74	0.06082 41838	-0.11543 869	0.13382 945	0.09506 206	-0.92961 727
.76	0.05854 22966	-0.11274 431	0.13554 741	0.07686 640	-0.88988 829
.78	0.05631 46165	-0.11001 916	0.13690 942	0.05946 846	-0.84986 942
.80	0.05414 16888	-0.10727 020	0.13793 149	0.04287 262	-0.80970 080
.82	0.05202 39229	-0.10450 406	0.13862 969	0.02708 053	-0.76951 553
.84	0.04996 15987	-0.10172 706	0.13902 007	+0.01209 127	-0.72943 954
.86	0.04795 48727	-0.09894 520	0.13911 867	-0.00209 857	-0.68959 143
.88	0.04600 37850	-0.09616 416	0.13894 142	-0.01549 465	-0.65008 248
.90	0.04410 82652	-0.09338 928	0.13850 412	-0.02810 482	-0.61101 661
.92	0.04226 81389	-0.09062 562	0.13782 240	-0.03993 892	-0.57249 036
.94	0.04048 31340	-0.08787 791	0.13691 166	-0.05100 863	-0.53459 292
.96	0.03875 28865	-0.08515 058	0.13578 706	-0.06132 737	-0.49740 627
.98	0.03707 69473	-0.08244 776	0.13446 347	-0.07091 012	-0.46100 520
.00	0.03545 47873	-0.07977 327	0.13295 545	-0.07977 327	-0.42545 745
	$\left[\begin{matrix}(-5)1\\6\end{matrix}\right]$	$\left[\begin{matrix}(-5)5\\6\end{matrix}\right]$	$\left[\begin{matrix}(-5)7\\6\end{matrix}\right]$	$\left[\begin{matrix}(-4)2\\6\end{matrix}\right]$	$\left[\begin{matrix}(-4)7\\6\end{matrix}\right]$

$P(-x)=1-P(x)$ $\quad Z(-x)=Z(x)$ $\quad Z^{(n)}(-x)=(-1)^n Z^{(n)}(x)$

Table 1

NORMAL PROBABILITY FUNCTION AND DERIVATIVES

x	$P(x)$	$Z(x)$	$Z^{(1)}(x)$
3.00	0.99865 01020	(-3) 4.43184 8412	(-2) -1.32955 45
3.05	0.99885 57932	(-3) 3.80976 2098	(-2) -1.16197 74
3.10	0.99903 23968	(-3) 3.26681 9056	(-2) -1.01271 39
3.15	0.99918 36477	(-3) 2.79425 8415	(-3) -8.80191 40
3.20	0.99931 28621	(-3) 2.38408 8201	(-3) -7.62908 22
3.25	0.99942 29750	(-3) 2.02904 8057	(-3) -6.59440 62
3.30	0.99951 65759	(-3) 1.72256 8939	(-3) -5.68447 75
3.35	0.99959 59422	(-3) 1.45873 0805	(-3) -4.88674 82
3.40	0.99966 30707	(-3) 1.23221 9168	(-3) -4.18954 52
3.45	0.99971 97067	(-3) 1.03828 1296	(-3) -3.58207 05
3.50	0.99976 73709	(-4) 8.72682 6950	(-3) -3.05438 94
3.55	0.99980 73844	(-4) 7.31664 4628	(-3) -2.59740 88
3.60	0.99984 08914	(-4) 6.11901 9301	(-3) -2.20284 69
3.65	0.99986 88798	(-4) 5.10464 9743	(-3) -1.86319 72
3.70	0.99989 22003	(-4) 4.24780 2706	(-3) -1.57168 70
3.75	0.99991 15827	(-4) 3.52595 6824	(-3) -1.32223 38
3.80	0.99992 76520	(-4) 2.91946 9258	(-3) -1.10939 83
3.85	0.99994 09411	(-4) 2.41126 5802	(-4) -9.28337 33
3.90	0.99995 19037	(-4) 1.98655 4714	(-4) -7.74756 34
3.95	0.99996 09244	(-4) 1.63256 4088	(-4) -6.44862 81
4.00	0.99996 83288	(-4) 1.33830 2258	(-4) -5.35320 90
4.05	0.99997 43912	(-4) 1.09434 0434	(-4) -4.43207 88
4.10	0.99997 93425	(-5) 8.92616 5718	(-4) -3.65972 79
4.15	0.99998 33762	(-5) 7.26259 3030	(-4) -3.01397 61
4.20	0.99998 66543	(-5) 5.89430 6776	(-4) -2.47560 88
4.25	0.99998 93115	(-5) 4.77186 3654	(-4) -2.02804 21
4.30	0.99999 14601	(-5) 3.85351 9674	(-4) -1.65701 35
4.35	0.99999 31931	(-5) 3.10414 0706	(-4) -1.35030 12
4.40	0.99999 45875	(-5) 2.49424 7129	(-4) -1.09746 87
4.45	0.99999 57065	(-5) 1.99917 9671	(-5) -8.89634 95
4.50	0.99999 66023	(-5) 1.59837 4111	(-5) -7.19268 35
4.55	0.99999 73177	(-5) 1.27473 3238	(-5) -5.80003 62
4.60	0.99999 78875	(-5) 1.01408 5207	(-5) -4.66479 20
4.65	0.99999 83403	(-6) 8.04718 2456	(-5) -3.74193 98
4.70	0.99999 86992	(-6) 6.36982 5179	(-5) -2.99381 78
4.75	0.99999 89829	(-6) 5.02950 7289	(-5) -2.38901 60
4.80	0.99999 92067	(-6) 3.96129 9091	(-5) -1.90142 36
4.85	0.99999 93827	(-6) 3.11217 5579	(-5) -1.50940 52
4.90	0.99999 95208	(-6) 2.43896 0746	(-5) -1.19509 08
4.95	0.99999 96289	(-6) 1.90660 0903	(-6) -9.43767 45
5.00	0.99999 97133	(-6) 1.48671 9515	(-6) -7.43359 76
	$\left[{(-6)3 \atop 7}\right]$		

NORMAL PROBABILITY FUNCTION FOR LARGE ARGUMENTS

x	$-\log Q(x)$	x	$-\log Q(x)$	x	$-\log Q(x)$
5	6.54265	15	50.43522	25	137.51475
6	9.00586	16	57.19458	26	148.60624
7	11.89285	17	64.38658	27	160.13139
8	15.20614	18	72.01140	28	172.09024
9	18.94746	19	80.06919	29	184.48283
10	23.11805	20	88.56010	30	197.30921
11	27.71882	21	97.48422	31	210.56940
12	32.75044	22	106.84167	32	224.26344
13	38.21345	23	116.63253	33	238.39135
14	44.10827	24	126.85686	34	252.95315
	$\left[{(-2)5 \atop 5}\right]$		$\left[{(-2)5 \atop 4}\right]$		$\left[{(-2)5 \atop 3}\right]$

Table 1

NORMAL PROBABILITY FUNCTION AND DERIVATIVES

x	$Z^{(2)}(x)$	$Z^{(3)}(x)$	$Z^{(4)}(x)$	$Z^{(5)}(x)$	$Z^{(6)}(x)$
3.00	(-2) 3.54547 87	(-2) -7.97732 71	(-1) 1.32955 45	(-2) -7.97732 71	(-1) -4.25457 45
3.05	(-2) 3.16305 50	(-2) -7.32336 28	(-1) 1.28470 92	(-2) -9.89017 82	(-1) -3.40704 15
3.10	(-2) 2.81273 12	(-2) -6.69403 89	(-1) 1.23133 27	(-1) -1.13951 58	(-1) -2.62416 45
3.15	(-2) 2.49317 71	(-2) -6.09312 50	(-1) 1.17138 12	(-1) -1.25260 09	(-1) -1.91121 33
3.20	(-2) 2.20289 75	(-2) -5.52345 55	(-1) 1.10663 65	(-1) -1.33185 47	(-1) -1.27124 77
3.25	(-2) 1.94027 72	(-2) -4.98701 97	(-1) 1.03869 82	(-1) -1.38096 14	(-2) -7.05366 66
3.30	(-2) 1.70362 07	(-2) -4.48505 27	(-2) 9.68981 20	(-1) -1.40361 69	(-2) -2.12970 34
3.35	(-2) 1.49118 76	(-2) -4.01812 87	(-2) 8.98716 85	(-1) -1.40345 00	(-2) +2.07973 11
3.40	(-2) 1.30122 34	(-2) -3.58625 07	(-2) 8.28958 19	(-1) -1.38395 76	(-2) 5.60664 85
3.45	(-2) 1.13198 62	(-2) -3.18893 82	(-2) 7.60587 84	(-1) -1.34845 27	(-2) 8.49222 78
3.50	(-3) 9.81768 03	(-2) -2.82531 02	(-2) 6.94328 17	(-1) -1.30002 45	(-1) 1.07844 49
3.55	(-3) 8.48913 69	(-2) -2.49416 18	(-2) 6.30753 35	(-1) -1.24150 96	(-1) 1.25359 25
3.60	(-3) 7.31834 71	(-2) -2.19403 56	(-2) 5.70302 39	(-1) -1.17547 44	(-1) 1.38019 58
3.65	(-3) 6.29020 46	(-2) -1.92328 53	(-2) 5.13292 98	(-1) -1.10420 53	(-1) 1.46388 44
3.70	(-3) 5.39046 16	(-2) -1.68013 34	(-2) 4.59935 51	(-1) -1.02970 80	(-1) 1.51024 21
3.75	(-3) 4.60578 11	(-2) -1.46272 12	(-2) 4.10347 00	(-2) -9.53712 78	(-1) 1.52468 79
3.80	(-3) 3.92376 67	(-2) -1.26915 17	(-2) 3.64564 64	(-2) -8.77684 95	(-1) 1.51237 96
3.85	(-3) 3.33297 22	(-2) -1.09752 68	(-2) 3.22558 66	(-2) -8.02840 11	(-1) 1.47814 11
3.90	(-3) 2.82289 42	(-3) -9.45977 49	(-2) 2.84244 39	(-2) -7.30162 14	(-1) 1.42641 04
3.95	(-3) 2.38395 17	(-3) -8.12688 36	(-2) 2.49493 35	(-2) -6.60423 39	(-1) 1.36120 56
4.00	(-3) 2.00745 34	(-3) -6.95917 17	(-2) 2.18143 27	(-2) -5.94206 20	(-1) 1.28610 85
4.05	(-3) 1.68555 79	(-3) -5.94009 36	(-2) 1.90007 05	(-2) -5.31924 82	(-1) 1.20426 03
4.10	(-3) 1.41122 68	(-3) -5.05408 43	(-2) 1.64880 65	(-2) -4.73847 30	(-1) 1.11837 07
4.15	(-3) 1.17817 42	(-3) -4.28662 75	(-2) 1.42549 82	(-2) -4.20116 64	(-1) 1.03073 50
4.20	(-4) 9.80812 65	(-3) -3.62429 14	(-2) 1.22795 86	(-2) -3.70770 95	(-2) 9.43258 69
4.25	(-4) 8.14199 24	(-3) -3.05473 83	(-2) 1.05400 40	(-2) -3.25762 18	(-2) 8.57487 24
4.30	(-4) 6.73980 59	(-3) -2.56671 38	(-3) 9.01492 78	(-2) -2.84973 34	(-2) 7.74638 98
4.35	(-4) 5.56339 62	(-3) -2.15001 71	(-3) 7.68355 55	(-2) -2.48233 98	(-2) 6.95640 04
4.40	(-4) 4.57943 77	(-3) -1.79545 89	(-3) 6.52618 76	(-2) -2.15333 90	(-2) 6.21159 79
4.45	(-4) 3.75895 76	(-3) -1.49480 91	(-3) 5.52421 34	(-2) -1.86035 13	(-2) 5.51645 66
4.50	(-4) 3.07687 02	(-3) -1.24073 79	(-3) 4.66025 95	(-2) -1.60082 16	(-2) 4.87356 75
4.55	(-4) 2.51154 32	(-3) -1.02675 14	(-3) 3.91825 60	(-2) -1.37210 59	(-2) 4.28395 39
4.60	(-4) 2.04439 58	(-4) -8.47126 22	(-3) 3.28346 19	(-2) -1.17154 20	(-2) 3.74736 21
4.65	(-4) 1.65953 02	(-4) -6.96842 75	(-3) 2.74245 97	(-3) -9.96506 67	(-2) 3.26252 61
4.70	(-4) 1.34339 61	(-4) -5.71519 82	(-3) 2.28312 43	(-3) -8.44460 51	(-2) 2.82740 22
4.75	(-4) 1.08448 75	(-4) -4.67351 25	(-3) 1.89457 22	(-3) -7.12981 28	(-2) 2.43937 50
4.80	(-5) 8.73070 32	(-4) -3.81045 28	(-3) 1.56709 63	(-3) -5.99788 09	(-2) 2.09543 47
4.85	(-5) 7.00939 74	(-4) -3.09767 67	(-3) 1.29209 13	(-3) -5.02757 21	(-2) 1.79232 68
4.90	(-5) 5.61204 87	(-4) -2.51088 57	(-3) 1.06197 25	(-3) -4.19931 11	(-2) 1.52667 62
4.95	(-5) 4.48098 88	(-4) -2.02933 60	(-4) 8.70091 63	(-3) -3.49521 92	(-2) 1.29508 77
5.00	(-5) 3.56812 68	(-4) -1.63539 15	(-4) 7.10651 93	(-3) -2.89910 31	(-2) 1.09422 56

NORMAL PROBABILITY FUNCTION FOR LARGE ARGUMENTS

x	$-\log Q(x)$	x	$-\log Q(x)$	x	$-\log Q(x)$
35	267.94888	45	441.77568	100	2173.87154
36	283.37855	46	461.54561	150	4888.38812
37	299.24218	47	481.74964	200	8688.58977
38	315.53979	48	502.38776	250	13574.49960
39	332.27139	49	523.45999	300	19546.12790
40	349.43701	50	544.96634	350	26603.48018
41	367.03664	60	783.90743	400	34746.55970
42	385.07032	70	1066.26576	450	43975.36860
43	403.53804	80	1392.04459	500	54289.90830
44	422.43983	90	1761.24604		

$$Q(x)=1-P(x)=\frac{1}{\sqrt{2\pi}}\int_x^\infty e^{-\frac{1}{2}t^2}dt \qquad Z(x)=\frac{1}{\sqrt{2\pi}}e^{-\frac{1}{2}x^2} \qquad P(x)=\int_{-\infty}^x Z(t)dt \qquad Z^{(n)}(x)=\frac{d^n}{dx^n}Z(x)$$

$$He_n(x)=(-1)^n Z^{(n)}(x)/Z(x) \qquad P(-x)=1-P(x) \qquad Z(-x)=Z(x) \qquad Z^{(n)}(-x)=(-1)^n Z^{(n)}(x)$$

x	$Z^{(7)}(x)$	$Z^{(8)}(x)$	$Z^{(9)}(x)$	$Z^{(10)}(x)$	$Z^{(11)}(x)$	$Z^{(12)}(x)$
0.0	0.00000 00	(1) 4.18889 39	0.00000 00	(2)-3.77000 46	0.00000 00	(3) 4.14700 50
0.1	(0) 4.12640 51	(1) 4.00211 42	(1)-3.70133 55	(2)-3.56488 94	(2) 4.05782 44	(3) 3.88080 01
0.2	(0) 7.88604 35	(1) 3.46206 56	(1)-7.00124 79	(2)-2.97583 41	(2) 7.59641 48	(3) 3.12148 92
0.3	(1) 1.09518 61	(1) 2.62702 42	(1)-9.54959 57	(2)-2.07783 39	(3) 1.01729 46	(3) 1.98042 89
0.4	(1) 1.30711 60	(1) 1.58584 37	(2)-1.10912 65	(1)-9.83608 69	(3) 1.14847 09	(2)+6.22581 20
0.5	(1) 1.40908 65	(0)+4.46820 41	(2)-1.14961 02	(1)+1.72666 73	(3) 1.14097 69	(2)-7.60421 83
0.6	(1) 1.39704 30	(0)-6.75565 29	(2)-1.07710 05	(2) 1.25426 91	(3) 1.00184 44	(3)-1.98080 26
0.7	(1) 1.27812 14	(1)-1.67416 58	(1)-9.05305 52	(2) 2.14046 31	(2) 7.55473 11	(3)-2.88334 06
0.8	(1) 1.06929 69	(1)-2.46111 11	(1)-6.58548 60	(2) 2.74183 89	(2) 4.39201 49	(3)-3.36738 39
0.9	(0) 7.94982 72	(1)-2.97666 59	(1)-3.68086 24	(2) 3.01027 69	(1)+9.71613 18	(3)-3.39874 98
1.0	(0) 4.83941 45	(1)-3.19401 36	(0)-6.77518 03	(2) 2.94236 40	(2)-2.26484 60	(3)-3.01011 58
1.1	(0)+1.65937 85	(1)-3.11962 40	(1)+2.10408 36	(2) 2.57621 24	(2)-4.93791 72	(3)-2.29066 27
1.2	(0)-1.31434 07	(1)-2.78951 64	(1) 4.39889 22	(2) 1.98269 77	(2)-6.77812 94	(3)-1.36759 19
1.3	(0)-3.85379 20	(1)-2.26227 70	(1) 6.02399 37	(2) 1.25293 01	(2)-7.65280 28	(2)-3.83358 74
1.4	(0)-5.79719 45	(1)-1.61006 61	(1) 6.89184 82	(1)+4.84200 76	(2)-7.56972 92	(2)+5.27141 25
1.5	(0)-7.05769 71	(0)-9.09001 03	(1) 7.00965 92	(1)-2.33347 96	(2)-6.65963 73	(3) 1.25562 83
1.6	(0)-7.62276 66	(0)-2.30231 44	(1) 6.46658 36	(1)-8.27445 07	(2)-5.14267 14	(3) 1.73301 70
1.7	(0)-7.54545 38	(0)+3.67230 07	(1) 5.41207 19	(2)-1.25055 93	(2)-3.28612 11	(3) 1.93425 58
1.8	(0)-6.92967 04	(0) 8.41240 26	(1) 4.02950 39	(2)-1.48242 69	(2)-1.36113 54	(3) 1.87567 40
1.9	(0)-5.91207 57	(1) 1.16856 49	(1) 2.50938 72	(2)-1.52849 20	(1)+3.94747 58	(3) 1.60633 92
2.0	(0)-4.64322 31	(1) 1.34437 51	(1)+1.02582 84	(2)-1.41510 32	(2) 1.80437 81	(3) 1.19573 79
2.1	(0)-3.27029 67	(1) 1.37966 95	(0)-2.81068 72	(2)-1.18267 82	(2) 2.76469 29	(2) 7.20360 48
2.2	(0)-1.92318 65	(1) 1.29729 67	(1)-1.31550 35	(1)-8.78156 27	(2) 3.24744 73	(2)+2.51533 48
2.3	(-1)-7.04932 91	(1) 1.12731 97	(1)-2.02888 89	(1)-5.47943 26	(2) 3.28915 84	(2)-1.53768 85
2.4	(-1)+3.13162 82	(0) 9.02423 01	(1)-2.41634 55	(1)-2.32257 79	(2) 2.97376 42	(2)-4.58219 83
2.5	(0) 1.09209 53	(0) 6.53922 01	(1)-2.50848 12	(0)+3.85905 05	(2) 2.41200 50	(2)-6.45450 80
2.6	(0) 1.62218 61	(0) 4.08745 39	(1)-2.36048 69	(1) 2.45855 73	(2) 1.72126 20	(2)-7.17969 42
2.7	(0) 1.91766 20	(0) 1.87558 77	(1)-2.04053 83	(1) 3.82142 44	(2) 1.00875 37	(2)-6.92720 18
2.8	(0) 2.00992 65	(-2)+4.01113 24	(1)-1.61917 24	(1) 4.49758 25	(1)+3.59849 29	(2)-5.95491 88
2.9	(0) 1.94057 71	(0)-1.35055 73	(1)-1.16080 01	(1) 4.58182 18	(1)-1.67928 25	(2)-4.55301 20
3.0	(0) 1.75501 20	(0)-2.28683 38	(0)-7.17959 44	(1) 4.21202 87	(1)-5.45649 18	(2)-2.99628 41
3.1	(0) 1.49720 05	(0)-2.80440 64	(0)-3.28394 42	(1) 3.54198 84	(1)-7.69621 99	(2)-1.51035 91
3.2	(0) 1.20591 21	(0)-2.96904 52	(-1)-1.46351 84	(1) 2.71897 33	(1)-8.55436 26	(1)-2.53474 56
3.3	(-1) 9.12450 33	(0)-2.86200 69	(0)+2.14502 00	(1) 1.86794 96	(1)-8.30925 36	(1)+6.87309 15
3.4	(-1) 6.39748 51	(0)-2.56761 03	(0) 3.61188 70	(1) 1.08280 77	(1)-7.29343 32	(2) 1.28867 88
3.5	(-1) 4.02558 98	(0)-2.16386 79	(0) 4.35306 57	(0)+4.23908 09	(1)-5.83674 40	(2) 1.57656 15
3.6	(-1) 2.08414 13	(0)-1.71642 80	(0) 4.51182 76	(-1)-7.94727 62	(1)-4.22572 56	(2) 1.60868 13
3.7	(-2)+5.90352 21	(0)-1.27559 98	(0) 4.24743 76	(0)-4.23512 06	(1)-2.68044 29	(2) 1.45762 72
3.8	(-2)-4.80932 87	(-1)-8.75911 24	(0) 3.71320 90	(0)-6.22699 31	(1)-1.34695 16	(2) 1.19681 09
3.9	(-1)-1.18202 76	(-1)-5.37496 49	(0) 3.04185 84	(0)-7.02577 94	(0)-3.01804 44	(1) 8.90539 46
4.0	(-1)-1.57919 67	(-1)-2.68597 26	(0) 2.33774 64	(0)-6.93361 02	(0)+4.35697 68	(1) 5.88418 05
4.1	(-1)-1.74223 60	(-2)-6.85427 28	(0) 1.67481 40	(0)-6.24985 27	(0) 8.87625 64	(1) 3.23557 28
4.2	(-1)-1.73706 08	(-2)+6.92844 60	(0) 1.09865 39	(0)-5.23790 66	(1) 1.10126 69	(1)+1.13637 65
4.3	(-1)-1.62110 76	(-1) 1.54828 96	(-1) 6.31121 50	(0)-4.10728 31	(1) 1.13501 02	(0)-3.62532 62
4.4	(-1)-1.44109 96	(-1) 1.99272 00	(-1) 2.76082 94	(0)-3.00821 29	(1) 1.04753 07	(1)-1.30010 10
4.5	(-1)-1.23261 24	(-1) 2.13525 86	(-2)+2.52235 61	(0)-2.03523 88	(0) 8.90633 89	(1)-1.76908 98
4.6	(-1)-1.02086 14	(-1) 2.07280 89	(-1)-1.36802 99	(0)-1.23623 43	(0) 7.05470 76	(1)-1.88530 78
4.7	(-2)-8.22202 74	(-1) 1.88517 13	(-1)-2.28268 33	(-1)-6.23793 04	(0) 5.21451 06	(1)-1.76464 76
4.8	(-2)-6.45935 81	(-1) 1.63368 76	(-1)-2.67421 39	(-1)-1.86696 14	(0) 3.57035 54	(1)-1.50840 48
4.9	(-2)-4.96112 66	(-1) 1.36227 87	(-1)-2.70626 44	(-1)+1.00018 72	(0) 2.21617 27	(1)-1.19594 52
5.0	(-2)-3.73166 60	(-1) 1.09987 51	(-1)-2.51404 27	(-1) 2.67133 76	(0) 1.17837 39	(0)-8.83034 08

$Z(x)=\frac{1}{\sqrt{2\pi}}e^{-\frac{1}{2}x^2}$ $\quad Z^{(n)}(x)=\frac{d^n}{dx^n}Z(x)$ $\quad He_n(x)=(-1)^nZ^{(n)}(x)/Z(x)$ $\quad Z^{(n)}(-x)=(-1)^nZ^{(n)}(x)$

NORMAL PROBABILITY FUNCTION—VALUES OF $Z(x)$ IN TERMS OF $P(x)$ AND $Q(x)$

$Q(x)$	0.000	0.001	0.002	0.003	0.004	0.005	0.006	0.007	0.008	0.009	0.010	
0.00	0.00000	0.00337	0.00634	0.00915	0.01185	0.01446	0.01700	0.01949	0.02192	0.02431	0.02665	0.99
0.01	0.02665	0.02896	0.03123	0.03348	0.03569	0.03787	0.04003	0.04216	0.04427	0.04635	0.04842	0.98
0.02	0.04842	0.05046	0.05249	0.05449	0.05648	0.05845	0.06040	0.06233	0.06425	0.06615	0.06804	0.97
0.03	0.06804	0.06992	0.07177	0.07362	0.07545	0.07727	0.07908	0.08087	0.08265	0.08442	0.08617	0.96
0.04	0.08617	0.08792	0.08965	0.09137	0.09309	0.09479	0.09648	0.09816	0.09983	0.10149	0.10314	0.95
0.05	0.10314	0.10478	0.10641	0.10803	0.10964	0.11124	0.11284	0.11442	0.11600	0.11756	0.11912	0.94
0.06	0.11912	0.12067	0.12222	0.12375	0.12528	0.12679	0.12830	0.12981	0.13130	0.13279	0.13427	0.93
0.07	0.13427	0.13574	0.13720	0.13866	0.14011	0.14156	0.14299	0.14442	0.14584	0.14726	0.14867	0.92
0.08	0.14867	0.15007	0.15146	0.15285	0.15423	0.15561	0.15698	0.15834	0.15970	0.16105	0.16239	0.91
0.09	0.16239	0.16373	0.16506	0.16639	0.16770	0.16902	0.17033	0.17163	0.17292	0.17421	0.17550	0.90
0.10	0.17550	0.17678	0.17805	0.17932	0.18057	0.18184	0.18309	0.18433	0.18557	0.18681	0.18804	0.89
0.11	0.18804	0.18926	0.19048	0.19169	0.19290	0.19410	0.19530	0.19649	0.19768	0.19886	0.20004	0.88
0.12	0.20004	0.20121	0.20238	0.20354	0.20470	0.20585	0.20700	0.20814	0.20928	0.21042	0.21155	0.87
0.13	0.21155	0.21267	0.21379	0.21490	0.21601	0.21712	0.21822	0.21932	0.22041	0.22149	0.22258	0.86
0.14	0.22258	0.22365	0.22473	0.22580	0.22686	0.22792	0.22898	0.23003	0.23108	0.23212	0.23316	0.85
0.15	0.23316	0.23419	0.23522	0.23625	0.23727	0.23829	0.23930	0.24031	0.24131	0.24232	0.24331	0.84
0.16	0.24331	0.24430	0.24529	0.24628	0.24726	0.24823	0.24921	0.25017	0.25114	0.25210	0.25305	0.83
0.17	0.25305	0.25401	0.25495	0.25590	0.25684	0.25778	0.25871	0.25964	0.26056	0.26148	0.26240	0.82
0.18	0.26240	0.26331	0.26422	0.26513	0.26603	0.26693	0.26782	0.26871	0.26960	0.27049	0.27137	0.81
0.19	0.27137	0.27224	0.27311	0.27398	0.27485	0.27571	0.27657	0.27742	0.27827	0.27912	0.27996	0.80
0.20	0.27996	0.28080	0.28164	0.28247	0.28330	0.28413	0.28495	0.28577	0.28658	0.28739	0.28820	0.79
0.21	0.28820	0.28901	0.28981	0.29060	0.29140	0.29219	0.29298	0.29376	0.29454	0.29532	0.29609	0.78
0.22	0.29609	0.29686	0.29763	0.29840	0.29916	0.29991	0.30067	0.30142	0.30216	0.30291	0.30365	0.77
0.23	0.30365	0.30439	0.30512	0.30585	0.30658	0.30730	0.30802	0.30874	0.30945	0.31016	0.31087	0.76
0.24	0.31087	0.31158	0.31228	0.31298	0.31367	0.31436	0.31505	0.31574	0.31642	0.31710	0.31778	0.75
0.25	0.31778	0.31845	0.31912	0.31979	0.32045	0.32111	0.32177	0.32242	0.32307	0.32372	0.32437	0.74
0.26	0.32437	0.32501	0.32565	0.32628	0.32691	0.32754	0.32817	0.32879	0.32941	0.33003	0.33065	0.73
0.27	0.33065	0.33126	0.33187	0.33247	0.33307	0.33367	0.33427	0.33486	0.33545	0.33604	0.33662	0.72
0.28	0.33662	0.33720	0.33778	0.33836	0.33893	0.33950	0.34007	0.34063	0.34119	0.34175	0.34230	0.71
0.29	0.34230	0.34286	0.34341	0.34395	0.34449	0.34503	0.34557	0.34611	0.34664	0.34717	0.34769	0.70
0.30	0.34769	0.34822	0.34874	0.34925	0.34977	0.35028	0.35079	0.35129	0.35180	0.35230	0.35279	0.69
0.31	0.35279	0.35329	0.35378	0.35427	0.35475	0.35524	0.35572	0.35620	0.35667	0.35714	0.35761	0.68
0.32	0.35761	0.35808	0.35854	0.35900	0.35946	0.35991	0.36037	0.36082	0.36126	0.36171	0.36215	0.67
0.33	0.36215	0.36259	0.36302	0.36346	0.36389	0.36431	0.36474	0.36516	0.36558	0.36600	0.36641	0.66
0.34	0.36641	0.36682	0.36723	0.36764	0.36804	0.36844	0.36884	0.36923	0.36962	0.37001	0.37040	0.65
0.35	0.37040	0.37078	0.37116	0.37154	0.37192	0.37229	0.37266	0.37303	0.37340	0.37376	0.37412	0.64
0.36	0.37412	0.37447	0.37483	0.37518	0.37553	0.37588	0.37622	0.37656	0.37690	0.37724	0.37757	0.63
0.37	0.37757	0.37790	0.37823	0.37855	0.37888	0.37920	0.37951	0.37983	0.38014	0.38045	0.38076	0.62
0.38	0.38076	0.38106	0.38136	0.38166	0.38196	0.38225	0.38254	0.38283	0.38312	0.38340	0.38368	0.61
0.39	0.38368	0.38396	0.38423	0.38451	0.38478	0.38504	0.38531	0.38557	0.38583	0.38609	0.38634	0.60
0.40	0.38634	0.38659	0.38684	0.38709	0.38734	0.38758	0.38782	0.38805	0.38829	0.38852	0.38875	0.59
0.41	0.38875	0.38897	0.38920	0.38942	0.38964	0.38985	0.39007	0.39028	0.39049	0.39069	0.39089	0.58
0.42	0.39089	0.39109	0.39129	0.39149	0.39168	0.39187	0.39206	0.39224	0.39243	0.39261	0.39279	0.57
0.43	0.39279	0.39296	0.39313	0.39330	0.39347	0.39364	0.39380	0.39396	0.39411	0.39427	0.39442	0.56
0.44	0.39442	0.39457	0.39472	0.39486	0.39501	0.39514	0.39528	0.39542	0.39555	0.39568	0.39580	0.55
0.45	0.39580	0.39593	0.39605	0.39617	0.39629	0.39640	0.39651	0.39662	0.39673	0.39683	0.39694	0.54
0.46	0.39694	0.39703	0.39713	0.39723	0.39732	0.39741	0.39749	0.39758	0.39766	0.39774	0.39781	0.53
0.47	0.39781	0.39789	0.39796	0.39803	0.39809	0.39816	0.39822	0.39828	0.39834	0.39839	0.39844	0.52
0.48	0.39844	0.39849	0.39854	0.39858	0.39862	0.39866	0.39870	0.39873	0.39876	0.39879	0.39882	0.51
0.49	0.39882	0.39884	0.39886	0.39888	0.39890	0.39891	0.39892	0.39893	0.39894	0.39894	0.39894	0.50
	0.010	0.009	0.008	0.007	0.006	0.005	0.004	0.003	0.002	0.001	0.000	$P(x)$

Linear interpolation yields an error no greater than 5 units in the fifth decimal place.

$$Z(x)=\frac{1}{\sqrt{2\pi}}e^{-\frac{1}{2}x^2} \qquad P(x)=1-Q(x)=\int_{-\infty}^{x} Z(t)\,dt$$

NORMAL PROBABILITY FUNCTION—VALUES OF x IN TERMS OF $P(x)$ AND $Q(x)$

$Q(x)$	0.000	0.001	0.002	0.003	0.004	0.005	0.006	0.007	0.008	0.009	0.010	
0.00	∞	3.09023	2.87816	2.74778	2.65207	2.57583	2.51214	2.45726	2.40892	2.36562	2.32635	0.99
0.01	2.32635	2.29037	2.25713	2.22621	2.19729	2.17009	2.14441	2.12007	2.09693	2.07485	2.05375	0.98
0.02	2.05375	2.03352	2.01409	1.99539	1.97737	1.95996	1.94313	1.92684	1.91104	1.89570	1.88079	0.97
0.03	1.88079	1.86630	1.85218	1.83842	1.82501	1.81191	1.79912	1.78661	1.77438	1.76241	1.75069	0.96
0.04	1.75069	1.73920	1.72793	1.71689	1.70604	1.69540	1.68494	1.67466	1.66456	1.65463	1.64485	0.95
0.05	1.64485	1.63523	1.62576	1.61644	1.60725	1.59819	1.58927	1.58047	1.57179	1.56322	1.55477	0.94
0.06	1.55477	1.54643	1.53820	1.53007	1.52204	1.51410	1.50626	1.49851	1.49085	1.48328	1.47579	0.93
0.07	1.47579	1.46838	1.46106	1.45381	1.44663	1.43953	1.43250	1.42554	1.41865	1.41183	1.40507	0.92
0.08	1.40507	1.39838	1.39174	1.38517	1.37866	1.37220	1.36581	1.35946	1.35317	1.34694	1.34076	0.91
0.09	1.34076	1.33462	1.32854	1.32251	1.31652	1.31058	1.30469	1.29884	1.29303	1.28727	1.28155	0.90
0.10	1.28155	1.27587	1.27024	1.26464	1.25908	1.25357	1.24808	1.24264	1.23723	1.23186	1.22653	0.89
0.11	1.22653	1.22123	1.21596	1.21072	1.20553	1.20036	1.19522	1.19012	1.18504	1.18000	1.17499	0.88
0.12	1.17499	1.17000	1.16505	1.16012	1.15522	1.15035	1.14551	1.14069	1.13590	1.13113	1.12639	0.87
0.13	1.12639	1.12168	1.11699	1.11232	1.10768	1.10306	1.09847	1.09390	1.08935	1.08482	1.08032	0.86
0.14	1.08032	1.07584	1.07138	1.06694	1.06252	1.05812	1.05374	1.04939	1.04505	1.04073	1.03643	0.85
0.15	1.03643	1.03215	1.02789	1.02365	1.01943	1.01522	1.01103	1.00686	1.00271	0.99858	0.99446	0.84
0.16	0.99446	0.99036	0.98627	0.98220	0.97815	0.97411	0.97009	0.96609	0.96210	0.95812	0.95416	0.83
0.17	0.95416	0.95022	0.94629	0.94238	0.93848	0.93458	0.93072	0.92686	0.92301	0.91918	0.91537	0.82
0.18	0.91537	0.91156	0.90777	0.90399	0.90023	0.89647	0.89273	0.88901	0.88529	0.88159	0.87790	0.81
0.19	0.87790	0.87422	0.87055	0.86689	0.86325	0.85962	0.85600	0.85239	0.84879	0.84520	0.84162	0.80
0.20	0.84162	0.83805	0.83450	0.83095	0.82742	0.82390	0.82038	0.81687	0.81338	0.80990	0.80642	0.79
0.21	0.80642	0.80296	0.79950	0.79606	0.79262	0.78919	0.78577	0.78237	0.77897	0.77557	0.77219	0.78
0.22	0.77219	0.76882	0.76546	0.76210	0.75875	0.75542	0.75208	0.74876	0.74545	0.74214	0.73885	0.77
0.23	0.73885	0.73556	0.73228	0.72900	0.72574	0.72248	0.71923	0.71599	0.71275	0.70952	0.70630	0.76
0.24	0.70630	0.70309	0.69988	0.69668	0.69349	0.69031	0.68713	0.68396	0.68080	0.67764	0.67449	0.75
0.25	0.67449	0.67135	0.66821	0.66508	0.66196	0.65884	0.65573	0.65262	0.64952	0.64643	0.64335	0.74
0.26	0.64335	0.64027	0.63719	0.63412	0.63106	0.62801	0.62496	0.62191	0.61887	0.61584	0.61281	0.73
0.27	0.61281	0.60979	0.60678	0.60376	0.60076	0.59776	0.59477	0.59178	0.58879	0.58581	0.58284	0.72
0.28	0.58284	0.57987	0.57691	0.57395	0.57100	0.56805	0.56511	0.56217	0.55924	0.55631	0.55338	0.71
0.29	0.55338	0.55047	0.54755	0.54464	0.54174	0.53884	0.53594	0.53305	0.53016	0.52728	0.52440	0.70
0.30	0.52440	0.52153	0.51866	0.51579	0.51293	0.51007	0.50722	0.50437	0.50153	0.49869	0.49585	0.69
0.31	0.49585	0.49302	0.49019	0.48736	0.48454	0.48173	0.47891	0.47610	0.47330	0.47050	0.46770	0.68
0.32	0.46770	0.46490	0.46211	0.45933	0.45654	0.45376	0.45099	0.44821	0.44544	0.44268	0.43991	0.67
0.33	0.43991	0.43715	0.43440	0.43164	0.42889	0.42615	0.42340	0.42066	0.41793	0.41519	0.41246	0.66
0.34	0.41246	0.40974	0.40701	0.40429	0.40157	0.39886	0.39614	0.39343	0.39073	0.38802	0.38532	0.65
0.35	0.38532	0.38262	0.37993	0.37723	0.37454	0.37186	0.36917	0.36649	0.36381	0.36113	0.35846	0.64
0.36	0.35846	0.35579	0.35312	0.35045	0.34779	0.34513	0.34247	0.33981	0.33716	0.33450	0.33185	0.63
0.37	0.33185	0.32921	0.32656	0.32392	0.32128	0.31864	0.31600	0.31337	0.31074	0.30811	0.30548	0.62
0.38	0.30548	0.30286	0.30023	0.29761	0.29499	0.29237	0.28976	0.28715	0.28454	0.28193	0.27932	0.61
0.39	0.27932	0.27671	0.27411	0.27151	0.26891	0.26631	0.26371	0.26112	0.25853	0.25594	0.25335	0.60
0.40	0.25335	0.25076	0.24817	0.24559	0.24301	0.24043	0.23785	0.23527	0.23269	0.23012	0.22754	0.59
0.41	0.22754	0.22497	0.22240	0.21983	0.21727	0.21470	0.21214	0.20957	0.20701	0.20445	0.20189	0.58
0.42	0.20189	0.19934	0.19678	0.19422	0.19167	0.18912	0.18657	0.18402	0.18147	0.17892	0.17637	0.57
0.43	0.17637	0.17383	0.17128	0.16874	0.16620	0.16366	0.16112	0.15858	0.15604	0.15351	0.15097	0.56
0.44	0.15097	0.14843	0.14590	0.14337	0.14084	0.13830	0.13577	0.13324	0.13072	0.12819	0.12566	0.55
0.45	0.12566	0.12314	0.12061	0.11809	0.11556	0.11304	0.11052	0.10799	0.10547	0.10295	0.10043	0.54
0.46	0.10043	0.09791	0.09540	0.09288	0.09036	0.08784	0.08533	0.08281	0.08030	0.07778	0.07527	0.53
0.47	0.07527	0.07276	0.07024	0.06773	0.06522	0.06271	0.06020	0.05768	0.05517	0.05266	0.05015	0.52
0.48	0.05015	0.04764	0.04513	0.04263	0.04012	0.03761	0.03510	0.03259	0.03008	0.02758	0.02507	0.51
0.49	0.02507	0.02256	0.02005	0.01755	0.01504	0.01253	0.01003	0.00752	0.00501	0.00251	0.00000	0.50
	0.010	0.009	0.008	0.007	0.006	0.005	0.004	0.003	0.002	0.001	0.000	$P(x)$

For $Q(x)>0.007$, linear interpolation yields an error of one unit in the third decimal place; five-point interpolation is necessary to obtain full accuracy.

$$P(x)=1-Q(x)=\int_{-\infty}^{x} Z(t)dt$$

NORMAL PROBABILITY FUNCTION—VALUES OF x FOR EXTREME VALUES OF $P(x)$ AND $Q(x)$

$Q(x)$	0.0000	0.0001	0.0002	0.0003	0.0004	0.0005	0.0006	0.0007	0.0008	0.0009	0.0010	
0.000	∞	3.71902	3.54008	3.43161	3.35279	3.29053	3.23888	3.19465	3.15591	3.12139	3.09023	0.999
0.001	3.09023	3.06181	3.03567	3.01145	2.98888	2.96774	2.94784	2.92905	2.91124	2.89430	2.87816	0.998
0.002	2.87816	2.86274	2.84796	2.83379	2.82016	2.80703	2.79438	2.78215	2.77033	2.75888	2.74778	0.997
0.003	2.74778	2.73701	2.72655	2.71638	2.70648	2.69684	2.68745	2.67829	2.66934	2.66061	2.65207	0.996
0.004	2.65207	2.64372	2.63555	2.62756	2.61973	2.61205	2.60453	2.59715	2.58991	2.58281	2.57583	0.995
0.005	2.57583	2.56897	2.56224	2.55562	2.54910	2.54270	2.53640	2.53019	2.52408	2.51807	2.51214	0.994
0.006	2.51214	2.50631	2.50055	2.49488	2.48929	2.48377	2.47833	2.47296	2.46765	2.46243	2.45726	0.993
0.007	2.45726	2.45216	2.44713	2.44215	2.43724	2.43238	2.42758	2.42283	2.41814	2.41350	2.40891	0.992
0.008	2.40891	2.40437	2.39989	2.39545	2.39106	2.38671	2.38240	2.37814	2.37392	2.36975	2.36562	0.991
0.009	2.36562	2.36152	2.35747	2.35345	2.34947	2.34553	2.34162	2.33775	2.33392	2.33012	2.32635	0.990
0.010	2.32635	2.32261	2.31891	2.31524	2.31160	2.30798	2.30440	2.30085	2.29733	2.29383	2.29037	0.989
0.011	2.29037	2.28693	2.28352	2.28013	2.27677	2.27343	2.27013	2.26684	2.26358	2.26034	2.25713	0.988
0.012	2.25713	2.25394	2.25077	2.24763	2.24450	2.24140	2.23832	2.23526	2.23223	2.22921	2.22621	0.987
0.013	2.22621	2.22323	2.22028	2.21734	2.21442	2.21152	2.20864	2.20577	2.20293	2.20010	2.19729	0.986
0.014	2.19729	2.19449	2.19172	2.18896	2.18621	2.18349	2.18078	2.17808	2.17540	2.17274	2.17009	0.985
0.015	2.17009	2.16746	2.16484	2.16224	2.15965	2.15707	2.15451	2.15197	2.14943	2.14692	2.14441	0.984
0.016	2.14441	2.14192	2.13944	2.13698	2.13452	2.13208	2.12966	2.12724	2.12484	2.12245	2.12007	0.983
0.017	2.12007	2.11771	2.11535	2.11301	2.11068	2.10836	2.10605	2.10375	2.10147	2.09919	2.09693	0.982
0.018	2.09693	2.09467	2.09243	2.09020	2.08798	2.08576	2.08356	2.08137	2.07919	2.07702	2.07485	0.981
0.019	2.07485	2.07270	2.07056	2.06843	2.06630	2.06419	2.06208	2.05998	2.05790	2.05582	2.05375	0.980
0.020	2.05375	2.05169	2.04964	2.04759	2.04556	2.04353	2.04151	2.03950	2.03750	2.03551	2.03352	0.979
0.021	2.03352	2.03154	2.02957	2.02761	2.02566	2.02371	2.02177	2.01984	2.01792	2.01600	2.01409	0.978
0.022	2.01409	2.01219	2.01029	2.00841	2.00653	2.00465	2.00279	2.00093	1.99908	1.99723	1.99539	0.977
0.023	1.99539	1.99356	1.99174	1.98992	1.98811	1.98630	1.98450	1.98271	1.98092	1.97914	1.97737	0.976
0.024	1.97737	1.97560	1.97384	1.97208	1.97033	1.96859	1.96685	1.96512	1.96340	1.96168	1.95996	0.975
	0.0010	0.0009	0.0008	0.0007	0.0006	0.0005	0.0004	0.0003	0.0002	0.0001	0.0000	$P(x)$

For $Q(x)>0.0007$, linear interpolation yields an error of one unit in the third decimal place; five-point interpolation is necessary to obtain full accuracy.

$Q(x)$	x	$Q(x)$	x	$Q(x)$	x	$Q(x)$	x
(-4) 1.0	3.71902	(- 9) 1.0	5.99781	(-14) 1.0	7.65063	(-19) 1.0	9.01327
(-5) 1.0	4.26489	(-10) 1.0	6.36134	(-15) 1.0	7.94135	(-20) 1.0	9.26234
(-6) 1.0	4.75342	(-11) 1.0	6.70602	(-16) 1.0	8.22208	(-21) 1.0	9.50502
(-7) 1.0	5.19934	(-12) 1.0	7.03448	(-17) 1.0	8.49379	(-22) 1.0	9.74179
(-8) 1.0	5.61200	(-13) 1.0	7.34880	(-18) 1.0	8.75729	(-23) 1.0	9.97305

$$P(x)=1-Q(x)=\int_{-\infty}^{x} Z(t)dt$$

Table 2

PROBABILITY INTEGRAL OF χ^2-DISTRIBUTION, INCOMPLETE GAMMA FUNCTION CUMULATIVE SUMS OF THE POISSON DISTRIBUTION

ν	χ^2=0.001	0.002	0.003	0.004	0.005	0.006	0.007	0.008	0.009	0.010
	m=0.0005	0.0010	0.0015	0.0020	0.0025	0.0030	0.0035	0.0040	0.0045	0.0050
1	0.97477	0.96433	0.95632	0.94957	0.94363	0.93826	0.93332	0.92873	0.92442	0.92034
2	0.99950	0.99900	0.99850	0.99800	0.99750	0.99700	0.99651	0.99601	0.99551	0.99501
3	0.99999	0.99998	0.99996	0.99993	0.99991	0.99988	0.99984	0.99981	0.99977	0.99973
4							0.99999	0.99999	0.99999	0.99999
ν	χ^2=0.01	0.02	0.03	0.04	0.05	0.06	0.07	0.08	0.09	0.10
	m=0.005	0.010	0.015	0.020	0.025	0.030	0.035	0.040	0.045	0.050
1	0.92034	0.88754	0.86249	0.84148	0.82306	0.80650	0.79134	0.77730	0.76418	0.75183
2	0.99501	0.99005	0.98511	0.98020	0.97531	0.97045	0.96561	0.96079	0.95600	0.95123
3	0.99973	0.99925	0.99863	0.99790	0.99707	0.99616	0.99518	0.99412	0.99301	0.99184
4	0.99999	0.99995	0.99989	0.99980	0.99969	0.99956	0.99940	0.99922	0.99902	0.99879
5			0.99999	0.99998	0.99997	0.99995	0.99993	0.99991	0.99987	0.99984
6							0.99999	0.99999	0.99999	0.99998
ν	χ^2=0.1	0.2	0.3	0.4	0.5	0.6	0.7	0.8	0.9	1.0
	m=0.05	0.10	0.15	0.20	0.25	0.30	0.35	0.40	0.45	0.50
1	0.75183	0.65472	0.58388	0.52709	0.47950	0.43858	0.40278	0.37109	0.34278	0.31731
2	0.95123	0.90484	0.86071	0.81873	0.77880	0.74082	0.70469	0.67032	0.63763	0.60653
3	0.99184	0.97759	0.96003	0.94024	0.91889	0.89643	0.87320	0.84947	0.82543	0.80125
4	0.99879	0.99532	0.98981	0.98248	0.97350	0.96306	0.95133	0.93845	0.92456	0.90980
5	0.99984	0.99911	0.99764	0.99533	0.99212	0.98800	0.98297	0.97703	0.97022	0.96257
6	0.99998	0.99985	0.99950	0.99885	0.99784	0.99640	0.99449	0.99207	0.98912	0.98561
7		0.99997	0.99990	0.99974	0.99945	0.99899	0.99834	0.99744	0.99628	0.99483
8			0.99998	0.99994	0.99987	0.99973	0.99953	0.99922	0.99880	0.99825
9				0.99999	0.99997	0.99993	0.99987	0.99978	0.99964	0.99944
10					0.99999	0.99998	0.99997	0.99994	0.99989	0.99983
11							0.99999	0.99998	0.99997	0.99995
12									0.99999	0.99999
ν	χ^2=1.1	1.2	1.3	1.4	1.5	1.6	1.7	1.8	1.9	2.0
	m=0.55	0.60	0.65	0.70	0.75	0.80	0.85	0.90	0.95	1.00
1	0.29427	0.27332	0.25421	0.23672	0.22067	0.20590	0.19229	0.17971	0.16808	0.15730
2	0.57695	0.54881	0.52205	0.49659	0.47237	0.44933	0.42741	0.40657	0.38674	0.36788
3	0.77707	0.75300	0.72913	0.70553	0.68227	0.65939	0.63693	0.61493	0.59342	0.57241
4	0.89427	0.87810	0.86138	0.84420	0.82664	0.80879	0.79072	0.77248	0.75414	0.73576
5	0.95410	0.94488	0.93493	0.92431	0.91307	0.90125	0.88890	0.87607	0.86280	0.84915
6	0.98154	0.97689	0.97166	0.96586	0.95949	0.95258	0.94512	0.93714	0.92866	0.91970
7	0.99305	0.99093	0.98844	0.98557	0.98231	0.97864	0.97457	0.97008	0.96517	0.95984
8	0.99753	0.99664	0.99555	0.99425	0.99271	0.99092	0.98887	0.98654	0.98393	0.98101
9	0.99917	0.99882	0.99838	0.99782	0.99715	0.99633	0.99537	0.99425	0.99295	0.99147
10	0.99973	0.99961	0.99944	0.99921	0.99894	0.99859	0.99817	0.99766	0.99705	0.99634
11	0.99992	0.99987	0.99981	0.99973	0.99962	0.99948	0.99930	0.99908	0.99882	0.99850
12	0.99998	0.99996	0.99994	0.99991	0.99987	0.99982	0.99975	0.99966	0.99954	0.99941
13	0.99999	0.99999	0.99998	0.99997	0.99996	0.99994	0.99991	0.99988	0.99983	0.99977
14			0.99999	0.99999	0.99999	0.99998	0.99997	0.99996	0.99994	0.99992
15						0.99999	0.99999	0.99999	0.99998	0.99997
16									0.99999	0.99999

$$Q(\chi^2|\nu)=1-P(\chi^2|\nu)=\left[2^{\frac{\nu}{2}}\Gamma\left(\frac{\nu}{2}\right)\right]^{-1}\int_{\chi^2}^{\infty}e^{-\frac{t}{2}}t^{\frac{\nu}{2}-1}dt=\left[\Gamma\left(\frac{\nu}{2}\right)\right]^{-1}\int_{\frac{1}{2}\chi^2}^{\infty}e^{-t}t^{\frac{\nu}{2}-1}dt=\sum_{j=0}^{c-1}e^{-m}m^j/j!\ (\nu \text{ even},\ c=\tfrac{1}{2}\nu,\ m=\tfrac{1}{2}\chi^2)$$

Table 2

PROBABILITY INTEGRAL OF χ^2-DISTRIBUTION, INCOMPLETE GAMMA FUNCTION CUMULATIVE SUMS OF THE POISSON DISTRIBUTION

ν	$\chi^2=2.2$ $m=1.1$	2.4 1.2	2.6 1.3	2.8 1.4	3.0 1.5	3.2 1.6	3.4 1.7	3.6 1.8	3.8 1.9	4.0 2.0
1	0.13801	0.12134	0.10686	0.09426	0.08327	0.07364	0.06520	0.05778	0.05125	0.04550
2	0.33287	0.30119	0.27253	0.24660	0.22313	0.20190	0.18268	0.16530	0.14957	0.13534
3	0.53195	0.49363	0.45749	0.42350	0.39163	0.36181	0.33397	0.30802	0.28389	0.26146
4	0.69903	0.66263	0.62682	0.59183	0.55783	0.52493	0.49325	0.46284	0.43375	0.40601
5	0.82084	0.79147	0.76137	0.73079	0.69999	0.66918	0.63857	0.60831	0.57856	0.54942
6	0.90042	0.87949	0.85711	0.83350	0.80885	0.78336	0.75722	0.73062	0.70372	0.67668
7	0.94795	0.93444	0.91938	0.90287	0.88500	0.86590	0.84570	0.82452	0.80250	0.77978
8	0.97426	0.96623	0.95691	0.94628	0.93436	0.92119	0.90681	0.89129	0.87470	0.85712
9	0.98790	0.98345	0.97807	0.97170	0.96430	0.95583	0.94631	0.93572	0.92408	0.91141
10	0.99457	0.99225	0.98934	0.98575	0.98142	0.97632	0.97039	0.96359	0.95592	0.94735
11	0.99766	0.99652	0.99503	0.99311	0.99073	0.98781	0.98431	0.98019	0.97541	0.96992
12	0.99903	0.99850	0.99777	0.99680	0.99554	0.99396	0.99200	0.98962	0.98678	0.98344
13	0.99961	0.99938	0.99903	0.99856	0.99793	0.99711	0.99606	0.99475	0.99314	0.99119
14	0.99985	0.99975	0.99960	0.99938	0.99907	0.99866	0.99813	0.99743	0.99655	0.99547
15	0.99994	0.99990	0.99984	0.99974	0.99960	0.99940	0.99913	0.99878	0.99832	0.99774
16	0.99998	0.99996	0.99994	0.99989	0.99983	0.99974	0.99961	0.99944	0.99921	0.99890
17	0.99999	0.99999	0.99998	0.99996	0.99993	0.99989	0.99983	0.99975	0.99964	0.99948
18			0.99999	0.99998	0.99997	0.99995	0.99993	0.99989	0.99984	0.99976
19				0.99999	0.99999	0.99998	0.99997	0.99995	0.99993	0.99989
20						0.99999	0.99999	0.99998	0.99997	0.99995
21								0.99999	0.99999	0.99998
22										0.99999

ν	$\chi^2=4.2$ $m=2.1$	4.4 2.2	4.6 2.3	4.8 2.4	5.0 2.5	5.2 2.6	5.4 2.7	5.6 2.8	5.8 2.9	6.0 3.0
1	0.04042	0.03594	0.03197	0.02846	0.02535	0.02259	0.02014	0.01796	0.01603	0.01431
2	0.12246	0.11080	0.10026	0.09072	0.08209	0.07427	0.06721	0.06081	0.05502	0.04979
3	0.24066	0.22139	0.20354	0.18704	0.17180	0.15772	0.14474	0.13278	0.12176	0.11161
4	0.37962	0.35457	0.33085	0.30844	0.28730	0.26739	0.24866	0.23108	0.21459	0.19915
5	0.52099	0.49337	0.46662	0.44077	0.41588	0.39196	0.36904	0.34711	0.32617	0.30622
6	0.64963	0.62271	0.59604	0.56971	0.54381	0.51843	0.49363	0.46945	0.44596	0.42319
7	0.75647	0.73272	0.70864	0.68435	0.65996	0.63557	0.61127	0.58715	0.56329	0.53975
8	0.83864	0.81935	0.79935	0.77872	0.75758	0.73600	0.71409	0.69194	0.66962	0.64723
9	0.89776	0.88317	0.86769	0.85138	0.83431	0.81654	0.79814	0.77919	0.75976	0.73992
10	0.93787	0.92750	0.91625	0.90413	0.89118	0.87742	0.86291	0.84768	0.83178	0.81526
11	0.96370	0.95672	0.94898	0.94046	0.93117	0.92109	0.91026	0.89868	0.88637	0.87337
12	0.97955	0.97509	0.97002	0.96433	0.95798	0.95096	0.94327	0.93489	0.92583	0.91608
13	0.98887	0.98614	0.98298	0.97934	0.97519	0.97052	0.96530	0.95951	0.95313	0.94615
14	0.99414	0.99254	0.99064	0.98841	0.98581	0.98283	0.97943	0.97559	0.97128	0.96649
15	0.99701	0.99610	0.99501	0.99369	0.99213	0.99029	0.98816	0.98571	0.98291	0.97975
16	0.99851	0.99802	0.99741	0.99666	0.99575	0.99467	0.99338	0.99187	0.99012	0.98810
17	0.99928	0.99902	0.99869	0.99828	0.99777	0.99715	0.99639	0.99550	0.99443	0.99319
18	0.99966	0.99953	0.99936	0.99914	0.99886	0.99851	0.99809	0.99757	0.99694	0.99620
19	0.99985	0.99978	0.99969	0.99958	0.99943	0.99924	0.99901	0.99872	0.99836	0.99793
20	0.99993	0.99990	0.99986	0.99980	0.99972	0.99962	0.99950	0.99934	0.99914	0.99890
21	0.99997	0.99995	0.99993	0.99991	0.99987	0.99982	0.99975	0.99967	0.99956	0.99943
22	0.99999	0.99998	0.99997	0.99996	0.99994	0.99991	0.99988	0.99984	0.99978	0.99971
23	0.99999	0.99999	0.99999	0.99998	0.99997	0.99996	0.99994	0.99992	0.99989	0.99986
24			0.99999	0.99999	0.99999	0.99998	0.99997	0.99996	0.99995	0.99993
25					0.99999	0.99999	0.99999	0.99998	0.99998	0.99997
26								0.99999	0.99999	0.99998
27									0.99999	0.99999

$\phi=\frac{1}{2}\left(\chi^2-\chi_0^2\right)$ $\quad w=\nu-\nu_0>0$

Interpolation on χ^2

$$Q(\chi^2|\nu)=Q\left(\chi_0^2\middle|\nu_0-4\right)\left[\tfrac{1}{2}\phi^2\right]+Q\left(\chi_0^2\middle|\nu_0-2\right)\left[\phi-\phi^2\right]+Q\left(\chi_0^2\middle|\nu_0\right)\left[1-\phi+\tfrac{1}{2}\phi^2\right]$$

Double Entry Interpolation

$$Q\left(\chi^2\middle|\nu\right)=Q\left(\chi_0^2\middle|\nu_0-4\right)\left[\tfrac{1}{2}\phi^2\right]+Q\left(\chi_0^2\middle|\nu_0-2\right)\left[\phi-\phi^2-w\phi\right]+Q\left(\chi_0^2\middle|\nu_0-1\right)\left[\tfrac{1}{2}w^2-\tfrac{1}{2}w+w\phi\right]$$

$$+Q\left(\chi_0^2\middle|\nu_0\right)\left[1-w^2-\phi+\tfrac{1}{2}\phi^2+w\phi\right]+Q\left(\chi_0^2\middle|\nu_0+1\right)\left[\tfrac{1}{2}w^2+\tfrac{1}{2}w-w\phi\right]$$

Table 2

PROBABILITY INTEGRAL OF χ^2-DISTRIBUTION, INCOMPLETE GAMMA FUNCTION CUMULATIVE SUMS OF THE POISSON DISTRIBUTION

	$\chi^2=6.2$	6.4	6.6	6.8	7.0	7.2	7.4	7.6	7.8	8.0
ν	$m=3.1$	3.2	3.3	3.4	3.5	3.6	3.7	3.8	3.9	4.0
1	0.01278	0.01141	0.01020	0.00912	0.00815	0.00729	0.00652	0.00584	0.00522	0.00468
2	0.04505	0.04076	0.03688	0.03337	0.03020	0.02732	0.02472	0.02237	0.02024	0.01832
3	0.10228	0.09369	0.08580	0.07855	0.07190	0.06579	0.06018	0.05504	0.05033	0.04601
4	0.18470	0.17120	0.15860	0.14684	0.13589	0.12569	0.11620	0.10738	0.09919	0.09158
5	0.28724	0.26922	0.25213	0.23595	0.22064	0.20619	0.19255	0.17970	0.16761	0.15624
6	0.40116	0.37990	0.35943	0.33974	0.32085	0.30275	0.28543	0.26890	0.25313	0.23810
7	0.51660	0.49390	0.47168	0.45000	0.42888	0.40836	0.38845	0.36918	0.35056	0.33259
8	0.62484	0.60252	0.58034	0.55836	0.53663	0.51522	0.49415	0.47349	0.45325	0.43347
9	0.71975	0.69931	0.67869	0.65793	0.63712	0.61631	0.59555	0.57490	0.55442	0.53415
10	0.79819	0.78061	0.76259	0.74418	0.72544	0.70644	0.68722	0.66784	0.64837	0.62884
11	0.85969	0.84539	0.83049	0.81504	0.79908	0.78266	0.76583	0.74862	0.73110	0.71330
12	0.90567	0.89459	0.88288	0.87054	0.85761	0.84412	0.83009	0.81556	0.80056	0.78513
13	0.93857	0.93038	0.92157	0.91216	0.90215	0.89155	0.88038	0.86865	0.85638	0.84360
14	0.96120	0.95538	0.94903	0.94215	0.93471	0.92673	0.91819	0.90911	0.89948	0.88933
15	0.97619	0.97222	0.96782	0.96296	0.95765	0.95186	0.94559	0.93882	0.93155	0.92378
16	0.98579	0.98317	0.98022	0.97693	0.97326	0.96921	0.96476	0.95989	0.95460	0.94887
17	0.99174	0.99007	0.98816	0.98599	0.98355	0.98081	0.97775	0.97437	0.97064	0.96655
18	0.99532	0.99429	0.99309	0.99171	0.99013	0.98833	0.98630	0.98402	0.98147	0.97864
19	0.99741	0.99679	0.99606	0.99521	0.99421	0.99307	0.99176	0.99026	0.98857	0.98667
20	0.99860	0.99824	0.99781	0.99729	0.99669	0.99598	0.99515	0.99420	0.99311	0.99187
21	0.99926	0.99905	0.99880	0.99850	0.99814	0.99771	0.99721	0.99662	0.99594	0.99514
22	0.99962	0.99950	0.99936	0.99919	0.99898	0.99873	0.99843	0.99807	0.99765	0.99716
23	0.99981	0.99974	0.99967	0.99957	0.99945	0.99931	0.99913	0.99892	0.99867	0.99837
24	0.99990	0.99987	0.99983	0.99978	0.99971	0.99963	0.99953	0.99941	0.99926	0.99908
25	0.99995	0.99994	0.99991	0.99989	0.99985	0.99981	0.99975	0.99968	0.99960	0.99949
26	0.99998	0.99997	0.99996	0.99994	0.99992	0.99990	0.99987	0.99983	0.99978	0.99973
27	0.99999	0.99999	0.99998	0.99997	0.99996	0.99995	0.99993	0.99991	0.99989	0.99985
28		0.99999	0.99999	0.99999	0.99998	0.99998	0.99997	0.99996	0.99994	0.99992
29				0.99999	0.99999	0.99999	0.99998	0.99998	0.99997	0.99996
30						0.99999	0.99999	0.99999	0.99999	0.99998

	$\chi^2=8.2$	8.4	8.6	8.8	9.0	9.2	9.4	9.6	9.8	10.0
ν	$m=4.1$	4.2	4.3	4.4	4.5	4.6	4.7	4.8	4.9	5.0
1	0.00419	0.00375	0.00336	0.00301	0.00270	0.00242	0.00217	0.00195	0.00175	0.00157
2	0.01657	0.01500	0.01357	0.01228	0.01111	0.01005	0.00910	0.00823	0.00745	0.00674
3	0.04205	0.03843	0.03511	0.03207	0.02929	0.02675	0.02442	0.02229	0.02034	0.01857
4	0.08452	0.07798	0.07191	0.06630	0.06110	0.05629	0.05184	0.04773	0.04394	0.04043
5	0.14555	0.13553	0.12612	0.11731	0.10906	0.10135	0.09413	0.08740	0.08110	0.07524
6	0.22381	0.21024	0.19736	0.18514	0.17358	0.16264	0.15230	0.14254	0.13333	0.12465
7	0.31529	0.29865	0.28266	0.26734	0.25266	0.23861	0.22520	0.21240	0.20019	0.18857
8	0.41418	0.39540	0.37715	0.35945	0.34230	0.32571	0.30968	0.29423	0.27935	0.26503
9	0.51412	0.49439	0.47499	0.45594	0.43727	0.41902	0.40120	0.38383	0.36692	0.35049
10	0.60931	0.58983	0.57044	0.55118	0.53210	0.51323	0.49461	0.47626	0.45821	0.44049
11	0.69528	0.67709	0.65876	0.64035	0.62189	0.60344	0.58502	0.56669	0.54846	0.53039
12	0.76931	0.75314	0.73666	0.71991	0.70293	0.68576	0.66844	0.65101	0.63350	0.61596
13	0.83033	0.81660	0.80244	0.78788	0.77294	0.75768	0.74211	0.72627	0.71020	0.69393
14	0.87865	0.86746	0.85579	0.84365	0.83105	0.81803	0.80461	0.79081	0.77666	0.76218
15	0.91551	0.90675	0.89749	0.88774	0.87752	0.86683	0.85569	0.84412	0.83213	0.81974
16	0.94269	0.93606	0.92897	0.92142	0.91341	0.90495	0.89603	0.88667	0.87686	0.86663
17	0.96208	0.95723	0.95198	0.94633	0.94026	0.93378	0.92687	0.91954	0.91179	0.90361
18	0.97551	0.97207	0.96830	0.96420	0.95974	0.95493	0.94974	0.94418	0.93824	0.93191
19	0.98454	0.98217	0.97955	0.97666	0.97348	0.97001	0.96623	0.96213	0.95771	0.95295
20	0.99046	0.98887	0.98709	0.98511	0.98291	0.98047	0.97779	0.97486	0.97166	0.96817
21	0.99424	0.99320	0.99203	0.99070	0.98921	0.98755	0.98570	0.98365	0.98139	0.97891
22	0.99659	0.99593	0.99518	0.99431	0.99333	0.99222	0.99098	0.98958	0.98803	0.98630
23	0.99802	0.99761	0.99714	0.99659	0.99596	0.99524	0.99442	0.99349	0.99245	0.99128
24	0.99888	0.99863	0.99833	0.99799	0.99760	0.99714	0.99661	0.99601	0.99532	0.99455
25	0.99937	0.99922	0.99905	0.99884	0.99860	0.99831	0.99798	0.99760	0.99716	0.99665
26	0.99966	0.99957	0.99947	0.99934	0.99919	0.99902	0.99882	0.99858	0.99830	0.99798
27	0.99981	0.99977	0.99971	0.99963	0.99955	0.99944	0.99932	0.99917	0.99900	0.99880
28	0.99990	0.99987	0.99984	0.99980	0.99975	0.99969	0.99962	0.99953	0.99942	0.99930
29	0.99995	0.99993	0.99991	0.99989	0.99986	0.99983	0.99979	0.99973	0.99967	0.99960
30	0.99997	0.99997	0.99996	0.99994	0.99993	0.99991	0.99988	0.99985	0.99982	0.99977

$$Q(\chi^2|\nu)=1-P(\chi^2|\nu)=\left[2^{\frac{\nu}{2}}\Gamma\left(\frac{\nu}{2}\right)\right]^{-1}\int_{\chi^2}^{\infty}e^{-\frac{t}{2}}t^{\frac{\nu}{2}-1}dt=\left[\Gamma\left(\frac{\nu}{2}\right)\right]^{-1}\int_{\frac{1}{2}\chi^2}^{\infty}e^{-t}t^{\frac{\nu}{2}-1}dt=\sum_{j=0}^{c-1}e^{-m}m^j/j!\;(\nu\text{ even},\ c=\tfrac{1}{2}\nu,\ m=\tfrac{1}{2}\chi^2)$$

Table 2

PROBABILITY INTEGRAL OF χ^2-DISTRIBUTION, INCOMPLETE GAMMA FUNCTION CUMULATIVE SUMS OF THE POISSON DISTRIBUTION

	χ^2=10.5	11.0	11.5	12.0	12.5	13.0	13.5	14.0	14.5	15.0
ν	m= 5.25	5.5	5.75	6.0	6.25	6.5	6.75	7.0	7.25	7.5
1	0.00119	0.00091	0.00070	0.00053	0.00041	0.00031	0.00024	0.00018	0.00014	0.00011
2	0.00525	0.00409	0.00318	0.00248	0.00193	0.00150	0.00117	0.00091	0.00071	0.00055
3	0.01476	0.01173	0.00931	0.00738	0.00585	0.00464	0.00367	0.00291	0.00230	0.00182
4	0.03280	0.02656	0.02148	0.01735	0.01400	0.01128	0.00907	0.00730	0.00586	0.00470
5	0.06225	0.05138	0.04232	0.03479	0.02854	0.02338	0.01912	0.01561	0.01273	0.01036
6	0.10511	0.08838	0.07410	0.06197	0.05170	0.04304	0.03575	0.02964	0.02452	0.02026
7	0.16196	0.13862	0.11825	0.10056	0.08527	0.07211	0.06082	0.05118	0.04297	0.03600
8	0.23167	0.20170	0.17495	0.15120	0.13025	0.11185	0.09577	0.08177	0.06963	0.05915
9	0.31154	0.27571	0.24299	0.21331	0.18657	0.16261	0.14126	0.12233	0.10562	0.09094
10	0.39777	0.35752	0.31991	0.28506	0.25299	0.22367	0.19704	0.17299	0.15138	0.13206
11	0.48605	0.44326	0.40237	0.36364	0.32726	0.29333	0.26190	0.23299	0.20655	0.18250
12	0.57218	0.52892	0.48662	0.44568	0.40640	0.36904	0.33377	0.30071	0.26992	0.24144
13	0.65263	0.61082	0.56901	0.52764	0.48713	0.44781	0.40997	0.37384	0.33960	0.30735
14	0.72479	0.68604	0.64639	0.60630	0.56622	0.52652	0.48759	0.44971	0.41316	0.37815
15	0.78717	0.75259	0.71641	0.67903	0.64086	0.60230	0.56374	0.52553	0.48800	0.45142
16	0.83925	0.80949	0.77762	0.74398	0.70890	0.67276	0.63591	0.59871	0.56152	0.52464
17	0.88135	0.85656	0.82942	0.80014	0.76896	0.73619	0.70212	0.66710	0.63145	0.59548
18	0.91436	0.89436	0.87195	0.84724	0.82038	0.79157	0.76106	0.72909	0.69596	0.66197
19	0.93952	0.92384	0.90587	0.88562	0.86316	0.83857	0.81202	0.78369	0.75380	0.72260
20	0.95817	0.94622	0.93221	0.91608	0.89779	0.87738	0.85492	0.83050	0.80427	0.77641
21	0.97166	0.96279	0.95214	0.93962	0.92513	0.90862	0.89010	0.86960	0.84718	0.82295
22	0.98118	0.97475	0.96686	0.95738	0.94618	0.93316	0.91827	0.90148	0.88279	0.86224
23	0.98773	0.98319	0.97748	0.97047	0.96201	0.95199	0.94030	0.92687	0.91165	0.89463
24	0.99216	0.98901	0.98498	0.97991	0.97367	0.96612	0.95715	0.94665	0.93454	0.92076
25	0.99507	0.99295	0.99015	0.98657	0.98206	0.97650	0.96976	0.96173	0.95230	0.94138
26	0.99696	0.99555	0.99366	0.99117	0.98798	0.98397	0.97902	0.97300	0.96581	0.95733
27	0.99815	0.99724	0.99598	0.99429	0.99208	0.98925	0.98567	0.98125	0.97588	0.96943
28	0.99890	0.99831	0.99749	0.99637	0.99487	0.99290	0.99037	0.98719	0.98324	0.97844
29	0.99935	0.99899	0.99846	0.99773	0.99672	0.99538	0.99363	0.99138	0.98854	0.98502
30	0.99963	0.99940	0.99907	0.99860	0.99794	0.99704	0.99585	0.99428	0.99227	0.98974

	χ^2=15.5	16.0	16.5	17.0	17.5	18.0	18.5	19.0	19.5	20.0
ν	m= 7.75	8.0	8.25	8.5	8.75	9.0	9.25	9.5	9.75	10.0
1	0.00008	0.00006	0.00005	0.00004	0.00003	0.00002	0.00002	0.00001	0.00001	0.00001
2	0.00043	0.00034	0.00026	0.00020	0.00016	0.00012	0.00010	0.00008	0.00006	0.00005
3	0.00144	0.00113	0.00090	0.00071	0.00056	0.00044	0.00035	0.00027	0.00022	0.00017
4	0.00377	0.00302	0.00242	0.00193	0.00154	0.00123	0.00099	0.00079	0.00063	0.00050
5	0.00843	0.00684	0.00555	0.00450	0.00364	0.00295	0.00238	0.00192	0.00155	0.00125
6	0.01670	0.01375	0.01131	0.00928	0.00761	0.00623	0.00510	0.00416	0.00340	0.00277
7	0.03010	0.02512	0.02092	0.01740	0.01444	0.01197	0.00991	0.00819	0.00676	0.00557
8	0.05012	0.04238	0.03576	0.03011	0.02530	0.02123	0.01777	0.01486	0.01240	0.01034
9	0.07809	0.06688	0.05715	0.04872	0.04144	0.03517	0.02980	0.02519	0.02126	0.01791
10	0.11487	0.09963	0.08619	0.07436	0.06401	0.05496	0.04709	0.04026	0.03435	0.02925
11	0.16073	0.14113	0.12356	0.10788	0.09393	0.08158	0.07068	0.06109	0.05269	0.04534
12	0.21522	0.19124	0.16939	0.14960	0.13174	0.11569	0.10133	0.08853	0.07716	0.06709
13	0.27719	0.24913	0.22318	0.19930	0.17744	0.15752	0.13944	0.12310	0.10840	0.09521
14	0.34485	0.31337	0.28380	0.25618	0.23051	0.20678	0.18495	0.16495	0.14671	0.13014
15	0.41604	0.38205	0.34962	0.31886	0.28986	0.26267	0.23729	0.21373	0.19196	0.17193
16	0.48837	0.45296	0.41864	0.38560	0.35398	0.32390	0.29544	0.26866	0.24359	0.22022
17	0.55951	0.52383	0.48871	0.45437	0.42102	0.38884	0.35797	0.32853	0.30060	0.27423
18	0.62740	0.59255	0.55770	0.52311	0.48902	0.45565	0.42320	0.39182	0.36166	0.33282
19	0.69033	0.65728	0.62370	0.58987	0.55603	0.52244	0.48931	0.45684	0.42521	0.39458
20	0.74712	0.71662	0.68516	0.65297	0.62031	0.58741	0.55451	0.52183	0.48957	0.45793
21	0.79705	0.76965	0.74093	0.71111	0.68039	0.64900	0.61718	0.58514	0.55310	0.52126
22	0.83990	0.81589	0.79032	0.76336	0.73519	0.70599	0.67597	0.64533	0.61428	0.58304
23	0.87582	0.85527	0.83304	0.80925	0.78402	0.75749	0.72983	0.70122	0.67185	0.64191
24	0.90527	0.88808	0.86919	0.84866	0.82657	0.80301	0.77810	0.75199	0.72483	0.69678
25	0.92891	0.91483	0.89912	0.88179	0.86287	0.84239	0.82044	0.79712	0.77254	0.74683
26	0.94749	0.93620	0.92341	0.90908	0.89320	0.87577	0.85683	0.83643	0.81464	0.79156
27	0.96182	0.95295	0.94274	0.93112	0.91806	0.90352	0.88750	0.87000	0.85107	0.83076
28	0.97266	0.96582	0.95782	0.94859	0.93805	0.92615	0.91285	0.89814	0.88200	0.86446
29	0.98071	0.97554	0.96939	0.96218	0.95383	0.94427	0.93344	0.92129	0.90779	0.89293
30	0.98659	0.98274	0.97810	0.97258	0.96608	0.95853	0.94986	0.94001	0.92891	0.91654

$\phi=\frac{1}{2}(\chi^2-\chi_0^2)$ $\qquad$ $w=\nu-\nu_0>0$

Interpolation on χ^2

$$Q(\chi^2|\nu)=Q(\chi_0^2|\nu_0-4)\left[\tfrac{1}{2}\phi^2\right]+Q(\chi_0^2|\nu_0-2)\left[\phi-\phi^2\right]+Q(\chi_0^2|\nu_0)\left[1-\phi+\tfrac{1}{2}\phi^2\right]$$

Double Entry Interpolation

$$Q(\chi^2|\nu)=Q(\chi_0^2|\nu_0-4)\left[\tfrac{1}{2}\phi^2\right]+Q(\chi_0^2|\nu_0-2)\left[\phi-\phi^2-w\phi\right]+Q(\chi_0^2|\nu_0-1)\left[\tfrac{1}{2}w^2-\tfrac{1}{2}w+w\phi\right]$$
$$+Q(\chi_0^2|\nu_0)\left[1-w^2-\phi+\tfrac{1}{2}\phi^2+w\phi\right]+Q(\chi_0^2|\nu_0+1)\left[\tfrac{1}{2}w^2+\tfrac{1}{2}w-w\phi\right]$$

Table 2

PROBABILITY INTEGRAL OF χ^2-DISTRIBUTION, INCOMPLETE GAMMA FUNCTION CUMULATIVE SUMS OF THE POISSON DISTRIBUTION

ν	$\chi^2 = 21$ $m = 10.5$	22 11.0	23 11.5	24 12.0	25 12.5	26 13.0	27 13.5	28 14.0	29 14.5	30 15.0
1	0.00001									
2	0.00003	0.00002	0.00001	0.00001						
3	0.00011	0.00007	0.00004	0.00003	0.00002	0.00001	0.00001			
4	0.00032	0.00020	0.00013	0.00008	0.00005	0.00003	0.00002	0.00001	0.00001	0.00001
5	0.00081	0.00052	0.00034	0.00022	0.00014	0.00009	0.00006	0.00004	0.00002	0.00002
6	0.00184	0.00121	0.00080	0.00052	0.00034	0.00022	0.00015	0.00009	0.00006	0.00004
7	0.00377	0.00254	0.00171	0.00114	0.00076	0.00050	0.00033	0.00022	0.00015	0.00010
8	0.00715	0.00492	0.00336	0.00229	0.00155	0.00105	0.00071	0.00047	0.00032	0.00021
9	0.01265	0.00888	0.00620	0.00430	0.00297	0.00204	0.00140	0.00095	0.00065	0.00044
10	0.02109	0.01511	0.01075	0.00760	0.00535	0.00374	0.00260	0.00181	0.00125	0.00086
11	0.03337	0.02437	0.01768	0.01273	0.00912	0.00649	0.00460	0.00324	0.00227	0.00159
12	0.05038	0.03752	0.02773	0.02034	0.01482	0.01073	0.00773	0.00553	0.00394	0.00279
13	0.07293	0.05536	0.04168	0.03113	0.02308	0.01700	0.01244	0.00905	0.00655	0.00471
14	0.10163	0.07861	0.06027	0.04582	0.03457	0.02589	0.01925	0.01423	0.01045	0.00763
15	0.13683	0.10780	0.08414	0.06509	0.04994	0.03802	0.02874	0.02157	0.01609	0.01192
16	0.17851	0.14319	0.11374	0.08950	0.06982	0.05403	0.04148	0.03162	0.02394	0.01800
17	0.22629	0.18472	0.14925	0.11944	0.09471	0.07446	0.05807	0.04494	0.03453	0.02635
18	0.27941	0.23199	0.19059	0.15503	0.12492	0.09976	0.07900	0.06206	0.04838	0.03745
19	0.33680	0.28426	0.23734	0.19615	0.16054	0.13019	0.10465	0.08343	0.06599	0.05180
20	0.39713	0.34051	0.28880	0.24239	0.20143	0.16581	0.13526	0.10940	0.08776	0.06985
21	0.45894	0.39951	0.34398	0.29306	0.24716	0.20645	0.17085	0.14015	0.11400	0.09199
22	0.52074	0.45989	0.40173	0.34723	0.29707	0.25168	0.21123	0.17568	0.14486	0.11846
23	0.58109	0.52025	0.46077	0.40381	0.35029	0.30087	0.25597	0.21578	0.18031	0.14940
24	0.63873	0.57927	0.51980	0.46160	0.40576	0.35317	0.30445	0.26004	0.22013	0.18475
25	0.69261	0.63574	0.57756	0.51937	0.46237	0.40760	0.35588	0.30785	0.26392	0.22429
26	0.74196	0.68870	0.63295	0.57597	0.51898	0.46311	0.40933	0.35846	0.31108	0.26761
27	0.78629	0.73738	0.68501	0.63032	0.57446	0.51860	0.46379	0.41097	0.36090	0.31415
28	0.82535	0.78129	0.73304	0.68154	0.62784	0.57305	0.51825	0.46445	0.41253	0.36322
29	0.85915	0.82019	0.77654	0.72893	0.67825	0.62549	0.57171	0.51791	0.46507	0.41400
30	0.88789	0.85404	0.81526	0.77203	0.72503	0.67513	0.62327	0.57044	0.51760	0.46565

ν	$\chi^2 = 31$ $m = 15.5$	32 16.0	33 16.5	34 17.0	35 17.5	36 18.0	37 18.5	38 19.0	39 19.5	40 20.0
5	0.00001	0.00001								
6	0.00003	0.00002	0.00001	0.00001						
7	0.00006	0.00004	0.00003	0.00002	0.00001	0.00001				
8	0.00014	0.00009	0.00006	0.00004	0.00003	0.00002	0.00001	0.00001		
9	0.00030	0.00020	0.00013	0.00009	0.00006	0.00004	0.00003	0.00002	0.00001	0.00001
10	0.00059	0.00040	0.00027	0.00019	0.00012	0.00008	0.00006	0.00004	0.00003	0.00002
11	0.00110	0.00076	0.00053	0.00036	0.00025	0.00017	0.00012	0.00008	0.00005	0.00004
12	0.00197	0.00138	0.00097	0.00068	0.00047	0.00032	0.00022	0.00015	0.00011	0.00007
13	0.00337	0.00240	0.00170	0.00120	0.00085	0.00059	0.00041	0.00029	0.00020	0.00014
14	0.00554	0.00401	0.00288	0.00206	0.00147	0.00104	0.00074	0.00052	0.00036	0.00026
15	0.00878	0.00644	0.00469	0.00341	0.00246	0.00177	0.00127	0.00090	0.00064	0.00045
16	0.01346	0.01000	0.00739	0.00543	0.00397	0.00289	0.00210	0.00151	0.00109	0.00078
17	0.01997	0.01505	0.01127	0.00840	0.00622	0.00459	0.00337	0.00246	0.00179	0.00129
18	0.02879	0.02199	0.01669	0.01260	0.00945	0.00706	0.00524	0.00387	0.00285	0.00209
19	0.04037	0.03125	0.02404	0.01838	0.01397	0.01056	0.00793	0.00593	0.00442	0.00327
20	0.05519	0.04330	0.03374	0.02613	0.02010	0.01538	0.01170	0.00886	0.00667	0.00500
21	0.07366	0.05855	0.04622	0.03624	0.02824	0.02187	0.01683	0.01289	0.00981	0.00744
22	0.09612	0.07740	0.06187	0.04912	0.03875	0.03037	0.02366	0.01832	0.01411	0.01081
23	0.12279	0.10014	0.08107	0.06516	0.05202	0.04125	0.03251	0.02547	0.01984	0.01537
24	0.15378	0.12699	0.10407	0.08467	0.06840	0.05489	0.04376	0.03467	0.02731	0.02139
25	0.18902	0.15801	0.13107	0.10791	0.08820	0.07160	0.05774	0.04626	0.03684	0.02916
26	0.22827	0.19312	0.16210	0.13502	0.11165	0.09167	0.07475	0.06056	0.04875	0.03901
27	0.27114	0.23208	0.19707	0.16605	0.13887	0.11530	0.09507	0.07786	0.06336	0.05124
28	0.31708	0.27451	0.23574	0.20087	0.16987	0.14260	0.11886	0.09840	0.08092	0.06613
29	0.36542	0.31987	0.27774	0.23926	0.20454	0.17356	0.14622	0.12234	0.10166	0.08394
30	0.41541	0.36753	0.32254	0.28083	0.24264	0.20808	0.17714	0.14975	0.12573	0.10486

Table 2

PROBABILITY INTEGRAL OF χ^2-DISTRIBUTION, INCOMPLETE GAMMA FUNCTION CUMULATIVE SUMS OF THE POISSON DISTRIBUTION

ν	$\chi^2=42$ $m=21$	44 22	46 23	48 24	50 25	52 26	54 27	56 28	58 29	60 30
10	0.00001									
11	0.00002	0.00001								
12	0.00003	0.00002	0.00001							
13	0.00006	0.00003	0.00001	0.00001						
14	0.00012	0.00006	0.00003	0.00001	0.00001					
15	0.00023	0.00011	0.00005	0.00003	0.00001	0.00001				
16	0.00040	0.00020	0.00010	0.00005	0.00002	0.00001	0.00001			
17	0.00067	0.00034	0.00017	0.00009	0.00004	0.00002	0.00001	0.00001		
18	0.00111	0.00058	0.00030	0.00015	0.00008	0.00004	0.00002	0.00001		
19	0.00177	0.00094	0.00050	0.00026	0.00013	0.00007	0.00003	0.00002	0.00001	
20	0.00277	0.00151	0.00081	0.00043	0.00022	0.00011	0.00006	0.00003	0.00001	0.00001
21	0.00421	0.00234	0.00128	0.00069	0.00036	0.00019	0.00010	0.00005	0.00003	0.00001
22	0.00625	0.00355	0.00198	0.00109	0.00059	0.00031	0.00016	0.00009	0.00004	0.00002
23	0.00908	0.00526	0.00299	0.00167	0.00092	0.00050	0.00027	0.00014	0.00007	0.00004
24	0.01291	0.00763	0.00443	0.00252	0.00142	0.00078	0.00043	0.00023	0.00012	0.00006
25	0.01797	0.01085	0.00642	0.00373	0.00213	0.00120	0.00066	0.00036	0.00020	0.00011
26	0.02455	0.01512	0.00912	0.00540	0.00314	0.00180	0.00102	0.00056	0.00031	0.00017
27	0.03292	0.02068	0.01272	0.00768	0.00455	0.00265	0.00152	0.00086	0.00048	0.00026
28	0.04336	0.02779	0.01743	0.01072	0.00647	0.00384	0.00224	0.00129	0.00073	0.00041
29	0.05616	0.03670	0.02346	0.01470	0.00903	0.00545	0.00324	0.00189	0.00109	0.00062
30	0.07157	0.04769	0.03107	0.01983	0.01240	0.00762	0.00460	0.00273	0.00160	0.00092

ν	$\chi^2=62$ $m=31$	64 32	66 33	68 34	70 35	72 36	74 37	76 38
21	0.00001							
22	0.00001	0.00001						
23	0.00002	0.00001	0.00001					
24	0.00003	0.00002	0.00001					
25	0.00006	0.00003	0.00002	0.00001				
26	0.00009	0.00005	0.00003	0.00001	0.00001			
27	0.00014	0.00008	0.00004	0.00002	0.00001	0.00001		
28	0.00023	0.00012	0.00007	0.00004	0.00002	0.00001	0.00001	
29	0.00035	0.00019	0.00011	0.00006	0.00003	0.00002	0.00001	
30	0.00052	0.00029	0.00016	0.00009	0.00005	0.00003	0.00001	0.00001

$$Q(\chi^2|\nu)=1-P(\chi^2|\nu)=\left[2^{\frac{\nu}{2}}\Gamma\left(\frac{\nu}{2}\right)\right]^{-1}\int_{\chi^2}^{\infty}e^{-\frac{t}{2}}t^{\frac{\nu}{2}-1}dt=\left[\Gamma\left(\frac{\nu}{2}\right)\right]^{-1}\int_{\frac{1}{2}\chi^2}^{\infty}e^{-t}t^{\frac{\nu}{2}-1}dt=\sum_{j=0}^{c-1}e^{-m}m^j/j!\,(\nu\text{ even},\ c=\tfrac{1}{2}\nu,\ m=\tfrac{1}{2}\chi^2)$$

$$\phi=\tfrac{1}{2}\left(\chi^2-\chi_0^2\right)\qquad w=\nu-\nu_0>0$$

Interpolation on χ^2

$$Q(\chi^2|\nu)=Q\left(\chi_0^2|\nu_0-4\right)\left[\tfrac{1}{2}\phi^2\right]+Q\left(\chi_0^2|\nu_0-2\right)\left[\phi-\phi^2\right]+Q\left(\chi_0^2|\nu_0\right)\left[1-\phi+\tfrac{1}{2}\phi^2\right]$$

Double Entry Interpolation

$$Q\left(\chi^2|\nu\right)=Q\left(\chi_0^2|\nu_0-4\right)\left[\tfrac{1}{2}\phi^2\right]+Q\left(\chi_0^2|\nu_0-2\right)\left[\phi-\phi^2-w\phi\right]+Q\left(\chi_0^2|\nu_0-1\right)\left[\tfrac{1}{2}w^2-\tfrac{1}{2}w+w\phi\right]$$
$$+Q\left(\chi_0^2|\nu_0\right)\left[1-w^2-\phi+\tfrac{1}{2}\phi^2+w\phi\right]+Q\left(\chi_0^2|\nu_0+1\right)\left[\tfrac{1}{2}w^2+\tfrac{1}{2}w-w\phi\right]$$

Table 3

PERCENTAGE POINTS OF THE χ^2-DISTRIBUTION—VALUES OF χ^2 IN TERMS OF Q AND ν

$\nu \backslash Q$	0.995	0.99	0.975	0.95	0.9	0.75	0.5	0.25
1	(−5) 3.92704	(−4) 1.57088	(−4) 9.82069	(−3) 3.93214	0.0157908	0.101531	0.454937	1.32330
2	(−2) 1.00251	(−2) 2.01007	(−2) 5.06356	0.102587	0.210720	0.575364	1.38629	2.77259
3	(−2) 7.17212	0.114832	0.215795	0.351846	0.584375	1.212534	2.36597	4.10835
4	0.206990	0.297110	0.484419	0.710721	1.063623	1.92255	3.35670	5.38527
5	0.411740	0.554300	0.831211	1.145476	1.61031	2.67460	4.35146	6.62568
6	0.675727	0.872085	1.237347	1.63539	2.20413	3.45460	5.34812	7.84080
7	0.989265	1.239043	1.68987	2.16735	2.83311	4.25485	6.34581	9.03715
8	1.344419	1.646482	2.17973	2.73264	3.48954	5.07064	7.34412	10.2188
9	1.734926	2.087912	2.70039	3.32511	4.16816	5.89883	8.34283	11.3887
10	2.15585	2.55821	3.24697	3.94030	4.86518	6.73720	9.34182	12.5489
11	2.60321	3.05347	3.81575	4.57481	5.57779	7.58412	10.3410	13.7007
12	3.07382	3.57056	4.40379	5.22603	6.30380	8.43842	11.3403	14.8454
13	3.56503	4.10691	5.00874	5.89186	7.04150	9.29906	12.3398	15.9839
14	4.07468	4.66043	5.62872	6.57063	7.78953	10.1653	13.3393	17.1170
15	4.60094	5.22935	6.26214	7.26094	8.54675	11.0365	14.3389	18.2451
16	5.14224	5.81221	6.90766	7.96164	9.31223	11.9122	15.3385	19.3688
17	5.69724	6.40776	7.56418	8.67176	10.0852	12.7919	16.3381	20.4887
18	6.26481	7.01491	8.23075	9.39046	10.8649	13.6753	17.3379	21.6049
19	6.84398	7.63273	8.90655	10.1170	11.6509	14.5620	18.3376	22.7178
20	7.43386	8.26040	9.59083	10.8508	12.4426	15.4518	19.3374	23.8277
21	8.03366	8.89720	10.28293	11.5913	13.2396	16.3444	20.3372	24.9348
22	8.64272	9.54249	10.9823	12.3380	14.0415	17.2396	21.3370	26.0393
23	9.26042	10.19567	11.6885	13.0905	14.8479	18.1373	22.3369	27.1413
24	9.88623	10.8564	12.4011	13.8484	15.6587	19.0372	23.3367	28.2412
25	10.5197	11.5240	13.1197	14.6114	16.4734	19.9393	24.3366	29.3389
26	11.1603	12.1981	13.8439	15.3791	17.2919	20.8434	25.3364	30.4345
27	11.8076	12.8786	14.5733	16.1513	18.1138	21.7494	26.3363	31.5284
28	12.4613	13.5648	15.3079	16.9279	18.9392	22.6572	27.3363	32.6205
29	13.1211	14.2565	16.0471	17.7083	19.7677	23.5666	28.3362	33.7109
30	13.7867	14.9535	16.7908	18.4926	20.5992	24.4776	29.3360	34.7998
40	20.7065	22.1643	24.4331	26.5093	29.0505	33.6603	39.3354	45.6160
50	27.9907	29.7067	32.3574	34.7642	37.6886	42.9421	49.3349	56.3336
60	35.5346	37.4848	40.4817	43.1879	46.4589	52.2938	59.3347	66.9814
70	43.2752	45.4418	48.7576	51.7393	55.3290	61.6983	69.3344	77.5766
80	51.1720	53.5400	57.1532	60.3915	64.2778	71.1445	79.3343	88.1303
90	59.1963	61.7541	65.6466	69.1260	73.2912	80.6247	89.3342	98.6499
100	67.3276	70.0648	74.2219	77.9295	82.3581	90.1332	99.3341	109.141
X	−2.5758	−2.3263	−1.9600	−1.6449	−1.2816	−0.6745	0.0000	0.6745

$$Q(\chi^2|\nu) = \left[2^{\frac{\nu}{2}}\Gamma\left(\frac{\nu}{2}\right)\right]^{-1} \int_{\chi^2}^{\infty} e^{-\frac{t}{2}} t^{\frac{\nu}{2}-1}\, dt$$

Table 3

PERCENTAGE POINTS OF THE χ^2-DISTRIBUTION—VALUES OF χ^2 IN TERMS OF Q AND ν

$\nu \backslash Q$	0.1	0.05	0.025	0.01	0.005	0.001	0.0005	0.0001
1	2.70554	3.84146	5.02389	6.63490	7.87944	10.828	12.116	15.137
2	4.60517	5.99147	7.37776	9.21034	10.5966	13.816	15.202	18.421
3	6.25139	7.81473	9.34840	11.3449	12.8381	16.266	17.730	21.108
4	7.77944	9.48773	11.1433	13.2767	14.8602	18.467	19.997	23.513
5	9.23635	11.0705	12.8325	15.0863	16.7496	20.515	22.105	25.745
6	10.6446	12.5916	14.4494	16.8119	18.5476	22.458	24.103	27.856
7	12.0170	14.0671	16.0128	18.4753	20.2777	24.322	26.018	29.877
8	13.3616	15.5073	17.5346	20.0902	21.9550	26.125	27.868	31.828
9	14.6837	16.9190	19.0228	21.6660	23.5893	27.877	29.666	33.720
10	15.9871	18.3070	20.4831	23.2093	25.1882	29.588	31.420	35.564
11	17.2750	19.6751	21.9200	24.7250	26.7569	31.264	33.137	37.367
12	18.5494	21.0261	23.3367	26.2170	28.2995	32.909	34.821	39.134
13	19.8119	22.3621	24.7356	27.6883	29.8194	34.528	36.478	40.871
14	21.0642	23.6848	26.1190	29.1413	31.3193	36.123	38.109	42.579
15	22.3072	24.9958	27.4884	30.5779	32.8013	37.697	39.719	44.263
16	23.5418	26.2962	28.8454	31.9999	34.2672	39.252	41.308	45.925
17	24.7690	27.5871	30.1910	33.4087	35.7185	40.790	42.879	47.566
18	25.9894	28.8693	31.5264	34.8053	37.1564	42.312	44.434	49.189
19	27.2036	30.1435	32.8523	36.1908	38.5822	43.820	45.973	50.796
20	28.4120	31.4104	34.1696	37.5662	39.9968	45.315	47.498	52.386
21	29.6151	32.6705	35.4789	38.9321	41.4010	46.797	49.011	53.962
22	30.8133	33.9244	36.7807	40.2894	42.7956	48.268	50.511	55.525
23	32.0069	35.1725	38.0757	41.6384	44.1813	49.728	52.000	57.075
24	33.1963	36.4151	39.3641	42.9798	45.5585	51.179	53.479	58.613
25	34.3816	37.6525	40.6465	44.3141	46.9278	52.620	54.947	60.140
26	35.5631	38.8852	41.9232	45.6417	48.2899	54.052	56.407	61.657
27	36.7412	40.1133	43.1944	46.9630	49.6449	55.476	57.858	63.164
28	37.9159	41.3372	44.4607	48.2782	50.9933	56.892	59.300	64.662
29	39.0875	42.5569	45.7222	49.5879	52.3356	58.302	60.735	66.152
30	40.2560	43.7729	46.9792	50.8922	53.6720	59.703	62.162	67.633
40	51.8050	55.7585	59.3417	63.6907	66.7659	73.402	76.095	82.062
50	63.1671	67.5048	71.4202	76.1539	79.4900	86.661	89.560	95.969
60	74.3970	79.0819	83.2976	88.3794	91.9517	99.607	102.695	109.503
70	85.5271	90.5312	95.0231	100.425	104.215	112.317	115.578	122.755
80	96.5782	101.879	106.629	112.329	116.321	124.839	128.261	135.783
90	107.565	113.145	118.136	124.116	128.299	137.208	140.782	148.627
100	118.498	124.342	129.561	135.807	140.169	149.449	153.167	161.319
	1.2816	1.6449	1.9600	2.3263	2.5758	3.0902	3.2905	3.7190

$$Q(\chi^2|\nu) = \left[2^{\frac{\nu}{2}}\Gamma\left(\frac{\nu}{2}\right)\right]^{-1} \int_{\chi^2}^{\infty} e^{-\frac{t}{2}} t^{\frac{\nu}{2}-1}\, dt$$

Table 4

PERCENTAGE POINTS OF THE F-DISTRIBUTION—VALUES OF F IN TERMS OF Q, ν_1, ν_2

$Q(F|\nu_1,\nu_2)=0.5$

$\nu_2\backslash\nu_1$	1	2	3	4	5	6	8	12	15	20	30	60	∞
1	1.00	1.50	1.71	1.82	1.89	1.94	2.00	2.07	2.09	2.12	2.15	2.17	2.20
2	0.667	1.00	1.13	1.21	1.25	1.28	1.32	1.36	1.38	1.39	1.41	1.43	1.44
3	0.585	0.881	1.00	1.06	1.10	1.13	1.16	1.20	1.21	1.23	1.24	1.25	1.27
4	0.549	0.828	0.941	1.00	1.04	1.06	1.09	1.13	1.14	1.15	1.16	1.18	1.19
5	0.528	0.799	0.907	0.965	1.00	1.02	1.05	1.09	1.10	1.11	1.12	1.14	1.15
6	0.515	0.780	0.886	0.942	0.977	1.00	1.03	1.06	1.07	1.08	1.10	1.11	1.12
7	0.506	0.767	0.871	0.926	0.960	0.983	1.01	1.04	1.05	1.07	1.08	1.09	1.10
8	0.499	0.757	0.860	0.915	0.948	0.971	1.00	1.03	1.04	1.05	1.07	1.08	1.09
9	0.494	0.749	0.852	0.906	0.939	0.962	0.990	1.02	1.03	1.04	1.05	1.07	1.08
10	0.490	0.743	0.845	0.899	0.932	0.954	0.983	1.01	1.02	1.03	1.05	1.06	1.07
11	0.486	0.739	0.840	0.893	0.926	0.948	0.977	1.01	1.02	1.03	1.04	1.05	1.06
12	0.484	0.735	0.835	0.888	0.921	0.943	0.972	1.00	1.01	1.02	1.03	1.05	1.06
13	0.481	0.731	0.832	0.885	0.917	0.939	0.967	0.996	1.01	1.02	1.03	1.04	1.05
14	0.479	0.729	0.828	0.881	0.914	0.936	0.964	0.992	1.00	1.01	1.03	1.04	1.05
15	0.478	0.726	0.826	0.878	0.911	0.933	0.960	0.989	1.00	1.01	1.02	1.03	1.05
16	0.476	0.724	0.823	0.876	0.908	0.930	0.958	0.986	0.997	1.01	1.02	1.03	1.04
17	0.475	0.722	0.821	0.874	0.906	0.928	0.955	0.983	0.995	1.01	1.02	1.03	1.04
18	0.474	0.721	0.819	0.872	0.904	0.926	0.953	0.981	0.992	1.00	1.02	1.03	1.04
19	0.473	0.719	0.818	0.870	0.902	0.924	0.951	0.979	0.990	1.00	1.01	1.02	1.04
20	0.472	0.718	0.816	0.868	0.900	0.922	0.950	0.977	0.989	1.00	1.01	1.02	1.03
21	0.471	0.716	0.815	0.867	0.899	0.921	0.948	0.976	0.987	0.998	1.01	1.02	1.03
22	0.470	0.715	0.814	0.866	0.898	0.919	0.947	0.974	0.986	0.997	1.01	1.02	1.03
23	0.470	0.714	0.813	0.864	0.896	0.918	0.945	0.973	0.984	0.996	1.01	1.02	1.03
24	0.469	0.714	0.812	0.863	0.895	0.917	0.944	0.972	0.983	0.994	1.01	1.02	1.03
25	0.468	0.713	0.811	0.862	0.894	0.916	0.943	0.971	0.982	0.993	1.00	1.02	1.03
26	0.468	0.712	0.810	0.861	0.893	0.915	0.942	0.970	0.981	0.992	1.00	1.01	1.03
27	0.467	0.711	0.809	0.861	0.892	0.914	0.941	0.969	0.980	0.991	1.00	1.01	1.03
28	0.467	0.711	0.808	0.860	0.892	0.913	0.940	0.968	0.979	0.990	1.00	1.01	1.02
29	0.466	0.710	0.808	0.859	0.891	0.912	0.940	0.967	0.978	0.990	1.00	1.01	1.02
30	0.466	0.709	0.807	0.858	0.890	0.912	0.939	0.966	0.978	0.989	1.00	1.01	1.02
40	0.463	0.705	0.802	0.854	0.885	0.907	0.934	0.961	0.972	0.983	0.994	1.01	1.02
60	0.461	0.701	0.798	0.849	0.880	0.901	0.928	0.956	0.967	0.978	0.989	1.00	1.01
120	0.458	0.697	0.793	0.844	0.875	0.896	0.923	0.950	0.961	0.972	0.983	0.994	1.01
∞	0.455	0.693	0.789	0.839	0.870	0.891	0.918	0.945	0.956	0.967	0.978	0.989	1.00

$Q(F|\nu_1,\nu_2)=0.25$

$\nu_2\backslash\nu_1$	1	2	3	4	5	6	8	12	15	20	30	60	∞
1	5.83	7.50	8.20	8.58	8.82	8.98	9.19	9.41	9.49	9.58	9.67	9.76	9.85
2	2.57	3.00	3.15	3.23	3.28	3.31	3.35	3.39	3.41	3.43	3.44	3.46	3.48
3	2.02	2.28	2.36	2.39	2.41	2.42	2.44	2.45	2.46	2.46	2.47	2.47	2.47
4	1.81	2.00	2.05	2.06	2.07	2.08	2.08	2.08	2.08	2.08	2.08	2.08	2.08
5	1.69	1.85	1.88	1.89	1.89	1.89	1.89	1.89	1.89	1.88	1.88	1.87	1.87
6	1.62	1.76	1.78	1.79	1.79	1.78	1.78	1.77	1.76	1.76	1.75	1.74	1.74
7	1.57	1.70	1.72	1.72	1.71	1.71	1.70	1.68	1.68	1.67	1.66	1.65	1.65
8	1.54	1.66	1.67	1.66	1.66	1.65	1.64	1.62	1.62	1.61	1.60	1.59	1.58
9	1.51	1.62	1.63	1.63	1.62	1.61	1.60	1.58	1.57	1.56	1.55	1.54	1.53
10	1.49	1.60	1.60	1.59	1.59	1.58	1.56	1.54	1.53	1.52	1.51	1.50	1.48
11	1.47	1.58	1.58	1.57	1.56	1.55	1.53	1.51	1.50	1.49	1.48	1.47	1.45
12	1.46	1.56	1.56	1.55	1.54	1.53	1.51	1.49	1.48	1.47	1.45	1.44	1.42
13	1.45	1.55	1.55	1.53	1.52	1.51	1.49	1.47	1.46	1.45	1.43	1.42	1.40
14	1.44	1.53	1.53	1.52	1.51	1.50	1.48	1.45	1.44	1.43	1.41	1.40	1.38
15	1.43	1.52	1.52	1.51	1.49	1.48	1.46	1.44	1.43	1.41	1.40	1.38	1.36
16	1.42	1.51	1.51	1.50	1.48	1.47	1.45	1.43	1.41	1.40	1.38	1.36	1.34
17	1.42	1.51	1.50	1.49	1.47	1.46	1.44	1.41	1.40	1.39	1.37	1.35	1.33
18	1.41	1.50	1.49	1.48	1.46	1.45	1.43	1.40	1.39	1.38	1.36	1.34	1.32
19	1.41	1.49	1.49	1.47	1.46	1.44	1.42	1.40	1.38	1.37	1.35	1.33	1.30
20	1.40	1.49	1.48	1.47	1.45	1.44	1.42	1.39	1.37	1.36	1.34	1.32	1.29
21	1.40	1.48	1.48	1.46	1.44	1.43	1.41	1.38	1.37	1.35	1.33	1.31	1.28
22	1.40	1.48	1.47	1.45	1.44	1.42	1.40	1.37	1.36	1.34	1.32	1.30	1.28
23	1.39	1.47	1.47	1.45	1.43	1.42	1.40	1.37	1.35	1.34	1.32	1.30	1.27
24	1.39	1.47	1.46	1.44	1.43	1.41	1.39	1.36	1.35	1.33	1.31	1.29	1.26
25	1.39	1.47	1.46	1.44	1.42	1.41	1.39	1.36	1.34	1.33	1.31	1.28	1.25
26	1.38	1.46	1.45	1.44	1.42	1.41	1.38	1.35	1.34	1.32	1.30	1.28	1.25
27	1.38	1.46	1.45	1.43	1.42	1.40	1.38	1.35	1.33	1.32	1.30	1.27	1.24
28	1.38	1.46	1.45	1.43	1.41	1.40	1.38	1.34	1.33	1.31	1.29	1.27	1.24
29	1.38	1.45	1.45	1.43	1.41	1.40	1.37	1.34	1.32	1.31	1.29	1.26	1.23
30	1.38	1.45	1.44	1.42	1.41	1.39	1.37	1.34	1.32	1.30	1.28	1.26	1.23
40	1.36	1.44	1.42	1.40	1.39	1.37	1.35	1.31	1.30	1.28	1.25	1.22	1.19
60	1.35	1.42	1.41	1.38	1.37	1.35	1.32	1.29	1.27	1.25	1.22	1.19	1.15
120	1.34	1.40	1.39	1.37	1.35	1.33	1.30	1.26	1.24	1.22	1.19	1.16	1.10
∞	1.32	1.39	1.37	1.35	1.33	1.31	1.28	1.24	1.22	1.19	1.16	1.12	1.00

Table 4

PERCENTAGE POINTS OF THE *F*-DISTRIBUTION—VALUES OF *F* IN TERMS OF *Q*, ν_1, ν_2

$Q(F|\nu_1,\nu_2)=0.1$

$\nu_2\backslash\nu_1$	1	2	3	4	5	6	8	12	15	20	30	60	∞
1	39.86	49.50	53.59	55.83	57.24	58.20	59.44	60.71	61.22	61.74	62.26	62.79	63.33
2	8.53	9.00	9.16	9.24	9.29	9.33	9.37	9.41	9.42	9.44	9.46	9.47	9.49
3	5.54	5.46	5.39	5.34	5.31	5.28	5.25	5.22	5.20	5.18	5.17	5.15	5.13
4	4.54	4.32	4.19	4.11	4.05	4.01	3.95	3.90	3.87	3.84	3.82	3.79	3.76
5	4.06	3.78	3.62	3.52	3.45	3.40	3.34	3.27	3.24	3.21	3.17	3.14	3.10
6	3.78	3.46	3.29	3.18	3.11	3.05	2.98	2.90	2.87	2.84	2.80	2.76	2.72
7	3.59	3.26	3.07	2.96	2.88	2.83	2.75	2.67	2.63	2.59	2.56	2.51	2.47
8	3.46	3.11	2.92	2.81	2.73	2.67	2.59	2.50	2.46	2.42	2.38	2.34	2.29
9	3.36	3.01	2.81	2.69	2.61	2.55	2.47	2.38	2.34	2.30	2.25	2.21	2.16
10	3.29	2.92	2.73	2.61	2.52	2.46	2.38	2.28	2.24	2.20	2.16	2.11	2.06
11	3.23	2.86	2.66	2.54	2.45	2.39	2.30	2.21	2.17	2.12	2.08	2.03	1.97
12	3.18	2.81	2.61	2.48	2.39	2.33	2.24	2.15	2.10	2.06	2.01	1.96	1.90
13	3.14	2.76	2.56	2.43	2.35	2.28	2.20	2.10	2.05	2.01	1.96	1.90	1.85
14	3.10	2.73	2.52	2.39	2.31	2.24	2.15	2.05	2.01	1.96	1.91	1.86	1.80
15	3.07	2.70	2.49	2.36	2.27	2.21	2.12	2.02	1.97	1.92	1.87	1.82	1.76
16	3.05	2.67	2.46	2.33	2.24	2.18	2.09	1.99	1.94	1.89	1.84	1.78	1.72
17	3.03	2.64	2.44	2.31	2.22	2.15	2.06	1.96	1.91	1.86	1.81	1.75	1.69
18	3.01	2.62	2.42	2.29	2.20	2.13	2.04	1.93	1.89	1.84	1.78	1.72	1.66
19	2.99	2.61	2.40	2.27	2.18	2.11	2.02	1.91	1.86	1.81	1.76	1.70	1.63
20	2.97	2.59	2.38	2.25	2.16	2.09	2.00	1.89	1.84	1.79	1.74	1.68	1.61
21	2.96	2.57	2.36	2.23	2.14	2.08	1.98	1.87	1.83	1.78	1.72	1.66	1.59
22	2.95	2.56	2.35	2.22	2.13	2.06	1.97	1.86	1.81	1.76	1.70	1.64	1.57
23	2.94	2.55	2.34	2.21	2.11	2.05	1.95	1.84	1.80	1.74	1.69	1.62	1.55
24	2.93	2.54	2.33	2.19	2.10	2.04	1.94	1.83	1.78	1.73	1.67	1.61	1.53
25	2.92	2.53	2.32	2.18	2.09	2.02	1.93	1.82	1.77	1.72	1.66	1.59	1.52
26	2.91	2.52	2.31	2.17	2.08	2.01	1.92	1.81	1.76	1.71	1.65	1.58	1.50
27	2.90	2.51	2.30	2.17	2.07	2.00	1.91	1.80	1.75	1.70	1.64	1.57	1.49
28	2.89	2.50	2.29	2.16	2.06	2.00	1.90	1.79	1.74	1.69	1.63	1.56	1.48
29	2.89	2.50	2.28	2.15	2.06	1.99	1.89	1.78	1.73	1.68	1.62	1.55	1.47
30	2.88	2.49	2.28	2.14	2.05	1.98	1.88	1.77	1.72	1.67	1.61	1.54	1.46
40	2.84	2.44	2.23	2.09	2.00	1.93	1.83	1.71	1.66	1.61	1.54	1.47	1.38
60	2.79	2.39	2.18	2.04	1.95	1.87	1.77	1.66	1.60	1.54	1.48	1.40	1.29
120	2.75	2.35	2.13	1.99	1.90	1.82	1.72	1.60	1.55	1.48	1.41	1.32	1.19
∞	2.71	2.30	2.08	1.94	1.85	1.77	1.67	1.55	1.49	1.42	1.34	1.24	1.00

$Q(F|\nu_1,\nu_2)=0.05$

$\nu_2\backslash\nu_1$	1	2	3	4	5	6	8	12	15	20	30	60	∞
1	161.4	199.5	215.7	224.6	230.2	234.0	238.9	243.9	245.9	248.0	250.1	252.2	254.3
2	18.51	19.00	19.16	19.25	19.30	19.33	19.37	19.41	19.43	19.45	19.46	19.48	19.50
3	10.13	9.55	9.28	9.12	9.01	8.94	8.85	8.74	8.70	8.66	8.62	8.57	8.53
4	7.71	6.94	6.59	6.39	6.26	6.16	6.04	5.91	5.86	5.80	5.75	5.69	5.63
5	6.61	5.79	5.41	5.19	5.05	4.95	4.82	4.68	4.62	4.56	4.50	4.43	4.36
6	5.99	5.14	4.76	4.53	4.39	4.28	4.15	4.00	3.94	3.87	3.81	3.74	3.67
7	5.59	4.74	4.35	4.12	3.97	3.87	3.73	3.57	3.51	3.44	3.38	3.30	3.23
8	5.32	4.46	4.07	3.84	3.69	3.58	3.44	3.28	3.22	3.15	3.08	3.01	2.93
9	5.12	4.26	3.86	3.63	3.48	3.37	3.23	3.07	3.01	2.94	2.86	2.79	2.71
10	4.96	4.10	3.71	3.48	3.33	3.22	3.07	2.91	2.85	2.77	2.70	2.62	2.54
11	4.84	3.98	3.59	3.36	3.20	3.09	2.95	2.79	2.72	2.65	2.57	2.49	2.40
12	4.75	3.89	3.49	3.26	3.11	3.00	2.85	2.69	2.62	2.54	2.47	2.38	2.30
13	4.67	3.81	3.41	3.18	3.03	2.92	2.77	2.60	2.53	2.46	2.38	2.30	2.21
14	4.60	3.74	3.34	3.11	2.96	2.85	2.70	2.53	2.46	2.39	2.31	2.22	2.13
15	4.54	3.68	3.29	3.06	2.90	2.79	2.64	2.48	2.40	2.33	2.25	2.16	2.07
16	4.49	3.63	3.24	3.01	2.85	2.74	2.59	2.42	2.35	2.28	2.19	2.11	2.01
17	4.45	3.59	3.20	2.96	2.81	2.70	2.55	2.38	2.31	2.23	2.15	2.06	1.96
18	4.41	3.55	3.16	2.93	2.77	2.66	2.51	2.34	2.27	2.19	2.11	2.02	1.92
19	4.38	3.52	3.13	2.90	2.74	2.63	2.48	2.31	2.23	2.16	2.07	1.98	1.88
20	4.35	3.49	3.10	2.87	2.71	2.60	2.45	2.28	2.20	2.12	2.04	1.95	1.84
21	4.32	3.47	3.07	2.84	2.68	2.57	2.42	2.25	2.18	2.10	2.01	1.92	1.81
22	4.30	3.44	3.05	2.82	2.66	2.55	2.40	2.23	2.15	2.07	1.98	1.89	1.78
23	4.28	3.42	3.03	2.80	2.64	2.53	2.37	2.20	2.13	2.05	1.96	1.86	1.76
24	4.26	3.40	3.01	2.78	2.62	2.51	2.36	2.18	2.11	2.03	1.94	1.84	1.73
25	4.24	3.39	2.99	2.76	2.60	2.49	2.34	2.16	2.09	2.01	1.92	1.82	1.71
26	4.23	3.37	2.98	2.74	2.59	2.47	2.32	2.15	2.07	1.99	1.90	1.80	1.69
27	4.21	3.35	2.96	2.73	2.57	2.46	2.31	2.13	2.06	1.97	1.88	1.79	1.67
28	4.20	3.34	2.95	2.71	2.56	2.45	2.29	2.12	2.04	1.96	1.87	1.77	1.65
29	4.18	3.33	2.93	2.70	2.55	2.43	2.28	2.10	2.03	1.94	1.85	1.75	1.64
30	4.17	3.32	2.92	2.69	2.53	2.42	2.27	2.09	2.01	1.93	1.84	1.74	1.62
40	4.08	3.23	2.84	2.61	2.45	2.34	2.18	2.00	1.92	1.84	1.74	1.64	1.51
60	4.00	3.15	2.76	2.53	2.37	2.25	2.10	1.92	1.84	1.75	1.65	1.53	1.39
120	3.92	3.07	2.68	2.45	2.29	2.17	2.02	1.83	1.75	1.66	1.55	1.43	1.25
∞	3.84	3.00	2.60	2.37	2.21	2.10	1.94	1.75	1.67	1.57	1.46	1.32	1.00

Table 4

PERCENTAGE POINTS OF THE *F*-DISTRIBUTION—VALUES OF *F* IN TERMS OF Q, ν_1, ν_2

$Q(F|\nu_1,\nu_2)=0.025$

$\nu_2 \backslash \nu_1$	1	2	3	4	5	6	8	12	15	20	30	60	∞
1	647.8	799.5	864.2	899.6	921.8	937.1	956.7	976.7	984.9	993.1	1001	1010	1018
2	38.51	39.00	39.17	39.25	39.30	39.33	39.37	39.41	39.43	39.45	39.46	39.48	39.50
3	17.44	16.04	15.44	15.10	14.88	14.73	14.54	14.34	14.25	14.17	14.08	13.99	13.90
4	12.22	10.65	9.98	9.60	9.36	9.20	8.98	8.75	8.66	8.56	8.46	8.36	8.26
5	10.01	8.43	7.76	7.39	7.15	6.98	6.76	6.52	6.43	6.33	6.23	6.12	6.02
6	8.81	7.26	6.60	6.23	5.99	5.82	5.60	5.37	5.27	5.17	5.07	4.96	4.85
7	8.07	6.54	5.89	5.52	5.29	5.12	4.90	4.67	4.57	4.47	4.36	4.25	4.14
8	7.57	6.06	5.42	5.05	4.82	4.65	4.43	4.20	4.10	4.00	3.89	3.78	3.67
9	7.21	5.71	5.08	4.72	4.48	4.32	4.10	3.87	3.77	3.67	3.56	3.45	3.33
10	6.94	5.46	4.83	4.47	4.24	4.07	3.85	3.62	3.52	3.42	3.31	3.20	3.08
11	6.72	5.26	4.63	4.28	4.04	3.88	3.66	3.43	3.33	3.23	3.12	3.00	2.88
12	6.55	5.10	4.47	4.12	3.89	3.73	3.51	3.28	3.18	3.07	2.96	2.85	2.72
13	6.41	4.97	4.35	4.00	3.77	3.60	3.39	3.15	3.05	2.95	2.84	2.72	2.60
14	6.30	4.86	4.24	3.89	3.66	3.50	3.29	3.05	2.95	2.84	2.73	2.61	2.49
15	6.20	4.77	4.15	3.80	3.58	3.41	3.20	2.96	2.86	2.76	2.64	2.52	2.40
16	6.12	4.69	4.08	3.73	3.50	3.34	3.12	2.89	2.79	2.68	2.57	2.45	2.32
17	6.04	4.62	4.01	3.66	3.44	3.28	3.06	2.82	2.72	2.62	2.50	2.38	2.25
18	5.98	4.56	3.95	3.61	3.38	3.22	3.01	2.77	2.67	2.56	2.44	2.32	2.19
19	5.92	4.51	3.90	3.56	3.33	3.17	2.96	2.72	2.62	2.51	2.39	2.27	2.13
20	5.87	4.46	3.86	3.51	3.29	3.13	2.91	2.68	2.57	2.46	2.35	2.22	2.09
21	5.83	4.42	3.82	3.48	3.25	3.09	2.87	2.64	2.53	2.42	2.31	2.18	2.04
22	5.79	4.38	3.78	3.44	3.22	3.05	2.84	2.60	2.50	2.39	2.27	2.14	2.00
23	5.75	4.35	3.75	3.41	3.18	3.02	2.81	2.57	2.47	2.36	2.24	2.11	1.97
24	5.72	4.32	3.72	3.38	3.15	2.99	2.78	2.54	2.44	2.33	2.21	2.08	1.94
25	5.69	4.29	3.69	3.35	3.13	2.97	2.75	2.51	2.41	2.30	2.18	2.05	1.91
26	5.66	4.27	3.67	3.33	3.10	2.94	2.73	2.49	2.39	2.28	2.16	2.03	1.88
27	5.63	4.24	3.65	3.31	3.08	2.92	2.71	2.47	2.36	2.25	2.13	2.00	1.85
28	5.61	4.22	3.63	3.29	3.06	2.90	2.69	2.45	2.34	2.23	2.11	1.98	1.83
29	5.59	4.20	3.61	3.27	3.04	2.88	2.67	2.43	2.32	2.21	2.09	1.96	1.81
30	5.57	4.18	3.59	3.25	3.03	2.87	2.65	2.41	2.31	2.20	2.07	1.94	1.79
40	5.42	4.05	3.46	3.13	2.90	2.74	2.53	2.29	2.18	2.07	1.94	1.80	1.64
60	5.29	3.93	3.34	3.01	2.79	2.63	2.41	2.17	2.06	1.94	1.82	1.67	1.48
120	5.15	3.80	3.23	2.89	2.67	2.52	2.30	2.05	1.94	1.82	1.69	1.53	1.31
∞	5.02	3.69	3.12	2.79	2.57	2.41	2.19	1.94	1.83	1.71	1.57	1.39	1.00

$Q(F|\nu_1,\nu_2)=0.01$

$\nu_2 \backslash \nu_1$	1	2	3	4	5	6	8	12	15	20	30	60	∞
1	4052	4999.5	5403	5625	5764	5859	5982	6106	6157	6209	6261	6313	6366
2	98.50	99.00	99.17	99.25	99.30	99.33	99.37	99.42	99.43	99.45	99.47	99.48	99.50
3	34.12	30.82	29.46	28.71	28.24	27.91	27.49	27.05	26.87	26.69	26.50	26.32	26.13
4	21.20	18.00	16.69	15.98	15.52	15.21	14.80	14.37	14.20	14.02	13.84	13.65	13.46
5	16.26	13.27	12.06	11.39	10.97	10.67	10.29	9.89	9.72	9.55	9.38	9.20	9.02
6	13.75	10.92	9.78	9.15	8.75	8.47	8.10	7.72	7.56	7.40	7.23	7.06	6.88
7	12.25	9.55	8.45	7.85	7.46	7.19	6.84	6.47	6.31	6.16	5.99	5.82	5.65
8	11.26	8.65	7.59	7.01	6.63	6.37	6.03	5.67	5.52	5.36	5.20	5.03	4.86
9	10.56	8.02	6.99	6.42	6.06	5.80	5.47	5.11	4.96	4.81	4.65	4.48	4.31
10	10.04	7.56	6.55	5.99	5.64	5.39	5.06	4.71	4.56	4.41	4.25	4.08	3.91
11	9.65	7.21	6.22	5.67	5.32	5.07	4.74	4.40	4.25	4.10	3.94	3.78	3.60
12	9.33	6.93	5.95	5.41	5.06	4.82	4.50	4.16	4.01	3.86	3.70	3.54	3.36
13	9.07	6.70	5.74	5.21	4.86	4.62	4.30	3.96	3.82	3.66	3.51	3.34	3.17
14	8.86	6.51	5.56	5.04	4.69	4.46	4.14	3.80	3.66	3.51	3.35	3.18	3.00
15	8.68	6.36	5.42	4.89	4.56	4.32	4.00	3.67	3.52	3.37	3.21	3.05	2.87
16	8.53	6.23	5.29	4.77	4.44	4.20	3.89	3.55	3.41	3.26	3.10	2.93	2.75
17	8.40	6.11	5.18	4.67	4.34	4.10	3.79	3.46	3.31	3.16	3.00	2.83	2.65
18	8.29	6.01	5.09	4.58	4.25	4.01	3.71	3.37	3.23	3.08	2.92	2.75	2.57
19	8.18	5.93	5.01	4.50	4.17	3.94	3.63	3.30	3.15	3.00	2.84	2.67	2.49
20	8.10	5.85	4.94	4.43	4.10	3.87	3.56	3.23	3.09	2.94	2.78	2.61	2.42
21	8.02	5.78	4.87	4.37	4.04	3.81	3.51	3.17	3.03	2.88	2.72	2.55	2.36
22	7.95	5.72	4.82	4.31	3.99	3.76	3.45	3.12	2.98	2.83	2.67	2.50	2.31
23	7.88	5.66	4.76	4.26	3.94	3.71	3.41	3.07	2.93	2.78	2.62	2.45	2.26
24	7.82	5.61	4.72	4.22	3.90	3.67	3.36	3.03	2.89	2.74	2.58	2.40	2.21
25	7.77	5.57	4.68	4.18	3.85	3.63	3.32	2.99	2.85	2.70	2.54	2.36	2.17
26	7.72	5.53	4.64	4.14	3.82	3.59	3.29	2.96	2.81	2.66	2.50	2.33	2.13
27	7.68	5.49	4.60	4.11	3.78	3.56	3.26	2.93	2.78	2.63	2.47	2.29	2.10
28	7.64	5.45	4.57	4.07	3.75	3.53	3.23	2.90	2.75	2.60	2.44	2.26	2.06
29	7.60	5.42	4.54	4.04	3.73	3.50	3.20	2.87	2.73	2.57	2.41	2.23	2.03
30	7.56	5.39	4.51	4.02	3.70	3.47	3.17	2.84	2.70	2.55	2.39	2.21	2.01
40	7.31	5.18	4.31	3.83	3.51	3.29	2.99	2.66	2.52	2.37	2.20	2.02	1.80
60	7.08	4.98	4.13	3.65	3.34	3.12	2.82	2.50	2.35	2.20	2.03	1.84	1.60
120	6.85	4.79	3.95	3.48	3.17	2.96	2.66	2.34	2.19	2.03	1.86	1.66	1.38
∞	6.63	4.61	3.78	3.32	3.02	2.80	2.51	2.18	2.04	1.88	1.70	1.47	1.00

Table 4

PERCENTAGE POINTS OF THE F-DISTRIBUTION—VALUES OF F IN TERMS OF Q, ν_1, ν_2

$Q(F|\nu_1,\nu_2)=0.005$

ν_2 \ ν_1	1	2	3	4	5	6	8	12	15	20	30	60	∞
1	16211	20000	21615	22500	23056	23437	23925	24426	24630	24836	25044	25253	25465
2	198.5	199.0	199.2	199.2	199.3	199.3	199.4	199.4	199.4	199.4	199.5	199.5	199.5
3	55.55	49.80	47.47	46.19	45.39	44.84	44.13	43.39	43.08	42.78	42.47	42.15	41.83
4	31.33	26.28	24.26	23.15	22.46	21.97	21.35	20.70	20.44	20.17	19.89	19.61	19.32
5	22.78	18.31	16.53	15.56	14.94	14.51	13.96	13.38	13.15	12.90	12.66	12.40	12.14
6	18.63	14.54	12.92	12.03	11.46	11.07	10.57	10.03	9.81	9.59	9.36	9.12	8.88
7	16.24	12.40	10.88	10.05	9.52	9.16	8.68	8.18	7.97	7.75	7.53	7.31	7.08
8	14.69	11.04	9.60	8.81	8.30	7.95	7.50	7.01	6.81	6.61	6.40	6.18	5.95
9	13.61	10.11	8.72	7.96	7.47	7.13	6.69	6.23	6.03	5.83	5.62	5.41	5.19
10	12.83	9.43	8.08	7.34	6.87	6.54	6.12	5.66	5.47	5.27	5.07	4.86	4.64
11	12.23	8.91	7.60	6.88	6.42	6.10	5.68	5.24	5.05	4.86	4.65	4.44	4.23
12	11.75	8.51	7.23	6.52	6.07	5.76	5.35	4.91	4.72	4.53	4.33	4.12	3.90
13	11.37	8.19	6.93	6.23	5.79	5.48	5.08	4.64	4.46	4.27	4.07	3.87	3.65
14	11.06	7.92	6.68	6.00	5.56	5.26	4.86	4.43	4.25	4.06	3.86	3.66	3.44
15	10.80	7.70	6.48	5.80	5.37	5.07	4.67	4.25	4.07	3.88	3.69	3.48	3.26
16	10.58	7.51	6.30	5.64	5.21	4.91	4.52	4.10	3.92	3.73	3.54	3.33	3.11
17	10.38	7.35	6.16	5.50	5.07	4.78	4.39	3.97	3.79	3.61	3.41	3.21	2.98
18	10.22	7.21	6.03	5.37	4.96	4.66	4.28	3.86	3.68	3.50	3.30	3.10	2.87
19	10.07	7.09	5.92	5.27	4.85	4.56	4.18	3.76	3.59	3.40	3.21	3.00	2.78
20	9.94	6.99	5.82	5.17	4.76	4.47	4.09	3.68	3.50	3.32	3.12	2.92	2.69
21	9.83	6.89	5.73	5.09	4.68	4.39	4.01	3.60	3.43	3.24	3.05	2.84	2.61
22	9.73	6.81	5.65	5.02	4.61	4.32	3.94	3.54	3.36	3.18	2.98	2.77	2.55
23	9.63	6.73	5.58	4.95	4.54	4.26	3.88	3.47	3.30	3.12	2.92	2.71	2.48
24	9.55	6.66	5.52	4.89	4.49	4.20	3.83	3.42	3.25	3.06	2.87	2.66	2.43
25	9.48	6.60	5.46	4.84	4.43	4.15	3.78	3.37	3.20	3.01	2.82	2.61	2.38
26	9.41	6.54	5.41	4.79	4.38	4.10	3.73	3.33	3.15	2.97	2.77	2.56	2.33
27	9.34	6.49	5.36	4.74	4.34	4.06	3.69	3.28	3.11	2.93	2.73	2.52	2.29
28	9.28	6.44	5.32	4.70	4.30	4.02	3.65	3.25	3.07	2.89	2.69	2.48	2.25
29	9.23	6.40	5.28	4.66	4.26	3.98	3.61	3.21	3.04	2.86	2.66	2.45	2.21
30	9.18	6.35	5.24	4.62	4.23	3.95	3.58	3.18	3.01	2.82	2.63	2.42	2.18
40	8.83	6.07	4.98	4.37	3.99	3.71	3.35	2.95	2.78	2.60	2.40	2.18	1.93
60	8.49	5.79	4.73	4.14	3.76	3.49	3.13	2.74	2.57	2.39	2.19	1.96	1.69
120	8.18	5.54	4.50	3.92	3.55	3.28	2.93	2.54	2.37	2.19	1.98	1.75	1.43
∞	7.88	5.30	4.28	3.72	3.35	3.09	2.74	2.36	2.19	2.00	1.79	1.55	1.00

Table 4

PERCENTAGE POINTS OF THE F-DISTRIBUTION—VALUES OF F IN TERMS OF Q, ν_1, ν_2

$Q(F|\nu_1, \nu_2) = 0.005$

$\nu_2 \backslash \nu_1$	1	2	3	4	5	6	8	12	15	20	30	60	∞
1	(5)4.053	(5)5.000	(5)5.404	(5)5.625	(5)5.764	(5)5.859	(5)5.981	(5)6.107	(5)6.158	(5)6.209	(5)6.261	(5)6.313	(5)6.366
2	998.5	999.0	999.2	999.2	999.3	999.3	999.4	999.4	999.4	999.4	999.5	999.5	999.5
3	167.0	148.5	141.1	137.1	134.6	132.8	130.6	128.3	127.4	126.4	125.4	124.5	123.5
4	74.14	61.25	56.18	53.44	51.71	50.53	49.00	47.41	46.76	46.10	45.43	44.75	44.05
5	47.18	37.12	33.20	31.09	29.75	28.84	27.64	26.42	25.91	25.39	24.87	24.33	23.79
6	35.51	27.00	23.70	21.92	20.81	20.03	19.03	17.99	17.56	17.12	16.67	16.21	15.75
7	29.25	21.69	18.77	17.19	16.21	15.52	14.63	13.71	13.32	12.93	12.53	12.12	11.70
8	25.42	18.49	15.83	14.39	13.49	12.86	12.04	11.19	10.84	10.48	10.11	9.73	9.33
9	22.86	16.39	13.90	12.56	11.71	11.13	10.37	9.57	9.24	8.90	8.55	8.19	7.81
10	21.04	14.91	12.55	11.28	10.48	9.92	9.20	8.45	8.13	7.80	7.47	7.12	6.76
11	19.69	13.81	11.56	10.35	9.58	9.05	8.35	7.63	7.32	7.01	6.68	6.35	6.00
12	18.64	12.97	10.80	9.63	8.89	8.38	7.71	7.00	6.71	6.40	6.09	5.76	5.42
13	17.81	12.31	10.21	9.07	8.35	7.86	7.21	6.52	6.23	5.93	5.63	5.30	4.97
14	17.14	11.78	9.73	8.62	7.92	7.43	6.80	6.13	5.85	5.56	5.25	4.94	4.60
15	16.59	11.34	9.34	8.25	7.57	7.09	6.47	5.81	5.54	5.25	4.95	4.64	4.31
16	16.12	10.97	9.00	7.94	7.27	6.81	6.19	5.55	5.27	4.99	4.70	4.39	4.06
17	15.72	10.66	8.73	7.68	7.02	6.56	5.96	5.32	5.05	4.78	4.48	4.18	3.85
18	15.38	10.39	8.49	7.46	6.81	6.35	5.76	5.13	4.87	4.59	4.30	4.00	3.67
19	15.08	10.16	8.28	7.26	6.62	6.18	5.59	4.97	4.70	4.43	4.14	3.84	3.51
20	14.82	9.95	8.10	7.10	6.46	6.02	5.44	4.82	4.56	4.29	4.00	3.70	3.38
21	14.59	9.77	7.94	6.95	6.32	5.88	5.31	4.70	4.44	4.17	3.88	3.58	3.26
22	14.38	9.61	7.80	6.81	6.19	5.76	5.19	4.58	4.33	4.06	3.78	3.48	3.15
23	14.19	9.47	7.67	6.69	6.08	5.65	5.09	4.48	4.23	3.96	3.68	3.38	3.05
24	14.03	9.34	7.55	6.59	5.98	5.55	4.99	4.39	4.14	3.87	3.59	3.29	2.97
25	13.88	9.22	7.45	6.49	5.88	5.46	4.91	4.31	4.06	3.79	3.52	3.22	2.89
26	13.74	9.12	7.36	6.41	5.80	5.38	4.83	4.24	3.99	3.72	3.44	3.15	2.82
27	13.61	9.02	7.27	6.33	5.73	5.31	4.76	4.17	3.92	3.66	3.38	3.08	2.75
28	13.50	8.93	7.19	6.25	5.66	5.24	4.69	4.11	3.86	3.60	3.32	3.02	2.69
29	13.39	8.85	7.12	6.19	5.59	5.18	4.64	4.05	3.80	3.54	3.27	2.97	2.64
30	13.29	8.77	7.05	6.12	5.53	5.12	4.58	4.00	3.75	3.49	3.22	2.92	2.59
40	12.61	8.25	6.60	5.70	5.13	4.73	4.21	3.64	3.40	3.15	2.87	2.57	2.23
60	11.97	7.76	6.17	5.31	4.76	4.37	3.87	3.31	3.08	2.83	2.55	2.25	1.89
120	11.38	7.32	5.79	4.95	4.42	4.04	3.55	3.02	2.78	2.53	2.26	1.95	1.54
∞	10.83	6.91	5.42	4.62	4.10	3.74	3.27	2.74	2.51	2.27	1.99	1.66	1.00

PERCENTAGE POINTS OF THE t-DISTRIBUTION—VALUES OF t IN TERMS OF A AND ν

$\nu \backslash A$	0.2	0.5	0.8	0.9	0.95	0.98	0.99	0.995	0.998	0.999	0.9999	0.99999	0.999999
1	0.325	1.000	3.078	6.314	12.706	31.821	63.657	127.321	318.309	636.619	6366.198	63661.977	636619.772
2	0.289	0.816	1.886	2.920	4.303	6.965	9.925	14.089	22.327	31.598	99.992	316.225	999.999
3	0.277	0.765	1.638	2.353	3.182	4.541	5.841	7.453	10.214	12.924	28.000	60.397	130.155
4	0.271	0.741	1.533	2.132	2.776	3.747	4.604	5.598	7.173	8.610	15.544	27.771	49.459
5	0.267	0.727	1.476	2.015	2.571	3.365	4.032	4.773	5.893	6.869	11.178	17.897	28.477
6	0.265	0.718	1.440	1.943	2.447	3.143	3.707	4.317	5.208	5.959	9.082	13.555	20.047
7	0.263	0.711	1.415	1.895	2.365	2.998	3.499	4.029	4.785	5.408	7.885	11.215	15.764
8	0.262	0.706	1.397	1.860	2.306	2.896	3.355	3.833	4.501	5.041	7.120	9.782	13.257
9	0.261	0.703	1.383	1.833	2.262	2.821	3.250	3.690	4.297	4.781	6.594	8.827	11.637
10	0.260	0.700	1.372	1.812	2.228	2.764	3.169	3.581	4.144	4.587	6.211	8.150	10.516
11	0.260	0.697	1.363	1.796	2.201	2.718	3.106	3.497	4.025	4.437	5.921	7.648	9.702
12	0.259	0.695	1.356	1.782	2.179	2.681	3.055	3.428	3.930	4.318	5.694	7.261	9.085
13	0.259	0.694	1.350	1.771	2.160	2.650	3.012	3.372	3.852	4.221	5.513	6.955	8.604
14	0.258	0.692	1.345	1.761	2.145	2.624	2.977	3.326	3.787	4.140	5.363	6.706	8.218
15	0.258	0.691	1.341	1.753	2.131	2.602	2.947	3.286	3.733	4.073	5.239	6.502	7.903
16	0.258	0.690	1.337	1.746	2.120	2.583	2.921	3.252	3.686	4.015	5.134	6.330	7.642
17	0.257	0.689	1.333	1.740	2.110	2.567	2.898	3.223	3.646	3.965	5.044	6.184	7.421
18	0.257	0.688	1.330	1.734	2.101	2.552	2.878	3.197	3.610	3.922	4.966	6.059	7.232
19	0.257	0.688	1.328	1.729	2.093	2.539	2.861	3.174	3.579	3.883	4.897	5.949	7.069
20	0.257	0.687	1.325	1.725	2.086	2.528	2.845	3.153	3.552	3.850	4.837	5.854	6.927
21	0.257	0.686	1.323	1.721	2.080	2.518	2.831	3.135	3.527	3.819	4.784	5.769	6.802
22	0.256	0.686	1.321	1.717	2.074	2.508	2.819	3.119	3.505	3.792	4.736	5.694	6.692
23	0.256	0.685	1.319	1.714	2.069	2.500	2.807	3.104	3.485	3.768	4.693	5.627	6.593
24	0.256	0.685	1.318	1.711	2.064	2.492	2.797	3.090	3.467	3.745	4.654	5.566	6.504
25	0.256	0.684	1.316	1.708	2.060	2.485	2.787	3.078	3.450	3.725	4.619	5.511	6.424
26	0.256	0.684	1.315	1.706	2.056	2.479	2.779	3.067	3.435	3.707	4.587	5.461	6.352
27	0.256	0.684	1.314	1.703	2.052	2.473	2.771	3.057	3.421	3.690	4.558	5.415	6.286
28	0.256	0.683	1.313	1.701	2.048	2.467	2.763	3.047	3.408	3.674	4.530	5.373	6.225
29	0.256	0.683	1.311	1.699	2.045	2.462	2.756	3.038	3.396	3.659	4.506	5.335	6.170
30	0.256	0.683	1.310	1.697	2.042	2.457	2.750	3.030	3.385	3.646	4.482	5.299	6.119
40	0.255	0.681	1.303	1.684	2.021	2.423	2.704	2.971	3.307	3.551	4.321	5.053	5.768
60	0.254	0.679	1.296	1.671	2.000	2.390	2.660	2.915	3.232	3.460	4.169	4.825	5.449
120	0.254	0.677	1.289	1.658	1.980	2.358	2.617	2.860	3.160	3.373	4.025	4.613	5.158
∞	0.253	0.674	1.282	1.645	1.960	2.326	2.576	2.807	3.090	3.291	3.891	4.417	4.892

$$A = A(t|\nu) = \left[\sqrt{\nu}\, B\left(\frac{1}{2}, \frac{\nu}{2}\right)\right]^{-1} \int_{-t}^{t} \left(1 + \frac{x^2}{\nu}\right)^{-\left(\frac{\nu+1}{2}\right)} dx$$

2500 FIVE DIGIT RANDOM NUMBERS

53479	81115	98036	12217	59526	40238	40577	39351	43211	69255
97344	70328	58116	91964	26240	44643	83287	97391	92823	77578
66023	38277	74523	71118	84892	13956	98899	92315	65783	59640
99776	75723	03172	43112	83086	81982	14538	26162	24899	20551
30176	48979	92153	38416	42436	26636	83903	44722	69210	69117
81874	83339	14988	99937	13213	30177	47967	93793	86693	98854
19839	90630	71863	95053	55532	60908	84108	55342	48479	63799
09337	33435	53869	52769	18801	25820	96198	66518	78314	97013
31151	58295	40823	41330	21093	93882	49192	44876	47185	81425
67619	52515	03037	81699	17106	64982	60834	85319	47814	08075
61946	48790	11602	83043	22257	11832	04344	95541	20366	55937
04811	64892	96346	79065	26999	43967	63485	93572	80753	96582
05763	39601	56140	25513	86151	78657	02184	29715	04334	15678
73260	56877	40794	13948	96289	90185	47111	66807	61849	44686
54909	09976	76580	02645	35795	44537	64428	35441	28318	99001
42583	36335	60068	04044	29678	16342	48592	25547	63177	75225
27266	27403	97520	23334	36453	33699	23672	45884	41515	04756
49843	11442	66682	36055	32002	78600	36924	59962	68191	62580
29316	40460	27076	69232	51423	58515	49920	03901	26597	33068
30463	27856	67798	16837	74273	05793	02900	63498	00782	35097
28708	84088	65535	44258	33869	82530	98399	26387	02836	36838
13183	50652	94872	28257	78547	55286	33591	61965	51723	14211
60796	76639	30157	40295	99476	28334	15368	42481	60312	42770
13486	46918	64683	07411	77842	01908	47796	65796	44230	77230
34914	94502	39374	34185	57500	22514	04060	94511	44612	10485
28105	04814	85170	86490	35695	03483	57315	63174	71902	71182
59231	45028	01173	08848	81925	71494	95401	34049	04851	65914
87437	82758	71093	36833	53582	25986	46005	42840	81683	21459
29046	01301	55343	65732	78714	43644	46248	53205	94868	48711
62035	71886	94506	15263	61435	10369	42054	68257	14385	79436
38856	80048	59973	73368	52876	47673	41020	82295	26430	87377
40666	43328	87379	86418	95841	25590	54137	94182	42308	07361
40588	90087	37729	08667	37256	20317	53316	50982	32900	32097
78237	86556	50276	20431	00243	02303	71029	49932	23245	00862
98247	67474	71455	69540	01169	03320	67017	92543	97977	52728
69977	78558	65430	32627	28312	61815	14598	79728	55699	91348
39843	23074	40814	03713	21891	96353	96806	24595	26203	26009
62880	87277	99895	99965	34374	42556	11679	99605	98011	48867
56138	64927	29454	52967	86624	62422	30163	76181	95317	39264
90804	56026	48994	64569	67465	60180	12972	03848	62582	93855
09665	44672	74762	33357	67301	80546	97659	11348	78771	45011
34756	50403	76634	12767	32220	34545	18100	53513	14521	72120
12157	73327	74196	26668	78087	53636	52304	00007	05708	63538
69384	07734	94451	76428	16121	09300	67417	68587	87932	38840
93358	64565	43766	45041	44930	69970	16964	08277	67752	60292
38879	35544	99563	85404	04913	62547	78406	01017	86187	22072
58314	60298	72394	69668	12474	93059	02053	29807	63645	12792
83568	10227	99471	74729	22075	10233	21575	20325	21317	57124
28067	91152	40568	33705	64510	07067	64374	26336	79652	31140
05730	75557	93161	80921	55873	54103	34801	83157	04534	81368

2500 FIVE DIGIT RANDOM NUMBERS

26687	74223	43546	45699	94469	82125	37370	23966	68926	37664
60675	75169	24510	15100	02011	14375	65187	10630	64421	66745
45418	98635	83123	98558	09953	60255	42071	40930	97992	93085
69872	48026	89755	28470	44130	59979	91063	28766	85962	77173
03765	86366	99539	44183	23886	89977	11964	51581	18033	56239
84686	57636	32326	19867	71345	42002	96997	84379	27991	21459
91512	49670	32556	85189	28023	88151	62896	95498	29423	38138
10737	49307	18307	22246	22461	10003	93157	66984	44919	30467
54870	19676	58367	20905	38324	00026	98440	37427	22896	37637
48967	49579	65369	74305	62085	39297	10309	23173	74212	32272
91430	79112	03685	05411	23027	54735	91550	06250	18705	18909
92564	29567	47476	62804	73428	04535	86395	12162	59647	97726
41734	12199	77441	92415	63542	42115	84972	12454	33133	48467
25251	78110	54178	78241	09226	87529	35376	90690	54178	08561
91657	11563	66036	28523	83705	09956	76610	88116	78351	50877
00149	84745	63222	50533	50159	60433	04822	49577	89049	16162
53250	73200	84066	59620	61009	38542	05758	06178	80193	26466
25587	17481	56716	49749	70733	32733	60365	14108	52573	39391
01176	12182	06882	27562	75456	54261	38564	89054	96911	88906
83531	15544	40834	20296	88576	47815	96540	79462	78666	25353
19902	98866	32805	61091	91587	30340	84909	64047	67750	87638
96516	78705	25556	35181	29064	49005	29843	68949	50506	45862
99417	56171	19848	24352	51844	03791	72127	57958	08366	43190
77699	57853	93213	27342	28906	31052	65815	21637	49385	75406
32245	83794	99528	05150	27246	48263	62156	62469	97048	16511
12874	72753	66469	13782	64330	00056	73324	03920	13193	19466
63899	41910	45484	55461	66518	82486	74694	07865	09724	76490
16255	43271	26540	41298	35095	32170	70625	66407	01050	44225
75553	30207	41814	74985	40223	91223	64238	73012	83100	92041
41772	18441	34685	13892	38843	69007	10362	84125	08814	66785
09270	01245	81765	06809	10561	10080	17482	05471	82273	06902
85058	17815	71551	36356	97519	54144	51132	83169	27373	68609
80222	87572	62758	14858	36350	23304	70453	21065	63812	29860
83901	88028	56743	25598	79349	47880	77912	52020	84305	02897
36303	57833	77622	02238	53285	77316	40106	38456	92214	54278
91543	63886	60539	96334	20804	72692	08944	02870	74892	22598
14415	33816	78231	87674	96473	44451	25098	29296	50679	07798
82465	07781	09938	66874	72128	99685	84329	14530	08410	45953
27306	39843	05634	96368	72022	01278	92830	40094	31776	41822
91960	82766	02331	08797	33858	21847	17391	53755	58079	48498
59284	96108	91610	07483	37943	96832	15444	12091	36690	58317
10428	96003	71223	21352	78685	55964	35510	94805	23422	04492
65527	41039	79574	05105	59588	02115	33446	56780	18402	36279
59688	43078	93275	31978	08768	84805	50661	18523	83235	50602
44452	10188	43565	46531	93023	07618	12910	60934	53403	18401
87275	82013	59804	78595	60553	14038	12096	95472	42736	08573
94155	93110	49964	27753	85090	77677	69303	66323	77811	22791
26488	76394	91282	03419	68758	89575	66469	97835	66681	03171
37073	34547	88296	68638	12976	50896	10023	27220	05785	77538
83835	89575	55956	93957	30361	47679	83001	35056	07103	63072

2500 FIVE DIGIT RANDOM NUMBERS

55034	81217	90564	81943	11241	84512	12288	89862	00760	76159
25521	99536	43233	48786	49221	06960	31564	21458	88199	06312
85421	72744	97242	66383	00132	05661	96442	37388	57671	27916
61219	48390	47344	30413	39392	91365	56203	79204	05330	31196
20230	03147	58854	11650	28415	12821	58931	30508	65989	26675
95776	83206	56144	55953	89787	64426	08448	45707	80364	60262
07603	17344	01148	83300	96955	65027	31713	89013	79557	49755
00645	17459	78742	39005	36027	98807	72666	54484	68262	38827
62950	83162	61504	31557	80590	47893	72360	72720	08396	33674
79350	10276	81933	26347	08068	67816	06659	87917	74166	85519
48339	69834	59047	82175	92010	58446	69591	56205	95700	86211
05842	08439	79836	50957	32059	32910	15842	13918	41365	80115
25855	02209	07307	59942	71389	76159	11263	38787	61541	22606
25272	16152	82323	76718	98081	38631	91956	49909	76253	33970
73003	29058	17605	49298	47675	90445	68919	05676	23823	84892
81310	94430	22663	06584	38142	00146	17496	51115	61458	65790
10024	44713	59832	80721	63711	67882	25100	45345	55743	67618
84671	52806	89124	37691	20897	82339	22627	06142	05773	03547
29296	58162	21858	33732	94056	88806	54603	00384	66340	69232
51771	94074	70630	41286	90583	87680	13961	55627	23670	35109
42166	56251	60770	51672	36031	77273	85218	14812	90758	23677
78355	67041	22492	51522	31164	30450	27600	44428	96380	26772
09552	51347	33864	89018	73418	81538	77399	30448	97740	18158
15771	63127	34847	05660	06156	48970	55699	61818	91763	20821
13231	99058	93754	36730	44286	44326	15729	37500	47269	13333
50583	03570	38472	73236	67613	72780	78174	18718	99092	64114
99485	57330	10634	74905	90671	19643	69903	60950	17968	37217
54676	39524	73785	48864	69835	62798	65205	69187	05572	74741
99343	71549	10248	76036	31702	76868	88909	69574	27642	00336
35492	40231	34868	55356	12847	68093	52643	32732	67016	46784
98170	25384	03841	23920	47954	10359	70114	11177	63298	99903
02670	86155	56860	02592	01646	42200	79950	37764	82341	71952
36934	42879	81637	79952	07066	41625	96804	92388	88860	68580
56851	12778	24309	73660	84264	24668	16686	02239	66022	64133
05464	28892	14271	23778	88599	17081	33884	88783	39015	57118
15025	20237	63386	71122	06620	07415	94982	32324	79427	70387
95610	08030	81469	91066	88857	56583	01224	28097	19726	71465
09026	40378	05731	55128	74298	49196	31669	42605	30368	96424
81431	99955	52462	67667	97322	69808	21240	65921	12629	92896
21431	59335	58627	94822	65484	09641	41018	85100	16110	32077
95832	76145	11636	80284	17787	97934	12822	73890	66009	27521
99813	44631	43746	99790	86823	12114	31706	05024	28156	04202
77210	31148	50543	11603	50934	02498	09184	95875	85840	71954
13268	02609	79833	66058	80277	08533	28676	37532	70535	82356
44285	71735	26620	54691	14909	52132	81110	74548	78853	31996
70526	45953	79637	57374	05053	31965	33376	13232	85666	86615
88386	11222	25080	71462	09818	46001	19065	68981	18310	74178
83161	73994	17209	79441	64091	49790	11936	44864	86978	34538
50214	71721	33851	45144	05696	29935	12823	01594	08453	52825
97689	29341	67747	80643	13620	23943	49396	83686	37302	95350

2500 FIVE DIGIT RANDOM NUMBERS

12367	23891	31506	90721	18710	89140	58595	99425	22840	08267
38890	30239	34237	22578	74420	22734	26930	40604	10782	80128
80788	55410	39770	93317	18270	21141	52085	78093	85638	81140
02395	77585	08854	23562	33544	45796	10976	44721	24781	09690
73720	70184	69112	71887	80140	72876	38984	23409	63957	44751
61383	17222	55234	18963	39006	93504	18273	49815	52802	69675
39161	44282	14975	97498	25973	33605	60141	30030	77677	49294
80907	74484	39884	19885	37311	04209	49675	39596	01052	43999
09052	65670	63660	34035	06578	87837	28125	48883	50482	55735
33425	24226	32043	60082	20418	85047	53570	32554	64099	52326
72651	69474	73648	71530	55454	19576	15552	20577	12124	50038
04142	32092	83586	61825	35482	32736	63403	91499	37196	02762
85226	14193	52213	60746	24414	57858	31884	51266	82293	73553
54888	03579	91674	59502	08619	33790	29011	85193	62262	28684
33258	51516	82032	45233	39351	33229	59464	65545	76809	16982
75973	15957	32405	82081	02214	57143	33526	47194	94526	73253
90638	75314	35381	34451	49246	11465	25102	71489	89883	99708
65061	15498	93348	33566	19427	66826	03044	97361	08159	47485
64420	07427	82233	97812	39572	07766	65844	29980	15533	90114
27175	17389	76963	75117	45580	99904	47160	55364	25666	25405
32215	30094	87276	56896	15625	32594	80663	08082	19422	80717
54209	58043	72350	89828	02706	16815	89985	37380	44032	59366
59286	66964	84843	71549	67553	33867	83011	66213	69372	23903
83872	58167	01221	95558	22196	65905	38785	01355	47489	28170
83310	57080	03366	80017	39601	40698	56434	64055	02495	50880
64545	29500	13351	78647	92628	19354	60479	57338	52133	07114
39269	00076	55489	01524	76568	22571	20328	84623	30188	43904
29763	05675	28193	65514	11954	78599	63902	21346	19219	90286
06310	02998	01463	27738	90288	17697	64511	39552	34694	03211
97541	47607	57655	59102	21851	44446	07976	54295	84671	78755
82968	85717	11619	97721	53513	53781	98941	38401	70939	11319
76878	34727	12524	90642	16921	13669	17420	84483	68309	85241
87394	78884	87237	92086	95633	66841	22906	64989	86952	54700
74040	12731	59616	33697	12592	44891	67982	72972	89795	10587
47896	41413	66431	70046	50793	45920	96564	67958	56369	44725
87778	71697	64148	54363	92114	34037	59061	62051	62049	33526
96977	63143	72219	80040	11990	47698	95621	72990	29047	85893
43820	13285	77811	81697	29937	70750	02029	32377	00556	86687
57203	83960	40096	39234	65953	59911	91411	55573	88427	45573
49065	72171	80939	06017	90323	63687	07932	99587	49014	26452
94250	84270	95798	13477	80139	26335	55169	73417	40766	45170
68148	81382	82383	18674	40453	92828	30042	37412	43423	45138
12208	97809	33619	28868	41646	16734	88860	32636	41985	84615
88317	89705	26119	12416	19438	65665	60989	59766	11418	18250
56728	80359	29613	63052	15251	44684	64681	42354	51029	77680
07138	12320	01073	19304	87042	58920	28454	81069	93978	66659
21188	64554	55618	36088	24331	84390	16022	12200	77559	75661
02154	12250	88738	43917	03655	21099	60805	63246	26842	35816
90953	85238	32771	07305	36181	47420	19681	33184	41386	03249
80103	91308	12858	41293	00325	15013	19579	91132	12720	92603

2500 FIVE DIGIT RANDOM NUMBERS

92630	78240	19267	95457	53497	23894	37708	79862	76471	66418
79445	78735	71549	44843	26104	67318	00701	34986	66751	99723
59654	71966	27386	50004	05358	94031	29281	18544	52429	06080
31524	49587	76612	39789	13537	48086	59483	60680	84675	53014
06348	76938	90379	51392	55887	71015	09209	79157	24440	30244
28703	51709	94456	48396	73780	06436	86641	69239	57662	80181
68108	89266	94730	95761	75023	48464	65544	96583	18911	16391
99938	90704	93621	66330	33393	95261	95349	51769	91616	33238
91543	73196	34449	63513	83834	99411	58826	40456	69268	48562
42103	02781	73920	56297	72678	12249	25270	36678	21313	75767
17138	27584	25296	28387	51350	61664	37893	05363	44143	42677
28297	14280	54524	21618	95320	38174	60579	08089	94999	78460
09331	56712	51333	06289	75345	08811	82711	57392	25252	30333
31295	04204	93712	51287	05754	79396	87399	51773	33075	97061
36146	15560	27592	42089	99281	59640	15221	96079	09961	05371
29553	18432	13630	05529	02791	81017	49027	79031	50912	09399
23501	22642	63081	08191	89420	67800	55137	54707	32945	64522
57888	85846	67967	07835	11314	01545	48535	17142	08552	67457
55336	71264	88472	04334	63919	36394	11196	92470	70543	29776
10087	10072	55980	64688	68239	20461	89381	93809	00796	95945
34101	81277	66090	88872	37818	72142	67140	50785	21380	16703
53362	44940	60430	22834	14130	96593	23298	56203	92671	15925
82975	66158	84731	19436	55790	69229	28661	13675	99318	76873
54827	84673	22898	08094	14326	87038	42892	21127	30712	48489
25464	59098	27436	89421	80754	89924	19097	67737	80368	08795
67609	60214	41475	84950	40133	02546	09570	45682	50165	15609
44921	70924	61295	51137	47596	86735	35561	76649	18217	63446
33170	30972	98130	95828	49786	13301	36081	80761	33985	68621
84687	85445	06208	17654	51333	02878	35010	67578	61574	20749
71886	56450	36567	09395	96951	35507	17555	35212	69106	01679
00475	02224	74722	14721	40215	21351	08596	45625	83981	63748
25993	38881	68361	59560	41274	69742	40703	37993	03435	18873
92882	53178	99195	93803	56985	53089	15305	50522	55900	43026
25138	26810	07093	15677	60688	04410	24505	37890	67186	62829
84631	71882	12991	83028	82484	90339	91950	74579	03539	90122
34003	92326	12793	61453	48121	74271	28363	66561	75220	35908
53775	45749	05734	86169	42762	70175	97310	73894	88606	19994
59316	97885	72807	54966	60859	11932	35265	71601	55577	67715
20479	66557	50705	26999	09854	52591	14063	30214	19890	19292
86180	84931	25455	26044	02227	52015	21820	50599	51671	65411
21451	68001	72710	40261	61281	13172	63819	48970	51732	54113
98062	68375	80089	24135	72355	95428	11808	29740	81644	86610
01788	64429	14430	94575	75153	94576	61393	96192	03227	32258
62465	04841	43272	68702	01274	05437	22953	18946	99053	41690
94324	31089	84159	92933	99989	89500	91586	02802	69471	68274
05797	43984	21575	09908	70221	19791	51578	36432	33494	79888
10395	14289	52185	09721	25789	38562	54794	04897	59012	89251
35177	56986	25549	59730	64718	52630	31100	62384	49483	11409
25633	89619	75882	98256	02126	72099	57183	55887	09320	73463
16464	48280	94254	45777	45150	68865	11382	11782	22695	41988

CHAPTER 16

SPECIAL FUNCTIONS

THE GAMMA FUNCTION

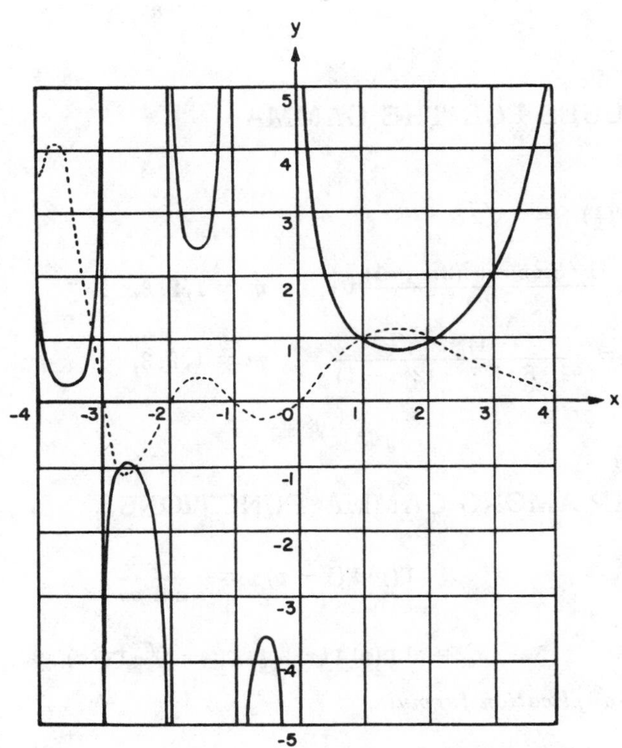

Fig. 1 *Gamma function.*

————, $y = \Gamma(x)$, - - - -, $y = 1/\Gamma(x)$

DEFINITION OF THE GAMMA FUNCTION $\Gamma(n)$ FOR $n>0$

16.1 $$\Gamma(n) = \int_0^\infty t^{n-1}e^{-t}\,dt \qquad n > 0$$

RECURSION FORMULA

16.2 $$\Gamma(n+1) = n\,\Gamma(n)$$

$$\Gamma(n+1) = n! \quad \text{if } n = 0, 1, 2, \ldots \text{ where } 0! = 1$$

THE GAMMA FUNCTION FOR $n<0$

For $n < 0$ the gamma function can be defined by using 16.2, i.e.

$$\Gamma(n) = \frac{\Gamma(n+1)}{n}$$

SPECIAL VALUES FOR THE GAMMA FUNCTION

$$\Gamma(\tfrac{1}{2}) = \sqrt{\pi}$$

$$\Gamma(m+\tfrac{1}{2}) = \frac{1 \cdot 3 \cdot 5 \cdots (2m-1)}{2^m}\sqrt{\pi} \qquad m = 1, 2, 3, \ldots$$

$$\Gamma(-m+\tfrac{1}{2}) = \frac{(-1)^m 2^m \sqrt{\pi}}{1 \cdot 3 \cdot 5 \cdots (2m-1)} \qquad m = 1, 2, 3, \ldots$$

RELATIONSHIP AMONG GAMMA FUNCTIONS

$$\Gamma(p)\,\Gamma(1-p) = \frac{\pi}{\sin p\pi}$$

16.3 $$2^{2x-1}\,\Gamma(x)\,\Gamma(x+\tfrac{1}{2}) = \sqrt{\pi}\,\Gamma(2x)$$

This is called the *duplication formula.*

$$\Gamma(x)\,\Gamma\left(x+\frac{1}{m}\right)\Gamma\left(x+\frac{2}{m}\right)\cdots\Gamma\left(x+\frac{m-1}{m}\right) = m^{1/2-mx}\,(2\pi)^{(m-1)/2}\,\Gamma(mx)$$

For $m = 2$ this reduces to 16.3

OTHER DEFINITIONS OF THE GAMMA FUNCTION

$$\Gamma(x+1) = \lim_{k\to\infty} \frac{1\cdot 2\cdot 3\cdots k}{(x+1)(x+2)\cdots(x+k)}\, k^x$$

$$\frac{1}{\Gamma(x)} = xe^{\gamma x}\prod_{m=1}^{\infty}\left\{\left(1+\frac{x}{m}\right)e^{-x/m}\right\}$$

This is an infinite product representation for the gamma function where γ is Euler's constant.

DERIVATIVES OF THE GAMMA FUNCTION

$$\Gamma'(1) = \int_0^\infty e^{-x}\ln x\, dx = -\gamma$$

$$\frac{\Gamma'(x)}{\Gamma(x)} = -\gamma + \left(\frac{1}{1}-\frac{1}{x}\right) + \left(\frac{1}{2}-\frac{1}{x+1}\right) + \cdots + \left(\frac{1}{n}-\frac{1}{x+n-1}\right) + \cdots$$

ASYMPTOTIC EXPANSIONS FOR THE GAMMA FUNCTION

16.4 $$\Gamma(x+1) = \sqrt{2\pi x}\, x^x e^{-x}\left\{1 + \frac{1}{12x} + \frac{1}{288x^2} - \frac{139}{51{,}840x^3} + \cdots\right\}$$

This is called *Stirling's asymptotic series.*

If we let $x = n$ a positive integer in 16.4, then a useful approximation for $n!$ where n is large [e.g. $n > 10$] is given by *Stirling's formula*

$$n! \sim \sqrt{2\pi n}\, n^n e^{-n}$$

where $\sim$ is used to indicate that the ratio of the terms on each side approaches 1 as $n \to \infty$.

MISCELLANEOUS RESULTS

$$|\Gamma(ix)|^2 = \frac{\pi}{x\sinh \pi x}$$

GAMMA FUNCTION

$$\Gamma(x) = \int_0^\infty t^{x-1}e^{-t}\,dt \quad \text{for } 1 \leqq x \leqq 2$$

[For other values use the formula $\Gamma(x+1) = x\,\Gamma(x)$]

x	$\Gamma(x)$	x	$\Gamma(x)$
1.00	1.00000	1.50	.88623
1.01	.99433	1.51	.88659
1.02	.98884	1.52	.88704
1.03	.98355	1.53	.88757
1.04	.97844	1.54	.88818
1.05	.97350	1.55	.88887
1.06	.96874	1.56	.88964
1.07	.96415	1.57	.89049
1.08	.95973	1.58	.89142
1.09	.95546	1.59	.89243
1.10	.95135	1.60	.89352
1.11	.94740	1.61	.89468
1.12	.94359	1.62	.89592
1.13	.93993	1.63	.89724
1.14	.93642	1.64	.89864
1.15	.93304	1.65	.90012
1.16	.92980	1.66	.90167
1.17	.92670	1.67	.90330
1.18	.92373	1.68	.90500
1.19	.92089	1.69	.90678
1.20	.91817	1.70	.90864
1.21	.91558	1.71	.91057
1.22	.91311	1.72	.91258
1.23	.91075	1.73	.91467
1.24	.90852	1.74	.91683
1.25	.90640	1.75	.91906
1.26	.90440	1.76	.92137
1.27	.90250	1.77	.92376
1.28	.90072	1.78	.92623
1.29	.89904	1.79	.92877
1.30	.89747	1.80	.93138
1.31	.89600	1.81	.93408
1.32	.89464	1.82	.93685
1.33	.89338	1.83	.93969
1.34	.89222	1.84	.94261
1.35	.89115	1.85	.94561
1.36	.89018	1.86	.94869
1.37	.88931	1.87	.95184
1.38	.88854	1.88	.95507
1.39	.88785	1.89	.95838
1.40	.88726	1.90	.96177
1.41	.88676	1.91	.96523
1.42	.88636	1.92	.96877
1.43	.88604	1.93	.97240
1.44	.88581	1.94	.97610
1.45	.88566	1.95	.97988
1.46	.88560	1.96	.98374
1.47	.88563	1.97	.98768
1.48	.88575	1.98	.99171
1.49	.88595	1.99	.99581
1.50	.88623	2.00	1.00000

THE BETA FUNCTION

DEFINITION $B(m,n) = \int_0^1 t^{m-1}(1-t)^{m-1}\,dt \quad m > 0, n > 0$

RELATIONSHIP WITH GAMMA FUNCTION $B(m,n) = \dfrac{\Gamma(m)\Gamma(n)}{\Gamma(m+n)}$

PROPERTIES

$$B(m,n) = B(n,m)$$

$$B(m,n) = 2\int_0^{\pi/2} \sin^{2m-1}\theta \cos^{2n-1}\theta\, d\theta$$

$$B(m,n) = \int_0^{\infty} \frac{t^{m-1}}{(1+t)^{m+n}}\,dt$$

$$B(m,n) = r^n(r+1)^m \int_0^1 \frac{t^{m-1}(1-t)^{n-1}}{(r+t)^{m+n}}\,dt$$

•

THE BESSEL FUNCTIONS

BESSEL'S DIFFERENTIAL EQUATION

$$x^2y'' + xy' + (x^2 - n^2)y = 0 \qquad n \geqq 0$$

Solutions of this equation are called *Bessel functions of order n.*

BESSEL FUNCTIONS OF THE FIRST KIND OF ORDER n

$$\begin{aligned} J_n(x) &= \frac{x^n}{2^n\,\Gamma(n+1)}\left\{1 - \frac{x^2}{2(2n+2)} + \frac{x^4}{2\cdot 4(2n+2)(2n+4)} - \cdots\right\} \\ &= \sum_{k=0}^{\infty} \frac{(-1)^k (x/2)^{n+2k}}{k!\,\Gamma(n+k+1)} \end{aligned}$$

$$J_{-n}(x) = \frac{x^{-n}}{2^{-n}\,\Gamma(1-n)}\left\{1 - \frac{x^2}{2(2-2n)} + \frac{x^4}{2\cdot 4(2-2n)(4-2n)} - \cdots\right\}$$

$$= \sum_{k=0}^{\infty} \frac{(-1)^k (x/2)^{2k-n}}{k!\,\Gamma(k+1-n)}$$

$$J_{-n}(x) = (-1)^n J_n(x) \qquad n = 0, 1, 2, \ldots$$

If $n \neq 0, 1, 2, \ldots,$ $J_n(x)$ and $J_{-n}(x)$ are linearly independent.

If $n \neq 0, 1, 2, \ldots,$ $J_n(x)$ is bounded at $x = 0$ while $J_{-n}(x)$ is unbounded.

For $n = 0, 1$ we have

$$J_0(x) = 1 - \frac{x^2}{2^2} + \frac{x^4}{2^2 \cdot 4^2} - \frac{x^6}{2^2 \cdot 4^2 \cdot 6^2} + \cdots$$

$$J_1(x) = \frac{x}{2} - \frac{x^3}{2^2 \cdot 4} + \frac{x^5}{2^2 \cdot 4^2 \cdot 6} - \frac{x^7}{2^2 \cdot 4^2 \cdot 6^2 \cdot 8} + \cdots$$

$$J_0'(x) = -J_1(x)$$

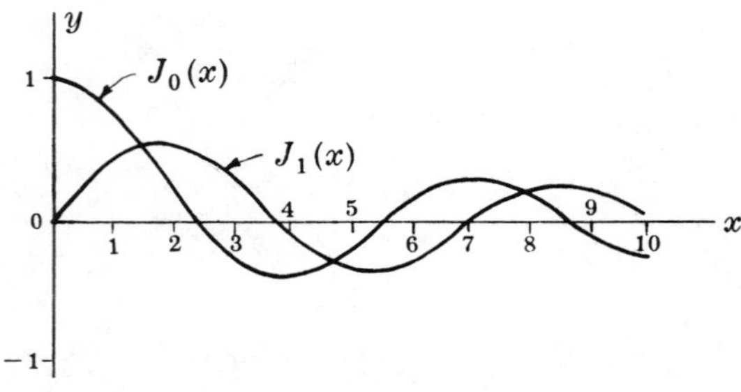

GENERATING FUNCTION FOR $J_n(x)$

$$e^{x(t-1/t)/2} = \sum_{n=-\infty}^{\infty} J_n(x) t^n$$

BESSEL FUNCTIONS OF THE SECOND KIND OF ORDER n

16.5 $$Y_n(x) = \begin{cases} \dfrac{J_n(x)\cos n\pi - J_{-n}(x)}{\sin n\pi} & n \neq 0, 1, 2, \ldots \\ \lim\limits_{p \to n} \dfrac{J_p(x)\cos p\pi - J_{-p}(x)}{\sin p\pi} & n = 0, 1, 2, \ldots \end{cases}$$

This is also called *Weber's function* or *Neumann's function* [also denoted by $N_n(x)$].

For $n = 0, 1, 2, \ldots,$ L'Hospital's rule yields

$$Y_n(x) = \frac{2}{\pi}\{\ln(x/2) + \gamma\} J_n(x) - \frac{1}{\pi}\sum_{k=0}^{n-1}\frac{(n-k-1)!}{k!}(x/2)^{2k-n} - \frac{1}{\pi}\sum_{k=0}^{\infty}(-1)^k\{\Phi(k) + \Phi(n+k)\}\frac{(x/2)^{2k+n}}{k!\,(n+k)!}$$

where $\gamma = .5772156\ldots$ is Euler's constant and

16.6 $$\Phi(p) = 1 + \frac{1}{2} + \frac{1}{3} + \cdots + \frac{1}{p}, \quad \Phi(0) = 0$$

For $n = 0$,

$$Y_0(x) = \frac{2}{\pi}\{\ln(x/2) + \gamma\} J_0(x) + \frac{2}{\pi}\left\{\frac{x^2}{2^2} - \frac{x^4}{2^2 4^2}(1 + \tfrac{1}{2}) + \frac{x^6}{2^2 4^2 6^2}(1 + \tfrac{1}{2} + \tfrac{1}{3}) - \cdots\right\}$$

$$Y_{-n}(x) = (-1)^n Y_n(x) \qquad n = 0, 1, 2, \ldots$$

For any value $n \geqq 0$, $J_n(x)$ is bounded at $x = 0$ while $Y_n(x)$ is unbounded.

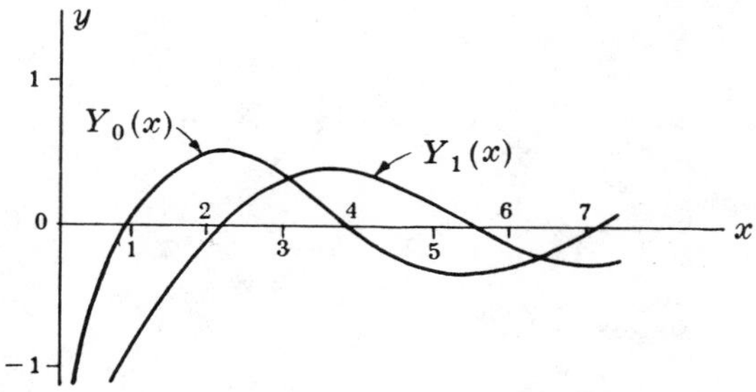

GENERAL SOLUTION OF BESSEL'S DIFFERENTIAL EQUATION

$$y = AJ_n(x) + BJ_{-n}(x) \qquad n \neq 0, 1, 2, \ldots$$

$$y = AJ_n(x) + BY_n(x) \qquad \text{all } n$$

$$y = AJ_n(x) + BJ_n(x)\int\frac{dx}{xJ_n^2(x)} \qquad \text{all } n$$

where A and B are arbitrary constants.

RECURRENCE FORMULAS FOR BESSEL FUNCTIONS

$$J_{n+1}(x) = \frac{2n}{x}J_n(x) - J_{n-1}(x)$$

$$J'_n(x) = \tfrac{1}{2}\{J_{n-1}(x) - J_{n+1}(x)\}$$

$$xJ'_n(x) = xJ_{n-1}(x) - nJ_n(x)$$

$$xJ'_n(x) = nJ_n(x) - xJ_{n+1}(x)$$

$$\frac{d}{dx}\{x^n J_n(x)\} = x^n J_{n-1}(x)$$

$$\frac{d}{dx}\{x^{-n} J_n(x)\} = -x^{-n} J_{n+1}(x)$$

The functions $Y_n(x)$ satisfy identical relations.

BESSEL FUNCTIONS OF ORDER EQUAL TO HALF AN ODD INTEGER

In this case the functions are expressible in terms of sines and cosines.

$$J_{1/2}(x) = \sqrt{\frac{2}{\pi x}}\sin x$$

$$J_{-1/2}(x) = \sqrt{\frac{2}{\pi x}}\cos x$$

$$J_{3/2}(x) = \sqrt{\frac{2}{\pi x}}\left(\frac{\sin x}{x} - \cos x\right)$$

$$J_{-3/2}(x) = \sqrt{\frac{2}{\pi x}}\left(\frac{\cos x}{x} + \sin x\right)$$

$$J_{5/2}(x) = \sqrt{\frac{2}{\pi x}}\left\{\left(\frac{3}{x^2} - 1\right)\sin x - \frac{3}{x}\cos x\right\}$$

$$J_{-5/2}(x) = \sqrt{\frac{2}{\pi x}}\left\{\frac{3}{x}\sin x + \left(\frac{3}{x^2} - 1\right)\cos x\right\}$$

For further results use the recurrence formula. Results for $Y_{1/2}(x), Y_{3/2}(x), \ldots$ are obtained from 16.5.

HANKEL FUNCTIONS OF FIRST AND SECOND KINDS OF ORDER n

$$H_n^{(1)}(x) = J_n(x) + iY_n(x) \qquad H_n^{(2)}(x) = J_n(x) - iY_n(x)$$

BESSEL'S MODIFIED DIFFERENTIAL EQUATION

$$x^2y'' + xy' - (x^2 + n^2)y = 0 \qquad n \geqq 0$$

Solutions of this equation are called *modified Bessel functions of order n.*

MODIFIED BESSEL FUNCTIONS OF THE FIRST KIND OF ORDER n

$$I_n(x) = i^{-n} J_n(ix) = e^{-n\pi i/2} J_n(ix)$$

$$= \frac{x^n}{2^n \Gamma(n+1)} \left\{ 1 + \frac{x^2}{2(2n+2)} + \frac{x^4}{2 \cdot 4(2n+2)(2n+4)} + \cdots \right\}$$

$$= \sum_{k=0}^{\infty} \frac{(x/2)^{n+2k}}{k!\, \Gamma(n+k+1)}$$

$$I_{-n}(x) = i^n J_{-n}(ix) = e^{n\pi i/2} J_{-n}(ix)$$

$$= \frac{x^{-n}}{2^{-n} \Gamma(1-n)} \left\{ 1 + \frac{x^2}{2(2-2n)} + \frac{x^4}{2 \cdot 4(2-2n)(4-2n)} + \cdots \right\}$$

$$= \sum_{k=0}^{\infty} \frac{(x/2)^{2k-n}}{k!\, \Gamma(k+1-n)}$$

$$I_{-n}(x) = I_n(x) \qquad n = 0, 1, 2, \ldots$$

If $n \neq 0, 1, 2, \ldots$, then $I_n(x)$ and $I_{-n}(x)$ are linearly independent.

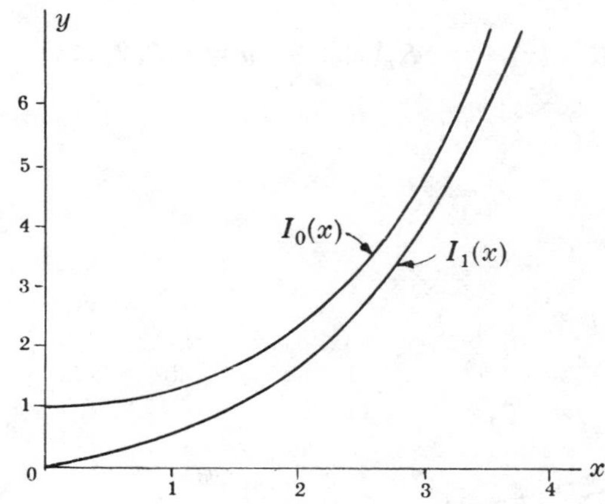

For $n = 0, 1,$ we have

$$I_0(x) = 1 + \frac{x^2}{2^2} + \frac{x^4}{2^2 \cdot 4^2} + \frac{x^6}{2^2 \cdot 4^2 \cdot 6^2} + \cdots$$

$$I_1(x) = \frac{x}{2} + \frac{x^3}{2^2 \cdot 4} + \frac{x^5}{2^2 \cdot 4^2 \cdot 6} + \frac{x^7}{2^2 \cdot 4^2 \cdot 6^2 \cdot 8} + \cdots$$

$$I_0'(x) = I_1(x)$$

GENERATING FUNCTION FOR $I_n(x)$

$$e^{x(t+1/t)/2} = \sum_{n=-\infty}^{\infty} I_n(x) t^n$$

MODIFIED BESSEL FUNCTIONS OF THE SECOND KIND OF ORDER n

16.7 $$K_n(x) = \begin{cases} \dfrac{\pi}{2 \sin n\pi} \{I_{-n}(x) - I_n(x)\} & n \neq 0, 1, 2, \ldots \\ \lim\limits_{p \to n} \dfrac{\pi}{2 \sin p\pi} \{I_{-p}(x) - I_p(x)\} & n = 0, 1, 2, \ldots \end{cases}$$

For $n = 0, 1, 2, \ldots,$ L'Hospital's rule yields

$$K_n(x) = (-1)^{n+1}\{\ln(x/2) + \gamma\} I_n(x) + \frac{1}{2} \sum_{k=0}^{n-1} (-1)^k (n-k-1)!\,(x/2)^{2k-n}$$
$$+ \frac{(-1)^n}{2} \sum_{k=0}^{\infty} \frac{(x/2)^{n+2k}}{k!\,(n+k)!} \{\Phi(k) + \Phi(n+k)\}$$

where $\Phi(p)$ is given by 16.6.

For $n = 0$,

$$K_0(x) = -\{\ln(x/2) + \gamma\} I_0(x) + \frac{x^2}{2^2} + \frac{x^4}{2^2 \cdot 4^2}(1 + \tfrac{1}{2}) + \frac{x^6}{2^2 \cdot 4^2 \cdot 6^2}(1 + \tfrac{1}{2} + \tfrac{1}{3}) + \cdots$$

$$K_{-n}(x) = K_n(x) \qquad n = 0, 1, 2, \ldots$$

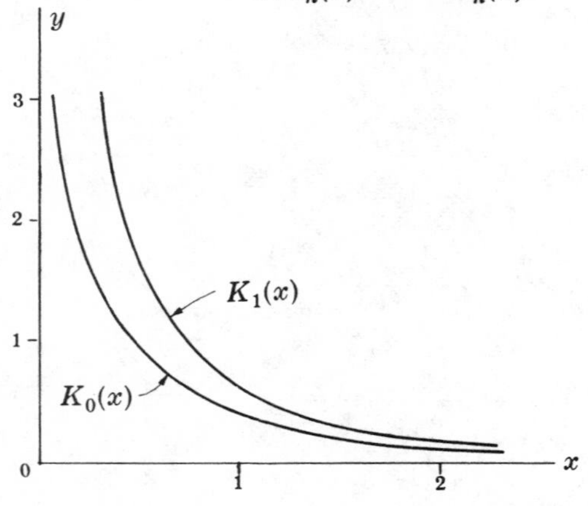

GENERAL SOLUTION OF BESSEL'S MODIFIED EQUATION

$$y = AI_n(x) + BI_{-n}(x) \qquad n \neq 0, 1, 2, \ldots$$

$$y = AI_n(x) + BK_n(x) \qquad \text{all } n$$

$$y = AI_n(x) + BI_n(x)\int\frac{dx}{xI_n^2(x)} \qquad \text{all } n$$

where A and B are arbitrary constants.

RECURRENCE FORMULAS FOR MODIFIED BESSEL FUNCTIONS

$$I_{n+1}(x) = I_{n-1}(x) - \frac{2n}{x}I_n(x)$$

$$I_n'(x) = \tfrac{1}{2}\{I_{n-1}(x) + I_{n+1}(x)\}$$

$$xI_n'(x) = xI_{n-1}(x) - nI_n(x)$$

$$xI_n'(x) = xI_{n+1}(x) + nI_n(x)$$

$$\frac{d}{dx}\{x^nI_n(x)\} = x^nI_{n-1}(x)$$

$$\frac{d}{dx}\{x^{-n}I_n(x)\} = x^{-n}I_{n+1}(x)$$

$$K_{n+1}(x) = K_{n-1}(x) + \frac{2n}{x}K_n(x)$$

$$K_n'(x) = -\tfrac{1}{2}\{K_{n-1}(x) + K_{n+1}(x)\}$$

$$xK_n'(x) = -xK_{n-1}(x) - nK_n(x)$$

$$xK_n'(x) = nK_n(x) - xK_{n+1}(x)$$

$$\frac{d}{dx}\{x^nK_n(x)\} = -x^nK_{n-1}(x)$$

$$\frac{d}{dx}\{x^{-n}K_n(x)\} = -x^{-n}K_{n+1}(x)$$

MODIFIED BESSEL FUNCTIONS OF ORDER EQUAL TO HALF AN ODD INTEGER

In this case the functions are expressible in terms of

hyperbolic sines and cosines.

$$I_{1/2}(x) = \sqrt{\frac{2}{\pi x}} \sinh x$$

$$I_{-1/2}(x) = \sqrt{\frac{2}{\pi x}} \cosh x$$

$$I_{3/2}(x) = \sqrt{\frac{2}{\pi x}}\left(\cosh x - \frac{\sinh x}{x}\right)$$

$$I_{5/2}(x) = \sqrt{\frac{2}{\pi x}}\left\{\left(\frac{3}{x^2}+1\right)\sinh x - \frac{3}{x}\cosh x\right\}$$

$$I_{-3/2}(x) = \sqrt{\frac{2}{\pi x}}\left(\sinh x - \frac{\cosh x}{x}\right)$$

$$I_{-5/2}(x) = \sqrt{\frac{2}{\pi x}}\left\{\left(\frac{3}{x^2}+1\right)\cosh x - \frac{3}{x}\sinh x\right\}$$

For further results use the recurrence formula. Results for $K_{1/2}(x), K_{3/2}(x), \ldots$ are obtained from 16.7.

Ber AND Bei FUNCTIONS

The real and imaginary parts of $J_n(xe^{3\pi i/4})$ are denoted by $\text{Ber}_n(x)$ and $\text{Bei}_n(x)$ where

$$\text{Ber}_n(x) = \sum_{k=0}^{\infty} \frac{(x/2)^{2k+n}}{k!\,\Gamma(n+k+1)} \cos\frac{(3n+2k)\pi}{4}$$

$$\text{Bei}_n(x) = \sum_{k=0}^{\infty} \frac{(x/2)^{2k+n}}{k!\,\Gamma(n+k+1)} \sin\frac{(3n+2k)\pi}{4}$$

If $n = 0$,

$$\text{Ber}(x) = 1 - \frac{(x/2)^4}{2!^2} + \frac{(x/2)^8}{4!^2} - \cdots$$

$$\text{Bei}(x) = (x/2)^2 - \frac{(x/2)^6}{3!^2} + \frac{(x/2)^{10}}{5!^2} - \cdots$$

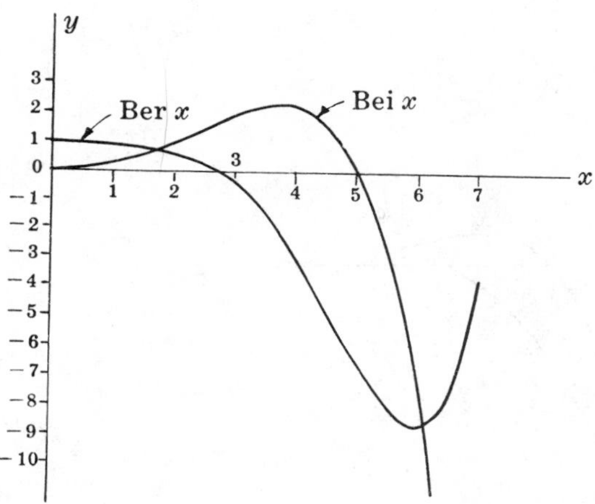

Ker AND Kei FUNCTIONS

The real and imaginary parts of $e^{-n\pi i/2}\,K_n(xe^{\pi i/4})$ are denoted by $\mathrm{Ker}_n\,(x)$ and $\mathrm{Kei}_n\,(x)$ where

$$\begin{aligned}\mathrm{Ker}_n\,(x) \;=\;& -\{\ln\,(x/2)+\gamma\}\,\mathrm{Ber}_n\,(x) + \tfrac{1}{4}\pi\,\mathrm{Bei}_n\,(x) \\ &+ \frac{1}{2}\sum_{k=0}^{n-1}\frac{(n-k-1)!\,(x/2)^{2k-n}}{k!}\cos\frac{(3n+2k)\pi}{4} \\ &+ \frac{1}{2}\sum_{k=0}^{\infty}\frac{(x/2)^{n+2k}}{k!\,(n+k)!}\{\Phi(k)+\Phi(n+k)\}\cos\frac{(3n+2k)\pi}{4}\end{aligned}$$

$$\begin{aligned}\mathrm{Kei}_n\,(x) \;=\;& -\{\ln\,(x/2)+\gamma\}\,\mathrm{Bei}_n\,(x) - \tfrac{1}{4}\pi\,\mathrm{Ber}_n\,(x) \\ &- \frac{1}{2}\sum_{k=0}^{n-1}\frac{(n-k-1)!\,(x/2)^{2k-n}}{k!}\sin\frac{(3n+2k)\pi}{4} \\ &+ \frac{1}{2}\sum_{k=0}^{\infty}\frac{(x/2)^{n+2k}}{k!\,(n+k)!}\{\Phi(k)+\Phi(n+k)\}\sin\frac{(3n+2k)\pi}{4}\end{aligned}$$

and Φ is given by 16.6.

If $n = 0$,

$$\begin{aligned}\mathrm{Ker}\,(x) \;=\;& -\{\ln\,(x/2)+\gamma\}\,\mathrm{Ber}\,(x) + \frac{\pi}{4}\,\mathrm{Bei}\,(x) + 1 - \frac{(x/2)^4}{2!^2}(1+\tfrac{1}{2}) \\ &+ \frac{(x/2)^8}{4!^2}(1+\tfrac{1}{2}+\tfrac{1}{3}+\tfrac{1}{4}) - \cdots\end{aligned}$$

$$\begin{aligned}\mathrm{Kei}\,(x) \;=\;& -\{\ln\,(x/2)+\gamma\}\,\mathrm{Bei}\,(x) - \frac{\pi}{4}\,\mathrm{Ber}\,(x) + (x/2)^2 \\ &- \frac{(x/2)^6}{3!^2}(1+\tfrac{1}{2}+\tfrac{1}{3}) + \cdots\end{aligned}$$

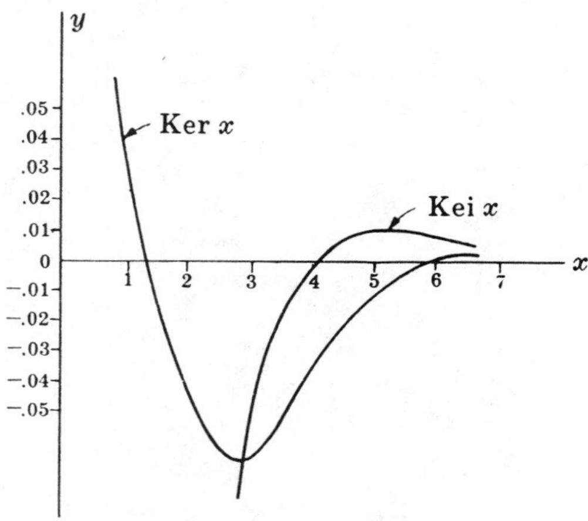

DIFFERENTIAL EQUATION FOR Ber, Bei, Ker, Kei FUNCTIONS

$$x^2y'' + xy' - (ix^2 + n^2)y = 0$$

The general solution of this equation is

$$y = A\{\text{Ber}_n(x) + i\,\text{Bei}_n(x)\} + B\{\text{Ker}_n(x) + i\,\text{Kei}_n(x)\}$$

•

ELLIPTIC INTEGRALS[1]

DEFINITION OF ELLIPTIC INTEGRALS

If $R(x, y)$ is a rational function of x and y, where y^2 is equal to a cubic or quartic polynomial in x, the integral

$$\int R(x, y)dx$$

is called an *elliptic integral.*

The elliptic integral just defined can not, in general, be expressed in terms of elementary functions.

Exceptions to this are

(i) when $R(x, y)$ contains no odd powers of y.
(ii) when the polynomial y^2 has a repeated factor.

We therefore exclude these cases.

[1]From "Handbook of Mathematical Functions", National Bureau of Standards

By substituting for y^2 and denoting by $p_s(x)$ a polynomial in x we get [2]

$$R(x,y)=\frac{p_1(x)+yp_2(x)}{p_3(x)+yp_4(x)}$$

$$=\frac{[p_1(x)+yp_2(x)][p_3(x)-yp_4(x)]y}{\{[p_3(x)]^2-y^2[p_4(x)]^2\}y}$$

$$=\frac{p_5(x)+yp_6(x)}{yp_7(x)}=R_1(x)+\frac{R_2(x)}{y}$$

where $R_1(x)$ and $R_2(x)$ are rational functions of x. Hence, by expressing $R_2(x)$ as the sum of a polynomial and partial fractions

$$\int R(x,y)dx=\int R_1(x)dx+\Sigma_s A_s\int x^s y^{-1}dx$$
$$+\Sigma_s B_s\int[(x-c)^s y]^{-1}dx$$

REDUCTION FORMULAE

Let

$$y^2=a_0x^4+a_1x^3+a_2x^2+a_3x+a_4 \qquad (|a_0|+|a_1|\neq 0)$$
$$=b_0(x-c)^4+b_1(x-c)^3+b_2(x-c)^2+b_3(x-c)+b_4$$
$$(|b_0|+|b_1|\neq 0)$$

$$I_s=\int x^s y^{-1}dx,\ J_s=\int[y(x-c)^s]^{-1}dx$$

By integrating the derivatives of yx^s and $y(x-c)^{-s}$ we get the reduction formulae

$$(s+2)a_0I_{s+3}+\tfrac{1}{2}a_1(2s+3)I_{s+2}+a_2(s+1)I_{s+1}$$
$$+\tfrac{1}{2}a_3(2s+1)I_s+sa_4I_{s-1}=x^sy \quad (s=0,\ 1,\ 2,\ \ldots)$$

$$(2-s)b_0J_{s-3}+\tfrac{1}{2}b_1(3-2s)J_{s-2}+b_2(1-s)J_{s-1}$$
$$+\tfrac{1}{2}b_3(1-2s)J_s-sb_4J_{s+1}=y(x-c)^{-s}$$
$$(s=1,\ 2,\ 3,\ \ldots)$$

By means of these reduction formulae and certain transformations every elliptic integral can be brought to depend on the integral of a rational function and on three canonical forms for elliptic integrals.

Elliptic integrals of the *first kind* are represented by

$$F(k, \phi) = \int_0^\phi \frac{d\Phi}{\sqrt{1 - k^2 \sin^2 \Phi}}$$
$$= \int_0^x \frac{d\xi}{\sqrt{(1 - \xi^2)(1 - k^2\xi^2)}}, \quad x = \sin \phi,\ k^2 < 1.$$

Elliptic integrals of the second kind are represented by

$$E(k, \phi) = \int_0^\phi \sqrt{1 - k^2 \sin^2 \Phi}\, d\Phi$$
$$= \int_0^x \frac{\sqrt{1 - k^2\xi^2}}{\sqrt{1 - \xi^2}}\, d\xi, \quad x = \sin \phi,\ k^2 < 1.$$

Elliptic integrals of the third kind are represented as

$$\pi(k, n, \phi) = \int_0^\phi \frac{d\Phi}{(1 + n \sin^2 \Phi)\sqrt{1 - k^2 \sin^2 \Phi}},$$
$$k^2 < 1,\ n \text{ an integer.}$$

Elliptic integrals of the third kind are also presented as

$$\pi_1(k, n, x) = \int_0^x \frac{d\xi}{(1 + n\xi^2)\sqrt{(1 - \xi^2)(1 - k^2\xi^2)}},$$
$$x = \sin \phi,\ k^2 < 1,\ n \text{ an integer.}$$

The complete integrals are

$$K = F\left(k, \frac{\pi}{2}\right) = \frac{\pi}{2}\left[1 + \left(\frac{1}{2}\right)^2 k^2 + \left(\frac{3}{2 \cdot 4}\right)^2 k^4 + \left(\frac{3 \cdot 5}{2 \cdot 4 \cdot 6}\right)^2 k^6 + \cdots\right]$$

$$E = E\left(k, \frac{\pi}{2}\right) = \frac{\pi}{2}\left[1 - \left(\frac{1}{2^2}\right) k^2 - \left(\frac{3^2}{2^2 \cdot 4^2}\right)\frac{k^4}{3} - \left(\frac{3^2 \cdot 5^2}{2^2 \cdot 4^2 \cdot 6^2}\right)\frac{k^6}{5} - \left(\frac{3^2 \cdot 5^2 \cdot 7^2}{2^2 \cdot 4^2 \cdot 6^2 \cdot 8^2}\right)\frac{k^8}{7} - \cdots\right].$$

$$K' = F\left(\sqrt{1 - k^2}, \frac{\pi}{2}\right),\quad E' = E\left(\sqrt{1 - k^2}, \frac{\pi}{2}\right)$$

LEGENDRE'S RELATION

$$EK' + E'K - KK' = \pi/2$$

where

$$E = \int_0^{\pi/2} \sqrt{1 - k^2 \sin^2 \theta}\, d\theta \qquad K = \int_0^{\pi/2} \frac{d\theta}{\sqrt{1 - k^2 \sin^2 \theta}}$$

$$E' = \int_0^{\pi/2} \sqrt{1 - k'^2 \sin^2 \theta} \qquad K' = \int_0^{\pi/2} \frac{d\theta}{\sqrt{1 - k'^2 \sin^2 \theta}}$$

LANDEN'S TRANSFORMATION

$$\tan \phi = \frac{\sin 2\phi_1}{k + \cos 2\phi_1} \quad \text{or} \quad k \sin \phi = \sin (2\phi_1 - \phi)$$

This yields

$$F(k, \phi) = \int_0^{\phi} \frac{d\theta}{\sqrt{1 - k^2 \sin^2 \theta}} = \frac{2}{1 + k} \int_0^{\phi_1} \frac{d\theta_1}{\sqrt{1 - k_1^2 \sin^2 \theta_1}}$$

where $k_1 = 2\sqrt{k}/(1 + k)$. By successive applications, sequences $k_1, k_2, k_3, \ldots$ and $\phi_1, \phi_2, \phi_3, \ldots$ are obtained such that $k < k_1 < k_2 < k_3 < \cdots < 1$ where $\lim_{n \to \infty} k_n = 1$. It follows that

$$F(k, \Phi) = \sqrt{\frac{k_1 k_2 k_3 \ldots}{k}} \int_0^{\Phi} \frac{d\theta}{\sqrt{1 - \sin^2 \theta}} = \sqrt{\frac{k_1 k_2 k_3 \ldots}{k}} \ln \tan \left(\frac{\pi}{4} + \frac{\Phi}{2} \right)$$

where

$$k_1 = \frac{2\sqrt{k}}{1 + k}, \quad k_2 = \frac{2\sqrt{k_1}}{1 + k_1}, \quad \ldots \quad \text{and} \quad \Phi = \lim_{n \to \infty} \phi_n$$

JACOBI'S ELLIPTIC FUNCTION

$$x = \sin (\text{am } u) = \text{sn } u$$

$$\sqrt{1 - x^2} = \cos (\text{am } u) = \text{cn } u$$

$$\sqrt{1 - k^2 x^2} = \sqrt{1 - k^2 \text{sn}^2 u} = \text{dn } u$$

We can also define the inverse functions $\text{sn}^{-1} x$, $\text{cn}^{-1} x$, $\text{dn}^{-1} x$ and the following

$$\text{ns } u = \frac{1}{\text{sn } u} \qquad \text{cd } u = \frac{\text{cn } u}{\text{dn } u}$$

$$\text{nc } u = \frac{1}{\text{cn } u} \qquad \text{cs } u = \frac{\text{cn } u}{\text{sn } u}$$

$$\text{nd } u = \frac{1}{\text{dn } u} \qquad \text{dc } u = \frac{\text{dn } u}{\text{cn } u}$$

$$\text{sc } u = \frac{\text{sn } u}{\text{cn } u} \qquad \text{ds } u = \frac{\text{dn } u}{\text{sn } u}$$

$$\text{sd } u = \frac{\text{sn } u}{\text{dn } u}$$

SPECIAL VALUES

$\mathrm{sn}\,0 = 0$ $\quad$ $\mathrm{sc}\,0 = 0$ $\quad$ $\mathrm{am}\,0 = 0$

$\mathrm{cn}\,0 = 1$ $\quad$ $\mathrm{dn}\,0 = 1$

ADDITION FORMULAS

$$\mathrm{sn}\,(u+v) = \frac{\mathrm{sn}\,u\,\mathrm{cn}\,v\,\mathrm{dn}\,v + \mathrm{cn}\,u\,\mathrm{sn}\,v\,\mathrm{dn}\,u}{1 - k^2\,\mathrm{sn}^2\,u\,\mathrm{sn}^2\,v}$$

$$\mathrm{cn}\,(u+v) = \frac{\mathrm{cn}\,u\,\mathrm{cn}\,v - \mathrm{sn}\,u\,\mathrm{sn}\,v\,\mathrm{dn}\,u\,\mathrm{dn}\,v}{1 - k^2\,\mathrm{sn}^2\,u\,\mathrm{sn}^2\,v}$$

$$\mathrm{dn}\,(u+v) = \frac{\mathrm{dn}\,u\,\mathrm{dn}\,v - k^2\,\mathrm{sn}\,u\,\mathrm{sn}\,v\,\mathrm{cn}\,u\,\mathrm{cn}\,v}{1 - k^2\,\mathrm{sn}^2\,u\,\mathrm{sn}^2\,v}$$

IDENTITIES INVOLVING ELLIPTIC FUNCTIONS

$$\mathrm{sn}^2\,u + \mathrm{cn}^2\,u = 1$$

$$\mathrm{dn}^2\,u - k^2\,\mathrm{cn}^2\,u = k'^2 \quad \text{where } k' = \sqrt{1-k^2}$$

$$\mathrm{cn}^2\,u = \frac{\mathrm{dn}\,2u + \mathrm{cn}\,2u}{1 + \mathrm{dn}\,2u}$$

$$\sqrt{\frac{1 - \mathrm{cn}\,2u}{1 + \mathrm{cn}\,2u}} = \frac{\mathrm{sn}\,u\,\mathrm{dn}\,u}{\mathrm{cn}\,u}$$

$$\mathrm{dn}^2\,u + k^2\,\mathrm{sn}^2\,u = 1$$

$$\mathrm{sn}^2\,u = \frac{1 - \mathrm{cn}\,2u}{1 + \mathrm{dn}\,2u}$$

$$\mathrm{dn}^2\,u = \frac{1 - k^2 + \mathrm{dn}\,2u + k^2\,\mathrm{cn}\,u}{1 + \mathrm{dn}\,2u}$$

$$\sqrt{\frac{1 - \mathrm{dn}\,2u}{1 + \mathrm{dn}\,2u}} = \frac{k\,\mathrm{sn}\,u\,\mathrm{cn}\,u}{\mathrm{dn}\,u}$$

PERIODS OF ELLIPTIC FUNCTIONS

Let

$$K = \int_0^{\pi/2} \frac{d\theta}{\sqrt{1 - k^2 \sin^2\theta}}, \qquad K' = \int_0^{\pi/2} \frac{d\theta}{\sqrt{1 - k'^2 \sin^2\theta}}$$

where $k' = \sqrt{1 - k^2}$

Then

$\operatorname{sn} u$ has periods $4K$ and $2iK'$

$\operatorname{cn} u$ has periods $4K$ and $2K + 2iK'$

$\operatorname{dn} u$ has periods $2K$ and $4iK'$

CATALAN'S CONSTANT

$$\frac{1}{2}\int_0^1 K\,dk = \frac{1}{2}\int_{k=0}^1 \int_{\theta=0}^{\pi/2} \frac{d\theta\,dk}{\sqrt{1 - k^2 \sin^2\theta}}$$

$$= \frac{1}{1^2} - \frac{1}{3^2} + \frac{1}{5^2} - \cdots = .915965594\ldots$$

SERIES EXPANSIONS

$$\operatorname{sn} u = u - (1 + k^2)\frac{u^3}{3!} + (1 + 14k^2 + k^4)\frac{u^5}{5!} - (1 + 135k^2 + 135k^4 + k^6)\frac{u^7}{7!} + \cdots$$

$$\operatorname{cn} u = 1 - \frac{u^2}{2!} + (1 + 4k^2)\frac{u^4}{4!} - (1 + 44k^2 + 16k^4)\frac{u^6}{6!} + \cdots$$

$$\operatorname{dn} u = 1 - k^2\frac{u^2}{2!} + k^2(4 + k^2)\frac{u^4}{4!} - k^2(16 + 44k^2 + k^4)\frac{u^6}{6!} + \cdots$$

DERIVATIVES

$$\frac{d}{du}\operatorname{sn} u = \operatorname{cn} u \operatorname{dn} u \qquad \frac{d}{du}\operatorname{dn} u = -k^2 \operatorname{sn} u \operatorname{cn} u$$

$$\frac{d}{du}\operatorname{cn} u = -\operatorname{sn} u \operatorname{dn} u \qquad \frac{d}{du}\operatorname{sc} u = \operatorname{dc} u \operatorname{nc} u$$

INTEGRALS

$$\int \operatorname{sn} u\,du = \frac{1}{k}\ln(\operatorname{dn} u - k \operatorname{cn} u)$$

$$\int \operatorname{cn} u\,du = \frac{1}{k}\cos^{-1}(\operatorname{dn} u)$$

$$\int \mathrm{dn}\, u\, du = \sin^{-1}(\mathrm{sn}\, u)$$

$$\int \mathrm{sc}\, u\, du = \frac{1}{\sqrt{1-k^2}} \ln\left(\mathrm{dc}\, u + \sqrt{1-k^2}\, \mathrm{nc}\, u\right)$$

$$\int \mathrm{cs}\, u\, du = \ln(\mathrm{ns}\, u - \mathrm{ds}\, u)$$

$$\int \mathrm{cd}\, u\, du = \frac{1}{k} \ln(\mathrm{nd}\, u + k\, \mathrm{sd}\, u)$$

$$\int \mathrm{dc}\, u\, du = \ln(\mathrm{nc}\, u + \mathrm{sc}\, u)$$

$$\int \mathrm{sd}\, u\, du = \frac{-1}{k\sqrt{1-k^2}} \sin^{-1}(k\, \mathrm{cd}\, u)$$

$$\int \mathrm{ds}\, u\, du = \ln(\mathrm{ns}\, u - \mathrm{cs}\, u)$$

$$\int \mathrm{ns}\, u\, du = \ln(\mathrm{ds}\, u - \mathrm{cs}\, u)$$

$$\int \mathrm{nc}\, u\, du = \frac{1}{\sqrt{1-k^2}} \ln\left(\mathrm{dc}\, u + \frac{\mathrm{sc}\, u}{\sqrt{1-k^2}}\right)$$

$$\int \mathrm{nd}\, u\, du = \frac{1}{\sqrt{1-k^2}} \cos^{-1}(\mathrm{cd}\, u)$$

ELLIPTIC INTEGRALS OF THE FIRST KIND: $F(k, \phi)$

$$F(k, \phi) = \int_0^\phi \frac{d\Phi}{\sqrt{1 - k^2 \sin^2 \Phi}}, \qquad \theta = \sin^{-1} k$$

ϕ \ θ	5°	10°	15°	20°	25°	30°	35°	40°	45°
1°	0.0175	0.0175	0.0175	0.0175	0.0175	0.0175	0.0175	0.0175	0.0175
2°	0.0349	0.0349	0.0349	0.0349	0.0349	0.0349	0.0349	0.0349	0.0349
3°	0.0524	0.0524	0.0524	0.0524	0.0524	0.0524	0.0524	0.0524	0.0524
4°	0.0698	0.0698	0.0698	0.0698	0.0698	0.0698	0.0698	0.0698	0.0698
5°	0.0873	0.0873	0.0873	0.0873	0.0873	0.0873	0.0873	0.0873	0.0873
6°	0.1047	0.1047	0.1047	0.1047	0.1048	0.1048	0.1048	0.1048	0.1048
7°	0.1222	0.1222	0.1222	0.1222	0.1222	0.1222	0.1223	0.1223	0.1223
8°	0.1396	0.1396	0.1397	0.1397	0.1397	0.1397	0.1398	0.1398	0.1399
9°	0.1571	0.1571	0.1571	0.1572	0.1572	0.1572	0.1573	0.1573	0.1574
10°	0.1745	0.1746	0.1746	0.1746	0.1747	0.1748	0.1748	0.1749	0.1750
11°	0.1920	0.1920	0.1921	0.1921	0.1922	0.1923	0.1924	0.1925	0.1926
12°	0.2095	0.2095	0.2095	0.2096	0.2097	0.2098	0.2099	0.2101	0.2102
13°	0.2269	0.2270	0.2270	0.2271	0.2272	0.2274	0.2275	0.2277	0.2279
14°	0.2444	0.2444	0.2445	0.2446	0.2448	0.2450	0.2451	0.2453	0.2456
15°	0.2618	0.2619	0.2620	0.2621	0.2623	0.2625	0.2628	0.2630	0.2633
16°	0.2793	0.2794	0.2795	0.2797	0.2799	0.2802	0.2804	0.2808	0.2811
17°	0.2967	0.2968	0.2970	0.2972	0.2975	0.2978	0.2981	0.2985	0.2989
18°	0.3142	0.3143	0.3145	0.3148	0.3151	0.3154	0.3159	0.3163	0.3167
19°	0.3317	0.3318	0.3320	0.3323	0.3327	0.3331	0.3336	0.3341	0.3347
20°	0.3491	0.3493	0.3495	0.3499	0.3503	0.3508	0.3514	0.3520	0.3526
21°	0.3666	0.3668	0.3671	0.3675	0.3680	0.3685	0.3692	0.3699	0.3706
22°	0.3840	0.3842	0.3846	0.3851	0.3856	0.3863	0.3871	0.3879	0.3887
23°	0.4015	0.4017	0.4021	0.4027	0.4033	0.4041	0.4049	0.4059	0.4068
24°	0.4190	0.4192	0.4197	0.4203	0.4210	0.4219	0.4229	0.4239	0.4250
25°	0.4364	0.4367	0.4372	0.4379	0.4387	0.4397	0.4408	0.4420	0.4433
26°	0.4539	0.4542	0.4548	0.4556	0.4565	0.4576	0.4588	0.4602	0.4616
27°	0.4714	0.4717	0.4724	0.4732	0.4743	0.4755	0.4769	0.4784	0.4800
28°	0.4888	0.4893	0.4899	0.4909	0.4921	0.4934	0.4950	0.4967	0.4985
29°	0.5063	0.5068	0.5075	0.5086	0.5099	0.5114	0.5132	0.5150	0.5170
30°	0.5238	0.5243	0.5251	0.5263	0.5277	0.5294	0.5313	0.5334	0.5356
31°	0.5412	0.5418	0.5427	0.5440	0.5456	0.5475	0.5496	0.5519	0.5543
32°	0.5587	0.5593	0.5603	0.5617	0.5635	0.5656	0.5679	0.5704	0.5731
33°	0.5762	0.5769	0.5780	0.5795	0.5814	0.5837	0.5862	0.5890	0.5920
34°	0.5937	0.5944	0.5956	0.5973	0.5994	0.6018	0.6046	0.6077	0.6109
35°	0.6111	0.6119	0.6133	0.6151	0.6173	0.6200	0.6231	0.6264	0.6300
36°	0.6286	0.6295	0.6309	0.6329	0.6353	0.6383	0.6416	0.6452	0.6491
37°	0.6461	0.6470	0.6486	0.6507	0.6534	0.6565	0.6602	0.6641	0.6684
38°	0.6636	0.6646	0.6662	0.6685	0.6714	0.6749	0.6788	0.6831	0.6877
39°	0.6810	0.6821	0.6839	0.6864	0.6895	0.6932	0.6975	0.7021	0.7071
40°	0.6985	0.6997	0.7016	0.7043	0.7076	0.7116	0.7162	0.7213	0.7267
41°	0.7160	0.7173	0.7193	0.7222	0.7258	0.7301	0.7350	0.7405	0.7463
42°	0.7335	0.7348	0.7370	0.7401	0.7440	0.7486	0.7539	0.7598	0.7661
43°	0.7510	0.7524	0.7548	0.7580	0.7622	0.7671	0.7728	0.7791	0.7859
44°	0.7685	0.7700	0.7725	0.7760	0.7804	0.7857	0.7918	0.7986	0.8059
45°	0.7859	0.7876	0.7903	0.7940	0.7987	0.8044	0.8109	0.8181	0.8260

ELLIPTIC INTEGRALS OF THE FIRST KIND: $F(k, \phi)$ (Continued)

$$F(k, \phi) = \int_0^{\phi} \frac{d\Phi}{\sqrt{1 - k^2 \sin^2 \Phi}}, \qquad \theta = \sin^{-1} k$$

ϕ \ θ	50°	55°	60°	65°	70°	75°	80°	85°	90°
1°	0.0175	0.0175	0.0175	0.0175	0.0175	0.0175	0.0175	0.0175	0.0175
2°	0.0349	0.0349	0.0349	0.0349	0.0349	0.0349	0.0349	0.0349	0.0349
3°	0.0524	0.0524	0.0524	0.0524	0.0524	0.0524	0.0524	0.0524	0.0524
4°	0.0698	0.0699	0.0699	0.0699	0.0699	0.0699	0.0699	0.0699	0.0699
5°	0.0873	0.0873	0.0873	0.0874	0.0874	0.0874	0.0874	0.0874	0.0874
6°	0.1048	0.1048	0.1049	0.1049	0.1049	0.1049	0.1049	0.1049	0.1049
7°	0.1224	0.1224	0.1224	0.1224	0.1224	0.1225	0.1225	0.1225	0.1225
8°	0.1399	0.1399	0.1400	0.1400	0.1400	0.1401	0.1401	0.1401	0.1401
9°	0.1575	0.1575	0.1576	0.1576	0.1577	0.1577	0.1577	0.1577	0.1577
10°	0.1751	0.1751	0.1752	0.1753	0.1753	0.1754	0.1754	0.1754	0.1754
11°	0.1927	0.1928	0.1929	0.1930	0.1930	0.1931	0.1931	0.1932	0.1932
12°	0.2103	0.2105	0.2106	0.2107	0.2108	0.2109	0.2109	0.2110	0.2110
13°	0.2280	0.2282	0.2284	0.2285	0.2286	0.2287	0.2288	0.2288	0.2289
14°	0.2458	0.2460	0.2462	0.2464	0.2465	0.2466	0.2467	0.2468	0.2468
15°	0.2636	0.2638	0.2641	0.2643	0.2645	0.2646	0.2647	0.2648	0.2648
16°	0.2814	0.2817	0.2820	0.2823	0.2825	0.2827	0.2828	0.2829	0.2830
17°	0.2993	0.2997	0.3000	0.3003	0.3006	0.3008	0.3010	0.3011	0.3012
18°	0.3172	0.3177	0.3181	0.3185	0.3188	0.3191	0.3193	0.3194	0.3195
19°	0.3352	0.3357	0.3362	0.3367	0.3371	0.3374	0.3377	0.3378	0.3379
20°	0.3533	0.3539	0.3545	0.3550	0.3555	0.3559	0.3561	0.3563	0.3564
21°	0.3714	0.3721	0.3728	0.3734	0.3740	0.3744	0.3747	0.3749	0.3750
22°	0.3896	0.3904	0.3912	0.3919	0.3926	0.3931	0.3935	0.3937	0.3938
23°	0.4078	0.4088	0.4097	0.4105	0.4113	0.4119	0.4123	0.4126	0.4127
24°	0.4261	0.4272	0.4283	0.4292	0.4301	0.4308	0.4313	0.4316	0.4317
25°	0.4446	0.4458	0.4470	0.4481	0.4490	0.4498	0.4504	0.4508	0.4509
26°	0.4630	0.4645	0.4658	0.4670	0.4681	0.4690	0.4697	0.4701	0.4702
27°	0.4816	0.4832	0.4847	0.4861	0.4873	0.4884	0.4891	0.4896	0.4897
28°	0.5003	0.5021	0.5038	0.5053	0.5067	0.5079	0.5087	0.5092	0.5094
29°	0.5190	0.5210	0.5229	0.5247	0.5262	0.5275	0.5285	0.5291	0.5293
30°	0.5379	0.5401	0.5422	0.5442	0.5459	0.5474	0.5484	0.5491	0.5493
31°	0.5568	0.5593	0.5617	0.5639	0.5658	0.5674	0.5686	0.5693	0.5696
32°	0.5759	0.5786	0.5812	0.5837	0.5858	0.5876	0.5889	0.5898	0.5900
33°	0.5950	0.5980	0.6010	0.6037	0.6060	0.6080	0.6095	0.6104	0.6107
34°	0.6143	0.6176	0.6208	0.6238	0.6265	0.6287	0.6303	0.6313	0.6317
35°	0.6336	0.6373	0.6408	0.6441	0.6471	0.6495	0.6513	0.6525	0.6528
36°	0.6531	0.6571	0.6610	0.6647	0.6679	0.6706	0.6726	0.6739	0.6743
37°	0.6727	0.6771	0.6814	0.6854	0.6890	0.6919	0.6941	0.6955	0.6960
38°	0.6925	0.6973	0.7019	0.7063	0.7102	0.7135	0.7159	0.7175	0.7180
39°	0.7123	0.7176	0.7227	0.7275	0.7318	0.7353	0.7380	0.7397	0.7403
40°	0.7323	0.7380	0.7436	0.7488	0.7535	0.7575	0.7604	0.7623	0.7629
41°	0.7524	0.7586	0.7647	0.7704	0.7756	0.7799	0.7831	0.7852	0.7859
42°	0.7727	0.7794	0.7860	0.7922	0.7979	0.8026	0.8062	0.8084	0.8092
43°	0.7931	0.8004	0.8075	0.8143	0.8204	0.8256	0.8295	0.8320	0.8328
44°	0.8136	0.8215	0.8293	0.8367	0.8433	0.8490	0.8533	0.8560	0.8569
45°	0.8343	0.8428	0.8512	0.8592	0.8665	0.8727	0.8774	0.8804	0.8814

ELLIPTIC INTEGRALS OF THE FIRST KIND: $F(k, \phi)$ (Continued)

$$F(k, \phi) = \int_0^{\phi} \frac{d\Phi}{\sqrt{1 - k^2 \sin^2 \Phi}}, \qquad \theta = \sin^{-1} k$$

ϕ \ θ	5°	10°	15°	20°	25°	30°	35°	40°	45°
46°	0.8034	0.8052	0.8080	0.8120	0.8170	0.8230	0.8300	0.8378	0.8462
47°	0.8209	0.8227	0.8258	0.8300	0.8353	0.8418	0.8492	0.8575	0.8666
48°	0.8384	0.8403	0.8436	0.8480	0.8537	0.8606	0.8685	0.8773	0.8870
49°	0.8559	0.8579	0.8614	0.8661	0.8721	0.8794	0.8878	0.8972	0.9076
50°	0.8734	0.8756	0.8792	0.8842	0.8905	0.8982	0.9072	0.9173	0.9283
51°	0.8909	0.8932	0.8970	0.9023	0.9090	0.9172	0.9267	0.9374	0.9491
52°	0.9084	0.9108	0.9148	0.9204	0.9275	0.9361	0.9462	0.9575	0.9701
53°	0.9259	0.9284	0.9326	0.9385	0.9460	0.9551	0.9658	0.9778	0.9912
54°	0.9434	0.9460	0.9505	0.9567	0.9646	0.9742	0.9855	0.9982	1.0124
55°	0.9609	0.9637	0.9683	0.9748	0.9832	0.9933	1.0052	1.0187	1.0337
56°	0.9784	0.9813	0.9862	0.9930	1.0018	1.0125	1.0250	1.0393	1.0552
57°	0.9959	0.9989	1.0041	1.0112	1.0204	1.0317	1.0449	1.0600	1.0768
58°	1.0134	1.0166	1.0219	1.0295	1.0391	1.0509	1.0648	1.0807	1.0985
59°	1.0309	1.0342	1.0398	1.0477	1.0578	1.0702	1.0848	1.1016	1.1204
60°	1.0484	1.0519	1.0577	1.0660	1.0766	1.0896	1.1049	1.1226	1.1424
61°	1.0659	1.0695	1.0757	1.0843	1.0953	1.1089	1.1250	1.1436	1.1646
62°	1.0834	1.0872	1.0936	1.1026	1.1141	1.1284	1.1452	1.1648	1.1868
63°	1.1009	1.1049	1.1115	1.1209	1.1330	1.1478	1.1655	1.1860	1.2093
64°	1.1184	1.1225	1.1295	1.1392	1.1518	1.1674	1.1859	1.2073	1.2318
65°	1.1359	1.1402	1.1474	1.1575	1.1707	1.1869	1.2063	1.2288	1.2545
66°	1.1534	1.1579	1.1654	1.1759	1.1896	1.2065	1.2267	1.2503	1.2773
67°	1.1709	1.1756	1.1833	1.1943	1.2085	1.2262	1.2472	1.2719	1.3002
68°	1.1884	1.1932	1.2013	1.2127	1.2275	1.2458	1.2678	1.2936	1.3232
69°	1.2059	1.2109	1.2193	1.2311	1.2465	1.2655	1.2885	1.3154	1.3464
70°	1.2234	1.2286	1.2373	1.2495	1.2655	1.2853	1.3092	1.3372	1.3697
71°	1.2410	1.2463	1.2553	1.2680	1.2845	1.3051	1.3299	1.3592	1.3931
72°	1.2585	1.2640	1.2733	1.2864	1.3036	1.3249	1.3507	1.3812	1.4167
73°	1.2760	1.2817	1.2913	1.3049	1.3226	1.3448	1.3715	1.4033	1.4403
74°	1.2935	1.2994	1.3093	1.3234	1.3417	1.3647	1.3924	1.4254	1.4640
75°	1.3110	1.3171	1.3273	1.3418	1.3608	1.3846	1.4134	1.4477	1.4879
76°	1.3285	1.3348	1.3454	1.3603	1.3800	1.4045	1.4344	1.4700	1.5118
77°	1.3460	1.3525	1.3634	1.3788	1.3991	1.4245	1.4554	1.4923	1.5359
78°	1.3636	1.3702	1.3814	1.3974	1.4183	1.4445	1.4765	1.5147	1.5600
79°	1.3811	1.3879	1.3995	1.4159	1.4374	1.4645	1.4976	1.5372	1.5842
80°	1.3986	1.4056	1.4175	1.4344	1.4566	1.4846	1.5187	1.5597	1.6085
81°	1.4161	1.4234	1.4356	1.4530	1.4758	1.5046	1.5399	1.5823	1.6328
82°	1.4336	1.4411	1.4536	1.4715	1.4950	1.5247	1.5611	1.6049	1.6572
83°	1.4512	1.4588	1.4717	1.4901	1.5143	1.5448	1.5823	1.6276	1.6817
84°	1.4687	1.4765	1.4897	1.5086	1.5335	1.5649	1.6035	1.6502	1.7062
85°	1.4862	1.4942	1.5078	1.5272	1.5527	1.5850	1.6248	1.6730	1.7308
86°	1.5037	1.5120	1.5259	1.5457	1.5720	1.6052	1.6461	1.6957	1.7554
87°	1.5212	1.5297	1.5439	1.5643	1.5912	1.6253	1.6673	1.7184	1.7801
88°	1.5388	1.5474	1.5620	1.5829	1.6105	1.6454	1.6886	1.7412	1.8047
89°	1.5563	1.5651	1.5801	1.6015	1.6297	1.6656	1.7099	1.7640	1.8294
90°	1.5738	1.5828	1.5981	1.6200	1.6490	1.6858	1.7312	1.7863	1.8541

ELLIPTIC INTEGRALS OF THE FIRST KIND: $F(k, \phi)$ (Continued)

$$F(k, \phi) = \int_0^{\phi} \frac{d\Phi}{\sqrt{1 - k^2 \sin^2 \Phi}}, \qquad \theta = \sin^{-1} k$$

ϕ \ θ	50°	55°	60°	65°	70°	75°	80°	85°	90°
46°	0.8552	0.8643	0.8734	0.8821	0.8900	0.8968	0.9019	0.9052	0.9063
47°	0.8761	0.8860	0.8958	0.9053	0.9139	0.9212	0.9269	0.9304	0.9316
48°	0.8973	0.9079	0.9185	0.9287	0.9381	0.9461	0.9523	0.9561	0.9575
49°	0.9186	0.9300	0.9415	0.9525	0.9627	0.9714	0.9781	0.9824	0.9838
50°	0.9401	0.9523	0.9647	0.9766	0.9876	0.9971	1.0044	1.0091	1.0107
51°	0.9617	0.9748	0.9881	1.0010	1.0130	1.0233	1.0313	1.0364	1.0381
52°	0.9835	0.9976	1.0118	1.0258	1.0387	1.0499	1.0587	1.0642	1.0662
53°	1.0055	1.0205	1.0359	1.0509	1.0649	1.0771	1.0866	1.0927	1.0948
54°	1.0277	1.0437	1.0602	1.0764	1.0915	1.1048	1.1152	1.1219	1.1242
55°	1.0500	1.0672	1.0848	1.1022	1.1186	1.1331	1.1444	1.1517	1.1542
56°	1.0725	1.0908	1.1097	1.1285	1.1462	1.1619	1.1743	1.1823	1.1851
57°	1.0952	1.1147	1.1349	1.1551	1.1743	1.1914	1.2049	1.2136	1.2167
58°	1.1180	1.1389	1.1605	1.1822	1.2030	1.2215	1.2362	1.2458	1.2492
59°	1.1411	1.1632	1.1864	1.2097	1.2321	1.2522	1.2684	1.2789	1.2826
60°	1.1643	1.1879	1.2125	1.2376	1.2619	1.2837	1.3014	1.3129	1.3170
61°	1.1877	1.2128	1.2392	1.2660	1.2922	1.3159	1.3352	1.3480	1.3524
62°	1.2113	1.2379	1.2661	1.2949	1.3231	1.3490	1.3701	1.3841	1.3890
63°	1.2351	1.2633	1.2933	1.3242	1.3547	1.3828	1.4059	1.4214	1.4268
64°	1.2591	1.2890	1.3209	1.3541	1.3870	1.4175	1.4429	1.4599	1.4659
65°	1.2833	1.3149	1.3489	1.3844	1.4199	1.4532	1.4810	1.4998	1.5065
66°	1.3076	1.3411	1.3773	1.4153	1.4536	1.4898	1.5203	1.5411	1.5485
67°	1.3321	1.3675	1.4060	1.4467	1.4880	1.5274	1.5610	1.5840	1.5923
68°	1.3568	1.3942	1.4351	1.4786	1.5232	1.5661	1.6030	1.6287	1.6379
69°	1.3817	1.4212	1.4646	1.5111	1.5591	1.6059	1.6466	1.6752	1.6856
70°	1.4068	1.4484	1.4944	1.5441	1.5959	1.6468	1.6918	1.7237	1.7354
71°	1.4320	1.4759	1.5246	1.5777	1.6335	1.6891	1.7388	1.7745	1.7877
72°	1.4574	1.5036	1.5552	1.6118	1.6720	1.7326	1.7876	1.8277	1.8427
73°	1.4830	1.5315	1.5862	1.6465	1.7113	1.7774	1.8384	1.8837	1.9008
74°	1.5087	1.5597	1.6175	1.6818	1.7516	1.8237	1.8915	1.9427	1.9623
75°	1.5345	1.5882	1.6492	1.7176	1.7927	1.8715	1.9468	2.0050	2.0276
76°	1.5606	1.6168	1.6812	1.7540	1.8347	1.9207	2.0047	2.0711	2.0973
77°	1.5867	1.6457	1.7136	1.7909	1.8777	1.9716	2.0653	2.1414	2.1721
78°	1.6130	1.6748	1.7462	1.8284	1.9215	2.0240	2.1288	2.2164	2.2528
79°	1.6394	1.7040	1.7792	1.8664	1.9663	2.0781	2.1954	2.2969	2.3404
80°	1.6660	1.7335	1.8125	1.9048	2.0119	2.1339	2.2653	2.3836	2.4362
81°	1.6926	1.7631	1.8461	1.9438	2.0584	2.1913	2.3387	2.4775	2.5421
82°	1.7193	1.7929	1.8799	1.9831	2.1057	2.2504	2.4157	2.5795	2.6603
83°	1.7462	1.8228	1.9140	2.0229	2.1537	2.3110	2.4965	2.6911	2.7942
84°	1.7731	1.8528	1.9482	2.0630	2.2024	2.3731	2.5811	2.8136	2.9487
85°	1.8001	1.8830	1.9826	2.1035	2.2518	2.4366	2.6694	2.9487	3.1313
86°	1.8271	1.9132	2.0172	2.1442	2.3017	2.5013	2.7612	3.0978	3.3547
87°	1.8542	1.9435	2,0519	2.1852	2.3520	2.5670	2.8561	3.2620	3.6425
88°	1.8813	1.9739	2.0867	2.2263	2.4026	2.6336	2.9537	3.4412	4.0481
89°	1.9084	2.0043	2.1216	2.2675	2.4535	2.7007	3.0530	3.6328	4.7413
90°	1.9356	2.0347	2.1565	2.3088	2.5046	2.7681	3.1534	3.8317	——

ELLIPTIC INTEGRALS OF THE SECOND KIND: $E(k, \phi)$

$$E(k, \phi) = \int_0^\phi \sqrt{(1 - k^2 \sin^2 \Phi)}\, d\Phi, \quad \theta = \sin^{-1} k$$

ϕ \ θ	5°	10°	15°	20°	25°	30°	35°	40°	45°
1°	0.0175	0.0175	0.0175	0.0175	0.0175	0.0175	0.0175	0.0175	0.0175
2°	0.0349	0.0349	0.0349	0.0349	0.0349	0.0349	0.0349	0.0349	0.0349
3°	0.0524	0.0524	0.0524	0.0524	0.0524	0.0524	0.0524	0.0523	0.0523
4°	0.0698	0.0698	0.0698	0.0698	0.0698	0.0698	0.0698	0.0698	0.0698
5°	0.0873	0.0873	0.0873	0.0873	0.0872	0.0872	0.0872	0.0872	0.0872
6°	0.1047	0.1047	0.1047	0.1047	0.1047	0.1047	0.1047	0.1046	0.1046
7°	0.1222	0.1222	0.1222	0.1221	0.1221	0.1221	0.1221	0.1220	0.1220
8°	0.1396	0.1396	0.1396	0.1396	0.1395	0.1395	0.1395	0.1394	0.1394
9°	0.1571	0.1571	0.1570	0.1570	0.1570	0.1569	0.1569	0.1568	0.1568
10°	0.1745	0.1745	0.1745	0.1744	0.1744	0.1743	0.1742	0.1742	0.1741
11°	0.1920	0.1920	0.1919	0.1918	0.1918	0.1917	0.1916	0.1915	0.1914
12°	0.2094	0.2094	0.2093	0.2093	0.2092	0.2091	0.2089	0.2088	0.2087
13°	0.2269	0.2268	0.2268	0.2267	0.2265	0.2264	0.2263	0.2261	0.2259
14°	0.2443	0.2443	0.2442	0.2441	0.2439	0.2437	0.2436	0.2433	0.2431
15°	0.2618	0.2617	0.2616	0.2615	0.2613	0.2611	0.2608	0.2606	0.2603
16°	0.2792	0.2791	0.2790	0.2788	0.2786	0.2784	0.2781	0.2778	0.2775
17°	0.2967	0.2966	0.2964	0.2962	0.2959	0.2956	0.2953	0.2949	0.2946
18°	0.3141	0.3140	0.3138	0.3136	0.3133	0.3129	0.3125	0.3121	0.3116
19°	0.3316	0.3314	0.3312	0.3309	0.3305	0.3301	0.3296	0.3291	0.3286
20°	0.3490	0.3489	0.3486	0.3483	0.3478	0.3473	0.3468	0.3462	0.3456
21°	0.3665	0.3663	0.3660	0.3656	0.3651	0.3645	0.3639	0.3632	0.3625
22°	0.3839	0.3837	0.3834	0.3829	0.3823	0.3817	0.3809	0.3802	0.3793
23°	0.4013	0.4011	0.4007	0.4002	0.3996	0.3988	0.3980	0.3971	0.3961
24°	0.4188	0.4185	0.4181	0.4175	0.4168	0.4159	0.4150	0.4139	0.4129
25°	0.4362	0.4359	0.4354	0.4348	0.4339	0.4330	0.4319	0.4308	0.4296
26°	0.4537	0.4533	0.4528	0.4520	0.4511	0.4500	0.4488	0.4475	0.4462
27°	0.4711	0.4707	0.4701	0.4693	0.4682	0.4670	0.4657	0.4643	0.4628
28°	0.4886	0.4881	0.4874	0.4865	0.4854	0.4840	0.4825	0.4809	0.4793
29°	0.5060	0.5055	0.5048	0.5037	0.5025	0.5010	0.4993	0.4975	0.4957
30°	0.5234	0.5229	0.5221	0.5209	0.5195	0.5179	0.5161	0.5141	0.5120
31°	0.5409	0.5403	0.5394	0.5381	0.5366	0.5348	0.5327	0.5306	0.5283
32°	0.5583	0.5577	0.5567	0.5553	0.5536	0.5516	0.5494	0.5470	0.5446
33°	0.5757	0.5751	0.5740	0.5725	0.5706	0.5684	0.5660	0.5634	0.5607
34°	0.5932	0.5924	0.5912	0.5896	0.5876	0.5852	0.5826	0.5797	0.5768
35°	0.6106	0.6098	0.6085	0.6067	0.6045	0.6019	0.5991	0.5960	0.5928
36°	0.6280	0.6272	0.6258	0.6238	0.6214	0.6186	0.6155	0.6122	0.6087
37°	0.6455	0.6445	0.6430	0.6409	0.6383	0.6353	0.6319	0.6283	0.6245
38°	0.6629	0.6619	0.6602	0.6580	0.6552	0.6519	0.6483	0.6444	0.6403
39°	0.6803	0.6792	0.6775	0.6750	0.6720	0.6685	0.6646	0.6604	0.6559
40°	0.6977	0.6966	0.6947	0.6921	0.6888	0.6851	0.6808	0.6763	0.6715
41°	0.7152	0.7139	0.7119	0.7091	0.7056	0.7016	0.6970	0.6921	0.6870
42°	0.7326	0.7313	0.7291	0.7261	0.7224	0.7180	0.7132	0.7079	0.7024
43°	0.7500	0.7486	0.7463	0.7431	0.7391	0.7345	0.7293	0.7237	0.7178
44°	0.7674	0.7659	0.7634	0.7600	0.7558	0.7508	0.7453	0.7393	0.7330
45°	0.7849	0.7832	0.7806	0.7770	0.7725	0.7672	0.7613	0.7549	0.7482

ELLIPTIC INTEGRALS OF THE SECOND KIND: $E(k, \phi)$ (Continued)

$$E(k, \phi) = \int_0^{\phi} \sqrt{(1 - k^2 \sin^2 \Phi)}\, d\Phi, \quad \theta = \sin^{-1} k$$

ϕ \ θ	50°	55°	60°	65°	70°	75°	80°	85°	90°
1°	0.0175	0.0175	0.0175	0.0175	0.0175	0.0175	0.0175	0.0175	0.0175
2°	0.0349	0.0349	0.0349	0.0349	0.0349	0.0349	0.0349	0.0349	0.0349
3°	0.0523	0.0523	0.0523	0.0523	0.0523	0.0523	0.0523	0.0523	0.0523
4°	0.0698	0.0698	0.0698	0.0698	0.0698	0.0698	0.0698	0.0698	0.0698
5°	0.0872	0.0872	0.0872	0.0872	0.0872	0.0872	0.0872	0.0872	0.0872
6°	0.1046	0.1046	0.1046	0.1046	0.1046	0.1045	0.1045	0.1045	0.1045
7°	0.1220	0.1220	0.1219	0.1219	0.1219	0.1219	0.1219	0.1219	0.1219
8°	0.1394	0.1393	0.1393	0.1393	0.1392	0.1392	0.1392	0.1392	0.1392
9°	0.1567	0.1566	0.1566	0.1566	0.1565	0.1565	0.1565	0.1564	0.1564
10°	0.1740	0.1739	0.1739	0.1738	0.1738	0.1737	0.1737	0.1737	0.1736
11°	0.1913	0.1912	0.1911	0.1910	0.1909	0.1909	0.1908	0.1908	0.1908
12°	0.2085	0.2084	0.2083	0.2082	0.2081	0.2080	0.2080	0.2079	0.2079
13°	0.2258	0.2256	0.2254	0.2253	0.2252	0.2251	0.2250	0.2250	0.2250
14°	0.2429	0.2427	0.2425	0.2424	0.2422	0.2421	0.2420	0.2419	0.2419
15°	0.2601	0.2598	0.2596	0.2594	0.2592	0.2590	0.2589	0.2588	0.2588
16°	0.2771	0.2768	0.2765	0.2763	0.2761	0.2759	0.2757	0.2757	0.2756
17°	0.2942	0.2938	0.2935	0.2932	0.2929	0.2927	0.2925	0.2924	0.2924
18°	0.3112	0.3107	0.3103	0.3099	0.3096	0.3094	0.3092	0.3091	0.3090
19°	0.3281	0.3276	0.3271	0.3267	0.3263	0.3260	0.3258	0.3256	0.3256
20°	0.3450	0.3444	0.3438	0.3433	0.3429	0.3425	0.3422	0.3421	0.3420
21°	0.3618	0.3611	0.3604	0.3598	0.3593	0.3589	0.3586	0.3584	0.3584
22°	0.3785	0.3777	0.3770	0.3763	0.3757	0.3752	0.3749	0.3747	0.3746
23°	0.3952	0.3943	0.3935	0.3927	0.3920	0.3915	0.3911	0.3908	0.3907
24°	0.4118	0.4108	0.4098	0.4090	0.4082	0.4076	0.4071	0.4068	0.4067
25°	0.4284	0.4272	0.4261	0.4251	0.4243	0.4236	0.4230	0.4227	0.4226
26°	0.4449	0.4436	0.4423	0.4412	0.4402	0.4394	0.4389	0.4385	0.4384
27°	0.4613	0.4598	0.4584	0.4572	0.4561	0.4552	0.4545	0.4541	0.4540
28°	0.4776	0.4760	0.4744	0.4730	0.4718	0.4708	0.4701	0.4696	0.4695
29°	0.4938	0.4920	0.4903	0.4887	0.4874	0.4863	0.4855	0.4850	0.4848
30°	0.5100	0.5080	0.5061	0.5044	0.5029	0.5016	0.5007	0.5002	0.5000
31°	0.5261	0.5239	0.5218	0.5199	0.5182	0.5169	0.5159	0.5152	0.5150
32°	0.5421	0.5396	0.5373	0.5352	0.5334	0.5319	0.5308	0.5301	0.5299
33°	0.5580	0.5553	0.5528	0.5505	0.5485	0.5468	0.5456	0.5449	0.5446
34°	0.5738	0.5709	0.5681	0.5656	0.5634	0.5616	0.5603	0.5595	0.5592
35°	0.5895	0.5863	0.5833	0.5806	0.5782	0.5762	0.5748	0.5739	0.5736
36°	0.6051	0.6017	0.5984	0.5954	0.5928	0.5907	0.5891	0.5881	0.5878
37°	0.6207	0.6169	0.6134	0.6101	0.6073	0.6050	0.6032	0.6022	0.6018
38°	0.6361	0.6321	0.6282	0.6247	0.6216	0.6191	0.6172	0.6160	0.6157
39°	0.6515	0.6471	0.6429	0.6391	0.6357	0.6330	0.6310	0.6297	0.6293
40°	0.6667	0.6620	0.6575	0.6533	0.6497	0.6468	0.6446	0.6432	0.6428
41°	0.6818	0.6767	0.6719	0.6674	0.6636	0.6604	0.6580	0.6566	0.6561
42°	0.6969	0.6914	0.6862	0.6814	0.6772	0.6738	0.6712	0.6697	0.6691
43°	0.7118	0.7059	0.7003	0.6952	0.6907	0.6870	0.6843	0.6826	0.6820
44°	0.7266	0.7204	0.7144	0.7088	0.7040	0.7000	0.6971	0.6953	0.6947
45°	0.7414	0.7346	0.7282	0.7223	0.7171	0.7129	0.7097	0.7078	0.7071

ELLIPTIC INTEGRALS OF THE SECOND KIND: $E(k, \phi)$ (Continued)

$$E(k, \phi) = \int_0^{\phi} \sqrt{(1 - k^2 \sin^2 \Phi)}\,d\Phi, \quad \theta = \sin^{-1} k$$

ϕ \ θ	5°	10°	15°	20°	25°	30°	35°	40°	45°
46°	0.8023	0.8006	0.7977	0.7939	0.7891	0.7835	0.7772	0.7704	0.7633
47°	0.8197	0.8179	0.8149	0.8108	0.8057	0.7998	0.7931	0.7858	0.7782
48°	0.8371	0.8352	0.8320	0.8277	0.8223	0.8160	0.8089	0.8012	0.7931
49°	0.8545	0.8525	0.8491	0.8446	0.8389	0.8322	0.8247	0.8165	0.8079
50°	0.8719	0.8698	0.8663	0.8614	0.8554	0.8483	0.8404	0.8317	0.8227
51°	0.8894	0.8871	0.8834	0.8783	0.8719	0.8644	0.8560	0.8469	0.8373
52°	0.9068	0.9044	0.9004	0.8951	0.8884	0.8805	0.8716	0.8620	0.8518
53°	0.9242	0.9217	0.9175	0.9119	0.9048	0.8965	0.8872	0.8770	0.8663
54°	0.9416	0.9389	0.9345	0.9287	0.9212	0.9125	0.9026	0.8919	0.8806
55°	0.9590	0.9562	0.9517	0.9454	0.9376	0.9284	0.9181	0.9068	0.8949
56°	0.9764	0.9735	0.9687	0.9622	0.9540	0.9443	0.9335	0.9216	0.9091
57°	0.9938	0.9908	0.9858	0.9789	0.9703	0.9602	0.9488	0.9363	0.9232
58°	1.0112	1.0080	1.0028	0.9956	0.9866	0.9760	0.9641	0.9510	0.9372
59°	1.0286	1.0253	1.0198	1.0123	1.0029	1.9918	0.9793	0.9656	0.9511
60°	1.0460	1.0426	1.0368	1.0290	1.0191	1.0076	0.9945	0.9801	0.9650
61°	1.0634	1.0598	1.0538	1.0456	1.0354	1.0233	1.0096	0.9946	0.9787
62°	1.0808	1.0771	1.0708	1.0623	1.0516	1.0389	1.0246	1 0090	0.9924
63°	1.0982	1.0943	1.0878	1.0789	1.0678	1.0546	1.0397	1.0233	1.0060
64°	1.1156	1.1115	1.1048	1.0955	1.0839	1.0702	1.0547	1.0376	1.0195
65°	1.1330	1.1288	1.1218	1.1121	1.1001	1.0858	1.0696	1.0518	1.0329
66°	1.1504	1.1460	1.1387	1.1287	1.1162	1.1013	1.0845	1.0660	1.0463
67°	1.1678	1.1632	1.1557	1.1453	1.1323	1.1168	1.0993	1.0801	1.0596
68°	1.1852	1.1805	1.1726	1.1618	1.1483	1.1323	1.1141	1.0941	1.0728
69°	1.2026	1.1977	1.1896	1.1784	1.1644	1.1478	1.1289	1.1081	1.0859
70°	1.2200	1.2149	1.2065	1.1949	1.1804	1.1632	1.1436	1.1221	1.0990
71°	1.2374	1.2321	1.2234	1.2114	1.1964	1.1786	1.1583	1.1359	1.1120
72°	1.2548	1.2493	1.2403	1.2280	1.2124	1.1939	1.1729	1.1498	1.1250
73°	1.2722	1.2666	1.2573	1.2445	1.2284	1.2093	1.1875	1.1636	1.1379
74°	1.2896	1.2838	1.2742	1.2609	1.2443	1.2246	1.2021	1.1773	1.1507
75°	1.3070	1.3010	1.2911	1.2774	1.2603	1.2399	1.2167	1.1910	1.1635
76°	1.3244	1.3182	1.3080	1.2939	1.2762	1.2552	1.2312	1.2047	1.1762
77°	1.3418	1.3354	1.3249	1.3104	1.2921	1.2704	1.2457	1.2183	1.1889
78°	1.3592	1.3526	1.3417	1.3268	1.3080	1.2856	1.2601	1.2319	1.2015
79°	1.3765	1.3698	1.3586	1.3432	1.3239	1.3009	1.2746	1.2454	1.2141
80°	1.3939	1.3870	1.3755	1.3597	1.3398	1.3161	1.2890	1.2590	1.2266
81°	1.4113	1.4042	1.3924	1.3761	1.3556	1.3312	1.3034	1.2725	1.2391
82°	1.4287	1.4214	1.4093	1.3925	1.3715	1.3464	1.3177	1.2859	1.2516
83°	1.4461	1.4386	1.4261	1.4090	1.3873	1.3616	1.3321	1.2994	1.2640
84°	1.4635	1.4558	1.4430	1.4254	1.4032	1.3767	1.3464	1.3128	1.2765
85°	1.4809	1.4729	1.4598	1.4418	1.4190	1.3919	1.3608	1.3262	1.2889
86°	1.4983	1.4901	1.4767	1.4582	1.4348	1.4070	1.3751	1.3396	1.3012
87°	1.5156	1.5073	1.4936	1.4746	1.4507	1.4221	1.3894	1.3530	1.3136
88°	1.5330	1.5245	1.5104	1.4910	1.4665	1.4372	1.4037	1.3664	1.3260
89°	1.5504	1.5417	1.5273	1.5074	1.4823	1.4523	1.4180	1.3798	1.3383
90°	1.5678	1.5589	1.5442	1.5238	1.4981	1.4675	1.4323	1.3931	1.3506

ELLIPTIC INTEGRALS OF THE SECOND KIND: $E(k, \phi)$ (Continued)

$$E(k, \phi) = \int_0^\phi \sqrt{(1 - k^2 \sin^2 \Phi)}\, d\Phi, \quad \theta = \sin^{-1} k$$

ϕ \ θ	50°	55°	60°	65°	70°	75°	80°	85°	90°
46°	0.7560	0.7488	0.7419	0.7356	0.7301	0.7255	0.7221	0.7200	0.7193
47°	0.7705	0.7628	0.7555	0.7488	0.7429	0.7380	0.7344	0.7321	0.7314
48°	0.7849	0.7768	0.7690	0.7618	0.7555	0.7502	0.7464	0.7440	0.7431
49°	0.7992	0.7905	0.7822	0.7746	0.7679	0.7623	0.7581	0.7556	0.7547
50°	0.8134	0.8042	0.7954	0.7872	0.7801	0.7741	0.7697	0.7670	0.7660
51°	0.8275	0.8177	0.8084	0.7997	0.7921	0.7858	0.7811	0.7781	0.7771
52°	0.8414	0.8311	0.8212	0.8120	0.8039	0.7972	0.7922	0.7891	0.7880
53°	0.8553	0.8444	0.8339	0.8241	0.8155	0.8084	0.8031	0.7998	0.7986
54°	0.8690	0.8575	0.8464	0.8361	0.8270	0.8194	0.8137	0.8102	0.8090
55°	0.8827	0.8705	0.8588	0.8479	0.8382	0.8302	0.8242	0.8204	0.8192
56°	0.8962	0.8834	0.8710	0.8595	0.8493	0.8408	0.8344	0.8304	0.8290
57°	0.9096	0.8961	0.8831	0.8709	0.8601	0.8511	0.8443	0.8401	0.8387
58°	0.9230	0.9088	0.8950	0.8822	0.8707	0.8612	0.8540	0.8496	0.8480
59°	0.9362	0.9213	0.9068	0.8932	0.8812	0.8711	0.8635	0.8588	0.8572
60°	0.9493	0.9336	0.9184	0.9042	0.8914	0.8808	0.8728	0.8677	0.8660
61°	0.9623	0.9459	0.9299	0.9149	0.9015	0.8903	0.8817	0.8764	0.8746
62°	0.9752	0.9580	0.9412	0.9254	0.9113	0.8995	0.8905	0.8849	0.8829
63°	0.9880	0.9700	0.9524	0.9358	0.9210	0.9085	0.8990	0.8930	0.8910
64°	1.0007	0.9818	0.9634	0.9460	0.9304	0.9173	0.9072	0.9009	0.8988
65°	1.0133	0.9936	0.9743	0.9561	0.9397	0.9258	0.9152	0.9086	0.9063
66°	1.0258	1.0052	0.9850	0.9659	0.9487	0.9341	0.9230	0.9159	0.9135
67°	1.0383	1.0167	0.9956	0.9756	0.9576	0.9422	0.9305	0.9230	0.9205
68°	1.0506	1.0281	1.0061	0.9852	0.9662	0.9501	0.9377	0.9299	0.9272
69°	1.0628	1.0394	1.0164	0.9946	0.9747	0.9578	0.9447	0.9364	0.9336
70°	1.0750	1.0506	1.0266	1.0038	0.9830	0.9652	0.9514	0.9427	0.9397
71°	1.0871	1.0617	1.0367	1.0129	0.9911	0.9724	0.9579	0.9487	0.9455
72°	1,0991	1.0727	1.0467	1.0218	0.9990	0.9794	0.9642	0.9544	0.9511
73°	1.1110	1.0836	1.0565	1.0306	1.0067	0.9862	0.9702	0.9599	0.9563
74°	1.1228	1.0944	1.0662	1.0392	1.0143	0.9928	0.9759	0.9650	0.9613
75°	1.1346	1.1051	1.0759	1.0477	1.0217	0.9992	0.9814	0.9699	0.9659
76°	1.1463	1.1158	1.0854	1.0561	1.0290	1.0053	0.9867	0.9745	0.9703
77°	1.1580	1.1263	1.0948	1.0643	1.0361	1.0113	0.9917	0.9789	0.9744
78°	1.1695	1.1368	1.1041	1.0724	1.0430	1.0171	0.9965	0.9829	0.9781
79°	1.1811	1.1472	1.1133	1.0805	1.0498	1.0228	1.0011	0.9867	0.9816
80°	1.1926	1.1576	1.1225	1.0884	1.0565	1.0282	1.0054	0.9902	0.9848
81°	1.2040	1.1678	1.1316	1.0962	1.0630	1.0335	1.0096	0.9935	0.9877
82°	1.2154	1.1781	1.1406	1.1040	1.0695	1.0387	1.0135	0.9965	0.9903
83°	1.2267	1.1883	1.1495	1.1116	1.0758	1.0437	1.0173	0.9992	0.9925
84°	1.2381	1.1984	1.1584	1.1192	1.0821	1.0486	1.0209	1.0017	0.9945
85°	1.2493	1.2085	1.1673	1.1267	1.0882	1.0534	1.0244	1.0039	0.9962
86°	1.2606	1.2186	1.1761	1.1342	1.0944	1.0581	1.0277	1.0060	0.9976
87°	1.2719	1.2286	1.1848	1.1417	1.1004	1.0628	1.0309	1.0078	0.9986
88°	1.2831	1.2386	1.1936	1.1491	1.1064	1.0673	1.0340	1.0095	0.9994
89°	1.2943	1.2487	1.2023	1.1565	1.1124	1.0719	1.0371	1.0111	0.9998
90°	1.3055	1.2587	1.2111	1.1638	1.1184	1.0764	1.0401	1.0127	1.0000

COMPLETE ELLIPTIC INTEGRALS

$$K = \int_0^{\pi/2} \frac{d\Phi}{\sqrt{1 - k^2 \sin^2 \Phi}} = F\left(k, \frac{\pi}{2}\right)$$

$\sin^{-1} k$	K	$\log K$	$\sin^{-1} k$	K	$\log K$
0°	1.5708	0.196120	**40°**	1.7868	0.252068
1	1.5709	0.196153	41	1.7992	0.255085
2	1.5713	0.196252	42	1.8122	0.258197
3	1.5719	0.196418	43	1.8256	0.261406
4	1.5727	0.196649	44	1.8396	0.264716
5	1.5738	0.196947	**45**	1.8541	0.268127
6	1.5751	0.197312	46	1.8691	0.271644
7	1.5767	0.197743	47	1.8848	0.275267
8	1.5785	0.198241	48	1.9011	0.279001
9	1.5805	0.198806	49	1.9180	0.282848
10	1.5828	0.199438	**50**	1.9356	0.286811
11	1.5854	0.200137	51	1.9539	0.290895
12	1.5882	0.200904	52	1.9729	0.295101
13	1.5913	0.201740	53	1.9927	0.299435
14	1.5946	0.202643	54	2.0133	0.303901
15	1.5981	0.203615	**55**	2.0347	0.308504
16	1.6020	0.204657	56	2.0571	0.313247
17	1.6061	0.205768	57	2.0804	0.318138
18	1.6105	0.206948	58	2.1047	0.323182
19	1.6151	0.208200	59	2.1300	0.328384
20	1.6200	0.209522	**60**	2.1565	0.333753
21	1.6252	0.210916	61	2.1842	0.339295
22	1.6307	0.212382	62	2.2132	0.345020
23	1.6365	0.213921	63	2.2435	0.350936
24	1.6426	0.215533	64	2.2754	0.357053
25	1.6490	0.217219	**65**	2.3088	0.363384
26	1.6557	0.218981	66	2.3439	0.369940
27	1.6627	0.220818	67	2.3809	0.376736
28	1.6701	0.222732	68	2.4198	0.383787
29	1.6777	0.224723	69	2.4610	0.391112
30	1.6858	0.226793	**70**	2.5046	0.398730
31	1.6941	0.228943	71	2.5507	0.406665
32	1.7028	0.231173	72	2.5998	0.414943
33	1.7119	0.233485	73	2.6521	0.423596
34	1.7214	0.235880	74	2.7081	0.432660
35	1.7312	0.238359	**75**	2.7681	0.442176
36	1.7415	0.240923	76	2.8327	0.452196
37	1.7522	0.243575	77	2.9026	0.462782
38	1.7633	0.246315	78	2.9786	0.474008
39	1.7748	0.249146	79	3.0617	0.485967
40	1.7868	0.252068	**80**	3.1534	0.498777

COMPLETE ELLIPTIC INTEGRALS (Continued)

$$K = \int_0^{\pi/2} \frac{d\Phi}{\sqrt{1 - k^2 \sin^2 \Phi}} = F\left(k, \frac{\pi}{2}\right)$$

$\sin^{-1} k$	K	$\log K$	$\sin^{-1} k$	K	$\log K$
80°	3.1534	0.498777	**85°**	3.8317	0.583396
81	3.2553	0.512591	86	4.0528	0.607751
82	3.3699	0.527613	87	4.3387	0.637355
83	3.5004	0.544120	88	4.7427	0.676027
84	3.6519	0.562514	89	5.4349	0.735192
85	3.8317	0.583396	**90**	∞	∞

Values of K for $\sin^{-1} k$ = 85° to 89° by 0.1° and 89° to 90° by minutes

$\sin^{-1} k$	K	$\log K$	$\sin^{-1} k$	K	$\log K$
85.0°	3.832	0.58343	**89° 0′**	5.435	0.73520
85.1	3.852	0.58569	89 2	5.469	0.73791
85.2	3.872	0.58794	89 4	5.504	0.74068
85.3	3.893	0.59028	89 6	5.540	0.74351
85.4	3.914	0.59262	89 8	5.578	0.74648
85.5	3.936	0.59506	**89 10**	5.617	0.74950
85.6	3.958	0.59748	89 12	5.658	0.75266
85.7	3.981	0.59999	89 14	5.700	0.75587
85.8	4.004	0.60249	89 16	5.745	0.75929
85.9	4.028	0.60509	89 18	5.791	0.76275
86.0	4.053	0.60778	**89 20**	5.840	0.76641
86.1	4.078	0.61045	89 22	5.891	0.77019
86.2	4.104	0.61321	89 24	5.946	0.77422
86.3	4.130	0.61595	89 26	6.003	0.77837
86.4	4.157	0.61878	89 28	6.063	0.78269
86.5	4.185	0.62170	**89 30**	6.128	0.78732
86.6	4.214	0.62469	89 32	6.197	0.79218
86.7	4.244	0.62778	89 34	6.271	0.79734
86.8	4.274	0.63083	89 36	6.351	0.80284
86.9	4.306	0.63407	89 38	6.438	0.80875
87.0	4.339	0.63739	**89 40**	6.533	0.81511
87.1	4.372	0.64068	89 41	6.584	0.81849
87.2	4.407	0.64414	89 42	6.639	0.82210
87.3	4.444	0.64777	89 43	6.696	0.82582
87.4	4.481	0.65137	89 44	6.756	0.82969
87.5	4.520	0.65514	**89 45**	6.821	0.83385
87.6	4.561	0.65916	89 46	6.890	0.83822
87.7	4.603	0.66304	89 47	6.964	0.84286
87.8	4.648	0.66727	89 48	7.044	0.84782
87.9	4.694	0.67154	89 49	7.131	0.85315
88.0	4.743	0.67605	**89 50**	7.226	0.85890
88.1	4.794	0.68070	89 51	7.332	0.86522
88.2	4.848	0.68556	89 52	7.449	0.87210
88.3	4.905	0.69064	89 53	7.583	0.87984
88.4	4.965	0.69592	89 54	7.737	0.88857
88.5	5.030	0.70157	**89 55**	7.919	0.89867
88.6	5.099	0.70749	89 56	8.143	0.91078
88.7	5.173	0.71374	89 57	8.430	0.92583
88.8	5.253	0.72041	89 58	8.836	0.94626
88.9	5.340	0.72754	89 59	9.529	0.97905
89.0	5.435	0.73520	**90 0**	∞	∞

COMPLETE ELLIPTIC INTEGRALS (Continued)

$$E = \int_0^{\pi/2} \sqrt{1 - k^2 \sin^2 \Phi} \cdot d\Phi = E\left(k, \frac{\pi}{2}\right)$$

$\sin^{-1} k$	E	$\log E$	$\sin^{-1} k$	E	$\log E$
0°	1.5708	0.196120	**45°**	1.3506	0.130541
1	1.5707	0.196087	46	1.3418	0.127690
2	1.5703	0.195988	47	1.3329	0.124788
3	1.5697	0.195822	48	1.3238	0.121836
4	1.5689	0.195591	49	1.3147	0.118836
5	1.5678	0.195293	**50**	1.3055	0.115790
6	1.5665	0.194930	51	1.2963	0.112698
7	1.5649	0.194500	52	1.2870	0.109563
8	1.5632	0.194004	53	1.2776	0.106386
9	1.5611	0.193442	54	1.2681	0.103169
10	1.5589	0.192815	**55**	1.2587	0.099915
11	1.5564	0.192121	56	1.2492	0.096626
12	1.5537	0.191362	57	1.2397	0.093303
13	1.5507	0.190537	58	1.2301	0.089950
14	1.5476	0.189646	59	1.2206	0.086569
15	1.5442	0.188690	**60**	1.2111	0.083164
16	1.5405	0.187668	61	1.2015	0.079738
17	1.5367	0.186581	62	1.1920	0.076293
18	1.5326	0.185428	63	1.1826	0.072834
19	1.5283	0.184210	64	1.1732	0.069364
20	1.5238	0.182928	**65**	1.1638	0.065889
21	1.5191	0.181580	66	1.1545	0.062412
22	1.5141	0.180168	67	1.1453	0.058937
23	1.5090	0.178691	68	1.1362	0.055472
24	1.5037	0.177150	69	1.1272	0.052020
25	1.4981	0.175545	**70**	1.1184	0.048589
26	1.4924	0.173876	71	1.1096	0.045183
27	1.4864	0.172144	72	1.1011	0.041812
28	1.4803	0.170348	73	1.0927	0.038481
29	1.4740	0.168489	74	1.0844	0.035200
30	1.4675	0.166567	**75**	1.0764	0.031976
31	1.4608	0.164583	76	1.0686	0.028819
32	1.4539	0.162537	77	1.0611	0.025740
33	1.4469	0.160429	78	1.0538	0.022749
34	1.4397	0.158261	79	1.0468	0.019858
35	1.4323	0.156031	**80**	1.0401	0.017081
36	1.4248	0.153742	81	1.0338	0.014432
37	1.4171	0.151393	82	1.0278	0.011927
38	1.4092	0.148985	83	1.0223	0.009584
39	1.4013	0.146519	84	1.0172	0.007422
40	1.3931	0.143995	**85**	1.0127	0.005465
41	1.3849	0.141414	86	1.0086	0.003740
42	1.3765	0.138778	87	1.0053	0.002278
43	1.3680	0.136086	88	1.0026	0.001121
44	1.3594	0.133340	89	1.0008	0.000326
45	1.3506	0.130541	**90**	1.0000	0.000000

SINE, COSINE, AND EXPONENTIAL INTEGRALS

$$Si(x) = \int_0^x \frac{\sin v}{v}\,dv; \qquad Ci(x) = \int_\infty^x \frac{\cos v}{v}\,dv;$$

$$Ei(x) = \int_{-\infty}^x \frac{e^v}{v}\,dv; \qquad -Ei(-x) = \int_x^\infty \frac{e^{-v}}{v}\,dv$$

x	$Si(x)$	$Ci(x)$	$Ei(x)$	$-Ei(-x)$
0.0	0.00000	$-\infty$	$-\infty$	$+\infty$
0.1	0.09994	-1.72787	-1.62281	1.82292
0.2	.19956	-1.04221	- .82176	1.22265
0.3	.29850	- .64917	- .30267	.90568
0.4	.39646	- .37881	.10477	.70238
0.5	.49311	- .17778	.45422	.55977
0.6	.58813	- .02227	.76988	.45438
0.7	.68122	.10051	1.06491	.37377
0.8	.77210	.19828	1.34740	.31060
0.9	.86047	.27607	1.62281	.26018
1.0	.94608	.33740	1.89512	.21938
1.1	1.02869	.38487	2.16738	.18599
1.2	1.10805	.42046	2.44209	.15841
1.3	1.18396	.44574	2.72140	.13545
1.4	1.25623	.46201	3.00721	.11622
1.5	1.32468	.47036	3.30129	.10002
1.6	1.38918	.47173	3.60532	.08631
1.7	1.44959	.46697	3.92096	.07465
1.8	1.50582	.45681	4.24987	.06471
1.9	1.55778	.44194	4.59371	.05620
2.0	1.60541	.42298	4.95423	.04890
2.1	1.64870	.40051	5.33324	.04261
2.2	1.68762	.37507	5.73261	.03719
2.3	1.72221	.34718	6.15438	.03250
2.4	1.75249	.31729	6.60067	.02844
2.5	1.77852	.28587	7.07377	.02491
2.6	1.80039	.25334	7.57611	.02185
2.7	1.81821	.22008	8.11035	.01918
2.8	1.83210	.18649	8.67930	.01686
2.9	1.84219	.15290	9.28602	.01482
3.0	1.84865	.11963	9.93383	.01305
3.1	1.85166	.08699	10.6263	.01149
3.2	1.85140	.05526	11.3673	.01013
3.3	1.84808	.02468	12.1610	.00894
3.4	1.84191	- .00452	13.0121	.00789
3.5	1.83313	- .03213	13.9254	.00697
3.6	1.82195	- .05797	14.9063	.00616
3.7	1.80862	- .08190	15.9606	.00545
3.8	1.79339	- .10378	17.0948	.00482
3.9	1.77650	- .12350	18.3157	.00427
4.0	1.75820	- .14098	19.6309	.00378
4.1	1.73874	- .15617	21.0485	.00335
4.2	1.71837	- .16901	22.5774	.00297
4.3	1.69732	- .17951	24.2274	.00263
4.4	1.67583	- .18766	26.0090	.00234

SINE, COSINE, AND EXPONENTIAL INTEGRALS (Continued)

x	$Si(x)$	$Ci(x)$	$Ei(x)$	$-Ei(-x)$
4.5	1.65414	− .19349	27.9337	.00207
4.6	1.63246	− .19705	30.0141	.00184
4.7	1.61100	− .19839	32.2639	.00164
4.8	1.58998	− .19760	34.6979	.00145
4.9	1.56956	− .19478	37.3325	.00129
5.0	1.54993	− .19003	40.1853	.00115
5.1	1.53125	− .18348	43.2757	.00102
5.2	1.51367	− .17525	46.6249	.00091
5.3	1.49732	− .16551	50.2557	.00081
5.4	1.48230	− .15439	54.1935	.00072
5.5	1.46872	− .14205	58.4655	.00064
5.6	1.45667	− .12867	63.1018	.00057
5.7	1.44620	− .11441	68.1350	.00051
5.8	1.43736	− .09944	73.6008	.00045
5.9	1.43018	− .08393	79.5382	.00040
6.0	1.42469	− .06806	85.9898	.00036
6.1	1.42087	− .05198	93.0020	.00032
6.2	1.41871	− .03587	100.626	.00029
6.3	1.41817	− .01989	108.916	.00026
6.4	1.41922	− .00418	117.935	.00023
6.5	1.42179	+ .01110	127.747	.00020
6.6	1.42582	+ .02582	138.426	.00018
6.7	1.43121	.03986	150.050	.00016
6.8	1.43787	.05308	162.707	.00014
6.9	1.44570	.06539	176.491	.00013
7.0	1.45460	.07670	191.505	.00012
7.1	1.46443	.08691	207.863	.00010
7.2	1.47509	.09596	225.688	.00009
7.3	1.48644	.10379	245.116	.00008
7.4	1.49834	.11036	266.296	.00007
7.5	1.51068	.11563	289.388	.00007
7.6	1.52331	.11960	314.572	.00006
7.7	1.53611	.12225	342.040	.00005
7.8	1.54894	.12359	372.006	.00005
7.9	1.56167	.12364	404.701	.00004
8.0	1.57419	.12243	440.380	.00004
8.1	1.58637	.12002	479.322	.00003
8.2	1.59810	.11644	521.831	.00003
8.3	1.60928	.11177	568.242	.00003
8.4	1.61981	.10607	681.919	.00002
8.5	1.62960	.09943	674.264	.00002
8.6	1.63857	.09194	734.714	.00002
8.7	1.64665	.08368	800.749	.00002
8.8	1.65379	.07476	872.895	.00002
8.9	1.65993	.06528	951.728	.00001
9.0	1.66504	.05535	1037.88	.00001
9.1	1.66908	.04507	1132.04	.00001
9.2	1.67205	.03455	1234.96	.00001
9.3	1.67393	.02391	1347.48	.00001
9.4	1.67473	.01325	1470.51	.00001

SINE, COSINE, AND EXPONENTIAL INTEGRALS (Continued)

x	$Si(x)$	$Ci(x)$	$Ei(x)$	$-Ei(-x)$
9.5	1.67446	.00268	1605.03	.00001
9.6	1.67316	− .00771	1752.14	.00001
9.7	1.67084	− .01780	1913.05	.00001
9.8	1.66757	− .02752	2089.05	.00001
9.9	1.66338	− .03676	2281.58	.00000
10.0	1.65835	− .04546	2492.23	.00000
10.5	1.62294	− .07828	3883.74	.00000
11.0	1.57831	− .08956	6071.41	.00000
11.5	1.53572	− .07857	9518.20	.00000
12.0	1.50497	− .04978	14959.5	.00000
12.5	1.49234	− .01141	23565.1	.00000
13.0	1.49936	+ .02676	37197.7	.00000
13.5	1.52291	+ .05576	58827.0	.00000
14.0	1.55621	.06940	93193.0	.00000
14.5	1.59072	.06554	147866.	.00000
15.0	1.61819	.04628	234955.	.00000

ORTHOGONAL POLYNOMIALS

LEGENDRE POLYNOMIALS

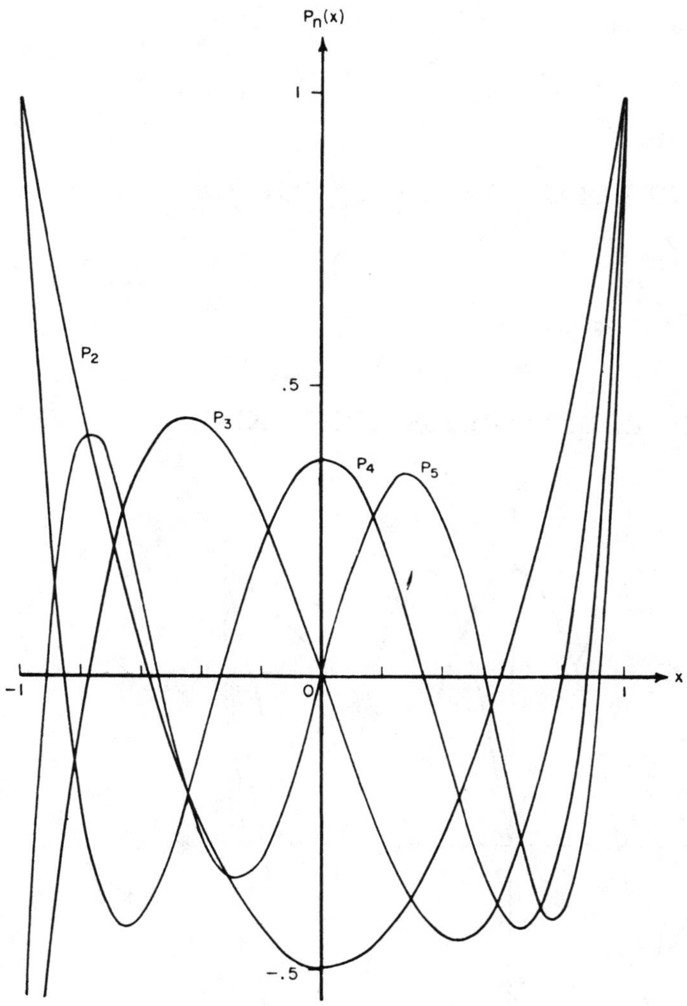

Legendre Polynomials $P_n(x)$, $n=2(1)5$.

SYMBOL: $P_n(x)$

INTERVAL: $[-1, 1]$

DIFFERENTIAL EQUATION: $(1 - x^2)y'' - 2xy' + n(n + 1)y = 0$

$$y = P_n(x)$$

EXPLICIT EXPRESSION: $P_n(x) = \frac{1}{2^n} \sum_{m=0}^{[n/2]} (-1)^m \binom{n}{m} \binom{2n - 2m}{n} x^{n-2m}$

RECURRENCE RELATION: $(n+1)P_{n+1}(x) = (2n+1)xP_n(x) - nP_{n-1}(x)$

WEIGHT: 1

STANDARDIZATION: $P_n(1) = 1$

NORM: $\int_{-1}^{+1} [P_n(x)]^2\,dx = \frac{2}{2n+1}$

RODRIGUES' FORMULA: $P_n(x) = \frac{(-1)^n}{2^n n!}\frac{d^n}{dx^n}\{(1-x^2)^n\}$

GENERATING FUNCTION: $R^{-1} = \sum_{n=0}^{\infty} P_n(x)z^n; \; -1 < x < 1, \; |z| < 1,$

$$R = \sqrt{1 - 2xz + z^2}.$$

INEQUALITY: $|P_n(x)| \le 1, \; -1 \le x \le 1.$

●

CHEBYSHEV POLYNOMIALS, FIRST KIND

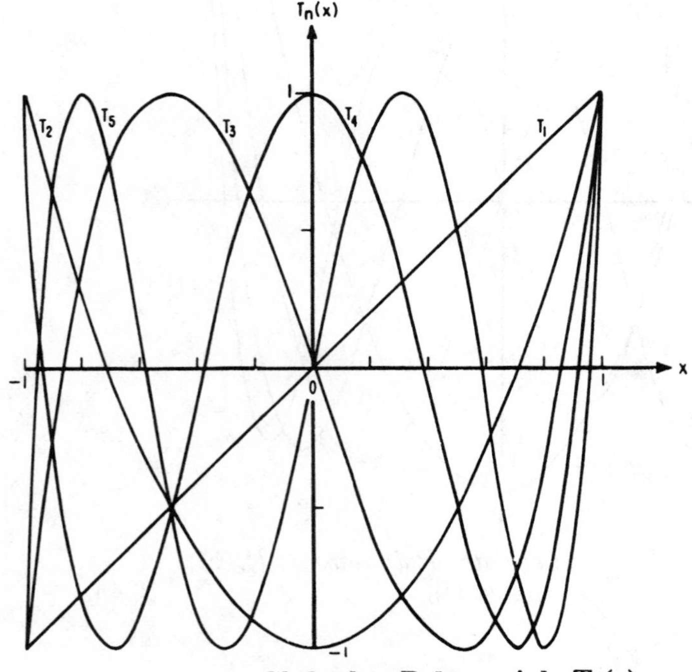

Chebyshev Polynomials $T_n(x)$,
$n=1(1)5$.

SYMBOL: $T_n(x)$

INTERVAL: $[-1, 1]$

DIFFERENTIAL EQUATION: $(1-x^2)y'' - xy' + n^2y = 0$

$$y = T_n(x)$$

EXPLICIT EXPRESSION: $\frac{n}{2}\sum_{m=0}^{[n/2]} (-1)^m \frac{(n-m-1)!}{m!(n-2m)!}(2x)^{n-2m} = \cos(n \arccos x)$
$= T_n(x)$

RECURRENCE RELATION: $T_{n+1}(x) = 2xT_n(x) - T_{n-1}(x)$

WEIGHT: $(1-x^2)^{-1/2}$

STANDARDIZATION: $T_n(1) = 1$

NORM: $\int_{-1}^{+1} (1-x^2)^{-1/2}[T_n(x)]^2\,dx = \begin{cases} \pi/2, & n \neq 0 \\ \pi, & n = 0 \end{cases}$

RODRIGUES' FORMULA: $\frac{(-1)^n(1-x^2)^{1/2}\sqrt{\pi}}{2^{n+1}\Gamma(n+\frac{1}{2})}\frac{d^n}{dx^n}\{(1-x^2)^{n-(1/2)}\} = T_n(x)$

GENERATING FUNCTION: $\frac{1-xz}{1-2xz+z^2} = \sum_{n=0}^{\infty} T_n(x)z^n, \; -1 < x < 1, \; |z| < 1.$

INEQUALITY: $|T_n(x)| \leq 1, \; -1 \leq x \leq 1.$

•

CHEBYSHEV POLYNOMIALS, SECOND KIND

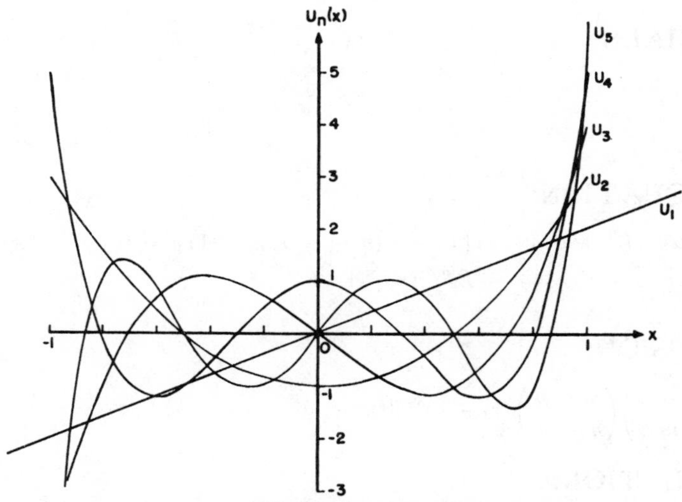

Chebyshev Polynomials $U_n(x)$, $n=1(1)5$.

SYMBOL: $U_n(x)$

INTERVAL: $[-1, 1]$

DIFFERENTIAL EQUATION: $(1-x^2)y'' - 3xy' + n(n+2)y = 0$
$y = U_n(x)$

EXPLICIT EXPRESSION:

$$U_n(x) = \sum_{m=0}^{[n/2]} (-1)^m \frac{(m-n)!}{m!(n-2m)!}(2x)^{n-2m}$$

$$U_n(\cos\theta) = \frac{\sin[(n+1)\theta]}{\sin\theta}$$

RECURRENCE RELATION: $U_{n+1}(x) = 2xU_n(x) - U_{n-1}(x)$

WEIGHT: $(1-x^2)^{1/2}$

STANDARDIZATION: $U_n(1) = n+1$

NORM: $\displaystyle\int_{-1}^{+1} (1-x^2)^{1/2}[U_n(x)]^2\,dx = \frac{\pi}{2}$

RODRIGUES' FORMULA: $\displaystyle U_n(x) = \frac{(-1)^n(n+1)\sqrt{\pi}}{(1-x^2)^{1/2}2^{n+1}\Gamma(n+\frac{3}{2})}\frac{d^n}{dx^n}\{(1-x^2)^{n+(1/2)}\}$

GENERATING FUNCTION: $\displaystyle\frac{1}{1-2xz+z^2} = \sum_{n=0}^{\infty} U_n(x)z^n,\ -1 < x < 1,\ |z| < 1.$

INEQUALITY: $|U_n(x)| \leq n+1,\ -1 \leq x \leq 1.$

•

JACOBI POLYNOMIALS

SYMBOL: $P_n^{(\alpha,\beta)}(x)$

INTERVAL: $[-1, 1]$

DIFFERENTIAL EQUATION:

$$(1-x^2)y'' + [\beta - \alpha - (\alpha+\beta+2)x]y' + n(n+\alpha+\beta+1)y = 0$$
$$y = P_n^{(\alpha,\beta)}(x)$$

EXPLICIT EXPRESSION:

$$P_n^{(\alpha,\beta)}(x) = \frac{1}{2^n}\sum_{m=0}^{n}\binom{n+\alpha}{m}\binom{n+\beta}{n-m}(x-1)^{n-m}(x+1)^m$$

RECURRENCE RELATION:

$$\begin{aligned}2(n+1)(n+\alpha+\beta+1)(2n+\alpha+\beta)P_{n+1}^{(\alpha,\beta)}(x) \\ = (2n+\alpha+\beta+1)[(\alpha^2-\beta^2)+(2n+\alpha+\beta+2) \\ \times(2n+\alpha+\beta)x]P_n^{(\alpha,\beta)}(x) \\ -2(n+\alpha)(n+\beta)(2n+\alpha+\beta+2)P_{n-1}^{(\alpha,\beta)}(x)\end{aligned}$$

WEIGHT: $(1-x)^\alpha(1+x)^\beta;\ \alpha,\beta > 1$

STANDARDIZATION: $P_n^{(\alpha,\beta)}(x) = \binom{n+\alpha}{n}$

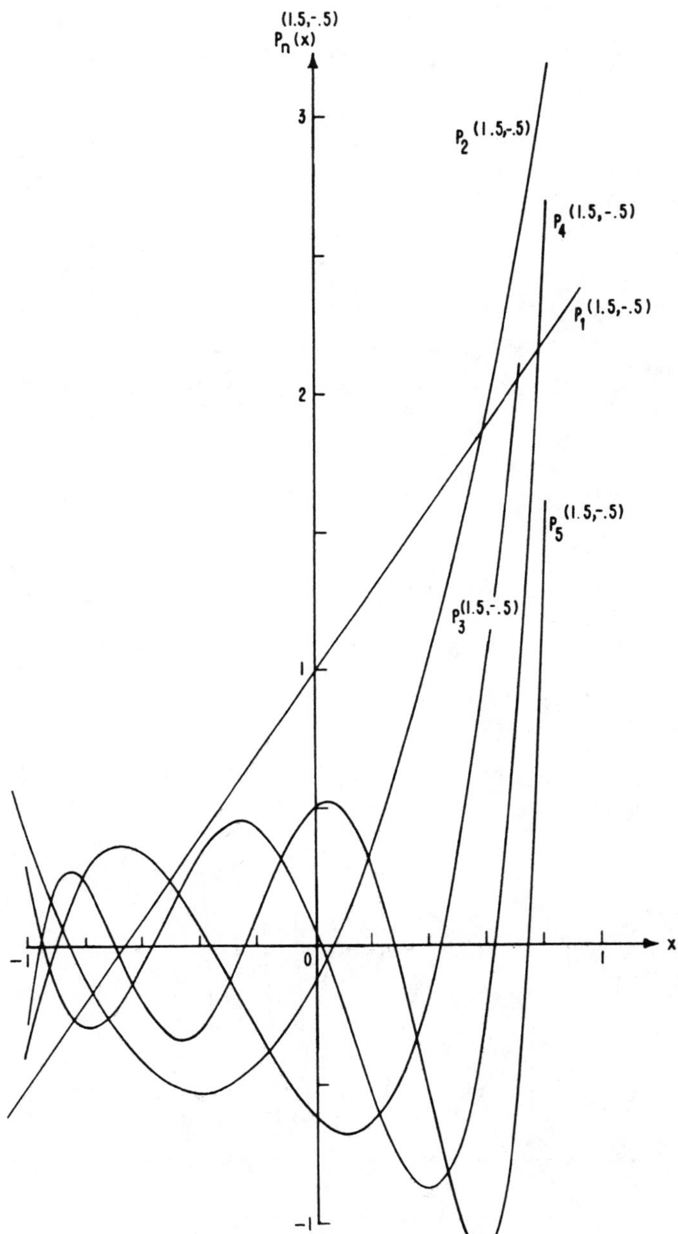

Jacobi Polynomials $P_n^{(\alpha,\beta)}(x)$,
$\alpha=1.5$, $\beta=-.5$, $n=1(1)5$.

NORM: $$\int_{-1}^{+1} (1-x)^\alpha(1+x)^\beta[P_n^{(\alpha,\beta)}(x)]^2\,dx = \frac{2^{\alpha+\beta+1}\Gamma(n+\alpha+1)\Gamma(n+\beta+1)}{(2n+\alpha+\beta+1)n!\,\Gamma(n+\alpha+\beta+1)}$$

RODRIGUES' FORMULA:

$$P_n^{(\alpha,\beta)}(x) = \frac{(-1)^n}{2^n n!(1-x)^\alpha(1+x)^\beta}\frac{d^n}{dx^n}\{(1-x)^{n+\alpha}(1+x)^{n+\beta}\}$$

GENERATING FUNCTION:

$$R^{-1}(1 - z + R)^{-\alpha}(1 + z + R)^{-\beta} = \sum_{n=0}^{\infty} 2^{-\alpha-\beta} P_n^{(\alpha,\beta)}(x) z^n,$$

$$R = \sqrt{1 - 2xz + z^2}, \; |z| < 1$$

INEQUALITY:

$$\max_{-1 \le x \le 1} |P_n^{(\alpha,\beta)}(x)| = \begin{cases} \binom{n+q}{n} \sim n^q \text{ if } q = \max(\alpha, \beta) \ge -\frac{1}{2} \\ |P_n^{(\alpha,\beta)}(x')| \sim n^{-1/2} \text{ if } q < -\frac{1}{2} \\ x' \text{ is one of the two maximum points nearest} \\ \dfrac{\beta - \alpha}{\alpha + \beta + 1} \end{cases}$$

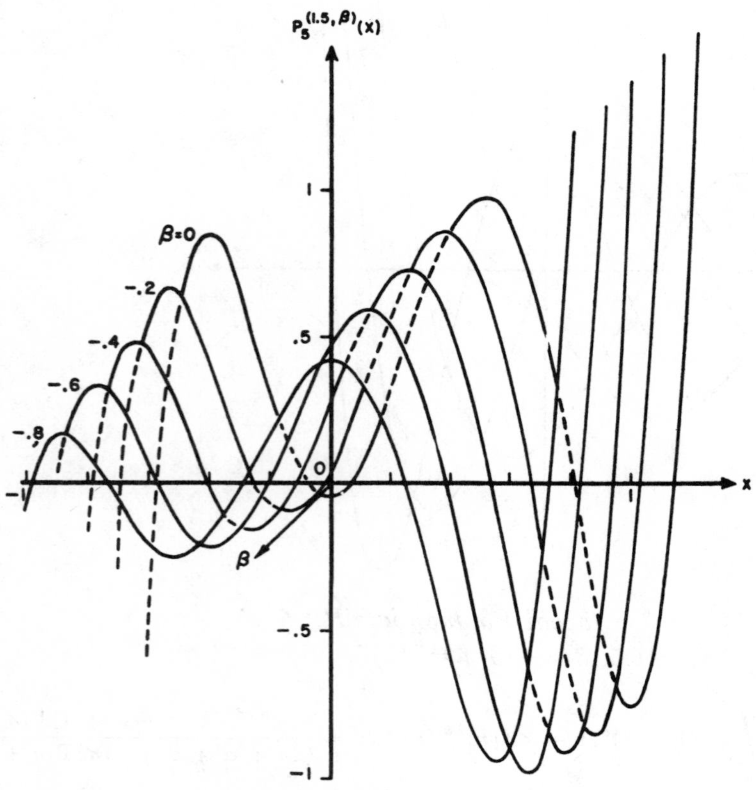

Jacobi Polynomials $P_n^{(\alpha,\beta)}(x)$,
$\alpha=1.5$, $\beta=-.8(.2)0$, $n=5$.

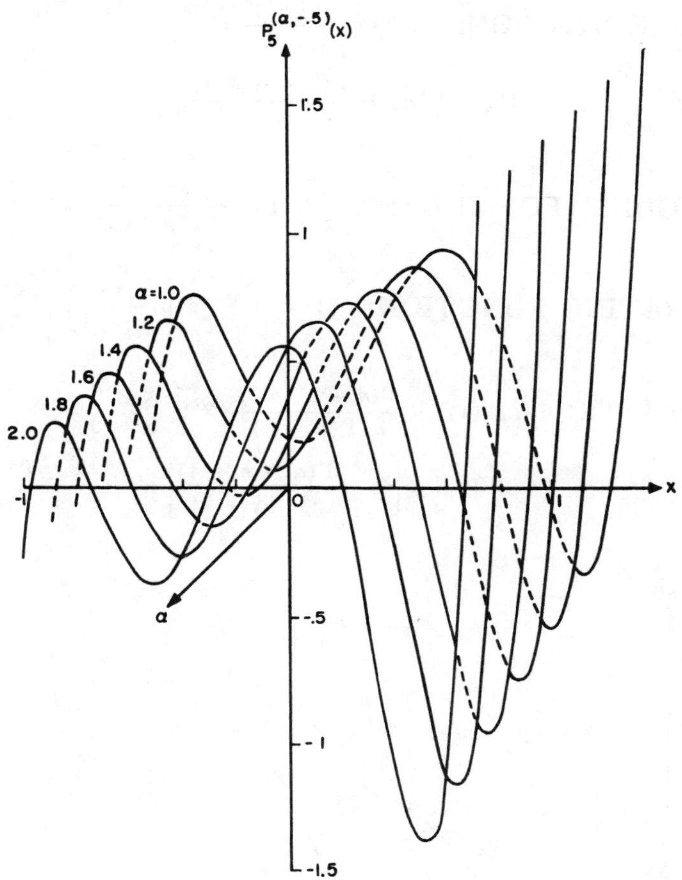

Jacobi Polynomials $P_n^{(\alpha,\beta)}(x)$,
$\alpha=1(.2)2,\ \beta=-.5,\ n=5.$

•

GENERALIZED LAGUERRE POLYNOMIALS

SYMBOL: $L_n^{(\alpha)}(x)$

INTERVAL: $[0, \infty]$

DIFFERENTIAL EQUATION: $xy'' + (\alpha + 1 - x)y' + ny = 0$

$$y = L_n^{(\alpha)}(x)$$

EXPLICIT EXPRESSION: $L_n^{(\alpha)}(x) = \sum_{m=0}^{n} (-1)^m \binom{n+\alpha}{n-m} \frac{1}{m!} x^m$

RECURRENCE RELATION:

$$(n+1)L_{n+1}^{(\alpha)}(x) = [(2n+\alpha+1) - x]L_n^{(\alpha)}(x) - (n+\alpha)L_{n-1}^{(\alpha)}(x)$$

WEIGHT: $x^{\alpha}e^{-x},\ \alpha > -1$

STANDARDIZATION: $L_n^{(\alpha)}(x) = \frac{(-1)^n}{n!} x^n + \cdots$

NORM: $\int_0^\infty x^\alpha e^{-x} [L_n^{(\alpha)}(x)]^2 \, dx = \frac{\Gamma(n + \alpha + 1)}{n!}$

RODRIGUES' FORMULA: $L_n^{(\alpha)}(x) = \frac{1}{n! \, x^\alpha e^{-x}} \frac{d^n}{dx^n} \{x^{n+\alpha} e^{-x}\}$

GENERATING FUNCTION: $(1 - z)^{-\alpha-1} \exp\left(\frac{xz}{z - 1}\right) = \sum_{n=0}^{\infty} L_n^{(\alpha)}(x) z^n$

INEQUALITY: $|L_n^{(\alpha)}(x)| \le \frac{\Gamma(n + \alpha + 1)}{n! \, \Gamma(\alpha + 1)} e^{x/2}; \quad x \ge 0, \ \alpha > 0$

$$|L_n^{(\alpha)}(x)| \le \left[2 - \frac{\Gamma(\alpha + n + 1)}{n! \, \Gamma(\alpha + 1)}\right] e^{x/2}; \quad x \ge 0, \ -1 < \alpha < 0$$

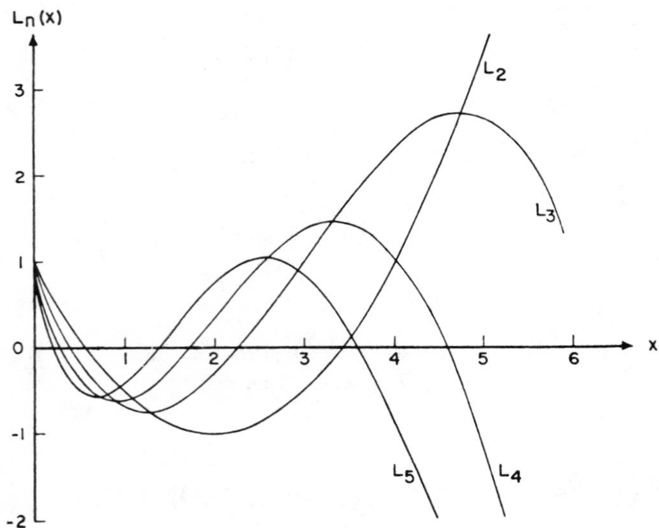

Laguerre Polynomials $L_n(x)$, $n = 2(1)5$.

•

HERMITE POLYNOMIALS

SYMBOL: $H_n(x)$

INTERVAL: $[-\infty, \infty]$

DIFFERENTIAL EQUATION: $y'' - 2xy' + 2ny = 0$

EXPLICIT EXPRESSION: $H_n(x) = \sum_{m=0}^{[n/2]} \frac{(-1)^m \, n! \, (2x)^{n-2m}}{m! \, (n - 2m)!}$

RECURRENCE RELATION: $H_{n+1}(x) = 2xH_n(x) - 2nH_{n-1}(x)$

WEIGHT: e^{-x^2}

STANDARDIZATION: $H_n(1) = 2^n x^n + \ldots$

NORM: $\int_{-\infty}^{\infty} e^{-x^2} [H_n(x)]^2 \, dx = 2^n n! \sqrt{\pi}$

RODRIGUES' FORMULA: $H_n(x) = (-1)^n e^{x^2} \frac{d^n}{dx^n}(e^{-x^2})$

GENERATING FUNCTION: $e^{-z^2 + 2zx} = \sum_{n=0}^{\infty} H_n(x) \frac{z^n}{n!}$

INEQUALITY: $|H_n(x)| < e^{\frac{x^2}{2}} k \, 2^{n/2} \sqrt{n!} \quad k \approx 1.086435$

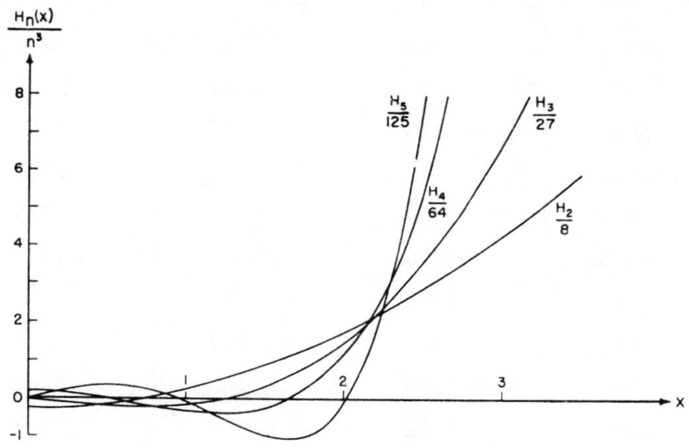

Hermite Polynomials $\frac{H_n(x)}{n^3}$,
$n=2(1)5$.

LEGENDRE POLYNOMIALS $P_n(x)$
$[P_0(x) = 1, \quad P_1(x) = x]$

x	$P_2(x)$	$P_3(x)$	$P_4(x)$	$P_5(x)$
.00	−.5000	.0000	.3750	.0000
.05	−.4963	−.0747	.3657	.0927
.10	−.4850	−.1475	.3379	.1788
.15	−.4663	−.2166	.2928	.2523
.20	−.4400	−.2800	.2320	.3075
.25	−.4063	−.3359	.1577	.3397
.30	−.3650	−.3825	.0729	.3454
.35	−.3163	−.4178	−.0187	.3225
.40	−.2600	−.4400	−.1130	.2706
.45	−.1963	−.4472	−.2050	.1917
.50	−.1250	−.4375	−.2891	.0898
.55	−.0463	−.4091	−.3590	−.0282
.60	.0400	−.3600	−.4080	−.1526
.65	.1338	−.2884	−.4284	−.2705
.70	.2350	−.1925	−.4121	−.3652
.75	.3438	−.0703	−.3501	−.4164
.80	.4600	.0800	−.2330	−.3995
.85	.5838	.2603	−.0506	−.2857
.90	.7150	.4725	.2079	−.0411
.95	.8538	.7184	.5541	.3727
1.00	1.0000	1.0000	1.0000	1.0000

LEGENDRE POLYNOMIALS $P_n(\cos\theta)$
$[P_0(\cos\theta) = 1]$

θ	$P_1(\cos\theta)$	$P_2(\cos\theta)$	$P_3(\cos\theta)$	$P_4(\cos\theta)$	$P_5(\cos\theta)$
0°	1.0000	1.0000	1.0000	1.0000	1.0000
5°	.9962	.9886	.9773	.9623	.9437
10°	.9848	.9548	.9106	.8532	.7840
15°	.9659	.8995	.8042	.6847	.5471
20°	.9397	.8245	.6649	.4750	.2715
25°	.9063	.7321	.5016	.2465	.0009
30°	.8660	.6250	.3248	.0234	−.2233
35°	.8192	.5065	.1454	−.1714	−.3691
40°	.7660	.3802	−.0252	−.3190	−.4197
45°	.7071	.2500	−.1768	−.4063	−.3757
50°	.6428	.1198	−.3002	−.4275	−.2545
55°	.5736	−.0065	−.3886	−.3852	−.0868
60°	.5000	−.1250	−.4375	−.2891	.0898
65°	.4226	−.2321	−.4452	−.1552	.2381
70°	.3420	−.3245	−.4130	−.0038	.3281
75°	.2588	−.3995	−.3449	.1434	.3427
80°	.1737	−.4548	−.2474	.2659	.2810
85°	.0872	−.4886	−.1291	.3468	.1577
90°	.0000	−.5000	.0000	.3750	.0000

BERNOULLI AND EULER NUMBERS

There are numerous sets of defined numbers and polynomials, among which the more important ones are those classified under the names of Bernoulli and Euler.

The *Bernoulli Polynomials* are generated by the defining condition

$$\frac{te^{tx}}{e^t - 1} = B_0(x) + B_1(x)t + B_2(x)\frac{t^2}{2!} + B_3(x)\frac{t^3}{3!} + \cdots$$

$$B_0(x) = 1, \quad B_1(x) = x - \tfrac{1}{2}, \quad B_2(x) = x^2 - x + \tfrac{1}{6}$$

$$B_3(x) = x^3 - \frac{3}{2}x^2 + \frac{x}{2}, \quad B_4(x) = x^4 - 2x^3 + x^2 - \tfrac{1}{30}, \ldots$$

The following useful relations should be noted

$$B_n'(x) = nB_{n-1}(x), \quad \int_a^x B_n(x)\,dx = \frac{1}{n+1}[B_{n+1}(x) - B_{n+1}(a)]$$

The Bernoulli numbers $B_0, B_1, B_2, \ldots,$ can be obtained by putting $x = 0$ in the respective polynomials. A simpler method is to use the generating function

$$\frac{x}{e^x - 1} = \sum_{n=0}^{\infty} \frac{B_n x^n}{n!}$$

and read the coefficients from the expansion. Here,

$$B_0 = 1, \quad B_1 = -\tfrac{1}{2}, \quad B_2 = \tfrac{1}{6}, \quad B_4 = -\tfrac{1}{30},$$

$$B_6 = \tfrac{1}{42}, \quad B_8 = -\tfrac{1}{30}, \ldots B_{2n+1} = 0 \qquad (n \geq 1)$$

An important application for the Bernoulli coefficients is in its use involved for the Euler-Maclaurin Sum Formula:

$$\sum_{x=1}^{n-1} f(x) = \int_1^n f(x)\,dx + \left[\sum_{i=1}^{\infty} \frac{B_i}{i!} f^{(i-1)}(x)\right]_{x=1}^{x=n}$$

$$= \left[\int f(x)\,dx - \frac{1}{2}f(x) + \frac{1}{12}f'(x) - \frac{1}{720}f'''(x) + \frac{1}{30{,}240}f^{(V)}(x) - \frac{1}{1{,}209{,}600}f^{(VII)}(x)\right]_{x=1}^{x=n}$$

Examples:

$$\sum_{x=1}^{n-1} \sqrt{x} = \sqrt{n}\left\{\frac{2}{3}n - \frac{1}{2} + \frac{1}{24n} - \frac{1}{1{,}920n^3} + \frac{1}{9{,}216n^5} - \frac{11}{163{,}840n^7}\right\} - 0.207{,}886{,}224{,}977{,}355$$

correct to 12 places for $n \geq 10$.

$\log_e(x!) = \log_e\Gamma(x + 1)$

$$= \left(x + \frac{1}{2}\right)\log_e x - x + \frac{1}{12x} - \frac{1}{360x^3} + \frac{1}{1{,}260x^5} - \frac{1}{1{,}680x^7} + 0.918{,}938{,}533{,}205$$

accurate to 12 places for $x \geq 10$.

$$f(x)\text{(digamma function)} = \frac{d \log \Gamma(x)}{dx}$$

$$= 1 + \frac{1}{2} + \frac{1}{3} + \cdots + \frac{1}{x-1} - \gamma$$

(Euler Constant) for x integer.

By Euler-Maclaurin

$$f(x) = \log_e x - \frac{1}{2x} - \frac{1}{12x^2} + \frac{1}{120x^4} - \frac{1}{252x^6} + \frac{1}{240x^8} - \frac{5}{660x^{10}} + \frac{691}{32{,}760x^{12}}$$

correct to 12 places for $x \geq 10$.

ASYMPTOTIC FORMULA FOR BERNOULLI NUMBERS

$$B_n \sim 4n^{2n}(\pi e)^{-2n}\sqrt{\pi n}$$

The Euler numbers together with their respective polynomials can be generated from

$$\frac{2e^{tx}}{e^t + 1} = \sum_{i=0}^{\infty} E_i(x)\frac{t^i}{i!}$$

and the relation

$$x^n = \frac{1}{2}[E_n(x+1) + E_n(x)]$$

The Euler polynomials are

$$E_0(x) = 1, \quad E_1(x) = x - \frac{1}{2}, \quad E_2(x) = x^2 - x$$

$$E_3(x) = x^3 - \frac{3}{2}x^2 + \frac{1}{4}, \quad E_4(x) = x^4 - 2x^3 + x$$

$$E_5(x) = x^5 - \frac{5}{2}x^4 + \frac{5}{2}x^2 - \frac{1}{2}$$

Tables which follow record the first fifteen polynomials of $B_k(x)$ and $E_k(x)$. The first sixty Bernoulli and Euler numbers are given in a separate table. The value of $x^n/n!$ is also important and this is given in a succeeding table for values of x from 1 to 9 and n from 1 to 50.

RELATIONSHIPS OF BERNOULLI AND EULER NUMBERS

$$\binom{2n+1}{2}2^2B_1 - \binom{2n+1}{4}2^4B_2 + \binom{2n+1}{6}2^6B_3 - \cdots (-1)^{n-1}(2n+1)2^{2n}B_n = 2n$$

$$E_n = \binom{2n}{2}E_{n-1} - \binom{2n}{4}E_{n-2} + \binom{2n}{6}E_{n-3} - \cdots (-1)^n$$

$$B_n = \frac{2n}{2^{2n}(2^{2n}-1)}\left\{\binom{2n-1}{1}E_{n-1} - \binom{2n-1}{3}E_{n-2}\right.$$

$$\left. + \binom{2n-1}{5}E_{n-3} - \cdots (-1)^{n-1}\right\}$$

SERIES INVOLVING BERNOULLI AND EULER NUMBERS

$$B_n = \frac{(2n)!}{2^{2n-1}\pi^{2n}}\left\{1 + \frac{1}{2^{2n}} + \frac{1}{3^{2n}} + \cdots\right\}$$

$$B_n = \frac{2(2n)!}{(2^{2n}-1)\pi^{2n}}\left\{1 + \frac{1}{3^{2n}} + \frac{1}{5^{2n}} + \cdots\right\}$$

$$B_n = \frac{(2n)!}{(2^{2n-1}-1)\pi^{2n}}\left\{1 - \frac{1}{2^{2n}} + \frac{1}{3^{2n}} - \cdots\right\}$$

$$E_n = \frac{2^{2n+2}(2n)!}{\pi^{2n+1}}\left\{1 - \frac{1}{3^{2n+1}} + \frac{1}{5^{2n+1}} - \cdots\right\}$$

BERNOULLI NUMBERS

$B_n = N/D$

n	N	D	B_n
0	1	1	(0) 1.0000 00000
1	−1	2	(−1) −5.0000 00000*
2	1	6	(−1) 1.6666 66667
4	−1	30	(−2) −3.3333 33333
6	1	42	(−2) 2.3809 52381
8	−1	30	(−2) −3.3333 33333
10	5	66	(−2) 7.5757 57576
12	−691	2730	(−1) −2.5311 35531
14	7	6	(0) 1.1666 66667
16	−3617	510	(0) −7.0921 56863
18	43867	798	(1) 5.4971 17794
20	−1 74611	330	(2) −5.2912 42424
22	8 54513	138	(3) 6.1921 23188
24	−2363 64091	2730	(4) −8.6580 25311
26	85 53103	6	(6) 1.4255 17167
28	−2 37494 61029	870	(7) −2.7298 23107
30	861 58412 76005	14322	(8) 6.0158 08739
32	−770 93210 41217	510	(10) −1.5116 31577
34	257 76878 58367	6	(11) 4.2961 46431
36	−26315 27155 30534 77373	19 19190	(13) −1.3711 65521
38	2 92999 39138 41559	6	(14) 4.8833 23190
40	−2 61082 71849 64491 22051	13530	(16) −1.9296 57934
42	15 20097 64391 80708 02691	1806	(17) 8.4169 30476
44	−278 33269 57930 10242 35023	690	(19) −4.0338 07185
46	5964 51111 59391 21632 77961	282	(21) 2.1150 74864
48	−560 94033 68997 81768 62491 27547	46410	(23) −1.2086 62652
50	49 50572 05241 07964 82124 77525	66	(24) 7.5008 66746
52	−80116 57181 35489 95734 79249 91853	1590	(26) −5.0387 78101
54	29 14996 36348 84862 42141 81238 12691	798	(28) 3.6528 77648
56	−2479 39292 93132 26753 68541 57396 63229	870	(30) −2.8498 76930
58	84483 61334 88800 41862 04677 59940 36021	354	(32) 2.3865 42750
60	−121 52331 40483 75557 20403 04994 07982 02460 41491	567 86730	(34) −2.1399 94926

*The floating decimal point notation is used here. For example for $n = 1$, $B_1 = -\frac{1}{2} = -.500000 = (-5.00000)(10^{-1}) = (-1) - 5.0000,0000$.

EULER NUMBERS

n	En
0	1
2	−1
4	5
6	−61
8	1385
10	−50521
12	27 02765
14	−1993 60981
16	1 93915 12145
18	−240 48796 75441
20	37037 11882 37525
22	−69 34887 43931 37901
24	15514 53416 35570 86905
26	−40 87072 50929 31238 92361
28	12522 59641 40362 98654 68285
30	−44 15438 93249 02310 45536 82821
32	17751 93915 79539 28943 66647 89665
34	−80 72329 92358 87898 06216 82474 53281
36	41222 06033 95177 02122 34707 96712 59045
38	−234 89580 52704 31082 52017 82857 61989 47741
40	1 48511 50718 11498 00178 77156 78140 58266 84425
42	−1036 46227 33519 61211 93979 57304 74518 59763 10201
44	7 94757 94225 97592 70360 80405 10088 07061 95192 73805
46	−6667 53751 66855 44977 43502 84747 73748 19752 41076 84661
48	60 96278 64556 85421 58691 68574 28768 43153 97653 90444 35185
50	−60532 85248 18862 18963 14383 78511 16490 88103 49822 51468 15121
52	650 61624 86684 60884 77158 70634 08082 29834 83644 23676 53855 76565
54	−7 54665 99390 08739 09806 14325 65889 73674 42122 40024 71169 98586 45581
56	9420 32189 64202 41204 20228 62376 90583 22720 93888 52599 64600 93949 05945
58	−126 22019 25180 62187 19903 40923 72874 89255 48234 10611 91825 59406 99649 20041
60	181089 11496 57923 04965 45807 74165 21586 88733 48734 92363 14106 00809 54542 31325

COEFFICIENTS b_k OF THE BERNOULLI POLYNOMIALS $B_n(x) = \sum_{k=0}^{n} b_k x^k$

$n \backslash k$	0	1	2	3	4	5	6	7	8	9	10	11	12	13	14	15
0	1															
1	$-\frac{1}{2}$	1														
2	$\frac{1}{6}$	-1	1													
3	0	$\frac{1}{2}$	$-\frac{3}{2}$	1												
4	$-\frac{1}{30}$	0	1	-2	1											
5	0	$-\frac{1}{6}$	0	$\frac{5}{3}$	$-\frac{5}{2}$	1										
6	$\frac{1}{42}$	0	$-\frac{1}{2}$	0	$\frac{5}{2}$	-3	1									
7	0	$\frac{1}{6}$	0	$-\frac{7}{6}$	0	$\frac{7}{2}$	$-\frac{7}{2}$	1								
8	$-\frac{1}{30}$	0	$\frac{2}{3}$	0	$-\frac{7}{3}$	0	$\frac{14}{3}$	-4	1							
9	0	$-\frac{3}{10}$	0	2	0	$-\frac{21}{5}$	0	6	$-\frac{9}{2}$	1						
10	$\frac{5}{66}$	0	$-\frac{3}{2}$	0	5	0	-7	0	$\frac{15}{2}$	-5	1					
11	0	$\frac{5}{6}$	0	$-\frac{11}{2}$	0	11	0	-11	0	$\frac{55}{6}$	$-\frac{11}{2}$	1				
12	$-\frac{691}{2730}$	0	5	0	$-\frac{33}{2}$	0	22	0	$-\frac{33}{2}$	0	11	-6	1			
13	0	$-\frac{691}{210}$	0	$\frac{65}{3}$	0	$-\frac{429}{10}$	0	$\frac{286}{7}$	0	$-\frac{143}{6}$	0	13	$-\frac{13}{2}$	1		
14	$\frac{7}{6}$	0	$-\frac{691}{30}$	0	$\frac{455}{6}$	0	$-\frac{1001}{10}$	0	$\frac{143}{2}$	0	$-\frac{1001}{30}$	0	$\frac{91}{6}$	-7	1	
15	0	$\frac{35}{2}$	0	$-\frac{691}{6}$	0	$\frac{455}{2}$	0	$-\frac{429}{2}$	0	$\frac{715}{6}$	0	$-\frac{91}{2}$	0	$\frac{35}{2}$	$-\frac{15}{2}$	1

COEFFICIENTS e_k OF THE EULER POLYNOMIALS $E_n(x) = \sum_{k=0}^{n} e_k x^k$

$n \backslash k$	0	1	2	3	4	5	6	7	8	9	10	11	12	13	14	15
0	1															
1	$-\frac{1}{2}$	1														
2	0	-1	1													
3	$\frac{1}{4}$	0	$-\frac{3}{2}$	1												
4	0	1	0	-2	1											
5	$-\frac{1}{2}$	0	$\frac{5}{2}$	0	$-\frac{5}{2}$	1										
6	0	-3	0	5	0	-3	1									
7	$\frac{17}{8}$	0	$-\frac{21}{2}$	0	$\frac{35}{4}$	0	$-\frac{7}{2}$	1								
8	0	17	0	-28	0	14	0	-4	1							
9	$-\frac{31}{2}$	0	$\frac{153}{2}$	0	-63	0	21	0	$-\frac{9}{2}$	1						
10	0	-155	0	255	0	-126	0	30	0	-5	1					
11	$\frac{691}{4}$	0	$-\frac{1705}{2}$	0	$\frac{2805}{4}$	0	-231	0	$\frac{165}{4}$	0	$-\frac{11}{2}$	1				
12	0	2073	0	-3410	0	1683	0	-396	0	55	0	-6	1			
13	$-\frac{5461}{2}$	0	$\frac{26949}{2}$	0	$-\frac{22165}{2}$	0	$\frac{7293}{2}$	0	$-\frac{1287}{2}$	0	$\frac{143}{2}$	0	$-\frac{13}{2}$	1		
14	0	-38227	0	62881	0	-31031	0	7293	0	-1001	0	91	0	-7	1	
15	$\frac{929569}{16}$	0	$-\frac{573405}{2}$	0	$\frac{943215}{4}$	0	$-\frac{155155}{2}$	0	$\frac{109395}{8}$	0	$-\frac{3003}{2}$	0	$\frac{455}{4}$	0	$-\frac{15}{2}$	1

BERNOULLI AND EULER POLYNOMIALS

$x^n/n!$

$n\backslash x$	2	3	4	5
1	(0) 2.0000 00000	(0) 3.0000 00000	(0) 4.0000 00000	(0) 5.0000 00000
2	(0) 2.0000 00000	(0) 4.5000 00000	(0) 8.0000 00000	(1) 1.2500 00000
3	(0) 1.3333 33333	(0) 4.5000 00000	(1) 1.0666 66667	(1) 2.0833 33333
4	(−1) 6.6666 66667	(0) 3.3750 00000	(1) 1.0666 66667	(1) 2.6041 66667
5	(−1) 2.6666 66667	(0) 2.0250 00000	(0) 8.5333 33333	(1) 2.6041 66667
6	(−2) 8.8888 88889	(0) 1.0125 00000	(0) 5.6888 88889	(1) 2.1701 38889
7	(−2) 2.5396 82540	(−1) 4.3392 85714	(0) 3.2507 93651	(1) 1.5500 99206
8	(−3) 6.3492 06349	(−1) 1.6272 32143	(0) 1.6253 96825	(0) 9.6881 20040
9	(−3) 1.4109 34744	(−2) 5.4241 07143	(−1) 7.2239 85891	(0) 5.3822 88911
10	(−4) 2.8218 69489	(−2) 1.6272 32144	(−1) 2.8895 94356	(0) 2.6911 44455
11	(−5) 5.1306 71797	(−3) 4.4379 05844	(−1) 1.0507 61584	(0) 1.2232 47480
12	(−6) 8.5511 19662	(−3) 1.1094 76461	(−2) 3.5025 38614	(−1) 5.0968 64499
13	(−6) 1.3155 56871	(−4) 2.5603 30295	(−2) 1.0777 04189	(−1) 1.9603 32500
14	(−7) 1.8793 66959	(−5) 5.4864 22060	(−3) 3.0791 54825	(−2) 7.0011 87499
15	(−8) 2.5058 22612	(−5) 1.0972 84412	(−4) 8.2110 79534	(−2) 2.3337 29166
16	(−9) 3.1322 78264	(−6) 2.0574 08272	(−4) 2.0527 69883	(−3) 7.2929 03644
17	(−10) 3.6850 33252	(−7) 3.6307 20481	(−5) 4.8300 46785	(−3) 2.1449 71660
18	(−11) 4.0944 81391	(−8) 6.0512 00801	(−5) 1.0733 43730	(−4) 5.9582 54611
19	(−12) 4.3099 80412	(−9) 9.5545 27582	(−6) 2.2596 71011	(−4) 1.5679 61740
20	(−13) 4.3099 80413	(−9) 1.4331 79137	(−7) 4.5193 42021	(−5) 3.9199 04350
21	(−14) 4.1047 43250	(−10) 2.0473 98768	(−8) 8.6082 70516	(−6) 9.3331 05595
22	(−15) 3.7315 84772	(−11) 2.7919 07410	(−8) 1.5651 40093	(−6) 2.1211 60362
23	(−16) 3.2448 56324	(−12) 3.6416 18361	(−9) 2.7219 82772	(−7) 4.6112 18179
24	(−17) 2.7040 46937	(−13) 4.5520 22952	(−10) 4.5366 37953	(−8) 9.6067 04540
25	(−18) 2.1632 37550	(−14) 5.4624 27543	(−11) 7.2586 20726	(−8) 1.9213 40908
26	(−19) 1.6640 28884	(−15) 6.3028 01010	(−11) 1.1167 10881	(−9) 3.6948 86362
27	(−20) 1.2326 13988	(−16) 7.0031 12233	(−12) 1.6543 86490	(−10) 6.8423 82151
28	(−22) 8.8043 85630	(−17) 7.5033 34535	(−13) 2.3634 09271	(−10) 1.2218 53956
29	(−23) 6.0719 90089	(−18) 7.7620 70209	(−14) 3.2598 74857	(−11) 2.1066 44751
30	(−24) 4.0479 93393	(−19) 7.7620 70209	(−15) 4.3464 99810	(−12) 3.5110 74585
31	(−25) 2.6116 08641	(−20) 7.5116 80847	(−16) 5.6083 86851	(−13) 5.6630 23524
32	(−26) 1.6322 55401	(−21) 7.0422 00795	(−17) 7.0104 83564	(−14) 8.8484 74257
33	(−28) 9.8924 56972	(−22) 6.4020 00722	(−18) 8.4975 55834	(−14) 1.3406 77918
34	(−29) 5.8190 92337	(−23) 5.6488 24167	(−19) 9.9971 24511	(−15) 1.9715 85173
35	(−30) 3.3251 95620	(−24) 4.8418 49286	(−19) 1.1425 28515	(−16) 2.8165 50246
36	(−31) 1.8473 30900	(−25) 4.0348 74405	(−20) 1.2694 76128	(−17) 3.9118 75343
37	(−33) 9.9855 72436	(−26) 3.2715 19788	(−21) 1.3724 06625	(−18) 5.2863 18032
38	(−34) 5.2555 64439	(−27) 2.5827 78779	(−22) 1.4446 38552	(−19) 6.9556 81619
39	(−35) 2.6951 61251	(−28) 1.9867 52908	(−23) 1.4816 80567	(−20) 8.9175 40539
40	(−36) 1.3475 80626	(−29) 1.4900 64681	(−24) 1.4816 80567	(−20) 1.1146 92567
41	(−38) 6.5735 64028	(−30) 1.0902 91230	(−25) 1.4455 42017	(−21) 1.3593 81180
42	(−39) 3.1302 68584	(−32) 7.7877 94498	(−26) 1.3767 06682	(−22) 1.6183 10928
43	(−40) 1.4559 38876	(−33) 5.4333 44999	(−27) 1.2806 57379	(−23) 1.8817 56893
44	(−42) 6.6179 03983	(−34) 3.7045 53408	(−28) 1.1642 33981	(−24) 2.1383 60106
45	(−43) 2.9412 90659	(−35) 2.4697 02271	(−29) 1.0348 74650	(−25) 2.3759 55673
46	(−44) 1.2788 22026	(−36) 1.6106 75395	(−31) 8.9989 09998	(−26) 2.5825 60514
47	(−46) 5.4417 95855	(−37) 1.0280 90677	(−32) 7.6586 46807	(−27) 2.7474 04803
48	(−47) 2.2674 14940	(−39) 6.4255 66736	(−33) 6.3822 05674	(−28) 2.8618 80003
49	(−49) 9.2547 54855	(−40) 3.9340 20450	(−34) 5.2099 63815	(−29) 2.9202 85717
50	(−50) 3.7019 10942	(−41) 2.3604 12270	(−35) 4.1679 71052	(−30) 2.9202 85717

BERNOULLI AND EULER POLYNOMIALS

$x^n/n!$

$n \backslash x$	6	7	8	9
1	(0) 6.0000 00000	(0) 7.0000 00000	(0) 8.0000 00000	(0) 9.0000 00000
2*	(1) 1.8000 00000	(1) 2.4500 00000	(1) 3.2000 00000	(1) 4.0500 00000
3	(1) 3.6000 00000	(1) 5.7166 66667	(1) 8.5333 33333	(2) 1.2150 00000
4	(1) 5.4000 00000	(2) 1.0004 16667	(2) 1.7066 66667	(2) 2.7337 50000
5	(1) 6.4800 00000	(2) 1.4005 83333	(2) 2.7306 66667	(2) 4.9207 50000
6	(1) 6.4800 00000	(2) 1.6340 13889	(2) 3.6408 88889	(2) 7.3811 25000
7	(1) 5.5542 85714	(2) 1.6340 13889	(2) 4.1610 15873	(2) 9.4900 17857
8	(1) 4.1657 14286	(2) 1.4297 62153	(2) 4.1610 15873	(3) 1.0676 27009
9	(1) 2.7771 42857	(2) 1.1120 37230	(2) 3.6986 80776	(3) 1.0676 27009
10	(1) 1.6662 85714	(1) 7.7842 60610	(2) 2.9589 44621	(2) 9.6086 43080
11	(0) 9.0888 31169	(1) 4.9536 20388	(2) 2.1519 59724	(2) 7.8616 17066
12	(0) 4.5444 15584	(1) 2.8896 11893	(2) 1.4346 39816	(2) 5.8962 12799
13	(0) 2.0974 22577	(1) 1.5559 44865	(1) 8.8285 52715	(2) 4.0819 93476
14	(− 1) 8.9889 53903	(0) 7.7797 24327	(1) 5.0448 87266	(2) 2.6241 38663
15	(− 1) 3.5955 81561	(0) 3.6305 38019	(1) 2.6906 06542	(2) 1.5744 83198
16	(− 1) 1.3483 43085	(0) 1.5883 60383	(1) 1.3453 03271	(1) 8.8564 67988
17	(− 2) 4.7588 57949	(− 1) 6.5403 07461	(0) 6.3308 38921	(1) 4.6887 18347
18	(− 2) 1.5862 85983	(− 1) 2.5434 52902	(0) 2.8137 06187	(1) 2.3443 59173
19	(− 3) 5.0093 24157	(− 2) 9.3706 15954	(0) 1.1847 18395	(1) 1.1104 85924
20	(− 3) 1.5027 97247	(− 2) 3.2797 15584	(− 1) 4.7388 73579	(0) 4.9971 86660
21	(− 4) 4.2937 06421	(− 2) 1.0932 38528	(− 1) 1.8052 85173	(0) 2.1416 51426
22	(− 4) 1.1710 10841	(− 3) 3.4784 86224	(− 2) 6.5646 73356	(− 1) 8.7613 01286
23	(− 5) 3.0548 10892	(− 3) 1.0586 69721	(− 2) 2.2833 64645	(− 1) 3.4283 35286
24	(− 6) 7.6370 27230	(− 4) 3.0877 86685	(− 3) 7.6112 15485	(− 1) 1.2856 25732
25	(− 6) 1.8328 86535	(− 5) 8.6458 02719	(− 3) 2.4355 88956	(− 2) 4.6282 52637
26	(− 7) 4.2297 38158	(− 5) 2.3277 16117	(− 4) 7.4941 19863	(− 2) 1.6020 87451
27	(− 8) 9.3994 18129	(− 6) 6.0348 19562	(− 4) 2.2204 79959	(− 3) 5.3402 91503
28	(− 8) 2.0141 61028	(− 6) 1.5087 04890	(− 5) 6.3442 28454	(− 3) 1.7165 22269
29	(− 9) 4.1672 29712	(− 7) 3.6417 01460	(− 5) 1.7501 31987	(− 4) 5.3271 38075
30	(−10) 8.3344 59424	(− 8) 8.4973 03406	(− 6) 4.6670 18634	(− 4) 1.5981 41423
31	(−10) 1.6131 21179	(− 8) 1.9187 45930	(− 6) 1.2043 91905	(− 5) 4.6397 65421
32	(−11) 3.0246 02211	(− 9) 4.1972 56723	(− 7) 3.0109 79764	(− 5) 1.3049 34025
33	(−12) 5.4992 76746	(−10) 8.9032 71836	(− 8) 7.2993 44881	(− 6) 3.5589 10976
34	(−13) 9.7046 06022	(−10) 1.8330 26555	(− 8) 1.7174 92913	(− 7) 9.4206 46701
35	(−13) 1.6636 46746	(−11) 3.6660 53108	(− 9) 3.9256 98086	(− 7) 2.4224 52008
36	(−14) 2.7727 44578	(−12) 7.1284 36600	(−10) 8.7237 73527	(− 8) 6.0561 30022
37	(−15) 4.4963 42559	(−12) 1.3486 23141	(−10) 1.8862 21303	(− 8) 1.4731 12708
38	(−16) 7.0994 88250	(−13) 2.4843 05785	(−11) 3.9709 92217	(− 9) 3.4889 51151
39	(−16) 1.0922 28962	(−14) 4.4590 10384	(−12) 8.1456 25061	(−10) 8.0514 25733
40	(−17) 1.6383 43443	(−15) 7.8032 68172	(−12) 1.6291 25012	(−10) 1.8115 70790
41	(−18) 2.3975 75770	(−15) 1.3322 65298	(−13) 3.1787 80512	(−11) 3.9766 18807
42	(−19) 3.4251 08241	(−16) 2.2204 42162	(−14) 6.0548 20021	(−12) 8.5213 26014
43	(−20) 4.7792 20803	(−17) 3.6146 73288	(−14) 1.1264 78144	(−12) 1.7835 33352
44	(−21) 6.5171 19276	(−18) 5.7506 16594	(−15) 2.0481 42079	(−13) 3.6481 36401
45	(−22) 8.6894 92369	(−19) 8.9454 03592	(−16) 3.6411 41473	(−14) 7.2962 72804
46	(−22) 1.1334 12048	(−19) 1.3612 57068	(−17) 6.3324 19955	(−14) 1.4275 31635
47	(−23) 1.4469 08998	(−20) 2.0274 04144	(−17) 1.0778 58716	(−15) 2.7335 71217
48	(−24) 1.8086 36247	(−21) 2.9566 31045	(−18) 1.7964 31193	(−16) 5.1254 46033
49	(−25) 2.2146 56629	(−22) 4.2237 58634	(−19) 2.9329 48887	(−17) 9.4140 84548
50	(−26) 2.6575 87955	(−23) 5.9132 62088	(−20) 4.6927 18219	(−17) 1 6945 35219

*The floating decimal point notation is used here. For example (1) 1.8000,000 = (1.8000,000) 10.

STIRLING NUMBERS

The Stirling numbers are used for reducing factorials such as $x^{(n)}$ to polynomials in x and vice versa.

STIRLING NUMBERS OF THE FIRST KIND

The factorial polynomial $x^{(n)}$ is defined and represented by

$$x^{(n)} = x(x-1)(x-2)\cdots(x-n+1)$$

where $x^{(0)}$ is 1 by definition.

If n is a non-negative integer, then

$$x^{(n)} = s_{n1}x + s_{n2}x^2 + \cdots + s_{nn}x^n.$$

Here the numbers $s_{n1}, s_{n2}, s_{n3}, \ldots,$ are Stirling numbers of the first kind.

EXAMPLE

Express $3x^{(3)} + 2x^{(1)}$ using Stirling's numbers of the first kind

$$\begin{aligned} 3x^{(3)} &= 3(2x - 3x^2 + x^3) \\ &= 6x - 9x^2 + 3x^3 \\ 2x^{(1)} &= 2x \\ \therefore\ 3x^{(3)} + 2x^{(1)} &= 8x - 9x^2 + 3x^3 \end{aligned}$$

This may be verified by carrying out the indicated operations as follows

$$\begin{aligned} 3x^{(3)} + 2x^{(1)} &= 3(x)(x-1)(x-2) + 2x \\ &= 3x^3 - 9x^2 + 6x + 2x \\ &= 8x - 9x^2 + 3x^3 \end{aligned}$$

STIRLING NUMBERS OF THE SECOND KIND

For every non-negative integer n the function defined by x^n can be expressed as a linear combination of factorial powers of x not higher than the n'th. In other words

$$x^n = t_{n1}x^{(1)} + t_{n2}x^{(2)} + \cdots + t_{nn}x^{(n)}.$$

The numbers $t_{n1}, t_{n2}, \ldots, t_{nn}$ are called Stirling's numbers of the second kind.

EXAMPLE

Express $2x^3 - 3x^2 + x + 2$ by use of factorial powers

$$\begin{aligned} 2x^3 &= 2[x^{(1)} + 3x^{(2)} + x^{(3)}] &&= 2x^{(1)} + 6x^{(2)} + 2x^{(3)} \\ -3x^2 &= -3[x^{(1)} + x^{(2)}] &&= -3x^{(1)} - 3x^{(2)} \\ x &= x^{(1)} &&= x^{(1)} \\ +2 &= +2x^{(0)} &&= 2x^{(0)} \end{aligned}$$

$$\therefore 2x^3 - 3x^2 + x + 2 = 2 + 3x^{(2)} + 2x^{(3)}$$

[1]STIRLING NUMBERS OF THE FIRST KIND $S_n^{(m)}$

$n\backslash m$	1	2	3
1	1		
2	−1	1	
3	2	−3	1
4	−6	11	−6
5	24	−50	35
6	−120	274	−225
7	720	−1764	1624
8	−5040	13068	− 13132
9	40320	−1 09584	1 18124
10	−3 62880	10 26576	−11 72700
11	36 28800	−106 28640	127 53576
12	−399 16800	1205 43840	−1509 17976
13	4790 01600	− 14864 42880	19315 59552
14	− 62270 20800	1 98027 59040	−2 65967 17056
15	8 71782 91200	−28 34656 47360	39 21567 97824
16	−130 76743 68000	433 91630 01600	−616 58176 14720
17	2092 27898 88000	−7073 42823 93600	10299 22448 37120
18	− 35568 74280 96000	1 22340 55905 79200	−1 82160 24446 24640
19	6 40237 37057 28000	−22 37698 80585 21600	34 01224 95938 22720
20	−121 64510 04088 32000	431 56514 68176 38400	−668 60973 03411 53280
21	2432 90200 81766 40000	−8752 94803 67616 00000	13803 75975 36407 04000
22	− 51090 94217 17094 40000	1 86244 81078 01702 40000	−2 98631 90286 32163 84000
23	11 24000 72777 76076 80000	−41 48476 77933 54547 20000	67 56146 67377 09306 88000
24	−258 52016 73888 49766 40000	965 38966 65249 30662 40000	−1595 39850 27606 68605 44000
25	6204 48401 73323 94393 60000	−23427 87216 39871 85664 00000	39254 95373 27809 77192 96000

[1]From "Handbook of Mathematical Functions", National Bureau of Standards

STIRLING NUMBERS OF THE FIRST KIND $S_n^{(m)}$

$n\backslash m$	4	5	6
4	1		
5	−10	1	
6	85	−15	1
7	−735	175	−21
8	6769	−1960	322
9	− 67284	22449	−4536
10	7 23680	−2 69325	63273
11	−84 09500	34 16930	−9 02055
12	1052 58076	−459 95730	133 39535
13	− 14140 14888	6572 06836	−2060 70150
14	2 03137 53096	− 99577 03756	33361 18786
15	−31 09892 60400	15 97216 05680	−5 66633 66760
16	505 69957 03824	−270 68133 45600	100 96721 07080
17	−8707 77488 75904	4836 60092 33424	−1886 15670 58880
18	1 58331 39757 27488	− 90929 99058 44112	36901 26492 34384
19	−30 32125 40077 19424	17 95071 22809 21504	−7 55152 75920 63024
20	610 11607 57404 91776	−371 38478 73452 28000	161 42973 65301 18960
21	− 12870 93124 51509 88800	8037 81182 26450 51776	−3599 97951 79476 07200
22	2 84093 31590 18114 68800	−1 81664 97952 06970 76096	83637 38169 95448 02976
23	−65 48684 85270 30686 97600	42 80722 86535 71471 42912	−20 21687 37691 06827 41568
24	1573 75898 28594 15107 32800	−1050 05310 75591 74529 84576	507 79532 53430 28501 98976
25	−39365 61409 13866 31181 31200	26775 03356 42796 03823 62624	−13237 14091 57918 58577 60000

STIRLING NUMBERS OF THE FIRST KIND $S_n^{(m)}$

$n\backslash m$	7	8	9
7	1		
8	−28	1	
9	546	−36	1
10	−9450	870	−45
11	1 57773	− 18150	1320
12	−26 37558	3 57423	− 32670
13	449 90231	−69 26634	7 49463
14	−7909 43153	1350 36473	−166 69653
15	1 44093 22928	− 26814 53775	3684 11615
16	−27 28032 10680	5 46311 29553	− 82076 28000
17	537 45234 77960	−114 69012 83528	18 59531 77553
18	− 11022 84661 84200	2487 18452 97936	−430 81053 01929
19	2 35312 50405 49984	− 55792 16815 47048	10241 77407 32658
20	−52 26090 33625 12720	12 95363 69899 43896	−2 50385 87554 67550
21	1206 64780 37803 73360	−311 33364 31613 90640	63 03081 20992 94896
22	− 28939 58339 73354 47760	7744 65431 01695 76800	−1634 98069 72465 83456
23	7 20308 21644 09246 53696	−1 99321 97822 10661 37360	43714 22964 95944 12832
24	−185 88776 35505 19497 76576	53 04713 71552 54458 12976	−12 04749 26016 17376 32496
25	4969 10165 05554 96448 36800	−1459 01905 52766 26492 88000	342 18695 95940 71489 92880

$n\backslash m$	10	11	12
10	1		
11	−55	1	
12	1925	−66	1
13	− 55770	2717	−78
14	14 74473	− 91091	3731
15	−373 12275	27 49747	−1 43325
16	9280 95740	−785 58480	48 99622
17	−2 30571 59840	21850 31420	−1569 52432
18	57 79248 94833	−6 02026 93980	48532 22764
19	−1471 07534 08923	166 15733 86473	−14 75607 03732
20	38192 20555 02195	−4628 06477 51910	446 52267 57381
21	−10 14229 98655 11450	1 30753 50105 40395	− 13558 51828 99530
22	276 01910 92750 35346	−37 60053 50868 59745	4 15482 38514 30525
23	−7707 40110 12973 61068	1103 23088 11859 49736	−129 00665 98183 31295
24	2 20984 45497 94337 17396	− 33081 71136 85742 04996	4070 38405 70075 69521
25	−65 08376 17966 81468 50000	10 14945 52782 52146 37300	−1 30770 92873 67558 73500

STIRLING NUMBERS OF THE FIRST KIND $S_n^{(m)}$

$n\backslash m$	13	14	15	16
13	1			
14	−91	1		
15	5005	−105	1	
16	−2 18400	6580	−120	1
17	83 94022	−3 23680	8500	−136
18	−2996 50806	138 96582	−4 68180	10812
19	1 02469 37272	−5497 89282	223 23822	−6 62796
20	−34 22525 11900	2 06929 33630	−9739 41900	349 16946
21	1131 02769 95381	−75 61111 84500	4 01717 71630	− 16722 80820
22	− 37310 09998 02531	2718 86118 69881	−159 97183 88730	7 52896 68850
23	12 36304 58470 86207	− 97125 04609 39913	6238 24164 21941	−325 60911 03430
24	−413 35671 43013 14056	34 70180 64487 04206	−2 40604 60386 44556	13727 25118 00831
25	13990 94520 02391 06865	−1246 20006 90702 15000	92 44691 13761 73550	−5 70058 63218 64500

$n\backslash m$	17	18	19	20	21	22	23	24	25
17	1								
18	−153	1							
19	13566	−171	1						
20	−9 20550	16815	−190	1					
21	533 27946	−12 56850	20615	−210	1				
22	− 27921 67686	797 21796	−16 89765	25025	−231	1			
23	13 67173 57942	− 45460 47198	1168 96626	−22 40315	30107	−253	1		
24	−640 05903 36096	24 12764 43496	− 72346 69596	1684 23871	−29 32776	35926	−276	1	
25	29088 66798 67135	−1219 12249 80000	41 49085 13800	−1 12768 42500	2388 10495	−37 95000	42550	−300	1

STIRLING NUMBERS OF THE SECOND KIND $\mathfrak{S}_n^{(m)}$

n\m	1	2	3	4	5	6
1	1					
2	1	1				
3	1	3	1			
4	1	7	6	1		
5	1	15	25	10	1	
6	1	31	90	65	15	1
7	1	63	301	350	140	21
8	1	127	966	1701	1050	266
9	1	255	3025	7770	6951	2646
10	1	511	9330	34105	42525	22827
11	1	1023	28501	1 45750	2 46730	1 79487
12	1	2047	86526	6 11501	13 79400	13 23652
13	1	4095	2 61625	25 32530	75 08501	93 21312
14	1	8191	7 88970	103 91745	400 75035	634 36373
15	1	16383	23 75101	423 55950	2107 66920	4206 93273
16	1	32767	71 41686	1717 98901	10961 90550	27349 26558
17	1	65535	214 57825	6943 37290	56527 51651	1 75057 49898
18	1	1 31071	644 39010	27988 06985	2 89580 95545	11 06872 51039
19	1	2 62143	1934 48101	1 12596 66950	14 75892 84710	69 30816 01779
20	1	5 24287	5806 06446	4 52321 15901	74 92060 90500	430 60788 95384
21	1	10 48575	17423 43625	18 15090 70050	379 12625 68401	2658 56794 62804
22	1	20 97151	52280 79450	72 77786 23825	1913 78219 12055	16330 53393 45225
23	1	41 94303	1 56863 35501	291 63425 74750	9641 68881 84100	99896 98579 83405
24	1	83 88607	4 70632 00806	1168 10566 34501	48500 07834 95250	6 09023 60360 84530
25	1	167 77215	14 11979 91025	4677 12897 38810	2 43668 49741 10751	37 02641 70000 02430

n\m	7	8	9	10
7	1			
8	28	1		
9	462	36	1	
10	5880	750	45	1
11	63987	11880	1155	55
12	6 27396	1 59027	22275	1705
13	57 15424	18 99612	3 59502	39325
14	493 29280	209 12320	51 35130	7 52752
15	4087 41333	2166 27840	671 28490	126 62650
16	32818 82604	21417 64053	8207 84250	1937 54990
17	2 57081 04786	2 04159 95028	95288 22303	27583 34150
18	19 74624 83400	18 90360 65010	10 61753 95755	3 71121 63803
19	149 29246 34839	170 97510 03480	114 46146 26805	47 72970 33785
20	1114 35540 45652	1517 09326 62679	1201 12826 44725	591 75849 64655
21	8231 09572 14948	13251 10153 47084	12327 24764 65204	7118 71322 91275
22	60276 23799 67440	1 14239 90799 91620	1 24196 33035 33920	83514 37993 77954
23	4 38264 19991 17305	9 74195 50199 00400	12 32006 88117 96900	9 59340 12973 13460
24	31 67746 38518 04540	82 31828 21583 20505	120 62257 43260 72500	108 25408 17849 31500
25	227 83248 29987 16310	690 22372 11183 68580	1167 92145 10929 73005	1203 16339 21753 87500

n\m	11	12	13	14
11	1			
12	66	1		
13	2431	78	1	
14	66066	3367	91	1
15	14 79478	1 06470	4550	105
16	289 36908	27 57118	1 65620	6020
17	5120 60978	620 22324	49 10178	2 49900
18	83910 04908	12563 28866	1258 54638	84 08778
19	12 94132 17791	2 34669 51300	28924 39160	2435 77530
20	190 08424 29486	41 10166 33391	6 10686 60380	63025 24580
21	2682 68516 89001	683 30420 30178	120 49092 18331	14 93040 04500
22	36628 25008 70286	10882 33560 51137	2249 68618 68481	329 51652 81331
23	4 86425 13089 51100	1 67216 27734 83930	40128 25603 41390	6862 91758 07115
24	63 10016 56957 75560	24 93020 45907 58260	6 88883 60579 22000	1 36209 10216 41000
25	802 35590 44384 62660	362 26262 07848 74680	114 48507 33437 44260	25 95811 03608 96000

n\m	15	16	17	18	19
15	1				
16	120	1			
17	7820	136	1		
18	3 67200	9996	153	1	
19	139 16778	5 27136	12597	171	1
20	4523 29200	223 50954	7 41285	15675	190
21	1 30874 62580	8099 44464	349 52799	10 23435	19285
22	34 56159 43200	2 60465 74004	14041 42047	533 74629	13 89850
23	847 94044 29331	76 23611 27264	4 99169 88803	23648 85369	797 81779
24	19582 02422 47080	2067 71824 65555	161 09499 36915	9 24849 25445	38807 39170
25	4 29939 46553 47200	52665 51616 95960	4806 33313 93110	327 56785 94925	16 62189 69675

n\m	20	21	22	23	24	25
20	1					
21	210	1				
22	23485	231	1			
23	18 59550	28336	253	1		
24	1169 72779	24 54606	33902	276	1	
25	62201 94750	1685 19505	32 00450	40250	300	1

FOURIER SERIES

1. If $f(x)$ is a bounded periodic function of period $2L$ (i.e. $f(x + 2L) = f(x)$), and satisfies the *Dirichlet conditions*:

   a) In any period $f(x)$ is continuous, except possibly for a finite number of jump discontinuities

   b) In any period $f(x)$ has only a finite number of maxima and minima.

Then $f(x)$ may be represented by the *Fourier series*

$$\frac{a_0}{2} + \sum_{n=1}^{\infty}\left(a_n \cos\frac{n\pi x}{L} + b_n \sin\frac{n\pi x}{L}\right)$$

where

$$\begin{cases} a_n = \dfrac{1}{L}\displaystyle\int_c^{c+2L} f(x)\cos\frac{n\pi x}{L}\,dx \\ b_n = \dfrac{1}{L}\displaystyle\int_c^{c+2L} f(x)\sin\frac{n\pi x}{L}\,dx \end{cases}$$

2. If in addition to the above restrictions, $f(x)$ is even (i.e. $f(-x) = f(x)$), the Fourier series reduces to

$$\frac{a_0}{2} + \sum_{n=1}^{\infty} a_n \cos\frac{n\pi x}{L}.$$

That is, $b_n = 0$. In this case, a simpler formula for a_n is

$$a_n = \frac{2}{L}\int_0^L f(x)\cos\frac{n\pi x}{L}\,dx, \quad n = 0, 1, 2, 3, \ldots$$

3. If in addition to the restrictions in (1), $f(x)$ is an odd function (i.e. $f(-x) = -f(x)$), then the Fourier series reduces to

$$\sum_{n=1}^{\infty} b_n \sin\frac{n\pi x}{L}.$$

That is, $a_n = 0$. In this case, a simpler formula for the b_n is

$$b_n = \frac{2}{L}\int_0^L f(x)\sin\frac{n\pi x}{L}\,dx, \quad n = 1, 2, 3, \ldots$$

4. If in addition to the restrictions in (2) above, $f(x) = -f(L - x)$, then a_n will be 0 for all even values of n, including $n = 0$. Thus in this case, the expansion reduces to

$$\sum_{m=1}^{\infty} a_{2m-1} \cos\frac{(2m-1)\pi x}{L}.$$

5. If in addition to the restrictions in (3) above, $f(x) = f(L - x)$, then b_n will be 0 for all even values of n. Thus in this case, the expansion reduces to

$$\sum_{m=1}^{\infty} b_{2m-1} \sin \frac{(2m - 1)\pi x}{L}.$$

(The series in (4) and (5) are known as *odd-harmonic series*, since only the odd harmonics appear. Similar rules may be stated for even-harmonic series, but when a series appears in the even-harmonic form, it means that $2L$ has not been taken as the smallest period of $f(x)$. Since any integral multiple of a period is also a period, series obtained in this way will also work, but in general computation is simplified if $2L$ is taken to be the smallest period.)

6. Assuming that the series 16.8 converges to $f(x)$, we have

$$f(x) = \sum_{n=-\infty}^{\infty} c_n e^{in\pi x/L}$$

where

$$c_n = \frac{1}{L}\int_c^{c+2L} f(x)e^{-in\pi x/L}\,dx = \begin{cases} \frac{1}{2}(a_n - ib_n) & n > 0 \\ \frac{1}{2}(a_{-n} + ib_{-n}) & n < 0 \\ \frac{1}{2}a_0 & n = 0 \end{cases}$$

7. **If both sine and cosine terms are present and if $f(x)$ is of period $2L$ and expandable by a Fourier series, it can be represented as**

$$f(x) = \frac{a_0}{2} + \sum_{n=1}^{\infty} c_n \cos\left(\frac{n\pi x}{L} + \phi_n\right),$$

where $a_n = c_n \cos \phi_n$, $b_n = c_n \cos \phi_n$, $c_n = \sqrt{a_n^2 + b_n^2}$, $\phi_n = \arctan\left(\frac{a_n}{b_n}\right)$

It can also be represented as

$$f(x) = \frac{a_0}{2} + \sum_{n=1}^{\infty} c_n \sin\left(\frac{n\pi x}{L} + \phi_n\right),$$

where $a_n = c_n \sin \phi_n$, $b_n = -c_n \sin \phi_n$, $c_n = \sqrt{a_n^2 + b_n^2}$, $\phi_n = \arctan\left(-\frac{b_n}{a_n}\right)$

where the quadrant of ϕ_n is chosen so as to make the formulas for a_n, b_n, and c_n hold.

8. The following table of trigonometric identities should be helpful for developing Fourier Series.

	n**	n even	n odd	$n/2$ odd	$n/2$ even
$\sin n\pi$	0	0	0	0	0
$\cos n\pi$	$(-1)^n$	+1	−1	+1	+1
$\sin \frac{n\pi}{2}$ *		0	$(-1)^{(n-1)/2}$	0	0
$\cos \frac{n\pi}{2}$ *		$(-1)^{n/2}$	0	−1	+1
$\sin \frac{n\pi}{4}$			$\frac{\sqrt{2}}{2}(-1)^{(n^2+4n+11)/8}$	$(-1)^{(n-2)/4}$	0

*A useful formula for $\sin \frac{n\pi}{2}$ and $\cos \frac{n\pi}{2}$ is given by

$$\sin \frac{n\pi}{2} = \frac{(i)^{n+1}}{2}[(-1)^n - 1] \text{ and } \cos \frac{n\pi}{2} = \frac{(i)^n}{2}[(-1)^n + 1], \text{ where } i^2 = -1.$$

** n any integer.

AUXILIARY FORMULAS FOR FOURIER SERIES

$$1 = \frac{4}{\pi}\left[\sin \frac{\pi x}{k} + \frac{1}{3} \sin \frac{3\pi x}{k} + \frac{1}{5} \sin \frac{5\pi x}{k} + \cdots\right] \qquad [0 < x < k]$$

$$x = \frac{2k}{\pi}\left[\sin \frac{\pi x}{k} - \frac{1}{2} \sin \frac{2\pi x}{k} + \frac{1}{3} \sin \frac{3\pi x}{k} - \cdots\right] \qquad [-k < x < k]$$

$$x = \frac{k}{2} - \frac{4k}{\pi^2}\left[\cos \frac{\pi x}{k} + \frac{1}{3^2} \cos \frac{3\pi x}{k} + \frac{1}{5^2} \cos \frac{5\pi x}{k} + \cdots\right] \qquad [0 < x < k]$$

$$x^2 = \frac{2k^2}{\pi^3}\left[\left(\frac{\pi^2}{1} - \frac{4}{1}\right) \sin \frac{\pi x}{k} - \frac{\pi^2}{2} \sin \frac{2\pi x}{k} + \left(\frac{\pi^2}{3} - \frac{4}{3^3}\right) \sin \frac{3\pi x}{k} - \frac{\pi^2}{4} \sin \frac{4\pi x}{k} + \left(\frac{\pi^2}{5} - \frac{4}{5^3}\right) \sin \frac{5\pi x}{k} + \cdots\right] \qquad [0 < x < k]$$

$$x^2 = \frac{k^2}{3} - \frac{4k^2}{\pi^2}\left[\cos \frac{\pi x}{k} - \frac{1}{2^2} \cos \frac{2\pi x}{k} + \frac{1}{3^2} \cos \frac{3\pi x}{k} - \frac{1}{4^2} \cos \frac{4\pi x}{k} + \cdots\right] \qquad [-k < x < k]$$

$$1 - \frac{1}{3} + \frac{1}{5} - \frac{1}{7} + \cdots = \frac{\pi}{4}$$

$$1 + \frac{1}{2^2} + \frac{1}{3^2} + \frac{1}{4^2} + \cdots = \frac{\pi^2}{6}$$

$$1 - \frac{1}{2^2} + \frac{1}{3^2} - \frac{1}{4^2} + \cdots = \frac{\pi^2}{12}$$

$$1 + \frac{1}{3^2} + \frac{1}{5^2} + \frac{1}{7^2} + \cdots = \frac{\pi^2}{8}$$

$$\frac{1}{2^2} + \frac{1}{4^2} + \frac{1}{6^2} + \frac{1}{8^2} + \cdots = \frac{\pi^2}{24}$$

FOURIER EXPANSIONS FOR BASIC PERIODIC FUNCTIONS

1

$$f(x) = \frac{c}{L} + \frac{2}{\pi} \sum_{n=1}^{\infty} \frac{(-1)^n}{n} \sin \frac{n\pi c}{L} \cos \frac{n\pi x}{L}$$

2

$$f(x) = \frac{4}{\pi} \sum_{n=1,3,5,\ldots} \frac{1}{n} \sin \frac{n\pi x}{L}$$

3

$$f(x) = \frac{2}{\pi} \sum_{n=1}^{\infty} \frac{(-1)^n}{n} \left(\cos \frac{n\pi c}{L} - 1 \right) \sin \frac{n\pi x}{L}$$

4

$$f(x) = \frac{2}{L} \sum_{n=1}^{\infty} \sin \frac{n\pi}{2} \frac{\sin (\frac{1}{2} n\pi c/L)}{\frac{1}{2} n\pi c/L} \sin \frac{n\pi x}{L}$$

5

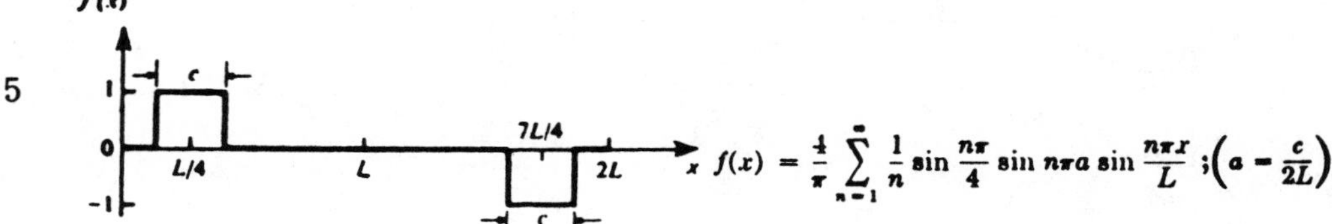

$$f(x) = \frac{4}{\pi} \sum_{n=1}^{\infty} \frac{1}{n} \sin \frac{n\pi}{4} \sin n\pi a \sin \frac{n\pi x}{L}; \left(a = \frac{c}{2L} \right)$$

6

$$f(x) = \frac{8}{\pi^2} \sum_{n=1,3,5,\ldots} \frac{(-1)^{(n-1)/2}}{n^2} \sin \frac{n\pi x}{L}$$

7

$$f(x) = \frac{32}{3\pi^2} \sum_{n=1}^{\infty} \frac{1}{n^2} \sin \frac{n\pi}{4} \sin \frac{n\pi x}{L}; \left(a = \frac{c}{2L}\right)$$

8

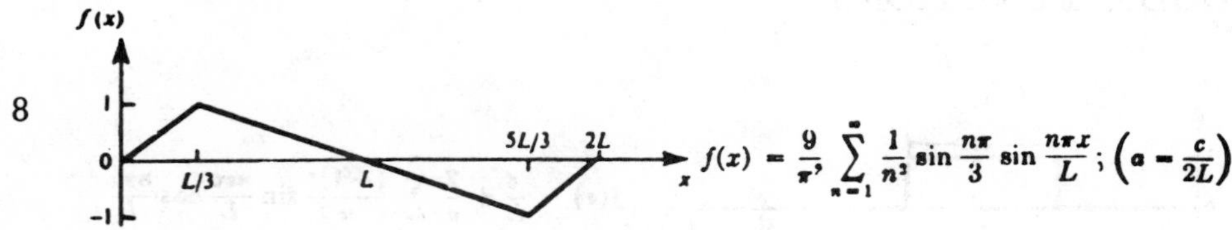

$$f(x) = \frac{9}{\pi^2} \sum_{n=1}^{\infty} \frac{1}{n^2} \sin \frac{n\pi}{3} \sin \frac{n\pi x}{L}; \left(a = \frac{c}{2L}\right)$$

9

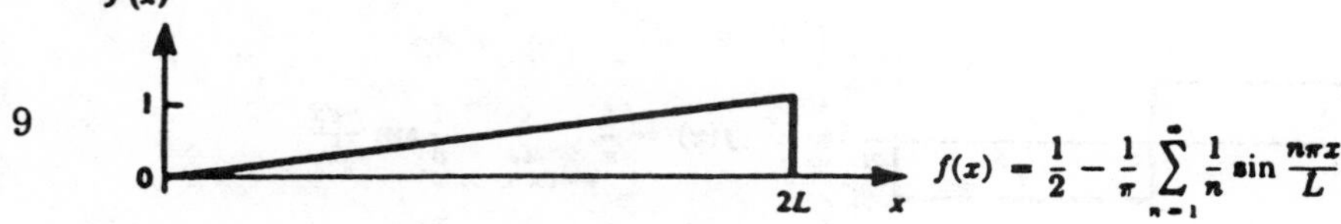

$$f(x) = \frac{1}{2} - \frac{1}{\pi} \sum_{n=1}^{\infty} \frac{1}{n} \sin \frac{n\pi x}{L}$$

10

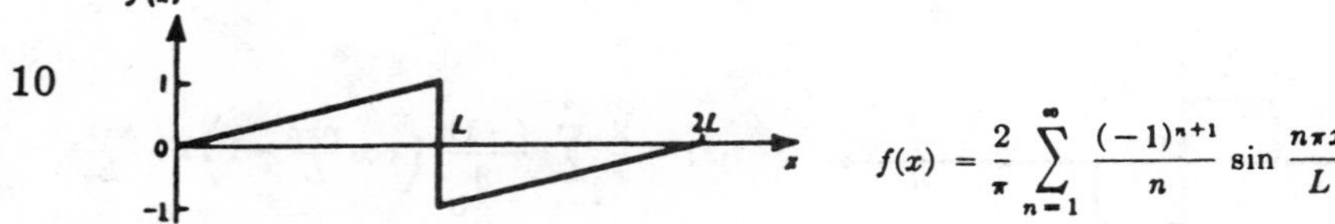

$$f(x) = \frac{2}{\pi} \sum_{n=1}^{\infty} \frac{(-1)^{n+1}}{n} \sin \frac{n\pi x}{L}$$

11

$$f(x) = \frac{1}{2} - \frac{4}{\pi^2} \sum_{n=1,3,5\ldots} \frac{1}{n^2} \cos \frac{n\pi x}{L}$$

12

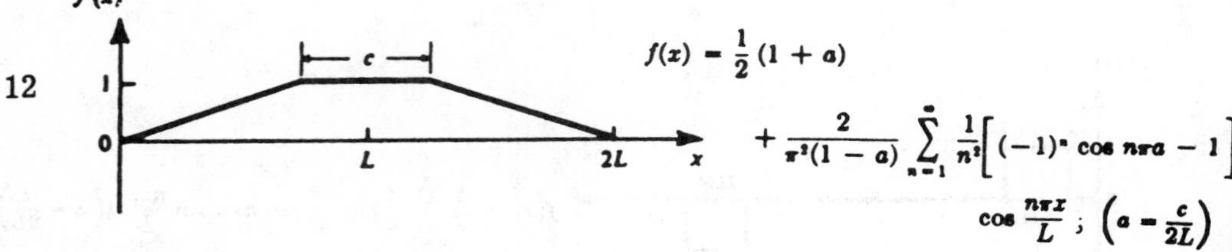

$$f(x) = \frac{1}{2}(1 + a) + \frac{2}{\pi^2(1-a)} \sum_{n=1}^{\infty} \frac{1}{n^2}\left[(-1)^n \cos n\pi a - 1\right] \cos \frac{n\pi x}{L}; \left(a = \frac{c}{2L}\right)$$

13

$$f(x) = \frac{1}{2} - \frac{4}{\pi^2(1-2a)} \sum_{n=1,3,5,\ldots} \frac{1}{n^2} \cos n\pi a \cos \frac{n\pi x}{L}; \quad \left(a = \frac{c}{2L}\right)$$

14

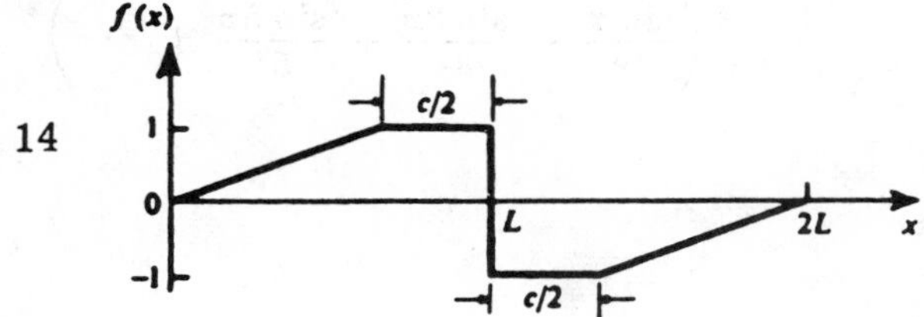

$$f(x) = \frac{2}{\pi} \sum_{n=1}^{\infty} \frac{(-1)^{n-1}}{n} \left[1 + \frac{\sin n\pi a}{n\pi(1-a)}\right] \sin \frac{n\pi x}{L}; \quad \left(a = \frac{c}{2L}\right)$$

15

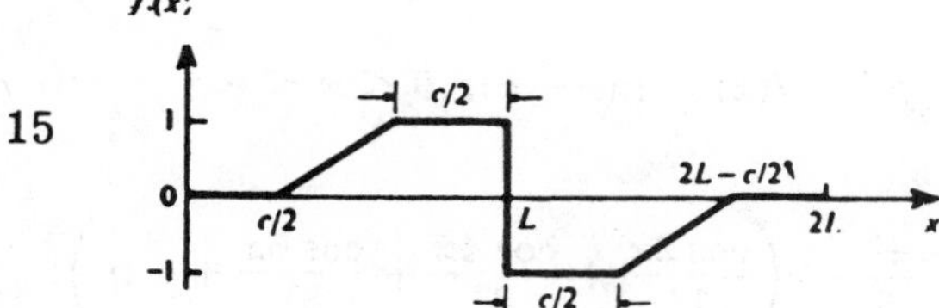

$$f(x) = -\frac{2}{\pi} \sum_{n=1}^{\infty} \frac{(-1)^{n}}{n} \left[1 + \frac{1+(-1)^n}{n\pi(1-2a)} \sin n\pi a\right] \sin \frac{n\pi x}{L}; \quad \left(a = \frac{c}{2L}\right)$$

16

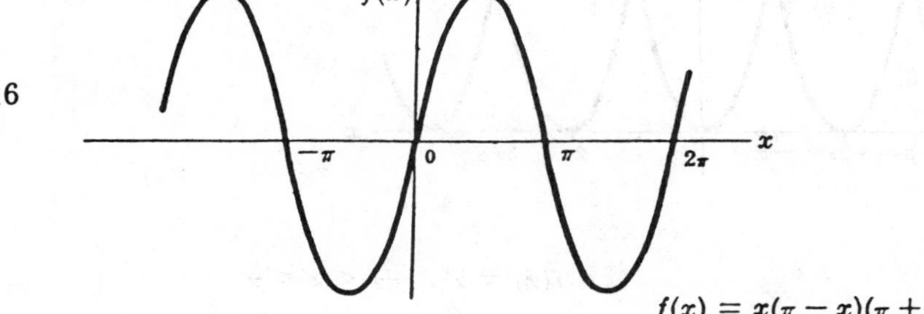

$$f(x) = x(\pi - x)(\pi + x), \quad -\pi < x < \pi$$

$$12\left(\frac{\sin x}{1^3} - \frac{\sin 2x}{2^3} + \frac{\sin 3x}{3^3} - \cdots\right)$$

17

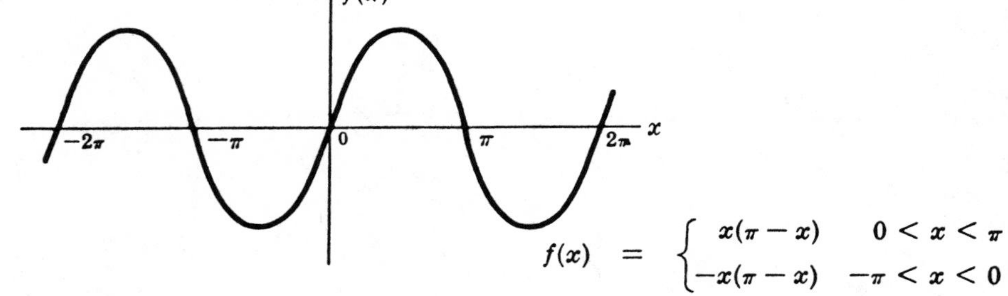

$$f(x) = \begin{cases} x(\pi - x) & 0 < x < \pi \\ -x(\pi - x) & -\pi < x < 0 \end{cases}$$

$$\frac{8}{\pi}\left(\frac{\sin x}{1^3} + \frac{\sin 3x}{3^3} + \frac{\sin 5x}{5^3} + \cdots\right)$$

18

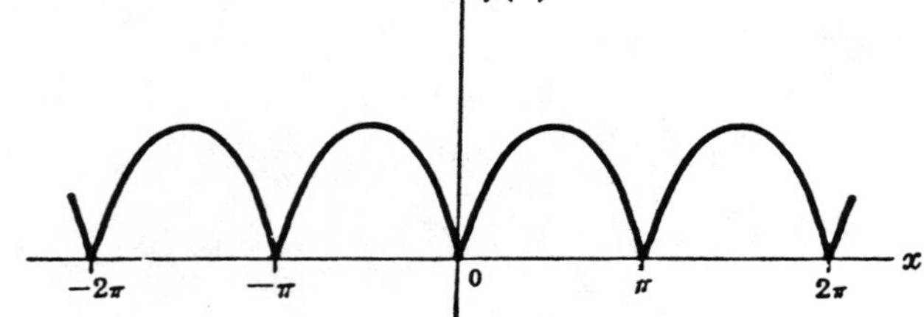

$$f(x) = x(\pi - x), \quad 0 < x < \pi$$

$$\frac{\pi^2}{6} - \left(\frac{\cos 2x}{1^2} + \frac{\cos 4x}{2^2} + \frac{\cos 6x}{3^2} + \cdots\right)$$

19

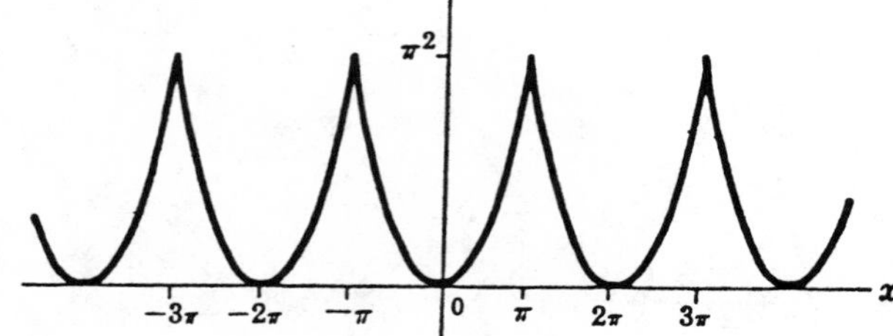

$$f(x) = x^2, \quad -\pi < x < \pi$$

$$\frac{\pi^2}{3} - 4\left(\frac{\cos x}{1^2} - \frac{\cos 2x}{2^2} + \frac{\cos 3x}{3^2} - \cdots\right)$$

20

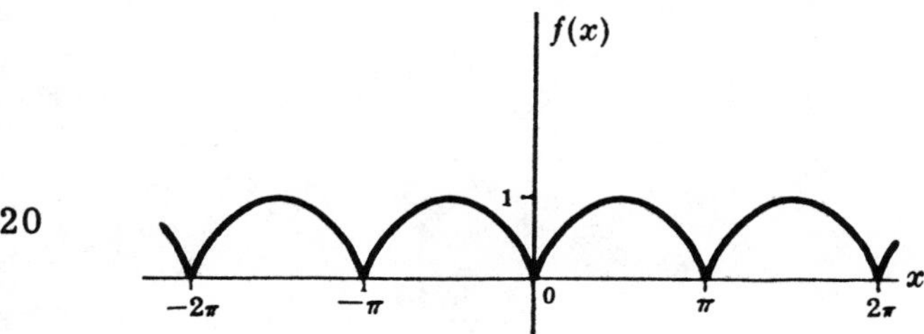

$$f(x) = |\sin x|, \quad -\pi < x < \pi$$

$$\frac{2}{\pi} - \frac{4}{\pi}\left(\frac{\cos 2x}{1 \cdot 3} + \frac{\cos 4x}{3 \cdot 5} + \frac{\cos 6x}{5 \cdot 7} + \cdots\right)$$

21

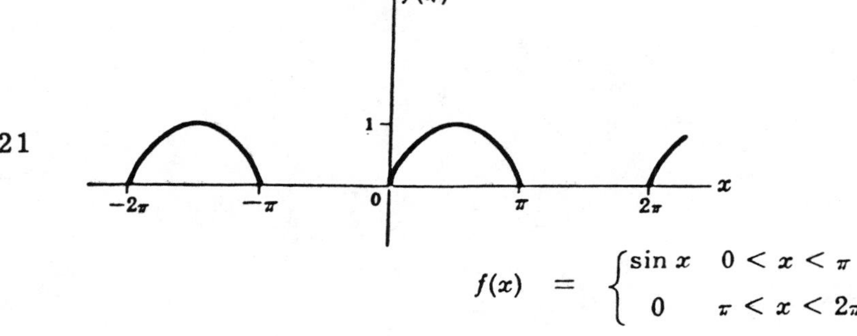

$$f(x) = \begin{cases} \sin x & 0 < x < \pi \\ 0 & \pi < x < 2\pi \end{cases}$$

$$\frac{1}{\pi} + \frac{1}{2}\sin x - \frac{2}{\pi}\left(\frac{\cos 2x}{1 \cdot 3} + \frac{\cos 4x}{3 \cdot 5} + \frac{\cos 6x}{5 \cdot 7} + \cdots\right)$$

22

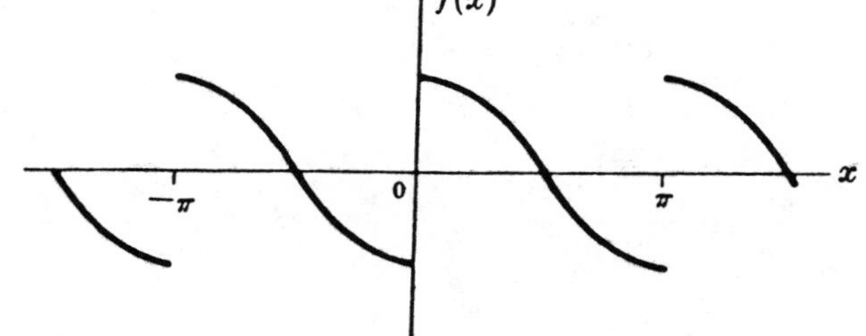

$$f(x) = \begin{cases} \cos x & 0 < x < \pi \\ -\cos x & -\pi < x < 0 \end{cases}$$

$$\frac{8}{\pi}\left(\frac{\sin 2x}{1 \cdot 3} + \frac{2 \sin 4x}{3 \cdot 5} + \frac{3 \sin 6x}{5 \cdot 7} + \cdots\right)$$

23 $f(x) = \frac{1}{12}x(x-\pi)(x-2\pi), \quad 0 \leqq x \leqq 2\pi$

$$\frac{\sin x}{1^3} + \frac{\sin 2x}{2^3} + \frac{\sin 3x}{3^3} + \cdots$$

24 $f(x) = \frac{1}{6}\pi^2 - \frac{1}{2}\pi x + \frac{1}{4}x^2, \quad 0 \leqq x \leqq 2\pi$

$$\frac{\cos x}{1^2} + \frac{\cos 2x}{2^2} + \frac{\cos 3x}{3^2} + \cdots$$

25 $f(x) = \frac{1}{90}\pi^4 - \frac{1}{12}\pi^2 x^2 + \frac{1}{12}\pi x^3 - \frac{1}{48}x^4, \quad 0 \leqq x \leqq 2\pi$

$$\frac{\cos x}{1^4} + \frac{\cos 2x}{2^4} + \frac{\cos 3x}{3^4} + \cdots$$

26 $f(x) = \sin \mu x, \quad -\pi < x < \pi, \quad \mu \neq \text{integer}$

$$\frac{2 \sin \mu\pi}{\pi}\left(\frac{\sin x}{1^2-\mu^2} - \frac{2\sin 2x}{2^2-\mu^2} + \frac{3 \sin 3x}{3^2-\mu^2} - \cdots\right)$$

27 $f(x) = \cos \mu x, \quad -\pi < x < \pi, \quad \mu \neq \text{integer}$

$$\frac{2\mu \sin \mu\pi}{\pi}\left(\frac{1}{2\mu^2} + \frac{\cos x}{1^2-\mu^2} - \frac{\cos 2x}{2^2-\mu^2} + \frac{\cos 3x}{3^2-\mu^2} - \cdots\right)$$

28 $f(x) = \tan^{-1}[(a \sin x)/(1 - a\cos x)], \quad -\pi < x < \pi, \quad |a| < 1$

$$a \sin x + \frac{a^2}{2}\sin 2x + \frac{a^3}{3}\sin 3x + \cdots$$

29 $f(x) = \frac{1}{2}\tan^{-1}[(2a \sin x)/(1-a^2)], \quad -\pi < x < \pi, \quad |a| < 1$

$$a \sin x + \frac{a^3}{3}\sin 3x + \frac{a^5}{5}\sin 5x + \cdots$$

30 $f(x) = \frac{1}{2}\tan^{-1}[(2a \cos x)/(1-a^2)], \quad -\pi < x < \pi, \quad |a| < 1$

$$a \cos x - \frac{a^3}{3}\cos 3x + \frac{a^5}{5}\cos 5x - \cdots$$

31 $f(x) = \ln|\sin \tfrac{1}{2}x|, \quad 0 < x < \pi$

$$-\left(\ln 2 + \frac{\cos x}{1} + \frac{\cos 2x}{2} + \frac{\cos 3x}{3} + \cdots\right)$$

32 $f(x) = \ln|\cos \tfrac{1}{2}x|, \quad -\pi < x < \pi$

$$-\left(\ln 2 - \frac{\cos x}{1} + \frac{\cos 2x}{2} - \frac{\cos 3x}{3} + \cdots\right)$$

33 $f(x) = e^{\mu x}, \quad -\pi < x < \pi$

$$\frac{2\sinh \mu\pi}{\pi}\left(\frac{1}{2\mu} + \sum_{n=1}^{\infty} \frac{(-1)^n(\mu\cos nx - n\sin nx)}{\mu^2 + n^2}\right)$$

34 $f(x) = \ln(1 - 2a\cos x + a^2), \quad -\pi < x < \pi, \quad |a| < 1$

$$-2\left(a\cos x + \frac{a^2}{2}\cos 2x + \frac{a^3}{3}\cos 3x + \cdots\right)$$

35 $f(x) = \sinh \mu x, \quad -\pi < x < \pi$

$$\frac{2\sinh \mu\pi}{\pi}\left(\frac{\sin x}{1^2 + \mu^2} - \frac{2\sin 2x}{2^2 + \mu^2} + \frac{3\sin 3x}{3^2 + \mu^2} - \cdots\right)$$

36 $f(x) = \cosh \mu x, \quad -\pi < x < \pi$

$$\frac{2\mu\sinh \mu\pi}{\pi}\left(\frac{1}{2\mu^2} - \frac{\cos x}{1^2 + \mu^2} + \frac{\cos 2x}{2^2 + \mu^2} - \frac{\cos 3x}{3^2 + \mu^2} + \cdots\right)$$

•

FOURIER TRANSFORMS

FOURIER'S INTEGRAL THEOREM

$$f(x) = \int_0^{\infty} \{A(\alpha)\cos \alpha x + B(\alpha)\sin \alpha x\}\, d\alpha$$

where

$$\begin{cases} A(\alpha) = \dfrac{1}{\pi}\displaystyle\int_{-\infty}^{\infty} f(x)\cos\alpha x\,dx \\ B(\alpha) = \dfrac{1}{\pi}\displaystyle\int_{-\infty}^{\infty} f(x)\sin\alpha x\,dx \end{cases}$$

Sufficient conditions under which this theorem holds are:

(i) $f(x)$ and $f'(x)$ are piecewise continuous in every finite interval $-L < x < L$;

(ii) $\int_{-\infty}^{\infty} |f(x)|\,dx$ converges;

(iii) $f(x)$ is replaced by $\frac{1}{2}\{f(x+0) + f(x-0)\}$ if x is a point of discontinuity.

FOURIER TRANSFORMS

The Fourier transform of $f(x)$ is defined as

$$\mathcal{F}\{f(x)\} = F(\alpha) = \int_{-\infty}^{\infty} f(x)\,e^{-i\alpha x}\,dx$$

Then the inverse Fourier transform of $F(\alpha)$ is

$$\mathcal{F}^{-1}\{F(\alpha)\} = f(x) = \frac{1}{2\pi}\int_{-\infty}^{\infty} F(\alpha)\,e^{i\alpha x}\,d\alpha$$

We call $f(x)$ and $F(\alpha)$ *Fourier transform pairs.*

CONVOLUTION THEOREM FOR FOURIER TRANSFORMS

If $F(\alpha) = \mathcal{F}\{f(x)\}$ and $G(\alpha) = \mathcal{F}\{g(x)\}$, then

$$\frac{1}{2\pi}\int_{-\infty}^{\infty} F(\alpha)\,G(\alpha)\,e^{i\alpha x}\,d\alpha = \int_{-\infty}^{\infty} f(u)\,g(x-u)\,du = f*g$$

where $f*g$ is called the *convolution* of f and g. Thus

$$\mathcal{F}\{f*g\} = \mathcal{F}\{f\}\,\mathcal{F}\{g\}$$

PARSEVAL'S IDENTITY

If $F(\alpha) = \mathcal{F}\{f(x)\}$, then

$$\int_{-\infty}^{\infty} |f(x)|^2\,dx = \frac{1}{2\pi}\int_{-\infty}^{\infty} |F(\alpha)|^2\,d\alpha$$

More generally if $F(\alpha) = \mathcal{F}\{f(x)\}$ and $G(\alpha) = \mathcal{F}\{g(x)\}$, then

$$\int_{-\infty}^{\infty} f(x)\,\overline{g(x)}\,dx = \frac{1}{2\pi}\int_{-\infty}^{\infty} F(\alpha)\,\overline{G(\alpha)}\,d\alpha$$

where the bar denotes complex conjugate.

FOURIER TRANSFORMS

If $F(x)$ is defined for $x \geqq 0$ and is piecewise continuous over any finite interval, and if

$$\int_0^{\infty} F(x)\,dx$$

is absolutely convergent, then

$$f_c(\alpha) = \sqrt{\frac{2}{\pi}}\int_0^{\infty} F(x)\cos(\alpha x)\,dx$$

is the *Fourier cosine transform* of $F(x)$. Furthermore,

$$\bar{F}(x) = \sqrt{\frac{2}{\pi}}\int_0^{\infty} f_c(\alpha)\cos(\alpha x)\,d\alpha.$$

If $\lim\limits_{x \to \infty} \dfrac{d^n F}{dx^n} = 0$, an important property of the Fourier cosine transform

$$f_c^{(2r)}(\alpha) = \sqrt{\frac{2}{\pi}}\int_0^{\infty}\left(\frac{d^{2r}F}{dx^{2r}}\right)\cos(\alpha x)\,dx$$

$$= -\sqrt{\frac{2}{\pi}}\sum_{n=0}^{r-1}(-1)^n a_{2r-2n-1}\alpha^{2n} + (-1)^r \alpha^{2r} f_c(\alpha)$$

where $\lim\limits_{x \to 0} \dfrac{d^r F}{dx^r} = a_r$, makes it useful in the solution of many problems.

Under the same conditions,

$$f_s(\alpha) = \sqrt{\frac{2}{\pi}}\int_0^{\infty} F(x)\sin(\alpha x)\,dx$$

defines the *Fourier sine transform* of $F(x)$, and

$$\bar{F}(x) = \sqrt{\frac{2}{\pi}}\int_0^{\infty} f_s(\alpha)\sin(\alpha x)\,d\alpha.$$

SPECIAL FOURIER TRANSFORM PAIRS

$f(x)$	$F(\alpha)$
$\begin{cases} 1 & \lvert x\rvert < b \\ 0 & \lvert x\rvert > b \end{cases}$	$\dfrac{2 \sin b\alpha}{\alpha}$
$\dfrac{1}{x^2 + b^2}$	$\dfrac{\pi e^{-b\alpha}}{b}$
$\dfrac{x}{x^2 + b^2}$	$-\dfrac{\pi i \alpha}{b} e^{-b\alpha}$
$f^{(n)}(x)$	$i^n \alpha^n F(\alpha)$
$x^n f(x)$	$i^n \dfrac{d^n F}{d\alpha^n}$
$f(bx)e^{itx}$	$\dfrac{1}{b} F\left(\dfrac{\alpha - t}{b}\right)$

FOURIER TRANSFORMS

	$F(x)$	$f(\alpha)$
1	$\dfrac{\sin ax}{x}$	$\begin{cases} \sqrt{\dfrac{\pi}{2}} & \lvert\alpha\rvert < a \\ 0 & \lvert\alpha\rvert > a \end{cases}$
2	$\begin{cases} e^{iwx} & (p < x < q) \\ 0 & (x < p,\ x > q) \end{cases}$	$\dfrac{i}{\sqrt{2\pi}} \dfrac{e^{ip(w+\alpha)} - e^{iq(w+\alpha)}}{(w+\alpha)}$
3	$\begin{cases} e^{-cx+iwx} & (x > 0) \\ 0 & (x < 0) \end{cases} \quad (c > 0)$	$\dfrac{i}{\sqrt{2\pi}(w + \alpha + ic)}$
4	$e^{-px^2} \quad R(p) > 0$	$\dfrac{1}{\sqrt{2p}} e^{-\alpha^2/4p}$
5	$\cos px^2$	$\dfrac{1}{\sqrt{2p}} \cos\left[\dfrac{\alpha^2}{4p} - \dfrac{\pi}{4}\right]$
6	$\sin px^2$	$\dfrac{1}{\sqrt{2p}} \cos\left[\dfrac{\alpha^2}{4p} + \dfrac{\pi}{4}\right]$
7	$\lvert x\rvert^{-p} \quad (0 < p < 1)$	$\sqrt{\dfrac{2}{\pi}} \dfrac{\Gamma(1-p)\sin\dfrac{p\pi}{2}}{\lvert\alpha\rvert^{(1-p)}}$
8	$\dfrac{e^{-a\lvert x\rvert}}{\sqrt{\lvert x\rvert}}$	$\dfrac{\sqrt{\sqrt{(a^2+\alpha^2)} + a}}{\sqrt{a^2+\alpha^2}}$
9	$\dfrac{\cosh ax}{\cosh \pi x} \quad (-\pi < a < \pi)$	$\sqrt{\dfrac{2}{\pi}} \dfrac{\cos\dfrac{a}{2}\cosh\dfrac{\alpha}{2}}{\cosh\alpha + \cos a}$
10	$\dfrac{\sinh ax}{\sinh \pi x} \quad (-\pi < a < \pi)$	$\dfrac{1}{\sqrt{2\pi}} \dfrac{\sin a}{\cosh\alpha + \cos a}$
11	$\begin{cases} \dfrac{1}{\sqrt{a^2-x^2}} & (\lvert x\rvert < a) \\ 0 & (\lvert x\rvert > a) \end{cases}$	$\sqrt{\dfrac{\pi}{2}} J_0(a\alpha)$
12	$\dfrac{\sin[b\sqrt{a^2+x^2}]}{\sqrt{a^2+x^2}}$	$\begin{cases} 0 & (\lvert\alpha\rvert > b) \\ \sqrt{\dfrac{\pi}{2}} J_0(a\sqrt{b^2-\alpha^2}) & (\lvert\alpha\rvert < b) \end{cases}$
13	$\begin{cases} P_n(x) & (\lvert x\rvert < 1) \\ 0 & (\lvert x\rvert > 1) \end{cases}$	$\dfrac{i^n}{\sqrt{\alpha}} J_{n+\frac{1}{2}}(\alpha)$
14	$\begin{cases} \dfrac{\cos[b\sqrt{a^2-x^2}]}{\sqrt{a^2-x^2}} & (\lvert x\rvert < a) \\ 0 & (\lvert x\rvert > a) \end{cases}$	$\sqrt{\dfrac{\pi}{2}} J_0(a\sqrt{a^2+b^2})$
15	$\begin{cases} \dfrac{\cosh[b\sqrt{a^2-x^2}]}{\sqrt{a^2-x^2}} & (\lvert x\rvert < a) \\ 0 & (\lvert x\rvert > a) \end{cases}$	$\sqrt{\dfrac{\pi}{2}} J_0(a\sqrt{\alpha^2-b^2})$

SPECIAL FOURIER SINE TRANSFORMS

$f(x)$	$F_C(\alpha)$
$\begin{cases} 1 & 0 < x < b \\ 0 & x > b \end{cases}$	$\dfrac{1 - \cos b\alpha}{\alpha}$
x^{-1}	$\dfrac{\pi}{2}$
$\dfrac{x}{x^2 + b^2}$	$\dfrac{\pi}{2} e^{-b\alpha}$
e^{-bx}	$\dfrac{\alpha}{\alpha^2 + b^2}$
$x^{n-1} e^{-bx}$	$\dfrac{\Gamma(n) \sin (n \tan^{-1} \alpha/b)}{(\alpha^2 + b^2)^{n/2}}$
xe^{-bx^2}	$\dfrac{\sqrt{\pi}}{4b^{3/2}} \alpha e^{-\alpha^2/4b}$
$x^{-1/2}$	$\sqrt{\dfrac{\pi}{2\alpha}}$
x^{-n}	$\dfrac{\pi\alpha^{n-1} \csc (n\pi/2)}{2\,\Gamma(n)} \qquad 0 < n < 2$
$\dfrac{\sin bx}{x}$	$\dfrac{1}{2} \ln \left(\dfrac{\alpha + b}{\alpha - b}\right)$
$\dfrac{\sin bx}{x^2}$	$\begin{cases} \pi\alpha/2 & \alpha < b \\ \pi b/2 & \alpha > b \end{cases}$
$\dfrac{\cos bx}{x}$	$\begin{cases} 0 & \alpha < b \\ \pi/4 & \alpha = b \\ \pi/2 & \alpha > b \end{cases}$
$\tan^{-1} (x/b)$	$\dfrac{\pi}{2\alpha} e^{-b\alpha}$
$\csc bx$	$\dfrac{\pi}{2b} \tanh \dfrac{\pi\alpha}{2b}$
$\dfrac{1}{e^{2x} - 1}$	$\dfrac{\pi}{4} \coth \left(\dfrac{\pi\alpha}{2}\right) - \dfrac{1}{2\alpha}$

SPECIAL FOURIER COSINE TRANSFORM

$f(x)$	$F_C(\alpha)$
$\begin{cases} 1 & 0 < x < b \\ 0 & x > b \end{cases}$	$\dfrac{\sin b\alpha}{\alpha}$
$\dfrac{1}{x^2 + b^2}$	$\dfrac{\pi e^{-b\alpha}}{2b}$
e^{-bx}	$\dfrac{b}{\alpha^2 + b^2}$
$x^{n-1} e^{-bx}$	$\dfrac{\Gamma(n) \cos (n \tan^{-1} \alpha/b)}{(\alpha^2 + b^2)^{n/2}}$
e^{-bx^2}	$\dfrac{1}{2}\sqrt{\dfrac{\pi}{b}} e^{-\alpha^2/4b}$
$x^{-1/2}$	$\sqrt{\dfrac{\pi}{2\alpha}}$
x^{-n}	$\dfrac{\pi\alpha^{n-1} \sec (n\pi/2)}{2\,\Gamma(n)}, \quad 0 < n < 1$
$\ln\left(\dfrac{x^2 + b^2}{x^2 + c^2}\right)$	$\dfrac{e^{-c\alpha} - e^{-b\alpha}}{\pi\alpha}$
$\dfrac{\sin bx}{x}$	$\begin{cases} \pi/2 & \alpha < b \\ \pi/4 & \alpha = b \\ 0 & \alpha > b \end{cases}$
$\sin bx^2$	$\sqrt{\dfrac{\pi}{8b}}\left(\cos \dfrac{\alpha^2}{4b} - \sin \dfrac{\alpha^2}{4b}\right)$
$\cos bx^2$	$\sqrt{\dfrac{\pi}{8b}}\left(\cos \dfrac{\alpha^2}{4b} + \sin \dfrac{\alpha^2}{4b}\right)$
$\operatorname{sech} bx$	$\dfrac{\pi}{2b} \operatorname{sech} \dfrac{\pi\alpha}{2b}$
$\dfrac{\cosh (\sqrt{\pi}\, x/2)}{\cosh (\sqrt{\pi}\, x)}$	$\sqrt{\dfrac{\pi}{2}} \dfrac{\cosh (\sqrt{\pi}\, \alpha/2)}{\cosh (\sqrt{\pi}\, \alpha)}$
$\dfrac{e^{-b\sqrt{x}}}{\sqrt{x}}$	$\sqrt{\dfrac{\pi}{2\alpha}}\{\cos (2b\sqrt{\alpha}) - \sin (2b\sqrt{\alpha})\}$

LAPLACE TRANSFORM

DEFINITION OF THE LAPLACE TRANSFORM

ONE-DIMENSIONAL LAPLACE TRANSFORM

16.9 $$f(s)=\mathscr{L}\{F(t)\}=\int_0^\infty e^{-st}F(t)dt$$

$F(t)$ is a function of the real variable t and s is a complex variable. $F(t)$ is called the original function and $f(s)$ is called the image function. If the integral in **16.9** converges for a real $s=s_0$, i.e.,

$$\lim_{\substack{A\to 0\\ B\to\infty}}\int_A^B e^{-s_0t}F(t)dt$$

exists, then it converges for all s with $\mathscr{R}s>s_0$, and the image function is a single valued analytic function of s in the half-plane $\mathscr{R}s>s_0$.

DEFINITION OF THE INVERSE LAPLACE TRANSFORM OF f (s)

If $\mathcal{L}\{F(t)\} = f(s)$, then we say that $F(t) = \mathcal{L}^{-1}\{f(s)\}$ is the *inverse Laplace transform* of $f(s)$. $\mathcal{L}^{-1}$ is called the *inverse Laplace transform operator.*

COMPLEX INVERSION FORMULA

The inverse Laplace transform of $f(s)$ can be found directly by methods of complex variable theory. The result is

$$F(t) = \frac{1}{2\pi i}\int_{c-i\infty}^{c+i\infty} e^{st} f(s)\, ds = \frac{1}{2\pi i}\lim_{T\to\infty}\int_{c-iT}^{c+iT} e^{st} f(s)\, ds$$

where c is chosen so that all the singular points of $f(s)$ lie to the left of the line $\operatorname{Re}\{s\} =$ in the complex s plane.

The important property

$$L\{F^{(r)}(t)\} = \int_0^\infty e^{-st}\left(\frac{d^rF}{dt^r}\right)dt = s^rf(s) - \sum_{n=0}^{r-1} s^{r-1-n}F^{(n)}(+0)$$

makes the Laplace transform very useful for solving linear differential equations with constant coefficients, and many boundary value problems.

OPERATIONS FOR THE LAPLACE TRANSFORM

ORIGINAL FUNCTION F(t)	IMAGE FUNCTION f(s)
$F(t)$	$\int_0^\infty e^{-st}F(t)dt$
Inversion Formula	
$\frac{1}{2\pi i}\int_{c-i\infty}^{c+i\infty} e^{ts}f(s)ds$	$f(s)$
Linearity Property	
$AF(t)+BG(t)$	$Af(s)+Bg(s)$
Differentiation	
$F'(t)$	$sf(s)-F(+0)$
$F^{(n)}(t)$	$s^nf(s)-s^{n-1}F(+0)-s^{n-2}F'(+0)-\dots-F^{(n-1)}(+0)$
Integration	
$\int_0^t F(\tau)d\tau$	$\frac{1}{s}f(s)$
$\int_0^t\int_0^\tau F(\lambda)d\lambda d\tau$	$\frac{1}{s^2}f(s)$

CONVOLUTION (FALTUNG) THEOREM

$\int_0^t F_1(t-\tau)F_2(\tau)d\tau=F_1*F_2$	$f_1(s)f_2(s)$	
$-tF(t)$	$f'(s)$	**Differentiation**
$(-1)^nt^nF(t)$	$f^{(n)}(s)$	
	Integration	
$\frac{1}{t}F(t)$	$\int_s^\infty f(x)dx$	
	Linear Transformation	
$e^{at}F(t)$	$f(s-a)$	
$\frac{1}{c}F\left(\frac{t}{c}\right) \quad (c>0)$	$f(cs)$	

ORIGINAL FUNCTION F(t)	IMAGE FUNCTION f(s)
$\frac{1}{c} e^{(b/c)t} F\left(\frac{t}{c}\right) \quad (c>0)$	$f(cs-b)$
Translation	
$F(t-b)u(t-b) \quad (b>0)$	$e^{-bs}f(s)$
Periodic Functions	
$F(t+a)=F(t)$	$\dfrac{\int_0^a e^{-st}F(t)dt}{1-e^{-as}}$
16.10 $\quad F(t+a)=-F(t)$	$\dfrac{\int_0^a e^{-st}F(t)dt}{1+e^{-as}}$
Half-Wave Rectification of *F*(*t*) in 16.10	
$F(t) \sum_{n=0}^{\infty} (-1)^n u(t-na)$	$\dfrac{f(s)}{1-e^{-as}}$
Full-Wave Rectification of *F*(*t*) in 16.10	
$\lvert F(t)\rvert$	$f(s) \coth \frac{as}{2}$

HEAVISIDE EXPANSION THEOREM

$\sum_{n=1}^{m} \frac{p(a_n)}{q'(a_n)} e^{a_n t}$	$\frac{p(s)}{q(s)}, \; q(s)=(s-a_1)(s-a_2)\ldots(s-a_m)$ $p(s)$ a polynomial of degree $<m$
$e^{at} \sum_{n=1}^{r} \frac{p^{(r-n)}(a)}{(r-n)!} \frac{t^{n-1}}{(n-1)!}$	$\frac{p(s)}{(s-a)^r}$ $p(s)$ a polynomial of degree $<r$

TABLE OF LAPLACE TRANSFORMS[1]

$f(s)$	$F(t)$
$\frac{1}{s^n} \quad (n=1, 2, 3, \ldots)$	$\frac{t^{n-1}}{(n-1)!}$
$\frac{1}{\sqrt{s}}$	$\frac{1}{\sqrt{\pi t}}$
$s^{-3/2}$	$2\sqrt{t/\pi}$
$s^{-(n+\frac{1}{2})} \quad (n=1, 2, 3, \ldots)$	$\frac{2^n t^{n-\frac{1}{2}}}{1\cdot 3\cdot 5 \ldots (2n-1)\sqrt{\pi}}$

[1] From "Handbook of Mathematical Functions", National Bureau of Standards

$f(s)$	$F(t)$
$\dfrac{\Gamma(k)}{s^k}\qquad (k>0)$	t^{k-1}
$\dfrac{1}{s+a}$	e^{-at}
$\dfrac{1}{(s+a)^2}$	te^{-at}
$\dfrac{1}{(s+a)^n}\qquad (n=1, 2, 3, \ldots)$	$\dfrac{t^{n-1}e^{-at}}{(n-1)!}$
$\dfrac{\Gamma(k)}{(s+a)^k}\qquad (k>0)$	$t^{k-1}e^{-at}$
$\dfrac{1}{(s+a)(s+b)}\qquad (a\neq b)$	$\dfrac{e^{-at}-e^{-bt}}{b-a}$
$\dfrac{s}{(s+a)(s+b)}\qquad (a\neq b)$	$\dfrac{ae^{-at}-be^{-bt}}{a-b}$
$\dfrac{1}{(s+a)(s+b)(s+c)}$ (a, b, c distinct constants)	$-\dfrac{(b-c)e^{-at}+(c-a)e^{-bt}+(a-b)e^{-ct}}{(a-b)(b-c)(c-a)}$
$\dfrac{1}{s^2+a^2}$	$\dfrac{1}{a}\sin at$
$\dfrac{s}{s^2+a^2}$	$\cos at$
$\dfrac{1}{s^2-a^2}$	$\dfrac{1}{a}\sinh at$
$\dfrac{s}{s^2-a^2}$	$\cosh at$
$\dfrac{1}{s(s^2+a^2)}$	$\dfrac{1}{a^2}(1-\cos at)$
$\dfrac{1}{s^2(s^2+a^2)}$	$\dfrac{1}{a^3}(at-\sin at)$
$\dfrac{1}{(s^2+a^2)^2}$	$\dfrac{1}{2a^3}(\sin at-at\cos at)$

$f(s)$	$F(t)$
$\dfrac{s}{(s^2+a^2)^2}$	$\dfrac{t}{2a}\sin at$
$\dfrac{s^2}{(s^2+a^2)^2}$	$\dfrac{1}{2a}(\sin at+at\cos at)$
$\dfrac{s^2-a^2}{(s^2+a^2)^2}$	$t\cos at$
$\dfrac{s}{(s^2+a^2)(s^2+b^2)} \quad (a^2\neq b^2)$	$\dfrac{\cos at-\cos bt}{b^2-a^2}$
$\dfrac{1}{(s+a)^2+b^2}$	$\dfrac{1}{b}e^{-at}\sin bt$
$\dfrac{s+a}{(s+a)^2+b^2}$	$e^{-at}\cos bt$
$\dfrac{3a^2}{s^3+a^3}$	$e^{-at}-e^{\frac{1}{2}at}\left(\cos\dfrac{at\sqrt{3}}{2}-\sqrt{3}\sin\dfrac{at\sqrt{3}}{2}\right)$
$\dfrac{4a^3}{s^4+4a^4}$	$\sin at\cosh at-\cos at\sinh at$
$\dfrac{s}{s^4+4a^4}$	$\dfrac{1}{2a^2}\sin at\sinh at$
$\dfrac{1}{s^4-a^4}$	$\dfrac{1}{2a^3}(\sinh at-\sin at)$
$\dfrac{s}{s^4-a^4}$	$\dfrac{1}{2a^2}(\cosh at-\cos at)$
$\dfrac{8a^3s^2}{(s^2+a^2)^3}$	$(1+a^2t^2)\sin at-at\cos at$
$\dfrac{1}{s}\left(\dfrac{s-1}{s}\right)^n$	$L_n(t)$
$\dfrac{s}{(s+a)^{\frac{3}{2}}}$	$\dfrac{1}{\sqrt{\pi t}}e^{-at}(1-2at)$
$\sqrt{s+a}-\sqrt{s+b}$	$\dfrac{1}{2\sqrt{\pi t^3}}(e^{-bt}-e^{-at})$
$\dfrac{1}{\sqrt{s}+a}$	$\dfrac{1}{\sqrt{\pi t}}-ae^{a^2t}\operatorname{erfc} a\sqrt{t}$

$f(s)$	$F(t)$
$\dfrac{\sqrt{s}}{s-a^2}$	$\dfrac{1}{\sqrt{\pi t}}+ae^{a^2t}\operatorname{erf} a\sqrt{t}$
$\dfrac{\sqrt{s}}{s+a^2}$	$\dfrac{1}{\sqrt{\pi t}}-\dfrac{2a}{\sqrt{\pi}}\, e^{-a^2t}\displaystyle\int_0^{a\sqrt{t}} e^{\lambda^2}d\lambda$
$\dfrac{1}{\sqrt{s}(s-a^2)}$	$\dfrac{1}{a}\, e^{a^2t}\operatorname{erf} a\sqrt{t}$
$\dfrac{1}{\sqrt{s}(s+a^2)}$	$\dfrac{2}{a\sqrt{\pi}}\, e^{-a^2t}\displaystyle\int_0^{a\sqrt{t}} e^{\lambda^2}d\lambda$
$\dfrac{b^2-a^2}{(s-a^2)(b+\sqrt{s})}$	$e^{a^2t}[b-a\operatorname{erf} a\sqrt{t}]-be^{b^2t}\operatorname{erfc} b\sqrt{t}$
$\dfrac{1}{\sqrt{s}(\sqrt{s}+a)}$	$e^{a^2t}\operatorname{erfc} a\sqrt{t}$
$\dfrac{1}{(s+a)\sqrt{s+b}}$	$\dfrac{1}{\sqrt{b-a}}\, e^{-at}\operatorname{erf}(\sqrt{b-a}\sqrt{t})$
$\dfrac{b^2-a^2}{\sqrt{s}(s-a^2)(\sqrt{s}+b)}$	$e^{a^2t}\left[\dfrac{b}{a}\operatorname{erf}(a\sqrt{t})-1\right]+e^{b^2t}\operatorname{erfc} b\sqrt{t}$
$\dfrac{(1-s)^n}{s^{n+\frac{1}{2}}}$	$\dfrac{n!}{(2n)!\sqrt{\pi t}}\, H_{2n}(\sqrt{t})$
$\dfrac{(1-s)^n}{s^{n+\frac{3}{2}}}$	$\dfrac{n!}{(2n+1)!\sqrt{\pi}}\, H_{2n+1}(\sqrt{t})$
$\dfrac{\sqrt{s+2a}}{\sqrt{s}}-1$	$ae^{-at}[I_1(at)+I_0(at)]$
$\dfrac{1}{\sqrt{s+a}\sqrt{s+b}}$	$e^{-\frac{1}{2}(a+b)t}I_0\left(\dfrac{a-b}{2}\, t\right)$
$\dfrac{\Gamma(k)}{(s+a)^k(s+b)^k}\quad (k>0)$	$\sqrt{\pi}\left(\dfrac{t}{a-b}\right)^{k-\frac{1}{2}} e^{-\frac{1}{2}(a+b)t}I_{k-\frac{1}{2}}\left(\dfrac{a-b}{2}\, t\right)$
$\dfrac{1}{(s+a)^{\frac{1}{2}}(s+b)^{\frac{3}{2}}}$	$te^{-\frac{1}{2}(a+b)t}\left[I_0\left(\dfrac{a-b}{2}\, t\right)+I_1\left(\dfrac{a-b}{2}\, t\right)\right]$
$\dfrac{\sqrt{s+2a}-\sqrt{s}}{\sqrt{s+2a}+\sqrt{s}}$	$\dfrac{1}{t}\, e^{-at}I_1(at)$

$f(s)$	$F(t)$
$\dfrac{(a-b)^k}{(\sqrt{s+a}+\sqrt{s+b})^{2k}} \quad (k>0)$	$\dfrac{k}{t}\, e^{-\frac{1}{2}(a+b)t} I_k\left(\dfrac{a-b}{2}\, t\right)$
$\dfrac{(\sqrt{s+a}+\sqrt{s})^{-2\nu}}{\sqrt{s}\sqrt{s+a}} \quad (\nu>-1)$	$\dfrac{1}{a^\nu}\, e^{-\frac{1}{2}at} I_\nu(\tfrac{1}{2}at)$
$\dfrac{1}{\sqrt{s^2+a^2}}$	$J_0(at)$
$\dfrac{(\sqrt{s^2+a^2}-s)^\nu}{\sqrt{s^2+a^2}} \quad (\nu>-1)$	$a^\nu J_\nu(at)$
$\dfrac{1}{(s^2+a^2)^k} \quad (k>0)$	$\dfrac{\sqrt{\pi}}{\Gamma(k)}\left(\dfrac{t}{2a}\right)^{k-\frac{1}{2}} J_{k-\frac{1}{2}}(at)$
$(\sqrt{s^2+a^2}-s)^k \quad (k>0)$	$\dfrac{ka^k}{t}\, J_k(at)$
$\dfrac{(s-\sqrt{s^2-a^2})^\nu}{\sqrt{s^2-a^2}} \quad (\nu>-1)$	$a^\nu I_\nu(at)$
$\dfrac{1}{(s^2-a^2)^k} \quad (k>0)$	$\dfrac{\sqrt{\pi}}{\Gamma(k)}\left(\dfrac{t}{2a}\right)^{k-\frac{1}{2}} I_{k-\frac{1}{2}}(at)$
$\dfrac{1}{s}\, e^{-ks}$	$u(t-k)$
$\dfrac{1}{s^2}\, e^{-ks}$	$(t-k)u(t-k)$
$\dfrac{1}{s^\mu}\, e^{-ks} \quad (\mu>0)$	$\dfrac{(t-k)^{\mu-1}}{\Gamma(\mu)}\, u(t-k)$
$\dfrac{1-e^{-ks}}{s}$	$u(t)-u(t-k)$

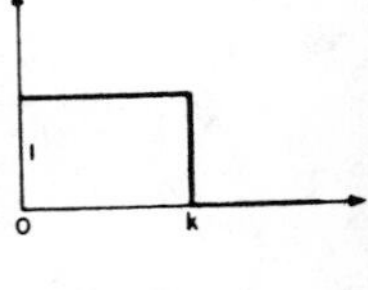

$f(s)$	$F(t)$
$\dfrac{1}{s(1-e^{-ks})}=\dfrac{1+\coth \frac{1}{2}ks}{2s}$	$\sum_{n=0}^{\infty} u(t-nk)$

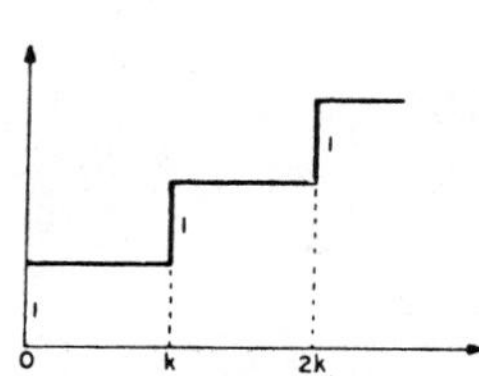

$\dfrac{1}{s(e^{ks}-a)}$	$\sum_{n=1}^{\infty} a^{n-1}u(t-nk)$

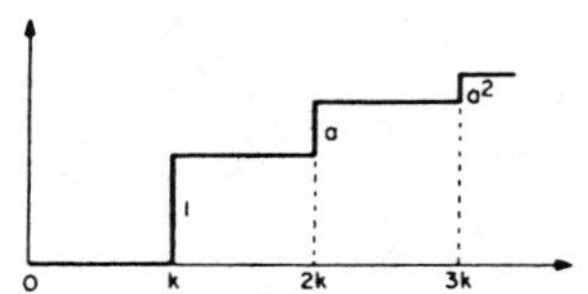

$\dfrac{1}{s}\tanh ks$	$u(t)+2\sum_{n=1}^{\infty} (-1)^n u(t-2nk)$

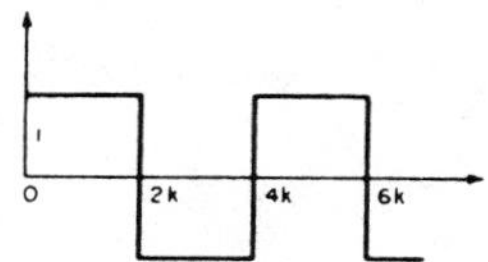

$\dfrac{1}{s(1+e^{-ks})}$	$\sum_{n=0}^{\infty} (-1)^n u(t-nk)$

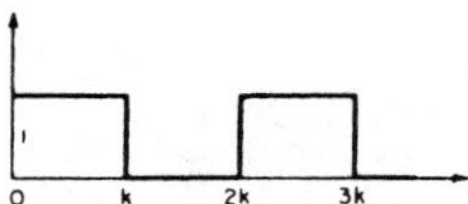

$\dfrac{1}{s^2}\tanh ks$	$tu(t)+2\sum_{n=1}^{\infty} (-1)^n (t-2nk)u(t-2nk)$

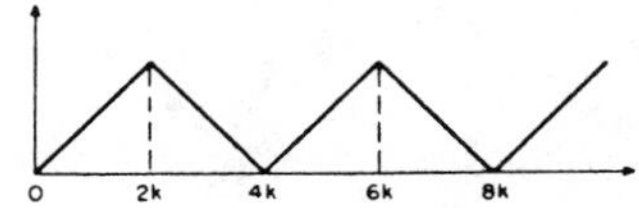

$\dfrac{1}{s\sinh ks}$	$2\sum_{n=0}^{\infty} u[t-(2n+1)k]$

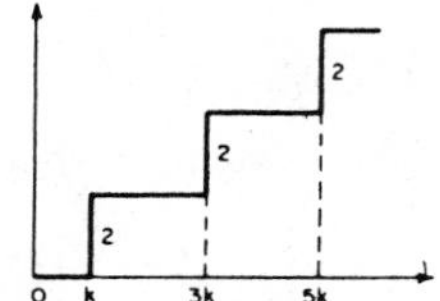

$f(s)$	$F(t)$
$\dfrac{1}{s \cosh ks}$	$2 \sum_{n=0}^{\infty} (-1)^n u[t-(2n+1)k]$ 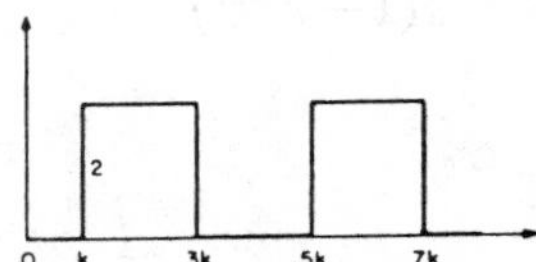
$\dfrac{1}{s} \coth ks$	$u(t)+2 \sum_{n=1}^{\infty} u(t-2nk)$ 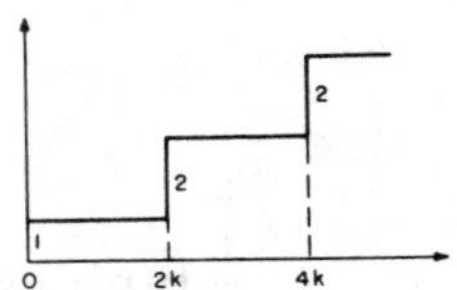
$\dfrac{k}{s^2+k^2} \coth \dfrac{\pi s}{2k}$	$\lvert\sin kt\rvert$ 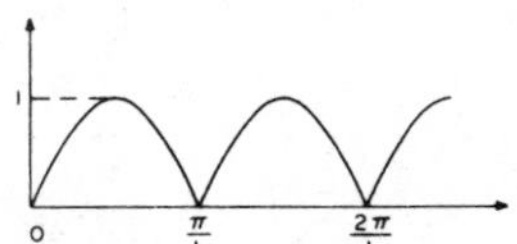
$\dfrac{1}{(s^2+1)(1-e^{-\pi s})}$	$\sum_{n=0}^{\infty} (-1)^n u(t-n\pi) \sin t$ 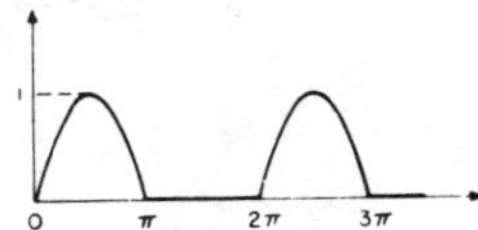
$\dfrac{1}{s} e^{-\frac{k}{s}}$	$J_0(2\sqrt{kt})$
$\dfrac{1}{\sqrt{s}} e^{-\frac{k}{s}}$	$\dfrac{1}{\sqrt{\pi t}} \cos 2\sqrt{kt}$
$\dfrac{1}{\sqrt{s}} e^{\frac{k}{s}}$	$\dfrac{1}{\sqrt{\pi t}} \cosh 2\sqrt{kt}$
$\dfrac{1}{s^{3/2}} e^{-\frac{k}{s}}$	$\dfrac{1}{\sqrt{\pi k}} \sin 2\sqrt{kt}$
$\dfrac{1}{s^{3/2}} e^{\frac{k}{s}}$	$\dfrac{1}{\sqrt{\pi k}} \sinh 2\sqrt{kt}$
$\dfrac{1}{s^{\mu}} e^{-\frac{k}{s}} \quad (\mu>0)$	$\left(\dfrac{t}{k}\right)^{\frac{\mu-1}{2}} J_{\mu-1}(2\sqrt{kt})$

$f(s)$		$F(t)$
$\frac{1}{s^{\mu}} e^{\frac{k}{s}}$	$(\mu>0)$	$\left(\frac{t}{k}\right)^{\frac{\mu-1}{2}} I_{\mu-1}(2\sqrt{kt})$
$e^{-k\sqrt{s}}$	$(k>0)$	$\frac{k}{2\sqrt{\pi t^3}} \exp\left(-\frac{k^2}{4t}\right)$
$\frac{1}{s} e^{-k\sqrt{s}}$	$(k\geq 0)$	$\operatorname{erfc} \frac{k}{2\sqrt{t}}$
$\frac{1}{\sqrt{s}} e^{-k\sqrt{s}}$	$(k\geq 0)$	$\frac{1}{\sqrt{\pi t}} \exp\left(-\frac{k^2}{4t}\right)$
$\frac{1}{s^{\frac{3}{2}}} e^{-k\sqrt{s}}$	$(k\geq 0)$	$2\sqrt{\frac{t}{\pi}} \exp\left(-\frac{k^2}{4t}\right)-k \operatorname{erfc} \frac{k}{2\sqrt{t}}$ $=2\sqrt{t}\ \mathrm{i}\operatorname{erfc} \frac{k}{2\sqrt{t}}$
$\frac{1}{s^{1+\frac{1}{2}n}} e^{-k\sqrt{s}}$	$(n=0, 1, 2, \ldots; k\geq 0)$	$(4t)^{\frac{1}{2}n}\ \mathrm{i}^n \operatorname{erfc} \frac{k}{2\sqrt{t}}$
$s^{\frac{n-1}{2}} e^{-k\sqrt{s}}$	$(n=0, 1, 2, \ldots; k>0)$	$\frac{\exp\left(-\frac{k^2}{4t}\right)}{2^n\sqrt{\pi t^{n+1}}} H_n\left(\frac{k}{2\sqrt{t}}\right)$
$\frac{e^{-k\sqrt{s}}}{a+\sqrt{s}}$	$(k\geq 0)$	$\frac{1}{\sqrt{\pi t}} \exp\left(-\frac{k^2}{4t}\right)$ $-ae^{ak}e^{a^2t} \operatorname{erfc}\left(a\sqrt{t}+\frac{k}{2\sqrt{t}}\right)$
$\frac{ae^{-k\sqrt{s}}}{s(a+\sqrt{s})}$	$(k\geq 0)$	$-e^{ak}e^{a^2t} \operatorname{erfc}\left(a\sqrt{t}+\frac{k}{2\sqrt{t}}\right)+\operatorname{erfc} \frac{k}{2\sqrt{t}}$
$\frac{e^{-k\sqrt{s}}}{\sqrt{s}(a+\sqrt{s})}$	$(k\geq 0)$	$e^{ak}e^{a^2t} \operatorname{erfc}\left(a\sqrt{t}+\frac{k}{2\sqrt{t}}\right)$
$\frac{e^{-k\sqrt{s(s+a)}}}{\sqrt{s(s+a)}}$	$(k\geq 0)$	$e^{-\frac{1}{2}at}I_0(\frac{1}{2}a\sqrt{t^2-k^2})u(t-k)$
$\frac{e^{-k\sqrt{s^2+a^2}}}{\sqrt{s^2+a^2}}$	$(k\geq 0)$	$J_0(a\sqrt{t^2-k^2})u(t-k)$

$f(s)$		$F(t)$
$\dfrac{e^{-k\sqrt{s^2-a^2}}}{\sqrt{s^2-a^2}}$	$(k\geq 0)$	$I_0(a\sqrt{t^2-k^2})u(t-k)$
$\dfrac{e^{-k(\sqrt{s^2+a^2}-s)}}{\sqrt{s^2+a^2}}$	$(k\geq 0)$	$J_0(a\sqrt{t^2+2kt})$
$e^{-ks}-e^{-k\sqrt{s^2+a^2}}$	$(k>0)$	$\dfrac{ak}{\sqrt{t^2-k^2}}J_1(a\sqrt{t^2-k^2})u(t-k)$
$e^{-k\sqrt{s^2-a^2}}-e^{-ks}$	$(k>0)$	$\dfrac{ak}{\sqrt{t^2-k^2}}I_1(a\sqrt{t^2-k^2})u(t-k)$
$\dfrac{a^\nu e^{-k\sqrt{s^2+a^2}}}{\sqrt{s^2+a^2}(\sqrt{s^2+a^2}+s)^\nu}$	$(\nu>-1, k\geq 0)$	$\left(\dfrac{t-k}{t+k}\right)^{\frac{1}{2}\nu}J_\nu(a\sqrt{t^2-k^2})u(t-k)$
$\dfrac{1}{s}\ln s$		$-\gamma-\ln t$ (γ=.57721 56649 . . . Euler's constant)
$\dfrac{1}{s^k}\ln s$	$(k>0)$	$\dfrac{t^{k-1}}{\Gamma(k)}[\psi(k)-\ln t]$
$\dfrac{\ln s}{s-a}$	$(a>0)$	$e^{at}[\ln a+E_1(at)]$
$\dfrac{\ln s}{s^2+1}$		$\cos t\ \mathrm{Si}\ (t)-\sin t\ \mathrm{Ci}\ (t)$
$\dfrac{s\ln s}{s^2+1}$		$-\sin t\ \mathrm{Si}\ (t)-\cos t\ \mathrm{Ci}\ (t)$
$\dfrac{1}{s}\ln(1+ks)$	$(k>0)$	$E_1\left(\dfrac{t}{k}\right)$
$\ln\dfrac{s+a}{s+b}$		$\dfrac{1}{t}(e^{-bt}-e^{-at})$
$\dfrac{1}{s}\ln(1+k^2s^2)$	$(k>0)$	$-2\ \mathrm{Ci}\left(\dfrac{t}{k}\right)$
$\dfrac{1}{s}\ln(s^2+a^2)$	$(a>0)$	$2\ln a-2\ \mathrm{Ci}\ (at)$

$f(s)$	$F(t)$
$\frac{1}{s^2}\ln(s^2+a^2) \qquad (a>0)$	$\frac{2}{a}[at\ln a+\sin at-at\,\mathrm{Ci}(at)]$
$\ln\frac{s^2+a^2}{s^2}$	$\frac{2}{t}(1-\cos at)$
$\ln\frac{s^2-a^2}{s^2}$	$\frac{2}{t}(1-\cosh at)$
$\arctan\frac{k}{s}$	$\frac{1}{t}\sin kt$
$\frac{1}{s}\arctan\frac{k}{s}$	$\mathrm{Si}(kt)$
$e^{k^2s^2}\operatorname{erfc} ks \qquad (k>0)$	$\frac{1}{k\sqrt{\pi}}\exp\left(-\frac{t^2}{4k^2}\right)$
$\frac{1}{s}e^{k^2s^2}\operatorname{erfc} ks \qquad (k>0)$	$\operatorname{erf}\frac{t}{2k}$
$e^{ks}\operatorname{erfc}\sqrt{ks} \qquad (k>0)$	$\frac{\sqrt{k}}{\pi\sqrt{t}(t+k)}$
$\frac{1}{\sqrt{s}}\operatorname{erfc}\sqrt{ks} \qquad (k\geq 0)$	$\frac{1}{\sqrt{\pi t}}u(t-k)$
$\frac{1}{\sqrt{s}}e^{ks}\operatorname{erfc}\sqrt{ks} \qquad (k\geq 0)$	$\frac{1}{\sqrt{\pi(t+k)}}$
$\operatorname{erf}\frac{k}{\sqrt{s}}$	$\frac{1}{\pi t}\sin 2k\sqrt{t}$
$\frac{1}{\sqrt{s}}e^{\frac{k^2}{s}}\operatorname{erfc}\frac{k}{\sqrt{s}}$	$\frac{1}{\sqrt{\pi t}}e^{-2k\sqrt{t}}$
$K_0(ks) \qquad (k>0)$	$\frac{1}{\sqrt{t^2-k^2}}u(t-k)$
$K_0(k\sqrt{s}) \qquad (k>0)$	$\frac{1}{2t}\exp\left(-\frac{k^2}{4t}\right)$
$\frac{1}{s}e^{ks}K_1(ks) \qquad (k>0)$	$\frac{1}{k}\sqrt{t(t+2k)}$
$\frac{1}{\sqrt{s}}K_1(k\sqrt{s}) \qquad (k>0)$	$\frac{1}{k}\exp\left(-\frac{k^2}{4t}\right)$

$f(s)$	$F(t)$
$\frac{1}{\sqrt{s}} e^{\frac{k}{s}} K_0\left(\frac{k}{s}\right) \quad (k>0)$	$\frac{2}{\sqrt{\pi t}} K_0(2\sqrt{2kt})$
$\pi e^{-ks} I_0(ks) \quad (k>0)$	$\frac{1}{\sqrt{t(2k-t)}} [u(t)-u(t-2k)]$
$e^{-ks} I_1(ks) \quad (k>0)$	$\frac{k-t}{\pi k\sqrt{t(2k-t)}} [u(t)-u(t-2k)]$
$e^{as} E_1(as) \quad (a>0)$	$\frac{1}{t+a}$
$\frac{1}{a} - s e^{as} E_1(as) \quad (a>0)$	$\frac{1}{(t+a)^2}$
$a^{1-n} e^{as} E_n(as) \quad (a>0; n=0, 1, 2, \ldots)$	$\frac{1}{(t+a)^n}$
$\left[\frac{\pi}{2} - \mathrm{Si}(s)\right] \cos s + \mathrm{Ci}(s) \sin s$	$\frac{1}{t^2+1}$

Saw tooth wave function

$$\frac{1}{as^2} - \frac{e^{-as}}{s(1-e^{-as})}$$

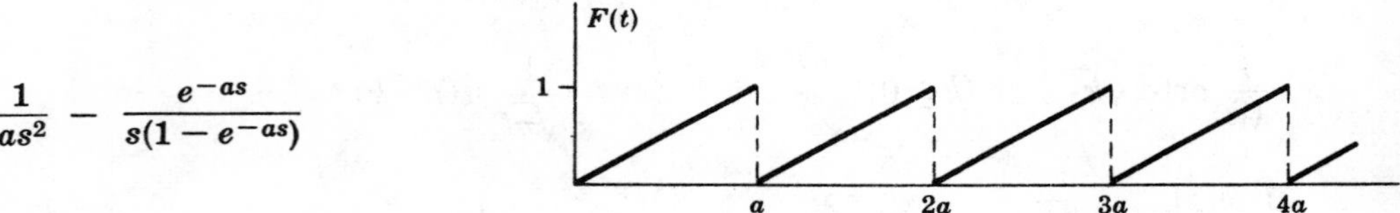

Pulse function

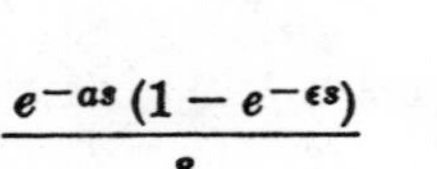

$$\frac{e^{-as}(1-e^{-\epsilon s})}{s}$$

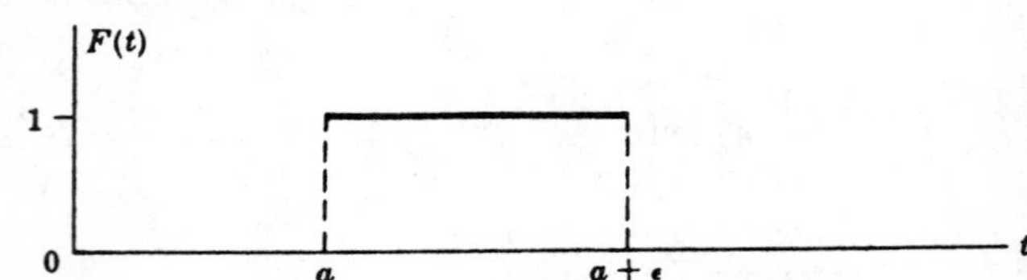

$F(t) = n^2, \; n \leqq t < n+1, \; n = 0, 1, 2, \ldots$

$$\frac{e^{-s} + e^{-2s}}{s(1-e^{-s})^2}$$

F(t)
4
3
2
1
0
1
2
3
t

$$\frac{\pi a(1 + e^{-as})}{a^2 s^2 + \pi^2}$$

$$F(t) = \begin{cases} \sin(\pi t/a) & 0 \leqq t \leqq a \\ 0 & t > a \end{cases}$$

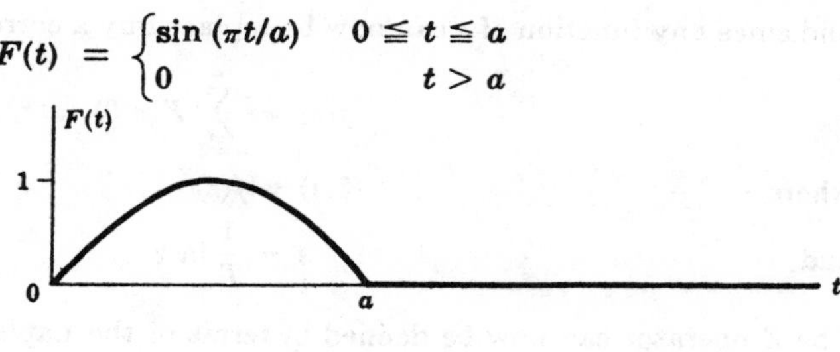

THE Z TRANSFORM

When $F(t)$, a continuous function of time, is sampled at regular intervals of period T the usual Laplace transform techniques are modified. The diagramatic form of a simple sampler together with its associated input-output waveforms is shown below

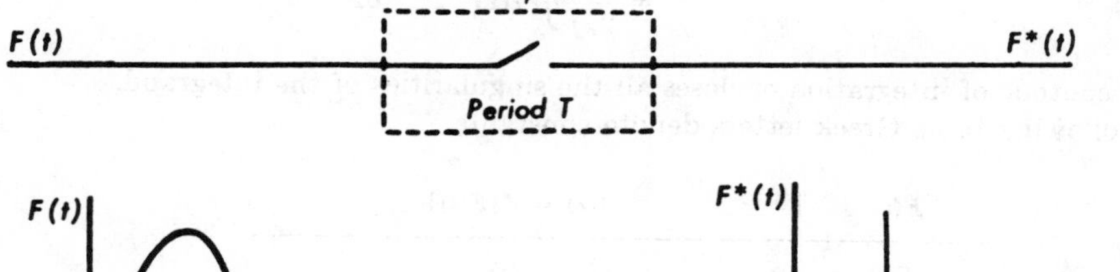

$$\frac{1}{T} \equiv F_s \qquad \text{the sampling frequency}$$

Defining the set of impulse functions $\delta_T(t)$ by

$$\delta_T(t) \equiv \sum_{n=0}^{\infty} \delta(t - nT)$$

the input-output relationship of the sampler becomes

$$\begin{aligned} F^*(t) &= F(t) \cdot \delta_T(t) \\ &= \sum_{n=0}^{\infty} F(nT) \cdot \delta(t - nT). \end{aligned}$$

While for a given $F(t)$ and T the $F^*(t)$ is unique, the converse is not true.

The Laplace transform can be used to define $F^*(s)$ as follows

$$\begin{aligned} L\{F^*(t)\} &\equiv f^*(s) \\ &= \sum_{n=0}^{\infty} F(nT) \cdot e^{-nTs}. \end{aligned}$$

The variable 'z' is introduced by means of the transformation

$$z = e^{Ts}$$

and since any function of s can now be replaced by a corresponding function of z we have

$$f(z) = \sum_{n=0}^{\infty} F(nT) \cdot z^{-n}$$

where
$$f^*(s) \equiv f(z)$$

and
$$s = \frac{1}{T} \ln z$$

The Z operator can now be defined in terms of the Laplace operator by the relationship

$$Z\{F(t)\} \equiv L\{F^*(t)\}$$

An alternative definition (quoted without proof) is

$$Z\{F(t)\} = \sum \text{residues of} \left[\left(\frac{1}{1 - e^{Tz}z^{-1}}\right) \cdot f(z)\right]$$

The inverse z transform

$$Z^{-1}\{f(z)\} \equiv F^*(t)$$
$$= \frac{1}{2\pi j} \oint f(z) \cdot z^{n-1}\, dz$$

where the contour of integration encloses all the singularities of the integrand.

In the following table Greek letters denote constants.

$F(t)$	$f(z) = Z\{F(t)\}$
$\alpha F(t)$	$\alpha f(z)$
$F(t) + G(t)$	$f(z) + g(z)$
$F(t + T)$	$zf(z) - zF(0)$
$F(t + 2T)$	$z^2 f(z) - z^2 F(0) - zF(T)$
$F(t + mT)$	$z^m f(z) - \sum_{r=0}^{m-1} z^{m-r} F(rT)$
	$= z^m f(z)$ when $F(rT) = 0,\ 0 \le r \le m - 1$
$F(t - mT)$	$z^{-m} f(z)$
$e^{\alpha t} F(t)$	$f(e^{-\alpha T} z)$
$e^{-\alpha t} F(t)$	$f(e^{\alpha T} z)$
$t \cdot F(t)$	$-Tz \frac{d}{dz} f(z)$
$t^{-1} F(t)$	$-\frac{1}{T} \int_0^z \frac{f(z)}{z}\, dz$
$\sum_{m=0}^{T/t} F(mT)$	$\left(\frac{z}{z-1}\right) f(z)$

The following limits are also valid

$$\lim_{t \to 0} F(t) = \lim_{z \to \infty} f(z)$$

$$\lim_{t \to \infty} F^*(t) = \lim_{z \to 1} \left[\left(\frac{z-1}{z}\right) f(z)\right]$$

In the table which follows, the Heavyside unit step function is defined by

$$H(t - nT) \equiv \begin{cases} 1;\ t \ge nT \\ 0;\ t < nT. \end{cases}$$

THE Z TRANSFORM

$F(t)$	$f(z)$
$\delta(t)$	1
$\delta(t - mT)$	$\frac{1}{z^m}$
$H(t)$	$\frac{z}{z-1}$
$H(t - T)$	$\frac{1}{z-1}$
$H(t - mT)$	$\frac{z}{z^m \cdot (z-1)}$
$H(t) - H(t - T)$	1
$H(t) - H(t - 2T)$	$1 + \frac{1}{z}$
$H(t - mT) - H(t - \overline{m+1}T)$	$\frac{1}{z} m$
$\frac{T}{t} H(t - T)$	$\ln\left(\frac{z}{z-1}\right)$
t	$\frac{Tz}{(z-1)^2}$
t^2	$\frac{T^2 z(z+1)}{(z-1)^3}$
t^3	$\frac{T^3 z(z^2 + 4z + 1)}{(z-1)^4}$
t^n	$(-1)^n \lim_{\chi \to 0} \frac{\partial^n}{\partial \chi^n}\left(\frac{z}{z - e^{-\chi T}}\right)$
$1 - a^{\omega t}$	$\frac{z(1 - a^{\omega T})}{(z-1)(z - a^{\omega T})}$
$a^{\omega t}$	$\frac{z}{(z - a^{\omega T})}$
$ta^{\omega t}$	$\frac{Tza^{\omega T}}{(z - a^{\omega T})^2}$
$t^2 a^{\omega t}$	$\frac{T^2 a^{\omega T} z(z + a^{\omega T})}{(z - a^{\omega T})^3}$
$\sin \omega t$	$\frac{z \sin \omega T}{z^2 - 2z \cos \omega T + 1}$
$\cos \omega t$	$\frac{z(z - \cos \omega T)}{z^2 - 2z \cos \omega T + 1}$
$\sinh \omega t$	$\frac{z \sinh \omega T}{z^2 - 2z \cosh \omega T + 1}$
$\cosh \omega t$	$\frac{z(z - \cosh \omega T)}{z^2 - 2z \cosh \omega T + 1}$
$e^{-\alpha t} \sin \omega t$	$\frac{ze^{-\alpha T} \sin \omega T}{z^2 - 2ze^{-T} \cos \omega T + e^{-2\alpha T}}$
$e^{-\alpha t} \cos \omega t$	$\frac{z(z - e^{-\alpha T} \cos \omega T)}{z^2 - 2ze^{-T} \cos \omega T + e^{-2\alpha T}}$
$e^{-\alpha t} \sinh \omega t$	$\frac{ze^{-\alpha T} \sinh \omega T}{z^2 - 2ze^{-\alpha T} \cosh \omega T + e^{-2\alpha T}}$
$e^{-\alpha t} \cosh \omega t$	$\frac{z(z - e^{-\alpha T} \cosh \omega T)}{z^2 - 2ze^{-\alpha T} \cosh \omega T + e^{-2\alpha T}}$

$F(t)$	$f(z)$
$-\frac{1}{a}[\delta(t) - a^{t/T}]$	$\frac{1}{z-a}$
$\frac{1}{(a-b)}\left[a^{\left(\frac{t}{T}-1\right)} - b^{\left(\frac{t}{T}-1\right)}\right]$	$\frac{1}{(z-a)(z-b)}$
$\frac{1}{(a-b)}\left[a^{\frac{t}{T}} - b^{\frac{t}{T}}\right]$	$\frac{z}{(z-a)(z-b)}$
$\frac{1}{(a-b)}\left[(a-c)a^{\left(\frac{t}{T}-1\right)} - (b-c)b^{\left(\frac{t}{T}-1\right)}\right]$	$\frac{z-c}{(z-a)(z-b)}$
$\frac{1}{(a-b)}\left[a^{\left(\frac{t}{T}+1\right)} - b^{\left(\frac{T}{t}+1\right)}\right]$	$\frac{z^2}{(z-a)(z-b)}$
$\frac{1}{\left(\frac{t}{T}\right)!}$	$e^{1/z}$
$\frac{1}{\left(\frac{2t}{T}\right)!}$	$\cosh(z^{-\frac{1}{2}})$

MEHTODS OF EVALUATING INVERSE z TRANSFORMS

(1) Cauchy's residue theorem.

For $t = nT$,

$$G(nT) = \sum_{\text{all } z_k} [\text{residues of } g(z)z^{n-1} \text{ at } z_k]$$

where the z_k define all the poles of $g(z)z^{n-1}$.

(2) Partial fractions.

Expand $g(z)/z$ into partial fractions. The product of z with each of the partial fractio will then be recognizable from the standard forms in the table of z transforms. Note how ever that the continuous functions obtained are only valid at the sampling instants.

(3) Power series expansion by long division using detached coefficients.

$g(z)$ is expanded into a power series in z^{-1} and the coefficient of the term in z^{-n} is th value of $g(nT)$ i.e. the value of $G(t)$ at the nth sampling instant.

The z transform as a means of determining approximately the inverse Laplace transforn

Since

$$z \equiv e^{Ts}$$

$$s^{-1} = \frac{T}{2}\left[\frac{1}{v} - \frac{v}{3} - \frac{4v^3}{45} - \frac{44v^5}{945} - \cdots\right]$$

where

$$v \equiv \frac{1 - z^{-1}}{1 + z^{-1}},$$

the series being very rapid in its convergence. Given $g(s)$, to find its inverse Laplac transform the following operations are carried out:-

i Divide the numerator and denominator of $g(s)$ by the highest power of s yielding as a alternative form for $g(s)$ the quotient of two polynomials in s^{-1}.

ii Chose as a numerical value of T, that which makes $2\pi/T$ much larger than the imagi nary part of the poles of $G(s)$.

iii Substitute into the alternative form for $g(s)$ obtained in i above the expansion fo s^{-n} determined from the following short table of approximations.

Do not at this stage insert the numerical value for T as tabulations with different intervals may be required.

iv Divide by T.

v Insert the chosen value for T and divide the numerator by the denominator.

vi The coefficient of z^{-n} is the required value of the function at $t = nT$.

s^{-n}	z transform (approximate)
s^{-1}	$\frac{T}{2}\left[\frac{1 + z^{-1}}{1 - z^{-1}}\right]$
s^{-2}	$\frac{T^2}{12}\left[\frac{1 + 10z^{-1} + z^{-2}}{(1 - z^{-1})^2}\right]$
s^{-3}	$\frac{T^3}{3}\left[\frac{z^{-1} + z^{-2}}{(1 - z^{-1})^3}\right]$
s^{-4}	$\frac{T^4}{144}\left[\frac{1 + 20z^{-1} + 102z^{-2} + 20z^{-3} + z^{-4}}{(1 - z^{-1})^4}\right]$
s^{-5}	$\frac{T^5}{24}\left[\frac{z^{-1} + 11z^{-2} + 11z^{-3} + z^{-4}}{(1 - z^{-1})^5}\right]$
s^{-6}	$\frac{T^6}{4}\left[\frac{z^{-2} + 2z^{-3} + z^{-4}}{(1 - z^{-1})^6}\right]$
s^{-7}	$\frac{T^7}{8}\left[\frac{z^{-2} + 3z^{-3} + 3z^{-4} + z^{-5}}{(1 - z^{-1})^7}\right]$

SCIENCE AND ENGINEERING DATA

CHAPTER 17

PHYSICAL CONSTANTS AND UNITS CONVERSION

Physical Constants I

Quantity	*Symbol*	*Value*	*Uncertainty* (ppm)	*Units* SI
Speed of light in vacuum	c	2.997924580(12)	0.004	10^8 m sec^{-1}
Fine-structure constant, $[\mu_0 c^2/4\pi](e^2/\hbar c)$	α	7.2973506(60)	0.82	10^{-3}
	α^{-1}	137.03604(11)	0.82	–
Electron charge	e	1.6021892(46)	2.9	10^{-19}C
		4.803242(14)	2.9	–
Planck constant	h	6.626176(36)	5.4	10^{-34} J sec
	$\hbar = h/2\pi$	1.0545887(57)	5.4	10^{-34} J sec
Avogadro constant	N	6.022045(31)	5.1	10^{26} kmol^{-1}
Atomic mass unit	u	1.6605655(86)	5.1	10^{-27} kg
Electron rest mass	m_e	9.109534 (47)	5.1	10^{-31} kg
		5.110034 (14)†	2.8†	–
$Nm_e = M_e$	M_e	5.4858026 (21)	0.38	10^{-4} u
Proton rest mass	m_p	1.6726485 (86)	5.1	10^{-27} kg
		9.382796 (27)†	2.8†	–
$Nm_p = M_p$	M_p	1.007276471 (11)	0.001	u
Ratio of proton mass to electron mass	m_p/m_e	1836.15152(70)	0.38	–
Neutron rest mass	m_n	1.6749543 (86)	5.1	10^{-27} kg
		9.395731 (27)†	2.8†	–
$Nm_p = M_n$	M_n	1.008665012(37)	0.037	u
Electron charge to mass ratio	e/m_e	1.7588047(49)	2.8	10^{11} C kg^{-1}
		5.272764(15)	2.8	–
Magnetic flux quantum, $[c]^{-1}(hc/2e)$	Φ_0	2.0678506(54)	2.6	10^{-15} Wb
	h/e	4.135701(11)	2.6	10^{-15} J sec C^{-1}
		1.3795215(36)	2.6	–
Josephson frequency-voltage ratio	$2e/h$	4.835939(13)	2.6	10^{14} Hz V^{-1}
Quantum of circulation	$h/2m_e$	3.6369455(60)	1.6	10^{-4} J sec kg^{-1}
	h/m_e	7.273891(12)	1.6	10^{-4} J sec kg^{-1}
Faraday constant	$\mathcal{F}$	9.648456(27)	2.8	10^7 C kmol^{-1}
		2.8925342(82)	2.8	–
Rydberg constant, $[\mu_0 c^2/4\pi]^2(m_e e^4/4\pi\hbar^3 c)$	R_∞	1.097373177(83)	0.075	10^7 m^{-1}
Bohr radius, $[\mu_0 c^2/4\pi]^{-1}(\hbar^2/m_e e^2) = \alpha/4\pi R_\infty$	a_0	5.2917706(44)	0.82	10^{-11} m
Classical electron radius, $[\mu_0 c^2/4\pi](e^2/m_e c^2) = \alpha^3/4\pi R_\infty$	r_0	2.8179380(70)	2.5	10^{-15} m
Free electron g-factor, or electron magnetic moment in Bohr magnetons	$g_j/2 = \mu_e/\mu_B$	1.0011596567(35)	0.0035	–
Free muon g-factor, or muon magnetic moment in units of $[c](e\hbar/2m_\mu c)$	$g_\mu/2$	1.00116616(31)	0.31	–
Bohr magneton, $[c](e\hbar/2m_e c)$	μ_B	9.274078(36)	3.9	10^{-24} J T^{-1}
Electron magnetic moment	μ_e	9.284832(36)	3.9	10^{-24} J T^{-1}
Gyromagnetic ratio of protons in H_2O	γ_p'	2.6751301(75)	2.8	10^8 rad sec^{-1} T^{-1}
	$\gamma_p'/2\pi$	4.257602(12)	2.8	10^7 Hz T^{-1}
γ_p' corrected for diamagnetism of H_2O	γ_p	2.6751987(75)	2.8	10^8 rad sec^{-1} T^{-1}
	$\gamma_p/2\pi$	4.257711(12)	2.8	10^7 Hz T^{-1}
Magnetic moment of protons in H_2O in Bohr magnetons	μ_p'/μ_B	1.52099322(10)	0.066	10^{-3}

Physical Constants I (continued)

Quantity	*Symbol*	*Value*	*Uncertainty* (ppm)	*Units* SI
Proton magnetic moment in Bohr magnetons	μ_p/μ_B	1.521032209(16)	0.011	10^{-3}
Ratio of electron and proton magnetic moments	μ_e/μ_p	658.2106880(66)	0.010	–
Proton magnetic moment	μ_p	1.4106171(55)	3.9	10^{-26} J T^{-1}
Magnetic moment of protons in H_2O in nuclear magnetons	μ'_p/μ_N	2.7927740(11)	0.38	–
μ'_p/μ_N corrected for diamagnetism of H_2O	μ_p/μ_N	2.7928456(11)	0.38	–
Nuclear magneton, [c] ($e\hbar/2m_pc$)	μ_N	5.050824(20)	3.9	10^{-27} J T^{-1}
Ratio of muon and proton magnetic moments	μ_μ/μ_p	3.1833402(72)	2.3	–
Muon magnetic moment	μ_μ	4.490474(18)	3.9	10^{-26} J T^{-1}
Ratio of muon mass to electron mass	m_μ/m_e	206.76865(47)	2.3	–
Muon rest mass	m_μ	1.883566(11)	5.6	10^{-28} kg
	M_μ	0.11342920(26)	2.3	u
Compton wavelength of the electron, h/m_ec	λ_C	2.4263089(40)	1.6	10^{-12} m
	$\lambda_C/2\pi$	3.8615905(64)	1.6	10^{-13} m
Compton wavelength of the proton, h/m_pc	$\lambda_{C,p}$	1.3214099(22)	1.7	10^{-15} m
	$\lambda_{C,p}/2\pi$	2.1030892(36)	1.7	10^{-16} m
Compton wavelength of the neutron, h/m_nc	$\lambda_{C,n}$	1.3195909(22)	1.7	10^{-15} m
	$\lambda_{C,n}/2\pi$	2.1001941(35)	1.7	10^{-16} m
Standard volume of ideal gas	V_0	22.71081(71)	31	10^5 J $kmol^{-1}$
		22.41383(70)	31	m^3 atm $kmol^{-1}$
Gas constant, V_0/T_0 (T_0 = 273.15 K)	R	8.31441(26)	31	10^3 J $kmol^{-1}$ K^{-1}
		8.20568(26)	31	10^{-2} m^3 atm $kmol^{-1}$ K^{-1}
Boltzmann constant, R/N	k	1.380662(44)	32	10^{-23} J K^{-1}
Stefan-Boltzmann constant, $\pi^2k^4/60\hbar^3c^2$	σ	5.67032(71)	125	10^{-8} W m^{-2} K^{-4}
First radiation constant, $2\pi hc^2$	c_1	3.741832(20)	5.4	10^{-16} W m^2
Second radiation constant, hc/k	c_2	1.438786(45)	31	10^{-2} m K
Gravitational constant	G	6.6720(41)	615	10^{-11} N m^2 kg^{-2}
kT for T = 300 K	kT	4.14199 (13)†	32	10^{-21} J
Energy-wave number conversion	hc	1.986478 (11)†	5.4	10^{-25} J m
		1.2398520 (32)†	2.6†	–
Energy-mass conversion	kg c^2	8.987551786 (72)†	0.008	10^{16} J
		5.609545 (16)†	2.9†	–
Energy-frequency conversion	1 eV h^{-1}†	2.4179696 (63)†	2.6†	–

Barry N. Taylor, National Bureau of Standards

Physical Constants II

Equatorial radius of the earth = 6378.388 km = 3963.34 miles (statute).
Polar radius of the earth, 6356.912 km = 3949.99 miles (statute).
1 degree of latitude at 40° = 69 miles.
1 international nautical mile = 1.15078 miles (statute) = 1852 m = 6076.115 ft.
Mean density of the earth = 5.522 g/cm^3 = 344.7 lb/ft^3.
Constant of gravitation, (6.673 ± 0.003) × 10^3 cm^3 gm^{-1} s^{-2}.
Acceleration due to gravity at sea level, latitude 45° = 980.665cm/s^2 = 32.1740 ft/sec^2
Length of seconds pendulum at sea level, latitude 45° = 99.3574 cm = 39.1171 in.
1 knot (international) = 101.269 ft/min = 1.6878 ft/sec = 1.1508 miles (statute)/hr.
1 micron = 10^{-4} cm.
1 angstrom = 10^{-8} cm.
Mass of hydrogen atom = (1.67339 ± 0.0031) × 10^{-24} g.
Density of mercury at 0°C = 13.5955 g/ml.
Density of water at 3.98°C = 1.000000 g/ml.
Density, maximum, of water, at 3.98°C = 0.999973 g/cm^3.
Density of dry air at 0°C, 760 mm = 1.2929 g/liter.
Velocity of sound in dry air at 0°C = 331.36 m/s - 1087.1 ft/sec.
Velocity of light in vacuum = (2.997925 ± 0.000002) × 10^{10} cm/s.
Heat of fusion of water 0°C = 79.71 cal/g.
Heat of vaporization of water 100°C = 539.55 cal/g.
Electrochemical equivalent of silver 0.001118 g/sec international amp.
Absolute wave length of red cadmium light in air at 15°C, 760 mm pressure = 6438.4696 A.
Wave length of orange-red line of krypton 86 = 6057.802 A.

DECIMAL EQUIVALENTS OF FRACTIONS OF AN INCH

		1/64	= 0.015 625
	1/32	2/64	= .031 25
		3/64	= .046 875
1/16	2/32	4/64	= .062 5
		5/64	= .078 125
	3/32	6/64	= .093 75
		7/64	= .109 375
1/8	4/32	8/64	= .125
		9/64	= .140 625
	5/32	10/64	= .156 25
		11/64	= .171 875
3/16	6/32	12/64	= .187 5
		13/64	= .203 125
	7/32	14/64	= .218 75
		15/64	= .234 375
1/4	8/32	16/64	= .25
		17/64	= .265 625
	9/32	18/64	= .281 25
		19/64	= .296 875
5/16	10/32	20/64	= .312 5
		21/64	= .328 125
	11/32	22/64	= 0.343 75
		23/64	= .359 375
3/8	12/32	24/64	= .375
		25/64	= .390 625
	13/32	26/64	= .406 25
		27/64	= .421 875
7/16	14/32	28/64	= .437 5
		29/64	= .453 125
	15/32	30/64	= .468 75
		31/64	= .484 375
1/2	16/32	32/64	= .50
		33/64	= .515 625
	17/32	34/64	= .531 25
		35/64	= .546 875
9/16	18/32	36/64	= .562 5
		37/64	= .578 125
	19/32	38/64	= .593 75
		39/64	= .609 375
5/8	20/32	40/64	= .625
		41/64	= .640 625
	21/32	42/64	= .656 25
		43/64	= 0.671 875
11/16	22/32	44/64	= .687 5
		45/64	= .703 125
	23/32	46/64	= .718 75
		47/64	= .734 375
3/4	24/32	48/64	= .75
		49/64	= .765 625
	25/32	50/64	= .781 25
		51/64	= .796 875
13/16	26/32	52/64	= .812 5
		53/64	= .828 125
	27/32	54/64	= .843 75
		55/64	= .859 375
7/8	28/32	56/64	= .875
		57/64	= .890 625
	29/32	58/64	= .906 25
		59/64	= .921 875
15/16	30/32	60/64	= .937 5
		61/64	= .953 125
	31/32	62/64	= .968 75
		63/64	= .984 375

UNITS CONVERSION FACTORS

I. The Metric System of Measurement

A. SI base units

The SI is constructed from seven base units for independent quantities plus two supplementary units for plane angle and solid angle.

Quantity	Name	Symbol
SI base units:		
length	meter	m
mass	kilogram	kg
time	second	s
electric current	ampere	A
thermodynamic temperature	kelvin	K
amount of substance	mole	mol
luminous intensity	candela	cd
SI supplementary units:		
plane angle	radian	rad
solid angle	steradian	sr

B. SI derived units

Quantity	SI Unit		
	Name	Symbol	Expression in terms of other units
frequency	hertz	Hz	1/s
force	newton	N	kg·m/s^2
pressure, stress	pascal	Pa	N/m^2
energy, work, quantity of heat	joule	J	N·m
power, radiant flux	watt	W	J/s
quantity of electricity, electric charge	coulomb	C	A·s
electric potential, potential difference, electromotive force	volt	V	W/A
capacitance	farad	F	C/V
electric resistance	ohm	Ω	V/A
conductance	siemens	S	A/V
magnetic flux	weber	Wb	V·s
magnetic flux density	tesla	T	Wb/m^2
inductance	henry	H	Wb/A
luminous flux	lumen	lm	cd·sr
illuminance	lux	lx	lm/m^2
activity (of ionizing radiation source)	becquerel	Bq	1/s
absorbed dose[a]	gray	Gy	J/kg

[a]Absorbed dose in rads (symbol rd) is the most often utilized quantity. In this handbook rad is also used as a unit symbol, following common usage (10^{-2} Gy = 1 rad). Note that rad is the accepted SI unit symbol for plane angle.

C. Multiplier prefixes for SI units

For use with SI units there is a set of 16 prefixes to form multiples and submultiples of the units. It is important to note that the kilogram is the only SI base unit with a prefix. Because double prefixes are not to be used, the prefixes, in the case of mass, are to be used with gram (symbol g) and not with kilogram (symbol kg).

SI prefixes

Factor	Prefix	Symbol
10^{18}	exa	E
10^{15}	peta	P
10^{12}	tera	T
10^{9}	giga	G
10^{6}	mega	M
10^{3}	kilo	k
10^{2}	hecto	h
10^{1}	deka	da
10^{-1}	deci	d
10^{-2}	centi	c
10^{-3}	milli	m
10^{-6}	micro	μ
10^{-9}	nano	n
10^{-12}	pico	p
10^{-15}	femto	f
10^{-18}	atto	a

II. Conversion Tables

A. Length

	cm	m	km	in	ft	mi
1 centimeter=	1	10^{-2}	10^{-5}	0.3937	3.281×10^{-2}	6.214×10^{-6}
1 METER=	100	1	10^{-3}	39.3	3.281	6.214×10^{-4}
1 kilometer=	10^{5}	1000	1	3.937×10^{4}	3281	0.6214
1 inch=	2.540	2.540×10^{-2}	2.540×10^{-5}	1	8.333×10^{-2}	1.578×10^{-5}
1 foot=	30.48	0.3048	3.048×10^{-4}	12	1	1.894×10^{-4}
1 mile=	1.609×10^{5}	1609	1.609	6.336×10^{4}	5280	1

1 angstrom = 10^{-10} m
1 nautical mile = 1852 m
= 1.151 miles = 6076 ft

1 light year = 9.4600×10^{12} km
1 parsec = 3.084×10^{13} km
1 fathom = 6 ft

1 yard = 3 ft
1 rod = 16.5 ft
1 mil = 10^{-3} in

PHYSICAL CONSTANTS AND CONVERSION FACTORS

Length Conversions

Unit	*Symbol (or Abbrevia-tion)*	*Relationship*		*Conversion to Meter* (m)
Fermi	–	–		10^{-15}
Angstrom	Å	–		10^{-10}
Micron	μm	10^4 Å	= 1 μm	10^{-6}
Inch	in.	2.54 cm	= 1 in.	0.0254*
Foot	ft	12 in.	= 1 ft	0.3048*
Yard	yd	3 ft (36 in.)	= 1 yd	0.9144*
Fathom	fath	2 yd (6 ft.)	= 1 fath	1.8288*
Link	–	7.92 in.	= 1 link	0.201168*
Rod	–	25 link (5.5 yd)	= 1 rod	5.0292
Chain	–	4 rod (22 yd)	= 1 chain	20.1168*
Furlong	fur	10 chain (220 yd)	= 1 fur	201.168*
Statute mile	mi	8 fur (1760 yd)	= 1 mi	1609.344
Nautical mile	nmi	1.15078 mi	= 1 nmi	1852.0*
Astronomical unit	AU	8.07775×10^7 nmi	= 1 AU	1.49600×10^{11}
Light year	light yr	6.3239×10^4 AU	= 1 light yr	9.46055×10^{15}
Parsec	pc	3.2562 light yr	= 1 pc	3.0857×10^{16}

***Defined value (1 in. = 2.54 cm exactly).**

Area Conversions

Unit	*Symbol*	*Relationship*	*Conversion to Square Meter* (m^2)*
Barn	–	–	1.0×10^{-28}
Are	–	–	1.0×10^{2}
Hectare	–	–	1.0×10^{4}
Square inch	in.^{2}	–	6.4516×10^{-4}
Square foot	ft^2	144 in.^{2} = 1 ft^2	9.290304×10^{-2}
Square yard	yd^2	9 ft^2 = 1 yd^2	8.3612736×10^{-1}
Acre	–	43,560 ft^2 = 1 acre	$4.0468564224 \times 10^{3}$
Square statute mile	mi^2	640 acre = 1 mi^2	$2.589988110336 \times 10^{6}$
Section	–	1 mi^2 = 1 section	$2.589988110336 \times 10^{6}$
Township	–	36 section = 1 township	9.3239572×10^{7}

*Defined values.

Volume Conversions

Unit	*Symbol*	*Relationship*	*Conversion to Cubic Meter* (m^3)
Cubic centimeter	cm^3	–	10^{-6}
Liter	l	10^3 cm^3 = 1 l	10^{-3}
Cubic inch	in.^{3}	1.6387064×10 cm^3 = 1 in.^{3}	1.6387064×10^{-5}*
Fluid ounce	fl oz	1.80469 in.^{3} = 1 fl oz	2.9573530×10^{-5}
Pint (liquid)	pt	16 fl oz = 1 pt	$4.73176473 \times 10^{-4}$*
Quart (liquid)	qt	2 pt = 1 qt	$9.46352946 \times 10^{-4}$*
Gallon (liquid)	gal	4 qt = 1 gal	$3.785411784 \times 10^{-3}$*
Cubic foot	ft^3	7.481 gal = 1 ft^3	$2.8316846592 \times 10^{-2}$*
Cubic yard	yd^3	27 ft^3 = 1 yd^3	$7.6455485844 \times 10^{-1}$*
Acre foot	–	–	1.2334818×10^{3}

*Defined values.

Angle Conversions

The radian, rad, is the basic unit of a plane angle. The steradian, sr, is the base unit of a solid angle.

Unit	*Symbol*	*Relationship*	*Conversion to Radian* (rad)
Milliradian	mrad	–	10^{-3}
Second	sec	–	$4.848136811 \times 10^{-6}$
Minute	min	60 sec = 1 min	$2.908882087 \times 10^{-4}$
Degree	° or deg	60 min = 1 deg	$1.745329252 \times 10^{-2}$
Quadrant	–	90 deg = 1 quadrant	1.570796327
Centesimal second	centesimal sec	10^{-6} quadrant = 1 centesimal sec	$1.570796327 \times 10^{-6}$
Centesimal minute	centesimal min	10^{-4} quadrant = 1 centesimal min	$1.570796327 \times 10^{-4}$
Grad	–	10^{-2} quadrant = 1 grad	$1.570796327 \times 10^{-2}$
Circumference	–	4 quadrant = 2π rad = 1 circumference	6.283185308
Mil (military)	–	1/6400 circumference = 1 mil	$9.817477044 \times 10^{-4}$

Mass Conversions, Avoirdupois, Apothecaries, Troy

	Unit	Symbol	Mass Conversion	Conversion to Kilograms (kg)
	Tonne (i.e., metric ton)	t	–	10^3
	Slug	–	–	1.45939029×10^1
Avoirdupois	Grain	–	–	6.479891×10^{-5}*
	Dram	dr	27.34375 grain = 1 dr	$1.771845195 \times 10^{-3}$
	Ounce	oz	16 dr = 1 oz	$2.834952313 \times 10^{-2}$
	Pound	lb	16 oz = 1 lb	4.5359237×10^{-1}*
	Ton (short)	tn	2000 lb = 1 tn	9.0718474×10^2
Apothecaries	Grain (see Avoirdupois)	–	–	–
	Scruple	s ap	20 grain = 1 s ap	1.2959782×10^{-3}*
	Dram	dr	3 s ap = 1 dr	3.8879346×10^{-3}
	Ounce	oz	8 dr = 1 oz	$3.11034768 \times 10^{-2}$
	Pound	lb	12 oz = 1 lb	$3.732317216 \times 10^{-1}$*
Troy	Grain (see Avoirdupois)	–	–	–
	Pennyweight	dwt	24 grain = 1 dwt	$1.555174384 \times 10^{-3}$*
	Ounce	oz	20 dwt = 1 oz	$3.11034768 \times 10^{-2}$
	Pound (see Apothecaries)	–	–	–

*Defined value.

Density Conversions

Unit	Symbol	Conversion to Kilogram/Cubic Meter (kg m^{-3})
Gram/liter	g l^{-1}	1
Gram/cubic centimeter	g cm^{-3}	10^3
Gram/milliliter	g ml^{-1}	10^3
Pound mass/cubic inch	lbm in.$^{-3}$	2.7679905×10^4
Pound mass/cubic foot	lbm ft^{-3}	1.6018463×10^1
Slug/cubic foot	slug ft^{-3}	5.15379×10^2

Time Conversion

Unit	*Symbol*	*Relationship*		*Conversion to Mean Solar Second** (sec)
Second (sidereal)	sec	–		9.9726957×10^{-1}
Minute (sidereal)	min	60 sec (sidereal)	= 1 min (sidereal)	5.9836174×10
Hour (sidereal)	h	60 min (sidereal)	= 1 h (sidereal)	3.5901704×10^{3}
Day (sidereal)	d	24 h (sidereal)	= 1 d (sidereal)	8.6164090×10^{4}
Year (sidereal)	yr	366.2564 d (sidereal)	= 1 yr (sidereal)	3.1558150×10^{7}
Minute (mean solar)	–	60 sec (mean solar)	= 1 min (mean solar)	6.0×10
Hour (mean solar)	–	60 min (mean solar)	= 1 h (mean solar)	3.60×10^{3}
Day (mean solar)	–	24 h (mean solar)	= 1 d (mean solar)	8.64×10^{4}
Month (mean calendar)	–	30.41667 d (mean solar)	= 1 month (mean calendar)	2.628×10^{6}
Year (calendar)	–	365 d (mean solar)	= 1 yr (calendar)	3.1536×10^{7}
Year (leap)		366 d (mean solar)	= 1 yr (leap)	3.16224×10^{7}
Year (tropical)	–	365.24219 d (mean solar)	= 1 yr (tropical)	3.1556926×10^{7}
Second		–		Consult *1977 American Ephemeris and Nautical Almanac*

*The unit of time (mean solar second) is based on the transition between two hyperfine levels of the ground state of the Cesium – 133 atom, with the value 9,192,631,770 cycles (Hz) as one second.

Force Conversions

Unit	*Symbol*	*Conversion to Newton* (N)
Dyne	–	10^{-5}
Kilogram force	kgf	9.80665*
Kip	–	$4.448221615 \times 10^{3}$
Ounce force (avoirdupois)	ozf	2.7801385×10^{-1}
Pound force (avoirdupois)	lbf	4.448221615
Poundal	–	$1.382549543 \times 10^{-1}$

*Defined value.

Speed Conversion

Unit	*Symbol*	*Relationship*		*Conversion to Meter/Second* ($m\ sec^{-1}$)
Centimeter/second	$cm\ sec^{-1}$	–		10^{-2}
Kilometer/second	$km\ sec^{-1}$	–		10^{3}
Kilometer/hour	$km\ h^{-1}$	–		$2.7777777778 \times 10^{-1}$
Inch/second	$in.\ sec^{-1}$	–		2.54×10^{-2}*
Foot/second	$ft\ sec^{-1}$	$12\ in.\ sec^{-1}$	$= 1\ ft\ sec^{-1}$	3.048×10^{-1}*
Statute mile/second	$mi\ sec^{-1}$	$5280\ ft\ sec^{-1}$	$= 1\ mi\ sec^{-1}$	1.609344×10^{3}*
Statute mile/minute	$mi\ min^{-1}$	$88\ ft\ sec^{-1}$	$= 1\ mi\ min^{-1}$	2.68224×10*
Statute mile/hour	mph	$1.4666667\ ft\ sec^{-1}$	= 1 mph	4.4704×10^{-1}*
Nautical mile/second	$nmi\ sec^{-1}$	$6076.1033\ ft\ sec^{-1}$	$= 1\ nmi\ sec^{-1}$	1.852×10^{3}
Nautical mile/hour or knot (international)	$nmi\ h^{-1}$	$1.6878099\ ft\ sec^{-1}$	= 1 knot (international)	5.1444444×10^{-1}

*Defined value.

Acceleration Conversions

Unit	*Symbol*	*Conversion to Meter/Second Squared* ($m\ sec^{-2}$)
Meter/minute squared	$m\ min^{-2}$	2.777778×10^{-4}
Inch/second squared	$in.\ sec^{-2}$	2.54×10^{-2}*
Inch/minute squared	$in.\ min^{-2}$	7.0555556×10^{-6}*
Feet/second squared	$ft\ sec^{-2}$	3.048×10^{-1}*
Feet/minute squared	$ft\ min^{-2}$	8.4666667×10^{-5}
Statute mile/second squared	$mi\ sec^{-2}$	1.609344×10^{3}
Statute mile/minute squared	$mi\ min^{-2}$	4.47040×10^{-1}
Statute mile/hour squared	$mi\ h^{-2}$	1.241778×10^{-4}
Gal (Galileo)	–	10^{-2}*
Free fall standard	–	9.80665*

*Defined value.

Torque Conversions

Unit	*Symbol*	*Conversion to Newton-Meter* (N m)
Dyne-centimeter	dyne cm	1.0×10^{-7}
Kilogram force-meter	kgf m	9.806650
Ounce force-inch	ozf in.	7.061552×10^{-3}
Pound force-inch	lbf in.	1.129848×10^{-1}
Pound force-foot	lbf ft	1.355818

Pressure Conversions

Unit	*Symbol*	*Conversion to Pascal (Pa) or Newton/Square-Meter* ($N\ m^{-2}$)
Barye	–	10^{-1}
Millibar	mbar	10^{2}
Bar	–	10^{5}
Dyne/square centimeter	dyne cm^{-2}	10^{-1}
Atmosphere	atm	1.01325×10^{5}
Kilogram force/square meter	kgf m^{-2}	9.80665
Kilogram force/square centimeter	kgf cm^{-2}	9.80665×10^{4}
Pound force/square foot	lbf ft^{-2}	4.7880258×10
Pound force/square inch	psi, lbf $in.^{-2}$	6.8947572×10^{3}
Centimeter of water (4°C)	–	9.80638×10
Inch of water (60°F)	–	2.4884×10^{2}
Inch of water (6°C)	–	2.49082×10^{2}
Foot of water (4°C)	–	2.98898×10^{3}
Millimeter of mercury (0°C)	mm Hg	1.333224×10^{2}
Inch of mercury (60°F)	–	3.37685×10^{3}
Inch of mercury (0°C)	–	3.386389×10^{3}
Torr	–	1.333223×10^{2}

Work and Energy Conversions

Unit	*Symbol*	*Conversion to Joule* (J)
Foot-pound force	ft lbf	1.3558179
Foot-poundal	–	4.2140110×10^{-2}
Ton (nuclear equivalent of TNT)	tn	4.20×10^{9}
British thermal unit (IST current)	Btu	1.055056×10^{3}*
Btu, IST before 1956	–	1.05504×10^{3}
Btu, mean	–	1.05587×10^{3}
Btu, thermochemical	–	1.054350×10^{3}
Btu (39°F)	–	1.05967×10^{3}
Btu (60°F)	–	1.05468×10^{3}
Calorie	cal	4.1868*
Calorie, mean	–	4.19002
Calorie, thermochemical	–	4.184**
Calorie (15°C)	–	4.18580
Calorie (20°C)	–	4.18190
Calorie kilogram	–	4.1868×10^{3}*
Calorie kilogram, mean	–	4.19002×10^{3}
Calorie kilogram, thermochemical	–	4.184×10^{3}**
Kilocalorie	kcal	4.1868×10^{3}*
Kilocalorie, mean	–	4.19002×10^{3}
Kilocalorie, thermochemical	–	4.184×10^{3}**
Electron volt	eV	$1.6021917 \times 10^{-19}$
Erg	–	1.00×10^{-7}**
Joule (International of 1948)	J	1.000165
Watt hour	W h	3.60×10^{3}**
Kilowatt hour	kW h	3.60×10^{6}**
Kilowatt hour (International of 1948)	–	3.60059×10^{6}

*International steam table.
**Defined value.

Power Conversions

Unit	*Symbol*	*Conversion to Watt* (W)
Foot pound force/hour	ft lbf h^{-1}	3.7661610×10^{-4}
Foot pound force/minute	ft lbf min^{-1}	2.2596966×10^{-2}
Foot pound force/second	ft lbf sec^{-1}	1.3558179
Horsepower*	hp	7.4569987×10^{2}
Btu (thermochemical)/second	–	1.054350×10^{3}
Btu (thermochemical)/minute	–	1.757250×10
Calorie (thermochemical)/second	cal sec^{-1}	4.184**
Calorie (thermochemical)/minute	cal min^{-1}	6.9733333×10^{-2}
Kilocalorie (thermochemical)/minute	kcal min^{-1}	6.9733333×10
Kilocalorie (thermochemical)/second	kcal sec^{-1}	4.184×10^{3}**
Watt (International of 1948)	W	1.000165

*1 hp = 550 ft lbf sec^{-1}.
**Defined value.

Electrical Unit Conversions

Quantity	*System International*	*Symbol*	*Electrostatic*	*Electromagnetic*
Charge	1 coulomb	C	2.9979×10^{9} statcoulomb	10^{-1} abcoulomb
Current	1 ampere	A	2.9979×10^{9} statampere	10^{-1} abampere
Potential	1 volt	V	3.3356×10^{-3} statvolt	10^{8} abvolt
Capacity	1 farad	F	8.9878×10^{11} statfarad	10^{-9} abfarad
Resistance	1 ohm	Ω	1.1126×10^{-12} statohm	10^{9} abohm
Conductance	1 siemen	S	8.987956×10^{11} statsiemen	10^{-9} absiemen
Inductance	1 henry	H	1.1126×10^{-12} stathenry	10^{9} abhenry

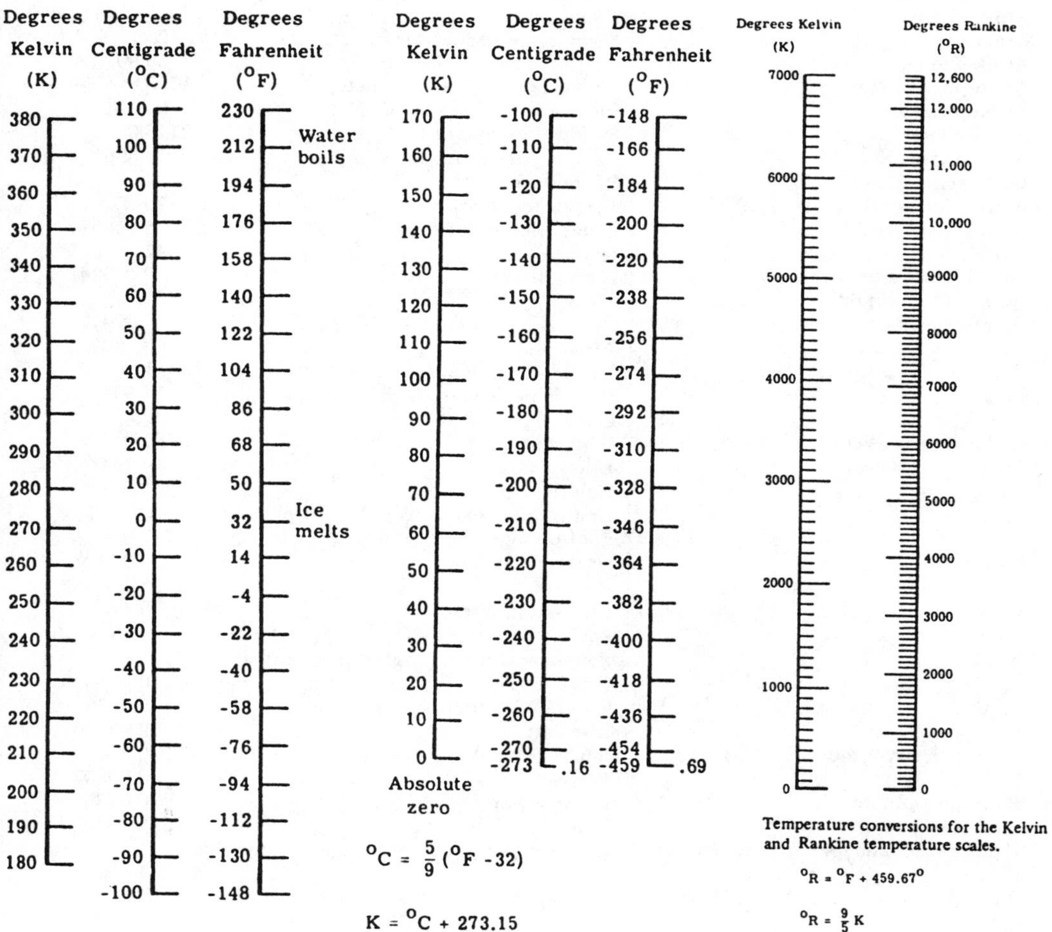

Temperature conversions for the Kelvin and Rankine temperature scales.

Temperature conversions for the Kelvin, Celsius and Farenheit temperature scales.

Temperature is a measure of the average, translational kinetic energy of the molecules of a substance; it has SI units of Kelvin, K.

CONVERSION FACTORS - GENERAL

To convert from	To	Multiply by
Acres	Square feet	43,560
Acres	Square meters	4074
Acres	Square miles	0.001563
Acre-feet	Cubic meters	1233
Ampere-hours (absolute)	Coulombs (absolute)	3600
Angstrom units	Inches	3.937×10^{-9}
Angstrom units	Meters	1×10^{-10}
Angstrom units	Microns	1×10^{-4}
Atmospheres	Millimeters of mercury at 32°F.	760
Atmospheres	Dynes per square centimeter	1.0133×10^{6}
Atmospheres	Newtons per square meter	101,325
Atmospheres	Feet of water at 39.1°F.	33.90
Atmospheres	Grams per square centimeter	1033.3
Atmospheres	Inches of mercury at 32°F.	29.921
Atmospheres	Pounds per square foot	2116.3
Atmospheres	Pounds per square inch	14.696
Bags (cement)	Pounds (cement)	94
Barrels (cement)	Pounds (cement)	376
Barrels (oil)	Cubic meters	0.15899
Barrels (oil)	Gallons	42
Barrels (U.S. liquid)	Cubic meters	0.11924
Barrels (U.S. liquid)	Gallons	31.5
Barrels per day	Gallons per minute	0.02917
Bars	Atmospheres	0.9869
Bars	Newtons per square meter	1×10^{5}
Bars	Pounds per square inch	14.504
Board feet	Cubic feet	$\frac{1}{12}$
Boiler horsepower	B.t.u. per hour	33,480
Boiler horsepower	Kilowatts	9.803
B.t.u.	Calories (gram)	252
B.t.u.	Centigrade heat units (c.h.u. or p.c.u.)	0.55556
B.t.u.	Foot-pounds	777.9
B.t.u.	Horsepower-hours	3.929×10^{-4}
B.t.u.	Joules	1055.1
B.t.u.	Liter-atmospheres	10.41
B.t.u.	Pounds carbon to CO_2	6.88×10^{-5}
B.t.u.	Pounds water evaporated from and at 212°F.	0.001036
B.t.u.	Cubic foot-atmospheres	0.3676
B.t.u.	Kilowatt-hours	2.930×10^{-4}
B.t.u. per cubic foot	Joules per cubic meter	37,260
B.t.u. per hour	Watts	0.29307
B.t.u. per minute	Horsepower	0.02357
B.t.u. per pound	Joules per kilogram	2326
B.t.u. per pound per degree Fahrenheit	Calories per gram per degree centigrade	1
B.t.u. per pound per degree Fahrenheit	Joules per kilogram per degree Kelvin	4186.8
B.t.u. per second	Watts	1054.4
B.t.u. per square foot per hour	Joules per square meter per second	3.1546
B.t.u. per square foot per minute	Kilowatts per square foot	0.1758
B.t.u. per square foot per second for a temperature gradient of 1°F. per inch	Calories, gram (15°C.), per square centimeter per second for a temperature gradient of 1°C. per centimeter	1.2405

(continued)

To convert from	To	Multiply by
B.t.u. (60°F.) per degree Fahrenheit	Calories per degree centigrade	453.6
Bushels (U.S. dry)	Cubic feet	1.2444
Bushels (U.S. dry)	Cubic meters	0.03524
Calories, gram	B.t.u.	3.968×10^{-3}
Calories, gram	Foot-pounds	3.087
Calories, gram	Joules	4.1868
Calories, gram	Liter-atmospheres	4.130×10^{-2}
Calories, gram	Horsepower-hours	1.5591×10^{-6}
Calories. gram, per gram per degree C.	Joules per kilogram per degree Kelvin	4186.8
Calories, kilogram	Kilowatt-hours	0.0011626
Calories, kilogram per second	Kilowatts	4.185
Candle power (spherical)	Lumens	12.556
Carats (metric)	Grams	0.2
Centigrade heat units	B.t.u.	1.8
Centimeters	Angstrom units	1×10^{8}
Centimeters	Feet	0.03281
Centimeters	Inches	0.3937
Centimeters	Meters	0.01
Centimeters	Microns	10,000
Centimeters of mercury at 0°C.	Atmospheres	0.013158
Centimeters of mercury at 0°C.	Feet of water at 39.1°F.	0.4460
Centimeters of mercury at 0°C.	Newtons per square meter	1333.2
Centimeters of mercury at 0°C.	Pounds per square foot	27.845
Centimeters of mercury at 0°C.	Pounds per square inch	0.19337
Centimeters per second	Feet per minute	1.9685
Centimeters of water at 4°C.	Newtons per square meter	98.064
Centistokes	Square meters per second	1×10^{-6}
Circular mils	Square centimeters	5.067×10^{-6}
Circular mils	Square inches	7.854×10^{-7}
Circular mils	Square mils	0.7854
Cords	Cubic feet	128
Cubic centimeters	Cubic feet	3.532×10^{-5}
Cubic centimeters	Gallons	2.6417×10^{-4}
Cubic centimeters	Ounces (U.S. fluid)	0.03381
Cubic centimeters	Quarts (U.S. fluid)	0.0010567
Cubic feet	Bushels (U.S.)	0.8036
Cubic feet	Cubic centimeters	28,317
Cubic feet	Cubic meters	0.028317
Cubic feet	Cubic yards	0.03704
Cubic feet	Gallons	7.481
Cubic feet	Liters	28.316
Cubic foot-atmospheres	Foot-pounds	2116.3
Cubic foot-atmospheres	Liter-atmospheres	28.316
Cubic feet of water (60°F.)	Pounds	62.37
Cubic feet per minute	Cubic centimeters per second	472.0
Cubic feet per minute	Gallons per second	0.1247
Cubic feet per second	Gallons per minute	448.8
Cubic feet per second	Million gallons per day	0.64632
Cubic inches	Cubic meters	1.6387×10^{-5}
Cubic yards	Cubic meters	0.76456
Curies	Disintegrations per minute	2.2×10^{12}
Curies	Coulombs per minute	1.1×10^{12}
Degrees	Radians	0.017453
Drams (apothecaries' or troy)	Grams	3.888

(continued)

To convert from	To	Multiply by
Drams (avoirdupois)	Grams	1.7719
Dynes	Newtons	1×10^{-5}
Ergs	Joules	1×10^{-7}
Faradays	Coulombs (abs.)	96,500
Fathoms	Feet	6
Feet	Meters	0.3048
Feet per minute	Centimeters per second	0.5080
Feet per minute	Miles per hour	0.011364
Feet per $(\text{second})^2$	Meters per $(\text{second})^2$	0.3048
Feet of water at 39.2°F.	Newtons per square meter	2989
Foot-poundals	B.t.u.	3.995×10^{-5}
Foot-poundals	Joules	0.04214
Foot-poundals	Liter-atmospheres	4.159×10^{-4}
Foot-pounds	B.t.u.	0.0012856
Foot-pounds	Calories, gram	0.3239
Foot-pounds	Foot-poundals	32.174
Foot-pounds	Horsepower-hours	5.051×10^{-7}
Foot-pounds	Kilowatt-hours	3.766×10^{-7}
Foot-pounds	Liter-atmospheres	0.013381
Foot-pounds force	Joules	1.3558
Foot-pounds per second	Horsepower	0.0018182
Foot-pounds per second	Kilowatts	0.0013558
Furlongs	Miles	0.125
Gallons (U.S. liquid)	Barrels (U.S. liquid)	0.03175
Gallons	Cubic meters	0.003785
Gallons	Cubic feet	0.13368
Gallons	Gallons (Imperial)	0.8327
Gallons	Liters	3.785
Gallons	Ounces (U.S. fluid)	128
Gallons per minute	Cubic feet per hour	8.021
Gallons per minute	Cubic feet per second	0.002228
Grains	Grams	0.06480
Grains	Pounds	$\frac{1}{7000}$
Grains per cubic foot	Grams per cubic meter	2.2884
Grains per gallon	Parts per million	17.118
Grams	Drams (avoirdupois)	0.5644
Grams	Drams (troy)	0.2572
Grams	Grains	15.432
Grams	Kilograms	0.001
Grams	Pounds (avoirdupois)	0.0022046
Grams	Pounds (troy)	0.002679
Grams per cubic centimeter	Pounds per cubic foot	62.43
Grams per cubic centimeter	Pounds per gallon	8.345
Grams per liter	Grains per gallon	58.42
Grams per liter	Pounds per cubic foot	0.0624
Grams per square centimeter	Pounds per square foot	2.0482
Grams per square centimeter	Pounds per square inch	0.014223
Hectares	Acres	2.471
Hectares	Square meters	10,000
Horsepower (British)	B.t.u. per minute	42.42
Horsepower (British)	B.t.u. per hour	2545
Horsepower (British)	Foot-pounds per minute	33,000
Horsepower (British)	Foot-pounds per second	550
Horsepower (British)	Watts	745.7
Horsepower (British)	Horsepower (metric)	1.0139
Horsepower (British)	Pounds carbon to CO_2 per hour	0.175

(continued)

To convert from	To	Multiply by
Horsepower (British)	Pounds water evaporated per hour at 212°F	2.64
Horsepower (metric)	Foot-pounds per second	542.47
Horsepower (metric)	Kilogram-meters per second	7.5
Hours (mean solar)	Seconds	3600
Inches	Meters	0.0254
Inches of mercury at 60°F	Newtons per square meter	3376.9
Inches of water at 60°F	Newtons per square meter	248.84
Joules (absolute)	B.t.u. (mean)	9.480×10^{-4}
Joules (absolute)	Calories, gram (mean)	0.2389
Joules (absolute)	Cubic foot-atmospheres	0.3485
Joules (absolute)	Foot-pounds	0.7376
Joules (absolute)	Kilowatt-hours	2.7778×10^{-7}
Joules (absolute)	Liter-atmospheres	0.009869
Kilocalories	Joules	4186.8
Kilograms	Pounds (avoirdupois)	2.2046
Kilograms force	Newtons	9.807
Kilograms per square centimeter	Pounds per square inch	14.223
Kilometers	Miles	0.6214
Kilowatt-hours	B.t.u.	3414
Kilowatt-hours	Foot-pounds	2.6552×10^{6}
Kilowatts	Horsepower	1.3410
Knots (international)	Meters per second	0.5144
Knots (nautical miles per hour)	Miles per hour	1.1516
Lamberts	Candles per square inch	2.054
Liter-atmospheres	Cubic foot-atmospheres	0.03532
Liter-atmospheres	Foot-pounds	74.74
Liters	Cubic feet	0.03532
Liters	Cubic meters	0.001
Liters	Gallons	0.26418
Lumens	Watts	0.001496
Micromicrons	Microns	1×10^{-6}
Microns	Angstrom units	1×10^{4}
Microns	Meters	1×10^{-6}
Miles (nautical)	Feet	6080
Miles (nautical)	Miles (U.S. statute)	1.1516
Miles	Feet	5280
Miles	Meters	1609.3
Miles per hour	Feet per second	1.4667
Miles per hour	Meters per second	0.4470
Milliliters	Cubic centimeters	1
Millimeters	Meters	0.001
Millimeters of mercury at 0°C	Newtons per square meter	133.32
Millimicrons	Microns	0.001
Mils	Inches	0.001
Mils	Meters	2.54×10^{-5}
Minims (U.S.)	Cubic centimeters	0.06161
Minutes (angle)	Radians	2.909×10^{-4}
Minutes (mean solar)	Seconds	60
Newtons	Kilograms	0.10197
Ounces (avoirdupois)	Kilograms	0.02835
Ounces (avoirdupois)	Ounces (troy)	0.9115
Ounces (U.S. fluid)	Cubic meters	2.957×10^{-5}
Ounces (troy)	Ounces (apothecaries')	1.000
Pints (U.S. liquid)	Cubic meters	4.732×10^{-4}
Poundals	Newtons	0.13826

(continued)

To convert from	To	Multiply by
Pounds (avoirdupois)	Grains	7000
Pounds (avoirdupois)	Kilograms	0.45359
Pounds (avoirdupois)	Pounds (troy)	1.2153
Pounds per cubic foot	Grams per cubic centimeter	0.016018
Pounds per cubic foot	Kilograms per cubic meter	16.018
Pounds per square foot	Atmospheres	4.725×10^{-4}
Pounds per square foot	Kilograms per square meter	4.882
Pounds per square inch	Atmospheres	0.06805
Pounds per square inch	Kilograms per square centimeter	0.07031
Pounds per square inch	Newtons per square meter	6894.8
Pounds force	Newtons	4.4482
Pounds force per square foot	Newtons per square meter	47.88
Pounds water evaporated from and at 212°F.	Horsepower-hours	0.379
Pound-centigrade units (p.c.u.)	B.t.u.	1.8
Quarts (U.S. liquid)	Cubic meters	9.464×10^{-4}
Radians	Degrees	57.30
Revolutions per minute	Radians per second	0.10472
Seconds (angle)	Radians	4.848×10^{-6}
Slugs	Gee pounds	1
Slugs	Kilograms	14.594
Slugs	Pounds	32.17
Square centimeters	Square feet	0.0010764
Square feet	Square meters	0.0929
Square feet per hour	Square meters per second	2.581×10^{-5}
Square inches	Square centimeters	6.452
Square inches	Square meters	6.452×10^{-4}
Square yards	Square meters	0.8361
Stokes	Square meters per second	1×10^{-4}
Tons (long)	Kilograms	1016
Tons (long)	Pounds	2240
Tons (metric)	Kilograms	1000
Tons (metric)	Pounds	2204.6
Tons (metric)	Tons (short)	1.1023
Tons (short)	Kilograms	907.18
Tons (short)	Pounds	2000
Tons (refrigeration)	B.t.u. per hour	12,000
Tons (British shipping)	Cubic feet	42.00
Tons (U.S. shipping)	Cubic feet	40.00
Torr (mm. mercury, 0°C.)	Newtons per square meter	133.32
Watts	B.t.u. per hour	3.413
Watts	Joules per second	1
Watts	Kilogram-meters per second	0.10197
Watt-hours	Joules	3600
Yards	Meters	0.9144

TEMPERATURE FACTORS

$$°F = 9/5\ (°C) + 32$$

Fahrenheit temperature = 1.8 (temperature in kelvins) −459.67

$$°C = 5/9\ [(°F) - 32]$$

Celsius temperature = temperature in kelvins −273.15

Fahrenheit temperature = 1.8 (Celsius temperature) +32

PHYSICAL CONCEPTS IN RATIONALIZED MKS UNITS

Concept	Symbol	Name of Unit	Abbreviation of Unit Name	Definition or Defining Equation	Explanations; Equivalent Units; Alternative Definitions; etc.
Permeability of vacuum	μ_0	Henry/metre	H m^{-1}	$\mu_0 = 4\pi \times 10^{-7}$ H/m	Defined value to give coherent rationalized electrical units.
Permittivity of vacuum	ϵ_0	Farad/metre	F m^{-1}	$\epsilon_0 = 1/\mu_0 c^2$	Derived in Maxwell's theory of EM radiation.
Electric charge	Q	Coulomb	C	$F = (Q_1 Q_2)/(4\pi\epsilon_0 r^2)$	A s Coulomb's Law Also $Q = \int I dt$
Electric current	I	Ampere	A	$F = (2\mu_0 I_1 I_2)/(4\pi r^2)$	Also $I = dQ/dt$
Electric potential (Potential difference)	V	Volt	V	$V_r = \int_{\infty} Fdl (V_{ab} = \int_a Fdl)$	N m C^{-1} or J C^{-1}
Electric field-strength (Electric force)	E	Volt/metre	V m^{-1}	$E = dV/dl$	N C^{-1} E = Force on unit point charge
Electric resistance	R	Ohm	Ω	$R = V/I$	J/A
Electric conductance	S	Siemens	Ω^{-1}	$\frac{I}{R} = \frac{I}{V}$	
Electric flux	Ψ	Coulomb	$\Psi = Q$		
Electric flux density (Displacement)	D	Coulomb/metre^2	C m^{-2}	$D = d\Psi/dA$	
Permittivity	ϵ	Farad/metre	F m^{-1}	$\epsilon = D/E$	
Relative permittivity	ϵ_r			$\epsilon_r = \epsilon/\epsilon_0$	A numeric.
Magnetic field-strength (Magnetic force)	H	Amp. turn/metre	AT m^{-1}	$dH = Idl \sin\theta/4\pi r^2$	The *turn* is a numeric not a unit.
Magnetic flux	Φ	Weber	Wb	$\Phi = -\int e dt$	V s Faraday-Lenz Law.
Magnetic flux density	B	Weber/metre^2, Tesla	Wb/m^2, T	$B = d\Phi/dA$	V s m^{-1}
Permeability	μ	Henry/metre	H/m	$\mu = B/H$	
Relative permeability	μ_r			$\mu_r = \mu/\mu_0$	A numeric
Coefficient of mutual induction	M	Henry	H	$e_2 = -M dI_1/dt$	Wb/A
Coefficient of self-induction	L	Henry	H	$e = -L dI/dt$	Wb/A
Capacitance	C	Farad	F	$C = Q/V$	CV^{-1}
Reactance	X	Equivalent ohm	Ω	$X = \omega L$ or $\frac{1}{\omega C}$	Sinusoidal a.c. Also
Impedance	Z	Equivalent ohm	Ω	$Z = \sqrt{R^2 + X^2}$	$\omega = 2\pi \times$ frequency

Names in parentheses with upper case initial, e.g., (Energy), are alternatives. Words in parentheses with l.c. initial, e.g., (dynamic), are adjectival.
Symbols and names are in accordance with BS 1991, BS 350, and BS 1637.
(Reproduced from Chambers's *Six-Figure Mathematical Tables*.)

(continued)

Concept	Symbol	Name of Unit	Abbreviation of Unit Name	Definition or Defining Equation	Explanations; Equivalent Units; Alternative Definitions; etc.
Length	l	metre	m	1 m = 1 650 763·73 wavelengths of radiation ($2p_{10}-5d_5$) of Kr 86.	
Mass	m	kilogram	kg	International Prototype Kilogram 1 kg pure water at 4°C and 760 mm pressure occupies 1 litre.	
Time	t	second	s	Mean solar second.	
The above are internationally agreed basic units.					
Area	A, a	square metre	m^2	$a = l^2$	(All concepts except the electrical concepts are derived from the three above. All concepts are capable of being written in terms of the first three; but electrical quantities need an additional fundamental concept.)
Volume	V, v	cubic metre	m^3	$V = l^3$	
Velocity	v, u	metre/second	$m\ s^{-1}$	$v = dl/dt$	
Acceleration	a	metre/second²	$m\ s^{-2}$	$a = d^2l/dt^2$	
Density	ρ	kilogram/metre³	$kg\ m^{-3}$	$\rho = m/V$	
Mass rate of flow		kilogram /sec	$kg\ s^{-1}$	dm/dt	
Volume rate of flow		cubic metre/sec	$m^3\ s^{-1}$	dV/db	
Moment of inertia	I	kilogram metre²	$kg\ m^2$	$I = Mk^2$	
Momentum	p	kilogram metre/sec	$kg\ m\ s^{-1}$	$p = mv$	
Angular momentum	$I\omega$	kilogram metre²/sec	$kg\ m^2\ s^{-1}$	dI/dt	
Kinetic energy	T, (W)	kilogram metre²/sec²	$kg\ m^2\ s^{-2}$	$T = \frac{1}{2}mv^2$	Newton metre (N m)
Force	F	Newton	N	$F = ma$	$kg\ m^2\ s^{-2}$
Torque (Moment of force)	T, (M)	Newton metre	N m	$T = Fl$	$T = P/2\pi n$. n in rev/s
Potential energy	V, (w)	Newton metre	N m	$V = \int F dl$	$kg\ m^2\ s^{-2}$
Work (Energy, Heat)	W, (U)	Joule	J	$W = \int F dl$	N m (1 J = 1 N m by definition)
Heat (Enthalpy)	Q, (H)	Joule	J	$H = U + p\ v$	Definition for a fluid U = Internal energy
Power	P	Watt	W	$P = dW/dt$	1 W = 1 J s^{-1} by definition
Pressure (Stress)	p (σ, f)	Newton/metre²	$N\ m^{-2}$	$p = F/A$	Usually pressure in fluid, stress in solids.
Surface tension	γ (σ)	Newton/metre²	$N\ m^{-2}$	$\gamma = F/l$	Free surface energy. $10^{-1}\ N\ s\ m^{-2}$
Viscosity, dynamic	η, μ	Poise	P	$\frac{P}{A} = \eta dv/dl$	Defined in CGS units $10^{-1}\ N\ s\ m\ kg^{-1}$
Viscosity, kinematic	ν	Stokes	S	$\nu = \eta/\rho$	Defined in CGS units
Action			J s	$\int W dt$	
Temperature	θ, T	degree C, degree K	°C, °K	T°K = $(\theta + 273{\cdot}16)$°C	International Temperature Scale
Velocity of light	c	metre/second	$m\ s^{-1}$	Fundamental, measured, constant	

UNIT CONVERSION TABLE

Quantity	Practical Units	Absolute Systems of Units								
		Rationalized MKSA Units		CGS Electromagnetic Units			CGS Electrostatic Units			
	Unit	*Unit*	*Multiple*	*Unit*		*Multiple*	*Unit*		*Multiple*
Energy	Joule	Joule	$\times 1$	Erg	(G)	$\times 10^{-7}$	Erg	(G)	$\times 10^{-7}$
Power	Watt	Watt	$\times 1$	Erg/s	(G)	$\times 10^{-7}$	Erg/s	(G)	$\times 10^{-7}$
Electric Charge	Coulomb	Coulomb	$\times 1$	Abcoulomb		$\times 10$	Statcoulomb	(G)	$\div 3 \times 10^{9}$
Polarization	Coulomb/cm²	Coulomb/m²	$\times 10^{-4}$	Abcoulomb/cm²		$\times 10$	Statcoulomb/cm²	(G)	$\div 3 \times 10^{9}$
Electric Potential	Volt	Volt	$\times 1$	Abvolt		$\times 10^{-8}$	Statvolt	(G)	$\times 300$
Electric Field	Volt/cm	Volt/m	$\times 10^{-2}$	Abvolt/cm		$\times 10^{-8}$	Statvolt/cm	(G)	$\times 300$
Permittivity		Farad/m	$\times 4\pi \times 10^{-9}$			$\times 10^{2}$		(G)	$\div 9 \times 10^{18}$
Capacitance	Farad	Farad	$\times 1$	Abfarad		$\times 10^{9}$	Cm or Statfarad	(G)	$\div 9 \times 10^{11}$
Displacement		Coulomb/m²	$\times 4\pi \times 10^{-4}$			$\times 10$		(G)	$\div 3 \times 10^{9}$
Electric Flux		Coulomb	$\times 4\pi$			$\times 10$		(G)	$\div 3 \times 10^{9}$
Current	Ampere	Ampere	$\times 1$	Abampere		$\times 10$	Statampere	(G)	$\div 3 \times 10^{9}$
Resistance	Ohm	Ohm	$\times 1$	Abohm		$\times 10^{-9}$	Statohm	(G)	$\times 9 \times 10^{11}$
Resistivity	Ohm/cm	Ohm/m	$\times 10^{2}$	Abohm/cm		$\times 10^{-9}$	Statohm/cm	(G)	$\times 9 \times 10^{11}$
Inductance	Henry	Henry	$\times 1$	Abhenry	(G)	$\times 10^{-9}$	Stathenry		$\times 9 \times 10^{11}$
Magnetic Pole		Weber	$\div 4\pi \times 10^{-8}$		(G)	$\times 1$			$\times 3 \times 10^{10}$
Magnetization		Weber/m²	$\div 4\pi \times 10^{-4}$		(G)	$\times 1$			$\times 3 \times 10^{10}$
Magnetic Field	Oersted	Ampere turn/m	$\times 4\pi \times 10^{-3}$	Oersted	(G)	$\times 1$			$\div 3 \times 10^{10}$
Permeability	Gauss/Oersted	Henry/m	$\div 4\pi \times 10^{-7}$	Gauss/Oersted	(G)	$\times 1$			$\times 9 \times 10^{20}$
Magnetic Induction	Gauss	Weber/m² (or Tesla)	$\times 10^{4}$	Gauss	(G)	$\times 1$			$\times 3 \times 10^{10}$
Magnetic Flux	Maxwell	Weber	$\times 10^{8}$	Maxwell	(G)	$\times 1$			$\times 3 \times 10^{10}$
Magnetic Potential	Gilbert	Ampere turn	$\times 4\pi \times 10^{-1}$	Gilbert	(G)	$\times 1$			$\div 3 \times 10^{10}$
Reluctance	Gilbert/Maxwell	Ampere turn/Weber	$\times 4\pi \times 10^{-9}$	Gilbert/Maxwell	(G)	$\times 1$			$\div 9 \times 10^{20}$
Electric Susceptibility	} Dimensionless		$\div 4\pi$			$\times 1$			$\times 1$
Magnetic Susceptibility			$\div 4\pi$			$\times 1$			$\times 1$

Note. The multiple column gives the ratio of the size of the absolute unit to that of the corresponding practical unit. To convert the numerical value of a quantity expressed in practical units to absolute units it must be multiplied by the reciprocal of this factor.

Where the practical unit is not the same as the corresponding MKSA absolute unit the ratio of the size of a CGS absolute unit to the MKSA absolute unit can be obtained by *dividing* the multiple given in the CGS column by that given in the MKSA column.

The conversion factors between the two CGS systems of units have been calculated by taking the velocity of electromagnetic waves in free space as 3×10^{10} cm/sec.

The symbol G indicates a Gaussian unit.

No generally accepted names exist for some of the units in this table.

CHAPTER 18

PHYSICAL BASIC PARAMETERS

Equivalent Pressures, Air Altitudes and Sea Depths

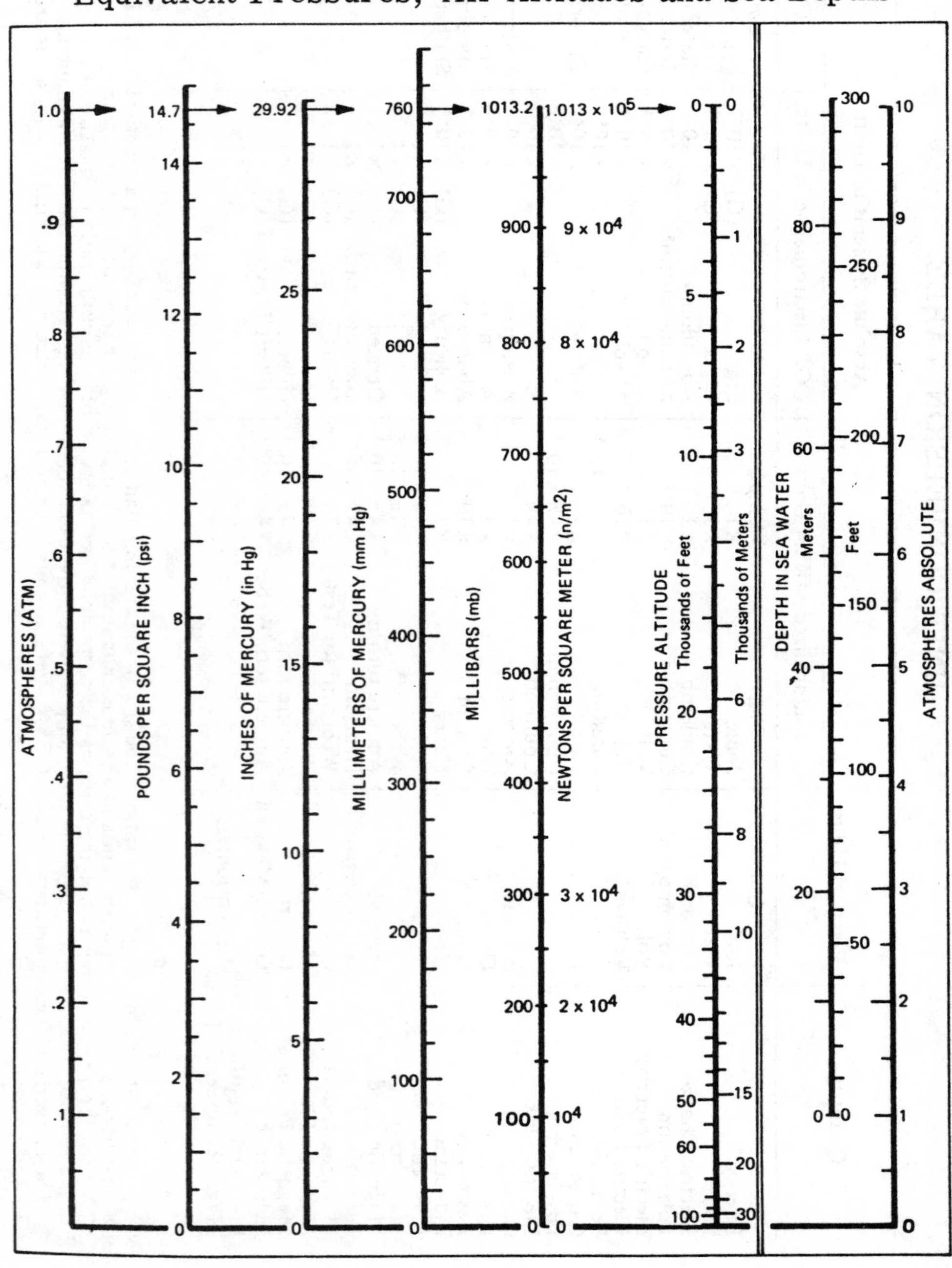

Conversion Table for Barometric Pressure Units

	Atm	N/M^2	bars	mb	kg/cm^2	gm/cm^2 ($cm\ H_2O$)	mm Hg	in. Hg (" Hg)	lb/in^2 (psi)
1 Atmosphere =	1	1.013×10^5	1.013	1013	1.033	1033	760	29.92	14.70
1 Newton/M^2 (N/M^2) =	$.9869 \times 10^{-5}$	1	10^{-5}	.01	1.02×10^{-5}	.0102	.0075	$.2953 \times 10^{-3}$	$.1451 \times 10^{-3}$
1 bar =	.9869	10^5	1	1000	1.02	1020	750.1	29.53	14.51
1 millibar (mb) =	$.9869 \times 10^{-3}$	100	.001	1	.00102	1.02	.7501	.02953	.01451
1 kg/cm^2 =	.9681	$.9807 \times 10^5$	.9807	980.7	1	1000	735	28.94	14.22
1 gm/cm^2 (1 $cm\ H_2O$) =	968.1	98.07	$.9807 \times 10^{-3}$	.9807	.001	1	.735	.02894	.01422
1 mm Hg =	.001316	133.3	.001333	1.333	.00136	1.36	1	.03937	.01934
1 in. Hg (" Hg) =	.0334	3386	.03386	33.86	.03453	34.53	25.4	1	.4910
1 lb/in^2 (psi) =	.06804	6895	.06895	68.95	.0703	70.3	51.70	2.035	1

Factors for Conversion of Volumes from ATPS to STPD and BTPS

Ambient Temperature °C	Aqueous Vapor Pressure (mmHg) at Saturation	Factor to Convert to: STPD	BTPS
14	12.0	0.936	1.133
15	12.8	0.932	1.128
16	13.6	0.928	1.123
17	14.5	0.924	1.118
18	15.5	0.920	1.113
19	16.5	0.916	1.108
20	17.5	0.911	1.102
21	18.7	0.906	1.096
22	19.8	0.902	1.091
23	21.1	0.897	1.085
24	22.4	0.893	1.080
25	23.8	0.888	1.075
26	25.2	0.883	1.069
27	26.7	0.878	1.063
28	28.3	0.874	1.057
29	30.0	0.869	1.051
30	31.8	0.864	1.045
31	33.7	0.859	1.039
32	35.7	0.853	1.032
33	37.7	0.848	1.026
34	39.9	0.843	1.020
35	42.2	0.838	1.014
36	44.6	0.832	1.007
37	47.1	0.826	1.000
38	49.7	0.821	0.994
39	52.4	0.816	0.987
40	55.3	0.810	0.980

DENSITY OF VARIOUS LIQUIDS

(From Smithsonian Tables.)

Liquid	Grams per cu. cm	Pounds per cu. ft.	Temp °C
Acetone	0.792	49.4	20°
Alcohol, ethyl	0.791	49.4	20
methyl	0.810	50.5	0
Benzene	0.899	56.1	0
Carbolic acid	0.950—0.965	59.2—60.2	15
Carbon disulfide	1.293	80.7	0
tetrachloride	1.595	99.6	20
Chloroform	1.489	93.0	20
Ether	0.736	45.9	0
Gasoline	0.66—0.69	41.0—43.0	
Glycerin	1.260	78.6	0
Kerosene	0.82	51.2	
Mercury	13.6	849.0	
Milk	1.028—1.035	64.2—64.6	
Naphtha, petroleum ether	0.665	41.5	15
wood	0.848—0.810	52.9—50.5	0
Oils:			
castor	0.969	60.5	15
cocoanut	0.925	57.7	15
cotton seed	0.926	57.8	16
creosote	1.040—1.100	64.9—68.6	15
linseed, boiled	0.924	58.8	15
olive	0.918	57.3	15
Sea water	1.025	63.99	15
Turpentine (spirits)	0.87	54.3	
Water	1.00	62.43	4

Densities of Miscellaneous Materials

Approximate Values at Ordinary Temperature

Name	Sp gr	Lb/ft³
Aluminum bronze	7.7	481
Anthracite	1.4–1.7	97
Asbestos*	2.1–2.8	153
Asphalt	1.1–1.5	81
Ashes (cinders)	....	45
Barytes*	4.5	281
Bituminous	1.2–1.5	84
Bluestone*	2.5–2.6	159
Borax*	1.7–1.8	109
Brass (70 Cu, 30 Zn)	8.53	532
Brick, common	1.8–2.0	120
Bronze (90 Cu, 10 Zn)	8.80	550
Cast iron	7.2	450
Cement, loose	....	90
Cement, set	2.7–3.2	183
Charcoal	0.4	25
Clay:		
Dry	1.0	63
Damp, plastic	1.8	110
Coke	1.0–1.4	75
Concrete:		
Plain	....	144
Reinforced	....	150
Cinder	....	100
Cork	0.24	15
Corn, bulk	0.73	45
Cotton, flax, hemp	1.47–1.50	93
Earth:		
Dry, loose	1.2	76
Dry, packed	1.5	95
Moist, loose	1.3	78
Moist, packed	1.6	96
Flour, loose	0.4–0.5	28
Gasoline	0.75	46.8
Glass, common	2.4–2.8	162
Granite*	2.6–2.7	165
Gravel:		
Dry	....	100
Wet	....	120
Greenstone (trap)*	2.8–3.2	187
Gypsum*	2.3–2.8	159
Hay and straw (bales)	0.32	20
Hematite (iron ore)	5.2	325
Leather	0.86–1.02	59
Lignite	1.1–1.4	78
Limestone*	2.1–2.86	155
Limonite (iron ore)	3.6–4.0	237
Magnesite*	3.0	187
Magnetite (iron ore)	4.9–5.2	315
Marble*	2.6–2.86	170
Mica	2.65–3.2	182
Oats, bulk	0.51	32
Oil:		
Vegetable	0.91–0.94	58
Fuel	1.0	63
Lubricant	0.9	56
Paper	0.70–1.15	58
Paraffin	0.87–0.91	56
Phosphate rock (apatite)*	3.2	200
Pitch	1.07–1.15	69
Plaster:		
On lath	....	100
On masonry	....	60
Potatoes, piled	0.67	44
Pumice, natural*	0.37–0.9	40
Rubber:		
Goods	1.0–2.0	93
Raw	0.92–0.96	59
Salt, granulated, piled	0.77	48
Sand:		
Dry	....	100
Wet	....	120
Sandstone*	2.0–2.6	143
Shale and slate*	2.6–2.9	172
Slag, blast furnace	2.5–3.0	172
Snow, loose	0.125	8
Stainless steel (18:8)	7.93	493
Steel	7.87	490
Sugar	1.61	100
Tile, hollow	....	55
Water:		
Fresh	1.00	62.3
Salt	1.02	64
Wheat, bulk	0.77	48
Wood, seasoned:		
Birch	0.71	44
Cedar	0.35	22
Cypress	0.48	30
Elm	0.56	35
Mahogany	0.56–0.85	44
Maple, white	0.53	33
Oak, red or black	0.64–0.71	42
Oak, white	0.77	48
Pine, white	0.43	27
Pine, yellow	0.71	44
Redwood	0.42	26
Spruce	0.45	28
Walnut	0.59	37

* Density for the mineral is specified. Most minerals, when quarried and piled are about 35 to 45 per cent less dense. Masonry is generally 5 to 10 per cent less dense than the mineral.

Densities of Selected Materials**

(From Smithsonian Tables.)

Material	Grams per cm^3	Pounds per $ft.^3$
Agate	2.5–2.7	156–168
Alabaster, carbon-ate	2.69–2.78	168–173
sulfate	2.26–2.32	141–145
Albite	2.62–2.65	163–165
Amber	1.06–1.11	66–69
Amphiboles	2.9–3.2	180–200
Anorthite	2.74–2.76	171–172
Asbestos	2.0–2.8	125–175
Asbestos slate	1.8	112
Asphalt	1.1–1.5	69–94
Basalt	2.4–3.1	150–190
Beeswax	0.96–0.97	60–61
Beryl	2.69–2.7	168–169
Biotite	2.7–3.1	170–190
Bone	1.7–2.0	106–125
Brick	1.4–2.2	87–137
Butter	0.86–0.87	53–54
Calamine	4.1–4.5	255–280
Calcspar	2.6–2.8	162–175
Camphor	0.99	62
Caoutchouc	0.92–0.99	57–62
Cardboard	0.69	43
Celluloid	1.4	87
Cement, set	2.7–3.0	170.190
Chalk	1.9–2.8	118–175
Charcoal, oak	0.57	35
pine	0.28–0.44	18–28
Cinnabar	8.12	507
Clay	1.8–2.6	112–162
Coal, anthracite	1.4–1.8	87–112
bituminous	1.2–1.5	75–94
Cocoa butter	0.89–0.91	56–57
Coke	1.0–1.7	62–105
Copal	1.04–1.14	65–71
Cork	0.22–0.26	14–16
Cork linoleum	0.54	34
Corundum	3.9–4.0	245–250
Diamond	3.01–3.52	188–220
Dolomite	2.84	177
Ebonite	1.15	72
Emery	4.0	250
Epidote	3.25–3.50	203–218
Feldspar	2.55–2.75	159–172
Flint	2.63	164
Fluorite	3.18	198
Galena	7.3–7.6	460–470
Gamboge	1.2	75
Garnet	3.15–4.3	197–268
Gas carbon	1.88	117
Gelatin	1.27	79

Material	Grams per cm^3	Pound per $ft.^3$
Glass, common	2.4–2.8	150–175
flint	2.9–5.9	180–370
Glue	1.27	79
Granite	2.64–2.76	165–172
Graphite*	2.30–2.72	144–170
Gum arabic	1.3–1.4	81–87
Gypsum	2.31–2.33	144–145
Hematite	4.9–5.3	306–330
Hornblende	3.0	187
Ice	0.917	57.2
Ivory	1.83–1.92	114–120
Leather, dry	0.86	54
Lime, slaked	1.3–1.4	81–87
Limestone	2.68–2.76	167–171
Linoleum	1.18	74
Magnetite	4.9–5.2	306–324
Malachite	3.7–4.1	231–256
Marble	2.6–2.84	160–177
Meerschaum	0.99–1.28	62–80
Mica	2.6–3.2	165–200
Muscovite	2.76–3.00	172–187
Ochre	3.5	218
Opal	2.2	137
Paper	0.7–1.15	44–72
Paraffin	0.87–0.91	54–57
Peat blocks	0.84	52
Pitch	1.07	67
Porcelain	2.3–2.5	143–156
Porphyry	2.6–2.9	162–181
Pressed wood		
pulp board	0.19	12
Pyrite	4.95–5.1	309–318
Quartz	2.65	165
Resin	1.07	67
Rock salt	2.18	136
Rubber, hard	1.19	74
Rubber, soft		
commercial	1.1	69
pure gum	0.91–0.93	57–58
Sandstone	2.14–2.36	134–147
Serpentine	2.50–2.65	156–165
Silica, fused trans-		
parent	2.21	138
translucent	2.07	129
Slag	2.0–3.9	125–240
Slate	2.6–3.3	162–205
Soapstone	2.6–2.8	162–175
Spermacéti	0.95	59
Starch	1.53	95
Sugar	1.59	99
Talc	2.7–2.8	168–174

Material	Grams per cm^3	Pounds per $ft.^3$
Tallow, beef	0.94	59
mutton	0.94	59
Tar	1.02	66
Topaz	3.5–3.6	219–223
Tourmaline	3.0–3.2	190–200
Wax, sealing	1.8	112
Wood (seasoned)		
alder	0.42–0.68	26–42
apple	0.66–0.84	41–52
ash	0.65–0.85	40–53
balsa	0.11–0.14	7–9
bamboo	0.31–0.40	19–25
basswood	0.32–0.59	20–37
beech	0.70–0.90	43–56
birch	0.51–0.77	32–48
blue gum	1.00	62
box	0.95–1.16	59–72
butternut	0.38	24
cedar	0.49–0.57	30–35
cherry	0.70–0.90	43–56
dogwood	0.76	47
ebony	1.11–1.33	69–83
elm	0.54–0.60	34–37
hickory	0.60–0.93	37–58
holly	0.76	47
juniper	0.56	35
larch	0.50–0.56	31–35
lignum vitae	1.17–1.33	73–83
locust	0.67–0.71	42–44
logwood	0.91	57
mahogany		
Honduras	0.66	41
Spanish	0.85	53
maple	0.62–0.75	39–47
oak	0.60–0.90	37–56
pear	0.61–0.73	38–45
pine, pitch	0.83–0.85	52–53
white	0.35–0.50	22–31
yellow	0.37–0.60	23–37
plum	0.66–0.78	41–49
poplar	0.35–0.5	22–31
satinwood	0.95	59
spruce	0.48–0.70	30–44
sycamore	0.40–0.60	24–37
teak, Indian	0.66–0.88	41–55
African	0.98	61
walnut	0.64–0.70	40–43
water gum	1.00	62
willow	0.40–0.60	24–37

* May be as low as 1.6.

** Densities are given at ordinary atmospheric temperature.

ELECTRICAL RESISTIVITY AND TEMPERATURE COEFFICIENTS AT GIVEN TEMPERATURES OF ELEMENTS

Element	Temperature °C	Microhm-Cm	Temperature Coefficient per °C	Element	Temperature °C	Microhm-Cm	Temperature Coefficient per °C
Aluminum, 99.996%	20	2.6548	0.00429^{20} [i]	Nickel	20	6.84	0.0069^{0-100}
Antimony	0	39.0		Niobium (Columbium)[g]	0	12.5	
Arsenic	20	33.3		Osmium	20	9.5	0.0042^{0-100}
Beryllium[a]	20	4.0	0.025^{20} [i]	Palladium	20	10.54	0.00374^{0-60} [g]
Bismuth	0	106.8		Phosphorus, white	11	1×10^{17}	
Boron	0	1.8×10^{12}		Platinum, 99.85%	20	10.6	0.003927^{0-100}
Cadmium	0	6.83	0.0042^{0} [i]	Plutonium	107	141.4	
Calcium	0	3.91	0.00416^{0} [i]	Potassium	0	6.15	
Carbon[b]	0	1375.0		Praseodymium	25	68	0.00171^{0-25}
Cerium	25	75.0	0.00087^{0-25}	Rhenium	20	19.3	0.00395^{0-100}
Cesium	20	20		Rhodium	20	4.51	0.0042^{0-100}
Chromium	0	12.9	0.003^{0} [i]	Rubidium	20	12.5	
Cobalt	20	6.24	0.00604^{0-100}	Ruthenium	0	7.6	
Copper	20	1.678	0.0068^{0-500} [g]	Samarium	25	88.0	0.00184^{0-25}
Dysprosium[c]	25	57.0	0.00119^{0-25}	Scandium[h]	22	61.0	0.00282^{0-25}
Erbium	25	107.0	0.00201^{0-25}	Selenium[k]	0	10^{6}	
Europium	25	90.0		Silicon	0	$3\text{-}4 \times 10^{6}$ [j]	
Gadolinium	25	140.5	0.00176^{0-25}	Silver	20	1.586	0.0061^{0-100} [g]
Gallium[d]	20	17.4		Sodium	0	4.2	
Germanium[e]	22	46×10^{6}		Strontium	20	23.0	
Gold	20	2.24	0.0083^{0-100} [g]	Sulfur, yellow	20	2×10^{23}	
Hafnium	25	35.1	0.0038^{25} [i]	Tantalum	25	12.45	0.00383^{0-100}
Holmium	25	87.0	0.00171^{0-25}	Tellurium	25	4.36×10^{6}	
Indium	20	8.37		Thallium	0	18.0	
Iodine	20	1.3×10^{15}		Thorium	0	13.0	0.0038^{0-100}
Iridium	20	5.3	0.003925^{0-100}	Thulium	25	79.0	0.00195^{0-25}
Iron, 99.99%	20	9.71	0.00651^{20} [i]	Tin	0	11.0	0.0047^{0-100}
Lanthanum	25	5.70	0.00218^{0-25}	Titanium	20	42.0	
Lead	20	20.648	0.00336^{20-40}	Tungsten	27	5.65	
Lithium	0	8.55		Uranium		30.0	
Lutetium	25	79.0	0.00240^{0-25}	Vanadium	20	24.8-26.0	
Magnesium[f]	20	4.45	0.0165^{20} [i]	Ytterbium	25	29.0	0.0013^{0-25}
Manganese α	23-100	185.0		Yttrium	25	57.0	0.0027^{0-25}
Mercury	50	98.4		Zinc	20	5.916	0.00419^{0-100}
Molybdenum	0	5.2		Zirconium	20	40.0	0.0044^{20} [i]
Neodymium	25	64.0	0.00164^{0-25}				

[a] Annealed, comm. pure.
[b] Graphite.
[c] Polycrystalline.
[d] Hard Wire.
[e] Intrinsic Ge.
[f] Polycrystalline.
[g] High Purity.
[h] Zone refined bar.
[i] Data not available to indicate range over which coefficient is valid.
[j] Very sensitive to purity.
[k] Crystalline.

SURFACE TENSION OF SELECTED LIQUIDS

LIQUIDS		In contact with	Temperature °C	Surface tension dynes/cm
Name	Formula			
Acetaldehyde	C_2H_4O	- -vapor	20	21.2
Acetaldoxime	C_2H_5NO	- -vapor	35	30.1
Acetamide	C_2H_5NO	- -vapor	85	39.3
Acetanilide	C_2H_5NO	- -vapor	120	35.6
Acetic acid	$C_2H_4O_2$	- -vapor	10	28.8
	C_2H_4O	- -vapor	20	27.8
	C_2H_4O	- -vapor	50	24.8
Acetic anhydride	$C_4H_6O_3$	- -vapor	20	32.7
Acetone	C_3H_6O	- -air or vapor	0	26.21
	C_3H_6O	- -air or vapor	20	23.70
	C_3H_6O	- -air or vapor	40	21.16
Acetonitrile	C_2H_3N	- -vapor	20	29.30
Acetophenone	C_8H_8O	- -vapor	20	39.8
Acetyl chloride	C_2H_3ClO	- -vapor	14.8	26.7
Acetylene	C_2H_2	- -vapor	−70.5	16.4
Acetylsalicylic acid (in aq. sol.)	$C_9H_8O_4$	- -vapor	25.9	60.06
Allyl alcohol	C_3H_6O	- -air or vapor	20	25.8
Allyl isothiocyanate	C_4H_5NS	- -air or vapor	20	34.5
Ammonia	NH_3	- -vapor	11.1	23.4
	NH_3	- -vapor	34.1	18.1
Aniline	C_6H_7N	- -air	10	44.10
	C_6H_7N	- -vapor	20	42.9
	C_6H_7N	- -air	50	39.4
Argon	A	- -vapor	−188	13.2
Azoxybenzene	$C_{12}H_{10}N_2O$	- -vapor	51	43.34
Benzaldehyde	C_7H_6O	- -air	20	40.04
Benzene	C_6H_6	- -air	10	30.22
	C_6H_6	- -air	20	28.85
	C_6H_6	- -saturated with vapor	20	28.89
	C_6H_6	- -air	30	27.56
Benzonitrile	C_7H_5N	- -air	20	39.05
Benzophenone	$C_{13}H_{10}O$	- -air or vapor	20	45.1
Benzylamine	C_7H_9N	- -vapor	20	39.5
Benzyl alcohol	C_7H_8O	- -air or vapor	20	39.0
Bromine	Br_2	- -air or vapor	20	41.5
Bromobenzene	C_6H_5Br	- -air	20	36.5
Bromoform	$CHBr_3$	- -vapor	20	41.53
p-Bromophenol	C_6H_5BrO	- -vapor	74.4	42.36
d-sec-Butyl alcohol	$C_4H_{10}O$	- -vapor	10	23.5
n-Butyl alcohol	$C_4H_{10}O$	- -air or vapor	0	26.2
	$C_4H_{10}O$	- -air or vapor	20	24.6
	$C_4H_{10}O$	- -air or vapor	50	22.1
tert-Butyl alcohol	$C_4H_{10}O$	- -air or vapor	20	20.7
n-Butylamine	$C_4H_{11}N$	- -nitrogen	41	19.7
n-Butyric acid	$C_4H_8O_2$	- -air	20	26.8
Carbon bisulfide	CS_2	- -vapor	20	32.33
Carbon dioxide	CO_2	- -vapor	20	1.16
	CO_2	- -vapor	−25	9.13
Carbon tetrachloride	CCl_4	- -vapor	20	26.95
	CCl_4	- -vapor	100	17.26
	CCl_4	- -vapor	200	6.53
Carbon monoxide	CO	- -vapor	−193	9.8
	CO	- -vapor	−203	12.1
Chloral	C_2HCl_3O	- -vapor	19.4	25.34
Chlorine	Cl_2	- -vapor	20	18.4
	Cl_2	- -vapor	−30	25.4
	Cl_2	- -vapor	−40	27.3
	Cl_2	- -vapor	−50	29.2
	Cl_2	- -vapor	−60	31.2
Chloroacetic acid	$C_2H_2Cl_2O_2$	- -nitrogen	25.7	35.4
Chlorobenzene	C_4H_5Cl	- -vapor	20	33.56
Chloroform	$CHCl_3$	- -air	20	27.14
o-Chlorophenol	C_6H_5ClO	- -vapor	12.7	42.25
Cyclohexane	C_6H_{12}	- -air	20	25.5
Dichloroacetic acid	$C_2H_2Cl_2O_2$	- -nitrogen	25.7	35.4
Dichloroethane	$C_2H_4Cl_2$	- -air	35.0	23.4
Diethylamine	$C_4H_{11}N$	- -air	56	16.4
Diethylaniline	$C_{10}H_{15}N$	- -vapor	20	34.2
Diethyl carbonate	$C_5H_{10}O$	- -air	20	26.31
Diethyl oxalate	$C_6H_{10}O_4$	- -vapor	20	32.0
Diethyl phthalate	$C_{12}H_{14}O_4$	- -vapor	20	37.5
Diethyl sulfate	$C_4H_{12}O_4S$	- -air	13	34.61
Dimethylamine	C_2H_7N	- -nitrogen	0	18.1
	C_2H_7N	- -nitrogen	5	17.7
Dimethylaniline	C_8H_{11}	- -air or vapor	20	36.6

(Continued)

LIQUIDS Name	Formula	In contact with	Temperature °C	Surface tension dynes/cm
1,5-Dimethyl-2-phenyl-3-pyrazolone	$C_{11}H_{12}N_2O$	--vapor	25.9	63.63
Dimethyl sulfate	$C_2H_6O_4S$	--air	18	40.12
Diphenylamine	$C_{12}H_{11}N$	--air or vapor	80	37.7
Ethyl acetate	$C_4H_8O_2$	--air	0	26.5
	$C_4H_8O_2$	--air	20	23.9
	$C_4H_8O_2$	--air	50	20.2
Ethyl acetoacetate	$C_6H_{10}O_3$	--air or vapor	20	32.51
Ethyl alcohol	C_2H_6O	--air	0	24.05
	C_2H_6O	--vapor	10	23.61
	C_2H_6O	--vapor	20	22.75
	C_2H_6O	--vapor	30	21.89
Ethylamine	C_2H_7N	--nitrogen	0	21.3
	C_2H_7N	--nitrogen	9.9	20.4
Ethylaniline	$C_8H_{11}N$	--air or vapor	20	36.6
Ethylbenzene	C_8H_{10}	--vapor	20	29.20
Ethylbenzoate	$C_9H_{10}O_2$	--vapor	20	35.5
Ethyl bromide	C_2H_5Br	--vapor	20	24.15
Ethyl chloroformate	$C_3H_5ClO_2$	--vapor	15.1	27.5
Ethyl Cinnamate	$C_{11}H_{12}O_2$	--air	20	38.37
Ethylene bromide	$C_2H_4Br_2$	--vapor	20	38.37
Ethylene chloride	$C_2H_4Cl_2$	--air	20	24.15
Ethylene oxide	C_2H_4O	--vapor	−20	30.8
	C_2H_4O	--vapor	0.0	27.6
	C_2H_4O	--vapor	20	24.3
Ethyl ether	$C_4H_{10}O$	--vapor	20	17.01
	$C_4H_{10}O$	--vapor	50	13.47
Ethyl format	$C_3H_6O_2$	--air or vapor	20	23.6
Ethyl iodide	C_2H_5I	--vapor	20	29.4
Ethyl nitrate	$C_2H_5NO_3$	--air or vapr	20	28.7
dl-Ethyl lactate	$C_5H_{10}O_3$	--air	20	29.9
Ethyl mercaptan	C_2H_6S	--air or vapor	20	22.5
Ethyl salicylate	$C_9H_{10}O_3$	--vapor	20.5	38.33
Formamide	CH_3NO	--vapor	20	58.2
Formic acid	CH_2O_2	--air	20	37.6
Furfural	$C_5H_4O_2$	--air or vapor	20	43.5
Gelatin solution (1%)		--water	2.85	8.3
Glycerol	$C_3H_8O_3$	--air	20	63.4
	$C_3H_8O_3$	--air	90	58.6
	$C_3H_8O_3$	--air	150	51.9
Glycol	$C_2H_6O_2$	--air or vapor	20	47.7
Helium	He	--vapor	−269	.12
	He	--vapor	−270	.239
	He	--vapor	−271.5	.353
n-Hexane	C_6H_{14}	--air	20	18.43
Hydrazine	N_2H_4	--vapor	25	91.5
Hydrogen	H_2	--vapor	−255	2.31
Hydrogen cyanide	HCN	--vapor	17	18.2
Hydrogen peroxide	H_2O_2	--vapor	18.2	76.1
Isobutyl alcohol	$C_4H_{10}O$	--vapor	20	23.0
Isobutylamine	$C_4H_{11}N$	--air	68	17.6
Isobutyl chloride	C_4H_9Cl	--air	20	21.94
Isobutyric acid	$C_4H_8O_2$	--air or vapor	20	25.2
Isopentane	C_5H_{12}	--air	20	13.72
Isopropyl alcohol	C_3H_8O	--air or vapor	20	21.7
Methyl acetate	$C_3H_6O_2$	--air or vapor	20	24.6
Methyl alcohol	CH_4O	--air	0	24.49
	CH_4O	--air	20	22.61
	CH_4O	--vapor	50	20.14
Methylamine	CH_3NH_2	--nitrogen	−12	22.2
	CH_3NH_2	--vapor	−20	23.0
	CH_3NH_2	--nitrogen	−70	29.2
N-Methylaniline	C_7H_9N	--air or vapor	20	39.6
Methyl benzoate	$C_8H_8O_2$	--air or vapor	20	37.6
Methyl chloride	CH_3Cl	--air	20	16.2
Methyl ether	C_2H_6O	--vapor	−10	16.4
	C_2H_6O	--vapor	−40	21
Methylene chloride	CH_2Cl_2	--air	20	26.52
Methylene iodide	CH_2I_2	--air	20	50.76
Methyl ethyl ketone	C_4H_8O	--air or vapor	20	24.6
Methyl formate	$C_2H_4O_2$	--vapor	20	25.08
Methyl iodide	CH_3I	--air	43.5	25.8
Methyl propionate	$C_4H_8O_2$	--air or vapor	20	24.9
Methyl salicylate	$C_8H_8O_3$	--nitrogen	94	31.9
Methyl sulfide	C_2H_6S	--vapor	11.1	26.50

(Continued)

LIQUIDS Name	Formula	In contact with	Temperature °C	Surface tension dynes/cm
Naphthalene	$C_{10}H_8$	- -air or vapor	127	28.8
Neon	Ne	- -vapor	−248	5.50
Nitric acid (98.8%)	HNO_3	- -air	11.6	42.7
Nitrobenzene	$C_6H_5NO_2$	- -air or vapor	20	43.9
Nitroethane	$C_2H_5NO_2$	- -air or vapor	20	32.2
Nitrogen	N_2	- -vapor	−183	6.6
	N_2	- -vapor	−193	8.27
	N_2	- -vapor	−203	10.53
Nitrogen tetra oxide	N_2O_4	- -vapor	19.8	27.5
Nitromethane	CH_3NO_2	- -vapor	20	36.82
Nitrous oxide	N_2O	- -vapor	20	1.75
n-Octane	C_8H_{18}	- -vapor	20	21.80
n-Octyl alcohol	$C_8H_{18}O$	- -air	20	27.53
Oleic acid	$C_{18}H_{34}O_2$	- -air	20	32.50
Oxygen	O_2	- -vapor	−183	13.2
Oxygen (65%)	O_2	- -air	−190.5	12.2
	O_2	- -vapor	−193	15.7
	O_2	- -vapor	−203	18.3
Paraldehyde	$C_6H_{12}O_3$	- -air	20	25.9
Phenetole	$C_8H_{10}O$	- -vapor	20	32.74
Phenol	C_6H_6O	- -air or vapor	20	40.9
	C_6H_6O	- -air or vapor	30	39.88
Phenylhydrazine	$C_6H_8N_2$	- -vapor	20	46.1
Phosphorus tribromide	PBr_3	- -air	24	45.8
Phosphorus trichloride	PCl_3	- -vapor	20	29.1
Phosphorus triiodide	PI_3	- -vapor	75.3	56.5
Propionic acid	$C_3H_6O_2$	- -vapor	20	26.7
n-Propyl acetate	$C_5H_{10}O_2$	- -air or vapor	20	24.3
n-Propyl alcohol	C_3H_8O	- -vapor	20	23.78
n-Propylamine	C_3H_9N	- -air	20	22.4
n-Propyl bromide	C_3H_7Br	- -vapor	71	19.65
n-Propyl chloride	C_3H_7Cl	- -air	47	18.2
n-Propyl formate	$C_4H_8O_2$	- -vapor	20	24.5
Pyridine	C_5H_5N	- -air	20	38.0
Quinoline	C_9H_7N	- -air	20	45.0
Ricinoleic acid	$C_{18}H_{34}O_3$	- -air	16	35.81
Selenium	Se	- -air	217	92.4
Styrene	C_8H_8	- -air	19	32.14
Sulfuric acid (98.5%)	H_2SO_4	- -air or vapor	20	55.1
Tetrabromoethane 1,1,2,2-	$C_2H_2Br_4$	- -air	20	49.67
Tetrachloroethane 1,1,2,2-	$C_2H_2Cl_4$	- -air	22.5	36.03
Tetrachloroethylene	C_2Cl_4	- -vapor	20	31.74
Toluene	C_7H_8	- -vapor	10	27.7
	C_7H_8	- -vapor	20	28.5
	C_7H_8	- -vapor	30	27.4
m-Toluidine	C_7H_9N	- -vapor	20	36.9
o-Toluidine	C_7H_9N	- -air or vapor	20	40.0
p-Toluidine	C_7H_9N	- -air	50	34.6
Trichloroacetic acid	$C_2HCl_3O_2$	- -nitrogen	80.2	27.8
Trichloroethane 1,1,2-	$C_2H_3Cl_3$	- -air	114	22.0
Triethyl phosphate	$C_6H_{15}O_4P$	- -air	15.5	30.61
Trimethylamine	C_3H_9N	- -nitrogen	−4	17.3
Triphenylcarbinol	$C_{19}H_{16}O$	- -vapor	165.8	30.38
Vinyl acetate	$C_4H_6O_2$	- -vapor	20	23.95
	$C_4H_6O_2$	- -vapor	25	23.16
	$C_4H_6O_2$	- -vapor	30	22.54
Water	H_2O	- -air	18	73.05
m-Xylene	C_8H_{10}	- -vapor	20	28.9
o-Xylene	C_8H_{10}	- -air	20	30.10
p-Xylene	C_8H_{10}	- -vapor	20	28.37

›URFACE TENSIONS OF SELECTED MATERIALS (in grouped orders)

JOSEPH J. JASPER

Polysiloxances (25°C)	dynes/cm
›methyltrisiloxane	67.56
›amethyltetrasiloxane	86.20
›ecamethylpentasiloxane	17.08
›adecamethylhexasiloxane	17.42
›adecamethylheptasiloxane	17.61
›adecamethyloctasiloxane	18.03

Linear Polymethylsiloxances (20°C)	
›tadecamer	19.9
›ecamer	19.6
›amer	19.2
›tamer	18.6
›amer	18.5
›tamer	18.1
›amer	17.6
›ner	17.0

Chloromethylsilances (20°C)	
›hloromethylsilane	20.3
›hlorodimethylsilane	20.1
›methylsilanol	18.4

Esters (20°C)	
›cresyl phosphate	40.9
›zyl phenylundecanoate	37.7
›2-ethylhexyl) phthalate	31.2
›2-ethylhexyl) sebacate	31.1
›taerythritol tetracaproate	30.4
›-Hexamethylene glycol di-2-ethylhexanoate	30.2
›2-ethylhexyl)adipate	30.2
›(2-ethylhexyl) tricarballylate	29.6
›ylphenylhendecanoate	38.0
›xa(2-ethyl-1-hexoxy) disiloxane	26.8
›xa(2-ethyl-1-butoxy) disiloxane	26.1

Acetals and ethers (20°C)	
›-Diisopropoxyethane	20.97
›-Dipentyloxyethane	25.67
›-Diisopentyloxyethane	23.98
›-Di(2-chloroethoxy)ethane	36.22
›-Di(2-ethoxyethyloxy)ethane	27.44
›utoxy-1-ethoxyethane	23.26
›utoxy-1-*tert*-butoxyethane	22.65
›-Butyl vinyl ether	20.76
›ntyl vinyl ether	23.44
›tyl vinyl ether	26.32

Alkyl glycolates (20°C)	dynes/cm
Propyl glycolate	31.55
Butyl glycolate	30.43
Hexyl glycolate	29.66
Heptyl glycolate	29.40

Alkyl glycol monoethers (20°C)	
2-Pentyloxyethanol	27.6
2-(2-Pentyloxyethoxy)ethanol	29.9
2-Hexyloxyethanol	27.7
2-(2-Hexyloxyethoxy)ethanol	29.6
1-(2-Hexyloxyethoxy)-2-(2-hydroxyethoxy)ethane	31.0
Tetraethylene glycol monohexyl ether	32.1
2-Octyloxyethanol	29.8
2-(2-Octyloxyethoxy)ethanol	30.7
1-(2-Hydroxyethoxy)-2-(2-octyloxyethoxy)ethane	31.5
Tetraethylene glycol mono-octyl ether	32.0
Pentaethylene glycol mono-octyl ether	32.17
Hexaethylene glycol mono-octyl ether	32.85
2-Dodecyloxyethanol	29.6
2-(2-Dodecyloxyethoxy)ethanol	30.6
1-(2-Dodecyloxyethoxy)-2-(2-hydroxyethoxy)ethane	31.3
Tetraethylene glycol mono-dodecyl ether	31.7
Pentaethylene glycol mono-dodecyl ether	32.0

Alkyl derivatives of benzene and styrene (20°C)	
Isopentylbenzene	28.34
tert-Pentylbenzene	29.02
p-Cymene	29.44
p-Di-*sec*-butylbenzene	29.98
β-Methylstyrene	34.12
Styrene	32.00
Allylbenzene	30.37
α-Methylstyrene	32.56
β-Ethylstyrene	32.01
1-Phenyl-2-butene	30.48
2-Phenyl-2-butene	33.61
(2-Methylpropenyl)benzene	31.86
β-Propylstyrene	33.23
β-Isopropylstyrene	31.40
β-Ethyl-β-methylstyrene	31.80
α,β,β-Trimethylstyrene	32.04
β-Ethyl-α-methylstyrene	32.25
α-Ethyl-β-methylstyrene	33.87
β,β-Diethylstyrene	32.22
β-Isobutyl-α-methylstyrene	31.38
β-Ethyl-α-propylstyrene	34.20
cis-β-Vinylstyrene	36.15
trans-β-Vinylstyrene	36.21
Isopropyl-1-pentylstyrene	36.59

Nonaromatic cyclic compounds (20°C)

Compound	dynes/cm
2-Methoxycyclopentanol	38.1
2-Ethoxycyclopentanol	35.0
2-Methoxycyclohexanol	37.7
2-Ethoxycyclohexanol	34.7
2-Propoxycyclohexanol	34.7
2-Methoxy-1-methylcyclohexanol	36.4
1-Ethyl-2-methoxycyclohexanol	36.1
1,2-Dimethoxycyclopentane	31.9
1,2-Diethoxycyclopentane	31.9
1,2-Dimethoxycyclohexane	34.2
1,2-Diethoxycyclohexane	33.9
1,2-Dipropoxycyclohexane	33.5
1,2-Diisopropoxycyclohexane	33.4
1,2-Dibutoxycyclohexane	33.1
1,2-Dicyclohexyloxycyclohexane	36.5
Bicyclopentyl[a] (30 °C)	29.8

Halogenated hydrocarbons (non-fluoro) (20°C)

Compound	dynes/cm
Trichlorobiphenyl	45.3
Tetrachlorobiphenyl	44.2
Perchlorocyclopentadiene	37.5
Hexachloro-1,3-butadiene	36.0

Fluorocarbons (25°C)

Compound	dynes/cm
Perfluoro-1,2-dimethylcyclohexane	15.5
Perfluoro-1,3,5-trimethylcyclohexane	17.2
Perfluoro-1,2,4-trimethylcyclohexane	17.3
Perfluoro-octane	13.7
Perfluorododecane (113.5 °C)	10.6
Perfluoro-2-methyldecahydronaphthalene	19.2
Perfluoropropylcyclohexane	17.2
Perfluorobutylcyclohexane	17.7
Perfluoro-4-isopropyl-1-methylcyclohexane	17.5
Perfluorodecahydronaphthalene	18.3
Perfluorononane	14.4
Perfluorodecane (45 °C)	13.5
Perfluoroundecane (70 °C)	12.7
Perfluoro-1,4-dimethylcyclohexane[a] (20 °C)	16.3

Fluorocarbons and derivatives (20°C)

Compound	dyn[illegible] cm
p-Difluorobenzene	27
Perfluorotrihexylamine	18
Perfluorotributylamine	16
Perfluorotripropylamine	15
Perfluorodibutyl ether	13
1,1,1,3,3,3-Hexachloro-2,2-difluoropropane	32
1,1,1,3,3-Pentachloro-2,2,3-trifluoropropane	27
1,2,2,3-Tetrachloro-1,1,3,3-tetrafluoropropane	22
2,2,3,3-Tetrafluoro-1-propanol	27
2,2,3,3,4,4,5,5-Octafluoro-1-pentanol	24
2,2,3,3,4,4,5,5,6,6,7,7-Dodecafluoro-1-heptanol	23

Fluorinated esters and ethers (20°C)

Compound	dynes/cm
Hexyl heptafluorobutyrate	19
Octadecyl heptafluorobutyrate	25
Butyl pentadecafluorooctanoate	18
1,6-Hexanediol bis(heptafluorobutyrate)	21
1,10-Decanediol bis(heptafluorobutyrate)	23
1,6-Hexanediol bis(7H-dodecafluoroheptanoate)	25
1,6-Hexanediol bis(pentadecafluorooctanoate)	20
2-Ethyl-2-(hydroxymethyl)-1,3-propanediol heptafluorobutyrate	21
1H,1H,9H-Hexadecafluorononyl 2-ethylhexanoate	23
Bis(1H,1H,5H-octafluoropentyl) glutarate	27
Bis(1H,1H-heptafluorobutyl) 3-methylglutarate	20
Bis(1H,1H,5H-octafluoropentyl) 3-methylglutarate	26
Bis(1H,1H-undecafluorohexyl) 3-methylglutarate	19
Bis(1H,1H,7H-dodecafluoroheptyl) 3-methylglutarate	2[illegible]
Bis(1H,1H-pentadecafluorooctyl) 3-methylglutarate	1[illegible]
Bis(1H,1H,9H-hexadecafluorononyl) 3-methylglutarate	2[illegible]
Bis(1H,1H,5H-octafluoropentyl) adipate	2[illegible]
Bis(1H,1H,7H-dodecafluoroheptyl) adipate	2[illegible]
Bis(1H,1H,9H-hexadecafluorononyl) 3-*tert*-butyl-adipate	2[illegible]
Bis(1H,1H-heptafluorobutyl) sebacate	2[illegible]
Bis(1H,1H,5H-octafluoropentyl) sebacate	2[illegible]
Bis(1H,1H,7H-dodecafluoroheptyl) pinate	2[illegible]
Bis(1H,1H,5H-octafluoropentyl) phthalate	2[illegible]
Tris(1H,1H,5H-octafluoropentyl) tricarballylate	2[illegible]
1H,1H,7H-Dodecafluoroheptyl methyl ether	2[illegible]
Bis(1H,1H,7H-dodecafluoroheptoxy)hexane	2[illegible]

Fluorinated esters

Compound	dynes/cm
Diethyl perfluoroadipate (27.0)	2[illegible]
Dibutyl perfluoroadipate (26.0)	2
Bis(1H,1H-heptafluorobutyl) adipate (25.8)	2[illegible]
2,2,3,3,4,4,5,5-Octafluoro-1,6-hexanediol dibutyrate	2
1,5-Pentanediol di(trifluoroacetate) (26.2)	2
1,5-Pentanediol di(perfluorobutyrate) (26.2)	2[illegible]
1,6-Hexanediol di(perfluorobutyrate) (26.2)	2
1,5-Pentanediol di(perfluorooctanoate) (26.2)	1
Pentaerythritol tetra(perfluorobutyrate) (26.0)	1[illegible]
m-(Trifluoromethyl)benzyl perfluorobutyrate (26.5)	2

Organic phosphoryl compounds(20°C)

	dynes/cm
Diisopropyl phosphonate	26.4
Diethyl phosphorofluoridate	25.9
Diisopropyl phosphorofluoridate	24.5
Diethyl phosphorochloridate	32.0
Diisopropyl phosphorochloridate	28.3
Methyl phosphorodichloridate	34.9
Ethyl phosphorodichloridate	32.8
Diethyl dimethylphosphoramidate	30.3
Ethyl dimethylphosphoramidochloridate	34.9
Dimethylphosphoramidic difluoridate	25.1
Dimethyphosphoramidic dichloride	36.1
Diethylphosphoramidic dichloride	35.7
Tetramethylphosphordiamidic fluoride	33.4

Esters of boric acid(20°C)

Propyl borate	22.48
Isopropyl borate	19.02
Butyl borate	24.42
Heptyl borate	26.15
Octyl borate	28.18
Decyl borate	29.38

Germanium compounds

Germanium tetrachloride (30 °C)	22.44
Germanium tetrabromide (30 °C)	35.51
Germanium tetrabromide (50 °C)	33.70
Tetraethylgermanium (30 °C)	22.96
Tetraethoxygermanium (30 °C)	23.00

Sulfur halides(20°C)

	dynes/cm
Sulfur dichloride	37.2
Trisulfur dichloride	47.8
Tetrasulfur dichloride	52.5
Sulfur monobromide	39.1
Trisulfur dibromide	41.0
Tetrasulfur dibromide	42.8

Miscellaneous compounds

Compound	Temp. °C	Surface tension
2,2′-Thiodiethanol	20	54.0
4-Hydroxy-4-methyl-2-pentanone	20	31.0
4-Methyl-3-penten-2-one	20	28.4
Triethylboron	30	19.84
Disilthiane	18	22.31
Sulfuryl chlorofluoride	0	17.2
Chlorotrinitromethane	20	34.2
1,1-Di-*p*-tolylethane	30	34.5
1-*o*-Tolyl-1-*p*-tolylethane	30	35.5
1-*m*-Tolyl-1-*o*-tolylethane	30	36.1
2,2-Di-*p*-tolylbutane	30	33.6
m,*m*′-Bitolyl	20	39.0
Cyclo-octadiene	20	31.46
Diolein succinate	70.2	25.35
Distearin succinate	94.6	28.01
Boron trifluoride—ethyl methyl etherate	25	30.8
Ozone	−182.7	38.1
1,1-Dichloro-2-propanone	20	31.91
2,2-Dichloroethyl sulfide	20	42.82
4-Methyl-4-penten-2-one	20	23.0
1-*o*-Allylglycerol	20	33.3

VISCOSITY OF WATER 0°C TO 100°C

°C	η(cp)	°C	η(cp)	°C	η(cp)	°C	η(cp)
0	1.787	26	0.8705	52	0.5290	78	0.3638
1	1.728	27	.8513	53	.5204	79	.3592
2	1.671	28	.8327	54	.5121	80	.3547
3	1.618	29	.8148	55	.5040	81	.3503
4	1.567	30	.7975	56	.4961	82	.3460
5	1.519	31	.7808	57	.4884	83	.3418
6	1.472	32	.7647	58	.4809	84	.3377
7	1.428	33	.7491	59	.4736	85	.3337
8	1.386	34	.7340	60	.4665	86	.3297
9	1.346	35	.7194	61	.4596	87	.3259
10	1.307	36	.7052	62	.4528	88	.3221
11	1.271	37	.6915	63	.4462	89	.3184
12	1.235	38	.6783	64	.4398	90	.3147
13	1.202	39	.6654	65	.4335	91	.3111
14	1.169	40	.6529	66	.4273	92	.3076
15	1.139	41	.6408	67	.4213	93	.3042
16	1.109	42	.6291	68	.4155	94	.3008
17	1.081	43	.6178	69	.4098	95	.2975
18	1.053	44	.6067	70	.4042	96	.2942
19	1.027	45	.5960	71	.3987	97	.2911
20	1.002	46	.5856	72	.3934	98	.2879
21	0.9779	47	.5755	73	.3882	99	.2848
22	.9548	48	.5656	74	.3831	100	.2818
23	.9325	49	.5561	75	.3781		
24	.9111	50	.5468	76	.3732		
25	.8904	51	.5378	77	.3684		

VISCOSITIES OF SELECTED LIQUIDS

Viscosity of liquids in centipoises (cp) including elements, inorganic and organic compounds and mixtures.

Liquid	Temp. °C	Viscosity cp
Acetaldehyde	0	.2797
	10	.2557
	20	.22
Acetanilide	120	2.22
	130	1.90
Acetic acid	15	1.31
	18	1.30
	25.2	1.155
	30	1.04
	41	1.00
	59	.70
	70	.60
	100	.43
anhydride	0	1.24
	15	.971
	18	.90
	30	.783
	100	.49
Acetone	-92.5	2.148
	-80.0	1.487
	-59.6	.932
	-42.5	.695
	-30.0	.575
	-20.9	.510
	-13.0	.470
	-10.0	.450
	0	.399
	15	.337
	25	.316
	30	.295
	41	.280
Acetonitrile	0	.442
	15	.375
	25	.345
Acetophenone	11.9	2.28
	23.5	1.59
	25.0	1.617
	50.0	1.246
	80.0	.734
Air, liq	-192.3	.172
Alcohol. See *Ethyl*, *Methyl*, etc.		
Allyl alcohol	0	2.145
	15	1.49
	20	1.363
	30	1.07
	40	.914
	70	.553
Allylamine	130	.506
Allyl chloride	15	.347
	30	.300
Ammonia	-69	.475
	-50	.317
	-40	.276
	-33.5	.255
n-Amyl acetate	11	1.58
	45	.805
alcohol	15	4.65
	30	2.99
ether	15	1.188
Aniline	-6	13.8
	0	10.2
	5	8.06
	10	6.50
	15	5.31
	20	4.40
	25	3.71
	30	3.16
	35	2.71
	40	2.37
	50	1.85
	60	1.51
	70	1.27
	80	1.09
	90	.935
	100	.825
Anisol	0	1.78
	20	1.32
	40	1.12
Antimony, liq	645	1.55
	700	1.26
	800	1.08
	850	1.05
Benzaldehyde	25	1.39
Benzene	0	.912
	10	.758
	20	.652
	30	.564
	40	.503
	50	.442
	60	.392
	70	.358
	80	.329
Benzonitrile	25	1.24
Benzophenone	55	4.79
	120	1.38
Benzyl alcohol	20	5.8
Benzylamine	25	1.59
Benzylaniline	33	2.18
	130	1.20
Benzyl ether	0	10.5
	20	5.33
	40	3.21
Bismuth	285	1.61
	304	1.662
	365	1.46
	451	1.280
	600	.998
Bromine, liq	-4.3	1.31
	0	1.241
	12.6	1.07
	16	1.0
	19.5	.995
	28.9	.911
o-Bromoaniline	40	3.19
m-Bromoaniline	20	6.81
	40	3.70
	80	1.70
p-Bromoaniline	80	1.81
Bromobenzene	15	1.196
	30	.985
Bromoform	15	2.152
	25	1.89
	30	1.741
Butyl acetate	0	1.004
	20	.732
	40	.563
n-Butyl alcohol	-50.9	36.1
	-30.1	14.7
	-22.4	11.1
	-14.1	8.38
	0	5.186
	15	3.379
	20	2.948
	30	2.30
	40	1.782
	50	1.411
	70	.930
	100	.540
sec-Butyl alcohol	15	4.21
n-Butyl bromide	15	.626
n-Butyl chloride	15	.469

(Continued)

Liquid	Temp. °C	Viscosity cp
Butyl chloride, tertiary	15	.543
n-Butyl formate	0	.940
	20	.689
Butyric acid	0	2.286
	15	1.81
	20	1.540
	40	1.120
	50	.975
	70	.760
	100	.551
Cadmium, liq	349	1.44
	506	1.18
	603	1.10
Carbolic acid. See *Phenol*.		
Carbon dioxide, liq., pressure that of saturated vapor	0	.099
	10	.085
	20	.071
	30	.053
disulfide	−13	.514
	−10	.495
	0	.436
	5	.380
	20	.363
	40	.330
Carbon tetrachloride	0	1.329
	15	1.038
	20	.969
	30	.843
	40	.739
	50	.651
	60	.585
	70	.524
	80	.468
	90	.426
	100	.384
Cetyl alcohol	50	13.4
Chlorine, liq	−76.5	.729
	−70.5	.680
	−60.2	.616
	−52.4	.566
	−35.4	.494
	0	.385
Chlorobenzene	15	.900
	20	.799
	40	.631
	80	.431
	100	.367
Chloroform	−13	.855
	0	.700
	8.1	.643
	15	.596
	20	.58
	25	.542
	30	.514
	39	.500
o-Chlorophenol	25	4.11
	50	2.015
m-Chlorophenol	25	11.55
p-Chlorophenol	50	4.99
Copper, liq	1,085	3.36
	1,100	3.33
	1,150	3.22
	1,200	3.12
o-Cresol	40	4.49
m-Cresol	10	43.9
	20	20.8
	40	6.18
p-Cresol	40	7.00
Creosote	20	12.0
Cycloheptane	13.5	1.64
Cyclohexane	17	1.02
Cyclohexanol	20	68
Cyclohexene	13.5	.696
	20	.66

Liquid	Temp. °C	Viscosity cp
Cyclooctane	13.5	2.35
Cyclopentane	13.5	.493
n Decane	20	.92
Diethylamine	25	.346
	25	.367
Diethylaniline	.5	3.84
	20.0	2.18
	25.0	1.95
Diethylcarbinol	15.0	7.34
Diethylketone	15	.493
Dimethylaniline	10	1.69
	20	1.41
	25	1.285
	30	1.17
	40	1.04
	50	.91
Dimethyl-α-naphthylamine	130	.868
Dimethyl-β-naphthylamine	130	.952
Diphenyl	70	1.49
	100	.97
Diphenylamine	130	1.04
Dodecane	25	1.35
Ether (diethyl-)	−100	1.69
	−80	.958
	−60	.637
	−40	.461
	−20	.362
	0	.2842
	17	.240
	20	.2332
	25	.222
	40	.197
	60	.166
	80	.140
	100	.118
Ethyl acetate	0	.582
	8.96	.516
	10	.512
	15	.473
	20	.455
	25	.441
	30	.400
	50	.345
	75	.283
Ethyl alcohol	−98.11	44.0
	−89.8	28.4
	−71.5	13.2
	−59.42	8.41
	−52.58	6.87
	−32.01	3.84
	−17.59	2.68
	−.30	1.80
	0	1.773
	10	1.466
	20	1.200
	30	1.003
	40	.834
	50	.702
	60	.592
	70	.504
Ethyl alcohol, anh.	−148	8,470
	−146	5,990
	−130	467
Ethyl aniline	25	2.04
Ethylbenzene	17	.691
Ethyl benzoate	20	2.24
Ethyl bromide	−120	5.6
	−100	2.89
	−80	1.81
	0	.487
	10	.441
	15	.418
	20	.402
	30	.348

(Continued)

Liquid	Temp. °C	Viscosity cp
n-Ethyl butyrate	15	.711
Ethyl carbonate	15	.868
Ethylene bromide	0	2.438
	17	1.95
	20	1.721
	40	1.286
	67.3	.922
	70	.903
	82.2	.750
	99.0	.648
chloride	0	1.077
	15	.887
	19.4	.800
	40	.652
	50	.565
	70	.479
glycol	20	19.9
	40	9.13
	60	4.95
	80	3.02
	100	1.99
oxide	−49.8	.577
	−38.2	.488
	−21.0	.394
	0	.320
Ethyl formate	20	.402
iodide	0	.727
	15	.617
	20	.592
	40	.495
	70	.391
malate	24.7	3.016
oxalate	15	2.31
propionate	15	.564
Eugenol	0	29.9
	20	9.22
	40	4.22
Fluorobenzene	20	.598
	40	.478
	60	.389
	80	.329
	100	.275
Formamide	0	7.55
	25	3.30
Formic acid	7.59	2.3868
	10	2.262
	20	1.804
	30	1.465
	40	1.219
	70	.780
	100	.549
Furfural	0	2.48
	25	1.49
Glucose	22	9.1×10^{15}
	30	6.6×10^{13}
	40	2.8×10^{11}
	60	9.3×10^{7}
	80	6.6×10^{5}
	100	2.5×10^{4}
Glycerin	−42	6.71×10^{6}
	−36	2.05×10^{6}
	−25	2.62×10^{5}
	−20	1.34×10^{5}
	−15.4	6.65×10^{4}
	−10.8	3.55×10^{4}
	−4.2	1.49×10^{4}
	0	12,110
	6	6,260
	15	2,330
	20	1,490
	25	954
	30	629
Glycerin trinitrate	10	69.2
	20	36.0
	30	21.0

Liquid	Temp. °C	Viscosity cp
	40	13.6
	60	6.8
Heptane	0	.524
	17	.461
	20	.409
	25	.386
	40	.341
	70	.262
n-Heptyl alcohol	15	8.53
Hexadecane	20	3.34
Hexane	0	.401
	17	.374
	20	.326
	25	.294
	40	.271
	50	.248
Hydrazine	1	1.29
	10	1.12
	20	.97
Hydrogen, liq		.011
Iodine, liq	116	2.27
Iodobenzene	15	1.74
Iron, 2.5% carbon, liq	1,400	2.25
Isoamyl acetate	8.97	1.030
	19.91	.872
alcohol	10	6.20
amine	25	.724
Isobutyl alcohol	15	4.703
amine	25	.553
Isobutyric acid	15	1.44
	30	1.13
Isoeugenol	25	26.72
Isoheptane	0	.481
	20	.384
	40	.315
Isohexane	0	.376
	20	.306
	40	.254
Isopentane	0	.273
	20	.223
Isopropyl alcohol	15	2.86
	30	1.77
Isoquinoline	25	3.57
Isosafrol	25	3.981
Lead, liq	350	2.58
	400	2.33
	441	2.116
	500	1.84
	551	1.70
	600	1.38
	703	1.349
	844	1.185
Menthol, liq	55.6	6.29
	74.6	2.47
	99.0	1.04
Mercury	−20	1.855
	−10	1.764
	0	1.685
	10	1.615
	19.02	1.56
	20	1.554
	20.2	1.55
	30	1.499
	40	1.450
	40.8	1.45
	41.86	1.44
	50	1.407
	60	1.367
	70	1.331
	80	1.298
	90	1.268
	100	1.240
	150	1.130
	200	1.052
	250	.995

(Continued)

Liquid	Temp. °C	Viscosity cp
	300	.950
	340	.921
Methyl acetate	0	.484
	20	.381
	40	.320
Methyl alcohol	−98.30	13.9
(Methanol)	−84.23	6.8
	−72.55	4.36
	−44.53	1.98
	−22.29	1.22
	0	.82
	15	.623
	20	.597
	25	.547
	30	.510
	40	.456
	50	.403
Methyl amine	0	.236
aniline	25	2.02
	30	1.55
chloride	20	.1834
Methylene bromide	15	1.09
	30	0.92
chloride	15	.449
	30	.393
Methyl iodide	0	.606
	15	.518
	20	.500
	30	.460
	40	.424
Naphthalene	80	.967
	100	.776
Nitric acid	0	2.275
	10	1.770
Nitrobenzene	2.95	2.91
	5.69	2.71
	5.94	2.71
	9.92	2.48
	14.94	2.24
	20.00	2.03
Nitromethane	0	.853
	25	.620
o-Nitrotoluene	0	3.83
	20	2.37
	40	1.63
	60	1.21
m-Nitrotoluene	20	2.33
	40	1.60
	60	1.18
p-Nitrotoluene	60	1.20
n-Nonane	20	.711
n-Octane	0	.706
	16	.574
	20	.542
	40	.433
Octodecane	40	2.86
n-Octylalcohol	15	10.6
Oil, castor	10	2,420
	20	986
	30	451
	40	231
	100	16.9
cottonseed	20	70.4
cylinder, filtered	37.8	240.6
	100	18.7
cylinder, dark	37.8	422.4
	100	24.0
linseed	30	33.1
	50	17.6
	90	7.1
machine, light	15.6	113.8
	37.8	34.2
	100	4.9
machine, heavy	15.6	660.6
	37.8	127.4
Oil, olive	10	138.0
	20	84.0
	40	36.3
	70	12.4
rape	0	2,530
	10	385
	20	163
	30	96
soya bean	20	69.3
	30	40.6
	50	20.6
	90	7.8
sperm	15.6	42.0
	37.8	18.5
	100.0	4.6
Oleic acid	30	25.6
Pentadecane	22	2.81
Pentane	0	.289
	20	.240
o-Phenetidine	0	16.5
	20	6.08
	30	4.22
m-Phenetidine	30	12.9
p-Phenetidine	20	12.9
	30	8.3
Phenol	18.3	12.7
	50	3.49
	60	2.61
	70	2.03
	90	1.26
Phenylcyanide	.28	1.96
	20.0	1.33
Phosphorus, liq	21.5	2.34
	31.2	2.01
	43.2	1.73
	50.5	1.60
	60.2	1.45
	69.7	1.32
	79.9	1.21
Potassium bromide, liq	745	1.48
	775	1.34
	805	1.19
nitrate, liq	334	2.1
	358	1.7
	333	2.97
	418	2.00
Propionic acid	10	1.289
	15	1.18
	20	1.102
	40	.845
Propyl acetate	10	.66
	20	.59
	40	.44
n-Propyl alcohol	0	3.883
	15	2.52
	20	2.256
	30	1.72
	40	1.405
n-Propyl alcohol	50	1.130
	70	.760
Propyl aldehyde	10	.47
	20	.41
	40	.33
bromide	0	.651
	20	.524
	40	.433
chloride	0	.436
	20	.352
	40	.291
n-Propyl ether	15	.448
Pyridine	20	.974
Salicylic acid	10	3.20
	20	2.71

(Continued)

Liquid	Temp. °C	Viscosity cp
	40	1.81
Salol	45	.746
Sodium bromide	762	1.42
	780	1.28
chloride, liq	841	1.30
	896	1.01
	924	.97
nitrate, liq	308	2.919
	348	2.439
	398	1.977
	418	1.828
Stearic acid	70	11.6
Sucrose (cane sugar)	109	2.8×10^6
	124.6	1.9×10^5
Sulfur (gas free)	123.0	10.94
	135.5	8.66
	149.5	7.09
	156.3	7.19
	158.2	7.59
	159.2	9.48
	159.5	14.45
	160.0	22.83
	160.3	77.32
	165.0	500.0
	171.0	4,500.0
	184.0	16,000.00
	190.5	19,700.0
	197.5	21,300.0
	200.0	21,500.0
	210.0	20,500.0
	217.0	19,100.0
	220.0	18,600.0
Sulfur dioxide, liq	−33.5	.5508
	−10.5	.4285
	0.1	.3936
Sulfuric acid	0	48.4
	15	32.8
	20	25.4
	30	15.7
	40	11.5
Sulfuric acid	50	8.82
	60	7.22
	70	6.09
	80	5.19
Tetrachloroethane	15	1.844
Tetradecane	20	2.18
Tin, liq	240	2.12

Liquid	Temp. °C	Viscosity cp
	280	1.678
	300	1.73
	301	1.680
	400	1.43
	450	1.270
	500	1.20
	600	1.08
	604	1.045
	750	.905
Toluene	0	.772
	17	.61
	20	.590
	30	.526
	40	.471
	70	.354
o-Toluidine	20	4.39
m-Toluidine	20	3.81
p-Toluidine	50	1.80
Triacetin	17	28.0
Tributyrin	20	11.6
Trichlorethane	20	1.2
Tridecane	23.3	1.55
Triethylcarbinol	20	6.75
Tripalmitin	70	16.8
Tristearin	75	18.5
Turpentine	0	2.248
	10	1.783
	20	1.487
	30	1.272
	40	1.071
	70	.728
Turpentine, Venice	17.3	1.3×10^5
n-Undecane	20	1.17
o-Xylene (xylol)	0	1.105
	16	.876
	20	.810
	40	.627
m-Xylene (xylol)	0	.806
	15	.650
	20	.620
	40	.497
p-Xylene (xylol)	16	.696
	20	.648
	40	.513
Zinc, liq	280	1.68
	357	1.42
	389	1.31

VISCOSITY OF GASES

Gas or vapor	Temp. °C	Viscosity micro-poises
Acetic acid, vap	119.1	107.0
Acetone, vap	100	93.1
	119.0	99.1
	190.4	118.6
	247.7	133.4
	306.4	148.1
Acetylene	0	93.5
Air	−194.2	55.1
	−183.1	62.7
	−104.0	113.0
	−69.4	133.3
	−31.6	153.9
	0	170.8
	18	182.7
	40	190.4
	54	195.8
	74	210.2

Gas or vapor	Temp. °C	Viscosity micro-poises
	229	263.8
	334	312.3
	357	317.5
	409	341.3
	466	350.1
	481	358.3
	537	368.6
	565	375.0
	620	391.6
	638	401.4
	750	426.3
	810	441.9
	923	464.3
	1034	490.6
	1134	520.6
Alcohol. See *Ethyl*, *Methyl*, etc.		

Gas or vapor	Temp. °C	Viscosity micro-poises
Ammonia	−78.5	67.2
	0	91.8
	20	98.2
	50	109.2
	100	127.9
	132.9	139.9
	150	146.3
	200	164.6
	250	181.4
	300	198.7
Argon	0	209.6
	20	221.7
	100	269.5
	200	322.3
	302	368.5
	401	411.5
	493	448.4
	584	481.5
	714	525.7
	827	563.2
Arsenic hydride (Arsine)	0	145.8
	15	114.0
	100	198.1
Benzene, vap	14.2	73.8
	131.2	103.1
	194.6	119.8
	252.5	134.3
	312.8	148.4
Bromine, vap	12.8	151
	65.7	170
	99.7	188
	139.7	208
	179.7	227
	220.3	248
Bromoform, vap	151.2	253.0
Butyl alcohol, *n*, vap	116.9	143
tert, vap	82.9	160
chloride, *n*, vap	78	149.5
iodide, vap	130	202
β-Butylene	18.8	74.4

Gas or vapor	Temp. °C	Viscosity micro-poises
	100.4	94.5
	200	119.2
Butyric acid, vap	161.7	130.0
Carbon dioxide	−97.8	89.6
	−78.2	97.2
	−60.0	106.1
	−40.2	115.5
	−21	129.4
	−19.4	126.0
	0	139.0
	15	145.7
	19	149.9
	20	148.0
	30	153
	32	155
	35	156
	40	157
	99.1	186.1
	104	188.9
	182.4	222.1
	235	241.5
	302.0	268.2
	490	330.0
	685	380.0
	850	435.8
	1052	478.6
disulfide, vap	0	91.1
	14.2	96.4
	114.3	130.3
	190.2	156.1
	309.8	196.6
monoxide	−191.5	56.1
	−78.5	127
	0	166
	15	172
	21.7	175.3
	126.7	218.3
	227.0	254.8
	276.9	271.4
tetrachloride, vap	76.7	195.0

MOHS HARDNESS SCALE

Hardness number	Original scale	Modified scale
1	Talc	Talc
2	Gypsum	Gypsum
3	Calcite	Calcite
4	Fluorite	Fluorite
5	Apatite	Apatite
6	Orthoclase	Orthoclase
7	Quartz	Vitreous silica
8	Topaz	Quartz or Stellite
9	Corundum	Topaz
10	Diamond	Garnet
11	....	Fused Zirconia
12	....	Fused Alumina
13	....	Silicon Carbide
14	....	Boron Carbide
15	....	Diamond

HARDNESS OF SELECTED MATERIALS

Material	Hardness	Material	Hardness
Agate	6–7	Indium	1.2
Alabaster	1.7	Iridium	6–6.5
Alum	2–2.5	Iridosmium	7
Aluminum	2–2.9	Iron	4–5
Alundum	9+	Kaolinite	2.0–2.5
Amber	2–2.5	Lead	1.5
Andalusite	7.5	Lithium	0.6
Anthracite	2.2	Loess (0°)	0.3
Antimony	3.0–3.3	Magnesium	2.0
Apatite	5	Magnetite	6
Aragonite	3.5	Manganese	5.0
Arsenic	3.5	Marble	3–4
Asbestos	5	Meerschaum	2–3
Asphalt	1–2	Mica	2.8
Augite	6	Opal	4–6
Barite	3.3	Orthoclase	6
Bell-metal	4	Osmium	7.0
Beryl	7.8	Palladium	4.8
Bismuth	2.5	Phosphorus	0.5
Boric acid	3	Phosphorbronze	4
Boron	9.5	Platinum	4.3
Brass	3–4	Plat-iridium	6.5
Cadmium	2.0	Potassium	0.5
Calamine	5	Pumice	6
Calcite	3	Pyrite	6.3
Calcium	1.5	Quartz	7
Carbon	10.0	Rock salt (halite)	2
Carborundum	9–10	Ross' metal	2.5–3.0
Cesium	0.2	Rubidium	0.3
Chromium	9.0	Ruthenium	6.5
Copper	2.5–3	Selenium	2.0
Corundum	9	Serpentine	3–4
Diamond	10	Silicon	7.0
Diatomaceous earth	1–1.5	Silver	2.5–4
Dolomite	3.5–4	Silver chloride	1.3
Emery	7–9	Sodium	0.4
Feldspar	6	Steel	5–8.5
Flint	7	Stibnite	2
Fluorite	4	Strontium	1.8
Galena	2.5	Sulfur	1.5–2.5
Gallium	1.5	Talc	1
Garnet	6.5–7	Tellurium	2.3
Glass	4.5–6.5	Tin	1.5–1.8
Gold	2.5–3	Topaz	8
Graphite	0.5–1	Tourmaline	7.3
Gypsum	1.6–2	Wax (0°)	0.2
Hematite	6	Wood's metal	3
Hornblende	5.5	Zinc	2.5

VAPOR PRESSURE OF WATER BELOW 100°C

in mm of Hg for temperatures from −15.8 to 100°C.

Temp. °C	0.0	0.2	0.4	0.6	0.8	Temp. °C	0.0	0.2	0.4	0.6	0.8
−15	1.436	1.414	1.390	1.368	1.345	18	15.477	15.673	15.871	16.071	16.2
−14	1.560	1.534	1.511	1.485	1.460	19	16.477	16.685	16.894	17.105	17.3
−13	1.691	1.665	1.637	1.611	1.585	20	17.535	17.753	17.974	18.197	18.4
−12	1.834	1.804	1.776	1.748	1.720	21	18.650	18.880	19.113	19.349	19.5
−11	1.987	1.955	1.924	1.893	1.863	22	19.827	20.070	20.316	20.565	20.8
−10	2.149	2.116	2.084	2.050	2.018	23	21.068	21.324	21.583	21.845	22.1
−9	2.326	2.289	2.254	2.219	2.184	24	22.377	22.648	22.922	23.198	23.4
−8	2.514	2.475	2.437	2.399	2.362	25	23.756	24.039	24.326	24.617	24.9
−7	2.715	2.674	2.633	2.593	2.553	26	25.209	25.509	25.812	26.117	26.4
−6	2.931	2.887	2.843	2.800	2.757	27	26.739	27.055	27.374	27.696	28.0
−5	3.163	3.115	3.069	3.022	2.976	28	28.349	28.680	29.015	29.354	29.6
−4	3.410	3.359	3.309	3.259	3.211	29	30.043	30.392	30.745	31.102	31.4
−3	3.673	3.620	3.567	3.514	3.461	30	31.824	32.191	32.561	32.934	33.3
−2	3.956	3.898	3.841	3.785	3.730	31	33.695	34.082	34.471	34.864	35.2
−1	4.258	4.196	4.135	4.075	4.016	32	35.663	36.068	36.477	36.891	37.3
−0	4.579	4.513	4.448	4.385	4.320	33	37.729	38.155	38.584	39.018	39.4
0	4.579	4.647	4.715	4.785	4.855	34	39.898	40.344	40.796	41.251	41.7
1	4.926	4.998	5.070	5.144	5.219	35	41.175	42.644	43.117	43.595	44.0
2	5.294	5.370	5.447	5.525	5.605	36	44.563	45.054	45.549	46.050	46.5
3	5.685	5.766	5.848	5.931	6.015	37	47.067	47.582	48.102	48.627	49.1
4	6.101	6.187	6.274	6.363	6.453	38	49.692	50.231	50.774	51.323	51.8
5	6.543	6.635	6.728	6.822	6.917	39	52.442	53.009	53.580	54.156	54.7
6	7.013	7.111	7.209	7.309	7.411	40	55.324	55.91	56.51	57.11	57.7
7	7.513	7.617	7.722	7.828	7.936	41	58.34	58.96	59.58	60.22	60.8
8	8.045	8.155	8.267	8.380	8.494	42	61.50	62.14	62.80	63.46	64.1
9	8.609	8.727	8.845	8.965	9.086	43	64.80	65.48	66.16	66.86	67.5
10	9.209	9.333	9.458	9.585	9.714	44	68.26	68.97	69.69	70.41	71.14
11	9.844	9.976	10.109	10.244	10.380	45	71.88	72.62	73.36	74.12	74.88
12	10.518	10.658	10.799	10.941	11.085	46	75.65	76.43	77.21	78.00	78.80
13	11.231	11.379	11.528	11.680	11.833	47	79.60	80.41	81.23	82.05	82.8
14	11.987	12.144	12.302	12.462	12.624	48	83.71	84.56	85.42	86.28	87.14
15	12.788	12.953	13.121	13.290	13.461	49	88.02	88.90	89.79	90.69	91.5
16	13.634	13.809	13.987	14.166	14.347	50	92.51	93.5	94.4	95.3	96.3
17	14.530	14.715	14.903	15.092	15.284	51	97.20	98.2	99.1	100.1	101.1

(continued)

Temp. °C	0.0	0.2	0.4	0.6	0.8
52	102.09	103.1	104.1	105.1	106.2
53	107.20	108.2	109.3	110.4	111.4
54	112.51	113.6	114.7	115.8	116.9
55	118.04	119.1	120.3	121.5	122.6
56	123.80	125.0	126.2	127.4	128.6
57	129.82	131.0	132.3	133.5	134.7
58	136.08	137.3	138.5	139.9	141.2
59	142.60	143.9	145.2	146.6	148.0
60	149.38	150.7	152.1	153.5	155.0
61	156.43	157.8	159.3	160.8	162.3
62	163.77	165.2	166.8	168.3	169.8
63	171.38	172.9	174.5	176.1	177.7
64	179.31	180.9	182.5	184.2	185.8
65	187.54	189.2	190.9	192.6	194.3
66	196.09	197.8	199.5	201.3	203.1
67	204.96	206.8	208.6	210.5	212.3
68	214.17	216.0	218.0	219.9	221.8
69	223.73	225.7	227.7	229.7	231.7
70	233.7	235.7	237.7	239.7	241.8
71	243.9	246.0	248.2	250.3	252.4
72	254.6	256.8	259.0	261.2	263.4
73	265.7	268.0	270.2	272.6	274.8
74	277.2	279.4	281.8	284.2	286.6
75	289.1	291.5	294.0	296.4	298.8
76	301.4	303.8	306.4	308.9	311.4
77	314.1	316.6	319.2	322.0	324.6
78	327.3	330.0	332.8	335.6	338.2
79	341.0	343.8	346.6	349.4	352.2
80	355.1	358.0	361.0	363.8	366.8
81	369.7	372.6	375.6	378.8	381.8
82	384.9	388.0	391.2	394.4	397.4
83	400.6	403.8	407.0	410.2	413.6
84	416.8	420.2	423.6	426.8	430.2
85	433.6	437.0	440.4	444.0	447.5
86	450.9	454.4	458.0	461.6	465.2
87	468.7	472.4	476.0	479.8	483.4
88	487.1	491.0	494.7	498.5	502.2
89	506.1	510.0	513.9	517.8	521.8
90	525.76	529.77	533.80	537.86	541.95
91	546.05	550.18	554.35	558.53	562.75
92	566.99	571.26	575.55	579.87	584.22
93	588.60	593.00	597.43	601.89	606.38
94	610.90	615.44	620.01	624.61	629.24
95	633.90	638.59	643.30	648.05	652.82
96	657.62	662.45	667.31	672.20	677.12
97	682.07	687.04	692.05	697.10	702.17
98	707.27	712.40	717.56	722.75	727.98
99	733.24	738.53	743.85	749.20	754.58
100	760.00	765.45	770.93	776.44	782.00
101	787.57	793.18	798.82	804.50	810.21

VAPOR PRESSURE OF WATER ABOVE 100°C

Temp. °C	Pressure mm
100	760.
101	787.51
102	815.86
103	845.12
104	875.06
105	906.07
106	937.92
107	970.60
108	1004.42
109	1038.92
110	1074.56
111	1111.20
112	1148.74
113	1187.42
114	1227.25
115	1267.98
116	1309.94
117	1352.95
118	1397.18
119	1442.63
120	1489.14
121	1536.80
122	1586.04
123	1636.36
124	1687.81
125	1740.93
126	1795.12
127	1850.83
128	1907.83
129	1966.35
130	2026.10
131	2087.42
132	2150.42
133	2214.64
134	2280.76
135	2347.26
136	2416.34
137	2488.16
138	2560.67
139	2634.84
140	2710.92
141	2788.44
142	2867.48
143	2948.80
144	3031.64
145	3116.76
146	3203.40
147	3292.32
148	3382.76
149	3476.24
150	3570.48
151	3667.00
152	3766.56
153	3866.88
154	3970.24
155	4075.88
156	4183.80
157	4293.24
158	4404.96
159	4519.72
160	4636.00
161	4755.32
162	4876.92
163	5000.04
164	5126.96
165	5256.16
166	5386.88
167	5521.40
168	5658.20
169	5798.04
170	5940.92
171	6085.32
172	6233.52
173	6383.24
174	6538.28
175	6694.08
176	6852.92
177	7015.56
178	7180.48
179	7349.20
180	7520.20
181	7694.24
182	7872.08
183	8052.96
184	8236.88
185	8423.84
186	8616.12
187	8809.92
188	9007.52
189	9208.16
190	9413.36
191	9620.08
192	9831.36
193	10047.20
194	10265.32
195	10488.76
196	10715.24
197	10944.76
198	11179.60
199	11417.48
200	11659.16
201	11905.40
202	12155.44
203	12408.52
204	12666.16
205	12929.12
206	13197.40
207	13467.96
208	13742.32
209	14022.76
210	14305.48
211	14595.04
212	14888.40
213	15184.80
214	15488.04
215	15792.80
216	16104.40
217	16420.56
218	16742.04
219	17067.32
220	17395.64
221	17731.56
222	18072.80
223	18417.84
224	18766.68
225	19123.12
226	19482.60
227	19848.92
228	20219.80
229	20596.76
230	20978.28
231	21365.12
232	21757.28
233	22154.00
234	22558.32
235	22967.96
236	23382.92
237	23802.44
238	24229.56
239	24661.24
240	25100.52
241	25543.60
242	25994.28
243	26449.52
244	26912.36
245	27381.28
246	27855.52
247	28335.84
248	28823.76
249	29317.00
250	29817.84
251	30324.00
252	30837.76
253	31356.84
254	31885.04
255	32417.80
256	32957.40
257	33505.36
258	34059.40
259	34618.76
260	35188.00
261	35761.80
262	36343.20
263	36932.20
264	37529.56
265	38133.00
266	38742.52
267	39361.92
268	39986.64
269	40619.72
270	41261.16
271	41910.20
272	42566.08
273	43229.56
274	43902.16
275	44580.84
276	45269.40
277	45964.04
278	46669.32
279	47382.20
280	48104.20
281	48833.80
282	49570.24
283	50316.56
284	51072.76
285	51838.08
286	52611.76
287	53395.32
288	54187.24
289	54989.04
290	55799.20
291	56612.40
292	57448.40
293	58284.40
294	59135.60
295	59994.40
296	60860.80
297	61742.40
298	62624.00
299	63528.40
300	64432.80
301	65352.40
302	66279.60
303	67214.40
304	68156.80
305	69114.40
306	70072.00
307	71052.40
308	72048.00
309	73028.40
310	74024.00
311	75042.40
312	76076.00
313	77117.20
314	78166.00
315	79230.00
316	80294.00
317	81373.20
318	82467.60
319	83569.60
320	84686.80
321	85819.20
322	86959.20
323	88114.40
324	89277.20
325	90447.60
326	91633.20
327	92826.40
328	94042.40
329	95273.60
330	96512.40
331	97758.80
332	99020.40
333	100297.20
334	101581.60
335	102881.20
336	104196.00
337	105526.00
338	106871.20
339	108224.00
340	109592.00
341	110967.60
342	112358.40
343	113749.20
344	115178.00
345	116614.40
346	118073.60
347	119532.80
348	121014.80
349	122504.40
350	124001.60
351	125521.60
352	127049.20
353	128599.60
354	130157.60
355	131730.80
356	133326.80
357	134945.60
358	136579.60
359	138228.80
360	139893.20
361	141572.80
362	143275.20
363	144992.80
364	146733.20
365	148519.20
366	150320.40
367	152129.20
368	153960.80
369	155815.20
370	157692.40
371	159584.80
372	161507.60
373	163468.40
374	165467.20

STATIC DIELECTRIC CONSTANT OF WATER AND STEAM

M. Uematsu and E. U. Franck

Units: t°C, P MPa

P	t									t					t			
	0.0	25.0	50.0	75.0	100.0	125.0	150.0	175.0	200.0	225.0	250.0	275.0	300.0	350.0	400.0	450.0	500.0	550.0
0.1	87.81	78.46	69.91	62.24	1.00	1.00	1.00	1.00	1.00	1.00	1.00	1.00	1.00	1.00	1.00	1.00	1.00	1.00
0.5	87.83	78.47	69.92	62.25	55.43	49.36	43.95	1.01	1.01	1.01	1.01	1.01	1.01	1.01	1.01	1.00	1.00	1.00
1.0	87.86	78.49	69.94	62.27	55.44	49.37	43.96	39.11	1.02	1.02	1.02	1.02	1.02	1.01	1.01	1.01	1.01	1.01
2.5	87.93	78.55	69.99	62.33	55.50	49.43	44.02	39.17	34.79	1.07	1.06	1.05	1.04	1.04	1.03	1.03	1.02	1.02
5.0	88.05	78.65	70.09	62.42	55.59	49.52	44.12	39.28	34.90	30.89	27.15	1.13	1.11	1.08	1.07	1.06	1.05	1.04
10.0	88.28	78.85	70.27	62.59	55.76	49.70	44.30	39.47	35.11	31.13	27.43	23.90	20.39	1.23	1.17	1.14	1.11	1.10
20.0	88.75	79.24	70.63	62.94	56.11	50.05	44.66	39.85	35.52	31.58	27.95	24.54	21.24	14.07	1.64	1.42	1.32	1.26
30.0	89.20	79.63	70.98	63.28	56.44	50.39	45.01	40.22	35.91	32.01	28.43	25.11	21.95	15.66	5.91	2.07	1.68	1.51
40.0	89.64	80.00	71.32	63.61	56.77	50.72	45.34	40.56	36.28	32.40	28.87	25.61	22.56	16.72	10.46	3.84	2.34	1.90
50.0	90.07	80.36	71.66	63.93	57.08	51.03	45.67	40.89	36.63	32.78	29.28	26.08	23.10	17.55	12.16	6.57	3.45	2.48
60.0	90.49	80.72	71.98	64.24	57.39	51.34	45.98	41.21	36.96	33.13	29.67	26.50	23.58	18.24	13.28	8.53	4.90	3.26
70.0	90.90	81.07	72.30	64.54	57.69	51.64	46.28	41.52	37.28	33.47	30.03	26.90	24.02	18.84	14.16	9.87	6.31	4.20
80.0	91.29	81.42	72.62	64.84	57.98	51.93	46.57	41.82	37.59	33.79	30.37	27.27	24.43	19.37	14.88	10.88	7.50	5.16
90.0	91.67	81.75	72.92	65.13	58.27	52.21	46.86	42.11	37.89	34.10	30.70	27.62	24.81	19.85	15.50	11.70	8.47	6.06
100.0	92.04	82.08	73.22	65.42	58.55	52.49	47.14	42.39	38.17	34.40	31.01	27.95	25.17	20.29	16.05	12.39	9.29	6.88
125.0	92.89	82.84	73.93	66.09	59.19	53.12	47.78	43.05	38.86	35.13	31.78	28.76	26.03	21.26	17.21	13.77	10.88	8.53
150.0	93.71	83.57	74.62	66.74	59.82	53.75	48.40	43.68	39.50	35.78	32.46	29.47	26.77	22.09	18.16	14.85	12.07	9.80
175.0	94.48	84.28	75.27	67.36	60.42	54.34	48.98	44.27	40.10	36.39	33.09	30.12	27.45	22.83	18.98	15.74	13.04	10.81
200.0	95.20	84.94	75.89	67.95	61.00	54.90	49.54	44.83	40.66	36.97	33.67	30.72	28.07	23.49	19.69	16.51	13.86	11.65
225.0	95.87	85.58	76.50	68.53	61.55	55.44	50.08	45.36	41.20	37.51	34.22	31.28	28.64	24.09	20.33	17.19	14.56	12.38
250.0	96.51	86.20	77.08	69.08	62.08	55.96	50.59	45.87	41.70	38.02	34.74	31.81	29.17	24.65	20.91	17.80	15.19	13.01
300.0	97.69	87.34	78.17	70.14	63.10	56.94	51.55	46.82	42.65	38.97	35.69	32.77	30.15	25.65	21.94	18.85	16.25	14.07
350.0	98.75	88.40	79.19	71.12	64.05	57.86	52.45	47.70	43.52	39.83	36.56	33.64	31.02	26.53	22.83	19.74	17.14	14.93
400.0	99.72	89.39	80.13	72.03	64.94	58.74	53.30	48.53	44.33	40.64	37.36	34.43	31.81	27.32	23.62	20.52	17.89	15.66
450.0	100.60	90.30	81.02	72.89	65.78	59.56	54.10	49.31	45.10	41.38	38.09	35.16	32.54	28.04	24.32	21.20	18.55	16.28
500.0	101.42	91.16	81.84	73.69	66.57	60.33	54.85	50.05	45.82	42.09	38.78	35.84	33.21	28.70	24.96	21.82	19.14	16.83

STATIC DIELECTRIC CONSTANT OF SATURATED WATER AND STEAM

Units: t°C

t	ε_L	ε_V	t	ε_L	ε_V
0.0	87.81	1.00	200.0	34.74	1.04
10.0	83.99	1.00	210.0	33.11	1.05
20.0	80.27	1.00	220.0	31.53	1.06
30.0	76.67	1.00	230.0	30.01	1.07
40.0	73.22	1.00	240.0	28.53	1.09
50.0	69.90	1.00	250.0	27.08	1.11
60.0	66.73	1.00	260.0	25.68	1.13
70.0	63.70	1.00	270.0	24.30	1.15
80.0	60.81	1.00	280.0	22.94	1.18
90.0	58.05	1.00	290.0	21.60	1.22
100.0	55.41	1.00	300.0	20.26	1.27
110.0	52.90	1.01	310.0	18.92	1.33
120.0	50.50	1.01	320.0	17.56	1.40
130.0	48.22	1.01	330.0	16.17	1.50
140.0	46.03	1.01	340.0	14.72	1.64
150.0	43.94	1.01	350.0	13.16	1.85
160.0	41.95	1.02	360.0	11.36	2.19
170.0	40.03	1.02	370.0	8.70	3.00
180.0	38.20	1.03			
190.0	36.44	1.03			

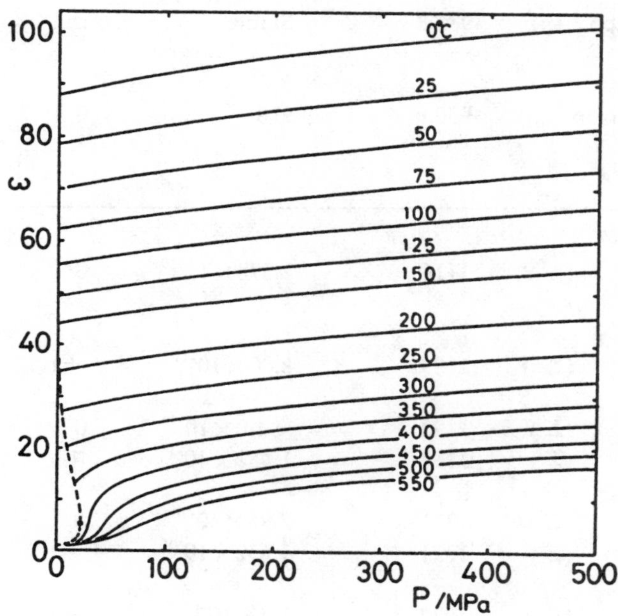

Static dielectric constant as a function of pressure.

Properties of Selected Elementary Particles

Family name		Particle name	Rest mass (MeV)	Mean life (seconds)	Charge (electron)	Typical decay mode
		Photon (γ)	0	Stable	0	
LEPTONS		Electron (e)	0.511	Stable	± 1	
		Muon (μ)	105.7	2.197×10^{-6}	± 1	$e + \nu + \overline{\nu}$
		Electron's neutrino (ν_e)	0	Stable	0	
		Muon's neutrino (ν_μ)	0	Stable	0	
HADRONS	MESONS	Pion (π)	139.6	2.603×10^{-8}	± 1	$\mu + \nu$
		(π^0)	135.0	8.28×10^{-17}	0	$\gamma + \gamma$
		K-meson (K)	493.7	1.237×10^{-8}	± 1	$\mu + \nu$
		(K^0)	497.7	8.930×10^{-11}	0	$\pi^+ + \pi^-$
				5.181×10^{-8}	0	$\pi^0 + \pi^0 + \pi^0$
		Eta-meson (η^0)	548.8	?	0	$\gamma + \gamma$
	BARYONS	NUCLEON Proton (p)	938.3	Stable	± 1	
		NUCLEON Neutron (n)	939.6	918	0	$p + e^- + \nu$
		Lambda particle (Λ^0)	1116	2.578×10^{-10}	0	$p + \pi^-$
		Sigma particle (Σ^+)	1189	8.00×10^{-11}	± 1	$p + \pi^0$
		(Σ^0)	1192	$< 1.0 \times 10^{-14}$	0	$\Lambda^0 + \gamma$
		(Σ^-)	1197	1.482×10^{-10}	∓ 1	$n + \pi^-$
		Xi particle (Ξ^0)	1315	2.96×10^{-10}	0	$\Lambda^0 + \pi^0$
		(Ξ^-)	1321	1.652×10^{-10}	∓ 1	$\Lambda^0 + \pi^-$
		Omega particle (Ω^-)	1672	1.3×10^{-10}	∓ 1	$\Xi^0 + \pi^-$

Summary of Nuclear Moment Value

Nucleus	Level (keV)	$T_{1/2}$	I $(h/2\pi)$	μ (nm)	Diam. Corr.	Q (b)	Ω (nmb)	Q_4 (b^2)
$^{1}_{0}n_{1}$		12m	1/2	−1.91312				
$^{1}_{1}\bar{H}_{0}$				−1.8				
$^{1}_{1}H_{0}$			1/2	+2.79278[j]	8			
$^{2}_{1}H_{1}$			1	+0.85742	2	+0.0028		
$^{3}_{1}H_{2}$		12y	1/2	+2.9789	1			
$^{3}_{2}He_{1}$			1/2	−2.1276	1			
$^{3}_{2}He^{+}_{1}$			1/2					
$^{4}_{2}He_{2}$			0					
$^{6}_{2}He_{4}$		0.8s	0♦					
$^{6}_{3}Li_{3}$			1	+0.82203	8	−0.0008[r]		
$^{6}_{3}Li^{+}_{3}$			1					
$^{7}_{3}Li_{4}$			3/2	+3.25636	33	−0.04[S]	[+0.09]	
$^{8}_{3}Li_{5}$		0.8s	2	+1.6532	2			
$^{9}_{4}Be_{5}$			3/2	−1.17745	18	+0.05	[−0.04]	
$^{8}_{5}B_{3}$		770ms	[2]	±1.0355[p]	2			
$^{10}_{5}B_{5}$			3	+1.8006	4	+0.085*[r]	[<±0.03]	
$^{10}_{5}B_{5}$	720	0.7ns	[1]	+0.6				
$^{11}_{5}B_{6}$			3/2	+2.6885	5	+0.041*[S]	[+0.08]	
$^{12}_{5}B_{7}$		20.4ms	1	+1.0028[g]		+0.018		
$^{13}_{5}B_{8}$		19ms	[3/2]	±3.1771[g]		±0.05		
$^{11}_{6}C_{5}$		21m	3/2	(−?)0.99		(+?)0.031*		
$^{12}_{6}C_{6}$			0					
$^{13}_{6}C_{7}$			1/2	+0.7024	2			
$^{14}_{6}C_{8}$		5.6ky	0					
$^{12}_{7}N_{5}$		12ms	1	±0.457				
$^{13}_{7}N_{6}$		10m	1/2	±0.3221	1			
$^{14}_{7}N_{7}$			1	+0.40375	13	+0.01		
$^{14}_{7}N_{7}$	5830	12.4ps	[3]	±1.5 to 2.6				
$^{15}_{7}N_{8}$			1/2	−0.2831	1			
$^{15}_{8}O_{7}$		2.1m	1/2	±0.7189	3			
$^{16}_{8}O_{8}$			0					
$^{17}_{8}O_{9}$			5/2	−1.8937	7	−0.026*		
$^{18}_{8}O_{10}$			0					
$^{18}_{8}O_{10}$	1980	3.3ps	[2]	±0.4 to 0.7				
$^{17}_{9}F_{8}$		66s	[5/2]	±4.722	2			
$^{18}_{9}F_{9}$	1125	153ns	[5]	+2.85				
$^{19}_{9}F_{10}$			1/2	+2.6288	12			

Summary of Nuclear Moment Value (continued)

Nucleus	Level (keV)	$T_{1/2}$	I $(h/2\pi)$	μ (nm)	Diam. Corr.	Q (b)	Ω (nmb)	Q_4 (b^2)
$^{19}_{9}F_{10}$	197	89ns	[5/2]	+3.60		±0.11[S]		
$^{20}_{9}F_{11}$		11s	[2]	+2.094	1	±0.06[r]		
$^{19}_{10}Ne_{9}$		18s	1/2	−1.887	1			
$^{19}_{10}Ne_{9}$	238	17.7ns	[5/2]	−0.74				
$^{20}_{10}Ne_{10}$			0♦					
$^{20}_{10}Ne_{10}$	1630	0.7ps	[2]			−0.25		
$^{21}_{10}Ne_{11}$			3/2	−0.66176	36	+0.09		
$^{22}_{10}Ne_{12}$			0♦					
$^{22}_{10}Ne_{12}$	1275	3ps	[2]			−0.21		
$^{23}_{10}Ne_{13}$		38s	[5/2]	−1.08				
$^{20}_{11}Na_{9}$		408ms	2	±0.369				
$^{21}_{11}Na_{10}$		23s	3/2	+2.3861	15			
$^{22}_{11}Na_{11}$		2.6y	3	+1.746	1			
$^{22}_{11}Na_{11}$	583	243ns	[1]	+0.55				
$^{23}_{11}Na_{12}$			3/2	+2.21740[f] (+2.21755)	139	+0.10*		
$^{24}_{11}Na_{13}$		15h	4	+1.690	1			
$^{24}_{12}Mg_{12}$			0♦					
$^{24}_{12}Mg_{12}$	1368	1ps	[2]			−0.27		
$^{25}_{12}Mg_{13}$			5/2	−0.8554	6	+0.22		
$^{26}_{12}Mg_{14}$			0♦					
$^{27}_{13}Al_{14}$			5/2	+3.6413	29	+0.15*[b]	[±0.3]	
$^{28}_{14}Si_{14}$			0♦					
$^{28}_{14}Si_{14}$	1779	0.5ps	[2]				+0.17	
$^{29}_{14}Si_{15}$			1/2	−0.55526	49			
$^{30}_{14}Si_{16}$			0♦					
$^{29}_{15}P_{14}$		4.2s	[1/2]	±1.235	1			
$^{30}_{15}P_{15}$		2.6m	1					
$^{31}_{15}P_{16}$			1/2	+1.1317	11			
$^{32}_{15}P_{17}$		14d	1	−0.2523	2			
$^{32}_{16}S_{16}$			0					
$^{32}_{16}S_{16}$	2237	0.25ps	[2]			−0.2		
$^{33}_{16}S_{17}$			3/2	+0.6435	7	−0.055[S]		
$^{34}_{16}S_{18}$			0♦					
$^{35}_{16}S_{19}$		87d	3/2	+1.00 or −1.07		+0.038[r]		
$^{36}_{16}S_{20}$			0♦					
$^{35}_{17}Cl_{18}$			3/2	+0.82181	94	−0.10*[S]	−0.016*[S]	
$^{36}_{17}Cl_{19}$		0.3My	2	+1.2853	14	−0.021*[r]		
$^{37}_{17}Cl_{20}$			3/2	+0.68407	78	−0.079*[r]	−0.013*[r]	

Summary of Nuclear Moment Value(continued)

Nucleus	Level (keV)	$T_{1/2}$	I ($h/2\pi$)	μ (nm)	Diam. Corr.	Q (b)	Ω (nmb)	Q_4 (b^2)
$^{35}_{18}Ar_{17}$		1.8s	[3/2]	+0.632[s]				
$^{36}_{18}Ar_{18}$			0♦					
$^{36}_{18}Ar_{18}$	1980	?	[2]			+0.11		
$^{37}_{18}Ar_{19}$		34d	3/2	+0.95				
$^{37}_{18}Ar_{19}$	1610	4.5ns	[7/2]	−1.33				
$^{38}_{18}Ar_{20}$			0♦					
$^{39}_{18}Ar_{21}$		265y	7/2	−1.3				
$^{40}_{18}Ar_{22}$			0♦					
$^{40}_{18}Ar_{22}$	1460	0.8ps	[2]			~0		
$^{36}_{19}K_{17}$		245ms	2	±0.548	1			
$^{37}_{19}K_{18}$		1.2s	3/2	+0.2032	3			
$^{37}_{19}K_{18}$	1380	10.5ns	[7/2]	+5.2				
$^{38}_{19}K_{19}$		7.7m	3	+1.374	2			
$^{39}_{19}K_{20}$			3/2	+0.39143[f] (+0.39147)	52	+0.049*[s]		
$^{40}_{19}K_{21}$		1.3Gy	4	−1.2981	17	−0.061*[r]		
$^{41}_{19}K_{22}$			3/2	(+0.21487)[f]	28	+0.060*[r]		
$^{41}_{19}K_{22}$	1290	7.3ns	[7/2]	+4.41				
$^{42}_{19}K_{23}$		12h	2	−1.1424	15			
$^{43}_{19}K_{24}$		22h	3/2	±0.163				
$^{45}_{19}K_{26}$		20m	3/2	±0.1734	2			
$^{40}_{20}Ca_{20}$			0♦					
$^{40}_{20}Ca_{20}$	3740	41ps	[3]	+0.4[p]				
$^{40}_{20}Ca_{20}$	4490	272ps	[5]	+1.6[p]				
$^{41}_{20}Ca_{21}$		110ky	7/2	−1.5946	23			
$^{42}_{20}Ca_{22}$	3190	5.5ns	[6]	−2.8				
$^{43}_{20}Ca_{23}$			7/2	−1.3172	19	<±0.2		
$^{41}_{21}Sc_{20}$		0.59s	[7/2]	±5.43[p]	1			
$^{43}_{21}Sc_{22}$		3.9h	7/2	+4.62	1	−0.26[r]		
$^{43}_{21}Sc_{22}$	3123	450ns	[19/2]	+3.14				
$^{44}_{21}Sc_{23}$		3.9h	2	+2.56		+0.11[r]		
$^{44}_{21}Sc_{23}$	69	153ns	[1]	+0.34		±0.18*		
$^{44}_{21}Sc_{23}$	270	2.4d	6	+3.88	1	−0.20[r]		
$^{45}_{21}Sc_{24}$			7/2	+4.7559	72	−0.22[s]		
$^{46}_{21}Sc_{25}$		84d	4	+3.03		+0.12[r]		
$^{47}_{21}Sc_{26}$		3.4d	7/2	+5.34	1	−0.22[r]		
$^{47}_{21}Sc_{26}$	767	274ns	[3/2]	±0.35				
$^{48}_{21}Sc_{27}$		1.8d	6					
$^{45}_{22}Ti_{23}$		3.1h	7/2	±0.095		~±0.02[r] μ/Q positive		
$^{46}_{22}Ti_{24}$	889	7ps	[2]			−0.2		
$^{47}_{22}Ti_{25}$			5/2	−0.78846	127	+0.29[s]		
$^{48}_{22}Ti_{26}$	983	3.6ps	[2]			−0.20		
$^{49}_{22}Ti_{27}$			7/2	−1.10414	177	+0.24[r]		

Summary of Nuclear Moment Value (continued)

Nucleus	Level (keV)	$T_{1/2}$	I $(h/2\pi)$	μ (nm)	Diam. Corr.	Q (b)	Ω (nmb)	Q_4 (b^2)
$^{50}_{22}Ti_{28}$	1550	1ps	[2]			~0		
$^{47}_{23}V_{24}$		31m	3/2					
$^{48}_{23}V_{25}$		16d	4	±1.6				
$^{48}_{23}V_{25}$	306	7.09ns	[2]	+0.38				
$^{49}_{23}V_{26}$		330d	7/2	±4.5[P]				
$^{50}_{23}V_{27}$		>40Jy	6	+3.3470	57	±0.06		
$^{51}_{23}V_{28}$			7/2	+5.1485	88	−0.05[b]		
$^{51}_{23}V_{28}$	320	173ps	[5/2]	+4.0				
$^{49}_{24}Cr_{25}$		42m	5/2	±0.476	1			
$^{50}_{24}Cr_{26}$	783	8.4ps	[2]			−0.3		
$^{51}_{24}Cr_{27}$		28d	7/2	±0.934	2			
$^{51}_{24}Cr_{27}$	749	7.5ns	[3/2]	±1.1				
$^{52}_{24}Cr_{28}$	1434	0.90ps	[2]			[−0.08]		
$^{53}_{24}Cr_{29}$			3/2	−0.4735[E]	9	+0.03		
				(−0.4744)				
$^{54}_{24}Cr_{30}$	834	8.9ps	[2]			−0.1		
$^{51}_{25}Mn_{26}$		45m	5/2	±3.56[E]				
$^{52}_{25}Mn_{27}$		5.7d	6	+3.059[E]	6	+0.6[r]		
$^{52}_{25}Mn_{27}$	383	21m	2	±0.0076				
$^{53}_{25}Mn_{28}$		2My	7/2	±5.02[E]	1			
$^{54}_{25}Mn_{29}$		312d	3	+3.278[E]	6	+0.4[r]		
$^{55}_{25}Mn_{30}$			5/2	+3.449[E]	7	+0.4[S]		
				(+3.4680)	66			
$^{56}_{25}Mn_{31}$		2.6h	3	+3.223[E]	6			
$^{54}_{26}Fe_{28}$	1408	1.0ps	[2]	+2.9				
$^{54}_{26}Fe_{28}$	2950	1.22ns	[6]	±8.2				
$^{56}_{26}Fe_{30}$	847	6.9ps	[2]	+1.2		−0.26		
$^{57}_{26}Fe_{31}$			1/2	+0.09042[E]	18			
				(+0.09060)				
$^{57}_{26}Fe_{31}$	14.4	~10ns	[3/2]	−0.1550	3	+0.19*		
$^{57}_{26}Fe_{31}$	136	8.8ns	[5/2]	+0.92				
$^{57}_{26}Fe_{31}$	367	7ps	[3/2]	<0.5				
$^{57}_{26}Fe_{31}$	707	3ps	[5/2]	<±1				
$^{58}_{26}Fe_{32}$	811	6.4ps	[2]	+1.1				
$^{59}_{26}Fe_{33}$		45d	3/2	±1.1				
$^{55}_{27}Co_{28}$		18h	[7/2]	±4.5				
$^{56}_{27}Co_{29}$		77d	4	±3.83	1			
$^{57}_{27}Co_{30}$		270d	7/2	+4.72	1	+0.5[r]		
$^{57}_{27}Co_{30}$	1378	19.4ps	[3/2]	+3				
$^{58}_{27}Co_{31}$		71.3d	2	+4.04	1	+0.2[r]		
$^{58}_{27}Co_{31}$	54	10.2μs	[4]	+4.18				
$^{59}_{27}Co_{32}$			7/2	+4.616	10	+0.38[S]		
$^{59}_{27}Co_{32}$	1292	564ps	[3/2]	+1.8				
$^{60}_{27}Co_{33}$		5.26y	5	+3.79	1	+0.4[r]		

Summary of Nuclear Moment Value (continued)

Nucleus	Level (keV)	$T_{1/2}$	I ($h/2\pi$)	μ (nm)	Diam. Corr.	Q (b)	Ω (nmb)	Q_4 (b^2)
$^{60}_{27}Co_{33}$	58	10.5m	2	+4.4		+0.3[r]		
$^{58}_{28}Ni_{30}$	1450	0.67ps	[2]			−0.14		
$^{60}_{28}Ni_{32}$	1330	0.80ps	[2]			[−0.10]		
$^{61}_{28}Ni_{33}$			3/2	−0.7498	17	+0.16*		
$^{61}_{28}Ni_{33}$	68	5.2ns	[5/2]	+0.42				
$^{62}_{28}Ni_{34}$	1170	1.57ps	[2]			+0.2		
$^{63}_{28}Ni_{35}$	87.2	1.72μs	[5/2]	+0.752[g]				
$^{64}_{28}Ni_{36}$	1350	0.78ps	[2]			+0.3		
$^{60}_{29}Cu_{31}$		24m	2	+1.219	3			
$^{61}_{29}Cu_{32}$		3.3h	3/2	+2.13				
$^{62}_{29}Cu_{33}$		9.9m	1	−0.380	1			
$^{62}_{29}Cu_{33}$	41	4.80ns	[2]	±1.3				
$^{62}_{29}Cu_{33}$	390	11.5ns	[3]	±1.9				
$^{63}_{29}Cu_{34}$			3/2	+2.2228[f]	53	−0.211*[r]		
				(+2.2262)				
$^{64}_{29}Cu_{35}$		13h	1	−0.216				
$^{64}_{29}Cu_{35}$	1590	20.4ns	[6]	+1.04				
$^{65}_{29}Cu_{36}$			3/2	+2.3812[f]	57	−0.195*[S]		
				(+2.3849)				
$^{66}_{29}Cu_{37}$		5.2m	1	−0.281	1			
$^{66}_{29}Cu_{37}$	1154	596ns	[6]	+1.04				
$^{63}_{30}Zn_{33}$		38m	3/2	−0.2816	7	+0.29[r]		
$^{64}_{30}Zn_{34}$			0♦					
$^{64}_{30}Zn_{34}$	992	2.7ps	[2]			[−0.14]		
$^{65}_{30}Zn_{35}$		245d	5/2	+0.7692	19	−0.024[r]		
$^{66}_{30}Zn_{36}$			0♦					
$^{67}_{30}Zn_{37}$			5/2	+0.87524[d]	218	+0.16[S]		
				(+0.8756)	22			
$^{67}_{30}Zn_{37}$	185	1.01ns	[3/2]	+0.4				
$^{67}_{30}Zn_{37}$	605	340ns	[9/2]	−1.09				
$^{68}_{30}Zn_{38}$			0♦					
$^{70}_{30}Zn_{40}$	884	3ps	[2]			[−0.2]		
$^{66}_{31}Ga_{35}$		9.5h	0♦					
$^{67}_{31}Ga_{36}$		78h	3/2	+1.849[f]	5	+0.22[r]		
$^{68}_{31}Ga_{37}$		68m	1	±0.0117		±0.031[r]		
$^{69}_{31}Ga_{38}$			3/2	+2.0145[f]	53	+0.19[S]	+0.14	
				(+2.0161)				
$^{70}_{31}Ga_{39}$		21m	1					
$^{71}_{31}Ga_{40}$			3/2	+2.5597[f]	67	+0.12[r]	+0.18	
				(+2.5617)				
$^{72}_{31}Ga_{41}$		14h	3	−0.1321[f]	3	+0.59[r]		
$^{67}_{32}Ge_{35}$	734	70ns	[9/2]	−0.94		Q/Q^{69}_{398}=1.22		
$^{69}_{32}Ge_{37}$		38h	5/2	±0.73		±0.03[r]		
						μ/Q positive		

Summary of Nuclear Moment Values (continued)

Nucleus	Level (keV)	$T_{1/2}$	I ($h/2\pi$)	μ (nm)	Diam. Corr.	Q (b)	Ω (nmb)	Q_4 (b^2)
$^{69}_{32}\mathrm{Ge}_{37}$	398	3μs	[9/2]	−1.001	3			
$^{70}_{32}\mathrm{Ge}_{38}$			0♦					
$^{70}_{32}\mathrm{Ge}_{38}$	1040	1.3ps	[2]	+1.8		~0		
$^{71}_{32}\mathrm{Ge}_{39}$		11d	1/2	+0.546	1			
$^{71}_{32}\mathrm{Ge}_{39}$	175	79ns	[5/2]	+1.02		Q/Q^{69}_{398}=0.22		
$^{71}_{32}\mathrm{Ge}_{39}$	198	20.2ms	[9/2]	−1.040	3	±0.3		
$^{72}_{32}\mathrm{Ge}_{40}$			0♦					
$^{72}_{32}\mathrm{Ge}_{40}$	835	3.14ps	[2]	+1.2				
$^{73}_{32}\mathrm{Ge}_{41}$			9/2	−0.87918	240	−0.18		
$^{74}_{32}\mathrm{Ge}_{42}$			0♦					
$^{74}_{32}\mathrm{Ge}_{42}$	596	12ps	[2]	+0.9		~0		
$^{75}_{32}\mathrm{Ge}_{43}$		82m	1/2	+0.51				
$^{76}_{32}\mathrm{Ge}_{44}$			0♦					
$^{76}_{32}\mathrm{Ge}_{44}$	563	17.6ps	[2]	+0.7		−0.2;~0		
$^{70}_{33}\mathrm{As}_{37}$		55m	4					
$^{72}_{33}\mathrm{As}_{39}$		26h	2	±2.2				
$^{72}_{33}\mathrm{As}_{39}$	215	80ns	[3]	+1.58				
$^{73}_{33}\mathrm{As}_{40}$	66.9	5.0ns	[5/2]	+1.6				
$^{73}_{33}\mathrm{As}_{40}$	427	5.8μs	[9/2]	+5.21	1			
$^{74}_{33}\mathrm{As}_{41}$	274	26.8ns	[3]	+2.43	1			
$^{75}_{33}\mathrm{As}_{42}$			3/2	+1.439	4	+0.29		
$^{75}_{33}\mathrm{As}_{42}$	265	11.9ps	[3/2]	+1.0				
$^{75}_{33}\mathrm{As}_{42}$	280	0.28ns	[5/2]	+0.9				
$^{76}_{33}\mathrm{As}_{43}$		26h	2	−0.905	2	±7 *8*		
$^{76}_{33}\mathrm{As}_{43}$	45	2.60μs	[1]	+0.559	1			
$^{77}_{33}\mathrm{As}_{44}$	473	116μs	[9/2]	±5.52	1			
$^{74}_{34}\mathrm{Se}_{40}$			0♦					
$^{75}_{34}\mathrm{Se}_{41}$		120d	5/2			+1.0[r]		
$^{76}_{34}\mathrm{Se}_{42}$			0♦					
$^{76}_{34}\mathrm{Se}_{42}$	559	11.1ps	[2]	+0.8				
$^{77}_{34}\mathrm{Se}_{43}$			1/2	+0.534	1			
$^{77}_{34}\mathrm{Se}_{43}$	249	9.4ns	[5/2]	+1.2				
$^{77}_{34}\mathrm{Se}_{43}$	440	24ps	[5/2]	+1.0				
$^{78}_{34}\mathrm{Se}_{44}$			0					
$^{78}_{34}\mathrm{Se}_{44}$	614	8.6ps	[2]	+0.8				
$^{79}_{34}\mathrm{Se}_{45}$		60ky	7/2	−1.02		+0.8[S]		
$^{80}_{34}\mathrm{Se}_{46}$			0					
$^{80}_{34}\mathrm{Se}_{46}$	666	8.05ps	[2]	+0.8				
$^{82}_{34}\mathrm{Se}_{48}$			0♦					
$^{82}_{34}\mathrm{Se}_{48}$	655	11.3ps	[2]	+0.9				
$^{76}_{35}\mathrm{Br}_{41}$		17h	1	±0.548	2	±0.30*[r]		
						μ/Q negative		
$^{77}_{35}\mathrm{Br}_{42}$		58h	3/2					
$^{78}_{35}\mathrm{Br}_{43}$	181	100μs	[4]	+4.11	1			
$^{79}_{35}\mathrm{Br}_{44}$			3/2	+2.1055	65	+0.37*[S]	+0.09*	
$^{80}_{35}\mathrm{Br}_{45}$		18m	1	±0.514	2	±0.22*[r]		

Summary of Nuclear Moment Values (continued)

Nucleus	Level (keV)	$T_{1/2}$	I ($h/2\pi$)	μ (nm)	Diam. Corr.	Q (b)	Ω (nmb)	Q_4 (b^2)
						μ/Q positive		
$^{80}_{35}Br_{45}$	85	4.5h	5	+1.317	4	+0.84^{*r}		
$^{81}_{35}Br_{46}$			3/2	+2.2696	70	+0.31^{*r}	+0.10^{*r}	
$^{81}_{35}Br_{46}$	540	35μs	[9/2]	±5.77	2			
$^{82}_{35}Br_{47}$		36h	5	+1.626	5	±0.84^{*r}		
$^{79}_{36}Kr_{43}$	148	77.7ns	[5/2]	+1.12				
$^{82}_{36}Kr_{46}$			0♦					
$^{83}_{36}Kr_{47}$			9/2	−0.9703	31	+0.26^{S}	−0.18	
$^{83}_{36}Kr_{47}$	9.3	143ns	[7/2]	−1.8		+0.44^{r}		
$^{84}_{36}Kr_{48}$			0♦					
$^{85}_{36}Kr_{49}$		11y	9/2	±1.005	3	+0.43^{r}		
$^{86}_{36}Kr_{50}$			0♦					
$^{81}_{37}Rb_{44}$		4.7h	3/2	+2.05				
$^{81}_{37}Rb_{44}$	85	32m	9/2					
$^{82}_{37}Rb_{45}$	30	6.3h	5	+1.643	6			
$^{83}_{37}Rb_{46}$		83d	5/2	+1.42				
$^{84}_{37}Rb_{47}$		33d	2	−1.32				
$^{85}_{37}Rb_{48}$			5/2	+1.3524^{df} (+1.3527)	45	+0.26^{*S}		
$^{86}_{37}Rb_{49}$		19d	2	−1.691	6			
$^{87}_{37}Rb_{50}$		47Gy	3/2	+2.7500^{df} (+2.7506)	92	+0.13^{*r}		
$^{88}_{37}Rb_{51}$		18m	2	±0.51				
$^{86}_{38}Sr_{48}$			0♦					
$^{86}_{38}Sr_{48}$	?	460ns	[8]	−1.9				
$^{87}_{38}Sr_{49}$			9/2	−1.093	4	+0.3		
$^{88}_{38}Sr_{50}$			0♦					
$^{86}_{39}Y_{47}$	243	28.5ns	[2]	−1.06				
$^{89}_{39}Y_{50}$			1/2	−0.13733	49			
$^{90}_{39}Y_{51}$		64h	2	−1.63	1	−0.15		
$^{91}_{39}Y_{52}$		58d	1/2	±0.164	1			
$^{90}_{40}Zr_{50}$	3590	130ns	[8]	±10.8				
$^{91}_{40}Zr_{51}$			5/2	−1.3028	48			
$^{91}_{40}Zr_{51}$	>2265	29.0ns	[15/2]	±5.3				
$^{91}_{41}Nb_{50}$	2378	10.0ns	[17/2]	±10.6				
$^{93}_{41}Nb_{52}$			9/2	+6.167	24	−0.22		
$^{95}_{41}Nb_{54}$		35d	[9/2]	±6.3				
$^{92}_{42}Mo_{50}$			0♦					
$^{92}_{42}Mo_{50}$	2761	190ns	[8]	±11.2				
$^{94}_{42}Mo_{52}$			0♦					
$^{94}_{42}Mo_{52}$	2953	97.7ns	[8]	+10.5				

Summary of Nuclear Moment Values (continued)

Nucleus	Level (keV)	$T_{1/2}$	I ($h/2\pi$)	μ (nm)	Diam. Corr.	Q (b)	Ω (nmb)	Q_4 (b^2)
$^{95}_{42}Mo_{53}$			5/2	−0.9135	36	±0.12		
$^{95}_{42}Mo_{53}$	204	760ps	[3/2]	−0.4				
$^{96}_{42}Mo_{54}$			0♦					
$^{97}_{42}Mo_{55}$			5/2	−0.9327	37	±1.1		
$^{98}_{42}Mo_{56}$			0♦					
$^{98}_{42}Mo_{56}$	787	3.5ps	[2]	+0.7				
$^{100}_{42}Mo_{58}$			0♦					
$^{100}_{42}Mo_{58}$	536	10ps	[2]	+0.7				
$^{96}_{43}Tc_{53}$		4.3d	6	±4.6				
$^{99}_{43}Tc_{56}$		210ky	9/2	+5.681	23	+0.3		
$^{99}_{43}Tc_{56}$	141	192ps	[7/2]	+5				
$^{99}_{43}Tc_{56}$	181	3.59ns	[5/2]	+3.3				
$^{98}_{44}Ru_{54}$	654	5.9ps	[2]	+0.8				
$^{99}_{44}Ru_{55}$			5/2	−0.62		$\lvert Q/Q_{90}\rvert \leq 0.3$		
$^{99}_{44}Ru_{55}$	90	20.7ns	3/2	−0.28		$\geq \pm 0.1$		
$^{100}_{44}Ru_{56}$	540	11.9ps	[2]	+1.0				
$^{101}_{44}Ru_{57}$			5/2	−0.68				
$^{101}_{44}Ru_{57}$	127	550ps	[3/2]	−0.31				
$^{102}_{44}Ru_{58}$	475	17.6ps	[2]	+0.74				
$^{104}_{44}Ru_{60}$	358	58ps	[2]	+0.8				
$^{100}_{45}Rh_{55}$	74.8	215ns	[2]	+4.32	2			
$^{103}_{45}Rh_{58}$			1/2	−0.0883	4			
$^{103}_{45}Rh_{58}$	93	1.13ns	[9/2]	±6.2				
$^{103}_{45}Rh_{58}$	298	6.3ps	[3/2]	+1[b]				
$^{103}_{45}Rh_{58}$	360	59ps	[5/2]	+1.2[b]				
$^{104}_{46}Pd_{58}$	556	9.7ps	[2]	+0.7		−0.3;~0		
$^{105}_{46}Pd_{59}$			5/2	−0.642	3	+0.8		
$^{106}_{46}Pd_{60}$	512	12.7ps	[2]	+0.73		−0.5;−0.3		
$^{106}_{46}Pd_{60}$	1128	2.5ps	[2]	+0.7				
$^{108}_{46}Pd_{62}$	434	23.8ps	[2]	+0.77		−0.6;−0.4		
$^{110}_{46}Pd_{64}$	374	45.8ps	[2]	+0.70		−0.3		
$^{101}_{47}Ag_{54}$		9m	9/2					
$^{102}_{47}Ag_{55}$		13m	5					
$^{102}_{47}Ag_{55}$	?	7m	2	+4.2				
$^{103}_{47}Ag_{56}$		66m	7/2	+4.45[a]				
$^{104}_{47}Ag_{57}$		1.2h	5	+4.0				
$^{104}_{47}Ag_{57}$	~20	27m	2	+3.7				
$^{105}_{47}Ag_{58}$		40d	1/2	±0.101				
$^{106}_{47}Ag_{59}$		24m	1	+2.9				
$^{106}_{47}Ag_{59}$	~300	8.3d	6					
$^{107}_{47}Ag_{60}$			1/2	−0.1135	5			
$^{107}_{47}Ag_{60}$	325	5.9ps	[3/2]	+0.7				
$^{107}_{47}Ag_{60}$	423	34ps	[5/2]	+0.9				

Summary of Nuclear Moment Values (continued)

Nucleus	Level (keV)	$T_{1/2}$	I ($h/2\pi$)	μ (nm)	Diam. Corr.	Q (b)	Ω (nmb)	Q_4 (b^2)
$^{108}_{47}Ag_{61}$		2.4m	1	+2.80	1			
$^{109}_{47}Ag_{62}$			1/2	−0.1305	6			
$^{109}_{47}Ag_{62}$	88	40s	7/2	±4.3				
$^{109}_{47}Ag_{62}$	309	5.2ps	[3/2]	+0.9				
$^{109}_{47}Ag_{62}$	414	33ps	[5/2]	+0.9				
$^{110}_{47}Ag_{63}$		24.4s	1	+2.72	1			
$^{110}_{47}Ag_{63}$	116	253d	6	+3.604	17			
$^{111}_{47}Ag_{64}$		7.5d	1/2	−0.145	1			
$^{112}_{47}Ag_{65}$		3.2h	2	±0.054				
$^{113}_{47}Ag_{66}$		5.3h	1/2	±0.159	1			
$^{105}_{48}Cd_{57}$		55m	5/2	−0.738	4	+0.43[r]		
$^{106}_{48}Cd_{58}$	633	6ps	[2]			−0.8		
$^{107}_{48}Cd_{59}$		6.7h	5/2	−0.61443[d]	294	+0.68[r]		
$^{108}_{48}Cd_{60}$	633	5ps	[2]			−0.8		
$^{109}_{48}Cd_{61}$		470d	5/2	−0.82701[d]	395	+0.69[S]		
$^{109}_{48}Cd_{61}$	469	8.9μs	[11/2]	−1.10	1			
$^{110}_{48}Cd_{62}$			0♦					
$^{110}_{48}Cd_{62}$	656	5.0ps	[2]	+0.7		−0.5;−0.3[b]		
$^{111}_{48}Cd_{63}$			1/2	−0.59428[d] (−0.59500)	284			
$^{111}_{48}Cd_{63}$	247	84ns	[5/2]	−0.793	4	+1		
$^{111}_{48}Cd_{63}$	397	49m	11/2	−1.1040	53	−0.85[r]		
$^{112}_{48}Cd_{64}$			0♦					
$^{112}_{48}Cd_{64}$	617	6.2ps	[2]	+0.7		−0.2		
$^{113}_{48}Cd_{65}$		>3Jy	1/2	−0.62167[d] (−0.62245)	297			
$^{113}_{48}Cd_{65}$	265	14y	11/2	−1.0871[d]	52	−0.71[r]		
$^{114}_{48}Cd_{66}$			0♦					
$^{114}_{48}Cd_{66}$	558	9.0ps	[2]	+0.8		−0.32		
$^{115}_{48}Cd_{67}$		2.3d	1/2	−0.6478[d]	31			
$^{115}_{48}Cd_{67}$	180	43d	11/2	−1.0400[d]	50	−0.55[r]		
$^{116}_{48}Cd_{68}$			0♦					
$^{116}_{48}Cd_{68}$	513	13.7ps	[2]	+0.8		−0.9[b]		
$^{109}_{49}In_{60}$		4.3h	9/2	+5.53	3	+0.85[r]		
$^{110}_{49}In_{61}$		66m	2	+4.36	2	+0.36[r]		
$^{110}_{49}In_{61}$	?	4.9h	7	+10.4 or −10.7		−0.21[r] or +0.22[r]		
$^{111}_{49}In_{62}$		2.8d	9/2	+5.53	3	+0.84[r]		
$^{112}_{49}In_{63}$		14m	1	+2.81	1	+0.089[r]		
$^{112}_{49}In_{63}$	155	21m	4					
$^{113}_{49}In_{64}$			9/2	+5.5229	271	+0.82[r]	+0.57	
$^{113}_{49}In_{64}$	393	1.7h	1/2	−0.210	1			
$^{114}_{49}In_{65}$		72s	[1]	≤±1.7				
$^{114}_{49}In_{65}$	190	50d	5	+4.7				
$^{115}_{49}In_{66}$		600Ty	9/2	+5.5348	272	+0.83[S]	+0.56	
$^{115}_{49}In_{66}$	335	4.5h	1/2	−0.244	1			
$^{116}_{49}In_{67}$		14s	[1]	±2.786	14	±0.1		

Summary of Nuclear Moment Values (continued)

Nucleus	Level (keV)	$T_{1/2}$	I ($h/2\pi$)	μ (nm)	Diam. Corr.	Q (b)	Ω (nmb)	Q_4 (b^2)
$^{116}_{49}In_{67}$	70	54m	5	+4.3				
$^{117}_{49}In_{68}$		45m	9/2					
$^{117}_{49}In_{68}$	310	1.9h	1/2	−0.2515	12			
$^{117}_{49}In_{68}$	660	60ns	[3/2]	+1.0		±0.64		
$^{112}_{50}Sn_{62}$	1257	0.3ps	[2]			~0		
$^{113}_{50}Sn_{63}$		118d	1/2	±0.88				
$^{114}_{50}Sn_{64}$	~3100	700ns	[9,7?]	g=−0.081				
$^{115}_{50}Sn_{65}$			1/2	−0.9178	46			
$^{115}_{50}Sn_{65}$	619	3.3μs	[7/2]	<±1.0				
$^{115}_{50}Sn_{65}$	726	159μs	[11/2]	−1.368	7	±0.8		
$^{116}_{50}Sn_{66}$			0•					
$^{116}_{50}Sn_{66}$	1290	0.4ps	[2]			[−0.1]		
$^{116}_{50}Sn_{66}$	2369	350ns	[5]	−0.32				
$^{117}_{50}Sn_{67}$			1/2	−0.9999	50			
$^{118}_{50}Sn_{68}$			0•					
$^{118}_{50}Sn_{68}$	1230	0.5ps	[2]			−0.2		
$^{118}_{50}Sn_{68}$	2320	21.7ns	[5]	−0.32				
$^{119}_{50}Sn_{69}$			1/2	−1.0461	53			
$^{119}_{50}Sn_{69}$	24	18.5ns	[3/2]	+0.68		−0.07		
$^{120}_{50}Sn_{70}$			0•					
$^{120}_{50}Sn_{70}$	1170	0.5ps	[2]			~0		
$^{120}_{50}Sn_{70}$	2300	5.5ns	[5]	−0.30		±0.02		
$^{121}_{50}Sn_{71}$		27h	3/2	±0.699	4	±0.08		
						μ/Q negative		
$^{122}_{50}Sn_{72}$	1140	0.6ps	[2]			~0.3		
$^{123}_{50}Sn_{73}$	~24	40m	3/2					
$^{124}_{50}Sn_{74}$	1130	0.8ps	[2]			−0.1		
$^{115}_{51}Sb_{64}$		31m	5/2	+3.46	2	−0.28[r]		
$^{116}_{51}Sb_{65}$		15m	3					
$^{117}_{51}Sb_{66}$		2.8h	5/2	+2.67	1	−0.42[r]		
$^{117}_{51}Sb_{66}$	3130	340μs	[21/2?]	+1.22				
$^{118}_{51}Sb_{67}$	~200	3.5m	1	±2.46	1			
$^{119}_{51}Sb_{68}$		38h	5/2	+3.45	2	−0.29[r]		
$^{120}_{51}Sb_{69}$		16m	1	±2.3				
$^{121}_{51}Sb_{70}$			5/2	+3.3592	174	−0.28[abS]		
$^{121}_{51}Sb_{70}$	37	3.5ns	[7/2]	+2.51	1	−0.4[r]		
$^{122}_{51}Sb_{71}$		2.73d	2	−1.90	1	+0.66[r]		
$^{122}_{51}Sb_{71}$	61	1.8μs	[3]	+2.98	2			
$^{123}_{51}Sb_{72}$			7/2	+2.5466	132	−0.36[r]		
$^{124}_{51}Sb_{73}$		60d	3	±1.3				
$^{125}_{51}Sb_{74}$		2.77y	7/2	±2.61	1			
$^{126}_{51}Sb_{75}$		12.5d	[8]	±1.3				
$^{127}_{51}Sb_{76}$		3.9d	[7/2]	±2.6				
$^{128}_{51}Sb_{77}$		8.6h	[8]	±1.3				
$^{116}_{52}Te_{64}$		2.5h	0•					

Summary of Nuclear Moment Values (continued)

Nucleus	Level (keV)	$T_{1/2}$	I ($h/2\pi$)	μ (nm)	Diam. Corr.	Q (b)	Ω (nmb)	Q_4 (b^2)
$^{117}_{52}Te_{65}$		61m	1/2					
$^{119}_{52}Te_{67}$		16h	1/2	±0.25				
$^{119}_{52}Te_{67}$	~300	4.5d	11/2					
$^{120}_{52}Te_{68}$	560	9.3ps	[2]	+0.6				
$^{122}_{52}Te_{70}$	564	7.6ps	[2]	+0.66				
$^{123}_{52}Te_{71}$		>50Ty	1/2	−0.7359	39			
$^{123}_{52}Te_{71}$	159	190ps	[3/2]	±0.7				
$^{123}_{52}Te_{71}$	248	117d	[11/2]	−1.00				
$^{123}_{52}Te_{71}$	440	?	?	g=+0.2				
$^{123}_{52}Te_{71}$	506	?	?	g=+0.03				
$^{124}_{52}Te_{72}$	603	6.6ps	[2]	+0.5[b]		−0.5;−0.3		
$^{125}_{52}Te_{73}$			1/2	−0.8872	47			
$^{125}_{52}Te_{73}$	35.5	1.6ns	3/2	+0.60		−0.2		
$^{125}_{52}Te_{73}$	145	58d	[11/2]	±0.9				
$^{125}_{52}Te_{73}$	321	695ps	[9/2]	−0.91				
$^{125}_{52}Te_{73}$	443	21ps	[3/2]	+0.5				
$^{125}_{52}Te_{73}$	463	13ps	[5/2]	+0.6				
$^{125}_{52}Te_{73}$	525	?	[7/2?]	negative				
$^{126}_{52}Te_{74}$			0♦					
$^{126}_{52}Te_{74}$	667	4.4ps	[2]	+0.6		−0.3;−0.1[b]		
$^{127}_{52}Te_{75}$		9.4h	[3/2]	±0.61				
$^{127}_{52}Te_{75}$	89	109d	[11/2]	−0.91				
$^{128}_{52}Te_{76}$			0♦					
$^{128}_{52}Te_{76}$	743	3.2ps	[2]	+0.5		−0.1;+0.1[b]		
$^{129}_{52}Te_{77}$		69m	[3/2]	±0.67				
$^{129}_{52}Te_{77}$	106	34d	[11/2]	−1.15				
$^{130}_{52}Te_{78}$			0♦					
$^{130}_{52}Te_{78}$	840	2.0ps	[2]	+0.6		−0.2;−0.1		
$^{123}_{53}I_{70}$		13h	5/2					
$^{124}_{53}I_{71}$		4.0d	2					
$^{125}_{53}I_{72}$		60d	5/2	+3		−0.89[r]		
$^{125}_{53}I_{72}$	188	35ns	[3/2]	±3				
$^{126}_{53}I_{73}$		13d	2					
$^{127}_{53}I_{74}$			5/2	+2.8091	153	−0.79[eS]	+0.18	
$^{127}_{53}I_{74}$	58	1.92ns	[7/2]	±2.2[b]		−0.71[r]		
$^{127}_{53}I_{74}$	203	330ps	[3/2]	≥±1.1				
$^{128}_{53}I_{75}$		25m	1					
$^{129}_{53}I_{76}$		16My	7/2	+2.6174	143	−0.55[r]		
$^{129}_{53}I_{76}$	27	15ns	[5/2]	+2.8		−0.68[r]		
$^{130}_{53}I_{77}$		12h	5					
$^{131}_{53}I_{78}$		8.1d	7/2	+2.738	15	−0.40[r]		
$^{131}_{53}I_{78}$	150	0.95ns	[5/2]	+2.8				
$^{131}_{53}I_{78}$	1797	5.9ns	[9/2	−0.7				
			or 11/2]	−0.9				
$^{132}_{53}I_{79}$		2.3h	4	±3.08	2	±0.08[r]		
						μ/Q negative		
$^{132}_{53}I_{79}$	49.7	0.95ns	[3]	+2.2				

Nucleus	Level (keV)	$T_{1/2}$	I ($h/2\pi$)	μ (nm)	Diam. Corr.	Q (b)	Ω (nmb)	Q_4 (b^2)
$^{133}_{53}I_{80}$		21h	7/2	+2.84	2	−0.26[r]		
$^{135}_{53}I_{82}$		6.7h	7/2					
$^{129}_{54}Xe_{75}$			1/2	−0.7768	43			
$^{129}_{54}Xe_{75}$	40	700ps	[3/2]			±0.41[r]		
$^{131}_{54}Xe_{77}$			3/2	+0.6908	39	−0.12[S]	+0.048	
$^{132}_{54}Xe_{78}$			0♦					
$^{132}_{54}Xe_{78}$	668	7ps	[2]	+0.9				
$^{134}_{54}Xe_{80}$			0♦					
$^{136}_{54}Xe_{82}$			0♦					
$^{125}_{55}Cs_{70}$		45m	1/2	+1.41	1			
$^{127}_{55}Cs_{72}$		6.2h	1/2	+1.45	1			
$^{129}_{55}Cs_{74}$		31h	1/2	+?1.479	8			
$^{130}_{55}Cs_{75}$		30m	1	+1.37 or −1.45	1			
$^{131}_{55}Cs_{76}$		10d	5/2	+3.54	2	−0.57*[r]		
$^{131}_{55}Cs_{76}$	133	9.3ns	[5/2]	+2.1				
$^{132}_{55}Cs_{77}$		6.2d	2	+2.22	1	+0.47*[r]		
$^{133}_{55}Cs_{78}$			7/2	+2.5779[d] (+2.5788)	148	−0.0030*[r]		<100
$^{133}_{55}Cs_{78}$	81	6.31ns	[5/2]	+3.44	2			
$^{133}_{55}Cs_{78}$	160	190ps	[5/2]	+1.5				
$^{134}_{55}Cs_{79}$		2.2y	4	+2.989	17	+0.36*[S]		
$^{134}_{55}Cs_{79}$	11.2	47.0ns	[5]	+3.34	2			
$^{134}_{55}Cs_{79}$	137	3.1h	8	+1.096	6			
$^{135}_{55}Cs_{80}$		2My	7/2	+2.7280[d] (+2.7289)	156	+0.044*[r]		
$^{136}_{55}Cs_{81}$		13d	5	+3.70	2			
$^{137}_{55}Cs_{82}$		30y	7/2	+2.8372[d] (+2.8382)	162	+0.045*[r]		
$^{138}_{55}Cs_{83}$		32m	3	±0.5				
$^{130}_{56}Ba_{74}$	356	63ps	[2]			+0.3[b]		
$^{134}_{56}Ba_{78}$			0♦					
$^{134}_{56}Ba_{78}$	605	7ps	[2]	~±0.006?		+0.1[b]		
$^{135}_{56}Ba_{79}$			3/2	+0.8365[d] (+0.8372)	49	+0.18[S]		
$^{135}_{56}Ba^{+}_{79}$			3/2					
$^{136}_{56}Ba_{80}$			0♦					
$^{136}_{56}Ba_{80}$	818	1.5ps	[2]			+0.3[b]		
$^{137}_{56}Ba_{81}$			3/2	+0.9357[d] (+0.9365)	55	+0.28[r]		
$^{137}_{56}Ba^{+}_{81}$			3/2					
$^{137}_{56}Ba_{81}$	662	2.55m	[11/2]			≈−0.05		
$^{138}_{56}Ba_{82}$			0♦					
$^{131}_{57}La_{74}$		59m	3/2					
$^{132}_{57}La_{75}$		4.5h	2					

Summary of Nuclear Moment Values (continued)

Nucleus	Level (keV)	$T_{1/2}$	I $(h/2\pi)$	μ (nm)	Diam. Corr.	Q (b)	Ω (nmb)	Q_4 (b^2)
$^{132}_{57}La_{75}$	?	25m	6					
$^{133}_{57}La_{76}$		4.0h	5/2					
$^{133}_{57}La_{76}$	535	49ns	[11/2]	±8				
$^{135}_{57}La_{78}$		19.4h	5/2					
$^{136}_{57}La_{79}$		9.9m	1					
$^{137}_{57}La_{80}$		60ky	7/2	+2.69	2	$+0.26*^{r}$		
$^{138}_{57}La_{81}$		112Gy	5	+3.704	22	$+0.51*^{r}$		
$^{139}_{57}La_{82}$			7/2	+2.778	17	$+0.22*^{S}$		
$^{140}_{57}La_{83}$		40h	3	+0.73		$+0.1*^{r}$		
$^{130}_{58}Ce_{72}$		25m	0♦					
$^{132}_{58}Ce_{74}$		4.2h	0♦					
$^{133}_{58}Ce_{75}$		5.4h	9/2					
$^{133}_{58}Ce_{75}$	?	97m	1/2					
$^{134}_{58}Ce_{76}$		72h	0♦					
$^{135}_{58}Ce_{77}$		17h	1/2					
$^{137}_{58}Ce_{79}$		9.0h	3/2	±0.7				
$^{137}_{58}Ce_{79}$	255	34.4h	11/2	±0.69				
$^{139}_{58}Ce_{81}$		140d	3/2	±0.9				
$^{140}_{58}Ce_{82}$	2083	3.41ns	[4]	+4.3		±0.40*		
$^{141}_{58}Ce_{83}$		33d	7/2					
$^{142}_{58}Ce_{84}$	650	6.2ps	[2]			−0.1		
$^{143}_{58}Ce_{85}$		34h	3/2	~±1				
$^{133}_{59}Pr_{74}$		7.5m	5/2					
$^{134}_{59}Pr_{75}$		18.5m	2					
$^{135}_{59}Pr_{76}$		24m	3/2					
$^{136}_{59}Pr_{77}$		13.5m	2					
$^{137}_{59}Pr_{78}$		1.28h	5/2					
$^{138}_{59}Pr_{79}$	?	2.0h	7					
$^{139}_{59}Pr_{80}$		4.5h	5/2					
$^{140}_{59}Pr_{81}$		3.4m	1					
$^{141}_{59}Pr_{82}$			5/2	+4.16	3	-0.058^{S}		
$^{142}_{59}Pr_{83}$		19h	2	−0.24		-0.034^{r}		
$^{142}_{59}Pr_{83}$	?	?	5					
$^{143}_{59}Pr_{84}$		14d	7/2					
$^{143}_{59}Pr_{84}$	57	4.17ns	[5/2]	+2.8				
$^{134}_{60}Nd_{74}$		8m	0♦					
$^{135}_{60}Nd_{75}$		15m	9/2					
$^{136}_{60}Nd_{76}$		55m	0♦					
$^{137}_{60}Nd_{77}$		37m	1/2					
$^{138}_{60}Nd_{78}$		5.2h	0♦					
$^{139}_{60}Nd_{79}$		29.7m	3/2					
$^{139}_{60}Nd_{79}$	232	5.5h	11/2					
$^{140}_{60}Nd_{80}$		3.4d	0♦					
$^{141}_{60}Nd_{81}$		2.4h	3/2					
$^{143}_{60}Nd_{83}$			7/2	−1.063	7	-0.48^{hS}		

Summary of Nuclear Moment Values (continued)

Nucleus	Level (keV)	$T_{1/2}$	I $(h/2\pi)$	μ (nm)	Diam. Corr.	Q (b)	Ω (nmb)	Q_4 (b^2)
$^{144}_{60}Nd_{84}$	695	4.2ps	[2]	+0.26		−0.2;−0.6		
$^{144}_{60}Nd_{84}$	1314	90ps	[4]	+0.2				
$^{145}_{60}Nd_{85}$			7/2	−0.654	4	−0.25[hr]		
$^{146}_{60}Nd_{86}$	454	21ps	[2]	+0.48		−0.7		
$^{147}_{60}Nd_{87}$		11d	5/2	±0.55		±0.7[r] μ/Q negative		
$^{148}_{60}Nd_{88}$	300	116ps	[2]	+0.45		−1.3		
$^{149}_{60}Nd_{89}$		1.9h	5/2	±0.35		±1.0[r]		
$^{150}_{60}Nd_{90}$	132	1.52ns	[2]	+0.64		−1.7		
$^{150}_{60}Nd_{90}$	397	55.9ps	[4]	+1.3				
$^{141}_{61}Pm_{80}$		20.9m	5/2					
$^{143}_{61}Pm_{82}$		265d	[5/2, or 7/2]	±3.8, ±3.9				
$^{144}_{61}Pm_{83}$		360d	[5, or 6]	±1.7, ±1.8				
$^{147}_{61}Pm_{86}$		2.6y	7/2	+2.62	2	+0.7		
$^{147}_{61}Pm_{86}$	91	2.55ns	[5/2]	+3.4				
$^{148}_{61}Pm_{87}$		5.4d	1	+2.0		+0.2		
$^{148}_{61}Pm_{87}$	137	43d	[6]	±1.8				
$^{149}_{61}Pm_{88}$		53h	7/2	±3.3				
$^{149}_{61}Pm_{88}$	114	2.58ns	[5/2]	+2.1				
$^{149}_{61}Pm_{88}$	188	3.24ns	[3/2]	+1.6				
$^{149}_{61}Pm_{88}$	211	80ps	[5/2]	+2.2				
$^{149}_{61}Pm_{88}$	270	2.59ns	[7/2]	+3				
$^{151}_{61}Pm_{90}$		28h	5/2	±1.6		±1.9 μ/Q positive		
$^{140}_{62}Sm_{78}$		15m	0*					
$^{141}_{62}Sm_{79}$		11.3m	1/2					
$^{141}_{62}Sm_{79}$	?	22.9m	11/2					
$^{142}_{62}Sm_{80}$		1.2h	0*					
$^{143}_{62}Sm_{81}$		8.8m	3/2					
$^{145}_{62}Sm_{83}$		340d	[7/2]	±0.92				
$^{147}_{62}Sm_{85}$		0.1Ty	7/2	−0.813	6	−0.18[s]		
$^{147}_{62}Sm_{85}$	121	780ps	[5/2]	−0.3				
$^{147}_{62}Sm_{85}$	198	1.31ns	[3/2]	−0.28				
$^{148}_{62}Sm_{86}$	551	7.35ps	[2]	+0.3		−0.8		
$^{149}_{62}Sm_{87}$			7/2	−0.670	5	+0.052[r]		
$^{149}_{62}Sm_{87}$	22	7.6ns	5/2	−0.61		+0.4		
$^{150}_{62}Sm_{88}$	334	48ps	[2]	+0.60		−1.3		
$^{151}_{62}Sm_{89}$	105	480ps	[5/2]	+0.5				
$^{151}_{62}Sm_{89}$	168	760ps	[3/2]	+0.6				
$^{152}_{62}Sm_{90}$	122	1.42ns	[2]	+0.69		−1.8		
$^{152}_{62}Sm_{90}$	366	57ps	[4]	+1.2				
$^{153}_{62}Sm_{91}$		47h	3/2	−0.0217	1	+0.9		
$^{154}_{62}Sm_{92}$	82	3.02n	[2]	+0.61				
$^{154}_{62}Sm_{92}$	267	165ps	[4]	+1.3				

Summary of Nuclear Moment Values (continued)

Nucleus	Level (keV)	$T_{1/2}$	I $(h/2\pi)$	μ (nm)	Diam. Corr.	Q (b)	Ω (nmb)	Q_4 (b^2)
$^{154}_{62}Sm_{92}$	549	23.5ps	[6]	+1.9				
$^{155}_{62}Sm_{93}$		24m	3/2			±0.8[r]		
$^{145}_{63}Eu_{82}$		5.9d	5/2					
$^{146}_{63}Eu_{83}$		4.65d	4					
$^{147}_{63}Eu_{84}$		22d	5/2					
$^{147}_{63}Eu_{84}$	625	765ns	[11/2]	+6.0				
$^{148}_{63}Eu_{85}$		54d	5					
$^{149}_{63}Eu_{86}$		93d	5/2					
$^{149}_{63}Eu_{86}$	497	2.43μs	[11/2]	+6.1				
$^{150}_{63}Eu_{87}$		12.5h	0♦					
$^{151}_{63}Eu_{88}$			5/2	+3.4631[h] [+3.466]	240	+1.1[S]		
$^{151}_{63}Eu_{88}$	21.7	9.4ns	7/2	+2.57	2	+1.8[r]		
$^{152}_{63}Eu_{89}$		13y	3	−1.937	13	+3.0[r]		
$^{152}_{63}Eu_{89}$	49	9.3h	0♦					
$^{153}_{63}Eu_{90}$			5/2	+1.530	11	+2.8[r]		
$^{153}_{63}Eu_{90}$	97	200ps	[5/2]	+3.2 or −0.5				
$^{153}_{63}Eu_{90}$	103	3.8ns	[3/2]	+1.5[b]				
$^{154}_{63}Eu_{91}$		16y	3	±2.001	14	+1.9[r]		
$^{155}_{63}Eu_{92}$	105	400ps	[5/2]	+2.5				
$^{145}_{64}Gd_{81}$		22.9m	1/2					
$^{147}_{64}Gd_{83}$		38.5h	7/2					
$^{149}_{64}Gd_{85}$		9.4d	7/2					
$^{151}_{64}Gd_{87}$		120d	7/2					
$^{152}_{64}Gd_{88}$	344	29ps	[2]	+1.0				
$^{153}_{64}Gd_{89}$		242d	3/2					
$^{154}_{64}Gd_{90}$	123	1.18ns	[2]	+0.84	1			
$^{154}_{64}Gd_{90}$	371	39ps	[4]					
$^{155}_{64}Gd_{91}$			3/2	−0.2584	18	+1.6[S]	−1.6	
$^{155}_{64}Gd_{91}$	87	6.66ns	5/2	−0.93[i]	1	~±0.2[r]		
$^{155}_{64}Gd_{91}$	105	1.1ns	3/2	+0.4[i]		~±1[r]		
$^{156}_{64}Gd_{92}$	89	2.22ns	[2]	+0.72[i]		±1.2[r]		
$^{156}_{64}Gd_{92}$	288	115ps	[4]	+1.4				
$^{156}_{64}Gd_{92}$	1513	190ps	[4]	+3.1				
$^{157}_{64}Gd_{93}$			3/2	−0.3388	24	+1.7[r]		
$^{157}_{64}Gd_{93}$	64	460ns	[5/2]			±3.0[r]		
$^{158}_{64}Gd_{94}$	79.5	2.49ns	[2]	+0.73[i]	1	±1.3[r]		
$^{159}_{64}Gd_{95}$		18h	3/2	±0.44				
$^{160}_{64}Gd_{96}$	75	2.7ns	[2]	+0.63		±1.3[r]		
$^{151}_{65}Tb_{86}$		18h	1/2					
$^{152}_{65}Tb_{87}$		18h	2					
$^{153}_{65}Tb_{88}$		2.3d	5/2					
$^{154}_{65}Tb_{89}$		21h	0♦					
$^{154}_{65}Tb_{89}$	?	8.5h	3					

Summary of Nuclear Moment Values (continued)

Nucleus	Level (keV)	$T_{1/2}$	I $(h/2\pi)$	μ (nm)	Diam. Corr.	Q (b)	Ω (nmb)	Q_4 (b^2)
$^{155}_{65}Tb_{90}$		5.6d	3/2					
$^{156}_{65}Tb_{91}$		5.4d	3	±1.4		+1.4		
$^{157}_{65}Tb_{92}$		>30y	[3/2]	±2.0				
$^{158}_{65}Tb_{93}$		150y	3	±1.75	1	+2.7*		
$^{159}_{65}Tb_{94}$			3/2	±2.008	14	+1.3*		
$^{159}_{65}Tb_{94}$	58	130ps	[5/2]	±2				
$^{160}_{65}Tb_{95}$		72d	3	±1.70	1	+2.3		
$^{161}_{65}Tb_{96}$		6.9d	3/2					
$^{151}_{66}Dy_{85}$		18m	7/2					
$^{152}_{66}Dy_{86}$		2.4h	0♦					
$^{153}_{66}Dy_{87}$		6.4h	7/2	±0.7		±0.14* μ/Q positive		
$^{155}_{66}Dy_{89}$		10h	3/2	±0.28		±0.9* μ/Q negative		
$^{157}_{66}Dy_{91}$		8.1h	3/2	±0.31		±1.2* μ/Q negative		
$^{158}_{66}Dy_{92}$	630	?	[6]	±2.2				
$^{159}_{66}Dy_{93}$		144d	3/2					
$^{160}_{66}Dy_{94}$	87	2.0ns	[2]	+0.73		−2		
$^{160}_{66}Dy_{94}$	966	2.2ps	[2]	+0.4				
$^{161}_{66}Dy_{95}$			5/2	−0.482	3	+2.4*[s]		~+0.6
$^{161}_{66}Dy_{95}$	26	28.4ns	5/2	+0.67[i]		+2.4*[r]		
$^{161}_{66}Dy_{95}$	75	3.4ns	3/2	−0.38		+1.3*[r]		
$^{162}_{66}Dy_{96}$	80.7	2.25ns	[2]	+0.72				
$^{163}_{66}Dy_{97}$			5/2	+0.676	5	+2.5*[r]		~+0.7
$^{164}_{66}Dy_{98}$	73.3	2.39ns	[2]	+0.70		−2.0*[r]		
$^{165}_{66}Dy_{99}$		2.3h	7/2	±0.52		+3.3*[r]		
$^{166}_{66}Dy_{100}$		82h	0♦					
$^{154}_{67}Ho_{87}$		12m	1					
$^{155}_{67}Ho_{88}$		50m	5/2					
$^{156}_{67}Ho_{89}$		55m	1					
$^{157}_{67}Ho_{90}$		14m	7/2					
$^{158}_{67}Ho_{91}$		11m	5					
$^{158}_{67}Ho_{91}$	67	29m	2					
$^{159}_{67}Ho_{92}$		33m	7/2					
$^{160}_{67}Ho_{93}$		26m	5					
$^{160}_{67}Ho_{93}$	60	5.0h	2					
$^{161}_{67}Ho_{94}$		2.5h	7/2					
$^{162}_{67}Ho_{95}$		15m	1					
$^{162}_{67}Ho_{95}$	~100	68m	6					
$^{164}_{67}Ho_{97}$		29m	1					
$^{164}_{67}Ho_{97}$	~46	38m	6					
$^{165}_{67}Ho_{98}$			7/2	+4.12	3	+2.7		~+0.8
$^{166}_{67}Ho_{99}$		27h	0♦					
$^{166}_{67}Ho_{99}$	9	1.2ky	[7]	±4.1				

Summary of Nuclear Moment Values (continued)

Nucleus	Level (keV)	$T_{1/2}$	I ($h/2\pi$)	μ (nm)	Diam. Corr.	Q (b)	Ω (nmb)	Q_4 (b^2)
$^{156}_{68}Er_{88}$	344	47.9ps	[2]	$g_{ave}\sim\pm0.4$				
	453	7.83ps	[4]					
$^{157}_{68}Er_{89}$		20m	3/2					
$^{158}_{68}Er_{90}$		2.3h	0*					
$^{158}_{68}Er_{90}$	193	433ps	[2]	$g_{ave}\sim\pm0.4$				
	356	20.8ps	[4]					
	434	4.04ps	[6]					
$^{159}_{68}Er_{91}$		36m	3/2					
$^{160}_{68}Er_{92}$		29h	0*					
$^{160}_{68}Er_{92}$	264	49.8ps	[4]	$g_{ave}\sim\pm0.3$				
	376	7.77ps	[6]					
	465	4.04ps	[8]					
$^{161}_{68}Er_{93}$		3.2h	3/2	−0.369	3	+1.2*		
$^{163}_{68}Er_{95}$		75m	5/2	+0.56		+2.2*		
$^{164}_{68}Er_{96}$	92	1.6ns	[2]	±0.71				
$^{165}_{68}Er_{97}$		10h	5/2	±0.65		±2.2^{r} μ/Q positive		
$^{166}_{68}Er_{98}$	80.6	1.82ns	[2]	+0.63		−2.0*		
$^{166}_{68}Er_{98}$	265	120ps	[4]	+1.2		−2.7		
$^{166}_{68}Er_{98}$	787	?	[2]			+2.0		
$^{167}_{68}Er_{99}$			7/2	−0.564	4	+2.83^{S}		
$^{168}_{68}Er_{100}$	80	1.91ns	[2]	+0.65				
$^{168}_{68}Er_{100}$	264	120ps	[4]	+1.2		−2		
$^{168}_{68}Er_{100}$	~7800	?	[4]	~0				
$^{168}_{68}Er_{100}$	~7800	?	[3]	±6				
$^{169}_{68}Er_{101}$		9.4d	1/2	+0.513	4			
$^{170}_{68}Er_{102}$	79	1.90ns	[2]	+0.64		−2.1*		
$^{170}_{68}Er_{102}$	261	135ps	[4]	±1.2		−2		
$^{171}_{68}Er_{103}$		7.5h	5/2	±0.70	1	±2.3^{r} μ/Q negative		
$^{159}_{69}Tm_{90}$		9m	5/2					
$^{160}_{69}Tm_{91}$		9m	1					
$^{161}_{69}Tm_{92}$		37m	7/2					
$^{162}_{69}Tm_{93}$		21m	1					
$^{163}_{69}Tm_{94}$		1.8h	1/2	±0.08				
$^{164}_{69}Tm_{95}$		2m	1					
$^{164}_{69}Tm_{95}$	?	5m	6					
$^{165}_{69}Tm_{96}$		29h	1/2	±0.138	1			
$^{166}_{69}Tm_{97}$		7.7h	2	±0.092	1	±1.9* μ/Q positive		
$^{167}_{69}Tm_{98}$		9.6d	1/2	−0.20				
$^{168}_{69}Tm_{99}$		85d	3					
$^{169}_{69}Tm_{100}$			1/2	−0.231	2			
$^{169}_{69}Tm_{100}$	8.4	4ns	[3/2]	+0.52		−1.3*		
$^{169}_{69}Tm_{100}$	118	62ps	[5/2]	+0.74	1			
$^{169}_{69}Tm_{100}$	139	320ps	[7/2]	+1.30	1	Q/Q_{118}=1.0		

Summary of Nuclear Moment Values (continued)

Nucleus	Level (keV)	$T_{1/2}$	I ($h/2\pi$)	μ (nm)	Diam. Corr.	Q (b)	Ω (nmb)	Q_4 (b^2)
$^{169}_{69}\mathrm{Tm}_{100}$	316	660ns	[7/2]	±0.15				
$^{169}_{69}\mathrm{Tm}_{100}$	379	36ns	[7/2]	±0.96	1			
$^{170}_{69}\mathrm{Tm}_{101}$		127d	1	±0.246	2	±0.59 μ/Q positive		
$^{171}_{69}\mathrm{Tm}_{102}$		1.9y	1/2	±0.229	2			
$^{171}_{69}\mathrm{Tm}_{102}$	117	55ps	[5/2]	+0.8				
$^{171}_{69}\mathrm{Tm}_{102}$	129	362ps	[7/2]	+1.2				
$^{169}_{70}\mathrm{Yb}_{99}$		32d	[7/2]	±0.6				
$^{170}_{70}\mathrm{Yb}_{100}$	84	1.58ns	[2]	+0.68	1	negative		
$^{171}_{70}\mathrm{Yb}_{101}$			1/2	+0.4919[d] (+0.4930)	40			
$^{171}_{70}\mathrm{Yb}_{101}$	67	900ps	[3/2]	±0.35				
$^{171}_{70}\mathrm{Yb}_{101}$	76	2ns	[5/2]	+1.01				
$^{172}_{70}\mathrm{Yb}_{102}$	78.7	1.6ns	[2]	+0.64[b]		+3		
$^{172}_{70}\mathrm{Yb}_{102}$	260	132ps	[4]			−2		
$^{172}_{70}\mathrm{Yb}_{102}$	1174	7.95ps	[3]	+0.66	1	±4		
$^{173}_{70}\mathrm{Yb}_{103}$			5/2	−0.6776[d] (−0.6791)	54	+3.0		
$^{173}_{70}\mathrm{Yb}_{103}$	79	38ps	[7/2]	−0.20				
$^{173}_{70}\mathrm{Yb}_{103}$	179	36ps	[9/2]	~+0.3				
$^{173}_{70}\mathrm{Yb}_{103}$	351	0.45ns	[11/2]	~−0.7				
$^{174}_{70}\mathrm{Yb}_{104}$	76.5	1.79ns	[2]	+0.68[b]	1			
$^{174}_{70}\mathrm{Yb}_{104}$	252	?	[4]			−2		
$^{175}_{70}\mathrm{Yb}_{105}$		4.2d	[7/2]	±0.3[b]		~±6		
$^{176}_{70}\mathrm{Yb}_{106}$	82	1.76ns	[2]	+0.76	1			
$^{176}_{70}\mathrm{Yb}_{106}$	270	?	[4]			~0		
$^{167}_{71}\mathrm{Lu}_{96}$		54m	7/2					
$^{169}_{71}\mathrm{Lu}_{98}$		1.5d	7/2					
$^{170}_{71}\mathrm{Lu}_{99}$		2.0d	0♦					
$^{171}_{71}\mathrm{Lu}_{100}$		8.3d	7/2					
$^{175}_{71}\mathrm{Lu}_{104}$			7/2	+2.230	18	+5.6[S]		
$^{175}_{71}\mathrm{Lu}_{104}$	114	100ps	[9/2]	+1.9				
$^{175}_{71}\mathrm{Lu}_{104}$	251	42ps	[11/2]	+1.9				
$^{176}_{71}\mathrm{Lu}_{105}$		20Gy	7	+3.18	3	+8.0[r]		
$^{176}_{71}\mathrm{Lu}_{105}$	~300	3.7h	1	+0.318	3	−2.3[r]		
$^{177}_{71}\mathrm{Lu}_{106}$		6.8d	7/2	+2.24	2	+5.4[r]		
$^{177}_{71}\mathrm{Lu}_{106}$	971	155d	[23/2]			+13		
$^{176}_{72}\mathrm{Hf}_{104}$	88.4	1.40ns	[2]	+0.53				
$^{177}_{72}\mathrm{Hf}_{105}$			7/2	+0.7902	66	+4.5*		
$^{177}_{72}\mathrm{Hf}_{105}$	113	500ps	[9/2]	+1.12	1			
$^{177}_{72}\mathrm{Hf}_{105}$	250	55ps	[11/2]	+2.6				
$^{177}_{72}\mathrm{Hf}_{105}$	321	660ps	[9/2]	−0.51				
$^{178}_{72}\mathrm{Hf}_{106}$			0♦					
$^{178}_{72}\mathrm{Hf}_{106}$	93	1.50ns	[2]	+0.58[b]				
$^{179}_{72}\mathrm{Hf}_{107}$			9/2	−0.638	5	+5.1*		
$^{180}_{72}\mathrm{Hf}_{108}$			0♦					

Summary of Nuclear Moment Values (continued)

Nucleus	Level (keV)	$T_{1/2}$	I $(h/2\pi)$	μ (nm)	Diam. Corr.	Q (b)	Ω (nmb)	Q_4 (b^2)
${}^{180}_{72}\mathrm{Hf}_{108}$	93	1.50ns	[2]	+0.64[b]				
${}^{180}_{72}\mathrm{Hf}_{108}$	309	71ps	[4]	+2.3				
${}^{181}_{73}\mathrm{Ta}_{108}$			7/2	+2.35	2	+3*[S]		
${}^{181}_{73}\mathrm{Ta}_{108}$	6.2	6.8μs	[9/2]	+5.1		+3*[r]		
${}^{181}_{73}\mathrm{Ta}_{108}$	482	10.8ns	[5/2]	+3.29	3	positive		
${}^{182}_{73}\mathrm{Ta}_{109}$		115d	[3]	±2.6				
${}^{183}_{73}\mathrm{Ta}_{110}$		5.0d	7/2					
${}^{182}_{74}\mathrm{W}_{108}$			0♦					
${}^{182}_{74}\mathrm{W}_{108}$	100	1.37ns	[2]	+0.51				
${}^{182}_{74}\mathrm{W}_{108}$	329	64ps	[4]	+0.7				
${}^{182}_{74}\mathrm{W}_{108}$	1289	1.04ns	[2]	+1.4				
${}^{182}_{74}\mathrm{W}_{108}$	1374	2.25ns	[3]	±0.10				
${}^{183}_{74}\mathrm{W}_{109}$			1/2	+0.1169	10			
${}^{183}_{74}\mathrm{W}_{109}$	46	180ps	[3/2]	−0.1				
${}^{183}_{74}\mathrm{W}_{109}$	99	700ps	[5/2]	+0.7				
${}^{184}_{74}\mathrm{W}_{110}$			0♦					
${}^{184}_{74}\mathrm{W}_{110}$	111	1.26ns	[2]	+0.56				
${}^{184}_{74}\mathrm{W}_{110}$	364	43.5ps	[4]	+1.2				
${}^{185}_{74}\mathrm{W}_{111}$		74d	3/2					
${}^{186}_{74}\mathrm{W}_{112}$			0♦					
${}^{186}_{74}\mathrm{W}_{112}$	123	1.01ns	[2]	+0.65	1			
${}^{186}_{74}\mathrm{W}_{112}$	399	25ps	[4]	+0.8		−3		
${}^{186}_{74}\mathrm{W}_{112}$	730	4.2ps	[2]			+0.7		
${}^{187}_{74}\mathrm{W}_{113}$		24h	3/2					
${}^{183}_{75}\mathrm{Re}_{108}$		70d	[5/2]	±3.1				
${}^{183}_{75}\mathrm{Re}_{108}$	496	7.89ns	[9/2]	±5.3				
${}^{184}_{75}\mathrm{Re}_{109}$		38d	[3]	±2.5				
${}^{184}_{75}\mathrm{Re}_{109}$	188	165d	[8]	±2.9				
${}^{185}_{75}\mathrm{Re}_{110}$			5/2	+3.172	28	+2.3[r]		
${}^{186}_{75}\mathrm{Re}_{111}$		90h	1	+1.73	1	~±0.4[r]		
${}^{187}_{75}\mathrm{Re}_{112}$		60Gy	5/2	+3.204	28	+2.2[S]		
${}^{187}_{75}\mathrm{Re}_{112}$	206	560ns	[9/2]	+4.8				
${}^{188}_{75}\mathrm{Re}_{113}$		17h	1	+1.78	1	~±0.4[r]		
${}^{184}_{76}\mathrm{Os}_{108}$	?	?	[2]			−2		
${}^{186}_{76}\mathrm{Os}_{110}$	137	840ps	[2]	+0.61	1	+1.5		
${}^{187}_{76}\mathrm{Os}_{111}$			1/2	+0.0643	6			
${}^{188}_{76}\mathrm{Os}_{112}$	155	710ps	[2]	+0.55		−0.4;−1.3		
${}^{188}_{76}\mathrm{Os}_{112}$	633	5.6ps	[2]	+0.9				
${}^{189}_{76}\mathrm{Os}_{113}$			3/2	+0.6565	59	+0.8*		
${}^{190}_{76}\mathrm{Os}_{114}$	187	350ps	[2]	+0.68	1	+0.3;−1.0		
${}^{190}_{76}\mathrm{Os}_{114}$	548	28ps	[4]	+0.9				
${}^{192}_{76}\mathrm{Os}_{116}$	206	280ps	[2]	+0.78	1	+1.2;−0.4		
${}^{192}_{76}\mathrm{Os}_{116}$	489	28ps	[2]	±0.7				
${}^{191}_{77}\mathrm{Ir}_{114}$			3/2	+0.1454	14	+1.1[r]		

Summary of Nuclear Moment Values (continued)

Nucleus	Level (keV)	$T_{1/2}$	I $(h/2\pi)$	μ (nm)	Diam. Corr.	Q (b)	Ω (nmb)	Q_4 (b^2)
$^{191}_{77}Ir_{114}$	82	3.8ns	[1/2]	+0.546	5			
$^{191}_{77}Ir_{114}$	129	131ps	[5/2]	+0.5				
$^{191}_{77}Ir_{114}$	171	4.9s	[11/2]	±6.1	1			
$^{192}_{77}Ir_{115}$		74d	4	+1.90	2			
$^{193}_{77}Ir_{116}$			3/2	+0.1583	15	+1.0[S]		
$^{193}_{77}Ir_{116}$	73	6.2ns	1/2	+0.468	4			
$^{193}_{77}Ir_{116}$	139	90ps	[5/2]	+0.6				
$^{194}_{77}Ir_{117}$		17h	1	±0.37				
$^{192}_{78}Pt_{114}$	316	35ps	[2]	+0.93[l]	1			
$^{192}_{78}Pt_{114}$	612	20ps	[2]	+1.0[k]				
$^{192}_{78}Pt_{114}$	785	12ps	[4]	±0.9[k]				
$^{194}_{78}Pt_{116}$			0♦					
$^{194}_{78}Pt_{116}$	328	35ps	[2]	+0.6		+0.6;+0.9		
$^{194}_{78}Pt_{116}$	622	44ps	[2]	+0.4				
$^{195}_{78}Pt_{117}$			1/2	+0.6022	56			
$^{195}_{78}Pt_{117}$	99	160ps	[3/2]	−0.60	1			
$^{195}_{78}Pt_{117}$	210	67ps	[3/2]	+0.3				
$^{195}_{78}Pt_{117}$	240	230ps	[5/2]	+0.22				
$^{195}_{78}Pt_{117}$	259	4.1d	[13/2]	±0.60	1			
$^{196}_{78}Pt_{118}$			0♦					
$^{196}_{78}Pt_{118}$	356	35ps	[2]	+0.55		+0.5;+0.6		
$^{197}_{78}Pt_{119}$		20h	1/2	±0.5[P]				
$^{198}_{78}Pt_{120}$	408	19ps	[2]	+0.5		+1.2		
$^{190}_{79}Au_{111}$		40m	1	±0.066	1			
$^{191}_{79}Au_{112}$		3.0h	3/2	±0.137	1			
$^{192}_{79}Au_{113}$		4.1h	1	±0.0079	1			
$^{193}_{79}Au_{114}$		18h	3/2	±0.139	1			
$^{194}_{79}Au_{115}$		39h	1	±0.074	1			
$^{195}_{79}Au_{116}$		192d	3/2	±0.147	1			
$^{196}_{79}Au_{117}$		6.2d	2	+0.588	6			
$^{196}_{79}Au_{117}$	596	9.7h	12	±5.4				
$^{197}_{79}Au_{118}$			3/2	+0.14486	137	+0.59	~+0.01	
$^{197}_{79}Au_{118}$	77	1.9ns	[1/2]	+0.42				
$^{198}_{79}Au_{119}$		2.7d	2	+0.590	6			
$^{198}_{79}Au_{119}$	367	123ns	[3]	±3.6				
$^{198}_{79}Au_{119}$	?	49h	[12?]	±5.6	1			
$^{199}_{79}Au_{120}$		3.2d	3/2	+0.270	3			
$^{200}_{79}Au_{121}$	?	18.7h	[12]	±6.1	1			
$^{183}_{80}Hg_{103}$		8.8s	1/2	+0.52				
$^{185}_{80}Hg_{105}$		50s	1/2	+0.50				
$^{187}_{80}Hg_{107}$		2.4m	3/2	−0.59	1	−0.3		
$^{193}_{80}Hg_{113}$		6h	3/2	−0.6236	60	−1[r]		
$^{193}_{80}Hg_{113}$	140	11h	13/2	−1.052	10	+1.2[r]		
$^{195}_{80}Hg_{115}$		9.5h	1/2	+0.538	5			
$^{195}_{80}Hg_{115}$	176	40h	13/2	−1.038	10	+1.2[r]		

Summary of Nuclear Moment Values (continued)

Nucleus	Level (keV)	$T_{1/2}$	I ($h/2\pi$)	μ (nm)	Diam. Corr.	Q (b)	Ω (nmb)	Q_4 (b^2)
$^{197}_{80}Hg_{117}$		65h	1/2	+0.5241	51			
$^{197}_{80}Hg_{117}$	134	7.3ns	[5/2]	+0.96	1			
$^{197}_{80}Hg_{117}$	299	24h	13/2	−1.0214	99	+1.4[r]		
$^{198}_{80}Hg_{118}$			0♦					
$^{198}_{80}Hg_{118}$	412	22.0ps	[2]	+1.1				
$^{199}_{80}Hg_{119}$			1/2	+0.50271[d] (+0.50415)	485			
$^{199}_{80}Hg_{119}$	158	2.32ns	[5/2]	+1.0		±0.7		
$^{199}_{80}Hg_{119}$	533	44m	13/2	±1.0083	97	+2		
$^{200}_{80}Hg_{120}$			0♦					
$^{200}_{80}Hg_{120}$	368	42ps	[2]	+0.9				
$^{201}_{80}Hg_{121}$			3/2	−0.55671[d] (−0.55830)	537	+0.44[S]	−0.13	
$^{202}_{80}Hg_{122}$			0♦					
$^{202}_{80}Hg_{122}$	439	26ps	[2]	+1.2				
$^{203}_{80}Hg_{123}$		47d	5/2	+0.86	1	+0.5		
$^{204}_{80}Hg_{124}$			0♦					
$^{204}_{80}Hg_{124}$	437	46ps	[2]	+0.8				
$^{205}_{80}Hg_{115}$		5.5m	1/2	+0.597	6			
$^{193}_{81}Tl_{112}$		23m	1/2					
$^{194}_{81}Tl_{113}$		33m	2	±0.135	1			
$^{195}_{81}Tl_{114}$		1.2h	1/2	+1.57	2			
$^{196}_{81}Tl_{115}$		1.8h	2	±0.0699	7			
$^{197}_{81}Tl_{116}$		2.7h	1/2	+1.56	2			
$^{198}_{81}Tl_{117}$		5.3h	2	±0.00121	1			
$^{198}_{81}Tl_{117}$	544	1.8h	7	±0.64	1			
$^{199}_{81}Tl_{118}$		7.4h	1/2	+1.62	2			
$^{200}_{81}Tl_{119}$		26h	2	±0.03568	35			
$^{201}_{81}Tl_{120}$		72h	1/2	+1.65	2			
$^{202}_{81}Tl_{121}$		12d	2	±0.0565	6			
$^{202}_{81}Tl_{121}$	950	560μs	[7]	±0.90	1			
$^{203}_{81}Tl_{122}$			1/2	+1.6115	158			
$^{203}_{81}Tl_{122}$	279	280ps	[3/2]	+0.16				
$^{204}_{81}Tl_{123}$		3.9y	2	±0.089	1			
$^{205}_{81}Tl_{124}$			1/2	+1.6274	160			
$^{206}_{81}Tl_{125}$		4.2m	0♦					
$^{204}_{82}Pb_{122}$	1274	260ns	[4]	+0.22		±0.3		
$^{205}_{82}Pb_{123}$	1014	5.55ns	[13/2]	−0.98	1			
$^{206}_{82}Pb_{124}$			0♦					
$^{206}_{82}Pb_{124}$	803	6ps	[2]	~0				
$^{206}_{82}Pb_{124}$	2200	123μs	[7]	−0.152	1			
$^{206}_{82}Pb_{124}$	2385	29ps	[6]	+0.8				
$^{206}_{82}Pb_{124}$	4027	200ns	[12]	−1.86	2			
$^{207}_{82}Pb_{125}$			1/2	+0.5783[d] (+0.5881)	58			
$^{207}_{82}Pb_{125}$	570	129ps	[5/2]	+0.8				

Summary of Nuclear Moment Values (continued)

Nucleus	Level (keV)	$T_{1/2}$	I ($h/2\pi$)	μ (nm)	Diam. Corr.	Q (b)	Ω (nmb)	Q_4 (b^2)
$^{208}_{82}Pb_{126}$			0♦					
$^{208}_{82}Pb_{126}$	2615	15ps	[3]	+1.8		−1.1		
$^{208}_{82}Pb_{126}$	3198	298ps	[5]	+0.10				
$^{199}_{83}Bi_{116}$		25m	9/2					
$^{200}_{83}Bi_{117}$		35m	7					
$^{201}_{83}Bi_{118}$		1.8h	9/2					
$^{202}_{83}Bi_{119}$		1.6h	5					
$^{203}_{83}Bi_{120}$		12h	9/2	+4.59	5	−0.71[r]		
$^{204}_{83}Bi_{121}$		12h	6	+4.25	4	−0.46[r]		
$^{205}_{83}Bi_{122}$		15d	9/2	~+5.5				
$^{206}_{83}Bi_{123}$		6.3d	6	+4.56	5	−0.21[r]		
$^{207}_{83}Bi_{124}$	2102	182μs	[21/2]	+3.4				
$^{209}_{83}Bi_{126}$		>2Ay	9/2	+4.080	41	−0.38[s]	+0.5	
$^{210}_{83}Bi_{127}$		5d	1	−0.0442	4	+0.14[r]		
$^{211}_{83}Bi_{128}$	405	318ps	[7/2]	+4.4				
$^{201}_{84}Po_{117}$		18m	3/2					
$^{202}_{84}Po_{118}$		51m	0♦					
$^{203}_{84}Po_{119}$		42m	5/2					
$^{204}_{84}Po_{120}$		3.5h	0♦					
$^{204}_{84}Po_{120}$	~1700	140ns	[8]	±7.8	1			
$^{205}_{84}Po_{121}$		1.8h	5/2	≈+0.26		+0.17		
$^{206}_{84}Po_{122}$		8.8d	0♦					
$^{206}_{84}Po_{122}$	?	212ns	[8]	±7.4	1			
$^{207}_{84}Po_{123}$		6.0h	5/2	≈+0.27		+0.28	+0.11	
$^{207}_{84}Po_{123}$	1115	47μs	[13/2]	−0.93	1			
$^{208}_{84}Po_{124}$	1530	380ns	[8]	±7.3	1			
$^{209}_{84}Po_{125}$		103y	1/2	+0.77	1			
$^{209}_{84}Po_{125}$	>1327	~100ns	[17/2?]	+7.5	1			
$^{210}_{84}Po_{126}$		138d	0♦					
$^{210}_{84}Po_{126}$	1472	38ns	[6]	±5.6	1			
$^{210}_{84}Po_{126}$	1552	110ns	[8]	+7.3	1			
$^{210}_{84}Po_{126}$	~2800	24ns	[11]	+12.0	1			
$^{210}_{84}Po_{126}$	4372	93ns	[13]	±7.1	1			
$^{211}_{84}Po_{127}$	1064	16ns	[15/2]	±0.4				
$^{211}_{85}At_{126}$		7.2h	9/2					
$^{211}_{85}At_{126}$	1416	50ns	[21/2]	±9.4	1			
$^{211}_{85}At_{126}$	4816	4.2μs	[39/2	±14				
			or 41/2]	±15				
$^{212}_{86}Rn_{126}$	~1700	1.0μs	[8]	±7.2	1			
$^{222}_{86}Rn_{136}$	186	320ps	[2]	+0.9				
$^{223}_{88}Ra_{135}$	50	630ps	[3/2]	+0.42				
$^{227}_{89}Ac_{138}$		22y	3/2	+1.1		+1.7		

Summary of Nuclear Moment Values (continued)

Nucleus	Level (keV)	$T_{1/2}$	I $(h/2\pi)$	μ (nm)	Diam. Corr.	Q (b)	Ω (nmb)	Q_4 (b^2)
$^{229}_{90}Th_{139}$		7.3ky	5/2	+0.38		~+4.6		
$^{231}_{91}Pa_{140}$		34ky	3/2	±1.98	2			
$^{233}_{91}Pa_{142}$		27d	3/2	+3.4		−3.0		
$^{233}_{92}U_{141}$		162ky	5/2	+0.64	1	+4.2[S]		
$^{235}_{92}U_{143}$		710My	7/2	−0.43		+4.9[r]		
$^{237}_{93}Np_{144}$		2.1My	5/2	+2.4[l]		positive		
$^{237}_{93}Np_{144}$	60	63ns	[5/2]	+1.3[k]		Q/Q_{gs}=+1.0		
$^{238}_{93}Np_{145}$		2.1d	2					
$^{239}_{93}Np_{146}$		2.3d	5/2					
$^{239}_{93}Np_{146}$	75	1.40ns	[5/2]	+1.3[k]				
$^{239}_{94}Pu_{145}$		24ky	1/2	+0.200[l]	2			
$^{241}_{94}Pu_{147}$		13y	5/2	−0.68[k]	1	+5.6		
$^{241}_{95}Am_{146}$		460y	5/2	+1.59	2	+4.9		
$^{242}_{95}Am_{147}$		16h	1	+0.383	5	−2.8		
$^{243}_{95}Am_{148}$		8ky	5/2	+1.59	2	+4.9		
$^{242}_{96}Cm_{146}$		160d	0♦					
$^{243}_{96}Cm_{147}$		28y	5/2	±0.4				
$^{245}_{96}Cm_{149}$		8.26ky	7/2	±0.5				
$^{247}_{96}Cm_{151}$		15.4My	9/2	±0.4				
$^{249}_{97}Bk_{152}$		314d	7/2	±5		±5		
$^{253}_{99}Es_{154}$		20.5d	7/2	+4.0		+6		

♦ No hyperfine structure observed

* Polarization or Sternheimer corrections included

[a] Weighted average

[b] Wide spread in tabulated values

[c] Atomic beam value of [59St46] adopted

[d] OP and NMR values discrepant. Values in ( )'s based on NMR values

[E] ENDOR and NMR values discrepant. Values in ( )'s based on NMR values

[f] ABMR and NMR values discrepant. Values in ( )'s based on NMR values

[g] No diamagnetic correction added. Not certain of corrections used by authors or if corrected.

[h] ABMR and ENDOR values discrepant. Values in [ ]'s based on ENDOR values

[i] Mössbauer and PAC values discrepant

[j] In the latest adjustment of fundamental constants [73CoTa], this value has been increased to 2.7928456 *11*.

[k] Relative value calculated from μ−ratio and μ^l

[l] Summary value upon which relative μ−values depend

[p] Preliminary value from meeting abstract, report, thesis or private communication

[r] Relative value calculated from Q−ratio and Q^S

[S] Summary value, average of tabulated values unless otherwise marked

Explanation of the Summary Table

Nucleus — Chemical symbol with Z−, A−, and N−numbers

Level — The energy of the nuclear level, in keV, given to identify the level for which nuclear moment information is presented
Values have been taken from the Nuclear Data Sheets (through B5), Table of Isotopes [67LeHo] or the experimenter's quoted value.

$T_{1/2}$ — The half-life of the radioactive nucleus or excited level
Values have been taken from Marelius [68Ma49], Nuclear Data Sheets (through B5), Table of Isotopes [67LeHo] or the experimenter's value.

I — Nuclear spin or angular momentum in units of $h/2\pi$
Values enclosed in brackets, [ ], were not determined by spectroscopic or resonance measurements but were assumed in order to interpret data.

μ — Nuclear magnetic moments in nuclear magnetons, with diamagnetic correction
When ratios are given, the level-energy of the reference isotope, in keV, is indicated by a subscript.

Diam. Corr. — The diamagnetic correction which was added to the last significant figure of the uncorrected magnetic dipole moment to get the value quoted in the previous column
For example, for Li^6, $\mu = \mu_{uncorrected}$ + Diam. corr. = +0.82195 + 0.00008 = +0.82203

Q — Nuclear electric quadrupole moment in barns
Values marked by "s" are averages of the Q−values listed in the individual tables.

Those marked by "r" have been calculated by use of Q^s and measured Q−ratios.

Values marked by "*" include polarization or Sternheimer corrections.

Values enclosed in brackets, [ ], are derived from electron−scattering experiments and are model-dependent.

When two values of Q are listed, the first refers to the value determined by Coulomb excitation reorientation assuming constructive interference and the second, destructive interference of the matrix elements.

When ratios are given, the level−energy of the reference isotope, in keV, is indicated by a subscript.

Ω — Nuclear magnetic octupole moment in nm−barns
Values enclosed in brackets, [ ], are derived from electron−scattering experiments and are model dependent.

Q_4 — Nuclear electric hexadecapole moment in $barns^2$

Neutron, Proton, and Anti-Proton Moments

I	μ	Quantity Measured	Method (Compound)
		Neutron	
1/2			Slow neutron scattering from ortho- and para- H_2
	−2 *1*		Non-adiabatic transitions in a rotating magnetic field
	±1.913002 *80*	ω_n/ω_p=0.685001 *30*	Neutron beam resonance; NMR (H_2O)

Neutron, Proton, and Anti-Proton Moments (continued)

I	μ	Quantity Measured	Method (Compound)
1/2			Reflection from magnetized iron
	negative	μ_p/μ_n negative	Neutron beam resonance; NMR (H_2O)
1/2			Neutron beam resonance
	−1.913159 *47*	ω_n/ω_p=0.685057 *17*	Neutron beam resonance and proton resonance over same field (H_2O)
		Proton	
1/2			Specific heat
1/2			Band spectra
1/2			Band spectra
	+2.785 *2*		Molecular beam magnetic resonance (H_2,HD)
	±2.79283 *11*	$g/g_J(\mathrm{Cs}, {}^2S_{1/2})=15.1911\times10^{-4}$ $g/g_J(\mathrm{In}, {}^2P_{1/2})=45.6877\times10^{-4}$ μ_p=(15.2106[a] *6*)$10^{-4}\mu_B$	Atomic and molecular beam magnetic resonance (NaOH)
	±2.79249 *20*	ω_p/ω_{cyc}=2.79242 *20*	NMR (H_2O); decelerating cyclotron
	positive		NMR (H_2O)
	±2.79288 *6* ±2.79292 *6* ±2.79292 *7* ±2.79291 *4*	γ_p=(2.67523 *6*)$10^8\,\mathrm{r{\cdot}s^{-1}T^{-1}}$ γ_p=(2.67534[aC] *6*)$10^8\,\mathrm{r{\cdot}s^{-1}T^{-1}}$ γ_p=(2.67527[C] *6*)$10^8\,\mathrm{r{\cdot}s^{-1}T^{-1}}$ γ_p=(2.675231[fgh] *26*)$10^8\,\mathrm{r{\cdot}s^{-1}T^{-1}}$	NMR in magnetic field determined by force on straight wire carrying known current (H_2O); strong field measurement
	±2.79274 *4* ±2.79275 *4*	ω_e/ω_p=657.475 *8* μ_p=(15.20970 *18*)$10^{-4}\mu_B$ μ_p=(15.21016[a] *19*)$10^{-4}\mu_B$	NMR (oil); measured cyclotron frequency of free electrons
	±2.792787 *17*	$g_J(\mathrm{H}, {}^2S_{1/2})/g_p$=658.2171 *6* μ_p=(15.210355[a] *13*)$10^{-4}\mu_B$	ABMR and NMR (oil) in same field
	±2.792764 *60* ±2.792763 *60*	ω_p/ω_{cyc}=2.792685 *60* ω_p/ω_{cyc}=2.792690[f] *60*	NMR (oil); omegatron
	±2.792784 *17*	$g_J(\mathrm{H})/g$=658.21734 *19* μ_p=(15.2103347[ai] *65*)$10^{-4}\mu_B$	Mic; NMR (cylinder of oil)
	±2.79281 *4*	ω_p/ω_{cyc}=2.79273 *4*	NMR (H_2O); decelerating cyclotron
	±2.79275 *10*	ω_p/ω_{cyc}=2.792675 *100*	NMR ($H_2O+FeCl_3$); decelerating cyclotron

I	μ	Quantity Measured	Method (Compound)
	±2.79315 *8*	γ_p=(2.67549 *8*)$10^8 r\cdot s^{-1}T^{-1}$	NMR in a field determined by the dimensions of and current in an iron−free coil (H_2O)
		g_J(D)/g=658.2162 *8*	Mic; NMR (cylinder of oil)
	±2.792788 *17*	μ_p=(15.210360[ai] *23*)$10^{-4}\mu_B$	
	±2.79277 *3*	γ_p=(2.67513 *2*)$10^8 r\cdot s^{-1}T^{-1}$	Free precession in the field of a
	±2.79277 *2*	γ_p=(2.675192[aC] *8*)$10^8 r\cdot s^{-1}T^{-1}$	standard solenoid (H_2O)
	±2.79278 *2*	γ_p=(2.675137[C] *11*)$10^8 r\cdot s^{-1}T^{-1}$	See 68Dr06 below
		ω_e/ω_p=657.4501	NMR (H_2)
	±2.79277 *3*	μ_p=(15.210280[ai] *12*)$10^{-4}\mu_B$	
		g_J(H)/g=658.215909[f] *44*	Mic (H+H_2+buffer gas)
	±2.792781 *17*	μ_p=(15.2099284[fj] *10*)$10^{-4}\mu_B$	
	±2.79280 *2*	ω_e/ω_p=657.462 *3*	NMR (oil); measured cyclotron
		μ_p=(15.21000 *7*)$10^{-4}\mu_B$	frequency of free electrons
	±2.79281 *2*	ω_e/ω_p=657.4620 *45*	
		μ_p=(15.21046[a] *10*)$10^{-4}\mu_B$	
	±2.79290 *6*	$\omega_{cyc}(H_2^+)/\omega_d$=1.65957 *28*	cyclotron frequency of H_2^+ and NMR ($D_2O+CuCl_2\cdot 2H_2O$) in same field;
		ω_p/ω_d=6.514411 *3*	NMR ($D_2O+CuCl_2\cdot 2H_2O$; $H_2O+CuCl_2$) Used $M_p/M(H_2^+)$=0.49986388 *50*
	±2.79288 *10*	γ_p=(2.67530[a] *10*)$10^8 r\cdot s^{-1}T^{-1}$	NMR in known magnetic field
	±2.79283 *10*	γ_p=(2.67525[aC] *10*)$10^8 r\cdot s^{-1}T^{-1}$	
	±2.79288 *10*	γ_p=(2.67523[C] *10*)$10^8 r\cdot s^{-1}T^{-1}$	
	±2.79277[b] *5*	ω_p/ω_{cyc}=2.792676 *50*	NMR ($H_2O+MnSO_4$); decelerating
	±2.79277 *7*	ω_p/ω_{cyc}=2.79270[f‡] *7*	cyclotron ‡Includes additional corrections
	±2.79280 *2*	ω_e/ω_p=657.4621 *25*	NMR (liquid paraffin); measured
		μ_p=(15.21043[a] *6*)$10^{-4}\mu_B$	cyclotron frequency of free electrons
	±2.79281 *15*	μ_p=(15.21046[a] *84*)$10^{-4}\mu_B$	
	±2.79277 *2*	γ_p=(2.67513 *1*)$10^8 r\cdot s^{-1}T^{-1}$	NMR in weak magnetic field (H_2O)
	±2.79276 *2*	γ_p=(2.675188[aC] *8*)$10^8 r\cdot s^{-1}T^{-1}$	
	±2.79278 *2*	γ_p=(2.675132[C] *8*)$10^8 r\cdot s^{-1}T^{-1}$	
	±2.79276 *3*	γ_p=(2.675144[fs] *16*)$10^8 r\cdot s^{-1}T^{-1}$	
	±2.79286 *2*	μ_p=2.79279 *2*	cyclotron frequency of $^4He^+$, $^{20}Ne^{2+}$.
		μ_p=2.792794[f] *17*	$^{20}Ne^+$; NMR ($H_2O+CuCl_2$)
	+3.0 *3*	γ_p=−(2.9 *3*)$10^8 r\cdot s^{-1}T^{-1}$	NMR in weak rotating rf field (H_2O)
	±2.792782 *17*	$g_J(H,{}^2S_{1/2})/g$=658.21049 *20*	MASER
		μ_p=(15.210326 *4*)$10^{-4}\mu_B$	
		g_J(H)/g=658.21053[d] *20*	See 70Wi22 below

Neutron, Proton, and Anti-Proton Moments (continued)

I	μ	Quantity Measured	Method (Compound)
	±2.79273	$\gamma_p=(2.675071)10^8 r\cdot s^{-1}T^{-1}$	NMR in strong field (H_2O)
	±2.79274[c]	$\gamma_p=(2.67510$[c]$)10^8 r\cdot s^{-1}T^{-1}$	
	±2.79277 *4*	$\gamma_p=(2.675105$[fg] *20*$)10^8 r\cdot s^{-1}T^{-1}$	
	±2.79267 *12*		
	±2.79267[d] *13*	$\mu_p=2.79260$[df] *13*	Determined from existing time–of–flight and magnetic analysis data on $^{27}Al(p,\gamma)^{28}Si$ resonanace and $^7Li(p,n)^7Be$ threshold
	±2.79281 *5*	$\omega_p/\omega_{cyc}=2.79274$ *5*	NMR (H_2O); omegatron (H_2^+; HD^+; D_2^+)
	±2.792773	$\gamma_p=(2.6751526)10^8 r\cdot s^{-1}T^{-1}$	Free precession in the field of a standard solenoid (H_2O)
		$\gamma_p=(2.6751465$[fg]$)10^8 r\cdot s^{-1}T^{-1}$	[Fredericksburg, Va. 1958]
		$\gamma_p=(2.6751555$[fg]$)10^8 r\cdot s^{-1}T^{-1}$	[Washington, D.C., 1960–1967]
		$\gamma_p=(2.6751526$[fg]$)10^8 r\cdot s^{-1}T^{-1}$	[Gaithersburg, Md. 1968]
	±2.792773 *30*	$\gamma_{p,ave}=(2.6751525$[fg]*99*$)10^8 r\cdot s^{-1}T^{-1}$	
	±2.79276 *3*	$\gamma_p=(2.675138$ *11*$)10^8 r\cdot s^{-1}T^{-1}$	Free precession in an air–core field–coil system (H_2O)
		$\gamma_p=(2.6751392$[fg]*86*$)10^8 r\cdot s^{-1}T^{-1}$	
	±2.792782 *18*	$\omega_p/\omega_e=(15.2099441$ *11*$)10^{-4}$	NMR (H_2O+0.2M $CuSO_2$ in cylinder); cyclotron resonance of electron
		$\omega_p/\omega_e=(15.210329$ *9*$)10^{-4}$	
	±2.79278 *3*	$\gamma_p=(2.675162$ *14*$)10^8 r\cdot s^{-1}T^{-1}$	Free precession in field of Helmholtz coils (H_2O)
	±2.79275	$\gamma_p=(2.6751349$[fg]$)10^8 r\cdot s^{-1}T^{-1}$	
		$\gamma_p=(2.67512)10^8 r\cdot s^{-1}T^{-1}$	Value adopted by BIPM for spherical sample of H_2O
	±2.792766 *15*	$g_J(H)/g_p(H)=658.210705$[j] *6*	MASER (atomic H)
		Anti–Proton	
	−1.8 *12*		Double scattering

[a] Includes diamagnetic correction
[b] Includes diamagnetic and paramagnetic corrections
[c] Corrected for standard BIPM ampere
[d] Includes corrections sent by experimenters to compilers 65Co20 or 69TaPa
[e] Value corrected for solution concentration, shape of holder, and shielding of electrons and neighboring molecules
[f] For spherical container of H_2O
[g] In terms of NBS as–maintained Ampere
[h] Includes corrections for better values of "g", the acceleration of gravity, and the diamagnetic correction
[i] Calculated using $\mu_e=1.001159639\mu_B$, $g_J=g_S(1-17.75$ ppm$)$, $\sigma(H_2)=26.6$ *3* ppm, $\sigma(H_20)=26.0$ *3* ppm, and σ(oil)=29.7 *6* ppm
[j] Calculated using $g_J(H)/g_p(H)=(g_S/g_p)(1-0.204$ ppm$)$.

I	Nuclear spin, in units of $h/2\pi$
μ	Nuclear magnetic dipole moment, in nuclear magnetons, given with diamagnetic correction
Quantity Measured	Directly measured frequency ratio, magnetic moment, or gyromagnetic ratio in appropriate units These values are given without diamagnetic correction unless otherwise indicated.
Method (Compound)	Method by which frequency ratio, moment, or gyromagnetic ratio was measured Compound used for measurement

MOSSBAUER MEASUREMENTS OF NUCLEAR MOMENTS

Summary of adopted values

Isotope	Energy	Spin	R_μ	μ	R_Q	Q
^{57}Fe	14.4	3/2	- 1.7142 (4)	- 0.15532 (4)		+ 0.209 (5)
^{61}Ni	67.4	5/2	- 0.637 (11)	+ 0.478 (7)	- 1.21 (13)	- 0.20 (3)
^{67}Zn	93.3	1/2	+ 0.66 (3)	+ 0.58 (3)		
^{83}Kr	9.4	7/2	+ 0.971 (2)	- 0.943 (2)	1.70 (2)	+ 0.430 (3)[U]
^{99}Ru	89.3	3/2	0.456 (2)	- 0.285 (5)	+ 2.88 (4)	0.3 (7[U]
^{99}Tc	140.5	7/2	+ 0.64 (16)	+ 3.6 (8)		
^{119}Sn	23.8	3/2	- 0.605 (17)	+ 0.633 (18)		- 0.065 (5)[U]
	89.0	11/2		1.40 (8)		- 0.14 (1)[U]
^{121}Sb	0	5/2				- 0.28 (6)[U]
	37.2	7/2	0.735 (9)	+2.47 (3)	1.34 (1)	- 0.38 (8)[U]
^{125}Te	35.5	3/2	- 0.69 (6)	+ 0.61 (5)		- 0.200 (23)[U]
^{127}I	57.6	7/2	0.905 (16)	2.54 (4)	- 0.896 (2)	- 0.71 (9)[U]
^{129}I	27.8	5/2	1.0687 (11)	+ 2.797 (3)	+ 1.2380 (15)	- 0 68 (6)[U]
^{129}Xe	39.6	3/2	- 0.75 (12)	+ 0.58 (9)		- 0.41 (4)
^{133}Cs	81.0	5/2	+ 1.335 (8)	+3.443 (21)		
^{141}Pr	145.2	7/2	+ 0.69(2)	+ 2.87 (9)		
^{145}Nd	72.5	5/2	+ 0.489 (4)	- 0.320 (3)		
^{147}Pm	91.0	5/2	1.373 (2)	3.60 (5)	0.8 (4)	0.6 (3)[U]
^{147}Sm	122.1	5/2	+ 0.551 (31)	- 0.448 (25)	+ 1.7 (7)	- 0.31(12)[U]
^{149}Sm	0	7/2				0.060 (15)[U]
	22.5	5/2	+ 0.929 (1)	- 0.623 (6)	8.3 (21)	0.50 (1)[U]
^{152}Sm	121.8	2		0.84 (5)		
^{154}Sm	81.9	2		0.78 (4)		- 1.3 (5)
^{151}Eu	21.6	7/2	0.7465 (6)	2.5865 (26)	1.312 (19)	+ 1.50 (7)[U]
^{153}Eu	83.4	7/2	+ 1.18 (4)	+ 1.80 (6)		
	97.4	5/2	+ 2.10 (15)	+ 3.21 (23)		
	103.2	3/2	+ 1.336 (3)	2.043 (5)	0.520 (3)	1.51 (6)[U]
^{155}Gd	86.5	5/2	+ 2.049 (20)	- 0.529 (5)	0.14 (5)	0.22 (8)[U]
	105.3	3/2	- 0.55 (5) or	+ 0.14 (4) or	0.90 (3)	1.57 (16)[U]
			+ 1.80 (18)	- 0.47 (7)		
^{156}Gd	88.9	2		0.778 (7)		- 2.40 (24)
^{157}Gd	0	3/2				2.00 (26)[U]
	64.0	5/2	+ 1.515 (7)	0.5133 (26)	1.79 (2)	3.6 (5)[U]
^{158}Gd	79.5	2		0.77 (4)		1.30 (14)
^{160}Gd	75.3	2				1.34 (14)
^{159}Tb	58.0	5/2	0.80 (5) or	+ 1.60 (10) or		
			1.15 (5)	+ 2.29 (10)		

Summary of adopted values-continued

Isotope	Energy	Spin	R_μ	μ	R_Q	Q
^{160}Dy	86.8	2		0.77(4)		
^{161}Dy	25.6	5/2	- 1.2368 (18)	+ 0.592 (6)	+ 0.9996 (4)	+ 2.34 (16)
	43.8	7/2	+ 0.293 (10)	- 0.140 (5)	0.21 (5)	0.49 (12)
	74.6	3/2	+ 0.840 (7)	- 0.403 (6)	+ 0.59 (3)	+ 1.36 (11)
^{162}Dy	80.7	2		+ 0.74 (8)		
^{164}Dy	73.4	2		0.69 (3)		- 1.95 (21)
^{165}Ho	94.7	9/2	0.99 (4)	4.08 (16)		
^{164}Er	91.5	2		0.694 (15)		
^{166}Er	80.6	2		0.629 (10)		- 1.59 (15)^{U}
^{168}Er	79.8	2		0.656 (13)		
^{170}Er	79.3	2		0.630 (13)		- 1.67 (30)^{U}
^{169}Tm	8.4	3/2	- 2.31(4)	+ 0.534 (10)		- 1.21(8)
^{170}Yb	84.3	2		0.669 (8)		
^{171}Yb	66.7	3/2	0.710 (5)	0.349 (2)		- 1.59 (3)^{U}
	75.9	5/2	2.055 (10)	1.011 (5)		- 2.21 (4)^{U}
^{172}Yb	78.7	2		0.664 (8)		- 2.17 (4)^{U}
^{174}Yb	76.5	2		0.672 (5)		- 2.14 (4)^{U}
^{176}Yb	82.1	2		0.76 (4)		- 2.24^{U}
^{176}Hf	88.4	2				- 2.08^{U}
^{178}Hf	93.2	2				- 1.94 (4)^{U}
^{180}Hf	93.3	2				- 1.92 (2)^{U}
^{181}Ta	6.2	9/2	+ 2.23 (3)	+ 5.24 (7)	+ 1.133 (10)	+ 4.4 (5)^{U}
^{180}W	103.0	2				- 1.82 (4)^{U}
^{182}W	100.1	2		0.512 (25)		
^{183}W	46.5	3/2		- 0.1 (1)		- 1.5
	99.1	5/2		0.92 (3)		- 1.63 (7)^{U}
^{184}W	111.2	2		0.58 (2)		- 1.72 (3)^{U}
^{186}W	122.5	2		0.62 (2)		- 1.65 (3)^{U}
^{186}Os	137.2	2		0.58 (3)		- 1.50 (10)^{U}
^{188}Os	155.0	2		0.63 (3)		- 1.36 (9)^{U}
^{189}Os	0	3/2				0.94 (7)^{U}
	36.3	1/2	0.34 (4)	0.23 (3)		
	69.6	5/2	1.502 (12)	0.986 (8)	- 0.735 (12)	- 0.6 (2)^{U}
^{190}Os	187.0	2				- 1.17 (11)^{U}
^{191}Ir	0	3/2				+ 0.78 (20)^{U}
	82.4	1/2	+ 3.71 (3)	+ 0.540 (5)		
	129.5	5/2	+ 3.8 (3)	+ 0.55 (5)		
^{193}Ir	73.1	1/2	2.958 (6)	0.4683 (20)		
^{195}Pt	98.8	3/2	- 1.01 (8)	- 0.61 (5)		
	129.8	5/2	1.35 (30)	0.854 (19)		

Summary of adopted values-continued

Isotope	Energy	Spin	R_μ	μ	R_Q	Q
^{197}Au	77.3	1/2	2.875 (22)	0.416 (3)		
^{236}U	45.2	2				-2.95 (14)[u]
^{238}U	44.9	2				-3.23 (17)[u]
^{237}Np	0	5/2				+4.1 (7)[u]
	59.5	5/2	+0.535 (4)	+1.34 (12)	0.99 (1)	+4.1 (7)[u]
^{243}Am	84.0	5/2			+0.962 (15)	+4.7[u]

[u]Uncorrected for the Sternheimer effect.

MICROWAVE SPECTRA OF DIATOMIC MOLECULES

Equilibrium Internuclear Distance

Isotopic Species	r_e(Å)[a]	r_e^{BO}(Å)[b]
AgBr	2.393138(3)	
AgCl	2.2808190(2)	2.280815(3)
AgF	1.983203(1)	
AgI	2.544651(3)	
AlBr	2.2948860(4)	2.294833(45)
AlCl	2.1301910(3)	2.130136(21)
AlF	1.6543883(4)	
AlI	2.5371326(3)	
BF	1.2623(1)	
BaO	1.9397119(2)	
BrCl	2.136091(21)	
BrCs	3.0722875(24)	
BrF	1.755747(85)	
BrGa	2.352519(32)	
^{79}Br^{1}H	1.4146569(40)	1.414490(5)
^{79}Br^{2}H	1.4144535(40)	1.414491(5)
BrI	2.484801(72)	
BrIn	2.543221(25)	
BrK	2.820809(3)	
BrLi	2.021491(3) U 2.021504(3) D	
BrNa	2.5020676(9)	
BrO	1.7172(25)	
BrRb	2.9447792(12)	
BrTl	2.6182148(80)	
CO	1.12833632(7)	1.1282427(6)
CS	1.534960(30)	
CSe	1.676198(64)	1.676086(64)
ClCs	2.9063065(9)	

Equilibrium Internuclear Distance

Isotopic Species	r_e(Å)[a]	r_e^{BO}(Å)[b]
ClF	1.6283323(22)	
ClGa	2.2017159(22)	
^{35}Cl^{1}H	1.2745717(4)	1.2746181[c]
^{35}Cl^{2}H	1.2745940(6)	1.2746149(9)[d]
^{35}Cl^{3}H	1.2746022(11)	
ClI	2.3209049(60)	
ClIn	2.4011967(17)	
ClK	2.6666830(7)	
ClLi	2.0206913(2)	2.020700(8)[d] 2.020705(10)[c]
ClNa	2.3608225(12)	
ClO	1.596(1)	
ClRb	2.7867690(14)	
ClTl	2.4848554(5)	
CsF	2.3453792(5)	
CsI	3.3152313(10)	
CuF	1.7449508(17)	
FGa	1.7743900(7)	
F^{1}H	0.91682	0.916905(X)
F^{2}H	0.91737(14)	
FI	1.9097813(38)	
FIn	1.9854199(7)	
FK	2.1714824(2) U 2.1714777(2) D	
FLi	1.5638785(3)	1.563884(15)
FNa	1.9259692(2) U 1.9259648(30) D	
FRb	2.2703609(8)	
FS	1.6006(20)[e]	

Equilibrium Internuclear Distance

Isotopic Species	r_e(Å)[a]	r_e^{BO}(Å)[b]
FTl	2.0844623(1) U 2.0844557(10) D	
GaI	2.574689(31)	
GeO	1.6246670(13)	
GeS	2.0120982(4)	
GeSe	2.1346561(8)	
GeTe	2.3401928(9)	
^{1}HI	1.609128(18)	1.609042(X)
^{2}H^{7}Li	1.595271(16)	1.59492(2)
^{1}HO	0.97998(X)	
IIn	2.753672(19)	
IK	3.0478794(11) U 3.0478801(11) D	
ILi	2.391944(16)	
INa	2.7114844(19)	
IRb	3.1769183(45)	
ITl	2.813709(7)	

Equilibrium Internuclear Distance

Isotopic Species	r_e(Å)[a]	r_e^{BO}(Å)[b]
NO	1.15074(10)	
NP	1.4908839(4)	
NS	1.4941(3)	
O_2	1.207546(60)	
OPb	1.9218359(12)	
OS	1.4811046(37)	1.480985(40)
OSi	1.5097560(9)	
OSn	1.8325271(16)	
PbS	2.2868898(8)	
PbSe	2.4022637(16)	
PbTe	2.595006(3)	
SSi	1.92934401(13)	1.9292866(30)
SSn	2.2090528(3)	
SeSi	2.0583513(30)	
SeSn	2.3256287(25)	
SnTe	2.5228436(11)	

[a] Value of r_e deremined from the uncorrected Y_{01}. In several cases the Dunham correction has been applied to Y_{01} in order to obtain B_e. When both r_e values are given the uncorrected value is labeled U and the value including the Dunham correction is labeled D.
[b] Value of r_e which includes all corrections to Y_{01} caused by the breakdown in the Born-Oppenheimer (BO) approximation.
[c] Obtained by the application of Bunker's technique.
[d] Obtained by Watson's method.
[e] The value shown is r_0.

Binding Energies of Electronic Shells of Selected Elements

Atomic number	Element	Binding energy of shell (keV) K	L_I	L_{II}	L_{III}
1	Hydrogen	0.0136			
6	Carbon	0.283			
8	Oxygen	0.531			
11	Sodium	1.08	0.055	0.034	0.034
13	Aluminum	1.559	0.087	0.073	0.072
14	Silicon	1.838	0.118	0.099	0.098
19	Potassium	3.607	0.341	0.297	0.294
20	Calcium	4.038	0.399	0.352	0.349
26	Iron	7.111	0.849	0.721	0.708
29	Copper	8.980	1.100	0.953	0.933
31	Gallium	10.368	1.30	1.134	1.117
32	Germanium	11.103	1.42	1.248	1.217
39	Yttrium	17.037	2.369	2.154	2.079
42	Molybdenum	20,002	2.884	2.627	2.523
47	Silver	25.517	3.810	3.528	3.352
53	Iodine	33.164	5.190	4.856	4.559
54	Xenon	34.570	5.452	5.104	4.782
56	Barium	37.410	5.995	5.623	5.247
57	Lanthanum	38.931	6.283	5.894	5.489
58	Cerium	40.449	6.561	6.165	5.729
74	Tungsten	69.508	12.090	11.535	10.198
79	Gold	80.713	14.353	13.733	11.919
82	Lead	88.001	15.870	15.207	13.044
92	Uranium	115.591	21.753	20.943	17.163

Photon Fluence, Energy Fluence, and Mass Energy-Absorption as a Function

Photon energy (MeV)	Photon fluence Φ/X (photons/(m^2·R))	Energy fluence Ψ/X (J/(m^2·R))	Mass energy absorption coefficient $(\mu_{en}/\rho)_{air}$ (cm^2/g)
0.010	$11.7\cdot10^{12}$	$18.7\cdot10^{-3}$	4.66
0.015	$28.1\cdot10^{12}$	$67.4\cdot10^{-3}$	1.29
0.020	$52.6\cdot10^{12}$	$169\cdot10^{-3}$	0.516
0.030	$123\cdot10^{12}$	$591\cdot10^{-3}$	0.147
0.040	$212\cdot10^{12}$	$1360\cdot10^{-3}$	0.0640
0.050	$283\cdot10^{12}$	$2270\cdot10^{-3}$	0.0384
0.060	$310\cdot10^{12}$	$2980\cdot10^{-3}$	0.0292
0.080	$288\cdot10^{12}$	$3690\cdot10^{-3}$	0.0236
0.100	$235\cdot10^{12}$	$3770\cdot10^{-3}$	0.0231
0.15	$144\cdot10^{12}$	$3470\cdot10^{-3}$	0.0251
0.20	$101\cdot10^{12}$	$3250\cdot10^{-3}$	0.0268
0.30	$62.8\cdot10^{12}$	$3020\cdot10^{-3}$	0.0288
0.40	$45.9\cdot10^{12}$	$2940\cdot10^{-3}$	0.0296
0.50	$36.6\cdot10^{12}$	$2930\cdot10^{-3}$	0.0297
0.60	$30.6\cdot10^{12}$	$2940\cdot10^{-3}$	0.0296
0.80	$23.5\cdot10^{12}$	$3010\cdot10^{-3}$	0.0289
1.00	$19.4\cdot10^{12}$	$3110\cdot10^{-3}$	0.0280
1.50	$14.2\cdot10^{12}$	$3410\cdot10^{-3}$	0.0255
2.00	$11.6\cdot10^{12}$	$3720\cdot10^{-3}$	0.0234
5.00	$6.28\cdot10^{12}$	$5030\cdot10^{-3}$	0.0173
10.00	$3.77\cdot10^{12}$	$6040\cdot10^{-3}$	0.0144

$$\frac{\Phi}{X} = \frac{5.43\cdot10^{14}}{(\mu_{en}/\rho)_{air}\cdot(h\nu)} \text{ photons/(m}^2\text{·R), } h\nu \text{ is in keV.} \qquad \frac{\Psi}{X} = \frac{86.9\cdot10^{-3}}{(\mu_{en}/\rho)_{air}} \text{ J/(m}^2\text{·R).}$$

Energy of K-edge and Fluorescent Yield as a Function of Atomic Number

Energy E_k, of K x-rays, and ω_k the fluorescent yield is presented as a function of atomic number Z. The fluorescent yield is the number of K x-rays per hole in the K shell; $1-\omega_k$ is the number of Auger electrons emitted per K shell vacancy.

Z	E_K (MeV)	ω_K	Z	E_K (MeV)	ω_K
10	0.0009	All energy is locally absorbed	50	0.025	0.85
15	0.002		55	0.031	0.87
20	0.004	0.15			
25	0.006	0.27			
30	0.009	0.43	70	0.052	0.94
35	0.012	0.63			
40	0.016	0.70	80	0.071	0.95
45	0.020	0.80	92		0.97

General equation governing fluorescent yield:

$$\frac{\omega_K}{1-\omega_K} = ((-6.4\cdot10^{-2}) + (3.4\cdot10^{-2}\cdot Z) - (1.03\cdot10^{-6}\cdot Z^3))^4,$$

where Z is the atomic number.

CHAPTER 19

HEAT AND THERMODYNAMICS

Expansion Coefficients of Solids*

Substance	Temperature or temp. range (°F)	CLE × 10^6/°F	CCE × 10^6/°F
Aluminum	68	12.4	
Aluminum	572	15.8	
Antimony	32–212	.....	17.6
Amber	32–194	34	
Bakelite, bleached	68–140	12	
Beryl	32–212	.....	0.58
Brass, cast	32–212	10.4	
Bronze(75Cu,25Sn)	62–212	10.2	
Carbon(graphite)	122	3.3	
Chromium	140	3.8	
Copper	32–212	9.3	27.8
Copper	392	9.4	
Fluorspar, CaF_2	32–212	10.8	
Glass:			
Flint	122–140	4.4	
Hard	32–212	.....	11.9
Plate	32–212	4.95	
Quartz	60–930	0.32	
Silica	32–176	.....	0.72
Tube	32–212	4.95	15.3
Ice	−15	28	62.5
Iron	32–212	6.6	19.7
Lead	32–212	.....	46.5
Lead	536	19.0	
Limestone	77–212	5	
Marble	59–212	6.5	
Monel metal	77–212	7.8	
Nickel	68	7.0	
Paraffin	32–60	59	
Paraffin	60–100	72	
Paraffin	100–120	265	
Platinum	68	4.9	
Quartz, parallel to axis	32–176	4.4	
Quartz, perpendicular to axis	32–176	7.4	21.4
Rubber, hard	32	38	
Silver	68	10.5	
Silver	32–212	12.8	38.2
Solder(2Pb,1Sn)	32–212	14	
Steel	0–400	6.8	
Tin	32–212	12.8	38.2
Wood, parallel to fiber:			
Mahogany	36	2.0	
Maple	36	3.5	
Oak	36	2.7	
Pine	36	3.0	
Wood, perpendicular to fiber:			
Mahogany	36	2.2	
Maple	36	2.7	
Oak	36	3.0	
Pine	36	1.9	
Zinc	32–212	.....	49.5
Zinc	68–482	22.1	

* CLE = coefficient of linear expansion, CCE = coefficient of cubical expansion. While simple theory predicts that the CCE should be three times the CLE, discrepancies may occur due to experimental error in determining one or both coefficients, isotropic effects, etc.

Cubical Expansion of Liquids

Liquid	Coef/°F × 10³	Liquid	Coef/°F × 10³
Acetic acid	0.595	Hydrochloric acid, 33.2%	0.253
Acetone	.825	Mercury	.109
Alcohol, ethyl	.622	Olive oil	.400
Alcohol, methyl	.666	Pentane (93°API)	.74
Benzene	.686	Petroleum:	
Bromine	.630	15°API	.35
Calcium chloride:		35°API	.44
5.8% solution	.139	Phenol	.605
40.9% solution	.260	Sodium chloride, 20.6% solution	.230
Carbon disulfide	.676	Sulfuric acid:	
Carbon tetrachloride	.686	10.9%	.215
Chloroform	.707	100.0%	.31
Ether	.920	Turpentine	.541
Glycerin	.280	Water	.115

Energy Conversion Devices

Device	Energy forms		Typical efficiency, %
	Input	Output	
Electric generator	M	E	98
Dry cell battery	C	E	90
Steam boiler	C	T	88
Fuel cell	C	E	60
Liquid-fuel rocket	T	K	48
Steam turbine	T	M	45
Steam power plant	C	E	40
Gas turbine	C	M	35
Solid-state laser	E	R	30
Automobile engine	C	M	25
Fluorescent lamp	E	R	20
Solar cell	R	E	10
Incandescent lamp	E	R	5

C—chemical, E—electrical, K—kinetic, M—mechanical, T—thermal, R—radiant.

Over-all Heat Transfer Coefficient U

Air-to-air heat transfer, Btu/(hr) (ft²) (°F)
Outside air 15-mph wind, inside still air

Example	Construction	*U*
Frame walls	Wood siding, building paper, air space, gypsum lath, plaster	0.24
	Wood siding, insulation board, air space, gypsum board	0.19
Frame partition	Gypsum board, air space, gypsum board	0.34
Frame construction ceilings and floors	Linoleum or tile, felt, plywood, wood subfloor, air space, metal lath, plaster	0.23
Pitched roofs	Asphalt shingles, building paper, wood sheathing, air space, gypsum lath, plaster	0.28
Masonry wall	Face brick 4″, common brick 4″	0.48
	Face brick 4″, common brick 4″, air space, gypsum lath, plaster	0.29
	Face brick 4″, concrete block 4″, air space, gypsum lath, plaster	0.26
Masonry partition	Cement block (cinder aggregate), plaster on both sides	0.31
Concrete floor and ceiling	Tile, felt, plywood ⅝″, air space, metal lath, plaster	0.23
Flat masonry roof	Built-up roofing, roof insulation 1″, concrete slab 4″, air space, metal lath, plaster	0.18

Viscosity of Air

Pressure, bars	Temp., °C.										
	0	50	100	150	200	250	300	400	500	600	700
1	17,500	5440	121	142	162	182	203	243	284	325	365
50	17,500	5450	2800	1820	1350	1070	206	250	289	329	369
100	17,500	5450	2810	1830	1360	1080	905	258	295	334	374
150	17,400	5460	2820	1840	1370	1100	917	269	302	340	379
200	17,400	5460	3830	1860	1380	1110	930	286	311	346	384
250	17,400	5470	2840	1870	1390	1120	943	321	321	353	389
300	17,400	5470	2850	1880	1400	1130	955	458	334	361	395
350	17,300	5480	2860	1890	1420	1150	968	573	349	369	401
400	17,300	5480	2870	1900	1430	1160	981	628	369	379	408
450	17,300	5490	2880	1910	1440	1170	993	664	393	389	415
500	17,200	5490	2890	1920	1450	1180	1010	693	421	401	423

Viscosity of Steam (Micropoises)

Viscosity, lb./(ft.)(hr.) × 10^{-2}

Pressure, lb./sq. in. abs.	Temp., °F.							
	−100	−50	0	50	100	150	200	250
200	3.27	3.64	3.98	4.29	4.57	4.78	5.12	5.45
400	3.39	3.73	4.06	4.36	4.63	4.86	5.19	5.51
600	3.54	3.83	4.14	4.43	4.69	4.94	5.25	5.56
800	3.72	3.95	4.22	4.50	4.76	5.02	5.31	5.61
1,000	3.90	4.07	4.31	4.58	4.84	5.10	5.38	5.67
1,200	4.08	4.20	4.42	4.66	4.92	5.16	5.44	5.72
1,400	4.26	4.35	4.54	4.77	5.00	5.24	5.50	5.77
1,600	4.47	4.55	4.68	4.87	5.08	5.31	5.57	5.84
1,800	4.70	4.75	4.83	5.00	5.17	5.39	5.63	5.90
2,000	5.10	4.95	4.97	5.10	5.27	5.47	5.70	5.97
2,500	6.05	5.52	5.36	5.38	5.51	5.68	5.87	6.07
3,000	6.82	6.14	5.77	5.70	5.76	5.91	6.06	6.19
3,500	7.62	6.76	6.23	6.06	6.06	6.12	6.25	6.43
4,000	8.35	7.34	6.65	6.42	6.38	6.42	6.43	6.63
4,500	9.10	7.91	7.09	6.76	6.68	6.69	6.71	6.82
5,000	9.88	8.49	7.55	7.16	6.99	6.99	6.97	7.02
6,000	11.35	9.66	8.39	7.90	7.66	7.52	7.43	7.60
7,000	12.83	10.78	9.17	8.61	8.26	8.03	7.92	8.10
8,000	14.56	11.94	10.16	9.42	8.89	8.56	8.39	8.52
9,000	16.09	12.94	11.08	10.11	9.46	9.07	8.83	8.90
10,000	17.70	14.03	11.85	10.79	10.10	9.65	9.37	9.17

SPECIFIC HEAT OF THE ELEMENTS AT 25°C

C_p = cal g^{-1} °K^{-1}

Element	Kelly: Bureau of Mines Bulletin 592 (1961)
Aluminum	0.215
Antimony	0.049
Argon	0.124
Arsenic	0.0785
Barium	0.046
Beryllium	0.436
Bismuth	0.0296
Boron	0.245
Bromine (Br_2)	0.113
Cadmium	0.0555
Calcium	0.156
Carbon (Diamond)	0.124
" (Graphite)	0.170
Cerium	0.049
Cesium	0.057
Chlorine (Cl_2)	0.114
Chromium	0.1073
Cobalt	0.109
Columbium	See Niobium
Copper	0.092
Dysprosium	0.0414
Erbium	0.0401
Europium	0.0421
Fluorine (F_2)	0.197
Gadolinium	0.055
Gallium	0.089
Germanium	0.077
Gold	0.0308
Hafnium	0.035
Helium	1.24
Holmium	0.0393
Hydrogen (H_2)	3.41
Indium	0.056
Iodine (I_2)	0.102
Iridium	0.0317
Iron (α)	0.106
Krypton	0.059
Lanthanum	0.047
Lead	0.038
Lithium	0.85
Lutetium	0.037
Magnesium	0.243
Manganese (α)	0.114
" (β)	0.119
Mercury	0.0331
Molybdenum	0.0599
Neodymium	0.049
Neon	0.246
Nickel	0.106
Niobium	0.064
Nitrogen (N_2)	0.249
Osmium	0.03127

SPECIFIC HEAT OF THE ELEMENTS AT 25°C (Continued)

Element	Kelly: Bureau of Mines Bulletin 592 (1961)
Oxygen (O_2)	0.219
Palladium	0.0584
Phosphorus, white	0.181
" red, triclinic	0.160
Platinum	0.0317
Polonium	0.030
Potassium	0.180
Praseodymium	0.046
Promethium	0.0442
Protactinium	0.029
Radium	0.0288
Radon	0.0224
Rhenium	0.0329
Rhodium	0.0583
Rubidium	0.0861
Ruthenium	0.057
Samarium	0.043
Scandium	0.133
Selenium (Se_2)	0.0767
Silicon	0.168
Silver	0.0566
Sodium	0.293
Strontium	0.0719
Sulfur, yellow	0.175
Tantalum	0.0334
Technetium	0.058
Tellurium	0.0481
Terbium	0.0437
Thallium	0.0307
Thorium	0.0271
Thulium	0.0382
Tin (α)	0.0510
Tin (β)	0.0530
Titanium	0.125
Tungsten	0.0317
Uranium	0.0276
Vanadium	0.116
Xenon	0.0378
Ytterbium	0.0346
Yttrium	0.068
Zinc	0.0928
Zirconium	0.0671

SPECIFIC HEAT OF WATER

Heat Capacity of Air-free Water 0°—100°C at 1 Atmosphere Pressure

Temp. °C	Thermal Capacity		Enthalpy		Temp. °C	Thermal Capacity		Enthalpy	
	Cal./g/°C	Joules/g/°C	Cal./g	Joules/g		Cal/g/°C	Joules/g/°C	Cal/g	Joules/g
0	1.00738	4.2177	0.0245	0.1026	50	.99854	4.1807	50.0079	209.3729
1	1.00652	4.2141	1.0314	4.3184	51	99862	4.1810	51.0065	213.5538
2	1.00571	4.2107	2.0376	8.5308	52	.99871	4.1814	52.0051	217.7350
3	1.00499	4.2077	3.0429	12.7400	53	.99878	4.1817	53 0039	221.9166
4	1.00430	4.2048	4.0475	16.9462	54	.99885	4.1820	54.0027	226.0984
5	1.00368	4.2022	5.0515	21.1498	55	.99895	4.1824	55.0016	230.2806
6	1.00313	4.1999	6.0549	25.3508	56	.99905	4.1828	56.0006	234.4632
7	1.00260	4.1977	7.0578	29.5496	57	.99914	4.1832	56.9997	238.6462
8	1.00213	4.1957	8.0602	33.7463	58	.99924	4.1836	57.9989	242.8296
9	1.00170	4.1939	9.0621	37.9410	59	.99933	4.1840	58.9982	247.0134
10	1.00129	4.1922	10.0636	42.1341	60	.99943	4.1844	59.9975	251.1976
11	1.00093	4.1907	11.0647	46.3255	61	.99955	4.1849	60.9970	255.3822
12	1.00060	4.1893	12.0654	50.5155	62	.99964	4.1853	61.9966	259.5673
13	1.00029	4.1880	13.0659	54.7041	63	.99976	4.1858	62.9963	263.7529
14	1.00002	4.1869	14.0660	58.8916	64	.99988	4.1863	63.9962	267.9390
15	.99976	4.1858	15.0659	63 0779	65	1.00000	4.1868	64.9961	272.1256
16	.99955	4.1849	16.0655	67.2632	66	1.00014	4.1874	65.9962	276.3127
17	.99933	4.1840	17.0650	71.4476	67	1.00026	4.1879	66.9964	280.5003
18	.99914	4.1832	18.0642	75.6312	68	1.00041	4.1885	67.9967	284.6885
19	.99897	4.1825	19.0633	79.8141	69	1.00053	4.1890	68.9972	288.8772
20	.99883	4.1819	20.0622	83.9963	70	1.00067	4.1896	69.9977	293.0665
21	.99869	4.1813	21.0609	88.1778	71	1.00081	4.1902	70.9985	297.2564
22	.99857	4.1808	22.0596	92.3589	72	1.00096	4.1908	71.9994	301.4469
23	.99847	4.1804	23.0581	96.5395	73	1.00112	4.1915	73.0004	305.6381
24	.99838	4.1800	24.0565	100.7196	74	1.00127	4.1921	74.0016	309.8299
25	.99828	4.1796	25.0548	104.8994	75	1.00143	4.1928	75.0030	314.0224
26	.99821	4.1793	26.0530	109.0788	76	1.00160	4.1935	76.0045	318.2155
27	.99814	4.1790	27.0512	113.2580	77	1.00177	4.1942	77.0062	322.4094
28	.99809	4.1788	28.0493	117.4369	78	1.00194	4.1949	78.0080	326.6039
29	.99804	4.1786	29.0474	121.6157	79	1.00213	4.1957	79.0101	330.7992
30	.99802	4 1785	30.0455	125 7943	80	1.00229	4.1964	80.0123	334.9952
31	.99799	4.1784	31.0435	129.9727	81	1.00248	4.1972	81.0147	339.1920
32	.99797	4.1783	32.0414	134.1510	82	1.00268	4.1980	82.0172	343.3897
33	.99797	4.1783	33.0394	138.3293	83	1.00287	4.1988	83.0200	347.5881
34	.99795	4.1782	34.0374	142.5076	84	1.00308	4.1997	84.0230	351.7873
35	.99795	4.1782	35.0353	146.6858	85	1.00327	4.2005	85.0262	355.9874
36	.99797	4.1783	36.0333	150.8641	86	1.00349	4.2014	86.0295	360.1883
37	.99797	4.1783	37.0312	155.0423	87	1.00370	4.2023	87.0331	364.3902
38	.99799	4.1784	38.0292	159.2207	88	1.00392	4.2032	88.0369	368.5929
39	.99802	4.1785	39.0272	163.3991	89	1.00416	4.2042	89.0410	372.7966
40	.99804	4.1786	40.0253	167.5777	90	1.00437	4.2051	90.0452	377.0012
41	.99807	4.1787	41.0233	171.7563	91	1.00461	4.2061	91.0497	381.2068
42	.99811	4.1789	42.0214	175.9351	92	1.00485	4.2071	92.0545	385.4135
43	.99816	4.1791	43.0195	180.1141	93	1.00509	4.2081	93.0594	389.6211
44	.99819	4.1792	44.0177	184.2933	94	1.00535	4.2092	94.0647	393.8297
45	.99826	4.1795	45.0159	188.4726	95	1.00561	4.2103	95.0701	398.0395
46	.99830	4.1797	46.0142	192.6522	96	1.00588	4.2114	96.0759	402.2503
47	.99835	4.1799	47.0125	196.8320	97	1.00614	4.2125	97.0819	406.4622
48	.99842	4.1802	48.0109	201.012[illegible]	98	1.00640	4.2136	98.0882	410.6753
49	.99847	4.1804	49.0094	205.1923	99	1.00669	4.2148	99.0947	414.8895
					100	1.00697	4.2160	100.1015	419.1049

Enthalpy of Air-saturated Water
1 Atmosphere Pressure 0-100°C

Temp. °C	Enthalpy		Temp.	Enthalpy	
	Cal/g	Joules/g	°C	Cal/g	Joules/g
0	0	0	50	49.9896	209.2964
5	5.0276	21.0496	55	54.9842	230.2077
10	10.0402	42.0363	60	59.9811	251.1289
15	15.0431	62.9826	65	64.9808	272.0619
20	20.0400	83.9034	70	69.9839	293.0087
25	25.0332	104.8089	75	74.9907	313.9712
30	30.0244	125.7063	80	80.0019	334.9519
35	35.0149	146.6003	85	85.0180	355.9532
40	40.0055	167.4949	90	90.0395	376.9773
45	44.9968	188.3928	95	95.0671	398.0270
			100	100.1016	419.1053

Specific Heats of Selected Liquids

Substance	*State*	c_P, *Btu/lbm-°R*	*Substance*	*State*	c_P, *Btu/lbm-°R*
Water	1 atm, 32°F	1.007	Glycerin	1 atm, 50°F	0.554
	1 atm, 77°F	0.998		1 atm, 120°F	0.617
	1 atm, 212°F	1.007			
			Bismuth	1 atm, 800°F	0.0345
Ammonia	sat., 0°F	1.08		1 atm, 1000°F	0.0369
	sat., 120°F	1.22		1 atm, 1400°F	0.0393
Freon-12	sat., −40°F	0.211	Mercury	1 atm, 50°F	0.033
	sat., 0°F	0.217		1 atm, 600°F	0.032
	sat., 120°F	0.244			
			Sodium	1 atm, 200°F	0.33
Benzene	1 atm, 60°F	0.43		1 atm, 1000°F	0.30
	1 atm, 150°F	0.46			
			n-Butane	1 atm, 32°F	0.550
Light oil	1 atm, 60°F	0.43			
	1 atm, 300°F	0.54	Propane	1 atm, 32°F	0.576

Specific Heats of Selected Solids P = 1 atm

Substance	*T*, °*C*	c_P, *cal/g-°K*	*Substance*	*T*, °*C*	c_P, *cal/g-°K*
Ice	−200	0.168	Lead	−270	0.00001
	−140	0.262		−259	0.0073
	−60	0.392		−100	0.0283
	−11	0.468		0	0.0297
	−2.6	0.500		+100	0.0320
				300	0.0356
Aluminum	−250	0.0039			
	−200	0.076	Iron	20	0.107
	−100	0.167	Silver	20	0.0558
	0	0.208			
	+100	0.225	Sodium	20	0.295
	300	0.248	Tungsten	20	0.034
	600	0.277			
			Graphite	20	0.17
Platinum	−256	0.00123			
	−152	0.0261	Wood	20	0.42
	0	0.0316	Rubber	20	0.44
	+500	0.0349			
	1000	0.0381	Mica	20	0.21

Specific Heat of Selected Gases

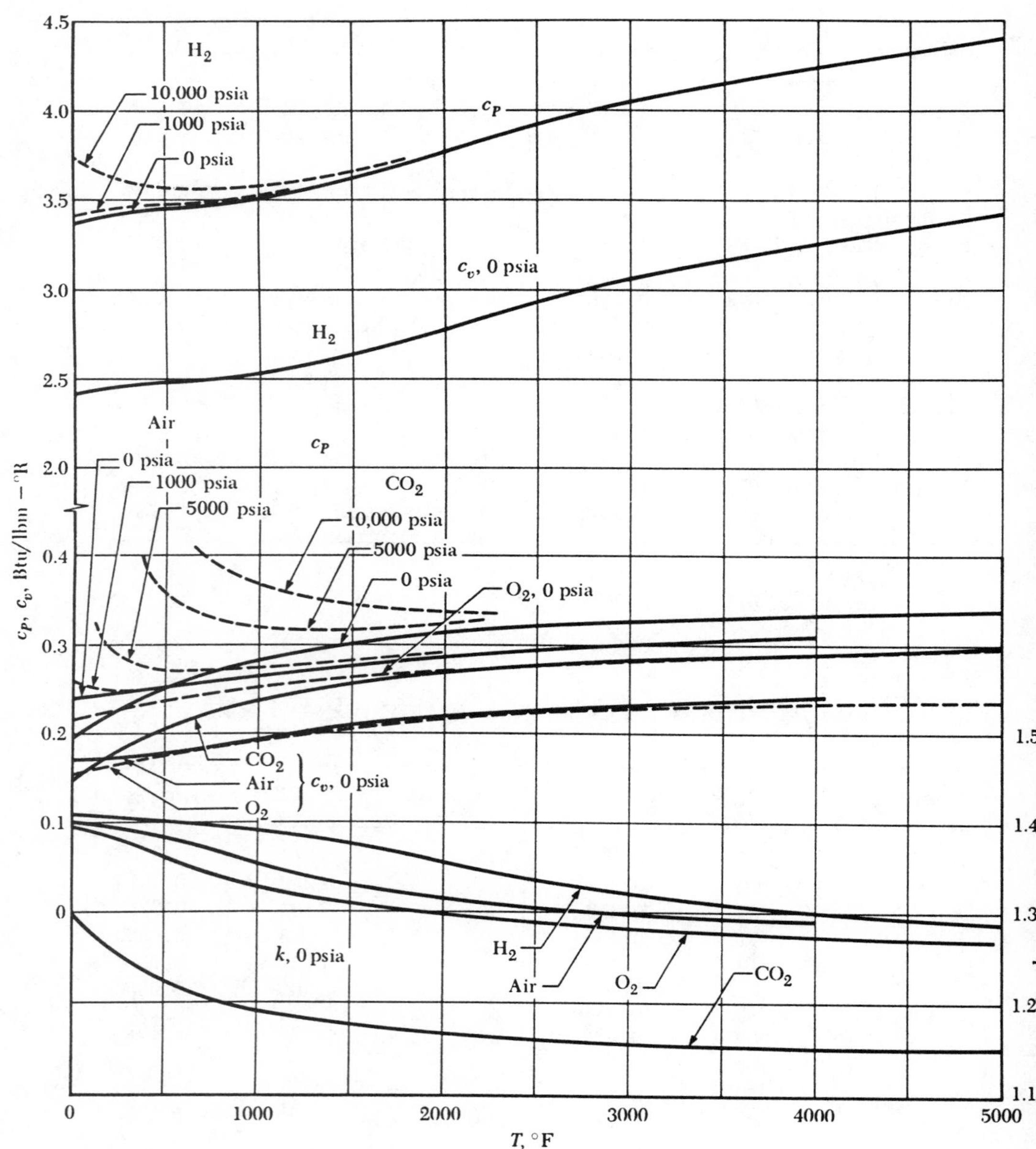

From National Bureau of Standards

Thermal Properties of Selected Gases at Low Pressures

Substance	$\hat{M}$, lbm/lbmole, g/gmole	c_P, Btu/lbm-°R	$\hat{c}_P$, Btu/lbmole-°R	c_v, Btu/lbm-°R	$\hat{c}_v$, Btu/lbmole-°R	R, ft-lbf/lbm-°R	R, Btu/lbm-°R	$k = c_P/c_v$
Argon, A	39.94	0.123	4.91	0.074	2.96	38.65	0.0496	1.67
Helium, He	4.003	1.25	5.00	0.75	3.000	386.3	0.4963	1.66
Hydrogen, H_2	2.016	3.42	6.89	2.43	4.90	767.0	0.9856	1.41
Nitrogen, N_2	28.02	0.248	6.95	0.177	4.96	55.13	0.0708	1.40
Oxygen, O_2	32.00	0.219	7.01	0.156	4.99	48.24	0.0620	1.40
Carbon monoxide, CO	28.01	0.249	6.97	0.178	4.98	55.13	0.0708	1.40
Air	28.97	0.240	6.95	0.171	4.95	53.34	0.0686	1.40
Water vapor, H_2O	18.016	0.446	8.07	0.336	6.03	85.58	0.1099	1.33
Methane, CH_4	16.04	0.532	8.53	0.403	6.46	96.4	0.1236	1.32
Carbon dioxide, CO_2	44.01	0.202	8.91	0.156	6.87	35.1	0.0451	1.30
Sulfur dioxide, SO_2	64.07	0.154	9.87	0.122	7.82	24.1	0.0310	1.26
Acetylene, C_2H_2	26.04	0.409	10.65	0.333	8.67	59.4	0.0761	1.23
Ethylene, C_2H_4	28.05	0.374	10.49	0.304	8.53	55.1	0.0707	1.23
Ethane, C_2H_6	30.07	0.422	12.69	0.357	10.73	51.3	0.0658	1.18
Propane, C_3H_8	44.09	0.404	17.81	0.360	15.87	35.0	0.0450	1.12
Isobutane, C_4H_{10}	58.12	0.420	24.41	0.387	22.49	26.6	0.0342	1.09

STEAM TABLES

Specific Heat at constant pressure of Steam and of Water

Temp. F	c_p, Btu/lbm × F														
Press., psia	1	1.5	2	3	4	6	8	10	15	20	30	40	60	80	100
Sat. Water	0.998	0.998	0.999	1.000	1.000	1.002	1.003	1.004	1.007	1.010	1.014	1.019	1.026	1.033	1.039
Sat. Steam	0.450	0.452	0.454	0.458	0.461	0.466	0.471	0.475	0.485	0.493	0.508	0.521	0.543	0.564	0.582
1500	0.559	0.559	0.559	0.559	0.559	0.559	0.559	0.559	0.559	0.559	0.560	0.560	0.560	0.561	0.561
1480	0.557	0.557	0.557	0.557	0.557	0.557	0.557	0.558	0.558	0.558	0.558	0.558	0.559	0.559	0.559
1460	0.556	0.556	0.556	0.556	0.556	0.556	0.556	0.556	0.556	0.556	0.556	0.557	0.557	0.557	0.558
1440	0.554	0.554	0.554	0.554	0.554	0.554	0.554	0.554	0.554	0.554	0.555	0.555	0.555	0.556	0.556
1420	0.552	0.552	0.552	0.552	0.552	0.552	0.553	0.553	0.553	0.553	0.553	0.553	0.554	0.554	0.555
1400	0.551	0.551	0.551	0.551	0.551	0.551	0.551	0.551	0.551	0.551	0.551	0.552	0.552	0.553	0.553
1380	0.549	0.549	0.549	0.549	0.549	0.549	0.549	0.549	0.549	0.549	0.550	0.550	0.550	0.551	0.551
1360	0.547	0.547	0.547	0.547	0.547	0.547	0.547	0.547	0.548	0.548	0.548	0.548	0.549	0.549	0.550
1340	0.546	0.546	0.546	0.546	0.546	0.546	0.546	0.546	0.546	0.546	0.546	0.547	0.547	0.548	0.548
1320	0.544	0.544	0.544	0.544	0.544	0.544	0.544	0.544	0.544	0.544	0.545	0.545	0.545	0.546	0.546
1300	0.542	0.542	0.542	0.542	0.542	0.542	0.542	0.542	0.542	0.543	0.543	0.543	0.544	0.544	0.545
1280	0.540	0.540	0.540	0.540	0.540	0.540	0.540	0.541	0.541	0.541	0.541	0.541	0.542	0.543	0.543
1260	0.538	0.539	0.539	0.539	0.539	0.539	0.539	0.539	0.539	0.539	0.539	0.540	0.540	0.541	0.541
1240	0.537	0.537	0.537	0.537	0.537	0.537	0.537	0.537	0.537	0.537	0.538	0.538	0.539	0.539	0.540
1220	0.535	0.535	0.535	0.535	0.535	0.535	0.535	0.535	0.535	0.536	0.536	0.536	0.537	0.537	0.538
1200	0.533	0.533	0.533	0.533	0.533	0.533	0.533	0.533	0.534	0.534	0.534	0.534	0.535	0.536	0.536
1180	0.531	0.531	0.531	0.531	0.531	0.531	0.532	0.532	0.532	0.532	0.532	0.533	0.533	0.534	0.535
1160	0.529	0.529	0.530	0.530	0.530	0.530	0.530	0.530	0.530	0.530	0.530	0.531	0.532	0.532	0.533
1140	0.528	0.528	0.528	0.528	0.528	0.528	0.528	0.528	0.528	0.528	0.529	0.529	0.530	0.531	0.531
1120	0.526	0.526	0.526	0.526	0.526	0.526	0.526	0.526	0.526	0.527	0.527	0.527	0.528	0.529	0.530
1100	0.524	0.524	0.524	0.524	0.524	0.524	0.524	0.524	0.525	0.525	0.525	0.526	0.526	0.527	0.528
1080	0.522	0.522	0.522	0.522	0.522	0.522	0.522	0.523	0.523	0.523	0.523	0.524	0.525	0.525	0.526
1060	0.520	0.520	0.520	0.520	0.520	0.521	0.521	0.521	0.521	0.521	0.522	0.522	0.523	0.524	0.524
1040	0.518	0.519	0.519	0.519	0.519	0.519	0.519	0.519	0.519	0.519	0.520	0.520	0.521	0.522	0.523
1020	0.517	0.517	0.517	0.517	0.517	0.517	0.517	0.517	0.517	0.518	0.518	0.518	0.519	0.520	0.521
1000	0.515	0.515	0.515	0.515	0.515	0.515	0.515	0.515	0.515	0.516	0.516	0.517	0.518	0.519	0.519
980	0.513	0.513	0.513	0.513	0.513	0.513	0.513	0.513	0.514	0.514	0.514	0.515	0.516	0.517	0.518
960	0.511	0.511	0.511	0.511	0.511	0.511	0.512	0.512	0.512	0.512	0.513	0.513	0.514	0.515	0.516
940	0.509	0.509	0.509	0.509	0.509	0.510	0.510	0.510	0.510	0.510	0.511	0.511	0.512	0.514	0.515
920	0.507	0.508	0.508	0.508	0.508	0.508	0.508	0.508	0.508	0.509	0.509	0.510	0.511	0.512	0.513
900	0.506	0.506	0.506	0.506	0.506	0.506	0.506	0.506	0.506	0.507	0.507	0.508	0.509	0.510	0.512
880	0.504	0.504	0.504	0.504	0.504	0.504	0.504	0.504	0.505	0.505	0.506	0.506	0.508	0.509	0.510
860	0.502	0.502	0.502	0.502	0.502	0.502	0.503	0.503	0.503	0.503	0.504	0.505	0.506	0.507	0.509
840	0.500	0.500	0.500	0.500	0.500	0.501	0.501	0.501	0.501	0.502	0.502	0.503	0.504	0.506	0.507
820	0.498	0.498	0.499	0.499	0.499	0.499	0.499	0.499	0.499	0.500	0.501	0.501	0.503	0.504	0.506
800	0.497	0.497	0.497	0.497	0.497	0.497	0.497	0.497	0.498	0.498	0.499	0.500	0.501	0.503	0.505
780	0.495	0.495	0.495	0.495	0.495	0.495	0.495	0.496	0.496	0.496	0.497	0.498	0.500	0.502	0.503
760	0.493	0.493	0.493	0.493	0.493	0.494	0.494	0.494	0.494	0.495	0.496	0.497	0.499	0.500	0.502
740	0.491	0.491	0.491	0.492	0.492	0.492	0.492	0.492	0.493	0.493	0.494	0.495	0.497	0.499	0.501
720	0.490	0.490	0.490	0.490	0.490	0.490	0.490	0.491	0.491	0.492	0.493	0.494	0.496	0.498	0.500
700	0.488	0.488	0.488	0.488	0.488	0.488	0.489	0.489	0.490	0.490	0.491	0.492	0.495	0.497	0.500
680	0.486	0.486	0.486	0.486	0.487	0.487	0.487	0.487	0.488	0.489	0.490	0.491	0.494	0.496	0.499
660	0.484	0.485	0.485	0.485	0.485	0.485	0.485	0.486	0.486	0.487	0.489	0.490	0.493	0.496	0.499
640	0.483	0.483	0.483	0.483	0.483	0.484	0.484	0.484	0.485	0.486	0.487	0.489	0.492	0.495	0.499
620	0.481	0.481	0.481	0.481	0.482	0.482	0.482	0.483	0.483	0.484	0.486	0.488	0.491	0.495	0.499
600	0.479	0.480	0.480	0.480	0.480	0.480	0.481	0.481	0.482	0.483	0.485	0.487	0.491	0.495	0.499
580	0.478	0.478	0.478	0.478	0.478	0.479	0.479	0.480	0.481	0.482	0.484	0.486	0.491	0.495	0.500
560	0.476	0.476	0.476	0.477	0.477	0.477	0.478	0.478	0.479	0.481	0.483	0.485	0.490	0.496	0.501
540	0.475	0.475	0.475	0.475	0.475	0.476	0.476	0.477	0.478	0.480	0.482	0.485	0.491	0.497	0.503
520	0.473	0.473	0.473	0.474	0.474	0.475	0.475	0.476	0.477	0.479	0.482	0.485	0.491	0.498	0.505
500	0.472	0.472	0.472	0.472	0.473	0.473	0.474	0.475	0.476	0.478	0.481	0.485	0.492	0.500	0.508
480	0.470	0.470	0.470	0.471	0.471	0.472	0.473	0.473	0.475	0.477	0.481	0.485	0.493	0.502	0.511
460	0.469	0.469	0.469	0.469	0.470	0.471	0.472	0.472	0.475	0.477	0.481	0.486	0.495	0.505	0.516
440	0.467	0.467	0.468	0.468	0.469	0.470	0.470	0.471	0.474	0.476	0.481	0.487	0.498	0.509	0.522
420	0.466	0.466	0.466	0.467	0.467	0.468	0.470	0.471	0.473	0.476	0.482	0.488	0.501	0.514	0.528
400	0.464	0.465	0.465	0.466	0.466	0.467	0.469	0.470	0.473	0.476	0.483	0.490	0.504	0.520	0.536
380	0.463	0.463	0.464	0.464	0.465	0.466	0.468	0.469	0.473	0.477	0.484	0.492	0.509	0.527	0.546
360	0.462	0.462	0.462	0.463	0.464	0.466	0.467	0.469	0.473	0.477	0.486	0.495	0.515	0.536	0.558
340	0.460	0.461	0.461	0.462	0.463	0.465	0.467	0.469	0.473	0.478	0.488	0.499	0.521	0.546	0.572
320	0.459	0.460	0.460	0.461	0.462	0.464	0.467	0.469	0.474	0.480	0.491	0.504	0.530	0.558	1.036
300	0.458	0.459	0.459	0.460	0.462	0.464	0.466	0.469	0.475	0.482	0.495	0.509	0.539	1.029	1.029
280	0.457	0.458	0.458	0.460	0.461	0.464	0.467	0.469	0.477	0.484	0.500	0.516	1.022	1.022	1.022
260	0.456	0.457	0.457	0.459	0.461	0.464	0.467	0.470	0.478	0.487	0.505	1.017	1.017	1.017	1.016
240	0.455	0.456	0.457	0.458	0.460	0.464	0.468	0.471	0.481	0.491	1.012	1.012	1.012	1.012	1.012
220	0.454	0.455	0.456	0.458	0.460	0.464	0.468	0.473	0.484	1.008	1.008	1.008	1.008	1.008	1.008
200	0.453	0.454	0.455	0.458	0.460	0.465	0.470	0.475	1.005	1.005	1.005	1.005	1.005	1.005	1.005
180	0.452	0.454	0.455	0.458	0.460	0.466	1.003	1.003	1.003	1.003	1.003	1.003	1.003	1.002	1.002
160	0.451	0.453	0.455	0.458	0.461	1.001	1.001	1.001	1.001	1.001	1.001	1.001	1.001	1.001	1.001
140	0.451	0.453	0.454	1.000	1.000	1.000	1.000	1.000	0.999	0.999	0.999	0.999	0.999	0.999	0.999
120	0.450	0.452	0.999	0.999	0.999	0.999	0.999	0.999	0.999	0.999	0.998	0.998	0.998	0.998	0.998
100	0.998	0.998	0.998	0.998	0.998	0.998	0.998	0.998	0.998	0.998	0.998	0.998	0.998	0.998	0.998
80	0.998	0.998	0.998	0.998	0.998	0.998	0.998	0.998	0.998	0.998	0.998	0.998	0.998	0.998	0.998
60	1.000	1.000	1.000	1.000	1.000	1.000	1.000	1.000	1.000	1.000	1.000	1.000	0.999	0.999	0.999
40	1.004	1.004	1.004	1.004	1.004	1.004	1.004	1.004	1.004	1.004	1.004	1.004	1.004	1.004	1.003
32	1.007	1.007	1.007	1.007	1.007	1.007	1.007	1.007	1.007	1.007	1.007	1.007	1.007	1.007	1.006

STEAM TABLES (Continued)

Specific Heat at constant pressure of Steam and of Water

Temp. F	c_p, Btu/lbm × F														
Press., psia	150	200	300	400	600	800	1000	1500	2000	3000	4000	6000	8000	10000	15000
Sat. Water	1.054	1.067	1.093	1.118	1.168	1.224	1.286	1.492	1.841	7.646	—	—	—	—	—
Sat. Steam	0.624	0.661	0.729	0.792	0.915	1.046	1.191	1.667	2.557	13.66	—	—	—	—	—
1500	0.562	0.563	0.565	0.567	0.571	0.576	0.580	0.590	0.601	0.623	0.645	0.691	0.737	0.780	0.868
1480	0.561	0.562	0.564	0.566	0.570	0.575	0.579	0.590	0.601	0.623	0.647	0.694	0.742	0.786	0.878
1460	0.559	0.560	0.562	0.565	0.569	0.573	0.578	0.589	0.601	0.624	0.648	0.698	0.747	0.793	0.888
1440	0.557	0.559	0.561	0.563	0.568	0.572	0.577	0.589	0.600	0.625	0.650	0.701	0.753	0.800	0.900
1420	0.556	0.557	0.559	0.562	0.566	0.571	0.576	0.588	0.600	0.625	0.651	0.705	0.759	0.808	0.909
1400	0.554	0.555	0.558	0.560	0.565	0.570	0.575	0.587	0.600	0.626	0.653	0.709	0.765	0.817	0.926
1380	0.553	0.554	0.556	0.559	0.564	0.569	0.574	0.587	0.600	0.627	0.655	0.714	0.773	0.827	0.939
1360	0.551	0.552	0.555	0.558	0.563	0.568	0.573	0.586	0.600	0.628	0.657	0.719	0.781	0.838	0.953
1340	0.549	0.551	0.553	0.556	0.561	0.567	0.572	0.586	0.600	0.629	0.660	0.725	0.790	0.850	0.968
1320	0.548	0.549	0.552	0.555	0.560	0.566	0.571	0.585	0.600	0.630	0.663	0.731	0.800	0.864	0.983
1300	0.546	0.548	0.550	0.553	0.559	0.565	0.570	0.585	0.600	0.632	0.666	0.738	0.811	0.879	0.998
1280	0.545	0.546	0.549	0.552	0.558	0.564	0.570	0.585	0.600	0.634	0.669	0.746	0.824	0.897	1.014
1260	0.543	0.544	0.547	0.550	0.556	0.563	0.569	0.585	0.601	0.636	0.673	0.755	0.838	0.918	1.033
1240	0.541	0.543	0.546	0.549	0.555	0.562	0.568	0.584	0.601	0.638	0.678	0.765	0.855	0.942	1.053
1220	0.540	0.541	0.544	0.548	0.554	0.561	0.567	0.584	0.602	0.641	0.683	0.777	0.875	0.969	1.072
1200	0.538	0.540	0.543	0.546	0.553	0.560	0.567	0.584	0.603	0.644	0.689	0.790	0.897	1.000	1.095
1180	0.536	0.538	0.541	0.545	0.552	0.559	0.566	0.584	0.604	0.647	0.696	0.805	0.922	1.033	1.117
1160	0.535	0.536	0.540	0.544	0.551	0.558	0.565	0.585	0.606	0.652	0.704	0.823	0.952	1.070	1.143
1140	0.533	0.535	0.539	0.542	0.550	0.557	0.565	0.585	0.607	0.656	0.713	0.843	0.986	1.107	1.167
1120	0.531	0.533	0.537	0.541	0.549	0.557	0.565	0.586	0.609	0.662	0.723	0.866	1.025	1.149	1.190
1100	0.530	0.532	0.536	0.540	0.548	0.556	0.564	0.587	0.612	0.668	0.735	0.893	1.070	1.193	1.220
1080	0.528	0.530	0.534	0.538	0.547	0.555	0.564	0.588	0.615	0.676	0.749	0.924	1.120	1.242	1.240
1060	0.527	0.529	0.533	0.537	0.546	0.555	0.564	0.590	0.618	0.685	0.765	0.960	1.176	1.295	1.260
1040	0.525	0.527	0.532	0.536	0.545	0.555	0.565	0.592	0.622	0.695	0.783	1.002	1.238	1.351	1.282
1020	0.523	0.526	0.530	0.535	0.545	0.555	0.565	0.594	0.627	0.707	0.804	1.051	1.306	1.399	1.298
1000	0.522	0.524	0.529	0.534	0.544	0.555	0.566	0.597	0.633	0.721	0.829	1.110	1.382	1.471	1.306
980	0.520	0.523	0.528	0.533	0.544	0.555	0.567	0.601	0.640	0.737	0.858	1.180	1.475	1.531	1.312
960	0.519	0.521	0.527	0.532	0.543	0.556	0.568	0.605	0.648	0.756	0.893	1.267	1.598	1.595	1.310
940	0.517	0.520	0.526	0.531	0.543	0.556	0.570	0.610	0.658	0.778	0.934	1.376	1.708	1.639	1.299
920	0.516	0.519	0.525	0.531	0.544	0.558	0.573	0.617	0.669	0.803	0.984	1.520	1.819	1.667	1.281
900	0.515	0.518	0.524	0.530	0.544	0.559	0.576	0.624	0.683	0.834	1.048	1.716	1.932	1.660	1.259
880	0.513	0.516	0.523	0.530	0.545	0.561	0.580	0.633	0.699	0.872	1.130	1.993	2.000	1.633	1.232
860	0.512	0.515	0.523	0.530	0.546	0.564	0.584	0.644	0.718	0.918	1.240	2.316	2.019	1.593	1.212
840	0.511	0.514	0.522	0.530	0.548	0.568	0.590	0.657	0.740	0.977	1.395	2.653	1.978	1.547	1.192
820	0.510	0.514	0.522	0.531	0.550	0.572	0.597	0.672	0.767	1.054	1.620	2.886	1.888	1.503	1.175
800	0.509	0.513	0.522	0.532	0.553	0.577	0.605	0.690	0.800	1.160	1.967	2.872	1.768	1.459	1.157
780	0.508	0.513	0.522	0.533	0.557	0.584	0.615	0.712	0.840	1.312	2.550	2.547	1.670	1.416	1.142
760	0.507	0.512	0.523	0.535	0.561	0.592	0.628	0.738	0.892	1.542	4.462	2.156	1.576	1.370	1.126
740	0.507	0.512	0.524	0.537	0.567	0.602	0.642	0.770	0.960	1.913	8.119	1.886	1.493	1.332	1.114
720	0.506	0.512	0.525	0.540	0.574	0.613	0.660	0.811	1.052	2.584	3.458	1.696	1.421	1.290	1.100
700	0.506	0.513	0.528	0.544	0.582	0.627	0.681	0.861	1.181	6.145⊙	2.237	1.557	1.358	1.250	1.089
680	0.506	0.514	0.530	0.549	0.592	0.644	0.707	0.927	1.365	2.469	1.789	1.450	1.303	1.217	1.079
660	0.507	0.515	0.534	0.555	0.604	0.665	0.738	1.015	1.639	1.851	1.587	1.369	1.256	1.187	1.071
640	0.507	0.517	0.538	0.562	0.619	0.690	0.777	1.135	2.219	1.601	1.454	1.303	1.216	1.157	1.063
620	0.509	0.519	0.543	0.571	0.637	0.720	0.826	1.308	1.614	1.455	1.362	1.252	1.184	1.136	1.056
600	0.510	0.522	0.550	0.582	0.659	0.757	0.888	1.586	1.453	1.358	1.295	1.211	1.157	1.118	1.052
580	0.513	0.526	0.558	0.595	0.685	0.804	0.969	1.393	1.351	1.289	1.243	1.178	1.134	1.102	1.046
560	0.516	0.531	0.568	0.611	0.717	0.862	1.079	1.309	1.281	1.237	1.202	1.151	1.115	1.087	1.039
540	0.519	0.538	0.580	0.630	0.756	0.937	1.272	1.249	1.229	1.196	1.169	1.128	1.098	1.074	1.031
520	0.524	0.545	0.594	0.653	0.804	1.035	1.221	1.204	1.189	1.164	1.142	1.109	1.083	1.062	1.024
500	0.530	0.554	0.611	0.680	0.865	1.187	1.181	1.169	1.157	1.137	1.120	1.092	1.069	1.051	1.017
480	0.537	0.565	0.632	0.714	1.159	1.154	1.150	1.140	1.131	1.115	1.101	1.077	1.057	1.041	1.010
460	0.545	0.578	0.657	0.755	1.132	1.128	1.125	1.117	1.110	1.096	1.084	1.064	1.047	1.033	1.004
440	0.556	0.594	0.687	1.113	1.110	1.107	1.104	1.098	1.092	1.080	1.070	1.052	1.038	1.025	0.999
420	0.568	0.614	0.724	1.094	1.091	1.089	1.087	1.081	1.076	1.067	1.058	1.042	1.029	1.018	0.994
400	0.583	0.636	1.079	1.078	1.076	1.074	1.072	1.067	1.063	1.055	1.047	1.034	1.022	1.011	0.990
380	0.601	1.066	1.065	1.065	1.063	1.061	1.059	1.056	1.052	1.044	1.038	1.026	1.015	1.006	0.986
360	0.622	1.054	1.054	1.053	1.052	1.050	1.049	1.045	1.042	1.036	1.030	1.019	1.009	1.001	0.982
340	1.045	1.044	1.044	1.043	1.042	1.040	1.039	1.036	1.033	1.028	1.022	1.013	1.004	0.996	0.979
320	1.036	1.036	1.035	1.034	1.033	1.032	1.031	1.028	1.026	1.021	1.016	1.007	0.999	0.992	0.976
300	1.028	1.028	1.028	1.027	1.026	1.025	1.024	1.022	1.019	1.015	1.010	1.002	0.995	0.988	0.973
280	1.022	1.022	1.021	1.021	1.020	1.019	1.018	1.016	1.014	1.009	1.005	0.998	0.991	0.985	0.971
260	1.016	1.016	1.016	1.015	1.014	1.013	1.013	1.011	1.009	1.005	1.001	0.994	0.988	0.982	0.968
240	1.012	1.011	1.011	1.011	1.010	1.009	1.008	1.006	1.004	1.001	0.997	0.991	0.985	0.979	0.966
220	1.008	1.008	1.007	1.007	1.006	1.005	1.005	1.003	1.001	0.998	0.994	0.988	0.982	0.977	0.964
200	1.005	1.004	1.004	1.004	1.003	1.002	1.002	1.000	0.998	0.995	0.992	0.986	0.980	0.975	0.963
180	1.002	1.002	1.002	1.001	1.001	1.000	0.999	0.998	0.996	0.993	0.989	0.983	0.978	0.973	0.961
160	1.000	1.000	1.000	0.999	0.999	0.998	0.997	0.996	0.994	0.991	0.987	0.981	0.976	0.971	0.959
140	0.999	0.999	0.998	0.998	0.997	0.997	0.996	0.994	0.992	0.989	0.986	0.980	0.974	0.969	0.958
120	0.998	0.998	0.997	0.997	0.996	0.996	0.995	0.993	0.991	0.988	0.984	0.978	0.972	0.967	0.957
100	0.997	0.997	0.997	0.996	0.996	0.995	0.994	0.992	0.990	0.986	0.983	0.976	0.970	0.965	0.955
80	0.998	0.997	0.997	0.996	0.995	0.994	0.994	0.991	0.989	0.985	0.981	0.974	0.968	0.962	0.951
60	0.999	0.999	0.998	0.997	0.996	0.995	0.994	0.991	0.989	0.984	0.979	0.970	0.963	0.956	0.942
40	1.003	1.003	1.002	1.001	1.000	0.998	0.997	0.993	0.989	0.983	0.976	0.965	0.954	0.945	0.920
32	1.006	1.006	1.005	1.004	1.002	1.000	0.999	0.994	0.990	0.983	0.975	0.962	0.949	0.937	0.904

Mollier Diagram for Steam

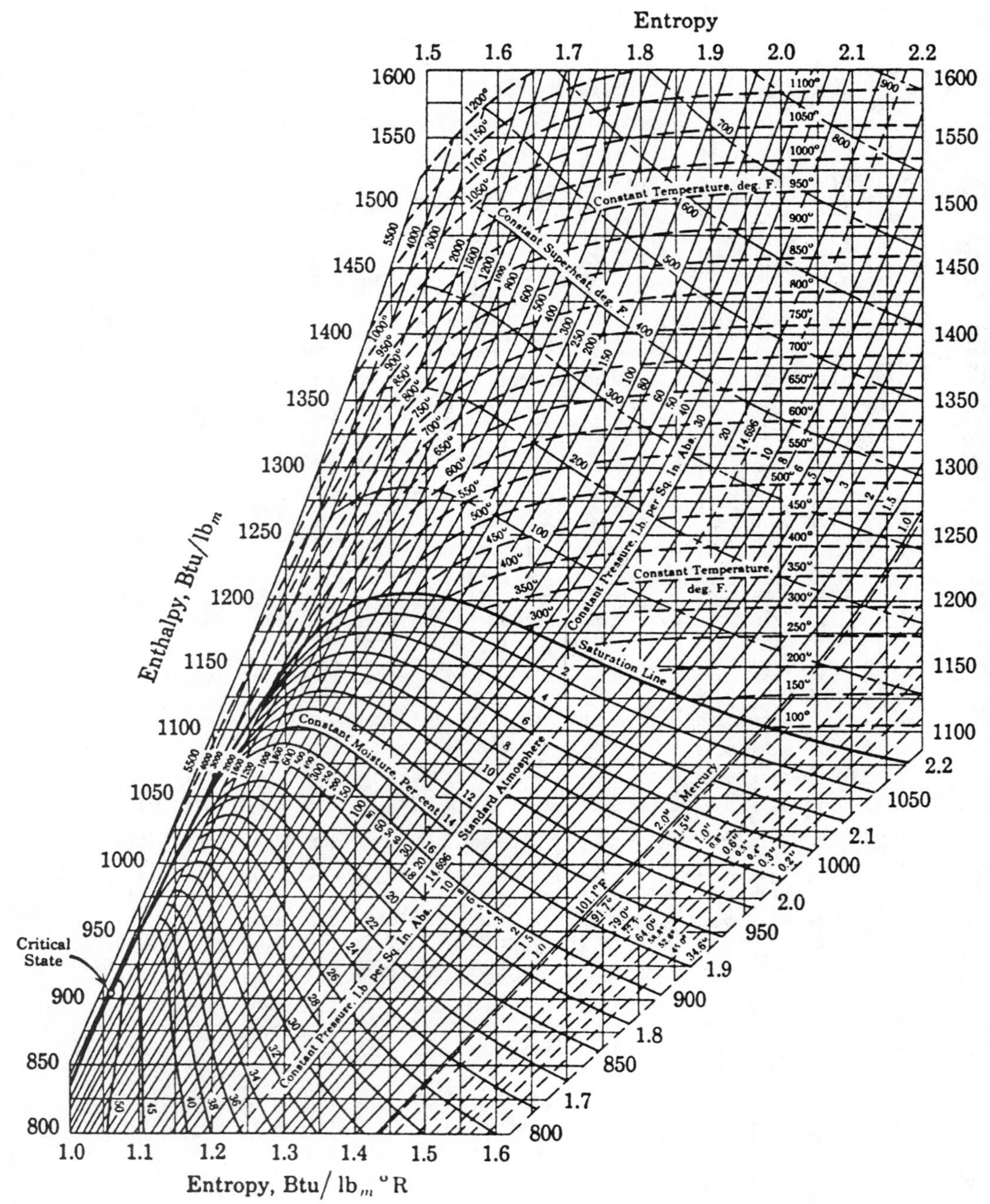

Properties of Saturated, Steam, Pressure Table

Abs press., lb/in.² p	Temp, °F t	Volume, ft³/lb		Enthalpy, Btu/lb		Entropy, Btu/(lb)(°R)		Internal energy, Btu/lb	
		Liquid v_f	Vapor v_g	Liquid h_f	Vapor h_g	Liquid s_f	Vapor s_g	Liquid u_f	Vapor u_g
1	101.74	0.01614	333.6	69.70	1,106.0	0.1326	1.9782	69.70	1,044.3
2	126.08	.01623	173.73	93.99	1,116.3	0.1749	1.9200	93.98	1,051.9
3	141.48	.01630	118.71	109.37	1,122.6	0.2008	1.8863	109.36	1,056.7
4	152.97	.01636	90.63	120.86	1,127.3	0.2198	1.8625	120.85	1,060.2
5	162.24	.01640	73.52	130.13	1,131.1	0.2347	1.8441	130.12	1,063.1
6	170.06	.01645	61.98	137.96	1,134.2	0.2472	1.8292	137.94	1,065.4
7	176.85	.01649	53.64	144.76	1,136.9	0.2581	1.8167	144.74	1,067.4
8	182.86	.01653	47.34	150.79	1,139.3	0.2674	1.8057	150.77	1,069.2
9	188.28	.01656	42.40	156.22	1,141.4	0.2759	1.7962	156.19	1,070.8
10	193.21	.01659	38.42	161.17	1,143.3	0.2835	1.7876	161.14	1,072.2
14.696	212.00	.01672	26.80	180.07	1,150.4	0.3120	1.7566	180.02	1,077.5
15	213.03	.01672	26.29	181.11	1,150.8	0.3135	1.7549	181.06	1,077.8
20	227.96	.01683	20.089	196.16	1,156.3	0.3356	1.7319	196.10	1,081.9
25	240.07	.01692	16.303	208.42	1,160.6	0.3533	1.7139	208.34	1,085.1
30	250.33	.01701	13.746	218.82	1,164.1	0.3680	1.6993	218.73	1,087.8
35	259.28	.01708	11.898	227.91	1,167.1	0.3807	1.6870	227.80	1,090.1
40	267.25	.01715	10.498	236.03	1,169.7	0.3919	1.6763	235.90	1,092.0
45	274.44	.01721	9.401	243.36	1,172.0	0.4019	1.6669	243.22	1,093.7
50	281.01	.01727	8.515	250.09	1,174.1	0.4110	1.6585	249.93	1,095.3
55	287.07	.01732	7.787	256.30	1,175.9	0.4193	1.6509	256.12	1,096.7
60	292.71	.01738	7.175	262.09	1,177.6	0.4270	1.6438	261.90	1,097.9
65	297.97	.01743	6.655	267.50	1,179.1	0.4342	1.6374	267.29	1,099.1
70	302.92	.01748	6.206	272.61	1,180.6	0.4409	1.6315	272.38	1,100.2
75	307.60	.01753	5.816	277.43	1,181.9	0.4472	1.6259	277.19	1,101.2
80	312.03	.01757	5.472	282.02	1,183.1	0.4531	1.6207	281.76	1,102.1
85	316.25	.01761	5.168	286.39	1,184.2	0.4587	1.6158	286.11	1,102.9
90	320.27	.01766	4.896	290.56	1,185.3	0.4641	1.6112	290.27	1,103.7
95	324.12	.01770	4.652	294.56	1,186.2	0.4692	1.6068	294.25	1,104.5
100	327.81	.01774	4.432	298.40	1,187.2	0.4740	1.6026	298.08	1,105.2
110	334.77	.01782	4.049	305.66	1,188.9	0.4832	1.5948	305.30	1,106.5
120	341.25	.01789	3.728	312.44	1,190.4	0.4916	1.5878	312.05	1,107.6
130	347.32	.01796	3.455	318.81	1,191.7	0.4995	1.5812	318.38	1,108.6
140	353.02	.01802	3.220	324.82	1,193.0	0.5069	1.5751	324.35	1,109.6
150	358.42	.01809	3.015	330.51	1,194.1	0.5138	1.5694	330.01	1,110.5
160	363.53	.01815	2.834	335.93	1,195.1	0.5204	1.5640	335.39	1,111.2
170	368.41	.01822	2.675	341.09	1,196.0	0.5266	1.5590	340.52	1,111.9
180	373.06	.01827	2.532	346.03	1,196.9	0.5325	1.5542	345.42	1,112.5
190	377.51	.01833	2.404	350.79	1,197.6	0.5381	1.5497	350.15	1,113.1
200	381.79	.01839	2.288	355.36	1,198.4	0.5435	1.5453	354.68	1,113.7
250	400.95	.01865	1.8438	376.00	1,201.1	0.5675	1.5263	375.14	1,115.8
300	417.33	.01890	1.5433	393.84	1,202.8	0.5879	1.5104	392.79	1,117.1
350	431.72	.01913	1.3260	409.69	1,203.9	0.6056	1.4966	408.45	1,118.0
400	444.59	.0193	1.1613	424.0	1,204.5	0.6214	1.4844	422.6	1,118.5
450	456.28	.0195	1.0320	437.2	1,204.6	0.6356	1.4734	435.5	1,118.7
500	467.01	.0197	0.9278	499.4	1,204.4	0.6487	1.4634	447.6	1,118.6
550	476.94	.0199	.8424	460.8	1,203.9	0.6608	1.4542	458.8	1,118.2
600	486.21	.0201	.7698	471.6	1,203.2	0.6720	1.4454	469.4	1,117.7
650	494.90	.0203	.7083	481.8	1,202.3	0.6826	1.4374	479.4	1,117.1
700	503.10	.0205	.6554	491.5	1,201.2	0.6925	1.4296	488.8	1,116.3
750	510.86	.0207	.6092	500.8	1,200.0	0.7019	1.4223	598.0	1,115.4
800	518.23	.0209	.5687	509.7	1,198.6	0.7108	1.4153	506.6	1,114.4
850	525.26	.0210	.5327	518.3	1,197.1	0.7194	1.4085	515.0	1,113.3
900	531.98	.0212	.5006	526.6	1,195.4	0.7275	1.4020	523.1	1,112.1
950	538.43	.0214	.4717	534.6	1,193.7	0.7355	1.3957	530.9	1,110.8
1,000	544.61	.0216	.4456	542.4	1,191.8	0.7430	1.3897	538.4	1,109.4
1,100	556.31	.0220	.4001	557.4	1,187.8	0.7575	1.3780	552.9	1,106.4
1,200	567.22	.0223	.3619	571.7	1,183.4	0.7711	1.3667	566.7	1,103.0
1,300	577.46	.0227	.3293	585.4	1,178.6	0.7840	1.3559	580.0	1,099.4
1,400	587.10	.0231	.3012	598.7	1,173.4	0.7963	1.3454	592.7	1,095.4
1,500	596.23	.0235	.2765	611.6	1,167.9	0.8082	1.3351	605.1	1,091.2
2,000	635.82	.0257	.1878	671.7	1,135.1	0.8619	1.2849	662.2	1,065.6
2,500	668.13	.0287	.1307	730.6	1,091.1	0.9126	1.2322	717.3	1,030.6
3,000	695.36	.0346	.0858	802.5	1,020.3	0.9731	1.1615	783.4	972.7
3,206.2	705.40	.0503	.0503	902.7	902.7	1.0580	1.0580	872.9	872.9

STEAM TABLES

Dry Saturated Steam: Temperature Table*

Temp., °F	Abs. Press., (lb/in²)	Specific volume: Sat. liquid	Specific volume: Evap.	Specific volume: Sat. vapor	Enthalpy: Sat. liquid	Enthalpy: Evap.	Enthalpy: Sat. vapor	Entropy: Sat. liquid	Entropy: Evap.	Entropy: Sat. vapor	Temp., °F
t	p	v_f	v_{fg}	v_g	h_f	h_{fg}	h_g	s_f	s_{fg}	s_g	t
32	0.08854	0.01602	3306	3306	0.00	1075.8	1075.8	0.0000	2.1877	2.1877	**32**
35	0.09995	0.01602	2947	2947	3.02	1074.1	1077.1	0.0061	2.1709	2.1770	**35**
40	0.12170	0.01602	2444	2444	8.05	1071.3	1079.3	0.0162	2.1435	2.1597	**40**
45	0.14752	0.01602	2036.4	2036.4	13.06	1068.4	1081.5	0.0262	2.1167	2.1429	**45**
50	0.17811	0.01603	1703.2	1703.2	18.07	1065.6	1083.7	0.0361	2.0903	2.1264	**50**
60	0.2563	0.01604	1206.6	1206.7	28.06	1059.9	1088.0	0.0555	2.0393	2.0948	**60**
70	0.3631	0.01606	867.8	867.9	38.04	1054.3	1092.3	0.0745	1.9902	2.0647	**70**
80	0.5069	0.01608	633.1	633.1	48.02	1048.6	1096.6	0.0932	1.9428	2.0360	**80**
90	0.6982	0.01610	468.0	468.0	57.99	1042.9	1100.9	0.1115	1.8972	2.0087	**90**
100	0.9492	0.01613	350.3	350.4	67.97	1037.2	1105.2	0.1295	1.8531	1.9826	**100**
110	1.2748	0.01617	265.3	265.4	77.94	1031.6	1109.5	0.1471	1.8106	1.9577	**110**
120	1.6924	0.01620	203.25	203.27	87.92	1025.8	1113.7	0.1645	1.7694	1.9339	**120**
130	2.2225	0.01625	157.32	157.34	97.90	1020.0	1117.9	0.1816	1.7296	1.9112	**130**
140	2.8886	0.01629	122.99	123.01	107.89	1014.1	1122.0	0.1984	1.6910	1.8894	**140**
150	3.718	0.01634	97.06	97.07	117.89	1008.2	1126.1	0.2149	1.6537	1.8685	**150**
160	4.741	0.01639	77.27	77.29	127.89	1002.3	1130.2	0.2311	1.6174	1.8485	**160**
170	5.992	0.01645	62.04	62.06	137.90	996.3	1134.2	0.2472	1.5822	1.8293	**170**
180	7.510	0.01651	50.21	50.23	147.92	990.2	1138.1	0.2630	1.5480	1.8109	**180**
190	9.339	0.01657	40.94	40.96	157.95	984.1	1142.0	0.2785	1.5147	1.7932	**190**
200	11.526	0.01663	33.62	33.64	167.99	977.9	1145.9	0.2938	1.4824	1.7762	**200**
210	14.123	0.01670	27.80	27.82	178.05	971.6	1149.7	0.3090	1.4508	1.7598	**210**
212	14.696	0.01672	26.78	26.80	180.07	970.3	1150.4	0.3120	1.4446	1.7566	**212**
220	17.186	0.01677	23.13	23.15	188.13	965.2	1153.4	0.3239	1.4201	1.7440	**220**
230	20.780	0.01684	19.365	19.382	198.23	958.8	1157.0	0.3387	1.3901	1.7288	**230**
240	24.969	0.01692	16.306	16.323	208.34	952.2	1160.5	0.3531	1.3609	1.7140	**240**
250	29.825	0.01700	13.804	13.821	218.48	945.5	1164.0	0.3675	1.3323	1.6998	**250**
260	35.429	0.01709	11.746	11.763	228.64	938.7	1167.3	0.3817	1.3043	1.6860	**260**
270	41.858	0.01717	10.044	10.061	238.84	931.8	1170.6	0.3958	1.2769	1.6727	**270**
280	49.203	0.01726	8.628	8.645	249.06	924.7	1173.8	0.4096	1.2501	1.6597	**280**
290	57.556	0.01735	7.444	7.461	259.31	917.5	1176.8	0.4234	1.2238	1.6472	**290**

300	67.013	0.01745	6.449	6.466	269.59	910.1	1179.7	0.4369	1.1980	1.6350	300
310	77.68	0.01755	5.609	5.626	279.92	902.6	1182.5	0.4504	1.1727	1.6231	310
320	89.66	0.01765	4.896	4.914	290.28	894.9	1185.2	0.4637	1.1478	1.6115	320
330	103.06	0.01776	4.289	4.307	300.68	887.0	1187.7	0.4769	1.1233	1.6002	330
340	118.01	0.01787	3.770	3.788	311.13	879.0	1190.1	0.4900	1.0992	1.5891	340
350	134.63	0.01799	3.324	3.342	321.63	870.7	1192.3	0.5029	1.0754	1.5783	350
360	153.04	0.01811	2.939	2.957	332.18	862.2	1194.4	0.5158	1.0519	1.5677	360
370	173.37	0.01823	2.606	2.625	342.79	853.5	1196.3	0.5286	1.0287	1.5573	370
380	195.77	0.01836	2.317	2.335	353.45	844.6	1198.1	0.5413	1.0059	1.5471	380
390	220.37	0.01850	2.0651	2.0836	364.17	835.4	1199.6	0.5539	0.9832	1.5371	390
400	247.31	0.01864	1.8447	1.8633	374.97	826.0	1201.0	0.5664	0.9608	1.5272	400
410	276.75	0.01878	1.6512	1.6700	385.83	816.3	1202.1	0.5788	0.9386	1.5174	410
420	308.83	0.01894	1.4811	1.5000	396.77	806.3	1203.1	0.5912	0.9166	1.5078	420
430	343.72	0.01910	1.3308	1.3499	407.79	796.0	1203.8	0.6035	0.8947	1.4982	430
440	381.59	0.01926	1.1979	1.2171	418.90	785.4	1204.3	0.6158	0.8730	1.4887	440
450	422.6	0.0194	1.0799	1.0993	430.1	774.5	1204.6	0.6280	0.8513	1.4793	450
460	466.9	0.0196	0.9748	0.9944	441.4	763.2	1204.6	0.6402	0.8298	1.4700	460
470	514.7	0.0198	0.8811	0.9009	452.8	751.5	1204.3	0.6523	0.8083	1.4606	470
480	566.1	0.0200	0.7972	0.8172	464.4	739.4	1203.7	0.6645	0.7868	1.4513	480
490	621.4	0.0202	0.7221	0.7423	476.0	726.8	1202.8	0.6766	0.7653	1.4419	490
500	680.8	0.0204	0.6545	0.6749	487.8	713.9	1201.7	0.6887	0.7438	1.4325	500
520	812.4	0.0209	0.5385	0.5594	511.9	686.4	1198.2	0.7130	0.7006	1.4136	520
540	962.5	0.0215	0.4434	0.4649	536.6	656.6	1193.2	0.7374	0.6568	1.3942	540
560	1133.1	0.0221	0.3647	0.3868	562.2	624.2	1186.4	0.7621	0.6121	1.3742	560
580	1325.8	0.0228	0.2989	0.3217	588.9	588.4	1177.3	0.7872	0.5659	1.3532	580
600	1542.9	0.0236	0.2432	0.2668	617.0	548.5	1165.5	0.8131	0.5176	1.3307	600
620	1786.6	0.0247	0.1955	0.2201	646.7	503.6	1150.3	0.8398	0.4664	1.3062	620
640	2059.7	0.0260	0.1538	0.1798	678.6	452.0	1130.5	0.8679	0.4110	1.2789	640
660	2365.4	0.0278	0.1165	0.1442	714.2	390.2	1104.4	0.8987	0.3485	1.2472	660
680	2708.1	0.0305	0.0810	0.1115	757.3	309.9	1067.2	0.9351	0.2719	1.2071	680
700	3093.7	0.0369	0.0392	0.0761	823.3	172.1	995.4	0.9905	0.1484	1.1389	700
705.4	3206.2	0.0503	0	0.0503	902.7	0	902.7	1.0580	0	1.0580	705.4

Note: Specific volume in ft^3/lbm; enthalpy in Btu/lbm; entropy in Btu/lbm·°F.

Enthalpies and Entropies of Gases

Gas	Symbol	Enthalpy, kcal/mole					Entropy, cal/mole				
		1000°K	1500°K	2000°K	2500°K	3000°K	1000°K	1500°K	2000°K	2500°K	3000°K
Acetylene	C_2H_2	317.9	326.5	336.0	346.0	356.3	64.32	71.34	76.78	81.22	84.97
Ammonia	NH_3	102.5	109.7	117.8	126.5	135.6	58.39	64.17	68.83	72.71	76.01
Argon	Ar	5.0	7.5	9.9	12.4	14.9	42.99	45.01	46.44	47.55	48.45
Benzene	C_6H_6	808.0	835.1	.....	.....	.....	106.73	128.68			
Butane	C_4H_{10}	704.1	712.7	.....	.....	.....	120.31	144.22			
Carbon dioxide	CO_2	10.2	17.0	24.1	31.4	38.8	64.34	69.82	73.90	77.15	79.85
Carbon monoxide	CO	73.0	77.1	81.4	85.8	90.2	56.03	59.35	61.81	63.76	65.37
Dichlorodifluoromethane	CCl_2F_2	134.4	146.8	159.5	172.2	185.1	97.92	107.98	115.27	120.96	125.63
Ethane	C_2H_6	381.2	397.8	416.4	436.2	456.6	79.39	92.46			
Ethylene	C_2H_4	347.9	360.3	374.0	388.5	403.4	72.06	82.02	89.90	96.36	101.79
Helium	He	4.97	7.45	9.94	12.4	14.9	36.15	38.16	39.59	40.70	41.61
Heptane	C_7H_{16}	1178.0	1192.3	.....	.....	.....	180.32	220.45			
Hydrogen	H_2	74.3	78.1	82.1	86.3	90.6	39.70	42.72	45.00	46.88	48.47
Methane	CH_4	222.3	231.9	242.7	254.3	266.3	59.14	66.84	73.08	78.23	82.60
Nitrogen	N_2	8.9	13.0	17.2	21.6	26.0	54.51	57.78	60.22	62.16	63.77
Oxygen	O_2	9.5	13.8	18.3	22.8	27.6	58.19	61.66	64.21	66.25	67.97
Propane	C_3H_8	543.4	550.2	.....	.....	.....	99.77	118.29			
Water	H_2O	19.9	25.2	31.1	37.4	43.9	55.59	59.86	63.23	66.03	68.42

Enthalpy of Combustion of Substances at 77°F(25°C) and 1 Atm

Substance	Formula	Molecular Weight	H_2O Appears as Liquid in Products of Combustion		H_2O Appears as Vapor in Products of Combustion	
			Btu/lb mol	Btu/lbm	Btu/lb mol	Btu/lbm
Hydrogen	$H_2(g)$	2.016	−122,970	−60,997	−104,040	−51,605
Carbon (graphite)	$C(s)$	12.011	−169,300	−14,095	−169,300	−14,095
Carbon monoxide	$CO(g)$	28.011	−121,750	−4,347	−121,750	−4,347
Methane	$CH_4(g)$	16.043	−383,030	−23,875	−345,160	−21,515
Acetylene	$C_2H_2(g)$	26.038	−559,110	−21,473	−540,170	−20,745
Ethylene	$C_2H(g)$	28.054	−607,110	−21,640	−569,240	−20,291
Ethane	$C_2H_6(g)$	30.070	−671,080	−22,317	−614,280	−20,428
Propane	$C_3H_8(g)$	44.097	−955,090	−21,659	−879,380	−19,941
Benzene	$C_6H_6(g)$	78.114	−1,420,400	−18,184	−1,363,600	−17,457
Octane	$C_8H_{18}(g)$	114.23	−2,371,400	−20,760	−2,201,000	−19,268
	$C_8H_{18}(l)$	114.23	−2,353,600	−20,604	−2,183,200	−19,112

SOURCES: 1. *JANAF Thermochemical Tables*, Second Edition, NSRDS-NBS-37 (Catalog No. C13.48:37), 1971.
2. *Circular No. 500*, National Bureau of Standards, 1952.

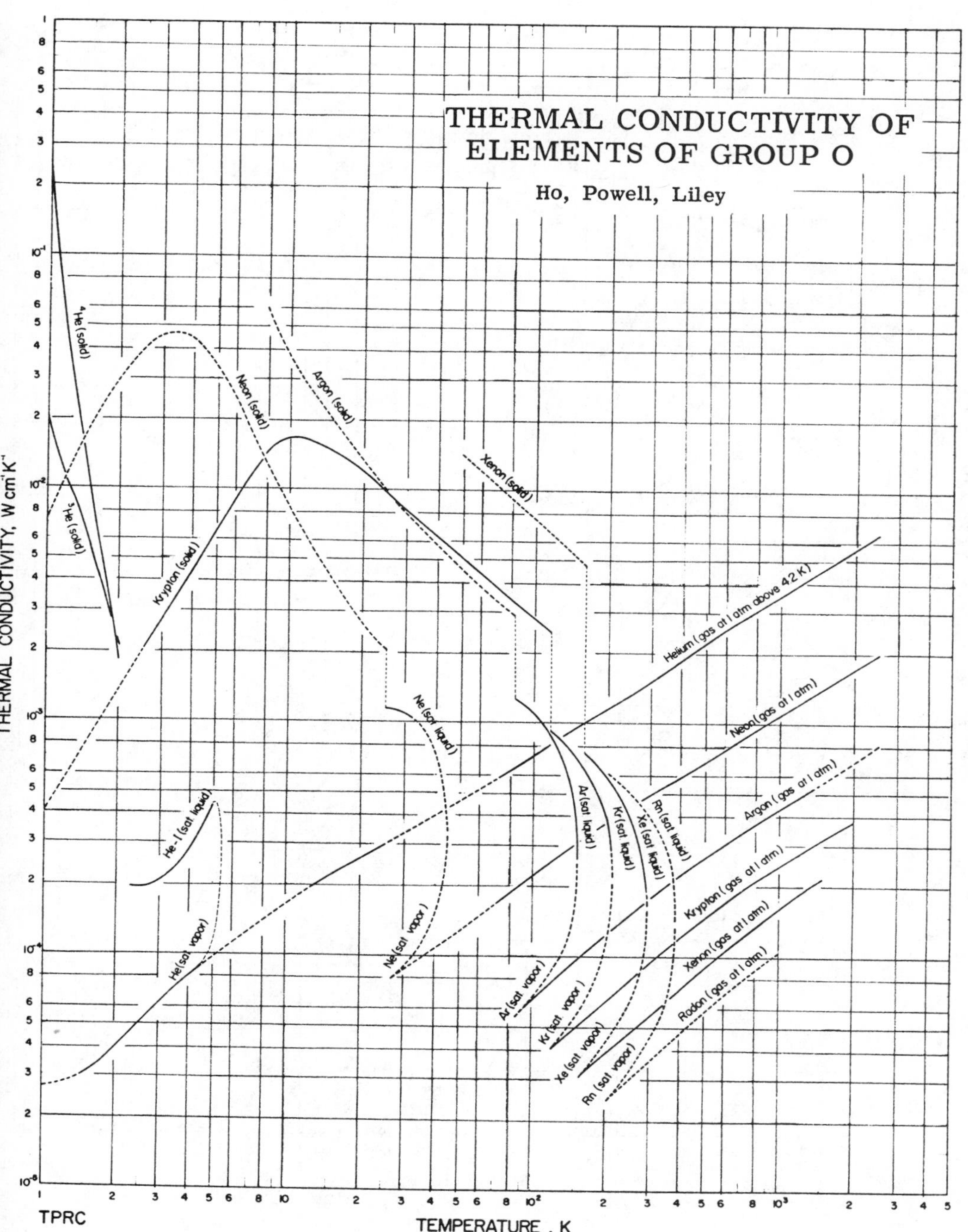

THERMAL CONDUCTIVITY OF ELEMENTS OF GROUP O
Ho, Powell, Liley
THERMAL CONDUCTIVITY, W cm⁻¹ K⁻¹
TEMPERATURE, K
TPRC
⁴He (solid)
³He (solid)
Neon (solid)
Argon (solid)
Krypton (solid)
Xenon (solid)
Helium (gas at 1 atm above 4.2 K)
Neon (gas at 1 atm)
Argon (gas at 1 atm)
Krypton (gas at 1 atm)
Xenon (gas at 1 atm)
Radon (gas at 1 atm)
He-I (sat liquid)
He (sat vapor)
Ne (sat liquid)
Ne (sat vapor)
Ar (sat liquid)
Ar (sat vapor)
Kr (sat liquid)
Kr (sat vapor)
Xe (sat liquid)
Xe (sat vapor)
Rn (sat liquid)
Rn (sat vapor)

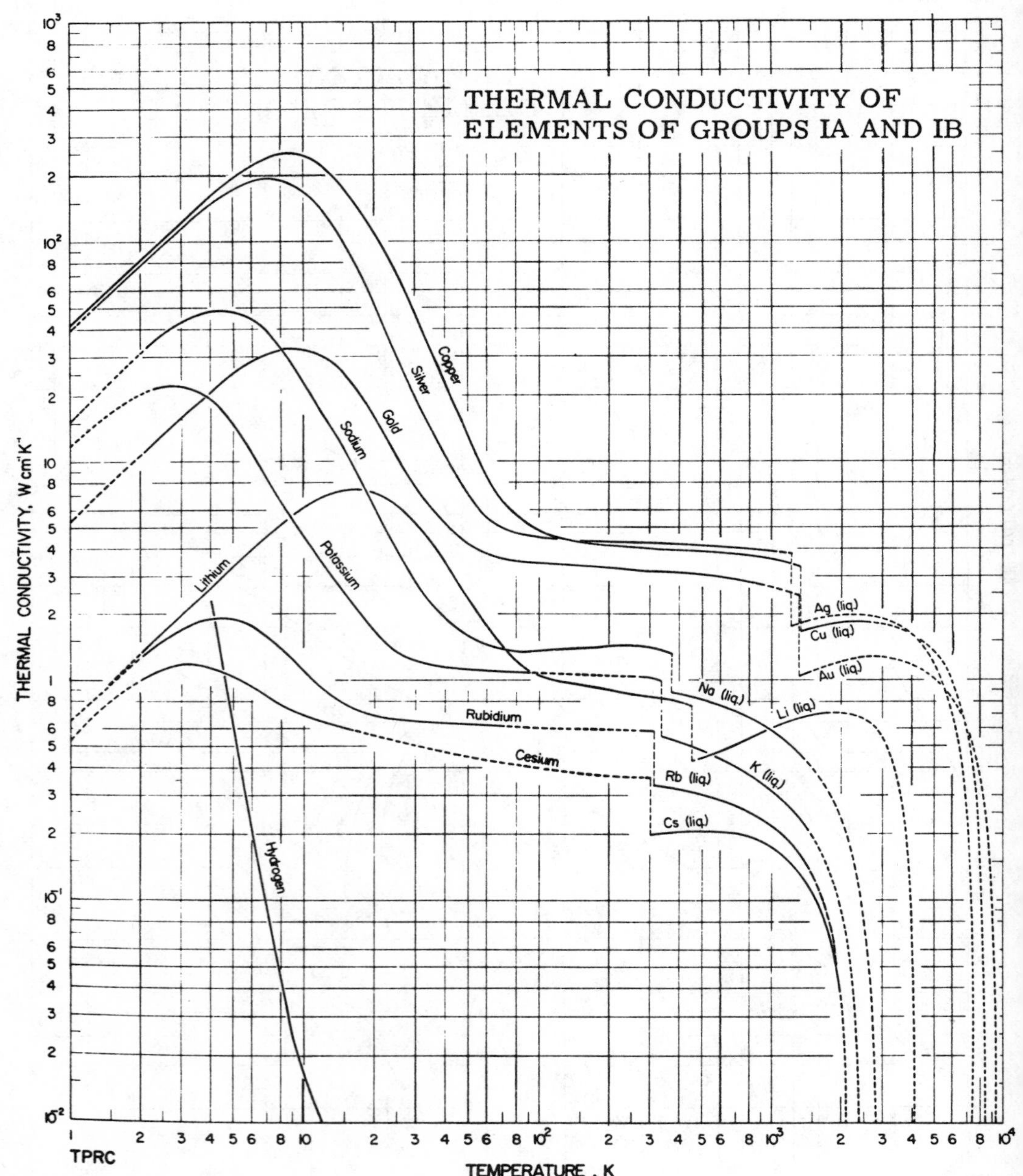

THERMAL CONDUCTIVITY OF ELEMENTS OF GROUPS IA AND IB
THERMAL CONDUCTIVITY, W cm⁻¹ K⁻¹
TEMPERATURE , K
TPRC
Copper
Silver
Gold
Sodium
Potassium
Lithium
Rubidium
Cesium
Hydrogen
Ag (liq.)
Cu (liq.)
Au (liq.)
Na (liq.)
Li (liq.)
K (liq.)
Rb (liq.)
Cs (liq.)

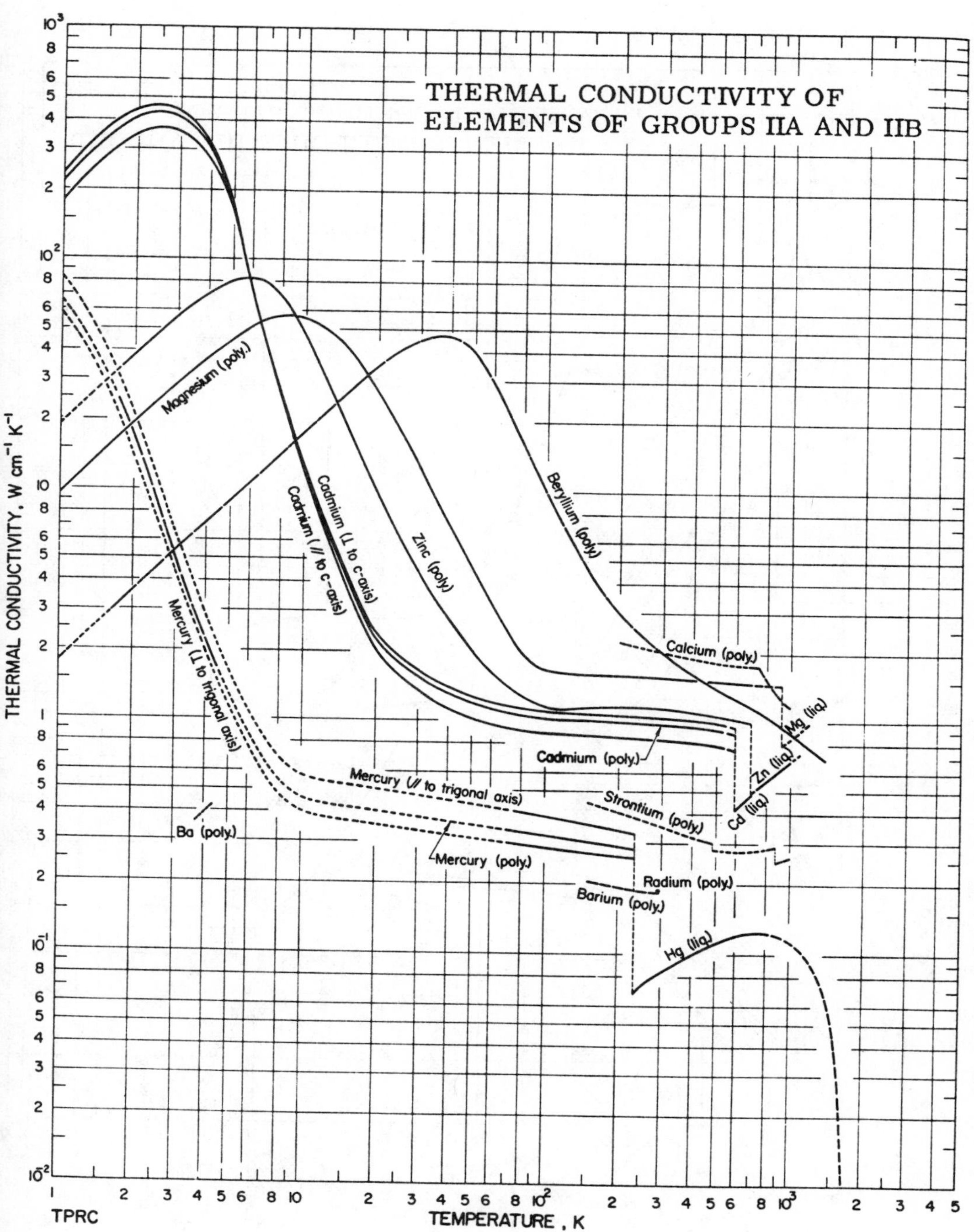
THERMAL CONDUCTIVITY OF
ELEMENTS OF GROUPS IIA AND IIB
THERMAL CONDUCTIVITY, W cm⁻¹ K⁻¹
TEMPERATURE, K
TPRC
Magnesium (poly.)
Beryllium (poly.)
Zinc (poly.)
Cadmium (⊥ to c-axis)
Cadmium (// to c-axis)
Cadmium (poly.)
Mercury (⊥ to trigonal axis)
Mercury (// to trigonal axis)
Mercury (poly.)
Ba (poly.)
Calcium (poly.)
Mg (liq.)
Zn (liq.)
Cd (liq.)
Strontium (poly.)
Radium (poly.)
Barium (poly.)
Hg (liq.)

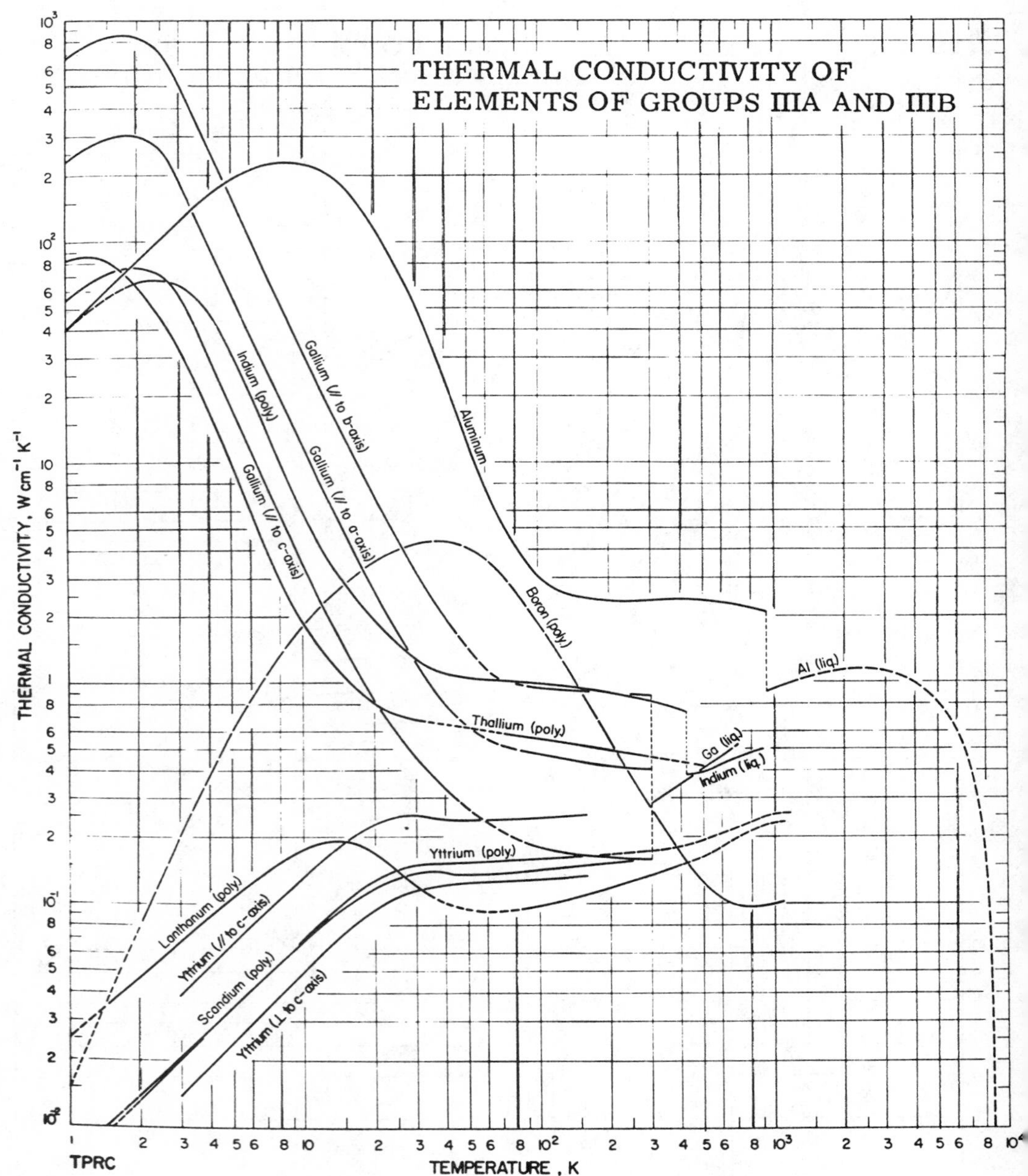
THERMAL CONDUCTIVITY OF
ELEMENTS OF GROUPS IIIA AND IIIB
THERMAL CONDUCTIVITY, W cm⁻¹ K⁻¹
TEMPERATURE, K
TPRC
Aluminum
Gallium (// to b-axis)
Gallium (// to a-axis)
Gallium (// to c-axis)
Indium (poly.)
Boron (poly.)
Thallium (poly.)
Al (liq.)
Ga (liq.)
Indium (liq.)
Yttrium (poly.)
Lanthanum (poly.)
Yttrium (// to c-axis)
Scandium (poly.)
Yttrium (⊥ to c-axis)

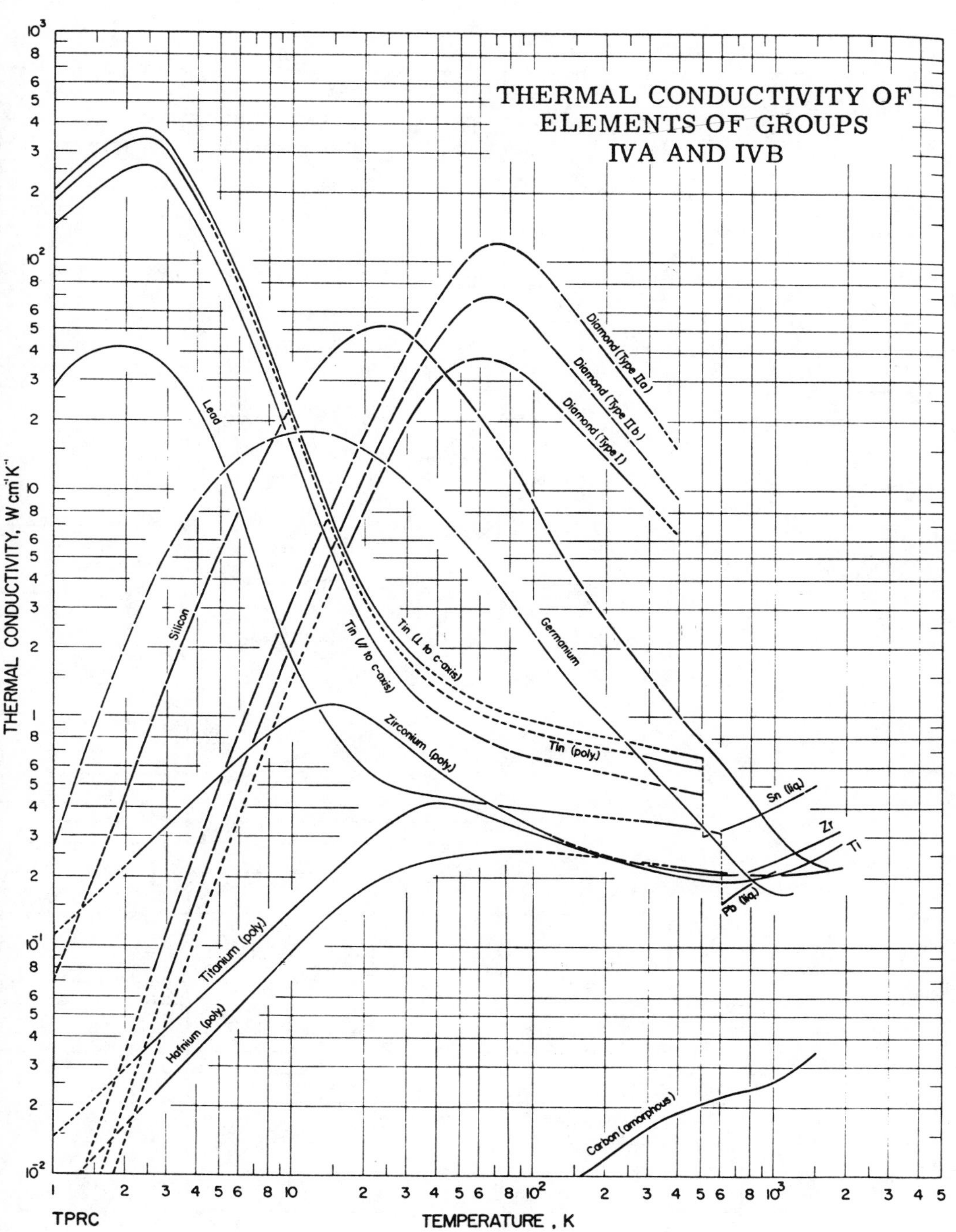
THERMAL CONDUCTIVITY OF ELEMENTS OF GROUPS IVA AND IVB
THERMAL CONDUCTIVITY, W cm⁻¹ K⁻¹
TEMPERATURE, K
TPRC
Diamond (Type IIa)
Diamond (Type IIb)
Diamond (Type I)
Lead
Silicon
Germanium
Tin (⊥ to c-axis)
Tin (// to c-axis)
Tin (poly.)
Zirconium (poly.)
Titanium (poly.)
Hafnium (poly.)
Carbon (amorphous)
Sn (liq.)
Pb (liq.)
Zr
Ti

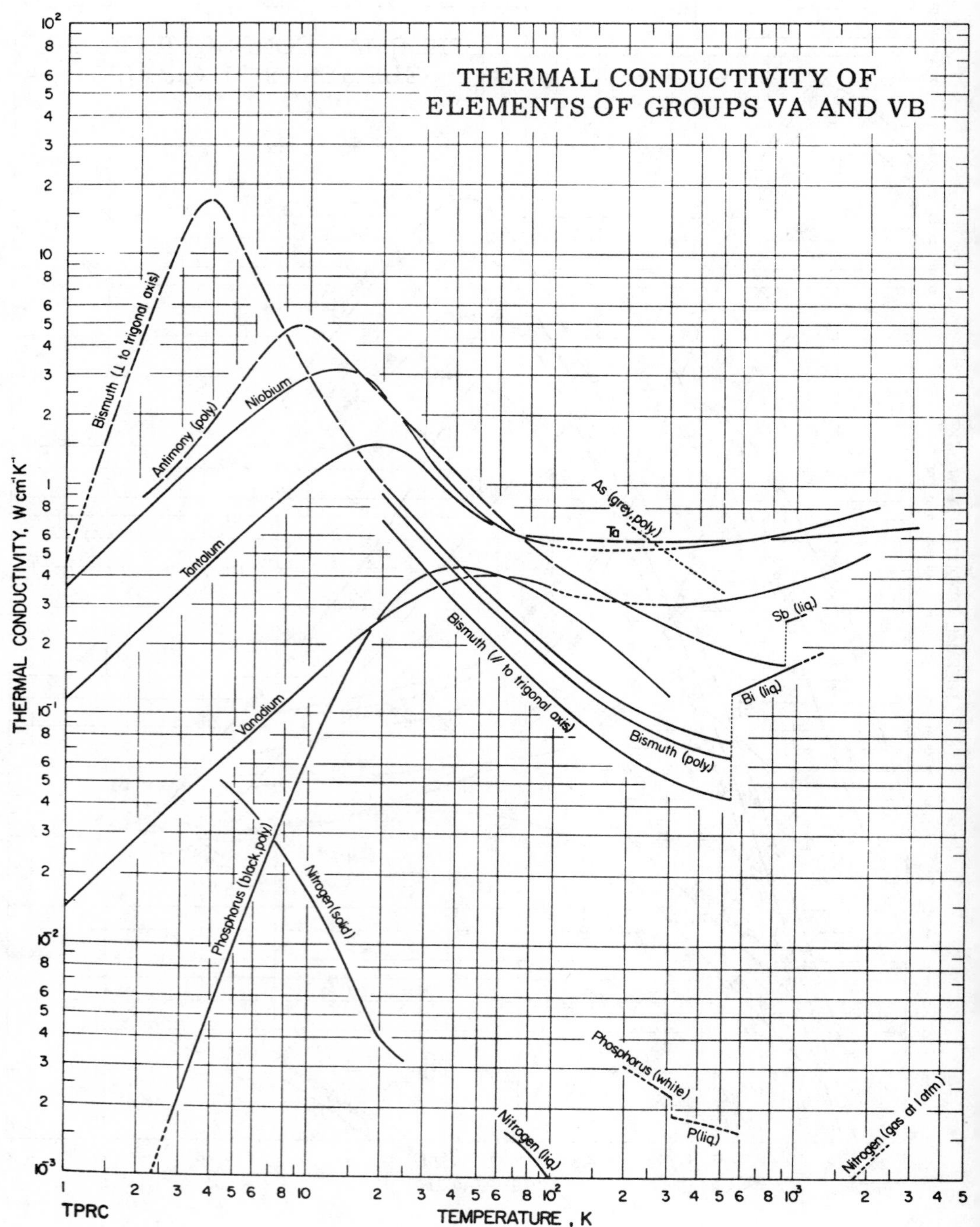

THERMAL CONDUCTIVITY OF ELEMENTS OF GROUPS VA AND VB
THERMAL CONDUCTIVITY, W cm⁻¹ K⁻¹
TEMPERATURE, K
TPRC
Bismuth (⊥ to trigonal axis)
Antimony (poly)
Niobium
Tantalum
Vanadium
Bismuth (// to trigonal axis)
Bismuth (poly)
As (grey, poly.)
Ta
Sb (liq)
Bi (liq)
Phosphorus (black, poly)
Nitrogen (solid)
Nitrogen (liq)
Phosphorus (white)
P(liq)
Nitrogen (gas at 1 atm)

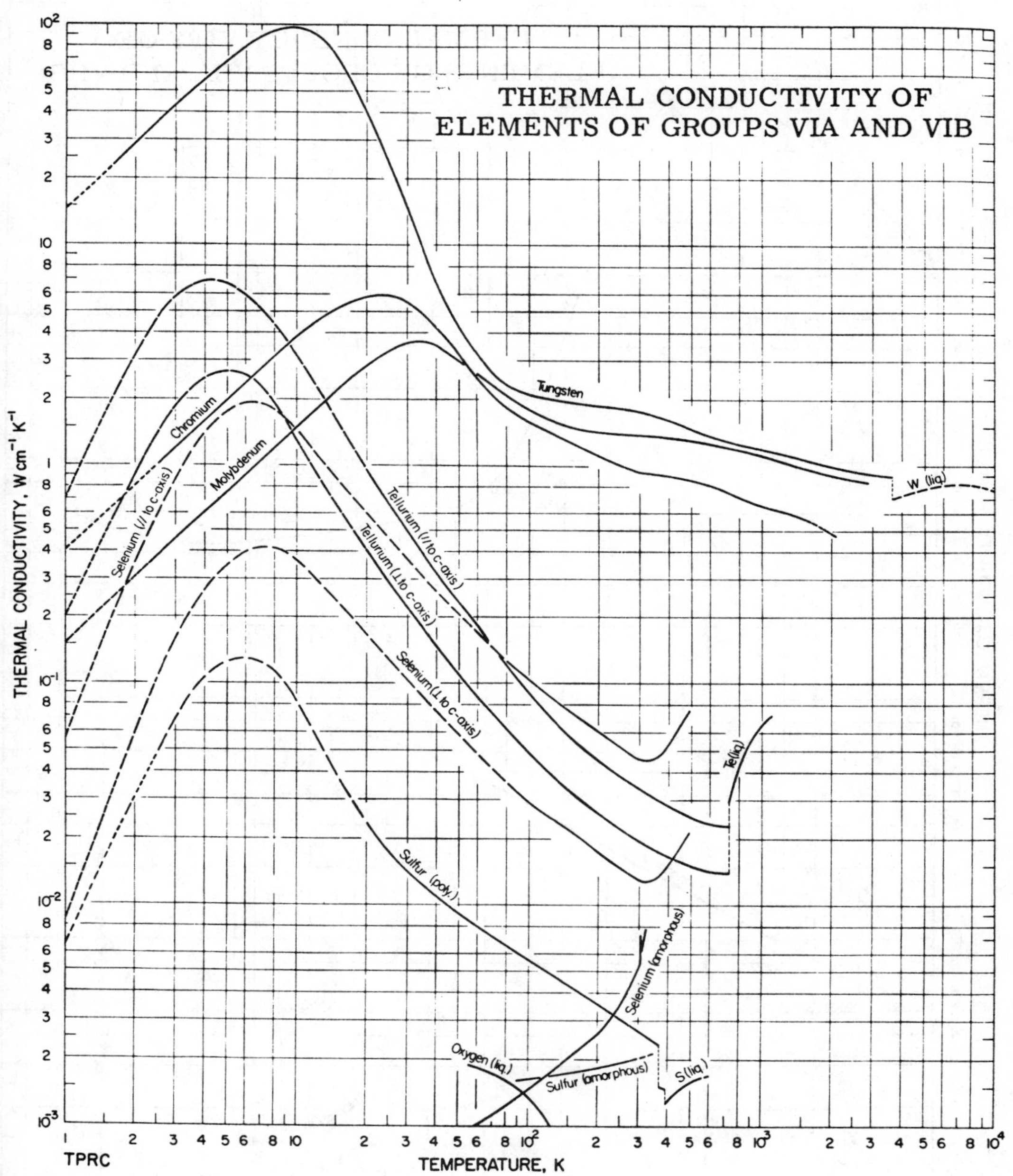

THERMAL CONDUCTIVITY OF
ELEMENTS OF GROUPS VIA AND VIB
Tungsten
W (liq.)
Chromium
Molybdenum
Selenium (// to c-axis)
Tellurium (// to c-axis)
Tellurium (⊥ to c-axis)
Selenium (⊥ to c-axis)
Te(liq.)
Sulfur (poly.)
Selenium (amorphous)
Oxygen (liq.)
Sulfur (amorphous)
S (liq.)
THERMAL CONDUCTIVITY, W cm⁻¹ K⁻¹
TEMPERATURE, K
TPRC

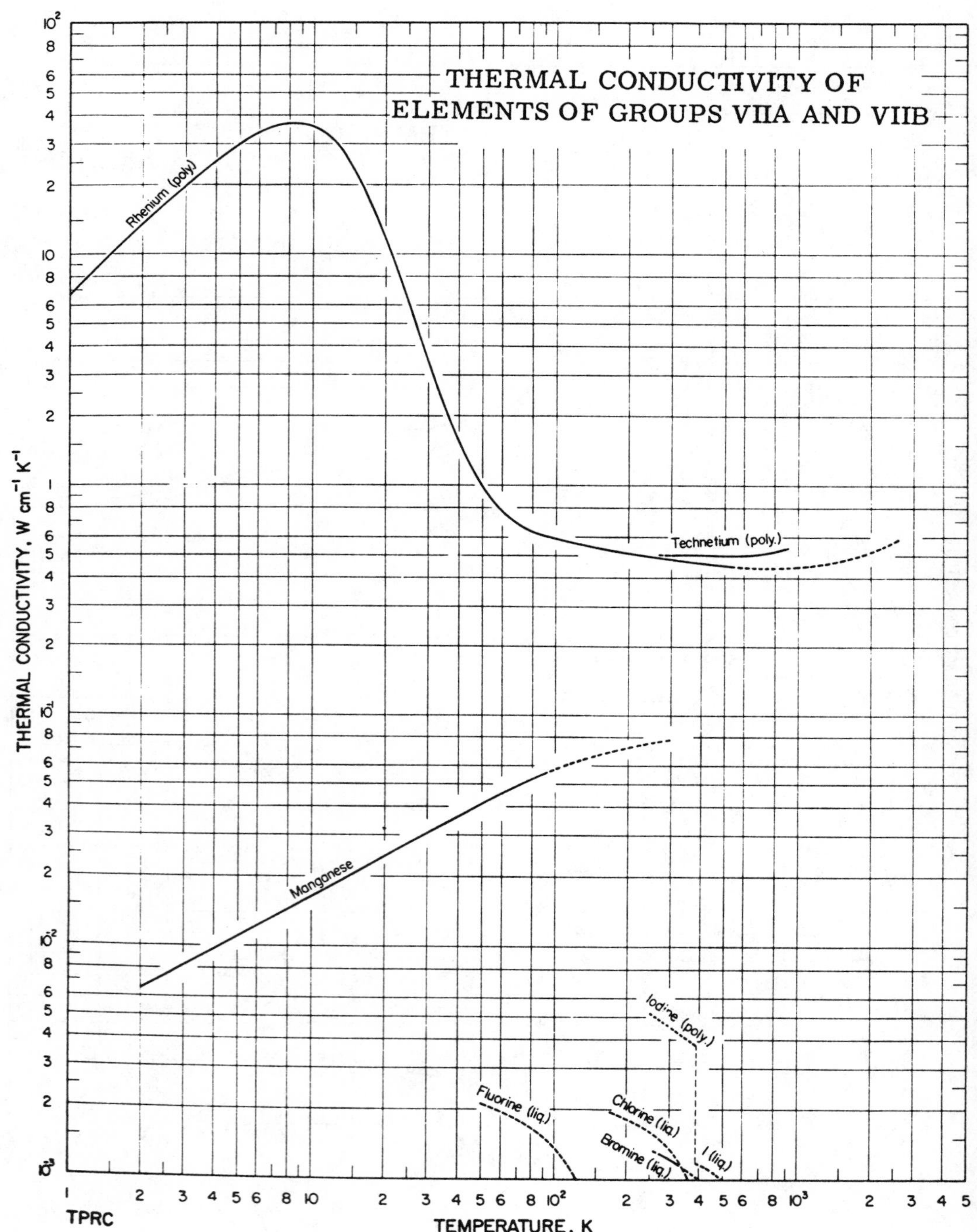
THERMAL CONDUCTIVITY OF
ELEMENTS OF GROUPS VIIA AND VIIB
Rhenium (poly.)
Technetium (poly.)
Manganese
Iodine (poly.)
Fluorine (liq.)
Chlorine (liq.)
Bromine (liq.)
I (liq.)
THERMAL CONDUCTIVITY, W cm⁻¹ K⁻¹
TEMPERATURE, K
TPRC

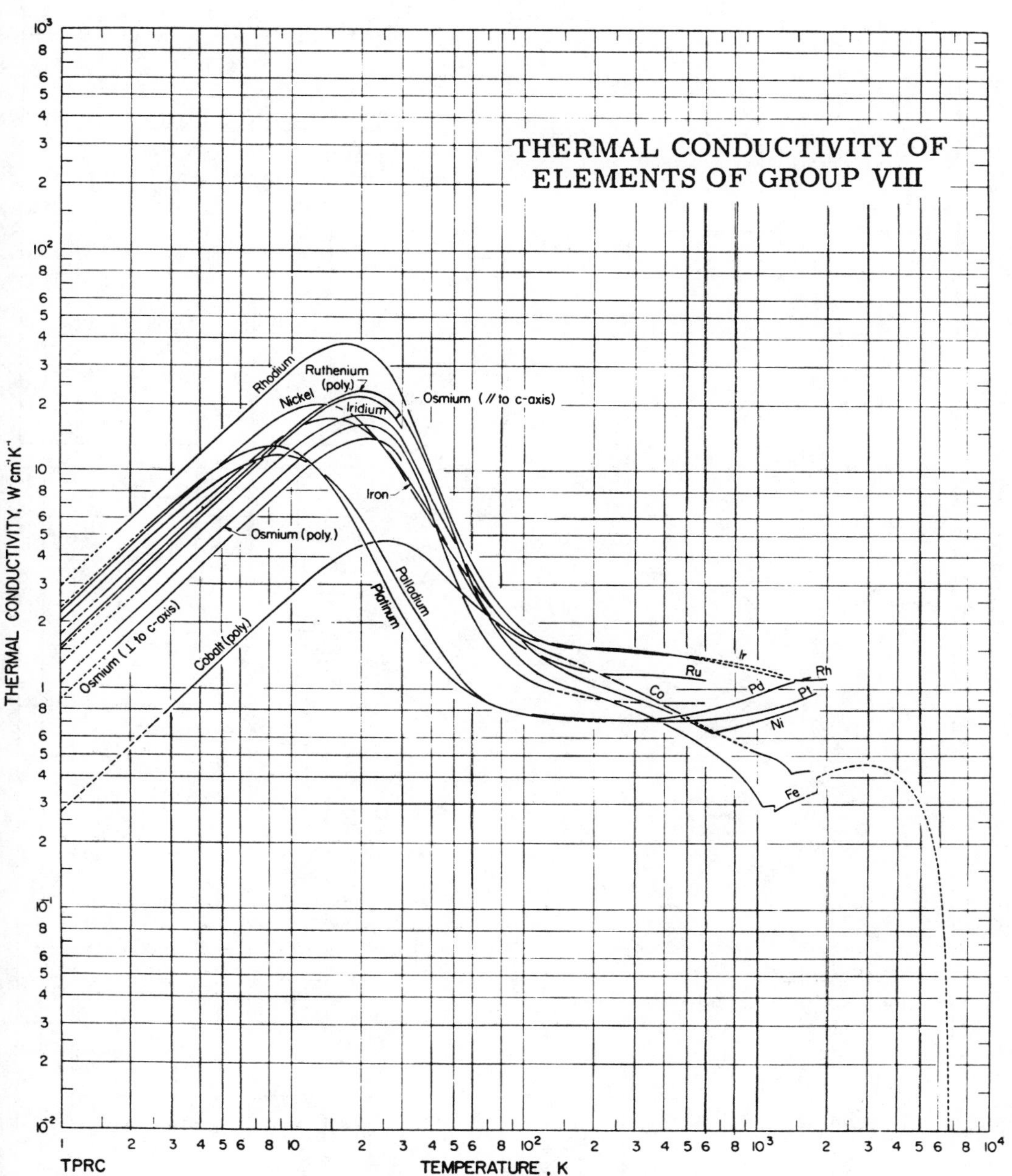

THERMAL CONDUCTIVITY OF
ELEMENTS OF GROUP VIII
THERMAL CONDUCTIVITY, W cm⁻¹ K⁻¹
TEMPERATURE, K
TPRC
Rhodium
Ruthenium
(poly)
Nickel
Iridium
Osmium (// to c-axis)
Iron
Osmium (poly.)
Osmium (⊥ to c-axis)
Cobalt (poly.)
Palladium
Platinum
Ir
Ru
Rh
Co
Pd
Pt
Ni
Fe

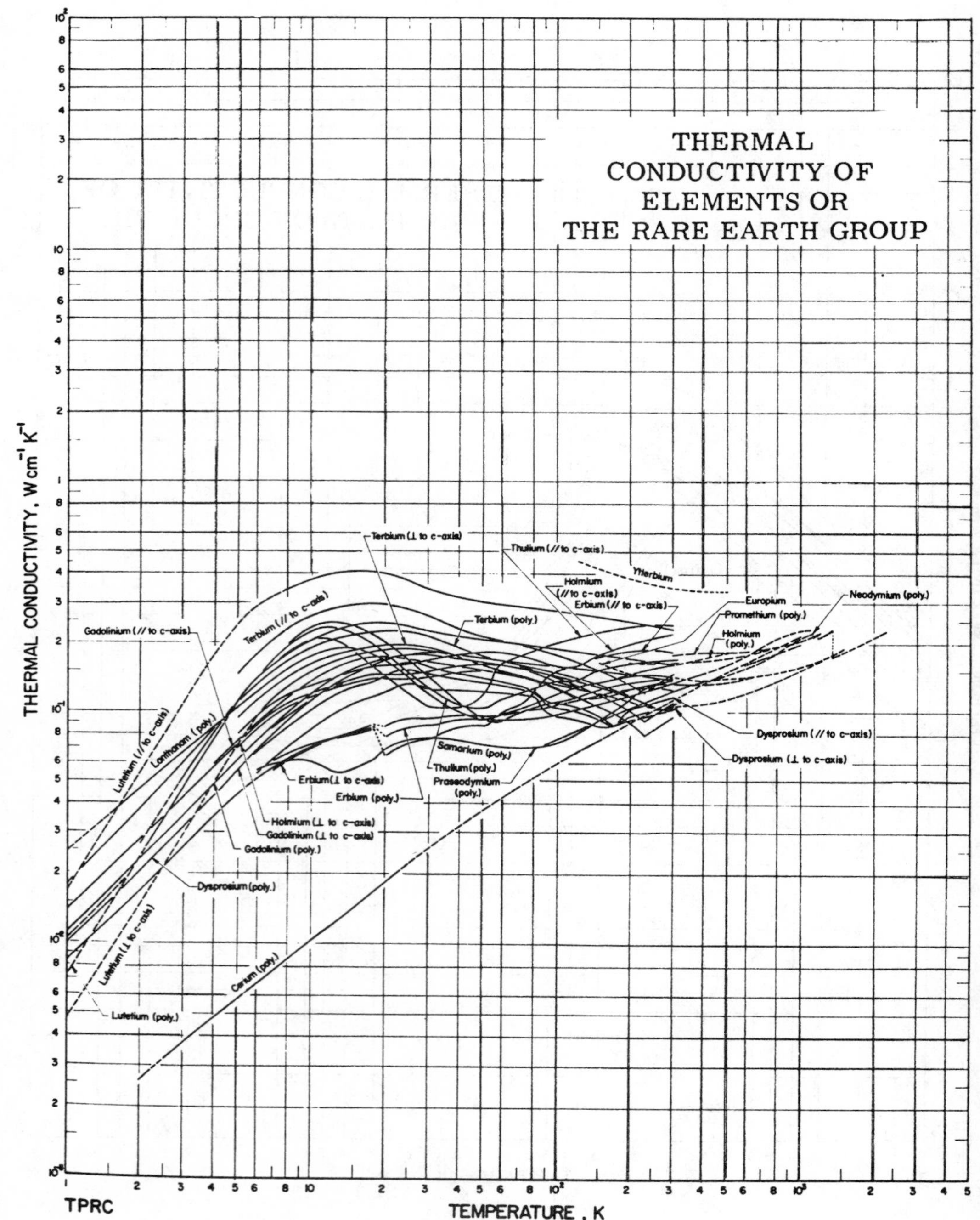
THERMAL CONDUCTIVITY OF ELEMENTS OR THE RARE EARTH GROUP
THERMAL CONDUCTIVITY, W cm⁻¹ K⁻¹
TEMPERATURE, K
TPRC
Terbium (⊥ to c-axis)
Thulium (// to c-axis)
Ytterbium
Holmium (// to c-axis)
Erbium (// to c-axis)
Europium
Promethium (poly.)
Neodymium (poly.)
Holmium (poly.)
Terbium (poly.)
Terbium (// to c-axis)
Gadolinium (// to c-axis)
Lutetium (// to c-axis)
Lanthanum (poly.)
Dysprosium (// to c-axis)
Dysprosium (⊥ to c-axis)
Samarium (poly.)
Thulium (poly.)
Praseodymium (poly.)
Erbium (⊥ to c-axis)
Erbium (poly.)
Holmium (⊥ to c-axis)
Gadolinium (⊥ to c-axis)
Gadolinium (poly.)
Dysprosium (poly.)
Lutetium (⊥ to c-axis)
Lutetium (poly.)
Cerium (poly.)

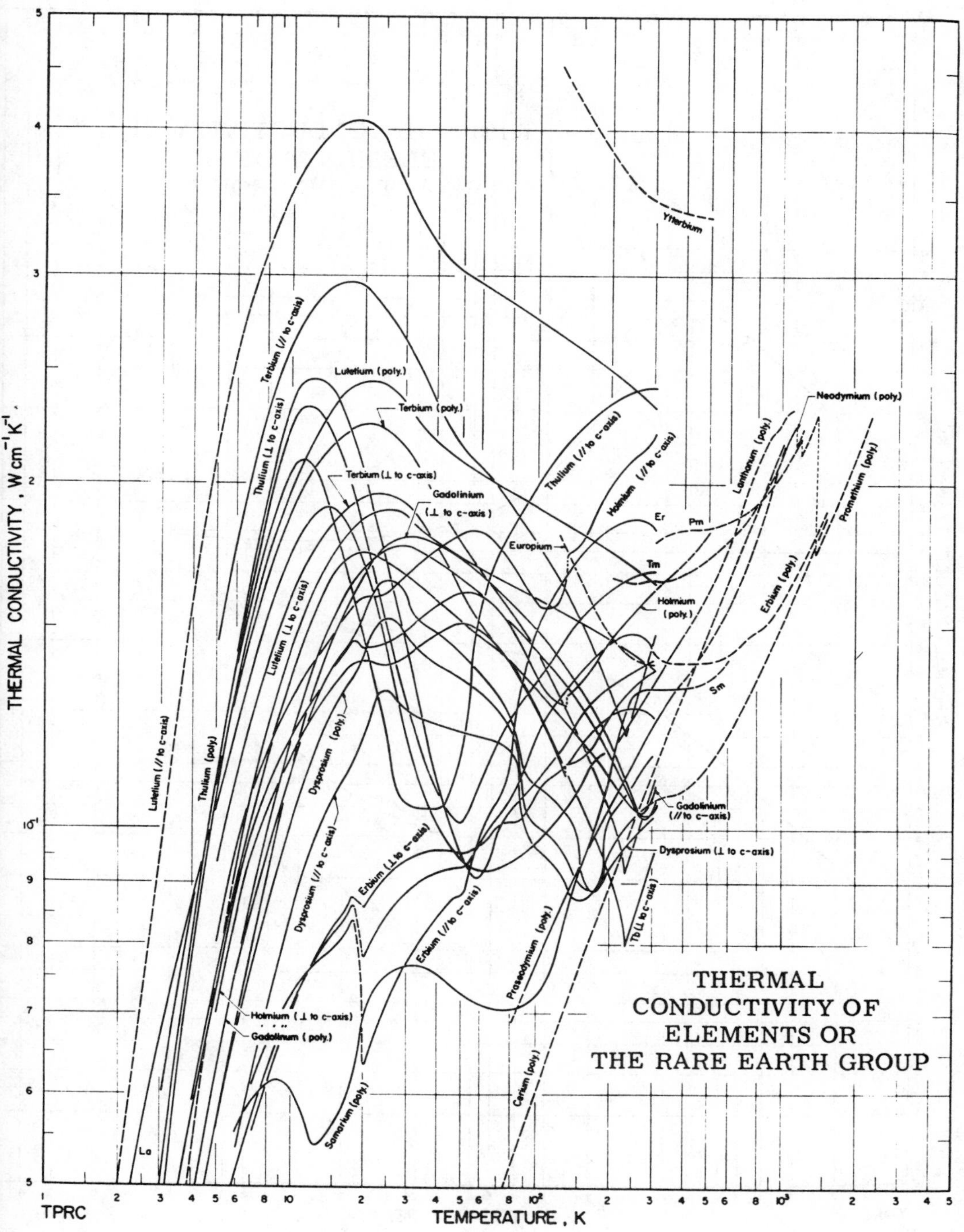

THERMAL CONDUCTIVITY OF ELEMENTS OR THE RARE EARTH GROUP

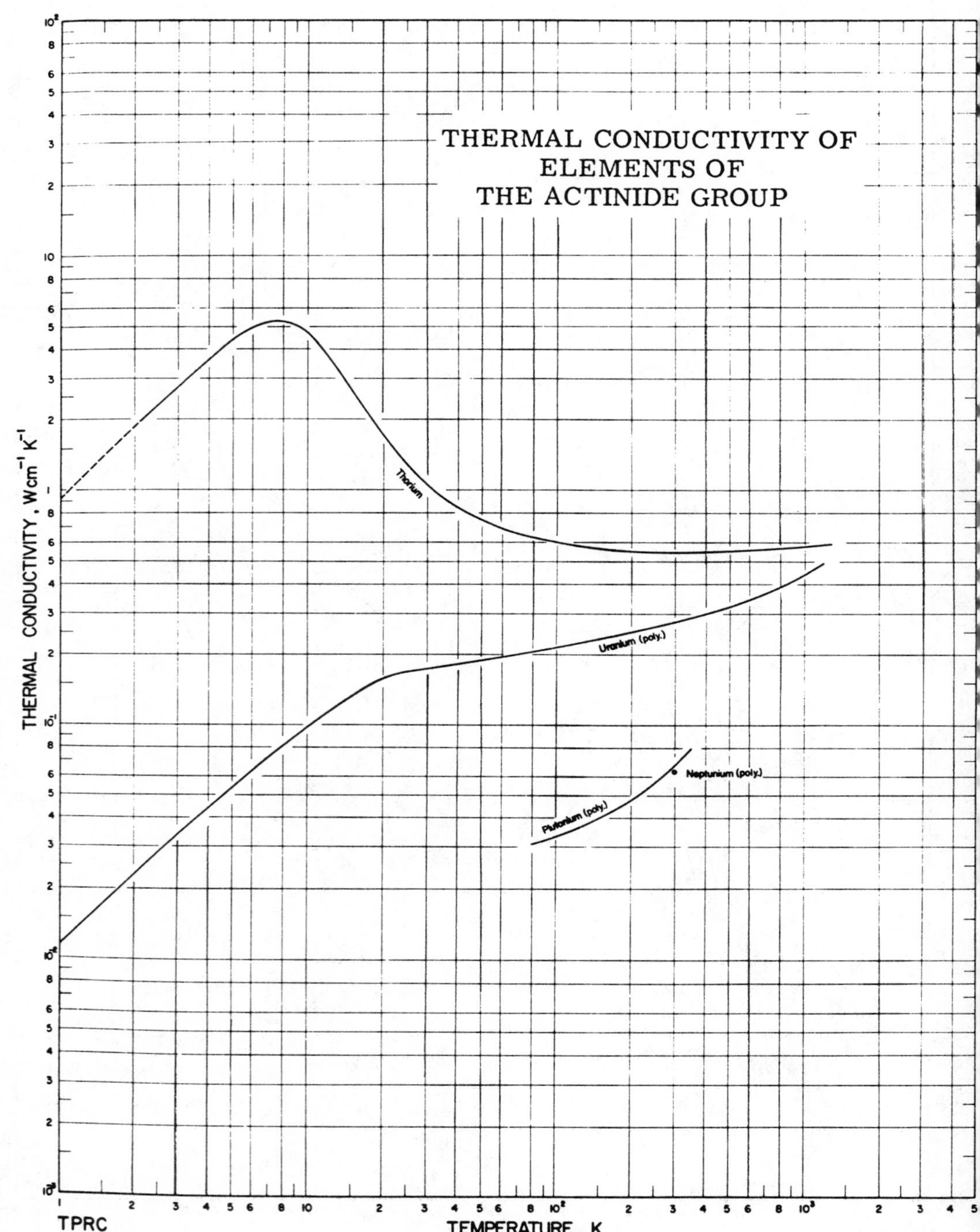
THERMAL CONDUCTIVITY OF
ELEMENTS OF
THE ACTINIDE GROUP
Thorium
Uranium (poly.)
Neptunium (poly.)
Plutonium (poly.)
THERMAL CONDUCTIVITY, $W cm^{-1} K^{-1}$
TEMPERATURE, K
TPRC

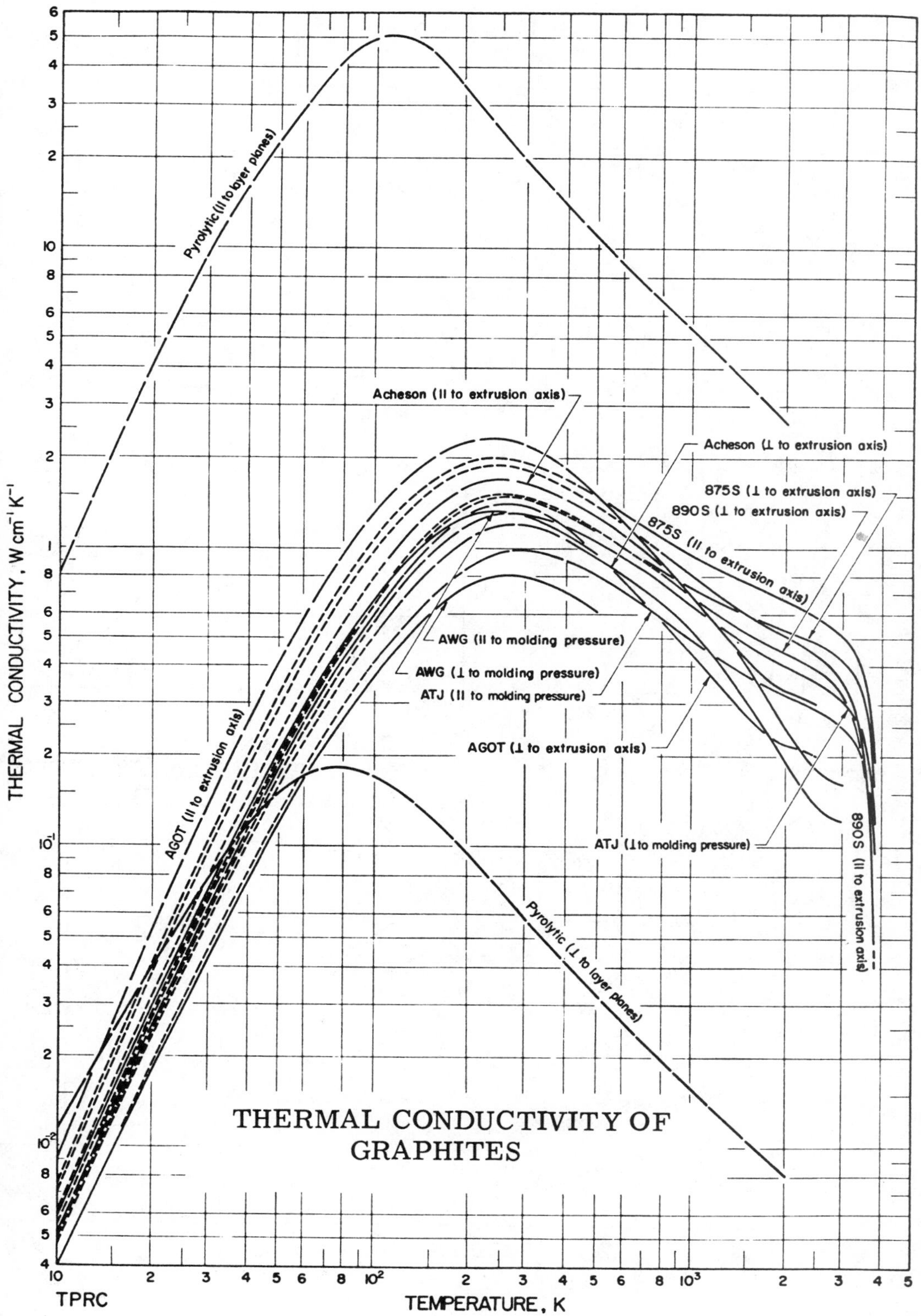
Pyrolytic (II to layer planes)
Acheson (II to extrusion axis)
Acheson (⊥ to extrusion axis)
875S (⊥ to extrusion axis)
890S (⊥ to extrusion axis)
875S (II to extrusion axis)
AWG (II to molding pressure)
AWG (⊥ to molding pressure)
ATJ (II to molding pressure)
AGOT (⊥ to extrusion axis)
ATJ (⊥ to molding pressure)
890S (II to extrusion axis)
AGOT (II to extrusion axis)
Pyrolytic (⊥ to layer planes)
THERMAL CONDUCTIVITY, W cm⁻¹ K⁻¹
TEMPERATURE, K
TPRC

THERMAL CONDUCTIVITY OF GRAPHITES

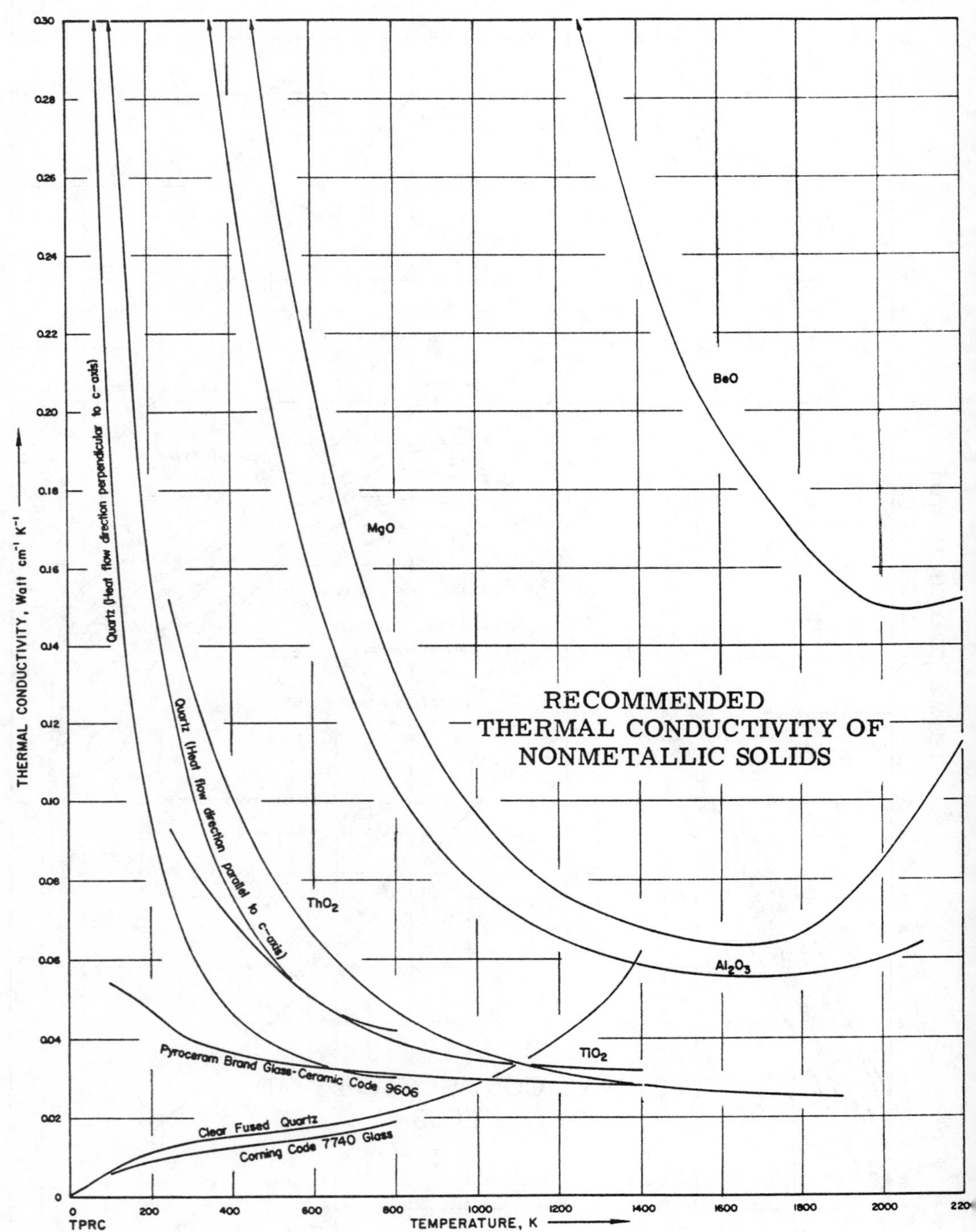

RECOMMENDED
THERMAL CONDUCTIVITY OF
NONMETALLIC SOLIDS
THERMAL CONDUCTIVITY, Watt cm⁻¹ K⁻¹
TEMPERATURE, K
TPRC
Quartz (Heat flow direction perpendicular to c-axis)
Quartz (Heat flow direction parallel to c-axis)
BeO
MgO
ThO2
Al2O3
TiO2
Pyroceram Brand Glass-Ceramic Code 9606
Clear Fused Quartz
Corning Code 7740 Glass
0.30
0.28
0.26
0.24
0.22
0.20
0.18
0.16
0.14
0.12
0.10
0.08
0.06
0.04
0.02
0
200
400
600
800
1000
1200
1400
1600
1800
2000

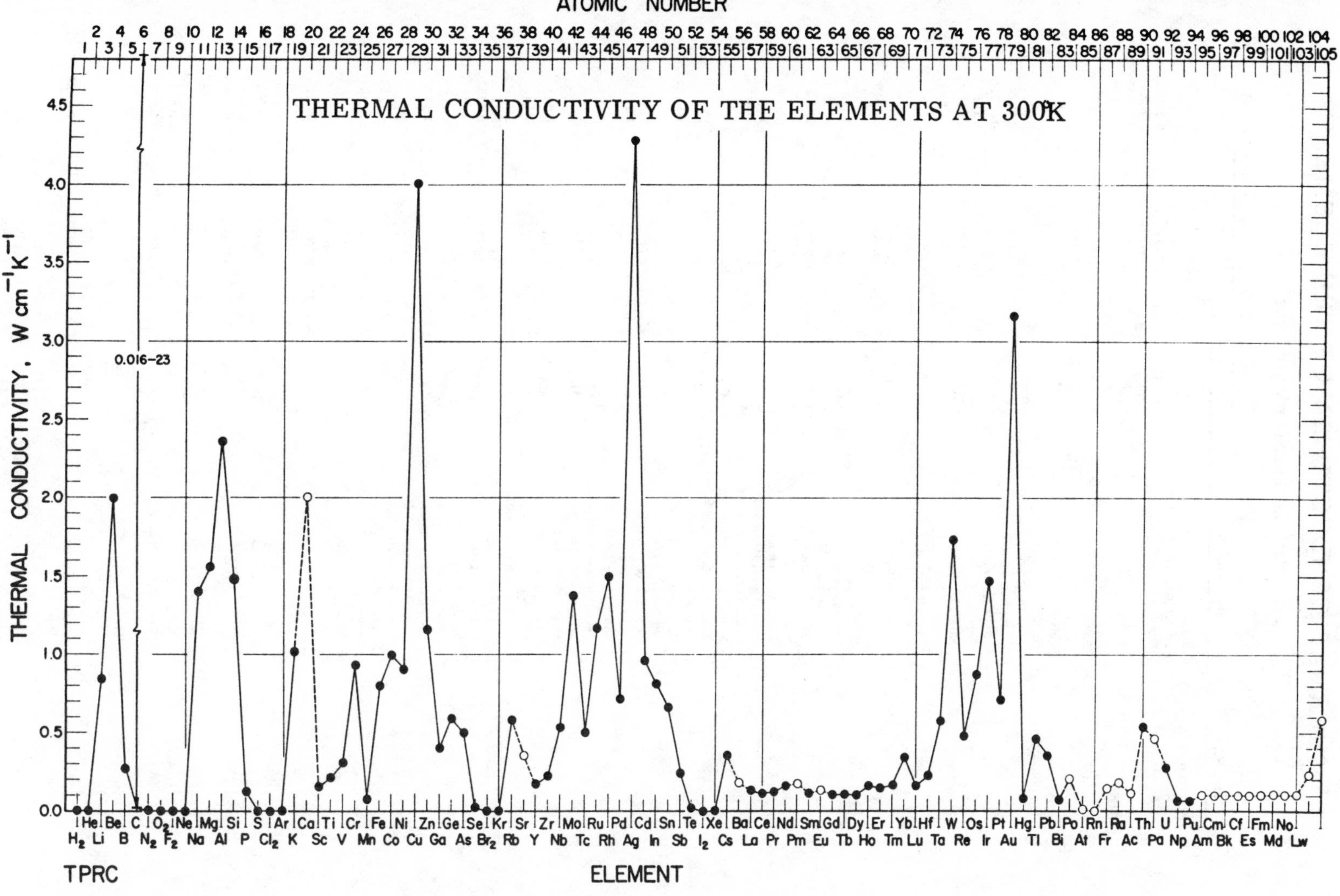
THERMAL CONDUCTIVITY OF THE ELEMENTS AT 300K
ATOMIC NUMBER
ELEMENT
THERMAL CONDUCTIVITY, W cm^{-1}K^{-1}
0.016–23
TPRC

THERMAL CONDUCTIVITY OF THE ELEMENTS

National Bureau of Standards

Element	State or Condition	Conductivity at (watt/cm/°K) 273.2K	298.2K	373.2K
Aluminum	Solid	2.36	2.37	2.40
Antimony	Polycrystalline	0.255	0.244	0.219
Argon	Gas at 1 atm.	0.1619[m]	0.1772[m]	0.2103[m]
		(270K)	(300K)	(370K)
Arsenic	Solid, Gray, Polycrystalline	0.539*	0.502*	0.427*
Barium	Solid	0.185*	0.184*	
			(295K)	
Beryllium	Polycrystalline	2.18*	2.01	1.68
Bismuth	Solid			
	∥ to triagonal axis	0.0554	0.0530	0.0481
	⊥ to triagonal axis	0.0953	0.0919	0.0844
	Polycrystalline	0.0822	0.0792	0.0722
Boron	Solid	0.318	0.274	0.188
Bromine	Saturated liquid	1.30[m]*	1.22[m]*	1.06[m]*
		(270K)	(300K)	(370K)
	Saturated vapor		0.048[m]*	
			(300K)	
	Gas	0.042[m]*	0.048[m]*	0.057[m]*
		(270K)	(300K)	(350K)
Cadmium	Solid			
	∥ to c-axis	0.835	0.830	0.816
	⊥ to c-axis	1.04	1.04	1.02
	Polycrystalline	0.975	0.969	0.953
Calcium	Solid	2.06*	2.01*	1.92*
Carbon	Solid, Amorphous	0.0150	0.0159	0.0182
	Solid, Type I (Diamond)	9.94	9.90	7.03*
	Solid, Type IIa (Diamond)	26.2	23.2	17.0*
	Solid, Type IIb (Diamond)	15.2	13.6	10.2*
	Solid, Acheson graphite			
	∥ to axis of extrusion	1.69	1.65	1.50
	⊥ to axis of extrusion	1.21	1.19	1.11
	Solid, AGOT graphite			
	∥ to axis of extrusion	2.28	2.21	1.95
	⊥ to axis of extrusion	1.41	1.38	1.22
	Solid, ATJ graphite			
	∥ to molding pressure	0.984	0.982	0.933
	⊥ to molding pressure	1.31	1.29	1.21
	Solid, AWG graphite			
	∥ to molding pressure	0.807	0.796	0.733
	⊥ to molding pressure	1.32	1.28	1.16
	Solid, Pyrolytic graphite			
	∥ to layer planes	21.3	19.6	15.1
	⊥ to layer planes	0.0636	0.0573	0.0442
	Solid, 875S graphite			
	∥ to axis of extrusion	1.97*	1.92*	1.75*
	⊥ to axis of extrusion	1.49*	1.46*	1.34*
	Solid, 890S graphite			
	∥ to axis of extrusion	1.87*	1.83*	1.66*
	⊥ to axis of extrusion	1.51*	1.48*	1.36*
Cerium	Solid, Polycrystalline	0.108*	0.113	0.128*
Cesium	Solid	0.361*	0.359	
			(301.9K)	
	Liquid		0.197	0.201
			(301.9K)	
Chlorine	Saturated liquid	1.49[m]*	1.34[m]*	0.95[m]*
		(270K)	(300K)	(370K)
	Saturated vapor	0.082[m]*	0.097[m]*	0.155[m]*
		(270K)	(300K)	(370K)
	Gas, 1 atm.	0.078[m]	0.089[m]	0.114[m]
		(270K)	(300K)	(370K)
Chromium	Solid, Polycrystalline	0.965	0.939	0.921
Cobalt	Solid, Polycrystalline	1.05	1.00	0.890
Copper	Solid	4.03	4.01	3.95
Dysprosium	Solid			
	∥ to c-axis	0.114*	0.117*	
	⊥ to c-axis	0.101*	0.103*	
	Polycrystalline	0.105*	0.107*	0.108*
Erbium	Solid			
	∥ to c-axis	0.187*	0.184*	
	⊥ to c-axis	0.127*	0.126*	
	Polycrystalline	0.147*	0.145*	0.140*

THERMAL CONDUCTIVITY OF THE ELEMENTS (continued)

Element	State or Condition	Conductivity at (watt/cm/°K) 273.2K	298.2K	373.2K
Europium	Solid	0.140*	0.139*	
Fluorine	Gas, 1 atm.	0.251^m^ (270K)	0.279^m^ (300K)	0.344^m^ (370K)
Gadolinium	Solid			
	// to c-axis	0.104*	0.108*	
	⊥ to c-axis	0.103*	0.103*	
	Polycrystalline	0.103*	0.105*	
Gallium	Solid			
	// to a-axis	0.410	0.408	
	// to b-axis	0.884	0.883	
	// to c-axis	0.160	0.159	
	Liquid		0.281 (302.93K)	0.328
Germanium	Solid	0.667	0.602	0.465
Gold	Solid	3.19	3.18	3.13
Hafnium	Solid, Polycrystalline	0.233*	0.230	0.224
Helium	Solid, ^{3}He	0.033 (0.9K)	0.020 (1K)	0.0021 (2K)
	Solid, ^{4}He	0.650 (0.9K)	0.245 (1K)	0.0018 (2K)
	Liquid, saturated: He-I	0.191^m^ (2.5K)	0.232^m^ (3.5K)	0.434^m^ (5K)
	Gas, 1 atm.	1.411^m^	1.520^m^	1.766^m^
Holmium	Solid			
	// to c-axis	0.215*	0.222*	
	⊥ to c-axis	0.136*	0.138*	
	Polycrystalline	0.159*	0.162*	0.170*
Hydrogen	Solid, Normal Hydrogen	2.30 (4K)	0.0158 (10K)	0.0090 (15K)
	Liquid, saturated: Normal Hydrogen	1.022^m^ (15K)	1.269^m^ (25K)	0.60^m^* (33K)
	Gas, 1 atm. Normal Hydrogen	1.665^m^	1.815^m^	2.106^m^
	Liquid, saturated; para-Hydrogen	0.824^m^ (14K)	0.998^m^ (25K)	0.58^m^ (32K)
	Vapor, saturated; para-Hydrogen	0.081^m^* (10K)	0.242^m^* (25K)	0.58^m^* (32K)
	Gas, 1 atm.; para-Hydrogen	1.768^m^*	1.880^m^*	2.126^m^*
	Deuterium:			
	Liquid, saturated	1.26^m^ (20K)	1.37^m^ (30K)	0.83^m^* (38K)
	Vapor, saturated	0.084^m^* (20K)	0.26^m^* (30K)	0.83^m^* (38K)
	Gas, 1 atm.	1.294^m^* (270K)	1.406^m^* (300K)	1.66^m^* (370K)
	Tritium:			
	Liquid, saturated	1.25 (21K)	1.34 (30K)	0.68 (44K)
Indium	Solid, Polycrystalline	0.837	0.818	0.762
Iodine	Solid	4.81^m^*	4.49^m^* (300K)	3.75^m^* (386.8K)
	Liquid, saturated			1.16^m^* (386.8K)
Iridium	Solid	1.48	1.47	1.45
Iron	Solid	0.865	0.804	0.720
	Armco Iron	0.747	0.728	0.676
Krypton	Solid	0.4^m^* (1K)	17^m^ (10K)	2.5^m^ (116K)
	Liquid, saturated			0.931^m^ (116K)
	Vapor, saturated	0.0406^m^* (120K)	0.0554^m^* (150K)	0.21^m^* (210K)
	Gas	0.0860^m^ (270K)	0.0949^m^ (300K)	0.1145^m^ (370K)
Lanthanum	Solid, Polycrystalline	0.131	0.134	0.145
Lead	Solid	0.356	0.353	0.344
Lithium	Solid	0.859	0.848	0.818
Lutetium	Solid			
	// to c-axis	0.236*	0.232*	
	⊥ to c-axis	0.140*	0.138*	
	Polycrystalline	0.167*	0.164*	
Magnesium	Solid, Polycrystalline	1.57	1.56	1.54

THERMAL CONDUCTIVITY OF THE ELEMENTS (continued)

Element	State or Condition	Conductivity at (watt/cm/°K) 273.2K	298.2K	373.2K
Manganese	Solid	0.0768*	0.0781*	
Mercury	Liquid	0.0782	0.0830	0.0947
Molybdenum	Solid	1.39*	1.38*	1.35*
Neodymium	Solid, Polycrystalline	0.165*	0.165*	0.167*
Neon	Gas	0.461^{m}* (270K)	0.493^{m}* (300K)	0.563^{m}* (370K)
Neptunium	Solid		0.063* (300K)	
Nickel	Solid	0.941	0.909	0.827
Niobium	Solid	0.533	0.537	0.548
Nitrogen	Solid	56^{m} (4K)	17^{m} (10K)	3.2^{m} (25K)
	Liquid, saturated	1.60^{m} (65K)	0.966^{m} (100K)	0.37^{m}* (126K)
	Vapor, saturated	0.061^{m}* (65K)	0.111^{m}* (100K)	0.37^{m}* (126K)
	Gas, 1 atm.	0.2374^{m} (270K)	0.2598^{m} (300K)	0.3065^{m} (370K)
Cesium	Solid			
	∥ to c-axis	2.93 (2K)	14.3 (10K)	15.4 (30K)
	⊥ to c-axis	1.76 (2K)	8.65 (10K)	11.1 (30K)
	Polycrystalline	2.09 (2K)	10.2 (10K)	12.4 (30K)
	Polycrystalline	0.880*	0.876*	0.870*
Oxygen	Liquid, saturated	1.501^{m} (90K)	1.023^{m} (125K)	0.41^{m} (155K)
	Vapor, saturated	0.081^{m}* (90K)	0.135^{m}* (125K)	0.41^{m}* (155K)
	Gas, 1 atm.	0.2424 (270K)	0.2674 (300K)	0.3204 (370K)
Palladium	Solid	0.716*	0.718	0.730
Phosphorus	Solid			
	Black (Polycrystalline)	0.132	0.121	
	White	0.00250*	0.00236*	
	Liquid, White			0.00181
Platinum	Solid	0.717	0.716	0.717
Plutonium	Solid, polycrystalline	0.0616*	0.0670*	0.0790* (350K)
Potassium	Solid	1.036*	1.025	
	Liquid			0.532
Praeseodymium	Solid, polycrystalline	0.120	0.125	0.134
Promethium	Solid, polycrystalline		0.179*	0.184*
Radium	Solid		0.186 (293.2K)	
Radon	Gas, 1 atm.	0.0327^{m}* (270K)	0.0364^{m}* (300K)	0.0445^{m}* (370K)
Rhenium	Solid, polycrystalline	0.486	0.480	0.466
Rhodium	Solid	1.51	1.50	1.47
Rubidium	Solid	0.583*	0.582	0.581 (312.04K)
	Liquid			0.333 (312.04K)
Ruthenium	Solid, polycrystalline	1.17	1.17	1.15
Samarium	Solid, polycrystalline	0.133*	0.133*	0.133*
Scandium	Solid, polycrystalline	0.157	0.158	
Selenium	Solid			
	∥ to c-axis	0.0481	0.0452	0.0483
	⊥ to c-axis	0.0137	0.0131	0.0139
	Amorphous	0.00428	0.00519	0.00818 (323.2K)
Silicon	Solid	1.68	1.49	1.08
Silver	Solid	4.29	4.29	4.26
Sodium	Solid	1.42	1.42	1.32 (371K)
Strontium	Solid, polycrystalline	0.364*	0.354*	0.325*
Sulfur	Solid, polycrystalline	0.00287	0.00270	0.00154
	Solid, amorphous	0.00200	0.00205	0.00216* (350K)
	Liquid			0.00129 (392.2K)
Tantalum	Solid	0.574	0.575	0.577

THERMAL CONDUCTIVITY OF THE ELEMENTS (continued)

Element	State or Condition	Conductivity at (watt/cm/°K) 273.2K	298.2K	373.2K
Technetium	Solid, polycrystalline	0.509*	0.506	0.501
Tellurium	Solid			
	∥ to c-axis	0.0360	0.0338	0.0292
	⊥ to c-axis	0.0208	0.197	0.173
Terbium	Solid			
	∥ to c-axis	0.138*	0.147*	
	⊥ to c-axis	0.0900*	0.0956*	
	Polycrystalline	0.104*	0.111*	
Thallium	Solid, polycrystalline	0.469	0.461	0.443
Thorium	Solid	0.540*	0.540*	0.543*
Thulium	Solid			
	∥ to c-axis	0.242*	0.242*	
	⊥ to c-axis	0.140*	0.141*	
	Polycrystalline	0.168*	0.169*	
Tin	Solid			
	∥ to c-axis	0.527	0.516	0.489
	⊥ to c-axis	0.759	0.743	0.704
	Polycrystalline	0.682	0.668	0.632
Titanium	Solid, polycrystalline	0.224	0.219	0.207
Tungsten	Solid	1.77	1.73	1.63
Uranium	Solid, polycrystalline	0.270	0.275	0.291
Vanadium	Solid	0.307*	0.307	0.310
Xenon	Liquid, saturated	0.31m (270K)	0.16m* (290K)	
	Vapor, saturated	0.084m*	0.16m*	
	Gas, 1 atm.	0.0514m (270K)	0.0569m (300K)	0.0695m (370K)
Ytterbium	Solid	0.354*	0.349*	0.343*
Yttrium	Solid, polycrystalline	0.170*	0.172*	0.177*
Zinc	Solid, polycrystalline	1.17	1.16	1.12
Zirconium	Solid, polycrystalline	0.232*	0.227 (300K)	0.218

Diffusivity of Selected Materials

Material	Diffusivity $\frac{cm^2}{s}$	Material	Diffusivity $\frac{cm^2}{s}$
Aluminium	0.853	Magnesium	0.93
Amber	—	Manganese	0.14
Antimony	1.56	Manganin	0.064
Arsenic	—	Mercury	0.044
Asbestos	0.00035 to 0.00046	Mica	0.0027 to 0.0034
Asphalt	0.005 to 0.007	Molybdenum	0.53
Barium	—	Nickel silver	0.074
Beryllium	0.526	Nickel	0.208
Bismuth	0.097	Palladium	0.198
Blast-furnace slag	0.0035	Paraffin	0.0013 to 0.0014
Brass	0.33 to 0.35	Phenol	—
Bronze	0.08 to 0.22	Phosphorus	—
Cadmium	0.455	Pigment (enamel)	—
Calcium	—	Platinum	0.248
Calcite	0.0034 to 0.0038	Potassium	1.51
Carbon		Rhenium	0.251
Diamond	—	Rhodium	0.27
Graphite	0.31	Rubber	0.0015 to 0.0016
Coke	0.022 to 0.026	Selenium (metallic)	—
Carbon	—	Silver	1.717
Coal	—	Silicon	—
Peat	—	Sodium	1.18
Cement	0.073	Soft Solder	—
Charcoal	0.0019	Sulphur (orthorhombic)	0.0014
Chromium	0.219	Tantalum	0.237
Concrete	0.004 to 0.008	Thorium	—
Constantan	0.063	Tin	0.400
Copper	1.12	Titanium	—
Cork	0.009 to 0.0019	Tungsten	0.646
Gallium	—	Uranium	—
Germanium	0.33	Vitreous quartz	0.0088
Glass (average)	0.0033 to 0.0044	Vulcanised fibre	0.0023 to 0.0030
Gold	1.24	Wax	0.00013 to 0.00014
Indium	—	Wood	
Iridium	0.195	Beech	—
Iron		Ebony	—
Cast iron	0.11	Fir	—
Chemically pure	0.195	Oak	0.0009 to 0.002
Invar	0.028	Pine	0.0007 to 0.0015
Steel	0.12	Wood's alloy	0.089
Lead	0.233	Zinc	0.434
Lime, caustic	—		

Thermal Conductivity--Temperature Table for Solids*

Substance	Temperature, K													
	10	20	40	60	80	100	200	300	400	500	600	800	1000	1200	1400
Alumina	7	32	121	174	160	125	55	36	26	20	16	10	8	7	6
Aluminum	38,000	13,500	2,300	850	380	300	237	237	240	237	232	220	93	99	105
Antimony	470	230	110	80	60	48	32	26	22	20					
Beryllium oxide	47	196	810	1,400	1,650	1,490	480	272	196	146	111	70	47	33	25
Bismuth	240	100	45	31	24	22	18	16	14	12					
Boron	165	305	400	327	230	170	45	25	15	12					
Cadmium	900	250	150	120	110	110	105	104	101	99					
Chromium	400	570	450	250	180	158	111	90	87	85	81	71	65	62	61
Cobalt	250	450	380	250	190	160	120	100	85	70					
Constantan	4	9	16	18	19	20	23	25	27	30					
Copper	19,000	10,700	2,100	850	570	483	413	398	392	388	383	371	357	342	
Gallium	2,200	640	250	200	170	140	100	85							
Gold	2,800	1,500	520	380	350	345	327	315	312	309	304	292	278	262	
Graphite[a]	27	108	135	81	54	39	15	10	7	5	4	3	3	2	2
Graphite[b]	81	420	1,630	2,980	4,290	4,980	3,250	2,000	1,460	1,140	930	680	530	440	370
Hastelloy	1	3	4	5	6	7	9	10	11	13					
Inconel	2	4	8	10	11	11	14	15							
Iridium	1,300	1,900	750	360	230	172	147	145	143	140					
Iron	710	1,000	560	270	170	132	94	80	69	61	55	43	33	28	31
Lead	175	57	43	42	41	40	37	35	34	33	31	19	22	24	26
Magnesium	1,200	1,300	620	290	190	169	159	156	153	151	149	146	84	98	112
Magnesium oxide	1,100	3,100	2,200	950	460	260	75	48	36	27	21	13	10	8	7
Manganese	2	2	4	5	5	6	7	8	9	9					
Manganin	2	4	9	11	13	13	17	22	28	34	40				
Mercury	54	40	35	33	33	32	32	8	10	11	12	13	14		
Molybdenum	150	280	350	250	210	179	143	138	134	130	126	118	112	105	100
Nickel	2,600	1,700	570	290	200	158	106	91	80	72	66	67	72	76	80
Nylon	0.04	0.10	0.17	0.20	0.23	0.25	0.28	0.30							
Palladium	1,200	610	160	100	88	80	78	78	79	80					
Platinum	1,200	490	130	92	82	79	75	73	72	72	72	73	78	78	81
PTFE[c]	0.94	1.43	1.94	2.1	2.15	2.16	2.20	2.25	2.3	2.5					
Pyrex	0.12	0.20	0.33	0.42	0.51	0.57	0.88	1.1	1.6	2.1					
Quartz	1,200	480	82	40	30										
Rhodium	2,900	3,900	1,000	370	250	190	160	150	145	140					
Rubber	....	....	0.13	0.15	0.16	0.17	0.20	0.22	0.24	0.25					
Selenium (axis)	140	57	25	15	10	8	6	4	3	2					
Silica	....	....	....	....	....	....	....	1.34	1.52	1.70	1.87	2.22	2.60		
Silver	16,500	5,200	1,100	630	500	430	425	424	420	413	405	389	374	358	
Tantalum	108	146	88	68	62	59	58	57	58	58	59	59	60	61	62
Tellurium	300	93	29	17	13	11	6	4	3	3					
Tin	....	320	130	101	90	84	72	67	62	60					
Titanium	14	28	39	37	33	31	26	21	20	20	19				
Tungsten	....	....	880	330	310	280	190	180	170	150	140				
Uranium	....	....	....	20	22	23	26	28	30	32					
Zinc	....	....	....	150	135	130	123	120	116	110	110				
Zirconium	100	110	59	42	38	34	25	23	22	21	21				

[a] Parallel to basal plane.
[b] Perpendicular to basal plane.
[c] Also known as "Teflon," etc.
* Especially at low temperatures, the thermal conductivity can often be markedly reduced by even small traces of impurities. These tables, for the highest purity specimens available, should thus be used with caution in applications with commercial materials. Thermal conductivities tabulated in W/(m · K).

THERMAL CONDUCTIVITIES OF SELECTED METALS (watt/cm/°k)

T,K	Aluminum 99.996+% ρ_0 = 0.00315 μohm cm	Copper 99.999+% ρ_0 = 0.000851 μohm cm	Gold 99.999+% ρ_0 = 0.0055 μohm cm	Iron 99.998+% ρ_0 = 0.0327 μohm cm	Manganin	Platinum 99.999% ρ_0 = 0.0106 μohm cm	Silver 99.999+% ρ_0 = 0.00062 μohm cm	Tungsten 99.99+% ρ_0 = 0.0017 μohm cm
0	0	0	0	0	0	0	0	0
1	7.8	28.7	4.4	0.75	0.0007	2.31	39.4	14.4
2	15.5	57.3	8.9	1.49	0.0018	4.60	78.3	28.7
3	23.2	85.5	13.1	2.24	0.0031	6.79	115	42.6
4	30.8	113	17.1	2.97	0.0046	8.8	147	55.6
5	38.1	138	20.7	3.71	0.0062	10.5	172	67.1
6	45.1	159	23.7	4.42	0.0078	11.8	187	76.2
7	51.5	177	26.0	5.13	0.0095	12.6	193	82.4
8	57.3	189	27.5	5.80	0.0111	12.9	190	85.3
9	62.2	195	28.2	6.45	0.0128	12.8	181	85.1
10	66.1	196	28.2	7.05	0.0145	12.3	168	82.4
11	69.0	193	27.7	7.62	0.0162	11.7	154	77.9
12	70.8	185	26.7	8.13	0.0180	10.9	139	72.4
13	71.5	176	25.5	8.58	0.0197	10.1	124	66.4
14	71.3	166	24.1	8.97	0.0215	9.3	109	60.4
15	70.2	156	22.6	9.30	0.0232	8.4	96	54.8
16	68.4	145	20.9	9.56	0.0250	7.6	85	49.3
18	63.5	124	17.7	9.88	0.0285	6.1	66	40.0
20	56.5	105	15.0	9.97	0.0322	4.9	51	32.6
25	40.0	68	10.2	9.36	0.0410	3.15	29.5	20.4
30	28.5	43	7.6	8.14	0.0497	2.28	19.3	13.1
35	21.0	29	6.1	6.81	0.0583	1.80	13.7	8.9
40	16.0	20.5	5.2	5.55	0.067	1.51	10.5	6.5
45	12.5	15.3	4.6	4.50	0.075	1.32	8.4	5.07
50	10.0	12.2	4.2	3.72	0.082	1.18	7.0	4.17
60	6.7	8.5	3.8	2.65	0.097	1.01	5.5	3.18
70	5.0	6.7	3.58	2.04	0.110	0.90	4.97	2.76
80	4.0	5.7	3.52	1.68	0.120	0.84	4.71	2.56
90	3.4	5.14	3.48	1.46	0.127	0.81	4.60	2.44
100	3.0	4.83	3.45	1.32	0.133	0.79	4.50	2.35
150	2.47	4.28	3.35	1.04	0.156	0.762	4.32	2.10
200	2.37	4.13	3.27	0.94	0.172	0.748	4.30	1.97
250	2.35	4.04	3.20	0.865	0.193	0.737	4.28	1.86
273	2.36	4.01	3.18	0.835	0.206	0.734	4.28	1.82
300	2.37	3.98	3.15	0.803	0.222	0.730	4.27	1.78
350	2.40	3.94	3.13	0.744	0.250	0.726	4.24	1.70
400	2.40	3.92	3.12	0.694	(0.279)	0.722	4.20	1.62
500	2.37	3.88	3.09	0.613	(0.338)	0.719	4.13	1.49
600	2.32	3.83	3.04	0.547	(0.397)	0.720	4.05	1.39
700	2.26	3.77	2.98	0.487		0.723	3.97	1.33
800	2.20	3.71	2.92	0.433		0.729	3.89	1.28
900	2.13	3.64	2.85	0.380		0.737	3.82	1.24
1000	[0.93]**	3.57	(2.78)	0.326		0.748	(3.74)	1.21
1100	[0.96]	3.50	(2.71)	0.297		0.760	(3.66)	1.18
1200	[0.99]	3.42	(2.62)	0.282		0.775	(3.58)	1.15
1300	[1.02]	(3.34)†	(2.51)	0.299		0.791		1.13
1400				0.309		0.807		1.11
1500				0.318		0.824		1.09
1600				(0.327)		0.842		1.07
						0.860		1.05
						0.877		1.03
						(0.895)		1.02
						(0.913)		1.00
								0.98
								0.96
								0.94
								0.925
								0.915
								0.905
								0.900
								(0.895)

* In the table the third significant figure is given only for the purpose of comparison and for smoothness and is not indicative of the degree of accuracy.

** Values in square brackets are for liquid state.

† Values in parentheses are extrapolated.

‡ Estimated.

THERMAL CONDUCTIVITY OF SELECTED NON-METALLIC MATERIALS

(Bureau of Standards Letter Circular No. 227)

D = Density in pound per cubic foot.
K = Thermal conductivity in B.T.U. per hour, square foot, and temperature gradient of 1 degree Fahrenheit per inch thickness. The lower the conductivity, the greater the insulating values.

Soft Flexible Materials in Sheet Form

		D	K
Dry zero	Kapok between burlap or paper	1.0	0.24
		2.0	0.25
Cabots quilt	Eel grass between kraft paper	3.4	0.25
		4.6	0.26
Hair felt	Felted cattle hair	11.0	0.26
		13.0	0.26
Balsam wool	Chemically treated wood fiber	2.2	0.27
Hairinsul	75% hair 25% jute	6.3	0.27
	50% hair 50% jute	6.1	0.26
Linofelt	Flax fibers between paper	4.9	0.28
Thermofelt	Jute and asbestos fibers, felted	10.0	0.37
	Hair and asbestos fibers, felted	7.8	0.28

Loose Materials

		D	K
Rock wool	Fibrous material made from rock	6.0	0.26
	also made in sheet form, felted and	10.0	0.27
	confined with wire netting	14.0	0.28
		18.0	0.29
Glass wool	Pyrex glass, curled	4.0	0.29
		10.0	0.29
Sil-O-Cel	Powdered diatomaceous earth	10.6	0.31
Regranulated	Fine particles	9.4	0.30
cork	about $^{3}/_{16}$ inch particles	8.1	0.31
Thermofill	Gypsum in powdered form	26	0.52
		34	0.60
Sawdust	Various	12.0	0.41
	redwood	10.9	0.42
Savings	Various, from planer	8.8	0.41
Charcoal	From maple, beech and birch, coarse	13.2	0.36
	6 mesh	15.2	0.37
	20 mesh	19.2	0.39

Semiflexible Materials in Sheet Form

		D	K
Flaxlinum	Flax fiber	13.0	0.31
Fibrofelt	Flax and rye fiber	13.6	0.32

Semiflexible Materials in Sheet Form

		D	K
Flaxlinum	Flax fiber	13.0	0.31
Fibrofelt	Flax and rye fiber	13.6	0.32

Semirigid Materials in Board Form

		D	K
Corkboard	No added binder; very low density	5.4	0.25
Corkboard	No added binder; low density	7.0	0.27
Corkboard	No added binder; medium density	10.6	0.30
Corkboard	No added binder; High density	14.0	0.34
Eureka	Corkboard with asphaltic binder	14.5	0.32
Rock Cork	Rock wood block with binder Also called "Tucork"	14.5	0.326
Lith	Board containing rock wool, flax and straw pulp	14.3	0.40

Stiff Fibrous Materials in Sheet Form

		D	K
Insulite	Wood pulp	16.2	0.34
		16.9	0.34
Celotex	Sugar cane fiber	13.2	0.34
		14.8	0.34
•Masonite		K =	0.33
•Inso-board			0.33
•Malzewood			0.33 to 0.39
•Cornstalk Pith Board			0.24 to 0.30
•Maftex			0.34

Cellular Gypsum

	D	K
Insulex or Pyrocell	8	0.35
	12	0.44
	18	0.59
	24	0.77
	30	1.00

Woods (Across Grain)

	D	K
Balsa	7.3	0.33
	8.8	0.38
	20	0.58
Cypress	29	0.67
White pine	32	0.78
Mahogany	34	0.90
Virginia pine	34	0.98
Oak	38	1.02
Maple	44	1.10

Miscellaneous Bulding Materials
(Data taken from various sources)

	K		K
Cinder concrete	2 to 3	Limestone	4 to 9
Building gypsum	About 3	Concrete	6 to 9
Plaster	2 to 5	Sandstone	8 to 16
Building brick	3 to 6	Marble	14 to 20
Glass	5 to 6	Granite	13 to 28

Thermal Conductivities of Liquids

Liquid	k, Btu/(hr)(ft²)(°F/ft)			
	50°F	100°F	200°F	300°F
Acetic acid:				
100%	0.099			
50%	0.20			
Ammonia	0.29			
Ammonia, aqueous, 26%	0.253	0.274	0.315	
Amyl acetate	0.083			
n-Amyl alcohol	0.096	0.094	0.0895	
Isoamyl alcohol	0.089	0.088	0.087	
Aniline	0.100			
Benzene	0.096	0.091	0.081	
Bromobenzene	0.075	0.074	0.071	
Butyl acetate	0.078	0.074	0.068	
n-Butyl alcohol	0.098	0.097	0.094	
Isobutyl alcohol	0.091			
Calcium chloride brine:				
30%	......	0.32		
15%	......	0.34		
Carbon disulfide	0.095	0.092	0.086	
Carbon tetrachloride	0.067	0.065	0.062	
Chloroform	0.073	0.071	0.067	
Decane	0.0865	0.0843	0.0805	0.0775
Dibutyl phthalate	0.079	0.078	0.073	
Dichlorodifluoromethane	0.054	0.048	0.035	
Dowtherm A	......	0.082	0.079	0.077
Dowtherm E	......	0.073	0.071	0.064
Ethane	0.056			
Ethyl acetate	0.085	0.080	0.068	
Ethyl alcohol	0.106	0.092	0.068	
Ethyl benzene	0.088	0.085	0.077	
Ethyl ether	......	0.080	0.077	
Ethylene glycol	0.145	0.144	0.142	0.140
Glycerine (USP)	0.156	0.159	0.164	
Heptane	0.083	0.0815	0.0765	0.071
Hexane	0.0820	0.0795	0.0745	0.068
Mercury	4.7	......	......	6.7
Methyl alcohol	0.128	0.117	0.095	
Nitrobenzene	0.097	0.093	0.089	
Nitromethane	0.128	0.124	0.115	
Octane	0.0845	0.0825	0.078	0.073
Olive oil (USP)	0.097	0.096	0.094	
Paraldehyde	0.086	0.083	0.0785	
Pentane	0.080	0.0775	0.071	0.0615
Propane	0.072	0.0675	0.056	
n-Propyl alcohol	0.101	0.099	0.094	
Propylene glycol	0.116	0.115	0.113	0.110
Sodium	......	......	14.8	15.0
Sodium chloride brine:				
25%	0.34			
12.5%	0.35			
Sulfuric acid:				
90%	0.22			
60%	0.26			
30%	0.31			
Sulfur dioxide	0.116	0.108		
Trichlorethylene	0.072	0.068	0.060	
Toluene	0.087	0.086	0.084	
Vinyl acetate	0.095	0.083	0.075	
Water	0.331	0.363	0.393	0.395
Dow Corning silicone DC-200 (500,000 centistokes)	......	0.090	0.085	
GE silicones SF-96:				
40 centistokes	......	0.085	0.080	
100 centistokes	......	0.086	0.081	
300 centistokes	......	0.090	0.084	
1,000 centistokes	......	0.091	0.086	

THERMAL CONDUCTIVITY OF SELECTED GASES

Gas °F / °C	−400 / −240	−300 / −184.4	−200 / −128.9	−100 / −73.3	−40 / −40	−20 / −28.9	0 / −17.8	20 / −6.7	40 / 4.4	60 / 15.6	80 / 26.7	100 / 37.8	120 / 48.9	200 / 93.3
Acetylene	....	....	....	28.10	34.71	37.19	39.67	42.15	45.04	47.94	50.83	53.72	56.62	69.43
Air	....	....	....	....	50.09	52.15	54.22	56.24	58.31	60.34	62.20	64.22	66.04	....
Ammonia	....	....	....	....	43.39	45.87	48.35	50.83	53.31	55.79	58.68	61.58	64.47	....
Argon	....	....	....	....	34.30	35.95	37.19	38.85	40.09	41.33	42.57	44.22	45.46	....
Bromine	....	....	....	....	....	....	9.09	....	....	....	....	11.57	....	....
n-Butane	....	....	....	....	....	....	....	30.99	33.06	35.54	38.02	40.91	43.39	54.14
i-Butane	....	....	....	....	....	....	....	32.65	33.89	36.37	38.85	41.74	44.22	55.79
Carbon dioxide	....	....	....	....	27.90	29.75	31.70	33.68	35.62	37.61	39.67	41.74	43.81	....
Carbon disulfide	....	....	....	....	....	....	14.05	15.29	16.53	17.77	19.01	19.84	....	....
Carbon monoxide	....	....	....	....	47.94	50.00	51.95	53.85	55.87	57.86	59.92	61.99	63.89	....
Chlorine	....	....	....	....	15.29	16.53	17.36	18.18	19.01	20.25	21.08	21.90	23.14	....
Deuterium	....	....	....	....	274.82	285.15	295.07	305.81	309.95	322.34	334.74	343.01	355.40	....
Ethane	....	....	....	23.97	32.65	35.54	38.43	41.33	44.63	47.94	51.24	54.55	58.27	74.39
Ethanol	....	....	....	....	....	....	....	29.34	30.99	32.65	34.71	36.78	....	....
Ethylamine	....	....	....	....	....	....	....	31.41	33.47	35.54	37.61	39.67	42.15	....
Ethylene	....	....	....	26.86	33.06	35.54	38.02	40.50	43.39	46.29	49.18	52.07	54.96	68.19
Fluorine	....	18.18	30.58	43.39	50.83	52.90	55.38	57.86	59.92	61.99	64.06	66.12	68.19	76.04
Helium	84.31	163.24	221.51	274.8	304.99	314.49	324.00	333.50	343.42	352.10	360.36	368.63	376.07	....
Hydrogen	59.92	142.57	227.29	308.7	357.47	371.93	388.46	405.00	417.39	433.92	446.32	458.72	471.11	....
Hydrogen bromide	....	....	....	....	15.29	16.11	16.49	17.77	18.60	19.84	20.66	21.49	....	....
Hydrogen chloride	....	....	....	....	25.62	26.86	28.51	29.75	30.99	32.23	33.89	35.12	....	....
Hydrogen cyanide	....	....	....	....	....	....	23.97	25.62	26.86	28.10	29.75	30.99	32.65	....
Hydrogen sulfide	....	....	....	....	....	....	28.10	29.75	31.41	33.47	....	36.78	....	....
Krypton	....	....	....	....	....	....	19.84	....	....	....	....	23.56	....	....
Methane	....	22.32	36.86	52.07	61.37	64.55	67.86	71.08	74.39	78.11	81.83	85.54	89.26	106.62
Neon	....	....	....	....	97.94	100.84	104.14	107.03	109.93	112.82	115.71	118.19	121.09	....
Nitric oxide	....	....	30.91	42.40	49.01	51.24	53.39	55.54	57.65	59.76	61.99	64.06	66.12	74.39
Nitrogen	....	20.25	33.06	44.22	50.42	52.48	54.55	56.20	58.27	60.34	62.40	64.06	65.71	....
Nitrous oxide	....	....	....	....	28.93	30.91	32.90	35.04	37.15	39.30	41.45	43.81	46.08	....
Oxygen	....	18.84	31.66	43.72	50.54	52.81	54.96	57.24	59.43	61.58	63.64	65.91	68.19	76.87
n-Propane	....	....	....	....	27.69	29.75	32.23	34.71	37.19	39.67	42.47	45.46	48.35	60.75
R-11(CCl_3F)	....	....	....	....	....	12.81	13.64	14.88	15.70	16.53	17.77	18.60	....	....
R-12(CCl_2F_2)	....	....	....	....	....	17.36	18.60	19.42	20.66	21.49	22.73	23.56	....	....
R-21($CHCl_2F$)	....	....	....	....	....	....	21.90	22.32	22.73	23.14	23.56	23.97	....	....
R-22($CHClF_2$)	....	....	....	....	....	....	24.80	25.62	26.45	27.28	28.10	28.93	....	....
Water	....	....	....	....	....	....	34.71	36.78	38.85	40.50	42.57	44.63	46.70	54.98

PROPERTIES OF SELECTED REFRIGERANTS

Refrigerant	Formula	Flash Point °F.	Ignition Temp. °F.	Explosive Limits % by Volume: Lower	Upper	Vapor Density (Air = 1)	Boiling Point °F	Threshold Limit Value* Parts per Million in Air	Water Soluble	Odor
Ammonia	NH_3	—	1204	16	25	0.59	−28	100	yes	yes
Bromotrifluoromethane (Kulene-131)	CF_3Br		nonflammable			5.25	−73.6	—	no	yes
Butane	C_4H_{10}	−76	806	1.8	8.4	2.04	33	—	no	no
Carbon dioxide	CO_2		nonflammable			1.53	−108	5000	yes	no
Carbon tetrachloride	CCL_4		nonflammable			5.32	170	25	no	yes
Dichlorodifluoromethane (Freon-12)	CCl_2F_2		nonflammable			4.17	−21.6	—	no	no
Dichlorodifluoromethane, 73.8% Ethylidene fluoride, 26.2% (Carrene-7)	CCl_2F_2 CH_3CHF_2		nonflammable			3.24	−28.0	—	no	yes
Dichloromonofluoromethane (Freon-21)	$CHCl_2F$		practically nonflammable			3.55	48	—	no	yes
Dichlorotetrafluoroethane (Freon-114)	$C_2Cl_2F_4$		practically nonflammable			5.89	38	—	no	no
Ethane	C_2H_6	<20	950	3.0	12.5	1.04	−128	—	no	no
Ethylene	C_2H_4	<20	842	3.1	32	0.972	−155	—	yes	yes
Isobutane	$(CH_3)_3CH$	<20	1010	1.8	8.4	2.01	14	—	no	no
Methyl chloride	CH_3Cl	632	1170	10.7	11.4	1.78	−11	100	yes	yes
Monochlorodifluoromethane (Freon-22)	$CHClF_2$		practically nonflammable			2.9	−41	—	yes	no
Monochlorotrifluoromethane (Freon-13)	$CClF_3$		nonflammable			3.6	−112	—	—	no
Propane	C_3H_8	<20	871	2.2	9.5	1.56	−45	—	no	no
Propylene	C_3H_6	<20	927	2.4	10.3	1.49	−53	—	yes	yes
Sulfur dioxide	SO_2		nonflammable			2.2	14	10	yes	yes
Tetrafluoromethane (Freon-14)	CF_4		nonflammable			3.0	−198	—	no	no
Trichloroethylene	C_2HCl_3		nonflammable at normal temperature			4.53	189	200	no	yes
Trichloromonofluoromethane (Freon-11)(Carrene-2)	CCL_3F		nonflammable			4.7	75.3	—	no	no
Trichlorotrifluoroethane (Freon-113)	$C_2Cl_3F_3$		practically nonflammable			6.4	118	—	no	no

* Maximum average atmospheric concentration of contaminants to which workers may be exposed for an eight-hour work day without injury to health. (American Conference of Governmental Industrial Hygienists: "Threshold Limit Values for 1954.)

Selected values of measured evaporation coefficients, α_r, for solids

Substance	Purity	Orientation	Surface imperfection	Surface clean iness	Vacuum	Temperature	Technique	Evaporation coefficient, α_r
Beryllium, solid.	Fairly high; vacuum cast; sintered.	Polycrystalline.	Unstated.	Unstated; probably fair.	Unstated; probably 10^{-6} torr.	1171–1552 K ± 1°.	Langmuir-Knudsen torsion balance.	1.0 ± 0.02.
Boron, solid; liquid.	Unstated; probably fair.	Polycrystalline; liquid (Still).		Unstated; probably fair.	Unstated; probably < 10^{-8} torr.	Melting Point, 2403 ± 40 K.	Mass spec. Langmuir-Knudsen on liq.	0.98 ± 0.02.
Cadmium, solid.	Unstated; probably fair.	Polycrystalline.	Polished.	Unstated; probably fair.	10^{-5} torr.	Just below mp, 586 ± 3 K.	Langmuir-Knudsen torsion balance.	0.996 ± 0.002.
Carbon.	Unstated; Acheson graphite.	Polycrystalline.	Unstated.	Unstated; probably fair.	10^{-8} torr.	2357–2870 K ± 10°.	Langmuir-Knudsen.	1 ± 0.1.
Cesium Bromide, solid.	Unstated; Harshaw.	Single crystal[a] but unknown orientation.	Unstated.	Unstated; probably fair.	Unstated; probably < 10^{-8} torr.	785–830 K ± 5°.	Mass spec. Langmuir-Knudsen.	0.27 ± 0.1.
Cesium Iodide, solid.	Unstated; Harshaw.	Single crystal[a] but unknown orientation.	Unstated.	Unstated; probably fair.	Unstated; probably < 10^{-8} torr.	757–772 K ± 5°.	Mass spec. Langmuir-Knudsen.	0.36 ± 0.1
Chromium, solid.	A. D. Mackay, Inc., 99.9%.	Polycrystalline.	Unstated.	Unstated; probably fair.	Unstated; probably < 10^{-7} torr.	1318–1563 K ± 3°.	Langmuir-Knudsen.	0.9 ± 0.1.
Ice.	Unstated; probably fair.	Polycrystalline.	Unstated.	Unstated; probably fair.	10^{-5} torr.	188–213 K ± 0.6°.	Langmuir-Equil. Vap.	0.9 ± 0.1.
Iron, solid.	Unstated; probably fair.	Polycrystalline.	Unstated.	Unstated; probably fair.	10^{-5} torr.	1540–1740 K ± 15°.	Langmuir-Knudsen.	1 ± 0.2.
Iron, solid.	Fisher Electrolytic; probably fair.	Polycrystalline.	Unstated.	Unstated; probably fair.	Unstated; probably < 10^{-7} torr.	1358–1520 K ± 3°.	Langmuir-Knudsen.	0.9 ± 0.1.
Lanthanum Fluoride, solid.	99.6%.	Macroscopically (0001).	Polished.	Unstated; probably fair.	10^{-8}–10^{-9} torr.	1340–1650 K ± 5°.	Torsion Knudsen-Torsion Langmuir.	0.95 ± 0.1.
n-$C_{17}H_{36}$.	Refractive indices and densities of liquid and X-ray diffraction of solid indicated good purity.	Polycrystalline.	Unstated.	Unstated; probably fair.	0.1 torr (of air).	15–22° C ± 0.01°.	Evap. of beads in air and Knudsen.	0.95 ± 0.05[b]
n-$C_{18}H_{38}$.	Refractive indices and densities of liquid and X-ray diffraction of solid indicated good purity.	Polycrystalline.	Unstated.	Unstated; probably fair.	0.1 torr (of air).	15–28° C ± 0.01°.	Evap. of beads in air and Knudsen.	1.00 ± 0.05[b].
Potassium Chloride.	Unstated; probably high.	Single crystal, macroscopically (100) and (100) plus (111).	Polished.	Unstated; probably fair.	Unstated; probably 10^{-7} torr.	672–788 K ± 0.1°.	Langmuir-Knudsen.	0.72 ± 0.15[c].

— continued —

Substance	Purity	Orientation	Surface imperfection	Surface cleanliness	Vacuum	Temperature	Technique	Evaporation coefficient, α_r
Potassium Chloride.	Unstated; probably high.	Single crystal, macroscopically (100) and (100) plus (111) and (100) plus (110).	Polished.	Unstated; probably fair.	Unstated; probably 10^{-7} torr.	672–788 K $\pm 0.1°$.	Langmuir-Knudsen.	0.63 ± 0.015 [c].
Potassium Perrhenate, solid.	Unstated.	Single crystal, macroscopically basal plane.	Irregular surface.	Unstated.	Unstated; probably 10^{-5} torr.	746–768 K $\pm 1°$.	Langmuir-Knudsen.	0.7 ± 0.1.
Silver, solid.	Unstated.	Polycrystalline.	Unstated.	Unstated; probably fair.	10^{-5} torr.	Just below mp, 1234 ± 3 K.	Torsion balance.	> 0.92.
Silver, solid.	Spectrographically pure.	Polycrystalline.	Unstated.	Unstated; probably fair.	5×10^{-6} torr.	1103 ± 1 K.	Langmuir-Knudsen.	0.9 ± 0.2.
Silver, solid.	Spectrographically pure.	Complex plane.	Unstated.	Unstated; probably fair.	5×10^{-6} torr.	1103 ± 1 K.	Langmuir-Knudsen.	0.8 ± 0.2.
Silver, solid.	Spectrographically pure.	Single crystal, (111) plane (epitaxial film).	Dislocation density $1.6 \pm 0.5 \times 10^7$/cm^2.	Unstated; probably fair.	5×10^{-6} torr.	1103 ± 1 K.	Langmuir-Knudsen.	0.85 ± 0.2.
Silver, solid.	Spectrographically pure.	Single crystal, (111) plane (epitaxial film).	Dislocation density $1.6 \pm 0.5 \times 10^7$/cm^2.	Unstated; probably fair.	5×10^{-6} torr.	1198 ± 1 K.	Langmuir-Knudsen.	0.54 ± 0.2.
Sodium Chloride.	Unstated; probably fair.	Single crystal, macroscopically (100).	Pitted, Increasingly so with increasing undersaturation.	Unstated; probably fair.	10^{-5} torr	853 ± 5 K.	Torsion balance with various effusion orifice sizes.	0.4 ± 0.05 at largest orifice size corresponding to free evaporation.
Sulfur, rhombic.	Unstated; probably high.	Single crystal, macroscopically bipyramidal planes.	Some pits.	Unstated; probably fair.	Unstated; probably 10^{-6} torr.	288.3–306.8 K ± 0.1.	Langmuir-Knudsen.	0.73 ± 0.05.
Sulfur, rhombic.	Unstated; probably high.	Polycrystalline.	Unstated.	Unstated; probably fair.	10^{-7} torr.	288.3–305.7 K ± 0.1.	Langmuir-Knudsen.	0.70 ± 0.05.

[a] Hirth and Pound [1963] later examined the specimens and observed that the surfaces had formed low-index planes in macroscopic steps or facets.

[b] The impedance due to diffusion through air was carefully considered. The evaporation coefficient in a higher vacuum should also be unity.

[c] Miller and Kusch [1956, 1957] report that approximately 10% dimer exists in the vapor over potassium chloride, and this could have an effect on the evaporation coefficient.

Substance	Purity	Surface condition	Surface cleanliness	Vacuum	Temperature	Technique	Evaporation coefficient, α_r
Boron, solid; liquid.	Unstated; probably fair.	Polycrystalline, liquid (Still).	Unstated; probably fair.	Unstated $< 10^{-8}$ torr.	Melting point 2403 ± 40 K	Mass spec. Langmuir-Knudsen on liq.	0.98 ± 0.02.
Carbon tetrachloride.	Unstated.	Still.	Unstated; probably fair.	~ 25 torr (of substance).	$0 \pm 0.1°$ C.	Nonequil. evap. and vapor pressure.	0.99 ± 0.02.
Di-*n*-butyl phthalate.	Washed with Na_2CO_3, vacuum distilled, molecular distillation. Refractive index and density given. Probably quite pure.	Still.	Unstated; probably fair.	0.14–84.6 torr (of gas).	15.00–35.00° C ± 0.01.	Evap. of drops in gases and Knudsen.	1.0 ± 0.05 [a].
Ethyl alcohol.	"Absolute".	Still.	Unstated; probably fair.	~ 20 torr (of substance).	12.40–15.50° C $\pm 0.03°$.	Nonequil. evap. and vapor pressure.	0.024 ± 0.002 [b].
Ethyl alcohol.	> 99.9%.	Still.	Unstated; probably fair.	~ 16 torr (of substance).	$0 \pm 0.1°$ C.	Nonequil. evap. and vapor pressure.	0.036 ± 0.003.
Glycerol.	Distilled.	Moving.	Unstated; probably fair.	10^{-5} torr residual gas.	18.0–70.0° C $\pm 0.1°$.	Langmuir and vapor pressure.	$1.0 + 0.15$.
Glycerol.	Well distilled.	Still and moving.	Unstated; probably fair.	Unstated; probably 10^{-5} torr residual gas (and 10^{-5} torr of substance).	13–25° C $\pm 0.1°$.	Nonquil. and equil. evaporations.	0.05–0.15 [c] ± 0.01.
Mercury.	Unstated; probably high.	Still.	Unstated; probably high.	Unstated; probably 10^{-5} torr.	$19.5 \pm 0.05°$ C.	Langmuir-Knudsen...	0.96 ± 0.01.
Mercury.	Unstated; probably high.	Still.	Fairly high.	10^{-5} torr.	-37 to $+59°$ C $\pm 0.2°$.	Langmuir-Knudsen...	1.0 ± 0.05.
n-$C_{17}H_{36}$.	Refractive indices, densities, and X-ray diffraction of solid indicated good purity.	Still.	Unstated; probably fair.	0.1 torr (of air).	22–40° C $\pm 0.01°$.	Evap. of drops in air and Knudsen.	0.95 ± 0.05 [a].
n-$C_{18}H_{38}$.	Refractive indices, densities, and X-ray diffraction of solid indicated good purity.	Still.	Unstated; probably fair.	0.1 torr (of air).	28–40° C $\pm 0.01°$	Evap. of drops in air and Knudsen.	0.95 ± 0.05 [a].
Potassium, liquid.	Unstated; probably high.	Still.	Unstated; probably fair.	Unstated; probably 10^{-6} torr.	66.7–119.3° C $\pm 0.1°$.	Langmuir-Knudsen...	0.95 ± 0.05.
Tin Chloride, $SnCl_2$.	Merck analytical, dehydrated with HCl gas. Distilled five times.	Still.	Unstated; probably fair.	~ 1 torr (of Substance).	$350 \pm 0.1°$ C.	Nonequil. evap. and vapor pressure.	0.96 ± 0.07.
Tridecyl Methane.	Refractive indices, densities, and dielectric constants indicated good purity.	Still.	Unstated; probably fair.	0.17 torr (of Air).	25–35° C $\pm 0.1°$.	Evap. of drops in air and Knudsen.	0.98 ± 0.03 [a].

-continued-

Substance	Purity	Surface condition	Surface cleanliness	Vacuum	Temperature	Technique	Evaporation coefficient, α_r
Triheptyl Methane.	Refractive indices, densities, and dielectric constants indicated good purity.	Still.	Unstated; probably fair.	0.17 torr (of Air).	25–35° C±0.1°.	Evap. of drops in air and Knudsen.	0.98 ± 0.03 [a].
Water.	Unstated.	Still.	Unstated; probably fair.	5–10 torr (of Substance).	14.24–19.04° C ±0.02°.	Nonequil. evap. and vapor pressure.	0.036 ± 0.002 [b].
Water.	Distilled.	Moving .	Unstated; probably fair.	1–7 torr (of Substance).	4.0±1° C.	Langmuir and vapor pressure.	> 0.24 [b].

[a] The impedance due to diffusion through gases was carefully considered. The evaporation coefficient at higher vacua should also be unity.

[b] Inadequate consideration was given to the problem of gaseous diffusion. Hence this value is probably too low.

[c] A function of system pressure, lower for higher partial pressures of glycerol.

Selected values of measured condensation coefficients, α_r, for solids

Condensing substance	Purity of condensing substance	Vapor temperature	Vapor beam flux	Substrate surface	Substrate surface orientation	Substrate surface cleanliness	Vacuum	Substrate temperature	Super-saturation ratio p/p_{eq}	Technique	Condensation coefficient
Cadmium .	99.9% .	Unstated.	1.6×10^{16} to 1.9×10^{16} atoms/cm^2-sec.	Cadmium .	Polycrystalline cadmium.	Unstated; probably good.	10^{-9} to 10^{-4} torr (residual gas).	291.1–296.1 K ±0.02°.	2×10^7.	Measurement of mass by chemical analysis.	1.00 ± 0.03.
Gold .	99.999% .	Unstated.	4×10^{-2} mono-layers/sec.	Gold single crystal and polycrystalline.	Macroscopically (100) ±2° and polycrystalline.	Unstated; probably good.	< 10^{-8} torr (residual gas).	375–900 K ±35°.	Unstated; probably high.	Measurement of mass by quartz micro-balance.	1.0 ± 0.1.
Gold .	Unstated.	Unstated.	3.13×10^{13} and 3.6×10^{13} atoms/cm^2-sec.	Gold on SiO.	Polycrystalline on SiO.	Unstated; probably good.	10^{-9} torr (residual gas).	297 and 452 K ± 1°	Unstated; probably high.	Ionization gauge for flux and crystal oscillator for deposit.	1.00 ± 0.05 (for greater than 20Å mean thickness).
Gold .	Unstated; probably fair.	Unstated.	Unstated .	Gold.	Polycrystalline.	Flashed by heating to high temp.	10^{-6} torr.	1100° ± 20 K.	Unstated; probably high.	Mass spectrometry.	> 0.99.
Platinum.	Unstated; probably fair.	Unstated.	Unstated .	Platinum .	Polycrystalline.	Flashed by heating to high temp.	10^{-6} torr.	1500° ± 20 K.	~ 10^4.	Mass spectrometry.	> 0.998.
Platinum.	Unstated; probably fair.	Unstated.	Unstated .	Platinum .	Polycrystalline.	Flashed by heating to high temp.	10^{-6} torr.	600° ± 20 K.	~ 10^3.	Mass spectrometry.	> 0.998.

-continued-

Condensing substance	Purity of condensing substance	Vapor temperature	Vapor beam flux	Substrate surface	Substrate surface orientation	Substrate surface cleanliness	Vacuum	Substrate temperature	Super-saturation Ratio p/p_{eq}	Technique	Condensation coefficient
Potassium.	Unstated.	334.1–340.1 K ±0.1°.	Corresponding to 6.9 $\times 10^{-7}$ to 13×10^{-7} torr.	Single crystals of potassium.	Unstated.	Unstated.	Unstated.	284.1 to 333.6 K±0.1.	1.4 to 520.	Measurement of crystal dimensions.	0.98 ± 0.05.
Rhodium.	Unstated; probably fair.	Unstated.	Unstated.	Rhodium.	Polycrystalline.	Flashed by heating to high temp.	10^{-6} torr.	1500° ± 20 K...	Unstated; probably high.	Mass spectrometry.	> 0.99.
Silver.	Unstated.	1173–1223 K.	1.3×10^{14} to 4.5×10^{14} atoms/cm²-sec.	Silver.	Polycrystalline.	Unstated; probably good.	7×10^{-5} to 6×10^{-10} torr (residual gas).	440° ± 30 K...	10^{6} to 10^{31}.	Measurement of mass by chemical analysis.	1.00 ± 0.02.
Silver.	Unstated.	Unstated.	3.8×10^{13} and 4.4×10^{13} atoms/cm²-sec.	Silver on SiO.	Polycrystalline on SiO.	Unstated; probably good.	10^{-9} torr (residual gas).	296 and 447 K ± 1°.	2×10^{18} and 3×10^{15}.	Ionization guage for flux and crystal oscillator for deposit.	1.00 ± 0.05 (for greater than 10Å mean thickness).
Tungsten.	Unstated; probably fair.	Unstated.	Unstated.	Tungsten.	Polycrystalline.	Flashed by heating to high temp.	10^{-6} torr.	2200° ± 20 K.	Unstated; probably high.	Mass spectrometry.	0.998 ± 0.0005.
Tungsten.	Unstated; probably fair.	Unstated.	Unstated.	Tungsten.	Polycrystalline.	Flashed by heating to high temp.	10^{-6} torr.	900° ± 20 K.	Unstated; probably high.	Mass spectrometry.	0.998 ± 0.0005.
Zinc.	99.98%.	Unstated.	8.3×10^{15} atoms/cm²-sec.	Zinc.	Polycrystalline zinc.	Unstated; probably good.	6×10^{-9} torr (residual gas).	341.1° ± 0.2 K.	8×10^{6}.	Measurement of mass by chemical analysis.	0.96 ± 0.03.

CRYOGENIC PROPERTIES OF SELECTED GASES

Property and conditions		He	Ne	Ar	Kr	Xe	H_2	CH_4	NH_3	N_2	O_2	F_2
Density	32°F, 1 atm, lb/ft³	0.01114	0.0562	0.1113	0.234	0.368	0.00561	0.0448	0.0481	0.0781	0.0892	0.106
	O°C, 1 atm, kg/m³	0.1784	0.9002	1.783	3.748	5.895	0.0899	0.718	0.770	1.251	1.429	1.698
Boiling point	°F, 1 atm	-452.08	-410.89	-302.3	-242.1	-160.8	-423.2	-263.2	-28.03	-320.4	-297.35	-306.7
	°C, 1 atm	-268.934	-246.048	-185.7	-152.90	-107.1	-252.87	-164.0	-33.35	-195.8	-182.97	-188.14
	°K, 1 atm	4.216	27.10	87.45	120.25	166.05	20.28	109.15	239.80	77.35	90.18	85.01
Melting point	°F, 1 atm	-458.0[a]	-415.6	-308.6	-249.9	-169.4	-434.5	-296.46	-107.9	-345.87	-361.1	-363.3
	°C, 1 atm	-272.2[a]	-248.67	-189.2	-156.6	-111.9	-259.14	-182.48	-77.7	-209.86	-218.4	-219.62
	°K, 1 atm	0.95[a]	24.48	83.95	116.55	161.25	14.01	90.67	195.45	63.29	54.75	53.53
Vapor density at boiling point	lb/ft³	0.999	0.593	0.368	0.518	0.606	0.0830	0.1124	0.0556	0.288	0.279	
	kg/m³	16.002	9.499	5.895	8.298	9.707	1.329	1.8004	0.8906	4.613	4.4692	
Liquid density at boiling point	lb/ft³	7.803	74.91	86.77	149.8	193.5	4.37	26.47	42.58	50.19	71.23	94.4
	kg/m³	125.	1200.	1390.	2400.	3100.	70.0	424.	682.1	804.	1142	1512
Vapor pressure of solid at melting point	lb/in.²		6.25	9.98	10.6	11.8	1.04	1.35	0.87	1.86	0.038	0.002
	kg/m²		323.	516	549.	612.	54.	70.	45.2	96.4	2.0	0.12
	(N/m²) × 10⁴		4.34	6.93	7.36	8.20	0.723	0.938	0.604	1.29	0.0026	0.00014
Heat of vaporization at boiling point	Btu/lb	10.3	37.4	70.0	46.4	41.4	194.4	248.4	588.6	85.7	91.588	73.7
	kcal/kg	5.72	20.8	38.9	25.8	23.0	108	138	327	47.6	50.88	40.9
	(J/kg) × 10³	23.932	87.027	162.76	107.95	96.23	451.9	577.4	1368.2	199.2	212.9	171.1
Heat of fusion at melting point	Btu/lb	1.8	7.2	12.1	7.0	5.9	25.2	26.1	152.1	11.0	5.9	5.8
	kcal/kg	1.0	4.0	6.7	3.9	3.3	14.0	14.5	84.0	6.1	3.27	3.2
	(J/kg) × 10³	4.184	16.74	28.03	16.3	13.8	58.6	60.7	351.5	25.5	13.7	13.4
C_p	59°F, 1 atm, Btu/lb-°F or 15°C, 1 atm, kcal/kg-°C	1.25[b]	0.25[c]	0.125	0.06[c]	0.04[c]	3.39	0.528	0.523	0.248	0.220	0.180
	288.15°K, 1 atm, (J/kg) × 10³	5.23[b]	1.05[c]	0.523	0.251[c]	0.167[c]	14.2	2.21	2.188	1.038	0.9205	0.753
C_p/C_v	15–20°C, 1 atm 288–293°K, 1 atm	1.66[b]	1.64[c]	1.67	1.68[c]	1.66[c]	1.41	1.31	1.31	1.40	1.40	
Critical temperature	°F	-450.2	-397.7	-188.5	-82.7	61.9	-399.8	-116.5	270.3	-232.8	-181.3	-200.2
	°C	-267.9	-228.7	-122.5	-63.7	16.6	-239.9	-82.5	132.4	-147.1	-118.57	-129.0
	°K	5.25	44.45	150.65	209.45	289.75	33.25	190.65	405.55	126.05	154.58	144.15
Critical pressure	lb/in.² (absolute)	33.2	394.6	705.2	798	855	188.1	672	1639	492.3	731.4	808.3
	kg/cm²	2.33	27.7	49.6	46.1	60.1	13.2	47.2	115.5	34.6	51.4	56.8
	(kg/m²) × 10³	23.3	277	496	561	601	132	472	1155	346	514	568
	(N/m²) × 10⁴	23.1	274.1	489.9	554.4	594	130.7	466.9	1139	342.	508.1	561.6

RECOMMENDED THERMAL CONDUCTIVITY OF METALS AT LOW TEMPERATURES*

Thermal Conductivity, k, Watt $cm^{-1} K^{-1}$

T, K	Aluminum 99.996⁺% pure ρ_0 = 0.00315 μohm cm	Copper 99.999⁺% pure ρ_0 = 0.000851 μohm cm	Gold 99.999⁺% pure ρ_0 = 0.0055 μohm cm	Iron 99.998⁺% pure ρ_0 = 0.0327 μohm cm	Manganin**	Platinum 99.999% pure ρ_0 = 0.0106 μohm cm	Silver 99.999⁺% pure ρ_0 = 0.00062 μohm cm	Tungsten 99.99⁺% pure ρ_0 = 0.0017 μohm cm	T, K
0	0	0	0	0	0	0	0	0	0
1	7.8	28.7	4.4	0.75	0.0007	2.31	39.4	14.4	1
2	15.5	57.3	8.9	1.49	0.0018	4.60	78.3	28.7	2
3	23.2	85.5	13.1	2.24	0.0031	6.79	115	42.6	3
4	30.8	113	17.1	2.97	0.0046	8.8	147	55.6	4
5	38.1	138	20.7	3.71	0.0062	10.5	172	67.1	5
6	45.1	159	23.7	4.42	0.0078	11.8	187	76.2	6
7	51.5	177	26.0	5.13	0.0095	12.6	193	82.4	7
8	57.3	189	27.5	5.80	0.0111	12.9	190	85.3	8
9	62.2	195	28.2	6.45	0.0128	12.8	181	85.1	9
10	66.1	196	28.2	7.05	0.0145	12.3	168	82.4	10
11	69.0	193	27.7	7.62	0.0162	11.7	154	77.9	11
12	70.8	185	26.7	8.13	0.0180	10.9	139	72.4	12
13	71.5	176	25.5	8.58	0.0197	10.1	124	66.4	13
14	71.3	166	24.1	8.97	0.0215	9.3	109	60.4	14
15	70.2	156	22.6	9.30	0.0232	8.4	96	54.8	15
16	68.4	145	20.9	9.56	0.0250	7.6	85	49.3	16
18	63.5	124	17.7	9.88	0.0285	6.1	66	40.0	18
20	56.5	105	15.0	9.97	0.0322	4.9	51	32.6	20
25	40.0	68	10.2	9.36	0.0410	3.15	29.5	20.4	25
30	28.5	43	7.6	8.14	0.0497	2.28	19.3	13.1	30
35	21.0	29	6.1	6.81	0.0583	1.80	13.7	8.9	35
40	16.0	20.5	5.2	5.55	0.067	1.51	10.5	6.5	40
45	12.5	15.3	4.6	4.50	0.075	1.32	8.4	5.07	45
50	10.0	12.2	4.2	3.72	0.082	1.18	7.0	4.17	50
60	6.7	8.5	3.8	2.65	0.097	1.01	5.5	3.18	60
70	5.0	6.7	3.58	2.04	0.110	0.90	4.97	2.76	70
80	4.0	5.7	3.52	1.68	0.120	0.84	4.71	2.56	80
90	3.4	5.14	3.48	1.46	0.127	0.81	4.60	2.44	90
100	3.0	4.83	3.45	1.32	0.133	0.79	4.50	2.35	100

* In the table the third significant figure is given only for the purpose of comparison and for smoothness and is not indicative of the degree of accuracy.

**Values for manganin are taken from the measurements of Zavaritskii and Zeldovich [44], since Equations (1) to (4) do not apply to an alloy.

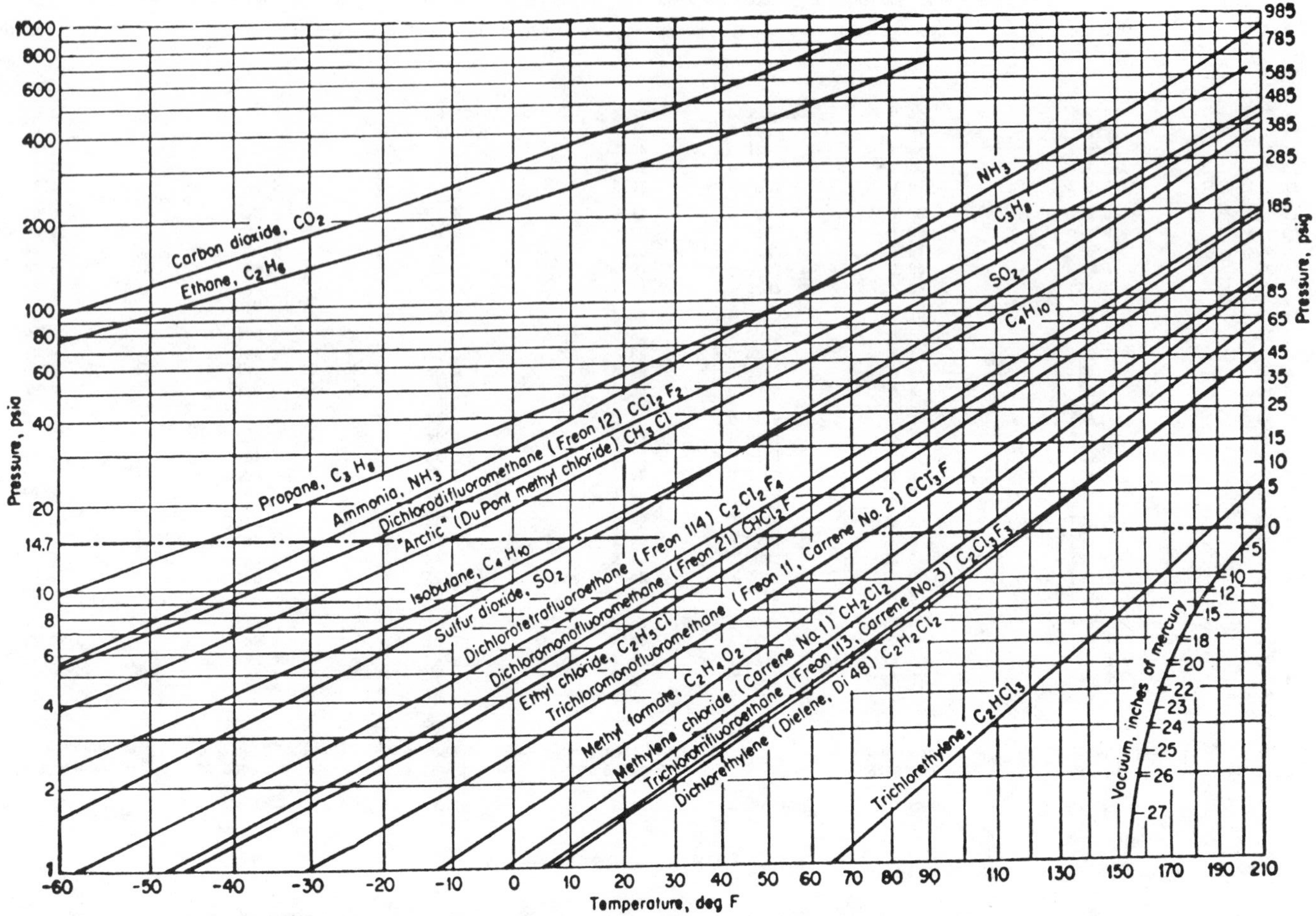

Pressure-Temperature characteristics of common refrigerants.

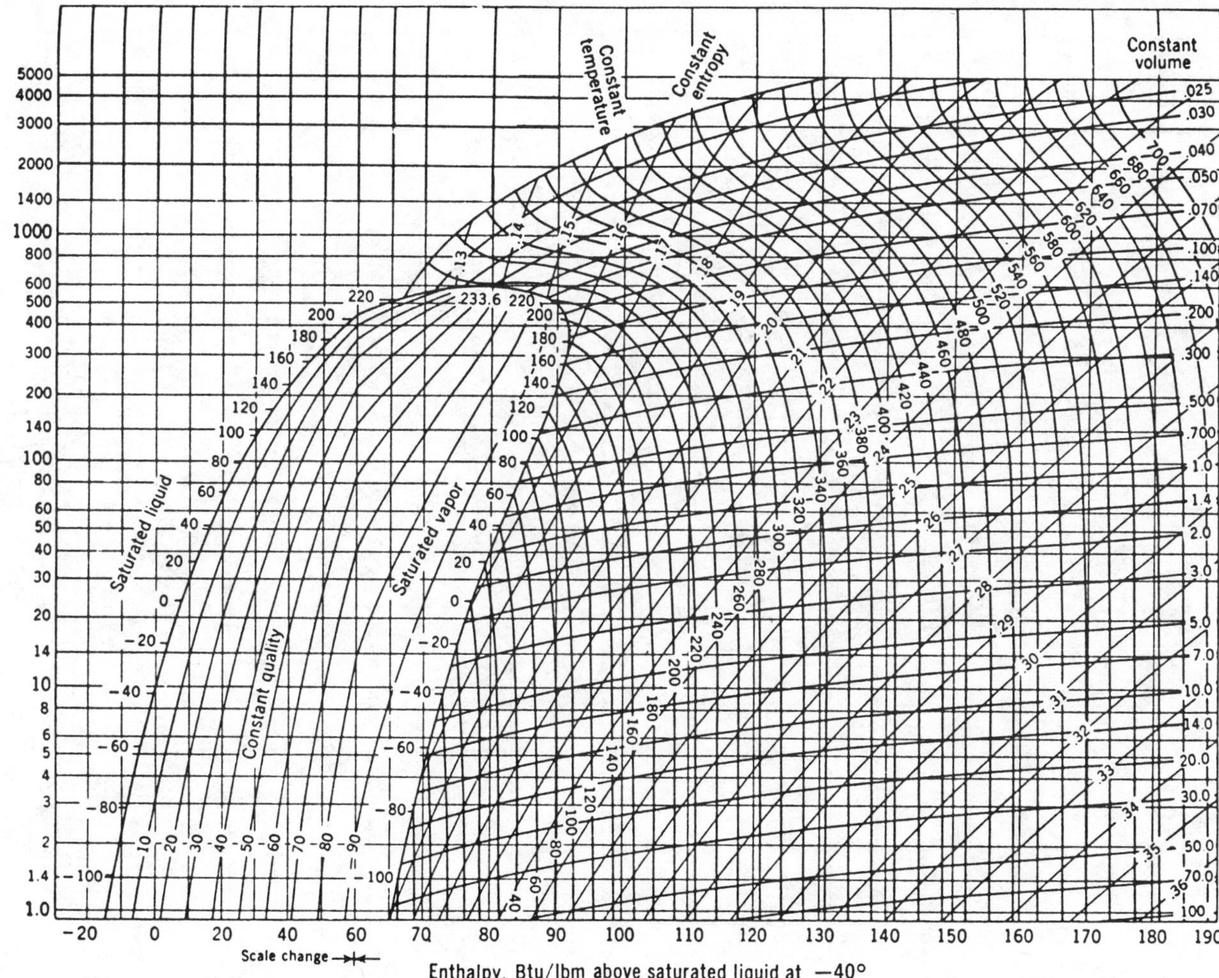

Pressure-enthalpy diagram for Freon-12. Courtesy of the E. I. du Pont de Nemours Company. T in °F, v in ft^3/lb_m, s in $Btu/lb_m{}^\circ R$, quality in percent.

Temperature-Entropy Chart for Air

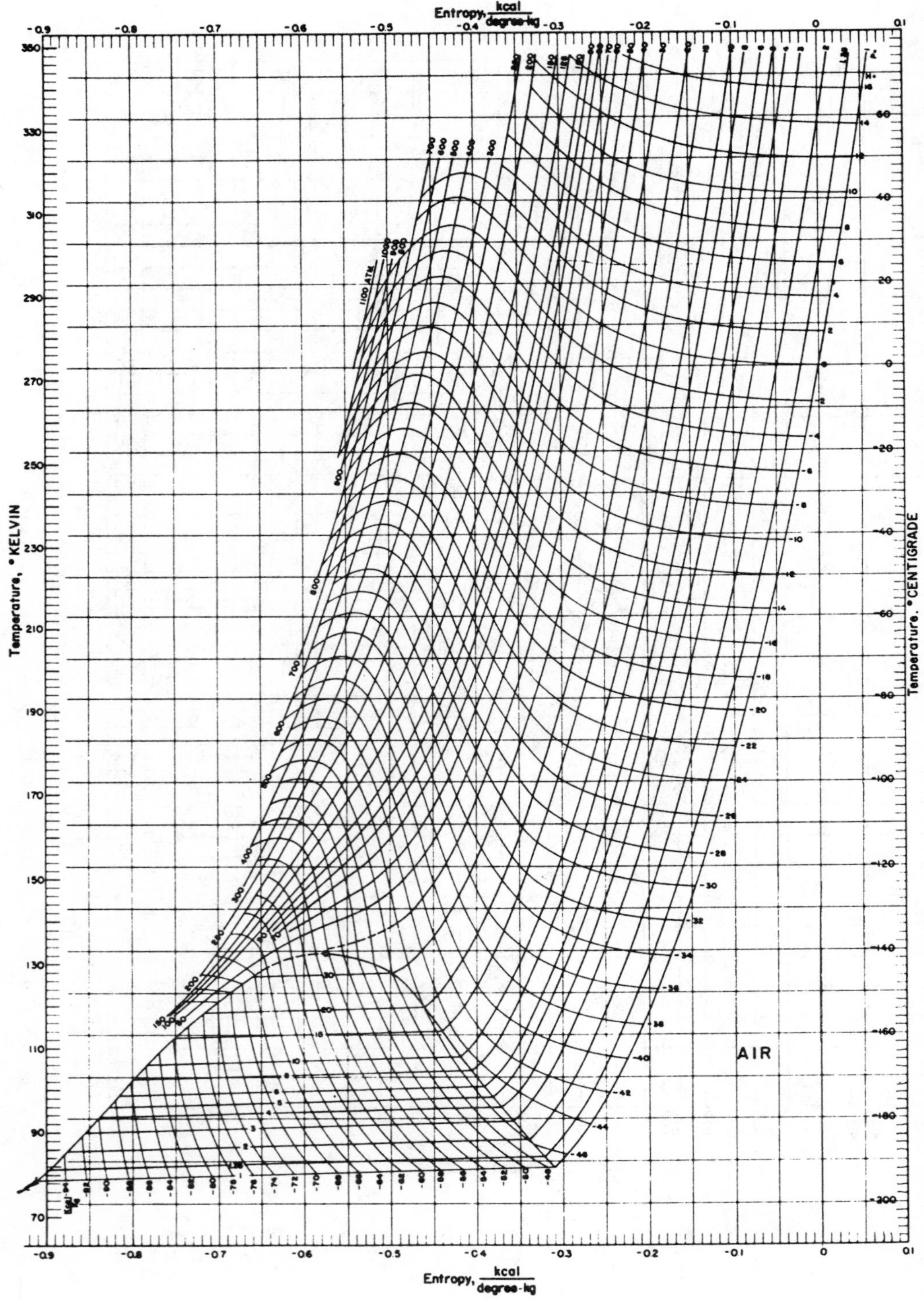

Temperature-Entropy Chart for Oxygen

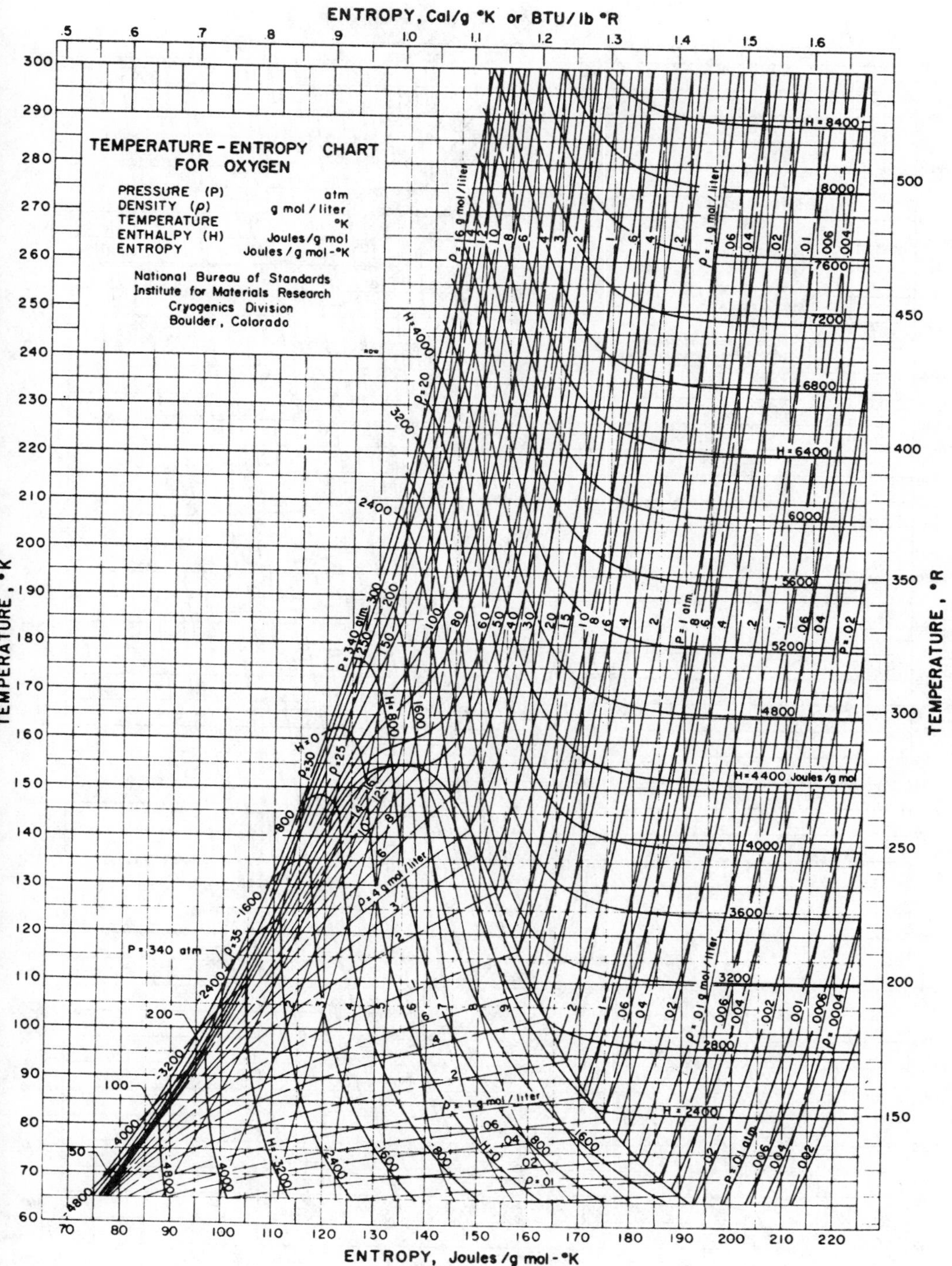
TEMPERATURE - ENTROPY CHART FOR OXYGEN
PRESSURE (P) atm
DENSITY (ρ) g mol / liter
TEMPERATURE °K
ENTHALPY (H) Joules/g mol
ENTROPY Joules / g mol - °K
National Bureau of Standards
Institute for Materials Research
Cryogenics Division
Boulder, Colorado
ENTROPY, Cal/g °K or BTU/lb °R
ENTROPY, Joules /g mol - °K
TEMPERATURE, °K
TEMPERATURE, °R
H = 4400 Joules/g mol
P = 340 atm
ρ = 4 g mol / liter
ρ = 16 g mol / liter
ρ = 1 g mol / liter
ρ = .01 g mol / liter

Temperature-Entropy Chart for Nitrogen

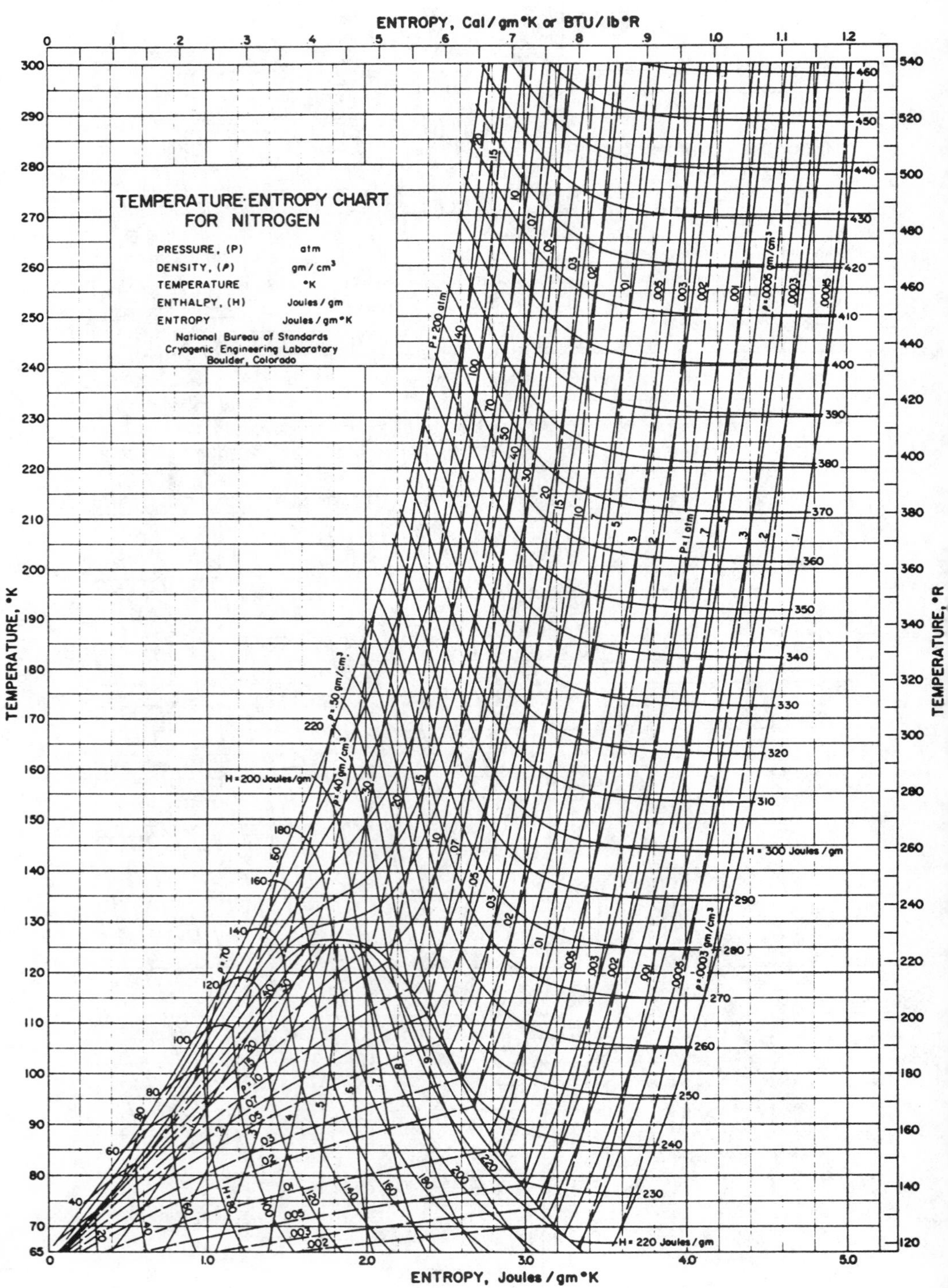

CHAPTER 20

OPTICS

REFLECTION COEFFICIENTS OF SELECTED MATERIAL SURFACES

Material	Wave lengths (μ)			
	0.400	0.500	0.600	0.700
Carbon black in oil	0.003	0.003	0.003	0.003
Clay				
Kaolin (treated)	0.82	0.81	0.82	0.82
Kaolin (untreated)	0.75	0.79	0.85	0.86
White georgia	0.94	0.92	0.93	0.94
Magnesium oxide	0.97	0.98	0.98	0.98
Paint				
Lithopone	0.95	0.98	0.98	0.98
$MgCO_3$-Vynal acetate lacquer	0.90	0.88	0.88	0.88
ZnO-Milk	0.74	0.84	0.85	0.86
Paper				
Blotting	0.64	0.72	0.79	0.79
Calendered	0.64	0.69	0.73	0.76
Crepe, green	0.23	0.49	0.19	0.48
Crepe, red	0.03	0.02	0.21	0.69
Crepe, yellow	0.17	0.44	0.75	0.79
News print stock	0.38	0.61	0.63	0.78
Peach				
Green	0.18	0.17	0.62	0.63
Ripe	0.10	0.10	0.41	0.42
Pear				
Green	0.04	0.12	0.29	0.41
Ripe	0.08	0.19	0.46	0.53
Pigment				
Chrome yellow	0.05	0.13	0.70	0.77
French ochre	0.06	0.14	0.50	0.56
Porcelain enamel				
Blue	0.44	0.10	0.05	0.23
Orange	0.09	0.09	0.59	0.69
Red	0.05	0.03	0.08	0.62
White	0.77	0.73	0.72	0.70
Yellow	0.11	0.46	0.62	0.62
Talcum, Italian	0.94	0.89	0.88	0.88
Wheat flour	0.75	0.87	0.94	0.97

INDEX OF REFRACTION OF GLASS

Relative to Air

Variety	Wave length in microns							
	.361	.434	.486	.589 (Na)	.656	.768	1.20	2.00
Zinc crown	1.539	1.528	1.523	1.517	1.514	1.511	1.505	1.497
Higher dispersion crown	1.546	1.533	1.527	1.520	1.517	1.514	1.507	1.497
Light flint	1.614	1.594	1.585	1.575	1.571	1.567	1.559	1.549
Heavy flint	1.705	1.675	1.664	1.650	1.644	1.638	1.628	1.617
Heaviest flint		1.945	1.919	1.890	1.879	1.867	1.848	1.832

INDEX OF REFRACTION OF WATER

Temp. °C	Water, pure relative to air	Temp. °C	Water, pure relative to air
14	1.33348	56	1.32792
15	1.33341	58	1.32755
16	1.33333	60	1.32718
18	1.33317	62	1.32678
20	1.33299	64	1.32636
22	1.33281	66	1.32596
24	1.33262	68	1.32555
26	1.33241	70	1.32511
28	1.33219	72	1.32466
30	1.33192	74	1.32421
32	1.33164	76	1.32376
34	1.33136	78	1.32332
36	1.33107	80	1.32287
38	1.33079	82	1.32241
40	1.33051	84	1.32195
42	1.33023	86	1.32148
44	1.32992	88	1.32100
46	1.32959	90	1.32050
48	1.32927	92	1.32000
50	1.32894	94	1.31949
52	1.32860	96	1.31897
54	1.32827	98	1.31842
		100	1.31783

INDEX OF REFRACTION OF AIR (15°C, 76 cm Hg)

Wave-length, λ angstroms	Dry air (n — 1) × 10⁷ 15°C 76 cm Hg	Vacuo correction for λ in air (nλ — λ) add	Frequency waves per cm 1/λ in air	Wave-length, λ angstroms	Dry air (n — 1) × 10⁷ 15°C 76 cm Hg	Vacuo correction for λ in air (nλ — λ) add	Frequency waves per cm 1/λ in air
2000	3256	.651	50,000	5500	2771	1.524	18,181
2100	3188	.670	47,619	5600	2769	1.551	17,857
2200	3132	.689	45,454	5700	2768	1.578	17,543
2300	3086	.710	43,478	5800	2766	1.604	17,241
2400	3047	.731	41,666	5900	2765	1.631	16,949
2500	3014	.754	40,000	6000	2763	1.658	16,666
2600	2986	.776	38,461	6100	2762	1.685	16,393
2700	2962	.800	37,037	6200	2761	1.712	16,129
2800	2941	.824	35,714	6300	2760	1.739	15,873
2900	2923	.848	34,482	6400	2759	1.766	15,625
3000	2907	.872	33,333	6500	2758	1.792	15,384
3100	2893	.897	32,258	6600	2757	1.819	15,151
3200	2880	.922	31,250	6700	2756	1.846	14,925
3300	2869	.947	30,303	6800	2755	1.873	14,705
3400	2859	.972	29,411	6900	2754	1.900	14,492
3500	2850	.998	28,571	7000	2753	1.927	14,285
3600	2842	1.023	27,777	7100	2752	1.954	14,084
3700	2835	1.049	27,027	7200	2751	1.981	13,888
3800	2829	1.075	26,315	7300	2751	2,008	13,698
3900	2823	1.101	25,641	7400	2750	2.035	13,513
4000	2817	1.127	25,000	7500	2749	2.062	13,333
4100	2812	1.153	24,390	7600	2749	2.089	13,157
4200	2808	1.179	23,809	7700	2748	2.116	12,987
4300	2803	1.205	23,255	7800	2748	2.143	12,820
4400	2799	1.232	22,727	7900	2747	2.170	12,658
4500	2796	1.258	22,222	8000	2746	2.197	12,500
4600	2792	1.284	21,739	8100	2746	2.224	12,345
4700	2789	1.311	21,276	8250	2745	2.265	12,121
4800	2786	1.338	20,833	8500	2744	2.332	11,764
4900	2784	1.364	20,406	8750	2743	2.400	11,428
5000	2781	1.391	20,000	9000	2742	2.468	11,111
5100	2779	1.417	19,607	9250	2741	2.536	10,810
5200	2777	1.444	19,230	9500	2740	2.604	10,526
5300	2775	1.471	18,867	9750	2740	2.671	10,256
5400	2773	1.497	18,518	10000	2739	2.739	10,000

Refractive index values of selected materials

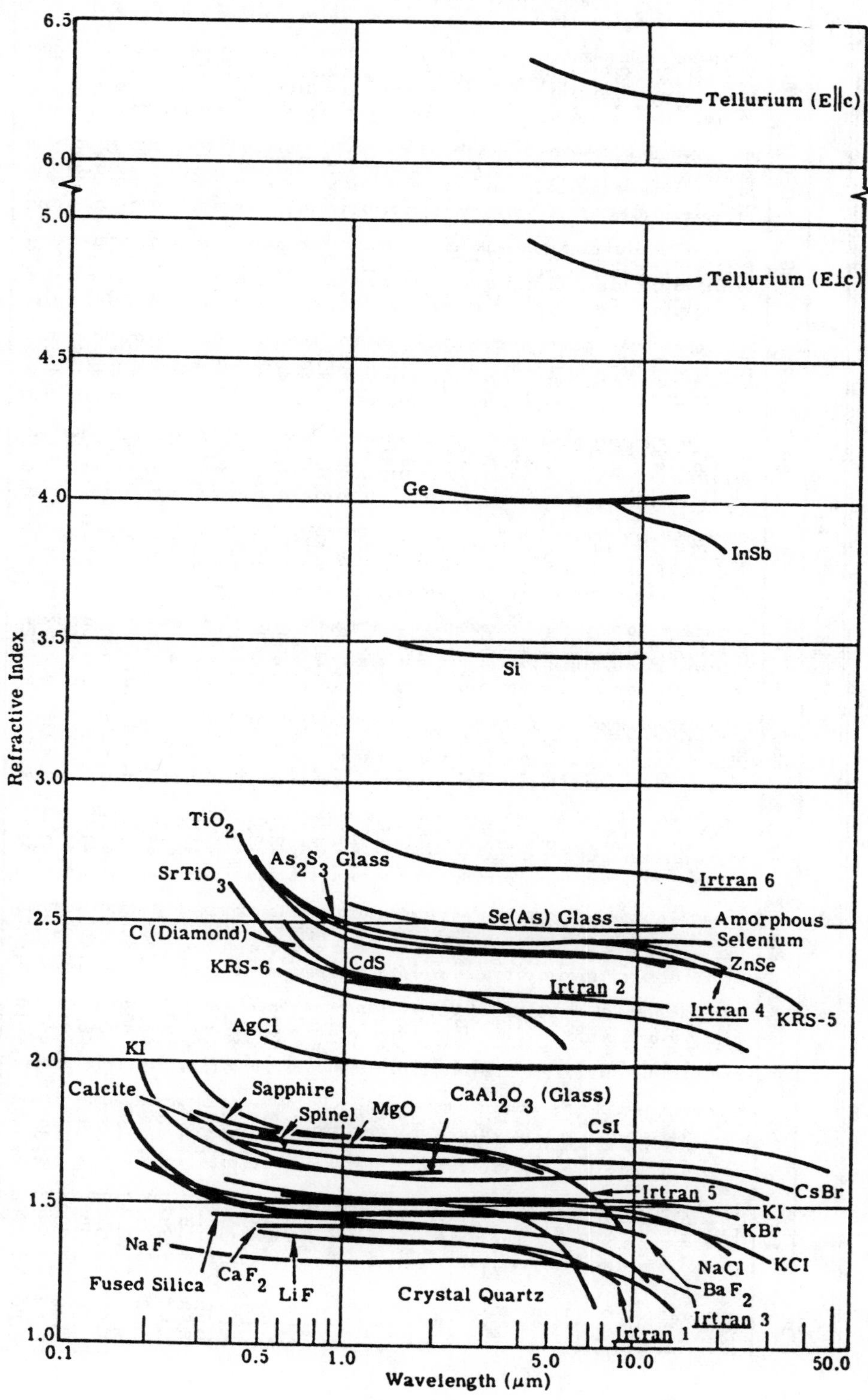

RECOMMENDED VALUES ON THE REFRACTIVE INDEX OF SILICON* (H. H. LI)

λ, μm	TEMPERATURE, K													
	100	150	200	250	293	350	400	450	500	550	600	650	700	750
1.20	3.4845	3.4915	3.4995	3.5084	3.5167	3.5284	3.5393	3.5508	3.5627	3.5749	3.5873	3.5999	3.6126	3.6252
1.22	3.4814	3.4884	3.4963	3.5051	3.5133	3.5250	3.5359	3.5473	3.5591	3.5712	3.5836	3.5961	3.6087	3.6213
1.24	3.4785	3.4854	3.4933	3.5020	3.5102	3.5218	3.5326	3.5439	3.5557	3.5678	3.5801	3.5925	3.6051	3.6176
1.26	3.4757	3.4826	3.4904	3.4991	3.5072	3.5187	3.5295	3.5407	3.5524	3.5644	3.5767	3.5891	3.6016	3.6140
1.28	3.4731	3.4799	3.4876	3.4963	3.5043	3.5158	3.5265	3.5377	3.5493	3.5613	3.5735	3.5858	3.5982	3.6106
1.30	3.4706	3.4773	3.4850	3.4936	3.5016	3.5130	3.5236	3.5348	3.5463	3.5582	3.5704	3.5827	3.5950	3.6074
1.32	3.4682	3.4748	3.4825	3.4910	3.4990	3.5103	3.5209	3.5320	3.5435	3.5554	3.5674	3.5797	3.5920	3.6043
1.34	3.4658	3.4725	3.4801	3.4885	3.4965	3.5077	3.5183	3.5293	3.5408	3.5526	3.5646	3.5768	3.5891	3.6013
1.36	3.4636	3.4702	3.4778	3.4862	3.4941	3.5053	3.5158	3.5268	3.5382	3.5500	3.5619	3.5741	3.5863	3.5985
1.38	3.4615	3.4681	3.4756	3.4839	3.4918	3.5029	3.5134	3.5243	3.5357	3.5474	3.5594	3.5715	3.5836	3.5957
1.40	3.4595	3.4660	3.4735	3.4818	3.4896	3.5007	3.5111	3.5220	3.5333	3.5450	3.5569	3.5689	3.5810	3.5931
1.45	3.4548	3.4612	3.4685	3.4768	3.4845	3.4955	3.5058	3.5166	3.5278	3.5394	3.5511	3.5631	3.5751	3.5871
1.50	3.4506	3.4568	3.4641	3.4722	3.4799	3.4908	3.5010	3.5117	3.5228	3.5343	3.5460	3.5578	3.5697	3.5816
1.55	3.4467	3.4529	3.4601	3.4681	3.4757	3.4865	3.4966	3.5072	3.5183	3.5297	3.5413	3.5530	3.5648	3.5766
1.60	3.4432	3.4493	3.4564	3.4644	3.4719	3.4826	3.4926	3.5032	3.5142	3.5255	3.5370	3.5487	3.5604	3.5721
1.65	3.4400	3.4461	3.4531	3.4610	3.4684	3.4791	3.4890	3.4995	3.5104	3.5216	3.5331	3.5447	3.5564	3.5680
1.70	3.4371	3.4431	3.4501	3.4579	3.4653	3.4758	3.4857	3.4961	3.5070	3.5181	3.5295	3.5411	3.5527	3.5643
1.80	3.4320	3.4379	3.4447	3.4524	3.4597	3.4701	3.4799	3.4902	3.5010	3.5120	3.5233	3.5347	3.5462	3.5577
1.90	3.4277	3.4334	3.4402	3.4478	3.4550	3.4653	3.4750	3.4852	3.4958	3.5068	3.5180	3.5293	3.5407	3.5521
2.00	3.4240	3.4297	3.4363	3.4439	3.4510	3.4612	3.4708	3.4809	3.4915	3.5023	3.5135	3.5247	3.5360	3.5473
2.25	3.4168	3.4223	3.4288	3.4362	3.4431	3.4532	3.4626	3.4726	3.4830	3.4937	3.5046	3.5157	3.5269	3.5380
2.50	3.4116	3.4170	3.4234	3.4306	3.4375	3.4474	3.4568	3.4666	3.4769	3.4875	3.4983	3.5093	3.5203	3.5313
2.75	3.4078	3.4131	3.4194	3.4266	3.4334	3.4432	3.4524	3.4622	3.4724	3.4829	3.4936	3.5045	3.5154	3.5264
3.00	3.4048	3.4101	3.4163	3.4234	3.4302	3.4399	3.4491	3.4588	3.4689	3.4794	3.4900	3.5009	3.5117	3.5226
4.00	3.3981	3.4032	3.4093	3.4163	3.4229	3.4325	3.4415	3.4510	3.4610	3.4713	3.4818	3.4925	3.5032	3.5139
5.00	3.3950	3.4000	3.4060	3.4129	3.4195	3.4290	3.4380	3.4474	3.4573	3.4675	3.4780	3.4886	3.4992	3.5099
6.00	3.3933	3.3983	3.4043	3.4111	3.4177	3.4271	3.4360	3.4455	3.4553	3.4655	3.4759	3.4865	3.4971	3.5077
7.00	3.3923	3.3973	3.4032	3.4100	3.4165	3.4260	3.4349	3.4443	3.4541	3.4643	3.4747	3.4852	3.4958	3.5063
8.00	3.3916	3.3966	3.4025	3.4093	3.4158	3.4252	3.4341	3.4435	3.4533	3.4635	3.4739	3.4844	3.4950	3.5055
9.00	3.3912	3.3961	3.4020	3.4088	3.4153	3.4247	3.4336	3.4430	3.4528	3.4629	3.4733	3.4838	3.4944	3.5049
10.00	3.3909	3.3958	3.4017	3.4085	3.4150	3.4244	3.4332	3.4426	3.4524	3.4625	3.4729	3.4834	3.4940	3.5045
11.00	3.3906	3.3955	3.4015	3.4082	3.4147	3.4241	3.4330	3.4423	3.4521	3.4623	3.4726	3.4831	3.4937	3.5042
12.00	3.3904	3.3954	3.4013	3.4080	3.4145	3.4239	3.4328	3.4421	3.4519	3.4620	3.4724	3.4829	3.4934	3.5039
13.00	3.3903	3.3952	3.4011	3.4079	3.4144	3.4237	3.4326	3.4420	3.4518	3.4619	3.4722	3.4827	3.4932	3.5037
14.00	3.3902	3.3951	3.4010	3.4078	3.4142	3.4236	3.4325	3.4418	3.4516	3.4617	3.4721	3.4826	3.4931	3.5036

* THE ESTIMATED UNCERTAINTY IN THE RECOMMENDED VALUES IS $\pm 2\times10^{-3}$. RECOMMENDED VALUES ARE GIVEN TO MORE DIGITS THAN WARRANTED MERELY FOR THE PURPOSE OF TABULAR SMOOTHNESS. THE INSIGNIFICANT DIGITS OF THE VALUES ARE INDICATED BY OVERSTRIKES.

RECOMMENDED n-λ-T DIAGRAM OF SILICON

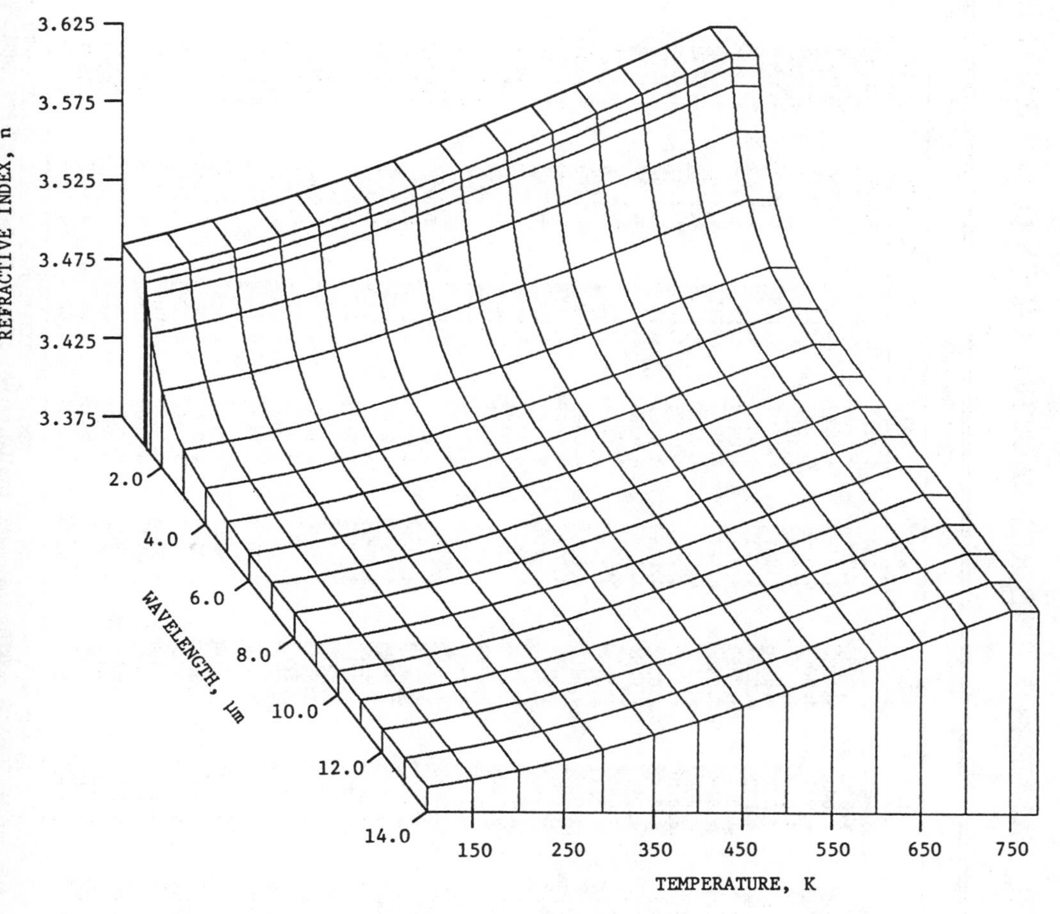

RECOMMENDED VALUES ON THE REFRACTIVE INDEX OF GERMANIUM* (H. H. LI)

λ, μm	TEMPERATURE, K 100	150	200	250	293	350	400	450	500	550
1.90	4.0290	4.0474	4.0680	4.0907	4.1117	4.1417	4.1697	4.1993	4.2302	4.2623
1.92	4.0270	4.0453	4.0659	4.0885	4.1094	4.1393	4.1672	4.1966	4.2274	4.2593
1.94	4.0251	4.0433	4.0638	4.0863	4.1072	4.1369	4.1647	4.1940	4.2246	4.2565
1.96	4.0232	4.0414	4.0618	4.0842	4.1050	4.1346	4.1623	4.1915	4.2220	4.2537
1.98	4.0214	4.0395	4.0598	4.0822	4.1029	4.1324	4.1599	4.1890	4.2194	4.2510
2.00	4.0197	4.0377	4.0579	4.0802	4.1008	4.1302	4.1577	4.1866	4.2169	4.2484
2.05	4.0156	4.0334	4.0534	4.0755	4.0959	4.1250	4.1523	4.1810	4.2110	4.2421
2.10	4.0117	4.0294	4.0493	4.0711	4.0914	4.1202	4.1472	4.1757	4.2054	4.2363
2.15	4.0081	4.0257	4.0454	4.0670	4.0872	4.1158	4.1426	4.1708	4.2003	4.2309
2.20	4.0048	4.0222	4.0417	4.0632	4.0832	4.1116	4.1382	4.1662	4.1954	4.2259
2.25	4.0017	4.0190	4.0383	4.0597	4.0795	4.1077	4.1341	4.1619	4.1909	4.2211
2.30	3.9987	4.0159	4.0352	4.0564	4.0761	4.1041	4.1303	4.1579	4.1867	4.2167
2.40	3.9934	4.0104	4.0294	4.0503	4.0698	4.0974	4.1233	4.1506	4.1791	4.2087
2.50	3.9887	4.0055	4.0243	4.0450	4.0642	4.0916	4.1172	4.1441	4.1723	4.2015
2.60	3.9845	4.0011	4.0197	4.0402	4.0593	4.0864	4.1117	4.1384	4.1663	4.1952
2.70	3.9808	3.9972	4.0157	4.0360	4.0549	4.0817	4.1068	4.1333	4.1609	4.1896
2.80	3.9775	3.9938	4.0121	4.0322	4.0509	4.0776	4.1025	4.1287	4.1561	4.1845
2.90	3.9745	3.9907	4.0088	4.0288	4.0474	4.0738	4.0985	4.1246	4.1518	4.1800
3.00	3.9718	3.9878	4.0059	4.0257	4.0442	4.0704	4.0950	4.1209	4.1479	4.1759
3.20	3.9671	3.9830	4.0008	4.0204	4.0387	4.0646	4.0889	4.1144	4.1411	4.1688
3.40	3.9632	3.9789	3.9966	4.0160	4.0341	4.0598	4.0838	4.1091	4.1355	4.1629
3.60	3.9600	3.9755	3.9930	4.0123	4.0302	4.0557	4.0795	4.1046	4.1308	4.1580
3.80	3.9572	3.9727	3.9900	4.0092	4.0270	4.0523	4.0759	4.1008	4.1268	4.1538
4.00	3.9549	3.9702	3.9875	4.0065	4.0242	4.0493	4.0728	4.0976	4.1234	4.1502
4.25	3.9524	3.9676	3.9848	4.0037	4.0212	4.0462	4.0696	4.0942	4.1198	4.1464
4.50	3.9503	3.9655	3.9825	4.0013	4.0188	4.0436	4.0668	4.0913	4.1168	4.1433
4.75	3.9485	3.9636	3.9806	3.9993	4.0167	4.0414	4.0645	4.0888	4.1142	4.1406
5.00	3.9470	3.9620	3.9789	3.9976	4.0149	4.0395	4.0625	4.0868	4.1121	4.1383
5.50	3.9446	3.9595	3.9763	3.9948	4.0120	4.0365	4.0594	4.0834	4.1086	4.1346
6.00	3.9428	3.9576	3.9743	3.9927	4.0098	4.0342	4.0569	4.0809	4.1059	4.1318
6.50	3.9413	3.9561	3.9727	3.9911	4.0081	4.0324	4.0550	4.0789	4.1038	4.1296
7.00	3.9402	3.9549	3.9715	3.9898	4.0068	4.0309	4.0536	4.0773	4.1021	4.1279
8.00	3.9385	3.9532	3.9697	3.9879	4.0048	4.0289	4.0514	4.0750	4.0997	4.1253
9.00	3.9374	3.9520	3.9684	3.9866	4.0034	4.0274	4.0498	4.0734	4.0981	4.1236
10.00	3.9365	3.9511	3.9675	3.9856	4.0025	4.0264	4.0488	4.0723	4.0969	4.1223
11.00	3.9359	3.9505	3.9669	3.9849	4.0017	4.0256	4.0480	4.0715	4.0960	4.1214
12.00	3.9355	3.9500	3.9664	3.9844	4.0012	4.0250	4.0474	4.0708	4.0953	4.1207
13.00	3.9351	3.9496	3.9660	3.9840	4.0008	4.0246	4.0469	4.0703	4.0948	4.1202
14.00	3.9348	3.9493	3.9657	3.9837	4.0004	4.0242	4.0465	4.0699	4.0944	4.1197
15.00	3.9346	3.9491	3.9654	3.9834	4.0001	4.0240	4.0462	4.0696	4.0941	4.1194
16.00	3.9344	3.9489	3.9652	3.9832	3.9999	4.0237	4.0460	4.0694	4.0938	4.1191
17.00	3.9342	3.9487	3.9650	3.9830	3.9997	4.0235	4.0458	4.0691	4.0936	4.1188
18.00	3.9341	3.9486	3.9649	3.9829	3.9996	4.0234	4.0456	4.0690	4.0934	4.1186

* THE ESTIMATED UNCERTAINTY IN THE RECOMMENDED VALUES IS $\pm 2\times10^{-4}$. RECOMMENDED VALUES ARE GIVEN TO MORE DIGITS THAN WARRANTED MERELY FOR THE PURPOSE OF TABULAR SMOOTHNESS. THE INSIGNIFICANT DIGITS OF THE VALUES ARE INDICATED BY OVERSTRIKES.

RECOMMENDED n-λ-T DIAGRAM OF GERMANIUM

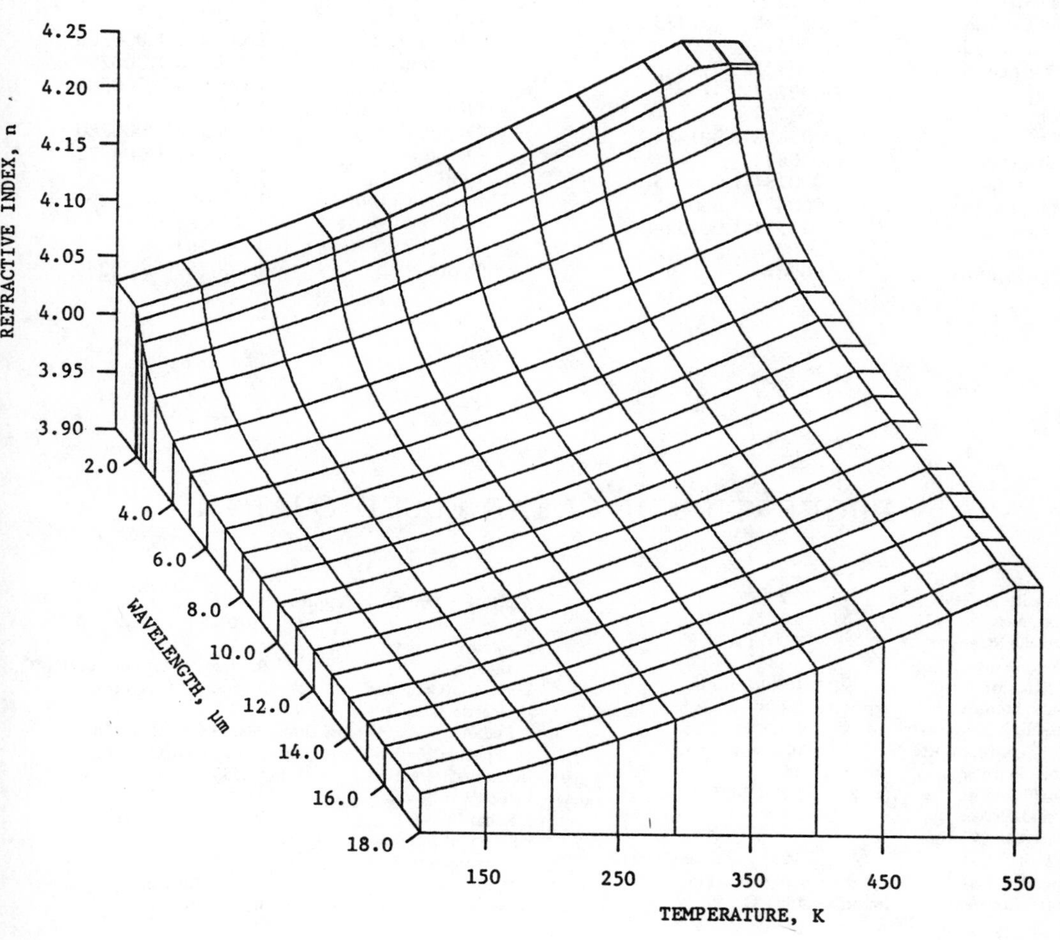

INDEX OF REFRACTION OF SELECTED GASES

Values are relative to a vacuum and for a Temp. of 0°C, and 760 mm pressure.
(From Smithsonian Tables)

Substance	Indices of refraction
Acetone	1.001079-1.001100
Air	1.0002926
Ammonia	1.000381-1.000385
Ammonia	1.000373-1.000379
Argon	1.000281
Benzene	1.001700-1.001823
Bromine	1.001132
Carbon dioxide	1.000449-1.000450
dioxide	1.000448-1.000454
disulfide	1.001500
disulfide	1.001478-1.001485
monoxide	1.000340
monoxide	1.000335
Chlorine	1.000772
Chlorine	1.000773
Chloroform	1.001436-1.001464
Cyanogen	1.000834
Cyanogen	1.000784-1.000825
Ethyl alcohol	1.000871-1.000885
ether	1.001521-1.001544
Helium	1.000036
Hydrochloric acid	1.000449

Substance	Indices of refraction
Hydrochloric acid	1.000447
Hydrogen	1.000138-1.000143
Hydrogen	1.000132
sulfide	1.000644
sulfide	1.000623
Methane	1.000443
Methane	1.000444
Methyl alcohol	1.000549-1.000623
Methyl ether	1.000891
Nitric oxide	1.000303
Nitric oxide	1.000297
Nitrogen	1.000295-1.000300
Nitrogen	1.000296-1.000298
Nitrous oxide	1.000503-1.000507
Nitrous oxide	1.000516
Oxygen	1.000272-1.000280
Oxygen	1.000271-1.000272
Pentane	1.001711
Sulfur dioxide	1.000665
Sulfur dioxide	1.000686
Water	1.000261
Water	1.000249-1.000259

PROPERTIES OF CLEAR FUSED QUARTZ

Density		2.2 g./c.c.
Hardness		4.9 (Mohs')
Tensile Strength		7,000 p.s.i.
Compressive Strength		> 160,000 p.s.i.
Bulk Modulus	(approx.)	5.3×10^6 p.s.i.
Rigidity Modulus		4.5×10^6 p.s.i.
Young's Modulus		10.4×10^6 p.s.i.
Poisson's Ratio		.16
Coefficient of Thermal Expansion	(av.)	5.5×10^{-7} cm./cm./°C {20°C 320°C
Thermal Conductivity		.0033 g. cal./cm.^{2}/sec./°C/cm.
Specific Heat		.18 g. cal./gm.
Softening Point	(approx.)	1665°C

Annealing Point	(approx.)	1140°C
Strain Point		1070°C
Electrical Resistance		9.5 $\log_{10}$ R for cm.^{3} at 350°C
Dielectric Constant		3.75 at 20°C. 1 Mc.
Dielectric Loss Factor	less than	.0004 at 20°C. 1 Mc.
Dissipation Factor	less than	.0001 at 20°C. 1 Mc.
Index of Refraction		1.4585
Velocity of Sound—Shear Wave		3.75×10^5 cm./sec.
Velocity of Sound—Compressional Wave		5.90×10^5 cm./sec.
Sonic Attenuation	less than	.033 db/ft./mc.

Transmission Regions of Optical Materials (2mm thickness)

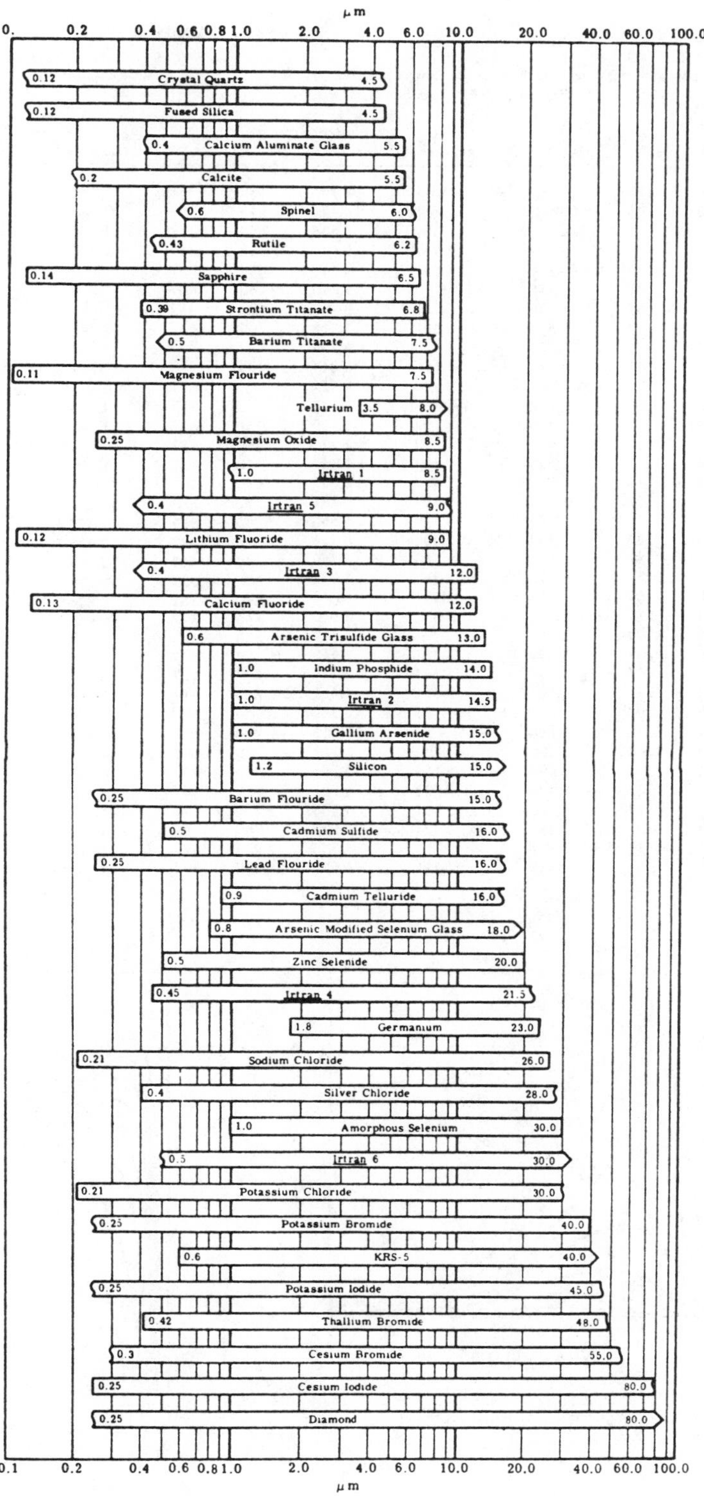

RADIATION FROM AN IDEAL BLACK BODY

From NASA TT-F-783

Temperature dependence of the specific power radiated, Q_T, and of λ_{max} for an ideal black body according to Kirchhoff's law ($\sigma_o = 5.68 \cdot 10^{-8}$ W/m²·deg⁴ = $4.88 \cdot 10^{-8}$ kcal/m²·deg⁴)

T, °K	t, °C	Q_T, W/cm²	Q_T kcal/m²·hr	λmax μ	T, °K	t, °C	Q_T W/cm²	Q_T kcal/m²·hr	λmax μ
100	−173	$5.680 \cdot 10^{-4}$	$4.880 \cdot 10^{0}$	28.96	630	357	8.948	7.687	4.597
200	−73	$9.088 \cdot 10^{-3}$	$7.808 \cdot 10^{1}$	14.48	640	367	9.529	8.187	4.525
273	0	$3.155 \cdot 10^{-2}$	$2.711 \cdot 10^{2}$	10.608	650	377	$1.014 \cdot 10^{0}$	8.711	4.455
300	27	4.601	3.953	9.655	660	387	1.078	9.260	4.388
310	37	5.246	4.507	9.342	670	397	1.145	9.831	4.322
320	47	5.956	5.117	9.050	680	407	1.214	$1.013 \cdot 10^{4}$	4.259
330	57	6.736	5.787	8.766	690	417	1.287	1.106	4.197
340	67	7.590	6.521	8.518	700	427	1.364	1.172	4.137
350	77	8.524	7.323	8.274	710	437	1.443	1.240	4.069
360	87	9.540	8.196	8.044	720	447	1.526	1.311	4.022
370	97	1.065	9.146	7.827	730	457	$1.613 \cdot 10^{0}$	$1.386 \cdot 10^{1}$	3.967
380	107	$1.184 \cdot 10^{-1}$	$1.018 \cdot 10^{3}$	7.621	740	467	1.703	1.463	3.914
390	117	1.314	1.128	7.426	750	477	1.797	1.544	3.861
400	127	1.454	1.249	7.270	760	487	1.895	1.628	3.811
410	137	1.605	1.379	7.053	770	497	1.997	1.715	3.761
420	147	1.76	1.519	6.865	780	507	2.102	1.806	3.713
430	157	1.942	1.668	6.735	790	517	2.212	1.901	3.666
440	167	2.129	1.829	6.562	800	527	2.327	1.999	3.620
450	177	2.329	2.001	6.436	810	537	2.445	2.101	3.565
460	187	2.543	2.185	6.266	820	547	2.568	2.206	3.532
470	197	2.772	2.381	6.162	830	557	2.696	2.316	3.489
480	207	3.015	2.591	6.033	840	567	2.828	2.430	3.448
490	217	3.274	2.813	5.910	850	577	2.965	2.547	3.407
500	227	3.550	3.05	5.792	860	587	3.107	2.670	3.367
510	237	3.843	3.301	5.668	870	597	3.254	2.796	3.329
520	247	4.163	3.568	5.559	880	607	3.406	2.927	3.291
530	257	4.482	3.851	5.454	890	617	3.564	3.062	3.254
540	267	4.830	4.150	5.363	900	627	3.727	3.202	3.218
550	277	5.198		5.255	910	637	3.895	3.346	3.162
560	287	5.586	4.799	5.161	920	647	4.069	3.496	3.148
570	297	5.996	5.151	5.061	930	657	4.249	3.650	3.114
580	307	6.428	5.522	4.963	940	667	4.435	3.810	3.081
590	317	6.883	5.913	4.908	950	677	4.626	3.975	3.048
600	327	7.361	6.324	4.827	960	687	4.824	4.145	3.017
610	337	7.864	6.757	4.748	970	697	5.028	4.320	2.986
620	347	8.393	7.211	4.671	980	707	5.239	4.501	2.955

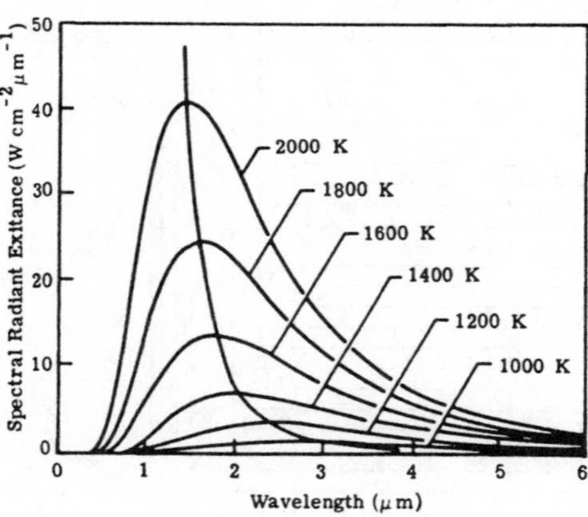

Blackbody curves, 1000 to 2000 K.

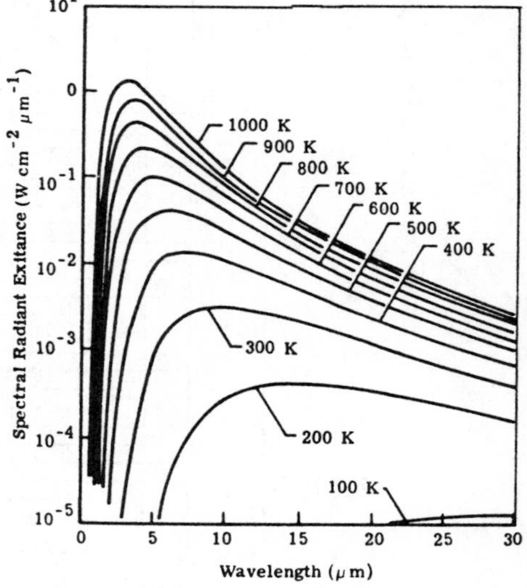

Blackbody curves, 100 to 1000 K.

The CIE chromaticity diagram

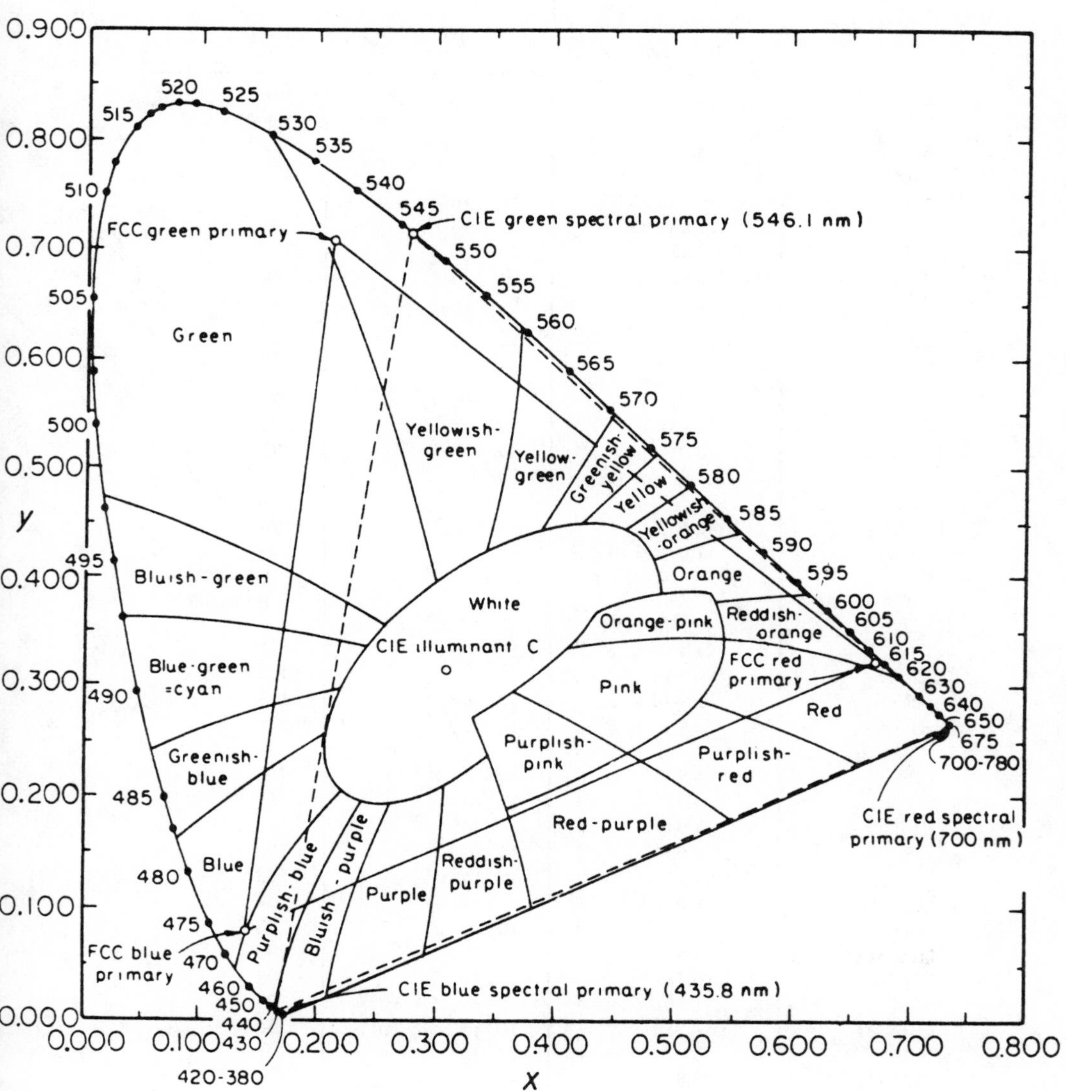

The CIE (Comité International d'Eclairage) diagram is the world-wide standard method of representing color. The x and y coordinates are transformations of three color primaries defined by CIE. The chromaticity diagram displays the hue and saturation of colors. The hue varies on the diagram with the angle measured with the white point (illuminant C) as the vertex. Saturation is measured by the radial distance from the white point at the center of the chart.

PERMITTIVITY OF OPTICAL MATERIALS
(Dielectric Constant)

Material	ϵ_r *(Relative dielectric constant)*	f (Hz)	T (K)
Al_2O_3	10.55^p	10^2 to 3×10^8	298
	8.6^s	10^2 to 2.5×10^{10}	298
As_2S_3	8.1	10^3 to 10^6	–
AgCl	12.3	10^6	293
ADP	56.4 to 55.9^s	10^2 to 10^8	–
	16.4 to 13.7^p	10^2 to 10^8	–
BaF_2	7.33	2×10^6	–
$BaTiO_3$	1240 to 1100	10^2 to 10^8	298
$CaCO_3$	8.5^s	10^4	290 to 295
	8.0^p	10^4	290 to 295
CuCl	10.0	5×10^5	293
CaF_2	6.76	10^5	–
CsBr	6.51	2×10^6	298
CsI	5.65	10^6	298
CuBr	8.0	3×10^6	293
CdTe	11.0	1 to 10^5	5.5×10^{13} carriers $(cc)^{-1}$
$CaTiO_3$	140.0	1.5×10^6	294
GaAs	11.06 ± 0.14	–	–
Ge	16.6	9.37×10^9	9.0 ohm cm resistivity
KDP	44.5 to 44.3^s	10^2 to 10^8	–
	21.4 to 20.2^p	10^2 to 10^8	–
KCl	4.64	10^6	302.5
KRS-5	32.9 to 32.5	10^2 to 10^7	298
KRS-6	32.9 to 31.8	10^2 to 10^5	298
KBr	4.90	10^2 to 10^{10}	298
	4.97	10^2 to 10^{10}	360
KI	4.94	2×10^6	–
LiF	9.00	10^2 to 10^{10}	298
	9.11	10^2 to 10^{10}	353
$MgO \cdot 3.5\ Al_2O_3$	8.0 to 9.0	–	–
MgO	9.65	10^2 to 10^8	298
Muscovite	5.4	10^2 to 3×10^9	299
$NaNO_3$	6.85	2×10^5	292
NaCl	5.90	10^2 to 2.5×10^{10}	298
	6.35 to 5.97	10^2 to 2.5×10^{10}	358
NaF	6.0	2×10^6	292
$PbMoO_4$	26.8	4×10^8	–
$PbCl_2$	33.5	5×10^5	293
PbS	17.9	10^6	288
PbF_2	3.6	10^6	–
Se	6.0	10^2 to 10^{10}	298
Si	13.0	9.37×10^9	–
SrF_2	7.69	2×10^6	–
SiO_2 (crystal)	4.34^s	3×10^7	290 to 295
	4.27^p	3×10^7	290 to 295
SiO_2 (fused)	3.78	10^2 to 10^{10}	300
$SrTiO_3$	306.0	10^2 to 10^5	298
Se(As)	234 to 230.0	10^2 to 10^{10}	298
TlCl	31.9	2×10^6	–
TlBr	30.3	10^3 to 10^7	298
TiO_2	200 to 160	10^4 to 10^7	298

THERMAL PROPERTIES OF OPTICAL MATERIALS

Materials	*Symbol*	*Melting Temperature* (K)	*Specific Heat*	*Specific Heat Temperature* (K)	*Thermal Conductivity* (cal cm^{-1} sec^{-1} K)	*Thermal Conductivity Temperature* (K)
Ammonium Dihydrogen Phosphate	ADP	463.0	–	–	1.7×10^{-3p}	315
		–	–	–	3.0×10^{-3s}	313
Silver	Ag	1233.8	0.0559	298	–	–
Silver Chloride	AgCl	730.7	0.0848	273	2.6×10^{-3}	295
		–	0.0906	323	2.71×10^{-3}	–
Silver Sulfide	Ag_2S	–	0.072	273	–	–
Aluminum	Al	933.2	0.218	298	–	–
Sapphire	Al_2O_3	2303.0	0.180	298	60×10^{-3p}	299
		–	0.174	273	55×10^{-3s}	296
Arsenic Trisulfide Glass	As_2S_3	483.0	–	–	4.0×10^{-4}	313
Gold	Au	1336.0	0.0309	298	–	–
Boron	B	2573.0	0.260	273	–	–
Barium Fluoride	BaF_2	1553.0	–	–	28×10^{-3}	286
		–	–	–	17×10^{-3}	311
Barium Titanate	$BaTiO_3$	1873.0	0.01799	55	1.6×10^{-3}	401
		–	0.03004	75	3.2×10^{-3}	273
		–	0.04471	100	–	–
		–	0.05709	125	–	–
		–	0.06868	150	–	–
		–	0.07813	175	–	–
Bismuth	Bi	544.0	0.02990	298	–	–
Carbon (diamond)	C	>3773.0	0.12000	298	34.0	76
		–	–	–	8.6	194
		–	–	–	6.59	273
Carbon (granular)	C	>3773.0	0.21600	298	–	–
Calcite	$CaCO_3$	1612.0	0.20300	273	1.32×10^{-2p}	273
		–	0.21400	373	1.11×10^{-2s}	273
Calcium Fluoride	CaF_2	1633.0	0.20400	273	9.32×10^{-2}	83
		–	0.21200	373	3.60×10^{-2}	200
		–	–	–	2.47×10^{-2}	273
		–	–	–	2.32×10^{-2}	298
		–	–	–	1.91×10^{-2}	373
Calcium Titanate	$CaTiO_3$	2248.0	–	–	–	–
Cadmium Fluoride	CdF_2	1047.0	–	–	–	–
Cadmium Sulfide	CdS	1560.0	0.08820	273	3.8×10^{-2}	287
		–	0.92200	323	–	–
Cadmium Sulfide (pressed)	CdS	–	–	–	1.0×10^{-1}	10 & 100
		–	–	–	4.0×10^{-1}	50
Cadmium Telluride	CdTe	1314 to 1323	0.01875	323	0.015	–
Cesium Bromide	CsBr	909.0	0.63000	293	2.3×10^{-3}	298
		–	–	–	2.2×10^{-3}	318
		–	–	–	2.6×10^{-3}	338
Cesium Iodide	CsI	894.0	0.04800	293	2.7×10^{-3}	298
Copper	Cu	1356.0	0.00717×10^{-3}	2	–	–
		–	0.038×10^{-3}	5	–	–
		–	0.21×10^{-3}	10	–	–
		–	1.8×10^{-3}	20	–	–
		–	24.0×10^{-3}	50	–	–
		–	61.0×10^{-3}	100	–	–
		–	92.0×10^{-3}	300	–	–
Copper Bromide	CuBr	777.0	–	–	–	–
Copper Chloride	CuCl	695.0	–	–	–	–
Copper Sulfide	CuS	–	0.12900	298	–	–
Iron	Fe	1808.0	0.11000	298	–	–
Iron Oxide	Fe_2O_3	–	0.17000	298	–	–
Gallium Arsenide	GaAs	1511.0	–	–	0.125	300
Gallium Phosphide	GaP	>773.0	–	–	0.13	300
Gallium Antimonide	GaSb	993.0	0.01828	–	0.105	300
Germanium	Ge	1209 to 1215	0.074	273	0.14	293
Indium Arsenide	InAs	1215	3.4	78 to 290	–	–
		–	5.2	290 to 573	–	–
		–	7.01	573 to 673	–	–
Indium Phosphide	InP	1323 to 1343	–	–	–	–
Indium Antimonide	InSb	796.0	0.0231	180	0.085	293
		–	0.248	300	–	–
<u>Irtran</u> 1		1528.0	–	–	0.035	329
			–	–	0.026	452
<u>Irtran</u> 2		2103.0	–	–	0.037	327
		(150 psi)	–	–	0.026	447
<u>Irtran</u> 3		1692.0	–	–	0.019	353
		–	–	–	0.015	449
<u>Irtran</u> 4		1788.0	–	–	0.031	327
		–	–	–	0.016	695
<u>Irtran</u> 5		3220.0	–	–	0.104	298
		–	–	–	0.070	441
<u>Irtran</u> 6*		1363.0	–	–	0.010	273
		–	–	–	0.0085	417
Potassium Bromide	KBr	1003.0	0.104	273	0.698×10^{-2}	299
		–	0.108	373	1.15×10^{-2}	319
Potassium Chloride	KCl	1049.0	0.162	273	1.56×10^{-2}	315
		–	0.168	373	–	–
Potassium Dihydrogen Phosphate	KDP	525.6	–	–	2.9×10^{-3p}	312
		–	–	–	3.2×10^{-3s}	319
Potassium Iodide	KI	996.0	0.73	200	$5.0(\pm 3\%) \times 10^{-3}$	299
		–	0.75	250	–	–
		–	0.75	270	–	–
Thallium Bromide Iodide	KRS-5	687.5	-	–	1.3×10^{-3}	293

***<u>Irtran</u> 6 is no longer manufactured by the Eastman Kodak Co. Irtran® is a registered trademark of the Eastman Kodak Co.**

THERMAL PROPERTIES OF OPTICAL MATERIALS (continued)

Materials	*Symbol*	*Melting Temperature* (K)	*Specific Heat*	*Specific Heat Temperature* (K)	*Thermal Conductivity* (cal cm^{-1} sec^{-1} K)	*Thermal Conductivity Temperature* (K)
Thallium Bromide Chloride	KRS-6	696.5	0.0482	293	17.1×10^{-4}	329
Lithium Fluoride	LiF	1143.0	0.373	283	2.70×10^{-2}	314
Magnesium Oxide	MgO	3073.0	0.209	273	262×10^{-2}	10
		–	–	–	756×10^{-2}	30
		–	–	–	638×10^{-2}	100
		–	–	–	14.0×10^{-2}	300
		–	–	–	7.7×10^{-2}	500
	$MgO \cdot 3.5\ Al_2O_3$	2303.0	0.03	308	3.3×10^{-2}	308
		–	0.026	441	–	–
		–	0.028	443	–	–
Mica	Muscovite	1473 to 1573	0.208	293 to 373	0.0006 to 0.0014	–
Sodium Chloride	NaCl	1074.0	0.204	273	1.55×10^{-2}	289
		–	0.217	373	–	–
Sodium Fluoride	NaF	1253.0	0.26	273	0.124	83
		–	–	–	0.0252	273
		–	–	–	0.0220	298
Sodium Nitrate	$NaNO_3$	579.8	0.247	273	–	
		–	0.270	373	–	–
Lead	Pb	600.5	0.031	273	–	–
Lead Chloride	$PbCl_2$	774.0	0.0649	273	–	–
		–	0.0681	373	–	–
Lead Fluoride	PbF_2	1128.0	–	–	–	–
Molybdate	$PbMoO_4$	1333 to 1343	0.100	288	–	–
Lead Sulfide	PbS	1387.0	0.0502	273	16×10^{-4}	–
		–	0.0511	373	–	–
Lead Selenide	PbSe	1338.0	–	–	100×10^{-4}	–
Lead Telluride	PbTe	1190.0	–	–	120×10^{-4}	–
		–	–	–	–	–
Polyethylene		–	0.3×10^{-3}	5	–	–
		–	2.3×10^{-3}	10	–	–
		–	16.1×10^{-3}	20	–	–
		–	78.8×10^{-3}	50	–	–
		–	157×10^{-3}	100	–	–
		–	566×10^{-3}	300	–	–
Platinum	Pt	2046.5	0.0318	273	–	–
Pyrex**		–	0.006×10^{-3}	2	–	–
		–	0.09×10^{-3}	5	–	–
		–	1.0×10^{-3}	10	–	–
		–	6.5×10^{-3}	20	–	–
Selenium	Se	308 (amorphous)	0.068	85 to 291	3.1×10^{-3}	308
		490 (crystal)	0.072	276	2.6×10^{-3}	–
		–	0.077	293.5	0.4×10^{-3}	–
		–	0.085	302.5	–	–
		–	0.127	305	–	–
		–	0.131	311	–	–
		–	0.95	291 to 311	–	–
	Se(As)	~343	–	–	3.3×10^{-4}	–
Silicon	Si	1693	0.177	298	0.39	313
Silicon Dioxide	SiO_2 (crystal)	2000	0.1657	273	2.82×10^{-3}	314
		–	0.201	373	2.64×10^{-3}	–
		–	–	–	4.50×10^{-3}	–
Silicon Dioxide	SiO_2 (quartz)	>1743	0.188	285 to 373	Parallel Perpend.	
		–	–	–	0.117 0.586	83
		–	–	–	0.0467 0.0249	195
		–	–	–	0.0273 0.0163	273
		–	–	–	0.0224 0.0135	323
		–	–	–	0.0190 0.0118	373
		–	–	–	0.0168 0.0160	423
		–	–	–	0.0151 0.00967	473
		–	–	–	0.0136 0.00895	523
		–	–	–	0.0123 0.0084	573
		–	–	–	0.0113 0.0079	623
Tin	Sn	504.85	0.0556	273	– –	–
Strontium Fluoride	SrF_2	1190.0	–	–	– –	–
Strontium Titanate	$SrTiO_3$	2353.0	–	–	– –	–
Tellurium	Te	722.8	0.0483	561 to 646	1.5×10^{-2}	–
Teflon†		–	0.07×10^{-3}	2	– –	–
		–	0.57×10^{-3}	5	– –	–
		–	4.3×10^{-3}	10	– –	–
		–	18×10^{-3}	20	– –	–
		–	48×10^{-3}	50	– –	–
		–	92×10^{-3}	100	– –	–
		–	241×10^{-3}	300	– –	–
Titanium Dioxide	TiO_2	2093.0	0.17	293	3.0×10^{-2}	309 (parallel)
		–	–	–	3.3×10^{-2}	340 (parallel)
		–	–	–	2.1×10^{-2}	317 (perpend.)
		–	–	–	1.7×10^{-2}	340 (perpend.)
Thallium Bromide	TlBr	733.0	0.045	293	1.4×10^{-3}	316
Thallium Chloride	TlCl	703.0	0.0520	273	1.9×10^{-3}	311
Zinc	Zn	692.4	0.0939	273	–	–

**Pyrex® is a registered trademark of Corning Glass Works.
†Teflon® is a registered trademark of the Dupont Corp.

THERMAL EXPANSION OF OPTICAL MATERIALS

$$\alpha = A \times 10^{-6} + B \times 10^{-8}\,T + C \times 10^{-11}\,T^2$$

Material	*T* (K)	*A*	*B*	*C*	*Remarks*
Al_2O_3	323	6.7	–	–	Parallel
	323	5.0	–	–	Perpendicular
As_2S_3	306-438	24.62	–	–	–
AgCl	298	30.01	–	–	–
	473	34.59	–	–	–
	623	52.09	–	–	–
	653	58.37	–	–	–
	673	63.19	–	–	–
	698	69.99	–	–	–
	293-333	30.00	–	–	–
ADP	297-407	39.3	–	–	–
		1.9	–	–	–
BaF_2	272-573	–	–	–	–
$BaTiO_3$	193-253	16.0	–	–	–
	283-343	19.0	–	–	–
	393-453	13.0	–	–	–
Borosilicate Crown Glass	295-771	9.0	–	–	–
$CaCO_3$	123-273	24.39	0.533	-30.7	Parallel
	123-273	-5.68	0.0333	-4.58	Perpendicular
	323	26.6	–	–	Parallel
	323	5.2	–	–	Parallel
	638	-3.8	–	–	Perpendicular
	348-673	24.71	3.775	-3.653	Parallel
CuCl	313-413	10.0	–	–	–
CaF_2	181-280	18.38	2.511	-21.10	–
	316-900	1.851	1.481	21.52	–
CsBr	293-323	47.9	–	–	–
	134-573	46.6	4.67	-1.78	–
CsI	298-323	50.0	–	–	–
CuBr	293-423	19.0	–	–	–
CdS	300-343	4.2	–	–	–
	323-773	3.5	–	–	Parallel
CdS Pressed	10	0	–	–	–
	50	-2.4	–	–	–
	110	0	–	–	–
	200	2.5	–	–	–
	300	4.2	–	–	–
CdTe	323	4.5	–	–	–
	873	5.9	–	–	–
GaAs	40	-0.5	–	–	–
	491	0.00	–	–	–
	78-290	3.64	–	–	–
	291-560	5.74	–	–	–
	560-680	7.44	–	–	–
CdF_2	293-393	27.0	–	–	–
GaSb	–	6.9	–	–	–
GaP	–	5.3	–	–	–
Ge	40	0.07	–	–	–
	50	0.20	–	–	–
	60	0.39	–	–	–
	70	0.67	–	–	–
	80	1.05	–	–	–
	90	1.54	–	–	–
	100	2.20	–	–	–
	110	2.79	–	–	–
	120	3.25	–	–	–
	130	3.62	–	–	–
	140	3.91	–	–	–
	150	4.12	–	–	–
	160	4.29	–	–	–
	170	4.45	–	–	–
	180	4.58	–	–	–
	190	4.70	–	–	–
	200	4.82	–	–	–
	210	4.93	–	–	–
	220	5.03	–	–	–
	230	5.13	–	–	–

(continued)

$$\alpha = A \times 10^{-6} + B \times 10^{-8}\, T + C \times 10^{-11}\, T^2$$

Material ·	*T* (K)	*A*	*B*	*C*	*Remarks*
GE (*Continued*)	240	5.23	–	–	–
	250	5.32	–	–	–
	260	5.42	–	–	–
	270	5.50	–	–	–
	280	5.59	–	–	–
	290	5.67	–	–	–
	300	5.75	–	–	–
InAs	–	5.3	–	–	–
InSb	10	-0.06	–	–	–
	30	-1.72	–	–	–
	50	-0.33	–	–	–
	70	0.89	–	–	–
	100	2.76	–	–	–
	160	4.08	–	–	–
	190	4.35	–	–	–
	220	4.58	–	–	–
	253	4.78	–	–	–
	270	4.89	–	–	–
	280	4.95	–	–	–
	300	5.04	–	–	–
InP	–	4.5	–	–	–
Irtran 1	298-573	11.0	–	–	–
Irtran 2	298-573	6.9	–	–	–
Irtran 3	298-573	20.0	–	–	–
Irtran 4	298-573	7.7	–	–	–
Irtran 5	298-573	12.0	–	–	–
Irtran 6*	298-573	5.9	–	–	–
KDP	123-293	21.6	–	–	–
KCl	293-333	36.0	–	–	–
KRS-5	223-293	61.0	–	–	–
	293-373	58.0	–	–	–
KRS-6	223	55.0	–	–	–
	233	56.0	–	–	–
	253	56.0	–	–	–
	273	55.0	–	–	–
	293	51.0	–	–	–
	313	48.0	–	–	–
	333	49.0	–	–	–
	353	51.0	–	–	–
	373	53.0	–	–	–
	393	56.0	–	–	–
	413	57.0	–	–	–
	433	58.0	–	–	–
	453	59.0	–	–	–
	473	59.0	–	–	–
KBr	113-573	27.6	4.1	–	–
	318-953	37.99	1.263	5.256	–
	293-333	43.0	–	–	–
KI	313	42.6	–	–	–
LiF	273-373	37.0	–	–	–
	123-273	31.95	5.049	-4.070	–
	320-1067	33.17	3.075	2.399	–
$MgOAl_2O_3$	313	5.9	–	–	–
MgO	323-988	10.98	–	–	–
	300	11.2	–	–	–
	481	12.3	–	–	–
	659	13.5	–	–	–
	825	14.6	–	–	–
	964	15.4	–	–	–
	1061	16.0	–	–	–
	293-1000	13.8	–	–	–
Muscovite	324	8.1	7.5	–	–
$NaNO_3$	323	12.0	–	–	–
	323	11.0	–	–	–
NaCl	223-473	44.0	–	–	–

(continued)

$$\alpha = A \times 10^{-6} + B \times 10^{-8}\, T + C \times 10^{-11}\, T^2$$

Material	*T* (K)	*A*	*B*	*C*	*Remarks*
NaF	Room temp.	36.0	–	–	–
PbTe	303	9.02	–	–	–
	313	12.08	–	–	–
	323	14.30	–	–	–
	333	15.57	–	–	–
	343	15.38	–	–	–
	353	16.42	–	–	–
	363	17.31	–	–	–
	373	17.70	–	–	–
	383	18.04	–	–	–
	393	18.33	–	–	–
	403	18.57	–	–	–
	413	18.78	–	–	–
	423	18.97	–	–	–
	433	19.42	–	–	–
	453	19.62	–	–	–
	473	19.74	–	–	–
	493	19.79	–	–	–
	513	19.80	–	–	–
	533	19.80	–	–	–
	553	19.80	–	–	–
	573	19.80	–	–	–
	593	19.80	–	–	–
	613	19.80	–	–	–
PbSe	303	7.65	–	–	–
	313	10.55	–	–	–
	323	12.92	–	–	–
	333	14.55	–	–	–
	343	15.63	–	–	–
	353	16.41	–	–	–
	363	16.97	–	–	–
	373	17.37	–	–	–
	383	17.66	–	–	–
	393	17.89	–	–	–
	403	18.09	–	–	–
	413	18.97	–	–	–
	423	18.43	–	–	–
	433	18.57	–	–	–
	433	18.57	–	–	–
	453	18.79	–	–	–
	473	18.94	–	–	–
	493	19.06	–	–	–
	513	19.16	–	–	–
	533	19.26	–	–	–
	553	19.34	–	–	–
	573	19.40	–	–	–
	593	19.46	–	–	–
	613	19.50	–	–	–
$PbCl_2$	293-393	31.0	–	–	–
Se	195-292	20.3	–	–	–
	195-273	42.7	–	–	–
	273-294	48.7	–	–	–
	293-373	22.9	–	–	–
	478	45.2	–	–	–
Si	50-100	2.5	–	–	<111>
	–	2.7	–	–	<110>
	100-200	3.1	–	–	<111>
	–	3.5	–	–	<110>
	200-300	3.9	–	–	<111>

(continued)

$$\alpha = A \times 10^{-6} + B \times 10^{-8}\, T + C \times 10^{-11}\, T^2$$

Material	T (K)	*A*	*B*	*C*	*Remarks*
Si (*Continued*)	–	3.8	–	–	<110>
	400-500	4.3	–	–	<111>
	–	4.1	–	–	<110>
	500-600	4.7	–	–	<111>
	–	4.4	–	–	<110>
	600-700	5.0	–	–	<111>
	–	4.5	–	–	<110>
	25-900	3.0024	0.1544	0.20576	<111>
SiO_2 (crystal)	173-310	7.067	2.11	–	Parallel
	283-607	–	–	–	Parallel
	273-633	7.067	1.6742	–	Parallel
	633-723	25.80	–	20.163	Parallel
	273-353	7.97	–	–	Parallel
	273-353	13.37	–	–	Perpendicular
SiO_2 (fused)	293-1173	0.5	–	–	–
$SrTiO_3$	–	9.4	–	–	–
Se(As)	–	34.0	–	–	–
Te	313	16.75	–	–	–
	293	-1.6	–	–	Parallel
	293	27.2	–	–	Perpendicular
	293-333	-1.7	–	–	Parallel
	293-333	27.0	–	–	Perpendicular
TlCl	293-333	53.0	–	–	–
TlBr	293-353	51.0	–	–	–
TiO_2	313	9.19	2.25	–	Parallel
	313	7.14	1.10	–	Perpendicular

ELASTIC COEFFICIENTS OF OPTICAL MATERIALS

Elastic coefficients can be thought of as the basis of the engineering moduli. They represent the stress-strain relationship along a particular direction in a crystal. Data are given in bar in this table.

Material	T (K)	c_{11} (Bar)	c_{12} (Bar)	c_{13} (Bar)	c_{33} (Bar)	c_{44} (Bar)
Al_2O_3	298	49.68	16.36	–	49.81	14.74
AgCl	–	6.01	3.62	–	–	0.625
ADP	–	617.00	0.72	1.94	3.28	0.85
BaF	–	9.01	4.03	–	–	2.49
$BaTiO_3$	298	8.18	2.98	1.95	6.76	18.30
$CaCO_3$	–	13.71	4.56	4.51	7.97	3.42
CaF_2	–	16.4 ± 0.1	5.3 ± 0.2	–	–	3.370 ± 0.01
CsBr	–	3.097	0.403	–	–	0.7500
CsI	–	2.46	0.67	–	–	0.624
CdS	–	8.432	5.212	4.638	9.397	1.489
CdTe	–	5.351	3.681	–	–	1.994
GaAs	–	1.192	0.5986	–	–	0.538
GaSb	–	8.849	4.037	–	–	4.325
GaP	–	14.7	–	–	–	–
Ge	–	1.29	4.83	–	–	6.71
InAs	–	8.329	4.526	–	–	3.959
InSb	300	6.472	3.625	–	–	3.071
InP	–	10.7	–	–	–	–

(continued)

Material	T (K)	c_{11} (Bar)	c_{12} (Bar)	c_{13} (Bar)	c_{33} (Bar)	c_{44} (Bar)
KDP	–	7.14	-0.49	1.29	5.62	1.27
KCl	–	3.98	0.62	–	–	–
KRS-5	–	3.31	1.32	–	–	0.597
KRS-6	–	3.85	1.49	–	–	0.737
KBr	–	3.45	0.54	–	–	0.508
KI	–	2.69	0.43	–	–	0.362
LiF	–	9.74	4.04	–	–	5.54
$MgAl_2O_3$	–	30.05	15.37	–	–	15.86
MgO	–	2.90	0.876	–	–	1.55
$NaNO_3$	–	8.67	1.63	1.60	3.74	2.13
NaCl	–	4.85	1.23	–	–	1.26
NaF	–	9.09	2.64	–	–	1.27
PbS	–	12.7	2.98	–	–	2.48
Si	–	1.67	0.65	–	–	0.80
SiO_2 (crystal)	–	8.675	0.687	1.13	10.68	5.786
$SrTiO_3$	–	31.56	10.27	–	–	12.15
Te	300	3.265	0.195	2.493	7.22	3.121
TlCl	–	4.01	1.53	–	–	0.760
TlBr	–	3.78	1.48	–	–	0.756
TiO_2	–	35.8	26.7	17.0	47.9	12.5
ZnS	–	9.45	5.70	–	–	4.36

ELASTIC MODULI OF OPTICAL MATERIALS

Material	*Young's* (10^6 psi)	*Rigidity* (10^6 psi)	*Bulk* (10^6 psi)	*Rupture* (psi)	*Apparent Elastic Limit* (psi)
Al_2O_3	50.00	21.50	0.30	–	–
As_2S_3	2.30	0.94	–	2.4×10^3	–
AgCl	0.02	1.03	6.39	–	3.8×10^3
	–	–	–	–	7.4×10^2
BaF_2	7.70	–	–	3.9×10^7	3.9×10^7
$BaTiO_3$	4.90	18.30	23.50	–	–
$CaCO_3$	10.50 parallel	–	18.80	–	–
	12.80 perpendicular	–	–	–	–
CaF_2	11.00	4.90	12.00	5.3×10^3	5.3×10^3
CsBr	2.30	–	–	23.9	12.2×10^2
CsI	0.769	–	–	–	8.1×10^2
CdTe	–	–	–	850.0	–
GaSb	9.19	6.28	8.19	–	–
Ge	14.90	9.73	11.30	–	–
InSb	6.21	4.45	6.28	–	–
Irtran 1	16.6×10^6 at 25°C	–	–	21,800.0 at 25°C	–
	16.6×10^6 at 500°C	–	–	10,000.0 at 500°C	–
Irtran 2	14.0×10^6 at 25°C	–	–	14,100.0 at 25°C	–
	10.6×10^6 at 250°C	–	–	13,500.0 at 250°C	–
Irtran 3	14.3×10^6 at 25°C	–	–	5,300.0 at 25°C	–
	14.0×10^6 at 500°C	–	–	9,000.0 at 500°C	–
Irtran 4	10.3×10^6 at 25°C	–	–	7,500.0 at 25°C	–
Irtran 5	48.2×10^6 at 25°C	–	–	19,200.0 at 25°C	–
	–	–	–	13,000.0 at 500°C	–
Irtran 6**	5.3×10^6 at 25°C	–	–	4,540.0 at 25°C	–
	4.5×10^6 at 100°C	–	–	5,880.0 at 100°C	–
KCl	4.30	0.906	2.52	6.4×10^2	3.3×10^2
KRS-5	2.30	0.840	2.87	1.81×10^4	3.8×10^4
KRS-6	3.00	1.230	3.31	–	3.05×10^3
KBr	3.90	0.737	2.18	4.8×10^2	1.6×10^2
KI	4.57	0.90	124.00	–	–
LiF	9.40	8.00	9.00	2.0×10^6	16.2×10^6
MgO	36.10	22.40	22.40	–	–

(continued)

Material	Young's (10^6 psi)	Rigidity (10^6 psi)	Bulk (10^6 psi)	Rupture (psi)	Apparent Elastic Limit (psi)
$NaNO_3$	–	–	3.80	–	–
NaCl	5.80	1.83	3.53	5.7×10^2	3.5×10^2
Si	1.9×10^7	1.16×10^7	1.48×10^7	–	–
SiO_2 (crystal)	11.10 perpendicular	5.28	–	–	–
	14.10 parallel	–	–	–	–
SiO_2 (quartz)	1.06×10^7	4.52	–	–	–
	$9.884 \pm 0.079 \times 10^7$	–	–	–	–
TlCl	4.60	1.10	3.42	–	–
TlBr	4.28	1.10	3.26	–	–

*To convert stresses and module from psi to dyne cm^{-2}, multiply the magnitude in psi by 6.90×10^4; to convert dyne cm^{-2} to psi, multiply the magnitude in dyne cm^{-2} by 1.45×10^{-5}.

**Irtran 6 is no longer manufactured by the Eastman Kodak Co. Irtran® is a registered trademark of the Eastman Kodak Co.

HARDNESS OF OPTICAL MATERIALS

Values of microhardness generally obtained by the Knoop test are tabulated in this table. The temperatures at which the measurement was taken as well as the orientation of the long direction of the diamond-shaped indenter and the load are given when available.

Material	Knoop (kg mm^{-2})	Temperature (K)	Load (g)
AgCl	9.5	–	200
AgTe	7.3	–	–
Al_2O_3	1370	–	1000
AlSb	400	–	–
As_2S_3	109	–	100
BaB_6	2900	–	120
BaF_2	82	–	500
C	8820	110	–
CaB_6	3150	–	120
CdS	55, 80	–	–
CdSe	90, 44, 66	–	–
CdTe	56	–	–
CeB_6	2350	–	120
Cr_2C	2160	–	120
Cr_2B_5	2150	–	120
CsBr	19.5	–	200
Cu	48, 17.5, 8	293, 773, 973	–
CuBr	21.2	–	–
CuTe	19.2	–	–
GaAs	721	–	–
GaSb	469	–	–
$GaSe_3$	316	–	–
$GaTe_3$	237	–	–
Ge	176, 83, 80, 24	873, 973, 1023, 1223	–
InAs	330	–	–
InSb	225	–	–
InP	430	–	–
In_2Te_3	180	–	–

(continued)

Material	*Knoop* (kg mm^{-2})	*Temperature* (K)	*Load* (g)
KBr	5.9, 7.0	–	200, 200
KCl	7.2, 9.3	–	200, 200
KRS-5	40.2, 39.8, 33.2	–	200, 500, 500
KRS-6	29.9, 38.5	–	500, 500
LaB_6	2500	–	120
LiF	102-113	–	600
MgO	692	–	600
$MgO\text{-}3.5Al_2O_3$	1140	–	1000
Mo_2B_5	2950	–	120
Mo_2C	1800	–	120
N_6B_2	2900	–	120
NaCl	15.2, 18.2	–	200, 200
$NaNO_3$	19.2	–	200
Si	1000, 500, 128	293, 773, 1273	–
SiO_2	461, 741	–	200, 500
$SrTiO_3$	595	–	–
TaB_2	2000	–	120
TaC	1629	–	120
$TaSi_2$	1200	–	120
TiB_2	3400	–	120
TiC	~2600	–	120
TiN	2100	–	120
TiO_2	879	–	500
TlBr	11.9, 11.9	–	500, 500
TlCl	12.8, 12.8	–	500, 500
W_2B_5	2500	–	120
WC	1800	–	120
WSi_2	1430	–	120
ZnS	178	–	–
ZnSe	137	–	–
ZnTe	82	–	–
ZrB_2	1500	–	120
ZrC	2400	–	120
ZrN	930	–	120

SOLUBILITY, MOLECULAR WEIGHT, AND SPECIFIC GRAVITY OF OPTICAL MATERIALS

Material	*Solubility* (g/100 g H_2O)	*Molecular Weight*	*Specific Gravity*
Al_2O_3	Insoluble	101.94	3.98 (3.95 to 4.10 for natural)
AgCl	8.9×10^{-5} at 283 K	143.34	5.589 at 273 K 5.56 at 293 K
As_2S_3	Insoluble	364.02	3.198
ADP	22.7 at 273 K	115.04	1.803 at 293 K
BaF_2	0.17	175.36	4.83 at 293 K
$BaTiO_3$	–	232.96	5.90 (single crystal)
$CaCO_3$	1.4×10^{-3} at 298 K 1.8×10^{-3} at 298 K	100.09	2.7102 at 293 K
CuCl	0.0062 at 293 K	99.00	3.53 at 293 K
CaF_2	0.0017 at 299 K Soluble in ammonia salt solutions	78.08	3.179 at 298 K

(continued)

Material	*Solubility* ($g/100\ g\ H_2O$)	*Molecular Weight*	*Specific Gravity*
CdF_2	–	150.41	6.382 ± 0.006 at 293 K
CsBr	124.3 at 298 K Soluble in acid	212.83	4.44 at 293 K
CuBr	Insoluble	143.46	4.718 at 293 K
CsI	–	259.83	4.526
CdS	Insoluble	144.48	4.82 at 293 K
CdTe	Probably insoluble	240.02	5.854
$CaTiO_3$	–	135.98	4.10 at 293 K
GaAs	Insoluble	144.63	5.3161 ± 0.0002 at 298 K
GaSb	Insoluble	191.48	–
GaP	–	100.70	–
Ge	Insoluble in water; soluble in hot sulfuric acid and aqua regia; etched in CP-4	72.60	5.327 at 298 K
InAs	Insoluble	189.73	5.66
InSb	Insoluble	237.0	5.78
InP	–	145.80	4.8
Irtran 1	Insoluble	62.32	3.18
Irtran 2	Insoluble	97.45	4.09
Irtran 3	Insoluble	78.08	3.18
Irtran 4	Insoluble	144.34	5.27
Irtran 5	0.00062	40.32	3.58
Irtran 6*	Insoluble	240.02	5.85
KDP	–	136.09	2.338
KCl	34.7 at 293 K	74.55	1.984 at 293 K
KRS-5	0.05 at room temperature	–	7.371 at 289 K
KRS-6	0.32 at 293 K The solubility of a micro crystal is that of the more soluble component, in this case TICI.	–	7.192 at 289 K
KBr	53.48 at 273 K Slightly hygroscopic 102 at 373 K	119.01	2.75 at 298 K
KI	127.5 at 273 K	116.02	3.13
LiF	0.27 at 291 K	25.94	2.639 at 298 K
$MgOAl_2O_3$	Insoluble in water; not attacked by common acid NaOH; slightly etched by HF	356.74	3.61
MgO	Insoluble in water; soluble in acids and ammonia salts	40.32	3.567 at 298 K
Muscovite	Insoluble	–	2.8-2.9
$NaNO_3$	73 at 273 K 180 at 373 K	85.01	2.261
NaCl	35.7 at 273 K 39.12 at 373 K Soluble in glycerine; slightly soluble in alcohol and liquid ammonia; insoluble in hydrochloric acid	58.45	2.164 at 293 K
NaF	4.22 at 291 K	42.00	2.79 at 293 K 2.558 at 314 K
$PbMoO_4$	Insoluble	367.16	6.03/7.01 at 293 K
PbTe	Insoluble	334.82	8.16
PbSe	Insoluble	286.17	8.10 at 288 K
$PbCl_2$	0.673 at 273 K 0.99 at 283 K 3.34 at 373 K	278.12	5.85 at 293 K
PbS	Insoluble	239.28	7.5
PbF_2	0.064 at 293 K	245.21	8.24 at 293 K 7.763 ± .001 at 291 K
Se	Insoluble in water	–	4.82 4.26
SrF_2	0.011 at 273 K 0.012 at 27 K	125.63	4.24 at 293 K
SiO_2 (crystal)	Insoluble in water	60.06	2.648 at 298 K
SiO_2 (fused)	Insoluble in water; very slightly soluble in alkalis; soluble in hydrofluoric acid	60.06	2.202 at 293 K
$SrTiO_3$	Insoluble	183.53	5.122 at 293 K
SeAs	Insoluble	Not applicable	Not applicable

(continued)

Material	Solubility (g/100 g H_2O)	Molecular Weight	Specific Gravity
Te	Insoluble	–	6.24 at 293 K
TlCl	0.32 at 293 K	238.85	7.018 at 298 K
TlBr	0.05 at 298 K 0.25 at 341 K	284.31	7.453 at 298 K
TlO_2	Insoluble in water; soluble in acid	79.90	4.25 (4.18-5.13)

COMMONLY USED OPTICAL SUBSTRATE MATERIALS

Material	Refractive Index	Transmission Range	Material	Refractive Index	Transmission Range
Irtran* 1	1.38-1.23	1.00- 9.00	Magnesium oxide	1.77-1.62	0.36- 5.35
Lithium fluoride	1.45-1.11	0.20- 9.80	Sapphire	1.83-1.59	0.27- 5.60
Calcium fluoride (also as Irtran 3)	1.44-1.32	0.20-12.00	Irtran 2	2.29-2.15	1.00-13.00
			Irtran 4	2.50-2.30	1.00-20.00
Vycor**	1.46	0.25- 3.50	Arsenic trisulfide glass	2.69-2.36	0.56-12.00
Fused quartz	1.48-1.41	0.20- 4.50			
Barium fluoride	1.51-1.40	0.26-10.35	Silicon	3.50-3.42	1.36- 7.00
Glass	1.70-1.51	0.32- 2.50	Germanium	4.10-4.00	1.80-23.00

*Irtran® is a registered trademark of Eastman Kodak Co.
**Vycor® is a registered trademark of Corning Glass Works.

COMMONLY USED OPTICAL FILM MATERIALS

Material	Refractive Index	Range of Transparency* from (nm)	to (μm)	Comments
Cryolite	1.35	<200	10	1
Chiolite	1.35	<200	10	1
Magnesium fluoride	1.38	230	5	2, 3
Thorium fluoride	1.45	<200	10	–
Cerium fluoride	1.62	300	>5	4
Silicon monoxide	1.45-1.90	350	8	5
Sodium chloride	1.54	180	>15	6
Zirconium dioxide	2.10	300	>7	2
Zinc sulfide	2.30	400	14	7
Titanium dioxide	2.40-2.90	400	>7	8
Cerium dioxide	2.30	400	5	2, 3
Silicon	3.50	900	8	–
Germanium	3.80-4.20	1400	>20	–
Lead telluride	5.10	3900	>20	–

1. Both materials are sodium-aluminum fluoride compounds, but differ in the ratio of Na to Al and have different crystal structure. Chiolite is preferable in the infrared, because it has less stress than cryolite.
2. These materials are hard and durable, especially when evaporated onto a hot substrate.
3. The long wavelength is limited by the fact that, when the optical thickness of the film is a quarter-wave at 5 μm, the film cracks because of the mechanical stress.
4. Other fluorides and oxides of rare earths have refractive indices in this range from 1.60 to 2.0
5. The refractive index of SiOx (called silicon monoxide) can vary from 1.45 to 1.90 depending upon the partial pressure of oxygen during the evaporation. Films with a refractive index of 1.75 and higher absorb at wave-lengths below 500 nm.
6. Sodium chloride is used in interference filters out to a wavelength of 20 μm. It has very little stress.
7. The refractive index of zinc sulfide is dispersive.
8. The refractive index of TiO_2 rises sharply in the blue spectral region.

*The range of transparency is for a film of quarter-wave optical thickness at this wavelength. These values are approximate and also depend quite markedly upon the conditions in the vacuum during the evaporation of the film.

Technologies Comparison

Device	*Comments*
Cathode ray tube (CRT)	A mature technology of high reliability in wide-spread use for black and white or color. Associated circuits and hardware readily available. Requires moderately high voltages for bright displays. Requires substantial depth behind display surfaces.
Video-driven image reproducers of pictorial or symbolic material	Basically a television-like reproducer of imagery which may be halftone or *only* black *or* white. Image usually refreshed at thirty frame, sixty field rate.
Extruded beam signal generators for alphanumeric symbols	Symbols limited to those built into the beam-shaping structure. Any shapes possible if built into the tube or formed by alternate super position of existing characters. Refresh necessary.
Stroke-driven symbol generation	Basic tube driven in direct strokes by program of beam addressing, beam blanking or brightening, and beam moving instructions. Repetitive programming necessary to maintain brightness of display.
Storage tubes	Large variety of methods for achieving storage of the functions discussed. Storage achieved by electrical charge stored on dielectrics within the tube in a manner to spatially modulate electron flow to the screen.
Digitally addressed flat-panel CRT	This quite recently developed nonconventional CRT is about 2 in. thick. It employs an area-cathode and dynode aperture plates. A multiplicity of electron beams is formed by the plates, one for each resolution element. A particular beam is selected by applying proper voltages to each plate in a binary selection scheme. The beam passing through the final plate impinges on a phosphor screen as in a conventional CRT. Resolution up to 80 lines per inch has been achieved. Viewing areas up to 7 X 7 in. are available.
Plasma panels	Transparent panel often containing a large array of discharge electrodes usually in a common gas cavity. Both ac and dc versions exist. The plasma display technology is being developed in many sizes and for many applications. For large graphic displays it is the only technology seriously challenging the cathode ray tube.
dc plasma panels	In one type of dc structure two sets of parallel electrodes oppose each other in the gas with one set directed orthogonally to the other. An aperture plate, placed between the electrode sets, confines the discharges. Appropriate addressing voltages on two intersecting electrodes cause an electrical breakdown and emission of light at the intersection; an addressing voltage on only one electrode is too small to ignite the discharge. In this matrix arrangement, the gas discharge cell, which is sufficiently nonlinear for the purpose, functions as a two input "and" circuit. This device is usually operated one row (or column) at a time with signals on the opposing electrodes determining which cells in the row (or column) will be on. Since the duty cycle in these devices becomes smaller as the device becomes larger, the peak currents limit the array size.
ac plasma panels	In the ac plasma display, as in the dc panel, two sets of electrodes oppose each other across a discharge gap and are directed orthogonally to one another. However, at each discharge site, the dielectric surfaces that isolate the electrodes from the gas define two capacitances which are in series with the discharge. An alternating voltage applied across the two electrode sets is too small by itself to ignite discharges. However, a pair of write voltages applied across two selected electrodes will ignite a pulsed discharge that extinguishes as ions and electrons flow to the dielectric surfaces and charge the capacitors. This charge augments the applied voltage on the next half cycle to ignite a second discharge, which then charges the capacitors in preparation for the third discharge. This sequence of pulsed discharges which characterizes the "on" state of a cell, terminates when an erase signal on the two intersecting electrodes produces a controlled discharge that reduces the charge on the series capacitance below the minimum required for ignition.
Electroluminescent (El) panels	Electroluminescent displays consist of an El powder or evaporated film between two electrodes, one of which is transparent. El displays can be made in many colors. However most displays are single-color, usually green or orange because of the higher efficiency achieved with copper-activated and manganese-activated materials. When a potential is applied across the El material, visible light is emitted. The potential may be ac or dc depending upon the specific structure, but El displays usually operate in the ac mode. The resolution or pattern is defined by the electrodes. Luminance is typically 5 to 30 ft L, although luminance in the thousands of ft L has been achieved and demonstrated in 1974.
Liquid crystals	A thin clear layer of a cholesteric material placed between transparent electrically conducting covers when excited by an electric field becomes turbulent and scatters ambient light in a manner that yields an apparent brightness related to the applied field. When an aggregate of such cells forms a two-dimensional array, a digitally-addressed display results. These displays can be small, light and relatively inexpensive. The driving circuitry for large arrays is the costly part, not the liquid crystalline materials.
Light emitting diodes (LEDs)	LED displays are now a mature technology for small scale displays such as in pocket calculators, small area indicators and related applications. Their utility for larger area or ambient brightness applications depends upon improved luminous efficiency, lower power dissipation in driving circuits, and costs to challenge other technologies.

(continued)

Device	Comments
Projection displays	Both high luminosity CRTs and light valves of the oil film type are available and useful. The projection CRTs fill the need for heads-up display and small screen systems. The oil film systems fill the need for small-to-large theater screen displays in both black and white or color. A typical device is the Eidophor. Projection CRT displays are finding increasing application in displays for tactical systems in sizes from 3 to 6 ft on a side. Current tubes, with typical F/0.9 optics, can develop 200 to 300 lm output after accounting for optical surface losses. Resolutions of 1,000 TV lines have been achieved on 5 in. projection CRTs. New longer-life and more efficient phospors are necessary to expand the application of projection CRTs. Oil-film light valves are currently being considered for a wide variety of command and control display applications of the fixed-site type. Devices can typically provide 525 line TV images with light outputs of 5,000 lm. They are in general large, complex, and expensive systems. Small sealed-off light-valves are available, but light-output and resolution are limited by light-source and cooling requirements.

Summary of Cathode Ray Tubes (CRTs) Display-Effectiveness Factors

Factor	*Required*	*Achieved*
Brightness	Average–min. of 50 ft L	Yes (far exceeded)
	Max.–3,000 ft L	Yes (far exceeded)
Contrast	Viewable in shade for stationary displays	Yes
	Viewable in direct sunlight	Only with direct view storage tubes (DVSTs)
Half-Tones	Radar displays–two-tone acceptable	Yes
	Television–5 or more required	Yes
Resolution	Min. size commensurate with eye acuity	Yes
	Size constant with brightness and position	No, but adequate for most purposes
Flicker	None present	Yes, for most applications
Distortion	Size constant with brightness and position	See "Resolution" above
Accuracy	Position linear with input voltage	System rather than device limited
Blemishes	Radar displays–min. loss of resolution elements	0.04 to 0.04% max. blemished area
	Television–indiscernible loss of picture detail	0.005 to 0.01% max. lost resolution elements
Volume-to-area	Overall volume small for desired viewing area	Poor; display device volume and shape may dictate equipment volume and shape
Power consumption	Negligible fraction of total equipment power	Yes, for TV No, for random access

Summary of LED and LCD Characteristics

Category	Comments	
	LEDs	LCDs
Visual appearance	Medium to wide viewing angle. Visible in dim ambient illumination but not as visible in bright ambient. All colors available except blue.	Medium viewing angle. Viewability insensitive to intensity of ambient illumination. All colors available.
Power dissipation	0.1 to 10 W cm^{-2} at 2 V	5 to 1.0 $\mu W\ cm^{-2}$ at 3 to 15 V
Response times	10 to 1,000 nsec	10 to 500 msec
Temperature dependencc	Unimportant. Operating range −40 to 100°C	The temperature dependence of the operating parameters can be significant. Operating range about 0 to 70°C
Circuit compatibility	High-current, low-voltage devices. Bipolar transistors usually required. Unipolar waveforms adequate for excitation. Easily multiplexed (>20 lines).	Low-current, low-to-medium voltage devices. CMOS IC compatible Bipolar waveforms necessary. Multiplexing is limited (~4 to 8 lines).
Packaging	Semiconductor processing techniques. Different structures are used to maximize light output with a minimum of LED material.	Glass and organic fluid technology. Need for hermiticity. Flexible with respect to size variations.
Reliability	>50,000 hours	>10 to 20,000 hours
Economic	Prices have dropped sharply in last few years. Well along learning curve. Cost is area-sensitive.	New relatively immature technology. Relatively low cost raw materials.

Characteristics of Typical Acousto-Optic Materials

Material (Approximate Range of IR Transmission)	*Index of Refraction, n(10.6 μm)*	*Acoustic Velocity,* $v \times 10^5$ cm sec^{-1}
Ge (2 to 20 μm)	4.0	5.5
CdS (0.5 to 11 μm)	2.22	4.32
GaAs (1 to 11 μm)	3.10	5.3
Si (1.5 to 10 μm)	3.42	9.85
Te (5 to 20 μm)	4.8	2.2
As_2S_3 Glass (0.6 to 11 μm)	2.38	2.6

CHAPTER 21

PROPERTIES OF SELECTED MATERIALS

ELEMENTS FOUND IN SEA WATER

Element	Average Concentration (mg/l)	Element	Average Concentration (mg/l)	Element	Average Concentration (mg/l)
Oxygen	8.57×10^{5}	Iron	1×10^{-2}	Selenium	9×10^{-5}
Hydrogen	1.08×10^{5}	Indium	$<2 \times 10^{-2}$	Germanium	7×10^{-5}
Chlorine	1.90×10^{4}	Molybdenum	1×10^{-2}	Xeon	5.2×10^{-5}
Sodium	1.05×10^{4}	Zinc	1×10^{-2}	Chromium	5×10^{-5}
Magnesium	1.35×10^{3}	Nickel	5.4×10^{-3}	Thorium	5×10^{-5}
Sulfur	8.85×10^{2}	Arsenic	3×10^{-3}	Gallium	3×10^{-5}
Calcium	4.00×10^{2}	Copper	3×10^{-3}	Mercury	3×10^{-5}
Potassium	3.80×10^{2}	Tin	3×10^{-3}	Lead	3×10^{-5}
Bromine	6.5×10^{1}	Uranium	3×10^{-3}	Zirconium	2.2×10^{-5}
Carbon	2.8×10^{1}	Krypton	2.5×10^{-3}	Bismuth	1.7×10^{-5}
Strontium	8.1×10^{0}	Manganese	2×10^{-3}	Lanthanum	1.2×10^{-5}
Boron	4.6×10^{0}	Vanadium	2×10^{-3}	Gold	1.1×10^{-5}
Silicon	3×10^{0}	Titanium	1×10^{-3}	Niobium	1×10^{-5}
Fluorine	1.3×10^{0}	Cesium	5×10^{-4}	Thallium	$<1 \times 10^{-5}$
Argon	6×10^{-1}	Cerium	4×10^{-4}	Hafnium	$<8 \times 10^{-6}$
Nitrogen	5×10^{-1}	Antimony	3.3×10^{-4}	Helium	6.9×10^{-6}
Lithium	1.8×10^{-1}	Silver	3×10^{-4}	Selenium	$<4 \times 10^{-6}$
Rubidium	1.2×10^{-1}	Yttrium	3×10^{-4}	Tantalum	$<2.5 \times 10^{-6}$
Phosphorus	7×10^{-2}	Cobalt	2.7×10^{-4}	Beryllium	6×10^{-7}
Iodine	6×10^{-2}	Neon	1.4×10^{-4}	Protoactinium	2×10^{-9}
Barium	3×10^{-2}	Cadmium	1.1×10^{-4}	Radium	6×10^{-11}
Aluminum	1×10^{-2}	Tungsten	1×10^{-4}	Radon	6×10^{-16}

* Concentration varies with geographic location.

PROPERTIES OF SELECTED METALS AND ALLOYS

Common name and classification	Thermal conductivity			Specific gravity	Coeff. of linear expansion, μ in./in. °F	Electrical resistivity, microhm-cm	Modulus of elasticity, millions of psi	Approximate melting point	
	J/sec cm °K	Btu/hr ft °F	kcal/sec cm °C					°F	°C
Ingot iron (included for comparison)	1.3	77	0.32	7.86	6.8	9	30	2800	1538
Plain carbon steel	1.0	56	0.23	7.86	6.7	10	30	2760	1515
Stainless steel type 304	0.3	19	0.08	8.02	9.6	72	28	2600	1427
Cast gray iron	0.8	48	0.20	7.2	6.7	67	13	2150	1177
Malleable iron				7.32	6.6	30	25	2250	1232
Ductile cast iron	0.6	34	0.14	7.2	7.5	60	25	2100	1149
Ni-resist cast iron, type 2	0.7	41	0.17	7.3	9.6	170	15.6	2250	1232
Cast 28-7 alloy (IID)	0.04	2	0.01	7.6	9.2	41	27	2700	1482
Hastelloy C	0.2	10	0.04	3.94	6.3	139	30	2350	1288
Inconel X, annealed	0.3	17	0.07	8.25	6.7	122	31	2550	1399
Haynes Stellite alloy 25 (L605)	0.2	10	0.04	9.15	7.61	88	34	2500	1371
Aluminum alloy 3003, rolled	2.8	164	0.68	2.73	12.9	4	10	1200	649
Aluminum alloy 2017, annealed	3.0	174	0.72	2.8	12.7	4	10.5	1185	641
Aluminum alloy 380	1.8	102	0.42	2.7	11.6	7.5	10.3	1050	566
Copper	7.1	411	1.70	8.91	9.3	1.7	17	1980	1082
Yellow brass (high brass)	2.2	126	0.52	8.47	10.5	7	15	1710	932
Aluminum bronze	1.3	75	0.31	7.8	9.2	12	17	1900	1038
Beryllium copper 25	0.2	12	0.05	8.25	9.3	-	19	1700	927
Nickel silver 18% alloy A (wrought)	0.6	34	0.14	8.8	9.0	29	18	2030	1110
Cupronickel 30%	0.5	31	0.13	8.95	8.5	35	22	2240	1227
Red brass (cast)	1.3	77	0.32	8.7	10	11	13	1825	996
Chemical lead	0.6	36	0.15	11.35	16.4	21	2	621	327
Antimonial lead (hard lead)	0.5	31	0.13	10.9	15.1	23	3	554	290
Solder 50-50	0.8	48	0.20	8.89	13.1	15		420	216
Magnesium alloy AZ31B	1.4	82	0.34	1.77	14.5	9	6.5	1160	627
K Monel	0.3	19	0.08	8.47	7.4	58	26	2430	1332
Nickel	1.1	63	0.26	8.89	6.6	10	30	2625	1441
Cupronickel 55-45 (Constantan)	0.4	24	0.10	8.9	8.1	49	24	2300	1260
Commercial titanium	0.3	19	0.08	5	4.9	80	16.5	3300	1816
Zinc	2.0	114	0.47	7.14	18	6	-	785	418
Zirconium, commercial	0.3	19	0.08	6.5	2.9	41	12	3350	1843

PROPERTIES OF HIGH-PERMEABILITY MAGNETIC MATERIALS

Material	Approximate composition (%) Fe	Ni	Co	Mo	Other	Typical heat treatment °C	Permeability at $B = 20$ gausses	Maximum permeability	Saturation flux density B gausses	Hysteresis ‡ loss, W_h ergs/cm	Coercive ‡ force H_c oersteds	Resistivity microhm cm	Density, g/cm³
Cold rolled steel	98.5	—	—	—	—	950 Anneal	180	2,000	21,000	—	1.8	10	7.88
Iron	99.91	—	—	—	—	950 Anneal	200	5,000	21,500	5,000	1.0	10	7.88
Purified iron	99.95	—	—	—	—	1480 H_2 + 880	5,000	180,000	21,500	300	.05	10	7.88
4% Silicon-iron	96	—	—	—	4 Si	800 Anneal	500	7,000	19,700	3,500	.5	60	7.65
Grain oriented*	97	—	—	—	3 Si	800 Anneal	1,500	30,000	20,000	—	.15	47	7.67
45 Permalloy	54.7	45	—	—	.3 Mn	1050 Anneal	2,500	25,000	16,000	1,200	.3	45	8.17
45 Permalloy †	54.7	45	—	—	.3 Mn	1200 H_2 Anneal	4,000	50,000	16,000	—	.07	45	8.17
Hipernik	50	50	—	—	—	1200 H_2 Anneal	4,500	70,000	16,000	220	.05	50	8.25
Monimax	—	—	—	—	—	1125 H_2 Anneal	2,000	35,000	15,000	—	.1	80	8.27
Sinimax	—	—	—	—	—	1125 H_2 Anneal	3,000	35,000	11,000	—	—	90	—
78 Permalloy	21.2	78.5	—	—	.3 Mn	1050 + 600 Q§	8,000	100,000	10,700	200	.05	16	8.60
4-79 Permalloy	16.7	79	—	4	.3 Mn	1100 + Q	20,000	100,000	8,700	200	.05	55	8.72
Mu metal	18	75	—	—	2 Cr, 5 Cu	1175 H_2	20,000	100,000	6,500	—	.05	62	8.58
Supermalloy	15.7	79	—	5	.3 Mn	1300 H_2 + Q	100,000	800,000	8,000		.002	60	8.77
Permendur	49.7	—	50	—	.3 Mn	800 Anneal	800	5,000	24,500	12,000	2.0	7	8.3
2V Permendur	49	—	49	—	2 V	800 Anneal	800	4,500	24,000	6,000	2.0	26	8.2
Hiperco	64	—	34	—	Cr	850 Anneal	650	10,000	24,200	—	1.0	25	8.0
2-81 Permalloy	17	81	—	2	—	650 Anneal	125	130	8,000	—	<1.0	10^6	7.8
Carbonyl iron	99.9	—	—	—	—	—	55	132	—	—	—	—	7.86
Ferroxcube III	$MnFe_2O_4$ + $ZnFe_2O_4$					—	1,000	1,500	2,500	—	.1	10^8	5.0

1 Materials are in sheet form except where indicated.

2 in powdered form

PROPERTIES OF PERMANENT MAGNET MATERIALS

Material	Percent composition (remainder Fe)	Heat treatment* (temperature, °C)	Magnetizing force H_{max} oersteds	Coercive force H_c oersteds	Residual induction B_r gausses	Energy product BH_{max} × 10^{-6}	Method of fabrication†	Mechanical properties‡	Weight lb/in.³
Carbon steel	1 Mn, 0.9 C	Q 800	300	50	10,000	.20	HR, M, P	H, S	.280
Tungsten steel	5 W, 0.3 Mn, 0.7 C	Q 850	300	70	10,300	.32	HR, M, P	H, S	.292
Chromium steel	3.5 Cr, 0.9 C, 0.3 Mn	Q 830	300	65	9,700	.30	HR, M, P	H, S	.280
17% Cobalt steel	17 Co, 0.75 C, 2.5 Cr, 8 W	—	1,000	150	9,500	.65	HR, M, P	H, S	—
36% Cobalt steel	36 Co, 0.7 C, 4 Cr, 5 W	Q 950	1,000	240	9,500	.97	HR, M, P	H, S	.296
Remalloy or Comol	17 Mo, 12 Co	Q 1200, B 700	1,000	250	10,500	1.1	HR, M, P	H	.295
Alnico I	12 Al, 20 Ni, 5 Co	A 1200, B 700	2,000	440	7,200	1.4	C, G	H, B	.249
Alnico II	10 Al, 17 Ni, 2.5 Co, 6 Cu	A 1200, B 600	2,000	550	7,200	1.6	C, G	H, B	.256
Alnico II (sintered)	10 Al, 17 Ni, 2.5 Co, 6 Cu	A 1300	2,000	520	6,900	1.4	Sn, G	H	.249
Alnico IV	12 Al, 28 Ni, 5 Co	Q 1200, B 650	3,000	700	5,500	1.3	Sn, C, G	H	.253
Alnico V	8 Al, 14 Ni, 24 Co, 3 Cu	AF 1300, B 600	2,000	550	12,500	4.5	C, G	H, B	.264
Alnico VI	8 Al, 15 Ni, 24 Co, 3 Cu, 1 Ti	—	3,000	750	10,000	3.5	C, G	H, B	.268
Alnico XII	6 Al, 18 Ni, 35 Co, 8 Ti	—	3,000	950	5,800	1.5	C, G	H, B	.26
Vicalloy I	52 Co, 10 V	B 600	1,000	300	8,800	1.0	C, CR, M, P	D	.295
Vicalloy II (wire)	52 Co, 14 V	CW + B 600	2,000	510	10,000	3.5	C, CR, M, P	D	.292
Cunife (wire)	60 Cu, 20 Ni	CW + B 600	2,400	550	5,400	1.5	C, CR, M, P	D, M	.311
Cunico	50 Cu, 21 Ni, 29 Co	—	3,200	660	3,400	.80	C, CR, M, P	D, M	.300
Vectolite	30 Fe_2O_3, 44 Fe_3O_4, 26 C_2O_3	—	3,000	1,000	1,600	.60	Sn, G	W	.113
Silmanal	86.8 Ag, 8.8 Mn, 4.4 Al	—	20,000	6,000*	550	.075	C, CR, M, P	D, M	.325
Platinum-cobalt	77 Pt, 23 Co	Q 1200, B 650	15,000	3,600	5,900	6.5	C, CR, M	D	—
Hyflux	Fine powder	—	2,000	390	6,600	.97	—	—	.176

* Value given is intrinsic H_c.

* Q—Quenched in oil or water. A—Air cooled. B—Baked. F—Cooled in magnetic field. CW—Cold worked.

† HR—Hot rolled or forged. CR—Cold rolled or drawn. M—Machined. G—Must be ground. P—Punched. C—Cast. Sn—Sintered.

‡ H—Hard. B—Brittle. S—Strong. D—Ductile. M—Malleable. W—Weak.

SELECTED PROPERTIES OF SOLID MATERIALS

Material	Density $\frac{g}{cm^3}$	Coefficient of linear expansion per deg between 0 and 100 °C $\times 10^{-6}$	Melting point °C	Boiling point °C	Latent heat of fusion $\frac{kcal}{kg}$	Specific heat capacity $\frac{kcal}{deg\ kg}$	Thermal conductivity (at 20 °C) $\frac{kcal \times 10^{-3}}{deg\ cm\ s}$	Heat of combustion $\frac{kcal}{kg}$	Tensile strength $\frac{kgf}{mm^2}$
Aluminium	2.70	23.86	659.7	2447	85	0.216	0.570	—	20 to 30
Amber	1.05 to 1.1	—	250 to 300	—	—	—	—	—	—
Antimony	6.69	12.8	630.5	1637	40	0.050	0.042	—	10
Asphalt	1.1 to 1.5	—	—	—	—	0.22	0.0017	9530	—
Barium	3.61	19	710	1637	—	0.068	—	—	—
Beryllium	1.86	12.3	1283	2477	250 to 275	0.277	0.43	—	—
Bismuth	9.79	13.5	271.3	1560	13	0.029	0.0194	—	—
Blast furnace slag	2.0 to 3.9	5.41	1300 to 1430	—	≈ 50	0.20	0.00025	—	—
Brass	8.1 to 8.6	18.75	900	≈ 2300	40	0.0917	0.26	—	40 to 60
Bronze	7.4 to 8.9	16.8 to 29.5	—	—	—	≈ 0.086	0.061 to 0.14	—	15 to 80
Cadmium	8.64	29.4	321	765	13	0.056	0.22	—	7.0 ∥ 10.0 ⊥
Calcium	1.54	25.2	850	1487	78,5	0.157	—	—	—
Carbon									
diamond	3.01 to 3.52	1.3	≈ 3600	4200	≈ 4000	0.121	—	8140 CO_2 2440 CO	—
graphite	2.3 to 2.72	7.86	Subl. 3652	4200	≈ 4000	0.170	0.012	≈ 7860	—
coke	1.0 to 1.7	5.40	—	—	≈ 4000	0.204	0.0085	≈ 7000	—
coal	1.2 to 1.5	—	—	—	—	—	0.00029	≈ 7500	—
peat	0.6 to 0.8	—	—	—	—	—	0.00015	≈ 3800	—
Cement	2.7 to 3.0	10 to 14	—	—	—	0.20	0.00071	—	—
Charcoal	0.3 to 0.5	—	—	—	—	0.16	0.000139	7260	—
Chromium	7.20	6.6	1903	2642	61.53	0.107	0.165 (18 °C)	—	10 to 40
Concrete	2.2 to 2.5	10 to 14	—	—	—	0.21	0.002 to 0.003	—	—
Constantan	8.9	14.5 to 17	1190	—	—	0.098	0.054	—	—
Copper	8.96	16.86	1083	2595	49 to 51	0.0921	0.941	590 (CuO)	40
Cork	0.2 to 0.35	—	—	—	—	0.4 to 0.5	0.00013 to 0.00017	—	—
Gallium	5.91	≈ 18	29.78	2227	—	0.089	—	—	—
Germanium	5.33	6	937.2	2830	—	0.077	0.148	—	—
Glass (average)	2.4 to 2.8	7 to 10	—	—	—	0.12 to 0.2	≈ 0.002	—	7 to 9
Gold	19.3	14.2	1064.76	2707	15.9	0.031	0.71	—	13.4
Indium	7.30	30	156.17	2047	—	0.0556	—	—	—
Iridium	22.42	6.5	2443	4350	—	0.0311	0.35	—	56
Iron, pure	7.87	11.5	1536	3070	67 to 94	0.107	0.161	1260 FeO	20 to 25
grey cast iron	7.6	10.6	1200	—	≈ 23	0.13 to 0.17	0.11	1680 Fe_3O_4	12 to 26
steel	7.7	10.5 to 13.2	1170 to 1530 according to C-content	—	—	0.11	0.108 to 0.115	1890 Fe_2O_3	37 to 64
Invar	7.9	0.9	—	—	—	0.120	0.0263	—	—
Lead	11.34	29.4	327.4	1751	5.6	0.031	0.0842	260	1.4
Lime (slaked)	1.3 to 1.4	—	—	—	—	0.214	—	—	—
Limestone	2.5 to 2.8	—	—	—	—	0.21	0.0020	—	—

(continued)

Magnesium	1.74	26.0	649.5	1120	88	0.215	0.40	6080	12
Manganese	7.43	23.0	1244	2095	63.7	0.114	0.12	—	50.6 γ-Mn
Manganin	8.4	17.5	910	—	—	0.097	0.052	—	—
Mercury	13.55	61	−38.86	356.73	2.8	0.0333	0.02	—	—
Molybdenum	10.22	5.1	2620	4800	69.8	0.059	0.33	—	100 to 250
Nickel silver	8.4	18.36	1100	—	—	0.094	0.055 to 0.115	—	—
Nickel	8.91	13.3	1455	2800	73.8	0.106	0.2	—	32.2
Niobium	8.55	7.31	2468	≈ 4900	68.9	0.0640	0.125 (0 °C)	—	27.3
Palladium	12.1	11.9	1550	3560	36.3	0.0584	0.17	—	19.8
Paraffin	0.86 to 0.93	107 to 477	42 to 75	300	35	0.694	0.0006	—	—
Phenol	1.071	—	40.6	182	29	0.39	—	—	—
Phosphorus (yellow)	1.82	124	44.1	280	5.2	0.189	—	5950	—
Platinum	21.5	9.09	1769	4300	27.2	0.0315	0.17	—	34
Plutonium	15.92 to 19.74	−120 to +75	640	3000 to 3200	3773	0.032	α-Phase 0.0105[1]	—	42 to 45 [2]
Potassium	0.86	84	63.2	753.8	13	0.18	0.232	—	—
Quartz glass	2.65	0.5	1700	2200	—	0.19	0.03 ∥ 0.016 ⊥	—	7 to 9
Rhenium	21.04	6.6	3180	≈ 5000	42.42	0.0326	0.17	—	sheet 105.5 to 268.6 wire 236.9 rod 115.3
Rhodium	12.5	8.5	1960	3960	—	0.0592	0.36	—	48.1
Rubber	0.92 to 0.99	77	125	—	—	0.502	0.00045	10 700	175 to 300
Selenium (metallic)	4.79	36.8 (40 °C)	217.4	684.9	16.4	0.0767	0.006	—	—
Silicon	2.33	7.63	1423	2355	—	0.168	0.347	7830	70
Silver	10.50	19.3	961.3	2180	26.0	0.0565	0.975	—	13 to 14.5
Sodium	0.971	71	97.82	890	27	0.292	0.33	—	—
Sulfur (rhombic)	2.07	64.13	115.18	444.6	9.4	0.168	0.0006	2200	—
Tantalum	16.65	6.5	2996	5400	41.5	0.0336	0.130	—	93
Thorium	11.7	10.5	1695	4200	19.82	0.0281	0.090 (100 °C)	—	38.4
Tin solder	—	—	100 to 210	—	—		—	—	—
Tin (white)	7.29	27	231.9	≈ 2687	14.2	0.0531	0.16	—	1.12
Titanium	4.51	8.35	1668	3280	10.43	0.1249	0.037 (50 °C)	510	—
Tungsten	19.27	4.5	3390	5500	—	0.0322	0.31	—	420
Uranium	19.1	15.3	1130	3930	19.74	0.0278	0.0574 (0 °C)	—	—
Vanadium	6.12	8.3 (23 to 100 °C)	1890	≈ 3380	78.5	0.115	0.084 (70 °C)	—	≈ 20
Vitreous enamel	2.4 to 5.0	—	960	—	—	0.0023 to 0.0028	0.002 to 0.003	—	30−32
Vulcanised fibre	1.1 to 1.45	63.6	—	—	—	0.3312	0.001	—	—
Wax	1.8	—	60 to 65	—	—	0.7	0.00009	9000	—
Wood, beech	0.7 to 0.9	2.57 ∥ 61.4 ⊥	—	—	—	≈ 0.42	≈ 0.0003 ∥	4100	13
ebony	1.11 to 1.33	3.6 ∥ 40.4 ⊥	—	—	—	≈ 0.42	≈ 0.0003 ∥	—	—
fir	0.37 to 0.85	5.41 ∥ 34.1 ⊥	—	—	—	≈ 0.42	≈ 0.0003 ∥	—	7.9 ∥
oak	0.6 to 0.9	4.92 ∥ 54.4 ⊥	—	—	—	≈ 0.42	≈ 0.0003 ∥	3990	9.5 ∥
pine	0.37 to 0.85	5.41 ∥ 34.1 ⊥	—	—	—	≈ 0.42	≈ 0.0003 ∥	4890	7.5 ∥
Woods metal	9.7	—	60 to 65	—	8	0.0352	0.031	—	4.5
Zinc	7.13	30.7	419.5	907	26	0.0931	0.27	1300	27.5
Zircaloy 2	6.55	5.49 (100 °C)			—	—	0.030	—	38 to 55
Zirconium	6.50	6.10 to 7.52	1855	≈ 4380	60.3	0.0659	0.050 (25° C)	—	15 to 45

[1]) At unspecified temperature. [2]) Depending upon rod-width and storage-time.

SELECTED PROPERTIES OF LIQUID MATERIALS

Material	Density $\frac{g}{cm^3}$	Viscosity at 20 °C in centipoise	Expansion coefficient at 18 °C $\times 10^{-3}$	Melting point at 760 mm Hg °C	Boiling point at 760 mm Hg °C	Latent heat of fusion $\frac{kcal}{kg}$	Specific heat capacity $\frac{kcal}{deg\ kg}$	Latent heat of evaporation $\frac{kcal}{kg}$	Critical temperature °C	Critical pressure atm	Relative permittivity at 20 °C	Chemical formula
Acetic acid	1.049	1.30	1.07	+16.7	118.5	46.4	0.491	96.8	321.6	57.11	6.15	$C_2H_4O_2$
Acetone	0.791	0.337	1.43	−94.6	56.2	23.0	0.514	124.5	236	62	20.7	C_3H_6O
Ammonia	0.771	0.010	—	−77.73	−33.41	108.1	1.0	283.6	132.4	115.2	17.8	NH_3
i-amyl- acetate	0.874	0.805	—	−70.8	124	—	0.459	—	326.18	—	4.8 (18 °C)	$C_7H_{14}O_2$
Aniline	1.022	4.40	0.84	−6.2	184.32	20.95	0.478	107	425.65	52.35	6.89	C_6H_7N
Benzene	0.879	0.652	1.16	+5.5	80.1	30.4	0.416	94.3	288.94	47.9	2.284	C_6H_6
Carbon disulfide	1.271	0.363	1.20	−112.1	46.25	17.7	0.237	87.83	277	77.5	2.64	CS_2
Castor oil	0.960 to 0.967	950	0.69	+13	—	—	0.46	—	—	—	4.6 (18 °C)	—
Chloroform	1.489	0.58	1.28	−63.5	61.2	19.0	0.234	59.0	260	56.7	4.806	$CHCl_3$
Ethyl alcohol	0.789	1.20	1.10	−112	78.5	25	0.578	204	243	65.1	24.3	C_2H_6O
Ethyl ether	0.714	0.233	1.62	−116.3	34.6	24	0.556	89.3	195	37.5	4.335	$C_4H_{10}O$
Glycerine	1.261	1490	0.50	+17.9	290	42	0.540	—	—	—	42.5	$C_3H_8O_3$
Heavy water	1.105	1.226	0.154 (20 °C)	+3.82	101.42	—	1.006	—	307.8 to 371.1	225.8	78.25 (25 °C)	D_2O
Kerosene	0.80	—	0.99	—	—	—	0.500	—	—	—	2.1 (18 °C)	—
Linseed oil	0.93 to 0.94	51.6	—	−20	316	—	—	—	—	—	2.2 (18 °C)	—
Machine oil	0.9 to 0.93	light 113.8 heavy 660.6 (15.6 °C)	—	—	—	—	0.40	—	—	—	—	—
Mercury	13.55	1.554	0.181	−38.86	356.73	2.8	0.0332	69.7	1460	1076	1.0074 (400 °C)	Hg
Methyl alcohol	0.792	0.597	1.19	−97.8	64.96	24	0.595	262.8	240	102.3	32.63	CH_4O
Naphthalene	1.145	—	—	+80.2	210.8	35.62	0.309	75.5	468.2	39.2	2.54	$C_{10}H_8$
Nitrobenzene	1.204	2.03	0.83	+5.7	210.8	23.5	0.339	95	—	—	35.74	$C_6H_5O_2N$
Sulfuric acid	1.834	25.4	0.57	+10.38	338	26.0	0.331	122.1	—	—	>84 (18 °C)	H_2SO_4
Sulfurous acid	1.03	—	—	−73	−10	—	0.32	94.9	—	—	13.8 (18 °C)	H_2SO_3
Toluene	0.866	0.590	110.90	−95	110.6	17.2	0.42	86	318.6	41.6	2.379	C_7H_8
Turpentine	0.855	1.46	9.7	−10	161	—	0.43	70	(376)	—	2.7	$C_{10}H_{16}$
Water	0.998	1.005	0.18	0	100	79.7	0.997	584.9	374.1	217.7	80.4	H_2O
m-xylene	0.864	0.620	0.99	−47.4	139	25.8	0.412	82	346	35.8	2.374	C_8H_{10}

Material	Density (at 0 °C and 760 mm Hg) $\frac{kg}{m^3}$	Thermal conductivity at 4.4 °C $\frac{cal \times 10^{-4}}{deg\ cm\ s}$	Specific heat capacity c_p at 15 °C $\frac{kcal}{deg\ kg}$	$\gamma = \frac{c_p}{c_v}$ (15 °C)	Critical pressure atm	Critical temperature °C	Melting point °C	Boiling point °C	Density as liquid at boiling point g/cm³	Relative permittivity at 0 °C and 760 mm Hg	1 l of water dissolves at 20 °C l	Chemical formula
Acetylene	1.175	0.45	0.3832	1.26	61.7	+35.5	−80.8	−84.03	0.613	—	1.03	C_2H_2
Air (CO_2-free)	1.293	0.58	0.219 (0 °C)	1.40 (0 °C)	38.49	−140.73	—	−191.4	0.875	1.000576	0.0187	—
Ammonia	0.771	0.53	0.5232	1.310	115.2	+132.4	−77.73	−33.41	0.68	1.0072	700	NH_3
Argon	1.784	0.42 [1]	0.1253	1.668	49.6	−122.44	−187.9	−185.88	1.404	1.00056	0.04	Ar
Butane	2.732	0.33	—	—	37.4	+152.01	−138.29	−0.50	0.600	—	—	C_4H_{10}
Carbon dioxide	1.977	0.356	0.1989	1.304	75.27	+31.0	−57	−78.45	1.53	1.000946	0.88	CO_2
Carbon monoxide	1.250	0.559	0.2478	1.404	35.68	−140.2	−207	−191.55	0.801	1.000695	0.023	CO
Chlorine	3.214	0.19	0.1149	1.355	78.5	+144	−101.5	−34.1	1.558	1.97	2.300	Cl_2
Deuterium	0.1796	3.09	—	—	16.98	−234.8	−245.6	−249.48	—	1.277 (20 °K)	—	D_2
Ethane	1.357	0.446	0.3861	1.22	48.4	+32.05	−183.3	−88.6	0.546	1.00150	0.047	C_2H_6
Ethylene	1.2604	0.434	0.3592	1.255	50.1	+9.50	−169.15	−103.78	0.568	1.00150	0.122	C_2H_4
Fluorine	1.696	0.599	—	—	56.8	−129	−217.9	−188.1	1.11	—	—	F_2
Helium	0.1785	3.43	1.25 (−180 °C)	1.66 (−180 °C)	2.336	−267.95	−272.1	−268.94	0.125	1.000074	0.0088	He
Hydrogen	0.0899	4.174	3.389	1.410	13.22	−239.9	−259.5	−252.77	0.0708	1.000264	0.0181	H_2
Hydrogen sulfide	1.536	0.314	0.2533	1.32	91.9	+100.38	−82.9	−60.2	0.92	1.00332	2.61	H_2S
Krypton	3.744	0.21 [1]	—	1.68 (19 °C)	56.0	−63.75	−156.6	−153.40	2.15	—	0.062	Kr
Methane	0.717	0.744	0.5284	1.31	47.2	−82.5	−182.52	−161.5	0.415	1.000944	0.0331	CH_4
Neon	0.900	1.10	—	1.64 (19 °C)	27.06	−228.75	−248.6	−246.06	1.207	1.000127	0.0104	Ne
Nitric oxide	1.340	0.555	0.2329	1.400	66.7	−92.9	−163	−151.75	—	1.00592	0.0471	NO
Nitrogen	1.250	0.583	0.2477	1.404	34.5	−146.9	−210.5	−195.82	0.81	1.000606	0.016	N_2
Oxygen	1.429	0.594	0.2178	1.401	51.8	−118.32	−219	−182.97	1.131	1.000547	0.0315	O_2
Ozone	2.142	—	—	1.29 (0 °C)	56.4	−12.1	−192.1	−111.9	—	—	0.450	O_3
Propane	2.0096	0.372	—	1.13 (16 °C 0.5 atm)	42.1	+96.8	−187.7	−42.1	0.585	—	—	C_3H_8
Propylene	1.915	—	—	—	45.6	+91.76	−185.25	−47.70	0.609	—	0.210	C_3H_6
Sulfur dioxide	2.926	0.20 [1]	0.1516	1.29	80.4	+157.5	−72.5	−10.02	1.460	1.00905	39.4	SO_2
Sulfur hexafluoride	6.602	0.336 (30 °C)	0.159	—	38.3	+45.58	−56	−63.8	1.91 (−50 °C)	1.00191 (27.5 °C, 708 mm Hg)	—	SF_6
Uranium hexafluoride	4.68	0.140 (0 °C)	0.108	1.063	≈ 50	≈ +240	+64.05	+56.5	3.624 (65 °C)	1.0038 (19.6 °C)	—	UF_6

[1]) At unspecified temperature.

Properties of Selected Thermoluminescent Materials

Property/Type	LiF	$Li_2B_4O_7$:Mn	CaF_2:Mn [b]	CaF_2 (TLD-200) [b]	$CaSO_4$:Mn [b]	$CaSO_4$:Dy [b]
Density (g/cm^3) (Powder ∼ ½ of Solid)	2.64	∼2.4	3.18	3.18	2.61	2.61
Effective Atomic No (Z) for photoelectric absorption	8.2	7.4	16.3	16.3	15.3	15.5
TI Emission Spectra	3500-6000Å (4000 max)	5300-6300Å (6050 max)	4400-6000Å (5000 max)	Peaks at 4835Å at 5765Å	4500-6000Å (5000 max)	4800Å; 5700Å
Temperature of main TL glow peak	195°C	200°C	260°C	180°C	110°C	220°C
Efficiency at ^{60}Co relative to LiF	1.0	0.15	10	30	70	20
Energy Response 30 Kev/^{60}Co	1.25	0.9	∼13	∼12.5	∼10	∼12.5
Useful Range	mR-3x10^5 R	50mR-10^6 + R	100μR-3x10^5 R	10μR-10^6 R	μR - 10^4 R	100μR - ∼10^5 R
Fading	Negligible [a] 5%/yr at 20°C	<5% in 3 months	10% in first 24 hours 15% total in 2 weeks	10% in first 24 hours [a] 16% total in 2 weeks	50% in first 24 hours	2% in 1 month 8% in 6 months
Physical Forms	TLD-100, 600, 700 Powder Ribbons Rods Bulbs Cards Cleaved Crystals	TLD-800 Powder Chips/Ribbons Bulbs Cards	TLD-400 on, Powder Chips/Ribbons Rods Bulbs	TLD-200 Powder Crystals Bulbs Cards	Powder	Powder

[a] Post-irradiation, pre-evaluation anneal for 10 minutes with LiF and 20 minutes with CaF_2 (TLD-200) at 100°C normalizes these materials and eliminates fading.

[b] The high sensitivity materials such as CaF_2 and $CaSO_4$ are extremely light (UV) sensitive, and fading is enhanced considerably. All of the high sensitivity materials should be handled, used and stored in opaque containers to prevent fading from light exposure.

Other TLD Materials include BeO:Mn, $CaSO_4$ (rare earth), Al_2O_3 (Mn), $CaSO_4$ (Tm), and Mg_2SiO_4 (Tb)

Mass Attenuation Coefficients of Selected Materials at Selected Energies

PHOTON ENERGY	ALUMINUM $Z=13$	SILICON $Z=14$	PHOSPHORUS $Z=15$	SULFUR $Z=16$	ARGON $Z=18$	POTASSIUM $Z=19$	CALCIUM $Z=20$	IRON $Z=26$
Mev				cm²/g				
1.00 − 02	2.58 + 01	3.36 + 01	4.02 + 01	5.03 + 01	6.38 + 01	8.01 + 01	9.56 + 01	1.72 + 02
1.50 − 02	7.66 + 00	9.97 + 00	1.20 + 01	1.52 + 01	1.95 + 01	2.46 + 01	2.96 + 01	5.57 + 01
2.00 − 02	3.24 + 00	4.19 + 00	5.10 + 00	6.42 + 00	8.27 + 00	1.05 + 01	1.26 + 01	2.51 + 01
3.00 − 02	1.03 + 00	1.31 + 00	1.55 + 00	1.94 + 00	2.48 + 00	3.14 + 00	3.82 + 00	7.86 + 00
4.00 − 02	5.14 − 01	6.35 − 01	7.31 − 01	8.91 − 01	1.11 + 00	1.39 + 00	1.67 + 00	3.46 + 00
5.00 − 02	3.34 − 01	3.96 − 01	4.44 − 01	5.27 − 01	6.30 − 01	7.77 − 01	9.25 − 01	1.84 + 00
6.00 − 02	2.55 − 01	2.92 − 01	3.18 − 01	3.67 − 01	4.20 − 01	5.12 − 01	5.95 − 01	1.13 + 00
8.00 − 02	1.89 − 01	2.07 − 01	2.15 − 01	2.38 − 01	2.52 − 01	2.96 − 01	3.34 − 01	5.50 − 01
1.00 − 01	1.62 − 01	1.73 − 01	1.75 − 01	1.89 − 01	1.89 − 01	2.16 − 01	2.37 − 01	3.42 − 01
1.50 − 01	1.34 − 01	1.40 − 01	1.38 − 01	1.45 − 01	1.36 − 01	1.50 − 01	1.59 − 01	1.84 − 01
2.00 − 01	1.20 − 01	1.25 − 01	1.22 − 01	1.27 − 01	1.17 − 01	1.28 − 01	1.33 − 01	1.39 − 01
3.00 − 01	1.03 − 01	1.07 − 01	1.04 − 01	1.08 − 01	9.79 − 02	1.06 − 01	1.09 − 01	1.07 − 01
4.00 − 01	9.22 − 02	9.54 − 02	9.28 − 02	9.58 − 02	8.68 − 02	9.38 − 02	9.66 − 02	9.21 − 02
5.00 − 01	8.41 − 02	8.70 − 02	8.46 − 02	8.72 − 02	7.90 − 02	8.52 − 02	8.78 − 02	8.29 − 02
6.00 − 01	7.77 − 02	8.05 − 02	7.82 − 02	8.06 − 02	7.29 − 02	7.87 − 02	8.09 − 02	7.62 − 02
8.00 − 01	6.83 − 02	7.06 − 02	6.86 − 02	7.08 − 02	6.40 − 02	6.90 − 02	7.09 − 02	6.65 − 02
1.00 + 00	6.14 − 02	6.35 − 02	6.17 − 02	6.36 − 02	5.75 − 02	6.20 − 02	6.37 − 02	5.96 − 02
1.50 + 00	5.00 − 02	5.18 − 02	5.03 − 02	5.19 − 02	4.69 − 02	5.06 − 02	5.20 − 02	4.87 − 02
2.00 + 00	4.32 − 02	4.48 − 02	4.36 − 02	4.49 − 02	4.07 − 02	4.39 − 02	4.52 − 02	4.25 − 02
3.00 + 00	3.54 − 02	3.68 − 02	3.59 − 02	3.71 − 02	3.38 − 02	3.66 − 02	3.78 − 02	3.62 − 02
4.00 + 00	3.11 − 02	3.24 − 02	3.17 − 02	3.29 − 02	3.02 − 02	3.28 − 02	3.40 − 02	3.31 − 02
5.00 + 00	2.84 − 02	2.97 − 02	2.92 − 02	3.04 − 02	2.80 − 02	3.06 − 02	3.17 − 02	3.14 − 02
6.00 + 00	2.66 − 02	2.79 − 02	2.75 − 02	2.87 − 02	2.67 − 02	2.91 − 02	3.03 − 02	3.05 − 02
8.00 + 00	2.44 − 02	2.57 − 02	2.55 − 02	2.68 − 02	2.51 − 02	2.76 − 02	2.89 − 02	2.98 − 02
1.00 + 01	2.31 − 02	2.46 − 02	2.45 − 02	2.58 − 02	2.44 − 02	2.70 − 02	2.83 − 02	2.98 − 02
1.50 + 01	2.19 − 02	2.34 − 02	2.36 − 02	2.51 − 02	2.41 − 02	2.68 − 02	2.83 − 02	3.07 − 02
2.00 + 01	2.16 − 02	2.33 − 02	2.35 − 02	2.52 − 02	2.44 − 02	2.73 − 02	2.89 − 02	3.21 − 02
3.00 + 01	2.19 − 02	2.38 − 02	2.42 − 02	2.61 − 02	2.55 − 02	2.86 − 02	3.05 − 02	3.45 − 02

PHOTON ENERGY	COPPER $Z=29$	MOLYBDENUM $Z=42$	TIN $Z=50$	IODINE $Z=53$	TUNGSTEN $Z=74$	LEAD $Z=82$	URANIUM $Z=92$	ABSORPTION EDGES
Mev				cm²/g				
1.00 − 02	2.23 + 02	8.40 + 01	1.39 + 02	1.58 + 02	9.12 + 01	1.28 + 02	1.73 + 02	
1.50 − 02	7.33 + 01	2.68 + 01	4.53 + 01	5.34 + 01	1.39 + 02	1.12 + 02	6.03 + 01	
2.00 − 02	3.30 + 01	1.17 + 01	2.02 + 01	2.47 + 01	6.51 + 01	8.34 + 01	6.85 + 01	L_{III} EDGE
3.00 − 02	1.06 + 01	2.83 + 01	4.07 + 01	7.98 + 00	2.18 + 01	2.84 + 01	3.96 + 01	L_{II}, L_I EDGES
4.00 − 02	4.71 + 00	1.30 + 01	1.89 + 01	2.23 + 01	9.97 + 00	1.31 + 01	1.87 + 01	
5.00 − 02	2.50 + 00	6.97 + 00	1.04 + 01	1.23 + 01	5.40 + 00	7.22 + 00	1.04 + 01	
6.00 − 02	1.52 + 00	4.25 + 00	6.32 + 00	7.55 + 00	3.28 + 00	4.43 + 00	6.45 + 00	
8.00 − 02	7.18 − 01	1.92 + 00	2.90 + 00	3.52 + 00	7.66 + 00	2.07 + 00	3.04 + 00	
1.00 − 01	4.27 − 01	1.05 + 00	1.60 + 00	1.91 + 00	4.29 + 00	5.23 + 00	1.71 + 00	
1.50 − 01	2.08 − 01	3.99 − 01	5.77 − 01	6.74 − 01	1.50 + 00	1.89 + 00	2.47 + 00	K EDGE
2.00 − 01	1.48 − 01	2.28 − 01	3.07 − 01	3.49 − 01	7.38 − 01	9.45 − 01	1.23 + 00	
3.00 − 01	1.08 − 01	1.31 − 01	1.55 − 01	1.68 − 01	3.02 − 01	3.83 − 01	4.85 − 01	
4.00 − 01	9.19 − 02	1.01 − 01	1.10 − 01	1.16 − 01	1.80 − 01	2.20 − 01	2.73 − 01	
5.00 − 01	8.22 − 02	8.59 − 02	9.11 − 02	9.36 − 02	1.29 − 01	1.54 − 01	1.85 − 01	
6.00 − 01	7.52 − 02	7.67 − 02	7.91 − 02	8.07 − 02	1.03 − 01	1.20 − 01	1.40 − 01	
8.00 − 01	6.55 − 02	6.52 − 02	6.55 − 02	6.61 − 02	7.73 − 02	8.56 − 02	9.64 − 02	
1.00 + 00	5.86 − 02	5.77 − 02	5.71 − 02	5.75 − 02	6.39 − 02	6.90 − 02	7.54 − 02	
1.50 + 00	4.79 − 02	4.68 − 02	4.59 − 02	4.60 − 02	4.88 − 02	5.10 − 02	5.39 − 02	
2.00 + 00	4.19 − 02	4.14 − 02	4.08 − 02	4.09 − 02	4.34 − 02	4.50 − 02	4.70 − 02	
3.00 + 00	3.59 − 02	3.66 − 02	3.67 − 02	3.69 − 02	4.01 − 02	4.16 − 02	4.35 − 02	
4.00 + 00	3.32 − 02	3.48 − 02	3.54 − 02	3.59 − 02	3.98 − 02	4.14 − 02	4.34 − 02	
5.00 + 00	3.18 − 02	3.43 − 02	3.53 − 02	3.59 − 02	4.06 − 02	4.24 − 02	4.44 − 02	
6.00 + 00	3.10 − 02	3.43 − 02	3.57 − 02	3.63 − 02	4.16 − 02	4.34 − 02	4.54 − 02	
8.00 + 00	3.06 − 02	3.50 − 02	3.69 − 02	3.78 − 02	4.39 − 02	4.59 − 02	4.79 − 02	
1.00 + 01	3.08 − 02	3.62 − 02	3.85 − 02	3.95 − 02	4.63 − 02	4.84 − 02	5.06 − 02	
1.50 + 01	3.23 − 02	3.93 − 02	4.25 − 02	4.38 − 02	5.24 − 02	5.48 − 02	5.73 − 02	
2.00 + 01	3.39 − 02	4.23 − 02	4.61 − 02	4.76 − 02	5.77 − 02	6.06 − 02	6.36 − 02	
3.00 + 01	3.68 − 02	4.70 − 02	5.17 − 02	5.36 − 02	6.59 − 02	6.96 − 02	7.33 − 02	

Properties of the Superconductive Elements

B. W. Roberts

Element	T_c(K)	H_o(oersted)	θ_D(K)	γ($mJ mol^{-1} K^{-1}$)
Al	1.175±0.002	104.9±0.3	420	1.35
Be	0.026			0.21
Cd	0.517±0.002	28±1	209	0.69
Ga	1.083±0.001	59.2±0.3	325	0.60
Ga (β)	5.9, 6.2	560		
Ga (γ)	7	950, HF[a]		
Ga (Δ)	7.85	815, HF		
Hf	0.128			
Hg (α)	4.154±0.001	411±2	87, 71.9	1.81
Hg (β)	3.949	339	93	1.37
In	3.408±0.001	281.5±2	109	1.672
Ir	0.1125±0.001	16±0.05	425	3.19
La (α)	4.88±0.02	800±10	151	9.8
La (β)	6.00±0.1	1096, 1600	139	11.3
Lu	0.1	<400		
Mo	0.915±0.005	96±3	460	1.83
Nb	9.25±0.02	2060±50, HF	276	7.80
Os	0.66±0.03	70	500	2.35
Pa	1.4			
Pb	7.196±0.006	803±1	96	3.1
Re	1.697±0.006	200±5	415	2.35
Ru	0.49±0.015	69±2	580	2.8
Sn	3.722±0.001	305±2	195	1.78
Ta	4.47±0.04	829±6	258	6.15
Tc	7.8±0.1	1410, HF	411	6.28
Th	1.38±0.02	160±3	165	4.32
Ti	0.40±0.04	56	415	3.3
Tl	2.38±0.04	178±5	78.5	1.47
V	5.40±0.05	1408	383	9.82
W	0.0154±0.0005	1.15±0.03	383	0.90
Zn	0.850±0.01	54±0.3	310	0.66
Zr	0.61±0.15	47	290	2.77
Zr (ω)	0.65, 0.95			

[a] HF denotes high field superconductive properties.

Range of Critical Temperatures Observed for Superconductive Elements in Thin Films Condensed Usually at Low Temperatures

Element	T_c Range (K)	H_o (oersted)
Al	1.15-~5.7	HF[a]
Ba	3.0	HF
Be	5-9.75	HF
(with KCl)	6.5-10.6	HF
(with zinc etio-porphyrin)	10.2	
Bi	6.17, 6.13-2.3, ~5-~2	
Ca	4.2	HF
Cd (Disordered)	0.79-0.91	
(Ordered)	0.53-0.59	
Ga	2.5-8.5	HF
In	3.43-4.65	HF
La	3.55 4.9, 5.0-6.74	
Mg	5.5	HF
Mo	3.3-3.8, 4-6.7	
Nb	6.3-10.1	
Pb	~2-7.5	
Re	1.7-~7	
Sn	3.5-~6	
Sr	3.6	HF
Ta	<1.7-4.51	HF
Tc	4.6-7.70	
Ti	1.3 Max	
Tl	2.33-2.96	
V	1.8-6.02	
W	<1.0-4.1	
Zn	0.77-1.70, ~1.9	

[a] HF denotes high magnetic field superconductive properties in Table 5.

Elements Exhibiting Superconductivity Under or After Application of High Pressure

Element	T_c Range(K)	Pressure (kbar)[b]
As	0.31-0.5	220-140
	0.2-0.25	~140-100
Ba II	~1-1.8	~55-85
III	1.8-5	~85-144
IV	4.5-5.4	144-190
Bi II	3.9	25-27
III	6.55, 7.25	~37, 27-28
IV	7.0, 8.7-6.0	43, 43-62
V	6.7, 8.3	68, 81
VI	8.55	90, 92-101
VII(?)	8.2	30
Ce	1.7	50
Cs V	~1.5	>125
Ga II	6.38	≥35
II'	7.5	≥35 then P removed
Ge	5.35	115
La	~5.5-11.93	0-~140
Lu	~0.6-<0.018	145-80
P	5.8	170
Pb II	3.55	160
Re II	2.3 Max.	"Plastic" compression
Sb (Prepared 120 kbar, held below 77K)	2.6-2.7	
Sb III	3.55-3.40	85 -~150
Se II	6.75, 6.95	~130
Si	6.7-7.1	120-130

(continued)

Element	T_c Range (K)	Pressure (kbar)[b]
Sn II	5.2-4.85	125-160
III	5.30	113
Te II	2.05	43
	3.4	50
III	4.28-4.15	68-80
IV	4.3-3.3	80-100
( )	3.3-2.8	100-260
Tl (cubic form)	1.45	35
(hexagonal form)	1.95	35
U	2.4-0.4	10-85
Y	2.3-1.7-2.5	110-125-160
Zr (omega form, metastable)	1-1.7	60-~130

[b] 1 kbar = 10^8 newton/meter^2 = 0.987 katm

Physical characteristics of copper, gold, palladium. and silver[a]

Element (chemical symbol)	Atomic number	Relative[b] atomic mass	Density,[c] Mg m^{-3}	Crystal structure	Debye[d] temperature, K		Melting point, K	Normal boiling point, K
					at 0 K	at 298 K		
Copper (Cu)	29	63.546	8.933	fcc	342±2	320	1357.6	2840
Gold (Au)	79	196.9665	18.88	fcc	165±1	178±8	1337.58	3135
Palladium (Pd)	46	106.4	12.02	fcc	283±16	275	1827	3243
Silver (Ag)	47	107.868	10.492	fcc	228±3	221	1235.08	2440

[a] Information taken from Touloukian, Kirby, Taylor, and Desai [228, pp. 39a, 41a, 42a] unless otherwise stated.
[b] Relative atomic masses are based on $^{12}C=12$ as adopted by the International Union of Pure and Appied Chemistry in 1971. Applies to material of terrestrial origin.
[c] Density values given for 293.2 K.
[d] From Gschneidner [209, table XV] as obtained from specific heat data.

Characteristic Properties of Thermal Insulation Materials

Insulation Type	Thermal Conductivity* k_{eff}		Density ρ		$\rho\, k_{eff}$		Gas Pressure (torr)
	(Btu-in. h^{-1} ft^{-2} $°F^{-1}$)	(mW cm^{-1} K^{-1})	(lb ft^{-3})	(g cm^{-3})			
Fibrous							
Fiberglass-ordered	0.0039	0.0056	14.98	0.24	0.0582	0.0013	760
Fiberglass-random	0.118	0.1702	3.12	0.05	0.0368	0.0085	760
Microspheres							
Al coated	0.0311	0.0448	4.99	0.08	0.2320	0.0036	10^{-6}
Al hemispherical	0.020	0.0288	4.99	0.08	0.1552	0.0023	10^{-6}
Uncoated	0.0465	0.0671	4.37	0.07	0.0878	0.0047	10^{-6}
Multilayer (50 layers/in.)							
**DAM/silk net	3.00×10^{-4}	4.33×10^{-4}	2.82	0.045	8.46×10^{-4}	19.49×10^{-6}	10^{-6}
DAM/nylon net	2.04×10^{-4}	2.94×10^{-4}	3.36	0.054	6.85×10^{-4}	15.88×10^{-6}	10^{-6}
NRC-2 †SAM crinkled	3.12×10^{-4}	4.50×10^{-4}	0.91	0.015	2.84×10^{-4}	6.75×10^{-6}	10^{-6}
††Superfloc	3.00×10^{-4}	4.33×10^{-4}	0.86	0.014	2.58×10^{-4}	6.06×10^{-6}	10^{-6}
Powder							
Perlite	0.0080	0.0115	6.00	0.096	0.048	0.0011	760
Santocel A (Monsanto)	0.0140	0.0202	6.00	0.096	0.084	0.0019	10^{-6}
Silica aerogel	0.0111	0.0160	5.00	0.080	0.0554	0.0013	10^{-3}
Preformed							
Expanded polystyrene	0.1665	0.2401	0.94	0.015	0.1560	0.0036	760
Polyurethane foam	0.1734	0.2500	3.06	0.049	0.530	0.0123	760
Resin bonded fiberglass	0.222	0.3201	1.87	0.030	0.4158	0.0096	760

*Approximate hot and cold boundary temperatures = 300K and 77K, respectively.
†SAM = Single-aluminized Mylar **DAM = Double-aluminized Mylar
††Superfloc = DAM/Dacron Tufts

Properties of Some Typical Plastics

Generic Name	Typical Tensile strength, ksi	Typical Impact Strength, ft-lb/in of notch	Typical Flexural Modulus, 10^5 psi	PV Rating, dry, continuous × 1000	Deflection Temperature °F		Abrasion Resist., mg. loss/1000 cycles	Endurance Limit ksi	Thermal Expansion, per °F × 10^5	Heat Resistance, Continuous, °F	Thermal Conductivity Btu/in/hr/ft²/°F	Water Absorption in 24 hr, %	Flammability	Formability	Machinability	Coefficient of Friction			Resistance to				Brittle Point, °F
					66 psi	264 psi										Slip-Stick	Dry	Lubricated	Acids	Alkalies	Solvents	Oils	
	Materials for Housings, Shrouds, Containers, Ducts																						
ABS	5	6	2.4	—	—	—	—	—	3.2-5.8	140-250	—	0.1-0.3	Slow	G	—	—	—	—	G	E	F	G	—
Styrenes, high-impact	4.3	1	2.3	—	—	—	—	—	2.2-5.6	126-165	—	0.03-0.2	Slow	G	—	—	—	—	G	E	P	F	—
Polypropylenes	5.5	1	1.75	—	—	—	—	—	3.4-6.2	230-320	—	0.01-0.03	Slow	G	—	—	—	—	E	E	G	E	—
Polyprohylenes high-density	4.2	12	2	—	—	—	—	—	6.5-16.7	170-260	—	<0.01	Very Slow	E	—	—	—	—	E	E	G	G	—
Cellulose Acetate Butyrates	5.5	2.1	1.3	—	—	—	—	—	6-10	140-220	—	0.9-2.8	Slow	E	—	—	—	—	P	P	P	G	—
Acrylics, modified	5.5	2	2.8	—	—	—	—	—	3-6	140-195	—	0.2-0.4	Slow	E	—	—	—	—	G	E	F	F	—
Acrylics-PVC alloy	6.5	15	4	—	—	—	—	—	3.5	165	—	0.06	Non-burn.	E	—	—	—	—	E	E	E	E	—
Polyester-Glass	16.5	15	15	—	—	—	—	—	1-1.4	200-550	—	0.1-2	Slow to nil	G	—	—	—	—	G	F	G	E	—
Epoxy-Glass	36	12.4	25	—	—	—	—	—	0.3-0.6	250-400	—	0.02-0.08	Slow to nil	G	—	—	—	—	G	E	E	E	—

(continued)

Generic Name	Typical Tensile strength, ksi	Typical Impact Strength, ft-lb/in of notch	Typical Flexural Modulus, 10^5 psi	PV Rating, dry, continuous × 1000	Deflection Temperature °F		Abrasion Resist., mg. loss/1000 cycles	Endurance Limit ksi	Thermal Expansion, per °F × 10^5	Heat Resistance, Continuous, °F	Thermal Conductivity Btu/in/hr/ft²/°F	Water Absorption in 24 hr, %	Flammability	Formability	Machinability	Coefficient of Friction			Resistance to				Brittle Point, °F
					66 psi	264 psi										Slip-Stick	Dry	Lubricated	Acids	Alkalies	Solvents	Oils	
	Materials for Housings, Shrouds, Containers, Ducts																						
Nylons	—	—	1.5–4	2-3	340-360	—	6-8	—	4.6-7.1	—	1.4-2	0.4-3.3	—	—	—	Yes	0.15-0.40	0.06	—	—	—	—	—
Acetals	—	—	3.1-4.1	2-3	316-338	—	6-20	—	4.5–6	—	1.6-1.9	0.12-0.41	—	—	—	No	0.15-0.35	0.1	—	—	—	—	—
Acetal, self-lubricating	—	—	—	18	302	—	5-12	—	—	—	—	—	—	—	—	No	0.10	0.05	—	—	—	—	—
Acetal, TFE-fiber filled.	—	—	4.14	7.5	329	—	—	—	4.6	—	1.7	0.6	—	—	—	No	0.12	0.07	—	—	—	—	—
Polyethylenes, high-density	—	—	1.3–2.2	—	140–180	—	6	—	6.5–16.7	—	3.4	<0.01	—	—	—	Yes	0.21	0.1	—	—	—	—	—
	Materials for Heavily Stressed Mechanical Components (e.g. Gears, Cams, Racks, Couplings, Rollers, etc.)																						
Nylons	7.1–12.6	0.6-4	1.5-4	—	340-360	140-165	6-8	3	—	—	—	—	—	—	E	—	—	—	F	E	E	E	—
Acetals	8.8-10	1.2–1.4	3.7–4.1	—	316-338	230-255	6-20	5	—	—	—	—	—	—	E	—	—	—	P	E	E	E	—
Acetal, TFE-fiber filled	6.9	0.86	4.14	—	329	212	—	—	—	—	—	—	—	—	E	—	—	—	P	P	E	E	—
Polycarb-onates	9-10.5	12-16	3.2-3.8	—	283-293	270-280	7-24	2	—	—	—	—	—	—	E	—	—	—	E	G	G	F	—
Phenolics, fabric filled	9-16	1-2.5	8-14	—	320	>320	—	—	—	—	—	—	—	—	F to E	—	—	—	F	F	E	E	—

(continued)

Generic Name	Typical Tensile strength, ksi	Typical Impact Strength, ft-lb/in of notch	Typical Flexural Modulus, 10^5 psi	PV Rating, dry, continuous × 1000	Deflection Temperature °F		Abrasion Resist., mg. loss/1000 cycles	Endurance Limit ksi	Thermal Expansion, per °F × 10^5	Heat Resistance, Continuous, °F	Thermal Conductivity Btu/in/hr/ft²/°F	Water Absorption in 24 hr, %	Flammability	Formability	Machinability	Coefficient of Friction			Resistance to				Brittle Point, °F
					66 psi	264 psi										Slip-Stick	Dry	Lubricated	Acids	Alkalies	Solvents	Oils	
	Materials for Low Friction Applications (e.g. Bearings, Slides, Guides, Valve Liners, Wear Surfaces, etc.)																						
TFE	—	—	—	1-2.5	250	—	7	—	5.5	—	1.7	None	—	—	—	No	0.04	0.04	—	—	—	—	—
FEP	—	—	0.95	.6-.9	<250	—	13.2	—	4.6	—	1.4	<0.01	—	—	—	No	0.08	0.08	—	—	—	—	—
TFE fabric	—	—	—	5-50	—	—	—	—	8	—	1.7	None	—	—	—	No	0.02–0.25	0.02-0.25	—	—	—	—	—
Filled TFE	—	—	1.2-2	5-35	>250	—	8–26	—	3-9.7	—	1.7-20	None	—	—	—	No	0.16-0.28	0.06	—	—	—	—	—
	Materials for Chemical and Thermal Equipment																						
TFE & FEP	1.5-4.5	2.5->16	—	—	250	—	—	—	—	—	—	—	None	—	—	—	—	—	E	E	E	—	-420
CTFE	4.6-5.6	3.1-7.3	—	—	265	—	—	—	—	400	—	—	None	—	—	—	—	—	E	E	G	—	-400
Polypropylenes	3.3-5.7	0.3-3	—	—	210	—	—	—	—	275	—	—	Slow	—	—	—	—	—	G	G	F	—	0
Polyphenylene Oxide	11.6	1.3	—	—	355	—	—	—	—	250	—	—	Self ext.	—	—	—	—	—	G	G	E	—	—
Epoxy-Glass	34–100	10-25	—	—	350-375	—	—	—	—	250-300	—	—	None	—	—	—	—	—	F	F	E	—	—

Properties of Thermoplastic Resins (insulating materials)

No.	Type	Material	Density	Flexural strength	Impact strength	Impact strength with notch	Tensile strength	Elastic modulus	Vicat soften-ing point	Thermal conduct-tivity	Thermal endurance	Electric strength	Volume resis-tivity	Permittivity (relative)		
			g/cm^3	kgf/cm^2	kgf cm/cm^2	kgf cm/cm^2	kgf/cm^2	kgf/cm^2 × 10^{-6}	°C	kcal/m h deg	°C	kV/mm	Ω cm	50 Hz	800 Hz	10^6 Hz
1	PE	High-pressure polyethylene	0.917	800 [1]	w. b. [2]	w. b. [2]	—	0.001	60	0.29	80	40	10^{17}	2.28	2.28	2.28
2	PE	Low-pressure polyethylene	0.960	800 [1]	w. b. [2]	w. b. [2]	250	0.014	65	0.41	95	45	10^{17}	2.30	2.30	2.30
3	PP	Polypropylene	0.906	450 [1]	w. b. [2]	5	330	0.012	85	0.19	105	40	10^{17}	2.27	2.26	2.25
4	PVC 642	Polyvinyl chloride high impact strength	1.35	200		>30	230	0.015	50	0.14	70	50	10^{13}	3.7	3.7	3.1
5	PS 502	Polystyrene standard	1.05	1000	22	2.5	550	0.032	100	0.12	60	50	10^{14}	2.5	2.5	2.5
6	AS	Polystyrene + acrylonitrile	1.07	1200	28	3.5	600	0.034	100	0.14	80	40	10^{16}	3.0	2.9	2.9
7	ABS	Polystyrene + acrylonitrile + butadiene	1.06	750 [1]	w. b. [2]	15	490	0.028	120	0.18	80	33	10^{14}	4.5	4.4	3.7
8	PA	6-polyamide	1.12	950 [1]	w. b. [2]	5	600	0.032	>200	0.3	90	34	10^{14}	4.3	4.1	3.6
9	PA	6.6-polyamide	1.14	850 [1]	w. b. [2]	5	570	0.017	250	0.22	110	45	10^{15}	3.7	3.7	3.6
10	PA	6.10-polyamide	1.08	650 [1]	w. b. [2]	10	400	0.020	>200	0.19	105	45	10^{15}	3.5	3.4	3.3
11	PA	11-polyamide	1.04	550 [1]	w. b. [2]	5	500	0.015	170	0.25	85	45	10^{14}	3.7	3.7	3.6
12	PA	12-polyamide	1.01	570 [1]	w. b. [2]	8	470	0.014	140	0.21	80	40	10^{15}	3.5	3.3	3.2
13	PUR	Polyurethane	1.21	600	w. b. [2]	2	600	0.012	180	0.3	90	35	10^{13}	3.6	3.5	3.4
14	PC 300	Polycarbonate	1.20	750 [1]	w. b. [2]	20	650	0.022	150	0.17	130	25	10^{16}	3.0	3.0	3.0
15		Polyacetal	1.41	1100 [1]	80	7	650	0.030	150	0.27	85	40	10^{15}	3.9	3.8	3.7
16	PMMA 528	Polymethyl-methacrylate	1.18	1150	18	2	760	0.032	110	0.16	100	30	10^{15}	3.6	3.1	2.6
17	PTFE	Polytetrafluoro-ethylene	2.20	190 [1]	w. b. [2]	13	200	0.004	110	0.21	250	30	10^{17}	2.0	2.0	2.0
18	PCTFE	Polytrifluoroethylene	2.10	540 [1]	w. b. [2]	8	320	0.009	80	0.10	155	40	10^{17}	2.7	2.6	2.4
19	CA 434	Cellulose acetate	1.29	600 [1]	75	5	460	0.016	80	0.22	65	33	10^{15}	5.1	4.8	4.1
20	CAB 411	Cellulose acetobutyrate	1.21	550 [1]	18	2	370	0.022	100	0.20	75	36	10^{16}	4.0	3.8	3.4
21	PPO	Polyphenylene oxide	1.06	1000 [1]	w. b. [2]	2	750	0.023	>200	0.28	110	40	10^{17}	2.6	2.6	2.6
22		Polysulphone	1.24	1200 [1]	w. b. [2]	3	750	0.023	195	0.21	155	35	10^{17}	3.0	2.9	2.9
23	PETP	Polyethylene terephthalate	1.38	1170 [1]	w. b. [2]	4	540	0.028	>250	0.25	120	30	10^{17}	3.5	3.4	3.4

[1]) Limiting flexural stress.
[2]) Without break.

(continued)

No.	Material	Trade name (examples)
1	High-pressure polyethylene	Lupolen 18; Trolen
2	Low-pressure polyethylene	Hostalen G; Vestolen A; Lupolen 60
3	Polypropylene	Hostalen PP, Vestolen P; Novolen
4	Polyvinyl chloride	Vestolit; Vinoflex; Vinnol; Hostalit
5	Standard polystyrene	Polystyrene VI; Vestyron; Trolitul VI
6	PS + acrylonitrile	Luran; Vestoran
7	ABS	Novodur; Vestodur; Cycolac
8	6-Polyamide	Ultramid B; Durethan BK
9	6.6-Polyamide	Ultramid A; Maranyl; Zytel
10	6.10-Polyamide	Ultramid S
11	11-Polyamide	Rilsan BM
12	12-Polyamide	Vestamid; Rilsan AM
13	Polyurethane	Durethan U
14	Polycarbonate	Makrolon; Lexan
15	Polyacetal	Delrin; Hostaform C
16	Polymethylmethacrylate	Plexigum; Resarit
17	Polytetrafluoroethylene	Hostaflon TF; Teflon
18	Polytrifluoroethylene	Hostaflon C2; KEL-F
19	Cellulose acetate	Cellidor A
20	Cellulose acetobutyrate	Cellidor B
21	Polyphenylene oxide	PPO
22	Polysulphone	
23	Polyethylene terephthalate	Arnite AR

CHAPTER 22

MECHANICAL TECHNOLOGY

Formulas of Motion

Nomenclature

t = time, sec
s = linear displacement, ft
v = linear velocity, fps
V_0 = linear velocity at time zero, fps
a = linear acceleration, ft/sec²
θ = angular displacement, radians
ω = angular velocity, radians/sec
ω_0 = angular velocity at time zero, radians/sec
α = angular acceleration, radians/sec²
w = weight of body, lb mass
f = force of acceleration, lb force
g_c = conversion factor = 32.2 (lb mass)(ft)/(lb force)(sec²)

v = constant	ω = constant	v = variable	ω = variable
$v = s/t$	$\omega = \theta/t$	$v = ds/dt$	$\omega = d\theta/dt$

a = constant	α = constant	a = variable	α = variable
$v = V_0 + at$	$\omega = \omega_0 + \alpha t$	$a = \frac{dv}{dt} = \frac{d^2s}{dt^2}$	$\alpha = \frac{d\omega}{dt} = \frac{d^2\theta}{dt^2}$
$s = V_0t + \frac{1}{2}at^2$	$\theta = \omega_0 t + \frac{1}{2}\alpha t^2$	$v = \int a\,dt$	$\omega = \int \alpha\,dt$
$v = \sqrt{V_0^2 + 2as}$	$\omega = \sqrt{\omega_0^2 + 2\alpha\theta}$	$s = \int v\,dt$	$\theta = \int \omega\,dt$

For uniform acceleration

$$f = (w/g_c)a$$

Coefficients of Static and Sliding Friction

Materials	Static coef		Sliding coef	
	Dry	Greasy	Dry	Greasy
Hard steel on hard steel	0.78	0.11	0.42	0.029
	....	.23	....	.081
	....	.15	....	.080
	....	.11	....	.058
	....	.0075	....	.084
	....	.0052	....	.105
	....	......	....	.096
	....	......	....	.108
	....	......	....	.12
Mild steel on mild steel	0.74	......	0.57	.09
	....	......	....	.19
Hard steel on graphite	0.21	.09		
Hard steel on babbitt (ASTM 1)	0.70	.23	0.33	.16
	....	.15	....	.06
	....	.08	....	.11
	....	.085		
Hard steel on babbitt (ASTM 8)	0.42	.17	0.35	.14
	....	.11	....	.065
	....	.09	....	.07
	....	.08	....	.08
Hard steel on babbitt (ASTM 10)	....	.25	....	.13
	....	.12	....	.06
	....	.10	....	.055
	....	.11		
Mild steel on cadmium silver	....	......	....	.097
Mild steel on phosphor bronze	....	......	0.34	.173
Mild steel on copper lead	....	......	....	.145
Mild steel on cast iron	....	.183	0.23	.133
Mild steel on lead	0.95	.5	0.95	.3
Nickel on mild steel	....	......	0.64	.178
Aluminum on mild steel	0.61	......	0.47	
Magnesium on mild steel	....	......	0.42	
Cadmium on mild steel	....	......	0.46	
Copper on mild steel	0.53	......	0.36	.18
Nickel on nickel	1.10	......	0.53	.12
Brass on mild steel	0.51	......	0.44	
Brass on cast iron	....	......	0.30	
Zinc on cast iron	0.85	......	0.21	
Magnesium on cast iron	....	......	0.25	
Copper on cast iron	1.05	......	0.29	
Tin on cast iron	....	......	0.32	
Lead on cast iron	....	......	0.43	
Aluminum on aluminum	1.05	......	1.4	
Glass on glass	0.94	.01	0.40	.09
	....	.005	....	.116
Carbon on glass	....	......	.18	
Garnet on mild steel	....	......	.39	
Glass on nickel	0.78	......	.56	
Copper on glass	0.68	......	.53	
Cast iron on cast iron	1.10	......	.15	.070
	....	......	....	.064
Bronze on cast iron	....	......	.22	.077
Oak on oak (parallel to grain)	0.62	......	.48	.164
	....	......	....	.067
Oak on oak (perpendicular to grain)	.54	......	.32	.072
Leather on oak (parallel)	.61	......	.52	
Cast iron on oak	....	......	.49	.075
Leather on cast iron	....	......	.56	.36
	....	......	....	.13
Laminated plastic on steel	....	......	.35	.05
Fluted rubber bearing on steel	....	......	....	.05

COEFFICIENT OF STATIC AND KINETIC (Sliding) FRICTION

Materials	Surface Condition	Temperature °C	μ (Static)
A. STATIC FRICTION			
Non Metals			
Glass on glass	clean	—	0.9–1.0
.. .. ..	lubricated with paraffin oil	—	0.5–0.6
.. .. ..	.. .. liquid fatty acids	—	0.3–0.6
.. .. ..	.. .. solid hydrocarbons, alcohols or fatty acids	—	0.1
.. .. metal	clean	—	0.5–0.7
.. .. ..	lubricated	—	0.2–0.3
Diamond on diamond	clean	—	0.1
.. .. ..	lubricated	—	0.05–0.1
.. .. metal	clean	—	0.1–0.15
.. .. ..	lubricated	—	0.1
Sapphire on sapphire	clean or lubricated	—	0.2
.. .. steel	.. .. ..	—	0.15
Hard carbon on carbon	clean	—	0.16
.. ..	lubricated	—	0.12–0.14
Graphite on graphite	clean or lubricated	—	0.1
.. .. ..	outgassed	—	0.5–0.8
.. .. steel	clean or lubricated	—	0.1
Mica on mica	freshly cleaved	—	1.0
.. .. ..	contaminated	—	0.2–0.4
Crystals of $NaNO_3$, KNO_3, NH_4Cl on self	clean	—	0.5
.. ..	lubricated with long chain polar compounds	—	0.12
Tungsten carbide on tungsten carbide	clean	room	0.17
Tungsten carbide on tungsten carbide	outgassed	room	0.58
.. ..	clean	820	0.35
.. ..	..	970	0.40
.. ..	..	1010	0.45
.. ..	..	1160	0.5
.. ..	..	1220	0.7
.. ..	..	1440	1.2
.. ..	..	1600	1.8
.. .. .. graphite	outgassed	room	0.62
.. ..	clean	..	0.15
.. ..	..	800	0.32
.. ..	..	910	0.30
.. ..	..	1000	0.25
.. ..	..	1120	0.29
.. ..	..	1220	0.26
.. ..	..	1300	0.25
.. ..	..	1410	0.25
.. ..	..	1800	0.24
.. ..	..	2030	0.25
.. .. .. steel	..	—	0.4–0.6
.. ..	lubricated	—	0.1–0.2
Polymethyl methacrylate on self	clean	—	0.8
.. .. .. steel	..	—	0.4–0.5
Polystyrene on self	..	—	0.5
.. .. steel	..	—	0.3–0.35
Polyethylene on self	..	—	0.2
.. .. steel	..	—	0.2
Polytetrafluoroethylene on self	..	—	0.04
.. .. steel	..	—	0.04
Nylon on nylon	..	—	0.15–0.25
Silk on silk	commercially clean	—	0.2–0.3
Cotton on cotton (thread)	.. ..	—	0.3
.. .. .. (from cotton wool)	.. ..	—	0.6
Rubber on solids	.. ..	—	1–4
Wood on wood	.. .. and dry	—	0.25–0.5
.. .. ..	.. .. .. wet	—	0.2
.. .. metals	.. .. .. dry	—	0.2–0.6
.. .. ..	.. .. .. wet	—	0.2
.. .. brick	.. ..	—	0.6
.. .. leather	.. ..	—	0.3–0.4
Leather on metal	.. ..	—	0.6
.. .. ..	.. .. and wet	—	0.4
.. .. ..	greasy	—	0.2
Brake material on cast iron	commercially clean	—	0.4
.. ..	.. .. and wet	—	0.2
.. ..	lubricated with mineral oil	—	0.1
Wool fiber on horn	clean (against scales)	—	0.8–1.0
.. ..	.. (with scales)	—	0.4–0.6
.. ..	greasy (against scales)	—	0.5–0.8
.. ..	.. (with scales)	—	0.3–0.4
Metals			
Steel on steel	clean	20	0.58
.. .. ..	vegetable oil lubricant		
.. .. ..	(a) castor oil	20	0.095
		100	0.105
.. .. ..	(b) rape	20	0.105
		100	0.105
.. .. ..	(c) olive	20	0.105
		100	0.105
.. .. ..	(d) coconut	20	0.08
		100	0.08

(continued)

Materials	Surface Condition	Temperature °C	μ (Static)
A. STATIC FRICTION (Cont.)			
Metals (Cont.)			
Steel on steel	Animal oil lubricant		
.. .. ..	(a) sperm	20	0.10
		100	0.10
.. .. ..	(b) pale whale	20	0.095
		100	0.095
.. .. ..	(c) neatsfoot	20	0.095
		100	0.095
.. .. ..	(d) lard	20	0.085
		100	0.085
.. .. ..	Mineral oil lubricant		
.. .. ..	(a) light machine	20	0.16
		100	0.19
.. .. ..	(b) thick gear	20	0.125
		100	0.15
.. .. ..	(c) solvent refined	20	0.15
		100	0.20
.. .. ..	(d) heavy motor	20	0.195
		100	0.205
.. .. ..	(e) extreme pressure	20	0.09–0.1
		100	0.09–0.1
.. .. ..	(f) graphited oil	20	0.13
		100	0.15
.. .. ..	(g) B.P. Paraffin	20	0.18
		100	0.22
.. .. ..	lubricated with trichloroethylene	20	0.33
.. .. ..	.. .. benzene	20	0.48
.. .. ..	.. .. glycerol	20	0.2
.. .. ..	.. .. ethyl alcohol	20	0.43
.. .. ..	.. .. butyl alcohol	room	0.3
.. .. ..	.. .. octyl	..	0.23
.. .. ..	.. .. decyl	..	0.16
.. .. ..	.. .. cetyl	..	0.10
.. .. ..	lubricated with nonane	room	0.26
.. .. ..	.. .. decane	..	0.23
.. .. ..	.. .. acetic acid	..	0.5
.. .. ..	.. .. proprionic acid	..	0.4
.. .. ..	.. .. valeric acid	..	0.17
.. .. ..	.. .. caproic acid	..	0.12
.. .. ..	.. .. pelargonic acid	..	0.11
.. .. ..	.. .. capric acid	..	0.11
.. .. ..	.. .. lauric acid	..	0.11
.. .. ..	.. .. myristic acid	..	0.11
.. .. ..	.. .. oleic acid	20–100	0.08
.. .. ..	.. .. palmitic acid	room	0.11
.. .. ..	.. .. stearic acid	..	0.10
.. .. hard steel	.. .. rape oil	—	0.14
.. ..	.. .. castor oil	—	0.12
.. ..	.. .. mineral oil	—	0.16
.. ..	.. .. long chain fatty acid	—	0.09
.. .. cast iron	.. .. rape oil	—	0.11
.. ..	.. .. castor oil	—	0.15
.. ..	.. .. mineral oil	—	0.21
.. ..	clean	—	0.4
.. .. gun metal	lubricated with rape oil	—	0.15
.. ..	.. .. castor oil	—	0.16
.. ..	.. .. mineral oil	—	0.21
.. .. bronze	.. .. rape oil	—	0.12
.. .. ..	.. .. caster oil	—	0.12
.. .. ..	.. .. mineral oil	—	0.16
.. .. lead	.. ..	—	0.5
.. .. ..	.. .. long chain fatty acid	—	0.22
.. .. base white metal	.. .. mineral oil	—	0.1
.. ..	.. .. long chain fatty acid	—	0.08
.. .. ..	clean	—	0.55
.. .. tin	lubricated with mineral oil	—	0.6
.. .. ..	.. .. long chain fatty acid	—	0.21
.. .. white metal, tin base	.. .. mineral oil	—	0.1
.. ..	.. .. long chain fatty acid	—	0.07
.. ..	clean	—	0.8
.. .. sintered bronze	lubricated with mineral oil	—	0.13
.. .. brass	.. ..	—	0.19
.. .. ..	.. .. castor oil	—	0.11
.. .. ..	.. .. long chain fatty acid	—	0.13
.. .. ..	clean	—	0.35
.. .. copper lead alloy	..	—	0.22
.. .. Wood's alloy	..	—	0.7
.. .. phosphor bronze	..	—	0.35
.. .. aluminum bronze	..	—	0.45
.. .. constantan	..	—	0.4
.. .. indium film deposited on steel	4 kg load, clean	—	0.08
.. ..	8 kg	—	0.04
.. .. silver	4 kg	—	0.1
.. ..	8 kg	—	0.07
.. .. lead film deposited on copper	4 kg	—	0.18
.. ..	8 kg	—	0.12
.. .. copper film deposited on steel	4 kg	—	0.3
.. ..	8 kg	—	0.2
Al on Al	in air or O_2	—	1.9
.. .. ..	.. H_2O vapor	—	1.1
Cu on Cu	.. H_2 or N_2	—	4.0
.. .. ..	.. air or O_2	—	1.6

(continued)

Materials	Surface Condition	Temperature °C	μ (Static)
A. STATIC FRICTION (Cont.)			
Metals (Cont.)			
Au on Au	in H_2 or N_2	—	4.0
.. .. ..	.. air or O_2	—	2.8
.. .. ..	.. H_2O vapor	—	2.5
Fe on Fe	.. in air or O_2	—	1.2
.. .. ..	.. H_2O vapor	—	1.2
Mo on Mo	.. air or O_2	—	0.8
.. .. ..	.. H_2O vapor	—	0.8
Ni on Ni	.. H_2 or N_2	—	5.0
.. .. ..	.. air or O_2	—	3.0
.. .. ..	.. H_2O vapor	—	1.6
Pt on Pt	.. air or O_2	—	3.0
.. .. ..	.. H_2O vapor	—	3.0
Ag on Ag	.. air or O_2	—	1.5
.. .. ..	.. H_2O vapor	—	1.5
Various Materials on Snow and Ice			
Ice on ice	clean	0	0.05–0.15
.. .. ..	..	−12	0.3
.. .. ..	..	−71	0.5
.. .. ..	..	−82	0.5
.. .. ..	..	−110	0.5
Polymethylmethylacrylate	on wet snow	0	0.5
..	.. dry ..	0	0.3
..	.. .. ..	−10	0.34
..	.. .. ..	−32	0.4
Polyester of teraphthalic acid and ethylene glycol	.. wet ..	0	0.5
Polyester of teraphthalic acid and ethylene glycol	.. dry ..	0	0.35
Polyester of teraphthalic acid and ethylene glycol	.. .. ..	−10	0.38
Nylon	on wet snow	0	0.4
..	.. dry ..	0	0.3
..	.. .. ..	−10	0.3
Polytetrafluoroethylene	.. wet ..	0	0.05
..	.. dry ..	0	0.02
..	.. .. ..	−10	0.08
..	.. .. ..	−32	0.1
Paraffin wax	.. wet ..	0	0.06
.. ..	.. dry ..	0	0.06
.. ..	.. .. ..	−10	0.35
.. ..	.. .. ..	−32	0.4
Swiss wax	.. wet ..	0	0.05
.. ..	.. dry ..	0	0.03
.. ..	.. .. ..	−10	0.2
.. ..	.. .. ..	−32	0.2
Ski wax	.. wet ..	0	0.1
.. ..	.. dry ..	0	0.04
.. ..	.. .. ..	−10	0.2
.. ..	.. .. ..	−32	0.2
.. laquer	.. wet ..	0	0.2
.. ..	.. dry ..	0	0.1
.. ..	.. .. ..	−10	0.4
.. ..	.. .. ..	−32	0.4
Aluminum	.. wet ..	0	0.4
..	.. dry ..	0	0.35
..	.. .. ..	−10	0.38

Materials	Surface Condition	Temperature °C	μ (Kinetic)
B. KINETIC FRICTION Various Materials			
Unwaxed hickory	4 m/sec on dry snow	−3	0.08
Waxed ..	0.1 m/sec .. wet ..	0	0.14
.. ..	0.1 m/sec .. dry ..	0	0.04
.. ..	0.1 m/sec	−3	0.09
.. ..	4 m/sec	−3	0.03
Waxed hickory	0.1 m/sec on dry snow	−10	0.18
.. ..	0.1 m/sec	−40	0.4
Ice on ice	4m/sec, clean	0	0.02
.. .. ..	.. . ..	−10	0.035
.. .. ..	.. . ..	−20	0.050
.. .. ..	.. . ..	−40	0.075
.. .. ..	.. . ..	−60	0.085
.. .. ..	.. . ..	−80	0.09
Ebonite	4m/sec on ice	0	0.02
..	.. .. ..	−10	0.05
..	.. .. ..	−20	0.065
..	.. .. ..	−40	0.085
..	.. .. ..	−60	0.10
..	.. .. ..	−80	0.11
Brass	4m/sec on ice	0	0.02
..	.. .. ..	−10	0.075
..	.. .. ..	−20	0.085
..	.. .. ..	−40	0.115
..	.. .. ..	−60	0.14
..	.. .. ..	−80	0.15
Natural rubber, vulcanized	100m/min on ground glass, clean	—	1.07
.. ..	100m/min .. .. .., wetted with water	—	0.94
.. ..	100m/min on concrete, clean	—	1.02
.. ..	100m/min .. .., wetted with water	—	0.97
.. ..	100m/min on bitumen, clean	—	1.07
.. ..	100m/min .. .., wetted with water	—	0.95
.. ..	100m/min on rubber flooring or rubber tread vulcanisate, clean	—	1.16
.. ..	100m/min on bitumen containing rubber powder, clean	—	1.15 (Varies with quantity of powder)
.. ..	100m/min on bitumen containing rubber powder, wetted with water	—	1.03

Friction factors for isothermal flow in round pipes and tubes

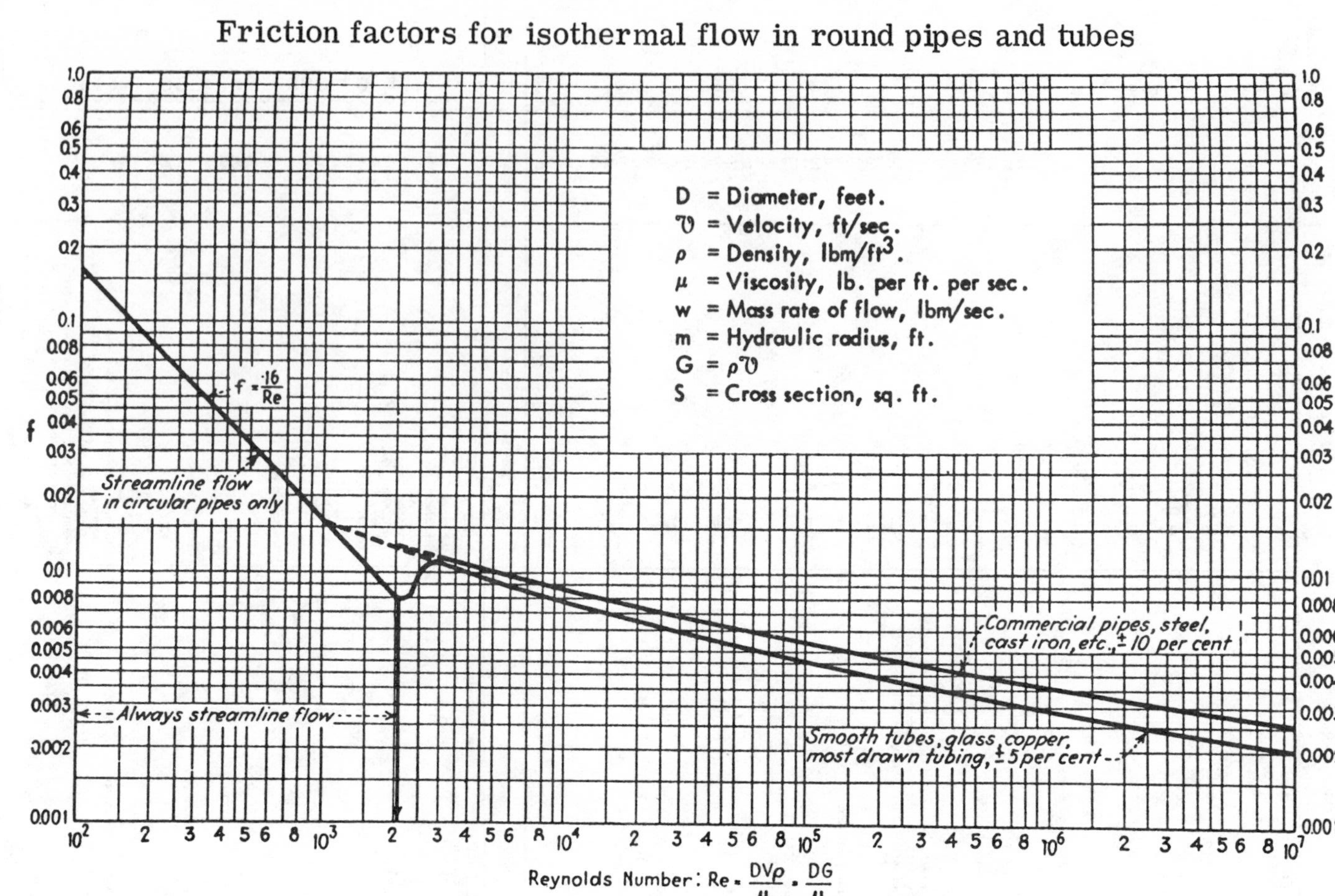

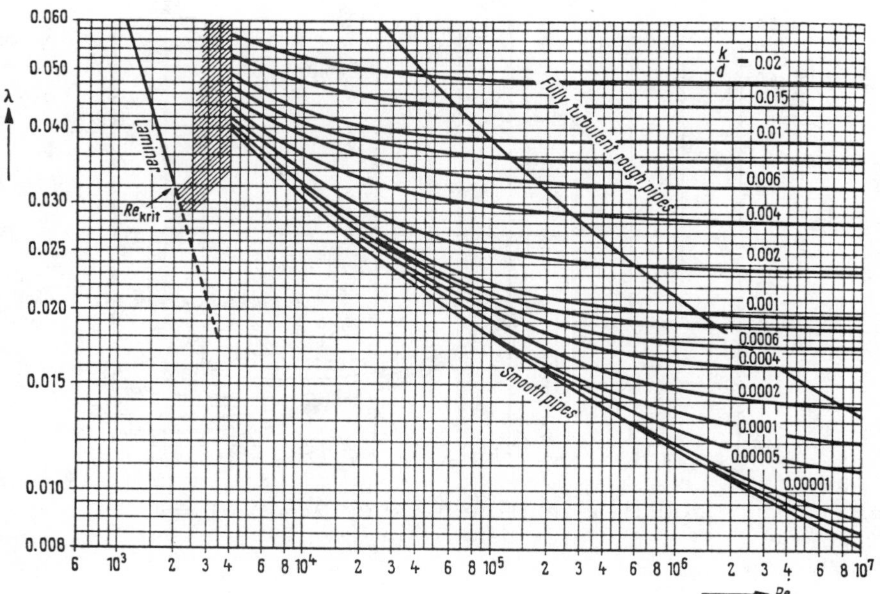

Pipe friction coeff. λ as a function of Reynolds number Re and relative roughness k/d.

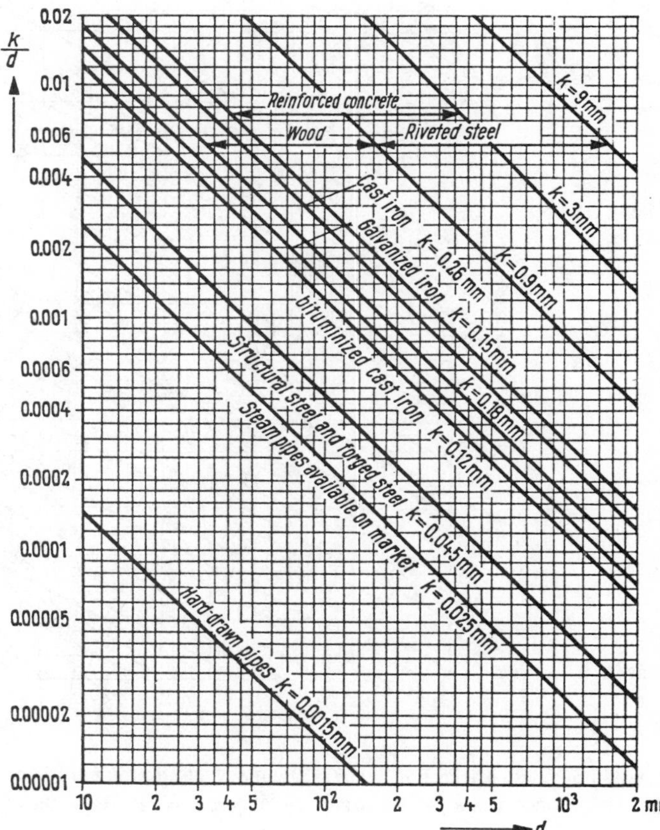

Relative pipe roughness k/d as a function of pipe diameter d.

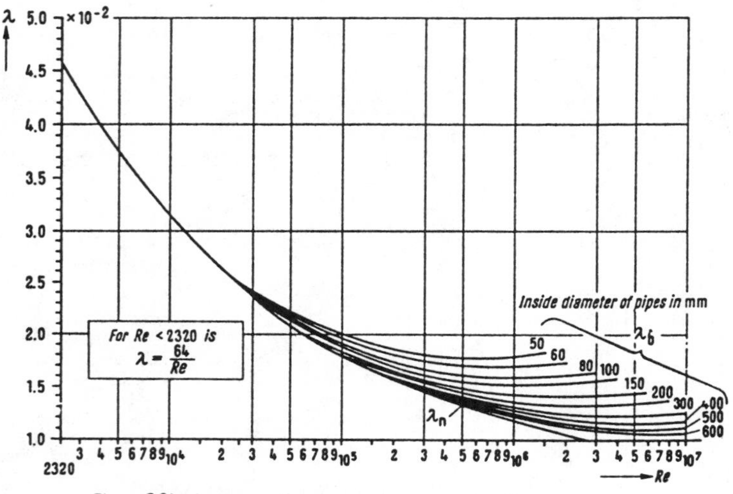

Coefficient of resistance λ for pipes

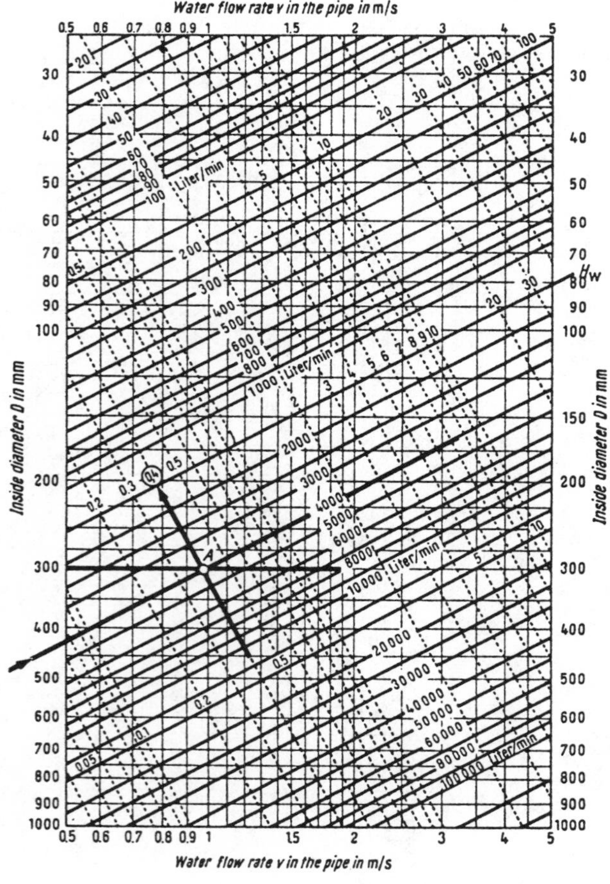

Determination of friction losses in standard pipes of 30 to 1000 mm inside diameter in the case of water
Example: At Q = 4000 1/min. and D = 300 mm; H_w = 0.4 m resistance head per 100 m pipe.

Coefficients of resistance (estimated) for round and rectangular pipes

No.	Mode of fluid flow	ζ	Remarks
A. Changes of direction			
1		1.5	Elbow joint, sharp corner, round or square cross section
2		0.5	Sharp elbow with deflector plates
3		0.2 to 0.25	Sharp elbow with fine mesh
4	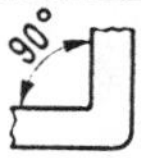	1.0	Elbow, corner slightly rounded off
5	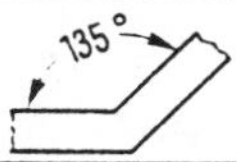	0.6	Elbow, corner left sharp
6	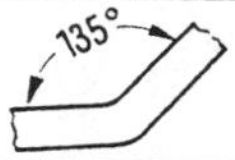	0.2	Elbow, corner rounded off
7	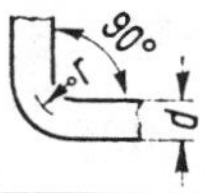	0.3 0.15 0.01	Angle of 90° $r=d$ $r\geqq 2d$ $r\geqq 6d$
8	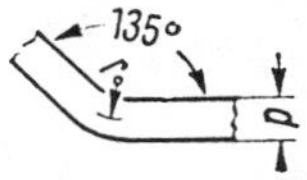	0.05	Angle 135° $r\geqq 2d$
9	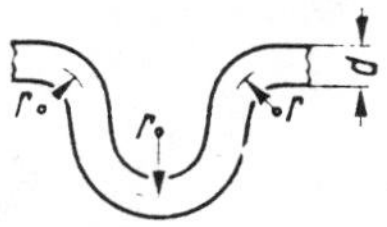	0.4 0	Bent-out section $r\geqq 3d$ $r\geqq 8d$
10	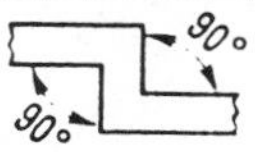	3.0	Two elbows, each 90°, closely adjacent
B. Branch pipes			
1	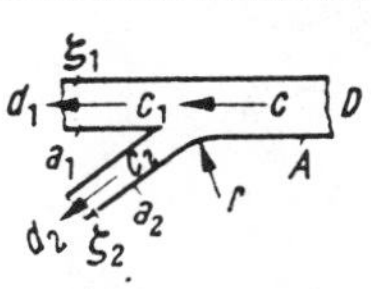		$\zeta_1=0$ $\approx$0.01 to 1.5, corresponding to cases A 1 to A 8 Only for branch pipes in which $c\approx c_1\approx c_2$ and $A\approx a_1+a_2$
2	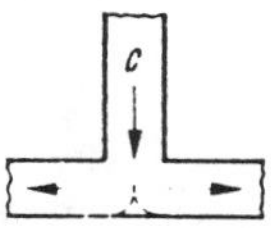	1.5 each	Branch pipe with air movement in two opposite directions

(continued)

No.	Mode of fluid flow	ζ	Remarks
B. Branch pipes (continued)			
3	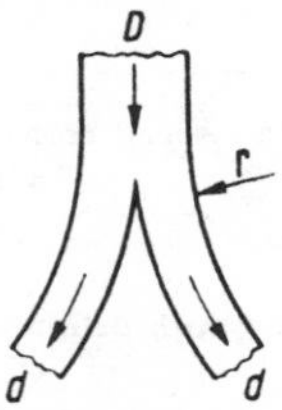	ζ corresponding to r from Case B 4	Y-piece
4	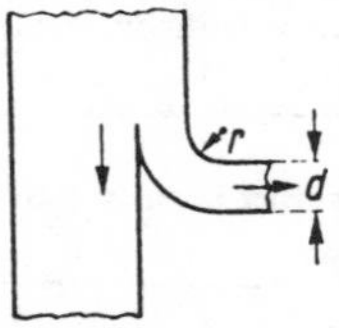		Rounded branch pipe $k = \frac{r}{d}$ $k=$ 0 / 0.5 / 2.0 / 4.0 / 6.0 $\zeta=$ 1.1 / 0.28 / 0.14 / 0.07 / 0.02
C. Variation in the cross section			
1	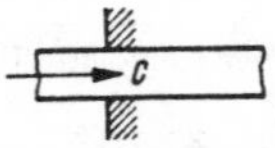	0.6	Sharp-edged inlet in pipe
2	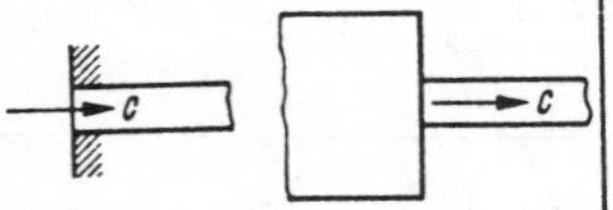	0.4	Sharp-edged inlet in pipe wall or change from large to small cross section
3	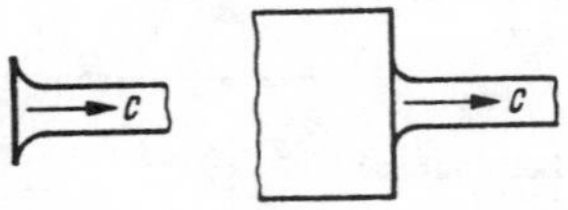	≈0	Inflow jet or change from large to small cross-section with inflow jet
4	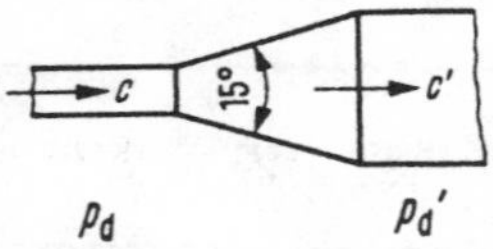	–	Diffuser loss $(p_d - p'_d)(1-\eta_D) = \frac{\gamma}{2g}(c^2 - c'^2)(1-\eta_D)$ Diffuser efficiency $\eta_D \approx 0.75$ for diffuser angle $\leq 15°$ with non-rotational flow
5	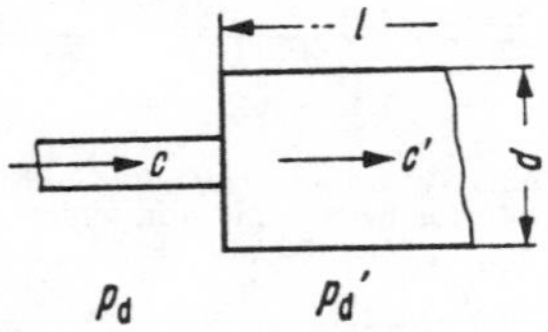		Impact loss $\frac{p}{2g}(c-c')^2$ where $l \leq 8d$, otherwise the loss will be greater

$k=$	0	0.5	2.0	4.0	6.0
$\zeta=$	1.1	0.28	0.14	0.07	0.02

Properties of Various Cross Sections

I = moment of inertia; I/c = section modulus; $r = \sqrt{I/A}$ = radius of gyration

Section	Moment of inertia	Section modulus	Radius of gyration
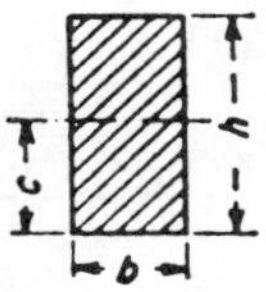 $I = \frac{bh^3}{12}$ $\frac{I}{c} = \frac{bh^2}{6}$ $r = \frac{h}{\sqrt{12}} = 0.289h$	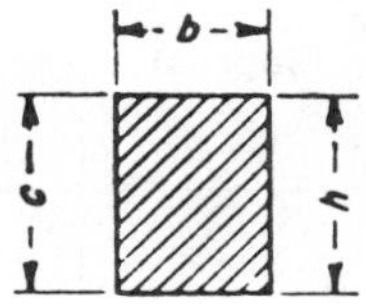 $\frac{bh^3}{3}$ $\frac{bh^2}{3}$ $\frac{h}{\sqrt{3}} = 0.577h$	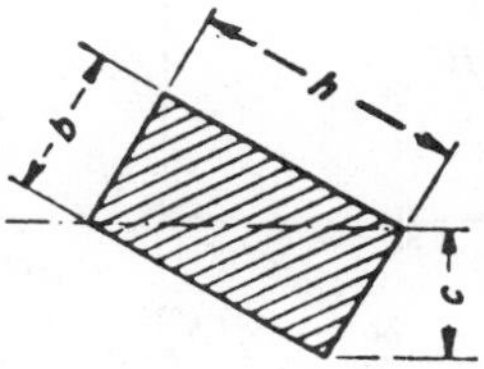 $\frac{b^3h^3}{6(b^2+h^2)}$ $\frac{b^2h^2}{6\sqrt{b^2+h^2}}$ $\frac{bh}{\sqrt{6(b^2+h^2)}}$	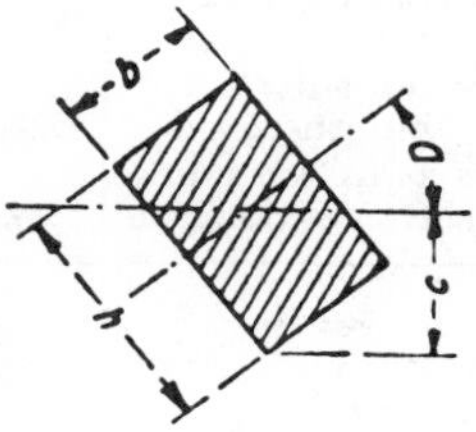 $\frac{bh}{12}(h^2\cos^2 a + b^2\sin^2 a)$ $\frac{bh}{6}\left(\frac{h^2\cos^2 a + b^2\sin^2 a}{h\cos a + b\sin a}\right)$ $\sqrt{\frac{h^2\cos^2 a + b^2\sin^2 a}{12}}$
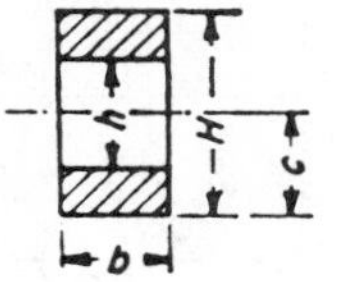 $I = \frac{b}{12}(H^3 - h^3)$ $\frac{I}{c} = \frac{b}{6}\frac{H^3 - h^3}{H}$ $r = \sqrt{\frac{H^3 - h^3}{12(H-h)}}$	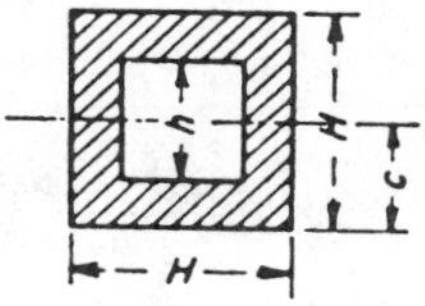 $\frac{H^4 - h^4}{12}$ $\frac{1}{6}\frac{H^4 - h^4}{H}$ $\sqrt{\frac{H^2 + h^2}{12}}$	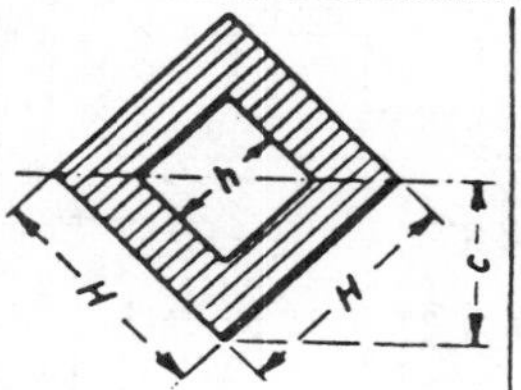 $\frac{H^4 - h^4}{12}$ $\frac{\sqrt{2}}{12}\frac{H^4 - h^4}{H}$ $\sqrt{\frac{H^2 + h^2}{12}}$	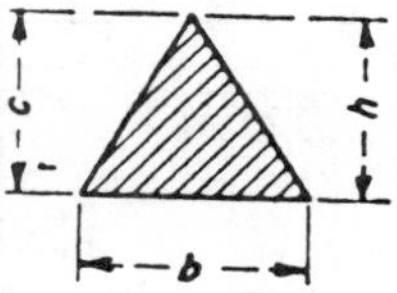 $\frac{bh^3}{36}$; $c = \frac{2}{3}h$ $\frac{bh^2}{24}$ $\frac{h}{\sqrt{18}}$
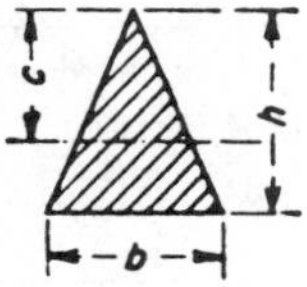 $I = \frac{bh^3}{12}$ $\frac{I}{c} = \frac{bh^2}{12}$ $r = \frac{h}{\sqrt{6}}$	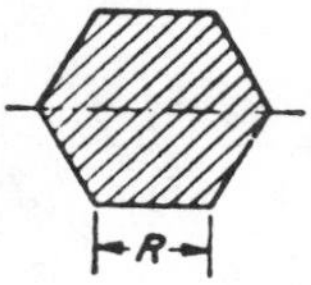 $\frac{5\sqrt{3}}{16}R^4$ $\frac{5}{8}R^3$	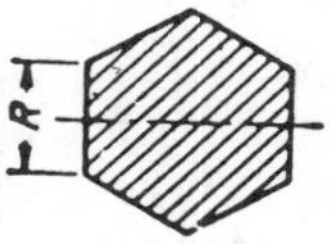 $\frac{5\sqrt{3}}{16}R^3$ $\sqrt{\frac{5}{24}}R$	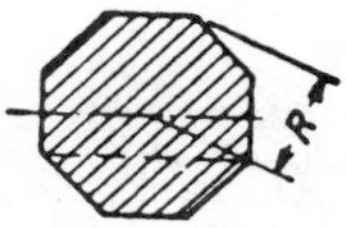 $\frac{1+2\sqrt{2}}{6}R^4$ $0.6906R^3$ $0.475R$

Properties of Various Cross Sections (continued)

Section	Moment of inertia	Section modulus	Radius of gyration
Equilateral polygon A = area R = radius circumscribed circle r = radius inscribed circle n = no. sides a = length of side Axis as in preceding section of octagon	$I = \frac{A}{24}(6R^2 - a^2)$ $= \frac{A}{48}(12r^2 + a^2)$ $= \frac{AR^2}{4}$ (approx)	$\frac{I}{c} = \frac{I}{r}$ $= \frac{I}{R \cos \frac{180°}{n}}$ $= \frac{AR}{4}$ (approx)	$\sqrt{\frac{6R^2 - a^2}{24}} \approx \frac{R}{2}$ $\sqrt{\frac{12r^2 + a^2}{48}}$
	$I = \frac{6b^2 + 6bb_1 + b_1^2}{36(2b + b_1)} h^3$ $c = \frac{1}{3} \frac{3b + 2b_1}{2b + b_1} h$	$\frac{I}{c} = \frac{6b^2 + 6bb_1 + b_1^2}{12(3b + 2b_1)} h^2$	$\frac{h\sqrt{12b^2 + 12bb_1 + 2b_1^2}}{6(2b + b_1)}$
	$I = \frac{BH^3 + bh^3}{12}$	$\frac{I}{c} = \frac{BH^3 + bh^3}{6H}$	$\sqrt{\frac{BH^3 + bh^3}{12(BH + bh)}}$
	$I = \frac{BH^3 - bh^3}{12}$	$\frac{I}{c} = \frac{BH^3 - bh^3}{6H}$	$\sqrt{\frac{BH^3 - bh^3}{12(BH - bh)}}$
	$I = \frac{1}{3}(Bc_1^3 - B_1h^3 + bc_2^3 - b_1h_1^3)$ $c_1 = \frac{1}{2} \frac{aH^2 + B_1d^2 + b_1d_1(2H - d_1)}{aH + B_1d + b_1d_1}$		$\sqrt{\frac{I}{(Bd + bd_1) + a(h + h_1)}}$
	$I = \frac{1}{3}(Bc_1^3 - bh^3 + ac_2^3)$ $c_1 = \frac{1}{2} \frac{aH^2 + bd^2}{aH + bd}$ $c_2 = H - c_1$ $r = \sqrt{\frac{I}{[Bd + a(H - d)]}}$		
	$I = \frac{\pi d^4}{64} = \frac{\pi r^4}{4} = \frac{A}{4} r^2$ $= 0.05d^4$ (approx)	$\frac{I}{c} = \frac{\pi d^3}{32} = \frac{\pi r^3}{4} = \frac{A}{4} r$ $= 0.1d^3$ (approx)	$\frac{r}{2} = \frac{d}{4}$

Properties of Various Cross Sections (continued)

Section	Moment of inertia	Section modulus	Radius of gyration
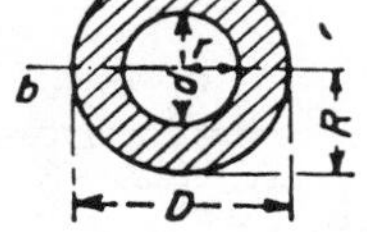 $d_m = \frac{1}{2}(D + d)$ $s = \frac{1}{2}(D - d)$	$I = \frac{\pi}{64}(D^4 - d^4)$ $= \frac{\pi}{4}(R^4 - r^4)$ $= \frac{1}{4}A(R^2 + r^2)$ $= 0.05(D^4 - d^4)$ (approx)	$\frac{I}{c} = \frac{\pi}{32}\frac{D^4 - d^4}{D}$ $= \frac{\pi}{4}\frac{R^4 - r^4}{R}$ $= 0.8d_m^2 s$ (approx) when $\frac{s}{d_m}$ is very small	$\frac{\sqrt{R^2 + r^2}}{2}$ $= \frac{\sqrt{D^2 + d^2}}{4}$
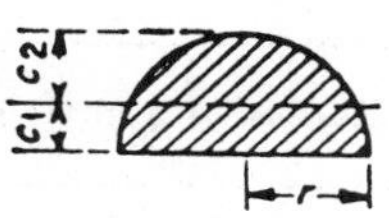	$I = r^4\left(\frac{\pi}{8} - \frac{8}{9\pi}\right)$ $= 0.1098r^4$	$\frac{I}{c_2} = 0.1908r^3$ $\frac{I}{c_1} = 0.2587r^3$ $c_1 = 0.4244r$	$\frac{\sqrt{9\pi^2 - 64}}{6\pi} r = 0.264r$
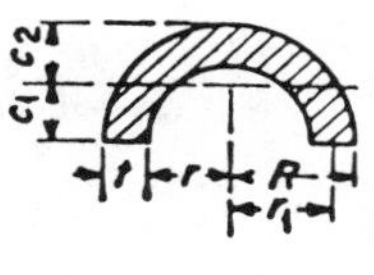	$I = 0.1098(R^4 - r^4)$ $- \frac{0.283R^2r^2(R - r)}{R + r}$ $= 0.3tr_1^3$ (approx) when $\frac{t}{r_1}$ is very small	$c_1 = \frac{4}{3\pi}\frac{R^2 + Rr + r^2}{R + r}$ $c_2 = R - c_1$	$\sqrt{\frac{2I}{\pi(R^2 - r^2)}}$ $= 0.31r_1$ (approx)
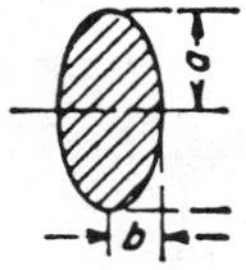	$I = \frac{\pi a^3 b}{4} = 0.7854a^3b$	$\frac{I}{c} = \frac{\pi a^2 b}{4} = 0.7854a^2b$	$\frac{a}{2}$
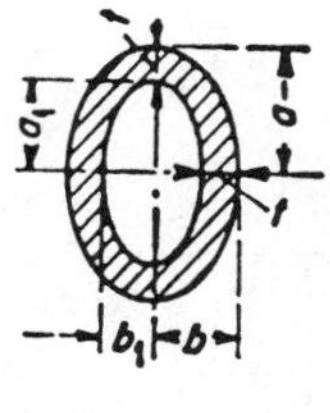	$I = \frac{\pi}{4}(a^3b - a_1^3b_1)$ $= \frac{\pi}{4}a^2(a + 3b)t$ (approx)	$\frac{I}{c} = \frac{\pi}{4}a(a + 3b)t$ (approx)	$\sqrt{\frac{I}{(\pi ab - a_1b_1)}} =$ $\frac{a}{2}\sqrt{\frac{a + 3b}{a + b}}$ (approx)
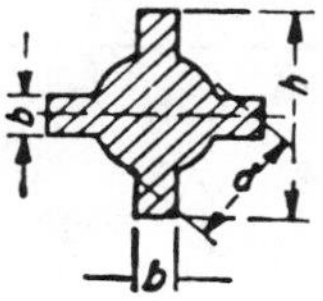	$I = \frac{1}{12}\left[\frac{3\pi}{16}d^4 + b(h^3 - d^3) + b^3(h - d)\right]$ $\frac{I}{c} = \frac{1}{6h}\left[\frac{3\pi}{16}d^4 + b(h^3 + d^3) + b^3(h - d)\right]$		$\sqrt{\frac{I}{\pi\frac{d^2}{4} + 2b(h - d)}}$ (approx)

Properties of Various Cross Sections (continued)

Section	Moment of inertia	Section modulus	Radius of gyration
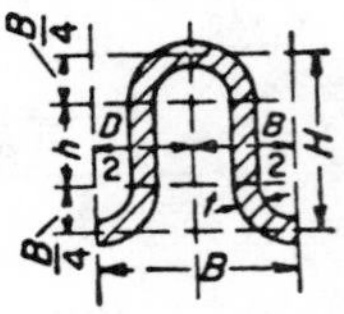	$I = \frac{t}{4}\left(\frac{\pi B^3}{16} + B^2h + \frac{\pi B h^2}{2} + \frac{2}{3}h^3\right)$ $h = H - \frac{1}{2}B$ $\frac{I}{c} = \frac{2I}{H+t}$		$\sqrt{\frac{I}{2\left(\frac{\pi B}{4} + h\right)t}}$

Section	Moment of inertia and section modulus	Radius of gyration
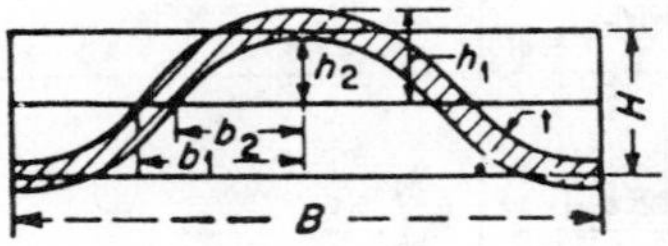 Corrugated sheet iron, parabolically curved	$I = \frac{64}{105}(b_1h_1^3 - b_2h_2^3)$, where $h_1 = \frac{1}{2}(H + t)$ | $b_1 = \frac{1}{4}(B + 2.6t)$ $h_2 = \frac{1}{2}(H - t)$ | $b_2 = \frac{1}{4}(B - 2.6t)$ $\frac{I}{c} = \frac{2I}{H+1}$	$r = \sqrt{\frac{3I}{t(2B + 5.2H)}}$

Approximate Values of Least Radius of Gyration r

	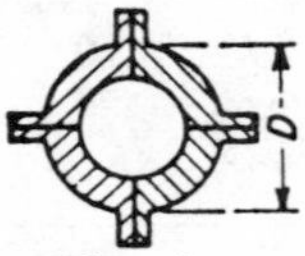 Phoenix column	 Carnegie Z-bar column	 I beam	 Channel	 Deck beam
$r =$	$0.3636D$	$0.295D$	$D/4.58$	$D/3.54$	$D/6$

	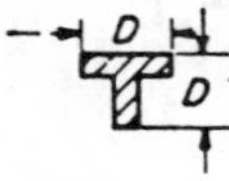 T beam	 Angle, equal legs	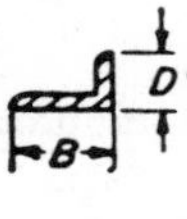 Angle, unequal legs	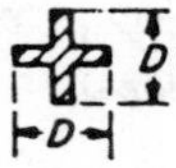 Cross
$r =$	$D/4.74$	$D/5$	$BD/2.6(B + D)$	$D/4.74$

NOTE: Square, axis same as first rectangle: side $= h$, $I = h^4/12$, $I/c = h^3/6$, $r = 0.289h$.
Square, diagonal taken as axis: $I = h^4/12$, $I/c = 0.1179h^3$, $r = 0.289h$.

Torsion of Shafts of Various Cross Sections

G = Shear Modulus of Elasticity, psi

Cross section	Torsional resisting moment M_t	Angular deflection, a_1 (length = 1 in., radius = 1 in.)		Work of torsion (V = volume)
		In terms of torsional moment	In terms of max shear	
d	$\frac{\pi}{16} d^3 S_v$	$\frac{M_t}{GJ} = \frac{32}{\pi d^4}\frac{M_t}{R}$	$2\frac{S_{v\max}}{G}\frac{1}{d}$	$\frac{1}{4}\frac{S^2_{v\max}}{G}V$ (Note 1)
d, D	$\frac{\pi}{16}\frac{D^4 - d^4}{D}S_v$	$\frac{32}{\pi(D^4 - d^4)}\frac{M_t}{G}$	$2\frac{S_{v\max}}{G}\frac{1}{D}$	$\frac{1}{4}\frac{S^2_{v\max}}{G}\frac{D^2 + d^2}{D^2}V$ (Note 2)
B, A, h, b	$\frac{\pi}{16} b^2hS_v$ ($h > b$)	$\frac{16}{\pi}\frac{b^2 + h^2}{b^3h^3}\frac{M_t}{G}$	$\frac{S_{v\max}}{G}\frac{b^2 + h^2}{bh^2}$	$\frac{1}{8}\frac{S^2_{v\max}}{G}\frac{b^2 + h^2}{h^2}V$ (Note 3)
h, b	$\frac{2}{9}b^2hS_v$ ($h > b$)	$3.6\frac{b^2 + h^2}{b^3h^3}\frac{M_t}{G}$ *	$0.8\frac{S_{v\max}}{G}\frac{b^2 + h^2}{bh^2}$ *	$\frac{4}{45}\frac{S^2_{v\max}}{G}\frac{b^2 + h^2}{h^2}V$ (Note 4)
h, h	$\frac{2}{9}h^3S_v$	$7.2\frac{1}{h^4}\frac{M_t}{G}$	$1.6\frac{S_{v\max}}{G}\frac{1}{h}$	$\frac{8}{45}\frac{S^2_{v\max}}{G}V$ (Note 5)
b, b, b	$\frac{b^3}{20}S_v$	$4.62\frac{1}{b^4}\frac{M_t}{G}$	$2.31\frac{S_{v\max}}{G}\frac{1}{b}$	
b, b	$\frac{b^3}{1.09}S_v$	$0.967\frac{1}{b^4}\frac{M_t}{G}$	$0.9\frac{S_{v\max}}{G}\frac{1}{b}$	

* When	h/b = 1	2	4	8
Coefficient 3.6 becomes	= 3.56	3.50	3.35	3.21
Coefficient 0.8 becomes	= 0.79	0.78	0.74	0.71

NOTES: (1) $S_{v\max}$ at circumference. (2) $S_{v\max}$ at outer circumference. (3) $S_{v\max}$ at A; $S_{vB} = 16M_t/\pi bh^2$. (4) $S_{v\max}$ at middle of side h; in middle of b, $S_v = 9M_t/2bh^2$. (5) $S_{v\max}$ at middle of side.

Compositions of Typical High-temperature Alloys

Alloy	C	Cr	Ni	Mo	Co	W	Cb	Ti	Al	Fe	Other
Ferritic Steels											
1.25 Cr, Mo	.10	1.25	....	0.50	....	...	....	....	....	Balance	
5 Cr, Mo	.20	5.00	....	0.50	....	...	....	....	....	Balance	
"17-22-A" S	.30	1.25	....	0.50	....	...	....	....	....	Balance	
410	.10	12.0	....	.....	....	...	....	....	....	Balance	
Austenitic Steels											
316	.08	17.0	12.0	2.50	....	...	....	....	....	Balance	
347	.06	18.0	12.0	.....	....	...	0.70	....	....	Balance	
16-25-6	.10	16.0	25.0	6.00	....	...	....	....	....	Balance	
A-286	.05	15.0	26.0	1.25	....	...	....	1.95	0.20	Balance	
Nickel-base Alloys											
Inconel	.04	15.5	76.0	.....	....	...	....	....	....	7.0	
Inconel X	.04	15.0	75.0	.....	....	...	....	2.5	0.6	7.0	
Nimonic 90	.08	20.0	58.0	.....	16	...	....	2.3	1.4	0.5	
Hastelloy B	.10	1.0	65.0	28	....	...	....	....	....	5.0	
René 41	.10	19.0	53.0	10	11	...	....	3.2	1.6	2.0	
Udimet 500	.10	19.4	55.6	4	14	...	....	2.9	2.9	0.6	
Cobalt-base Alloys											
Vitallium (HS-21)	.25	27.0	3.0	5	62	...	....	....	....	1.0	
X-40 (HS-31)	.40	25.0	10.0	.....	55	8	....	....	....	1.0	
Complex Superalloys											
N-155 (Multimelt)	.15	21.0	20.0	3	20	2.5	1.0	....	....	Balance	0.15N
S-590	.40	20.0	20.0	4	20	4.0	4.0	....	....	Balance	
S-816	.40	20.0	20.0	4	Bal.	4.0	4.0	....	....	3.0	
K 42 B	.05	18.0	43.0	.....	22	...	....	2.5	0.2	13	
Refractaloy 26	.05	18.0	37.0	3	20	...	....	2.8	0.2	18	

Relative Characteristics of Centrifugal Fans

Characteristic	Backward	Radial	Forward
First cost	High	Medium	Low
Efficiency	High	Medium	Poor
Stability of operation	Good	Good	Poor
Space required	Medium	Medium	Small
Tip speed	High	Medium	Low
Resistance to abrasion	Medium	Good	Poor

Wire and Sheet-metal Gauge Equivalents

Values in Approximate Decimals of an Inch.

As a number of gauges are in use for various shapes and metals, it is advisable to state the thickness in thousandths when specifying gauge number.

Metric wire gauge is ten times the diameter in millimeters.

Gauge no.	(1)*	(2)*	(3)*	(4)*	(5)*	(6)*	Gauge no.
0000000	.....	0.4900	.....	.....	0.6666	0.500	0000000
000000	.....	.4615	.....	.....	.6250	.464	000000
00000	.....	.4305	.....	.....	.5883	.432	00000
0000	0.460	.3938	0.454	.....	.5416	.400	0000
000	.410	.3625	.425	.....	.5000	.372	000
00	.365	.3310	.380	.....	.4452	.348	00
0	.325	.3065	.340	.....	.3964	.324	0
1	.289	.2830	.300	.....	.3532	.300	1
2	.258	.2625	.284	.....	.3147	.276	2
3	.229	.2437	.259	0.239	.2804	.252	3
4	.204	.2253	.238	.224	.2500	.232	4
5	.182	.2070	.220	.209	.2225	.212	5
6	.162	.1920	.203	.194	.1981	.192	6
7	.144	.1770	.180	.179	.1764	.176	7
8	.128	.1620	.165	.164	.1570	.160	8
9	.114	.1483	.148	.150	.1398	.144	9
10	.102	.1350	.134	.135	.1250	.128	10
11	.091	.1205	.120	.120	.1113	.116	11
12	.081	.1055	.109	.105	.0991	.104	12
13	.072	.0915	.095	.090	.0882	.092	13
14	.064	.0800	.083	.075	.0785	.080	14
15	.057	.0720	.072	.067	.0699	.072	15
16	.051	.0625	.065	.060	.0625	.064	16
17	.045	.0540	.058	.054	.0556	.056	17
18	.040	.0475	.049	.0478	.0495	.048	18
19	.036	.0410	.042	.0418	.0440	.040	19
20	.032	.0348	.035	.0359	.0392	036	20
21	.0285	.0317	.032	.0329	.0349	.032	21
22	.0253	.0286	.028	.0299	.0313	.028	22
23	.0226	.0258	.025	.0269	.0278	.024	23
24	.0201	.0230	.022	.0239	.0248	.022	24
25	.0179	.0204	.020	.0209	.0220	.020	25
26	.0159	.0181	.018	.0179	.0196	.018	26
27	.0142	.0173	.016	.0164	.0175	.0164	27
28	.0126	.0162	.014	.0149	.0156	.0148	28
29	.0113	.0150	.013	.0135	.0139	.0136	29
30	.0100	.0140	.012	.0120	.0123	.0124	30
31	.0089	.0132	.010	.0105	.0110	.0116	31
32	.0080	.0128	.009	.0097	.0098	.0108	32
33	.0071	.0118	.008	.0090	.0087	.0100	33
34	.0063	.0104	.007	.0082	.0077	.0092	34
35	.0056	.0095	.005	.0075	.0069	.0084	35
36	.0050	.0090	.004	.0067	.0061	.0076	36
37	.0045	.0085	.....	.0064	.0054	.0068	37
38	.0040	.0080	.....	.0060	.0048	.0060	38
39	.0035	.0075	.....	.....	.0043	.0052	39
40	.0031	.0070	.....	.....	.0039	.0048	40

* Gauges are arranged in columns as follows:

1. American (Awg) or Brown & Sharpe (B.&S.), for nonferrous wire and sheet; sometimes used for iron wire.
2. U.S. Steel Wire, Washburn & Moen, Roebling, or American Steel & Wire Co., for steel wire.
3. Birmingham, B.W.G., for steel wire and heat-exchanger tubing, or Stubs Iron Wire, for iron or brass wire; sometimes used for copper plate and for steel plate 12 gauge and heavier and for steel tubes.
4. U.S. Standard, for sheet and plate metal and wrought iron.
5. Standard Birmingham, B.G., for sheet and hoop metal.
6. Imperial Standard Wire Gauge, S.W.G., British legal standard for sheet metal.

Tap-drill Sizes for American Standard Screw Threads*

Size, no. or in.	Coarse-thread series		Fine-thread series		Size, no. or in.	Coarse-thread series		Fine-thread series	
	Thds/in.	Drill size	Thds/in.	Drill size		Thds/in.	Drill size	Thds/in.	Drill size
0	...	......	80	3/64	3/4	10	21/32	16	11/16
1	64	No. 53	72	No. 53	7/8	9	49/64	14	13/16
2	56	No. 50	64	No. 50	1	8	7/8	14	15/16
3	48	No. 47	56	No. 45	1 1/8	7	63/64	12	1 3/64
4	40	No. 43	48	No. 42	1 1/4	7	1 7/64	12	1 11/64
5	40	No. 38	44	No. 37	1 3/8	6	1 7/32	12	1 19/64
6	32	No. 36	40	No. 33	1 1/2	6	1 21/64	12	1 27/64
8	32	No. 29	36	No. 29	1 3/4	5	1 35/64		
10	24	No. 25	32	No. 21	2	4 1/2	1 25/32		
12	24	No. 16	28	No. 14	2 1/4	4 1/2	2 1/32		
1/4	20	No. 7	28	No. 3	2 1/2	4	2 1/4		
5/16	18	F	24	I	2 3/4	4	2 1/2		
3/8	16	5/16	24	Q	3	4	2 3/4		
7/16	14	U	20	25/64	3 1/4	4	3		
1/2	13	27/64	20	29/64	3 1/2	4	3 1/4		
9/16	12	31/64	18	33/64	3 3/4	4	3 1/2		
5/8	11	17/32	18	37/64	4	4	3 3/4		

* The sizes listed are the commercial tap drills to produce approx 75 per cent full thread.

Standard Pipe

Nominal diameter, in.	Actual external diameter, in.	Approximate internal diameter, in.	Nominal weight per foot, lb_m
1/8	0.405	0.27	0.24
1/4	0.540	0.36	0.42
1/2	0.840	0.62	0.85
1	1.315	1.05	1.68
1 1/2	1.900	1.61	2.72
2	2.375	2.07	3.65
4	4.500	4.03	10.79
8	8.625	8.07	24.69
10	10.75	10.19	31.20
12	12.75	12.09	43.77

Extra Strong Pipe

Nominal diameter, in.	Actual external diameter, in.	Approximate internal diameter, in.	Nominal weight per foot, lb_m
1/8	0.405	0.21	0.31
1/4	0.540	0.29	0.54
1/2	0.840	0.54	1.09
1	1.315	0.95	2.17
1 1/2	1.900	1.49	3.63
2	2.375	1.93	5.02
4	4.500	3.82	14.98
8	8.625	7.63	43.34
10	10.750	9.75	54.73
12	12.750	11.75	65.41

Common Configurations of Rolling Element Bearings

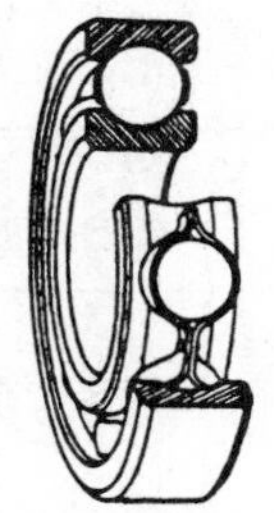

Single-row deep-groove ball bearing without filling slot

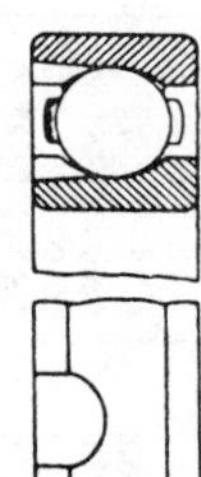

Single-row ball bearing with filling slot

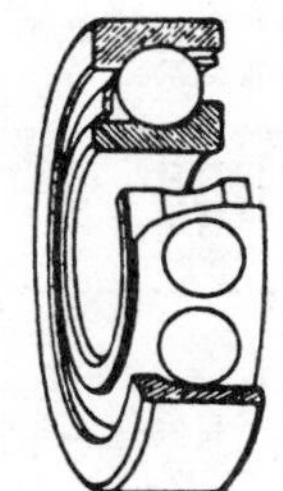

Single-row angular-contact ball bearing

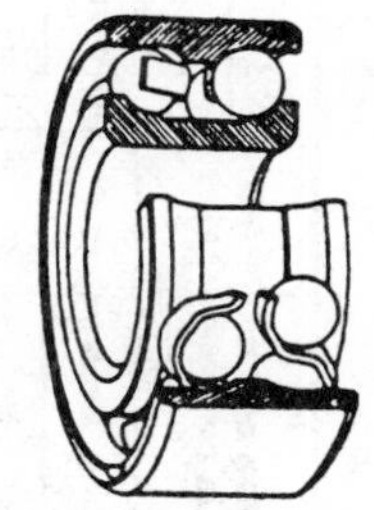

Double-row angular-contact ball bearing

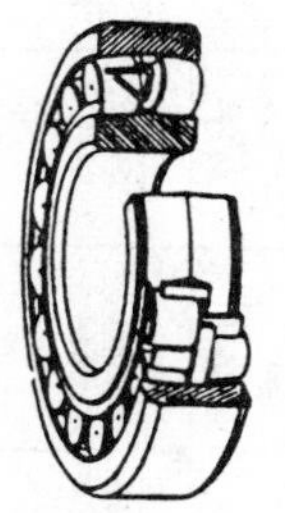

Spherical roller bearing with separate guide flange

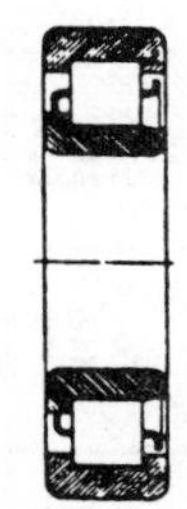

Type NJ Cylindrical-roller bearings

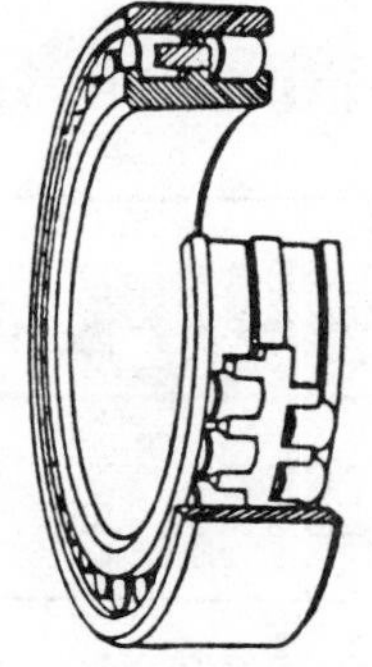

Double-row cylindrical-roller bearing

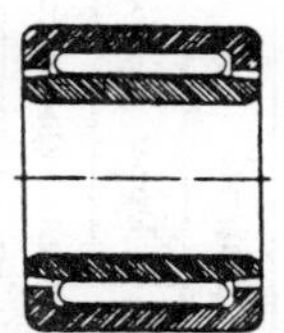

Needle bearing

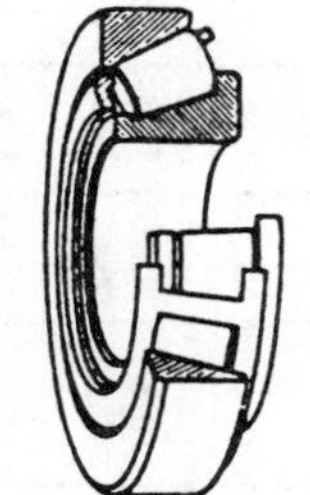

Tapered-roller bearing with large contact angle

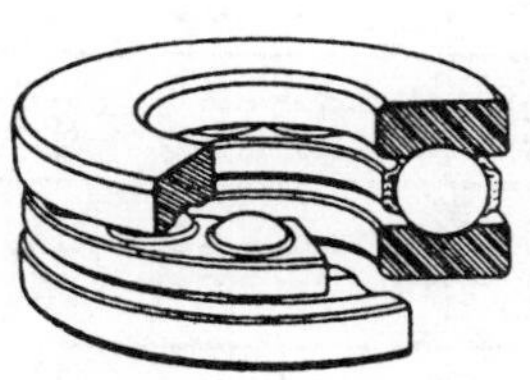

One direction thrust ball bearing

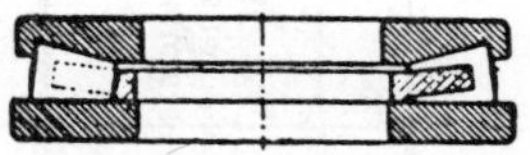

Tapered roller thrust bearing

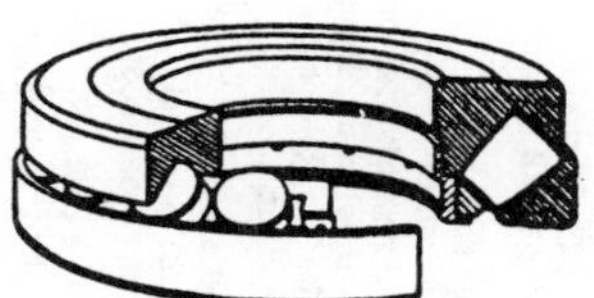

Spherical roller thrust bearing

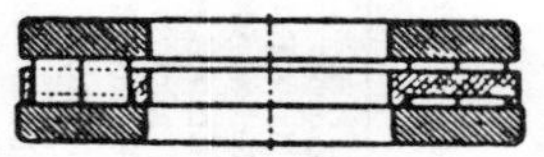

Cylindrical roller thrust bearing

Calorific Values of Selected Solid Fuels

Solide fuels	Referred to water and ashfree material								Water content in raw coal	Water ballast	Ash content in raw coal	Ash ballast	Upper	Lower
	Volatile constit-uents	C	H	O	N	S	Upper	Lower	Mean values		Mean values		cal. value in kcal/kg (in raw coal)	
							cal. value in kcal/kg							
	%	%	%	%	%	%	H_o	H_u	%	kg/kg	%	kg/kg	H_o	H_u
Lignite														
Halle-Bitterfeld	57.5	71.9	5.6	18.3	0.8	3.4	7125	6830	15.1	0.2057	11.5	0.1567	5230	4925
Schwandorf	55.25	63.63	5.10	26.08	1.27	3.92	6049	5780	53.1	1.5436	12.5	0.3634	2081	1678
Rhineland	55	68.3	5.0	27.5	0.5	0.5	6300	6037	60.0	1.6086	2.7	0.0724	2350	1900
Yallourn	51.4	67.58	4.76	26.86	0.54	0.26	6152	5900	66.5	2.0336	0.8	0.0245	2012	1540
Fohnsdorf	47.2	72.5	5.4	17.5	17.5	4.6	7270	6986	9.0	0.1268	20.0	0.2817	5162	4623
Moscow	46	67.5	5.2	19.5	1.3	6.5	6750	6476	33.0	0.7143	20.8	0.4502	3119	2799
Bituminous coal														
Bavaria	52	73.9	5.5	15.0	1.4	4.2	7365	7075	10.0	0.1299	13.0	0.1688	5671	5389
Open-burning coal														
Illinois, Madison	46.7	77.0	5.4	10.6	1.3	5.7	7850	7566	12.5	0.1673	12.8	0.1714	5864	5579
Scotland	41.5	81.4	5.4	10.3	2.1	0.8	8111	7827	13.8	0.1691	4.6	0.0564	6619	6306
W. Midlands	39.6	80.5	5.5	11.9	1.4	0.7	7883	7593	8.6	0.0994	4.9	0.0566	6819	6518
Ruhr nut coal	36.8	82.84	5.15	9.34	1.78	0.89	8203	7932	4.5	0.0503	6.0	0.0670	7342	7073
Saar small coal	36.6	85.00	5.42	7.24	1.22	1.21	8412	8127	9.0	0.1084	8.0	0.0964	6982	6693
Gas coal														
Yorkshire	34.4	82.3	5.2	8.0	1.7	0.8	8600	8326	2.0	0.0219	6.8	0.0746	7843	7581
Ruhr small coal	33.7	85.91	5.51	6.20	1.63	0.75	8417	8127	9.0	0.1084	8.0	0.0964	6986	6694
Pennsylvania, Fayette	33.4	85.55	5.23	6.24	1.52	1.46	8579	8304	3.9	0.0450	9.5	0.1097	7429	7168
Virginia, Dickenson	31.2	87.38	5.30	4.58	1.70	0.44	8608	8329	3.2	0.0351	5.6	0.0614	7850	7577
Durham	29.4	87.8	5.3	4.6	1.4	0.9	8700	8421	2.6	0.0287	6.9	0.0762	7874	7606
Fat coal														
Ruhr nut coal	27	86.88	4.80	5.82	1.58	0.92	8548	8295	4.5	0.0503	6.0	0.0670	7650	7398
Ruhr small coal	24.4	88.70	4.90	4.14	1.60	0.66	8636	8378	9.0	0.1084	8.0	0.0964	7168	6901
Forge coal														
Ruhr nut coal	15.4	90.20	4.33	3.22	1.58	0.67	8645	8417	4.8	0.0539	6.2	0.0697	7694	7463
Hard coal														
Ruhr nut coal	12.4	90.70	3.98	2.52	1.50	1.30	8644	8434	4.5	0.0503	6.0	0.0670	7736	7522
Ruhr small coal	10.5	90.85	3.84	2.73	1.74	0.84	8614	8412	9.0	0.1084	8.0	0.0964	7150	6929
Anthracite														
Ruhr nut coal	7.7	91.80	3.56	2.55	1.38	0.71	8576	8389	4.5	0.0503	6.0	0.0670	7676	7482
Pennsylvania (Schuylkill)	2.45	93.67	2.34	2.30	0.80	0.89	8314	8191	2.8	0.0324	10.8	0.1250	7183	7061

Calorific Values of Selected Liquid Fuels

Type	Density	Composition in %					Upper	Lower
							cal. value in kcal/kg	
	g/ml	C	H	O	N	S	H_o	H_u
Fuel oils								
Extra light (EL)	0.840	85.9	13.0	0.4	0.4	0.7	10 880	10 200
Light (L)	0.880	85.5	12.5	0.8	0.8	1.2	10 700	10 050
Medium (M)	0.920	85.3	11.6	0.6	0.6	2.5	10 350	9 725
Heavy (H)	0.970	84.0	11.0	1.11	0.39	3.5	10 200	9 600
Extra heavy (EH)	—	—	—	—	—	—	—	9 200
Coal-tar oil	1.02—1.10	89.8	6.5	1.7	1.2	0.8	9 300	9 000

Calorific Values of Selected Gaseous Fuels

Type	Density [1])	Composition in Vol. %									Upper	Lower
											cal. value in kcal/kg	
	kg/m³	CO	H_2	CH_4	C_2H_6	C_3H_8	C_2H_4	H_2S	CO_2	N_2	H_o	H_u
Natural gas												
USA (Calif.)	0.850	—	—	86.8	7.2	4.3	—	—	0.5	—	10 900	9860
USA (Texas, stripped) . .	0.775	—	—	89.8	2.3	—	—	—	0.2	7.7	8 930	8030
Germany (Bentheim) .	0.754	—	—	93.2	0.6	—	—	—	—	6.2	8 960	8060
Austria . .	0.751	—	0.7	94.7	1.8	0.2	—	—	1.4	1.2	9 380	8440
Italy (Corte-maggiore) .	0.766	—	—	91.8	5.1	—	—	—	—	3.1	9 590	8630
France (Lacq, raw)	1.034	—	—	69.52	3.20	1.42	—	15.30	9.60	—	8 740	7900
France (Lacq, pure)	0.746	—	—	95.9	3.2	0.5	—	—	—	0.4	9 780	8800
Sahara (Hassi R'Mel)	0.928	—	—	81.3	6.8	2.3	—	—	0.5	4.8	11 040	9990
USSR (Saratow) .	0.772	—	—	93.1	2.5	1.5	—	—	0.6	2.3	9 640	8680
Town gas (mixed) . .	0.591	21.5	51.5	17.0	—	—	2.0	—	4.0	4.0	4 140	3710
Water gas .	0.705	40.0	50.0	0.3	—	—	—	—	5.0	4.7	2 760	2520
Blast-furnace gas	1.287	31.0	2.3	0.3	—	—	—	—	9.0	57.4	1 035	1020

[1]) Volume in m³ at $t = 0$ °C and $p = 760$ mm Hg.

Heat-transfer Coefficients for Selected Exchangers

Designation	Heat transfer coeff. k kcal/h m² degree approx
De-superheater & superheater (steam/steam)	100 to 200
De-superheater (built in or separate) for cooling steam with feedwater	180 to 700
Heat-exchanger (vapour/liquid)	
h.p. heater	2500 to 3600
l.p. heater	1500 to 2500
vacuum heater.	900 to 1500
Heat-exchanger (liquid/liquid) h.p. condensate cooler	
built in	600 to 1500 acc. to design
separate.	1400 to 2100
L.p. condensate cooler	
built in	400 to 700
separate.	600 to 1400

CHAPTER 23

STRUCTURAL DESIGN

Yield Points for Structural Steels

ASTM Specification	Designation	Specified minimum yield point F_y, psi*
"Structural Steel"	A36	36,000
"Structural Steel"	A529	42,000
"High-strength Low-alloy Structural Steel"	A242	42,000
"High-strength Structural Steel"	A440	42,000
"High-strength Low-alloy Structural Magnesium Vanadium Steel"	A441	42,000
"High-strength Low-alloy Structural Columbium Vanadium Steel, Grade 42"	A572	42,000
"High-strength Low-alloy Structural Steel with 50,000 psi Minimum Yield Point"	A588	50,000†
"High-Yield Strength, Quenched and Tempered Alloy Steel Plate"	A514	90,000

* Values given are for heavy sections and for plates 1½ to 4 in. thick; higher values may be allowed for light sections and thinner plates and lower values for thicker plates.
† For sections weighing more than 600 lb/ft, F_y = 42,000 psi.

Average Properties of Structural Materials

Material	Modulus of Elasticity, E (lb/in.^{2} × 10^6)	Shear Modulus of Elasticity, G (lb/in.^{2} × 10^6)	Poisson's Ratio, μ	Weight Density (lb/in.^{3})
Aluminum alloys	10.2	3.9	0.33	0.10
Beryllium copper	18.0	7.0	0.29	0.30
Carbon steel	29.0	11.5	0.29	0.28
Cast iron	14.5	6.0	0.21	0.26
Inconel	31.0	11.5	0.29	0.31
Magnesium	6.5	2.4	0.35	0.07
Molybdenum	48.0	17.1	0.31	0.37
Monel metal	26.0	9.5	0.32	0.32
Nickel silver	18.5	7.0	0.32	0.32
Nickel steel	29.0	11.0	0.29	0.28
Nylon	1.5	0.6	—	0.04
Phosphor bronze	16.1	6.0	0.35	0.30
Stainless steel	27.6	10.6	0.31	0.28
Titanium	16.5	6.5	—	0.16

Strength Properties of Iron and Steel

Material	Ultimate Strength: Tension, Thousands of Pounds per Square Inch, *T*	Ultimate Strength: Compression, in terms of *T*	Ultimate Strength: Shear, in terms of *T*	Yield Point, Thousands of Pounds per Square Inch	Modulus of Elasticity, in Tension, Millions of psi, *E*	Modulus of Elasticity, in Shear,[b] in terms of *E*
Cast iron, gray, class 20..	20[a]	3.6 *T* to 4.4 *T*	1.6 *T*	.....	11.6	0.40 *E*
class 25..............	25[a]	3.6 *T* to 4.4 *T*	1.4 *T*	.....	14.2	0.40 *E*
class 30..............	30[a]	3.7 *T*	1.4 *T*	.....	14.5	0.40 *E*
class 35..............	35[a]	3.2 *T* to 3.9 *T*	1.4 *T*	.....	16.0	0.40 *E*
class 40..............	40[a]	3.1 *T* to 3.4 *T*	1.3 *T*	.....	17	0.40 *E*
class 50..............	50[a]	3.0 *T* to 3.4 *T*	1.3 *T*	.....	18	0.40 *E*
class 60..............	60[a]	2.8 *T*	1.0 *T*	.....	19.9	0.40 *E*
malleable............	40 to 100[c]	....	....	30 to 80[c]	25	0.43 *E*
nodular (ductile iron).	60 to 120[d]	....	....	40 to 90[d]	23	......
Cast steel, carbon.......	60 to 100	*T*	0.75 *T*	30 to 70	30	0.38 *E*
low alloy.............	70 to 200	*T*	0.75 *T*	45 to 170	30	0.38 *E*
Steel, SAE 950 (low alloy)	65 to 70	*T*	0.75 *T*	45 to 50	30	0.38 *E*
1025 (low carbon).....	60 to 103	*T*	0.75 *T*	40 to 90	30	0.38 *E*
1045 (medium carbon).	80 to 182	*T*	0.75 *T*	50 to 162	30	0.38 *E*
1095 (high carbon)....	90 to 213	*T*	0.75 *T*	20 to 150	30	0.39 *E*
1112 (free cutting)[e]....	60 to 100	*T*	0.75 *T*	30 to 95	30	0.38 *E*
1212 (free cutting).....	57 to 80	*T*	0.75 *T*	25 to 72	30	0.38 *E*
1330 (alloy)...........	90 to 162	*T*	0.75 *T*	27 to 149	30	0.38 *E*
2517 (alloy)[e]..........	88 to 190	*T*	0.75 *T*	60 to 155	30	0.38 *E*
3140 (alloy)...........	93 to 188	*T*	0.75 *T*	62 to 162	30	0.38 *E*
3310 (alloy)[e]..........	104 to 172	*T*	0.75 *T*	56 to 142	30	0.38 *E*
4023 (alloy)[e]..........	105 to 170	*T*	0.75 *T*	60 to 114	30	0.38 *E*
4130 (alloy)...........	81 to 179	*T*	0.75 *T*	46 to 161	30	0.38 *E*
4340 (alloy)...........	109 to 220	*T*	0.75 *T*	68 to 200	30	0.38 *E*
4640 (alloy)...........	98 to 192	*T*	0.75 *T*	62 to 169	30	0.38 *E*
4820 (alloy)[e]..........	98 to 209	*T*	0.75 *T*	68 to 184	30	0.38 *E*
5150 (alloy)...........	98 to 210	*T*	0.75 *T*	51 to 190	30	0.38 *E*
52100 (alloy)..........	100 to 238	*T*	0.75 *T*	81 to 228	30	0.38 *E*
6150 (alloy)...........	96 to 228	*T*	0.75 *T*	59 to 210	30	0.38 *E*
8650 (alloy)...........	110 to 228	*T*	0.75 *T*	69 to 206	30	0.38 *E*
8740 (alloy)...........	100 to 179	*T*	0.75 *T*	60 to 165	30	0.38 *E*
9310 (alloy)[e]..........	117 to 187	*T*	0.75 *T*	63 to 162	30	0.38 *E*
9840 (alloy)...........	120 to 285	*T*	0.75 *T*	45 to 50	30	0.38 *E*
Steel, stainless, SAE						
30302[f]................	85 to 125	*T*	.....	35 to 95	28	0.45 *E*
30321[f]................	85 to 95	*T*	.....	30 to 60	28	......
30347[f]................	90 to 100	*T*	.....	35 to 65	28	......
51420[g]................	95 to 230	*T*	.....	50 to 195	29	0.40 *E*
51430[h]................	75 to 85	*T*	.....	40 to 70	29	......
51446[h]................	80 to 85	*T*	.....	50 to 70	29	......
51501[g]................	70 to 175	*T*	.....	30 to 135	29	......
Steel, structural,						
common..............	60 to 75	*T*	0.75 *T*	33[a]	29	0.41 *E*
rivet................	52 to 62	*T*	0.75 *T*	28[a]	29	......
rivet, high strength...	68 to 82	*T*	0.75 *T*	38[a]	29	......
Wrought iron...........	34 to 54	*T*	0.83 *T*	23 to 32	28	......

[a] Minimum specified value of the American Society of Testing Materials. The specifications for the various materials are as follows: Cast iron, ASTM A48; structural steel for bridges and structures, ASTM A7; structural rivet steel, ASTM A141; high-strength structural rivet steel, ASTM A195.
[b] Synonomous in other literature to the modulus of elasticity in torsion and the modulus of rigidity, *G*.
[c] Range of minimum specified values of the ASTM (ASTM A47, A197, and A220).
[d] Range of minimum specified values of the ASTM (ASTM A339) and the Munitions Board Standards Agency (MIL-I-17166A and MIL-I-11466).
[e] Carburizing grades of steel.
[f] Non-hardenable nickel-chromium and chromium-nickel-manganese steel (austenitic).
[g] Hardenable chromium steel (martensitic).
[h] Non-hardenable chromium steel (ferritic).

Strength Properties of Non-Ferrous Metals*

Material	Ultimate Strength, Thousands of Pounds per Square Inch: in Tension	Ultimate Strength, Thousands of Pounds per Square Inch: in Shear	Yield Strength (0.2 per cent offset), Thousands of Pounds per Square Inch	Modulus of Elasticity, Millions of Pounds per Square Inch: in Tension, *E*	Modulus of Elasticity, Millions of Pounds per Square Inch: in Shear, *G*
Aluminum alloys, cast,					
sand cast,...............	19 to 35	14 to 26	8 to 25	10.3	...
heat-treated............	20 to 48	20 to 34	16 to 40	10.3	...
permanent mold cast,..	23 to 35	16 to 27	9 to 24	10.3	...
heat-treated.........	23 to 48	15 to 36	8.5 to 43	10.3	...
die-cast...............	30 to 46	19 to 29	16 to 27	10.3	...
Aluminum alloys, wrought,					
annealed..............	10 to 42	7 to 26	4 to 22	10.0 to 10.6	...
cold-worked...........	12 to 63	8 to 34	11 to 59	10.0 to 10.3	...
heat-treated..........	22 to 83	14 to 48	13 to 73	10.0 to 11.4	...
Aluminum bronze, cast,..	62 to 90	...	25 to 37	15 to 18	...
heat-treated...........	80 to 110	...	32 to 65	15 to 18	...

(continued)

Material	Ultimate Strength, Thousands of Pounds per Square Inch		Yield Strength (0.2 per cent offset), Thousands of Pounds per Square Inch	Modulus of Elasticity, Millions of Pounds per Square Inch	
	in Tension	in Shear		in Tension, *E*	in Shear, *G*
Aluminum bronze, wrought,					
annealed	55 to 80	...	20 to 40	16 to 19	...
cold-worked	71 to 110	...	62 to 66	16 to 19	...
heat-treated	101 to 151	...	48 to 94	16 to 19	...
Brasses, leaded, cast	32 to 40	29 to 31	12 to 15	12 to 14	...
flat products, wrought	46 to 85	31 to 45	14 to 62	14 to 17	5.3 to 6.4
wire, wrought	50 to 88	34 to 46	....	15	5.6
Brasses, non-leaded,					
flat products, wrought	34 to 99	28 to 48	10 to 65	15 to 17	5.6 to 6.4
wire, wrought	40 to 130	29 to 60	....	15 to 17	5.6 to 6.4
Copper, wrought,					
flat products	32 to 57	22 to 29	10 to 53	17	6.4
wire	35 to 66	24 to 33	....	17	6.4
Inconel, cast	70 to 95	...	30 to 45	23	...
flat products, wrought	80 to 170	...	30 to 160	31	11
wire, wrought	80 to 185	...	25 to 175	31	11
Lead	2.2 to 4.9	...	....	0.8 to 2.0	...
Magnesium, cast,					
sand & permanent mold	22 to 40	17 to 22	12 to 23	6.5	2.4
die-cast	33	20	22	6.5	2.4
Magnesium, wrought,					
sheet and plate	35 to 42	21 to 23	20 to 32	6.5	2.4
bars, rods, and shapes	37 to 55	19 to 27	26 to 44	6.5	2.4
Monel, cast	65 to 90	...	32 to 40	19	...
flat products, wrought	70 to 140	...	25 to 130	26	9.5
wire, wrought	70 to 170	...	25 to 160	26	9.5
Nickel, cast	45 to 60	...	20 to 30	21.5	...
flat products, wrought	55 to 130	...	15 to 115	30	11
wire, wrought	50 to 165	...	10 to 155	30	11
Nickel silver, cast	40 to 50	...	24 to 25	...	...
flat products, wrought	49 to 115	41 to 59	18 to 90	17.5 to 18	6.6 to 6.8
wire, wrought	50 to 145	...	25 to 90	17.5 to 18	6.6 to 6.8
Phosphor bronze, wrought,					
flat products	40 to 128	...	14 to 80	15 to 17	5.6 to 6.4
wire	50 to 147	...	20 to 80	16 to 17	6 to 6.4
Silicon bronze, wrought,					
flat products	56 to 110	42 to 63	21 to 62	15	5.6
wire	50 to 145	36 to 70	25 to 70	15 to 17	5.6 to 6.4
Tin bronze, leaded, cast	21 to 38	23 to 43	15 to 18	10 to 14.5	...
Titanium	50 to 135	...	40 to 120	15.0 to 16.5	...
Zinc, commercial rolled	19.5 to 31	...	....	....	...
Zirconium	22 to 83	...	....	9 to 14.5	4.8

• Consult the index for data on metals not listed and for more data on metals listed.

Influence of Temperature on the Strength of Metals

Material	210	400	570	750	930	1100	1300	1475
	Degrees Fahrenheit — Strength in Per Cent of Strength at 70 Degrees F.							
Wrought iron	104	112	116	96	76	42	25	15
Cast iron	....	100	99	92	76	42	....	....
Steel castings	109	125	121	97	57	....	....	....
Structural steel	103	132	122	86	49	28	....	....
Copper	95	85	73	59	42	....	....	....
Bronze	101	94	57	26	18	....	....	....

Strength of Copper-Zinc-Tin Alloys

(U. S. Government Tests)

Percentage of Copper	Percentage of Zinc	Percentage of Tin	Tensile Strength, Lbs. per Sq. In.	Percentage of Copper	Percentage of Zinc	Percentage of Tin	Tensile Strength, Lbs. per Sq. In.	Percentage of Copper	Percentage of Zinc	Percentage of Tin	Tensile Strength, Lbs. per Sq. In.
45	50	5	15,000	60	20	20	10,000	75	20	5	45,000
50	45	5	50,000	65	30	5	50,000	75	15	10	45,000
50	40	10	15,000	65	25	10	42,000	75	10	15	43,000
55	43	2	65,000	65	20	15	30,000	75	5	20	41,000
55	40	5	62,000	65	15	20	18,000	80	15	5	45,000
55	35	10	32,500	65	10	25	12,000	80	10	10	45,000
55	30	15	15,000	70	25	5	45,000	80	5	15	47,500
60	37	3	60,000	70	20	10	44,000	85	10	5	43,500
60	35	5	52,500	70	15	15	37,000	85	5	10	46,500
60	30	10	40,000	70	10	20	30,000	90	5	5	42,000

Average Ultimate Strength of Common Materials other than Metals

(Pounds per square inch)

Material	Compression	Tension
Bricks, best hard	12,000	400
Bricks, light red	1,000	40
Brickwork, common	1,000	50
Brickwork, best	2,000	300
Cement, Portland, one month old	2,000	400
Cement, Portland, one year old	3,000	500
Concrete, Portland	1,000	200
Concrete, Portland, one year old	2,000	400
Granite	19,000	700
Limestone and sandstone	9,000	300
Trap rock	20,000	800
Slate	14,000	500
Vulcanized Fiber	39,000	13,000

Permissible Working Stresses for Structural Timbers

(U. S. Government Tests)

Kind of Timber	Bending, Pounds per Sq. In.			Compression, Pounds per Sq. In.			
	Allowable Stress in Extreme Fiber		Allowable Horizontal Shear Stress	Allowable Stress Parallel to Grain "Short Columns"		Allowable Stress Perpendicular to Grain	
	Outside Location	Dry Location	All Locations	Outside Location	Dry Location	Outside Location	Inside Location
Cedar, western red	800	900	80	700	700	150	200
Cedar, northern white	650	750	70	500	550	140	175
Chestnut	850	950	90	700	800	200	300
Cypress	1100	1300	100	1100	1100	250	350
Douglas fir (No. 1 str'l) *	1400	1600	100	1100	1200	250	350
Douglas fir (No. 2 str'l)	1100	1300	90	900	1000	225	300
Fir, balsam	750	900	70	600	700	125	150
Gum, red	900	1100	100	750	800	200	300
Hemlock, western	1100	1300	75	900	900	225	300
Hemlock, eastern	900	1000	70	700	700	225	300
Hickory	1500	1900	140	1200	1500	400	600
Maple, sugar or hard	1300	1500	150	1100	1200	375	500
Maple, silver or soft	900	1000	100	700	800	250	350
Oak, white or red	1200	1400	125	900	1000	375	500
Pine, s. yellow (dense) †	1400	1600	125	1100	1200	250	350
Pine, s. yellow (sound)	1100	1300	105	900	1000	225	300
Pine, eastern white	800	900	85	750	750	150	250
Pine, western white	800	900	85	750	750	150	250
Pine, Norway	1000	1100	85	800	800	175	300
Redwood	1000	1200	70	900	1000	150	250
Spruce, red or white	900	1100	85	750	800	150	250
Spruce, Englemann	650	750	70	550	600	140	175

* The strength of large timbers depends chiefly upon the density or weight per cubic foot of the dry wood and upon the character, size, number and location of defects. "Dense" Douglas fir of the "No. 1 structural grade" shows on one end an average of at least six annual rings per inch and at least one-third "summer wood," measured over 3 inches on a line extending from the pith to the corner farthest from the pith when the least dimension of the timber is 5 inches or more. The point where the 3-inch line begins is found by the formula $A = \frac{1}{2} D - 2$, where A = distance in inches from pith to beginning of 3-inch line and D = minimum dimension of timber in inches. The "No. 2 structural grade" for Douglas fir includes timbers not passing the No. 1 grade, because (1) there is less density than required or (2) greater defects than are permitted.

† The term "southern yellow pine" includes the species known heretofore as long-leaf pine, short-leaf pine, loblolly pine, Cuban pine and pond pine. "Dense" southern yellow pine shows on either end an average of at least six annual rings per inch and at least one-third summer wood, or else the greater number of rings shows at least one-third summer wood all as measured over the third, fourth, and fifth inches of a radial line extending from the pith. Wide-ringed material, excluded by this rule, is acceptable, provided the amount of summer wood measured as previously specified is at least one-half. "Sound" southern yellow pine includes pieces without any ring or summer wood requirement.

Allowable Unit Stresses, Stress-grade Lumber for Normal Loading Conditions

Species, commercial grade, and modulus of elasticity	Allowable unit stresses, psi					
	Extreme fiber in bending, *f*, and tension parallel to grain, *t*		Horizontal shear, *H*	Compression perpendicular to grain, $c \perp$	Compression parallel to grain, *c*	
	J and P B and S	P and T			J and P P and T	B and S
Douglas fir						
E = 1,760,000 psi						
Dense select structural	2,050	1,900	120	455	1,650	1,500
Select structural	1,900	1,750	120	415	1,500	1,400
Dense construction	1,750	1,500	120	455	1,400	1,200
Construction	1,500	1,200	120	390	1,200	1,000
Standard (J and P only)	1,200	.....	95	390	1,000	
Pine, southern, 5 in. thick and up						
E = 1,760,000 psi						
Dense structural 86	2,400	2,400	150	455	1,800	1,800
Dense structural 72	2,000	2,000	135	455	1,550	1,550
Dense structural 65	1,800	1,800	120	455	1,400	1,400
Dense structural 58	1,600	1,600	105	455	1,300	1,300
No. 1 dense SR	1,600	1,600	120	455	1,500	1,500
No. 1 SR	1,400	1,400	120	390	1,300	1,300
No. 2 dense SR	1,400	1,400	105	455	1,050	1,050
No. 2 SR	1,200	1,200	105	390	900	900
Pine, Norway (J and P only)						
E = 1,320,000						
Prime structural	1,200	.....	75	360	900	
Common structural	1,100	.....	75	360	775	
Utility structural	950	.....	75	360	650	
Spruce, eastern (J and P only)						
E = 1,320,000						
1450 f structural grade	1,450	.....	110	300	1,050	
1300 f structural grade	1,300	.....	95	300	975	
1200 f structural grade	1,200	.....	95	300	900	
Redwood						
E = 1,320,000 psi						
Dense structural	1,700	.....	110	320	1,450	1,450
Heart structural	1,300	.....	95	320	1,100	1,100
Hemlock, eastern						
E = 1,210,000						
Select structural	1,300	.....	85	360	850	850
Prime structural (J and P only)	1,200	.....	60	360	775	
Common structural (J and P only)	1,100	.....	60	360	650	
Utility structural (J and P only)	950	.....	60	360	600	
Hemlock, western						
E = 1,540,000						
Select structural (J and P only)	1,600	.....	100	365	1,200	
Construction	1,500	1,200	100	365	1,100	1,000
Standard (J and P only)	1,200	.....	80	365	1,000	

NOTE: J and P = joists and planks, B and S = beams and stringers, P and T = posts and timbers.

Properties of Sections for Standard Lumber Sizes

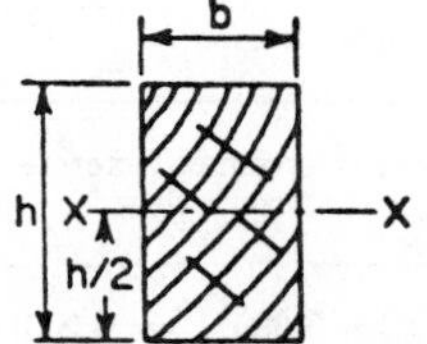

Dressed (S4S) Sizes

Moment of inertia and section modulus are given with respect to xx axis, with dimensions b and h as shown on sketch.

Nominal size b h	Standard dressed size S4S b h	Area of section $A = bh$	Moment of inertia $I = \frac{bh^3}{12}$	Section modulus $S = \frac{bh^2}{6}$	Board feet per linear foot of piece
2 × 4	1⅝ × 3⅝	5.89	6.45	3.56	⅔
2 × 6	1⅝ × 5½	8.93	22.53	8.19	1
2 × 8	1⅝ × 7½	12.19	57.13	15.23	1⅓
2 × 10	1⅝ × 9½	15.44	116.10	24.44	1⅔
2 × 12	1⅝ × 11½	18.69	205.95	35.82	2
3 × 4	2⅝ × 3⅝	9.52	10.42	5.75	1
3 × 6	2⅝ × 5½	14.43	36.40	13.23	1½
3 × 8	2⅝ × 7½	19.69	92.29	24.61	2
3 × 10	2⅝ × 9½	24.94	187.55	39.48	2½
3 × 12	2⅝ × 11½	30.19	332.69	57.86	3
4 × 6	3⅝ × 5½	19.95	50.25	18.28	2
4 × 8	3⅝ × 7½	27.19	127.44	33.98	2⅔
4 × 10	3⅝ × 9½	34.44	259.00	54.43	3⅓
4 × 12	3⅝ × 11½	41.69	459.43	79.90	4
4 × 14	3⅝ × 13½	48.94	743.24	110.11	4⅔
4 × 16	3⅝ × 15½	56.19	1,124.92	145.15	5⅓
6 × 6	5½ × 5½	30.25	76.26	27.73	3
6 × 8	5½ × 7½	41.25	193.36	51.56	4
6 × 10	5½ × 9½	52.25	392.96	82.73	5
6 × 12	5½ × 11½	63.25	697.07	121.23	6
6 × 14	5½ × 13½	74.25	1,127.67	167.06	7
6 × 16	5½ × 15½	85.25	1,705.78	220.23	8
6 × 18	5½ × 17½	96.25	2,456.38	280.73	9
8 × 8	7½ × 7½	56.25	263.67	70.31	5⅓
8 × 10	7½ × 9½	71.25	535.86	112.81	6⅔
8 × 12	7½ × 11½	86.25	950.55	165.31	8
8 × 14	7½ × 13½	101.25	1,537.73	227.81	9⅓
8 × 16	7½ × 15½	116.25	2,327.42	300.31	10⅔
8 × 18	7½ × 17½	131.25	3,349.61	382.81	12
8 × 20	7½ × 19½	146.25	4,634.30	475.31	13⅓
10 × 10	9½ × 9½	90.25	678.76	142.90	8⅓
10 × 12	9½ × 11½	109.25	1,204.03	209.40	10
10 × 14	9½ × 13½	128.25	1,947.80	288.56	11⅔
10 × 16	9½ × 15½	147.25	2,948.07	380.40	13⅓
10 × 18	9½ × 17½	166.25	4,242.84	484.90	15
10 × 20	9½ × 19½	185.25	5,870.11	602.06	16⅔
12 × 12	11½ × 11½	132.25	1,457.51	253.48	12
12 × 14	11½ × 13½	155.25	2,357.86	349.31	14
12 × 16	11½ × 15½	178.25	3,568.71	460.48	16
12 × 18	11½ × 17½	201.25	5,136.07	586.98	18
12 × 20	11½ × 19½	224.25	7,105.92	728.81	20
12 × 22	11½ × 21½	247.25	9,524.28	885.98	22
12 × 24	11½ × 23½	270.25	12,437.13	1,058.48	24

Decking (Based on Strip 1 Ft Wide and of Thickness Indicated)

1′0 × 2	12 × 1⅝	19.50	4.29	5.28	2
1′0 × 3	12 × 2⅝	31.50	18.00	13.76	3
1′0 × 4	12 × 3⅝	42.00	42.88	24.50	4

Allowable Stresses in Concrete

Description		Allowable stresses: For any strength of concrete	For strength of concrete shown below: f'_c = 2,500 psi	f'_c = 3,000 psi	f'_c = 4,000 psi	f'_c = 5,000 psi
Modulus of elasticity ratio: n..........		$\frac{29,000,000}{w^{1.5}33\sqrt{f'_c}}$				
For concrete weighing 145 lb per cu ft	n	....	10	9	8	7
Flexure: f_c						
Extreme fiber stress in compression...	f_c	$0.45f'_c$	1,125	1,350	1,800	2,250
Extreme fiber stress, in tension in plain concrete footings and walls........	f_c	$1.6\sqrt{f'_c}$	80	88	102	113
Shear: v (as a measure of diagonal tension at a distance d from the face of the support)						
Beams with no web reinforcement....	v_c	$1.1\sqrt{f'_c}$	55	60	70	78
Joists with no web reinforcement.....	v_c	$1.2\sqrt{f'_c}$	61	66	77	86
Members with vertical or inclined web reinforcement or properly combined bent bars and vertical stirrups.....	v	$5\sqrt{f'_c}$	250	274	316	354
Slabs and footings (peripheral shear)..	v_c	$2\sqrt{f'_c}$	100	110	126	141
Bearing: f_c						
On full area........................		$0.25f'_c$	625	750	1,000	1,250
On one-third area or less*...........		$0.375f'_c$	938	1,125	1,500	1,875

* This increase is permitted only when the least distance between the edges of the loaded and unloaded areas is a minimum of one-fourth of the parallel side dimension of the loaded area. The allowable bearing stress on a reasonably concentric area greater than one-third but less than the full area is to be interpolated between the values given.

NOTE: f'_c = compressive strength of concrete, psi; n = ratio of modulus of elasticity of steel to that of concrete; w = weight of concrete, lb/ft^3.

Maximum Permissible Water-cement Ratios for Concrete

Specified compressive strength at 28 days, psi f'_c	Maximum permissible water-cement ratio*: Non-air-entrained concrete: U.S. gal per 94-lb bag of cement	Non-air-entrained concrete: Absolute ratio by weight	Air-entrained concrete: U.S. gal per 94-lb bag of cement	Air-entrained concrete: Absolute ratio by weight
2,500	7.3	0.65	6.1	0.54
3,000	6.6	0.58	5.2	0.46
3,500	5.8	0.51	4.5	0.40
4,000	5.0	0.44	4.0	0.35

* Including free surface moisture on aggregates.

Bearing Capacities of Soils

The approximate ultimate bearing capacity under a long footing at the surface of a soil is given by Prandtl's equation as

$$q_u = (c/\tan\phi + \tfrac{1}{2}\gamma_{dry} b\sqrt{K_p})(K_p e^{\pi \tan\phi} - 1)$$

where q_u = ultimate bearing capacity of soil, lb/ft²
c = cohesion, lb/ft²
ϕ = angle of internal friction, deg
γ_{dry} = unit weight of dry soil, lb/ft³ (Sec. 6-14)
b = width of footing, ft
d = depth of footing below surface, ft
K_p = coefficient of passive pressure = $[\tan(45 + \phi/2)]^2$
e = 2.718 . . .

For footings below the surface, the ultimate bearing capacity of the soil may be modified by the factor $1 + Cd/b$. The coefficient C is about 2 for cohesionless soils and about 0.3 for cohesive soils. The increase in bearing capacity with depth for cohesive soils is often neglected.

Allowable Bearing Capacity of Soils

	Allowable Bearing Capacity, Tons/Ft²
Medium soft clay	1.5
Medium stiff clay	2.5
Sand, fine, loose	2
Sand, coarse, loose; compact fine sand; loose sand-gravel mixture	3
Gravel, loose; compact coarse sand	4
Sand-gravel mixture, compact	6
Hardpan and exceptionally compacted or partially cemented gravels or sands	10
Sedimentary rocks, such as hard shales, sandstones, limestones, and silt stones, in sound condition	15
Foliated rocks, such as schist or slate, in sound condition	40
Massive bedrock, such as granite, diorite, gneiss, and trap rock, in sound condition	100

CHAPTER 24

ELECTRICAL TECHNOLOGY

ELECTRICAL RESISTIVITY OF COPPER

(R. A. Matula)

[Temperature, T,°K; Total Resistivity, ρ, 10^{-8} Ω m; Intrinsic Resistivity, ρ_i, 10^{-8} Ω m]

T	ρ_i	ρ	Solid	T	ρ_i	ρ
1		0.00200		175	0.872	0.874
4		0.00200		200	1.044	1.046
7		0.00200		225	1.215	1.217
10		0.00202		250	1.385	1.387
15		0.00218		273.15	1.541	1.543
20	0.000798*	0.00280		293	1.676	1.678
25	0.00249*	0.00449		300	1.723	1.725
30	0.00628	0.00828		350	2.061	2.063
35	0.0127	0.0147		400	2.400	2.402
40	0.0219	0.0239		500	3.088	3.090
45	0.0338	0.0358		600	3.790	3.792
50	0.0498	0.0518		700	4.512	4.514
55	0.0707	0.0727		800	5.260	5.262
60	0.0951	0.0971		900	6.039	6.041
70	0.152	0.154		1000	6.856	6.858
80	0.213	0.215		1100	7.715	7.717
90	0.279	0.281		1200	8.624	8.626
100	0.346	0.348		1300	9.590	9.592
125	0.520	0.522		1357.6	10.169	10.171
150	0.697	0.699				

ELECTRICAL RESISTIVITY OF SILVER

(R. A. Matula)

[Temperature, T,°K; Total Resistivity, ρ, 10^{-8} Ω m; Intrinsic Resistivity, ρ_i, 10^{-8} Ω m]

T	ρ_i	ρ	Solid	T	ρ_i	ρ
1		0.00100		150	0.725	0.726
4		0.00100		175	0.877	0.878
7		0.00103		200	1.028	1.029
10		0.00115		225	1.178	1.179
15		0.00189		250	1.328	1.329
20	0.00322	0.00422		273.15	1.466	1.467
25	0.00855	0.00955		293	1.586	1.587
30	0.0184	0.0194		300	1.628	1.629

(continued)

T	ρ_i	ρ	T	ρ_i	ρ
35	0.0331	0.0341	350	1.931	1.932
40	0.0529	0.0539	400	2.240	2.241
45	0.0763	0.0773	500	2.874	2.875
50	0.103	0.104	600	3.530	3.531
55	0.131	0.132	700	4.208	4.209
60	0.161	0.162	800	4.911	4.912
70	0.224	0.225	900	5.637	5.638
80	0.288	0.289	1000	6.395	6.396
90	0.353	0.354	1100	7.214	7.215
100	0.417	0.418	1200	8.088	8.089
125	0.572	0.573	1235.08	8.414	8.415

ELECTRICAL RESISTIVITY OF GOLD

(R. A. Matula)

[Temperature, T, °K; Total Resistivity, ρ, 10^{-8} Ω m; Intrinsic Resistivity, ρ_i, 10^{-8} Ω m]

		Solid			
T	ρ_i	ρ	T	ρ_i	ρ
1		0.0220	175	1.240	1.262
4		0.0220	200	1.440	1.462
7		0.0221	225	1.640	1.662
10		0.0226	250	1.842	1.864
15	0.00376*	0.0258	273.15	2.029	2.051
20	0.0126*	0.0346*	293	2.192	2.214
25	0.0282*	0.0502*	300	2.249	2.271
30	0.0505*	0.0725*	350	2.663	2.685
35	0.0798*	0.1018*	400	3.085	3.107
40	0.119*	0.141*	500	3.952	3.974
45	0.159	0.181	600	4.853	4.875
50	0.199	0.221	700	5.794	5.816
55	0.248	0.270	800	6.786	6.808
60	0.286	0.308	900	7.840	7.862
70	0.373	0.395	1000	8.964	8.986
80	0.459	0.481	1100	10.169	10.191
90	0.544	0.566	1200	11.464	11.486
100	0.628	0.650	1300	12.832	12.854
125	0.835	0.857	1337.58	13.366	13.388
150	1.039	1.061			

ELECTRICAL RESISTIVITY OF PALLADIUM

(R. A. Matula)

[Temperature, T, K; Total Resistivity, ρ, 10^{-8} Ω m; Intrinsic Resistivity, ρ_i, 10^{-8} Ω m]

T	ρ_i	ρ	Solid	T	ρ_i	ρ
1	0.0000309*	0.0200		225	7.87	7.89
4	0.000505*	0.0205		250	8.86	8.88
7	0.00170*	0.0217		273.15	9.76	9.78
10	0.00421*	0.0242		293	10.52	10.54
15	0.0145	0.0345		300	10.78	10.80
20	0.0363	0.0563		350	12.65	12.67
25	0.0736	0.0936		400	14.46	14.48
30	0.130	0.150		500	17.92	17.94
35	0.210	0.230		600	21.16	21.18
40	0.314	0.334		700	24.21	24.23
45	0.440	0.460		800	27.05	27.07
50	0.586	0.606		900	29.72	29.74
55	0.745	0.765		1000	32.21	32.23
60	0.918	0.938		1100	34.52	34.54
70	1.30	1.32		1200	36.66	36.68
80	1.73	1.75		1300	38.64	38.66
90	2.17	2.19		1400	40.44	40.46
100	2.60	2.62		1500	42.08	42.10
125	3.71	3.73		1600	43.55	43.57
150	4.78	4.80		1700	44.86	44.88
175	5.83	5.85		1800	45.99	46.01
200	6.86	6.88		1827	46.27	46.29

Conversion Factors Between MKS (Practical), CGS Electrostatic (ESU), and CGS Electromagnetic (EMU) Systems of Units

Quantity	Symbol	Mks unit	Conversion factors		Cgs (esu) unit	Conversion factors		Cgs (emu) unit	Conversion factors	
			To cgs (esu)	To cgs (emu)		To cgs (emu)	To mks		To cgs (esu)	To mks
Acceleration	a	Meter per second per second	10^2	10^2	Centimeter per second per second	1	10^{-2}	Centimeter per second per second	1	10^{-2}
Area	A	Square meter	10^4	10^4	Square centimeter	1	10^{-4}	Square centimeter	1	10^{-4}
Capacitance	C	Farad	9×10^{11}	10^{-9}	Statfarad	$\frac{1}{9} \times 10^{-20}$	$\frac{1}{9} \times 10^{-11}$	Abfarad	9×10^{20}	10^9
Charge	Q	Coulomb	3×10^9	10^{-1}	Statcoulomb	$\frac{1}{3} \times 10^{-10}$	$\frac{1}{3} \times 10^{-9}$	Abcoulomb	3×10^{10}	10
Charge density:										
Linear	q	Coulomb per meter	3×10^7	10^{-3}	Statcoulomb per centimeter	$\frac{1}{3} \times 10^{-10}$	$\frac{1}{3} \times 10^{-7}$	Abcoulomb per centimeter	3×10^{10}	10^3
Area	σ	Coulomb per square meter	3×10^5	10^{-5}	Statcoulomb per square centimeter	$\frac{1}{3} \times 10^{-10}$	$\frac{1}{3} \times 10^{-5}$	Abcoulomb per square centimeter	3×10^{10}	10^5
Volume	ρ	Coulomb per cubic meter	3×10^3	10^{-7}	Statcoulomb per cubic centimeter	$\frac{1}{3} \times 10^{-10}$	$\frac{1}{3} \times 10^{-3}$	Abcoulomb per cubic centimeter	3×10^{10}	10^7
Conductance	G	Mho	9×10^{11}	10^{-9}	Statmho	$\frac{1}{9} \times 10^{-20}$	$\frac{1}{9} \times 10^{-11}$	Abmho	9×10^{20}	10^9
Conductivity	τ	Mho per meter	9×10^9	10^{-11}	Statmho per centimeter	$\frac{1}{9} \times 10^{-20}$	$\frac{1}{9} \times 10^{-9}$	Abmho per centimeter	9×10^{20}	10^{11}
Current	I	Ampere	3×10^9	10^{-1}	Statampere	$\frac{1}{3} \times 10^{-10}$	$\frac{1}{3} \times 10^{-9}$	Abampere	3×10^{10}	10
Current density	ϱ	Ampere per square meter	3×10^5	10^{-5}	Statampere per square centimeter	$\frac{1}{3} \times 10^{-10}$	$\frac{1}{3} \times 10^{-5}$	Abampere per square centimeter	3×10^{10}	10^5
Elastance	S	Daraf	$\frac{1}{9} \times 10^{-11}$	10^9	Statdaraf	9×10^{20}	9×10^{11}	Abdaraf	$\frac{1}{9} \times 10^{-20}$	10^{-9}
Electric intensity	E	Volt per meter	$\frac{1}{3} \times 10^{-4}$	10^6	Statvolt per centimeter	3×10^{10}	3×10^2	Abvolt	$\frac{1}{3} \times 10^{-10}$	10^{-6}
Electrostatic flux	χ	....	3×10^9	10^{-1}	....	$\frac{1}{3} \times 10^{-10}$	$\frac{1}{3} \times 10^{-9}$	....	3×10^{10}	10
Electrostatic flux density	D	....	3×10^5	10^{-5}	....	$\frac{1}{3} \times 10^{-10}$	$\frac{1}{3} \times 10^{-5}$	....	3×10^{10}	10^5
Energy	W	Joule	10^7	10^7	Erg	1	10^{-7}	Erg	1	10^{-7}
Force	f	Newton	10^5	10^5	Dyne	1	10^{-5}	Dyne	1	10^{-5}
Inductance	L	Henry	$\frac{1}{9} \times 10^{-11}$	10^9	Stathenry	9×10^{20}	9×10^{11}	Abhenry	$\frac{1}{9} \times 10^{-20}$	10^{-9}
Length	l	Meter	10^2	10^2	Centimeter	1	10^{-2}	Centimeter	1	10^{-2}
Magnetic flux	Φ	Weber	$\frac{1}{3} \times 10^{-2}$	10^8	....	3×10^{10}	3×10^2	Maxwell	$\frac{1}{3} \times 10^{-10}$	10^{-8}
Magnetic flux density	B	Weber per square meter	$\frac{1}{3} \times 10^{-6}$	10^4	....	3×10^{10}	3×10^6	Gauss	$\frac{1}{3} \times 10^{-10}$	10^{-4}
Magnetic intensity	H	Praoersted	3×10^7	10^{-3}	....	$\frac{1}{3} \times 10^{-10}$	$\frac{1}{3} \times 10^{-7}$	Oersted	3×10^{10}	10^3
Magnetic linkages	λ	Weber-turn	$\frac{1}{3} \times 10^{-2}$	10^8	....	3×10^{10}	3×10^2	Maxwell-turn	$\frac{1}{3} \times 10^{-10}$	10^{-8}
Magnetomotive force	F	Pragilbert	3×10^9	10^{-1}	....	$\frac{1}{3} \times 10^{-10}$	$\frac{1}{3} \times 10^{-9}$	Gilbert	3×10^{10}	10

Conversion Factors between MKS (Practical), CGS Electrostatic (ESU), and CGS Electromagnetic (EMU) Systems of Units (continued)

Quantity	Symbol	Mks unit	Conversion Factors		Cgs (esu) unit	Conversion factors		Cgs (emu) unit	Conversion factors:	
			To cgs (esu)	To cgs (emu)		To cgs (emu)	To mks		To cgs (esu)	To mks
Mass	m	Kilogram	10^3	10^3	Gram	1	10^{-3}	Gram	1	10^{-3}
Permeability	μ	....	$\frac{1}{9} \times 10^{-13}$	10^7	....	9×10^{20}	9×10^{13}	Gauss per oersted	$\frac{1}{9} \times 10^{-20}$	10^{-7}
Permeability of free space	μ_0	10^{-7}	$\frac{1}{9} \times 10^{-13}$	10^7	$\frac{1}{9} \times 10^{-20}$	9×10^{20}	9×10^{13}	1	$\frac{1}{9} \times 10^{-20}$	10^{-7}
Permeance	P	Weber per pragilbert	$\frac{1}{9} \times 10^{-11}$	10^9	....	9×10^{20}	9×10^{11}	Maxwell per gilbert	$\frac{1}{9} \times 10^{-20}$	10^{-9}
Permittivity	ϵ	Farads per meter	9×10^9	10^{-11}	....	$\frac{1}{9} \times 10^{-20}$	$\frac{1}{9} \times 10^{-9}$	....	9×10^{20}	10^{11}
Permittivity of free space	ϵ_0	$\frac{1}{9} \times 10^{-9}$	9×10^9	10^{-11}	....	$\frac{1}{9} \times 10^{-20}$	$\frac{1}{9} \times 10^{-9}$	$\frac{1}{9} \times 10^{-20}$	9×10^{20}	10^{11}
Potential difference	v	Volt	$\frac{1}{3} \times 10^{-2}$	10^8	Statvolt	3×10^{10}	3×10^2	Abvolt	$\frac{1}{3} \times 10^{-10}$	10^{-8}
Power	P	Watt	10^7	10^7	Erg per second	1	10^{-7}	Erg per second	1	10^{-7}
Reluctance	$\mathcal{R}$	Pragilbert weber	9×10^{11}	10^{-9}	....	$\frac{1}{9} \times 10^{-20}$	$\frac{1}{9} \times 10^{-11}$	Gilbert per maxwell	9×10^{20}	10^9
Reluctivity	ν	....	9×10^{13}	10^{-7}	....	$\frac{1}{9} \times 10^{-20}$	$\frac{1}{9} \times 10^{-13}$	Oersted per gauss	9×10^{20}	10^7
Resistance	R	Ohm	$\frac{1}{9} \times 10^{-11}$	10^9	Statohm	9×10^{20}	9×10^{11}	Abohm	$\frac{1}{9} \times 10^{-20}$	10^{-9}
Resistivity	ρ	Ohm-meter	$\frac{1}{9} \times 10^{-9}$	10^{11}	Statohm centimeter	9×10^{20}	9×10^9	Abohm centimeter	$\frac{1}{9} \times 10^{-20}$	10^{-11}
Time	t	Second	1	1	Second	1	1	Second	1	1

Nomenclature and Symbols for SI Units

Quantity	Symbol	Unit
Acceleration	m/s^2	meter per second per second
Area	m^2	square meter
Capacitance	F	farad
Charge	C	coulomb
Charge density:		
Linear	C/m	coulomb per meter
Area	C/m^2	coulomb per square meter
Volume	C/m^3	coulomb per cubic meter
Conductance	S	siemens
Conductivity	S/m	siemens per meter
Current	A	ampere
Current density	A/m^2	ampere per square meter
Electric intensity	V/m	volt per meter
Force	N	newton
Inductance	H	henry
Length	m	meter
Magnetic flux	Wb	weber
Magnetic flux density	T	tesla
Magnetomotive force	A	ampere (or ampere-turn)
Mass	kg	kilogram
Potential difference	V	volt
Power	W	watt
Resistance	Ω	ohm
Time	s	second

Characteristics of Copper Conductors

Hard-drawn, 97.3 Per Cent Conductivity

Conductor size		OD, in.	Wt, lb/mile	Capacity,* amp	x'_a†	r_a‡	x_a¶
Cir mils	Awg or B.&S.						
1,000,000	...	1.152	16,300	1,300	0.0901	0.0685	0.400
900,000	...	1.092	14,670	1,220	.0916	0.0752	.406
800,000	...	1.029	13,040	1,130	.0934	0.0837	.413
750,000	...	0.997	12,230	1,090	.0943	0.0888	.417
700,000	...	.963	11,410	1,040	.0954	0.0947	.422
600,000	...	.891	9,781	940	.0977	0.109	.432
500,000	...	.814	8,151	840	.1004	0.130	.443
450,000	...	.770	7,336	780	.1020	0.144	.451
400,000	...	.726	6,521	730	.1038	0.162	.458
350,000	...	.679	5,706	670	.1058	0.184	.466
300,000	...	.629	4,891	610	.1080	0.215	.476
250,000	...	.574	4,076	540	.1108	0.257	.487
211,600	4/0	.522	3,450	480	.1136	0.303	.503
167,800	3/0	.464	2,736	420	.1171	0.382	.518
133,100	2/0	.414	2,170	360	.1205	0.481	.532
105,500	1/0	.368	1,720	310	.1240	0.607	.546
83,690	1	.328	1,364	270	.1274	0.765	.560
66,370	2	.320	1,071	240	.1281	0.955	.571
52,630	3	.285	850	200	.1315	1.20	.585
41,740	4	.254	674	180	.1349	1.52	.599
33,100	5	.226	534	150	.1384	1.91	.613
26,250	6	.162	420	120	.1483	2.39	.637
20,800	7	.144	333	110	.1517	3.01	.651
16,510	8	.129	264	90	.1552	3.80	.665

* Approximate current-carrying capacity for conductor at 75°C, air at 25°C, wind 1.4 mph (2 fps), 60 cycles.
† x'_a = shunt capacitive reactance at 1 ft, megohms/mile.
‡ r_a = resistance at 50°C, 60 cycles, ohms/(conductor)(mile).
¶ x_a = reactance at 1-ft spacing, 60 cycles, ohms/(conductor)(mile).

Conversion Factors for Units of Electrical Resistivity

MULTIPLY VALUE by appropriate factor to OBTAIN →	(SI unit) Ω m	(Unit used in this work) 10^{-8} Ω m
10^{-8} Ω m	1×10^{-8}	1
$\mu\Omega$ cm	1×10^{-8}	1
Ω cm	1×10^{-2}	1×10^{6}
Ω m	1	1×10^{8}
Ω cmil ft^{-1}	1.66243×10^{-9}	1.66243×10^{-1}
Ω in	2.54×10^{-2}	2.54×10^{6}
Ω ft	3.048×10^{-1}	3.048×10^{7}
abohm—centimeter	1×10^{-11}	1×10^{-3}
emu	1×10^{-11}	1×10^{-3}
statohm—centimeter	8.98755×10^{9}	8.98755×10^{17}
esu	8.98755×10^{9}	8.98755×10^{17}
10^{-6} ohm per centimeter cubed	1×10^{-8}	1
σ(in units of $(\Omega\ cm)^{-1}$)	$(1\times10^{-2})/\sigma$	$(1\times10^{6})/\sigma$
Ω mm^2 m^{-1}	1×10^{-6}	1×10^{2}
percent IACS	1×10^{-4}/(58×percent IACS)	1×10^{4}/(58×percent IACS)

Dimensions, Weight, and Resistance of Pure Copper Wire

AWG	Diam, in.	Area, d^2, cir mils	Lb/1,000 ft (bare wire)	Ft length/lb	Resistance, 77°F, ohms/1,000 ft (bare wire)
	1.152	1,000,000	3,088	0.3238	0.0108
	1.031	800,000	2,470	0.4048	0.0135
	0.964	700,000	2,161	0.4627	0.0154
	.893	600,000	1,853	0.5397	0.0180
	.813	500,000	1,544	0.6477	0.0216
	.728	400,000	1,235	0.8 97	0.0270
	.575	250,000	772	1.30	0.0431
0000	.4600	211,600	653.3	1.53	0.0509
000	.4096	167,800	518.1	1.93	0.0642
00	.3648	133,100	410.9	2.43	0.0811
0	.3248	105,500	325.8	3.07	0.102
1	.2893	83,690	258.9	3.87	0.129
2	.2576	66,370	204.9	4.88	0.162
3	.2294	52,640	162.5	6.15	0.205
4	.2043	41,740	128.9	7.76	0.259
6	.1620	26,250	81.05	12.34	0.410
8	.1284	16,510	49.98	20.01	0.641
10	.1018	10,380	31.43	31.82	1.018
12	.0808	6,530	19.77	50.59	1.619
14	.0640	4,107	12.43	80.44	2.575
16	.0508	2,583	7.82	127.90	4.094
18	.0403	1,624	4.92	203.40	6.510
20	.0319	1,022	3.09	323.4	10.35
22	.0254	642	1.95	514.2	16.46
24	.0201	404	1.22	817.7	26.17
26	.0159	254	0.77	1,300	41.62
28	.0126	159.8	.48	2,067	66.17
30	.0100	100.5	.30	3,287	105.2
32	.0080	63.2	.19	5,227	167.3
34	.0063	39.7	.12	8,310	266.0
36	.0050	25.0	.076	13,210	423.0
38	.0040	15.7	.047	21,010	672.6
40	.0031	9.89	.030	33,410	1,069
42	.0025	6.22	.019	52,800	1,701
44	.0020	3.91	.012	82,500	2,703
46	.0016	2.46	.008	128,800	4,299
48	.0012	1.55	.004	229,600	6,836
50	.0010	0.97	.003	330,000	10,870

PLATINUM WIRE

Mass in Grams per Foot

B. & S. Gauge	Diameter, inches	Mass, g per ft.	B. & S. Gauge	Diameter, inches	Mass, g per ft.
10	.1019	37.5	23	.02257	1.8
11	.09074	28.0	24	.02010	1.4
12	.08081	22.0	25	.01790	1.1
13	.07196	17.5	26	.01594	0.9
14	.06408	14.0	27	.01420	0.7
15	.05707	11.0	28	.01264	0.6
16	.05082	9.0	29	.01126	0.45
17	.04526	7.0	30	.01003	0.35
18	.04030	5.7	31	.008928	0.28
19	.03589	4.4	32	.007950	0.22
20	.03196	3.4	33	.007080	0.17
21	.02846	2.9	34	.006305	0.15
22	.02535	2.3	35	.005615	0.11

RESISTANCE OF WIRES (at 20°C)

The following dimensions have been adopted in the computations.

B. & S. gauge	Diameter mm	Diameter mils 1 mil = .001 in.	B. & S. gauge	Diameter mm	Diameter mils 1 mil = .001 in.
10	2.588	101.9	26	0.4049	15.94
12	2.053	80.81	27	0.3606	14.20
14	1.628	64.08	28	0.3211	12.64
16	1.291	50.82	30	0.2546	10.03
18	1.024	40.30	32	0.2019	7.950
20	0.8118	31.96	34	0.1601	6.305
22	0.6438	25.35	36	0.1270	5.000
24	0.5106	20.10	40	0.07987	3.145

***Advance (0°C) $\varrho = 48. \times 10^{-6}$ ohm cm**

B. & S. No.	Ohms per cm
10	.000912
12	.00145
14	.00231
16	.00367
18	.00583
20	.00927
22	.0147
24	.0234
26	.0373
27	.0470
28	.0593
30	.0942
32	.150
34	.238
36	.379
40	.958

Brass $\varrho = 7.00 \times 10^{-6}$ ohm cm

B. & S. No.	Ohms per cm
10	.000133
12	.000212
14	.000336
16	.000535
18	.000850
20	.00135
22	.00215
24	.00342
26	.00543
27	.00686
28	.00864
30	.0137
32	.0219
34	.0348
36	.0552
40	.140

Constantan (0°C) $\varrho = 44.1 \times 10^{-6}$ ohm cm

B. & S. No.	Ohms per cm
10	.000838
12	.00133
14	.00212
16	.00337
18	.00536
20	.00852
22	.0135
24	.0215
26	.0342
27	.0432
28	.0545
30	.0866
32	.138
34	.219
36	.348
40	.880

Aluminum $\varrho = 2.828 \times 10^{-6}$ ohm cm

B. & S. No.	Ohms per cm
10	.0000538
12	.0000855
14	.000136
16	.000216
18	.000344
20	.000546
22	.000869
24	.00138
26	.00220
27	.00277
28	.00349
30	.00555
32	.00883
34	.0140
36	.0223
40	.0564

Climax $\varrho = 87. \times 10^{-6}$ ohm cm

B. & S. No.	Ohms per cm
10	.00165
12	.00263
14	.00418
16	.00665
18	.0106
20	.0168
22	.0267
24	.0425
26	.0675
27	.0852
28	.107
30	.171
32	.272
34	.432
36	.687
40	1.74

Copper, annealed $\varrho = 1.724 \times 10^{-6}$ ohm cm

B. & S. No.	Ohms per cm
10	.0000328
12	.0000521
14	.0000828
16	.000132
18	.000209
20	.000333
22	.000530
24	.000842
26	.00134
27	.00169
28	.00213
30	.00339
32	.00538
34	.00856
36	.0136
40	.0344

Eureka (0°C) $\varrho = 47. \times 10^{-6}$ ohm cm

B. & S. No.	Ohms per cm
10	.000893
12	.00142
14	.00226
16	.00359
18	.00571
20	.00908
22	.0144
24	.0230
26	.0365
27	.0460
28	.0580
30	.0923
32	.147
34	.233
36	.371
40	.938

German silver $\varrho = 33. \times 10^{-6}$ ohm cm

B. & S. No.	Ohms per cm
10	.000627
12	.000997
14	.00159
16	.00252
18	.00401
20	.00638
22	.0101
24	.0161
26	.0256
27	.0323
28	.0408
30	.0648
32	.103
34	.164
36	.260
40	.659

Iron $\varrho = 10. \times 10^{-6}$ ohm cm

B. & S. No.	Ohms per cm
10	.000190
12	.000302
14	.000481
16	.000764
18	.00121
20	.00193
22	.00307
24	.00489
26	.00776
27	.00979
28	.0123
30	.0196
32	.0312
34	.0497
36	0.789
40	.200

Excello $\varrho = 92. \times 10^{-6}$ ohm cm

B. & S. No.	Ohms per cm
10	.00175
12	.00278
14	.00442
16	.00703
18	.0112
20	.0178
22	.0283
24	.0449
26	.0714
27	.0901
28	.114
30	.181
32	.287
34	.457
36	.726
40	1.84

Gold $\varrho = 2.44 \times 10^{-6}$ ohm cm

B. & S. No.	Ohms per cm
10	.0000464
12	.0000737
14	.000117
16	.000186
18	.000296
20	.000471
22	.000750
24	.00119
26	.00189
27	.00239
28	.00301
30	.00479
32	.00762
34	.0121
36	.0193
40	.0487

Lead $\varrho = 22. \times 10^{-6}$ ohm cm

B. & S. No.	Ohms per cm
10	.000418
12	.000665
14	.00106
16	.00168
18	.00267
20	.00425
22	.00676
24	.0107
26	.0171
27	.0215
28	.0272
30	.0432
32	.0687
34	.109
36	.174
40	.439

RESISTANCE OF WIRES (continued)

Magnesium $\varrho = 4.6 \times 10^{-6}$ ohm cm

B. & S. No.	Ohms per cm	Ohms per ft.
10	.0000874	
12	.000139	
14	.000221	
16	.000351	
18	.000559	
20	.000889	
22	.00141	
24	.00225	
26	.00357	
27	.00451	
28	.00568	
30	.00903	
32	.0144	
34	.0228	
36	.0363	
40	.0918	

Molybdenum $\varrho = 5.7 \times 10^{-6}$ ohm cm

B. & S. No.	Ohms per cm	Ohms per ft.
10	.000108	
12	.000172	
14	.000274	
16	.000435	
18	.000693	
20	.00110	
22	.00175	
24	.00278	
26	.00443	
27	.00558	
28	.00704	
30	.0112	
32	.0178	
34	.0283	
36	.0450	
40	.114	

***Nichrome $\varrho = 150. \times 10^{-6}$ ohm cm**

B. & S. No.	Ohms per cm	Ohms per ft.
10	.0021281	
12	.0033751	
14	.0054054	
16	.0085116	
18	.0138383	
20	.0216218	
22	.0346040	
24	.0548088	
26	.0875760	
28	.1394328	
30	.2214000	
32	.346040	
34	.557600	
36	.885600	
38	1.383832	
40	2.303872	

Manganin $\varrho = 44. \times 10^{-6}$ ohm cm

B. & S. No.	Ohms per cm	Ohms per ft.
10	.000836	
12	.00133	
14	.00211	
16	.00336	
18	.00535	
20	.00850	
22	.0135	
24	.0215	
26	.0342	
27	.0431	
28	.0543	
30	.0864	
32	.137	
34	.218	
36	.347	
40	.878	

Monel Metal $\varrho = 42. \times 10^{-6}$ ohm cm

B. & S. No.	Ohms per cm	Ohms per ft.
10	.000798	
12	.00127	
14	.00202	
16	.00321	
18	.00510	
20	.00811	
22	.0129	
24	.0205	
26	.0326	
27	.0411	
28	.0519	
30	.0825	
32	.131	
34	.209	
36	.331	
40	.838	

Nickel $\varrho = 7.8 \times 10^{-6}$ ohm cm

B. & S. No.	Ohms per cm	Ohms per ft.
10	.000148	
12	.000236	
14	.000375	
16	.000596	
18	.000948	
20	.00151	
22	.00240	
24	.00381	
26	.00606	
27	.00764	
28	.00963	
30	.0153	
32	.0244	
34	.0387	
36	.0616	
40	.156	

Platinum $\varrho = 10. \times 10^{-6}$ ohm cm

B. & S. No.	Ohms per cm	Ohms per ft.
10	.000190	
12	.000302	
14	.000481	
16	.000764	
18	.00121	
20	.00193	
22	.00307	
24	.00489	
26	.00776	
27	.00979	
28	.0123	
30	.0196	
32	.0312	
34	.0497	
36	.0789	
40	.200	

Steel, piano wire (0°C) $\varrho = 11.8 \times 10^{-6}$ ohm cm

B. & S. No.	Ohms per cm	Ohms per ft.
10	.000224	
12	.000357	
14	.000567	
16	.000901	
18	.00143	
20	.00228	
22	.00363	
24	.00576	
26	.00916	
27	.0116	
28	.0146	
30	.0232	
32	.0368	
34	.0586	
36	.0931	
40	.236	

Tantalum $\varrho = 15.5 \times 10^{-6}$ ohm cm

B. & S. No.	Ohms per cm	Ohms per ft.
10	.000295	.00898
12	.000468	.0143
14	.000745	.0227
16	.00118	.0361
18	.00188	.0574
20	.00299	.0913
22	.00476	.145
24	.00757	.231
26	.0120	.367
27	.0152	.463
28	.0191	.583
30	.0304	.928
32	.0484	1.47
34	.0770	2.35
36	.122	3.73
40	.309	9.43

Silver (18°C) $\varrho = 1.629 \times 10^{-6}$ ohm cm

B. & S. No.	Ohms per cm	Ohms per ft.
10	.0000310	
12	.0000492	
14	.0000783	
16	.000124	
18	.000198	
20	.000315	
22	.000500	
24	.000796	
26	.00126	
27	.00160	
28	.00201	
30	.00320	
32	.00509	
34	.00809	
36	.0129	
40	.0325	

Steel, invar (35% Ni) $\varrho = 81. \times 10^{-6}$ ohm cm

B. & S. No.	Ohms per cm	Ohms per ft.
10	.00154	
12	.00245	
14	.00389	
16	.00619	
18	.00984	
20	.0156	
22	.0249	
24	.0396	
26	.0629	
27	.0793	
28	.100	
30	.159	
32	.253	
34	.402	
36	.639	
40	1.62	

Tin $\varrho = 11.5 \times 10^{-6}$ ohm cm

B. & S. No.	Ohms per cm	Ohms per ft.
10	.000219	
12	.000348	
14	.000553	
16	.000879	
18	.00140	
20	.00222	
22	.00353	
24	.00562	
26	.00893	
27	.0113	
28	.0142	
30	.0226	
32	.0359	
34	.0571	
36	.0908	
40	.230	

Tungsten $\varrho = 5.51 \times 10^{-6}$ ohm cm

B. & S. No.	Ohms per cm
10	.000105
12	.000167
14	.000265
16	.000421
18	.000669
20	.00106
22	.00169
24	.00269
26	.00428
27	.00540
28	.00680
30	.0108
32	.0172
34	.0274
36	.0435
40	.110

Zinc (0°C) $\varrho = 5.75 \times 10^{-6}$ ohm cm

B. & S. No.	Ohms per cm
10	.000109
12	.000174
14	.000276
16	.000439
18	.000699
20	.00111
22	.00177
24	.00281
26	.00446
27	.00563
28	.00710
30	.0113
32	.0180
34	.0286
36	.0454
40	.115

COMPARISON OF WIRE GAUGES

Wire Diameter (Inches)

Gauge No.	Brown & Sharpe	Birmingham or Stubs'	Washburn & Moen	Imperial or Brit. Std.	Stubs' Steel	U.S. Std. plate
00000000						
0000000				.500		
000000				.464		.46875
00000				.432		.4375
0000	.4600	.454	.3938	.400		.40625
000	.4096	.425	.3625	.372		.375
00	.3648	.380	.3310	.348		.34375
0	.3249	.340	.3065	.324		.3125
1	.2893	.300	.2830	.300	.227	.28125
2	.2576	.284	.2625	.276	.219	.26525
3	.2294	.259	.2437	.252	.212	.25
4	.2043	.238	.2253	.232	.207	.234375
5	.1819	.220	.2070	.212	.204	.21875
6	.1620	.203	.1920	.192	.201	.203125
7	.1443	.180	.1770	.176	.199	.1875
8	.1285	.165	.1620	.160	.197	.171875
9	.1144	.148	.1483	.144	.194	.15625
10	.1019	.134	.1350	.128	.191	.140625
11	.09074	.120	.1205	.116	.188	.125
12	.08081	.109	.1055	.104	.185	.109375
13	.07196	.095	.0915	.092	.182	.09375
14	.06408	.083	.0800	.080	.180	.078125
15	.05707	.072	.0720	.072	.178	.0703125
16	.05082	.065	.0625	.064	.175	.0625
17	.04526	.058	.0540	.056	.172	.05625
18	.04030	.049	.0475	.048	.168	.05
19	.03589	.042	.0410	.040	.164	.04375
20	.03196	.035	.0348	.036	.161	.0375
21	.02846	.032	.0318	.032	.157	.034375
22	.02535	.028	.0286	.028	.155	.03125
23	.02257	.025	.0258	.024	.153	.028125
24	.02010	.022	.0230	.022	.151	.025
25	.01790	.020	.0204	.020	.148	.021875
26	.01594	.018	.0181	.018	.146	.01875
27	.01419	.016	.0173	.0164	.143	.0171875
28	.01264	.014	.0162	.0149	.139	.015625
29	.01126	.013	.0150	.0136	.134	.0140625
30	.01003	.012	.0140	.0124	.127	.0125
31	.008928	.010	.0132	.0116	.120	.0109375
32	.007950	.009	.0128	.0108	.115	.01015625
33	.007080	.008	.0118	.0100	.112	.009375
34	.006304	.007	.0104	.0092	.110	.00859375
35	.005614	.005	.0095	.0084	.108	.0078125
36	.005000	.004	.0090	.0076	.106	.00703125
37	.004453		.0085	.0068	.103	.006640625
38	.003965		.0080	.0060	.101	.00625
39	.003531		.0075	.0052	.099	
40	.003145		.0070	.0048	.097	
41			.0066	.0044	.095	
42			.0062	.0040	.092	
43			.0060	.0036	.088	
44			.0058	.0032	.085	
45			.0055	.0028	.081	
46			.0052	.0024	.079	
47			.0050	.0020	.077	
48			.0048	.0016	.075	
49			.0046	.0012	.072	
50			.0044	.0010	.069	

Wire Diameter (Centimeters)

Gauge No.	Brown & Sharpe	Birmingham or Stubs'	Washburn & Moen	Imperial or Brit. Std.	Stubs' Steel	U.S. Std. plate
00000000						
0000000			1.245	1.27		1.27
000000			1.172	1.18		1.191
00000			1.093	1.10		1.111
0000	1.168	1.15	1.000	1.02		1.032
000	1.040	1.08	0.9208	0.945		0.9525
00	0.9266	0.965	0.8407	0.884		0.8731
0	0.8252	0.864	0.7785	0.823		0.7938
1	0.7348	0.762	0.7188	0.762	0.577	0.7144
2	0.6543	0.721	0.6668	0.701	0.556	0.6747
3	0.5827	0.658	0.6190	0.640	0.538	0.6350
4	0.5189	0.605	0.5723	0.589	0.526	0.5953
5	0.4620	0.559	0.5258	0.538	0.518	0.5556
6	0.4115	0.516	0.4877	0.488	0.511	0.5159
7	0.3665	0.457	0.4496	0.447	0.505	0.4763
8	0.3264	0.419	0.4115	0.406	0.500	0.4366
9	0.2906	0.376	0.3767	0.366	0.493	0.3969
10	0.2588	0.304	0.3429	0.325	0.485	0.3572
11	0.2305	0.305	0.3061	0.295	0.478	0.3175
12	0.2053	0.277	0.2680	0.264	0.470	0.2778
13	0.1828	0.241	0.232	0.234	0.462	0.2381
14	0.1628	0.211	0.203	0.203	0.457	0.1984
15	0.1450	0.183	0.183	0.183	0.452	0.1786
16	0.1291	0.165	0.159	0.163	0.445	0.1588
17	0.1150	0.147	0.137	0.142	0.437	0.1429
18	0.1024	0.124	0.121	0.122	0.427	0.1270
19	0.09116	0.107	0.104	0.102	0.417	0.1111
20	0.08118	0.089	0.0884	0.0914	0.409	0.09525
21	0.07229	0.081	0.0808	0.0813	0.399	0.08731
22	0.06439	0.071	0.0726	0.0711	0.394	0.07938
23	0.05733	0.064	0.0655	0.0610	0.389	0.07144
24	0.05105	0.056	0.0584	0.0559	0.384	0.06350
25	0.04547	0.051	0.0518	0.0508	0.376	0.05556
26	0.04049	0.046	0.0460	0.0457	0.371	0.04763
27	0.03604	0.041	0.0439	0.0417	0.363	0.04366
28	0.03211	0.036	0.0411	0.0378	0.353	0.03969
29	0.02860	0.033	0.0381	0.0345	0.340	0.03572
30	0.02548	0.030	0.0356	0.0315	0.323	0.03175
31	0.02268	0.025	0.0335	0.0295	0.305	0.02778
32	0.02019	0.023	0.0325	0.0274	0.292	0.02580
33	0.01798	0.020	0.0300	0.0254	0.284	0.02381
34	0.01601	0.018	0.0264	0.0234	0.279	0.02183
35	0.01426	0.013	0.024	0.0213	0.274	0.01984
36	0.01270	0.010	0.023	0.0193	0.269	0.01786
37	0.01131		0.022	0.0173	0.262	0.01687
38	0.01007		0.020	0.0152	0.257	0.01588
39	0.008969		0.019	0.0132	0.251	
40	0.007988		0.018	0.0122	0.246	
41			0.017	0.0112	0.241	
42			0.016	0.0102	0.234	
43			0.015	0.0091	0.224	
44			0.015	0.0081	0.216	
45			0.014	0.0071	0.206	
46			0.013	0.0061	0.201	
47			0.013	0.0051	0.196	
48			0.012	0.0041	0.191	
49			0.012	0.0030	0.183	
50			0.011	0.0025	0.175	

ALLOWABLE CARRYING CAPACITIES OF COPPER AND ALUMINUM CONDUCTORS[1]

Size A.W.G.	Area Circular Mils	Diameter of Solid Wires Mils	Rubber Insulation Amperes	Varnished Cambric Insulation Amperes	Other Insulations and Bare Conductors Amperes
18	1,624.	40.3	3*		6†
16	2,583.	50.8	6*		10†
14	4,107.	64.1	15	18	20
12	6,530.	80.8	20	25	30
10	10,380.	101.9	25	30	35
8	16,510.	128.5	35	40	50
6	26,250.	162.0	50	60	70
5	33,100.	181.9	55	65	80
4	41,740.	204.3	70	85	90
3	52,630.	229.4	80	95	100
2	66,370.	257.6	90	110	125
1	83,690.	289.3	100	120	150
0	105,500.	325.0	125	150	200
00	135,100.	364.8	150	180	225
000	167,800.	409.6	175	210	275
0000	211,600.	460	225	270	325

* The allowable carrying capacities of No. 18 and 16 are 5 and 7 amperes respectively, when in flexible cords.

† The allowable carrying capacities of No. 18 and 16 are 10 and 15 amperes respectively, when in cords for portable heaters. Types AFS, AFSJ, HC, HPD, and HSJ.

1 Ratings are given for copper wire. Ratings for aluminum wire are to be taken as 84% of the copper ratings.

WIRE TABLE, STANDARD ANNEALED COPPER

Gauge No.	Diameter in mm at 20°C	Cross section in mm² at 20°C	Ohms per kilometer* 0°C	20°C	50°C	75°C
0000	11.68	107.2	0.1482	0.1608	0.1798	0.1956
000	10.40	85.03	.1868	.2028	.2267	.2466
00	9.266	67.43	.2356	.2557	.2858	.3110
0	8.252	53.48	.2971	.3224	.3604	.3921
1	7.348	42.41	.3746	.4066	.4545	.4944
2	6.544	33.63	.4724	.5127	.5731	.6235
3	5.827	26.67	.5956	.6465	.7227	.7862
4	5.189	21.15	.7511	.8152	.9113	.9914
5	4.621	16.77	.9471	1.028	1.149	1.250
6	4.115	13.30	1.194	1.296	1.449	1.576
7	3.665	10.55	1.506	1.634	1.827	1.988
8	3.264	8.366	1.899	2.061	2.304	2.506
9	2.906	6.634	2.395	2.599	2.905	3.161
10	2.588	5.261	3.020	3.277	3.663	3.985
11	2.305	4.172	3.807	4.132	4.619	5.025
12	2.053	3.309	4.801	5.211	5.825	6.337
13	1.828	2.624	6.054	6.571	7.345	7.991
14	1.628	2.081	7.634	8.285	9.262	10.08
15	1.450	1.650	9.627	10.45	11.68	12.71
16	1.291	1.309	12.14	13.17	14.73	16.02
17	1.150	1.038	15.31	16.61	18.57	20.20
18	1.024	.8231	19.30	20.95	23.42	25.48
19	.9116	.6527	24.34	26.42	29.53	32.12
20	.8118	.5176	30.69	33.31	37.24	40.51
21	.7230	.4105	38.70	42.00	46.95	51.08
22	.6438	.3255	48.80	52.96	59.21	64.41
23	.5733	.2582	61.54	66.79	74.66	81.22
24	.5106	.2047	77.60	84.21	94.14	102.4
25	.4547	.1624	97.85	106.2	118.7	129.1
26	.4049	.1288	123.4	133.9	149.7	162.9
27	.3606	.1021	155.6	168.9	188.8	205.4
28	.3211	.08098	196.2	212.9	238.0	258.9
29	.2859	.06422	247.4	268.5	300.1	326.5
30	.2546	.05093	311.9	338.6	378.5	411.7
31	.2268	.04039	393.4	426.9	477.2	519.2
32	.2019	.03203	496.0	538.3	601.8	654.7
33	.1798	.02540	625.5	678.8	758.8	825.5
34	.1601	.02014	788.7	856.0	956.9	1041
35	.1426	.01597	994.5	1079	1207	1313
36	.1270	.01267	1254	1361	1522	1655
37	.1131	.01005	1581	1716	1919	2087
38	.1007	.007967	1994	2164	2419	2632
39	.08969	.006318	2514	2729	3051	3319
40	.07987	.005010	3171	3441	3847	4185

* Resistance at the stated temperatures of a wire whose length is 1 kilometer at 20°C.

Characteristics of Aluminum Cable, Steel-reinforced

Conductor Size, cir mils or Awg	Cu equiv, cir mils or Awg*	OD, in	Wt, lb/mile	Capacity amp†	x'_a‡	r_a§	x_a¶
1,590,000	1000,000	1.545	10,777	1,380	0.0814	0.0684	0.359
1,510,000	950,000	1.506	10,237	1,340	.0821	0.0720	.362
1,431,000	900,000	1.465	9,699	1,300	.0830	0.0760	.365
1,351,000	850,000	1.424	9,160	1,250	.0838	0.0803	.369
1,272,000	800,000	1.382	8,621	1,200	.0847	0.0851	.372
1,192,500	750,000	1.338	8,082	1,160	.0857	0.0906	.376
1,113,000	700,000	1.293	7,544	1,110	.0867	0.0969	.380
1,033,500	650,000	1.246	7,019	1,060	.0878	0.104	.385
954,000	600,000	1.196	6,479	1,010	.0890	0.113	.390
900,000	566,000	1.162	6,112	970	.0898	0.119	.393
874,500	550,000	1.146	5,940	950	.0903	0.123	.395
795,000	500,000	1.093	5,399	900	.0917	0.138	.401
666,000	419,000	1.000	4,527	800	.0943	0.160	.412
636,000	400,000	0.977	4,319	770	.0950	0.169	.414
605,000	380,500	.953	4,109	750	.0957	0.178	.417
556,500	350,000	.927	4,039	730	.0965	0.186	.420
477,000	300,000	.858	3,462	670	.0988	0.216	.430
397,500	250,000	.783	2,885	590	.1015	0.259	.441
336,400	4/0	.721	2,442	530	.1039	0.306	.451
266,800	3/0	.642	1,936	460	.1074	0.385	.465
4/0	2/0	.563	1,542	340	.1113	0.592	.581
3/0	1/0	.502	1,223	300	.1147	0.723	.621
2/0	1	.447	970	270	.1182	0.895	.641
1/0	2	.398	769	230	.1216	1.12	.656
1	3	.355	610	200	.1250	1.38	.665
2	4	.316	484	180	.1285	1.69	.665
4	6	.250	304	140	.1355	2.57	.659

* Based on copper 97 per cent; aluminum 61 per cent.
† Approximate current-carrying capacity for conductor at 75°C, air at 25°C, wind 1.4 mph (2 fps), 60 cycles.
‡ x'_a = capacitive reactance at 1 ft, megohms/mile.
§ Resistance, ohms/(conductor)(mile) at 50°C, 60 cycles.
¶ Reactance at 1-ft spacing, 60 cycles, ohms/(conductor)(mile).

Power, Voltage, and Current Ratios and Their Corresponding Values in Decibels

Power ratio	Voltage or current ratio	Decibels (db)	Efficiency, %
1.26	1.12	1.0	79.5
1.58	1.26	2.0	63.4
2.0	1.41	3.0	50.0
3.16	1.78	5.0	31.6
5.01	2.24	7.0	20.0
10.0	3.16	10.0	10.0
50.12	7.08	17.0	1.99
100.0	10.0	20.0	1.0
1,000.0	31.6	30.0	0.1
10^5	316.2	50.0	.001
10^8	10,000.0	80.0	.000001
10^{10}	100,000.0	100.0	.00000001

Color Code for Fixed Resistors

Values in Ohms

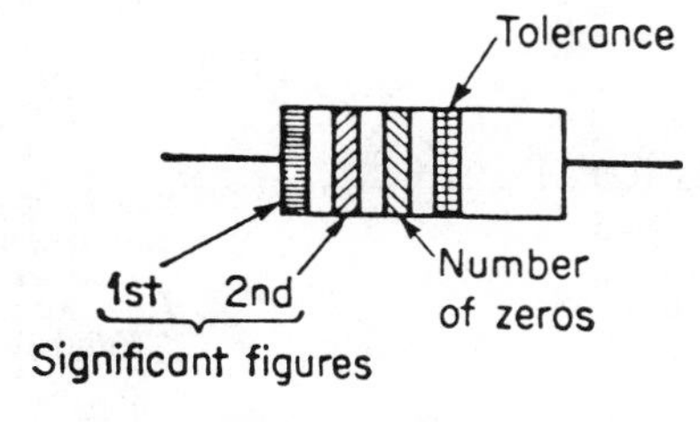

Resistor with axial wire leads

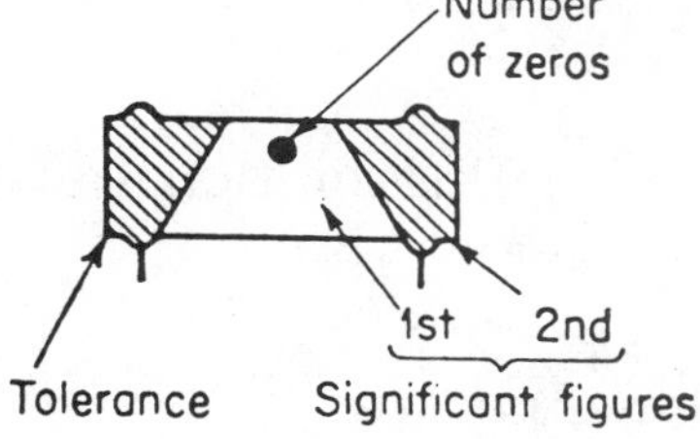

Resistor with radial wire leads

Resistor with radial wire leads....	Body	End	Dot or band	End	
Resistor with axial wire leads.....	1st band	2nd band	3rd band	End band	
Color	Value	Value	Value	Color	Tolerance, %
Black	0	0	None	Gold	± 5
Brown	1	1	0	Silver	± 10
Red	2	2	00	None	± 20
Orange	3	3	000		
Yellow	4	4	0000		
Green	5	5	00000		
Blue	6	6	000000		
Violet	7	7	0000000		
Gray	8	8	00000000		
White	9	9	000000000		

Color Code for JAN Fixed Micra Capacitors

Color-code scheme for JAN standard fixed mica capacitors. The significance of the letters denoting characteristic will be found in Specification JAN C-5.

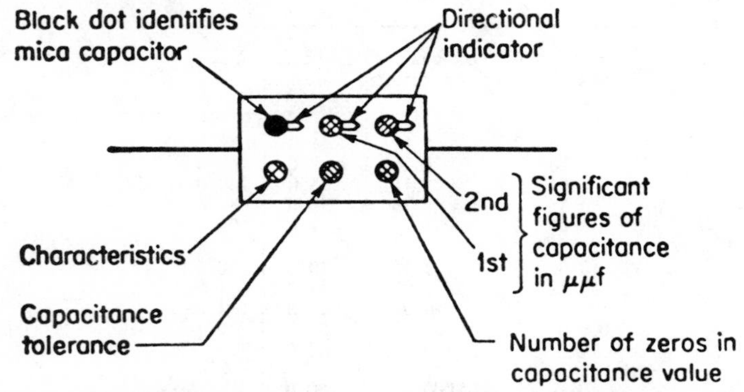

Color	Capacitance, μμf (significant fig)	Decimal multiplier	Tolerance, %	Characteristic
Black......	0	1	20 (M)*	A
Brown.....	1	10	......	B
Red.......	2	100	2 (G)*	C
Orange....	3	1,000	......	D
Yellow.....	4	....	......	E
Green.....	5	....	......	F
Blue.......	6	....	......	G
Violet.....	7			
Gray......	8			
White	9			
Gold......	...	0.1	5 (J)*	
Silver......	...	.01	10 (K)*	

* Code letter for indicated % tolerance.

Illumination Levels, Interior Lighting

	Foot-candles Maintained in Service (Not Initial Values)
Assembly (manufacturing):	
Rough	20
Medium	50
Fine	100
Extra fine	300*
Auditoriums:	
Assembly only	10
Exhibitions	30
Banks:	
Lobby	20
Cages and offices	50
Barber shops and beauty parlors	50
Bathrooms:	
General lighting	5
At mirror (on face)	40
Bedrooms:	
General lighting	5
At mirror (on face)	20
Churches:	
Auditorium	10
Sunday-school rooms	20
Pulpit	20
Classrooms, on desks and chalkboards:	
Typical	30
Sight-saving or special	50
Depots and stations:	
Waiting room	20
Ticket rack and counter	50
Concourse	5
Platforms	5

Illumination Levels, Interior Lighting (continued)

	Foot-candles Maintained in Service (Not Initial Values)
Dining rooms:	
Homes (general lighting)	5
Hotels and restaurants	10
Drafting rooms	50
Elevators	10
Garages:	
Storage	10
Repair and servicing	50
Gymnasiums:	
Exhibitions and matches	30
General exercise	20
Assemblies	10
Dances	5
Lockers and shower rooms	10
Halls and corridors	5
Homes (see specific rooms)	
Hospitals:	
Private rooms and wards:	
General lighting	5
Supplementary for reading	20
Surgery:	
General lighting	50
Operating table	1800
Obstetrical:	
Delivery room	50
Delivery table	200
Examination table	50
Inspection:	
Rough	20
Medium	50
Fine	100
Extra fine	200 or more*
Ironing	40
Kitchens:	
General lighting	10
Supplementary (at task)	40
Laboratories:	
General lighting	30
Work tables	50
Close work	100
Living rooms (see also specific visual task):	
General lighting	5
Lobbies	20
Machine shops:	
Rough bench and machine work	20
Medium bench and machine work	50
Fine bench and machine work	100
Extra fine bench and machine work	200 or more*
Mail rooms	30
Museums and art galleries:	
General lighting	10
On displays	50
Offices:	
Casual visual tasks: inactive file rooms, reception rooms, stairways, washrooms, and other service areas	10
Ordinary visual tasks: general office work (except for work classified as "difficult visual tasks"), private office work, general correspondence, conference rooms, active file rooms, mail rooms	30
Difficult visual tasks: auditing and accounting, business-machine operation, transcribing and tabulation, bookkeeping, drafting, designing	50
Reading:	
Short periods, material of reasonably good visibility	20
Prolonged periods or smaller type	40
Proofreading	100
Schools (see specific rooms)	
Sewing:	
Coarse work, high contrast between thread and fabric	20
Light fabrics, occasional periods	40
Light to medium fabrics, prolonged periods	80
Dark fabrics, fine detail, low contrast	150 or more*

Illumination Levels, Interior Lighting (continued)

	Foot-candles Maintained in Service (Not Initial Values)
Show windows:	
Low surrounding brightness:	
General displays	50
Feature displays	100
Medium surrounding brightness:	
General displays	100
Feature displays	200
High surrounding brightness:	
General displays	200
Feature displays	500
Stairways	10
Storage and stock rooms:	
Rough bulky material	5
Medium material	10
Fine material requiring care	20
Store interiors:	
Circulation areas	20
General merchandising areas	50
Showcases, wall cases, and open-counter displays	100*
Feature displays	200*
Theaters and motion-picture houses:	
Auditorium during intermission	5
Auditorium during picture	0.1
Foyer	5
Lobby	20
Toilets and washrooms	10
Waiting rooms	20
Woodworking:	
Rough sawing and bench work	30
Sizing, planing, rough sanding, veneering, medium machine and bench work	50
Fine bench and machine work, fine sanding and finishing	100
Writing	20

* Usually obtained by supplementary luminaires in combination with general lighting systems providing not less than one-tenth of the recommended value for the task.

Incandescent-lamp Data

Watts	Bulb	Base	Finish	Rated avg life, hr	Initial lumens
General-service Lamps					
100	A-21	Med.	I.f	750	1,620
150	A-23	Med.	I.f-cl.	750	2,600
200	PS-30	Med.	I.f.-cl.	750	3,700
300	PS-30	Med.	I.f.-cl.	750	5,900
300	PS-35	Mogul	I.f.-cl.	1,000	5,650
500	PS-40	Mogul	I.f.-cl.	1,000	9,900
750	PS-52	Mogul	I.f.-cl.	1,000	15,600
1,000	PS-52	Mogul	I.f.-cl.	1,000	21,500
1,500	PS-52	Mogul	I.f.-cl.	1,000	33,000
Projector and Reflector Lamps					
75	PAR-38	Med. skt.	Projector spot	1,000	450 (0–15°)
75	PAR-38	Med. skt.	Projector flood	1,000	550 (0–30°)
150	PAR-38	Med. skt.	Projector spot	1,000	1,150 (0–15°)
150	PAR-38	Med. skt.	Projector flood	1,000	1,400 (0–30°)
75	R-30	Med.	Reflector spot	1,000	220 (0–15°)
75	R-30	Med.	Reflector flood	1,000	300 (0–30°)
150	R-40	Med.	Reflector spot	1,000	600 (0–15°)
150	R-40	Med.	Reflector flood	1,000	800 (0–30°)
300	R-40	Med.	Reflector spot	1,000	1,350 (0–15°)
300	R-40	Med.	Reflector flood	1,000	1,600 (0–30°)

Mercury-lamp Data

Designation	Watts	Bulb	Base	Ballast loss/lamp, watts	Rated avg life, hr*	Initial lumens
A-H1	400	T-16	Mogul	40†	4,000	16,000
A-H12	1,000	T-28	Mogul	85†	3,000	60,000
A-H9	3,000	T-9½	S.C. term	165‡	5,000	120,000

* Rated average life under specified test conditions at 5 hr per start. At 10 hr/start, rated average life is 6,000 hr.

† Single lamp high PF 110 to 125-volt ballasts. Losses for two-lamp ballasts are generally lower.

‡ Single lamp high PF, 230-volt ballast.

Fluorescent-lamp Data

Bulb	Watts	Base	Rated avg life, hr	Rated initial lumens*		
				White	Std cool white	Std warm white
Preheat Lamps						
33″ T-12	25	Med. bipin	7,500†	1,430	1,370	1,440
48″ T-12	40	Med. bipin	7,500†	2,480	2,370	2,500
60″ T-17	90	Mog. bipin	7,500†	4,860	4,650	4,900
Instant-start Lamps						
48″ T-12	40	Med. bipin	6,000‡	2,480	2,370	2,500
60″ T-17	40	Mog. bipin	6,000‡	.....	2,300	
Slimline Lamps¶						
48″ T-12	38§	Single pin	6,000‡	2,320	2,200	2,340
	52			3,020	2,870	3,050
72″ T-12	59	Single pin	6,000‡	3,660	3,500	3,700
	72			4,300	4,100	4,340
96″ T-12	75	Single pin	6,000‡	4,800	4,575	4,850
	96			5,800	5,540	5,860
96″ T-8	34	Single pin	6,000‡	2,280	2,180	2,300
	51			3,300	3,150	3,330
	69			4,350	4,150	4,390

* Lumens measured after 100 hr burning at 80°F ambient and under specified test conditions. The lumen outputs of the de luxe cool white and de luxe warm white lamps are approximately 40 per cent less than those of the corresponding standard cool white and standard warm white. The lumen values of daylight and soft white lamps are 85 and 73 per cent, respectively, of the white values.

† Life under specified test conditions at 3 burning hours per start. Lamp life is slightly longer for more burning hours per start.

‡ Life (tentative) under specified test conditions at 12 burning hours per start. Lamp life is somewhat shorter for fewer burning hours per start.

¶ Slimlines may be operated at any current density within their design range. The figures listed for the 96″ T-8 Slimline are for 120, 200, and 300 ma. The data listed for the T-12 Slimlines are for 425 and 600 ma.

§ Operates on a standard 40-watt instant-start ballast at 420 ma.

Approximate Ballast Loss per Lamp, Watts

Bulb	Watts	Starter switch no. or current, ma	Approximate ballast loss per lamp, watts					
			110–125 volt				220–250 volt, high PF	
			Single lamp		Two-lamp high PF			
			Low PF	High PF	Series	Lead lag	Single lamp	Two lamp
Preheat Lamps								
48″ T-12	40	FS-4	8.5	11	...	7.8	8.5	9.3
60″ T-17	90	FS-85	....	25	...	19.5	...	16
Rapid-start Lamps								
48″ T-12	40	430	12	12	...	8.5		
Instant-start Lamps								
48″ and 60″	40	415	....	20	11	12		
Slimline Lamps								
72″ T-12	55	425	....	25	16	15		
	67	600	....	...	...	17.5		
96″ T-12	74	425	....	29	16	17.5		
	95	600	....	...	...	29		
96″ T-8	50	200	....	20	...	16		
	69	300	....	25	...	23		

Comparison of Electrical Properties and Thermal Endurance of Standard Insulating Materials

	Permittivity (relative) ε_r	Dissipation factor tan δ	Electric strength kV/mm	Thermal endurance °C
Transformer oil	2.2	0.004	20	95
Electrical insulating paper	2.4	0.002	8	105
Pressboard	3.6	0.003	12	105
Laminated paper	4.0	0.030	15	120
Phenolic moulded material	4.0	0.300	8	120
Melamine resin moulded material	6.0	0.500	8	105
Polyester resin moulded material	3.6	0.030	10	120
Polyethylene	2.3	0.0001	40	90
Polystyrene	2.5	0.001	50	60
Polyamide	3.7	0.025	40	100
Polycarbonate	3.0	0.001	25	130
Triacetate film	4.0	0.014	120	120
Polyester film	3.1	0.002	160	125
Polycarbonate film	3.0	0.003	170	130
Polyimide film	3.8	0.003	200	180
Epoxy cast resin	3.3	0.006	20	130
Polyester cast resin	3.4	0.006	25	105
Porcelain	6.0	0.017	35	>200
Air	1.00058	—	45	—

The above electrical properties are given at 20 °C and 50 Hz for a layer thickness of 1 mm or 0.04 mm in the case of film.

Dielectric Constants of Selected Materials (17 to 22°C)

Material	Dielectric constant	Material	Dielectric constant
Acetamide	4.0	Phenanthrene	2.80
Acetanilide	2.9	Phenol (10°C)	4.3
Acetic acid (2°C)	4.1	Phosphorus, red	4.1
Aluminum oleate	2.40	Phosphorus, yellow	3.6
Ammonium bromide	7.1	Potassium aluminum sulfate	3.8
Ammonium chloride	7.0	Potassium carbonate (15°C)	5.6
Antimony trichloride	5.34	Potassium chlorate	5.1
Apatite ⊥ optic axis	9.50	Potassium chloride	5.03
Apatite ∥ optic axis	7.41	Potassium chromate	7.3
Asphalt	2.68	Potassium iodide	5.6
Barium chloride (anhyd.)	11.4	Potassium nitrate	5.0
Barium chloride ($2H_2O$)	9.4	Potassium sulfate	5.9
Barium nitrate	5.9	Quartz ⊥ optic axis	4.34
Barium sulfate (15°C)	11.4	Quartz ∥ optic axis	4.27
Beryl ⊥ optic axis	7.02	Resorcinol	3.2
Beryl ∥ optic axis	6.08	Ruby ⊥ optic axis	13.27
Calcite ⊥ optic axis	8.5	Ruby ∥ optic axis	11.28
Calcite ∥ optic axis	8.0	Rutile ⊥ optic axis	86
Calcium carbonate	6.14	Rutile ∥ optic axis	170
Calcium fluoride	7.36	Selenium	6.6
Calcium sulfate ($2H_2O$)	5.66	Silver bromide	12.2
Cassiterite ⊥ optic axis	23.4	Silver chloride	11.2
Cassiterite ∥ optic axis	24	Silver cyanide	5.6
d-Cocaine	3.10	Smithsonite ⊥ optic axis	9.3
Cupric oleate	2.80	Smithsonite ∥ optic axis	9.4
Cupric oxide (15°C)	18.1	Sodium carbonate (anhyd.)	8.4
Cupric sulfate (anhyd.)	10.3	Sodium carbonate ($10H_2O$)	5.3
Cupric sulfate ($5H_2O$)	7.8	Sodium chloride	6.12
Diamond	5.5	Sodium nitrate	5.2
Diphenylmethane	2.7	Sodium oleate	2.75
Dolomite ⊥ optic axis	8.0	Sodium perchlorate	5.4
Dolomite ∥	6.8	Sucrose (mean)	3.32
Ferrous oxide (15°C)	14.2	Sulfur (mean)	4.0
Iodine	4	Thallium chloride	46.9
Lead acetate	2.6	*p*-Toluidine	3.0
Lead carbonate (15°C)	18.6	Tourmaline ⊥ optic axis	7.10
Lead chloride	4.2	Tourmaline ∥ optic axis	6.3
Lead monoxide (15°C)	25.9	Urea	3.5
Lead nitrate	37.7	Zircon ⊥, ∥	12
Lead oleate	3.27		
Lead sulfate	14.3		
Lead sulfide (15°)	17.9		
Malachite (mean)	7.2		
Mercuric chloride	3.2		
Mercurous chloride	9.4		
Naphthalene	2.52		

Applications	JFET	MOST	IGFET
Gate noise	Shot noise across isolation diode to gate.	Does not have shot noise at gate.	Does not have shot noise at gate.
Channel noise	Buried channel results in low $1/f$ noise.	Surface channel results in $1/f$ noise. Can be reduced by ion inplantation to bury channel.	Surface channel results in $1/f$ noise. Can be reduced by ion inplantation to bury channel.
Substrate noise	The gate is the substrate. Some channel modulation of noise is usually observable.	Up to 80% of substrate noise has been observed modulating the channel.	Up to 80% of substrate noise has been observed modulating the channel.
Transconductance	Rate of decrease in g_m with a reduction in temperature and bias current is greater at the low extreme than in MOSTs.	Rate of decrease in g_m with a reduction in temperature and bias current is less at low extremes than in JFETs.	–
Gate capacitance	Varies with bias.	Remains relatively constant.	Remains relatively constant.
Gate leakage current	Has leakage currents across the reverse-biased diode junctions which are usually an order of magnitude greater than the leakage of the insulation in a MOST or IGFET.	Leakage through the insulation is lower by an order of magnitude than the leakage currents in a JFET.	Leakage through the insulation is lower by an order of magnitude than the leakage currents in a JFET.
Temperature effects	Has a g_m of less than 10% of its room temperature value at 30 K.	Maintains greater than 50% of its room temperature g_m at 4 K.	–
Element density	Obtainable density of devices per chip size is less than for MOSTs and IGFETs.	Obtainable density of devices per chip size is greater than for a comparable JFET.	Obtainable density of devices per chip size is greater than for a comparable JFET.
Sensitive to change by static charge buildup	–	Sensitive to damage by static charge buildup.	Sensitive to damage by static charge buildup.

Types of Transistor Circuits

Transistor Type	*Amplifier Configuration*	*Input Z Range* (ohms)	*State of Art Input Noise Voltage* (rms nV) -55 to + 75°C 1 kHz	100 kHz	*Operating Temperature Range* (K)	*Comments*
Bipolars	Common-Emitter	20×10^3	0.5	0.25	220 - 398	–
	Emitter-Follower	20×10^3	0.7	0.35	220 - 398	–
	Common-Base	10^3	0.5	0.25	220 - 398	–
JFET	Common-Source	10^{10}	2.0	1.5	30 - 398	Silicon units
	Source-Follower	10^{10}	2.0	1.5	30 - 398	–
MOSTs and IGFETs	Common-Source	10^{12}	12.0	4.0	4 - 398	Input *Z* range for cooled device
	Source-Follower	10^{12}	12.0	4.0	4 - 398	

Relations for the resultant of several two-port networks

	Series connection	Parallel connection	Cascade connection	Series-parallel connection	Parallel-series connection
Arrangement	W_1, W_2	Y_1, Y_2	K_1, K_2	H_1, H_2	P_1, P_2
Resulting two-port network	W	Y	K	H	P
Resulting matrix	$W = W_1 + W_2$	$Y = Y_1 + Y_2$	$K = K_1 \cdot K_2$	$H = H_1 + H_2$	$P = P_1 + P_2$

Parameters of important symmetrical two-port networks

	T-network	*Π*-network	Mash network
	Z_r, Z_r, Z_p	Z_r, Z_p, Z_p	Z_r, Z_p, Z_p, Z_r
Open-circuit input impedance $\underline{W}_1$	$\underline{Z}_r + \underline{Z}_p$	$\frac{\underline{Z}_p(\underline{Z}_r + \underline{Z}_p)}{\underline{Z}_r + 2\underline{Z}_p}$	$\frac{1}{2}(\underline{Z}_r + \underline{Z}_p)$
Short-circuit input admittance $\underline{Y}_k$	$\frac{(\underline{Z}_r + \underline{Z}_p)}{\underline{Z}_r(\underline{Z}_r + 2\underline{Z}_p)}$	$\frac{1}{\underline{Z}_r} + \frac{1}{\underline{Z}_p}$	$\frac{1}{2}\left(\frac{1}{\underline{Z}_r} + \frac{1}{\underline{Z}_p}\right)$
Open-circuit transfer impedance $\underline{W}_m$	$\underline{Z}_p$	$\frac{\underline{Z}_p^2}{\underline{Z}_r + 2\underline{Z}_p}$	$\frac{1}{2}(\underline{Z}_p - \underline{Z}_r)$
Short circuit transfer admittance $\underline{Y}_m$	$\frac{\underline{Z}_p}{\underline{Z}_r(\underline{Z}_r + 2\underline{Z}_p)}$	$\frac{1}{\underline{Z}_r}$	$\frac{1}{2}\left(\frac{1}{\underline{Z}_r} - \frac{1}{\underline{Z}_p}\right)$
Open-circuit voltage, short-circuit current-transfer $\underline{\ddot{u}}_U$, $\underline{\ddot{u}}_I$	$1 + \frac{\underline{Z}_r}{\underline{Z}_p}$	$1 + \frac{\underline{Z}_r}{\underline{Z}_p}$	$\frac{\underline{Z}_p + \underline{Z}_r}{\underline{Z}_p - \underline{Z}_r}$
Characteristic impedance $\underline{Z}$	$\sqrt{\underline{Z}_r(\underline{Z}_r + 2\underline{Z}_p)}$	$\underline{Z}_p \cdot \sqrt{\frac{\underline{Z}_r}{\underline{Z}_r + 2\underline{Z}_p}}$	$\sqrt{\underline{Z}_r\underline{Z}_p}$

In the case of a symmetrical two-port network, the series equations can be written as follows:

$$\underline{U}_1 = \underline{K}_{11}\underline{U}_2 + \underline{K}_{12}\underline{I}_2 = \cosh \underline{g} \times \underline{U}_2 + \underline{Z} \sin h\, \underline{g} \times \underline{I}_2$$

where $\cosh \underline{g} = \underline{K}_{11} \stackrel{!}{=} \underline{K}_{22}$,

$$\underline{I}_2 = \underline{K}_{21}\underline{U}_2 + \underline{K}_{22}\underline{I}_2 = \cosh \underline{g} \times \underline{I}_2 + \frac{1}{\underline{Z}} \sinh \underline{g} \times \underline{U}_2$$

where $\underline{Z} \sinh \underline{g} = \underline{K}_{12}$; $\frac{1}{\underline{Z}} \sinh \underline{g} = \underline{K}_{21}$.

Frequency Response, Step Function Response and Characteristic Qualities of Controllers

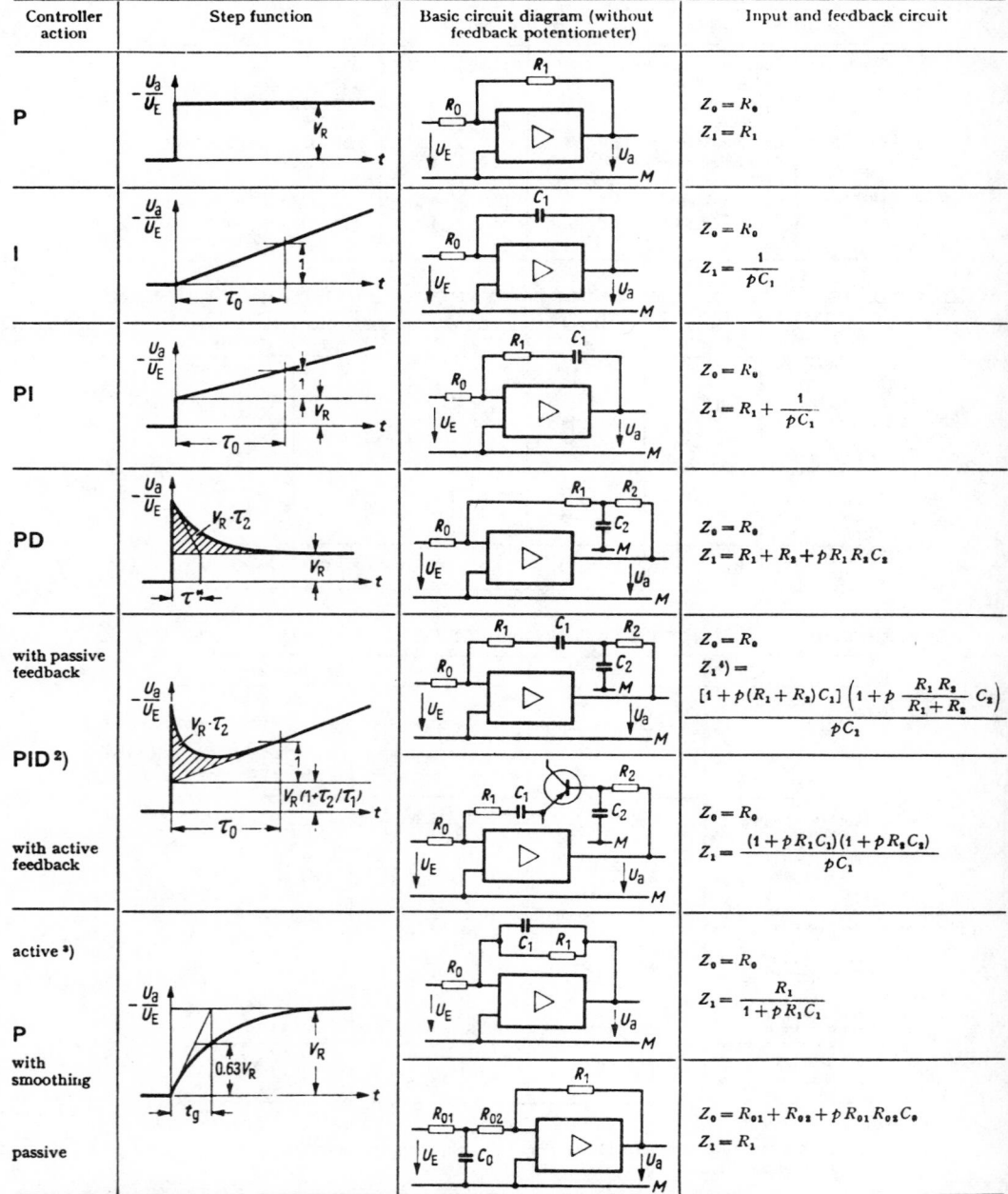

Controller action	Step function	Basic circuit diagram (without feedback potentiometer)	Input and feedback circuit
P		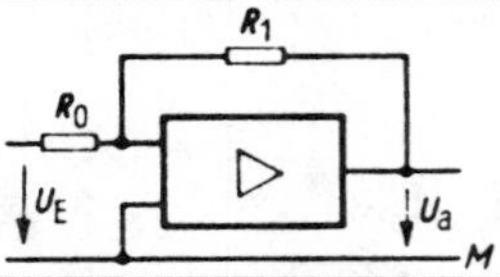	$Z_0 = R_0$ $Z_1 = R_1$
I	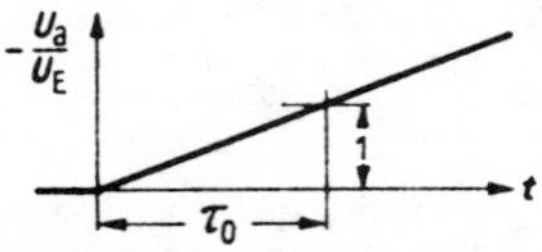	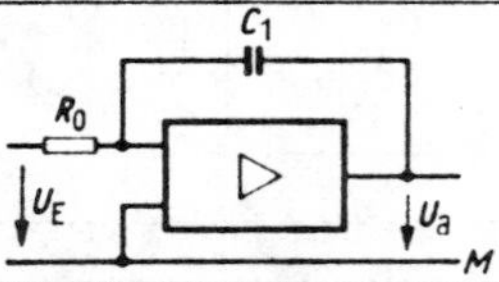	$Z_0 = R_0$ $Z_1 = \frac{1}{pC_1}$
PI	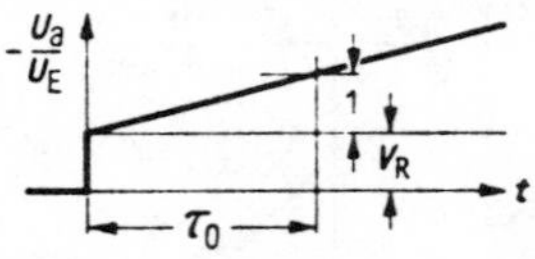	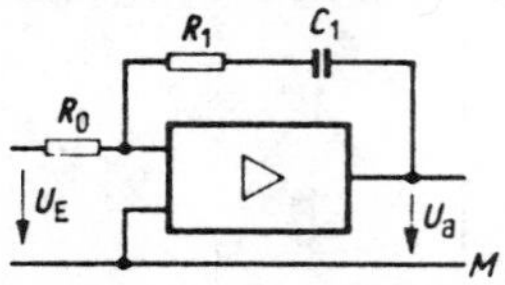	$Z_0 = R_0$ $Z_1 = R_1 + \frac{1}{pC_1}$
PD	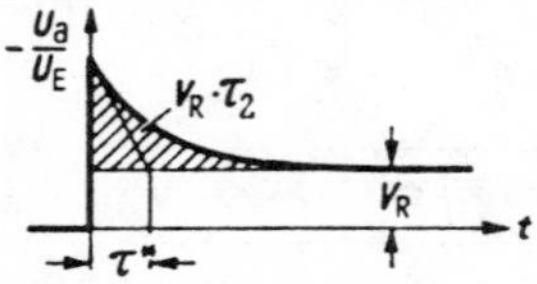	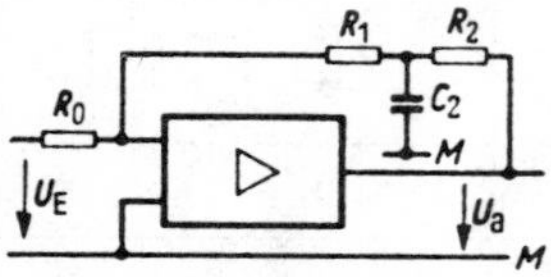	$Z_0 = R_0$ $Z_1 = R_1 + R_2 + pR_1R_2C_2$
PID [2]) with passive feedback	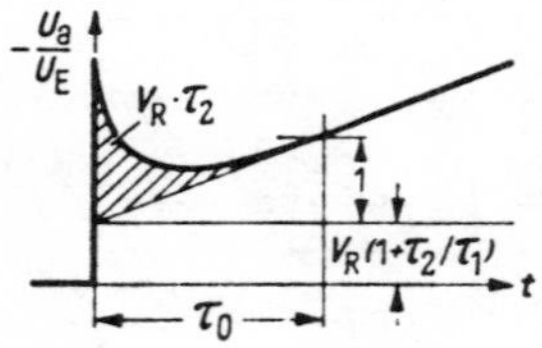	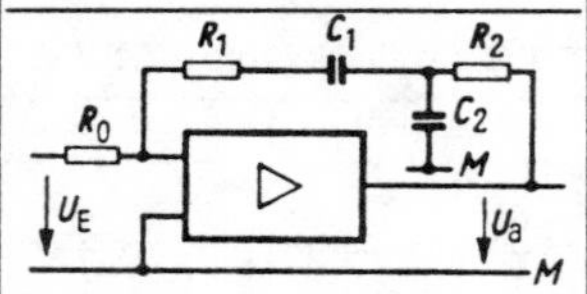	$Z_0 = R_0$ Z_1 [1]) $=$ $\frac{[1 + p(R_1 + R_2)C_1]\left(1 + p\frac{R_1R_2}{R_1 + R_2}C_2\right)}{pC_1}$
PID [2]) with active feedback		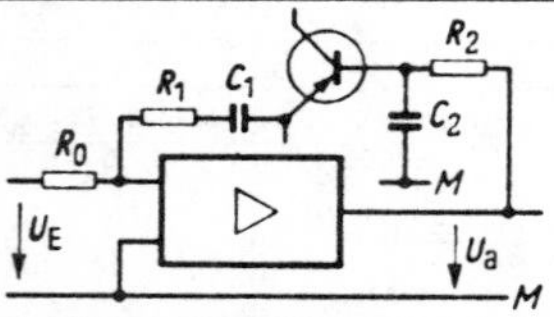	$Z_0 = R_0$ $Z_1 = \frac{(1 + pR_1C_1)(1 + pR_2C_2)}{pC_1}$
P with smoothing, active [3])	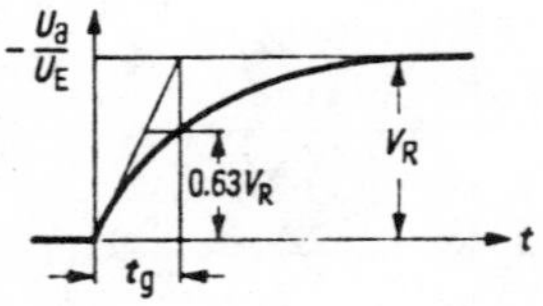	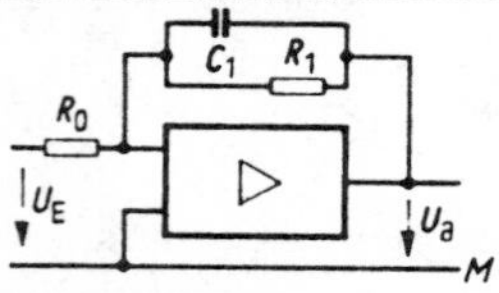	$Z_0 = R_0$ $Z_1 = \frac{R_1}{1 + pR_1C_1}$
P with smoothing, passive		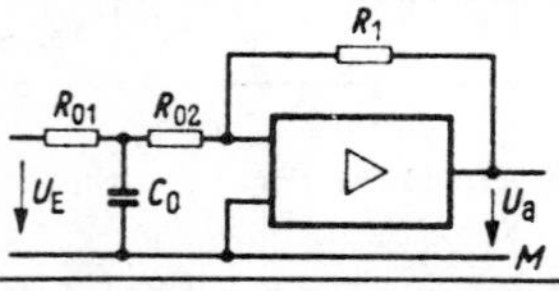	$Z_0 = R_{01} + R_{02} + pR_{01}R_{02}C_0$ $Z_1 = R_1$

[1]) Influence of a feedback potentiometer see page 721. [2]) In connection with this see page 733.
[3]) In connection with this see page 734.

(continued)

Controller action	Frequency response	Proportional amplification [1])	Time constants [1])
P	$F_R = V_R$	$V_R = \frac{R_1}{R_0}$	–
I	$F_R = \frac{1}{p\tau_0}$	–	$\tau_0 = R_0 C_1$
PI	$F_R = V_R \frac{1+p\tau_1}{p\tau_1}$	$V_R = \frac{R_1}{R_0}$	$\tau_1 = R_1 C_1 = V_R \tau_0$ $\tau_0 = R_0 C_1$
PD	$F_R = V_R \frac{1+p\tau_2}{1+p\tau^*}$	$V_R = \frac{R_1+R_2}{R_0}$	$\tau_2 = \frac{R_1 R_2}{R_1+R_2} C_2$ $\tau^* = \frac{R_1+R_2}{R_{ü}} \left(1 + \frac{R_e}{R_0} + \frac{R_e}{R_1}\right) \tau_2$
PID [2]) with passive feedback	$F_R = V_R \frac{(1+p\tau_1)(1+\tau_2)}{p\tau_1}$ [5])	$V_R = \frac{R_1+R_2}{R_0}$	$\tau_1 = (R_1+R_2) C_1$ $\tau_2 = \frac{R_1 R_2}{R_1+R_2} C_2$ $\tau_0 = R_0 C_1$
PID [2]) with active feedback		$V_R = \frac{R_1}{R_0}$	$\tau_1 = R_1 C_1$ $\tau_2 = R_2 C_2$ $\tau_0 = R_0 C_1$
P with smoothing, active [3])	$F_R = \frac{V_R}{1+pt_g}$	$V_R = \frac{R_1}{R_0}$	$t_g = R_1 C_1$
P with smoothing, passive		$V_R = \frac{R_1}{R_{01}+R_{02}}$	$t_g = \frac{R_{01} R_{02}}{R_{01}+R_{02}} C_0$

[4]) Approximation. [5]) Coupling time constant $\tau_{21} = R_2 C_1$ is neglected in the case of PID-controller with passive feedback.

CHAPTER 25

CHEMICAL TECHNOLOGY

Periodic Table of the Elements

The number above the symbol is the atomic weight, the numbers below are the atomic number and the density in g/cm^3 at room temperature (20°C).

I	II	III	IV	V	VI	VII	VIII		
1.01 H 1 0.0001									4.00 He 2 0.0002
6.94 Li 3 0.5	9.01 Be 4 1.8	10.81 B 5 2.5	12.01 C 6 2.3/3.5	14.01 N 7 0.0013	16.00 O 8 0.0014	19.00 F 9 0.0017			20.18 Ne 10 0.0009
22.99 Na 11 1.0	24.31 Mg 12 1.7	26.98 Al 13 2.7	28.09 Si 14 2.4	30.97 P 15 1.8/2.3	32.06 S 16 2.0/2.1	35.45 Cl 17 0.0032			39.95 Ar 18 0.0018
39.10 K 19 0.9	40.08 Ca 20 1.6	44.96 Sc 21 2.5	47.88 Ti 22 4.5	50.94 V 23 6.0	52.00 Cr 24 7.1	54.94 Mn 25 7.4	55.85 Fe 26 7.9	58.93 Co 27 8.9	58.69 Ni 28 8.9
63.55 Cu 29 8.9	65.38 Zn 30 7.1	69.72 Ga 31 5.9	72.59 Ge 32 5.9	74.92 As 33 5.7	78.96 Se 34 4.5/4.8	79.90 Br 35 3.1			83.80 Kr 36 0.0037
85.47 Rb 37 1.5	87.62 Sr 38 2.6	88.91 Y 39 5.5	91.22 Zr 40 6.5	92.91 Nb 41 8.5	95.94 Mo 42 10.2	(98) Tc 43 11.5	101.1 Ru 44 12.3	102.9 Rh 45 12.5	106.4 Pd 46 12.0
107.9 Ag 47 10.5	112.4 Cd 48 8.6	114.8 In 49 7.3	118.7 Sn 50 5.8/7.3	121.8 Sb 51 6.7	127.6 Te 52 6.2	126.9 I 53 4.9			131.3 Xe 54 0.0059
132.9 Cs 55 1.9	137.3 Ba 56 3.5	1)	178.5 Hf 72 13.3	180.9 Ta 73 16.6	183.9 W 74 19.3	186.2 Re 75 20.5	190.2 Os 76 22.5	192.2 Ir 77 22.4	195.1 Pt 78 21.4
197.0 Au 79 19.3	200.6 Hg 80 13.5	204.4 Tl 81 11.8	207.2 Pb 82 11.3	209.0 Bi 83 9.8	(209) Po 84 9.2	(210) At 85			(222) Rn 86 0.0099
(223) Fr 87	226.0 Ra 88 5.0	2)	(261) Unq 104	(262) Unp 105	(263) Unh 106				
1) Lanthanides:			138.9 La 57 6.2	140.1 Ce 58 6.8	140.9 Pr 59 6.5	144.2 Nd 60 6.9	(145) Pm 61	150.4 Sm 62 7.7	152.0 Eu 63 5.2
		157.3 Gd 64 7.9	158.9 Tb 65 8.3	162.5 Dy 66 8.6	164.9 Ho 67 10.1	167.3 Er 68 9.1	168.9 Tm 69 9.3	173.0 Yb 70 7.0	175.0 Lu 71 9.7
2) Actinides:			227.0 Ac 89	232.0 Th 90 11.6	231.0 Pa 91 15.4	238.0 U 92 18.7	237.0 Np 93	(244) Pu 94	(243) Am 95
		(247) Cm 96	(247) Bk 97	(251) Cf 98	(252) Es 99	(257) Fm 100	(258) Md 101	(258) No 102	(260) Lr 103

Names, Symbols, and Atomic Number of the Elements

Name	Symbol	Atomic number	Name	Symbol	Atomic number	Name	Symbol	Atomic number
Actinium	Ac	89	Gold (Aurum)	Au	79	Praseodymium	Pr	59
Aluminum	Al	13	Hafnium	Hf	72	Promethium	Pm	61
Americium	Am	95	Helium	He	2	Protactinium	Pa	91
Antimony	Sb	51	Holmium	Ho	67	Radium	Ra	88
Argon	Ar	18	Hydrogen	H	1	Radon	Rn	86
Arsenic	As	33	Indium	In	49	Rhenium	Re	75
Astatine	At	85	Iodine	I	53	Rhodium	Rh	45
Barium	Ba	56	Iridium	Ir	77	Rubidium	Rb	37
Berkelium	Bk	97	Iron (Ferrum)	Fe	26	Ruthenium	Ru	44
Beryllium	Be	4	Krypton	Kr	36	Samarium	Sm	62
Bismuth	Bi	83	Lanthanum	La	57	Scandium	Sc	21
Boron	B	5	Lead (Plumbum)	Pb	82	Selenium	Se	34
Bromine	Br	35	Lithium	Li	3	Silicon	Si	14
Cadmium	Cd	48	Lutetium	Lu	71	Silver (Argentum)	Ag	47
Calcium	Ca	20	Magnesium	Mg	12	Sodium	Na	11
Californium	Cf	98	Manganese	Mn	25	Strontium	Sr	38
Carbon	C	6	Mendelevium	Md	101	Sulfur	S	16
Cerium	Ce	58	Mercury	Hg	80	Tantalum	Ta	73
Cesium	Cs	55	Molybdenum	Mo	42	Technetium	Tc	43
Chlorine	Cl	17	Neodymium	Nd	60	Tellurium	Te	52
Chromium	Cr	24	Neon	Ne	10	Terbium	Tb	65
Cobalt	Co	27	Neptunium	Np	93	Thallium	Tl	81
Copper (Cuprum)	Cu	29	Nickel	Ni	28	Thorium	Th	90
Curium	Cm	96	Niobium	Nb	41	Thulium	Tm	69
Dysprosium	Dy	66	Nitrogen	N	7	Tin (Stannum)	Sn	50
Einsteinium	Es	99	Nobelium	No	102	Titanium	Ti	22
Erbium	Er	68	Osmium	Os	76	Tungsten (Wolfram)	W	74
Europium	Eu	63	Oxygen	O	8	Uranium	U	92
Fermium	Fm	100	Palladium	Pd	46	Vanadium	V	23
Fluorine	F	9	Phosphorus	P	15	Xenon	Xe	54
Francium	Fr	87	Platinum	Pt	78	Ytterbium	Yb	70
Gadolinium	Gd	64	Plutonium	Pu	94	Yttrium	Y	39
Gallium	Ga	31	Polonium	Po	84	Zinc	Zn	30
Germanium	Ge	32	Potassium	K	19	Zirconium	Zr	40

ELECTRONIC STRUCTURE OF THE ELEMENTS

Atomic No.	Element	K	L		M			N				O				P				Q			
		1	2		3			4				5				6				7			
		s	s	p	s	p	d	s	p	d	f	s	p	d	f	s	p	d	f	s	p	d	f
1	H	1																					
2	He	2																					
3	Li	2	1																				
4	Be	2	2																				
5	B	2	2	1																			
6	C	2	2	2																			
7	N	2	2	3																			
8	O	2	2	4																			
9	F	2	2	5																			
10	Ne	2	2	6																			
11	Na	2	2	6	1																		
12	Mg	2	2	6	2																		
13	Al	2	2	6	2	1																	
14	Si	2	2	6	2	2																	
15	P	2	2	6	2	3																	
16	S	2	2	6	2	4																	
17	Cl	2	2	6	2	5																	
18	Ar	2	2	6	2	6																	
19	K	2	2	6	2	6	..	1															
20	Ca	2	2	6	2	6	..	2															
21	Sc	2	2	6	2	6	1	2															
22	Ti	2	2	6	2	6	2	2															
23	V	2	2	6	2	6	3	2															
24	Cr	2	2	6	2	6	5*	1															
25	Mn	2	2	6	2	6	5	2															
26	Fe	2	2	6	2	6	6	2															
27	Co	2	2	6	2	6	7	2															
28	Ni	2	2	6	2	6	8	2															
29	Cu	2	2	6	2	6	10*	1															
30	Zn	2	2	6	2	6	10	2															
31	Ga	2	2	6	2	6	10	2	1														
32	Ge	2	2	6	2	6	10	2	2														
33	As	2	2	6	2	6	10	2	3														
34	Se	2	2	6	2	6	10	2	4														
35	Br	2	2	6	2	6	10	2	5														
36	Kr	2	2	6	2	6	10	2	6														
37	Rb	2	2	6	2	6	10	2	6	..		1											
38	Sr	2	2	6	2	6	10	2	6	..		2											
39	Y	2	2	6	2	6	10	2	6	1		2											

(continued)

Atomic No	Element	K	L	M	N	O	P	Q
		1	2	3	4	5	6	7
		s	s p	s p d	s p d f	s p d f	s p d f	s p d f
40	Zr	2	2 6	2 6 10	2 6 2 ..	2		
41	Nb	2	2 6	2 6 10	2 6 4* ..	1		
42	Mo	2	2 6	2 6 10	2 6 5 ..	1		
43	Tc	2	2 6	2 6 10	2 6 6 ..	1		
44	Ru	2	2 6	2 6 10	2 6 7 ..	1		
45	Rh	2	2 6	2 6 10	2 6 8 ..	1		
46	Pd	2	2 6	2 6 10	2 6 10* ..	0		
47	Ag	2	2 6	2 6 10	2 6 10 ..	1		
48	Cd	2	2 6	2 6 10	2 6 10 ..	2		
49	In	2	2 6	2 6 10	2 6 10 ..	2 1		
50	Sn	2	2 6	2 6 10	2 6 10 ..	2 2		
51	Sb	2	2 6	2 6 10	2 6 10 ..	2 3		
52	Te	2	2 6	2 6 10	2 6 10 ..	2 4		
53	I	2	2 6	2 6 10	2 6 10 ..	2 5		
54	Xe	2	2 6	2 6 10	2 6 10 ..	2 6		
55	Cs	2	2 6	2 6 10	2 6 10 ..	2 6	1	
56	Ba	2	2 6	2 6 10	2 6 10 ..	2 6	2	
57	La	2	2 6	2 6 10	2 6 10 ..	2 6 1 ..	2	
58	Ce	2	2 6	2 6 10	2 6 10 2*	2 6	2	
59	Pr	2	2 6	2 6 10	2 6 10 3	2 6	2	
60	Nd	2	2 6	2 6 10	2 6 10 4	2 6	2	
61	Pm	2	2 6	2 6 10	2 6 10 5	2 6	2	
62	Sm	2	2 6	2 6 10	2 6 10 6	2 6	2	
63	Eu	2	2 6	2 6 10	2 6 10 7	2 6	2	
64	Gd	2	2 6	2 6 10	2 6 10 7	2 6 1 ..	2	
65	Tb	2	2 6	2 6 10	2 6 10 9*	2 6	2	
66	Dy	2	2 6	2 6 10	2 6 10 10	2 6	2	
67	Ho	2	2 6	2 6 10	2 6 10 11	2 6	2	
68	Er	2	2 6	2 6 10	2 6 10 12	2 6	2	
69	Tm	2	2 6	2 6 10	2 6 10 13	2 6	2	
70	Yb	2	2 6	2 6 10	2 6 10 14	2 6	2	
71	Lu	2	2 6	2 6 10	2 6 10 14	2 6 1 ..	2	
72	Hf	2	2 6	2 6 10	2 6 10 14	2 6 2 ..	2	
73	Ta	2	2 6	2 6 10	2 6 10 14	2 6 3 ..	2	
74	W	2	2 6	2 6 10	2 6 10 14	2 6 4 ..	2	
75	Re	2	2 6	2 6 10	2 6 10 14	2 6 5 ..	2	
76	Os	2	2 6	2 6 10	2 6 10 14	2 6 6 ..	2	
77	Ir	2	2 6	2 6 10	2 6 10 14	2 6 7 ..	2	
78	Pt	2	2 6	2 6 10	2 6 10 14	2 6 9 ..	1	
79	Au	2	2 6	2 6 10	2 6 10 14	2 6 10 ..	1	
80	Hg	2	2 6	2 6 10	2 6 10 14	2 6 10 ..	2	
81	Tl	2	2 6	2 6 10	2 6 10 14	2 6 10 ..	2 1	
82	Pb	2	2 6	2 6 10	2 6 10 14	2 6 10 ..	2 2	
83	Bi	2	2 6	2 6 10	2 6 10 14	2 6 10 ..	2 3	
84	Po	2	2 6	2 6 10	2 6 10 14	2 6 10 ..	2 4	
85	At	2	2 6	2 6 10	2 6 10 14	2 6 10 ..	2 5	
86	Rn	2	2 6	2 6 10	2 6 10 14	2 6 10 ..	2 6	
87	Fr	2	2 6	2 6 10	2 6 10 14	2 6 10 ..	2 6.. ..	1
88	Ra	2	2 6	2 6 10	2 6 10 14	2 6 10 ..	2 6.. ..	2
89	Ac	2	2 6	2 6 10	2 6 10 14	2 6 10 ..	2 6 1 ..	2
90	Th	2	2 6	2 6 10	2 6 10 14	2 6 10 ..	2 6 2 ..	2
91	Pa	2	2 6	2 6 10	2 6 10 14	2 6 10 2*	2 6 1 ..	2
92	U	2	2 6	2 6 10	2 6 10 14	2 6 10 3	2 6 1 ..	2
93	Np	2	2 6	2 6 10	2 6 10 14	2 6 10 4	2 6 1 ..	2
94	Pu	2	2 6	2 6 10	2 6 10 14	2 6 10 6	2 6.. ..	2
95	Am	2	2 6	2 6 10	2 6 10 14	2 6 10 7	2 6.. ..	2
96	Cm	2	2 6	2 6 10	2 6 10 14	2 6 10 7	2 6 1 ..	2
97	Bk	2	2 6	2 6 10	2 6 10 14	2 6 10 9*	2 6.. ..	2
98	Cf	2	2 6	2 6 10	2 6 10 14	2 6 10 10	2 6.. ..	2
99	Es	2	2 6	2 6 10	2 6 10 14	2 6 10 11	2 6.. ..	2
100	Fm	2	2 6	2 6 10	2 6 10 14	2 6 10 12	2 6.. ..	2
101	Md	2	2 6	2 6 10	2 6 10 14	2 6 10 13	2 6.. ..	2
102	No	2	2 6	2 6 10	2 6 10 14	2 6 10 14	2 6.. ..	2
103	Lr	2	2 6	2 6 10	2 6 10 14	2 6 10 14	2 6 1 ..	2
104	—	2	2 6	2 6 10	2 6 10 14	2 6 10 14	2 6 2 ..	2

Gas-constant Values

Temp. Scale	Press. units	Vol. units	Wt. units	Energy units	R
Kelvin..........	....	....	g moles	calories	1.9872
	....	....	g moles	joules (abs)	8.3144
	....	....	g moles	joules (int)	8.3130
	atm	cm^3	g moles	atm-cm^3	82.057
	atm	liters	g moles	atm-liters	0.08205
	mm Hg	liters	g moles	mm Hg–liters	62.361
	bar	liters	g moles	bar-liters	0.08314
	kg/cm^2	liters	g moles	$kg/(cm^2)$(liters)	0.08478
	atm	ft^3	lb moles	atm-ft^3	1.314
	mm Hg	ft^3	lb moles	mm Hg–ft^3	998.9
	....	....	lb moles	chu or pcu	1.9872
Rankine.........	....	....	lb moles	Btu	1.9872
	....	....	lb moles	hp-hr	0.0007805
	....	....	lb moles	kw-hr	0.0005819
	atm	ft^3	lb moles	atm-ft^3	0.7302
	in. Hg	ft^3	lb moles	in. Hg–ft^3	21.85
	mm Hg	ft^3	lb moles	mm Hg–ft^3	555.0
	lb/in.^{2} abs	ft^3	lb moles	lb/(in.^{2})(ft^3)	10.73
	lb/ft^2 abs	ft^3	lb moles	ft-lb	1,545.0

Relations for ideal-gas Processes

Reversible Processes with Constant Specific Heats

Process..............	**Isothermal** $T = \text{const}$	**Constant pressure** $p = \text{const}$	**Constant volume** $v = \text{const}$	**Isentropic** $S = \text{const}$
pvT relations.........	$pv = \text{const}$	$\frac{v}{T} = \text{const}$	$\frac{p}{T} = \text{const}$	$pv^k = \text{const}$ $Tv^{k-1} = \text{const}$ $\frac{p^{(k-1)/k}}{T} = \text{const}$
Nonflow work $-\int p\,dv$..........	$pv \ln \frac{v_f}{v_i}$	$p\,\Delta v$	0	$nC_V\,\Delta T$
Steady-flow work $-\int v\,dp$.........	$pv \ln \frac{p_i}{p_f}$	0	$v\,\Delta p$	$nC_p\,\Delta T$
Heat $\int T\,dS$......	$pv \ln \frac{p_i}{p_f}$	$nC_p\,\Delta T$	$nC_V\,\Delta T$	0
ΔU................	0	$nC_V\,\Delta T$	$nC_V\,\Delta T$	$nC_V\,\Delta T$
ΔH................	0	$nC_p\,\Delta T$	$nC_p\,\Delta T$	$nC_p\,\Delta T$
ΔS................	$nR \ln \frac{p_i}{p_f}$	$nC_p \ln \frac{T_f}{T_i}$	$nC_V \ln \frac{T_f}{T_i}$	0

Properties of Gases

Gas	Symbol	Approx. mol. wt.	Critical pressure, psia	Critical temp., °F	Enthalpy of formation at 25°C, kcal/mole	Total enthalpy at 25°C, kcal/mole	Entropy at 25°C, cal/mole	Log of equil. const. formation at 25°C
Acetylene	C_2H_2	26	911	96.3	54.19	308.0	48.00	36.64
Air	......	29	546	−220.3	....	3.8		
Ammonia	NH_3	17	1640	270.3	−11.04	95.0	45.97	2.907
Argon	Ar	40	706	−187.7	0	1.5	36.98	0
Benzene	C_6H_6	78	702	551.4	19.82	781.2	64.34	−22.714
Butane	C_4H_{10}	58	530	307.4	−29.81	686.1	74.10	2.752
Carbon dioxide	CO_2	44	1073	88.0	−94.05	2.24	51.07	69.095
Carbon monoxide	CO	28	515	−220.3	−26.42	67.82	47.21	24.029
Dichlorodifluoromethane (R12)	CCl_2F_2	121	597	233.6	−112.00	118.83	71.92	75.280
Ethane	C_2H_6	30	718	90.0	−20.24	372.4	54.85	5.761
Ethylene	C_2H_4	28	748	49.3	12.50	335.7	52.45	11.934
Helium	He	4	33	−450.2	0	1.48	30.13	0
Heptane	C_7H_{16}	100	394	517.1	−44.89	1147	101.64	−1.532
Hydrogen	H_2	2	188	−399.8	0	69.4	31.21	0
Methane	CH_4	16	674	−116.5	−17.89	213.2	44.50	8.902
Nitrogen	N_2	28	493	−232.8	0	3.8	45.77	0
Oxygen	O_2	32	731	−181.8	0	4.11	49.00	0
Propane	C_3H_8	44	632	206.3	−24.82	529.5	64.51	4.037
Water	H_2O	18	3106	705.5	−57.80	13.7	45.11	40.048

Physical and Biochemical Properties of Inert Gases

ELEMENT	Helium	Nitrogen	Neon	Argon	Krypton	Xenon
Symbol	He	N_2	Ne	A	Kr	Xe
Atomic number	2	7	10	18	36	54
Molecular weight	4.00	28.00	20.18	39.94	83.80	131.3
Density at 0^oC, 1 atm, gm/l	0.1784	1.251	0.9004	1.784	3.708	5.851
Viscosity at 0^oC, 1 atm, micropoise	194.1	175.0	311.1	221.7	249.6	226.4
Thermal conductivity at 0^oC, 1 atm, cal/^oC-cm-sec	34.0×10^{-5}	5.66×10^{-5}	11.04×10^{-5}	3.92×10^{-5}	2.09×10^{-5}	1.21×10^{-5}
Bunsen solubility coefficients:						
in water at 38^oC	0.0086	0.013	0.0097	0.026	0.045	0.085
in olive oil at 38^oC	0.015	0.061	0.019	0.14	0.43	1.7
in human fat at 37^oC	?	0.062	0.020	?	0.41	1.6
oil:water solubility ratio	1.74	4.69	1.96	5.38	9.56	20.0

(Roth, 1967)

Ideal Gas Properties

Gas	*Chemical formula*	*M Molecular Weight*	*R (ft lbf/lbm°R)*	*c_p (Btu/lbm°R) at 77°F*	*c_v (Btu/lbm°R) at 77°F*	*c_p/c_v*
air		28.96	53.3	0.240	0.171	1.40
ammonia	NH_3	17.02	90.7	0.52	0.404	1.29
argon	A	39.90	38.73	0.124	0.074	1.68
butane	C_4H_{10}	58.08	26.61	0.406	0.372	1.09
carbon dioxide	CO_2	44.00	35.12	0.201	0.156	1.29
carbon monoxide	CO	28.00	55.19	0.248	0.177	1.40
ethane	C_2H_6	30.05	51.43	0.418	0.352	1.19
ethylene	C_2H_4	28.03	55.13	0.360	0.289	1.25
helium	He	4.00	386.33	1.25	0.75	1.67
hydrogen	H_2	2.016	766.53	3.416	2.431	1.41
methane	CH_4	16.03	96.40	0.532	0.408	1.30
nitrogen	N_2	28.02	55.15	0.248	0.177	1.40
octane	C_8H_{18}	114.14	13.54	0.407	0.390	1.04
oxygen	O_2	32.00	48.29	0.219	0.157	1.39
sulfur dioxide	SO_2	64.07	24.12	0.154	0.123	1.25
water vapor	H_2O	18.02	85.6	0.445	0.335	1.33

Generalized compressibility chart (low-pressure region)

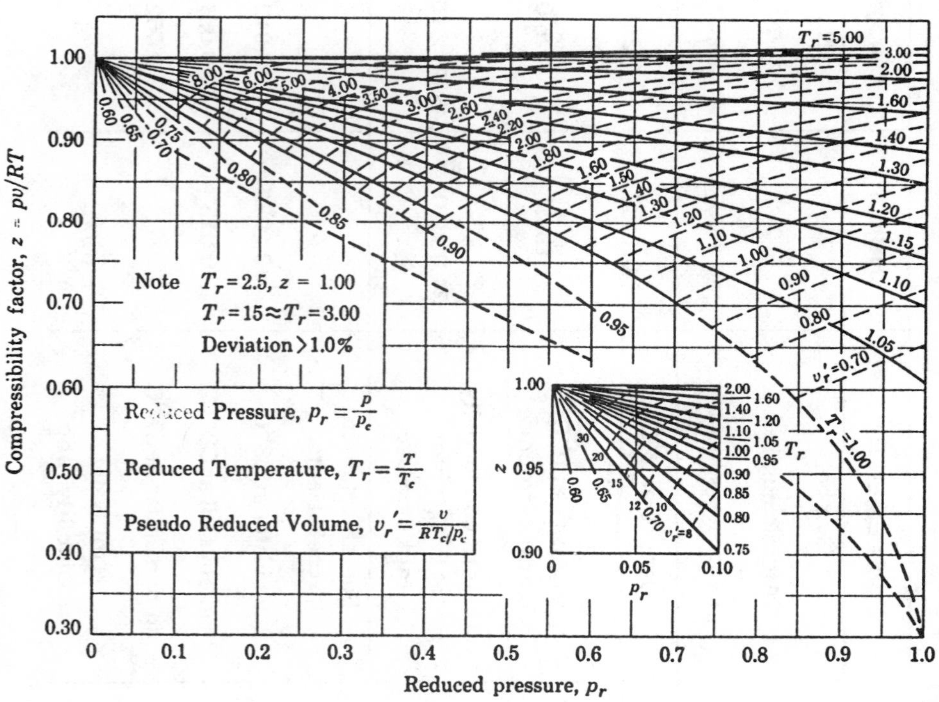

Generalized compressibility chart (medium-pressure region)

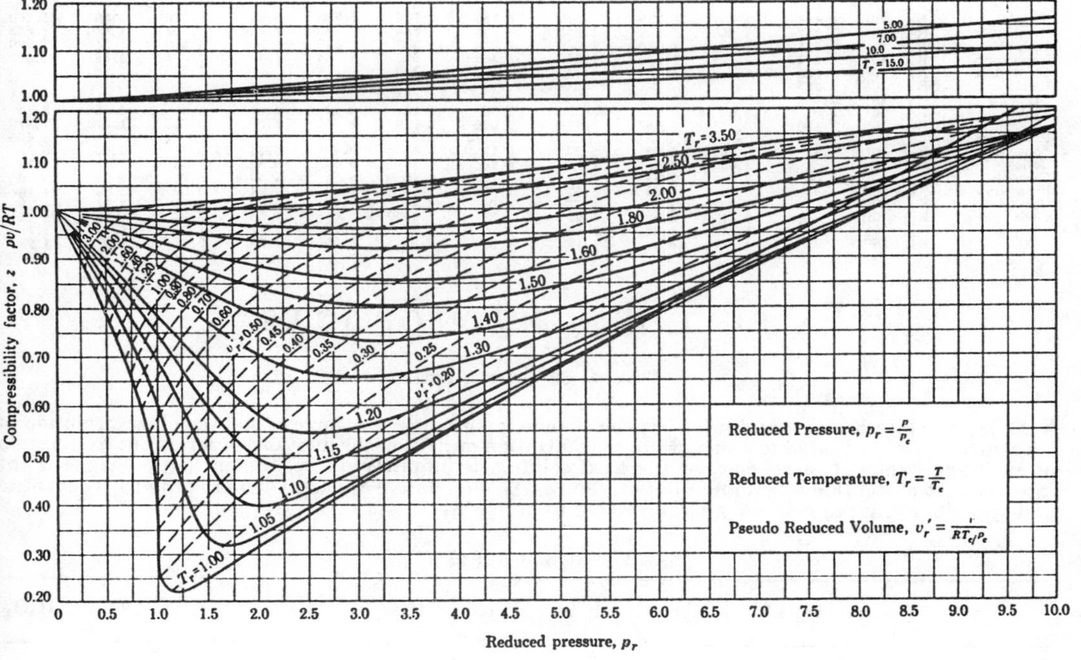

Generalized compressibility chart (high-pressure region)

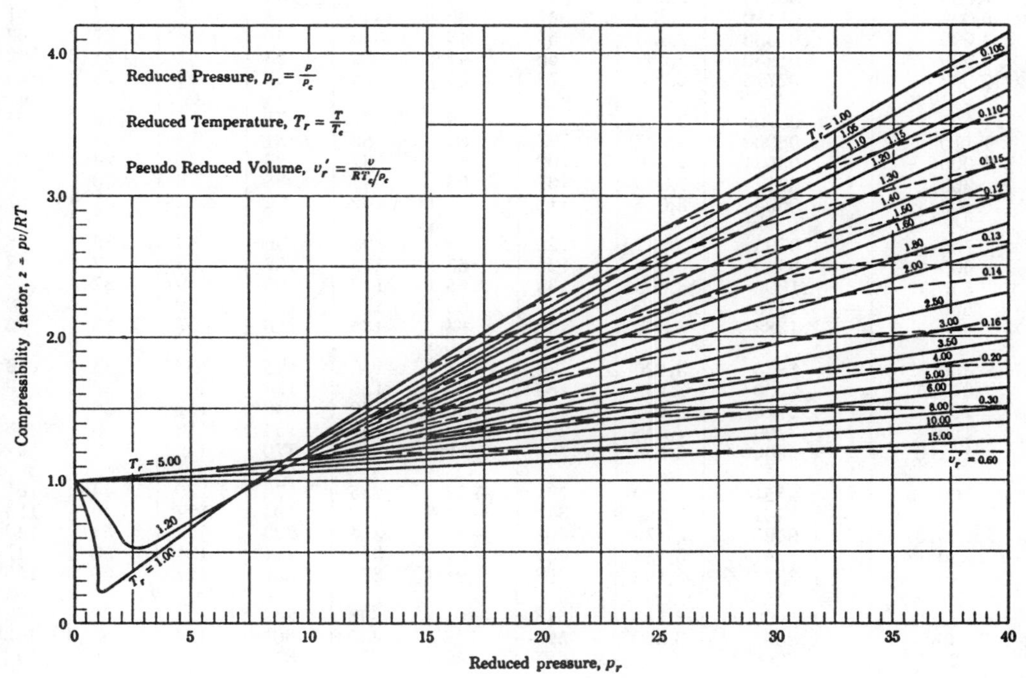

IONIZATION CONSTANTS OF SELECTED ACIDS IN WATER

Acids		0°	5°	10°	15°	20°	25°	30°	35°	40°	45°	50°
Formic	$K_A \cdot 10^4$	1.638	1.691	1.728	1.749	1.765	1.772	1.768	1.747	1.716	1.685	1.650
Acetic	$K_A \cdot 10^5$	1.657	1.700	1.729	1.745	1.753	1.754	1.750	1.728	1.703	1.670	1.633
Propionic	$K_A \cdot 10^5$	1.274	1.305	1.326	1.336	1.338	1.336	1.326	1.310	1.280	1.257	1.229
n-Butyric	$K_A \cdot 10^5$	1.563	1.574	1.576	1.569	1.542	1.515	1.484	1.439	1.395	1.347	1.302
Chloracetic	$K_A \cdot 10^3$	1.528	....	1.488	....	....	1.379	....	....	1.230	....	....
Lactic	$K_A \cdot 10^4$	1.287	....	....	....	....	1.374	....	....	....	....	1.270
Glycollic	$K_A \cdot 10^4$	1.334	....	....	....	....	1.475	....	....	....	....	1.415
Oxalic	$K_{2A} \cdot 10^5$	5.91	5.82	5.70	5.55	5.40	5.18	4.92	4.67	4.41	4.09	3.83
Malonic	$K_{2A} \cdot 10^6$	2.140	2.165	2.152	2.124	2.076	2.014	1.948	1.863	1,768	1.670	1.575
Phosphoric	$K_A \cdot 10^3$	8.968	....	....	....	....	7.516	....	....	....	....	5.495
Phosphoric	$K_{2A} \cdot 10^8$	4.85	5.24	5.57	5.89	6.12	6.34	6.46	6.53	6.58	6.59	6.55
Boric	$K_A \cdot 10^{10}$	....	3.63	4.17	4.72	5.26	5.79	6.34	6.86	7.38	....	8.32
Carbonic	$K_{1A} \cdot 10^7$	2.64	3.04	3.44	3.81	4.16	4.45	4.71	4.90	5.04	5.13	5.19
Phenol-sulfonic	$K_{2A} \cdot 10^{10}$	4.45	5.20	6.03	6.92	7.85	8.85	9.89	10.94	12.00	13.09	14.16
Glycine	$K_{1A} \cdot 10^7$	....	3.82	3.99	4.17	4.32	4.46	4.57	4.66	4.73	4.77	4.79
Citric	$K_{1A} \cdot 10^4$	6.03	6.31	6.69	6.92	7.21	7.45	7.66	7.78	7.96	7.99	8.04
	$K_{2A} \cdot 10^5$	1.45	1.54	1.60	1.65	1.70	1.73	1.76	1.77	1.78	1.76	1.75
	$K_{3A} \cdot 10^7$	4.05	4.11	4.14	4.13	4.09	4.02	3.99	3.78	3.69	3.45	3.28

HEAT OF DILUTION OF ACIDS

National Bureau of Standards

ΔH_{diln}, the integral heat of dilution, is the change in enthalpy, per mole of solute, when a solution of concentration m_1 is diluted to a final finite concentration m_2. When the dilution is carried out by addition of an infinite amount of solvent, so the final solution is infinitely dilute, the enthalpy change is the integral heat of dilution to infinite dilution. Since Φ_L, the relative apparent molal enthalpy, is equal to and opposite in sign to this, only Φ_L is referred to here.

Φ_L, cal/mole, at 25°C

n	m	HF	HCl	$HClO_4$	HBr	HI	HNO_3	CH_2O_2	$C_2H_4O_2$
∞	0.00	0	0	0	0	0	0	0	0
500,000	.000111	300	5	5	5	5	5	9	40
100,000	.000555	900	10	10	9	9	11	13	50
50,000	.00111	1,300	16	14	13	12	15	20	53
20,000	.00278	1,800	25	22	22	20	23	23	55
10,000	.00555	2,130	34	30	31	29	31	25	58
7,000	.00793	2,250	40	35	37	34	36	26	59
5,000	.01110	2,360	47	40	44	41	42	26	61
4,000	.01388	2,450	54	43	49	46	46	27	62
3,000	.01850	2,550	60	47	56	52	51	28	62
2,000	.02775	2,700	74	54	68	63	59	28	63
1,500	.03700	2,812	85	58	77	71	65	29	64
1,110	.05000	2,927	97	62	89	81	73	29	65
1,000	.05551	2,969	102	62	92	84	76	29	65
900	.0617	2,989	107	63	97	88	78	30	66
800	.0694	3,015	113	64	102	92	81	31	67
700	.0793	3,037	120	65	108	96	84	32	68
600	.0925	3,057	129	65	115	102	88	32	68
555.1	.1000	3,060	133	65	119	105	89	32	69
500	.1110	3,077	140	65	124	108	92	32	70
400	.1388	3,097	156	64	135	116	97	33	72
300	.1850	3,126	176	61	150	125	103	34	76
277.5	.2000	3,129	182	59	155	128	105	35	79
200	.2775	3,142	212	50	176	140	117	36	82
150	.3700	3,148	242	36	197	154	118	39	88
111.0	.5000	3,156	280	18	225	170	119	42	97
100	.5551	3,160	295	+12	235	176	120	44	101
75	.7401	3,167	343	−14	270	194	121	49	113
55.51	1.0000	3,179	405	−48	314	223	121	54	130
50	1.1101	3,184	431	−61	331	234	121	56	147
40	1.3877	3,192	493	−91	379	260	121	60	155
37.00	1.5000	3,194	518	−103	398	269	121	62	162
30	1.8502	3,200	595	−138	455	301	124	65	183
27.75	2.0000	3,203	627	−149	477	315	126	66	192

(continued)

Φ_L, cal/mole, at 25°C

n	m	HF	HCl	$HClO_4$	HBr	HI	HNO_3	CH_2O_2	$C_2H_4O_2$
25	2.2202	3,208	674	−162	510	336	130	67	204
22.20	2.5000	3,211	732	−173	550	365	139	68	218
20	2.7753	3,214	792	−182	590	396	149	69	233
18.50	3.0000	3,216	838	−187	624	427	159	69	245
15.86	3.500	3,221	946	−196	709	503	189	69	268
15	3.7004	3,227	988	−195	743	536	203	69	277
13.88	4.0000	3,234	1,052	−188	796	588	229	69	291
12.33	4.5000	3,246	1,171	−175	887	676	265	69	313
12	4.6255	3,249	1,190	−170	911	700	277	69	318
11.10	5.0000	3,256	1,271	−150	983	764	313	69	333
10	5.5506	3,265	1,396	−117	1,097	855	368	68	353
9.5	5.8427	3,269	1,462	−97	1,156	920	400	68	363
9.251	6.0000	3,272	1,498	−84	1,196	950	418	67	368
9.0	6.1674	3,274	1,535	−72	1,230	980	437	67	373
8.5	6.5301	3,278	1,618	−40	1,313	1,050	480	66	383
8.0	6.9383	3,282	1,710	+4	1,401	1,115	530	65	392
7.929	7.0000	3,283	1,725	11	1,416	1,130	538	65	394
7.5	7.4008	3,286	1,820	61	1,497	1,210	595	63	402
7.0	7.9295	3,290	1,942	135	1,608	1,325	661	61	411
6.938	8.0000	3,291	1,960	146	1,622	1,340	667	61	412
6.5	8.5394	3,296	2,090	229	1,738	1,450	745	58	420
6.167	9.0000	3,302	2,202	306	1,845	1,570	805	55	426
6.0	9.2510	3,305	2,265	348	1,903	1,630	840	53	429
5.551	10.0000	3,316	2,447	481	2,078	1,820	940	49	436
5.5	10.0920	3,317	2,472	499	2,102	1,850	950	49	437
5.0	11.1012	3,335	2,721	730	2,344	2,100	1,098	43	445
4.5	12.3346	3,362	3,025	1,144	2,655	2,460	1,270	37	453
4.0	13.8765	3,400	3,404	1,574	3,089	2,960	1,495	29	462
3.700	15.0000	3,428	3,680	1,893	3,415	3,350	1,645	26	469
3.5	15.8589	3,450	3,882	2,150	3,668	3,660	1,770	21	473
3.25	17.0788	3,483	4,160	2,460	4,005	4,110	1,920	17	481
3.0	18.5020	3,520	4,460	2,880	4,370	4,630	2,101	13	488
2.775	20.0000	3,557	4,750	3,300	4,760	5,190	2,270	9	496
2.5	22.2024	3,607	5,180	4,000	5,300	6,000	2,520	+4	506
2.0	27.7530	3,712	6,260	5,500	6,650	.....	3,060	−5	528
1.5	37.0040	.....	8,240	.....	8,530	.....	3,770	−13	532
1.0	55.506	.....	10,900	.....	11,670	.....	4,715	+11	518
0.5	111.012	.....	......	.....	......	.....	.....	77	495
0.25	222.02	.....	......	.....	......	.....	.....	129	......

HEATS OF SOLUTION

Vivian B. Parker

National Bureau of Standards

ΔH°_∞ 25°C for uni-univalent electrolytes in H_2O

Substance	State	ΔH°_∞	Substance	State	ΔH°_∞	Substance	State	ΔH°_∞
		cal/mole	NH_4ClO_4	c	8,000	$AgClO_4$	c	1,760
HF	g	−14,700	NH_4Br	c	4,010	$AgNO_2$	c	8,830
HCl	g	−17,888	NH_4I	c	3,280	$AgNO_3$	c	5,400
$HClO_4$	l	−21,215	NH_4IO_3	c	7,600			
$HClO_4 \cdot H_2O$	c	−7,875	NH_4NO_2	c	4,600	LiOH	c	−5,632
HBr	g	−20,350	NH_4NO_3	c	6,140	$LiOH \cdot H_2O$	c	−1,600
HI	g	−19,520	$NH_4C_2H_3O_2$	c	−570	LiF	c	1,130
HIO_3	c	2,100	NH_4CN	c	4,200	LiCl	c	−8,850
HNO_3	l	−7,954	NH_4CNS	c	5,400	$LiCl \cdot H_2O$	c	−4,560
HCOOH	l	−205	CH_3NH_3Cl	c	1,378	$LiClO_4$	c	−6,345
CH_3COOH	l	−360	$(CH_3)_3NHCl$	c	350	$LiClO_4 \cdot 3H_2O$	c	7,795
			$N(CH_3)_4Cl$	c	975	LiBr	c	−11,670
NH_3	g	−7,290	$N(CH_3)_4Br$	c	5,800	$LiBr \cdot H_2O$	c	−5,560
NH_4Cl	c	3,533	$N(CH_3)_4I$	c	10,055			

(continued)

Substance	State	ΔH°_∞	Substance	State	ΔH°_∞
		cal/mole			*cal/mole*
$LiBr\cdot 2H_2O$	c	−2,250	KCl	c	4,115
$LiBrO_3$	c	340	$KClO_3$	c	9,890
LiI	c	−15,130	$KClO_4$	c	12,200
$LiI\cdot H_2O$	c	−7,090	KBr	c	4,750
$LiI\cdot 2H_2O$	c	−3,530	$KBrO_3$	c	9,830
$LiI\cdot 3H_2O$	c	140	KI	c	4,860
$LiNO_2$	c	−2,630	KIO_3	c	6,630
$LiNO_2\cdot H_2O$	c	1,680	KNO_2	c	3,190
$LiNO_3$	c	−600	KNO_3	c	8,340
			$KC_2H_3O_2$	c	−3,665
$NaOH$	c	−10,637	KCN	c	2,800
$NaOH\cdot H_2O$	c	−5,118	$KCNO$	c	4,840
NaF	c	218	$KCNS$	c	5,790
$NaCl$	c	928	$KMnO_4$	c	10,410
$NaClO_2$	c	80			
$NaClO_2\cdot 3H_2O$	c	6,830	$RbOH$	c	−14,900
$NaClO_3$	c	5,191	$RbOH\cdot H_2O$	c	−4,310
$NaClO_4$	c	3,317	$RbOH\cdot 2H_2O$	c	210
$NaClO_4\cdot H_2O$	c	5,380	RbF	c	−6,240
$NaBr$	c	−144	$RbF\cdot H_2O$	c	−100
$NaBr\cdot 2H_2O$	c	4,454	$RbF\cdot 1\frac{1}{2}H_2O$	c	320
$NaBrO_3$	c	6,430	$RbCl$	c	4,130
NaI	c	−1,800	$RbClO_3$	c	11,410
$NaI\cdot 2H_2O$	c	3,855	$RbClO_4$	c	13,560
$NaIO_3$	c	4,850	$RbBr$	c	5,230
$NaNO_2$	c	3,320	$RbBrO_3$	c	11,700
$NaNO_3$	c	4,900	RbI	c	6,000
$NaC_2H_3O_2$	c	−4,140	$RbNO_3$	c	8,720
$NaC_2H_3O_2\cdot 3H_2O$	c	4,700			
$NaCN$	c	290	$CsOH$	c	−17,100
$NaCN\cdot \frac{1}{2}H_2O$	c	790	$CsOH\cdot H_2O$	c	−4,900
$NaCN\cdot 2H_2O$	c	4,440	CsF	c	−8,810
$NaCNO$	c	4,590	$CsF\cdot H_2O$	c	−2,500
$NaCNS$	c	1,632	$CsF\cdot 1\frac{1}{2}H_2O$	c	−1,300
			$CsCl$	c	4,250
KOH	c	−13,769	$CsClO_4$	c	13,250
$KOH\cdot H_2O$	c	−3,500	$CsBr$	c	6,210
$KOH\cdot 1\frac{1}{2}H_2O$	c	−2,500	$CsBrO_3$	c	12,060
KF	c	−4,238	CsI	c	7,970
$KF\cdot 2H_2O$	c	1,666	$CsNO_3$	c	9,560

PHASE TRANSITION CHARACTERISTICS OF THE ELEMENTS

From the U.S. Atomic Energy Commision Report ANL-5750

Element	Phase	Temperature of Transition (°K)	Heat of Transition (kcal/g mole)	Entropy of Transition (e.u.)	Entropy at 298°K (e.u.)
Ac	solid	(1090)	(2.5)	(2.3)	(13)
	liquid	(2750)	(70)	(25)	—
Ag	solid	1234	2.855	2.313	10.20
	liquid	2485	60.72	24.43	—
	gas	—	—	—	—
Al	solid	931.7	2.57	2.76	6.769
	liquid	2600	67.9	26	—
Am	solid	(1200)	(2.4)	(2.0)	(13)
	liquid	2733	51.7	18.9	—
As	solid	883	3¼	35.¼	8.4
Au	solid	1336.16	3.03	2.27	11.32
	liquid	2933	74.21	25.30	—
B	solid	2313	(3.8)	(1.6)	1.42
	liquid	2800	75	27	—
Ba	solid, α	648	0.14	0.22	16
	solid, β	977	1.83	1.87	—
	liquid	1911	35.665	18.63	—
	gas	—	—	—	—
Be	solid	1556	2.919	1.501	2.28
	liquid	—	—	—	—
Bi	solid	544.2	2.63	4.83	13.6
	liquid	1900	41.1	21.6	—
	gas	—	—	—	—
C	solid	—	—	—	1.3609
Ca	solid, α	723	0.24	0.33	9.95
	solid, β	1123	2.2	1.96	—
	liquid	1755	38.6	22.0	—
	gas	—	—	—	—
Cd	solid	594.1	1.46	2.46	12.3
	liquid	1040	23.86	22.94	—
	gas	—	—	—	—
Ce	solid	1048	2.1	2.0	13.8
	liquid	2800	73	26	—
Cl_2	gas	—	—	—	53.286
Co	solid, α	723	0.005	0.007	6.8
	solid, β	1398	0.095	0.068	—
	solid, γ	1766	3.7	2.1	—
	liquid	3370	93	28	—

Element	Phase	Temperature of Transition (°K)	Heat of Transition (kcal/g mole)	Entropy of Transition (e.u.)	Entropy at 298°K (e.u.)
Cr	solid	2173	3.5	1.6	5.68
	liquid	2495	72.97	29.25	—
	gas	—	—	—	—
Cs	solid	301.9	0.50	1.7	19.8
	liquid	963	16.32	17.0	—
	gas	—	—	—	—
Cu	solid	1356.2	3.11	2.29	7.97
	liquid	2868	72.8	25.4	—
F_2	gas	—	—	—	48.58
Fe	solid, α	1033	0.410	0.397	6.491
	solid, β	1180	0.217	0.184	—
	solid, γ	1673	0.15	0.084	—
	solid, d	1808	3.86	2.14	—
	liquid	3008	84.62	28.1	—
Ga	solid	302.94	1.335	4.407	9.82
	liquid	2700	—	—	—
Ge	solid	1232	8.3	6.7	10.1
	liquid	2980	68	23	—
H_2	gas	—	—	—	31.211
Hf	solid	(2600)	(6.0)	(2.3)	13.1
Hg	liquid	629.73	13.985	22.208	18.46
	gas	—	—	—	—
In	solid	430	0.775	1.80	13.88
	liquid	2440	53.8	22.0	—
	gas	—	—	—	—
Ir	solid	2727	6.6	2.4	8.7
K	solid	336.4	0.5575	1.657	15.2
	liquid	1052	18.88	17.95	—
	gas	—	—	—	—
La	solid	1153	(2.3)	(2.0)	13.7
	liquid	3000	80	27	—
Li	solid	459	0.69	1.5	6.70
	liquid	1640	32.48	19.81	—
	gas	—	—	—	—
Mg	solid	923	2.2	2.4	7.77
	liquid	1393	31.5	22.6	—
	gas	—	—	—	—

(continued)

Element	Phase	Temperature of Transition (°K)	Heat of Transition (kcal/g mole)	Entropy of Transition (e.u.)	Entropy at 298°K (e.u.)
Mn	solid, α	1000	0.535	0.535	7.59
	solid, β	1374	0.545	0.397	—
	solid, γ	1410	0.430	0.305	—
	solid, δ	1517	3.5	2.31	—
	liquid	2368	53.7	22.7	—
	gas	—	—	—	—
Mo	solid	2883	(5.8)	(2.0)	6.83
N_2	gas	—	—	—	45.767
Na	solid	371	0.63	1.7	12.31
	liquid	1187	23.4	20.1	—
	gas	—	—	—	—
Nb	solid	2760	(5.8)	(2.1)	8.3
Nd	solid	1297	(2.55)	(1.97)	13.9
	liquid	(2750)	(61)	(22)	—
Ni	solid, α	626	0.092	0.15	7.137
	solid, β	1728	4.21	2.44	—
	liquid	3110	90.48	29.0	—
Np	solid	913	(2.3)	(2.5)	(14)
	liquid	(2525)	(55)	(22)	—
O_2	gas	—	—	—	49.003
Os	solid	2970	(6.4)	(2.2)	7.8
P_4	solid, white	317.4	0.601	1.89	42.4
	liquid	553	11.9	21.5	—
	gas	—	—	—	—
Pa	solid	(1825)	(4.0)	(2.2)	(13.5)
	liquid	(4500)	(115)	(26)	—
Pb	solid	600.6	1.141	1.900	15.49
	liquid	2023	42.5	21.0	—
	gas	—	—	—	—
Pd	solid	1828	4.12	2.25	8.9
	liquid	3440	89	26	—
Po	solid	525	(2.4)	(4.6)	13
	liquid	(1235)	(24.6)	(19.9)	—
	gas	—	—	—	—
Pr	solid	1205	(2.5)	(2.1)	(13.5)
	liquid	3563	—	—	—
Pt	solid	2042.5	5.2	2.5	10.0
	liquid	4100	122	29.8	—
Pu	solid	913	(2.26)	(2.48)	(13.0)
	liquid	—	—	—	—
Ra	solid	1233	(2.3)	(1.9)	(17)
	liquid	(1700)	(35)	(21)	—
	gas	—	—	—	—
Rb	solid	312.0	0.525	1.68	16.6
	liquid	952	18.11	19.0	—
	gas	—	—	—	—
Re	solid	3440	(7.9)	(2.3)	(8.89)
Rh	solid	2240	(5.2)	(2.3)	7.6
	liquid	4150	127	30.7	—
Ru	solid, α	1308	0.034	0.026	6.9
	solid, β	1473	0	0	—
	solid, γ	1773	0.23	0.13	—
	solid, δ	2700	(6.1)	(2.3)	—

Element	Phase	Temperature of Transition (°K)	Heat of Transition (kcal/g mole)	Entropy of Transition (e.u.)	Entropy at 298°K (e.u.)
S	solid, α	368.6	0.088	0.24	7.62
	solid, β	392	0.293	0.747	—
	liquid	717.76	2.5	3.5	—
½S_2	gas	—	—	—	—
Sb	solid (α,β,γ)	903.7	4.8	5.3	10.5
	liquid	1713	46.665	27.3	—
½Sb_2	gas	—	—	—	—
Sc	solid	1670	(4.0)	(2.4)	(9.0)
	liquid	3000	80	27	—
Se	solid	490.6	1.25	2.55	10.144
	liquid	1000	14.27	14.27	—
Si	solid	1683	11.1	6.60	4.50
	liquid	2750	71	26	—
Sm	solid	1623	3.7	2.3	(15)
	liquid	(2800)	(70)	(25)	—
Sn	solid (α,β)	505.1	1.69	3.35	12.3
	liquid	2473	(55)	(22)	—
	gas	—	—	—	—
Sr	solid	1043	2.2	2.1	13.0
	liquid	1657	33.61	20.28	—
	gas	—	—	—	—
Ta	solid	3250	7.5	2.3	9.9
Te	solid	(2400)	(5.5)	(2.3)	(8.0)
	liquid	(3800)	(120)	(32)	—
Te	solid, α	621	0.13	0.21	11.88
	solid, β	723	4.28	5.92	—
	liquid	1360	11.9	8.75	—
½Te_2	gas	—	—	—	—
Th	solid	2173	(4.6)	(2.1)	12.76
	liquid	4500	(130)	(29)	—
Ti	solid, α	1155	0.950	0.822	7.334
	solid, β	2000	(4.6)	(2.3)	—
	liquid	3550	(101)	(28)	—
Tl	solid, α	508.3	0.082	0.16	15.4
	solid, β	576.8	1.03	1.79	—
	liquid	1730	38.81	22.4	—
	gas	—	—	—	—
U	solid, α	938	0.665	0.709	12.03
	solid, β	1049	1.165	1.111	—
	solid, γ	1405	(3.0)	(2.1)	—
	liquid	3800	—	—	—
V	solid	2003	(4.0)	(2.0)	7.05
	liquid	3800	—	—	—
W	solid	3650	8.42	2.3	8.0
Y	solid	1750	(4.0)	(2.3)	(11)
	liquid	3500	(90)	(26)	—
Zn	solid	692.7	1.595	2.303	9.95
	liquid	1180	27.43	23.24	—
	gas	—	—	—	—
Zr	solid, α	1135	0.920	0.811	9.29
	solid, β	2125	(4.9)	(2.3)	—
	liquid	(3900)	(100)	(26)	—

PHASE TRANSITION CHARACTERISTICS OF SELECTED OXIDES

Oxide	Phase	Temperature of Transition (°K)	Heat of Transition kcal/mole	Entropy of Transition (e.u.)	Entropy at 298°K (e.u.)
Ac_2O_3	solid	(2250)	(20)	(8.9)	(36.5)
	liquid	—	—	—	—
Ag_2O	solid	dec. 460	—	—	29.09
Ag_2O_2	solid	dec	—	—	(20.4)
Al_2O_3	solid	2300	26	11	12.186
	liquid	dec.	—	—	—
Am_2O_3	solid	(2225)	(17)	(7.6)	(37)
	liquid	(3400)	(85)	(25)	—
AmO_2	solid	dec.	—	—	(20)
As_2O_3	solid, α	503	4.1	8.2	25.6
	solid, β	586	4.4	7.5	—
	liquid	730	7.15	9.79	—
	gas	—	—	—	—
AsO_2	solid	(1200)	(9.0)	(7.5)	(13)
	liquid	(dec.)	—	—	—
As_2O_5	solid	dec. >1100	—	—	25.2
Au_2O_3	solid	dec.	—	—	30
B_2O_3	solid	723	5.27	7.29	12.91
	liquid	2520	(55)	(22)	—
Ba_2O	solid	(880)	(5.2)	(5.9)	(23.5)
	liquid	(1040)	(20)	(19)	—
	gas	—	—	—	—
BaO	solid	2196	13.8	6.28	16.8
	liquid	3000	(62)	(21)	—
BaO_2	solid	723	(5.7)	(7.9)	(18.5)
	liquid	dec.1110	—	—	—
BeO	solid	dec.	—	—	3.37
BiO	solid	(1175)	(3.7)	(3.1)	(15)
	liquid	(1920)	(54)	(28)	—
	gas	—	—	—	—

Oxide	Phase	Temperature of Transition (°K)	Heat of Transition kcal/mole	Entropy of Transition (e.u.)	Entropy at 298°K (e.u.)
Bi_2O_3	solid	1090	6.8	6.2	36.2
	liquid	(dec.)	—	—	—
CO	gas	—	—	—	47.30
CO_2	gas	—	—	—	51.06
CaO	solid	2860	(18)	(6.3)	9.5
CdO	solid	dec.	—	—	13.1
Ce_2O_3	solid	1960	(20)	(10)	(33.5)
	liquid	(3500)	(80)	(23)	—
CeO_2	solid	3000	(19)	(6.3)	17.7
CoO	solid	2078	(12)	(5.8)	10.5
	liquid	(2900)	(61)	(21)	—
Co_3O_4	solid	dec., 1240	—	—	(35.5)
Cr_2O_3	solid	2538	(25)	(10)	19.4
CrO_2	solid	dec. 700	—	—	(11.5)
CrO_3	solid	460	(6.1)	(13)	(17.5)
	liquid	(1000)	(25)	(25)	—
	gas	—	—	—	—
Cs_2O	solid	763	(4.58)	(6.0)	(23)
	liquid	dec.	—	—	—
Cs_2O_2	solid	867	(5.5)	(6.3)	(40)
	liquid	dec.	—	—	—
Cs_2O_3	solid	775	(7.75)	(10)	(47)
	liquid	dec.	—	—	—
Cu_2O	solid	1503	13.4	8.92	22.44
	liquid	dec.	—	—	—
CuO	solid	1609	(8.9)	(5.5)	10.4
	liquid	dec.	—	—	—
FeO	solid	1641	7.5	4.6	12.9
	liquid	(2700)	(55)	(20)	—
Fe_3O_4	solid, α	900	(0)	(0)	35.0
	solid, β	dec.	—	—	—

(continued)

Oxide	Phase	Temperature of Transition (°K)	Heat of Transition kcal/mole	Entropy of Transition (e.u.)	Entropy at 298°K (e.u.)
Fe_2O_3	solid, α	950	0.16	0.17	21.5
	solid, β	1050	0	0	—
	solid, γ	dec.	—	—	—
Ga_2O	solid	(925)	(8.5)	(9.2)	(22.5)
	liquid	(1000)	(20)	(20.)	—
	gas	—	—	—	—
Ga_2O_3	solid	2013	(22)	(11.)	20.23
	liquid	(2900)	(75)	(26.)	—
GeO	solid	983	(50)	(51)	(12.5)
	gas	—	—	—	—
GeO_2	solid, (α,β)	1389	10.5	7.56	(12.5)
	liquid	(2625)	(61)	(23.)	—
H_2O	liquid	373.16	9.770	26.18	16.716
	gas	—	—	—	—
HfO_2	solid	3063	(17)	(5.6)	14.18
Hg_2O	solid	dec.	—	—	(30.)
HgO	solid	dec.	—	—	16.839
In_2O	solid	(600)	(4.5)	(7.5)	(28.)
	liquid	(800)	(16)	(20.)	—
	gas	—	—	—	—
InO	solid	(1325)	(4.0)	(3.0)	(14.5)
	liquid	(2000)	(60.)	(30)	—
	gas	—	—	—	—
In_2O_3	solid	(2000)	(20.)	(10.)	30.1
	liquid	(3600)	(85)	(24)	—
Ir_2O_3	solid	(1450)	(10)	(6.8)	(26.5)
	liquid	(2250)	(50)	(22)	—
	gas	—	—	—	—
IrO_2	solid	dec. 1373	—	—	(15.9)
K_2O	solid	(980)	(6.8)	(6.9)	(23)
	liquid	dec.	—	—	—
K_2O_2	solid	763	(7.0)	(9.2)	(27)
	liquid	(1800)	(45)	(25)	—
	gas	—	—	—	—
K_2O_3	solid	703	(6.1)	(8.7)	(33.5)
	liquid	(975)	(25)	(26)	—
	gas	—	—	—	—
KO_2	solid	653	(4.9)	(7.5)	27.9
	liquid	dec.	—	—	—
La_2O_3	solid	2590	(18)	(7)	(36.5)
Li_2O	solid	2000	(14)	(7)	9.06
	liquid	2600	(56)	(22)	—
Li_2O_2	solid	dec.470	—	—	(16.5)
MgO	solid	3075	18.5	5.8	6.4
MgO_2	solid	dec. 361	—	—	(20.5)
MnO	solid	2058	13.0	6.32	14.27
	liquid	dec.	—	—	—
Mn_3O_4	solid, α	1445	4.97	3.44	35.5
	solid, β	1863	(33)	(18)	—
	liquid	(2900)	(75)	(26)	—
Mn_2O_3	solid	dec. 1620	—	—	26.4
MnO_2	solid	dec. 1120	—	—	12.7
MoO_2	solid	(2200)	(16)	(7.3)	(14.5)
	liquid	dec. 2250	—	—	—
MoO_3	solid	1068	12.54	11.74	18.68
	liquid	1530	33	22	—
	gas	—	—	—	—
N_2O	gas	—	—	—	52.58
Na_2O	solid	1193	(7.1)	(6.0)	17.4
	liquid	dec.	—	—	—
Na_2O_2	solid	dec. 919	—	—	22.6
NaO_2	solid	(825)	(6.2)	(7.5)	27.7
	liquid	(1300)	(28)	(22)	—
	gas	—	—	—	—
NbO	solid	(2650)	(16)	(6.0)	(12)
NbO_2	solid	(2275)	(16)	(7.0)	(12.7)
	liquid	(3800)	(85)	(22)	—
Nb_2O_5	solid	1733	(28)	(16)	32.8
	liquid	(3200)	(80)	(25)	—
Nd_2O_3	solid	2545	(22)	(8.8)	(35.3)
NiO	solid	2230	(12.1)	(5.43)	9.22
	liquid	dec.	—	—	—
NpO_2	solid	(2600)	(15)	(5.7)	19.19
Np_2O_5	solid	dec. 800—900°K	—	—	(43)
OsO_2	solid	dec. 923	—	—	(14.5)
OsO_4	solid	813.3	3.41	10.9	34.7
	liquid	403	9.45	23.4	—
	gas	—	—	—	—
P_2O_3	liquid	448.5	4.5	10	(34)
	gas	—	—	—	—
PO_2	solid	(350)	(2.7)	(7.7)	(11.5)
	liquid	(dec)	—	—	—
P_2O_5	solid	631	8.8	13.9	33.5
	gas	—	—	—	—
PaO_2	solid	(2560)	(20)	(7.8)	(17.8)
Pa_2O_5	solid	(2050)	(26)	(13)	(37.5)
	liquid	(3350)	(95)	(28)	—
PbO	solid, red	762	(0.4)	(0.5)	16.2
	solid, yellow	1159	2.8	2.4	—
	liquid	1745	51	29	—
	gas	—	—	—	—
Pb_3O_4	solid	dec.	—	—	50.5
PbO_2	solid	dec.	—	—	18.3
PdO	solid	dec. 1150	—	—	(9.1)
PoO_2	solid	(825)	(5.5)	(6.7)	(17)
	liquid	(dec.)	—	—	—
Pr_2O_3	solid	(2200)	(22)	(10)	(35.5)
	liquid	(4000)	(90)	(23)	—
PrO_2	solid	dec. 700	—	—	(17)
PtO	solid	dec. 780	—	—	(13.5)
Pt_3O_4	solid	(dec.)	—	—	(41)
PtO_2	solid	723	(4.6)	(6.4)	(16.5)
	liquid	dec. 750	—	—	—
PuO	solid	(1290)	(7.2)	(5.6)	(20)
	liquid	(2325)	(47)	(20)	—
	gas	—	—	—	—
Pu_2O_3	solid	(1880)	(16)	(8.5)	(38)
	liquid	(3250)	(75)	(23)	—
PuO_2	solid	(2400)	(15)	(6.2)	19.7)
	liquid	(3500)	(90)	(26)	—
RaO	solid	(>2500)	—	—	(17)
Rb_2O	solid	(910)	(5.7)	(6.3)	(27)
	liquid	dec.	—	—	—
Rb_2O_2	solid	843	(7.3)	(8.7)	(27.5)
	liquid	(dec.)	—	—	—
Rb_2O_3	solid	762	(7.6)	(10)	(32.5)
	liquid	dec.	—	—	—
RbO_2	solid	685	(4.1)	(6.0)	(21.5)
	liquid	dec.	—	—	—
ReO_2	solid	(1475)	(12)	(8.1)	(15)
ReO_2	liquid	(3250)	(80)	(25)	—
ReO_3	solid	433	5.2	12	19.8
	liquid	dec.	—	—	—
Re_2O_7	solid	569	15.8	27.8	44
	liquid	635.5	17.7	27.9	—
	gas	—	—	—	—
ReO_4	solid	420	(4.2)	(10)	(34.5)
	liquid	(460)	(9.3)	(20)	—
	gas	—	—	—	—
Rh_2O	solid	dec. 1400	—	—	(25.5)
RhO	solid	dec. 1394	—	—	(12)
Rh_2O_3	solid	dec. 1388	—	—	(23)
RuO_2	solid	dec. 1400	—	—	(12.5)
RuO_4	solid	300	(3.2)	(11)	(32.5)
	liquid	dec.	—	—	—
SO_2	gas	—	—	—	59.40
Sb_2O_3	solid	928	14.74	15.88	29.4
	liquid	1698	8.92	5.25	—
	gas	—	—	—	—
SbO_2	solid	dec.	—	—	15.2
Sb_2O_5	solid	dec.	—	—	29.9
Sc_2O_3	solid	(2500)	(23)	(9.3)	24.8
SeO	solid	(1375)	(7.6)	(5.5)	(11)
	liquid	(2075)	(45)	(22)	—
	gas	—	—	—	—
SeO_2	solid	603	(24.5)	(40.6)	(15)
	gas	—	—	—	—
SiO	solid	(2550)	(12)	(4.7)	(6.5)
SiO_2	solid, β	856	0.15	0.18	10.06
	solid, α	1883	2.04	1.08	—
	liquid	dec. 2250	—	—	—
Sm_2O_3	solid	(2150)	(20)	(9.3)	(36.5)
	liquid	(3800)	(80)	(21)	—
SnO	solid	(1315)	(6.4)	(4.9)	13.5
	liquid	(1800)	(60)	(33)	—
	gas	—	—	—	—
SnO_2	solid	1898	(11.39)	(5.95)	12.5
	liquid	(3200)	(75)	(23)	—
SrO	solid	2703	16.7	6.2	13.0
SrO_2	solid	dec. 488	—	—	(14.8)
Ta_2O_5	solid	2150	(16)	(7.4)	34.2
	liquid	—	—	—	—
TcO_2	solid	(2400)	(18)	(7.5)	(13.5)
	liquid	(4000)	(105)	(26)	—
TcO_3	solid	(dec. <1200)	—	—	(19.5)
Tc_2O_7	solid	392.7	(11)	(28)	(42.5)
	liquid	583.8	(14)	(24)	—
	gas	—	—	—	—
TeO	solid	(1020)	(7.1)	(7.0)	(13)
	liquid	(1775)	(50)	(28)	—
	gas	—	—	—	—
TeO_2	solid	1006	3.2	3.2	16.99
	liquid	dec.	—	—	—
ThO	solid	(2150)	(13)	(6.0)	(16)
	liquid	(3250)	(65)	(20)	—
ThO_2	solid	3225	(18)	(5.6)	15.59
TiO	solid, α	1264	0.82	0.65	8.31
	solid, β	dec. 2010	—	—	—
Ti_2O_3	solid, α	473	0.215	0.455	18.83
	solid, β	2400	(24)	(10)	—
	liquid	3300	—	—	—
Ti_3O_5	solid, α	450	2.24	4.98	30.92
	solid, β	(2450)	(50)	(20)	—
	liquid	(3600)	(85)	(24)	—
TiO_2	solid	2128	(16)	(7.5)	12.01
	liquid	dec.3200	—	—	—

(continued)

Oxide	Phase	Temperature of Transition (°K)	Heat of Transition kcal/mole	Entropy of Transition (e.u.)	Entropy at 298°K (e.u.)
Ti_2O	solid	573	(5.0)	(8.7)	23.8
	liquid	773	(17)	(22)	—
	gas	—	—	—	—
Ti_2O_3	solid	990	(12.4)	(13)	(33.5)
	liquid	(dec)	—	—	—
UO	solid	(2750)	(14)	(5.1)	(16)
UO_2	solid	3000	—	—	18.63
U_3O_8	solid	dec.	—	—	(66)
UO_3	solid	dec. 925	—	—	23.57
VO	solid	(2350)	(15)	(6.4)	9.3
	liquid	(3400)	(70)	(21)	—
V_2O_3	solid	2240	(24)	(11)	23.58
	liquid	dec. 3300	—	—	—
V_3O_5	solid	(2100)	(42)	(20)	(32)
	liquid	(dec.)	—	—	—
VO_2	solid, α	345	1.02	2.96	12.32
	solid, β	1818	13.60	7.48	—
	liquid	dec. 3300	—	—	—
V_2O_5	solid	943	15.56	16.50	313
	liquid	(2325)	(63)	(27)	—
	gas	—	—	—	—
VO_2	solid	(1543)	(11.5)	(7.45)	(15)
	liquid	dec. 2125	—	—	—
WO_3	solid	1743	(17)	(9.8)	19.90
	liquid	(2100)	(43)	(20)	—
	gas	—	—	—	—
Y_2O_3	solid	(2500)	(25)	(10)	(29.5)
ZnO	solid	dec.	—	—	10.4
ZrO_2	solid, α	1478	1.420	0.961	12.03
	solid, β	2950	20.8	7.0	—

THERMODYNAMIC FUNCTIONS OF COPPER, SILVER AND GOLD

Furukawa, Saba, Reilly
National Bureau of Standards

1 cal = 4.1840*J* H°_0 is the enthalpy of the solid at 0°K and 1 atm pressure

T °K	C°_p J/deg-mol			$H^\circ_T - H^\circ_0$ J/mol			$(H^\circ_T - H^\circ_0)/T$ J/deg-mol		
	Cu	Ag	Au	Cu	Ag	Au	Cu	Ag	Au
1.00	0.000743	0.000818	0.00118	0.000359	0.000367	0.000478	0.000359	0.000367	0.000478
2.00	0.00177	0.00265	0.00504	0.00158	0.00197	0.00326	0.000790	0.000987	0.00163
3.00	0.00337	0.00650	0.0141	0.00409	0.00633	0.0123	0.00136	0.00211	0.00410
4.00	0.00582	0.0134	0.0306	0.00860	0.0160	0.0340	0.00215	0.00399	0.00849
5.00	0.00943	0.0243	0.0570	0.0161	0.0344	0.0768	0.00322	0.00689	0.0154
6.00	0.0145	0.0403	0.0955	0.0279	0.0663	0.152	0.00466	0.0110	0.0253
7.00	0.0213	0.0626	0.149	0.0456	0.117	0.273	0.00652	0.0167	0.0390
8.00	0.0301	0.0927	0.220	0.0712	0.194	0.456	0.00889	0.0243	0.0570
9.00	0.0414	0.132	0.313	0.107	0.306	0.720	0.0119	0.0340	0.0800
0.00	0.0555	0.183	0.431	0.155	0.462	1.090	0.0155	0.0462	0.109
.00	0.0727	0.247	0.577	0.219	0.676	1.592	0.0199	0.0614	0.145
2.00	0.0936	0.325	0.755	0.302	0.961	2.255	0.0251	0.0801	0.188
3.00	0.119	0.421	0.963	0.407	1.332	3.112	0.0313	0.102	0.239
.00	0.149	0.535	1.203	0.541	1.809	4.193	0.0386	0.129	0.299
5.00	0.184	0.670	1.474	0.706	2.409	5.529	0.0471	0.161	0.369
6.00	0.225	0.826	1.772	0.910	3.155	7.149	0.0569	0.197	0.447
7.00	0.273	1.002	2.096	1.158	4.067	9.081	0.0681	0.239	0.534
.00	0.328	1.199	2.442	1.458	5.166	11.35	0.0810	0.287	0.630
.00	0.390	1.414	2.807	1.816	6.471	13.97	0.0956	0.341	0.735
.00	0.462	1.647	3.187	2.242	8.001	16.97	0.112	0.400	0.848
5.00	0.963	3.066	5.245	5.703	19.62	37.97	0.228	0.785	1.519
.00	1.693	4.774	7.375	12.25	39.14	69.53	0.408	1.305	2.318
5.00	2.638	6.612	9.395	22.99	67.58	111.5	0.657	1.931	3.186
.00	3.740	8.419	11.22	38.89	105.2	163.2	0.972	2.630	4.079
5.00	4.928	10.11	12.86	60.54	151.6	223.4	1.345	3.368	4.965
.00	6.154	11.66	14.29	88.23	206.1	291.4	1.765	4.121	5.828
.00	7.385	13.04	15.52	122.1	267.9	366.0	2.220	4.871	6.654
.00	8.595	14.27	16.59	162.0	336.2	446.3	2.701	5.604	7.438
.00	9.759	15.35	17.51	208.0	410.4	531.6	3.199	6.313	8.179
.00	10.86	16.30	18.31	259.5	489.5	621.2	3.708	6.993	8.874
.00	11.89	17.14	19.01	316.4	573.2	714.6	4.219	7.642	9.528
.00	12.85	17.87	19.63	378.4	660.7	811.2	4.729	8.259	10.14
.00	13.74	18.53	20.17	444.9	751.8	910.7	5.234	8.844	10.71
.00	14.56	19.11	20.64	515.7	845.9	1013.	5.730	9.399	11.25
.00	15.31	19.63	21.06	590.4	942.8	1117.	6.215	9.924	11.76

(continued)

T °K	C_p° J/deg-mol			$H_T^\circ - H_0^\circ$ J/mol			$(H_T^\circ - H_0^\circ)/T$ J/deg-mol		
	Cu	Ag	Au	Cu	Ag	Au	Cu	Ag	Au
100.00	16.01	20.10	21.44	668.7	1042.	1223.	6.687	10.42	12.23
105.00	16.64	20.52	21.77	750.3	1144.	1331.	7.146	10.89	12.68
110.00	17.22	20.89	22.06	835.0	1247.	1441.	7.591	11.34	13.10
115.00	17.76	21.23	22.33	922.5	1353.	1552.	8.021	11.76	13.49
120.00	18.25	21.54	22.56	1013.	1460.	1664.	8.438	12.16	13.87
125.00	18.70	21.82	22.78	1105.	1568.	1777.	8.839	12.54	14.22
130.00	19.12	22.07	22.97	1199.	1678.	1892.	9.227	12.91	14.55
135.00	19.51	22.31	23.15	1296.	1789.	2007.	9.601	13.25	14.87
140.00	19.87	22.52	23.31	1395.	1901.	2123.	9.961	13.58	15.17
145.00	20.20	22.72	23.45	1495.	2014.	2240.	10.31	13.89	15.45
150.00	20.51	22.90	23.59	1597.	2128.	2358.	10.64	14.19	15.72
155.00	20.79	23.07	23.70	1700.	2243.	2476.	10.97	14.47	15.97
160.00	21.05	23.22	23.81	1804.	2358.	2595.	11.28	14.74	16.22
165.00	21.30	23.37	23.91	1910.	2475.	2714.	11.58	15.00	16.45
170.00	21.53	23.50	24.00	2017.	2592.	2834.	11.87	15.25	16.67
175.00	21.74	23.63	24.08	2125.	2710.	2954.	12.15	15.49	16.88
180.00	21.94	23.75	24.15	2235.	2828.	3075.	12.42	15.71	17.08
185.00	22.13	23.86	24.22	2345.	2947.	3196.	12.68	15.93	17.27
190.00	22.31	23.96	24.29	2456.	3067.	3317.	12.93	16.14	17.46
195.00	22.47	24.06	24.35	2568.	3187.	3438.	13.17	16.34	17.63
200.00	22.63	24.16	24.41	2681.	3308.	3650.	13.40	16.54	17.80
205.00	22.77	24.24	24.48	2794.	3429.	3683.	13.63	16.72	17.96
210.00	22.91	24.33	24.54	2908.	3550.	3805.	13.85	16.90	18.12
215.00	23.04	24.41	24.60	3023.	3672.	3928.	14.06	17.08	18.27
220.00	23.17	24.49	24.65	3139.	3794.	4051.	14.27	17.25	18.41
225.00	23.28	24.56	24.71	3255.	3917.	4174.	14.47	17.41	18.55
230.00	23.39	24.63	24.76	3372.	4040.	4298.	14.66	17.56	18.69
235.00	23.50	24.69	24.82	3489.	4163.	4422.	14.85	17.71	18.82
240.00	23.60	24.76	24.87	3607.	4287.	4546.	15.03	17.86	18.94
245.00	23.69	24.82	24.92	3725.	4411.	4671.	15.20	18.00	19.06
250.00	23.78	24.88	24.97	3844.	4535.	4796.	15.37	18.14	19.18
255.00	23.86	24.93	25.02	3963.	4659.	4921.	15.54	18.27	19.30
260.00	23.94	24.99	25.07	4082.	4784.	5046.	15.70	18.40	19.41
265.00	24.02	25.04	25.12	4202.	4909.	5171.	15.86	18.53	19.51
270.00	24.09	25.09	25.17	4322.	5035.	5297.	16.01	18.65	19.62
273.15	24.13	25.12	25.20	4398.	5114.	5376.	16.10	18.72	19.68
275.00	24.15	25.14	25.21	4443.	5160.	5423.	16.16	18.76	19.72
280.00	24.22	25.19	25.26	4564.	5286.	5549.	16.30	18.88	19.82
285.00	24.28	25.24	25.31	4685.	5412.	5676.	16.44	18.99	19.91
290.00	24.34	25.28	25.35	4807.	5538.	5802.	16.57	19.10	20.01
295.00	24.40	25.32	25.39	4929.	5665.	5929.	16.71	19.20	20.10
298.15	24.44	25.35	25.42	5005.	5745.	6009.	16.79	19.27	20.15
300.00	24.46	25.37	25.43	5051.	5792.	6056.	16.84	19.31	20.19

T °K	S_T° J/deg-mol			$-(G_T^\circ - H_0^\circ)$ J/mol			$-(G_T^\circ - H_0^\circ)/T$ J/deg-mol		
	Cu	Ag	Au	Cu	Ag	Au	Cu	Ag	Au
1.00	0.000711	0.000706	0.000880	0.000351	0.000339	0.000402	0.000351	0.000339	0.000402
2.00	0.00152	0.00175	0.00266	0.00145	0.00152	0.00206	0.000727	0.000762	0.00103
3.00	0.00251	0.00347	0.00620	0.00345	0.00406	0.00631	0.00115	0.00135	0.00210
4.00	0.00379	0.00619	0.0123	0.00657	0.00879	0.0153	0.00164	0.00220	0.00383
5.00	0.00546	0.0103	0.0218	0.0112	0.0169	0.0321	0.00223	0.00338	0.00641
6.00	0.00760	0.0160	0.0354	0.0176	0.0299	0.0603	0.00294	0.00498	0.0100
7.00	0.0103	0.0238	0.0539	0.0265	0.0496	0.104	0.00379	0.00709	0.0149
8.00	0.0137	0.0341	0.0782	0.0385	0.0783	0.170	0.00481	0.00979	0.0212
9.00	0.0179	0.0472	0.109	0.0542	0.119	0.263	0.00602	0.0132	0.0292
10.00	0.0229	0.0636	0.148	0.0746	0.174	0.391	0.00746	0.0174	0.0391
11.00	0.0290	0.0839	0.196	0.100	0.247	0.562	0.00913	0.0225	0.0511
12.00	0.0362	0.109	0.253	0.133	0.343	0.786	0.0111	0.0286	0.0655
13.00	0.0447	0.138	0.322	0.173	0.466	1.073	0.0133	0.0359	0.0825
14.00	0.0545	0.174	0.402	0.223	0.622	1.434	0.0159	0.0444	0.102
15.00	0.0660	0.215	0.494	0.283	0.815	1.880	0.0189	0.0544	0.125
16.00	0.0791	0.263	0.598	0.355	1.054	2.426	0.0222	0.0659	0.152
17.00	0.0941	0.318	0.715	0.442	1.344	3.081	0.0260	0.0790	0.181
18.00	0.111	0.381	0.845	0.544	1.693	3.861	0.0302	0.0940	0.214
19.00	0.131	0.452	0.987	0.665	2.109	4.775	0.0350	0.111	0.251
20.00	0.152	0.530	1.140	0.806	2.599	5.838	0.0403	0.130	0.292
25.00	0.305	1.043	2.069	1.917	6.446	13.76	0.0767	0.258	0.550
30.00	0.541	1.750	3.214	3.995	13.35	26.89	0.133	0.445	0.896
35.00	0.871	2.623	4.505	7.487	24.22	46.14	0.214	0.692	1.318
40.00	1.294	3.625	5.881	12.86	39.79	72.08	0.322	0.995	1.802
45.00	1.802	4.715	7.299	20.57	60.61	105.0	0.457	1.347	2.334
50.00	2.385	5.862	8.729	31.01	87.04	145.1	0.620	1.741	2.902
55.00	3.029	7.040	10.15	44.52	119.3	192.3	0.809	2.169	3.496
60.00	3.724	8.228	11.55	61.38	157.5	246.5	1.023	2.624	4.109
65.00	4.458	9.414	12.91	81.82	201.6	307.7	1.259	3.101	4.734
70.00	5.222	10.59	14.24	106.0	251.6	375.6	1.514	3.594	5.366
75.00	6.007	11.74	15.53	134.1	307.4	450.0	1.788	4.099	6.001
80.00	6.806	12.87	16.78	166.1	368.9	530.8	2.076	4.612	6.635
85.00	7.612	13.97	17.98	202.1	436.1	617.7	2.378	5.130	7.267

(continued)

T °K	S_T° J/deg-mol			$-(G_T^\circ - H_0^\circ)$ J/mol			$-(G_T^\circ - H_0^\circ)/T$ J/deg-mol		
	Cu	Ag	Au	Cu	Ag	Au	Cu	Ag	Au
0.00	8.421	15.05	19.15	242.2	508.6	710.6	2.691	5.652	7.895
5.00	9.229	16.10	20.28	286.4	586.5	809.1	3.014	6.174	8.517
0.00	10.03	17.12	21.37	334.5	669.6	913.3	3.345	6.696	9.133
5.00	10.83	18.11	22.42	386.7	757.7	1023.	3.683	7.216	9.740
0.00	11.62	19.07	23.44	442.8	850.6	1137.	4.025	7.733	10.34
5.00	12.39	20.01	24.43	502.8	948.3	1257.	4.372	8.246	10.93
0.00	13.16	20.92	25.38	566.7	1051.	1382.	4.723	8.755	11.51
5.00	13.91	21.80	26.31	634.4	1157.	1511.	5.075	9.260	12.09
0.00	14.66	22.66	27.20	705.8	1269.	1645.	5.429	9.759	12.65
5.00	15.39	23.50	28.07	780.9	1384.	1783.	5.785	10.25	13.21
0.00	16.10	24.32	28.92	859.7	1504.	1925.	6.140	10.74	13.75
5.00	16.80	25.11	29.74	941.9	1627.	2072.	6.496	11.22	14.29
0.00	17.49	25.88	30.54	1028.	1755.	2223.	6.851	11.70	14.82
5.00	18.17	26.64	31.31	1117.	1886.	2377.	7.206	12.17	15.34
0.00	18.84	27.37	32.07	1209.	2021.	2536.	7.559	12.63	15.85
5.00	19.49	28.09	32.80	1305.	2160.	2698.	7.910	13.09	16.35
0.00	20.13	28.79	33.52	1404.	2302.	2864.	8.260	13.54	16.85
5.00	20.75	29.47	34.21	1506.	2448.	3033.	8.608	13.99	17.33
0.00	21.37	30.14	34.89	1612.	2597.	3206.	8.954	14.43	17.81
5.00	21.97	30.79	35.55	1720.	2749.	3382.	9.298	14.86	18.28
0.00	22.57	31.43	36.20	1831.	2904.	3561.	9.639	15.29	18.74
5.00	23.15	32.05	36.83	1946.	3063.	3744.	9.978	15.71	19.20
0.00	23.72	32.66	37.45	2063.	3225.	3930.	10.31	16.12	19.65
5.00	24.28	33.26	38.05	2183.	3390.	4118.	10.65	16.54	20.09
0.00	24.83	33.85	38.64	2306.	3558.	4310.	10.98	16.94	20.52
5.00	25.37	34.42	39.22	2431.	3728.	4505.	11.31	17.34	20.95
0.00	25.90	34.98	39.79	2559.	3902.	4702.	11.63	17.74	21.37
5.00	26.42	35.53	40.34	2690.	4078.	4903.	11.96	18.12	21.79
0.00	26.94	36.07	40.89	2824.	4257.	5106.	12.28	18.51	22.20
5.00	27.44	36.60	41.42	2960.	4439.	5312.	12.59	18.89	22.60
0.00	27.94	37.12	41.94	3098.	4623.	5520.	12.91	19.26	23.00
5.00	28.42	37.63	42.46	3239.	4810.	5731.	13.22	19.63	23.39
0.00	28.90	38.14	42.96	3382.	4999.	5945.	13.53	20.00	23.78
5.00	29.37	38.63	43.46	3528.	5191.	6161.	13.83	20.36	24.16
0.00	29.84	39.11	43.94	3676.	5386.	6379.	14.14	20.71	24.53
5.00	30.30	39.59	44.42	3826.	5582.	6600.	14.44	21.07	24.91
0.00	30.75	40.06	44.89	3979.	5782.	6823.	14.74	21.41	25.27
.15	31.02	40.35	45.18	4076.	5908.	6965.	14.92	21.63	25.50
5.00	31.19	40.52	45.35	4134.	5983.	7049.	15.03	21.76	25.63
0.00	31.62	40.97	45.81	4291.	6187.	7277.	15.32	22.10	25.99
5.00	32.05	41.42	46.25	4450.	6393.	7507.	15.61	22.43	26.34
0.00	32.48	41.86	46.69	4611.	6601.	7739.	15.90	22.76	26.69
5.00	32.89	42.29	47.13	4775.	6811.	7974.	16.19	23.09	27.03
.15	33.15	42.56	47.40	4879.	6945.	8123.	16.36	23.29	27.24
0.00	33.30	42.72	47.56	4940.	7024.	8211.	16.47	23.41	27.37

Summary of Evaluated Photochemical Data for Atmospheric Chemistry

Reaction	Quantum yield, $\phi(\lambda)$	Wavelength λ, nm	Wavelength range nm, for absorption coefficients
$HNO_3 + h\nu \rightarrow HO + NO_2$	no recommendation*		190–370*
$H_2O_2 + h\nu \rightarrow 2\,HO$	1	200–300	185–225
			254
$O_3 + h\nu\,(vis) \rightarrow O + O_2$	1	450–750	440–850
$O_3 + h\nu\,(uv) \rightarrow O(^1D) + O_2(^1\Delta)$	1	250–310	200–360
	0	> 310	
$\rightarrow O(^1D) + O_2(^3\Sigma_g^-)$	0	< 350	
$\rightarrow O(^3P) + O_2$ (Singlet)	0	< 310	
	~ 1	310–350	
$\rightarrow O\,(total) + O_2$	1	250–350	
$\rightarrow O(^1D) + O_2(^1\Sigma_g^+)$	0	250–350	
$\rightarrow O(^3P) + O_2(^3\Sigma_g^-)$	0	250–350	

*Changed from value recommended in NBS Report 10828 (April 1972).

Summary of Evaluated Rate Data for Atmospheric Chemistry

Hampson, Et Al

Reaction		Rate constant k (cm^3 molecule^{-1} s^{-1})	Temperature range (K)	Uncertainty in log k
$H + HNO$	$\rightarrow H_2 + NO$	$> 5 \times 10^{-14}$	211–703	
		7×10^{-12}*	2000	±0.3
$H + HNO_2$	$\rightarrow$ Products	No recommendation		
$H + HNO_3$	$\rightarrow$ Products	$< 1 \times 10^{-13}$	300	
$H + H_2O_2$	$\rightarrow H_2 + HO_2$	$2.8 \times 10^{-12} \exp(-1900/T)$	300–800	±0.3
	$H_2O + HO$	No recommendation		
$H + NO_2$	$\rightarrow HO + NO$	4.8×10^{-11}	300	±0.1
$H + O_3$	$\rightarrow HO + O_2$	2.6×10^{-11}	300	±0.1
$HNO + HO$	$\rightarrow H_2O + NO$	7×10^{-11}*	1600–2100	±0.7
$HNO_2 + HO$	$\rightarrow H_2O + NO_2$	No recommendation*		
$HNO_2 + O$	$\rightarrow HO + NO_2$	No recommendation		
$HNO_3 + HO$	$\rightarrow H_2O + NO_3$	$6 \times 10^{-13} \exp(-400/T)$*	300–650	±0.5
$HNO_3 + O$	$\rightarrow HO + NO_3$	$< 1.5 \times 10^{-14}$	300	
$HO + H_2O_2$	$\rightarrow HO_2 + H_2O$	$1.7 \times 10^{-11} \exp(-910/T)$	300–800	±0.2
$HO + O_3$	$\rightarrow HO_2 + O_2$	$1.6 \times 10^{-12} \exp(-1000/T)$*	220–450	±0.3
$HO_2 + HO_2$	$\rightarrow H_2O_2 + O_2$	$3 \times 10^{-11} \exp(-500/T)$ (a)*	300–1000	±0.3 (b)
$H_2O + NO + NO_2$	$\rightarrow 2\ HNO_2$	$< 1.1 \times 10^{-55}$ (c)*	300	
$H_2O + N_2O_5$	$\rightarrow 2\ HNO_3$	$< 1 \times 10^{-20}$*	300	
$H_2O + O(^1D)$	$\rightarrow 2\ HO$	3.5×10^{-10}*	300	±0.1
$H_2O_2 + NO$	$\rightarrow HO + HNO_2$	$< 5 \times 10^{-20}$*	300	
		$\sim 2 \times 10^{-20}$	550	
$NO + O_3$	$\rightarrow NO_2 + O_2$	$9 \times 10^{-13} \exp(-1200/T)$	198–330	±0.11
$NO_2 + O_3$	$\rightarrow NO_3 + O_2$	5×10^{-17}*	298	±0.2
$N_2O + O(^1D)$	$\rightarrow N_2 + O_2$	1.1×10^{-10}*	300	±0.1
	$\rightarrow 2\ NO$	1.1×10^{-10}*	300	±0.1
$O + O_3$	$\rightarrow 2\ O_2$	$1.9 \times 10^{-11} \exp(-2300/T)$*	200–1000	±0.1
$O_2(^1\Delta) + M$	$\rightarrow O_2 + M$	$2.2(T/300)^{0.8} \times 10^{-18} (M = O_2)$	285–322	±0.1
		$< 2 \times 10^{-20}$ $(M = N_2)$	300	
$O_2(^1\Sigma) + M$	$\rightarrow O_2 + M$	$1.5 \times 10^{-16} (M = O_2)$	300	±0.12
		$2.0 \times 10^{-15} (M = N_2)$	300	±0.1
		$4 \times 10^{-12} (M = H_2O)$	300	±0.18

(a) $-d[HO_2]/dt = 2k[HO_2]^2$.

(b) Error in log k increases to ±1 at 1000 K.

(c) $-d[NO_2]/dt = k[NO][NO_2][H_2O]^2$. Value of k is for a surface reaction. This is adopted as the upper limit for he gas phase rate constant.

*Changed from value recommended in NBS Reports 10692 (Jan. 1972) and 10828 (April 1972).

Summary of Preferred Rate Data for Atmospheric Chemistry

BAULCH ET AL

Reaction	k_{298} cm^3 molecule^{-1} s^{-1}	$\Delta \log k_{298}$	Temp. dependence of k/cm^3 molecule^{-1} s^{-1}	Temp. range/K	$\Delta(E/R)$/ K
O_x Reactions					
$O + O_2 + M \rightarrow O_3 + M$	3.6×10^{-34} [Ar](k_0)	±0.1	$3.6 \times 10^{-34} (T/300)^{-1.83}$ [Ar]	200–1100	
	5.6×10^{-34} $[N_2](k_0)$	±0.1	$5.6 \times 10^{-34} (T/300)^{-2.36}$ $[N_2]$	220–300	
	2.8×10^{-12} (k_∞)	±0.3			
$O + O_3 \rightarrow 2O_2$	9.5×10^{-15}	±0.1	$2.0 \times 10^{-11} \exp(-2280/T)$	220–1000	±130
$O(^1D) + O_2 \rightarrow O(^3P) + O_2(^1\Sigma_g^+)$	3.7×10^{-11}	±0.15	$3.0 \times 10^{-11} \exp(67/T)$	200–350	±100
$\rightarrow O(^3P) + O_2(^3\Sigma_g^-)$	0.9×10^{-11}	±0.15	$0.7 \times 10^{-11} \exp(67/T)$	200–350	±100
$O_2 + h\nu \rightarrow 2O$	See data sheets				
$O_3 + h\nu \rightarrow O + O_2$	See data sheets				
HO_x Reactions					
$H + HO_2 \rightarrow H_2 + O_2$	1.4×10^{-11}	±0.4			
$\rightarrow 2HO$	3.2×10^{-11}	±0.4			
$\rightarrow H_2O + O$	$\leq 9.4 \times 10^{-13}$	+0.3 −?			

(continued)

Reaction	k_{298} cm^3 molecule^{-1} s^{-1}	$\Delta\log k_{298}$	Temp. dependence of k/cm^3 molecule^{-1} s^{-1}	Temp. range/K	$\Delta(E/R)$/K
$H + O_2 + M \rightarrow HO_2 + M$	1.8×10^{-32} [Ar](k_0)	±0.2	$1.8 \times 10^{-32}(T/300)^{-0.8}$ [Ar]	200–2000	
	5.9×10^{-32} [N_2](k_0)	±0.2	$5.9 \times 10^{-32}(T/300)^{-1.0}$ [N_2]	200–400	
$H + O_3 \rightarrow HO + O_2$	2.8×10^{-11}	±0.2	$1.4 \times 10^{-10} \exp(-480/T)$	220–360	±100
$O + HO \rightarrow O_2 + H$	3.8×10^{-11}	±0.3			
$O + HO_2 \rightarrow HO + O_2$	3.1×10^{-11}	±0.5			
$O + H_2O_2 \rightarrow HO + HO_2$ $\rightarrow O_2 + H_2O$	2.1×10^{-15}	±0.3	$2.7 \times 10^{-12} \exp(-2100/T)$	283–368	±500
$O(^1D) + H_2 \rightarrow HO + H$ $\rightarrow O(^3P) + H_2$	2.0×10^{-10}	±0.3	2.0×10^{-10}	200–350	±100
$O(^1D) + H_2O \rightarrow 2HO$ $\rightarrow O(^3P) + H_2O$	2.8×10^{-10}	±0.3	2.8×10^{-10}	200–350	±100
$HO + H_2 \rightarrow H_2O + H$	7.1×10^{-15}	±0.1	$1.8 \times 10^{-11} \exp(-2330/T)$	210–300	±300
$HO + HO \rightarrow H_2O + O$	1.8×10^{-12}	±0.2			
$HO + HO + M \rightarrow H_2O_2 + M$	6.5×10^{-31} [Ar](k_0)	±0.3	$6.5 \times 10^{-31}(T/300)^{-2.0}$ [Ar]	300–1500	
	6.5×10^{-31} [N_2](k_0)	±0.3			
$HO + HO_2 \rightarrow H_2O + O_2$	3.5×10^{-11}	±0.5			
$HO + H_2O_2 \rightarrow H_2O + HO_2$	8.0×10^{-13}	±0.3	$7.6 \times 10^{-12} \exp(-670/T)$	200–700	±200
$HO + O_3 \rightarrow HO_2 + O_2$	6.7×10^{-14}	±0.15	$1.9 \times 10^{-12} \exp(-1000/T)$	220–450	$^{+250}_{-100}$
$HO_2 + HO_2 \rightarrow H_2O_2 + O_2$	2.3×10^{-12}	±0.3			
$HO_2 + O_3 \rightarrow HO + 2O_2$	2.0×10^{-15}	±0.2	$1.4 \times 10^{-14} \exp(-600/T)$	250–400	±200
$H_2O + h\nu \rightarrow HO + H$	See data sheets				
$H_2O_2 + h\nu \rightarrow 2HO$	See data sheets				
NO_x Reactions					
$N + O_2 \rightarrow NO + O$	8.9×10^{-17}	±0.1	$4.4 \times 10^{-12} \exp(-3220/T)$	280–333	±350
$N + O_3 \rightarrow NO + O_2$	$\leq 5 \times 10^{-16}$				
$N + NO \rightarrow N_2 + O$	3.4×10^{-11}	±0.15	3.4×10^{-11}	200–400	±100
$N + NO_2 \rightarrow N_2O + O$	1.4×10^{-12}	±0.2			
$O + NO + M \rightarrow NO_2 + M$	6.4×10^{-32} [Ar](k_0)	±0.1	$6.4 \times 10^{-32}(T/300)^{-1.80}$ [Ar]	200–2000	
	1.2×10^{-31} [N_2](k_0)	±0.1	$1.2 \times 10^{-31}(T/300)^{-1.82}$ [N_2]	200–300	
	3.0×10^{-11} (k_∞)	±0.2	$3.0 \times 10^{-11}(T/300)^{+0.3}$	300–1500	
$O + NO_2 \rightarrow NO + O_2$	9.3×10^{-12}	±0.06	9.3×10^{-12}	230–340	$^{+0}_{-150}$
$O + NO_2 + M \rightarrow NO_3 + M$	9×10^{-32} [N_2](k_0)	±0.1			
	2.2×10^{-11} (k_∞)	±0.1			
$O + NO_3 \rightarrow O_2 + NO_2$	1×10^{-11}	±0.5			
$O + N_2O_5 \rightarrow$ products	$\leq 3 \times 10^{-16}$		$\leq 3 \times 10^{-16}$	220–300	
$O(^1D) + N_2 \rightarrow O(^3P) + N_2$	4.5×10^{-11}	±0.15	$3.2 \times 10^{-11} \exp(107/T)$	200–350	±100
$O(^1D) + N_2O \rightarrow N_2 + O_2$	7.4×10^{-11}	±0.15	7.4×10^{-11}	200–350	±100
$\rightarrow 2NO$	8.6×10^{-11}	±0.15	8.6×10^{-11}	200–350	±100
$\rightarrow O(^3P) + N_2O$	No recommendation (see data sheet)				
$HO + NO + M \rightarrow HONO + M$	6.5×10^{-31} [N_2](k_0)	±0.1	$6.5 \times 10^{-31}(T/300)^{-2.4}$ [N_2]	220–440	
	1.0×10^{-11} (k_∞)	±0.2	1.0×10^{-11}	220–440	
$HO + NO_2 + M \rightarrow HONO_2 + M$	1.0×10^{-30} [Ar](k_0)	±0.1	$1.0 \times 10^{-30}(T/300)^{-2.9}$ [Ar]	300–1200	
	2.6×10^{-30} [N_2](k_0)	±0.1	$2.6 \times 10^{-30}(T/300)^{-2.7}$ [N_2]	220–550	
	1.6×10^{-11} (k_∞)	±0.2	1.6×10^{-11}	200–1200	
$HO + HONO_2 \rightarrow H_2O + NO_3$	8.5×10^{-14}	±0.1	8.5×10^{-14}	240–470	±300
$HO + HO_2NO_2 \rightarrow$ products	No recommendation (see data sheet)				
$HO_2 + NO \rightarrow HO + NO_2$	8.4×10^{-12}	±0.08	$4.3 \times 10^{-12} \exp(200/T)$	230–425	±200
$HO_2 + NO_2 + M \rightarrow HO_2NO_2 + M$	2.1×10^{-31} [N_2](k_0)	±0.1			
	5×10^{-12} (k_∞)	±0.4			
$HO_2NO_2 + M \rightarrow HO_2 + NO_2 + M$	1.2×10^{-20} [N_2](k_0/s^{-1})	±0.1		298	
	0.09 (k_∞/s^{-1})	±0.6	$1.4 \times 10^{14} \exp(-10420/T)$ s^{-1}	250–300	±500
$NO + O_3 \rightarrow NO_2 + O_2$	1.8×10^{-14}	±0.06	$2.3 \times 10^{-12} \exp(-1450/T)$	200–360	±200
$NO + NO_3 \rightarrow 2NO_2$	2×10^{-11}	±0.5			
$NO_2 + NO_3 + M \rightarrow N_2O_5 + M$	1.5×10^{-30} [N_2](k_0)	±0.3	$1.5 \times 10^{-30}(T/300)^{-4.6}$ [N_2]	300–340	
	5×10^{-12} (k_∞)	±0.3	5×10^{-12}	200–400	
$N_2O_5 + M \rightarrow NO_2 + NO_3 + M$	6.4×10^{-20} [N_2](k_0/s^{-1})	±0.3	$8.8 \times 10^{-6} \exp(-9700/T)$ [N_2] s^{-1}	300–340	
	0.20 (k_∞/s^{-1})	±0.3	$5.7 \times 10^{14} \exp(-10600/T)$ s^{-1}	273–300	
$NO_2 + O_3 \rightarrow NO_3 + O_2$	3.2×10^{-17}	±0.06	$1.2 \times 10^{-13} \exp(-2450/T)$	230–360	±150
$NO + h\nu \rightarrow$ products	See data sheets				
$NO_2 + h\nu \rightarrow$ products	See data sheets				
$NO_3 + h\nu \rightarrow$ products	See data sheets				
$N_2O + h\nu \rightarrow$ products	See data sheets				
$N_2O_5 + h\nu \rightarrow$ products	See data sheets				
$HONO + h\nu \rightarrow$ products	See data sheets				
$HONO_2 + h\nu \rightarrow$ products	See data sheets				
$HO_2NO_2 + h\nu \rightarrow$ products	See data sheets				
CH_4 Reactions					
$O(^1D) + CH_4 \rightarrow HO + CH_3$	2.2×10^{-10}	±0.3	2.2×10^{-10}	200–300	±100
$\rightarrow HCHO + H_2$	2.4×10^{-11}		2.4×10^{-11}	200–300	±100
$HO + CH_4 \rightarrow H_2O + CH_3$	8.0×10^{-15}	±0.1	$2.4 \times 10^{-12} \exp(-1710/T)$	200–300	±200
$HO + CO \rightarrow H + CO_2$	1.5×10^{-13} (≤ 100 Torr)	±0.05	1.5×10^{-13}	200–300	
$HO + CO \rightarrow$ products	2.8×10^{-13} (1 atm air)	±0.1			
$HO + HCHO \rightarrow H_2O + HCO$ $\rightarrow H + HCOOH$	1.3×10^{-11}	±0.15	1.3×10^{-11}	200–400	±200
$HO_2 + CH_3O_2 \rightarrow O_2 + CH_3OOH$ $\rightarrow HO + O_2 + CH_3O$	6.5×10^{-12}	±0.3			
$HCO + O_2 \rightarrow HO_2 + CO$	5.1×10^{-12}	±0.1			
$HCO + O_2 + M \rightarrow HCO_3 + M$	No recommendation (see data sheet)				

(continued)

Reaction	k_{298} $cm^3\ molecule^{-1}\ s^{-1}$	$\Delta \log k_{298}$	Temp. dependence of $k/cm^3\ molecule^{-1}\ s^{-1}$	Temp. range/K	$\Delta(E/R)$/ K
$CH_3 + O_2 \rightarrow HCHO + HO$	No recommendation (see data sheet)				
$CH_3 + O_2 + M \rightarrow CH_3O_2 + M$	$2.6 \times 10^{-31}\ [N_2](k_0)$	±0.3	$2.6 \times 10^{-31}(T/300)^{-3}\ [N_2]$	260–340	
	$2 \times 10^{-12}\ (k_\infty)$	±0.3	2.0×10^{-12}	200–400	
$CH_3O + O_2 \rightarrow HCHO + HO_2$	6×10^{-16}	±0.6	$5 \times 10^{-13} \exp(-2000/T)$	300–450	±1000
$CH_3O_2 + CH_3O_2 \rightarrow CH_3OH + HCHO + O_2$; $\rightarrow 2\ CH_3O + O_2$; $\rightarrow CH_3OOCH_3 + O_2$	4.6×10^{-13}	±0.1			
$CH_3O_2 + NO \rightarrow CH_3O + NO_2$	7.5×10^{-12}	±0.3	7.5×10^{-12}	200–300	±500
$CH_3O_2 + NO_2 + M \rightarrow CH_3O_2NO_2 + M$	1.6×10^{-12} (1 atm)	±0.5	1.6×10^{-12} (1 atm)	200–300	±500
$HCHO + h\nu \rightarrow$ products	See data sheets				
$CH_3OOH + h\nu \rightarrow$ products	See data sheets				
SO_x Reactions					
$O + H_2S \rightarrow HO + HS$	2.7×10^{-14}	±0.1	$7.2 \times 10^{-12} \exp(-1660/T)$	250–500	±150
$O + CS \rightarrow CO + S$	2.1×10^{-11}	±0.1	$2.7 \times 10^{-10} \exp(-760/T)$	150–300	±250
$O + OCS \rightarrow SO + CO$	1.4×10^{-14}	±0.2	$2.6 \times 10^{-11} \exp(-2250/T)$	220–600	±150
$O + CS_2 \rightarrow SO + CS$; $\rightarrow CO + S_2$; $\rightarrow OCS + S$	5.5×10^{-12}	±0.2	$5.8 \times 10^{-11} \exp(-700/T)$	200–500	±100
$O + SO_2 + M \rightarrow SO_3 + M$	$1.4 \times 10^{-33}\ [N_2](k_0)$	±0.3	$4.0 \times 10^{-32} \exp(-1000/T)$	200–400	$^{+200}_{-100}$
$HO + H_2S \rightarrow H_2O + HS$	5.3×10^{-12}	±0.1	$1.1 \times 10^{-11} \exp(-225/T)$	250–400	±225
$HO + OCS \rightarrow$ products	$\leq 6 \times 10^{-14}$	$^{+0.7}_{-?}$			
$HO + CS_2 \rightarrow$ products	$\leq 2 \times 10^{-13}$	$^{+0.7}_{-?}$			
$HO + SO_2 + M \rightarrow HOSO_2 + M$	$3 \times 10^{-31}\ [N_2](k_0)$	±0.3	$3 \times 10^{-31}(T/300)^{-2.9}\ [N_2]$	200–400	
	$2 \times 10^{-12}\ (k_\infty)$	±0.4	2×10^{-12}	200–400	
$HO_2 + SO_2 \rightarrow$ products	No recommendation (see data sheet)				
$S + O_2 \rightarrow SO + O$	2.0×10^{-12}	±0.15	2.0×10^{-12}	230–400	±100
$S + O_3 \rightarrow SO + O_2$	1.2×10^{-11}	±0.3			
$HS + O_2 \rightarrow HO + SO$	No recommendation (see data sheet)				
$CS + O_2 \rightarrow CO + SO$; $\rightarrow OCS + O$	No recommendation (see data sheet)				
$SO + O_2 \rightarrow SO_2 + O$	9×10^{-18}	±0.5	$6 \times 10^{-13} \exp(-3300/T)$	300–1000	±500
$SO + O_3 \rightarrow SO_2 + O_2$	6×10^{-14}	±0.3	$2.5 \times 10^{-12} \exp(-1100/T)$	220–300	±400
$SO + NO_2 \rightarrow SO_2 + NO$	1.4×10^{-11}	±0.3			
$SO_3 + H_2O \rightarrow$ products	No recommendation (see data sheet)				
$CH_3O_2 + SO_2 \rightarrow CH_3O + SO_3$; $\rightarrow CH_3O_2SO_2$	No recommendation (see data sheet)				
$OCS + h\nu \rightarrow$ products	See data sheets				
$CS_2 + h\nu \rightarrow$ products	See data sheets				
FO_x Reactions					
$O + FO \rightarrow O_2 + F$	5×10^{-11}	±0.5			
$O + FO_2 \rightarrow O_2 + FO$	5×10^{-11}	±0.7			
$O(^1D) + HF \rightarrow HO + F$; $\rightarrow O(^3P) + HF$	1×10^{-10}	±0.5			
$F + H_2 \rightarrow HF + H$	2.5×10^{-11}	±0.2	$2.0 \times 10^{-10} \exp(-620/T)$	200–400	±250
$F + O_2 + M \rightarrow FO_2 + M$	$1.1 \times 10^{-32}\ [N_2](k_0)$	±0.3	$1.1 \times 10^{-32}(T/300)^{-2.0}\ [N_2]$	270–360	
	$3 \times 10^{-11}\ (k_\infty)$	±0.5	3×10^{-11}	200–400	
$F + O_3 \rightarrow FO + O_2$	1.3×10^{-11}	±0.3	$2.8 \times 10^{-11} \exp(-226/T)$	250–365	±200
$F + H_2O \rightarrow HF + HO$	1.1×10^{-11}	±0.5	$2.2 \times 10^{-11} \exp(-200/T)$	240–360	±200
$F + CH_4 \rightarrow HF + CH_3$	8×10^{-11}	±0.2	$3.0 \times 10^{-10} \exp(-400/T)$	250–450	±150
$FO + O_3 \rightarrow F + 2O_2$; $\rightarrow FO_2 + O_2$	No recommendation (see data sheet)				
$FO + NO \rightarrow F + NO_2$	2×10^{-11}	±0.5			
$FO + NO_2 + M \rightarrow FONO_2 + M$	$1.7 \times 10^{-31}\ [N_2](k_0)$	±0.7	$1.7 \times 10^{-31}(T/300)^{-3.0}\ [N_2]$	200–400	
	$1.2 \times 10^{-11}\ (k_\infty)$	±0.4	1.2×10^{-11}	200–400	
$FO + FO \rightarrow 2F + O_2$; $\rightarrow FO_2 + F$; $\rightarrow F_2 + O_2$	1.5×10^{-11}	±0.3			
$HF + h\nu \rightarrow$ products	See data sheets				
$COF_2 + h\nu \rightarrow$ products	See data sheets				
$FONO_2 + h\nu \rightarrow$ products	See data sheets				
ClO_x Reactions					
$O + HCl \rightarrow HO + Cl$	1.4×10^{-16}	±0.3	$1.1 \times 10^{-11} \exp(-3370/T)$	293–718	±350
$O + ClO \rightarrow O_2 + Cl$	5.0×10^{-11}	±0.1	$7.5 \times 10^{-11} \exp(-120/T)$	220–425	±120
$O + ClONO_2 \rightarrow ClO + NO_3$; $\rightarrow OClO + NO_2$; $\rightarrow O_2 + ClONO$	1.9×10^{-13}	±0.1	$3.0 \times 10^{-12} \exp(-808/T)$	213–295	±200
$O(^1D) + CF_2Cl_2 \rightarrow ClO + CF_2Cl$; $\rightarrow O(^3P) + CF_2Cl_2$	2.8×10^{-10}	±0.3			
$O(^1D) + CFCl_3 \rightarrow ClO + CFCl_2$; $\rightarrow O(^3P) + CFCl_3$	3.5×10^{-10}	±0.2			

(continued)

Reaction	k_{298} cm^3 molecule^{-1} s^{-1}	$\Delta \log k_{298}$	Temp. dependence of k/cm^3 molecule^{-1} s^{-1}	Temp. range/K	$\Delta(E/R)$/K
$O(^1D) + CCl_4 \rightarrow ClO + CCl_3$ $\rightarrow O(^3P) + CCl_4$	4.8×10^{-10}	± 0.2			
$Cl + H_2 \rightarrow HCl + H$	1.8×10^{-14}	± 0.2	$4.7 \times 10^{-11} \exp(-2340/T)$	210–1070	± 200
$Cl + HO_2 \rightarrow HCl + O_2$	4.1×10^{-11}	± 0.3			
$Cl + H_2O_2 \rightarrow HCl + HO_2$	4.3×10^{-13}	± 0.2	$1.1 \times 10^{-11} \exp(-980/T)$	265–424	± 500
$Cl + O_3 \rightarrow ClO + O_2$	1.2×10^{-11}	± 0.06	$2.7 \times 10^{-11} \exp(-257/T)$	205–298	± 100
$Cl + CH_4 \rightarrow HCl + CH_3$	1.04×10^{-13}	± 0.06	$9.9 \times 10^{-12} \exp(-1360/T)$	200–300	± 150
$Cl + C_2H_6 \rightarrow HCl + C_2H_5$	5.7×10^{-11}	± 0.06	$7.7 \times 10^{-11} \exp(-90/T)$	220–350	± 100
$Cl + HCHO \rightarrow HCl + HCO$	7.3×10^{-11}	± 0.06	$7.9 \times 10^{-11} \exp(-34/T)$	200–500	± 100
$Cl + HONO_2 \rightarrow HCl + NO_3$	$\leq 7 \times 10^{-15}$	$^{+0.3}_{-2.5}$			
$Cl + CH_3Cl \rightarrow HCl + CH_2Cl$	4.9×10^{-13}	± 0.1	$3.4 \times 10^{-11} \exp(-1260/T)$	233–350	± 200
$Cl + ClONO_2 \rightarrow Cl_2 + NO_3$ $\rightarrow ClONO + ClO$	2.2×10^{-13}	± 0.3	$1.7 \times 10^{-12} \exp(-610/T)$	224–273	± 400
$HO + HCl \rightarrow H_2O + Cl$	6.6×10^{-13}	± 0.06	$3.0 \times 10^{-12} \exp(-425/T)$	210–460	± 100
$HO + ClO \rightarrow HO_2 + Cl$ $\rightarrow HCl + O_2$	9.1×10^{-12}	± 0.3			
$HO + ClONO_2 \rightarrow HOCl + NO_3$ $\rightarrow HO_2 + ClONO$ $\rightarrow HNO_3 + ClO$	3.9×10^{-13}	± 0.2	$1.2 \times 10^{-12} \exp(-330/T)$	246–387	± 200
$HO + CH_3Cl \rightarrow H_2O + CH_2Cl$	4.1×10^{-14}	± 0.1	$2.2 \times 10^{-12} \exp(-1140/T)$	240–422	± 200
$HO + CHF_2Cl \rightarrow H_2O + CF_2Cl$	4.4×10^{-15}	± 0.1	$1.3 \times 10^{-12} \exp(-1670/T)$	240–400	± 200
$HO + CHFCl_2 \rightarrow H_2O + CFCl_2$	2.8×10^{-14}	± 0.1	$1.5 \times 10^{-12} \exp(-1180/T)$	240–400	± 200
$HO + CH_3CCl_3 \rightarrow H_2O + CH_2CCl_3$	1.2×10^{-14}	± 0.15	$5.1 \times 10^{-12} \exp(-1800/T)$	250–460	± 200
$ClO + HO_2 \rightarrow HOCl + O_2$ $\rightarrow HCl + O_3$	5.2×10^{-12}	± 0.2			
$ClO + NO \rightarrow Cl + NO_2$	1.8×10^{-11}	± 0.1	$8.9 \times 10^{-12} \exp(+210/T)$	227–415	± 100
$ClO + NO_2 + M \rightarrow ClONO_2 + M$	1.7×10^{-31} $[N_2]$(k_0)	± 0.1	$1.7 \times 10^{-31}(T/300)^{-3.0}$ $[N_2]$	250–400	
	1.2×10^{-11} (k_∞)	± 0.4	1.2×10^{-11}	200–400	
$HOCl + h\nu \rightarrow$ products	See data sheets				
$COFCl + h\nu \rightarrow$ products	See data sheets				
$ClONO_2 + h\nu \rightarrow$ products	See data sheets				
$COCl_2 + h\nu \rightarrow$ products	See data sheets				
$CF_2Cl_2 + h\nu \rightarrow$ products	See data sheets				
$CFCl_3 + h\nu \rightarrow$ products	See data sheets				
$CCl_4 + h\nu \rightarrow$ products	See data sheets				
BrO_x Reactions					
$O + HBr \rightarrow HO + Br$	3.9×10^{-14}	± 0.2	$7.0 \times 10^{-12} \exp(-1560/T)$	250–400	± 300
$O + BrO \rightarrow O_2 + Br$	3×10^{-11}	± 0.5			
$Br + HO_2 \rightarrow HBr + O_2$	1×10^{-11}	± 0.7			
$Br + H_2O_2 \rightarrow HBr + HO_2$	$\leq 2 \times 10^{-14}$	$^{+0.3}_{-1.7}$			
$Br + O_3 \rightarrow BrO + O_2$	1.1×10^{-12}	± 0.1	$1.4 \times 10^{-11} \exp(-760/T)$	220–360	± 200
$HO + HBr \rightarrow H_2O + Br$	8.5×10^{-12}	± 0.3	8.5×10^{-12}	249–416	± 250
$HO + CH_3Br \rightarrow H_2O + CH_2Br$	3.8×10^{-14}	± 0.1	$7.6 \times 10^{-13} \exp(-890/T)$	244–350	± 200
$BrO + HO_2 \rightarrow HOBr + O_2$ $\rightarrow HBr + O_3$	5×10^{-12}	± 0.5			
$BrO + NO \rightarrow Br + NO_2$	2.1×10^{-11}	± 0.1	$8.7 \times 10^{-12} \exp(+260/T)$	224–425	± 100
$BrO + NO_2 + M \rightarrow BrONO_2 + M$	3×10^{-31} $[N_2]$(k_0)	± 0.4	$3 \times 10^{-31}(T/300)^{-3.0}$ $[N_2]$	200–400	
	1.2×10^{-11} (k_∞)	± 0.4	1.2×10^{-11}	200–400	
$BrO + O_3 \rightarrow Br + 2O_2$	$< 5 \times 10^{-15}$	$^{+0.5}_{-?}$			
$BrO + ClO \rightarrow Br + OClO$	6.7×10^{-12}	± 0.3			
$\rightarrow Br + Cl + O_2$ $\rightarrow BrCl + O_2$	6.7×10^{-12}	± 0.3			
$BrO + BrO \rightarrow 2Br + O_2$ $\rightarrow Br_2 + O_2$	2.8×10^{-12}	± 0.1	2.8×10^{-12}	220–440	± 500
$BrO + h\nu \rightarrow$ products	See data sheets				
$HOBr + h\nu \rightarrow$ products	See data sheets				
$BrONO_2 + h\nu \rightarrow$ products	See data sheets				

Enthalpy Data

Substance	$\Delta H^\circ_f(298)$ kJ mol^{-1}	$\Delta H^\circ_f(0)$ kJ mol^{-1}
H	217.997	216.03
H_2	0	0
O	249.17	246.78
$O(^1D)$	438.9	436.6
O_2	0	0
$O_2(^1\Delta)$	94.3	94.3
$O_2(^1\Sigma)$	156.9	156.9
O_3	142.7	145.4
HO	39.0	38.7

(continued)

Substance	$\Delta H^\circ_f(298)$ kJ mol^{-1}	$\Delta H^\circ_f(0)$ kJ mol^{-1}
HO_2	2 ± 8	5 ± 8
H_2O	−241.81	−238.92
H_2O_2	−136.32	−130.04
N	472.68	470.82
N_2	0	0
NH	343	343
NH_2	185	188
NH_3	− 45.94	− 38.95
NO	90.25	89.75
NO_2	33.2	36.0
NO_3	71 ± 20	77 ± 20
N_2O	82.05	85.50
N_2O_4	9.1	18.7
N_2O_5	11.3	23.8
HNO	99.6	102.5
HNO_2	− 79.5	− 74
HNO_3	−135.06	−125.27
HO_2NO_2	− 54 ± 20	
CH	594.1	590.8
CH_2	386	386
CH_3	145.6	149.0
CH_4	− 74.81	− 66.82
CO	−110.53	−113.81
CO_2	−393.51	−393.14
HCO	37.6	37.2
CH_2O	−108.6	−104.7
HCOOH	−378.6	−371.6
CH_3O	14.6	22.6
CH_3O_2	7 ± 8	
CH_3OH	−200.7	−189.7
CH_3OOH	−131	
CH_3ONO	− 65.3	− 52.6
CH_3ONO_2	−119.7	−103.4
C_2H_5	107.5	
C_2H_6	− 83.8	− 68.3
CH_3OOCH_3	−125.5	
S	276.98	274.72
S_2	128.49	128.20
HS	146 ± 4	145 ± 4
H_2S	− 20.63	− 17.70
SO	5.0	5.0
SO_2	−296.81	−294.26
SO_3	−395.7	−390
SOH	21 ± 17	
HSO_3	−481 ± 25	
CS	272	268
CS_2	117.2	116.6
OCS	−142	−142
F	79.39	77.28
F_2	0	0
HF	−273.30	−273.26
HOF	− 98 ± 4	− 95 ± 4
FO	109 ± 8	109 ± 8
FO_2	50 ± 12	52 ± 12
FONO	67	

(continued)

Substance	$\Delta H^\circ_f(298)$ kJ mol^{-1}	$\Delta H^\circ_f(0)$ kJ mol^{-1}
$FONO_2$	10	18
CF_2	−182 ± 8	−182 ± 8
CF_3	−470 ± 4	−468 ± 4
CF_4	−933	−927
FCO	−170 ± 60	−170 ± 60
COF_2	−634.7	−631.6
Cl	121.30	119.62
Cl_2	0	0
HCl	− 92.31	− 92.13
ClO	102	102
ClOO	89 ± 5	91
OClO	97 ± 8	100 ± 8
Cl_2O	81.4	83.2
HOCl	− 78	− 75
ClNO	51.7	53.6
$ClNO_2$	12.5	18.0
ClONO	83	
$ClONO_2$	26.4	
FCl	− 50.7	− 50.8
CCl	502 ± 20	498 ± 20
CCl_2	238 ± 20	237 ± 20
CCl_3	79.5	80.1
CCl_4	− 95.8	− 93.6
$CHCl_3$	−102.9	− 98.0
CH_2Cl	125	
CH_2Cl_2	− 95.4	− 88.5
CH_3Cl	− 82.0	− 74.0
ClCO	− 17	
$COCl_2$	−220.1	−218.4
CFCl	30 ± 25	30 ± 25
$CFCl_2$	− 96	
$CFCl_3$	−284.9	−281.8
CF_2Cl	−269	
CF_2Cl_2	−493.3	−489.1
CF_3Cl	−707.9	−702.9
$CHFCl_2$	−284.9	−279.5
CHF_2Cl	−483.7	−477.4
COFCl	−427 ± 33	−423 ± 33
C_2Cl_4	− 12.4	− 11.9
C_2HCl_3	− 7.8	− 4.3
CH_2CCl_3	45 ± 30	
CH_3CCl_3	−142.3	−145.0
Br	111.86	117.90
Br_2	30.91	45.69
HBr	− 36.38	− 28.54
HOBr	− 80 ± 8	
BrO	125	133
BrNO	82.2	91.5
$BrONO_2$	20 ± 30	
BrCl	14.6	22.1
CH_2Br	163	
CH_3Br	− 37.7	− 22.3

Structure, operating data and characteristics of batteries

Structure and operating data	Lead battery	Nickel-cadmium battery (with pocket plates)	Nickel-cadmium battery (with sintered plates)	Edison battery (nickel-iron)	Silver-zinc battery	Silver-cadmium battery
Cell	Cell in plastic housing, glass or hard rubber with positive armoured and negative grid plates	Cell in nickel-coated steel housing, with positive and negative pocket plates	Cell with nickel-coated steel housing and sintered plates	Cell with steel housing and positive and negative plates	Cell with positive and negative solid plates in plastic housing	
Positive electrode	Lead dioxide (PbO_2)	Nickel-oxy-hydrate (NiOOH)		Nickel (III)-oxide (Ni_2O_3)	Silver (I)- and (II)-oxide $Ag_2O+Ag_2O_2$	
Negative electrode	Lead (Pb)	Cadmium (Cd)		Iron (Fe+HgO)	Zinc (Zn)	Cadmium (Cd)
Electrolyte	Dilute sulphuric acid H_2SO_4, 22 °Bé ($\gamma=1.18$)	20% caustic potash solution KOH ($\gamma=1.2$)			Alkaline solution	
No-load voltage fully charged	2.1 V	1.3 V	1.3 V	1.4 V	1.86 V	1.4 V
Rated voltage	2 V	1.2 V	1.2 V	1.2 V	1.5 V	1.1 V
Mean voltage after 1 hour discharge [1])	at +25 °C 1.86 V at −20 °C 1.84 V at −40 °C 1.72 V	at +25 °C 1.05 V at −20 °C 1.00 V at −40 °C 0.75 V	at +25 °C 1.18 V at −20 °C 1.15 V at −40 °C 1.08 V	at +25 °C 0.85 V at −20 °C — at −40 °C —	at +25 °C 1.43 V at −20 °C 1.27 V at −40 °C —	at +25 °C 0.99 V at −20 °C 0.95 V at −40 °C 0.77 V
Discharge figures Ah	200 to 2000 (approx.)	>2000 (approx.)	>2000 (approx.)	(>2000 approx.)	10 to 400 (approx.)	300 to 1000 (approx.)
Capacity W/kg	4 to 8 (11)	3 to 5	5 to 7	5 to 6	11 to 25	5 to 17
a) Advantages	Low cost; wide range of application, good service life, high voltage per cell, good capacity	Excellent service life, high reliability for vehicles, suitable for heavy duty operations	Excellent life, high reliability for vehicles, good charging and discharging characteristics at low temperatures, suitable for heavy duty operation, can be hermetically sealed	Excellent life, suitable for extremely heavy duty, not damaged by overcharging or intensive discharging	Excellent capacity in relation to weight and volume, excellent characteristics with regard to intensive discharge	Good delivery of power in relation to weight and volume, good service life
b) Disadvantages	Sensitivity to mishandling (mechanical shock, current surge, over-charging, failure to charge) poor discharge characteristics at low temperatures	High cost, poor discharge characteristics at low temperatures, poor discharge cycle	High cost	High cost, poor discharge reserve, poor discharge characteristics at low temperatures	High cost, poor service life, poor discharge characteristics at low temperatures	High cost

[1]) If 20% of the voltage is used for discharge.

CHAPTER 26

BIOMEDICAL TECHNOLOGY

MODEL OF MAN

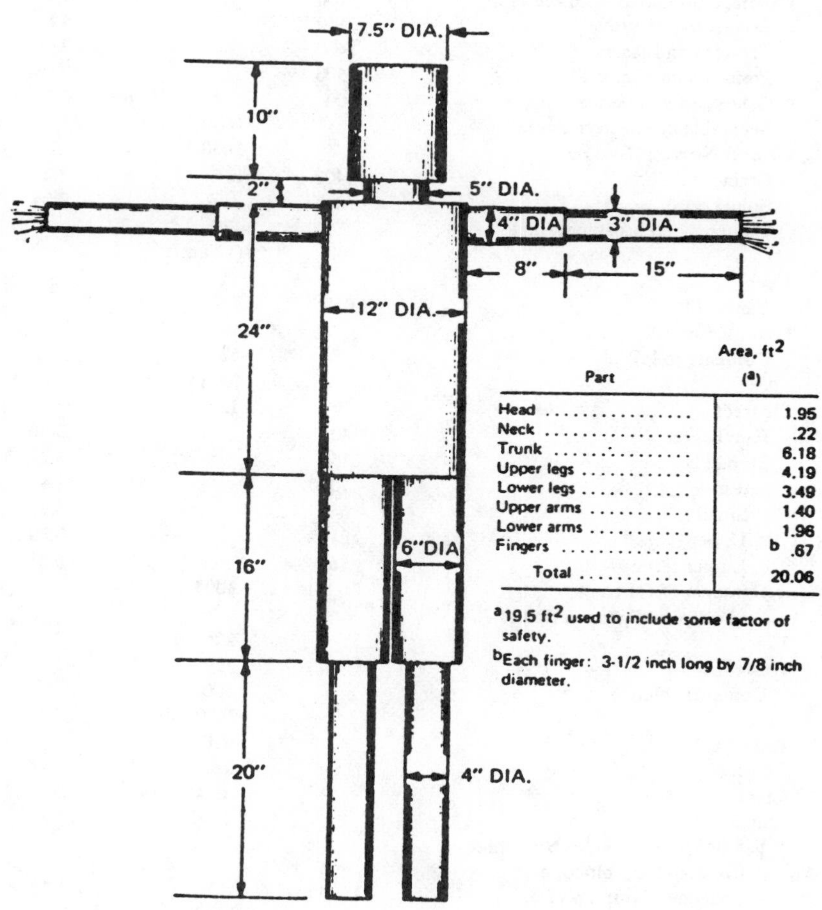

Part	Area, ft^2 ([a])
Head	1.95
Neck	.22
Trunk	6.18
Upper legs	4.19
Lower legs	3.49
Upper arms	1.40
Lower arms	1.96
Fingers	[b] .67
Total	20.06

[a] 19.5 ft^2 used to include some factor of safety.

[b] Each finger: 3-1/2 inch long by 7/8 inch diameter.

Mass of organs of the standard adult human body

Tissue or organ	Mass[a] (g)		% of total body	
Adipose tissue	15000		21	
Subcutaneous		7500		11
Other separable		5000		7.1
Interstitial	1000		1.4	
Yellow marrow (included with skeleton)	1500		2.1	
Adrenals (2)		14		0.02
Aorta		100		0.14
Contents (blood)		190 (180 ml)		0.27
Blood-total	5500 g (5200 ml)		7.8	
Plasma	3100 g (3000 ml)		4.4	
Erythrocytes	2400 g (2200 ml)		3.4	
Blood vessels (not including aorta and pulmonary)		200		0.29
Contents (blood)		3000 (2900 ml)		4.3
Cartilage (included with skeleton)	1100		1.6	
Connective tissue	3400		4.8	
Tendons and fascia	1400		2.0	
Periarticular tissue	1500		2.1	
Other connective tissue	500		0.7	
Separable connective tissue		1600		2.3
Central Nervous System		1430		2.04
Brain	1400		2.0	
Spinal cord	30		0.04	
Contents-cerebrospinal fluid		120 (120 ml)		0.17
Eyes		15		0.02
Lenses (2)	0.4			
Gall bladder		10		0.01
Contents (bile)		62 (60 ml)		0.09
GI tract		1200		1.7
Esophagus	40		0.06	
Stomach	150		0.21	
Intestine	1000		1.4	
Small	640		0.91	
Upper large	210		0.30	
Lower large	160		0.23	
Contents of GI tract (food plus digestive fluids)		1005		1.4
Hair		20		0.03
Heart		330		0.47
Contents (blood)		500 (470 ml)		0.71
Kidneys		310		0.44
Larynx		28		0.04
Liver		1800		2.6
Lungs		1000		1.4
Parenchyma (includes bronchial tree, capillary blood, and associated lymph nodes)	570		0.81	
Pulmonary blood	430 (400 ml)		0.61	

(continued)

Tissue or organ	Mass[a] (g)		% of total body	
Lymphocytes	1500		2.1	
Lymphatic tissue	700		1.0	
Lymph nodes (dissectible)		250		0.36
Miscellaneous (by difference)		2953.1		4.2
Soft tissue (nasopharynx, etc.)	300		0.43	
Fluids (synovial, pleural, etc.)	350		0.50	
Muscle (skeletal)		28,000		40.0
Nails		3		
Pancreas		100		0.14
Parathyroids		0.12		
Pineal		0.18		
Pituitary		0.6		
Prostate		16		0.023
Salivary glands		85		0.12
Skeleton		10,000		14
Bone	5000		7.2	
Cortical	4000		5.7	
Trabecular	1000		1.4	
Red Marrow	1500		2.1	
Yellow Marrow	1500		2.1	
Cartilage	1100		1.6	
Periarticular tissue (skeletal)	900		1.3	
Skin		2600		3.7
Epidermis	100		0.14	
Dermis	2500		3.6	
Hypodermis	7500		11	
Spleen		180		0.26
Teeth		46		0.066
Testes		35		0.05
Thymus		20		0.029
Thyroid		20		0.029
Tongue		70		0.10
Tonsils		4		0.006
Trachea		10		0.014
Ureters		16		0.023
Urethra		10		0.014
Urinary bladder		45		0.064
Contents (urine)		102 (100 ml)		0.15
Total body		70,000		100

[a] Values for organs and tissues listed in the right hand column under "Mass" make up the totality of Reference Man (70,000 g).

Chemical composition, adult human body

Element	Amount (g)	Percent of total body weight
Oxygen	43,000	61
Carbon	16,000	23
Hydrogen	7000	10
Nitrogen	1800	2.6
Calcium	1000	1.4
Phosphorus	780	1.1

(continued)

Element	Amount (g)	Percent of total body weight
Sulfur	140	0.20
Potassium	140	0.20
Sodium	100	0.14
Chlorine	95	0.12
Magnesium	19	0.027
Silicon	18	0.026
Iron	4.2	0.006
Fluorine	2.6	0.0037
Zinc	2.3	0.0033
Rubidium	0.32	0.00046.
Strontium	0.32	0.00046
Bromine	0.20	0.00029
Lead	0.12	0.00017
Copper	0.072	0.00010
Aluminum	0.061	0.00009
Cadmium	0.050	0.00007
Boron	<0.048	0.00007
Barium	0.022	0.00003
Tin	<0.017	0.00002
Manganese	0.012	0.00002
Iodine	0.013	0.00002
Nickel	0.010	0.00001
Gold	<0.010	0.00001
Molybdenum	<0.0093	0.00001
Chromium	<0.0018	0.000003
Cesium	0.0015	0.000002
Cobalt	0.0015	0.000002
Uranium	0.00009	0.0000001
Beryllium	0.000036	
Radium	3.1×10^{-11}	

Properties of the Skin

Approximate values of the physical dimensions of whole skin for the "average man": 154 lb, 5 ft 9 in.

Weight	8.8 lb	4 kg
Surface area	20 sq ft	1.8 sq m
Volume	3.7 qt	3.6 liters
Water content	70 to 75 percent	
Specific gravity	1.1	
Thickness	0.02 to 0.2 in.	0.5 to 5.0 mm

Approximate values for thermal properties of skin:

Heat production	240 kcal/day
Conductance	9 to 30 kcal/sq m-hr-°C
Thermal conductivity (k)	$(1.5 \pm 0.3) \times 10^{-3}$ cal/cm sec-°C, at 23° to 25°C ambient
Diffusivity ($k/\rho c$)	7×10^{-4} sq cm/sec (surface layer 0.26 mm thick)
Thermal inertia ($k\rho c$)	90 to 400 $\times 10^{-5}$ cal^2/cm^4 sec $(°C)^2$
Heat capacity	~ 0.8 cal/gm

Skin temperature and thermal sensation:

Pain threshold for any area of skin 113°F (45°C)

When mean weighted skin temperature is:	The typical sensation is:
above 95°F (35°C)	unpleasantly warm
93°F (34°C)	comfortably warm
below 88°F (31°C)	uncomfortably cold
86°F (30°C)	shivering cold
84°F (29°C)	extremely cold

(continued)

When the hands reach:	When the feet reach:	They feel:
68°F (20°C)	73.5°F (23°C)	uncomfortably cold
59°F (15°C)	64.5°F (18°C)	extremely cold
50°F (10°C)	55.5°F (13°C)	painful and numb

Approximate optical properties of skin:

Emissivity (infrared)	~ 0.99
Reflectance (wavelength dependent)	Maximum 0.6 to 1.1μ
	Minima < 0.3 and > 1.2μ
Transmittance (wavelength dependent)	Maxima 1.2, 1.7, 2.2, 6, 11μ
	Minima 0.5, 1.4, 1.9, 3, 7, 12μ
Solar reflectivity of surface	
Very white skin	42 percent
5 "white" subjects	28 to 40 percent, average 34 percent
6 "colored" subjects	19 to 24 percent, average 21 percent
Very black skin	10 percent
Solar penetration--very white skin	45.5 percent passes 0.1 mm depth
	39.6 percent passes 0.2 mm depth
	32.0 percent passes 0.4 mm depth
	19.0 percent passes 1.0 mm depth
	10.2 percent passes 2.0 mm depth
Solar penetration--very dark skin	75 percent passes 0.1 mm depth
	40 percent absorbed in the melanin layer
	35 percent passes 0.2 mm depth

Regional cooling requirements of the human body in air at sea level at rest

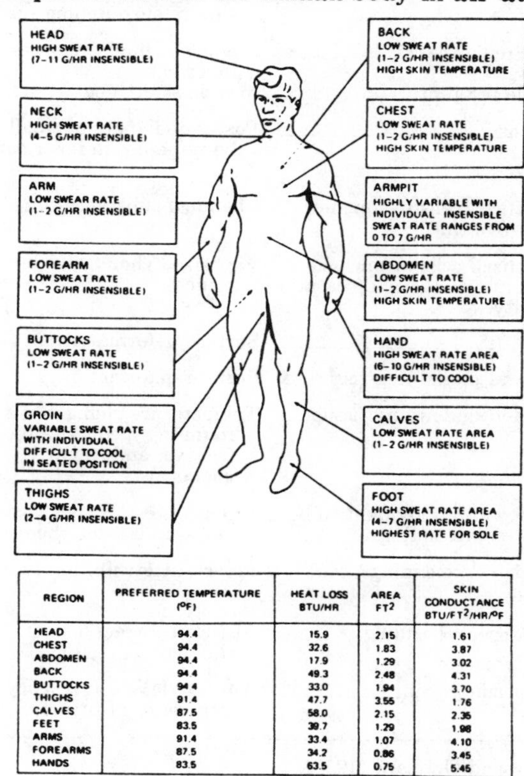

REGION	PREFERRED TEMPERATURE (°F)	HEAT LOSS BTU/HR	AREA FT²	SKIN CONDUCTANCE BTU/FT²/HR/°F
HEAD	94.4	15.9	2.15	1.61
CHEST	94.4	32.6	1.83	3.87
ABDOMEN	94.4	17.9	1.29	3.02
BACK	94.4	49.3	2.48	4.31
BUTTOCKS	94.4	33.0	1.94	3.70
THIGHS	91.4	47.7	3.55	1.76
CALVES	87.5	58.0	2.15	2.36
FEET	83.5	39.7	1.29	1.98
ARMS	91.4	33.4	1.07	4.10
FOREARMS	87.5	34.2	0.86	3.45
HANDS	83.5	63.5	0.75	5.45

Recruitment of Sweating

Area	Usual (But Not Invariable) Order of Recruitment
Dorsum foot	1
Lateral calf	2
Medial calf	3
Lateral thigh	4
Medial thigh	5
Abdomen	6
Dorsum hand	7 or 8
Chest	8 or 7
Ulnar forearm	9
Radial forearm	10
Medial arm	11
Lateral arm	12

(After Randall and Hertzman, 1953)

Man's senses and the energies that stimulate them

Sensation	Sense organ	Stimulation	Origin
Sight	Eye	Some electromagnetic waves.	External.
Hearing	Ear	Some amplitude and frequency variations of pressure in surrounding media.	External.
Rotation	Semicircular canals	Change of fluid pressures in inner ear.	Internal.
	Muscle receptors	Muscle stretching	Internal.
Falling and rectilinear movement.	Otoliths	Position changes of small, bony bodies in inner ear.	Internal.
Taste	Specialized cells in tongue and mouth.	Chemical substances	External on contact.
Smell	Specialized cells in mucous membrane at top of nasal cavity.	Vaporized chemical substances.	External.
Touch	Skin	Surface deformation	On contact.
Pressure	Skin and underlying tissue	Surface deformation	On contact.
Temperature	Skin and underlying tissue	Temperature changes of surrounding media or objects, friction, and some chemicals.	External on contact.
Pain	Unknown, but thought to be free nerve endings.	Intense pressure, heat, cold, shock, and some chemicals.	External on contact.
Position and movement (kinesthesis).	Muscle nerve endings	Muscle stretching	Internal.
	Tendon nerve endings	Muscle contraction	Internal.
	Joints	Unknown	Internal.
Mechanical vibration.	No specific organ	Amplitude and frequency variations of pressure.	External on contact.

Adapted from Mowbray and Gebhard (1958).

Stimulation-intensity ranges of man's senses

Sensation	Smallest detectable (threshold)	Largest tolerable or practical
Sight	10^{-6} mL.	10^{4} mL.
Hearing	2×10^{-4}dynes/cm^{2}	$<10^{3}$ dynes/cm^{2}.
Mechanical vibration	25×10^{-5} mm average amplitude at the fingertip (Maximum sensitivity 200 Hz).	Varies with size and location of stimulator. Pain likely 40 dB above threshold.
Touch (pressure)	Fingertips, 0.04 to 1.1 erg (One erg approx. kinetic energy of 1 mg dropped 1 cm.) "Pressure," 3 gm/mm^{2}.	Unknown.
Smell	Very sensitive for some substances, e.g., 2×10^{-7} mg/m^{3} of vanillin.	Unknown.
Taste	Very sensitive for some substances, e.g., 4×10^{-7} molar concentration of quinine sulfate.	Unknown.
Temperature	15×10^{-5} gm-cal/cm^{2}/sec. for 3 sec. exposure of 200 cm^{2} skin.	22×10^{-2} gm-cal/cm^{2}/sec. for 3 sec. exposure of 200 cm^{2} skin.
Position and movement	0.2–0.7 deg. at 10 deg./min. for joint movement.	Unknown.
Acceleration	0.02 g for linear acceleration 0.08 g for linear deceleration 0.12 deg./sec^{2} rotational acceleration for oculogyral illusion (apparent motion or displacement of viewed object).	5 to 8 g positive; 3 to 4 g negative. Disorientation, confusion, vertigo, blackout, or redout.

Frequency-sensitivity ranges of the senses

Stimulus	Lower Limit	Upper Limit
Color (hue)	300 nm (300 × 10^{-9} m.)	800 nm.
Interrupted white light	Unlimited	50 interruptions/sec. at moderate intensities and duty cycle of 0.5.
Pure tones	20 Hz	20,000 Hz.
Mechanical vibration	Unlimited	10,000 Hz at high intensities.

American experience table of mortality

(Based on 100,000 living at age of 10)

Age x	Number living l_x	Number dying d_x	Yearly probability of dying q_x	Yearly probability of living p_x
10	100 000	749	0.007 490	0.992 510
11	99 251	746	0.007 516	0.992 484
12	98 505	743	0.007 543	0.992 457
13	97 762	740	0.007 569	0.992 431
14	97 022	737	0.007 596	0.992 404
15	96 285	735	0.007 634	0.992 366
16	95 550	732	0.007 661	0.992 339
17	94 818	729	0.007 688	0.992 312
18	94 089	727	0.007 727	0.992 273
19	93 362	725	0.007 765	0.992 235
20	92 637	723	0.007 805	0.992 195
21	91 914	722	0.007 855	0.992 145
22	91 192	721	0.007 906	0.992 094
23	90 471	720	0.007 958	0.992 042
24	89 751	719	0.008 011	0.991 989
25	89 032	718	0.008 065	0.991 935
26	88 314	718	0.008 130	0.991 870
27	87 596	718	0.008 197	0.991 803
28	86 878	718	0.008 264	0.991 736
29	86 160	719	0.008 345	0.991 655
30	85 441	720	0.008 427	0.991 573
31	84 721	721	0.008 510	0.991 490
32	84 000	723	0.008 607	0.991 393
33	83 277	726	0.008 718	0.991 282
34	82 551	729	0.008 831	0.991 169
35	81 822	732	0.008 946	0.991 054
36	81 090	737	0.009 089	0.990 911
37	80 353	742	0.009 234	0.990 766
38	79 611	749	0.009 408	0.990 592
39	78 862	756	0.009 586	0.990 414
40	78 106	765	0.009 794	0.990 206
41	77 341	774	0.010 008	0.989 992
42	76 567	785	0 010 252	0.989 748
43	75 782	797	0.010 517	0.989 483
44	74 985	812	0.010 829	0.989 171
45	74 173	828	0.011 163	0.988 837
46	73 345	848	0.011 562	0.988 438
47	72 497	870	0.012 000	0.988 000
48	71 627	896	0.012 509	0.987 491
49	70 731	927	0.013 106	0.986 894
50	69 804	962	0.013 781	0.986 219
51	68 842	1 011	0.014 541	0.985 459
52	67 841	1 044	0.015 389	0.984 611
53	66 797	1 091	0.016 333	0.983 667
54	65 706	1 143	0.017 396	0.982 604

(continued)

Age x	Number living l_x	Number dying d_x	Yearly probability of dying q_x	Yearly probability of living p_x
55	64 563	1 199	0.018 571	0.981 429
56	63 364	1 260	0.019 885	0.980 115
57	62 104	1 325	0.021 335	0.978 665
58	60 779	1 394	0.022 936	0.977 064
59	59 385	1 468	0.024 720	0.975 280
60	57 917	1 546	0.026 693	0.973 307
61	56 371	1 628	0.028 880	0.971 120
62	54 743	1 713	0.031 292	0.968 708
63	53 030	1 800	0.033 943	0.966 057
64	51 230	1 889	0.036 873	0.963 127
65	49 341	1 980	0.040 129	0.959 871
66	47 361	2 070	0.043 707	0.956 293
67	45 291	2 158	0.047 647	0.952 353
68	43 133	2 243	0.052 002	0.947 998
69	40 890	2 321	0.056 762	0.943 238
70	38 569	2 391	0.061 993	0.938 007
71	36 178	2 448	0.067 665	0.932 335
72	33 730	2 487	0.073 733	0.926 267
73	31 243	2 505	0.080 178	0.919 822
74	28 738	2 501	0.087 028	0.912 972
75	26 237	2 476	0.094 371	0.905 629
76	23 761	2 431	0.102 311	0.897 689
77	21 330	2 369	0.111 064	0.888 936
78	18 961	2 291	0.120 827	0.879 173
79	16 670	2 196	0.131 734	0.868 266
80	14 474	2 091	0.144 466	0.855 534
81	12 383	1 964	0.158 605	0.841 395
82	10 419	1 816	0.174 297	0.825 703
83	8 603	1 648	0.191 561	0.808 439
84	6 955	1 470	0.211 359	0.788 641
85	5 485	1 292	0.235 552	0.764 448
86	4 193	1 114	0.265 681	0.734 319
87	3 079	933	0.303 020	0.696 980
88	2 146	744	0.346 692	0.653 308
89	1 402	555	0.395 863	0.604 137
90	847	385	0.454 545	0.545 455
91	462	246	0.532 466	0.467 534
92	216	137	0.634 259	0.365 741
93	79	58	0.734 177	0.265 823
94	21	18	0.857 143	0.142 857
95	3	3	1.000 000	0.000 000

Height and weight of white male and female Americans at different ages

	Male				Female			
	Height (in.)		Weight (lb)		Height (in.)		Weight (lb)	
Age (yr)	Mean	S.D.	Mean	S.D.	Mean	S.D.	Mean	S.D.
1	29.7	1.1	23	3	29.3	1.0	21	3
2	34.5	1.2	28	3	34.1	1.2	27	3
3	37.8	1.3	32	3	37.5	1.4	31	4
4	40.8	1.9	37	5	40.6	1.6	36	5
5	43.7	2.0	42	5	43.8	1.7	41	5
6	46.1	2.1	47	6	45.7	1.9	45	5
7	48.2	2.2	54	7	47.9	2.0	50	7
8	50.4	2.3	60	8	50.3	2.2	58	11
9	52.8	2.4	66	8	52.1	2.3	64	11
10	54.5	2.5	73	10	54.6	2.5	72	14
11	56.8	2.6	82	11	57.1	2.6	82	18
12	58.3	2.9	87	12	59.6	2.7	93	18
13	60.7	3.2	99	13	61.4	2.6	102	18
14	63.6	3.2	113	15	62.8	2.5	112	19
15	66.3	3.1	128	16	63.4	2.4	117	20
16	67.7	2.8	137	16	63.9	2.2	120	21
17	68.3	2.6	143	19	64.1	2.2	122	19
18	68.5	2.6	149	20	64.1	2.3	123	17
19	68.6	2.6	153	21	64.1	2.3	124	17
20–24	68.7	2.6	158	23	64.0	2.4	125	19
25–29	68.7	2.6	163	24	63.7	2.5	127	21
30–34	68.5	2.6	165	25	63.6	2.4	130	24
35–39	68.4	2.6	166	25	63.4	2.4	136	25
40–49	68.0	2.6	167	25	63.2	2.4	142	27
50–59	67.3	2.6	165	25	62.8	2.4	148	28
60–69	66.8	2.4	162	24	62.2	2.4	146	28
70–79	66.5	2.2	157	24	61.8	2.2	144	27
80–89	66.1	2.2	151	24				

VISION

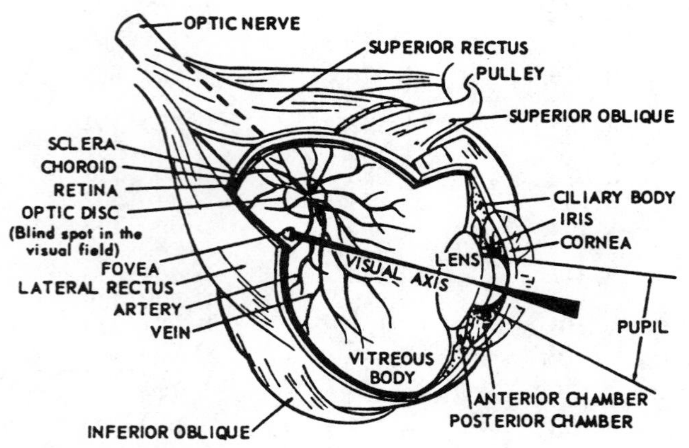

Right eye, viewed from outer side, showing visual axis passing through center of lens to point of sharpest vision at fovea, where cones are concentrated.

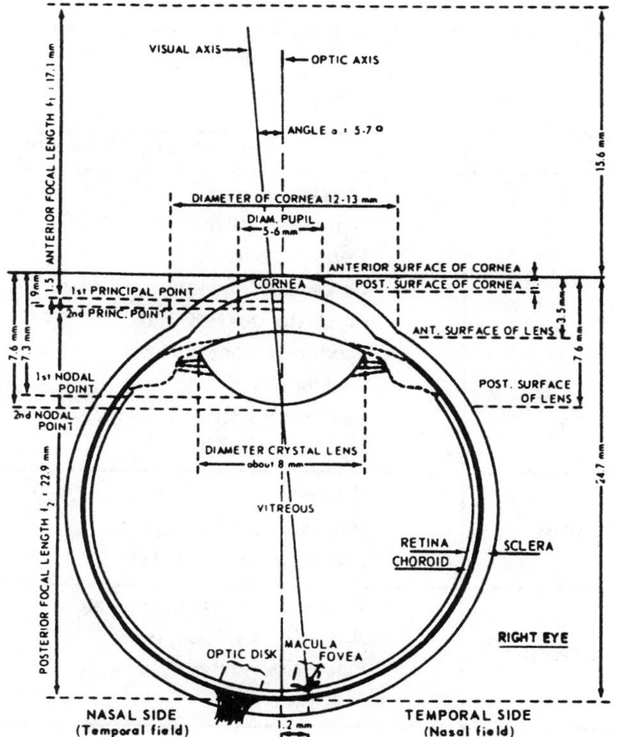

Dimensions of the human eye

Proportion of population 18-79 years old reaching or exceeding the test levels for binocular distance vision

Test Level	Proportion for Distance Vision	
	Un-Corrected	"Corrected"[a]
20/10 or better --	1.1	1.5
20/15 or better --	30.3	40.0
20/20 or better --	53.9	72.9
20/30 or better --	69.3	90.6
20/40 or better --	75.8	95.1
20/50 or better --	80.4	96.8
20/70 or better --	83.9	97.7
20/100 or better --	93.5	99.2
20/200 or better --	97.6	99.6

[a] Uncorrected testing was without glasses. "Corrected" testing was with glasses, if worn to the examination; otherwise, without them.

Optical Constants for the Human Eye

Constant	Eye Area or Measurement	
Refractive index	Cornea	1.37
	Aqueous humor	1.33
	Lens capsule	1.38*
	Outer cortex, lens	
	Anterior cortex, lens	
	Posterior cortex, lens	
	Center, lens	1.41
	Calculated total index	1.41
	Vitreous body	1.33
Radius of curvature, mm	Cornea	7.7
	Anterior surface, lens	9.2 - 12.2
	Posterior surface, lens	5.4 - 7.1
Distance from cornea, mm	Posterior surface, cornea	1.2
	Anterior surface, lens	3.5
	Posterior surface, lens	7.6
	Retina	24.8
Focal distance, mm	Anterior focal length	17.1
		[14.2] **
	Posterior focal length	22.8
		[18.9]
Position of cardinal points measured from corneal surface, mm	1. Focus	−15.7
		[−12.4]
	2. Focus	24.4
		[21.0]
	1. Principal point	1.5
		[1.8]
	2. Principal point	1.9
		[2.1]
	1. Nodal point	7.3
		[6.5]
	2. Nodal point	7.6
		[6.8]
Diameter, mm	Optic disk	2-5
	Macula	1-3
	Fovea	1.5
Depth, mm	Anterior chamber	2.7 - 4.2

*Cortex of lens and its capsule.

**Values in brackets refer to state of maximum accommodation.

A log-log plot of visual acuity against background luminance, with contrast as the parameter

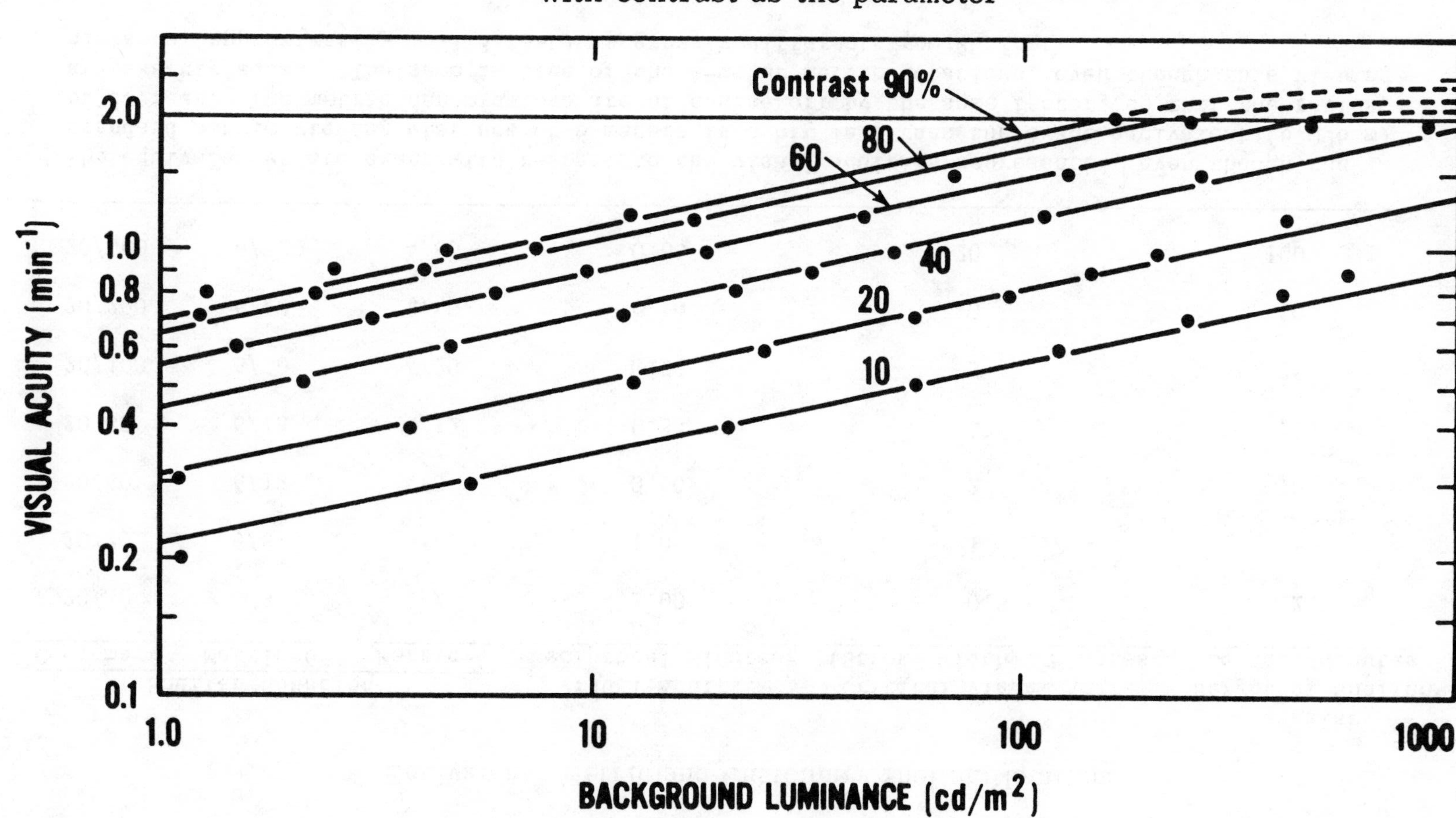

Equivalent[a] Metric and Customary Snellen Notations

Snellen Notation			Visual Acuity, Reciprocal Minutes[b]	Critical Visual Angle (Stroke Width), Minutes	Angular Height of Snellen Letter, Minutes
Customary	Metric-6	Metric-4			
20/10	6/3	4/2	2.00	0.5	2.5
20/20	6/6	4/4	1.00	1	5
20/40	6/12	4/8	0.50	2	10
20/60	6/18	4/12	0.33	3	15
20/100	6/30	4/20	0.20	5	25
20/200	6/60	4/40	0.10	10	50
20/400	6/120	4/80	0.05	20	100

a The equivalences are exact with respect to the visual acuities represented, even though the standard metric viewing distance of 6 meters is a bit less than the exact equivalent (6.096 m) of 20 feet. The metric denominators are of course off by the same factor, so that the ratios are exactly equal. The same is true of the 4-meter metric notations, even though this viewing distance (approximately 13.123 feet) is grossly different from 20 feet.

b Note that visual acuity is numerically equal to the Snellen fraction (customary or metric), if the latter is interpreted as an actual arithmetic quantity.

Log-log plot of minimum stroke width (mm) required for letter visibility as a function of viewing distance (m), with observer visual acuity as a parameter

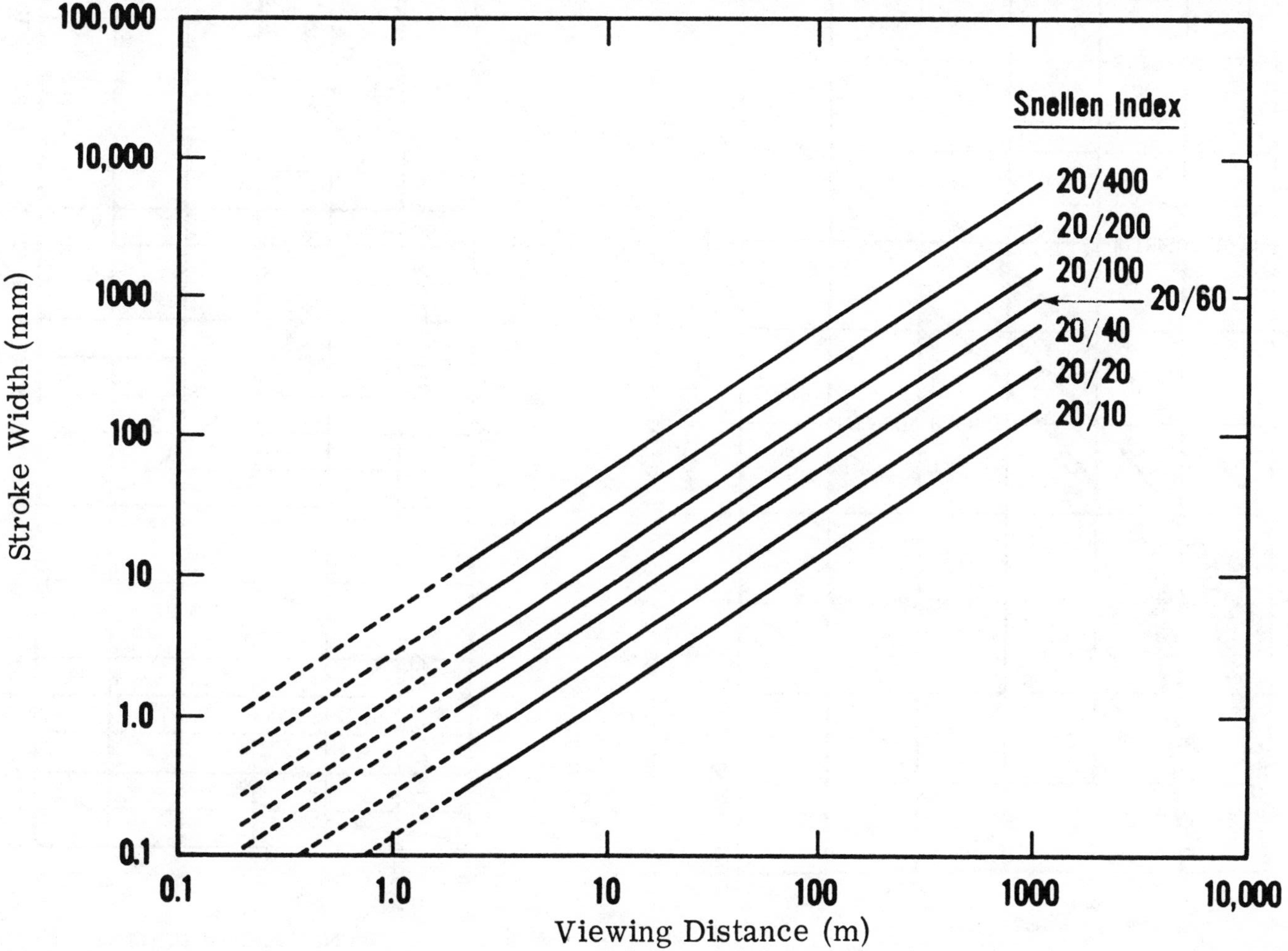

Snellen Index

20/400
20/200
20/100
20/60
20/40
20/20
20/10

Stroke Width (mm)

100,000
10,000
1000
100
10
1.0
0.1

0.1
1.0
10
100
1000
10,000

Viewing Distance (m)

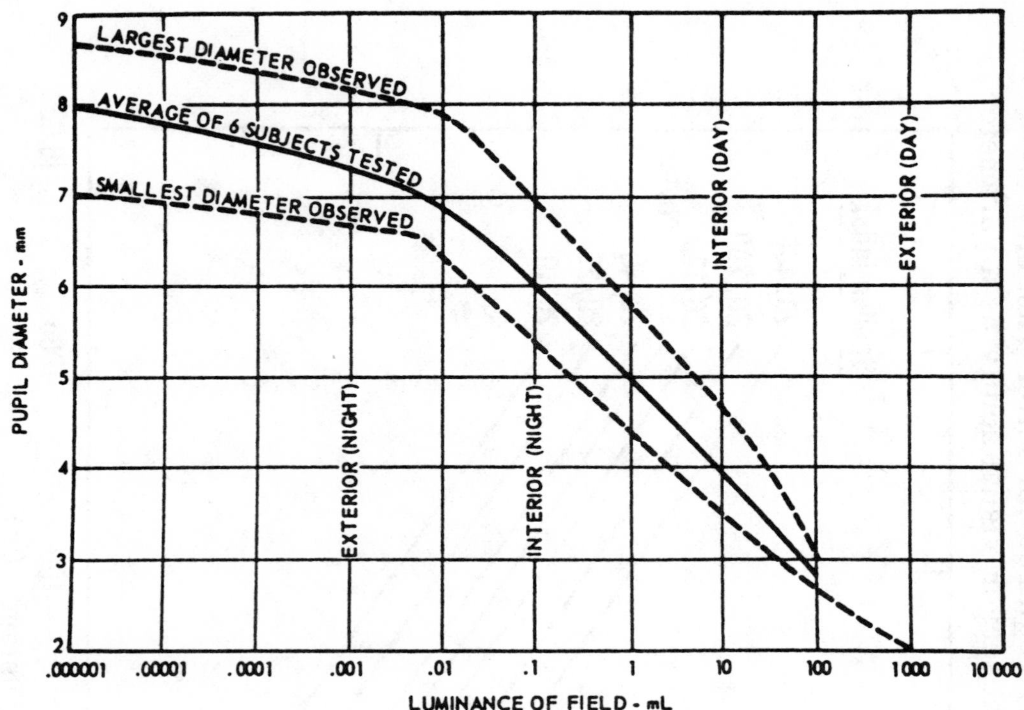

Pupil diameter as a function of luminance in adapting field

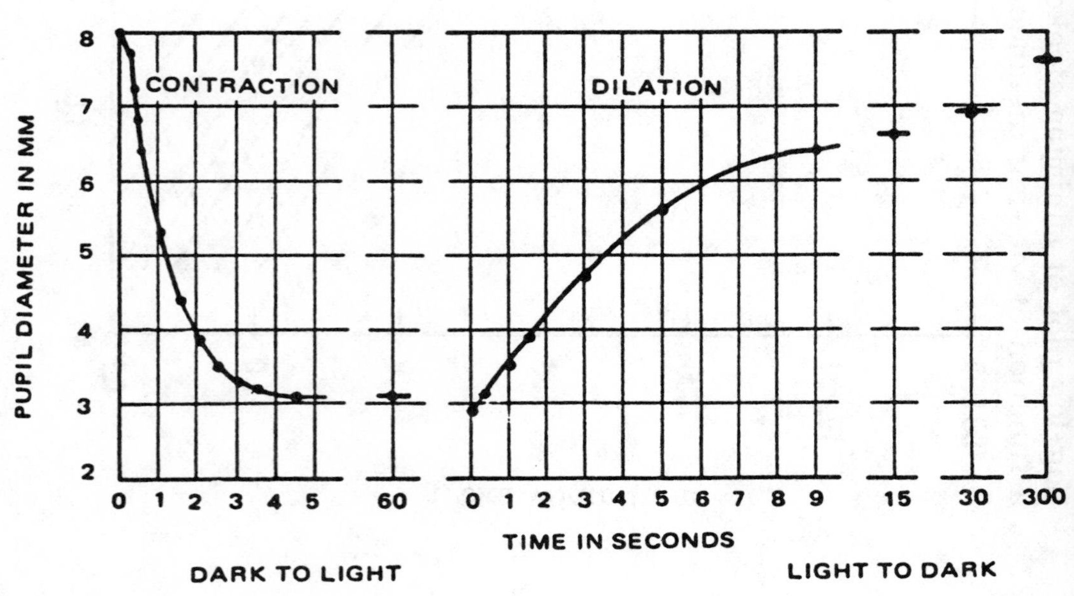

Pupillary response to light and darkness

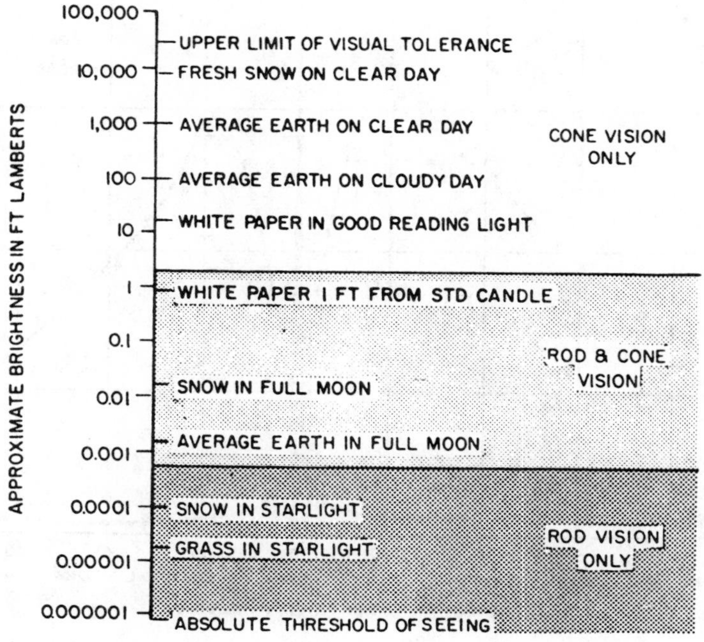

Examples of various levels of luminance

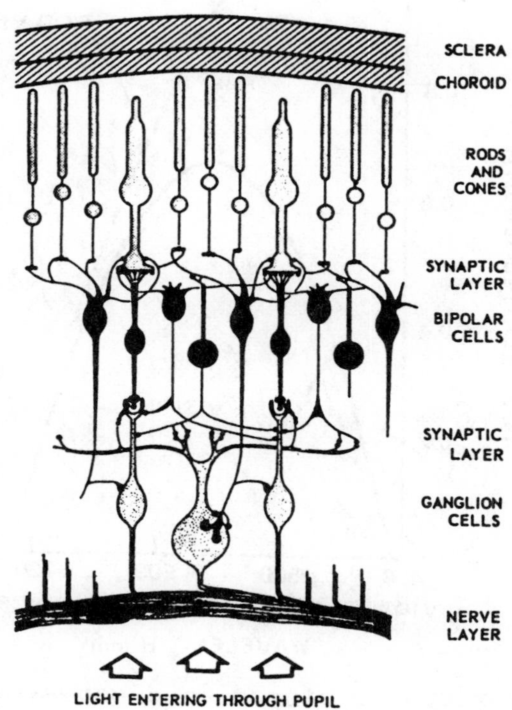

Simplified digram of retina

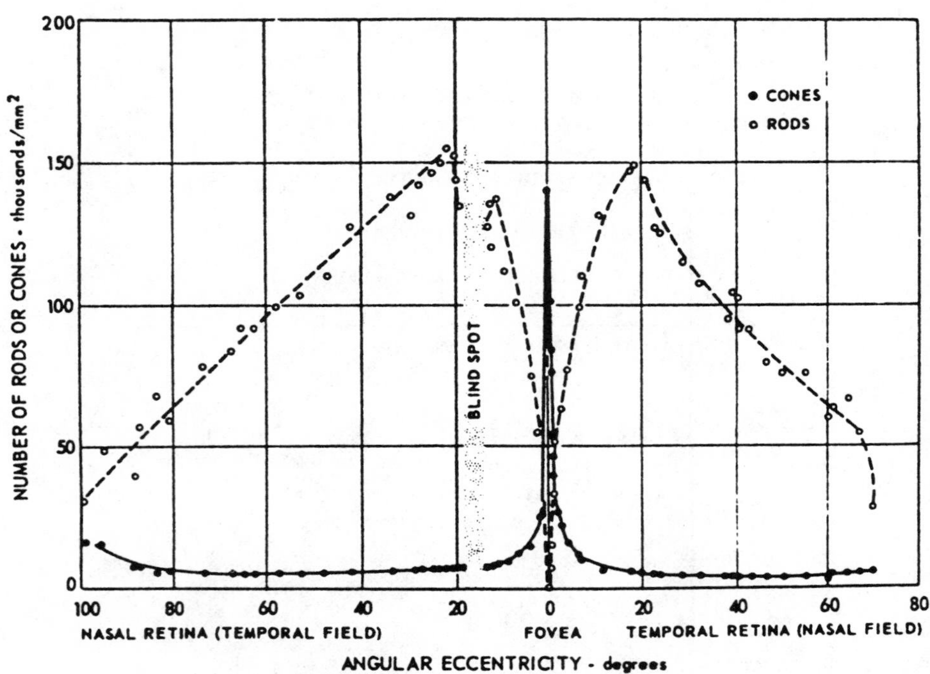

Typical distribution of rods and cones

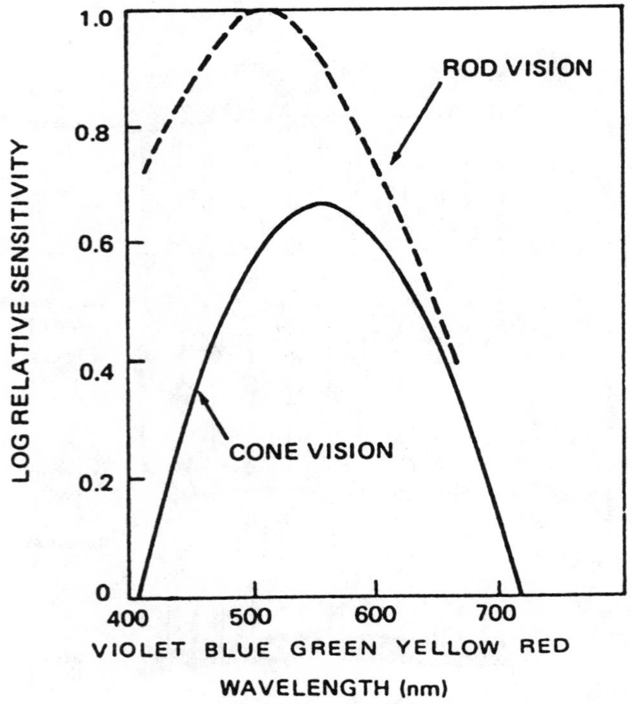

Spectral response of rods and cones, with relative amount of radiant energy for vision at absolute threshold shown as a function of wavelength of light

Human field of vision

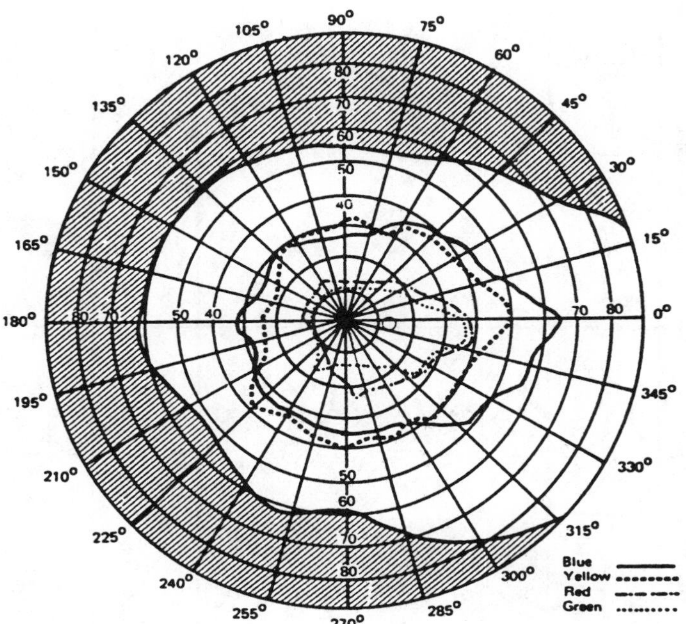

Field of view from right eye

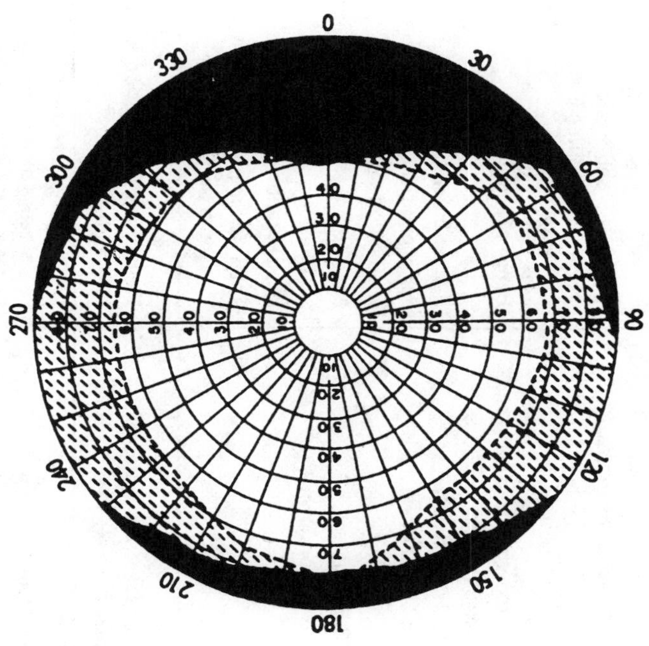

Binocular field of vision

The Limits of the Visual Field Under Various Kinds of Restraint

Movement Permitted	Type of Field and Factors Limiting Field	Horizontal Limits		Vertical Limits	
		Temporal Ambinocular Field (each side)	Nasal Binocular Field (each side)	Field Angle Up	Field Angle Down
Moderate movements of head and eyes assumed as:	Range of fixation	60°		45°	
Eyes: 15° right or left	Eye deviation (assumed)	15°	15°	15°	15°
15° up or down	Peripheral field from point of fixation	95°	(45°)	46°	67°
Head: 45° right or left	Net peripheral field from central fixation	110°	60°***	61°	82°
30° up or down	Head rotation (assumed)	45°	45°	30°*	30°*
	Total peripheral field (from central body line)	155°	105°	91°	112°**
Head fixed Eyes fixed (central position with respect to head)	Field of peripheral vision (central fixation)	95°	60°	46°	67°
Head fixed Eye maximum deviation	Limits of eye deviation (= range of fixation)	74°	55°	48°	66°
	Peripheral field (from point of fixation)	91°	Approx (5°)	18°	16°
	Total peripheral field (from central head line)	165°	60°***	66°	82°

(continued)

Movement Permitted	Type of Field and Factors Limiting Field	Horizontal Limits		Vertical Limits	
		Temporal Ambinocular Field (each side)	Nasal Binocular Field (each side)	Field Angle Up	Field Angle Down
		60°		45°	
Head maximum movement Eyes fixed (central with respect to head)	Limits of head motion (= range of fixation)	72°	72°	80°*	90°*
	Peripheral field (from point of fixation)	95°	60°	46°	67°
	Total peripheral field (from central body line)	167°	132°	126°	157°**
Maximum movement of head and eyes	Limits of head motion	72°	72°	80°*	90°*
	Maximum eye deviation	74°	55°	48°	66°
	Range of fixation (from central body line)	146°	127°	128°	156°**
	Peripheral field (from point of fixation)	91°	Approx (5°)	18°	16°
	Total peripheral field (from central body line)	237°	132°	146°	172°**

*Estimated by the authors on the basis of a single subject.

**Ignoring obstruction of body (and knees if seated). This obstruction would probably impose a maximum field of 90° (or less, seated) directly downward; however, this would not apply downward to either side.

***This is the maximum possible peripheral field; rotating the eye in the nasal direction will not extend it, because it is limited by the nose and other facial structures rather than by the optical limits of the eye. The figures in parentheses on the line above are calculated values, chosen to given the maximum limit thus indicated.

Notes: 1. All data except as noted are from Hall and Greenbaum (1950). 2. The ambinocular field is defined here as the total area that can be seen by either eye; it is not limited to the binocular field, which can be seen by both eyes at once. That is, at the sides, it includes monocular regions visible to the right eye but not to the left, and vice versa. 3. The term binocular is here restricted to the central region that can be seen by both eyes simultaneously (stereoscopic vision). It is bounded by the nasal field-limits of the eyes.

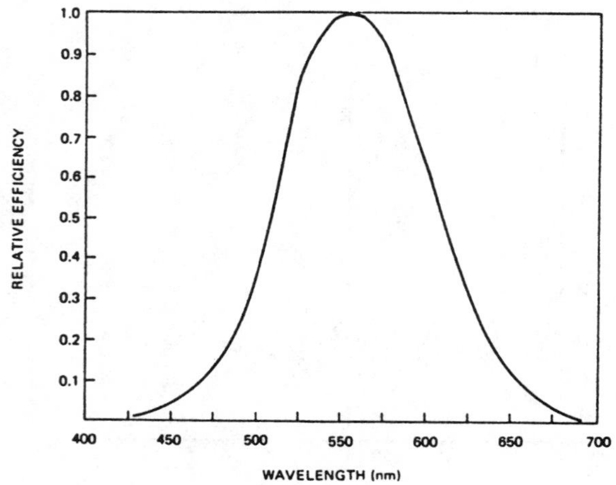

Spectral sensitivity to the human eye for daylight conditions

Conversion factors for units of luminance

Number of / Multiplied by / Equals Number of	cd/m^2 (nit)*	cd/cm^2 (stilb)	cd/ft^2	cd/in^2	apostilb (blondel)	millilambert	foot-lambert
cd/m^2 (nit)*	1	10 000	10.764	1550	0.3183	3.183	3.426
cd/cm^2 (stilb)	0.0001	1	0.001076	0.155	0.00003183	0.0003183	0.0003426
cd/ft^2	0.0929	929	1	144	0.02957	0.2957	0.3183
cd/in^2	0.000645	6.452	0.00694	1	0.0002054	0.002054	0.002211
apostilb (blondel)	3.1416	31 416	33.82	4869	1	10	10.764
millilambert	0.31416	3 141.6	3.382	486.9	0.1	1	1.0764
foot-lambert	0.2919	2919	3.1416	452.4	0.0929	0.929	1

*The name "nit" is not in widespread use.

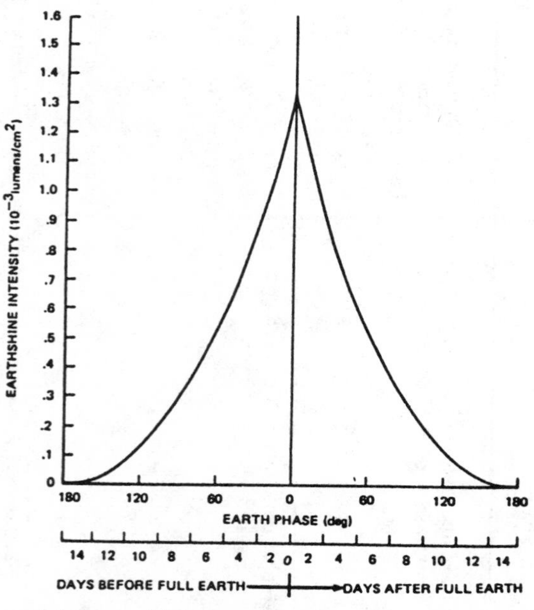

Earthshine intensity on the moon as a function of the Earth phase

Illumination Levels in Space

Planet	Mean Solar Illumination (Foot-Candles)
Mercury	84 600
Venus	24 400
Earth	12 700
Mars	5430
Jupiter	470
Saturn	138
Uranus	34
Pluto	8

HEARING

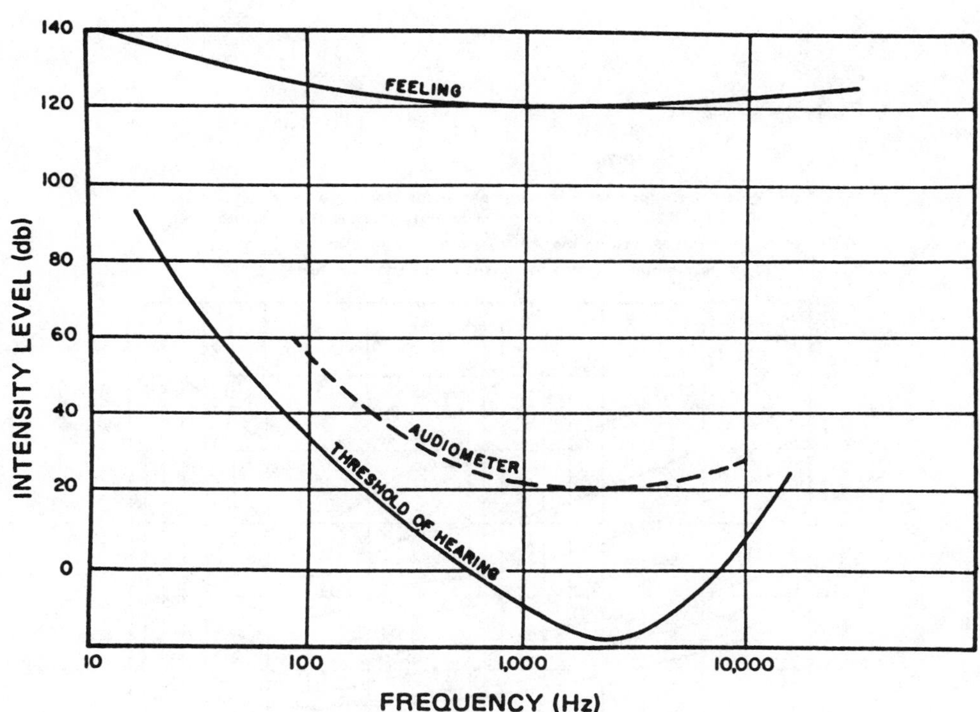

Audibility curve in man, showing thresholds of feeling and hearing

Relationship between Decibels, Newtons/Meter^2, Microbar*, and Pounds/Inch^2

dB	N/m^2	μbar	PSI
0	0.00002	0.0002	2.94 X 10^{-9}
14	0.0001	0.001	14.70 X 10^{-9}
34	0.001	0.01	147.0 X 10^{-9}
54	0.01	0.1	1.47 X 10^{-6}
74	0.1	1	14.70 X 10^{-6}
94	1	10	147.0 X 10^{-6}
114	10	100	1.47 X 10^{-3}
134	100	1000	14.70 X 10^{-3}
154	1000	10 000	147.0 X 10^{-3}
174	10 000	100 000	1.47

*Also note that 1 μbar = 1 dyn/cm^2

Occupational Safety and Health Act Permissible Daily Noise Exposure*

Duration (hr)	Sound Level (dBA)
8	90
6	92
4	95
3	97
2	100
1.5	102
1	105
0.5	110
0.25	115

*When the exposure is intermittent at different levels the fraction $C_1/T_1 + C_2/T_2 \ldots C_n/T_n$ should not exceed unity to meet the exposure limit.

C_n = total exposure time at the specified noise level.

T_n = total exposure time permitted at the specified level.

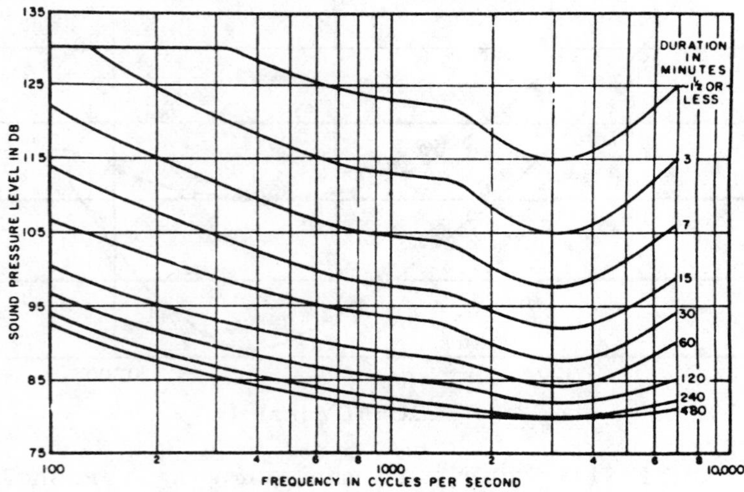

Damage risk contours for 1 exposure/day to pure tones

Typical SPL's for some common sounds

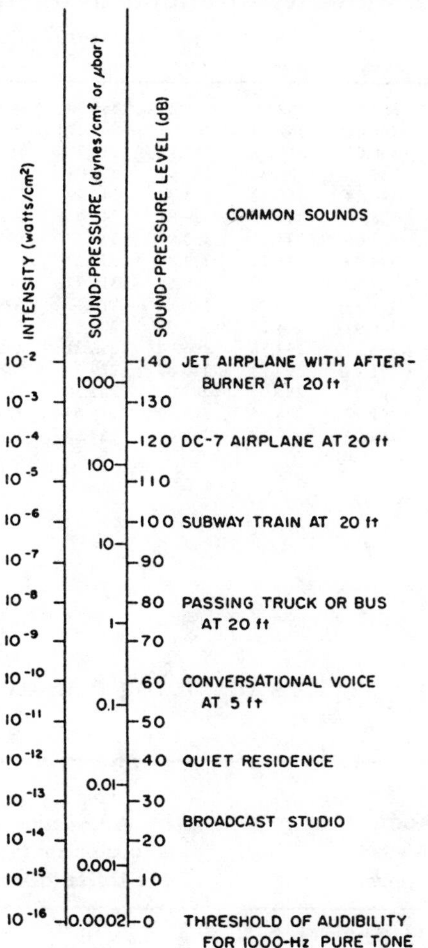

Typical R.M.S. pressure levels of the fundamental speech sounds

Key word	Sound*	Pressure level (dB)	Key word	Sound*	Pressure level (dB)
talk	o'	28.2	chat	ch	16.2
top	a	27.8	me	m	15.5
ton	o	27.0	jot	i	13.6
tap	a'	26.9	azure	zh	13.0
tone	o	26.7	zip	z	12.0
took	u	26.6	sit	s	12.0
tape	'a	25.7	tap	t	11.7
ten	e	25.4	get	g	11.7
tool	u	25.0	kit	k	11.1
tip	i	24.1	vat	v	10.8
team	e	23.4	that	th	10.4
err	r	23.2	bat	b	8.0
let	l	20.0	dot	d	8.0
shot	sh	19.0	pat	p	7.7
ring	ng	18.6	for	f	7.0
me	m	17.2	thin	th	0

* Spoken by an average talker at a normal level of effort (Fletcher, 1953).

Velocity of sound in dry air

Temp °C	0 m sec⁻¹	1 m sec⁻¹	2 m sec⁻¹	3 m sec⁻¹	4 m sec⁻¹	5 m sec⁻¹	6 m sec⁻¹	7 m sec⁻¹	8 m sec⁻¹	9 m sec⁻¹
60	366.05	366.60	367.14	367.69	368.24	368.78	369.33	369.87	370.42	370.96
50	360.51	361.07	361.62	362.18	362.74	363.29	363.84	364.39	364.95	365.50
40	354.89	355.46	356.02	356.58	357.15	357.71	358.27	358.83	359.39	359.95
30	349.18	349.75	350.33	350.90	351.47	352.04	352.62	353.19	353.75	354.32
20	343.37	343.95	344.54	345.12	345.70	346.29	346.87	347.44	348.02	348.60
10	337.46	338.06	338.65	339.25	339.84	340.43	341.02	341.61	342.20	342.78
0	331.45	332.06	332.66	333.27	333.87	334.47	335.07	335.67	336.27	336.87
-10	325.33	324.71	324.09	323.47	322.84	322.22	321.60	320.97	320.34	319.72
-20	319.09	318.45	317.82	317.19	316.55	315.92	315.28	314.64	314.00	313.36
-30	312.72	312.08	311.43	310.78	310.14	309.49	308.84	308.19	307.53	306.88
-40	306.22	305.56	304.91	304.25	303.58	302.92	302.26	301.59	300.92	300.25
-50	299.58	298.91	298.24	297.56	296.89	296.21	295.53	294.85	294.16	293.48
-60	292.79	292.11	291.42	290.73	290.03	289.34	288.64	287.95	287.25	286.55
-70	285.84	285.14	284.43	283.73	283.02	282.30	281.59	280.88	280.16	279.44
-80	278.72	278.00	277.27	276.55	275.82	275.09	274.36	273.62	272.89	272.15
-90	271.41	270.67	269.92	269.18	268.43	267.68	266.93	266.17	265.42	264.66

Acoustic properties of biologic materials (at 37°C)

Tissue	Acoustic Velocity (m/s)	Attenuation coefficient (α) at 1MHz (dB/cm)	Approximate frequency dependence of α
Amniotic fluid	1510±3	5.1×10^{-3}	1.6
Blood	1581 at 40% HMTC[a]	0.13	1.33
Brain, fetus	1520–1540	.63	1.27
Breast	1465±5 postmenopause 1529±5 premenopause		
Eye, lens	1638.4±3	.8	1.0
Eye, vitreous	1531.7±.9		
Fat	1479	.6	1.0
Liver	1540	.9	1.0
Muscle	1500–1610	1.3 for gastronemius muscle perpendicular to fibers	1.0

[a] Velocity in blood for a specified hematocrit (HMTC)=1541.8+(.98) (HMTC), (m/s).

Acoustic properties of non-biologic materials (at 20°C and 1 atmosphere unless noted)

Material	Chemical formula	Density g/cm³	Acoustic Velocity[a] m/s	Temperature coefficient of velocity (m/s)/°C	Attenuation coefficient at 1 MHz α, dB/cm	Approximate[b] frequency dependence of α
Pure water[c]	H_2O	0.9982	1482.343(5)	+3.071	0.0022	f^2
Carbon tetrachloride	CCl_4	1.5896	939(10)	-2.7	.047 at 25 °C	f^2
Acetone	C_3H_6O	.7899	1196(10)	-4.5	.0047 at 30 °C	f^2
Ethanol, 95%	C_2H_6O, H_2O	.7998	1227(10)	-4.0		
Ethanol	C_2H_6O	.7893	1161.8	-3.5	.0042 at 30 °C	f^2
Methanol	CH_4O	.7914	1121.2	-3.3	.0026 at 30 °C	f^2
Glycerol	$C_3H_8O_3$	1.2613	1997	-2.2	.16 at 26 °C	f^2
Ethylene glycol	$C_2H_6O_2$	1.1088	1667	-2.2		
Castor oil	$C_{11}H_{10}O_{10}$	.969	1495	-3.6	.95 at ?	f^2
Aluminum	Al	2.695	6420		.018	f
Brass	.7Cu, .3Zn	8.6	4700		.02	f
347 Stainless steel		7.91	5790			
Pyrex glass		2.32	5640			
Rubber gum		.95	1550			
Lucite		1.182	2680		2.0	f
Polyethylene		.90	1950	−8	4.7	$f^{1.1}$
Lexan polycarbonate		1.19	2280(.5)	-3.58		
Nylon		1.11	2620			

[a] The velocity is measured at the frequency noted in parentheses (MHz), however, for most materials little velocity dispersion is observed.

[b] Classical attenuation due to the effects of viscosity and heat conduction is observed in most liquids with a square dependence on frequency over a broad frequency range. Attenuation in solids is more complex. The frequency dependence noted is observed in the frequency range from 1 to 10 MHz.

[c] Doubly distilled water. The measured velocity is unaffected by dissolved gas.

Velocity of Sound Propagation in Selected Substances

Material	Temperature in °C	Sound velocity in m s⁻¹	
In gases and vapours			
Air	+500	558	
	+100	387.2	
	+20	343.8	
	0	331.8	
	−20	319.3	
Carbon dioxide	+100	297.2	
	+18	265.8	
Ethylene	0	317	
Helium	0	971	
Hydrogen	+100	1643	
	+18	1301	
	0	1286	
In liquids			
Ethyl alcohol	+20	1168	
Glycerin	+20	1923	
Mercury	+20	1451	
Paraffin oil	+33.5	1420	
Water (light)	+25	1497	
Water (heavy)	+25	1401	
Sea water (3.235% salt content)			
on the surface	0	1440	
	+15	1498	
at a depth of 1500 m	0	1456	
	+15	1511	
In solids		in bars $d \ll \lambda$	in unbounded media $d \gg \lambda$
a) Metals			
Aluminium	(ν=0.34)	5240	6400
Brass	(ν=0.35)	3420	4250
Copper	(ν=0.35)	3580	4606
Iron	(ν=0.27)	5170	5850
Lead	(ν=0.45)	1250	2400
Steel		5050	6100
b) Other materials			
Beechwood (along fibre)		3400	
Deal (along fibre)		5260	
Degussit (Degussa aluminium oxide)		9600	
Glass		3490 to 5300	3760 to 5660
Granite		3950	
Ice (−4 °C)		3232	
Quartz, crystalline (X-plane)		5440	5720
Rochelle salt (45° Y-plane)		2740	
Sand		100 to 300	
Vitreous quartz		5370	5570

In plastics the sound velocity is very highly dependent on the temperature and the frequency.

ENVIRONMENTAL ANALYSIS

National Ambient Air Quality Standards*

	Primary standard		Secondary standard	
	μg/m³	ppm	μg/m³	ppm
Sulfur oxides:				
Annual arithmetic mean	80	0.03		
24-hr concentration	365†	0.14†		
3-hr concentration	.....	.....	1.300†	0.5†
Suspended particulate matter:				
Annual geometric mean	75	.....	60‡	
24-hr concentration	260†	.....	150†	
Carbon monoxide:				
8-hr concentration	.....	9.0†	Same as primary	
1-hr concentration	.....	35.0†		
Photochemical oxidants:				
1-hr concentration	160†	0.08†	Same as primary	
Hydrocarbons (corrected for methane):				
3-hr concentration (6–9 am)	160†	0.24†	Same as primary	
Nitrogen oxides:				
Annual arithmetic mean	100	0.05	Same as primary	

* ppm = parts per million, μg/m³ = micrograms per cubic meter.
† Not to be exceeded more than once a year.
‡ A guide for assessing achievement of the 24-hr standard.

Outdoor Air Requirements

Application	Smoking	Cfm per person		Cfm/ft² of floor, min.
		Recommended	Min.	
Apartment:				
Average	Some	20	10	
Deluxe	Some	20	10	
Banking space	Occasional	10	7½	
Barber shops	Considerable	15	10	
Beauty parlors	Occasional	10	7½	
Brokers' board rooms	Very heavy	50	20	
Cocktail bars	....	40	25	
Corridors (supply or exhaust)	....	....	....	0.25
Department stores	None	7½	5	0.05
Directors' rooms	Extreme	50	30	
Drugstores	Considerable	10	7½	
Factories	None	10	7½	0.10
Five and ten cent stores	None	7½	5	
Funeral parlors	None	10	7½	
Garages	....	....	....	1.0
Hospitals:				
Operating rooms	None	....	....	2.0
Private rooms	None	30	25	0.33
Wards	None	20	10	
Hotel rooms	Heavy	30	25	0.33
Kitchens:				
Restaurant	....	....	....	4.0
Residence	....	....	....	2.0
Laboratories	Some	20	15	
Meeting rooms	Very heavy	50	30	1.25
Offices:				
General	Some	15	10	
Private	None	25	15	0.25
	Considerable	30	25	0.25
Restaurants:				
Cafeteria	Considerable	12	10	
Dining-room	Considerable	15	12	
Schoolrooms	None			
Shop, retail	None	10	7½	
Theater	None	7½	5	
	Some	15	10	
Toilets (exhaust)	....	....	....	2.0

Rates of Heat Gain from Occupants of Conditioned Spaces[a]

Degree of Activity	*Typical Application*	*Total Heat Adults, Male, Btu/Hr*	*Total Heat Adjusted,[b] Btu/Hr*	*Sensible Heat, Btu/Hr*	*Latent Heat, Btu/Hr*
Seated at rest	Theater—Matinee	390	330	225	105
	Theater—Evening	390	350	245	105
Seated, very light work	Offices, hotels, apartments	450	400	245	155
Moderately active office work	Offices, hotels, apartments	475	450	250	200
Standing, light work; or walking slowly	Department store, retail store, dime store	550	450	250	200
Walking; seated Standing; walking slowly	Drug store, Bank	550	500	250	250
Sedentary work	Restaurant[c]	490	550	275	275
Light bench work	Factory	800	750	275	475
Moderate dancing	Dance hall	900	850	305	545
Walking 3 mph; moderately heavy work	Factory	1000	1000	375	625
Bowling[d] Heavy work	Bowling alley Factory	1500	1450	580	870

[a] *Note:* Tabulated values are based on 75 F room dry-bulb temperature. For 80 F room dry-bulb, the total heat remains the same, but the sensible heat values should be decreased by approximately 20 percent, and the latent heat values increased accordingly.

[b] *Adjusted total heat gain* is based on normal percentage of men, women, and children for the application listed, with the postulate that the gain from an adult female is 85 percent of that for an adult male, and that the gain from a child is 75 percent of that for an adult male.

[c] Adjusted total heat value for *sedentary work, restaurant,* includes 60 Btu per hour for food per individual (30 Btu sensible and 30 Btu latent).

[d] For *bowling* figure one person per alley actually bowling, and all others as sitting (400 Btu per hour) or standing (550 Btu per hour).

Recommended Inside Design Conditions--Winter

Type of application	Winter				
	With humidification			Without humidification	
	Dry-bulb, F	Rel. hum., %	Temp. swing,* F	Dry-bulb, F	Temp. swing,* F
General Comfort Apartment, house, hotel, office, hospital, school, etc.	74-76	35-30	−3 to −4	75-77	−4
Retail Shops (Short-term occupancy) Bank, barber or beauty shop, department store, supermarket, etc.	72-74	35-30†	−3 to −4	73-75	−4
Low Sensible Heat Factor Applications (High latent load) Auditorium, church, bar, restaurant, kitchen, etc.	72-74	40-35	−2 to −3	74-76	−4
Factory Comfort Assembly areas, machining rooms etc.	68-72	35-30	−4 to −6	70-74	−6

* Temperature swing is below the thermostat setting at peak winter load conditions (no lights, people, or solar heat gain).

† Winter humidification in retail clothing shops is recommended to maintain the quality texture of goods.

Recommended Inside Design Conditions--Summer

Type of application	Summer				
	Deluxe		Commercial practice		
	Dry-bulb, °F	Rel. hum., %	Dry-bulb, °F	Rel. hum., %	Temp. swing,* °F
General Comfort Apartment, house, hotel, office, hospital, school, etc.	74-76	50-45	77-79	50-45	2-4
Retail Shops (Short-term occupancy) Bank, barber, or beauty shop, department store, supermarket, etc	76-78	50-45	78-80	50-45	2-4
Low Sensible Heat Factor Applications (High latent load) Auditorium, church, bar, restaurant, kitchen, etc.	76-78	55-50	78-80	60-50	1-2
Factory Comfort Assembly areas, machining rooms, etc	77-80	55-45	80-85	60-50	3-6

* Temperature swing is above the thermostat setting at peak summer load conditions.

LIMITS FOR HUMAN EXPOSURE TO AIR CONTAMINANTS

Substance	ppm[a]	mg/m³ [b]
Abate		15
Acetaldehyde	200	360
Acetic acid	10	25
Acetic anhydride	5	20
Acetone	1,000	2,400
Acetonitrile	40	70
Acetylene dichloride, see 1,2-Dichloroethylene		
Acetylene tetrabromide	1	14
Acrolein	0.1	0.25
Acrylamide–Skin		0.3
Acrylonitrile–Skin	20	45
Aldrin–Skin		0.25
Allyl alcohol–Skin	2	5
Allyl chloride	1	3
**C Allyl glycidyl ether (AGE)	10	45
Allyl propyl disulfide	2	12
2-Aminoethanol, see Ethanolamine		
2-Aminopyridine	0.5	2
**Ammonia	50	35
Ammonium sulfamate (Ammate)		15
n-Amyl acetate	100	525
sec-Amyl acetate	125	650
Aniline–Skin	5	19
Anisidine (*o,p*-isomers)–Skin		0.5
Antimony and compounds (as Sb)		0.5
ANTU (alpha naphthyl thiourea)		0.3
Arsenic and compounds (as As)		0.5
Arsine	0.05	0.2
Azinphos-methyl–Skin		0.2
Barium (soluble compounds)		0.5
p-Benzoquinone, see Quinone		
Benzoyl peroxide		5
Benzyl chloride	1	5
Biphenyl, see Diphenyl		
Bisphenol A, see Diglycidyl ether		
Boron oxide		15
Boron tribromide	1	10
C Boron trifluoride	1	3
Bromine	0.1	0.7
*Bromine pentafluoride	0.1	0.7
Bromoform–Skin	0.5	5
Butadiene (1,3-butadiene)	1,000	2,200
Butanethiol, see Butyl mercaptan		
2-Butanone	200	590
2-Butoxy ethanol (Butyl Cellosolve)–Skin	50	240
Butyl acetate (*n*-butyl acetate)	150	710
sec-Butyl acetate	200	950
tert-Butyl acetate	200	950
Butyl alcohol	100	300
sec-Butyl alcohol	150	450
tert-Butyl alcohol	100	300
C Butylamine–Skin	5	15
C *tert*-Butyl chromate (as CrO_3)–Skin		0.1
n-Butyl glycidyl ether (BGE)	50	270
*Butyl mercaptan	0.5	1.5
p-tert-Butyltoluene	10	60
Calcium arsenate		1
Calcium oxide		5
**Camphor (Synthetic)	2	
Carbaryl (Sevin®)		5
Carbon black		3.5
Carbon dioxide	5,000	9,000
Carbon monoxide	50	55
Chlordane–Skin		0.5
Chlorinated camphene–Skin		0.5
Chlorinated diphenyl oxide		0.5

Substance	ppm[a]	mg/m³ [b]
*Chlorine	1	3
Chlorine dioxide	0.1	0.3
C Chlorine trifluoride	0.1	0.4
C Chloroacetaldehyde	1	3
α-Chloroacetophenone (phenacylchloride)	0.05	0.3
Chlorobenzene (monochlorobenzene)	75	350
o-Chlorobenzylidene malononitrile (OCBM)	0.05	0.4
Chlorobromomethane	200	1,050
2-Chloro-1,3-butadiene, see Chloroprene		
Chlorodiphenyl (42 percent Chlorine)–Skin		1
Chlorodiphenyl (54 percent Chlorine)–Skin		0.5
1-Chloro-2,3-epoxypropane, see Epichlorhydrin		
2-Chloroethanol, see Ethylene chlorohydrin		
Chloroethylene, see Vinyl chloride		
C Chloroform (trichloromethane)	50	240
1-Chloro-1-nitropropane	20	100
Chloropicrin	0.1	0.7
Chloroprene (2-chloro-1,3-butadiene)–Skin	25	90
Chromium, sol. chromic, chromous salts as Cr		0.5
Metal and insol. salts		1
Coal tar pitch volatiles (benzene soluble fraction) anthracene, BaP, phenanthrene, acridine, chrysene, pyrene		0.2
Cobalt, metal fume and dust		0.1
Copper fume		0.1
Dusts and Mists		1
Cotton dust (raw)		1
Crag® herbicide		15
Cresol (all isomers)–Skin	5	22
Crotonaldehyde	2	6
Cumene–Skin	50	245
Cyanide (as CN)–Skin		5
*Cyanogen	100	
Cyclohexane	300	1,050
Cyclohexanol	50	200
Cyclohexanone	50	200
Cyclohexene	300	1,015
Cyclopentadiene	75	200
2,4-D		10
DDT–Skin		1
DDVP, see Dichlorvos		
Decaborane–Skin	0.05	0.3
Demeton®–Skin		0.1
Diacetone alcohol (4-hydroxy-4-methyl-2-pentanone)	50	240
1,2-Diaminoethane, see Ethylenediamine		
Diazomethane	0.2	0.4
Diborane	0.1	0.1
Dibutyl phosphate	1	5
Dibutylphthalate		5
*C Dichloroacetylene	0.1	0.4
C *o*-Dichlorobenzene	50	300
p-Dichlorobenzene	75	450
Dichlorodifluoromethane	1,000	4,950
1,3-Dichloro-5,5-dimethyl hydantoin		0.2

(continued)

Substance	ppm[a]	mg/m³ [b]
1,1-Dichloroethane	100	400
1,2-Dichloroethylene	200	790
C Dichloroethyl ether–Skin	15	90
Dichloromethane, see Methylenechloride		
Dichloromonofluoromethane	1,000	4,200
C 1,1-Dichloro-1-nitroethane	10	60
1,2-Dichloropropane, see Propylenedichloride		
Dichlorotetrafluoroethane	1,000	7,000
Dichlorvos (DDVP) Skin		1
Dieldrin–Skin		0.25
Diethylamine	25	75
Diethylamino ethanol–Skin	10	50
**C Diethylene triamine–Skin	10	42
Diethylether, see Ethyl ether		
Difluorodibromomethane	100	860
C Diglycidyl ether (DGE)	0.5	2.8
Dihydroxybenzene, see Hydroquinone		
Diisobutyl ketone	50	290
Diisopropylamine–Skin	5	20
Dimethoxymethane, see Methylal		
Dimethyl acetamide–Skin	10	35
Dimethylamine	10	18
Dimethylaminobenzene, see Xylidene		
Dimethylaniline (*N*-dimethylaniline)–Skin	5	25
Dimethylbenzene, see Xylene		
Dimethyl 1,2-dibromo-2,2-dichloroethyl phosphate, (Dibrom)		3
Dimethylformamide–Skin	10	30
2,6-Dimethylheptanone, see Diisobutyl ketone		
1,1-Dimethylhydrazine–Skin	0.5	1
Dimethylphthalate		5
Dimethylsulfate–Skin	1	5
Dinitrobenzene (all isomers)–Skin		1
Dinitro-*o*-cresol–Skin		0.2
Dinitrotoluene–Skin		1.5
Dioxane (Diethylene dioxide)–Skin	100	360
Diphenyl	0.2	1
Diphenylamine		10
Diphenylmethane diisocyanate (see Methylene bisphenyl isocyanate (MDI)		
Dipropylene glycol methyl ether–Skin	100	600
Di-*sec*, octyl phthalate (Di-2-ethylhexylphthalate)		5
*Endosulfan (Thiodan®)–Skin		0.1
Endrin–Skin		0.1
Epichlorhydrin–Skin	5	19
EPN–Skin		0.5
1,2-Epoxypropane, see Propyleneoxide		
2,3-Epoxy-1-propanol, see Glycidol		
Ethanethiol, see Ethylmercaptan		
Ethanolamine	3	6
2-Ethoxyethanol–Skin	200	740
2-Ethoxyethylacetate (Cellosolve acetate)–Skin	100	540
Ethyl acetate	400	1,400
Ethyl acrylate–Skin	25	100
Ethyl alcohol (ethanol)	1,000	1,900

Substance	ppm[a]	mg/m³ [b]
Ethylamine	10	18
Ethyl *sec*-amyl ketone (5-methyl-3-heptanone)	25	130
Ethyl benzene	100	435
Ethyl bromide	200	890
Ethyl butyl ketone (3-Heptanone)	50	230
Ethyl chloride	1,000	2,600
Ethyl ether	400	1,200
Ethyl formate	100	300
Ethyl mercaptan	0.5	1
Ethyl silicate	100	850
Ethylene chlorohydrin–Skin	5	16
Ethylenediamine	10	25
Ethylene dibromide, see 1,2-Dibromoethane		
Ethylene dichloride, see 1,2-Dichloroethane		
C Ethylene glycol dinitrate and/or Nitroglycerin–Skin	[d]0.2	
Ethylene glycol monomethyl ether acetate, see Methyl cellosolve acetate		
Ethylene imine–Skin	0.5	1
Ethylene oxide	50	90
Ethylidine chloride, see 1,1-Dichloroethane		
N-Ethylmorpholine–Skin	20	94
Ferbam		15
Ferrovanadium dust		1
Fluoride (as F)		2.5
Fluorine	0.1	0.2
Fluorotrichloromethane	1,000	5,600
Formic acid	5	9
Furfural–Skin	5	20
Furfuryl alcohol	50	200
Glycidol (2,3-Epoxy-1-propanol)	50	150
Glycol monoethyl ether, see 2-Ethoxyethanol		
Guthion®, see Azinphosmethyl		
Hafnium		0.5
Heptachlor–Skin		0.5
Heptane (*n*-heptane)	500	2,000
Hexachloroethane–Skin	1	10
Hexachloronaphthalene–Skin		0.2
Hexane (*n*-hexane)	500	1,800
2-Hexanone	100	410
Hexone (Methyl isobutyl ketone)	100	410
sec-Hexyl acetate	50	300
Hydrazine–Skin	1	1.3
Hydrogen bromide	3	10
C Hydrogen chloride	5	7
Hydrogen cyanide–Skin	10	11
Hydrogen peroxide	1	1.4
Hydrogen selenide	0.05	0.2
Hydroquinone		2
*Indene	10	45
Indium and compounds, as In		0.1
C Iodine	0.1	1
Iron oxide fume		10
Iron salts, soluble, as Fe		1
Isoamyl acetate	100	525
Isoamyl alcohol	100	360
Isobutyl acetate	150	700
Isobutyl alcohol	100	300
Isophorone	25	140
Isopropyl acetate	250	950
Isopropyl alcohol	400	980
Isopropylamine	5	12
Isopropylether	500	2,100
Isopropyl glycidyl ether (IGE)	50	240

(continued)

Substance	ppm[a]	mg/m³ [b]
Ketene	0.5	0.9
Lead arsenate		0.15
Lindane–Skin		0.5
Lithium hydride		0.025
L.P.G. (liquefied petroleum gas)	1,000	1,800
Magnesium oxide fume		15
Malathion–Skin		15
Maleic anhydride	0.25	1
C Manganese and compounds, as Mn		5
Mesityl oxide	25	100
Methanethiol, see Methyl mercaptan		
Methoxychlor		15
2-Methoxyethanol, see Methyl cellosolve		
Methyl acetate	200	610
Methyl acetylene (propyne)	1,000	1,650
Methyl acetylene-propadiene mixture (MAPP)	1,000	1,800
Methyl acrylate–Skin	10	35
Methylal (dimethoxymethane)	1,000	3,100
Methyl alcohol (methanol)	200	260
Methylamine	10	12
Methyl amyl alcohol, see Methyl isobutyl carbinol		
*Methyl isoamyl ketone	100	475
Methyl (*n*-amyl) ketone (2-Heptanone)	100	465
C Methyl bromide–Skin	20	80
Methyl butyl ketone, see 2-Hexanone		
Methyl cellosolve–Skin	25	80
Methyl cellosolve acetate–Skin	25	120
Methyl chloroform	350	1,900
Methylcyclohexane	500	2,000
Methylcyclohexanol	100	470
o-Methylcyclohexanone–Skin	100	460
Methyl ethyl ketone (MEK), see 2-Butanone		
Methyl formate	100	250
Methyl iodide–Skin	5	28
Methyl isobutyl carbinol–Skin	25	100
Methyl isobutyl ketone, see Hexone		
Methyl isocyanate–Skin	0.02	0.05
*Methyl mercaptan	0.5	1
Methyl methacrylate	100	410
Methyl propyl ketone, see 2-Pentanone		
C Methyl silicate	5	30
C α-Methyl styrene	100	480
C Methylene bisphenyl isocyanate (MDI)	0.02	0.2
Molybdenum:		
Soluble compounds		5
Insoluble compounds		15
Monomethyl aniline–Skin	2	9
C Monomethyl hydrazine–Skin	0.2	0.35
Morpholine–Skin	20	70
Naphtha (coaltar)	100	400
Naphthalene	10	50
Nickel carbonyl	0.001	0.007
Nickel, metal and soluble cmpds, as Ni		1
Nicotine–Skin		0.5
Nitric acid	2	5
Nitric oxide	25	30
p-Nitroaniline–Skin	1	6
Nitrobenzene–Skin	1	5

Substance	ppm[a]	mg/m³ [b]
p-Nitrochlorobenzene–Skin		1
Nitroethane	100	310
Nitrogen dioxide	5	9
Nitrogen trifluoride	10	29
Nitroglycerin–Skin	0.2	2
Nitromethane	100	250
1-Nitropropane	25	90
2-Nitropropane	25	90
Nitrotoluene–Skin	5	30
Nitrotrichloromethane, see Chloropicrin		
Octachloronaphthalene–Skin		0.1
*Octane	400	1,900
*Oil mist, particulate		5
Osmium tetroxide		0.00
Oxalic acid		1
Oxygen difluoride	0.05	0.1
Ozone	0.1	0.2
Paraquat–Skin		0.5
Silver, metal and soluble compounds		0.01
Sodium fluoroacetate (1080)–Skin		0.05
Sodium hydroxide		2
Stibine	0.1	0.5
*Stoddard solvent	200	1,150
Strychnine		0.15
Sulfur dioxide	5	13
Sulfur hexafluoride	1,000	6,000
Sulfuric acid		1
Sulfur monochloride	1	6
Sulfur pentafluoride	0.025	0.25
Sulfuryl fluoride	5	20
Systox, see Demeton®		
2,4,5T		10
Tantalum		5
TEDP–Skin		0.2
Tellurium		0.1
Tellurium hexafluoride	0.02	0.2
TEPP–Skin		0.05
C Terphenyls	1	9
1,1,1,2-Tetrachloro-2,2-difluoroethane	500	4,170
1,1,2,2-Tetrachloro-1,2-difluoroethane	500	4,170
1,1,2,2-Tetrachloroethane–Skin	5	35
Tetrachloroethylene, see Perchloroethylene		
Tetrachloromethane, see Carbon tetrachloride		
Tetrachloronaphthalene–Skin		2
Tetraethyl lead (as Pb)–Skin		[f]0.10
Tetrahydrofuran	200	590
Tetramethyl lead (as Pb)–Skin		[f]0.15
Tetramethyl succinonitrile–Skin	0.5	3
Tetranitromethane	1	8
Tetryl (2,4,6-trinitrophenylmethylnitramine)–Skin		1.5
Thallium (soluble compounds)–Skin as Tl		0.1
Thiram		5
Tin (inorganic cmpds, except SnH_4 and SnO_2)		2
Tin (organic cmpds)		0.1
C Toluene-2,4-diisocyanate	0.02	0.14
o-Toluidine–Skin	5	22
Toxaphene, see Chlorinated camphene		
Tributyl phosphate		5

(continued)

Substance	ppm[a]	mg/M³[b]
1,1,1-Trichloroethane (see Methyl chloroform		
1,1,2-Trichloroethane–Skin	10	45
Parathion–Skin		0.1
Pentaborane	0.005	0.01
Pentachloronaphthalene–Skin		0.5
Pentachlorophenol–Skin		0.5
*Pentane	500	1,500
2-Pentanone	200	700
Perchloromethyl mercaptan	0.1	0.8
Perchloryl fluoride	3	13.5
Phenol–Skin	5	19
p-Phenylene diamine–Skin		0.1
Phenyl ether (vapor)	1	7
Phenyl ether-biphenyl mixture (vapor)	1	7
Phenylethylene, see Styrene		
Phenyl glycidyl ether (PGE)	10	60
Phenylhydrazine–Skin	5	22
Phosdrin (Mevinphos®)–Skin		0.1
Phosgene (carbonyl chloride)	0.1	0.4
Phosphine	0.3	0.4
Phosphoric acid		1
Phosphorus (yellow)		0.1
Phosphorus pentachloride		1
Phosphorus pentasulfide		1
Phosphorus trichloride	0.5	3
Phthalic anhydride	2	12
Picric acid–Skin		0.1
Pival® (2-Pivalyl-1,3-indandione)		0.1
Platinum (Soluble Salts) as Pt		0.002
Propargyl alcohol–Skin	1	
n-Propyl acetate	200	840
Propyl alcohol	200	500
n-Propyl nitrate	25	110
Propylene dichloride	75	350
Propylene imine–Skin	2	5
Propylene oxide	100	240
Propyne, see Methylacetylene		
Pyrethrum		5
Pyridine	5	15
Quinone	0.1	0.4
RDX–Skin		1.5
Rhodium, Metal fume and dusts, as Rh		0.1
Soluble salts		0.001
Ronnel		10
Rotenone (commercial)		5
Selenium compounds (as Se)		0.2
Selenium hexafluoride	0.05	0.4
Trichloromethane, see Chloroform		
Trichloronaphthalene–Skin		5
1,2,3-Trichloropropane	50	300
1,1,2-Trichloro 1,2,2-trifluoroethane	1,000	7,600
Triethylamine	25	100
Trifluoromonobromomethane	1,000	6,100
*Trimethyl benzene	25	120
2,4,6-Trinitrophenol, see Picric acid		
2,4,6-Trinitrophenylmethylnitramine, see Tetryl		
Trinitrotoluene–Skin		1.5
Triorthocresyl phosphate		0.1
Triphenyl phosphate		3
Tungsten and compounds, as W:		
Soluble		1
Insoluble		5
Turpentine	100	560
Uranium (natural) sol. and insol. compounds as U		0.2
C Vanadium:		
V_2O_5 dust		0.5
V_2O_5 fume		0.1
Vinyl benzene, see Styrene		
**C Vinyl chloride	500	1,300
Vinylcyanide, see Acrylonitrile		
Vinyl toluene	100	480
Warfarin		0.1
Xylene (xylol)	100	435
Xylidine–Skin	5	25
Yttrium		1
Zinc chloride fume		1
Zinc oxide fume		5
Zirconium compounds (as Zr)		5

*1970 Addition.

[a]Parts of vapor or gas per million parts of contaminated air by volume at 25°C and 760 mm Hg pressure.

[b]Approximate milligrams of particulate per cubic meter of air.

(No footnote "c" is used to avoid confusion with ceiling value notations.)

[d]An atmospheric concentration of not more than 0.02 ppm, or personal protection may be necessary to avoid headache.

[e]As sampled method that does not collect vapor.

[f]For control of general room air, biologic monitoring is sential for personnel control.

	8-hour time weighted average
Benzene	10 ppm
Beryllium and beryllium compounds	0.002 mg/M³
Cadmium dust (as Cd)	0.2 mg/M³
Cadmium fume (as Cd)	0.1 mg/M³
Carbon disulfide	20 ppm
Carbon tetrachloride	10 ppm
Ethylene dibromide	20 ppm
Ethylene dichloride	50 ppm
Formaldehyde	3 ppm
Hydrogen fluoride	3 ppm
Fluoride as dust	2.5 mg/M³
Lead and its inorganic compounds	0.2 mg/M³
Methyl chloride	100 ppm
Methylene chloride	500 ppm
Organo (alkyl) mercury	0.01 mg/M³
Styrene	100 ppm
Tetrachloroethylene	100 ppm
Toluene	200 ppm

	Acceptable ceiling concentration
Hydrogen sulfide	20 ppm
Chromic acid and chromates	1 mg/10M³
Mercury	1 mg/10M³

(continued)

Substance	Mppcf[a]	Mg/M³
Silica:		
Crystalline:		
Quartz (respirable)	250[f] / %SiO_2+5	10mg/M³[m] / %SiO_2+2
Quartz (total dust)		30mg/M³ / %SiO_2+2
Cristobalite: Use ½ the value calculated from the count or mass formulae for quartz.		
Trioymite: Use ½ the value calculated from the formulae for quartz.		
Amorphous, including natural diatomaceous earth	20	80mg/M³ / %SiO_2
Tremolite	5	20mg/M³ / %SiO_2
Silicates (less than 1% crystalline silica):		
Asbestos – 12 fibers per milliliter greater than 5 microns in length,[i] or	2	
Mica	20	
Soapstone	20	
Talc	20	
Portland cement	50	
Graphite (natural)	15	
Coal dust (respirable fraction less than 5% SiO_2)		2.4mg/M³ or
For more than 5% SiO_2		10mg/M³ / %SiO_2+2
Inert or Nuisance Dust:		
Respirable fraction	15	5mg/M³
Total dust	50	15mg/M³

NOTE: Conversion factors–
mppcf x 35.3 = million particles per cubic meter
= particles per cc

[a]Millions of particles per cubic foot of air, based on impinger samples counted by light-field technics.

[f]The percentage of crystalline silica in the formula is the amount determined from air-borne samples, except in those instances in which other methods have been shown to be applicable.

[i]As determined by the membrane filter method at 430 x phase contrast magnification.

[m]Both concentration and percent quartz for the application of this limit are to be determined from the fraction passing a size-selector with the following characteristics:

Aerodynamic diameter (unit density sphere)	Percent passing selector
2	90
2.5	75
3.5	50
5.0	25
10	0

The measurements under this note refer to the use of an AEC instrument. If the respirable fraction of coal dust is determined with a MRE the figure corresponding to that of 2.4 Mg/M³ in the table for coal dust is 4.5 Mg/M³.

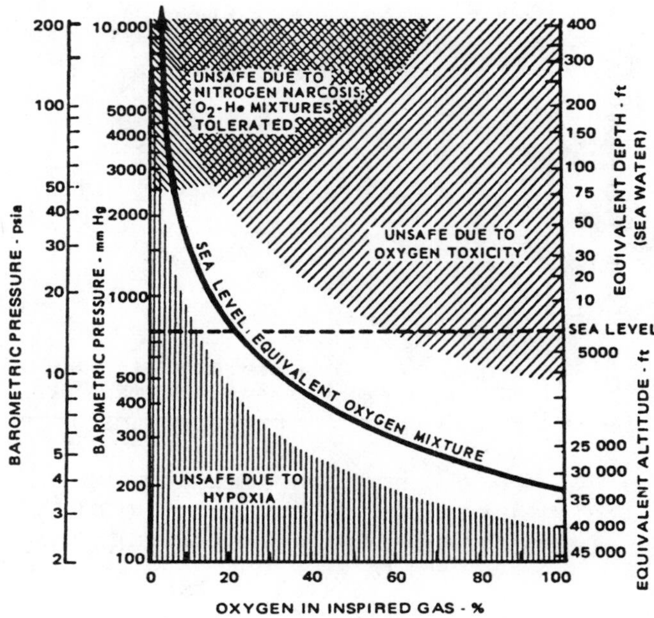

Approximate range of barometric pressure (above and below sea level) tolerated by humans breathing gas mixtures containing the indicated concentrations of O_2. Heavy curve indicates gas mixture which will maintain a sea level equivalent PO_2 in the lungs at the indicated barometric pressure

Type of Ear Complaints Encountered During Change in Barometric Pressure

Ascent (mm Hg)	Complaint	Descent (mm Hg)
0	No sensation; hearing is normal (level flight)	0
+ 3 – 5	Feeling of fullness in ears	– 3 – 5
+10 – 15	More fullness, lessened sound intensity	– 10 – 15
+15 – 30	Fullness, discomfort, tinnitus in ears:	– 15 – 30
	Ears usually "pop" as air leaves middle ear	
	Desire to clear ears; if this is done, symptoms stop	
+30 plus	Increasing pain, tinnitus, and dizziness	– 30 – 60
	Severe and radiating pain, dizziness, and nausea	– 60 – 80
	Voluntary clearing becomes difficult or impossible	–100
	Eardrum ruptures	200+

NOTE: During ascent pressure in middle ear is higher than ambient pressure; during descent, middle ear pressure is lower than ambient.

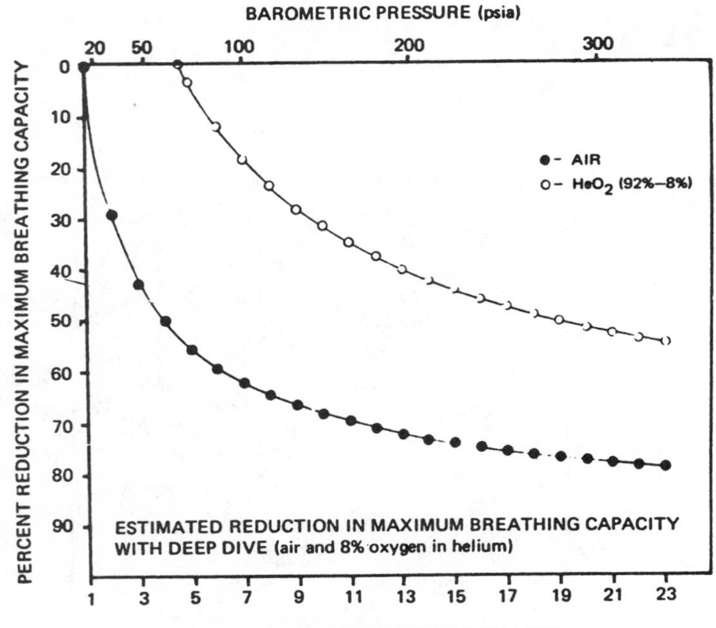

Calculated reductions in maximum ventilatory capacity with increasing depth breathing air or He-O_2.

TEMPERATURE REGULATION AND TOLERANCES

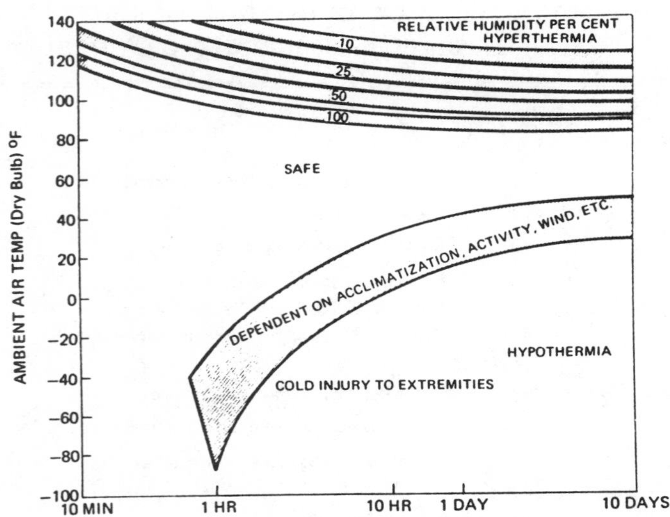

Approximate human time-tolerance temperature with optimum clothing

Approximate time-tolerance temperature with optimum clothing

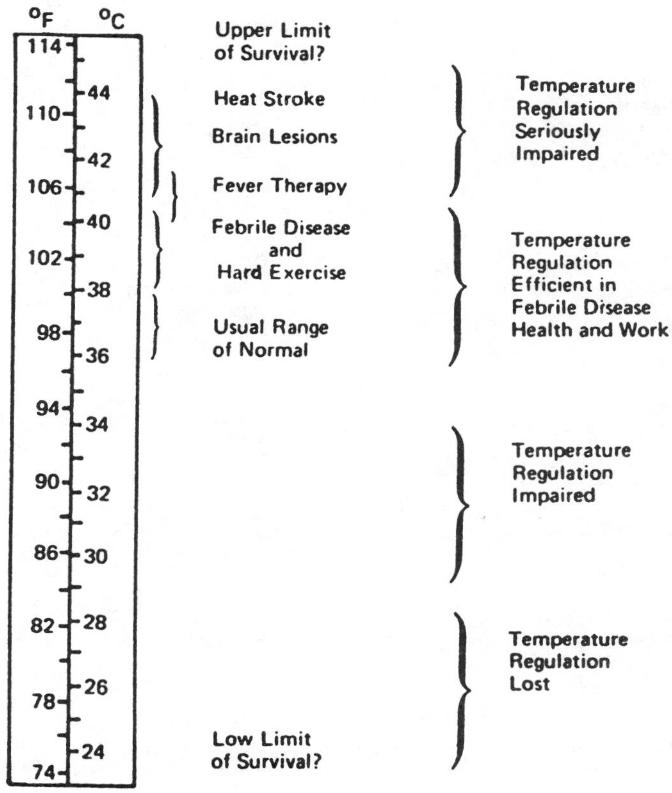

Oxygen Consumption and Heat Production of Organs in Man

Organ	Oxygen Consumption (ml/min – 100 gm)	Typical Organ Weight (gm)	Heat Production (kcal/min)
Heart	9.5 (7.8 – 10.5)	320	0.15
Brain	3.6 (3.3 – 3.9)	1380	0.25
Kidneys	10.2 (6.2 – 16.0)	300	0.15
Muscle			
Rest	0.17	28 000	0.24
Exercise	11.2	28 000	15.7

Oxygen Cost of Everyday Activities

Activity	Oxygen Consumption (liters/min)	Equivalent Heat Production (kcal/min)
Asleep		
Sleeping, men over 40	0.22	1.1
Sleeping, men aged 30 – 40	0.24	1.2
Sleeping, men aged 20 – 30	0.24	1.2
Sleeping, men aged 15 – 20	0.25	1.3
Resting		
Lying fully relaxed	0.24	1.2
Lying moderately relaxed	0.26	1.3
Lying awake, after meals	0.28	1.4
Sitting at rest	0.34	1.7
Very light activity—seated		
Writing	0.36	1.8
Riding in automobile	0.40	2.0
Typing	0.46	2.3
Polishing	0.48	2.4
Very light activity—standing		
Relaxed	0.36	1.8
Drafting	0.38	1.9
Taking lecture notes	0.40	2.0
Peeling potatoes	0.42	2.1
Light activity—seated		
Playing musical instruments	0.58	2.9
Repairing boots and shoes	0.60	3.0
At lecture	0.60	3.0
Assembling weapons	0.72	3.6
Light activity—standing		
Entering ledgers	0.52	2.6
Washing clothes	0.74	3.7
Ironing	0.88	4.4
Scrubbing	0.94	4.7
Light activity—moving		
Slow movement about room	0.50	2.5
Vehicle repairs	0.68	3.4
Slow walking	0.76	3.8
Washing	0.84	4.2
Moderate activity—lying		
Creeping, crawling, prone resting maneuvers	1.14	5.7
Crawling	1.22	6.1
Swimming breaststroke at 1 mph	1.36	6.8
Swimming crawl at 1 mph	1.40	7.0
Moderate activity—sitting		
Rowing for pleasure	1.00	5.0
Cycling at 8 – 11 mph	1.14	5.7
Cycling rapidly	1.38	6.9
Trotting on horseback	1.42	7.1

(continued)

Activities	Oxygen Consumption (liters/min)	Equivalent Heat Production (kcal/min)
Moderate activity—standing		
Gardening	1.16	5.8
Chopping wood	1.24	6.2
Baseball pitching	1.30	6.5
Shoveling sand	1.36	6.8
Moderate activity—moving		
Golf	1.08	5.4
Table tennis	1.16	5.8
Tennis	1.26	6.3
Army drill	1.42	7.1
Heavy activity—lying		
Leg exercises, average	1.50	7.5
Swimming breaststroke at 1.6 mph	1.64	8.2
Swimming backstroke at 1.0 mph	1.66	8.3
Lying on back, head raising	1.76	8.8
Heavy activity—sitting		
Cycling rapidly, own pace	1.66	8.3
Cycling at 10 mph, heavy bicycle	1.78	8.9
Cycling in race (100 mi in 4 hr 22 min)	1.96	9.8
Trotting on horseback	1.96	9.8
Heavy activity—standing		
Chopping wood	1.50	7.5
Shoveling sand	1.54	7.7
Sawing wood by hand	1.60	8.0
Digging	1.78	8.9
Heavy activity—moving		
Skating at 9 mph	1.56	7.8
Playing soccer	1.66	8.3
Skiing at 3 mph on level	1.80	9.0
Climbing stairs at 116 steps/min	1.96	9.8
Very heavy activity—sitting		
Cycling at 13.2 mph	2.00	10.0
Rowing with two oars at 3.5 mph	2.20	11.0
Galloping on horseback	2.28	11.4
Sculling (97 strokes/min)	2.52	12.6
Very heavy activity—moving		
Fencing	2.10	10.5
Playing squash	2.10	10.5
Playing basketball	2.28	11.4
Climbing stairs	2.40	12.0
Extreme activity		
Wrestling	2.60	13.0
Marching at double	2.66	13.3
Endurance marching	2.96	14.8
Harvard Step Test	3.22	16.1

Wind Chill Table

Source: National Weather Service, NOAA, U.S. Commerce Department

Both temperature and wind cause heat loss from body surfaces. A combination of cold and wind makes a body feel colder than the actual temperature. The table shows, for example, that a temperature of 20 degrees Fahrenheit, plus a wind of 20 miles per hour, causes a body heat loss equal to that in minus 10 degrees with no wind. In other words, the wind makes 20 degrees feel like minus 10.

Top line of figures shows actual temperatures in degrees Fahrenheit. Column at left shows wind speeds.

MPH	35	30	25	20	15	10	5	0	−5	−10	−15	−20	−25	−30	−35	−40	−45
5	33	27	21	19	12	7	0	−5	−10	−15	−21	−26	−31	−36	−42	−47	−52
10	22	16	10	3	−3	−9	−15	−22	−27	−34	−40	−46	−52	−58	−64	−71	−77
15	16	9	2	−5	−11	−18	−25	−31	−38	−45	−51	−58	−65	−72	−78	−85	−92
20	12	4	−3	−10	−17	−24	−31	−39	−46	−53	−60	−67	−74	−81	−88	−95	−103
25	8	1	−7	−15	−22	−29	−36	−44	−51	−59	−66	−74	−81	−88	−96	−103	−110
30	6	−2	−10	−18	−25	−33	−41	−49	−56	−64	−71	−79	−86	−93	−101	−109	−116
35	4	−4	−12	−20	−27	−35	−43	−52	−58	−67	−74	−82	−89	−97	−105	−113	−120
40	3	−5	−13	−21	−29	−37	−45	−53	−60	−69	−76	−84	−92	−100	−107	−115	−123
45	2	−6	−14	−22	−30	−38	−46	−54	−62	−70	−78	−85	−93	−102	−109	−117	−125

(Wind speeds greater than 45 mph have little additional chilling effect.)

Heat Stress Index

The overall effect of excessive heat on the body is known as heat stress. Important factors contributing to heat stress are: air temperature; humidity; air movements; radiant heat from incoming solar radiation (insolation), bright lights, an oven, stove, or other sources; atmospheric pressure; physiological factors which vary among people; physical activity; and clothing.

This index is a measure of what hot weather "feels like" to the average person for various temperatures and relative humidities.

	Air Temperature (°F)										
	70	**75**	**80**	**85**	**90**	**95**	**100**	**105**	**110**	**115**	**120**
Relative Humidity	Apparent Temperature (°F)										
0%	64	69	73	78	83	87	91	95	99	103	107
10%	65	70	75	80	85	90	95	100	105	111	116
20%	66	72	77	82	87	93	99	105	112	120	130
30%	67	73	78	84	90	96	104	113	123	135	148
40%	68	74	79	86	93	101	110	123	137	151	
50%	69	75	81	88	96	107	120	135	150		
60%	70	76	82	90	100	114	132	149			
70%	70	77	85	93	106	124	144				
80%	71	78	86	97	113	136					
90%	71	79	88	102	122						
100%	72	80	91	108							

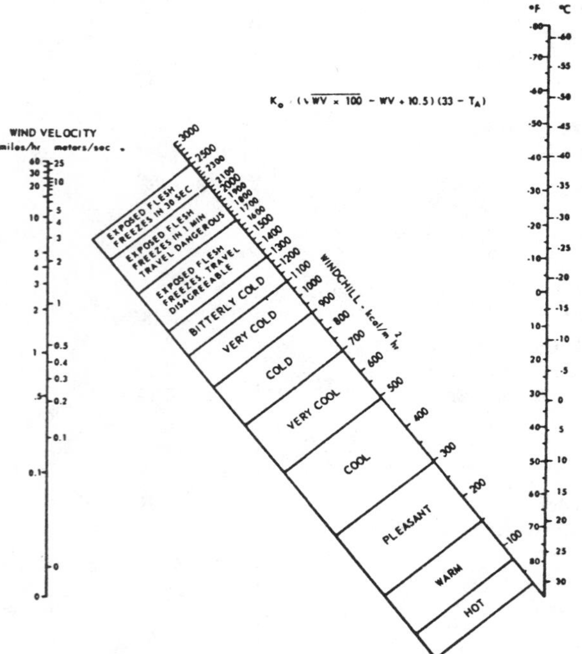

Wind Chill Nomogram

Nomogram for Estimating Tolerance Times to Cold Water Immersion

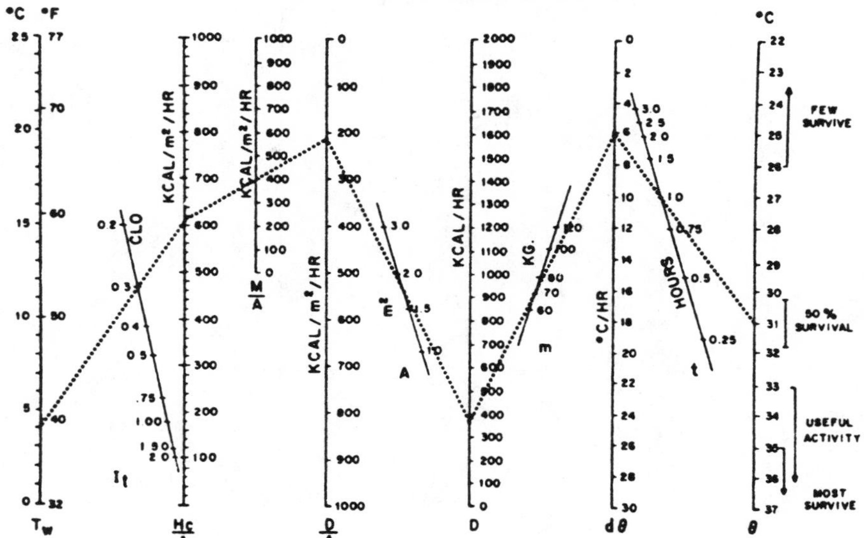

Evaporative Capacity Comfort Criterion

Percent of Maximum Evaporative Capacity	Comfort Level	Skin Temperature °F	Skin Temperature °C
0 to 10	Cold	<89	31.7
10 to 25	Comfortable	90 to 92	32.2 – 33.3
25 to 70	Tolerable	93 to 94	33.9 – 34.4
70 to 100	Hot	>95	35
Over 100	Dangerous		

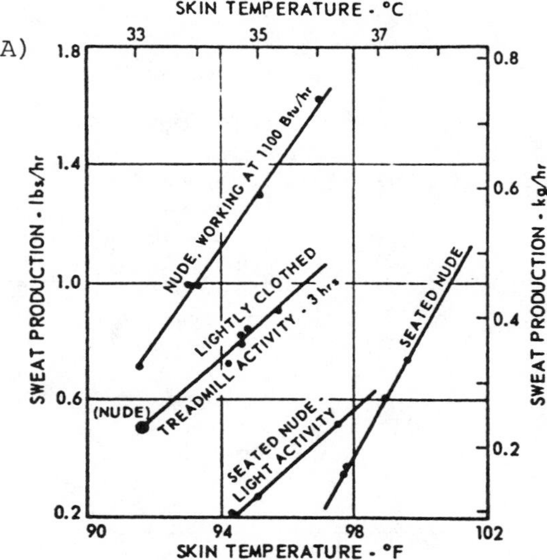

Water loss by sweating during different environmental conditions:
a) sweat rates during various laboratory procedures

(continued)

B)

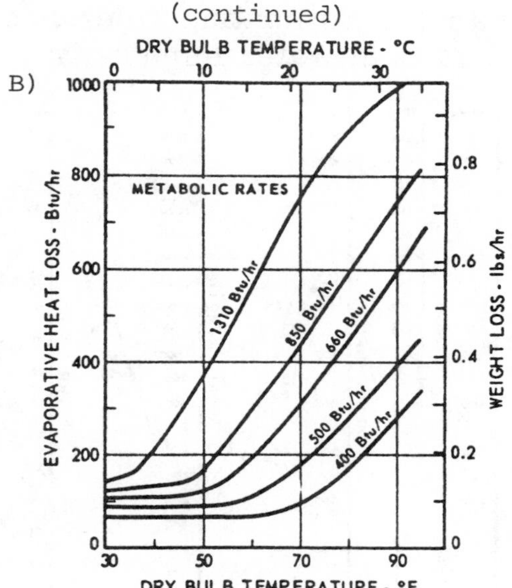

b) sweating and evaporative heat loss varying with air temperature and activity level

RADIATION SAFETY

License-exempt levels of activity

Radionuclide	Microcuries	Radionuclide	Microcuries
Calcium-45	10	Krypton-85	100
Calcium-47	10	Mercury-197	100
Carbon-14	100	Mercury-203	10
Cesium-137	10	Molybdenum-99	100
Chromium-51	1,000	Phosphorus-32	10
Cobalt-58	10	Potassium-42	10
Cobalt-60	1	Rubidium-86	10
Copper-64	100	Selenium-75	10
Fluorine-18	1,000	Sodium-24	10
Gold-198	100	Strontium-85	10
Hydrogen-3	1,000	Strontium-90	0.1
Indium-113m	100	Sulfur-35	100
Iodine-125	1	Technetium-99m	100
Iodine-129	0.1	Thallium-201	100
Iodine-131	1	Xenon-133	100
Iron-55	100	Zinc-65	10
Iron-59	10		

Any alpha emitting radionuclide not listed above or mixtures of alpha emitters of unknown composition, 0.01 μCi.

Any radionuclide other than alpha emitting radionuclides not listed above or mixtures of beta emitters of unknown composition, 0.1μCi.

Levels to be used as a guide in the establishment of contamination zones

Type of radiation	Airborne contamination ($\mu Ci/cm^3$ in air)	Direct reading surface contamination	Transferable surface contamination ($dpm/100\ cm^2$)
α	2×10^{-12}	300 $dpm/100\ cm^2$	30
β,γ	3×10^{-10}	0.25 mrad/h	1,000

Maximum permissible contamination guide for skin surfaces

	Direct survey		Transferable (smear)
Surface	α ($dpm/100\ cm^2$)	β,γ (mrad/h)	α,β,γ
General body	150	<0.06	None detectable
Hands	150	<0.3	

Maximum permissible contamination on clothing

	Direct survey		Transferable (smear)	
Item	α ($dpm/100\ cm^2$)	β,γ (mrad/h)	α ($dpm/100\ cm^2$)	β,γ ($dpm/100\ cm^2$)
Shoes, contamination zone:				
Inside	300	1.0	30	1,000
Outside	300	2.5	30	1,000
Shoes, personal:				
Inside	300	0.3	30	1,000
Outside	300	.6	30	1,000
Clothing, contamination zone	150	.75		
Clothing, other company issued, and personal	150	.25		

Permissible contamination for items given radiation clearance

Direct survey		Transferable (smear)	
α ($dpm/100\ cm^2$)	β,γ (mrad/h)	α ($dpm/100\ cm^2$)	β,γ ($dpm/100\ cm^2$)
<300	<0.05	<30	<200

Radiation coefficient of materials

Material	Surface condition	Radiation coefficient C $\frac{kcal}{m^2\,h\,(100\,deg)^4}$
a) At room temperature		
Black Body		4.96 ± 0.07
Bare metal: Aluminium	Polished	0.26
Copper	Polished	0.20
Brass	Polished	0.25
Nickel	Polished, nickel-plated sheet iron	0.27 to 0.30
Oxydised metals: Lead	Grey oxydised	1.4
Iron	Hot rolled	3.2 to 3.3
Iron	Smooth or rough cast	4.0 to 4.1
Copper	Black	3.9
Surface layers: Aluminium varnish	With varnish on rough sheet iron	1.9 to 2.0
Enamel varnish	Snow-white	4.5
Varnish	Black, glossy	4.4
Oil paint		3.9
Enamel	White	4.5
Building materials, synthetic and natural: Tar paper		4.5
Oak wood	Dressed	4.4
Gypsum	0.5 mm thick	4.5
Glass	Smooth	4.7
Lime mortar	Rough, white	4.6
Paper		4.0
Porcelain	Glazed	4.6
Wool, silk, and cotton materials		3.9
Bricks	Red, rough, but no very great unevenness	4.6
b) At higher temperatures		
Mild steel	Oxydised, smooth at 800°C	4.6
Carbon	Ground, 0.9% ash at 630°C	3.9
Fire-proof clay	Vitrified through practical usage at 1000°C	3.7
Silica brick	Rough at 1000°C	4.0

[1]) In accordance with SCHACK, A.: "Der industrielle Wärmeübergang", 5. Aufl., Düsseldorf 1957.

Permissible concentrations and pertinent radiologic data of selected radionuclides

Isotope	MPBB (μCi)	Critical organ[a]	MPC (air) workers (μCi/cm³)	MPC (air) public (μCi/cm³)	MPC (water) workers (μCi/cm³)	MPC (water) public (μCi/cm³)
Ca-47	5	Bone	2×10^{-7}	6×10^{-9}	1×10^{-3}	3×10^{-5}
C-14	300	Fat	4×10^{-6}	1×10^{-7}	2×10^{-2}	8×10^{-4}
Cs-137	30	T.B.	1×10^{-8}	5×10^{-10}	4×10^{-4}	2×10^{-5}
Cr-51	800	T.B.	2×10^{-6}	8×10^{-8}	5×10^{-2}	2×10^{-3}
Co-57	200	T.B.	2×10^{-7}	6×10^{-9}	1×10^{-2}	4×10^{-4}
Co-58	30	T.B.	5×10^{-8}	2×10^{-9}	3×10^{-3}	9×10^{-5}
Co-60	10	T.B.	9×10^{-9}	3×10^{-10}	1×10^{-3}	3×10^{-5}
Cu-64	10	Spl.	1×10^{-6}	4×10^{-8}	6×10^{-3}	2×10^{-4}
F-18	20	T.B.	3×10^{-6}	9×10^{-8}	1×10^{-2}	5×10^{-4}
Au-198	20	Kid.	2×10^{-7}	8×10^{-9}	1×10^{-3}	5×10^{-5}
H-3	1000	B.T.	5×10^{-6}	2×10^{-7}	1×10^{-1}	3×10^{-3}
In-113m	30	Kid.	7×10^{-6}	2×10^{-7}	4×10^{-2}	1×10^{-3}
I-125			5×10^{-9}	8×10^{-11}	4×10^{-5}	2×10^{-7}
I-129	3	Thy.	2×10^{-9}	2×10^{-11}	1×10^{-5}	6×10^{-8}
I-131	0.7	Thy.	9×10^{-9}	1×10^{-10}	6×10^{-5}	3×10^{-7}
Fe-55	1000	Spl.	9×10^{-7}	3×10^{-8}	2×10^{-2}	8×10^{-4}
Fe-59	20	Spl.	5×10^{-9}	2×10^{-9}	2×10^{-3}	5×10^{-5}
Kr-85			1×10^{-5}	3×10^{-7}		
Hg-197	20	Kid.	1×10^{-6}	4×10^{-8}	9×10^{-3}	3×10^{-4}
Hg-203	4	Kid.	7×10^{-8}	2×10^{-9}	5×10^{-4}	2×10^{-5}

(continued)

Isotope	MPBB (μCi)	Critical organ[a]	MPC (air) workers (μCi/cm³)	MPC (air) public (μCi/cm³)	MPC (water) workers (μCi/cm³)	MPC (water) public (μCi/cm³)
Mo-99	8	Kid.	2×10^{-7}	7×10^{-9}	1×10^{-3}	4×10^{-5}
P-32	6	Bone	7×10^{-8}	2×10^{-9}	5×10^{-4}	2×10^{-5}
K-42	10	T.B.	1×10^{-7}	4×10^{-9}	6×10^{-4}	2×10^{-5}
Rb-86	30	T.B.	7×10^{-8}	2×10^{-9}	7×10^{-4}	2×10^{-5}
Se-75	90	Kid.	1×10^{-7}	4×10^{-9}	8×10^{-3}	3×10^{-4}
Na-24	7	T.B.	1×10^{-7}	5×10^{-9}	8×10^{-4}	3×10^{-5}
Sr-85	60	T.B.	1×10^{-7}	4×10^{-9}	3×10^{-3}	1×10^{-4}
Sr-90	2	Bone	1×10^{-9}	3×10^{-11}	1×10^{-5}	3×10^{-7}
S-35	90	Test.	3×10^{-7}	9×10^{-9}	2×10^{-3}	6×10^{-5}
Tc-99m	200	T.B.	1×10^{-5}	5×10^{-7}	8×10^{-2}	3×10^{-3}
Tl-201	40	Kid.	9×10^{-7}	3×10^{-8}	5×10^{-3}	2×10^{-4}
Xe-133			1×10^{-5}	3×10^{-7}		
Zn-65	60	T.B.	6×10^{-8}	2×10^{-9}	3×10^{-3}	1×10^{-4}

[a] T.B.—total body; Spl—spleen; B.T.—body tissue; Kid.—kidney; Thy.—Thyroid; Test.—testes.
MPC—maximum permissible concentration.
MPBB—maximum permissible body burden.

RBE Values for Various Types of Radiation

Type of Radiation	RBE
X-rays	1
Gamma rays and bremsstrahlung	1
Beta particles, 1.0 MeV	1
" " 0.1 MeV	1
Neutrons, thermal energy	2.8
" 0.0001 MeV	2.2
" 0.005 MeV	2.4
" 0.02 MeV	5
" 0.5 MeV	10.2
" 1.0 MeV	10.5
" 10.0 MeV	6.4
Protons, greater than 100 MeV	1-2
" 1.0 MeV	8.5
" 0.1 MeV	10
Alpha particles, 5 MeV	15
" " 1 MeV	20

Instaneous and Mean RBE's for Protons of Various Energies

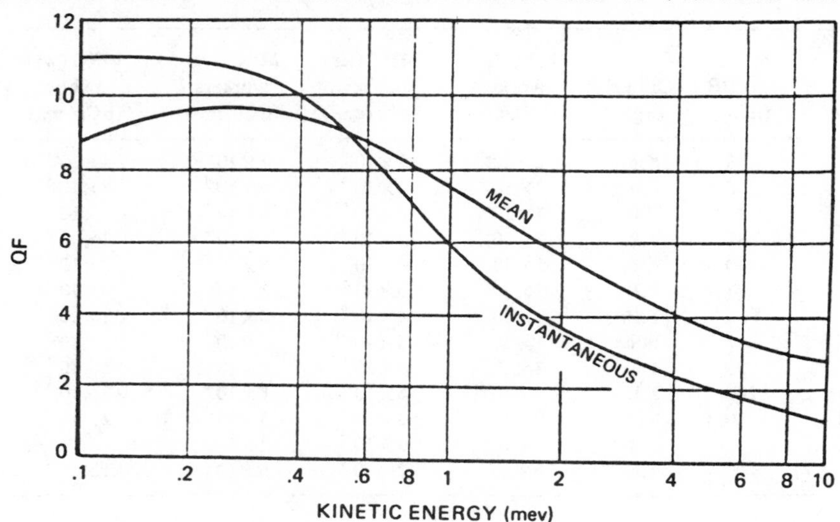

Physical Characteristics of Clinically Used Radionuclides

Radio-nuclide	Principal means of production	Half-life	Major radiations-energies (MeV)-abundance (%) Beta	Gamma	Γ, specific γ-ray constant (R·cm²/h·mCi)	Half-value layer in Pb (cm)
^{3}H	$^{6}Li(n,\alpha)^{3}H$	12.3 y	0.018(100)			
^{11}C	$^{11}B(p,n)^{11}C$	20.3 m	β^{+} 0.97(100)	0.511(200)	5.91	0.4
^{14}C	$^{14}N(n,p)^{14}C$	5.73×10^{3} y	0.16(100)			
^{13}N	$^{13}C(p,n)^{13}N$	10.0 m	β^{+} 1.20(100)	0.511(200)	5.91	0.4
^{15}O	$^{14}N(d,n)^{15}O$	2.07 m	β^{+} 1.74(100)	0.511(200)	5.91	0.4
^{18}F	$^{19}F(p,pn)^{18}F$	1.8 h	β^{+} 0.65(97)	0.511(194)	5.73	0.4
^{24}Na	$^{23}Na(n,\gamma)^{24}Na$	15.0 h	1.4(100)	1.369(100) 2.754(100)	18.40	1.5
^{32}P	$^{31}S(n,p)^{32}P$ $^{31}P(n,\gamma)^{32}P$	14.3 d	1.7(100)			
^{35}S	$^{35}Cl(n,p)^{35}S$ $^{34}S(n,\gamma)^{35}S$	88.0 d	0.17(100)			
^{42}K	$^{41}K(n,\gamma)^{42}K$	12.4 h	2.00(18) 3.52(82)	1.524(18)	1.35	
^{43}K	$^{40}Ar(\alpha,p)^{43}K$	22.4 h	0.83(87)	0.371(85) 0.338(07) 0.394(11) 0.590(13) 0.619(81)	5.60	
^{45}Ca	$^{44}Ca(n,\gamma)^{45}Ca$	165 d	0.25(100)			
^{47}Ca	$^{46}Ca(n,\gamma)^{47}Ca$	4.56 d	0.67(82) 1.98(18)	0.490(5) 0.815(5) 1.308(74)	5.70	

(continued)

Radio-nuclide	Principal means of production	Half-life	Major radiations-energies (MeV)-abundance (%) Beta	Gamma	Γ, specific γ-ray constant (R·cm²/h·mCi)	Half-value layer in Pb (cm)
^{51}Cr	^{50}Cr(n,γ)^{51}Cr	27.8 d	0.3e⁻ (trace)	0.320(9)	0.164	0.2
^{52}Fe	^{50}Cr(α,2n)^{52}Fe	8.3 h	β⁺ 0.80(56)	0.165(100) 0.511(112)	17.20	1.2
^{55}Fe	^{54}Fe(n,γ)^{55}Fe	2.7 y		0.006(13) Mn x-rays		
^{59}Fe	^{58}Fe(n,γ)^{59}Fe	45 d	0.48(99)	1.095(56) 1.292(44)	6.20	1.1
^{57}Co	^{60}Ni(p,α)^{57}Co	267 d		0.014(9) 0.122(87) 0.136(10)	0.93	0.03
^{58}Co	^{55}Mn(α,n)^{58}Co	71.3 d	β⁺ 0.47(15)	0.511(30) 0.811(99)	5.50	
^{60}Co	^{59}Co(n,γ)^{60}Co	5.26 y	0.31(99⁺)	1.173(100) 1.332(100)	13.20	1.2
^{64}Cu	^{63}Cu(n,γ)^{64}Cu	12.9 h	0.57(39) β⁺ 0.64(19)	0.511(38)	1.16	0.4
^{65}Zn	^{64}Zn(n,γ)^{65}Zn	245 d	1.1e⁻ (trace) β⁺ 0.33(2)	1.150(49)	2.70	1.0
^{67}Ga	^{67}Zn(p,n)^{67}Ga	78 h	0.09e⁻ (15)	0.093(40) 0.185(24) 0.296(22) 0.388(7)	~1.1	
^{68}Ga	^{68}Zn(p,n)^{68}Ga	1.13 h	β⁺ 1.90(87)	0.511(176)	5.37	0.4
^{75}Se	^{74}Se(n,γ)^{75}Se	120 d	e⁻ 0.08–0.25	0.121(16) 0.136(57) 0.265(60) 0.280(25) 0.401(13)	2.00	0.2
^{85m}Kr	^{84}Kr(n,γ)^{85m}Kr	4.4 h	0.82(77)	0.150(74) 0.305(16)	1.29	
^{85}Sr	^{84}Sr(n,γ)^{85}Sr	64 d	0.5(1)	0.514(100)	3.00	0.4
^{87m}Sr	^{87}Y daughter	2.8 h	0.4e⁻ (22)	0.388(80)	1.74	
^{90}Sr	Fission	28 y	0.55(100) ^{90}Y–2.30(100)			
^{86}Rb	^{85}Rb(n,γ)^{86}Rb	18.7 d	0.70(9) 1.78(9)	1.078(9)	0.49	1.0
^{99}Mo	^{98}Mo(n,γ)^{99}Mo	66.7 h	1.23(82) 0.45(17)	0.181(7) 0.740(14) 0.778(5)	~1.80	
^{99m}Tc	^{99}Mo daughter	6.04 h	0.12e⁻ (trace)	0.140(90)	0.70	0.03
^{111}In	^{111}Cd(p,n)^{111}In	2.8 d	0.15–0.24e⁻(15)	0.173(89) 0.247(94)		

(continued)

Radio-nuclide	Principal means of production	Half-life	Major radiations-energies (MeV)-abundance (%) Beta	Gamma	Γ, specific γ-ray constant (R·cm²/h·mCi)	Half-value layer in Pb (cm)
^{113m}In	Daughter ^{113}Sn	1.67 h		0.393(65)	0.32(x) 1.57(γ) ~1.7	0.03
^{123}I	$^{123}Te(p,n)^{123}I$	13.0 h	0.13e⁻ (trace)	0.159(97)	0.97(x) 0.66(γ)	0.04
^{125}I	$^{124}Xe(n,\gamma) \rightarrow$ $^{125}Xe^{ec} \rightarrow {}^{125}I$	60.2 d	0.03e⁻ (90)	0.036(7) 0.027 x-rays (90)	~0.70	
^{129}I	Fission	1.7×10^7 y	0.150(100)	0.04(9) Xe x-rays		
^{131}I	Fission	8.05 d	0.61(90)	0.284(5) 0.365(83) 0.637(7)	2.23	0.3
^{127}Xe	$^{127}I(p,n)^{127}Xe$	36.4 d		0.172(22) 0.203(65) 0.375(20)		
^{133}Xe	Fission	5.3 d	0.35(100)	0.08(35) Cs x-rays	0.10	
^{129}Cs	$^{127}I(\alpha,2n)^{129}Cs$	32.1 h		0.375(48) 0.416(25) 0.550(5)		
^{137}Cs	Fission	30 y	0.51(95) 1.18(7)	0.662(84)	3.32	0.6
^{198}Au	$^{197}Au(n,\gamma)^{198}Au$	2.7 d	0.97(99)	0.412(96)	2.34	0.3
^{197}Hg	$^{196}Hg(n,\gamma)^{197}Hg$	2.7 d		0.077(19)	~0.40	
^{203}Hg	$^{202}Hg(n,\gamma)^{203}Hg$	47 d	0.214(100)	0.279(82)	1.33	
^{169}Yb	$^{168}Yb(n,\gamma)^{169}Yb$	31.8 d		0.063(45) 0.177(22) 0.110(18) 0.197(35) 0.131(11) 0.308(10)		
^{201}Tl	Daughter ^{201}Pb	73 h		0.167(8) Hg x-rays	0.08	

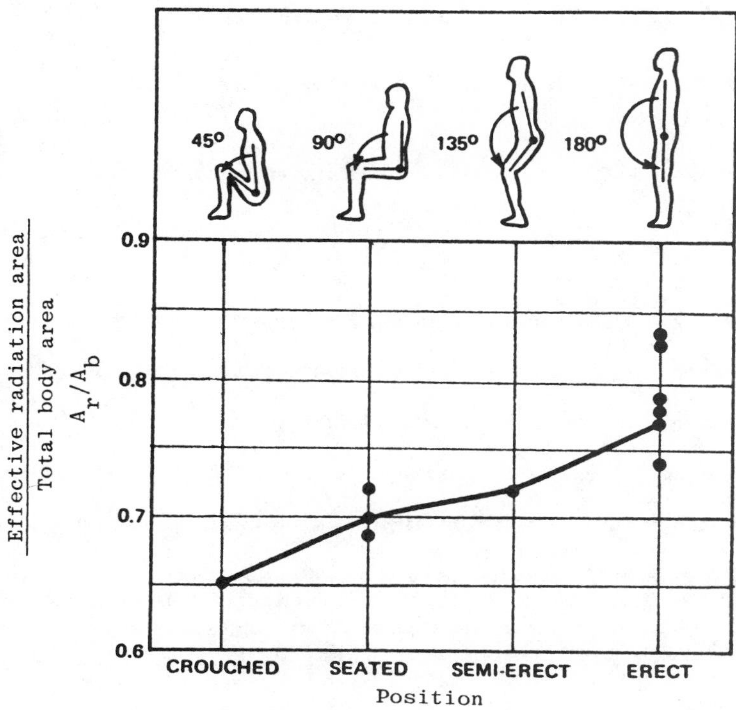

Body radiation area for various body positions

Nature and location of electromagnetic and particulate ionizing radiation in space

Name	Nature of Radiation	Charge	Mass	Where Found
Photon	Electromagnetic	0	0	Radiation belts, solar radiation (produced by nuclear reactions and by stopping electrons), and everywhere in space
X-ray	Electromagnetic	0	0	
Gamma ray	Electromagnetic	0	0	
Electron	Particle	−e	1 m_e	Radiation belt and elsewhere
Positron	Particle	+e	1 m_e	Cosmic rays, radiation belt, solar flares
Proton	Particle	+e	1840 m_e or 1 amu*	Primary cosmic rays, radiation belt, solar flares
Neutron	Particle	0	1841 m_e	Secondary particles produced by nuclear interactions involving primary particle flux
Pi meson	Particle	+,−, or 0	273 m_e	Cosmic rays, radiation belt, solar flares
Alpha particle	Particle	+2e	4 amu	Primary cosmic radiation (nucleus of helium atom)
Heavy primary nuclei	Particle	≥+3e	≥6 amu	Primary cosmic radiation (nuclei of heavier atoms)

*amu = atom mass unit

Selected Rules of Thumb for Radiation Safety

The following rules of thumb are only approximate, and should be treated as such!

Alpha Particles

Alpha particles of at least 7.5 MeV are required to penetrate the epidermis, the protective layer of skin, 0.07 mm thick.

Electrons

Electrons of at least 70 keV are required to penetrate the epidermis, the protective layer of skin, 0.07 mm thick.

The range (R) of electrons in g/cm^2 is approximately equal to the maximum energy (E) in MeV divided by 2 (i.e., $R \cong E/2$).

The range of electrons in air is about 3.65 m per MeV; for example, a 3 MeV electron has a range of about 11 m in air.

A chamber wall thickness of 30 mg/cm^2 will transmit 0.7 of the initial fluence of 1 MeV electrons and 0.2 of 0.4-MeV electrons.

When electrons of 1 to 2 MeV pass through light materials such as water, aluminum, or glass, less than 1% of their energy is dissipated as bremsstrahlung.

The bremsstrahlung from 1 Ci of ^{32}P aqueous solution in a glass bottle is about 1 mR/h at 1 meter.

When electrons from a 1 Ci source of ^{90}Sr–^{90}Y are absorbed, the bremsstrahlung hazard is approximately equal to that presented by the gamma radiation from 12 mg of radium. The average energy of the bremsstrahlung is about 300 keV.

Gamma Rays

The air-scattered radiation (sky-shine) from a 100-Ci ^{60}Co source placed 1 ft behind a 4-ft-high shield is about 100 mrad/h at 6 ft from the outside of the shield.

Within $\pm 20\%$ for point source gamma emitters with energies between 0.07 and 4 MeV, the exposure rate (R/h) at 1 ft is $6{\cdot}C{\cdot}E{\cdot}n$ where C is the activity in curies, E the energy in MeV, and n is the number of gammas per disintegration.

Neutrons

An approximate HVL for 1-MeV neutrons is 3.2 cm of paraffin; that for 5-MeV neutrons is 6.93 cm.

Miscellaneous

The activity of any radionuclide is reduced to less than 1% after 7 half-lives (i.e., $2^{-7} = 0.8\%$).

For material with a half-life greater than 6 days, the change in activity in 24 hours will be less than 10%.

There is 0.64 mm^3 of radon gas in transient equilibrium with 1 Ci of radium.

1 year $\cong \pi \times 10^7$ s.

10 HVL attenuates approximately by 10^{-3}.

CHAPTER 27

SOLAR SYSTEM

Perpetual Calendar

The number shown for each year indicates which Gregorian calendar to use. For 1583-1802, or for Julian calendar, see page 738. For years 1803-1820, use numbers for 1983-2000, respectively.

Year	No.	Year	No.	Year	No.	Year	No.	Year	No.	Year	No.	Year	No.	Year	No.	Year	No.	Year	No.
1821	2	1847	6	1873	4	1899	1	1925	5	1951	2	1977	7	2003	4	2029	2	2055	6
1822	3	1848	14	1874	5	1900	2	1926	6	1952	10	1978	1	2004	12	2030	3	2056	14
1823	4	1849	2	1875	6	1901	3	1927	7	1953	5	1979	2	2005	7	2031	4	2057	2
1824	12	1850	3	1876	14	1902	4	1928	8	1954	6	1980	10	2006	1	2032	12	2058	3
1825	7	1851	4	1877	2	1903	5	1929	3	1955	7	1981	5	2007	2	2033	7	2059	4
1826	1	1852	12	1878	3	1904	13	1930	4	1956	8	1982	6	2008	10	2034	1	2060	12
1827	2	1853	7	1879	4	1905	1	1931	5	1957	3	1983	7	2009	5	2035	2	2061	7
1828	10	1854	1	1880	12	1906	2	1932	13	1958	4	1984	8	2010	6	2036	10	2062	1
1829	5	1855	2	1881	7	1907	3	1933	1	1959	5	1985	3	2011	7	2037	5	2063	2
1830	6	1856	10	1882	1	1908	11	1934	2	1960	13	1986	4	2012	8	2038	6	2064	10
1831	7	1857	5	1883	2	1909	6	1935	3	1961	1	1987	5	2013	3	2039	7	2065	5
1832	8	1858	6	1884	10	1910	7	1936	11	1962	2	1988	13	2014	4	2040	8	2066	6
1833	3	1859	7	1885	5	1911	1	1937	6	1963	3	1989	1	2015	5	2041	3	2067	7
1834	4	1860	8	1886	6	1912	9	1938	7	1964	11	1990	2	2016	13	2042	4	2068	8
1835	5	1861	3	1887	7	1913	4	1939	1	1965	6	1991	3	2017	1	2043	5	2069	3
1836	13	1862	4	1888	8	1914	5	1940	9	1966	7	1992	11	2018	2	2044	13	2070	4
1837	1	1863	5	1889	3	1915	6	1941	4	1967	1	1993	6	2019	3	2045	1	2071	5
1838	2	1864	13	1890	4	1916	14	1942	5	1968	9	1994	7	2020	11	2046	2	2072	13
1839	3	1865	1	1891	5	1917	2	1943	6	1969	4	1995	1	2021	6	2047	3	2073	1
1840	11	1866	2	1892	13	1918	3	1944	14	1970	5	1996	9	2022	7	2048	11	2074	2
1841	6	1867	3	1893	1	1919	4	1945	2	1971	6	1997	4	2023	1	2049	6	2075	3
1842	7	1868	11	1894	2	1920	12	1946	3	1972	14	1998	5	2024	9	2050	7	2076	11
1843	1	1869	6	1895	3	1921	7	1947	4	1973	2	1999	6	2025	4	2051	1	2077	6
1844	9	1870	7	1896	11	1922	1	1948	12	1974	3	2000	14	2026	5	2052	9	2078	7
1845	4	1871	1	1897	6	1923	2	1949	7	1975	4	2001	2	2027	6	2053	4	2079	1
1846	5	1872	9	1898	7	1924	10	1950	1	1976	12	2002	3	2028	14	2054	5	2080	9

1

JANUARY

S	M	T	W	T	F	S
1	2	3	4	5	6	7
8	9	10	11	12	13	14
15	16	17	18	19	20	21
22	23	24	25	26	27	28
29	30	31				

FEBRUARY

S	M	T	W	T	F	S
			1	2	3	4
5	6	7	8	9	10	11
12	13	14	15	16	17	18
19	20	21	22	23	24	25
26	27	28				

MARCH

S	M	T	W	T	F	S
			1	2	3	4
5	6	7	8	9	10	11
12	13	14	15	16	17	18
19	20	21	22	23	24	25
26	27	28	29	30	31	

APRIL

S	M	T	W	T	F	S
						1
2	3	4	5	6	7	8
9	10	11	12	13	14	15
16	17	18	19	20	21	22
23	24	25	26	27	28	29
30						

MAY

S	M	T	W	T	F	S
	1	2	3	4	5	6
7	8	9	10	11	12	13
14	15	16	17	18	19	20
21	22	23	24	25	26	27
28	29	30	31			

JUNE

S	M	T	W	T	F	S
				1	2	3
4	5	6	7	8	9	10
11	12	13	14	15	16	17
18	19	20	21	22	23	24
25	26	27	28	29	30	

JULY

S	M	T	W	T	F	S
						1
2	3	4	5	6	7	8
9	10	11	12	13	14	15
16	17	18	19	20	21	22
23	24	25	26	27	28	29
30	31					

AUGUST

S	M	T	W	T	F	S
		1	2	3	4	5
6	7	8	9	10	11	12
13	14	15	16	17	18	19
20	21	22	23	24	25	26
27	28	29	30	31		

SEPTEMBER

S	M	T	W	T	F	S
					1	2
3	4	5	6	7	8	9
10	11	12	13	14	15	16
17	18	19	20	21	22	23
24	25	26	27	28	29	30

OCTOBER

S	M	T	W	T	F	S
1	2	3	4	5	6	7
8	9	10	11	12	13	14
15	16	17	18	19	20	21
22	23	24	25	26	27	28
29	30	31				

NOVEMBER

S	M	T	W	T	F	S
			1	2	3	4
5	6	7	8	9	10	11
12	13	14	15	16	17	18
19	20	21	22	23	24	25
26	27	28	29	30		

DECEMBER

S	M	T	W	T	F	S
					1	2
3	4	5	6	7	8	9
10	11	12	13	14	15	16
17	18	19	20	21	22	23
24	25	26	27	28	29	30
31						

2

JANUARY

S	M	T	W	T	F	S
	1	2	3	4	5	6
7	8	9	10	11	12	13
14	15	16	17	18	19	20
21	22	23	24	25	26	27
28	29	30	31			

FEBRUARY

S	M	T	W	T	F	S
				1	2	3
4	5	6	7	8	9	10
11	12	13	14	15	16	17
18	19	20	21	22	23	24
25	26	27	28			

MARCH

S	M	T	W	T	F	S
				1	2	3
4	5	6	7	8	9	10
11	12	13	14	15	16	17
18	19	20	21	22	23	24
25	26	27	28	29	30	31

APRIL

S	M	T	W	T	F	S
1	2	3	4	5	6	7
8	9	10	11	12	13	14
15	16	17	18	19	20	21
22	23	24	25	26	27	28
29	30					

MAY

S	M	T	W	T	F	S
		1	2	3	4	5
6	7	8	9	10	11	12
13	14	15	16	17	18	19
20	21	22	23	24	25	26
27	28	29	30	31		

JUNE

S	M	T	W	T	F	S
					1	2
3	4	5	6	7	8	9
10	11	12	13	14	15	16
17	18	19	20	21	22	23
24	25	26	27	28	29	30

JULY

S	M	T	W	T	F	S
1	2	3	4	5	6	7
8	9	10	11	12	13	14
15	16	17	18	19	20	21
22	23	24	25	26	27	28
29	30	31				

AUGUST

S	M	T	W	T	F	S
			1	2	3	4
5	6	7	8	9	10	11
12	13	14	15	16	17	18
19	20	21	22	23	24	25
26	27	28	29	30	31	

SEPTEMBER

S	M	T	W	T	F	S
						1
2	3	4	5	6	7	8
9	10	11	12	13	14	15
16	17	18	19	20	21	22
23	24	25	26	27	28	29
30						

OCTOBER

S	M	T	W	T	F	S
	1	2	3	4	5	6
7	8	9	10	11	12	13
14	15	16	17	18	19	20
21	22	23	24	25	26	27
28	29	30	31			

NOVEMBER

S	M	T	W	T	F	S
				1	2	3
4	5	6	7	8	9	10
11	12	13	14	15	16	17
18	19	20	21	22	23	24
25	26	27	28	29	30	

DECEMBER

S	M	T	W	T	F	S
						1
2	3	4	5	6	7	8
9	10	11	12	13	14	15
16	17	18	19	20	21	22
23	24	25	26	27	28	29
30	31					

3 — 1985

JANUARY

S	M	T	W	T	F	S
		1	2	3	4	5
6	7	8	9	10	11	12
13	14	15	16	17	18	19
20	21	22	23	24	25	26
27	28	29	30	31		

FEBRUARY

S	M	T	W	T	F	S
					1	2
3	4	5	6	7	8	9
10	11	12	13	14	15	16
17	18	19	20	21	22	23
24	25	26	27	28		

MARCH

S	M	T	W	T	F	S
					1	2
3	4	5	6	7	8	9
10	11	12	13	14	15	16
17	18	19	20	21	22	23
24	25	26	27	28	29	30
31						

APRIL

S	M	T	W	T	F	S
	1	2	3	4	5	6
7	8	9	10	11	12	13
14	15	16	17	18	19	20
21	22	23	24	25	26	27
28	29	30				

MAY

S	M	T	W	T	F	S
			1	2	3	4
5	6	7	8	9	10	11
12	13	14	15	16	17	18
19	20	21	22	23	24	25
26	27	28	29	30	31	

JUNE

S	M	T	W	T	F	S
						1
2	3	4	5	6	7	8
9	10	11	12	13	14	15
16	17	18	19	20	21	22
23	24	25	26	27	28	29
30						

JULY

S	M	T	W	T	F	S
	1	2	3	4	5	6
7	8	9	10	11	12	13
14	15	16	17	18	19	20
21	22	23	24	25	26	27
28	29	30	31			

AUGUST

S	M	T	W	T	F	S
				1	2	3
4	5	6	7	8	9	10
11	12	13	14	15	16	17
18	19	20	21	22	23	24
25	26	27	28	29	30	31

SEPTEMBER

S	M	T	W	T	F	S
1	2	3	4	5	6	7
8	9	10	11	12	13	14
15	16	17	18	19	20	21
22	23	24	25	26	27	28
29	30					

OCTOBER

S	M	T	W	T	F	S
		1	2	3	4	5
6	7	8	9	10	11	12
13	14	15	16	17	18	19
20	21	22	23	24	25	26
27	28	29	30	31		

NOVEMBER

S	M	T	W	T	F	S
					1	2
3	4	5	6	7	8	9
10	11	12	13	14	15	16
17	18	19	20	21	22	23
24	25	26	27	28	29	30

DECEMBER

S	M	T	W	T	F	S
1	2	3	4	5	6	7
8	9	10	11	12	13	14
15	16	17	18	19	20	21
22	23	24	25	26	27	28
29	30	31				

4

JANUARY

S	M	T	W	T	F	S
			1	2	3	4
5	6	7	8	9	10	11
12	13	14	15	16	17	18
19	20	21	22	23	24	25
26	27	28	29	30	31	

FEBRUARY

S	M	T	W	T	F	S
						1
2	3	4	5	6	7	8
9	10	11	12	13	14	15
16	17	18	19	20	21	22
23	24	25	26	27	28	

MARCH

S	M	T	W	T	F	S
						1
2	3	4	5	6	7	8
9	10	11	12	13	14	15
16	17	18	19	20	21	22
23	24	25	26	27	28	29
30	31					

APRIL

S	M	T	W	T	F	S
		1	2	3	4	5
6	7	8	9	10	11	12
13	14	15	16	17	18	19
20	21	22	23	24	25	26
27	28	29	30			

MAY

S	M	T	W	T	F	S
				1	2	3
4	5	6	7	8	9	10
11	12	13	14	15	16	17
18	19	20	21	22	23	24
25	26	27	28	29	30	31

JUNE

S	M	T	W	T	F	S
1	2	3	4	5	6	7
8	9	10	11	12	13	14
15	16	17	18	19	20	21
22	23	24	25	26	27	28
29	30					

JULY

S	M	T	W	T	F	S
		1	2	3	4	5
6	7	8	9	10	11	12
13	14	15	16	17	18	19
20	21	22	23	24	25	26
27	28	29	30	31		

AUGUST

S	M	T	W	T	F	S
					1	2
3	4	5	6	7	8	9
10	11	12	13	14	15	16
17	18	19	20	21	22	23
24	25	26	27	28	29	30
31						

SEPTEMBER

S	M	T	W	T	F	S
	1	2	3	4	5	6
7	8	9	10	11	12	13
14	15	16	17	18	19	20
21	22	23	24	25	26	27
28	29	30				

OCTOBER

S	M	T	W	T	F	S
			1	2	3	4
5	6	7	8	9	10	11
12	13	14	15	16	17	18
19	20	21	22	23	24	25
26	27	28	29	30	31	

NOVEMBER

S	M	T	W	T	F	S
						1
2	3	4	5	6	7	8
9	10	11	12	13	14	15
16	17	18	19	20	21	22
23	24	25	26	27	28	29
30						

DECEMBER

S	M	T	W	T	F	S
	1	2	3	4	5	6
7	8	9	10	11	12	13
14	15	16	17	18	19	20
21	22	23	24	25	26	27
28	29	30	31			

5

JANUARY

S	M	T	W	T	F	S
				1	2	3
4	5	6	7	8	9	10
11	12	13	14	15	16	17
18	19	20	21	22	23	24
25	26	27	28	29	30	31

FEBRUARY

S	M	T	W	T	F	S
1	2	3	4	5	6	7
8	9	10	11	12	13	14
15	16	17	18	19	20	21
22	23	24	25	26	27	28

MARCH

S	M	T	W	T	F	S
1	2	3	4	5	6	7
8	9	10	11	12	13	14
15	16	17	18	19	20	21
22	23	24	25	26	27	28
29	30	31				

APRIL

S	M	T	W	T	F	S
			1	2	3	4
5	6	7	8	9	10	11
12	13	14	15	16	17	18
19	20	21	22	23	24	25
26	27	28	29	30		

MAY

S	M	T	W	T	F	S
					1	2
3	4	5	6	7	8	9
10	11	12	13	14	15	16
17	18	19	20	21	22	23
24	25	26	27	28	29	30
31						

JUNE

S	M	T	W	T	F	S
	1	2	3	4	5	6
7	8	9	10	11	12	13
14	15	16	17	18	19	20
21	22	23	24	25	26	27
28	29	30				

JULY

S	M	T	W	T	F	S
			1	2	3	4
5	6	7	8	9	10	11
12	13	14	15	16	17	18
19	20	21	22	23	24	25
26	27	28	29	30	31	

AUGUST

S	M	T	W	T	F	S
						1
2	3	4	5	6	7	8
9	10	11	12	13	14	15
16	17	18	19	20	21	22
23	24	25	26	27	28	29
30	31					

SEPTEMBER

S	M	T	W	T	F	S
		1	2	3	4	5
6	7	8	9	10	11	12
13	14	15	16	17	18	19
20	21	22	23	24	25	26
27	28	29	30			

OCTOBER

S	M	T	W	T	F	S
				1	2	3
4	5	6	7	8	9	10
11	12	13	14	15	16	17
18	19	20	21	22	23	24
25	26	27	28	29	30	31

NOVEMBER

S	M	T	W	T	F	S
1	2	3	4	5	6	7
8	9	10	11	12	13	14
15	16	17	18	19	20	21
22	23	24	25	26	27	28
29	30					

DECEMBER

S	M	T	W	T	F	S
		1	2	3	4	5
6	7	8	9	10	11	12
13	14	15	16	17	18	19
20	21	22	23	24	25	26
27	28	29	30	31		

6

JANUARY

S	M	T	W	T	F	S
					1	2
3	4	5	6	7	8	9
10	11	12	13	14	15	16
17	18	19	20	21	22	23
24	25	26	27	28	29	30
31						

FEBRUARY

S	M	T	W	T	F	S
	1	2	3	4	5	6
7	8	9	10	11	12	13
14	15	16	17	18	19	20
21	22	23	24	25	26	27
28						

MARCH

S	M	T	W	T	F	S
	1	2	3	4	5	6
7	8	9	10	11	12	13
14	15	16	17	18	19	20
21	22	23	24	25	26	27
28	29	30	31			

APRIL

S	M	T	W	T	F	S
				1	2	3
4	5	6	7	8	9	10
11	12	13	14	15	16	17
18	19	20	21	22	23	24
25	26	27	28	29	30	

MAY

S	M	T	W	T	F	S
						1
2	3	4	5	6	7	8
9	10	11	12	13	14	15
16	17	18	19	20	21	22
23	24	25	26	27	28	29
30	31					

JUNE

S	M	T	W	T	F	S
		1	2	3	4	5
6	7	8	9	10	11	12
13	14	15	16	17	18	19
20	21	22	23	24	25	26
27	28	29	30			

JULY

S	M	T	W	T	F	S
				1	2	3
4	5	6	7	8	9	10
11	12	13	14	15	16	17
18	19	20	21	22	23	24
25	26	27	28	29	30	31

AUGUST

S	M	T	W	T	F	S
1	2	3	4	5	6	7
8	9	10	11	12	13	14
15	16	17	18	19	20	21
22	23	24	25	26	27	28
29	30	31				

SEPTEMBER

S	M	T	W	T	F	S
			1	2	3	4
5	6	7	8	9	10	11
12	13	14	15	16	17	18
19	20	21	22	23	24	25
26	27	28	29	30		

OCTOBER

S	M	T	W	T	F	S
					1	2
3	4	5	6	7	8	9
10	11	12	13	14	15	16
17	18	19	20	21	22	23
24	25	26	27	28	29	30
31						

NOVEMBER

S	M	T	W	T	F	S
	1	2	3	4	5	6
7	8	9	10	11	12	13
14	15	16	17	18	19	20
21	22	23	24	25	26	27
28	29	30				

DECEMBER

S	M	T	W	T	F	S
			1	2	3	4
5	6	7	8	9	10	11
12	13	14	15	16	17	18
19	20	21	22	23	24	25
26	27	28	29	30	31	

(continued)

7 1983

JANUARY

S	M	T	W	T	F	S
						1
2	3	4	5	6	7	8
9	10	11	12	13	14	15
16	17	18	19	20	21	22
23	24	25	26	27	28	29
30	31					

FEBRUARY

S	M	T	W	T	F	S
		1	2	3	4	5
6	7	8	9	10	11	12
13	14	15	16	17	18	19
20	21	22	23	24	25	26
27	28					

MARCH

S	M	T	W	T	F	S
		1	2	3	4	5
6	7	8	9	10	11	12
13	14	15	16	17	18	19
20	21	22	23	24	25	26
27	28	29	30	31		

APRIL

S	M	T	W	T	F	S
					1	2
3	4	5	6	7	8	9
10	11	12	13	14	15	16
17	18	19	20	21	22	23
24	25	26	27	28	29	30

MAY

S	M	T	W	T	F	S
1	2	3	4	5	6	7
8	9	10	11	12	13	14
15	16	17	18	19	20	21
22	23	24	25	26	27	28
29	30	31				

JUNE

S	M	T	W	T	F	S
			1	2	3	4
5	6	7	8	9	10	11
12	13	14	15	16	17	18
19	20	21	22	23	24	25
26	27	28	29	30		

JULY

S	M	T	W	T	F	S
					1	2
3	4	5	6	7	8	9
10	11	12	13	14	15	16
17	18	19	20	21	22	23
24	25	26	27	28	29	30
31						

AUGUST

S	M	T	W	T	F	S
	1	2	3	4	5	6
7	8	9	10	11	12	13
14	15	16	17	18	19	20
21	22	23	24	25	26	27
28	29	30	31			

SEPTEMBER

S	M	T	W	T	F	S
				1	2	3
4	5	6	7	8	9	10
11	12	13	14	15	16	17
18	19	20	21	22	23	24
25	26	27	28	29	30	

OCTOBER

S	M	T	W	T	F	S
						1
2	3	4	5	6	7	8
9	10	11	12	13	14	15
16	17	18	19	20	21	22
23	24	25	26	27	28	29
30	31					

NOVEMBER

S	M	T	W	T	F	S
		1	2	3	4	5
6	7	8	9	10	11	12
13	14	15	16	17	18	19
20	21	22	23	24	25	26
27	28	29	30			

DECEMBER

S	M	T	W	T	F	S
				1	2	3
4	5	6	7	8	9	10
11	12	13	14	15	16	17
18	19	20	21	22	23	24
25	26	27	28	29	30	31

8 1984

JANUARY

S	M	T	W	T	F	S
1	2	3	4	5	6	7
8	9	10	11	12	13	14
15	16	17	18	19	20	21
22	23	24	25	26	27	28
29	30	31				

FEBRUARY

S	M	T	W	T	F	S
			1	2	3	4
5	6	7	8	9	10	11
12	13	14	15	16	17	18
19	20	21	22	23	24	25
26	27	28	29			

MARCH

S	M	T	W	T	F	S
				1	2	3
4	5	6	7	8	9	10
11	12	13	14	15	16	17
18	19	20	21	22	23	24
25	26	27	28	29	30	31

APRIL

S	M	T	W	T	F	S
1	2	3	4	5	6	7
8	9	10	11	12	13	14
15	16	17	18	19	20	21
22	23	24	25	26	27	28
29	30					

MAY

S	M	T	W	T	F	S
		1	2	3	4	5
6	7	8	9	10	11	12
13	14	15	16	17	18	19
20	21	22	23	24	25	26
27	28	29	30	31		

JUNE

S	M	T	W	T	F	S
					1	2
3	4	5	6	7	8	9
10	11	12	13	14	15	16
17	18	19	20	21	22	23
24	25	26	27	28	29	30

JULY

S	M	T	W	T	F	S
1	2	3	4	5	6	7
8	9	10	11	12	13	14
15	16	17	18	19	20	21
22	23	24	25	26	27	28
29	30	31				

AUGUST

S	M	T	W	T	F	S
			1	2	3	4
5	6	7	8	9	10	11
12	13	14	15	16	17	18
19	20	21	22	23	24	25
26	27	28	29	30	31	

SEPTEMBER

S	M	T	W	T	F	S
						1
2	3	4	5	6	7	8
9	10	11	12	13	14	15
16	17	18	19	20	21	22
23	24	25	26	27	28	29
30						

OCTOBER

S	M	T	W	T	F	S
	1	2	3	4	5	6
7	8	9	10	11	12	13
14	15	16	17	18	19	20
21	22	23	24	25	26	27
28	29	30	31			

NOVEMBER

S	M	T	W	T	F	S
				1	2	3
4	5	6	7	8	9	10
11	12	13	14	15	16	17
18	19	20	21	22	23	24
25	26	27	28	29	30	

DECEMBER

S	M	T	W	T	F	S
						1
2	3	4	5	6	7	8
9	10	11	12	13	14	15
16	17	18	19	20	21	22
23	24	25	26	27	28	29
30	31					

9

JANUARY

S	M	T	W	T	F	S
	1	2	3	4	5	6
7	8	9	10	11	12	13
14	15	16	17	18	19	20
21	22	23	24	25	26	27
28	29	30	31			

FEBRUARY

S	M	T	W	T	F	S
				1	2	3
4	5	6	7	8	9	10
11	12	13	14	15	16	17
18	19	20	21	22	23	24
25	26	27	28	29		

MARCH

S	M	T	W	T	F	S
					1	2
3	4	5	6	7	8	9
10	11	12	13	14	15	16
17	18	19	20	21	22	23
24	25	26	27	28	29	30
31						

APRIL

S	M	T	W	T	F	S
	1	2	3	4	5	6
7	8	9	10	11	12	13
14	15	16	17	18	19	20
21	22	23	24	25	26	27
28	29	30				

MAY

S	M	T	W	T	F	S
			1	2	3	4
5	6	7	8	9	10	11
12	13	14	15	16	17	18
19	20	21	22	23	24	25
26	27	28	29	30	31	

JUNE

S	M	T	W	T	F	S
						1
2	3	4	5	6	7	8
9	10	11	12	13	14	15
16	17	18	19	20	21	22
23	24	25	26	27	28	29
30						

JULY

S	M	T	W	T	F	S
	1	2	3	4	5	6
7	8	9	10	11	12	13
14	15	16	17	18	19	20
21	22	23	24	25	26	27
28	29	30	31			

AUGUST

S	M	T	W	T	F	S
				1	2	3
4	5	6	7	8	9	10
11	12	13	14	15	16	17
18	19	20	21	22	23	24
25	26	27	28	29	30	31

SEPTEMBER

S	M	T	W	T	F	S
1	2	3	4	5	6	7
8	9	10	11	12	13	14
15	16	17	18	19	20	21
22	23	24	25	26	27	28
29	30					

OCTOBER

S	M	T	W	T	F	S
		1	2	3	4	5
6	7	8	9	10	11	12
13	14	15	16	17	18	19
20	21	22	23	24	25	26
27	28	29	30	31		

NOVEMBER

S	M	T	W	T	F	S
					1	2
3	4	5	6	7	8	9
10	11	12	13	14	15	16
17	18	19	20	21	22	23
24	25	26	27	28	29	30

DECEMBER

S	M	T	W	T	F	S
1	2	3	4	5	6	7
8	9	10	11	12	13	14
15	16	17	18	19	20	21
22	23	24	25	26	27	28
29	30	31				

10

JANUARY

S	M	T	W	T	F	S
		1	2	3	4	5
6	7	8	9	10	11	12
13	14	15	16	17	18	19
20	21	22	23	24	25	26
27	28	29	30	31		

FEBRUARY

S	M	T	W	T	F	S
					1	2
3	4	5	6	7	8	9
10	11	12	13	14	15	16
17	18	19	20	21	22	23
24	25	26	27	28	29	

MARCH

S	M	T	W	T	F	S
						1
2	3	4	5	6	7	8
9	10	11	12	13	14	15
16	17	18	19	20	21	22
23	24	25	26	27	28	29
30	31					

APRIL

S	M	T	W	T	F	S
		1	2	3	4	5
6	7	8	9	10	11	12
13	14	15	16	17	18	19
20	21	22	23	24	25	26
27	28	29	30			

MAY

S	M	T	W	T	F	S
				1	2	3
4	5	6	7	8	9	10
11	12	13	14	15	16	17
18	19	20	21	22	23	24
25	26	27	28	29	30	31

JUNE

S	M	T	W	T	F	S
1	2	3	4	5	6	7
8	9	10	11	12	13	14
15	16	17	18	19	20	21
22	23	24	25	26	27	28
29	30					

JULY

S	M	T	W	T	F	S
		1	2	3	4	5
6	7	8	9	10	11	12
13	14	15	16	17	18	19
20	21	22	23	24	25	26
27	28	29	30	31		

AUGUST

S	M	T	W	T	F	S
					1	2
3	4	5	6	7	8	9
10	11	12	13	14	15	16
17	18	19	20	21	22	23
24	25	26	27	28	29	30
31						

SEPTEMBER

S	M	T	W	T	F	S
	1	2	3	4	5	6
7	8	9	10	11	12	13
14	15	16	17	18	19	20
21	22	23	24	25	26	27
28	29	30				

OCTOBER

S	M	T	W	T	F	S
			1	2	3	4
5	6	7	8	9	10	11
12	13	14	15	16	17	18
19	20	21	22	23	24	25
26	27	28	29	30	31	

NOVEMBER

S	M	T	W	T	F	S
						1
2	3	4	5	6	7	8
9	10	11	12	13	14	15
16	17	18	19	20	21	22
23	24	25	26	27	28	29
30						

DECEMBER

S	M	T	W	T	F	S
	1	2	3	4	5	6
7	8	9	10	11	12	13
14	15	16	17	18	19	20
21	22	23	24	25	26	27
28	29	30	31			

11

JANUARY

S	M	T	W	T	F	S
			1	2	3	4
5	6	7	8	9	10	11
12	13	14	15	16	17	18
19	20	21	22	23	24	25
26	27	28	29	30	31	

FEBRUARY

S	M	T	W	T	F	S
						1
2	3	4	5	6	7	8
9	10	11	12	13	14	15
16	17	18	19	20	21	22
23	24	25	26	27	28	29

MARCH

S	M	T	W	T	F	S
1	2	3	4	5	6	7
8	9	10	11	12	13	14
15	16	17	18	19	20	21
22	23	24	25	26	27	28
29	30	31				

APRIL

S	M	T	W	T	F	S
			1	2	3	4
5	6	7	8	9	10	11
12	13	14	15	16	17	18
19	20	21	22	23	24	25
26	27	28	29	30		

MAY

S	M	T	W	T	F	S
					1	2
3	4	5	6	7	8	9
10	11	12	13	14	15	16
17	18	19	20	21	22	23
24	25	26	27	28	29	30
31						

JUNE

S	M	T	W	T	F	S
	1	2	3	4	5	6
7	8	9	10	11	12	13
14	15	16	17	18	19	20
21	22	23	24	25	26	27
28	29	30				

JULY

S	M	T	W	T	F	S
			1	2	3	4
5	6	7	8	9	10	11
12	13	14	15	16	17	18
19	20	21	22	23	24	25
26	27	28	29	30	31	

AUGUST

S	M	T	W	T	F	S
						1
2	3	4	5	6	7	8
9	10	11	12	13	14	15
16	17	18	19	20	21	22
23	24	25	26	27	28	29
30	31					

SEPTEMBER

S	M	T	W	T	F	S
		1	2	3	4	5
6	7	8	9	10	11	12
13	14	15	16	17	18	19
20	21	22	23	24	25	26
27	28	29	30			

OCTOBER

S	M	T	W	T	F	S
				1	2	3
4	5	6	7	8	9	10
11	12	13	14	15	16	17
18	19	20	21	22	23	24
25	26	27	28	29	30	31

NOVEMBER

S	M	T	W	T	F	S
1	2	3	4	5	6	7
8	9	10	11	12	13	14
15	16	17	18	19	20	21
22	23	24	25	26	27	28
29	30					

DECEMBER

S	M	T	W	T	F	S
		1	2	3	4	5
6	7	8	9	10	11	12
13	14	15	16	17	18	19
20	21	22	23	24	25	26
27	28	29	30	31		

12

JANUARY

S	M	T	W	T	F	S
				1	2	3
4	5	6	7	8	9	10
11	12	13	14	15	16	17
18	19	20	21	22	23	24
25	26	27	28	29	30	31

FEBRUARY

S	M	T	W	T	F	S
1	2	3	4	5	6	7
8	9	10	11	12	13	14
15	16	17	18	19	20	21
22	23	24	25	26	27	28
29						

MARCH

S	M	T	W	T	F	S
	1	2	3	4	5	6
7	8	9	10	11	12	13
14	15	16	17	18	19	20
21	22	23	24	25	26	27
28	29	30	31			

APRIL

S	M	T	W	T	F	S
				1	2	3
4	5	6	7	8	9	10
11	12	13	14	15	16	17
18	19	20	21	22	23	24
25	26	27	28	29	30	

MAY

S	M	T	W	T	F	S
						1
2	3	4	5	6	7	8
9	10	11	12	13	14	15
16	17	18	19	20	21	22
23	24	25	26	27	28	29
30	31					

JUNE

S	M	T	W	T	F	S
		1	2	3	4	5
6	7	8	9	10	11	12
13	14	15	16	17	18	19
20	21	22	23	24	25	26
27	28	29	30			

JULY

S	M	T	W	T	F	S
				1	2	3
4	5	6	7	8	9	10
11	12	13	14	15	16	17
18	19	20	21	22	23	24
25	26	27	28	29	30	31

AUGUST

S	M	T	W	T	F	S
1	2	3	4	5	6	7
8	9	10	11	12	13	14
15	16	17	18	19	20	21
22	23	24	25	26	27	28
29	30	31				

SEPTEMBER

S	M	T	W	T	F	S
			1	2	3	4
5	6	7	8	9	10	11
12	13	14	15	16	17	18
19	20	21	22	23	24	25
26	27	28	29	30		

OCTOBER

S	M	T	W	T	F	S
					1	2
3	4	5	6	7	8	9
10	11	12	13	14	15	16
17	18	19	20	21	22	23
24	25	26	27	28	29	30
31						

NOVEMBER

S	M	T	W	T	F	S
	1	2	3	4	5	6
7	8	9	10	11	12	13
14	15	16	17	18	19	20
21	22	23	24	25	26	27
28	29	30				

DECEMBER

S	M	T	W	T	F	S
			1	2	3	4
5	6	7	8	9	10	11
12	13	14	15	16	17	18
19	20	21	22	23	24	25
26	27	28	29	30	31	

13

JANUARY

S	M	T	W	T	F	S
					1	2
3	4	5	6	7	8	9
10	11	12	13	14	15	16
17	18	19	20	21	22	23
24	25	26	27	28	29	30
31						

FEBRUARY

S	M	T	W	T	F	S
	1	2	3	4	5	6
7	8	9	10	11	12	13
14	15	16	17	18	19	20
21	22	23	24	25	26	27
28	29					

MARCH

S	M	T	W	T	F	S
		1	2	3	4	5
6	7	8	9	10	11	12
13	14	15	16	17	18	19
20	21	22	23	24	25	26
27	28	29	30	31		

APRIL

S	M	T	W	T	F	S
					1	2
3	4	5	6	7	8	9
10	11	12	13	14	15	16
17	18	19	20	21	22	23
24	25	26	27	28	29	30

MAY

S	M	T	W	T	F	S
1	2	3	4	5	6	7
8	9	10	11	12	13	14
15	16	17	18	19	20	21
22	23	24	25	26	27	28
29	30	31				

JUNE

S	M	T	W	T	F	S
			1	2	3	4
5	6	7	8	9	10	11
12	13	14	15	16	17	18
19	20	21	22	23	24	25
26	27	28	29	30		

JULY

S	M	T	W	T	F	S
					1	2
3	4	5	6	7	8	9
10	11	12	13	14	15	16
17	18	19	20	21	22	23
24	25	26	27	28	29	30
31						

AUGUST

S	M	T	W	T	F	S
	1	2	3	4	5	6
7	8	9	10	11	12	13
14	15	16	17	18	19	20
21	22	23	24	25	26	27
28	29	30	31			

SEPTEMBER

S	M	T	W	T	F	S
				1	2	3
4	5	6	7	8	9	10
11	12	13	14	15	16	17
18	19	20	21	22	23	24
25	26	27	28	29	30	

OCTOBER

S	M	T	W	T	F	S
						1
2	3	4	5	6	7	8
9	10	11	12	13	14	15
16	17	18	19	20	21	22
23	24	25	26	27	28	29
30	31					

NOVEMBER

S	M	T	W	T	F	S
		1	2	3	4	5
6	7	8	9	10	11	12
13	14	15	16	17	18	19
20	21	22	23	24	25	26
27	28	29	30			

DECEMBER

S	M	T	W	T	F	S
				1	2	3
4	5	6	7	8	9	10
11	12	13	14	15	16	17
18	19	20	21	22	23	24
25	26	27	28	29	30	31

14

JANUARY

S	M	T	W	T	F	S
						1
2	3	4	5	6	7	8
9	10	11	12	13	14	15
16	17	18	19	20	21	22
23	24	25	26	27	28	29
30	31					

FEBRUARY

S	M	T	W	T	F	S
		1	2	3	4	5
6	7	8	9	10	11	12
13	14	15	16	17	18	19
20	21	22	23	24	25	26
27	28	29				

MARCH

S	M	T	W	T	F	S
			1	2	3	4
5	6	7	8	9	10	11
12	13	14	15	16	17	18
19	20	21	22	23	24	25
26	27	28	29	30	31	

APRIL

S	M	T	W	T	F	S
						1
2	3	4	5	6	7	8
9	10	11	12	13	14	15
16	17	18	19	20	21	22
23	24	25	26	27	28	29
30						

MAY

S	M	T	W	T	F	S
	1	2	3	4	5	6
7	8	9	10	11	12	13
14	15	16	17	18	19	20
21	22	23	24	25	26	27
28	29	30	31			

JUNE

S	M	T	W	T	F	S
				1	2	3
4	5	6	7	8	9	10
11	12	13	14	15	16	17
18	19	20	21	22	23	24
25	26	27	28	29	30	

JULY

S	M	T	W	T	F	S
						1
2	3	4	5	6	7	8
9	10	11	12	13	14	15
16	17	18	19	20	21	22
23	24	25	26	27	28	29
30	31					

AUGUST

S	M	T	W	T	F	S
		1	2	3	4	5
6	7	8	9	10	11	12
13	14	15	16	17	18	19
20	21	22	23	24	25	26
27	28	29	30	31		

SEPTEMBER

S	M	T	W	T	F	S
					1	2
3	4	5	6	7	8	9
10	11	12	13	14	15	16
17	18	19	20	21	22	23
24	25	26	27	28	29	30

OCTOBER

S	M	T	W	T	F	S
1	2	3	4	5	6	7
8	9	10	11	12	13	14
15	16	17	18	19	20	21
22	23	24	25	26	27	28
29	30	31				

NOVEMBER

S	M	T	W	T	F	S
			1	2	3	4
5	6	7	8	9	10	11
12	13	14	15	16	17	18
19	20	21	22	23	24	25
26	27	28	29	30		

DECEMBER

S	M	T	W	T	F	S
					1	2
3	4	5	6	7	8	9
10	11	12	13	14	15	16
17	18	19	20	21	22	23
24	25	26	27	28	29	30
31						

(continued)

7 — 1983

JANUARY

S	M	T	W	T	F	S
						1
2	3	4	5	6	7	8
9	10	11	12	13	14	15
16	17	18	19	20	21	22
23	24	25	26	27	28	29
30	31					

FEBRUARY

S	M	T	W	T	F	S
		1	2	3	4	5
6	7	8	9	10	11	12
13	14	15	16	17	18	19
20	21	22	23	24	25	26
27	28					

MARCH

S	M	T	W	T	F	S
		1	2	3	4	5
6	7	8	9	10	11	12
13	14	15	16	17	18	19
20	21	22	23	24	25	26
27	28	29	30	31		

APRIL

S	M	T	W	T	F	S
					1	2
3	4	5	6	7	8	9
10	11	12	13	14	15	16
17	18	19	20	21	22	23
24	25	26	27	28	29	30

MAY

S	M	T	W	T	F	S
1	2	3	4	5	6	7
8	9	10	11	12	13	14
15	16	17	18	19	20	21
22	23	24	25	26	27	28
29	30	31				

JUNE

S	M	T	W	T	F	S
			1	2	3	4
5	6	7	8	9	10	11
12	13	14	15	16	17	18
19	20	21	22	23	24	25
26	27	28	29	30		

JULY

S	M	T	W	T	F	S
					1	2
3	4	5	6	7	8	9
10	11	12	13	14	15	16
17	18	19	20	21	22	23
24	25	26	27	28	29	30
31						

AUGUST

S	M	T	W	T	F	S
	1	2	3	4	5	6
7	8	9	10	11	12	13
14	15	16	17	18	19	20
21	22	23	24	25	26	27
28	29	30	31			

SEPTEMBER

S	M	T	W	T	F	S
				1	2	3
4	5	6	7	8	9	10
11	12	13	14	15	16	17
18	19	20	21	22	23	24
25	26	27	28	29	30	

OCTOBER

S	M	T	W	T	F	S
						1
2	3	4	5	6	7	8
9	10	11	12	13	14	15
16	17	18	19	20	21	22
23	24	25	26	27	28	29
30	31					

NOVEMBER

S	M	T	W	T	F	S
		1	2	3	4	5
6	7	8	9	10	11	12
13	14	15	16	17	18	19
20	21	22	23	24	25	26
27	28	29	30			

DECEMBER

S	M	T	W	T	F	S
				1	2	3
4	5	6	7	8	9	10
11	12	13	14	15	16	17
18	19	20	21	22	23	24
25	26	27	28	29	30	31

8 — 1984

JANUARY

S	M	T	W	T	F	S
1	2	3	4	5	6	7
8	9	10	11	12	13	14
15	16	17	18	19	20	21
22	23	24	25	26	27	28
29	30	31				

FEBRUARY

S	M	T	W	T	F	S
			1	2	3	4
5	6	7	8	9	10	11
12	13	14	15	16	17	18
19	20	21	22	23	24	25
26	27	28	29			

MARCH

S	M	T	W	T	F	S
				1	2	3
4	5	6	7	8	9	10
11	12	13	14	15	16	17
18	19	20	21	22	23	24
25	26	27	28	29	30	31

APRIL

S	M	T	W	T	F	S
1	2	3	4	5	6	7
8	9	10	11	12	13	14
15	16	17	18	19	20	21
22	23	24	25	26	27	28
29	30					

MAY

S	M	T	W	T	F	S
		1	2	3	4	5
6	7	8	9	10	11	12
13	14	15	16	17	18	19
20	21	22	23	24	25	26
27	28	29	30	31		

JUNE

S	M	T	W	T	F	S
					1	2
3	4	5	6	7	8	9
10	11	12	13	14	15	16
17	18	19	20	21	22	23
24	25	26	27	28	29	30

JULY

S	M	T	W	T	F	S
1	2	3	4	5	6	7
8	9	10	11	12	13	14
15	16	17	18	19	20	21
22	23	24	25	26	27	28
29	30	31				

AUGUST

S	M	T	W	T	F	S
			1	2	3	4
5	6	7	8	9	10	11
12	13	14	15	16	17	18
19	20	21	22	23	24	25
26	27	28	29	30	31	

SEPTEMBER

S	M	T	W	T	F	S
						1
2	3	4	5	6	7	8
9	10	11	12	13	14	15
16	17	18	19	20	21	22
23	24	25	26	27	28	29
30						

OCTOBER

S	M	T	W	T	F	S
	1	2	3	4	5	6
7	8	9	10	11	12	13
14	15	16	17	18	19	20
21	22	23	24	25	26	27
28	29	30	31			

NOVEMBER

S	M	T	W	T	F	S
				1	2	3
4	5	6	7	8	9	10
11	12	13	14	15	16	17
18	19	20	21	22	23	24
25	26	27	28	29	30	

DECEMBER

S	M	T	W	T	F	S
						1
2	3	4	5	6	7	8
9	10	11	12	13	14	15
16	17	18	19	20	21	22
23	24	25	26	27	28	29
30	31					

9

JANUARY

S	M	T	W	T	F	S
	1	2	3	4	5	6
7	8	9	10	11	12	13
14	15	16	17	18	19	20
21	22	23	24	25	26	27
28	29	30	31			

FEBRUARY

S	M	T	W	T	F	S
				1	2	3
4	5	6	7	8	9	10
11	12	13	14	15	16	17
18	19	20	21	22	23	24
25	26	27	28	29		

MARCH

S	M	T	W	T	F	S
					1	2
3	4	5	6	7	8	9
10	11	12	13	14	15	16
17	18	19	20	21	22	23
24	25	26	27	28	29	30
31						

APRIL

S	M	T	W	T	F	S
	1	2	3	4	5	6
7	8	9	10	11	12	13
14	15	16	17	18	19	20
21	22	23	24	25	26	27
28	29	30				

MAY

S	M	T	W	T	F	S
			1	2	3	4
5	6	7	8	9	10	11
12	13	14	15	16	17	18
19	20	21	22	23	24	25
26	27	28	29	30	31	

JUNE

S	M	T	W	T	F	S
						1
2	3	4	5	6	7	8
9	10	11	12	13	14	15
16	17	18	19	20	21	22
23	24	25	26	27	28	29
30						

JULY

S	M	T	W	T	F	S
	1	2	3	4	5	6
7	8	9	10	11	12	13
14	15	16	17	18	19	20
21	22	23	24	25	26	27
28	29	30	31			

AUGUST

S	M	T	W	T	F	S
				1	2	3
4	5	6	7	8	9	10
11	12	13	14	15	16	17
18	19	20	21	22	23	24
25	26	27	28	29	30	31

SEPTEMBER

S	M	T	W	T	F	S
1	2	3	4	5	6	7
8	9	10	11	12	13	14
15	16	17	18	19	20	21
22	23	24	25	26	27	28
29	30					

OCTOBER

S	M	T	W	T	F	S
		1	2	3	4	5
6	7	8	9	10	11	12
13	14	15	16	17	18	19
20	21	22	23	24	25	26
27	28	29	30	31		

NOVEMBER

S	M	T	W	T	F	S
					1	2
3	4	5	6	7	8	9
10	11	12	13	14	15	16
17	18	19	20	21	22	23
24	25	26	27	28	29	30

DECEMBER

S	M	T	W	T	F	S
1	2	3	4	5	6	7
8	9	10	11	12	13	14
15	16	17	18	19	20	21
22	23	24	25	26	27	28
29	30	31				

10

JANUARY

S	M	T	W	T	F	S
		1	2	3	4	5
6	7	8	9	10	11	12
13	14	15	16	17	18	19
20	21	22	23	24	25	26
27	28	29	30	31		

FEBRUARY

S	M	T	W	T	F	S
					1	2
3	4	5	6	7	8	9
10	11	12	13	14	15	16
17	18	19	20	21	22	23
24	25	26	27	28	29	

MARCH

S	M	T	W	T	F	S
						1
2	3	4	5	6	7	8
9	10	11	12	13	14	15
16	17	18	19	20	21	22
23	24	25	26	27	28	29
30	31					

APRIL

S	M	T	W	T	F	S
		1	2	3	4	5
6	7	8	9	10	11	12
13	14	15	16	17	18	19
20	21	22	23	24	25	26
27	28	29	30			

MAY

S	M	T	W	T	F	S
				1	2	3
4	5	6	7	8	9	10
11	12	13	14	15	16	17
18	19	20	21	22	23	24
25	26	27	28	29	30	31

JUNE

S	M	T	W	T	F	S
1	2	3	4	5	6	7
8	9	10	11	12	13	14
15	16	17	18	19	20	21
22	23	24	25	26	27	28
29	30					

JULY

S	M	T	W	T	F	S
		1	2	3	4	5
6	7	8	9	10	11	12
13	14	15	16	17	18	19
20	21	22	23	24	25	26
27	28	29	30	31		

AUGUST

S	M	T	W	T	F	S
					1	2
3	4	5	6	7	8	9
10	11	12	13	14	15	16
17	18	19	20	21	22	23
24	25	26	27	28	29	30
31						

SEPTEMBER

S	M	T	W	T	F	S
	1	2	3	4	5	6
7	8	9	10	11	12	13
14	15	16	17	18	19	20
21	22	23	24	25	26	27
28	29	30				

OCTOBER

S	M	T	W	T	F	S
			1	2	3	4
5	6	7	8	9	10	11
12	13	14	15	16	17	18
19	20	21	22	23	24	25
26	27	28	29	30	31	

NOVEMBER

S	M	T	W	T	F	S
						1
2	3	4	5	6	7	8
9	10	11	12	13	14	15
16	17	18	19	20	21	22
23	24	25	26	27	28	29
30						

DECEMBER

S	M	T	W	T	F	S
	1	2	3	4	5	6
7	8	9	10	11	12	13
14	15	16	17	18	19	20
21	22	23	24	25	26	27
28	29	30	31			

11

JANUARY

S	M	T	W	T	F	S
			1	2	3	4
5	6	7	8	9	10	11
12	13	14	15	16	17	18
19	20	21	22	23	24	25
26	27	28	29	30	31	

FEBRUARY

S	M	T	W	T	F	S
						1
2	3	4	5	6	7	8
9	10	11	12	13	14	15
16	17	18	19	20	21	22
23	24	25	26	27	28	29

MARCH

S	M	T	W	T	F	S
1	2	3	4	5	6	7
8	9	10	11	12	13	14
15	16	17	18	19	20	21
22	23	24	25	26	27	28
29	30	31				

APRIL

S	M	T	W	T	F	S
			1	2	3	4
5	6	7	8	9	10	11
12	13	14	15	16	17	18
19	20	21	22	23	24	25
26	27	28	29	30		

MAY

S	M	T	W	T	F	S
					1	2
3	4	5	6	7	8	9
10	11	12	13	14	15	16
17	18	19	20	21	22	23
24	25	26	27	28	29	30
31						

JUNE

S	M	T	W	T	F	S
	1	2	3	4	5	6
7	8	9	10	11	12	13
14	15	16	17	18	19	20
21	22	23	24	25	26	27
28	29	30				

JULY

S	M	T	W	T	F	S
			1	2	3	4
5	6	7	8	9	10	11
12	13	14	15	16	17	18
19	20	21	22	23	24	25
26	27	28	29	30	31	

AUGUST

S	M	T	W	T	F	S
						1
2	3	4	5	6	7	8
9	10	11	12	13	14	15
16	17	18	19	20	21	22
23	24	25	26	27	28	29
30	31					

SEPTEMBER

S	M	T	W	T	F	S
		1	2	3	4	5
6	7	8	9	10	11	12
13	14	15	16	17	18	19
20	21	22	23	24	25	26
27	28	29	30			

OCTOBER

S	M	T	W	T	F	S
				1	2	3
4	5	6	7	8	9	10
11	12	13	14	15	16	17
18	19	20	21	22	23	24
25	26	27	28	29	30	31

NOVEMBER

S	M	T	W	T	F	S
1	2	3	4	5	6	7
8	9	10	11	12	13	14
15	16	17	18	19	20	21
22	23	24	25	26	27	28
29	30					

DECEMBER

S	M	T	W	T	F	S
		1	2	3	4	5
6	7	8	9	10	11	12
13	14	15	16	17	18	19
20	21	22	23	24	25	26
27	28	29	30	31		

12

JANUARY

S	M	T	W	T	F	S
				1	2	3
4	5	6	7	8	9	10
11	12	13	14	15	16	17
18	19	20	21	22	23	24
25	26	27	28	29	30	31

FEBRUARY

S	M	T	W	T	F	S
1	2	3	4	5	6	7
8	9	10	11	12	13	14
15	16	17	18	19	20	21
22	23	24	25	26	27	28
29						

MARCH

S	M	T	W	T	F	S
	1	2	3	4	5	6
7	8	9	10	11	12	13
14	15	16	17	18	19	20
21	22	23	24	25	26	27
28	29	30	31			

APRIL

S	M	T	W	T	F	S
				1	2	3
4	5	6	7	8	9	10
11	12	13	14	15	16	17
18	19	20	21	22	23	24
25	26	27	28	29	30	

MAY

S	M	T	W	T	F	S
						1
2	3	4	5	6	7	8
9	10	11	12	13	14	15
16	17	18	19	20	21	22
23	24	25	26	27	28	29
30	31					

JUNE

S	M	T	W	T	F	S
		1	2	3	4	5
6	7	8	9	10	11	12
13	14	15	16	17	18	19
20	21	22	23	24	25	26
27	28	29	30			

JULY

S	M	T	W	T	F	S
				1	2	3
4	5	6	7	8	9	10
11	12	13	14	15	16	17
18	19	20	21	22	23	24
25	26	27	28	29	30	31

AUGUST

S	M	T	W	T	F	S
1	2	3	4	5	6	7
8	9	10	11	12	13	14
15	16	17	18	19	20	21
22	23	24	25	26	27	28
29	30	31				

SEPTEMBER

S	M	T	W	T	F	S
			1	2	3	4
5	6	7	8	9	10	11
12	13	14	15	16	17	18
19	20	21	22	23	24	25
26	27	28	29	30		

OCTOBER

S	M	T	W	T	F	S
					1	2
3	4	5	6	7	8	9
10	11	12	13	14	15	16
17	18	19	20	21	22	23
24	25	26	27	28	29	30
31						

NOVEMBER

S	M	T	W	T	F	S
	1	2	3	4	5	6
7	8	9	10	11	12	13
14	15	16	17	18	19	20
21	22	23	24	25	26	27
28	29	30				

DECEMBER

S	M	T	W	T	F	S
			1	2	3	4
5	6	7	8	9	10	11
12	13	14	15	16	17	18
19	20	21	22	23	24	25
26	27	28	29	30	31	

13

JANUARY

S	M	T	W	T	F	S
					1	2
3	4	5	6	7	8	9
10	11	12	13	14	15	16
17	18	19	20	21	22	23
24	25	26	27	28	29	30
31						

FEBRUARY

S	M	T	W	T	F	S
	1	2	3	4	5	6
7	8	9	10	11	12	13
14	15	16	17	18	19	20
21	22	23	24	25	26	27
28	29					

MARCH

S	M	T	W	T	F	S
		1	2	3	4	5
6	7	8	9	10	11	12
13	14	15	16	17	18	19
20	21	22	23	24	25	26
27	28	29	30	31		

APRIL

S	M	T	W	T	F	S
					1	2
3	4	5	6	7	8	9
10	11	12	13	14	15	16
17	18	19	20	21	22	23
24	25	26	27	28	29	30

MAY

S	M	T	W	T	F	S
1	2	3	4	5	6	7
8	9	10	11	12	13	14
15	16	17	18	19	20	21
22	23	24	25	26	27	28
29	30	31				

JUNE

S	M	T	W	T	F	S
			1	2	3	4
5	6	7	8	9	10	11
12	13	14	15	16	17	18
19	20	21	22	23	24	25
26	27	28	29	30		

JULY

S	M	T	W	T	F	S
					1	2
3	4	5	6	7	8	9
10	11	12	13	14	15	16
17	18	19	20	21	22	23
24	25	26	27	28	29	30
31						

AUGUST

S	M	T	W	T	F	S
	1	2	3	4	5	6
7	8	9	10	11	12	13
14	15	16	17	18	19	20
21	22	23	24	25	26	27
28	29	30	31			

SEPTEMBER

S	M	T	W	T	F	S
				1	2	3
4	5	6	7	8	9	10
11	12	13	14	15	16	17
18	19	20	21	22	23	24
25	26	27	28	29	30	

OCTOBER

S	M	T	W	T	F	S
						1
2	3	4	5	6	7	8
9	10	11	12	13	14	15
16	17	18	19	20	21	22
23	24	25	26	27	28	29
30	31					

NOVEMBER

S	M	T	W	T	F	S
		1	2	3	4	5
6	7	8	9	10	11	12
13	14	15	16	17	18	19
20	21	22	23	24	25	26
27	28	29	30			

DECEMBER

S	M	T	W	T	F	S
				1	2	3
4	5	6	7	8	9	10
11	12	13	14	15	16	17
18	19	20	21	22	23	24
25	26	27	28	29	30	31

14

JANUARY

S	M	T	W	T	F	S
						1
2	3	4	5	6	7	8
9	10	11	12	13	14	15
16	17	18	19	20	21	22
23	24	25	26	27	28	29
30	31					

FEBRUARY

S	M	T	W	T	F	S
		1	2	3	4	5
6	7	8	9	10	11	12
13	14	15	16	17	18	19
20	21	22	23	24	25	26
27	28	29				

MARCH

S	M	T	W	T	F	S
			1	2	3	4
5	6	7	8	9	10	11
12	13	14	15	16	17	18
19	20	21	22	23	24	25
26	27	28	29	30	31	

APRIL

S	M	T	W	T	F	S
						1
2	3	4	5	6	7	8
9	10	11	12	13	14	15
16	17	18	19	20	21	22
23	24	25	26	27	28	29
30						

MAY

S	M	T	W	T	F	S
	1	2	3	4	5	6
7	8	9	10	11	12	13
14	15	16	17	18	19	20
21	22	23	24	25	26	27
28	29	30	31			

JUNE

S	M	T	W	T	F	S
				1	2	3
4	5	6	7	8	9	10
11	12	13	14	15	16	17
18	19	20	21	22	23	24
25	26	27	28	29	30	

JULY

S	M	T	W	T	F	S
						1
2	3	4	5	6	7	8
9	10	11	12	13	14	15
16	17	18	19	20	21	22
23	24	25	26	27	28	29
30	31					

AUGUST

S	M	T	W	T	F	S
		1	2	3	4	5
6	7	8	9	10	11	12
13	14	15	16	17	18	19
20	21	22	23	24	25	26
27	28	29	30	31		

SEPTEMBER

S	M	T	W	T	F	S
					1	2
3	4	5	6	7	8	9
10	11	12	13	14	15	16
17	18	19	20	21	22	23
24	25	26	27	28	29	30

OCTOBER

S	M	T	W	T	F	S
1	2	3	4	5	6	7
8	9	10	11	12	13	14
15	16	17	18	19	20	21
22	23	24	25	26	27	28
29	30	31				

NOVEMBER

S	M	T	W	T	F	S
			1	2	3	4
5	6	7	8	9	10	11
12	13	14	15	16	17	18
19	20	21	22	23	24	25
26	27	28	29	30		

DECEMBER

S	M	T	W	T	F	S
					1	2
3	4	5	6	7	8	9
10	11	12	13	14	15	16
17	18	19	20	21	22	23
24	25	26	27	28	29	30
31						

The Planets and the Solar System

Planet	Mean daily motion ″	Orbital velocity miles per sec.	Sidereal revolution days	Synodical revolution days	Dist. from sun in millions of mi. Max.	Min.	Dist. from Earth in millions of mi. Max.	Min.	Light at[1] perihelion	aphelion
Mercury . .	14732	29.75	88.0	115.9	43.4	28.6	136	50	10.58	4.59
Venus . . .	5768	21.76	224.7	583.9	67.7	66.8	161	25	1.94	1.89
Earth	3548	18.51	365.3	—	94.6	91.4	—	—	1.03	0.97
Mars	1887	14.99	687.0	779.9	155.0	128.5	248	35	0.524	0.360
Jupiter . . .	299	8.12	4332.1	398.9	507.0	460.6	600	368	0.0408	0.0336
Saturn . . .	120	5.99	10825.9	378.1	937.5	838.4	1031	745	0.01230	0.00984
Uranus . . .	42	4.23	30676.1	369.7	1859.7	1669.3	1953	1606	0.00300	0.00250
Neptune . .	21	3.38	59911.1	367.5	2821.7	2760.4	2915	2667	0.00114	0.00109
Pluto	14	2.95	90824.2	366.7	4551.4	2756.4	4644	2663	0.00114	0.00042

Light at perihelion and aphelion is solar illumination in units of mean illumination at Earth.

Planet	Mean longitude of:* ascending node °	′	″	perihelion °	′	″	Inclination* of orbit to ecliptic °	′	″	Mean* distance**	Eccentricity* of orbit	Mean longitude at the epoch* °	′	″
Mercury. . . .	48	08	17	77	12	04	7	01	22	0.387101	0.205638	45	06	51
Venus.	76	31	55	131	19	12	3	23	40	0.723335	0.006776	17	53	08
Earth		—	—	102	40	08				0.999997	0.016762	1	17	09
Mars	49	26	06	335	48	22	1	51	00	1.523662	0.093370	120	06	08
Jupiter	100	19	34	15	18	14	1	18	21	5.20303	0.048028	260	13	46
Saturn.	113	31	08	94	38	10	2	29	10	9.57255	0.051007	260	56	07
Uranus	73	59	20	177	24	04	0	46	20	19.3143	0.048926	243	16	55
Neptune . . .	131	35	28	354	02	24	1	46	23	30.2779	0.005172	269	00	18
Pluto	110	10	41	223	59	02	17	08	11	39.8282	0.255556	215	16	35

*Consistent for the standard Epoch: 1983 Sept. 23.0 Ephemeris Time **Astronomical units

Sun and planets	Semi-diameter at unit distance ′ ″	at mean least dist. ′ ″	in miles mean s.d.	Volume ⊕ = 1.	Mass. ⊕ = 1.	Density ⊕ = 1.	Axial rotation d.	h.	m.	s.	Gravity at surface ⊕ = 1.	Reflecting power Pct.	Probable temperature °F.
Sun.	15 59.62	___	432560	1303730	332830	0.26	24	16	48		27.9	___	+ 10,000
Mercury . . .	3.37	5.45	1505	0.054	0.0554	0.98	59				0.37	0.06	+ 620
Venus	8.46	30.50	3762	0.880	0.8150	0.94	244.3	(R)			0.88	0.72	+ 900
Earth	___	___	3960	1.000	1.000	1.00		23	56	4	1.00	0.39	+ 72
Moon	2.40	16 43.00	1080	0.020	0.0123	0.61	27	7	43	12	0.17	0.07	− 10
Mars	4.68	8.94	2107	0.149	0.1075	0.72		24	37	23	0.38	0.16	− 10
Jupiter	98.37	23.43	44270	1316.	317.84	0.24		9	50	30	2.64	0.70	− 240
Saturn	82.80	9.76	37300	755.	95.147	0.13		10	14		1.15	0.75	− 300
Uranus	32.90	1.80	15200	52.	14.54	0.29		10	49	(R)	1.15	0.90	− 340
Neptune . . .	31.10	1.06	15600	57.	17.23	0.30		18	12		1.12	0.82	− 370
Pluto*.	1.80	0.06	800	0.008	0.0016	0.19	6	9			0.04	0.14	? ?

*Much of this information is too new to be verified, but observers at the U.S. Naval Observatory have derived values similar to these after having discovered that Pluto has a satellite. It apparently revolves about Pluto in a period equal to Pluto's rotation period. (R) retrograde of Venus and Uranus.

Object	Average distance from sun (astronomical units)[a]	Radius (Earth = 1)[b]	Mass (Earth = 1)[b]	Density (g/cc) (water = 1)	Atmospheric constituents	Surface (or interior) pressure (bars)	Surface (or cloud top) temperature (K)	Number of satellites
Mercury	0.38	0.38	0.05	5.4	Helium, argon, neon, hydrogen	$>10^{-12}$	600—700 (day); ~95 (night)	0
Venus	0.72	0.96	0.82	5.2	Carbon dioxide (96%), nitrogen (~3%), helium, sulfur dioxide, water vapor, argon, neon, carbon monoxide, oxygen, sulfuric acid, hydrogen chloride, hydrogen fluoride, hydrogen sulfide	90	~730	0
Earth	1.00	1.00	1.00	5.5	Nitrogen (78%), oxygen (21%), argon (0.9%), water vapor, carbon dioxide, neon, methane, krypton, helium, xenon, hydrogen, nitrous oxide, carbon monoxide, nitrogen dioxide, sulfur dioxide, ozone	1.0	250—310	1
Mars	1.52	0.53	0.11	4.0	Carbon dioxide (95%), nitrogen (3%), argon (~2%), water vapor, oxygen, neon, carbon monoxide, krypton, xenon, ozone	0.005 to 0.008	~200—245	2
Jupiter	5.20	10.8	318	1.3	Hydrogen (89%), helium (11%), methane, ammonia, water, ethane, acetylene, phosphine, germanium tetrahydride, hydrogen cyanide?, ammonium hydrosulfide	$(<10^{6})$	(~140)	16
Saturn	9.5	9.0	95	0.7	Hydrogen (94%), helium (6%), methane, ammonia, ethane, phosphine, acetylene, propane?, methylacetylene?	$(>10^{6})$	(~100?)	17
Uranus	19.2	4.1	15	1.2	Hydrogen (<90%), methane, helium?	$(>10^{6})$	(~50—60)	5
Neptune	30.0	3.85	17	1.7	Hydrogen (<90%), methane, helium?	$(>10^{6})$	(~50—60)	2
Pluto	39.5	~0.22 to 0.25?	~0.002?	<2	Methane, argon?, neon?	$\sim10^{-3}$ to 10^{-4}?	~50—60	1
Moon	—	0.27	0.0123	3.3	Neon, argon, helium	$\sim2 \times 10^{-14}$	370 (day); 120 (night)	—

[a] Astronomical unit defined as average distance between the centers of the sun and Earth (1.496×10^{8} km)).
[b] Compared to Earth as unity: Earth radius is 6371 km; mass is 5.975×10^{24} kg.

CHARACTERISTICS OF THE EARTH

(NASA TT-F-533)

Quantity	Unit of Measurement	Symbol	Numerical Value
Mass	Proportion of the mass of the sun	M	1/331950
	gram		$5.9763 \cdot 10^{27}$
Major Orbital semi-axis	Astronomical unit	a_{orb}	1.000000
	km		149,457,000
Distance from sun at perihelion	a.u.	r_π	0.983298
Distance from sun at aphelion	a.u.	r_α	1.016744
Moment of perihelion passage		T_π	Jan. 2, $4^{hr}52^{m}$
Moment of aphelion passage		T_α	Jul 4 $5^{hr}05^{m}$
Siderial rotation period around sun	sec	P_{orb}	$31.558 \cdot 10^{6}$
Mean rotational velocity	km/sec	U_{orb}	29.8
Mean equatorial radius	km	$\bar{a}$	6,378.245 6,378.077
Mean polar compression		α	1/298.3 1/296.6 1/298.2
Difference in equatorial and polar semi-axes	km	$a - c$	21.382 21.500
Compression of meridian of major equatorial axis		α_a	1/295.2
Compression of meridian of minor equatorial axis		α_b	1/298.0
Equatorial compression		ϵ	1/30 000 1/32 000
Difference in equatorial semi-axes	m	$a - b$	213 199

(continued)

Quantity	Unit of Measurement	Symbol	Numerical Value
Meridian of longitude of minor equatorial semi-axis		λ_a	15°E - 6°W
Meridian of longitude of minor equatorial axis		λ_b	105°E - 75° W; 84°E - 96°W
Difference in polar semi-axes	m	$C_N - C_S$	~ 70 <100
Polar asymmetry		η	$\sim 1.10^{-5}$
Mean acceleration of gravity at equator		g_e	978,057.3
Mean acceleration of gravity at poles	milligals (mgl)	g_p	983,225.1
Difference in acceleration of gravity at pole and at equator		$g_p - g_e$	+5,167.8
Difference in acceleration of gravity at equator	mgl	$g_a - g_b$	+30.2
Difference in acceleration of gravity at poles		$g_N - g_S$	+30
Mean acceleration of gravity for entire surface of terrestrial ellipsoid	mgl	g	979,783.0
Mean radius	km	R	6,370.949
Area of surface	km^2	S	$510.0501 \cdot 10^6$
Volume	km^3	V	$1{,}083.1579 \cdot 10^9$
Mean density	gr/cm^3	δ	5.5170
Siderial rotational period	sec	P	86,164.09
Angular rotational velocity	rad/sec	ω	$7.292116 \cdot 10^{-5}$
Mean equatorial rotational velocity	km/sec	ν	0.465

(continued)

Quantity	Unit of Measurement	Symbol	Numerical Value
Ratio of centrifugal force to attractive force at equator		q	$\frac{1}{289}$
Ratio of centrifugal force to force of gravity at equator		q_c	$0.0034677 = \frac{1}{288}$
Coefficients characterizing the radial distribution of densities within the earth		κ_1 κ	0.966 0.331
Radius of inertia	km; proportion of mean radius	R_i	3,674.735 0.5768
Geocentric latitude of inertial parallel		ϕ_i	54°47′
Moment of inertia	$gr \cdot cm^2$	I	$8.070 \cdot 10^{44}$
Moment of rotation	$gr \cdot cm^2/sec$	L	$5.885 \cdot 10^{40}$
Relative true secular braking of earth's rotation due to tidal friction		$\frac{\Delta\omega_e}{\omega}$	$-4.2 \cdot 10^{-8}$ per century
Relative proper secular acceleration of earth's rotation		$\frac{\Delta\omega_i}{\omega}$	$+1.4 \cdot 10^{-8}$ per century
Relative observed secular braking of earth's rotation		$\frac{\Delta\omega}{\omega}$	$-2.8 \cdot 10^{-8}$ per century
Mean rotational velocity of terrestrial radius due to abyssal compression	cm/century	$\frac{\Delta R}{\Delta t}$	~5 Assumed invariability of mass (M) 4.5 and distribution of masses (κ)
Secular variation in potential gravitational energy of earth accompanying reduction of terrestrial radius by 5 cm and corresponding increase in earth's kinetic energy	erg/century	ΔE	$\sim 17 \cdot 10^{30}$

(continued)

Quantity	Unit of Measurement	Symbol	Numerical Value
Probable value of total energy of tectonic deformation of earth	erg/century	E_t	$\sim 1 \cdot 10^{30}$
Secular loss of heat of earth through radiation into space	erg/century cal/century	$\Delta' E_k$	$1 \cdot 10^{30}$ $2.4 \cdot 10^{22}$
Portion of earth's kinetic energy transformed into heat as a result of lunar and solar tides in the hydrosphere	erg/century cal/century	$\Delta'' E_k$	$0.11 \cdot 10^{30}$ $0.26 \cdot 10^{22}$
Difference in duration of days in March and August	sec	ΔP	0.0025 (March-Aug.)
Corresponding relative annual variation in earth's rotational velocity		$\frac{\Delta^* \omega}{\omega}$	$2.9 \cdot 10^{-8}$ (Aug.-March)
Presumed variation in earth's radius between August and March	cm	$\Delta^* R$	-9.2 (Aug.-March)
Annual variation in level of world ocean	cm	Δh_0	~10 (Sept.-March)

The Earth's Lithosphere, Hydrosphere, Atmosphere and Biosphere

Quantity	Unit of Measurement	Symbol	Numerical Value
Area of continents	km^2; in % of area of surface of earth	S_C	$149 \cdot 10^6$ 29.2
Area of world ocean	km^2; in % of area of surface of earth	S_o	$361 \cdot 10^6$ 70.8
Mean height of continents above sea level	m	h_C	875
Mean depth of world ocean	m	h_o	=3794
Mean position of earth's surface with respect to sea level	m	h_m	=2430

CHAPTER 28

COMPOUND INTEREST FACTORS

Examples in Use of Tables

GIVEN: \$2,500 is invested now at 5 per cent.
REQUIRED: Accumulated value in 10 years (i.e., the amount of a given principal).

SOLUTION:

$$S = P(1+i)^n = \$2{,}500 \times 1.05^{10}$$

$$\text{Compound-amount factor} = (1+i)^n = 1.05^{10} = 1.629$$

$$S = \$2{,}500 \times 1.629 = \$4{,}062.50$$

GIVEN: \$19,500 will be required in 5 years to replace equipment now in use.
REQUIRED: With interest available at 3 per cent, what sum must be deposited in the bank at present to provide the required capital (i.e., the principal which will amount to a given sum).

SOLUTION:

$$P = S\frac{1}{(1+i)^n} = \$19{,}500\,\frac{1}{1.03^5}$$

$$\text{Present-worth factor} = 1/(1+i)^n = 1/1.03^5 = 0.8626$$

$$P = \$19{,}500 \times 0.8626 = \$16{,}821$$

GIVEN: \$50,000 will be required in 10 years to purchase equipment.
REQUIRED: With interest available at 4 per cent, what sum must be deposited each year to provide the required capital (i.e., the annuity which will amount to a given fund).

SOLUTION:

$$R = S\frac{i}{(1+i)^n - 1} = \$50{,}000\,\frac{0.04}{1.04^{10} - 1}$$

$$\text{Sinking-fund factor} = \frac{i}{(1+i)^n - 1} = \frac{0.04}{1.04^{10} - 1} = 0.08329$$

$$R = \$50{,}000 \times 0.08329 = \$4{,}164$$

GIVEN: \$20,000 is invested at 10 per cent interest.
REQUIRED: Annual sum that can be withdrawn over a 20-year period (i.e., the annuity provided by a given capital).

SOLUTION:

$$R = P\frac{i(1+i)^n}{(1+i)^n - 1} = \$20{,}000\,\frac{0.10 \times 1.10^{20}}{1.10^{20} - 1}$$

$$\text{Capital-recovery factor} = \frac{i(1+i)^n}{(1+i)^n - 1} = \frac{0.10 \times 1.10^{20}}{1.10^{20} - 1} = 0.11746$$

$$R = \$20{,}000 \times 0.11746 = \$2{,}349.20$$

GIVEN: \$500 is invested each year at 8 per cent interest.
REQUIRED: Accumulated value in 15 years (i.e., amount of an annuity).

SOLUTION:

$$S = R\frac{(1+i)^n - 1}{i} = \$500\,\frac{1.08^{15} - 1}{0.08}$$

$$\text{Compound-amount factor} = \frac{(1+i)^n - 1}{i} = \frac{1.08^{15} - 1}{0.08} = 27.152$$

$$S = \$500 \times 27.152 = \$13{,}576$$

GIVEN: \$8,000 is required annually for 25 years.
REQUIRED: Sum that must be deposited now at 6 per cent interest.

SOLUTION:

$$P = R\frac{(1+i)^n - 1}{i(1+i)^n} = \$8{,}000\,\frac{1.06^{25} - 1}{0.06 \times 1.06^{25}}$$

$$\text{Present-worth factor} = \frac{(1+i)^n - 1}{i(1+i)^n} = \frac{1.06^{25} - 1}{0.06 \times 1.06^{25}} = 12.783$$

$$P = \$8{,}000 \times 12.78 = \$102{,}264$$

1% Compound Interest Factors

n	Single Payment: Compound Amount Factor	Single Payment: Present Worth Factor	Uniform Annual Series: Sinking Fund Factor	Uniform Annual Series: Capital Recovery Factor	Uniform Annual Series: Compound Amount Factor	Uniform Annual Series: Present Worth Factor	n
	Given P To find S $(1+i)^n$	Given S To find P $\frac{1}{(1+i)^n}$	Given S To find R $\frac{i}{(1+i)^n-1}$	Given P To find R $\frac{i(1+i)^n}{(1+i)^n-1}$	Given R To find S $\frac{(1+i)^n-1}{i}$	Given R To find P $\frac{(1+i)^n-1}{i(1+i)^n}$	
1	1.010	0.9901	1.00000	1.01000	1.000	0.990	**1**
2	1.020	0.9803	0.49751	0.50751	2.010	1.970	**2**
3	1.030	0.9706	0.33002	0.34002	3.030	2.941	**3**
4	1.041	0.9610	0.24628	0.25628	4.060	3.902	**4**
5	1.051	0.9515	0.19604	0.20604	5.101	4.853	**5**
6	1.062	0.9420	0.16255	0.17255	6.152	5.795	**6**
7	1.072	0.9327	0.13863	0.14863	7.214	6.728	**7**
8	1.083	0.9235	0.12069	0.13069	8.286	7.652	**8**
9	1.094	0.9143	0.10674	0.11674	9.369	8.566	**9**
10	1.105	0.9053	0.09558	0.10558	10.462	9.471	**10**
11	1.116	0.8963	0.08645	0.09645	11.567	10.368	**11**
12	1.127	0.8874	0.07885	0.08885	12.683	11.255	**12**
13	1.138	0.8787	0.07241	0.08241	13.809	12.134	**13**
14	1.149	0.8700	0.06690	0.07690	14.947	13.004	**14**
15	1.161	0.8613	0.06212	0.07212	16.097	13.865	**15**
16	1.173	0.8528	0.05794	0.06794	17.258	14.718	**16**
17	1.184	0.8444	0.05426	0.06426	18.430	15.562	**17**
18	1.196	0.8360	0.05098	0.06098	19.615	16.398	**18**
19	1.208	0.8277	0.04805	0.05805	20.811	17.226	**19**
20	1.220	0.8195	0.04542	0.05542	22.019	18.046	**20**
21	1.232	0.8114	0.04303	0.05303	23.239	18.857	**21**
22	1.245	0.8034	0.04086	0.05086	24.472	19.660	**22**
23	1.257	0.7954	0.03889	0.04889	25.716	20.456	**23**
24	1.270	0.7876	0.03707	0.04707	26.973	21.243	**24**
25	1.282	0.7798	0.03541	0.04541	28.243	22.023	**25**
26	1.295	0.7720	0.03387	0.04387	29.526	22.795	**26**
27	1.308	0.7644	0.03245	0.04245	30.821	23.560	**27**
28	1.321	0.7568	0.03112	0.04112	32.129	24.316	**28**
29	1.335	0.7493	0.02990	0.03990	33.450	25.066	**29**
30	1.348	0.7419	0.02875	0.03875	34.785	25.808	**30**
31	1.361	0.7346	0.02768	0.03768	36.133	26.542	**31**
32	1.375	0.7273	0.02667	0.03667	37.494	27.270	**32**
33	1.389	0.7201	0.02573	0.03573	38.869	27.990	**33**
34	1.403	0.7130	0.02484	0.03484	40.258	28.703	**34**
35	1.417	0.7059	0.02400	0.03400	41.660	29.409	**35**
40	1.489	0.6717	0.02046	0.03046	48.886	32.835	**40**
45	1.565	0.6391	0.01771	0.02771	56.481	36.095	**45**
50	1.645	0.6080	0.01551	0.02551	64.463	39.196	**50**
55	1.729	0.5785	0.01373	0.02373	72.852	42.147	**55**
60	1.817	0.5504	0.01224	0.02224	81.670	44.955	**60**
65	1.909	0.5237	0.01100	0.02100	90.937	47.627	**65**
70	2.007	0.4983	0.00993	0.01993	100.676	50.169	**70**
75	2.109	0.4741	0.00902	0.01902	110.913	52.587	**75**
80	2.217	0.4511	0.00822	0.01822	121.672	54.888	**80**
85	2.330	0.4292	0.00752	0.01752	132.979	57.078	**85**
90	2.449	0.4084	0.00690	0.01690	144.863	59.161	**90**
95	2.574	0.3886	0.00636	0.01636	157.354	61.143	**95**
100	2.705	0.3697	0.00587	0.01587	170.481	63.029	**100**

2½% Compound Interest Factors

n	Single Payment: Compound Amount Factor — Given P To find S $(1+i)^n$	Single Payment: Present Worth Factor — Given S To find P $\frac{1}{(1+i)^n}$	Uniform Annual Series: Sinking Fund Factor — Given S To find R $\frac{i}{(1+i)^n-1}$	Uniform Annual Series: Capital Recovery Factor — Given P To find R $\frac{i(1+i)^n}{(1+i)^n-1}$	Uniform Annual Series: Compound Amount Factor — Given R To find S $\frac{(1+i)^n-1}{i}$	Uniform Annual Series: Present Worth Factor — Given R To find P $\frac{(1+i)^n-1}{i(1+i)^n}$	n
1	1.025	0.9756	1.00000	1.02500	1.000	0.976	1
2	1.051	0.9518	0.49383	0.51883	2.025	1.927	2
3	1.077	0.9286	0.32514	0.35014	3.076	2.856	3
4	1.104	0.9060	0.24082	0.26582	4.153	3.762	4
5	1.131	0.8839	0.19025	0.21525	5.256	4.646	5
6	1.160	0.8623	0.15655	0.18155	6.388	5.508	6
7	1.189	0.8413	0.13250	0.15750	7.547	6.349	7
8	1.218	0.8207	0.11447	0.13947	8.736	7.170	8
9	1.249	0.8007	0.10046	0.12546	9.955	7.971	9
10	1.280	0.7812	0.08926	0.11426	11.203	8.752	10
11	1.312	0.7621	0.08011	0.10511	12.483	9.514	11
12	1.345	0.7436	0.07249	0.09749	13.796	10.258	12
13	1.379	0.7254	0.06605	0.09105	15.140	10.983	13
14	1.413	0.7077	0.06054	0.08554	16.519	11.691	14
15	1.448	0.6905	0.05577	0.08077	17.932	12.381	15
16	1.485	0.6736	0.05160	0.07660	19.380	13.055	16
17	1.522	0.6572	0.04793	0.07293	20.865	13.712	17
18	1.560	0.6412	0.04467	0.06967	22.386	14.353	18
19	1.599	0.6255	0.04176	0.06676	23.946	14.979	19
20	1.639	0.6103	0.03915	0.06415	25.545	15.589	20
21	1.680	0.5954	0.03679	0.06179	27.183	16.185	21
22	1.722	0.5809	0.03465	0.05965	28.863	16.765	22
23	1.765	0.5667	0.03270	0.05770	30.584	17.332	23
24	1.809	0.5529	0.03091	0.05591	32.349	17.885	24
25	1.854	0.5394	0.02928	0.05428	34.158	18.424	25
26	1.900	0.5262	0.02777	0.05277	36.012	18.951	26
27	1.948	0.5134	0.02638	0.05138	37.912	19.464	27
28	1.996	0.5009	0.02509	0.05009	39.860	19.965	28
29	2.046	0.4887	0.02389	0.04889	41.856	20.454	29
30	2.098	0.4767	0.02278	0.04778	43.903	20.930	30
31	2.150	0.4651	0.02174	0.04674	46.000	21.395	31
32	2.204	0.4538	0.02077	0.04577	48.150	21.849	32
33	2.259	0.4427	0.01986	0.04486	50.354	22.292	33
34	2.315	0.4319	0.01901	0.04401	52.613	22.724	34
35	2.373	0.4214	0.01821	0.04321	54.928	23.145	35
40	2.685	0.3724	0.01484	0.03984	67.403	25.103	40
45	3.038	0.3292	0.01227	0.03727	81.516	26.833	45
50	3.437	0.2909	0.01026	0.03526	97.484	28.362	50
55	3.889	0.2572	0.00865	0.03365	115.551	29.714	55
60	4.400	0.2273	0.00735	0.03235	135.992	30.909	60
65	4.978	0.2009	0.00628	0.03128	159.118	31.965	65
70	5.632	0.1776	0.00540	0.03040	185.284	32.898	70
75	6.372	0.1569	0.00465	0.02965	214.888	33.723	75
80	7.210	0.1387	0.00403	0.02903	248.383	34.452	80
85	8.157	0.1226	0.00349	0.02849	286.279	35.096	85
90	9.229	0.1084	0.00304	0.02804	329.154	35.666	90
95	10.442	0.0958	0.00265	0.02765	377.664	36.169	95
100	11.814	0.0846	0.00231	0.02731	432.549	36.614	100

3% Compound Interest Factors

n	Single payment		Uniform annual series				n
	Compound-amount factor	Present-worth factor	Sinking-fund factor	Capital-recovery factor	Compound-amount factor	Present-worth factor	
	Given P, to find S $(1+i)^n$	Given S, to find P $\frac{1}{(1+i)^n}$	Given S, to find R $\frac{i}{(1+i)^n-1}$	Given P, to find R $\frac{i(1+i)^n}{(1+i)^n-1}$	Given R, to find S $\frac{(1+i)^n-1}{i}$	Given R, to find P $\frac{(1+i)^n-1}{i(1+i)^n}$	
	3 per cent Compound Interest Factors						
1	1.030	0.9709	1.00000	1.03000	1.000	0.971	1
2	1.061	.9426	0.49261	0.52261	2.030	1.913	2
3	1.093	.9151	.32353	.35353	3.091	2.829	3
4	1.126	.8885	.23903	.26903	4.184	3.717	4
5	1.159	.8626	.18835	.21835	5.309	4.580	5
6	1.194	.8375	.15460	.18460	6.468	5.417	6
7	1.230	.8131	.13051	.16051	7.662	6.230	7
8	1.267	.7894	.11246	.14246	8.892	7.020	8
9	1.305	.7664	.09843	.12843	10.159	7.786	9
10	1.344	.7441	.08723	.11723	11.464	8.530	10
11	1.384	.7224	.07808	.10808	12.808	9.253	11
12	1.426	.7014	.07046	.10046	14.192	9.954	12
13	1.469	.6810	.06403	.09403	15.618	10.635	13
14	1.513	.6611	.05853	.08853	17.086	11.296	14
15	1.558	.6419	.05377	.08377	18.599	11.938	15
16	1.605	.6232	.04961	.07961	20.157	12.561	16
17	1.653	.6050	.04595	.07595	21.762	13.166	17
18	1.702	.5874	.04271	.07271	23.414	13.754	18
19	1.754	.5703	.03981	.06981	25.117	14.324	19
20	1.806	.5537	.03722	.06722	26.870	14.877	20
21	1.860	.5375	.03487	.06487	28.676	15.415	21
22	1.916	.5219	.03275	.06275	30.537	15.937	22
23	1.974	.5067	.03081	.06081	32.453	16.444	23
24	2.033	.4919	.02905	.05905	34.426	16.936	24
25	2.094	.4776	.02743	.05743	36.459	17.413	25
26	2.157	.4637	.02594	.05594	38.553	17.877	26
27	2.221	.4502	.02456	.05456	40.710	18.327	27
28	2.288	.4371	.02329	.05329	42.931	18.764	28
29	2.357	.4243	.02211	.05211	45.219	19.188	29
30	2.427	.4120	.02102	.05102	47.575	19.600	30
31	2.500	.4000	.02000	.05000	50.003	20.000	31
32	2.575	.3883	.01905	.04905	52.503	20.389	32
33	2.652	.3770	.01816	.04816	55.078	20.766	33
34	2.732	.3660	.01732	.04732	57.730	21.132	34
35	2.814	.3554	.01654	.04654	60.462	21.487	35
40	3.262	.3066	.01326	.04326	75.401	23.115	40
45	3.782	.2644	.01079	.04079	92.720	24.519	45
50	4.384	.2281	.00887	.03887	112.797	25.730	50
55	5.082	.1968	.00735	.03735	136.072	26.774	55
60	5.892	.1697	.00613	.03613	163.053	27.676	60
65	6.830	.1464	.00515	.03515	194.333	28.453	65
70	7.918	.1263	.00434	.03434	230.594	29.123	70
75	9.179	.1089	.00367	.03367	272.631	29.702	75
80	10.641	.0940	.00311	.03311	321.363	30.201	80
85	12.336	.0811	.00265	.03265	377.857	30.631	85
90	14.300	.0699	.00226	.03226	443.349	31.002	90
95	16.578	.0603	.00193	.03193	519.272	31.323	95
100	19.219	.0520	.00165	.03165	607.288	31.599	100

4% Compound Interest Factors

n	Single payment: Compound-amount factor — Given P, to find S — $(1+i)^n$	Single payment: Present-worth factor — Given S, to find P — $\frac{1}{(1+i)^n}$	Uniform annual series: Sinking-fund factor — Given S, to find R — $\frac{i}{(1+i)^n-1}$	Uniform annual series: Capital-recovery factor — Given P, to find R — $\frac{i(1+i)^n}{(1+i)^n-1}$	Uniform annual series: Compound-amount factor — Given R, to find S — $\frac{(1+i)^n-1}{i}$	Uniform annual series: Present-worth factor — Given R, to find P — $\frac{(1+i)^n-1}{i(1+i)^n}$	n
	4 per cent Compound Interest Factors						
1	1.040	0.9615	1.00000	1.04000	1.000	0.962	1
2	1.082	.9246	0.49020	0.53020	2.040	1.886	2
3	1.125	.8890	.32035	.36035	3.122	2.775	3
4	1.170	.8548	.23549	.27549	4.246	3.630	4
5	1.217	.8219	.18463	.22463	5.416	4.452	5
6	1.265	.7903	.15076	.19076	6.633	5.242	6
7	1.316	.7599	.12661	.16661	7.898	6.002	7
8	1.369	.7307	.10853	.14853	9.214	6.733	8
9	1.423	.7026	.09449	.13449	10.583	7.435	9
10	1.480	.6756	.08329	.12329	12.006	8.111	10
11	1.539	.6496	.07415	.11415	13.486	8.760	11
12	1.601	.6246	.06655	.10655	15.026	9.385	12
13	1.665	.6006	.06014	.10014	16.627	9.986	13
14	1.732	.5775	.05467	.09467	18.292	10.563	14
15	1.801	.5553	.04994	.08994	20.024	11.118	15
16	1.873	.5339	.04582	.08582	21.825	11.652	16
17	1.948	.5134	.04220	.08220	23.698	12.166	17
18	2.026	.4936	.03899	.07899	25.645	12.659	18
19	2.107	.4746	.03614	.07614	27.671	13.134	19
20	2.191	.4564	.03358	.07358	29.778	13.590	20
21	2.279	.4388	.03128	.07128	31.969	14.029	21
22	2.370	.4220	.02920	.06920	34.248	14.451	22
23	2.465	.4057	.02731	.06731	36.618	14.857	23
24	2.563	.3901	.02559	.06559	39.083	15.247	24
25	2.666	.3751	.02401	.06401	41.646	15.622	25
26	2.772	.3607	.02257	.06257	44.312	15.983	26
27	2.883	.3468	.02124	.06124	47.084	16.330	27
28	2.999	.3335	.02001	.06001	49.968	16.663	28
29	3.119	.3207	.01888	.05888	52.966	16.984	29
30	3.243	.3083	.01783	.05783	56.085	17.292	30
31	3.373	.2965	.01686	.05686	59.328	17.588	31
32	3.508	.2851	.01595	.05595	62.701	17.874	32
33	3.648	.2741	.01510	.05510	66.210	18.148	33
34	3.794	.2636	.01431	.05431	69.858	18.411	34
35	3.946	.2534	.01358	.05358	73.652	18.665	35
40	4.801	.2083	.01052	.05052	95.026	19.793	40
45	5.841	.1712	.00826	.04826	121.029	20.720	45
50	7.107	.1407	.00655	.04655	152.667	21.482	50
55	8.646	.1157	.00523	.04523	191.159	22.109	55
60	10.520	.0951	.00420	.04420	237.991	22.623	60
65	12.799	.0781	.00339	.04339	294.968	23.047	65
70	15.572	.0642	.00275	.04275	364.290	23.395	70
75	18.945	.0528	.00223	.04223	448.631	23.680	75
80	23.050	.0434	.00181	.04181	551.245	23.915	80
85	28.044	.0357	.00148	.04148	676.090	24.109	85
90	34.119	.0293	.00121	.04121	827.983	24.267	90
95	41.511	.0241	.00099	.04099	1,012.785	24.398	95
100	50.505	.0198	.00081	.04081	1,237.624	24.505	100

5% Compound Interest Factors

n	Single payment		Uniform annual series				n
	Compound-amount factor	Present-worth factor	Sinking-fund factor	Capital-recovery factor	Compound-amount factor	Present-worth factor	
	Given P, to find S $(1+i)^n$	Given S, to find P $\frac{1}{(1+i)^n}$	Given S, to find R $\frac{i}{(1+i)^n-1}$	Given P, to find R $\frac{i(1+i)^n}{(1+i)^n-1}$	Given R, to find S $\frac{(1+i)^n-1}{i}$	Given R, to find P $\frac{(1+i)^n-1}{i(1+i)^n}$	
5 per cent Compound Interest Factors							
1	1.050	0.9524	1.00000	1.05000	1.000	0.952	1
2	1.103	.9070	0.48780	0.53780	2.050	1.859	2
3	1.158	.8638	.31721	.36721	3.153	2.723	3
4	1.216	.8227	.23201	.28201	4.310	3.546	4
5	1.276	.7835	.18097	.23097	5.526	4.329	5
6	1.340	.7462	.14702	.19702	6.802	5.076	6
7	1.407	.7107	.12282	.17282	8.142	5.786	7
8	1.477	.6768	.10472	.15472	9.549	6.463	8
9	1.551	.6446	.09069	.14069	11.027	7.108	9
10	1.629	.6139	.07950	.12950	12.578	7.722	10
11	1.710	.5847	.07039	.12039	14.207	8.306	11
12	1.796	.5568	.06283	.11283	15.917	8.863	12
13	1.886	.5303	.05646	.10646	17.713	9.394	13
14	1.980	.5051	.05102	.10102	19.599	9.899	14
15	2.079	.4810	.04634	.09634	21.579	10.380	15
16	2.183	.4581	.04227	.09227	23.657	10.838	16
17	2.292	.4363	.03870	.08870	25.840	11.274	17
18	2.407	.4155	.03555	.08555	28.132	11.690	18
19	2.527	.3957	.03275	.08275	30.539	12.085	19
20	2.653	.3769	.03024	.08024	33.066	12.462	20
21	2.786	.3589	.02800	.07800	35.719	12.821	21
22	2.925	.3418	.02597	.07597	38.505	13.163	22
23	3.072	.3256	.02414	.07414	41.430	13.489	23
24	3.225	.3101	.02247	.07247	44.502	13.799	24
25	3.386	.2953	.02095	.07095	47.727	14.094	25
26	3.556	.2812	.01956	.06956	51.113	14.375	26
27	3.733	.2678	.01829	.06829	54.669	14.643	27
28	3.920	.2551	.01712	.06712	58.403	14.898	28
29	4.116	.2429	.01605	.06605	62.323	15.141	29
30	4.322	.2314	.01505	.06505	66.439	15.372	30
31	4.538	.2204	.01413	.06413	70.761	15.593	31
32	4.765	.2099	.01328	.06328	75.299	15.803	32
33	5.003	.1999	.01249	.06249	80.064	16.003	33
34	5.253	.1904	.01176	.06176	85.067	16.193	34
35	5.516	.1813	.01107	.06107	90.320	16.374	35
40	7.040	.1420	.00828	.05828	120.800	17.159	40
45	8.985	.1113	.00626	.05626	159.700	17.774	45
50	11.467	.0872	.00478	.05478	209.348	18.256	50
55	14.636	.0683	.00367	.05367	272.713	18.633	55
60	18.679	.0535	.00283	.05283	353.584	18.929	60
65	23.840	.0419	.00219	.05219	456.798	19.161	65
70	30.426	.0329	.00170	.05170	588.529	19.343	70
75	38.833	.0258	.00132	.05132	756.654	19.485	75
80	49.561	.0202	.00103	.05103	971.229	19.596	80
85	63.254	.0158	.00080	.05080	1,245.087	19.684	85
90	80.730	.0124	.00063	.05063	1,594.607	19.752	90
95	103.035	.0097	.00049	.05049	2,040.694	19.806	95
100	131.501	.0076	.00038	.05038	2,610.025	19.848	100

6% Compound Interest Factors

	Single payment		Uniform annual series				
n	Compound-amount factor	Present-worth factor	Sinking-fund factor	Capital-recovery factor	Compound-amount factor	Present-worth factor	n
	Given P, to find S $(1+i)^n$	Given S, to find P $\frac{1}{(1+i)^n}$	Given S, to find R $\frac{i}{(1+i)^n-1}$	Given P, to find R $\frac{i(1+i)^n}{(1+i)^n-1}$	Given R, to find S $\frac{(1+i)^n-1}{i}$	Given R, to find P $\frac{(1+i)^n-1}{i(1+i)^n}$	
			6 per cent Compound Interest Factors				
1	1.060	0.9434	1.00000	1.06000	1.000	0.943	1
2	1.124	.8900	0.48544	0.54544	2.060	1.833	2
3	1.191	.8396	.31411	.37411	3.184	2.673	3
4	1.262	.7921	.22859	.28859	4.375	3.465	4
5	1.338	.7473	.17740	.23740	5.637	4.212	5
6	1.419	.7050	.14336	.20336	6.975	4.917	6
7	1.504	.6651	.11914	.17914	8.394	5.582	7
8	1.594	.6274	.10104	.16104	9.897	6.210	8
9	1.689	.5919	.08702	.14702	11.491	6.802	9
10	1.791	.5584	.07587	.13587	13.181	7.360	10
11	1.898	.5268	.06679	.12679	14.972	7.887	11
12	2.012	.4970	.05928	.11928	16.870	8.384	12
13	2.133	.4688	.05296	.11296	18.882	8.853	13
14	2.261	.4423	.04758	.10758	21.015	9.295	14
15	2.397	.4173	.04296	.10296	23.276	9.712	15
16	2.540	.3936	.03895	.09895	25.673	10.106	16
17	2.693	.3714	.03544	.09544	28.213	10.477	17
18	2.854	.3503	.03236	.09236	30.906	10.828	18
19	3.026	.3305	.02962	.08962	33.760	11.158	19
20	3.207	.3118	.02718	.08718	36.786	11.470	20
21	3.400	.2942	.02500	.08500	39.993	11.764	21
22	3.604	.2775	.02305	.08305	43.392	12.042	22
23	3.820	.2618	.02128	.08128	46.996	12.303	23
24	4.049	.2470	.01968	.07968	50.816	12.550	24
25	4.292	.2330	.01823	.07823	54.865	12.783	25
26	4.549	.2198	.01690	.07690	59.156	13.003	26
27	4.822	.2074	.01570	.07570	63.706	13.211	27
28	5.112	.1956	.01459	.07459	68.528	13.406	28
29	5.418	.1846	.01358	.07358	73.640	13.591	29
30	5.743	.1741	.01265	.07265	79.058	13.765	30
31	6.088	.1643	.01179	.07179	84.802	13.929	31
32	6.453	.1550	.01100	.07100	90.890	14.084	32
33	6.841	.1462	.01027	.07027	97.343	14.230	33
34	7.251	.1379	.00960	.06960	104.184	14.368	34
35	7.686	.1301	.00897	.06897	111.435	14.498	35
40	10.286	.0972	.00646	.06646	154.762	15.046	40
45	13.765	.0727	.00470	.06470	212.744	15.456	45
50	18.420	.0543	.00344	.06344	290.336	15.762	50
55	24.650	.0406	.00254	.06254	394.172	15.991	55
60	32.988	.0303	.00188	.06188	533.128	16.161	60
65	44.145	.0227	.00139	.06139	719.083	16.289	65
70	59.076	.0169	.00103	.06103	967.932	16.385	70
75	79.057	.0126	.00077	.06077	1,300.949	16.456	75
80	105.796	.0095	.00057	.06057	1,746.600	16.509	80
85	141.579	.0071	.00043	.06043	2,342.982	16.549	85
90	189.465	.0053	.00032	.06032	3,141.075	16.579	90
95	253.546	.0039	.00024	.06024	4,209.104	16.601	95
100	339.302	.0029	.00018	.06018	5,638.368	16.618	100

7% Compound Interest Factors

n	Single Payment		Uniform Annual Series				n
	Compound Amount Factor	Present Worth Factor	Sinking Fund Factor	Capital Recovery Factor	Compound Amount Factor	Present Worth Factor	
	Given P To find S $(1+i)^n$	Given S To find P $\frac{1}{(1+i)^n}$	Given S To find R $\frac{i}{(1+i)^n-1}$	Given P To find R $\frac{i(1+i)^n}{(1+i)^n-1}$	Given R To find S $\frac{(1+i)^n-1}{i}$	Given R To find P $\frac{(1+i)^n-1}{i(1+i)^n}$	
1	1.070	0.9346	1.00000	1.07000	1.000	0.935	1
2	1.145	0.8734	0.48309	0.55309	2.070	1.808	2
3	1.225	0.8163	0.31105	0.38105	3.215	2.624	3
4	1.311	0.7629	0.22523	0.29523	4.440	3.387	4
5	1.403	0.7130	0.17389	0.24389	5.751	4.100	5
6	1.501	0.6663	0.13980	0.20980	7.153	4.767	6
7	1.606	0.6227	0.11555	0.18555	8.654	5.389	7
8	1.718	0.5820	0.09747	0.16747	10.260	5.971	8
9	1.838	0.5439	0.08349	0.15349	11.978	6.515	9
10	1.967	0.5083	0.07238	0.14238	13.816	7.024	10
11	2.105	0.4751	0.06336	0.13336	15.784	7.499	11
12	2.252	0.4440	0.05590	0.12590	17.888	7.943	12
13	2.410	0.4150	0.04965	0.11965	20.141	8.358	13
14	2.579	0.3878	0.04434	0.11434	22.550	8.745	14
15	2.759	0.3624	0.03979	0.10979	25.129	9.108	15
16	2.952	0.3387	0.03586	0.10586	27.888	9.447	16
17	3.159	0.3166	0.03243	0.10243	30.840	9.763	17
18	3.380	0.2959	0.02941	0.09941	33.999	10.059	18
19	3.617	0.2765	0.02675	0.09675	37.379	10.336	19
20	3.870	0.2584	0.02439	0.09439	40.995	10.594	20
21	4.141	0.2415	0.02229	0.09229	44.865	10.836	21
22	4.430	0.2257	0.02041	0.09041	49.006	11.061	22
23	4.741	0.2109	0.01871	0.08871	53.436	11.272	23
24	5.072	0.1971	0.01719	0.08719	58.177	11.469	24
25	5.427	0.1842	0.01581	0.08581	63.249	11.654	25
26	5.807	0.1722	0.01456	0.08456	68.676	11.826	26
27	6.214	0.1609	0.01343	0.08343	74.484	11.987	27
28	6.649	0.1504	0.01239	0.08239	80.698	12.137	28
29	7.114	0.1406	0.01145	0.08145	87.347	12.278	29
30	7.612	0.1314	0.01059	0.08059	94.461	12.409	30
31	8.145	0.1228	0.00980	0.07980	102.073	12.532	31
32	8.715	0.1147	0.00907	0.07907	110.218	12.647	32
33	9.325	0.1072	0.00841	0.07841	118.933	12.754	33
34	9.978	0.1002	0.00780	0.07780	128.259	12.854	34
35	10.677	0.0937	0.00723	0.07723	138.237	12.948	35
40	14.974	0.0668	0.00501	0.07501	199.635	13.332	40
45	21.002	0.0476	0.00350	0.07350	285.749	13.606	45
50	29.457	0.0339	0.00246	0.07246	406.529	13.801	50
55	41.315	0.0242	0.00174	0.07174	575.929	13.940	55
60	57.946	0.0173	0.00123	0.07123	813.520	14.039	60
65	81.273	0.0123	0.00087	0.07087	1146.755	14.110	65
70	113.989	0.0088	0.00062	0.07062	1614.134	14.160	70
75	159.876	0.0063	0.00044	0.07044	2269.657	14.196	75
80	224.234	0.0045	0.00031	0.07031	3189.063	14.222	80
85	314.500	0.0032	0.00022	0.07022	4478.576	14.240	85
90	441.103	0.0023	0.00016	0.07016	6287.185	14.253	90
95	618.670	0.0016	0.00011	0.07011	8823.854	14.263	95
100	867.716	0.0012	0.00008	0.07008	12381.662	14.269	100

8% Compound Interest Factors

n	Single payment: Compound-amount factor — Given P, to find S — $(1+i)^n$	Single payment: Present-worth factor — Given S, to find P — $\frac{1}{(1+i)^n}$	Uniform annual series: Sinking-fund factor — Given S, to find R — $\frac{i}{(1+i)^n-1}$	Uniform annual series: Capital-recovery factor — Given P, to find R — $\frac{i(1+i)^n}{(1+i)^n-1}$	Uniform annual series: Compound-amount factor — Given R, to find S — $\frac{(1+i)^n-1}{i}$	Uniform annual series: Present-worth factor — Given R, to find P — $\frac{(1+i)^n-1}{i(1+i)^n}$	n
	8 per cent Compound Interest Factors						
1	1.080	0.9259	1.00000	1.08000	1.000	0.926	1
2	1.166	.8573	0.48077	0.56077	2.080	1.783	2
3	1.260	.7938	.30803	.38803	3.246	2.577	3
4	1.360	.7350	.22192	.30192	4.506	3.312	4
5	1.469	.6806	.17046	.25046	5.867	3.993	5
6	1.587	.6302	.13632	.21632	7.336	4.623	6
7	1.714	.5835	.11207	.19207	8.923	5.206	7
8	1.851	.5403	.09401	.17401	10.637	5.747	8
9	1.999	.5002	.08008	.16008	12.488	6.247	9
10	2.159	.4632	.06903	.14903	14.487	6.710	10
11	2.332	.4289	.06008	.14008	16.645	7.139	11
12	2.518	.3971	.05270	.13270	18.977	7.536	12
13	2.720	.3677	.04652	.12652	21.495	7.904	13
14	2.937	.3405	.04130	.12130	24.215	8.244	14
15	3.172	.3152	.03683	.11683	27.152	8.559	15
16	3.426	.2919	.03298	.11298	30.324	8.851	16
17	3.700	.2703	.02963	.10963	33.750	9.122	17
18	3.996	.2502	.02670	.10670	37.450	9.372	18
19	4.316	.2317	.02413	.10413	41.446	9.604	19
20	4.661	.2145	.02185	.10185	45.762	9.818	20
21	5.034	.1987	.01983	.09983	50.423	10.017	21
22	5.437	.1839	.01803	.09803	55.457	10.201	22
23	5.871	.1703	.01642	.09642	60.893	10.371	23
24	6.341	.1577	.01498	.09498	66.765	10.529	24
25	6.848	.1460	.01368	.09368	73.106	10.675	25
26	7.396	.1352	.01251	.09251	79.954	10.810	26
27	7.988	.1252	.01145	.09145	87.351	10.935	27
28	8.627	.1159	.01049	.09049	95.339	11.051	28
29	9.317	.1073	.00962	.08962	103.966	11.158	29
30	10.063	.0994	.00883	.08883	113.283	11.258	30
31	10.868	.0920	.00811	.08811	123.346	11.350	31
32	11.737	.0852	.00745	.08745	134.214	11.435	32
33	12.676	.0789	.00685	.08685	145.951	11.514	33
34	13.690	.0730	.00630	.08630	158.627	11.587	34
35	14.785	.0676	.00580	.08580	172.317	11.655	35
40	21.725	.0460	.00386	.08386	259.057	11.925	40
45	31.920	.0313	.00259	.08259	386.506	12.108	45
50	46.902	.0213	.00174	.08174	573.770	12.233	50
55	68.914	.0145	.00118	.08118	848.923	12.319	55
60	101.257	.0099	.00080	.08080	1,253.213	12.377	60
65	148.780	.0067	.00054	.08054	1,847.248	12.416	65
70	218.606	.0046	.00037	.08037	2,720.080	12.443	70
75	321.205	.0031	.00025	.08025	4,002.557	12.461	75
80	471.955	.0021	.00017	.08017	5,886.935	12.474	80
85	693.456	.0014	.00012	.08012	8,655.706	12.482	85
90	1,018.915	.0010	.00008	.08008	12,723.939	12.488	90
95	1,497.121	.0007	.00005	.08005	18,701.507	12.492	95
100	2,199.761	.0005	.00004	.08004	27,484.516	12.494	100

10% Compound Interest Factors

n	Single payment: Compound-amount factor — Given P, to find S $(1+i)^n$	Single payment: Present-worth factor — Given S, to find P $\frac{1}{(1+i)^n}$	Uniform annual series: Sinking-fund factor — Given S, to find R $\frac{i}{(1+i)^n-1}$	Uniform annual series: Capital-recovery factor — Given P, to find R $\frac{i(1+i)^n}{(1+i)^n-1}$	Uniform annual series: Compound-amount factor — Given R, to find S $\frac{(1+i)^n-1}{i}$	Uniform annual series: Present-worth factor — Given R, to find P $\frac{(1+i)^n-1}{i(1+i)^n}$	n
10 per cent Compound Interest Factors							
1	1.100	0.9091	1.00000	1.10000	1.000	0.909	1
2	1.210	.8264	0.47619	0.57619	2.100	1.736	2
3	1.331	.7513	.30211	.40211	3.310	2.487	3
4	1.464	.6830	.21547	.31547	4.641	3.170	4
5	1.611	.6209	.16380	.26380	6.105	3.791	5
6	1.772	.5645	.12961	.22961	7.716	4.355	6
7	1.949	.5132	.10541	.20541	9.487	4.868	7
8	2.144	.4665	.08744	.18744	11.436	5.335	8
9	2.358	.4241	.07364	.17364	13.579	5.759	9
10	2.594	.3855	.06275	.16275	15.937	6.144	10
11	2.853	.3505	.05396	.15396	18.531	6.495	11
12	3.138	.3186	.04676	.14676	21.384	6.814	12
13	3.452	.2897	.04078	.14078	24.523	7.103	13
14	3.797	.2633	.03575	.13575	27.975	7.367	14
15	4.177	.2394	.03147	.13147	31.772	7.606	15
16	4.595	.2176	.02782	.12782	35.950	7.824	16
17	5.054	.1978	.02466	.12466	40.545	8.022	17
18	5.560	.1799	.02193	.12193	45.599	8.201	18
19	6.116	.1635	.01955	.11955	51.159	8.365	19
20	6.727	.1486	.01746	.11746	57.275	8.514	20
21	7.400	.1351	.01562	.11562	64.002	8.649	21
22	8.140	.1228	.01401	.11401	71.403	8.772	22
23	8.954	.1117	.01257	.11257	79.543	8.883	23
24	9.850	.1015	.01130	.11130	88.497	8.985	24
25	10.835	.0923	.01017	.11017	98.347	9.077	25
26	11.918	.0839	.00916	.10916	109.182	9.161	26
27	13.110	.0763	.00826	.10826	121.100	9.237	27
28	14.421	.0693	.00745	.10745	134.210	9.307	28
29	15.863	.0630	.00673	.10673	148.631	9.370	29
30	17.449	.0573	.00608	.10608	164.494	9.427	30
31	19.194	.0521	.00550	.10550	181.943	9.479	31
32	21.114	.0474	.00497	.10497	201.138	9.526	32
33	23.225	.0431	.00450	.10450	222.252	9.569	33
34	25.548	.0391	.00407	.10407	245.477	9.609	34
35	28.102	.0356	.00369	.10369	271.024	9.644	35
40	45.259	.0221	.00226	.10226	442.593	9.779	40
45	72.890	.0137	.00139	.10139	718.905	9.863	45
50	117.391	.0085	.00086	.10086	1,163.909	9.915	50
55	189.059	.0053	.00053	.10053	1,880.591	9.947	55
60	304.482	.0033	.00033	.10033	3,034.816	9.967	60
65	490.371	.0020	.00020	.10020	4,893.707	9.980	65
70	789.747	.0013	.00013	.10013	7,887.470	9.987	70
75	1,271.895	.0008	.00008	.10008	12,708.954	9.992	75
80	2,048.400	.0005	.00005	.10005	20,474.002	9.995	80
85	3,298.969	.0003	.00003	.10003	32,979.690	9.997	85
90	5,313.023	.0002	.00002	.10002	53,120.226	9.998	90
95	8,556.676	.0001	.00001	.10001	85,556.760	9.999	95
100	13,780.612	.0001	.00001	.10001	137,796.123	9.999	100

Capital Recovery Factors for High Interest Rates

n	10%	12%	15%	17%	20%	25%	30%	35%	40%	45%	50%
1	1.10000	1.12000	1.15000	1.17000	1.20000	1.25000	1.30000	1.35000	1.40000	1.45000	1.50000
2	0.57619	0.59170	0.61512	0.63083	0.65455	0.69444	0.73478	0.77553	0.81667	0.85816	0.90000
3	0.40211	0.41635	0.43798	0.45257	0.47473	0.51230	0.55063	0.58966	0.62936	0.66966	0.71053
4	0.31547	0.32923	0.35027	0.36453	0.38629	0.42344	0.46163	0.50076	0.54077	0.58156	0.62308
5	0.26380	0.27741	0.29832	0.31256	0.33438	0.37165	0.41058	0.45046	0.49136	0.53318	0.57583
6	0.22961	0.24323	0.26424	0.27861	0.30071	0.33882	0.37839	0.41926	0.46126	0.50426	0.54812
7	0.20541	0.21912	0.24036	0.25495	0.27742	0.31634	0.35687	0.39880	0.44192	0.48607	0.53108
8	0.18744	0.20130	0.22285	0.23769	0.26061	0.30040	0.34192	0.38489	0.42907	0.47427	0.52030
9	0.17364	0.18768	0.20957	0.22469	0.24808	0.28876	0.33124	0.37519	0.42034	0.46646	0.51335
10	0.16275	0.17698	0.19925	0.21466	0.23852	0.28007	0.32346	0.36832	0.41432	0.46123	0.50882
11	0.15396	0.16842	0.19107	0.20676	0.23110	0.27349	0.31773	0.36339	0.41013	0.45768	0.50585
12	0.14676	0.16144	0.18448	0.20047	0.22526	0.26845	0.31345	0.35982	0.40718	0.45527	0.50388
13	0.14078	0.15568	0.17911	0.19538	0.22062	0.26454	0.31024	0.35722	0.40510	0.45362	0.50258
14	0.13575	0.15087	0.17469	0.19123	0.21689	0.26150	0.30782	0.35532	0.40363	0.45249	0.50172
15	0.13147	0.14682	0.17102	0.18782	0.21388	0.25912	0.30598	0.35393	0.40259	0.45172	0.50114
16	0.12782	0.14339	0.16795	0.18500	0.21144	0.25724	0.30458	0.35290	0.40185	0.45118	0.50076
17	0.12466	0.14046	0.16537	0.18266	0.20944	0.25576	0.30351	0.35214	0.40132	0.45081	0.50051
18	0.12193	0.13794	0.16319	0.18071	0.20781	0.25459	0.30269	0.35158	0.40094	0.45056	0.50034
19	0.11955	0.13576	0.16134	0.17907	0.20646	0.25366	0.30207	0.35117	0.40067	0.45039	0.50023
20	0.11746	0.13388	0.15976	0.17769	0.20536	0.25292	0.30159	0.35087	0.40048	0.45027	0.50015
21	0.11562	0.13224	0.15842	0.17653	0.20444	0.25233	0.30122	0.35064	0.40034	0.45018	0.50010
22	0.11401	0.13081	0.15727	0.17555	0.20369	0.25186	0.30094	0.35048	0.40024	0.45013	0.50007
23	0.11257	0.12956	0.15628	0.17472	0.20307	0.25148	0.30072	0.35035	0.40017	0.45009	0.50004
24	0.11130	0.12846	0.15543	0.17402	0.20255	0.25119	0.30055	0.35026	0.40012	0.45006	0.50003
25	0.11017	0.12750	0.15470	0.17342	0.20212	0.25095	0.30043	0.35019	0.40009	0.45004	0.50002
30	0.10608	0.12414	0.15230	0.17154	0.20085	0.25031	0.30011	0.35004	0.40002	0.45001	0.50000
35	0.10369	0.12232	0.15113	0.17070	0.20034	0.25010	0.30003	0.35001	0.40000	0.45000	0.50000
40	0.10226	0.12130	0.15056	0.17032	0.20014	0.25003	0.30001	0.35000	0.40000	0.45000	0.50000
45	0.10139	0.12074	0.15028	0.17015	0.20005	0.25001	0.30000	0.35000	0.40000	0.45000	0.50000
50	0.10086	0.12042	0.15014	0.17007	0.20002	0.25000	0.30000	0.35000	0.40000	0.45000	0.50000
60	0.10033	0.12013	0.15003	0.17001	0.20000	0.25000	0.30000	0.35000	0.40000	0.45000	0.50000
70	0.10013	0.12004	0.15001	0.17000	0.20000	0.25000	0.30000	0.35000	0.40000	0.45000	0.50000
80	0.10005	0.12001	0.15000	0.17000	0.20000	0.25000	0.30000	0.35000	0.40000	0.45000	0.50000
90	0.10002	0.12000	9.15000	0.17000	0.20000	0.25000	0.30000	0.35000	0.40000	0.45000	0.50000
100	0.10001	0.12000	0.15000	0.17000	0.20000	0.25000	0.30000	0.35000	0.40000	0.45000	0.50000

Present Value of $1

Years Hence	1%	2%	4%	6%	8%	10%	12%	14%	15%	16%	18%	20%	22%	24%	25%	26%	28%	30%	35%	40%	45%	50%
1	0.990	0.980	0.962	0.943	0.926	0.909	0.893	0.877	0.870	0.862	0.847	0.833	0.820	0.806	0.800	0.794	0.781	0.769	0.741	0.714	0.690	0.667
2	0.980	0.961	0.925	0.890	0.857	0.826	0.797	0.769	0.756	0.743	0.718	0.694	0.672	0.650	0.640	0.630	0.610	0.592	0.549	0.510	0.476	0.444
3	0.971	0.942	0.889	0.840	0.794	0.751	0.712	0.675	0.658	0.641	0.609	0.579	0.551	0.524	0.512	0.500	0.477	0.455	0.406	0.364	0.328	0.296
4	0.961	0.924	0.855	0.792	0.735	0.683	0.636	0.592	0.572	0.552	0.516	0.482	0.451	0.423	0.410	0.397	0.373	0.350	0.301	0.260	0.226	0.198
5	0.951	0.906	0.822	0.747	0.681	0.621	0.567	0.519	0.497	0.476	0.437	0.402	0.370	0.341	0.328	0.315	0.291	0.269	0.223	0.186	0.156	0.132
6	0.942	0.888	0.790	0.705	0.630	0.564	0.507	0.456	0.432	0.410	0.370	0.335	0.303	0.275	0.262	0.250	0.227	0.207	0.165	0.133	0.108	0.088
7	0.933	0.871	0.760	0.665	0.583	0.513	0.452	0.400	0.376	0.354	0.314	0.279	0.249	0.222	0.210	0.198	0.178	0.159	0.122	0.095	0.074	0.059
8	0.923	0.853	0.731	0.627	0.540	0.467	0.404	0.351	0.327	0.305	0.266	0.233	0.204	0.179	0.168	0.157	0.139	0.123	0.091	0.068	0.051	0.039
9	0.914	0.837	0.703	0.592	0.500	0.424	0.361	0.308	0.284	0.263	0.225	0.194	0.167	0.144	0.134	0.125	0.108	0.094	0.067	0.048	0.035	0.026
10	0.905	0.820	0.676	0.558	0.463	0.386	0.322	0.270	0.247	0.227	0.191	0.162	0.137	0.116	0.107	0.099	0.085	0.073	0.050	0.035	0.024	0.017
11	0.896	0.804	0.650	0.527	0.429	0.350	0.287	0.237	0.215	0.195	0.162	0.135	0.112	0.094	0.086	0.079	0.066	0.056	0.037	0.025	0.017	0.012
12	0.887	0.788	0.625	0.497	0.397	0.319	0.257	0.208	0.187	0.168	0.137	0.112	0.092	0.076	0.069	0.062	0.052	0.043	0.027	0.018	0.012	0.008
13	0.879	0.773	0.601	0.469	0.368	0.290	0.229	0.182	0.163	0.145	0.116	0.093	0.075	0.061	0.055	0.050	0.040	0.033	0.020	0.013	0.008	0.005
14	0.870	0.758	0.577	0.442	0.340	0.263	0.205	0.160	0.141	0.125	0.099	0.078	0.062	0.049	0.044	0.039	0.032	0.025	0.015	0.009	0.006	0.003
15	0.861	0.743	0.555	0.417	0.315	0.239	0.183	0.140	0.123	0.108	0.084	0.065	0.051	0.040	0.035	0.031	0.025	0.020	0.011	0.006	0.004	0.002
16	0.853	0.728	0.534	0.394	0.292	0.218	0.163	0.123	0.107	0.093	0.071	0.054	0.042	0.032	0.028	0.025	0.019	0.015	0.008	0.005	0.003	0.002
17	0.844	0.714	0.513	0.371	0.270	0.198	0.146	0.108	0.093	0.080	0.060	0.045	0.034	0.026	0.023	0.020	0.015	0.012	0.006	0.003	0.002	0.001
18	0.836	0.700	0.494	0.350	0.250	0.180	0.130	0.095	0.081	0.069	0.051	0.038	0.028	0.021	0.018	0.016	0.012	0.009	0.005	0.002	0.001	0.001
19	0.828	0.686	0.475	0.331	0.232	0.164	0.116	0.083	0.070	0.060	0.043	0.031	0.023	0.017	0.014	0.012	0.009	0.007	0.003	0.002	0.001	
20	0.820	0.673	0.456	0.312	0.215	0.149	0.104	0.073	0.061	0.051	0.037	0.026	0.019	0.014	0.012	0.010	0.007	0.005	0.002	0.001	0.001	
21	0.811	0.660	0.439	0.294	0.199	0.135	0.093	0.064	0.053	0.044	0.031	0.022	0.015	0.011	0.009	0.008	0.006	0.004	0.002	0.001		
22	0.803	0.647	0.422	0.278	0.184	0.123	0.083	0.056	0.046	0.038	0.026	0.018	0.013	0.009	0.007	0.006	0.004	0.003	0.001	0.001		
23	0.795	0.634	0.406	0.262	0.170	0.112	0.074	0.049	0.040	0.033	0.022	0.015	0.010	0.007	0.006	0.005	0.003	0.002	0.001			
24	0.788	0.622	0.390	0.247	0.158	0.102	0.066	0.043	0.035	0.028	0.019	0.013	0.008	0.006	0.005	0.004	0.003	0.002	0.001			
25	0.780	0.610	0.375	0.233	0.146	0.092	0.059	0.038	0.030	0.024	0.016	0.010	0.007	0.005	0.004	0.003	0.002	0.001	0.001			
26	0.772	0.598	0.361	0.220	0.135	0.084	0.053	0.033	0.026	0.021	0.014	0.009	0.006	0.004	0.003	0.002	0.002	0.001				
27	0.764	0.586	0.347	0.207	0.125	0.076	0.047	0.029	0.023	0.018	0.011	0.007	0.005	0.003	0.002	0.002	0.001	0.001				
28	0.757	0.574	0.333	0.196	0.116	0.069	0.042	0.026	0.020	0.016	0.010	0.006	0.004	0.002	0.002	0.002	0.001	0.001				
29	0.749	0.563	0.321	0.185	0.107	0.063	0.037	0.022	0.017	0.014	0.008	0.005	0.003	0.002	0.002	0.001	0.001	0.001				
30	0.742	0.552	0.308	0.174	0.099	0.057	0.033	0.020	0.015	0.012	0.007	0.004	0.003	0.002	0.001	0.001	0.001					
40	0.672	0.453	0.208	0.097	0.046	0.022	0.011	0.005	0.004	0.003	0.001	0.001										
50	0.608	0.372	0.141	0.054	0.021	0.009	0.003	0.001	0.001	0.001												

Present Value of $1 Received Annually for N Years

Years (*N*)	1%	2%	4%	6%	8%	10%	12%	14%	15%	16%	18%	20%	22%	24%	25%	26%	28%	30%	35%	40%	45%	50%
1	0.990	0.980	0.962	0.943	0.926	0.909	0.893	0.877	0.870	0.862	0.847	0.833	0.820	0.806	0.800	0.794	0.781	0.769	0.741	0.714	0.690	0.667
2	1.970	1.942	1.886	1.833	1.783	1.736	1.690	1.647	1.626	1.605	1.566	1.528	1.492	1.457	1.440	1.424	1.392	1.361	1.289	1.224	1.165	1.111
3	2.941	2.884	2.775	2.673	2.577	2.487	2.402	2.322	2.283	2.246	2.174	2.106	2.042	1.981	1.952	1.923	1.868	1.816	1.696	1.589	1.493	1.407
4	3.902	3.808	3.630	3.465	3.312	3.170	3.037	2.914	2.855	2.798	2.690	2.589	2.494	2.404	2.362	2.320	2.241	2.166	1.997	1.849	1.720	1.605
5	4.853	4.713	4.452	4.212	3.993	3.791	3.605	3.433	3.352	3.274	3.127	2.991	2.864	2.745	2.689	2.635	2.532	2.436	2.220	2.035	1.876	1.737
6	5.795	5.601	5.242	4.917	4.623	4.355	4.111	3.889	3.784	3.685	3.498	3.326	3.167	3.020	2.951	2.885	2.759	2.643	2.385	2.168	1.983	1.824
7	6.728	6.472	6.002	5.582	5.206	4.868	4.564	4.288	4.160	4.039	3.812	3.605	3.416	3.242	3.161	3.083	2.937	2.802	2.508	2.263	2.057	1.883
8	7.652	7.325	6.733	6.210	5.747	5.335	4.968	4.639	4.487	4.344	4.078	3.837	3.619	3.421	3.329	3.241	3.076	2.925	2.598	2.331	2.108	1.922
9	8.566	8.162	7.435	6.802	6.247	5.759	5.328	4.946	4.772	4.607	4.303	4.031	3.786	3.566	3.463	3.366	3.184	3.019	2.665	2.379	2.144	1.948
10	9.471	8.983	8.111	7.360	6.710	6.145	5.650	5.216	5.019	4.833	4.494	4.192	3.923	3.682	3.571	3.465	3.269	3.092	2.715	2.414	2.168	1.965
11	10.368	9.787	8.760	7.887	7.139	6.495	5.937	5.453	5.234	5.029	4.656	4.327	4.035	3.776	3.656	3.544	3.335	3.147	2.757	2.438	2.185	1.977
12	11.255	10.575	9.385	8.384	7.536	6.814	6.194	5.660	5.421	5.197	4.793	4.439	4.127	3.851	3.725	3.606	3.387	3.190	2.779	2.456	2.196	1.985
13	12.134	11.343	9.986	8.853	7.904	7.103	6.424	5.842	5.583	5.342	4.910	4.533	4.203	3.912	3.780	3.656	3.427	3.223	2.799	2.468	2.204	1.990
14	13.004	12.106	10.563	9.295	8.244	7.367	6.628	6.002	5.724	5.468	5.008	4.611	4.265	3.962	3.824	3.695	3.459	3.249	2.814	2.477	2.210	1.993
15	13.865	12.849	11.118	9.712	8.559	7.606	6.811	6.142	5.847	5.575	5.092	4.675	4.315	4.001	3.859	3.726	3.483	3.268	2.825	2.484	2.214	1.995
16	14.718	13.578	11.652	10.106	8.851	7.824	6.974	6.265	5.954	5.669	5.162	4.730	4.357	4.033	3.887	3.751	3.503	3.283	2.834	2.489	2.216	1.997
17	15.562	14.292	12.166	10.477	9.122	8.022	7.120	6.373	6.047	5.749	5.222	4.775	4.391	4.059	3.910	3.771	3.518	3.295	2.840	2.492	2.218	1.998
18	16.398	14.992	12.659	10.828	9.372	8.201	7.250	6.467	6.128	5.818	5.273	4.812	4.419	4.080	3.928	3.786	3.529	3.304	2.844	2.494	2.219	1.999
19	17.226	15.678	13.134	11.158	9.604	8.365	7.366	6.550	6.198	5.877	5.316	4.844	4.442	4.097	3.942	3.799	3.539	3.311	2.848	2.496	2.220	1.999
20	18.046	16.351	13.590	11.470	9.818	8.514	7.469	6.623	6.259	5.929	5.353	4.870	4.460	4.110	3.954	3.808	3.546	3.316	2.850	2.497	2.221	1.999
21	18.857	17.011	14.029	11.764	10.017	8.649	7.562	6.687	6.312	5.973	5.384	4.891	4.476	4.121	3.963	3.816	3.551	3.320	2.852	2.498	2.221	2.000
22	19.660	17.658	14.451	12.042	10.201	8.772	7.645	6.743	6.359	6.011	5.410	4.909	4.488	4.130	3.970	3.822	3.556	3.323	2.853	2.498	2.222	2.000
23	20.456	18.292	14.857	12.303	10.371	8.883	7.718	6.792	6.399	6.044	5.432	4.925	4.499	4.137	3.976	3.827	3.559	3.325	2.854	2.499	2.222	2.000
24	21.243	18.914	15.247	12.550	10.529	8.985	7.784	6.835	6.434	6.073	5.451	4.937	4.507	4.143	3.981	3.831	3.562	3.327	2.855	2.499	2.222	2.000
25	22.023	19.523	15.622	12.783	10.675	9.077	7.843	6.873	6.464	6.097	5.467	4.948	4.514	4.147	3.985	3.834	3.564	3.329	2.856	2.499	2.222	2.000
26	22.795	20.121	15.983	13.003	10.810	9.161	7.896	6.906	6.491	6.118	5.480	4.956	4.520	4.151	3.988	3.837	3.566	3.330	2.856	2.500	2.222	2.000
27	23.560	20.707	16.330	13.211	10.935	9.237	7.943	6.935	6.514	6.136	5.492	4.964	4.524	4.154	3.990	3.839	3.567	3.331	2.856	2.500	2.222	2.000
28	24.316	21.281	16.663	13.406	11.051	9.307	7.984	6.961	6.534	6.152	5.502	4.970	4.528	4.157	3.992	3.840	3.568	3.331	2.857	2.500	2.222	2.000
29	25.066	21.844	16.984	13.591	11.158	9.370	8.022	6.983	6.551	6.166	5.510	4.975	4.531	4.159	3.994	3.841	3.569	3.332	2.857	2.500	2.222	2.000
30	25.808	22.396	17.292	13.765	11.258	9.427	8.055	7.003	6.566	6.177	5.517	4.979	4.534	4.160	3.995	3.842	3.569	3.332	2.857	2.500	2.222	2.000
40	32.835	27.355	19.793	15.046	11.925	9.779	8.244	7.105	6.642	6.234	5.548	4.997	4.544	4.166	3.999	3.846	3.571	3.333	2.857	2.500	2.222	2.000
50	39.196	31.424	21.482	15.762	12.234	9.915	8.304	7.133	6.661	6.246	5.554	4.999	4.545	4.167	4.000	3.846	3.571	3.333	2.857	2.500	2.222	2.000

8%

MONTHLY INSTALLMENTS TO REPAY OR AMORTIZE A LOAN

TERM / AMOUNT	5 YEARS	7 YEARS	10 YEARS	15 YEARS	20 YEARS	25 YEARS	29 YEARS	30 YEARS	35 YEARS
$ 50	1.02	.78	.61	.48	.42	.39	.37	.37	.36
100	2.03	1.56	1.22	.96	.84	.78	.74	.74	.72
200	4.06	3.12	2.43	1.92	1.68	1.55	1.48	1.47	1.43
300	6.09	4.68	3.64	2.87	2.51	2.32	2.22	2.21	2.14
400	8.12	6.24	4.86	3.83	3.35	3.09	2.96	2.94	2.85
500	10.14	7.80	6.07	4.78	4.19	3.86	3.70	3.67	3.56
600	12.17	9.36	7.28	5.74	5.02	4.64	4.44	4.41	4.27
700	14.20	10.92	8.50	6.69	5.86	5.41	5.18	5.14	4.98
800	16.23	12.47	9.71	7.65	6.70	6.18	5.92	5.88	5.69
900	18.25	14.03	10.92	8.61	7.53	6.95	6.66	6.61	6.40
1000	20.28	15.59	12.14	9.56	8.37	7.72	7.40	7.34	7.11
2000	40.56	31.18	24.27	19.12	16.73	15.44	14.80	14.68	14.21
3000	60.83	46.76	36.40	28.67	25.10	23.16	22.20	22.02	21.31
4000	81.11	62.35	48.54	38.23	33.46	30.88	29.60	29.36	28.42
5000	101.39	77.94	60.67	47.79	41.83	38.60	37.00	36.69	35.52
6000	121.66	93.52	72.80	57.34	50.19	46.31	44.40	44.03	42.62
7000	141.94	109.11	84.93	66.90	58.56	54.03	51.80	51.37	49.72
8000	162.22	124.69	97.07	76.46	66.92	61.75	59.20	58.71	56.83
9000	182.49	140.28	109.20	86.01	75.28	69.47	66.60	66.04	63.93
10000	202.77	155.87	121.33	95.57	83.65	77.19	74.00	73.38	71.03
15000	304.15	233.80	182.00	143.35	125.47	115.78	111.00	110.07	106.54
20000	405.53	311.73	242.66	191.14	167.29	154.37	147.99	146.76	142.06
25000	506.91	389.66	303.32	238.92	209.12	192.96	184.99	183.45	177.57
26000	527.19	405.25	315.46	248.47	217.48	200.68	192.39	190.78	184.67
27000	547.47	420.83	327.59	258.03	225.84	208.40	199.79	198.12	191.78
28000	567.74	436.42	339.72	267.59	234.21	216.11	207.19	205.46	198.88
29000	588.02	452.01	351.86	277.14	242.57	223.83	214.59	212.80	205.98
30000	608.30	467.59	363.99	286.70	250.94	231.55	221.99	220.13	213.08
31000	628.57	483.18	376.12	296.26	259.30	239.27	229.39	227.47	220.19
32000	648.85	498.76	388.25	305.81	267.67	246.99	236.79	234.81	227.29
33000	669.13	514.35	400.39	315.37	276.03	254.70	244.19	242.15	234.39
34000	689.40	529.94	412.52	324.93	284.39	262.42	251.59	249.48	241.49
35000	709.68	545.52	424.65	334.48	292.76	270.14	258.99	256.82	248.60
36000	729.96	561.11	436.78	344.04	301.12	277.86	266.39	264.16	255.70
37000	750.23	576.69	448.92	353.60	309.49	285.58	273.78	271.50	262.80
38000	770.51	592.28	461.05	363.15	317.85	293.30	281.18	278.84	269.90
39000	790.78	607.87	473.18	372.71	326.22	301.01	288.58	286.17	277.01
40000	811.06	623.45	485.32	382.27	334.58	308.73	295.98	293.51	284.11
41000	831.34	639.04	497.45	391.82	342.95	316.45	303.38	300.85	291.21
42000	851.61	654.63	509.58	401.38	351.31	324.17	310.78	308.19	298.31
43000	871.89	670.21	521.71	410.94	359.67	331.89	318.18	315.52	305.42
44000	892.17	685.80	533.85	420.49	368.04	339.60	325.58	322.86	312.52
45000	912.44	701.38	545.98	430.05	376.40	347.32	332.98	330.20	319.62
46000	932.72	716.97	558.11	439.60	384.77	355.04	340.38	337.54	326.73
47000	953.00	732.56	570.24	449.16	393.13	362.76	347.78	344.87	333.83
48000	973.27	748.14	582.38	458.72	401.50	370.48	355.18	352.21	340.93
49000	993.55	763.73	594.51	468.27	409.86	378.19	362.58	359.55	348.03
50000	1013.82	779.32	606.64	477.83	418.23	385.91	369.98	366.89	355.14
51000	1034.10	794.90	618.78	487.39	426.59	393.63	377.38	374.22	362.24
52000	1054.38	810.49	630.91	496.94	434.95	401.35	384.78	381.56	369.34
53000	1074.65	826.07	643.04	506.50	443.32	409.07	392.18	388.90	376.44
54000	1094.93	841.66	655.17	516.06	451.68	416.79	399.58	396.24	383.55
55000	1115.21	857.25	667.31	525.61	460.05	424.50	406.98	403.58	390.65
56000	1135.48	872.83	679.44	535.17	468.41	432.22	414.37	410.91	397.75
57000	1155.76	888.42	691.57	544.73	476.78	439.94	421.77	418.25	404.85
58000	1176.04	904.01	703.71	554.28	485.14	447.66	429.17	425.59	411.96
59000	1196.31	919.59	715.84	563.84	493.50	455.38	436.57	432.93	419.06
60000	1216.59	935.18	727.97	573.40	501.87	463.09	443.97	440.26	426.16
65000	1317.97	1013.11	788.63	621.18	543.69	501.69	480.97	476.95	461.67
70000	1419.35	1091.04	849.30	668.96	585.51	540.28	517.97	513.64	497.19
75000	1520.73	1168.97	909.96	716.74	627.34	578.87	554.96	550.33	532.70
80000	1622.12	1246.90	970.63	764.53	669.16	617.46	591.96	587.02	568.21
85000	1723.50	1324.83	1031.29	812.31	710.98	656.05	628.96	623.70	603.73
90000	1824.88	1402.76	1091.95	860.09	752.80	694.64	665.96	660.39	639.24
95000	1926.26	1480.70	1152.62	907.87	794.62	733.23	702.95	697.08	674.75
100000	2027.64	1558.63	1213.28	955.66	836.45	771.82	739.95	733.77	710.27

10% (continued)

TERM / AMOUNT	5 YEARS	7 YEARS	10 YEARS	15 YEARS	20 YEARS	25 YEARS	29 YEARS	30 YEARS	35 YEARS
$ 50	1.07	.84	.67	.54	.49	.46	.45	.44	.43
100	2.13	1.67	1.33	1.08	.97	.91	.89	.88	.86
200	4.25	3.33	2.65	2.15	1.94	1.82	1.77	1.76	1.72
300	6.38	4.99	3.97	3.23	2.90	2.73	2.65	2.64	2.58
400	8.50	6.65	5.29	4.30	3.87	3.64	3.53	3.52	3.44
500	10.63	8.31	6.61	5.38	4.83	4.55	4.42	4.39	4.30
600	12.75	9.97	7.93	6.45	5.80	5.46	5.30	5.27	5.16
700	14.88	11.63	9.26	7.53	6.76	6.37	6.18	6.15	6.02
800	17.00	13.29	10.58	8.60	7.73	7.27	7.06	7.03	6.88
900	19.13	14.95	11.90	9.68	8.69	8.18	7.95	7.90	7.74
1000	21.25	16.61	13.22	10.75	9.66	9.09	8.83	8.78	8.60
2000	42.50	33.21	26.44	21.50	19.31	18.18	17.65	17.56	17.20
3000	63.75	49.81	39.65	32.24	28.96	27.27	26.48	26.33	25.80
4000	84.99	66.41	52.87	42.99	38.61	36.35	35.30	35.11	34.39
5000	106.24	83.01	66.08	53.74	48.26	45.44	44.13	43.88	42.99
6000	127.49	99.61	79.30	64.48	57.91	54.53	52.95	52.66	51.59
7000	148.73	116.21	92.51	75.23	67.56	63.61	61.78	61.44	60.18
8000	169.98	132.81	105.73	85.97	77.21	72.70	70.60	70.21	68.78
9000	191.23	149.42	118.94	96.72	86.86	81.79	79.43	78.99	77.38
10000	212.48	166.02	132.16	107.47	96.51	90.88	88.25	87.76	85.97
15000	318.71	249.02	198.23	161.20	144.76	136.31	132.38	131.64	128.96
20000	424.95	332.03	264.31	214.93	193.01	181.75	176.50	175.52	171.94
25000	531.18	415.03	330.38	268.66	241.26	227.18	220.62	219.40	214.92
26000	552.43	431.64	343.60	279.40	250.91	236.27	229.45	228.17	223.52
27000	573.68	448.24	356.81	290.15	260.56	245.35	238.27	236.95	232.12
28000	594.92	464.84	370.03	300.89	270.21	254.44	247.10	245.73	240.71
29000	616.17	481.44	383.24	311.64	279.86	263.53	255.92	254.50	249.31
30000	637.42	498.04	396.46	322.39	289.51	272.62	264.75	263.28	257.91
31000	658.66	514.64	409.67	333.13	299.16	281.70	273.57	272.05	266.50
32000	679.91	531.24	422.89	343.88	308.81	290.79	282.40	280.83	275.10
33000	701.16	547.84	436.10	354.62	318.46	299.88	291.22	289.60	283.70
34000	722.40	564.45	449.32	365.37	328.11	308.96	300.05	298.38	292.29
35000	743.65	581.05	462.53	376.12	337.76	318.05	308.87	307.16	300.89
36000	764.90	597.65	475.75	386.86	347.41	327.14	317.70	315.93	309.49
37000	786.15	614.25	488.96	397.61	357.06	336.22	326.52	324.71	318.08
38000	807.39	630.85	502.18	408.35	366.71	345.31	335.35	333.48	326.68
39000	828.64	647.45	515.39	419.10	376.36	354.40	344.17	342.26	335.28
40000	849.89	664.05	528.61	429.85	386.01	363.49	353.00	351.03	343.87
41000	871.13	680.65	541.82	440.59	395.66	372.57	361.82	359.81	352.47
42000	892.38	697.25	555.04	451.34	405.31	381.66	370.65	368.59	361.07
43000	913.63	713.86	568.25	462.09	414.96	390.75	379.47	377.36	369.66
44000	934.87	730.46	581.47	472.83	424.61	399.83	388.29	386.14	378.26
45000	956.12	747.06	594.68	483.58	434.26	408.92	397.12	394.91	386.86
46000	977.37	763.66	607.90	494.32	443.91	418.01	405.94	403.69	395.45
47000	998.62	780.26	621.11	505.07	453.57	427.09	414.77	412.46	404.05
48000	1019.86	796.86	634.33	515.82	463.22	436.18	423.59	421.24	412.65
49000	1041.11	813.46	647.54	526.56	472.87	445.27	432.42	430.02	421.24
50000	1062.36	830.06	660.76	537.31	482.52	454.36	441.24	438.79	429.84
51000	1083.60	846.67	673.97	548.05	492.17	463.44	450.07	447.57	438.44
52000	1104.85	863.27	687.19	558.80	501.82	472.53	458.89	456.34	447.03
53000	1126.10	879.87	700.40	569.55	511.47	481.62	467.72	465.12	455.63
54000	1147.35	896.47	713.62	580.29	521.12	490.70	476.54	473.89	464.23
55000	1168.59	913.07	726.83	591.04	530.77	499.79	485.37	482.67	472.82
56000	1189.84	929.67	740.05	601.78	540.42	508.88	494.19	491.45	481.42
57000	1211.09	946.27	753.26	612.53	550.07	517.96	503.02	500.22	490.02
58000	1232.33	962.87	766.48	623.28	559.72	527.05	511.84	509.00	498.62
59000	1253.58	979.47	779.69	634.02	569.37	536.14	520.67	517.77	507.21
60000	1274.83	996.08	792.91	644.77	579.02	545.23	529.49	526.55	515.81
65000	1381.06	1079.08	858.98	698.50	627.27	590.66	573.62	570.43	558.79
70000	1487.30	1162.09	925.06	752.23	675.52	636.10	617.74	614.31	601.78
75000	1593.53	1245.09	991.14	805.96	723.77	681.53	661.86	658.18	644.76
80000	1699.77	1328.10	1057.21	859.69	772.02	726.97	705.99	702.06	687.74
85000	1806.00	1411.11	1123.29	913.42	820.27	772.40	750.11	745.94	730.73
90000	1912.24	1494.11	1189.36	967.15	868.52	817.84	794.23	789.82	773.71
95000	2018.47	1577.12	1255.44	1020.88	916.78	863.27	838.36	833.70	816.69
100000	2124.71	1660.12	1321.51	1074.61	965.03	908.71	882.48	877.58	859.68

(continued)

12%

TERM / AMOUNT	5 YEARS	7 YEARS	10 YEARS	15 YEARS	20 YEARS	25 YEARS	29 YEARS	30 YEARS	35 YEARS
$ 50	1.12	.89	.72	.61	.56	.53	.52	.52	.51
100	2.23	1.77	1.44	1.21	1.11	1.06	1.04	1.03	1.02
200	4.45	3.54	2.87	2.41	2.21	2.11	2.07	2.06	2.04
300	6.68	5.30	4.31	3.61	3.31	3.16	3.10	3.09	3.05
400	8.90	7.07	5.74	4.81	4.41	4.22	4.13	4.12	4.07
500	11.13	8.83	7.18	6.01	5.51	5.27	5.17	5.15	5.08
600	13.35	10.60	8.61	7.21	6.61	6.32	6.20	6.18	6.10
700	15.58	12.36	10.05	8.41	7.71	7.38	7.23	7.21	7.11
800	17.80	14.13	11.48	9.61	8.81	8.43	8.26	8.23	8.13
900	20.03	15.89	12.92	10.81	9.91	9.48	9.30	9.26	9.14
1000	22.25	17.66	14.35	12.01	11.02	10.54	10.33	10.29	10.16
2000	44.49	35.31	28.70	24.01	22.03	21.07	20.65	20.58	20.32
3000	66.74	52.96	43.05	36.01	33.04	31.60	30.98	30.86	30.47
4000	88.98	70.62	57.39	48.01	44.05	42.13	41.30	41.15	40.63
5000	111.23	88.27	71.74	60.01	55.06	52.67	51.62	51.44	50.78
6000	133.47	105.92	86.09	72.02	66.07	63.20	61.95	61.72	60.94
7000	155.72	123.57	100.43	84.02	77.08	73.73	72.27	72.01	71.09
8000	177.96	141.23	114.78	96.02	88.09	84.26	82.59	82.29	81.25
9000	200.21	158.88	129.13	108.02	99.10	94.80	92.92	92.58	91.40
10000	222.45	176.53	143.48	120.02	110.11	105.33	103.24	102.87	101.56
15000	333.67	264.80	215.21	180.03	165.17	157.99	154.86	154.30	152.34
20000	444.89	353.06	286.95	240.04	220.22	210.65	206.48	205.73	203.11
25000	556.12	441.32	358.68	300.05	275.28	263.31	258.09	257.16	253.89
26000	578.36	458.98	373.03	312.05	286.29	273.84	268.42	267.44	264.05
27000	600.61	476.63	387.38	324.05	297.30	284.38	278.74	277.73	274.20
28000	622.85	494.28	401.72	336.05	308.31	294.91	289.07	288.02	284.36
29000	645.09	511.93	416.07	348.05	319.32	305.44	299.39	298.30	294.51
30000	667.34	529.59	430.42	360.06	330.33	315.97	309.71	308.59	304.67
31000	689.58	547.24	444.76	372.06	341.34	326.50	320.04	318.87	314.83
32000	711.83	564.89	459.11	384.06	352.35	337.04	330.36	329.16	324.98
33000	734.07	582.55	473.46	396.06	363.36	347.57	340.68	339.45	335.14
34000	756.32	600.20	487.81	408.06	374.37	358.10	351.01	349.73	345.29
35000	778.56	617.85	502.15	420.06	385.39	368.63	361.33	360.02	355.45
36000	800.81	635.50	516.50	432.07	396.40	379.17	371.65	370.31	365.60
37000	823.05	653.16	530.85	444.07	407.41	389.70	381.98	380.59	375.76
38000	845.29	670.81	545.19	456.07	418.42	400.23	392.30	390.88	385.91
39000	867.54	688.46	559.54	468.07	429.43	410.76	402.62	401.16	396.07
40000	889.78	706.11	573.89	480.07	440.44	421.29	412.95	411.45	406.22
41000	912.03	723.77	588.24	492.07	451.45	431.83	423.27	421.74	416.38
42000	934.27	741.42	602.58	504.08	462.46	442.36	433.60	432.02	426.54
43000	956.52	759.07	616.93	516.08	473.47	452.89	443.92	442.31	436.69
44000	978.76	776.73	631.28	528.08	484.48	463.42	454.24	452.59	446.85
45000	1001.01	794.38	645.62	540.08	495.49	473.96	464.57	462.88	457.00
46000	1023.25	812.03	659.97	552.08	506.50	484.49	474.89	473.17	467.16
47000	1045.49	829.68	674.32	564.08	517.52	495.02	485.21	483.45	477.31
48000	1067.74	847.34	688.67	576.09	528.53	505.55	495.54	493.74	487.47
49000	1089.98	864.99	703.01	588.09	539.54	516.08	505.86	504.03	497.62
50000	1112.23	882.64	717.36	600.09	550.55	526.62	516.18	514.31	507.78
51000	1134.47	900.29	731.71	612.09	561.56	537.15	526.51	524.60	517.94
52000	1156.72	917.95	746.05	624.09	572.57	547.68	536.83	534.88	528.09
53000	1178.96	935.60	760.40	636.09	583.58	558.21	547.16	545.17	538.25
54000	1201.21	953.25	774.75	648.10	594.59	568.75	557.48	555.46	548.40
55000	1223.45	970.91	789.10	660.10	605.60	579.28	567.80	565.74	558.56
56000	1245.69	988.56	803.44	672.10	616.61	589.81	578.13	576.03	568.71
57000	1267.94	1006.21	817.79	684.10	627.62	600.34	588.45	586.31	578.87
58000	1290.18	1023.86	832.14	696.10	638.63	610.88	598.77	596.60	589.02
59000	1312.43	1041.52	846.48	708.10	649.65	621.41	609.10	606.89	599.18
60000	1334.67	1059.17	860.83	720.11	660.66	631.94	619.42	617.17	609.33
65000	1445.89	1147.43	932.57	780.11	715.71	684.60	671.04	668.60	660.11
70000	1557.12	1235.70	1004.30	840.12	770.77	737.26	722.66	720.03	710.89
75000	1668.34	1323.96	1076.04	900.13	825.82	789.92	774.27	771.46	761.67
80000	1779.56	1412.22	1147.77	960.14	880.87	842.58	825.89	822.90	812.44
85000	1890.78	1500.49	1219.51	1020.15	935.93	895.25	877.51	874.33	863.22
90000	2002.01	1588.75	1291.24	1080.16	990.98	947.91	929.13	925.76	914.00
95000	2113.23	1677.01	1362.98	1140.16	1046.04	1000.57	980.75	977.19	964.78
100000	2224.45	1765.28	1434.71	1200.17	1101.09	1053.23	1032.36	1028.62	1015.55

14% (continued)

TERM / AMOUNT	5 YEARS	7 YEARS	10 YEARS	15 YEARS	20 YEARS	25 YEARS	29 YEARS	30 YEARS	35 YEARS
$ 50	1.17	.94	.78	.67	.63	.61	.60	.60	.59
100	2.33	1.88	1.56	1.34	1.25	1.21	1.19	1.19	1.18
200	4.66	3.75	3.11	2.67	2.49	2.41	2.38	2.37	2.36
300	6.99	5.63	4.66	4.00	3.74	3.62	3.57	3.56	3.53
400	9.31	7.50	6.22	5.33	4.98	4.82	4.76	4.74	4.71
500	11.64	9.38	7.77	6.66	6.22	6.02	5.94	5.93	5.88
600	13.97	11.25	9.32	8.00	7.47	7.23	7.13	7.11	7.06
700	16.29	13.12	10.87	9.33	8.71	8.43	8.32	8.30	8.23
800	18.62	15.00	12.43	10.66	9.95	9.64	9.51	9.48	9.41
900	20.95	16.87	13.98	11.99	11.20	10.84	10.69	10.67	10.59
1000	23.27	18.75	15.53	13.32	12.44	12.04	11.88	11.85	11.76
2000	46.54	37.49	31.06	26.64	24.88	24.08	23.76	23.70	23.52
3000	69.81	56.23	46.58	39.96	37.31	36.12	35.63	35.55	35.28
4000	93.08	74.97	62.11	53.27	49.75	48.16	47.51	47.40	47.03
5000	116.35	93.71	77.64	66.59	62.18	60.19	59.39	59.25	58.79
6000	139.61	112.45	93.16	79.91	74.62	72.23	71.26	71.10	70.55
7000	162.88	131.19	108.69	93.23	87.05	84.27	83.14	82.95	82.30
8000	186.15	149.93	124.22	106.54	99.49	96.31	95.02	94.79	94.06
9000	209.42	168.67	139.74	119.86	111.92	108.34	106.89	106.64	105.82
10000	232.69	187.41	155.27	133.18	124.36	120.38	118.77	118.49	117.57
15000	349.03	281.11	232.90	199.77	186.53	180.57	178.15	177.74	176.36
20000	465.37	374.81	310.54	266.35	248.71	240.76	237.53	236.98	235.14
25000	581.71	468.51	388.17	332.94	310.89	300.95	296.91	296.22	293.92
26000	604.98	487.25	403.70	346.26	323.32	312.98	308.79	308.07	305.68
27000	628.25	505.99	419.22	359.58	335.76	325.02	320.67	319.92	317.44
28000	651.52	524.73	434.75	372.89	348.19	337.06	332.54	331.77	329.19
29000	674.78	543.47	450.28	386.21	360.63	349.10	344.42	343.62	340.95
30000	698.05	562.21	465.80	399.53	373.06	361.13	356.30	355.47	352.71
31000	721.32	580.95	481.33	412.84	385.50	373.17	368.17	367.32	364.46
32000	744.59	599.69	496.86	426.16	397.93	385.21	380.05	379.16	376.22
33000	767.86	618.43	512.38	439.48	410.37	397.25	391.93	391.01	387.98
34000	791.13	637.17	527.91	452.80	422.80	409.28	403.80	402.86	399.73
35000	814.39	655.91	543.44	466.11	435.24	421.32	415.68	414.71	411.49
36000	837.66	674.65	558.96	479.43	447.67	433.36	427.56	426.56	423.25
37000	860.93	693.39	574.49	492.75	460.11	445.40	439.43	438.41	435.00
38000	884.20	712.13	590.02	506.07	472.54	457.43	451.31	450.26	446.76
39000	907.47	730.87	605.54	519.38	484.98	469.47	463.18	462.10	458.52
40000	930.74	749.61	621.07	532.70	497.41	481.51	475.06	473.95	470.27
41000	954.00	768.35	636.60	546.02	509.85	493.55	486.94	485.80	482.03
42000	977.27	787.09	652.12	559.34	522.28	505.58	498.81	497.65	493.79
43000	1000.54	805.83	667.65	572.65	534.72	517.62	510.69	509.50	505.54
44000	1023.81	824.57	683.18	585.97	547.15	529.66	522.57	521.35	517.30
45000	1047.08	843.31	698.70	599.29	559.59	541.70	534.44	533.20	529.06
46000	1070.34	862.05	714.23	612.61	572.02	553.74	546.32	545.05	540.81
47000	1093.61	880.79	729.76	625.92	584.46	565.77	558.20	556.89	552.57
48000	1116.88	899.53	745.28	639.24	596.89	577.81	570.07	568.74	564.33
49000	1140.15	918.27	760.81	652.56	609.33	589.85	581.95	580.59	576.08
50000	1163.42	937.01	776.34	665.88	621.77	601.89	593.82	592.44	587.84
51000	1186.69	955.75	791.86	679.19	634.20	613.92	605.70	604.29	599.60
52000	1209.95	974.49	807.39	692.51	646.64	625.96	617.58	616.14	611.36
53000	1233.22	993.23	822.92	705.83	659.07	638.00	629.45	627.99	623.11
54000	1256.49	1011.97	838.44	719.15	671.51	650.04	641.33	639.84	634.87
55000	1279.76	1030.71	853.97	732.46	683.94	662.07	653.21	651.68	646.63
56000	1303.03	1049.45	869.50	745.78	696.38	674.11	665.08	663.53	658.38
57000	1326.30	1068.19	885.02	759.10	708.81	686.15	676.96	675.38	670.14
58000	1349.56	1086.93	900.55	772.42	721.25	698.19	688.84	687.23	681.90
59000	1372.83	1105.67	916.08	785.73	733.68	710.22	700.71	699.08	693.65
60000	1396.10	1124.41	931.60	799.05	746.12	722.26	712.59	710.93	705.41
65000	1512.44	1218.11	1009.24	865.64	808.29	782.45	771.97	770.17	764.19
70000	1628.78	1311.81	1086.87	932.22	870.47	842.64	831.35	829.42	822.98
75000	1745.12	1405.51	1164.50	998.81	932.65	902.83	890.73	888.66	881.76
80000	1861.47	1499.21	1242.14	1065.40	994.82	963.01	950.12	947.90	940.54
85000	1977.81	1592.91	1319.77	1131.99	1057.00	1023.20	1009.50	1007.15	999.33
90000	2094.15	1686.61	1397.40	1198.57	1119.17	1083.39	1068.88	1066.39	1058.11
95000	2210.49	1780.31	1475.04	1265.16	1181.35	1143.58	1128.26	1125.63	1116.89
100000	2326.83	1874.01	1552.67	1331.75	1243.53	1203.77	1187.64	1184.88	1175.68

16%

(continued)

TERM / AMOUNT	5 YEARS	7 YEARS	10 YEARS	15 YEARS	20 YEARS	25 YEARS	29 YEARS	30 YEARS	35 YEARS
$ 50	1.22	1.00	.84	.74	.70	.68	.68	.68	.67
100	2.44	1.99	1.68	1.47	1.40	1.36	1.35	1.35	1.34
200	4.87	3.98	3.36	2.94	2.79	2.72	2.70	2.69	2.68
300	7.30	5.96	5.03	4.41	4.18	4.08	4.05	4.04	4.02
400	9.73	7.95	6.71	5.88	5.57	5.44	5.39	5.38	5.36
500	12.16	9.94	8.38	7.35	6.96	6.80	6.74	6.73	6.70
600	14.60	11.92	10.06	8.82	8.35	8.16	8.09	8.07	8.04
700	17.03	13.91	11.73	10.29	9.74	9.52	9.43	9.42	9.37
800	19.46	15.89	13.41	11.75	11.14	10.88	10.78	10.76	10.71
900	21.89	17.88	15.08	13.22	12.53	12.23	12.13	12.11	12.05
1000	24.32	19.87	16.76	14.69	13.92	13.59	13.47	13.45	13.39
2000	48.64	39.73	33.51	29.38	27.83	27.18	26.94	26.90	26.77
3000	72.96	59.59	50.26	44.07	41.74	40.77	40.41	40.35	40.16
4000	97.28	79.45	67.01	58.75	55.66	54.36	53.87	53.80	53.54
5000	121.60	99.32	83.76	73.44	69.57	67.95	67.34	67.24	66.93
6000	145.91	119.18	100.51	88.13	83.48	81.54	80.81	80.69	80.31
7000	170.23	139.04	117.26	102.81	97.39	95.13	94.28	94.14	93.70
8000	194.55	158.90	134.02	117.50	111.31	108.72	107.74	107.59	107.08
9000	218.87	178.76	150.77	132.19	125.22	122.30	121.21	121.03	120.47
10000	243.19	198.63	167.52	146.88	139.13	135.89	134.68	134.48	133.85
15000	364.78	297.94	251.27	220.31	208.69	203.84	202.02	201.72	200.78
20000	486.37	397.25	335.03	293.75	278.26	271.78	269.35	268.96	267.70
25000	607.96	496.56	418.79	367.18	347.82	339.73	336.69	336.19	334.62
26000	632.27	516.42	435.54	381.87	361.73	353.32	350.16	349.64	348.01
27000	656.59	536.28	452.29	396.55	375.64	366.90	363.63	363.09	361.39
28000	680.91	556.14	469.04	411.24	389.56	380.49	377.09	376.54	374.78
29000	705.23	576.00	485.79	425.93	403.47	394.08	390.56	389.98	388.16
30000	729.55	595.87	502.54	440.62	417.38	407.67	404.03	403.43	401.55
31000	753.86	615.73	519.30	455.30	431.29	421.26	417.50	416.88	414.93
32000	778.18	635.59	536.05	469.99	445.21	434.85	430.96	430.33	428.32
33000	802.50	655.45	552.80	484.68	459.12	448.44	444.43	443.77	441.70
34000	826.82	675.32	569.55	499.36	473.03	462.03	457.90	457.22	455.08
35000	851.14	695.18	586.30	514.05	486.94	475.62	471.37	470.67	468.47
36000	875.46	715.04	603.05	528.74	500.86	489.20	484.83	484.12	481.85
37000	899.77	734.90	619.80	543.42	514.77	502.79	498.30	497.57	495.24
38000	924.09	754.76	636.55	558.11	528.68	516.38	511.77	511.01	508.62
39000	948.41	774.63	653.31	572.80	542.59	529.97	525.24	524.46	522.01
40000	972.73	794.49	670.06	587.49	556.51	543.56	538.70	537.91	535.39
41000	997.05	814.35	686.81	602.17	570.42	557.15	552.17	551.36	548.78
42000	1021.36	834.21	703.56	616.86	584.33	570.74	565.64	564.80	562.16
43000	1045.68	854.07	720.31	631.55	598.25	584.33	579.11	578.25	575.55
44000	1070.00	873.94	737.06	646.23	612.16	597.92	592.57	591.70	588.93
45000	1094.32	893.80	753.81	660.92	626.07	611.50	606.04	605.15	602.32
46000	1118.64	913.66	770.57	675.61	639.98	625.09	619.51	618.59	615.70
47000	1142.95	933.52	787.32	690.29	653.90	638.68	632.98	632.04	629.09
48000	1167.27	953.38	804.07	704.98	667.81	652.27	646.44	645.49	642.47
49000	1191.59	973.25	820.82	719.67	681.72	665.86	659.91	658.94	655.86
50000	1215.91	993.11	837.57	734.36	695.63	679.45	673.38	672.38	669.24
51000	1240.23	1012.97	854.32	749.04	709.55	693.04	686.84	685.83	682.62
52000	1264.54	1032.83	871.07	763.73	723.46	706.63	700.31	699.28	696.01
53000	1288.86	1052.69	887.82	778.42	737.37	720.22	713.78	712.73	709.39
54000	1313.18	1072.56	904.58	793.10	751.28	733.80	727.25	726.17	722.78
55000	1337.50	1092.42	921.33	807.79	765.20	747.39	740.71	739.62	736.16
56000	1361.82	1112.28	938.08	822.48	779.11	760.98	754.18	753.07	749.55
57000	1386.13	1132.14	954.83	837.16	793.02	774.57	767.65	766.52	762.93
58000	1410.45	1152.00	971.58	851.85	806.93	788.16	781.12	779.96	776.32
59000	1434.77	1171.87	988.33	866.54	820.85	801.75	794.58	793.41	789.70
60000	1459.09	1191.73	1005.08	881.23	834.76	815.34	808.05	806.86	803.09
65000	1580.68	1291.04	1088.84	954.66	904.32	883.28	875.39	874.10	870.01
70000	1702.27	1390.35	1172.60	1028.10	973.88	951.23	942.73	941.33	936.93
75000	1823.86	1489.66	1256.35	1101.53	1043.45	1019.17	1010.06	1008.57	1003.86
80000	1945.45	1588.97	1340.11	1174.97	1113.01	1087.12	1077.40	1075.81	1070.78
85000	2067.04	1688.28	1423.87	1248.40	1182.57	1155.06	1144.74	1143.05	1137.70
90000	2188.63	1787.59	1507.62	1321.84	1252.14	1223.00	1212.08	1210.29	1204.63
95000	2310.22	1886.90	1591.38	1395.27	1321.70	1290.95	1279.41	1277.52	1271.55
100000	2431.81	1986.21	1675.14	1468.71	1391.26	1358.89	1346.75	1344.76	1338.47

20%

TERM / AMOUNT	5 YEARS	7 YEARS	10 YEARS	15 YEARS	20 YEARS	25 YEARS	29 YEARS	30 YEARS	35 YEARS
$ 50	1.33	1.12	.97	.88	.85	.84	.84	.84	.84
100	2.65	2.23	1.94	1.76	1.70	1.68	1.68	1.68	1.67
200	5.30	4.45	3.87	3.52	3.40	3.36	3.35	3.35	3.34
300	7.95	6.67	5.80	5.27	5.10	5.04	5.02	5.02	5.01
400	10.60	8.89	7.74	7.03	6.80	6.72	6.69	6.69	6.68
500	13.25	11.11	9.67	8.79	8.50	8.40	8.36	8.36	8.35
600	15.90	13.33	11.60	10.54	10.20	10.08	10.04	10.03	10.01
700	18.55	15.55	13.53	12.30	11.90	11.75	11.71	11.70	11.68
800	21.20	17.77	15.47	14.06	13.60	13.43	13.38	13.37	13.35
900	23.85	19.99	17.40	15.81	15.29	15.11	15.05	15.04	15.02
1000	26.50	22.21	19.33	17.57	16.99	16.79	16.72	16.72	16.69
2000	52.99	44.42	38.66	35.13	33.98	33.57	33.44	33.43	33.37
3000	79.49	66.62	57.98	52.69	50.97	50.36	50.16	50.14	50.05
4000	105.98	88.83	77.31	70.26	67.96	67.14	66.88	66.85	66.74
5000	132.47	111.04	96.63	87.82	84.95	83.93	83.60	83.56	83.42
6000	158.97	133.24	115.96	105.38	101.93	100.71	100.32	100.27	100.10
7000	185.46	155.45	135.28	122.95	118.92	117.50	117.04	116.98	116.78
8000	211.96	177.65	154.61	140.51	135.91	134.28	133.76	133.69	133.47
9000	238.45	199.86	173.94	158.07	152.90	151.07	150.48	150.40	150.15
10000	264.94	222.07	193.26	175.63	169.89	167.85	167.20	167.11	166.83
15000	397.41	333.10	289.89	263.45	254.83	251.77	250.80	250.66	250.25
20000	529.88	444.13	386.52	351.26	339.77	335.70	334.40	334.21	333.66
25000	662.35	555.16	483.14	439.08	424.71	419.62	418.00	417.76	417.07
26000	688.85	577.37	502.47	456.64	441.70	436.40	434.72	434.47	433.76
27000	715.34	599.57	521.80	474.21	458.69	453.19	451.44	451.18	450.44
28000	741.83	621.78	541.12	491.77	475.68	469.97	468.16	467.89	467.12
29000	768.33	643.98	560.45	509.33	492.66	486.76	484.88	484.60	483.81
30000	794.82	666.19	579.77	526.89	509.65	503.54	501.60	501.31	500.49
31000	821.32	688.40	599.10	544.46	526.64	520.33	518.32	518.02	517.17
32000	847.81	710.60	618.42	562.02	543.63	537.11	535.04	534.73	533.85
33000	874.30	732.81	637.75	579.58	560.62	553.89	551.76	551.44	550.54
34000	900.80	755.02	657.07	597.15	577.61	570.68	568.48	568.15	567.22
35000	927.29	777.22	676.40	614.71	594.59	587.46	585.20	584.86	583.90
36000	953.78	799.43	695.73	632.27	611.58	604.25	601.92	601.57	600.59
37000	980.28	821.63	715.05	649.83	628.57	621.03	618.64	618.28	617.27
38000	1006.77	843.84	734.38	667.40	645.56	637.82	635.36	634.99	633.95
39000	1033.27	866.05	753.70	684.96	662.55	654.60	652.08	651.70	650.63
40000	1059.76	888.25	773.03	702.52	679.53	671.39	668.80	668.41	667.32
41000	1086.25	910.46	792.35	720.09	696.52	688.17	685.52	685.12	684.00
42000	1112.75	932.67	811.68	737.65	713.51	704.95	702.24	701.83	700.68
43000	1139.24	954.87	831.00	755.21	730.50	721.74	718.95	718.54	717.36
44000	1165.74	977.08	850.33	772.78	747.49	738.52	735.67	735.25	734.05
45000	1192.23	999.28	869.66	790.34	764.48	755.31	752.39	751.96	750.73
46000	1218.72	1021.49	888.98	807.90	781.46	772.09	769.11	768.67	767.41
47000	1245.22	1043.70	908.31	825.46	798.45	788.88	785.83	785.38	784.10
48000	1271.71	1065.90	927.63	843.03	815.44	805.66	802.55	802.09	800.78
49000	1298.21	1088.11	946.96	860.59	832.43	822.45	819.27	818.80	817.46
50000	1324.70	1110.31	966.28	878.15	849.42	839.23	835.99	835.51	834.14
51000	1351.19	1132.52	985.61	895.72	866.41	856.02	852.71	852.22	850.83
52000	1377.69	1154.73	1004.93	913.28	883.39	872.80	869.43	868.93	867.51
53000	1404.18	1176.93	1024.26	930.84	900.38	889.58	886.15	885.64	884.19
54000	1430.67	1199.14	1043.59	948.41	917.37	906.37	902.87	902.36	900.88
55000	1457.17	1221.35	1062.91	965.97	934.36	923.15	919.59	919.07	917.56
56000	1483.66	1243.55	1082.24	983.53	951.35	939.94	936.31	935.78	934.24
57000	1510.16	1265.76	1101.56	1001.09	968.34	956.72	953.03	952.49	950.92
58000	1536.65	1287.96	1120.89	1018.66	985.32	973.51	969.75	969.20	967.61
59000	1563.14	1310.17	1140.21	1036.22	1002.31	990.29	986.47	985.91	984.29
60000	1589.64	1332.38	1159.54	1053.78	1019.30	1007.08	1003.19	1002.62	1000.97
65000	1722.11	1443.41	1256.17	1141.60	1104.24	1091.00	1086.79	1086.17	1084.39
70000	1854.58	1554.44	1352.79	1229.41	1189.18	1174.92	1170.39	1169.72	1167.80
75000	1987.05	1665.47	1449.42	1317.23	1274.12	1258.84	1253.99	1253.27	1251.21
80000	2119.52	1776.50	1546.05	1405.04	1359.06	1342.77	1337.59	1336.82	1334.63
85000	2251.99	1887.53	1642.68	1492.86	1444.01	1426.69	1421.19	1420.37	1418.04
90000	2384.45	1998.56	1739.31	1580.67	1528.95	1510.61	1504.78	1503.92	1501.46
95000	2516.92	2109.59	1835.93	1668.49	1613.89	1594.53	1588.38	1587.47	1584.87
100000	2649.39	2220.62	1932.56	1756.30	1698.83	1678.46	1671.98	1671.02	1668.28

INDEX